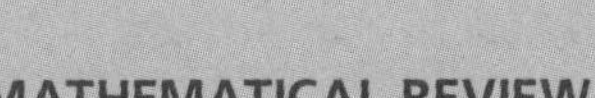

MATHEMATICAL REVIEW

Area of a circle of radius R — $A = \pi R^2$

Circumference of a circle — $C = 2\pi R$

Surface area of a sphere — $A = 4\pi R^2$

Volume of a sphere — $V = \frac{4}{3}\pi R^3$

Area of a triangle — $A = \frac{1}{2}bh$

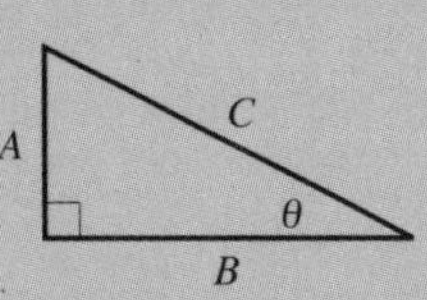

Volume of a circular cylinder of length l — $V = \pi R^2 l$

Pythagorean Theorem — $C^2 = A^2 + B^2$

$\sin\theta = A/C$

$\cos\theta = B/C$ $\qquad \tan\theta = \dfrac{\sin\theta}{\cos\theta}$

$\tan\theta = A/B$

Quadratic Equation: Where $ax^2 + bx + c = 0$

$$x = \frac{-b \pm \sqrt{b^2 - 4ac}}{2a}$$

PHYSICAL CONSTANTS

Quantity	Symbol	Value
Gravitation constant	G	$6.672\,59 \times 10^{-11}$ N·m²/kg²
Speed of light in vacuum	c	$2.997\,924\,58 \times 10^{8}$ m/s
Electron charge	e	$1.602\,18 \times 10^{-19}$ C
Planck's Constant	h	$6.626\,076 \times 10^{-34}$ J·s
		$4.135\,669 \times 10^{-15}$ eV·s
Universal gas constant	R	8.314 510 J/mol·K
Avogadro's number	N_A	$6.022\,137 \times 10^{23}$ mol^{-1}
Boltzmann Constant	k_B	$1.380\,66 \times 10^{-23}$ J/K
		$8.617\,39 \times 10^{-5}$ eV/K
Coulomb force constant	k	$8.987\,55 \times 10^{9}$ N·m²/C²
Permittivity of free space $(1/\mu_0 c^2)$	ϵ_0	$8.854\,19 \times 10^{-12}$ C²/N·m²
Permeability of free space	μ_0	$1.256\,64 \times 10^{-6}$ T·m/A
Permeability constant	$\mu_0/4\pi$	10^{-7} T·m/A
Electron mass	m_e	$9.109\,39 \times 10^{-31}$ kg
Electron rest energy	$m_e c^2$	0.510 999 MeV
Electron magnetic moment	μ_e	$9.284\,77 \times 10^{-24}$ J/T
Electron charge/mass ratio	e/m_e	$1.758\,82 \times 10^{11}$ C/kg
Electron Compton wavelength	λ_c	$2.426\,31 \times 10^{-12}$ m
Proton mass	m_p	$1.672\,623 \times 10^{-27}$ kg
		1.007 276 u
Proton rest energy	$m_p c^2$	938.272 MeV
Proton magnetic moment	μ_p	$1.410\,608 \times 10^{-26}$ J/T
Neutron mass	m_n	$1.674\,929 \times 10^{-27}$ kg
		1.008 66 u
Neutron rest energy	$m_n c^2$	939.566 MeV
Neutron magnetic moment	μ_n	$9.662\,37 \times 10^{-27}$ J/T
Bohr magneton	μ_B	$9.274\,015 \times 10^{-24}$ J/T
Stefan-Boltzmann Constant	σ	$5.670\,51 \times 10^{-8}$ W/m²·K⁴
Rydberg constant	R	$1.097\,373 \times 10^{7}$ m^{-1}
Bohr radius	r_1	$5.291\,77 \times 10^{-11}$ m
Faraday constant	F	$9.648\,53 \times 10^{4}$ C/mol

PHYSICS

PHYSICS

EUGENE HECHT
Adelphi University

Brooks/Cole Publishing Company
Pacific Grove, California

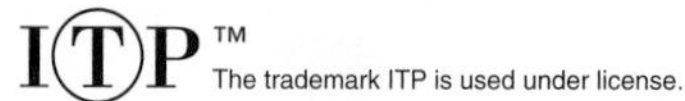

Brooks/Cole Publishing Company
A Division of Wadsworth, Inc.

Printed in the United States of America

10 9 8 7 6 5 4 3 2 1

Library of Congress Cataloging-in-Publication Data

Hecht, Eugene.
Physics / Eugene Hecht.
p. cm.
Includes index.
ISBN 0-534-09114-8
1. Physics. I. Title.
QC23.H3917 1994
530—dc20 93-43106
CIP

Sponsoring Editor: *Lisa J. Moller*
Project Development Editor: *Suzanne Ewing*
Editorial Assistant: *Beth Wilbur*
Production Editor: *Marjorie Z. Sanders*
Production Assistant: *Tessa A. McGlasson*
Manuscript Editor: *Stuart Kenter*
Permissions Editor: *May Clark*
Interior and Cover Design: *E. Kelly Shoemaker*
Cover Photo: *Tom Skrivan*
Art Coordinator: *Susan Haberkorn*
Interior Illustration: *Precision Graphics, Carl Brown, LM Graphics, Matrix Communications*
Photo Coordination and Digital Photo Design: *Larry Molmud*
Photo Researcher: *Stuart Kenter*
Typesetting: *Beacon Graphics*
Cover Printing: *Phoenix Color Corporation*
Printing and Binding: *R.R. Donnelley & Sons Company/Willard*

Credits continue on p. C1.

To Ca, b. w. l.

Preface

Physics is the study of the material Universe—all there *is*. And that's a bold and wonderful agenda. The Universe is incredibly awesome and tantalizingly mysterious, and we, after all, are just beginning to understand it. Almost 3000 years in the making, physics—incomplete as it is—stands as one of the great creations of the human intellect. It's been a privilege and an unending joy to have spent much of my life studying physics, and it is out of gratitude and admiration that this book takes its form. If this work, while insightfully teaching basic physics, transmits a sense of the grandeur, unity, and vitality of the subject, it will have met my primary objectives.

Philosophy and Goals

Designed for a one-year College Physics course, this text assumes the student comes with little more than a rudimentary knowledge of algebra and just the basics of geometry. There is an appendix that will help to develop all the requisite mathematical skills. Whatever trigonometry is required is taught in place, as the need arises. Special attention is always paid to carefully and slowly developing mathematical ideas that might be new to the student—graphs, vectors, logarithms, and so forth.

The central glory of twentieth-century physics is the discovery of an overarching unity in nature. There is revealed an internal simplicity that is well confirmed, even if its complete comprehension is, as yet, just beyond our reach. All matter is composed of myriad identical clones of a small number of fundamental interacting particles. Everything physical is presumably understandable within that context. Thus, to treat the subject as if it were an encyclopedic collection of unrelated ideas is to miss the point of the twentieth century. From the beginning of this text to its end, we will study the unity of natural phenomena, the various manifestations of matter interacting with matter via the fundamental Four Forces. Even as we explore concepts conceived centuries ago, we will bring to bear the perspective of contemporary physics.

Bulk matter is atomic, and to truly comprehend its behavior (mechanical, thermal, electrical, magnetic, acoustical, optical, etc.), we must appreciate how atoms interact to produce all the phenomena of nature. It's no longer appropriate to wait until Chapter 30 and the second semester of College Physics to learn that the objects of our everyday world are composed of atoms. Nor is it necessary to teach the initial 16 chapters of the first semester as if nineteenth-century thought was still valid in all regards—it is not. We have learned a great deal during this century, and that knowledge should inform everything taught, including Classical Physics.

Pacing

The standard College Physics course makes a whole range of challenging intellectual demands on the student, and the experienced instructor knows that the first few weeks of the semester can be crucial. It is then that some students are likely to become overwhelmed. With that in mind, years ago I began to explore the benefits of rearranging the material in kinematics in order to introduce new physical concepts more gradually and allow time for the ideas to be assimilated. Thus, Chapter 2 deals only with speed, displacement, and velocity. The explanations are elaborate, and there are many examples, graphs, and illustrations. The concept of vectors is introduced as it relates to displacement. It is justified, made logically appealing, and very carefully extended and applied to velocity. With the visual aid of multi-frame drawings, the concept of vector addition is next applied to relative velocities. Only then, after the student has presumably worked out dozens of problems and has begun to learn how to learn physics, do we turn to Chapter 3, acceleration. By the time the equations of uniform acceleration are reached, the typical student is much better able to deal with the logical complexities involved.

With the mathematics of trigonometry and rudimentary vector algebra in place, and with a more realistic understanding of the demands of the experience, the serious student is ready to move ahead more rapidly.

A Modern Perspective

The discoveries of this century have radically changed our perception of every aspect of the physical Universe. And yet Classical Physics is often taught as though the last hundred years of revelation had little or no effect on our thinking. This is certainly not the case. Just consider the impact the discovery of the photon has had. If radiant energy is particulate, if the electromagnetic field is quantized, then continuous electromagnetic waves lose their traditional reality, and electric and magnetic field lines become a metaphor for an alternative corpuscular vision.

One of the most far-reaching insights of twentieth-century physics (one still rarely discussed in books at this level) is the theoretical importance of the relationship between symmetry and conservation. This relationship, which culminates in the theory of the electroweak interaction, is a leitmotif that runs throughout the text. Students are usually fascinated to learn that symmetries of space and time manifest themselves in the conservation laws. The relationship is first discussed, albeit briefly, in Chapter 1, with an introduction to the concept of symmetry. It recurs in Chapter 4 with Conservation of Linear Momentum, in Chapter 8 with Conservation of Angular Momentum, in Chapter 9 with Conservation of Energy, and so on all the way up to gauge symmetry in Chapter 33. Further, modern insights are gently introduced whenever they are relevant and as early as possible. Chapter 1, for example, briefly summarizes and considers the interrelationships of the various theoretical formalisms (Classical Physics, Relativity, and Quantum Mechanics). It is important for the student to have an overview of the discipline, to know that this is an ongoing, unending study. Chapter 1 also introduces some of the vocabulary and ideas that will be considered throughout the development. In this way, the concept of quantization will be a familiar one long before it is dealt with formally in Chapter 30. Likewise, atoms, electrons, protons, neutrons, photons, gravitons, and quarks are all at least introduced early on.

The central role played by the Four Forces of nature is an ever-present theme, and an effort is made to see everyday phenomena from this unifying perspective. Students are made aware from the outset that contact forces, friction, cohesion, adhesion, etc. are the result of electromagnetic interaction, although they will have to wait until Chapter 17 to study the details.

Contemporary, Practical Applications

A driving force shaping this text is the belief that students are more easily motivated to learn about what directly affects their lives and concerns. The perspective of this book is therefore largely practical: How do we walk? Why do we get backaches? How does an airplane fly? The text is rich in applications of physics to the life sciences, but it strives to be broadly interdisciplinary, drawing from and reflecting back upon biology, geology, astronomy, architecture, medicine, and meteorology.

Problem Solving and Self-Study

Examples

The text contains a great many example calculations systematically worked out in detail, in order to guide students through the processes of analysis and numerical problem solving. All such examples are followed by a Quick Check, which teaches a wide variety of verification methods and helps to establish the habit of checking one's computations.

Problem-Solving Techniques

Every chapter includes a section of suggestions on problem solving that may also contain approximation techniques, as well as a discussion of the pitfalls and common errors specific to the material at hand. For example, a common error is to compute the average speed of a uniformly accelerating object using $v_{av} = \frac{1}{2}(v_f - v_i)$ rather than $v_{av} = \frac{1}{2}(v_f + v_i)$.

Core Material

Every chapter includes a distilled review of the core material for that chapter.

Discussion Questions

All chapters contain a selection of discussion questions designed to develop and extend the conceptual understanding of the material.

Multiple Choice Questions

A group of roughly 20 multiple choice questions similar to those found on national medical school entrance exams is included in each chapter. Among other types, these include single-concept calculational questions, as well as pointedly probing conceptual questions.

Numerical Problems

An extensive selection of problems, most built on real data and referring to actual situations, is provided at the end of each chapter. An instructor will find a wealth of choices from which to assign homework. The problem sets are arranged in three levels of difficulty and always include a large selection of single-concept problems that explore one idea at a time (from several perspectives) to help students establish a strong foundation of competence and confidence with the basics. Each set of problems is deployed as a carefully developed and integrated pedagogical unit that will carry the student from one level of mastery to the next. All students working diligently should soon be able to do grade-I problems without much trouble. Grade-II problems require a more developed competence, and a student who is comfortable with them should do well in the course. Approximately 15% of the end-of-chapter problems are worked out succinctly at the back of the book in order to encourage and strengthen independent study. These problems are indicated by a boldface numeral. Problem solving, like playing an instrument, can only be mastered by a combination of practice and guiding example. Students should be encouraged to spend several hours a day solving problems on their own, especially those with boldface numerals for which skeletal solutions are provided. Another 10% of the solutions are worked out in a student solutions manual, which is available at the instructor's request to the bookstore. These problems are indicated by an italic numeral.

Organization

A good working text has responsibilities to both the student and the instructor. It should, as much as possible, provide a complete resource to students who are studying on their own at home. At the same time, it must provide an instructor with a broad range of conceptual material from which to fashion a course that meets his or her specific requirements.

It is impossible to lecture in class on every topic covered in a text that deals with the whole of introductory physics. The book has been organized in a way that addresses this problem and should therefore be a more effective teaching instrument. All section headings are coded in three colors: brown, black, and purple. Topics bearing **brown headings** are intended to be read and understood by students working on their own prior to coming to class. This information can then be assumed as basic communal knowledge, and it thereafter requires little or no classroom time. Topics bearing **black headings** represent the basic material from which the instructor can fashion the specific course. Topics with **purple headings** are enrichment sections that can be used to enhance and strengthen the primary discourse.

Units

Units play a special role in physics and so represent an issue that requires considerable attention. Without exception, the proper units to be used in physics are those of the Système International (SI). Thus, SI units will be used almost exclusively throughout this text. The commitment to the Système International must, however, be flexible enough to contend with three outstanding challenges: (1) The United States is the only major country in which there still lingers an antique system of units. Consequently, the intuitive experiences of most students are framed in terms of feet, miles, quarts, and pounds. Moreover, the data of everyday life, which must be brought into the discussion, is invariably given in U.S. Customary units. Apparently baseball pitchers hurl balls in miles per hour exclusively. (2) In order to extend our analyses into the domains of other disciplines, we must deal realistically with their idiosyncratic unit preferences. Like it or not, the pressure in the lungs is measured in centimeters of water, and for good reason. (3) There is a vast body of extant scientific data in a variety of units; the well-trained student will need to have access to that treasure.

Thus, this text will gradually, compassionately, and yet unswervingly move toward the total adoption of SI units. Even as it does, it will attempt to deal effectively with the substantive issues considered here.

Art and Photography Program

The graphics and text for this book were conceived and developed from the very start as an integrated whole; every photograph and every drawing is justified in its inclusion. The result is an unparalleled level of pedagogical effectiveness. There are about 1300 fresh, insightful, and uncompromisingly accurate illustrations that make it possible for students to visualize a diversity of physical phenomena. Many of these are multi-frame sequential drawings that allow students to apprehend the temporal unfolding of complex events (e.g., Fig's. 2.27, 2.28, 13.38, and 13.39). There are nearly 500 photographs (many of them overlaid with explanatory graphics) that form an important part of the pedagogical program.

A Complete Ancillary Package

Accompanying the text is a complete ancillary package, which includes the following instructional aids:

Instructor's Solutions Manual
Includes answers to all discussion questions, answers to all multiple choice questions, and solutions to all problems.

Student Solutions Manual for Selected Problems
Includes answers to selected odd discussion questions, answers to odd multiple choice questions, and solutions to selected odd problems (approximately 10%) not already solved in the book.

Transparencies
Includes approximately 100 full-color transparencies illustrating key concepts.

Test Bank
Includes 1000 multiple choice and short-answer questions.

Electronic Testing (EXP-Test, MathWriter)
All test items are available in electronic format for IBM PCs and Macintosh computers.

BCX-Physics Software
An electronic study guide for the Macintosh and IBM PC, consisting of multiple choice and fill-in-the-blank questions with hints, selected from the multiple choice problems and numerical problems in the text.

Acknowledgments

Over the several years during which this work has developed, there have been many people who have been kind enough to share their thoughts about the book with me. Accordingly, I take this opportunity to express my appreciation to:

Reviewers

William Achor
Western Maryland College

Robert Boughton
Bowling Green State University

Michael Browne
University of Idaho

Anthony Buffa
California Polytechnic State University, San Luis Obispo

Gary Buckwalter
Daytona Beach Community College

Lawrence B. Coleman
University of California, Davis

Miles Dresser
Washington State University

David Ernst
Vanderbilt University

Roger Freedman
University of California, Santa Barbara

Sherman Frye
North Virginia Community College

Robert Hallock
University of Massachusetts, Amherst

Fred Inman
Mankato State University

Gordon Jones
College of Charleston

Sanford Kern
Colorado State University

James Kettler
Ohio University Eastern Campus

Paul Lee
Louisiana State University

David Markowitz
University of Connecticut

Robert March
University of Wisconsin

Reviewers (continued)

George Matous
Indiana University of Pennsylvania

Marvin Morris
San Jose State University

Melvin Oakes
University of Texas, Austin

Kurt Reibel
Ohio State University

Kandula Sastry
University of Massachusetts, Amherst

Jerald Tunheim
Dakota State University

Lonnie Van Zandt
Purdue University

Richard Whitlock
University of North Carolina

Answer Checkers

Sol Friedman
University of California, Santa Cruz

Petr Gaidarev
Cornell University

Chris Keller
University of California, Berkeley

Stefan Koch
Cornell University

Leo Krzewina
University of Wisconsin

Wei Liu
Adelphi University

Scott Milster
University of Wisconsin

Sangwook Park
Purdue University

Betty Richardson
Cornell University

Gerardo A. Rodriguez
Cornell University

V. K. Saxena
Purdue University

Robin Shelton
University of Wisconsin

Bingxi Sun
Cornell University

William Tiernan
Trinity College

Ashok Tripathi
Cornell University

Deborah Wolkovitch
Cornell University

In particular, I thank Professor G. N. Rao of Adelphi University for all the delightful time we have spent talking about physics. Anyone wishing to exchange ideas is welcome to write to the author c/o Physics Department, Adelphi University, Garden City, NY 11530.

The team that worked so mightily to produce this book had as its production editor, Marjorie Sanders; permissions editor, May Clark; supplements editor, Audra Silverie; marketing manager, Connie Jirovsky; and advertising manager, Margaret Parks. The design and layout were brilliantly done with an uncompromising commitment to excellence by Kelly Shoemaker; the art was skillfully managed by Susan Haberkorn, and the photos artfully rendered by Larry Molmud. I thank them for bearing with me for these many months. Each can be justly proud; they have produced a beautiful book. I also thank Sue Ewing, whose clear thinking kept the project on course; Beth Wilbur, whose good counsel sustained the effort; Stuart Kenter, whose experience and patience was so valuable; and Lisa Moller, whose devotion, energy, and wisdom guided the book into being.

My appreciation is also extended to Amy Allison, Jennifer Michal, and Jamey Adam for their assistance in the preparation of every aspect of this work. Finally, I nod appreciatively to my good friend, Carolyn Eisen Hecht, for tracking down hundreds of photographs and reading thousands of pages and, of course, for going through all of this, one more time.

Eugene Hecht

Brief Contents

Contents

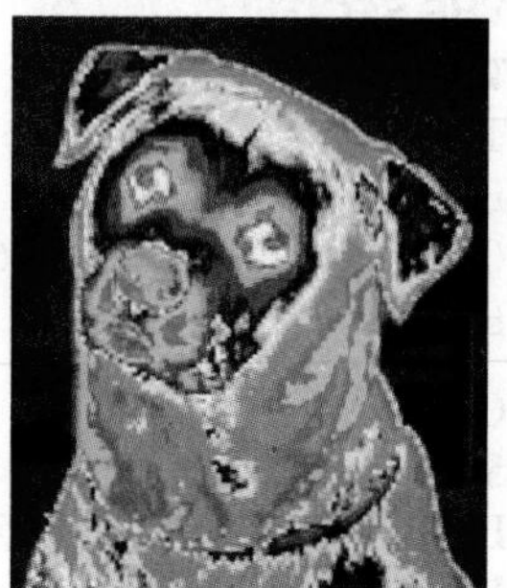

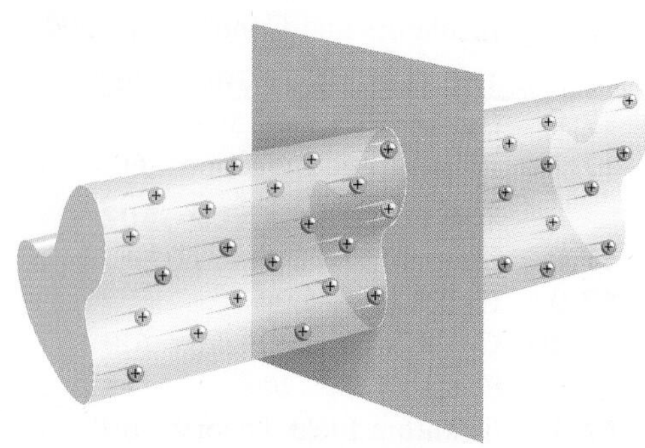

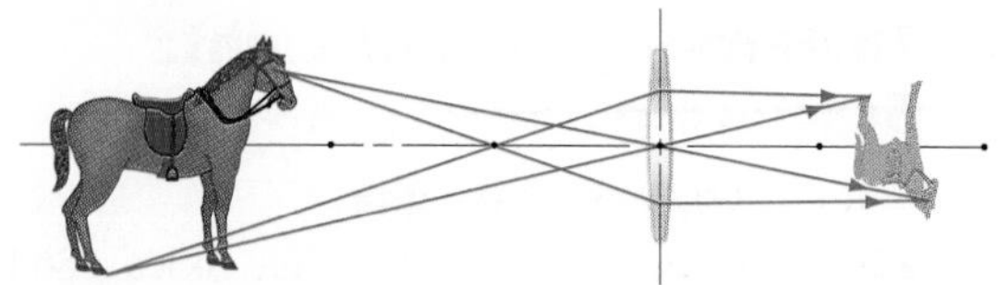

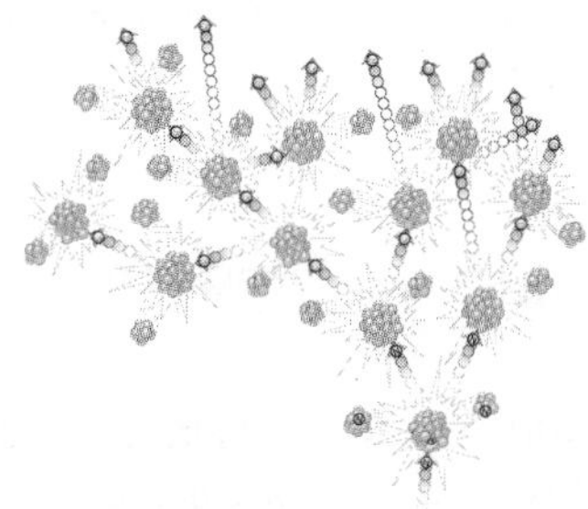

NOTE: The Table of Contents is color-coded as per p. x in the Preface.

1

An Introduction to Physics

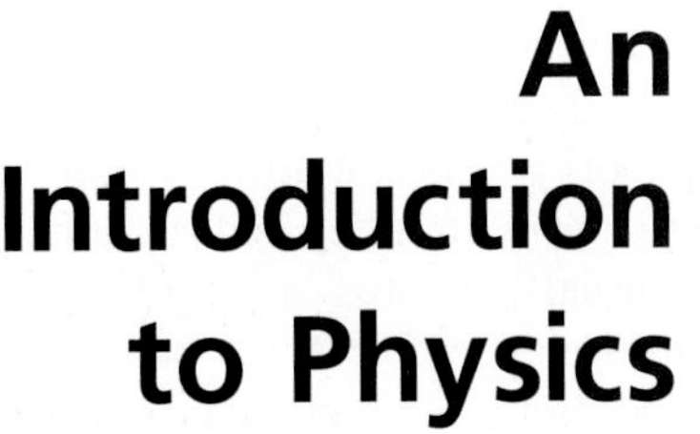

P*HYSICS IS THE STUDY of the physical Universe.* And though we all know what the words *physical Universe* refer to, it would take volumes to attempt to explain what they mean. The problem is that the basic underlying concepts—matter, space, and time—are difficult if not impossible to define conceptually, a shortcoming we share with the philosophers.

All the material things we are familiar with, from stars to toenails, are structures composed of ever smaller structures. Ultimately, the smallest indivisible subatomic grains of matter are the fundamental *real particles,* which compose all that exists. The primary property of matter is that it is

The question "What is matter?" is of the kind that is asked by metaphysicians, and answered in vast books of incredible obscurity.

BERTRAND RUSSELL (1959)
British philosopher

observable; it interacts in ways that result in changes we can detect. **Physics, then, is the study of matter, interaction, and change**.

Physicists try to describe phenomena in the simplest, most precise way, an approach that leads to a language of exact terminology. There are, for instance, concepts in physics that correspond to basic properties of matter (such as mass, charge, momentum, and energy); that describe the location of matter in space and time (such as displacement, velocity, and acceleration); or that pertain to the behavior of matter in bulk (such as heat, current, resistance, and pressure). All of these concepts have very specific scientific meanings.

In addition to conceptual or dictionary-like definitions, physicists rely on *operational definitions* to augment the working language of the discipline. These definitions are procedural, and include actual things: a standard mass, a thermometer, a stream of cesium atoms, and so on. Operational definitions demarcate a quantity by telling us how to measure it. For example, a *second* is the interval of time corresponding to exactly 9 192 631 770 oscillations of a cesium-133 atom. Not being able to conceptually define time in a completely satisfactory way doesn't prevent us from thinking about it, or from being able to measure it (or even from growing old).

The natural sciences, as distinct from the social sciences, have nature as their subject; that is, material phenomena. Physics, the premier natural science, deals with composition, structure, shape, creation, annihilation, motion, light and sound, atoms and molecules, fission and fusion, solids and liquids and gases—with the whole of things physical. Its purview is all that can be observed, all that can be measured.

1.1 Law and Theory

The aim of physics is to understand the natural events that we are party to, to understand the Universe—what it is, how it works, what it is doing, and maybe even why it exists. If that bold agenda is possible at all, it is only because natural phenomena occur in reproducible ways; there are rules and rhythms in the seeming chaos.

The doing of physics usually involves a record of phenomena: observation and the collection of **data**—information objectively perceived. An event is observed, either deliberately or not, and things are recorded, *measured* (How much? How long? How many? How big?). **Physics quantifies; it associates numbers with its concepts**. Within the body of observations, patterns are sought that reveal relationships among the data. A **law** is a description of a relationship in nature that manifests itself in recurring patterns of events. It can be a prescription for how things change, or it might be a statement of how things remain invariant to change.

How does our basic underlying knowledge, our world view, of space and time, of matter interacting with matter, account for the patterns of events? How can we understand what is observed? **Theory** is the explanation of phenomena in terms of more basic natural processes and relationships. To explain phenomena, we draw on intuition and imagination and guess at what is happening. We propose **hypotheses**, and leap beyond what we know to what might be. *A construct of definitions, hypotheses, and laws that explains some observed order in nature is the essence of theory.* A powerful theory allows us to deduce already known laws, as well as to predict new occurrences and relationships that, once tested and confirmed, may become new laws.

. . . by what words, yea and in what measures, I may avail to spread before your mind a bright light, whereby you may see to the heart of hidden things.

This terror, then, this darkness of the mind, must needs be scattered not by the rays of the sun and the gleaming shafts of day, but by the outer view and inner law of nature.

TITUS LUCRETIUS CARUS (99–55 B.C.)
De Rerum Natura

All theory is tentative. The formalisms of physics cannot be proven absolutely true, or even absolutely false. Our current world view may be wrong and all the proofs based on it equally wrong, our hypotheses may be in error, our laws may only be approximate. Newtonian Mechanics worked incredibly well for 200 years before it began to deviate even slightly from the most refined observations. Then Albert Einstein proved that the world view on which it was based is wrong; Newtonian Mechanics is a brilliant approximation of a more complete truth.

Physics must be continuously tested against nature. Our ideas must correspond in every detail to the observed workings of the Universe, but even that does not prove them true. The testing never ends. The process goes on—checked and rechecked—predictions, confirmations, discrepancies, refinements, new visions, new laws, new theories, evermore powerful understandings.

At this moment, there is no single significant theory in physics that can be said to be both finished and completely satisfactory. We have a magnificent understanding, but it's only the beginning. The game is wonderfully wide open and the real surprise, as Einstein suggested, is that we can do so much and know so little.

Insofar as the propositions of mathematics refer to reality they are not certain, and insofar as they are certain they do not refer to reality.

ALBERT EINSTEIN

1.2 The Modern Perspective

Physics has evolved to its present state as a result of about 2500 years of effort. Simple early theories were displaced by increasingly more effective later ones, and these, in turn, were subsumed by still more powerful, far-reaching, and complex contemporary treatments. A theory that has functioned well is assumed to possess some measure of truth, and so each new formulation must encompass the wisdom of its predecessors.

Classical Physics

The discipline, as it developed up until the 1920s, is known as Classical Physics. The classical period rests on three theoretical pillars (Fig. 1.1): Newtonian Mechanics, Electromagnetic Theory, and Thermodynamics. Within its boundaries are also several subdisciplines, such as acoustics, wave motion, condensed matter physics, and so on. The completion of Classical Physics was accomplished by Einstein. His Special Theory of Relativity (1905) reformulates our conception of space, time, and motion. His General Theory of Relativity (1915) considers gravity in terms of curved space-time. It subsumes the Special Theory, and contains and transcends Newton's Theory of Gravitation. The General Theory describes things on a grand scale, allowing us to begin to comprehend such cosmic phenomena as gravitational lenses, black holes, pulsars, and the Big Bang creation of the Universe.

Classical Physics produced a wonderful but limited picture of the physical world. Indeed, humanity was able to set foot on the Moon in 1969 thanks largely to Newton's theory. Except for the electronics, which would never have existed without our modern knowledge, the journey to the Moon was a classical trip. Classical Physics represents the basic conceptual material that should be understood if one is to interact effectively with the physical environment on an everyday macroscopic level.

Philosophically, Classical Physics is *deterministic*—it maintains that things can be known with certainty, that immutable law constrains nature to unfold in a totally predictable way. The cosmos is thus viewed as a great machine

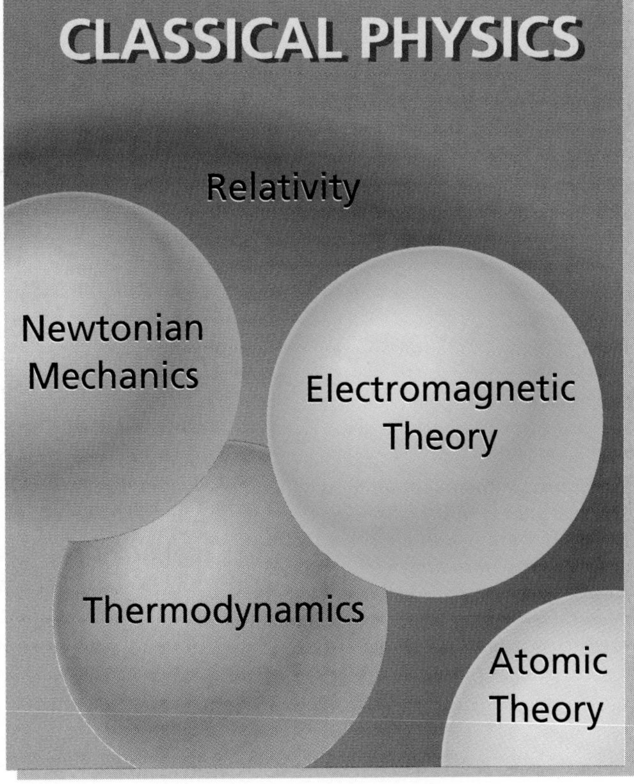

Figure 1.1 Classical Physics encompasses Newtonian Mechanics (the study of motion and gravity), Electromagnetic Theory (the synthesis of electricity and magnetism), Thermodynamics (the study of thermal energy), and Relativity. It corresponds to physics prior to the 1920s.

Light is, in short, the most refined form of matter.

LOUIS DE BROGLIE (1892–1987)
French physicist

that grinds along according to an intricate internal order. Once that order is known, we can, in principle at least, compute both the future and the past.

Physicists were supremely confident throughout the last decades of the 1800s, an era dominated by breakthroughs in early experimental atomic physics when everyone expected to successfully apply established theory to the atom. Surprisingly, as time went by, the atomic domain seemed to become more and more *inaccessible* to classical analysis.

Modern Physics

The reality of the atom was finally established only in the beginning of the twentieth century, in large measure due to the work of Einstein. Moreover, in direct conflict with classical theory, Einstein showed that light was emitted and absorbed in minute blasts, or *quanta.* These particles of light, the most tenuous form of matter, are called *photons.* Today, the material Universe is considered to be composed of myriads of identical clones of a handful of elementary particles—the *quarks, leptons,* and photons. All of us are basically organized clouds of quarks surrounded by leptons (electrons), absorbing and emitting photons.

Contemporary physics has extended its domain from the subnuclear to the entire Universe. Though it seems presumptuous, Genesis has become a branch of Modern Physics. Particle accelerators can replicate conditions that existed in the first moments of creation. All of this, and more, are the fruits of Quantum Mechanics. We know now that physical processes, although they appear to be continuous on a macroscopic level, are fundamentally discrete. The entities of the microworld exist and move in ways that are distinctly restricted. The continuum of motion is an illusion. *Change occurs in jumps;* the landscape is quantized.

Philosophically, it can be argued that Quantum Mechanics embraces determinism, but it certainly is not the all-knowing determinism of Classical Physics. The Universe may be a great machine, but its tiniest parts are mysteriously illusive. We cannot possibly know everything we would like to know. There is a soft edge to certainty, and the future has yet to be written. Despite its baffling text and strange version of reality, Quantum Mechanics has never once failed an experimental test. It is extremely reliable, though not transparently comprehendible. It is likely true that "no one understands Quantum Mechanics" (Sect. 31.9), although it is equally true that in some wonderful way Quantum Mechanics understands the Universe.

The Four Forces

Matter, whatever it is, interacts. The Universe is laced with a web of interactions that has evolved since the moment of its creation. Contemporary theoretical physics maintains that in the world as it exists today, there are four distinct basic interactions, or forces: gravitational, electromagnetic, strong, and weak. These influence the entire range of observable occurrences (Table 1.1).

To be sure, we say that an area of space is free of matter; we call it empty, if there is nothing present except a gravitational field. However, this is not found in reality, because even far out in the universe there is starlight, and that is matter.

ERWIN SCHRÖDINGER (1887–1961)
Austrian physicist

When a change in the state of something occurs, we assume a cause and logically presume that the change is effectuated by a force. When an apple falls, or a grasshopper leaps, or a supernova explodes, or a neutron decays, these very different events unfold because of the involvement of the Four Forces. Thus, we define **force** broadly as *the agent of change.*

The **gravitational force** keeps you, the atmosphere, and the seas fixed to the surface of the planet. Though gravity is the weakest of all the interactions, it is also the least selective—it acts between all particles. Because its range is unlimited and because it is only attractive, gravity rules the cosmos on a grand scale. It holds the Earth in orbit around the Sun, keeps the Sun locked within our galaxy of a hundred thousand million stars, and reaches all across the thousands of millions of galaxies that constitute the Universe. If it were much stronger than it is, it would quickly halt the present expansion of the Universe and send all the galaxies collapsing back down into oblivion (it may even be strong enough, as is, to inevitably do that anyway).

This galaxy is one of perhaps one hundred thousand million in the Universe, and it resembles our own Milky Way galaxy. Our galaxy is an island of about one hundred thousand million stars, as well as dust and gas, all pulled together by gravity. The Sun is but one rather ordinary star, around which orbit nine tiny planets, also held in place by the gravitational force.

The **electromagnetic force** binds together the smaller things, such as atoms, molecules, trees, buildings, and you. As with the gravitational force, we can perceive its consequences directly in the macroworld (just hold two magnets near one another). The electromagnetic force acting on a microscopic level is responsible for a variety of seemingly different macroscopic forces. It produces contact forces, as for example, between a fist and a punching bag, or a hamburger and teeth. It generates friction and drag, produces adhesion and cohesion, and is responsible for the elastic force. The range of the electromagnetic force is also unlimited, but it can be either attractive or repulsive, and that fact inhibits its influence over great distances. Electromagnetism governs chemistry and biology, rules life and death, and keeps the Earth and everything on it from being crushed by gravity.

The **strong force** binds quarks together to form neutrons and protons, and binds these together to form the nuclei of atoms (Fig. 1.2). It's an extremely powerful, exceedingly short-range force whose influence extends about as far as the diameter of a proton. That tiny range accounts for its not obtruding directly into our normal experience, and, as a result, not being discovered until this century. Without it, familiar matter, from planets to puppies, would all disintegrate into a fine quark dust.

The **weak force** is a million times fainter than the strong force and a hundred times shorter in range. Among other things, it transforms one flavor or type of quark into another. Thus, it can change protons into neutrons and is thereby responsible for the slow decay of certain radioactive atoms, such as uranium. A star like the Sun derives its energy from the thermonuclear furnace at its core. There, it "burns" hydrogen into helium, a process that relies on the gradual transformation of protons via the weak force. Hence, no weak force, no sunshine, no life on Earth.

Most physicists believe that the Four Forces are quantized. They do not act as continuous streamers, or invisible tentacles extending from one interacting body

TABLE 1.1 The Four Forces

Force	Couples with	Strength*	Range
Strong	Quarks and particles composed of them	10^4	$\approx 10^{-15}$ meter
Electromagnetic	Electrically charged particles	10^2	Unlimited
Weak	Most particles	10^{-2}	$\approx 10^{-17}$ meter
Gravitational	All particles	10^{-34}	Unlimited

*Strengths listed are the forces (in newtons) between two protons separated by a distance equal to their diameter ($\approx$2 femtometers). See Appendix A-2 for scientific notation.

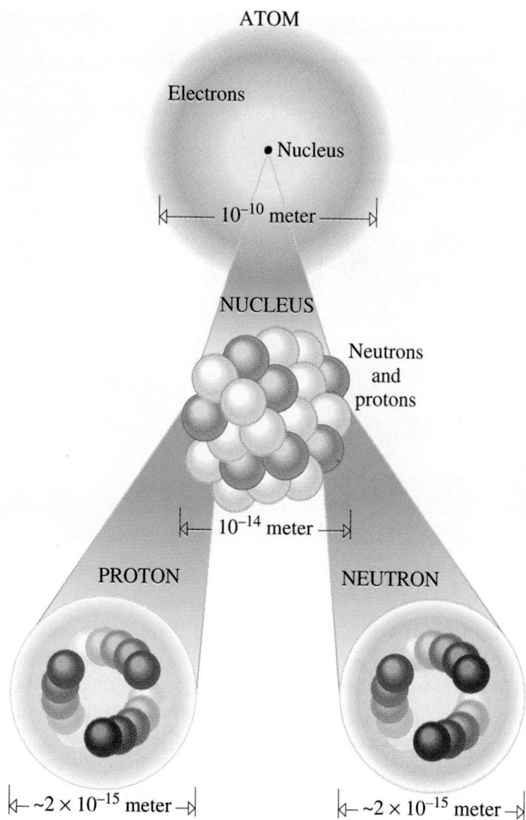

Figure 1.2 An atom is composed of a tiny nucleus surrounded by a cloud of fast-moving electrons. The nucleus is a minute cluster of neutrons and protons. Each neutron and proton consists of three quarks.

to the other. Instead, forces are caused by the back-and-forth exchange of a special class of force-mediating particles. These *virtual particles* only exist as transmitters of force and cannot be otherwise observed. This line of thinking has been quite successful, though it remains tantalizingly incomplete and philosophically problematic.

For the moment, the greatest impasse experienced by contemporary physics involves the inability to bring into harmony two fundamental theories: General Relativity (primarily its vision of gravity) and Quantum Mechanics. Relativity is an overview grounded in the analysis of space and time. All theory should be relativistically correct. Quantum Mechanics is an overview grounded in the analysis of energy and momentum, in the way change occurs. All theory should be quantum-mechanically correct. As yet, there is no satisfactory blending of the two. There is no adequate quantum theory of gravity. If researchers succeed, we will be well on the road to what is brashly called the Theory of Everything (Sect. 33.7).

Symmetry

Physics in the last quarter of this century has come to a remarkable revelation: on a fundamental level, nature possesses a range of symmetries that are associated with the most basic laws of physics. By understanding these symmetries, we can produce theories that presumably have the correct internal logical structure. Physicists now believe that for a basic theoretical construct to be right, it must have a subtle mathematical form called *gauge symmetry*. Such an incredible statement could not have been made a hundred years ago, nor could scientists have uttered it with much assurance even three or four decades back. Our confidence derives from an amazing series of discoveries that culminated in 1983.

Gauge symmetry was introduced as an abstract mathematical idea in the 1950s. By the mid-1960s, theoreticians were attempting to use it to produce a theory of the weak force. Gradually, the work began to suggest that the weak and electromagnetic interactions were actually two aspects of the same phenomenon. The treatment predicted the existence of three extraordinarily massive new particles. Although experiments in the 1970s confirmed a series of other predictions based on the theory, no laboratory in the world had equipment powerful enough to create the trio of hypothetical particles. When that feat was finally accomplished with a new generation of colliding-beam machines in 1983, the three particles created were found to have exactly the charges and almost exactly the masses predicted. Gauge symmetry was firmly established. We discuss symmetries and trace the development of these profound ideas throughout this text.

Figure 1.3 A uniform disc mounted on a central axle displays rotational symmetry.

Noether's Principle

There are many kinds of symmetry, the simplest of which is geometrical. Imagine a perfectly uniform disc mounted on a central axle (Fig. 1.3). Suppose you close your eyes and someone rotates the disc. When you open your eyes, everything is as it was before—you can't tell that anything has occurred. *When a change in a physical*

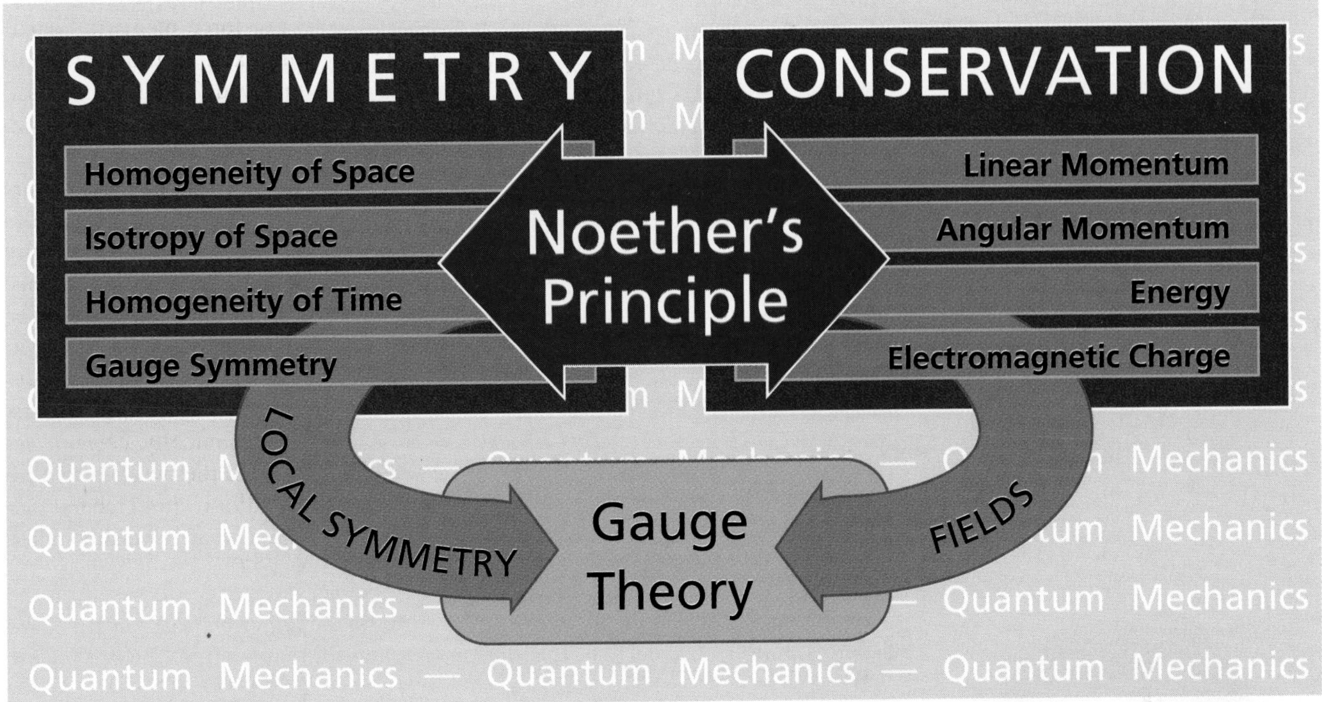

Figure 1.4 A summary of a number of interrelated ideas that we develop throughout this text. Noether's Principle relates symmetry and conservation.

system leaves some aspect of the system unchanged, the system possesses a corresponding symmetry. The existence of a symmetry means that a feature of the system is changeless, or *invariant.* In this instance, we say that the disc possesses rotational symmetry and that the system is invariant under rotation about the central axis. The amount of rotation can be varied smoothly, and the symmetry is said to be *continuous.*

The letter S is invariant under a rotation of 180°, just as a snowflake is invariant under a rotation of 60° (see illustration). In contrast, the letter A is symmetric under a mirror reflection along a vertical centerline. That kind of symmetry is *discontinuous;* that is, it jumps—all or nothing. These are all geometrical symmetries that we can appreciate by just looking at them, but, on a deeper level, any invariance suggests a symmetry.

In nature, there are important invariances associated with more subtle symmetries. The first person to grasp the physical implications of this was Amalie Noether. Among the central conceptual pillars of Classical Physics are three conservation laws, powerful insights that took hundreds of years to develop: Conservation of Linear Momentum, Conservation of Angular Momentum, and Conservation of Energy. Each maintains that a specific quantity is constant at every moment as an isolated system undergoes change. **Noether's Principle** states that *for every continuous symmetry there is a corresponding conservation law and vice versa.* In due course, we will learn how each of the great conservation laws turns out to be a direct consequence of a specific symmetry of space and time (Fig. 1.4). Noether's Principle is more general yet; it also applies to subtler abstract symmetries, such as gauge symmetry. In Chapter 18, we see how this principle relates to the conservation of electromagnetic charge.

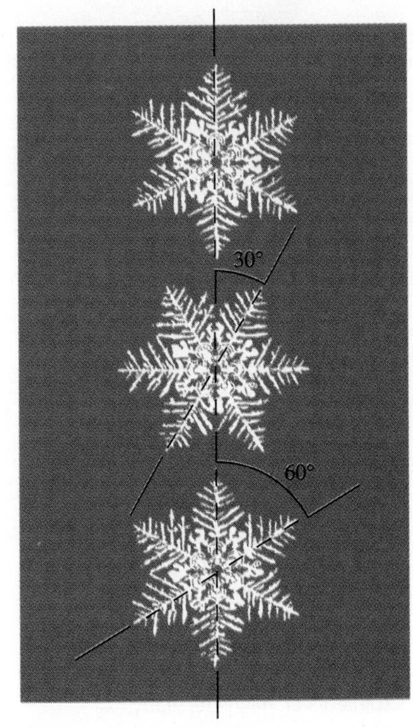

60° snowflake symmetry.

Now, suppose we paint the disc in Fig. 1.3 half red and half green (Fig. 1.5). If the viewers happen to be color-blind, nothing is altered, but if they are not, the

Figure 1.5 A uniform disc whose symmetry was broken by being painted.

symmetry is radically reduced—it's *broken.* In nature, symmetries are often *spontaneously broken.* Imagine a perfectly symmetrical ballpoint pen balanced on its tip. We might suppose that if there were no vibrations, winds, or other external influences the pen would remain upright indefinitely. But that's not likely. We know from Quantum Mechanics that things on an atomic level are always changing; the symmetry will spontaneously break, and the pen will fall.

In an analogous way, many physicists believe that the wildly energetic, incredibly dense infant Universe was a place of great symmetry and undifferentiated simplicity. The early Universe was ruled by a single superforce that ultimately broke into the Four Forces as the Universe cooled, calmed, and shattered into complexity (Fig. 33.23).

MEASUREMENT

Physics is founded on experimentation: we ground the science directly to nature through observations that entail the measurement of physical quantities. The basic notions of length, volume, weight, and time were quantified in antiquity and simply carried over into physics. Today, the scientific community follows the *Système International* (SI), a program in which the majority of measurements are based on the following units: for length, the *meter* (m); for mass, the *kilogram* (kg); for time, the *second* (s); for electric current, the *ampere* (A); and for temperature, the *kelvin* (K).

1.3 Length

Amalie (Emmy) Noether (1882–1935) was an outstanding mathematician who did most of her work in abstract algebra. After a long struggle she won the right as a woman to lecture, without pay, at Göttingen University in Germany. It was then, in 1918, that she presented the results of an analysis dealing with symmetry that became a guiding principle for contemporary physics. Noether taught at Göttingen until 1933 when she came to the United States after the Nazis learned that she was Jewish and expelled her from Germany.

Length is a distance or extent in space (whatever distance is, or for that matter, whatever space is). Given the limitations of language, it's easier to measure length precisely than to talk about it precisely. Stick out your arm and announce that the distance from elbow to farthest outstretched fingertip is a length hence to be known as a *cubit.* That's the way it was done over 4000 years ago in Egypt and Mesopotamia. *An agreed-upon measure by which we reckon any physical quantity is called a* **unit**. The Great Pyramid was built to cubit specifications, as was Noah's Ark. Of course, you can't build a large structure if everyone's forearms are different. An advanced society must evolve *an unchanging embodiment of each unit to serve as a primary reference, or* **standard**. The black granite master cubit was just such a standard against which cubit sticks in Egypt were regularly checked.

For centuries, primitive body measures were used throughout Europe: the width of a man's thumb is an inch; the length of a sandal is a foot. The rise of modern science painfully exposed the lack of a universally standardized system. In 1790, Thomas Jefferson proposed a decimal scheme for weights and measures, but the U.S. Congress would have nothing to do with it. At the same time in France, the climate of the Revolution was right for innovation. The French adopted a decimal approach for their *Metric System,* dividing all quantities into 10, 100, 1000, and so on, equal parts. The French decided on a new unit of length, the **meter** (or *metre,* from the Greek *metron,* a measure), abbreviated m. It was to be one ten-millionth of the distance from the North Pole to the Equator along a meridian line passing through Paris. However patriotic, errors in the difficult measurement made the whole exercise quite arbitrary. They could have just as well stayed home and used the length of Napoleon's sword instead. Be that as it may, the distance between two lines inscribed in a bar of platinum-iridium alloy became—in 1889—the world's standard meter (Fig. 1.6).

The **Metric System**, which is the basis of the SI, gradually spread across Europe, although it was shunned by the British for over a century. The United States, in happy distant oblivion, continued to use the dreadful British system and is still trying to make the transition. The system used in the United States is called *U.S. Customary units.* It is based on length, weight, and time expressed in feet, pounds, and seconds, respectively (Table 1.2).

Because a billion (1 000 000 000) in the United States is a thousand times smaller than a European billion (1 000 000 000 000), the common names for numbers beyond a million are rarely used in scientific work. Instead, prefixes (Table 1.3) are added to the units. Greek prefixes like kilo, mega, and giga are multiples; the Latin ones like centi, milli, and micro are subdivisions (see Appendix A-2 for a review of scientific notation).

A related approach that is used, especially where things are on a small scale, is the cgs (centimeter, gram, second) system. Here, a *centimeter* (cm) is a one-hundredth part of a meter (1 cm = 0.01 m) and 1 inch (in.) is defined to be exactly 2.54 cm. The centimeter, gram, and second are SI units, but all the others in the cgs system (the units for force, pressure, and density, for instance) are not.

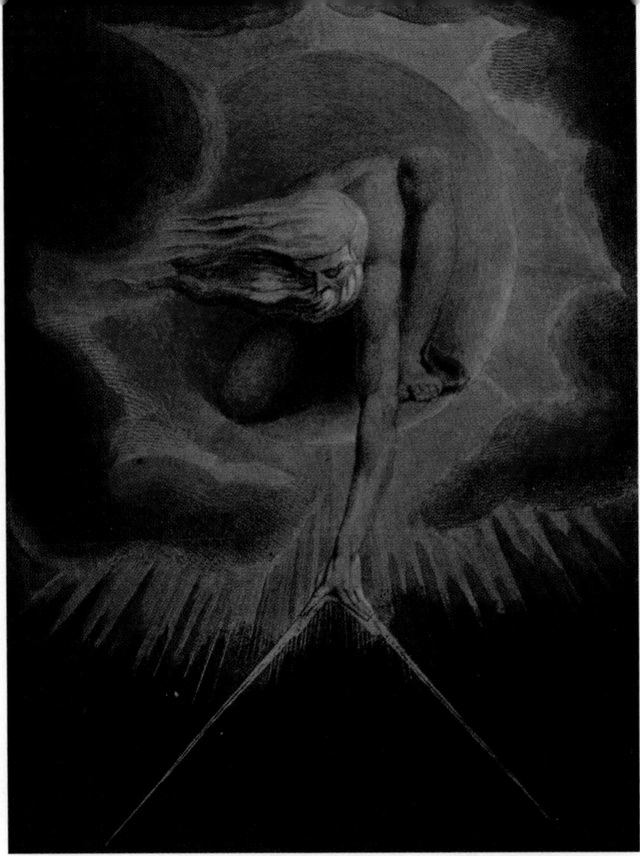

God taking the measure of the Universe, painted by the poet and artist William Blake (1757–1827).

Example 1.1 How many centimeters are there in 25.00 meters?

Solution: [Given: a length of 25.00 m. Find: the corresponding number of centimeters.] The most systematic way to do all unit conversions is to form a ratio equal to 1, with the unit you want on top and the one you want to replace on the bottom. Since 1 m is exactly 100 cm,

$$1 = \frac{100.0 \text{ cm}}{1.000 \text{ m}}$$

Then multiply the original number by the unit ratio

$$25.00 \text{ m} \times \frac{100.0 \text{ cm}}{1.000 \text{ m}} = \boxed{2500 \text{ cm}}$$

▶ **Quick Check:** There are 100 cm per 1 m, hence 25 m × 100 cm/m = 2500 cm.

Twenty-nine duplicates of the International Prototype Meter were made, and the major industrial nations still preserve their copies (the United States maintains Number 27). After the Nazis conquered France, it became obvious that the world's reliance on a single vulnerable set of standards was foolhardy—far better to redefine the meter in terms of some measurement that could be performed in any well-equipped laboratory. In 1960, the *Système International* was accepted, and the meter became 1 650 763.73 wavelengths of the orange-red light from a krypton-86 lamp.

In the decades that followed, with the advent of stable lasers, the speed of light was measured with extraordinary precision. In 1983, the decision was made to use it as the basis of a redefined meter that would be better than 10 times more precise.

In the scientific world, the kilogram is the basic unit of mass.

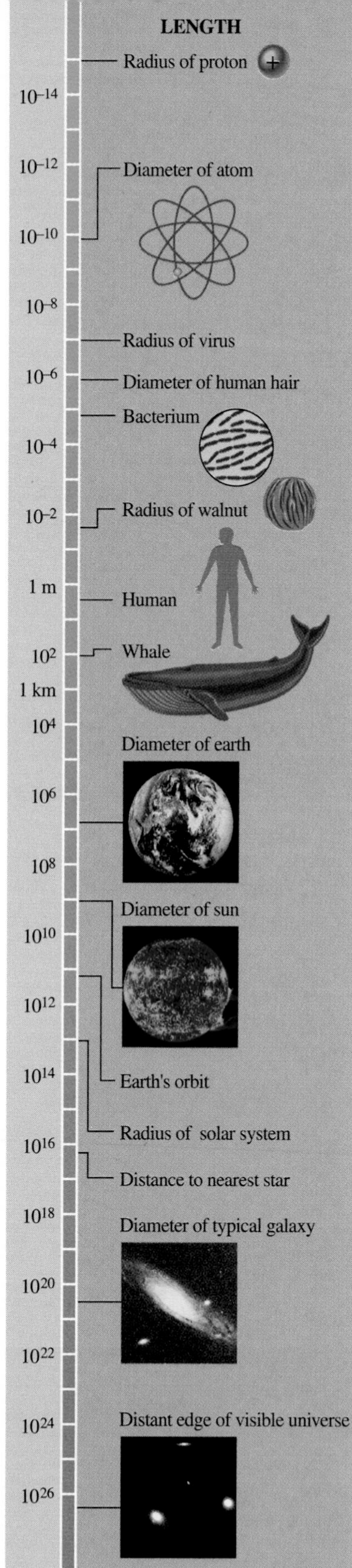

Figure 1.6 The sizes (in meters) of objects from protons to the Universe itself.

The speed of light in vacuum (c) is now defined to be exactly 299 792 458 meters per second, which is the best measured value. The meter is, then, *the distance light travels in vacuum during a time of exactly 1/299 792 458 second.* This new standard will never need to be revised. Improvements in measuring time will simply improve the measurement of the meter.

TABLE 1.2 Length and volume equivalents

1 light-year (5.88×10^{12} mi)	$9.460\,55 \times 10^{15}$ m
1 mile (5280 ft)	1.609 344 km
1 yard (yd; 3 ft)	0.914 4 m
1 foot	0.304 8 m
1 inch	2.54 cm
1 meter	3.281 ft
1 meter	39.37 in.
1 meter	1.094 yd
1 gallon U.S. (4 quarts)	$3.785\,412 \times 10^{-3}$ m^3
1 quart (2 pints)	$9.463\,530 \times 10^{-4}$ m^3
1 ounce	$2.957\,353 \times 10^{-5}$ m^3
1 tablespoon (1/2 oz)	$1.478\,676 \times 10^{-5}$ m^3
1 cubic foot	28 320 cm^3
1 cubic foot	0.028 32 m^3
1 cubic meter	1.308 yd^3
1 cubic meter	61 023 in.^{3}

TABLE 1.3 Common unit prefixes

Power of ten	Prefix	Pronunciation	Symbol	Example
10^{18}	exa-	ex'a	E	exajoule, EJ
10^{15}	peta-	pe'ta	P	petasecond, Ps
10^{12}	tera-	ter'a	T	terahertz, THz
10^{9}	giga-	jig'a[a]	G	gigavolts, GV
10^{6}	mega-	meg'a	M	megawatt, MW
10^{3}	kilo-	kil'oe	k	kilogram, kg
10^{2}	hecto-	hek'toe[b]	h	
10	deka-	dek'a[b]	da	
10^{-1}	deci-	des'i	d	decibel, dB
10^{-2}	centi-	sen'ti	c	centimeter, cm
10^{-3}	milli-	mil'i	m	millimeter, mm
10^{-6}	micro-	my'kroe	μ	microgram, μg
10^{-9}	nano-	nan'oe	n	nanometer, nm
10^{-12}	pico-	pee'koe	p	picofarad, pF
10^{-15}	femto-	fem'toe	f	femtometer, fm
10^{-18}	atto-	at'toe	a	attocoulomb, aC

[a]Although the given pronunciation is supposed to be standard, many physicists use gig'a instead.
[b]The prefixes hecto- and deka- are avoided in physics.

Example 1.2 How many cubic centimeters are in a cubic meter?

Solution: [Given: 1 cubic meter. Find: the number of cubic centimeters.] We know that 1 m = 100 cm exactly, hence a cube 1 m by 1 m by 1 m is 100 cm by 100 cm by 100 cm. From Fig. 1.7 and Appendix B, the volume (V) is given by the length times width times height:

$$V = (100 \text{ cm})(100 \text{ cm})(100 \text{ cm}) = \boxed{1\,000\,000 \text{ cm}^3}$$

▶ **Quick Check:** Imagine a big cube made up of 1 cm^3 blocks. There will be 100 layers, each 1 cm tall and 100 cm long by 100 cm wide. Each layer contains 100 × 100 or 10 000 little cubes, and there are 100 such layers, yielding 100 × 10 000, which equals 10^6 cm^3.

1.4 Mass and Weight

When the French set out to construct the Metric System, they followed the ancient Babylonian example of defining weight in terms of a fixed volume of water. The gram (g), originally a unit of weight, was specified as "the absolute weight of a volume of pure water equal to a cube of one-hundredth part of a meter [that is, a cube one centimeter on a side]."

Historically, weight was the downward force acting on an object at the surface of the Earth, and it was generally assumed to be constant. The fact that the weight of an object varies with location was discovered inadvertently in 1671 by J. Richer, who found that his pendulum clock, when brought from Europe to French Guiana, ran slow, losing $2\frac{1}{2}$ minutes each day. Newton promptly explained this surprising effect by clearly distinguishing between *weight* and *mass,* two ideas that had been around for centuries.

Mass is a property of a sample of matter immersed in the Universe-at-large (p. 92). It is apparently independent of the local distribution of matter and is constant everywhere on the Earth. Indeed, the mass of an object in a laboratory will be measured to be constant no matter where in the Universe the laboratory is. Masses interact gravitationally, and this causes the force we call **weight**. What gives rise to mass is still an unsettled issue.

The weight of an object at the surface of our planet is due to the gravitational interaction between that object and the Earth. Speaking practically, weight arises out of a local gravitational interaction between the sample and any massive object in its immediate vicinity, such as a planet, moon, or asteroid. At a far distance from all such objects, the weight of a sample effectively vanishes. Mass never vanishes. Richer's pendulum bob, its mass constant, was moved to a place where gravity was reduced, and consequently its weight was reduced. Thus, the pendulum was pulled down on each swing with less force, and it swung more slowly.

Ultimately, physicists properly ceased considering weight to be a fundamental property of matter, and in 1889,

Figure 1.7 The areas and volumes of several geometric figures. See Fig. B4 in Appendix B.

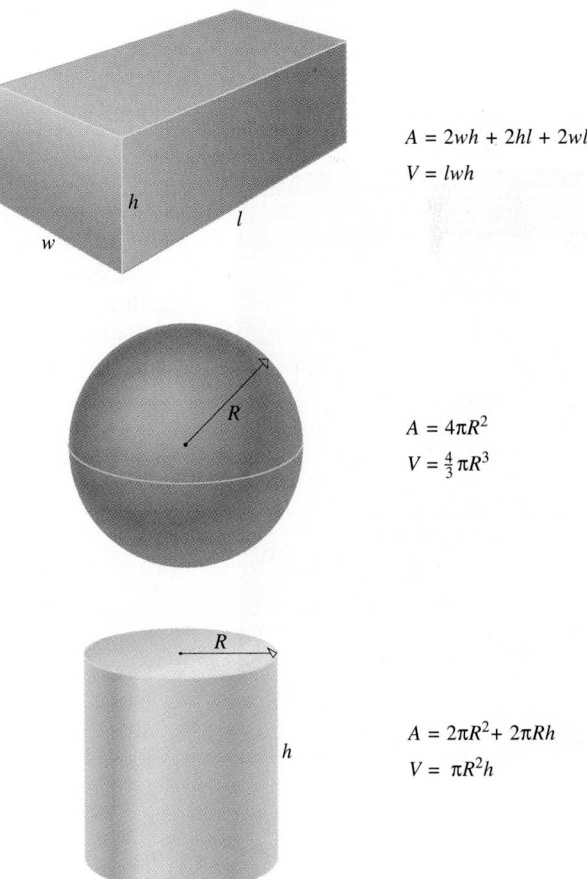

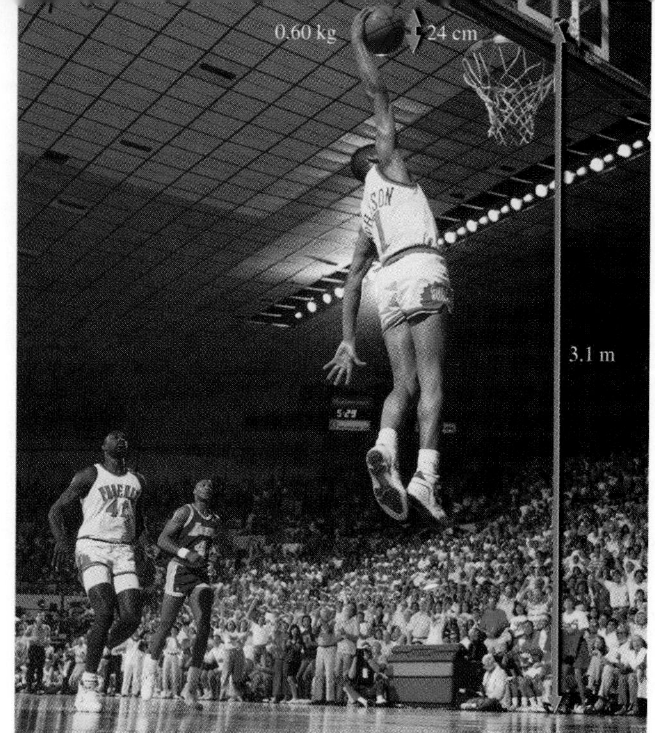

A basketball must weigh between 20 and 22 ounces. It must be no less than $29\frac{1}{2}$ in. and no more than 30 in. in circumference ($2\pi R$). The basket must be hung 10 ft above the floor. Check that the diameter of the ball is about 24 cm.

the **kilogram** (kg) was redefined as the unit of mass. The standard kilogram is a platinum-iridium cylinder watched over by the International Bureau of Weights and Measures. It's a little thing about the size of a whisky shot glass, 39 mm (1.5 in.) in both height and diameter. By definition, a gram is a one-thousandth part of a kilogram. Once the International Kilogram was agreed upon and copies made, a pan balance could then be used anywhere on the planet to compare the mass of an object with a set of standard masses. Since gravity is the same at each pan, if the balance is level the masses are equally independent of what they might weigh at that location.

Forty nearly identical masses were made in London in 1884. The standard for the United States is Kilogram Number 20. It and a clone, Kilogram Number 4, sit next to each other under sealed glass domes in a basement vault at the National Institute of Standards and Technology (NIST). Kilogram Number 20 is rarely handled, for fear that the slightest trace of metal may be worn away. Even so, it has traveled to Paris on four occasions in its lifetime of over one hundred years. The kilogram is the only remaining unit in the SI that is still based on an actual artifact; this brings special problems. Exposing the mass to air or touching it even with tongs can change it, but if it's to be used, it must be handled, however carefully. Although an atomic standard for mass is probably inevitable, the cylinder survives because it's easy to compare objects of that size to within an accuracy of several parts in a thousand million.

Because weight depends on mass, we explore the physics of converting from one to the other in Chapter 4. Whenever you buy anything by the gram, kilogram, ounce, grain, carat, or pound, that measure must be traceable directly back through a long list of calibrations to the primary standard (Table 1.4).

Mass is a fundamental property of matter that we can attempt to define via three seemingly different approaches: (1) it is that which participates in the gravitational interaction; (2) it is a measure of an object's resistance to a change in motion; and (3) it is a measure of the quantity of matter possessed by an object. So primary is the concept of mass that we return to it many times in this text (Fig. 1.8).

Kilogram Number 20 is the standard mass of the United States. It is a copy of the International Standard Kilogram housed at Sèvres, France.

TABLE 1.4 Weight-mass correspondence on Earth*

1 gram	0.035 273 40 ounce
1 kilogram	2.204 622 pounds
1 ounce	28.349 52 grams
1 pound	0.453 592 4 kilogram

*Mass and weight are very different features of matter. The values given here are the weights that would be measured for the corresponding masses at the surface of the Earth.

1.5 Time

The rhythms of life—the passing of successive days, the pounding of a heart—suggest a progression, a sequencing of occurrences. Time is a measure of the unfolding of events. Space is revealed by the juxtaposition of matter; time is revealed by the change in that juxtaposition. Whatever time *is,* it can be measured against uniform change. That's often done with some cyclic phenomenon like the rising of the Sun. More than 3000 years ago the Egyptians divided the day and night into 12 equal hours. Babylonian arithmetic had 60 as its number base, and there subsequently developed a tradition of subdividing things into 60 parts. It was probably sometime in the fourteenth century, after the arrival of the mechanical clock, that the hour (h) got divided into 60 minutes (min). The word minute was abbreviated from the medieval Latin for *first minūte part* (which is why the words minute and minūte are spelled the same). Later, when the so-called *second minūte part* could be measured, it was natural to take 60 seconds (s) to a minute.

As a young pre-med undergraduate in 1582, Galileo noticed that a swinging pendulum kept a steady beat and suggested that it might serve to measure time. By the early 1600s, astronomers were using pendulums 39.1 in. long (0.994 m), pushed every now and then by hand, to count out seconds. A tremendous improvement in accuracy was achieved in 1656 when Huygens introduced the pendulum clock, which had an error of only about 10 s a day. Minute hands, which had been rare, became the rule, and clocks were soon fitted with second hands for the first time. By the 1920s, the free pendulum clock was keeping time to better than 0.005 s per day.

The day, high noon to high noon, has been the benchmark of time for millennia, but the Earth's spin rate is slowing down, and the days grow longer. In 1967, the SI unit of time, the *second,* was defined as the interval required for 9 192 631 770 vibrations of the cesium-133 atom measured via an atomic beam clock. At present, such clocks are in error by about 1 second in 3 000 000 years. Their remarkable constancy is a manifestation of the precisely quantized energy states of the cesium atom.

So accurate is modern timekeeping, that the atomic clocks of the NIST at Boulder, Colorado, have been found, as expected from the General Theory of Relativity, to run 15.5 nanoseconds (ns) faster per day than those in Paris and Washington. Einstein showed that time depends on gravity: the city of Boulder is a mile above sea level; at that altitude, time runs a little faster.

Operationally, time is that which is measured by a clock. Conceptually, time is a measure of the rate at which change occurs. Facetiously, time is that which keeps everything from happening all at once (Fig. 1.9).

If we ever communicate with extraterrestrials, they surely won't use meters, kilograms, and seconds, but they will certainly measure length, mass, and time. We should have no trouble making the necessary conversions. Our physics, to the extent that it's correct, must be their physics. And that's what gauge symmetry is all about.

1.6 Significant Figures

Measurement is very different from counting, even though both associate numbers with notions. We can count the number of beans in a jar and know it exactly. But we cannot measure the height of the jar exactly. *There is no such thing as an exact measurement.* Practically, measurements are made to some desired precision that suits

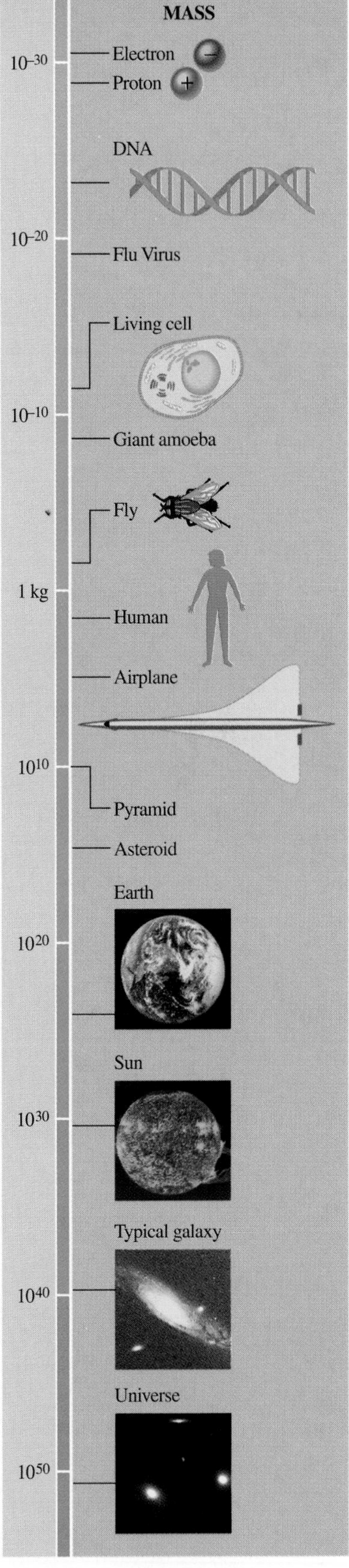

Figure 1.8 Masses (in kg) of objects from the electron to the Universe.

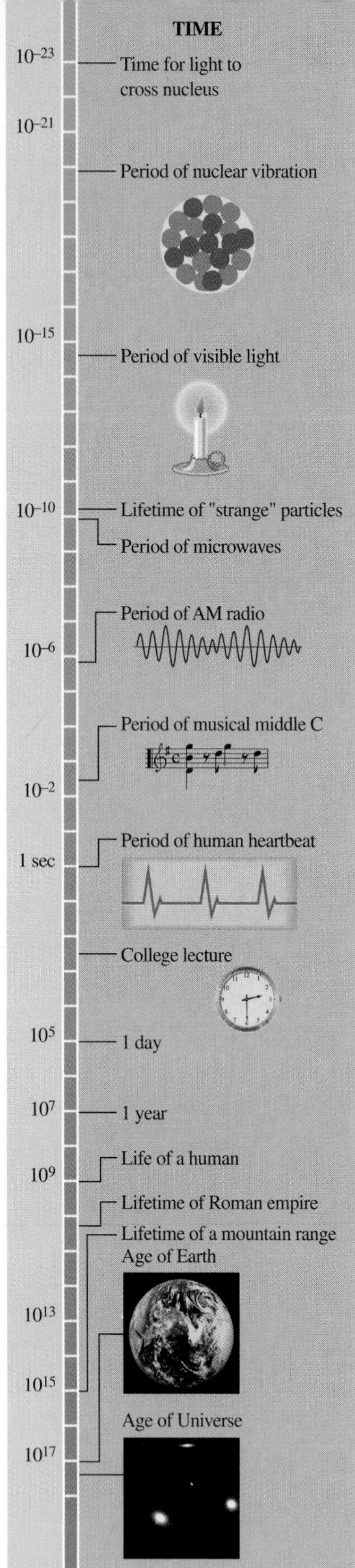

Figure 1.9 Intervals (in seconds) from the time it takes light to traverse a nucleus to the age of the Universe.

the experimenter's purposes (often determined by the limitations of the available instruments).

Figure 1.10 shows a rod whose length is being measured with a ruler. Assuming one end of the ruler and rod are lined up precisely, the object is between 4.1 and 4.2 cm long. Because there are no finer divisions than 1 mm on the scale, fractions of a millimeter will have to be estimated. Doing that, we conclude that the rod is 4.15 cm long. It's inappropriate to say that the length is exactly 4.15 cm. The length is closer to 4.15 cm than to either 4.14 cm or 4.16 cm, which is what we mean when we write 4.15, although it can be stated explicitly as 4.15 ± 0.01. The 4 is certain, as is the 1, but the 5 might be in error by as much as ± 1, and trying to arrive at any more figures would be meaningless. A digit in a number is said to be a **significant figure** when it is known with some reliability, like the 4, the 1, and even the 5. When a result is presented, such as 4.15 cm, it can be assumed that all the figures are significant and that the 5 is the least significant. Note that to specify the rod's length exactly would require an infinite number of significant figures.

To three significant figures, the length of the rod is 4.15 cm, or 41.5 mm, or 0.041 5 m, or 0.000 041 5 km. Each of these is equivalent, and each has three significant figures. Moving the decimal point and changing the unit prefix has no effect on the accuracy of the number. Hence, zeros to the left of the 4 only locate the decimal point and are not significant figures. Zeros to the right, however, are significant. If you use a micrometer and measure the rod to have a length of 4.150 cm, there are four significant figures. Needless to say, carelessness with these right-hand zeros can cause ambiguities. For example, the distance to the Sun is 146 million kilometers, but writing it as 146 000 000 km implies 9 significant figures, which is not the case. Only significant figures should be given and the decimal should be located using scientific notation. The distance to the Sun is then 146×10^9 m, or 146×10^6 km, or 14.6×10^7 km, and so forth.

When we conduct a careful measurement, a number of determinations are made and averaged. The deviations are examined, all possible errors analyzed, and a numerical uncertainty is associated with the measurement. Thus, it's found that the mass of an electron is $9.109\,389\,7 \times 10^{-31}$ kg, with an uncertainty of 0.59 *parts per million* (ppm). That latter phrase is translated: $(0.59/1\,000\,000) \times (9.109\,389\,7) = 0.000\,005\,4$. Hence, the mass is $(9.109\,389\,7 \pm 0.000\,005\,4) \times 10^{-31}$ kg.

Processing Significant Figures

Computations frequently involve measured quantities, and we need to establish how to deal with the multiplication, division, addition, and subtraction of significant figures. Consider the area of a rectangle with sides of 10.7**5** cm and 3.5**4** cm (to help keep track of things, the doubtful digits are boldfaced). The area (Appendix B) is the product of length and width, (10.7**5** cm)(3.5**4** cm) = 38.**055 0** cm^2. As a rule, we *keep only one doubtful digit, and round off the answer accordingly. If the first of the insignificant figures being dropped is 5 or more, raise the last significant figure being retained by 1; if it's less than 5, leave it unchanged.* Here 38.**055 0** cm^2 is rounded up to 38.1 cm^2. Rather than computing the number of significant figures in each case, we use this rule:

> The result of multiplication and/or division should be rounded off so that it has as many significant figures as the least precise quantity used in the calculation.

This rule is the equivalent of saying that making a calculation cannot improve on the precision of the numbers used. The rule is rough; it can occasionally reduce the precision. For example, it's possible, using the same instrument, to measure a flat plate to have sides of 0.**91** m, and 1.**51** m, with 2 and 3 significant figures, respectively. The area is (0.**91** m)(1.**51** m) = 1.**374**1 m². According to the rule, that's 1.4 m², even though the uncertainty is in the 7 and 1.37 m² is a better answer. Regardless of the rare exceptions, we will stick with the rule.

```
   10.75          1.51
    3.54          0.91
    4300           151
   5 375         1 359
  32 25         1.3741
  38.0550
```

Example 1.3 Someone is 2.00 yd tall. Using the fact that 1 in. is exactly 2.54 cm, how tall is the person in centimeters?

Solution: [Given: height = 2.00 yd, and 1.00 in. = 2.54 cm. Find: height in centimeters.] Note that the data, the height, is given to three significant figures. We have a relationship between inches and centimeters, but the height is in yards. If we had the height in inches, it would be easy to find it in centimeters. So, we compute it in inches. Since there are 3.00 feet per yard,

$$1 = \frac{3.00\text{ ft}}{1.00\text{ yd}}$$

The height, call it h, in feet, is

$$h = 2.00\text{ yd}\,\frac{3.00\text{ ft}}{1.00\text{ yd}} = 6.00\text{ ft}$$

Similarly, since there are 12.0 inches per foot,

$$1 = \frac{12.0\text{ in.}}{1.00\text{ ft}}$$

and $$h = 6.00\text{ ft}\,\frac{12.0\text{ in.}}{1.00\text{ ft}} = 72.0\text{ in.}$$

Inasmuch as there are 2.54 centimeters per inch,

$$1 = \frac{2.54\text{ cm}}{1.00\text{ in.}}$$

hence $$h = 72.0\text{ in.}\,\frac{2.54\text{ cm}}{1.00\text{ in.}} = 182.88\text{ cm}$$

and to three significant figures $\boxed{h = 183\text{ cm}}$.

▶ **Quick Check:** 1 yd ≈ 0.9 m; 2 yd ≈ 1.8 m ≈ 180 cm.

The significance of digits in addition and subtraction is determined in relation to the location of the decimal point. Suppose a rod of length 140 mm is added to one 2.0 m long, and we want the total length. It might be assumed that (0.140 m) + (2.0 m) = 2.140 m, but we know nothing about the digits in the second and third decimal places of the 2.0**-m rod and cannot know the sum accurate to three decimal places. It makes sense to round off the sum to the least number of decimal places of any number appearing in the summation: the length of the rod is 2.1 m and both the 2 and the 1 are significant figures. As a rule:

```
  0.140
 +2.0
  2.140
```

> The result of addition and or subtraction should be rounded off so that it has the same number of decimal places (to the right of the decimal point) as the quantity in the calculation having the least number of decimal places.

Thus, (275 s − 270 s) = 5 s and (120 kg − 40.0 kg) = 80 kg.

In time, we will work with several transcendental functions: sines, cosines, tangents, logarithms, and exponentials. All of these can be determined using a scientific calculator; for example, sin 35.1° = 0.575 005 252. Here, the number whose sine is being taken, the *argument* of the function, is 35.1°. The calculator, trying to please, provides

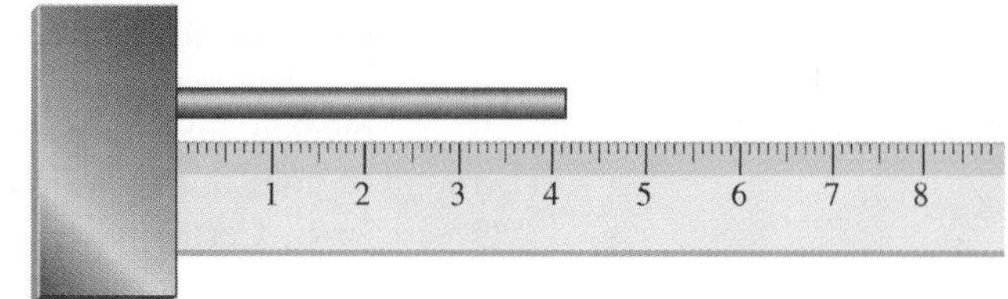

Figure 1.10 This rod measures somewhere between 4.1 cm and 4.2 cm, closest to 4.15 cm, but that last digit is not certain.

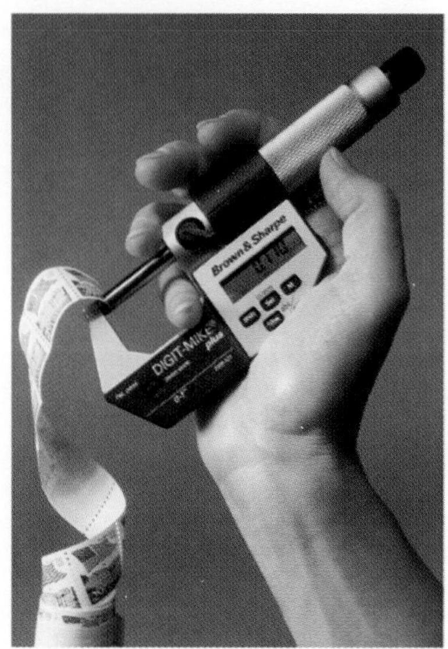
A fine micrometer can measure distances in divisions of 0.001 mm.

us with a full register of figures, most of which are meaningless. What it actually gives us is sin 35.100 000 00°. We shall follow the rule that:

> The values of transcendental functions have the same number of significant figures as their arguments.

Thus, sin 35.**1**° = 0.57**5**. The 35 in the argument is certain, but there is some doubt about the .1°, and so on the right we are certain about the 0.57 and have doubt about the .005. In agreement with the rule, sin 35.0° = 0.574 and sin 35.2° = 0.576.

Because rounding off intermediate numbers within a calculation can produce cumulative errors, *one or two insignificant figures should be carried through intermediate calculations and only the final answers should be properly rounded off.* It's appropriate to run through an entire analysis on a calculator and round off only the final answer.

THE LANGUAGE OF PHYSICS

Physics is done in almost every language on Earth, and yet it has a symbolic dialect of its own that is understandable to all its practitioners. Learning that vocabulary is the most basic task confronting the student. Being able to make the translation from everyday discourse to the language of physics and on to the symbolic statements of mathematics is part of the process of mastering the subject. As you move through this text, *commit to memory the definitions of physical concepts* like displacement, velocity, acceleration, and so on. Physics is progressively elaborated in terms of quantities that have previously been defined: you cannot understand acceleration without remembering velocity, and you cannot understand velocity without remembering displacement. Physics, perhaps more than any other discipline you have studied before, builds on definitions and basic principles.

Once we learn the scientific vocabulary, we can transpose the experiences of everyday life into the carefully defined terminology of physics. For example, How much space is occupied by a big round red balloon 10 meters across? This question translates into, What is the volume of a sphere 10 m in diameter? The scientific words *volume, sphere,* and *diameter* have exact meanings, and there is a specific relationship between them—we know how to calculate the volume of a sphere in terms of its diameter (Fig. 1.7). No such relationship exists for the ordinary words *space, round,* and *across.*

1.7 Equations

Newton was the last great physicist to do his work without routinely representing physical quantities symbolically and writing equations describing their interrelationships (see Appendix A-4). Today, that approach would be unthinkable. We assign each physical quantity a symbol; t for time, a for acceleration, V for volume, R for radius, and so forth. Of course, we try to pick symbols that will be easy to remember, though that's not always possible. Moreover, we have standardized mathematical signs that act as a conceptual shorthand (Table 1.5).

The sight of day and night, of months and the revolving years, of equinox and solstice has caused the invention of number and bestowed on us the notion of time.

PLATO
Timaeus

The reason for transforming verbal scientific statements into symbolic ones is twofold. First, they are concise. A sentence such as "The volume of a sphere is equal to four-thirds of the product of pi times the radius cubed" becomes $V = \frac{4}{3}\pi R^3$. Certainly you have to know the definitions of volume, radius, and pi for this equation to make any sense. Once the technique is mastered, though, a whole page of symbols can be read like a story. Second, there exists a mathematical symbolic logic called algebra that's immediately applicable to physics. If you know the volume of a sphere and want to find its radius, the rules of algebra tell us that $3V = 4\pi R^3$, $R^3 = 3V/4\pi$, and $R = \sqrt[3]{3V/4\pi}$. Try thinking that through in sentences! Using algebra, we can easily combine several physical equations to arrive at new relationships that would otherwise be extremely difficult to deduce.

TABLE 1.5 Mathematical symbols

Symbol	Meaning
$=$	equal to
$\neq$	not equal to
$\equiv$	defined as
$\pm$	plus or minus
$\mp$	minus or plus
$\parallel$	parallel
$\perp$	perpendicular
$\div$	divide by
$\rightarrow$	approaches
$\approx$	approximately equal to
$<$	less than
$>$	greater than
$\ll$	much less than
$\gg$	much greater than
$\leq$	less than or equal to
$\geq$	greater than or equal to
$\lvert z \rvert$	absolute value of z
$\propto$	proportional to
∞	infinity
Σ	sum of

1.8 Graphs

Complex relationships can be represented algebraically, but it's often hard to get a feeling for the interplay of the variables in an equation. That's a great virtue of graphs; we can plot an equation and literally see what happens to one quantity as another varies, and take it all in at a glance. Futhermore, physicists usually have collections of data, and by graphing them they can get a handle on the interdependencies of the measured quantities.

The simplest situation is when one variable (y) depends *directly* on another (x): $y = mx$, where m is a constant. The larger m is, the faster y changes with changes in x. For example, if you get a 10% discount ($m = 0.10$) on the price of any shirt (x) in the store, you save ($y = mx$) dollars. Figure 1.11 is a plot of y versus x, and m is the slope of the line (see Appendix A-4). Because the graph is a straight line, $y = mx$ is called a *linear* equation. There are many physical relationships, such as Ohm's Law and Hooke's Law, that are linear. Here, the line passes through zero so that at $x = 0$, $y = 0$, but that needn't be the case. Figure 1.12 shows a graph of the function $y = mx + b$, where m is again the slope and b is the point where the line now crosses the y-axis when $x = 0$.

Frequently, a quantity depends on the square of some other parameter. Mathematically, that situation appears as $y = kx^2$, where k is a constant. For example, the area (A) of a square whose sides are of length l is given by $A = l^2$ (where $k = 1$). Double the side length and the area of the square increases by a factor of four. A second degree equation of this kind is said to be *quadratic,* and the graph of y versus x is a *parabola* (Fig. 1.13). As we will see, a falling object descending along a straight line drops a distance (y) that is proportional to the time of fall (x) squared. A plot of distance versus time is parabolic. In a very different way, a ball thrown into the air follows an arcing path in space that's essentially parabolic (p. 77). Equations can describe relationships between physical parameters, but they can also represent actual flight paths through space, and both can be graphed.

Another common situation occurs when a quantity (y) depends *inversely* on some other quantity (x): $y = k/x$ or $yx = k$, where k is constant. The bigger x gets, the smaller y becomes, but yx is constant. For example, suppose you are designing a rectangular floor that must have a fixed area (k). The longer (x) the floor is made, the narrower (y) it must be, since $yx = k$. The graph of y versus x (Fig. 1.14) is a hyperbola that gradually approaches zero as x increases. At a constant temperature, a plot of pressure against volume for a variety of gases is hyperbolic.

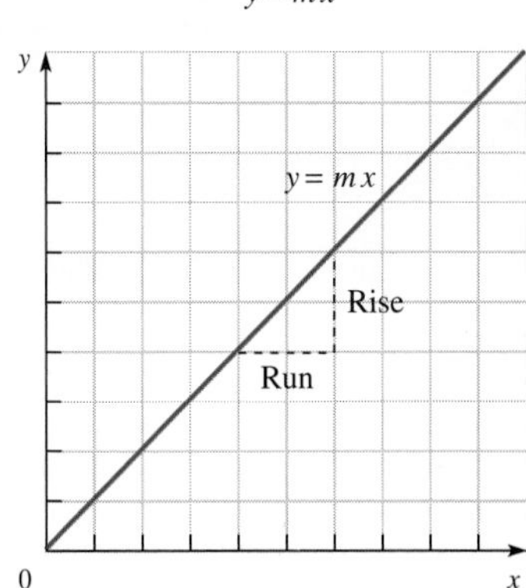

Figure 1.11 A curve is linear when it plots as a straight line, which means that the slope, the rise over the run, is constant and equal to m. This plot passes through the origin when $x = 0$, $y = 0$.

1.9 Approximations and Checks

There are about 10^{11} galaxies in the Universe. No one has ever counted them, but we have photographed the sky, counted the number of galaxies in a small region, and then estimated the total. Making approximations like that are part of the everyday doing of physics. We approximate how much material we need for an experiment, how much time it takes, how much power to expend, and so forth.

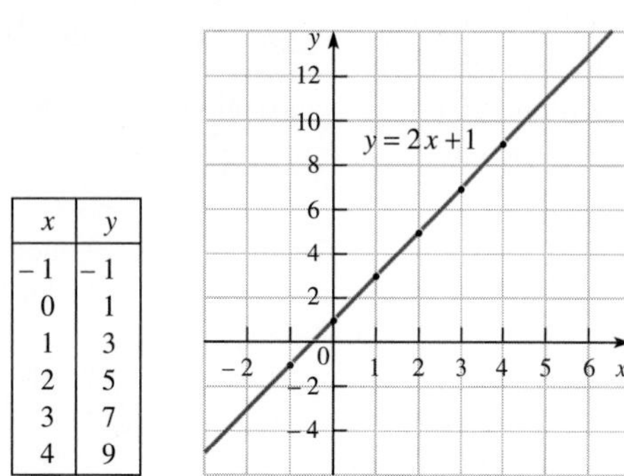

x	y
−1	−1
0	1
1	3
2	5
3	7
4	9

Figure 1.12 Here $y = 2x + 1$, the slope is $+2$, and the y-intercept (b) is 1. When $x = 0$, $y = 1$. Between $x = 0$ and $x = 4$, the rise is 8, the run is 4, and the slope, 8/4, is 2.

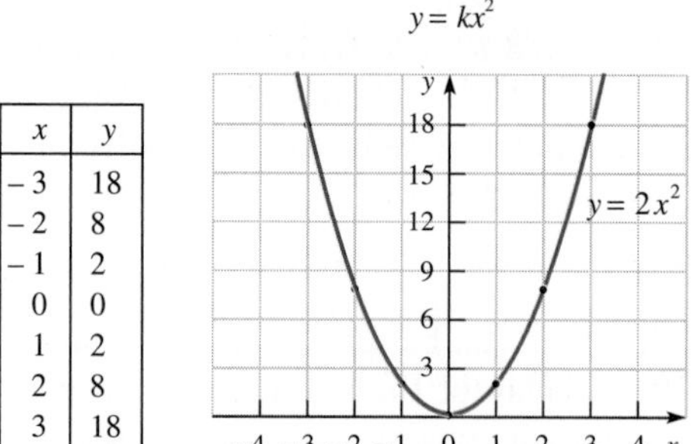

x	y
−3	18
−2	8
−1	2
0	0
1	2
2	8
3	18

Figure 1.13 A plot of $y = kx^2$ (where k is a constant) is a curve that is symmetric about the y-axis called a *parabola.* At $x = 0$, $y = 0$, and the curve passes through the origin. There are many physical relationships that are parabolic.

The volume of gasoline pumped into the tank is directly proportional to the time (t) during which it is pumped. If V is the volume of gas in the tank at any moment, b is the amount in at the start, and m is the constant flow rate, then $V = mt + b$.

How accurate an approximation is depends on the accuracy of the data that goes into it and the level of detail we devote to the calculation. For example, if all the people in the United States were placed head-to-foot, how far would they extend? There are about 250×10^6 people, each roughly 2 m tall; that's $\approx 5 \times 10^8$ m, which is greater than the distance to the Moon (3.8×10^8 m). That's not a very good approximation because 2 m is a poor guess for the average height—it's too tall. So what we calculated is an extreme. We know that the exact answer (there probably isn't one, with people being born and dying) is less than 5×10^8 m. We could improve things by assuming there are $\approx 186 \times 10^6$ adults ≈ 1.6 m tall and $\approx 64 \times 10^6$ children averaging ≈ 1 m tall. That yields a distance of $\approx 4 \times 10^8$ m, which can't be too far off—after all, if everyone in the country averaged only 1 m tall, the distance would still be $\approx 2.5 \times 10^8$ m. Notice how looking at the extremes of a situation puts things in perspective.

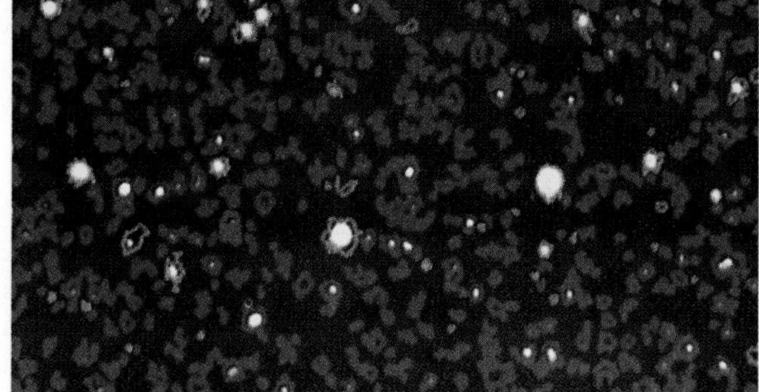

A long-exposure image of a small region of the southern sky. Here galaxies appear red and stars green. Clearly, there are a lot of galaxies in the Universe.

It's also common practice to check the solutions of problems by making quick approximations (see Appendix A-5). For example, suppose we had a swimming pool 3.1079 m deep, 4.921 m wide, and 18.689 m long. What's its volume (V)? Since the volume is the length times width times depth (Fig. 1.7):

$$V = (18.689 \text{ m})(4.921 \text{ m})(3.1079 \text{ m}) = 285.8 \text{ m}^3$$

To check this result, round each number off to 1 significant figure and redo the calculation, as follows:

$$V = (2 \times 10^1 \text{ m})(5 \text{ m})(3 \text{ m}) = 3 \times 10^2 \text{ m}^3$$

Because we rounded up more than down, this approximation can be expected to be on the large side. Still, it provides a quick check to show us that 285.8 m^3 is roughly the right size.

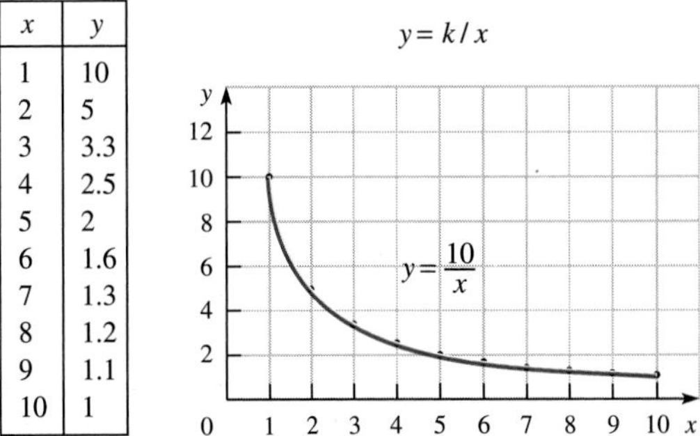

x	y
1	10
2	5
3	3.3
4	2.5
5	2
6	1.6
7	1.3
8	1.2
9	1.1
10	1

Figure 1.14 A plot of $y = k/x$ (where k is constant) is a hyperbola that approaches $y = 0$ as $x \to \infty$. For the gas in an air pump, the pressure times the volume is constant at a fixed temperature. Double the pressure, and you halve the volume.

Core Material

Physics is the study of the physical Universe. **Data** is the objective record of our observations. A **law** is a statement of a recurring pattern of events. **Theory** is the explanation of phenomena.

An agreed-upon measure of a physical quantity is a **unit**. An unchanging embodiment of a unit that serves as a primary reference is a **standard**. The SI is based on the meter, kilogram, and second. U.S. Customary units use the foot, the pound, and the second.

The result of multiplication and/or division should be rounded off so that it has as many **significant figures** as the least precise quantity used in the calculation. The result of addition and/or subtraction should be rounded off so that it has the same number of decimal places as the quantity in the calculation having the least number of decimal places. The values of transcendental functions have the same number of significant figures as their arguments.

Suggestions on Problem Solving

1. Problems are usually posed in nonscientific prose, for example, "A young man runs 20 meters from his front door to the street in 10 seconds. On average, how fast was he going?" We develop the physics to handle this problem in Chapter 2; here, we are only concerned with general principles of problem solving. Clearly, the question is full of extraneous information—it doesn't matter that it's a young man, and it doesn't matter that he goes from door to street. The following is a good way to proceed. (a) Decide on what's happening as far as the physics is concerned and lift that information from the inconsequential background. Object travels 20 m in 10 s—how fast? (b) Translate the statement into the language of physics, as you continue to learn that language. Object travels a *distance* (20 m) in a *time* (10 s) at what *average speed*? (c) Symbolically represent the variables: Object travels a distance ($d = 20$ m) in a time ($t = 10$ s); find average speed (v_{av}). Once the problem has been so restated, write down what was *given* ($d = 20$ m, and $t = 10$ s) and what you must *find* (v_{av}). That completes the crucial translation phase of the solution.

2. Determine a relationship between the given quantities and the one(s) to be found. It is the business of physics to produce such relationships in the form of laws and definitions. For example, a 10.0-meters tall cylinder contains 50.0 cubic meters of water. What is its diameter? First, translate it: A cylinder of height 10.0 m has a volume of 50.0 m^3; find its diameter. Notice that the original statement did not contain the crucial word *volume.* Without that conceptual translation we could not proceed. Now write down what is given and what's to be found. *Given:* cylinder, $h = 10.0$ m, $V = 50.0$ m^3; *Find:* D. This problem is more one of math than of physics, but the procedures are the same. Next, use an equation that provides a relationship between the quantities (Fig. 1.7), namely, $V = \pi R^2 h$. We want the diameter ($D = 2R$), not the radius. Solve for $R = \sqrt{V/\pi h} = 1.26$ m and double it.

3. As a rule, it's always a good idea to draw a diagram. The process helps to organize your thoughts.

4. Wherever possible, check your calculation. We all make numerical mistakes, it's perfectly normal. The thing you must do is catch them! With the above cylinder, take $R = 1.26$ m and compute something we know, namely V: $V = \pi R^2 h = \pi(1.26 \text{ m})^2(10.0 \text{ m}) = 49.9 \text{ m}^3$—close enough, since we rounded off R to three figures.

5. Show all of your work when making a calculation. Doing so will allow you, and anyone else attempting to understand your efforts at some later date, to follow your logic.

6. When solving a problem involving quantities with units having prefixes such as ms, or kg, or μm, replace the prefixes by scientific notation before carrying out the calculation. Notice that $(1.0 \text{ km})^3 = (1.0 \times 10^3 \text{ m})^3 = 1.0 \times 10^9 \text{ m}^3$, not 1.0×10^9 m.

Discussion Questions

1. There are people who define science as that which scientists do. Why isn't that a reasonable approach?

2. Deliberate hoaxes occur occasionally in science. A good example is the famous Piltdown man (around the turn of the century, a human skull fragment and an apelike jaw were combined and touted as the "missing link"). How do you think such hoaxes get revealed for what they are?

3. Given that a law says that some relationship among physical quantities will be found to *always* exist, how does that transcend experience? Implicit in the statement of law is the understanding that it applies *everywhere.* Comment on the practicalities of this notion.

4. What do you think the French mathematician H. Poincaré meant when he wrote, "Nature can only be governed by obeying her"?

5. What did Einstein mean when he wrote:

> The aim of science is, on the one hand, a comprehension, as *complete* as possible, of the connection between the sense experiences in their totality, and, on the other hand, the accomplishment of this aim *by the use of a minimum of primary concepts and relations.*

What's the point of specifying "a minimum of primary concepts"?

6. When W. K. Clifford, more than a century ago, said "scientific thought is the guide to action," what might he have had in mind?

7. Discuss the distinction between *data* and *fact.* When Einstein was confronted by some reporters who excitedly proclaimed that the facts of a new experiment had proven Relativity to be incorrect, he simply replied, "The facts are wrong." Months later, the data was reinterpreted and the theory confirmed. Fact is not immutable. Indeed, a good trial lawyer knows that facts can often be cast within a shadow of doubt.

8. The poet Samuel Taylor Coleridge said that beauty was "unity in variety." How is that applicable to physical theory?

9. Having spoken for an hour or so about his recent theoretical work on elementary particles, Wolfgang Pauli settled down for the usual discussion involving the audience. It was New York, 1958, and the great Niels Bohr was there. After some heated remarks, Bohr rose and said, "We are all agreed that your theory is crazy. The question which divides us is whether it is crazy enough to have a chance of being correct." How would they ultimately find out if it was "crazy enough"?

10. Robert Millikan, the precise experimentalist and Nobel laureate, wrote in 1936, "In science, truth once discovered always remains truth." What can you say about this idea in light of the tentative nature of science? In what way was Millikan naive?

Multiple Choice Questions

1. Which of the following is longest? (a) 10^4 cm (b) 10^4 mm (c) 10^6 μm (d) 10^9 nm (e) none of these.
2. Which length is the largest? (a) 10^1 cm (b) 10^{-10} m (c) 1×10^2 mm (d) 1 m (e) none of these.
3. Which mass is the smallest? (a) 10^5 μg (b) 10^2 g (c) 1 kg (d) 10^3 mg (e) none of these.
4. Given are four masses: (1) 10 mg (2) 1000 μg (3) 10^2 kg (4) 10^{-4} kg. These are ordered in ascending size as (a) 1, 2, 3, 4 (b) 2, 1, 4, 3 (c) 4, 3, 2, 1 (d) 2, 1, 3, 4 (e) none of these.
5. A day has roughly (a) 86×10^2 s (b) 8640 s (c) 9×10^4 s (d) 1.44×10^3 s (e) none of these.
6. A year has roughly (a) 8.77×10^2 h (b) 5×10^5 min (c) 3.7×10^3 days (d) 32×10^5 s (e) none of these.
7. A cube 1000 cm on a side has a volume of (a) 10^2 cm^2 (b) 10^2 cm^3 (c) 10^6 cm^3 (d) 10^9 cm^3 (e) none of these.
8. A rectangular floor is 6.6 m by 12 m. Its area is (a) 79 m^2 (b) 18.6 m^2 (c) 7.92 m^2 (d) 79.2 m (e) none of these.
9. A femtosecond is (a) 10^{-12} s (b) -15 s (c) 10^{15} s (d) 10^{-15} s (e) none of these.
10. A 20.0-in. bar is (a) 20.0 cm long (b) 508 mm long (c) 51 m long (d) 2.54/20 cm long (e) none of these.
11. One pound has an equivalent mass of exactly 453.592 37 g. To four significant figures, that's (a) 453.5 g (b) 453.592 3 g (c) 400.0 g (d) 453.6 g (e) none of these.
12. The product of 12.4 m and 2 m should be written as (a) 24.8 m (b) 24.8 (c) 25 m^2 (d) 0.2×10^2 m (e) none of these.
13. The product of 15.0 cm and 5 cm should be written as (a) 75 cm^2 (b) 7.5×10^1 cm^2 (c) 0.75×10^2 cm^2 (d) 0.8×10^2 cm^2 (e) none of these.
14. The weight of 1 kg on Earth is about (a) 1 lb (b) 1000 g (c) 2-1/4 lb (d) 0 (e) none of these.
15. If a bag of screws costs 10¢ per pound, a kilogram of them will cost about (a) 100¢ (b) 22¢ (c) 4.5¢ (d) $22 (e) none of these.
16. If coffee is $12 a kilogram, roughly how much will it be by the pound? (a) $26 (b) $12 (c) $5.5 (d) 55¢ (e) none of these.
17. A liter is 1000 cm^3, which means that a cube 100 cm on a side has a volume of (a) 1000 liters (b) 0.001 m^3 (c) 100 liters (d) 1000 liters^3 (e) none of these.
18. A kilometer is (a) just under half a mile (b) just over half a mile (c) about 1000 ft (d) roughly 5280 ft (e) none of these.
19. A dime is about (a) 1 cm thick (b) 1 kg in mass (c) 1 mm thick (d) 10^{-3} m in diameter (e) none of these.

Problems

MEASUREMENT

1. [I] The human brain has about 10 (U.S.) billion nerve cells, more tightly packed than in any other tissue. Write that number out and then put it in scientific notation.

2. [I] It's been estimated that the human brain can store 100 (U.S.) trillion bits of information. Write that number out, put it in scientific notation, and then express it by adding the proper prefix to the word *bits*.

3. [I] Express each of the following quantities in seconds in decimal form: (a) 10.0 ms; (b) 1000 μs; (c) 10.000 ks; (d) 100 Ms; (e) 1000 ns. (Don't worry about significant figures in this problem.)

4. [I] Express each of the following quantities in micrograms: (a) 10.0 mg; (b) 10^4 g; (c) 10.000 kg; (d) 100×10^3 g; (e) 1000 ng. (Don't worry about significant figures.)

5. [I] How many millimeters are in 10.0 km?

6. [I] A red blood cell lives for about four months and travels roughly 1.0×10^3 mi through the body. How far is that in kilometers?

7. [I] How many millimeters are in 3.00 in.?

8. [I] The unit known as an angstrom (Å), exactly equal to 0.1 nm, is still widely used. It's named after the nineteenth century physicist Anders J. Ångstrom (Å is the letter that comes after Z in the Swedish alphabet). Given that an atom is about 1 Å in diameter, how much is that in centimeters?

9. [I] The length of a lightwave is about 5×10^3 Å (see previous problem). How much is that in nanometers?

10. [I] About how many centimeters tall is a stack of 25 dimes?

11. [I] A nickel (5-cent coin) has a mass of about 5 g. How many nickels correspond to a kilogram?

12. [I] It takes your brain about one five-hundredth of a second to recognize a familiar object once the light from that object enters your eye. Express that time interval in milliseconds, microseconds, and nanoseconds, each to one significant figure.

13. [I] How many seconds are there in an exactly 24-h day?

14. [I] Each second the human brain undergoes 1×10^5 different chemical reactions. At that rate, how many will it experience in 10 h?

15. [I] The distance light travels in a year is called a *light-year* (ly). Given that 1.00 ly = 5.88×10^{12} mi, how far is that in meters?

16. [I] By what number would you multiply a given number of centimeters to convert it to millimeters?

17. [I] There are an average of 32 million bacteria on each square inch of the human body. Given a total skin area of 1.7 m^2, how many bacteria are you carrying around (exclude the bacteria that are internal)?

18. [I] How many square centimeters are in a square inch?

19. [I] In a human nose, the total area of the region that detects odors is about 3/4 in.^{2}. Compare that to the sensory organ of a hunting dog, whose nose has an active area of about 65 cm^2.

20. [I] Can the quantity 100.0 kg be expressed in grams in decimal form with the correct number of significant figures? Explain.

21. [I] Express each of the following in grams in decimal form with the correct number of significant figures: (a) 1.00 μg; (b) 0.001 ng; (c) 100.0 mg; (d) 10 000 μg; and (e) 10.000 kg.

22. [I] Determine the number of significant figures for each of the following: (a) 0.002 0; (b) 0.99; (c) 1.75 ± 0.02; (d) 1.001; (e) 4.44×10^4; and (f) 0.01×10^{34}.

23. [I] Write the speed of light in vacuum (299 792 458 m/s) to one, three, four, and eight significant figures.

24. [I] State the number π = 3.141 592 65 . . . to one, three, four, and five significant figures.

25. [I] The electromagnetic charge of an electron is $1.602\,177\,33 \times 10^{-19}$ coulombs. If the uncertainty is 0.30 ppm, which figures are in doubt and by how much?

26. [I] One of the most important numbers in physics is Planck's constant: $6.626\,075\,5 \times 10^{-34}$ kg·m^2/s ± 0.60 ppm. (a) Write Planck's constant out to four significant figures. (b) Write it out to two significant figures. (c) Write it out to one significant figure. (d) Write it out to five significant figures.

27. [I] To 4 significant figures, how many square meters correspond to a square foot?

28. [I] Liters are sometimes used for liquid and gas measurement, although the preferred unit is the cubic meter. How much is 1.000 liter in m^3?

29. [I] The radius of the Moon ($R_☾$) is 1.738×10^3 km. Assume the Moon is a sphere and compute its volume to two significant figures.

30. [II] A GM diesel locomotive pulling 40 to 50 loaded freight cars at 70 mi/h uses 1.0 gallon of fuel every 632 yd. How many meters can it travel on 10 gallons of fuel?

31. [II] By what number would you multiply a given number of centimeters to convert it into inches? Give the answer to 4 significant figures and show your work.

32. [II] A Boeing 747 jumbo jet carrying 385 people while cruising at 39 000 ft travels 280 yd on a gallon of aviation fuel. How many gallons will it need to travel 2000 km?

33. [II] Most people lose about 45 hairs per day out of a typical headful of 125 000. Suppose each hair averages 10 cm long. If you placed a year's lost hairs end-to-end, how far would they extend?

34. [II] What's the SI equivalent of 1.000 in.^{2}?

35. [II] Express the equivalent of 1.000 ft^2 in SI units.

36. [II] In an average lifetime, a human inhales roughly 5.0×10^5 yd^3 of air. How many cubic meters is that?

37. [II] One cup is equivalent to 237 milliliters. How much is that in the preferred SI unit of cubic meters?

38. [II] Roughly what size cube of water would have a mass of 1.0 kg?

39. [II] Each hour a large man sheds about 6×10^5 particles of skin. In a year, that amounts to about 1.5 lb (on Earth). Roughly how much is the mass of each such particle? What mass of skin will be shed in 50 years of adulthood?

40. [II] We are advised to keep our total blood cholesterol level at less than 200. What that actually means is 200 mg of cholesterol per deciliter of blood. Physicists totally shun the *deci* prefix and avoid the liter as well, so convert this quantity to SI units.

41. [II] Compute the sum of the following quantities: 0.066 m, 1.132 m, 200.1 m, 5.3 m, and 1600.22 m.

42. [II] Add the following quantities: 0.10 ms, 20.2 s, 6.33 s, 18 μs, and 200.55 ms.

43. [II] Add the following quantities: 1.00 g, 1.00 mg, 1.00 kg, and 1.00 μg.

44. [II] Compute the circumferences of circles having diameters of (a) 5.42 μm, (b) 0.5420 nm, (c) 5.420 000 mm, (d) 542.0 m, (e) 0.542 km.

45. [II] What is the product of the following quantities: 0.002 1 g, 655.1 kg, and 4.41 μg?

46. [III] A *pipe* is 2 *hogsheads* and a hogshead is 63 gallons (or 2 barrels). Given that 1 U.S. gallon is 231 in.^{3}, what's the SI equivalent of a pipe to three significant figures?*

***Answers** to all odd-numbered Multiple Choice Questions and all odd-numbered Problems are in the Answers in the back of the book. Numbers in boldface indicate that a **solution** is provided in the Answers in the back of the book. Numbers in italic indicate that a **solution** is provided in the Student Solutions Manual.

2

Kinematics: Speed and Velocity

EVERY ATOM IN EVERY solid object, including the human body, is jiggling wildly. The Earth is spinning and revolving around a star that is itself whirling along with the galaxy. The entire Universe is expanding. Everything is in motion. When anything happens, when an event occurs, there is change, and change on a fundamental level is predicated on motion. The idea of motion is so basic that it has occupied human thought for millennia, and we still struggle with the subtleties of space and time. Physics *quantifies*—it associates numbers with concepts in a way that renders them *measurable.* In particular, the branch of physics that describes motion, that

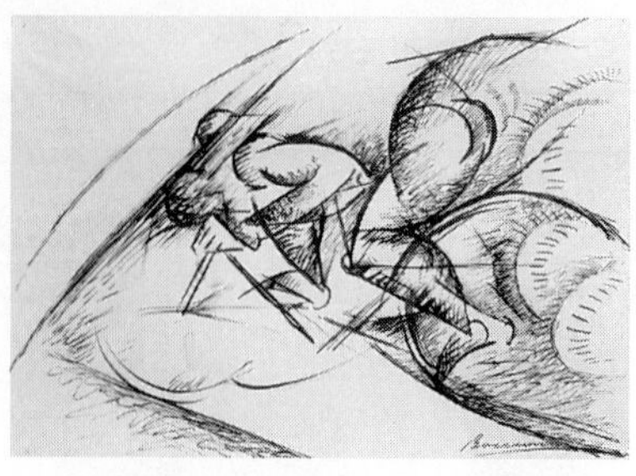

Study for Dynamic Force of a Cyclist I by the Italian futurist Umberto Boccioni (1882–1916). The artist attempts to capture the feeling of motion; the physicist attempts to quantify motion and thereby measure it.

defines the needed ideas and develops their interrelationships *with no concern for what is causing the motion,* is known as **kinematics**. This chapter develops the concepts of speed, displacement, velocity, and relative velocity. Once these ideas are in place, we go on to study changes in velocity—acceleration (Chapter 3). Thus, armed with kinematics (the vocabulary of motion), we can begin to understand dynamics—what drives the motion, what drives the Universe.

SPEED

The faster a thing moves, the farther it goes in any interval of time. We have meter sticks to determine *distance* and clocks for *time;* we only need to relate these two fundamental ideas to quantify speed, to measure "how fast." Among the oldest existing thoughts on the subject are those of Aristotle (384–322 B.C.) and he, like his fellow Greeks of the era, specified speed as the distance traveled in a given amount of time. That's the way it was treated for well over 1000 years—one traveled with a speed of "so many miles in so many hours."

2.1 Average Speed

Today, we define the **average speed** *as the distance traveled divided by the time it took to do the traveling,* or

$$\text{Average speed} = \frac{\text{distance traveled}}{\text{time elapsed}}$$

Dividing by the time in specific units—seconds or minutes or hours—yields the *distance covered per unit time.* The Greeks never performed that division, apparently because they could not bring themselves philosophically to divide the unlike quantities of space and time. The division creates a new concept: a single quantity with a single numerical value and its own unit.

The above definition can be expressed symbolically as

$$v_{av} = \frac{l}{t} \tag{2.1}$$

Figure 2.1 The rocket travels a distance measured along its actual path through space. After a time t, having gone a distance l, its average speed is $v_{av} = l/t$.

where v_{av} is the average speed,* l the length of the path traveled, and t the time it took to travel it (Fig. 2.1). The units of speed are always those of *length over time* (kilometers per hour, meters per second, and so on). An automobile traversing a total along-the-road distance of 30 km in 2.0 h has an average speed of 15 km/h. Evidently, it could have stopped a few times on the way and then sped up and still covered the 30 km in the same 2.0 h—that's why we call this concept *average* speed.

A wooden horse on a carousel can swing around on its pedestal through 40 m in 20 s, and though it is forever coming back to the same spot in space, it still travels with an average speed of (40 m)/(20 s) = 2.0 m/s. *Speed is independent of the direction of motion,* and like all such quantities that have nothing to do with spatial orientation (such as length, temperature, time, mass, density, charge, and volume), it has only a *scale* or size and no associated direction. Accordingly, it's called a **scalar quantity**.

*We do not use s for speed because it has often been associated with the displacement in "space" (see p. 32). The symbol v derives from the word velocity (see p. 31).

Example 2.1 The Moon moves in a nearly circular orbit around the Earth with an average radius R of 3.84×10^8 m. If it takes 27.3 days for it to complete one revolution, determine its average orbital speed in m/s.

Solution: [Given: $R = 3.84 \times 10^8$ m and a once-around time of $t = 27.3$ days. Find: v_{av}.] We compute the motion over one complete orbit, which takes 27.3 days (that is, 27.3 d × 24 h/d × 60 min/h × 60 s/min = 2.359×10^6 s). The Moon's orbital path has an overall length equal to the circumference, namely $2\pi R$ (Appendix B). From Eq. (2.1), $v_{av} = l/t = 2\pi R/t$, or

$$v_{av} = \frac{2\pi(3.84 \times 10^8 \text{ m})}{2.359 \times 10^6 \text{ s}} = \boxed{1.02 \times 10^3 \text{ m/s}}$$

▶ **Quick Check:** *Always make sure your answer is the correct order-of-magnitude* (Appendix A-5). The quantity $2\pi \approx 6$ and the circumference is about $6 \times 4 \times 10^8 \text{ m} \approx 24 \times 10^8$ m; that value divided by the time (about 24×10^5 s) yields 10^3 m/s.

Example 2.2 A student drives from home to school at an average speed of 25.3 km/h. If it takes 4.72 h to get to her destination, how far did she travel?

Solution: [Given: $v_{av} = 25.3$ km/h and $t = 4.72$ h. Find: l.] We are provided the average speed and time, and are asked to find distance traveled. Equation (2.1) contains all of these quantities, and solving it for distance we get

$$l = v_{av}t = (25.3 \text{ km/h})(4.72 \text{ h})$$

Since the least number of significant figures is three, $\boxed{l = 119 \text{ km}}$.

▶ **Quick Check:** Recalculate one of the given terms using l, say v_{av}: $v_{av} = l/t = (119.4 \text{ km})/(4.72 \text{ h}) = 25.297 \text{ km/h} = 25.3$ km/h.

Table 2.1 shows some typical speeds and Table 2.2 compares mi/h with several other unit representations. An average speed in one set of units can be expressed in any other set of units. To convert, we use multiplicative factors equal to 1 so that the

TABLE 2.1 Typical speeds

Speed, m/s	Motion	Speed, mi/h
300 000 000	Light, radio waves, X-rays, microwaves (in vacuum)	669 600 000
210 000	Earth-Sun travel around the galaxy	481 000
29 600	Earth around the Sun	66 600
1 000	Moon around the Earth	2300
980	SR-71 reconnaissance jet	2200
333	Sound (in air)	750
267	Commercial jet airliner	600
62	Commercial automobile (max.)	140
37	Falcon in a dive	82
29	Running cheetah	65
10	100-yard dash (max.)	22
9	Porpoise swimming	20
5	Flying bee	12
4	Human running	10
2	Human swimming	4.5
0.01	Walking ant	0.03
0.000 045	Swimming sperm	0.000 1

TABLE 2.2 A rough comparison of speeds in different units

mi/h	ft/s	km/h	m/s
0	0	0	0
10	15	16	4
20	29	32	9
30	44	48	13
40	59	64	18
50	73	80	22
60	88	97	27
70	103	113	31
80	117	129	36
90	132	145	40
100	147	161	45
200	293	322	89
500	734	805	224
1000	1467	1609	447

TABLE 2.3 Speed conversion table*

m/s	km/h	mi/h	ft/s	
1	0.277 8	0.447 0	0.304 8	**m/s**
3.600	1	1.609	1.097	**km/h**
2.237	0.621 4	1	0.681 8	**mi/h**
3.281	0.911 3	1.467	1	**ft/s**

*The units being converted head up the columns. The units into which they are transformed via multiplication are on the right of each row. Example: To go from km/h (second column) to mi/h (third row), multiply by 0.621 4. Thus, 1.00 km/h = (1.00)(0.621 4) mi/h.

size of the physical quantity never changes. For example, to change 15 km/h into m/s, first convert kilometers to meters by multiplying by (1000 m/1 km), which equals 1 exactly:

$$\left(\frac{15\ \text{km}}{1\ \text{h}}\right)\left(\frac{1000\ \text{m}}{1\ \text{km}}\right) = \frac{15\,000\ \text{m}}{1\ \text{h}}$$

To convert this expression to m/s, multiply by whatever it takes to transform the hours into seconds, namely (1 h/3600 s) = 1:

$$\left(\frac{15\,000\ \text{m}}{1\ \text{h}}\right)\left(\frac{1\ \text{h}}{3600\ \text{s}}\right) = 4.2\ \text{m/s}$$

Converting from one system of units to another is a practical necessity, especially in the United States, where quantities of interest are not always given in SI units. To go from U.S. Customary units to SI, or vice versa, requires the knowledge of a numerical constant that relates the two systems for each quantity involved. For example, for distance 1.000 in. = 2.540 cm, or 1.000 0 mi = 1.609 3 km. Thus, the conversion of 60 mi/h to SI is

$$60\ \text{mi/h} = \left(\frac{60\ \text{mi}}{1\ \text{h}}\right)\left(\frac{1.609\ \text{km}}{1\ \text{mi}}\right)\left(\frac{1000\ \text{m}}{1\ \text{km}}\right)\left(\frac{1\ \text{h}}{3600\ \text{s}}\right) = 27\ \text{m/s}$$

The three terms multiplying 60 mi/h can be compressed into a single conversion factor; that is, (0.447 0 m/s) = (1.000 mi/h), (see Table 2.3). *It is not recommended that such specific conversion factors be memorized;* 1.000 mi ≈ 1.609 km should do nicely.

Example 2.3 Roger Bannister was the first person to run the mile in less than 4 min. He accomplished this feat in a time of 3.00 min 59.4 s in 1954. Compute his average speed in both miles per hour and meters per second.

Solution: [Given: l = 1.00 mi and t = 3.00 min 59.4 s. Find: v_{av}.] First, convert the time interval into some fraction of an hour because we want the answer in mi/h. (59.4 s)/(60 s/1 min) = 0.990 min and adding this to 3.00 min yields a total time of 3.990 min, or (3.990 min)/(60 min/1 h) = 0.066 5 h. From Eq. (2.1)

$$v_{av} = \frac{l}{t} = \frac{1.00\ \text{mi}}{0.066\,5\ \text{h}} = \boxed{15.0\ \text{mi/h}}$$

Using Table 2.3 to convert the units, we obtain

$$v_{av} = (15.04\ \text{mi/h})\left(0.4470\ \frac{\text{m/s}}{\text{mi/h}}\right) = \boxed{6.72\ \text{m/s}}$$

▶ **Quick Check:** 1.00 mi = 1.609 3 km, hence v_{av} = (1.609 3 km)/(3.00 min × 60 s/1 min + 59.4 s) = 6.72 m/s. Or (1 mi)/(4 min) = 15 mi/h.

Example 2.4 The 1980 speed record for human-powered vehicles was set on a measured 200-m run by a sleek machine called Vector. Pedaling back-to-back, its two drivers averaged 62.92 mi/h. This awkward mix of units is the way the data appeared in an article reporting the event. Determine the speed in m/s and then compute how long the record run lasted.

Solution: [Given: $l = 200$ m and $v_{av} = 62.92$ mi/h. Find: v_{av} in m/s and t.] Changing miles to feet, $v_{av} = (62.92 \text{ mi/h})(5280 \text{ ft/mi}) = 3.322 \times 10^5$ ft/h. Since 1.00 m = 3.281 ft, $v_{av} = (3.322 \times 10^5 \text{ ft/h})/(3.281 \text{ ft/m}) = 10.12 \times 10^4$ m/h. To convert hours to seconds, divide by (3600 s/h); so $(10.12 \times 10^4 \text{ m/h})/(3600 \text{ s/h}) = \boxed{28.1 \text{ m/s}}$. Equation (2.1) relates l, v_{av}, and t, hence

$$t = \frac{l}{v_{av}} = \frac{200 \text{ m}}{28.1 \text{ m/s}} = \boxed{7.12 \text{ s}}$$

▶ **Quick Check:** 200 m = $(200 \text{ m})(6.214 \times 10^{-4} \text{ mi/m}) = 0.124$ mi. $t = (0.124 \text{ mi})/(62.92 \text{ mi/h}) = 1.98 \times 10^{-3}$ h = 7.1 s. Or, with $v_{av} \approx 60$ mi/h = 88 ft/s ≈ 29 m/s; (200 m)/(29 m/s) = 7 s.

Example 2.5 A 600-km cross-country automobile race is won by a team of two drivers, each of whom had the wheel for half the distance of the trip. If one averaged 60 km/h and the other 20 km/h, what was their overall average speed?

Solution: Here two distances and average speeds are involved, so introduce a subscript notation that will keep track of things. [Given: total length $l = 600$ km; distances traveled by drivers 1 and 2, $l_1 = l_2 = 300$ km (half the total distance); average speeds $(v_{av})_1 = 60$ km/h and $(v_{av})_2 = 20$ km/h. Find: v_{av}.] The overall average speed, as stated in Eq. (2.1) is the total length l divided by the total time t. We have l but need $t = t_1 + t_2$. Problems will often be encountered that deal with portions of a total distance or time—*the sum of the parts equals the whole* is one more equation to be used in the analysis. The individual times can be computed from the individual journeys using Eq. (2.1), as

$$t_1 = \frac{l_1}{(v_{av})_1} = \frac{300 \text{ km}}{60 \text{ km/h}} = 5.0 \text{ h}$$

and

$$t_2 = \frac{l_2}{(v_{av})_2} = \frac{300 \text{ km}}{20 \text{ km/h}} = 15 \text{ h}$$

The total time elapsed is 20 h, and therefore the average speed for the total trip is

$$v_{av} = \frac{l}{t} = \frac{600 \text{ km}}{20 \text{ h}} = \boxed{30 \text{ km/h}}$$

Notice that this is *not* equal to the average of the two average speeds, which is

$$\frac{(v_{av})_1 + (v_{av})_2}{2} = \frac{60 \text{ km/h} + 20 \text{ km/h}}{2} = 40 \text{ km/h}$$

The two are not the same because one person drove three times longer than the other and has a correspondingly greater influence on v_{av}. If that bothers you, imagine two rooms, one containing 100 basketball players all 6 ft tall and the other containing 2 kids each 3 ft tall. Find the average height of the people in each room and then find the average height of all the people. If you got 4.5 ft, you missed the point.

▶ **Quick Check:** Since the second driver traveled three times longer, that person's speed must be three times more weighty in the average. One person drives for 5 h out of 20 h, and the other for 15 h out of 20 h, hence

$$v_{av} = \frac{5(60 \text{ km/h}) + 15(20 \text{ km/h})}{20} = \frac{120 \text{ km/h}}{4}$$

and $v_{av} = 30$ km/h

2.2 Constant Speed

All of us have traveled in automobiles and have some intuitive sense of what **constant** or **uniform speed** is—just keep the speedometer locked at, say, 55 km/h and you have it. The scholars of the Middle Ages produced the first workable definition: *Uniform speed corresponds to the traversal of equal distances in equal intervals of time—of any fixed duration, however small* (Fig. 2.2). Galileo, 300 years later, was

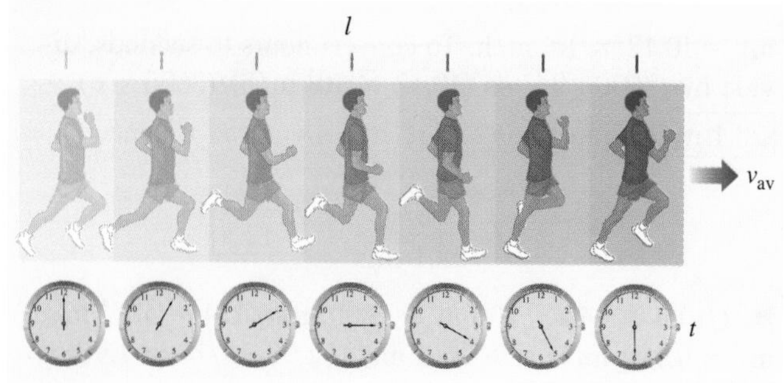

Figure 2.2 An object traveling at a constant speed covers equal distances in equal intervals of time.

very careful to stress that word *any* in his definition. If the interval is large, a body moving alternately faster and then slower can still have the same average speed over each such interval of time and so seem to be moving with a constant speed.

If v_{const} is the constant speed with which some object moves, then that is clearly also its average speed, and Eq. (2.1) becomes $v_{const} = l/t$. The path length traveled in a given time at a fixed speed is:

$$l = v_{const}t \tag{2.2}$$

When the speed is constant (or if we know the average speed), the distance traversed during any time period can be determined.

A modern speedometer in the United States shows speed in both mph (miles per hour) and kph (kilometers per hour).

Figure 2.3 is a **distance-time graph** for a bumblebee flying at a constant speed of 10 m/s. Starting at $t = 0$ and $l = 0$, the bee traverses 10 m during each second of motion; its *speed* or *rate-of-change of distance* is constant, and so the plot is a straight line. The *slope* of the line is the ratio of its *rise* to its *run* (the vertical distance between any two points on the line divided by their horizontal separation—see Appendix A-4). For the two points shown in the figure, the rise is (30 m − 10 m) = 20 m, and the run is (3.0 s − 1.0 s) = 2.0 s. Thus, the slope is 20 m/2.0 s = 10 m/s. *The slope of the straight line numerically equals the value of the constant speed:* a faster object produces a greater slope (Fig. 2.4).

Another important plot is the **speed-time graph**, which in this case of constant speed is simple. At each of the 1st, 2nd, 3rd, 4th, and so on, seconds, the speed of the bee is always 10 m/s. Hence, the curve is a horizontal straight line (Fig. 2.5). The rectangular area contained under the line within, say, the first 3.0 s of flight equals the height (that is, the speed, 10 m/s), multiplied by the length (that is, the time, 3.0 s), which yields 30 m, the distance. The bee traveled 30 m in those 3.0 s. *The area under the speed-time curve bounded by any two moments in time equals the distance traveled during that interval.* This statement may not seem like much of a revelation in this instance, but it applies to all speed-time graphs. In fact, it was in order to determine just such areas that integral calculus was devised (p. 65).

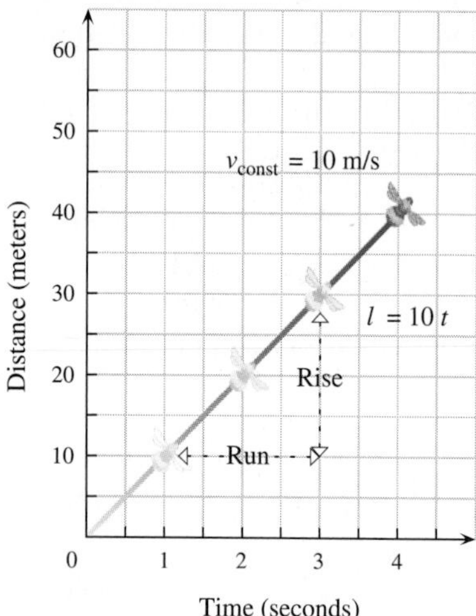

Figure 2.3 Distance-time graph of a bumblebee traveling at a constant speed (v_{const}). Take the distance to be the vertical axis and the time to be the horizontal axis. Note that the angle (although not the slope) of the line depends on the scale used to plot the curve.

2.3 Delta Notation: The Change in a Quantity

Imagine now that you decide to measure the average speed of your new Ferrari (Fig. 2.6). As you pass some arbitrary initial point, P_i, on the road, your watch reads an initial time, t_i, of 12:04. Just at that moment, the odometer shows an initial distance, l_i, of 16 275.50 km. A far-off tree marks the final point of the course, P_f. As the car passes it, the odometer reads a final distance, l_f, of 16 285.50 km and the clock a final time, t_f, of 12:14. Here, $l_f - l_i = 10$ km and $t_f - t_i = 10$ min.

The Greek capital letter delta, Δ, placed before the symbol for some quantity means *the change in that quantity.* Thus, Δl (read delta "el") is the *change in the distance* or the length traversed: $\Delta l = l_f - l_i$. Similarly, Δt (read delta "tee") is the *change in time,* or the duration elapsed: $\Delta t = t_f - t_i$. Consequently,

$$v_{av} = \frac{\Delta l}{\Delta t} = \frac{l_f - l_i}{t_f - t_i} \tag{2.3}$$

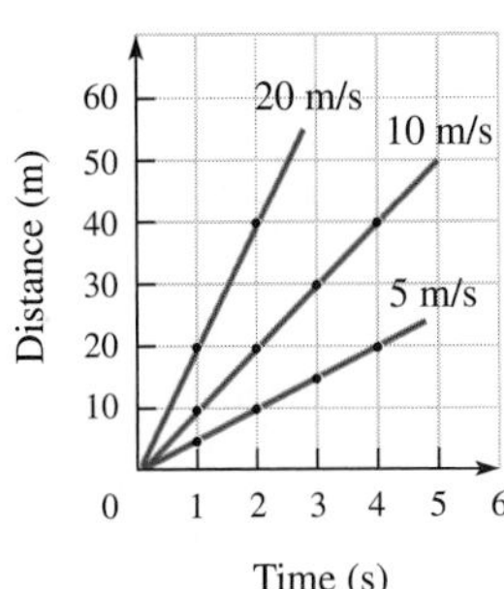

Figure 2.4 For a uniformly moving object, the distance-time graph is a straight line, and its slope is the speed. The greater the slope, the greater the speed.

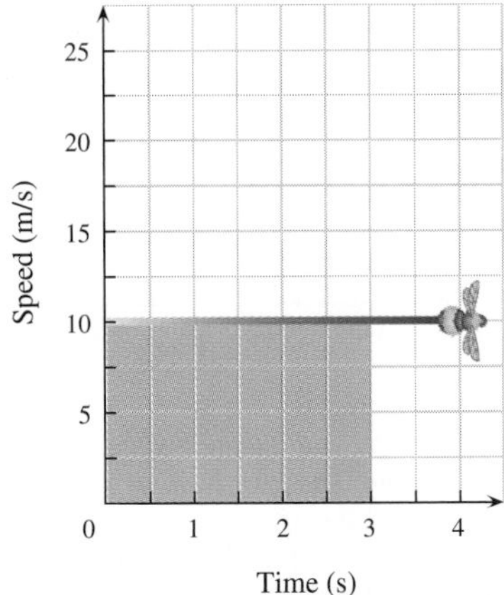

Figure 2.5 Plotting speed along the vertical axis and time along the horizontal axis gives us a speed-time graph. The area under the curve between any two times is the height (that is, the speed) multiplied by the length (that is, the time), and that result equals the distance traveled.

In this case, where $\Delta l = 10$ km and $\Delta t = 10$ min or 1/6 h, $v_{av} = 60$ km/h. This form of Eq. (2.1) is useful when the data are presented in terms of initial and final values, as opposed to overall lengths and times. Had you used a stopwatch and a resettable odometer or "tripmeter," you could have started the run at $t_i = 0$ and $l_i = 0$. In that instance, $\Delta l = l_f = l = 10$ km, the whole distance, and $\Delta t = t_f = t = 10$ min, the whole time. Eq. (2.3) becomes identical to Eq. (2.1).

2.4 Instantaneous Speed

Scholars of the mid-1300s dealt successfully with the idea of uniform speed. But when they tried to take the next logical step, to define the speed at any moment, they failed. They lacked the mathematical imagery of motion that Newton would create centuries later—the calculus of change. While in a car or an airplane, if you ask "How fast am I going NOW?" you want the *instantaneous speed*—the speed at that very moment. Today, that's usually read from a speedometer. Still, what are we really talking about?

The bumblebee flying in Figs. 2.3 and 2.5 has a constant speed, and it does not matter when you yell NOW, its speed at that moment (at any moment) would be 10 m/s. More realistically, Fig. 2.7a depicts a nonlinear distance-time plot. The bee leaps into the air at $t = 0$, lands on a flower after flying 2.0 m, rests, and takes off again at $t = 3.0$ s. It flies past your nose at $t = 5.0$ s, and alights on a branch at $t = 7.0$ s, ending the journey. Its average speed over any portion of the trip is the slope of the line from the beginning to the end of that portion. In Fig. 2.7b, for the interval from say $t = 2.0$ s to $t = 7.0$ s, $v_{av} = \Delta l/\Delta t = 8.0\text{ m}/5.0\text{ s} = 1.6$ m/s.

How do we specify the speed at some instant—NOW—like the one when the bee flew past the point P on the tip of your nose at $t = 5.0$ s? To that end, evaluate v_{av} using as an interval Δl, a small test course that extends along the flight path from a little before to a little after point P (Fig. 2.8a). Because it takes a time Δt for

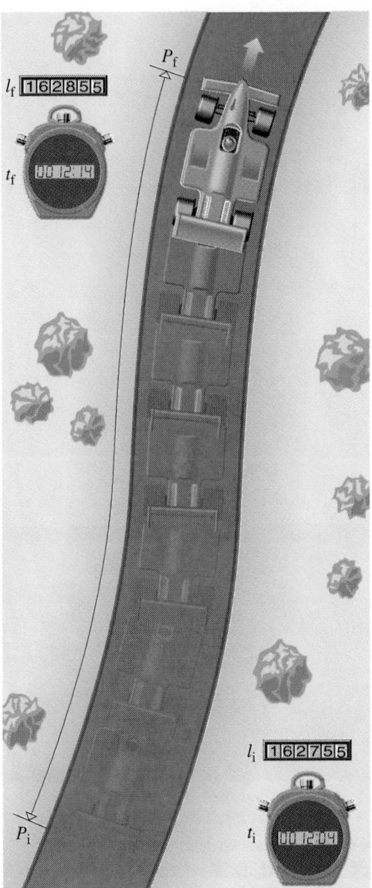

Figure 2.6 Measuring the average speed of a car over a run from point P_i to point P_f. The distance traveled is $l_f - l_i = 10$ km and the time elapsed is $t_f - t_i = 10$ min. The average speed is: $v_{av} = 10\text{ km}/10\text{ min} = 60$ km/h.

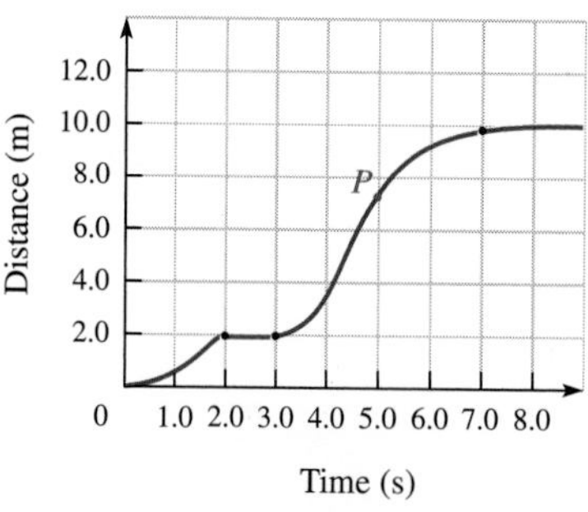

Figure 2.7 (a) The distance-time curve for a bee. The bee flew to a flower at $t = 2.0$ s, rested until $t = 3.0$ s, and then flew past your nose at point P, alighting on a branch at $t = 7.0$ s.

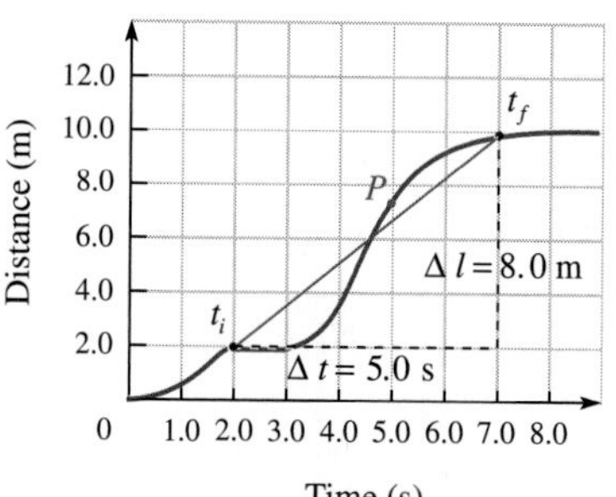

(b) The average speed of the bee during the interval from $t = 2.0$ s to $t = 7.0$ s is $v_{av} = \Delta l/\Delta t = 8.0$ m/5.0 s $= 1.6$ m/s.

the bee to traverse this Δl, the average speed is again $\Delta l/\Delta t$. Our scheme will become increasingly better as the length interval Δl straddling P is narrowed, and that narrowing will occur if we reduce the time interval of the measurement (Fig. 2.8b). For example, suppose the bee flew from 0.000 005 m before P to 0.000 005 m beyond P in 0.000 002 s. Its average speed over that tiny distance is equal to 0.000 010 m divided by 0.000 002 s, or 5 m/s. This is the average speed measured during a small but finite time N-O-W as opposed to an instantaneous NOW. It is not necessarily the speed exactly at P, but it is very close to it. By further shrinking Δt, and thereby Δl, we can get as close as we like to the speed NOW.

There is a geometrical way to envision the instantaneous speed of the bee at P. As Δt is made ever smaller, Δl also shrinks, and $\Delta l/\Delta t$ (the average speed) approaches the slope of the distance-time curve at P. The slope of any curve at any point is the slope of the *tangent* to the curve at that point (Fig. 2.9a). The slope precisely at P is the instantaneous speed at P, just as the slope anywhere along the curve of Fig. 2.3 is the constant instantaneous speed 10 m/s. From Fig. 2.9b, we see that the bee flew by your nose at 2.9 m/s.

The central notion here is that Δl will become infinitesimally small as Δt becomes infinitesimally small, and yet their ratio, $\Delta l/\Delta t$, will approach a finite limiting value: a tiny number divided by another tiny number need not itself be tiny. This limiting value approached by $\Delta l/\Delta t$ as $\Delta t \rightarrow 0$ (that is, as Δt gets infinitesimally small) is called the **instantaneous speed**, v. Mathematically that's written as:

A 0.30-caliber bullet "frozen" in an exposure lasting less than a millionth of a second. During that time, the bullet certainly moved, but not enough to blur its image. It had already cut the card in half before the picture was taken. The average speed of the bullet during the exposure is a practical value for its instantaneous speed at the "moment" shown.

$$v = \lim_{\Delta t \rightarrow 0} \left[\frac{\Delta l}{\Delta t}\right] \tag{2.4}$$

The instantaneous speed is the limiting value of the average speed ($\Delta l/\Delta t$) *determined as the interval over which the averaging takes place* (Δt) *approaches zero.* Henceforth, when we talk about "speed," this concept is what we have in mind. Notice that $\Delta l/\Delta t$ is to be determined as Δt *approaches,* rather than reaches, zero. It would make no sense to ask for the average speed over a zero time interval; nothing moves a real distance in no time, and the definition of average speed falls apart when $\Delta t = 0$. This subtlety was a point of confusion for centuries. It was to solve the very problem of instantaneous speed that Newton devised the calculus. Indeed,

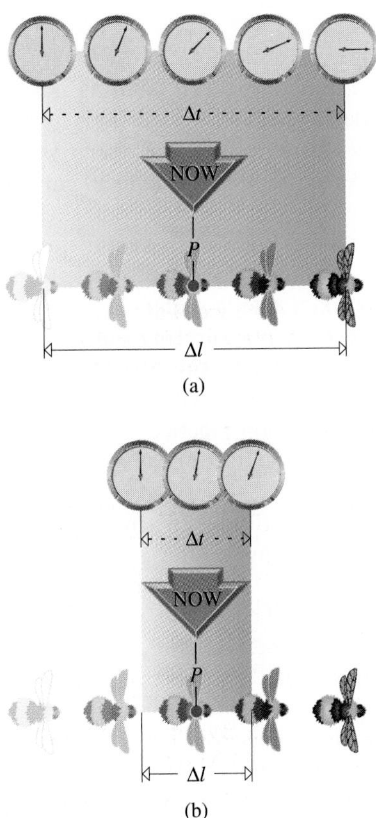

Figure 2.8 To find the instantaneous speed of the bee at point P, determine the average speed over a tiny interval straddling P. Then shrink Δl until v_{av} remains constant at the instantaneous speed, v.

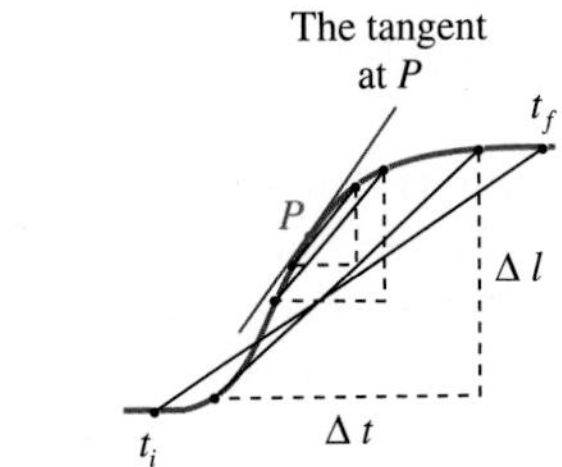

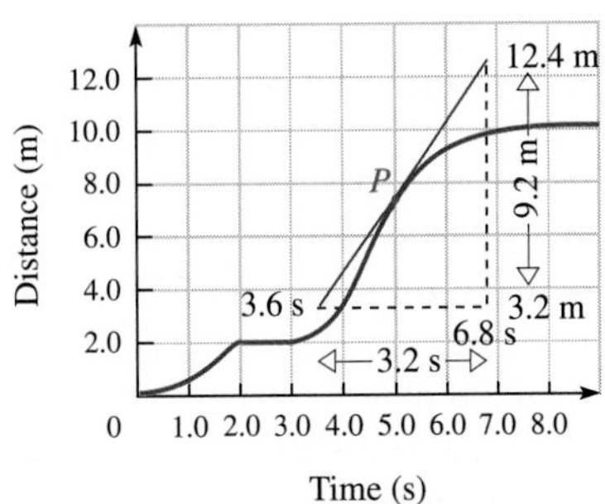

Figure 2.9 (a) As Δt shrinks, Δl shrinks and the line from t_i to t_f approaches the tangent to the curve at P. In this case, both t_i and t_f approach $t = 5.0$ s. (b) The slope of the tangent at P is $v = 9.2$ m/3.2 s $= 2.9$ m/s, and this value is the instantaneous speed at $t = 5.0$ s.

Eq. (2.4) serves as the definition of the *derivative,* written as $v = (dl/dt)$. Speed is the derivative of l with respect to t. Not to worry, we will not be using calculus. For us, it will suffice to think of v as v_{av} taken over a tiny interval Δt centered on NOW, where Δt is small enough so that v is constant within it.

On Newton, Leibniz and Limits. *Neither man [the two founders of the calculus] succeeded in doing more with the limit concept proper than confusing himself, his contemporaries, and even his successors. At one point in his* Principles, *Newton does state the correct version of the notion of instantaneous rate of change but apparently he did not recognize this fact, for in later writings he gave poorer explanations of the logic of his procedure.*

MORRIS KLINE (1953)
Mathematics in Western Culture

VELOCITY

To describe the motion of an object completely, we must specify both the rate of travel (that is, the speed) and its direction. The single concept that at once embraces both speed and direction is called **velocity**. Although we have managed thus far with just the notion of speed, more complex circumstances will require us to consider direction as well. A situation in which several motions occur simultaneously in different directions can only be analyzed using velocities. For example, a boy running east at 5 m/s throws a ball north at 10 m/s with respect to him. In what direction and at what speed relative to the ground will the ball travel? We now develop the mathematics needed to treat all such directional quantities (velocity, acceleration, force, momentum, and so on), the mathematics of *vectors* (Appendix D).

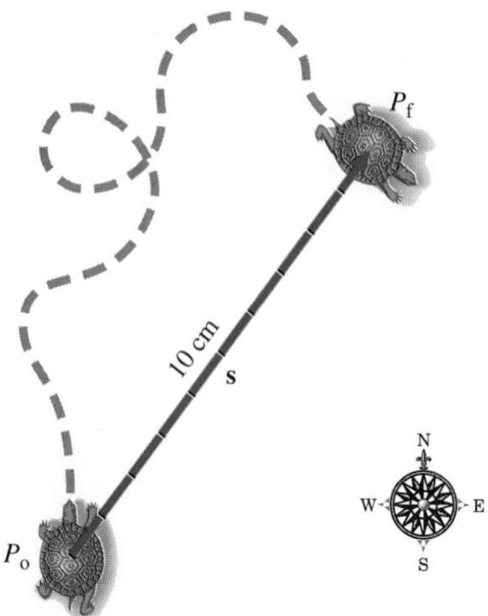

Figure 2.10 The displacement vector **s** of a turtle that meandered from point P_o to point P_f. The vector quantity **s** tells us where the turtle ended up in relation to where it started.

2.5 The Displacement Vector

Suppose we put a little turtle on a piece of paper at some point P_o and leave for a while. On our return, the turtle is discovered at a new spot, P_f. Nothing is known about the route it took, but we do know the turtle has been displaced. This *displacement is designated as the straight-line shift in position from P_o to P_f specifying both its length and direction.* In Fig. 2.10, the meandering reptile ended up 10 cm northeast of where it started. Any concept such as this that can be stated completely only if we provide both its magnitude (size) and direction is known as a **vector quantity**. It's standard in textbooks to designate vector quantities by boldfaced letters, reminding us that we are dealing with something other than a scalar.

To be precise, **the displacement of an object from a starting point P_o is the vector extending from that point to the position where the object is finally.** Displacement is often of considerable interest, as is the case with the displacement of a football from the person who threw it. It is common practice to represent the *displacement vector* by **s**, a symbol that comes from the idea of the space between two points.

In Fig. 2.10, the vector **s** can be represented pictorially by an arrow 10 cm long pointing northeast. And if 10 cm is inconveniently long to draw, simply scale it down to 10 "anythings" and just label it 10 cm. The *length* of the arrow representing a vector corresponds to what is called the **magnitude** (the positive numerical value) of that vector, and it is symbolized by an italic light typeface letter. Thus, the magnitude of the displacement vector **s** is s, which is a scalar specified by a single number, in this case equal to 10 cm.

Imagine now that the turtle takes off on a new trip across the page, this time traveling in a straight line east for 3 cm and thereby undergoing its first displacement $\mathbf{s}_1$, arriving at point A in Fig. 2.11. After an idle moment, it turns 90-degrees left, going from A straight north for 4 cm to point P_f, thereby undergoing a displacement $\mathbf{s}_2$. The *resultant* displacement with respect to P_o is the vector **s**, which is independent of the actual path taken in getting to P_f. Its magnitude can be computed from the right triangle using the famous theorem of Pythagoras (Appendix B), $s = \sqrt{3^2 + 4^2} = \sqrt{25}$. Fortunately, the turtle walked two sides of a 3-4-5 right triangle, so s, the hypotenuse, comes out neatly as 5 cm long—P_f is 5 cm from P_o in a direction somewhat east of north.

There is a considerable difference between *distance* and *displacement.* For instance, imagine yourself throwing a football to someone running in zigzags. He started at your side and is now far down the field, dodging the opposition. You certainly might worry about the total distance traveled by the runner; surely he is considering it. But you are more concerned with the moment-by-moment *displacement* of the potential receiver—with the straight-line arrow from you to him.

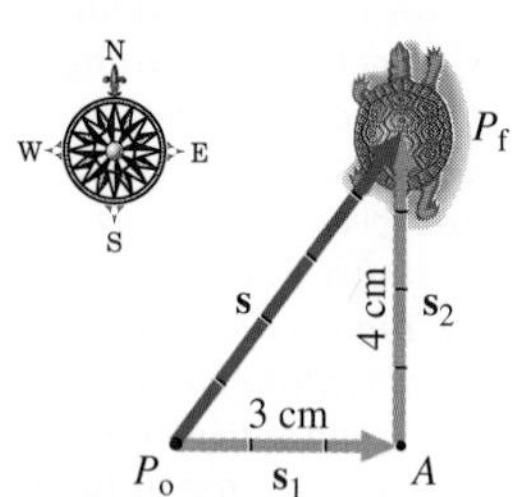

Figure 2.11 The total displacement **s** of a turtle that traveled from P_o to A to P_f via displacements $\mathbf{s}_1$ and $\mathbf{s}_2$. The total displacement equals the vector sum of the individual displacements: $\mathbf{s} = \mathbf{s}_1 + \mathbf{s}_2$.

2.6 Some Vector Algebra

We have not completely defined what a vector is until we establish some rules for its behavior. One such needed operation is **vector addition**. We do not have a complete idea of what numbers, or scalars, are until we know how to manipulate them, and the same is true with vectors.

Referring to Fig. 2.11, we can envision somehow adding the intermediate displacements ($\mathbf{s}_1 + \mathbf{s}_2$) to arrive at the total displacement **s**, but it's evident that we cannot simply add in the usual algebraic manner. The total path-length traveled is $s_1 + s_2 = 7$ cm, while the magnitude of the total displacement is $s = 5$ cm, and obviously $s \neq s_1 + s_2$. The illustration itself provides a general rule for adding vectors known as the **tip-to-tail method**. *Two vectors can be added together by arranging them such that the tail of the second is at the tip of the first; the sum of these, also* *called the* **resultant vector**, *extends from the tail of the first to the tip of the second.* In the process of adding two vectors, the sum will not change if either one (or both) is moved around, provided it is kept parallel to its original direction.

Notice that had the turtle walked north first and then east (inverting the order of the displacements), it would still have arrived at P_f, which means that $\mathbf{s} = \mathbf{s}_1 + \mathbf{s}_2 = \mathbf{s}_2 + \mathbf{s}_1$. Just as with ordinary scalars, the order of addition is irrelevant, and you can put either tip to either tail when adding two vectors.

Figure 2.12 makes the same point, this time with vectors that form an oblique triangle. It suggests another technique for graphically adding two vectors. Beginning with the tip-to-tail scheme and realizing that the order of addition is irrelevant, we can construct the resultant vector in two ways that, taken together, form a parallelogram. In other words, draw the two vectors we wish to add, $\mathbf{s}_1$ and $\mathbf{s}_2$, emerging from a *common origin.* Using these as adjacent sides, construct a parallelogram. *The diagonal emerging from the origin is then the resultant of the two.* This procedure is called the **parallelogram method**.

Once we have a rule for adding two vectors, we have a rule for adding any number of vectors. Thus in Fig. 2.13, we can add the four vectors, $\mathbf{A} + \mathbf{B} + \mathbf{D} + \mathbf{F}$,

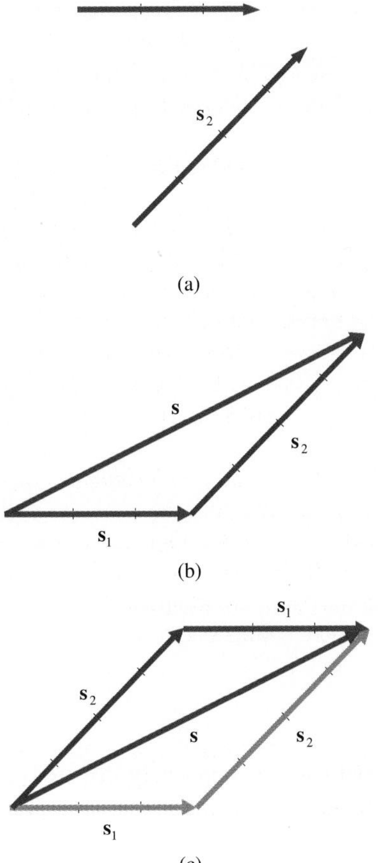

Figure 2.12 (a) Two vectors that are to be added in two ways. (b) The tip-to-tail addition of two vectors ($\mathbf{s}_1 + \mathbf{s}_2 = \mathbf{s}_2 + \mathbf{s}_1 = \mathbf{s}$). (c) The parallelogram formed with $\mathbf{s}_1$ and $\mathbf{s}_2$ as its sides.

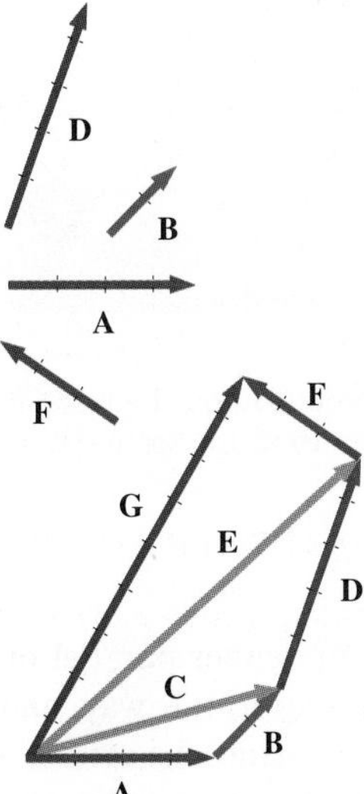

Figure 2.13 The addition of several vectors ($\mathbf{A} + \mathbf{B} + \mathbf{D} + \mathbf{F}$) via the tip-to-tail method. Here $\mathbf{C} = \mathbf{A} + \mathbf{B}$, $\mathbf{E} = \mathbf{C} + \mathbf{D}$, and $\mathbf{G} = \mathbf{E} + \mathbf{F} = \mathbf{A} + \mathbf{B} + \mathbf{D} + \mathbf{F}$. Knowing how to add two vectors allows us to add any number of vectors.

two at a time, following the tip-to-tail method. Accordingly, **A** plus **B** is **C**, to which we add **D**, such that **C** + **D** = **E**, to which we add **F**, such that **E** + **F** = **G**, so that **A** + **B** + **D** + **F** = **G**. Evidently, it is not really necessary to bother with the intermediate sums of **C** and **E**. The figure shows that we could have just as well tipped-to-tailed **A**, **B**, **D**, and **F** directly and, for that matter, we could have done it in any order at all.

Example 2.6 A treasure map was torn into six pieces by its maker, a clever one-eyed gentleman with a parrot. The first piece shows a large recognizable tree and simply reads "start here and go." Although in no particular order, each of the other five fragments reads, in turn: "3 m–east," "5 m–north," "5 m–southeast," "2.5 m–south," and "6.5 m–west." Find the treasure.

Solution: Each of the five instructions is a displacement vector and vector addition is independent of order. It follows from Fig. 2.14 that the treasure is buried 1 m due south of the old tree or, if you like, the resultant displacement vector has a magnitude of 1 m and points south. Try this example with cut matchsticks; so long as their lengths and orientations remain constant, the resultant will be the same for a fixed starting point independent of the order in which the sticks are put down.

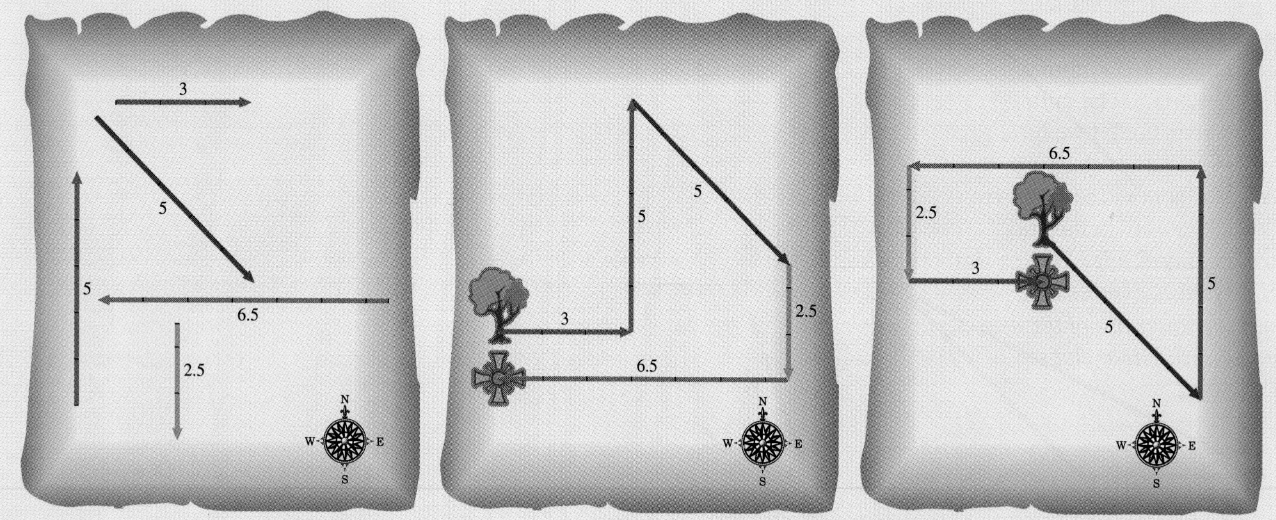

Figure 2.14 The five vectors in (a) can be added in any order to find the treasure. In (b), we start at the tree and first move east 3 m. In (c), that 3-m displacement is last, but we still arrive at the same spot where the treasure is buried.

Vectors that are either parallel or antiparallel (that is, oppositely directed) are added in the usual tip-to-tail way, and the results are simple and useful (Fig. 2.15). In the case of two parallel vectors, $\mathbf{s}_1$ and $\mathbf{s}_2$, the magnitude of the resultant, s, equals the sum of the magnitudes of the scalar distances.

Suppose now that we add to a vector **s** an identical vector **s**. The resultant of (**s** + **s**) lies in the same direction as either constituent **s** and is twice as long as either one. It would seem reasonable from the usual procedures of algebra to write the resultant as 2**s** = **s** + **s**, which defines the process of **multiplication of a vector by a scalar**. Accordingly, if **s** is a displacement of 10 cm–east, 6**s** is a displacement of

60 cm–east. *Multiplying a vector by a positive scalar multiplies the vector's magnitude in the usual way, leaving its direction unchanged.*

If two vectors are antiparallel (Fig. 2.15b), the magnitude of the resultant equals the difference of the individual magnitudes, $s_1 - s_2$. *Vectors that are oppositely directed subtract algebraically.* A displacement of 10 m–east, followed by a displacement of 10 m–west, yields a total displacement of zero. If we interpret the symbol $(-\mathbf{s})$ to be a vector equal in magnitude but antiparallel to $\mathbf{s}$, then the sum $\mathbf{s} + (-\mathbf{s}) = \mathbf{s} - \mathbf{s} = 0$, so the two cancel one another in the usual algebraic way. Accordingly, **vector subtraction** can be defined as follows: any vector $\mathbf{B}$ can be subtracted from any vector $\mathbf{A}$ by reversing its direction to form $-\mathbf{B}$ and then adding them in the usual tip-to-tail way (Fig. 2.16).

Multiplying a vector by a negative scalar multiplies the vector's magnitude in the usual way but also reverses its direction. If $\mathbf{s}$ is a displacement of 10 cm–east, $-6\mathbf{s}$ is a displacement of 60 cm–west. Note that the magnitude is always positive.

Each of these operations (vector addition, subtraction, and multiplication by a scalar) has its direct application in our description of the physical Universe.

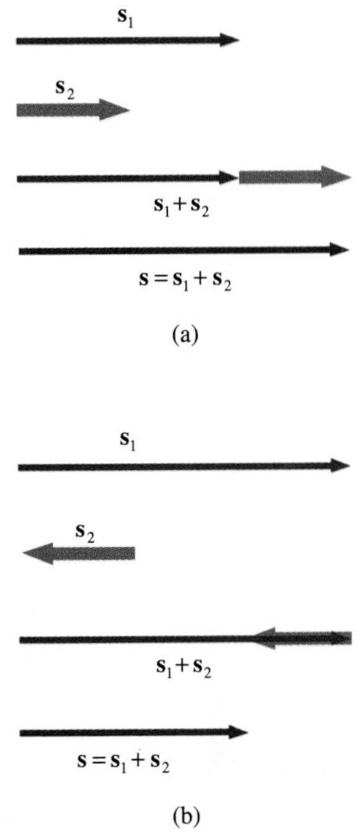

Figure 2.15 (a) The sum of two parallel vectors. Vectors that point in the same direction can be added like scalars in that $s = s_1 + s_2$. (b) The sum of two antiparallel vectors. Vectors that are oppositely directed subtract as do scalars, in that $s = s_1 - s_2$.

2.7 Average Velocity

Speed is a scalar that tells us "how fast" without dealing with "in what direction." But more often than not, direction is an essential aspect of motion. There is a difference between the command "March at 2 km/h" and "March at 2 km/h–south, toward the cliff." The culminating concept of this chapter is *instantaneous velocity,* and to understand it we have to examine what happens to the average velocity vector as $\Delta t \to 0$. As we shall see, the instantaneous velocity is the vector whose direction at any moment is the direction of motion and whose magnitude is the speed.

Back to our bumblebee, which in a time t now flies from P_o to P_f in an arc shown in Fig. 2.17a. We see from Eq. (2.1) that the bee's average speed is $v_{av} = l/t$, and, with this expression as a guide, we define the **average velocity** to be the *vector formed by the ratio of the displacement, measured from the starting point of the motion, over the time elapsed,* or

$$\mathbf{v}_{av} = \frac{\mathbf{s}}{t} \tag{2.5}$$

Since t is a positive scalar, multiplying $\mathbf{s}$ by $1/t$ has no effect on its direction; $\mathbf{v}_{av}$ is parallel to $\mathbf{s}$. Average velocity is a necessary idea on the way to considering instantaneous velocity, although by itself $\mathbf{v}_{av}$ is of limited usefulness—a racecar roaring around a track will have zero displacement and zero average velocity whenever it returns to its starting point, no matter how fast it is going.

Example 2.7 A pulse from a laser travels at a constant speed in a straight line reaching a detector 3×10^4 m due north in 100 μs. Calculate the average velocity of the beam.

Solution: [Given: speed is constant; $\mathbf{s} = 3 \times 10^4$ m–north; $t = 100 \times 10^{-6}$ s. Find: $\mathbf{v}_{av}$.] The beam travels with a constant speed in a straight line, so it has a constant velocity. This constant velocity is also equal to its average velocity, namely

$$\mathbf{v}_{av} = \frac{\mathbf{s}}{t} = \frac{3 \times 10^4 \text{ m–north}}{100 \times 10^{-6} \text{ s}} = \boxed{3 \times 10^8 \text{ m/s–north}}$$

The displacement is due north, and so the average velocity is due north.

▶ **Quick Check:** $v_{av}t = s = (3 \times 10^8 \text{ m/s})(10^{-4} \text{ s}) = 3 \times 10^4$ m.

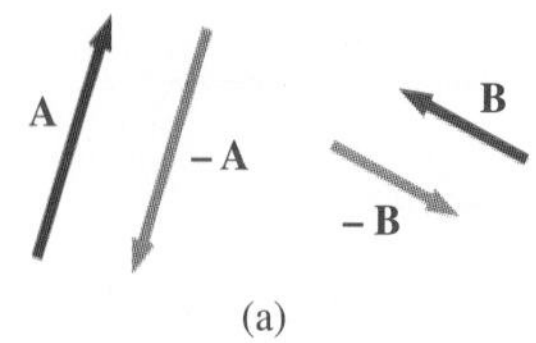

(a)

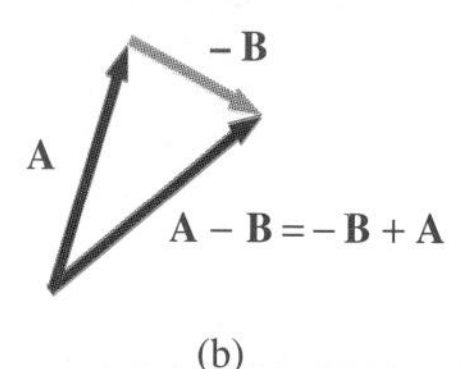

(b)

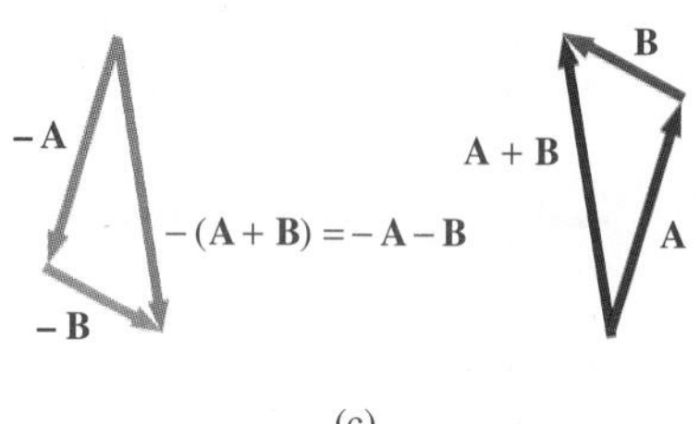

(c)

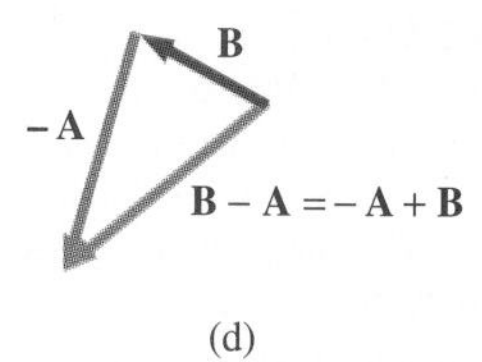

(d)

Figure 2.16 The sum and differences of two vectors. In (a) we see vectors **A** and **B** and their negatives −**A** and −**B**; (b) shows **A** − **B**; and (c) shows −**A** − **B** and **A** + **B**; (d) adds −**A** and **B** to get **B** − **A**. Notice that −**B** + **A** is the negative of −**A** + **B**.

Before the idea of instantaneous speed was introduced (p. 29) we studied the average speed in terms of intervals, $\Delta l/\Delta t$. The same thinking applies to the average velocity where we now focus on a portion of the flight of length Δl. As the bee in Fig. 2.18 flies the segment of its path from P_i to P_f in a time Δt, the vector **s**, measured from P_o, changes as it tracks the bee instant-by-instant. The displacement with respect to P_o initially, $\mathbf{s}_i$, plus the change in the displacement, $\Delta\mathbf{s}$, equals the final displacement with respect to P_o, $\mathbf{s}_f$ (Fig. 2.19). The symbolic form is ($\mathbf{s}_i + \Delta\mathbf{s} = \mathbf{s}_f$) and, therefore, $\Delta\mathbf{s} = \mathbf{s}_f - \mathbf{s}_i$, which is reminiscent of Eq. (2.3). Note that $\Delta\mathbf{s}$ is also the displacement of the bee at P_f with respect to its starting point P_i.

Using the definition as stated in Eq. (2.5), we can now express the average velocity over the curved route from P_i to P_f as

$$\mathbf{v}_{av} = \frac{\Delta\mathbf{s}}{\Delta t} = \frac{\mathbf{s}_f - \mathbf{s}_i}{t_f - t_i} \tag{2.6}$$

Notice that when P_i and P_f are coincident, $\Delta\mathbf{s} = 0$ and $\mathbf{v}_{av} = 0$.

2.8 Instantaneous Velocity

Earlier, we transformed the notion of average speed into instantaneous speed by causing the time interval over which the averaging was happening to shrink toward zero. We now apply the same limiting process to the average velocity, Eq. (2.6), to get the *instantaneous velocity* **v** (or the velocity, for short).

Figure 2.17 The distance traveled (l) and the displacement (**s**) in the two different cases of (a) motion along a curve, and (b) motion along a straight line. For the special case of straight-line motion in a fixed direction, the magnitude of the average velocity s/t equals the average speed, l/t. That is generally not the case.

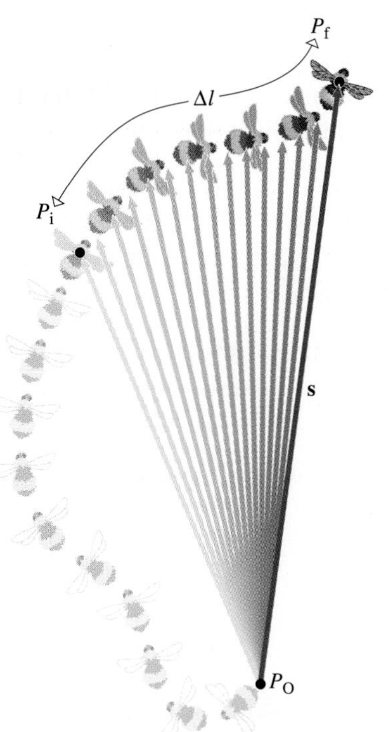

Figure 2.18 The bumblebee flies from P_i to P_f always tracked by the vector **s**. The distance it travels in the time Δt is Δl.

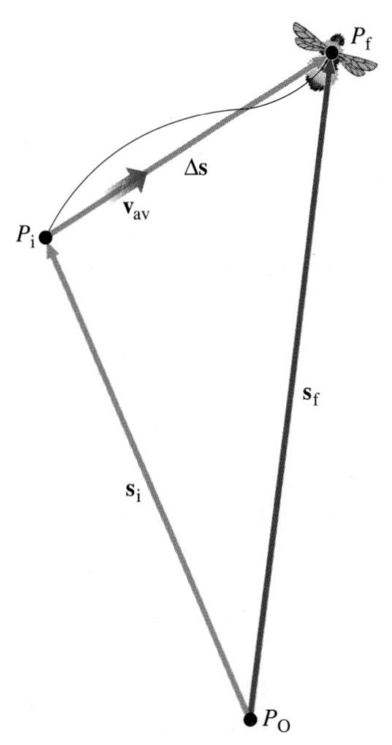

Figure 2.19 The change in the displacement of the bumblebee from P_i to P_f occurring in a time Δt is Δ**s**. Notice that $\mathbf{v}_{av}$ is in the direction of Δ**s**.

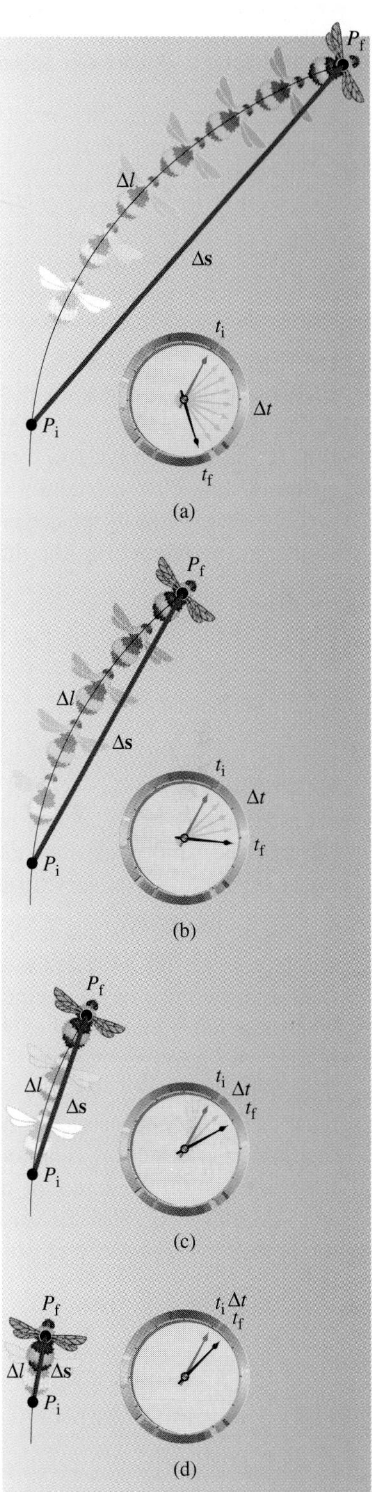

Figure 2.20 The bee flies from point P_i to point P_f. In successive parts of the figure, from (a) to (d), less and less time passes, and point P_f comes closer and closer to P_i.As Δt shrinks, Δs approaches Δl, and the magnitude of **v** approaches v.

The bee in Fig. 2.19 flies from P_i to P_f in a time Δt, experiencing a displacement in the process, Δ**s**. As the time, Δt, allowed for the bee to fly shrinks, the insect cannot travel as far; P_f moves nearer to P_i and Δ**s** diminishes as well (Fig. 2.20). Accordingly, we define the **instantaneous velocity** *at any point* (P_i) *as the limit* (*as* $\Delta t \rightarrow 0$) *of the ratio of the displacement of the object from that starting point over the time interval,* or

$$\mathbf{v} = \lim_{\Delta t \rightarrow 0} \left[\frac{\Delta \mathbf{s}}{\Delta t} \right] \tag{2.7}$$

Now that we have **v** mathematically, let's see what it is physically. There are two crucial insights here: first, as P_f approaches P_i, the magnitude of Δ**s**, that is Δs, gradually shrinks and the change in path-length, Δl, shrinks right along with it (Fig. 2.20). For extremely small values of Δt, Δs and Δl become increasingly indistinguishable. We can conclude that as $\Delta t \rightarrow 0$, the two merge and become equal, which means that in the limit $\Delta s/\Delta t = \Delta l/\Delta t$; *the magnitude of the instantaneous velocity vector,* **v**, *equals the instantaneous speed* v. That is why we used the symbol v for speed in the first place—*speed is the magnitude of the velocity.*

The second important quality of instantaneous velocity as defined by Eq. (2.7), is its direction. As the vector Δ**s** diminishes in size (Fig. 2.20) with decreasing Δt, its end point glides along the curve toward P_i. As $\Delta t \rightarrow 0$, P_f moves closer and closer to sitting on top of P_i and $\Delta\mathbf{s}$, however small, becomes tangent to the path at P_i. Since **v** is defined by the limit of this shrinking process, *the velocity vector at any point is tangent to the path, pointing in the direction of motion* (Fig. 2.21). The magnitude of the velocity vector is the speed, and its direction is always tangent to the path in space.

Example 2.8 A particle-beam pulse fired due north, 45° up from the horizon strikes a distant orbiting satellite. If the beam travels in a straight line at a constant speed of 2.5×10^8 m/s and it takes 1.0 ms to reach the target, what is the displacement of the pulse from the source at the moment it hits the satellite?

Solution: [Given: $\mathbf{v} = 2.5 \times 10^8$ m/s–north-45° up, and $t = 1.0$ ms. Find: **s**.] The instantaneous velocity of the beam is constant, hence

$$\mathbf{s} = \mathbf{v}t = (2.5 \times 10^8 \text{ m/s–north-45° up})(1.0 \times 10^{-3} \text{ s})$$

and so $\boxed{\mathbf{s} = 2.5 \times 10^5 \text{ m–north-45° up}}$

▶ **Quick Check:** because the motion is in a straight line $s = vt$; $v = 1/4 \times 10^9$ m/s, $t = 10^{-3}$ s, $s = 1/4 \times 10^6$ m.

Example 2.9 A large clock mounted on a wall in a railroad station has a second hand 1.0 m long. Assuming that the hand sweeps smoothly around, find the velocity of a point on its very tip at exactly 15 s after noon.

Solution: [Given: the point moves around on a circle of radius $r = 1.0$ m, its speed is uniform, and it traverses the circumference in 60 s. Find: **v** at $t = 45$ s.] Most of what is given is not stated explicitly—you need to recognize the implied. Nature is rarely explicit. *Draw a diagram* (Fig. 2.22). We have to find the point's instantaneous velocity—speed and direction—when the second hand is horizontal. The total distance swept out by the very tip of the hand in one revolution is $2\pi r = 2(3.14)1.0 \text{ m} = 6.28$ m. Because the motion is uniform, the point's average speed equals its instantaneous speed, thus

$$v = \frac{l}{t} = \frac{6.28 \text{ m}}{60 \text{ s}} = 0.10 \text{ m/s}$$

The velocity 15 s after noon is then $\boxed{0.10 \text{ m/s–straight downward}}$, while 15 s before noon it's 0.10 m/s–straight upward.

Figure 2.22 A clockface at 15 s after 12:00. At the instant shown, the point on the very end of the second hand is moving downward with a velocity **v**. At 15 s before noon, the hand will be pointing at 9 and **v** will be directed straight upward.

▶ **Quick Check:** The velocity is always tangent to the circle (that is, perpendicular to the radius), pointing in a direction determined by the clockwise motion. $2\pi R \approx 6$ m, $v \approx 6 \text{ m}/60 \text{ s} \approx 0.1$ m/s.

Instantaneous velocity cannot be measured with the same accuracy that it can be thought about. In the end, Δt will be as tiny as you can measure it, but finite, meaning that we will always be restricted to measuring average velocity. This restriction holds even if we measure so rapidly there is no practical distinction between average velocity and the ideal of instantaneous velocity.

One-Dimensional Motion

An especially simple graphical relationship exists between displacement and velocity when the motion is confined to one dimension. To illustrate, suppose an object moves on a straight line (Fig. 2.23) along the x-axis. Let the magnitude of the displacement from the origin O be s_x (Fig. 2.24). This object is already moving with a velocity of 0.6 m/s in the positive x-direction at $t = 0$, and it continues at that rate (constant slope) until $t = 9$ s, when the speed slowly begins to decrease (Fig. 2.24). At $t = 11.5$ s, the tangent to the displacement-time curve has zero slope and the speed is zero. Because we are plotting displacement from O and not the total distance traveled, s_x can decrease and the curve will have a negative slope whenever the object moves back toward O. At $t = 19.6$ s, the object momentarily stops and thereafter moves in the positive x-direction.

The velocity is always in the direction of $\Delta\mathbf{s}_x$ which is the direction of motion. When the slope of Fig. 2.24 is positive, **v** points away from the origin, and when the slope is negative, **v** points toward the origin. Thus, since we know the direction of motion, we can talk about velocity and not just speed. *For 1-dimensional motion, the slope of the displacement-time curve at every point in time is the velocity at that instant.* Fig. 2.25 is a plot of the instantaneous velocity determined from the tangents to the curve in Fig. 2.24.

Figure 2.21 The velocity is everywhere tangent to the path, pointing in the direction of motion.

RELATIVE MOTION

Most people believe that they can know with certainty that either a thing is moving or it is not; they think that motion is absolute. But the Earth is orbiting the Sun (at about 66 000 mi/h), and the Sun is whirling along with the galaxy (at about 480 000 mi/h), and the whole Universe is expanding. Nothing is at rest; besides, and even more importantly, there is no way to measure the difference between rest and uniform motion. There is no physical difference between rest and uniform motion. That's why it feels as though you are at rest at this very moment, sitting reading this book.

We learn from the Special Theory of Relativity that motion is relative, not absolute. Motion is always measured with respect to something. We usually hold the Earth to be at "rest," and think about objects moving relative to it. Even so, there are countless situations not quite that simple: two things moving with respect to the Earth may surely be moving relative to each other, and that latter motion might even

The plane moving at a constant speed is an inertial reference frame. The rockets are traveling at relatively low speeds with respect to it. By comparison, the speed of the rockets with respect to the ground is much higher.

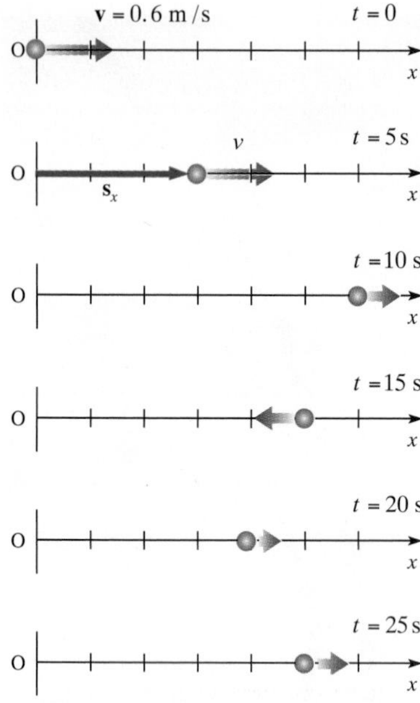

Figure 2.23 The motion of an object in one dimension at successive 5-s intervals. Notice that sometime between $t = 10$ s and $t = 15$ s, the object stops and reverses direction. It can't move along a line and change direction without stopping, if only for an instant.

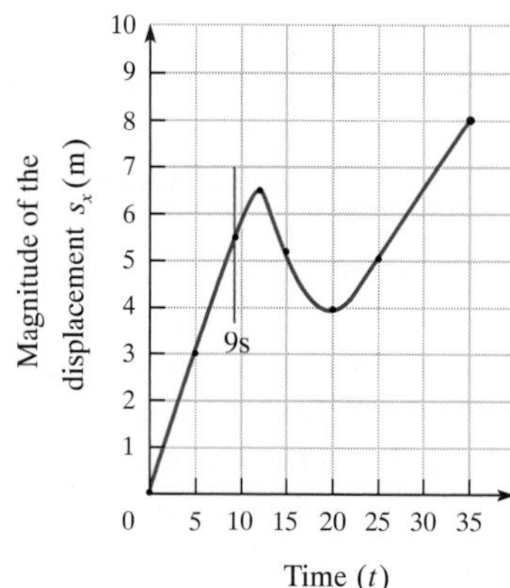

Figure 2.24 A plot of the magnitude of the displacement versus time for that object as it moves along its straight path.

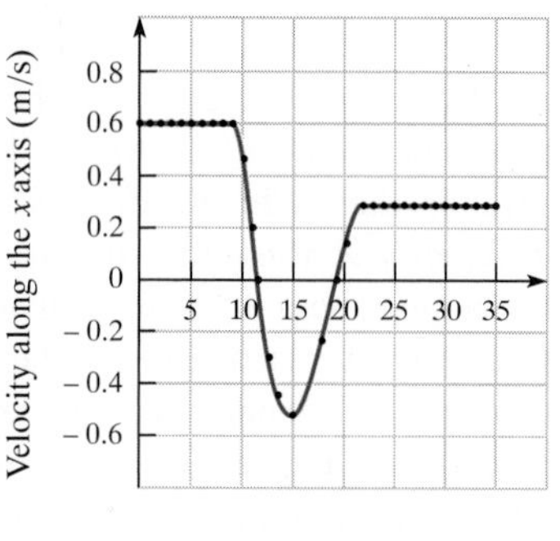

Figure 2.25 The velocity along the x-axis determined each second from the slope of the curve of Fig. 2.24. At $t = 11.5$ s and $t = 19.6$ s, the slope of the displacement versus time curve is zero and the velocity is zero.

be the more important one. A front-runner in a race should be as concerned about her velocity relative to the person in second place as with her velocity relative to the finish line. A pilot landing a plane on the deck of a moving aircraft carrier has relatively similar concerns.

Consider a laboratory that is either at rest or moving with a constant velocity. Such a system is called an *inertial frame of reference* or an *inertial system.* In the remainder of this chapter, we study uniformly moving objects traveling with respect to inertial systems. In most cases of motion occurring on the surface of the Earth, the planet itself (that is, the ground) serves as just such an inertial frame (Fig. 2.26).

The central idea in relative motion is *velocity.* To combine the effects of several simultaneous motions, we need only to add the corresponding velocity vectors. The procedure is an elegant example of the power of the vector formulation.

2.9 Velocity with Respect to . . .

Consider the journey of a little turtle meandering eastward along the moving ruler of Fig. 2.27. At the last moment, frozen in the diagram, the *r*uler has been displaced with respect to the *E*arth by $\Delta\mathbf{s}_{\mathrm{RE}}$. The *t*urtle, in turn, has undergone a displacement with respect to the end of the *r*uler of $\Delta\mathbf{s}_{\mathrm{TR}}$. The total displacement of the *t*urtle with respect to the *E*arth (that is, with respect to the starting point P_{o} fixed on the ground) is given by the vector sum

$$\Delta\mathbf{s}_{\mathrm{TE}} = \Delta\mathbf{s}_{\mathrm{TR}} + \Delta\mathbf{s}_{\mathrm{RE}} \tag{2.8}$$

An observer sitting stationary on the Earth would see the turtle travel from P_o to P_f in some time interval Δt. Consequently, if we divide both sides of the equation by Δt and take the limit as $\Delta t \rightarrow 0$, we get, via Eq. (2.7), the corresponding velocity expression, namely

$$\mathbf{v}_{\mathrm{TE}} = \mathbf{v}_{\mathrm{TR}} + \mathbf{v}_{\mathrm{RE}} \qquad (2.9)$$

If, for example, the velocity of the *t*urtle with respect to the *r*uler is 0.01 m/s–east, and the velocity of the *r*uler with respect to the *E*arth is 0.50 m/s–east, then $\mathbf{v}_{\mathrm{TE}} =$ 0.51 m/s–east. The velocity vectors add to give the resultant, just as the displacement vectors do. Remember that the equation is really quite general even though the subscripts may change; it applies as well to *t*urtles, *r*ulers, and *E*arths as to *t*rains, *r*ubberbands, and *e*lephants.

The order of the subscripts in Eq. (2.9) is crucial to the logic, even though the order of the vector addition is not. When setting up expressions such as this one, arrange the subscripts so that the first and last on one side of the equation are the same as the first and last on the other side. It is convenient to imagine that identical letters cancel when they come one after the other in the sequence of subscripts on the right. For instance, the vector addition effectively removes the R subscripts in Eq. (2.9)—the intermediate role of the moving ruler is subsumed.

Suppose that the turtle in Fig. 2.27 turns around and begins heading *west* at a speed with respect to the ruler of 0.01 m/s. Equation (2.9) can still be used. In fact, it would apply regardless of the directions of the two vectors. In the present instance, $\mathbf{v}_{\mathrm{TR}}$ and $\mathbf{v}_{\mathrm{RE}}$ are antiparallel. The resulting speed is the *difference* between their speeds, and subtracting the smaller from the larger yields $\mathbf{v}_{\mathrm{TE}} = 0.49$ m/s–east. The turtle approaches the left edge of the ruler but, because v_{RE} is greater than v_{TR}, it is nonetheless carried east of the fixed point P_f in Fig. 2.28.

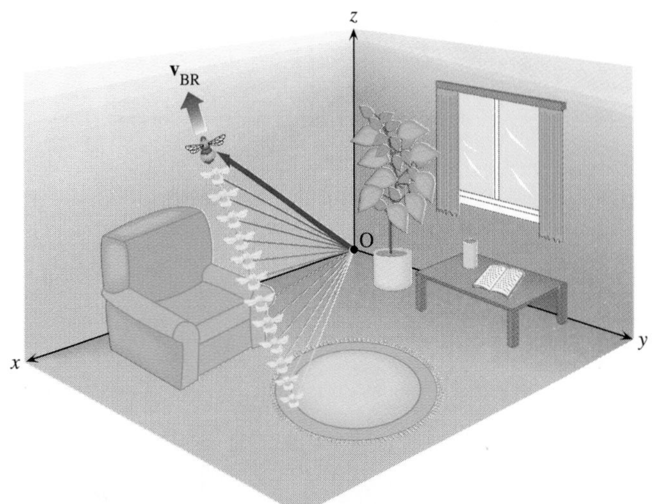

Figure 2.26 A bee flying a straight path in a room. The room is fixed to the ground and represents an inertial reference frame. A vector from the origin, O, to the bee tracks the insect's motion. The velocity of the bee with respect to the room is $\mathbf{v}_{\mathrm{BR}}$. If the room happens to be in a mobile home, the bee's motion with respect to the ground can get rather complicated.

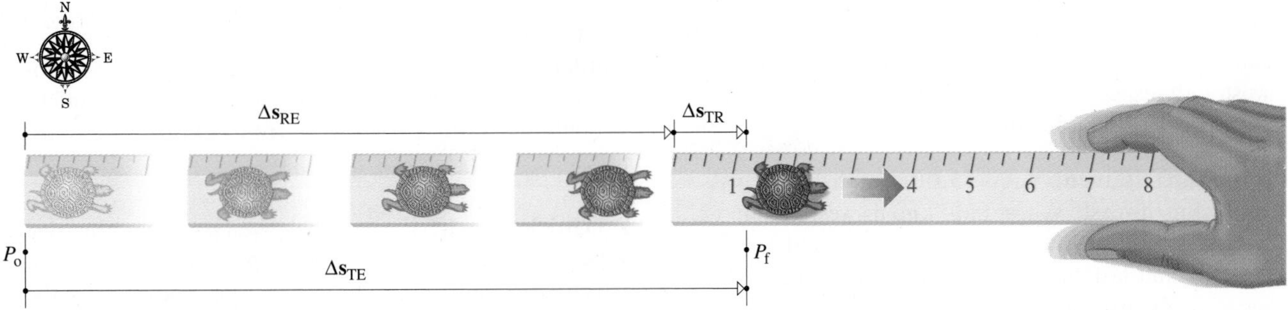

Figure 2.27 The displacement of a turtle walking on a moving ruler. The turtle walks to the right with respect to the ruler, while the ruler moves right with respect to the Earth. As a result, the little traveler ends up a rather large distance Δs_{TE}, from the point on Earth, P_o, where it started.

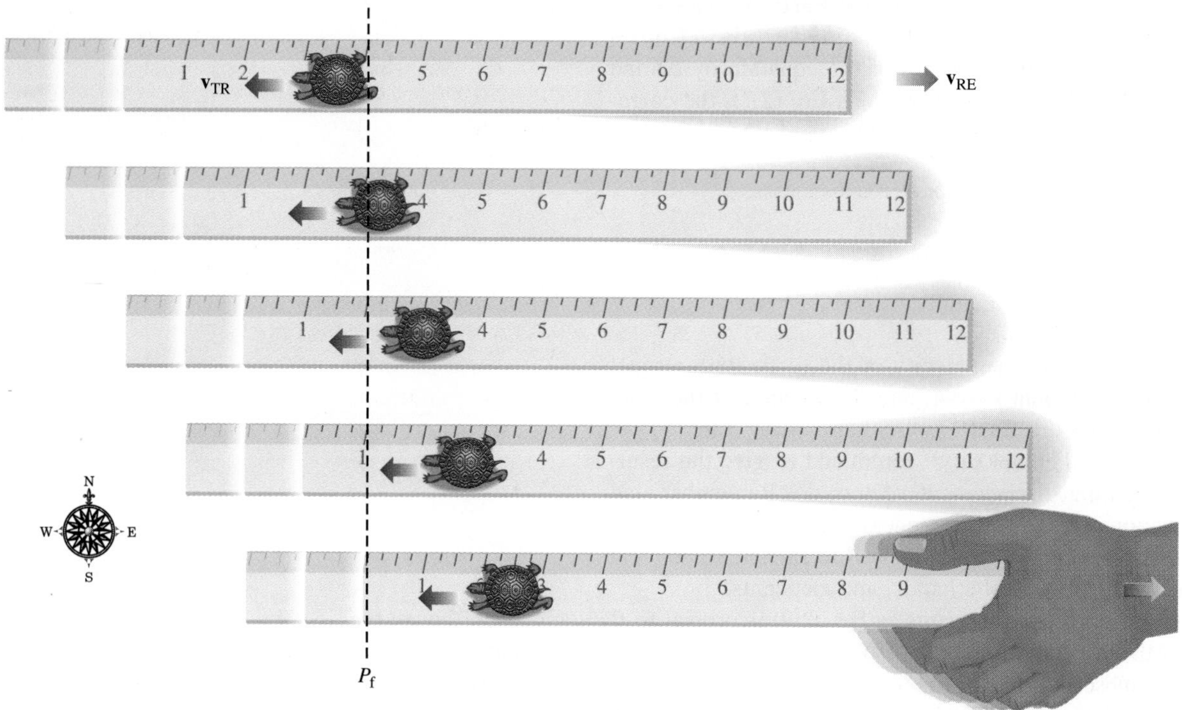

Figure 2.28 With respect to the Earth, the turtle travels east even though it's heading toward the west end of the ruler. The situation is much like that of a person walking to the back of a jet airliner while flying at 600 mi/h. If the walk takes 1 min, the plane travels 10 mi in the process.

On the other hand, if the turtle began moving west at P_f with a constant speed equal to that of the ruler (0.50 m/s), it would again march to the left end of the ruler and yet still be trudging along "in place" at P_f. Its speed relative to the Earth would be zero. This is what occurs when someone aboard a train pulling out of a station runs to the rear of the car trying to linger another moment face-to-face with a friend standing on the platform. It is the classic treadmill scheme well known to rodents and cardiologists.

Example 2.10 A big, black dog saunters down the aisle of an Amtrak Pullman at a velocity of 5.00 km/h–east with respect to the car. The train, all the while, is traveling at 10.00 km/h–east. A resident flea heading rumpward along the hound's back moves at 0.01 km/h–west with respect to the dog. Find the velocity of the flea with respect to the Earth.

Solution: [Given: $\mathbf{v}_{DT}$ = 5.00 km/h–east; $\mathbf{v}_{TE}$ = 10.00 km/h–east; $\mathbf{v}_{FD}$ = 0.01 km/h–west. Find: $\mathbf{v}_{FE}$.] Extending Eq. (2.9) to a three-vector sum and arranging the first and last subscripts on the right to be F and E, respectively, we get

$$\mathbf{v}_{FE} = \mathbf{v}_{FD} + \mathbf{v}_{DT} + \mathbf{v}_{TE} \tag{2.10}$$

Notice that the pairs of Ds and Ts will effectively cancel, yielding the desired resultant. In particular, $\mathbf{v}_{FE}$ = (0.01 km/h–west) + (5.00 km/h–east) + (10.00 km/h–east). Because parallel and antiparallel vectors can be dealt with algebraically, we simply add the speeds of similarly directed velocities and subtract those of oppositely directed ones. The easterly speed is greater, and so we subtract 0.01 km/h–west from 15.00 km/h–east to get $\boxed{\mathbf{v}_{FE} = 14.99 \text{ km/h–east}}$.

▶ **Quick Check:** Even though the flea is running due west, it is going east with respect to the ground—still, it is getting right to the end it wants. Both the dog and the train are going east, so $\mathbf{v}_{FE} \approx 15$ km/s is reasonable. Some keen-eyed observer outside at rest watching this circus will see the train, the dog, and the flea all moving east at different speeds.

Now, for a slight variation, imagine the midnight mail train carrying a safe full of gold heading due north at v_{TE} = 20 km/h relative to the Earth. A robber, riding hard at $\mathbf{v}_{RE}$ = 25 km/h–north next to the caboose, is about to leap onto the train. What is his speed relative to the train (v_{RT})? His speed is his main concern because he wants to land on the train, not the ground. Assuredly, unless he misses, he couldn't care less about either $\mathbf{v}_{RE}$ or $\mathbf{v}_{TE}$ individually.

We want $\mathbf{v}_{RT}$ and can construct an appropriate vector equation for it, namely

$$\mathbf{v}_{RT} = \mathbf{v}_{RE} + \mathbf{v}_{ET} \tag{2.11}$$

which is fine except we don't have $\mathbf{v}_{ET}$. Notice that *reversing the order of the subscripts reverses the direction of the vector.* For example, $\mathbf{v}_{ET} = -\mathbf{v}_{TE}$; that is, the motion of the train with respect to the sheriff standing still on the Earth is 20 km/h–north (that is, $\mathbf{v}_{TE}$), but to the robber, once he makes his jump onto the caboose, the sheriff will be receding evermore *southward* at 20 km/h from him (that is, $\mathbf{v}_{ET}$). Therefore Eq. (2.11) can be rewritten in terms of vectors whose values we already know, namely

$$\mathbf{v}_{RT} = \mathbf{v}_{RE} + (-\mathbf{v}_{TE})$$

$$\mathbf{v}_{RT} = (25\text{ km/h–north}) + (-20\text{ km/h–north}) \tag{2.12}$$

This gives the robber a velocity just before he leaps of 5.0 km/h–north with respect to the train. The analysis applies as well to spaceships docking, airplanes midair-refueling, relay runners baton-passing, and police cars hotly pursuing.

It does not appear to me that there can be any motion other than relative; *so that to conceive motion there must be at least conceived two bodies, whereof the distance or position in regard to each other is varied. Hence, if there was only one body in being it could not possibly be moved. This seems evident, in that the idea I have of motion does necessarily include relation.*

GEORGE BERKELEY
The Principles of Human Knowledge (1710)

Example 2.11 Two knights with lances and banners are about to joust. Sir John the Slow gallops south at 5.0 km/h, while Sir Peter the Fast races north at 25 km/h. At what speed do they close the gap between one another?

Solution: [Given: $\mathbf{v}_{JE}$ = (5.0 km/h–south); $\mathbf{v}_{PE}$ = (25 km/h–north). Find: $\mathbf{v}_{JP}$.]

$$\mathbf{v}_{JP} = \mathbf{v}_{JE} + \mathbf{v}_{EP}$$

We don't have $\mathbf{v}_{EP}$, but it equals $(-\mathbf{v}_{PE})$, and we have that. Thus,

$$\mathbf{v}_{JP} = \mathbf{v}_{JE} - \mathbf{v}_{PE}$$

$$\mathbf{v}_{JP} = (5.0\text{ km/h–south}) - (25\text{ km/h–north})$$

The minus sign means that, if we do the vector addition graphically, we must draw the second vector reversed; that is, pointing south. The resultant is then

$$\boxed{(30\text{ km/h–south})}$$

▶ **Quick Check:** Peter sees John coming south at 30 km/h and John sees Peter ($\mathbf{v}_{PJ}$) galloping north toward him at 30 km/h, where $\mathbf{v}_{JP} = -\mathbf{v}_{PJ}$. The distance between them is being covered from both ends at a net speed of 30 km/h.

Figure 2.29 shows an ant scampering with a velocity of $\mathbf{v}_{AP}$ east across a piece of paper that is simultaneously being pulled north at a velocity of $\mathbf{v}_{PE}$. The ant, as seen from above by an observer at rest on the Earth, has a diagonal velocity of $\mathbf{v}_{AE}$. In exactly the same way, we can think of this situation in terms of a stream rushing north ($\mathbf{v}_{SE}$) being crossed by a boat bearing due east ($\mathbf{v}_{BS}$) with respect to the stream. The resultant motion of the boat with respect to the inertial frame of the Earth (that is, the shore) is diagonal ($\mathbf{v}_{BE}$)—the stream carries the boat somewhat northward even though its prow is always heading east.

The same vector relationships in Fig. 2.29 will exist for many situations. For example, imagine a jet plane heading east through the air ($\mathbf{v}_{JA}$) which, in turn, is

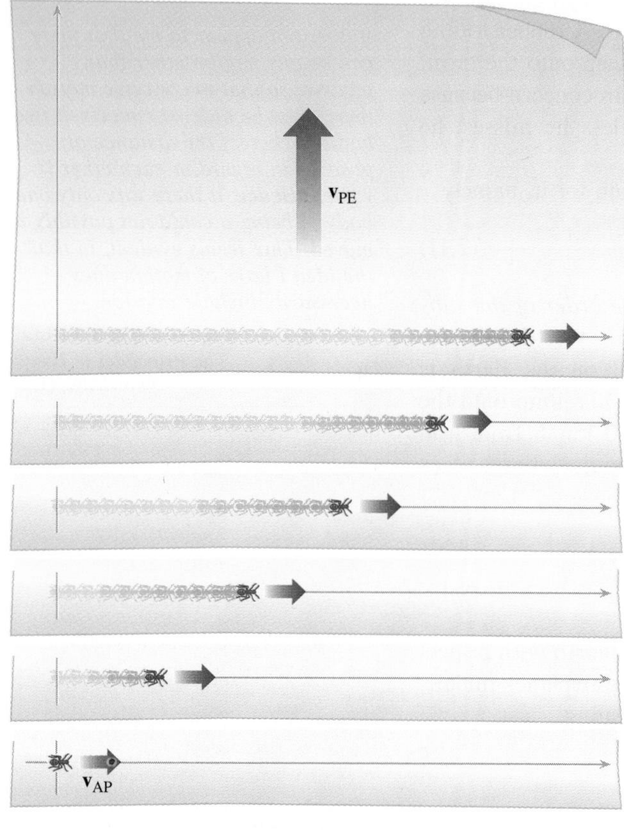

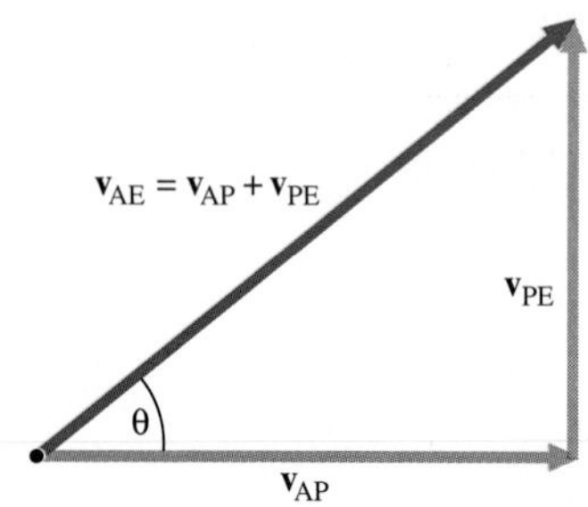

Figure 2.29 An ant walking across a sheet of paper that is itself being moved at a speed $\mathbf{v}_{PE}$. The bug is carried along with the paper so that it moves northeast with respect to the Earth.

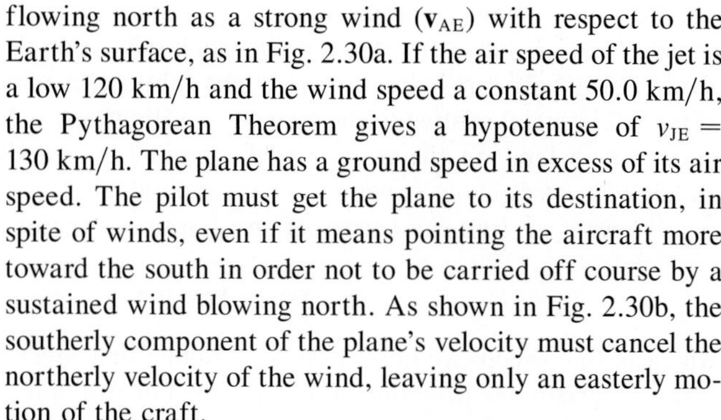

flowing north as a strong wind ($\mathbf{v}_{AE}$) with respect to the Earth's surface, as in Fig. 2.30a. If the air speed of the jet is a low 120 km/h and the wind speed a constant 50.0 km/h, the Pythagorean Theorem gives a hypotenuse of $v_{JE} =$ 130 km/h. The plane has a ground speed in excess of its air speed. The pilot must get the plane to its destination, in spite of winds, even if it means pointing the aircraft more toward the south in order not to be carried off course by a sustained wind blowing north. As shown in Fig. 2.30b, the southerly component of the plane's velocity must cancel the northerly velocity of the wind, leaving only an easterly motion of the craft.

The angle of flight, θ, can be determined via the definition of the tangent function for right triangles (Appendix C), as

$$\tan\theta = \frac{\text{opposite side}}{\text{adjacent side}}$$

The tangent of either acute angle of a right triangle is the ratio of the side opposite the angle to the side adjacent to that angle. In the case of Fig. 2.30a, (50.0 km/h)/(120 km/h) = 0.416 7 = tan θ. We want the angle (θ) whose tangent is known, and that angle is obtained using the *inverse* or *arc tangent* function; $\tan^{-1}(0.417) = \theta = 22.6°$, which here is north of east. On a calculator, the sequence is [0] [·] [4] [1] [6] [7] [$\tan^{-1}$].

2.10 Vector Components

We know how to add two vectors to form a third *resultant* vector that produces the same physical effects as the original two. The reverse process is also possible and also useful. Figure 2.31 shows two perpendicular vectors (acting along the x and y directions) $\mathbf{A}_x$ and $\mathbf{A}_y$. Using the parallelogram method (p. 33), it is evident that $\mathbf{A} = \mathbf{A}_x + \mathbf{A}_y$. The same diagram can be interpreted as showing that vector $\mathbf{A}$ may be *resolved into two perpendicular components:* $\mathbf{A}_x$ and $\mathbf{A}_y$. Given a specific value of $\mathbf{A}$ (that is, given its magnitude, A, and direction, θ), it is possible to measure these two components graphically. It will often be more convenient and accurate, however, to use the sine and cosine trigonometric functions instead. If you know any two sides, or any side and either of the acute angles of a right triangle, everything else about it can be determined easily.

Note (see Appendix C) that, for a right triangle, the sine of either acute angle is the ratio formed of the side opposite that angle over the hypotenuse:

$$\sin\theta = \frac{\text{opposite side}}{\text{hypotenuse}}$$

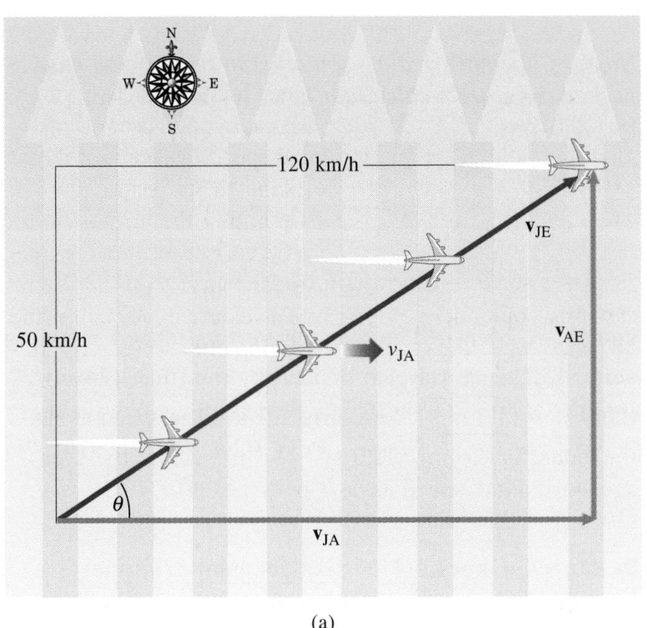

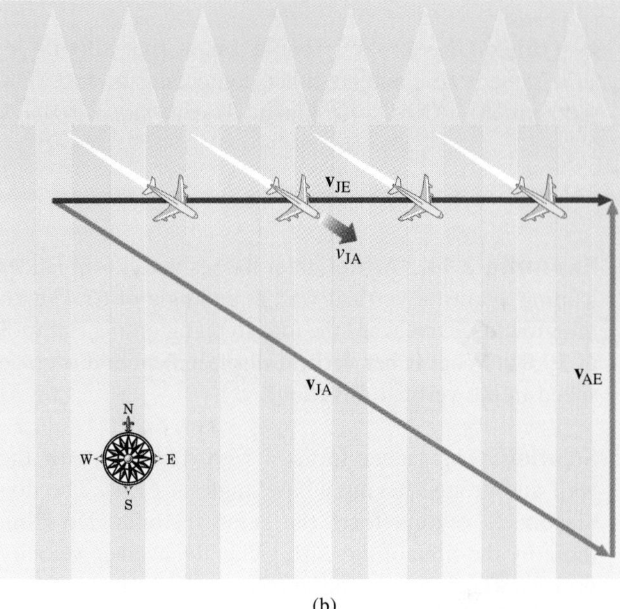

Figure 2.30 (a) A plane heading east across a wind blowing north. The plane is carried N of E at an angle θ. (b) To cross the air stream due east, the plane must point south-of-east such that its southerly velocity component effectively cancels the northerly velocity of the wind. Someone hitting a golf ball in a crosswind has the same problem as the pilot. When you are moving across a medium that's moving, you can't head straight toward your destination.

Similarly, the cosine of either acute angle of a right triangle is the ratio formed of the side adjacent to that angle over the hypotenuse:

$$\cos\theta = \frac{\text{adjacent side}}{\text{hypotenuse}}$$

The triangle in Fig. 2.31 has sides equal to the lengths (or magnitudes) of the vectors $\mathbf{A}_x$, $\mathbf{A}_y$, and $\mathbf{A}$. Accordingly, the appropriate ratios of these are

$$\sin\theta = A_y/A \qquad \text{and} \qquad \cos\theta = A_x/A$$

Multiplying both sides of both these equations by A yields

$$A_y = A\sin\theta \qquad \text{and} \qquad A_x = A\cos\theta \qquad (2.13)$$

These then, are the scalar x- and y-components of $\mathbf{A}$.

Example 2.12 An aircraft in a 45.0° dive is traveling at a constant 800 km/h. Determine its speed in the vertical direction—that is, the rate at which the plane is losing altitude.

Solution: The plane's *constant* velocity vector **v** points downward at an angle θ. [Given: $v = 800$ km/h and $\theta = 45.0°$. Find: v_y.] Equation (2.13) provides the needed relationships between any vector and its two perpendicular components.

Therefore

$$v_y = v\sin\theta = (800\text{ km/h})\sin 45.0°$$

Because both acute angles of this velocity triangle are equal (both are 45°), it is isosceles and the two sides, v_x and v_y, are equal. The plane's horizontal and vertical speeds are therefore both $\boxed{566\text{ km/h}}$. As the aircraft advances horizontally, it simultaneously descends at the same speed vertically.

(continued)

(continued)

▶ **Quick Check:** You should know that sin 45° = 0.707 = cos 45°, and so either component is ≈(0.7) × (800 km/h) ≈ 5.6 × 10^2 km/h. *Always make a rough check of the numerical computation to be sure you have pushed the correct calculator keys:* [4] [5] [·] [0] [sin] [×] [8] [0] [0] [=].

Example 2.13 A firefighter dashes up a 26-m ladder leaning against a vertical wall at an angle of 67.4° with the ground. She reaches the top, stepping onto a flat roof in 20.0 s. What is her vertical displacement and average speed in the vertical direction?

Solution: The ladder forms a right triangle with the wall and ground having a base angle of 67.4°. The displacement vectors form the same triangle. Dividing these by the time of the dash yields the average velocity vectors, which form a similar triangle. [Given: s = 26 m, t = 20.0 s and θ = 67.4°. Find: s_y the vertical displacement and the average vertical speed.] From Eq. (2.13)

$$s_y = s \sin\theta = (26\text{ m})(\sin 67.4°) = (26\text{ m})(0.923)$$

and $\boxed{s_y = 24\text{ m}}$. The average vertical speed is the vertical distance traveled divided by the time, (24 m)/(20.0 s) = $\boxed{1.2\text{ m/s}}$. Similarly, the average speed along the ladder is (26 m)/(20.0 s) = 1.3 m/s while the average horizontal speed is $s_x/t = (s \cos\theta)/t$ = (10 m)/(20.0 s) = 0.50 m/s.

▶ **Quick Check:** The displacement triangle is a special whole-number right triangle (Appendix B), a 5-12-13. Its sides are 10-24-26. The same must be true for the velocity triangle.

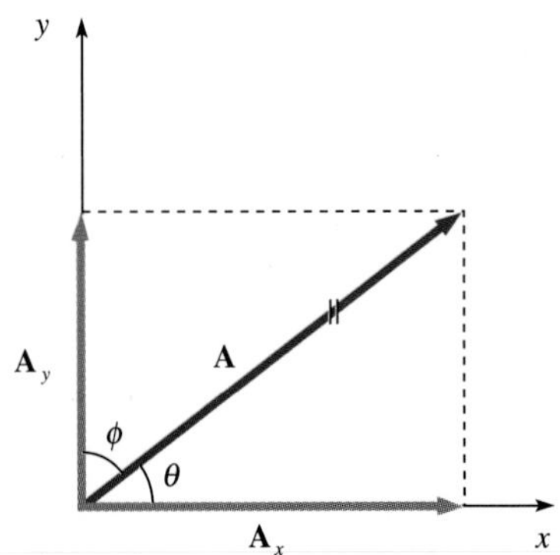

Figure 2.31 Any vector **A** can be thought of as equivalent to its two perpendicular components $\mathbf{A}_x$ and $\mathbf{A}_y$. Notice that $A_x = A\cos\theta$ and $A_y = A\sin\theta$, but if angle ϕ is used, $A_x = A\sin\phi$ and $A_y = A\cos\phi$. Be careful with those angles.

Now let's develop the most powerful analytic scheme for summing any number of vectors. Figure 2.32a shows two arbitrary vectors **A** and **B**, which we wish to add. The usual graphical tip-to-tail technique is illustrated in Fig. 2.32b, and the resultant vector **C** is constructed in Fig. 2.32c. Notice that each of the components of the resultant are equal to the sums of the components of the constituents. In other words,

$$C_x = A_x + B_x \quad \text{and} \quad C_y = A_y + B_y \tag{2.14}$$

Find the scalar x-components of the vectors to be summed, give them proper signs (Fig. 2.33), and then add them algebraically. That total is the scalar x-component of the resultant and the same procedure applies to the y-component. To reconstruct **C** analytically, use the Pythagorean Theorem

$$C = \sqrt{C_x^2 + C_y^2} \tag{2.15}$$

to compute the magnitude of the resultant. The most convenient way to find its orientation is to use

$$\theta = \tan^{-1}\frac{|C_y|}{|C_x|} \tag{2.16}$$

where $|C_y|$ and $|C_x|$ are the positive values of the components (Fig. 2.34), as discussed in detail in Appendix D.

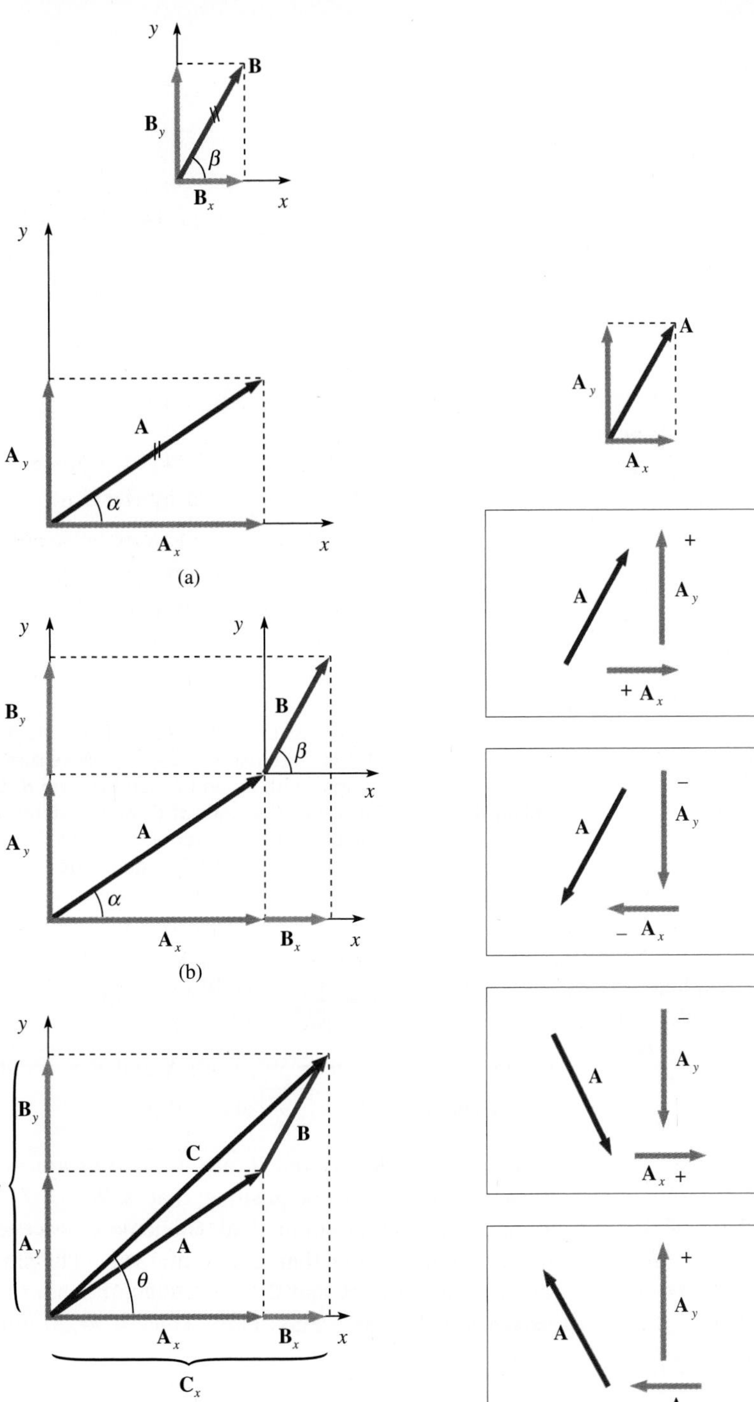

Figure 2.32 The summation of two vectors using their perpendicular components. (a) Those x- and y-components are $A_x = A \cos \alpha$, $A_y = A \sin \alpha$, $B_x = B \cos \beta$, and $B_y = B \sin \beta$. In (b) we place **A** and **B** tip-to-tail. In (c) we see how this arrangement is equivalent to adding their components in the x- and y-directions: $C_x = A_x + B_x$ and $C_y = A_y + B_y$.

Figure 2.33 The vector **A** has components in the x- and y-directions of $\mathbf{A}_x$ and $\mathbf{A}_y$. Here we see the signs that must be associated with the scalar quantities A_x and A_y if they are to be correctly added algebraically to the scalar components of other vectors.

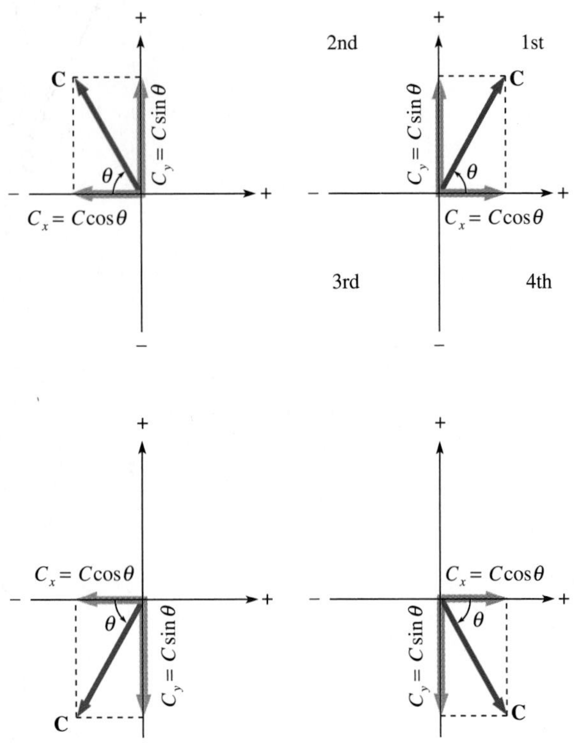

Figure 2.34 In the process of adding any two vectors **A** and **B** to get **C** use Eq. (2.14) to determine C_x and C_y. Make sure to associate the proper signs with A_x and B_x, and A_y and B_y before adding them (see Fig. 2.33). The signs of C_x and C_y determine in which quadrant **C** will be and also determine θ. Thus, if C_x is negative and C_y is positive, **C** is in the 2nd quadrant.

Example 2.14 Two vectors lie in the x-y plane, as shown in Fig. 2.32. Vector **A** is 10 units long and points upward 30° above the positive x-axis. Vector **B** is 15 units long and points 45° above the positive x-axis. What is the resultant of the two vectors?

Solution: [Given: **A** and **B**. Find: **C** = **A** + **B**.] From Eq. (2.13) $A_x = A \cos \alpha = 10 \cos 30° = 8.66$ and $A_y = A \sin \alpha = 10 \sin 30° = 5.00$; $B_x = B \cos \beta = 15 \cos 45° = 10.61$ and $B_y = B \sin \beta = 15 \sin 45° = 10.61$. Hence, $C_x = 8.66 + 10.61 = 19.27$ and $C_y = 5.00 + 10.61 = 15.61$. Using Eq. (2.15),

$$C = \sqrt{C_x^2 + C_y^2} = \sqrt{(19.27)^2 + (15.61)^2} = 24.8$$

To two significant figures, $\boxed{C = 25 \text{ units}}$. To find the angle θ that **C** makes with the x-axis, we use Eq. (2.16):

$$\tan \theta = \frac{|C_y|}{|C_x|} = \frac{15.61}{19.27} = 0.8101$$

and $\tan^{-1} 0.8101 = \theta = 39°$. Both **A** and **B** point above the x-axis and so **C** is $\boxed{39°}$ above it too.

▶ **Quick Check:** A_x and B_x are positive as are A_y and B_y, so C_x and C_y must be positive; that is, $90° \geq \theta \geq 0°$. From the parallelogram method, **C** should lie between **A** and **B** and be greater than either, and it is. The smaller the angle between **A** and **B** (45° − 30°), the more C approaches $A + B$ as a maximum. Thus, $C = 20$ units is reasonable.

Example 2.15 A blimp cruising with a constant air speed of 180 km/h heads due north against a steady 100 km/h wind blowing out of the northwest. Compute its ground speed, its actual direction of flight, and the distance traveled after 3.00 h of cruising.

Solution: This problem concerns relative velocity. [Given: $\mathbf{v}_{BA} = 180$ km/h–north; $\mathbf{v}_{AG} = 100$ km/h–southeast; $t = 3.00$ h. Find: $\mathbf{v}_{BG}$ and **s**.] Inasmuch as we need $\mathbf{v}_{BG}$, set up the appropriate expression for it, using Eq. (2.9) as a guide:

(continued)

(continued)

$$\mathbf{v}_{BG} = \mathbf{v}_{BA} + \mathbf{v}_{AG}$$

Figure 2.35 represents this equation graphically. To carry out the addition, we will first resolve $\mathbf{v}_{BA}$ and $\mathbf{v}_{AG}$ into their respective x- and y-components, which is easily done for $\mathbf{v}_{BA}$; it points along the positive y-axis and is the entire y-component—it has no x-component. The scalar x-component of $\mathbf{v}_{AG}$ is just v_{AG} cos 45° and the y-component is $-v_{AG}$ sin 45°. The minus sign is used here because the vector points downward in the negative y-direction. It has to be subtracted from v_{BA}, which points in the positive y-direction. For the moment, let's represent vectors pointing either left in the negative x-direction (west), or down in the negative y-direction (south), as negative. Similarly, those pointing either right in the positive x-direction (east), or up in the positive y-direction (north) are positive. In that way the algebraic subtraction of oppositely directed vectors takes place systematically as one adds all the components in a given direction. With the signs taken care of, every numerical value is entered as a positive number.

The scalar y-component of $\mathbf{v}_{BG}$ is then

$$v_{BA} - v_{AG} \sin 45° = 180 \text{ km/h} - (100 \text{ km/h})(0.707) = 109 \text{ km/h}$$

while the scalar x-component is

$$v_{AG} \cos 45° = (100 \text{ km/h})(0.707) = 70.7 \text{ km/h}$$

The magnitude of $\mathbf{v}_{BG}$ can now be gotten from the Pythagorean Theorem:

$$v_{BG} = \sqrt{(109 \text{ km/h})^2 + (70.7 \text{ km/h})^2} = \boxed{130 \text{ km/h}}$$

The angle $\mathbf{v}_{BG}$ makes with the x-axis (or east-west line) is θ where

$$\tan \theta = \frac{109 \text{ km/h}}{70.7 \text{ km/h}} = 1.54$$

and so $\tan^{-1}(1.54) = \boxed{\theta = 57.0° \text{ N of E}}$.

In 3.00 h, the blimp travels a distance

$$s = v_{BG}t = (130 \text{ km/h})(3.00 \text{ h}) = 390 \text{ km}$$

▶ **Quick Check:** There are many problems that can be checked using the Law of Sines (Appendix C). This law maintains that for any triangle the ratio of any one side to the sine of the angle opposite that side equals the ratio of any other side to the sine of the angle opposite it. For the triangle in Fig. 2.35, $v_{BG}/\sin 45° = 184$ m/s and this quantity should equal $v_{AG}/\sin(90° - 57°) = 184$ m/s, and all is well.

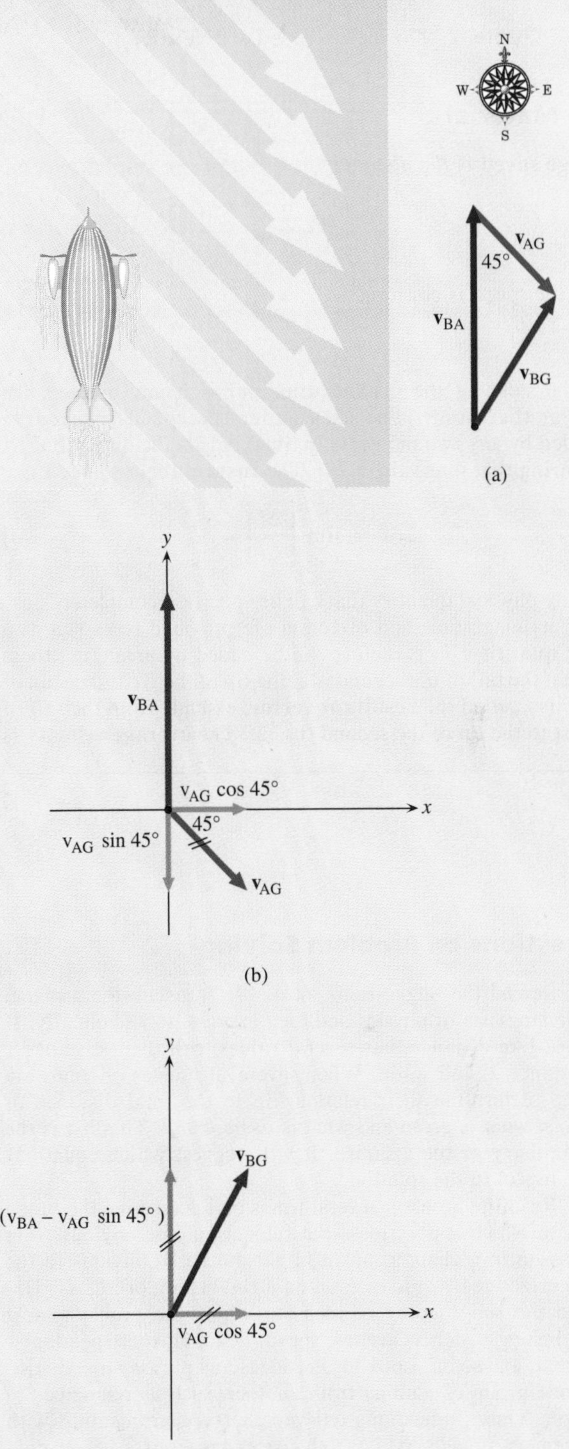

Figure 2.35 (a) A blimp heads north through the air with a velocity $\mathbf{v}_{BA}$ while the air is moving with a velocity $\mathbf{v}_{AG}$ with respect to the ground. Thus, the blimp is blown off course and travels northeast with a velocity $\mathbf{v}_{BG}$ with respect to the ground. (b) To find its velocity $\mathbf{v}_{BG}$, vectors $\mathbf{v}_{BA}$ and $\mathbf{v}_{AG}$ are resolved into their components along the x- and y-axes. (c) All the components along each axis are added, yielding two perpendicular vectors that are themselves the x- and y-components of the resultant $\mathbf{v}_{BG}$.

Core Material

Average speed is *the distance traversed in one unit of time,* or

$$v_{\text{av}} = \frac{l}{t} \qquad [2.1]$$

When the speed is constant

$$l = v_{\text{const}} t \qquad [2.2]$$

The slope of the distance-time curve at any point is the speed at that point. The area under the speed-time curve (bounded by any two moments in time) equals the distance traveled during that interval (p. 28). The **instantaneous speed** is

$$v = \lim_{\Delta t \to 0} \left[\frac{\Delta l}{\Delta t}\right] \qquad [2.4]$$

Any physical quantity that can be specified completely only if both its magnitude and direction are provided is known as a **vector quantity**. Two vectors can be added by arranging them such that the tail of the second is at the tip of the first; the sum of these, also called the **resultant vector**, extends from the tail of the first to the tip of the second (p. 32). The **average velocity** is

$$\mathbf{v}_{\text{av}} = \frac{\mathbf{s}}{t} \qquad [2.5]$$

and the **instantaneous velocity** is

$$\mathbf{v} = \lim_{\Delta t \to 0} \left[\frac{\Delta \mathbf{s}}{\Delta t}\right] \qquad [2.7]$$

Figure 2.26 illustrates the concept of relative motion. In this case, the velocity of the turtle with respect to the Earth is

$$\mathbf{v}_{\text{TE}} = \mathbf{v}_{\text{TR}} + \mathbf{v}_{\text{RE}} \qquad [2.9]$$

The scalar x- and y-components of any vector **A** are

$$A_y = A \sin \theta \quad \text{and} \quad A_x = A \cos \theta \qquad [2.13]$$

If **C** is the sum of the two vectors **A** and **B**, then

$$C_x = A_x + B_x \quad \text{and} \quad C_y = A_y + B_y \qquad [2.14]$$

and

$$C = \sqrt{C_x^2 + C_y^2} \qquad [2.15]$$

where

$$\theta = \tan^{-1}\frac{|C_y|}{|C_x|} \qquad [2.16]$$

Suggestions on Problem Solving

1. Reread the Suggestions on p. 19. Translate the problem into the language of physics and then express it symbolically. If one idea, like distance, has several values, call them distance 1 and distance 2, and so on. When several distances or times are involved, determine their relationship to the total distance or time. List what is *given* and what is to be *found.* This list is the best summary of the exercise. It will suggest which equations will be useful in the solution.

2. Read the problem several times and determine the broad subject to which it pertains—the subject may be obvious from the location in a chapter, but it might not be so obvious in the uncategorized real world or even on a final exam or MCAT. Determine the subgrouping of ideas to which the problem most clearly belongs, such as average speed, instantaneous speed, displacement, and so on. Look for key ideas and phrases that distinguish one group of notions from another, such as references to direction. These indications tell you that you are dealing with displacement or velocity. Watch for expressions like *straight road, steady motion, constant speed,* and so forth—they provide vital information.

3. Draw sketches that illustrate the problem being analyzed. Put into these all the known information and, where appropriate, draw in the vectors to represent directional quantities.

4. For most simple single-concept problems there will be one unknown that can be determined from one formula. Write down the equation that contains the desired unknown (and only that one unknown) and solve for it algebraically in terms of the remaining knowns. Problems involving two unknowns require solving two equations simultaneously (Appendix A-4). Remember you will need the same number of equations as there are unknowns to be found.

5. **Unless otherwise required, work all problems in SI units**. Convert the given numerical information into appropriate units *before* substituting into equations. It's best to convert as soon as possible to avoid accidentally using the wrong values (10 cm is *not* 10 m). Carry the units for each term through the whole calculation as a double check that you haven't messed up the algebra: an answer of 10 km/h for how *far* the chicken traveled suggests something is amiss; an answer of just 10 tells you a good deal less and is incomplete. **Always** include units with your numerical answers.

6. Always try to check your answers by recomputing the problem in a different way.

7. Watch out for the number of significant figures. If you travel 10.2 m (three significant figures) in 2.1 s (two significant figures), your average speed is 4.9 m/s (two significant figures) and *not* 4.857 142 9 m/s.

8. To foster some independence from the calculator, a number of problems are given with "nice" numbers so they are easier to do by hand than by machine. Avoid using the calculator to divide 100 by 2, wondering all the while why you bothered. Some important national examinations do not allow calculators to be used. If you plan to take such exams, you must know what the curves of the sine, cosine, and tangent functions look like. Also know their values for 0°, 30°, 45°, 60°, and 90°. Become familiar with the 3-4-5 and 5-12-13 right triangles.

Discussion Questions

1. Suppose that the average speed over some time interval is zero. Is it possible for the average speed over a still smaller segment of that interval to be nonzero? Suppose that the average velocity over some time interval is zero. Is it possible for the average velocity over a still smaller segment of that interval to be nonzero?

2. If the tangent at a given point on the distance-time graph of some object is horizontal, what is the object's instantaneous speed at that moment? What does it mean if the tangent to the graph is vertical? Can that actually occur?

3. Is it possible for a plot of l versus t to have a negative slope at any point? Explain.

4. Is it possible to travel from one place to another with some average speed without ever having had an instantaneous speed equal to it somewhere along the journey? Explain.

5. Imagine that you leave the base of a mountain at 9:00 A.M. and ascend along a path reaching the top at 11:00 A.M. You sleep there and descend along the same path starting at 9:00 A.M. the next day. Will there be some point along the trail that you reach at exactly the same time you were there the day before, regardless of the variations in your speeds during the two journeys?

6. Can the change in displacement of an object in some interval of time exceed the distance traveled in that time?

7. Is it possible at any moment to be moving with a constant velocity and not a constant speed? How about the reverse of this situation—a constant speed but not a constant velocity?

8. When we talk about displacement vectors for travel on our planet, we assume a flat Earth, or at least trips short enough so it's approximately flat. Just for fun, suppose you were standing at the North Pole and walked on the surface (no burrowing) 10 km–south, 20 km–east, and 10 km–north—where would you end up?

9. Figure Q9 is a plot of the total distance traveled by a dog running in a field. Describe the motion. What can be said about the dog's displacement at any time? What was the total distance traveled? When was he moving the fastest?

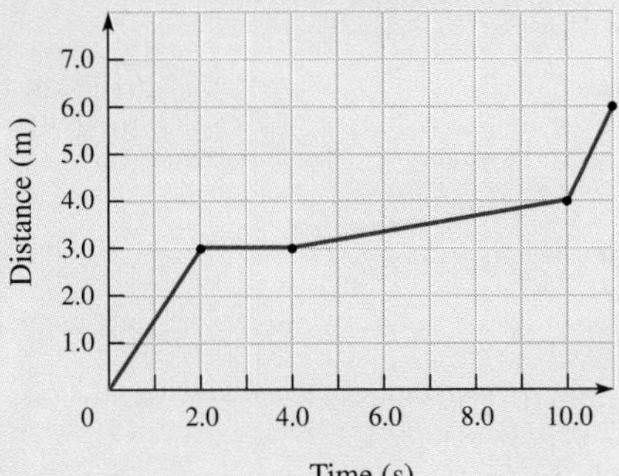

Figure Q9

10. Is it possible during a given interval for a graph of *distance (l) versus time* for the same object to be different from a graph of *the magnitude of the displacement (s) versus time?* Explain.

11. Suppose we have a vector representing some physical quantity. Will the numerical value of the magnitude change if we express it in different units? Will its orientation change?

12. After reaching cruising altitude, a jet can fly along at a constant 750 km/h, over the "shortest path" from New York to Tokyo. What dynamical quantity would you say was constant for that portion of the flight? Explain.

13. Figure Q13 is a plot of the displacement of a mouse running through a straight, narrow tunnel, the entrance to which is taken as the origin. Describe the mouse's motion in general terms throughout its journey. Where and when did it start moving? Where did it finish? Did it ever rest?

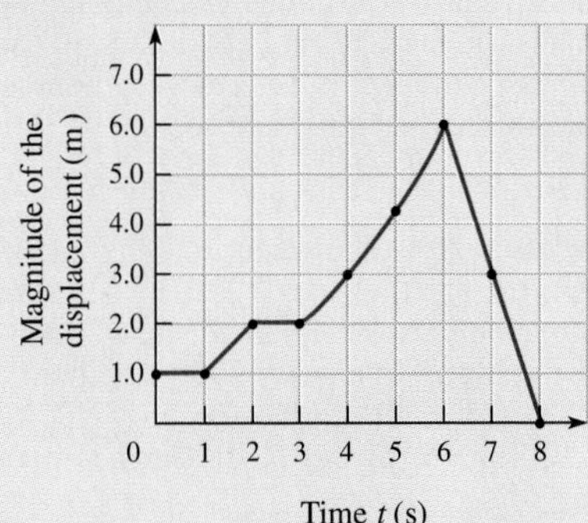

Figure Q13

14. A plane dives in a straight line at a fixed angle. With the Sun right above it, the plane's shadow moves across the ground at a constant speed. What can you say about the plane's rate of descent?

15. If two vectors have magnitudes of 4 units and 6 units, what are the largest and the smallest possible resultants? Can the vector **A** have a component with a magnitude that is larger than A? If two vectors are to produce a resultant equal to zero, what can you say about them?

16. Suppose we multiply a vector **M** representing some physical quantity, by a scalar n representing yet another physical quantity. Will n**M** have the same units as **M**?

17. If the magnitude of the displacement of a body is given by $s = (At^2 + Bt)/D(C + t)$ where A, B, C, and D are constants, (a) determine the displacement at $t = 0$; (b) find the approximate value of s when t is very much larger than C—that is, when $t \gg C$. What is the approximate displacement when $t \ll C$?

18. Determine the resultant of the two vectors in Fig. Q18 where $A_1 = A_2 = A = 25$. Show that the resultant has a magnitude of $2A \cos 30°$. Now try the last part again, this time rotating the coordinate system along which you resolve the vectors so that the y-axis bisects the 60° angle. Notice how much easier the problem is now!

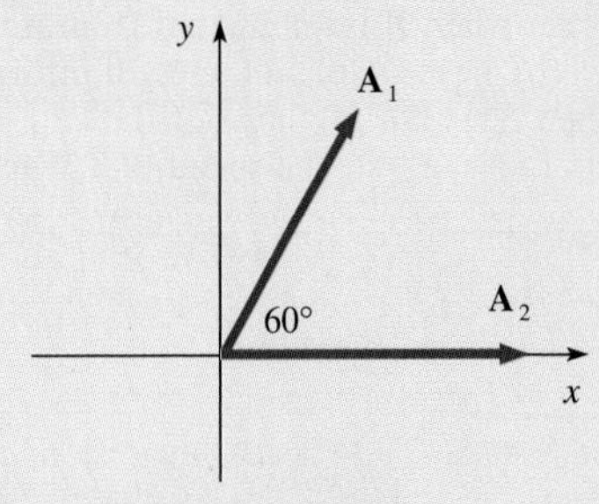

Figure Q18

Multiple Choice Questions

1. If $[L]$ represents the dimensions of length and $[T]$ that of time, then the dimensions of speed are (a) $[L + T]$ (b) $[T/L]$ (c) $[L/T^2]$ (d) $[L/T]$ (e) none of these.
2. When converting a given number of kilometers per hour to kilometers per second, you can expect the resulting number to be (a) always smaller (b) the same (c) sometimes smaller (d) never smaller (e) none of these.
3. How far does light travel in one year moving at 3×10^8 m/s? (a) 3×10^8 m (b) 9.5×10^{15} m (c) 3×10^8 km (d) 9.5×10^8 km (e) none of these.
4. A speed of 300 m/s is equivalent to (a) 0.300 km/h (b) 984 ft/s (c) 134 mi/h (d) 83.3 km/h (e) none of these.
5. A speed of 60 mi/h is equivalent to (a) 44 ft/s (b) 22 m/s (c) 88 ft/s (d) 11 km/h (e) none of these.
6. The three-person crew of *Apollo 11* made the 3.8×10^5-km journey to the Moon in exactly 3 days with an average speed of (a) 5.3×10^3 km/d (b) 53×10^3 km/h (c) 5.3×10^3 km/s (d) 5.3×10^3 km/h (e) none of these.
7. If a body moves at a constant speed v along a closed path, its average speed in comparison to v is (a) greater (b) less (c) the same (d) sometimes greater, sometimes less (e) none of these.
8. On a distance-time graph, the slope at any point is (a) the distance traveled (b) the time elapsed (c) the instantaneous speed (d) the average speed (e) none of these.
9. During a journey over which the speed varies, the average speed in comparison to the maximum attained speed is always (a) one-half of it (b) greater than it (c) less than it (d) equal to it (e) none of these.
10. A horse on a continuously and uniformly turning carousel is carried through three rotations every 6 min. Its average speed over each and any 18-min interval is (a) a constant (b) undetermined and we can not say anything about it (c) always different (d) may be different (e) none of these.
11. The *Mariner 2* spacecraft lifted off from Cape Canaveral on August 27, 1962. After a successful flight, it passed out of tracking range at about 87 million kilometers (54×10^6 mi) on January 4, 1963. The magnitude of its average velocity up until that point was roughly (a) not enough information to calculate (b) 1.7×10^4 km/h (c) equal to the average speed (d) 28×10^3 km/h (e) none of these.
12. When is the resultant that arises when **A** is subtracted from **B** equal to that when **B** is subtracted from **A**? (a) always (b) never (c) not enough information (d) only when **A** = **B** (e) none of these.
13. If, just as its fuel ran out, the speed of a World War II German V-2 rocket was 5633 km/h–straight up, what was the velocity of the Earth with respect to the rocket at that moment? (a) zero (b) 5633 km/h–downward (c) 5633 km/h–upward (d) 11 266 km/h (e) none of these.
14. Two vectors of magnitude 8.0 and 3.0 can be added so as to produce a third vector of magnitude (a) 12 (b) 15 (c) 8.0 (d) 3.0 (e) none of these.
15. A vector 10 units long pointing northeast is added to a vector 24 units long pointing northwest. The magnitude of the resultant is (a) 26 units (b) 14 units (c) 34 units (d) 0 units (e) not enough information.
16. The inhabitants of the mythical planet Mongo measure distance in units of "glongs," one of which is the length of their leader's breathing tube. A Mongoian displacement vector 90.0 glongs long points south from the truffle tree to the methane fountain, while a vector 120 glongs long points west from the fountain to the main reflector. The displacement from tree to reflector is (a) 150 glongs, 49° S of W (b) 150 m, 37° S of W (c) 150 glongs, 37° S of W (d) 80.0 glongs, 41° S of E (e) not enough information.
17. In general, if a vector **A** is to be added to a vector **B** the magnitude of the resultant when $A \geq B$ must be between (a) $2A$ and B (b) A and $A + B$ (c) $A + B$ and $A - B$ (d) $B - A$ and B (e) none of these.
18. Figure MC18 is a portion of a distance-time curve for a jogger. At $t = 14$ s the instantaneous speed is about (a) 0.50 m/s (b) 0.27 m/s (c) 6.5 m/s (d) 0.46 m/s (e) none of these.

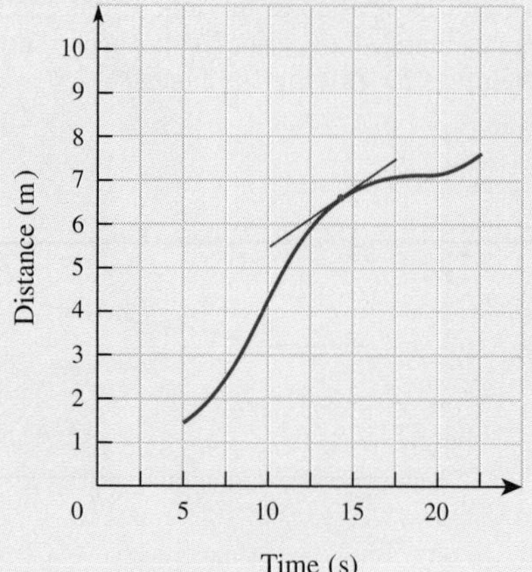

Figure MC18

Problems

SPEED

1. [I] In the next 30.0 s, the Earth will travel roughly 885 km (550 mi) along its orbit around the Sun. Compute its average orbital speed in km/s.

2. [I] You know how you close your eyes when you sneeze? Suppose you are driving along at a constant 96.5 km/h (60 mi/h) and you experience a 1.00-s-long, eyes-closed, giant sneeze. How many meters does the car travel while you are out of control?

3. [I] Light travels in vacuum at a fixed speed of roughly 2.998×10^8 m/s and its speed in air is only negligibly (0.03%) slower than that. (a) How long does it take light to traverse 1.0 ft? (b) That means that when you look at something 1000 m away, you are seeing it as it was ________ second(s) back in time.

4. [I] If an alien power straight out of a comic book were to cause the Sun to vanish *now,* we would still be bathed in sunshine for the next 8.3 min. We would see our star blazing in the sky as usual for all that time. Using that information and taking the speed of light to be roughly 3.0×10^8 m/s, compute the average Earth–Sun distance.

5. [I] Referring to Table 2.3 (p. 26), derive each conversion factor in the first column.

6. [I] Referring to Table 2.3 (p. 26), derive each conversion factor in the second column.

7. [I] Figure P7 is a plot of the speed of a cat versus time. How far did the cat travel during the third second of its journey? What were its maximum and minimum speeds? When, if ever, was its speed nonzero and constant?

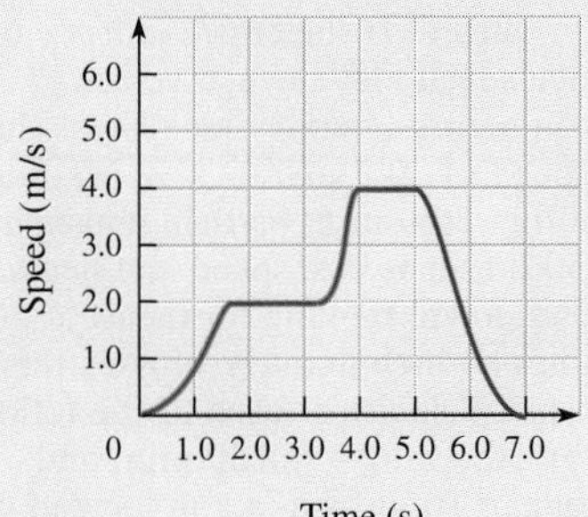

Figure P7

8. [I] The Earth rotates once around its spin axis in 23 h 56 min and its equatorial diameter is 1.276×10^7 m (7927 mi). At what speed would you be traveling with respect to the stationary stars while sitting still in Mbandaka, Zaire, on the Equator?

9. [I] A 30-km trip is negotiated in two equal-length segments. The first, on highways taking 15 min; the second, through heavy traffic and taking 45 min. Compute (a) the average speed along each leg, and (b) the average speed over the whole trip.

10. [I] The SR-71 strategic reconnaissance aircraft, the *Blackbird,* set a world speed record by flying from London to Los Angeles (5463 mi) in 3 h 47 min 36 s. (a) Compute its average speed in m/s. (b) It recaptured the 1000-km closed-circuit-course record (previously held by the Russian MIG-25 *Foxbat*) at an average speed of 2092 mi/h. How long in time was that flight?

11. [I] Use Fig. 2.5 (p. 29) to calculate the distance traveled by the bee (whose speed-time graph is plotted) in the time interval from 1.33 s to 2.83 s.

12. [I] The driver of a car sets a tripmeter to zero and starts a stopwatch while en route along a highway. The following are a series of pairs of readings taken during the run: 6.0 min, 2.75 km; 30 min, 13.8 km; 45 min, 20.6 km; 1.0 hr, 27.5 km; 1.00 h 15 min, 34.4 km; 1.00 h 30 min, 41.3 km. Draw a distance-time graph and describe the car's speed.

13. [I] Figure P13 shows the distance traveled versus time curve for a toy car. What was the toy car's average speed during the time interval from t = 2.0 s to 8.0 s?

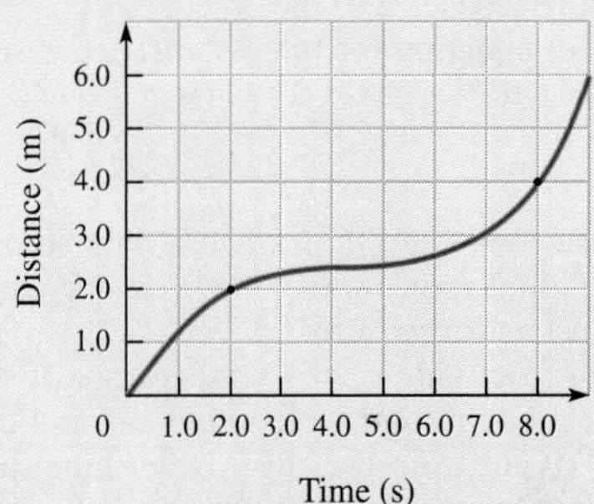

Figure P13

14. [II] Suppose you fire a rifle bullet (1600 m/s) in a shooting gallery and hear the gong on the target ring 0.731 s later. Taking the speed of sound to be 330 m/s and assuming the bullet travels straight downrange at a constant speed, how far away is the target?

15. [II] The speed of sound in air (at 0°C) is a constant 3.3×10^2 m/s, while the speed of light is about 3.0×10^8 m/s. If you see a flash of lightning and then 5.0 s later hear its roll of thunder, how far away did the bolt strike?

16. [II] The evil Dr. X secretly leaves Space Port L4 in a warpship capable of traveling at an average speed of $(v_{av})_X$. Two hours later her escape is noticed and our hero blasts off after her at a speed that will effect rendezvous in

6.4 hours. Write an expression for his speed, $(v_{av})_H$, in terms of hers, $(v_{av})_X$.

17. [II] The bad guys come roaring onto a highway at 100 km/h headed for the Mexican border 300 km away. The cops, in hot pursuit, arrive at the highway entrance one-half hour later. What must be their minimum average speed if they are to intercept the crooks this side of the border?

18. [III] Having traveled halfway to the end of a journey at an average speed of 15 km/h, how fast must you traverse the rest of the trip in order to average 20 km/h?

19. [III] A swimmer crosses a channel at an average speed of 10 km/h and returns at half that rate. What was his overall average speed for the round trip?

20. [III] If the equation describing the motion of a rocket in SI units is $l = 10 + 5t^2$, find its average speed during the first 5.0 s of flight. (Hint: At a time $(t + \Delta t)$, the object is at $(l + \Delta l)$.) Write an expression for the instantaneous speed of the rocket [see Eq. (2.4)].

21. [III] Given the equation describing the motion of a falling object $l = Ct^2$ (where C is a constant), show that its average speed during the interval from t to $(t + \Delta t)$ is

$$v_{av} = 2Ct + C\Delta t$$

(Hint: At a time $(t + \Delta t)$, the object is at $(l + \Delta l)$.) Now determine an equation for the instantaneous speed from Eq. (2.4).

VELOCITY

22. [I] A mouse runs straight north 1.414 m, stops, turns right through 90°, and runs another 1.414 m due east. Through what distance is it displaced?

23. [I] A green frog with a body temperature of 18°C at rest at the edge of a 1.0 m high table jumps off and lands on the ground 1.0 m out from the edge. Determine the length of its displacement vector.

24. [I] A toy electric train runs along a straight length of track. Its displacement versus time curve is shown in Fig. P24. Is the train's velocity ever constant, and if so, when? What is its instantaneous velocity at $t = 2.0$ s and at $t = 6.5$ s? Did it change direction, and if so when?

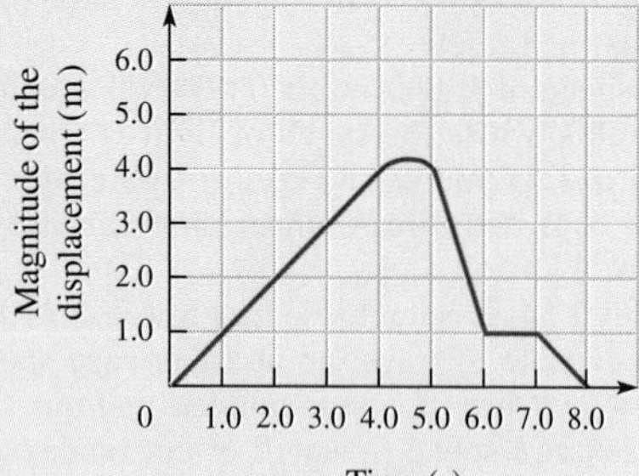

Figure P24

25. [I] A robot on the planet Mongo leaves its storage closet and heads due east for 9.0 km. It then turns south for 12 km and stops. Neglecting the curvature of the planet, what is the magnitude of the robot's displacement?

26. [I] A jogger in the city runs 4 blocks north, 2 blocks east, 1 block south, 4 blocks west, 1 block north, 1 block east, and collapses. Determine the magnitude of the jogger's displacement.

27. [I] A stationary wooden horse on a merry-go-round is carried along a 31.4-m circumference each time the machine turns once around. Determine the magnitude of the horse's displacement for half a turn.

28. [I] Draw the vector **A** 10 units long pointing due east and graphically add to it the vector **B** 15 units long pointing northeast at 45°. Measure both the magnitude of the resultant and its direction.

29. [I] A boy scout troop marches 10 km east, 5.0 km south, 4.0 km west, 3.0 km south, 6.0 km west, and 8.0 km north. What is their total displacement from their starting point? How far have they marched?

30. [I] Graphically determine **C**, which is the resultant of **A** = 20 units–30° S of E and **B** = 10 units–30° N of E.

31. [I] Someone runs up four zigzagging flights of stairs, each one 10 m long and rising 6.0 m high. At the end of the climb, the person stands directly above the starting point. Determine the runner's total displacement.

32. [I] A marble, on the first shot of a tournament game in 1942 in Brooklyn, was displaced by **s** equal to 3.0 m in a direction 60° N of E. On the next shot, it was displaced from its new location by −4.0 **s**. Where did it end up?

33. [I] Graphically add **A** (5.0 units–east) to **B** (7.1 units–45° N of W) to get **C**.

34. [I] Graphically subtract **A** (5.0 units–east) from **B** (7.1 units–45° N of E) to get **C**.

35. [I] A bumblebee flew 43 m along a twisting path only to land on a flower 3.0 m due south of the point on its hive from which it started. If the entire journey took 10 s, what was its average speed and average velocity?

36. [I] A driver in a cross-country race heads due east, attaining a speed of 30 km/h in 10 s. Maintaining that speed for another 30.0 min, he then makes a hard 90° left turn. After 50.0 min at that speed and heading, he veers right and slows down, moving northeast at 20 km/h for the next 40.0 min, at which point he crosses the finish line. What is the car's velocity at each of the following times: (a) 40 s, (b) 30.0 min, (c) 50 min, and (d) 100 min?

37. [II] A cannonball fired from a gun located 20 m away from a castle rises high into the air in a smooth arc and sails down, crashing into the wall 60 m up from the ground. Determine the projectile's displacement.

38. [II] Figure P38 shows the displacement-time graph of a

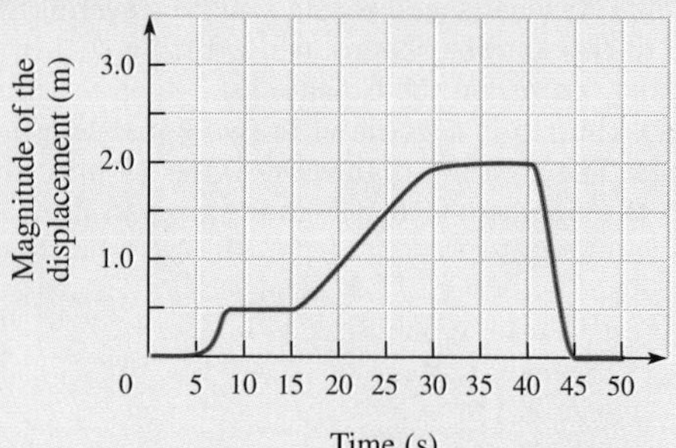

Figure P38

gerbil running inside a straight length of clear plastic tubing. (a) How far did it travel during the interval from 15 s to 20 s? (b) What distance did it traverse in the first 35 s? (c) What is its average speed during the first 35 s? (d) What is its instantaneous speed at $t = 20$ s?

39. [II] Refer to Fig. P38 and assume the gerbil starts out heading north. (a) What is his velocity at $t = 12$ s; at $t = 38$ s? (b) What is his velocity at 42.5 s?

40. [II] An athlete dashes past the starting point in a straight-line run at a fairly constant speed of 5 m/s for the first 2 s and then at 10 m/s for the next 3 s. At that moment she abruptly stops, turns around, waits 1 s, and runs back to the start with a constant speed in 3 s. Draw a graph of the magnitude of her displacement versus time.

41. [II] Ideally, a bullet fired straight up at 300 m/s will constantly slow down as it rises to a height of 4588 m in 30.6 s, at which point it will come to a midair stop. It then plummets back down again (and overlooking friction losses) and will speed up, reaching the gun barrel at the original 300 m/s after falling 30.6 s. Determine its average speed and average velocity on both the upward leg and on the round trip.

42. [II] Referring to the previous problem: (a) What is the velocity of the bullet just as it leaves the gun? (b) What is the bullet's velocity at $t = 30.6$ s; at $t = 61.2$ s?

43. [II] A chicken is resting at a location 3.0 m north of a stationary farmer who weighs 185 lb. The chicken then meanders to a new location 3.0 m east of the farmer in a time of 2.0 s. Compute the bird's average velocity during its little journey.

44. [II] A trolley travels along a straight run of track and Fig. P44 is a plot of its velocity versus time. Approximately how far did it travel in the first 3.0 s of its journey? How far from its starting point is it at $t = 6.0$ s?

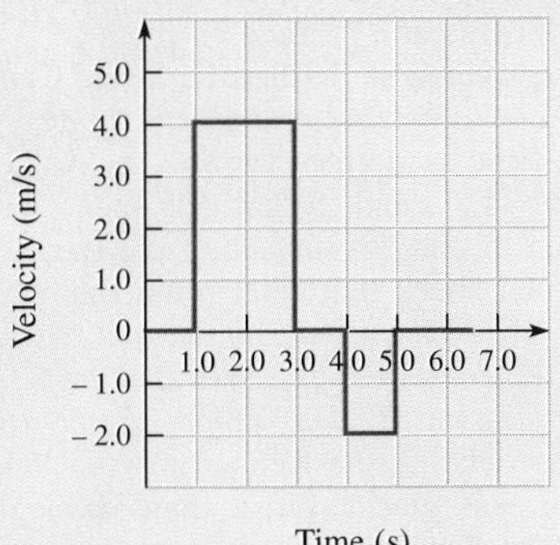

Figure P44

45. [III] A cannon is fired due north with an elevation of 45°. The projectile arcs into the air and then descends, crashing to the ground 600 m downrange 3 s later. With the Sun directly overhead, the projectile's shadow races along the flat stretch of Earth at a fairly constant speed. (a) Compute the instantaneous velocity of the shadow at the moment of impact. (b) What was the average velocity of the shell for the entire flight?

46. [III] An observer on a golf course, at 2:00 in the afternoon, stands 60 m west of a player who drives a ball due north down the fairway. If the ball lands 2.0 s later, 156 m from the observer, what was its average velocity?

RELATIVE MOTION

47. [I] Two runners at either end of a 1000-m straight track each jogs toward the other at a constant 5.00 m/s. How long will it take before they meet?

48. [I] The jet stream is a narrow current of air that flows from west to east in the stratosphere above the temperate zone. Suppose a passenger plane capable of cruising at an air speed of 965 km/h (600 mi/h) rode the jet stream on a day when that current remained at a constant 483 km/h (300 mi/h) with respect to the Earth. Find the plane's ground speed.

49. [I] Suppose that a fly were to go back and forth, essentially without a pause, from one runner to the other in Problem 47 at an average speed of 10 m/s. How far will it have traveled in total by the time the athletes meet? (Hint: How long is the fly in the air?)

50. [I] A deckhand on a ship steaming north walks toward the rear of the vessel at 5.00 km/h carrying a horizontal wooden plank. A ladybug on the plank scampers away from the human at 0.01 km/h–south. If the ship cuts through the calm sea at 15.00 km/h: (a) What is the speed of the ladybug with respect to the ship? (b) What is the speed of the plank with respect to the Earth? (c) What is the velocity of the bug with respect to the shore?

51. [I] Determine the two acute angles of an exactly 5-12-13 right triangle accurate to two decimal places (4 significant figures).

52. [I] A pack of hounds running at 15.0 m/s is 50.0 m behind a mechanical rabbit, itself sailing along at 10.0 m/s. (a) How long will it take before the rabbit is caught? (b) How far will it travel before being overtaken?

53. [I] A rope tied to the top of a flagpole makes an angle of 45° with the ground. If a circus performer walks up the $10\sqrt{2}$-m rope, what is her altitude at the top?

54. [I] What is the ratio of the sides of a right triangle whose acute angles are 30° and 60°? *This ratio is a good one to remember.* Which side is opposite the 30° angle?

55. [I] A telescope pointed directly at the top of a distant flagpole makes an angle of 25° with the ground. If the scope is low to the ground and 25 m from the base of the pole, how tall is the pole?

56. [I] A Roman catapult fires a boulder with a launch speed of 20 m/s at an angle of 60° with the ground. What are the projectile's horizontal and vertical speeds? (See the previous problem.)

57. [I] A skier is sailing straight down a 3000-ft-tall mountain with a 60.0° slope at 85 km/h. At what rate is his altitude decreasing?

58. [I] A pintail duck is flying northwest at 10 m/s. (a) At what rate is it progressing north? (b) What is the magnitude of its westward velocity component?

59. [I] A bullet is fired from a 1905 9-mm Luger pointed up at 32.0° with respect to the horizontal. If the muzzle speed—that is, the speed at which the projectile leaves the gun—is 300 m/s, find the magnitudes of: (a) the horizontal, and (b) the vertical, components of the bullet's velocity. (c) Check your results using the Pythagorean Theorem.

60. [II] A jet fighter heading directly north toward a flying target drone fires an air-to-air missile at it from 1000 m away. The drone has a constant speed (due north) with respect to the ground of 700 km/h. The fighter maintains a ground speed of 800 km/h, and the missile travels at 900 km/h with respect to it. (a) What is the speed of the plane with respect to the drone? (b) What is the ground speed of the missile? (c) How long will it take the missile to overtake the target?

61. [II] At the moment this sentence was written, I was in New York at a latitude of just about 41° N. Using Fig. P61, compute the circumference of the circle (the *parallel* of latitude) swept out by a point at that location as the planet rotates. Take the Earth to be a sphere of 6400 km radius turning once in 24 h. At what speed was I moving about the spin axis while sitting here at "rest"? Where are you and how fast are you moving?

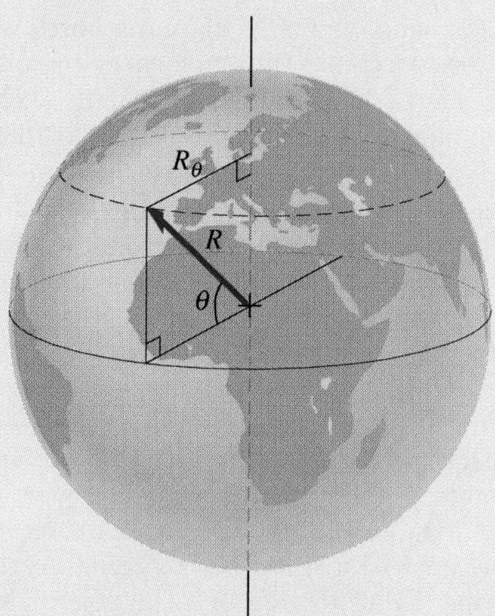

Figure P61

62. [II] A ferryboat has a still-water speed of 10 km/h. Its helmsman steers a course due north straight across a river that runs with a current of 5.0 km/h–east. If the trip lasts 2.0 h, how wide is the river and where does the ferry dock?

63. [II] A hawk 50 m above ground sees a mouse directly below running due north at 2.0 m/s. If it reacts immediately, at what angle and speed must the hawk dive in a straight line, keeping a constant velocity, to intercept its prey in 5.0 s? Incidentally, the mouse escaped by jumping in a hole.

64. [II] A ship, in an old black-and-white war movie, is steaming 45° S of W at 30 km/h. At that moment 50 km due south of it, an enemy submarine heading 30° W of N, spots it. At what speed must a torpedo from the sub travel if it's to strike the ship? Assume that the sub fires straight ahead from its forward tubes, and overlook the fact that this problem is a bit unrealistic. (Hint: What can be said about both westerly displacements?)

65. [III] At times of flood, the Colorado River reaches a speed of 30 mi/h near the Lava Falls. Suppose that you wanted to cross it there perpendicularly, in a motorboat capable of traveling at 30 mi/h in still water. (a) Is that possible? Explain. (b) If on a quieter day the water flows at 20 mi/h, at what angle would you head the boat to cut directly across the river? (c) At what speed (in SI units) with respect to the shore would you be traveling?

66. [III] A train is traveling east along a straight run of track at 60 km/h (37 mi/h). Inside, two children 2.0 m apart are playing catch directly across the aisle. The kid wearing a Grateful Dead T-shirt throws the ball horizontally north. The ball crosses the train and is caught in 1.0 s by her little brother. Ignoring any effects of gravity or friction, (a) find the ball's velocity with respect to the little brother. (b) What is the velocity of the ball as seen by someone outside standing still?

67. [III] A cruise ship heads 30° W of N at 20 km/h in a still sea. Someone in a sweatsuit dashes across the deck traveling 60° E of N at 10 km/h. What is the jogger's velocity with respect to the Earth?

68. [III] A ship bearing 60° N of E is making 25 km/h while negotiating a 10-km/h current heading 30° N of E. (a) What is the velocity of the ship with respect to the Earth? (b) How fast is it traveling northward?

69. [III] The Kennedy Space Center on the east coast of Florida is located at a latitude of about 28.5°. Using Problem 61, we can show that the Space Center is moving eastward at 1470 km/h as the Earth rotates. (a) Suppose a rocket is fired so it arcs over and travels horizontally westward at 1470 km/h. What's its speed with respect to both the center and the surface of the Earth? (Incidentally, it's the former speed that determines the orbit of a satellite.) (b) What would be the rocket's speed with respect to the center of the Earth if it were launched at 1470 km/h eastward? (c) Which way do you think we usually fire rockets from Kennedy? (d) Why, overlooking the weather, was Florida rather than Maine originally chosen for the Kennedy Space Center?*

---

*__Answers__ to all odd-numbered Multiple Choice Questions and all odd-numbered Problems are in the Answers in the back of the book. Numbers in boldface indicate that a **solution** is provided in the Answers in the back of the book. Numbers in italic indicate that a **solution** is provided in the Student Solutions Manual.

3

Kinematics: Acceleration

VARIATIONS IN MOTION ARE commonplace. Change is the rule rather than the exception. We travel from one place to the next, speed up, slow down, turn, pause, and move again. How do we quantify this kind of complex motion? In the previous chapter, we learned about velocity, the time-rate-of-change of displacement, but were careful to avoid such complexities. This chapter goes a step further and evolves the concept of *acceleration,* the time-rate-of-change of velocity. To keep things manageable, the discussion will be limited to objects undergoing constant accelerations. But even confining ourselves in this manner will allow us to develop a set of

At present it is the purpose of our Author merely to investigate and to demonstrate some of the properties of accelerated motion (whatever the cause of this acceleration may be).

GALILEO
Dialogues Concerning Two New Sciences (1638)

equations that describes the motion of a great range of occurrences, from braking automobiles to falling apples.

THE CONCEPT OF ACCELERATION

Aristotle hinted at the idea of acceleration, although he never grasped it clearly. Not long afterwards, the philosopher-physicist Strato (ca. 300 B.C.) rightly suggested that a body is accelerating when *equal increments of distance are traversed in shorter and shorter times.* An alternative formulation appeared during the twelfth century in Europe: *acceleration occurs when, during equal periods of time, a body travels greater and greater distances* (Fig. 3.1). What is missing in both of these definitions is the realization that velocity is a distinct concept that may vary from moment to moment. This point was first appreciated in the fourteenth century.

The change in velocity must be part of our definition of acceleration, but it's only a part. The *rate* at which change occurs is crucial. Waiting in a car, at rest on the entrance ramp to a highway, we know what is needed. The car must move from an initial speed $v_i = 0$ to a final speed $v_f = 88.5$ km/h (that is, 55 mi/h) to merge into a gap in the flowing ribbon of traffic. Clearly, the change in velocity ($\Delta \mathbf{v}$) must occur in a specific amount of time (Δt)—it cannot take 10 h, or 10 min, but more like 10 s. **Acceleration is the time-rate-of-change of velocity.**

3.1 Average Acceleration

The **average acceleration ($\mathbf{a}_{av}$)** of a body is defined as *the ratio of the change in its velocity over the time elapsed in the process,* thus

$$\text{Average acceleration} = \frac{\text{change in velocity}}{\text{time elapsed}}$$

$$\mathbf{a}_{av} = \frac{\Delta \mathbf{v}}{\Delta t} = \frac{\mathbf{v}_f - \mathbf{v}_i}{t_f - t_i} \tag{3.1}$$

A vector divided by a scalar is a vector; both velocity and acceleration are vector quantities. Something accelerates in a specific direction—it changes its velocity in that direction.

Figure 3.1 Uniform acceleration. Notice how the distances traveled in equal time intervals increase. As we'll see, those distances go as the odd integers 1, 3, 5, 7, ...

Speed skiing can take a person from 0 to 60 mi/h in 3 s and up to 100 mi/h in 6 s. The men's world record is 139 mi/h. The special rubber suit, helmet, and leg fins have been wind-tunnel tested to cut air resistance.

Acceleration is a two-part concept that arises because of a change in the *direction* of the velocity or because of a change in the *magnitude* of the velocity (that is, speed) or both. A racehorse running faster and faster on a straightaway is accelerating because its speed is changing. A merry-go-round stallion revolving at a constant speed is also accelerating, because the direction of its motion (the direction its nose is pointing) is forever changing. A colt chasing butterflies in a field is likely to be accelerating because of changes in both speed and direction.

Whenever an object moves along a curved path, its velocity vector changes direction and its acceleration will not be parallel to the direction of motion. By contrast, much of this chapter will deal with the simpler situation of rectilinear motion, where velocities and accelerations occur along the same straight line. Even so, there are a great many practical situations that meet this condition (a falling rock is just one of them). Accordingly, $\mathbf{v}_f$ and $\mathbf{v}_i$ are either parallel or antiparallel and $(\mathbf{v}_f - \mathbf{v}_i)$ is in the direction of one or the other. Putting it slightly differently, $\mathbf{v}_i$, $\mathbf{v}_f$, and $\mathbf{a}_{av}$ are *colinear;* that is, they all act along the same line. As we have seen, colinear vectors can be treated algebraically by assigning them appropriate signs. *Let's take the direction in which the object first moves to be positive.* Hence, if $\mathbf{v}_f$ is in the same direction as $\mathbf{v}_i$ and is larger than it, $\mathbf{a}_{av}$ is also in that direction (the time difference $t_f - t_i$ must be a positive number). When the speed increases in the positive direction, it corresponds to a positive acceleration.

By contrast, if $\mathbf{v}_f$ turns out to be smaller than $\mathbf{v}_i$ (or oppositely directed), $\mathbf{a}_{av}$ is negative. A decrease in speed in the positive direction is a negative acceleration, often called a *deceleration.* Many people, however, avoid that term, not wanting to suggest that two different ideas are at work. Regardless of whether velocity is increasing or decreasing, its time-rate-of-change is the acceleration, be it positive or negative.

With an appropriate sign convention, Eq. (3.1) can be written as a scalar, thus

$$a_{av} = \frac{\Delta v}{\Delta t} = \frac{v_f - v_i}{t_f - t_i} \tag{3.2}$$

In the same way, Eq. (2.3) was formulated. The dimensions of a_{av} are those of *speed over time* or, equivalently, *distance over time, over time*—meters per second, per second (m/s^2); feet per second, per second (ft/s^2), and so on. The former is the preferred SI unit of acceleration, and it is read "meters per second squared" (although (m/s)/s is more informative—an object changes speed by so many meters per second, every second).

Example 3.1 A yellow messenger robot is traveling at 1.00 m/s in a straight line along a ramp in a spaceship. If it speeds up to 2.50 m/s in a time of 0.50 s, what is the magnitude of its average acceleration?

Solution: [Given: $v_i = 1.00$ m/s, $v_f = 2.50$ m/s, and $\Delta t = 0.50$ s. Find: a_{av}]. From the definition

$$a_{av} = \frac{\Delta v}{\Delta t} = \frac{(2.50 \text{ m/s} - 1.00 \text{ m/s})}{0.50 \text{ s}}$$

and

$$a_{av} = \boxed{3.0 \text{ m/s}^2}$$

▶ **Quick Check:** $\Delta v = a_{av}\Delta t = (3.0 \text{ m/s}^2)(0.50 \text{ s}) = 1.5$ m/s.

Example 3.2 A red, 12-cylinder Jaguar XJ12L sedan (getting a dreadful 13 miles per gallon) can travel a straight line from a dead stop to 48.3 km/h (30.0 mi/h) in 3.80 s. It takes an additional 3.00 s to reach 80.5 km/h (50.0 mi/h) and 16.7 more seconds to attain 161 km/h (100 mi/h). What is the magnitude of its average acceleration in m/s^2 during each of these intervals?

Solution: We take the direction in which the car is traveling as positive. [Given: for the three intervals $v_{i1} = 0$, $v_{f1} = 48.3$ km/h, $\Delta t_1 = 3.80$ s; $v_{i2} = 48.3$ km/h, $v_{f2} = 80.5$ km/h, $\Delta t_2 = 3.00$ s; $v_{i3} = 80.5$ km/h, $v_{f3} = 161$ km/h, $\Delta t_3 = 16.7$ s. Find: the corresponding values of a_{av}.] From the definition of a_{av} we'll need Δv over Δt. The three successive values of $\Delta v = v_f - v_i$ are 48.3 km/h, 32.2 km/h, and 80.5 km/h. To convert these to m/s, multiply each by [(1000 m/km)/(3600 s/h)], which equals 0.2778 according to Table 2.3 (p. 26), thereby yielding 13.42 m/s, 8.945 m/s, and 22.36 m/s. During the first 3.80 s

$$(a_{av})_1 = \frac{13.42 \text{ m/s}}{3.80 \text{ s}} = 3.53 \text{ m/s}^2$$

The car moves faster and faster, increasing its speed on the average 3.53 m/s each second. During the next 3.00 s

$$(a_{av})_2 = \frac{8.945 \text{ m/s}}{3.00 \text{ s}} = 2.98 \text{ m/s}^2$$

And over the last 16.7 s

$$(a_{av})_3 = \frac{22.36 \text{ m/s}}{16.7 \text{ s}} = 1.34 \text{ m/s}^2$$

▶ **Quick Check:** As expected, the vehicle, encountering increasing air resistance, has a decreasing acceleration. From Table 2.2 (p. 25), we can get some fast rough answers: 30 mi/h ≈ 13 m/s, 50 mi/h ≈ 22 m/s, and 100 mi/h ≈ 45 m/s. Hence, (13 m/s)/4 s = 3 m/s^2, (22 m/s − 13 m/s)/3 s = 3 m/s^2, and (45 m/s − 22 m/s)/17 s = 1.4 s.

3.2 Instantaneous Acceleration

The people who made the great advances in kinematics in the early fourteenth century were scientists at Oxford with marvelous names like Bradwardine, Dumbleton, Heytesbury, and Swineshead. From their work, we understand that constant or **uniform acceleration** happens *when equal changes in speed occur during equal intervals of time, of any duration.*

For example, if an object at rest increases its speed precisely by 5 m/s during each 1 s of travel, the magnitude of its acceleration will be constant at 5 m/s^2. Because the motion is along a straight line, we graph velocity versus time (Fig. 3.2a), with the initial direction of motion positive. The plot is a diagonal *straight line,* where $v = 0$ at $t = 0$. At the end of successive 1-s intervals, each point is plotted 5 m/s higher than the previous point, and so the graph has to be a straight line, rising at a constant rate. The slope, the rise (Δv) over the run (Δt), is the same everywhere; it's constant at 5 m/s over 1 s and *equal to the scalar value of the average acceleration.* The greater the constant acceleration, the greater the slope of the line. If the speed was *decreasing* ($\Delta v < 0$), the rise over the run would be *negative.* A negative slope, one progressing downward, corresponds to a negative acceleration (Fig. 3.2b). By contrast, a constant speed corresponds to a horizontal line, a slope of zero, and no acceleration (Fig. 2.5, p. 29).

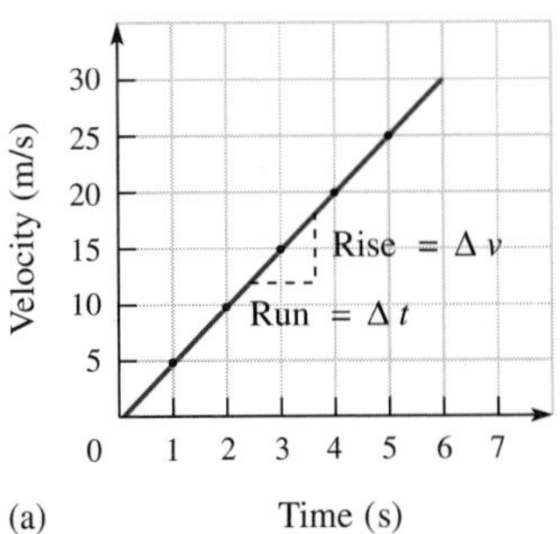

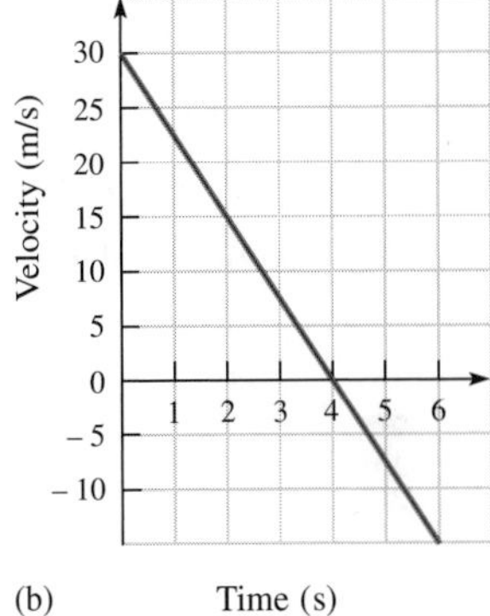

Figure 3.2 Graphs of the motions of two objects moving along straight lines. The slope of the velocity-time plot is the average acceleration over the interval. In (a), the slope is positive and constant, as is the acceleration. In (b), the slope is constant and negative. The body decelerates to $v = 0$ and continues to accelerate in the negative direction such that, after $t = 4$ s, it has a negative velocity.

Figure 3.3 is a plot of the velocity-time curve for the Jaguar XJ12L in the previous example. Although it is fairly straight for the first 3 or 4 s, the curve leans over more and more, leveling off as time goes on. Its slope (that is, the slope of the tangent at any point) decreases, slowly at first but more so as time progresses—the car's speed is increasing, but ever more slowly. Evidently, it cannot accelerate forever because of increasing air drag. The Jaguar will never hit 1000 km/h, even with the gas pedal floored for a week.

Again, because the motion varies in time, we need to consider what is happening at each moment. It's not very meaningful to ask about the vehicle's average acceleration without specifying the interval over which we want it. Furthermore, the larger that time interval, the more obscured the details of the motion become. Thus, the car ends up with an average acceleration of 1.90 m/s^2 for the whole run. That single number is poor in information, and has less meaning than the graph that makes it clear there is a range of acceleration. What we need is the average acceleration at *any moment* over a *tiny interval.* The limiting value approached by the average acceleration—Eq. (3.1)—as the time interval (over which the averaging occurs) shrinks toward zero is known as the **instantaneous acceleration**, and is expressed as

$$\mathbf{a} = \lim_{\Delta t \to 0} \left[\frac{\Delta \mathbf{v}}{\Delta t} \right] \tag{3.3}$$

The limiting value has the same form as Eq. (2.4). This expression represents the moment-by-moment value of the acceleration of an object. *The slope of the tangent*

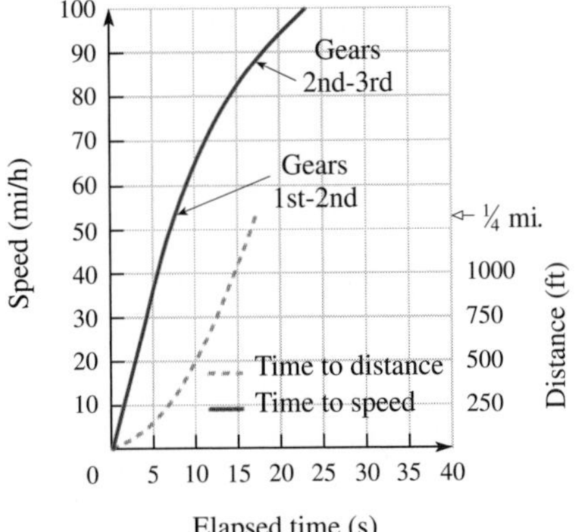

Figure 3.3 The speed-time graph for a classic Jaguar XJ12L sedan traveling along a straight line.

to the velocity-time curve at any moment is the instantaneous acceleration at that moment. The Jaguar's velocity-time curve shows a gradual decrease in slope, and therefore a gradual decrease in acceleration, as time goes on.

UNIFORMLY ACCELERATED MOTION

Although in the real world, acceleration is rarely if ever constant, there are many circumstances where it can be treated that way, at least over some limited time interval. The first few seconds of an automobile's acceleration is one such case (Fig. 3.3). Without calculus, we have to be content considering only situations in which **a** may be taken to be constant, and therefore equal to $\mathbf{a}_{av}$. Thus, no matter how large Δt is

$$\mathbf{a} = \frac{\Delta \mathbf{v}}{\Delta t} = \frac{\mathbf{v}_f - \mathbf{v}_i}{t_f - t_i} \tag{3.4}$$

Continuing to restrict the discussion to straight-line motion, we can treat the acceleration as a scalar. Furthermore, if we start our clocks so that $t_i = 0$, the final time t_f is the entire trip-time, which we represent simply as t. Hence, Eq. (3.4) becomes

$$a = \frac{v_f - v_i}{t} \tag{3.5}$$

Multiplying both sides by t and shifting v_i over to the other side, leaving v_f to stand alone, we arrive at one of the equations governing *uniformly accelerated motion,* namely

$$v_f = v_i + at \tag{3.6}$$

Four quantities are interrelated by Eq. (3.6) and given any three, the remaining one can be calculated. *Taking the direction of initial motion as positive, numerical values entered into Eq. (3.6), or any other equation in this chapter, can be positive or negative.* For example, if **a** is in the negative direction opposing the initial motion, the numerical value of a must be negative when put in Eq. (3.6), which will ensure that $v_f < v_i$.

Example 3.3 A red Jaguar XJ12L can screech to a straight-line stop from 96.54 km/h (60.00 mi/h) in about 3.7 s. Compute the magnitude of its acceleration, assuming it to be constant.

Solution: [Given: $v_i = 96.54$ km/h, $v_f = 0$, and $t = 3.7$ s. Find: a.] We have v_i, v_f, and t and need a, which suggests Eq. (3.6), or Eq. (3.5). In either case, we will require the initial speed in m/s: $v_i = (96.54 \text{ km/h}) \times (1000 \text{ m}/1 \text{ km})(1 \text{ h}/3600 \text{ s}) = 26.82$ m/s. Since

$$a = \frac{v_f - v_i}{t}$$

we have

$$a = \frac{0 - 26.82 \text{ m/s}}{3.7 \text{ s}} = \boxed{-7.2 \text{ m/s}^2}$$

▶ **Quick Check:** 60 mi/h = 88 ft/s and $a = -23.78$ ft/s², multiplying by 0.3048 m/ft, we get -7.2 m/s².

Example 3.4 A bicyclist pedaling along a straight road at 25.0 km/h uniformly accelerates at 3.00 m/s^2 for 3.00 s. Find her final speed.

Solution: The words *uniformly accelerating* immediately classify this as a problem for which $a =$ constant and to which Eq. (3.6) applies. We select the direction of motion as positive. [Given: $v_i = +25.0$ km/h, $a = +3.00$ m/s^2, and $t = 3.00$ s. Find: v_f.] Put everything in the same SI units:

$$v_i = \frac{(25.0 \text{ km/h})(1000 \text{ m/km})}{3600 \text{ s/h}} = 6.94 \text{ m/s}$$

Hence
$$v_f = v_i + at$$

$$v_f = (6.94 \text{ m/s}) + (3.00 \text{ m/s}^2)(3.00 \text{ s})$$

and to three significant figures, $\boxed{v_f = 15.9 \text{ m/s}}$.

▶ **Quick Check:** $\Delta v/\Delta t = (15.9 \text{ m/s} - 6.94 \text{ m/s})/3 \text{ s} = 3 \text{ m/s}^2$.

3.3 The Mean Speed

Perhaps the most important contribution to the theory of motion that came from the Middle Ages was the Mean-Speed Theorem—a gift from Oxford, around 1335. Having successfully defined uniform acceleration, scholars next wanted to relate a to the distance traveled. They knew that the magnitude of the displacement s traversed in rectilinear motion at some average speed was given by $s = v_{av}t$ (though all of this work was done long before any algebraic expressions were ever written out). Could it then be possible to determine an *average speed* that would produce the same displacement in the same time interval as would the constant acceleration in question? (See Fig. 3.4.) In other words, does an average value of the speed exist for an object that is uniformly accelerating along a straight line from an initial speed v_i to a final speed v_f? The answer is yes, and that **mean speed** is given by

$$v_{av} = \tfrac{1}{2}(v_i + v_f) \tag{3.7}$$

Today, we often contend that a constant acceleration produces a uniformly changing speed, and so the average value is midway between the initial and final speeds (Fig. 3.5). There were proofs like that given in the fourteenth century, but let's look at a different approach that is quite fruitful. Around 1360, Nicolas Oresme used a primitive speed-time graph to assert that the area under the sloping straight line representing uniform acceleration corresponded to the magnitude of the displacement (Fig. 3.6). We considered something similar in Chapter 2.

The area in question has two distinct pieces and can be thought of as the sum of the areas of the rectangle (base t times height v_i) and the triangle (one-half the base t times the altitude $v_f - v_i$). The total area is therefore given by

$$v_i t + \tfrac{1}{2}(v_f - v_i)t$$

or
$$(v_i - \tfrac{1}{2}v_i + \tfrac{1}{2}v_f)t = \tfrac{1}{2}(v_i + v_f)t$$

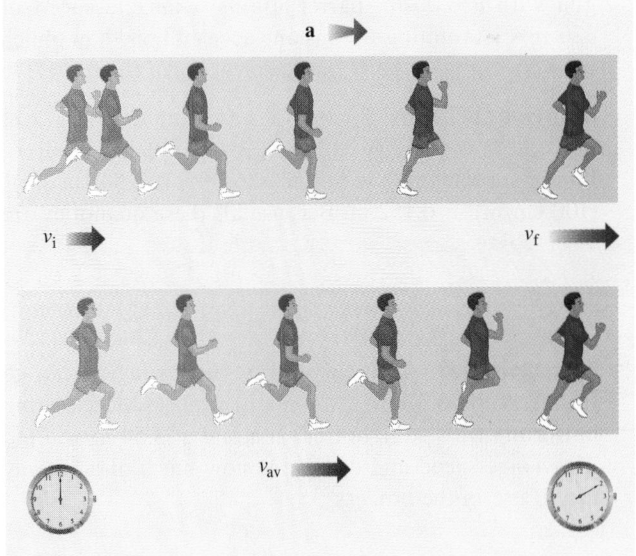

Figure 3.4 A body that accelerates uniformly from v_i to v_f over a certain straight-line distance will cover exactly the same distance in the same time traveling at a fixed speed of v_{av}, known as the *mean speed.*

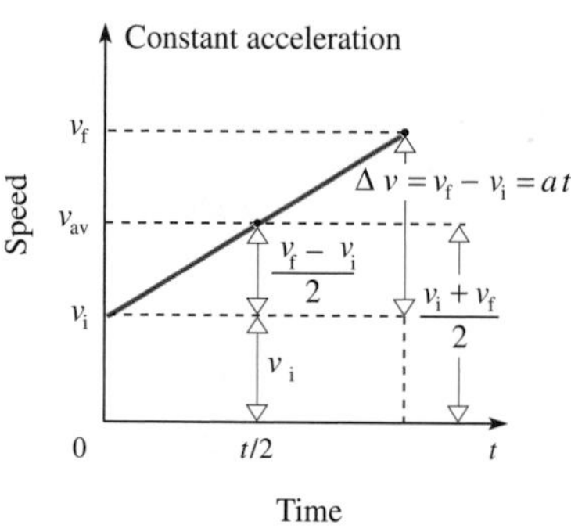

Figure 3.5 For a body that is uniformly accelerating, the mean-speed, v_{av}, is the height of the midpoint of the straight-line speed-time graph.

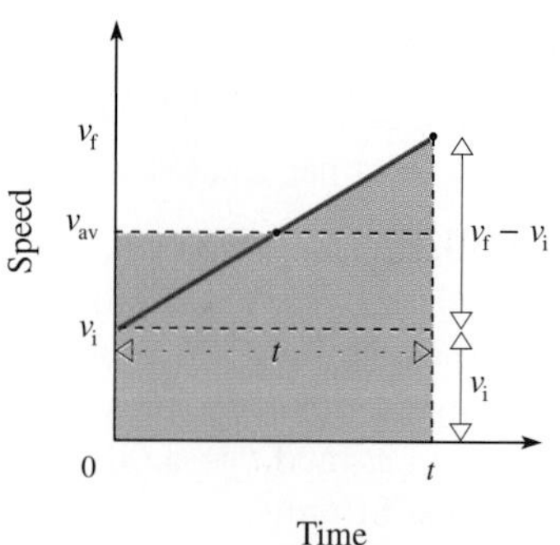

Figure 3.6 The area under the speed-time curve is the distance traveled. The area under the curve (that is, the tilted line) equals the area under the v_{av} line. In other words, $v_{av}t = v_it + \frac{1}{2}(v_f - v_i)t$.

Setting this equal to the magnitude of the displacement (s) experienced during the uniform accelerated motion, we arrive at what was called the *Mean-Speed Theorem*:

$$s = \tfrac{1}{2}(v_i + v_f)t \qquad (3.8)$$

This equation is simply $s = v_{av}t$ with v_{av} given by Eq. (3.7). Notice that the mean speed is attained at the journey's halfway point in time (Fig. 3.5) and not at the halfway point in space. Also remember that Eq. (2.1) is always true by definition, whereas Eqs. (3.7) and (3.8) are true only when a is constant.

Example 3.5 A bullet fired from a 38-caliber handgun with a 6.00-in. barrel attains a muzzle speed of 330 m/s. Assuming a constant acceleration, how much time does it take the bullet to travel down the barrel?

Solution: This is a case of constant acceleration. [Given: $v_i = 0$, $v_f = 330$ m/s, $s = 6.00$ in. Find: t.] First, put everything in SI units: (6.00 in.)(2.54 cm/in.)/(100 cm/m) = 0.152 m. Because all these quantities are related by $s = \frac{1}{2}(v_i + v_f)t$ we have

$$s = 0.152 \text{ m} = \tfrac{1}{2}(0 + 330 \text{ m/s})t$$

and $t = 2(0.152 \text{ m})/(330 \text{ m/s}) = \boxed{9.21 \times 10^{-4} \text{ s}}$.

▶ **Quick Check:** In U.S. Customary units, $\frac{1}{2}$ ft $= \frac{1}{2}(1083 \text{ ft/s})t$; $t = 0.92$ ms.

Example 3.6 That classic 12-cylinder Jaguar can go from rest to 48.3 km/h (30 mi/h) in 3.8 s, accelerating uniformly (in a straight line) at a rate of 3.54 m/s². Find its average speed and determine how much of a runway it will use in the process.

Solution: The key words here are *accelerating uniformly,* which means a is constant. Take the direction of motion as positive. [Given: $v_i = 0$, $v_f = +48.3$ km/h, $t = 3.8$ s and $a = +3.54 \text{ m/s}^2$. Find: v_{av} and s.] First we convert v_f to m/s, so

(continued)

(continued)

$$v_f = \frac{(48.3 \text{ km/h})(1000 \text{ m/km})}{3600 \text{ s/h}} = 13.4 \text{ m/s}$$

Equation (3.7) gives us

$$v_{av} = \tfrac{1}{2}(v_i + v_f) = \tfrac{1}{2}(0 + 13.4 \text{ m/s}) = \boxed{6.7 \text{ m/s}}$$

which equals 15 mi/h. The distance traveled [Eq. (3.8)] is that average speed multiplied by the time of travel, or

$$s = (6.7 \text{ m/s})(3.8 \text{ s}) = \boxed{25 \text{ m}}$$

or about 84 ft.

▶ **Quick Check:** From Table 2.3, $v_f = (48.3) \times (0.2778)$ m/s $= 13.4$ m/s. Since 60 mi/h $= 88$ ft/s, 30 mi/h $= 44$ ft/s; hence, $v_{av} = 22$ ft/s $= 6.7$ m/s.

3.4 Integrals and Intervals

Oresme's area-under-the-curve analysis introduced the use of graphs to represent motion and served as an early cornerstone for the integral calculus of Newton (1665–1666) and Leibniz (1673–1676).

Thus far, we have computed areas under curves, or portions of curves, that were simple straight lines. Now, we want to generalize the technique to include nonuniform motion, such as that represented in Fig. 3.7. Since the object depicted in Fig. 3.7a is moving along a straight path, we can call the diagram a velocity-time curve. We would like to be certain that the area under it, bounded by the times t_i and t_f, is the scalar value of the displacement of the body. In this case, where the velocity is always positive, the magnitude of the displacement equals the distance traveled. Divide the region into several subintervals, each corresponding to a small rectangular area. During any one of these finite subintervals, the body will travel a finite distance.

Let's look at the second subinterval for which the distance traveled is Δs_2. We fix the height of the rectangle at a value of speed equal to $(v_{av})_2$ such that $\Delta s_2 = (v_{av})_2 \Delta t_2$. In so doing, we make its area exactly equal to Δs_2, the appropriate subinterval displacement. If we add up all the subareas, the sum equals the total distance traveled by the body, but we still don't know if that is exactly the area *under the curve.*

The tops of all the rectangles can be made to more closely coincide with the curve by increasing the number of subintervals, making each one correspondingly narrower. If we allow the number of subintervals to approach infinity while the

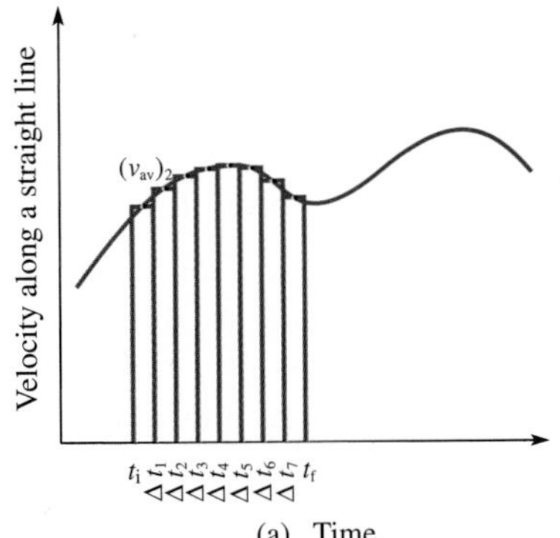

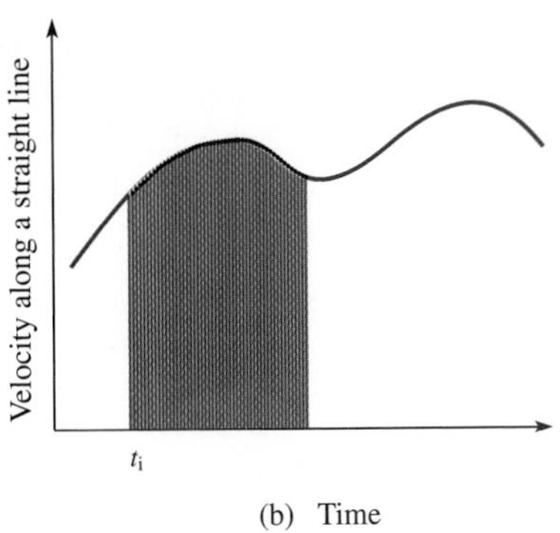

Figure 3.7 (a) The velocity-time curve for an object moving in a straight path at a speed that changes from moment to moment. The second area element is $\Delta s_2 = (v_{av})_2 \Delta t_2$, and the total area under the curve between t_i and t_f is the distance traveled in that time interval. (b) As each time interval Δt shrinks, and the number of such intervals increases, the uppermost points form the curve. The total area of the rectangles approaches the area under the curve.

width of each one (that is, Δt) approaches zero, the tops of the rectangles will blend smoothly into the curve. Those topmost speeds on the rectangles are the various v_{av} values and, in the limit as $\Delta t \rightarrow 0$, they become the instantaneous speeds [Eq. (2.4)] that correspond to points on the curve. Thus, the area under the curve is the distance traveled (Fig. 3.7b).

Isaac Beeckman in 1618 divided time into "indivisible moments," and areas into infinitesimal rectangles, long before the work of Newton and Leibniz. The scheme for adding up an infinity of infinitesimals is the heart of integral calculus. For us, it will suffice to draw the velocity-time curves on graph paper and count the little boxes. Keep in mind that if the object moves in the negative direction (as in Fig. P44, p. 55), the area will then be negative, and any such contribution must be subtracted from the total positive area. An object that moves outward, stops, turns around, and comes all the way back, will have zero displacement and, properly, zero net area under its velocity-time curve.

Example 3.7 Figure 3.8 depicts a speed-time curve for a toy compressed-gas rocket car that has enough fuel for 3.0 s of full power. Approximately how far did the toy car travel between the second and fourth seconds?

Solution: We want the area under the curve between $t = 2.0$ s and $t = 4.0$ s. In the interval from 2.0 to 3.0 s, the area is roughly that of a triangle above a rectangle: $\frac{1}{2}(30\text{ m/s} - 20\text{ m/s})1.0\text{ s} + (20\text{ m/s})1.0\text{ s} = 25\text{ m}$. Between 3.0 and 4.0 s, the speed is fairly constant at 30 m/s and the area is that of a rectangle, about $(30\text{ m/s})1.0\text{ s} = 30\text{ m}$. Therefore, the total distance is $\boxed{\approx 55\text{ m}}$.

▶ **Quick Check:** The average speed is >25 m/s and <30 m/s so for 2 s, 55 m is reasonable.

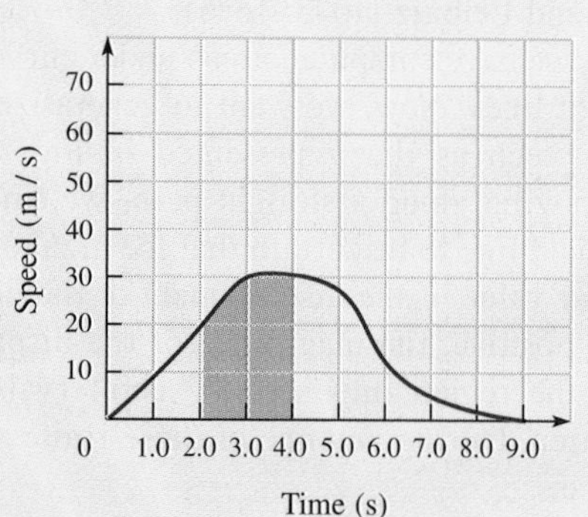

Figure 3.8 The speed-time curve for a toy car traveling in a straight line. The acceleration is positive and nearly constant for the first 3 s.

3.5 The Equations of Constant Acceleration

Usually, acceleration changes with time, and the calculus is necessary to describe the motion. Nonetheless, many situations occur with uniform, or almost uniform, acceleration, and we already have the kinematics to deal with those. To that end, we will develop five constant-a equations that interrelate the five motion variables (v_i, v_f, a, s, t). These derive from the two basic definitions of speed and acceleration, and so there are only two independent variables that can be solved for. First, from the definition of acceleration, we have

$$v_f = v_i + at \qquad [3.6]$$

The average speed is

$$v_{av} = \tfrac{1}{2}(v_i + v_f) \qquad [3.7]$$

although it does not shed any new light on the relationship among v_i, v_f, a, s, and t. The Mean-Speed Theorem gives us

$$s = v_{av}t = \tfrac{1}{2}(v_i + v_f)t \qquad [3.8]$$

It would be useful to have an expression for s that is not explicitly dependent on v_f but is written in terms of v_i, t, and a. Often, objects begin their motion from rest, so that v_i (equal to zero) is known. Furthermore, although s and t are easy to measure, v_f is frequently not available because it's much more difficult to determine directly. Replace v_f in Eq. (3.8) by Eq. (3.6):

$$s = \tfrac{1}{2}[v_i + (v_i + at)]t = \tfrac{1}{2}(2v_it + at^2)$$

and

$$s = v_it + \tfrac{1}{2}at^2 \qquad (3.9)$$

The first term (v_it) is the scalar value of the body's displacement as it coasts along at its initial speed, even if it is not accelerating ($a = 0$). To the first term is added another displacement ($\tfrac{1}{2}at^2$) traversed as the speed increases beyond v_i. If a is negative, the term ($\tfrac{1}{2}at^2$) is also negative, and the object is slowing down—it will not travel as far as (v_it). In fact, since **s** is the displacement, it's possible for the negative acceleration to carry the object back to where it started such that $s = 0$, or even to where $s < 0$. Using Eq. (3.9) is equivalent to using both Eqs. (3.6) and (3.8), and that's sometimes more convenient.

Example 3.8 According to *Road & Track* magazine, a Lotus Esprit super-coupe can travel 100 ft from rest in 3.30 s. Assume the acceleration is constant, as it nearly is, and determine it in m/s^2.

Solution: This is a uniform-acceleration problem for which we currently have four applicable equations. Take the direction of motion as positive. [Given: $v_i = 0$, $s = +100$ ft, and $t = 3.30$ s. Find: a.] This list (v_i, s, t, and a) suggests Eq. (3.9), which contains all, and only, these quantities. Hence:

$$s = v_it + \tfrac{1}{2}at^2$$

$$(100 \text{ ft})(0.3048 \text{ m/ft}) = 0 + \tfrac{1}{2}a(3.30 \text{ s})^2$$

Multiplying both sides by 2 and dividing by $(3.30 \text{ s})^2$ yields

$$a = \frac{60.96 \text{ m}}{(3.30 \text{ s})^2} = 5.598 \text{ m/s}^2$$

and to three figures, $\boxed{a = 5.60 \text{ m/s}^2}$. Note that s and a are both positive quantities.

▶ **Quick Check:** Using Eq. (3.6), $v_f = v_i + at = 0 + (5.6 \text{ m/s}^2)(3.3 \text{ s}) = 18.5$ m/s, so the average speed is 9.25 m/s, which in 3.30 s means a distance traveled of 30.5 m = 100 ft.

The last of the five expressions we want relates the displacement, speeds, and acceleration *independent of time*. First, solve Eq. (3.6) for the time, $t = (v_f - v_i)/a$.

Then substitute this along with v_{av} from Eq. (3.7) into Eq. (3.8) so that t no longer appears explicitly:

$$s = \left(\frac{v_i + v_f}{2}\right)\left(\frac{v_f - v_i}{a}\right) = \frac{v_f^2 - v_i^2}{2a}$$

Multiplying both sides by $2a$ and adding v_i^2 to both sides yields

$$v_f^2 = v_i^2 + 2as \tag{3.10}$$

Whenever we have a problem concerning uniform acceleration in which time does not explicitly appear, Eq. (3.10) will likely be the key formula in the analysis.

Example 3.9 The fastest animal sprinter is the cheetah, reaching speeds in excess of 113 km/h (70 mi/h). These animals have been observed to bound from a standing start to 72 km/h in 2.0 s. Compute the cheetah's maximum acceleration, assuming it to be constant. What minimum distance is required for the cheetah to go from rest to 17.9 m/s (40 mi/h)?

Solution: [Given: for the first part, $v_i = 0$, $v_f =$ 72 km/h, $t = 2.0$ s. Find: a. In the second part, $v_i = 0$, $v_f = 17.9$ m/s, and a is assumed constant. Find: s.] Eq. (3.6) is perfect for the first part: $v_f = 72$ km/h $=$ 20 m/s

$$a = \frac{v_f - v_i}{t} = \frac{20\ \text{m/s}}{2.0\ \text{s}} = \boxed{10\ \text{m/s}^2}$$

The second part calls for s without specifying the corresponding time and that suggests Eq. (3.10). Using $v_f =$ 17.9 m/s and rearranging ($v_f^2 = v_i^2 + 2as$) where $v_i = 0$, we get

$$s = \frac{v_f^2}{2a} = \frac{(17.9\ \text{m/s})^2}{2(10\ \text{m/s}^2)} = \boxed{16\ \text{m}}$$

The cheetah can hit 17.9 m/s from a dead stop in a mere 16 m.

▶ **Quick Check:** $v_{av} = \frac{1}{2}\,72$ km/h $= 10$ m/s; hence, it takes a distance $s = (10\ \text{m/s})(2.0\ \text{s}) = 20$ m to reach 20 m/s, and so 16 m for a top speed of 17.9 m/s is reasonable.

Suppose you are driving a car and decide to stop. Assuming you apply steady pressure to properly working brakes that do not lock, and assuming good road conditions, the vehicle should experience a uniform negative acceleration. How far can you expect to travel before coming to rest? According to Eq. (3.10), realizing that $v_f = 0$, we have

$$0 = v_i^2 + 2as$$

and

$$-v_i^2 = 2as$$

Thus, the *stopping distance* is

$$s = -\frac{v_i^2}{2a} \tag{3.11}$$

Because a is a negative number, the presence of the minus sign ensures that s will be positive. You brake, and the car continues to be displaced a distance s in the positive direction of the initial motion.

The stopping distance depends on v_i *squared*—double the driving speed, and you will need four times the space to stop. The stopping distance also depends on the negative acceleration of your car (the condition and type of its brakes and tires), as well as the road surface. Under good conditions, a superb automobile like a Lamborghini has a negative acceleration of around $-9.3\ \text{m/s}^2$, whereas a less than superb compact might slow down at around $-6.6\ \text{m/s}^2$. If you brake while traveling 36 m/s (80 mi/h) in that compact on a sunny day and manage not to skid, you will coast for 98 m before the thing stops. (Which is one good reason to never drive that fast.)

Our description overlooks the fact that ordinarily it takes some time (t_R) for the driver to react. That reaction time might be as little as 0.2 s, but it's likely to be more. Thus, if you are traveling at a speed v_i, a distance $s = v_i t_R$ will be traversed before you even activate the brakes. Adding this term to Eq. (3.11) means that stopping will take a distance of

$$s = v_i t_R - \frac{v_i^2}{2a} \tag{3.12}$$

which should make the point that tailgating is a bad idea (Fig. 3.9).

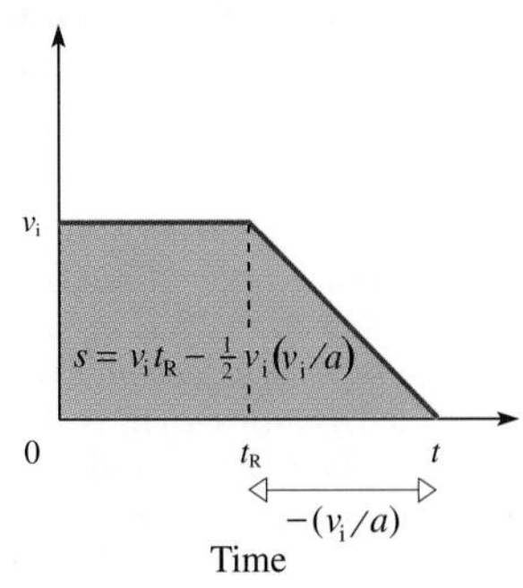

Figure 3.9 The speed-time curve for a driver who, while traveling at v_i, waits a time t_R before slamming on the brakes. The area under the curve is the distance traveled, s. The slope of the curve becomes negative once the brakes engage at $t = t_R$.

Example 3.10 It is a typical sunny day. Assume your reaction time is 0.50 s and the acceleration of your car is $-8.2\ \text{m/s}^2$. What is the no-skid stopping distance when the car's speed is initially 26.8 m/s (60 mi/h)?

Solution: [Given: $t_R = 0.50$ s, $a = -8.2\ \text{m/s}^2$, $v_i = 26.8$ m/s. Find: s.] This specialized problem relates to the analysis that produced Eq. (3.12):

$$s = v_i t_R - \frac{v_i^2}{2a}$$

$$s = (26.8\ \text{m/s})(0.50\ \text{s}) - \frac{(26.8\ \text{m/s})^2}{2(-8.2\ \text{m/s}^2)}$$

and $s = 13.4\ \text{m} + 43.8\ \text{m} = \boxed{57\ \text{m}} \approx 188\ \text{ft}$

▶ **Quick Check:** From Eq. (3.6), the time to decelerate is $(-26.8\ \text{m/s})/(-8.2\ \text{m/s}^2) = 3.3$ s, and from Eq. (3.9) the distance traveled with the brakes engaged is 43.8 m.

FREE-FALL

Probably the most important situation where acceleration is effectively constant is that of free-fall. Drop a heavy object at the Earth's surface and it will descend in such a way that Eqs. (3.6) through (3.10) describe the motion quite accurately.

A falling object ordinarily moves through the surrounding air, so the essential nature of the motion can be obscured by friction. We now know that, in a vacuum, all bodies uniformly accelerate downward at the same rate, regardless of their weight. This crucial insight reveals the hidden nature of gravity, but the realization of the role played by gravity was a long time coming. Aristotle created a theory of the Universe that dominated the science of Muslims and Christians alike, right through the Renaissance. He maintained that a vacuum could not exist (p. 90), and so objects always fell through some medium.

This Aristotelian notion produced the misleading conclusion that all falling bodies descend at different rates proportional to their individual weights, which is a fair picture of what happens to extremely light objects, such as feathers and snowflakes, in air. It applies, as well, for heavy bodies descending through thick media, such as oil or honey, but it misses the central point. For centuries, people wrongly thought that a stone 10 times heavier than another would fall through air 10 times faster—even today many believe this to be the case.

If Galileo Galilei were alive he probably would be tugging at his beard with pleasure about the elementary physics experiment Apollo 15 *astronaut David R. Scott performed today on the moon.*

Scott dropped a hammer and a feather from waist high to illustrate that both objects accelerated equally by the moon's gravity and that both would hit the surface at the same time despite their differences in mass or weight.

The experiment, similar to those Galileo did 300 years ago from the top of the Leaning Tower of Pisa in Italy, was performed as Scott and James B. Irwin were finishing up their final lunar excursion.

"In my left hand I have a feather. In the right hand a hammer," Scott said, standing in front of the camera mounted on the lunar Rover.

"The reason I have these here today is because of Galileo's discourse on falling bodies in gravity fields. Where better to confirm his findings than on the moon."

Then he dropped both objects and, sure enough, they struck the lunar surface simultaneously.

Associated Press News Release, July, 1971

The person who contributed most effectively to the modern understanding of falling bodies was Galileo Galilei, the hero of free-fall. Galileo was born in Italy in 1564. He was vigorous and robust, built short and solid with wavy red hair, a man who fathered three more children than perhaps was proper for a lifelong bachelor. He was an arrogant, argumentative, self-assured spokesman of modernity.

Galileo really was not saying anything new when he asserted that, **in a vacuum, where there is no air resistance, all bodies fall at the same rate**. Others (John Philoponus in the sixth century, for instance) had guessed as much before. By contrast, when an object falls, moving ever more rapidly through the air, the resistance to its motion, the *drag,* also increases. (Riding on a motorcycle gives a sense of how formidable the effect can be—at 100 mi/h the force exerted on a 1.0 ft^2 perpendicular area is about 25 lb, and it increases to 100 lb at 200 mi/h.) The greater the rate of descent, the greater the resistance, as more air must be pushed aside per second (roughly speaking, for small objects at low speeds, resistance goes up with v; for large objects at high speeds, it goes up with v^2). Finally, a balance is reached where no further increase in falling speed can occur. This point is called the **terminal speed**, and it depends on the shape, surface, and weight of the object (see Table 3.1).

A sky diver, after dropping about 620 m (in the thick atmosphere below an altitude of 3000 m or so) with arms and legs extended spread-eagle, will reach a top speed of roughly 200 km/h (120 mi/h).* In a head-down dive, the same person will have a terminal speed up around 300 km/h (185 mi/h). With an open parachute, drag increases tremendously, and the terminal speed drops to around 30 km/h (20 mi/h). On the other hand, a compact heavy object, a smooth stone or an aerial bomb, might drop 150 m to 200 m before even beginning to deviate from free-fall.

All else being constant, terminal speed decreases as the size of an object decreases. The terminal speed of a BB ($\approx$9 m/s) is lower than that of a cannonball ($\approx$250 m/s). The terminal speed of a child is less than that of an adult. This fact accounts, in part, for small animals and insects surviving falls that are immense in proportion to their size. Air friction spares us from being hurt by raindrops and hailstones, which ordinarily strike the Earth at only about 25 km/h to 30 km/h, but might otherwise fall at several hundreds of kilometers per hour (Problem 49, p. 87).

Light objects with a lot of surface area have low terminal speeds that are reached quickly even in air, which is why they seem to conform to Aristotle's notion of weight-dependent fall (why a piece of paper drops slower than a penny). But crumple the paper into a ball, cutting drag without changing weight, and it will fall together with the penny. Galileo knew that for all falling bodies, air resistance would "render the motion uniform," provided the descent was long enough. His ideas were confirmed some 30 years later, when Robert Boyle pumped the air out of a long cylinder and, for the first time, dropped various objects in a vacuum. No amount of theorizing is as convincing as seeing a feather and a lead ball come slamming down together.

TABLE 3.1 Approximate terminal speeds

Object	Speed (m/s)
Fluffy feather	0.4
Sheet of paper (flat)	0.5
Snowflake	1
Parachutist	7
Penny	9
No. 6 shot	9
Mouse	13
Sky diver spread-eagle (reached in 10 to 12 s)	58
Bullet (high caliber)	100
Large rock	200
Cannonball	250

*Five seconds is the record for a 310-ft free-fall onto an air bag. The stunt man, who reached 80 mi/h and burst the bag on impact, walked away smiling.

The terminal speed of each person in this position is about 200 km/h. Notice how this flat, belly-down orientation is stable. Throw some playing cards into the air. Do they fall edge-first?

3.6 Acceleration Due to Gravity

Aristotle implied that falling bodies accelerate, but his remarks were ambiguous. His successor Strato, however, was perfectly clear: *objects in free-fall accelerate.* Rainwater running off a roof begins its descent in a continuous stream that ultimately breaks up into a flutter of separate splashes. "This could never happen," Strato maintained, "unless the water was falling more swiftly through each successive distance than it had through the earlier ones." The droplets do accelerate, and the ones falling longer move faster and pull away from the slower ones behind them. On some level, you already knew that falling bodies accelerate—although you might volunteer to catch a baseball dropped from a few meters, you would surely decline if it was to come down from the top of the Sears Tower in Chicago (443 m). The ball would then descend at a tremendous rate, which could only mean that it accelerated.

Leonardo da Vinci maintained that a body in free-fall moves at a speed proportional to its time of flight, which is equivalent to saying that a body accelerates uniformly, since, by Eq. (3.6), when $v_i = 0$, $v_f = at$. But it was Galileo who first designed and carried out an experiment to confirm that the acceleration is indeed constant. Two practical obstacles made it impossible for him to simply drop an object, measure v_f at several values of t, and then compute a for each. The first was that free-fall happens rapidly and clocks in his era were extremely crude. He surmounted this problem by slowing things down—instead of dropping balls, he rolled

On free-fall. *For all things that fall through the water and thin air, these things must needs quicken their fall in proportion to their weights, just because the body of water and the thin nature of air cannot check each thing equally, but give place more quickly when overcome by heavier bodies. But, on the other hand, the empty void cannot on any side, at any time, support anything, but rather, as its own nature desires, it continues to give place; wherefore all things must needs be borne on through the calm void, moving at equal rate with unequal weights.*

LUCRETIUS (ca. 94–55 B.C.)
On the Nature of Things

THE RECORD FREE-FALL

Lieutenant I. M. Chisov of the former Soviet Union was flying his *Ilyushin 4* on a bitter cold day in January, 1942, when it was attacked by 12 German *Messerschmitts.* Convinced that he had no chance of surviving if he stayed with his badly battered plane, Chisov bailed out at 21 980 ft. With the fighters still buzzing around, Chisov cleverly decided to fall freely out of the arena. It was his plan not to open his chute until he was down to only about 1000 ft above the ground. Unfortunately, he lost consciousness en route. As luck would have it, he crashed at the edge of a steep ravine covered with 3 ft of snow. Hitting at about 120 mi/h, he plowed along its slope until he came to rest at the bottom. Chisov awoke 20 minutes later, bruised and sore, but miraculously he had suffered only a concussion of the spine and a fractured pelvis. Three and one-half months later he was back at work as a flight instructor.

them along inclined planes (Fig. 3.10). He argued that if a was constant for each incline, it would continue to be constant, even when the track was vertical and the ball fell freely.

The second difficulty Galileo encountered was that he had no way of measuring the instantaneous final speed at the end of each run. To handle this obstacle, he reformulated the problem, replacing v_f by quantities that were directly observable. Thus, he arrived at a verbal statement of Eq. (3.9), $s = v_i t + \frac{1}{2}at^2$. *If the acceleration was constant, starting from rest* ($v_i = 0$), *the distance traveled by the ball would depend on the time squared* ($s \propto t^2$). Doubling the time a ball was allowed to move would quadruple the distance it traveled, and that situation would be an easy one to observe. Galileo's results were remarkably good (even though he overlooked the fact that the balls were rolling, not sliding, which decreases the constant acceleration—see Fig. 8.16). "In such experiments," he wrote, "repeated a full hundred times, we always found that the spaces traversed were to each other as the squares of the times, and this was true for all inclinations of the plane."

In the absence of air friction, all bodies, regardless of their weight, fall with the same uniform acceleration. Well, not quite. The acceleration is the same for all bodies, but it is not exactly uniform. Newton pointed out that the acceleration gradually decreases with increasing height above the Earth (Eq. 7.3). At an altitude of 1000 m, the acceleration is about 0.03% less than it is at the planet's surface. Over the small distances objects fall in the laboratory, this altitude dependence is insignificant, and it's not surprising that Galileo missed it.

A cannon ball weighing one or two hundred pounds, or even more, will not reach the ground by as much as a span ahead of a musket ball weighing only half a pound, provided both are dropped from a height of 200 cubits.

GALILEO GALILEI

The acceleration due to gravity is represented by its own symbol, g, traditionally taken to be the net *measured* downward acceleration (Table 3.2). In nearby space there is no ambiguity, but *on* Earth that acceleration does not depend only on gravity. There is a small negative contribution (see Fig. 7.7) arising from the spin of the planet; objects at rest at the surface (a rock in your hand, for example) are revolving around the center of the Earth. If the planet stopped spinning, g would

Regarding terminal speeds. *You can drop a mouse down a thousand-yard mine shaft and, on arriving at the bottom, it gets a slight shock and walks away. A rat is killed, a man is broken, a horse splashes.*

J. B. S. HALDANE
British geneticist, 1892–1964

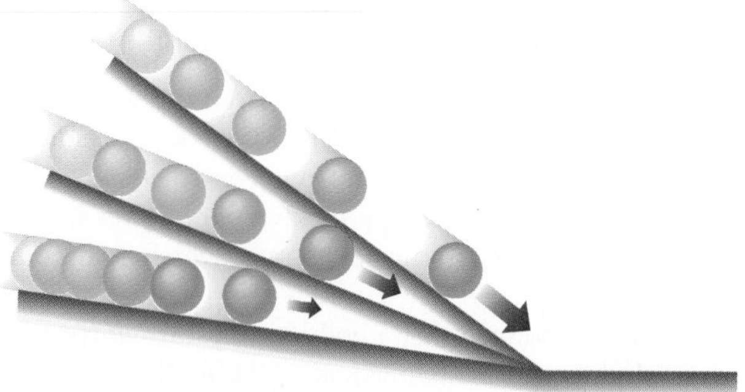

Figure 3.10 To see if the acceleration of a falling body was constant, Galileo rolled balls down inclined planes of increasing tilt. As the angle increased, the acceleration increased, but for each position it was constant. He argued that at a tilt of 90° the object would free-fall, also with a constant acceleration.

TABLE 3.2 Acceleration due to gravity*

Location	g (m/s^2)	g (ft/s^2)
Equator	9.780	32.09
Panama	9.782	32.09
Honolulu	9.789	32.12
Key West	9.790	32.12
Austin	9.793	32.13
Tokyo	9.798	32.15
San Francisco	9.800	32.15
Princeton	9.802	32.16
New York	9.803	32.16
Chicago	9.803	32.16
Munich	9.807	32.18
Winnipeg	9.810	32.18
Leningrad	9.819	32.22
Arctic Red River	9.824	32.23
North Pole	9.832	32.26

*The measured values of g are actually due to both gravity and the rotation of the Earth.

Galileo Galilei at the age of about 60, drawn by his contemporary Ottavio Leoni. Galileo was born in 1564, just three days before Michelangelo died, and within months of Shakespeare's birth.

increase slightly everywhere but at the poles. Because the effect is small (<0.4 %), it's usually ignored. We can generally take g to be a constant equal to its average value of $9.806\,65\ m/s^2$ ($32.2\ ft/s^2$). Galileo never did determine g—he did not have a clock accurate enough for the purpose. C. Huygens, who invented the pendulum clock in 1656, was the first to measure g. He showed that it could be adduced from the swing of a pendulum using only a ruler and a good timepiece (see Sect. 12.3).

3.7 All Fall Down: Constant Acceleration

If an object is moving freely near the surface of the Earth, its gravitational acceleration is along a vertical line and is assumed to be constant. All the equations of uniformly accelerated motion—Eqs. (3.6), (3.7), (3.8), (3.9), and (3.10)—apply where $a = g$. **In free-fall the acceleration is always g and always straight downward regardless of the motion.**

Example 3.11 A salmon is dropped by a hovering eagle. How far will the fish fall in 2.5 s before the bird catches it again? Neglect air drag.

Solution: The initial motion is down, so take down as positive. [Given: $a = g = +9.81\ m/s^2$, $v_i = 0$, and $t = 2.5$ s. Find: the distance fallen, s.] We have a, v_i, and t, and need s, which suggests Eq. (3.9). Thus

$$s = v_i t + \tfrac{1}{2}gt^2 = 0 + \tfrac{1}{2}(+9.81\ m/s^2)(2.5\ s)^2$$

The fish falls a distance $s = +30.6$ m or to two figures, $\boxed{s = 31\ m}$.

▶ **Quick Check:** Its final speed is $v_f = v_i + gt = 24.5$ m/s, and since $v_f^2 = v_i^2 + 2gs$, $s = (24.5\ m/s)^2/2g = 30.6$ m.

Example 3.12 A ball is thrown straight down from the roof of a dormitory at 10.0 m/s. If the building is 100 m tall, at what speed will the ball hit the ground? How long will the trip take?

Solution: Since the initial motion is down, take that as the positive direction. [Given: $v_i = +10.0$ m/s, $a = g = +9.81$ m/s^2, and $s = +100$ m. Find: v_f and t.] Using Eq. (3.10), we have

$$v_f^2 = v_i^2 + 2gs = (+10.0 \text{ m/s})^2 + 2(+9.81 \text{ m/s}^2)(100 \text{ m})$$

and $\boxed{v_f = 45.4 \text{ m/s}}$. The time can most easily be gotten from Eq. (3.6), $v_f = v_i + at$ where

$$t = \frac{v_f - v_i}{g} = \frac{45.4 \text{ m/s} - 10.0 \text{ m/s}}{+9.81 \text{ m/s}^2} = \boxed{3.61 \text{ s}}$$

▶ **Quick Check:** Using $s = v_i t + \frac{1}{2}at^2$, $\frac{1}{2}(9.81 \text{ m/s}^2)t^2 + (10.0 \text{ m/s})t - 100 \text{ m} = 0$. From the quadratic equation (Appendix A-4)

$$t = \frac{-10.0 \text{ m/s} \pm 45.4 \text{ m/s}}{9.81 \text{ m/s}^2}$$

ignoring the negative solution $t = 3.6$ s.

If a ball is thrown upward, its speed continuously diminishes at a rate of −9.81 m/s per second as it climbs. Always acting downward, g has the effect of increasing the speed of descending objects and decreasing the speed of ascending objects. The rising ball, moving slower and slower, will momentarily stop (the acceleration is still g) and then plunge downward faster and faster. After all, a body cannot move in opposite directions along a straight line without stopping first. The height of that stopping-point, the maximum or **peak altitude** (s_p), is the vertical equivalent of the stopping distance for a uniformly decelerating car, Eq. (3.11)—the distance it takes to go from v_i to $v_f = 0$.

The up-and-down journey is symmetrical in space and time, around the peak altitude (Fig. 3.11). Ignoring air friction, the object will be moving at the same speed when at the same altitude, whether it's on the way up or down. Moveover, the speeds will be equal for any amount of time before and after the peak altitude is reached. The ascent unfolds as if it were a movie of the descent run backward. And the descent is exactly the same as if the object were dropped from the peak altitude. At high projectile speeds, air friction limits the motion causing the trajectory to be lower and asymmetrical (see p. 93). For example, typical antiaircraft fire rises to less than about 1500 m (see Example 3.13).

For a projectile shot into the air, at any instant on the way up or down, s is the scalar value of the displacement: *it is the distance the object is from the starting point.* It is not the total distance traversed to get there. When a projectile is thrown straight up, reaches peak altitude, and comes halfway back down, the displacement ($\frac{1}{2}s_p$) at that moment is one third of the distance traveled.

Example 3.13 A .32-caliber bullet fired from a revolver with a 3-in. long barrel will have a relatively low muzzle speed of about 200 m/s. If it's shot straight up, neglecting air resistance, (a) What is the peak height the bullet will reach? (b) How fast will it be moving when it returns to the height of the gun? (c) How long will the whole trip take?

Solution: Select up as the positive direction. [Given: $v_i = +200$ m/s. Find: (a) s_p; (b) final speed; (c) total time t_T.] For the upward leg of the trip use Eq. (3.10), where $a = g$, to get the height at which the speed goes to zero ($v_f = 0$). Thus

$$0 = v_i^2 + 2gs_p$$

and

$$s_p = -\frac{v_i^2}{2g}$$

which is essentially Eq. (3.11). The bullet is decelerating, *g is downward, and so we enter it as a negative number.*

(continued)

(continued)

Hence the answer to part (a) is

$$s_p = -\frac{(200 \text{ m/s})^2}{2(-9.81 \text{ m/s}^2)} = \boxed{+2.04 \times 10^3 \text{ m}}$$

(b) Because the flight is symmetrical, the bullet ideally returns to the height of the gun at a speed of $\boxed{200 \text{ m/s}}$. Table 3.1 tells us that the actual speed is closer to 100 m/s, which is still very dangerous. (c) The time for the trip *up* can be computed using Eqs. (3.6), (3.8), or (3.9)—Eq. (3.6) is the easiest. Accordingly

$$v_f = v_i + gt$$
$$0 = 200 \text{ m/s} + (-9.81 \text{ m/s}^2)t_p$$

and $t_p = -(200 \text{ m/s})/(-9.80 \text{ m/s}^2) = 20.4$ s, which is half the total flight time and thus $\boxed{t_T = 40.8 \text{ s}}$.

▶ **Quick Check:** Compute the downtrip time starting with Eq. (3.9), $s = v_i t + \frac{1}{2}gt^2$. Now $v_i = 0$, $g = -9.80 \text{ m/s}^2$, and the displacement is −2040 m measured from the top of the trajectory downward to the initial height of firing. Hence

$$t_p = \left(\frac{2s}{g}\right)^{1/2} = \left[\frac{2(-2040 \text{ m})}{-9.81 \text{ m/s}^2}\right]^{1/2} = 20.4 \text{ s}$$

Alternatively, factor Eq. (3.9) and set $s = 0$; $t(v_i + \frac{1}{2}gt) = 0$, hence $t = 0$ and $(v_i + \frac{1}{2}gt) = 0$, and so $t = t_T = -2v_i/g = 40.8$ s.

Example 3.14 A ball is hurled straight up at a speed of 15.0 m/s, leaving the hand of the thrower 2.00 m above ground. Compute the times and the ball's speeds when it passes an observer sitting at a window in line with the throw 10.0 m above the point of release.

Solution: Because nothing about ground level is to be computed, the height of the hand at launch is taken as the origin. Take *up* as positive throughout; $a = g$, which is always downward and a negative number. [Given: $v_i = +15.0$ m/s, $a = -9.81 \text{ m/s}^2$, and $s = +10.0$ m. Find: the *two* times and the *two* final speeds.] We could use $s = v_i t + \frac{1}{2}at^2$ to find t directly, which yields a quadratic equation (Problem 74). Instead we will find v_f first, via $v_f^2 = v_i^2 + 2as$. Consequently

$$v_f^2 = (15.0 \text{ m/s})^2 + 2(-9.81 \text{ m/s}^2)(+10.0 \text{ m})$$
$$v_f^2 = 28.9 \text{ m}^2/\text{s}^2$$

and $$\boxed{v_f = \pm 5.37 \text{ m/s}}$$

The ball passes the window traveling at +5.37 m/s on the way up and −5.37 m/s on the way down. The corresponding times (t_u and t_d) can be gotten from $v_f = v_i + at$. On the way up

$$+5.37 \text{ m/s} = +15.0 \text{ m/s} + (-9.81 \text{ m/s}^2)t_u$$
$$\boxed{t_u = 0.982 \text{ s}}$$

On the way down, $v_f = -5.37$ m/s, and so

$$-5.37 \text{ m/s} = +15.0 \text{ m/s} + (-9.81 \text{ m/s}^2)t_d$$
$$\boxed{t_d = 2.08 \text{ s}}$$

Notice that the peak altitude was reached when $v_f = 0 = v_i + (-9.81 \text{ m/s}^2)t_p$; that is, when $t_p = 1.53$ s. Hence, the ball passed the window on the way up 0.55 s earlier than t_p and on the way down, 0.55 s later. This symmetry around the peak can be useful in problem solving.

▶ **Quick Check:** $s = v_i t + \frac{1}{2}at^2 = (15.0 \text{ m/s}) \times (0.982 \text{ s}) + \frac{1}{2}(-9.81 \text{ m/s}^2)(0.982 \text{ s})^2 = 10.0$ m.

3.8 Projectile Motion

Suppose that we throw an object into the air at some angle other than straight up. Galileo recognized that the projectile's unpowered movement can be imagined as if it were two simultaneous *independent* motions: a horizontal flight (ideally at constant speed) and a vertical gravity fall (ideally at constant acceleration). The gravitational interaction causes a vertical acceleration, but no such influence exists horizontally, and the horizontal acceleration (overlooking air drag) is zero. We will

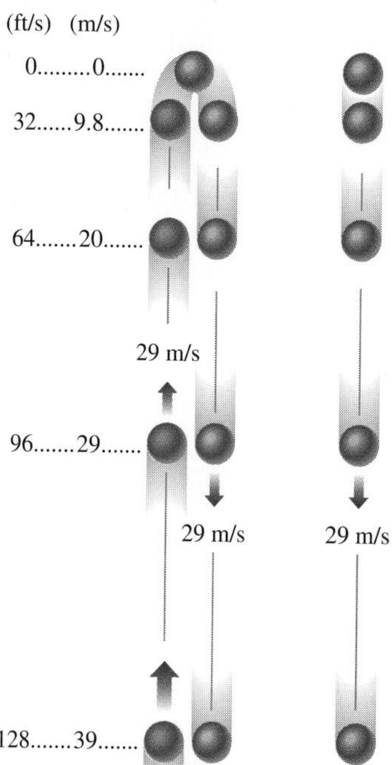

Figure 3.11 A ball thrown straight up at some speed—in this case 39 m/s (128 ft/s)—will (neglecting air friction) return to its launch height at that same speed. In fact, the speeds moving up and down are equal at equal heights. In this instance, the ball takes 4.0 s to reach maximum altitude and 4.0 s to get back down.

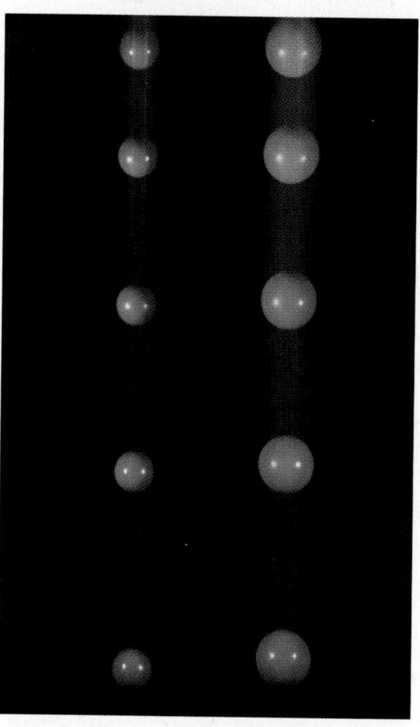

At the surface of the Earth, in situations where air friction is negligible, objects fall with the same acceleration regardless of their weights.

study the forces at work in this situation in the next chapter; here, we concentrate on the kinematics, the description of the motion.

Imagine a ball thrown horizontally. There is then no initial component of the velocity in the vertical direction ($v_{iV} = 0$). The ball sails straight off horizontally at a *constant speed* equal to its initial speed, v_{iH}. But on Earth, heavy objects fall freely downward with a constant acceleration. Accordingly, the projectile will continuously descend, faster and faster, as it progresses laterally, sweeping out a smooth arc that curves increasingly downward (Fig. 3.12).

The shape of the arc traveled by a projectile can be approximated by forming an equation relating its horizontal and vertical displacements, an equation that is independent of time since the path is fixed, even if the object moving along it is not. In the case of a projectile launched with an initial horizontal speed (v_{iH}), which is assumed constant, the horizontal distance traveled is proportional to the time of flight ($s_H = v_{iH}t$). The vertical free-fall distance ($s_V = \frac{1}{2}gt^2$) varies with the time squared. These two distance equations can be combined to produce an expression for s_V in

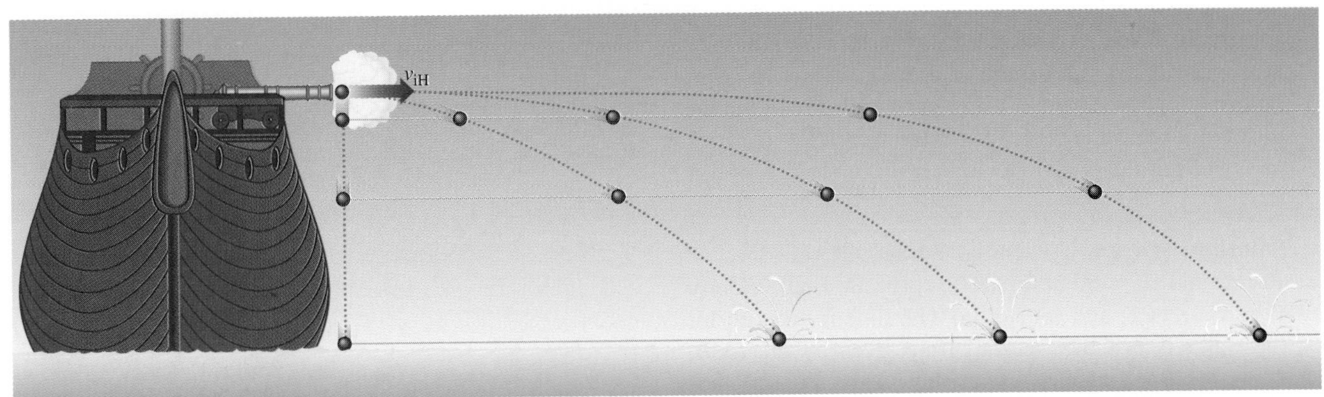

Figure 3.12 The faster each cannonball is fired, the farther it will go, but all fall at the same rate and hit the water after the same flight times. The horizontal speed, v_{iH}, is constant, provided air friction is negligible. The vertical speed increases at the rate g.

terms of s_H, which is independent of time. Rearranging the first expression to get $t = s_H/v_{iH}$ and substituting this into the second yields

$$s_V = \left(\frac{g}{2v_{iH}^2}\right)s_H^2 \tag{3.13}$$

The term in parentheses is constant for any given launch and that means that Eq. (3.13) is the equation of a *parabola*—one space variable proportional to the square of another space variable (Fig. 3.13). To the extent that both the curvature of the Earth and friction can be neglected, the path of a ballistic projectile is parabolic.

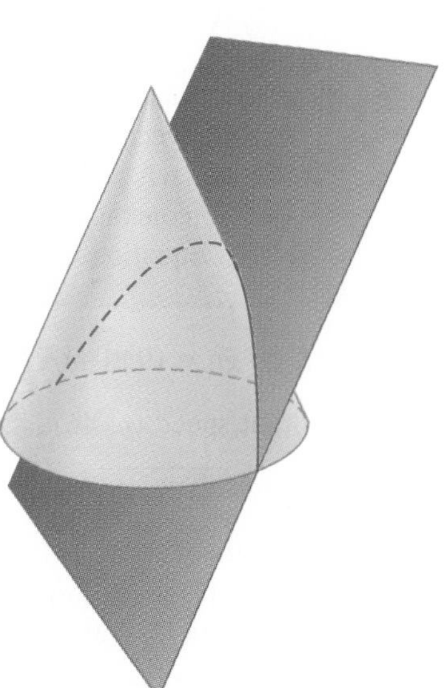

Figure 3.13 The parabola can be visualized as the intersection of a plane and a cone. The plane is tilted parallel to the edge of the cone.

Sparks fly into the air, arcing along parabolic paths.

Example 3.15 A youngster hurls a rock horizontally at a speed of 10 m/s from a bridge 50 m above a river. Ignoring air resistance: (a) How long will it take for the rock to hit the water? (b) What is the velocity of the rock just before it lands? (c) How far from the bridge will it strike?

Solution: [Given: The initial horizontal speed $v_{iH} =$ 10 m/s and $s_V = 50$ m. Find: (a) t; (b) the final velocity $\mathbf{v}_f = \mathbf{v}_{fH} + \mathbf{v}_{fV}$, and (c) s_H.] (a) The time it takes to hit the water is just the free-fall time given by Eq. (3.9), $s = v_i t + \frac{1}{2}at^2$, applied exclusively to the vertical motion as if it alone were occurring. With $a = g$ and no initial *vertical* speed, either up or down, this equation becomes

$$s_V = \tfrac{1}{2}gt^2$$

Taking down as positive

$$(50 \text{ m}) = \tfrac{1}{2}(9.81 \text{ m/s}^2)t^2$$

and $$t = \sqrt{\frac{2(50 \text{ m})}{(9.81 \text{ m/s}^2)}} = 3.19 \text{ s} = \boxed{3.2 \text{ s}}$$

(b) The velocity at which the rock hits the water (Fig. 3.14) is the vector resultant of its final horizontal, $\mathbf{v}_{fH}$, and vertical, $\mathbf{v}_{fV}$, motions. We know that the horizontal speed is constant at $v_{iH} = 10$ m/s and so must now find $\mathbf{v}_{fV}$. Equation (3.6), with no initial vertical speed, yields

$$v_{fV} = gt = (9.81 \text{ m/s}^2)(3.19 \text{ s}) = 31.3 \text{ m/s}$$

The resultant speed (p. 46) is then

$$v_f = \sqrt{v_{fH}^2 + v_{fV}^2} = \sqrt{1079} \text{ m/s} = 33 \text{ m/s}$$

Notice that this quantity is larger than the largest of the two components, as it should be. From Fig. 3.14, we see that the angle, θ, made by $\mathbf{v}_f$ and the horizontal is such that

$$\tan\theta = \frac{v_{fV}}{v_{fH}} = \frac{31.3 \text{ m/s}}{10 \text{ m/s}} = 3.13$$

and therefore $\theta = 72°$ below the horizontal. We could use either $\sin\theta$ or $\cos\theta$ instead of $\tan\theta$ but that would involve using v_f, and if an error was made in computing v_f, it would make θ wrong, too. It's best, if all else is equal, to use quantities that are least likely to be in error. Thus, $\boxed{\mathbf{v}_f = 33 \text{ m/s–}72° \text{ below horizontal}}$. (c) Because the rock's horizontal speed is constant and because it is in the air for a time t

$$s_H = v_{iH}t = (10 \text{ m/s})(3.2 \text{ s}) = \boxed{32 \text{ m}}$$

▶ **Quick Check:** The average vertical speed is $\frac{1}{2}v_{fV} = 15.65$ m/s; the time-of-flight is then $t =$ 50 m/(15.65 m/s) = 3.19 s. Observe that v_{fV} is about 3 times v_{fH}, so a fairly large value of θ is reasonable.

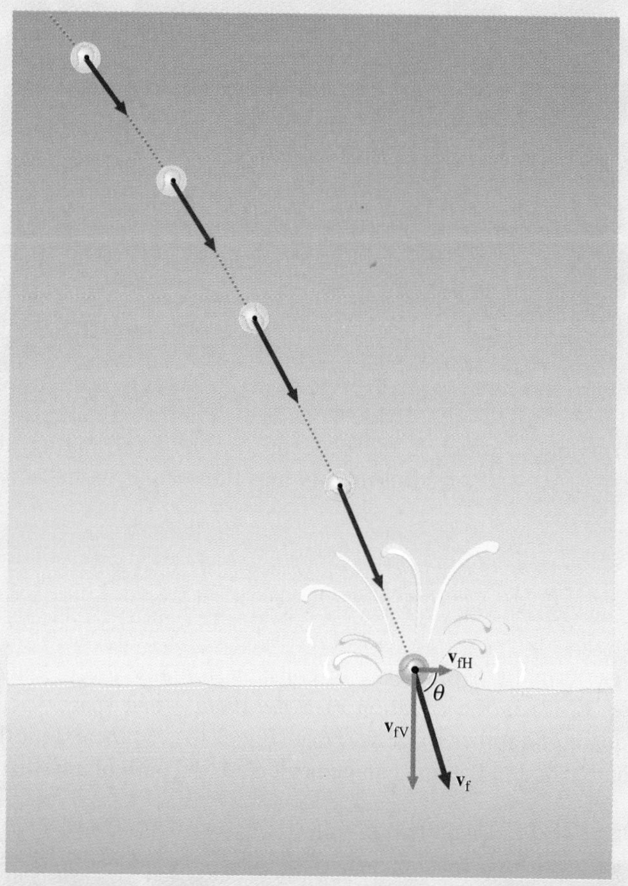

Figure 3.14 A projectile thrown horizontally from a bridge hits the water with a velocity $\mathbf{v}_f$ that has both horizontal, $\mathbf{v}_{fH}$, and vertical $\mathbf{v}_{fV}$, components. With negligible air-friction losses, the initial horizontal speed v_{iH} equals v_{fH}.

Now for the more general case. Imagine a cork popping out of a champagne bottle into the air at some angle θ with an initial speed v_i, as depicted in Fig. 3.15a. (This could just as well be a leaping frog, an athlete executing a broad jump, or a kicked football—any unpowered flight.) We deal with this situation practically

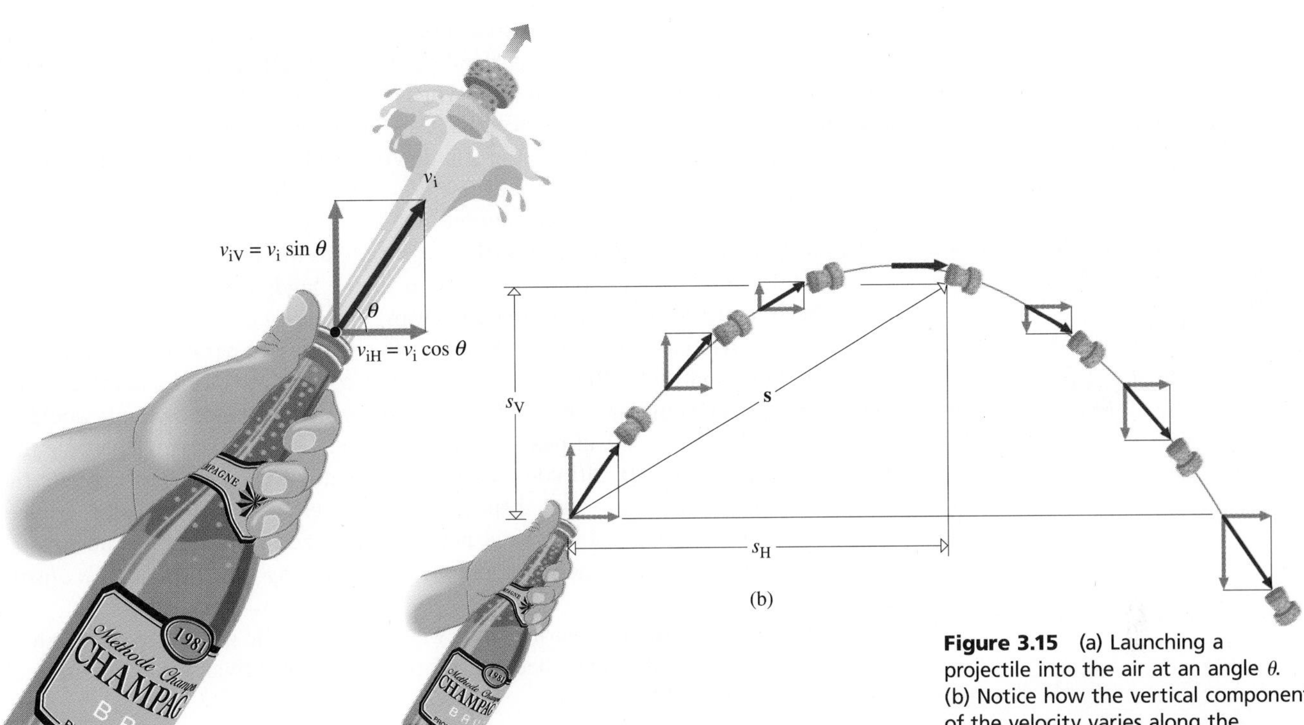

Figure 3.15 (a) Launching a projectile into the air at an angle θ. (b) Notice how the vertical component of the velocity varies along the trajectory. A longstanding record was set in Reno, Nevada on July 4, 1981, when a cork from a champagne bottle was "fired" 105 ft 9 in. Because of air friction, corks are launched at 40° to cover more distance than height.

by realizing that the motion unfolds as if it were two independent superimposed motions—one vertical, with an initial speed of

$$v_{iV} = v_i \sin\theta$$

the other horizontal, with an initial speed of

$$v_{iH} = v_i \cos\theta$$

That's equivalent to saying that **s**, the displacement vector drawn from the point of launch to the projectile at any moment, has a scalar horizontal component of

$$s_H = v_{iH}t$$

since $a_H = 0$, and a scalar vertical component of

$$s_V = v_{iV}t + \tfrac{1}{2}gt^2$$

since $a_V = g$.

As before, when dealing with the scalar components of vectors, we must select certain directions to be positive so that the components along and opposite to those directions can be added algebraically, preserving the directional nature of the process. Accordingly, g can be either a positive or negative number (± 9.81 m/s^2). If the initial motion has an upward component, we might choose *up* to be positive, in which case v_{iV} is positive, but g, because it is downward as always, is then negative.

The cork rises (in the positive direction) accelerating at -9.81 m/s^2, slowing down under the influence of a downward acceleration, until it reaches its peak altitude. At that point, its vertical speed is zero although it continues uninterruptedly to

From accounts given by gunners, I was already aware of the fact that in the use of cannon and mortars, the maximum range, that is the one in which the shot goes farthest, is obtained when the elevation is 45° . . . but to understand why this happens far outweighs the mere information obtained by the testimony of others or even by repeated experiment.

GALILEO GALILEI
Dialogues Concerning Two New Sciences (1638)

move horizontally (Fig. 3.15b). Thereafter, the cork descends along an arc, as ever under the influence of gravity. Accelerating all the while, it returns to the launch elevation at very nearly the speed with which it was first fired. During the descent (because *up* is positive), the velocity increases in the negative direction.*

The equation for s_V describes the trip down as well as the trip up; you need not change signs midway through the journey. We are describing the *displacement* from the launch point, *not* the distance traveled. On the way up, it increases; on the way down, it decreases. When the projectile drops back to the launch height, s_V is zero. If it should fall into a hole or over a cliff so that it lands beneath launch level, s_V would become negative (being in the downward direction).

Provided friction losses are negligible, the path is a parabola, symmetrical around the peak altitude. But if you ran along at a constant speed staying directly under the cork, it would have no horizontal velocity with respect to you; it would be seen simply to rise up, stop at peak altitude, and come straight down. The time that it takes (t_p) to reach peak altitude (s_p) is again obtained from Eq. (3.6), where the final vertical speed is zero and, hence, $0 = v_{iV} + gt_p$. Accordingly, $t_p = -v_{iV}/g$ (which is simple to derive and should not be memorized). When *up* is positive so that v_{iV} is positive, g will be negative and vice versa. In any event, *the time is always positive.*

In the simplest case, the journey ends when the projectile returns to the height at which it was launched. Then, because of the symmetry, twice the time-to-peak equals the **total-flight time** t_T:

$$t_T = -\frac{2v_{iV}}{g} = -\frac{2v_i \sin\theta}{g} = \frac{2v_i \sin\theta}{9.8\ \mathrm{m/s^2}} \tag{3.14}$$

The **peak height** attained is

$$s_p = -\frac{v_{iV}^2}{2g} = -\frac{(v_i \sin\theta)^2}{2g} = \frac{(v_i \sin\theta)^2}{2(9.8\ \mathrm{m/s^2})} \tag{3.15}$$

which applies as well to basketballs or water fountains. As long as *up* is positive, $g = -9.8\ \mathrm{m/s^2}$ and s_p is positive.

Example 3.16 The person in the gondola shown in Fig. 3.16 throws a ball at 40.0° to the horizon at 10.0 m/s. If the ball is launched at a height of 100 m, where will it land? Ignore air friction.

Solution: [Given: $v_i = 10.0$ m/s, $\theta = 40.0°$, and the launch height is 100 m. Find: s_H.] We need the total time of flight, but this path is not symmetrical and Eq. (3.14) does not apply. As a rule, we can get the total-flight time by finding how long the vertical motion lasts. With the launch point as the origin

$$s_V = v_{iV}t + \tfrac{1}{2}gt^2$$

and if we take up positive, the journey ends when $s_V = -100$ m, at which point $t = t_T$. Using $v_{iV} = (10.0\ \mathrm{m/s}) \sin 40.0° = 6.428$ m/s in the expression for s_V, we get

(continued)

*When the Palestinian Liberation Organization left Beirut, Lebanon, in 1982, they fired so many shots into the air to celebrate that at least half a dozen people were seriously wounded by falling bullets.

(continued)

$$-100 \text{ m} = (6.428 \text{ m/s})t_T + \tfrac{1}{2}(-9.807 \text{ m/s}^2)t_T^2$$

To solve for t_T, rearrange this equation so it has the standard form of the quadratic equation (Appendix A-4), namely

$$(4.904 \text{ m/s}^2)t_T^2 + (-6.428 \text{ m/s})t_T + (-100 \text{ m}) = 0$$

whereupon t_T then equals

$$\frac{6.428 \text{ m/s} \pm \sqrt{(-6.428 \text{ m/s})^2 - 4(4.904 \text{ m/s}^2)(-100 \text{ m})}}{2(4.904 \text{ m/s})}$$

There is only one positive solution, namely, $t_T = 5.22$ s. Hence, the ball hits the ground at a horizontal distance from the launch point equal to

$$s_H = v_{iH}t_T = (10.0 \text{ m/s})(\cos 40.0°)(5.22 \text{ s}) = \boxed{40.0 \text{ m}}$$

▶ **Quick Check:** The time it takes to get back down to the launch height is $2t_p = 1.31$ s. At that point, the ball is moving down at 6.428 m/s and if the answer is right, it will take (5.22 s − 1.31 s) = 3.91 s to descend the 100 m. Hence, $s_V = v_{iV}t + \frac{1}{2}gt^2 = (6.428 \text{ m/s}) \times (3.91 \text{ s}) + \frac{1}{2}(9.81 \text{ m/s}^2)(3.91 \text{ s})^2 = 100$ m and all's well.

Figure 3.16 A ball is thrown at 40.0° from a gondola 100 m above the ground. Where will it land?

Range s_R *is the total horizontal distance that a projectile travels returning to the same height from which it was launched.* The only way this notion makes sense is if the object is fired in some direction upward. Again *up* is positive and g is negative. Since v_{iH} is constant

$$s_R = v_{iH}t_T = (v_i \cos\theta)t_T$$

Using Eq. (3.14) for the total-flight time, the range becomes

$$s_R = -\frac{2v_i^2}{g}\cos\theta\sin\theta = \frac{2v_i^2}{9.8 \text{ m/s}^2}\cos\theta\sin\theta \qquad (3.16)$$

This horizontal displacement is nonzero and positive (to the right of the launch point) when $\theta < 90°$ whereupon both $\sin\theta$ and $\cos\theta$ are positive. Notice that when $\theta = 90°$, the projectile goes straight up and down, $\cos 90° = 0$, and $s_R = 0$; moreover, when $\theta = 0°$, $s_R = 0$. Neglecting aerodynamic effects, maximum range occurs when $(\cos\theta\sin\theta)$ has its maximum value. A simple plot of that product reveals that it reaches a maximum when $\theta = 45°$. Altering the launch angle by the same amounts, either increasing or decreasing it from 45°, produces the same range (Fig. 3.17). *The projected horizontal distance is the same for each member of any pair of firing angles that add up to 90°.*

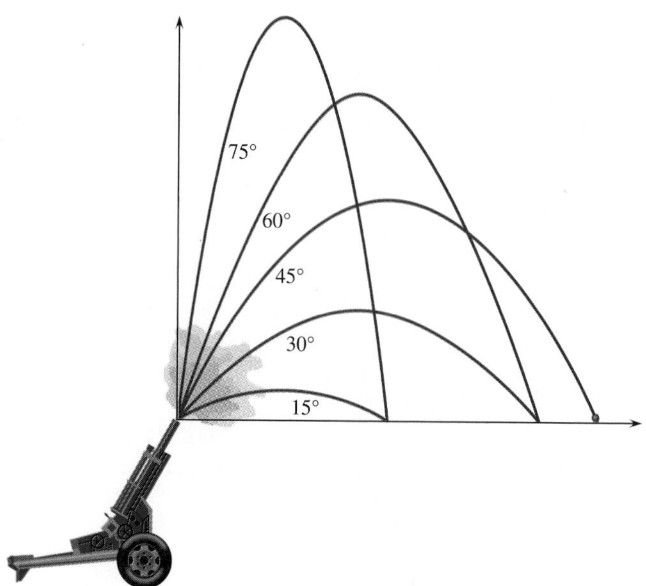

Figure 3.17 Ideally, the maximum range of an artillery shell corresponds to a firing angle of 45°. Notice that if we change the angle by going up or down some amount from 45°, the range is the same. For example, firing at 45° + 15° or 45° − 15° yields the same range.

Example 3.17 A baseball recoiling from a bat soars into the air at an angle of 40.0° above the ground traveling at 45.7 m/s (150 ft/s). Calculate its range ignoring aerodynamic effects.

Solution: [Given: $\theta = 40.0°$, and $v_i = 45.7$ m/s. Find: s_R.] It is not necessary to memorize Eq. (3.16) when it's so easily derived from $s_R = v_{iH}(2t_p)$. In any event

$$s_R = -\frac{2v_i^2}{g}\cos\theta\sin\theta$$

$$s_R = \frac{2(45.7\text{ m/s})^2}{9.81\text{ m/s}^2}\cos 40.0°\sin 40.0°$$

or $\boxed{s_R = 210\text{ m}}$ (688 ft). In reality, a baseball experiences both drag and lift. At the speed given, it's not likely to go much farther than about 500 ft. Moreover, maximum range is actually attained when the ball is hit at 40° rather than 45°.

▶ **Quick Check:** $t_T = 2t_p = -2v_{iV}/g = -2(45.7\text{ m/s}) \times \sin 40°/9.8\text{ m/s}^2 = 6.0$ s; $s_R = v_{iH}t_T = (45.7\text{ m/s})\cos 40° \times (6.0\text{ s}) = 210$ m.

Air friction (p. 93) has a great effect on projectiles, especially on those that are light and fast. Accordingly, cannon fire can travel much farther than bullets. For example, the 210-mm (diameter) howitzer (used by Iraq in the 1991 Gulf War) has the greatest range of all field artillery, namely 57000 m. That is more than ten times the range of a high-speed rifle bullet.

Core Material

The **average acceleration** is defined as:

$$\mathbf{a}_{av} = \frac{\Delta\mathbf{v}}{\Delta t} = \frac{\mathbf{v}_f - \mathbf{v}_i}{t_f - t_i} \qquad [3.1]$$

As $\Delta t \to 0$, this ratio approaches the instantaneous acceleration

$$\mathbf{a} = \lim_{\Delta t \to 0}\left[\frac{\Delta\mathbf{v}}{\Delta t}\right] \qquad [3.3]$$

Restricting the treatment to *straight-line motion at constant acceleration,* there are five useful equations:

$$v_f = v_i + at \qquad [3.6]$$

$$v_{av} = \tfrac{1}{2}(v_i + v_f) \qquad [3.7]$$

$$s = \tfrac{1}{2}(v_i + v_f)t \qquad [3.8]$$

$$s = v_i t + \tfrac{1}{2}at^2 \qquad [3.9]$$

and $$v_f^2 = v_i^2 + 2as \qquad [3.10]$$

Objects in free-fall at the surface of the Earth (for which air friction is inconsequential) descend with a nearly *uniform acceleration* (g) equal to 9.81 m/s^2 or 32.2 ft/s^2. The above set of equations can be applied to projectiles (p. 80) where the constant-speed horizontal motion and the constant-acceleration vertical motion occur independently. Such quantities as the time to peak altitude (t_p), the total time of flight (t_T), and the range (s_R) are important ideas that can be easily calculated when friction is negligible.

Suggestions on Problem Solving

1. Watch units. When making conversions anticipate whether the answer will be larger or smaller than the original number *before* doing the calculation. Never mix different sets of units in the same problem, let alone the same equation. Be careful not to take 32 as the numerical value of g in a problem using SI units.

2. Be consistent with the signs of the vector quantities (**s**, **v** and **a**), particularly in free-fall and projectile problems. Take the initial direction of motion as positive and stick with it throughout the problem. This approach is not always necessary, but in the beginning it is certainly advisable. *Draw diagrams.*

3. Consider Eq. (3.9), which varies with t and t^2 both. When t is given, it's easy to find s, v, or a. When t is the unknown to be determined, the most straightforward approach is to use the quadratic equation. Rearranging Eq. (3.9) yields $\frac{1}{2}at^2 + v_it - s = 0$, which has the general form $At^2 + Bt + C = 0$, where $A = \frac{1}{2}a$, $B = v_i$, and $C = -s$. The solution is then $t = (-B \pm \sqrt{B^2 - 4AC})/2A$ (see Appendix A-4). The quadratic equation can be avoided by using combinations of other equations: Eq. (3.10) to find v_f, and then Eq. (3.6) or (3.8) to get t. The wise choice of equations will save time and effort.

4. Think about your answers. Often an error will produce a result that is unrealistically large or small. If you think the problem through so that you can anticipate the size, direction, sign, and so on, most computational errors will be spotted. We all make computational errors; not all of us find them. A reindeer falling off the roof of a one-story building is not going to land on the ground at 5000 km/h. Given the real-world physical situation of a problem, ask yourself if your answer is reasonable.

5. Check your answers. The best way to do that (although it may not always be practical) is to recalculate the answer using a different approach. At least go over your calculations several times.

6. When a problem requests the distance traveled during a certain time interval, such as the fifth second of motion, this is not the same as asking how far the body moved in five seconds. The fifth second is the one that begins at $t = 4$ s and ends at $t = 5$ s.

7. It is not really necessary to remember the whole set of equations for projectile motion: t_p is easily derived from the definition of a, where $v_V = 0$; t_T is just $2t_p$ and s_R is simply v_Ht_T. The fact that $v_V = 0$ (and $v_H \neq 0$) at the peak altitude is important and should always be kept in mind.

8. The Mean-Speed Theorem applies to straight-line motion in one direction, and so is appropriate for dealing with the uniformly accelerated vertical component of a projectile's movement. *Do not apply it to the overall motion* (as for example, in an attempt to determine the total final speed v_f knowing s and t)— the projectile travels along an arc and $l \neq s$.

Discussion Questions

1. Does a car's speedometer measure a vector or a scalar? How about the odometer? What does it measure? Is it possible for the car to accelerate and yet have the speedometer read a constant value? Is it possible that at some moment two airplanes have the same acceleration and different speeds? The same speeds and different accelerations? Can either plane at a given instant have a large acceleration and a small velocity? A small acceleration and a large velocity? Explain.

2. Is an ant stuck to a glob of honey fixed on a spinning phonograph record accelerating (a) with respect to the record? (b) with respect to the room? Explain.

3. Galileo used a "thought experiment" from the Middle Ages to justify the conclusion that all things fall at the same rate in vacuum. See if you can reconstruct it from the following: consider three identical objects (three balls of clay, for instance). How would they fall separately in vacuum? Stick two together. How would the pair of clay masses fall now? Explain.

4. Overlooking the planet's own motion, why is it likely that everything moving for any appreciable distance on the surface of the Earth is accelerating?

5. An airplane in a dive increases its speed every second by one-tenth its original value. What, if anything, can you say about its acceleration? Is it possible for a rocket to have an acceleration due east and a velocity due west? Explain.

6. Can the average acceleration of a body over a finite time ever equal the instantaneous acceleration for more than an instant during the journey? Can it ever not equal the instantaneous acceleration for at least an instant? Explain.

7. Two identical cars are in a drag race. The first starts and continues to uniformly accelerate up to the finish line some time later. The second car follows the first out of the starting box 1 s afterward traveling exactly as it did. Will the cars be the same distance apart throughout the race? If not, when will they be closest to each other? How much time will there be between their successive crossings of the finish line?

8. Is it accurate to say that the total distance traveled is the product of the time and the average speed, even though the instantaneous speed varies from moment to moment?

9. A rocket can accelerate at an increasing rate as its fuel is consumed and it gets lighter and lighter. If the rocket starts from

rest, what can we say about the *average speed* as compared with the *final speed?*

10. Explain why the distance traversed by a body during its tenth second of fall is greater than the distance covered during its first second—if, in fact, it is. What's the average speed of an apple during its first second of fall from a tree?

11. Suppose you are in a free-falling elevator and you hold your keys motionless right in front of your face and then let go. What will happen to them? Explain.

12. What is the average speed and acceleration of a banana at maximum altitude when hurled straight up at the surface of the Earth? A pea is fired straight up through a drinking straw by an undergrad in the cafeteria. When is it moving at its greatest speed? Imagine that you have taken a motion picture of a ball fired straight up, reaching maximum altitude and then falling straight down. What would it look like run backwards? Explain.

13. Consider a free-falling sky diver descending straight down. Make a graph of v versus t showing (a) the straight-line terminal speed v_T; (b) the zero-drag free-fall with $v_i = 0$; and (c) your approximation of the actual speed-time descent curve.

14. If you throw a ball straight up next to something whose height you already know (like a building) and time the round trip, you can determine g. How would you proceed?

15. To get a feeling for how extremes of a variable affect an equation, suppose that the acceleration of a body is given by $a = [(At^2 - Bt)/(t + C)D] + Et^2/(t - C)D$ where A, B, C, D, and E are constants: (a) Determine the value of a when $t = 0$. (b) Find the value of a when $t \gg C$. (c) Find the value of a when $t \ll C$.

16. Imagine an object thrown straight upward into the air. Ignoring friction, draw a rough plot of the vertical velocity versus time. Does the velocity ever become negative, and if so, when? What can you say about the object's acceleration?

17. A portion of the flight of a toy rocket that travels straight up from the surface of the Earth is depicted in Fig. Q17. Describe the motion as completely as you can.

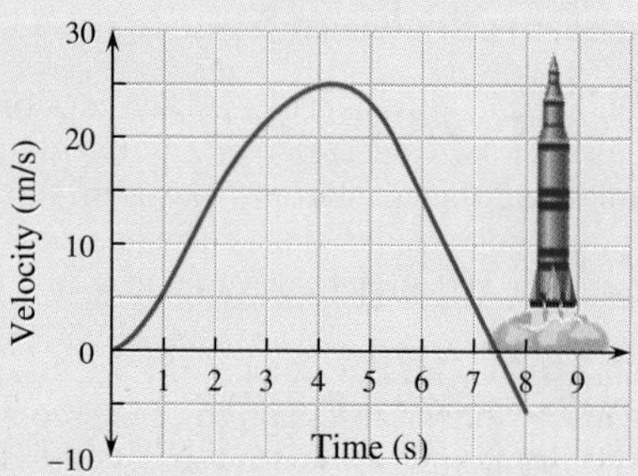

Figure Q17

18. A football is kicked into the air at some angle θ and lands down field. Ignoring air friction, draw a rough plot of its speed versus time. Is the slope of the curve the acceleration? Explain.

Multiple Choice Questions

1. The expression $s = vt$ is applicable when (a) the speed is constant (b) the acceleration is constant (c) the distance is constant (d) the acceleration is linear (e) none of these.
2. A body moving with an acceleration having a constant magnitude must experience a change in (a) velocity (b) speed (c) acceleration (d) weight (e) none of these.
3. The term "10 m/s²–south" specifies (a) displacement (b) speed (c) acceleration (d) velocity (e) none of these.
4. Which of the following pairs of concepts cannot both simultaneously be constant and nonzero for a given body? (a) the speed and velocity (b) the distance and displacement (c) the magnitude of the acceleration and the acceleration (d) the velocity and acceleration (e) none of these.
5. A mouse runs along a straight narrow tunnel. If its velocity-time curve is a straight line parallel to the time axis, then the acceleration is (a) a nonzero constant (b) zero (c) varying linearly (d) quadratic (e) none of these.
6. Which one of the following must be held constant if *uniformly accelerated motion* is likely to occur? The (a) speed (b) direction (c) acceleration (d) velocity (e) none of these.
7. The speed of a body traveling in a straight line with a constant positive acceleration increases linearly with (a) distance (b) time (c) displacement (d) distance squared (e) none of these.
8. If $[L]$ represents the dimension of length and $[T]$ that of time, then the dimensions of acceleration are (a) $[L + T^2]$ (b) $[L/T]$ (c) $[L^2/T]$ (d) $[L/T^2]$ (e) none of these.
9. At any moment the slope of the velocity-time graph of a body moving along a straight line is its (a) displacement (b) distance traveled (c) speed (d) time (e) none of these.
10. Taking the direction of the initial motion to be positive, the slope of the velocity-time curve of a body moving in a straight line is negative when the (a) acceleration is in the opposite direction to the initial motion (b) acceleration increases the speed (c) acceleration decreases in time (d) acceleration is constant and in the same direction as the initial motion (e) none of these.
11. If the displacement of a body is a quadratic function of time, the body is moving with (a) a uniform acceleration (b) a nonconstant acceleration (c) a uniform speed (d) a uniform velocity (e) none of these.
12. The expression $s = v_i t + \frac{1}{2}at^2$ is applicable only when (a) v is constant (b) t is constant (c) a is constant (d) s is constant (e) none of these.
13. A driver traveling at 20 km/h abruptly stops the car in 3 m. Later, while moving at 40 km/h, the driver again stops the car with the same deceleration, bringing it to a halt after (a) 6 m (b) 9 m (c) 60 m (d) 12 m (e) none of these.
14. Two statues of Kermit the Frog, one made of aluminum and one of brass, are the same size, although the former is 3.2 times lighter than the latter. Both are dropped at the

same moment from the same height of 2 m. They hit the ground (a) at nearly the same time with very different speeds (b) at very different times with nearly the same speeds (c) at nearly the same times with nearly the same speeds (d) at very different times with very different speeds (e) none of these.

15. A ballast bag of sand dropped from a hot air balloon hits the ground at a certain speed and the craft slowly rises and soon comes to a stop. If a second identical bag is then dropped and it hits the ground twice as fast as the first, how high was the balloon when it was dropped in comparison to when the first bag fell? (a) 1/2 as high (b) 2 times as high (c) 4 times as high (d) 8 times as high (e) none of these.

16. The average speed of a coconut during a 2-s fall from a tree, starting at rest, is (a) 19.6 m/s (b) 9.8 m/s^2 (c) 39.2 m/s (d) 9.8 m/s (e) none of these.

17. An alien space traveler exploring Earth records that his phasor pistol, dropped from a high cliff, fell a distance of 1 glong in a time of 1 tock. How far will it fall in two tocks? (a) 1.5 glongs (b) 2 glongs (c) 3 glongs (d) 4 glongs (e) none of these.

18. A body moving with a uniform acceleration in a straight line has an acceleration-time curve (a vertical against t horizontal) that is (a) a vertical straight line (b) a horizontal straight line (c) a vertical parabola (d) unable to be plotted (e) none of these.

Problems

THE CONCEPT OF ACCELERATION

1. [I] A rocket lifts off its launch pad and travels straight up attaining a speed of 100 m/s in 10 s. Calculate its average acceleration.

2. [I] A canvasback duck heading south at 50 km/h at 2:02 A.M. is spotted at 2:06 A.M. still traveling south but at 40 km/h. Calculate its average acceleration over that interval—magnitude and direction.

3. [I] An android on guard duty in front of the Institute of Robotics is heading due south at 1:07 P.M. at a speed of 10 m/s when it receives a command to alter course. At 1:09 P.M. it is recorded to be moving at 10 m/s due north. Compute its average acceleration over that interval—magnitude and direction.

4. [I] At time $t = 0$, a body located at $s = 0$ has a speed of $v = 0$ and an acceleration of 1 m/s^2. It travels in a straight line maintaining that constant acceleration for 4 s and then immediately ceases accelerating. That condition lasts for 1 s at which point it decelerates at 2 m/s^2 for 2 s and then again ceases accelerating. Draw graphs of a versus t, and v versus t.

5. [I] A finalist in the Soap Box Derby starts with a push down a long hill at an initial speed of 1.0 m/s. At the bottom, 1.0 min 2.0 s later, it reaches a speed of 15.0 m/s. Find its average acceleration.

6. [I] A car goes from rest to 10 km/h in 10 s; at the end of 20 s it's moving at 20 km/h; at the end of 30 s it reaches 30 km/h. What can you say about its acceleration? Make a graph of a.

7. [I] A piston-engine dragster set a world record by starting from rest and hitting a top speed of 244 mi/h over a measured track of 440 yd. Compute its average acceleration in m/s^2.

8. [I] A VW Rabbit can go from rest to 80.5 km/h (50.0 mi/h) in a modest 8.20 s. Assuming its acceleration is precisely constant (which, of course, it isn't), how long will it take to speed up from 48.3 km/h to 64.4 km/h?

9. [I] What is the instantaneous acceleration of the object whose motion is depicted in Fig. 3.2a (p. 61) at a time of 1.58 s?

10. [I] Figure P10 is a velocity-time graph for a test car on a straight track. The test car initially moved backward in the negative x-direction at 20 m/s. It slowed, came to a stop, and then moved off in the positive x-direction at $t = 2.0$ s. What was its average acceleration during each of the time intervals 0 to 0.5 s, 1.5 s to 2.0 s, and 2.0 s to 2.5 s?

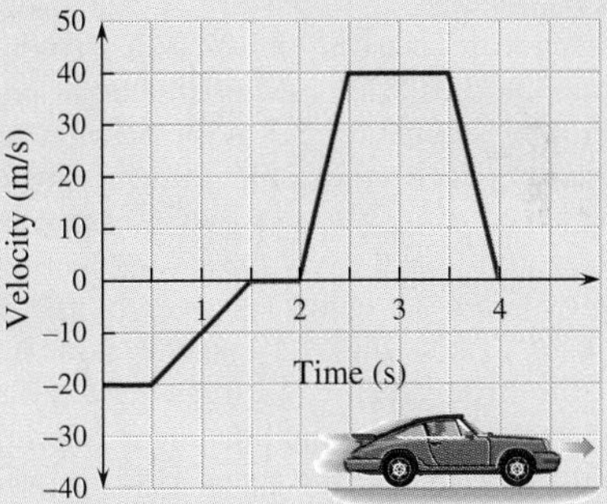

Figure P10

11. [II] Referring back to Fig. P10, what was the car's average acceleration during the interval from $t = 0$ to $t = 3.0$ s?

12. [III] Figure P12 shows the speed-time curves of three cyclists. Describe their motions and compute their average accelerations over the entire interval shown.

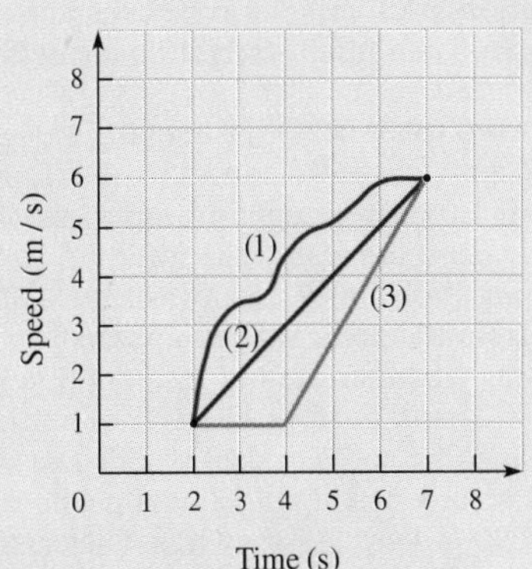

Figure P12

13. [II] Using Fig. 3.3 (p. 61), graphically determine the car's approximate acceleration at 3.8 s into the run.
14. [II] A motorboat starting from a dead stop accelerates at a constant 2.0 m/s^2 for 3.0 s, then very rapidly roars up to 4.0 m/s^2 and holds it constant for 4.0 s. What is its approximate average acceleration over the first 5.0 s of motion?
15. [II] Suppose that in the previous problem, after holding at 4.0 m/s^2 for 4.0 s, the boat next decelerates to a stop 20 s later at a rate of 1.1 m/s^2. (a) What is its average acceleration 27 s after the start of the run? (b) What is its instantaneous acceleration 10 s into the trip?
16. [II] A rocket accelerates straight up at 10 m/s^2 toward a helicopter that is descending uniformly at 5.0 m/s^2. What is the relative acceleration of the missile with respect to the aircraft as seen by the pilot who is about to bail out?
17. [II] Superman slams head-on into a locomotive speeding along at 60 km/h, bringing it smoothly to rest in an amazing 1/1000 s and saving Lois Lane, who was tied to the tracks. Calculate the constant deceleration of the train.
18. [III] Two motorcycle stuntpersons are driving directly toward one another, each having started at rest and each accelerating at a constant 5.5 m/s^2. At what speed will they be approaching each other 2.0 s into this lunacy?
19. [III] The speed of a flying saucer in ascent mode-4 as reported by an alleged eyewitness (who got it straight from the talkative alien navigator) is given by the expression $v = At + Bt^2$. Amazingly, A and B are constants in SI units but the little green navigator would not reveal their values. Determine the instantaneous acceleration of the craft. (Hint: at $(t + \Delta t)$ the speed is $(v + \Delta v)$.)

UNIFORMLY ACCELERATED MOTION

20. [I] Ask a friend to hold his or her thumb and forefinger parallel to each other in a horizontal plane. The fingers should be about an inch apart. Now you hold a 1-ft ruler vertically in the gap just above and between these fingers so it can be dropped between them. Have your friend look at the ruler and catch it when you, without warning, let it fall. Calculate the corresponding response time. Now position a dollar bill vertically so that Washington's face is between your accomplice's fingers—is it likely to be caught when dropped?
21. [I] A locust, extending its hind legs over a distance of 4 cm, leaves the ground at a speed of 340 cm/s. Determine its acceleration, presuming it to be constant.
22. [I] A kangaroo can jump straight up about 2.5 m—what is its takeoff speed?
23. [I] A wayward robot, R2D3, is moving along at 1.5 m/s when it suddenly shifts gears and roars off, accelerating uniformly at 1.0 m/s^2 straight toward a wall 10 m away. At what speed will it crash into the wall?
24. [I] Regarding Problem 17, Lois was a mere 10 cm down the tracks from the point where Superman struck the train. How far away from her did the "Man of Steel" finally stop the engine? (Thank goodness he arrived in time!)
25. [I] A Jaguar in an auto accident in England in 1960 left the longest recorded skid marks on a public road: an incredible 290 m long. As we will see later, the friction force between the tires and the pavement varies with speed, producing a deceleration that increases as the speed decreases. Assuming an average acceleration of -3.9 m/s^2 (that is, -0.4 g), calculate the Jag's speed when the brakes locked.
26. [I] A modern supertanker is gigantic: 1200 to 1300 ft long with a 200-ft beam. Fully loaded, it chugs along at about 16 knots (30 km/h or 18 mi/h). It can take 20 min to bring such a monster to a full stop. Calculate the corresponding deceleration in m/s^2 and determine the stopping distance.
27. [I] A bullet traveling at 300 m/s slams into a block of moist clay, coming to rest with a fairly uniform acceleration after penetrating 5 cm. Calculate its acceleration.
28. [I] The velocity-time graph for a rocket is shown in Fig. Q17 (p. 84). Roughly what was its peak altitude?
29. [II] A driver traveling at 60 km/h sees a chicken dash out onto the road and slams on the brakes. Accelerating at -7 m/s^2, the car stops just in time 23.3 m down the road. What was the driver's reaction time?
30. [II] The longest passenger liner ever built was the *France,* at 66 348 tons and 315.5 m long. Suppose its bow passes the edge of a pier at a speed of 2.50 m/s while the ship is accelerating uniformly at 0.01 m/s^2. At what speed will the stern of the vessel pass the pier?
31. [II] A swimmer stroking along at a very fast 2.2 m/s ceases all body movement and uniformly coasts to a dead stop in 10 m. Determine how far she moved during her third second of unpowered drift.
32. [II] A little electric car having a maximum speed of 40.0 km/h can speed up uniformly at any rate from 1.00 to 4.00 m/s^2 and slow down uniformly at any rate from 0.00 to -6.00 m/s^2. What is the shortest time in which such a vehicle can traverse a distance of 1.00 km starting and ending at rest?
33. [II] Two trains heading straight for each other on the same track are 250 m apart when their engineers see each other and hit the brakes. The Express, heading west at 96 km/h, slows down, accelerating at an average of -4 m/s^2, while the eastbound Flyer, traveling at 110 km/h, slows down, accelerating at an average of -3 m/s^2. Will they collide?
34. [II] The drivers of two cars in a demolition derby are at rest 100 m apart. A clock on a billboard reads 12:17:00 at the moment they begin heading straight toward each other. If both are accelerating at a constant 2.5 m/s^2, at what time will they collide?
35. [II] A rocket-launching device contains several solid-propellant missiles that are successively fired horizontally at 1-s intervals. (a) What is the separation of the first and second missiles just at the moment the second one is fired, if each has an initial speed of 60 m/s and a constant acceleration of 20 m/s^2 lasting for 10 s? (b) What is the separation of the first and second just as the third is launched?
36. [II] Referring to the previous problem, what is the relative acceleration of the first missile with respect to the second one once both are in the air and accelerating? What will be their separation 6 s into the flight?
37. [II] Imagine that you are driving toward an intersection at a speed v_i just as the light changes from green to yellow. Assuming a response time of 0.6 s and an acceleration of -6.9 m/s^2, write an expression for the smallest distance (s_S) from the corner in which you could stop in time. How much is that if you are traveling 35 km/h?

38. [II] Considering the previous problem, it should be clear that the yellow light might reasonably be set for a time t_Y, which is long enough for a car to traverse the distance equal to both s_S and the width of the intersection s_I. Assuming a constant speed v_i equal to the legal limit, write an equation for t_Y, which is independent of s_S.

39. [II] The driver of a pink Cadillac traveling at a constant 60 mi/h in a 55 mi/h zone is being chased by the law. The police car is 20 m behind the perpetrator when it too reaches 60 mi/h, and at that moment the officer floors the gas pedal. If her car roars up to the rear of the Cadillac 2.0 s later, what was her acceleration, assuming it to be constant?

40. [II] With Problems 37 and 38 in mind, how long should the yellow light stay lit if we assume a driver-response time of 0.6 s, an acceleration of -6.9 m/s^2, a speed of 35 km/h, and an intersection 25 m wide? Which of the several contributing aspects requires the greatest time?

41. [II] In Problem 39, how far does the police car travel in the process of closing the Cadillac's 20-m lead?

42. [III] Having taken a nap under a tree only 20 m from the finish line, Rabbit wakes up to find Turtle 19.5 m beyond him, grinding along at 1/4 m/s. If the bewildered hare can accelerate at 9 m/s^2 up to his top speed of 18 m/s (40 mi/h) and sustain that speed, will he win?

43. [III] Superman is jogging alongside the railroad tracks on the outskirts of Metropolis, at 100 km/h. He overtakes the caboose of a 500-m-long freight train traveling at 50 km/h. At that moment he begins to accelerate at $+10$ m/s^2. How far will the train have traveled before Superman passes the locomotive?

44. [III] A motorcycle cop, parked at the side of a highway reading a magazine, is passed by a woman in a red Ferrari 308 GTS doing 90.0 km/h. After a few attempts to get his cycle started, the officer roars off 2.00 s later. At what average rate must he accelerate if 110 km/h is his top speed and he is to catch her just at the state line 2.00 km away?

FREE-FALL

45. [I] At what speed would you hit the floor if you stepped off a chair 0.50 m high? Ignore friction. Express your answer in m/s, ft/s, and mi/h.

46. [I] If a stone dropped (not thrown) from a bridge takes 3.7 s to hit the water, how high is the rock-dropper? Ignore friction.

47. [I] Ignoring air friction, how fast will an object be moving and how far will it have fallen after dropping from rest for 1.0 s, 2.0 s, 5.0 s, and 10 s?

48. [I] A cannonball is fired straight up at a rather modest speed of 9.8 m/s. Compute its maximum altitude and the time it takes to reach that height (ignoring air friction).

49. [I] Calculate the speed at which a hailstone, falling from 3.00×10^4 ft out of a cumulonimbus cloud, would strike the ground, presuming air friction is negligible (which it certainly is not). Give your answer in mi/h and m/s.

50. [I] A circus performer juggling while standing on a platform 15.0 m high tosses a ball directly upward into the air at a speed of 5.0 m/s. If it leaves his hand 1.0 m above the platform, what is the ball's maximum altitude? If the juggler misses the ball, at what speed will it hit the floor? Ignore air friction.

51. [I] Draw a curve of s in meters versus t in seconds for a free-falling body dropped from rest. Restrict the analysis to the first second of fall, and make up a table of values of t, t^2, and s at 1/10 s intervals. Draw a curve of s versus t^2. Explain your findings.

52. [I] A shoe is flung into the air such that at the end of 2.0 s it is at its maximum altitude, moving at 6.0 m/s. How far away from the thrower will it be when it returns to the height it was tossed from? Ignore air friction.

53. [I] Show that the range of a projectile can be expressed as

$$s_R = -\frac{2v_{iH}v_{iV}}{g}$$

Ignore air friction.

54. [I] A lit firecracker is hurled upward from a height of 2.00 m at a speed of 20.0 m/s at an angle above the ground of 60°. If it explodes 2.00 s later, how high above the ground is the blast? Ignore air friction.

55. [I] Suppose you point a rifle horizontally directly at the center of a paper target 100 m away from you. If the muzzle speed of the bullet is 1000 m/s, where will it strike the target? Assume aerodynamic effects are negligible.

56. [I] A rifle must always be pointed slightly above the target—the bullet travels an arc. The peak height it reaches (at midrange) above the horizontal straight line from rifle to target is called the *mid-range trajectory,* and the smaller that is, the better. A 243 Winchester with a muzzle speed of 3500 ft/s has a mid-range trajectory of 4.7 in. when zeroed in on a target 300 yd away. Ignoring air effects, at what angular elevation must the rifle be pointed to hit that target? A bullet like this will lose about 30 to 40% of its initial speed over such a trip, so the above approximation is tolerable, but not great.

57. [I] A raw egg is thrown horizontally straight out of the open window of a fraternity house. If its initial speed is 20 m/s and it hits ground 2.0 s later, at what height was it launched? At such low speeds air friction is negligible.

58. [I] A golfer wishes to chip a shot into a hole 50 m away on flat level ground. If the ball sails off at 45°, what speed must it have initially? Ignore aerodynamic effects.

59. [I] Check the dimensions of both sides of the equation

$$v_i s_V^2 = t v_i^2 \cos\theta + \tfrac{1}{2} g v_i t^2$$

to see if they are the same; if not, the equation is wrong. This technique was introduced by the French mathematician and physicist, J. B. J. Fourier, around 1822. Is it possible that the above equation is correct?

60. [I] The acceleration due to gravity on the surface of the Moon is about $g/6$. If you can throw a ball straight up to a height of 25 m on Earth, how high would it reach on the Moon when launched at the same speed? Ignore the minor effects of air friction.

61. [II] A young kid with a huge baseball cap is playing catch with himself by throwing a ball straight up. How fast does he throw it if the ball comes back to his hands a second later? At low speeds air friction is negligible.

62. [II] An arrow is launched vertically upward from a crossbow at 98.1 m/s. Ignoring air friction, what is its instantaneous speed at the end of 10.0 s of flight? What is its average speed up to that moment? How high has it

risen? What is its instantaneous acceleration 4.20 s into the flight?

63. [II] A lit firecracker is shot straight up into the air at a speed of 50.00 m/s. How high is it above ground level 5.000 s later when it explodes? How fast is it moving when it blows up? How far has it fallen, if at all, from its maximum height? Ignore drag and take g equal to 9.800 m/s^2.

64. [II] According to their literature, the observation deck of the World Trade Center in New York is 1377 ft above ground. Ignoring air friction, how long will it take to free-fall that far? Take $g = 9.81$ m/s^2. Considering the discussion in the text, is it reasonable to ignore air drag in this problem? Explain.

65. [II] A human being performing a vertical jump generally squats and then springs upward, accelerating with feet touching the Earth through a distance s_a. Once fully extended, the jumper leaves the ground and glides upward, decelerating until the feet are a maximum height s_{max} off the floor. Assuming the jumping acceleration a to be constant, derive an expression for it in terms of s_a and s_{max}. Ignore friction.

66. [II] A small rocket is launched vertically, attaining a maximum speed at burnout of 1.0×10^2 m/s and thereafter coasting straight up to a maximum altitude of 1510 m. Assuming the rocket accelerated uniformly while the engine was on, how long did it fire and how high was it at engine cutoff? Ignore air friction.

67. [II] Using Fig. 3.17 (p. 82) and extending it where necessary, show that an angle less than 45° with respect to the ground will result in a greater horizontal displacement when the landing point is lower than the launch point. Neglect friction.

68. [II] A flea jumps into the air and lands about 8.0 in. away, having risen to an altitude of about 130 times its own height (that's comparable to you jumping 650 ft up). Assuming a 45° launch, compute the flea's takeoff speed. Make use of the fact that $2 \sin\theta \cos\theta = \sin 2\theta$ and ignore air friction.

69. [II] Two diving platforms 10 m high terminate just at the edge of each end of a swimming pool 30 m long. How fast must two clowns run straight off their respective boards if they are to collide at the surface of the water midpool? Ignore friction.

70. [II] A golf ball hit with a 7-iron soars into the air at 40.0° with a speed of 54.86 m/s (that is, 180 ft/s). Overlooking the effect of the atmosphere on the ball, determine (a) its range and (b) when it will strike the ground.

71. [II] A silver dollar is thrown downward at an angle of 60.0° below the horizontal from a bridge 50.0 m above a river. If its initial speed is 40.0 m/s, where and at what speed will it strike the water? Ignore the effects of the air.

72. [II] Salmon, swimming to their spawning grounds, leap over all sorts of obstacles. The unofficial salmon-altitude record is an amazing 3.45 m (that is, 11 ft 4 in.) jump. Assuming the fish took off at 45.0°, what was its speed on emerging from the water? Ignore friction.

73. [II] Show that at any time t, where friction is negligible, the velocity of a projectile (launched at an angle θ with a speed v_i) makes an angle θ_t with the horizontal given by the expression

$$\theta_t = \tan^{-1}\left(\tan\theta + \frac{gt}{v_i \cos\theta}\right)$$

where $g < 0$.

74. [II] The illustrative Example 3.14 deferred from using the equation $s = v_i t + \frac{1}{2}at^2$ because it required the use of the quadratic formula. Carry out that calculation, solving it for the two values of t. Notice the symmetry around the time of peak altitude.

75. [III] A bag of sand dropped by a would-be assassin from the roof of a building just misses Tough Tony, a gangster 2 m tall. The missile traverses the height of Tough Tony in 0.20 s, landing with a thud at his feet. How high was the building? Ignore friction.

76. [III] Imagine that someone dropped a firecracker off the roof of a building and heard it explode exactly 10 s later. Ignoring air friction, taking $g = 9.81$ m/s^2, and using 330 m/s as the speed of sound, calculate how far the cherry bomb had fallen at the very moment it blew up.

77. [III] A burning firecracker is tossed into the air at an angle of 60° up from the horizon. If it leaves the hand of the hurler at a speed of 30 m/s, how long should the fuse be set to burn if the explosion is to occur 20 m away? Ignoring friction, just set up the equation for t.

78. [III] A baseball is hit as it comes in, 1.30 m over the plate. The blast sends it off at an angle of 30° above the horizontal with a speed of 45.0 m/s. The outfield fence is 100 m away and 11.3 m high. Ignoring aerodynamic effects, will the ball clear the fence?

4

Newton's Three Laws: Momentum

THIS CHAPTER AND THE next are about Newton's Three Laws—(1) the Law of Inertia; (2) the relationship between force and the accompanying change in motion; (3) and the Law of Interaction. These insights initiated the study of how motion changes, a discipline called **dynamics**. Traditionally, there are two different ways to develop this material: One is through the concept of momentum and how momentum is changed by force, which we will concentrate on in this chapter; the other uses acceleration and how it is produced by force, discussed in Chapter 5. Superficially, the two seem equally insightful, but they are not. Newton's own formulation was in terms of momentum, and we now know from Einstein's Special Theory of Relativity that Newton's Second Law is perfectly correct as he gave it. However, the alternative formulation in terms of acceleration ($F = ma$) is actually only a low-speed approximation—a very useful one, but an approximation nonetheless.

We concentrate first on momentum because it has been learned in this century that momentum is one of the premier ideas of physics. When there is no net external force acting on a system, its linear momentum remains unchanged and we have the all-important law of Conservation of Linear Momentum. We now know that this powerful principle arises from the homogeneity of space itself (p. 116). As we shall see, in all of physics there are only a handful of these wonderful conservation laws, and each is related to some fundamental symmetry of the Universe.

Though their perspectives are different, Chapters 4 and 5 both use Newton's laws to develop *dynamics;* that is, what happens when momentum changes. Chapter 6 uses those same laws to elaborate *statics;* that is, what happens when momentum remains constant.

THE LAW OF INERTIA

Sir Isaac Newton, 1642–1727.

Isaac Newton was born on Christmas day, 1642, within a year of Galileo's death. Newton was a short man, nearsighted, already silver-gray in his thirties, inattentive to personal appearance, incredibly forgetful, and probably virginal. This nervous, hypochondriacal, sensitive, pious, vulnerable individual was one of the greatest geniuses to ever live. Among Newton's many achievements was a theory of motion that withstood every test for over 200 years.

The first of Newton's Three Laws was actually formulated much earlier by Galileo. The **Law of Inertia** states that *a body once set in motion and thereafter undisturbed will continue in uniform motion forever, all by itself.* Some scholars thought along similar lines in Galileo's day, but this concept was completely opposed to the prevailing Aristotelian view. To Aristotle and his followers, any motion of a heavy object other than free-fall required the exertion of a force. They argued—wrongly—that without an impelling force, there can be no sustained motion. End that force, and the motion in progress spontaneously ceases. Aristotle's position appears to agree with experience, but it is wrong. We live on a planet where gravity and friction obscure what is really going on. So much so that it took 18 centuries before Galileo saw to the heart of the matter.

The experiment that led Galileo to the Law of Inertia used two inclined planes set end-to-end, one tilted down, the other up (Fig. 4.1). A ball was rolled, from rest, down the first incline. It swept past the bottom and ascended the second incline, gradually losing speed until it momentarily came to rest. Regardless of the tilt of that second surface, the ball always stopped ascending at a vertical height a little lower than that from which it was first released. Rather than overlook that slight difference, Galileo searched for its cause. He polished the ball and smoothed the track, and finally concluded that the loss was due to friction between the track and the ball. Friction was a crucial piece to the puzzle, a piece Aristotle never appreciated.

Next, Galileo did something quite clever: He gradually reduced the angle of the second plane so that the ball rolled farther and farther before coming to rest. Then he asked the right question: What would happen if the second incline were made *perfectly frictionless* and *perfectly horizontal?* The ball would roll on forever, neither speeding up nor slowing down, with "a motion which was uniform and perpetual." His imagination had gone beyond the limitations of the experiment, from

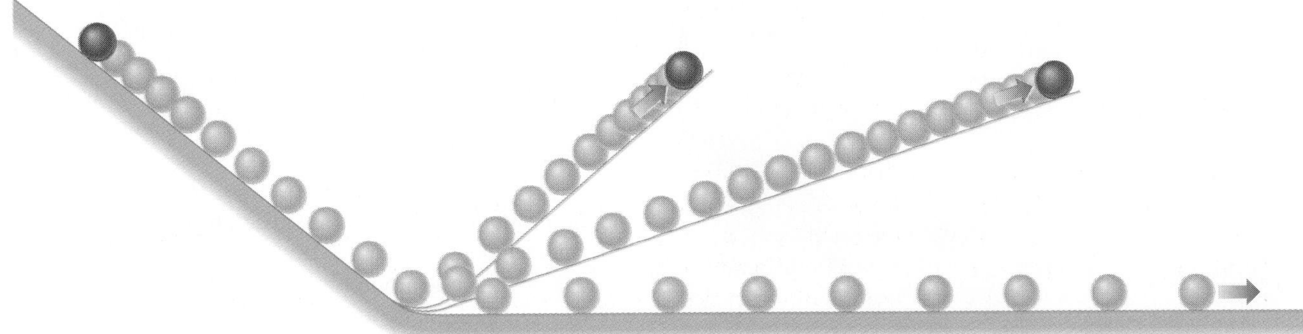

Figure 4.1 A ball rolled down an inclined plane always climbs back to nearly the same height at which it was released before it stops. Galileo thus argued that if the second plane was horizontal and frictionless, the ball would roll forever.

which he could never have totally removed friction. We ordinarily must keep pushing objects to maintain them in motion because of friction (p. 146), not because of the fundamental nature of the process. Had Aristotle lived aboard a spaceship, the Law of Inertia would have been obvious.

Despite the brilliant insight, Galileo soon got lost worrying about how an infinite horizontal plane could exist on a spherical Earth. In the end, he erroneously decided that the unimpeded motion of objects had to be in great circles, in this case around the planet. The law was soon properly linearized by a professional soldier turned philosopher and mathematician—René Descartes (pronounced Day-cart). Guided by conjecture rather than experiment, Descartes maintained, in 1644, "that every body which moves tends to continue its movement in a straight line."

In 1687, Isaac Newton set out the "axioms or laws of motion" in his great work, *Mathematical Principles of Natural Philosophy* (nowadays often called the *Principia* from the original Latin title, and pronounced prin-sip-′pee-uh). His First Law was the Law of Inertia: **Every body continues in a state of rest or of uniform motion in a straight line except insofar as it is compelled to change that state by forces impressed upon it**—and he gave Galileo credit for the insight.

A new equivalence is drawn here between rest and uniform motion. Altering either requires a force, but once either state is established, it persists forever in the absence of force. Rest and uniform motion are only "relatively distinguished," as Newton put it. For example, a coin sits at "rest" in your open hand even though the car you are in is speeding along at a constant 50 km/h. Someone standing outside would see the car, you, and the coin streak past, but the Law of Inertia holds from both perspectives; that is, it holds in both inertial frames. Measured with regard to your hand, the coin's speed is zero and will remain so, just as it will remain 50 km/h with respect to the outside observer, so long as no applied force alters the motion.

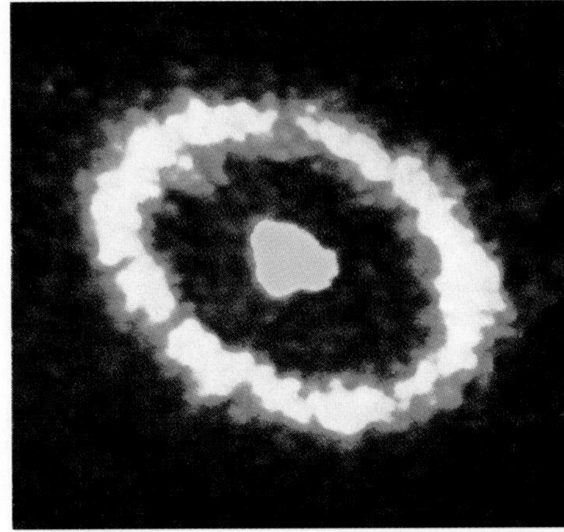

The pink central region is the remnant of a star that was seen to explode in 1987 (supernova 1987A). This 1990 photo shows a yellow ring of glowing gas about 12×10^{12} mi across. It was blown off by the star some 20 000 years ago and is still hurtling out into space in accord with Newton's First Law. The central debris of the supernova is moving outward much faster, and will overtake and tear apart the ring within the next 100 years.

The *Apollo* astronauts, in 1969, shut down their engines and coasted in unpowered flight for much of the way to the Moon. At this very instant, *Pioneer 10* is hurtling toward the stars at tens of thousands of kilometers per hour, its rockets burned out decades ago. The famous trick of whisking a tablecloth out from under a setting of dishes is a grandiose First-Law gesture. And the purpose of seat belts becomes clear when the body in motion tending to stay in motion after slamming on the brakes is yours.

Be aware that the Law of Inertia, the first basic insight into the behavior of nature that we have come upon so far, is an idealization. There is no place in this Universe where an object can be completely free of external influences, and the idea of infinite straight-line motion is unrealistic, especially in a cosmos cluttered with

The First Law: A body in motion tends to stay in motion.

galaxies. So, we will not be able to confirm directly the full implications of the First Law, namely that an object free of external influences will move with a constant velocity forever. Still, the Law of Inertia allows us to understand a tremendous range of observed phenomena—and that's its real power.

Newton felt that **inertia**—that is, *the resistance to a change in motion*—was an intrinsic property of matter itself, independent of the environment. Yet, within a few decades of his publication of the Law of Inertia, that opinion was challenged by the philosopher George Berkeley. Many today believe, like Berkeley, that inertia arises from the interaction (p. 208) of all matter in the Universe, an idea that is called Mach's Principle. As we will see, this concept is quite appealing. Still, there are other possibilities. Physicists have certainly not yet resolved the fundamental question of the origin of inertia.

4.1 Ballistic Motion

The unpowered flight of an object launched into the air at the Earth's surface is a fine example of the primary role played by the Law of Inertia. Moreover, it demonstrates how a body is caused to deviate from its straight-line, First-Law inertial path by forces applied to it, in this case gravity acting downward and frictional drag acting tangent to the path of the motion.

The first person to comprehend ballistics was Galileo, precisely because he knew about both free-fall and the Law of Inertia. Galileo understood that an unpowered projectile moves in a way that is the result of the superposition of two independent motions: a straight-line inertial flight in the direction of launch (ideally at constant speed) and a vertical gravity fall (ideally at constant acceleration). The Law of Inertia maintains that left on its own, a ball should sail straight off in the direction thrown, at a *constant speed* equal to its initial speed, forever (and this phenomenon is exactly what would happen in empty space in the absence of air friction and gravity). On a sizeable body such as the Earth, gravity will exert a substantial downward force. From the First Law, we know that such a force will alter the motion, in this instance turning that straight path into a curved trajectory (p. 75). Newton showed that on a spherical planet, with gravity pulling down toward its center, a dragless ballistic trajectory should actually be elliptical. (The parabolic path discussed in Chapter 3 was a flat-Earth approximation, excellent for short ranges.)

An unappreciated anticipation. *No body begins to move or comes to rest of itself.*

ABU ALI IBN SINA, known as Avicenna (980–1037)

If we fire a cannonball horizontally and at that very instant drop a pebble from the same height, *the ball and pebble will hit the ground at very nearly the same moment*—at least to the degree that air effects are negligible. Upon emerging from the cannon, the ball is in free-fall and it immediately accelerates downward at g. Despite the fact that the ball may land quite a distance away, under the influence of gravity it will drop at 9.8 m/s^2, in step with the descent of the free-falling pebble. Figure 4.2 shows how the ball's flight path at any instant t is composed of a horizontal inertial displacement, $s_{\mathrm{H}} = v_{\mathrm{iH}}t$, from which the projectile continuously free-falls a distance $s_{\mathrm{V}} = \frac{1}{2}gt^2$. You can do a similar experiment by placing a dime at the

On the motion of bodies before Galileo. *Once moving they are never at rest unless impeded. Once in motion only an outside force can restrain them.*

G. B. BENEDETTI (1585)

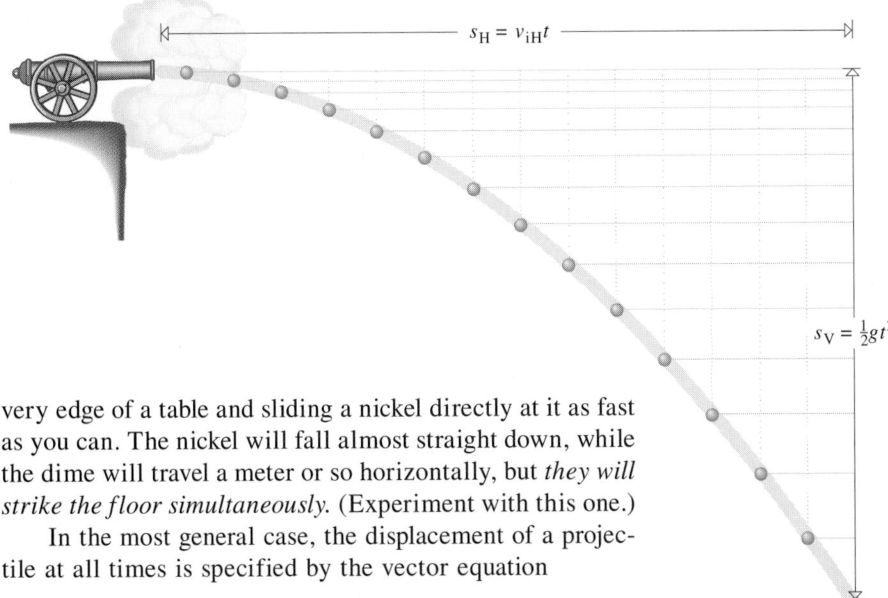

Figure 4.2 Without friction, a projectile fired horizontally falls as it progresses, sustaining its constant horizontal inertial motion while experiencing a vertical acceleration of g.

very edge of a table and sliding a nickel directly at it as fast as you can. The nickel will fall almost straight down, while the dime will travel a meter or so horizontally, but *they will strike the floor simultaneously.* (Experiment with this one.)

In the most general case, the displacement of a projectile at all times is specified by the vector equation

$$\mathbf{s} = \mathbf{v}_i t + \tfrac{1}{2}\mathbf{g}t^2$$

wherein $\mathbf{g}$ is the vector acceleration due to gravity: 9.81 m/s^2 pointing downward. The equation provides the displacement from the point of liftoff at each and every moment. Notice that the first term is the straight-line inertial path, whereas the second is the downward free-fall contribution. Without gravity, the ball in Fig. 4.3 would have followed the straight line, being displaced in a time t by an amount $\mathbf{v}_i t$. With gravity, the ball falls continuously, and is further displaced by an amount $\frac{1}{2}\mathbf{g}t^2$ downward. The vector sum of these two displacements is $\mathbf{s}$.

The concern is sometimes voiced that the effect of gravity on a projectile might be different because of the lateral motion, but it isn't. If we perform the free-fall experiment inside a uniformly moving train, the results are unaltered. A dropped bundle of keys does not slam into the rear wall of the car, but instead falls straight down as usual (Fig. 4.4a). To someone outside at "rest" looking in, the keys would have the constant horizontal motion of the train in addition to the downward fall, and so would be seen to move along a parabolic arc (Fig. 4.4b).

Although we will not be able to analyze aerodynamic effects, people concerned with accurate ballistics have to. They consider the complexities of air friction, lift (Sect. 11.12), and wind (Fig. 4.5). In their own way, so do people trying to catch a "high fly" or hit a golf ball to a distant green. Since the retarding force generated on a fast projectile by the air (the **drag**) is proportional to the *square of the speed,* that force becomes increasingly problematic as speeds go up, usually in an attempt to extend the range. The analysis of projectiles in Chapter 3 is fine for cannoneers on the Moon where there is no atmosphere, but on Earth it fails noticeably when we go beyond a gentle game of catch.

A well-struck baseball, in the air for a rather long time, can lose as much as half its initial speed due to aerodynamic drag, and so travel only a bit more than half as far as it would have, had there been no friction. A basketball is

The flight of exploding mortar shells, drawn by Leonardo da Vinci about 500 years ago. Leonardo knew that on a flat Earth the trajectory would ideally be parabolic.

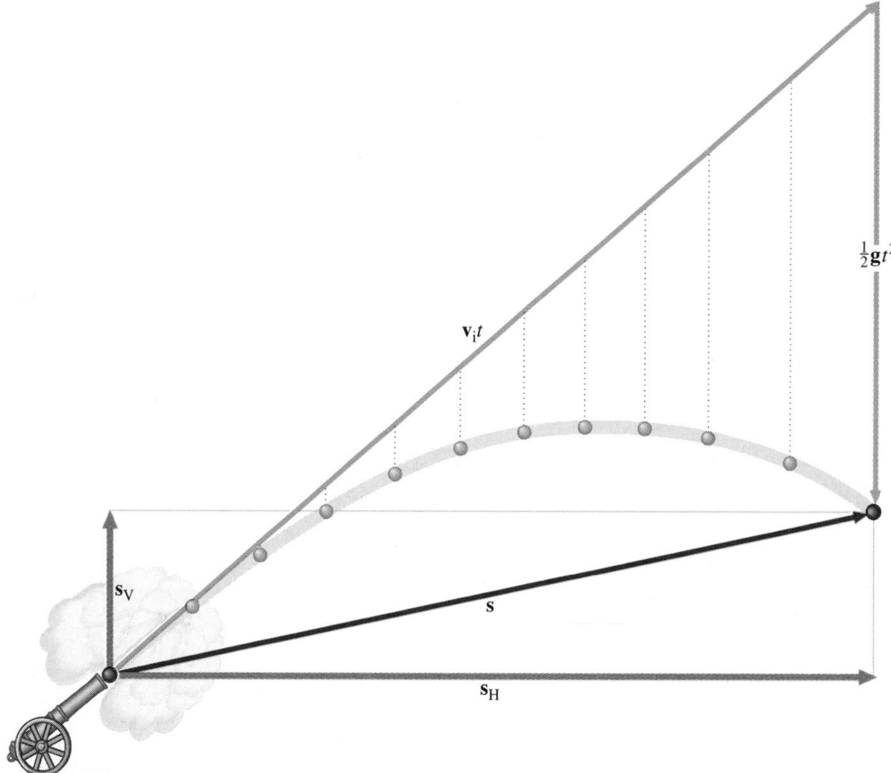

Figure 4.3 At every instant, the motion of the ball is the composite of a straight-line inertial path ($\mathbf{v}_i t$) in the direction of launch, and a vertical free-fall ($\frac{1}{2}\mathbf{g}t^2$) drop down from it.

usually thrown at a fairly low speed (around 7.6 m/s, or 25 ft/s), which keeps drag down, but the ball's light weight and large surface area nonetheless make air resistance appreciable. The drag force results in a deceleration ($\approx$1 m/s^2) tangent to the flight path. A player must compensate for that by launching the ball with a speed about 5% greater than would be necessary if the court were in vacuum. By contrast, a rifle bullet fired at 0.6 km/s will experience a good deal of drag. Although, if unimpeded, it would have a tremendous maximum range of about 40 km according to Eq. (3.16, p. 81), the bullet is not likely to travel much beyond 4 km (Fig. 4.5).

Were it not for the accidental impediment of the air, I verily believe that, if at the time of the ball's going out of the piece [that is, a cannon] another were let fall from the same height directly downwards, they would both come to the ground at the same instant.

GALILEO

4.2 Force

The First Law tells us that a body at rest will remain at rest and a body in uniform motion will remain in uniform motion *except insofar as it is compelled to change that state by a force impressed upon it.* But what is a force? Force is often naively thought of as pushes and pulls (reasonably enough, we tend to go back to our physiological experiences as a starting point), but we have to be more precise than that. The sensation of force is as subjective as the sensations of taste, smell, and color. To say that force is your muscle sensation of force is simplistic. If we were ghost-like and viewed the world remotely without being able to feel, what would we say about force? Forces are not directly observable. What is perceived is the effect, the change produced by a force.

If you insist upon a precise definition of force, you will never get it!

RICHARD FEYNMAN
Nobel Laureate (1963)

Figure 4.4 An experiment performed inside a uniformly moving train. (a) As seen from inside, the dropped keys fall straight down. (b) As seen from outside, the keys, tossed straight up, move along a parabola.

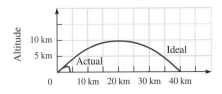

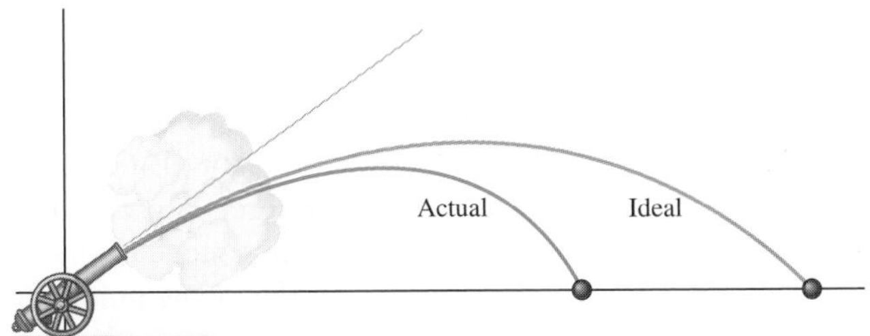

Figure 4.5 Actual and ideal paths of a relatively slow cannonball in free flight through the air, and a high-speed projectile like a rifle bullet with a muzzle speed of 0.6 km. To see the effect, hurl something lightweight, like a wad of tissue paper. Use a piece about an inch or two square and loosely balled up.

In the beginning of the *Principia,* Newton defined force as "an action exerted upon a body, in order to change its state, either of rest or of moving uniformly. . . ." **Force is the agent of change**. In the particular case of dynamics, it is *that which alters motion*. In the Universe as it exists today, there are four distinct fundamental forces: gravitational, electromagnetic, strong, and weak. In one way or another, all of physics is concerned with the Four Forces.*

*The concept of force or interaction has taken on a broadened meaning in contemporary physics. Thus, for example, the weak force mediates the transmutation of quarks and the decay of the neutron; force not only changes motion, it transforms matter.

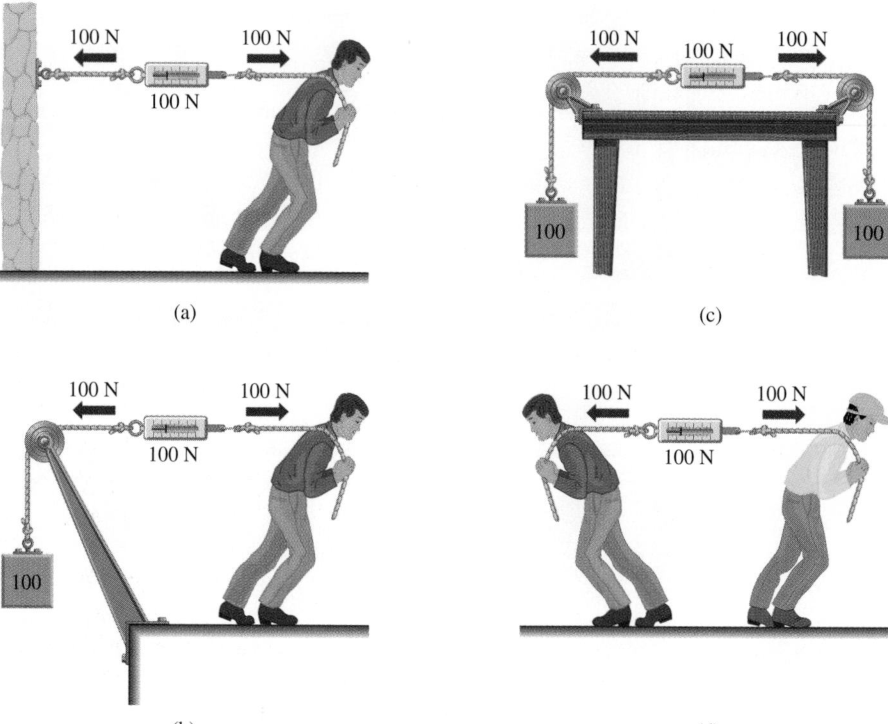

Figure 4.6 A spring scale being used to measure force. Shown are several situations that look different but are equivalent as far as the scale is concerned.

An apple of medium size weighs about 1 N.

As with most scientific concepts, we try to define force in a way that allows us to measure it. The easiest procedure to follow is to take a standard mass (p. 11), a kilogram, hang it on a spring scale and call whatever the arrow points to "so many units of force." After all, ordinary weight is the downward gravitational force. With a beam balance, we can divide a kilogram mass into fractional pieces that could then be used to calibrate the spring scale. A 0.453 592 37-kg piece, at a place where $g = 9.806\,65\ \text{m/s}^2$, has a weight equal to 1 lb. Once appropriately labeled, the spring scale then provides a convenient, reproducible way to measure forces (Fig. 4.6). Unfortunately, this scheme uses gravity and has little to do with Newton's definition of force as the changer of motion.

Later in the chapter, when we formally define the unit of force, the *newton* (N), in the SI system, it will be by way of the change in motion of an object, independent of gravity. For now, 1 N is the weight on Earth of a mass of about 0.10 kg. In terms of the outmoded pound, $1\ \text{N} = 0.224\,809$ lb. From now on, we will treat force in newtons, keeping in mind that $1\ \text{N} \approx \frac{1}{4}$ lb—the weight of a medium-sized apple.

Example 4.1 The *Apollo Command Module* on the way to the Moon was *effectively weightless* (p. 136) after being placed in what was called a parking orbit around the Earth. On the other hand, its mass (5896 kg) was unchanged by the ride into space. Compute the craft's weight on the launch pad in both pounds and newtons.

Solution: [Given: $m_A = 5896$ kg. Find: its Earth-weight.] Because a mass of 0.453 59 kg weighs 1.000 0 lb on Earth, the *Apollo* weighed

$$\frac{5896 \text{ kg}}{0.45359 \text{ kg/lb}} = \boxed{13.00 \times 10^3 \text{ lb}}$$

Because 1 N = 0.224 809 lb, the inverse of this is 4.448 22 N = 1 lb, and the *Apollo* weighed

$$(13.00 \times 10^3 \text{ lb})(4.44822 \text{ N/lb}) = \boxed{57.83 \text{ kN}}$$

▶ **Quick Check:** As we will see [Eq. (5.5)], the proportionality constant between weight and mass is g; that is, the weight of an object divided by its mass equals $9.8 \text{ m/s}^2 = (57.83 \times 10^3 \text{ N})/(5896 \text{ kg})$.

Force as a Vector

Hold a book upright, placing a hand flat on either side (Fig. 4.7). Now press your hands together exerting an equal force on each cover. Regardless of how great the force applied by each hand, as long as both are equal, the book remains at rest. It seems that the *net applied force* is zero—no change in the motion, no net force according to the First Law. Two oppositely directed forces acting on the same object work against each other, partially or totally canceling one another. Moreover, several forces applied to a point on a body act as if they were a single force. The man on the left in Fig. 4.8 experiences one large force pulling him backward even though he has four individual adversaries. With your eyes closed, there is no way to tell how many people are pulling against you in a tug-of-war; you experience the one resultant force.

Figure 4.7 Force is a vector quantity, which means that a force to the left tends to cancel a force to the right. If the two forces on the book are equal, it remains at rest.

If we say "push on the person next to you with a force of 10 N," it could be done in a number of different ways: up, down, right, left, and so on. The statement is ambiguous. Force is specified completely only when both its magnitude and direction are given. Like displacement and velocity, *force is a vector quantity*. We draw this conclusion from observations. And *it's the net force, the resultant of all the forces acting on a body, that is manifested in a change in motion.*

Figure 4.8 The four forces to the right (which all pass through a single point in the ring) act as a single net force equal to their vector sum.

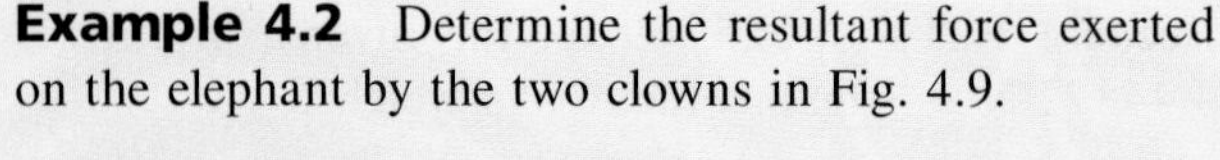

Example 4.2 Determine the resultant force exerted on the elephant by the two clowns in Fig. 4.9.

Solution: [Given: $F_1 = 300$ N and $F_2 = 400$ N. Find: the resultant **F**.] We want the vector sum

$$\mathbf{F} = \mathbf{F}_1 + \mathbf{F}_2$$

and since the forces form a 3-4-5 right triangle $\boxed{F = 500 \text{ N}}$. To find the angle **F** makes with the west-to-east axis, we use

$$\theta = \tan^{-1}\frac{300 \text{ N}}{400 \text{ N}} = \boxed{36.9° \text{ north of east}}$$

▶ **Quick Check:** (500 N) cos 36.9° = 400 N.

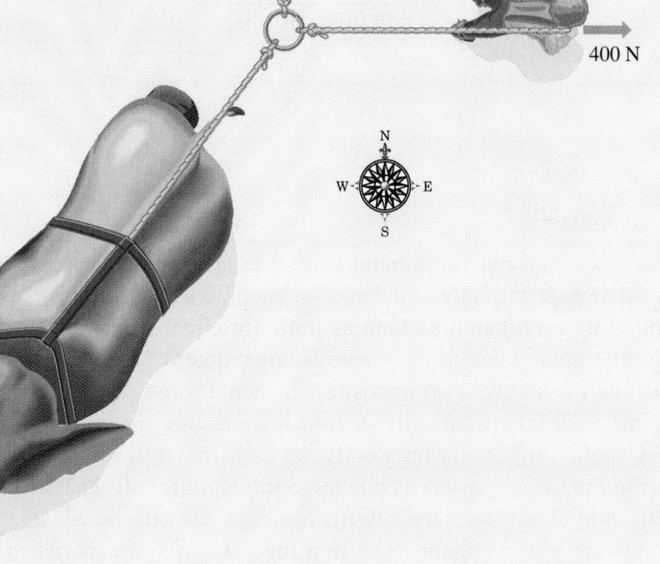

Figure 4.9 Each clown pulls on the ring with a force as shown. The elephant feels a net force pulling him north of east.

Example 4.3 Determine the net force exerted on the ring by the three people in Fig. 4.10.

Solution: [Given: $F_1 = 707$ N, $F_2 = 500$ N, and $F_3 = 966$ N. Find: $\mathbf{F} = \mathbf{F}_1 + \mathbf{F}_2 + \mathbf{F}_3$.] Resolve the three vectors into their components to get the resultant (see Appendix D). The choice of the two perpendicular axes along which to do the resolving is arbitrary, but often one set yields simpler computations, and it's usually along one of the force vectors. Accordingly, resolve the forces along and perpendicular to the y-axis down which $\mathbf{F}_3$ acts, as in Fig. 4.10b. The y-component of the resultant is then

$$F_y = F_{1y} + F_{2y} - F_3 = F_1 \cos 30° + F_2 \cos 45° - F_3$$

where F_3 is entered with a negative sign because it points in the negative y-direction. Hence

$$F_y = (707 \text{ N})(0.866) + (500 \text{ N})(0.707) - (966 \text{ N})$$

and so to 3 significant figures

$$F_y = 0$$

(continued)

(continued)

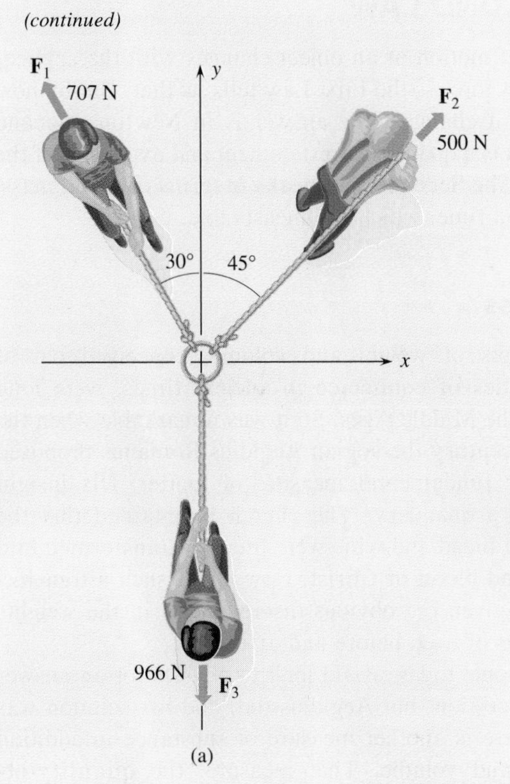

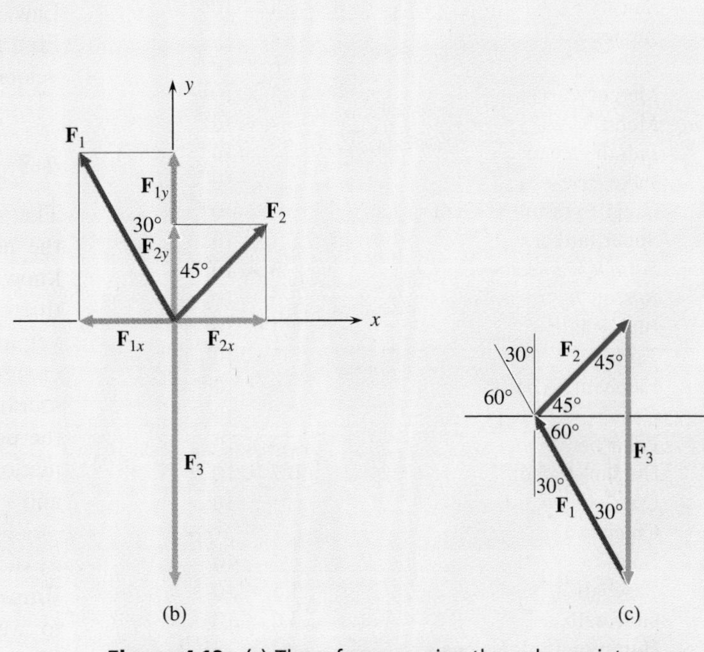

Figure 4.10 (a) Three forces passing through a point are applied in different directions. (b) The forces are resolved in the two perpendicular directions x and y. (c) Because the net force is zero, the three force vectors form a closed triangle.

Similarly, in the x-direction

$$F_x = F_{2x} - F_{1x} = F_2 \sin 45° - F_1 \sin 30°$$

and therefore

$$F_x = (500 \text{ N})(0.707) - (707 \text{ N})(0.500) = 0$$

The people heave and strain, but if the situation began with the ring at rest, it remains at rest because the net force acting on it is zero!

▶ **Quick Check:** If the three vectors actually add up to zero, they must form a closed triangle, as in Fig. 4.10c. The angles inside the triangle do add up to 180°, which is a good sign. To be sure things are right, we can use the Law of Sines:

$$\frac{707}{\sin 45°} = \frac{500}{\sin 30°} = \frac{966}{\sin 105°} = 10^3$$

Everyday ideas about force are often erroneous. That only living creatures can exert force is a commonly held (invalid) belief. Moreover, there are people who wrongly insist that only self-propelled things, living or not, are force makers. Surely a 5000-kg safe resting on your foot exerts a force. The gravitational interaction with the planet pulls the safe downward with a crushing force (its weight), and both Earth and safe are inanimate.

TABLE 4.1 A comparison of masses

Approximate masses	kg
Our galaxy	2×10^{41}
Sun	2×10^{30}
Jupiter	2×10^{27}
Earth	6×10^{24}
Mercury	3×10^{23}
Moon	7×10^{22}
Indian Ocean	10^{20}
Lake Erie	10^{15}
Great Pyramid at Giza	10^{10}
Super tanker	10^{8}
Saturn V rocket	3×10^{6}
Boeing 747	3.5×10^{5}
Blue whale	1.4×10^{5}
Tank	5×10^{4}
Elephant, African	6×10^{3}
Car	10^{3}
Polar bear	3×10^{2}
Human—adult	0.7×10^{2}
Dog	10^{1}
Chicken	3
Book	10^{0}
Baseball	1.5×10^{-1}
Golf ball	4.6×10^{-2}
Hummingbird, Helena's	2×10^{-3}
Paperclip	0.5×10^{-3}
House spider	10^{-4}
Ant	10^{-5}
Giant amoeba	10^{-8}
Pencil mark	10^{-10}
Bacterium	10^{-13} to 10^{-15}
DNA molecule	10^{-20}
Insulin molecule	10^{-23}
Oxygen atom	3×10^{-26}
Proton	2×10^{-27}
Electron	9×10^{-31}

THE SECOND LAW

The state of motion of an object changes with the application of a net force—the First Law tells us that. But by how much does it change? The answer is in Newton's Second Law, which is a quantified restatement and extension of the First Law. The Second Law speaks in terms of force, mass, velocity, and time, which are measurable.

4.3 Mass

The concepts of weight and volume, conceived out of the necessities of commerce in ancient times, were long known by the Middle Ages. So it was remarkable when the thirteenth-century theologian Aegidius Romanus proposed yet another fundamental measure of matter. His insight came from a quandary: The church maintained that the sacramental bread and wine were literally transformed into the body and blood of Christ. How could such a transformation be, given the obvious discrepancies in the weights and volumes of each before and after?

Few people today would look to physics for an answer to such a question, but Aegidius did, and his solution was simple: There is another measure of substance in addition to weight and volume. That measure, the **quantity-of-matter**, indicates *how much matter* there is in a material object. However unlikely, this was the origin of one of the most important ideas in physics, the concept of **mass** (m). Although the innovation of Aegidius had no effect on the course of theology, it did emerge as a useful, though undefined, concept in medieval science. Almost 400 years later, Newton used the terms *mass* and *quantity-of-matter* interchangeably, after providing a rather unsatisfactory definition of his own.

Mass manifests itself in two seemingly distinct ways: gravitationally and inertially. We now know from the General Theory of Relativity that these are equivalent properties of matter, that mass is mass, that there is only one kind of mass. In fact, if inertia actually is the result of the gravitational interaction of any one object with all other objects, this conclusion is quite natural. No matter what mass is conceptually (and science is certainly not finished trying to define it), we can measure it *relative to a standard kilogram*. We need only counterpoise any object against an appropriate selection of standard masses on a simple beam balance. Gravity presumably acts equally on each side, and so once balanced, the masses are equal as well. In effect, this definition of mass is an operational one that scrupulously avoids any conceptual description of what is being measured. The mass of an object is whatever we measure with a beam balance, but that still requires gravity (Table 4.1).

If you were to go off into the distant reaches of space, far from anything else, your weight would all but vanish—without gravity, there is no weight. In outer space, there is no such thing as standing *on* a scale; you would simply float above it while it reads zero. Other than that dramatic weight loss, nothing else about you would change in the journey. You would certainly look the same, even if you were overweight. Moreover, the force it takes to push you is the same whether you are hovering weightlessly in space or standing on slick ice in Brooklyn. That's the crucial point and it reveals a need for a concept, associated with matter, that remains constant—namely, *mass*.

4.4 Linear Momentum

The realization that speed alone fails to give us some essential quality of the motion of an object goes back to the Middle Ages. A Parisian, Jean Buridan, working around 1330, elaborated on an idea that had been around for centuries. When a body is thrown, it continues to move because it supposedly captured some of the original push, or, as he called it, impetus. That explanation was naive, but it led, 300 years later, to the Law of Inertia.

Buridan's most important contribution came from his attempt to understand why an iron ball of the same size as a wooden one would travel much farther when both were launched at the same speed. Or, equivalently, why you would rather be hit by a firefly traveling at 60 km/h than a fire engine at the same speed? Buridan reasoned that the crucial concept was the product of the quantity-of-matter and the speed. The "drive," the "motive power," the "impetus"—all those ideas that convey the sense of something unalterably plowing along—cannot be described by speed alone. The moving body possesses more than speed (the firefly, after all, has the same speed as the fire engine). The essential measure of motion was a new basic quantity, the product of *mass* and *speed:* (mv).

By the time Galileo's work matured, his *impeto* had an inertial quality related to what he called *momento* (weight times velocity). Descartes spoke of the measure of motion as the *quantity-of-motion,* equating it to the product of "matter" and speed. When Sir Isaac Newton set out the central concept of his theory—the measure of motion—he defined the quantity-of-motion, or the **momentum (p)**, as *the product of the object's mass and velocity:*

$$\mathbf{p} = m\mathbf{v} \tag{4.1}$$

and because velocity is a vector quantity, so too is momentum. The direction of **p** is the direction of **v**; the magnitude of **p** is mv with SI units of kg·m/s. When **p** is constant, there is straight-line motion, and so it's more precisely called *linear momentum.*

A critical feature of momentum is that oppositely directed momenta can cancel, which is a key quality of vectors. If a train travels with a velocity of 20 km/h–east, and you jog along on its flatcars with a velocity of 20 km/h–west, you can run right off the end, motionless with respect to the ground. You will drop *straight down* as if you had stepped off a chair in your living room. Your horizontal velocity with respect to the Earth up there in midair will be zero, and your momentum will be zero as well. The momentum imparted in the eastward direction by the train will be canceled by the westward momentum gained in the running. Like velocity, linear momentum is relative. The momentum of an object moving at **v** with respect to some observer is $m\mathbf{v}$ with respect to that observer. The momentum of a pilot flying at 800 km/h is zero with respect to the plane.

The modern physicist may rightfully be proud of his spectacular achievements in science and technology. However, he should always be aware that the foundations of his imposing edifice, the basic notions of his discipline, such as the concept of mass, are entangled with serious uncertainties and perplexing difficulties that have as yet not been resolved.

MAX JAMMER (1961)

Example 4.4 Rich Gossage set a fastball record by hurling a 0.14-kg baseball at a speed of 46.3 m/s (that is, 153 ft/s). What was the magnitude of the ball's momentum as it left his hand? Here everything is measured with respect to the Earth.

Solution: [Given: $m = 0.14$ kg and $v_B = 46.3$ m/s. Find: p_B.] The momentum is provided by the scalar form of Eq. (4.1):

$$p_B = m_B v_B = (0.14 \text{ kg})(46.3 \text{ m/s}) = \boxed{6.5 \text{ kg·m/s}}$$

▶ **Quick Check:** A baseball thrown at a casual 7.2 m/s has a momentum of (0.14 kg)(7.2 m/s) = 1 kg·m/s, so the above momentum at a speed ≈6.4 times faster is correct.

4.5 Impulse and Momentum Change

> The change of motion is ever proportional to the motive force impressed; and is made in the direction of the right [straight] line in which that force is impressed.

This translation from the Latin is the Second Law as Newton stated it. It is not the version you see in textbooks because the language is archaic. In spite of that, it is worth the effort to appreciate Newton's thinking because it's much more reasonable than most modern versions. We know from the First Law that force is the changer of motion—if there is no applied force, there is no change in straight-line uniform motion.

When a force is applied to a body, there will be a resulting proportional change in its motion. Moreover, the measure of that motion is the quantity-of-motion, the momentum. In fact, when Newton spoke about "the motion," he meant the momentum. Thus, in the definition of quantity-of-motion, he wrote, ". . . in a body double in quantity [that is, mass], with equal velocity, the motion is double." In other words, if you double m, keeping v constant, then p doubles. The Second Law therefore states that an applied force (F) produces a proportional change in the momentum of the body (Δp): $F \propto \Delta p$. Since momentum is relative, the significant quantity is not p, which depends on the motion of the observer, but Δp, which does not.

Example 4.5 A 0.149-kg baseball traveling at 28 m/s due south approaches a waiting batter. The ball is hit and momentarily crushed to almost half its original size in a collision with the bat lasting a few milliseconds. The ball springs back, sailing away at 46 m/s due north. Determine the magnitudes of its initial and final momenta and the change in its momentum.

Solution: [Given: $m_B = 0.149$ kg, $\mathbf{v}_i = 28$ m/s–south, $\mathbf{v}_f = 46$ m/s–north. Find: p_i, p_f, and Δp.]

$$p_i = m_B v_i = (0.149 \text{ kg})(28 \text{ m/s}) = \boxed{4.2 \text{ kg·m/s}}$$

Similarly

$$p_f = m_B v_f = (0.149 \text{ kg})(46 \text{ m/s}) = \boxed{6.9 \text{ kg·m/s}}$$

Since

$$\Delta \mathbf{p} = \mathbf{p}_f - \mathbf{p}_i$$

we can either visualize the subtraction vectorially, reversing $\mathbf{p}_i$ and adding it tip-to-tail to $\mathbf{p}_f$, or let north be positive and subtract scalar values, whereupon

$$\Delta p = (+6.9 \text{ kg·m/s}) - (-4.2 \text{ kg·m/s})$$

and

$$\Delta p = +11 \text{ kg·m/s}$$

Either way, $\boxed{\Delta \mathbf{p} = 11 \text{ kg·m/s–north}}$.

▶ **Quick Check:** $p_i/p_f = 28/46 = 4.2/6.9 = 0.61$, so the numbers look right. The ball has had its direction reversed, as if a momentum of 11 kg·m/s–north was added to its original 4.2 kg·m/s–south by the impact. This kind of large change always occurs when the motion of an object is turned around in a collision.

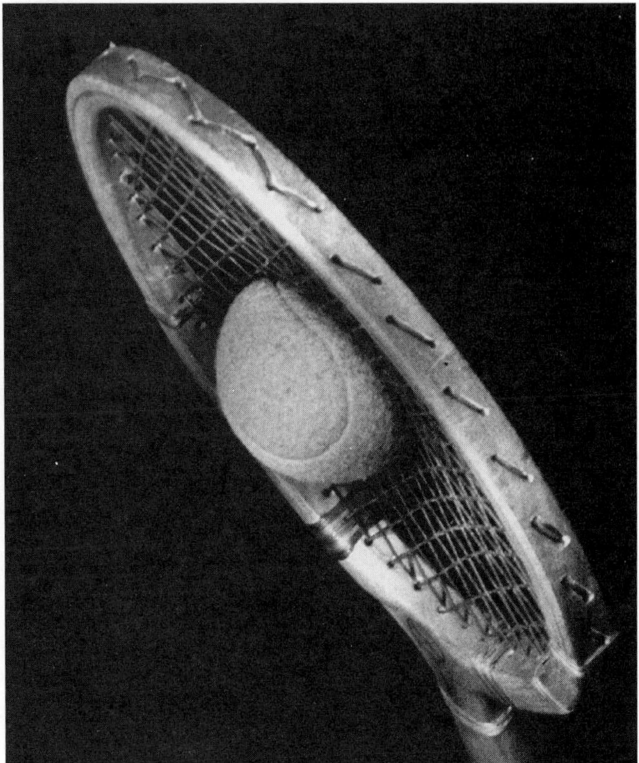

The force of the impact crushes a tennis ball. The racket and ball are in contact for a duration known as the *collision time.*

We know from experience that time is involved here, too. If several people push on a stalled car, the longer they push, the more swiftly it will roll away, and the greater will be Δp. Interestingly, there is no explicit statement about time in the original version of the Second Law. But it is clear from Newton's discussion that he was speaking about the change in motion *per unit time.* Double the force exerted on a body and the resulting change in momentum will double (assuming the forces act for the same time); triple it, and the change in momentum will triple. *The force applied equals the corresponding change in momentum per unit time.* Thus, if a certain change in motion (Δp) is to occur quickly, the applied force must be large. (Have twice as many people push, and the car will get up to speed in half the time.) To get a desired momentum change, the force applied and the interval of time over which it must be applied (Δt) vary inversely: $F \propto 1/\Delta t$. Bring to bear a large force and that force need be applied for only a short time.

If Δt is substantial, it is certainly possible that the force might vary during that interval (someone pushing behind the car could get tired). So, to be precise, we should either let Δt approach zero and work with the instantaneous force or let Δt be appreciable and use the average force over the interval (as we used v_{av} and a_{av} earlier). For the moment, we will do the latter, which will be sufficient for many practical situations, including those where the force is constant and therefore always equal to F_{av}. Accordingly, a modern version of the Second Law reads:

> *The average force exerted on a body equals the resulting change in momentum divided by the time elapsed in the process.*

Newton was quite aware of the directional nature of the phenomenon—the change in motion is in the direction of the force impressed, which is something we can deal with rather nicely using vectors. Hence, $\Delta\mathbf{p}$ is parallel to $\mathbf{F}_{av}$, and putting it all together symbolically, we have

$$\mathbf{F}_{av} = \frac{\Delta\mathbf{p}}{\Delta t} = \frac{\Delta(m\mathbf{v})}{\Delta t} \tag{4.2}$$

The **average net force equals the time-rate-of-change of momentum**.

Example 4.6 A rocket fires its engine, exerting an average force of 1000 N for 40 s. By how much will its momentum change?

Solution: [Given: $F_{av} = 1000$ N and $\Delta t = 40$ s. Find: Δp.] The momentum change occurs in the direction of the force. Hence, we can use the scalar form of Eq. (4.2), namely

$$F_{av} = \frac{\Delta p}{\Delta t}$$

Thus $$\Delta p = F_{av}\Delta t = (1000\text{ N})(40\text{ s})$$

and so $$\boxed{\Delta p = 40 \times 10^3\text{ N·s}}$$

▶ **Quick Check:** 1 kN × 40 s = 40 kN·s.

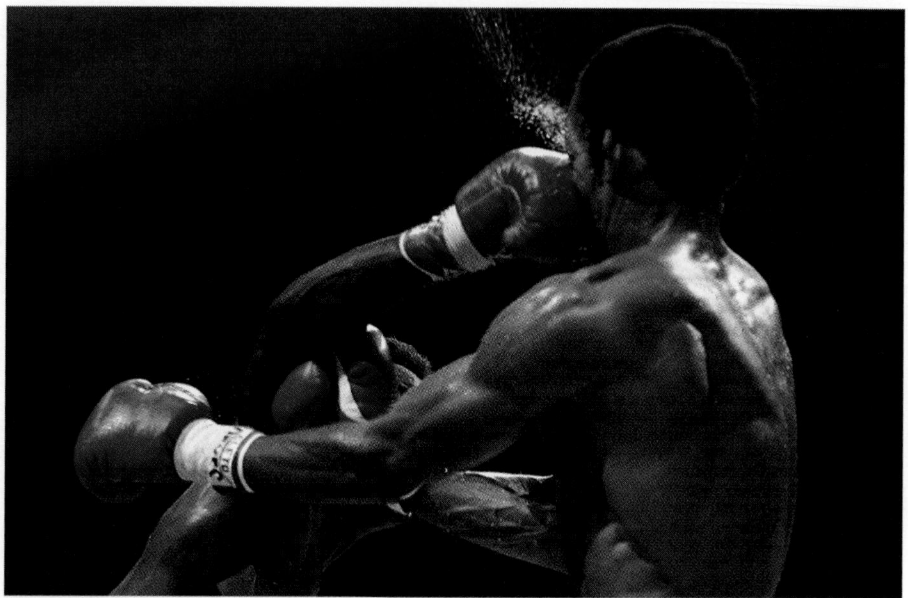

The impulse exerted by the punch changes the momentum of the recipient's head. The brain, which is surrounded by a liquid, slams into the skull. The gloves are heavily padded to increase Δt thereby decreasing F_{av}, but it often doesn't help.

Equation (4.2) suggests a dynamic means of quantifying force since we can measure m, v, and t. Consequently, we now define 1 *newton* (N) to be the amount of force that will produce a change in the momentum of *any* object equal to 1 kg·m/s in 1 s. That is

$$1 \text{ N} = 1 \text{ kg·m/s}^2$$

Newton's early studies had been done around the same time as important work by Hooke, Wren, Wallis, and Huygens was accomplished on collisions. Billiards was very popular at the time, and there was a lot of interest in momentary impacts. In such cases, neither $\mathbf{F}_{av}$ nor Δt is known very well, but the change in momentum is easily determined. Consequently, it is useful to combine the force and time into a single notion that equals the known momentum change. That single concept ($\mathbf{F}_{av}\,\Delta t$) is called the **impulse** of the force, and an alternative statement of the Second Law becomes

$$\text{Impulse} = \text{change in momentum}$$

$$\mathbf{F}_{av}\,\Delta t = \Delta \mathbf{p} \tag{4.3}$$

A given impulse will produce a specific change in momentum no matter what the mass or speed of the recipient body. An object originally at rest will take off in the direction of the net applied force, acquiring a momentum $\Delta(m\mathbf{v}) = \mathbf{F}_{av}\,\Delta t$, which is what happens when you throw a dart, discharge a syringe, or strike a golf ball. As long as the club is in contact with the ball, applying force, there is an ongoing gain in momentum in the direction of the force (Table 4.2). Once the ball leaves the club, it flies off under the First Law, forceless and changeless. The longer the barrel of a gun, or the greater the pitcher's throwing motion—the larger will be the time (Δt) during which the propelling force acts and the more the change in momentum (Δp) of the projectile (Fig. 4.11).

Force applied to a body already in motion may either increase or decrease its momentum depending on whether $\mathbf{F}_{av}$ acts parallel or antiparallel to the initial ve-

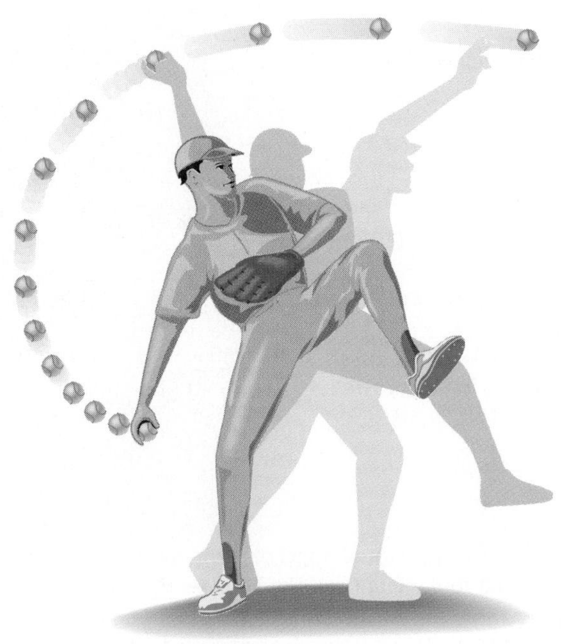

Figure 4.11 The pitcher exerts a force on the ball over as long a distance and for as long a time as possible. The greater is $F\Delta t$, the larger is Δp, and the higher the release speed. Here the ball is shown every 1/100th of a second. The greater the distance between images of the ball, the greater the speed.

TABLE 4.2 Typical parameters for balls hit from rest by the appropriate object

Ball	Mass (kg)	Speed imparted (m/s)	Impact Time (ms)
Baseball	0.149	39	1.25
Football (punt)	0.415	28	8
Golf ball (drive)	0.047	69	1
Handball (serve)	0.061	23	12.5
Soccer ball (kick)	0.425	26	8
Tennis ball (serve)	0.058	51	4

locity ($\mathbf{v}_f = \mathbf{v}_i + \Delta\mathbf{v}$). To slow the *Lunar Excursion Module* as it plunged toward the Moon's surface in 1969, a downward-pointing retrorocket was fired that exerted an upward force on the craft, thereby decreasing its downward momentum and speed.

Example 4.7 On September 12, 1966, a *Gemini* spacecraft piloted by astronauts Conrad and Gorden met and docked with an orbiting *Agena* launch vehicle. With plenty of fuel left in the spacecraft, NASA decided to determine the mass of the *Agena*. While coupled, *Gemini*'s motor was fired, exerting a constant thrust of 890 N for 7.0 s. As a result of that little nudge, the *Gemini-Agena* sped up by 0.93 m/s. Assuming *Gemini*'s mass as a constant 34×10^2 kg, compute the mass of the *Agena*.

Solution: [Given: $F_{av} = 890$ N, $\Delta t = 7.0$ s, $\Delta v = +0.93$ m/s, and $m_G = 34 \times 10^2$ kg. Find: m_A.] This calculation will be inexact to the degree that both the

(continued)

(continued)

thrust and m_G are assumed constant. The basic relationship is

$$\mathbf{F}_{av}\,\Delta t = \Delta\mathbf{p} \qquad [4.3]$$

Because m is constant

$$\Delta\mathbf{p} = \Delta(m\mathbf{v}) = m\,\Delta\mathbf{v}$$

where $m = m_A + m_G$

Since the direction of motion is of no concern, we can use a scalar form

$$F_{av}\,\Delta t = (m_A + m_G)\Delta v$$

The rocket's impulse produces a well-defined *change* in speed, regardless of the initial speed. Substituting in the numbers yields

$$(890\text{ N})(7.0\text{ s}) = (34 \times 10^2\text{ kg} + m_A)(0.93\text{ m/s})$$

Solving for m_A

$$m_A = \frac{(890\text{ N})(7.0\text{ s})}{(0.93\text{ m/s})} - (34 \times 10^2\text{ kg})$$

and $m_A = 3299$ kg, or to two figures, $\boxed{33 \times 10^2\text{ kg}}$.

▶ **Quick Check:** Assume the answer is correct and calculate the force. $F_{av} = (67 \times 10^2\text{ kg})(0.93\text{ m/s})/(7.0\text{ s}) = 890$ N, as given.

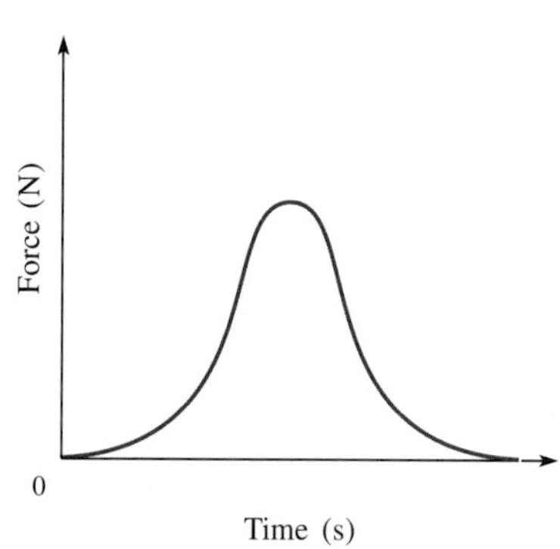

Figure 4.12 The force-time graph for a body undergoing an impact. The force rises, peaks, and falls to zero.

It follows from Eq. (4.3) that for a given change in momentum (either an increase or a decrease), the longer the force acts (Δt), the smaller $\mathbf{F}_{av}$ will have to be. Your hands provide the stopping force for an incoming ball when you catch it; that is, when you change $\mathbf{p}_i$ to $\mathbf{p}_f = 0$. If the ball is soft and deforms as it comes to rest, the stopping time will be large and the stopping force will be comparatively small. Even with the same incoming momentum, so that $\Delta\mathbf{p}$ is the same when the balls are brought to rest, a soft ball will take longer to stop and so require less force to catch than a hard ball. Furthermore, if you wear a padded mitt (which will extend Δt) and draw the glove back toward you just as the ball strikes it, the stopping force will be still smaller. Automobile bumpers that compress on impact extend the stopping time (Δt) and thereby decrease the force of collision even though the car undergoes the same change in momentum.

4.6 Varying Force

Force may change from moment to moment (the force on your feet as you walk around certainly does). To obtain a complete picture, we must envision a concept that varies in time. At each and every moment, there is a value of the **instantaneous force**; that is, *the limit approached by the average force as the time interval (over which the averaging is taking place) shrinks to zero*:

$$\mathbf{F} = \lim_{\Delta t \to 0}\left[\frac{\Delta\mathbf{p}}{\Delta t}\right] \qquad (4.4)$$

Push on something for a moment. The force exerted initially builds, the flesh on your hand compresses and distorts, the force peaks, and then decreases to zero. That's what happens when you shove a book or hurl a shot put. Similarly, a quick blow with a hammer or the blast of an explosive charge will usually produce a force that even more rapidly rises, peaks, and falls off to zero.

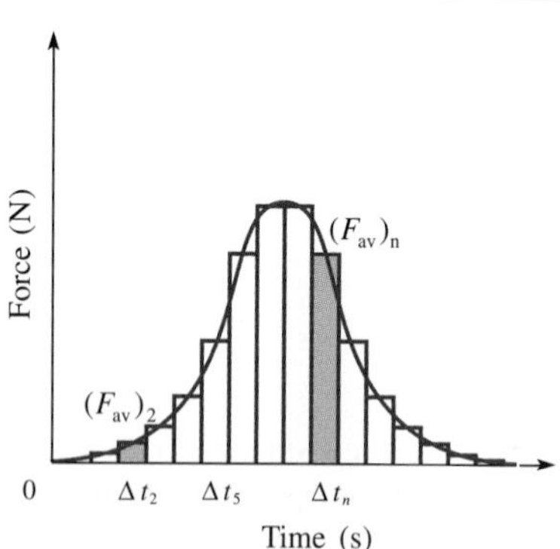

Figure 4.13 The area under the force-time graph is the net impulse exerted, and that equals the net change in momentum imparted to the body in question.

Figure 4.12 is a force-time graph for an impact of some sort, a plot of the instantaneous force versus time. The width of the curve and the details of its shape will vary from one situation to another, but the area under the curve is the net impulse imparted to the body. To quantify the impulse, divide the region under the curve into a number of subintervals, each with a little rectangular area whose base is an interval of time: Δt_1, Δt_2, Δt_3, and so forth (Fig. 4.13). Next, fix the height of each rectangle at a value equal to the average force exerted over that particular time interval. Thus, for example, the area of the second rectangle is $(F_{av})_2\,\Delta t_2$, which corresponds to the *impulse* associated with that subinterval. The sum of all the little subareas is exactly equal to the total impulse exerted by the force.

If the number of subintervals is made increasingly large while the width of each approaches zero ($\Delta t \rightarrow 0$), the tops of the rectangles will blend ever more smoothly into the curve. In other words, the value of the average force for each subinterval will approach the value of the instantaneous force and these, in turn, are the corresponding points on the curve. Just as the area under any portion of the speed-time curve is the distance traveled during a given period, **the area under the force-time curve is the impulse exerted during that interval and it therefore equals the resulting change in momentum**.

Example 4.8 A golfer's club hits a 47.0-g golf ball from rest to a speed of 60.0 m/s in a collision lasting 1.00 ms. The force on the ball rises to a peak value of F_{max} and then drops to zero as it leaves the club. Compute a rough value for this maximum force by approximating the force-time curve, with a triangle of altitude F_{max}.

Solution: [Given: $m = 47.0 \times 10^{-3}$ kg, $v_i = 0$, $v_f = 60.0$ m/s, and $\Delta t = 1.00 \times 10^{-3}$ s. Find: F_{max}.] First we will find the change in momentum, and that equals the net impulse. The net impulse, in turn, equals the area under the force-time curve, which can be approximated by a triangle with an altitude of F_{max}. Thus

$$F_{av}\,\Delta t = \Delta(mv) = (47.0 \times 10^{-3}\text{ kg})(60.0\text{ m/s})$$

and the impulse is 2.82 kg·m/s, or 2.82 N·s. Representing the force-time curve by a triangle of altitude F_{max} and base Δt, the area encompassed is $\frac{1}{2}\Delta t(F_{max})$, which equals the impulse; that is

$$2.82\text{ N·s} = \tfrac{1}{2}(1.00 \times 10^{-3}\text{ s})F_{max}$$

Hence, $\boxed{F_{max} = 5.64\text{ kN}}$ or 1.27×10^3 lb.

▶ **Quick Check:** Whenever the force-time curve is approximated by a triangle, $F_{av}\,\Delta t = \frac{1}{2}(\Delta t)F_{max}$ and the peak force is twice the average force. As a consequence, F_{max} may break bones, even though F_{av} is not large enough to do damage. $F_{av} = (2.82\text{ N·s})/(1.00\text{ ms})$ and $2F_{av} = 5.6 \times 10^3\text{ N} = F_{max}$.

A real force-time curve can have a complex shape (Fig. 4.14). A spaceship with a variable-thrust engine can generate a force-time curve with many bumps and wiggles, but the area encompassed in a given time interval will always be equal to the resulting change in momentum. If the ship turns 180° around, the same engine can fire in the opposite direction, against the motion. That would generate a negative impulse, a negative change in momentum, and a decrease in speed. The total impulse is, then, the area of the force-time curve above the axis minus the area below the axis. If those two portions happen to be equal, the net change in momentum will be zero. When you throw a punch, your fist starts with $p_i = 0$ and ends with $p_f = 0$. It accelerates up to a maximum speed and then decelerates to zero when your arm is extended. Given this fact, a karate blow is always aimed at a point inside the target so that it makes contact when p is maximum.

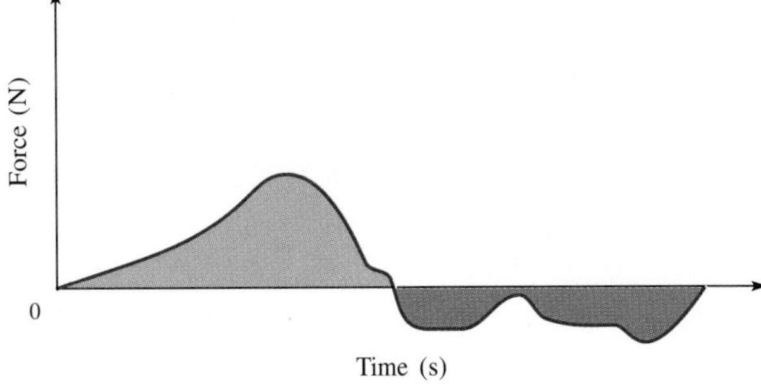

Figure 4.14 Instantaneous force versus time. The total impulse is the net area under the curve; take the area above the axis as positive and that below as negative. Oppositely directed forces produce impulses with opposite signs and adding the areas above and below the axis takes this opposition into account.

When a car crashes into a brick wall, its front end deforms as it slows to a crushing stop. Today's average production vehicle compresses roughly 1 inch for every mile per hour of speed just prior to impact; smaller cars typically collapse a

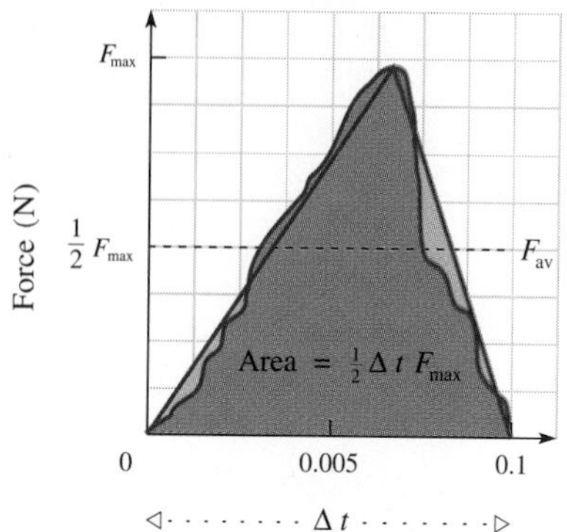

Figure 4.15 A triangular approximation to the area under the force-time curve.

little less than this, larger cars a little more. The process is similar to a struck baseball, except the car does not spring back to its original shape.

If we assume that the collision-deceleration is fairly uniform, the crush-distance s_c divided by $v_{av} = \frac{1}{2}(v_i + v_f)$, with $v_f = 0$, is the impact time. Thus

$$\Delta t = \frac{s_c}{v_{av}} = \frac{2s_c}{v_i}$$

Since tests show that s_c is proportional to v_i, we conclude that the impact time for cars of comparable stiffness should be independent of the speed of impact. A typical head-on, brick-wall collision lasts around 100 milliseconds. By the way, a driver would not be much better off hitting an oncoming car. In fact, crashing head-on into an identical car traveling toward you at the same speed is effectively the same as hitting a stationary brick wall. To appreciate that, just imagine a large movable sheet of steel mounted on wheels. Now, if both cars plow into this steel sheet on opposite sides, the autos will be quite destroyed, but the sheet will remain in the middle of the melee unmoved, as if it were unmovable, just like the brick wall.

We can expect the force-time curve for a car colliding with a stationary barrier to have a base width of about 100 ms. An 1800-kg car impacting at 26.8 m/s (60 mi/h) will be crushed about 1.5 m (60 in.). The engine will end up in the passenger compartment as the remainder of the car moves forward, past it. The momentum change of the car is $\Delta p = 48.2 \times 10^3$ kg·m/s, which equals the area under the force-time curve in Fig. 4.15. If we again approximate that curve by a triangle, the peak force exerted on the car is approximately 0.96×10^6 N, or over 100 tons.

Example 4.9 A 70-kg passenger riding in a typical automobile is involved in a 17.9 m/s (40 mi/h) head-on collision with a concrete barrier. Taking the stopping time as 100 ms, compute the average force exerted by the seat belt and shoulder strap on the person.

Solution: [Given: $m_p = 70$ kg; taking the initial direction of motion as positive, $\Delta v = -17.9$ m/s, $\Delta t = 0.100$ s. Find: F_{av}.] This collision problem involves mass, speed, time, and force, and suggests impulse-momentum, that is

$$F_{av}\,\Delta t = m_p \Delta v$$

$$F_{av} = \frac{(70 \text{ kg})(-17.9 \text{ m/s})}{0.100 \text{ s}}$$

Hence, $\boxed{F_{av} = -1.3 \times 10^4 \text{ N}}$ or -2.8×10^3 lb. The minus sign means the force is in the opposite direction to the initial motion.

▶ **Quick Check:** Assume the force is correct and calculate the time; $\Delta t = (70 \text{ kg})(-17.9 \text{ m/s})/(-1.3 \times 10^4 \text{ N}) = 0.096 \text{ s} = 0.10$ s.

INTERACTION

Newton's Third Law completes the logical picture of the concept of force. A change of motion arises out of a net impressed force. Whenever we talked about a force applied to a body, it was an **external force**, one whose source was external to the body. Newton defined force as any action that causes an object to change its momentum. But there is some circular logic here. By definition, whenever a body's momentum is altered, the cause must be a force acting upon it. Something is still missing. What ties the concept of force back to the observable Universe? What links the cause of the force to the effect, the change of momentum?

An isolated body obeys the Law of Inertia. It cannot alter its own uniform straight-line motion because that requires some outside intervention via one or more external entities. We can assume that when two bodies interact, *both* will be affected by the encounter; the motions of both will be altered, both will experience an external force. Two identical bodies moving toward each other at the same speed must experience the same alteration of motion on impact, and therefore must exert the same force on each other.

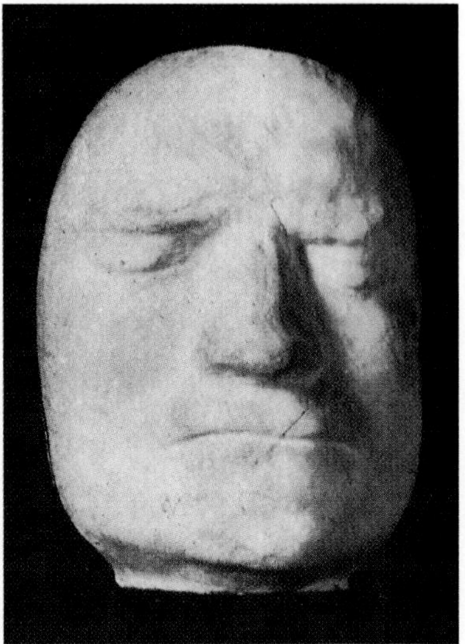

Isaac Newton's death mask (a plaster casting of his face), made soon after he died at about 1:00 A.M. on March 20, 1727.

Newton had other things to draw insight from, in addition to billiards. Experiments on gases, for example, fashionable when he was a young man, demonstrated that when a gas is compressed within a container by a piston pushing down on it, there is a backward reaction—the piston tends to pop back up. The "spring of the air" was a common descriptive phrase coming out of those studies and, indeed, a spring was a perfect image to use—push on it and it pushes back on you; pull it in one direction and it pulls you in the opposite direction.

Newton provided the final link in the credo of the Three Laws. In his own words:

> To every action there is always opposed an equal reaction: or the mutual actions of two bodies upon each other are always equal, and directed to contrary parts.

The interaction of two bodies always occurs via a force and an equal-magnitude oppositely directed counterforce. There is no such thing as a single force exerted by some active entity on a passive one; there is only **interaction**. A hand cannot simply push on a chair—the chair must always push back on the hand. Entities interact, one upon the other: *force is a thing of pairs.* Whenever we see some object deviate from the Law of Inertia, we must be able to find another interacting with it.

> *An object offers as much resistance to the air as the air does to the object. You may see that the beating of an eagle's wings against the air supports its heavy body in the highest and rarest atmosphere . . . .*
>
> LEONARDO DA VINCI (1452–1519)
> *Codex Atlanticus*

Consider, for example, a balloon being pressed into a wall. You could argue that the balloon is squashed by the pusher alone, without any help from the wall; that the person's hand is doing the pushing, not the wall. Yet the balloon is flattened on both sides just as it would be if the wall itself had moved to where the balloon was being held fixed in hand and proceeded to squash it. If you came along at precisely that

These firefighters are struggling to control a blasting hose. The water is propelled forward with great force and it reacts back with an equal-magnitude force.

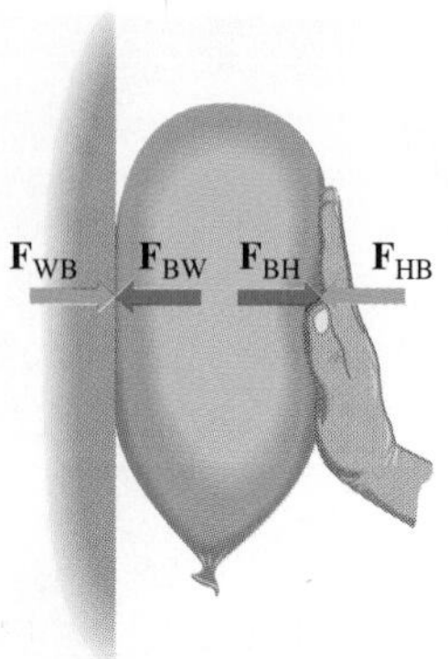

Figure 4.16 A balloon pressed against a wall. $\mathbf{F}_{HB}$ and $\mathbf{F}_{BH}$ are an interaction pair, as are $\mathbf{F}_{WB}$ and $\mathbf{F}_{BW}$. All forces are the result of interactions and all come in pairs comprising equal and opposite influences.

moment and saw a wall, a flattened balloon, and a hand, you could not possibly tell whether the wall or the hand had moved. The hand pushes on the balloon (Fig. 4.16) with a force $\mathbf{F}_{HB}$, and the balloon pushes back with an equal and opposite force $\mathbf{F}_{BH}$. The balloon, acting as if it were an extension of the hand, transmits the force that is applied to it to the wall ($\mathbf{F}_{BW} = \mathbf{F}_{HB}$). The wall, in turn, reacts with an equal-magnitude, oppositely directed counterforce $\mathbf{F}_{WB}$. The balloon is squashed in on both sides by the forces *applied to it,* $\mathbf{F}_{WB}$ and $\mathbf{F}_{HB}$.

Interaction between entities means *two equal-magnitude, oppositely directed forces (one on each participant) forming the pair.* Here, each interaction pair comprises two forces with reverse-ordered subscripts. **The forces of such a pair are always in opposite directions and never act on the same body**. Thus, $\mathbf{F}_{HB}$ and $\mathbf{F}_{BH}$ are an interaction pair; $\mathbf{F}_{HB}$ is exerted on the balloon while $\mathbf{F}_{BH}$ is exerted on the hand. Two external horizontal forces are acting on the balloon, $\mathbf{F}_{WB}$ and $\mathbf{F}_{HB}$. These are not an interaction pair even though they happen to be equal in magnitude and opposite in direction. Hence, as far as the motion of the balloon as a whole is concerned, the two external horizontal forces cancel, yielding a net applied force of zero. Momentum remains unchanged—the balloon stays put at rest, squashed up against the wall.

A chisel struck by a mallet transmits a force to a piece of wood, just as the balloon transmitted a force from the hand to the wall. One layer of atoms pushes (electromagnetically) on the next layer down to the end of the chisel. During such a process some distortion occurs, even in a hard steel chisel. By contrast, the distortion of the balloon was obvious. In the real world, a perfectly rigid body does not exist. Even a flea deforms the rock it lands on (granted, it's only by a miniscule amount, but to some extent the rock *is* squashed). The shifting of atoms within the rock and the subsequent change in their electrical interaction supports the flea.

Example 4.10 One tough killer bee has a mass of 0.10 g. Traveling due south at 5.0 m/s, it crashes into the windshield of a 10-ton truck heading due north at 60 km/h. The bee bounces off the window at 35 m/s–north (with respect to the ground) after an impact lasting 0.090 s. (a) Compute the average force exerted by the bee on the truck. (b) What was the change in momentum of the truck as a result of the collision? (c) The bee started out moving south and ended up moving north, which means it must have come to rest, with respect to the Earth, during the collision. Does that mean the truck, with which it was in contact, also stopped for that instant?

Solution: [Given: $m_b = 0.10$ g, $v_{bi} = -5.0$ m/s, $v_{bf} = +35$ m/s, the weight of the truck is 10 tons or 20 000 lb, $v_T = +60$ km/h, and $t = 0.090$ s. Find: (a) the average force exerted by the bee on the truck. Because there are so many subscripts, let's use pointed brackets to indicate average value and write it as $\langle\mathbf{F}_{bT}\rangle$; and (b) $\Delta\mathbf{p}_T$.] Take north as the positive direction. (a) We know the change in velocity of the bee, and so can compute its change in momentum and the corresponding force exerted on it. The force it exerts on the truck is just equal and opposite to that. The truck exerts a force on the bee $\langle\mathbf{F}_{Tb}\rangle$, which alters the bee's momentum such that

$$\langle\mathbf{F}_{Tb}\rangle\Delta t = \Delta\mathbf{p}_b = m_b\,\Delta\mathbf{v}_b$$

Here, $\Delta v_b = (+35 \text{ m/s}) - (-5.0 \text{ m/s}) = +40$ m/s. The fact that the bee was turned around adds to the change in velocity and momentum. The bee experiences a force pointing north that first brings it to rest and then reverses its direction. Hence

$$\langle F_{Tb}\rangle\Delta t = m_b(40 \text{ m/s})$$

and

$$\langle F_{Tb}\rangle = \frac{(0.10 \times 10^{-3}\text{ kg})(40 \text{ m/s})}{0.090 \text{ s}} = 0.044 \text{ N}$$

The bee exerts an equal and opposite force

$$\boxed{\langle\mathbf{F}_{bT}\rangle = 4.4 \times 10^{-2}\text{ N–south}}$$

(continued)

(continued)

(b) The impulse acting on the bee was the same in magnitude and opposite in direction to the impulse acting on the truck, and so $\Delta\mathbf{p}_T = -\Delta\mathbf{p}_b$. Accordingly, $\Delta p_b = m_b \Delta v_b$, hence

$$\Delta p_b = (0.10 \times 10^{-3}\ \text{kg})(40\ \text{m/s})$$

and $$\Delta p_b = 4.0 \times 10^{-3}\ \text{kg·m/s}$$

The bee's initial momentum points south; its momentum *change* is positive and points north ($\mathbf{p}_{bi} + \Delta\mathbf{p}_b = \mathbf{p}_{bf}$). Hence, the truck's momentum change is

$$\boxed{4.0 \times 10^{-3}\ \text{kg·m/s–south}}$$

(c) The bee crashed into the window moving southward at a relative speed of 78 km/h. The window deformed ever so slightly, slowing the bee and bringing it to rest for an instant with respect to the ground. At that moment, the bee was moving southward at 60 km/h with respect to the truck. The bit of glass in contact with the bee did stop for an instant with respect to the ground, but the truck kept right on rolling.

▶ **Quick Check:** As seen by the driver, the bee comes in with a speed of (5.0 m/s + 16.66 m/s) and bounces off with a speed of (35 m/s − 16.66 m/s), giving a net momentum change of $[(+21.66\ \text{m/s}) - (-18.33\ \text{m/s})] \times (0.10 \times 10^{-3}\ \text{kg}) = 4.0 \times 10^{-3}\ \text{kg·m/s}$ and, dividing by 0.090 s, $\langle F_{Tb}\rangle = 0.044$ N.

An X-ray photo of a 45-caliber bullet fired from a pistol. In effect, the gun pushes on the bullet and the bullet pushes back on the gun. Some of that reaction goes into operating the slide exposing the barrel and ejecting the spent shell.

4.7 Rockets

Imagine yourself on roller skates holding a bag of oranges. Now throw one of them due north, and away you go due south. You are thrust backward just as a gun recoils backward when it fires a projectile. You push on the orange in the forward direction during the throw, and it pushes back with the same impulse on you, and back you go. Hurl several oranges and you have an orange-rocket, because that's exactly the way a rocket works. Rather than throwing a few large objects slowly, a rocket engine hurls out a tremendous number of tiny high-speed objects—molecules. For example, during launch, the solid-fuel boosters for the Space Shuttle expel 8.5 tons of fiery exhaust each second. The engines blast exhaust downward and the escaping gas, in turn, pushes up on the rocket (interaction). That's why rockets can be used in space. *Rockets are not propelled by pushing against either the ground or the atmosphere,*

A jet plane sucks in air and blasts out a high-speed exhaust that drives the craft forward. The engines push the exhaust backward, and the exhaust pushes the engines forward.

On flying rockets in space. *That Professor Goddard . . . does not know the relation of action to reaction, and of the need to have something better than a vacuum against which to react—to say that would be absurd. Of course, he only seems to lack the knowledge ladled out daily in high schools.*

(The *New York Times,* Jan. 13, 1920)

Further investigation and experimentation have confirmed the findings of Isaac Newton in the 17th century, and it is now definitely established that a rocket can function in a vacuum as well as in an atmosphere. The Times *regrets the error.*

(The *New York Times,* July 17, 1969)

Octopuses and squid can accelerate rapidly by squirting a stream of water backward, just like a jet plane. The animal pushes the water stream, the stream pushes back on the animal.

though that erroneous opinion is widely held even today. In fact, a *New York Times'* editorial of 1920 advised Robert H. Goddard, the American who launched the first liquid-fuel rocket, to give up any thoughts of space travel. After all, even a schoolboy knows that rockets obviously cannot fly in space because a vacuum is devoid of anything to push on.

Example 4.11 A small rocket engine expels 10.0 kg of exhaust gas per second. If these molecules are ejected at an average speed of 600 m/s, what is the thrust of the engine?

Solution: [Given: for the exhaust gas, $v_i = 0$, $v_f = 600$ m/s, and 10.0 kg are ejected per 1.00 s. Find: the thrust, F_{av}, exerted by the motor.] We know that each second, 10.0 kg of exhaust gas experience an increase in speed of $\Delta v = 600$ m/s. Hence, there's an average force exerted on the gas given by

$$F_{av} = \frac{m\Delta v}{\Delta t} = \frac{(10.0 \text{ kg})(600 \text{ m/s})}{1.00 \text{ s}}$$

and $$F_{av} = 6.00 \times 10^3 \text{ kg·m/s}$$

The engine pushes the gas backward with a force of $\boxed{F_{av} = 6.0 \times 10^3 \text{ N}}$; that is, 1.3×10^3 lb, and the gas exerts an equal reaction force forward on the rocket.

▶ **Quick Check:** Assume the force is correct and calculate the mass: $m = (F_{av})\,\Delta t/\Delta v = 10$ kg.

4.8 Conservation of Linear Momentum

One of the great guiding ideas of physics is the law of Conservation of Linear Momentum: **When the resultant of all the external forces acting on a system is zero, the linear momentum of the system remains constant**. Descartes, long before Newton had refined dynamics, wrote of his Creator of the Universe:

> He set in motion in many different ways the parts of matter when He created them, and since He maintained them with the same behavior and with the same laws as He laid upon them in their creation, He conserves continually in this matter an equal quantity-of-motion.

In other words, the total momentum of the Universe persists unchanged and will continue to be preserved forever.

In the process of throwing a bouquet or catching a cigar, a force is applied by the hand to the projectile. Simultaneously, there is an equal and opposite force back on the hand. The *impulse* exerted on the cigar, bringing it to rest, is matched by an equal and oppositely directed impulse on the hand that catches it. The interaction times are the same, and the forces are equal in magnitude and opposite in direction. Because interacting objects exert such forces on each other, the resulting momentum changes must be equal and opposite.

A particle or a group of particles constitutes a *system*. Whenever forces act to change the location of material objects, it's wise to have a clear idea of the extent of the system being dealt with. That's true now when we are concerned with changes in momentum resulting from the application of force, and it will be true later when we deal with changes in energy. Forces whose sources are within the system are said to be **internal forces** (Fig. 4.17a). Hence, *the total momentum of a system of interacting masses must remain unaltered, provided that no net external force is applied.* Just try pushing a car off the road while sitting inside it (p. 128).

The momenta of individual interacting members of a system may certainly change, but each change is accompanied by an equal and opposite change in momentum of the interaction partner. If the whole of our Universe is taken as the system, there are no *external forces* whose sources lie outside the system, and the total momentum must be conserved forever (Fig. 4.17b). If a system is not isolated, and external forces do act on it (Fig. 4.17c), its momentum will change, say, by an

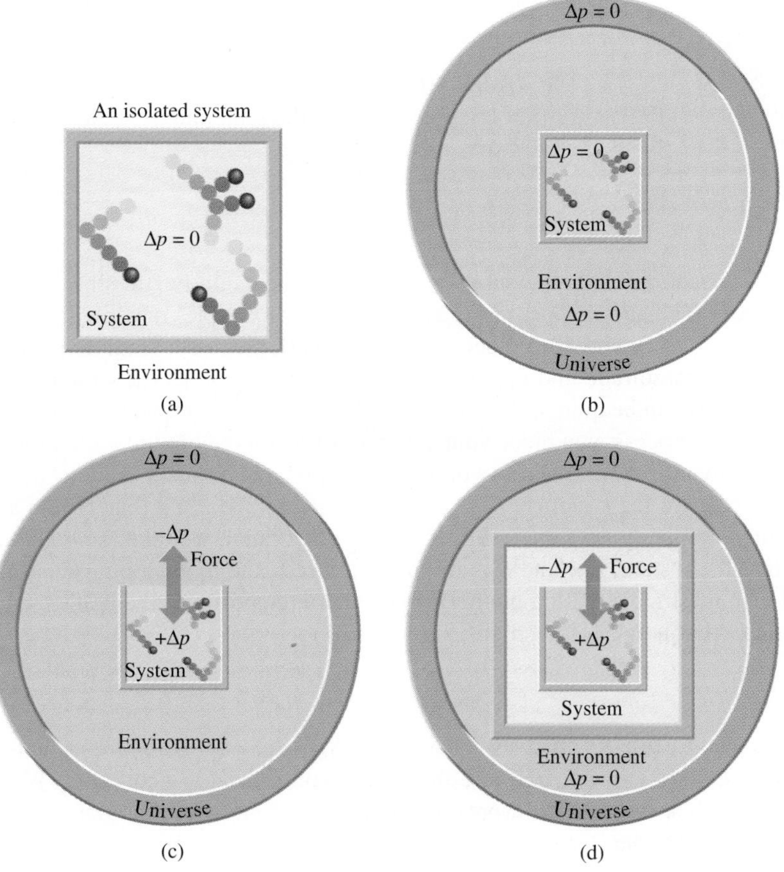

Figure 4.17 (a) A system isolated from the environment conserves momentum; that is, $\Delta p = 0$. (b) Here, we see an isolated system within the Universe. In all cases, Δp for the Universe is zero. (c) When the system communicates with its environment, it may experience a net momentum change, $+\Delta p$, but the environment will then experience an equal and opposite change, $-\Delta p$. (d) If the system is enlarged and yet isolated, its parts can experience momentum changes, but its net change will be zero.

amount $+\Delta p$. But the momentum of the communicating environment will then change by $-\Delta p$. If that system is part of a still larger isolated system (Fig. 4.17d), then the change in momentum of everything going on inside the large system will be zero, as if it were a little universe unto itself.

Figure 4.18, an example of how this phenomenon works, shows two people and a small moon, all floating in deep space and all at relative rest. Let's first take the three bodies to be the system. The total initial momentum is zero. The moon is subsequently thrown back and forth, momentum is imparted to the individual members of the system, but the forces are all internal and the net momentum is always zero, as it was at the outset. If we now take the moon as the system, its momentum changes every time someone applies an external force to it. By contrast, take the moving moon and the person on the right to be the system in Fig. 4.18b and thereafter. There is initially a net momentum of 10 kg·m/s to the right, and it will remain constant (Fig. 4.18c and d) unless the person on the left, who is not in the system, exerts an external force.

Example 4.12 According to published figures, a bullet fired from a 9-mm Luger pistol weighs 124 grains, has a mass of 8.0 g, and a muzzle speed of 1155 ft/s. If the mass of the gun is 0.90 kg, what is its recoil speed when the gun is fired horizontally? Ignore the escaping gases.

Solution: [Given: $m_B = 8.0$ g, $m_G = 0.90$ kg, $v_{Bi} = v_{Gi} = 0$, $v_{Bf} = 1155$ ft/s. Find: v_{Gf}.] The muzzle speed in SI is (1155 ft/s)(0.3048 m/ft) = 352 m/s. Taking the gun and bullet as the system, there are no external horizontal forces, and so the initial momentum must equal the final momentum, therefore

$$m_B v_{Bi} + m_G v_{Gi} = 0 = m_B v_{Bf} + m_G v_{Gf}$$

hence

$$v_{Gf} = -\frac{m_B}{m_G} v_{Bf} = -\frac{0.0080 \text{ kg}}{0.90 \text{ kg}}(352 \text{ m/s})$$

and $\boxed{v_{Gf} = -3.1 \text{ m/s}}$. The minus sign means that the gun moves in the opposite direction to the bullet.

▶ **Quick Check:** $p_{Gf} = p_{Bf}$, the bullet is roughly 100 times less massive than the gun, and its speed is roughly 100 times greater.

The planet often plays a subtle part in experiments on Earth, since it is usually connected to what is happening in some way. For example, if while hovering motionlessly ($p_i = 0$) far off in space you throw a ball, then your final momentum will be equal in magnitude and opposite in direction to the projectile's momentum. On Earth, things can be quite different. The ball gets thrown, but you usually just stand there. In the process, you brace yourself by pushing on the ground with a force (F_{YG}) and the ground pushes back on you with an equal and opposite force (F_{GY}). If you and the ball are the system, F_{GY} is an *external* force acting on that system and momentum is properly not conserved. If, on the other hand, we take the ball, you, and Earth as the system, F_{GY} and F_{YG} are internal forces and momentum is conserved—the Earth must move with a momentum equal and opposite to the ball's momentum (an effect that is much too small to measure directly).

Example 4.13 Imagine two masses, m_1 and m_2, resting on a horizontal frictionless surface. The masses are tied together with a string so that they compress a spring located between them. Now cut the string and let the masses fly apart. (a) Write an expression for m_1 in terms of m_2 and the final recoil speeds v_1 and v_2. (b) What is the expression when m_2 is a standard mass of 1 kg?

(continued)

(continued)

Solution: [Given: m_1, m_2, and the final speeds v_{f1} and v_{f2}. Find: m_1.] There being no external horizontal forces, momentum is conserved, and since it was zero initially

$$0 = m_1 v_{f1} + m_2 v_{f2}$$

and $m_1 = -m_2 v_{f2}/v_{f1}$, When $m_2 = 1$ kg

$$\boxed{m_1 = -\left(\frac{v_{f2}}{v_{f1}}\right)(1 \text{ kg})}$$

This procedure can serve as an operational definition of mass that does not require gravity. Given a standard mass (m_2), we can determine the mass of any object without ever defining mass conceptually.

▶ **Quick Check:** v_{f2}/v_{f1} is unitless, so the units are fine. When $m_1 > m_2$, we expect $v_{f2} > v_{f1}$, and the equation agrees with that.

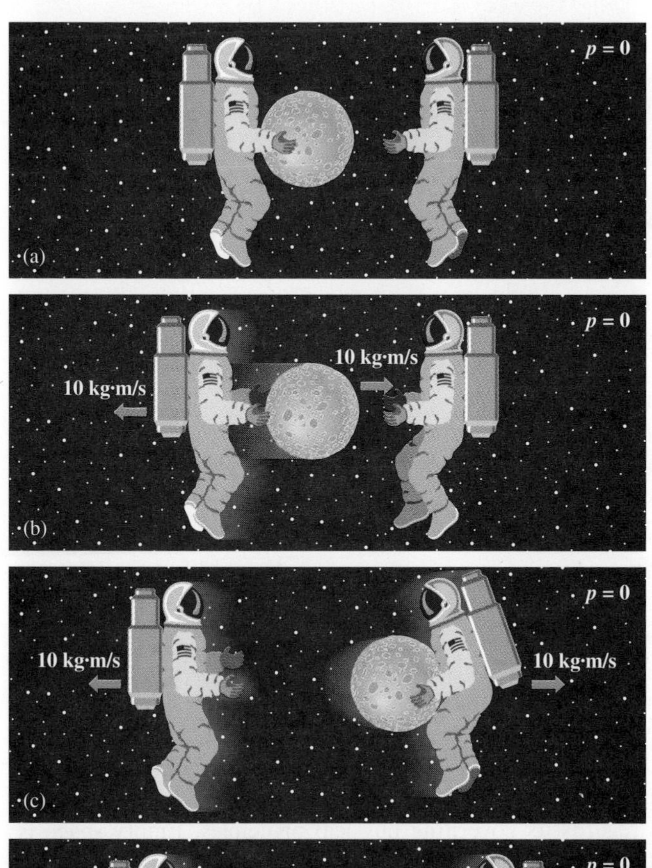

Figure 4.18 Take the two astronauts playing catch with a little moon as the system. There will then be no external forces acting, and momentum will be conserved. (a) Everything starts out at rest with a net momentum of zero: $p = 0$. (b) After the moonlet is thrown, the net momentum of the system is still zero. (c) And it's zero after the moonlet is caught and (d) thrown again. Alternatively, take the moon and/or either astronaut as the system. External forces will now be applied, and the momentum of this less inclusive system will change. Note that we do not know how much momentum the moonlet has in (c).

No experiment can prove that the law of Conservation of Linear Momentum is true, although observations confirm it all the time: There are no places in this Universe free of outside influences. Still, subatomic particles appear to conserve momentum scrupulously. Indeed, Conservation of Momentum is more fundamental

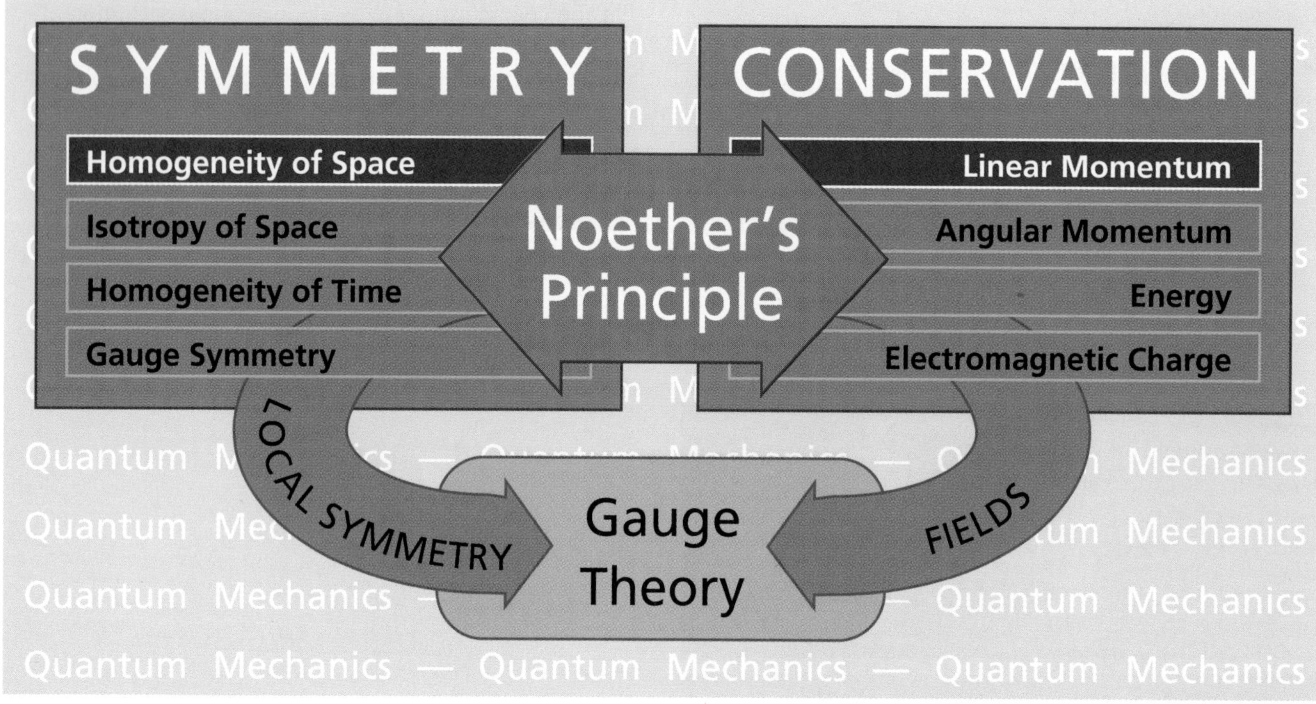

Figure 4.19 Until now we examined symmetry, conservation, and Noether's Principle in general terms. Here, we learn that the homogeneity of space gives rise to Conservation of Linear Momentum.

than the Third Law. That's true because interactions between separated objects take time to propagate, and there can be an appreciable delay between the observed action and the observable reaction. Modern Quantum Field Theory (Ch. 33) maintains that all the basic interactions (gravitational, electromagnetic, strong, and weak) arise from the exchange of momentum-carrying particles that travel at finite speeds. Although momentum is conserved at all times, we may have to wait a while after the action for the reaction.

Linear Momentum and Symmetry

Recall the discussion of symmetry and conservation in Chapter 1. When a change in a physical system leaves it essentially unchanged, that system possesses a corresponding symmetry. Figure 1.4 (p. 6) showed a disc that appears uniform and could therefore be rotated to a new position without producing any observable change—the disc possesses a symmetry. In the same way, we believe that the Universe is homogeneous as regards its governing physical principles. Empty space is the same everywhere. The same laws of physics apply everywhere, and the fundamental constants (for example, the speed of light, the quantum of charge, the gravitational constant) are independent of position. We could relocate our laboratory anywhere in the Universe and the laws of physics, the way things happen, will not change—the Universe possesses a spatial symmetry; there is an invariance in its space-time behavior.

Noether's Principle maintains that for every conservation law there is a corresponding symmetry, and vice versa. In particular, Noether showed that Conservation of Linear Momentum is a consequence of the invariance of nature with respect to place. Insofar as no conceivable experiment can distinguish between one point in empty space and another (on the basis of variations in physical behavior), space is homogeneous and linear momentum must be conserved (Fig. 4.19). Noether's proof of this is beyond our means, but we can show the interrelationship.

Assume that the physical Universe is *not* homogeneous, that the laws of physics are *not* the same everywhere. Now hold two identical objects at separate distant points (A and B) in space. Suppose that they attract each other either electrically, magnetically, or gravitationally. Given the "inhomogeneity," assume the force law is different at the two locations. Then the force on object A due to object B will be different from the force on object B due to object A. These forces are not an equal-and-opposite interaction pair (which conflicts with the Third Law). Suppose the objects are now released with a total initial momentum of zero. They will immediately travel toward each other, increasing in momentum but at different rates, thereby violating Conservation of Linear Momentum. The converse is as follows: space is homogeneous, the force laws are the same everywhere, the two forces are equal in magnitude and opposite in direction, the change in momentum of the two objects is equal in magnitude and opposite in direction, and linear momentum is conserved throughout the Universe.

Momentum is a vector in space, and it is conserved because of the homogeneity of space. That homogeneity is manifest in the equal-and-opposite nature of interaction. Hence, wherever there is a $\Delta\mathbf{p}$, there must, somewhere else in space, be an equal-magnitude, oppositely directed $\Delta\mathbf{p}$.

Core Material

Newton's First Law is the **Law of Inertia**: Every body continues in a state of rest, or in uniform motion in a straight line except insofar as it is compelled to change that state by forces impressed upon it. Force is the agent of change. The *weight* of an object on Earth is the downward force arising from the gravitational interaction between the object and the planet. The SI unit of force, the **newton** (N), is approximately equal on Earth to the weight of a mass of 0.10 kg, or $1\text{ N} \approx \frac{1}{4}$ lb. It is the *net external force* acting on a body that changes the motion (p. 97).

The measure of motion in Newton's theory is **momentum (p)**, defined as

$$\mathbf{p} = m\mathbf{v} \qquad [4.1]$$

Newton's Second Law says that

$$\mathbf{F}_{\text{av}} = \frac{\Delta\mathbf{p}}{\Delta t} = \frac{\Delta(m\mathbf{v})}{\Delta t} \qquad [4.2]$$

and

$$1\text{ N} = 1\text{ kg}\cdot\text{m/s}^2$$

Alternatively, *the change in the momentum of a body equals the net applied impulse,* thus

$$\mathbf{F}_{\text{av}}\,\Delta t = \Delta\mathbf{p} \qquad [4.3]$$

The **Third Law** maintains that two objects interact such that for every force exerted on one, there is an equal and oppositely directed force exerted on the other.

The law of **Conservation of Linear Momentum** states that *when the resultant of all the external forces acting on a system is zero, the momentum of the system remains constant* (p. 112). Conservation of Linear Momentum arises from the homogeneity of empty space.

Suggestions on Problem Solving

1. Begin problem solving with a sketch. Watch out for the signs. Take the direction of initial motion as positive and stick with it for the whole analysis. Put the signs in your diagram at the outset.

2. When adding force vectors (Appendix D), keep in mind that the x- and y-axes may not be the simplest choice of directions along which to resolve your vectors. Watch for symmetries that tell you force components are equal or perhaps cancel. Two equal and oppositely directed vectors cancel each other no matter what the angle of their common line-of-action is.

3. Do not mix up 32.2 ft/s^2 and 9.81 m/s^2—that is, be sure to use the proper numerical value of g for any given set of units in a problem. Always work in SI units. Most data tabulated for common consumption in this country are still in U. S. Customary units—you should learn how to deal with them.

4. Think in terms of Conservation of Momentum whenever you have and want masses and speeds before and after some interaction. Newton's laws will usually provide an alternative means of solving such problems, and either approach can be used to check the other.

5. The more experience you have, the further along you will go in the analysis before substituting in numbers. It makes no sense to multiply by some number only to find that you have to divide by it later on. Also, do not forget to square the quantity when you have a problem involving t^2 or v^2.

Discussion Questions

1. In a science-fiction TV show several years ago, the hero spent the entire 40 min fretting about running out of fuel before he could pilot his spaceship back to base. Time and again the camera cut from his worried face to the blasting rockets, to the trembling fuel gauge and back to his pained expression as the engines roared and the fuel diminished. What should he have done and why would you say the writers were Aristotelians?

2. A car stopped at a red light is hit in the rear by another car. Describe what will happen to the two drivers as a result of the impact. What kind of injuries might each driver sustain?

3. Some seat belts have a movable chest strap that locks whenever the brakes are hit sharply while the car is in motion. The belt takeup spool has a tooth wheel on it that is engaged by the top of a little pendulum whenever the latter swings forward. Describe how the First Law rules this arrangement.

4. If, while having lunch aboard a plane flying at 600 km/h, you happen to drop a pea right above your plate, where will it land? Explain. Is that why first class is in the front of the plane?

5. Suppose a hunter fires a blast of No. 6 shot straight up into the air. Given that the pellets have a terminal speed of about 9 m/s, describe the flight of a typical shot and comment on the danger to the hunter. The terminal speed of a .22-caliber bullet (nose first) is about 67 m/s. What does that suggest about shooting into the air anywhere near people? (See the footnote on p. 80 and comment on the physics.)

6. The bombardier in the World War II B-17 shown in Fig. Q6 has evidently just dropped a "stick" of bombs. Explain why they form a nice neat column in the picture.

Figure Q6

7. During the First World War (1918), the Germans surprised the French by bombarding Paris from 115 km away with a super-long-range gun, the *Kaiser Wilhelm Geschütz*. The 120-kg shells, fired with a muzzle speed of 2 km/s, took 3.5 min to reach the city. They were launched at an elevation of 52° and climbed to a height of 40 km. The Germans had found the technique of extending the range of cannon fire quite by accident. What had they discovered that made the bombardment possible?

8. If you dive into the water and land flat on your chest in what is referred to in some circles as a "belly whop," you are likely to remember it for a while. Use the Second Law to explain what happens.

9. Hold a ball high in your right hand and position your left hand a few feet directly below it. Now walk with a constant velocity and, somewhere along the way, drop the ball without breaking stride. Where will it land? Where will the ball land if, just as you drop it, you begin to decelerate? Try it.

10. Suppose you are riding a bike at a constant leisurely rate. Is it possible to bounce a ball (thrown straight downward) off the ground and catch it coming straight up, on the move? Draw a diagram of the ball's trajectory as seen by a stationary bystander as you ride past.

11. It is fairly easy to catch a 20-N steel weight dropped from a height of a $\frac{1}{4}$ m or so. But if you attempt to catch the weight while resting your hand flat on a table, you are likely to be injured. Explain.

12. Why should a batter pay more attention to follow-through when hitting a softball as opposed to a hardball? Every now and then a player will hit a ball so hard that the bat itself shatters. Explain how this happens in terms of Newton's laws. Drop a ball and watch it hit the floor and rebound back upward. According to the First Law, there must have been an externally applied upward force on the ball. What was it? Explain.

13. Which would be more difficult to throw at a given speed while hovering in space: a "weightless" tennis ball or a "weightless" bowling ball? Explain. Can astronauts floating in orbit tell which objects around them would be heavy or light on Earth even though everything is effectively weightless? How? Suppose that while working out in space an astronaut accidentally hits her thumb with a "weightless" hammer. Will it hurt?

14. A human being can survive a feet-first impact at speeds up to roughly 12 m/s (27 mi/h) on concrete; 15 m/s (34 mi/h) on soil; and 34 m/s (75 m/h) on water—explain the spread in these values.

15. There is a classic demonstration in which a heavy mass is suspended from a fairly fine thread and another length of the same thread is attached to its underside and allowed to hang beneath it. Someone then gives a sharp tug downward on the dangling thread. Which string will break and why? What would happen if the tug were gradual and strong? Explain.

16. Which stings the hand more (if either)—catching a ball in flight while running toward it or away from it? Explain.

17. What would happen to two astronauts if, while floating stationary with respect to one another, they played catch with a baseball?

Multiple Choice Questions

1. If we represent the dimensions of mass, length, and time by M, L, and T, respectively, then the dimensions of impulse are (a) $[ML/T]$ (b) $[ML^2/T^2]$ (c) $[ML/T^2]$ (d) $[LT/M]$ (e) none of these.

2. The blue team in the Pentagon's annual charity tug-of-war is being dragged toward the losing line by a superior red team. The eight blue pullers are each hauling horizontally with 400 N, but the eight red players are each tugging with 460 N. You can save the day for the blue team by joining them and pulling with a force: (a) of 400 N (b) in excess of 480 N (c) of 460 N (d) of greater than 400 N but less than 460 N (e) none of these.
3. Two identical rockets are dropped simultaneously from a plane. One is faulty and does not ignite at all; the other fires horizontally for 1 ms and then shuts off. (a) They hit the ground at about the same time. (b) The faulty one hits the ground first. (c) The one that fired sped up and so hits the ground first. (d) Not enough information is given to reach a conclusion. (e) None of these.
4. Figure MC4 shows a ball rolling down a curved incline. Which diagram depicts the resulting motion correctly? (a), (b), (c), (d), or (e) none of these.

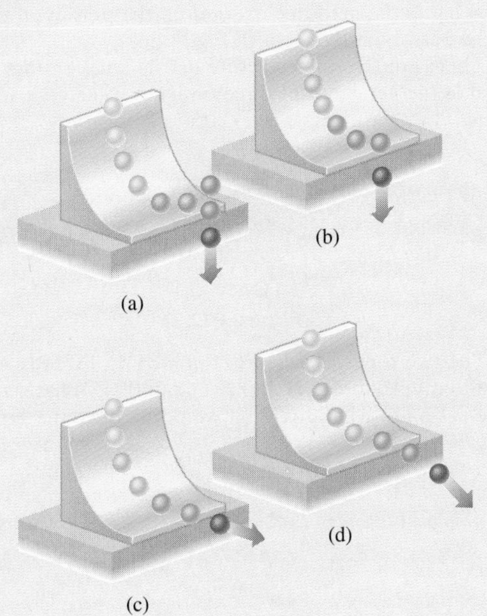

Figure MC4

5. Suppose that while walking in a straight line at a uniform speed, you throw a ball straight up about 2 or 3 m without breaking stride. Where will it land? (a) behind you (b) in front of you (c) right back in your hand (d) to the right of you (e) none of these. Try it!
6. Figure MC6 depicts a spiral tube lying at rest horizontally on a table, as viewed from above. A ball is rolled into one end of the tube and emerges from the other end. Ignoring friction, which diagram correctly represents the motion: (a), (b), (c), (d), or (e) none of these.
7. Suppose a projectile's speed and mass are both doubled. Its momentum will then be (a) doubled (b) unchanged (c) quadrupled (d) none of these.
8. A 1.0-kg body at rest experiences a force of 5.0 N exerted in the positive x-direction for 2.0 s, followed by a force of

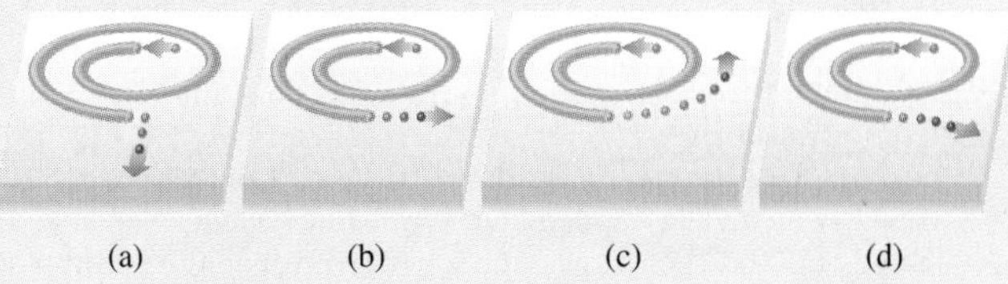

Figure MC6

10 N exerted in the negative x-direction for 1.0 s. Its resulting speed will be (a) +10 m/s (b) −10 m/s (c) +20 m/s (d) −20 m/s (e) none of these.

9. Refer to the force-time curve for an object of constant mass m in Fig. MC9. When was the object at rest? (a) 0 s to 1 s (b) 4 s to 5 s (c) 2 s to 3 s (d) 3 s to 4 s (e) not enough information given.

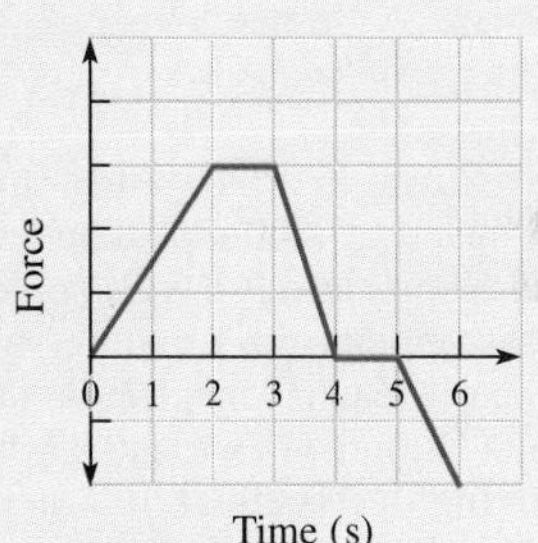

Figure MC9

10. When did the object in Fig. MC9 have a constant velocity? (a) 3 s to 4 s (b) 2 s to 3 s (c) 4 s to 5 s (d) 5 s to 6 s (e) not enough information given.
11. Which interval in Fig. MC9 corresponds to the greatest change in the speed of the body? (a) 0 s to 1 s (b) 1 s to 2 s (c) 2 s to 3 s (d) 3 s to 4 s (e) 5 s to 6 s.
12. During which time interval in Fig. MC9 did the body decelerate? (a) 0 s to 1 s (b) 2 s to 3 s (c) 3 s to 4 s (d) 5 s to 6 s (e) none of these.
13. Light carries momentum so when a beam impinges on a surface, it will exert a force on that surface. If the light is reflected rather than absorbed, the force will be (a) less (b) greater (c) equal either way (d) not enough information given (e) none of these.
14. A car traveling at a certain speed undergoes a head-on collision with an identical vehicle traveling at the same speed, and both come to rest in 10 ms. Now suppose, instead, that the car crashes into a brick wall and again comes to a rest in 10 ms. From the perspective of the original driver, which would be less dangerous? (a) the two-car crash (b) the car-wall crash (c) both are identical (d) not enough information (e) none of these.
15. A garbage truck crashes head-on into a Volkswagen and the two come to rest in a cloud of flies. Which experiences the greater impact force? (a) the truck (b) the Volkswagen (c) both experience the same force (d) not enough information given (e) none of these.
16. Two equal-mass bullets traveling with the same speed strike a target. One of the bullets is rubber and bounces off: the other is metal and penetrates, coming to rest in the target. Which exerts the greater impulse on the target?

(a) the rubber bullet (b) the metal bullet (c) both exert the same (d) not enough information (e) none of these.

17. An open railroad car filled with coal is coasting frictionlessly. A girl on board starts throwing the coal horizontally backward straight off the car, one chunk at a time. The car (a) speeds up (b) slows down (c) first speeds up and then slows down (d) travels at constant speed (e) none of the above.
18. A tank car coasting frictionlessly horizontally along the rails has a leak in its bottom and dribbles several thousand gallons of water onto the roadbed. In the process it (a) speeds up (b) slows down (c) gains momentum (d) loses momentum (e) none of the above.
19. A bomb hanging from a string explodes into pieces of different sizes and shapes. After the explosion (a) the vector momentum of each piece is identical (b) the total momentum is increased (c) the momentum of all the pieces, exhaust, and smoke adds up to zero (d) not enough information to comment (e) none of the above.
20. A can of whipped cream floating in space develops a hole in the bottom from which it squirts backward a mess of gas and cream at a constant speed with respect to the can. The can thereupon (a) accelerates forward throughout the squirting (b) moves forward at a constant speed (c) remains at rest (d) first speeds up, and then slows down when the gas runs out (e) none of the above.

Problems

THE LAW OF INERTIA

1. [I] A youngster sitting in a bus gently throws a ball straight up at a speed of 1.00 m/s. The bus is traveling at a constant rate of 10 m/s and it passes someone sitting on a fence. With what velocity does that person outside see the ball moving when it reaches maximum altitude?

2. [I] The hero in a Western movie is standing on top of a train that is moving due north at 30 km/h, while a pack of bad guys chases in hot pursuit. Cleverly, the hero tosses a lit bomb straight up at a speed of 9.8 m/s. When and where will it land (neglecting air friction)?

3. [I] Marc Antony, stretched out on a couch, is waiting, mouth upward and open. Cleopatra, carrying a bunch of grapes, is dashing across the palace floor straight toward him at a speed of 2.2136 m/s. She holds one grape 1.0000 m above his face. How far from him should she release her projectile if it is to land in the middle of his mouth? Neglect air friction and take the acceleration of gravity at the palace to be 9.8000 m/s^2.

4. [I] A girl running at a constant speed of 2.0 m/s in a straight line throws a ball directly upward at 4.9 m/s. How far will she travel before the ball drops back into her hands? Ignore air friction.

5. [I] When a 12-bore shotgun using No. 6 shot is fired straight up, the pellets attain a height of about 110 m. How does this situation compare with the ideal no-friction approximation, given that the muzzle speed is 405 m/s? Explain the difference.

6. [I] If you use the same setup as in Problem 5 but fire horizontally from a height of 1.22 m (with no air drag), what maximum horizontal distance should the pellets reach? (In actuality, they travel about 85 m.)

7. [I] What is the ideal maximum range you can expect from the shotgun in Problem 5? Maximum range is more nearly 200 m and is attained by firing at a reduced elevation of about 33°. Explain why this might be the case.

8. [I] Calculate your Earth-weight in newtons and then determine your mass in kilograms. What is the combined mass of $\frac{1}{2}$ lb of hard salami, 3 oz of Swiss cheese, 1 oz of mustard, all on a 4-oz roll "fresh this morning" on Earth?

9. [I] What is the force exerted on the wall by the person in Fig. 4.6a (p. 96)?

10. [I] What is the net force exerted on the scale in Fig. 4.6a?

11. [I] What is the net force exerted on the person on the left by the person on the right in Fig. 4.6d?

12. [I] There are five forces acting on the truck in Fig. P12. What is the net external force on it?

Figure P12

13. [I] The two ropes attached to the hook in Fig. P13 are pulled on with forces of 100 N and 200 N. What size single force acting in what direction would produce the same effect?

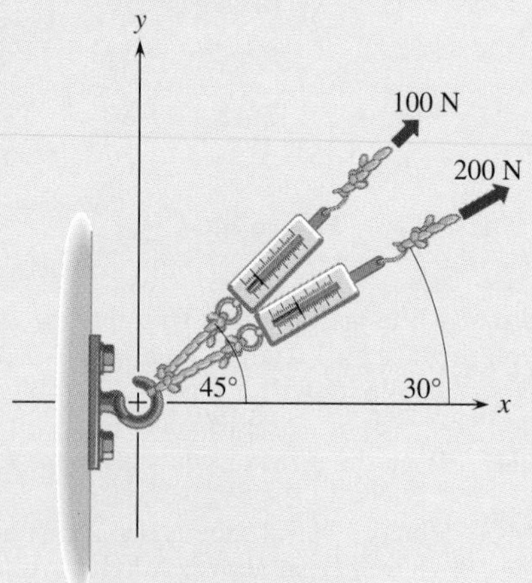

Figure P13

14. [I] The scale in Fig. P14 is being pulled on via three ropes. What net force does the scale read?

15. [I] What is the net force acting on the ring in Fig. P15?

16. [I] What is the net force acting on the ring in Fig. P16?

17. [II] A baseball is hurled at 30.7 m/s at an angle up from the ground of exactly 45°. Neglecting air friction, compute

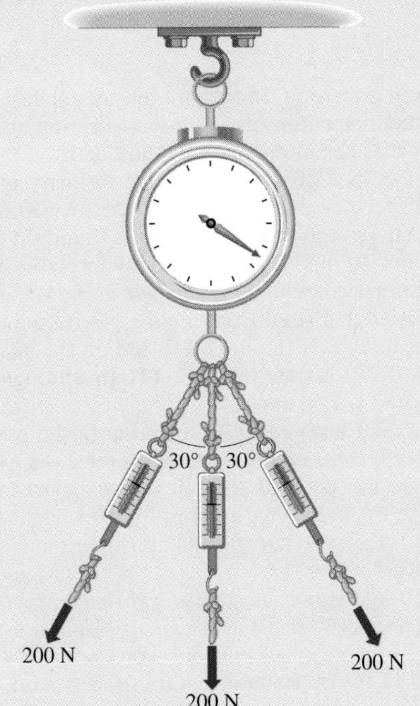

Figure P14

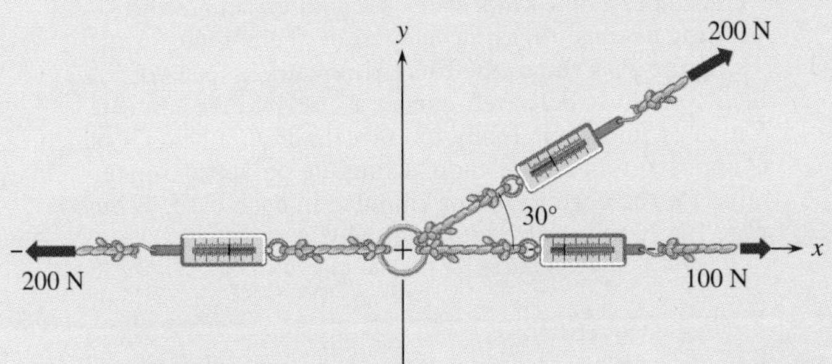

Figure P15

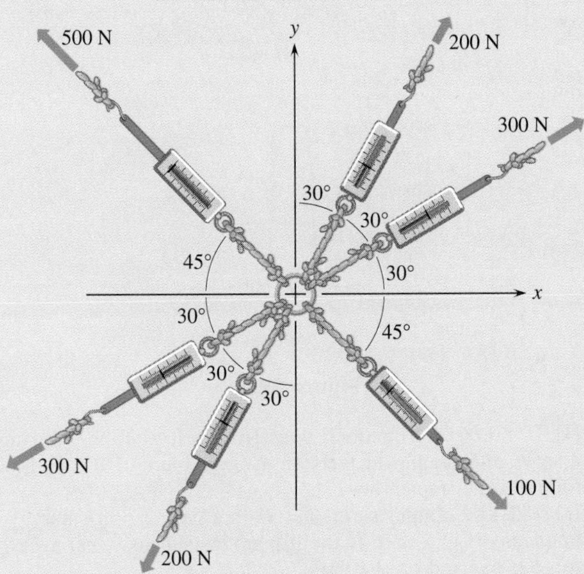

Figure P16

(a) its peak altitude, and (b) how far downrange it will be when it returns to the height at which it was thrown. In actuality, because of air friction, the ball will rise to only about 20 m and return to launch level less than 70 m downrange, descending more steeply than it was thrown by about 10° (see Fig. 4.5, p. 95).

18. [II] According to the manufacturer, a 256 Winchester Magnum cartridge shot by a revolver with an 8-in. barrel results in a bullet with a muzzle speed of 2200 ft/s. Fired horizontally, that bullet drops 0.90 in. after traveling 50 yd, 4.4 in. after 100 yd, and 10.9 in. after 150 yd. Compare these figures with those obtained theoretically, neglecting aerodynamic effects. Make a plot of both sets of values. What's happening?

19. [II] Here's a classic: Suppose a conservationist points a dart gun directly at a monkey in a tree. Show that if the monkey drops from its branch just when it sees the gun go off, it will still get hit.

20. [II] Write expressions for the net horizontal and vertical forces acting on the block in Fig. P20.

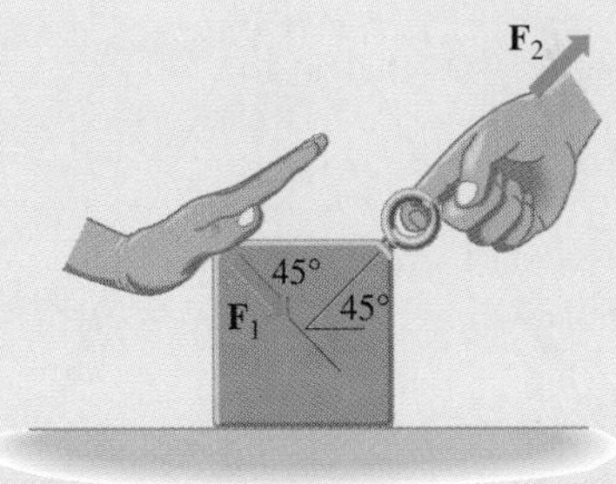

Figure P20

21. [II] What is the net force acting on the ring in Fig. P21? Try to solve this one in an elegant way.

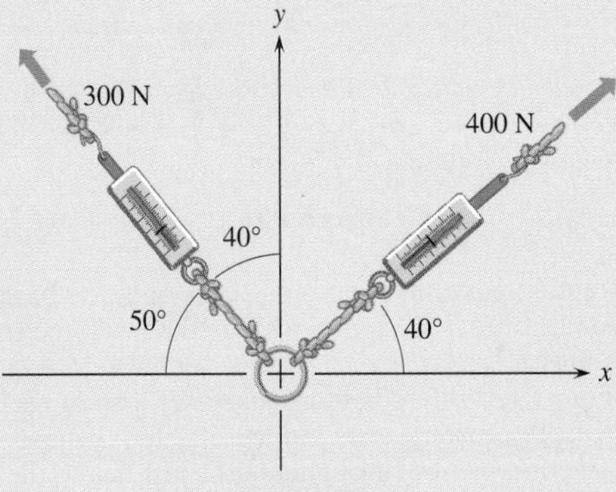

Figure P21

22. [II] Determine the net force acting on the ring in Fig. P22. Again, try to do this one elegantly.

23. [II] A ring is fixed via a harness to a block of stone and three ropes are attached to the ring. Each rope is to be pulled in the same horizontal plane with a force of 2 kN. How should they be arranged so as to produce a net force due east of 4 kN on the block?

24. [III] Figure P24 shows three people pulling down via ropes, each with 200 N, on the top of a mast 20 m tall. If

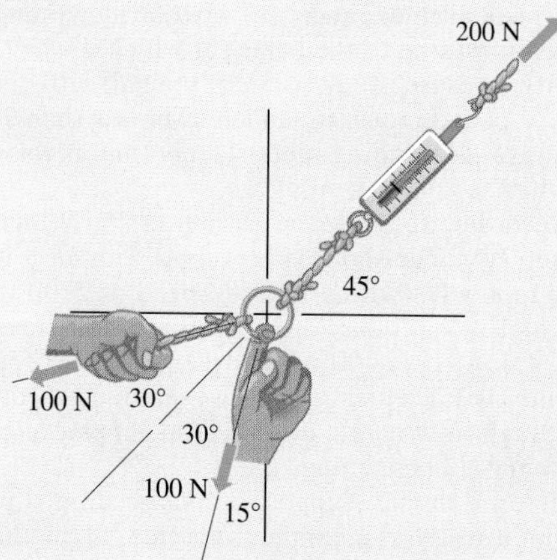

Figure P22

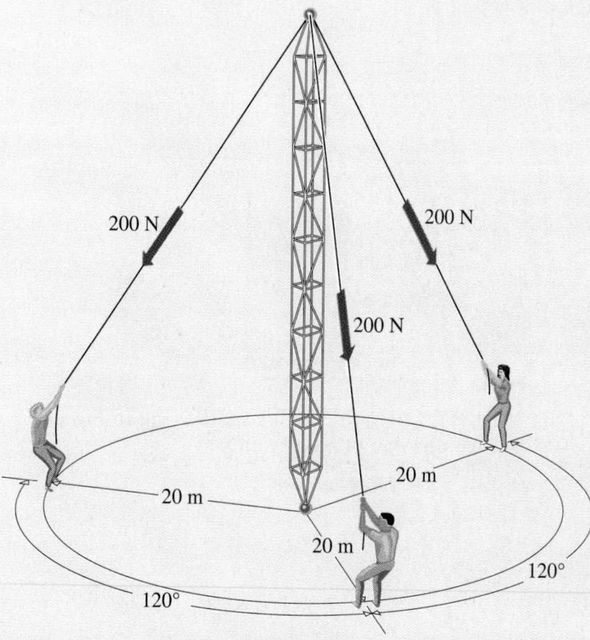

Figure P24

they stand at equal distances of 20 m from the base of the mast along lines 120° apart, what is the net force they exert on the mast?

25. [III] Four firefighters hold a square net, one at each corner. Each person exerts a force of 200 N whose line-of-action passes through a point just below the center of the net and makes an angle with the vertical of 60°. What is the net force the firefighters exert on the net?

THE SECOND LAW
INTERACTIONS

26. [I] Compute the mass of your head in kilograms, given that its Earth-weight in pounds is *very roughly* equal to $(0.028\ F_W + 6.35)$, where F_W is your total body weight in pounds.

27. [I] From how high (in meters) must a car fall if it's to have the same momentum it would have driving at 60 mi/h?

28. [I] What is the momentum of the Earth ($m_\oplus = 5.975 \times 10^{27}$ g) as it moves through space if its orbital speed is about 6.66×10^4 mi/h? (SI units please.)

29. [I] A 1.0-kg wad of clay is slammed straight into a wall at a speed of 10 m/s. If the clay sticks in place, what was the impulse that acted on it via the wall?

30. [I] A dried pea fired from a plastic drinking straw has a mass of 0.50 g. If the force exerted on the pea is an average 0.070 lb over the 0.10-s flight through the straw, at what speed will it emerge?

31. [I] A 1.0-kg body initially traveling in the positive x-direction at 10 m/s is acted upon for 2.0 s by a force in that same direction of 20 N. It then experiences a force acting in the negative direction for 20 s equal to 2.0 N. Draw a force-time curve and determine the final momentum.

32. [I] A 10-kg asteroid is traveling through space at 2 m/s toward a spaceship. To avoid a collision, an astronaut, with the help of a backpack rocket, exerts a force of 20 N on the asteroid in the opposite direction for 5 s. (a) Find the asteroid's final speed. (b) How long should she have pushed in order to just stop the asteroid?

33. [I] A rocket engine blasts out 1000 kg of exhaust gas at an average speed of 2 km/s every second. Calculate the resulting average thrust developed by the engine.

34. [I] Figure P34 shows the force-time curves for two idealized blows delivered during a martial arts contest. Which of the two is likely to correspond to a boxer's punch as opposed to a karate chop? Compute an approximate value for the corresponding impulse in each case. Which is more likely to break bones, and why?

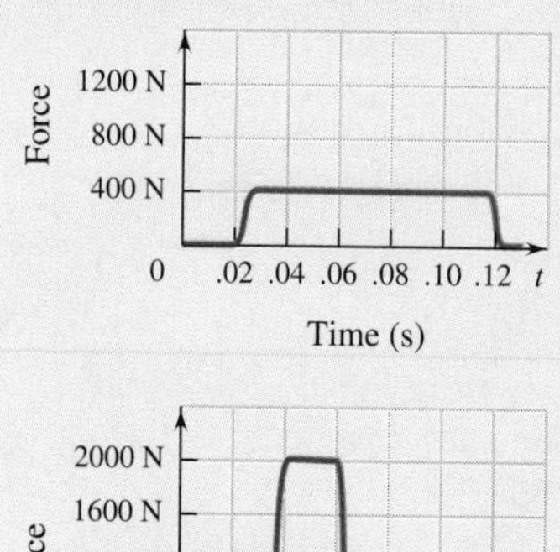

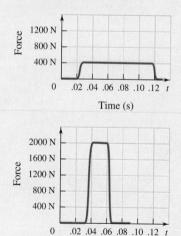

Figure P34

35. [I] If a 0.061-kg handball goes from rest to 20 m/s during a serve and the player feels an average force of 100 N, how long did the impact last?

36. [I] A 1-kg hammer slams down on a nail at 5 m/s and bounces off at 1 m/s. If the impact lasts 1 ms, what average force is exerted on the nail?

37. [I] A 50-kg person at the southernmost end of a 150-kg rowboat at rest in the water walks to the northern end. If, at a given instant, her speed with respect to the water is 10 m/s, what is the velocity of the boat with respect to the water at that moment?

38. [I] Water squirting from a hose emerges with a speed of 10 m/s at a rate of 100 kg/s. Compute the average reaction force exerted on the hose. Have you ever seen firefighters working to control a large hose?

39. [I] A 60-kg cosmonaut jumps from her 5000-kg spaceship in order to meet her copilot who is floating at rest a few hundred meters away. If she sails toward him at a speed of 10 m/s, what is the resulting motion of the ship?

40. [II] A golf ball with a mass of 47.0 g can be blasted from rest to a speed of 70.0 m/s during the impact with a clubhead. Taking that impact to last only about 1.00 ms, calculate the change in momentum of the ball and the average force applied to it.

41. [II] A bullet is fired from a handgun with a barrel 24.0 cm long. Its muzzle speed is 350 m/s and its mass is 6.00 g. Compute the average force exerted on the bullet by the expanding gas in the barrel.

42. [II] During a particular rocket-powered sled run, Colonel J. P. Stapp of the U.S. Air Force decelerated from a speed of 940 ft/s to a dead stop in 1.40 s. Assuming he weighed 175 lb, compute the average force, in newtons, exerted on him. Incidentally, the primary dangers in this sort of test are things like blood vessels tearing away from the lungs or the heart, and even retinal detachment, via the First Law.

43. [II] The human heart pumps about 2 ounces of blood into the aorta on each stroke, which lasts roughly 0.1 s. During that time, the pulse of blood is accelerated from rest to about 50 cm/s. Compute the corresponding average propulsion force exerted on the blood by the heart.

44. [II] Suppose a car stopped on the road is hit from behind by a bus so that it accelerates up to 4.47 m/s (10 mi/h) in 0.10 s. If the driver of the car has a mass of 50 kg, and her front-seat passenger has a mass of 80 kg, what average force will the seat have to exert on them? (Car seats have been known to collapse under this sort of treatment, though they obviously shouldn't.)

45. [II] In Problem 44, how much force must the driver's neck exert to keep her head in line with her body during the collision, if her head weighs 44.5 N (10 lb)? Think about "whiplash."

46. [II] A person can just survive a full-body collision (either to the front, back, or side) at roughly 9 m/s (20 mi/h) with an impact time of approximately 10 ms. At greater speeds or shorter times, fatal brain damage will likely occur. Could someone survive a fall from 4.0 m landing flat on his back on soft soil, so that he decelerates to rest through a distance of 10 cm (that's the total compression of body and soil)? If his mass is 70 kg, what's the impulse exerted on his body by the ground? Assume the deceleration is constant.

47. [II] Suppose that a person of mass m jumps off a ladder at a height s_h and lands on the ground without bouncing. If the total compression of the body and the soil during impact is s_c, and if the deceleration is assumed constant, show that the force exerted by the ground is given by

$$F_{av} = mg\left(\frac{s_h}{s_c}\right)$$

Notice that bending the knees extends s_c considerably, decreasing F_{av} accordingly.

48. [II] During a 40 mi/h car crash, a 50-kg passenger comes to rest in 100 ms. What is the area under the force-time curve for that person? Assuming a seat belt was worn, which would smooth out the curve so there were no devastating peaks; what average force would be experienced?

49. [II] When a person jumps and lands stiff-legged on the heels of the feet, a considerable force can be exerted on the long leg bones. The greatest stress occurs in the tibia, or shin bone, a bit above the ankle where the bone has its smallest cross section. If a force in excess of about 50 000 N is applied upward on the heel, the tibia will probably fracture. Keeping in mind the results of Problem 47, what is the minimum height of a fall above which a 60-kg person is likely to suffer a tibial fracture? Assume the body decelerates uniformly through a distance of 1.0 cm and the landing occurs squarely on both feet.

50. [II] An ion engine is a rocket designed to produce small thrusts for very long periods of time. The SERT 2 engine develops 30 mN of thrust by expelling mercury ions at speeds of around 22 000 m/s (about 50 000 mi/h). How many kilograms of exhaust emerge per second?

51. [II] A modern nuclear aircraft carrier weighs in at around 90 000 tons and is capable of traveling in excess of 30 knots (15.4 m/s). Suppose one of these giants plowed into a pier at 30 knots, coming to rest in 0.50 min, what average force would it exert in the process?

52. [II] A beam consisting of around 1.0×10^{15} electrons per second, each of mass 9.1×10^{-31} kg, paints pictures on the face of a TV tube. Assuming that the electrons are absorbed by the screen and exert an average net force on it of 1.0×10^{-7} N, what is their average speed at impact?

53. [II] A bullet fired into wet clay will decelerate fairly uniformly. If a 10-g slug hits a block of clay at 200 m/s and comes to rest in 20 cm, what average force does it exert on the block?

54. [II] Imagine a car involved in a head-on crash. The driver, whose mass is m, is to be brought uniformly to rest within the passenger compartment by compressing an inflated air bag through a distance s_c. Write an expression for the average force exerted on the bag in terms of m, v_i, and s_c. Compute that average force for a 60 mi/h collision, where the driver's mass is 60 kg and the allowed stopping distance in the bag is 30 cm. Assume the car deforms only negligibly.

55. [II] Suppose a 6.00-g bullet traveling at 100 m/s strikes a bulletproof vest and comes to rest in about 600 μs. What average force will it impart to the happy wearer?

56. [II] A 2785-lb Triumph TR-8 traveling south at 40 km/h (25 mi/h) crashes head-on into a 3840-lb Checker cab moving north at 90 km/h (56 mi/h). If the two cars remain tangled together but free to coast, describe their motion immediately after the collision. What is their final speed?

57. [II] A railroad flatcar having a mass of 10 000 kg is coasting along at 20 m/s. As it passes under a bridge, 10 men (having an average mass of 90 kg) drop straight down onto the car. What is its speed as it emerges with its new passengers from beneath the bridge?

58. [II] A 90-kg astronaut floating out in space is carrying a 1.0-kg TV camera and a 10-kg battery pack. He's drifting toward his ship but, in order to get back faster, he hurls the camera out into space at 15 m/s and then throws the battery at 10 m/s in the same direction. What's the resulting increase in his speed after each throw?

59. [II] An ice skater (with a mass of 55.0 kg) throws a snowball while standing at rest. The ball has a mass of 200 g and moves straight out with a horizontal speed of 20.0 km/h. Neglecting friction, and assuming the skate blades are parallel to the direction of the throw, describe the skater's resulting motion in detail.

60. [II] A 4265-lb Jaguar XJ12L requires 164 ft of runway as a minimum stopping distance from a speed of 60.0 mi/h. Assuming its deceleration is uniform, compute the average stopping force exerted on the car.

61. [II] An F-14A jet weighing 70 000 lb has a typical lift-off distance of 3000 ft and a corresponding lift-off speed of 135 knots (69.5 m/s). Assuming its acceleration is constant, calculate the net average thrust of its engines in newtons.

62. [III] Navy jets are hurled off the deck of a modern carrier by a combination of catapult and engine thrust. The catapult (the C-7, for example) has a 250-ft stroke (that is, runway) and will yank a 70 000-lb F-14A jet, with its engines developing a net average thrust of 16 000 lb, from rest to a speed of 200 ft/s in just 2.4 s. (a) Compute the average force exerted by the catapult. (b) Is the acceleration of the plane uniform? (Hint: Try several of the equations for constant a.)

63. [III] During a parachute jump over Alaska in 1955, a United States trooper jumped from a C-119 at 1200 feet, but his chute failed to open. He was found flat on his back at the bottom of a $3\frac{1}{2}$-ft deep crater in the snow, alive and with only an incomplete fracture of the clavicle. Compute the average force that acted on him as he plowed into the snow. Assume the deceleration was constant, take his mass to be 90 kg and his terminal speed to be 120 mi/h.

64. [III] A typical tubular aluminum arrow 70 cm long has a mass of 25.0 g. The force exerted by a bow varies in a fairly complicated fashion from the initial draw force of, say, 175 N (39 lb) to zero, as the arrow leaves the string. Assume a full draw (70 cm) and a launch time of 16.0 ms. If the arrow when fired straight up reaches a maximum height of 143.3 m above the point of launch, then (a) what is the average force exerted on it by the bow? (b) What's the area under the force-time curve? (c) What impulse is applied to the shooter?

The catapult of a modern aircraft carrier hurls planes off the deck.

65. [III] Two astronauts floating at rest with respect to their ship in space decide to play catch with a 0.500-kg asteroid. Neil (whose mass is 100 kg) heaves the asteroid at 20.0 m/s toward Sally (whose mass is 50.0 kg). She catches it and heaves it back at 20.0 m/s (with respect to the ship). Before Neil catches it a second time, how fast is each person moving, and in what direction?

5

Dynamics: Force and Acceleration

WHEN ALL THE EXTERNAL forces exerted on a system act in such a way that there is no change in momentum, no acceleration, the system is in *equilibrium*—there is a balance, the study of which is called *statics* (p. 163). Alternatively, the discipline dealing with the nonequilibrium behavior of objects is *dynamics.* A body experiencing a net force will accelerate in the direction of that force. It is the relationship between various kinds of forces and the resulting accelerations that will be examined in this chapter.

Figure 5.1 The application of a net force F_{av} for a time Δt results in a change in momentum and an average acceleration, $a_{av} = (v_f - v_i)/\Delta t$.

FORCE, MASS, AND ACCELERATION

There is an alternative formulation of Newton's Second Law that, although less general, is very useful. Recall that the average force on an object equals its time-rate-of-change of momentum (Fig. 5.1), expressed as

$$\mathbf{F}_{av} = \frac{\Delta \mathbf{p}}{\Delta t} = \frac{\Delta(m\mathbf{v})}{\Delta t} \quad [4.2]$$

The change in its momentum $\Delta \mathbf{p} = (\mathbf{p}_f - \mathbf{p}_i)$ equals $(m_f\mathbf{v}_f - m_i\mathbf{v}_i)$, but *if we take the mass of the object to be constant* $m_i = m_f = m$, then $\Delta \mathbf{p} = m(\mathbf{v}_f - \mathbf{v}_i) = m\,\Delta\mathbf{v}$. Consequently, Eq. (4.2) becomes

$$\mathbf{F}_{av} = \frac{m\,\Delta \mathbf{v}}{\Delta t} \quad (5.1)$$

With Eq. (3.4), $\mathbf{a}_{av} = \Delta\mathbf{v}/\Delta t$, for the average acceleration, the average force becomes

$$\mathbf{F}_{av} = m\mathbf{a}_{av} \quad (5.2)$$

An average force of 1.0 N will cause a mass of 1.0 kg to accelerate at an average rate of 1.0 m/s^2, and as we saw before (p. 104), 1 N = 1 kg·m/s^2. Table 5.1 lists the units of force in the several systems that are likely to be encountered.

The force in Fig. 5.1 is applied to a rope and not directly to the object. A rope, chain, cable, or tendon transmits the applied force to the object via electromagnetic interactions between its intervening atoms. Force is a vector quantity, it acts along a line, and the line of the taut rope is the **line-of-action** of the force. Because that force tends to stretch the rope, it is known as a **tensile force**. Imagine that we cut the rope and splice in a spring scale of negligible mass. The scale will read out the magnitude of the applied tensile force. That scalar quantity is called the **tension** (see Fig. 4.6, p. 96).

Equation (5.2) provides a means of computing average quantities when you know the overall change in motion but not the moment-by-moment details. If we can measure the initial and final speeds, we can compute the average accelerating forces acting on baseballs, cars, rockets, whatever.

Example 5.1 A baseball (0.142 kg) leaving a pitcher's hand at a speed of 45.7 m/s is clocked using gun-radar. If the throw lasted 0.12 s, what was the magnitude of the average force exerted on the ball?

Solution: [Given: m = 0.142 kg, $v_i = 0$, v_f = 45.7 m/s, and Δt = 0.12 s. Find: F_{av}.] First determine the average acceleration using

$$a_{av} = \frac{\Delta v}{\Delta t} = \frac{+45.7\text{ m/s}}{0.12\text{ s}} = 380.8\text{ m/s}^2$$

From Eq. (5.2)

$$F_{av} = (0.142\text{ kg})(380.8\text{ m/s}^2) = \boxed{54\text{ N}}$$

This quantity is about 12 lb, which is reasonable given that it corresponds to an average force (see Fig. 4.12).

▶ **Quick Check:** The impulse, $F_{av}\,\Delta t$, is about 6.5 N·s, as is Δp, so it looks okay.

TABLE 5.1 Units of force, mass, and acceleration in various systems

	SI	cgs*	U.S. Customary units
Force	$1\ \text{N} \equiv 1\ \text{kg·m/s}^2$	$1\ \text{dyne} \equiv 1\ \text{g·cm/s}^2$	1 lb
Mass	1 kg	1 g	$1\ \text{slug}^{\dagger} \equiv 1\ \text{lb·s}^2/\text{ft}$
Acceleration	$1\ \text{m/s}^2$	$1\ \text{cm/s}^2$	$1\ \text{ft/s}^2$

*The centimeter-gram-second system, which we will avoid, uses the dyne as the unit of force. $1\ \text{N} = 10^5$ dynes = 0.225 lb.
†Although the slug is still used in engineering as the unit of mass, we will make no use of it.

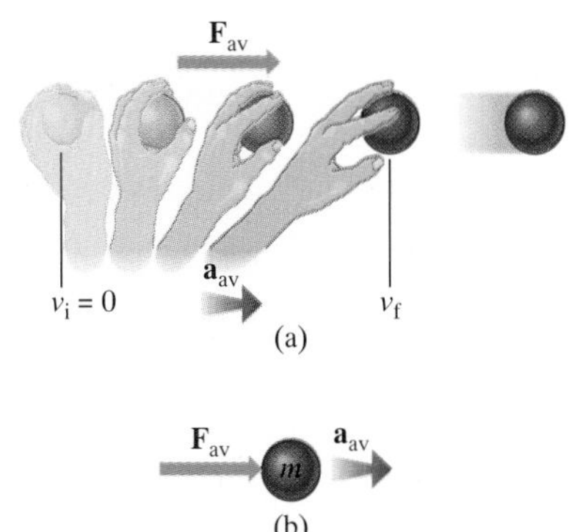

Figure 5.2 A ball being launched out in space where the only force acting on it is the pushing hand. (a) The average force accelerates the ball from $v_i = 0$ to v_f at an average rate $\mathbf{a}_{av}$. (b) Displayed is a free-body diagram of the hand pushing the ball. Notice how $\mathbf{a}_{av}$ is in the direction of $\mathbf{F}_{av}$, which in this case is in the direction of motion.

As we progress in our analysis, we will encounter situations in which several forces act on a body in different directions. Even though things can get complicated, there is a wonderful conceptual device called a **free-body diagram** that will help. To create one, isolate the object of concern from the rest of the system; that is, (1) remove anything in contact with it and (2) replace each such source of interaction, be it a pushing hand or a pulling string, by a force vector. Following these procedures leaves us with a picture of the object all by itself with a bunch of force vectors acting on it. Figure 5.2a shows a ball being launched in space, where the only force acting arises from the interaction with the thrower. The free-body diagram (Fig. 5.2b) is the simplest possible one, namely a single force, accelerating a single mass.

Example 5.2 An old Checker cab weighing 17.08 kN cruises along a level road at 35.8 m/s. The driver lets the car coast in neutral, and air drag decelerates it at a nonconstant rate down to 22.4 m/s in 24 s. (a) Calculate the average deceleration during the period. (b) Determine the average retarding force acting on the cab.

Solution: [Given: $F_W = 17.08$ kN, $v_i = 35.8$ m/s, $v_f = 22.4$ m/s, and $\Delta t = 24$ s. Find: a_{av} and F_{av}.] Figure 5.3a shows the cab moving to the right, with an initial speed of 35.8 m/s. Ignoring gravity, Figs. 5.3b and c are the free-body diagrams showing the average drag force acting to the left. It opposes the motion and slows the vehicle. (a) From the definition

$$a_{av} = \frac{\Delta v}{\Delta t} = \frac{(22.4\ \text{m/s} - 35.8\ \text{m/s})}{24\ \text{s}} = -0.558\ \text{m/s}^2$$

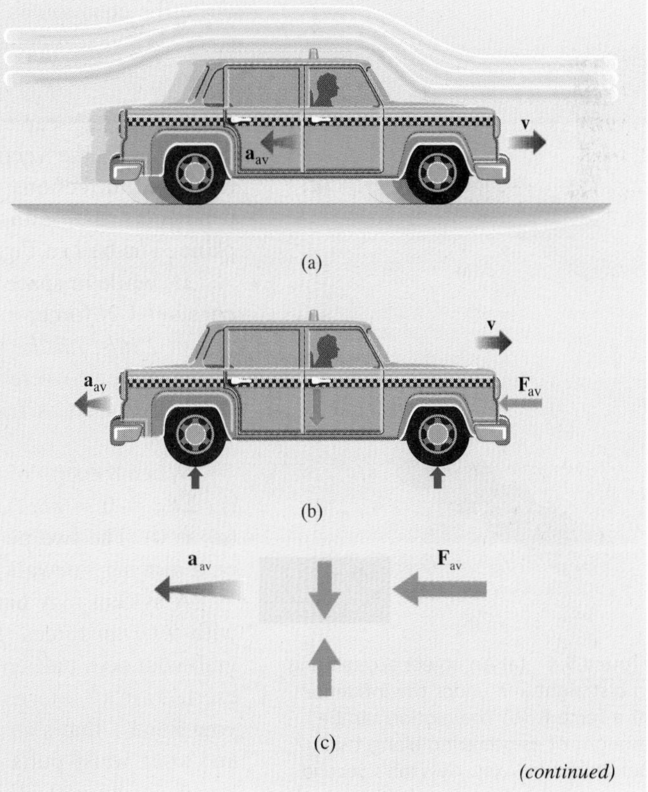

Figure 5.3 (a) An old cab coasting down from a speed v_i to a speed v_f. Air drag produces an average force $\mathbf{F}_{av}$ and the cab decelerates. Overlooking vertical forces for the moment, (b) is a free-body diagram of the cab. Notice how $\mathbf{a}_{av}$ is in the same direction as $\mathbf{F}_{av}$ but opposite to the motion. (c) When we begin to do a problem, the free-body diagram can be simple.

(continued)

(continued)

or $\boxed{-0.56 \text{ m/s}^2}$. (b) Consequently

$$F_{\text{av}} = ma_{\text{av}} = (F_{\text{W}}/g)a_{\text{av}} = (1741.7 \text{ kg})(-0.558 \text{ m/s}^2)$$

and so $F_{\text{av}} = -971.9$ N or $\boxed{-0.97 \text{ kN}}$. The minus sign indicates deceleration.

▶ **Quick Check:** From Eq. (4.3), $F_{\text{av}}\Delta t = \Delta p$, (0.97 kN)(24 s) = 23.3 kN·s, while $\Delta p = (1742 \text{ kg}) \times (13.4 \text{ m/s}) = 23.3 \times 10^3$ kg·m/s.

5.1 Instantaneous Motion

It is possible to go from average force to instantaneous force in the following way. Return to Eq. (5.1), $\mathbf{F}_{\text{av}} = m\,\Delta\mathbf{v}/\Delta t$, and take the limit of both sides as $\Delta t \to 0$. On the left, the average force determined over an infinitesimal interval is the *instantaneous force* (**F**), via Eq. (4.4). On the right, we have the product of the constant mass (m) and the *instantaneous acceleration* (**a**), via Eq. (3.3). At each and every moment, the *net force applied to a body equals the mass of the body multiplied by the resulting acceleration,* that is

$$\mathbf{F} = m\mathbf{a} \tag{5.3}$$

Evidently, the net applied force **F** and the resulting acceleration **a** are always parallel.

Remember that **F** represents the *net externally applied force.* It's helpful to write the equation as

$$\Sigma\mathbf{F} = m\mathbf{a} \tag{5.4}$$

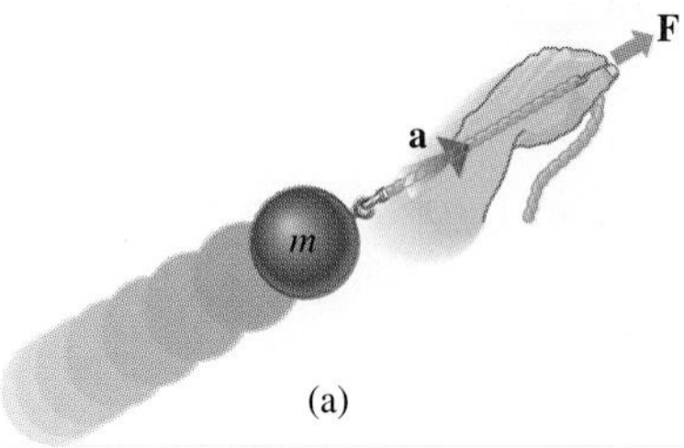

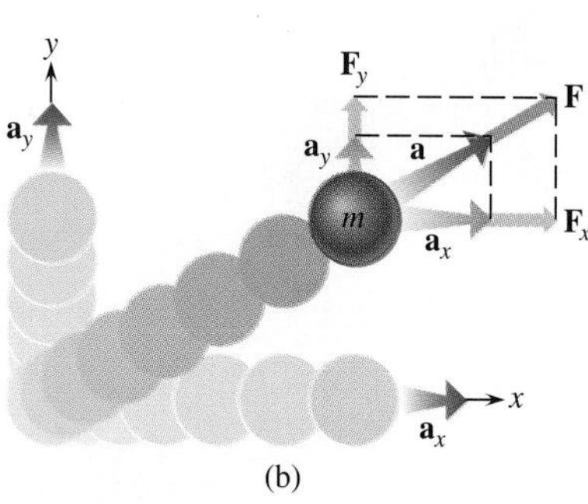

Figure 5.4 (a) An object accelerating in a straight line under the influence of a force **F**. (b) The motion can be thought of as occurring along two perpendicular axes. Newton's Second Law applies independently along both axes.

where Σ (Greek capital sigma) stands for "the sum of" and reminds us that we are dealing with the vector sum of all the forces acting on the body. The equation is true in all directions, and **F** can be resolved into components along any three perpendicular axes, although we restrict ourselves to the simpler case of motion in a plane. The ball in Fig. 5.4 moves in the xy-plane, so $\Sigma F_x = ma_x$ and $\Sigma F_y = ma_y$.

If, while in space (Fig. 5.5), we push on a floating weightless $\frac{1}{2}$-kg mass with a constant 1-N force, it will immediately accelerate in the direction of the force, at a rate of 2 m/s² *all the while the force acts.* Similarly, if we push on a 100-kg refrigerator (Fig. 5.6) here on Earth, with a horizontal force of 100 N (and if all other influences such as friction are negligible), away it will go with an acceleration of 1.00 m/s².

When the sum of the external horizontal forces acting on an object is zero (that is, $\Sigma F_{\text{H}} = 0 = ma_{\text{H}}$), as in Fig. 5.7, there will be no horizontal acceleration ($a_{\text{H}} = 0$). The two people and the "fridge" are not necessarily at rest (they may be on a moving sidewalk); they are just not accelerating.

A system may have several interacting parts that push and pull on each other with *internal* forces that do not affect the overall motion. The reason you cannot grab your belt, pull up, and fly, is that you and the belt are the system and the forces exerted are all internal. Your hand pulls up on your belt and your belt pulls down on your hand—that's an interaction pair. Meanwhile, your belt pulls up on your waist and your waist pulls down on your belt—that's an interaction pair. And all the forces are numerically equal. Hence, the net force on the belt is zero, and the net force on you is zero—you go nowhere. You have to get out and push if you want to

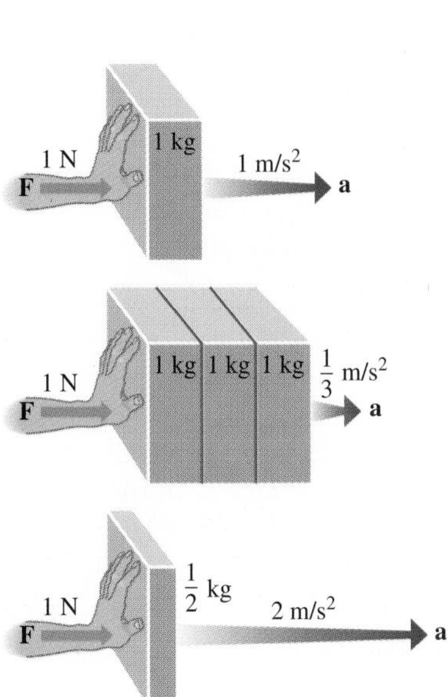

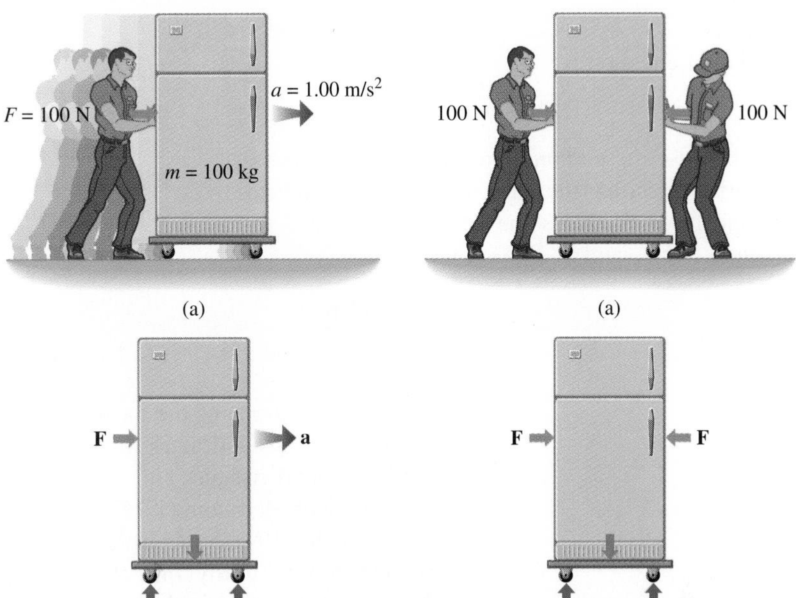

Figure 5.5 Several examples of how $\mathbf{F} = m\mathbf{a}$ works. In each case the force—at the moment shown—produces a proportional instantaneous acceleration. For as long as the force is applied, the body accelerates in the direction of **F**.

Figure 5.6 (a) If the only horizontal force acting on a body of mass m is F, then it will accelerate horizontally such that $F = ma$. This motion is independent of the vertical forces that we'll deal with shortly. (b) The free-body diagram shows the net horizontal force.

Figure 5.7 (a) Two people apply equal, oppositely directed forces to an object. Because force is a vector quantity, these two forces cancel each other and $\Sigma F_{\mathrm{H}} = 0$, which means there cannot be any horizontal acceleration. (b) The free-body diagram makes the same point more directly.

move a stalled car: inside, you push forward on the dashboard and it pushes you backward against the seat. Your back pushes to the rear of the car with the same force your hands push forward with—the net force you exert on the car is zero, and the net force it exerts on you is zero.

Example 5.3 The person in Fig. 5.8 pulls a loaded wagon having a total mass of 100 kg. A force of 100 N is applied through the handle at 30.0°. Ignoring friction, compute the horizontal force on the wagon and the resulting horizontal acceleration.

Solution: [Given: $m = 100$ kg, $F = 100$ N, and $\theta = 30.0°$. Find: F_{H} and a_{H}.] The free-body diagram (Fig. 5.8b) shows **F**, acting up along the handle, resolved into two components, one vertical and one horizontal.

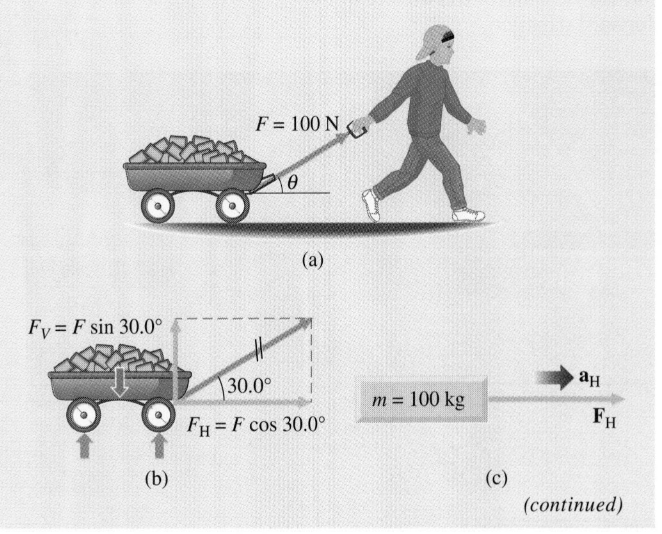

Figure 5.8 (a) The wagon experiences a force directed upward along the handle at an angle θ. (b) Its horizontal component $\mathbf{F}_{\mathrm{H}}$ is the only horizontal force, and it will determine the horizontal acceleration. (c) Ignoring the vertical forces for the moment, the free-body diagram is simple.

(continued)

(continued)

The guiding equation is

$$\overset{+}{\rightarrow}\Sigma F_{\mathrm{H}} = ma_{\mathrm{H}}$$

where, because the motion is to the right, we make that the positive direction (the arrow and + sign so indicate). Only one force produces the acceleration and it's the horizontal component of **F**. That is,

$$F_{\mathrm{H}} = F\cos\theta = F\cos 30.0° = \boxed{+86.6\ \mathrm{N}}$$

Thus $\overset{+}{\rightarrow}\Sigma F_{\mathrm{H}} = +86.6\ \mathrm{N} = ma_{\mathrm{H}}$

and $\boxed{a_{\mathrm{H}} = +0.866\ \mathrm{m/s^2}}$

▶ **Quick Check:** The total 100-N force acting horizontally would produce an acceleration of 1.00 m/s^2, so the magnitude of this result is appropriate.

$F = ma$ is one of the most famous equations in science. To the modern physicist it's Newton's hallmark, his signature. Yet it does not appear in the *Principia*—and with good reason. This restricted formulation of the Second Law is the later work of Jacob Hermann (1716), refined by the Swiss mathematician Leonhard Euler. It's a tribute to Newton's insight that his analysis, framed in terms of momentum, carries over naturally into the more precise Theory of Relativity. The formula $F = ma$ is the low-speed approximation of a more accurate velocity-dependent expression. Still, Eq. (5.3) applies at ordinary speeds (much lower than the speed of light), and we will make good use of it.

5.2 Weight: Gravitational Force

Weight, anywhere near the Earth, is the downward force toward the center of the planet experienced by an object as a result of the Earth-object gravitational interaction. Because the Earth is revolving, a scale in your bathroom is actually moving in a circle and therefore accelerating (p. 140). Let's call the "weight" it reads, the **effective weight**. Effective weight is slightly lower than the true weight as defined above, but the differences are usually negligible.

The Second Law provides a relationship between the concepts of weight ($\mathbf{F}_{\mathrm{W}}$) and mass (m). Because the object's Earth-weight is pulling down, causing it to accelerate at g, $\mathbf{F} = m\mathbf{a}$ becomes

$$\mathbf{F}_{\mathrm{W}} = m\mathbf{g} \tag{5.5}$$

By pushing back on a fixed block the runner causes the block to push forward on her. It is this force that initially accelerates the athlete in the forward direction.

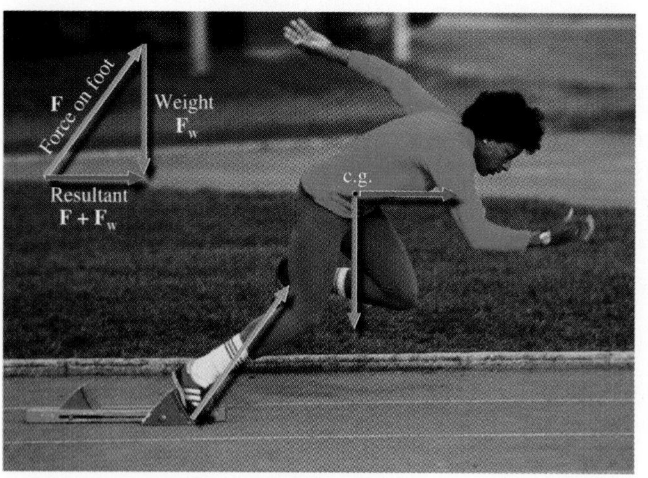

that is, the more mass, the more weight. Here, **g** is again the gravitational acceleration vector pointing downward. The weight of an object acting down numerically equals the force that must be exerted up on it (for example, by the floor or a scale) in order to keep the object from accelerating because of gravity (p. 71). The weight on Earth of a 2.0-kg chicken is just (2.0 kg)(9.8 m/s^2) = 20 N. Conversely, an apple weighing 1 N on our planet has a mass of (1 N)/(9.8 m/s^2) = 0.1 kg. That apple, anywhere in the Universe, has a mass of 0.1 kg, although its weight depends on the local gravity.

We can demonstrate some of these ideas using a quasipractical example. Suppose you meet a 54.4-kN (12 000-lb) bull elephant wearing ice skates and stranded in the middle of a frozen pond, and you decide to push him to shore. Neglecting friction, how hard must you shove to uniformly accelerate the pachyderm from rest up to 6.7 m/s (15 mi/h) in 10 s? The required acceleration ($a = \Delta v/\Delta t$) is 0.67 m/s^2, which is not very much. Thus

$$F = ma = \frac{F_W}{g}a$$

and
$$F = \frac{(54.4 \times 10^3 \text{ N})}{9.81 \text{ m/s}^2}(0.67 \text{ m/s}^2) = 3.7 \text{ kN}$$

or 0.84×10^3 lb. Well, forget that idea. If you were even more ambitious and wanted the elephant to accelerate over the ice at $a = 9.8 \text{ m/s}^2 = g$, a would cancel g in the equation and $F = F_W$. You would have to push with a force equal to the elephant's weight. That's true on a frictionless ice pond or out in space far from any celestial body where gravity is negligible and the object is essentially weightless—the needed force depends on the mass and the acceleration. The moral of this tale is that you *can* get squashed out in space between two colliding weightless elephants.

Weights on Ropes

Let's now examine the forces acting in the system depicted in Fig. 5.9a. A 1-kg brass mass on a rope hangs from a hook. Suppose we want to study the forces on both the rope and brass mass. Accordingly, draw a free-body diagram of each object (Figs. 5.9b and c). Doing so will allow us to write equations for those objects and thereby determine the forces. Gravity tugs on every atom of a body, but it will suffice (for the moment) to draw the overall weight vector acting straight down from the center of each object. The weight of the rope is F_{Wr}, and the weight of the brass mass is F_{Wb}. Gravity pulls down on the brass with F_{Wb}, and the brass pulls down on the rope (F_{br}). The rope, in turn, pulls up on the brass with an equal and opposite interaction (F_{rb}). The hook pulls up on the rope (F_{hr}), and the rope pulls back down on the hook (F_{rh})—another interaction pair ($F_{hr} = F_{rh}$). Applying the Second Law in the vertical direction to the brass object (Fig. 5.9b) and taking up as positive, we have

$$+\uparrow \Sigma F_{Vb} = F_{rb} - F_{Wb} = m_b a_b = 0$$

and $F_{rb} = F_{Wb}$. Three forces act on the rope and, since it is not accelerating,

$$+\uparrow \Sigma F_{Vr} = F_{hr} - F_{Wr} - F_{br} = 0$$

And with $F_{rb} = F_{Wb}$

$$F_{hr} = F_{Wr} + F_{br} = F_{Wr} + F_{Wb}$$

Not surprisingly, the hook (pulling on the rope) holds up all the weight. The tension at the top of the rope (F_{hr}) is greater than the tension on the bottom (F_{br}) by an amount equal to the weight of the rope. From now on, we shall deal only with ropes having negligible weight compared to the loads they support. In that case, *the tension (F_T) at any two points in a rope will be the same (provided no tangential forces act on the rope between those two points).* Henceforth we can assume that **the tensile force exerted by any rope has a scalar value F_T and is directed along the rope's length toward its center**.

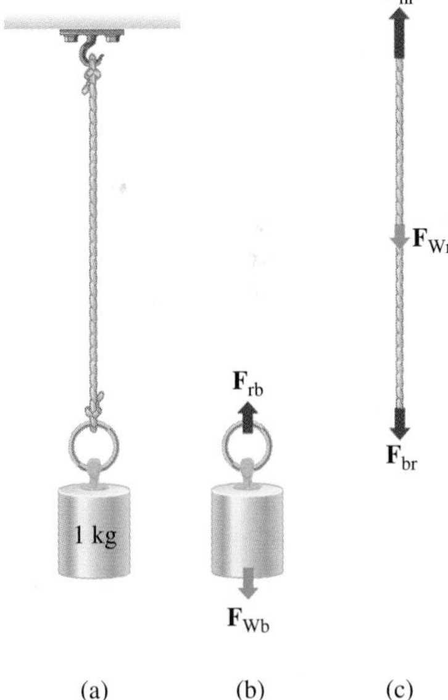

Figure 5.9 (a) A hook supporting a rope and a 1-kg brass mass. Suppose we wish to study the forces acting on both the mass and the rope. In that case, draw a free-body diagram of each object of interest, (b) and (c). The weights act down from the middles of the mass ($\mathbf{F}_{Wb}$) and the rope ($\mathbf{F}_{Wr}$). All the forces are colinear, and the Second Law applies to each separate object.

The Physics of Standing Still

The force exerted by a surface on an object is called the **reaction force**. It may have both a tangential component, known as the **force of friction** ($\mathbf{F}_f$) and a perpendicular component (that is, normal to the surface) known as the **normal force** ($\mathbf{F}_N$).

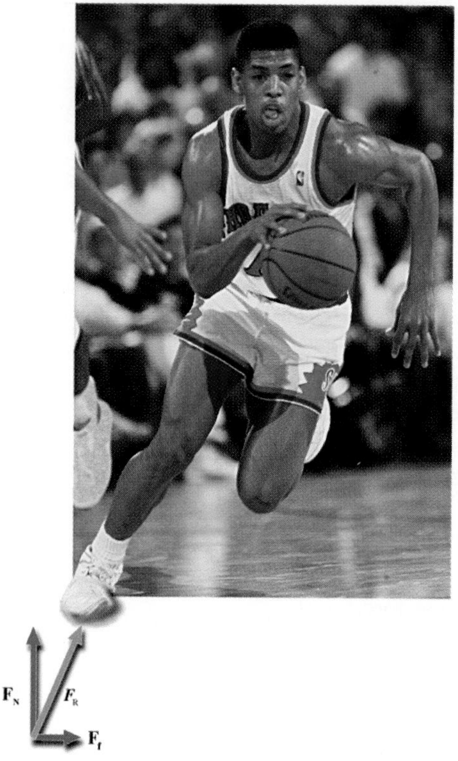

The floor pushes up and to the left with a reaction force $\mathbf{F}_R$. It has components $\mathbf{F}_f$ (a friction force keeping his foot from sliding) and $\mathbf{F}_N$ (the normal force equal to his weight).

Both these so-called *contact forces* are due to the electromagnetic interaction between atoms of the surface and atoms of the object when they come into proximity. Let's first deal with the normal force, leaving friction for later. Because we live on a planet where gravity is appreciable, objects most often rest on horizontal surfaces and $\mathbf{F}_N$ usually points straight up (Fig. 5.10). That need not be the case: $\mathbf{F}_N$ must be perpendicular to the surface, but the latter could be tilted in any direction (a Ping-Pong ball flattened against a vertical paddle experiences a horizontal normal force).

Weight acts down, it behaves as an *external force* acting *on* an object. In effect, the Earth pulls down on the object and the object pulls up on the Earth—that's an interaction pair. As a rule, we do not go plummeting down to the center of the planet, so there must be some other force stopping us. Consider the person in Fig. 5.10. Her weight causes her feet to press on the ground and the ground pushes back on her—that's also an interaction pair. This contact-force pair, which is basically electromagnetic, is the same phenomenon that keeps you from walking through walls.

Call the force exerted perpendicularly on a supporting surface (such as a floor, pillar, or roadway) the **load**. The weight of a brick resting on a floor numerically equals the downward load *on the floor,* and the equal-and-opposite force *on the brick* exerted by the floor is the upward normal force. If you lift up slightly on the brick, decreasing the load, then the normal force will be reduced accordingly. If instead you step onto the brick, the normal force will numerically equal the combined weight, the total load.

When you stand still, the net vertical force *on* you (weight down, normal force up) equals zero. The Second Law then demands that you continue to remain at rest in the vertical direction. If, for structural reasons, the floor cannot exert an upward force equal to your weight, the nonzero net force acting on you would be down, and down you would accelerate through the floor, just as if you were trying to stand on a paper box or walk on water (Table 5.2).

Floors do not usually collapse underfoot. They exactly match your weight with a reaction force, and one might wonder how they manage that. Materials, no matter how rigid, give under the influence of a body pushing on them, regardless of how lightly. When two objects exert a contact force on each other, atoms in both must be

It is the same for me with mechanics as it is with languages. I understand the mathematical laws, but the simplest technical reality demanding perception is harder to me than to the biggest blockheads.

KARL MARX (1818–1883)

TABLE 5.2 The physiological effects of acceleration

Acceleration	Body orientation	Effect
$2g$	Upright parallel to a	Walking becomes strenuous
$3g$	Upright parallel to a	Walking is impossible
$4g$–$6g$	Upright parallel to a	Progressive dimming of vision due to decrease of blood to retina, ultimate blackout
$9g$–$12g$	Reclining perpendicular to a	Chest pain, fatigue, some loss of peripheral vision, but one is still conscious and can move hands and fingers

somewhat displaced by the interaction. The greater the load, the more a floor stretches, building up a counterforce via springlike atomic interactions (that are electromagnetic) within the floor. The material ceases to distort any further when the upward reaction of the floor exactly cancels the load on it. Wooden floors often betray the sagging by squeaking as one board rubs past another, but all floors sag underfoot.

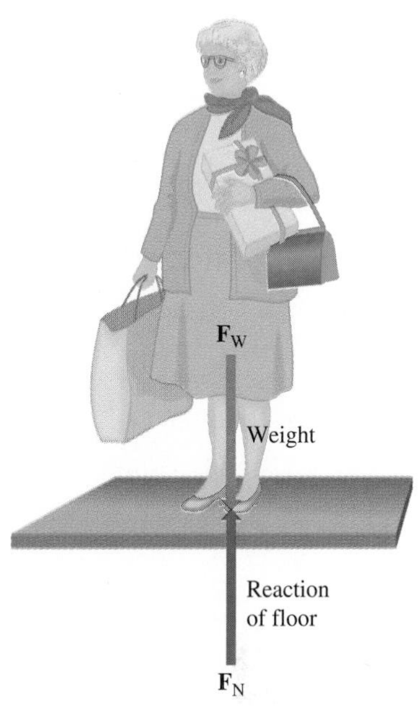

Figure 5.10 The weight of an object is supported by an equal normal force exerted upward by the floor. The free-body diagram shows the two forces acting on the standing person. The normal force is actually distributed over both feet.

The Inclined Plane

Although the force of gravity acts straight down, when an object is supported by an inclined plane (Fig. 5.11), components of its weight exist both parallel ($\mathbf{F}_{W\parallel}$) and perpendicular ($\mathbf{F}_{W\perp}$) to the surface. As you can see in Fig. 5.11b, the only possible motion is along the incline, and it is $F_{W\parallel}$ that drives that motion. As we will learn presently, $F_{W\perp}$ influences friction and will be of interest as well. The angle between the line-of-action of the weight vector ($\mathbf{F}_W$) and the normal to the incline equals θ, the incline angle. Consequently

$$F_{W\parallel} = F_W \sin\theta \tag{5.6}$$

and

$$F_{W\perp} = F_W \cos\theta \tag{5.7}$$

$F_W \sin\theta$ pushes the object down the incline, and $F_W \cos\theta$ pushes it into the surface. The Second Law applies independently in each perpendicular direction. As shown in the free-body diagram (Fig. 5.11b), there is a force $F_{W\parallel}$ that will accelerate the object down the slope; $\Sigma F_\parallel = ma_\parallel$, and we can compute $a_\parallel$. There is no acceleration perpendicular to the slope; $\Sigma F_\perp = ma_\perp = 0$, and $F_{W\perp}$ must be balanced by an equal and oppositely directed reaction force F_N.

Example 5.4 The 50.0-kg skier in Fig. 5.11 coasts along the surface of a snow-covered slope (assumed to be frictionless) tilted at an angle of 30.0°. Compute (a) the magnitude of the normal force acting on her; (b) the magnitude of the force tending to drive her down the inclined plane; and (c) the resulting acceleration ignoring air drag.

Solution: [Given: $m = 50.0$ kg and $\theta = 30.0°$. Find: (a) F_N; (b) the force parallel to the incline; and (c) $a_\parallel$.] The component of the skier's weight pressing her into the surface is

$$F_{W\perp} = mg\cos\theta = (50.0 \text{ kg})(9.81 \text{ m/s}^2)\cos 30.0°$$

and so

$$F_{W\perp} = 425 \text{ N}$$

We know that the skier does not leave the surface of the incline, so $a_\perp = 0$. Taking up (normal to the surface) as positive, the sum of the forces perpendicular to the incline is zero, thus

$$+\nwarrow \Sigma F_\perp = 0 = F_N + (-F_{W\perp})$$

$F_{W\perp}$ points into the surface in the negative direction—we enter it with a minus sign. Hence, $\boxed{F_N = 425 \text{ N}}$.

(b) The magnitude of the driving force down the incline is

$$F_{W\parallel} = mg\sin\theta = (50.0 \text{ kg})(9.81 \text{ m/s}^2)\sin 30.0°$$

and

$$\boxed{F_{W\parallel} = 245 \text{ N}}$$

(c) To compute the down-plane acceleration (taking the direction of motion down the slope as positive), we write

$$+\swarrow \Sigma F_\parallel = ma_\parallel = F_{W\parallel}$$

and, hence

$$a_\parallel = \frac{(mg\sin\theta)}{m} = g\sin 30.0° \tag{5.8}$$

or $\boxed{a_\parallel = \frac{1}{2}g}$. This result is independent of the mass, and applies to any body sliding down a frictionless incline at $\theta = 30°$.

▶ **Quick Check:** We can think of the inclined plane as a kind of "gravity reducer." With $\sin 30° = \frac{1}{2}$ in Eq. (5.6), the body behaves as if it were in free-fall (down the slope) at $\frac{1}{2}g$. Since the skier's weight is 490 N, in this tilted world it's driven down by $\frac{1}{2}$ 490 N. A normal force of 425 N at a small incline is also reasonable.

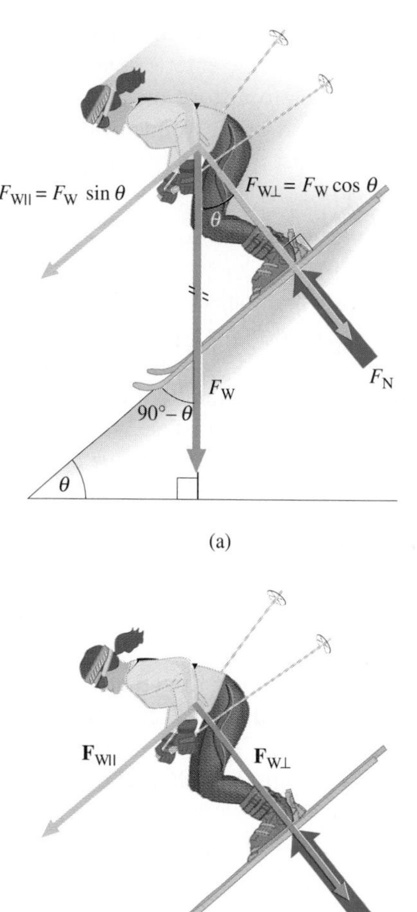

Figure 5.11 A body on an inclined plane. (a) The component of weight acting down the inclined plane drives the body downhill. The component of weight acting perpendicular to the plane is matched by the normal force. (b) Applying the Second Law down the incline yields $\Sigma F_{\parallel} = F_{W\parallel} = ma_{\parallel}$.

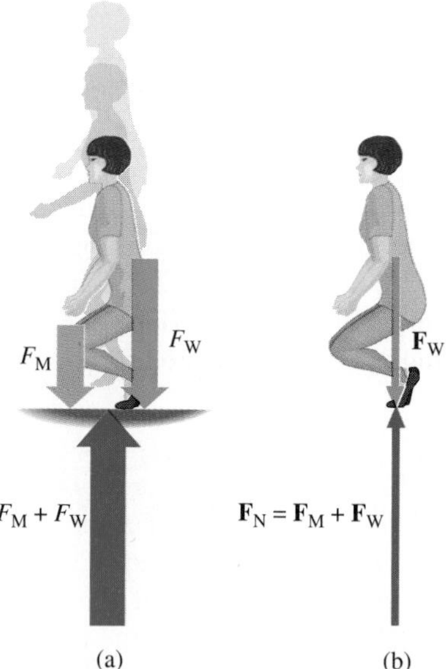

Figure 5.12 (a) When an individual jumps, two forces act *on* the person, F_W down and the normal force F_N, acting up. (b) The floor experiences a net downward force of ($F_M + F_W$), and it exerts an upward equal and opposite reaction F_N. The net force on the jumper is F_M upward.

The roof of this building in the Northeast is constructed as a strongly tilted inclined plane to allow snow to slide off rather than pile up and load the structure.

5.3 Vertical Accelerations

Have you ever seen $\approx 1\frac{1}{2}$-year-old kids (who have not yet learned to jump) trying desperately to leap into the air? What they do not know is that a net external vertical force must be applied if a mass is to accelerate upward. The Third Law is the key: *push down on the floor and it will push up on you.* Push with your leg muscles downward (F_M) on the floor (Fig. 5.12a). The total load is then ($F_M + F_W$), down. The upward reaction, the normal force (F_N), is then numerically equal to ($F_M + F_W$). But only two forces act *on you,* F_W down and F_N up. F_M is a force exerted by you *on the floor* (Fig. 5.12b). The sum of the vertical forces on you equals your mass times your acceleration. Taking up as positive

$$+\uparrow\Sigma F = ma = (-F_W) + (F_N) = -F_W + (F_M + F_W) = F_M$$

which is a net upward force, and up you go ($a \neq 0$). The harder you push down, the greater the resulting acceleration. Skeptics are encouraged to leap up from a bathroom scale to see it read $(F_W + F_M)$ during the accelerated ascent.

Example 5.5 A graduate student (having a mass of 40 kg) is standing inside an elevator on a scale that reads in newtons (Fig. 5.13). (a) If the elevator is at rest, what does the scale read? (b) Now suppose the elevator accelerates upward at one-half the acceleration of gravity ($a = \frac{1}{2}g$)—what does the scale read?

Solution: [Given: $m = 40$ kg, $a = \frac{1}{2}g$. Find: (a) F_W when $a = 0$; (b) F_N when $a = 0.50g$.] (a) Take up as positive. When $a = 0$

$$+\uparrow\Sigma F_V = F_N + (-F_W) = 0$$

The directions of the vectors are taken care of with the signs, so all quantities are entered as positive numbers. Since $F_W = mg$, the load read by the scale is

$$F_W = F_N = (40\text{ kg})(9.8\text{ m/s}^2) = \boxed{0.39\text{ kN}}$$

(Fig. 5.13b), and this is also the case even when the elevator moves at a constant speed. (b) When the elevator accelerates upward at a, it must carry the student up with it (Fig. 5.13c). Taking up as positive

$$+\uparrow\Sigma F_V = F_N + (-F_W) = ma$$

$$F_N - mg = ma$$

and $$F_N = ma + mg = m(a + g)$$

The floor must push up on the passenger with a total force F_N that equals the sum of her original weight (mg) plus an upward accelerating force (ma); this sum equals the load, and it is the force the scale will read. Since $a = 0.50g$

$$F_N = m(1.5\,g) = (40\text{ kg})(1.5)(9.8\text{ m/s}^2) = 0.59\text{ kN}$$

The scale reads an "effective weight" of $\boxed{0.59\text{ kN}}$. Everything within this accelerating environment behaves as if gravity were increased. That's why packages always seem heavier during the brief upward accelerating phase of an elevator ride.

▶ **Quick Check:** Her weight is around 10 × 40 or 400 N, and with an increase of $\frac{1}{2}g$ the effective weight should be roughly 600 N, so 0.59 kN is reasonable.

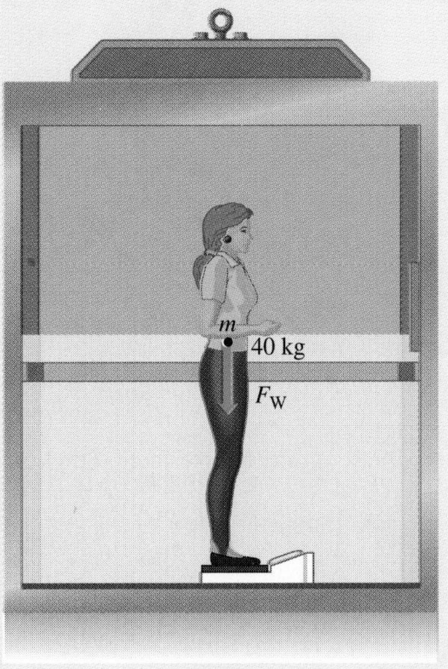

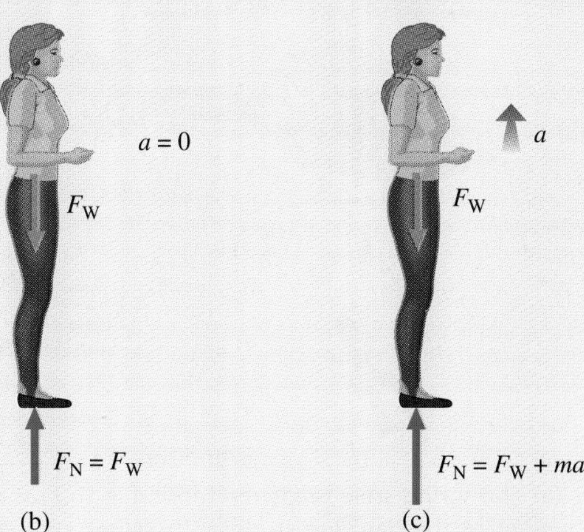

Figure 5.13 (a) A student standing in an elevator on a bathroom scale. (b) When $a = 0$, the normal force equals the weight, and the scale reads F_W. (c) When the elevator accelerates up at a, the normal force increases beyond mg by ma. The scale reads $F_W + ma$.

A 12-cylinder Jaguar can accelerate from rest at roughly $g/3$. The driver is then pushed forward by the seat in proportion to the acceleration (regardless of the speed). Much the same thing happens to spacebound astronauts. The space shuttle crews are accelerated skyward by an up-pointing normal force. In reaction, they press down against their reclining couches with a peak force three times as much as they felt under normal gravity. Yet even that was a pleasant change from the $6g$ accelerations experienced by the *Apollo* crews during their Saturn-rocket lift-offs to the Moon (Table 5.2, p. 132).

Example 5.6 Jane, a 50-kg undergrad, is rescued from a burning building by a police helicopter (Fig. 5.14a). She hangs at the end of a rope dangling beneath the chopper. (a) Compute the tension in the rope as read by the spring scale when the craft is hovering. (b) Suppose the helicopter accelerates straight downward with respect to the ground at 3.0 m/s². What is the tension in the rope now?

Solution: [Given: $m_J = 50$ kg; a_J is Jane's acceleration with respect to the ground. Find: the tension F_T (a) when $a_J = 0$ and (b) when $a_J = 3.0\text{ m/s}^2$.] (a) Figure 5.14b is Jane's free-body diagram where $a_J = 0$. The tensile force $\mathbf{F}_T$ (that is, the pull of the rope) acts up, while her weight acts down. It doesn't matter, but since the motion will be down later, take down as positive, and so

$$+\downarrow\Sigma F_V = (-F_T) + F_W = m_J a_J = 0$$

Accordingly, the tension equals her weight, $F_T = F_W = m_J g = (50\text{ kg})(9.8\text{ m/s}^2)$, and $\boxed{F_T = 0.49\text{ kN}}$. (b) When the chopper drops downward at a rate less than g, she drops with the same acceleration. The magnitude of the net force *on* her (Fig. 5.13c) equals her mass times that acceleration

$$+\downarrow\Sigma F_V = (-F_T) + F_W = m_J a_J$$

and

$$F_T = F_W - m_J a_J$$

Hence

$$F_T = 0.49\text{ kN} - (50\text{ kg})(3.0\text{ m/s}^2) = \boxed{0.34\text{ kN}}$$

▶ **Quick Check:** The chopper accelerates down at roughly $g/3$. That puts her in a $\frac{1}{3}$-reduced gravity environment. Her scale-measured "effective weight" is reduced by $\frac{1}{3}$, hence the tension should be about $\frac{2}{3}gm \approx \frac{2}{3}500\text{ N} \approx 0.33\text{ kN}$.

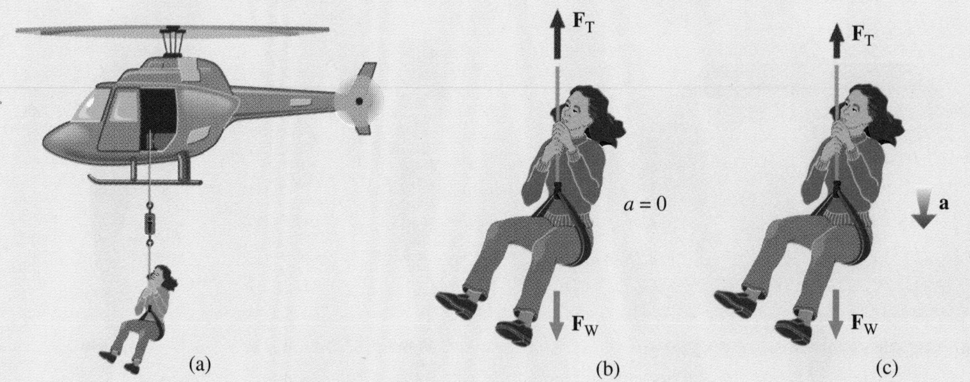

Figure 5.14 (a) The spring scale reads the tension in the rope produced by weight of the dangling student. (b) When there's no acceleration, the sum of the forces acting on her is zero, and $F_T = F_W$. (c) When the person accelerates down, $F_T < F_W$.

5.4 Effectively Weightless

Nothing in this Universe of $\approx 10^{22}$ stars is gravityless. Yet a body falling freely behaves as if it were without weight, and so we say it is *effectively weightless.* Hold

This diver, high in the air over Barcelona at the 1992 Olympic games, is effectively weightless in free-fall.

Figure 5.15 Floating around within a free-falling elevator after pushing off from its floor. The elevator moves downward more rapidly than the person who slowed his descent by pushing on the floor.

your palm facing skyward, and place a bunch of keys in it, keeping your hand open. Now drop your arm, accelerating it at 9.8 m/s^2. The keys descend at this rate all by themselves—let them fall freely and simply have your hand precede them down. At the point where your hand reaches $a = g$, you will no longer feel the keys pressing on your palm; your hand will fall away as the keys fall, and the keys will become effectively weightless.

Now imagine that you are in an elevator falling freely, accelerating at g. Not only would the keys again hover "weightlessly" in your hand, but if you were standing on a scale it would read zero. The floor, dropping out from beneath your feet, would exert no upward response unless you pushed down on it with some muscle-generated force (Fig. 5.15). To be precise, the gravitational force still acts; in fact, *it's the only force that does act.* According to our definition of weight, an object in free-fall is not weightless; $F_W = mg$ is not zero. However, if we define the **effective weight** of an object as *the force it exerts on a scale,* then bodies in free-fall have an effective weight of zero, which is what the popular literature calls "zero-g" or "zero gravity." As we will see, your bathroom scale is also accelerating downward, so it too reads effective weight.

Suppose that our elevator is now hurled upward, at first accelerating at a rate a and then let loose to continue up in free unpowered flight. If you stepped off a chair inside the elevator during the powered ascent, you would accelerate down at g while the floor accelerated up to meet you at a. The combined result would be a relative acceleration of you with respect to the floor of $(g + a)$. Once you alight on the floor, you would have an effective weight of $m(g + a)$. During the unpowered portion of the ascent, our elevator-turned-rocketship will continue to climb, slowing down as it rises because of the ever-present tug of gravity. Step off the chair now and you will accelerate Earthward at g as usual, but so will the elevator, even as it continues to climb. Both passenger and craft are in free-fall although heading upward, and you again hover effectively weightless in midair right where you left the chair.

The *Apollo* astronauts floated around in their spaceship on the way to and from the Moon. They were not gravityless, or weightless, just effectively weightless in free-fall. The National Aeronautic and Space Administration (NASA) routinely flies a padded research plane called the "Vomit Comet" (see photo, p. 141) in arcs so that trainee astronauts can float effectively weightless. Gravity acts uniformly on all parts of the body, and so in free-fall one feels nothing—no perception of endless descent—just effortless floating.

5.5 Coupled Motions

Figure 5.16 shows several different situations in which two masses are attached together by an unstretchable rope. The pulleys are weightless and frictionless, so there are no tangential forces, and the tensions are therefore constant throughout each rope. For the moment, the surfaces are frictionless, as well. Suppose the motion takes place in the direction shown by the arrow in each case. The leading mass m_1 pulls the connecting rope along, and the rope pulls the trailing mass m_2 along. The trailing mass can never overtake and slacken the rope, nor can it lag behind, accelerating slower than the rope. Each mass must accelerate at the same rate. With two masses, we can write two coupled Second-Law equations and solve for two unknowns, usually F_T and a, although any two parameters could be unknown.

If there is any ambiguity about how the system moves, *guess at the direction of the overall motion and make that positive,* even if it means down is positive for one part and up for another. If you guess wrong, the values of the unknowns will just come out negative.

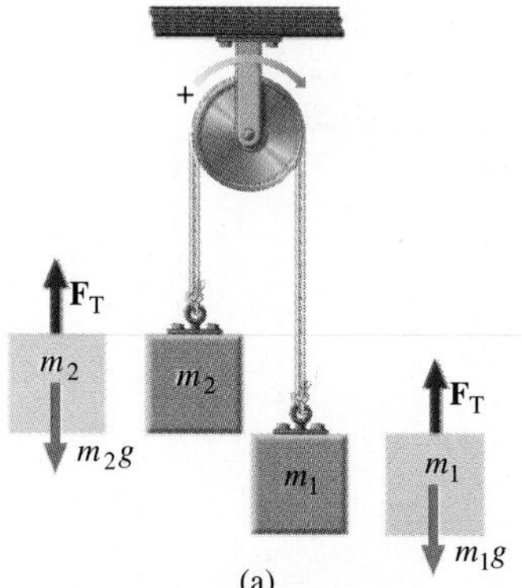

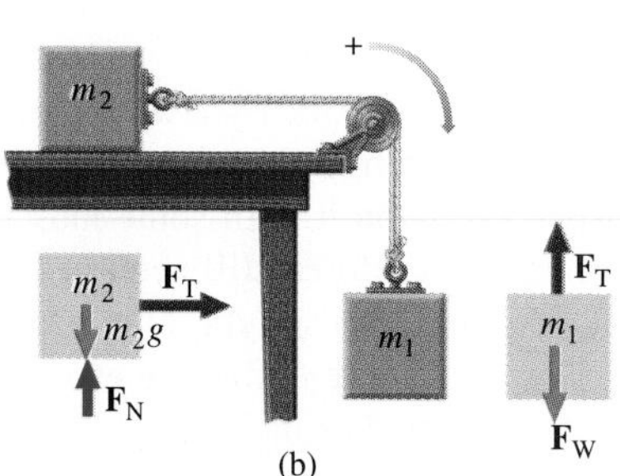

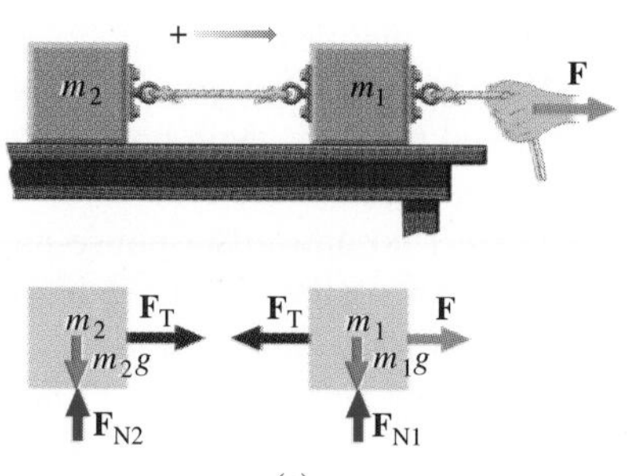

Figure 5.16 Three examples of coupled arrangements where two masses are attached together by a rope. The free-body diagram for every mass is drawn. In each of the three examples the same tension acts on both masses, and both accelerate at the same rate.

Example 5.7 Marylou ($m_M = 50$ kg) and her boyfriend, Doug ($m_D = 70$ kg), are tied together by a rope of negligible mass. She is standing on a frictionless horizontal sheet of wet ice when Doug accidentally steps off a cliff (Fig. 5.17a). Assume the unlikely possibility that the tree limb is frictionless and that her length of the rope is horizontal. Determine (a) the tension in the rope, and (b) the accelerations of the lovers.

Solution: [Given: $m_M = 50$ kg and $m_D = 70$ kg. Find: F_T, a_M and a_D.] First, draw a free-body diagram of each person, as in Fig. 5.17b and c. As long as the rope does not stretch, $a_D = a_M = a$. Draw a curved arrow in the direction of motion of the system from Marylou over the branch to Doug. For him, down is positive, while for her, to the right is positive. There are two unknowns, a and F_T, and we have to construct two equations. Marylou

(continued)

(continued)

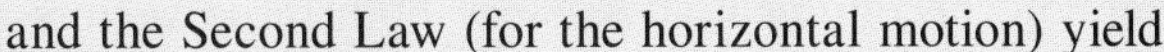

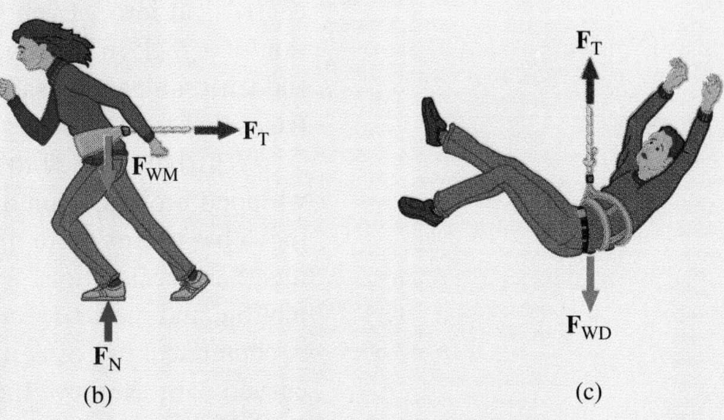

Figure 5.17 (a) Marylou sliding along the ice, after Doug falls off the cliff. Free-body diagrams of (b) Marylou, and (c) Doug.

and the Second Law (for the horizontal motion) yield

$$\overset{+}{\rightarrow}\Sigma F_{HM} = F_T = m_M a \tag{5.9}$$

while Doug's vertical contribution provides

$$+\downarrow\Sigma F_{VD} = F_{WD} - F_T = m_D a$$

We solve these two equations simultaneously (Appendix A-4) for either unknown. Let's do it for a first. Substituting $F_T = m_M a$ from Eq. (5.9) into the vertical summation leads to

$$F_{WD} - m_M a = m_D a$$

We do not have Doug's weight, but Eq. (5.5) lets us write it in terms of his mass: $F_{WD} = m_D g$, hence

$$m_D g = m_D a + m_M a$$

and
$$a = \frac{m_D}{(m_D + m_M)} g$$

Putting this expression into Eq. (5.9) produces the desired equation for the tension, namely

$$F_T = \frac{m_M m_D}{(m_D + m_M)} g$$

The numerical values are $\boxed{a = 0.58\ \text{g}}$ and $\boxed{F_T = 0.29\ \text{kN}}$. The tension is the magnitude of the force accelerating dear Marylou. If the rope were to be cut, the tension would drop to zero, and she would coast at a fixed speed. Doug would descend in free-fall.

▶ **Quick Check:** When $m_D \gg m_M$, $(m_D + m_M) \approx m_D$, $a \approx g$, which means Doug is approximately in free-fall, and that's reasonable. Moreover, $F_T \approx m_M g$, and that too is to be expected. When $m_M \gg m_D$, $a \approx 0$ and $F_T \approx F_{WD}$, as it would be if Doug were tied to a massive boulder. There is then no acceleration and he just hangs over the edge. Both equations produce sensible results at the extremes, suggesting that they may be correct. Moreover, both have correct units.

CURVILINEAR MOTION

The central problem of Newton's *Principia,* indeed the central question of the age, was "How do the planets move?" The motion of the heavens had long been one of the great issues of Western science, philosophy, and theology. It was Newton who finally welded dynamics and astronomy into a mathematical understanding that swept away Aristotelian scholarship as if that genius had been little more than an amusing schoolboy. At the heart of the Newtonian synthesis is the concept of a center-seeking gravitational force holding the planets in their orbits. We introduce one of the crucial ideas here, leaving a discussion of the role of gravity for Chapter 7.

5.6 Centripetal Acceleration

Newton probably came to the correct understanding of motion along a curve in the early 1680s, no doubt with the inadvertent help of his chief antagonist, the brilliant, argumentative, lascivious Robert Hooke. The common wisdom, then and now (though it is wrong!), was that revolving objects are thrown radially outward by a centrifugal force. Even so, back in 1585, G. B. Benedetti, while considering the motion of a rock in a sling, rightly pointed out that an object revolving in a circle and suddenly unleashed sails off in a straight line *tangent* to the curve at the point of release. (David knew as much intuitively. Perhaps Goliath did, too.)

Whirl a ball tied to a string in a circle. Its velocity continuously changes as it is whipped around, even at a constant speed, because the direction of $\mathbf{v}$ keeps changing. That means there must be an acceleration and, by the Second Law, a force causing that acceleration. The hand pulls the string inward toward the center of the motion, and the string continuously pulls the ball off its straight-line inertial course. A center-seeking or **centripetal force** must be exerted if any object is to move in a curved path. Remove the inward centripetal force and the motion instantly becomes straight-line inertial, and that's tangential, *not* outwardly radial.

We can derive an expression for the *centripetal acceleration,* $\mathbf{a}_C$, for an object that moves uniformly in a circle (Fig. 5.18). The body travels from A to B in a time interval Δt, and as it does, the radius r sweeps through an angle θ. The velocity $\mathbf{v}_i$ at A subsequently changes direction, becoming $\mathbf{v}$ at B. Since both these vectors are perpendicular to r, $\mathbf{v}_i$ must also sweep through an angle θ as it moves into $\mathbf{v}$, and that is shown in the velocity triangle of Fig. 5.18b. The latter diagram shows that the initial velocity plus the change $\Delta\mathbf{v}$ equals the final velocity ($\mathbf{v} = \mathbf{v}_i + \Delta\mathbf{v}$). Inasmuch as $v = v_i$, since the speed is constant, the velocity triangle is isosceles, just as is triangle ABO. Indeed, these two are similar triangles, both having the same vertex angle θ. It follows that the ratios of any two of their corresponding sides are equal, thus

$$\frac{\Delta v}{v} = \frac{\overline{AB}}{r}$$

The side $\overline{AB}$ is the magnitude of the displacement Δs, as the body moves from A to B, and so

$$\Delta v = \frac{v}{r}\Delta s$$

Dividing by Δt gives us the average acceleration ($\Delta v/\Delta t$) in terms of the magnitude of the average velocity ($\Delta s/\Delta t$), hence

$$\frac{\Delta v}{\Delta t} = \frac{v}{r}\frac{\Delta s}{\Delta t}$$

In the limit as Δt approaches zero, (Fig. 2.20, p. 37), A comes ever closer to B, $\Delta s/\Delta t$ becomes v, by way of Eq. (2.7), and $\Delta v/\Delta t$ becomes a, by way of Eq. (3.3). This particular instantaneous acceleration is called the **centripetal acceleration**. For a body in circular motion, a_C is given by

$$a_C = \frac{v^2}{r} \tag{5.10}$$

where the units are m/s^2.

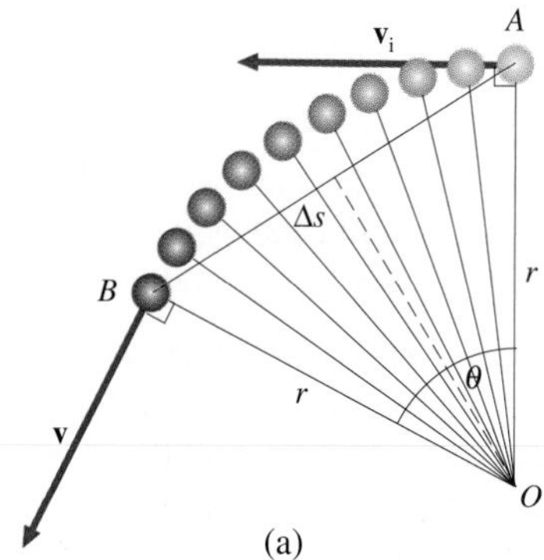

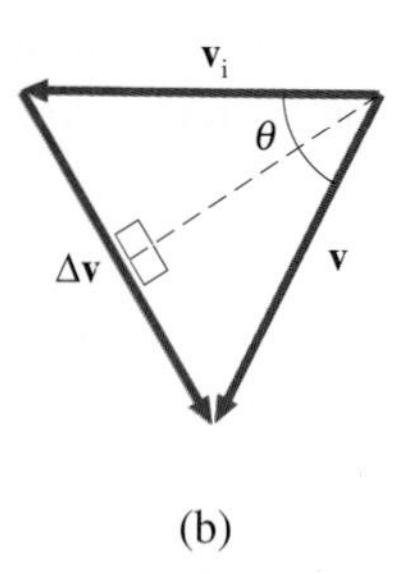

Figure 5.18 The geometry of circular motion. In this simple derivation the speed is constant, but that need not be the case. At any instant, $a_C = v^2/r$.

The direction of $\mathbf{a}_C$ is the direction of $(\Delta\mathbf{v}/\Delta t)$ in the limit as $\Delta t \to 0$. There are several ways to see that $\Delta\mathbf{v}$ points inward perpendicular to Δs—for instance, in the two similar triangles, $\overline{OA}$ is perpendicular to $\mathbf{v}_i$, $\overline{OB}$ is perpendicular to $\mathbf{v}$, and therefore Δs must be perpendicular to $\Delta\mathbf{v}$. Alternatively, the dashed angle-bisector in Fig. 5.18b is parallel to Δs and also perpendicular to $\Delta\mathbf{v}$. Since $\mathbf{v}_{av} = \Delta\mathbf{s}/\Delta t$, the average acceleration $(\Delta\mathbf{v}/\Delta t)$ must always be perpendicular to the average velocity. In the limit as $\Delta t \to 0$, the instantaneous acceleration $\mathbf{a}_C$ is everywhere perpendicular to the instantaneous velocity $\mathbf{v}$, which is always tangent to the path; $\mathbf{a}_C$ *points radially inward toward the center.* It's also common practice to refer to a_C as the *radial acceleration.* Interestingly, Huygens was the first to derive Eq. (5.10) but he, like many others, thought it was outward, away from the center.

Example 5.8 A beetle standing on the edge of a 12-in. recording of "Sergeant Pepper's Lonely Hearts Club Band" is whirling around at 33-$\frac{1}{3}$ rotations per minute. Compute the magnitude of the creature's centripetal acceleration.

Solution: [Given: $r = 6.0$ in. $= 0.152$ m and a rotation rate of 33-$\frac{1}{3}$ rpm. Find: a_C.] There is only one expression for a_C, and that is

$$a_C = \frac{v^2}{r} \qquad [5.10]$$

so v must be determined first. The total distance traveled in 1 min is $33\text{-}\frac{1}{3} \times 2\pi r = 31.9$ m; hence $v = (31.9\text{ m})/(60\text{ s}) = 0.532$ m/s. Consequently

$$a_C = \frac{(0.532\text{ m/s})^2}{(0.152\text{ m})} = \boxed{1.9\text{ m/s}^2}$$

▶ **Quick Check:** At a speed of 1 m/s and a radius of 1 m, a_C would be 1 m/s^2. Here, v is more than one-half that and r about one-sixth, so the answer should be a bit more than $(6/4)(1\text{ m/s}^2) \approx 3/2$ m/s^2.

Equation (5.10) applies to paths that are not completely circular (Fig. 5.19) provided that, at any instant, we use for r the radius of the circular segment being traveled, the *radius of curvature.*

5.7 Center-Seeking Forces

It follows from the Second Law that if a body of mass m is accelerating, it must be experiencing a net force given by $F = ma$. Any object moving in a circle must therefore be restrained by an inwardly directed centripetal force $\mathbf{F}_C$, with a magnitude of

$$F_C = ma_C = \frac{mv^2}{r} \qquad (5.11)$$

This is the force tugging the body off its otherwise straight-line inertial course. And the tighter the curve (the smaller is r), the more the path deviates from straight-line motion and the greater the needed force (Fig. 5.20). Centripetal force is not a new physical interaction, it is simply the name given to any force directed toward the center of the motion. For the Moon revolving about the Earth, $\mathbf{F}_C$ is gravitational. For a car rounding a turn on a flat highway, $\mathbf{F}_C$ is frictional. There are trains derailed all over the world simply because they are driven too fast around curves; Amtrak's Silver Meteor was derailed in December 1991 by an engineer oblivious to Eq. (5.11).

By flying along large parabolas in the sky, the so-called Vomit Comet can provide 40 s of effective weightlessness.

It was Newton who first showed that the bulge of the Earth (about 20 km at the equator) was due to the planet's rotation. Any piece of spinning Earth tends to fly

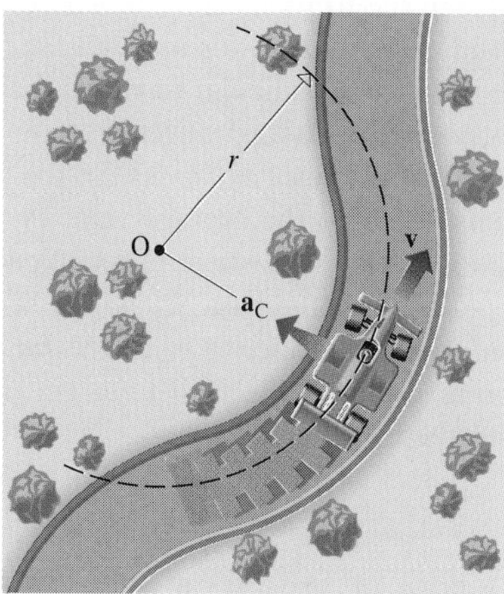

Figure 5.19 The centripetal acceleration of a car on a curved road at the moment when the radius of curvature is *r*.

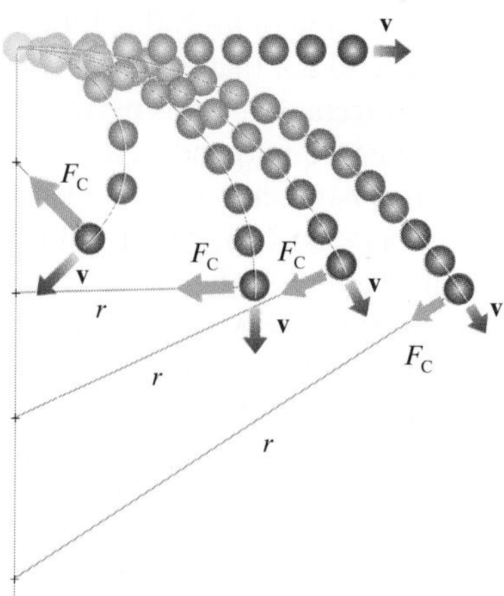

Figure 5.20 It takes no force for an object to move along a straight line at a uniform speed. Such a path can be thought of as a circle having an infinite radius. If the object is made to follow an arc of finite radius *r*, a centripetal force must be applied. The smaller the radius, the tighter the curve, and the greater is the force needed to pull the object away from its inertial path.

off tangentially and the planet essentially stretches out—the elastic forces holding it together provide the needed centripetal force. Failing that (given a high enough v), the dirtball would rip apart. It's the equatorial region that stretches out because v is greatest there.

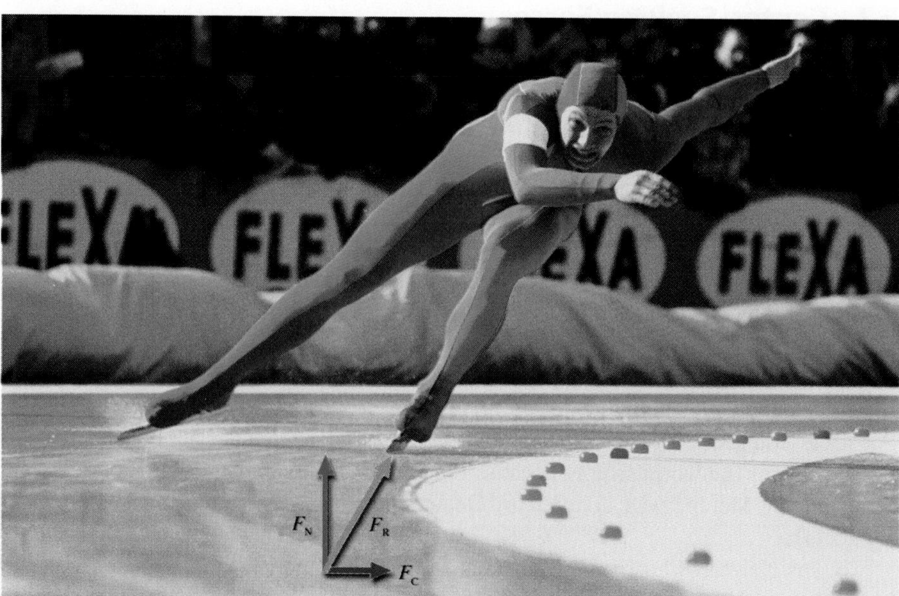

The reaction force F_R acting up along the line of the skate blade has two components: F_N (which is vertical and equal to the weight, and supports the skater) and F_C (which is the needed centripetal force).

Example 5.9 Compute the centripetal acceleration of an object located on the Earth's equator. The period of rotation, or time to spin once around, is 23 h 56 min 4 s (not 24 h; in fact, it's increasing even as you read this—see Problem 8.41) and the equatorial radius is approximately 6378 km. Let g_0 represent the local value of the acceleration due to gravity for a nonspinning Earth. Given that $g_0 = 9.81\ \text{m/s}^2$ at the equator, at what rate would a stone fall there, and what would be its weight on a scale compared to its nonrotating-Earth weight?

Solution: [Given: $T = 23$ h 56 min 4 s, or 86 164 s, and $r_\oplus = 6378$ km. Find: a_C.] First, compute v:

$$v = \frac{(2\pi r_\oplus)}{T} = 465.09\ \text{m/s}$$

and so

$$a_C = \frac{v^2}{r_\oplus} = \boxed{3.391 \times 10^{-2}\ \text{m/s}^2}$$

Any object at rest on the equator is already accelerating downward at $0.0339\ \text{m/s}^2$, because it is circling around with the planet. A stone would fall with an acceleration of $g_0 - a_C = \boxed{9.78\ \text{m/s}^2}$. Its measured weight would be $\boxed{99.65\%}$ of its nonrotating-Earth weight (see Fig. 7.6, p. 216).

▶ **Quick Check:** Look at Table 3.2, p. 73.

5.8 Circular Motion

It is possible to bank a roadway so that the normal force provides the centripetal force rather than relying on friction to provide it. Figure 5.21 shows how the normal force $\mathbf{F}_N$ (perpendicular to the track's surface) acting on a runner has components both vertical (supporting the weight) and horizontal (providing the centripetal force). With no vertical acceleration

$$+\uparrow \Sigma F_V = F_N \cos\theta - F_W = 0$$

Notice that $F_N > F_W$. There is a horizontal or radial acceleration, and so

$$\overset{+}{\rightarrow} \Sigma F_H = m a_H$$

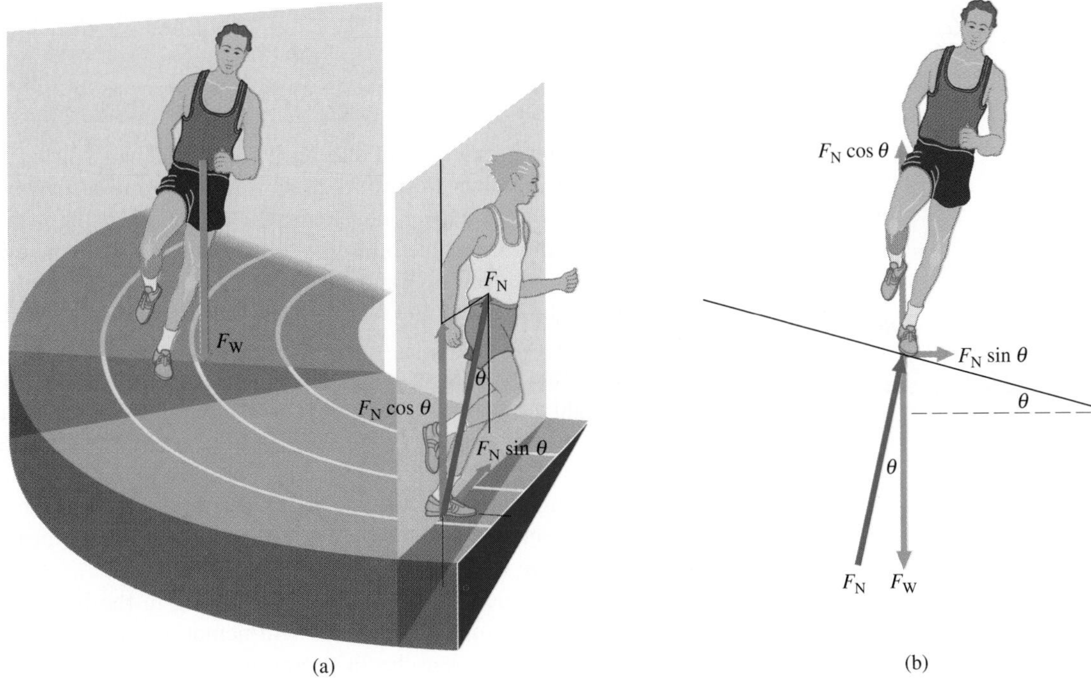

Figure 5.21 On a banked road the normal force F_N, which is always perpendicular to the road, is not aligned with the weight F_W, which is always straight down. The normal force has a radial component, $F_N \sin\theta$, that can be made equal to F_C.

A banked luge track. The athlete takes a position that will minimize air resistance.

with $a_H = a_C$ $$F_N \sin\theta = F_C = \frac{mv^2}{r}$$

The first equation contains cos θ and the second, sin θ. One way to solve for θ is to combine the two into a single expression in terms of tan θ. Accordingly, write the first equation as $F_N \cos\theta = F_W = mg$ and divide the second one by it, thus

$$\frac{F_N \sin\theta}{F_N \cos\theta} = \frac{mv^2/r}{mg}$$

and

$$\tan\theta = \frac{v^2}{gr} \tag{5.12}$$

This expression gives the proper banking angle for any one speed, and it applies equally well to cars on roads, trains on tracks, and even to birds and planes banking during turns in level flight. Any car—Eq. (5.12) is independent of mass—traveling at the posted speed can safely make the tilted turn even if the road is slippery, which is the reason highway ramps are banked. It follows from Eq. (5.12) that the smaller the radius and the greater the speed, the larger is tan θ and, hence, the greater the banking angle. Motorcyclists and tobogganers are forever careening at high speeds around incredibly tilted tracks.

Example 5.10 A circular track with a 20-m radius is to be banked at an angle θ appropriate for a "4.0-min mile." Compute θ.

Solution: [Given: $r = 20$ m and v corresponding to a speed of 1 mi per 4 min. Find: θ.] Using Eq. (5.12), we fix θ for a 4.0-min mile; that is, a 240-s (1.609×10^3 m) run, for which

$$v = \frac{1.609 \times 10^3 \text{ m}}{240 \text{ s}} = 6.71 \text{ m/s}$$

Hence

$$\tan\theta = \frac{v^2}{gr} = \frac{(6.71 \text{ m/s})^2}{(9.81 \text{ m/s}^2)(20 \text{ m})} = 0.229$$

and

$$\boxed{\theta = 13^\circ}$$

▶ **Quick Check:** The runner will lean into the curve balancing automatically and quite comfortably at a 13° tilt: $v = \sqrt{gr\tan\theta} = \sqrt{(9.8 \text{ m/s}^2)(20 \text{ m})(\tan 13^\circ)} = 6.7$ m/s.

Imagine a frog sitting in the bottom of a bucket that is being whirled in a horizontal plane (Fig. 5.22). The frog, left to its own, would move in a straight tangential line, but the bottom of the bucket, constrained to a circular path, advances and continuously intercepts that line. The bucket is forever nudging its passenger out of the way, thereby exerting an inward centripetal force on it. This is a normal force exerted *on the frog* by the bucket. The frog, in turn, exerts a radially outward reaction force *on the bucket.*

Figure 5.22 A frog being whirled around in a horizontal circle. The bottom of the bucket exerts a normal force that is the centripetal force.

Suppose this situation were happening in deep space, where there is no appreciable gravity. We could adjust the radius and speed so that $a_C = v^2/r = 9.81$ m/s^2, whereupon the "frogonaut" would experience a normal force equal to its Earth-weight, as though it were sitting on the ground back at the old mudhole. It has been suggested that great doughnut-shaped, spinning space stations (where "up" is radially inward toward the central axis) could provide a more natural long-term environment for humans who were used to living only on the planet.

The *centrifuge* is a practical device using the same frog-in-the-bucket principle. The idea is simple—for a sample at rest, the sedimentation rate of particles suspended in a liquid (like blood corpuscles in serum) is proportional to g. The centrifuge revolves the sample, creating a simulated gravity of as much as 500 000g at perhaps 60 000 rpm. Particles denser than the surrounding liquid end up in the bottom of a test tube, much as the frog ended up pressing on the bottom of the bucket.

A driver colliding with the door in a turning car (Fig. 5.23) experiences the same thing as a cell in a centrifuge or a frog in a pail. The wall of the car intercepts the driver's linear motion, forcing her radially inward toward the center of the turn. The wall pushes on her, and she pushes back on the wall—with no friction from the seat, there is only one force acting horizontally *on* her and it's the normal force of the wall, which equals $F_C = ma_C$. For ordinary turns, there is usually enough friction via the seats to supply F_C and keep people from crashing into doors.

An amusement-park ride that makes use of gravity to supply centripetal force.

Some interesting physics occurs when we take the rope and bucket and swing the frog in a vertical plane (Fig. 5.24). Now gravity and the tension in the rope work alternatively with and against each other to supply the centripetal force. When the rope is at some arbitrary angle θ, the combined weight of the frog and bucket (total mass m) is the vector $\mathbf{F}_W = m\mathbf{g}$. Always acting down, it can be resolved into two components, one radial $F_W \cos\theta$, the other tangential, $F_W \sin\theta$. The weight provides the only tangential force. Because there is a radial acceleration (a_C), there must be a net radial force inward toward O, namely F_C, which we take as positive. This centripetal force must be provided by the only two physical forces acting, the tensile force and the weight, hence

$$F_C = F_T - F_W \cos\theta$$

and since $F_C = ma_C$

$$a_C = \frac{F_T - mg\cos\theta}{m} = \frac{v^2}{r}$$

Solving for the tension,

$$F_T = m\left(\frac{v^2}{r} + g\cos\theta\right)$$

At the bottom of the circle $\theta = 0$, $\cos\theta = 1$ and

$$F_T = m\left(\frac{v^2}{r} + g\right)$$

At the top, $\theta = 180°$, $\cos\theta = -1$ and

$$F_T = m\left(\frac{v^2}{r} - g\right)$$

The fact that the bucket is tied to a rope rather than a rigid rod has not been introduced into the analysis. That is why this last equation allows negative solutions when $g > v^2/r$ at the top of the path. Then the bucket would be whirling too slowly to keep the rope taut, which is fine for a supporting rod that can push or pull on the bucket but not for a rope that can only pull. When $g > v^2/r$ or $v < \sqrt{rg}$, the weight will exceed the needed centripetal force, the rope will slacken, and the bucket and frog will drop out of their circular orbit into a free-fall parabola that never reaches the top. Notice that the tension can be zero if the system is moving such that $v^2/r = g$ at the very top. At that moment, the only force acting is gravity and the system is in free-fall. That's the fun part of those amusement-park rides where the

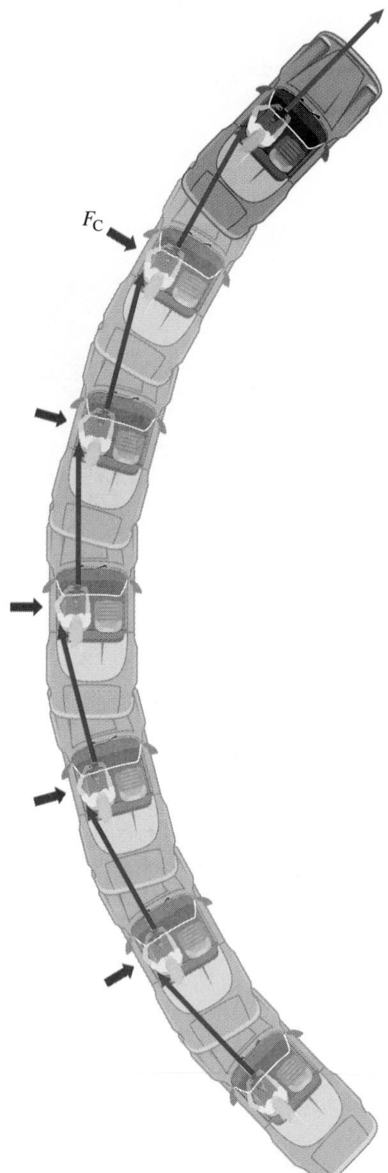

Figure 5.23 A car moving along a curve intercepts the straight-line inertial path of the driver, exerting an inward centripetal force on her.

car takes you upside down at the top of a loop. When $v > \sqrt{rg}$, the bucket and its passenger are actually pulled down by the rope with a force greater than in free-fall. If the frog was on a scale on the bottom of the bucket, it would have an effective weight upward, even while it was upside down at the top of the circle. That is why the frog does not drop out of the bucket—the bucket is accelerating down faster than the frog can fall.

FRICTION AND MOTION

Our common experience is that objects in motion do not often stay in motion, despite the unrefuted wisdom of the First Law. In practice we move, or try to move, one thing against another, and there are interactions that resist the motion. A force of this sort that *opposes an impending or actual motion* is said to be a *frictional force*. There are several different mechanisms that retard movement, and all produce what is broadly called **friction**. When friction impedes a motion that is in progress, it's called *kinetic friction;* when it prevents motion from occurring altogether, it's called *static friction.*

Friction arises via the electromagnetic interaction between atoms that can be in any bulk state—solid, liquid, or gas. *Liquid-gas friction* saves us from being pelted by high-speed raindrops; it allows the wind to make the sea choppy and retards the rising bubbles in a glass of beer. *Solid-liquid friction* slows the flow of petroleum in pipelines and blood in vessels; it helps to make bullets fairly useless under water and oil slicks fairly deadly on highways. The friction of the ocean tides as the waters rub across the land decreases the planet's spin and increases the length of the day. *Solid-gas friction* hinders the flow of air in pumps, pipes, and lungs; it slows skaters, cyclists, baseballs, and fire engines (at 70 mi/h a typical car expends 70% of its fuel just pushing air out of its way); it burns up meteors, threatens returning spaceships, and is the unflagging joy of parachutists. *Solid-solid friction* stops your car whenever you use your brakes, and it dissipates about 20% of the engine's power; without it, you could neither walk nor hold up your socks; it keeps cloth, carpets, ropes, baskets, and wicker furniture from unraveling; it allows us to write with pencils and turn pages. Friction is the great two-faced force of nature, at once the bane and the backbone of modern terrestrial technology.

Our present concern is with dry friction and its influence on motion. We limit the discussion to solids sitting, sliding, and rolling on other solids. Even so, our knowledge is rather rudimentary, the analysis is incomplete, and the experimental data are often imprecise and difficult to reproduce. Dry friction is dominated by the presence of contaminants that make the behavior erratic and somewhat unpredictable.

5.9 Static Friction

Figure 5.25 shows a block of weight F_W on a table that supports it with a normal force F_N. A small horizontal force F is applied to the right—the spring scale registers that force, but the block does not accelerate—it does not move at all. The credo of the Three Laws explains what is happening: Because $a = 0$, the net force is zero, and there must be a force to the left equal to F. That force acting parallel to the surface, resisting the motion, is the force of **static friction**, F_f. Increase F and if the object remains at rest, F_f must also have increased. If F is made larger and larger, the block will ultimately break loose and move when F just exceeds the maximum possible value of the static friction F_f(max). When there is no motion, the static friction force is equal and opposite to the driving force F, and that equality continues right up to F_f(max), just as a rope can react with any size tension right up to the moment when it breaks.

Three basic insights regarding F_f(max) come from simple observations (Leonardo da Vinci was aware of them in the mid-fifteenth century). (1) *F_f(max) is proportional to the normal force, F_N*: $F_f(\max) \propto F_N$. Push on an empty chair, first gently, then harder and harder until it slides. Now have someone sit in it and try again. With the added load, F_f(max) is increased tremendously. (2) The same block in Fig. 5.25 sitting on another surface will experience an F_f(max) proportional to F_N, but it will likely be different. Some surfaces are more slick than others. The statement $F_f(\max) \propto F_N$ can be turned into an equality (Appendix A-4) using a constant of proportionality, namely μ_s. *This* **coefficient of static friction** *depends on the two materials in contact.*

We get around our theoretical limitations by compressing the details of what is happening on an atomic level into a single coefficient that, because it cannot yet be calculated from theory, must be determined experimentally (Table 5.3). Accordingly

$$F_f(\max) = \mu_s F_N \tag{5.13}$$

A 40-N dictionary on a surface where $\mu_s = 0.3$ will require a minimum horizontal force of 12 N to overcome static friction and make it slide.

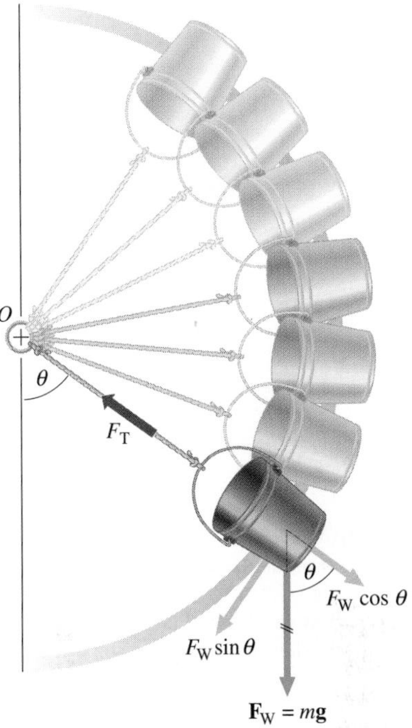

Figure 5.24 The forces on a bucket revolving in a vertical circle. The tension in the rope depends on θ and the speed at that instant v. If you do this with a bucket full of water, make sure that $v > \sqrt{rg}$ at the top ($\theta = 180°$), or you're likely to get wet.

TABLE 5.3 Approximate friction coefficients*

Material	μ_s	μ_k
Steel on ice	0.1	0.05
Steel on steel—dry	0.6	0.4
Steel on steel—greased	0.1	0.05
Rope on wood	0.5	0.3
Teflon on steel	0.04	0.04
Shoes on ice	0.1	0.05
Climbing boots on rock	1.0	0.8
Leather-soled shoes on carpet	0.6	0.5
Leather-soled shoes on wood	0.3	0.2
Rubber-soled shoes on wood	0.9	0.7
Auto tires on dry concrete	1.0	0.7–0.8
Auto tires on wet concrete	0.7	0.5
Auto tires on icy concrete	0.3	0.02
Rubber on asphalt	0.60	0.40
Teflon on Teflon	0.04	0.04
Wood on wood	0.5	0.3
Ice on ice	0.05–0.15	0.02
Glass on glass	0.9	0.4

*The first column lists values of various coefficients of static friction. The second gives the corresponding values of the kinetic coefficients of friction, a concept that will be discussed presently.

An inverted airfoil on the back of this race car pushes down on the rear to increase tire traction without increasing weight.

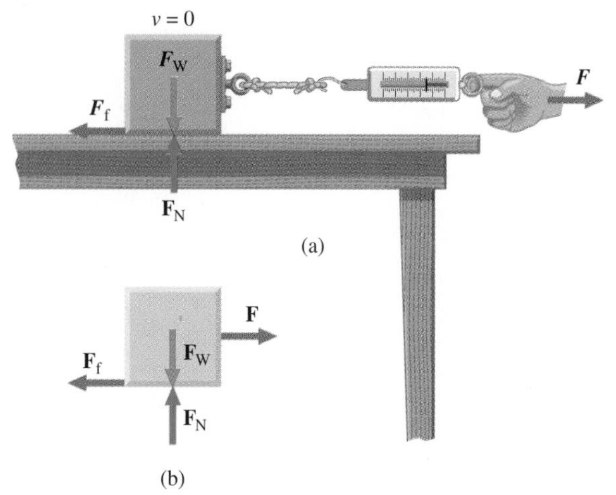

Figure 5.25 (a) A horizontal force F acting through a spring scale is applied to a motionless block of weight F_W. Static friction opposes the impending motion with a force F_f that can be as much as $F_f(\max) = \mu_s F_N$. (b) The free-body diagram indicates that no horizontal acceleration will occur until $F > F_f$.

Teams of horses drag heavy loads of concrete across the ground. They strain to overcome static friction.

Example 5.11 A high point of a country fair is the horse-pulling contest. Teams of horses compete to haul the largest weight using a flat-bottomed sled. How much force must two horses exert to get the sled moving if the total load is 50 kN (or 11 × 10³ lb) and the coefficient of static friction is 0.60?

Solution: [Given: $F_N = 50$ kN and $\mu_s = 0.60$. Find: $F_f(\max)$.] Any horizontal force in excess of $F_f(\max)$ will set the sled moving, thus

$$F_f(\max) = \mu_s F_N = 0.60(50 \text{ kN})$$

and

$$\boxed{F_f(\max) = 30 \text{ kN}}$$

▶ **Quick Check:** The force to make it start to slide is 60% of the load, or 30 kN.

The values in Table 5.3 should not be taken literally—μ_s for, say, glass on glass is given as 0.9, but what is really meant is glass on contaminants on glass. Unless special precautions are taken, oxides, dust, oils, and so forth, will be present on the surfaces and will generally reduce the friction. Nonetheless, in most practical situations, we deal with materials covered with foreign matter, so the table is useful, even if a little misleading.

(3) The third insight is the surprising observation that $F_f(\max)$ *is independent of the apparent size of the contact area between the two solid surfaces.* The dictionary mentioned earlier can be resting on either its narrow edge or its broad face, and the maximum-static-friction force will be the same.

Example 5.12 A climber stands on the rock face of a mountain. The soles and heels of her boots have a high static-friction coefficient equal to 1.0. (a) What is the steepest slope she can stand on without slipping? (b) Assuming her pants have a static-friction coefficient for rock of 0.3, what happens if she sits down to rest?

(continued)

(continued)

Solution: [Given: $\mu_s = 1.0$. Find: the corresponding maximum slope θ_{max}.] We need to formulate an expression for θ in terms of known quantities. Figure 5.26 shows the climber standing still on the mountain. Her weight, acting straight downward, has two components: $F_W \cos\theta$ perpendicular to the tilted surface, and $F_W \sin\theta$ parallel to the surface. It is $F_W \sin\theta$ that causes her to slide down the slope and it is this component that is resisted by the static friction. Because she is standing still, $a_\perp = 0$ and

$$+\nearrow\Sigma F_\perp = 0 = F_N - F_W \cos\theta$$

or

$$F_N = F_W \cos\theta$$

Static friction always acts parallel to the surface, opposing impending motion. Inasmuch as $a_\parallel = 0$

$$+\nwarrow\Sigma F_\parallel = 0 = F_f - F_W \sin\theta$$

or

$$F_f = F_W \sin\theta$$

To combine the two functions of θ into one tangent expression, divide F_f by F_N, yielding

$$\frac{F_f}{F_N} = \tan\theta$$

This ratio has its maximum value when $F_f = F_f(\text{max})$, whereupon $\theta = \theta_{max}$, thus

$$\frac{F_f(\text{max})}{F_N} = \tan\theta_{max} = \mu_s$$

and

$$\theta_{max} = \tan^{-1}\mu_s \qquad (5.14)$$

Measuring the maximum tilt angle at which a body starts to slide down an incline is a convenient way to determine μ_s. With $\mu_s = 1$, $\boxed{\theta = 45^\circ}$. If she were to sit, tan (0.3) = $\boxed{17^\circ}$, and she would slide down any hill tilted at more than 17°.

▶ **Quick Check:** Both sides of Eq. (5.14) are unitless, which is okay. At the extremes, when $\mu_s = 0$, $\theta_{max} = 0$; when $\mu_s = \infty$, $\theta_{max} = 90°$, and that makes sense.

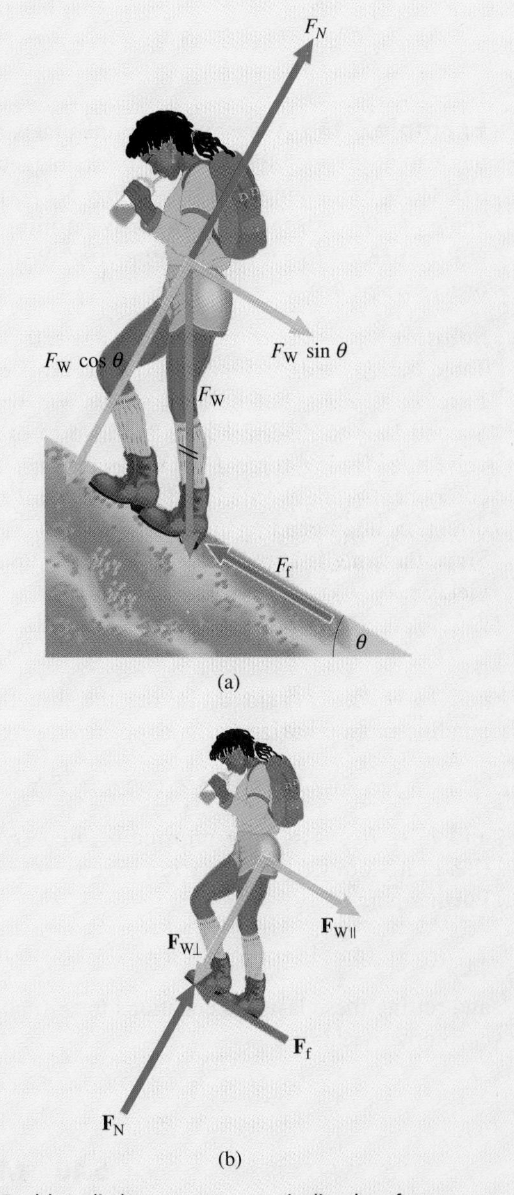

Figure 5.26 (a) A climber at rest on an inclined surface, where friction opposes her impending downward motion. (b) The free-body diagram is similar to Fig. 5.25b except that it's tilted.

Both the proportionality between F_f and F_N, and the area independence of F_f, are working principles rather than fundamental relationships. They generally hold true, and even when they are off, it's usually by no more than about 10%. As we will see, however, there are a few exceptions. Similarly, the idea that μ_s is always constant is not exactly true either—it can change as the contact time increases. Have you ever tried to unscrew a nut that has been in place for a few years or move a vase that has been sitting on a painted surface for several months?

Example 5.13 A washing machine in a wooden crate has a total mass of 100 kg. It is to be dragged across an oak floor by tugging on a rope (Fig. 5.27), making an angle of 30° with the horizontal. What minimum force will be needed to get the thing moving? Will it be more or less when $\theta = 0$?

Solution: [Given: $m = 100$ kg, $\theta = 30°$, and from Table 5.3, $\mu_s = 0.5$. Find: F.] The central equation is $F_f(\max) = \mu_s F_N$, but here $F_N \neq F_W$. We will use the Second Law to determine F_N and also to explicitly involve the driving force F in the equations. F has two components (one parallel and one perpendicular to the direction of impending motion): $F \cos \theta$ and $F \sin \theta$. Since the crate is at rest, taking vertically up as positive yields

$$+\uparrow \Sigma F_V = 0 = F_N + F \sin \theta - F_W$$

and $F_N = F_W - F \sin \theta$. Taking the direction of impending motion horizontally as positive, we get

$$\overset{+}{\rightarrow} \Sigma F_H = 0 = F \cos \theta - F_f$$

and $F_f = F \cos \theta$. The minimum force (F) to move the crate obeys the equation $F \cos \theta = F_f(\max)$. Furthermore

$$F_f(\max) = \mu_s F_N = \mu_s(F_W - F \sin \theta)$$

and setting these last two equations for $F_f(\max)$ equal to each other yields

Figure 5.27 (a) The washing machine crate experiences an applied force **F** that both lifts upward, lightening the load, and pulls forward to overcome static friction. (b) The normal force equals the load, which now is ($F_W - F_V$). The crate will move when $F_H > F_f(\max)$, where $F_H = F \cos \theta$.

$$F \cos \theta = \mu_s F_W - \mu_s F \sin \theta$$

and

$$F = \frac{\mu_s mg}{\cos \theta + \mu_s \sin \theta}$$

Thus

$$F = \frac{490.3 \text{ N}}{(0.866 + 0.250)} = \boxed{0.4 \text{ kN}}$$

When $\theta = 0$, $F = \mu_s F_W = 490$ N, or $\boxed{0.5 \text{ kN}}$.

▶ **Quick Check:** If F was horizontal, then $\theta = 0$, $\cos 0 = 1$, $\sin 0 = 0$, and $F = \mu_s mg$, which equals $\mu_s F_N$, and that suggests that the equation for F might be right.

5.10 Moving via Static Friction

We walk by cleverly arranging for an external force to propel us. Newton's Third Law is at the center of the process—push backward on the floor and it will react with a forward force that propels you. But only if there is friction can you push tangentially backward on the floor. And that push cannot exceed $F_f(\max)$ or your foot will slide out from under you. Friction, opposing the backward motion of your foot, propels you forward—*friction is the driving force* (Fig. 5.28).

It takes just a few ounces of force to overcome air drag and keep someone walking at a constant ($a = 0$) pace. To see that, have a person walk in place and as she stands there marking time, push on her with a continuous force—once started, it takes very little effort to keep up a constant forward motion. When you walk at a uniform rate, you gently push backward, but your feet do not move horizontally while in contact with the ground. That's why you make neat footprints trotting on

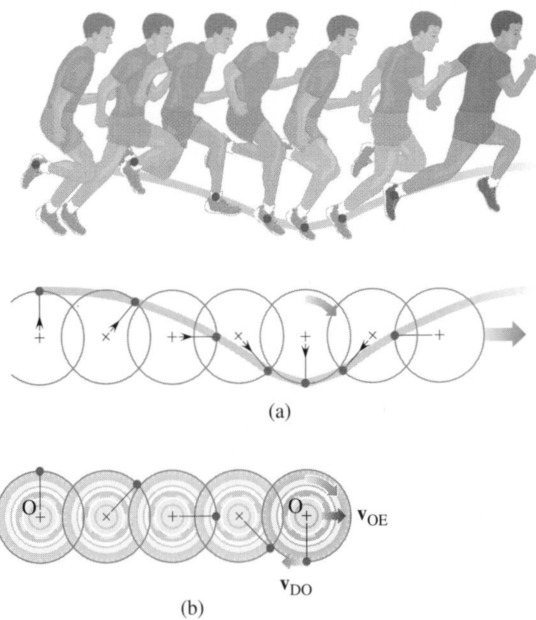

Figure 5.28 (a) The motion of a running foot resembles that of a point on a uniformly rolling wheel. (b) At the instant shown, the red dot is moving backward at a velocity with respect to the center of the wheel of $\mathbf{v}_{DO}$. This is equal in magnitude and opposite in direction to the wheel's velocity with respect to the Earth, $\mathbf{v}_{OE}$. Hence $\mathbf{v}_{DE} = \mathbf{v}_{DO} + \mathbf{v}_{OE} = 0$ and the dot is at rest, at that instant, with respect to the Earth.

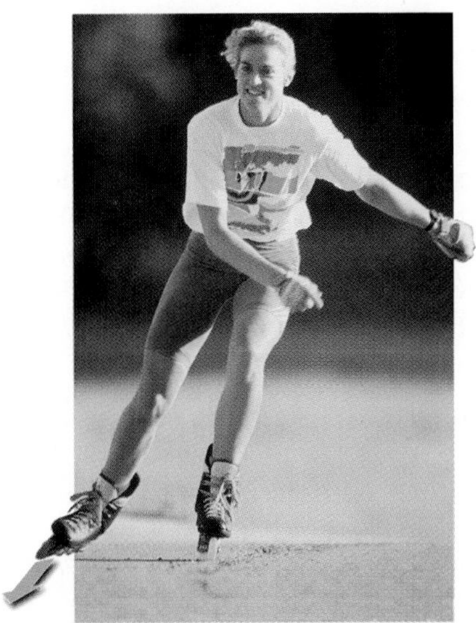

A skater pushes down, to the side, and back, transverse to the skate's rolling axis. The ground pushes forward on the skate, and that frictional reaction propels the person.

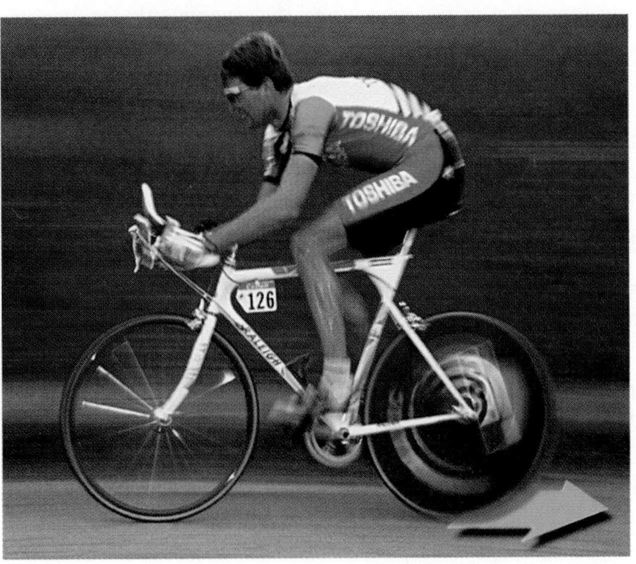

The bicycle moves forward when the rear tire pushes back on the ground via friction. The ground pushes the tire, and hence the bike, forward.

damp sand. While traveling at a constant speed, a runner's foot actually moves through space with a motion that closely matches that of a rolling wheel (Fig. 5.28), and *static* friction propels both!

Imagine a dot of red paint on the rim of a uniformly rolling wheel. As the dot comes around, it will reside for an instant on the forward-most portion of the wheel.

It gradually moves down and *back*. At the instant the dot hits the ground, it will be moving backward with respect to the center at the same speed the entire wheel is moving forward. With respect to the ground, just as the dot comes down and touches, it will be motionless.

Example 5.14 The weight distribution of a Datsun 510 with a driver is 55% in front and 45% on the rear wheels. With $\mu_s = 0.9$ for tires-on-road, what is the maximum acceleration this rear-wheel-drive vehicle can attain? It actually goes from 0 to 30 mi/h in 4.1 s. How does that compare?

Solution: [Given: 45% F_W on the rear wheels and $\mu_s = 0.9$. Find: a_{max}.] The back two tires propel the car, and the force they exert on the ground determines the motion. The total load on both drive wheels is 45% mg. The maximum acceleration occurs at the point just before slipping, when

$$F_f(\max) = \mu_s(0.45)mg = ma_{max}$$

and

$$a_{max} = 0.45(0.9)g$$

or

$$\boxed{a_{max} = 0.4g}$$

The car actually accelerates at $a = \Delta v/\Delta t = (13.4 \text{ m/s})/(4.1 \text{ s}) = 3.27 \text{ m/s}^2 = \boxed{0.3g}$, which is a bit less than the predicted maximum and suggests that a more powerful engine could improve the acceleration.

▶ **Quick Check:** All ordinary cars accelerate at a fraction of g, so the units are right and the order-of-magnitude is okay, too.

A car traveling at a constant speed leaves unsmeared tire marks in damp sand because there is no horizontal motion of the tread with respect to the ground. Ideally, the region in contact with the road does not push forward or backward; it does that only when braking or accelerating. Provided there is no slipping, it's the static-friction force between tires and road that propels a car, even when it is in motion. It's the same static-friction force that accelerates it and that supplies the centripetal force on turns (although there is a small amount of slipping that occurs during each one of these maneuvers). Skidding with the brakes locked and the tires motionless or attempting to accelerate so rapidly that the wheels spin (burning rubber), results in something different that is called *kinetic friction.*

The skier moves ahead by pushing back on the ground with the pole. The ground pushes forward on the pole, which transmits the force to the person.

5.11 Kinetic Friction

Please look at Fig. 5.25, where we pulled on a body at rest to observe its static-friction force. As the driving force increased, F_f increased until it reached a maximum value of $F_f(\max)$. Applying a still greater force will cause the body to break loose and move in the direction of the applied force. The **kinetic-friction force**, *the retarding force exerted on a sliding body in contact with a surface, is equal and opposite to the driving force that is needed to maintain the body in uniform motion.* Once again, experiments show that the three principles of friction apply, this time with a *coefficient of kinetic friction* μ_k, where

$$F_f = \mu_k F_N \tag{5.15}$$

Table 5.3 lists experimentally determined values of μ_k. For example, the force F needed to push a 600-N (135-lb) person wearing leather-soled shoes across a wood floor at a constant speed ($\Sigma F = ma = 0$) is $F = F_f = \mu_k F_N = (0.2)(600 \text{ N}) = 30$ N. Table 5.3 makes it clear that static friction is usually greater than kinetic friction (a fact apparent to anyone dragging a heavy mass).

In practice, kinetic friction decreases as the sliding speed increases. Drivers know from experience that they must ease off the brakes if they are to bring an automobile to a smooth stop—the kinetic friction on the brake shoes increases as the wheels slow, and after a few jolting stops, the new driver learns to lighten the pressure appropriately. Thus, while μ_k for a sliding tire is roughly 0.8 at 8 km/h, it is likely to be less than about 0.5 at 130 km/h, and that can be crucial if your car is skidding. The best way to stop a car is to engage the brakes so the wheels are just on the threshold of locking up but are still turning. This technique will ensure a maximum retardation via *static* friction, which is greater than the kinetic friction of skidding rubber. The process can be done manually—you can feel the wheels locking—but it takes practice, especially if it is to be accomplished under stress. That's why computer-operated antilocking brakes are a good idea.

On friction. *Generally it is easier to further the motion of a moving body than to move a body at rest.*

THEMISTIUS (390–320 B.C.)
Physica

Example 5.15 A camp is on a mountainside that slopes at 30° (Fig. 5.29). Someone is hired to pull a 200-kg sled up the slippery incline for which the sled-slope coefficient of kinetic friction is 0.1. If the person can exert a continuous force parallel to the incline of 200 N, compute the acceleration of the sled.

Solution: [Given: $\theta = 30°$, $m = 200$ kg, and $\mu_k = 0.1$. Find: a.] We will need F_f and therefore F_N. From the figure, we have

$$+\nwarrow \Sigma F_\perp = 0 = F_N - F_W \cos\theta$$

and again $F_N = F_W \cos\theta$. Taking the direction of motion as that of the applied force **F**, friction must oppose it, and so

$$+\nearrow \Sigma F_\parallel = F - F_W \sin\theta - F_f = ma$$

Hence $\quad ma = F - mg\sin\theta - \mu_k mg\cos\theta$

and

$$ma = 200\text{ N} - 981\text{ N} - 170\text{ N} = -951\text{ N}$$

STOP!

The component of the weight down the plane (981 N) far exceeds the upward driving force (200 N)—the sled slides downhill and a is negative. That's fine, but we must reverse F_f because it must oppose the motion (Fig. 5.30b)

$$ma = 200\text{ N} - 981\text{ N} + 170\text{ N}$$

and $\boxed{a = -3\text{ m/s}^2}$. The sled cannot be hauled uphill; the person can only slow the acceleration downhill.

▶ **Quick Check:** Since $\sin\theta = \frac{1}{2}$, the force down the hill ($mg\sin\theta$) is $\frac{1}{2}$(200 kg)g = (100 kg)g = 981 N. The normal force is $mg\cos\theta$ = 1699 N, and the coefficient of friction is 0.1, so F_f = 170 N.

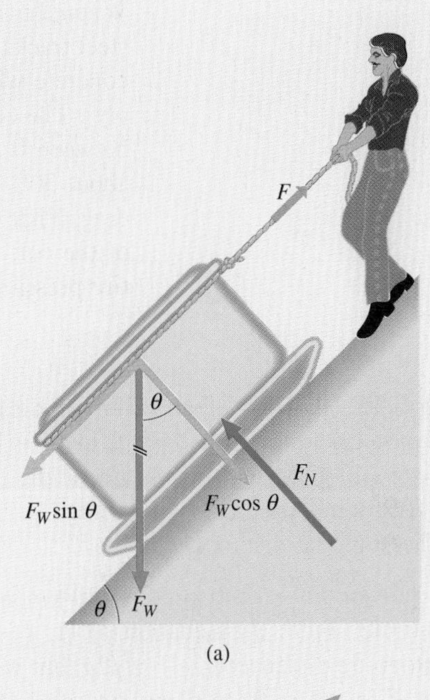

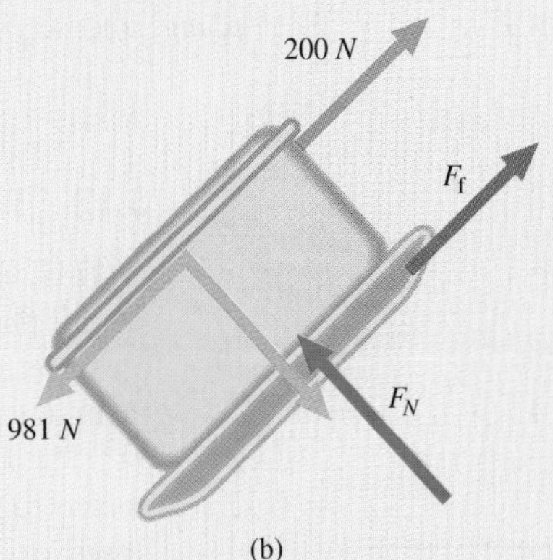

Figure 5.29 (a) A person pulling upward on a sled with a force $F = 200$ N. Since $F_W \sin\theta$ exceeds F, the sled slides downhill, but you might not have noticed that at the start of the analysis.

5.12 Rolling Friction

Wheels were used in transportation 5000 years ago, and it has been obvious for all that time that it is easier to roll a load than slide it. Ideally, if a rigid wheel is rolled along a flat inflexible surface and nothing stops the motion, it should go on forever. That does not happen, and we call the operative retarding force **rolling friction**. Even though the primary mechanism is quite different, the effect can be described approximately in the same form as were static and kinetic friction, thus

$$F_f = \mu_r F_N \tag{5.16}$$

where μ_r is the *coefficient of rolling friction.* The friction force is equal in magnitude to the force required to keep the object rolling at a uniform speed.

The effect arises from the deformation of both the wheel and the surface, which creates an obstacle to free rolling that must be overcome by the exertion of a force. When the surfaces are hard, there is little deformation, as when steel wheels roll on steel tracks, whereupon μ_r is only about 0.001. By contrast, a flexible rubber tire on concrete will have a coefficient around 0.01 to 0.02 (with radials lower than cross-ply). The coefficient for tires actually varies with speed, increasing as much as 15% between 0 and 120 km/h. However small μ_r is, the average auto at 80 km/h uses about 30% of its power to overcome rolling friction. At low speeds it exceeds air drag. That situation changes at around 55 km/h (35 mi/h), depending on the shape of the car, after which air drag predominates. The admonition to maintain proper tire pressure translates into lower rolling friction and better fuel economy.

Example 5.16 A driver in an aerodynamically trim, 1200-kg sports car traveling at 20 km/h shifts into neutral and coasts to a stop uniformly under the influence of rolling friction. Assuming an average $\mu_r = 0.013$, compute the stopping distance.

Solution: [Given: $m = 1200$ kg, $v_i = 20$ km/h, $v_f = 0$, and $\mu_r = 0.013$. Find: s.] Equation (3.11, p. 68) provides the stopping distance: $s = -v_i^2/2a$, but we need a. Notice that $F = ma = F_f = \mu_r mg$, and so $a = \mu_r g$. If we take the direction of motion as positive, a (which is a deceleration) must be entered as a negative quantity. Hence, $s = v_i^2/2\mu_r g$, and

$$s = \frac{(5.56 \text{ m/s})^2}{2(0.013)(9.81 \text{ m/s}^2)} = \boxed{0.12 \text{ km}}$$

▶ **Quick Check:** The time it takes to stop is $v_i/a = 43.6$ s, and at an average speed of $\frac{1}{2}(5.56 \text{ m/s})$, $s = 121$ m.

5.13 The Causes of Friction

In 1734, John Desaguliers (an acquaintance of Newton's) rightly concluded that *adhesion* played a role in producing friction. He took two freshly cut pieces of lead, pressed them together by hand and found that it could take as much as 50 lb to pull them apart.

The attraction between the atoms and molecules that holds solid matter together is electromagnetic in nature. Without it, there would be no such thing as textbooks, noses, or teacups, nor would there be any friction. But only when the atoms are close to one another will they experience an appreciable attractive force; the effect is short-ranged. Such forces become negligible at distances equivalent to only about 4 or 5 atomic diameters (each of which is $\approx 1 \times 10^{-10}$ m to 2×10^{-10} m). Thus, if a substantial amount of friction is to exist, the interacting surfaces must be in close contact.

Common objects may seem smooth to the eye and touch, but on a microscopic level they are jagged and rough. Even a polished surface is a craggy affair with features about 3×10^{-8} m across, each hill corresponding to perhaps 100 atomic diameters. When two ordinary surfaces are in contact, they only touch at a few high spots. Any material will deform if the *force per unit area,* or *pressure,* on it is great enough. With the load supported on a few tiny prominences, each experiences a crushing pressure and subsequently flattens out. The contact area thereby increases, decreasing the pressure until the material squashes no further. Still, only a small fraction of the total surface area is in contact, while the remainder is separated by gaps of 10 to 50 atomic diameters. Clearly, friction should be proportional to this *actual* contact area, not the apparent contact area.

Suppose a book is oriented to give the greatest apparent contact area with a surface (Fig. 5.30). Resting on a flat face may mean more contact points, but each will experience less pressure and will deform less. The overall effect is that the actual total contact area is the same, independent of the apparent area. That situation is equivalent to the third principle of friction. The actual contact area depends on the physical properties of the materials. Moreover, the interatomic forces, and thereby the adhesion and the friction, also depend on the kinds of atoms involved, which is equivalent to the second principle of friction.

Static friction arises out of the need to rip apart the areas of bonded contact. Two clean pieces of metal, such as gold, pressed together, will literally fuse at the contact points, a process known as *cold welding.* To move the two surfaces, the welds must be torn. On the other hand, keeping the surfaces separated reduces adhesion and friction—that's what oil and grease do, and why baby powder protects the skin against chafing (Fig. 5.31).

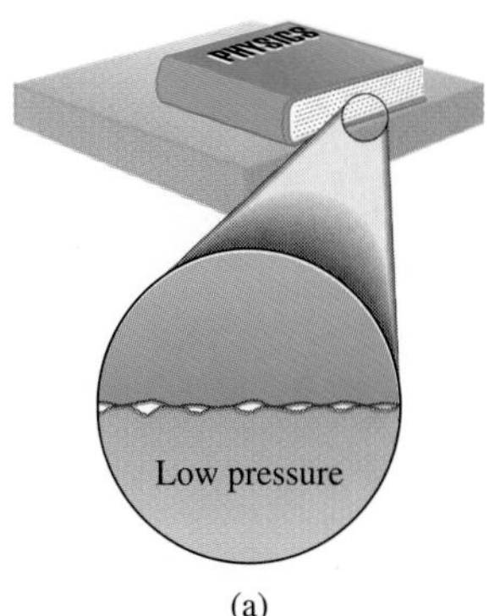

(a)

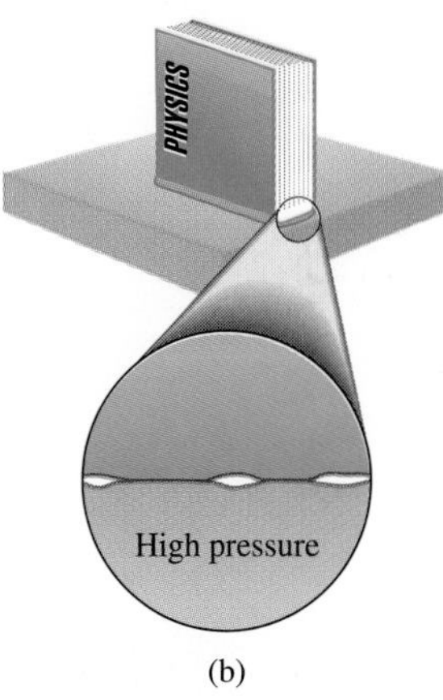

(b)

Figure 5.30 (a) When the load is spread out over a large area, the pressure is low, and there are many small contact areas. (b) When the load is concentrated, the pressure is greater and each place of contact is larger, but there are fewer of them. Thus, the total contact area is the same in (a) as in (b). These contact areas literally weld together and must be pulled apart.

Burned rubber due to rapid deceleration. A tire's surface heats up and the rubber burns, leaving marks on the road as the vehicle accelerates. These tracks were left by the two tires on the right side of a truck. Variations show that the driver pumped the brakes.

The flattening of the tire contributes to rolling friction. Every time the wheel turns, energy is lost, reshaping the rubber, and the tire heats up.

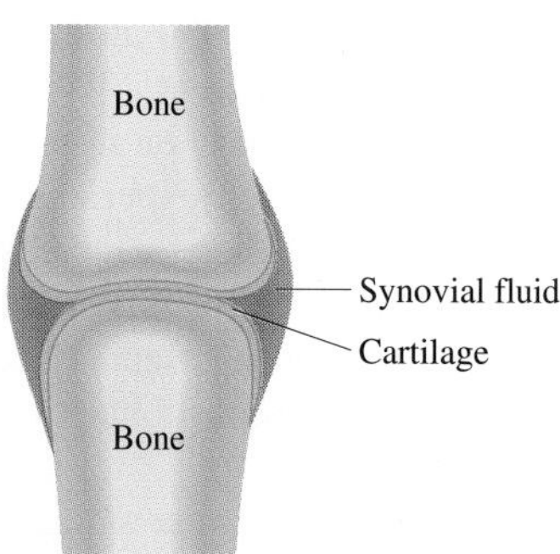

Figure 5.31 A human bone joint lubricated with synovial fluid. As the bones move, the contact area shifts, squeezing fluid out of the porous cartilage. The result is a remarkably small coefficient of friction of about 0.000 3.

Gauge blocks are precise, smooth steel blocks that are used to calibrate measuring tools. When they are squeezed together they interact so strongly that they stick to each other. The adhesive force at work here is the same force that gives rise to friction.

Adhesion is lessened if the two surfaces are already moving past one another, and so usually $\mu_s > \mu_k$. When no relative motion occurs, the surfaces enmesh, coming into close contact. When there is sliding, the surfaces ride across each other mostly on the high prominences. By contrast, in rolling friction the welds are peeled apart rather than sheared or ripped, and that takes less effort. Moreover, during sliding, surface layers of dirt, oxides, and grease will be cut through and good contacts can be made. This action does not happen with rolling objects, where adhesion is still lower and $\mu_k > \mu_r$.

Core Material

The net force (**F**) applied to a body equals the mass (m) of the body multiplied by the resulting instantaneous acceleration (**a**):

$$\mathbf{F} = m\mathbf{a} \quad [5.3]$$

The weight of an object is

$$\mathbf{F}_W = m\mathbf{g} \quad [5.5]$$

The force exerted by a surface perpendicular to itself and equal to the load supported is the normal force (F_N).

When the only force acting on an object is gravity, the object is in free-fall, and it is **effectively weightless**.

A centripetal force must be exerted on an object if it is to move in a curved path. The *centripetal acceleration,* a_C, which occurs when any object uniformly moves in a circle, is

$$a_C = \frac{v^2}{r} \quad [5.10]$$

Such an object must experience an inwardly directed centripetal force given by

$$F_C = ma_C = \frac{mv^2}{r} \quad [5.11]$$

The maximum static friction force is

$$F_f(\max) = \mu_s F_N \quad [5.13]$$

where μ_s is the *coefficient of static friction.* For surfaces in relative motion, kinetic friction is given by

$$F_f = \mu_k F_N \quad [5.15]$$

where μ_k is the *coefficient of kinetic friction.* For the friction encountered when one object rolls across another

$$F_f = \mu_r F_N \quad [5.16]$$

where μ_r is the *coefficient of rolling friction.*

Suggestions on Problem Solving

1. Draw a diagram. Guess at the direction of the overall motion of the system (actual or impending) and call that positive. If a complicated system is moving clockwise, make that the positive direction and draw a big arrow in that direction from one end of the system to the other. Part of the system may be moving up, part right, and part down. Provided each portion follows the overall clockwise motion, up, right, and down are positive for the corresponding portions of the system. (See, for example, Fig. P22.)

2. Isolate each object of interest via its own free-body diagram. Then apply $\Sigma F = ma$ to each body, remembering that if the body is at rest or moving at a constant velocity, $a = 0$. Apply the Second Law in the direction of actual or impending motion and (if you need normal forces) perpendicular to it. Each such application will provide an independent equation, and the number of equations must equal the number of unknowns.

3. If forces act that are neither perpendicular nor parallel to the actual or impending motion, they must be resolved into components in those directions. The motions in these two directions are independent. For an inclined plane, we apply $\Sigma F = ma$ both down the incline and perpendicular to it—the former tells us about the acceleration, the latter about the normal force.

4. The tensions acting at each end of a rope are equal and opposite (provided no tangential forces act). Thus, if two masses are connected by a rope, the same tension will act in opposite directions on the two bodies.

5. When doing centripetal force problems we often encounter revolving objects for which we must find the speed. Thus, if a body turns at 1000 rotations per minute, a point on its surface a distance r from the axis moves in 1 min through 1000 circular paths, each equal to a circumference of $2\pi r$, for a total distance of $1000(2\pi r)$. The speed is then $v = (1000\ \text{min}^{-1})(2\pi r)/(60\ \text{s/min})$.

6. Don't forget that $F_W = mg$, which means that if you know either the weight or the mass, you can calculate the other.

7. For a body at rest, make sure that the net force can overcome the static friction before you go on to calculate the effects of kinetic friction.

Discussion Questions

1. Can we shoot a hole through a silver dollar floating out in space where the bullet is weightless? An old circus trick is to fire a bullet at a target while holding a pure silk handkerchief so it hangs freely directly in the line of fire. The gun is shot, the target hit, but no hole is made in the handkerchief. Explain.

2. Describe the process of swimming in terms of Newton's laws. Swimmers sometimes use rubber fins on their feet. How do the fins work? While floating in space, can you propel yourself by moving your hands and feet as if swimming? Explain how one moves a boat by rowing.

3. Explain *in terms of Newton's laws* each of the following: How does a rug get cleaned by beating it? What happens when we shake down a mercury thermometer used for measuring body temperatures? How does a hummingbird, a helicopter, or a bumblebee hover in midair? What makes an unsealed, air-filled balloon fly around when it's released? What are the external forces acting and their interrelationships when an automobile is traveling on a level road at a constant speed?

4. A weightlifter and a barbell are both on a large floor scale. While the weightlifter lifts and holds the barbell, will the scale ever momentarily exceed and/or be less than the combined weight of the two bodies? Explain.

5. Can an object ever move in a direction other than that of the net applied force? Can it accelerate in a direction different from that of the force?

6. Two equal-weight monkeys each hang from the ends of a rope passing over a weightless, frictionless pulley. If one accelerates up the rope, what happens to the other?

7. A truck carrying a load of caged canaries pulls onto the weighing scales at a highway toll station. While no one was looking, the driver banged on the side of the cab so the birds would fly around inside. What happens to the weight of the load?

8. A 160-lb gymnast doing a giant swing on a high bar experiences a force directed inward toward the bar of about 120 lb at the top and roughly 790 lb at the bottom. Explain what is happening.

9. Knowing that the mass of muscle tissue increases more rapidly than does its strength, provide a reasonable physical explanation for the fact that dogs and deer run faster than people.

10. What physical mechanism accounts for the "stickiness" of adhesive tape? Attach a piece of tape a few inches long to a solid horizontal surface leaving a little tab to hold on to. Now try to pull the tape off horizontally (that is, shear it off). Next, peel it off with a vertical force. Compare the results and discuss the significance as it applies to rolling friction.

11. Ice skates melt the ice beneath them so that they ride on a layer of water—provided it's not too cold. What role does that process play in skating? Can you skate on dry glass? Why are skis waxed? Suppose you wanted to slide a 4-kN block of marble into a narrow space—why might it be advantageous to slip several ice cubes under the block? There is a rumor that mausoleum workers sometimes use a handful of expendable marbles to move heavy coffins into narrow vaults. Explain the physics behind such a procedure.

12. Why is it advisable to avoid rapid acceleration, "deceleration," and sharp cornering while driving a car on a rainy day? The main purpose of the treads on an automobile tire is to provide a space into which water on the road can be squeezed. Why is that a good idea? What is the purpose of spikes on golf shoes and cleats on football shoes?

13. Take a toy car and wedge the rear wheels so they lock. Now roll it down an inclined plane, front first—what happens and why? Try it again with the front wheels locked. How should the brakes on a real car be engaged? Many cars are biased in the opposite way, which accounts for some of the broadside sluing on icy roads. Why is it recommended to pump the brakes when a car goes into a skid while making a rapid stop?

14. Draw a picture of a horse pulling a wagon over a smooth road. What are all the forces acting on the horse? On the wagon?

15. What is the smallest force with which you can raise an object on Earth into the air? Explain your ideas in detail.

16. What will happen when the string holding the wrench handle in Fig. Q16 burns through?

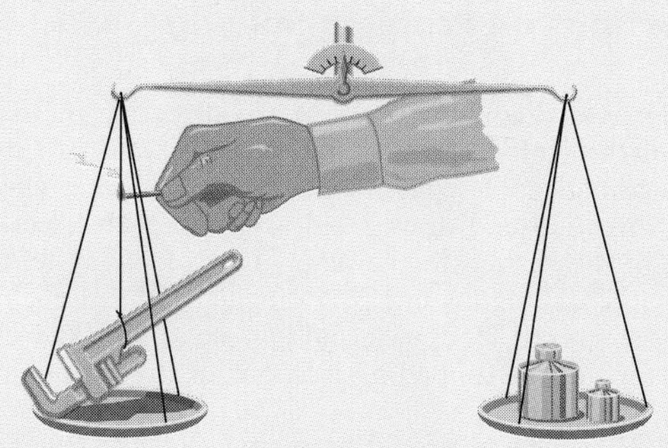

Figure Q16

Multiple Choice Questions

1. If L stands for length, T for time, and M for mass, the dimensions of force are (a) $[ML^2]$ (b) $[ML/T]$ (c) $[ML/T^2]$ (d) $[LT/M]$ (e) none of these.
2. One *newton* is the force (a) needed to move a 1-kg mass at 1 m/s (b) needed to accelerate a 1-kg mass at 1 m/s^2 (c) needed to accelerate a 1-g mass at 1 cm/s^2 (d) equal to the Earth-weight of a 1-kg mass (e) none of these.
3. Two forces, one 10 N and the other 40 N, cannot be applied to a body so that they produce the same acceleration as a single force of (a) 35 N (b) 50 N (c) 40 N (d) 10 N (e) none of these.
4. The weight of a TV star on Earth is 686 N. Out in deep space his mass is (a) 686 kg (b) 6723 kg (c) 0 kg (d) 70 kg (e) none of these.
5. The mass of an astronaut on a planet where gravity is 10 times greater than Earth's gravity is (a) 10 times smaller (b) 10 times larger (c) $10g$ times greater (d) $10g$ times smaller (e) none of these.
6. Is it possible to devise a technique to push on a table without it pushing back on you? (a) Yes, out in space. (b) Yes, if someone else also pushes on it. (c) A table never pushes in the first place. (d) No. (e) None of these.
7. If a nonzero constant net horizontal force is acting on a body sitting at rest on a frictionless table, the body will (a) sometimes accelerate (b) always move off at a constant speed (c) always accelerate at a constant rate (d) accelerate whenever the force exceeds its weight (e) none of these.
8. If (with no friction) a force F results in an acceleration a when acting on a mass m, then tripling the mass and increasing the force sixfold will result in an acceleration of (a) a (b) $a/2$ (c) $2a$ (d) $a/6$ (e) none of these.
9. A bubble level can be used as an accelerometer. If the level is accelerated due east while in normal operating position aligned east-west, the bubble will (a) move west (b) move east (c) move north (d) remain at rest (e) none of these. Try it.
10. A 250-lb man holding a 30-lb bag of potatoes is standing on a scale in an amusement park. He heaves the bag straight up into the air, and before it leaves his hands, a card pops out of a slot with his weight and fortune. It reads (a) 250 lb (b) 280 lb (c) less than 250 lb (d) more than 280 lb (e) none of these.
11. Why does it take more force to start walking than to continue walking? (a) It doesn't. (b) Because kinetic friction is less than static friction. (c) Because air drag at walking speeds is much less than the needed inertial force, *ma*. (d) People get tired walking and prefer to stand still. (e) None of these.
12. Imagine that you are standing on a cardboard box that just supports you. What would happen to it if you jumped into the air? It would (a) collapse (b) be unaffected (c) spring up as well (d) move sidewise (e) none of these.
13. Someone weighing 600 N jumps onto a bathroom scale. In the process, the scale will read (a) a constant 600 N (b) 0 and then rise to 600 N (c) 0 then rise to over 600 N and stay there (d) 0 then rise to over 600 N, then drop down to 600 N (e) none of these.
14. Why do thin people with skinny legs make the best runners? (a) They tend to be strong. (b) Thin legs are longer, so people with them can take longer strides. (c) They have less massive legs to accelerate. (d) Heavy people just don't like to run. (e) None of these.
15. Imagine a flat, lightweight wheeled cart that is low to the ground and has well-oiled bearings. What will happen to it if, while standing at rest on it, you begin to walk along its length? It will (a) remain stationary (b) advance along with you (c) not enough information to say (d) move in the opposite direction (e) none of these.
16. With the previous question in mind, what would happen if you approached the cart, stepped onto it, and walked its length at a constant speed? It would (a) remain nearly stationary (b) advance along with you (c) move rapidly in the opposite direction (d) move forward then backward (e) none of these.
17. For an object moving along a curved path at a constant speed (a) its acceleration is perpendicular to its instantaneous velocity (b) its velocity is constant (c) its acceleration vector is constant (d) its acceleration is zero (e) none of these.
18. Mica is a crystalline material that can be sheared along atomic planes. It is the only material that can be prepared

in a form that is smooth on a molecular level over sizable areas. Place two sheets of mica such that an area A on both are in contact under a 1-kg load, then measure $F_f(\max)$ for mica on mica. If that load is then doubled, the maximum static friction force will likely (a) remain unchanged (b) double (c) drop to zero (d) quadruple (e) none of these.

19. With the previous question in mind, suppose the overlapping area of the two sheets is made $2A$. The maximum friction force will thereby (a) remain unchanged (b) decrease by half (c) increase proportionately (d) quadruple (e) none of these.

20. A man is walking with a cane, which at some instant makes an angle θ with the ground. The reaction force on the cane is (a) zero (b) $\mathbf{F}_N$ (c) $-(\mathbf{F}_N + \mathbf{F}_f)$ (d) $+(\mathbf{F}_N + \mathbf{F}_f)$ (e) none of these.

Problems

FORCE, MASS, AND ACCELERATION

1. [I] A person pushes a shopping cart with a force of 100 N acting down at 45°. If the cart travels at a constant speed of 1.0 m/s, what is the magnitude of the combined retarding force acting horizontally on the cart due to friction and air drag?

2. [I] Two bar magnets of mass 1.0 kg and 2.0 kg have their like poles pressed together so that there is a repulsive interaction between them. In a frictionless environment the two fly apart when released such that the more massive one has an initial acceleration of 10.0 m/s^2 due north. What is the initial acceleration of the other magnet?

3. [I] What is the smallest force needed to lift a 0.50-kg bullfrog up from the ground and under what circumstances would that answer be applicable?

4. [I] A 100-kg gentleman standing on slippery grass is pulled by his two rather unruly children. One tugs him with 50 N north, toward the ice cream stand, while the other hauls with 120 N toward the bathroom, due east. Overlooking friction, compute Pop's resulting acceleration.

5. [I] During training on Earth, an astronaut is accelerated upward at $0.50g$. If her mass is m, what is her effective weight?

6. [I] The gravitational acceleration on the surface of Mercury is 0.38 times its value on Earth. What is the weight of a 1.0-kg mass on that planet?

7. [I] A youngster of mass 40.0 kg jumps into the air with an average vertical acceleration of 5.00 m/s^2. Determine the average force exerted by the ground on the jumper during the leap, and compare it to the youngster's weight.

8. [I] A maintenance robot 1.70 m tall, weighing 752 N when constructed on Earth, is found floating 400 m from an orbiting power station. The human crew decides to tie a light rope to it and haul it aboard. How much force must be exerted if the robot is to be uniformly accelerated for 1.00 s at 9.81 m/s^2?

9. [I] An *F-14* twin engine supersonic fighter weighs 69 000 lb. Its two Pratt and Whitney TF30-P-412 engines can each deliver a peak thrust, using afterburners, of 20 200 lb. Compute its maximum acceleration (in terms of g) during horizontal flight (neglect air friction).

10. [I] A rope of mass 10.0 kg is tied to a 1.00×10^3-kg signal relay in intergalactic space. The device is to be returned for repairs as rapidly as possible by a relay-tender android capable of exerting a maximum sustained force on the rope of 10.0 kN. Compute the resulting acceleration.

11. [II] A spread-eagled sky diver whose mass is 65.0 kg is falling straight down at a speed in excess of 150 km/h with an acceleration of 2.00 m/s^2. Compute the total friction force due to air drag acting at that moment.

12. [II] A parachutist of mass m lands with legs bent, coming to rest with an upwardly directed average acceleration of $4.0g$. What is the average force exerted on him by the ground?

13. [II] Someone of mass 100 kg is standing on top of a steep cliff with only an old rope that he knows will support no more than 500 N. His plan is to slide down the rope using friction to keep from falling freely. At what minimum rate can he accelerate down the rope in order not to break it?

14. [II] The ballistocardiograph is a device used to assess the pumping action of the human heart. A patient lies on a horizontal 2.0-kg platform "frictionlessly" suspended on air-bearings. The rush of blood pumped in one direction will be accompanied by a counterforce on the body and table in the opposite direction. The resulting acceleration of the platform is recorded by an extremely sensitive accelerometer able to measure values as small as 10^{-5} m/s^2. For a young, healthy adult the acceleration over an interval of 0.10 s during the pumping cycle may be as great as 0.06 m/s^2. (a) Compute the force exerted by the heart if the patient has a mass of 70 kg. (b) What is the corresponding change in momentum?

15. [II] A pink pom-pom of mass m hangs on a chain from the rearview mirror of a Corvette L82. That machine can go from rest to 26.8 m/s (60 mi/h) in 6.8 s. The pom-pom serves as a pendulum accelerometer. Assuming a constant maximum acceleration, compute the angle the pom-pom makes with the vertical.

16. [II] On its first flight in 1981, the *Columbia* spaceship was part of a 4.5×10^6-lb, 18-story high launch assembly that developed a total thrust of about 6.4×10^6 lb. (a) What was its initial acceleration at full power? (b) According to newspaper reports, it was traveling at 33.5 m/s (75 mi/h) 6.0 s after blast-off when it cleared the 347-ft support tower. Are these observations consistent? What average acceleration does that correspond to? (c) Compare these values of acceleration and explain any differences.

17. [II] A tug pulls two small barges (of mass $m_1 = 4.00 \times 10^3$ kg and $m_2 = 3.50 \times 10^3$ kg) tied together, one behind the other. The tug exerts 1.00 kN on the line to the first barge (m_1) and accelerates at 0.100 m/s^2. Compute the tension in the two ropes and, knowing that friction opposes the motion, determine the friction on each barge, assuming it's the same for both.

18. [II] Refer to Fig. P18. Compute the acceleration of the 1.3×10^3-N weight. Neglect friction and the mass of the

Figure P18

rope and pulleys. Now remove the 1.8×10^3-N weight, replace it by a constant downward tug of 1.8×10^3 N acting on the end of the rope, and recalculate the acceleration of the first weight.

19. [II] A 2000-kg car in neutral at the top of a 20° inclined driveway 20 m long slips its parking brake and rolls down. At what speed will it hit the garage door at the bottom of the incline? Neglect all retarding forces.

20. [II] A rescue helicopter lifts two people from the sea on an essentially weightless rope. Jamey (100 kg) hangs 15 m below Amy (50.0 kg), who is 5.0 m below the aircraft. What is the tension (a) on the topmost end of the rope, and (b) at its middle while the helicopter hovers? (c) Compute those answers again, this time with the aircraft accelerating upward at 9.8 m/s^2.

21. [II] The all-time champion jumper is the flea. This tiny pest can attain a range of up to 12 in. or about 200 times its own body length. (That's the same as you jumping five city blocks.) Assuming a 45° lift-off angle, a push-off time of 1.0 ms, and a mass of 4.5×10^{-7} kg, compute (a) the initial acceleration of the flea, assuming it to be constant, and (b) the force it exerts on the floor.

22. [II] In 1784, George Atwood published the description of a device for "diluting" the effect of gravity, thereby facilitating the accurate measurement of g. Figure P22 shows the apparatus: two masses tied together by a length of essentially massless rope slung over an essentially massless free-turning pulley. With $m_2 > m_1$, prove that both masses accelerate at a rate of

$$a = \frac{(m_2 - m_1)}{(m_2 + m_1)} g$$

Show that the tension in the rope is

$$F_T = \frac{2m_1 m_2}{(m_2 + m_1)} g$$

If $m_2 = 2m_1$, what is a? When is a equal to zero? When $m_2 \gg m_1$, find a.

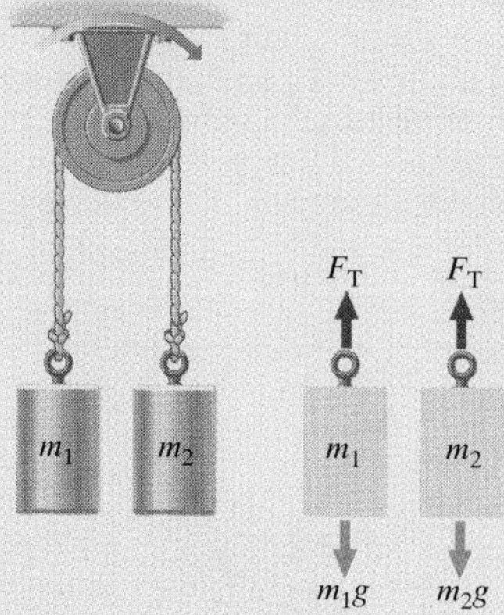

Figure P22

23. [III] Figure P23 shows three masses attached via massless ropes over weightless, frictionless pulleys on a frictionless surface. Compute (a) the tension in the ropes and (b) the acceleration of the system. Draw all appropriate free-body diagrams. (c) Discuss your results in terms of internal and external forces.

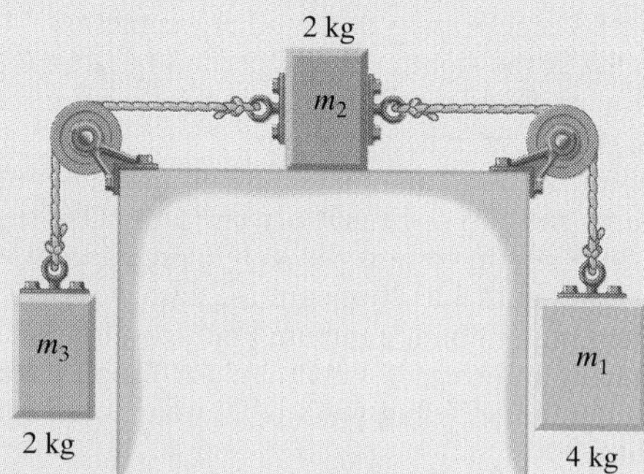

Figure P23

24. [III] Two identical blocks, each of mass 10.0 kg, are to be used in an experiment on a frictionless surface. The first is held at rest on a 20.0° inclined plane 10.0 m from the second, which is at rest at the foot of the plane. The one descends the incline, slams into and sticks to the second, and they sail off together horizontally. Calculate their final speed.

25. [III] A physicist on planet Mongo is using a device equivalent to Atwood's machine to measure the Mongoian gravitational acceleration g_M. He fixes one of the two 0.25-kg masses at each end of the rope. While both are at rest, he places a 0.025-kg gronch (a toadlike creature) on one of the masses. That body and its wart-covered passenger descend 0.50 m before the gronch hops off. The body continues traveling downward another 1.2 m in the next 3.0 s. Compute g_M. See Problem 22.

CURVILINEAR MOTION

26. [I] Determine the acceleration of a kid on a bike traveling at a constant 10.0 m/s around a flat circular track of radius 200 m.

27. [I] A 25-g pebble is stuck in the tread of a 28-in. diameter tire. If the tire can exert an inward radial friction force of up to 20 N on the pebble, how fast will the pebble be traveling with respect to the center of the wheel when it flies out tangentially?

28. [I] A 10.0-kg mass is tied to a 3/16-in. Manila line, which has a breaking strength of 1.80 kN. What is the maximum speed the mass can have if it is whirled around in a horizontal circle with a 1.0-m radius and the rope is not to break?

29. [I] A baseball player rounds second base in an arc with a radius of curvature of 4.88 m at a speed of 6.1 m/s. If he weighs 845 N, what is the centripetal force that must be acting on him? Notice how this limits the tightness of the turn. What provides the centripetal force?

30. [I] Compute the Earth's centripetal acceleration toward the Sun. Take the time it takes to go once around as 365 d and the radius on average as 1.50×10^8 km.

31. [I] A hammer thrower at a track-and-field meet whirls around at a rate of 2.0 revolutions per second, revolving a 16-lb ball at the end of a cable that gives it a 6.0-ft effective radius. Compute the inward force that must be exerted on the ball.

32. [I] A youngster on a carousel horse 5.0 m from the center revolves at a constant rate, once around in 15.0 s. What is her acceleration?

33. [I] A supersonic jet diving at 290 m/s pulls out into a circular loop of radius R. If the craft is designed to withstand forces accompanying accelerations of up to 9.0 g, compute the minimum value of R.

34. [I] A test tube in a centrifuge is pivoted so that it swings out horizontally as the machine builds up speed. If the bottom of the tube is 150 mm from the central spin axis, and if the machine hits 50 000 revolutions per minute, what would be the centripetal force exerted on a giant amoeba of mass 1.0×10^{-8} kg at the bottom of the tube?

35. [I] A test tube 100 mm long is held rigidly at 30° with respect to the vertical in a centrifuge with its top lip 5.0 cm from the central spin axis of the machine. If it rotates at 40 000 revolutions per minute, what is the centripetal acceleration of a cell at the bottom of the tube?

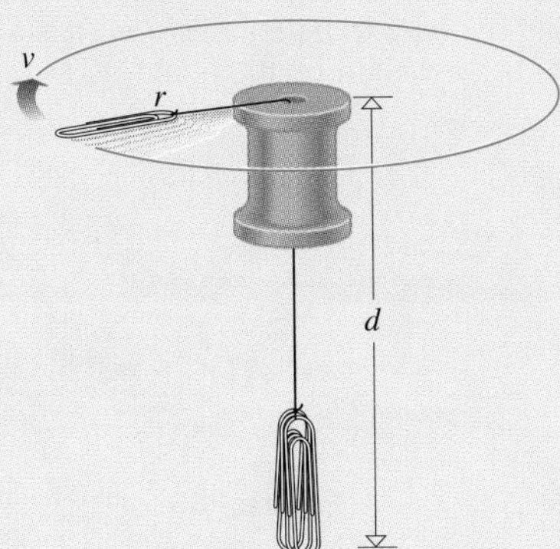

Figure P36

36. [II] Drop a string of length L through the hole in a vertically held spool of thread (Fig. P36). Tie 10 paper clips to each end and twirl the string so that one bunch of clips moves in a horizontal circle at a speed v, while the other hangs vertically. Neglecting friction, (a) write an expression for the distance d as a function of v and L where $L = (r + d)$. (b) Write an expression for g in terms of v and r, the radius of the circle.

37. [II] A front-loading clothes washer has a horizontal drum that is thoroughly perforated with small holes. Assuming it to spin dry at 1 rotation per second, have a radius of 40 cm, and contain a 4.5-kg wet Teddy bear, what maximum force is exerted by the wall on the bear? What happens to the water?

38. [II] A circular automobile racetrack is banked at an angle θ such that no friction between road and tires is required when a car travels at 30.0 m/s. If the radius of the track is 400 m, determine θ.

39. [II] A skier, with a speed v and a mass m, comes down a slope that has the shape of a vertical segment of a circle (of radius r), ending in a tangential, flat, horizontal run. Write an expression in terms of v, m, r, and g for the normal force exerted by the snow on the skis at the bottom just before the skier leaves the circular portion.

40. [II] A 1000-kg car traveling on a road that runs straight up a hill reaches the rounded crest at 10.0 m/s. If the hill at that point has a radius of curvature (in a vertical plane) of 50 m, what is the effective weight of the car at the instant it is horizontal at the very peak?

41. [II] At a given instant, someone strapped into a roller coaster car hangs upside down at the very top of the circle (of radius 25.0 m) while executing a so-called loop-the-loop. At what speed must he be traveling if at that moment the force exerted by his body on the seat is half his actual weight?

42. [II] Suppose you wish to whirl a pail full of water in a vertical circle without spilling any of its contents. If your arm is 0.90 m long (shoulder to fist) and the distance from the handle to the surface of the water is 20.0 cm, what minimum speed is required?

43. [II] A cylindrically shaped space station 1500 m in diameter is to revolve about its central symmetry axis to provide a simulated 1.0-g environment at the periphery. (a) Compute the necessary spin rate. (b) How would "g" vary with altitude up from the floor (which is the inside curved wall of the cylinder)?

44. [II] A stunt pilot flying an old biplane climbs in a vertical circular loop. While upside down, the force acting upward normally on the seat is $\frac{1}{3}$ of her usual weight. If the plane is traveling at that moment at 300 km/h, what is the radius of the loop? (A World War I pilot did this little trick without fastening his seat belt and—you guessed it—fell out. What can you say about his speed at the top of the loop?)

45. [III] Take the Earth to be a perfect sphere of diameter 1.274×10^7 m. If an object has a weight of 100 N while on a scale at the south pole, how much will it weigh at the equator? Take the equatorial spin speed to be $v = 465$ m/s.

46. [III] Design a carnival ride on which standing passengers are pressed against the inside curved wall of a rotating vertical cylinder. It is to turn at most at $\frac{1}{2}$ revolution per second. Assuming a minimum coefficient of friction of 0.20 between clothing and wall, what diameter should the

ride have if we can safely make the floor drop away when it reaches running speed?

FRICTION

47. [I] A dog weighing 300 N harnessed to a sled can exert a maximum horizontal force of 160 N without slipping. What is the coefficient of static friction between the dog's foot pads and the road?

48. [I] Assume that your mass is 70.0 kg and that you are wearing leather-soled shoes on a wooden floor. Now walk over to a wall and push horizontally on it. How much force can you exert before you start sliding away?

49. [I] What is the maximum acceleration attainable by a 4-wheel-drive vehicle with a tires-on-the-road static coefficient of μ_s?

50. [I] Mass m_1 sits on top of mass m_2, which is pulled along at a constant speed by a horizontal force F. If $m_1 = 10.0$ kg, $m_2 = 5.0$ kg, and μ_k for all surfaces is 0.30, find F.

51. [I] Someone wearing leather shoes is standing in the middle of a wooden plank. One end of the board is gradually raised until it makes an angle of 17° with the floor, at which point the person begins to slide down the incline. Compute the coefficient of static friction.

52. [I] A 30.0-kg youngster is dragged around the living room floor at a constant speed (giggling all the while) via a 60-N horizontal force. What was the appropriate pants-carpet friction coefficient?

53. [I] A wooden crate containing old $1000 bank notes has a total mass of 50.0 kg. It is transported on a flatbed truck to a facility to be burned. If the coefficient of static friction between the bed and the crate is 0.3 and the truck begins to climb a 20° incline at a constant speed, will the crate begin to slide?

54. [I] A garbage can partly filled with coal ash weighs 100 N, and it takes a force of 40 N to drag it down to the street at a uniform speed. How much force will it take to drag a can full of ash weighing 150 N?

55. [I] A 60.0-kg woman and a 20.0-kg child are both standing still in a moving passenger car when the engineer hits the brakes decelerating the train at $-1.0\ \text{m/s}^2$. What must be the appropriate coefficients of friction between floor and shoes if neither of them is to slip?

56. [II] What is the steepest incline that can be climbed at a constant speed by a 4-wheel-drive vehicle having a tires-on-the-road coefficient of static friction of 0.90?

57. [II] A crate is being transported on a flatbed truck. The coefficient of static friction between the crate and the bed is 0.50. What is the minimum stopping distance if the truck, traveling at 50.0 km/h, is to decelerate uniformly and the crate is not to slide forward on the bed?

58. [II] A yellow Triumph TR8 automobile with a total weight, including the driver, of 2850 lb is turning a flat curve of radius 200 m at 30 m/s. (a) What is the minimum value of μ_s that must exist between the tires and the road if no skidding is to occur? (b) Will this be possible on icy concrete?

59. [II] Assume an even weight distribution on all four tires of a car with 4-wheel antilock brakes. What is the minimum stopping time from 27 m/s if the coefficients of static and kinetic friction are 0.9 and 0.8, respectively?

60. [II] Place a book flat on a table and press down on it with your hand. Now suppose the hand-book and table-book values of μ_k are 0.50 and 0.40, respectively; the book's mass is 1.0 kg and your downward push on it is 10 N. How much horizontal force is needed to keep the book moving at a constant speed if the hand is stationary?

61. [II] A skier on a 4.0° inclined, snow-covered run skied downhill at a constant speed. Compute the coefficient of friction between the waxed skis and snow on that day when the temperature was around 0°C. Why is it that if the temperature were to drop to −10°C, μ_k would rise to around 0.22? Neglect air friction.

62. [II] A youngster shoots a bottle cap up a 20° inclined board at 2.0 m/s. The cap slides in a straight line, slowing to 1.0 m/s after traveling some distance. If $\mu_k = 0.4$, find that distance.

63. [II] A 100-kg bale of dried hay falls off a truck traveling on a level road at 88.0 km/h. It lands flat on the blacktop and skids 100 m before coming to rest. Assuming a uniform deceleration, compute the coefficient of kinetic friction.

64. [II] A 100-kg trunk loaded with old books is to be slid across a floor by a young woman who exerts a force of 300 N down and forward at 30° with the horizontal. If $\mu_k = 0.4$ and $\mu_s = 0.5$, compute the resulting acceleration.

65. [II] Suppose the woman in Problem 64 puts aside some of the books so that the mass of the load is 50 kg. (a) What will the acceleration be now? (b) A bit annoyed, she squirts some oil under the trunk so that $\mu_s = 0.4$ and $\mu_k = 0.3$. What now?

66. [II] One of two 200-kg crates is dragged along the floor at a constant speed by a horizontal force of 200 N. How much force would it take to pull them both at some constant speed if (a) they are tied together one behind the other? (b) the second is stacked on top of the first?

67. [III] What is the steepest incline that can be climbed at a constant speed by a rear-wheel-drive auto having a tires-on-the-road μ_s of 0.9 and 57% of its weight on the front axle?

68. [III] Determine the force (F) needed to keep the blocks depicted in Fig. P68 moving at a constant speed—m_2 to the right and m_1 to the left. Assume both the frictionless pulley and the rope to be massless: m_1 has a weight of 5.0 N; m_2 has a weight of 10 N; and $\mu_k = 0.40$ for all surfaces.

Figure P68

69. [III] A greasy flatbed truck is carrying a crate weighing 2.0 kN. The truck accelerates uniformly from rest to a speed of 30 km/h in a distance of 30.0 m. In that time, the crate slides 1.0 m back toward the end of the truck. Compute the coefficient of friction between bed and box.

6

Equilibrium: Statics

WE NOW FOCUS ON a special case where, even though two or more external forces act on a body, their effects cancel and there is no change in motion. Such a body is said to be in **equilibrium**—it may or may not be moving, but its motion is unchanging ($\mathbf{a} = 0$). The associated discipline is called **statics**. In Chapter 29, we will learn that absolute rest is meaningless. Still, it is convenient for planet-bound observers to consider a thing to be at rest when it is not moving with respect to the Earth (the Golden Gate Bridge is, in that manner of speaking, at rest). An object at rest is in **static equilibrium**.

In practical terms, statics is the study of structures—the analysis of the

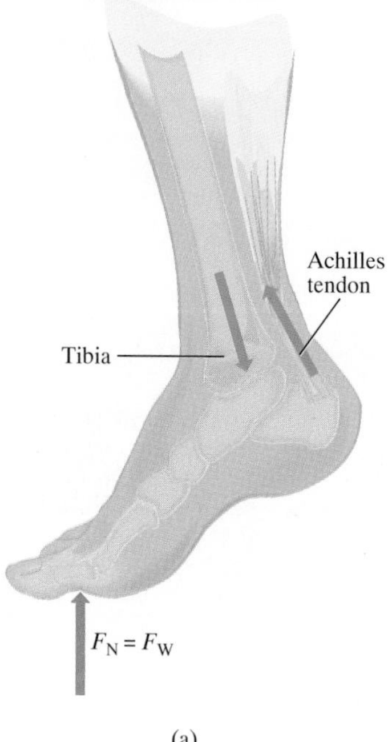

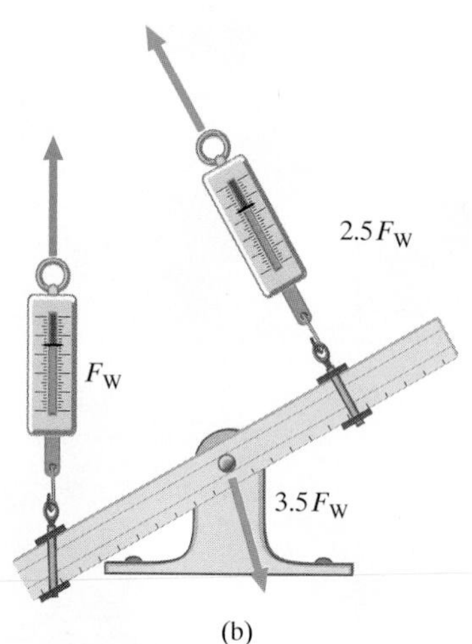

Figure 6.1 (a) The foot on tiptoe. Ligaments hold the bones together, and tendons attach the muscles to the bone. Think of the foot pivoted at the ankle joint, with the Achilles tendon pulling upward. (b) The foot can be modeled, and, as we'll see presently, the forces in the tibia and tendon can be computed.

forces acting on and within skyscrapers, towers, wrenches, crutches, and even people. Structurally, the human body shares remarkable similarities with high-rise buildings and trestle bridges. Moreover, about 500 years ago, Leonardo da Vinci discovered that the bones and muscles of vertebrates form a system of levers. We move most of our body parts (fingers, arms, neck, back, legs, feet, and so on) using muscles to pivot one bone against another. For example, when the calf muscles contract, the Achilles tendon pulls up on the heel bone (Fig. 6.1). The foot pivots in the ankle joint, forcing the toes down, and up you go. As we will see, this mechanism is identical to a wide range of systems, from the crowbar to the seesaw. Among other things, knowing the statics of the human body allows us to design efficient sports gear, enhance the techniques of physical therapy, better repair broken bones and damaged teeth, and make improved prostheses.

There are two governing principles in statics, both of which derive from Newton's Second Law: the *sum-of-the-forces equals zero,* and the *sum-of-the-torques equals zero* (p. 174). The first, by itself, can deal with situations where the lines-of-action of all the forces on a body intersect at a single point and are said to be *concurrent.* The second principle must be used when the forces are not concurrent. Concurrent-force systems are easier to analyze, and so after some preliminaries we will focus on them, develop technique, and go on to nonconcurrent systems.

TRANSLATIONAL EQUILIBRIUM

When every part of a body travels with the same speed in the same fixed direction, we have *translational motion,* as compared to *rotational motion,* where the speeds are different and the direction changes from moment to moment. *Translational equilibrium* corresponds to straight-line motion along a fixed direction at a constant speed (which, in this chapter, will be taken to be zero).

If the lines-of-action of all the forces acting on a body lie in a single plane, the system of forces is *coplanar.* Many practical arrangements of cables, pulleys, muscles, and machines correspond to coplanar force systems. Noncoplanar systems are also commonplace (in fact, you are probably sitting on one—the four forces supporting the legs of a chair are not coplanar). Forces not confined to a common plane require more effort to analyze, and so this treatment is limited to the study of *equilibrium under coplanar concurrent forces.*

It follows from Newton's Second Law [Eq. (5.4)] that translational equilibrium ($\mathbf{a} = 0$) occurs when

$$\Sigma\mathbf{F} = 0 \tag{6.1}$$

and this is called the **First Condition of Equilibrium**. If we resolve all the forces acting on a body into their components along *any* two perpendicular axes (for example, horizontal and vertical), the equivalent scalar statement is

$$\Sigma F_H = 0 \quad \text{and} \quad \Sigma F_V = 0 \tag{6.2}$$

These analytic tools are all we will require to deal with concurrent-force systems. What we need now is to learn how they are applied in practice.

6.1 Parallel and Colinear Force Systems

Figure 6.2a shows a pot of begonias resting on an essentially weightless pedestal. The Earth-plant gravitational interaction causes the begonias to have a weight

These people have reached terminal speed. The weight of each diver acting down exactly equals the drag force acting up, or $\Sigma F_V = 0$. Hence, $a = 0$ and each person is in translational equilibrium even though he or she is falling at about 160 km/h.

$F_{Wb} = 300$ N, and the several resulting forces that arise in response are depicted in the diagram. If the potted plant were weightless, those forces would be zero. As is, they all act along the same line and are said to be **colinear**.

As shown in the free-body diagram, only two forces act on the flowers. Given that the begonias are in static equilibrium ($a = 0$), and taking *up* as positive, we have

$$+\uparrow \Sigma F_V = (+F_{PB}) + (-F_{Wb}) = 0$$

and

$$F_{PB} = F_{Wb}$$

The same logic for the pedestal yields

$$+\uparrow \Sigma F_V = (+F_{GP}) + (-F_{BP}) = 0$$

and

$$F_{GP} = F_{BP}$$

Hence $F_{GP} = F_{BP} = F_{PB} = F_{Wb} = 300$ N. The ground holds up the begonias.

When two colinear forces acting on a body point toward each other, the body is under compression—the forces on the pedestal tend to compress or squash it. An object in tension pulls, an object in compression pushes. The external forces that support a system are called **reaction forces**: Here, the ground exerts a vertical reaction force on the pedestal, the normal force.

Let's take things one step further and add to a system of vertical forces a horizontal force (Fig. 6.3). Each tower of the World Trade Center weighs about 1.2×10^9 N (1.4×10^5 tons). Suppose a horizontal wind averaging 50 mi/h blows against a face of one of the towers, having an area of roughly 1.63×10^4 m^2 (1.75×10^5 ft^2). From measurements, we know that such a wind produces a pressure of around 3×10^2 N/m^2 (6 lb/ft^2) and would result in a total horizontal force on the tower of 5×10^6 N ($\approx 1 \times 10^6$ lb). It follows from $\Sigma F_H = 0$ that the ground must exert a horizontal reaction force of 5×10^6 N on the tower in the opposite direction to the wind, or else the building would begin to slide across town (Fig. 6.3b). So a system can experience both horizontal and vertical reaction forces.

Most buildings are constructed on a frame of wood, reinforced concrete, or steel, and that structure carries the loads of weight and wind down to the ground. Skyscrapers were once made rigid with heavy masonry walls, but today they are constructed on a lighter frame, reinforced with steel stiffening members. The human body is constructed on a similar frame of bones, ligaments, and muscles. Loads in the upper body are carried by the spine to the pelvis and then down

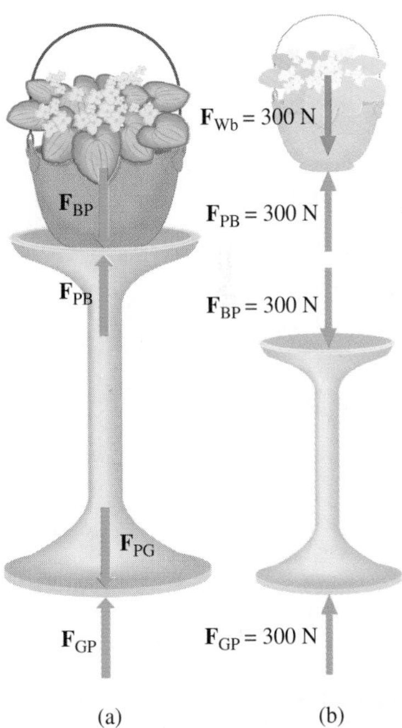

Figure 6.2 (a) A 300-N pot of begonias rests on an essentially weightless pedestal. The gravitational force on the pot gives rise to two interaction pairs ($\mathbf{F}_{BP} = -\mathbf{F}_{PB}$ and $\mathbf{F}_{PG} = -\mathbf{F}_{GP}$). (b) These both equal 300 N because the pedestal's weight is negligible.

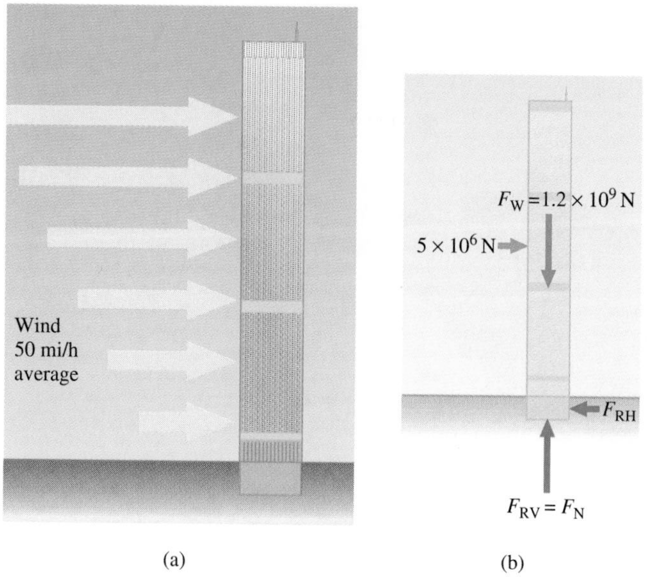

Figure 6.3 The force exerted by wind increases with height and is a major consideration with buildings over 20 or 30 stories. About 10% of the structural weight of a high-rise building goes into wind bracing. (a) One of the 1350-ft-tall towers of the World Trade Center in New York City. A wind with an average speed of 50 mi/h blows against one face and pushes with 5×10^6 N. (b) The free-body diagram shows the needed reaction forces.

through the legs to the ground. The bones of the skeleton are primarily compressive members that carry both body weight and the forces exerted by the muscles. The skeletal muscles are attached at both ends via tendons to different bones. These muscles can only contract, pulling on the bones and putting them in compression.

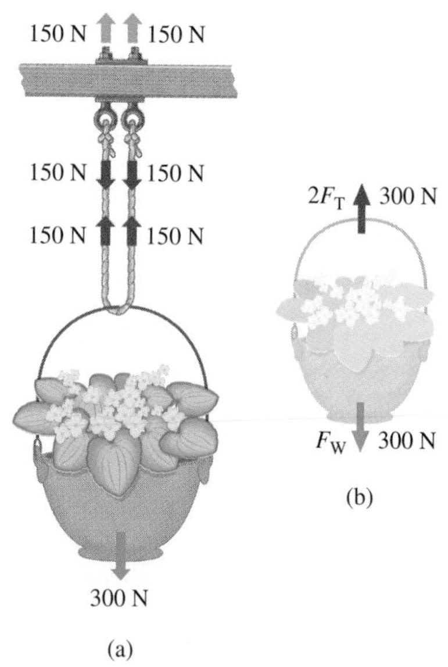

Figure 6.4 (a) A load supported by two ropes. Since the system is in equilibrium, $\Sigma F_V = 0$. (b) The net upward force ($2F_T$) equals the net downward force (F_W). Here you count the number of ropes, just as you would count the number of legs supporting a table.

Ropes and Pulleys

Return to Fig. 5.9, which depicts a brass mass attached to an essentially weightless rope hanging from a ceiling hook. The mass, the rope, and the hook are separately in equilibrium and, for each, $\Sigma F_V = 0$. The hook holds up the rope, and the rope holds up the mass. A single length of rope supporting a load is called a *hanger.* **When there is only one hanger, the tension in it equals the load**: $F_T = F_{Wb}$.

If two lengths of light rope support a load, such as the 300-N pot in Fig. 6.4, the tension in each length is 150 N and the force on each hook is 150 N. If three ropes equally share the 300-N load, the tension in each is 100 N and the force on each ceiling hook to which a rope is attached is 100 N. In each case, the tension in every hanger supporting the load is equal. Figure 6.5 incorporates an essentially weightless, frictionless pulley that distributes the load equally on the two rope segments; other than that, the situation is identical to Fig. 6.4. We apply a force of 150 N, and the system delivers a force of 300 N. This arrangement is a *force-multiplier,* one of the most important simple machines ever devised. It's often set up in a horizontal plane (anchored to a tree to haul heavy loads along the ground, for example).

Figure 6.6 indicates how any number of pulleys can be used to multiply the applied force; in this case by a factor of 4. Several pulleys can be mounted next to, or just below, one another (Fig. 6.7). This ancient device is still used today extensively in applications such as elevators, scaffolds, tow trucks, cranes, and pile drivers. A typical derrick can hoist a 25-ton load into the air via pulleys with 10 or 11 lengths of steel cable sharing the effort.

Each of the pulleys in Fig. 6.8 serves to change the direction of the applied force rather than act as a multiplier. A pulley used in this way is a convenient means of

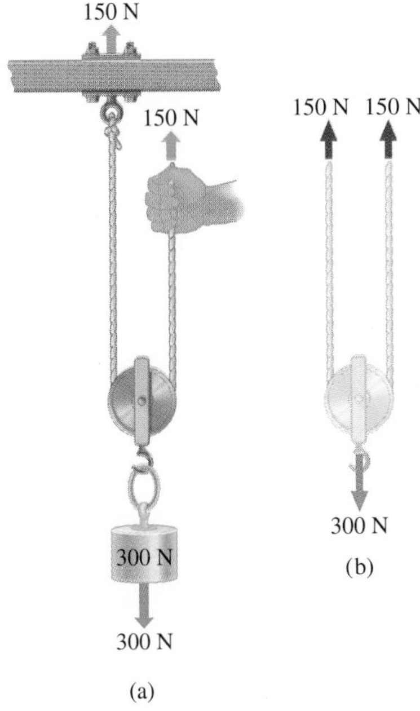

Figure 6.5 (a) This pulley arrangement acts as a force multiplier. Count the number of ropes supporting the load. In this case, there are two, and the system applies twice the force applied to it. The hand lifts with 150 N and the system lifts with 300 N. (b) The tension in each supporting length of rope is 150 N. The total upward force is therefore 300 N.

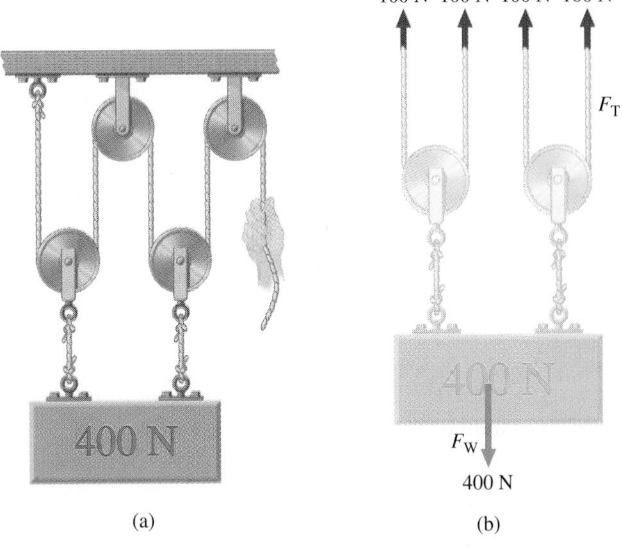

Figure 6.6 (a) When four ropes support the load in equilibrium, each carries an equal tension of $\frac{1}{4}$ that load. (b) Here, the tension is 100 N and the hand need only apply 100 N to hold the load, or a tiny bit more to start it accelerating upward.

applying a constant force (equal to the hanging weight) to a system, whether it's an accelerating mass in a physics lab or a broken leg in a hospital.

6.2 Concurrent Force Systems

A system acted on by concurrent forces (forces that all pass through a single point) is in equilibrium when

$$\Sigma F_x = 0 \qquad \text{and} \qquad \Sigma F_y = 0$$

where x and y are any two perpendicular directions. There are therefore two independent equations, and no more than two unknowns can be determined. Each force vector in the plane of the system corresponds to two quantities: either a magnitude and the angle at which it acts, or any two of its perpendicular components. In general, two equations will be needed to determine a force vector. In practice, however, we often know the line-of-action of the force and require only the magnitude. The nice thing about ropes, cables, and muscles is that they always act in tension, and the forces applied to and by them act along them.

Figure 6.7 (a) This configuration is equivalent to that of Fig. 6.6. (b) Four ropes support the load, and the hand need only pull down with 100 N.

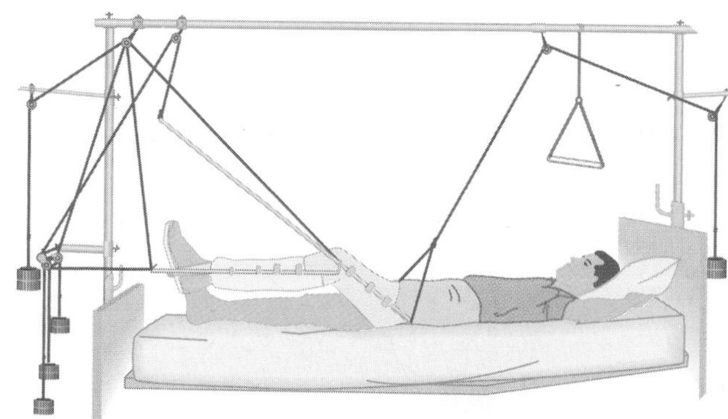

Figure 6.8 It's been common practice since the fourteenth century to use ropes and pulleys to put broken limbs in traction. Shown here is a Thomas leg splint with a Pearson attachment.

Example 6.1 The engine hanging motionlessly in Fig. 6.9 weighs 800 N. If θ is 20.0°, compute (a) the tension in each length of rope and (b) the horizontal force tending to pull out the support pins. Use Table 6.1

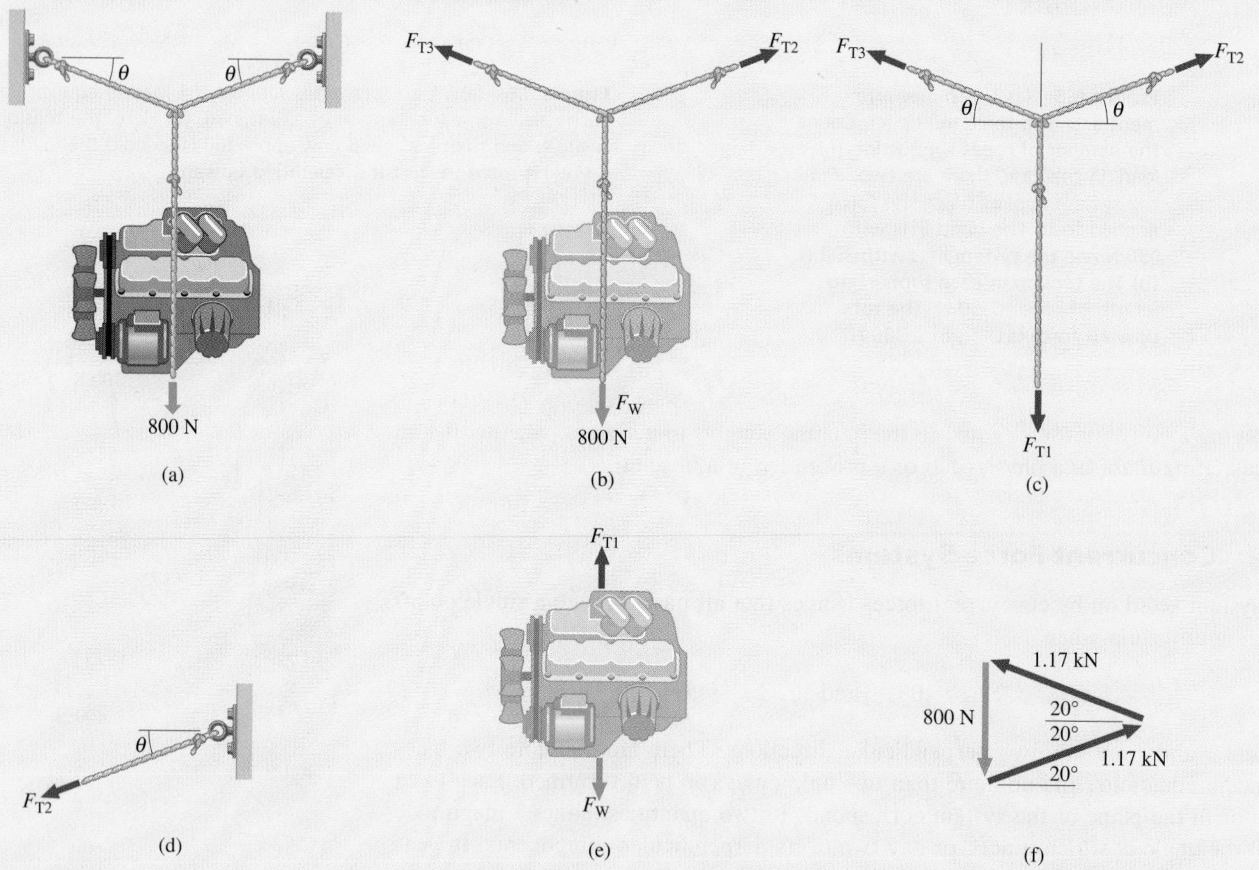

Figure 6.9 (a) The engine is supported by a system of ropes, and all the forces acting on that arrangement are concurrent. (b), (c), (d), and (e) are free-body diagrams. (f) Because $\Sigma\mathbf{F} = 0$, the three forces must form a closed triangle such that the resultant is zero.

(continued)

(continued)

to pick an appropriate rope for the job. To avoid failure from high momentary forces (sudden jerks, for example), it's customary to use ropes at no more than about one-sixth their breaking strength.

Solution: [Given: $F_W = 800$ N and $\theta = 20.0°$. Find: tensions F_{T1}, F_{T2}, F_{T3}, and the pull-out force.] The system consists of the three ropes and the load—its free-body diagram is Fig. 6.9b. The force vectors acting on the system—$\mathbf{F}_{T2}$, $\mathbf{F}_{T3}$, and $\mathbf{F}_W$—are concurrent, and pass through the knot. A nice place to start is with $\mathbf{F}_{T1}$; the engine is supported by a single hanger, so the tension in that rope equals the load

$$\boxed{F_{T1} = F_W = 800 \text{ N}}$$

Because of the symmetry, the tensions on either segment of the V-shaped support rope must be equal ($F_{T2} = F_{T3}$). Whatever happens, happens identically to both segments. Let's not take advantage of that fact now, but just press on with the analysis step by step. We need F_{T2} and F_{T3}, and could use either Fig. 6.9b for the whole system or Fig. 6.9c for just the knot. Each must be in equilibrium, and each contains the two forces we want. Take up and to the right as positive. The sum-of-the-horizontal-forces equals zero on the knot, and so

$$\overset{+}{\rightarrow}\Sigma F_H = F_{T2} \cos\theta - F_{T3} \cos\theta = 0$$

confirming that $F_{T2} = F_{T3}$. The sum-of-the-vertical-forces equals zero on the knot, and so

$$+\uparrow\Sigma F_V = F_{T2} \sin\theta + F_{T2} \sin\theta - F_W = 0$$

Hence $$2F_{T2} \sin 20.0° = 800 \text{ N}$$

and $\boxed{F_{T2} = 1.17 \text{ kN}}$. Note that the tension in either supporting rope is greater than the load.

At each wall pin (Fig. 6.9d), the horizontal force is

$$F_{T2} \cos 20.0° = \boxed{1.10 \text{ kN}}$$

Since $F_{T1} = 800$ N, and $F_{T2} = 1.17$ kN, none of the 3/16-in. ropes will suffice. The braided nylon and Dacron™ 1/4-in. lines will do nicely.

▶ **Quick Check:** Make a sketch of the three force vectors acting at the knot. They must add up to zero and form a closed triangle, and they do (Fig. 6.9f).

TABLE 6.1 Various line strengths

	Breaking strength*					
Diameter (inches)	Manila hemp (kN)	Nylon filament (kN)	Dacron™ filament (kN)	Nylon braid (kN)	Dacron™ braid (kN)	Cable (kN)
3/16	1.8	4.0	4.0	4.6	4.6	19
1/4	2.4	6.6	6.7	7.1	8.2	31
5/16	4.0	9.6	10.3	12.0	12.8	44
3/8	5.4	14.9	15.4	16.9	18.4	64
7/16	7.0	20.0	20.0	23.1	25.0	78
1/2	10.6	27.1	28.3	32.0	32.7	101

*As a rule, working loads should not exceed one-sixth of the breaking strength for ropes and one-fifth for cable.

A spider web is a total tension structure. Every fiber is in tension.

Notice in Fig. 6.9 that, if F_W has any nonzero value, θ cannot equal zero because sin 0 equals zero, and an undeflected rope exerts no upward force. The rope must sag as it interacts with the object being supported. In fact, the rope will sag under its own weight. No amount of horizontal force can pull the rope so that it straightens and hangs totally horizontal.

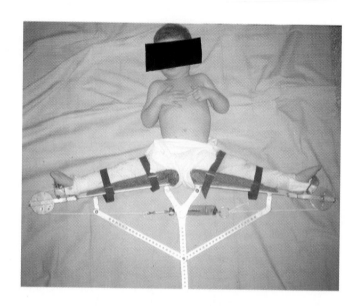

An interesting application of statics for the treatment of congenital hip dislocation. The spring scale reads the tension applied to each leg. The little turnbuckle (on the left of the scale) allows the tension to be adjusted as needed. The double-Y-shaped frame keeps the legs apart and allows $\Sigma \mathbf{F} = 0$.

Example 6.2 Determine the angle θ in the arrangement shown in Fig. 6.10. Assume the pulley is weightless and frictionless.

Solution: [Given: the angle is 20.0°, the leg weight 150 N, and the counterforce 200 N. Find: θ.] With no friction on the weightless pulley, the tension in the rope on the right is uniform throughout. Draw the free-body diagram of the ring (Fig. 6.10b). It should be evident that $F_{T2} = 200$ N, $F_{T1} = 150$ N, and, because θ may not equal 20°, F_{T3} may not equal F_{T2}. At the ring

$$\overset{+}{\rightarrow}\Sigma F_H = (200\text{ N})\cos 20.0° - F_{T3}\cos\theta = 0$$

and

$$+\uparrow\Sigma F_V = (200\text{ N})\sin 20.0° + F_{T3}\sin\theta - 150\text{ N} = 0$$

Hence, $F_{T3}\cos\theta = 187.9$ N and $F_{T3}\sin\theta = 81.60$ N. These are the horizontal and vertical components of F_{T3}, and if we divide the second by the first, the tension cancels. We get

$$\frac{\sin\theta}{\cos\theta} = \tan\theta = 0.434\,274$$

and

$$\boxed{\theta = 23.5°}$$

▶ **Quick Check:** $F_{T3} = (187.9\text{ N})/\cos\theta = 204.9$ N, and $F_{T3} = (81.60\text{ N})/\sin\theta = 204.9$ N. Figure 6.10c is the force triangle, and it correctly has a zero resultant.

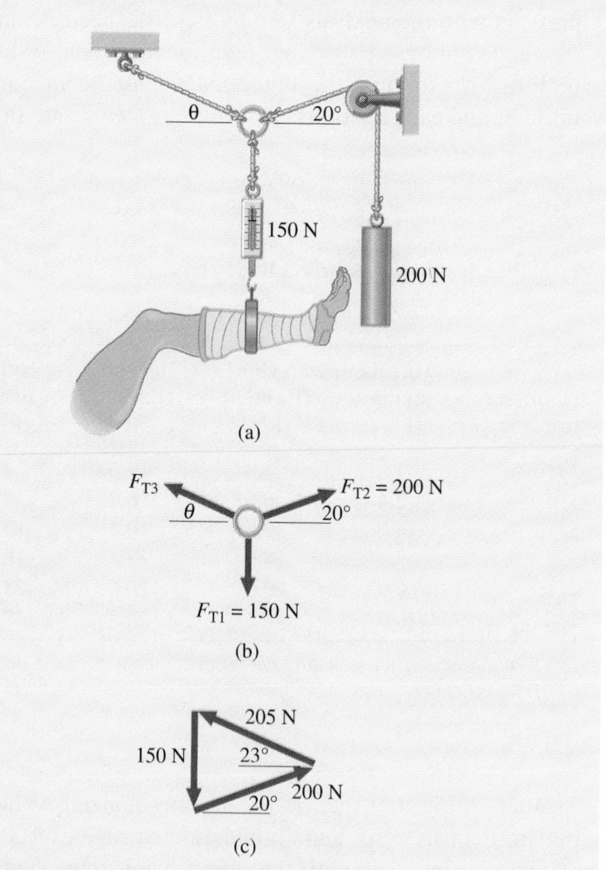

Figure 6.10 (a) A concurrent force system. (b) The free-body diagram of the ring showing the three forces that meet at its center. (c) The force triangle corresponding to equilibrium.

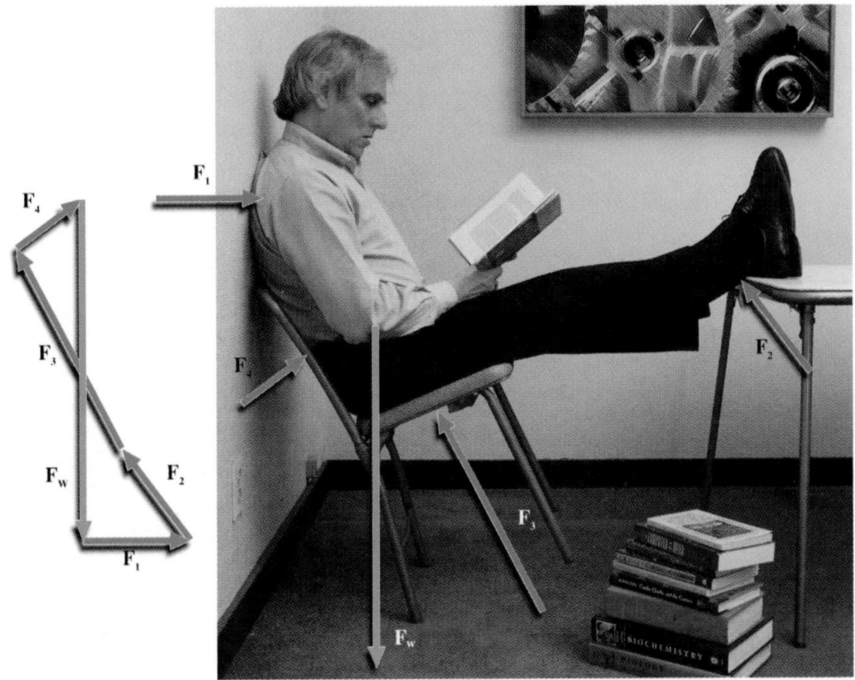

This person is acted upon by five forces. His gravitational interaction with the Earth produces a downward force $\mathbf{F}_W$. The wall ($\mathbf{F}_1$), table ($\mathbf{F}_2$), and chair ($\mathbf{F}_3$ and $\mathbf{F}_4$) each exert normal forces on him. Since he's in equilibrium, the sum-of-the-forces *on him* equals zero.

Bars, Bones, and Trusses

Many modern structures, from billboards to bridges, are supported by a rigid framework of slender, short, interconnected members called *bars* (which are usually made of steel, aluminum, or wood). We will restrict our treatment to rigid structural members that are light compared to the supported loads. These are attached together by smooth pins (bolts or rivets) at their ends at what are called *joints.* All loads are applied at the joints. This construction has the effect of allowing *only forces that act along the length of each bar.* The bars are either in pure tension or compression, and no bending occurs. To prove that contention, imagine a free-body diagram of a typical member, say bar 1 in Fig. 6.11. Bar 1 is at rest, and if it is to stay at rest, the net force acting on each of its two ends must be equal in magnitude, oppositely directed, and colinear.

Example 6.3 Two lightweight, pinned bars support a 300-N sign (Fig. 6.11). If the angle θ is 30.0°, compute the magnitudes of the forces acting on each member and determine if they are in tension or compression.

Solution: [Given: $F_W = 300$ N and $\theta = 30.0°$. Find: the forces on the bars, F_1 and F_2.] The weight of the sign pulls downward on the joint, and that force must be countered entirely by bar 2 because bar 1 is horizontal. *The internal forces can only act along the line of either pinned member.* As shown in Fig. 6.11c, the free-body diagram of the pin supporting the sign, bar 2 pulls upward and to the left on the joint and is in tension. Bar 2 exerts a force to the left (tending to squash bar 1). This force must be countered by bar 1, which is in compression. If, in your diagram, you put the forces in the wrong directions, they will simply come out negative, so don't worry about that. Thus, for the pin, where the forces are concurrent

$$+\uparrow \Sigma F_V = F_2 \sin 30.0° - F_W = 0$$

and

$$\overset{+}{\rightarrow} \Sigma F_H = F_1 - F_2 \cos 30.0° = 0$$

(continued)

(continued)

Hence, $F_2 = (300\ \text{N})/\sin 30.0° = \boxed{600\ \text{N}}$ and $F_1 = (600\ \text{N}) \cos 30.0° = \boxed{520\ \text{N}}$; both are positive, so they are correct as drawn.

▶ **Quick Check:** Since the system is in equilibrium, the force vectors must add up to zero and so form a closed triangle. Hence, F_2^2 must equal $F_1^2 + F_W^2$, which it does (Fig. 6.11d).

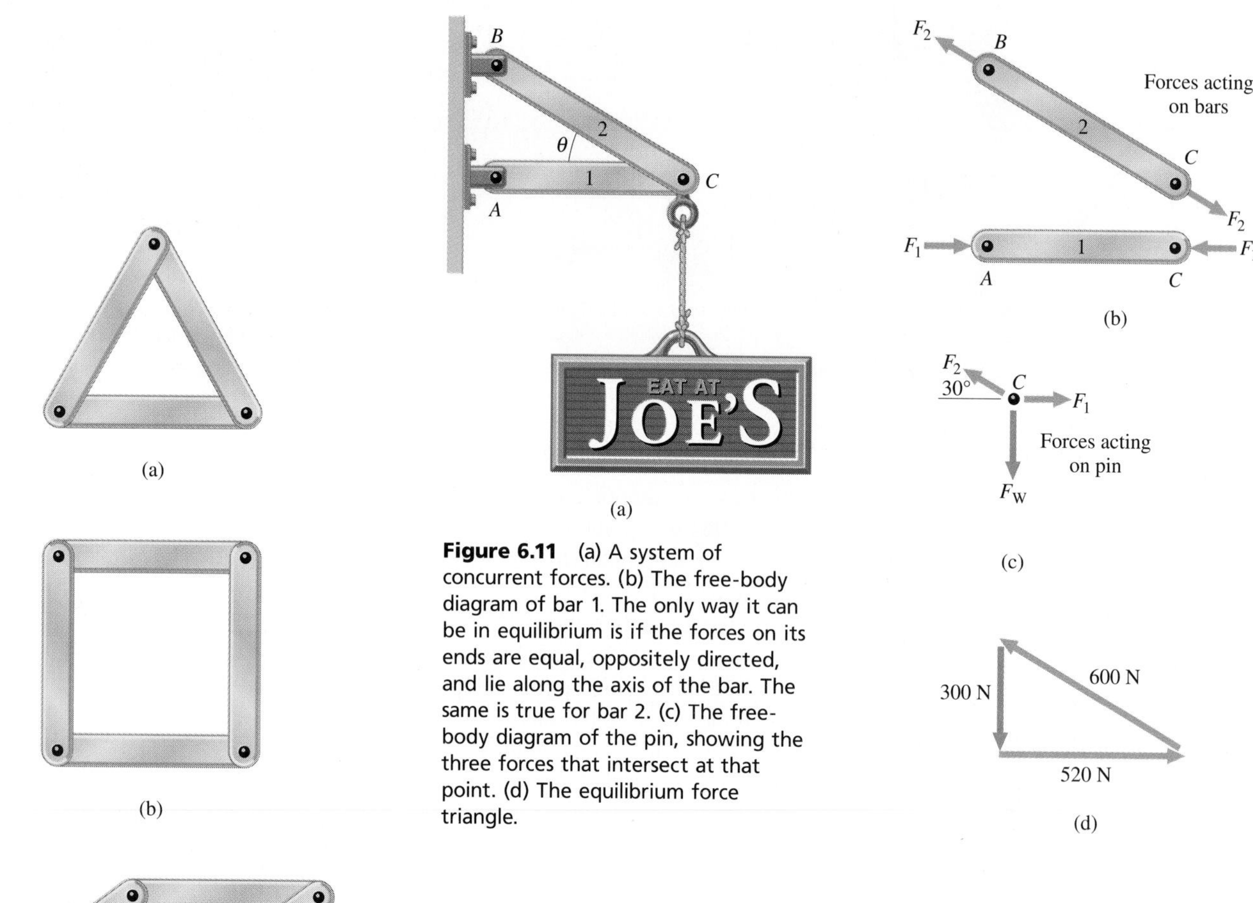

Figure 6.11 (a) A system of concurrent forces. (b) The free-body diagram of bar 1. The only way it can be in equilibrium is if the forces on its ends are equal, oppositely directed, and lie along the axis of the bar. The same is true for bar 2. (c) The free-body diagram of the pin, showing the three forces that intersect at that point. (d) The equilibrium force triangle.

Figure 6.12 (a) The three-bar frame is stable. Push on it, and it will not collapse. The bars cannot stretch or compress to allow the three-bar frame to fold up. (b) The four-bar frame is unstable. Push on that frame, load it with a horizontal force, and it will collapse (c).

A simple *truss* is a structural framework of members joined together at their ends in triangular patterns lying in a plane. Trusses are used in aircraft, space stations, antennas, TV towers, bridges, and buildings. Three-sided component frames are the rule because they are inherently rigid (Fig. 6.12). The truss, using lightweight components of moderate length, provides strength and great space-spanning capability.

In the early days of the railroads, a variety of truss bridges were developed. These were economical, easy to build at remote sites, and fairly rigid under heavy loads. In the United States, where iron was expensive, timber beams were used in compression, with thin iron rods as the tension members. Figure 6.13 shows one of these truss bridges, and it immediately brings to mind the musculoskeletal system of quadrupedal vertebrates, with the spine spanning the space between the shoulder girdle and the pelvis.

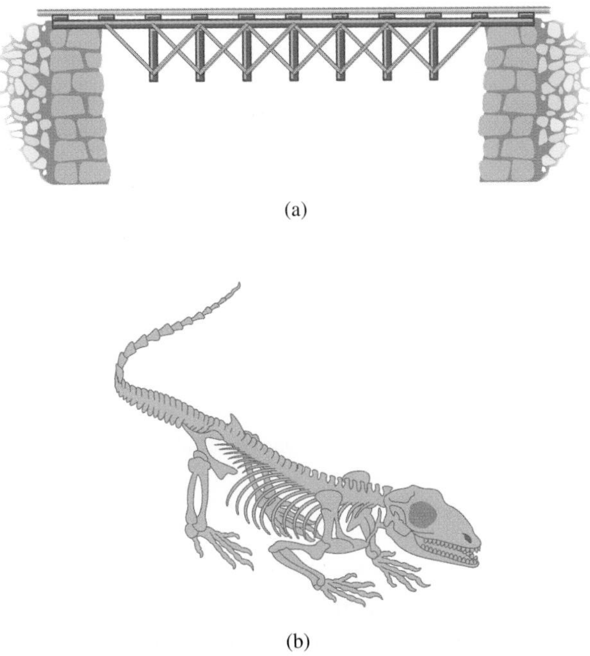

Figure 6.13 (a) An old railroad trusswork bridge made of timber and braced with iron tension rods. (b) Vertebrates, such as this lizard, are supported by a similar system of bone in compression, crisscrossed by muscles in tension.

A clutter of steel trusses supporting an overhead railroad track. Today, high-strength bolts have replaced the rivets that can be seen here.

In humans and other animals, the compression members—vertebrae and ribs—are made of bone. The tension members are the intercostal muscles that crisscross between the ribs, pulling them together. Similarly, if the bridge were to sag appreciably under a load, the vertical beams would have to spread apart at their lower ends, an action prevented by the tension crossbars that pull the beams together. The spine-rib-muscle structure is the bridge supporting the weight of the trunk of the body. The four legs are the compression columns on which the entire load rests. So effective is the musculoskeletal system that camels, elephants, donkeys, and horses can easily carry additional loads on their backs.

Archimedes, who was kinsman and a friend of King Hieron of Syracuse, wrote to him that with any given force it was possible to move any given weight, and emboldened, as we were told, by the strength of his demonstration, he declared that, if there were another world, and he could go to it, he could move this. Hieron was astonished, and begged him to put his proposition into execution, and show him some great weight moved by a slight force. Archimedes therefore fixed upon a three-master merchantman of the royal fleet, which had been dragged ashore by the great labors of many men, and after putting on board many passengers and the customary freight, he seated himself at a distance from her, and without any great effort, but quietly setting in motion with his hand a system of compound pulleys, drew her towards him smoothly and evenly, as though she were gliding through the water.

PLUTARCH (ca. 46–120 A.D.)
Life of Marcellus

ROTATIONAL EQUILIBRIUM

The equal-arm balance was already in widespread use well over 4000 years ago. It is not surprising, then, that Aristotle, in his book *Mechanics,* attempted to analyze the practical problem represented by balances, levers, and seesaws. When equal downward forces are applied to each end of a centrally pivoted rod, the system remains balanced horizontally. The beam does not revolve (or, more generally, there is no *change* in its rotational motion), and we say that it is in *rotational equilibrium.* The word *equilibrium* comes from the Latin words *aequus,* meaning even or equal, and *libra,* meaning balance.

The general problem of the lever, where the weights and arm lengths are different on each side of the pivot, was treated (though not altogether satisfactorily) by Aristotle. The job was completed by that genius of Greek science, Archimedes of Syracuse (287 B.C.–212 B.C.). Their rather rudimentary *Law of the Lever* states that

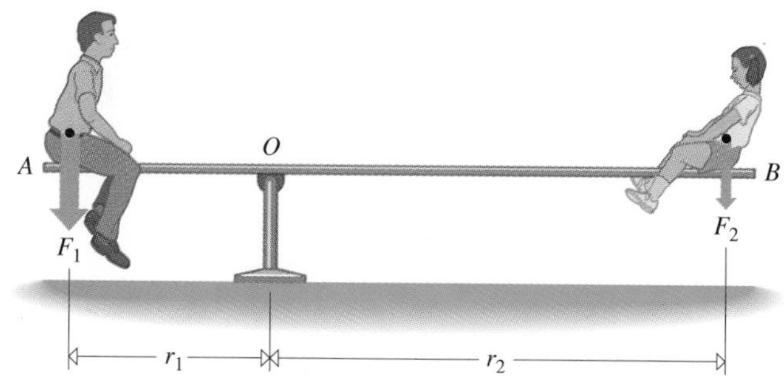

Figure 6.14 Unequal parallel forces acting at unequal distances from the pivot. The system is in rotational equilibrium.

unequal forces, acting perpendicularly on a pivoted bar, balance each other, provided $F_1r_1 = F_2r_2$ (Fig. 6.14). One must consider the sizes of the forces, as well as the distances at which they act from the pivot—that much can easily be confirmed experimentally.

Example 6.4 A boy (mass 30 kg) wishes to play on a centrally pivoted seesaw with his dog Irving (mass 10 kg). When the dog sits 3.0 m from the pivot, where must the boy sit if the 6.5-m-long board is to be balanced horizontally?

Solution: [Given: $m_B = 30$ kg, $m_D = 10$ kg, the dog-pivot distance $r_D = 3.0$ m, and the extraneous board length is 6.5 m. Find: the boy-pivot distance r_B.] The forces are the weights $F_{WB} = m_B g$ and $F_{WD} = m_D g$; therefore, from the Law of the Lever

$$(m_B g)r_B = (m_D g)r_D$$

Hence $$r_B = \frac{(10\text{ kg})(3.0\text{ m})}{(30\text{ kg})} = 1.0\text{ m}$$

The boy must sit 1.0 m from the pivot.

▶ **Quick Check:** $m_B r_B = m_D r_D = 30$ kg·m.

The Roman balance, or steelyard (Fig. 6.15), is a practical example of the unequal-arm balance. The steelyard has been used for at least 2000 years to measure heavy objects by counterpoising them against small standard weights. You have probably been weighed in a physician's office on a platform scale that is essentially a steelyard. Extending the idea, da Vinci realized that the human body achieved its mobility via various kinds of levers. Both the human foot (Fig. 6.1a) and the steelyard are levers pivoted at a point between the two forces. da Vinci was also aware of the significance of the *moment of a force,* a concept called the *torque,* which is central to the modern treatment of statics.

6.3 Torque

Figure 6.16 shows situations in which a force **F** acts on a wrench along different directions and at different points. The axis about which rotation will occur runs down the center of the bolt through point O perpendicular to the plane of the diagram. The point of application of the force is marked by the end of the *position vector* **r** drawn from O. In each case, the ability of **F** to produce a rotation of the wrench will be different. It is this turning effect, or *moment,* that da Vinci appreciated as crucial. When the line-of-action of the force passes through the pivot, no rotation will occur at all (Fig. 6.16a). You cannot close a door by pushing on its edge *toward* the

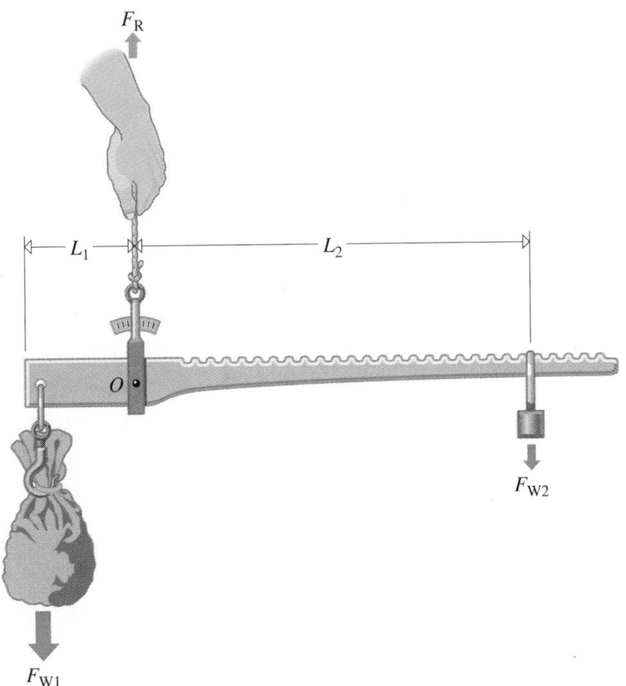

Figure 6.15 An unequal-arm balance, or steelyard. The small weight, the *poise,* is moved down the arm until a balance is reached with the heavy object being weighed. The balanced steelyard is a system of separated forces whose lines-of-action are parallel and therefore nonconcurrent.

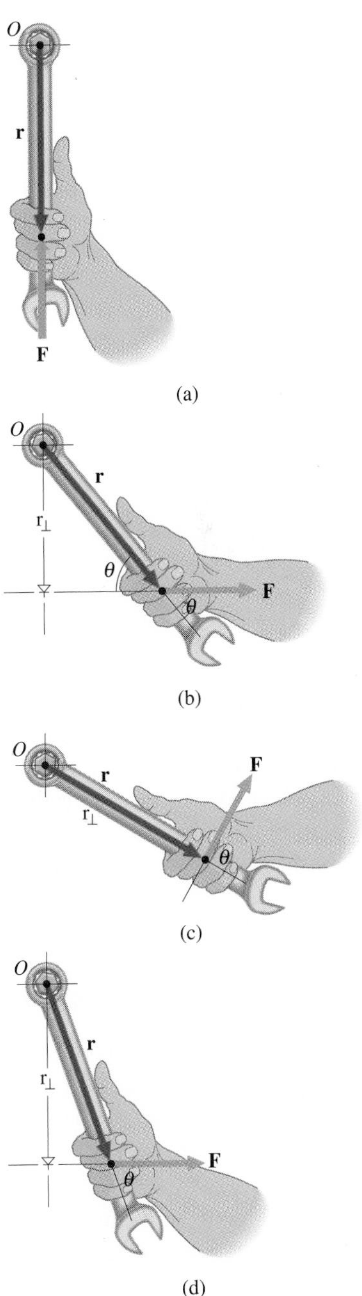

Figure 6.16 The amount of twist, or torque, one can get from a wrench depends on how big the force is, in what direction it acts, and where it is applied. (a) No torque is exerted when the force goes through the pivot point O. (b) Here, torque about O exists. It can be increased (c) by increasing θ or decreased (d) by decreasing r.

hinge. By contrast, the force applied at the very end of the wrench, where the line-of-action is far from O, as in Fig. 6.16c, produces the greatest twist. In fact, it's precisely for that reason that the wrench was invented.

The **moment-arm** *of the force with respect to the axis passing through O is defined as the perpendicular distance* ($r_\perp$) *drawn from O to the line-of-action of* **F**. In each part of Fig. 6.16, notice that, although the magnitude of the force is unchanged, the moment-arm increases from $r_\perp = 0$ in (a), to $r_\perp = r \sin \theta$ in (b), to $r_\perp = r$ in (c). The **moment of the force** *about O is defined as the product of F and the moment-arm.* We symbolize the moment of the force by the Greek letter *tau* (τ) and particularize it with respect to point O as τ_0. Thus

$$\tau_0 = r_\perp F$$

This quantity is called the **torque** (from the Latin *torquere,* to twist). The term "moment of the force" is also used. The torque is a measure of the twist produced by a force about a particular axis that may be located *anywhere.* Torque has the dimensions of force multiplied by distance and the SI units of N·m.

In Fig. 6.17a, the line-of-action of the force vector **F** makes an angle of θ with the line-of-action of the position vector **r**. Here, $r_\perp = r \sin \theta$, and so

$$\tau_0 = F r_\perp = F r \sin \theta$$

Alternatively, resolve **F** into its components ($\mathbf{F}_\parallel$, parallel to **r**, and $\mathbf{F}_\perp$, perpendicular to **r**). $\mathbf{F}_\parallel$ passes through O, its moment-arm is zero, and its torque about O is also zero. It follows that, with respect to O, the torque produced by $\mathbf{F}_\perp$ must equal the

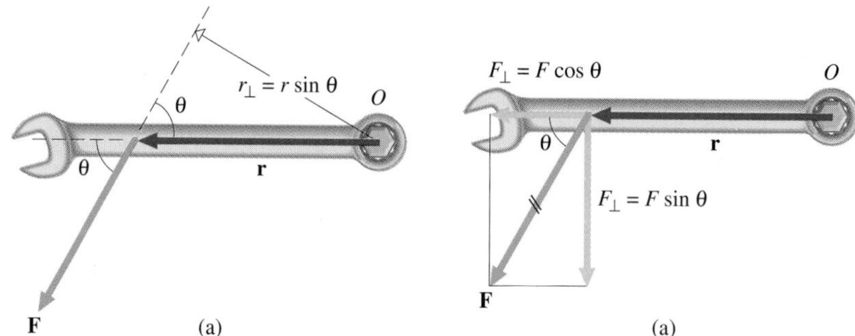

Figure 6.17 The torque can be thought of as $Fr_\perp$ or $F_\perp r$, since $Fr_\perp = F_\perp r = Fr \sin \theta$.

torque produced by **F** as a whole. Since $F_\perp$ (which equals $F \sin \theta$) has a moment-arm of r, its torque about O is

$$\tau_O = rF_\perp = rF \sin \theta \tag{6.3}$$

which is the same result we obtained earlier. The torque equals either $r_\perp F$ or $rF_\perp$, and we use whichever form is the more convenient.

Example 6.5 The exercise illustrated in Fig. 6.18 consists of swinging the lower leg out until it's horizontal and then lowering it, all the while supporting a mass m. Write an expression for the torque about O as a function of θ, m, g, and the knee-heel length r. For the time being, ignore the mass of the leg.

Solution: [Given: m, r, and θ. Find: τ_O where O is at the knee joint.] By definition

$$\tau_O = r_\perp F$$

The downward force is the weight mg. The moment-arm is the perpendicular distance from O to the line-of-action of the force, that is

$$r_\perp = r \sin \theta$$

where θ is the angle between **r** and **F**. Hence

$$\boxed{\tau_O = mgr \sin \theta}$$

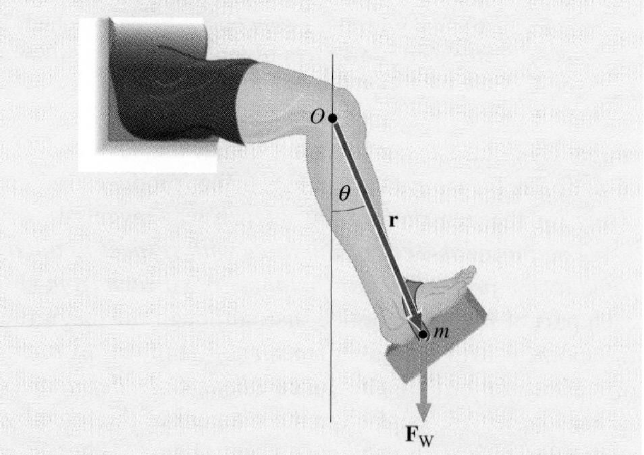

Figure 6.18 The weight exerts a torque about O that is countered by an equal and opposite muscle-generated torque.

▶ **Quick Check:** The torque varies with $\sin \theta$; reasonably, it's zero when $\theta = 0$ and maximum, $\tau_O(\max) = mgr$, when $\theta = 90°$.

When you look down on the diagram in Fig. 6.17, you can see that the force rotates the wrench counterclockwise about O. Some other force could act on the wrench in a different direction to produce an opposing clockwise twist (as when two people arm wrestle, for instance). The resulting twist produced by opposing torques would then be the sum of the two: $\Sigma\tau_O$, taking one positive and the other negative.

Example 6.6 At the moment depicted in Fig. 6.19, the hand exerts a 200-N force at the end of a pivoted 1.0-m-long lever. The return spring fixed to the lever's midpoint pulls back on it with a horizontal force of 80 N. What is the net torque acting on the lever about the pivot?

Solution: [Given: referring to Fig. 6.19b, $F_h = 200$ N, $r_h = 1.0$ m, $F_s = 80$ N, and $r_s = 0.50$ m. Find: τ_O where O is at the pivot.] The moment-arm of the hand-force (the perpendicular distance from O to the line-of-action of F_h) is just r_h, while the moment-arm of the spring-force is $r_s \cos 60° = (0.50 \text{ m})(0.50) = 0.25$ m. Arbitrarily taking clockwise as positive, we have

$$\circlearrowright + \Sigma\tau_O = r_h F_h + (-r_s F_s \cos 60°)$$

and so

$$\circlearrowright + \Sigma\tau_O = (1.0 \text{ m})(200 \text{ N}) - (0.50 \text{ m})(80 \text{ N})(0.50)$$

and $$\boxed{\circlearrowright + \Sigma\tau_O = +0.18 \text{ kN·m}}$$

▶ **Quick Check:** The angle between $\mathbf{r}_s$ and $\mathbf{F}_s$ is 150°; hence, the torque $\mathbf{F}_s$ contributes is $-r_s F_s \sin 150°$, which equals $-r_s F_s \cos 60°$.

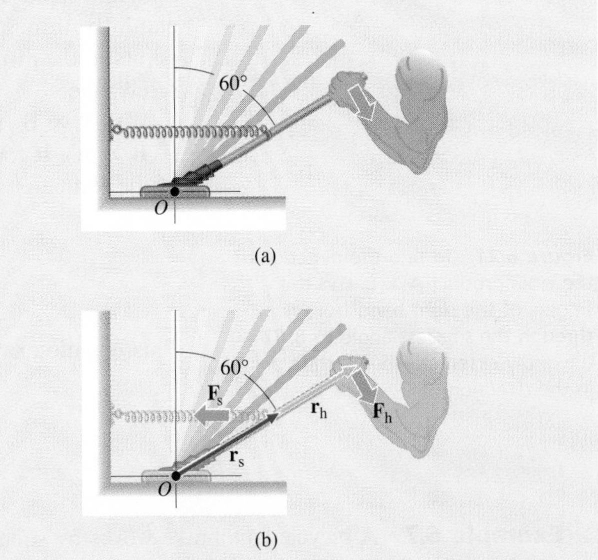

Figure 6.19 (a) A pivoted bar with two forces acting on it. (b) The forces and their moment-arms about point O.

Inasmuch as both **F** and **r** are vectors, the torque has a vector interpretation that includes a meaningful direction. It is customary to envision the torque vector, $\boldsymbol{\tau}$, lying along the axis of the rotation it tends to cause. That is, it passes through O and is *perpendicular to both* **r** *and* **F**. The sense of the vector, which way it points, is suggested by the *right-hand-screw rule.* When the naturally curved fingers of the right hand point in the direction of the force (Fig. 6.20), the outstretched thumb points in the direction of $\boldsymbol{\tau}$. In Fig. 6.16, the wrench tends to rotate counterclockwise about O under the influence of the torque, and the thumb and $\boldsymbol{\tau}$ point up out of the plane of the paper.

This description can be summarized using the mathematical idea of the **cross product**, which is a procedure for multiplying two coplanar vectors in such a way as to produce a third vector. Accordingly, define **C** as the cross product (also called the vector product) of **A** and **B**, such that

$$\mathbf{C} = \mathbf{A} \times \mathbf{B} \tag{6.4a}$$

which is read: **C** equals **A** cross **B**. The symbol × stands for a set of specific requirements: First, **C** is to be perpendicular to both **A** and **B** (that is, perpendicular to the plane containing **A** and **B**). Second, the magnitude of **C** is given by

$$C = AB \sin \theta \tag{6.4b}$$

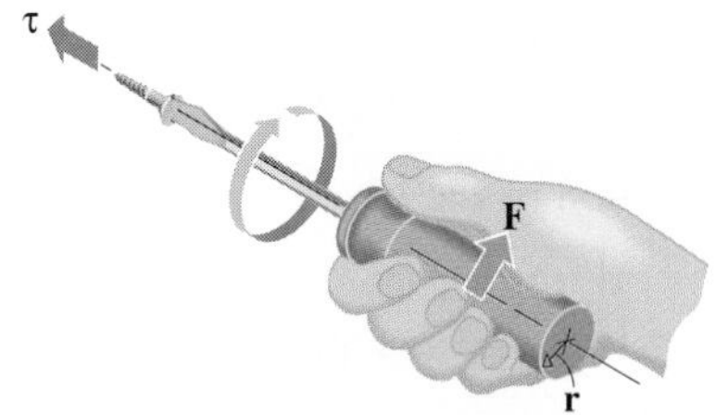

Figure 6.20 The right-hand-screw rule. An ordinary screw advances when it's turned in the direction in which the fingers on the right hand point. You will probably use this diagram more often to figure out how to turn screws than find torques.

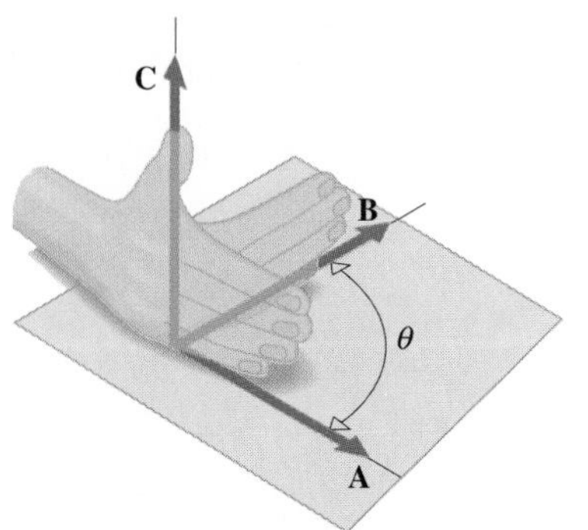

Figure 6.21 To find the direction of the cross product **A** × **B**, curl the fingers of the right hand from **A** through the smallest angle to **B**. The upwardly extended thumb then points in the direction of **C** = **A** × **B**.

Here, θ is the *smaller* of the two angles between **A** and **B**—each vector can be slid along its respective line-of-action until the two are tail-to-tail (Fig. 6.21) and separated by θ. Third, the direction or sense of **C** is determined by placing the fingers of the right hand along the first vector (**A**) in such a way that the fingers can close naturally through θ as they curl toward the second vector (**B**). The thumb then points in the direction of **C**, and the procedure is an embodiment of the right-hand-screw rule.

If $\mathbf{A} \times \mathbf{B}$ points upward, $\mathbf{B} \times \mathbf{A}$ points downward; that is, $(\mathbf{A} \times \mathbf{B}) = -(\mathbf{B} \times \mathbf{A})$. It follows, then, from Eq. (6.3) that, with respect to O, the scalar value of the torque is $\tau_0 = rF \sin \theta$, and

$$\boldsymbol{\tau}_0 = \mathbf{r} \times \mathbf{F} \tag{6.5}$$

This equation says it all.

Example 6.7 A bicycle handbrake works by squeezing the metal rim of the wheel between two rubber pads. Suppose that the maximum inward force on each brake pad is 50 N and the metal-rubber kinetic coefficient of friction is 0.70. Taking the rim-to-axle distance as 30 cm, determine the maximum torque produced by a set of brakes about the axis of rotation.

Solution: [Given: $F_N = 50\ N$, $\mu_k = 0.70$, and $r = 0.30$ m. Find: τ_0, taking O at the axle.] The friction force F_f applied to the rim by each pad (Fig. 6.22) is

$$F_f = \mu_k F_N = 0.70(50\text{ N}) = 35\text{ N}$$

Since the friction is tangential ($\mathbf{r}$ is $\perp$ to $\mathbf{F}_f$), the moment-arm equals 0.30 m. Hence, for each pad

$$\tau_0 = (0.30\text{ m})(35\text{ N}) = 10.5\text{ N·m}$$

and so the total torque is $\boxed{21\text{ N·m}}$, opposing the motion of the wheel.

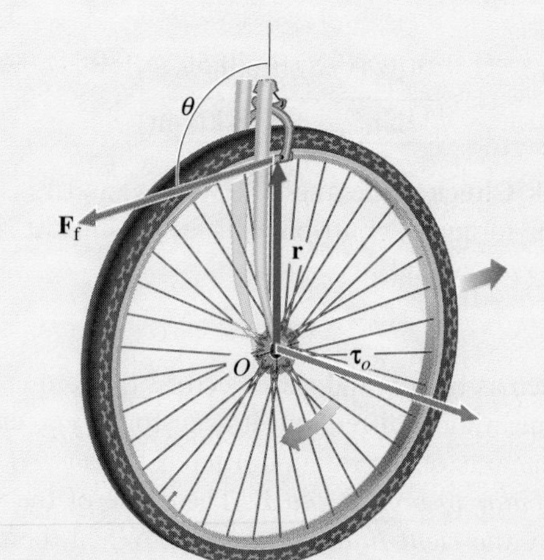

Figure 6.22 The torque about O exerted by a brake on a bicycle wheel.

▶ **Quick Check:** If the wheel is seen revolving clockwise as in Fig. 6.22, **r** lies along a vertical line, $\mathbf{F}_f$ is horizontal, $\theta = 90°$, and $\mathbf{r} \times \mathbf{F}_f = \boldsymbol{\tau}_0$ extends as shown. The twist of $\boldsymbol{\tau}_0$ (via the right-hand-screw rule) is counterclockwise against the rotation.

6.4 Rigid Bodies

Real bodies deform to some extent under the influence of applied forces, but many change so little that we can take them to be totally rigid for all practical purposes. Assume such a body is made up of a large number of separate, stationary, interacting particles (atoms or molecules). If the body is in equilibrium under the action of several externally applied forces, all of its constituent particles are also in equilibrium. The net force (F) acting on any one particle must be zero. Now, choose some point O—anywhere you like—and determine the resultant torque about O due to the net force acting on that arbitrary particle; that is, $\boldsymbol{\tau}_0 = \mathbf{r} \times \mathbf{F}$. But here, $\mathbf{F} = 0$, so the torque about O is zero, too. That's true for each and every particle in the body, so the net torque on the body due to all the internal and external forces is zero. Because of the colinear equal-and-opposite nature of the internal interaction pairs holding the body together, the torque about any point due to each pair must cancel. Consequently, the sum of all the torques due to the internal forces must be zero. Thus **the sum-of-the-torques about any point due to all the externally applied forces acting on a rigid body in equilibrium must be zero**. Here we have only the suggestion of a derivation, but it makes the point that this powerful insight is based squarely on Newton's laws and is not some new physical principle.

The wind blowing to the right on the sail produces a clockwise torque that must be balanced by a net counterclockwise torque produced by the sailors.

The fact that

[*translational equilibrium*] $$\Sigma\mathbf{F} = 0 \qquad [6.1]$$

is the First Condition of Equilibrium. In addition, for a rigid body

[*rotational equilibrium*] $$\Sigma\boldsymbol{\tau}_0 = 0 \qquad (6.6)$$

which is known as the **Second Condition of Equilibrium**. Note that the point O about which the torques are taken is arbitrary.

When the forces acting on a body pass through a common point, the First Condition suffices for equilibrium. (The Second Condition is then automatically true because the torque produced by a system of *concurrent forces* equals the torque produced by its resultant, which is zero.) *When dealing with concurrent forces, there will be only two independent equations (either sum-of-the-forces or sum-of-the-torques), and no more than two unknowns can be determined.* Figures 6.9, 6.10, and 6.11 depict concurrent force systems, and although we solved those problems using the sum-of-the-forces, we could have used the sum-of-the-torques instead.

Example 6.8 Figure 6.23 recalls Example 6.3, but this time let's solve for the forces acting on bar 2, using the Second Condition of Equilibrium.

Solution: [Given: $F_W = 300$ N, $\overline{AC} = 1.00$ m, and $\theta = 30.0°$. Find: the force on bar 2, that is, F_2.] We can use either free-body diagram—Fig. 6.23b or 6.23c—since both contain F_2 and the load. Let's select Fig. 6.23c. The torques can be taken about any point, so we should choose cleverly. $\mathbf{F}_1$ passes through A, and since we do not need to find F_1, taking torques about A will exclude it from the equation. The moment-arm of $\mathbf{F}_2$ about point A is the perpendicular distance from A to the line-of-action of $\mathbf{F}_2$, and it equals (1.00 m) sin 30.0°. Similarly, the moment-arm of the load about A is the perpendicular distance from A to the line-of-action of the 300-N load, and it equals 1.00 m. Taking clockwise as positive, the sum-of-the-torques about A equals zero, thus

$$\circlearrowright\!+\Sigma\tau_A = 0 = (300 \text{ N})(1.00 \text{ m}) - F_2(1.00 \text{ m})\sin 30.0°$$

$$F_2(0.500 \text{ m}) = 300 \text{ N·m}$$

and $$\boxed{F_2 = 600 \text{ N}}$$

▶ **Quick Check:** F_2 is the same as in Example 6.3.

(continued)

(continued)

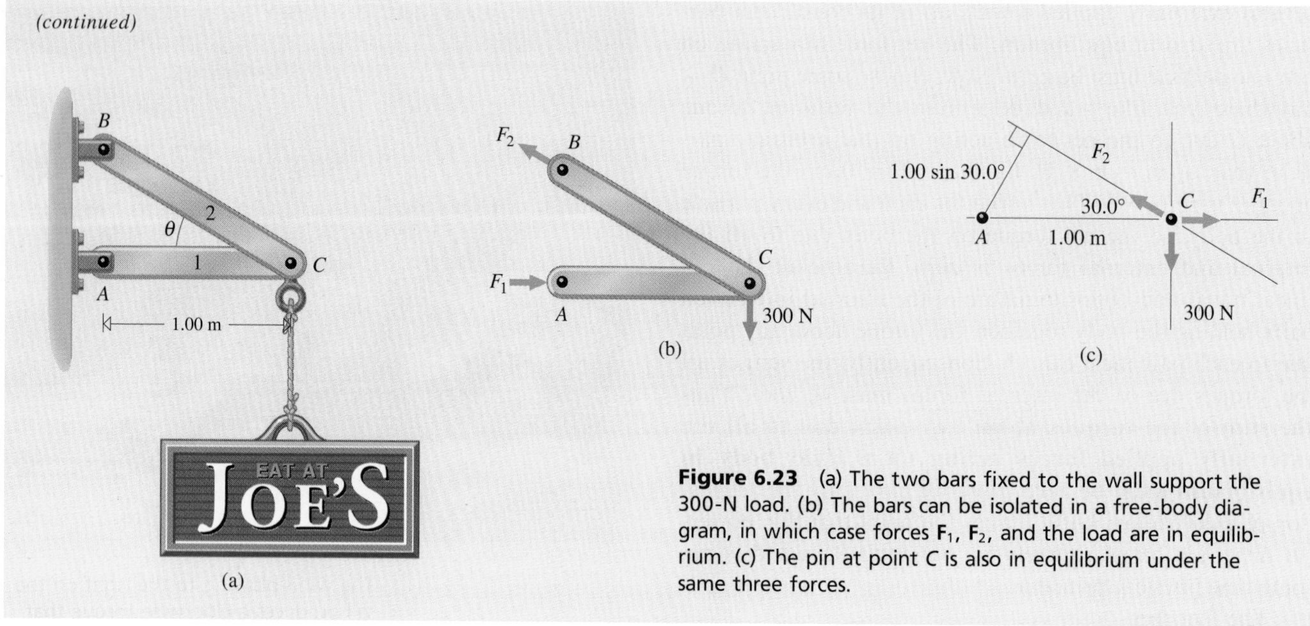

Figure 6.23 (a) The two bars fixed to the wall support the 300-N load. (b) The bars can be isolated in a free-body diagram, in which case forces $\mathbf{F}_1$, $\mathbf{F}_2$, and the load are in equilibrium. (c) The pin at point *C* is also in equilibrium under the same three forces.

Example 6.9 Prove that any three nonparallel, coplanar forces resulting in equilibrium must be concurrent.

Solution: Suppose the three forces acting are $\mathbf{F}_1$, $\mathbf{F}_2$, and $\mathbf{F}_3$, where $\mathbf{F}_1 + \mathbf{F}_2 + \mathbf{F}_3 = 0$ (Fig. 6.24). Label as *O* the point of intersection of $\mathbf{F}_1$ and $\mathbf{F}_2$. Now, if $\mathbf{F}_3$ does not pass through *O*, it will have a moment-arm with respect to it and therefore a torque about it, $\tau_O \neq 0$. But this possibility violates the Second Condition of Equilibrium. Hence, $\mathbf{F}_3$ must pass through *O*.

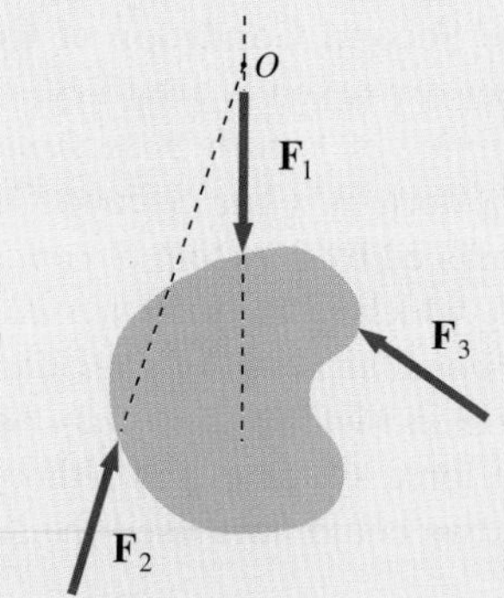

Figure 6.24 Shown are three nonparallel, coplanar forces acting on a body. If the body is in equilibrium under the influence of such forces, then the forces must meet at a point.

Nonconcurrent Forces

When two equal but antiparallel forces act at different points on a body (Fig. 6.25), equilibrium will not be established despite the fact that $\Sigma\mathbf{F} = 0$. There will be a net torque, and the body will revolve increasingly rapidly—it will accelerate. Although the body is in translational equilibrium, it will only be in rotational equilibrium when $\Sigma\tau = 0$. *A nonconcurrent system of any number of parallel separated forces in equilibrium can give rise to only two independent equations (either one force and one torque, or two torque), and only two unknowns can be determined.*

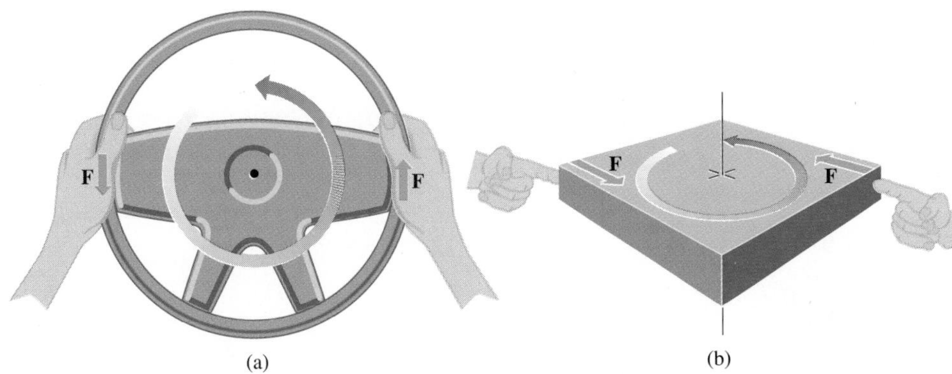

Figure 6.25 (a) Two equal, oppositely directed forces acting at different points on a body do not produce equilibrium even though $\Sigma \mathbf{F} = 0$. There is a torque, and the system cannot be in rotational equilibrium. (b) The block will rotate without translating.

Example 6.10 The steelyard in Fig. 6.15 consists of a nonuniform horizontal bar that is balanced without any added weights when supported at point O. If the weights F_{W1} and F_{W2} are attached, and the system is balanced as shown, write an expression for L_2 in terms of L_1, F_{W1}, and F_{W2}. What is the magnitude of the reaction force F_R supporting the system?

Solution: [Given: F_{W1}, F_{W2}, and the moment-arm L_1. Find: the moment-arm L_2 and the reaction F_R.] The forces are nonconcurrent, and we will have to apply both conditions of equilibrium. The sum-of-the-torques about *any* point must be zero, but not all such points are equally convenient. It would be advantageous not to have to deal with the reaction force in the first part of the problem. Since F_R passes through O, it has no torque around O and consequently does not enter the equation, which therefore contains only one unknown. Weight 2 tends to swing the system clockwise around O, whereas weight 1 tends to rotate it counterclockwise. Hence, with clockwise as positive, we have

$$\circlearrowright\!+\Sigma\tau_O = 0 = F_{W2}L_2 - F_{W1}L_1$$

and

$$\boxed{L_2 = \frac{F_{W1}L_1}{F_{W2}}}$$

As for the reaction force

$$+\!\uparrow\Sigma F_V = F_R - F_{W1} - F_{W2} = 0$$

and, not surprisingly

$$\boxed{F_R = F_{W1} + F_{W2}}$$

▶ **Quick Check:** The problem can be redone using the Law of the Lever, which is a special case of Eq. (6.6). Note that the units check.

Figure 6.26 illustrates a tilted variation of the previous example. The moment-arms are $r_{1\perp} = r_1 \cos 30°$ and $r_{2\perp} = r_2 \sin 60°$; hence, taking clockwise as positive, equilibrium occurs when

$$\circlearrowright\!+\Sigma\tau_O = 0 = F_{W2}r_2 \sin 60° - F_{W1}r_1 \cos 30°$$

The steelyard and the seesaw, the crowbar and the scissor (Fig. 6.27), and the human foot are all the same kind of simple machine—a lever of the class of levers that have the fulcrum, or pivot, located *between* the two forces acting on it.

When a system of nonconcurrent forces (that are not all parallel and perpendicular) produces equilibrium, there will be three independent equations, from which no more than three unknowns may be determined.

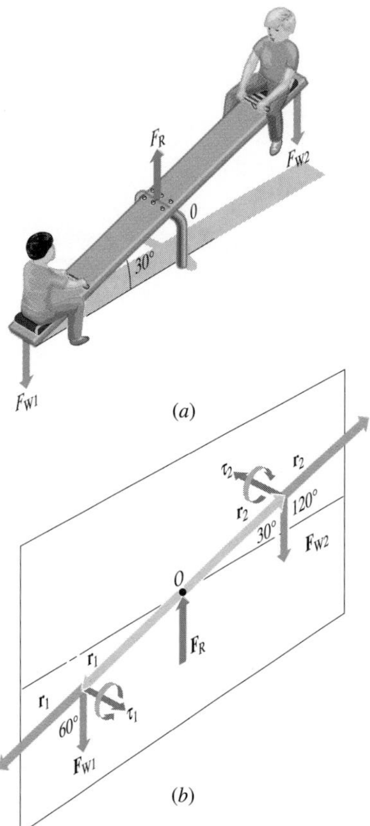

Figure 6.26 (a) The seesaw. (b) The torque around *O* due to the weight F_{W1} is counterclockwise. The torque around *O* due to the weight F_{W2} is clockwise.

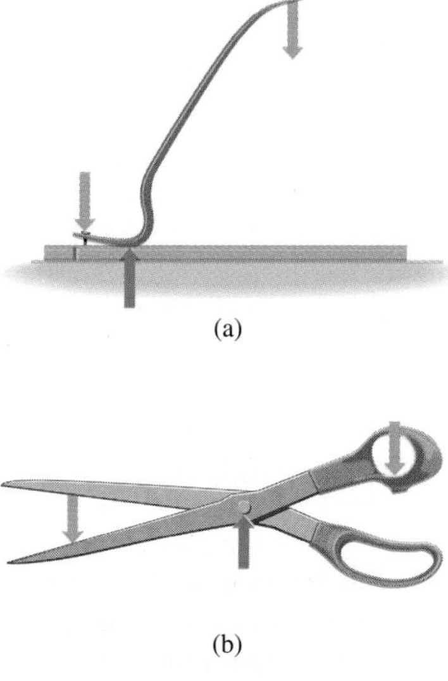

Figure 6.27 Two tools that are the same kind of lever as the seesaw. Both have fulcrums or pivot points between the applied force and the force exerted by the device. Shown are the forces acting on the tools.

Example 6.11 A 400-N sign is suspended from the end of a 2.0-m-long, very light horizontal bar, which is attached to a wall via a pivot (Fig. 6.28). The bar is held up by a rope attached one-quarter of the way from its end, making an angle of 50°. Find the reaction force on the pivot and select the lightest possible support rope from Table 6.1.

Solution: [Given: $F_W = 400$ N, $L = 2.0$ m, and F_T acts 0.50 m from the end point *E* at 50°. Find: F_T and the force on the pivot $\mathbf{F}_R$.] The force acting at *O* is not at a joint. Therefore, this is not a simple case in which the bar experiences forces only along its length. Moreover, it might seem that we have two unknowns, but including F_T there are actually three: $\mathbf{F}_R$ is equivalent to F_{RH} and F_{RV}, or F_R and θ—two unknowns. First draw the free-body diagram of the bar, Fig. 6.28b with $\mathbf{F}_R$ in any direction you like: since $\mathbf{F}_T$ is up and to the left, it's clear $\mathbf{F}_R$ is to the right. Moreover, $\mathbf{F}_R$ must act downward if $\Sigma\tau_0$ is to equal zero. Had you drawn it incorrectly, it would simply turn out negative when you solved for it. Now, to determine F_T, we focus on point *P* so that $\mathbf{F}_R$ is absent from the torque equation

$$\circlearrowleft + \Sigma\tau_P = F_W(2.0 \text{ m}) - (F_T \sin 50°)(1.50 \text{ m}) = 0$$

Here, only the vertical component of the tension, $F_T \times$

(continued)

(continued)
sin 50°, contributes to the torque because the horizontal one goes through P. And solving this yields $F_T = 0.70$ kN. The lightest possible rope on which to hang the sign is 3/16-in. hemp, but since that material tends to rot in damp weather, do not stand under the sign.

We have two remaining unknowns, $\mathbf{F}_R$ and θ, and thus require two equations. Let's first try using two torque equations about points O and E

$$\circlearrowleft+\Sigma\tau_O = F_W(0.50\text{ m}) - (F_R \sin\theta)(1.50\text{ m}) = 0$$

$$\circlearrowleft+\Sigma\tau_E = (F_T \sin 50°)(0.50\text{ m}) - (F_R \sin\theta)(2.0\text{ m}) = 0$$

STOP!

It's clear that these two equations cannot be solved for F_R and θ because they are not independent. We have determined that $F_T = F_W(2.0\text{ m})/(\sin 50°)(1.50\text{ m})$, and as soon as we substitute that expression in the second torque equation, it becomes identical to the first equation.

Taking the torques about a point off the line of the bar (as, for example, where the rope attaches to the wall) would give us the needed independent equation (see Problem 40). Applying the torque equation more than once requires care. By contrast, the force equations are far more straightforward, and it's advisable to use them in preference wherever appropriate. From Fig. 6.28c

$$+\uparrow\Sigma F_V = F_T \sin 50° - F_W - F_{RV} = 0$$

$$\overset{+}{\rightarrow}\Sigma F_H = F_{RH} - F_T \cos 50° = 0$$

or $F_{RV} = 0.13$ kN and $F_{RH} = 0.45$ kN. Therefore

$$F_R = \sqrt{(0.13\text{ kN})^2 + (0.45\text{ kN})^2} = \boxed{0.47\text{ kN}}$$

The angle that $\mathbf{F}_R$ makes with the horizontal is then

$$\theta = \tan^{-1}\left(\frac{F_{RV}}{F_{RH}}\right) = \boxed{17°}$$

▶ **Quick Check:** Taking torques about O, $F_{RV} = (400\text{ N})(0.50\text{ m})/(1.50\text{ m}) = 0.13$ kN and about A, $F_{RH} = (400\text{ N})(2.0\text{ m})/(1.50\text{ m})(\tan 50°) = 0.45$ kN.

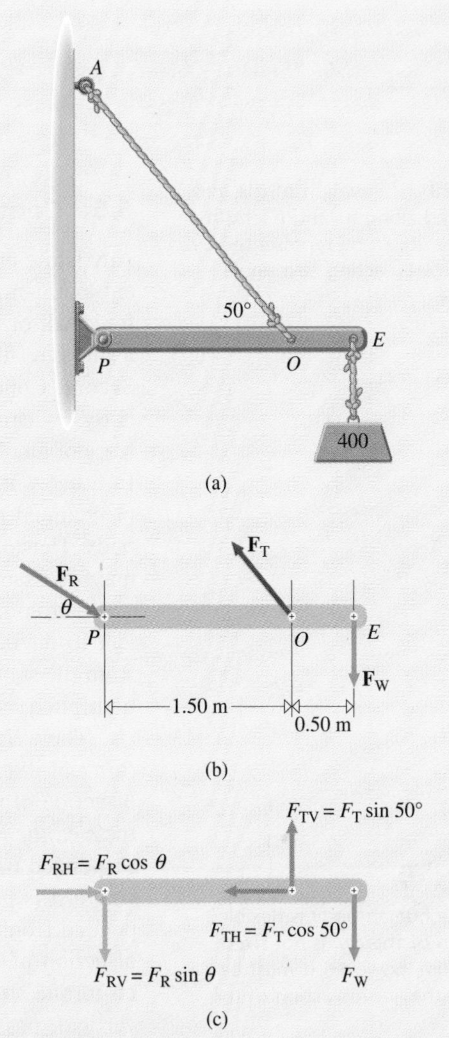

Figure 6.28 (a) The problem is to calculate the reaction at P and the tension in the rope from O to A. Unlike Fig. 6.11, the forces are nonconcurrent, and we cannot learn much from the free-body diagrams of points O or E. (b) Since the bar must be in equilibrium and all the unknown forces act on it, we draw a free-body diagram of the bar. (c) The forces resolved into horizontal and vertical components.

If you want to use three torque equations, the points about which they are taken must not lie on the same line (as do P, O, and E in Example 6.11). *If you want to use one force and two torque equations, the points about which the latter are taken must not lie on a line perpendicular to the force-summation axis.*

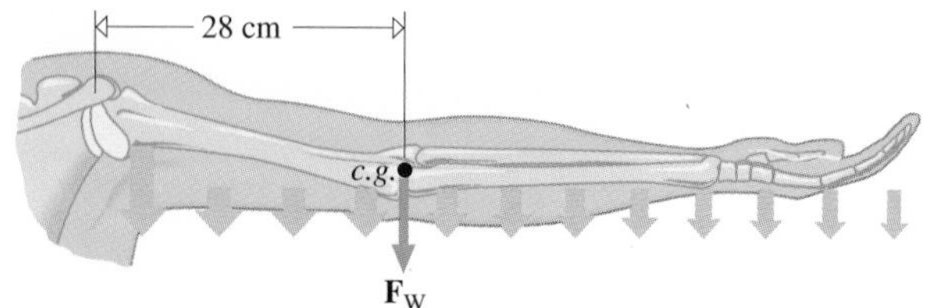

Figure 6.29 The weight of the arm is distributed along its entire length. Nonetheless, it can be represented by a single force F_W acting at a single point, the *c.g.*

EQUILIBRIUM OF EXTENDED BODIES

On planet Earth, gravity imparts weight, and weight affects equilibrium. You, right there, are an extended body probably sitting in equilibrium. How do we deal with gravity acting on you?

6.5 Center-of-Gravity

Any finite object can be thought of as composed of a very large number of constituent point-masses. Each atom within you experiences a downward gravitational force, all of which can be considered parallel, and all of which combine to produce a single resultant force, the *weight* of the body. Besides knowing the net weight, the Second Condition of Equilibrium suggests that we must also know where on the body $\mathbf{F}_W$ can be understood to act. Consequently, we define the **center-of-gravity** (*c.g.*) of an object to be *that point where the total weight,* $\mathbf{F}_W$, *can be imagined to act.* In other words, having a single force $\mathbf{F}_W$ act at the *c.g.* produces exactly the same mechanical result as having gravity act on all the point-masses that constitute the body. Hence, the torque about any point arising from the weight of a distribution of mass is exactly the same as if the net weight acted only at the *c.g.* Gravity acts on every atom of the outstreched arm (Fig. 6.29), and yet the effect is exactly equivalent to having the total weight act at the *c.g.* located near the elbow, about 28 cm from the shoulder joint. The torque produced at the shoulder is the weight of the arm multiplied by the distance from the joint to the *c.g.*

The weight of the balanced athlete in the photograph is straight down, and if we think of it as acting through a single point, the *c.g.* must be somewhere on the line-of-action of the normal force. Since the line-of-action of the weight passes through the *c.g.*, *the force of gravity generates no torque about the center-of-gravity.* A body suspended from its *c.g.* is in equilibrium in any orientation whatsoever and will stay where it is put. *The c.g. is the balance point.* If a body of any shape is freely suspended from *any* point, it will so orient itself in reaching equilibrium that the line-of-action of $\mathbf{F}_W$ will pass through the support point (Fig. 6.30), thus producing a no-torque situation. In that case, $\mathbf{F}_R = -\mathbf{F}_W$ and the line-of-action of $\mathbf{F}_R$, which is vertical, passes through the *c.g.* To find the center-of-gravity experimentally, one need only suspend the object in turn from two different points. The lines-of-action of the two successive reaction forces will intersect at the *c.g.*

Because the human body is flexible, the location of the *c.g.* is not fixed. In equilibrium, however, it must be located on the line-of-action of the normal force.

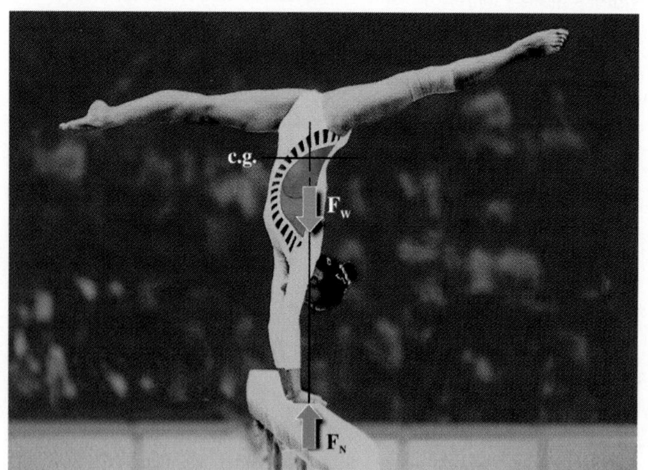

For simplicity, suppose we have a flat body made up of a large number of tiny masses: $m_1, m_2, m_3, \ldots$, as shown in Fig. 6.31. We can compute the location of its center-of-gravity in the following way. Each particle experiences a torque (due to its weight) about the *arbitrary* point O. To find the moment-arm for each of these tiny torques, extend the line-of-action of the weight of each particle and drop a perpendicular to it from O—that x-distance is the moment-arm. The torques have the form $F_{W1}x_1$, $F_{W2}x_2$, and so forth. The total weight-generated torque about O produced by all the individual particles, taking clockwise as positive, is

$$F_{W1}x_1 + F_{W2}x_2 + \cdots = \Sigma F_W x$$

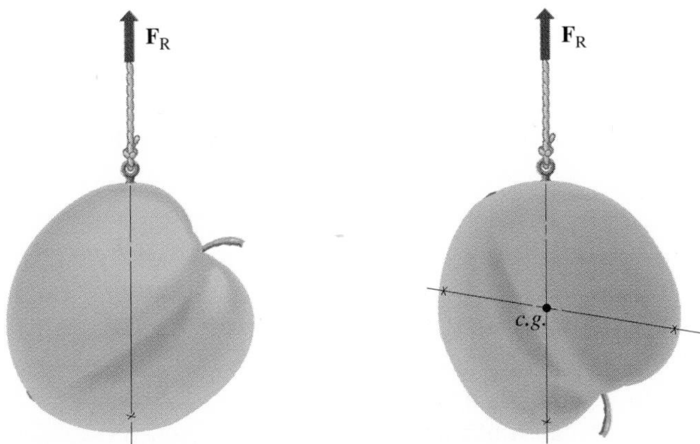

Figure 6.30 The *c.g.* of an object can be located by suspending it from several different points. The *c.g.* is always on the line-of-action of the force supporting the object.

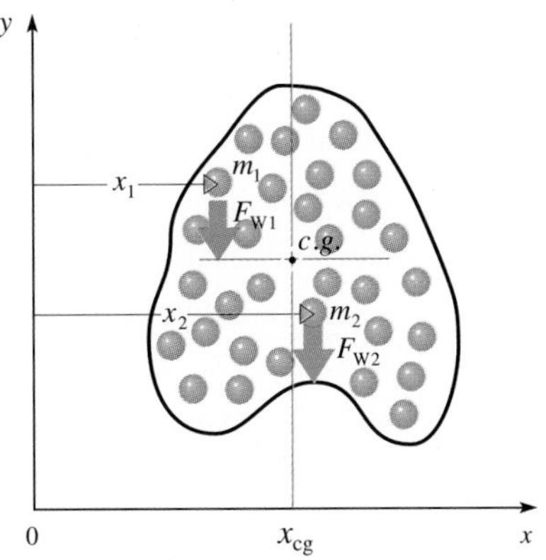

Figure 6.31 The sum of all the torques about any point 0 caused by the weights of the point-masses constituting a body is equal to the torque produced by the total weight of the body acting at the *c.g.* Measured along the *x*-axis, the *c.g.* is at x_{cg}. Measured along the *y*-axis, it's at y_{cg}.

It follows from the definition of the *c.g.* that the total weight, written as

$$F_{W1} + F_{W2} + \cdots = \Sigma F_W$$

must produce that same torque when it acts as a single force at the *c.g.*; that is

$$(\Sigma F_W)x_{cg} = \Sigma F_W x$$

and

$$x_{cg} = \frac{\Sigma F_W x}{\Sigma F_W} \tag{6.7}$$

Take each point-mass, multiply its weight by its distance from the $x = 0$ axis, add them, divide by the total weight, and you obtain x_{cg}, measured from the axis no matter where the origin was chosen to be. If the object and the coordinate system are both rotated 90° so that the *y*-axis is horizontal, we get an expression of exactly the same form for y_{cg}.

The c.g. *of a regularly shaped body of uniform composition lies at its geometric center.* This statement applies to spheres, cubes, cylinders, rings, and so forth. Envision the object in Fig. 6.31 to be a sphere and slide it over so that the *y*-axis, which is quite arbitrary, passes through its geometric center. Then, by symmetry, $\Sigma F_W x = 0$, and Eq. (6.7) yields $x_{cg} = 0$; the *c.g.* is indeed *on* the central symmetry axis. When a body is composed of several parts, the overall *c.g.* can be located by treating each part as a point-mass located at its own *c.g.*, acted upon by its own weight.

Example 6.12 For a man of mass *m* and height *h*, the *c.g.* of the lower leg and foot, and the upper leg are shown in Fig. 6.32 along with the corresponding weights. Find the *c.g.* of the whole leg in the extended position measured from the sole of the foot. This kind of information is important in physical therapy.

Solution: [Given: $F_{W1} = 0.059mg$, $x_1 = 0.19h$, $F_{W2} = 0.097mg$, and $x_2 = 0.42h$. Find: x_{cg}.] Treating the two

(continued)

(continued)
leg parts as point-masses located at their respective *c.g.*s, we find that

$$x_{cg} = \frac{F_{W1}x_1 + F_{W2}x_2}{F_{W1} + F_{W2}}$$

$$x_{cg} = \frac{(0.059mg)(0.19h) + (0.097mg)(0.42h)}{0.097mg + 0.059mg}$$

thus $$x_{cg} = \frac{0.052mgh}{0.156mg} = \boxed{0.33\ h}$$

▶ **Quick Check:** The torque about the heel due to the whole leg is $(0.33h)(0.156mg) = (0.05mgh)$; the torque due to the separate parts $\approx (0.06mg \times 0.2h) + (0.1mg \times 0.4h) \approx 0.05mgh$ and they *are* equal.

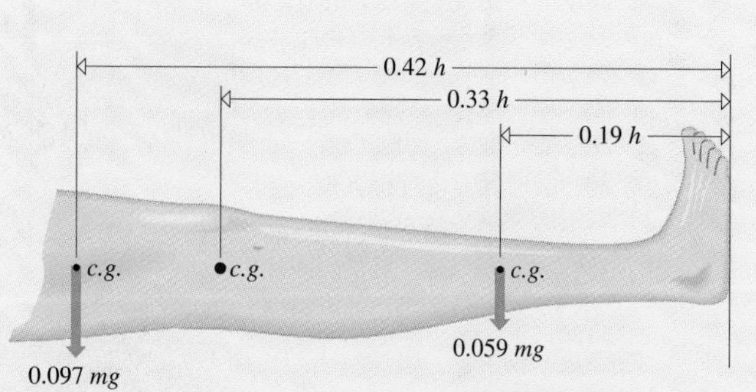

Figure 6.32 The locations of the various *c.g.*s for several parts of a man's leg.

Bear in mind that the *c.g.* of an object need not reside within the space occupied by material. The center-of-gravity of a doughnut is at the center of the hole.

Example 6.13 The gadget depicted in Fig. 6.33 consists of a 30-N metal block 3.0 cm thick, epoxied to a 16-cm-long rod weighing 50 N, on whose end is affixed a 6.0-cm-diameter sphere weighing 10 N. The whole thing is to be hung on a thin wire so that it stays horizontal. Find the point where the wire must be attached.

Solution: [Given: $F_{W1} = 30$ N, thickness 0.030 m; $F_{W2} = 50$ N, length 0.16 m, and $F_{W3} = 10$ N, diameter 0.060 m. Find: x_{cg}.] Each piece can be envisioned as replaced by a point-mass located at its geometric center. Taking $x = 0$ at the very left end, the individual *c.g.*s are at $x_1 = 0.015$ m, $x_2 = 0.11$ m, and $x_3 = 0.22$ m for block, rod, and sphere, respectively. Hence

$$x_{cg} = \frac{(30\text{ N})(0.015\text{ m}) + (50\text{ N})(0.11\text{ m}) + (10\text{ N})(0.22\text{ m})}{30\text{ N} + 50\text{ N} + 10\text{ N}}$$

and $$\boxed{x_{cg} = 0.091\text{ m}}$$

The *c.g.* is 9.1 cm to the right of the left end.

▶ **Quick Check:** The net torque ↺+ about the *c.g.* is $(30\text{ N})(7.555\text{ cm}) - (50\text{ N})(1.944\text{ cm}) - (10\text{ N}) \times (12.944\text{ cm}) \approx 0$.

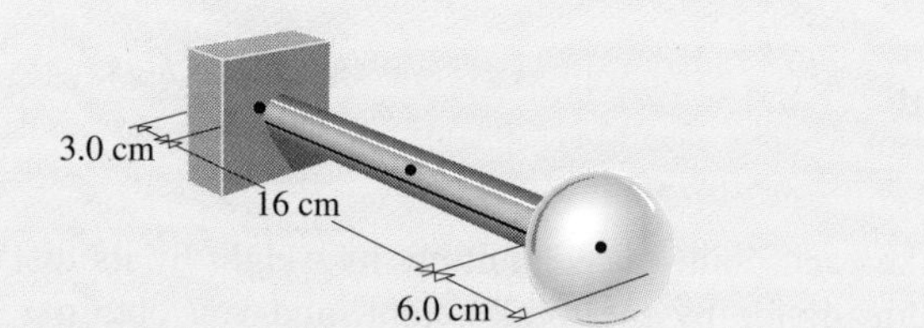

Figure 6.33 The *c.g.* of a composite object can be found by representing each constituent body by a point-mass at its *c.g.* For example, the sphere here weighs 10 N and can be replaced by a 10-N point at its center.

For a complicated inhomogeneous body, like that of a human being, it's far easier to determine the *c.g.* in the laboratory (Problem 63) than it is to calculate it directly. Moreover, the location of the *c.g.* varies with posture and limb position. For a person standing at attention, it lies on a vertical centerline about 3 cm forward

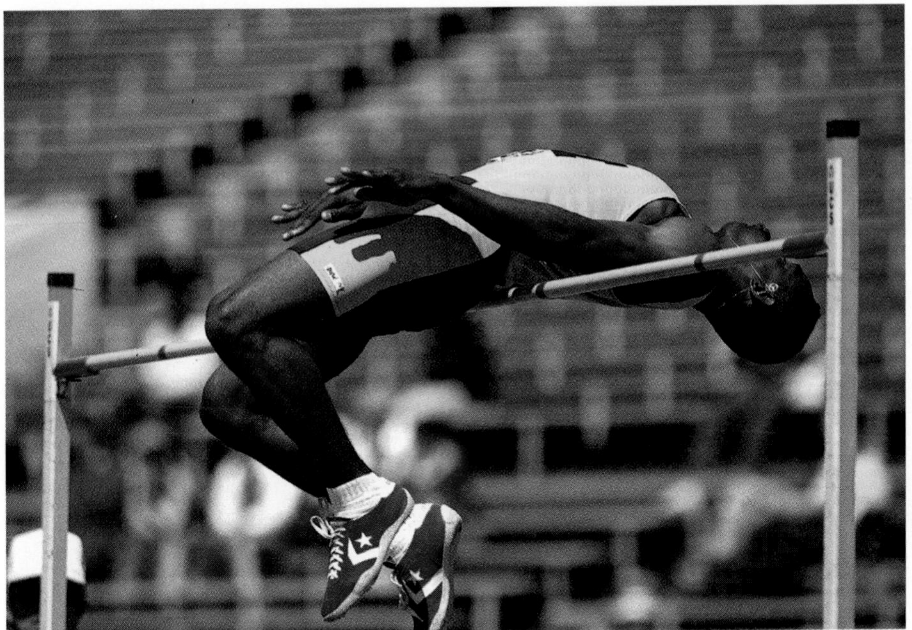

The Fosbury flop. The jumper extends the head and legs downward so the *c.g.* actually passes below the bar. As we'll see, raising the *c.g.* takes energy, and having it slip under the bar means the bar can be higher.

This mechanical arm is driven by two hydraulic pistons that can both push and pull, so each does the work of a pair of muscles.

of the ankle joint and up around the height of the second sacral vertebra, or roughly a few inches behind the navel. The *c.g.* also varies from person to person—in men with large shoulders, it's usually a little higher than in women with large pelvises. The location of the body's *c.g.* is of interest to everyone who stands or walks, but it's of particular concern to dancers and athletes. For example, Dick Fosbury devised a backward high-jump technique in which his body, face upward, arcs over the bar, while his *c.g.* passes beneath it. How high the *c.g.* is raised determines the amount of effort that will have to be expended in the jump (p. 291).

Example 6.14 The biceps muscle (Fig. 6.34) is connected from the shoulder (scapula) to the radius bone at a point 5.0 cm from the elbow. Its contractions flex the arm. Taking the mass of the hand and forearm together to be 5.5% of the total body mass (with a *c.g.* as shown), compute the force exerted by the biceps of a 70-kg man holding a 2.0-kg sphere. Assume the force of the biceps acts vertically.

Solution: [Given: the mass of the arm $m_a = 5.5\% \times$ (70 kg), $m_s = 2.0$ kg, and all the appropriate distances. Find: F_B, the force exerted by the biceps.] Assuming F_B acts vertically, take the torque about the elbow joint to avoid dealing with the force exerted by the humerus (Fig. 6.34b), whereupon

$$\circlearrowleft + \Sigma\tau_0 = 0 = m_s g(0.34\text{ m}) + m_a g(0.16\text{ m}) - F_B(0.050\text{ m})$$

Accordingly,

$$F_B = g\frac{(2.0\text{ kg})(0.34\text{ m}) + (3.85\text{ kg})(0.16\text{ m})}{(0.050\text{ m})}$$

and $\boxed{F_B = 2.5 \times 10^2\text{ N}}$, or about 57 lb.

(continued)

(continued)

▶ **Quick Check:** Rounding off the numbers, the clockwise torque is roughly $g(2\text{ kg})(0.3\text{ m}) + g(4\text{ kg}) \times (0.2\text{ m}) \approx 1.4g \approx 14\text{ m·N}$, and dividing by 0.05 m yields $F_B \approx 0.28$ kN.

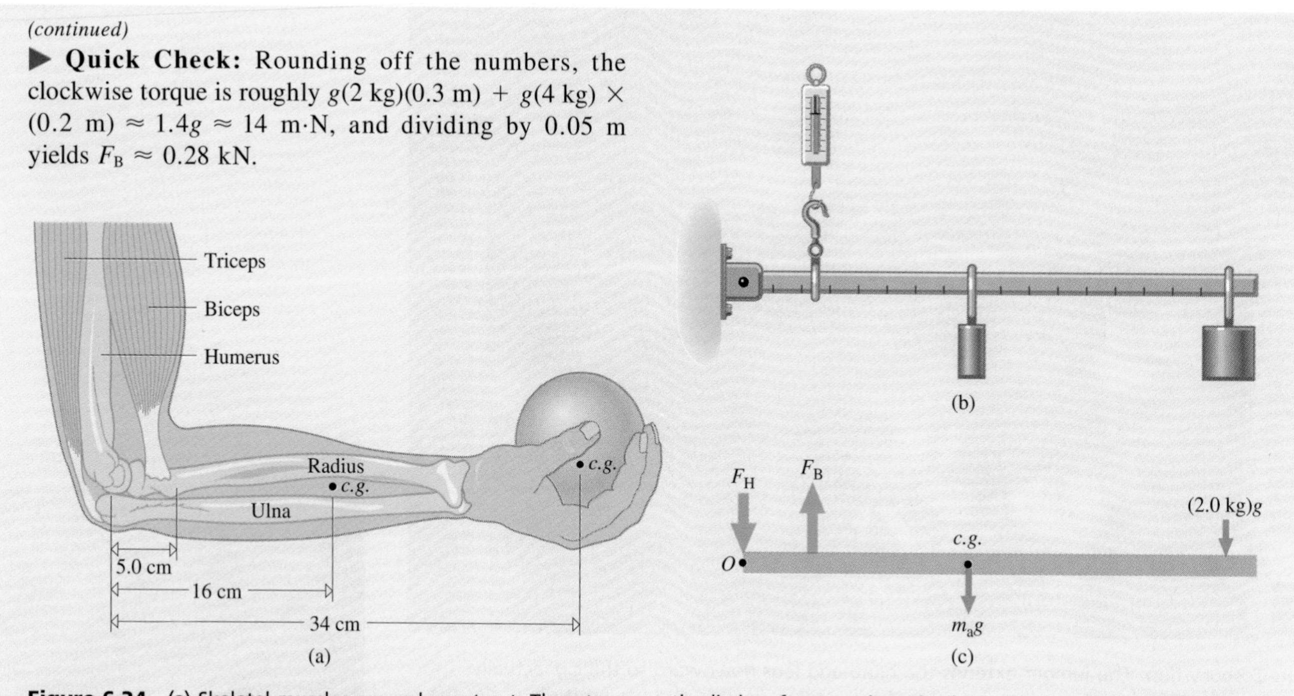

Figure 6.34 (a) Skeletal muscles can only contract. Thus, to raise (flex) the arm, the biceps is contracted. The pivot is at one end, the load at the other, and the applied force is between the two. This kind of end-pivoted lever is common in the limbs of mammals and is basically the same as encountered in the use of the shovel and rake, or a pair of tweezers. (b) A model of the arm. (c) The four forces acting are shown in the free-body diagram.

6.6 Stability and Balance

As bipedal beings, we cope with the problem of stability every day. The common wisdom holds that stability is about objects not toppling over, and that's a good place to start. When a thing falls over, the driving force is gravity and the pivotal concept is the *c.g.* Out in space, where everything is essentially weightless, there's no need to worry about objects tipping over.

Figure 6.35 Bodies such as these are in stable equilibrium because they fall back to their original orientations after a large displacement.

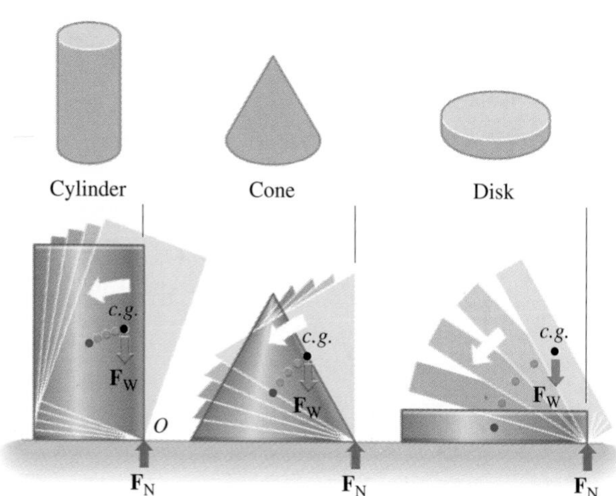

An object that is supported against gravity is in equilibrium when $\Sigma\mathbf{F} = 0$ and $\Sigma\tau = 0$, but we can distinguish three degrees of equilibrium. *A body is said to be in* **stable equilibrium** *if, after receiving a small momentary displacement, it returns to its original condition.* Each object in Fig. 6.35 is in stable equilibrium, because once slightly tilted, the resulting torque produced by its weight acting through the *c.g.* tends to rotate the body about the pivot, back to its original orientation. Notice that in each such case the initial displacement causes a raising of the *c.g.*, which, when released, tends naturally to fall back downward.

A body is said to be in **unstable equilibrium** *if, after receiving a small momentary displacement, it moves farther away from its original condition.* A cone balanced on its point is in equilibrium when the normal force and the weight are colinear, $\mathbf{F}_N = -\mathbf{F}_W$ and $\Sigma\tau = 0$. But if the slightest tilt disrupts that condition, $\mathbf{F}_W$ produces a net torque about the pivot point and the cone rotates from the vertical. Again the *c.g.* descends, but this time the fall carries it away from the original location.

A body is said to be in **neutral equilibrium** *if, after receiving a small momentary displacement, it remains in equilibrium, tending neither to move away any farther from, nor to return to, its original condition.* Each of the objects in Fig. 6.36 has its *c.g.* directly over the point or line-of-contact with the surface, and this situation holds when it is rolled into some new position. The sum-of-the-torques produced by $\mathbf{F}_N$ and $\mathbf{F}_W$ is always zero. Moreover, the height of the *c.g.* remains unchanged.

As a rule, *an object supported at a point above the c.g. will be stable, while one supported at a point below the c.g. will be unstable.* A cane hanging from its crook on a finger is stable, but one balanced upright on a fingertip is unstable.

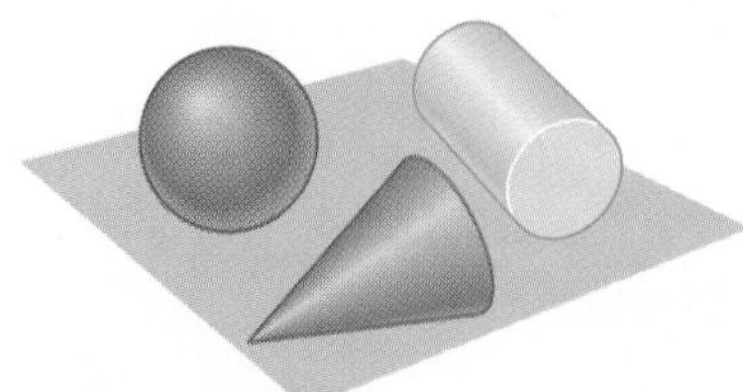

Figure 6.36 These objects are in neutral equilibrium. A small shift of the *c.g.* leaves them unchanged—they stay put, with the *c.g.* always above the point or line of support.

Example 6.15 The 0.25-kg toy man on the platform in Fig. 6.37 holds a rod of negligible weight. Considering him alone, his *c.g.* is 8.00 cm up from his feet. The two counterweights, each of mass 0.50 kg, hang on a horizontal level 12.0 cm beneath the top of the platform. Are the counterweights, the toy man, and the rod in stable equilibrium?

Solution: [Given: $m_1 = 0.25$ kg, $m_2 = m_3 = 0.50$ kg, and taking $y = 0$ at the level of the counterweights, $y_1 = 20.0$ cm, $y_2 = y_3 = 0$. Find: the *c.g.* for the total system.] From Eq. (6.7)

$$y_{cg} = \frac{m_1gy_1 + m_2gy_2 + m_3gy_3}{m_1g + m_2g + m_3g} = \frac{m_1y_1}{m_1 + m_2 + m_3}$$

and $$y_{cg} = \frac{(0.25\text{ kg})(0.20\text{ m})}{1.25\text{ kg}} = \boxed{0.040\text{ m}}$$

The system's *c.g.* is 4.0 cm up from the *x*-axis and 8.0 cm below the point of support on the platform, and so it is stable.

▶ **Quick Check:** The toy man has 1/5th of the mass of the system and therefore must be 4 times farther above the *c.g.*, (20 cm × 4/5 = 16 cm) than are the counterweights beneath the *c.g.*, (20 cm × 1/5 = 4.0 cm).

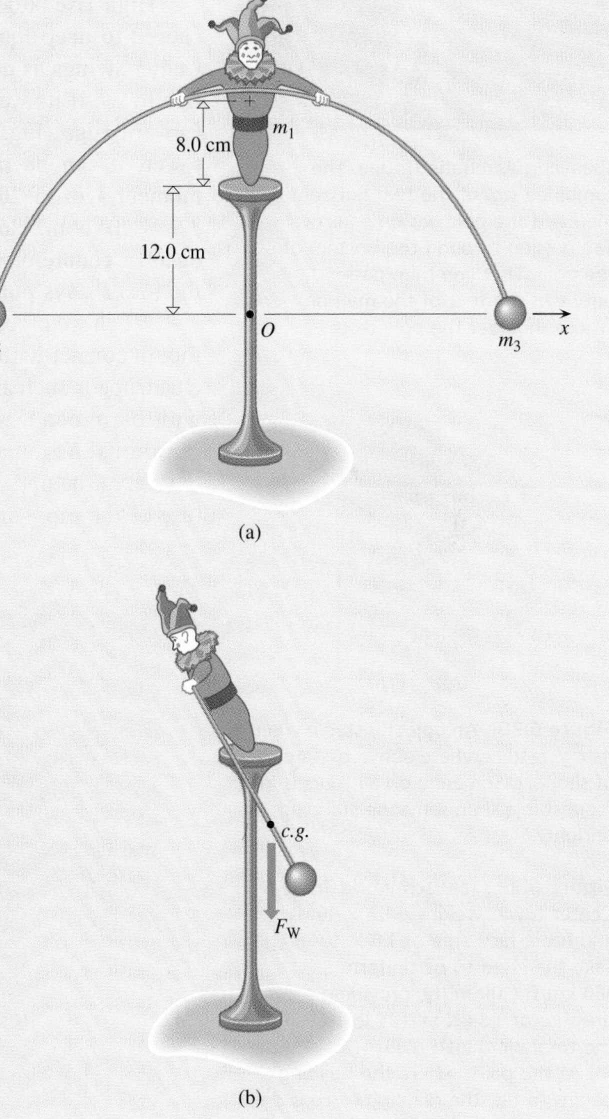

Figure 6.37 (a) The two counterweights shift the *c.g.* well below the platform. (b) Side view. Any shift of the little man raises the *c.g.*, giving F_W a moment-arm and causing a torque that drives the system back to its original orientation.

Shenyang Acrobatic Troupe. The combined *c.g.* of the two horizontal men and the pole lies on a vertical line passing through the bottom of the pole. That line then passes between the feet of the man on whose shoulder the pole rests.

An object resting on a horizontal surface will overturn once its *c.g.* passes beyond the farthest extent of its base of support (whereupon $\mathbf{F}_N$ and $\mathbf{F}_W$ are no longer colinear). Lean forward, hands at your sides, until your *c.g.* is just above the line of your toes—any farther, and over you go. We can imagine an *area of support* bounded by the perimeter of lines connecting the outermost contact points, as shown in Fig. 6.38. *So long as the c.g. is directly above some point within this area of support, the system will be stable.*

Four things affect the ease or difficulty with which the *c.g.* will pass beyond its support: the weight of the object, the height of the *c.g.*, its location within the area of support, and the size of that area. A light object will overturn more easily than a heavy one, all else being equal. So an empty paper cup will blow over in the wind before a full one. On a lurching bus, we instinctively increase the area of support by assuming a wider than normal stance. As a rule, a low *c.g.* in the middle of a large area is stable. The boxer crouches, the sumo wrestler squats, and the football guard gets down on all fours, each in order to lower the *c.g.*

High-rise buildings can experience tremendous wind loads and are therefore anchored to deep foundations. In a strong hurricane, each tower of the World Trade Center swings as much as 7 ft on either side of vertical. Let's make a rough calculation to see if its weight alone would keep the tower from toppling around its bottom leeward edge (Fig. 6.39). Assuming the wind load is uniform over the face of the tower, it can be taken to act at the center (where the *c.g.* ought to be) with a moment-arm of 206 m. The wind produces a torque of (19×10^6 N $\times$ 206 m $=$ 3.9×10^9 N·m) clockwise about the back edge. On the other hand, the weight produces a counterclockwise torque of (1.2×10^9 N $\times$ 18 m $= 2.3 \times 10^{10}$ N·m), and the tower stays put.

When we bipeds walk, our *c.g.* must remain somewhere over the one supporting foot in contact with the floor, and that requires us to do a little side-to-side wiggling. Quadrupeds such as dogs and horses can walk slowly with three feet always in touch with the ground, which maintains a large triangular area of support, even though the animal has small paws or hooves. We are the creatures with the big feet. In fact, the *c.g.* of the upper portion of the human body resides well above the support structure of the hips, and that's not a particularly good design. Birds, which have been

Figure 6.38 An object is stable when its *c.g.* is somewhere above the area of support. Someone on all fours is far more stable than someone standing upright.

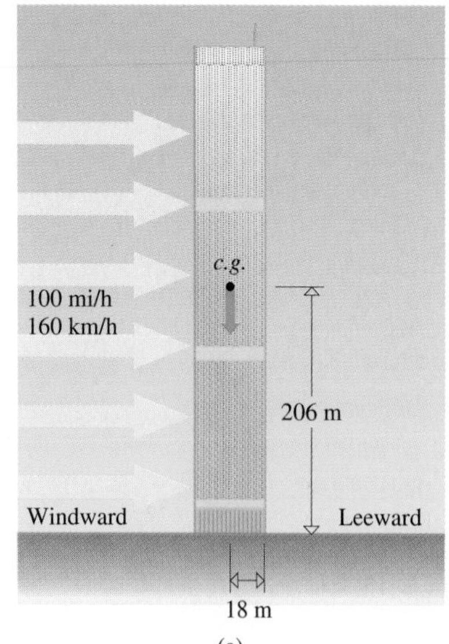

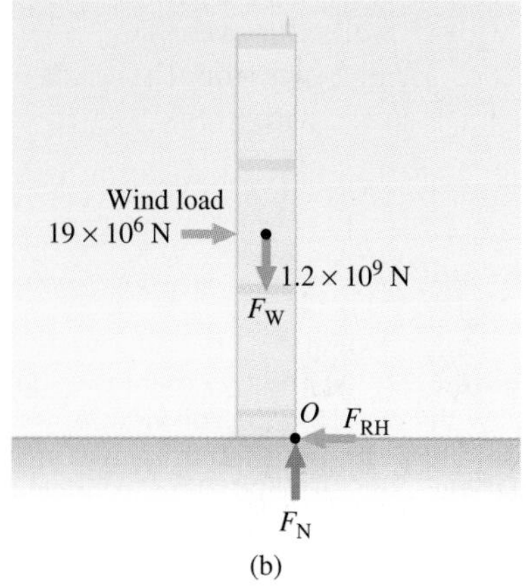

Figure 6.39 (a) Each World Trade Center tower weighs $\approx 1.2 \times 10^9$ N and has a face area of 1.6×10^4 m^2. Take the wind to be uniform at 160 km/h (100 mi/h), providing a pressure of 1.2 kN/m^2 (25 lb/ft^2) so the total wind load is 19×10^6 N. (b) At the point where the building begins to tilt, the reaction force is on its leeward edge. The tower is actually well anchored in the ground.

bipedal a lot longer than us, are far better suited to that purpose, having their *c.g.* hanging *below* the hip line (Fig. 6.40). It's like comparing a Model-T and a modern sports car (Fig. 6.41). One teeters above the axles, the other hangs below them. Despite Hollywood, we are not likely to see a terrestrial robot designed in the likeness of a human—it would require too much circuitry just to keep the thing in balance as it walked.

STABILITY AND BALANCE

According to Walter Reed Army Medical Center, between 1975 and 1977, 3 young men were killed and 12 severely injured by falling soda vending machines at U.S. army bases. The victims had apparently deliberately tipped the machines, which, when fully loaded, weigh between 800 and 1000 pounds. Since that study was published, another 48 victims (of whom 11 died) were reported. All were male, ranging in age from 10 to 33; 7 were civilians.

6.7 Structures, Walls, and Arches

Children build towers of wooden blocks that invariably come tumbling down. That's not because of some structural failure (the blocks themselves do not crumble). Toy towers topple because they are unstable—the weight of the blocks, instead of holding them together, acts to pull them apart. Much of the ancient stone architecture that still survives was similarly constructed by stacking blocks without any mortar or cement between them. The primary function of mortar is to bring the layers of blocks into good contact, so that the compressive loads are distributed uniformly over the whole surface of the joint. Mortar has comparatively little strength in tension and is not used to bond the blocks together, a point well illustrated every time an automobile crashes through a brick wall. What holds together the cathedral of Notre Dame in Paris is the weight of its parts and the friction between them.

We can expect a stone wall to compress slightly under the combined action of its own weight and whatever load it's carrying, such as the weight of a roof. (A typical modern building 300 m tall compresses its supporting columns by 2 or 3 cm. When the resultant, the so-called *thrust,* acts down the center of the wall, the force spreads out evenly, and the compression is nearly uniform. If the load shifts, the thrust is made to move closer to the wall's edge, and things distort (Fig. 6.42). Once the line-of-action of the thrust passes beyond what is known as the "middle third" of the wall, the masonry experiences compression on that side and tension on the other. The result is usually a crack in the joint where the mortar tears loose. We see that kind of thing all the time in modern buildings. If the condition becomes extreme, and the line-of-action moves beyond the surface of the wall, beyond the area of support, the wall topples over.

The character of a presteel building was determined to a large extent by the kind of roof that it supported. The simplest approach to roofing was to lay a beam from wall to wall. But even King Solomon could not buy roof beams much longer than about 8 m, hewn from the "cedar-trees out of Lebanon." The best solution to spanning a space with stone was the arch. The downward load (the weight of a

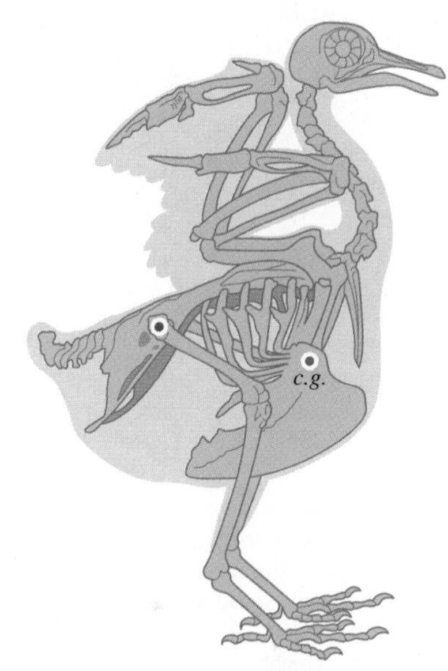

Figure 6.40 Birds are stable on two legs because those legs pivot at the pelvis above the *c.g.,* which is low because of the massive breast bone that anchors the wing muscles. In flight, the *c.g.* is supported by, and located under, the wings.

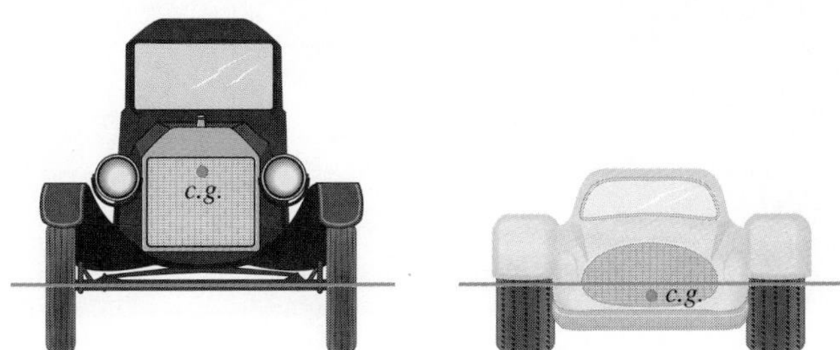

Figure 6.41 Modern automobiles have their *c.g.*s as low as possible. When the *c.g.* is high above the axles, the vehicle is unstable. Large trailer trucks, foolishly loaded, are constantly toppling over in high winds. Many roads are particularly hazardous in this regard and are often closed to commercial traffic whenever the wind comes up.

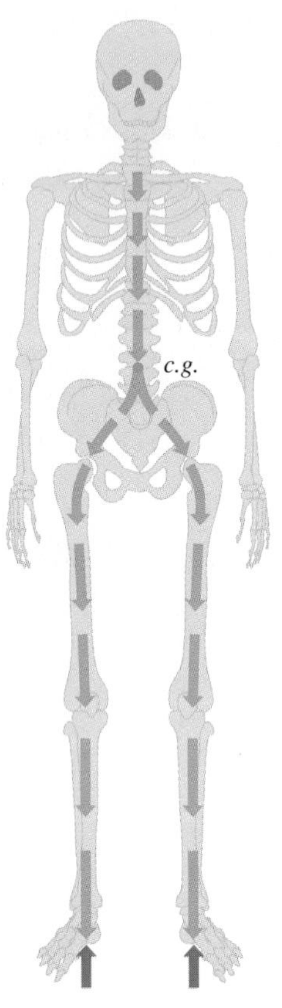

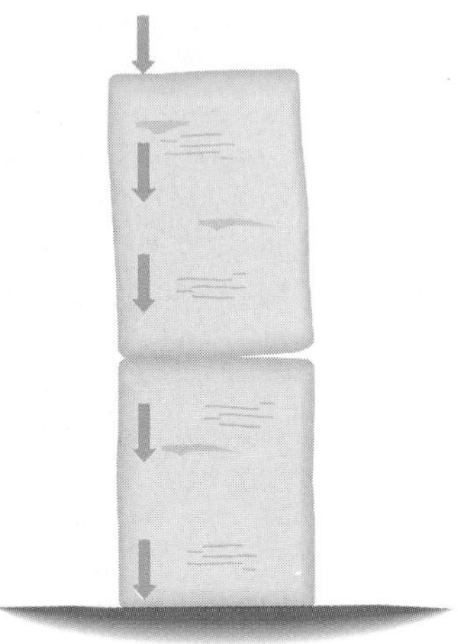

Figure 6.42 The "middle third" rule maintains that an elastic block will be entirely in compression, provided the load remains in the middle third. Masonry structures are weak in tension and are designed to keep the load in the middle third.

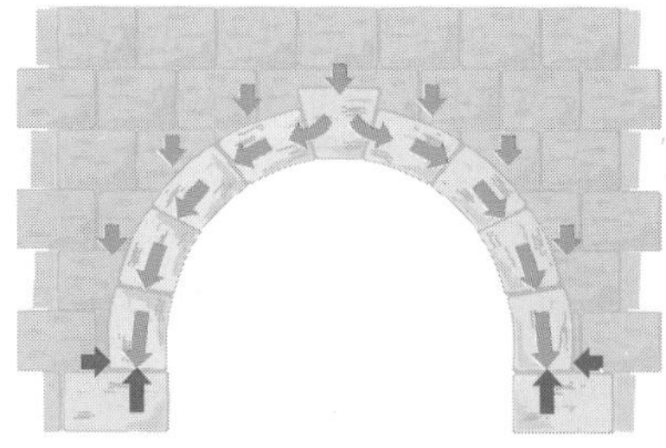

Figure 6.43 The arch channels the load down to the abutments. In the process, each block is compressed in two directions, which is fine since stone has great strength in compression. A masonry arch can readily span 60 m or so. Such arches are also very durable. Many Roman arches from antiquity still stand in excellent condition.

Figure 6.44 The weights of the various parts of the human body are supported by the musculoskeletal system. The hanging arms are held in place by tendons and muscles in the shoulders. That load is carried back to the spine, which transmits the weight of the entire upper body to the pelvis. The bone arch of the pelvis splits the downward force, directing it through the legs to the ground.

bridge or building) squeezes the separate wedge-shaped stones against each other, redirecting the force (Fig. 6.43). In effect, the thrust flows down along the arch-ring and out into the abutments at the base. Each stone acts mainly in compression, where it is strongest. If all goes well, the thrust lines will remain within the arch-ring no matter how the load shifts. If not, cracks will develop just as in the wall, but there is a significant difference. Arches are, by their very nature, stable (without crumbling, the wedge-shaped stones cannot move past each other).

Today, concrete arched highway overpasses are a common sight. In the longest arch bridges, spanning upwards of 1700 ft, steel truss frames replace blocks of stone. The human pelvis (Fig. 6.44) is a bone arch that carries the load of the spine,

A flying buttress of the cathedral of Notre Dame de Paris. The buttress is a half-arch that keeps the walls of the building from moving outward.

dividing it in two and channeling it down to the legs. Not surprisingly, the pelvis would work even better if we walked on all fours.

One of the great inventions of the Middle Ages (done by trial and error long before the development of structural analysis) was the Gothic arch that crowned the cathedrals of Europe. By spanning the same space with a high pointed arch, the horizontal reaction forces were greatly reduced and the support walls made much thinner. The thrust flowed more nearly vertically and stayed within a thinner supporting wall, which could then be ornamented with splendid windows. Where the thrust line might stray out of the bounds of a thin wall (as in a strong wind), a flying buttress channeled the force safely downward (see photo on p. 192).

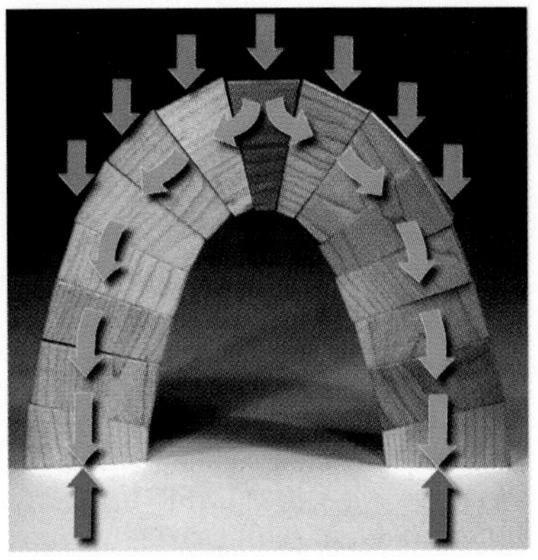

The pointed arch channels the load straight down to the ground. There is no sideways outward thrusting, as there is with the circular arch. As a result, the walls can be much lighter. That allowed for the soaring architecture of the Gothic cathedral.

Core Material

For *coplanar concurrent forces,* translational equilibrium occurs when (p. 164)

$$\Sigma \mathbf{F} = 0 \qquad [6.1]$$

which is the First Condition of Equilibrium. A system acted upon by forces that all pass through a single point is in equilibrium when

$$\Sigma F_{\mathrm{H}} = 0 \quad \text{and} \quad \Sigma F_{\mathrm{V}} = 0 \qquad [6.2]$$

The **torque** about O is defined as

$$\tau_O = Fr_\perp = F_\perp r = Fr \sin\theta \qquad [6.3]$$

The torque vector $\boldsymbol{\tau}_O$ is *perpendicular to both* **r** *and* **F**, and

$$\boldsymbol{\tau}_O = \mathbf{r} \times \mathbf{F} \qquad [6.5]$$

For a rigid body in equilibrium, the sum of the torques about any point O must be zero, that is

$$\Sigma \tau_O = 0 \qquad [6.6]$$

which is the Second Condition of Equilibrium. When we deal with concurrent forces, there will be only two independent equations (either sum-of-the-forces *or* sum-of-the-torques) and no more than two unknowns can be determined. A nonconcurrent system of any number of parallel, separated forces in equilibrium can give rise to only two independent equations (either one force and one torque, or two torque), and only two unknowns can be determined. When a system of nonconcurrent forces (those that are not all parallel and perpendicular) produces equilibrium,

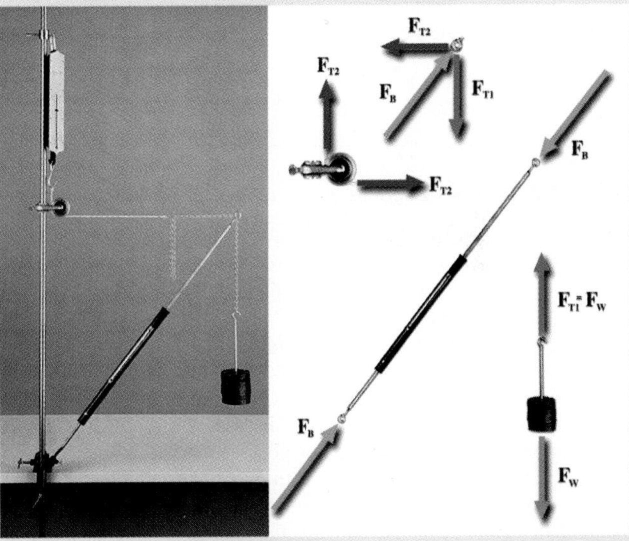

Figure 6.45 An experimental crane boom. The pulley redirects the tension force in the rope (F_{T2}) without altering it. The vertical length of chain is a hanger, and the tension in it (F_{T1}) equals the load (F_W) it supports. The boom is made of a compression spring scale, and it reads the force (F_B) tending to squash it. The bottom of the boom is pivoted on a pin; therefore, the forces (F_B) at its ends act inward along the scale's length. This arrangement is a concurrent force system.

there will be three independent equations, from which no more than three unknowns may be determined.

Figure 6.45 serves as a summary of some of the conceptual details that are involved in analyzing a statics problem.

The **center-of-gravity** *(c.g.) of an object is that point where the total weight of an object can be imagined to act* (p. 184), thus the equation

$$x_{cg} = \frac{\Sigma F_W x}{\Sigma F_W} \qquad [6.7]$$

Suggestions on Problem Solving

1. When all the forces acting on a body pass through the same point, we solve the problem by applying the sum-of-the-forces equation (in each of two perpendicular directions if necessary). When the forces are nonconcurrent, at least one application of the sum-of-the-torques equation will be required. Use Eq. (6.6) sparingly to avoid nonindependent torque equations.

2. Always draw a picture of the entire system. Then draw a free-body diagram of each segment of the setup to be analyzed. Guess at the directions of the internal forces, and if you are wrong, they will turn out negative. In time, you should be able to guess correctly by recognizing what the directions of the forces and torques must be to sustain equilibrium. If you are to solve for the force in a given rope or bar, part of that member must be included in one of these diagrams. You cannot solve for a quantity unless it appears in one of the equations. If possible, start the analysis in the simplest place where there is only one unknown. Then move on to the more complicated regions, using what you have just learned.

3. Consider the geometrical symmetries of a given setup. Such symmetries usually mean that the internal forces will be symmetrical if the load is symmetrical, as in Figs. 6.9 and P15. If the arrangement is asymmetrical, the internal forces are also likely to be different (Fig. P11).

4. When finding torques, consider the point about which they will be computed. Avoid the unknown forces you are not interested in by taking torques about points through which those forces pass. By computing the net torque about point P in Fig. 6.28, you preclude the unknown reaction force from entering the equation and can immediately solve for the tension.

5. When dealing with pulleys, determine how many lengths or segments of a single continuous rope support the load. A set of pulleys may be fixed to a load and move with it; if so, determine the number of rope segments supporting it (as in Fig. 6.6), because they support the load. If the rope is continuous and there are no tangential forces on it, the tension is the same at every point in it. If there are N supporting rope segments, the tension in each is $1/N$th of the load. When several different ropes support the load (Fig. P14), each may have a different tension in it.

6. To determine the signs in a sum-of-the-torques equation, assign a direction (clockwise or counterclockwise) as positive. Then imagine an axis perpendicular to your drawing running through the point about which you are taking the torques. For each force, determine how it, acting alone, would rotate the system about that axis. For example, in Fig. 6.28, taking $\circlearrowleft + \Sigma\tau_P$, the weight produces a positive torque, and the tension, a negative torque.

Discussion Questions

1. A body is in equilibrium under the influence of three equal-magnitude nonparallel forces. Determine their relative orientations. Can concurrent forces of 30 N, 40 N, and 80 N produce equilibrium?

2. A rock thrown straight up stops at its peak altitude. Is it in equilibrium at the instant it comes to rest? A ball thrown into a wall on a spaceship deforms and comes to rest for an instant before bouncing back, off the wall. Is it in static equilibrium at that moment? Once the ball springs back to shape and leaves the wall (assuming zero gravity), is it in equilibrium?

3. What is the tension in the rope on the right in Fig. Q3a? Each of the pulley systems in Fig. Q3b is supposed to be in equilibrium. Redraw any of them that may be in error.

4. Imagine a hammock hung by lengths of rope between two trees. Are the ropes more or less likely to break when someone sits in the hammock if it is initially pulled tight or allowed to sag? Explain.

5. Figure Q5 shows a forearm and hand exerting a downward force on a book. Explain the physics of what's happening. How would it be different if the arm were supporting the book? Try it yourself, checking your own muscles.

6. Why is it easier to hold a long stick horizontally at its midpoint rather than at its end? Position a stick horizontally so it

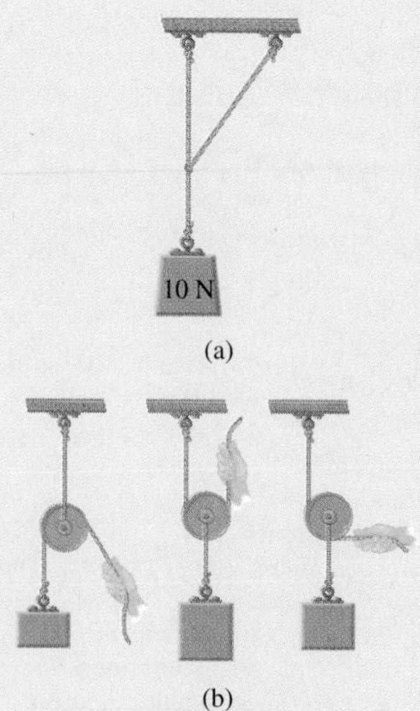

Figure Q3

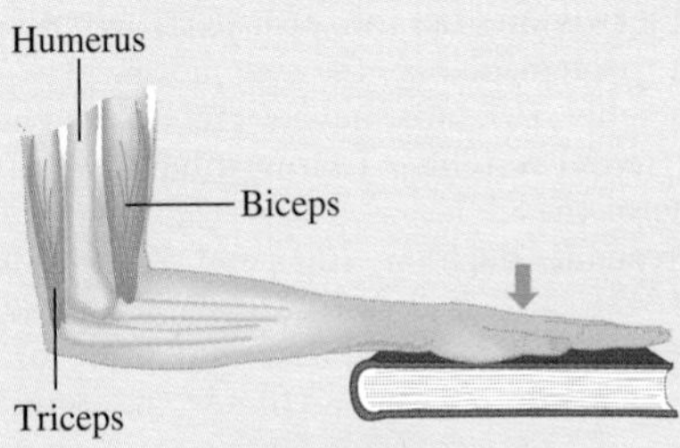

Figure Q5

rests on your two outstretched index fingers held about 2 ft apart. Now slowly move your fingers toward one another. Observe the movement of the stick over your fingers: Why does it alternate, sliding first over one finger and then the other? What is the significance of the point where your fingers finally meet?

7. Why do people lean forward when they ascend an inclined ramp? Why do you think pregnant women often suffer backaches in their third trimester? Why does a swimmer or a sprinter crouched at a starting line lean way forward?

8. About 100 years ago, there was a popular toy called a tumbler (Fig. Q8). When toppled, the toy immediately righted itself. The modern equivalents are great inflatable plastic things that kids punch. How do such toys work?

Figure Q8

9. Stand, feet flat on the ground, with your back and heels against a wall. Try to touch your toes. Explain why you cannot do it. Now face the edge of an open door. Place one foot on either side of the door, so that both feet extend somewhat beyond the edge on which you are resting your nose. Now try to stand up on your toes. Explain what is happening.

10. Try to balance an ordinary kitchen bowl on the eraser end of a pencil, first at its middle on the inside, and then on the bottom outside of the bowl. Explain the difference.

11. Because insects are so light, it would seem that they would be blown over by even a slight wind. What anatomical difference between them and mammals prevents this from happening?

12. Describe the equilibrium state of each configuration shown in Fig. Q12.

Figure Q12

13. What, if anything, can you say about the location of the *c.g.* of the chair in Fig. Q13?

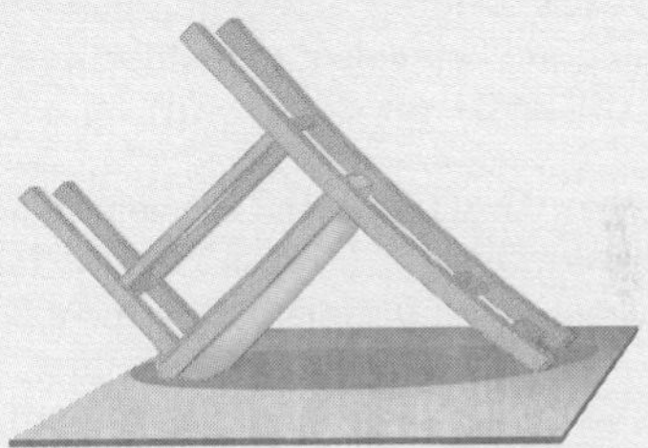
Figure Q13

14. Why is it more strenuous to do situps with your hands behind your head than with them out in front of you?

15. It's common during demonstrations of karate for the practitioner to strike several pine boards held tightly together, one behind the other. It often happens that the back board will break in two, while the front one, which receives the blow, is little more than cracked. Explain.

16. How would you make a pair of "loaded" dice? Explain.

17. Imagine that you are standing straight, hands at sides, holding a weight in each hand. Draw a free-body diagram of your arms and legs and describe the forces acting on them. Are they in tension or compression?

18. Is a truck loaded with several tons of feathers more likely to tip over in a high wind than one carrying the same weight of bricks? Explain. Occasionally, one sees a narrow van filled with cargo carrying a heavy load on its roof. Why is that dangerous, and how can the risk be minimized? Explain the significance of the narrowness of the van.

Multiple Choice Questions

1. Ten scales, each weighing 10 N, are in equilibrium, stacked one on top of the other. What are the readings of the top and bottom scales, and what is the reaction force exerted by the ground on the stack? (a) 0 N, 100 N, 100 N (b) 10 N, 90 N, 100 N (c) 10 N, 100 N, 100 N (d) 0 N, 90 N, 100 N (e) none of these.

2. An 800-N acrobat holds a 40-N chicken in one hand while his 480-N assistant Jane sits on his shoulders holding a 4.0-N box of cigars. The acrobat stands motionlessly on one foot on a 36-N bathroom scale. The scale reads (a) 1360 N (b) 1324 N (c) 662 N (d) 680 N (e) none of these.
3. A woman holds a 20-N shovel containing 16 N of dirt at rest horizontally. She grips it with one hand at the end of the handle and one at the middle. The net vertical force she exerts on the shovel is (a) not enough information to know (b) 20 N upward (c) zero (d) 36 N upward (e) none of these.
4. The essential aspect of a body in equilibrium is that it has (a) zero speed (b) zero mass (c) zero momentum (d) zero acceleration (e) none of these.
5. It is possible to apply forces of 6 N and 10 N to a body in such a way as to counter the effect of a single additional force of (a) 18 N (b) 20 N (c) 7 N (d) 3 N (e) none of these.
6. Two forces, each of 100 N acting at a point such that they are 120° apart, are equivalent to a single force of (a) 100 N (b) zero (c) 200 N (d) 86.6 N (e) none of these.
7. No matter what the setup is, the tension in a rope used to support a load is always (a) less than the load (b) equal to the load (c) zero (d) more than the load (e) none of these.
8. The lightweight pivoted bar in Fig. MC8 will be in rotational equilibrium when a 200-N force acts (a) down at *D* (b) up at *B* (c) down at *E* or up at *C* (d) up at *C* only (e) none of these.

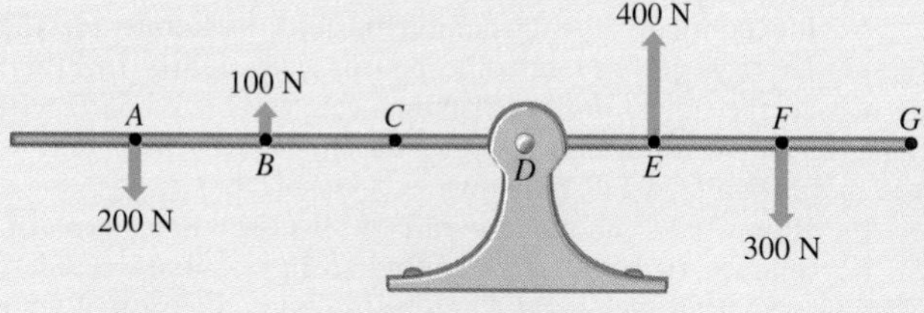

Figure MC8

9. An egg balanced on an end is (a) in stable equilibrium (b) in unstable equilibrium (c) in neutral equilibrium (d) not in equilibrium (e) none of these.
10. Given a body acted upon by several forces, if the sum of those forces is exactly zero, the body (a) may be in both translational and rotational equilibrium (b) must be in both translational and rotational equilibrium (c) cannot be in both translational and rotational equilibrium (d) not enough information to come to any conclusion (e) none of these.
11. In the photo of Chinese acrobats on p. 190, the combined *c.g.* of the pole and the two suspended people (a) is on a vertical line passing through the third man's left shoulder (b) is to the left of the third man (c) is to the right of the third man (d) is on a horizontal line passing through the third man's shoulder (e) none of these.
12. If an object is divided into two parts by slicing it vertically straight through its *c.g.*, the two pieces (a) must have the same weight (b) must have different weights (c) can have different weights (d) must have the same mass (e) none of these.
13. Two astronauts push on either end of a satellite floating in space. If the craft is then in translational and rotational equilibrium, it is required that (a) both forces must pass through the satellite's *c.g.* (b) the two forces be equal, coplanar, and pass through the *c.g.* (c) there be two forces, they need not be equal, but they must be antiparallel (d) the forces must be equal, colinear, antiparallel, and pass through the *c.g.* (e) none of these.
14. A light rope is looped over a weightless pulley and tied to a 50-N rock. The other end is held by a 50-N monkey who climbs up the rope a bit and then stops. The tension in the rope at that moment is (a) 100 N (b) zero (c) 50 N (d) 25 N (e) none of these.
15. In analyzing the condition of rotational equilibrium, one takes the sum of the torques about an axis (a) which must pass through the body's *c.g.* (b) through which the line-of-action of all the forces must pass (c) passing through the center of the body (d) located anywhere (e) none of these.
16. Someone places a 6.0-m-long steel rod on a rock so that one end is under a baby moose weighing 2.0 kN. The person pushes down on the other end of the rod with a force of 400 N, and the moose is held in the air at rest. The rock was (a) 1.0 m from the moose (b) 5.0 m from the moose (c) 1.0 m from the person (d) 6.0 m from the person (e) none of these.
17. A 100-N load hangs from the ceiling on a rope. Someone pushes on the load with a continuing horizontal force, and it moves off to the side a little, coming to rest at a bit of an angle with the vertical. As a result, the tension in the rope (a) is unchanged (b) is zero (c) is decreased (d) is increased (e) none of these.
18. A 10-m length of rope, weighing 10 N per meter, is hanging straight down from a ceiling hook. The tensions in the rope at the free end, at 5.0 m up from the free end, and at the hook are, respectively (a) 0, 100 N, and 100 N (b) 0, 50 N, and 100 N (c) 100 N, 100 N, and 100 N (d) 100 N, 50 N, and 0 (e) none of these.
19. In order to hold the weight in Fig. MC19 in equilibrium, the hand and the ceiling exert forces, respectively, of (a) 100 N and 300 N (b) 300 N and 100 N (c) 100 N and 400 N (d) 300 N and 600 N (e) none of these.

Figure MC19

Problems

1. [I] A 200-N chandelier is hung from a ceiling hook by a chain weighing 50 N. (a) What force is exerted on the hook? (b) What force acts on the bottom of the chain?
2. [I] What are the tension forces acting on each of the weightless ropes in Fig. P2?

Figure P2

3. [I] A 20-m length of chain weighing 2.0 N per meter is hung vertically from one end on a hook. (a) What is the tension three-quarters of the way up? (b) What is the tension 1.0 m from the top and bottom?
4. [I] Using a ruler and a compass, determine if forces of 11 N, 16 N, and 20 N acting at a point can result in equilibrium. Explain your answer.
5. [I] A 10-kg child and a 100-kg adult both stand upright and barefooted at rest on a smooth platform sloping at some angle θ. Who is more likely to slide down the incline?
6. [I] A 200-N child stands at rest on an inclined metal slide tilted at an angle of 24° up from the ground. Compute the friction force on her shoes. What minimum value of μ_s must exist?
7. [I] The 10-kg block in Fig. P7 is held at rest by the four ropes shown. If the tension in the ropes on the right and top are each 98 N, what is the tension in the remaining two ropes?

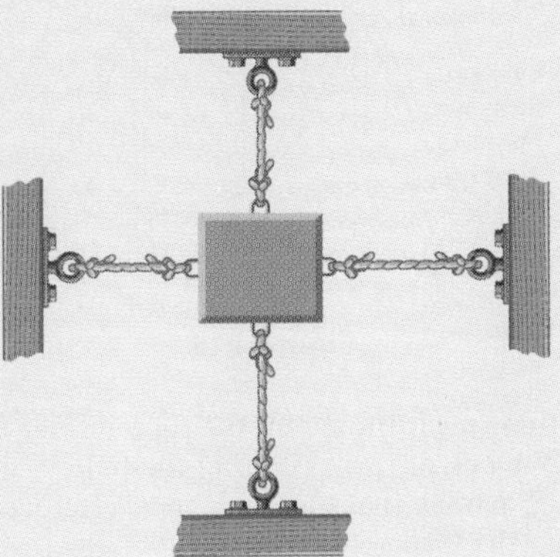

Figure P7

8. [I] The orthodontal wire brace in Fig. P8 makes an angle of 80.0° with the perpendicular to the protruding tooth. If the tension in the wire is 10.0 N, what force is exerted on the tooth by the brace?

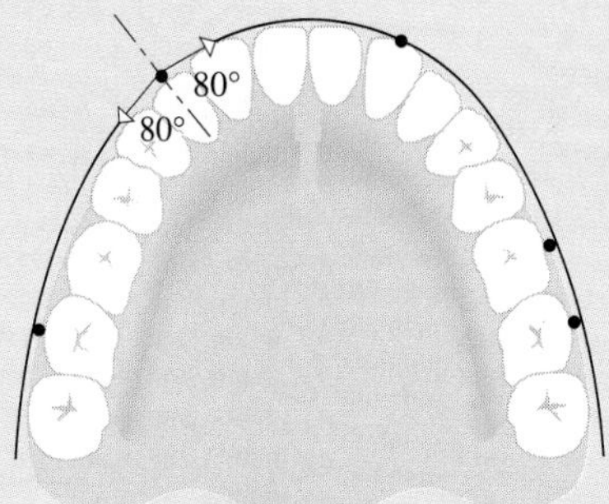

Figure P8

9. [I] Referring to Fig. P9, determine the scale reading in the right arm of the suspension and the angle θ.

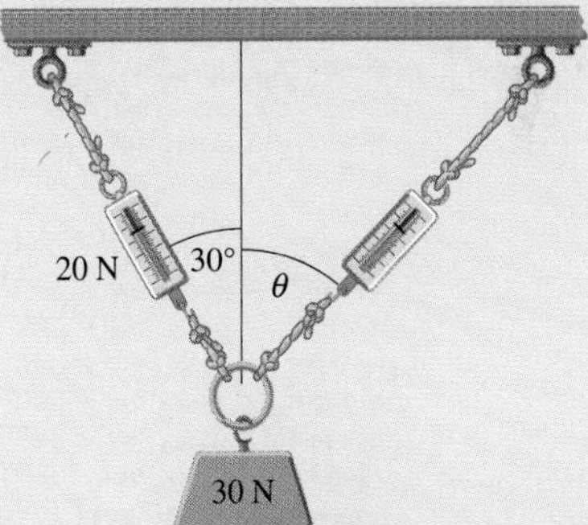

Figure P9

10. [I] The steel beam that hangs horizontally in Fig. P10 weighs 8.00 kN, and the lower cables hook on to it at angles of 70.0°. What are the tensions in the lower cables?

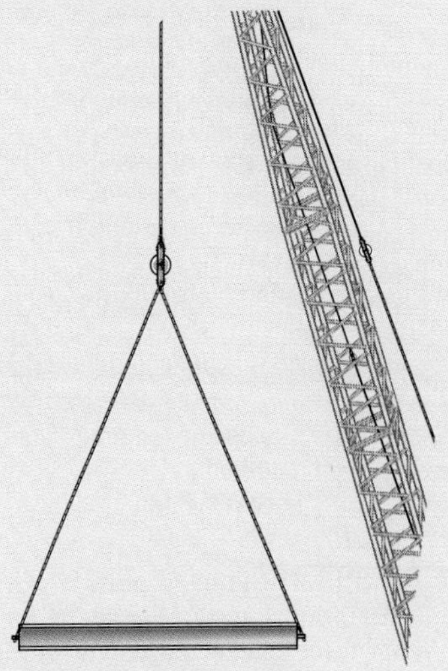

Figure P10

11. [I] In the static arrangement shown in Fig. P11, the pulleys and ropes are essentially weightless. If weight 1 is 15.0 N and weight 2 is 31.0 N, and the two angles are measured to be $\theta = 45.0°$ and $\phi = 20.0°$, determine the value of weight 3. Check that the system is in equilibrium.

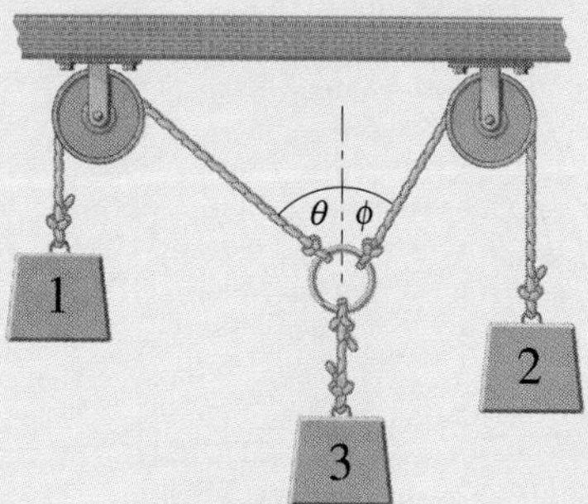

Figure P11

12. [I] The hand in Fig. P12 exerts a downward force that holds the length of pipe at rest. (a) Compute that force. (b) What is the tension in the rope that is strung over the pulley? (c) What force supports the upper hook?

Figure P12

13. [I] Look at Fig. P13. If the hand pulls down so that the 100-N weight remains at rest, what is the tension in the long rope? What force is exerted down on the right-most ceiling bracket?

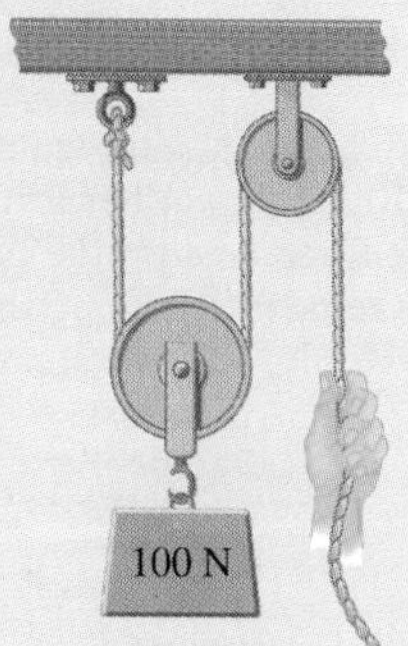

Figure P13

14. [I] What is the tension in each length of rope in Fig. P14, given that the load is at rest?

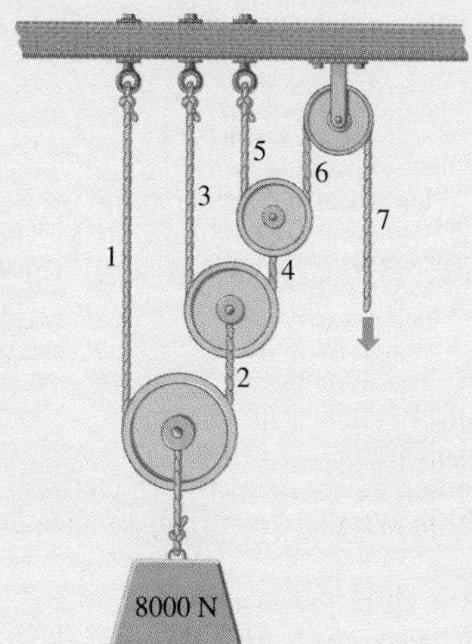

Figure P14

15. [I] Referring to the pin-jointed truss in Fig. P15, compute the force in bars *BC*, *DE*, and *FG*.

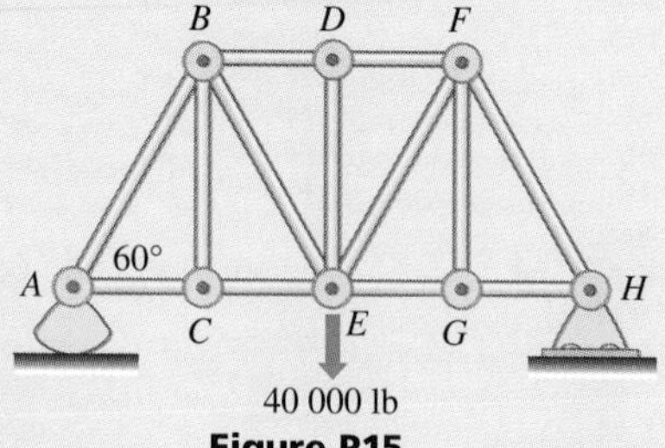

Figure P15

16. [I] What downward force on the rope in Fig. P16 will hold the 300-N block at rest assuming the pulleys are essentially weightless? Notice that there are two different ropes holding up the load.

17. [II] Determine the weight of the mass *m* in Fig. P17. Assume the pulleys and ropes are all essentially weightless.

18. [II] An essentially weightless rope is strung nearly horizontally over a light pulley, as shown in Fig. P18. The

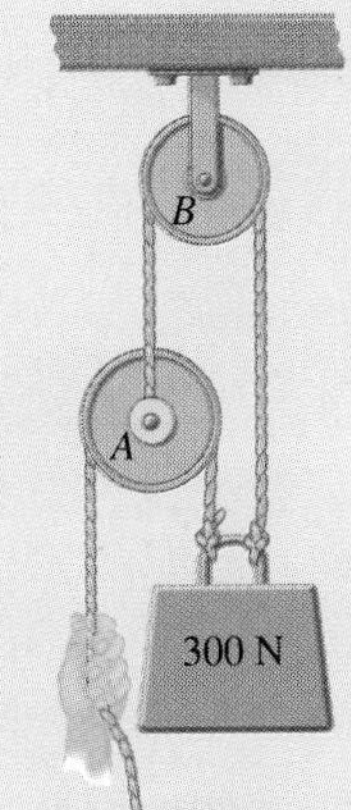

Figure P16

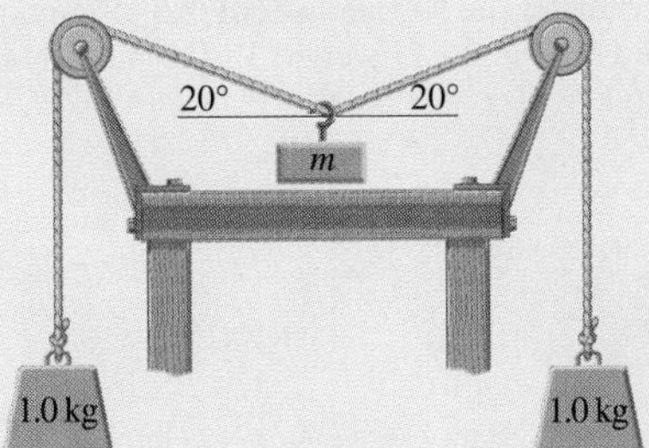

Figure P17

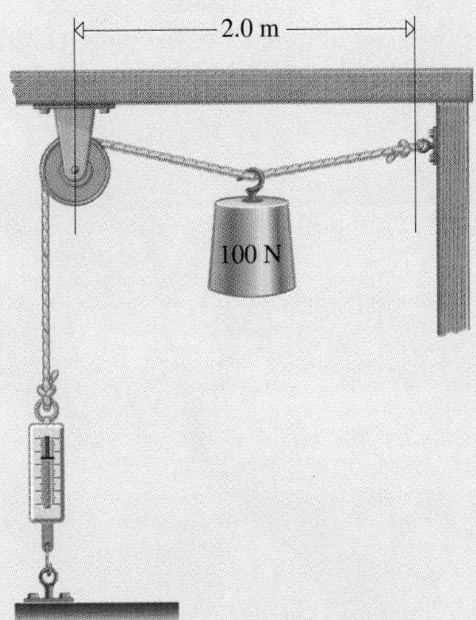

Figure P18

spring balance is adjusted to read zero. A 100-N weight is then hung at the midpoint of the span, and the rope sags, descending 10 cm from the horizontal. Determine the scale reading.

19. [II] A 533.8-N tightrope walker dances out to the middle of a 20-m-long wire stretched parallel to the ground between two buildings. She is wearing a pink tutu, of negligible weight; the wire sags, making a 5.0° angle on both sides of her feet with the horizontal. Find the tension.

20. [II] If a force of 100 N in Fig. P20 holds the weight motionless, what is its mass? What is the tension in rope 1?

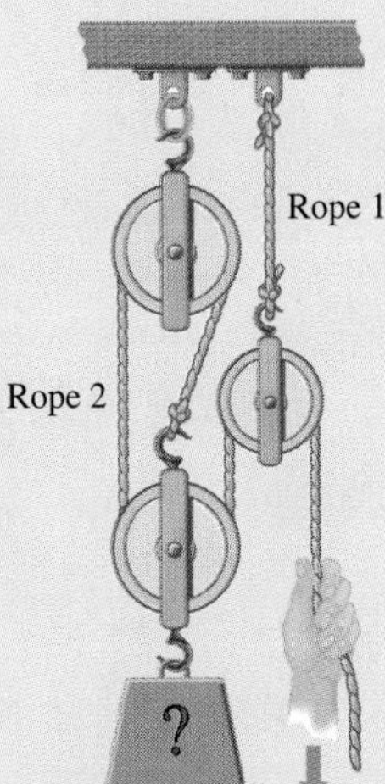

Figure P20

21. [II] Determine both the angle at which the pulley hangs and the tension in the hook supporting it in Fig. P21.

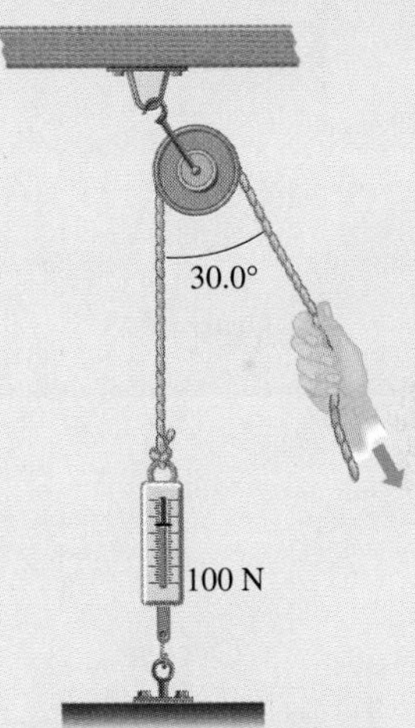

Figure P21

22. [II] The pin-connected symmetrical structure in Fig. P22 (which is used to support floodlights at a rock concert) is known as a Fink truss. If the lamps on top weigh 1.00 kN and the ones on the bottom weigh 2.00 kN each, compute the reaction forces exerted by the towers.

23. [II] The three bodies in Fig. P23 are at rest. Compute the static friction force between the box and the surface. If a fly lands on the 80.0-kg mass, and the system then begins to move, what is the static coefficient of friction for the box and surface?

24. [III] Consider the arrangement depicted in Fig. P24. Now, compute the internal forces in each of the ropes and in the boom.

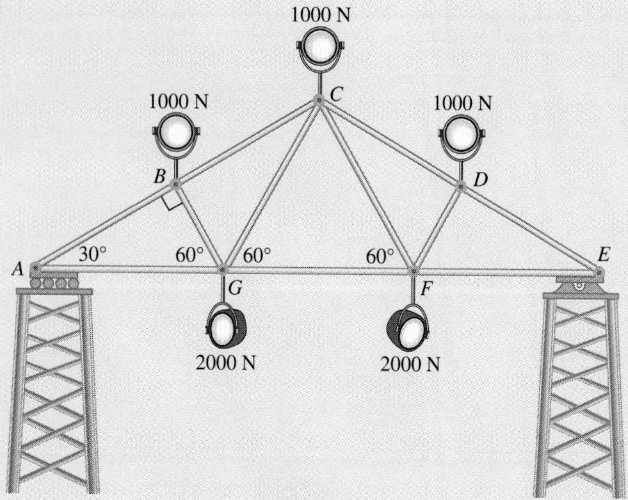

Figure P22

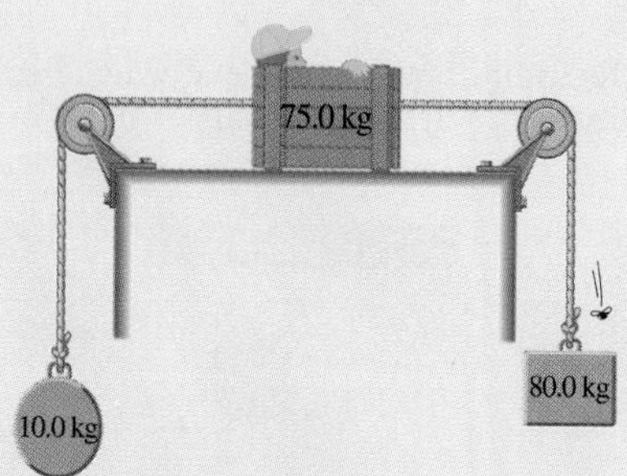

Figure P23

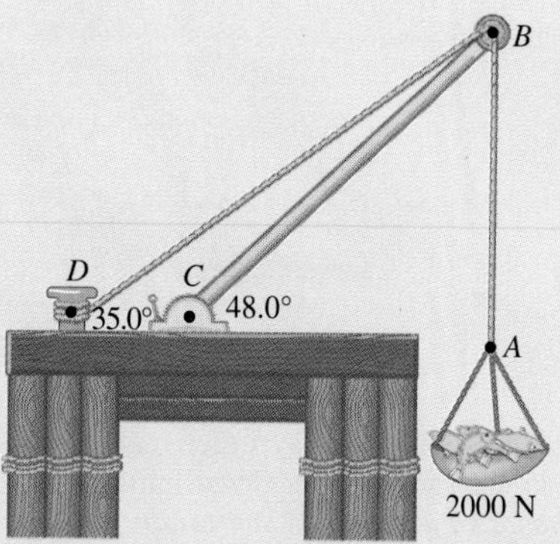

Figure P24

25. [III] Determine the reaction forces at the four contact points A, B, C, and D, in Fig. P25. The hollow balls, which are homogeneous and made of the same material, have masses of 102 kg and 204 kg, and the chamber they sit in is 55.98 cm wide. Everything is smooth, so that friction is negligible.

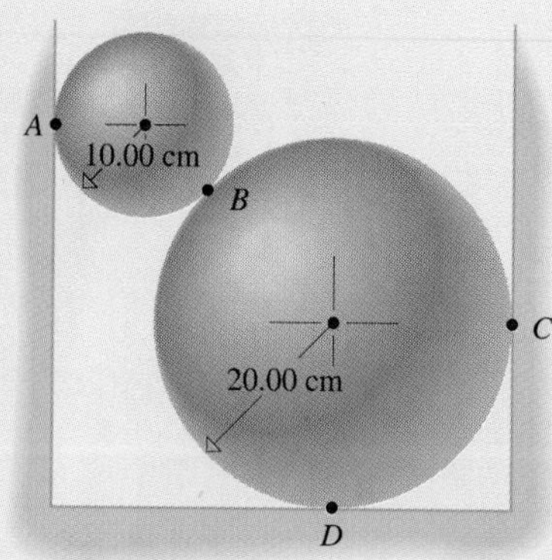

Figure P25

ROTATIONAL EQUILIBRIUM

26. [I] Determine the torques about the elbow and shoulder produced by the 20-N weight in the outstretched hand in Fig. P26.

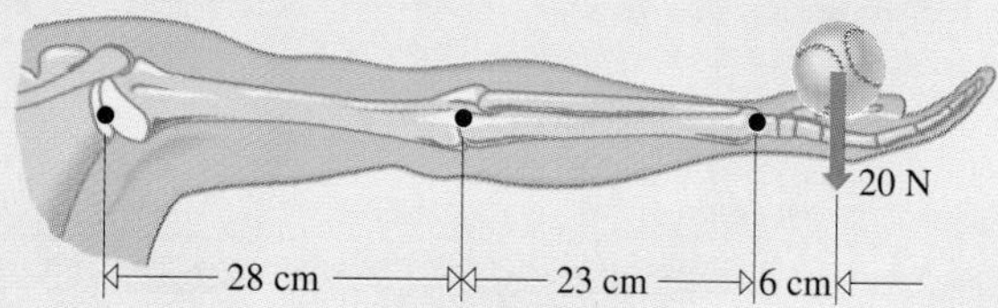

Figure P26

27. [I] Compute the torque about the pivot O, generated by the 100-N force on the gearshift lever in Fig. P27.

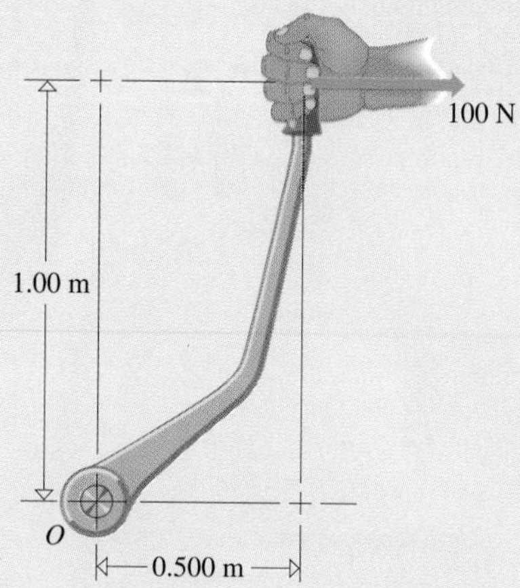

Figure P27

28. [I] Figure P28 shows a tendon exerting an 80-N force on bones in the lower leg. Assume the tendon acts horizontally 0.06 m from the knee pivot. Draw a simplified version of the system and compute the tendon's torque about the pivot.
29. [I] While working with precision components, especially out in space, it is often necessary to use a torque wrench, a device that allows the user to exert only a pre-set amount

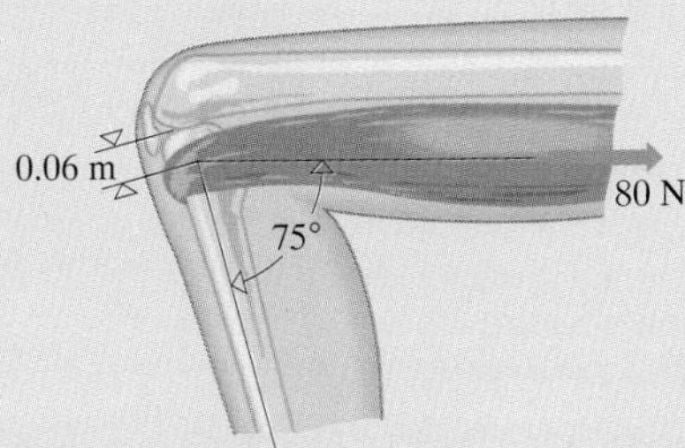

Figure P28

of torque. Having dialed in a value of 35.0 N·m, what maximum perpendicular force should be exerted on the handle of a wrench 25.0 cm from the bolt?

30. [I] Referring to the *xyz*-coordinate system, imagine a vector **A** lying on the positive *x*-axis and a vector **B** on the positive *y*-axis. Determine the direction of each of the following: **C**, where $\mathbf{C} = \mathbf{A} \times \mathbf{B}$; **D**, where $\mathbf{D} = \mathbf{A} \times \mathbf{C}$; **E**, where $\mathbf{E} = \mathbf{C} \times \mathbf{B}$; and **F**, where $\mathbf{F} = \mathbf{E} \times \mathbf{D}$.

31. [I] Two campers carry their gear (90.72 kg) on a light, rigid horizontal pole whose ends they support on their shoulders 1.829 m apart. If Selma experiences a compressive force of 533.8 N, where is the load hung on the pole, and what will Rocko feel?

32. [I] Harry, who weighs 320 N, and 200-N Gretchen are about to play on a 5.00-m-long seesaw. He sits at one end and she at the other. Where should the pivot be located, if they are to be balanced? Neglecting the weight of the seesaw beam, what is the reaction force exerted by the support on it?

33. [I] An 8000-N automobile is stalled one-quarter of the way across a bridge (see Fig. P33). Compute the additional reaction forces at supports *A* and *B* due to the presence of the car. Take the length of the bridge to be $\overline{AB}$.

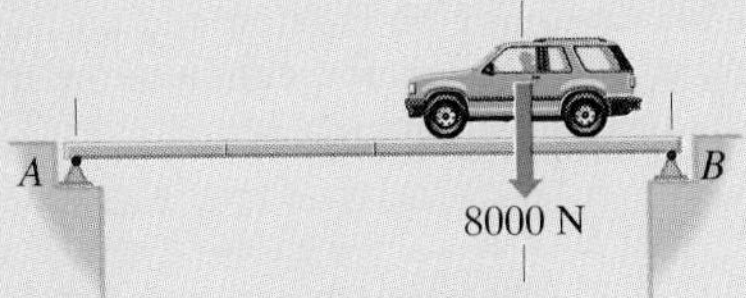

Figure P33

34. [I] The weightless ruler in Fig. P34 is in equilibrium. Determine both the unknown mass *M* and the left-hand scale reading.

35. [I] A 10.0-kg brass sphere rests in a 90.0° groove as shown in Fig. P35. Assuming that there is no friction and that the weight of the sphere acts at its center, determine the two reaction forces.

36. [I] The little bridge in Fig. P36 has a weight of 20.0 kN, which acts at its center. Calculate the reaction forces at points *A* and *B*.

37. [II] If the beam in Fig. P37 is of negligible mass, what value does the scale read (in kg)?

38. [II] Figure P38 shows a laboratory model of a roof truss constructed of two very light rods pinned at points *B* and *C*. A light cable tightened by a turnbuckle runs from *A* to *C* to complete the truss. What do the two spring scales read in newtons? Determine the compression in rod *AB*. (Hint: Find out the reading of scale 1 and then draw a free-body diagram of *A*. Remember the forces exerted by the pinned rods act along their lengths.)

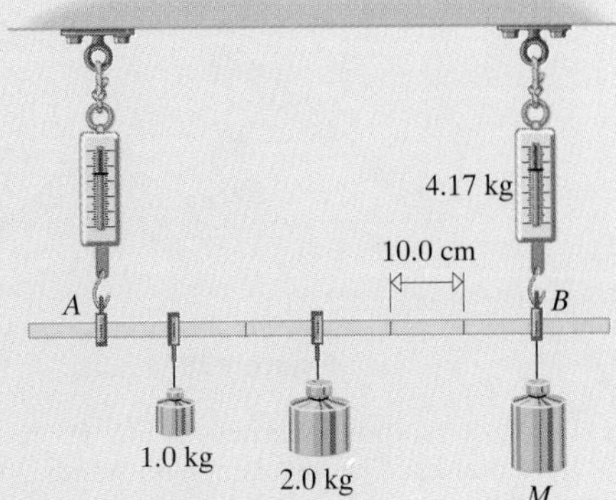

Figure P34

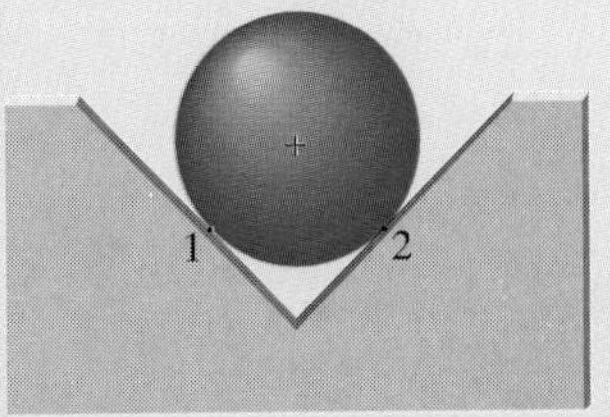

Figure P35

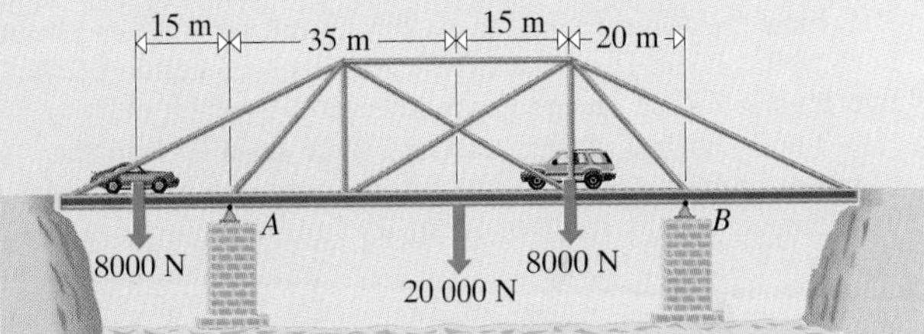

Figure P36

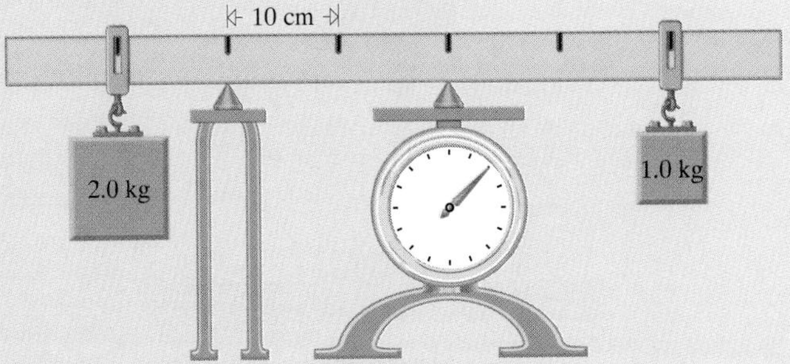

Figure P37

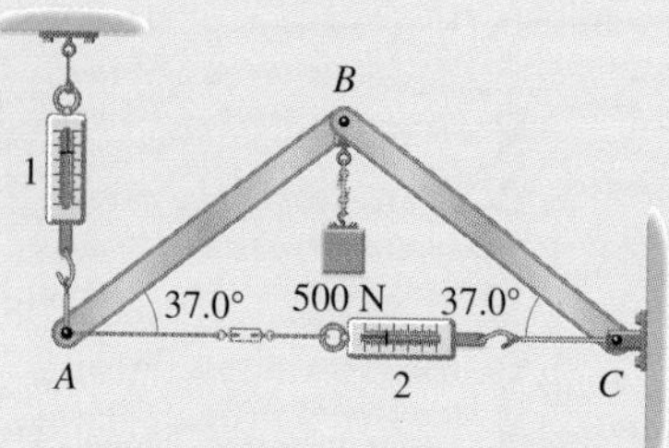

Figure P38

39. [II] The little temporary bridge in Fig. P39 consists of two very light planks. The 480-N person stands 3.00 m from the end of the upper board, which itself rests on rollers. What are the reaction forces at points A, C, and D?

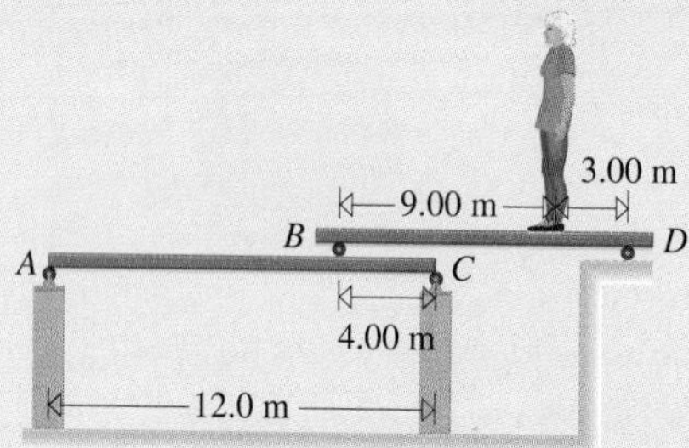

Figure P39

40. [II] Find the force F_R in Fig. 6.28b by taking torques about both point A and the point where the lines-of-action of the two tensile forces on the ropes intersect.

41. [II] The hand in Fig. P41 exerts a force of 300 N on the hammer handle. Compute the force acting on the nail, just as the head begins to pivot about its front edge in contact with the board.

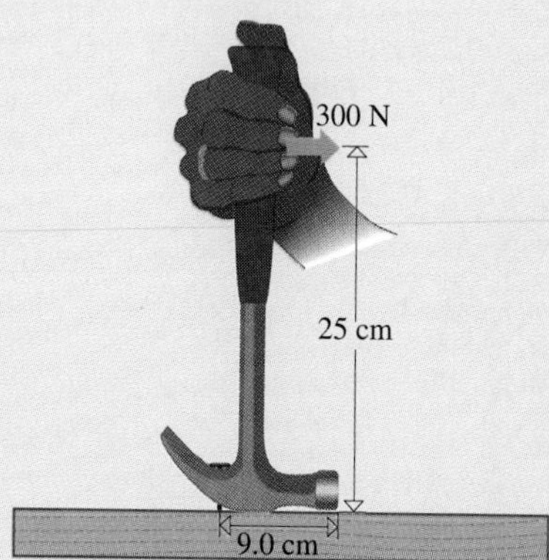

Figure P41

42. [II] A forceps is used to pinch off a piece of rubber tubing, as shown in Fig. P42. What is the force exerted on the rubber if each finger squeezes with 10.0 N? What is the force exerted on the pivot by each half of the forceps?

43. [III] Suppose a downward force of 80 N acts on the pedal of the bike shown in Fig. P43. The chain goes around the front chain wheel (of radius 10 cm) and the rear sprocket wheel (of radius 3.0 cm). The pedal is attached to a 17-cm crank arm that turns around point A. What impelling force will the scale read (the bike, chain, and wheel are motionless at the moment depicted)? What provides that force?

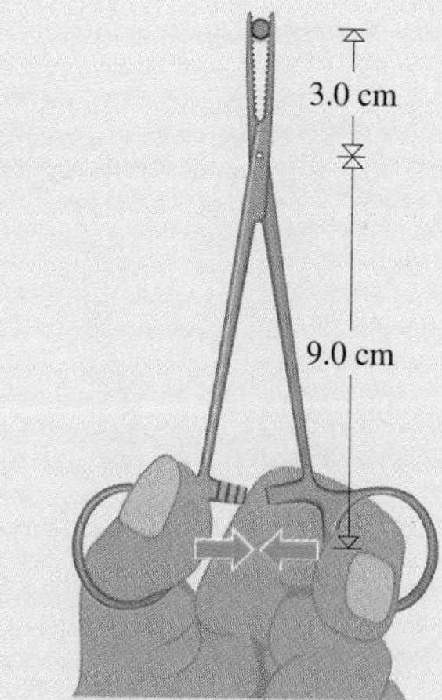

Figure P42

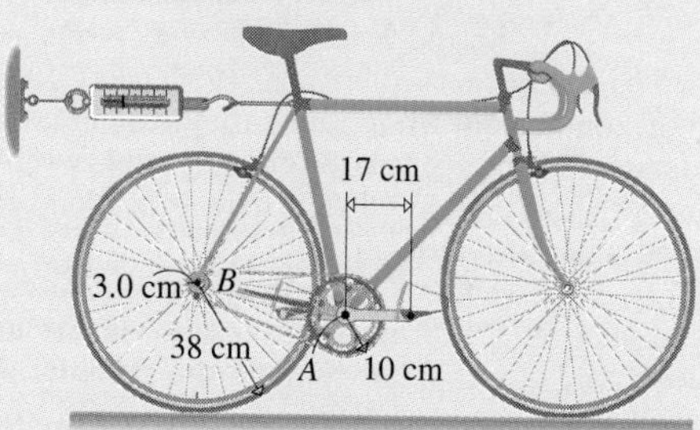

Figure P43

44. [III] A sphere of mass 10.0 kg rests in a groove, as shown in Fig. P44. Assuming no friction and taking the weight of the sphere to act at its center, compute the reaction forces exerted by the two surfaces.

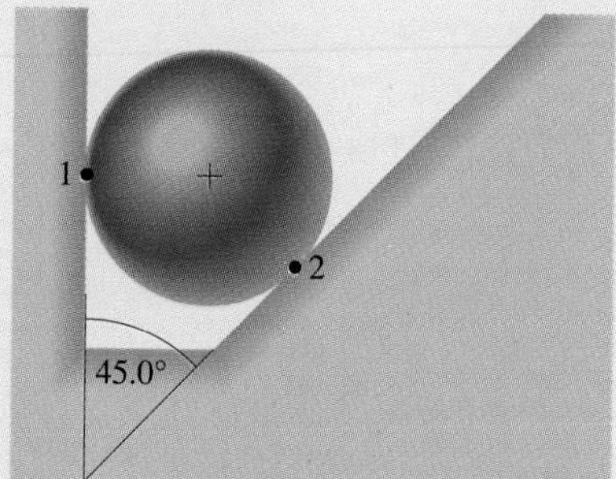

Figure P44

45. [III] Determine the internal forces acting at point A in Fig. P45 when a 1000-N weight is hung from the hook. All of the structural members can be taken as weightless.

46. [III] Figure P46 shows a smooth rod resting horizontally inside a bowl. Compute the force F that will maintain the rod in position, given negligible friction.

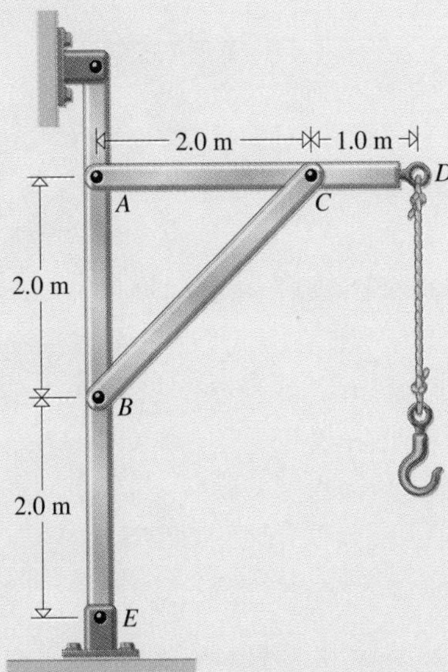

Figure P45

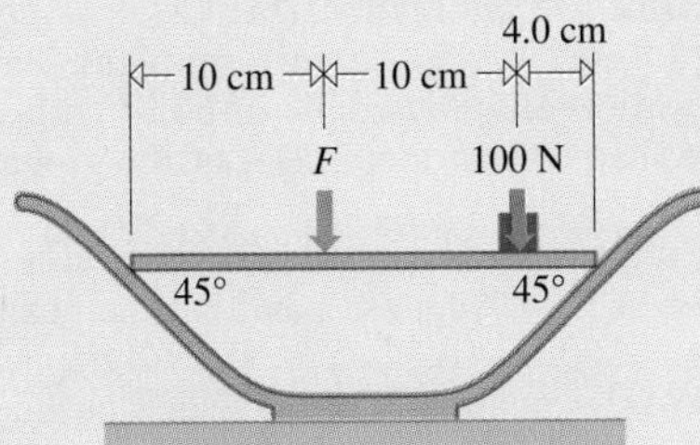

Figure P46

EQUILIBRIUM OF EXTENDED BODIES

47. [I] Two uniform blocks of wood 4 cm × 4 cm × 12 cm, each weighing 1.0 N, are glued together as shown in Fig. P47. Find the center-of-gravity of the system.

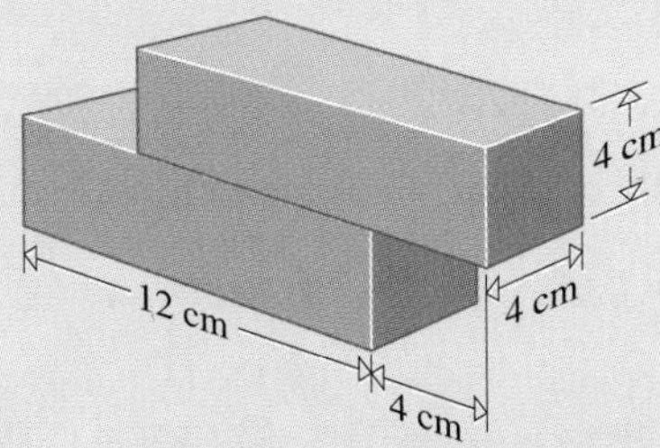

Figure P47

48. [I] A uniform piece of soft, thin copper wire 60 cm long and weighing 1.2 N is bent into three equal length segments to form two right angles, so that it has the shape of an angular letter "C." Locate the folded wire's center-of-gravity.

49. [I] A 0.50-m-long thin steel rod with a mass per unit length of 0.50 kg/m is bent in half (making a right angle) into the shape of a letter "L." Locate its *c.g.*

50. [I] The glider depicted in Fig. P50 is descending at a constant speed. If the drag (F_D) is 600 N and the plane weighs 6000 N, determine both the angle of descent, θ, and the lift, F_L.

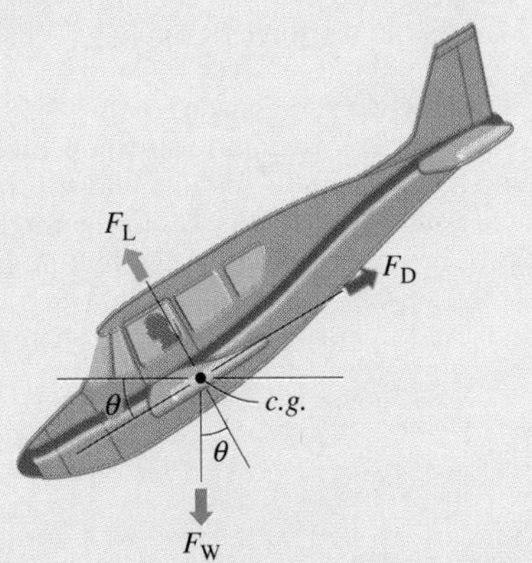

Figure P50

51. [I] Scale 1 in Fig. P51 reads 100 N. What does scale 2 read? How much does the suspended body weigh? Discuss the location of the vertical line passing through the *c.g.*

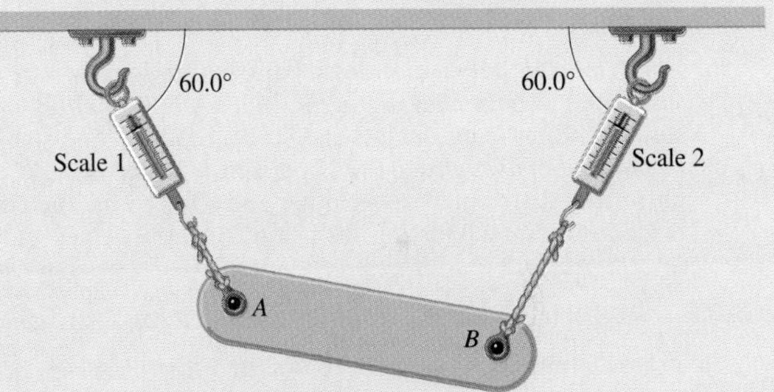

Figure P51

52. [I] Refer to the mobile in Fig. P52. If the star weighs 10.0 N, how much does the sphere weigh?

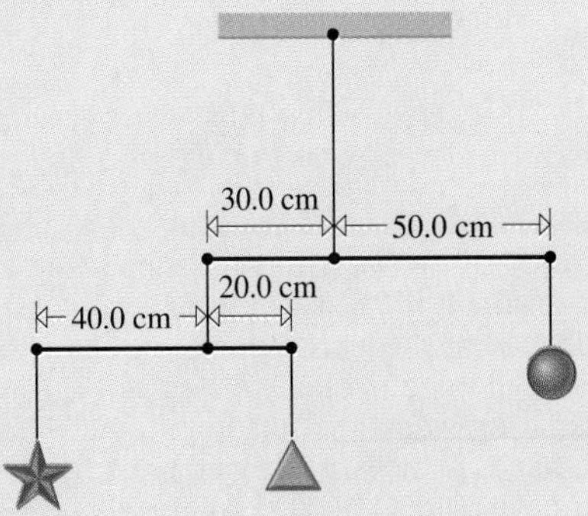

Figure P52

53. [I] Now suppose we displace the sphere in Problem 52 downward, whereupon the top rod tilts down at 30.0° with the horizontal. Will the mobile be in equilibrium?

54. [II] A pole vaulter carries a 6.0-m uniform pole horizontally. He holds one end with his right hand pushing downward, and 1.0 m away pushes upward with his left hand. If the pole weighs 40 N, compute both forces he exerts. What is the net force he applies? What would these forces be if the pole were vertical? What does this answer suggest about the carrying angle?

55. [II] We wish to stack five uniform wooden blocks so that they extend as far right as possible and still remain stable. How should each be positioned? Can the top block have its entire length beyond the edge of the bottom block (Fig. P55)?

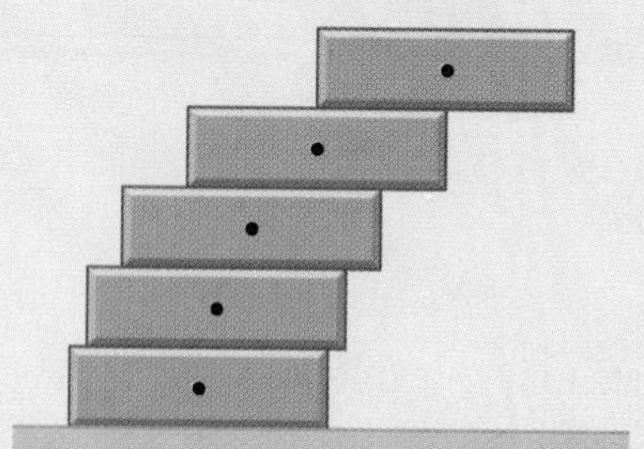

Figure P55

56. [II] An essentially weightless beam 10.0 m long is supported at both ends, as in Fig. P56. A 300-N child stands 2.00 m from the left end, and a 6.00-m-long stack of newspapers weighing 100 N per linear meter is uniformly distributed at the other end. Determine the two reaction forces supporting the beam.

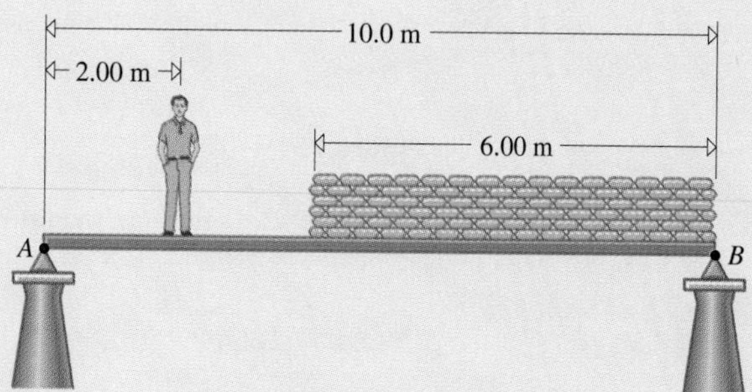

Figure P56

57. [II] Figure P57 shows two ropes supporting a body. Prove that the resultant of the two tensile forces, as indicated by the scale readings, equals the weight. Show that the plumb line passes through the body's *c.g.*

58. [II] A 55.0-kg woman is standing "rigid"—straight upright, hands at sides, feet together. At their widest, her two feet have a breadth of 20.0 cm and her *c.g.* is 95.0 cm above the ground. What minimum horizontal force in the plane of her body (that is, perpendicular to the forward direction) will tilt her over if it acts along a shoulder-to-shoulder line 145 cm high?

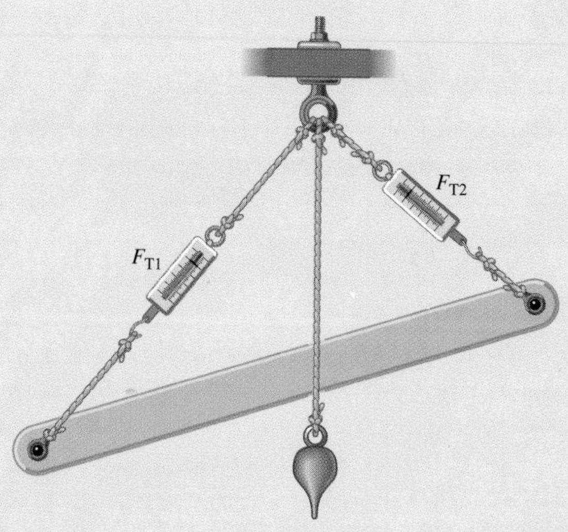

Figure P57

59. [II] An 8.0-m-long, 30-kg uniform plank leans against a smooth wall, making an angle of 60.0° with the ground. Compute the reaction force, $\mathbf{F}_{RG}$, of the ground on the plank.

60. [II] In Problem 32, Harry and Gretchen were sitting on a weightless seesaw. Suppose now that the beam is uniform and weighs 200 N—find the new pivot point.

61. [II] A 65-kg woman is horizontal in a pushup position in Fig. P61. What are the forces acting on her hands and feet?

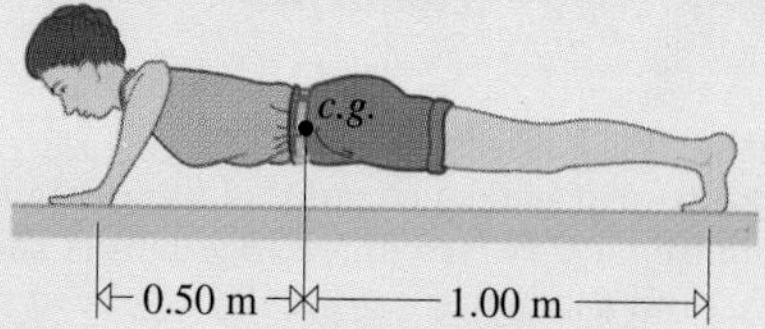

Figure P61

62. [II] A large box of breakfast cereal is 6.0 cm × 35 cm × 46 cm high and has a net mass of 1.0 kg. The contents are a uniformly distributed material. The box is next to an open window. Determine the speed of a uniform wind hitting the large surface if it just starts tilting the box over. The force exerted by a perpendicular wind in newtons on each square meter is given roughly by $0.6v^2$, wherein v is expressed in m/s. Assume the distributed force acts at the geometrical center of the face.

63. [III] One technique for measuring the *c.g.* of a person is illustrated in Fig. P63. The board is positioned according to the person's height (h), and the scales are then reset to zero. The participant lies down, and the scale readings are taken. Determine an expression for x_{cg} in terms of the measured quantities.

64. [III] A uniform ladder leans against a smooth wall, making an angle of θ with respect to the ground. The

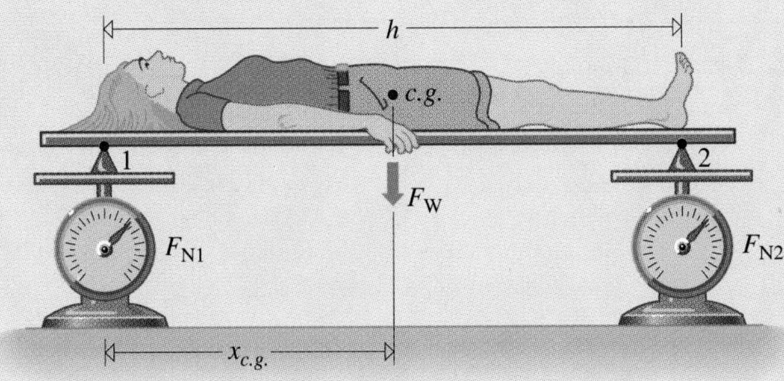

Figure P63

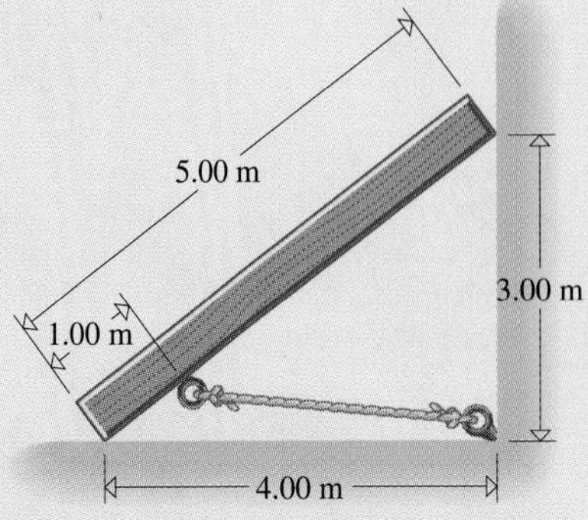

Figure P65

ground's reaction force on the ladder is at an angle of ϕ with the horizontal. Determine the relationship between θ and ϕ. Which is larger?

65. [III] The uniform plank in Fig. P65 is 5.00 m long and weighs 100 N. The cord that attaches to the plank 1.00 m from the bottom end holds it from sliding, there being no friction. Find the tension in the cord.

7

Gravity, According to Newton

AFTER 2000 YEARS OF thinking about gravity, we have a powerful, albeit still incomplete, understanding of the phenomenon. That understanding comes from several different theoretical perspectives. First, there is Newtonian Theory, which is reliable, practical, and simple. It was the formulation that guided us to a precise landing on the Moon (1969) and yet, on a grand scale, we know that it is inadequate. Then there is Einstein's General Theory of Relativity, which relates gravity to the fabric of space and time (p. 229). It subsumes Newton's Theory and goes far beyond it, allowing us to analyze events on a vast cosmic scale. But Einstein's formulation is

Nicolaus Copernicus (1473–1543), pictured here on a stamp as a young man in Torun, Poland.

highly mathematical and quite unnecessary for most earthly considerations. Today, many theoreticians believe that all interactions arise from the continuous exchange of force-carrying particles (Chapter 33) and are therefore quantum mechanical in nature. They argue that General Relativity, which is not quantized, must be incomplete. This chapter focuses mainly on Newtonian theory, exploring how it applies on Earth and beyond.

THE LAW OF UNIVERSAL GRAVITATION

Newton's *Law of Universal Gravitation* states that the gravitational force F_G between any two bodies of mass m and M, separated by a distance r, is described by

$$F_G \propto \frac{mM}{r^2} \tag{7.1}$$

Gravity varies directly in proportion to the product of the masses of the interacting objects and inversely with their separation squared. Moreover, *it is a center-to-center attraction between all forms of matter.* Yet Newton's achievement was not so much in formulating the law as it was in applying it with supreme mathematical skill. He established that the formula was correct mainly by using it to explain the motions of the Moon and planets. As we will see, the pieces to the gravity puzzle were already there before Newton's great work of 1685.

On gravity. *For my part I think that gravity is nothing but a certain natural striving with which parts have been endowed . . . so that by assembling in the form of a sphere they may join together in their unity and wholeness.*

NICOLAUS COPERNICUS

7.1 Evolution of the Law

The word *gravity* derives from the Latin *gravitas,* meaning weight or heaviness. And since a force equal and opposite to a body's weight must be applied to keep it from falling, the tendency for things to drop toward the center of the Earth was recognized early on as part of the mystery of gravity.

Copernicus (1543) challenged the Earth-centered Aristotelian world view, maintaining instead that the Sun was at the center of the Universe. With that, the Earth became a spinning planet rushing through space. The dominant gravitational theory at the time was Aristotle's—things fall because they seek their "natural place" at the center of the Universe, the center of the Earth. But the new vision of planet Earth orbiting the Sun made that age-old wisdom implausible and, within two centuries, the Newtonian formulation replaced it.

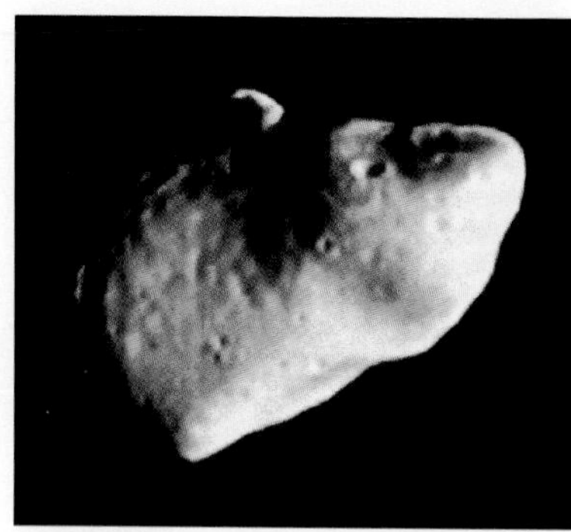
The asteroid Gaspra as photographed by the space probe *Galileo.* It's only 12 km by 20 km by 11 km, and is much too small to gravitationally crush itself into a sphere.

The "universal" aspect of Newton's law alludes to the fact that *all* matter interacts via gravity. That insight, which still surprises some people today, developed gradually. Copernicus suggested that gravity was the tendency of *similar* matter to come together. He explained that the Sun, Moon, and Earth were all spherical because each kind of substance tends to pull itself inward into a symmetrical shape. That idea is in part correct, but there is no suggestion here that Earth-stuff might attract Moon-stuff. A century later, Galileo was still committed to the same parochial gravity, in which only like substances attract.

In 1600, William Gilbert published *De Magnete,* a book of remarkable insight, although he wrongly attributed the primary functions of gravity to magnetism. (Hold two powerful magnets close together and you immediately know the wonder of an invisible force reaching across space.) For a while, magnetism and gravity were a single mystery of matter joining matter.

Johannes Kepler, influenced by Gilbert, made the proper distinction between gravity and magnetism. Kepler argued correctly that both the Sun and Moon attract the "waters of the sea," causing the rising of the tides. The Earth, the Moon, the Sun, the waters—all apparently interact gravitationally. Extending this concept, Roberval (1644) hinted at an attraction between all "worldly matter." In 1670, Robert Hooke stated that gravity was an interaction between "All Celestial Bodies." And, finally, Newton proposed that the gravitational interaction exists between all material objects: "We must . . . universally allow that all bodies whatsoever are endowed with a principle of mutual gravitation."

The Earth pulls you downward with a gravitational force called *weight,* just as it pulls downward on apples, rocks, and frogs giving them weight as well. You, in turn, gravitationally attract the Earth with an equal and opposite force; that's the interaction pair. Matter somehow draws on matter, and you attract (weakly, to be sure) all the apples, rocks, and even the frogs—everything attracts everything else.

The realization that the force of gravity, Eq. (7.1), is proportional to the interacting masses comes from a suggestion by Gilbert, even though it was little more than a lucky guess. He observed that the force between two magnets depended on their sizes and weights—which just happened to be the case for the natural magnets available at the time. Carrying the idea over to gravity, Kepler (1609) maintained that "two stones . . . placed anywhere in space" would gravitationally attract and "come together . . . at an intermediate point [the center-of-gravity], each approaching the other in proportion to the other's mass."

The force of gravity decreases as the separation increases, Eq. (7.1), and we might also wonder how that insight came to be. In fact, it was Kepler who initiated it, albeit in a confused context. Kepler wrongly envisioned the planets propelled in orbit by a solar force that pushed them along. Since the farther a planet is from the Sun, the slower it moves, he proposed that this sweeping force decreased with distance. After all, the magnetic force between two bodies decreases as their separation increases. The concept of a force extending from the Sun to the planets would soon be more correctly applied to gravity itself. But how was the specific $1/r^2$ dependence arrived at?

When anything diffuses outward uniformly from a *point* source (Fig. 7.1), it spreads out, becoming less concentrated as it advances. We can imagine all of the flow passing successively through surrounding concentric spheres centered on the source, any one of which has a radius r and an area $4\pi r^2$. The same quantity of emanation that initially passes densely through a tiny spherical surface will later pass tenuously through a much larger one. The concentration of the emanation (the amount per unit area) decreases inversely with the area of the sphere—it *decreases inversely with the distance squared* ($1/r^2$). Kepler certainly knew that this phenomenon was true for light; everyone knew you can only read by a candle flame if you stand near it. And he came close to embracing an inverse-square law for his fictitious solar force, once even writing that it "weakened through spreading from the Sun in the same manner as light." If gravity spreads out from a body uniformly in all directions, an inverse-square dependence would be reasonable to suppose.

In 1645, the astronomer Ismaël Bullialdus asserted that the Sun's attractive action on each planet *was along the center-to-center line joining the two and dropped off inversely with the distance squared.* Newton was aware of this work; in fact, he explicitly mentions the author by name. The idea that gravity acts along a line connecting the "centers" of the interacting bodies was not startling either. The Medieval scholar Albert of Saxony had defined weight as the tendency of the center-of-gravity of a body to approach the center-of-gravity of the Earth. Ever since Archimedes (Sect. 11.6), it had been common to think of weight acting at the *c.g.* Thus, the vision of a center-to-center gravitational interaction was quite natural.

Johannes Kepler (1571–1630).

The Moon causes the movement of the waters and the tides of the ocean.

The Sun (chief inciter of action in nature) . . . causes the planets to advance in their course.

The Moon alone of all planets directs its movements as a whole toward the Earth's centre, and is near of kin to Earth, and as it were held by ties to Earth.

The force which emanates from the Moon reaches to the Earth, and, in like manner, the magnetic virtue of the Earth pervades the region of the Moon.

WILLIAM GILBERT (1540–1603)

On gravity. *If the attractive force of the Moon reaches down to the Earth, it follows that the attractive force of the Earth, all the more, extends to the Moon and even farther.*

J. KEPLER

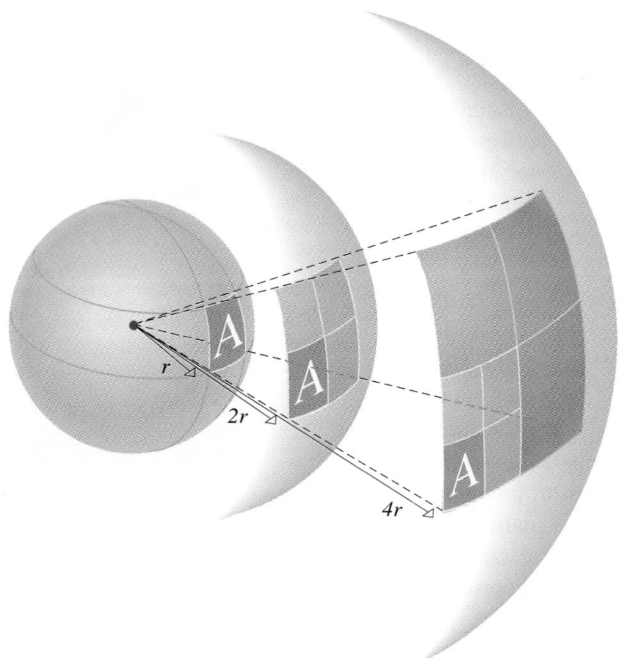

Figure 7.1 When the radius of a sphere r increases, the area of the sphere increases as r^2. Thus, if we double the radius, the little area A projects into an area of $4A$. Similarly, if the radius goes from r to $4r$, the area A projects out to an area of $16A$. The total area of the sphere, $4\pi r^2$, expands to $4\pi(4r)^2 = 16(4\pi r^2)$, or 16 times the original area.

7.2 Newton's Confirmation of $1/r^2$

Early in the 1660s, Hooke, already convinced of the variation of gravity with distance, attempted to arrive at its precise form experimentally. Being a practical man, he carried out a series of measurements to determine the minute differences in the weight of a body at various altitudes, within deep wells and high "on Paul's steeple and Westminster Abbey, but none that were fully satisfactory." Years later, Newton claimed to have made his own first test of the inverse-square dependence just around that same time, himself a young man of twenty-four. As he put it, I "compared the force requisite to keep the Moon in her orb with the force of gravity at the surface of the Earth; and found them answer pretty nearly."

Although its orbit is slightly elliptical we can envision the Moon circling the Earth at a center-to-center distance of 38.4×10^7 m under the influence of a centripetal force provided by gravity (Fig. 7.2). Its centripetal acceleration $a_C = v^2/r$, Eq. (5.10), can be easily determined numerically. The lunar orbital speed (v) is the distance traveled in one revolution ($2\pi r$) divided by the time it takes—the *period.* The orbit's circumference equals $2\pi(38.4 \times 10^7$ m), while the period is 27.32 days (or 2.360×10^6 s), which yields a speed of 1.02 km/s and an acceleration of

$$a_C = \frac{v^2}{r} = 0.0027 \text{ m/s}^2$$

Any object at that distance from Earth, whether orbiting or not, should experience this same downward acceleration via gravity.

Newton next compared this value with the acceleration at the planet's surface (6.371×10^6 m from its center). An object *on* the Earth experiences an acceleration due to gravity of 9.81 m/s^2. Assuming that the downward gravitational force trails off with distance from the planet as $1/r^2$, the resulting acceleration will do the same ($F = ma$). Let $g_\oplus$ be this distance-dependent acceleration associated with the Earth ($\oplus$ is the usual symbol for our planet). The Moon is 60.33 Earth-radii away, so it (and anything else at that distance) should experience a gravitational acceleration toward Earth that is about 60^2 times smaller than does an object at the planet's surface; thus,

$$g_\oplus = \frac{9.81 \text{ m/s}^2}{3600} = 0.0027 \text{ m/s}^2$$

Amazing! The Moon's centripetal acceleration equals the gravitational acceleration due to the Earth at that altitude: $a_C = g_\oplus$. The whole calculation, simple and profound, is wonderful.

Example 7.1 An astronaut out in space descends straight down toward the Earth in free-fall. Compute her acceleration at the moment she reaches an altitude of 1.911×10^4 km from the planet's surface. Neglect the presence of any other celestial object.

Solution: [Given: $h = 1.911 \times 10^4$ km, measured from the *surface.* Find: $g_\oplus$.] For the moment, we don't have an expression for $g_\oplus$ in terms of r (p. 216). We'll have to arrive at a value using the acceleration at the surface and the fact that $g_\oplus$ drops off as $1/r^2$. Remember that r is the

(continued)

(continued)
center-to-center distance. Letting $R_{\oplus}$ be the Earth's radius (6.371×10^3 km), $h = 3R_{\oplus}$, and the astronaut is at a distance $r = h + R_{\oplus} = 4R_{\oplus}$ from the planet's center. The gravitational acceleration at that altitude must be 4^2 or 16 times less than the surface value. Hence

$$g_{\oplus} = \frac{9.81\ \text{m/s}^2}{16} = \boxed{0.613\ \text{m/s}^2}$$

▶ **Quick Check:** $r = 25\,481$ km and $g_{\oplus}/(9.8\ \text{m/s}^2) = (6.4 \times 10^3\ \text{km})^2/(25 \times 10^3\ \text{km})^2$; hence, $g_{\oplus} = 0.6\ \text{m/s}^2$.

It would certainly seem that the force of gravity acting between two bodies (F_G) should be inversely proportional to their center-to-center (*c.g.*-to-*c.g.*) separation squared, or

$$F_G \propto \frac{1}{r^2}$$

7.3 The Product of the Masses

That all terrestrial objects at the surface of the planet accelerate downward at the same rate of $g = 9.81\ \text{m/s}^2$ is both remarkable and suggestive. From Newton's Second Law ($F = ma$), we know that a body of mass m propelled by a force equal to its own weight F_W accelerates at a rate g such that $F_W = mg$. Consequently, $F_W/m = g$, but this equivalence is true for *all* masses, which can only be the case if F_W is proportional to m so that the ratio can be independent of m. Inasmuch as the weight of an object is the force of gravity pulling it down toward the planet ($F_W = F_G$), it follows that $F_G \propto m$ and

By all their influences, you may as well Forbid the sea for to obey the moon . . .

W. SHAKESPEARE (1564–1616)
The Winter's Tale

$$F_G \propto \frac{m}{r^2}$$

Arguing from Newton's Third Law, we know that, if the Earth pulls down on an object, the object must tug *up* on the Earth with an equal force. Imagine two interacting masses m and M. Since the force of gravity acting on any body depends on its mass, and since there is an equal and opposite force F_G on each of them, $F_G \propto m$ and $F_G \propto M$, implying that the gravitational interaction on either one of two objects is proportional to both m and M, represented mathematically as

$$F_G \propto \frac{mM}{r^2}$$

The force being proportional to the product mM agrees with our everyday experiences. We know that if one apple of mass m in a bag on a scale at the market has a weight of 1 N, then two identical apples will have a weight of 2 N. Doubling the mass in the bag (m) doubles the weight ($F_W = mg$) despite the immense mass of the Earth ($M_{\oplus} = 5.975 \times 10^{24}$ kg). This would not occur, for instance, if F_G were in proportion to the sum of the masses ($m + M_{\oplus}$) rather than their product ($mM_{\oplus}$).

To be rigorous, Newton's Law—Eq. (7.1)—ought to be applied only to idealized point masses. We would then, by mathematical means, determine the gravitational force due to a body of any shape by adding up the forces from all its constituent particles. Developing the ability to perform that calculation and then proving that the law as it stands applies between spherical masses (such as planets) was one of the main problems Newton had to overcome.

Universal Gravitation on a Celestial Scale. *All Celestial Bodies whatsoever have an attraction or gravitating power towards their own Centres, whereby they attract not only their own parts . . . but that they do also attract all the other Celestial Bodies that are within the sphere of their activity.*

ROBERT HOOKE (1670)

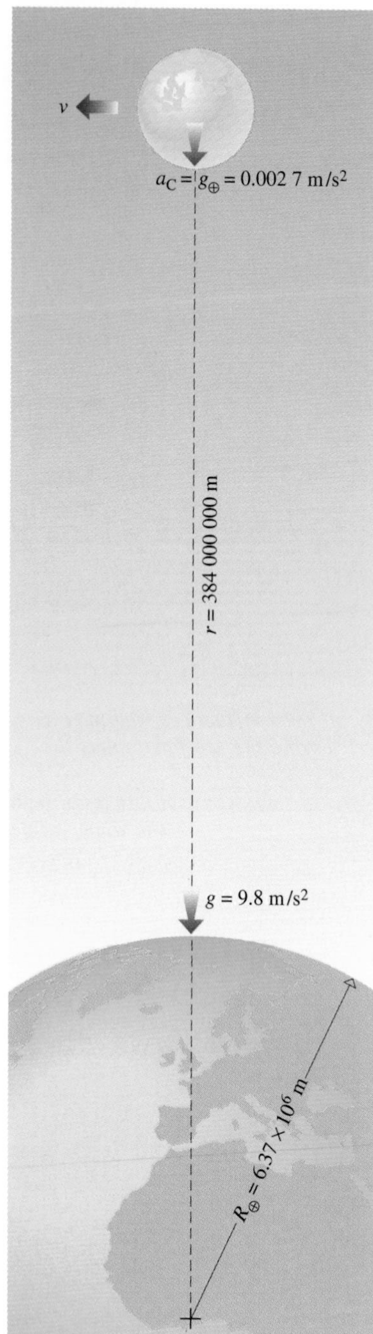

Figure 7.2 The acceleration due to gravity drops off inversely with distance squared. Since g at the Earth's surface is 9.8 m/s^2, it should be equal to 0.0027 m/s^2 at the distance of the Moon. And that's precisely equal to the centripetal acceleration we know the Moon must have to maintain its orbit.

7.4 Gravitational Constant

The relationship between F_G, m, M, and r^2 can be recast as an equality (something Newton never had the need to do) by introducing a constant of proportionality G (see Appendix A-4). This constant also serves to make the resulting equation dimensionally correct—the units on both sides of the proportionality are not the same. Newton's Law of Universal Gravitation takes its final form as

$$F_G = G\frac{mM}{r^2} \tag{7.2}$$

Here, G has the units of N·m^2/kg^2, and the force is properly in newtons. The numerical value of this *Universal Gravitational Constant* has to be determined experimentally.

That measurement was first done in 1798 by Lord Henry Cavendish, one of the richest men of his age and one of the most peculiar. Incredibly withdrawn, he rarely spoke to men and never, if he could help it, interacted with women. It's said that he would dismiss a female servant if she so much as entered a room he was in.

Eccentricities aside, determining G was a formidable technical accomplishment. To do it, Cavendish attached a small lead sphere about 2 in. in diameter to each end of a 6-ft-long, light, rigid rod, which was suspended horizontally from its middle on a thin vertical wire (Fig. 7.3). He then brought two lead balls, roughly 8 in. in diameter, into stationary positions, one in front of and one behind the small spheres. The resulting gravitational attraction subsequently caused the suspended rod-assembly to turn. The smaller balls swung very slowly toward the larger ones until the torque produced by the twisting wire brought them to rest. Having previously determined the force that would twist the wire through any angle, Cavendish promptly arrived at a value for F_G. Measuring m, M, r, and F_G, the constant G could then be computed. The presently accepted value is

$$G = 6.672\,59 \times 10^{-11}\ \text{N·m}^2/\text{kg}^2$$

(Cavendish obtained the equivalent of 6.75×10^{-11} N·m^2/kg^2). It's interesting that of all the fundamental constants we will encounter, G is the least accurately known.

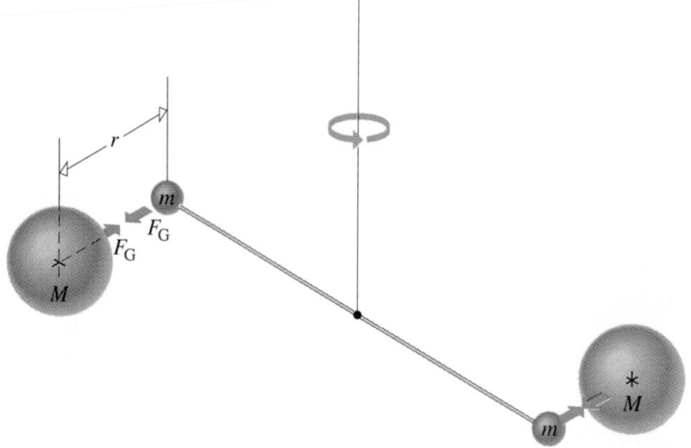

Figure 7.3 The Cavendish torsion pendulum for determining G. Drawn by their gravitational attraction, the two small balls move toward the large stationary balls, thereby twisting the wire.

Example 7.2 Compute the gravitational attraction between two 100-kg uniform spheres separated center-to-center by 1.00 m. (That is the rough equivalent of two 220-lb football players at arm's length.)

Solution: [Given: $m = 100$ kg, $M = 100$ kg, and $r = 1.00$ m. Find: F_G.]

$$F_G = G\frac{mM}{r^2}$$

$$F_G = (6.67 \times 10^{-11}\ \text{N}\cdot\text{m}^2/\text{kg}^2)\frac{(100\ \text{kg})(100\ \text{kg})}{(1.00\ \text{m})^2}$$

and

$$\boxed{F_G = 6.67 \times 10^{-7}\ \text{N}}$$

which equals 0.1×10^{-6} lb, or just about the weight of a baby flea. No wonder we're not noticeably drawn to each other via gravity.

▶ **Quick Check:** The force between two 1-kg masses separated by 1 m is 6.7×10^{-11} N, and the force in our problem is 100×100 times greater.

Example 7.3 The mass of the Moon is 7.35×10^{22} kg and its distance from Earth is 384×10^3 km. Taking the planet's mass to be 5.98×10^{24} kg, compute what Newton called "the force requisite to keep the Moon in her orb."

Solution: [Given: $m_☾ = 7.35 \times 10^{22}$ kg, $M_\oplus = 5.98 \times 10^{24}$ kg, and $r_{\oplus☾} = 384 \times 10^3$ km. Find: F_G.]

$$F_G = G\frac{m_☾ M_\oplus}{r_{\oplus☾}^2}$$

$$F_G = (6.67 \times 10^{-11}\ \text{N}\cdot\text{m}^2/\text{kg}^2) \times \frac{(7.35 \times 10^{22}\ \text{kg})(5.98 \times 10^{24}\ \text{kg})}{(384 \times 10^6\ \text{m})^2}$$

and

$$\boxed{F_G = 1.99 \times 10^{20}\ \text{N}}$$

Incidentally, this value is approximately 4.5×10^{19} lb, or the equivalent of the thrust of 6 million million *Saturn 5* rockets.

▶ **Quick Check:** From p. 210, $g_\oplus = 0.0027\ \text{m/s}^2$; hence, $F_G = ma = (7.4 \times 10^{22}\ \text{kg})(0.0027\ \text{m/s}^2) = 2 \times 10^{20}$ N.

The quantity G is a measure of the strength of the gravitational force. Since it's an extremely small number, F_G will be appreciable only when at least one of the interacting masses is very large. Still, because it has indefinite range (dropping to zero when $r = \infty$), and because it is purely attractive (not being weakened by an opposing repulsive aspect), gravity rules the Universe on a grand scale. Consequently, gravity is central to all cosmological theories and it's from that background that have come, in recent times, some interesting suggestions that perhaps G is not a constant after all.

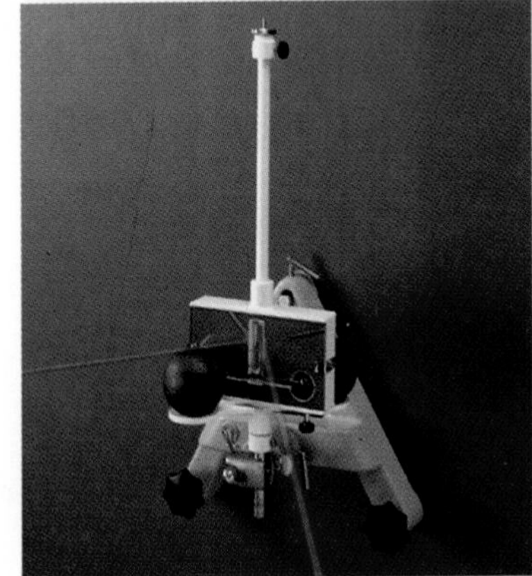

A modern version of the Cavendish experiment. The two large lead balls are easily visible, one in front, one in back. One of the small balls can be seen just to the right of the red laser beam.

7.5 Terrestrial Gravity

Cavendish had an ulterior purpose when he performed his famous experiment—he wanted to determine the density of the Earth and ended up being the first person to effectively "weigh" the planet, as he put it. An apple of mass m, hanging on a scale at the Earth's surface a distance $R_\oplus = 6.37 \times 10^6$ m from its center, weighs $F_W = mg$. The scale and the apple are motionless on the Earth and therefore both are revolving with the planet. If, for a moment, we remove the small effect of that

TABLE 7.1 Field events in the Solar System: long jump, high jump, and shot put on the Moon and nine planets*

Planet	Gravitational acceleration (m/s^2)	Long jump[1] distance (m)	High jump[2] height (m)	Shot put[3] distance (m)
Mercury	3.7	24	4.7	56
Venus	8.9	9.9	2.5	25
Earth	9.8	8.9	2.4	22
Moon	1.6	54	9.4	126
Mars	3.7	23	4.6	56
Jupiter	26	3.3	1.5	9.4
Saturn	12	7.5	2.2	19
Uranus	11	7.6	2.2	19
Neptune	12	7.3	2.1	19
Pluto	2	45	7.9	105

1. The athlete runs at 10 m/s and then jumps at 71° at 4.08 m/s.
2. The athlete jumps at 5.20 m/s.
3. The athlete throws the shot at 14.2 m/s at slightly less than 45°.

rotation, we must replace g by g_0, the nonspinning, *absolute acceleration due to gravity,* whereupon $F_W = mg_0$. This equation represents the "true" weight of an object due only to its gravitational interaction with the Earth and is therefore identical to $F_W = GmM_\oplus/R_\oplus^2$. Setting the two weight expressions equal and canceling m, we get *the absolute gravitational acceleration at the Earth's surface*

$$g_0 = \frac{GM_\oplus}{R_\oplus^2} \tag{7.3}$$

In the vicinity of the planet, the acceleration due to gravity is the same for all bodies independent of their masses. By appropriately substituting for M and R, the acceleration due to gravity can be determined for any celestial body (Table 7.1).

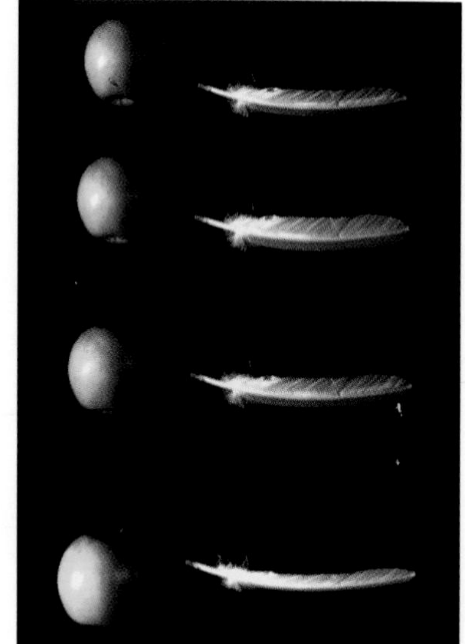

In vacuum, a feather and an apple fall at the same rate. The apple is more massive and must experience a greater force ($F_W = mg$) if it's to accelerate at a rate equal to that of the feather. But it has more mass and does weigh more ($F_W \propto mM_\oplus/R_\oplus^2$). Accordingly, the acceleration due to gravity is independent of the mass of the falling object.

7.6 Earth's Density

Knowing both the value of g_0 (which differs from g by less than 0.4%) and $R_\oplus$, and having just determined G experimentally, Eq. (7.3) can be solved for the mass of the Earth

$$M_\oplus = \frac{R_\oplus^2 g_0}{G} = 6.0 \times 10^{24}\ \text{kg}$$

and that's a splendid thing to be able to compute. Taking the planet to be a sphere of volume

$$V_\oplus = \tfrac{4}{3}\pi R_\oplus^3 = 1.1 \times 10^{21}\ \text{m}^3$$

Cavendish found that the **average density**, or *mass per unit volume,* was

$$\frac{M_\oplus}{V_\oplus} = 5.5 \times 10^3\ \text{kg/m}^3$$

which is 5.5 times greater than that of water (whose density is 1.0×10^3 kg/m^3).

We now believe that the Earth is a layered structure. The thin, cool surface *crust* composed of the lightest materials has an average density (Sect. 10.3) of only 2.5×10^3 kg/m^3. It surrounds a *mantle* of hot rock that in turn encloses a dense *core* of iron and nickel.

Newton never had access to a measured value of G, but he could have guessed at the Earth's density and then just run the above calculation backwards to find G, had he so wished. We know that he made just such an extraordinary guess in the *Principia,* where he wrote: "Since . . . the common matter of our Earth on the surface thereof is about twice as heavy as water, and a little lower, in mines, is found about three or four, or even five times more heavy, it is probable that the quantity of the whole matter of the Earth may be five or six times greater than if it consisted all of water . . .".

The Apple Myth. *One day in the year 1666 Newton had gone to the country, and seeing the fall of an apple, as his niece told me, let himself be led into a deep meditation on the cause which thus draws every object along a line whose extension would pass almost through the center of the Earth.*

VOLTAIRE (1738)

7.7 Gravity of a Sphere

Let's look now at the one sticky problem sidestepped so far. An apple near the surface of the planet is apparently drawn downward by a gravity force that arises much as if the entire 6.0×10^{24} kg of Earth were located at a point 6.4×10^6 m away at the planet's center. And why that should be is certainly not obvious. Newton solved the problem in a complicated geometrical way, although it's suspected that he had already analyzed it using his own invention, the integral calculus. Sir Isaac showed that *a particle located outside of a very thin uniform spherical shell will be drawn toward the sphere's center as if all the shell's mass were concentrated at that point* (Fig. 7.4). The shell is imagined to be constructed of numerous circular strips that in turn are divided into minute segments of mass. The Law of Universal Gravitation holds exactly for each of these tiny mass segments interacting with the outside particle. If we pair up segments on diametrically opposite sides of any one strip, it's easy to see that for each pair of equal forces acting on the particle, only the two components along the center line will not cancel one another. By symmetry, the total force on the outside particle due to all mass segments—that is, due to the whole spherical shell—is directed inward along the center-to-center line. The rest of the analysis, which is too involved to do here, leads to the conclusion that the spherical shell behaves gravitationally like a central point-mass, behavior that is a consequence of the inverse-square dependence. No other force law would combine with the symmetry of the sphere to produce so simple a result.

Figure 7.4 The gravitational attraction of a hollow spherical shell on a point-mass m. Note that because of the symmetry, F_G is directly inward toward the center of the sphere (the transverse components of the forces toward the two area elements cancel).

The same combination of influences, of geometry and force, leads to another surprise when the particle is placed *anywhere inside* the spherical shell. In that case, the particle is pulled equally in all directions by the shell, so that it experiences no resultant force whatsoever. That much can been seen from Fig. 7.5, using an analysis very like Newton's original treatment. Shown are two narrow cones bounded by straight lines passing through the particle. The mass of the region of the shell corresponding to the base of each cone is proportional to the area thereof. Thus, each produces an oppositely directed gravitation force on the particle such that $F_{G1} = A_1/r_1^2$ and $F_{G2} = A_2/r_2^2$. Hence

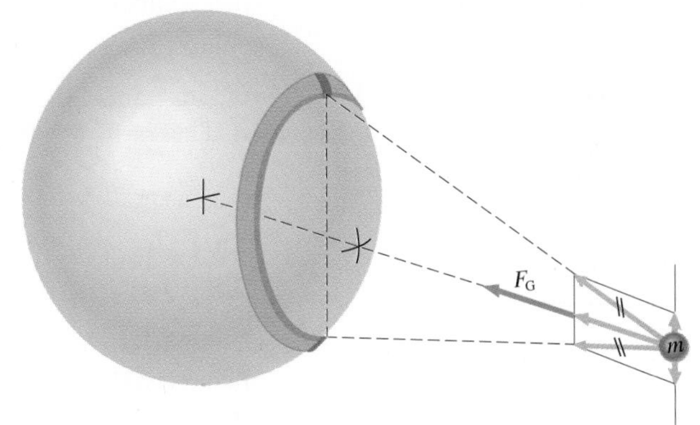

$$\frac{F_{G1}}{F_{G2}} = \frac{A_1 r_2^2}{A_2 r_1^2}$$

The cones have the same apex angles and are similar; their corresponding sides are in the same ratios. Consequently,

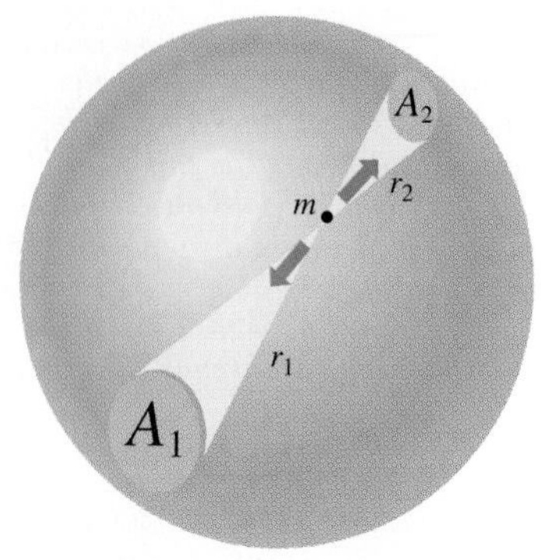

Figure 7.5 A point-mass, m, inside a spherical shell experiences a net gravitational force of zero. The little mass discs of area A_1 and A_2 produce equal and opposite forces on the point mass.

the base areas are in proportion to the squares of the sides

$$\frac{A_1}{A_2} = \frac{r_1^2}{r_2^2}$$

Substituting this expression into the preceding, it follows that $F_{G1}/F_{G2} = 1$: the forces are equal and exactly cancel. The small area nearby produces the same force as the large area far away—the areas go as r^2, and the forces as $1/r^2$. The whole surface can be divided into similar pairs of area elements with the overall effect that *no force acts on the internal particle.*

As a first approximation, we can imagine the Earth to be composed of a cluster of contiguous, uniform, spherical, thin shells (not necessarily with the same density) resembling a huge onion. On interacting with an external object such as an apple, each shell, no matter what its density, acts like a central point-mass. Accordingly, the whole cluster must behave as if the total mass $M_\oplus$ was at its center, pulling on the apple such that

$$g_\oplus = \frac{GM_\oplus}{r^2} \tag{7.4}$$

is the *absolute gravitational acceleration beyond the Earth.* Note that this expression becomes Eq. (7.3) at the planet's surface, where $r = R_\oplus$.

If the apple were to descend beneath the surface, into a deep vertical tunnel, we can imagine it traversing successive concentric shells. Once having passed through a shell, the latter's influence on the falling apple would vanish—the only thing pulling down on the apple at any moment is the spherical mass "beneath" it. When the apple is at a distance $r < R_\oplus$ from the planet's center, the volume V "beneath" it is $\frac{4}{3}\pi r^3$. The apple is outside of this spherical region, being drawn toward its center. If, for simplicity's sake, we take the density of the planet to be a constant ($\rho_\oplus$) then *the absolute gravitational acceleration beneath the Earth's surface is*

$$g_\oplus = \frac{G(V\rho_\oplus)}{r^2} = \tfrac{4}{3}\pi\rho_\oplus Gr \tag{7.5}$$

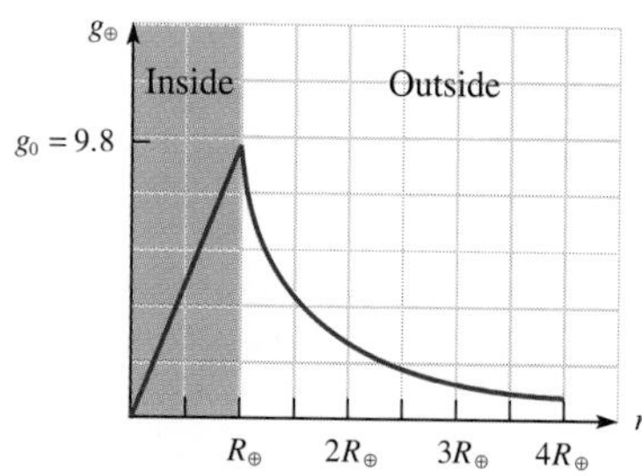

Figure 7.6 Absolute gravitational acceleration of the Earth, inside and outside, assuming the planet to have a uniform density.

The acceleration decreases linearly as the apple plunges downward, becoming zero at the Earth's center (Fig. 7.6) where it would truly be weightless. Notice that it's only at $r = 0$ and $r = \infty$ that the planet's gravity vanishes.

Example 7.4 By what percentage would your weight change if you ascended a height equivalent to that of the tallest building in the world, the Sears Tower in Chicago (1454 ft)? Take the Earth's diameter to be $1.274\,246 \times 10^7$ m.

Solution: [Given: $h = 4.432 \times 10^2$ m and $R_\oplus = 6.371\,23 \times 10^6$ m. Find: F_{Wh}/F_W.] The ratio of the weights, from Eq. (7.2), is

$$\frac{F_{Wh}}{F_W} = \frac{1/(R_\oplus + h)^2}{1/R_\oplus^2} = \frac{(6.371\,23 \times 10^6\ \text{m})^2}{(6.371\,67 \times 10^6\ \text{m})^2}$$

which to four significant figures is $\boxed{99.99\%}$. The reduction in the weight, and therefore the reduction in $g_\oplus$, is a mere 0.01%.

▶ **Quick Check:** The ratio equals $[R_\oplus/(R_\oplus + h)]^2 = [1/(1 + h/R_\oplus)]^2 = 0.999\,9$.

7.8 An Imperfect Spinning Earth

The Earth is not a sphere and it is not uniform. Excluding its hills and valleys, the planet is slightly flattened at the poles, with a little bulge at the north giving the whole thing a pear shape ($R_\oplus$ at the equator is about 21.5 km greater than at the poles). At close range, the deviation in the force from an ideal $1/r^2$ dependence is small (about 0.1%), and yet it markedly affects the orbits of artificial Earth satellites. Farther out, at the distance of the Moon, the interaction with the Earth deviates from the inverse square by only about 0.000 03%. In general, *any two mass configurations will interact in an approximately* $1/r^2$ *fashion when r is large in comparison to the sizes of the interacting objects.*

Guidance systems for missiles and aircraft require a detailed knowledge of the Earth's gravity. Here U.S. Air Force technicians very precisely measure variations in F_G as the altitude increases.

The planet's oblate shape, combined with the fact that it is rotating, produces a variation in the measured surface value of the gravitation acceleration (g) that is about five parts in a thousand (0.5%) as one moves from pole to equator (Table 3.2). To determine the effect of the Earth's rotation on g, consider an object of mass m supported by a scale situated at the *equator*. The two external forces acting on the object are its downward interaction with the planet (F_G) and the upward scale-force equal in magnitude to the object's measured weight (F_W). But since the object rotates as the planet turns, it must experience a centripetal acceleration. Hence, $(+\downarrow \Sigma F = ma)$

$$F_G - F_W = ma_C$$

$$\frac{GmM_\oplus}{R_\oplus^2} - mg = ma_C$$

and so

$$g = \frac{GM_\oplus}{R_\oplus^2} - a_C = g_0 - a_C$$

At the poles, $a_C = 0$, $g = g_0$, and the measured weight mg equals the *true weight* mg_0. As we saw in Example 5.9, $a_C = 0.0339\ \text{m/s}^2$ at the equator. The absolute value of the gravitational acceleration gradually varies from $g_0 = 9.814\ \text{m/s}^2$ at the equator to $9.832\ \text{m/s}^2$ at the poles, while $g = 9.780\ \text{m/s}^2$ and $9.832\ \text{m/s}^2$, respectively. The planet's spin produces almost two-thirds of the observed change in g with latitude (Fig. 7.7).

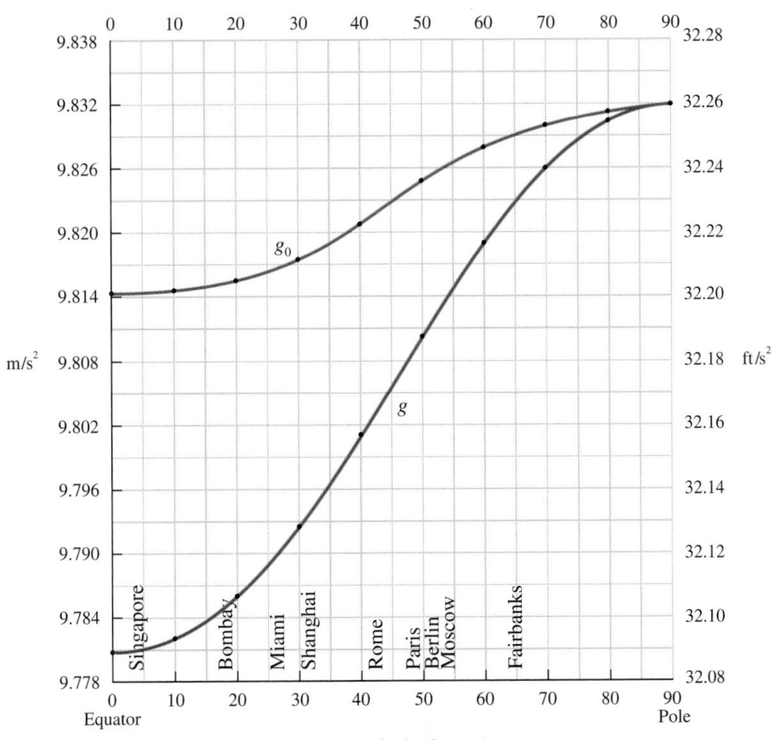

Figure 7.7 A plot of both the measured (g) and the absolute (g_0) values of the acceleration due to gravity on Earth. The former includes the rotational effects of the spinning planet, the latter does not.

Superimposed on the planetwide variation in gravity due to asphericity are innumerable small local alterations arising from geological details in the surface of the Earth. A modern instrument known as a gravimeter, usually consisting of a sensitive quartz spring balance, can measure minute changes in g of 1 part in 10^8 (or 0.000 001%), which is equivalent to a shift in elevation of only about 3 cm. Gravimeters are commonly used to survey the surface of the planet in order to explore geological structures and locate deposits of ore, gas, water, coal, and oil.

THE COSMIC FORCE

Brahe and Kepler met for the first time in Prague in 1600. Renowned as the leading astronomer of Europe, Tycho Brahe was 53. Self-conscious about his silver nose (the reminder of a college duel), he was aristocratic, suspicious, and overbearing. But he needed a theoretician to prove his own anti-Copernican model of the solar system. His opposite in every regard, Kepler, at twenty-nine, was sensitive, provincial, and an ardent Copernican. The young mystic, who came from a family of misfits and witches, was a little-known teacher of mathematics, a pauper, a caster of uncanny horoscopes, and a genius. Kepler went to work for Tycho to get his hands on the treasure of planetary data the latter had spent his life amassing. For 18 months they quarreled and reconciled and fought again, each trying to use the other to his own ends.

In 1601, Tycho, a dinner guest amongst illustrious companions, was drinking heavily and, being reluctant to leave the table, he (as Kepler put it) "held back his water beyond the demand of courtesy." Racked by pain, fever, and delirium, the greatest naked-eye astronomer of all time died, no doubt of uremia. Kepler quickly usurped the precious data and refused to hand it over to Tycho's rightful heirs who continued, with less than high motives, to harass him for years.

Finding patterns in the volumes of data, Kepler ultimately formulated his famous Three Laws of Planetary Motion. In time, Newton would come to rely on them as the touchstone of his theory of gravity.

7.9 The Laws of Planetary Motion

Unlike almost everyone before him* who insisted that the planetary paths must have the mystic perfection of the circle, *Kepler's First Law* maintains that *the planets move in elliptical orbits with the Sun at one focus* (Fig. 7.8). Granted, the paths are nearly circular—but the difference is crucial.

Kepler also noticed that each planet progressed at varying speeds over the course of its journey around the Sun. This would have been strange for a circular orbit; it wasn't for an elliptical one. Relying on the mathematics of his era—geometry—he succeeded in describing the motion in detail. *Kepler's Second Law* states that *as a planet orbits the Sun it moves in such a way that a line drawn from the Sun to the planet sweeps out equal areas in equal time intervals* (Fig. 7.9). For example, the Earth-Sun line will sweep out the same area in a week no matter where it is in orbit, whether it's a week in June or a week in January. Near the Sun the speed is great, the arc is long, and the slice of orbital pie is short and broad. Far from the Sun, the speed is low, the slice is narrow but tall, and still the area swept out is the same.

*Perhaps the one exception is the Muslim al-Zarqali (1081), "the blue-eyed one" who wrote about the "oval" orbit of Mercury.

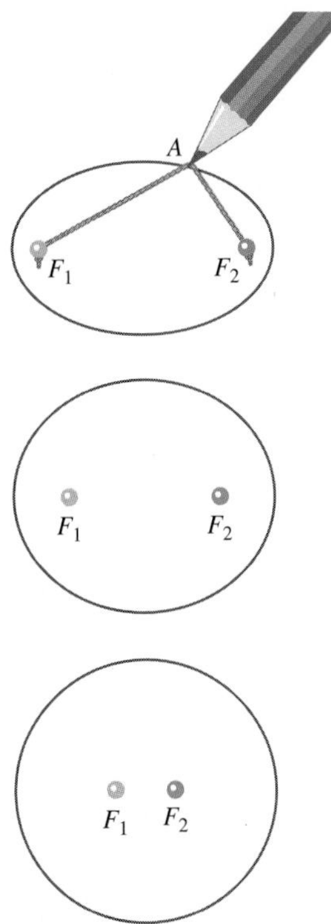

Figure 7.8 The ellipse. Point A moves along an ellipse when the sum of the distances $\overline{F_1 A}$ and $\overline{F_2 A}$ is constant. Imagine a slack string tacked down at its ends, F_1 and F_2. A pencil holding the string taut will sweep out an ellipse, since $(\overline{F_1 A} + \overline{F_2 A})$ always equals the constant length of the string. As the two foci, F_1 and F_2, are set closer together, the figure more and more resembles a circle.

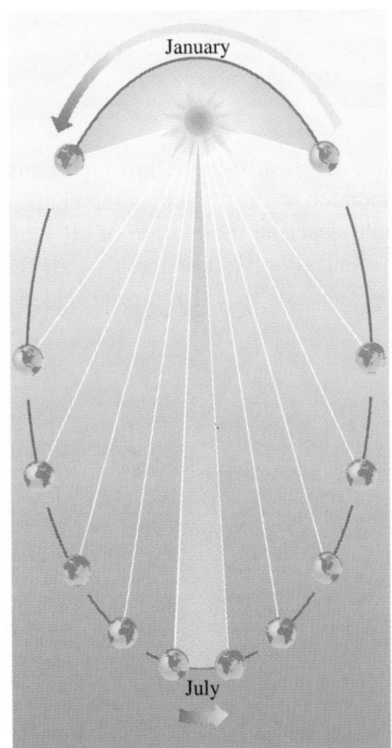

Figure 7.9 A line drawn from the Earth to the Sun sweeps out equal areas in equal times. Here each segment corresponds to a one-month interval and all have the same area. The Earth reaches its perihelion in January where it travels at maximum speed.

Armed with the physics of gravity, we can appreciate what's happening on another level. The planet, attracted to the Sun, swoops "down," accelerating as it rushes in and, as it does, the attraction increases even more. Reaching a maximum speed at its point of closest approach, the *perihelion,* it sails around and past. As it "climbs" against the Sun's gravity-pull, it decelerates. The force gradually weakens, and the planet continues to slow down until it reaches its minimum speed at the point farthest away, the *aphelion,* only to sweep around and begin the plunge once again.

As a rule, the more distant a body is from the Sun, the more slowly it moves in orbit—that much was already known (certainly Copernicus knew it). But unlike most of his predecessors, Kepler looked for precise law in such patterns. In 1618, after nearly 20 years of work he found the relationship between a planet's orbital period (T), the time it takes to travel once around the Sun, and its average distance from the Sun ($r_\odot$). *Kepler's Third Law* states that *the ratio of the average distance from the Sun cubed to the period squared is the same constant value for all planets.*

More succinctly

$$\frac{r_\odot^3}{T^2} = C_\odot \tag{7.6}$$

where $C_\odot$ is a to-be-determined constant that depends on the specific units, and is associated with the Sun as the central body. This equation with a different constant would describe the motion of all satellites about the Earth, and the same is true for any central body. The fact that all the planets orbiting the Sun share the same constant suggests that they are all ruled by the same underlying mechanism: gravity.

Example 7.5 Given that the mean solar distance for the Earth is 149.6×10^6 km, compute the corresponding value for Venus. The periods (in Earth days) of the Earth (⊕) and Venus (♀) are 365.256 days and 224.701 days, respectively.

Solution: [Given: $r_{\odot\oplus} = 149.6 \times 10^6$ km, $T_\oplus = 365.256$ d, $T_♀ = 224.701$ d. Find: $r_{\odot♀}$.] From Eq. (7.6)

$$\frac{r_{\odot\oplus}^3}{T_\oplus^2} = \frac{r_{\odot♀}^3}{T_♀^2} = C_\odot$$

Hence

$$r_{\odot♀} = \left(\frac{T_♀^2 r_{\odot\oplus}^3}{T_\oplus^2}\right)^{\frac{1}{3}} = (149.6 \times 10^6\ \text{km})\left(\frac{224.701\ \text{d}}{365.256\ \text{d}}\right)^{\frac{2}{3}}$$

and $\boxed{r_{\odot♀} = 108.2 \times 10^6\ \text{km}}$. Notice that the division canceled the units of time, so no conversions were needed.

▶ **Quick Check:** We use the data for the Earth to determine $C_\odot = 2.5096 \times 10^{19}\ \text{km}^3/\text{d}^2$ and from that value and $T_♀$ find $r_{\odot♀} = \sqrt[3]{C_\odot T_♀^2} = 108.2 \times 10^6$ km, which at least shows that we did the math correctly.

Example 7.6 Knowing the period of the Moon ($27\frac{1}{3}$ days) and its distance from Earth, compute the Keplerian constant for our planet ($C_\oplus$) in units of m^3/d^2. This same value is applicable to all artificial Earth satellites (see Problem 51).

Solution: [Given: $r_{\oplus☾} = 3.84 \times 10^8$ m, and $T_☾ = 27\frac{1}{3}$ d. Find: $C_\oplus$.]

$$C_\oplus = \frac{r_{\oplus☾}^3}{T_☾^2} = \frac{5.66 \times 10^{25}\ \text{m}^3}{746.9\ \text{d}^2}$$

and $\boxed{C_\oplus = 7.58 \times 10^{22}\ \text{m}^3/\text{d}^2}$

Various numerical values of this constant result when different sets of units are used.

▶ **Quick Check:** $r_{\oplus☾} = \sqrt[3]{C_\oplus T_☾^2} = 0.384 \times 10^9$ m, so the math is okay. The best check would be to use Eq. (7.7), but we'll have to wait for that.

Kepler's solar constant $C_\odot$ has its simplest value, namely 1.00, when we measure distance in *astronomical units* (1 AU $= r_{\odot\oplus} \approx 1.5 \times 10^8$ km $\approx 93 \times 10^6$ mi) and time in *Earth-years*. Then, using Eq. (7.6), we can evaluate the constant for Earth, and that same value applies to all the planets orbiting the Sun; thus, $C_\odot = (1\ \text{AU})^3/(1\ \text{Earth-year})^2$. Table 7.2 gives the results for the planets known to Kepler.

7.10 Gravity and Kepler's Laws

Newton took the Sun as an immovable *center-of-force* toward which a planet is continuously impelled. He then showed that *an orbit will be elliptical if and only if the*

TABLE 7.2 Kepler's Third Law*

Planet	Average distance to Sun (astonomical units) $r_\odot$	Period (Earth-years) T	Period squared T^2	Distance cubed $r_\odot^3$
Mercury	0.39	0.24	0.058	0.059
Venus	0.72	0.62	0.38	0.37
Earth	1.00	1.00	1.00	1.00
Mars	1.53	1.88	3.53	3.58
Jupiter	5.21	11.9	142	141
Saturn	9.55	29.5	870	871

*Compare the last two columns.

centripetal force varies as the inverse square of the distance to the center-of-force, which is located at one of the two foci (Fig. 7.8). This statement is essentially Kepler's First Law. The needed **central force**, *a force always directed toward the same point,* is gravity and it must vary as $1/r^2$. Newton proved that a variety of orbits (closed and open) were possible under the influence of an inverse-square force. Thus, any object interacting gravitationally with some other stationary body, such as a star, moves through space along either a circular, elliptical, parabolic, or hyperbolic path.

As for Kepler's Second Law: Imagine an object moving at a constant speed in a straight line past some fixed point O, as in Fig. 7.10. In equal time intervals, Δt, a line from O to the object sweeps out areas A_1, A_2, A_3, and so on. Since each of these has the same base (Δt) and the same altitude, each area is equal. Suppose the object having reached point B in Fig. 7.11 receives a sudden momentary blow, a pulse of force directed toward O. That burst briefly accelerates the object, adding a component of velocity along the $\overline{OB}$ line to the original velocity that was along $\overline{BC}$. The result is that, in a time Δt, the object arrives at point D, having traversed the line $\overline{BD}$. The length $\overline{BE}$ is the distance the object would have traveled in Δt had it initially been at rest at B. Since $\overline{BE}$ is parallel to $\overline{CD}$, it follows that triangles BOD and BOC, which have the same base ($\overline{OB}$) and equal altitudes, are equal in area. The area A_2, which would have been swept out without the action of a centripetal force,

Halley's comet as it appeared in its most recent approach in 1985. The comet is held by gravity in an elongated elliptical orbit that causes it to swing near the Sun roughly every 75 years.

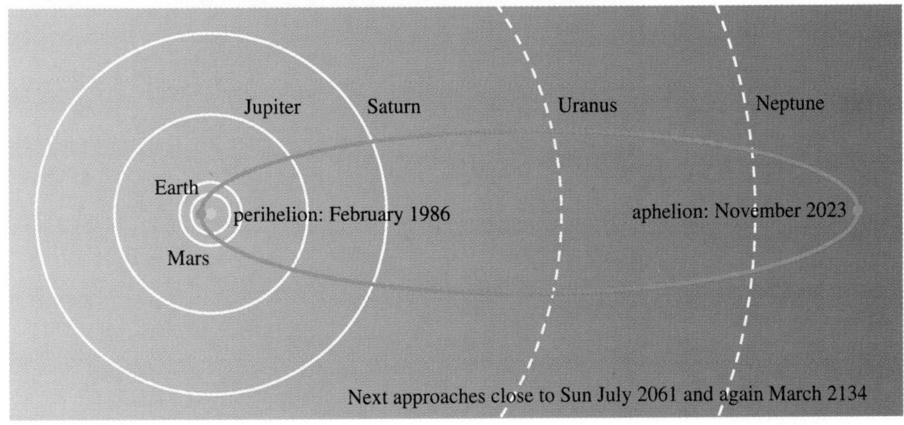

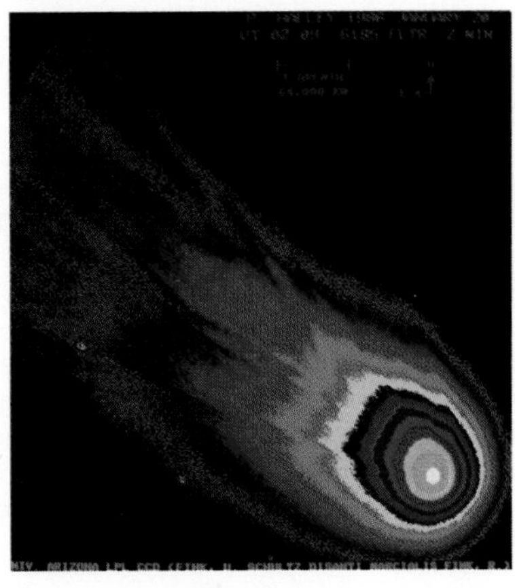

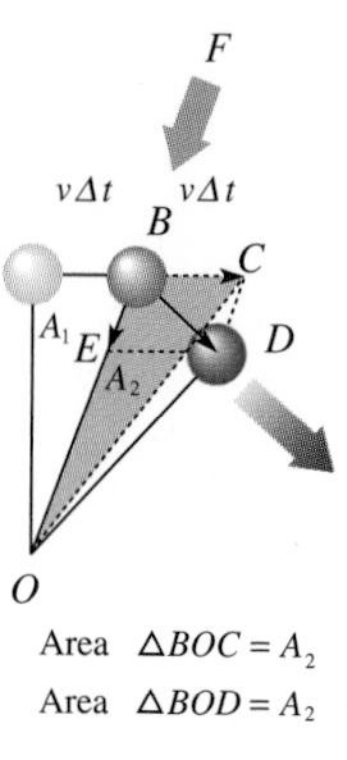

(a)

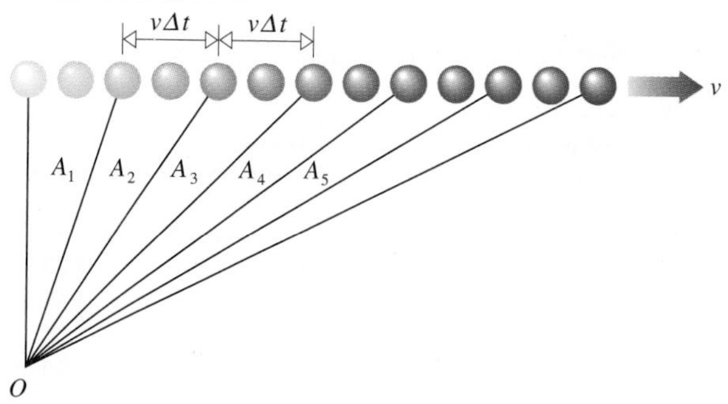

Figure 7.10 An object moving in a straight line past a point O sweeps out equal areas in equal intervals of time Δt.

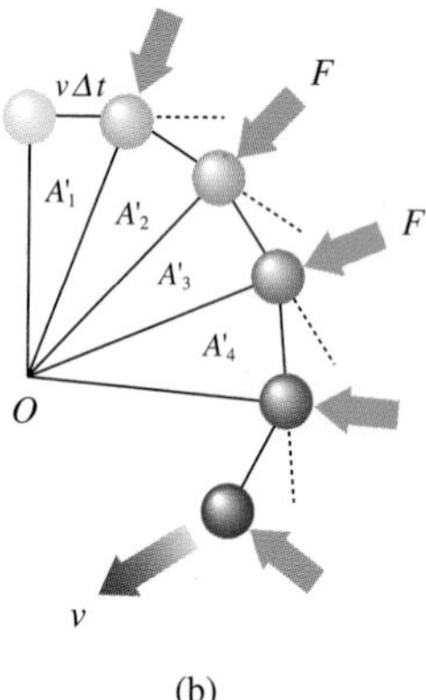

(b)

Figure 7.11 A planet moving in orbit under the influence of a central force sweeps out equal areas in equal times. (a) The altitude of a triangle BOD is obtained by dropping a perpendicular from D to $\overline{BO}$. Because $\overline{CD}$ is parallel to $\overline{BO}$, that altitude is equal to the altitude of triangle BOC, which is obtained by dropping a perpendicular from C to $\overline{BO}$. Thus, the area of triangle BOC equals the area of BOD, equals A_2. (b) A succession of pushes, all toward point O.

equals the area A_2', and so $A_1 = A_2$. Similiarly, successive blows directed toward O and occurring regularly at time intervals of Δt will cause the object to sweep out areas of A_3', A_4', and so on, each of which will be equal. Note that these momentary forces need not be the same size for the swept areas to be constant. All that's required is that they be directed toward O.

And now for the final touch. Let the interval Δt be made vanishingly small as the number of blows becomes infinitely many, until the flood of distinct impulses blends into a sustained continuous center-seeking force and the segmented path dissolves into a smooth curve. *Under the influence of a centripetal force, an object moves about a center-of-force sweeping out equal areas in equal times.* The enigmatic machinery underlying Kepler's Second Law is revealed: gravity is the central force.

The last of Kepler's relationships is easy to understand, particularly in the special case of a circular orbit. A planet of mass m moving in a circle at a distance $r_\odot$ from the Sun must be under the influence of a centripetal force given by

$$F_C = ma_C = \frac{mv^2}{r_\odot} \qquad [5.11]$$

And if gravity is to provide such a force ($F_G = F_C$), then

$$F_G = G\frac{mM_\odot}{r_\odot^2} = \frac{mv^2}{r_\odot} = F_C$$

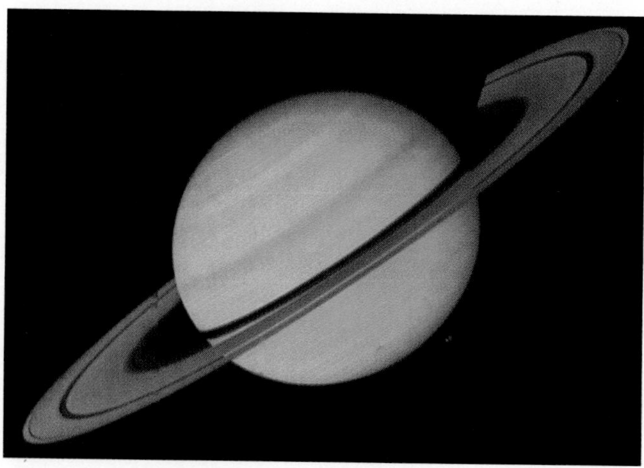

Saturn and its ring system. The rings are composed of billions of individual particles that typically range in size from pebbles to boulders. Gravity holds them in orbit and also, by its tidal action, keeps them from coalescing. Any moon orbiting closer than a certain distance will be ripped into pieces by its gravitation interaction with Saturn.

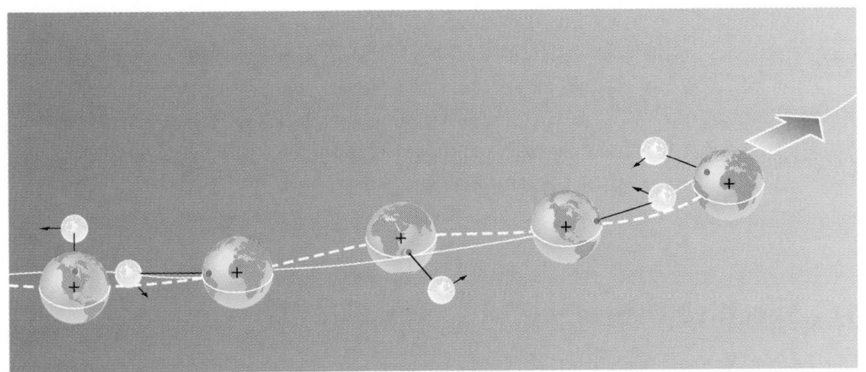

Figure 7.12 Just as a planet orbiting a star will displace the star, the Moon causes the Earth to snake along its orbit. It's the barycenter (the red dot) that moves in an elliptical orbit, as the Earth and Moon both weave in and out. The barycenter is 4672 km from the Earth's center (about 1000 miles beneath the planet's surface).

where $M_\odot$ is the mass of the central body. Since $2\pi r_\odot$ is the circumference, dividing this by the period $(2\pi r_\odot/T)$ yields the orbital speed. Accordingly, canceling m and substituting for v, we obtain

$$G\frac{M_\odot}{r_\odot^2} = \frac{(2\pi r_\odot/T)^2}{r_\odot}$$

and

$$\frac{r_\odot^3}{T^2} = \frac{GM_\odot}{4\pi^2} \tag{7.7}$$

which is Kepler's Third Law [Eq. (7.6)], where the constants on the right side combine to equal $C_\odot$. Notice that *only the mass of the central body remains in the expression;* the mass of the orbiting object cancels out (see Problem 44 for a slightly more accurate version). Exchange $M_\oplus$ for $M_\odot$, and replace $r_\odot$ with $r_\oplus$ (the distance to our planet's center), and the same equation applies to the circling Moon or to any artificial satellite in Earth-orbit.

Newton soon realized that Kepler's Laws could only be approximately true. For an object to orbit the exact *c.g.* of a larger body, the latter's mass must be infinite. In real life, when a planet orbits a star, both revolve around a common center-of-gravity, the **barycenter** (see Problem 43). That means that a distant star orbited by an otherwise invisible planet should move through space with a slight wiggle (Fig. 7.12). Jupiter (♃), the largest planet ($m_♃ = 318\ m_\oplus$) in the Solar System, displaces the Sun by almost a million miles. Half a dozen nearby stars show this telltale planetary wiggle.

When Newton saw an apple fall, he
found
In that slight startle from his
contemplation—
'Tis said (for I'll not answer above
ground
For any sage's creed or
calculation)—
A mode of proving that the earth
turn'd round
In a most natural whirl, called
"gravitation";
And this is the sole mortal who
could grapple,
Since Adam, with a fall, or with an
apple.

LORD BYRON (1788–1824)

Example 7.7 The Sun is one of an immense number of stars in a huge rotating pinwheel system known as the Milky Way galaxy. The Sun is located about 3×10^{20} m (or 3×10^4 light-years) from the center of the spherical hub about which it orbits once every 250 million years. Make a *very crude estimate* of the equivalent mass of galactic material acting on the Sun and compare that with its own mass of $M_\odot \approx 2 \times 10^{30}$ kg. Assume all the mass within the confines of the Sun's orbit can be taken to act as if it were at the center of the galaxy.

Solution: [Given: $r = 3 \times 10^{20}$ m and $T_\odot = 250 \times 10^6$ years $= 7.88 \times 10^{15}$ s. Find: M, the mass of the galaxy.] From Eq. (7.7), where now the central body corresponds to the galactic mass, we have

$$\frac{r^3}{T_\odot^2} = \frac{GM}{4\pi^2}$$

Solving this equation, we get

$$M = \frac{4\pi^2 r^3}{GT_\odot^2} = \frac{4\pi^2(3 \times 10^{20}\ \text{m})^3}{(6.7 \times 10^{-11}\ \text{N·m}^2/\text{kg}^2)(7.9 \times 10^{15}\ \text{s})^2}$$

(continued)

(continued)

and $M \approx 3 \times 10^{41}\ \text{kg} \approx \boxed{10^{11}\ M_\odot}$

This result approximately agrees with estimates that our galaxy contains about 10^{11} stars. Incidentally, there are likely some 10^{11} galaxies in the Universe.

▶ **Quick Check:** There are $\approx\pi \times 10^7$ seconds per year, so the period is right. As given above, the units of M are $\text{m}^3/(\text{N}\cdot\text{m}^2/\text{kg}^2)\text{s}^2 = \text{m}/(\text{N}\cdot\text{s}^2/\text{kg}^2)$; since m/s^2 is acceleration and N is force, $1\ \text{N}/(\text{m/s}^2) = 1$ kg; hence, M has the units of $\text{kg}^2/\text{kg} = \text{kg}$.

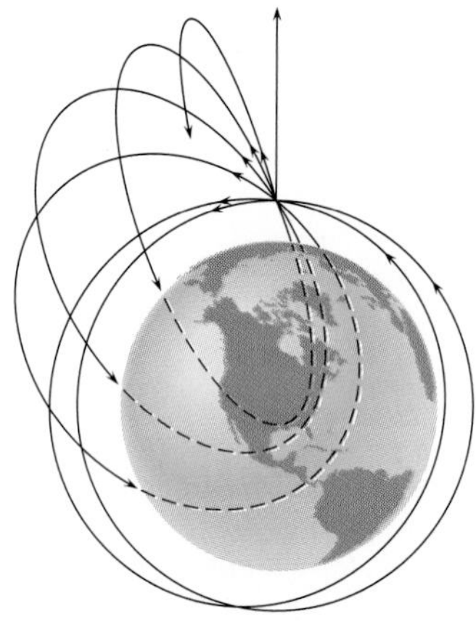

Figure 7.13 Various orbits for projectiles with the same speeds launched in different directions. In each case, the major or long axis is unchanged.

7.11 Satellite Orbits

Long before *Sputnik* and its thousands of progeny, Newton had observed that any projectile launched horizontally is, in a sense, an Earth *satellite* (a word coined by Kepler). A rock thrown from a tall building sails in a modest orbit that soon intersects the Earth not far from its point of launch (Fig. 7.13). Galileo maintained that this trajectory should be parabolic, but that assumed a flat world. On a spherical Earth, a horizontally hurled baseball (neglecting friction) theoretically arcs along an elliptical orbit, with its far focus at the planet's center. All the planet's mass acts as if it were at its *c.g.*, and the ball attempts to orbit that point. If the Earth were squashed down small enough, the baseball would become a moonlet. In real life, the Earth looms up in the way, and the motion is abruptly interrupted when the projectile crashes into the ground.

If the ball were fired more swiftly to start with, it would travel further; the ellipse would flatten out, becoming less elongated (Fig. 7.14). Further increasing the speed would result in ever larger, rounder elliptical paths and more distant impact points. Finally, at one particular launch speed, the ball would glide out just above

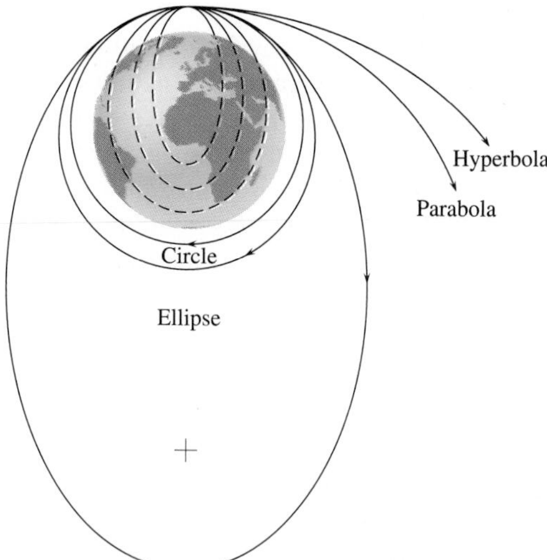

Figure 7.14 The orbits of projectiles fired horizontally close to the Earth's surface at various speeds (based on a drawing by Newton). The circular orbit occurs at a launch speed of about 8 km/s. From 8 to 11.2 km/s, the orbit is elliptical. At 11.2 km/s, it's parabolic and, beyond that, it's hyperbolic.

Y^e suns conversion doth turn the planet out of this line framing its motion into a circular, but the former desire of ye planet to move in a streight line hinders the full conquest of ye Sun, and forces it into an Ellipticke figure.

JEREMIAH HORROCKS (1618–1641)

the planet's surface all the way around to the other side without ever striking the ground. Like a little leather moon, it would wheel around the globe in a circular orbit until something (air friction, collision with birds or buildings) brought it down. At successively greater launch speeds, the ball would revolve in ever-increasing elliptical orbits until it moved so fast initially that it sailed off in an open parabolic or (if even faster) into a still flatter, hyperbolic orbit, never to come back to its starting point.

A low, circular flight path means that the satellite follows the planet's curvature. In effect, it continually drops toward the center-of-force, but since the ground "falls away" at the same rate, it never gets any closer to the surface. Figure 7.15 shows how the curvature of the Earth's surface drops a distance h beneath a tangent of length L such that the Pythagorean theorem yields

$$(r + h)^2 = r^2 + L^2$$

$$r^2 + 2rh + h^2 = r^2 + L^2$$

and

$$2rh + h^2 = L^2$$

Here we are looking at a very small segment of a large curve, and so h is tiny and h^2 by comparison with $2rh$ (since r is big) is quite negligible. Hence

$$2rh \approx L^2$$

and

$$L \approx \sqrt{2rh}$$

Now, suppose we are dealing with an orbit just above the planet's surface and the small segment corresponds in time to 1 s of motion. Then, L is the straight-line distance the object would travel if there were no gravity and h is the height it would fall due to gravity alone (see Fig. 4.2). From Eq. (3.9), $h = \frac{1}{2}g(1\text{ s})^2 = 4.9$ m and, with $r \approx R_\oplus$, we have

$$L \approx \sqrt{2R_\oplus h} \approx 7.9 \text{ km}$$

This result means that if an object were initially fired horizontally with a speed such that it traveled 7.9 km in 1 s (that is, 18×10^3 mi/h), it would subsequently orbit the Earth at the *lowest possible altitude,* effectively dropping 4.9 m each second.

An alternative way of seeing this, which is less graphic but more general, comes from equating F_G and F_C. An object of mass m at a distance r from the center of a large body of mass M will travel a circular path with a *tangential orbital speed* v_o when the gravitational force exactly equals the centripetal force, whereupon

$$G\frac{mM}{r^2} = \frac{mv_o^2}{r}$$

and so

$$v_o = \sqrt{\frac{GM}{r}} \tag{7.8}$$

Any satellite *regardless of its mass* (provided $M \gg m$) will move in a circular orbit of radius r, so long as its speed obeys Eq. (7.8), and it does so *independently of any rotation of the central body* (Fig. 7.16).

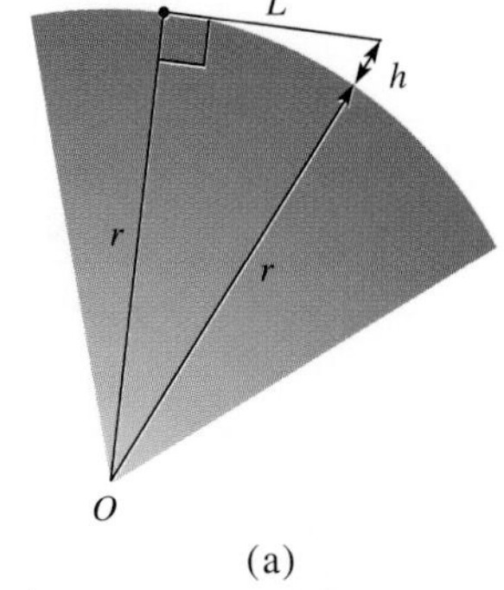

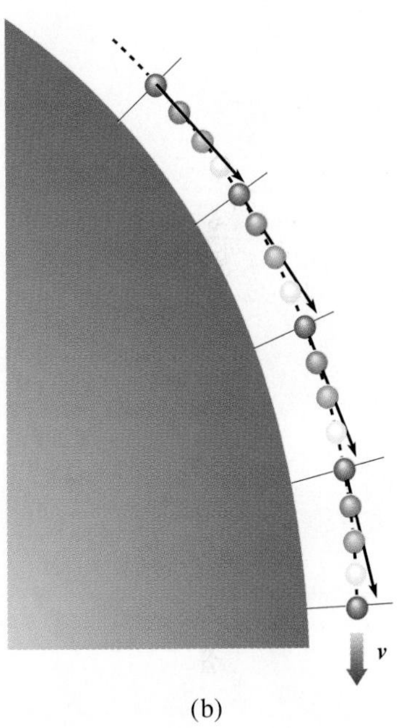

Figure 7.15 (a) The Earth drops away by an amount h below the tangent line of length L. (b) If a projectile is fired with a speed such that it travels a distance L in the time it falls a height h, it will remain in orbit.

Example 7.8 Use Eq. (7.8) to compute the necessary launch speed for an impractical treetop-height, circular Earth-orbit.

Solution: [Given: $r = R_\oplus = 6.4 \times 10^6$ m and $M = M_\oplus = 6.0 \times 10^{24}$ kg. Find: v_o.] If the object is to orbit just above ground level, it must have a horizontal speed of

(continued)

(continued)

$$v_o = \left(\frac{GM_\oplus}{R_\oplus}\right)^{\frac{1}{2}}$$

$$v_o = \left[\frac{(6.67 \times 10^{-11}\ \text{N}\cdot\text{m}^2/\text{kg}^2)(6.0 \times 10^{24}\ \text{kg})}{(6.4 \times 10^6\ \text{m})}\right]^{\frac{1}{2}}$$

and $\boxed{v_o = 7.9 \times 10^3\ \text{m/s}}$

▶ **Quick Check:** This speed is the same one we obtained from the analysis where we used Fig. 7.15 and the Earth's curvature.

Astronauts orbiting Earth are effectively weightless in free-fall.

The independence of v_o on m makes the point that an astronaut inside an orbiting spaceship is also orbiting at that same speed. And so, too, is all the loose paraphernalia in the vehicle—all of which is in orbit within the confines of the craft independent of it. The pilot will float around inside the ship just as she would float around outside of it, moving along at speed v_o in orbit. And since the only force acting is gravity, the crew will be *effectively weightless.* Indeed, orbiting astronauts are in *free-fall* toward the Earth at a rate $g_\oplus = a_C$. The concept of "falling" is generally thought of from a flat-Earth perspective, where the thing that falls gets closer to the surface in the process, which need not be the case on a spherical planet.

A launch vehicle initially rises vertically, gradually rolling over. If on separation the payload is moving tangentially at a speed v_o, satisfying Eq. (7.8), it will be injected into a circular orbit. If the speed is less than v_o, the craft will descend in an elliptical orbit (Fig. 7.14). If the speed is greater than v_o but less than $\sqrt{2}v_o$, it will ascend into a large elliptical orbit. At $\sqrt{2}v_o$, it will escape into an open parabolic path that carries it out and away forever (p. 301). At speeds in excess of $\sqrt{2}v_o$, the escape trajectory becomes hyperbolic and increasingly flattened (Fig. 7.17).

One of the most desirable orbital arrangements, especially for communications purposes, is the *geostationary* or *geosynchronous* orbit. While parked in one of these, a satellite, having been lofted directly above the equator, will circle once around in a day, revolving in synchronization with the rotating Earth. As seen from anywhere on the globe, the craft will remain in a fixed location in the sky and therefore in continuous line-of-sight communication with a ground station. Well over 100 military and civilian spacecraft are crowded into the geosynchronous band with more being added every year.

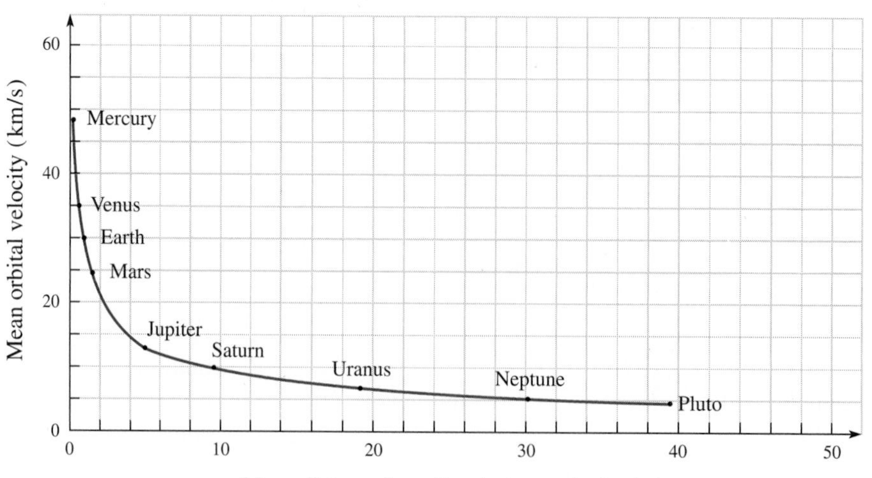

Figure 7.16 Because the Sun represents 99.9% of the mass of the Solar System, each planet essentially "sees" only the Sun as the central gravitating body. Consequently, for all planets $v_o \propto 1/\sqrt{r}$; the larger the orbit, the slower the orbital speed.

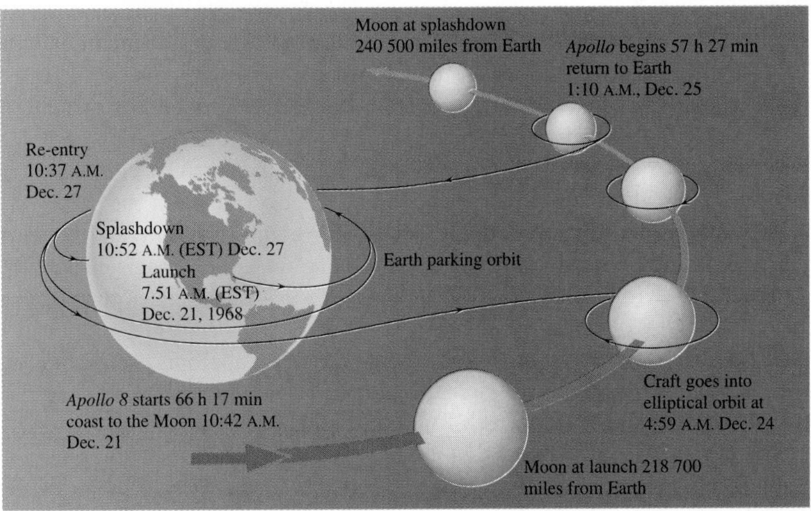

Figure 7.17 This drawing summarizes the *Apollo 8* journey to the Moon in 1968. The spacecraft first went into a parking orbit around the Earth. After the crew briefly fired its rockets and sped it up, the ship coasted off to the Moon.

Example 7.9 Compute the altitude above the planet's surface and the necessary injection speed for a geostationary orbit.

Solution: [Given: $M_\oplus = 5.98 \times 10^{24}$ kg and the most important fact (though not stated explicitly) $T = 24$ h. Find: the altitude h.] Remember that $h = r - R_\oplus$, the difference between the distance from the center of the Earth to the satellite and the Earth's radius. Setting $F_G = F_C$, canceling m, and using $v = 2\pi r/T$, we get, once again

$$\frac{GM_\oplus}{r^2} = \frac{(2\pi r/T)^2}{r}$$

and so

$$r = \left(\frac{M_\oplus T^2 G}{4\pi^2}\right)^{\frac{1}{3}}$$

hence $r = (7.544 \times 10^{22}\ \text{m}^3)^{\frac{1}{3}} = 42.25 \times 10^6$ m

We want $h = r - R_\oplus = \boxed{3.59 \times 10^7\ \text{m}} \approx 5.6R_\oplus \approx 22 \times 10^3$ mi. Do you suppose this is why the nation of Colombia, which straddles the equator, claims air space above it up to 23 000 miles? As for the speed—in one transit of the orbit, the satellite travels a distance of $2\pi r_\oplus$ in a time T. Hence

$$v_o = \frac{2\pi r}{T} = \frac{2\pi(42.25 \times 10^6\ \text{m})}{(24\ \text{h})(60\ \text{min/h})(60\ \text{s/min})}$$

and

$$\boxed{v_0 = 3.1 \times 10^3\ \text{m/s}}$$

or 6.9×10^3 mi/h. By the way, *Syncom II* was the first of the species successfully launched in July 1963.

▶ **Quick Check:** Using Eq. (7.8), $v_o = [(6.67 \times 10^{-11}\ \text{N}\cdot\text{m}^2/\text{kg}^2)(5.975 \times 10^{24}\ \text{kg})/(42.25 \times 10^6\ \text{m})]^{\frac{1}{2}} = 3.1 \times 10^3$ m/s.

7.12 The Gravitational Field

All things by immortal power,
Near or far,
Hiddenly
To each other linked are,
That thou canst not stir a flower
Without troubling of a star.

FRANCIS THOMPSON (1859–1907)

An apple loosed from a branch drops earthward under the influence of gravity. But how does the apple get the message? How does it know which way is down and how hard it's supposedly being pulled?

We are content with the idea of *direct contact*—two things pressed against one another, force transmitted by touching. But matter is composed of separate, spaced atoms and the idea of *direct* contact becomes meaningless. Atoms certainly interact, yet they never touch in the familiar way we think of macroscopic objects touching each other. Even here, in the very essence of contact, there is *action-at-a-distance,* the idea of objects exerting influences across an intervening void. When your finger "touches" this book, it approaches close enough for the electromagnetic repulsion between the electron clouds of finger atoms and book atoms to become palpable.

Each particle of matter has associated with it a surrounding field of influence, and it is that field that carries the interaction between separated bits of matter. Euler, in the mid-eighteenth century, devised a mathematical description of fluid flow that assigned a velocity to every point in a moving liquid, thereby creating a field of motion. Euler's concept was the origin of Field Theory. In much the same way, we could associate a thermometer reading with every point in a room and speak about the corresponding scalar temperature field. The idea became indispensable to physics in the late nineteenth century with the conceptualization of the electric and magnetic force fields (Chs. 17 and 21).

A **field of force** exists in a region of space when an appropriate object placed at any point therein experiences a force. Apparently, we can envision a gravitational force field surrounding any object of mass M. Suppose that we hang a small test mass m on a scale and take measurements moving from one point to another throughout the space surrounding the object in question. At each point, we could read off the force F_G (that is, the weight of the test mass), record its magnitude and direction, and thereby map the vector force field of the object.

One disadvantage to this scheme is that all the field values depend on the size of the test mass. It would be far more useful to conceive a new field quantity varying only with the mass of the object that was creating it, and not with the probe. Thus, we define the **gravitational field strength** $\mathfrak{g}$ as

$$\mathfrak{g} = \frac{\mathbf{F}_G}{m} \tag{7.9}$$

This is the force on the test mass, divided by the test mass, and it's called the *gravitational field intensity,* or just the *gravitational field* of the mass distribution

Figure 7.18 This summary of theoretical considerations gives us a picture of where we have been and where we are going. Thus far, we have considered the relationship between symmetry and conservation as put forth in Noether's Principle, and we have learned about Conservation of Linear Momentum. Now, we've encountered the concept of the force field in the form of gravity.

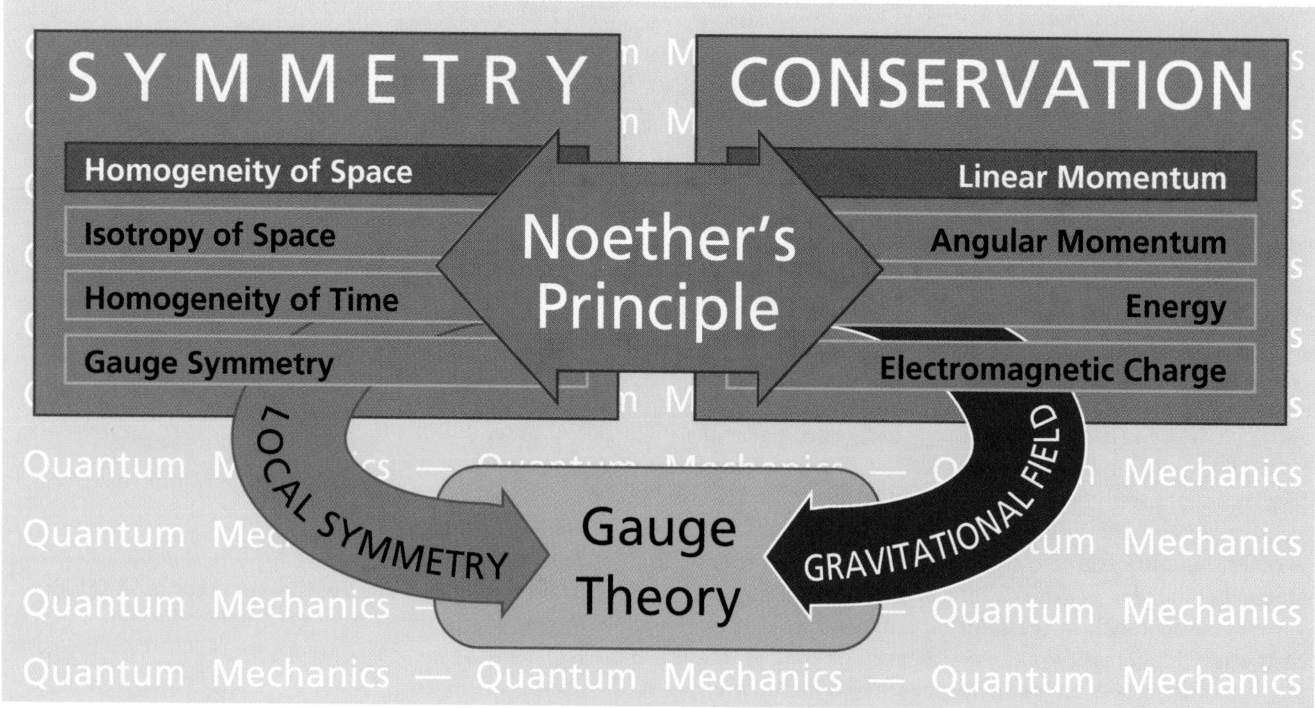

M. The gravitational field at any point is equal to the force exerted on a unit test mass located at that point (Fig. 7.18). Inasmuch as the "true" weight of an object is proportional to the absolute gravitational acceleration ($F_W = mg_0$), we can equate the field with the free-fall acceleration of the test mass.

From the definition of field strength, the appropriate units are N/kg, or, equivalently, m/s^2. In the workaday world of measuring gravitational fields, the principal unit used is the *gal,* named after Galileo. Defined as 1 cm/s^2 or 10^{-2} m/s^2, 1 gal is about one-thousandth of the Earth's surface field strength (9.8 m/s^2). Determinations of $\mathfrak{g}$ can be made accurate to 1×10^{-6} gal by measuring the free-fall time of a body over a distance fixed via laser interferometery.

Example 7.10 Derive an expression for the magnitude of the gravitational field strength, $\mathfrak{g}$, at any distance r from a point mass M.

Solution: [Given: M, the mass of a particle. Find: $\mathfrak{g}$ at any r.] We start with the definition of $\mathfrak{g}$ [Eq. (7.9)] in terms of the force F_G on a test mass m. But we already know that for a point mass M interacting with a point mass m

$$F_G = G\frac{mM}{r^2} \qquad [7.2]$$

Taking m as our test mass and substituting it into Eq. (7.9), the field strength anywhere in the surrounding space beyond M becomes

$$\boxed{\mathfrak{g} = \frac{GM}{r^2}} \qquad (7.10)$$

▶ **Quick Check:** According to Eq. (7.10), $\mathfrak{g}$ has the units of GM/r^2 or (N·m^2/kg^2)(kg)/m^2 = N/kg, which is what it should be.

To get a picture of the field of an object, we could draw a vector representing $\mathfrak{g}$ at a number of points in the surrounding region, as in Fig. 7.19. *All such vectors are directed toward the object,* since each one points in the direction of the force on a test mass and that force is always attractive. Alternatively, it's far more convenient to construct *lines-of-force,* called *field lines.* At each point in space, the direction of the field is tangent to the line-of-force (Fig. 7.20). Visualized as a three-dimensional distribution, the number of lines per unit area, the density of lines, is proportional to the field strength.

Figure 7.19 The gravitational field of mass M. At every point in space surrounding M we draw a vector corresponding to the force on a test mass m divided by m.

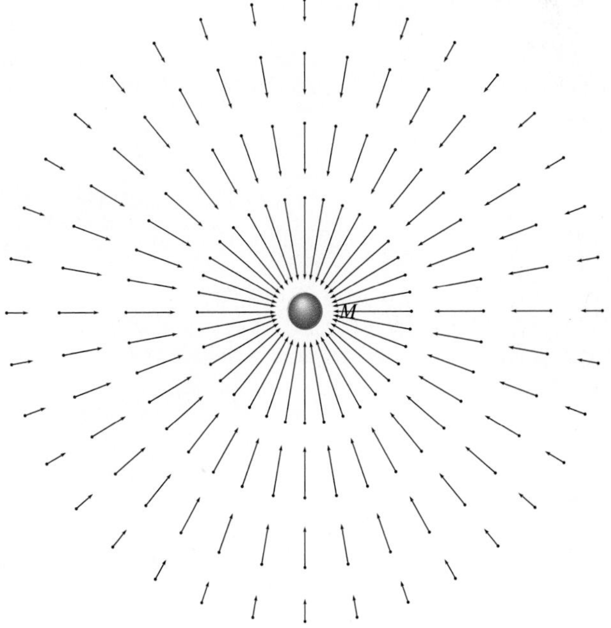

We think of every object that has mass permeating the surrounding space with a gravitational field that extends out indefinitely, its strength dropping off with distance. The field is a physical entity, and any other object immersed in it will interact directly with it, experiencing an attractive force. This explanation of action-at-a-distance raises fascinating questions that still await definitive answers. Is the field separate from M or is it the nonlocalized aspect of a single thing called mass? Does a particle of mass fill the surrounding space and exist everywhere its field exists? What, if anything, does the particle continue to give off as it sustains the expanding field?

7.13 New Directions

In early conceptions, space was a vessel and the field was strung across it like a web. But Einstein's General Relativ-

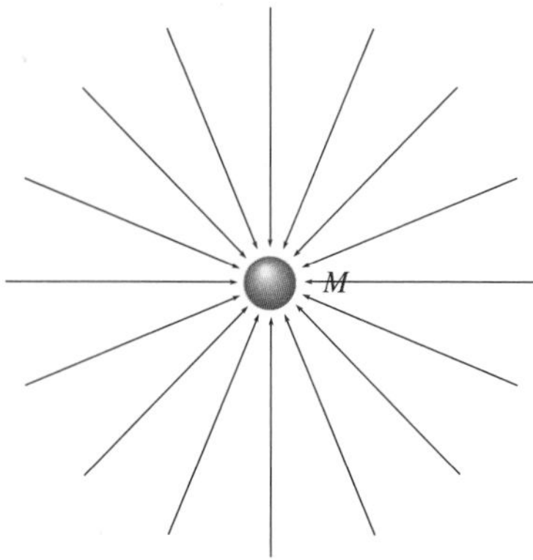

Figure 7.20 The gravitational field of a mass M. Rather than drawing a vector field, it's more convenient to draw the lines to which the vectors are tangent. These are the lines-of-force.

ity has space-time itself as the medium of gravity. Matter is viewed as a localized deformation of the space, distortions of which propagate out into the surroundings as the field. A material object simply moves within and across the warped space created by other objects. There is no attractive force per se; there is only *free* motion along contours in space-time. Since the introduction of Relativity Theory, it has become widely accepted that the speed of light in vacuum (c) represents the upper bound to motion. This suggests that changes in the gravitational field of an object move out into space at that speed, too.

That finite propagation rate leads to an apparent conflict with Newton's Third Law whenever there's a change in the relationship between any two interacting objects. For example, suppose you jump into the air, moving in the Sun's gravity field in such a way that the solar tug on you changes. Yet your field, in which the Sun is immersed and with which *it* interacts, will remain unaltered in the vicinity of that star for about eight minutes. For all that time, the interaction pair between you and the Sun will be out of tune.

It is from the study of the microworld that we have learned to expect that *all* forces are mediated by streams of special particles. Contemporary Quantum Field Theory maintains that the forces between objects arise from the exchange of these field particles. The carrier of the gravitational interaction is the still hypothetical massless **graviton**. These particles, streaming at the speed of light from Earth to apple and vice versa, carry the interaction from one to the other. And with the reasonable assumption that gravitons possess momentum (the field transmits momentum), the law of Conservation of Momentum (p. 112) holds exactly at all times.

By comparison to the other kinds of forces to be encountered in later chapters (namely, electromagnetic, weak, and strong), gravity is by far the faintest. Not surprisingly, then, the graviton remains elusive, its existence neither confirmed nor denied. For the time being, the graviton is a theoretical entity in a marvelous scheme that has replaced invisible tentacles with invisible particles. On the frontier of modern research are theories that attempt to unite the force particles of Quantum Field Theory with the geometrical space-time perspective of General Relativity.

Variations in the strength of the gravitational field across North America. The units are milligals. The field is strongest (red) in the Atlantic basins off the continental shelf and weakest in the high reaches of the Rocky Mountains (purple).

We now believe that the Universe was created some 17 thousand million years ago in a violent burst known as the Big Bang. That primeval space-time fireball grew and evolved into the cosmos as we know it, still expanding today. It has been suggested (by A. H. Guth, 1979) that within the very first moments of time ($\approx 10^{-35}$ s to $\approx 10^{-33}$ s), the Universe experienced an *inflationary era* during which it underwent an immensely rapid expansion. Under bizarre circumstances General Relativity allows the possibility of negative gravity. And so, driven by a momentary spasm of repulsive gravitation, the Universe may have expanded, perhaps by a factor of as much as 10^{90}. Reaching a kind of saturation, a dramatic transition occurred and the motherlode of material particles condensed into existence. The simplest of all substances, hydrogen, soon formed as particles captured one another. And $\approx 17 \times 10^9$ years later, the very same primordial atoms, dating back almost to the beginning of time, are there in every cell of your body.

Core Material

The primary insight of this chapter is Newton's **Law of Universal Gravitation**, that is,

$$F_G = G\frac{mM}{r^2} \quad [7.2]$$

where $G = 6.672\,59 \times 10^{-11}$ N·m²/kg². This equation represents the attractive interaction of gravity between two masses M and m separated by a center-to-center distance of r. The expression holds exactly for point masses or uniform spheres. In general, any two mass configurations will interact in an approximately $1/r^2$ fashion when r is large in comparison to the sizes of the interacting objects.

On Earth's surface, the *absolute acceleration due to gravity* is

$$g_0 = \frac{GM_\oplus}{R_\oplus^2} \quad [7.3]$$

Kepler's *Three Laws of Planetary Motion* are

1. The planets move in elliptical orbits with the Sun at one focus (and nothing at the other).
2. Any planet moves in such a way that a line drawn from the Sun to its center sweeps out equal areas in equal time intervals.
3. The ratio of the average distance from the Sun cubed to the period squared is the same constant value for all planets. That is

$$\frac{r_\odot^3}{T^2} = C_\odot \quad [7.6]$$

Newton's formulation immediately leads to an expression for this constant, namely

$$\frac{r_\odot^3}{T^2} = \frac{GM_\odot}{4\pi^2} \quad [7.7]$$

A satellite in a circular orbit about a large mass M, at a distance r from its center, moves with a speed

$$v_0 = \sqrt{\frac{GM}{r}} \quad [7.8]$$

It's useful to imagine that any mass is accompanied by a *gravitational field* that extends out into the surrounding space. The *field intensity,* or *strength,* is defined as

$$\mathfrak{g} = \frac{\mathbf{F}_G}{m} \quad [7.9]$$

and is represented pictorially by a series of lines converging down to the source mass.

Suggestions on Problem Solving

1. Make sure that your answers are reasonable. Compare your results with known values to see if they fit within recognized extremes. Accordingly, a value of $g_\oplus$ computed anywhere inside or outside the Earth should be between 0 and roughly 9.8 m/s². A value of 98 m/s² must be wrong!

2. As ever, check your answers whenever possible by recomputing using a different approach.

3. Don't confuse G, g, g_0, $g_\oplus$ and $\mathfrak{g}$—they are all different.

4. Consider units—don't enter a number in *kilometers* when you want *meters,* or use 32 m/s² when you should use 9.8 m/s².

5. Remember that F_G varies inversely with the square of the distance. An object moved from the surface of the Earth to a point 100 Earth-radii from the planet's *center* drops in weight to 1/10 000th its original value, *not* 1/100th.

6. Many of the equations in the chapter call for the radial distance. When given the numerical value of a diameter remember to divide by 2 before substituting. Keep in mind that the altitude above the surface of a planet is not the radial distance to its center.

Discussion Questions

1. The following quote is from *Bioastronautics Data Book,* 2nd ed., NASA SP-3006 (1973), p. 149. Why is it unmitigated nonsense?

> In its third form, acceleration occurs as a component of the attraction between masses. The resulting force is directly proportional to the product of the masses and inversely proportional to the square of the distance between them. The proportionality constant is the gravitational constant g which represents an acceleration of 32.24 feet per second (fps) within the terrestrial field of reference. This is the accepted unit of measurement of acceleration.

2. Why do you lose weight when you enter a tunnel passing horizontally through a mountain, or walk inside a skyscraper?

3. A comic book space colony occupies a huge central cavity symmetrically hollowed out of a spherical planetoid. The idea is to provide a home for any space traveler not wishing to return to a traditional gravity environment on the surface of some ordinary planet. Is such a weightless heaven possible? Explain. Can the lighthearted inhabitants of the colony detect the arrival of a spacecraft landing on the planet's surface via their gravitational interactions, or are they *shielded* inside the sphere?

4. Jupiter has a mass 318 times that of the Earth and yet its "surface" acceleration due to gravity is roughly only 26 m/s^2. Explain how this could be.

5. It's convenient, particularly with weather and spy satellites, to have them always point radially downward toward the planet. One imaginative *passive* system for accomplishing that is to extend from the bottom of the craft a long boom with a small mass at the end and let the Earth's gravity do the rest. Explain qualitatively what effects might be at work here. (Hint: Think of the boom swinging slightly about its *c.g.* away from verticality.) The operative mechanism is called *differential gravity.* The satellite is also stable with the boom pointing up and, believe it or not, that has actually happened—somewhere in space a spy satellite is upside down and useless.

6. Suppose that an astronaut in space is in a high circular Earth orbit moving eastward (that is, counterclockwise looking down on the North Pole). Imagine that she releases an apple into space. What will be the flight path of the apple? What if the apple is hurled directly forward (eastward)? What if it's thrown in the backward direction? What do you think would happen to it if it were thrown radially toward the planet? (This last question goes beyond the material we have studied so far, and is only meant as a test of your intuition.)

7. Why is it easier to launch Earth satellites in an easterly direction than in a westerly direction? The John F. Kennedy Space Center is on which coast of Florida? Why? Why is it in Florida?

8. The Sun seems to move across the star field more rapidly in the winter than it does in the summer. What does this phenomenon imply about the Earth-Sun distances at these two times?

9. A number of spacecraft, including the several *Apollo* Command Modules, have been put into low lunar orbits. Why are such low paths possible around the Moon but not around the Earth?

10. Is it possible to have one object orbit another in a closed path if the mutual interaction is repulsive?

11. Figure Q11 shows a spaceship changing from a low to a high circular orbit. Explain how it's being done. Compare the initial and final orbital speeds. Assume an ordinary chemical rocket that can be fired in limited duration bursts.

12. Figure Q12 shows a maneuver that a rocket can use to escape from a circular orbit about a planet into a hyperbolic unpowered flight path. What must happen at point *A*? If the process is run backward and the craft free-falls toward the planet, what must happen at *A* to effect capture?

13. Figure Q13 depicts a double-thrust hyperbolic departure of a rocket from a circular orbit. Explain how it happens. Incidentally, for high speeds at great distances, this maneuver is more fuel-efficient than the one considered in the previous question.

14. The Sun gravitationally attracts the Moon. Does the Moon orbit the Sun? Explain.

15. It has become commonplace to see large (≈9-ft-diameter) parabolic TV antennas on the roofs of buildings, especially bars and hotels. In what direction do they point, and at what are they pointed?

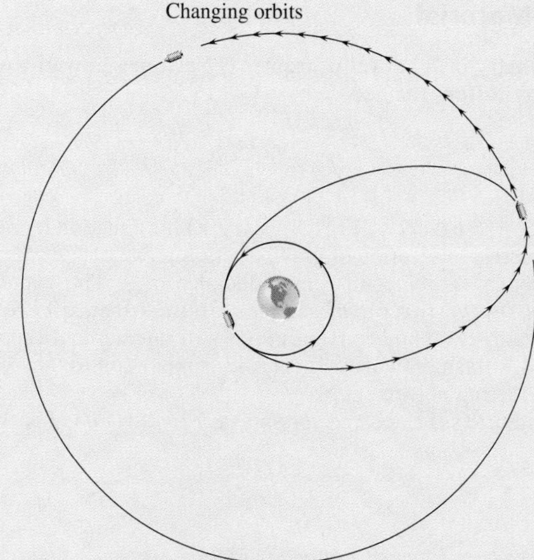

Figure Q11

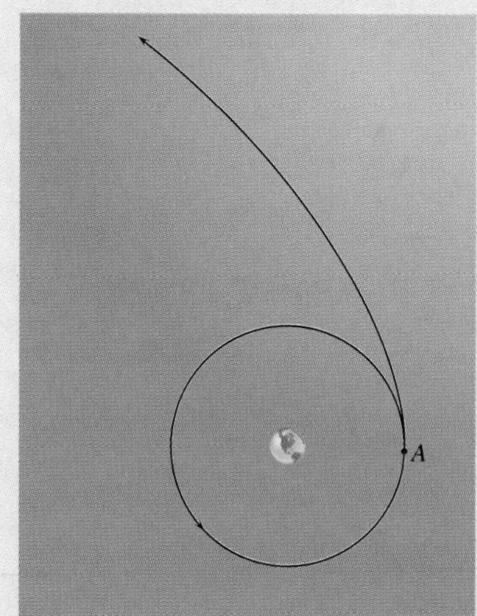

Figure Q12

16. Imagine that all the mass of the Earth were somehow compressed into a sphere < 200 m in diameter. It can be shown that a *solid steel* satellite 1.0 m in diameter descending toward the planet would be pulled apart as soon as it approached to within 100 m of the center of this miniworld. Similarly, if the Moon were to orbit Earth at a center-to-center distance of less than $2.9R_{\oplus}$, it, too, would be torn apart. Explain.

17. The third outermost moon of Saturn is Hyperion, an irregular object 400 km long by 220 km wide. It is one of the largest lopsided bodies in the Solar System. Why do you think Hyperion isn't spherical like most other moons?

18. Figure Q18 shows the Earth and Moon and a number of lines forming some sort of pattern. Identify the pattern and discuss its important features. (See Problem 18.)

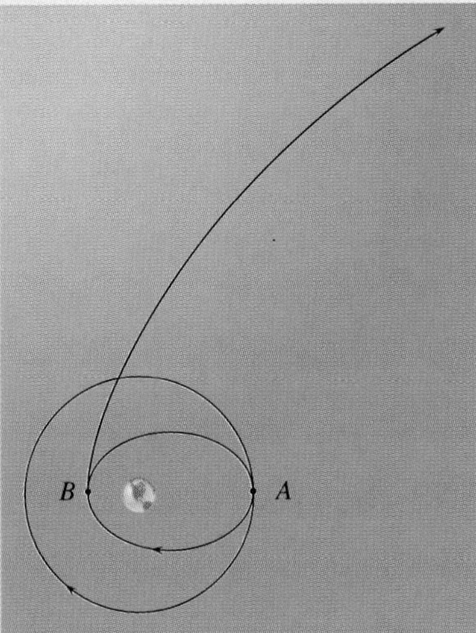

Figure Q13

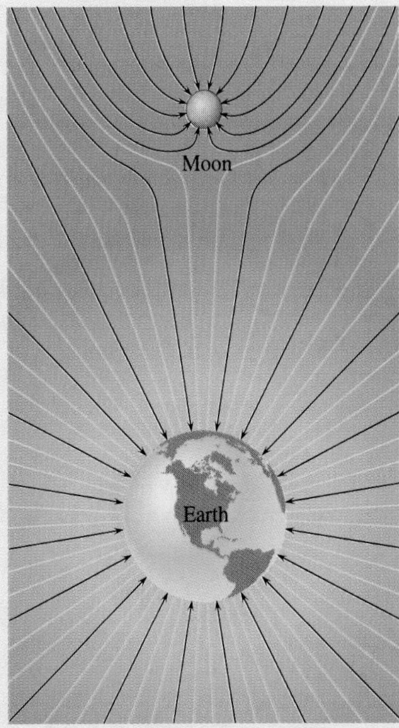

Figure Q18

19. In 1978, the *International Sun-Earth Explorer 3* satellite (*ISEE-3*) was launched into a remarkable orbit. What's so extraordinary is that it orbits a *libration point* in space where no central mass exists. This point lies on the Earth-Sun center-to-center line and is the place where the gravitational forces of these two bodies exactly counterbalance each other. In other words, the gravitational field there is zero. The plane of the orbit is perpendicular to the center line and contains the libration point. Explain, qualitatively, how *ISEE-3* could be held in orbit this way.

20. Make a sketch of the gravitational field both inside and outside a hollow spherical shell of mass m, assuming no other masses are present. Why is it necessary to add that last requirement?

Multiple Choice Questions

1. If the dimensions of length, time, and mass are L, T, and M, respectively, then the dimensions of G are (a) $[L^2/MT^2]$ (b) $[L^3/MT]$ (c) $[L^3/MT^2]$ (d) $[L^3/M^2T]$ (e) $[L^2/M^3T]$.
2. A 1.00-kg chicken weighs 9.8 N on the surface of the Earth. At a distance of one Earth-radius above the planet's surface (a) its weight is 4.9 N (b) its mass is 0.50 kg (c) its weight is 19.6 N (d) its mass is 2.00 kg (e) none of these.
3. Call the gravitational attraction between you and the planet, your *true* weight. Because of the rotation of Earth, the reading that you get when you stand on a scale (a) will be more than your *true* weight (b) will everywhere be less than your *true* weight (c) will always be your *true* weight (d) will only be your *true* weight at the two poles (e) will only be your *true* weight at the equator.
4. Suppose you were transported to the mythical planet Mongo, which is four times as massive as Earth and has twice the diameter. Your Mongoian weight as compared to your present weight would be (a) 4 times larger (b) the same (c) 2 times smaller (d) 4 times smaller (e) none of these.
5. Mars has a mass of 0.1074 $M_\oplus$ and is at a mean distance from the Sun, which is 1.52 times larger than that of Earth. By comparison to the gravitational force exerted on Mars by our world, the force exerted on Earth by Mars is (a) 0.1074 times smaller (b) 0.1074 times larger (c) the same (d) 1.52 times less (e) none of these.
6. The acceleration of a meteor at a height above the Earth of $1R_\oplus$ is (a) about 2.5 m/s^2 (b) 9.8 m/s^2 (c) not enough information to say (d) can be anything from 9.8 m/s^2 to 0 (e) none of these.
7. The asteroid Geographos (one of the Apollo group, each of which crosses the Earth's orbit on the way around the Sun) has a radius of $2.4 \times 10^{-4}R_\oplus$ and a mass of $8.4 \times 10^{-12}M_\oplus$. How does the gravitational acceleration on its surface compare to the corresponding value g on the Earth? It equals (a) $2.4 \times 10^{-4}g$ (b) $8.4 \times 10^{-12}g$ (c) $1.5 \times 10^{-4}g$ (d) $3.5 \times 10^{-8}g$ (e) none of these.
8. Suppose a deep vertical shaft is drilled into the Moon and a rock is slowly lowered down. It will be found that the rock's (a) mass and weight both decrease (b) mass decreases but the weight increases (c) mass and weight both remain unchanged (d) mass is constant but its weight decreases gradually (e) none of these.
9. The Earth's value of the acceleration due to gravity (a) is constant inside (b) varies by about 0.5% over the surface

(c) is constant over the surface (d) is constant beyond the surface (e) none of these.

10. An astronaut on the Moon has a mass that by comparison to his mass on Earth is (a) unchanged (b) six times greater (c) six times less (d) not enough information to say (e) none of these.
11. The acceleration due to gravity, as measured by a spring balance determination of the weight of an object ($F_W = mg$), varies from place to place on Earth because (a) the mass changes (b) g is affected by the rotation of the planet only (c) g depends on the shape of the planet only (d) g depends on both the rotation and shape of the planet (e) none of these.
12. If *Martian Orbiter 1* is sailing about the planet in a circle with an orbital radius nine times that of *Orbitor 2,* whose speed is v_2, what is the speed of *Orbitor 1?* (a) $\frac{1}{3}v_2$ (b) $3v_2$ (c) v_2 (d) $81v_2$ (e) none of these.
13. Figure MC13 shows a spaceship in orbit about a star. If its speeds at the four points shown are v_A, v_B, v_C, and v_D, respectively, then (a) $v_A < v_B < v_C < v_D$ (b) $v_A > v_B > v_C > v_D$ (c) $v_A > v_B = v_D > v_C$ (d) $v_A < v_B = v_D < v_C$ (e) none of these.

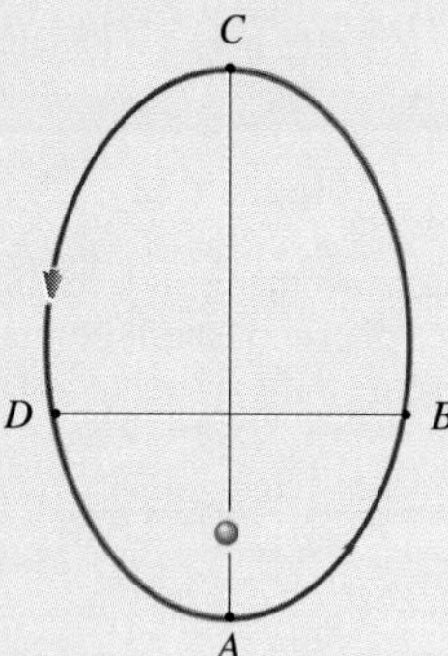

Figure MC13

14. A spacecraft is in a circular orbit about a planet located at point O in Fig. MC14. When it reaches point A on its orbit, the First Mate throws a canister out the front port straight ahead; the canister goes into a new orbit shown in (a) part a (b) part b (c) part c (d) part d (e) none of these.

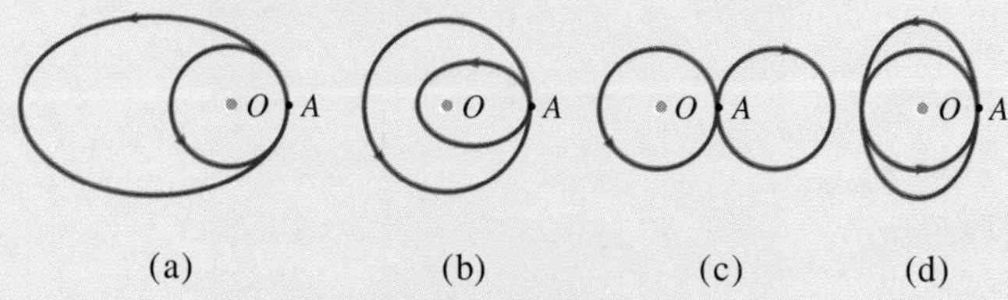

Figure MC14

15. It is desired that a spacecraft initially in a circular orbit drop straight down to the planet below. Figure MC15 shows the planned double-thrust maneuver, which supposedly will save fuel. The craft (a) slows down at B and brakes to zero speed at D (b) speeds up at A and brakes to zero speed at E (c) brakes slightly at A and then again at E (d) the maneuver is impossible (e) none of these.

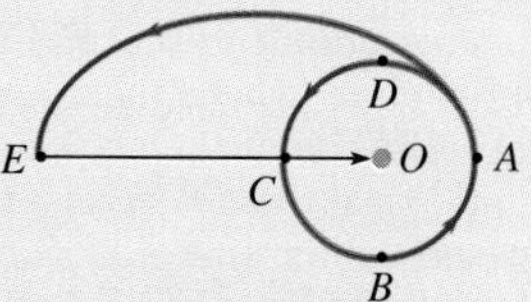

Figure MC15

16. The gravitational field intensity at a point 1.00 m away from the center of a small 1.00-kg spherical mass (assuming no other sources exist) is (a) 1.00 N/kg (b) 1.00 gal (c) 6.67×10^{-11} m/s^2 (d) 9.8 m/s^2 (e) none of these.

Problems

Miscellaneous data: $M_\oplus = 5.975 \times 10^{24}$ kg, $M_\odot = 1.987 \times 10^{30}$ kg, $M_☾ = 7.35 \times 10^{22}$ kg, $R_\odot = 6.97 \times 10^8$ m, $R_\oplus = 6371.23$ km, $R_☾ = 1.74 \times 10^6$ m, $r_{\odot\oplus} = 1.495 \times 10^{11}$ m, $r_{\oplus☾} = 3.844 \times 10^8$ m.

THE UNIVERSAL LAW OF GRAVITATION

1. [I] What would happen to the weight of an object if its mass was doubled while its distance from the center of the Earth was also doubled?

2. [I] The gravitational attraction between a 20-kg cannonball and a marble separated center-to-center by 30 cm is 1.48×10^{-10} N. Compute the mass of the marble.

3. [I] Suppose that two identical spheres, separated center-to-center by 1.00 m, experience a mutual gravitational force of 1.00 N. Compute the mass of each sphere.

4. [I] At what center-to-center distance from the Earth would a 1.0-kg mass weigh 1.0 N?

5. [I] Suppose the Earth were compressed to half its diameter. What would happen to the acceleration due to the gravity at its surface?

6. [I] Compare the gravitational force of the Earth on the Moon to that of the Sun on the Moon.

7. [I] If the average distance between Uranus (♅) and Neptune (♆) is 4.9×10^9 km, and $M_♅ = 14.6M_\oplus$ while $M_♆ = 17.3M_\oplus$, compute their average gravitational interaction.

8. [I] Imagine two uniform spheres of radius R and density ρ in contact with each other. Write an expression for their mutual gravitational interaction as a function of R, ρ, and G.

9. [I] If you can jump 1.00 m high on Earth, how high can you jump on Venus, where $g_♀ = 0.88g_\oplus$? Assume the same takeoff speed.

10. [I] What fraction of what you weigh on Earth would you weigh in a rocket ship firing its rockets so that it was stationary with respect to the planet $4R_\oplus$ from its surface?

11. [I] Consider two subatomic particles, an electron and a proton, which have masses of 9.1×10^{-31} kg and 1.7×10^{-27} kg, respectively. When separated by a distance of 5.3×10^{-11} m, as they are in a hydrogen atom, the electrical attraction (F_E) between them is 8.2×10^{-8} N. Compare this with the corresponding gravitational interaction. How many times larger is F_E than F_G?
12. [I] Show that Eq. (7.3) is equivalent to Eq. (7.4) when $r = R_\oplus$.
13. [I] The acceleration due to gravity on the surface of Mars is 3.7 m/s². If the planet's diameter is 6.8×10^6 m, determine the mass of the planet and compare it to Earth.
14. [II] Imagine an astronaut having a mass of 70 kg floating in space 10.0 m away from the *c.g.* of an *Apollo* Command Module whose mass is 6.00×10^3 kg. Determine the gravitational force acting on, and the resulting accelerations (at that instant) of, both the ship and the person.
15. [II] Venus has a diameter of 12.1×10^3 km and a mean density of 5.2×10^3 kg/m³. How far would an apple fall in one second at its surface?
16. [II] Given that $M_☾/M_\oplus = 0.012\,30$ and $R_☾/R_\oplus = 0.2731$, compute the ratio of an astronaut's Moon-weight ($F_{W☾}$) to Earth-weight ($F_{W\oplus}$).
17. [II] Taking the surface value of $g_\oplus$ to be g_0, show that

$$g_\oplus = g_0(R_\oplus/r)^2 \quad \text{for } r \geq R_\oplus$$

18. [II] Locate the position of a spaceship on the Earth-Moon center line such that, at that point, the tug of each celestial body exerted on it would cancel and the craft would literally be weightless. (See Fig. Q18.)
19. [II] Mars has a mass of $M_♂ = 0.108M_\oplus$ and a mean radius $R_♂ = 0.534R_\oplus$. Find the acceleration of gravity at its surface in terms of $g_0 = 9.8$ m/s².
20. [II] Three very small spheres of mass 2.50 kg, 5.00 kg and 6.00 kg are located on a straight line in space away from everything else. The first one is at a point between the other two, 10.0 cm to the right of the second and 20.0 cm to the left of the third. Compute the net gravitational force it experiences.
21. [II] It is believed that during the gravitational collapse of certain stars, such great densities and pressures will be reached that the atoms themselves will be crushed, leaving only a residual core of neutrons. Such a *neutron star* is, in some respects, very much like a giant atomic nucleus with a tremendous density of roughly about 3×10^{17} kg/m³. Compute the surface acceleration due to gravity for a one-solar-mass neutron star.
22. [II] Fig. P22 shows two concentric, thin, uniform spherical shells of mass m_1 and m_2, at the center of which is a small ball of lead of mass m_3. Write an expression for the gravitational force exerted on a particle of mass m at each point A, B, and C located at distances r_A, r_B, and r_C from the very center.
23. [III] Write an expression for the gravitational force on a small mass m imbedded in a uniform spherical cloud of mass M and radius R. Take the particle to be at $r < R$.
24. [III] Two 2.0-kg crystal balls are 1.0 m apart. Compute the magnitude and direction of the gravitational force they exert on a 10-g marble located 0.25 m from the center-to-center line as shown in Fig. P24.

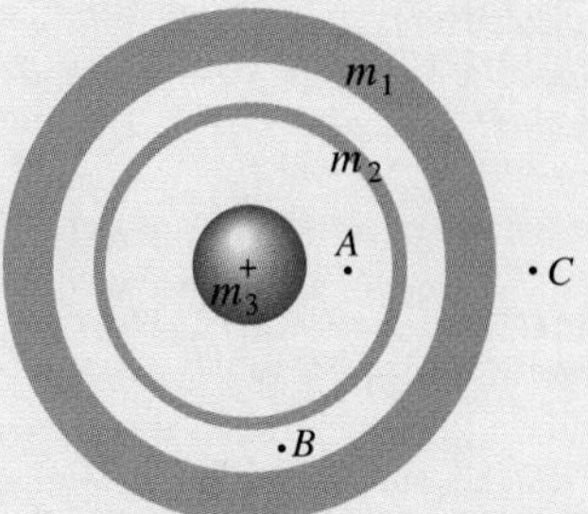

Figure P22

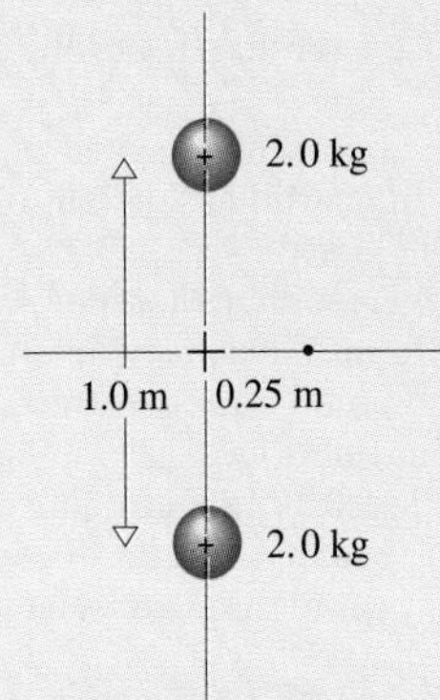

Figure P24

25. [III] A neutron star (see Problem 21) can be envisioned as an immense nucleus held together by its self-gravitation. What is the shortest period with which such a star could rotate, if it's not to lose mass flying off at the equator? Take $\rho = 1 \times 10^{17}$ kg/m³. It is widely believed that *pulsars,* strange celestial emitters of precisely pulsating radiation, are rapidly rotating neutron stars (see p. 260, Chapter 8).
26. [III] Draw a graph of the weight of an object of mass m due to the Earth versus the height above the surface, out to roughly 700 km. What can you say about the curve (so long as $R \gg h$)?
27. [III] Let M be the mass of a uniform spherical planet of radius R. If h is the height above its surface, show that the absolute gravitational acceleration g_p varies with h as

$$g_p = \frac{GM}{R^2}\left(1 + \frac{h}{R}\right)^{-2}$$

This expression can be approximated using the binomial expansion. (See Appendix A-5.)

$$(a + x)^n = a^n + na^{n-1}x + \tfrac{1}{2}n(n-1)a^{n-2}x^2 + \cdots$$

where $x^2 < a^2$. Here, $a = 1$, $x = h/R$, $n = -2$, and we limit the calculation to the case where $h \ll R$. Show that

$$g_p \approx \frac{GM}{R^2}\left(1 - \frac{2h}{R}\right)$$

Notice that GM/R^2 is the surface value of g_p occurring when $h = 0$, as in Eq. (7.3).

28. [III] Calculate the acceleration due to gravity 10 000 m above the Earth's surface in the following two ways: (1) using Eq. (7.3); and (2) using the approximation of Problem 27. Compare your results.
29. [III] In light of Problem 27, determine the acceleration due to gravity experienced by the Lunar Module when it was 100 m above the Moon's surface. Is it appreciably different from the surface value?

THE COSMIC FORCE

30. [I] Using the data for the Earth's orbit, compute the mass of the Sun.
31. [I] Determine the approximate speed of a Lunar Orbiter revolving in a circular orbit at a height of 62 km. Take the Moon's radius as 1738 km.
32. [I] Each of the *Apollo* Lunar Modules was in a very low orbit around the Moon. Given a typical mass of 14.7×10^3 kg, assume an altitude of 60.0 km and determine the orbital period.
33. [I] For any Earth satellite in a circular orbit, show that both its period (in seconds) and its distance from the center of the planet (in meters) are related by way of

$$T = 3.15 \times 10^{-7}\,(r_{\oplus})^{\frac{3}{2}}$$

34. [I] *Sputnik I,* the first artificial satellite to circle the planet (October 1957) had a mean orbital radius of 6950 km. Compute its period.
35. [I] A satellite is to be raised from one circular orbit to another twice as large. What will happen to its period?
36. [I] Referring to Problem 35, compare the two orbital speeds v_1 and v_2.
37. [I] What is the acceleration due to the gravity of the Moon at the Earth?
38. [I] Find the gravitational field strength 1.00 m from a small sphere whose mass is 1.00 kg.
39. [I] What is the gravitational field strength of the Sun at the Earth?
40. [II] By definition, the Earth is a distance of 1.000 AU from the Sun. Using the fact that Jupiter is, on average, 5.202 8 AU from the Sun, compute its period in Earth-years.
41. [II] Determine the period in Earth-years of a satellite placed in a circular solar orbit with a radius of 371.6 million miles.
42. [II] Imagine a central body of mass M_B (be it a star, planet, or moon) about which another object is orbiting such that its Keplerian constant is C_B. Show that C_B/M_B is a universal constant, the same for all bodies.
43. [II] Imagine the two comparable masses m_1 and m_2 of Fig. P43 orbiting their barycenter O at distances r_1 and r_2, respectively, with a common period T. Since their mutual gravitational interaction provides their individual centripetal forces F_{C1} and F_{C2}, these must be equal. Show that

$$\frac{r_1}{r_2} = \frac{m_2}{m_1}$$

and compare this with the definition of the *c.g.*

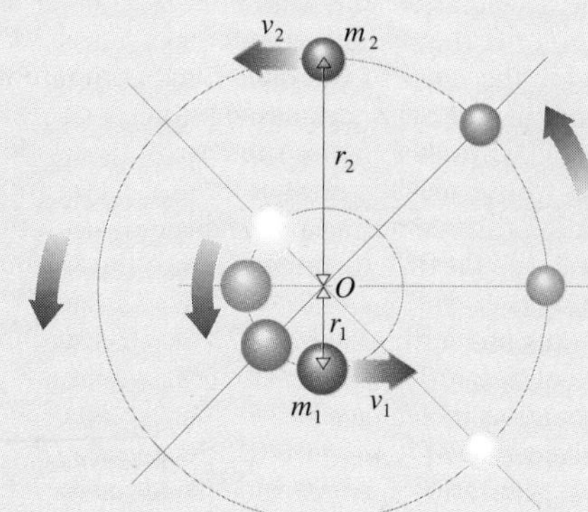

Figure P43

44. [II] Referring to Problem 43 and Fig. P43, write explicit expressions for $F_G = F_{C1}$ and $F_G = F_{C2}$, cancel out whatever you can, and then combine them to get Kepler's Third Law in the more accurate Newtonian version, namely

$$\frac{(r_1 + r_2)^3}{T^2} = \frac{G(m_1 + m_2)}{4\pi^2}$$

Unlike Eq. (7.7), which applies when the central body (here the Sun) has essentially infinite mass, this expression is more realistic. Notice that the equations are identical when $m_1 \gg m_2$, so that m_2 is negligible; of course, $r = r_1 + r_2$.

45. [II] For an elliptical orbit, the *mean* or *average distance* from the central body to the orbiter is just half the main symmetry axis. This *semi-major axis* is also equal to the average of the aphelion and perihelion distances. The tiny asteroid Icarus (which has a mass of 5.0×10^{12} kg and a radius of just 0.7 km) orbits the Sun in an elongated elliptical path. It crosses the Earth's orbit as it sweeps in to a perihelion of only 0.186 AU and out to an aphelion at 1.97 AU. Compute its mean distance from the Sun and its orbital period in Earth-years.
46. [II] The Andromeda galaxy, known as M31, is a great spiral star-island 2.2×10^6 light-years away. Measurements show that a star out at its extremities—5×10^9 AU from the center—orbits the nucleus at a speed of 200 km/s. Approximate the mass of M31.
47. [II] What is the value of the gravitational field intensity halfway down inside of a solid uniform sphere of mass M and radius R? (See Problem 23.)
48. [II] Determine the gravitational field at the center of a uniform ring of mass M and radius R.

49. [II] Galileo discovered the four major moons of the planet Jupiter in 1610. The nearest one, Io, has a period of 1.769 9 d and is 5.578 Jovian radii ($R_{♃}$) from the center of the planet. Use this information to calculate the mean density of Jupiter.
50. [III] The binary stars Sirius A and Sirius B each orbit their barycenter with a period of 50 years. Their separation is measured to be 20.0 AU (or 2.99×10^{12} m) with the much fainter star, Sirius B, being twice as far from the barycenter as is Sirius A. Compute both their masses. (Hint: Go back and look at Problems 43 and 44.)
51. [III] We wish to put an artificial Earth satellite in a circular orbit halfway out to the Moon. Compute its period and the necessary orbital speed.
52. [III] Two identical masses of value m are each located on the y-axis a distance d above and below the origin. Show that the gravitational field intensity at any point P on the x-axis at a distance from the origin of x is given by

$$g = \frac{2Gmx}{(x^2 + d^2)^{\frac{3}{2}}}$$

Notice that the distance from either mass to P is $r = (x^2 + d^2)^{\frac{1}{2}}$.

8

Rotational Motion

MOTION APPEARS IN TWO basic types: **translational** and **rotational** (Fig. 8.1). If a body is moving such that a line drawn between any two of its internal points remains parallel to itself, the body is *translating*—it may be traveling either in a straight line (*rectilinear translation*) or along an arc (*curvilinear translation*). If, on the other hand, that line between any two points does not remain parallel to itself, the body is *rotating*.

At this very instant, the Earth is rotating and revolving around the Sun, which is itself spinning and revolving around the center of the galaxy, which is itself spinning through space. At the other extreme, the atoms that form

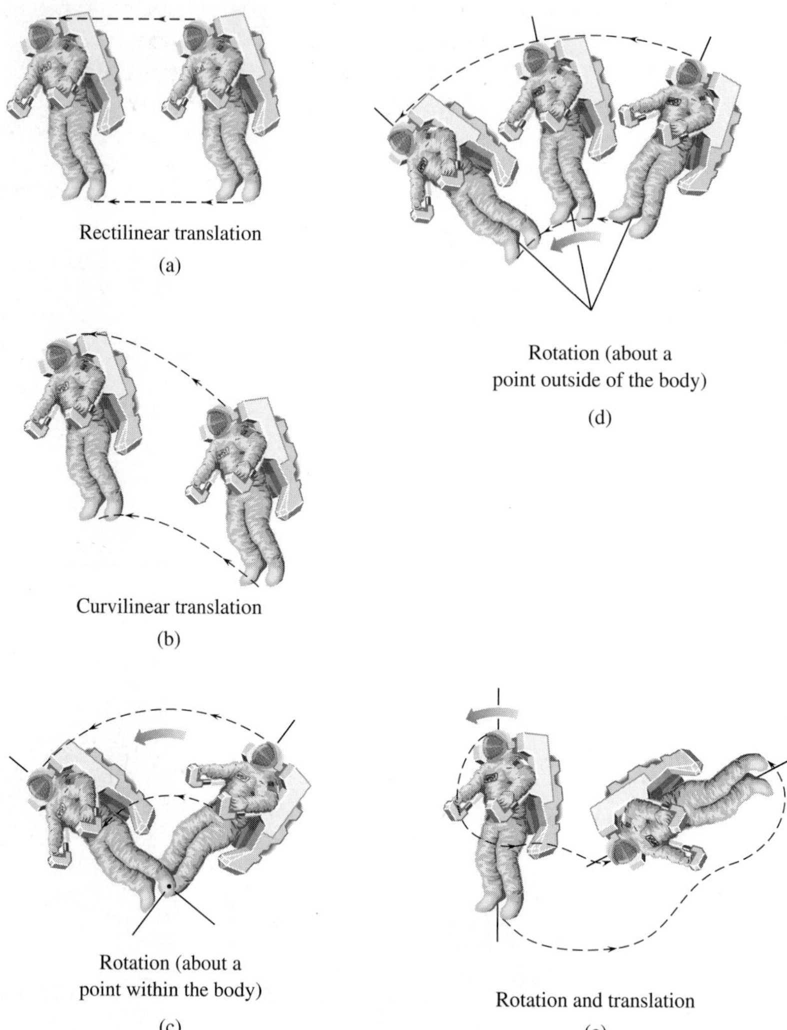

Figure 8.1 The movement of all rigid bodies in space can be described in terms of rotational and translational motion.

all matter are composed of far smaller parts that are in perpetual rotation.

On a more pragmatic level, many of our machines, from washing machines, tape recorders, CD players, jet engines, and pencil sharpeners to clocks and carousels, involve rotation. After all, the most basic means of mechanically transmitting power is by way of a rotating drive shaft.

This chapter focuses on the kinematics and dynamics of rotational motion. Happily, the conceptual framework is very similar to that of translational motion. There will be a few fresh definitions, a new circularization of imagery, but you have already mastered the essence of the treatment (see Chapters 2–6). Just as linear momentum is the measure of translational motion, *angular momentum* is the measure of rotational motion.

At the center of this treatment is a grand insight: Conservation of Angular Momentum. This is another of those rare principles that reveals the inner symmetry of the Universe (p. 267)—empty space is the same in all directions, and so the laws of physics do not depend on direction.

THE KINEMATICS OF ROTATION

Imagine an object revolving about some point; it might, for instance, be a wheel, a line of hand-holding skaters, or a string of beads as in Fig. 8.2. Whirled in a circle, each bead sails along a different arc, but each sweeps through the same angle θ. Up to now, we have measured angles in *degrees;* 360° to a circle. Yet, there is nothing special about "degrees"—they are a remnant of the ancient Babylonian number system, whose base was 60 rather than 10.

The angle θ can be specified in a less arbitrary way by relating it back to an essential aspect of the circle, its circumference equal to $2\pi r$. Each bead in Fig. 8.2 moves an arc-length (l) at some radial distance from the center (r). The larger is r, the larger is l. In each case, θ is the same, suggesting that we might formulate θ in terms of the ratio of l to r. Thus, let

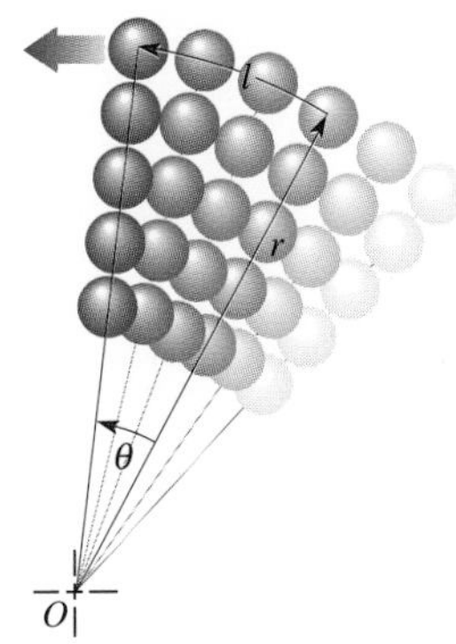

Figure 8.2 A revolving string of beads all moving around their respective circles in step with each other.

$$\theta = \frac{l}{r} \quad \text{or} \quad l = r\theta \tag{8.1}$$

The angle θ is now framed as a distance over a distance and is *unitless*—it's a pure number. The degree is not a physical unit in the strict sense either—it is not referenced back to meters, kilograms, or seconds. When $l = r$, $\theta = 1$ and as a reminder that we are measuring angles via Eq. (8.1), we call this 1 *radian* or 1 rad. If we ever meet up with aliens from some other world, they will probably use the equivalent of radians, too.

Draw a circle of radius r (Fig. 8.3a). Now lay off arc-lengths equal to r around the circle. You will end up with six such segments plus an additional fraction of one. That's because the circumference equals $2\pi r$, which is $(6.283\,185\ldots)r$. Figure 8.3b shows the relationship between an arc-length l and the angle θ subtended by it at point O. When $l = r$, $\theta = 1$ rad, but it also equals $1/2\pi$ of the angle corresponding to a whole turn (360°), which means that

$$1 \text{ rad} = \frac{360°}{2\pi} = \frac{360°}{(6.283\ldots)} = 57.3° \tag{8.2}$$

as in Fig. 8.3c. Compare this angle to that formed in an equilateral triangle, which has to be a little wider (Fig. 8.3d). Clearly, 2π rad $= 360°$. Any angle θ in radians can be transformed into degrees by simply taking proportions, such that

$$\frac{\theta(\text{radians})}{2\pi(\text{radians})} = \frac{\theta(\text{degrees})}{360(\text{degrees})} \tag{8.3}$$

Thus, π rad $= 180°$, $\pi/2$ rad $= 90°$, and so forth (see Table 8.1).

The Moon has a diameter $D_☾$ equal to about 3.4×10^6 m, and it is a distance $r = 3.8 \times 10^8$ m from the Earth's surface. If we approximate its straight-line diameter as an arc-length, then the angle θ subtended at the Earth (Fig. 8.4) by the Moon is

$$\theta = \frac{l}{r} = \frac{D_☾}{r} = \frac{3.4 \times 10^6 \text{ m}}{3.8 \times 10^8 \text{ m}} = 0.009 \text{ rad}$$

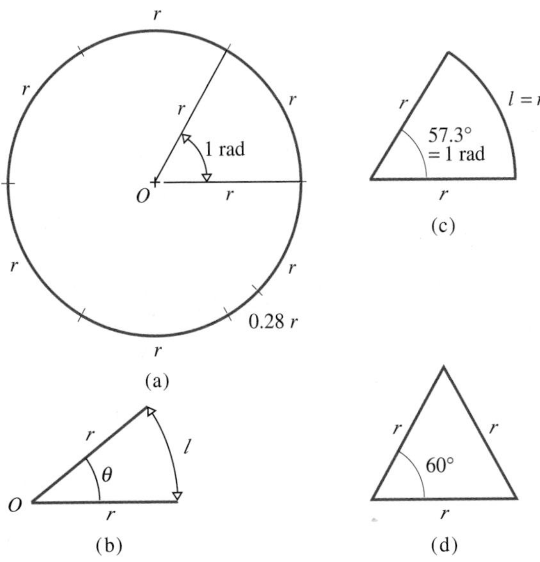

Figure 8.3 The relationship between arc-length, radius, and angle. (a) The circumference of a circle equals $2\pi r$. (b) As r moves through the angle θ, its end point sweeps out a length l. (c) When $l = r$, $\theta = 1$ rad. This shape is close to that of an equilateral triangle where $\theta = 60°$ (d).

TABLE 8.1 Angles in degrees and radians

Degrees	Radians in terms of π	Radians in decimals
0	0	0
15	$\pi/12$	0.261
30	$\pi/6$	0.524
45	$\pi/4$	0.785
57.3	1.0	1.0
60	$\pi/3$	1.047
90	$\pi/2$	1.571
120	$2\pi/3$	2.094
180	π	3.142
270	$3\pi/2$	4.712
360	2π	6.283

or 0.5°. By comparison, the diameter of the Sun (1.4×10^9 m) is immensely larger, but its distance to the Earth (1.5×10^{11} m) is also much greater. Coincidentally, the Sun also subtends an angle of 0.009 rad. That's why the Sun and Moon seem to be the same size.

Example 8.1 The Andromeda galaxy is a giant spiral star-island whose mass is that of 300 thousand million suns. You can see it with the naked eye as a faint elongated cloud in the night sky. Inasmuch as it subtends an angle of 4.1° and is known to be larger than our own galaxy [163×10^3 light-years (ly) in diameter for Andromeda as compared to 100×10^3 ly for our galaxy], how far away is it?

Solution: [Given: the diameter of Andromeda $D_A = 163 \times 10^3$ ly and its subtended angle $\theta = 4.1°$. Find: r_A.] First, convert θ into radians, which yields 0.072 rad. Then, approximate the arc-length by the diameter. Hence, $\theta \approx D_A/r_A$. Consequently

$$r_A = \frac{D_A}{\theta} = \frac{163 \times 10^3 \text{ ly}}{0.072} = \boxed{2.3 \times 10^6 \text{ ly}}$$

▶ **Quick Check:** When the arm r_A-long sweeps through θ, its end point traces an arc of length $(4.1°\pi/180°)(2.3 \times 10^6 \text{ ly}) = 1.6 \times 10^5$ ly.

Example 8.2 A decent spy-satellite camera can distinguish objects that subtend angles of as little as 0.5×10^{-6} rad. How small an object could such a device detect from 200 mi away?

Solution: [Given: $\theta = 0.5 \times 10^{-6}$ rad and $r = 200$ mi. Find: l.] This analysis involves a straightforward substitution into Eq. (8.1). Using

$$r = (200 \text{ mi})(1.609 \text{ km/mi}) = 321.8 \text{ km}$$

$$l = r\theta = (321.8 \times 10^3 \text{ m})(0.5 \times 10^{-6} \text{ rad}) = 0.160\,9 \text{ m}$$

or, to one significant figure, $\boxed{l = 0.2 \text{ m}}$.

▶ **Quick Check:** $l = (200 \text{ mi})(0.5 \times 10^{-6} \text{ rad}) = 100 \times 10^{-6}$ mi, or $l = 0.5$ ft.

Incidentally, the Muslim Alhazen (A.D. 1000) correctly suggested that visual perspective arises because the angle an object subtends gets smaller as it recedes from us—a nearby tree appears large, whereas a distant one, the same size, seems minute.

8.1 Angular Speed

Imagine the last bead in Fig. 8.2 moving in a circle of radius r. As it travels along the circular segment l, the radius sweeps through angle θ, the **angular displacement**. If the bead revolves through six complete counterclockwise turns and comes back to its starting point, we will take $\theta = 6 \times (2\pi)$ rad rather than $\theta = 0$. Define θ swept out counterclockwise as positive and clockwise as negative, and the same for l. Consequently, four turns in one direction followed by four turns in the other brings the bead back to where it started, $\theta = 0$ ($l = 0$); in that sense, we are dealing with angular displacements.

If the bead travels the distance $l = r\theta$ in a total time t, divide both sides to get

$$\frac{l}{t} = r\frac{\theta}{t} \tag{8.4}$$

The ratio on the left is evidently the average speed (that is, $v_{av} = l/t$). The ratio on the right is the **average angular speed**, denoted by the lowercase Greek letter *omega* (ω) such that

$$\omega_{av} = \frac{\theta}{t} \tag{8.5}$$

The units of angular speed are rad/s although *degrees per second, revolutions per second,* and *rotations per minute* (rpm) are commonly used. Notice that each revolution swings through 2π rad, so that 1 rev/s $= 2\pi$ rad/s. Only when the angular speed is specified in rad/s will

$$v_{av} = r\omega_{av} \tag{8.6}$$

Example 8.3 On August 24, 1968, a very fast horse named Dr. Fager finished a 1.00-mi race in 1.00 min 32.2 s. Assuming he ran once around a circular track, what was his average speed and average angular speed?

Solution: [Given: $l = 1.00$ mi on a circular path and $t = 1.00$ min 32.2 s. Find: v_{av} and ω_{av}.] By Eq. (2.1)

$$v_{av} = \frac{l}{t} = \frac{1.00 \text{ mi}}{1.537 \text{ min}} = 0.651 \text{ mi/min}$$

or $\boxed{v_{av} = 62.8 \text{ km/h}}$. To find the radius, we use the fact that $2\pi r = 1.00$ mi; hence, $r = 0.159$ mi $= 0.256$ km. From Eq. (8.6)

$$\omega_{av} = \frac{v_{av}}{r} = \frac{0.0174 \text{ km/s}}{0.256 \text{ km}} = \boxed{0.0682 \text{ rad/s}}$$

▶ **Quick Check:** You could use Eq. (8.5) instead. Since $\theta = 2\pi$ rad

$$\omega_{av} = \frac{\theta}{t} = \frac{2\pi \text{ rad}}{1.537 \text{ min}} = \frac{6.28 \text{ rad}}{1.537 \text{ min}} = 4.09 \text{ rad/min}$$

Figure 8.5 depicts the slightly different circumstance in which both θ and l are measured from some reference line; in this case, the x-axis. Here, the finite intervals $\Delta l = l_f - l_i$ and $\Delta\theta = \theta_f - \theta_i$, as before, are related by

$$\Delta l = r\Delta\theta \tag{8.7}$$

If these are traversed in a time duration Δt, then

$$\frac{\Delta l}{\Delta t} = r\frac{\Delta\theta}{\Delta t} \tag{8.8}$$

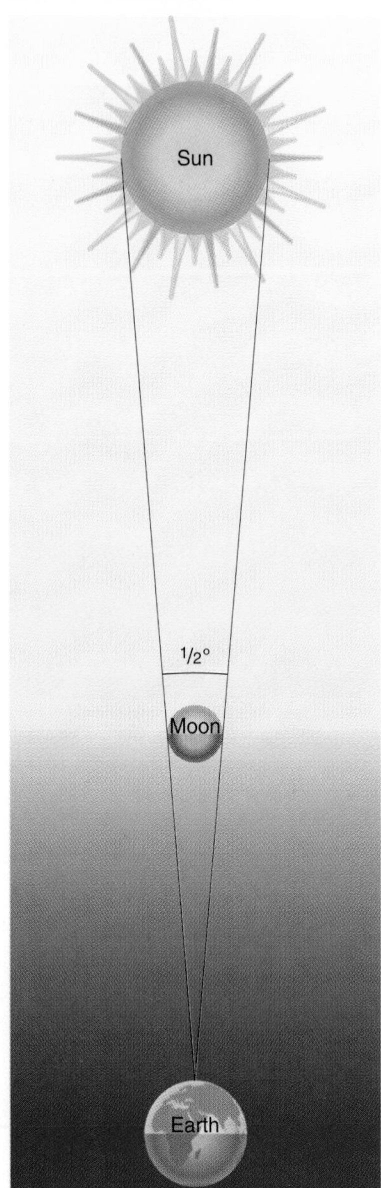

Figure 8.4 The Moon is closer to us than the Sun. Therefore, even though it is much smaller than the Sun, it happens to subtend very nearly the same angle when viewed from Earth.

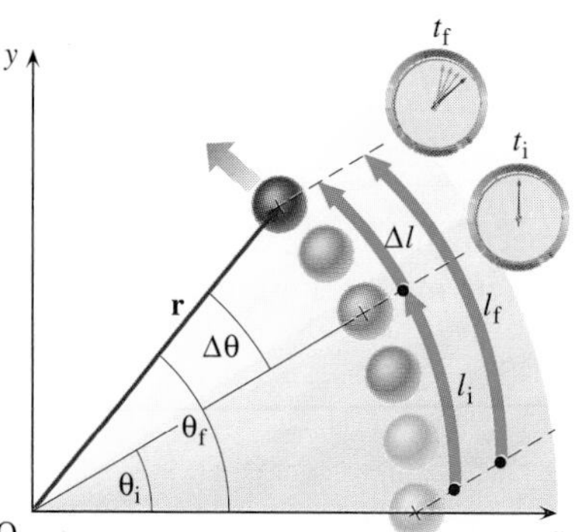

Figure 8.5 As the ball moves from angular position θ_i at l_i to angular position θ_f at l_f on the arc, the angle changes by $\Delta\theta$ and the arc-length measured up from the x-axis changes by Δl.

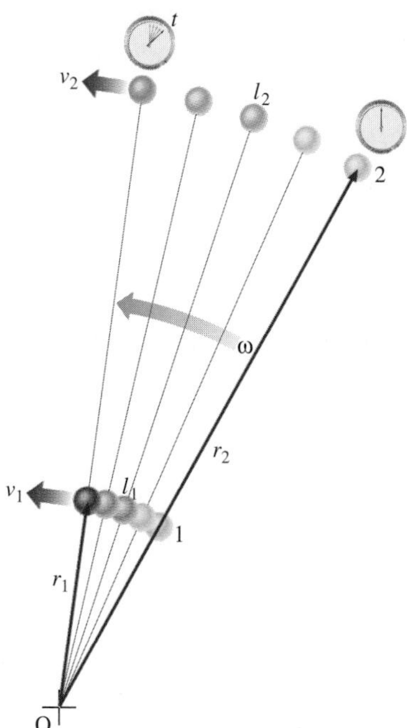

Figure 8.6 Here, beads 1 and 2 move in step, swinging around their respective circles together. Bead 2 travels a larger path and must therefore be moving faster ($v_2 > v_1$), even though both beads have the same ω.

which provides an alternative statement for the average speeds

$$v_{av} = \frac{\Delta l}{\Delta t} \quad \text{and} \quad \omega_{av} = \frac{\Delta\theta}{\Delta t} \tag{8.9}$$

These reduce to Eqs. (2.1) and (8.5) when $\theta_i = 0$, $\theta_f = \theta$, $t_i = 0$, and $t_f = t$. The advantage to this notation is that we can define the **instantaneous angular speed** (or just the angular speed ω) by finding the limit of the average angular speed as the averaging time interval becomes smaller and smaller. In other words

$$\omega = \lim_{\Delta t \to 0} \left[\frac{\Delta\theta}{\Delta t}\right] \tag{8.10}$$

Thus, ω *is the time-rate-of-change of* θ. Taking the limit as $\Delta t \to 0$ of Eq. (8.8) yields

$$v = r\omega \tag{8.11}$$

where ω is in rad/s. Keep in mind that the velocity is always tangent to the curved path. At any instant, if we know ω, we can find v, and vice versa.

It's evident from Fig. 8.6 that the more distant a bead is from the center of rotation, the more rapidly it must be moving. Once around for bead 1 is a small circle, while bead 2 travels an appreciably larger path in the same time. Thus, if they stay abreast, $v_2 > v_1$ even though ω is the same for each of them. That's why the outermost skater in a long rotating line has to be the fastest, and why the inner rail position in a race is most advantageous.

Example 8.4 A phonograph record of Beethoven's Ninth Symphony is whirling around at $33\frac{1}{3}$ rpm. Two 1.0×10^{-5}-kg lint balls are at rest on the disk, one on the end of the paper label 5.0 cm from the central spin axis and the other 15 cm out on the very edge of the record. Determine their respective linear speeds and the corresponding friction forces acting on them.

Solution: [Given: ω equivalent to $33\frac{1}{3}$ rpm, $m_1 = m_2 = 1.0 \times 10^{-5}$ kg, $r_1 = 0.050$ m and $r_2 = 0.15$ m. Find: v_1, v_2, F_{f1}, and F_{f2}.] Convert $33\frac{1}{3}$ rpm to rad/s: $\omega = (33\frac{1}{3}\text{ rpm})(2\pi\text{ rad/rot})/(60\text{ s/min}) = 3.491$ rad/s. Hence, from Eq. (8.11), the lint balls are moving tangentially at speeds of

$$v_1 = r_1\omega = (0.050\text{ m})(3.491\text{ rad/s}) = \boxed{0.17\text{ m/s}}$$

and $$v_2 = r_2\omega = (0.15\text{ m})(3.491\text{ rad/s}) = \boxed{0.52\text{ m/s}}$$

The rad unit is dropped when dealing with linear quantities.

The balls are held in place by friction, which must supply the needed centripetal force. It follows from Eq. (5.11) that

$$F_{f1} = F_{C1} = \frac{m_1 v_1^2}{r_1} = \frac{(1.0 \times 10^{-5}\text{ kg})(0.17\text{ m/s})^2}{0.050\text{ m}}$$

and $$\boxed{F_{f1} = 5.8 \times 10^{-6}\text{ N}}$$

Moreover

$$F_{f2} = F_{C2} = \frac{m_2 v_2^2}{r_2} = \frac{(1.0 \times 10^{-5}\text{ kg})(0.52\text{ m/s})^2}{0.15\text{ m}}$$

and $$\boxed{F_{f2} = 1.8 \times 10^{-5}\text{ N}}$$

▶ **Quick Check:** $F_C = mr\omega^2$; hence, $F_{C1} = (1 \times 10^{-5}\text{ kg})(0.050\text{ m})(3.5\text{ rad/s})^2 \approx 6\ \mu\text{N}$ and $F_{C2} = (1 \times 10^{-5}\text{ kg})(0.15\text{ m})(3.5\text{ rad/s})^2 \approx 20\ \mu\text{N}$.

Example 8.5 A horizontal cylinder (Fig. 8.7) with a radius of 50 cm has a rope wrapped around it that supports a weight at its end. At what constant angular speed must the shaft revolve if the weight is to descend at 1.0 m/s?

Solution: [Given: $R = 0.50$ m and $v = 1.0$ m/s. Find: ω.] As the shaft turns through an angle θ, a length of rope l unwinds such that $l = R\theta$, which *is* the distance the weight drops. Hence, $v = R\omega$ and

$$\omega = \frac{v}{R} = \frac{1.0\text{ m/s}}{0.50\text{ m}} = \boxed{2.0\text{ rad/s}}$$

▶ **Quick Check:** $v = R\omega = (0.5\text{ m})(2\text{ rad/s}) = 1$ m/s.

Figure 8.7 As the cylinder revolves, the rope wrapped around it unwinds, and the weight descends.

When a rigid body rotates about an axis at some angular speed ω, *that rate characterizes the motion and is the same for the entire object.* By contrast, each constituent point of the body moves with a linear speed determined by its perpendicular distance from the spin axis.

8.2 Angular Acceleration

Variations in ω are as commonplace as are variations in v. Just imagine a rock whirling faster and faster on the end of a string, or a carousel slowing down as it comes to rest. If a rotating body changes its angular speed by an amount $\Delta\omega$ in a time interval Δt, its **average angular acceleration** (denoted by the lowercase Greek letter *alpha*) is

$$\alpha_{\text{av}} = \frac{\Delta\omega}{\Delta t} = \frac{\omega_{\text{f}} - \omega_{\text{i}}}{t_{\text{f}} - t_{\text{i}}} \tag{8.12}$$

The unit of angular acceleration is radians per second-per second or rad/s^2. Once again, if we take the limit of this expression as $\Delta t \to 0$, the average angular acceleration approaches the **instantaneous angular acceleration** and

$$\alpha = \lim_{\Delta t \to 0} \left[\frac{\Delta\omega}{\Delta t}\right] \tag{8.13}$$

For simplicity, we restrict our study to situations where α is constant.

From Eq. (8.11), it follows that a change in angular speed will be accompanied by a change in linear speed, $\Delta v = r\Delta\omega$. And Eq. (8.12) can be rewritten as

$$\alpha_{\text{av}} = \frac{\Delta v}{r\Delta t} = \frac{1}{r} a_{\text{av}} \tag{8.14}$$

Taking the limit as $\Delta t \to 0$ of this equation provides the link between the linear and angular acceleration, namely

$$\alpha = \frac{a_{\text{T}}}{r} \tag{8.15}$$

as long as α is in rad/s^2 (Table 8.2). Here, a_{T} is a *tangential acceleration arising only from a change in the tangential speed*—a_{T} doesn't exist if v is constant; that is, if $\Delta v = 0$. It is the scalar value of an acceleration vector drawn from the object tangent to the path of motion at any instant (Fig. 8.8):

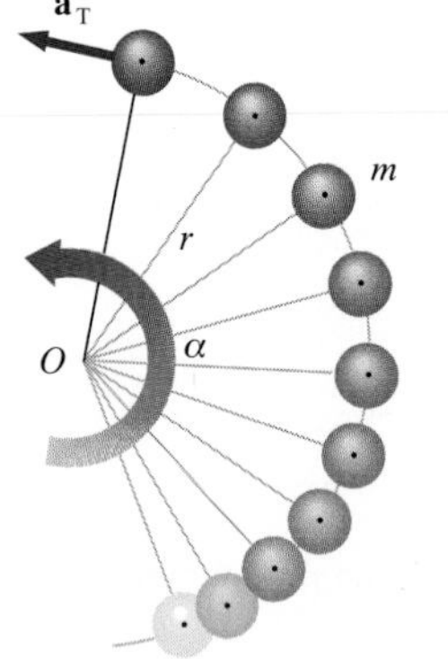

Figure 8.8 When an object moving in a circle goes faster (or slower), it accelerates tangentially. The angle that marks its position (θ) changes at a rate (ω) that is not constant. The object has an angular acceleration (α) such that $a_{\text{T}} = r\alpha$.

$$a_{\text{T}} = r\alpha \tag{8.16}$$

Centripetal acceleration a_{C} arises from a change in the direction of the motion (not a change in speed): v may or may not be constant. So long as the direction of $\mathbf{v}$ changes, there will be an a_{C}, and that always occurs for circular motion. When both the direction and magnitude of $\mathbf{v}$ vary in time, a_{T} will exist acting perpendicular to a_{C}. This is true for a runner who speeds up around a turn, or a monkey who slows while swinging on a vine.

TABLE 8.2 Linear and angular kinematic parameters

Equation	Units	
$l = r\theta$	l (m)	θ (rad)
$v = r\omega$	v (m/s)	ω (rad/s)
$a_{\text{T}} = r\alpha$	a_{T} (m/s^2)	α (rad/s^2)

Example 8.6 At a particular moment in a race, a car roaring around a turn with a radius of 50 m had an angular speed of 0.60 rad/s and an angular acceleration of 0.20 rad/s^2. Compute its linear speed, its centripetal acceleration, its tangential acceleration, and its *total* linear acceleration (Fig. 8.9).

Solution: [Given: $r = 50$ m, $\omega = 0.60$ rad/s, and $\alpha = 0.20$ rad/s^2. Find: v, a_C, a_T, and the total acceleration a.] The linear speed is

$$v = r\omega = (50\text{ m})(0.60\text{ rad/s}) = \boxed{30\text{ m/s}}$$

From which it follows that

$$a_C = \frac{v^2}{r} = \frac{(30\text{ m/s})^2}{50\text{ m}} = \boxed{18\text{ m/s}^2}$$

The tangential acceleration is

$$a_T = r\alpha = (50\text{ m})(0.20\text{ rad/s}^2) = \boxed{10\text{ m/s}^2}$$

To find the total linear acceleration a, refer to the figure, which shows the relationship between $\mathbf{a}_C$, $\mathbf{a}_T$, and $\mathbf{a} = \mathbf{a}_C + \mathbf{a}_T$. Accordingly,

$$a = \sqrt{a_C^2 + a_T^2} = \sqrt{(18\text{ m/s}^2)^2 + (10\text{ m/s}^2)^2}$$

and $$a = 21\text{ m/s}^2$$

at an angle of

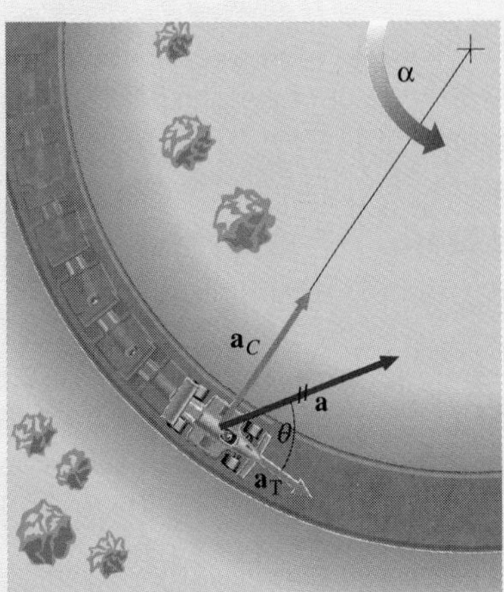

Figure 8.9 A car moving faster and faster around a curve has both a tangential ($\mathbf{a}_T$) and a centripetal ($\mathbf{a}_C$) acceleration. Its total acceleration is the vector $\mathbf{a} = \mathbf{a}_T + \mathbf{a}_C$.

$$\phi = \tan^{-1}\frac{a_C}{a_T} = \boxed{61°}$$

▶ **Quick Check:** It's always best to avoid making a calculation that depends on a previous one. Since we computed v, we really should have used

$$a_C = r\omega^2 = (50\text{ m})(0.60\text{ rad/s})^2 = 18\text{ m/s}^2$$

Moreover, $a_C = a \sin\phi = (21\text{ m/s}^2)(\sin 61°) = 18\text{ m/s}^2$.

8.3 Equations of Constant Acceleration

In Chapter 3, we considered linear motion at constant acceleration and arrived at five important equations. With the proper interpretation, these same expressions also apply to curvilinear motion. Hence, if we take l to be the arc-length traveled and require a *constant* tangential acceleration a_T, then

$$v_f = v_i + a_T t \tag{8.17}$$

$$v_{av} = \tfrac{1}{2}(v_i + v_f) \tag{8.18}$$

$$l = v_{av}t = \tfrac{1}{2}(v_i + v_f)t \tag{8.19}$$

$$l = v_i t + \tfrac{1}{2}a_T t^2 \tag{8.20}$$

and

$$v_f^2 = v_i^2 + 2a_T l \tag{8.21}$$

As usual, the positive direction for l determines the positive directions of both v and a_T.

Example 8.7 A space station in the shape of a 100-m-diameter wheel is spinning so as to impart a linear tangential speed of 22.1 m/s to a point on its outermost wall, thereby providing an artificial gravity of 1g. A

(continued)

(continued)

pair of thrusters is to be fired to decelerate that point at a rate of 1.00 m/s^2. How long should the burn last to bring the station to rest? How large an arc-length will a reference point on the rim traverse during the process? What will happen if the rockets burn too long?

Solution: [Given: Take the initial direction of rotation as positive; thus, $v_i = +22.1$ m/s, $R = 50$ m and, because we are dealing with a deceleration here, $a_T = -1.00$ m/s^2. Find: t and l for $v_f = 0$.] Let the origin ($l = 0$) correspond to the location of the reference point at the moment the thrusters fire ($t = 0$). From Eq. (8.17), we obtain t, as

$$t = \frac{(v_f - v_i)}{a_T} = \frac{(0 - 22.1\text{ m/s})}{(-1.00\text{ m/s}^2)} = \boxed{22.1\text{ s}}$$

Knowing t, we can use either one of Eqs. (8.19), (8.20), or (8.21) to find l, thus

$$l = \frac{(v_f^2 - v_i^2)}{2a_T} = -\frac{-(22.1\text{ m/s})^2}{2(-1.00\text{ m/s}^2)} = \boxed{244\text{ m}}$$

If the rockets burn longer than 22.1 s, the station will begin to rotate in the opposite direction, picking up speed such that v_f becomes negative and l decreases, ultimately becoming negative too.

▶ **Quick Check:** $l = v_{av}t = \frac{1}{2}(22.1\text{ m/s} + 0) \times (22.1\text{ s}) = 244$ m.

These equations of curvilinear motion can most easily be transformed into an equivalent angular description when the path is circular. We know that

$$l = r\theta \qquad [8.1]$$

$$v = r\omega \qquad [8.11]$$

$$a_T = r\alpha \qquad [8.16]$$

It is then a simple matter of substitution into Eqs. (8.17) through (8.21) to arrive at

$$\omega_f = \omega_i + \alpha t \qquad (8.22)$$

$$\omega_{av} = \tfrac{1}{2}(\omega_i + \omega_f) \qquad (8.23)$$

$$\theta = \omega_{av}t = \tfrac{1}{2}(\omega_i + \omega_f)t \qquad (8.24)$$

$$\theta = \omega_i t + \tfrac{1}{2}\alpha t^2 \qquad (8.25)$$

and

$$\omega_f^2 = \omega_i^2 + 2\alpha\theta \qquad (8.26)$$

where the direction in which θ is positive is the direction in which ω and α are both positive.

Example 8.8 Mounted in a bus is a 2.0-m-diameter flywheel, a massive disk often used to store rotational energy. If it is accelerated from rest at a constant rate of 2.0 rpm per second, what will be the angular speed of a point on the rim of the flywheel after 5.0 s? Through what angle will that point have rotated?

(continued)

(continued)

Solution: [Given: $r = 1.0$ m, $\alpha = +2.0$ rpm/s and $\omega_i = 0$. Find: ω_f and θ at $t = 5.0$ s.] Let's first get α in rad/s^2; that is,

$$\alpha = \frac{(2.0 \text{ rpm/s})(2\pi \text{ rad/rot})}{(60 \text{ s/min})} = 0.209 \text{ rad/s}^2$$

To find the final angular speed, use Eq. (8.22)

$$\omega_f = \omega_i + \alpha t = 0 + (0.209 \text{ rad/s}^2)(5.0 \text{ s})$$

and
$$\omega_f = 1.045 \text{ rad/s}$$

or, to two figures, $\boxed{\omega_f = 1.0 \text{ rad/s}}$. Equation (8.25) provides θ:

$$\theta = \tfrac{1}{2}\alpha t^2 = \tfrac{1}{2}(0.209 \text{ rad/s}^2)(5.0 \text{ s})^2 = \boxed{2.6 \text{ rad}}$$

▶ **Quick Check:** Notice that at $\alpha = 2.0$ rpm/s, the flywheel's angular speed should pick up 10 rpm in 5 s; that is, $(2\pi 10 \text{ rad/min})/(60 \text{ s/min})$, or $\omega_f = 1.05$ rad/s, which gives the flywheel an ω_{av} of (1.05/2) rad/s, so it will revolve through $\theta = \omega_{av} t = (1.05/2)(5 \text{ rad}) = 2.6$ rad.

The situation of a freely rolling disk, cylinder, or sphere (Fig. 8.10) merits a closer study. The term *freely rolling* means there is no slipping at the point of contact with the ground—no skidding and no spinning in place. The wheel in the figure rolls to the right and point O on the axis moves to O' as A moves to A', and B moves to B'. The arc-length from B to A equals l just as length $\overline{BA'} = \overline{OO'} = l$. Thus, the rectilinear distance traveled by the centerpoint is $l = R\theta$, and so $v = R\omega$ and $a = R\alpha$. *The center of the wheel has a linear speed and acceleration equal to that of any point on its rim.* Notice that the velocity of point B, while in contact with the ground, measured with respect to O, namely $\mathbf{v}_{BO}$, is to the left. The velocity of O with respect to the ground, $\mathbf{v}_{OG}$, is to the right. Furthermore, $v_{BO} = v_{OG} = R\omega$; hence

$$\mathbf{v}_{BG} = \mathbf{v}_{BO} + \mathbf{v}_{OG} = 0$$

The velocity of point B with respect to the ground at the moment of touching is zero—the contact point doesn't slip across the surface. (See Fig. 5.28.)

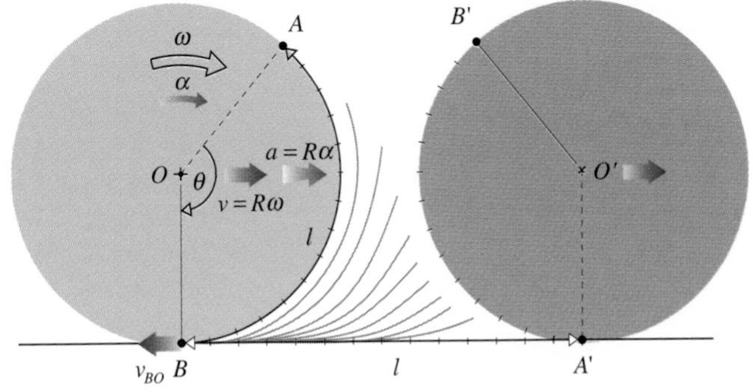

Figure 8.10 As a body rolls, revolving about its center at a rate ω, with an angular acceleration α, point O moves at a speed $v = R\omega$, with an acceleration at that instant of $a = R\alpha$.

Example 8.9 A cyclist traveling at 5.0 m/s uniformly accelerates up to 10.0 m/s in 2.0 s. Each tire of the bike has a 35.0-cm radius, and a small pebble is caught in the tread of one of them. (a) What is the angular acceleration of the pebble during those two seconds? (b) Through what angle does the pebble revolve? (c) How far does the pebble travel during that accelerating interval?

Solution: [Given: $v_i = 5.0$ m/s, $v_f = 10.0$ m/s, $t = 2.0$ s and $R = 0.35$ m. Find: (a) α, (b) θ, and (c) l.] The bike moves along with the center of each wheel, accelerating at a linear rate of

$$a = \frac{(v_f - v_i)}{t} = \frac{(10 \text{ m/s}) - (5.0 \text{ m/s})}{2.0 \text{ s}} = 2.5 \text{ m/s}^2$$

This, in turn, equals $R\alpha$ and therefore

$$\alpha = \frac{a}{R} = \frac{2.5 \text{ m/s}^2}{0.35 \text{ m}} = \boxed{7.1 \text{ rad/s}^2}$$

which is the angular acceleration of each wheel and, hence, of the pebble. Inasmuch as the linear speed of the pebble equals the linear speed of the bike, $v_i = r\omega_i = 5.0$ m/s, $\omega_i = 14.29$ rad/s and so, from Eq. (8.25)

$$\theta = \omega_i t + \tfrac{1}{2}\alpha t^2$$

(continued)

(continued)

Therefore

$$\theta = (14.29\ \text{rad/s})(2.0\ \text{s}) + \tfrac{1}{2}(7.1\ \text{rad/s}^2)(2.0\ \text{s})^2$$

and $\boxed{\theta = 43\ \text{rad}}$. Both bike and pebble travel distances of $l = R\theta = \boxed{15\ \text{m}}$.

▶ **Quick Check:** At an average speed of 7.5 m/s, moving for 2.0 s, the cyclist traverses 15 m, and everything checks.

THE DYNAMICS OF ROTATION

Newton's Three Laws are the foundation of our understanding of dynamics, both linear and rotational. A body revolves as a single entity when its various parts are constrained to stay together by internal forces. That being the case, we may describe the overall rotation in terms of our new kinematic concepts θ, ω, and α and introduce new dynamical quantities, as well.

Newton's First Law has its rotational equivalent in the concept that *a body at rest tends to stay at rest while a body in uniform rotational motion tends to stay in such motion, except insofar as it is acted upon by a torque.* Remember the discussion in the section "Rotational Equilibrium" (p. 173), where we considered the two determining conditions of nonaccelerating systems: $\Sigma\mathbf{F} = 0$ and $\Sigma\tau = 0$. Force is the changer of all motion. Specifically, the *moment of the force* is the *torque,* and *torque is the changer of rotational motion.*

The resistance to the change in motion of an object is called *inertia,* and that resistance is physically embodied in the *inertial mass* (or just *mass,* for short). This same property of matter can appear as a resistance to the change in rotational motion, where it is called *rotational inertia.* This rotational resistance is associated with both the amount of mass and its distribution with respect to the axis of rotation. Euler called that composite concept the *moment of inertia.*

8.4 Rotational Inertia

In swinging closed a heavy door or turning a massive wheel, it's evident that if the thing is to accelerate rotationally, a force must be applied, and it must be applied with some moment-arm. There must be a net torque.

Envision a very small mass m constrained to move on a circle of radius r, about an axis passing through O, under the influence of an applied tangential force F (Fig. 8.8). The particle will experience a tangential acceleration via the Second Law

$$F = ma_{\text{T}}$$

or, using Eq. (8.16)

$$F = mr\alpha$$

The torque about O arising from F is rF, and multiplying both sides of the above equation by r yields

$$\tau_0 = rF = mr^2\alpha \tag{8.27}$$

The equation suggests that the rotational equivalent of mass (of *inertia*) is the quantity mr^2. This is the **moment of inertia of a point-mass** about a given axis, and it is designated as $I. = mr^2$. Hence, $\tau_0 = I.\alpha$ just as $F = ma$.

A rigid body consists of a great many interacting particles. When such a body is set into rotation, Eq. (8.27) describes any one of those component particles. The sum of all the torques acting on all the constituent point-masses results in an overall angular acceleration of the body such that

$$\Sigma\tau_0 = (\Sigma mr^2)\alpha \tag{8.28}$$

The summation on the right is the moment of inertia of the body about the axis of rotation passing through O, hence

$$I = \Sigma mr^2 \tag{8.29}$$

Each particle (1, 2, 3, . . .) with its own mass and perpendicular distance from O has its own moment of inertia. All these can be added so that $I = m_1r_1^2 + m_2r_2^2 + m_3r_3^2 + \ldots$.

Calculus is usually used to compute moments of inertia, but we'll rely on Table 8.3, which lists values of I for some uniform symmetrical bodies about various axes. *The more mass and the farther it is from the axis, the greater will be I, and the greater will be the resistance to the change in the rotational motion* (Fig. 8.11). That's why the thin walled ring in Table 8.3 has twice the moment of inertia of the

TABLE 8.3 Moments of inertia

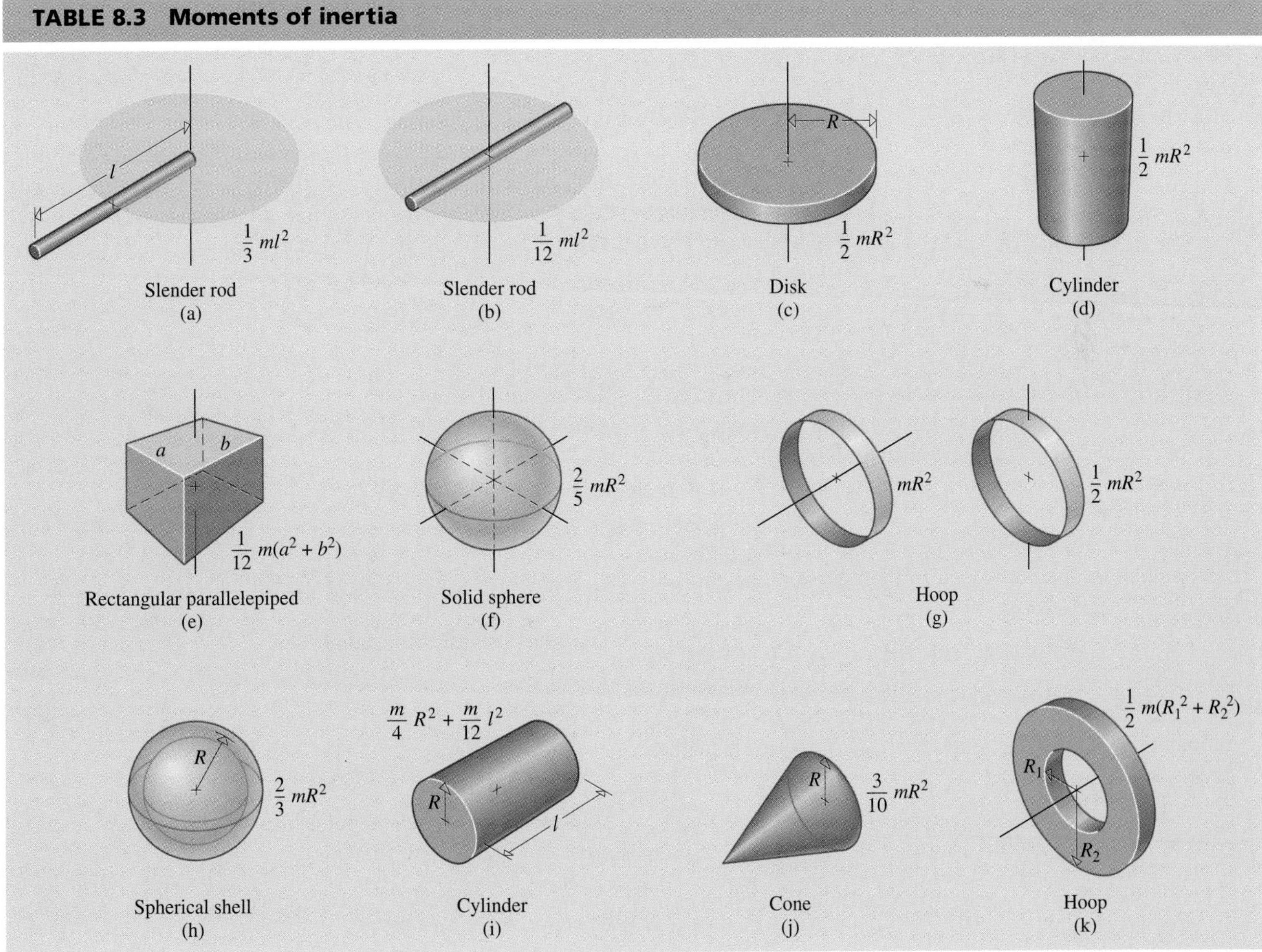

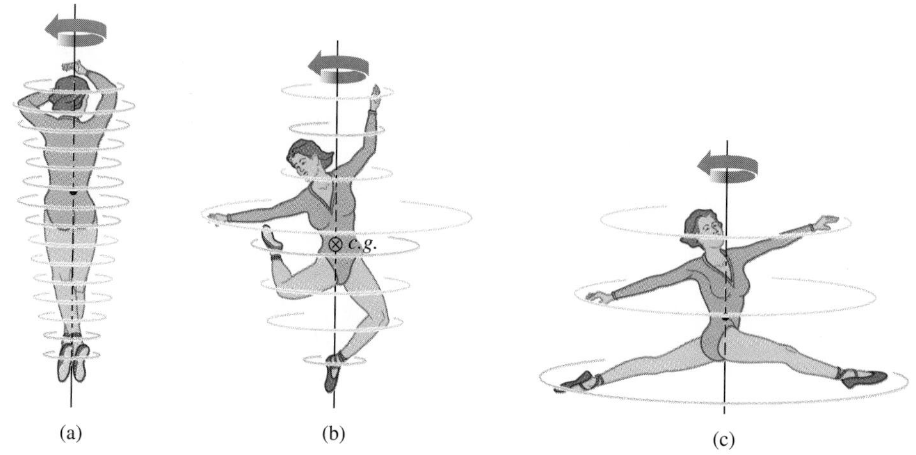

Figure 8.11 With the limbs drawn in near the spin axis as in (a), the moment of inertia is comparatively small. With bent knees I increases, as in (b), and reaches a maximum with arms and legs extended perpendicular to the axis (c).

solid disk with the same mass, and why the moment of inertia of the hollow sphere is $\frac{5}{3}$ that of the solid sphere. The expression

$$\Sigma\tau_0 = I\alpha \tag{8.30}$$

is the primary dynamical equation of rotating systems. The farther a bit of mass is from the axis, the larger will be I, and the more torque must be applied to attain a given α. "Choking up" on a baseball bat (that is, gripping it in from the end) decreases I about the pivot and allows the same torque to produce a larger α. If a youngster can't swing the bat fast enough to meet the incoming pitch, a larger α will help.

Example 8.10 The evil Dr. Doom, involved in an extortion threat to de-spin the Earth, plans to mount a series of surplus rockets tangentially all along the Equator. Taking the planet to be a uniform sphere of radius 6.37×10^6 m and mass 5.98×10^{24} kg, how much continuous total thrust would the rockets need to apply to accomplish the heinous deed in the course of 12 hours?

Solution: [Given: $R_\oplus = 6.37 \times 10^6$ m and $M_\oplus = 5.98 \times 10^{24}$ kg. Find: F such that $\omega_f = 0$ when $\Delta t = 24$ h.] The necessary torque is given by Eq. (8.30), so we must first determine I and α. According to Table 8.3, the moment of inertia of a solid sphere with the mass of the Earth is

$$I = \tfrac{2}{5}mR_\oplus^2 = \tfrac{2}{5}(5.98 \times 10^{24}\text{ kg})(6.37 \times 10^6\text{ m})^2$$

or $I = 9.71 \times 10^{37}$ kg·m². It must decelerate from its present angular rate to zero in 12 h. The former (ω_i) can be computed accurately enough assuming the planet now spins through 2π rad in 24 h; that is, $\omega_i = 0.727 \times 10^{-4}$ rad/s, and

$$\alpha = \frac{\omega_f - \omega_i}{\Delta t} = \frac{0 - 0.727 \times 10^{-4}\text{ rad/s}}{12\text{ h} \times 60\text{ min/h} \times 60\text{ s/min}}$$

and $$\alpha = -1.68 \times 10^{-9}\text{ rad/s}^2$$

The total required torque is then

$$\Sigma\tau_0 = I\alpha$$

$$\Sigma\tau_0 = (9.71 \times 10^{37}\text{ kg·m}^2)(-1.68 \times 10^{-9}\text{ rad/s}^2)$$

and $$\Sigma\tau_0 = -1.63 \times 10^{29}\text{ N·m}$$

Since each rocket acts about the spin axis with a moment arm of $R_\oplus$, it follows that

$$\Sigma\tau_0 = R_\oplus\Sigma F$$

(continued)

(continued)

and the total thrust must be

$$\Sigma F = \frac{\Sigma \tau_O}{R_{\oplus}} = \frac{-1.63 \times 10^{29}\ \text{N}\cdot\text{m}}{6.37 \times 10^{6}\ \text{m}} = \boxed{-2.6 \times 10^{22}\ \text{N}}$$

The minus sign shows that the thrust must oppose the positive ω_i. The value 2.6×10^{22} N is a realistically large force. Compare it to the 3.4×10^{7} N thrust of a *Saturn* Moon rocket. Dr. Doom would have to fire over 700 million million such rockets continuously over the whole 12 hours! So, not to worry.

▶ **Quick Check:** $I \approx 0.4(6 \times 10^{24}\ \text{kg})(36 \times 10^{12}\ \text{m}^2) \approx 9 \times 10^{37}\ \text{kg}\cdot\text{m}^2$; $\omega_i(24\ \text{h} \times 60\ \text{min/h} \times 60\ \text{s/min}) = 6.28\ \text{rad} = 2\pi$, so $\omega_i = 0.727 \times 10^{-4}$ rad/s is okay. Using Eq. (8.24), $\theta = (3.6 \times 10^{-5}\ \text{rad/s}) \times (4.3 \times 10^{4}\ \text{s}) = 1.6$ rad. From Eq. (8.26) $\alpha = -\omega_i^2/2\theta = -1.7 \times 10^{-9}\ \text{rad/s}^2$.

Equation (8.30) governs all sorts of angular motion including, for example, that of a pole-carrying tightrope walker. The long balance pole has a large moment of inertia about its center. Since the acrobat cannot topple over sideways off the wire without rotating the pole, and since such angular acceleration is resisted by the large rotational inertia of the pole, it can be quite handy. Any kid walking on a narrow curb knows to stick both arms out for "balance."

Legs swinging about hip joints also have rotational inertia. Long, heavy legs require large torques to accelerate them. That's why animals with four lightweight legs can usually run faster than erect creatures with two necessarily big feet and relatively massive legs. One thing to do to reduce the rotational inertia I about the hip joint when high accelerations are needed is to bend the legs. Most people must know that, since few run stiff-legged.

Example 8.11 A 10.0-kg mass hangs on a rope wrapped around a freely rotating 2.00-kg cylinder of radius 10.0 cm as shown in Fig. 8.12. Determine the tension in the rope and the accelerations of both the cylinder and the mass. Use $g = 9.81\ \text{m/s}^2$.

Solution: [Given: $m = 10.0$ kg, $M_c = 2.00$ kg, and $R = 10.0$ cm $= 0.100$ m. Find: F_T, α, and a.] There are three unknowns, and we'll need three equations (the sum-of-the-torques, the sum-of-the-forces, and the relation between a and α). The tension of the rope produces a torque on the cylinder such that

$$\circlearrowright^{+}\Sigma \tau_O = F_T R = I\alpha$$

and so $F_T = I\alpha/R$. Furthermore, for the descending mass traveling in the vertical direction

$$\downarrow^{+}\Sigma F_V = mg - F_T = ma$$

As we saw earlier, the rope accelerates at a rate equal to that of a point on the cylinder, $a = R\alpha$. All three equations yield

$$mg - F_T = ma = mR\alpha$$

$$mg - \frac{I\alpha}{R} = mR\alpha$$

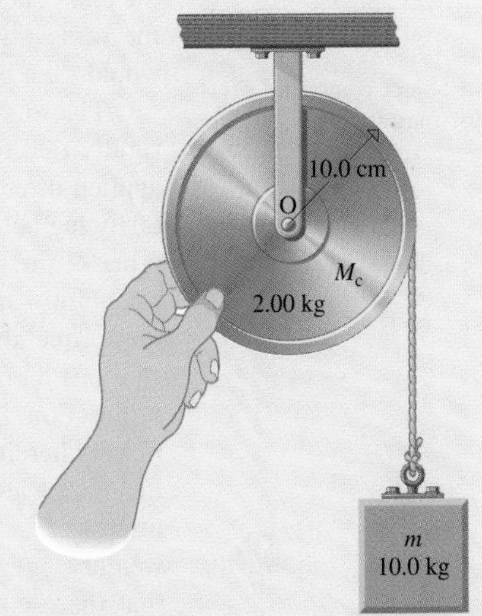

Figure 8.12 Up to this point, we neglected the motion of the pulley in this sort of problem. Now, however, we can take into account that it really isn't massless. The pulley has a moment of inertia and will reduce the acceleration of m.

(continued)

(continued)

and so
$$\alpha = \frac{mg}{mR + \dfrac{I}{R}}$$

Since

$$I = \tfrac{1}{2}M_cR^2 = \tfrac{1}{2}(2.00\text{ kg})(0.100\text{ m})^2 = 0.010\text{ kg}\cdot\text{m}^2$$

$$\alpha = \frac{98.1\text{ kg}\cdot\text{m/s}^2}{(1.00\text{ kg}\cdot\text{m} + 0.100\text{ kg}\cdot\text{m})} = \boxed{89.2\text{ rad/s}^2}$$

This result corresponds to $a = R\alpha = \boxed{8.92\text{ m/s}^2}$ and a tension of $F_T = m(g - a) = \boxed{8.90\text{ N}}$.

▶ **Quick Check:** Let's examine the expression for α at its extremes; namely, when $m \gg M_c$ and $m \ll M_c$. In the first case ($m \gg M_c$), I/R would be negligible compared to mR. Thereupon $\alpha \approx g/R$; that is, $a \approx g$; $F_T = 0$, and we would have free-fall. When $m \ll M_c$, $I/R \gg mR$, and $\alpha \approx mgR/I$, which is vanishingly small—hardly any motion occurs. Both conclusions are reasonable.

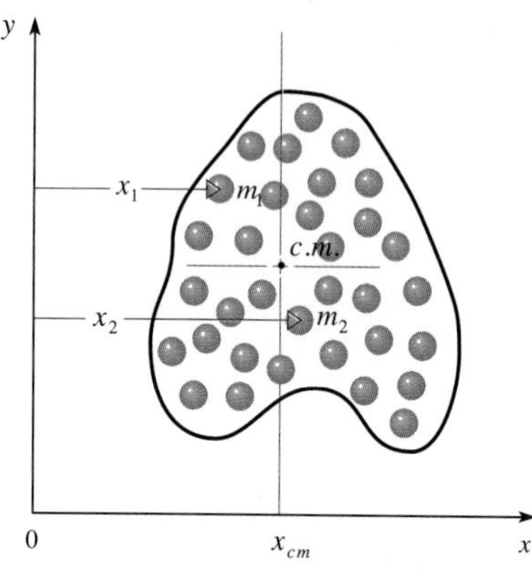

Figure 8.13 An object composed of a number of point-masses m_1, m_2, ...

8.5 The Center-of-Mass

Given an extended object, it is possible to find a point, the **center-of-mass** (*c.m.*), *where all the mass (M) of the object can be thought of as concentrated,* as the object translates in compliance with Newton's Second Law: $\Sigma\mathbf{F} = M\mathbf{a}_{cm}$.

To locate the *c.m.*, envision an object of mass M composed of particles bound together via internal interactions (Fig. 8.13). Suppose the body, initially at rest, is acted upon by several externally applied forces that *do not cause it to rotate*—$\Sigma\tau = 0$. If the sum of those forces is nonzero, the body must translate in the direction of the resultant. Assume that the net force is in the y-direction. An acceleration (a) of the body in that direction will then occur. What we have to find is the point, the *c.m.*, such that if the net force vector acted through that point it would produce the same translational motion of the object as did the group of forces. The body would then behave as if its entire mass was at the *c.m.*

The answer is evident for the simple case of two identical spheres of mass m rigidly tied to one another by a massless rod, as in Fig. 8.14. If the net force $\Sigma\mathbf{F}$ is applied directly to one of the spheres causing it to accelerate, the other will momentarily lag behind and swing around. If the net force is now applied near to one or the other of the spheres, the system will again rotate. Only when $\Sigma\mathbf{F}$ passes through the midpoint—the *c.m.*—will both spheres undergo the same acceleration. The resulting torque about the *c.m.* will then be zero, and no rotation will occur.

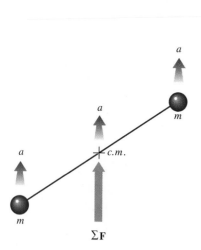

Figure 8.14 The center-of-mass of two identical spherical masses.

The more complex object in Fig. 8.14, uniformly accelerating at a upward in the y-direction, behaves exactly as if it were at rest in a *uniform* downward gravitational field (wherein $g = a$). We established that behavior much earlier in problems dealing with elevators. In these two equivalent pictures, the upward net accelerating force ($\Sigma F = ma$) is equal in magnitude and opposite in direction to the effective weight (mg), which acts through the *c.g.* If $\Sigma\mathbf{F}$ is to be applied through the *c.m.* so that there is no rotation, the net torque about the *c.m.* must equal zero. We know from Section 6.5 that in a gravitational field the torque due to the weight of each particle will be zero about the *c.g.*, where

$$x_{cg} = \frac{\Sigma F_W x}{\Sigma F_W} = \frac{\Sigma mgx}{\Sigma mg} \qquad [6.7]$$

In the present case, the acceleration due to gravity is equated with a, so that

$$x_{cm} = \frac{\Sigma max}{\Sigma ma}$$

Since the body accelerates as a whole at a single rate, a is the same on the top and bottom of this equation, and cancels, yielding

$$x_{\mathrm{cm}} = \frac{\Sigma m x}{\Sigma m} \tag{8.31}$$

Similar expressions hold in three dimensions for y_{cm} and z_{cm}.

In the event that the gravitational field is precisely uniform over the body, g can be canceled in Eq. (6.7), and the *c.g.* is then identical to the *c.m.* For a symmetric homogeneous body (for example, a sphere, cylinder, cube, disk, hoop, doughnut, or rectangular parallelepiped), the *c.m.* is at the geometric center. Usually, the field is adequately uniform for the *c.g.* and *c.m.* to be at essentially the same point. A rigid body has a fixed *c.m.* independent of the gravitational environment; the location of the *c.g.*, however, depends on the gravitational field. Indeed, several artificial satellites have maintained their orientations in space using the $1/r^2$ drop in the Earth's gravity. A long tube extending down from a satellite experiences a nonuniform field—the *c.g.* is displaced downward from the *c.m.*, allowing torques to develop that maintain a stable alignment (Discussion Question 7.5, p. 232).

We started this discussion by assuming that the applied torque was zero and the motion was purely translational. If the opposite is true—namely, $\Sigma\mathbf{F} = 0$ and $\Sigma\tau \neq 0$—the body will rotate accelerating about the *c.m.*, which cannot itself accelerate—the motion will be purely rotational. Adding in an additional net force will superimpose a translational acceleration of the *c.m.* onto the rotation about the *c.m.* A wrench or a baseball bat hurled through the air may tumble around (if $\Sigma\mathbf{F}$ does not pass through its *c.m.* at launch), but the center-of-mass will sail along a parabola as if it marked the motion of a point-mass. Only if you hit a cue ball along a line through its *c.m.* will it move away spinless, and only if you push this book along a line through its *c.m.* will it slide without rotating.

A body in motion does not have to be rigid for its mass to behave as if it were at the *c.m.* A diver in the air (Fig. 8.15) can tumble and twist, changing the location of her *c.m.* with respect to her body from one instant to the next via *internal forces,* and the *c.m.* will still free-fall along a parabola. Her *c.m.* will move in a ballistic arc as she rotates around it.

Two unequal-mass balls fastened together with a string are free to slide along the central horizontal rod. The system can be spun about the vertical shaft. Only when that vertical axis passes through the *c.m.* of the two balls will they remain where they are as the system rotates. A slight displacement to either side will unbalance the centripetal forces and cause the balls to slide to that side.

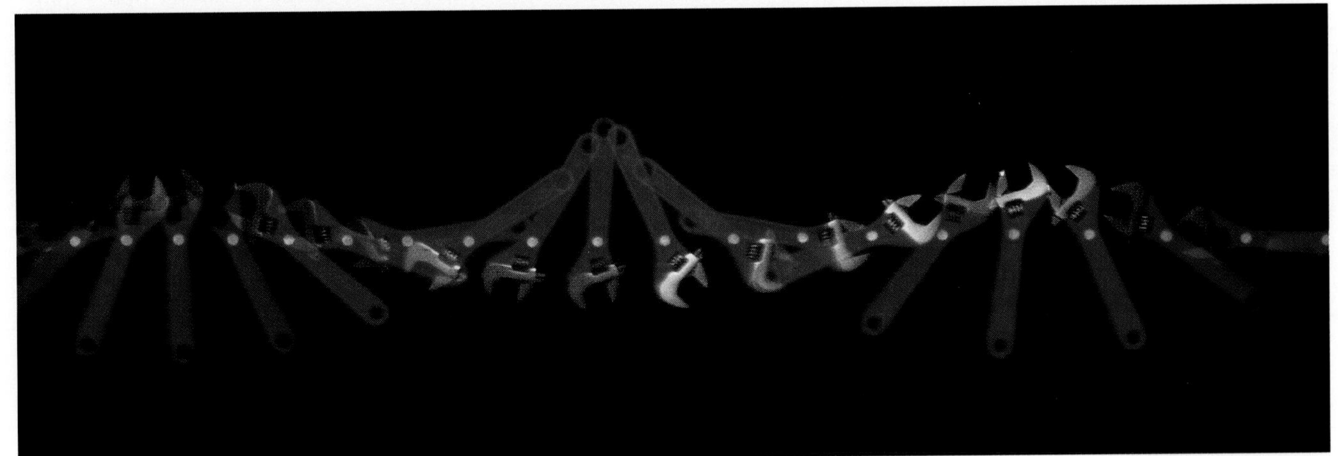

A tumbling wrench shown every 1/30 of a second. The center-of-mass (indicated by a spot of light paint) moves along a straight path, while the wrench revolves around it.

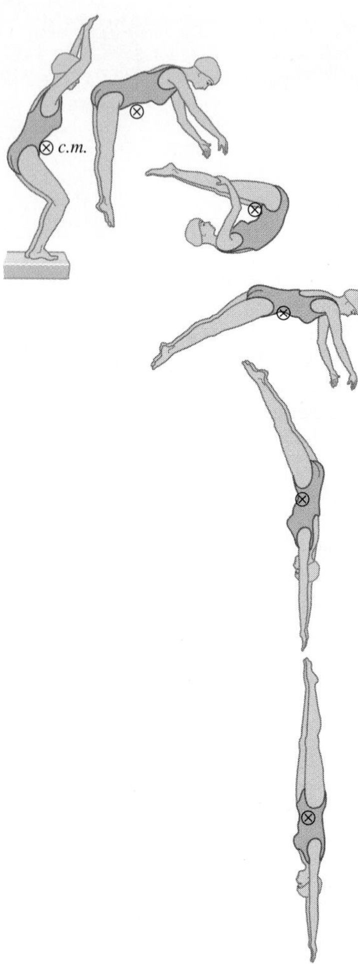

Figure 8.15 Since the motion of the *c.m.* can only change with the application of an external force, and no new force is applied, the *c.m.* moves in a ballistic arc regardless of the diver's twisting.

8.6 Rolling Down an Incline

The solid sphere (of radius R) in Fig. 8.16 is rolling freely, *without slipping* down an inclined plane. The point (A) on the sphere in contact, for an instant, with the incline would tend to slide down the plane were it not for the friction force. Thus, F_f points up the incline, resisting the downward motion of A and thereby generating a torque about the *c.m.*, which turns the sphere. *It is* F_f *that causes the rotation of the ball;* without friction, it would slide downward with point A in constant contact with the ramp.

The motion can be analyzed as a translation of the *c.m.* coupled with a rotation about the *c.m.* As ever, $\Sigma\mathbf{F} = m\mathbf{a}$, where m is the mass of the sphere and $\mathbf{a}$ is the linear acceleration of its center-of-mass. Consequently

$$\searrow\!+\Sigma F_{\parallel} = F_W \sin\theta - F_f = ma \tag{8.32}$$

Because the sphere is rotating, we also have $\Sigma\tau = I\alpha$. Taking the torque with reference to the *c.m.*, we get

$$\curvearrowright\!+\Sigma\tau_{cm} = F_f R = I_{cm}\alpha$$

for the rotation about the *c.m.* Now, from Table 8.3, $I_{cm} = \frac{2}{5}mR^2$. Moreover, we saw earlier that $a = R\alpha$; the two motions are interrelated. Substituting into the torque equation, we get

$$F_f R = \tfrac{2}{5}mR^2(a/R) \tag{8.33}$$

Using the fact that $F_W = mg$, Eqs. (8.32) and (8.33) can be solved simultaneously to yield

$$a = \tfrac{5}{7}g\sin\theta \tag{8.34}$$

and

$$F_f = \tfrac{2}{7}mg\sin\theta \tag{8.35}$$

The acceleration is independent of both m and R. It's *not* $g\sin\theta$ as it would have been if the sphere had *slid* down the incline; instead, it's appreciably smaller because of the rotational inertia.

The requirement that no slipping occurs demands that the frictional force be *static* friction, but that doesn't mean that the sphere is just about to slip. It cannot be assumed that $F_f = \mu_s F_N$. In Problem 63, you will find that the minimum coefficient of friction needed to prevent slipping is $\mu_s = \frac{2}{7}\tan\theta$. Tilt the plane beyond that, and the sphere slides.

Notice that, had the sphere been hollow, its moment of inertia ($I_{cm} = \frac{2}{3}mR^2$) would have been larger and the resulting acceleration smaller. But again a would be

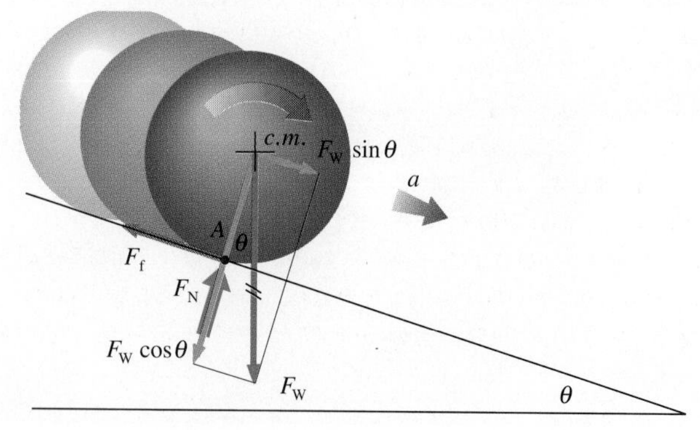

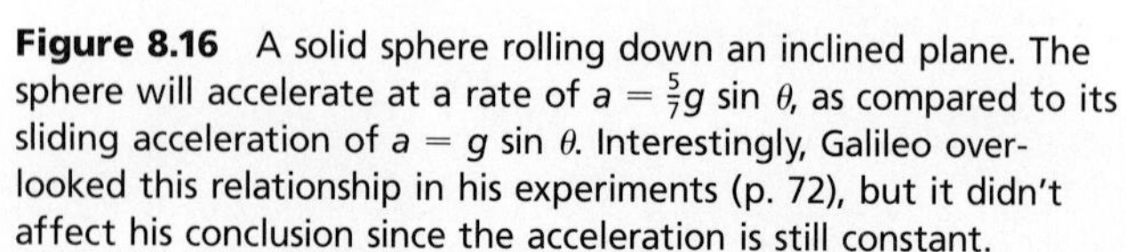
Figure 8.16 A solid sphere rolling down an inclined plane. The sphere will accelerate at a rate of $a = \frac{5}{7}g\sin\theta$, as compared to its sliding acceleration of $a = g\sin\theta$. Interestingly, Galileo overlooked this relationship in his experiments (p. 72), but it didn't affect his conclusion since the acceleration is still constant.

independent of both R and m. In a no-slip-rolling-down-an-incline race, the solid sphere will always beat the hollow one, while on a *frictionless* surface, it will be a tie every time (Problem 54). The acceleration depends only on the shape of the rolling body.

8.7 Angular Momentum

The rotational equivalent of force is torque, the *moment of the force.* Likewise, the rotational equivalent of linear momentum ($m\mathbf{v}$) is **angular momentum** ($\mathbf{L}$), the *moment of momentum.*

Figure 8.17 depicts a particle of mass m and speed v moving along a straight line past a point O. We define the magnitude of the angular momentum with respect to O as

$$L_0 = r_\perp p = r_\perp mv \tag{8.36}$$

with units of kg·m²/s. Note that as long as $\mathbf{v}$ is constant, $r_\perp$ is constant and L_0 is constant. It might seem surprising that a particle moving in a straight line has *angular* momentum. Yet experience tells us that a straight-blowing wind can impart angular motion to a windmill and that a door can be swung closed (imparting angular momentum to it) by bouncing a ball straight off it.

Just as linear momentum is inherently a vector quantity, angular momentum is also. Because angular momentum is a measure of the "quantity-of-angular-motion" of a body, it is reasonable that oppositely acting contributions must tend to cancel. For instance, a system of two identical particles heading toward each other at the same speed along a straight line has no net linear momentum and should similarly have no net angular momentum about any reference point.

Equation (8.36) is reminiscent of the relationship between force and torque, Eq. (6.3). If we recognize that $r_\perp = r \sin \phi$, then

$$L_0 = rp \sin \phi$$

where ϕ is the smallest angle between the lines-of-action of $\mathbf{r}$ and $\mathbf{p}$. And, as with the torque (the moment of the force, $\boldsymbol{\tau}_0 = \mathbf{r} \times \mathbf{F}$), this suggests a vector cross product via Eq. (6.5). In other words, if we define the angular momentum vector (the moment of the momentum) as

$$\mathbf{L}_0 = \mathbf{r} \times \mathbf{p} \tag{8.37}$$

we specify both the required magnitude and a meaningful direction.

The direction of $\mathbf{L}_0$ is found by first drawing $\mathbf{r}$ from the axial point of rotation O to the mass point, as in Fig. 8.17b. Next, slide $\mathbf{r}$ along its line-of-action so that it has a common origin at m with $\mathbf{p}$. Then cross $\mathbf{r}$ into $\mathbf{p}$ through the smallest angle (via the method of p. 178) to get $\mathbf{L}_0 = \mathbf{r} \times \mathbf{p}$.

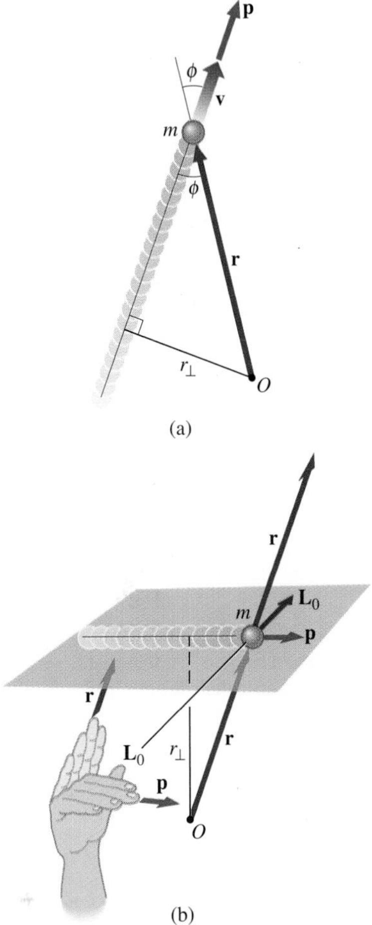

Figure 8.17 (a) A particle of mass m has an angular momentum L_0 with respect to point O equal to $r_\perp mv$. (b) The direction of $\mathbf{L}_0$ is perpendicular to the plane of $\mathbf{r}$ and $\mathbf{p}$ such that $\mathbf{L}_0 = \mathbf{r} \times \mathbf{p}$. Slide $\mathbf{r}$ along itself until it's tail-to-tail with $\mathbf{p}$ and then form $\mathbf{r} \times \mathbf{p}$. Determine the smallest angle between $\mathbf{r}$ and $\mathbf{p}$ by placing your fingers along $\mathbf{r}$ so they close naturally into $\mathbf{p}$. Your thumb will point in the direction of $\mathbf{L}_0$.

Now, examine the flat extended object made up of point masses in Fig. 8.18. It's revolving about a perpendicular axis through point O. Each mass point has a distinctive angular momentum even though they all have the same angular speed ω. In this case of an essentially two-dimensional body, all the constituent angular momenta are directed perpendicular to the plane of the body, all are parallel to each other, and their magnitudes can be added algebraically. Since, for every particle, $\mathbf{r}$ is perpendicular to $\mathbf{v}$, the magnitude of the body's total angular momentum L_0 about O is

$$L_0 = \Sigma rmv$$

Because $v = r\omega$,

$$L_0 = (\Sigma mr^2)\omega$$

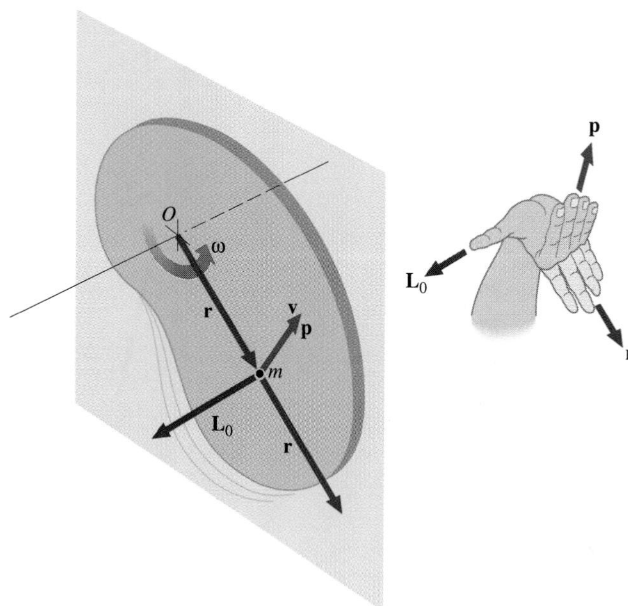

Figure 8.18 To find $\mathbf{L}_0$ for point mass m, slide $\mathbf{r}$ so it's tail-to-tail with $\mathbf{p}$, and then construct $\mathbf{r} \times \mathbf{p}$.

The term in parentheses is the body's total moment of inertia about O [Eq. (8.29)], and so

$$L_0 = I_0\omega$$

Here point O, which locates the axis of rotation, happened to lie on the body, but it need not have. The O subscript tells us that L is referenced to that point; it's really not crucial and is often omitted. The last equation shows angular momentum to be analogous to linear momentum: $p = mv$. In transforming to the rotational regime, m becomes I, v becomes ω, and p becomes L.

The similarities between the linear and angular formalisms can be made complete by assigning a direction to the angular velocity, recognizing its vector nature. To that end, we take $\boldsymbol{\omega}$ to lie *on the axis of rotation* and define its direction via the right-hand-rule; namely, the thumb on the right hand points in the direction of the vector $\boldsymbol{\omega}$ when the fingers curl around in the direction of rotation of the body. Thus, in Fig. 8.2, $\boldsymbol{\omega}$ is out of the page as it is in Fig. 8.9, whereas in Figs. 8.7, 8.10, 8.12, and 8.16, it is into the page.

For the flat object illustrated in Fig. 8.19, $\boldsymbol{\omega}$ is parallel to $\mathbf{L}_0$, and so we can extend the last equation, writing it as

$$\mathbf{L}_0 = I_0\boldsymbol{\omega} \tag{8.38}$$

This important result is a little misleading and commonly misunderstood. In general, for a three-dimensional body (Fig. 8.20), the angular momentum of each point-mass is along a direction that need not correspond to $\boldsymbol{\omega}$. Thus, *the net angular momentum vector of an arbitrary revolving body need not be parallel to* $\boldsymbol{\omega}$ and Eq. (8.38) must be modified; I_0 cannot be a scalar. If the body is symmetrical about the axis of rotation, then for every point-mass on one side there will be one diametrically opposite it, the same distance away from the axis. For every such pair of point-masses, the two components of $\mathbf{L}_0$ perpendicular to the axis will cancel, leaving a resultant angular momentum parallel to $\boldsymbol{\omega}$. So Eq. (8.38) is true, as is, for the important case of symmetrical objects revolving about their symmetry axes, and we will limit our treatment to these.

Example 8.12 Taking the Earth to be a uniform sphere of radius 6.37×10^6 m and mass 5.98×10^{24} kg, compute its angular momentum about its spin axis. What are the directions of $\mathbf{L}$ and $\boldsymbol{\omega}$?

Solution: [Given: $R_\oplus = 6.37 \times 10^6$ m and $M_\oplus = 5.98 \times 10^{24}$ kg. Find: $\mathbf{L}$.] As we saw earlier, $I = \frac{2}{5} \times mR_\oplus^2 = 9.71 \times 10^{37}$ kg·m^2, while $\omega = 0.727 \times 10^{-4}$ rad/s; hence

$$L = I\omega = (9.71 \times 10^{37} \text{ kg·m}^2)(7.27 \times 10^{-5} \text{ rad/s})$$

and $\boxed{L = 7.06 \times 10^{33} \text{ kg·m}^2\text{/s}}$

Looking down onto the North Pole, the Earth rotates counterclockwise. Hence, $\boldsymbol{\omega}$ points up northward along the spin axis and $\mathbf{L}$ and $\boldsymbol{\omega}$ are parallel.

▶ **Quick Check:** Let's see if the units are right: kg·m^2/s = m(kg·m/s), which are the units of $L = rmv \sin\phi$. $L \approx (10^{38}$ kg·m$^2)(7 \times 10^{-5}$ rad/s$) \approx 7 \times 10^{33}$ kg·m^2/s.

On an Atomic Level

A variety of experiments performed during this century have revealed a surprising fact: *angular momentum is an intrinsic property of matter on the atomic and subatomic level.* Each of the elementary particles that constitutes ordinary matter (the leptons and quarks—see Chapter 33) has a characteristic angular momentum, just as it has a characteristic mass and electric charge. The building blocks of the atom—electrons, protons, and neutrons—each have an intrinsic angular momentum. The discrete measure of angular momentum associated with these particles turns out to be the smallest observed amount—namely, $h/4\pi = 5.27 \times 10^{-35}$ kg·m²/s. Because it comes in specific doses, we say that angular momentum is *quantized*—it exists only in whole-number multiples of $h/4\pi$. Even the particles of light—photons—each have an angular momentum of $h/2\pi$. The quantity h is the fundamental constant of Quantum Mechanics known as **Planck's Constant**. The angular momentum of macroscopic objects is so large compared to $h/4\pi$ that we never have to worry about its inherently quantized nature.

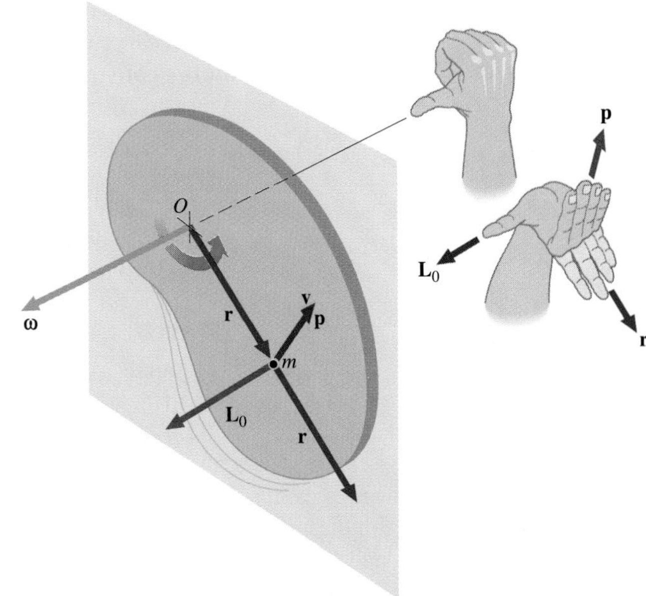

Figure 8.19 For this flat object, $\mathbf{L}_0$ is in the same direction as $\boldsymbol{\omega}$ for all point-masses.

8.8 Conservation of Angular Momentum

Suppose that a constant net torque τ acts on a body for a time interval Δt. There will then be a constant angular acceleration such that

$$\tau = I\alpha = I\frac{\Delta\omega}{\Delta t}$$

Forasmuch as I too is constant, any change in the angular momentum is a result of a change in ω; that is, $\Delta L = I\Delta\omega$, and therefore

$$\tau = \frac{\Delta L}{\Delta t}$$

The torque equals the time-rate-of-change of the angular momentum. To be more inclusive, we can write this expression as a vector equation and so

$$\boldsymbol{\tau} = \frac{\Delta \mathbf{L}}{\Delta t} \tag{8.39}$$

where the change in $\mathbf{L}$ may be in magnitude, direction, or both.

Remember that we are limiting the discussion to symmetrical objects revolving around symmetry axes. And so, when the torque is parallel to the angular momentum (when $\boldsymbol{\tau}$ is in the same direction as $\boldsymbol{\omega}$), $\Delta\mathbf{L}$ is parallel to $\mathbf{L}$, and the body speeds up. In the same way, when $\boldsymbol{\tau}$ and $\mathbf{L}$ are antiparallel, $\Delta\mathbf{L}$ is opposite to $\mathbf{L}$, and the body slows down. This behavior occurs in both instances without a change in the direction of $\mathbf{L}$; the body does not change its orientation. By contrast, if $\boldsymbol{\tau}$ is not colinear with $\mathbf{L}$, then the direction of $\mathbf{L}$, and therefore of $\boldsymbol{\omega}$, must change.

Equation (8.39) is the rotational equivalent of Newton's Second Law and, as before, it points to a grand underlying insight. If the net torque applied to a body is

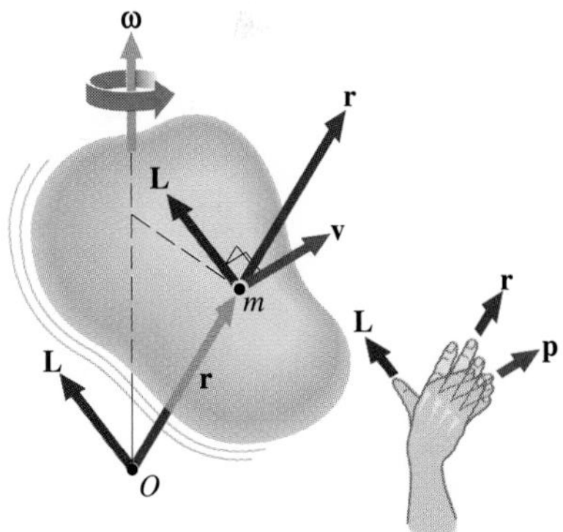

Figure 8.20 An asymmetrical body may revolve such that $\mathbf{L}$ and $\boldsymbol{\omega}$ are not parallel. Here, we see how that can happen for a single point-mass m. In a symmetrical body revolving about a symmetry axis, there will always be a point-mass on the opposite side of the axis so that the net $\mathbf{L}$ will be parallel to $\boldsymbol{\omega}$.

zero, there will be no change in its angular momentum: $\boldsymbol{\tau} = 0$ and $\Delta\mathbf{L} = 0$. **In the absence of a net external torque acting on a system, its angular momentum remains constant in magnitude and direction.** This is the law of **Conservation of Angular Momentum**. And it applies when $\Delta\mathbf{L}$ is due to a change in $\boldsymbol{\omega}$, or I, or both.

The tendency for matter to sustain rotational motion is utilized in a number of practical devices. The modern automobile engine is fitted with a flywheel that is set in high-speed rotation by the engine. Mounted on the end of the crankshaft, it stores angular momentum and keeps the shaft turning smoothly between successive strokes of the pistons.*

Example 8.13 By pushing on the ground, a skater with arms extended (as in Fig. 8.21a) manages to whirl around at a maximum angular speed of 1.0 rev/s. In that configuration, her moment of inertia about the spin axis is 3.5 kg·m^2. The point-masses making up her arms and outstretched leg are, for the most part, fairly far from the vertical spin axis, and therefore her moment of inertia is relatively large. What will happen to her spin rate as she draws her arms in and stands upright in a pirouette (Fig. 8.21b), whereupon her moment of inertia is only 1.0 kg·m^2? Assume the friction forces acting on her are negligible.

Solution: [Given: $\omega_i = 1.0$ rev/s, $I_i = 3.5$ kg·m^2, and $I_f = 1.0$ kg·m^2. Find: ω_f.] Since $\Sigma\boldsymbol{\tau} = 0$, $\mathbf{L} =$ constant; hence, from Eq. (8.38)

$$L_i = L_f$$

$$I_i\omega_i = I_f\omega_f$$

and so

$$\omega_f = \frac{I_f}{I_i}\omega_i = \frac{3.5 \text{ kg}\cdot\text{m}^2}{1.0 \text{ kg}\cdot\text{m}^2}(1.0 \text{ rev/s})$$

yielding $\boxed{\omega_f = 3.5 \text{ rev/s}}$.

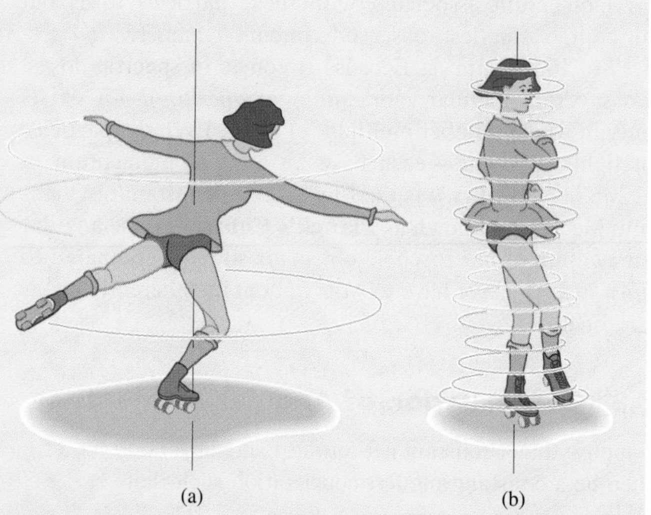

Figure 8.21 (a) The skater revolves as rapidly as possible, maximizing ω_i while maintaining a large I about the spin axis, which produces a substantial $L_i = I_i\omega_i$. (b) By pulling in her arms and legs, her moment of inertia decreases ($I_f < I_i$). Since there are no appreciable external torques, $L_i = L_f$ and $\omega_f > \omega_i$; her spin rate increases.

▶ **Quick Check:** The skater speeds up as her moment of inertia decreases, all the while conserving her angular momentum of 3.5 kg·m^2/s.

The same mechanism that governs a pirouetting skater governs a whirling hurricane or a spinning star. A fuel-spent star will collapse as it dies, gravitationally crushing itself to immense densities. The tiny remnant of this cosmic compression is a neutron star. If the thing is spinning at the outset, as it collapses, I decreases tremendously and ω increases in proportion. The end result is a rapidly rotating source of radiant energy—a *pulsar.* One such object, perhaps only 25 km (15 mi) in diameter and having about the same mass as the Sun, is at the center of the Crab Nebula. Whirling around 30 times each second it emits regular pulses of light and X-rays that sweep the sky every 33 milliseconds.

*The flywheel is geared via a disengaging mechanism to the starting motor, and if you've ever heard a horrible grinding sound when turning the ignition key for the second or third time, it's because the flywheel was still spinning.

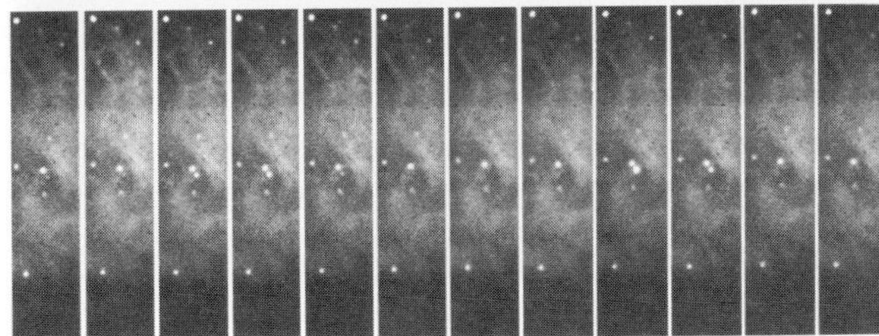

The Crab Nebula is the remnant of a star that exploded (a supernova) in A.D. 1054. At its center are the crushed remains now in the form of a spinning neutron star. It's called a pulsar because it emits radiant energy like a lighthouse beam that sweeps past, blinking on and off. You can see it appear and disappear in the center of this sequence of closeup photos.

An upwelling of air, here above warm water, causes a drop in pressure that draws in air from all sides. If there is any circulation of the surrounding air, it spirals inward, moving faster and faster as its distance from the center decreases. Thus, a great vortex forms, conserving angular momentum all the while.

Once the jumper is airborne, no external torques are applied and angular momentum is conserved. Since he wants to land legs extended, he jumps with arms up. Then, while in midair, he pulls his arms down (producing angular momentum into the page) as he draws his legs up (producing an equal amount of angular momentum out of the page). The net change in **L** is zero, and the jumper lands feet first.

In this old-fashioned long jump, the athlete rotates his legs forward (with an angular momentum vector out of the page) while pulling his arms backward (with an angular momentum vector into the page). Once airborne, his net angular momentum cannot change because there are no external torques on him.

On a far less grand scale, a high diver leaping into a pool can exert a torque about his *c.m.* while still in contact with the board and so begin to rotate. Once in the air and "torqueless," **L** is constant no matter how he tumbles, and tumble he must. Just as a translating body will continue to move with a constant **v** unless acted upon by a force, a rotating body will continue to rotate forever except insofar as it is acted upon by an *external* torque.

Example 8.14 A very small sphere of mass 0.20 kg is being whirled in a horizontal circle at the end of a 2.0-m-long string at a constant speed of 1.0 m/s. (a) Determine its orbital angular momentum about the axis of rotation. Suppose the string is quickly reeled in leaving a radius of only 1.0 m. (b) How fast will the sphere be moving then? (c) Talk about the tension in the rope.

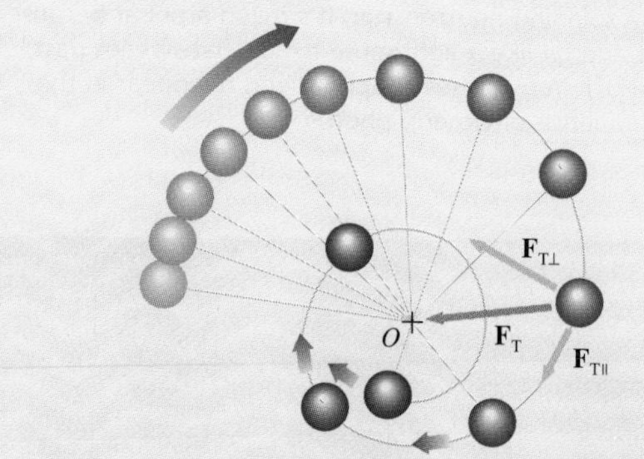

Figure 8.22 A body spiraling inward due to the tension in an attached string.

Solution: [Given: $m = 0.20$ kg, $v_i = 1.0$ m/s, $r_i = 2.0$ m, and $r_f = 1.0$ m. Find: L_O taking O at the axis of rotation, and v_f.] The angular momentum can be obtained by approximating the sphere as a point located at its *c.m.* whereupon, via Eq. (8.36), we have

(a) $$L_O = rmv = (2.0\text{ m})(0.20\text{ kg})(1.0\text{ m/s})$$

and $$\boxed{L_O = 0.40\text{ kg}\cdot\text{m}^2/\text{s}}$$

The string exerts a centripetal force on the ball that passes through O and, therefore, no torque about the axis is produced. No net torque means no change in the ball's angular momentum—the initial value of the angular momentum (mr_iv_i) before the string is reeled in must equal the final value (mr_fv_f) afterwards, consequently

$$mr_iv_i = mr_fv_f$$

or $$I_i\omega_i = I_f\omega_f$$

In either case

(b) $$v_f = \frac{v_ir_i}{r_f} = \frac{(1.0\text{ m/s})(2.0\text{ m})}{(1.0\text{ m})} = \boxed{2.0\text{ m/s}}$$

(continued)

(continued)

(c) One might wonder which force causes the sphere to accelerate from v_i to v_f. Figure 8.22 shows that as the body is pulled out of its circular orbit by an increased tensile force $\mathbf{F}_T$, the latter can be resolved into two components, one parallel ($\mathbf{F}_{T\parallel}$) and one perpendicular ($\mathbf{F}_{T\perp}$) to the motion. While $\mathbf{F}_{T\perp}$ changes the direction, it is $\mathbf{F}_{T\parallel}$ that tangentially accelerates the body. The *net* torque about O is always zero, and $\mathbf{L}$ is constant throughout.

▶ **Quick Check:** The moment of inertia decreases and as a result the angular speed increases, as does the linear speed. The radius is halved; the speed must double.

Planetary Motion

The string in the above example behaves like the gravitational field acting on an orbiting planet. Each—the string and the field—is the instrument of a centripetal force that passes through O and therefore produces no net torque $\boldsymbol{\tau}_O$. Thus, *a planet moving under the influence of a gravitational force directed toward the center of the Sun will maintain a constant orbital angular momentum:* r may change and v may change, but $\mathbf{L}$ will be constant. Since the force on the planet is always pointing toward the Sun, the planet cannot acquire any velocity out of the plane in which it was originally moving. Hence, *each planetary orbit lies within a plane.* That is, since $\mathbf{L}$ is constant, its direction is constant and therefore perpendicular to a plane, the plane formed by $\mathbf{r}$ and $\mathbf{v}$, the plane containing the motion.

Figure 8.23 depicts a planet in orbit moving a distance Δl in an interval of time Δt. Take Δt to be so small that Δl is essentially a straight line. The area of the little triangular section (equal to one-half the base Δl times the altitude $r_\perp$) swept out by the radius is

$$\Delta A \approx \tfrac{1}{2}\Delta l\, r_\perp$$

an approximation that gets better as Δt gets smaller. Dividing by Δt and taking the limit of both sides as $\Delta t \to 0$ leads to

$$\lim_{\Delta t \to 0}\left[\frac{\Delta A}{\Delta t}\right] = \frac{1}{2} v r_\perp = \frac{1}{2m}(r_\perp m v) = \frac{1}{2m} L$$

The left side of the equation is the time-rate-of-change of the area swept out by the radius vector drawn from the Sun to the planet. We have already seen, as a result of Conservation of Angular Momentum, that L is constant for central forces. Thus, the right side of the equation is constant, and so the left must be—the radius sweeps out equal areas in equal times. Of course, this is Kepler's Second Law.

Example 8.15 The Earth's nearest (perihelion) and farthest (aphelion) distances from the Sun are 1.47×10^8 km and 1.52×10^8 km, respectively. If its orbital speed at its closest approach is 30.3 km/s, how fast is the planet moving when most distant from the Sun?

Solution: [Given: $r_a = 1.52 \times 10^8$ km, $r_p = 1.47 \times 10^8$ km and $v_p = 30.3$ km/s. Find: v_a.] Since the angular momentum of orbiting planets is constant, the values of L at aphelion and perihelion are equal; accordingly

$$M_\oplus r_a v_a = M_\oplus r_p v_p$$

where, at both these locations, $\mathbf{r}$ is perpendicular to $\mathbf{v}$. Hence

$$v_a = \frac{r_p}{r_a} v_p = \frac{1.47 \times 10^8 \text{ km}}{1.52 \times 10^8 \text{ km}} 30.3 \text{ km/s}$$

and $\boxed{v_a = 29.3 \text{ km/s}}$.

▶ **Quick Check:** The Earth, reasonably enough, slows down as it pulls away from the Sun. The radius increases by a factor of 1.034, so the speed must decrease by the same factor: $v_a = (30.3 \text{ km/s})/1.034 = 29.3$ km/s.

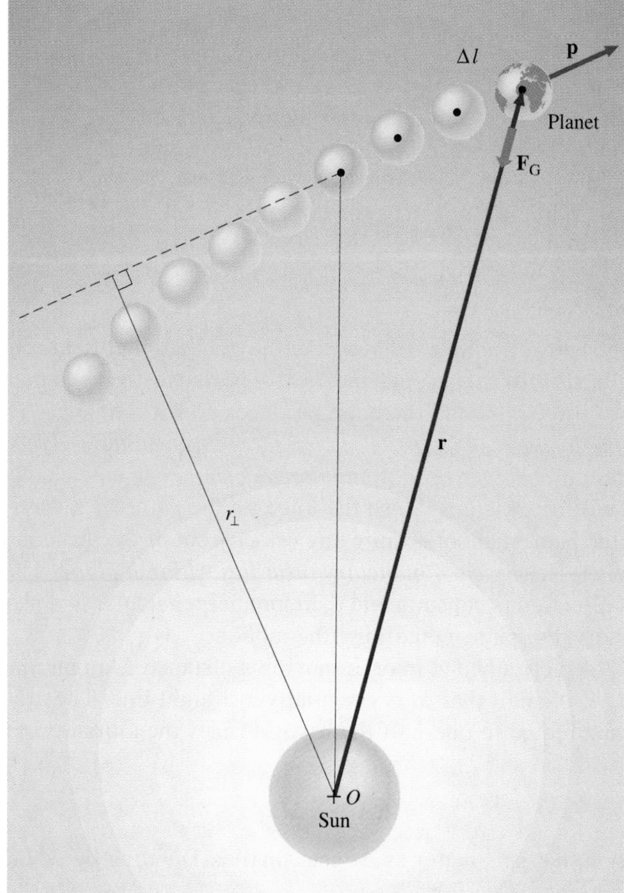

Figure 8.23 A planet orbiting the Sun. The direction of $\mathbf{L}_O$ is into the page parallel to $\mathbf{r} \times \mathbf{p}$ and perpendicular to the plane of the orbit. Since $\mathbf{F}_G$ passes through O, there is no torque about O, and $\mathbf{L}_O$ cannot change.

The Earth and Moon form a single system bound together by gravity. This photograph was taken by the *Galileo* spacecraft from a distance of 6.2×10^6 km.

The Moon gravitationally draws the Earth's crust, waters, and atmosphere, generating tidal motions in each. These, in turn, create frictional forces that gradually slow the planet's spin rate. The effect is quite small, an increase in period of roughly 25×10^{-9} s each day. Still, one thousand million years from now, a day may be approximately 3 h longer as a result (see Problem 41). What makes these predictions a bit unreliable is that the rate of de-spin is varying. Nonetheless, it's been suggested that 1.5×10^9 years ago a day may have been only 9 or 10 hours long, with roughly 900 such days per year.

The source of the torque causing the de-spin is the gravitational interaction of the Moon, which is *external* to the Earth; hence, the planet's angular momentum must change. Alternatively, if this same torque is envisioned as *internal* to the Earth-Moon system, the angular momentum of that larger system must then be conserved. This means that as the planet loses angular momentum, its moon, to which it is coupled gravitationally, must gain angular momentum, and it does. The Moon is receding from the Earth and thereby increasing its distance, r, at a rate of 1 foot in about 30 years and, consequently, increasing its orbital angular momentum ($F_C = F_G$, therefore $mv^2/r \propto m/r^2$ and $v^2 \propto 1/r$, so $mvr = L \propto m\sqrt{r}$).

Gyroscopic Stability

If there is no net torque acting on a revolving body, then its angular momentum vector will be fixed in space, in both magnitude and orientation. There is no more important example of this phenomenon than the spinning Earth itself, whose tilted axis of rotation points very nearly in the same direction all the while the planet orbits the Sun. It's this effect that gives rise to the seasons and that keeps the star Polaris due north all year long.

The tendency of the spin axis to remain fixed has important implications, even when a torque is present. We know, from Eq. (8.39) wherein I is constant, that

$$\boldsymbol{\tau}\Delta t = \Delta\mathbf{L} = I\boldsymbol{\omega}_{\rm f} - I\boldsymbol{\omega}_{\rm i}$$

and hence

$$I\boldsymbol{\omega}_{\rm f} = I\boldsymbol{\omega}_{\rm i} + \boldsymbol{\tau}\Delta t \tag{8.40}$$

Imagine that we have a massive body spinning with a large initial angular speed $\boldsymbol{\omega}_{\rm i}$. A torque subsequently applied produces a term, $\boldsymbol{\tau}\Delta t$, in Eq. (8.40). Suppose this contribution is small, either because the torque itself or the time of application is small. Adding it vectorially to the large value of $I\boldsymbol{\omega}_{\rm i}$ can only result in a final $I\boldsymbol{\omega}_{\rm f}$ that is not very different from its initial value. By comparison, if the body isn't initially rotating ($\boldsymbol{\omega}_{\rm i} = 0$), the entire final rotation ($I\boldsymbol{\omega}_{\rm f}$) will be due to the action of the torque.

Thus, a spinning body has a kind of **gyroscopic stability**. A casually tossed football might easily be made to tumble about its *c.m.* from the slightest torque, but one first set spinning around its long axis maintains its orientation throughout the flight (so does a Frisbee and a discus). A modern rifle or pistol fires spinning bullets to resist tumbling that would otherwise drastically increase air resistance and diminish accuracy. The *Space Shuttle* has a turntable in its cargo bay that is used to spin up satellites just prior to releasing them into space. Gyroscopic stability also accounts for the wonderful way a spinning top stays upright on its pointed end, just as it accounts for the way a twirling quarter, balanced on edge, seems to defy gravity. The fact that a freely spinning mass keeps its axis fixed in space allows us to use that mechanical reference line as the heart of guidance systems for missiles, airliners, submarines, and spaceships.

This photo, made at a location well below the equator, was taken by holding the camera's shutter open for several hours. As the Earth revolved, the stars left circular streaks about the spin axis. A similar photo, taken in the Northern Hemisphere, would show Polaris, the North Star, as a bright, tiny arc very near the center of the star traces.

Figure 8.24 shows a top spinning about its symmetry axis with an angular velocity $\boldsymbol{\omega}_{\rm s}$. The spin axis can simultaneously swing around, or **precess** ($\omega_{\rm p}$), and it can also nod up and down, or *nutate* ($\omega_{\rm n}$). Still, by making $\boldsymbol{\omega}_{\rm s}$ much larger than either of the other two angular speeds, the analysis becomes tremendously simplified. In that case, $\mathbf{L}$ is very nearly fixed in the body along $\boldsymbol{\omega}_{\rm s}$, and Eq. (8.38) again applies.

The tilted top in Fig. 8.24 (which we require to be rapidly spinning) experiences a constant torque due to its own weight acting downward at its *c.g.* As shown in Fig. 8.24b, the torque $\boldsymbol{\tau} = \mathbf{r}_{\rm cg} \times \mathbf{F}_{\rm W}$ is perpendicular to the spin axis, along which $\mathbf{L}$ lies. In a small time Δt, it will produce an angular momentum change parallel to itself, $\Delta\mathbf{L} = \boldsymbol{\tau}\Delta t$, as in Fig. 8.24d. The angular momentum will go from some

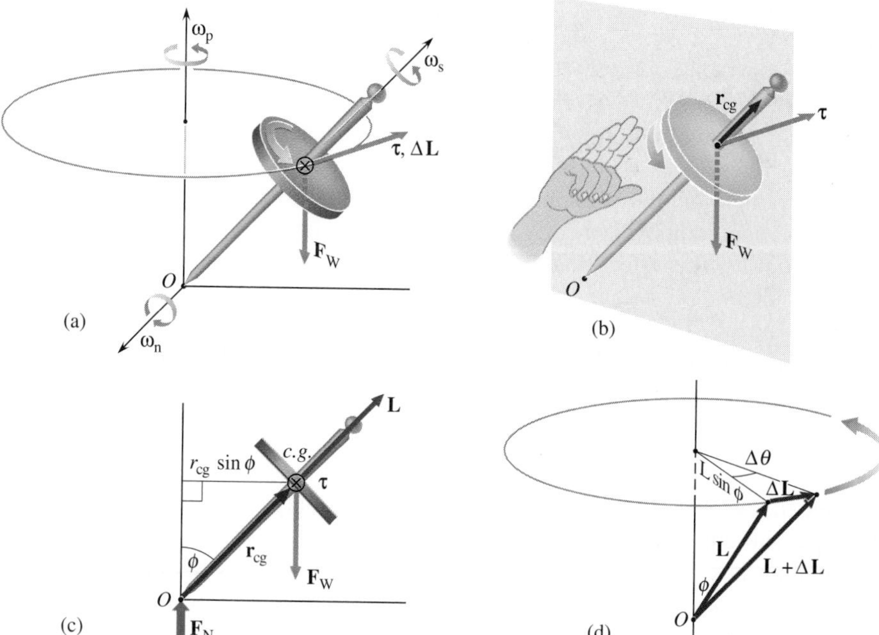

Figure 8.24 (a) A top spinning about a tilted axis. (b) The weight acts down at the *c.g.* producing a torque $\boldsymbol{\tau} = \mathbf{r}_{cg} \times \mathbf{F}_W$. (c) The torque $\tau = r_{cg}F_W \sin\phi$ produces a change in the angular momentum. (d) The $\Delta\mathbf{L}$ that results is parallel to $\boldsymbol{\tau}$.

initial value **L** to some final value **L** + Δ**L**, only to continuously change again as the torque continues to act. There is no component of torque in the direction of **L**, and so L is constant. Without falling, the spin axis of the top, tilted at some constant angle ϕ, will sweep around in a uniform conical wobble, or *precession*. Only when the spin rate becomes small will the top flop over.

The actual precessional angular velocity ω_p can be determined with the help of Fig. 8.24d and the realization that $\omega_p = \Delta\theta/\Delta t$ in the limit as $\Delta t \to 0$. From the geometry and Eq. (8.1), $\Delta\theta = \Delta L/(L \sin\phi)$, but since $\Delta L = \tau\Delta t$ and $\tau = mgr_{cg}\sin\phi$, then $\Delta L = (mgr_{cg}\sin\phi)\Delta t$ and so $\Delta\theta = (mgr_{cg}\sin\phi)\Delta t/(L\sin\phi)$. Hence, $\Delta\theta/\Delta t = mgr_{cg}/L$ and in the limit as $\Delta t \to 0$, using $L = I\omega_s$ and assuming $\omega_s \gg \omega_p$, we get

$$\omega_p = \frac{mgr_{cg}}{I\omega_s} \tag{8.41}$$

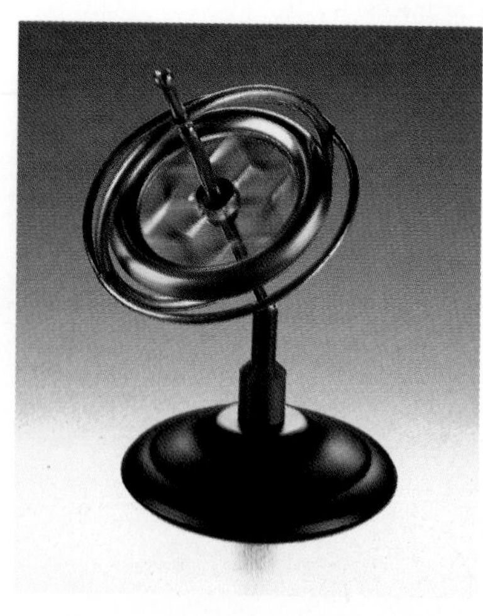

A spinning gyroscope, balanced and precessing in the manner depicted in Fig. 8.24.

Of course, a nonspinning top would have to be supported by an appropriate, additional vertical force in order to be held at an angle ϕ. That is to say (when $\omega_s = 0$), for nonzero ϕ, $F_W > F_N$, and without an additional force $\Sigma F_V \neq 0$, and down goes the *c.g.* The net torque produces a parallel Δ**L** and a parallel rotation ($\boldsymbol{\omega}$) about O. By contrast, for a spinning top $F_N = F_W$, $\Sigma F_V = 0$ and the *c.g.* cannot descend. There is still a torque, but it produces a ΔL that now is not parallel to the existing **L**. In general, **L** tends to move toward $\boldsymbol{\tau}$ and the top stays at ϕ while it precesses. Clearly, a torque does not have the same effect on both a spinning and a nonspinning object, and that's what makes the gyroscope seem so magical.

It is common to demonstrate the effect with a wheel whose spin axis is initially horizontal (Fig. 8.25). When let loose, the wheel precesses as expected, with its axis seemingly floating in air pivoting about its supported end. But something is being overlooked—where does the precessional angular momentum come from? There are no torques about the vertical axis. What happens is this: the spin axis actually tilts down a little below the horizontal. This gives the spin angular momentum a downward vertical component that cancels the upward precessional angular momentum, thereby conserving vertical angular momentum at a value of zero.

The phenomenon of precession isn't limited to tops: we steer bicycles and motorcycles by tilting them so they precess. For a guidance gyro we want to minimize ω_p, which means making I and ω_s large and r_{cg} small. The best guidance gyros are military secrets that probably have precession rates around 0.02° per day (the Earth as a gyro is over a thousand times better than that).

Because of its bulging middle, the Earth experiences an unbalanced attraction to both the Sun and Moon and a corresponding torque. That, in turn, gives rise to a precession of the spin axis that takes 25.78×10^3 years to sweep once around. Thus, the pole of the heavens slowly shifts, and when the ancient Chinese wrote about the "celestial pivot," their pole star was not Polaris. Approximately 4000 years ago, Thuban in the constellation Draco shone as the North Star, and in A.D. 14,000 bright Vega will have that small honor.

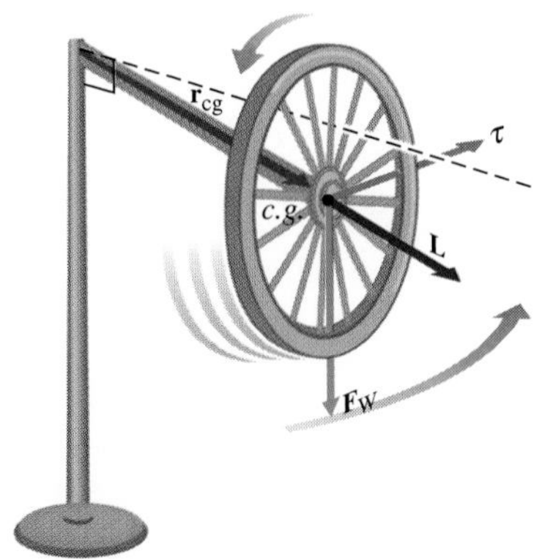

Figure 8.25 A spinning wheel, initially horizontal when let loose, drops slightly and precesses.

Angular Momentum and Symmetry

Noether's Principle (p. 6) maintains that every conservation law is linked to a symmetry of some kind (Fig. 8.26). We change an aspect of the physical system and some characteristic remains unchanged or conserved—that's what a symmetry is. Rotate the disk of Fig. 1.3 and its appearance does not change. But the same logic also applies to subtle conceptual symmetries. Conservation of Linear Momentum arises from the spatial homogeneity of the fabric of the Universe: the physical nature of empty space is the same everywhere (p. 116). We can move our laboratory to any place in the Universe and the laws of physics would apply unaltered—they are invariant under translation through a homogeneous space. That symmetry is equivalent to Conservation of Linear Momentum.

Figure 8.26 Conservation of Angular Momentum arises from the isotropy of free space. The laws and constants of physics are independent of orientation in space. The four fundamental forces of nature are isotropic, and so the net angular momentum of an isolated system cannot change.

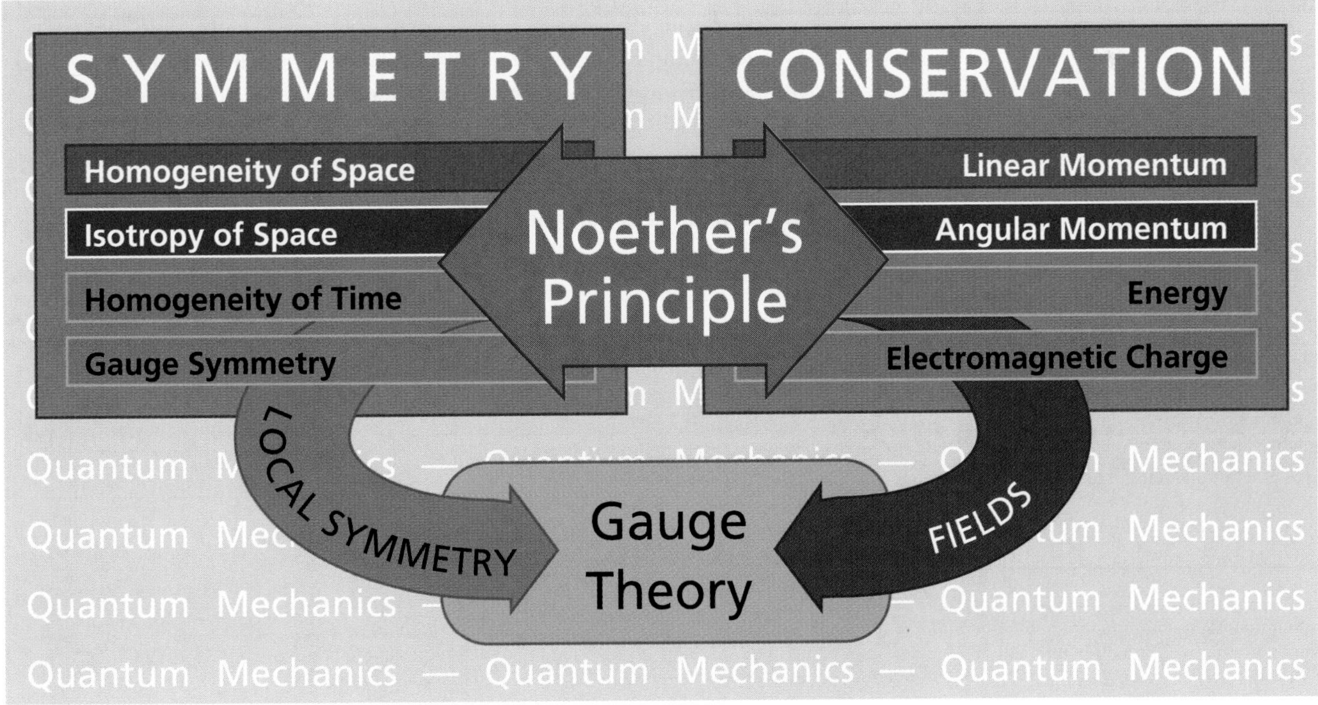

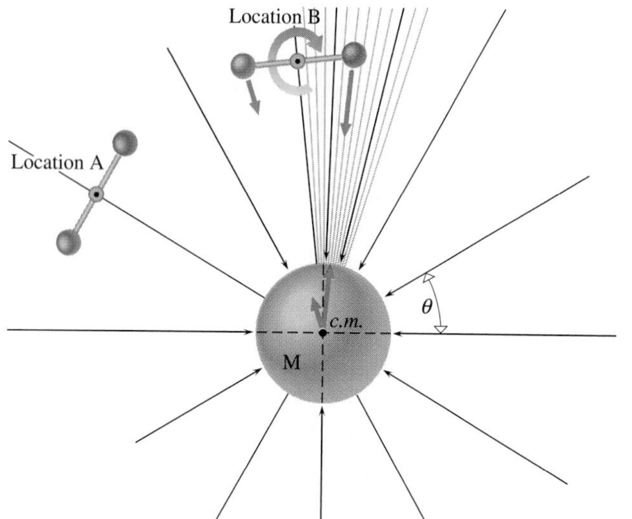

Figure 8.27 A dumbbell at two locations in the anisotropic gravity field of a uniform sphere in an imaginary anisotropic space. At location B, the dumbbell will have an angular acceleration and a nonzero **L**. The torque on the dumbbell is now not equal to the torque on the sphere, and, as a result, Conservation of Angular Momentum will be violated.

Similarly, we find that Conservation of Angular Momentum is a consequence of the isotropy of empty space. If space is *isotropic* (that is, the same in all directions), the constants and the laws of physics must be independent of orientation. Imagine a laboratory in a spaceship far from any other objects. If the ship is rotated through some arbitrary angle, none of the experiments performed in the lab will change. A magnet will attract a piece of iron oriented one way or another. The gravitational field of a uniform sphere is the same in every direction, and the speed of light is independent of the way the source is pointing. Natural law is invariant under rotation within an isotropic space. That this invariance demands Conservation of Angular Momentum is not easy to prove, but we can at least make it reasonable.

Suppose space were anisotropic and therefore the laws of physics were dependent on spatial orientation. For example, imagine a uniform spherical mass M having a hypothetical gravitational field that varies with angle (that is, θ in Fig. 8.27). A dumbbell with two identical balls, one red and one blue, is pivoted at its center so that it can turn freely. Holding it at rest at location A in a region of uniform field, the gravitational attraction on both small masses is equal, and the dumbbell remains motionless. But holding it at location B is like putting a paddlewheel in a waterfall. Although initially $L = 0$, as soon as the dumbbell is allowed to, it will start to revolve clockwise. The red ball in the strong-field region will descend, swing past the vertical and enter the weak-field region, whereupon the blue ball will fall through the strong field, and the dumbbell will accelerate around. On the other hand, the forces acting on the large spherical mass pass through its center. They do produce a torque about the *c.m.* of the system, but because the *c.m.* is so close to the sphere's center, that torque is negligible. Thus, the dumbbell experiences a large torque and a large $\Delta\mathbf{L}$, while the sphere experiences a small oppositely directed torque and a small, oppositely directed $\Delta\mathbf{L}$. The system has a net $\Delta\mathbf{L}$ and Conservation of Angular Momentum is violated.

We assume that in our Universe the converse is true: in all cases, the interactions between two bodies are such that the torque on body 1 due to body 2 is equal in magnitude and opposite in direction to the torque on body 2 due to body 1. **Empty space is isotropic (physics is indifferent to orientation) and therefore angular momentum is conserved.**

Core Material

The basic measure of **angular displacement** θ is defined by

$$\theta = \frac{l}{r} \quad \text{or} \quad l = r\theta \qquad [8.1]$$

where l is the *arc-length.* The units of θ are *radians* such that 1 rad = 57.3° and π rad = 180°. The *average angular speed* ω_{av} equals $\Delta\theta/\Delta t$ and in the limit as $\Delta t \to 0$ that ratio approaches ω, the **instantaneous angular speed**, where

$$v = r\omega \qquad [8.11]$$

and ω has units of rad/s. The **instantaneous angular acceleration**, α, equals the limit of $\Delta\omega/\Delta t$ as $\Delta t \to 0$, and

$$a_T = r\alpha \qquad [8.16]$$

The units of a_T are m/s^2, whereas those of α are rad/s^2 (p. 246). When α *is constant,*

$$\omega_f = \omega_i + \alpha t \qquad [8.22]$$

$$\omega_{av} = \tfrac{1}{2}(\omega_i + \omega_f) \qquad [8.23]$$

$$\theta = \omega_{av}t = \tfrac{1}{2}(\omega_i + \omega_f)t \qquad [8.24]$$

$$\theta = \omega_i t + \tfrac{1}{2}\alpha t^2 \qquad [8.25]$$

$$\omega_f^2 = \omega_i^2 + 2\alpha\theta \qquad [8.26]$$

These equations are identical to those of translational motion in Chapter 3, where $s \to \theta$, $v \to \omega$, $a \to \alpha$, and $t \to t$.

For a body composed of a number of particles of mass m, the moment of inertia about an axis passing through some point O is

$$I = \Sigma mr^2 \qquad [8.29]$$

Table 8.3 lists many common shapes and the associated moments of inertia for homogeneous bodies. The rotational equivalent of $F = ma$ is

$$\Sigma\tau_0 = I\alpha \qquad [8.30]$$

It is possible to find a point, the **center-of-mass**, where, as far as its motion is concerned, all of the mass of a body can be thought of as concentrated, thus

$$x_{cm} = \frac{\Sigma mx}{\Sigma m} \quad \text{and} \quad y_{cm} = \frac{\Sigma my}{\Sigma m} \qquad [8.31]$$

Angular momentum, the moment of momentum, is defined as

$$L_0 = r_\perp p = r_\perp mv \qquad [8.36]$$

and

$$\mathbf{L}_0 = \mathbf{r} \times \mathbf{p} \qquad [8.37]$$

For symmetrical bodies spinning around symmetry axes

$$\mathbf{L} = I_0\boldsymbol{\omega} \qquad [8.38]$$

The rotational equivalent of the Second Law is

$$\boldsymbol{\tau} = \frac{\Delta\mathbf{L}}{\Delta t} \qquad [8.39]$$

In the absence of a net torque, the angular momentum of the system remains constant; this is the law of **Conservation of Angular Momentum** (p. 259), and it has its origins in the isotropy of free space.

Suggestions on Problem Solving

1. Make a quick sketch of what is happening and always use free-body diagrams if forces are involved.
2. Watch out for those unit conversions! Do not confuse rpm with rev/s or rad/s. Remember the factor of 2π.
3. Keep your left hand in your pocket—the vector cross products are done only with the right hand.
4. Remember that m, I, $\mathbf{F}$, $\boldsymbol{\tau}$, $\mathbf{p}$, and $\mathbf{L}$ are dynamical variables; seeing any of these in a problem should call to mind certain specific equations from the second half of the chapter. For example, it's not very likely that you will ever come across a numerical problem that talks about angular momentum and does not require either Eq. (8.38) or Eq. (8.39). Similarly, the mere mention of the word *torque* suggests angular acceleration and vice versa, although a time-varying angular momentum should come to mind as well. It's certainly possible to have a torque problem without ever seeing the word explicitly. Such "disguised" torque problems work by giving $\mathbf{F}$ and r instead—when you see these, think torque. The crucial step is to relate the "Given" and "Find" via one or more familiar equations once you recognize the general type of problem. The simple problems require one application of one of the basic equations. More difficult problems may require solving for things that weren't asked for.

Discussion Questions

1. When a car turns, do the wheels on its right side have the same angular velocity as those on its left side? What does the answer suggest about mounting the wheels rigidly on a common axle?

2. People have been predicting that a rise in the temperature of the planet would result in the melting of the polar ice caps. Would such massive melting, in turn, have any effect on the length of the day? Explain.

3. Figure Q3 shows a gymnast in three postures, spinning about the frontal axis passing through her *c.m.* Which has the greatest moment of inertia about that axis and which the least? Explain. Draw a rectangular parallelepiped in the shape of a deck of playing cards. Fix the *c.m.* at its geometric center. Draw, passing through the *c.m.*, the three axes perpendicular to the three pairs of faces, numbering them 1, 2, and 3 as the area of each increases. About which axis is the moment of inertia a maximum? A minimum?

4. A diver shown in Fig. Q4 leaves the springboard an instant after being projected by it into the air. Resolve the normal force on his feet into two components, one along the line through his *c.m.* and one perpendicular to it. Discuss the motion that results from each component.

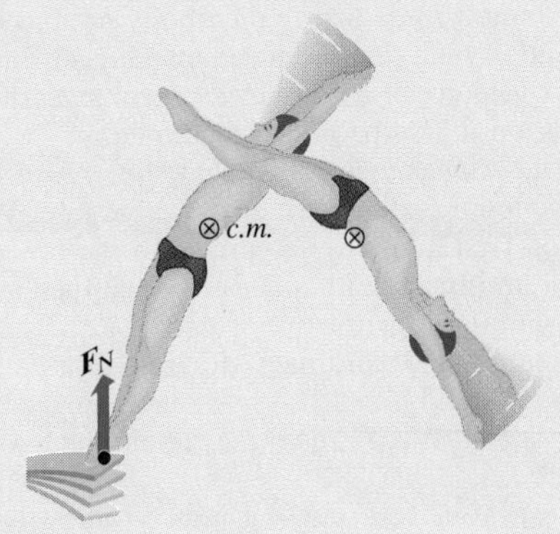

Figure Q4

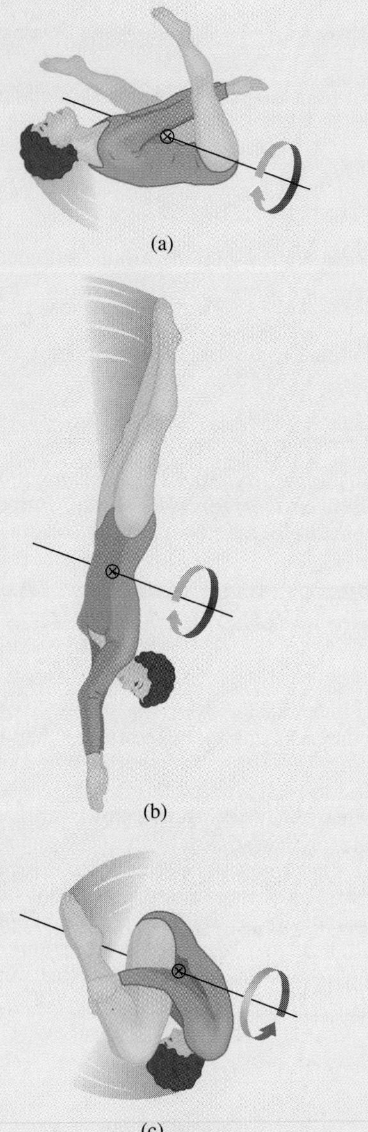

Figure Q3

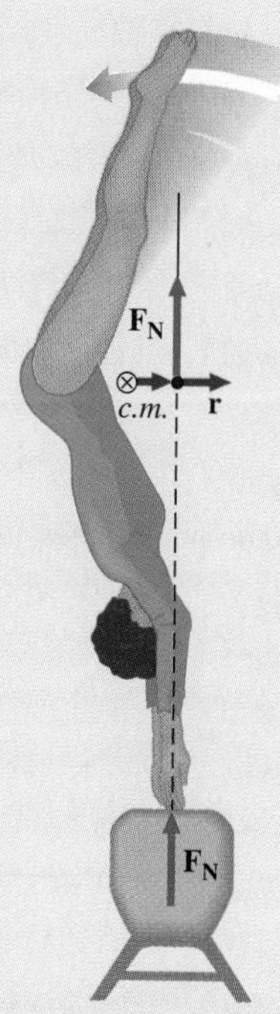

Figure Q9

5. Can a small force acting on a body produce a larger angular acceleration than that produced by a bigger force? Explain. If the angular velocity of a body is nonzero, must the sum of the torques acting on it also be nonzero? Explain.

6. Imagine that you spin two eggs, one raw and one hard-boiled, on a smooth flat surface at the same rate. Which would come to a stop first and why? Again, spin the raw egg but now stop it just for an instant with a fingertip and then let it go. Why does it start spinning again?

7. Consider the extraordinary situation of a polished sphere rolling without slipping on a flat horizontal surface with a constant angular speed. What can you say about the force of friction on the sphere?

8. Suppose you were playing pool with balls floating in space. How would you hit a motionless cue ball to propel it forward in a straight line, making it rotate as if it were in contact with a table? (By the way, on Earth where the game is only played *on* a table, this is known as a "follow shot" because the cue ball follows after the ball it hits.)

9. The vaulting gymnast executing a push-off at the moment shown in Fig. Q9 experiences a vertical normal force acting at her fingertips and somewhat exceeding her weight. Determine the resulting direction of $\mathbf{L}$, $\boldsymbol{\omega}$, and $\boldsymbol{\tau}$ about her *c.m.* Describe her motion soon after.

10. Explain why most helicopters have a small additional propeller at the tail end that revolves in a vertical plane.

11. Imagine Superman in a black leather jacket sitting on a motorcycle floating out in space. Describe what would happen as he gunned the engine (presuming it was modified to run out there).

12. An astronaut floating in space has zero angular momentum initially. He then proceeds to rotate both his extended arms around an axis passing across his chest through his two shoulders. While looking at his right side, you see his arms revolving counterclockwise. What motion, if any, will be executed by the rest of his body? What will happen when he stops?

13. If a person throws a ball, its angular momentum is increased, but the person doesn't rotate. Is angular momentum conserved in the ball-person system? What, if any, external torque acts on the person? Is the angular momentum of the ball-person-Earth system conserved? What, if anything, happens to the motion of the planet? What would happen to someone floating in space who threw a ball in an overhand pitch?

14. One of the best ways of transporting the magnetic tape in a recorder is by a direct-drive system where a motor turns the capstan that moves the tape (see Problem 28). Such motors are often fitted with heavy flywheel housings that revolve with the shaft. Why do you think this might be done? (See Fig. Q14.)

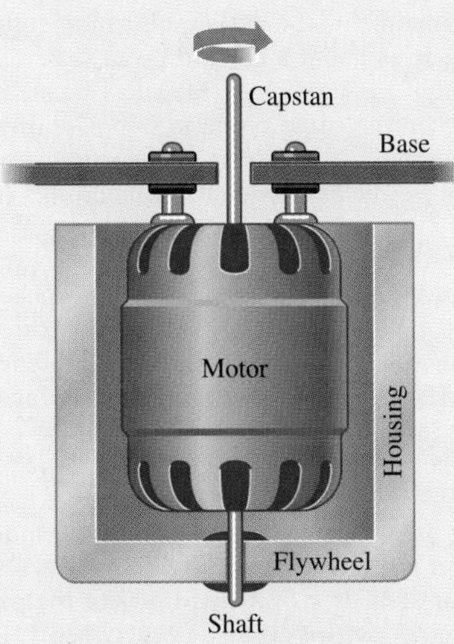

Figure Q14

15. A motorboat generally noses up out of the water, and the more it does so, the faster it travels. Make a sketch locating the *c.m.* of the motorboat and the level of its propeller. Draw in all the forces acting on the boat while it is traveling at a constant speed—don't forget the upward force of the water on the boat and the friction on the hull bottom. Explain what's happening in terms of $\Sigma\mathbf{F}$ and $\Sigma\boldsymbol{\tau}_{\text{cm}}$.

16. When a rear-wheel drive car or motorcycle rapidly accelerates, it tends to rotate about its *c.m.* so that the front end noses upward. What forces are acting *on* the vehicle? Make a sketch and explain what is happening. What would happen if the front brakes engaged first and locked during a quick stop?

17. Imagine a gradual, inwardly spiraling groove cut just into the surface of a flat tabletop. Now, suppose we propel a small sphere into the channel with a speed v so that it rolls toward the center point O (with no appreciable losses due to friction). What, if anything, will happen to its speed and why?

18. In the past, many satellites have had to be put into orbit spinning more rapidly than was operationally desirable. A cute de-spin system consists of two masses each (known as a yo-yo) tied to a long cable. These are wound around the craft and the weights are held in place on opposite sides. On command, the weights are released and they unwind as the satellite spins until the ends of the cables unhook and each weight, trailing its cable, flies away into space. Explain how this system works to reduce the satellite's spin rate—usually by a factor of about 10.

19. A physics student, with arms stretched out to either side, holds a weight in each hand. While sitting on a bar stool, he is swung around by a helpful friend. What will happen to his angular speed, if anything, when he lets the weights fall to the floor? Think this one through carefully.

Multiple Choice Questions

1. A flying saucer in a TV show banks and just blocks out the disk of the Moon (which subtends an angle of 0.009 rad) when a nearby radar device determines its range to be 10 km away. How big was the craft? (a) 9 m (b) 0.09 km (c) 0.9 km (d) 0.9 m (e) none of these.
2. Every point-mass on a rigid body that is rotating has (a) the same angular speed and angular acceleration (b) a different angular speed but the same angular acceleration (c) the same linear and angular speed (d) the same angular speed and the same linear acceleration (e) none of these.
3. At what latitude on Earth would you have the greatest angular velocity? (a) 0° (b) 45° N (c) 45° S (d) 60° N (e) none of these.
4. A body free of support and experiencing a zero net applied force (a) must be accelerating (b) must only be translating at a constant speed (c) must only be rotating about its *c.m.* (d) may be at rest or moving at a constant speed and may be rotating about its *c.m.* as well (e) none of these.
5. Given that object 1 exerts a torque on object 2 about some axis. Then with respect to that same axis, object 2 (a) exerts an identical torque on object 1 (b) exerts an equal but oppositely directed torque on object 1 (c) exerts no torque on object 1 (d) may or may not exert a torque on object 1 (e) none of these.
6. Two slender uniform rods have the same cross section and the same mass. If one is light plastic and the other lead, what can you say regarding their respective moments of inertia about an axis through the very center perpendicular to each rod's long axis? (a) Nothing. (b) The moments of inertia of the rods are equal. (c) The lead rod must have a greater moment of inertia. (d) The plastic rod must have a greater moment of inertia. (e) None of these.
7. The moment of inertia of a spinning body about its spin axis depends on its (a) angular speed, shape, and mass (b) angular acceleration, mass, and spin-axis position (c) mass, shape, spin-axis position, and size (d) mass, size, shape, and speed (e) none of these.
8. When a single force is applied to a free body floating in space, it is possible that there will be a resulting change in (a) only its translational motion (b) both its translational and rotational motions (c) both (a) and (b) are possible (d) only its rotational motion (e) none of these.
9. For the net force acting on a body to result in purely linear motion (a) it must be zero (b) it must pass through the *c.m.* (c) it must be less than the weight of the body (d) it must not pass through the *c.m.* (e) none of these.
10. Two straight inclined grooved tracks are lined up next to one another. Number 1 track is coated with a slippery film and number 2 is not; otherwise, they are identical. Two steel balls are placed one on each track and released so that ball 1 slides down and ball 2 rolls all the way. (a) Both reach the bottom at the same time. (b) What happens depends on the exact tilt of the tracks and so we cannot say who wins. (c) Ball 1 wins always. (d) Ball 2 wins always. (e) Ball 1 wins as often as ball 2.
11. A solid cylinder, a solid sphere, and a hoop all of the same mass but different radii, roll without sliding down an inclined plane. The body that gets to the bottom first will invariably be (a) the hoop (b) the sphere (c) the

cylinder (d) they all arrive together (e) not enough information.

12. The turntable platter, the large circular revolving disk of a record player, is designed to resist small variations in speed that would occur with slight fluctuations in the drive motor's output. This design is accomplished by (a) making the platter as small as possible (b) using a platter as heavy as possible (c) using a very light platter (d) making the platter's moment of inertia as large as possible (e) none of these.
13. A body isolated from all other objects can alter, by internal interactions among its various parts, its (a) angular momentum (b) linear momentum (c) both (a) and (b) (d) neither (a) nor (b) (e) none of these.
14. The dimensions of angular momentum are (a) $[ML^2T]$ (b) $[M^{-1}L^2T]$ (c) $[ML^{-2}T]$ (d) $[ML^2T^{-1}]$ (e) $[ML^{-2}T^{-1}]$.
15. Low-altitude artificial Earth satellites traversing wisps of atmosphere are initially slowed down by air-drag so that they cannot remain in their original orbits and inevitably fall back to the planet. Spiraling down ever closer to the Earth, they usually break up and burn up before hitting ground. The speed of such a device (a) increases as it descends and its angular momentum is not conserved (b) decreases as it descends and its angular momentum is conserved (c) increases and then decreases, conserving picture is its angular momentum (d) must remain constant, conserving its angular momentum (e) none of these.
16. At a given moment, the net angular momentum of a system of two particles about a certain point is zero. Accordingly, we know that the particles (a) must be at rest (b) must be moving on the same circle in opposite directions (c) must be moving on straight perpendicular paths (d) must have the same linear momentum (e) none of these.
17. A particle of mass m and speed v has zero angular momentum with respect to a point A in space. We know that the particle (a) has already passed through A (b) will never pass through A (c) is moving along a line that is perpendicular to the plane containing it and A (d) is moving along a line that passes through A (e) will go somewhere, but we don't have enough information to comment.
18. Imagine a particle traveling with a linear momentum **p** past a point O with respect to which it has an angular momentum of **L**. If **r** drawn from O to the particle makes an angle of 45° with **v**, the angle between **L** and **p** is (a) 0° (b) 45° (c) 90° (d) 180° (e) none of these.
19. When a rigid body rotates about a symmetry axis with no external torque being applied (a) the angular acceleration is a nonzero constant (b) the angular velocity is a nonzero constant (c) the angular speed changes uniformly (d) the angular acceleration changes uniformly (e) none of these.

Problems

THE KINEMATICS OF ROTATION

1. [I] What does 1.000 0° equal in radians?
2. [I] How many radians are swept out by the minute hand of a clock as it goes from 12:00 noon to each of the following times: 1:45 P.M., 2:30 P.M., and 3:17 P.M.?
3. **3.** [I] Determine, via geometry, the angle in radians subtended by a 2.00-m-wide car seen head-on from a distance of 1.00 m away. What would that angle be as seen standing 100 m away? Compute this last part exactly (using the straight-line width) and approximately (taking the width as an arc-length). Now, go back to the first part and do it with the same approximation. Compare your results.
4. [I] A 25-cent coin is roughly 2.5 cm in diameter. How far away should it be held if it is to subtend the same angle as does the Moon?
5. **5.** [I] A well-adjusted laser beam is emitted as an exceedingly narrow diverging cone of light with a typical spread angle of about 8×10^{-4} rad. How large a circle will be illuminated on the surface of the Moon by such a device ($r_{☾} = 3.8 \times 10^8$ m)?
6. [I] The 10-point center-circle bull's eye on the official slow fire pistol target is $\frac{7}{8}$ in. in diameter. What angle does it subtend as seen by a shooter 50 ft away? How does this compare with the apparent size of the Moon?
7. *7.* [I] Assume the human eye can resolve details separated by about 1.0 minute of arc (where there are 60 minutes per degree). How far away must you sit from a TV set (whose picture is 30 cm high) in order to no longer distinguish the horizontal scan lines? (Typically, 525 horizontal lines make up a modest TV picture.)
8. [I] An ant positioned on the very edge of a Beatles record that is 13 cm in radius revolves through an angle of 100° as the disk turns. How far does he travel?
9. **9.** [I] A ball 30 cm in diameter rolls 65 m without slipping. How many revolutions did it make in the process?
10. [I] What is the equivalent of 1.00 rev/min in rad/s?
11. [I] A well-trained runner can swing a leg forward in the vertical plane around a horizontal axis through the knee at a rate of about 700° per second. How much is that in rad/s?
12. [I] What is the angular speed of the 10-cm-long second hand of a clock? What is ω for the minute hand?
13. **13.** [I] A videodisc revolves at 1800 rpm beneath a laser read-out head. If the beam is 12 cm from the center of the disc, how many meters of data pass beneath it in 0.10 s?
14. [I] A variable speed electric drill motor turning at 100 rev/s is uniformly accelerated at 50.0 rev/s^2 up to 200 rev/s. How many turns does it make in the process?
15. *15.* [I] Thirty seconds after the start button is pressed on a big electric motor, its shaft is whirling around at 500 rev/s. Determine its average acceleration.
16. [I] A large motor-driven grindstone is spinning at 4 rad/s when the power is turned off. If it rotates through 100 rad as it uniformly comes to rest, what is its angular acceleration?
17. **17.** [I] A steam engine is running at 200 rpm when the

engineer shuts it off. The friction of its various parts produces torques that combine to decelerate the machine at 5.0 rad/s/s. How long will it take to come to rest?

18. [I] What angular acceleration is required to stop a turbine blade spinning at 20 000 rpm in 50.0 s?
19. [II] Take the Earth to be moving in a circular orbit ($r_{\oplus\odot} = 1.495 \times 10^{11}$ m) about the Sun. Compute both its linear and angular speeds.
20. [II] A bicycle with 24-in.-diameter wheels is traveling at 10 mi/h. At what angular speed do the wheels turn? How long do they take to turn once around?
21. [II] A 1.0-m-diameter nontranslating disk is made to accelerate from rest up to 20 rpm at a rate of 5 rad/s^2. Through how many turns will it revolve in the process? How far will a point on its rim travel while all this is happening?
22. [II] The motor in Fig. P22 is mounted with a 20-cm-diameter pulley and revolves at 100 rpm. We would like to drive the attic fan so that no point on each of its four 1.0-m-long blades exceeds a speed of 7.0 m/s. What size pulley should it have?

Figure P22

23. [II] A train rounds a turn 1000 ft in radius while traveling at a speed of 20 mi/h. Determine its angular speed and centripetal acceleration. Give the answer in SI units.
24. [II] A bicycle has its gears set so that the front sprocket, which is driven by the pedals, has 52 teeth whereas the rear one, to which it is attached via the chain, has 16 teeth. If it's pedaled at a most efficient 50 rpm and has 24-in. wheels, how fast is the bicycle moving? Give the answer in SI units.
25. [II] A chimp sitting on a yellow unicycle with a wheel diameter of 20 in. is pedaling away at 100 rpm. How fast does it travel? Give the answer in SI units.
26. [II] Imagine two ordinary gears of different diameters meshed together, with the larger being the driver. If the larger gear has 100 teeth around its circumference and rotates at 5.0 rad/s, the smaller gear, which has only 25 teeth, will rotate at what speed?
27. [II] The pulley on the left of Fig. P27 is 0.6 m in diameter and rotates at 1.0 rpm. It is attached by a twisted belt to the 0.2-m-diameter hub of a compound pulley whose outer diameter is 0.8 m. If the driver turns clockwise, find the velocity of the suspended body.
28. [II] Figure P28 shows the structure of a magnetic tape cassette. The tape is transported across the heads at a constant speed of $1\frac{7}{8}$ inches per second. Contrary to popular belief, this is not done by driving the feed and take-up hubs. Why not? Actually, the tape is moved by being squeezed between a precisely rotating shaft—the capstan—and a soft pinch roller. If the capstan has a diameter of 3/32 in., how fast must it turn? Give the answer in SI units.

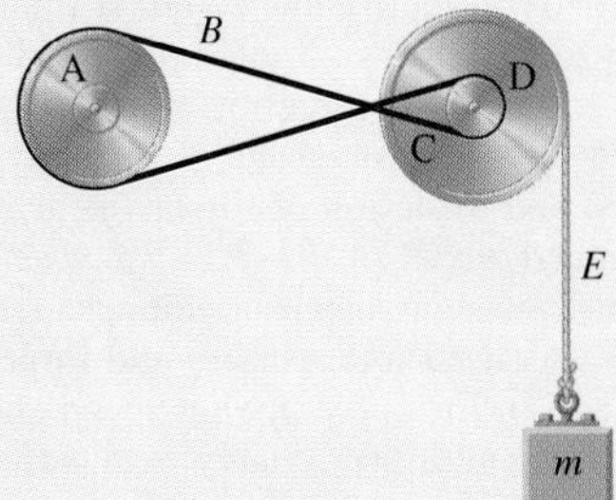

Figure P27

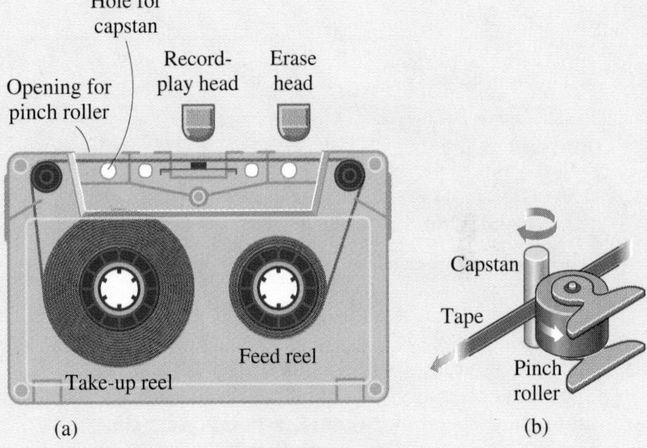

Figure P28

29. [II] An electric circular saw reaches an operating speed of 1500 rpm in the process of revolving through 200 turns. Assuming the angular acceleration is constant, determine its value. How long does it take to get up to speed? Now redo the problem using the average angular speed.
30. [II] A wheel of radius R is freely rolling along a flat surface at an angular rate of ω. Determine, analytically, the linear speed with respect to the ground at any instant of (a) the *c.m.*, (b) the topmost point, and (c) the point in contact with the ground.
31. [II] New York City is traveling around at a tangential speed of about 790 mi/h as the Earth spins. Assuming the planet is a sphere of radius 6371 km, how long is the perpendicular from the city to the spin axis? Compute the city's latitude.
32. [II] Two spacecraft are freely coasting in circular orbits, one 2 km beneath the other. Referring to the conclusions of Chapter 7, compare their angular speeds and show that the lower ship will pull ahead of the upper one.
33. [II] A gear train consists of five meshed gears arranged so that the first (20 teeth) drives the second (80 teeth), which

drives the third (40 teeth), and so on. If the first is mounted on the shaft of a 1500-rpm motor rotating clockwise, at what speed and in what direction will the fourth (25 teeth) and fifth (75 teeth) gears rotate? Figure out a general rule for the speed and direction of the last gear in such a train.

34. [II] The pegged wheel in Fig. P34 was a precursor of the modern cam. Here it's shown operating a trip hammer the same way it was used to crush ore and forge metal 1000 years ago in Europe. At what speed should the central drive shaft be turned, presumably by a waterwheel, for the hammer to pound away at one smash every 3 s?

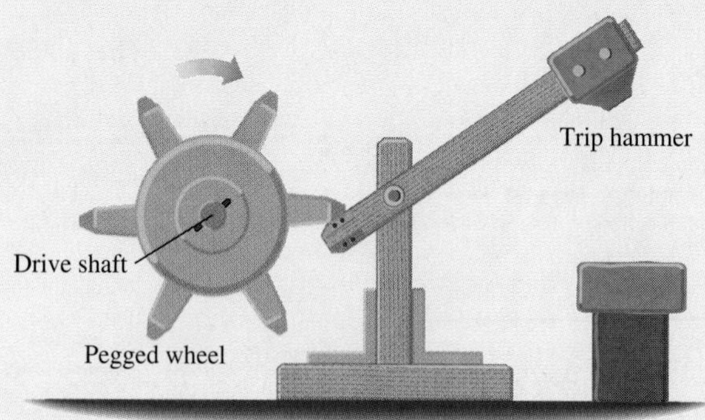

Figure P34

35. [III] A cylindrical wheel is rotating with a constant angular acceleration of 4.0 rad/s^2. At the instant it reaches an angular speed of 2.0 rad/s, a point on its rim experiences a total acceleration of 8.0 m/s^2. What is the radius of the wheel?

36. [III] A cord is wrapped around a 0.24-m-diameter pulley and the free end is tied to a weight. That weight is allowed to descend at a constant acceleration. A stopwatch is started at the instant when the weight reaches a speed of 0.02 m/s. At a time of $t = 2.0$ s, the weight has dropped an additional 0.10 m. Take $y = 0$ to occur at $t = 0$ and show that

$$y = (0.02t + 15 \times 10^{-3}t^2)\text{ m}$$

Prove that for any point on the circumference

$$a_C = (33 \times 10^{-4} + 10 \times 10^{-3}t + 75 \times 10^{-4}t^2)\text{ m/s}^2$$

and $$a_T = 0.03\text{ m/s}^2$$

37. [III] If the hand in Fig. P37 pulls down on the rope moving it at a speed of 2.0 m/s, what will be the resulting velocity of the hanging mass?

38. [III] If the hand in Problem 37 uniformly accelerates the rope downward from rest at $1\frac{1}{3}$ m/s^2, how fast will the hanging mass be moving 1.0 s later?

39. [III] A 30-cm-diameter turntable platter of a record player is attached, by a belt (to reduce vibrations), to a motor-driven pulley 2.0 cm in diameter (Fig. P39). Determine the motor's angular acceleration if the platter is to reach $33\text{-}\frac{1}{3}$ rpm in 6.0 s. How many rotations will it make before reaching operating speed?

40. [III] An inexpensive rim-drive record player has a 6.0-cm-diameter rubber wheel rotating in contact with the

Figure P37

Figure P39

Figure P40

inner surface of the 30-cm-diameter platter (Fig. P40). If the turntable is to go from $33\frac{1}{3}$ rpm to 78 rpm in 3.0 s, what must be the angular acceleration of the drive wheel?

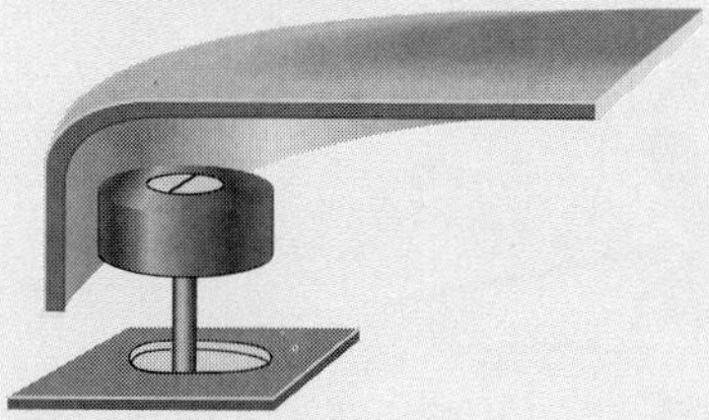

41. [III] Because of tidal friction, the Earth is de-spinning. Its period is increasing at a rate of roughly 25×10^{-9} s per day. Approximate the corresponding angular acceleration in rad/s^2. Compute the decrease in the Earth's angular speed that would exist one thousand million years from now, assuming α is constant. What will its new period be?

THE DYNAMICS OF ROTATION

42. [I] Figure P42 depicts a tone arm, turntable, and record showing the friction force exerted by the record *on the stylus* (the needle). Determine the corresponding frictional torque on the arm when $F_f = 0.008$ N, $\theta = 4.5°$, and $l = 20$ cm. This torque tends to make the arm "skate" across the record whenever the stylus happens to jump out of the groove. Which way will it skate?

43. [I] Consider a thin ring or hoop of mass M and radius R of negligible thickness. Compute the moment of inertia about its *c.m.* by imagining it to be made up of a large number of small segments.

44. [I] Two solid 3.0-kg cones, each with a base radius of 0.30 m, are glued with their flat faces together. Determine the moment of inertia of this device about the symmetry

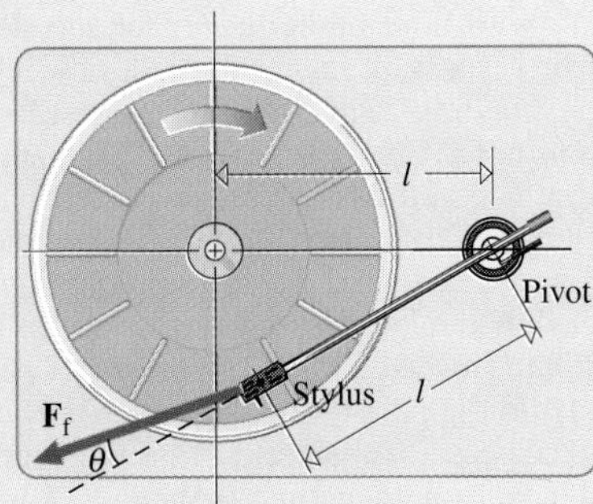

Figure P42

axis passing through both vertices.

45. [I] A gymnast in the process of performing a forward somersault in midair increases his angular velocity by 450% while going from an arms-overhead layout position to a tight tuck. What can you say about the change, if any, in the moment of inertia about the frontal axis (parallel to the shoulder-to-shoulder line) passing through his *c.m.* (in one side and out the other)?

46. [I] An astronaut of mass 70 kg rotates his arms backward about a shoulder-to-shoulder axis at a rate of 1 rev/s. *Both of his arms together* have a mass of 12.5% of his body mass and a total moment of inertia about the shoulders of 1 kg·m^2. Compute the angular momentum imparted to the rest of his body in the process.

47. [I] The **Parallel-Axis Theorem** states that if I_{cm} is the moment of inertia of a body about an axis through the *c.m.*, then I, the moment of inertia about any axis parallel to that first one, is given by

$$I = I_{cm} + md^2$$

Here, m is the object's mass and d is the perpendicular distance through which the axis is displaced. Accordingly, use the theorem for a thin rod of length l; $I_{cm} = \frac{1}{12}ml^2$ to compute I about one end.

48. [I] Show that a small object (such as a sphere) that orbits a distant axis can be approximated by a point mass when you want to compute its moment of inertia. Begin with the theorem of the previous problem.

49. [I] A tiny hummingbird with a mass of only 2×10^{-3} kg is circling around a flower in a 1.0-m-diameter orbit. If it travels once around in 1.0 s, approximate its angular momentum with respect to the blossom (look at Problem 48).

50. [I] Envision a solid cylindrical wheel of radius r and mass m resting upright on a flat plane. With Problem 47 in mind, compute the wheel's moment of inertia about the line of contact with the surface.

51. [I] Two small rockets are mounted tangentially on diametrically opposite sides of a cylindrically shaped artificial satellite. The spacecraft has a 1.0-m diameter and a moment of inertia about its central symmetry axis of 25 kg·m^2. The rockets each develop a thrust of 5.0 N, are oppositely directed, and are aligned to produce a maximum spin-up of the craft. What's the resulting angular acceleration when they are both fired?

52. [II] Compute the orbital angular momentum of Jupiter ($M_{♃} = 1.9 \times 10^{27}$ kg, $r_{♃} = 7.8 \times 10^{11}$ m, and $v_{av} = 13.1 \times 10^3$ m/s) and then compare it to the spin angular momentum of the Sun ($M_\odot = 1.99 \times 10^{30}$ kg, $R_\odot = 6.96 \times 10^8$ m). Assume the Sun, whose equator rotates once in about 26 days, is a rigid sphere of uniform density. It would seem that most of the angular momentum of the Solar System is out there with the giant planets that also rotate quite rapidly.

53. [II] A 26-in.-diameter bicycle wheel supported vertically at its center is spun about a horizontal axis by a torque of 50.0 ft·lb. If the rim and tire together weigh 3.22 lb, determine the approximate angular acceleration of the wheel. Ignore the contribution of the spokes. Of course, these ugly units should cancel out.

54. [II] Derive an expression for the acceleration of a hollow sphere of mass m and radius R rolling, without slipping, down an incline of angle θ. Compare this result with Eq. (8.34) and discuss the implications of the absence of m and R in both those formulas.

55. [II] A block of mass m is tied to a light cord that is wrapped around a vertical pulley of radius R and moment of inertia I. Assuming the bearings are essentially frictionless, write an expression for the rate at which the block will accelerate when released.

56. [II] A 10-kg solid steel cylinder with a 10-cm radius is mounted on bearings so that it rotates freely about a horizontal axis. Around the cylinder is wound a number of turns of a fine gold thread. A 1.0-kg monkey named Fred holds on to the loose end and descends on the unwinding thread as the cylinder turns. Compute Fred's acceleration and the tension in the thread.

57. [II] How would Fred manage in Problem 56 if the bearings exerted a frictional torque of 0.060 N· m?

58. [II] A small mass m is tied to a string and swung in a horizontal plane. The string winds around a vertical rod as the mass revolves, like a length of jewelry chain wrapping around an outstretched finger. Given that the initial speed and length are v_i and r_i, compute v_f when $r_f = r_i/10$.

59. [II] A very thin 1.0-kg disk with a diameter of 80 cm is mounted horizontally to rotate freely about a central vertical axis. On the edge of the disk, sticking out a little, is a small, essentially massless, tab or "catcher." A 1.0-g wad of clay is fired at a speed of 10.0 m/s directly at the tab perpendicular to it and tangent to the disk. The clay sticks to the tab, which is initially at rest, at a distance of 40 cm from the axis. What is the moment of inertia (a) of the clay about the axis? (b) of the disk about the axis? (c) of both clay and disk about the axis? (d) What is the linear momentum of the clay before impact? (e) What is the angular momentum of the clay with respect to the axis just before impact? (f) What is the angular speed of the disk after impact?

60. [II] A water turbine (having a moment of inertia of 1000 kg·m^2) is essentially a wheel with a lot of blades attached to it much like its predecessor, the waterwheel. When a high-speed blast of water hits the blades, it drives the turbine into rotation. Suppose the input valve is closed and the turbine winds down from its operating speed of 200 rpm, coming to rest in 30 minutes. Determine the frictional torque acting.

61. [II] A hoop of mass m and radius R rolls without slipping down an incline at an angle of θ. Write an expression for the acceleration of its *c.m.* and find its numerical value when $m = 1.0$ kg, $R = 1.0$ m, and $\theta = 30°$.

62. [II] Write an expression for the orbital angular momentum of a small artificial satellite of mass m in a circular flight-path of radius r about the Earth. Check the units. What happens to angular momentum as the orbit increases?

63. [II] A solid sphere of mass m and radius r rolls without sliding down an inclined plane tilted at an angle θ. Show that the minimum coefficient of static friction needed to prevent slipping is $\mu_s = (2/7) \tan \theta$.

64. [III] Figure P64 shows two masses strung over a 0.50-m-diameter pulley whose moment of inertia is 0.035 kg·m². If there is a constant frictional torque of 0.25 N·m at the bearing, compute the acceleration of either mass. (The tension in the rope is not the same on either side of the pulley.)

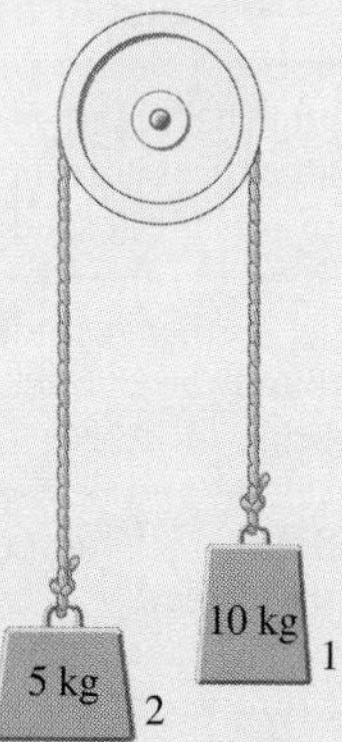

Figure P64

65. [III] An astronaut is working out at the far reach of a 100-m tether, the opposite end of which is attached to a pivoting ring on the nose of the space station. She does not notice that an air hose on her backpack has developed a leak. The little blast of gas produces a tangential thrust and a corresponding acceleration of $a_T = 1.0 \times 10^{-3}\, g$. After two minutes, she realizes what has happened and shuts off the leak. At that point, she and her life support system (total mass of 150 kg) are sailing along at a tangential speed of ________ with an angular momentum of ________. Annoyed, she decides to return to the craft by pulling hand-over-hand on the tether. If she manages to get 5.0 m from the ring on the ship, her tangential speed would be ________. Considering that the centripetal force she would have to hold against is then ________, it's unlikely she would ever get that close.

66. [III] A solid cylindrical pulley of mass m and radius R is held upright and a number of circular turns of light cord are wrapped around it within the groove. The loose end of the cord is held vertically and the pulley is released so it falls straight downward, unwinding like a yo-yo. Please compute its linear acceleration, in terms of g, and the tension in the cord, in terms of m and g, before it reaches the end of its rope.

67. [III] A solid cylinder is placed at rest on an inclined plane at a vertical height h and allowed to roll down without slipping. Show that its speed at the bottom is given by $v = 2\sqrt{\frac{1}{3}gh}$.

68. [III] A cylinder of radius R and mass m mounted with an axle is placed on a horizontal surface. It's pulled forward via a mechanical arrangement so that the applied horizontal force F passes through the *c.m.* Assuming the cylinder rolls without slipping, determine both its acceleration and the friction force at the surface in terms of F.

69. [III] Consider a long slender rod of mass M and length l pivoted at its far end at point O. Now compute I_0 directly by dividing the rod into 100 equal-mass parts and summing up the moments of inertia for all of these. You will need to know that

$$1^2 + 3^2 + 5^2 + \cdots + (2n - 1)^2 = \frac{n}{3}(2n + 1)(2n - 1)$$

9

Energy

THUS FAR WE HAVE studied Newtonian Mechanics, the description of motion in terms of momentum. This was the unrivaled theoretical formalism into the 1800s. But by the beginning of that century, a powerful alternative, predicated on the concept of *energy,* was already taking shape. One of the great achievements of that era was the formulation of the Law of Conservation of Energy (p. 297). In modern times, the Special Theory of Relativity and the work of Amalie Noether have made it clear that momentum and energy are complementary ideas rooted in the nature of space and time.

We now begin a treatment of energy, first looking at how it is trans-

It is important to realize that in physics today, we have no knowledge of what energy is.

R. P. FEYNMAN
Nobel laureate

ferred through *work* and then at how it is manifest in its basic forms—*kinetic* and *potential*. The emphasis is on mechanics, although the concept of energy is all-embracing. In one way or another, it influences our thinking about every branch of physics.

THE TRANSFER OF ENERGY

Over the centuries, the word *energy* has had different meanings. The nontechnical usage derives from the Greek *en* (which means *in*) and *ergon* (which means *work*). Hence, energy is the capacity to do work, an inherent vigor. The word has been used in this way at least since the late 1500s. Galileo (1638) made use of the term *l'energia,* though he never defined it. Only in the last 200 years has the idea taken on a scientific meaning, albeit one not altogether satisfactorily defined.

In very general terms, *energy describes the state of a system in relation to the action of the Four Forces.* It is a property of all matter and is observed indirectly through changes in speed, mass, position, and so forth. There is no universal energy meter that measures energy directly. The change in the energy of a system, which is all we can ever determine experimentally, is a measure of the physical change in that system. **Force is the agent of change; energy is a measure of change**. Because a system can change through the action of different forces in different ways, there are several distinct manifestations of energy.

We buy *electrical energy* "packaged" from the hardware store and "on tap" from the power company. *Chemical energy* can be gotten from a slice of pizza or a tank of gasoline. Rubber bands, ligaments, and girdles all store *elastic energy.* Snowflakes fall because they have *gravitational energy.* We bake apple pies with *thermal energy* and defend them with *nuclear energy.* You don't play on railroad tracks because trains have lots of *kinetic energy. Radiant energy* floods in from the Sun to warm us and from TV stations to entertain us.

The most important characteristic of energy is that it is transferred from one configuration of interactions to another, such that the total amount of energy remains unchanged. Thermal energy can be converted into electrical energy, and some of that turned into light, and back again into thermal energy, but the net amount of energy is always the same—energy is conserved.

Energy is a scalar quantity associated in various amounts with all the "things" that exist, from minute massless particles to immense whirling galaxies. By observing the changing behavior of matter, we infer the presence of one form or another of energy. Like linear and angular momentum, energy is not an entity in and of itself—there is no such thing as pure energy.

9.1 Work

The ancient Greeks seem to have had a vague concept of work; it appears just beneath the surface in their explanations of how a large weight could be lifted by exerting a small force on a lever. In the early 1600s, Galileo was only beginning to grope toward the essential idea. He considered the behavior of a pile driver and recognized that the combination of the weight of the hammer and the distance through which it fell determined its effectiveness. Distance and force are related in some crucial way. Newton, in his *Principia,* doesn't mention energy, but it is clear that he had a strong intuitive sense of the concept.

We learned earlier that the product of *force* and the *time* over which it acts is the *change in momentum.* Now, we will see that *the product of force and the distance over which it acts is a measure of the change in energy.* When that idea was finally formalized by Gaspard Coriolis (1829), he called the product of force and distance **work**. That word is still used, although *mechanical change in energy,* or *energy mechanically transferred,* would be more to the point. **Work is the change in the energy of a system resulting from the application of a force acting over a distance**.

At the heart of the concept of work is the notion of *movement against resistance,* be it the resistance produced by gravity, or friction, or inertia, or whatever. Work can be done on a body to get it moving, to keep it moving, or to change the way it is moving. Work is done to overcome, to move against the resistance of, some force.

> *As we cannot give a general definition of energy, the principle of the conservation of energy simply signifies that there is* something *which remains constant. Well, whatever new notions of the world future experiments may give us, we know beforehand that there will be something which remains constant and which we shall be able to call* energy.
>
> HENRI POINCARÉ (1854–1912)

Work Along a Straight Line: F ∥ s

Let's begin with the simplest case of a point-mass, or equivalently, a finite object that is perfectly rigid so it acts as though it were a point-mass located at its center-of-mass. The idea is to avoid the complexities that exist with deformable bodies. Figure 9.1 shows a constant external force (F) exerted on a rigid object, moving it through a horizontal displacement (s). The applied force is parallel to the straight-line motion and, in that case, a preliminary definition of work (W) is

$$W = \pm Fs \tag{9.1}$$

As long as the object moves along the line-of-action of the force while the force acts on it, work is being done. Here $\mathbf{F}$ is parallel to the displacement $\mathbf{s}$; it sustains the motion and does positive work on the object. When $\mathbf{F}$ is antiparallel to $\mathbf{s}$, it opposes the motion and does negative work on the object.

The SI unit of work is the *newton-meter.* To be more concise and to honor J. P. Joule, the work done by a 1-newton force moving a body through 1 meter is defined as 1 *joule* (J), or

$$1\ \text{J} = 1\ \text{N}\cdot\text{m}$$

Thus, if the force in Fig. 9.1 is 20 N and the displacement 2.0 m, the work done on the body is 40 N·m, or 40 J. In U.S. Customary units, still used in engineering, work is expressed in *foot-pounds,* where

$$1\ \text{J} = 0.7376\ \text{ft}\cdot\text{lb} \qquad \text{and} \qquad 1\ \text{ft}\cdot\text{lb} = 1.356\ \text{J}$$

Positive work is done on an object when the point of application of the force moves in the direction of the force. If the object experiencing a force does not move,

Figure 9.1 (a) A man dragging a crate uses a force F to pull it along the ground. (b) When F acts over a distance s, an amount of work (W) is done by the man.

there is no work done. Raise a 100-kg barbell into the air against its downward weight and you do work. Hold it above your head, and you do none.

Example 9.1 A locomotive exerts a constant forwardly directed force of 400 kN on a train that it pulls for 500 m. (a) How much work does the engine do on the train (W_{et}) during that period? Coming toward a station, the locomotive slows the train, applying a constant force of 100 kN in the opposite direction. (b) How much work does it then do on the train over a distance of 1000 m?

Solution: [Given: (a) $F = +400$ kN, $s = 500$ m, (b) $F = -100$ kN, $s = 1000$ m. Find: W_{et} in both cases.] (a) **F** and **s** are parallel and so $F > 0$; hence

$$W_{et} = +Fs = +(400 \text{ kN})(500 \text{ m}) = \boxed{+200 \times 10^6 \text{ J}}$$

The locomotive does 200 MJ of positive work *on* the train. (b) Now **F** and **s** are antiparallel and so

$$W_{et} = -Fs = -(100 \text{ kN})(1000 \text{ m}) = \boxed{-100 \times 10^6 \text{ J}}$$

The engine opposes the motion and does negative work on the train. Equivalently, the train does work ($W_{te} = -W_{et}$) on the engine.

▶ **Quick Check:** (a) $W_{et} = +4 \times 5 \times 10^7$ J. (b) $W_{et} = -1 \times 1 \times 10^8$ J.

Work in General

Imagine a rigid crate being pulled along by a person tugging on a rope that makes an angle θ, as in Fig. 9.2a. Part of the force **F** acts upward (tending to lift the body and lessen the load), and part acts in the direction of the motion and does work. The force is a vector; it is equivalent to two perpendicular component vectors, one vertical ($F \sin \theta$) and the other horizontal ($F \cos \theta$). The situation is identical to two people pulling on the crate at the same time via two ropes, one up and one forward. Assuming it doesn't lift off the floor, the crate only moves horizontally and only the horizontal puller does work.

Equation (9.1) can be generalized by requiring that *the work done on a body by an applied force is the product of the component of the force in the direction of the displacement multiplied by the magnitude of the displacement*:

$$W = (F \cos \theta)s = Fs \cos \theta \tag{9.2}$$

A working force must to some extent be in, or opposite to, the direction of the motion. While walking across a room at a constant speed with a barbell above your head, you do no work on it because the supporting force, counteracting gravity, is vertical and the displacement is horizontal ($\theta = 90°$).

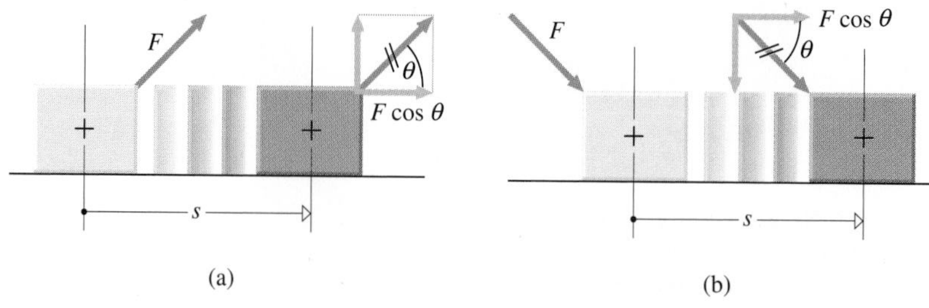

Figure 9.2 Often the force propelling an object is not parallel to the displacement—a stick pushing on a hockey puck or a foot pressing on the pedal of a bike are familiar examples. Only the component of the force parallel to the displacement does work. In both (a) and (b), that component is $F \cos \theta$.

Example 9.2 The truck in Fig. 9.3 is dragging a stalled car up a 20° incline. The tensile force on the tow-line is constant, and the two vehicles accelerate at a constant rate. If the chain makes an angle of 30° with the road and the tension is 1600 N, how much work was done by the truck on the car in pulling it 0.50 km up the incline?

Solution: [Given: $F_T = 1600$ N, $\theta = 30°$, and $s = 500$ m; the angle of the incline is irrelevant here. Find: W_{tc}.] The motion is along the incline, and the component of the force acting in the direction of the displacement is therefore $F_T \cos 30°$, no matter what the tilt of the hill. Hence

$$W_{tc} = (F_T \cos 30°)s = (1600 \text{ N})(0.866)(500 \text{ m})$$

and

$$\boxed{W_{tc} = 6.9 \times 10^5 \text{ J}}$$

▶ **Quick Check:** The working force is 1386 N acting over 500 m; hence, $W_{tc} \approx (1.4 \times 10^3 \text{ N})(5 \times 10^2 \text{ m}) \approx 7 \times 10^5$ N·m.

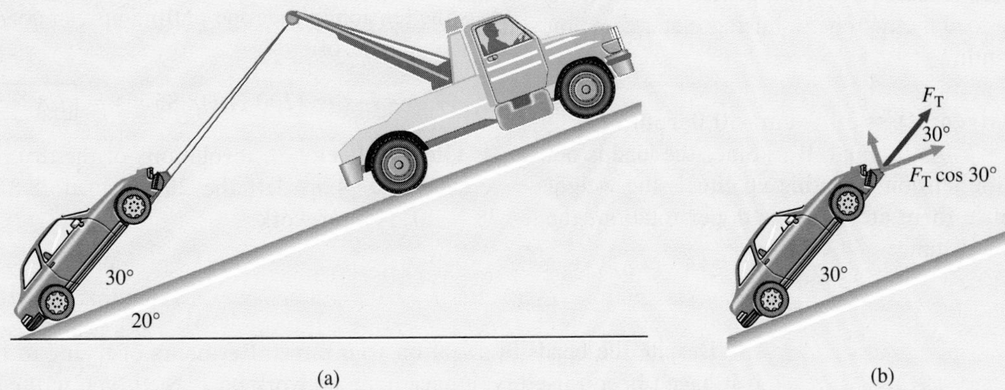

Figure 9.3 (a) A tow truck pulling a car up an inclined plane. Only the component of the force parallel to the displacement does work. (b) That working component is $F_T \cos 30°$.

Example 9.3 Figure 9.4 depicts someone very slowly raising a rigid 1.0-kg block at a constant speed through a vertical distance of 2.0 m at the Earth's surface. The center-of-mass rises from point *A* to point *B*. Neglecting the tiny additional force needed to begin the motion and assuming *g* to be constant, determine the work done on the block by the hand.

Solution: [Given: $m = 1.0$ kg, $s = 2.0$ m, and $g = 9.8$ m/s^2. Find: W_{hb}.] To raise the block at a constant speed ($a = 0$), the sum-of-the-forces acting vertically must be zero ($\Sigma F = ma = 0$), meaning that the applied force **F** must be equal in magnitude and opposite in direction to the weight of the block $\mathbf{F}_W$. The force is straight up, as is the displacement, so that in Eq. (9.2) $\theta = 0$, $\cos\theta = 1$, and

$$W_{hb} = Fs = F_W s = mgs = (1.0 \text{ kg})(9.8 \text{ m/s}^2)(2.0 \text{ m})$$

and

$$\boxed{W_{hb} = 20 \text{ J}}$$

This is the work done in overcoming gravity.

▶ **Quick Check:** $W_{hb}/F = s = 20$ N·m/9.8 N ≈ 2 m.

In Example 9.3, the hand pushes up on the block and the block pushes down on the hand. The latter force comes from the block-Earth gravitational interaction. The work done by the hand ($W_{hb} = +20$ J) on the block is positive because the force is in the direction of the displacement. By contrast, while the block is moved upward, the force it exerts on the hand acts downward; $\theta = 180°$, $\cos\theta = -1$, and so the work done on the hand is $W_{bh} = -20$ J. It is negative because the weight of the block resists the upward motion. When the work done *on* system 1 (the block) *by* system 2 (the hand) is positive, system 1 receives energy from system 2. Thus, while one of the *interacting* systems gains energy, the other loses an equal amount of energy; while one has positive work done on it, the other does the positive work, and has negative work done on it.

Example 9.4 A 30.0-cm-diameter horizontal drum capable of rotating at 10.0 rpm has a rope wound around it. A 200-N load is hung from the rope. How much work is done on the load by the rope in raising it at a constant speed for 2.0 min?

Solution: [Given: $D = 30.0$ cm, 10.0 rpm, $F_W = 200$ N, and $t = 2.0$ min. Find: W_{r1}.] Since the load is not accelerating, the tension F_T acting up equals the weight F_W, down. Since there are πD meters per rotation, the load is raised a distance

$$s = (\pi D \text{ m/rot})(10 \text{ rpm})(2.0 \text{ min})$$

and

$$s = \pi(0.30 \text{ m})20 = 18.85 \text{ m}$$

It is the tension in the rope pulling up that does the work in the amount of

$$W_{r1} = F_T s = (200 \text{ N})(18.85 \text{ m}) = \boxed{3.8 \times 10^3 \text{ J}}$$

▶ **Quick Check:** 20 revolutions of the drum (circumference = 0.94 m) lift the 200-N load 18.8 m, doing 38×10^2 N·m of work.

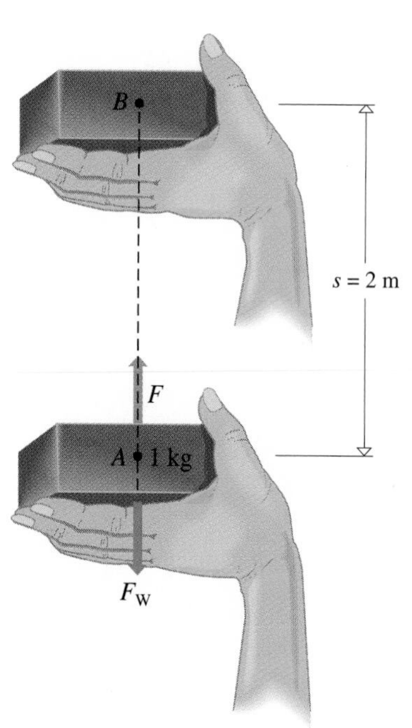

Figure 9.4 The hand slowly raises the block by exerting an upward force F negligibly greater than its weight F_W. Work is done to overcome gravity—that is, to move against the block-Earth gravitational interaction.

Despite the beads of sweat on your brow after hours of trying to lift a barbell, if it hasn't been raised, you have done no work *on it.* Still, you will certainly have generated a good deal of "heat"; your metabolic rate and oxygen consumption will both have gone up. This seeming contradiction can be understood by examining how muscles function. When a muscle contracts, exerting a force over a distance, it does work, but a muscle maintaining a constant tension is doing something as well. The skeletal muscles are composed of bundles of elongated cells known as *fibers.* In response to nerve signals, these cells individually contract (generating a brief pulse of tension) and then relax, all in a matter of a few milliseconds. Each time a fiber contracts it does work on the muscle. The net result of all of these tiny, short-lived contributions is the apparently constant muscle force that may or may not be doing work on some other body.

Deformable Objects and Friction

Until now, s was the displacement of every point-mass constituting a rigid object, and it was the same for all of them. Thus, s was also the displacement of the center-of-mass, s_{cm}. But not all statements about work that are true for a point-mass are necessarily true for a deformable object. Portions of a real extended object can move in ways different from that of its *c.m.* The concept of work has to be applied to extended bodies more cautiously than was sometimes done in the past. We will come back to this later when we examine the energetics of walking, swimming, and so forth. For the moment, let's consider the example of friction.

Kinetic friction arises from the electromagnetic interaction at the interface between two objects in relative motion. Imagine a block being pulled at a constant speed along the floor by an applied force F equal to the force of kinetic friction F_f (Fig. 9.5). On a macroscopic level, we deal with the friction force as if it were continuous, and it's tempting to assume that the work done *on the block* by F_f as it moves a distance s_{cm} is simply $W_f = -F_f s_{cm}$ *but it is not!* The friction force is an average effect arising from a complex of occurrences on the atomic level. Neither the body nor the floor is rigid. Regions in contact abrade each other, microscopic welds are formed and pulled apart, and work is done. In the process, atoms are jostled and energy is transferred to *both* the block and the floor. *There is no way to calculate the work done on just the block by the friction force:* F_f is not the force on each weld and it does not act over a distance s_{cm}. However, the net work done on the block-floor system is Fs_{cm}. This is the total mechanical energy delivered to the system and, since $F = F_f$, it does equal $F_f s_{cm}$. As we will soon see, the energy transferred into the system by the action of F appears primarily as thermal energy in *both* the block and the floor.

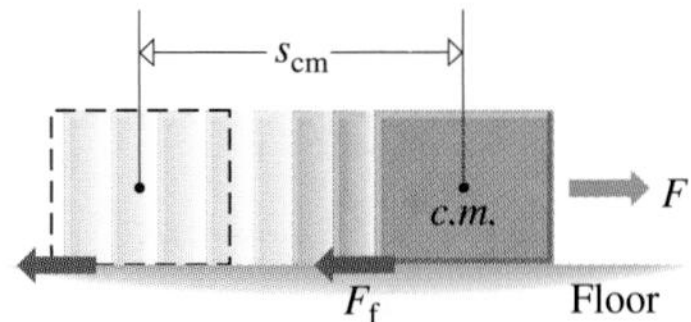

Figure 9.5 A force F is applied to a block that also experiences a friction force F_f. An amount of work $W_f = F_f s_{cm}$ goes into overcoming friction in the block-floor system, and that energy ultimately appears as thermal energy.

Example 9.5 A youngster weighing 250 N is sitting on the grass holding on to a large dog via a leash stretched horizontally. The dog pulls on the leash with a force of 100 N and drags the kid at a slow, constant speed 20 m straight across the yard and then stops. How much work did the dog do on the child and the ground in overcoming kinetic friction?

Solution: [Given: $F = F_f = 100$ N and $s = 20$ m. Find: W_f.] The force F exerted by the dog was constant and in the direction of the displacement; hence, from Eq. (9.1)

$$W_f = F_f s = (100 \text{ N})(20 \text{ m}) = \boxed{2.0 \times 10^3 \text{ N·m}}$$

▶ **Quick Check:** 10^2 N × 2 × 10^1 m = 2 × 10^3 N·m.

When there is a displacement and positive work is done on one system by another, an amount of energy equal to W is transferred to the first from the second. In Example 9.5, the dog was the source of 2.0 kJ that ended up warming the child and the ground. If it takes +2.0 kJ to drag the youngster 20 m to the right, it will take another +2.0 kJ to slide him back. The friction force changes direction as the motion changes, and the work done against friction is always positive. *The net work done equals the force applied to overcome friction times the total path-length traveled.* Thus, when a body returns to its starting point so that its displacement is zero, the work done in overcoming friction is not zero!

Example 9.6 The force needed to overcome air friction for a person walking at a constant normal rate is only about 1 ounce (0.3 N). How much work does someone do against air friction in walking twice around a circular track having a radius of $\frac{1}{4}$ km?

Solution: [Given: $F = 0.3$ N, $R = \frac{1}{4}$ km, $s =$ twice around $= 2 \times 2\pi R$. Find: W_f.] Since F is in the direction of the motion

$$W_f = Fs = (0.3 \text{ N})(2 \times 2\pi R) = \boxed{9 \times 10^2 \text{ J}}$$

▶ **Quick Check:** The path is $2 \times 2\pi R = 2 \times 2(3.1416)(250 \text{ m})$, or $10^3\pi$ m ≈ 3×10^3 m, and $W \approx (0.3 \text{ N})(3 \times 10^3 \text{ m}) \approx 0.9$ kJ.

Changing Forces

When we defined work ($W = Fs \cos\theta$), it was assumed that the driving force F was constant over the displacement. Although there are many important situations where this *is* true, there are many more where it is not. For instance, the force of gravity

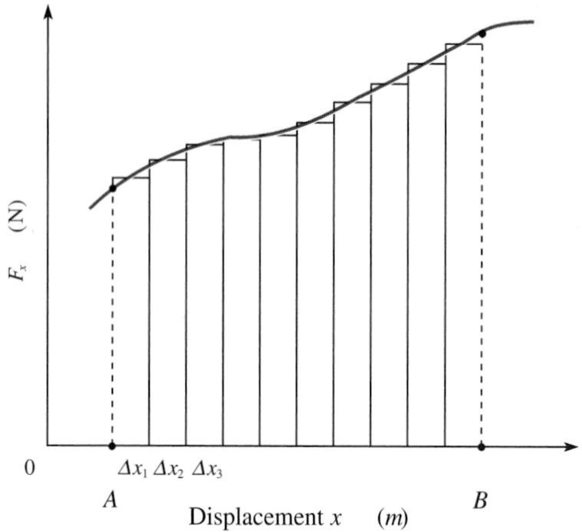

Figure 9.6 A force that varies from point to point in space plotted against distance. The area under the curve is the work done in moving from *A* to *B*.

actually varies as $1/r^2$; the force exerted by a hammer on a nail or a fist on a nose will vary with contact distance.

Imagine an object being displaced along the x-axis by a varying force $\mathbf{F}$ that has a component F_x that does the work. Figure 9.6 is a plot of F_x versus x. (Recall the treatment that led to Fig. 4.13, where the momentum change associated with a time-varying force was determined.) We cannot simply plug into Eq. (9.2) to find the work done because the force varies from point to point, but we can use Fig. 9.6, which is a plot of exactly how F_x changes. The area under the curve (F_x times x) is the desired work. Therefore, divide that curve into a number of very narrow intervals: Δx_1, Δx_2, Δx_3, and so forth. Assume that the force is constant at some average value over each of these segments.

We have a definition for work done by a constant force; hence, for any segment $\Delta W = F_x \Delta x$. The net work done in going from $x = A$ to $x = B$ is approximately the sum of all these contributions. And that approximation improves as the Δx intervals get narrower and narrower because the average force in each case then more closely approaches the instantaneous force; that is, the corresponding point on the F_x-curve. As these intervals approach zero in width and the number of them increases toward infinity, *the sum approaches the area under the curve between the two points, which is the total work done by the variable force in moving the object from A to B*. The work done is the area under the force-displacement curve.

If we can plot the curve, the *work diagram*, we can determine the area and therefore the work. The area can be found directly, by counting boxes on graph paper; mechanically, by tracing the curve with a gadget called a planimeter; or electronically, using a microcomputer. It's common practice when evaluating all kinds of devices, from mechanical hearts to steam engines, to automatically produce work diagrams.

Example 9.7 Figure 9.7a shows a mass attached to a spring in a zero-gravity environment. The mass is initially at a height of 0.3 m. It is subsequently pulled down to 0.0 m and released. The force-displacement curve for the spring is shown in Fig. 9.7b. Describe the varying force and compute the work done by the spring on the mass in moving from 0.0 m to 0.3 m. How much work is done by the spring as the mass goes from 0.0 m to 0.6 m?

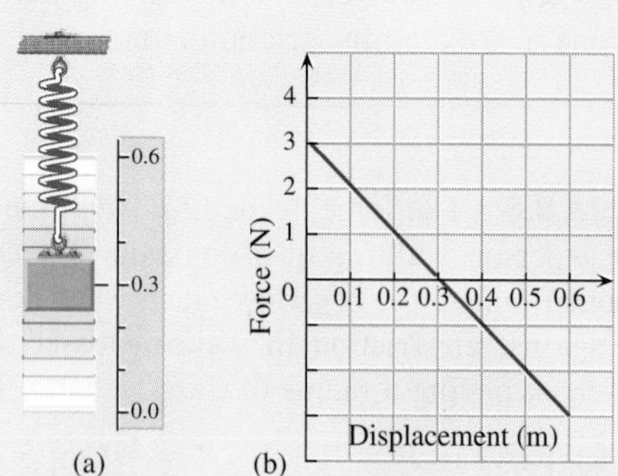

Figure 9.7 (a) A mass in a zero-gravity environment attached to a vertical spring. It is pulled down to point 0.0 and released. (b) The plot is a graph of the variable force exerted *by the spring* versus displacement. We'll study the spring-force in detail later; for the moment, notice that it's linear.

Solution: [Given: the curve. Find: the work done.] The spring-force acts along a vertical line such that it is parallel to the displacement ($\theta = 0$). It follows from the curve that the force varies linearly with the distance from the unstretched equilibrium position (0.3 m). The force is up and maximum (3.0 N) at the very bottom (0.0 m); zero at equilibrium (0.3 m); and down and maximum (-3.0 N) at the top (0.6 m). From 0.0 m to

(continued)

(continued)

0.3 m, the spring-force causes the motion and, from 0.3 m to 0.6 m, it opposes it. The triangular area under the curve from 0.0 m to 0.3 m equals $\frac{1}{2}$(base)(altitude); hence, the work done by the spring on the mass is

$$W_{sm} = \tfrac{1}{2}(0.3 \text{ m})(3.0 \text{ N}) = 0.45 \text{ N·m}$$

or

$$\boxed{W_{sm} = +0.5 \text{ J}}$$

Once the mass passes the equilibrium point on the way up, the spring-force opposes the motion. The work done by the spring is negative; the mass does work on the spring. In this case, the area between 0.3 m and 0.6 m is negative and $W_{sm} = -0.5$ J. Thus, the total amount of work done by the spring in the complete upward journey is zero.

▶ **Quick Check:** The total area under the curve is zero.

Meandering Paths

Imagine a person holding a bowling ball with a constant upward force F. Suppose that the ball is to be slowly brought up a hill along a smooth path (Fig. 9.8a) so that it is at rest at both A and B. After each step, the ball will have been raised a small amount vertically ($\Delta\mathbf{s}_V$) and translated a small amount horizontally ($\Delta\mathbf{s}_H$). The ball is displaced at each step by $\Delta\mathbf{s} = \Delta\mathbf{s}_V + \Delta\mathbf{s}_H$. Since the weight always acts downward and $\mathbf{F}$, accordingly, always acts upward, positive work is done ($\Delta W = F\Delta s_V$) only during the rising part of the motion—no work is done on the ball in moving it horizontally. Although the force doing the lifting is constant from point to point along the route, the angle between $\mathbf{F}$ and the tangent to the path changes. Hence, we divide the route into tiny intervals, Δs, that are so small that θ can be taken as constant over each one (Fig. 9.8b). In any one such interval, $\Delta W = (F \cos \theta)\, \Delta s$. The total work done in going from A to B is

$$W(A \rightarrow B) = \Sigma\, \Delta W = \Sigma F(\Delta s \cos \theta)$$

Because F is constant, it can be factored out of the summation; thus

$$W(A \rightarrow B) = F \Sigma\, \Delta s \cos \theta = F \Sigma\, \Delta s_V$$

The sum of all the little vertical displacements is the total vertical displacement ($s_V = \Sigma\, \Delta s_V$) and

$$W(A \rightarrow B) = F s_V \tag{9.3}$$

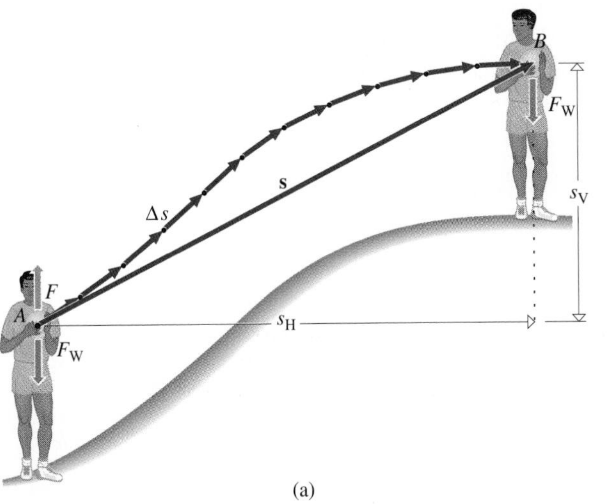

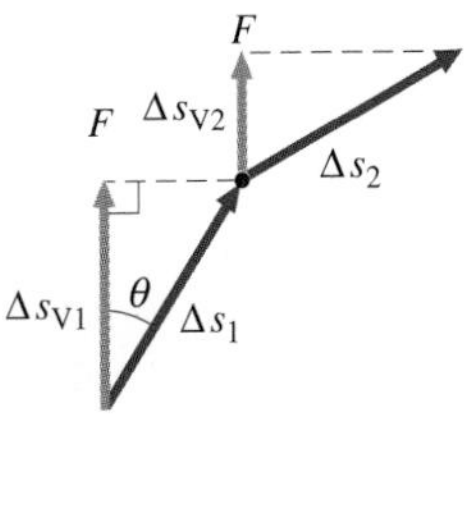

Figure 9.8 The curved path from A to B is divided into tiny straight segments $\Delta s_1, \Delta s_2, \Delta s_3, \cdots$. The small amount of work done during each minute step has the form $\Delta W = F\Delta s \cos \theta$; hence, $W(A \rightarrow B) = F\Delta s_{V1} + F\Delta s_{V2} + F\Delta s_{V3} + \cdots = Fs_V$, where $\Delta s_{V1} + \Delta s_{V2} + \Delta s_{V3} + \cdots = s_V$.

The work depends only on the net vertical distance traversed. **In going from *A* to *B* in a gravitational field, the work done is independent of the path taken—it is determined solely by the weight of the body and the vertical displacement**. The ball could be carried from *A* up, down, and around the hill in any zigzag path to *B*; negative work will subtract from positive work and, in the end, the net work will always be the same.

9.2 Power

When you want a day's labor and are concerned about getting every minute's worth, you buy a certain amount of work per hour. That's the practical language of industry, and it was in the beginning of the Industrial Revolution that the idea of *power* became quantified. **Power** is *the rate of doing work,* that is

$$\text{Power} = \frac{\text{work done}}{\text{time interval}}$$

More generally, power is the rate at which energy is transferred into or out of a system. An engine that will do the same amount of work as another in half the time is twice as powerful. If the time interval Δt is finite, we are talking about the *average power,* or

$$\mathrm{P_{av}} = \frac{\Delta W}{\Delta t} \tag{9.4}$$

To express the moment-by-moment power, the **instantaneous power**, take the limit of this ratio as the time interval goes toward zero, whereupon

$$\mathrm{P} = \lim_{\Delta t \to 0} \left(\frac{\Delta W}{\Delta t}\right) \tag{9.5}$$

In many cases, the power will be constant and Eq. (9.4) will be P.

The idea of standardizing the measure of power using the output of a horse had been around for a while before James Watt (1783) formalized it. Watt found that a large draft horse could exert a pull of about 150 lb while walking at about $2\frac{1}{2}$ mi/h for a considerable amount of time. That rate of doing work is equivalent to 33 000 foot-pounds per minute or *550 ft·lb/s,* which he called 1 **horsepower** (hp). Even though *horsepower* is now an outdated measure, its use persists in the United States.

In the SI system, the unit of power is the **watt** (W), equal to 1 joule of work done per 1 second: 1 W = 1 J/s. This unit is familiar because almost all electrical devices are rated in watts to indicate how much power they require. Since a watt is a fairly small amount of power, the more convenient kilowatt (1 kW = 1000 W) is often used. Thus,

$$1 \text{ hp} = 550 \text{ ft·lb/s} = 746 \text{ W} = 0.746 \text{ kW}$$

Presumably, someday all forms of power will be specified in SI units so that automobile engines will be rated in kilowatts along with toasters (Table 9.1).

Example 9.8 The express elevator in the Sears Tower in Chicago averages a speed of 1800 ft/min in its climb to the 103rd floor, 1340 ft above ground. Assuming a total load of 1.0×10^3 kg, what average power must the lifting motor supply? Always use SI units.

(continued)

(continued)

Solution: [Given: $v_{av} = 30.00$ ft/s $= 9.144$ m/s, $\Delta s =$ 1340 ft $= 408.4$ m, and $m = 1.0 \times 10^3$ kg. Find: P_{av}.] From Eq. (9.4) it's clear that we need ΔW, which, in turn, equals the weight (F_W) times the displacement (Δs). Hence

$$\Delta W = F_W \Delta s = mg\Delta s$$

$$\Delta W = (1.0 \times 10^3 \text{ kg})(9.8 \text{ m/s}^2)(408.4 \text{ m})$$

and
$$\Delta W = 4.0 \times 10^6 \text{ J}$$

To find Δt, remember that $v_{av} = \Delta s/\Delta t$ and so

$$\Delta t = \frac{408.4 \text{ m}}{9.144 \text{ m/s}} = 44.66 \text{ s}$$

Finally

$$P_{av} = \frac{\Delta W}{\Delta t} = \frac{4.0 \times 10^6 \text{ J}}{44.66 \text{ s}} = 89.57 \text{ kW} = \boxed{90 \text{ kW}}$$

▶ **Quick Check:** 90 kW will lift the 9.8×10^6-N load a distance of 408 m in a time of about $(9.8 \times 10^3 \text{ N}) \times (408 \text{ m})/(90 \text{ kW}) = 44$ s.

A person in good physical condition can deliver work at a fairly continuous rate of roughly $\frac{1}{10}$ hp, or about 75 W. Working at a steady pace, the human body develops power in proportion to the amount of oxygen it consumes. A rate of consumption of about 1 liter (10^3 cm^3) per minute is what's required for an output of 75 W. An athlete can raise that to about 5.5 liters per minute and develop continuous power at close to 400 W. A sustained power level of about twice this amount is the limit that a human can maintain for a period in excess of a minute or so. Over short time intervals (for example, the time it takes to leap into the air or throw a ball), much higher power levels can be attained. When muscles are rested, they have an additional short-term supply of oxygen that may also be called upon. This limited reserve can supply bursts of power of, say, 450 W for a minute or so, or as much as several kilowatts for a fraction of a second. Once exhausted, that reserve supply takes some time to replenish, and any athlete is aware of its one-shot nature.

TABLE 9.1 Power equivalencies

Quantity	Equivalent
1 Btu per hour	0.293 W
1 joule per second	1 W
1 foot-pound per second	1.356 W
1 kilocalorie per hour	1.16 W
1 kilowatt-hour per day	41.7 W
1 kilocalorie per minute	69.77 W
1 horsepower	745.7 W
1 kilowatt	1000 W
1 Btu per second	1054 W
1 gallon of gasoline per hour	39 kW
1 gallon of oil per minute	2.5 MW
1 million barrels of oil per day	73 GW

Power and Motion

When a constant force F acts on a body which, in the process, moves through a displacement Δs, we know from Eq. (9.2) that an amount of work is done, $\Delta W = (F \cos \theta)\Delta s$. If this occurs in a time interval Δt

$$P_{av} = \frac{\Delta W}{\Delta t} = (F \cos \theta)\frac{\Delta s}{\Delta t} = (F \cos \theta)v_{av}$$

or, in the limit, as $\Delta t \to 0$

$$P = Fv \cos \theta \qquad (9.6)$$

Power equals the product of the component of the force in the direction of motion and the speed. At any instant, it depends on the instantaneous speed at which the point of application of the force is moving. In many cases, **F** and **v** are

In 1979, the human-powered Gossamer Albatross flew across the English Channel. The pedaling pilot averaged 190 W, or about 0.25 horsepower.

parallel, $\theta = 0$, and the expression becomes

$$[\theta = 0] \qquad P = Fv \qquad (9.7)$$

Example 9.9 In 1935, R. H. Goddard, the American rocket pioneer, launched several A-Series sounding rockets. Given that the engine had a constant thrust of 200 lb, how much power did it transfer to the rocket traveling at a maximum speed of 1130 km/h?

Solution: [Given: $F = 200$ lb and $v = 1130$ km/h. Find: P.] Putting things in SI units, we obtain $F = (200 \text{ lb})(4.448 \text{ N/lb}) = 889.6$ N and 1130 km/h = 313.9 m/s. Using Eq. (9.7)

$$P = Fv = (889.6 \text{ N})(313.9 \text{ m/s}) = \boxed{279 \text{ kW}}$$

▶ **Quick Check:** Let's recalculate using U.S. Customary units; (1130 km/h)(0.9113 ft·h/km·s) = 1029 ft/s; hence, P = (200 lb)(1029 ft/s) = 20.6×10^4 ft·lb/s. Using Table 9.1, we multiply this value by 1.356 W/ft·lb/s, yielding 279 kW.

MECHANICAL ENERGY

There are four basic forces in nature as it exists today, and when work is done by or against any of them—and that's the only way work can be done—change occurs and energy is transferred.*

All matter interacts gravitationally, resulting in a web of force that binds every entity in the Universe together. It's been suggested that inertia arises from that invisible web (Mach's Principle, 1872). If that's true (and the idea is quite speculative), we can anticipate that a change in the distribution of matter effected by a moving body will correspond to an energy of motion.

9.3 Kinetic Energy

Huygens never cared for the idea that the basic measure of motion was momentum. Nonetheless, that view was widely held in the seventeenth century by the followers of Descartes. To be meaningful, momentum had to be directional. A bomb at rest could explode into two pieces, rapidly flying apart in opposite directions, and yet the total momentum of the system would remain zero—that's Conservation of Momentum. But what kind of essential quality of motion equals zero when the bodies being described are hurtling through the air? That thought did not sit well with Huygens, who searched to find a different measure independent of direction, one that would vanish only when motion ceased. His studies of rigid colliding balls led him to conclude that there was something special about the product of mass and speed-squared. Remarkably, adding the values of mv^2 for each ball prior to collision yielded a total that was essentially the same after collision, even though all the velocities had changed. Squaring the speed removed any dependence on the sign of v (that is, on the direction of $\mathbf{v}$); mv^2 is always positive and only vanishes when v vanishes.

Leibniz picked up on the idea, calling mv^2 the *vis viva.* He showed that, for a falling body, mv^2 was proportional to Galileo's product of weight and height. It was not until 1807 that Thomas Young, an English physicist and physician, shifted the

Christian Huygens (1629–1695), Dutch mathematician, physicist, and astronomer.

*The common dictionary definition, *energy is the ability to do work,* is at best misleading (see Discussion Question 9). We will learn that the First Law of Thermodynamics maintains that energy is conserved, while the Second Law tells us that the availability of energy to do work tends to decrease as processes occur.

imagery and spoke of mv^2 for the first time as *energy*. He concluded that "labour expended in producing any motion, is proportional . . . to the energy which is obtained." In other words, work that causes motion equals the resulting change in energy.

Work and the Change in KE

Today, we call the energy associated with motion **kinetic energy** (KE), a term introduced in 1849 by one of the great figures of the nineteenth century, Lord Kelvin. Imagine a *rigid* body moving under the influence of a constant net force, such as a bullet decelerating as it penetrates soft clay. In each case, $W = Fs$ and, since $F = ma$, $W = mas$. From Eq. (3.10), $v_f^2 = v_i^2 + 2as$; hence, $as = \frac{1}{2}v_f^2 - \frac{1}{2}v_i^2$, and multiplying both sides by m yields

$$W = \tfrac{1}{2}mv_f^2 - \tfrac{1}{2}mv_i^2 \tag{9.8}$$

Here, s is the displacement of the *c.m.* Actually, the net force need not be constant. We would get the same result using the average force, producing an average acceleration. Under the influence of a net force ($\theta = 0$), a body accelerates as positive work is done on it to overcome inertia. It increases its speed and gains kinetic energy (KE) in the process. We define the kinetic energy of any object of mass m traveling at speed v as

$$\text{KE} = \tfrac{1}{2}mv^2 \tag{9.9}$$

such that Eq. (9.8) is the difference between the final and the initial KE:

$$W = \text{KE}_f - \text{KE}_i = \Delta\text{KE} \tag{9.10}$$

The net work done accelerating a rigid object equals its change in kinetic energy (Fig. 9.9). From the derivation, the units of KE are the units of W; namely, *joules.* That can be confirmed by the following: $F = ma$, $1\text{ N} = 1\text{ kg·m/s}^2$, and $W = Fs$; hence, $1\text{ J} = 1\text{ N·m} = 1\text{ (kg·m/s}^2\text{)m} = 1\text{ kg(m/s)}^2$. All forms of energy will be measured in joules (Table 9.2).

Equation (9.10) cannot always be applied to deformable bodies. For example, throw a wad of soft clay against a stone wall. The clay comes to a stop, ΔKE can be quite large, but W_{wc} is, for all practical purposes, zero. The wall exerts forces on the clay across the essentially stationary clay-wall interface. The wall is not perfectly rigid, and there is a tiny motion of the points of application of force as the wall distorts, but $W_{wc} \approx 0$. This sort of **nonworking force** will become important to us later in the discussion.

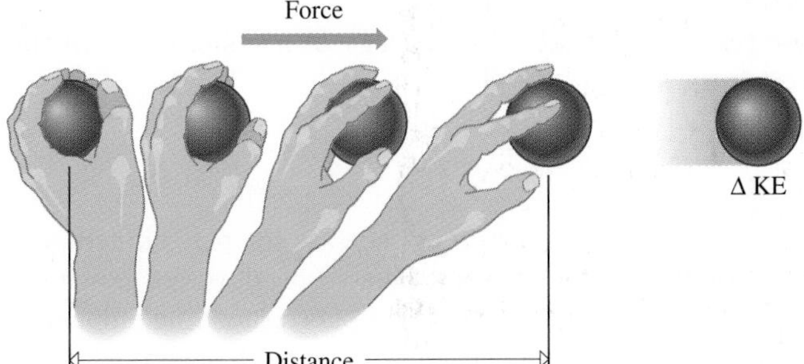

Figure 9.9 A force acting over a distance results in a change in kinetic energy.

TABLE 9.2 Rough values of the energies of various occurrences

Occurrence	Energy (J)	Occurrence	Energy (J)
Creation of the Universe	10^{68}	Burning a cord of wood	10^{10}
Emission from a radio galaxy	10^{55}	One gallon of gasoline	10^{8}
Supernova explosion	10^{44}	One day of heavy manual labor	10^{7}
Yearly solar emission	10^{34}	Explosion of 1 kg of TNT	10^{6}
Earth moving in orbit	10^{33}	Woman running for 1 h	10^{6}
Earth spinning	10^{29}	Burning match	10^{3}
Earth's annual sunshine	10^{25}	Hard-hit baseball	10^{3}
Yearly U.S. sunshine	10^{22}	Apple falling 1 m	1
Annual tidal friction	10^{20}	Human heartbeat	0.5
Exploding volcano (Krakatoa)	10^{19}	Depressing typewriter key	10^{-2}
Severe earthquake (Richter 8)	10^{18}	Cricket chirrup	10^{-3}
100-megaton H-bomb	10^{17}	Hopping flea	10^{-7}
Burning a million tons of coal	10^{16}	Proton in supercollider	10^{-7}
Hurricane	10^{15}	Fission of 1 uranium nucleus	10^{-11}
Thunderstorm	10^{15}	Electron mass-energy	10^{-13}
Atomic bomb (Hiroshima)	10^{14}	Electron in atom	10^{-18}
Saturn V rocket	10^{11}	Photon of light	10^{-19}
Lightning bolt	10^{10}	KE of room-temperature air molecule	10^{-21}

Example 9.10 A Boeing *747* airliner weighing 2.2×10^6 N (5.0×10^5 lb) at takeoff is cruising at a ground speed of 268 m/s (600 mi/h). Compute its kinetic energy. If 1 kg of TNT yields 4.6×10^6 J, how much TNT is the KE equivalent to?

Solution: [Given: $F_W = 2.2 \times 10^6$ N and $v = 268$ m/s. Find: KE.] We need the mass to compute KE; hence, $m = F_W/g = (2.2 \times 10^6\ \text{N})/(9.8\ \text{m/s}^2) = 2.24 \times 10^5$ kg. Consequently

$$\text{KE} = \tfrac{1}{2}mv^2 = \tfrac{1}{2}(2.24 \times 10^5\ \text{kg})(268\ \text{m/s})^2$$

and

$$\boxed{\text{KE} = 8.0 \times 10^9\ \text{J}}$$

This is equivalent to the amount of energy given off when $\boxed{1.7 \times 10^3\ \text{kg}}$ of TNT explodes.

▶ **Quick Check:** $8.0 \times 10^9\ \text{J} = \frac{1}{2}mv^2$; $v^2 = 2(8.04 \times 10^9\ \text{J})/(2.24 \times 10^5\ \text{kg})$; $v = 268$ m/s.

Example 9.10 suggests a fundamental point: KE *is a relative quantity;* we effectively choose the zero of KE by selecting the coordinate system with respect to which v is measured. Here the Earth was the motionless reference frame, and ground speed was used in the calculation. If you were in that jet sitting at "rest" next to your 20-kg suitcase, it would have no KE with respect to you and yet, to someone at "rest" on the ground watching, your bag would flash past with almost 3×10^3 J. **As a rule, we are not interested in the absolute energy of a system, but in the energy added to or taken from it.**

Work → KE → Work

Your arm can do work on a bowling ball and thereby increase the ball's kinetic energy ($W = \Delta\text{KE}$). The ball, going from rest to some launch speed, travels down the lane, crashes into the pins, does work on them, and loses a corresponding amount of KE. A moving object can do a maximum amount of work equal to $\frac{1}{2}mv^2$. Work →

KE $\rightarrow$ work, the transformation goes either way. When work is done on the system, its KE can increase; when work is done by the system, its KE can decrease. Every system that moves transports energy in the form of KE from one place to another.

Example 9.11 A man of average size having a mass of 70 kg can run at a maximum speed of about 10 m/s. Compute his maximum KE.

Solution: [Given: $m = 70$ kg and $v_{max} = 10$ m/s. Find: KE_{max}.] His maximum kinetic energy is

$$KE_{max} = \tfrac{1}{2}mv_{max}^2 = \tfrac{1}{2}(70 \text{ kg})(10 \text{ m/s})^2$$

and

$$\boxed{KE_{max} = 3.5 \times 10^3 \text{ J}}$$

▶ **Quick Check:** $v_{max}^2 = 2KE_{max}/m = 100 \text{ m}^2/\text{s}^2$; $v = 10$ m/s.

Example 9.12 According to the record books, Aleksandr Zass (known to his admirers as "Samson") would, when he wasn't bending iron bars, catch a 104-lb (463-N) woman fired from a cannon at around 45 mi/h. Assuming that Samson brought her to rest uniformly in a distance of 1.00 m, compute the force he exerted on our heroine. As a guess, take her "landing speed" to be 8.94 m/s (20 mi/h).

Solution: [Given: $F_W = 463$ N, $v = 8.94$ m/s, and $s = 1.00$ m. Find: F.] To find the force, we need the work done in stopping her, and that work equals her KE at impact. To find her KE, we need her mass: $m = F_W/g = (463 \text{ N})/(9.81 \text{ m/s}^2) = 47.2$ kg. Hence

$$KE = \tfrac{1}{2}(47.2 \text{ kg})(8.94 \text{ m/s})^2 = 1.89 \text{ kJ}$$

Samson had to do 1.89 kJ of work on her to bring her to rest; hence, $W = Fs = \Delta KE$ and

$$F = \frac{\Delta KE}{s} = \frac{1.89 \times 10^3 \text{ J}}{1.00 \text{ m}} = \boxed{1.89 \text{ kN}}$$

▶ **Quick Check:** It is sometimes useful to check a problem by reworking it in different units. KE = $\tfrac{1}{2}(3.3 \text{ lb/ft/s}^2)(29 \text{ ft/s})^2 = 1.4 \times 10^3$ ft·lb, $(1.4 \times 10^3 \text{ ft·lb})(1.356 \text{ J/ft·lb}) = 1.9$ kJ.

Note that doubling the speed of an object (the young lady fired from the cannon or perhaps the family car on the highway) quadruples the kinetic energy and also quadruples the work needed either to get it up to speed or to stop it. Insofar as the braking force on a car is fairly constant, once the brakes are engaged, it will take four times the distance to stop an auto traveling at 40 km/h than it takes for one moving at 20 km/h.

9.4 Potential Energy

Imagine a force that is *continuously* exerted on a body; for example, the gravitational interaction producing the body's weight. To move upward against that downward pull requires the application of a counterforce (provided by leg muscles or elevators) and the doing of work. The crucial point is that the force will continue to act after the displacement. A body raised in a gravitational field still experiences a downward force while held up there at rest. When let loose, it will fall; it will be driven back down toward where it came from and kinetic energy will be imparted to it in the process.

All of that is fine; work done on the system is ultimately converted into KE. But what is happening while the body is held motionless, high in the gravitational field? KE is not "liberated" until the body is released. Apparently, it is possible to do

Wind blows against the sails, exerting a force F and driving the ship forward. Air and water resistance produce a net drag F_f on the craft. When $F > F_f$, $\Sigma F = F - F_f = ma$, the ship accelerates and its KE increases.

work on a system and not have it immediately appear as KE. Yet the potential for generating that energy is there because of the ever-present force. *Energy is stored in the system of interacting objects,* waiting to be let loose. This *retrievable stored energy, energy by virtue of position or configuration in relation to a force,* is known as **potential energy** (PE), a name suggested by W. Rankine in the mid-1800s. The concept of PE gives continuity to the idea of energy; W done on a body against a force such as gravity goes into changing its PE, which may later go into changing its KE.

The change in the potential energy of a body incurred in moving from one point to another equals the work done in overcoming the interaction that stores the energy. PE exists only in relation to systems of interacting objects.

Gravitational-PE

When a painter climbs a flagpole, she does work to overcome the downward pull of gravity. Work corresponds to a change in energy, and we associate this change with the *gravitational*-PE; inasmuch as she is motionless, high on her perch, $W = \Delta \text{PE}_G$. That work equals the product of the force she exerts (that is, a force insignificantly greater than her weight F_W) and the vertical height through which she ascends, Δh. Thus

$$\Delta \text{PE}_G = F_W \Delta h \tag{9.11}$$

Given that the painter has a mass m, we can write Eq. (9.11) in terms of her initial (h_i) and final (h_f) heights independent of the path taken (p. 285); accordingly

$$\Delta \text{PE}_G = mg\Delta h = mg(h_f - h_i) \tag{9.12}$$

For the time being, we limit our activities to the Earth and assume g is constant. Potential energy, like KE, is a relative quantity; even the idea of "height" is relative. How high is your own nose? Above what—your lip, the ground, or perhaps sea level? Any of these would do; the zero-reference level of PE_G is arbitrary and can be taken wherever it's convenient. We simply say that the height at some level is zero and take the PE_G with respect to that level. Usually, we are concerned only with changes in PE_G due to changes in altitude, which then removes the need to deal with the zero-reference level. In fact, **the only thing we are able to measure is the change in PE_G, the change in energy**. That tells us something very profound about the nature of the Universe (p. 308).

The dogs pull on the sled and do work on it, W_{ds}, over some displacement. This energy goes into overcoming drag due to both the air and snow, W_f. If the sled accelerates over the run, $W_{ds} = W_f + \Delta\text{KE}$.

It is useful, nonetheless, to have an expression for the PE_G of a body at any height h above the chosen zero-reference, especially if that reference is a commonly used one, such as the surface of the planet. Here, what is meant by "the height of a body" is the height of its center-of-gravity. Accordingly, Eq. (9.12) becomes

$$\text{PE}_{Gf} - \text{PE}_{Gi} = mg(h_f - h_i)$$

which suggests that

$$\text{PE}_{Gf} = mgh_f \qquad \text{and} \qquad \text{PE}_{Gi} = mgh_i$$

If h_i is the reference level (that is, $h_i = 0$), $\text{PE}_{Gi} = 0$, and we set $\text{PE}_G = \text{PE}_{Gf}$ (at any elevation $h = h_f$ measured from $h = 0$). Hence

$$\mathrm{PE_G} = mgh \qquad (9.13)$$

The work done raising the mass a height h (namely, $F_W h$) equals the increase in potential energy. Our planet is spinning; consequently, the g here should represent the absolute gravitational acceleration. It's common practice, however, to overlook the small discrepancy and just take the acceleration as 9.81 m/s^2.

Example 9.13 The flagpole painter in Fig. 9.10 is carrying a 2.00-kg can of white paint. She has climbed (raising her can's *c.g.* 10.0 m) up the pole from its base, where she just finished having lunch on the roof of a 30.0-m-tall tower. (a) What is the increase in the potential energy of the can as a result of that last ascent? (b) What is the total increase in the can's potential energy above the value it had while it sat on the ground? (c) If we fix $\mathrm{PE_G} = 0$ at the top of the roof, what was the $\mathrm{PE_G}$ of the paint can while it sat on the ground? Assume the can's *c.g.* is 10.0 cm above its bottom.

Solution: [Given: heights of 30.0 m and 10.0 m, m = 2.00 kg, and the *c.g.* 10.0 cm above the can's bottom. Find: appropriate PEs.] (a) Using Eq. (9.12)

$$\Delta\mathrm{PE_G} = mg\Delta h = (2.00\ \mathrm{kg})(9.81\ \mathrm{m/s^2})(10.0\ \mathrm{m})$$

and so $\boxed{\Delta\mathrm{PE_G} = 196\ \mathrm{J}}$

(b) Its *c.g.* rose 40.0 m; hence

$$\Delta\mathrm{PE_G} = mg\Delta h = (2.00\ \mathrm{kg})(9.81\ \mathrm{m/s^2})(40.0\ \mathrm{m})$$

therefore $\boxed{\Delta\mathrm{PE_G} = 785\ \mathrm{J}}$

(c) The *c.g.* is 0.100 m above the ground when the can is standing upright. At such a time, the *c.g.* would be (30.0 − 0.100) m *below* the zero level, and so from Eq. (9.13)

$$\mathrm{PE_G} = mgh = (2.00\ \mathrm{kg})(9.81\ \mathrm{m/s^2})(-29.9\ \mathrm{m})$$

and $\boxed{\mathrm{PE_G} = -586\ \mathrm{J}}$

▶ **Quick Check:** If we take the roof as the zero of $\mathrm{PE_G}$, raising the can from the ground to the top of the pole will change the $\mathrm{PE_G}$ by 196 J − (−586 J) = 785 J.

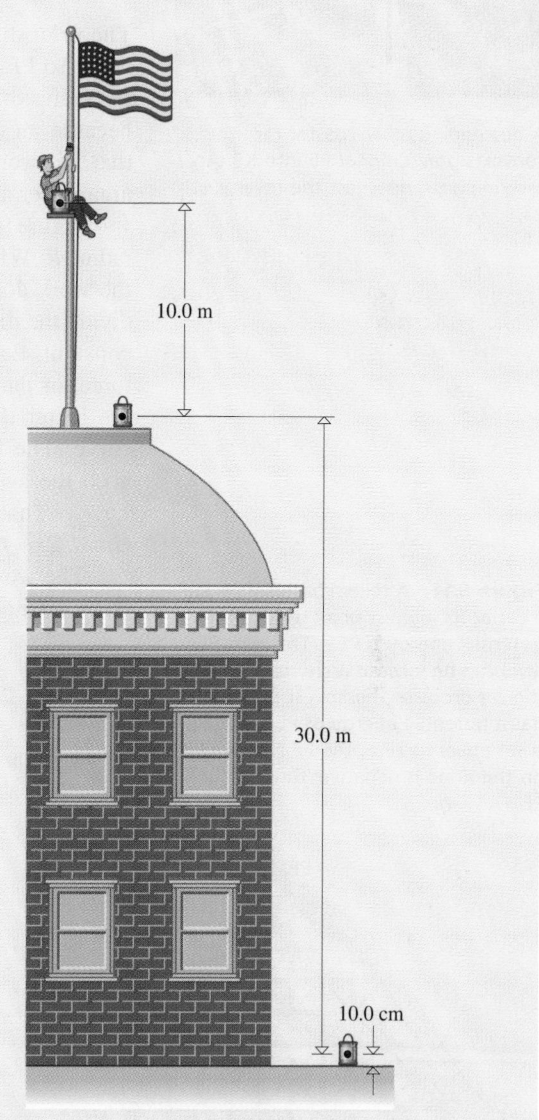

Figure 9.10 A flagpole painter has *gravitational*-PE given by $\mathrm{PE_G} = mgh$. Here, h is measured with respect to any level taken as $h = 0$. Only the change in $\mathrm{PE_G}$ is measurable, so the location of the $h = 0$ level is arbitrary.

A descending roller-coaster car converts *gravitational*-PE into KE. An ascending car does just the reverse.

Depending on where the zero-reference is chosen to be, PE_G can be positive or negative. Figure 9.11 shows a car on a roller coaster. It starts at the top and drops to the first low point. If we take the ground level as the zero of *gravitational*-PE, $\Delta PE_G = (mgh_f - mgh_i) = -mg\Delta h$, where Δh is positive. PE_G decreases, ΔPE_G is negative, but each value of PE_G is positive. If, however, we take the very top to be the zero-PE_G level, then $PE_{Gi} = 0$ and $\Delta PE_G = PE_{Gf} = -mg\Delta h$. The PE_G is negative everywhere on the ride, but the change in potential energy is the same as before, when the ground was the zero level.

Forces Varying with Distance

The prior discussion is true provided that g is constant, but as seen in Eq. (7.4), it isn't! So $PE_G = mgh$ is an approximation, a fine one when h is small compared to the radius of the Earth, but an approximation all the same. This circumstance arises because an object's weight is a function of distance ($F_G = GmM/r^2$). That means that the work done in moving against gravity must be determined by finding the area under an appropriate portion of the force-displacement curve.

Figure 9.12a is a plot of the external, inverse-square gravity-force for a planet of radius R. When an object of mass m is raised from the surface R to some distance r, the work done on it will equal the area beneath the curve from R to r. And so we divide the distance into short segments Δr, over each of which we can assume F_G is constant. Each little rectangular region is topped off at a value equal to the average force for that region. The smaller is Δr, the greater the number of little regions, and the better the average force approximates the actual force at every point on the curve. The first interval is bounded by the forces GmM/R^2 and GmM/r_1^2, and we need the average force somewhere between. On the left is $1/RR$ and on the right $1/r_1r_1$. That suggests using $1/Rr_1$ as the average. The average force is then GmM/Rr_1. As $\Delta r \to 0$, $r_1 \to R$ and $1/R^2 \to 1/r_1^2 \to 1/Rr_1$, which is fine.

The work done in traversing the first interval is the average force times $\Delta r = (r_1 - R)$, namely

$$\Delta W_1 = G\frac{mM}{Rr_1}(r_1 - R) = GmM\left(\frac{1}{R} - \frac{1}{r_1}\right)$$

In the same way

$$\Delta W_2 = GmM\left(\frac{1}{r_1} - \frac{1}{r_2}\right)$$

and so on, until

$$\Delta W_n = GmM\left(\frac{1}{r_{n-1}} - \frac{1}{r_n}\right)$$

Figure 9.11 A roller coaster showing a car at its highest point, where the potential energy is PE_{Gi}. The zero-PE_G level can be located anywhere. When the car drops to the lowest point, it has a potential energy of PE_{Gf}. If PE_{Gi} is set equal to zero, the PE_G anywhere on the slope is negative; thus, $PE_{Gf} = -mg\Delta h$.

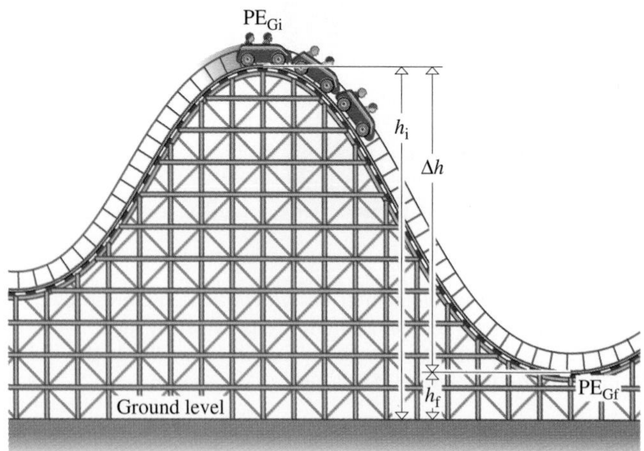

where the number of little segments is n, and $r = r_n$ finishes the trip. In the end, we will cause $\Delta r \to 0$, while $n \to \infty$. Now add all of these work contributions, whereupon $W = \Delta W_1 + \Delta W_2 + \cdots + \Delta W_n$. Realize that each ΔW expression contains a $1/r$ term that cancels a corresponding identical term in the next expression. Thus, the $1/r_1$, $1/r_2$, $1/r_3$, $\cdots$, $1/r_{n-1}$ terms all cancel, leaving only

$$W = GmM\left(\frac{1}{R} - \frac{1}{r}\right)$$

no matter how many segments there are (even as $n \to \infty$). In general then, *the change in the* PE *experienced by a body of mass m relocating from* r_i *out to* r_f *in the gravita-*

tional field of a body of mass M is

$$\Delta PE_G = GmM\left(\frac{1}{r_i} - \frac{1}{r_f}\right) \tag{9.14}$$

Since $\Delta PE_G = PE_{Gf} - PE_{Gi} = (-GmM/r_f) - (-GmM/r_i)$, if at $r_f = \infty$ we choose PE_{Gf} to be zero, the $1/r_f$ term goes to zero and the i subscript can be dropped. Then $PE_{Gi} = PE_G$ and

$$PE_G = -\frac{GmM}{r} \tag{9.15}$$

is the *gravitational*-PE of a mass m at any distance r from the center of a mass M. The farther m is from M, the larger is r and the larger is the PE_G, inasmuch as it gets to be a less negative number. At $r = \infty$, the PE_G has its greatest value: zero.

We can apply Eq. (9.14) to the surface of the Earth (Fig. 9.13) to show that it agrees with the more limited statement of Eq. (9.13). Neglecting spin, Eq. (7.3) tells us that $gR_\oplus^2 = GM_\oplus$. Using that, we can find the change in potential energy of a mass m raised to a low altitude ($h = r - R_\oplus$) for which $r \approx R_\oplus$. Equation (9.14) becomes

$$\Delta PE_G = mgR_\oplus^2\left(\frac{1}{R_\oplus} - \frac{1}{r}\right) = mgR_\oplus^2\left(\frac{r - R_\oplus}{rR_\oplus}\right) = mgR_\oplus\frac{h}{r} \approx mgh$$

Raising his *c.m.*, this climber increases the *gravitational*-PE stored via the interaction between him and the Earth. Considering the gravitational field as a mapping of the interaction, the PE can be thought of as stored in the field.

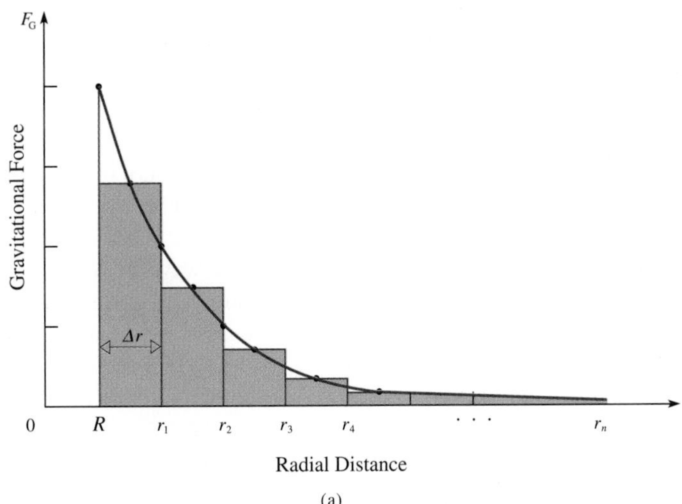

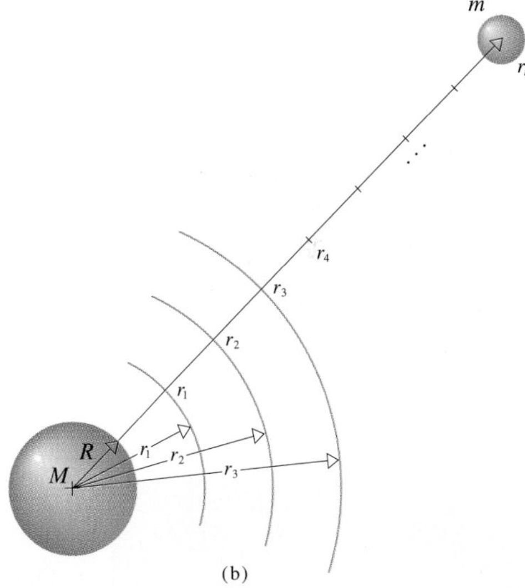

Figure 9.12 (a) A spherical mass of radius R is surrounded by a gravitational force field that drops off inversely with the square of the distance r. (b) The distance from the surface out to a mass m is divided into a large number n of segments.

Example 9.14 What is the change in potential energy of a 100-kg meteorite if it free-falls from an altitude of 1000 km down to the surface of the Earth? Take the mass of the planet to be 6.0×10^{24} kg and the diameter to be 1.28×10^7 m.

Solution: [Given: $m = 100$ kg, $h = 1000$ km, $M_\oplus = 6.0 \times 10^{24}$ kg, $R_\oplus = 0.64 \times 10^7$ m, and $G = 6.67 \times 10^{-11}$ N·m²/kg². Find: ΔPE_G.] Falling from r to $R_\oplus$, the PE_G decreases and so ΔPE_G must be negative:

(continued)

(continued)

$$\Delta PE_G = GmM_\oplus\left(\frac{1}{r} - \frac{1}{R_\oplus}\right)$$

With $r = R_\oplus + h = 7.4 \times 10^6$ m

$$\Delta PE_G = (4.0 \times 10^{16}\ \text{N}\cdot\text{m}^2) \times (1.35 \times 10^{-7}\ \text{m}^{-1} - 1.56 \times 10^{-7}\ \text{m}^{-1})$$

Hence $\boxed{\Delta PE_G = -8.5 \times 10^8\ \text{J}}$

▶ **Quick Check:** $\Delta PE_G \approx mg\Delta h = (100\ \text{kg}) \times (9.81\ \text{m/s}^2)(-10^6\ \text{m}) = -9.8 \times 10^8$ J.

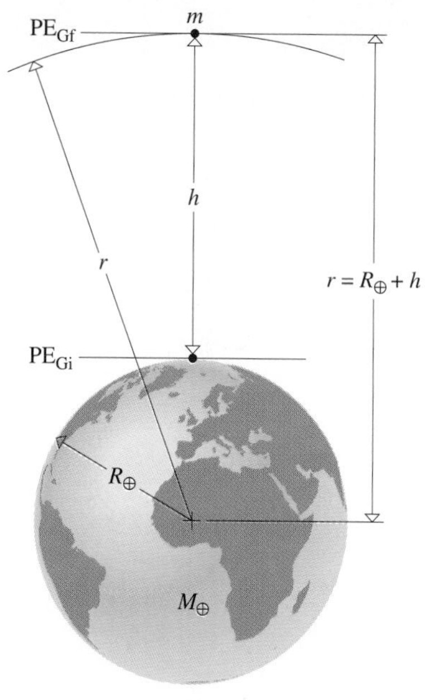

Figure 9.13 A mass m at a distance r from the center of the Earth. Given that the altitude of m is h, it follows that $r = R_\oplus + h$. When h is small, $r \approx R_\oplus$ and $\Delta PE_G = PE_{Gf} - PE_{Gi} \approx mgh$.

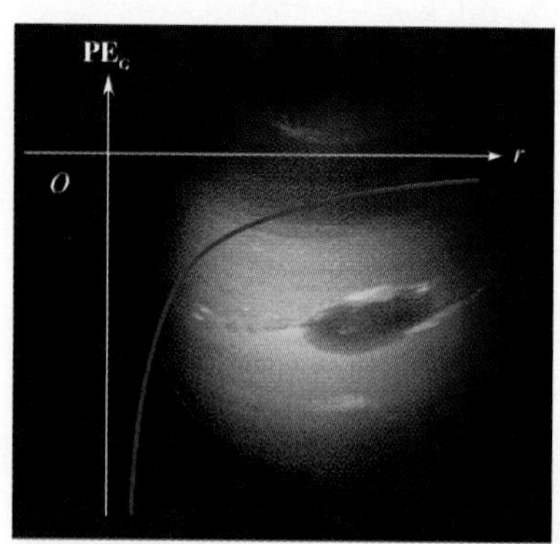

Potential energy is often formulated so that it is negative when the interaction is attractive.

9.5 Internal Energy

It is convenient to distinguish two basic forms of energy, KE (energy of motion) and PE (energy of position). For objects that have structure (that is, interacting parts), it's helpful to further recognize two manifestations of each of these forms called *internal* and *external.* A bowling ball at rest has all its atoms randomly vibrating about their equilibrium positions at fairly high speeds (typically ≈0.4 km/s). We refer to that disorganized internal kinetic energy as *thermal energy.* A bowling ball flying through the air has an organized motion superimposed on all its jiggling atoms, and we have called this external energy *kinetic energy.* Similarly, a bowling ball in the air has external potential energy by virtue of its interaction with the Earth. When the ball is distorted (for example, during a collision) it has *elastic energy,* one of several kinds of internal PE.

Imagine two structureless particles interacting gravitationally. As we have seen, the work done in bringing a mass from any point A to any point B is independent of the path taken. It follows that moving a particle from A to B and back to A results in zero work done. That feature, namely, that *zero work is done in moving a particle around a closed path,* defines a **conservative force**. To be conservative, the force on a particle at any given location in space must be constant, no matter how or *when* the particle gets there. Whatever energy goes into such an interaction on moving from A to B, comes out on moving from B to A; the process is reversible. The conservative interaction stores PE.

As we continue our study, we will encounter a number of ways in which the Four Forces act to store PE both internally and externally. A stretched garter retains *elastic*-PE; a bright red stick of dynamite is gift-wrapped *chemical*-PE; a battery supplies *electrical*-PE; two magnets pulled apart wait to reunite with *magnetic*-PE; and all those neatly painted warheads hiding in their silos threaten with *nuclear*-PE.

Internal Energy Sources and Work

The work done on passive objects (crates, blocks, balls, and sailboats) by externally applied forces has been our primary concern thus far. Now, consider self-propelled objects (cars, people, helicopters, frogs, etc.) that have their own internal energy sources. Each can be accelerated by a net external force ($F = ma$) arising from its interaction with the environment. *As a rule, such a force does no net positive work on the active nonrigid body, and* $W \neq \Delta KE$. No energy is transferred to the body from the environment via that reaction force, even though that force accelerates the body.

A swimmer's hand (Fig. 9.14a) pushes back on the water and the water pushes forward on the hand, accelerating the person. But the hand is moving backward all the while. The hand does positive work on the water; the force it exerts is in the same direction as the displacement. By contrast, the water pushes in the forward direction on the hand. It does negative work on the swimmer: $W_{ws} < 0$ even though $\Delta KE > 0$. The water gets energy from the swimmer—it gets warmer. The swimmer is self-powered and uses the water to generate a reaction force so she can accelerate. Just imagine a motorboat suspended above the water with its engine running at full speed, going nowhere. All the energy needed is provided by the fuel, but without the water to push on, the boat cannot accelerate.

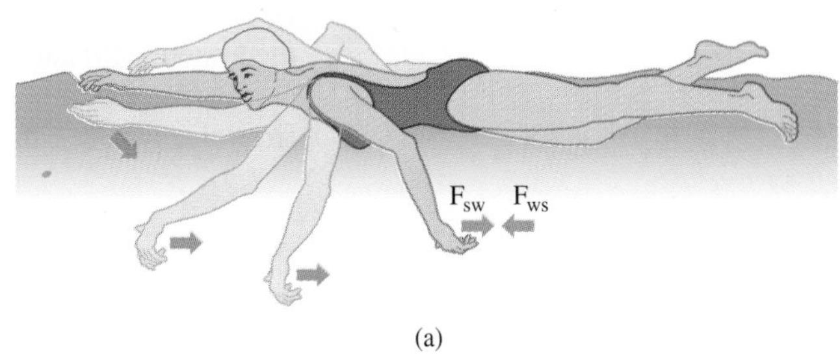

(a)

The energy to walk, climb (Fig. 9.14b), skate, or jump comes from internal energy stored in the person (see Discussion Questions 2 and 3). When you jump (Fig. 5.12), the upward reaction force that accelerates you acts at the stationary foot-floor interface. If the floor is rigid, *there is no motion of the point of application of the force, and no work is done by the floor on you:* $W_{fu} = 0$ even though $\Delta KE > 0$. Actually, the floor will sag slightly while it exerts a normal force on you and W_{fu} is positive though very small, since s is very small.

When a car accelerates, the tires on the drive wheels push back on the ground; the ground pushes forward on the tires and the car accelerates forward. But the region of contact between tire and road is motionless, and no work is done on the car by the ground. Of course, the energy equivalent to ΔKE comes from the fuel via the engine—a car doesn't drain energy out of the ground, it just pushes off it.

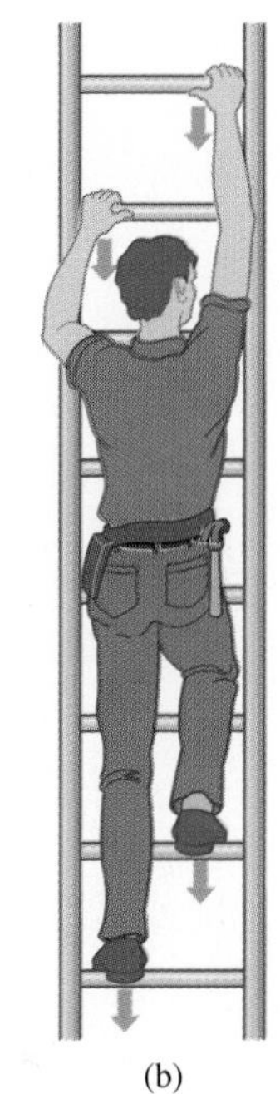

(b)

Figure 9.14 (a) A swimmer pulls on the water with a force F_{sw}, drawing it back a distance x and doing positive work on it. The forward reaction of the water on the swimmer's hand (F_{ws}) is opposite to the displacement of the hand, and the work done on the swimmer is negative. Energy, supplied by her food, is transferred to the water. (b) A person climbing a ladder exerts forces down on the rungs, but the point of application does not move and no work is done on the ladder or on the climber. The increase in PE_G comes from internal energy.

CONSERVATION OF MECHANICAL ENERGY

It was Descartes who proclaimed the Law of Conservation of Momentum (1644). And so when Huygens sought to replace mv as the essential measure of motion, it was natural for him to look for a quantity that was likewise conserved (that is, constant in time). The realization that there were various interchangeable forms of an otherwise unchanging total energy had already been proposed by several people before Rankine (1853) provided us with the ringing phrase "Conservation of Energy." That era gave rise to the study of Thermodynamics with its premier insight that "heat" was another form of energy. Yet it was not until 1905 that Einstein added the final piece to the construct. He showed that mass can be thought of as equivalent to energy; that by virtue of its very existence, a particle of mass possesses a store of energy.

The grandest generalization in all of physics is the fully developed statement of the Law of Conservation of Energy:

> **The total energy of any system that is isolated from the rest of the Universe remains constant, even though energy may be transformed from one kind to another within the system.**

Assuming that *the energy of the Universe is constant,* the energy of any portion of the Universe isolated from the rest must also be constant. If there is no flow of energy into or out of a system (and that's what is meant by *isolated*), then there can be no change in energy either there or in the remainder of the Universe.

9.6 Mechanical Energy

Let's begin with the case of a system acted on by both externally applied contact forces and gravity. To keep things simple, we eliminate any possible changes in internal energy—no friction, no deformation. The effect of the contact forces is reflected in the net work they perform on the system: W_{net}. The *gravitational*-PE of the system delineates the influence of the gravitational field. And the energy associated with inertia is accounted for via ΔKE. Thus, energy delivered to or removed from the system can go into changing the *gravitational*-PE and/or the KE, hence

$$W_{net} = \Delta PE_G + \Delta KE \tag{9.16}$$

There are other conservative forces besides gravity, and we can add appropriate PE terms to Eq. (9.16) later.

We define the **mechanical energy** (E) of a system as *the sum of the* KE *and gravitational*-PE *of all its parts.* Therefore

$$W_{net} = \Delta E = E_f - E_i$$

and

$$E_f = E_i + W_{net}$$

If no contact forces are applied W_{net} is zero, no energy is transferred into or out of the system, $E_f = E_i$, and *mechanical energy is conserved.* This is an early, limited formulation of what would later develop into the concept of conservation of all forms of energy.

Assuming that g is constant, Eq. (9.16) becomes

$$W_{net} = (\tfrac{1}{2}mv_f^2 - \tfrac{1}{2}mv_i^2) + (mgh_f - mgh_i)$$

When the only force acting is gravity, $W_{net} = 0$ and

$$\tfrac{1}{2}mv_f^2 + mgh_f = \tfrac{1}{2}mv_i^2 + mgh_i \tag{9.17}$$

At every instant in the motion, the total mechanical energy is constant; if the body's KE increases, its PE_G must decrease, and vice versa. Fire a ball straight up into the

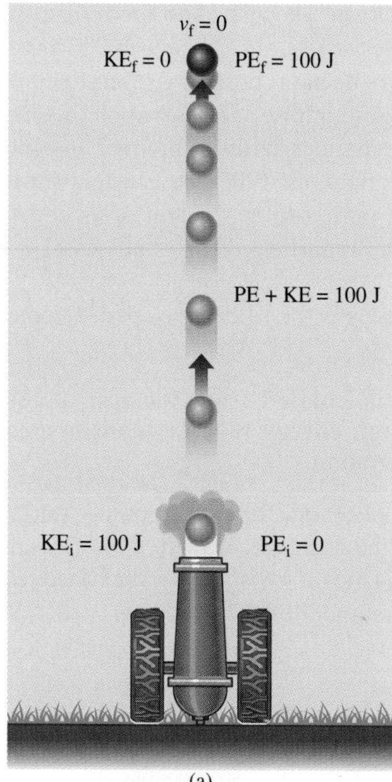

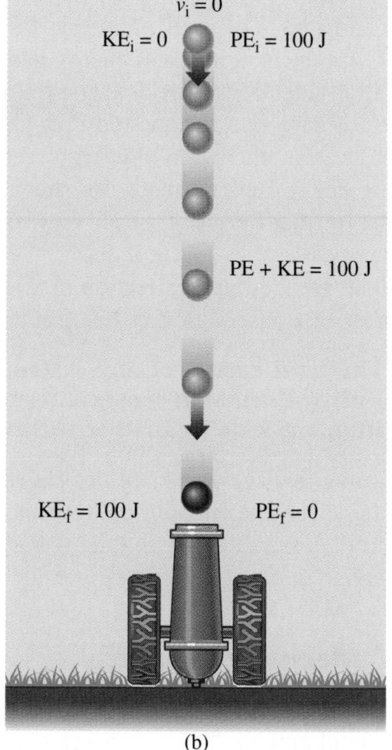

Figure 9.15 (a) A cannonball fired upward loses KE as it rises and gains PE_G. Fired with a $KE_i = 100$ J, it will always have a total energy of 100 J. Of course, in real life, some energy will be lost to air friction. (b) When the ball drops from rest, it loses PE_G and gains KE.

air and in the process do work on it, giving it a kinetic energy of 100 J (Fig. 9.15a). Let's focus on the launched projectile. Once fired (ignore air friction), no forces other than gravity act and $W_{net} = 0$. At $h_i = 0$, $PE_G = 0$, while the KE, equal to 100 J, is a maximum (equal to E). As the ball rises, increasing in PE_G, it slows down, decreasing in KE such that at any moment $KE + PE_G = E = 100$ J. At peak altitude, where the ball momentarily comes to rest, KE = 0, and $PE_G = 100$ J is a maximum (equal to E). As it falls back (Fig. 9.15b), PE_G decreases to zero, its speed increases, and the KE attains a maximum value of 100 J (equal to E) as it returns to the point from which it was launched. The similar situation of a swinging pendulum is depicted in Fig. 9.16.

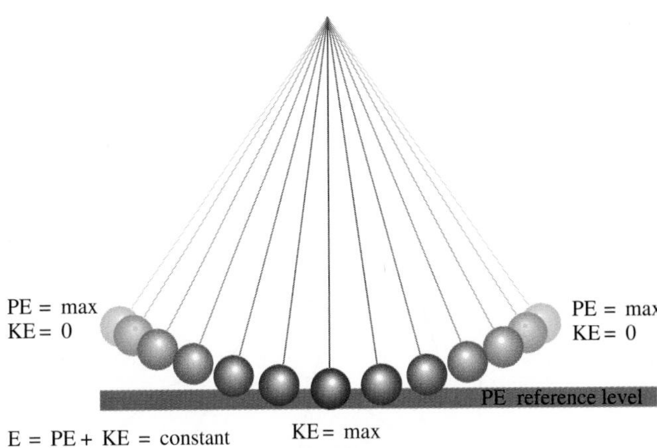

Figure 9.16 The transfer of energy from one form to another in the swinging pendulum.

During a running jump, a person can only get as high as his vertical velocity will carry him. However, using a pole to vault into the air provides a means of transforming nearly all of a runner's KE into PE_G. At the start, the person runs as fast as possible to gain as much KE as possible. Once the athlete jumps off the ground and is pole-borne, his run-up KE is gradually converted into both *gravitational*-PE and *elastic*-PE (p. 329) as the pole continues to bend while he rises. It soon straightens out, gives up its *elastic*-PE, and hurls him even higher.

Example 9.15 A vaulter carrying a graphite-fiberglass pole (a thin-walled tube weighing around 4 lb) is about to make a jump. At what speed must he run in order to clear the 20.0-ft (6.096-m) mark? Neglect all possible energy losses, take his *c.g.* to be 1.00 m above the floor while standing, and presume that it just clears the bar.

Solution: [Given: $h_i = 1.00$ m, $\Delta h_{cg} + h_i = 6.10$ m. Find: v_i when $v_f = 0$.] Neglect friction. It follows from Eq. (9.17) that the final PE_G equals the initial KE. Take the zero-PE level at the initial height of the *c.g.;* thus, if the *c.g.* rises an additional $\Delta h_{cg} = 6.096 \text{ m} - 1.00 \text{ m}$, then

$$0 + mg\Delta h_{cg} = \tfrac{1}{2}mv_i^2 + 0$$

The mass cancels, yielding

$$v_i^2 = 2g\Delta h_{cg} = 2(9.81 \text{ m/s}^2)(5.096 \text{ m}) = 99.949 \text{ m}^2/\text{s}^2$$

and $\boxed{v_i = 10.0 \text{ m/s}}$. This is a high speed for someone running with a pole. Actually, at the last moment, high in the air, as he approaches the bar, the athlete will straighten his arms, lifting his *c.g.* above his hand position and even above the end of the pole.

▶ **Quick Check:** With a launch speed v_i, the peak altitude (p. 80) reached by a projectile is $v_i^2/2g \approx 10^2/(<20)$ m $\approx >5$ m above the 1-m lift-off height.

9.7 The Energy Perspective

Unlike Newton's Laws, which require us to know the details of the motion in order to carry out the analysis, the energy formulation provides an overview. With it we can determine the final state of a system from its initial configuration, without looking at the intermediate stages it passes through. For instance, knowing that the identical cannonballs in Fig. 9.17 were launched with the same initial speed, and therefore with the same KE, we can be sure that at any one height (where their potential energies are the same), they will all be moving with equal speed regardless of the path taken. Similarly, knowing the height of a snowy slope, we can compute a skier's speed at the bottom of a run without any knowledge of the route she took, or even the shape of the mountain, so long as friction is negligible.

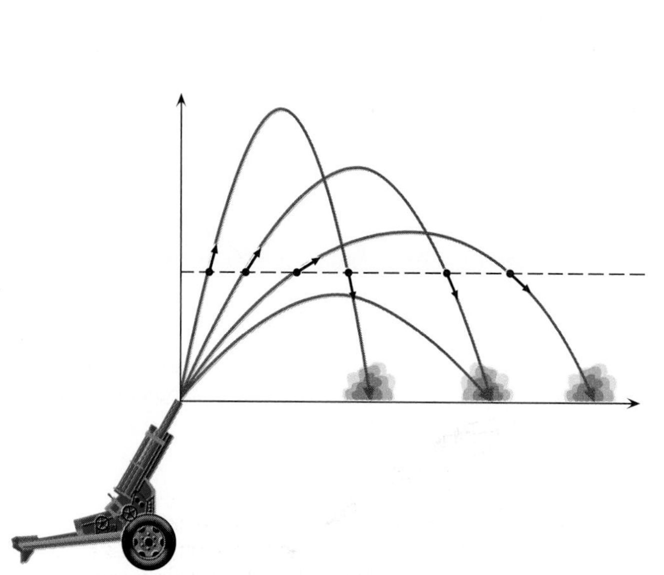

Figure 9.17 Cannonballs fired with the same muzzle speed will all be traveling at the same speed at the same altitude.

A vaulter transfers his run-up KE into *elastic*-PE as the pole bends. When the pole straightens out, it imparts energy to the vaulter. At maximum altitude, he's motionless and subsequently drops straight down.

Example 9.16 A 60-kg skier starts from rest at the top of a 60-m-high slope. She descends without using her poles. (a) What is her initial *gravitational*-PE with respect to the level ground at the bottom? (b) Assume friction is negligible and compute at what speed she will ideally arrive at the bottom. (c) She reaches the bottom of the run, traveling at 25 m/s. What is the net energy transferred via friction to her, the skis, the slope, and the air?

Solution: [Given: $m = 60$ kg, $h_i = 60$ m, $v_i = 0$, $v_f = 25$ m/s, and $s = 10$ m. Find: (a) PE_G, (b) v_f, no friction, and (c) W_F.] (a) Taking zero PE at the bottom, at the top of the hill

$$\text{PE}_{Gi} = mgh_i = (60\text{ kg})(9.8\text{ m/s}^2)(60\text{ m})$$

and

$$\boxed{\text{PE}_{Gi} = 3.5 \times 10^4\text{ J}}$$

(b) With no work done against friction, $\text{E}_f = \text{E}_i$ and from Eq. (9.17), with $v_i = 0$ and $h_f = 0$, $\text{KE}_f = \text{PE}_{Gi}$, and

$$\tfrac{1}{2}mv_f^2 + 0 = 0 + mgh_i = 3.5 \times 10^4\text{ J}$$

Hence

$$v_f = \sqrt{\frac{2(3.5 \times 10^4\text{ J})}{60\text{ kg}}} = \boxed{34\text{ m/s}} = 76\text{ mi/h}$$

(c) With friction acting on the system, $\text{E}_f = \text{E}_i + W_F$; hence

$$\tfrac{1}{2}mv_f^2 + 0 = 0 + mgh_i + W_F$$

and

$$W_F = \tfrac{1}{2}mv_f^2 - mgh_i = \tfrac{1}{2}(60\text{ kg})(25\text{ m/s})^2 - (60\text{ kg})(9.8\text{ m/s}^2)(60\text{ m})$$

Hence, $W_F = 18\,750\text{ J} - 35\,280\text{ J} = \boxed{-1.7 \times 10^4\text{ J}}$. The work done on the system by friction is negative because it opposes the motion.

▶ **Quick Check:** Working backwards, $\tfrac{1}{2}mv_f^2 = \tfrac{1}{2}(60\text{ kg})(34\text{ m/s})^2 = 35\text{ kJ} = \text{PE}_{Gi}$.

Satellites and Spaceships

As a further example of the power of the energy formulation, consider the following: Why does a planet in orbit speed up as it approaches the Sun? We know how to find the answer from Newton's force picture, but now we have an alternative. A planet's speed and KE both increase because its *gravitational*-PE decreases as it approaches the Sun. Any energy losses are negligible and the total mechanical energy is everywhere constant. At the point of closest approach and smallest PE, the planet must have its maximum KE and be traveling with maximum speed (Fig. 9.18).

Reconsider Eq. (9.15), $\text{PE}_\text{G} = -GmM/r$. At $r = \infty$, PE_G is zero, while everywhere else it's negative. A negative PE is characteristic of attractive forces. Had the interaction been repulsive, as between two magnets with like poles, the PE would have been positive at finite distances. When the force is repulsive, work must be done on one body to bring it from infinity "down" to the vicinity of the other body, and so the change in PE (from zero) must be positive. By contrast, if the force is attractive, the bodies will come together on their own. The force field will do the work, and the change in PE (from zero) will be negative. *If a system can return to its zero-PE configuration by doing positive work, then the* PE *stored in the system is positive.* This is the case with two magnets with like poles repelling and separating out to infinity. *If positive work must be done on the system to bring it back to the zero-*PE *state, then the* PE *stored is negative.* This is the case with two attracting masses being pulled apart.

It's been suggested that since *gravitational*-PE is negative, if the total of all other forms of energy is equal in magnitude to the total PE_G, then the net energy of the entire Universe is zero. It follows that such a system could have spontaneously appeared out of "nothingness" without violating energy conservation.

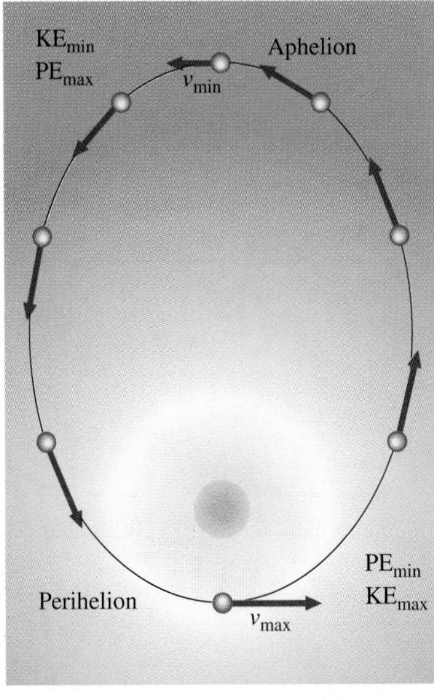

Figure 9.18 A planet in an elliptical orbit around the Sun. As it gets closer to the Sun, it speeds up. The speeds are the same for equal distances, either moving toward or away from the Sun, since there the PEs must be the same and therefore the KEs must also be the same.

Example 9.17 How much work must be done on a projectile of mass m to slowly raise it from the surface of a motionless planet (of radius R and mass M) out to an altitude that is essentially infinitely far away? Neglect all frictional losses.

Solution: [Given: $v \approx 0$, $r_\text{i} = R$ and $r_\text{f} = \infty$. Find: W, the work done on the rocket.] We know that

$$\text{E}_\text{f} = \text{E}_\text{i} + W$$

hence $$\text{KE}_\text{f} + \text{PE}_\text{Gf} = \text{KE}_\text{i} + \text{PE}_\text{Gi} + W$$

Using Eq. (9.15),

$$W = \text{PE}_\text{Gf} - \text{PE}_\text{Gi} = 0 - \left(-\frac{GmM}{R}\right)$$

And so the work done, the energy that must be supplied, is

$$\boxed{W = \frac{GmM}{R}}$$

▶ **Quick Check:** The units of W are those of force (GmM/R^2) times distance (R), as it should be. As $R \to \infty$, $W \to 0$, which is appropriate. The smaller R is, the larger W gets, which is also reasonable.

With the results of the Example 9.17, suppose we wish to fire a craft into deep space with a single short-lived burst of energy. The minimum blast-off speed, the lowest speed with which we could fire a projectile straight up and never have it fall back to the planet from which it was launched, is the **escape velocity, $\mathbf{v}_\text{esc}$**. When given that speed measured with respect to the planet's *c.m.*, the object will have enough KE to overcome the planet's gravity. It will sail on forever, gradually slow-

TABLE 9.3 Escape speeds from the surfaces of celestial bodies

Body	Escape speed (km/h)
Eros asteroid	55
Moon	8×10^3
Mercury	15×10^3
Mars	18×10^3
Venus	37×10^3
Earth	40×10^3
Neptune	83×10^3
Saturn	130×10^3
Jupiter	216×10^3
Neutron star*	$\approx 0.8 \times 10^9$

*That's about 80% c.

ing down, but nonetheless arriving at infinity (whatever that means) as the final speed becomes zero. At launch, at the planet's surface (a distance R from its center)

$$\mathrm{E_i} = \mathrm{KE_i} + \mathrm{PE_{Gi}} = \tfrac{1}{2}mv_{\mathrm{esc}}^2 - \frac{GmM}{R}$$

and no matter what this equals, it must be forever constant, provided nothing external acts on the craft. Ideally, the object will reach infinity, where $\mathrm{PE_{Gf}} = 0$, traveling at zero speed such that $\mathrm{KE_f} = 0$. Hence, $\mathrm{E_f} = 0$ at $r = \infty$, and since E is everywhere constant, it must be that $\mathrm{E_i} = 0$. Thus

$$\tfrac{1}{2}mv_{\mathrm{esc}}^2 - \frac{GmM}{R} = 0$$

and

$$v_{\mathrm{esc}} = \sqrt{\frac{2GM}{R}} \tag{9.18}$$

At the surface of the Earth, where, from Eq. (7.4), $mg_\oplus = GmM_\oplus/R_\oplus^2$

$$v_{\mathrm{esc}} = \sqrt{2g_\oplus R_\oplus} \tag{9.19}$$

Although the escape speed is independent of the mass of the projectile, the needed KE certainly is not. That's why it takes big rockets to raise big payloads. Notice that this is also the speed with which an object would collide with the planet, having fallen from a very great distance. Recall Eq. (7.8) for the speed of a body in a circular orbit; namely, $v_0 = \sqrt{GM/r}$. The escape speed from *any altitude* (where $r = R$) is equal to $\sqrt{2}$ times the orbital speed at that altitude. The higher the orbit, the slower the speed, and the easier the escape (Table 9.3).

Example 9.18 In 1865, Jules Verne published a story called *From the Earth to the Moon.* In it, a gigantic cannon fires a capsule carrying three men to the Moon. The cannon was located at "27°7′N. lat. 5°7′W. long. of the meridian of Washington." (That, incidentally, turns out to be in Florida about 100 mi south of the Kennedy Spaceflight Center.) Compute the muzzle speed needed to escape the Earth. Use $R_\oplus = 6.4 \times 10^6$ m.

Solution: [Given: $R_\oplus = 6.4 \times 10^6$ m and $g_\oplus = 9.8$ m/s^2. Find: v_{esc}.] Using

$$v_{\mathrm{esc}} = \sqrt{2g_\oplus R_\oplus} \quad [9.19]$$

$$v_{\mathrm{esc}} = [2(9.8\ \mathrm{m/s^2})(6.4 \times 10^6\ \mathrm{m})]^{\frac{1}{2}} = \boxed{1.1 \times 10^4\ \mathrm{m/s}}$$

This is equal to 25×10^3 mi/h, which is formidable. In fact, enough friction would be encountered in the dense lower atmosphere at that speed to burn up the ship. This is one reason why we use rockets that build up speed gradually, only reaching v_{esc} in the thin upper atmosphere.

▶ **Quick Check:** From Table 9.3, $v_{\mathrm{esc}} = 40 \times 10^3$ km/h $= (40 \times 10^6/3600)$ m/s $= 1.1 \times 10^4$ m/s.

Because gravity drops off inversely with r^2, most of the energy required to travel away from a large body like the Earth is needed to get beyond the first few hundred thousand miles. While the engine-cutoff speed to escape to deep space from Earth is 25×10^3 mi/h, it takes a launch at about 99% of that speed just to get as far as the Moon. The Earth orbits the Sun, and attempting to leave our star requires a still greater speed. By taking advantage of the planet's orbital motion and sailing off in a judicious direction, this speed can be reduced to 16.4 km/s or 37×10^3 mi/h. The *Pioneer 10* space probe passed beyond the orbit of Pluto in

1983 (traveling at 30 613 mi/h) to become the first humanmade object to leave the Solar System.

Example 9.19 In his *Exposition of the System of the World* (1798) the great French mathematician Laplace pointed out that an object of the same density as the Earth with a diameter 250 times that of the Sun would have so large a gravitational attraction that its escape speed would exceed the speed of light. He concluded that such a body, which could prevent light from leaving its surface, would be invisible to us. (a) Show that for a body of a certain density ρ the escape speed is proportional to its radius R. (b) Verify Laplace's assertion.

Solution: [Given: $R = 250\ R_{\odot}$. Find: v_{esc}.] (a) Mass equals density times volume, $M = V\rho = \frac{4}{3}\pi R^3\rho$. It follows from Eq. (9.18) that

$$v_{esc} = \left(\frac{2G\frac{4}{3}\pi R^3\rho}{R}\right)^{\frac{1}{2}} = \boxed{R(\tfrac{8}{3}G\pi\rho)^{\frac{1}{2}}}$$

(b) The ratio of the diameter of the Sun to the Earth is $(1.393 \times 10^9\ \text{m})/(1.274 \times 10^7\ \text{m}) = 109$, which means that an object 250 times the size of the Sun is (109)(250), or 2.7×10^4 times the size of the Earth. Hence, its escape speed is 2.7×10^4 times 11.2 km/s, or $\boxed{3.1 \times 10^8\ \text{m/s}}$ and that slightly exceeds c. Collapsed stars that are so dense and so powerful gravitationally that light cannot escape from them seem to exist. Nowadays, they are known as **black holes**. Although Laplace had the wrong theory for both gravity and light, he ended up predicting what we think is a real entity.

▶ **Quick Check:** The units of $R(\frac{8}{3}G\pi\rho)^{\frac{1}{2}}$ are $m[(\text{N}\cdot\text{m}^2/\text{kg}^2)(\text{kg/m}^3)]^{\frac{1}{2}} = \text{m}[(\text{kg}\cdot\text{m/s}^2)(1/\text{kg}\cdot\text{m})]^{\frac{1}{2}} = \text{m/s}$.

The Universe is an explosion that began $\approx 17 \times 10^9$ years ago. Its galactic star-islands are still rushing outward to this day; the Universe—all there is—is expanding. If it has enough KE to override its own gravitational field, if the galaxies, and whatever else is out there, have the needed escape speed, the Universe will go on expanding forever. If not, it will ultimately stop and fall back upon itself thousands of millions of years from now.

9.8 Collisions

Much of what we know about the atomic and subatomic domains has been learned from the observation of collisions. Beams of particles are hurled against targets, and the motion of the scattered fragments is analyzed. Similar relationships govern both these exotic events and the more mundane ones, such as clubs hitting golfballs. *A collision is marked by the transfer of momentum between objects in relative motion resulting from their interaction via at least one of the Four Forces.* **In all cases where there are no external forces, the total momentum of the colliding objects is conserved**.

Inelastic Collisions

An **inelastic collision** *is one where the final* KE *of the system is different from the initial* KE. A tennis ball dropped against a concrete floor distorts, momentarily comes to rest, and then springs back, popping into the air. But the squashing produces some internal heating as the molecules shift position (and even the crack of sound carries off a little energy) and so the ball only returns about two-thirds of the way back up. One-third of the initial KE has been converted into other energy forms. If, after an impact between two macroscopic objects, either one increases in temperature or remains distorted, the collision is inelastic.

Most of the KE of the moving car is transformed into thermal energy during the crash. The yellow plastic bumper extends the collision time and therefore decreases the force exerted on the car, as indicated in Eq. (4.3), p. 104.

Macroscopic bodies always collide more or less inelastically. What usually happens is that the system either changes its configuration, or transforms some of the impact KE into thermal energy, or both. Even the most rigid macroscopic objects convert some small amount of KE into thermal energy. To be precise, only subatomic particles collide elastically.

The *completely inelastic* collision is at one extreme—where the impacting objects stick together—and the maximum amount of KE is transformed (Fig. 9.19). A wad of oatmeal landing on a table, a bug hitting the windshield of a moving bus, or a neutron absorbed by a uranium nucleus are all representative of inelastic collisions. Less drastic is the case where the colliding objects bounce apart and yet lose KE in the process. A lead ball rebounding from a steel anvil will not even retain 1% of its original KE, whereas a soft cork sphere might retain around 35%. The harder the colliders, the less KE is transformed, which is why billiard balls are not soft. A glass marble will bounce off the anvil with about 95% of its initial KE. Our purposes will be served by examining only the two extremes of elastic and completely inelastic collisions.

Early in its existence (1663), the British Royal Society sponsored a scientific study of collisions. Among the leading scholars who participated were Huygens and the English mathematician John Wallis. Wallis, in a paper of 1668, was probably the first to apply the Law of Conservation of Momentum correctly. When two objects of mass m_1 and m_2, initially moving with speeds v_{1i} and v_{2i}, collide head-on and fly off with speeds v_{1f} and v_{2f}, initial momentum equals final momentum, and

$$m_1 v_{1i} + m_2 v_{2i} = m_1 v_{1f} + m_2 v_{2f} \tag{9.20}$$

The head-on restriction requires that the motion of both bodies at the moment of impact be along a common straight line connecting the two centers-of-mass. As a result, the final motion is also along that line. Hitting off-center or with noncolinear velocities results in the objects moving away at angles to the original direction. Any pool player knows how to strike a ball on the side so it sails away at an angle. Momentum is still conserved, but now the components in the x and y directions will require separate applications of Eq. (9.20).

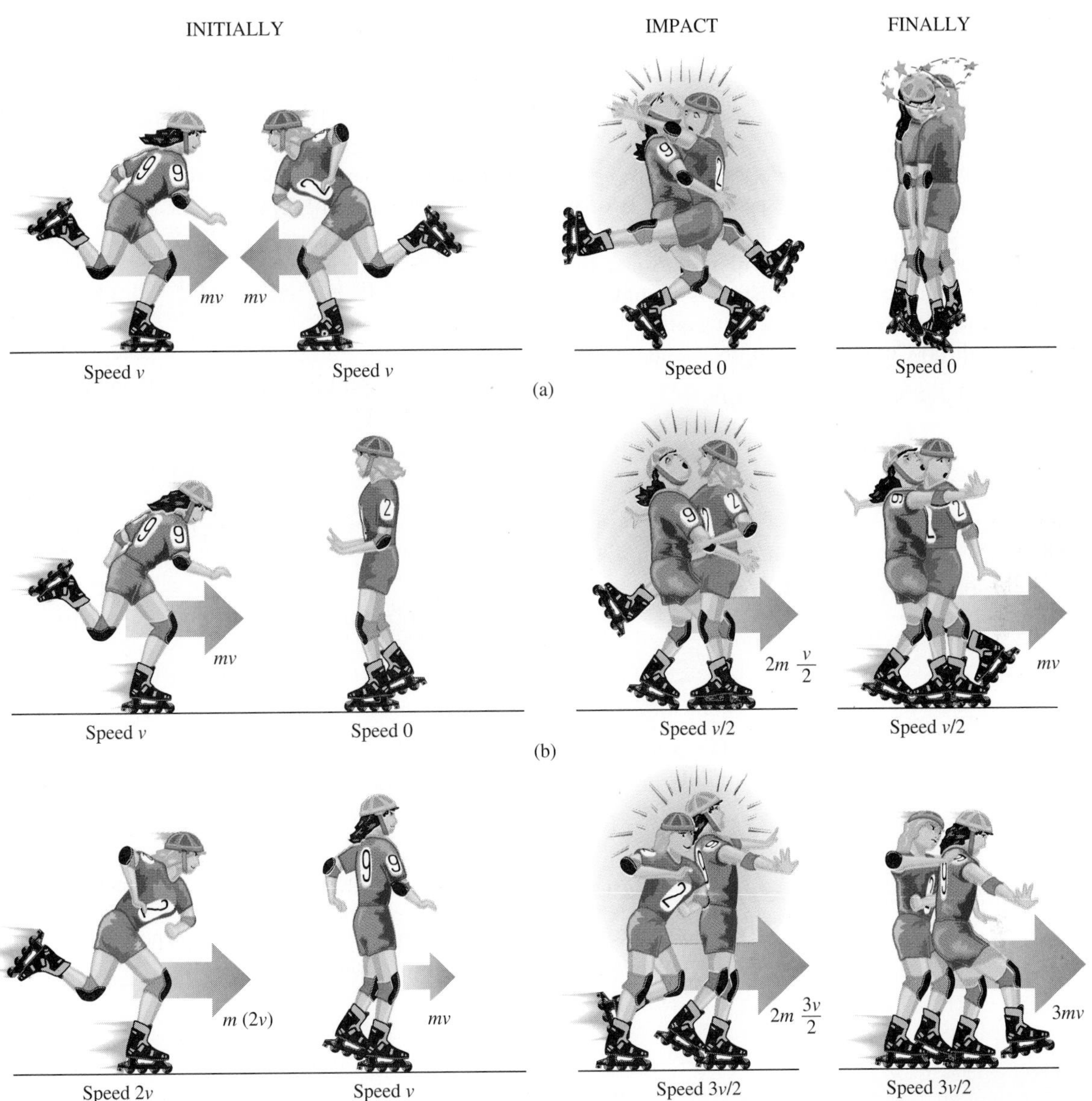

Figure 9.19 Inelastic collisions. (a) Here the skaters flying toward each other have equal masses and speeds. The total momentum is initially zero, and it remains zero after the collision, as they simply come to rest and stay that way. (b) Here number 9 initially has all of the system's momentum. After impact, since the total momentum is constant and they stay together, the moving mass is twice what it was, and therefore the speed must be halved. The two sail off at half the speed number 9 came in with. (c) Taking revenge, number 2 comes flying at number 9, at twice 9's speed. The total momentum is $m(2v) + mv$. After the collision, the two skaters sail off together ($2m$) with a speed ($3v/2$) that conserves the initial momentum ($3mv$).

Example 9.20 During a rainy day football game, a 192-lb quarterback is standing holding the ball looking for a receiver when he's unkindly hit by a 288-lb tackle charging in at 6.1 m/s (20 ft/s). (a) At what speed do the two men, tangled together, initially sail off onto the wet field? Assume friction is negligible and that the impact

(continued)

(continued)
is head-on. (b) How much mechanical energy is lost to friction? As always, work the problem through in SI units only.

Solution: [Given: $F_{Wq} = 192$ lb, $F_{Wt} = 288$ lb, $v_{qi} = 0$ and $v_{ti} = 20$ ft/s. Find: the final speed of the two, v_f and ΔKE.] First change everything to SI: $F_{Wq} = 854$ N, $F_{Wt} = 1281$ N, $v_{qi} = 0$, and $v_{ti} = 6.10$ m/s. Since the two slide away together, this collision is completely inelastic—momentum is conserved, but neither E nor KE is conserved. (a) Consequently, $p_i = p_f$ and

$$m_q v_{qi} + m_t v_{ti} = (m_q + m_t)v_f$$

Using $m = F_W/g$, and taking the direction of the initial motion as positive

$$0 + \frac{1281 \text{ N}}{9.81 \text{ m/s}^2} 6.10 \text{ m/s} = \left(\frac{854 \text{ N}}{9.81 \text{ m/s}^2} + \frac{1281 \text{ N}}{9.81 \text{ m/s}^2}\right) v_f$$

$$796.8 \text{ N/s} = (217.7 \text{ N·s}^2\text{/m})\, v_f$$

and $\boxed{v_f = 3.7 \text{ m/s}}$

The two players move off in the positive direction. Without ever coming to rest, the tackle captures the quarterback and carries him off. This, of course, is why tackles are designed with so much mass. (b) No change in potential energy occurs provided no one falls, but there is a change in KE, since

$$\text{KE}_i = \tfrac{1}{2} m_t v_{ti}^2 = 2.4 \text{ kJ}$$

while $$\text{KE}_f = \tfrac{1}{2}(m_q + m_t)v_f^2 = 1.5 \text{ kJ}$$

Hence, $\boxed{0.9 \text{ kJ}}$ was transformed, mostly to thermal energy. This problem has one unknown and can be solved using only Conservation of Momentum.

▶ **Quick Check:**

$$0 + \frac{288 \text{ lb}}{32 \text{ ft/s}^2} 20 \text{ ft/s} = \left(\frac{192 \text{ lb}}{32 \text{ ft/s}^2} + \frac{288 \text{ lb}}{32 \text{ ft/s}^2}\right) v_f$$

and $v_f = 12$ ft/s. Similarly, $\text{KE}_i = 1.8 \times 10^3$ ft·lb and $\text{KE}_f = 1.1 \times 10^3$ ft·lb; $\Delta\text{KE} = 0.7 \times 10^3$ ft·lb $=$ 0.9 kJ.

For completely inelastic collisions between two bodies, only one of which (m_1) is initially moving, the ratio of the total final KE to the total initial KE is

$$\frac{\text{KE}_f}{\text{KE}_i} = \frac{\frac{1}{2}(m_1 + m_2)v_f^2}{\frac{1}{2}m_1 v_{1i}^2}$$

Using momentum conservation, $m_1 v_{1i} = (m_1 + m_2)v_f$; hence, $v_f = m_1 v_{1i}/(m_1 + m_2)$, and so squaring and substituting yields

$$\text{KE}_f = \left(\frac{m_1}{m_1 + m_2}\right)\text{KE}_i$$

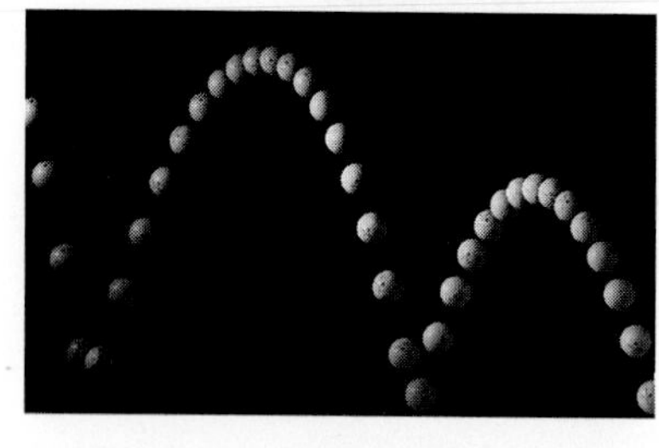

The loss of energy generally runs a system down—in this case, a bouncing golf ball. With each impact, a certain amount of energy goes into random thermal motion, warming the ball and floor; less is returned as organized KE to the ball, so it rises more slowly and doesn't bounce as high. The closer the images of the ball, the slower it is moving.

which suggests that $\text{KE}_f \approx \text{KE}_i$ when m_1, the moving mass, is much greater than m_2. Then $(m_1 + m_2) \approx m_1$ and the term in parentheses becomes 1. If a sports car locks bumpers with the back of a stopped garbage truck, the two will roll slowly ahead—KE_i will be greater than KE_f, and the difference will go into mangling the participants. On the other hand, if the truck did the crashing with the same KE, the two would roll away at nearly the same speed as the truck came in at, and the collision would be less destructive. A small, high-speed projectile in an inelastic collision with a movable target will do more damage than a massive, slow one with the same KE.

Elastic Collisions

A collision is said to be **elastic** *when* KE *is conserved.* Particles on the submicroscopic level collide elastically, presumably because they don't suffer any permanent rearrangement of their component parts (if there are any). Thus, suppose a head-on collision takes place between mass 1 and mass 2 on a horizontal plane with m_2 at

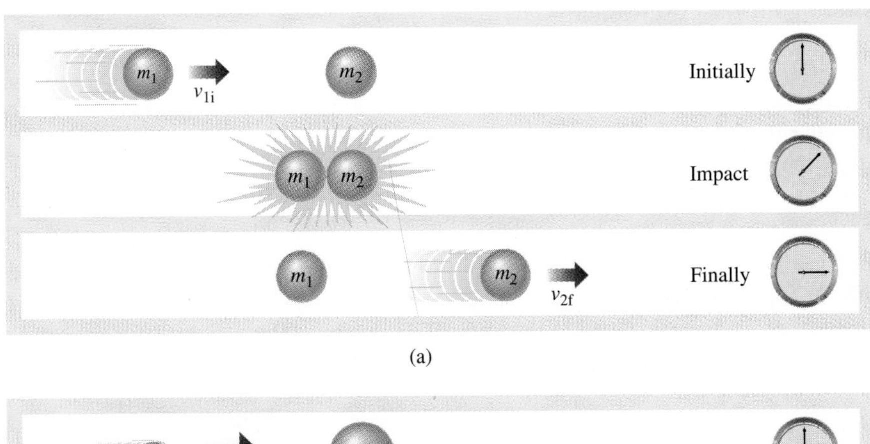

(a)

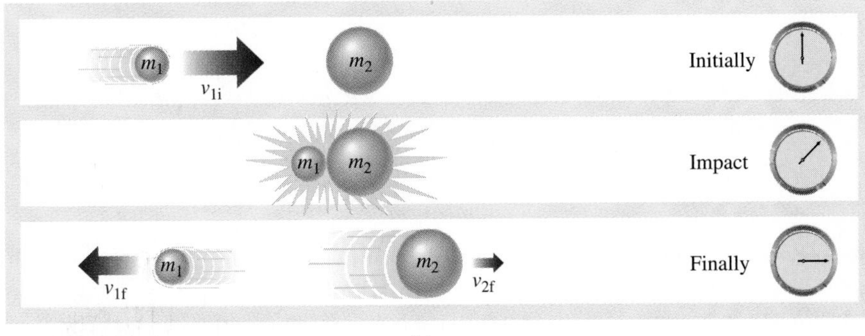

(b)

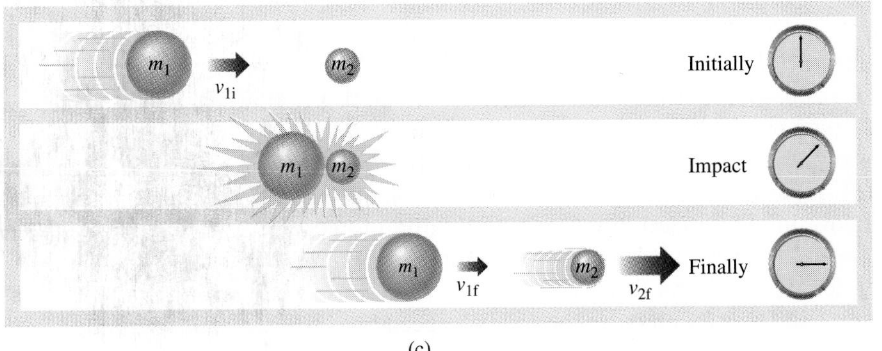

(c)

Figure 9.20 Three elastic collisions in which a moving mass m_1 crashes into a stationary mass m_2. In (a) $m_1 = m_2$ and they essentially exchange speeds. In (b) $m_1 < m_2$, and the balls move off in opposite directions. In (c) $m_1 > m_2$, and both balls move off in the direction in which m_1 was originally traveling.

rest, where this time the impact is completely elastic (Fig. 9.20). Conservation of Momentum provides

$$m_1 v_{1i} + 0 = m_1 v_{1f} + m_2 v_{2f}$$

but now the participants move off independently. Since $\Delta PE_G = 0$, Conservation of Energy provides

$$KE_{1i} + KE_{2i} = KE_{1f} + KE_{2f}$$

$$\tfrac{1}{2} m_1 v_{1i}^2 + 0 = \tfrac{1}{2} m_1 v_{1f}^2 + \tfrac{1}{2} m_2 v_{2f}^2$$

Now there can be two unknowns, usually the two final speeds, because there are two independent equations. Accordingly, we can solve for v_{1f} and v_{2f} by first rearranging these two equations as

$$m_1(v_{1i} - v_{1f}) = m_2 v_{2f} \qquad (9.21)$$

and

$$m_1(v_{1i}^2 - v_{1f}^2) = m_2 v_{2f}^2$$

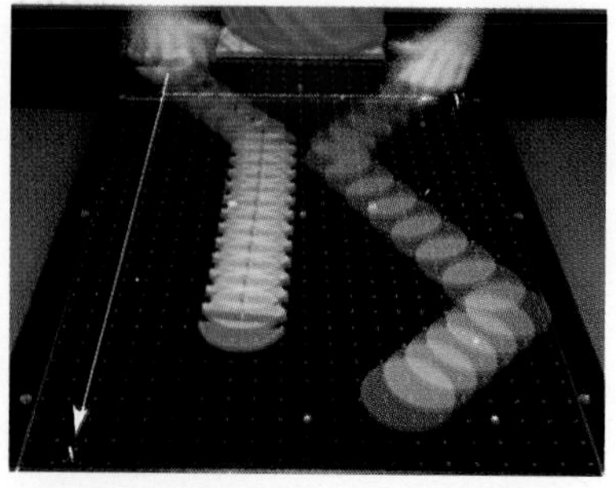

Two identical pucks on an air table. The blue puck has a positive y-component of momentum mv_{By}. The yellow puck has a negative y-component of momentum $-mv_{Yy}$, where $mv_{By} > mv_{Yy}$. After colliding, the blue puck has no y-component and the yellow puck moves off with a positive y-component equal to the initial net y-component $(mv_{By} - mv_{Yy})$. Both the x- and y-components of momentum are conserved independently.

Factoring this last expression yields

$$m_1(v_{1i} + v_{1f})(v_{1i} - v_{1f}) = m_2 v_{2f}^2 \tag{9.22}$$

Dividing the right side of Eq. (9.22) by the right side of Eq. (9.21) and the left by the left gets rid of the masses, and

$$v_{1i} + v_{1f} = v_{2f} \tag{9.23}$$

Hence

$$v_{1i} = v_{2f} - v_{1f}$$

The "relative speeds" of the two bodies before and after impact are equal: they initially approach one another at the same speed with which they finally separate. Substituting v_{2f} from Eq. (9.23) into Eq. (9.21) leads to the sought-after expression for the final speed of mass 1, which is

$$v_{1f} = \frac{m_1 - m_2}{m_1 + m_2} v_{1i}$$

Substituting this expression into Eq. (9.23), we get $v_{2f} = [v_{1i}(m_1 - m_2)/(m_1 + m_2)] + v_{1i}$, and the expression for the final speed of mass 2 becomes

$$v_{2f} = \frac{2m_1}{m_1 + m_2} v_{1i}$$

If we take the direction of $\mathbf{v}_{1i}$ as positive, the last two equations describe the resulting motion for the three possible cases.

(1) When $m_1 = m_2$, as with two billiard balls, $v_{1f} = 0$ and $v_{2f} = v_{1i}$. The moving ball slams to a stop and the struck ball slides (no rolling, please) away with the initial speed. The balls, in effect, exchange velocities (Fig. 9.20a). (2) When $m_1 < m_2$, v_{1f} is negative, and the incoming object rebounds in the opposite direction while the initially stationary object moves off in the positive direction, as in Fig. 9.20b. Indeed, when $m_1 \ll m_2$, $v_{1f} \approx -v_{1i}$ and $v_{2f} \approx 0$, which is like bouncing a ball (m_1) off the back of a stationary bus (m_2). (3) Finally, when $m_1 > m_2$, v_{1f} is positive as is v_{2f}; both move off in the initial direction of motion (Fig. 9.20c). When a tennis racket slams into an essentially motionless ball during a serve, it continues to move in the forward direction but at a diminished speed. In the extreme, when $m_1 \gg m_2$, $v_{1f} \approx v_{1i}$, and since m_2 is negligible in the denominator of the last equation for v_{2f}, it follows that $v_{2f} \approx 2v_{1i}$. Hit a Ping Pong ball with a massive paddle and it will fly off with twice the paddle's speed (*and still the relative speeds will be unchanged*).

9.9 Energy Conservation and Symmetry

Linear momentum is a vector quantity and Conservation of Linear Momentum arises from the homogeneity of free space (p. 116). The laws and physical constants of nature are the same everywhere—they are invariant with respect to place. Position is relative in the sense that no experiment performed in an isolated laboratory can produce results that depend on its location in empty space. Thus, there is spatial displacement symmetry. Similarly, angular momentum is a vector quantity and Conservation of Angular Momentum arises from the isotropy of free space. There is no absolute zero angle of orientation, no preferred direction. The laws and constants of nature are the same at every orientation.

Noether showed that Conservation of Energy is a logical consequence of the temporal homogeneity of the Universe (Fig. 9.21). Energy is a scalar quantity; in an isolated laboratory, it is constant in time, independent of the lab's position and orientation in free space. Time is relative in the sense that no experiment performed

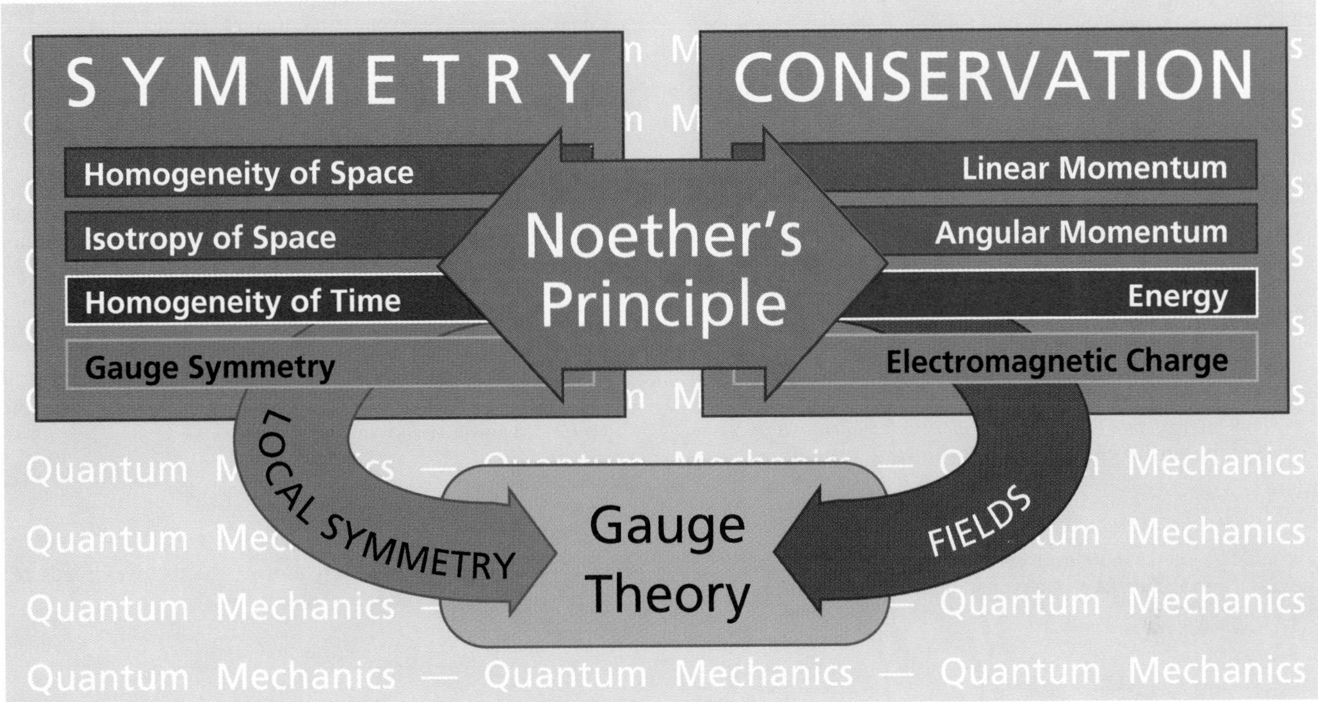

Figure 9.21 Thus far, we have examined the ideas of symmetry, conservation, and Noether's Principle. We have seen that linear and angular momentum are conserved. Now we know that energy is also conserved. Our Universe exhibits temporal displacement symmetry, which requires Conservation of Energy.

in the laboratory can produce results that depend on its location in time. All experiments repeated exactly will behave in precisely the same way today as they did yesterday. If, as we believe is the case, physical law is timeless throughout the Universe, then energy must be conserved.

Though the rigorous proof of this assertion is beyond the level of our treatment, we can make the conclusion plausible. To that end, imagine an isolated lab in which you separate two gravitationally interacting masses, thereby doing work and storing a certain amount of PE_G in that system. Assume, for the sake of argument, that the laws of physics change in time. If you hold the objects motionless and simply wait for gravity to change, a time will come when the PE_G is either larger or smaller than it was at the start. At that moment, you let the objects fall back to where they were originally. The lab is returned to its original configuration, but there is now an energy imbalance—Conservation of Energy is violated.

The nature of the fundamental interactions (Newton's Third Law) ensures that the transfer of energy everywhere, while the laws of physics are constant, will be such as to maintain a fixed total energy. When any two systems interact, both will change; one will lose the energy the other gains. Insofar as our Universe exhibits temporal displacement symmetry, energy is and will be conserved.

Core Material

The measure of the amount of energy transferred into or out of a system via the action of an applied force is known as **work**, where

$$W = Fs \cos \theta \qquad [9.2]$$

The unit of work is the *newton-meter* or *joule*.

A force is **conservative** if *the work done in moving an object from one point to another against the force depends only on the locations of the points and not on the path taken* (p. 296).

Power is *the rate of doing work*, the rate at which energy is transferred. *Average power* is

$$P_{av} = \frac{\Delta W}{\Delta t} \qquad [9.4]$$

The SI unit of power is the **watt**: 1 W = 1 J/s. When an object is in motion at a speed v under the influence of a force

$$P = Fv \cos \theta \qquad [9.6]$$

When work is done on a system to overcome inertia, it goes into **kinetic energy** (KE), where

$$KE = \tfrac{1}{2}mv^2 \qquad [9.9]$$

The energy *stored in a system of interacting objects* is known as **potential energy** (PE). Assuming g is constant

$$PE_G = mgh \qquad [9.13]$$

The change in the potential energy experienced by a body in moving from the surface of a planet out to some distance r from the center is

$$\Delta PE_G = GmM\left(\frac{1}{R} - \frac{1}{r}\right) \qquad [9.14]$$

The Law of *Conservation of Energy* (p. 297) maintains that

The total energy of any system that is isolated from the rest of the Universe remains constant, even though energy may be transformed from one kind to another within that system.

In the absence of friction and any other force but gravity, mechanical energy is conserved; thus

$$\tfrac{1}{2}mv_f^2 + mgh_f = \tfrac{1}{2}mv_i^2 + mgh_i \qquad [9.17]$$

The lowest speed at which we could fire a projectile so that it never falls back to the place from which it was launched is the **escape speed**

$$v_{esc} = \sqrt{\frac{2GM}{R}} \qquad [9.18]$$

For *inelastic* collisions between two bodies, only one of which (m_1) is initially moving

$$KE_f = \left(\frac{m_1}{m_1 + m_2}\right)KE_i$$

When the collision is completely *elastic*, the relative speeds of the two bodies before and after impact are equal, and

$$v_{1f} = \frac{m_1 - m_2}{m_1 + m_2}v_{1i}$$

and

$$v_{2f} = \frac{2m_1}{m_1 + m_2}v_{1i}$$

Suggestions on Problem Solving

1. A common error in *work* problems involving forces that are not parallel to the displacement is to forget to use the component of F, namely ($F \cos \theta$), in computing W. If the force is not parallel to the displacement, check whether or not the perpendicular component ($F \sin \theta$) is affecting the system. For example, ($F \sin \theta$) can increase or decrease the normal force and thus affect the friction and W.

2. When a body moves under the influence of a constant force, the power developed is given by $P = Fv$. If you know the power provided by a rocket motor traveling at a given speed, you can determine its thrust. Or if you know the maximum power of a locomotive and the force exerted on it, you can calculate its maximum speed.

3. Remember to *square* the speed when doing a KE calculation. When considering the change in KE, bear in mind that $v_f^2 - v_i^2$ is *not* equal to $(v_f - v_i)^2$.

4. Momentum is a vector quantity. If the velocity vectors of two bodies lie along a line connecting their centers, a collision problem can be treated via a single scalar momentum expression. This is the case provided that you first assign a sign to the appropriate directions. Then colinear vectors will properly add (or subtract) as scalars. If the initial velocity vectors are not colinear, their separate component momenta in, say, the x- and y-directions will be conserved individually.

5. We are restricted to the two extreme classes of collision problems, and that in itself is helpful to remember. Thus, if the colliding bodies stick together, the collision involves only Conservation of Momentum, and there will be one equation and only one unknown. When the two bodies move off together, there will be one final velocity for both. By contrast, if the collision is elastic, we can also use Conservation of Energy, and there will be two equations and there can be two unknowns. The final velocities of the bodies will be different.

6. Proceed cautiously when confronted with a collision problem where it's not explicitly stated that mechanical energy is conserved. A bullet that squashes against a steel plate or imbeds in a phone book or passes right through a wooden board will "lose" an appreciable amount of energy. Nonetheless, provided no external forces are acting on the bullet-board system, momentum will be conserved. Watch out for situations where any of the objects taking part suffer deformation or generate "heat" via friction.

Discussion Questions

1. A rocket is held in place as its engines are fired up. Is any work done on the rocket? Once released, the rocket rises into the air. What force accelerates it? Considering both the vehicle and the exhaust, talk about the KE of each and the nature of the forces doing work. Where does the energy that ends up as KE come from? How is this different from swimming? Even though the thrust is constant, the longer the engine fires, the faster the rocket goes. What does that say about the power transferred to the craft as a function of time?

2. In the process of rowing, what external force propels the boat? Describe the work done on the boat and on the water. Where does the change in KE of the boat come from? How is this similar to the operation of a propeller on an airplane?

3. A person walks at a constant speed by pushing backward on the ground with a force F. If he travels a distance s in a straight line on level ground, (a) what force is he overcoming? (b) Suppose his feet slide a little as he pushes back—does the reaction force of the ground do work on him? (c) What if his feet don't slip as he walks; is work done on him by the floor? (d) Where does the energy expended come from?

4. A lump of soft clay of mass m falls from a height h above the floor, where it crashes and comes to rest. What force brings the clay to a stop? How much work is done by the floor on the clay? What is the net amount of work done on the clay during the fall? Where does that energy end up?

5. A youngster on a bicycle starts at rest and pedals a straight path on level ground at a constant speed, coming to a stop some distance away. (a) Is it reasonable to assume that he is the energy source for the journey? (b) What *external* force propels the bicycle and rider? (c) Does the ground do positive work on the bike?

6. Work is done against resistance. In each of the following cases, describe the forces overcome: (a) hammering a nail into a piece of wood; (b) stretching a piece of rubber; (c) cutting a loaf of bread; (d) putting a book on a high shelf; (e) opening the door on a modern refrigerator; (f) pulling a piece of masking tape off a roll.

7. Two identical balls travel along frictionless grooves in a flat surface. One groove is straight and runs horizontally along the surface, while the other dips down below the surface and up again smoothly in a vertical plane. The two runways begin and end parallel and adjacent to each other. Assume that neither ball leaves its track and that both are given the same initial speed. Describe their motions in a race where the starting and finish lines are across horizontal portions of the grooves.

8. A rock at the bottom of a well has a negative amount of *gravitational*-PE with respect to the ground level. What does that mean? What will happen if it's launched upward with an equal positive amount of KE? Can a body have a negative KE? Discuss the significance of the situation in which a body has a positive KE and a negative PE, where $|PE| > KE$. What happens when $|PE| < KE$?

9. The late physicist and Nobel laureate Richard P. Feynman once wrote, "It is important to realize that in physics today, we have no knowledge of what energy *is*." What do you think he meant by that? The usual definition of energy is *the ability to do work*. Discuss the circular nature of that definition. Consider the alternative definition, energy is *the ability to impart vis viva* (mv^2); is this definition any better?

10. Potential energy belongs to the entire system of interacting objects and not to any one of them. Explain this statement and discuss how it applies to the *gravitational*-PE of a frog sitting on a rock on the planet Earth.

11. A ball on a string whirls around in a circle at a constant speed. Is there a net force on the ball? Does it do work? Is there a corresponding change in either PE or KE? How can the ball be made to accelerate tangentially? Will there then be work done on it and a change in either PE or KE?

12. The number One billard ball moving with a speed v_{1i} strikes the Two ball, which is at rest, hitting it off-center and sailing away in an elastic collision. With respect to the initial line of motion, the two balls move off at angles θ and ϕ, respectively. Draw a diagram showing that $\mathbf{p}_{1i}$, $\mathbf{p}_{1f}$, and $\mathbf{p}_{2f}$ form a closed triangle. Explain your reasoning.

13. The cars of a roller coaster are usually tugged up to the top of the first hill by a chain drive and then released to coast the rest of the way around. Consider that friction is acting all along the journey. (a) Where is the *gravitational*-PE maximum? (b) How should the ride be configured to produce the greatest value of the maximum speed? (c) Having once attained its maximum *gravitational*-PE, can it reach that value again? (d) Having once attained its maximum KE (which may well not be the maximum possible KE for a structure of that height), can it reach that value again? (e) Must each successive hill be lower than the previous one? (f) Must the second hill be lower than the first?

14. It is possible to hold a large rock or a massive lump of soft clay and slam it with a hammer and yet not hurt the hand holding the object. Explain what is happening in terms of energy and momentum.

15. A glass marble is dropped from some height onto wet sand and penetrates to a depth d. The experiment is repeated, but this time from twice the original height. How deep will the sphere penetrate and why? What assumptions are made?

16. You wish to hit a nail into a board with a hammer. It is your intention to slam the nail so the hammer remains in contact with the nail without rebounding. Would you do better with a light or a heavy hammer? Explain.

17. A crate of mass m rests on a flatbed truck. The truck accelerates from rest to a speed v, during which time the crate does not slide on the bed. How much work is done on the crate by friction? Is that work positive or negative? What is the work done against? In what final form is the energy that was provided to the crate? Now suppose the truck comes to a gradual stop. Given this situation, answer the same questions again.

18. During the first Moon-landing flight (*Apollo 11,* July 1969), the spacecraft was put into a circular parking orbit at an altitude of 119 mi above the Earth. The engines were later fired, sending the ship off to the Moon at 24 245 mi/h. Compare this speed to the escape speed and explain the difference. When we calculated the formula for the escape speed for a planet, we neglected the fact that our particular dirtball is spinning. Discuss the effect, if any, of the spin if launch is from some parking orbit about the Earth. What if lift-off is from the surface?

Multiple Choice Questions

1. A 10-kg mass is held 1.0 m above a table for 25 s. How much work is done during that period? (a) none (b) 10 J (c) 250 J (d) 9.8 J (e) none of these.
2. The net work done on an object moving along a closed path in a force field is zero when it returns to the origin. The force is (a) conservative (b) nonconservative (c) impossible (d) liberal (e) none of these.
3. A typical adult male's heart pumps about 160 milliliters of blood per beat. It beats around 70 times per minute and does roughly 1 J of work per beat. How much work does it do in a day? (a) 10^5 J (b) 10^6 J (c) 70 J (d) 70×10^5 J (e) none of these.
4. Which of the following is not a measure of the same quantity as the others? (a) newton-meter per second (b) kilogram-meter^2 per second^3 (c) joule per second (d) watt (e) none of these.
5. If a 20-kW engine can raise a load 50 m in 10 s, how long will it take for it to raise that same load 100 m? (a) 20 s (b) 40 s (c) 5.0 s (d) not enough information (e) none of these.
6. If a 25-hp motor can raise an elevator 10 floors in 20 s, how long will it take a 50-hp motor to do the same? (a) 40 s (b) 10 s (c) 20 s (d) 5.0 s (e) not enough information.
7. Which of the following is not a measure of the same quantity as the others? (a) foot-pound (b) newton-meter (c) watt (d) joule (e) none of these.
8. If this book is placed on an ordinary table and slid along a path that brings it back to where it started, (a) no net power will have been required (b) work will certainly have been done (c) assuming a conservative gravitational field, no net work will be done (d) not enough information is given to say anything about the work done (e) none of these.
9. A fairly small asteroid (1000 kg) out in deep space is to be accelerated from rest up to 10 m/s. Inasmuch as it is weightless, will work have to be done on it during the acceleration and, if so, how much? (a) no (b) yes, 10 000 J (c) yes, 50×10^3 J (d) yes, 10 000 N (e) yes, 50×10^3 N
10. Work is done on an object far out in space where it has negligible *gravitational*-PE. If in the process there is no net change in its KE, we can conclude (a) that friction *may* have been operative (b) that this situation is impossible (c) that the energy of the object has decreased (d) that the object's speed decreased (e) none of these.
11. A rocket coasting along in space at some speed v fires its engines thereupon doubling its speed, but at the same time it jettisons some cargo, reducing its mass to half its previous value. In the process, its KE is (a) doubled (b) tripled (c) quadrupled (d) unchanged (e) none of these.
12. A kid in a wagon rolls from rest down a hill reaching the bottom at 12 m/s. On the next run, she gets a push and starts down at 5.0 m/s. At what speed does she now arrive at the bottom? (a) 12 m/s (b) 17 m/s (c) 7 m/s (d) 13 m/s (e) none of these.
13. An arrow is fired, via a bow, straight up. It rises for a while and then drops back to the ground. The process, taking the arrow from loading to just prior to touch down, can best be described by a series of energy transformations corresponding to (a) work, *elastic*-PE, KE, *gravitational*-PE, KE (b) work, KE, *elastic*-PE, KE (c) KE, *gravitational*-PE, work (d) *elastic*-PE, *gravitational*-PE, KE (e) none of these.
14. How fast must a 1.0-kg mass be moving if its kinetic energy is 1.0 J? (a) 1.0 m/s (b) 9.8 m/s (c) 1.414 m/s (d) 10 m/s (e) 0.102 m/s.
15. An open wagon is rolling along frictionlessly on a level road when it begins to snow. After a load of snow has fallen vertically into the wagon, its speed, kinetic energy, and momentum will each, respectively (a) increase, increase, stay the same (b) decrease, stay the same, decrease (c) decrease, decrease, stay the same (d) stay the same, increase, increase (e) decrease, stay the same, increase.
16. How high above the Earth's surface must a 1.0-kg mass be for it to have a *gravitational*-PE of 1.0 J with respect to that surface? (a) 9.8 m (b) 1.0 m (c) 0.10 m (d) 0.01 m (e) 32 m.
17. While a ball rolls down the circularly curved track shown in Fig. MC17, its speed, acceleration, and kinetic energy, respectively (a) increase, increase, increase (b) decrease, decrease, decrease (c) increase, decrease, decrease (d) decrease, increase, increase (e) none of these.

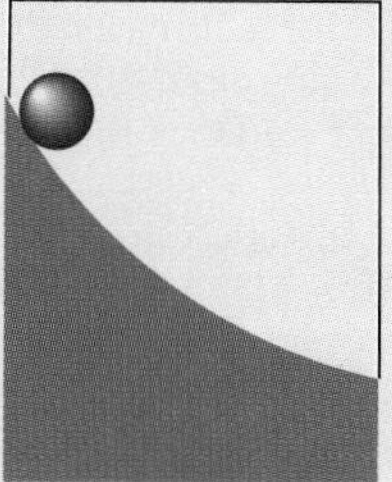

Figure MC17

18. A 458 Winchester magnum cartridge has a 500-grain (1 gr = 0.0648 g) bullet that attains a muzzle speed of 644.9 m/s (2120 ft/s). After traveling 456 m (500 yd), it would still be moving at 365.6 m/s (1202 ft/s). What fraction of its original KE does it have at that distance? (a) 23% (b) 50% (c) 32% (d) 68% (e) 77%.
19. Taking infinity as the zero of *gravitational*-PE, through what height at the Earth's surface must a 1.0-N body be raised if its PE_G is to be increased by 1.0 J? (a) ∞ (b) 10 m (c) 1.0 m (d) −1.0 m (e) none of these.
20. If Superman really is "more powerful than a locomotive," then with regard to freight trains (a) he can pull more cars at the same speed (b) he can pull the same number of cars faster (c) he can pull the same number of cars up steeper hills (d) all of these (e) none of these.
21. A 9.5-g wad of chewing gum traveling at 1.0 m/s horizontally crashes into and sticks to a 9.5-g puck floating frictionlessly on an air table. They stick together and sail away at (a) 1.0 m/s (b) 2.0 m/s (c) 0.0 m/s (d) 0.50 m/s (e) none of these.

Problems

THE TRANSFER OF ENERGY

1. [I] A book is slowly slid at a constant speed 1.5 m across a level table by a 15-N horizontal force. How much work is done on the book and what is it done against?
2. [I] A 1.0-kg mass is raised 10 m into the air by a 10-N force exerted vertically. Determine the amount of work done by the force. Now suppose a 10-N force pulling horizontally moves a 1.0-kg mass through 10 m over a frictionless floor. How much work is done here? What is the work done against in both instances?
3. [I] A delicatessen owner holds a knife horizontally above a 10-cm-thick cheese. With one smooth motion, exerting a downward force of 20 N, the cheese is cut in two. How much work was done by the owner?
4. [I] A 2224-N piano is rolled on little wheels 3.1 m across a horizontal floor by a 890-N piano mover. If, while lifting upward on one end with 111 N, he pushes horizontally with 445 N, how much work will he do?
5. [I] When a solid-fuel rocket burns, the flame front advances into the material in a direction perpendicular to the ignited surface. By configuring the fuel core in various ways and lighting it along the entire length, one can obtain all sorts of performance characteristics. Figure P5 shows a starred-cavity motor. How much work was done by the device in the first 90 s of flight if the rocket rose vertically to a height of 2.0 km in that time?

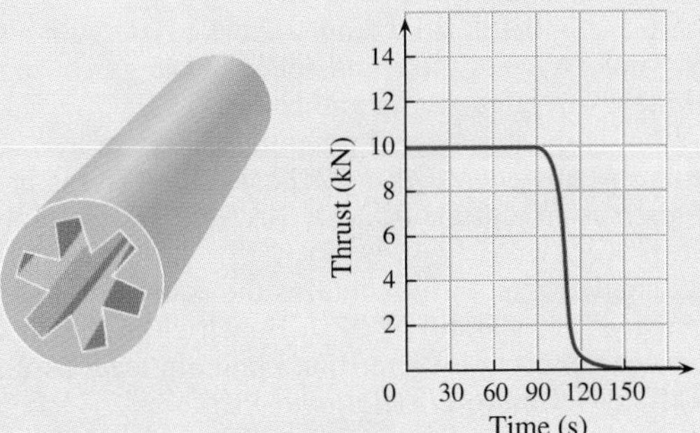

Figure P5

6. [I] A newspaper delivery boy pulls a rubber-wheeled cart that has a coefficient of rolling friction of 0.02 and a mass of 25 kg. It travels 10 km on level streets. How much work does he do in overcoming friction between the cart and the ground?
7. [I] A nurse pushes someone in a wheelchair 100 m doing 400 J of work in the process. What average force did the nurse exert in the direction of motion?
8. [I] Prove that 1 hp = 746 W.
9. [I] A runner traverses a 50-m-long stretch on a horizontal track in 10 s at a fairly constant speed. All the while she experiences a retarding force of 1.0 N due to air friction. What power was developed in overcoming that friction?
10. [I] A small hoist can raise 100 kg of bricks to the top of a construction project 30 m above the street in half a minute. Determine the power provided.
11. [I] The per capita power consumption in the United States is around 10 kJ each and every second. At what speed would you have to push a car exerting a force of 1.0 kN on it all year, day in and day out, to be equivalent to your share?
12. [I] A rigid steel crowbar is rested on an upright brick standing near the back of a car. The end of the bar touches the bottom of the bumper 30 cm from its pivot point on the brick. Someone pushes straight down on the other end of the bar, 270 cm from the pivot point, and raises the car's chassis 5.0 cm. Given ideally that the work-in equals the work-out, if it takes 3200 N to raise the car, how much force did the person exert on the bar?
13. [I] Assuming no friction and weightless pulleys, how much rope will have to be drawn off if the weight in Fig. P13 is to be raised 1.0 m? Remember: ideally, work-in equals work-out.

Figure P13

14. [II] A pickup truck is hauling a barge along a canal at a constant speed. The truck, driving parallel to the waterway, is attached to the barge by a cable tied to the bow making a 30° angle with the forward direction. If the truck exerts a force of 1000 N on the cable, how much work is done in overcoming friction as the barge is moved 10 km?
15. [II] A 100-N box filled with books is slid along a 13-m-long ramp up to a platform 5.0 m above the ground. How much work is done if friction is negligible? How much work would have been done if the box were lifted straight up to the platform?
16. [II] A 50-kg keg of beer slides upright down a 3.0-m-long plank leading from the back of a truck 1.5 m high to the ground. Determine the amount of work done on the keg by gravity.
17. [II] The newspaper boy in Problem 6 pushes his cart 25 m along a road inclined at 10°. How much work does he do to overcome road friction?

18. [II] Five fat dictionaries, each 10 cm thick and each 2.5 kg, are resting side by side flat on a table 1.0 m high. How much work would it take to stack them one atop the other? How high would that raise the center-of-mass of all the books?
19. [II] A constant force of 100 N is applied to an object over a distance of 2.0 m parallel to the displacement. Draw a work diagram (F vs. s) with a scale of 1.0 cm = 10 N and 1.0 cm = 0.25 m. (a) What was the total work done? (b) How much work does an area of 1.0 cm^2 equal? (c) What is the area under the curve?
20. [II] The Zambesi River in Africa rushes over Victoria Falls at a rate of 25×10^6 gallons per minute. The falls are 108 meters high, 1 gallon equals 0.133 7 cubic foot, and each cubic foot weighs 62.4 pounds. Determine the power developed by the water in SI units.
21. [II] A 2.5-hp motor drives a hoist that can raise a load of 50 kg to a height of 20 m. At full power, how long will the hoist take to do it?
22. [II] The belt connecting an auto engine to an air conditioner is moving at 40 m/min and has an effective tension of 20 N. How much work does it do in an hour? How much power does it transmit?
23. [II] An animal's power level is known as its **metabolic rate**. If a man typically uses 10 MJ in the course of a day, what is his metabolic rate? While resting but awake, an average person has a *basal* power level of about 90 W. This figure, however, depends on the size of the person, being larger for larger people, who have greater surface areas and thus lose more energy. And so to standardize the notion, making it independent of body size, it is customary to divide by the skin surface area (typically 1.5 m^2)—the number thus arrived at is the *basal metabolic rate* or BMR. What is the BMR for our average person?
24. [II] The oxygen taken in by the body reacts with fats, carbohydrates, and protein, liberating energy *internally* at a rate of about 2.0×10^4 J per liter. If a 70-kg man requires 77 W of power even while sleeping, what is his rate of oxygen consumption?
25. [II] Refer to Problems 23 and 24. What is the BMR of a person who (according to the standard charts) has a surface area of 1.8 m^2 and is consuming oxygen at a rate of 0.40 liters per minute?
26. [II] Efficiency is defined as work-out/work-in. Muscles operate with an efficiency of about 20% in converting energy internally into work externally. Accordingly, how much energy will be expended by a 60-kg person in the process of ascending several flights of stairs to a height of 25 m?
27. [II] A variable force, depicted in Fig. P27, acts on a 10-kg body. How much work does it do in moving the body from $x = 2$ to $x = 6$? What is the total amount of work done in going from $x = 0$ to $x = 10$? Over which 1-m interval was the smallest amount of work done? How much work is done in going from $x = 2$ to $x = 12$?

MECHANICAL ENERGY

28. [I] A major concern in the design of spacecraft is the presence in space of tiny, high-speed meteoroids. Micrometeoroids, as they're called, have been detected traveling as fast as 70 km/s. Compute the kinetic energy of a 1.0-g mass moving at that rate.

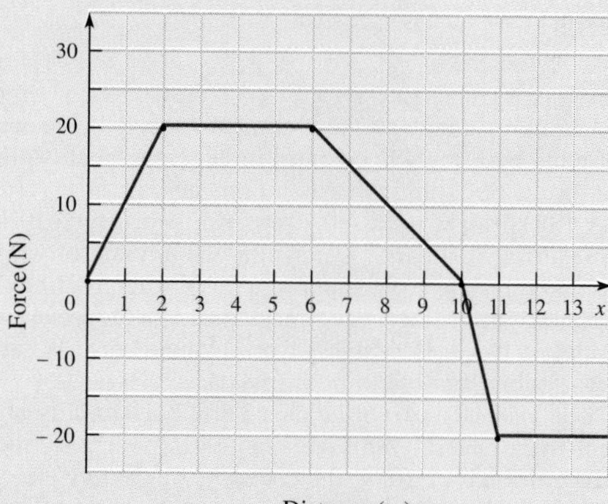

Figure P27

29. [I] It has been suggested that a controlled fusion reaction (a mini-H-bomb) could be achieved by causing a small (on the order of 1 mm), extremely high-speed projectile to hit a stationary target, both made of the appropriate materials. Determine the kinetic energy of a 0.5-g mass traveling at 200 km/s. If the collision takes 10 ns, how much power would be provided to the target?
30. [I] One ton of uranium-235 can provide about 7.4×10^{16} J of nuclear energy. If that much energy went into accelerating a 3.5×10^6-kg spaceship (that's the size of a fully loaded *Saturn V* Moon rocket) from rest, what would its final speed be?
31. [I] A 6.5-g bullet is fired from a 2.0-kg rifle with a speed of 300 m/s. What is the recoil speed of the gun? Compare the kinetic energies of gun and bullet.
32. [I] What is the kinetic and potential energy of a Boeing *747* airliner weighing 2.22×10^6 N, flying at 600 mi/h at an altitude of 20×10^3 ft? Work the problem in SI, please.
33. [I] A single barrel of oil contains the equivalent *chemical*-PE of about 6×10^9 J. How high in the air could that much energy raise a million kilogram load, assuming it is all converted to *gravitational*-PE?
34. [I] A uniform rod 6.0 m long, weighing 60 N, is pivoted about a horizontal axis 1.0 m up from its *c.m.* so that it hangs vertically. Neglecting friction, (a) what's the net work done on the rod by gravity in the process of raising its end, making it horizontal? (See if you can come up with two ways to do this.) (b) What change in potential energy, if any, has the rod experienced?
35. [I] A car with a mass of 1000 kg is at rest at the base of a hill. It accelerates up the incline reaching a speed of 20 m/s at a height of 100 m. What is the total increase in mechanical energy at that point?
36. [I] A 17.8-kN car is to be accelerated from 0 to 96.5 km/h (60 mi/h) in 10 s. Neglecting all frictional losses, how much power must be supplied in the process (independent of the exact nature of the acceleration)? Incidentally, air friction increases rapidly with speed, becoming a major concern above 64 km/h (40 mi/h).
37. [II] (a) What is the KE of a 17.8-kN car traveling at 64 km/h? (b) Assuming the car gets 20 miles to the gallon at that speed, and each gallon of gasoline provides

1.3×10^8 J of chemical energy, what's its power consumption? Incidentally, of this total, only about 20% gets into the mechanical system (pumps, transmission, etc.). About 10^4 J/s goes into propelling the car, of which half is dissipated in overcoming tire and road friction and half in air friction. Older cars typically lose about 10^3 J/s in unburned fuel that evaporates from the carborator and never even gets into the engine.

38. [II] A 60-kg person is slowly doing chin-ups, during each one of which her *c.g.* rises just about $\frac{1}{2}$ m. If the biceps contract roughly 4.0 cm in the process of each lift, how much tension is there, on average, in the muscles of each arm?

39. [II] A 2×10^3-kg spacecraft has an ion engine that delivers 150 kW of propulsive power via its six thrusters. Assuming it to be far from any massive object, how much time would it take to accelerate from "rest" to 268 m/s (600 mi/h)?

40. [II] A 1000-kg car at rest at the top of a hill accelerates down the road, reaching a speed of 20 m/s after descending a height of 100 m. What was its total change in mechanical energy as of that moment?

41. [II] A 100-N weight sitting on the floor is tied to a light rope that passes over a very light frictionless pulley 20 m above it. The other end of the rope hangs down to the floor where it is being held by a 10-N monkey named George. Suppose George now climbs 10 m above the floor. (a) How much work does he do? (b) How much rope ends up on the floor? (c) What is the total change in *gravitational*-PE of the system?

42. [II] During a vertical jump, a person crouches down to lower his or her *c.m.* and then leaps straight up. The leg muscles essentially do all the work, accelerating the body over a push-off distance of about $\frac{1}{3}$ m. Typically, people can support an additional load using the legs equal to their own weight, but only with considerable effort. Let's suppose then that our person can push off with a muscle-force equal to 1.5 times body-weight. Neglecting losses due to friction, how high can our friend jump; that is, how much above the approximately 1.0-m standing position can we expect the *c.m.* to rise?

43. [II] (a) Determine the value of the gravitational acceleration at the surface of the Moon. (b) What is the change in the *gravitational*-PE of a 1000-kg spacecraft when it has risen 100 m above the surface? Take the radius of the Moon to be $\frac{1}{4}$ that of the Earth and the mass to be about 100 times less.

44. [II] An object explodes into two pieces as, for example, when an atom emits an α-particle. Show that the ratio of their energies equals the inverse ratio of their masses—the smaller particle carries off the larger share of KE.

45. [III] A locomotive exerts a maximum force of 22.7×10^4 N while pulling a freight train up a $\frac{1}{2}$% grade (that is, the rise in the roadbed is $\frac{1}{2}$ m per 100 m). The train is 45 cars long, and each one loaded weighs 4.0×10^5 N. The total friction varies with the load and speed, but in this case, 35 N per 10^4 N of weight is a reasonable number. Using energy considerations, how far will the train travel while accelerating from a speed of 6.7 m/s to twice that?

46. [III] A runner accelerates to a top speed of 9.9 m/s taking a number of strides (N). If the average horizontal force exerted on the ground is 1.5 times body-weight, and if it acts during each stride over a length of $\frac{1}{3}$ m, determine the value of N.

47. [III] The second half of a smooth horizontal track is bent upward to form an incline at 20°. A solid homogeneous sphere 20 cm in diameter with a mass of 2.0 kg is rolled along the track so that it reaches the sloped portion at a speed of 5.0 m/s. Part of the sphere's energy of motion is associated with its rotation about its *c.m.*, and that quantity is given by the expression $\frac{1}{2}I\omega^2$. In addition, there is the usual KE $= \frac{1}{2}mv^2$ term. Neglecting friction, how far along the incline will the sphere progress before momentarily coming to rest?

CONSERVATION OF MECHANICAL ENERGY

48. [I] A 60-kg stuntperson runs off a cliff at 5.0 m/s and lands safely in the river 10.0 m below. What was the splashdown speed?

49. [I] Two automobiles of weight 7.12 kN and 14.24 kN are traveling along horizontally at 96 km/h when they both run out of gas. Luckily, there is a town in a valley not far off, but it's just beyond a 33.5-m-high hill. Assuming that friction can be neglected, which of the cars will make it to town?

50. [I] A ball having a mass of 0.50 kg is thrown straight up at a speed of 25.0 m/s. (a) How high will it go if there is no friction? (b) If it rises 22 m, what was the average force due to air friction?

51. [I] An athlete whose mass is 55.0 kg steps off a 10.0-m-high platform and drops onto a trampoline, which, while stretching, brings her to a stop 1.00 m above the ground. Assuming no losses, how much energy must have momentarily been stored in the trampoline as she came to rest? How high will she rise?

52. [I] While traveling along at 96 km/h, a 14.2-kN auto runs out of gas 16 km from a service station. Neglecting friction, if the station is on a level 15.2 m above the elevation where the car stalled, how fast will the car be going when it rolls into the station, if in fact it gets there?

53. [I] A 90-kg signal relay floating in space is struck by a 1000-g meteoroid. The latter imbeds itself in the craft and the two sail away at 5.0 m/s. What was the initial speed of the meteoroid?

54. [I] What is the potential energy of a 1.0-kg mass sitting on the surface of the Earth if we take the zero of PE_G at infinity?

55. [II] Referring to Problem 54, compute the *gravitational*-PE of a 1.0-kg object at distances of 1, 2, 3, 4, 5, and 10 Earth-radii. Draw a plot of PE_G against r measured from the center of the planet in units of Earth-radii. Now suppose the object is fired upward with a KE of 5.2×10^7 J. What is its initial total mechanical energy? Draw a horizontal line on your diagram representing E. Where does it cross your curve and what is the significance of that point?

56. [II] A 1000-kg car racing up a mountain road runs out of gas at a height of 35 m while traveling at 22 m/s. Cleverly, the driver shifts into neutral and coasts onward. Neglecting all friction losses, will he clear the 65-m peak? Would it help to throw out any extra weight or even jump out and run alongside the car? Not having any brakes, at what speed will he reach the bottom of the mountain?

57. [II] If the mass of the Moon is 7.4×10^{22} kg and its radius is 1.74×10^6 m, compute the speed with which an object would have to be fired in order to sail away from it, completely overcoming the Moon's gravity pull.

58. [II] A pendulum consists of a spherical mass attached to a rope so that its *c.g.* hangs down a distance L beneath the suspension point. The bob is displaced so that the taut string makes an angle of θ_i with the vertical, whereupon it is let loose and swings downward. Show that its maximum speed is given by

$$v_{max}^2 = 2gL(1 - \cos\theta_i)$$

59. [II] It's been suggested that we mine either the Moon or some of the asteroids in order to get raw material from which to build space stations. The idea is that removing material from the Moon "would consume only five percent of the energy needed to lift the same payload off Earth." Show that this conclusion is roughly true.

60. [II] At what speed should a space probe be fired from the Earth if it is required to still be traveling at a speed of 5.00 km/s, even after coasting to an exceedingly great distance from the planet (a distance that is essentially infinite)?

61. [II] A satellite is in a circular orbit about the Earth moving at a speed of 1500 m/s. It is desired that by firing its rocket, the craft attain a speed that will allow it to escape the planet. What must that speed be?

62. [II] In an arrangement for measuring the muzzle velocity of a rifle or pistol, the bullet is fired up at a wooden mass, into which it imbeds. The wood is blasted straight up into the air to a measured height h. Assuming negligible losses to friction, write an expression for the velocity in terms of the known masses and height. If a 100-grain (6.48-g) 25-06 Remington rifle bullet is fired into a 5.00-kg block that then rises 4.0 cm into the air, what was the muzzle speed of that bullet?

63. [II] An 8.0-kg puck floating on an air table is traveling east at 15 cm/s. Coming the other way at 25 cm/s is a 2.0-kg puck on which is affixed a wad of bubblegum. The two slam head-on into each other and stick together. Find their velocity after the impact.

64. [II] As seen from the window of a space station, a 100-kg satellite sailing along at 10.0 m/s collides head-on with a small 300-kg asteroid, which was initially at rest. Taking the collision to be elastic, and neglecting their mutual gravitational interaction, what are the final velocities of the two bodies? What were their "relative speeds" before and after the collision?

65. [II] Two billiard balls, one heading north at 15.0 m/s and one heading south at 10 m/s, collide head-on. Take the collision to be perfectly elastic. What is the post-impact speed of each ball? (Hint: It will help if you remember that for completely elastic collisions, the "relative speeds" before and after are equal.)

66. [II] A skater with a mass of 75 kg is traveling east at 5.0 m/s when he collides with another skater of mass 45 kg heading 60° south of west at 15 m/s. If they stay tangled together, what is their final velocity?

67. [III] A light rope is passed over a weightless, frictionless pulley and masses m_1 and m_2 are attached to its ends. The arrangement is called Atwood's Machine. The starting configuration corresponds to both masses held at rest at the same height. The two are then released. During some arbitrary interval of time, the heavier one falls a distance y, while the lighter one rises a distance y. Derive an expression for the speed of either body in terms of g, y, and the masses.

68. [III] The coefficient of restitution of two colliding bodies is defined as the ratio of their "relative speeds" (after the impact to before the impact). Imagine that we drop a sphere made of some material of interest from an initial height onto a test anvil and measure the final height to which the ball bounces. Derive an expression for the coefficient of restitution in terms of these two heights. If the coefficient for glass on steel is 0.96, to what height will a glass marble bounce off a steel plate when dropped from 1.0 m?

69. [III] A soft clay block is suspended so as to form a so-called ballistic pendulum, as shown in Fig. P69. A bullet is fired point-blank into the block, imbedding itself therein and raising the latter to a height h. Write an expression for the muzzle speed of the bullet in terms of the masses g and h. Remember that although friction on the pendulum is negligible, clay-bullet friction losses are not, and *mechanical energy is not conserved* in the collision.

Figure P69

70. [III] Referring to Problem 69, derive an expression for the percentage of the kinetic energy converted into internal energy during the bullet-clay impact.

71. [III] Two identical hard spheres, one at rest and the other moving, slam into one another in a *non-head-on* elastic collision, as shown in Fig. P71. Prove that they will always fly apart at 90° to each other.

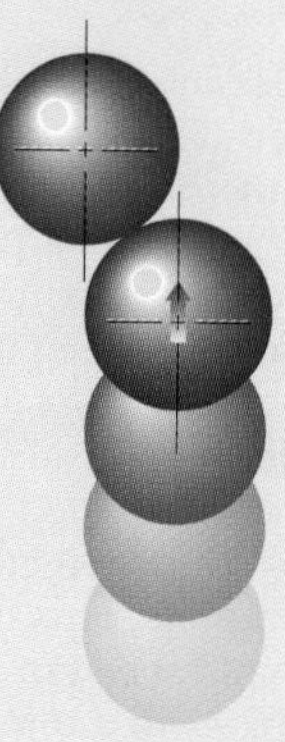

Figure P71

10

Solids

ONE OF OUR CONTINUING concerns in studying the physical Universe is *matter:* What is it? How does it behave, and why? This chapter is the beginning of an exploration into the mechanical, thermal, electrical, magnetic, and optical properties of matter. It develops a sketch of the atomic landscape, introducing ideas that will be elaborated throughout the remainder of the text.

The behavior of materials is of great practical interest. After all, few of us go through life without ever breaking a bone, chipping a tooth, or tearing a tendon. And although collapsing buildings, falling bridges, and toppling tow-

The fuselage of this Aloha Airlines jet ripped off during a flight in 1988. Experts attribute the failure to metal fatigue.

ers are a rarity, these things happen too, especially with some provocation like an earthquake. Boats still snap in half on the high seas and airplanes occasionally come apart in awkward ways. Our world is filled with material structures (trees, cars, people, skyscrapers, heart valves, and even the Earth itself), all of which we rely on to function in predictable ways. To understand why falling off a ladder might shatter a leg bone, we have to know a little about the materials. This chapter is a practical one, dealing as it does with the mechanical properties of solids: how they stretch and compress, fatigue, break, and shear. It examines an idea—Hooke's Law—that will have important ramifications throughout our study. That law describes elastic systems, and the atom itself is the most fundamental of all such systems. Atoms vibrate much as springs vibrate, and so as we study the common spring, we are laying the groundwork for far grander things to come.

ATOMS AND MATTER

It might seem easy to hold a thing in hand and know what matter is—it's stuff, substance, the material aspect of the Universe. But language fails us if we try to go beyond that intuitive apprehension. Matter cannot satisfactorily be defined, certainly not in terms of more fundamental concepts.

The observable Universe is composed of material particles and electromagnetic radiation, but these are the same, as much as they are different. Radiant energy may be thought of as "the most refined form of matter." A century ago, one could have responded that matter is that which possesses mass. But we now know that there are things called *neutrinos* (Sect. 33.5)—tiny, chargeless particles that carry momentum and energy, but which are nonetheless apparently massless.

It is probably as meaningless to discuss how much room an electron takes up as to discuss how much room a fear, an anxiety, or an uncertainty takes up.

SIR JAMES JEANS (1877–1946)

The old maxim that "matter is that which is impenetrable and occupies space" is also not very useful. A steel cannonball may appear solid enough, but it's mostly empty space. Bulk matter is composed of separated atoms and does not fill space in any ordinary sense. The particles that make up the atoms, that make up the cannonball, and everything else, are fluttery, elusive things that recoil from localization quite unlike the imagery of tiny billiard balls. The idea of atoms as ultimate impenetrable specks of solid matter is an ancient mirage.

Roughly 110 basic materials are known, and they are called **elements** (oxygen, iron, hydrogen, and so on). *These cannot be further simplified or broken down into another substance by any chemical or physical process* short of nuclear meddling. A sample of a given element is composed of identical submicroscopic entities called **atoms**, each of which is the smallest representative sample of the element. Chemical **compounds** are formed when elements combine. The basic unit of a compound, the smallest piece that behaves chemically like the whole, is a **molecule**, itself made up of atoms.

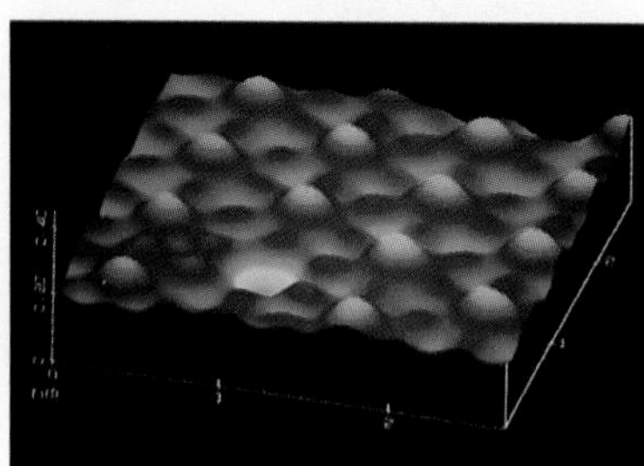
This regular pattern of bumps is formed by iodine atoms absorbed on a platinum surface.

An adult human being is a collection of over 10^{27} atoms bound together as $\approx 10^{14}$ cells. We are a constantly changing aggregate of atoms, themselves created thousands of millions of years ago. The living cells are recent enough, though almost without exception the atoms that form us are at least as old as the Solar System and often much older than that. They have circulated around the Earth for the last four-and-a-half thousand-million years, through air and water, through fish and fowl and trees and burgers and dung, and back to the soil and then, for the moment, to you. Formed in the thermonuclear fires of stars long gone, the atoms you ate for breakfast extend your lineage back in time to the dawn of creation. We *are* stardust, borrowers in the ancient ritual of "ashes to ashes."

10.1 Atomism

The concept of the *atom* is an old idea that dates back to the Greek Leucippus (ca. 450 B.C.). If a substance, a piece of gold, say, is divided over and over again into ever smaller portions, could the process go on forever, or would some last indivisible fragment—the *atomos (that which cannot be further cut)*—be reached? The most influential early opponent of atomism was Aristotle, and not until his hold on Western thought started to collapse in the seventeenth century was atomism again taken seriously. Robert Boyle embraced the concept of the atom as an explanation of the "springiness of air," (in a balloon, for example). In 1665, Robert Hooke proposed that certain crystalline structures resulted from the close packing of tiny spherical particles—atoms stacked like cannonballs. Still others were convinced of the reality of atoms by the nature of the mixing process. That salt dissolves into water suggests that the liquid, instead of being continuous, is rife with empty spaces. A cup of alcohol and a cup of water will combine to form less than two cups of "good cheer"—the atoms seem to be sliding in between one another, filling holes. A droplet of ink released into a beaker of water will spread out uniformly, just as the smell of perfume can fill a room in minutes, suggesting a mingling of colliding particles.

A few droplets of ink in a beaker of water. The warmer the water, the faster the ink diffuses.

Despite the long list of believers—Galileo, Bacon, Boyle, Hooke, Newton, and Leibniz, to name a few—the quantitative evidence in favor of the atom was far from impressive. The breakthrough came in the early 1800s at the hands of an unknown country schoolmaster, a gruff, awkward man named John Dalton. It had already been shown that *elements combine to form compounds in certain definite proportions.* One gram of hydrogen burned with eight grams of oxygen combine to produce nine grams of water (H_2O). Dalton recognized that these definite proportions were just the observed manifestations of an invisible atomic structure. He argued that if water is formed of equal numbers of hydrogen and oxygen atoms (and here he was wrong) in a mass ratio of 1 to 8, then each O atom must be 8 times as massive as an H atom. Suspecting, and rightly so, that hydrogen was the lightest of the atoms, he set its mass equal to 1. The mass of an oxygen atom became 8, and so on down the list of elements involved in the known chemical reactions.

The table of *relative atomic masses* that Dalton drew up had many errors, but they were soon corrected. Water was electrically broken down and its true formula established: one oxygen atom has 8 times the mass of *two* hydrogens—that is, 8×2, or 16. Nowadays, we take carbon (^{12}C) as the standard instead of hydrogen, set its mass equal to exactly 12.000 00 *unified atomic mass units* (u), and measure all the other atoms relative to it (Table 32.1). Thus, a hydrogen atom has a mass of 1.007 825 u.

10.2 Avogadro's Number

In 1811, Amedeo Avogadro, Count of Quaregna, set forth two brilliant proposals (coining the word *molecule* in the process): (1) *The gaseous elements can exist in molecular form.* (2) *Equal volumes of molecular gases (under the same conditions of temperature and pressure) contain the same number of molecules.*

Molecular hydrogen (H_2) has a relative mass of 2. Suppose we admit 2 g of it into an expandable chamber that can be maintained at **standard temperature** (0°C)

A gram-mole of mercury, water, glass, salt, and iron. The glass marble can be used for establishing the scale.

and pressure (1 atm)—STP. We would find that the gas occupied a volume of 22.4 liters (that is, 22.4×10^3 cm^3). According to Avogadro, 22.4 liters of oxygen, or of any gas, would contain the same number of molecules. Since each O_2 molecule has a mass of 2×16 u, or 32 u, we expect any volume of oxygen to be 16 times more massive than the same volume of hydrogen. Consequently, 22.4 liters of oxygen should have a mass of 32 g. Indeed, *22.4 liters of any gas at STP has a mass in grams numerically equal to its molecular mass.* This highly useful measure is called a *gram-molecular mass,* and it's often spoken of as a **mole**. For instance, one molecule of carbon dioxide (CO_2) has a relative mass of 12 u + 2×16 u or 44 u; therefore, 1 mole of it has a mass of 44 grams and, as a gas at STP, occupies 22.4 liters.

Similarly, *a* **kilomole** *is the mass of any substance in kilograms that is numerically equal to the molecular mass of that substance.* Thus, a kilomole (abbreviated kmol) of any gas at STP occupies 22.4 m^3. Many elements exist as collections of separate atoms rather than molecules, and the word *kilomole* then stands for a *kilogram-atomic mass.*

We could start out with 22.4 m^3 of a gas and cool it down to a liquid or freeze it into a solid and still have a kilomole. Thus, a kilomole of uranium, a shiny 238-kg block, has the same number of atoms as there are molecules in 44 kg of carbon dioxide, whether it is gaseous or solid (dry ice). To honor the good Count, the number of molecules in a mole is known as **Avogadro's number** (N_A), though he himself had no idea of its numerical value. The best current measure of that quantity is

$$N_A = (6.022\,136\,7 \pm 0.000\,003\,6) \times 10^{23} \text{ molecules/mole}$$

Example 10.1 Determine (a) the mass of a carbon atom in kilograms and (b) the equivalent of one atomic mass unit in kilograms. Use both sets of units.

Solution: [Given: the mass of a carbon atom is 12.00 u, and a mole or 12.00 g of carbon contains N_A atoms. Find: the mass of a carbon atom in kg, and the equivalent of 1.000 u in kg.] The mass of a carbon atom equals the mass of a mole of carbon divided by the number of atoms in a mole; thus

$$m_C = \frac{12.00 \text{ g}}{6.022 \times 10^{23}} = 1.993 \times 10^{-23} \text{ g}$$

and $$\boxed{m_C = 1.993 \times 10^{-26} \text{ kg}}$$

Since 12.00 u = $m_C = 1.993 \times 10^{-26}$ kg

$$1.000 \text{ u} = \frac{1.993 \times 10^{-26} \text{ kg}}{12} = \boxed{1.661 \times 10^{-27} \text{ kg}}$$

▶ **Quick Check:** In SI units, $N_A = 6.022 \times 10^{26}$ molecules/kilomole and

$$m_C = \frac{12.00 \text{ kg}}{6.022 \times 10^{26}} = 1.993 \times 10^{-26} \text{ kg}$$

10.3 Density

One of the more evident features of bulk matter is that different substances are not packed with equal density—a kilogram of lead has a much smaller volume than a kilogram of feathers. When dealing with the behavior of materials rather than individual objects, it's often more convenient to use quantities that are *independent of volume.* Thus, we define the **density** of any material *as the mass per unit volume:*

$$\text{Density} = \frac{\text{mass}}{\text{volume}}$$

TABLE 10.1 Densities of some materials

Substance	Density (kg/m^3)	Substance	Density (kg/m^3)
Interstellar space	10^{-18} to 10^{-21}	Chloroform	1.53×10^3
Hydrogen*	0.090	Sugar	1.6×10^3
Oxygen	1.43	Magnesium	1.7×10^3
Helium	0.178	Bone	$(1.5–2.0) \times 10^3$
Air, dry (30°C)	1.16	Clay	$(1.8–2.6) \times 10^3$
Air, dry (0°C)	1.29	Ivory	$(1.8–1.9) \times 10^3$
Styrofoam	0.03×10^3	Glass	$(2.4–2.8) \times 10^3$
Balsa wood	0.12×10^3	Cement	$(2.7–3.0) \times 10^3$
Cork	$(0.2–0.3) \times 10^3$	Aluminum	2.7×10^3
Pine wood	$(0.4–0.6) \times 10^3$	Diamond	$(3.0–3.5) \times 10^3$
Oak wood	$(0.6–0.9) \times 10^3$	Moon	3.34×10^3
Ether	0.74×10^3	Planet Earth, average	5.25×10^3
Ethyl alcohol	0.79×10^3	Iron	7.9×10^3
Acetone	0.79×10^3	Nickel	8.8×10^3
Terpentine	0.87×10^3	Copper	8.9×10^3
Benzene	0.88×10^3	Silver	10.5×10^3
Butter	0.9×10^3	Lead	11.3×10^3
Olive oil	0.92×10^3	Mercury	13.6×10^3
Ice	0.92×10^3	Uranium	18.7×10^3
Water (0°C)	$0.999\,87 \times 10^3$	Gold	19.3×10^3
Water (3.98°C)	$1.000\,00 \times 10^3$	Tungsten	19.3×10^3
Water (20°C)	$1.001\,80 \times 10^3$	Platinum	21.5×10^3
Tar	1.02×10^3	Osmium	22.5×10^3
Seawater	1.025×10^3	Pulsar	$10^8–10^{11}$
Blood plasma	1.03×10^3	Nuclear matter	$\approx 10^{17}$
Blood, whole	1.05×10^3	Neutron star, core	$\approx 10^{18}$
Ebony wood	$(1.1–1.3) \times 10^3$	Black hole (1 solar mass)	$\approx 10^{19}$
Rubber, hard	1.2×10^3		
Brick	$(1.4–2.2) \times 10^3$		
Sun, average	1.41×10^3		

*Gases are at 0°C and 1 atm unless otherwise indicated.

The symbol for density is the lowercase Greek letter ρ (rho), and so

$$\rho = \frac{m}{V} \tag{10.1}$$

Density is a characteristic of a substance; mass is a characteristic of an object (Table 10.1).

At one time, the centimeter-gram-second (cgs) unit of g/cm^3 was widely used; it still persists where small quantities of substances are dealt with, as in medicine. Notice that since

$$1 \text{ g/cm}^3 = \frac{1 \times 10^{-3} \text{ kg}}{(1 \times 10^{-2} \text{ m})(1 \times 10^{-2} \text{ m})(1 \times 10^{-2} \text{ m})} = 10^3 \text{ kg/m}^3$$

the numerical values only differ by a factor of 10^3. Because the gram was defined as the mass of a 1-cm^3 volume of water, the density of water is 1 g/cm^3, or 10^3 kg/m^3. The pungent-smelling metal osmium, whose atoms are massive at 190 u (though not the most massive), is so closely packed that it is the densest substance on Earth.

The pale blue, almost transparent solid resting on a mound of shaving cream is silica aerogel. Its density is only about 3 times that of air and yet it can support a load ≈1600 times its own weight.

Example 10.2 Determine the mass of a gold sphere 0.10 m in diameter. How many atoms are in it? The atomic mass of gold is 197 u.

Solution: [Given: $D = 0.10$ m, from Table 10.1 $\rho = 19.3 \times 10^3$ kg/m^3. Find: m and the number of atoms N.] The volume of the sphere

$$V = \tfrac{4}{3}\pi R^3 = \tfrac{4}{3}3.1416\ (0.050\ \text{m})^3 = 0.524 \times 10^{-3}\ \text{m}^3$$

together with the density provides the mass, which is

$$m = \rho V = (19.3 \times 10^3\ \text{kg/m}^3)(0.524 \times 10^{-3}\ \text{m}^3)$$

and $\boxed{m = 10\ \text{kg}}$

A kilomole of gold is 197 kg, so this sphere is the equivalent of 0.051 27 kmol and therefore

$$N = N_A\, 0.051\,27 = (6.022 \times 10^{26}\ \text{atoms/kmol}) \times (0.051\,27\ \text{kmol}) = \boxed{3.1 \times 10^{25}}$$

▶ **Quick Check:** $(3.1 \times 10^{25}$ atoms$)(197$ kg·kmol$^{-1})/(6.022 \times 10^{26}$ atoms/kmol$) = 10$ kg.

10.4 The Size of Atoms

By the 1880s, a number of experiments had established a rough measure of the size of atoms and molecules. One of the easiest to appreciate takes a known volume of oil and allows it to spread out on the surface of water. There are good reasons to assume that the resulting film would thin out until it was only one molecule thick. Measuring the area of the slick gives us the thickness of the film, and therefore the diameter of the oil molecule. Nowadays, we know that *atoms of every kind are nearly the same size, varying in diameter from roughly 1×10^{-10} m to about 3×10^{-10} m.*

We can get a rough estimate of the dimensions of atoms and molecules in a solid or liquid by ignoring the space between them, assuming that they "fill" the volume of the material, however tenuously. They do not, but there cannot be a great deal of interatomic space since neither solids nor liquids can be compressed very much. If we take each atom as a little cube of volume L^3, the volume V of a mole would then be $N_A L^3$. If the mass of that mole is M_m

$$\rho = \frac{M_m}{V} = \frac{M_m}{(N_A L^3)}$$

and

$$L = \left(\frac{M_m}{N_A \rho}\right)^{\frac{1}{3}} \tag{10.2}$$

This means that a water molecule should have a size of very roughly

$$L = \left[\frac{18\ \text{g}}{(6.022 \times 10^{23}\ \text{atoms/mole})(1.0\ \text{g/cm}^3)}\right]^{\frac{1}{3}}$$

which is 3×10^{-10} m, or 0.3 nm. Interestingly, if we do this same calculation for most simple solids and liquids, they generally turn out to have atomic dimensions within a factor of 2 or so of water (a result confirmed by X-ray scattering). For example, $L = 0.5$ nm for ethyl alcohol.

Example 10.3 Assuming that the atoms in a solid are packed shoulder to shoulder, and taking a typical diameter to be 0.2 nm, (a) how big a cube corresponds to a mole of almost any solid? (b) What is a typical value for the molar volume (that is, the volume of a mole) of a solid?

(continued)

(continued)

Solution: [Given: $D = 0.2$ nm, and N_A. Find: the volume of a 1-mole cube.] (a) The cube root of 6.022×10^{23} is 8.445×10^7, and this must be the number of atoms along each edge of the cubic volume. Each atom is 0.2 nm in diameter, so the edges must be $(8.445 \times 10^7)(0.2 \times 10^{-9}\text{ m}) = 16\text{ mm} \approx 2\text{ cm}$ long. (b) And the molar volume is approximately 8 cm^3.

▶ **Quick Check:** 1 mole of gold has a mass of 197 g and a volume of

$$V = \frac{m}{\rho} = \frac{(197\text{ g})}{(19.3\text{ g/cm}^3)} = 10\text{ cm}^3$$

which is close to our approximation.

The numbers of atoms in even modest amounts of material are incredibly vast. If we took all the molecules in a tablespoon of water, about 18 g or 1 mole, and put them end to end, the column would be $(6.022 \times 10^{23})(3 \times 10^{-10}\text{ m}) = 1.8 \times 10^{14}$ m long and would stretch from here to the Sun (1.495×10^{11} m) and back over 600 times! There are more molecules in a tablespoon of water than there are stars ($\approx 10^{22}$) in all the Universe.

The atom is actually a complex structure, a thing of interacting, whirling parts, many of whose intricacies evade us still. Each atom consists of a dense, minute core—the *nucleus*—surrounded by a tenuous mist of electrons. The nucleus, whose diameter is only about one ten-thousandth the diameter of the atom, is a ball of rapidly moving, positively charged protons and neutral neutrons bound together by the strong force (Sect. 32.3). The electrons, which tumble around at tremendous speeds in the outer reaches of the atom, are retained by the attractive electromagnetic force (Fig. 10.1).

Since each proton and neutron is roughly 2000 times more massive than an electron, a typical nucleus is many thousands of times more massive than the accompanying electrons. The tiny atomic core where practically all the mass resides is surrounded by a far-distant, relatively massless, negative wisp of charge. Thus, the space occupied by the atom as a whole is mostly devoid of matter.

We no longer envision neat little well-defined electron orbits. Instead, the atomic landscape is calculated in terms of the patterns in which the electrons are likely to reside, patterns known as electron clouds. The diameter of an atom is not perfectly defined because its electrons are in constant shifting motion. Indeed, an atom's size varies slightly with the different kinds of experiments used to measure it.

10.5 The States of Matter

Atoms in solids and liquids can be imagined as tightly packed (that is, cloud to cloud), whereas in gases they are quite a bit farther apart—gases under ordinary conditions are about a million times more compressible than solids. The regions between the nuclei of adjacent atoms in all forms of bulk matter are comparatively immense, and comparatively empty; that's true even within solids. If you could drink Alice in Wonderland's growth potion and swell up until the core of an atom in your body were the size of a cherry pit, you would stand about a thousand million kilometers tall! And most of your huge volume would be void, with hundreds of meters of emptiness separating the pit-sized concentrations of matter. Atoms combine under the influence of the electromagnetic force to produce all the

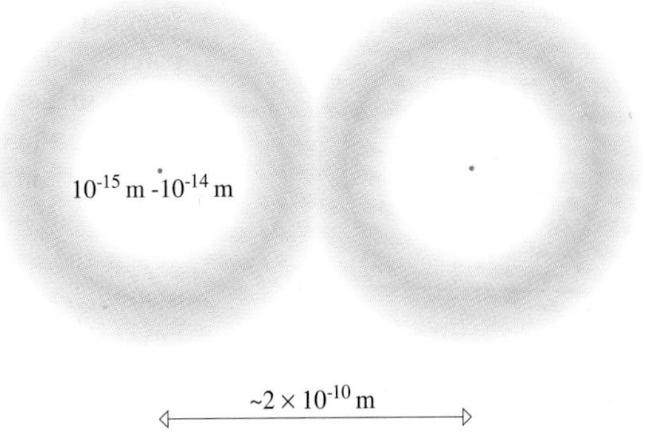

Figure 10.1 The atoms in a solid are roughly 0.2 nm apart, center-to-center, and have almost all their mass concentrated in the tiny 10^{-15} m to 10^{-14} m nuclei.

10^{-15} m - 10^{-14} m

~2×10^{-10} m

TABLE 10.2 Molecular number-densities

Substance	State	Number-density (molecules/m^3)
Gold	solid	5.9×10^{28}
Ice	solid	3.1×10^{28}
Water	liquid	3.3×10^{28}
Mercury	liquid	4.2×10^{28}
Oxygen	gas	27×10^{25}
Air	gas	27×10^{25}

things of our world. The kinds of atoms grouping together determine the strengths of their electrical interactions, which define the ultimate structure of the substance. Under ordinary planet-Earth conditions, matter exists as solid, liquid, gas, or plasma. If the interatomic forces are strong enough, the collection of particles independently *maintains its shape and volume.* Along with its elastic properties, that ability to maintain shape is a distinguishing feature of a **solid**. A **liquid** is characterized by weaker binding, so that *it flows, assuming the shape of its container, although it maintains a constant volume regardless of that shape.* If the forces are still weaker, the material exists as a **gas**; the atoms or molecules tend to disperse, and *the substance flows, assuming both the shape and volume of its container.* If some of the atoms of a gas are *ionized* (that is, electrons are removed or added) *the mix of atoms, ions, and electrons is a conducting fluid* called a **plasma**.

Some materials can exist as solid, liquid, or gas and readily shift from one form, or **phase**, to another and back, depending on the prevailing conditions of temperature and pressure. Both internal motion and disorder progressively increase as a sample transforms from the solid to the liquid and then to the gaseous state. Most solids become less dense by a few percent on melting (ice is a notable exception, being less dense than water). The space between molecules thereby increases, but not by very much. On the other hand, a tremendous decrease in density results when a liquid is transformed to a gas. Table 10.2 shows the effect in terms of *number-density,* the number of molecules per cubic meter, which provides a sense of the intermolecular spacing.

10.6 Solids

Many more materials in our moderate-temperature environment are in the solid state than in either the liquid or gaseous states. One way to classify these solids is as either *crystalline, quasi-crystalline, amorphous,* or *composite.* As with liquids and gases, the atoms of a solid are in motion, but it's a small jiggling, vibratory motion about fixed, closely spaced positions. When the atoms arrange themselves in *an orderly, three-dimensional pattern that repeats itself over and over again,* the solid is **crystalline**. Most minerals and all metals and salts exist in at least one crystalline configuration.

If a substance is composed of atoms that are *not arranged in any orderly and repetitive array,* the solid is said to be **amorphous**. Rubber, pitch, resin, plastics, and various glassy materials are examples. The atoms of glasses are strung out in an irregular mesh that lacks any *long-range* order (Fig. 10.2). What order there is, is patchy, localized, and nothing like the neat repetitive array of row upon row of atoms in a crystal. In contrast, the long-chain molecules of rubber and many plastics resemble the entwined, but still not total, chaos of a bowl of spaghetti.

Between these two long-recognized traditional configurations—crystalline and amorphous—there is a new form of solid matter that was only discovered in 1983. When the atoms arrange themselves in *an orderly three-dimensional pattern that nonetheless does not repeat itself with a regular periodicity,* the solid is **quasi-crystalline**.

And lastly, **composite** solids such as wood, concrete, fiberglass, bone, and blood vessel walls are composed of several different materials bonded together.

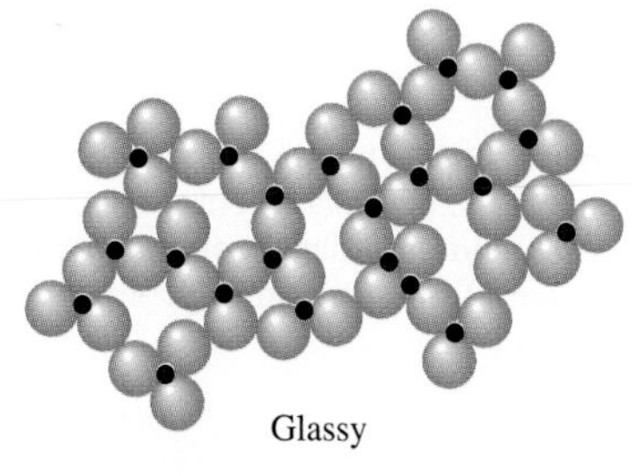

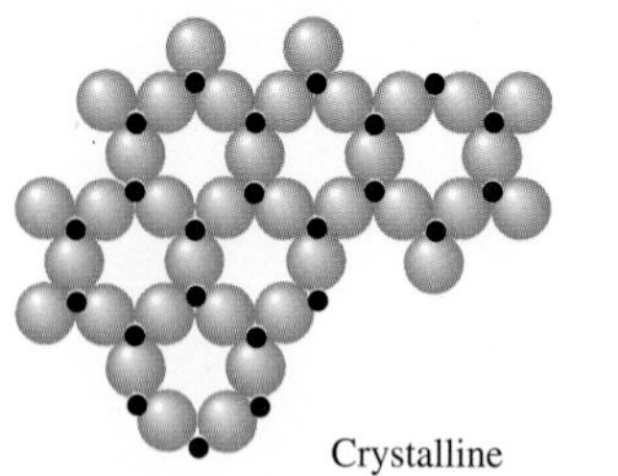

Figure 10.2 Glassy and crystalline solids—short- and long-range order.

Crystals

Many materials are made up of tiny crystal grains, each no more than a fraction of a millimeter in size. Indeed, ordinary metal objects (knives, forks, and hubcaps) are all **polycrystalline** (many-crystalled).

A distinguishing property of most crystalline solids is their sharp transition to the liquid state at a well-defined constant temperature. Crystals melt suddenly. Because the distances between atoms are well defined, the interatomic bonds that hold the crystal together are all nearly equal in strength, and all rupture almost simultaneously. This behavior is unlike that of a noncrystalline material (for example, glass or tar) that melts gradually over a broad range of increasing temperature.

Zinc crystals on a galvanized parking lot rail guard. Crystals like these can be seen on streetlamps and garbage cans. Old brass doorknobs etched by sweat and years of handling also show a fine-grain polycrystalline structure.

The smooth, flat, angular faces of crystals reflect their internal order, an order that can be appreciated by imagining the atoms arranged in minute, identical building blocks. A particular basic grouping of atoms forms a type of building block. Countless such blocks are then stacked one upon another, filling the space of the sample and forming its overall geometrical design. There are only seven basic structures (Fig. 10.3) from which all crystals can be constructed. The key point here is that crystals are filled, their atoms are packed in, leaving no large empty regions. These imagined blocks must also completely fill the space when they are stacked, and not every nice geometrical figure does that. The idea is easy to see in two dimensions by just considering a tiled floor. Without leaving empty spaces between

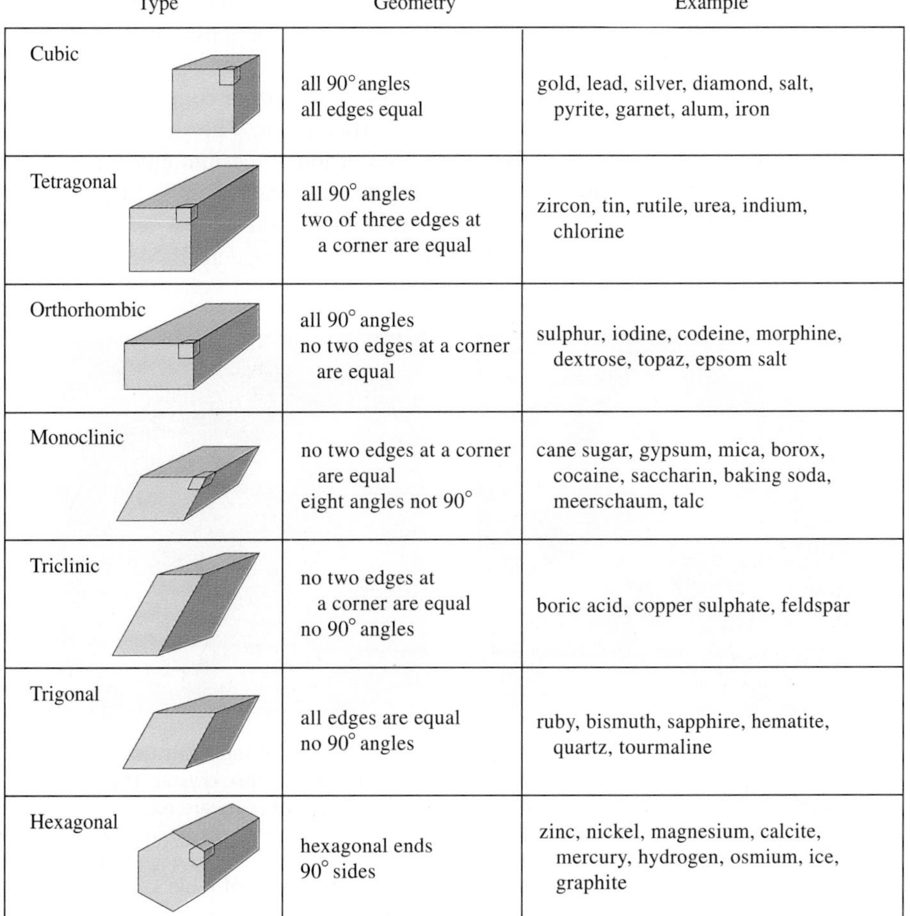

Type	Geometry	Example
Cubic	all 90° angles all edges equal	gold, lead, silver, diamond, salt, pyrite, garnet, alum, iron
Tetragonal	all 90° angles two of three edges at a corner are equal	zircon, tin, rutile, urea, indium, chlorine
Orthorhombic	all 90° angles no two edges at a corner are equal	sulphur, iodine, codeine, morphine, dextrose, topaz, epsom salt
Monoclinic	no two edges at a corner are equal eight angles not 90°	cane sugar, gypsum, mica, borox, cocaine, saccharin, baking soda, meerschaum, talc
Triclinic	no two edges at a corner are equal no 90° angles	boric acid, copper sulphate, feldspar
Trigonal	all edges are equal no 90° angles	ruby, bismuth, sapphire, hematite, quartz, tourmaline
Hexagonal	hexagonal ends 90° sides	zinc, nickel, magnesium, calcite, mercury, hydrogen, osmium, ice, graphite

Figure 10.3 Crystal building blocks.

Figure 10.4 (a) The packing of regular objects with 3, 4, 5, 6, 7, and 8 sides in two dimensions. (b) Regular hexagons fill the flat two-dimensional space of this floor.

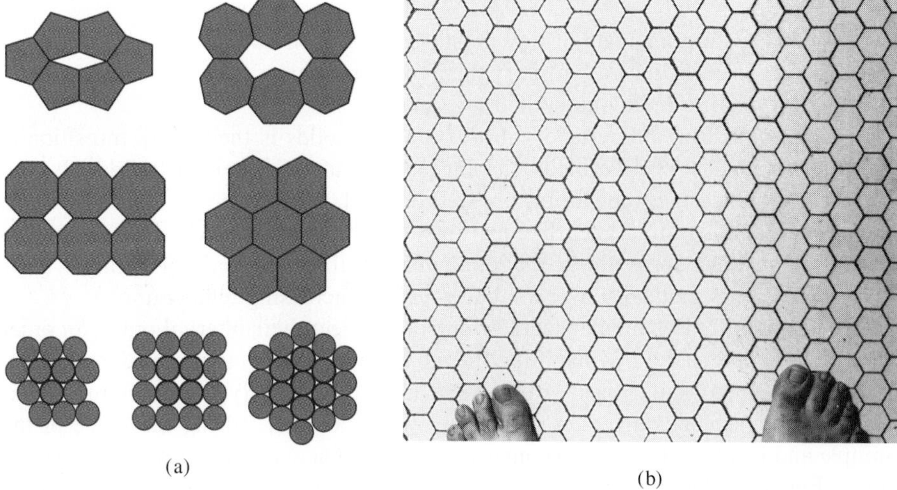

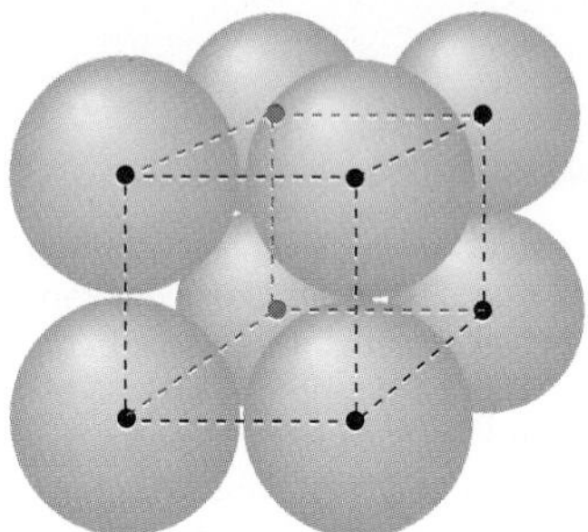

Simple cubic (SC)

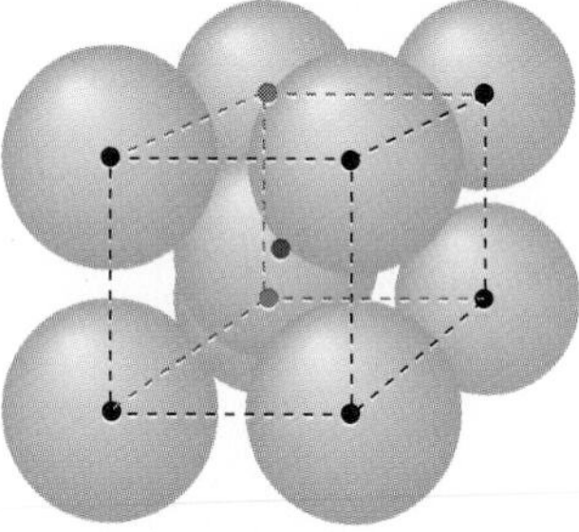

Body-centered cubic (BCC)

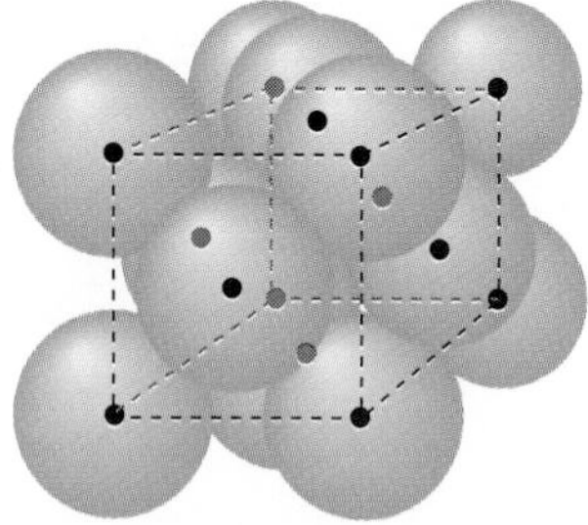

Face-centered cubic (FCC)

Figure 10.5 The three basic variations of the cubic crystal form.

adjacent tiles, we can cover the floor with triangles, squares, rectangles, parallelograms, or hexagons. But it cannot be done with regular pentagons or any other shape (Fig. 10.4). There isn't a washroom in the Universe with a flat floor tiled solid in white pentagons.

The simplest building block is cubic, and there are three variations on the theme (Fig. 10.5). The points at the corners, faces, and center of these cells are *lattice points.* They locate the positions of the atoms or molecules of the crystal. In Fig. 10.5 a single atom is seen at each lattice point, although in complex materials a cluster of hundreds of atoms may occupy each lattice point. The simple cubic (SC) structure is not uncommon with solids composed of ions—cesium chloride (CsCl), for instance—and metallic alloys. Of the elements, only polonium is SC, and then only in a particular temperature range. If we imagine the atoms as "touching" spheres, the SC structure can fill only about 52% of the space available. More effi-

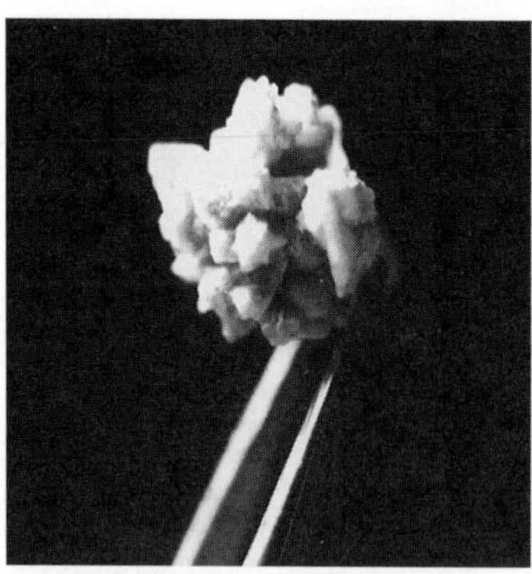

This polycrystalline mass of calcium carbonate ($CaCO_3$) grew in a human kidney until it rather painfully found its way down and out. Fortunately this photo is about six times life-size.

A single crystal of table salt (NaCl). It's a face-centered cubic crystal. The atomic structure (a square roughly 0.56 nm on a side) has two ions, a sodium and a chlorine, at each lattice point. Most brands of table salt come as tiny cubes.

cient, at 68% packing, is the body-centered cubic (BCC) structure utilized by several metals, including iron and sodium. The densest packing, 74%, is achieved in the face-centered cubic (FCC) arrangement assumed by many elemental metals including silver, gold, nickel, copper, aluminum, and lead.

There are two ways to stack spheres (or atoms) so as to minimize the unfilled, or interstitial, volume. One arrangement is the face-centered cubic, and the other, which also fills 74% of the volume (Fig. 10.6), is the hexagonal close-packed (HCP). There is nothing very exotic about this arrangement; it's the way many things are usually stacked—from plums to cannonballs. We can expect that atoms that are attracted uniformly in all directions will have the spherical symmetry of cannonballs and so pack relatively densely. Those that do not combine into asymmetrical molecules usually form densely packed crystals; atoms of the metals and rare gases (helium, neon, and argon) are typical. Thus, argon is face-centered cubic, while zinc is hexagonal close-packed.

Figure 10.6 The hexagonal close-packed (HCP) arrangement of atoms.

ELASTICITY

Galileo began his book, *Two New Sciences* (1638), with a discussion "treating of the resistance which solid bodies offer to fracture." By the middle of the 1600s, there were some other tentative studies of the strengths of wires and rods, but it remained for that remarkably homely, lascivious genius, Robert Hooke, to really get things going in what they called "*ye* science of *elasticity*"—that is, the analysis of the behavior of materials and structures under the influence of applied forces.

10.7 Hooke's Law

In 1676, wishing to establish priority for some new discoveries, Hooke published two cryptic anagrams. Two years later he decoded them, announcing the invention of the spring balance and the discovery of what has since come to be known as **Hooke's Law**. When an object is acted on by a force, it may be compressed, stretched, or bent. If, after the force is removed, the object rapidly returns to its original configuration, we say that it is *elastic*. If the force is not too large, materials such as steel, bone, rock, tendon, glass, flesh, and rubber are elastic. The metal hair-spring in a pocket watch coils and uncoils elastically over 170 000 times a day. By contrast, solids or quasi-solids, such as chewing gum, lead, moist clay, and putty, do not return to their original configurations once distorted. These materials are categorized as *plastic*.

An HCP stack of plums. Notice how each plum is surrounded by six nearest neighbors.

What Hooke found was that, besides being elastic, **many materials deform in proportion to the load they support**. That's Hooke's Law, and it applies to *linearly elastic* or *Hookean* objects—to steel bars, rods, and wires, to springs, diving boards, and more or less to rubber bands, among other things. Thus, if an ordinary coiled (helical) spring in equilibrium ($\Sigma F = 0$) is either stretched or compressed by an applied force F, the spring will be deformed (extended or contracted) by a distance s that is proportional to the force: $F \propto s$ (Fig. 10.7). Doubling the stretching force doubles the extension of the spring.

When one parameter is proportional to another, the relationship can be transformed into an equality by introducing a constant-of-proportionality. Here, this constant, k, takes care of the differences between units on both sides of the equation. It introduces the physical characteristics of that particular spring into the formula. Presumably, if we knew more physics, we could express the constant in terms

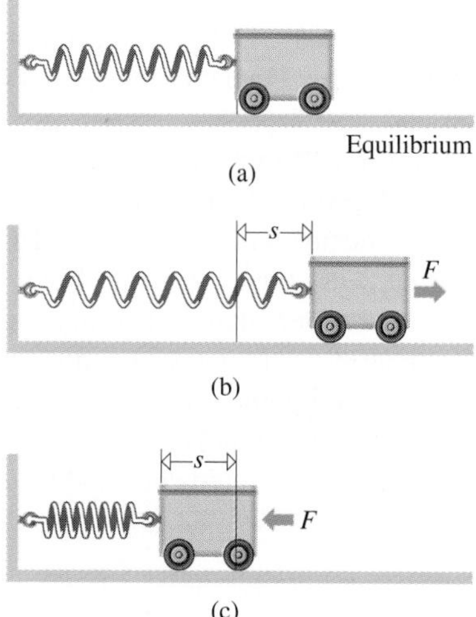

Figure 10.7 A force F applied to a spring (a) can either (b) stretch or (c) compress it by a distance s. Provided s is not large, $F = ks$.

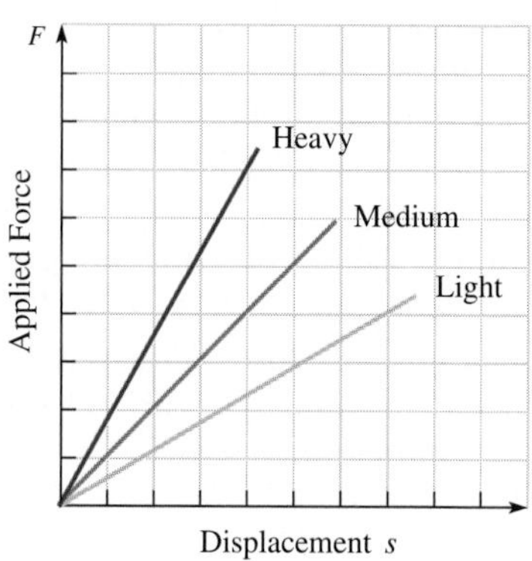

Figure 10.8 The force F applied to a spring stretches it a distance s. This figure is a plot of F versus s for three different springs. The slope of each curve is the spring's elastic constant, k. The less rigid the spring (heavy, medium, and light), the smaller the value of k.

Hooke's Law. *The Power [that is, force] of any Spring is in the same proportion with the Tension [meaning, at the time,* extension*] thereof; That is, if one power [force] stretch or bend it one space, two will bend it two, three will bend it three and so forward.*

ROBERT HOOKE

of the atomic parameters that give rise to it, but that's still beyond our means. Hooke's Law is then

$$F = ks \tag{10.3}$$

This is the expression for the force that must be exerted *on* the spring to produce a corresponding displacement: when F is positive, s is positive. From Newton's Third Law, just the opposite would be true for the force exerted *by* the spring (Fig. 9.7, p. 284). Because this force always tends to return the system to its equilibrium configuration, it's called a *restoring force.*

The **spring constant** or **elastic force constant** k is defined by Eq. (10.3). It has the dimensions of force over distance and the units of N/m. *For a given force, the larger the value of k, the smaller the extension of the spring—that is, the more rigid the spring* (Fig. 10.8). The greater the slope (k) of the force-displacement curve, the stiffer the spring, rod, or diving board. The elastic constant varies with the size and shape of the object as well as with the material: the element, compound or alloy, its impurities, heat treatment, crystal structure, and so on.

Example 10.4 A spring balance (Fig. 10.9) has a 10-cm graduated scale that can read mass from zero to 500 g. How does the device work? Determine the spring constant. What displacement would the pointer experience if you pulled down on the support hook with 10 N? Would this balance be useful on the Moon?

(continued)

(continued)

Solution: [Given: mass range 0–500 g, displacement 0–10 cm, and a load of 10 N. Find: k and then s for 10 N.] The spring scale indicates s, which is proportional to F, or the weight of the load. On Earth, that force is more or less constant and proportional to the constant mass of the load and so, if properly calibrated on this planet, such a scale can be used to measure mass. If the scale was set to read newtons directly, it would function perfectly on the Moon. However, when set to read grams, it is assumed the device is on Earth, where $F_W/(9.8\ \text{m/s}^2) = m$; hence, it would have to be recalibrated at 1/6th scale.

The weight of a 500-g mass is

$$F_W = mg = (0.500\ \text{kg})(9.8\ \text{m/s}^2) = 4.9\ \text{N}$$

producing a displacement of 10 cm, or 0.10 m; hence

$$k = \frac{F}{s} = \frac{4.9\ \text{N}}{0.10\ \text{m}} = \boxed{49\ \text{N/m}}$$

If you pulled down on the balance with a force of 10 N, a displacement

$$s = \frac{F}{k} = \frac{10\ \text{N}}{49\ \text{N/m}} = \boxed{0.20\ \text{m}}$$

would be necessary, which for this device is physically impossible because it's only 10 cm long.

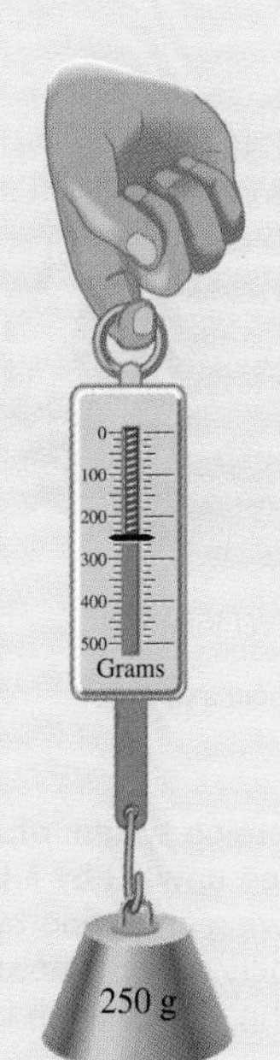

Figure 10.9 A spring scale much like the one invented by Robert Hooke. A load either stretches or compresses a spring, changing its length proportionately.

▶ **Quick Check:** Each 1 cm corresponds to 500 g/10 = 0.050 kg; hence, $k = F/s = mg/s = (0.050\ \text{kg}) \times (9.8\ \text{m/s}^2)/(0.010\ \text{m}) = 49\ \text{N/m}$.

You may have seen a Slinky mercilessly stretched until portions of it became distorted, unable to return to its original unstretched condition. Any rod, girder, spring, or wire can be bent or stretched beyond its **elastic limit** so that it becomes permanently deformed, losing its springiness. In such a sorry state, it will no longer obey Eq. (10.3); it will no longer be *linearly* elastic. Accordingly, those nice straight-line curves in Fig. 10.8 don't go on forever.

Elastic Potential Energy

One of the practical functions served by springs is to store energy (Fig. 10.10). The mainspring in a clock (a device invented by Hooke) replaces the *gravitational*-PE of a falling weight or pendulum with the *elastic*-PE of a spring. As we saw in Chapter 9, *the work done on a system against a nonconstant force, such as the force of a spring, is the area under the force-displacement curve which, if the force is conservative, equals the change in potential energy of the system.* Thus, for any of the plots in Fig. 10.8, the *elastic*-PE (PE_e) stored in stretching the spring a distance s is the area of the triangle under the line—namely, $\frac{1}{2}$ the base (s), times the corresponding altitude ($F = ks$):

$$\Delta\text{PE}_e = \tfrac{1}{2}s(ks)$$

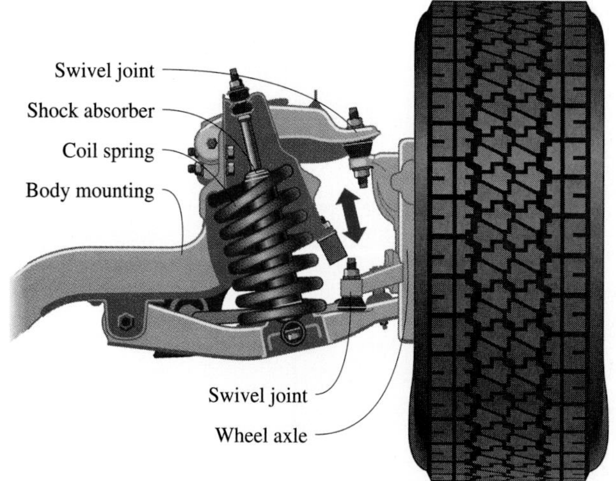

Figure 10.10 The suspension system of an automobile. When the wheel of a car is jolted upward by a bump in the road, a coil spring is compressed, slowing and smoothing out the vertical motion. The shock absorbers convert the stored mechanical energy to thermal energy, keeping the car from bouncing on the springs.

Heavy trucks use leaf springs and shock absorbers to smooth out the ride. The leaf spring is a stack of thick steel bars of decreasing length strapped together. They are slightly bent so that an impulsive vertical force tends to straighten them out.

and so

$$\Delta PE_e = \tfrac{1}{2}ks^2 \tag{10.4}$$

This is the same as saying that the work done by the displacement-dependent spring-force is its average value, $F_{av} = \frac{1}{2}(F_f + F_i) = \frac{1}{2}(F_f + 0) = \frac{1}{2}ks$ times the displacement s: $W = F_{av}s = \frac{1}{2}ks^2 = \Delta PE_e$.

Example 10.5 Four people each with a mass of 70 kg enter a car of mass 1100 kg—two in front and two in back. The vehicle settles down in a little while on its four identical springs, compressing them a distance of 2.5 cm. Determine the elastic constant of each spring and the amount of energy stored in each. What is the equivalent elastic constant of the suspension taken as a single unit? Where does the energy come from?

Solution: [Given: mass of each person, 70 kg; mass of car, 1100 kg; and s = 2.5 cm. Find: each k, each ΔPE_e, and the net spring constant k'.] The spring system can be taken to be in equilibrium when the car is empty, so its mass is irrelevant. The total mass of people supported by the four springs is 4 × 70 kg and this weighs

$$F_W = (280 \text{ kg})(9.8 \text{ m/s}^2) = 2744 \text{ N}$$

Each spring compresses the same amount (0.025 m) and supports the same load; hence, the spring-force exerted upward by each is $\frac{1}{4}$ the total load

$$F = \tfrac{1}{4}(2744 \text{ N}) = ks = k(0.025 \text{ m})$$

and

$$\boxed{k = 2.7 \times 10^4 \text{ N/m}}$$

Taking the springs together as one, exerting 2744 N upward, the equivalent force constant is

$$k' = \frac{2744 \text{ N}}{0.025 \text{ m}} = \boxed{1.1 \times 10^5 \text{ N/m}}$$

which is 4 times the individual spring constant.

The *elastic*-PE stored in each spring is

$$\Delta PE_e = \tfrac{1}{2}ks^2 = \tfrac{1}{2}(2.7 \times 10^4 \text{ N/m})(0.025 \text{ m})^2 = \boxed{8.4 \text{ J}}$$

while the whole system stores 4 times that. The energy stored elastically is a fraction of the decrease in the *gravitational*-PE—the remainder is lost as thermal energy.

▶ **Quick Check:** The total stored PE_e is $\frac{1}{2}k's^2$ = 34 J, which equals 4(8.4 J).

The elastic energy of a spring arises from pulling against the interatomic bonds holding the material together—it's really *electromagnetic*-PE. In fact, whenever a contact force acts on any kind of body, the interaction is accompanied by a distortion of the body and the resulting electrical restoring forces among the many millions of atoms involved. There is no such thing as an absolutely rigid body.

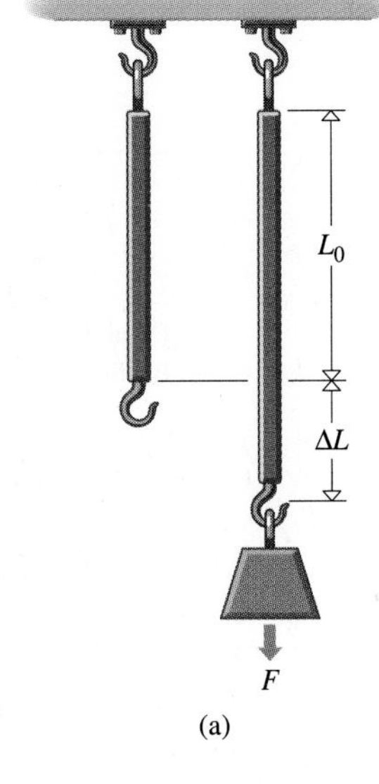

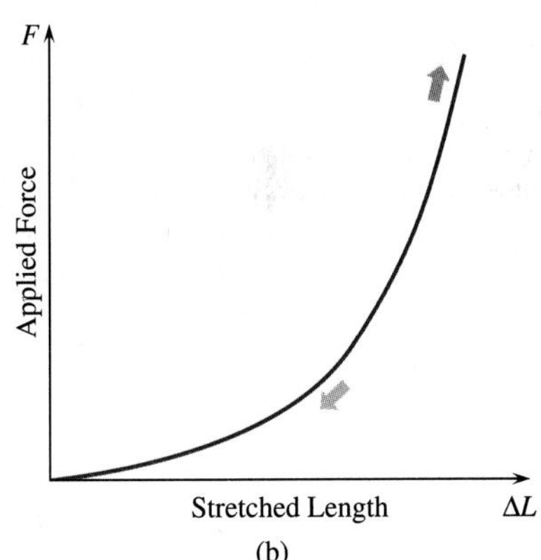

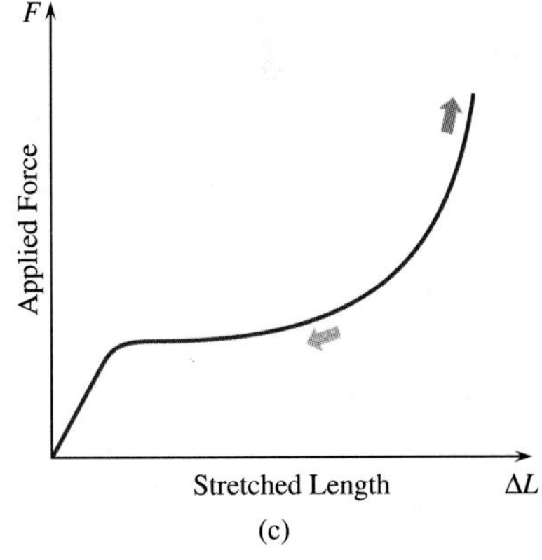

Figure 10.11 (a) A length (L_0) of material stretches an amount ΔL under a load F. (b) Soft animal tissue stretches elastically, forming a J-shaped curve and contracting back along the same arc as the force is reduced. (c) Rubbery synthetic solids stretch elastically along an S-shaped curve. They also return to $\Delta L = 0$ along the curve as F goes to zero.

Elastic Materials

Suppose we take a sample of some material, apply an increasing force F to it, and plot the resulting change in length ΔL, as in Fig. 10.11. If the sample is animal tissue, the graph will usually be a curve like that of Fig. 10.11b, whereas if it's a synthetic stretchy polymer, it will look more like Fig. 10.11c. For moderate extensions, the data goes up and back along the two curves, and both represent elastic behavior. By contrast, some elastic materials manifest a linear relationship between F and ΔL that corresponds to Hooke's Law

$$F = k\Delta L \tag{10.5}$$

Typically, however, a linear response only occurs over a limited range of applied force and that's determined by the nature of the interatomic bonding being stretched. Removing a small load from a sample allows the atoms, which have only been slightly shifted, to return to their equilibrium configuration, and the material comes back to its original length.

Consider what happens to a sample as the load on it gets very large. Depicted in Fig. 10.12 is the behavior of several different solid rods under increasing tension. The longest curve with the greatest value of ΔL corresponds to a material like pure iron or aluminum. Such a sample will elongate a little linearly until its **elastic limit**, or **yield point**, is reached and the bonds between atoms begin to be overcome. Rows of atoms then slide past one another in a process known as *slip,* and the material plastically deforms in an unrecoverable way. Once the plastic mode is initiated, the elongation occurs more dramatically until ultimately the specimen ruptures at the **breaking point**. You have probably never stretched a steel rod to the breaking point, but the same thing happens with a piece of taffy or a Tootsie Roll.

As a rule, removing the load anywhere between the elastic limit and the breaking point will leave the sample permanently longer than it was originally. A material that can be stretched thin or drawn without breaking even while cold (like chewing gum) is *ductile.* Metals such as copper, gold, and silver, are quite ductile at STP and elongate considerably via slipping. Such substances are *tough* and resist cracking. The more carbon is fused into iron, the less ductile is the resulting alloy. Steel is iron with less than about 1% carbon. Pure iron has a low elastic limit, low-carbon steel a moderate one, and high-carbon steel a still greater one (Fig. Q3, p. 347). If we built a bridge out of pure iron, it would permanently sag the first time it was heavily loaded.

Carbon embedded among iron atoms serves as a kind of grit that impedes slipping. Brittle substances such as glass, stone, ceramics, and cast iron (which contains about as much carbon as it can, $\approx$4%) break close to their elastic limits, without appreciable deformation. Even though glass is stronger than steel in both compression and tension, you are not likely to see glass houses because glass does not yield rapidly at STP—it's just too brittle to be a major structural material. If you want to shatter an old cast iron household radiator, just hit it with a sledge hammer.

The degree to which a material will display ductile behavior depends on several factors: temperature, the type of loading, and the rate of loading. A substance that is

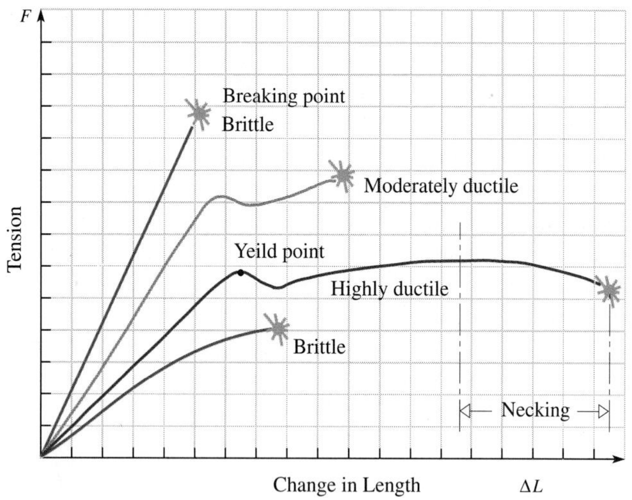

Figure 10.12 Load or tension versus elongation for several different engineering materials. Brittle substances rupture suddenly without much elongating. Ductile substances elongate via plastic flow. A ductile metal might be Hookean up to 0.25%–0.5% elongation and fracture at around 50%. Some nonmetallic materials such as wood and fiberglass fracture after elongations of 1%–3%.

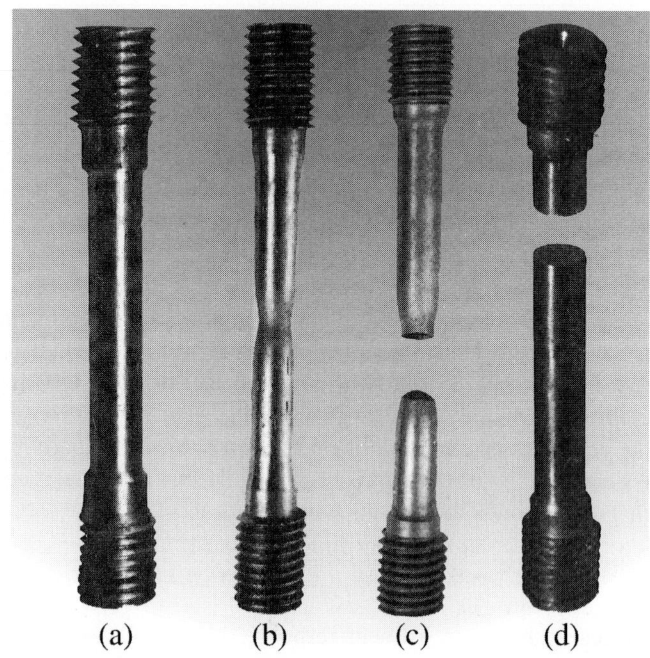

(a) When a sample is put under tension, it elongates linearly by less than ≈0.5%. If it's ductile, it continues to elongate several hundred times as much under slightly increased loads. (b) After a maximum load is reached, the sample begins to neck down until (c) it fails. (d) If the material is brittle, it fractures without deforming plastically.

brittle at −200°F may well be ductile at room temperature, and red-hot glass stretches like chewing gum. All across the northern United States, cheap plastic outdoor garbage pails bounce in July and shatter in January. The strongest structural material used today is steel, but because it becomes plastic at 1200°F, steel columns and beams in buildings have to be fireproofed. At the other extreme, steel becomes brittle at around −30°F, so much so that a railroad bridge in northern Canada broke up when a heavy train rattled over it on an especially cold winter evening.

For the most part, *brittle materials tend to rupture sooner in tension than compression* because of the presence of flaws (for example, microscopic cavities and cracks) that weaken the specimen far more in tension than in compression.

Bone is a composite structure made of crystals of hydroxyapatite (a calcium compound good in compression) and collagen fibers (good in tension). Bone, which is varyingly brittle, remains elastic almost to the point of rupture and cannot be either permanently compressed or elongated unless the loads are applied for a very long time. Infants' bones are high in collagen and very flexible. By young adulthood, hydroxyapatite has built up to the point (≈67%) where bones are both flexible and strong. As adults age, collagen becomes less flexible and calcium tends to be reabsorbed from the hydroxyapatite, making bones brittle and more breakable.

10.8 Stress

Remarkably, about a century and a half elapsed before Hooke's ideas were further developed by the French mathematician Augustin Louis Cauchy. It is likely that this long period of idleness was in no small part due to Newton, who detested Hooke and, outliving him by 25 years, had ample opportunity to denigrate the latter's practical science. Sir Isaac was a loner—chaste, sensitive, and vulnerable—while Hooke was argumentative, vain, and a womanizer, a testy, earthy sort who took pleasure in controversy as much as Newton shrank from it. It can be argued that Hooke contributed to Newton's nervous breakdown, and it can also be argued that Sir Isaac had his revenge, at least on the theory of elasticity.

In any event, Cauchy (who himself was widely held to be selfish and narrow-minded) introduced the central notions of *stress* and *strain* (1822). What severely limits Hooke's Law is that it deals with the specimen or spring or structure in its entirety. The elastic constant depends on the details of the object under test and not just on the material of which it's made. When we bend a tree and compare the applied force and resulting deflection, we arrive at the elastic constant of that particular tree and not some clever insight about the behavior of oak or pine. Recall Example 10.5. The elastic constant of the four springs acting as one was 4 times as great as the constant for any one of the springs, even though nothing different was happening inside them in either case. What we need is a reformulation of Hooke's Law that focuses not on the forces and deformations of the thing as a whole, but on the internal processes at work that are independent of the external dimensions.

When a force acts over a finite area, it's called a *distributed force.* Pressure then becomes a very useful concept. Here, the pressure exerted on the platform is the net load divided by the total contact area. By contrast, the stress anywhere within either leg is the magnitude of the force on that leg divided by the cross-sectional area at that point. Where the leg is narrow, the stress is large. And that's one reason elephants have stout legs.

Cauchy introduced the notion of *stress* to describe the distribution of force within a solid acted upon by a load. Ordinary interactions between bodies are always distributed over some contact area; the idea of a force acting at a point is an idealization. **The stress experienced within a solid is defined as the magnitude of the force acting, divided by the area over which it acts**:

$$\text{Stress} = \frac{\text{force}}{\text{area}}$$

Using the Greek letter sigma (σ) for stress, this equation becomes

$$\sigma = \frac{F}{A} \qquad (10.6)$$

The idea is very much like the concept of pressure, and like pressure it has the SI units of $\mathrm{N/m^2}$, or the U.S. Customary units of $\mathrm{lb/in.^2}$ (psi). To make matters a bit more concise, $1\ \mathrm{N/m^2}$ is called a *pascal* (Pa) honoring the French physicist by that name. Since this quantity is actually very small, stress is more often given in megapascals (MPa $= 10^6$ Pa) or even gigapascals (GPa $= 10^9$ Pa).

The hanging bar in Fig. 10.13b experiences both a downward force and the upward supporting force provided by the ceiling. The line-of-action of these two equal-magnitude forces is along the rod's central axis, and so the loading is said to be *axial.* Imagine a plane cutting perpendicularly across the bar. The corresponding stress in that cross section is given by Eq. (10.6) and it's known as the *normal stress* and, in particular, since the bar is in tension, it's called the **normal tensile stress** (σ_t). Similarly, under an axial compressive load, Eq. (10.6) would provide the **normal compressive stress** (σ_c). Notice in Fig. 10.13 that while a load F is supported by a single rod (which stretches a distance ΔL in the process), two such rods will support twice that load ($2F$), each elongating by that same ΔL, a situation that corresponds to Hooke's Law. Now if each rod has a cross-sectional area A, the tensile stress (F/A) in the single rod exactly equals the stress in the double-rod suspension ($2F/2A$), even though the elastic constants are not the same. That's the great virtue of the idea of stress—it reflects how strongly the atoms are being forced to interact.

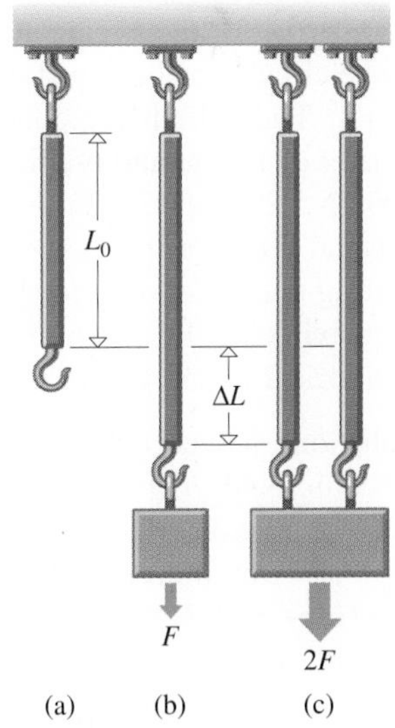

Figure 10.13 (a) A sample of an elastic material. (b) Under a load F, it elongates by an amount ΔL. (c) Two identical samples loaded with $2F$ also elongate by the same ΔL.

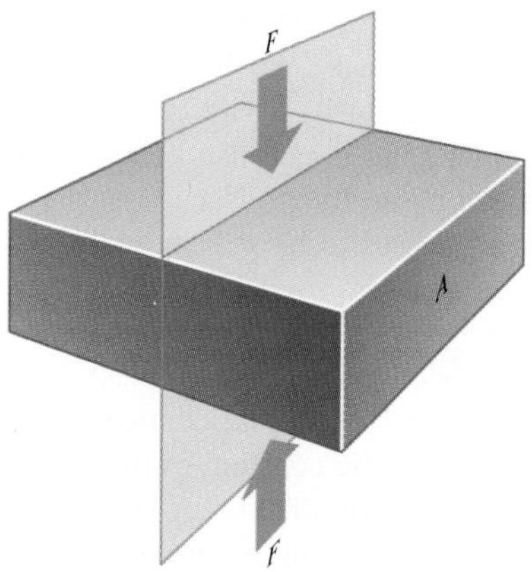

Figure 10.14 A pair of equal-magnitude, parallel forces shearing a sample. Think of how you tear a piece of cardboard or snip a twig or wire.

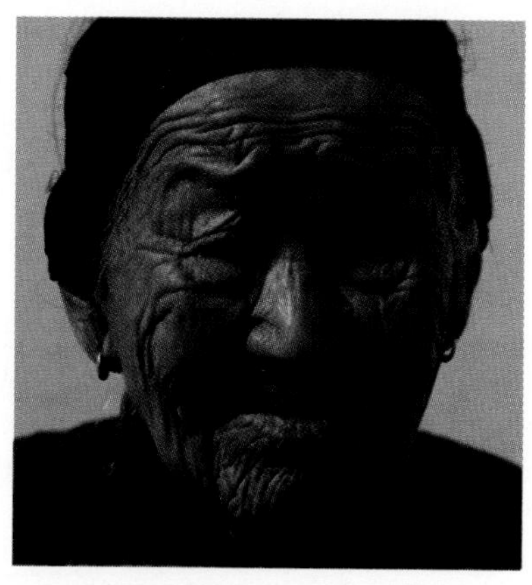

Human skin is under tension like a rubber glove. As a person ages, the skin wrinkles because it loses its elasticity.

In addition to these normal stresses, there is also something known as **shear stress** (σ_s) that exists when the force-pair acts parallel to the area A (Fig. 10.14). A scissors cuts paper and a pruning clipper cuts twigs by applying a pair of equal-magnitude, oppositely directed forces parallel to the cross section to be severed. These instruments produce a shearing stress that is great enough to sever the material. Thus, the shear stress, again given by Eq. (10.6), is a measure of the effort needed to cause the atoms in one part of a solid to slide past an adjacent part—like a pushed deck of cards that edges over at a slant or one that is fanned out. The latter reflects the fact that shear stress also occurs when the sample (a wrench, wrist, or screwdriver) is twisted.

Example 10.6 Figure 10.15 depicts a punch used to make circular holes in a stack of paper or a sheet of soft metal. The cylindrical punch has a cross-sectional diameter of 1.0 cm and is driven downward with a force of 100 kN. The punch and the die beneath produce a ring of vertical force on the paper that will ultimately shear it, cutting out a disk-shaped slug. Compute the shear stress that exists in the 2.0-mm-thick stack along a vertical cylindrical region beneath the edge of the punch.

Solution: [Given: $D = 1.0$ cm, $F = 100$ kN, and $d = 2.0$ mm. Find: σ_s.] The area of shear is that of the vertical cylindrical surface (corresponding to the edge of the die) of circumference $2\pi R$ and height d; $A = 2\pi \times (D/2)d = \pi(1.0 \times 10^{-2}\text{ m})(2.0 \times 10^{-3}\text{ m}) = 6.283 \times 10^{-5}\text{ m}^2$. Hence

$$\sigma_s = \frac{F}{A} = \frac{10 \times 10^4\text{ N}}{6.283 \times 10^{-5}\text{ m}^2} = \boxed{1.6\text{ GPa}}$$

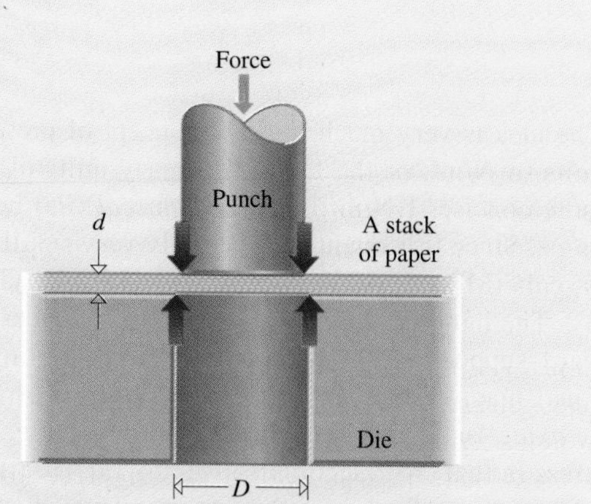

Figure 10.15 A paper punch. The cylindrical punch is forced down through the stack of paper above the die hole. It shears the paper along the circular outline of the punch.

which is about 2.3×10^5 psi.

▶ **Quick Check:** $F = \sigma_s A = (1.6 \times 10^9\text{ Pa})2\pi \times (0.50 \times 10^{-2}\text{ m})(2.0 \times 10^{-3}\text{ m}) = 10 \times 10^4\text{ N}$.

Several decades ago, women's high-heeled shoes with a bare steel shank in the heel were fashionable. Ordinarily, a shoe might have roughly 20 in.^{2} of contact area with the ground and a 140-lb woman, her weight shifted to one foot, would produce a meager stress of about 48 kN/m^2 (7 psi). But when that weight was concentrated on the spiked heel of one shoe (with an area of about 0.12 in.^{2}), the stress would soar to 8 MN/m^2 (1.1×10^3 psi). The resulting dents, holes, and chipped tiles prompted some shopkeepers to ban the menace.

Floor tiles sheared by spiked heels. The downward force increases as a person shifts back onto the heels just before sitting.

10.9 Strain

Until now, we have talked about rods and the way they elongate or shorten by some amount ΔL under a tensile or compressive load. But that change in length depends on the original dimensions of the rod. Imagine that a tensile force pulling on the atoms stretches their bonds and increases the interatomic separation by, say, 0.01%. Then, with all its atoms 0.01% farther apart axially, the rod must be 0.01% longer, no matter how long it was originally. Different length rods will deform by different amounts, but the percentage change will be the same, and it's the change in the interatomic spacing that is crucial here. It's that change that generates the counterforce to the load. Hence, we define the **normal strain under an axial load as the change in the length over the original length**:

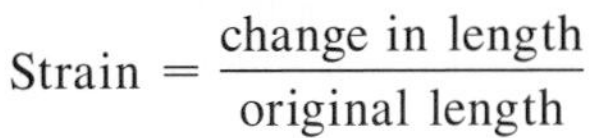

$$\text{Strain} = \frac{\text{change in length}}{\text{original length}}$$

which can be compressive or tensile. Commonly, the Greek letter epsilon (ε) represents the strain, which symbolically becomes

$$\varepsilon = \frac{\Delta L}{L_0} \tag{10.7}$$

where it is assumed that the sample has a constant cross-sectional area. Strain is a measure of by how much the atoms are shifted and, therefore, by how much the bonds are being stretched or compressed. It is a *dimensionless quantity.* In engineering applications, strain is generally kept within ±1.0% and rarely even exceeds ±0.1%. The columns holding up a high-rise building 300 m tall, whether made of concrete or steel, will be designed to compress by little more than about 3 cm. A 100-m elevator cable that stretches 10 cm under a load has a strain of 10×10^{-2} m/100 m $= 1.0 \times 10^{-3} = 0.1\%$.

The threadlike structures of collagen magnified tens of thousands of times.

Clearly, there must be a strain as a solid—any solid—yields under the influence of a stress, however small. That's true for normal stress, as well as shear stress, and there must be a **shear strain** (ε_s), a corresponding change in shape. The solid block shown in Fig. 10.16 is being twisted

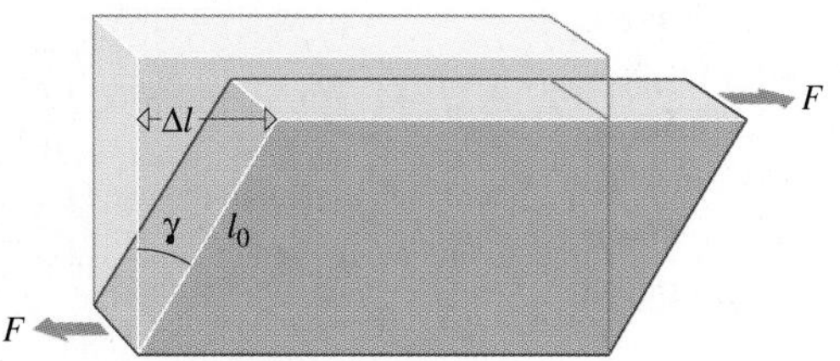

Figure 10.16 Two equal-magnitude, oppositely directed forces twist the solid block through an angle γ that corresponds to the shear strain.

by a force-couple. The larger the side l_0, the more the edge is displaced, Δl, but the shifting on an atomic level will be the same no matter what l_0 is, provided the angle γ is constant. Hence, *we define the shear strain as the angle of distortion,* γ (Greek lowercase gamma), measured in radians, which makes it dimensionless. Since the angle of distortion is generally small, tan $\gamma \approx \gamma$ and

$$\varepsilon_s = \gamma \approx \tan \gamma \approx \frac{\Delta l}{l_0} \tag{10.8}$$

An elastic material is one that returns to its original shape after an initial deformation, and that applies to shear strain as well. Metals, bone, stone, and concrete display *elastic* shear strain of less than about 1°, which is typical of many hard solids. Still, stone bends more than one might imagine; indeed, every building bends in the wind. The towers of the World Trade Center heave over as much as 3 ft in a strong wind (you can even hear them creak) and will bend away from the vertical by 6 or 7 ft in a hurricane. By comparison, stretchy things such as soft biological tissue can recover from shear strains of upwards of 40°. Take hold of the skin on your belly and pull; the way it returns depends on such things as dehydration and, of course, age.

Figure 10.17 is a *stress-strain diagram* for several materials. It's similar to the force-displacement graph of Fig. 10.12. Here, provided the strains are never very large, the test pieces behave elastically, returning to their original shapes in a Hookean fashion. In Fig. 10.17b, the curve is the idealized S-shape associated with

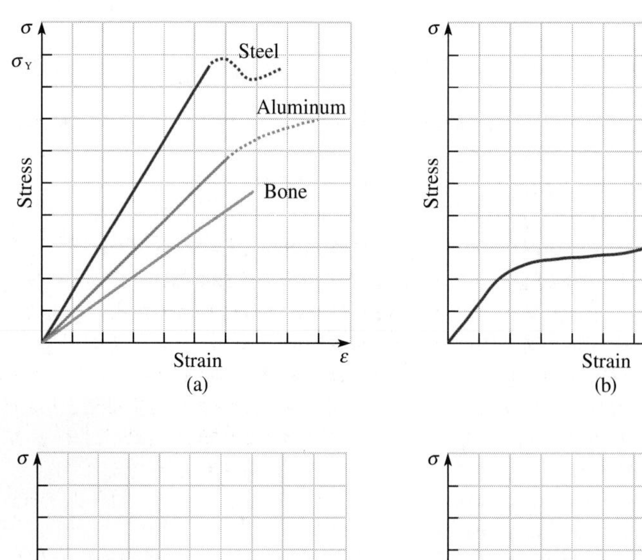

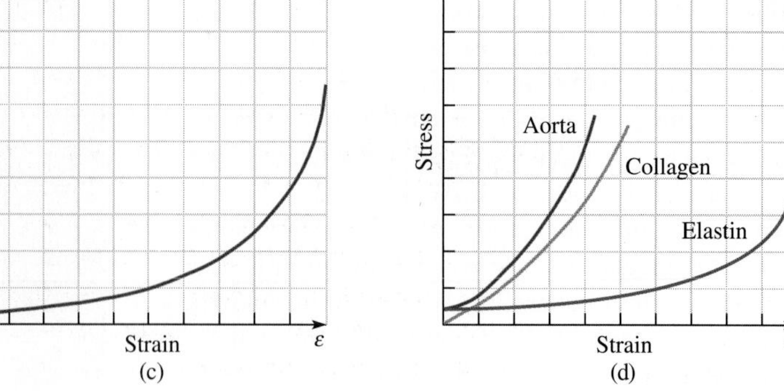

Figure 10.17 (a) The stress-strain curves for several Hookean materials. The ductile samples, such as soft steel, stop behaving linearly at their yield points (σ_Y). (b) When stressed, rubbery polymers first elongate by straightening out their molecules and thereafter by tugging on the chemical bonds. (c) Most biological materials are under tension even when not strained. Your skin is like a rubber glove stretched over your body. (d) Elastin is usually reinforced with collagen in biological systems such as arteries. Tendon is made principally of collagen.

elastomers, stretchy polymers such as Spandex. Their giant molecules are intertwined, coiled strands that straighten out when stretched and spring back when released. They are elastic, even out to strains of several hundred percent, but not Hookean.

Figure 10.17c depicts the behavior of a typical biological tissue sample—pull on your lip or ear lobe. Soft connective animal tissue often contains a mix of two different protein materials: elastin and collagen. The elastin fibers, abundant in the vessels of the human circulatory system, are easily stretched, requiring little stress to produce considerable strain. These are frequently reinforced with a network of quasi-Hookean collagen fibers, as in a human artery. Collagen (p. 335) is the strongest of the connective tissues. It's the collagen that gives flesh its toughness so it does not tear easily. Collagen makes meat tough, although at high temperatures (≈100°C) it breaks down to soft gelatin, leaving the other more tender constituents like elastin behind—that's one reason why we cook meat.

Some biological tissue can stretch elastically well beyond the 800% value that's typical of soft materials such as rubber.

10.10 Strength

In broad terms, *strength* means the ability to withstand a load without failure, but there is a difference in the definition of this term when referring to structures as opposed to materials. When applied to structures, the strength of a truss, building, bridge, or tower equals the load, in newtons, that will initiate collapse. On the other hand, the **strength** *of a material is the stress, in newtons per meter-squared, that will cause failure in a sample of that substance.* Strength is inherently linked to the atomic behavior of the material and tells us how any sample of the stuff should behave, all else being equal (Table 10.3).

In the case of ductile materials, several different strengths are of interest. Figure 10.18 is the stress-strain curve for a sample of an aluminum alloy. *The stress at which slip is first observed, at which the curve stops being linear, is the* **yield strength**. Any mechanical part that will have to support a load without plastically deforming will have to have an appropriately high yield strength—we can't make the girders supporting a building out of soft steel. Yet, if a sheet of aluminum is to be squashed into a mold so it comes out looking like a beer can, its yield strength needs to be lower than the applied stress.

The **tensile strength**, *or* **ultimate strength**, *is the stress corresponding to the greatest applied force*—it's the high point on the stress-strain curve. Usually a ductile sample will deform nonuniformly over its cross section, tending to narrow or **neck** down. The start of that process also marks the ultimate strength (see photo on p. 332). This concept can be misleading, since by the time a piece of material reaches its point of ultimate strength, it has already been irreversibly elongated. We don't often want a ladder that will permanently sag when the first person climbs it.

The stress corresponding to the rupture of the sample is the **breaking strength**, and that applies whether the sample is being stretched, compressed, or sheared. Brittle materials show little or no necking, and their ultimate strengths equal their breaking strengths (see photo on p. 332). Notice in Fig. 10.19 that the stress-strain curve can be quite different in tension than in compression, and the ultimate strengths can be appreciably different too.

TABLE 10.3 Approximate ultimate strengths of various materials

Material	Tensile (MPa)	Compressive (MPa)	Shear (MPa)
Natural rubber	0.03		
Muscle	0.1		
Bladder wall	0.2		
Intestine	0.5		
Cartilage	3		
Concrete	4	30–40	
Brick	5.5	10–21	
Skin	10		
Marble	10	110	22
Lead	12		12
Granite	20	240	35
Ceramics	35–350		
Glass	40–175	50	
Leather	41		
Polystyrene	48	90	55
Fir, Douglas	50	50	8
Tendon	65–82		
Pine	100	27–50	9
Bronze	100–600		
Bone, compact	110	150	
Brass	120–400		
Femur, horse	121	145	
Femur, human	124	170	
Iron, cast-gray	170	650	240
Hair	190		
Iron, wrought	200		
Spider's silk	240		
Magnesium alloy	280		150
Catgut	350		
Cotton	350		
Aluminum	300–570		200–330
Silk	350		
Steel, structural	400		
Nylon thread	1100		
Kevlar	2760		
Steel piano wire	3100		

Example 10.7 The human thigh bone or femur at its narrowest point resembles a hollow cylinder with an outer radius of roughly 1.1 cm and an inner radius of just about half of that. Taking the compressive strength of the bone to be 170 MPa, how much force will be required to rupture it?

Solution: [Given: $R_o = 1.1$ cm, $R_i = 0.5R_o$, and $\sigma_c = 170$ MPa. Find: F.] The cross-sectional area of bone matter is

$$A = \pi(R_o^2 - R_i^2) = \pi R_o^2(1 - 0.5^2) = \tfrac{3}{4}\pi(1.1 \times 10^{-2}\text{ m})^2$$

therefore $A = 2.85 \times 10^{-4}\text{ m}^2$

Hence, since $\sigma_c = F/A$

$$F = \sigma_c A = (170 \times 10^6\text{ Pa})(2.85 \times 10^{-4}\text{ m}^2)$$

and so $\boxed{F = 4.8 \times 10^4\text{ N}}$

(continued)

(continued)

or about 11 × 10^3 lb. An adult who steps off a kitchen table and lands on the heel of one bare foot with the knee locked will experience a force large enough to shatter the femur.

▶ **Quick Check:** $A = 3.8 \times 10^{-4}\ \text{m}^2 - 0.79 \times 10^{-4}\ \text{m}^2 = 3 \times 10^{-4}\ \text{m}^2$; $F/A = \sigma_c \approx (50\ \text{kN})/(3 \times 10^{-4}\ \text{m}^2) \approx 0.17$ GPa.

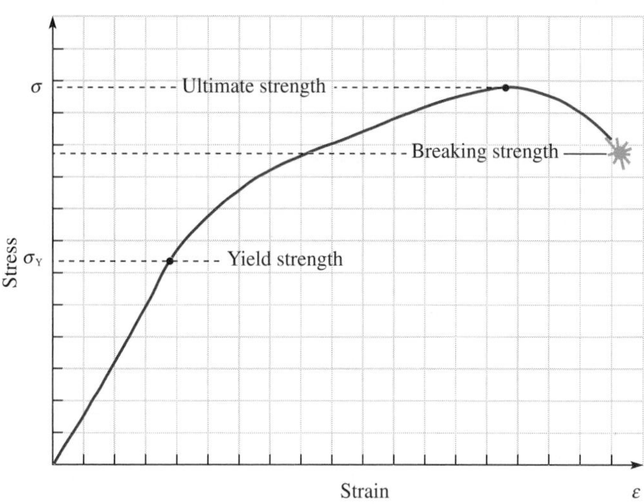

Figure 10.18 The stress-strain curve for an aluminum alloy. There are several different strengths that are of interest to someone using the material: (1) The yield strength tells us when it will cease being elastic and begin to permanently deform. (2) The ultimate strength is the maximum stress it can experience. (3) The breaking strength is the stress at which it fails.

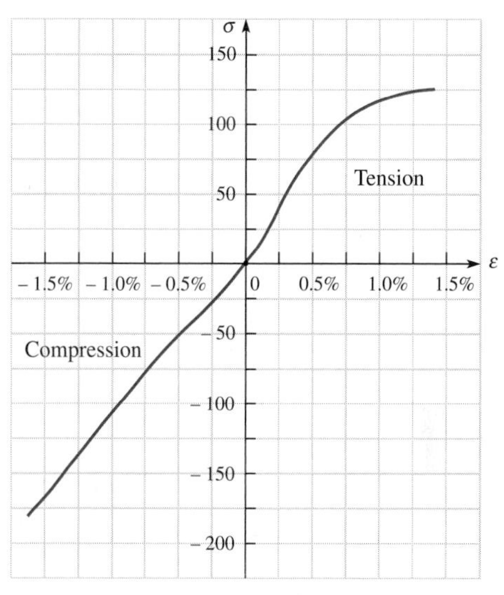

Figure 10.19 Stress-strain for compact (dense) bone.

Because strength depends on many factors (the presence of impurities, holes, surface scratches, cracks, and so on), engineers tend to be cautious in their designs. They usually incorporate a "safety factor" of 3 or 4, or if the situation is potentially dangerous, perhaps 10 or more. Thus, while a large structure might typically experience a stress of 80 MN/m^2 to 95 MN/m^2, mild steel (over 95% of all steel manufactured is of this type) has a strength of around 450 MN/m^2, affording a safety factor of about 4 to 5. In the early 1800s, one out of every four bridges built collapsed. We've come a long way from that sorry record, but earthquakes still play havoc with our cities.

The groundwork for a modern building. Reinforced concrete (concrete with rods of steel embedded in it) has added tensile strength. This is important because concrete is strong in compression and weak in tension.

Our forearms and calves house massive concentrations of muscle. Muscle tissue has a very low strength (0.1 MPa), and there has to be a good deal of it to maintain the needed cross-sectional area. Hands and feet, which must move around quickly, need to be light, agile, and still strong. The solution is to bunch our muscles fairly far away and link them to our fingers and toes by strong (82 MPa), and therefore affordably thin, light tendons. Tendon, though its strength is less than that of mild steel (≈450 MPa), is about 7 times less dense and so is actually stronger pound for pound.

Some modern plastics such as Kevlar (1972) are considerably stronger, per unit weight, and stiffer than steel. This strength derives from the highly ordered arrangement of its long rodlike molecules. With a tensile strength of 2.8 GPa (4 ×

10^5 psi), Kevlar has been used for such things as reinforcing tires, making special mooring cables, and for bulletproof vests.

10.11 Fatigue

Many ordinary situations exist in which a metal component can experience a cyclical application of stress. Even though the applied stress is well below the yield strength of the metal, it is still possible for the component to rupture after repeated applications of stress. This failure mode, induced by a relatively small, cyclic stress, is known as **fatigue**. It is characterized by brittle behavior even in metals that are ordinarily ductile, and so the effect must always be considered in dealing with varying loads.

The traffic on a bridge creates fluctuating stress levels, but the complete reversal of the loading is an even more severe condition. Picture the axle of a railroad car supporting the downward weight of the vehicle and cargo. Every time the axle turns through half a rotation, the portion that was on top in compression is now on the bottom in tension, and vice versa—the bending of the axle completely reverses as it rotates. That's why the axles on trains are inspected regularly, and it's also why cracks appear in the structural members of buses bouncing along potholed city streets. Most of the United States' B-1B bomber fleet was grounded in 1992 because of cracks in the landing gear assembly. Similarly, a paper clip, which certainly would be impossible to pull apart by hand, ruptures after about ten 180°-bending cycles.

Figure 10.20 is a plot of stress (σ) versus number-of-cycles-to-failure (N) for two typical metals, tool steel and aluminum. Accordingly, a machine part designed to last through 10^5 cycles will rupture at a stress of 620 MPa (90×10^3 psi) if made of tool steel, and at about half that if made of aluminum. At higher stress levels, the part will survive fewer operating cycles before failing. Steel generally shows a behavior that levels off at what is called the *endurance,* or *fatigue, limit.* That's where the curve becomes horizontal. Presumably, a cyclic stress at a level less than the endurance limit will "never" result in fatigue failure regardless of the number of stress cycles. Incidentally, N is ordinarily likely to be quite large. Even a slow-moving crane might be loaded and unloaded a few hundred times a day and, at 5 days a week

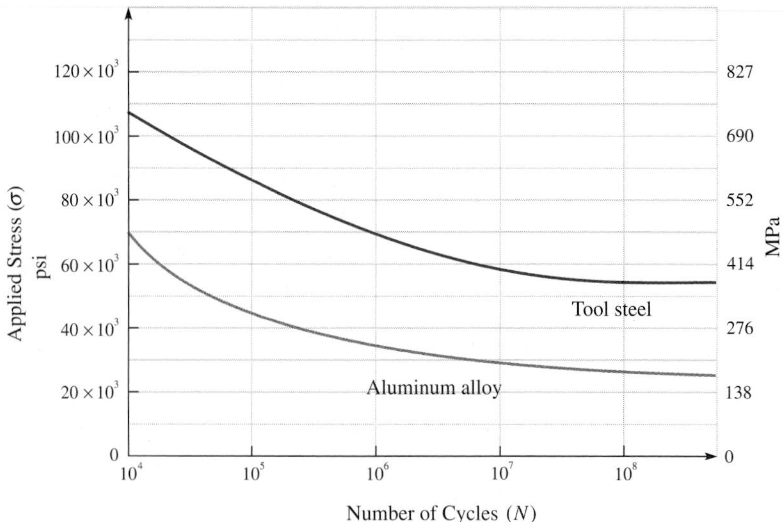

Figure 10.20 Fatigue curves for tool steel and aluminum alloy. At any particular value of stress (σ), there is a corresponding number-of-cycles-to-failure (N). Steel below a certain stress never fails. The curve for aluminum continues down and can fail at any level.

for 30 years, that is over a million and a half cycles. By comparison, parts in the propulsion system of an automobile can move through 500 times as many cycles in their much shorter lifetimes.

In contrast to steel, nonferrous metals and alloys, aluminum and copper among them, have σ-versus-N curves that give no indication of ever leveling off, and so no fatigue limit exists for these materials. This means that an aluminum part can be expected to fail at any applied stress sooner or later—the idea is to make the "later" long after the designed lifetime.

Figure 10.21a is an X-ray of a broken ulna in the forearm of a man showing the steel brace used to hold the two segments together. Unfortunately, the bone did not knit and so whenever the patient used his arm, especially during physical therapy, the brace carried the load and experienced cyclic stresses, often reversing completely. Of course it should have been expected that the fairly thin brace would fail sooner or later due to fatigue, but no one knew the bone had not fused. Finally, the brace ruptured at the place where the steel was narrowest and the stress the greatest, right across a screw hole.

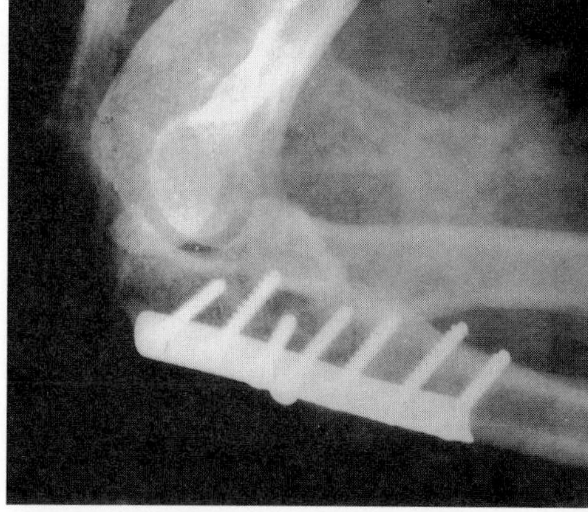

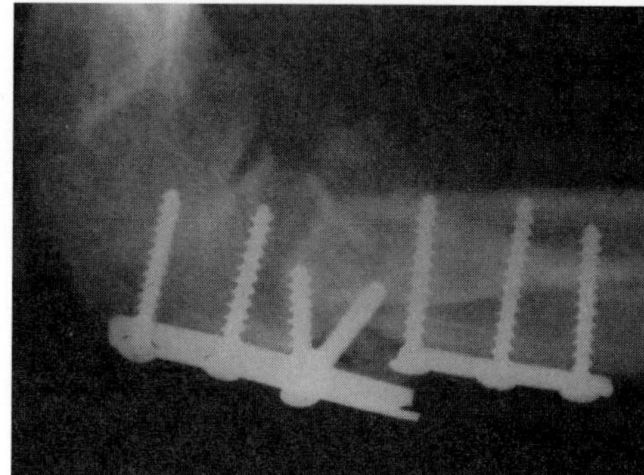

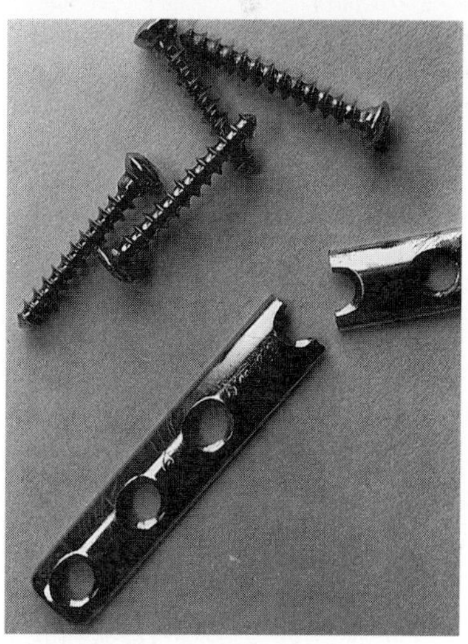

Figure 10.21 An X-ray of a steel plate used to bridge a broken bone. Months later, the plate fatigued and broke. The break occurred at a hole, right where you would expect.

MODULI AND ENERGY

In 1807, Dr. Thomas Young, one of the most brilliant minds of the era, introduced a notion he called "the modulus of the elasticity" (*modulus* is Latin for "a small measure"). But it was the French engineer Claude Navier who in 1826 provided the modern formulation of what has come to be known as **Young's Modulus** (Y), or the **Modulus of Elasticity**. What Young had found was a measure of the stiffness of materials.

10.12 Young's Modulus

The normal stress-strain graphs for many materials begin with a linear region where the substance, at least at low strain, is Hookean (Fig. 10.17a). This means that the slope of the curve is a constant, which reflects how the material elastically strains under an increasing normal stress. This **elastic stress per unit strain** defines Young's Modulus:

$$\text{Young's Modulus} = \frac{\text{stress}}{\text{strain}}$$

or symbolically

$$Y = \frac{\sigma}{\varepsilon} \tag{10.9}$$

which corresponds to either tension or compression (Table 10.4). The units of Y are the same as those for stress—namely, N/m^2 or Pa. The steeper the slope, the less distortion the sample experiences for a given stress, the stiffer it is, and the greater is Y. A stiff material more nearly maintains its size and shape under a load in the elastic regime.

Referring to the loaded rod in Fig. 10.13b, since $\sigma = F/A$ and $\varepsilon = \Delta L/L_0$, Eq. (10.9) rearranged, $\sigma = Y\varepsilon$, leads to $F/A = Y\varepsilon$, and so

$$F = \frac{YA}{L_0}\Delta L \tag{10.10}$$

which is the equivalent of Hooke's Law, $F = k\Delta L$; hence, $k = YA/L_0$.

Example 10.8 Return to Example 10.7, which deals with a broken femur, and calculate the fractional change in length of the bone, assuming Y is constant right up to failure. Use Table 10.4.

Solution: [Given: $\sigma_c = 170$ MPa. Find: ε_c.] From Eq. (10.9) and Table 10.4

$$\varepsilon_c = \frac{\sigma_c}{Y} = \frac{170 \text{ MPa}}{10 \text{ GPa}} = 17 \times 10^{-3} = \boxed{1.7\%}$$

▶ **Quick Check:** Figure 10.19 agrees with 1.7%.

TABLE 10.4 Approximate values of Young's Modulus for various solids

Material	Young's Modulus (GPa)	Material	Young's Modulus (GPa)
Biological tissue, soft	0.0002	Compact bone, compression	10
Rubber	0.007	tension	22
Egg membrane	0.008	Concrete	25–30
Cartilage, human	0.024	Magnesium	42
Collagen	0.6	Marble	50
Tendon, human	0.6	Aluminum	56–77
Nylon	2	Glass	65
Polystyrene	3	Iron, cast, gray	70–145
Spider thread	3	Granite	70
Catgut	3	Gold	79
Nylon fiber	5.5	Brass, yellow	105
Plywood	7	Copper	120
Hair	10	Bronze	120
Pine, southern	11	Iron, wrought	180–210
Fir, Douglas	13	Steel, stainless	190
Oak	14	Steel, structural	200
Brick	14	Tungsten	360
Lead	16	Sapphire	420
		Osmium	560
		Diamond	1200

The stronger the interatomic bonding, the greater the force that must be applied to separate the atoms and, in turn, to stretch the material. Thus, we should expect that for materials with larger and larger bonding forces, as indicated by increasing melting points, Young's Modulus would increase as well (Table 10.5).

Notice that when a substance elongates to twice its original length $\Delta L = L_0$, whereupon the strain is 100%. Young's Modulus then equals the corresponding stress (presuming the thing does not break before then). *We can estimate the stress needed to double the separation of the atoms in a molecule, and that should provide something resembling Y.*

The energy needed to pull a molecule apart, its **binding energy**, is determined by the specific bonding mechanisms at work. Table 10.6 provides a range of values for the energies associated with the various types of atomic bonding. Thus, while it takes about 9×10^{-19} J to separate the sodium from the chlorine in a molecule of table salt, just about half as much energy will break up a molecule of water into hy-

TABLE 10.5 Young's Modulus and the melting point for various solids

Material	Young's Modulus (GPa)	Melting point (°C)
Lead	16	327
Magnesium	42	650
Aluminum	56–77	660
Copper	120	1083
Tungsten	360	3410

TABLE 10.6 Binding energies

Bond mechanism	Energy (eV/particle)*
Weak	<0.4
Metallic	1–4.5
Covalent	0.6–8.4
Ionic	6–11

*1 eV = 1.6×10^{-19} J.

drogen and oxygen. In fact, the majority of atoms are bound to their respective molecules and neighboring atoms in a solid with an energy, within a factor of 10 or so, of 1.6×10^{-19} J. This particular quantity is called an **electron volt** (eV) and it's especially convenient in the microworld: $1.6 \times 10^{-19}\ \text{J} = 1\ \text{eV}$.

To approximate Young's Modulus, let's take the interatomic separation of the material as 0.2 nm and the binding energy as 2 eV = 3.2×10^{-19} J. If we assume that pulling the atoms completely apart is not going to require appreciably more energy than separating them by an additional 0.2 nm, then the work done in so doing is approximately the binding energy. Since $\Delta W = F_{av}\Delta r$, for each atom

$$F_{av} = \frac{\Delta W}{\Delta r}$$

and

$$F_{av} = \frac{3.2 \times 10^{-19}\ \text{J}}{0.2 \times 10^{-9}\ \text{m}} = 16 \times 10^{-10}\ \text{N}$$

Because the atoms are originally separated by 0.2 nm, there are $(0.2\ \text{nm})^{-1}$ atoms per meter, and about $(0.2\ \text{nm})^{-2}$, or roughly 2.5×10^{19}, atoms per square meter. Thus, the average stress is $F_{av}/A = (16 \times 10^{-10}\ \text{N/atom})(2.5 \times 10^{19}\ \text{atoms/m}^2) = 40$ GPa, which is a fine value of Young's Modulus for a metal.

As something of a review, keep in mind that nylon fishing line and catgut are both *flexible* (as exemplified by low Y values), and yet both are quite *strong.* Cartilage, cooked spaghetti, and Jello are also all *flexible,* but each is relatively *weak* (they have low breaking strengths). On the other hand, a pretzel or a dried noodle is *stiff* (large Y) but *weak,* breaking easily, whereas a steel crowbar is *stiff* and *strong.*

10.13 Shear and Bulk Moduli

The stress-strain curves for shear are also often linear initially and therefore an elastic **Shear Modulus** (S), or **Modulus of Rigidity**, can be introduced:

$$\text{Shear Modulus} = \frac{\text{shear stress}}{\text{shear strain}}$$

that is

$$S = \frac{\sigma_s}{\varepsilon_s} = \frac{F/A}{\gamma} \tag{10.11}$$

where the force acts parallel to the area, and we have made use of Eq. (10.8). The Shear Modulus is typically 2 or 3 times less than the Young's Modulus for the same material (Table 10.7).

Thomas Young (1773–1829), physicist and physician, has a long list of accomplishments to his credit, including his introduction of the Modulus of Elasticity. He is best known for his work in optics.

The top of a failed extension ladder. Each side had a steel holding plate fastened to the body with three aluminum rivets. The steel plates are much harder than the rivets, and when there was enough torque, they sheared the rivets in half. The plate twisted upward, still holding its half of the three severed rivets. One disembodied head has dropped out of its hole in the body and rests on the flange. The ladder collapsed and the man on it luckily broke only his wrist.

TABLE 10.7 Shear and bulk moduli

Material	Shear Modulus (GPa)	Bulk Modulus (GPa)
Rubber	0.000 8–0.001 6	
Ethyl alcohol		0.9
Water		2
Seawater		2.1
Pine	0.6	3
Glycerin		4.5
Lead	5	8
Mercury		25
Quartz	8–30	35
Compact bone	3.5	12
Marble		70
Aluminum	30	70
Glass	8–30	40
Iron, cast	50	60
Granite		50
Brass	40	70
Copper	50	140
Iron, wrought	85	60
Steel, stainless	85	70
Steel, structural	80	160
Tungsten	150	200
Diamond		620

When an ordinary helical spring is "stretched," the coiled wire itself does not elongate. Rather than being in tension or compression, the wire is actually twisted by a load that tends to pull it out into a straight length. Thus, the spring constant is actually dependent on the Shear Modulus.

Figure 10.22 shows the geometry for a tube, resembling a hollow human bone, being twisted by a torque τ. The resulting angle $\gamma \approx (R_0\phi)/l$ is proportional to τ/S. If some structure or component is twisted (a bone, for example) through a large enough angle ϕ, it may fracture, which is what often happens to the tibia when a skier's leg is twisted in a fall (Fig. 10.23). Table 10.8 lists a few values of both the breaking torque and the corresponding value of ϕ for some human bones.

Similarly, we can define a **Bulk Modulus** (B) as the ratio of the volumetric stress (the pressure P) to the volumetric strain:

$$\text{Bulk Modulus} = -\frac{\text{volumetric stress}}{\text{volumetric strain}}$$

that is

$$B = -\frac{\sigma_\text{v}}{\varepsilon_\text{v}} = -\frac{F/A}{\Delta V/V_0} \qquad (10.12)$$

The negative sign makes the modulus positive since an increase in stress causes a decrease in volume and therefore a negative strain (Table 10.7). Notice that the larger the value of B, the smaller the change in the volume ΔV, the less compressible is the material.

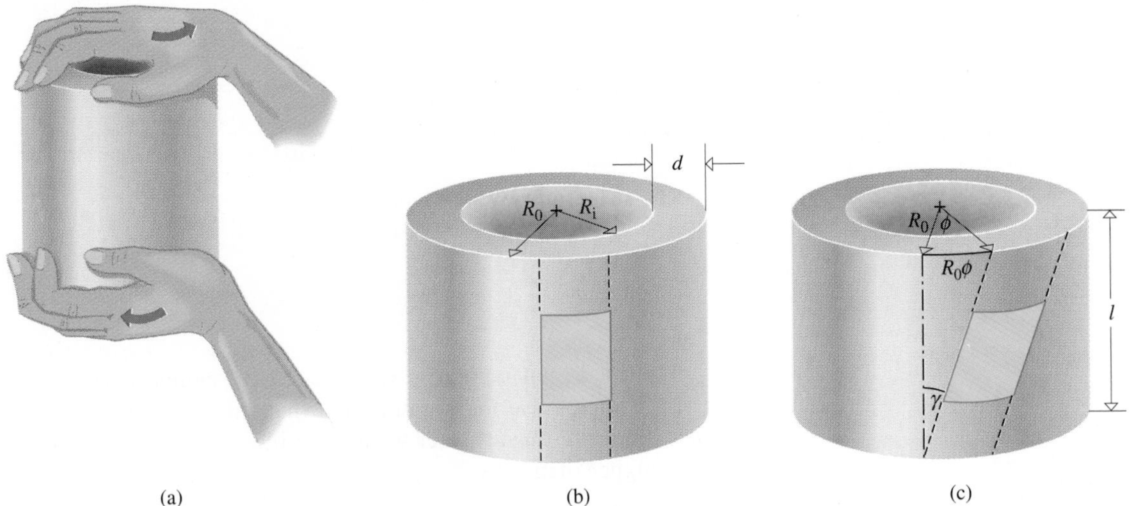

Figure 10.22 As an example of the theoretical Shear Modulus, consider a hollow thin-walled tube: $S = (F/A)/\gamma$. (a) A torque is applied and the rod is twisted through an angle γ. A small region in the untwisted state (b) experiences shear (c) in the twisted state.

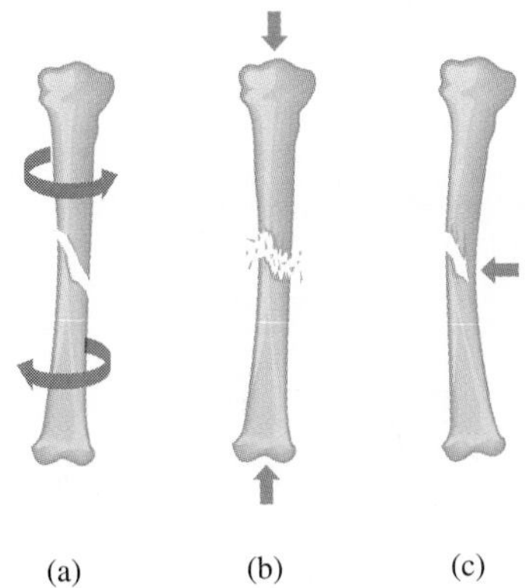

Figure 10.23 Common bone breaks. (a) A spiral fracture results from a torque that twists the bone. This type of break is often associated with skiing accidents. (b) Under compression, the bone may fail and the fractured ends break into pieces. (c) An applied transverse force bends the bone. The side away from the force is stretched, and since bone is weak in tension, it shatters. The near side is in compression, and if the bone is flexible, it might not break there at all.

TABLE 10.8 Torques and breaking angles for human bones

Bone	τ (N·m)	ϕ (degrees)
Ulna or radius (forearm)	20	15
Humerus (upper arm)	60	6
Femur (thigh)	140	1.5
Tibia (lower leg)	100	3

Example 10.9 Determine the change in the volume of a 1.00-m^3 block of granite when it's submerged about 3 km in the ocean, where the pressure on all its surfaces is about 300 times ordinary atmospheric pressure (1.013×10^5 Pa).

Solution: [Given: $P = F/A = 300(1.013 \times 10^5 \text{ Pa})$, $V_0 = 1.00 \text{ m}^3$. Find: ΔV.] From Table 10.7, $B = 50$ GPa; hence from Eq. (10.12)

$$\Delta V = -\frac{V_0 F/A}{B}$$

$$\Delta V = -\frac{(1.00 \text{ m}^3)(3.039 \times 10^7 \text{ Pa})}{50 \times 10^9 \text{ Pa}} = \boxed{-6.1 \times 10^{-4} \text{ m}^3}$$

▶ **Quick Check:** $B = -(F/A)/(\Delta V/V_0) \approx (-3 \times 10^7)/(-6 \times 10^{-4}) \text{ Pa} \approx 0.5 \times 10^{11} \text{ Pa}$.

Clearly, solids do not compress very much. By comparison, water is about 25 times more compressible than granite, but that still isn't a great deal. Nonetheless, estimates indicate that if water were truly incompressible, sea level would be roughly 30 m higher than it is.

Core Material

An **element** is a homogeneous material that cannot be broken down into another substance by any chemical process. Carbon (^{12}C) is the mass standard, equal to 12.000 00 *unified atomic mass units* (u). A volume of 22.4 liters of any gas at STP has a mass in grams numerically equal to its molecular mass and that constitutes a **mole** (p. 320). The number of molecules per mole is Avogadro's number: $N_A \approx 6.022 \times 10^{23}$ molecules/mole.

Density is defined as

$$\text{Density} = \frac{\text{mass}}{\text{volume}}$$

$$\rho = \frac{m}{V} \qquad [10.1]$$

Solids can be classified as either *crystalline, quasi-crystalline, amorphous,* or *composite* (p. 324).

Hooke's Law for a spring is

$$F = ks \qquad [10.3]$$

where s is either the distance it is elongated or compressed and k is the elastic constant with units of N/m. The elastic potential energy stored in such a spring is

$$\Delta \text{PE}_e = \tfrac{1}{2}ks^2 \qquad [10.4]$$

Many materials manifest a linear relationship between the load F they support and the resulting change in length of the sample (ΔL):

$$F = k\Delta L \qquad [10.5]$$

Stress is defined as the force acting, divided by the area over which it acts:

$$\sigma = \frac{F}{A} \qquad [10.6]$$

Strain is the fractional change in some geometrical measure of a sample under stress. Thus, for a specimen experiencing an axial load, the *normal strain,* compressive or tensile, is

$$\varepsilon = \frac{\Delta L}{L_0} \qquad [10.7]$$

Similarly, the *shear strain* equals the angle of distortion γ:

$$\varepsilon_s = \gamma \approx \tan\gamma \approx \frac{\Delta l}{l_0} \qquad [10.8]$$

The **strength** of a material is the stress, in N/m^2, that will cause failure of some sort in a sample.

The *elastic stress per unit strain* defines **Young's Modulus**:

$$Y = \frac{\sigma}{\varepsilon} \qquad [10.9]$$

in either tension or compression. The units of Y are the same as those for stress—namely, Pa. Alternatively

$$F = \frac{YA}{L_0}\Delta L \qquad [10.10]$$

which is the equivalent of Hooke's Law, $F = k\Delta L$.

The elastic **Shear Modulus**, or **Modulus of Rigidity**, is the shear stress over the shear strain:

$$S = \frac{\sigma_s}{\varepsilon_s} = \frac{F/A}{\gamma} \qquad [10.11]$$

Similarly, the **Bulk Modulus** is the ratio of the volumetric stress (the pressure P) to the volumetric strain:

$$B = -\frac{\sigma_V}{\varepsilon_V} = -\frac{F/A}{\Delta V/V_0} \qquad [10.12]$$

Suggestions on Problem Solving

1. Area conversions (for example, cm^2 to m^2) are common in this chapter, so be careful with them. Don't mix up diameters and radii when calculating areas.

2. We deal extensively with the abbreviated notation of mega- and giga-, as for example MPa and GPa, which is often the way data is given in the literature. Get in the habit of entering the numbers in your calculations in scientific notation right away. It's easier to substitute the numbers in MPa, but it's also easier to lose the factor of 10^6 that way.

3. Young's Modulus provides an insight into the meaning of the spring constant by way of Eq. (10.10). Hence, when confronted with a problem dealing with the physical parameters that describe the object being loaded (length, cross section, etc.) think in terms of Y and that equation.

Discussion Questions

1. Where in your home might you find each of the following?

aluminum	tungsten	carbon	oxygen
mercury	zinc	chlorine	chromium
copper	sodium	iron	nickel
lead	magnesium	cobalt	

2. Prestressed concrete structural forms are made by setting up steel rods within a mold and then pouring in concrete around the rods, which are kept stretched until the concrete sets. Why is this manner of construction advantageous?

3. Figure Q3 shows the stress-strain curves for several kinds of steel. (a) Which is the most ductile? (b) Which is the most brittle? (c) What can be said about their Young's Moduli? (d) Which has the greatest ultimate strength? (e) Which has the greatest rupture strength? (f) Which has the least yield strength? (g) Which has the greatest stiffness in the elastic region? (h) Which has the greatest strain at rupture?

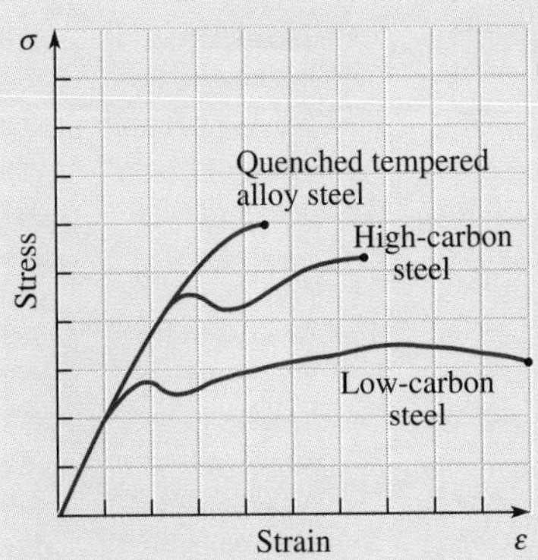

Figure Q3

4. Refer to Fig. Q3. If a high-strength steel is substituted for a low-strength steel and all else is kept constant, what will happen to the load-carrying capacity and stiffness of the structure?

5. The area under the force-displacement curve is called the **strain energy**. (a) Discuss the physical meaning of this quantity. Similarly, the area under a stress-strain curve is called the **strain-energy density**. (b) Discuss its physical meaning. What are the units of strain-energy density? (c) Figure Q5 shows a σ-ε curve for a material strained to a value of ε_F beyond the yield point. It is unloaded and returns to zero stress, but not zero strain. There is a permanent strain ε_p. What is the significance of the colored area?

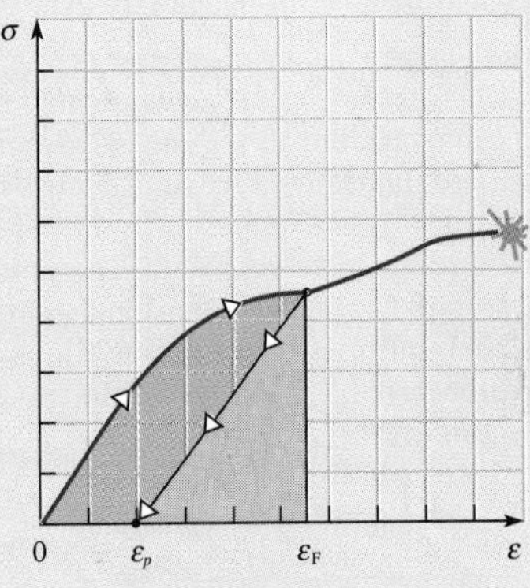

Figure Q5

6. Why are "steel-belted" tires belted with steel? Or, alternatively, why is it advisable to cover a recurring crack in a ceiling with cloth mesh before replastering it?

7. Imagine two vertical hanging ropes that are identical except that one is considerably longer than the other. A gradually increasing load is applied to each of their bottom ends until one or both break. (a) What can you say about the two breaking loads? (b) Compare the amounts of strain energy stored—that is, the amount of energy pumped in—in order to break the ropes. (See Question 5.)

8. Human bones in the embryo stage are mostly collagen and only later are they gradually reinforced with calcium and phosphorus. Compare the corresponding values of Young's Modulus and discuss the implications in regard to the mechanical properties of children.

9. Mountains are typically shaped like a cone. Why is that configuration more able to reach great heights than a straight vertical columnlike structure? Explain why a column of marble about 4.2 km high would collapse under its own weight. Given that granite has about the same density as marble, how tall a column could be made of it before it crushed itself?

10. There is a general principle (attributed to Galileo) that deals with scaling structures—that is, increasing or decreasing their sizes proportionately. For example, King Kong was a scaled-up gorilla. The *square-cube rule* points out that while the volume of a thing being scaled-up increases as the cube of the dimensions, the cross-sectional area of the load-supporting members only increases as the square of the dimensions. How does the weight of the structure increase? What happens to the stress in the load-bearing members? Assuming King Kong was

made out of normal gorilla-stuff, why would he have to be very careful when standing up?

11. The strengths and rigidities of animal bones do not vary all that much from one type of mammal to the next. Human bone (femur), for instance, has a relatively high compressive strength of 170 MPa; at the other extreme, pig bone has a low compressive strength of 100 MPa—not that much different. With Question 10 in mind, what does this suggest about the relative safety of hopping rabbits as compared to jumping elephants? What advice would you give to a horse about prancing around?

12. Rubber at a working elastic strain of 300% has a stress of 7 MPa, and the area under its nonlinear stress-strain curve is 10 MJ/m^3. What is the significance of this area (which, by the way, is about 10 times the corresponding value for spring steel)? Why do we make the heels of shoes out of rubber? (See Question 5.)

13. The stress-strain diagram for vulcanized rubber is shown in Fig. Q13. The sample returns to its original length when the load is removed, but not along the same curve—this phenomenon is called **elastic hysteresis**. Discuss the work done *on* the sample to deform it as compared to the work done *by* the rubber on returning to zero strain. What happens to the difference? Why do we put rubber legs under vibrating pieces of equipment or rubber bumpers on automobiles and tugboats? Stretch a heavy rubber band quickly a few times—what will happen to its temperature?

14. The inverse of the Bulk Modulus is known as the *compressibility*. Is this name appropriate? Explain.

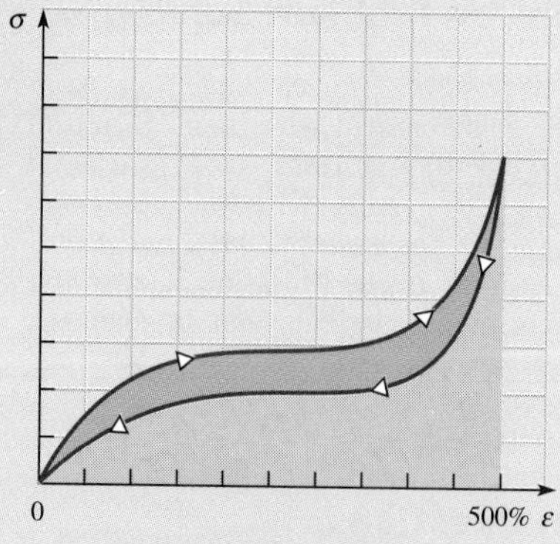

Figure Q13

15. For a Hookean material, the area under the linear portion of the stress-strain curve is called the **resilience**. It's a measure of how much energy the material or structure can store without permanently distorting. The area under the curve all the way out to rupture is known as the **toughness**. It's a measure of the ability of a material or structure to withstand a blow without rupturing. In other words, it's a measure of the energy it takes to break the object. What can you say about the resilience of a human tendon or a bridge or a steel spring? Is a glass bowl or a china plate tough? Explain. Suppose you were designing a bow for shooting arrows or a pole for vaulting. Do you want these to be tough or resilient?

Multiple Choice Questions

1. The dimensions of density are (a) $[L/M]$ (b) $[L^3/M]$ (c) $[M^3/L]$ (d) $[M/L^3]$ (e) $[M/L]$.
2. If we totally decompose 11 g of carbon dioxide, how many grams of oxygen would be liberated? (a) 11 (b) 8 (c) 3 (d) 44 (e) none of these.
3. Figure MC3 shows a nail that was left in a concentrated solution of sugar for several days. What has grown on the nail can best be described as (a) a single crystal (b) an amorphous mass (c) a glassy solid (d) a polycrystalline mass (e) none of these.
4. Methane melts at −184°C, whereas table salt melts at 801°C. From this we can conclude that the crystal's cohesive forces, the bonding forces, (a) are much larger in sodium chloride than methane (b) have nothing to do with melting (c) are both ionic (d) are both van der Waals (e) none of these.
5. Figure MC5 shows a stack of cotton swabs. In two dimensions, it suggests a crystal system known as (a) FCC (b) HCP (c) SC (d) LSD (e) none of these.
6. A simple molecule such as water has a size of about (a) 0.3 nm (b) 3.0 nm (c) 3×10^{-6} m (d) 0.3 mm (e) none of these.
7. The dimensions of stress are (a) $[L/M]$ (b) $[L^3T/M]$ (c) $[M^3T/L]$ (d) $[M/T^2L]$ (e) $[M/T^3L]$.
8. A metal part fails via fatigue (a) when it gets tired (b) as a result of the repeated application of a cyclic stress (c) when the stress exceeds the breaking strength (d) when the metal is very old and the load very large (e) none of these.

Figure MC3

9. An ordinary piece of blackboard chalk is (a) ductile (b) quite tough (c) brittle (d) highly resilient (e) none of these.

Figure MC5

10. A rod of tungsten is quite stiff, and so one can say that tungsten (a) is soft (b) is tough (c) has a large conductivity (d) has a large Young's Modulus (e) none of these.
11. A spring with a large elastic force constant can be said to be fairly (a) soft (b) rigid (c) long (d) short (e) none of these.
12. A spring balance with a linear scale can be used to measure mass because (a) springs are strong (b) metal springs are nearly weightless (c) a spring-force is linear because it has a large Young's Modulus (d) the spring-force is linear in displacement via Hooke's Law (e) none of these.
13. A piece of wire is stretched by a certain amount and then allowed to return to its original configuration. It is then stretched twice that initial amount, without exceeding its elastic limit. Compared to the first stretching, the second elongation stored (a) twice as much energy (b) four times as much energy (c) half as much energy (d) no additional energy (e) none of these.
14. A concrete block supported horizontally at its two ends can be broken with a downward karate chop to the middle because (a) the bottom of the block will be in tension and concrete is weak in tension so it will crack on the bottom face (b) the top will be in compression and concrete is weak in compression so it will crack at the top (c) the bottom will be in compression so it will crack on the bottom (d) the top will be in tension so it will crack on top (e) none of these.
15. It is possible, while tightening the nuts on the wheel of your car, to use too much torque and break off one of the bolts. This happens when the (a) shear stress exceeds the breaking strength (b) compressive stress becomes too large (c) screw has too low an elastic constant (d) volumetric stress on the bolt is too great (e) none of these.
16. The units of strain are (a) m/s (b) N/m (c) m^2 (d) N/m^2 (e) none of these.
17. The breaking strength of a ductile material is generally (a) lower than the yield strength (b) very low (c) higher than the yield strength (d) higher than the ultimate strength (e) none of these.
18. A steel wire stretches by 10 cm when a load is hung from it. If a new wire made of the same material with twice the cross-sectional area supports the same load, it will stretch (a) the same amount (b) not at all (c) half as much (d) twice as much (e) not enough information given.
19. A dress made of cloth will hang nicely, conforming to the shape of the body beneath, because cloth has both (a) a large Young's Modulus and a low Bulk Modulus (b) a low Shear Modulus and a low Modulus of Elasticity (c) a high Young's Modulus and a low Modulus of Toughness (d) a low Bulk Modulus and a high Shear Modulus (e) none of these.
20. Of the several moduli, the only one applicable to liquids and even to gases is (a) Young's Modulus (b) the Modulus of Toughness (c) the Shear Modulus (c) the Elastic Modulus (d) the Bulk Modulus (e) none of these.

Problems

ATOMS AND MATTER

1. [I] How many molecules of trichloromonofluoromethane are there in 22.4×10^3 cm^3 of the gas at STP?

2. [I] What quantity of water will be produced when 5.0 g of hydrogen are burned in the presence of 40 g of oxygen? What if 6.0 g of hydrogen are burned in the presence of the same amount of oxygen?

3. [I] White dwarf stars are very small and very massive. A 1-in.^{3} chunk of such star-stuff brought to Earth would weigh about 1 ton (2000 lb). Determine its density in SI units.

4. [I] The mass of the Earth is 5.975×10^{24} kg, and its mean diameter is $1.274\,246 \times 10^7$ m. Determine its average density. Given that the density of nuclear matter is about 2×10^{17} kg/m^3, if the Earth were squashed into a small sphere, how big would it have to be to have a comparable density?

5. [I] A typical human being contains roughly 13 gallons of water, where 1.00 gallon = 3.785 liters. What is the corresponding mass of water?

6. [I] How many grams of salt are in a gram-mole of NaCl?

7. [I] Imagine that a kilomole of a substance undergoes some chemical change that liberates 2.0 eV per reaction, per molecule. How much energy will be released by the entire kilomole?

8. [I] A cubic centimeter of water has a mass of 1.00 g. How many *molecules* does it contain? How many atoms are in a 1.00-g ice cube?

9. [I] How many molecules of pure sugar ($C_{12}H_{22}O_{11}$) are in 684 g of the stuff?

10. [I] A ring of pure gold has a mass of 19.7 g. How many atoms are in it?

11. [I] Benzoylmethylecgonine (cocaine) is $C_8H_{13}N(OOCC_6H_5)(COOCH_3)$. What is its atomic weight?

12. [I] Assuming that you weigh 150 lb and that your body is roughly 70% water, how many H_2O molecules are you?

13. [I] Make a rough estimate of the number of atoms in all the stars in the Universe. (Hint: Assume an average atomic mass of, say, 10 u; a star-mass of 10^{33} g; 10^{11} stars in a galaxy; and 10^{11} galaxies in the Universe.)

14. [II] Derive a general expression for the number of molecules there are in a cubic meter (the number-density) of any substance at STP.

15. [II] Using the results of Problem 14, determine the number-density for water—that is, the number of molecules there are per cubic meter. (Check your answer with Table 10.2.)

16. [II] The mass of a proton is 1.673×10^{-27} kg, and it can be considered to be a sphere of roughly 1.35×10^{-15}-m radius. Determine its density and compare it with that of most solids. How much is that in tons per cubic inch (1 ton = 2000 lb)?

17. [II] A cubic foot of water weighs 62.4 lb. How many kilograms of hydrogen are in that much water?

18. [II] What is the mass of a single water molecule, and how many of them are in 1.000 g of pure water? The atomic masses of hydrogen and oxygen are 1.008 u and 16.00 u, respectively.

19. [II] Ethyl alcohol, C_2H_5OH, has a density of 0.789×10^3 kg/m^3. Estimate the size of the molecule.

20. [II] Assume that a water molecule is a sphere of diameter 0.3 nm. Approximately how many of these spherical molecules will be in 18 cm^3 of water? How does this result compare with the correct answer? Why is it in error?

21. [III] What is the approximate separation between gas molecules at STP? How does this compare to the separation between molecules in a solid?

ELASTICITY MODULI AND ENERGY

22. [I] Given a spring with an elastic constant of 500 N/m, how much will it contract when pushed on by an axial force of 10 N?

23. [I] A helical spring 20 cm long extends to a length of 25 cm when it supports a load of 50 N. Determine the spring constant.

24. [I] How much energy is stored in a helical spring with an elastic constant of 50 N/m when compressed 0.05 m?

25. [I] An axial force of 200 N is applied to a rod with a cross-sectional area of 10^{-4} m^2. What is the normal stress within the rod?

26. [I] A man leans down on a cane with a vertical axial force of 100 N. The narrowest part of the cane has a cross-sectional area of 1.0 cm^2. What is the maximum compressive stress in the cane?

27. [I] A 10-m-long wire strung between two trees has a "Curb Your Dog" sign weighing 10 N hung from it. If the wire stretches 1.5 cm, what strain is it under?

28. [I] A rubber band is stretched to twice its original length by a kid who is going to shoot it across a room. What strain is it experiencing?

29. [I] If the normal strain experienced by a rod in compression is 5.000×10^{-4} and its original length is 40.00 cm, what's its final length?

30. [I] Given that collagen has an ultimate tensile strength of 60 MPa, what maximum load could be supported by a fiber with a cross-sectional area of 10^{-6} m^2?

31. [I] A length of tendon L_0 mm long stretches an amount ΔL under a load F. A second piece twice as long, with twice the cross-sectional area, experiences the same load. By how much will it elongate in comparison?

32. [I] Pure natural rubber has a tensile strength of 21 MPa. When carbon particles are added to reinforce it, the tensile strength rises to 31 MPa (which is one reason why you see so much black rubber around). If a rod of natural rubber can support a maximum load of 1000 N, how much will an identical rod of carbon-reinforced rubber support?

33. [I] A rod made of structural steel, 4 m long and 2 mm in diameter, supports a 1.0-lb microphone hanging 3 m above the floor of a gym. A 181.4-kg (400-lb) Sumo wrestler reaches up, grabs the mike, and hangs from it. How much will the rod stretch, assuming it doesn't break?

34. [I] A piece of fresh human cartilage, with a 1.0 cm^2 cross-sectional area, is being studied in a laboratory. It is found that when loaded with 100 N, its length increases by 4.2%. Determine its Elastic Modulus.

35. [I] A dictionary weighing 10 N rests face down on a table. The book's cover is 20 cm by 30 cm. What is the pressure it exerts on the table?

36. [I] A cube of marble 1.00 m on each edge is submerged in the sea so that there is a distributed force F acting perpendicularly inward on each of its faces. What must that force be if the block is to decrease in volume by 1.00%?

37. [II] A helical spring is 55 cm long when a load of 100 N is hung from it and 57 cm long when the load is 110 N. Find its spring constant.

38. [II] Two identical helical springs are attached to one another, end-to-end, making one long spring. If $k = 500$ N/m is the elastic constant of each separate (essentially weightless) spring, what is the constant for the combination?

39. [II] Estimate the value of the spring constant for an atom bound in a molecule. Take the potential energy stored when the displacement from equilibrium is 0.2 nm to be 2 eV $= 3.2 \times 10^{-19}$ J.

40. [II] Two identical wires each of length L_0 are attached to one another producing a single long wire. If $k = 500$ N/m was the elastic constant of each separate (essentially weightless) wire, what is the constant for the combination?

41. [II] A brass wire 1.5 m long is to support a 200-N "Eat At Joe's" sign without permanently elongating. Given that the yield strength of this variety of yellow cold-rolled brass is 435 MPa, what minimum diameter wire is called for?

42. [II] A thin rod of structural steel 5.0 m long is to carry a tensile load such that it's just about to begin to deform plastically. Given that the yield strength in tension of steel is 250 MPa, how much will the rod stretch?

43. [II] A rubber motor belt with a circular cross section of radius 1.25 cm breaks at a tensile load of 7363 N. If a new belt was made of the same material with twice the cross-sectional radius, at what load would it break?

44. [II] What is the largest working strain you should design for when dealing with a structural steel member that is not to be permanently deformed by the tensile load? The yield

strength of the steel in tension is 250 MPa as compared to its tensile strength of 400 MPa.

45. [II] Marble crushes under a compressive stress of 110 MPa. What maximum strain will it experience?

46. [II] The compressive strength and maximum strain of the crown of a human tooth are 150 MPa and 2.3%, respectively. Assuming the crown to behave elastically until it ruptures, determine its Young's Modulus and (remembering Discussion Question 15) its resilience.

47. [II] The cable on a crane is to be made of high-strength, low-alloy steel with a yield strength in tension of 345 MPa. If the maximum load is to be 25 000 lb, how big in diameter should the cable be? What maximum strain can it experience given that its Modulus of Elasticity is 200 GPa?

48. [II] A certain kind of brick has a density of $2 \times 10^3\ \mathrm{kg/m^3}$ and a compressive strength of 40 MPa. What's the highest vertical parallel-walled tower that can be built that will not crush the bottom bricks?

49. [II] A young pine tree experiences a horizontal force of 200 N and bends through an angle of 0.001 rad. Approximate its diameter.

50. [II] A structural steel bar 4.0 cm in diameter juts horizontally 50 cm straight out of a wall. A 90-kg (200-lb) person reaches up, grabs the bar at its end, and hangs from it. By how much will the bar deflect? (Hint: Use the Shear Modulus.)

51. [II] Two 1.00-cm-thick steel plates on a bridge are riveted to one another with five 4.00-cm-diameter steel rivets. Given that the yield strength in shear of structural steel is 145 MPa, what maximum force parallel to the plates can they be expected to handle without permanent deformation?

52. [II] Show that the *elastic*-PE stored in a normally strained material of uniform cross-sectional area A and original length L_0 is given by

$$\mathrm{PE_e} = \frac{F^2 L_0}{2AY}$$

53. [II] The bar in Fig. P53 is made from a single piece of material. It consists of two segments of equal length $\frac{1}{2}L_0$ and cross-sectional areas of A and $2A$. Show that the *elastic*-PE stored under the action of an axial force F is

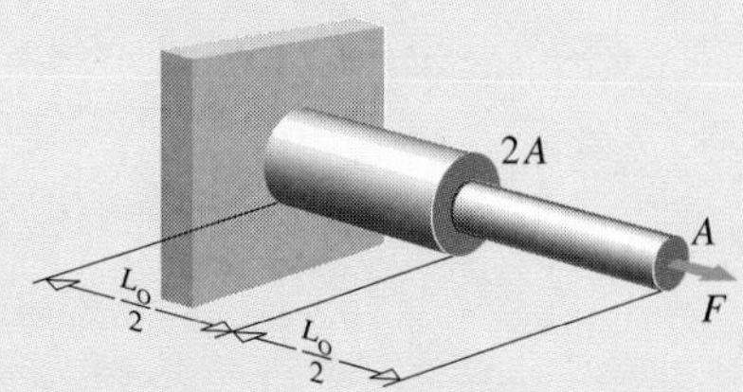

Figure P53

$$\mathrm{PE_e} = \frac{3F^2 L_0}{8AY}$$

(Hint: Look at Problem 52.)

54. [III] A solid lead sphere is placed in a pressure chamber so that the force per unit area acting on the sphere is hundreds of times atmospheric pressure (1 atm = 1.013×10^5 Pa). What pressure will produce a 1% increase in the density of the lead?

55. [III] If a uniform rod of cross-sectional area A and length L_0 can sustain a maximum stress of σ_R without rupture, show that it stores an amount of elastic energy given by

$$\mathrm{PE_e} = \frac{\frac{1}{2}AL_0\sigma_R^2}{Y}$$

in getting to that stress level.

56. [III] A robot's main body is supported on two legs, each containing two vertical rods one above the other jointed at the knee in human fashion. Each rod has a cross-sectional area of 5 cm² and a length of 50 cm and is made of a bonelike material that can sustain a maximum stress of 150 MPa without rupturing and has a Young's Modulus (in compression) of 10 GPa. How much elastic energy can be stored in these two leg supports (as a result of jumping in a gravity field) without having to go back to the shop for repair? If the robot has a mass of 70 kg, how high a drop can it sustain on Earth?

11

Fluids

LIQUIDS AND GASES FLOW, so we call both *fluids.* Their atoms and/or molecules can move around fairly freely, and that contributes to a range of shared properties. Still, there are difficulties in defining the term "fluid" precisely. Even diamond can be made to flow like soft wax at high enough pressures ($>1.7 \times 10^{11}$ Pa). It might be more practical to say that a fluid is an aggregate of atoms that does not significantly resist shear.

The mobility inherent in fluids makes them crucial to all known life forms. The human body itself is a fluid-dynamical system; we breathe, drink, bleed, and excrete fluids. This chapter is about the physics of gases and liquids. It first treats characteristics that are independent of any bulk motion: hydrostatic pressure, hydraulics, and surface tension, which are all examples of *fluid statics.* The chapter then deals with behavior that derives from bulk motion of the medium: fluid flow, Bernoulli's Equation, and vortices, all of which are examples of *fluid dynamics.*

FLUID STATICS

When most solids (such as ice or steel) are heated, their atoms and/or molecules pick up KE in the form of random vibrational motion: they oscillate more vigorously about their equilibrium locations. Once enough energy is obtained to overcome the intermolecular forces, a solid melts. Small groupings of associated molecules still persist in the *liquid*, but as the patterns shift, they link and unlink. There is order, but it's local and changing. A web of cohesive, fairly long-ranged, intermolecular force remains, though the molecules are now energetic enough to easily move against it. They stay relatively near to each other, still interacting appreciably, but the locked, powerful bonding of the solid is gone.

Raising the temperature further brings the liquid to the boiling point, where bonds between molecules are overcome. For some molecules, their random thermal KE exceeds the cohesive potential energy, and they escape the liquid altogether. The short-range groupings disperse and the liquid vaporizes—it turns into a *gas.*

11.1 Liquids

Very few liquids occur naturally in any great quantity. Apart from water and petroleum, most others, such as gasoline, are manufactured. Not surprisingly, living organisms are typically composed of between 65 and 95% water. In fact, many of the familiar liquids—from blood and beer to berry juice—are primarily water.

The balance between bonding energy and thermal energy determines the state of an atomic system. The liquid is a transition state, existing between the random violence of the gas and the relative calm of the solid. At ordinary temperature and pressure, the liquid state is most effectively formed of either highly polarized molecules with electrically positive and negative ends (like water) or large heavy molecules (like petroleum). As with Goldilocks's porridge, the temperature of the liquid has to be "just right" with respect to the intermolecular forces—too hot and it's a gas, too cold and it's a solid.

The ability to flow, the hallmark of the fluid, varies with the cohesive force from one substance to the next—from acetone to water to motor oil to molasses to tar. This notion was first treated analytically by Newton (ca. 1687). *Viscosity* is the internal resistance, or friction, offered to an object moving through a fluid. Usually small molecules, such as water and benzene, move around easily and manifest little viscosity compared to large complex molecules, such as tar. Much of our discussion will be simplified by treating *ideal liquids* that are incompressible and nonviscous—characteristics well approximated by water and many other real liquids under ordinary conditions.

No physicist would have predicted the existence of a liquid state from our present knowledge of atomic properties.

VICTOR WEISSKOPF (1972)
Physicist

11.2 Gases

Despite their basic mechanical similarity, gases display a diversity of characteristics. The majority of them (nitrogen, oxygen, helium, carbon dioxide, and so forth) are transparent. A few are colorful: chlorine is pale yellow, bromine is orange-brown, and iodine vapor is violet. Radon is radioactive and nasty, oxygen is the breath of life, carbon monoxide is deadly, and nitrous oxide is amusing. Natural gas burns, hydrogen explodes, and carbon dioxide extinguishes. Ozone has a sweet metallic odor, hydrogen sulfide smells rotten-egg vile, and water vapor has no scent at all.

The alchemist van Helmont (ca. 1620) transmuted the Greek word for *chaos* into Flemish and came up with the term *gas*—and a raging chaos it is. At this very instant, each square inch of your face, and every other exposed surface at sea level, is being bombarded by 2×10^{24} air molecules every second! They pelt, ricochet, hit other molecules, and rebound, over and over in a tireless fury (powered, for the most part, by energy from the Sun). At its greatest, the density of the atmosphere at our planet's surface is only 1/800 that of water. Still, at STP there are roughly 3×10^{19} molecules careening around in every cubic centimeter. Though the distance between them is constantly changing, a typical separation is about 10 times the size of the molecules themselves or around 3 nm, which is relatively large. Except for a very slight electromagnetic cohesion and an even slighter attraction due to gravity, these widely spaced molecules behave independently.

Nitrogen, the main constituent of air, becomes a liquid at a very low temperature (−196°C). It looks and pours like water but evaporates much more quickly.

Aside from breezes, an average air molecule (Table 11.1) travels at a remarkable 450 m/s, or about 1000 mi/h! Usually a molecule will get no farther than 8×10^{-6} cm before crashing into some other one, only to recoil and collide again. This aimless bombardment goes on for each molecule in a gas at a rate of roughly 6×10^9 collisions per second. Such a jostling mass of molecules will spread out in a container until it encounters and rebounds from the walls. Swelling of its own internal violence, a gas more or less uniformly fills whatever vessel it's in (provided gravity does not play a significant role). The room you are in at this moment is happily "filled" with some 10^{26} to 10^{30} molecules of air, so you needn't worry about walking into an empty region.

Gravity can function on a large scale to restrain a gas much as a container does. It is the internal gravity that causes an interstellar gas cloud to contract into a star, and it is gravity that holds down the Earth's atmosphere. A balance of molecular speed against strength of attraction keeps that thin layer of air (99.999% lies below 90 km) from escaping and leaving us all breathless.

TABLE 11.1 The composition of the Earth's atmosphere

Gas	Percent by volume
Dry air	
Nitrogen	78.08
Oxygen	20.95
Argon	0.9
Carbon dioxide	0.03
Neon	0.002
Helium	0.000 5
Methane	0.000 1
Krypton	0.000 1
Nitrous oxide	0.000 05
Hydrogen	0.000 05
Ozone	0.000 007
Xenon	0.000 009
Typical air	
Nitrogen	76.9
Oxygen	20.7
Water vapor	1.4
All others	1.0

11.3 Hydrostatic Pressure

Rather than forces that act at a few points on a body, we now treat *distributed forces* that act over an extended area. Accordingly, **pressure** (P) is defined as *the magnitude of the force acting perpendicular to, and distributed over, a surface divided by the area of that surface:*

$$P = \frac{F_\perp}{A} \tag{11.1}$$

Pressure is a scalar quantity; at any point, it has magnitude but no direction (Table 11.2). The units of P are N/m^2 or pascals (Pa).

TABLE 11.2 A sampling of pressures

Source	Pressure (Pa)
Sun, at the center	2×10^{16}
Earth, at the center	4×10^{11}
Laboratory, highest sustained	1.5×10^{10}
Ocean, deepest point	1.1×10^{8}
Spiked heels	$\approx 10^{7}$
Atmospheric, Venus	90×10^{5}
Earth	1.0×10^{5}
Mars	≈ 700
Air, Earth 1-km altitude	90×10^{3}
10-km altitude	26×10^{3}
100-km altitude	≈ 0.1
Air, top of Sears Tower, Chicago	95×10^{3}
Air, top of Mt. Everest	30×10^{3}
Air, commercial jet altitude (11 km)	23×10^{3}
Solar radiation, at Earth	5×10^{-6}
Laboratory vacuum (very good)	10^{-12}

Imagine a container filled with a liquid—a cup of tea, say, or a barrel of wine. Gravity acts on the substance and very slightly compresses the fluid, pushing molecule against molecule and ultimately transmitting the force to the bottom of the container, which pushes upward on the liquid. Like a mound of dry sand, the liquid would run off sideways were it not for the walls of the tank. A fluid exerts an outward push on the walls of the container, which restrains it with a counterforce—a counterpressure. This outward pushing of the liquid is observed when water spurts from a hole in the side of a container.

The force exerted by a fluid at rest acting on any rigid surface is always everywhere perpendicular to that surface. It cannot be otherwise since the fluid has no rigidity and will not sustain shear stress. A wooden stick certainly could be used to push on a wall in a direction that is not perpendicular, and the same thing could also be done with a stream of water from a hose, but a fluid *at rest* pushes only perpendicularly. If it somehow managed to exert a tangential force on the wall, the wall would react with a counter tangential force (via Newton's Third Law), and the fluid, unable to resist shear stress, would flow.

A fluid at rest within a container is typically in static equilibrium under the perpendicular compressive forces exerted by the walls. It is important to realize, however, that a liquid, by virtue of its internal (electromagnetic) cohesive force, can also sustain tensile stress (Fig. 11.1). Although liquids usually push outward, they can also pull inward. *The pressure exerted by a fluid will be taken as positive when the fluid is under compression,* as is most often the case (Discussion Question 11).

The mobility of the molecules is the mechanism that transmits pressure *independent of direction.* The value of P measured at any point in a liquid is independent of the orientation of the measuring gauge. Underwater or in a jet plane you can turn your head any way you like and your ears will still feel stuffy.

By supporting a load at a number of "points," the normal force at each is small. The pressure at each "point" is then also relatively small.

Gravity and Hydrostatic Pressure

Gravity is the ultimate cause of hydrostatic pressure. To see this, suppose there is a tank of liquid with a postage stamp in it having an area A parallel to and below the surface at a depth h (Fig. 11.2). The downward normal force on the top face of the stamp due to the liquid must be equal to the weight of the column of fluid above the stamp. Assuming the density ρ is constant (because the fluid is essentially incompressible), the mass of the column is its volume $V = Ah$ times its density; that is, $m = \rho Ah$. Hence, the weight ($F_W = mg$) of the column is ρAhg, and the average pressure on the stamp due to the liquid *only* is

$$P_l = \frac{F_W}{A} = \frac{\rho Ahg}{A}$$

or

$$P_l = \rho g h \tag{11.2}$$

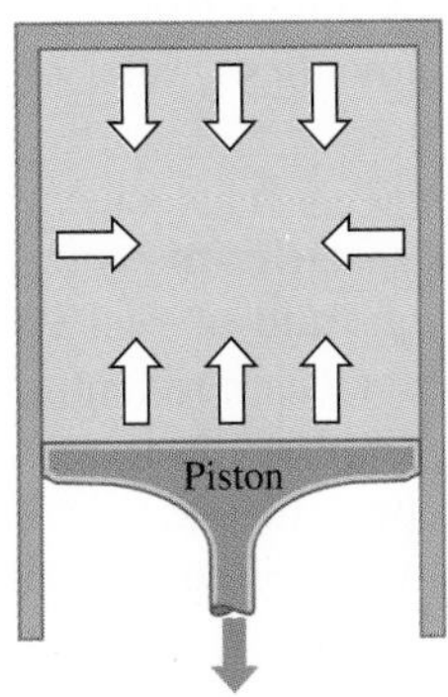

Figure 11.1 A liquid in a cylinder being stretched as the tight-fitting piston is pulled down. This is resisted by the liquid's intermolecular (electromagnetic) cohesive forces pulling inward. Pure water has a tensile strength of about 30 MN/m^2.

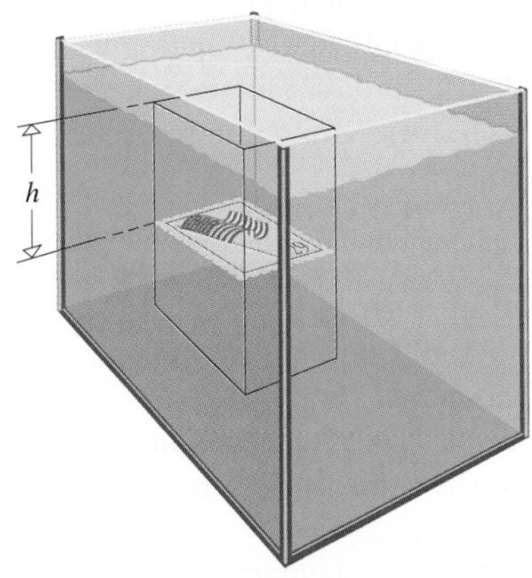

Figure 11.2 The weight of the column of liquid above the stamp, of height h, determines the pressure at that level.

Example 11.1 What pressure (due to only the water) will a swimmer 20 m below the surface of the ocean experience?

Solution: [Given: $h = 20$ m and (from Table 10.1) for seawater $\rho = 1.025 \times 10^3$ kg/m^3. Find: P_l.] From Eq. (11.2)

$$P_l = \rho g h = (1.025 \times 10^3 \text{ kg/m}^3)(9.8 \text{ m/s}^2)(20 \text{ m})$$

and $$\boxed{P_l = 2.0 \times 10^5 \text{ Pa}}$$

or, since 1 Pa $= 1.45 \times 10^{-4}$ lb/in.^{2}, this is 29 lb/in.^{2}, or 29 psi, which is about the pressure in an automobile tire.

▶ **Quick Check:** The pressure increase per 1 m of depth is roughly $(1 \times 10^3 \text{ kg/m}^3)(10 \text{ m/s}^2)(1 \text{ m}) = 10^4$ Pa; at a depth of 20 m, the pressure must be $\approx 20 \times 10^4$ Pa.

For a "weightless" fluid, one far out in space or, more practically, one in free-fall, no internal pressure exists (other than the trivial amount caused by the surface tension). That's true provided the amount of fluid is modest, which means its self-gravity is negligible. This is not the case for something huge like a star. Generally, the force on the postage stamp in any fluid, and therefore the pressure at that level, is determined by the weight of the column above, which is the case whether or not the density varies, as it does in the atmosphere. If the fluid is not homogeneous, Eq. (11.2) will not apply. Still, it can be used successively if there are several homogeneous layers of fluid floating one above the other. When a uniform pressure P_s exists on the surface of the liquid (Fig. 11.3), the total pressure at any point within that liquid is the sum of the contributing pressures:

$$P = P_s + P_l = P_s + \rho g h \tag{11.3}$$

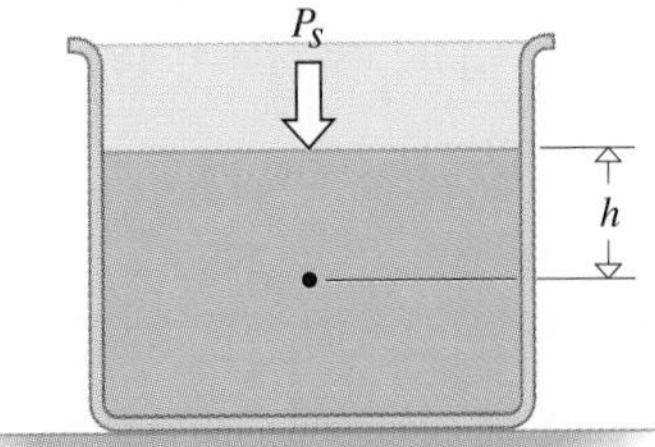

Figure 11.3 In a single liquid, the total pressure at a point a distance h beneath the surface is the sum of the pressure on the surface P_s and $\rho g h$.

The Dutch boy who held back the North Sea by sticking his finger in the dike could really have accomplished the feat. The pressure that he had to overcome de-

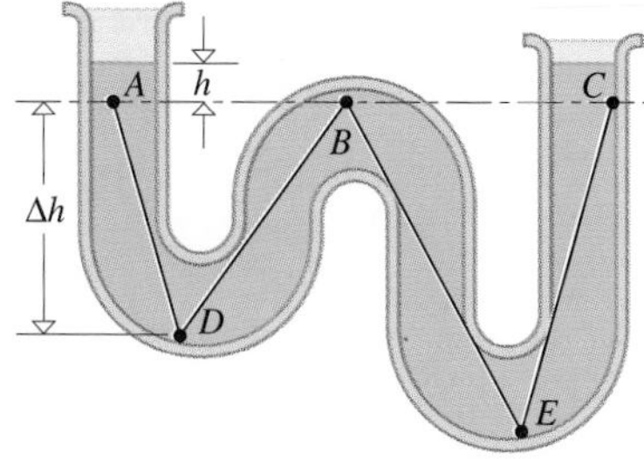

Figure 11.4 The pressure at point A is $P_s + \rho gh$. It increases in going down to D by $\rho g \Delta h$, but decreases by the same amount rising back up to B. Again, it increases even more in descending to E, but drops back that same amount in rising the same vertical distance to point C. Hence, the pressure at A, B, and C is the same.

pended on h, the depth in the liquid. The force on the end of his legendary finger acted perpendicularly, tending to pop it out of the hole like a cork in a champagne bottle. But it didn't matter whether he was holding back a leak from an ocean or a bathtub; there is nothing in Eq. (11.3) about the total amount of water present.

Pressure in a Container

It follows from successive applications of Eq. (11.3) that **the pressure at every point at a given horizontal level in a single body of fluid at rest is the same** (Fig. 11.4). Consequently, the pressure at any point on the flat horizontal bottom of a container of fluid is the same, regardless of how weirdly shaped the container may be, since P only depends on the depth beneath the open surface (Fig. 11.5). The validity of this conclusion was proven experimentally by Simon Stevin in 1586 (Fig. 11.6).

What's happening in Fig. 11.6 is that the walls of the container are interacting with the liquid and participating in the process. Examine the long, narrow-necked vessel with the wide shoulders shown in Fig. 11.7. The pressure at point A differs from that at point B by an amount determined by the weight of the column of liquid between A and B, namely $\rho g \Delta h$. The pressure at B equals that at D and so the pressure at C, also a height Δh above D, must equal that at A, even though C is under the glass shoulder. Anywhere at the level of point A, the pressure in the fluid is the same whether there is liquid or glass directly above the point. The pressure generates an upward normal force on each unit area of the horizontal shoulder, and the glass, in turn, pushes down on the liquid with the same force per unit area. The pressure is precisely equal to that produced by the weight of the column of fluid above A in the neck of the vessel. A guppy swimming around could not tell from the pressure whether it was under the shoulder or the neck.

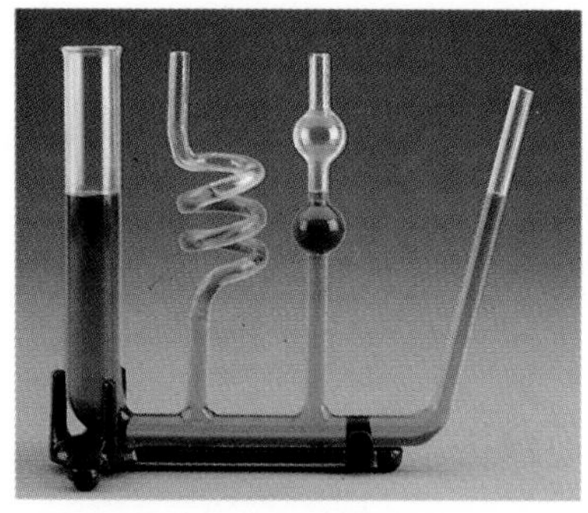

The pressure at the surface of each liquid column must be the same. Therefore, the level of each surface must be independent of the shape of the column and all must stand at the same height.

Atmospheric Pressure and the Barometer

A formidable problem plaguing the European mining community in the seventeenth century was the seepage of groundwater into the deep shafts of iron, silver, alum, and salt mines. The real frustration arose from the mysterious inability of even the best suction pumps to raise the water any more than 34 ft above the flood level. In Italy, Galileo was consulted about the strange phenomenon. The maestro knew that air had weight. He had determined that much experimentally by weighing a glass bulb sealed under room conditions and then reweighing it after forcing in air under pressure. Yet it was only after Galileo died that his assistant, Evangelista Torricelli, was able to recognize that the two apparently unrelated ideas were cause and effect (1643).

Being 13.6 times denser than water, mercury would presumably rise only about 30 in. instead of 34 ft and so would be far more convenient to work with in the laboratory. Cleverly, Torricelli sealed one end of a glass tube about 2 m long, filled it with mercury, and covering the opening with a finger, upended the tube into a bowl of mercury (Fig. 11.8). On removing his finger from the submerged end, some of the liquid poured out of the tube into the bowl, and the shiny column dropped to a height of about 76 cm (30 in.) above the open mercury level. Since no air had entered the tube, the space above the mercury column was apparently empty—the vacuum, forbidden by Aristotelian dogma, was there in the tube, at last.

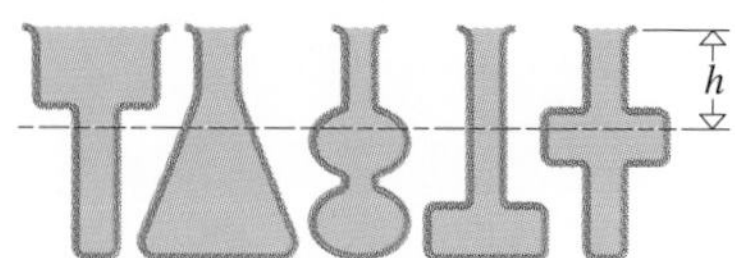

Figure 11.5 The pressure at a given depth is independent of the shape of the vessel. It depends only on the density of the liquid and the depth h.

"On the surface of the liquid which is in the bowl," wrote Torricelli, "there rests the weight of a height of fifty miles of air." Air pressure produces a normal force that pushes down on the open liquid surface, driving the mercury up the tube. The column settles with a height such that the downward force it exerts, at the level of the air-mercury interface, equals the upward force beneath and supporting the column. In other words, *the pressure is equal everywhere at the level of the open surface, both inside the liquid in the tube and out* (Fig. 11.9). The column of mercury effectively takes the place of a much taller, though equally weighty, column of air, producing the same pressure inside as the air produces outside. Clearly, the height of the mercury is a direct measure of the **atmospheric pressure** acting on the device. Torricelli had constructed the first mercury *barometer.*

"We live immersed at the bottom of a sea of . . . air," he observed—a fluid sea wherein pressures are generated just as they are in a liquid. Approximately 5×10^{18} kg of air press down on the planet, giving rise to an average sea-level atmospheric pressure (P_A) of 1.013×10^5 N/m^2 or, in proper SI units, 1.013×10^5 Pa. The study of pressure is particularly rife with antiquated units that are still in common usage (Table 11.3). Thus, an *atmosphere* (abbreviated atm) is defined as 1 atm $= 1.01325 \times 10^5$ Pa, which equals 14.7 lb/in.^{2}. The *torr* (for Torricelli) was formerly the "millimeter of mercury" (or mm Hg), a pressure corresponding to a barometric mercury height of 1 mm and equal to 133.3 Pa. The mm Hg still survives, however, especially in medical literature.

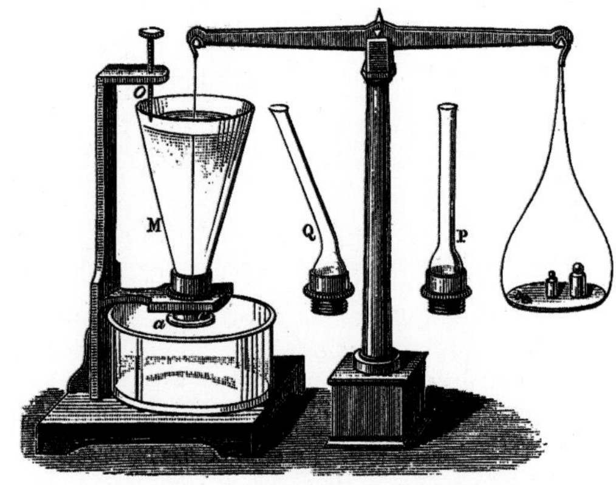

Figure 11.6 Stevin's experiment to determine the pressure on the bottoms of differently shaped containers. The balance holds up the bottom of each test vessel with a force equal to the load on the right-hand pan. When the weight of fluid poured into the left exceeds the load on the right, the bottom drops out on the left.

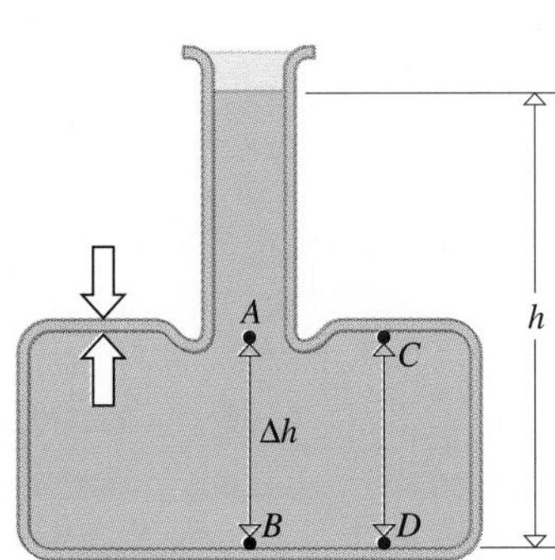

Figure 11.7 The walls of the container contribute to the process that leads to the pressure being dependent only on the depth in the liquid.

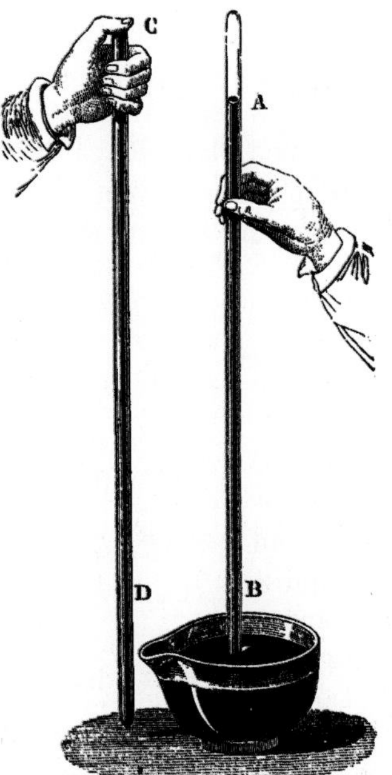

Figure 11.8 The atmosphere pressing down on the open mercury surface can support a 30-in. column. The tube is first filled and then upended in the bowl of mercury. The column drops a little, leaving a partial vacuum in the space above it.

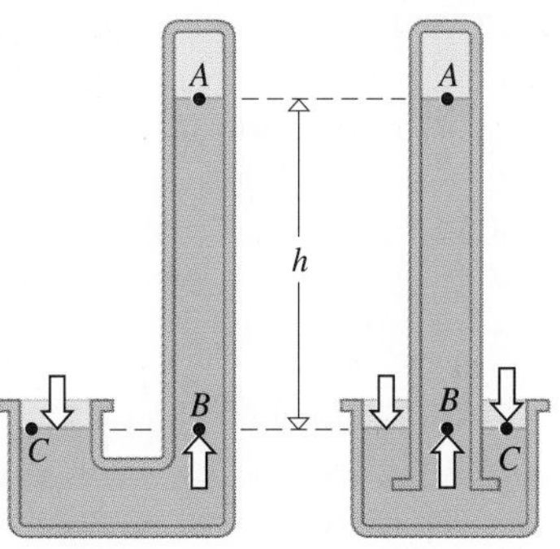

Figure 11.9 Two equivalent versions of the barometer. The pressure at *A* is essentially zero. The pressure at *B* equals that at *C*, which is atmospheric. Hence ρgh must equal atmospheric pressure.

TABLE 11.3 Pressure conversions

Quantity	Equivalent in pascals
1 atm	1.013×10^5
1 lb/in.^{2} (psi)	6.895×10^3
1 mm Hg (torr)	1.333×10^2
1 in. Hg	3.386×10^3
1 in. H_2O	2.491×10^2
1 bar	1.000×10^5
Equivalent of 1 atmosphere	
1.013×10^5 Pa	
14.70 lb/in.^{2}	
7.60×10^2 mm Hg (0°C)	
29.92 in. Hg (0°C)	
1.013 bar	
2.117×10^3 lb/ft^2	
4.068×10^2 in. H_2O	

One standard *atmosphere* corresponds to a barometric reading of 29.9 in. of mercury (76 cm), although air pressure ordinarily varies from 29 to 30 in. of mercury. Incidentally, water vapor is less dense than air (a liter of moist air has a mass of 1.18 g, whereas a liter of dry air has a mass of 1.29 g). As the air becomes drier, it becomes heavier, and a rising barometer usually portends fair weather, whereas falling pressure heralds foul weather.

In 1644, Torricelli wrote a letter to a friend describing the barometer, explaining the effect of the atmosphere, and discussing the vacuum created above the mercury. In the next few years, that letter was circulated around Europe. A copy went to a young man named Blaise Pascal who subsequently (1648) proved that the level of a barometer dropped as it went higher in the atmosphere. At the top of a 3200-ft mountain, the mercury column was about 3 in. lower than before the ascent (p. 394).

Example 11.2 One *atmosphere* is defined as the pressure equivalent to that produced at 0°C by exactly 76 cm of mercury of density 13.5950×10^3 kg/m^3 under standard gravitational conditions, where $g = 9.80665$ m/s^2. Show that a mercury barometer column 76.00 cm (29.921 in.) high corresponds to an air pressure of 1.013×10^5 Pa.

Solution: [Given: Hg column 76.00 cm high. Find: pressure beneath it.] Consider the barometer in Fig. 11.9. The height of the column above the open-surface level is $h = 0.7600$ m. The pressure at point B is

$$P = P_s + P_l = 0 + \rho g h$$

Since there is vacuum above the mercury in the tube ($P_s = 0$), the pressure at B, at the open-surface level, is

$$P = \rho g h = (13.5950 \times 10^3 \text{ kg/m}^3)(9.80665 \text{ m/s}^2) \times (0.7600 \text{ m}) = \boxed{1.013 \times 10^5 \text{ Pa}}$$

▶ **Quick Check:** This is also the pressure outside at the open surface of the mercury—namely, at point C where $P_l = 0$. Since $h = 0$, $P = P_s = P_A = 1.013 \times 10^5$ Pa.

The medicine dropper and its larger culinary cousin, the meat baster, are devices on the family tree of the barometer. The low-pressure region created by squeezing the rubber bulb and allowing it to expand back to size corresponds to the near vacuum in the top of the barometer. Air acting on the open surface of the liquid forces it up the medicine dropper tube. A good bit of residual air left in the bulb pushes down on the top of the liquid column, effectively adding to its weight. The result is that a much smaller amount of liquid is forced up the tube, which is fine, since one rarely wants a 34-ft-long eye dropper. The pipette and the drinking straw are the same machine, with the rubber bulb replaced by a mouth. A drop in mouth pressure of a few percent will operate the thing nicely, while an excess of mouth pressure will force the liquid in the straw below the open-surface level, allowing the operator to "make bubbles."

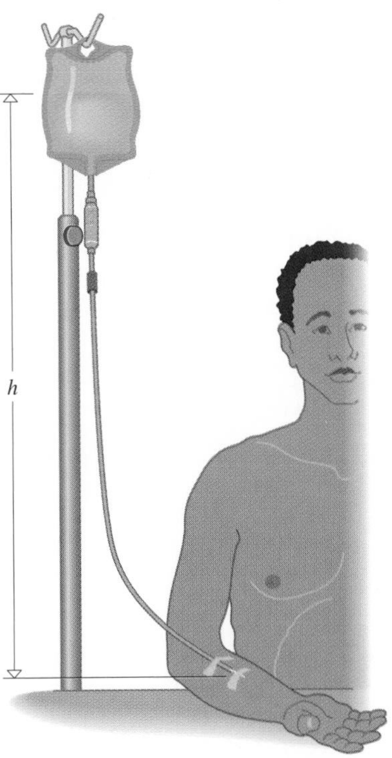

Figure 11.10 The liquid's gauge pressure at the needle is $P_G = \rho g h$, where ρ is the density of the fluid being injected.

Gauge Pressure

In the practical world of pumps, automobile tires, and compressed gas tanks, pressure is measured with a gauge that is set to read zero in the open air and thus neglects the effect of the atmosphere. This pressure above or below atmospheric is the **gauge pressure** (we will distinguish it with a subscript G). The pressure referenced to the perfect vacuum as zero is the **absolute pressure** (P). Until now we have only considered absolute pressure. It's given as the sum of the gauge and atmospheric pressures:

$$P = P_A + P_G \tag{11.4}$$

There is nothing profound about this relationship. If we agreed to limit ourselves to absolute pressure, P_G would be superfluous. But people measure the pressure in veins, chests, and scuba tanks using gauge pressure, which makes Eq. (11.4) a practical necessity.

For the fluid in Fig. 11.10 to flow from the collapsible pouch into the vein, the gauge pressure at the needle ($P_G = \rho g h$) must exceed the gauge blood pressure in the arm. The gauge pressure in a vein (Table 11.4) is less than about 2 kPa (15 mm Hg), and so the pouch must be raised at least 20 cm above the arm.

TABLE 11.4 Some fluid pressures in the human body

Location	Gauge pressure mm Hg	kPa
Brain, surrounding fluid	5–12	0.7–1.6
Eye, aqueous humor	12–24	1.6–3.2
Gastrointestinal	10–20	1.3–2.7
Lungs		
inhalation	−2	−0.3
exhalation	3	0.4
Chest (intrathoracic)		
inhalation	−6	−0.8
exhalation	−2.5	−0.3
Venous blood		
venules	8–15	1–2
veins	4–8	0.5–1
major veins	4	0.5
Arterial blood at heart level		
maximum (systolic)	100–140	13–19
minimum (diastolic)	60–90	8–12
Bladder		
average	0–25	0–3
during urination	15–30	2–4

Example 11.3 A straw is to be used to raise water from a glass to a mouth 15.0 cm above the liquid's surface. Determine the absolute pressure that must be maintained in the mouth. What is the corresponding gauge pressure?

Solution: [Given: $h = 15.0$ cm and $\rho = 1.00 \times 10^3$ kg/m^3. Find: P_s in the mouth at the surface of the water at the top of the straw.] The column is to rise 0.150 m above the open-surface level where the pressure is P_A. Then in the liquid within the straw at that open-surface level

$$P = P_l + P_s = P_A$$

Hence

$$P_s = P_A - \rho g h = 1.013 \times 10^5 \text{ Pa} - (1.00 \times 10^3 \text{ kg/m}^3)(9.81 \text{ m/s}^2)(0.150 \text{ m})$$

and $\boxed{P_s = 0.998 \times 10^5 \text{ Pa}}$. That's 98.5% of atmospheric pressure. To find the gauge pressure just subtract atmospheric pressure—that is

$$P_G = P_s - P_A = -1.53 \times 10^3 \text{ Pa}$$

▶ **Quick Check:** The weight of the water column of cross-sectional area A is $\rho g h A \approx (10^3 \text{ kg/m}^3) \times (10 \text{ m/s}^2)(0.15 \text{ m})A$ and the pressure is ≈ 1.5 kPa: air pressure must exceed mouth pressure by this amount.

For several centuries the only accurate practical pressure gauge was the mercury manometer (Fig. 11.11), a U-shaped open tube filled with mercury. The gauge pressure at point A equals the pressure at point B equals the chamber pressure. When h is above or below point B, the chamber pressure is above or below atmospheric: $P_G = \pm \rho g h$. Because of the manometer, blood pressure is measured in mm Hg to this day.

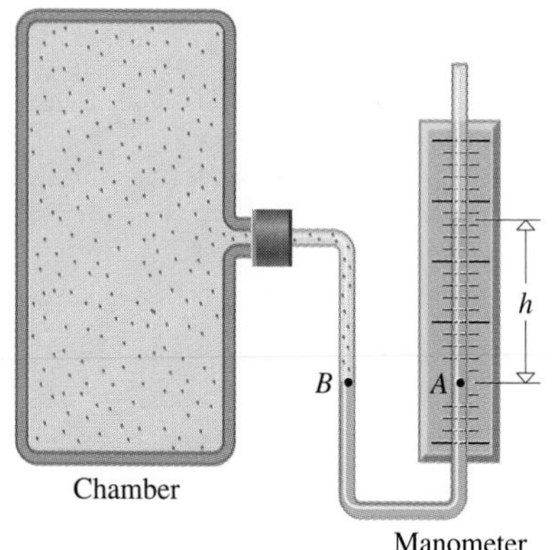

Figure 11.11 A manometer. It is usually filled with mercury, but water is also used in medical applications, such as measuring lung pressure, because mercury vapor is poisonous. Here, the gas pressure in the chamber is being measured. It is assumed that the weight of the column of gas above B is negligible (that is, the pressure in the chamber is uniform). As shown, the chamber pressure exceeds atmospheric by $\rho g h$.

Vacuum

After centuries of philosophical insistence on Aristotle's idea that an empty void cannot exist, the "Torricellian vacuum" was seen by many as miraculous. The great Descartes, who debated the existence of vacuum with Pascal, never accepted the notion. In fact, just before his death (1650), he suggested that Pascal had a void in his head.

The next advance in vacuum technology was made by the German engineer Otto von Guericke. In 1650, he succeeded in constructing the first air-extraction pump, which was a modified version of a commonplace fire extinguisher—"a brass squirt-pump" that operated with a movable piston like a large hypodermic syringe. Later, von Guericke constructed a pair of bronze hemispheres, 22 in. in diameter, between which was placed a soft ring gasket. These were then evacuated with his pump until the outside air pressure forced the hemispheres together (Fig. 11.12a). The normal force is everywhere radially inward (Fig. 11.12b), but the symmetry is such that for every component acting downward above the central z-axis, there is an equal upward component below that axis. The end result is that all force components in the x- and y-directions must cancel on each half-sphere. Only a net z-axis force remains inward on each, driving the hemispheres together.

If each half-sphere was independently sealed with a flat plate of radius R and separately re-evacuated, each would again be in equilibrium (Fig. 11.12c). There would be the same net axial force F acting on the curved surface as before, but now it would be balanced by an atmospheric-pressure force, $(\pi R^2)P_A$, acting in opposition on the flat face. It follows, assuming the pressure inside is zero, that the net atmospheric force on each hemispherical surface is

$$F = \pi R^2 P_A$$

With $R = 11$ in. and $P_A = 14.7$ lb/in.^{2}, it would take a force of 5.6×10^3 lb pulling on each half to yank them apart. On May 18, 1654, with Emperor Ferdinand III

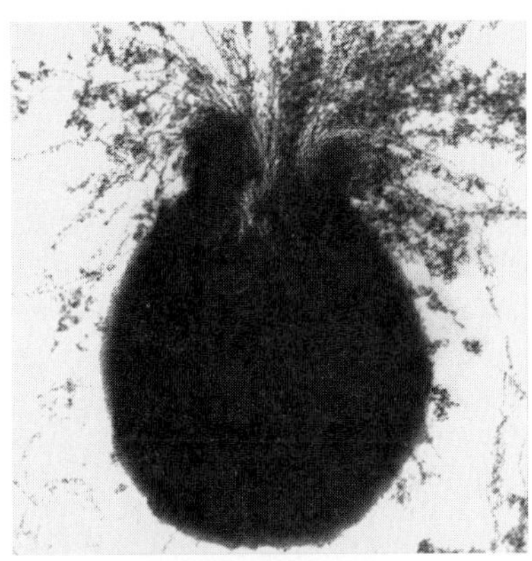

Magnified about 1.5×10^5 times, this staphylococcus bacterium explodes after being exposed to a low-level dose of antibiotic. With an internal pressure of between 25 and 30 times atmospheric pressure, the cell wall ruptured at its weakest point.

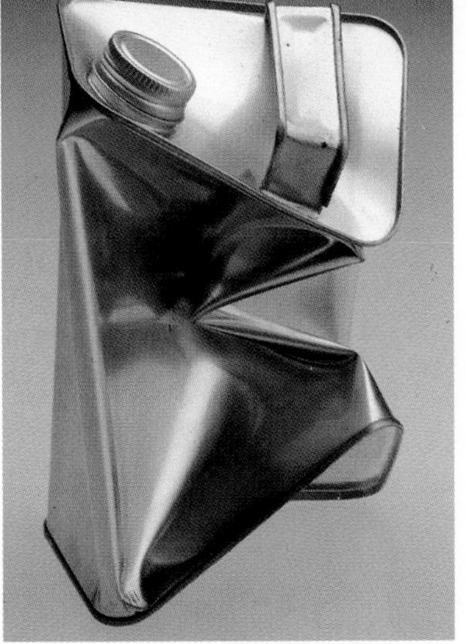

Put a little water in an open empty gallon can, and gently heat it until the escaping steam displaces most of the air that was inside. Cap it tightly, cool it, and as the vapor condenses and the internal pressure drops, it will gradually collapse under the weight of the atmosphere. Notice how the can is pushed in on all sides and not just flattened top down. The pressure is in all directions—it's a scalar.

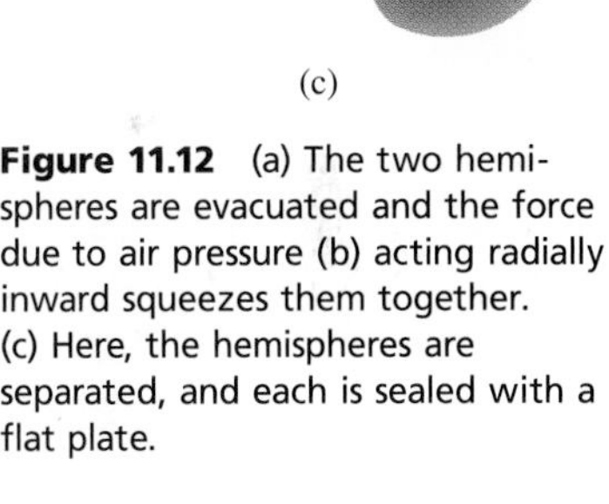

Figure 11.12 (a) The two hemispheres are evacuated and the force due to air pressure (b) acting radially inward squeezes them together. (c) Here, the hemispheres are separated, and each is sealed with a flat plate.

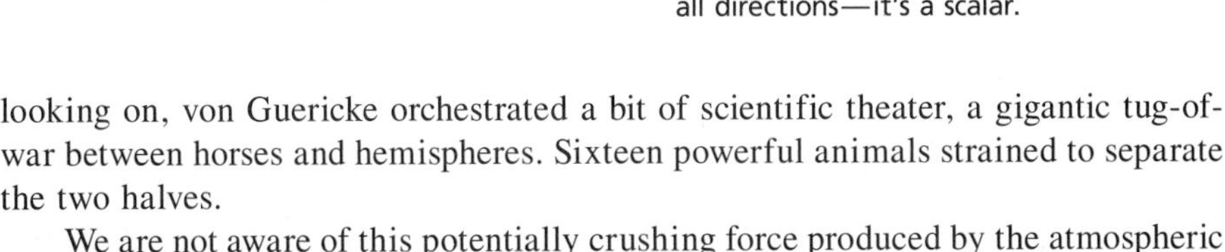

looking on, von Guericke orchestrated a bit of scientific theater, a gigantic tug-of-war between horses and hemispheres. Sixteen powerful animals strained to separate the two halves.

We are not aware of this potentially crushing force produced by the atmospheric sea because we are filled with, and surrounded by, air. Were the air to be drawn out

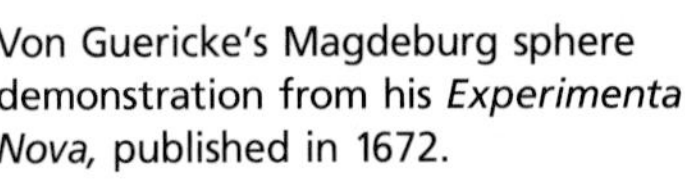

Von Guericke's Magdeburg sphere demonstration from his *Experimenta Nova,* published in 1672.

This house exploded during a hurricane. The pressure outside dropped below the inside pressure, which raised the roof. Had the occupants left several windows open, the explosion would not have happened.

of a vessel (as with a vacuum-sealed jar), that container had better be able to withstand 1.013×10^5 N/m^2, or it will be crushed. Youngsters often squash paper cups on their faces by sucking the air out of them, wearing the crumpled remains as a muzzle. An ordinary window (1 m by 1.5 m) experiences an inward atmospheric force of 1.5×10^5 N (34×10^3 lb), which is usually counterbalanced by an equal outward force since the building is filled with air at the same pressure. Sometimes, for example, during a tornado, the pressure outside can suddenly drop and, if the windows are closed, a house can explode.

The Honourable Robert Boyle, the seventh son of the Earl of Cork, read about von Guericke's work in 1657 and immediately set out to better it. By 1659, Boyle and his assistant Hooke had built an extraction air-pump (a so-called vacuum pump, though it certainly doesn't pump vacuum) far superior to any in existence at the time. They soon established that air was necessary to both combustion and respiration (at least for flies, bees, birds, and mice). They even put a barometer in a chamber and confirmed that the mercury column descended as the pressure dropped. Today, vacuum systems are commonplace in the production of a variety of familiar items, from TV picture tubes and light bulbs to computer circuits, freeze-dried food, and coated sunglasses.

11.4 Boyle's Law

Boyle's early work with vacuum pumps led him to the study of gases, a research that culminated in the recognition of a simple interdependence between pressure and volume. Although R. Towneley and H. Power actually published the relationship a year earlier, it is still most often referred to as Boyle's Law.

Boyle's approach was simple and elegant. He bent a glass tube about 2 m long into the shape of a J and sealed off the short arm (Fig. 11.13a). Then some mercury was poured in, usually trapping a quantity of air under pressure in the short segment. By tipping the device "so that air might freely pass from one leg into the other," the pressure was equalized at atmospheric and the mercury level made the same on both sides (Fig. 11.13b). The device, turned back upright, was ready to use. Adding more mercury caused the liquid level to rise unequally in both sections until the volume of gas in the smaller arm was halved and the open column had risen another 30 in. (Fig. 11.13c). The pressure in the trapped gas was then twice atmospheric—the volume of the gas had been halved and the pressure doubled. Similarly, at 3 atm of pressure the volume was further reduced to $\frac{1}{3}$, and so forth. **Keeping the temperature constant, the volume of a gas varies inversely with the pressure**, which is equivalent to saying that the **pressure times the volume is constant. Boyle's Law** is then

$$PV = \text{constant} \tag{11.5}$$

30"

(a) (b) (c)

Figure 11.13 The apparatus used by Boyle to determine the law bearing his name. The gas captured in the space on the left in (b) is at atmospheric pressure. By adding 30 in. of mercury, the pressure is doubled and the volume is halved, as in (c).

Consider a balloon floating upward in the air or a bubble rising in a liquid. As the external pressure drops, the internal counterpressure must also decrease. If PV is to be constant, the volume must increase; the balloon grows larger as the pressure decreases. Similarly, Fig. 11.14 shows how Boyle's Law applies to the way we breathe.

The imagery of a gas as a swarm of tiny bouncing, colliding atoms or molecules is called the Kinetic Theory. It was first applied analytically by Daniel Bernoulli (1738), who succeeded in deriving Boyle's Law using the concept. A molecule impacting on, and ricocheting off, a chamber wall changes its momentum in the process. That, in turn, corresponds to a force on the molecule and an equal-magnitude oppositely directed force on the wall. The ongoing bombardment of millions of molecules blurs into a seemingly continuous force and pressure.

Suppose we have a gas in an upright cylindrical chamber fitted with a vertical descending piston (Fig. 11.15). Compressing the gas to half its original volume doubles the density, giving the same number of molecules half the room to fly around in. As the piston slowly drops, halving the top-to-bottom distance, the average time to traverse the chamber vertically is halved, since the speed of the particles is unchanged. Therefore, the number of impacts per second is doubled, and so is the pressure. Similarly, lowering the piston halves the exposed side-wall area without altering the average time to traverse the chamber horizontally. That, in turn, results in the same number of impacts per second happening on half as much area; that is, the pressure on the walls also doubles—halving the volume doubles the pressure.

Boyle's Law is an idealization that works over a wide range of temperature and pressure. It applies especially well when the intermolecular forces are negligible, as they are when the molecular separations are large. It is most in error at very high pressures and at temperatures near liquefaction, where the molecules are relatively close and do attract one another. Carbon dioxide (CO_2), a gas that is fairly easily liquefied, deviates appreciably from Boyle's Law, whereas helium and hydrogen comply with it rather well under ordinary conditions. The make-believe stuff for which Boyle's Law holds exactly would be a *noninteracting* collection of independent molecules known as an **ideal gas**. Insofar as all real gases resemble the ideal gas to some extent, Boyle's Law is quite useful.

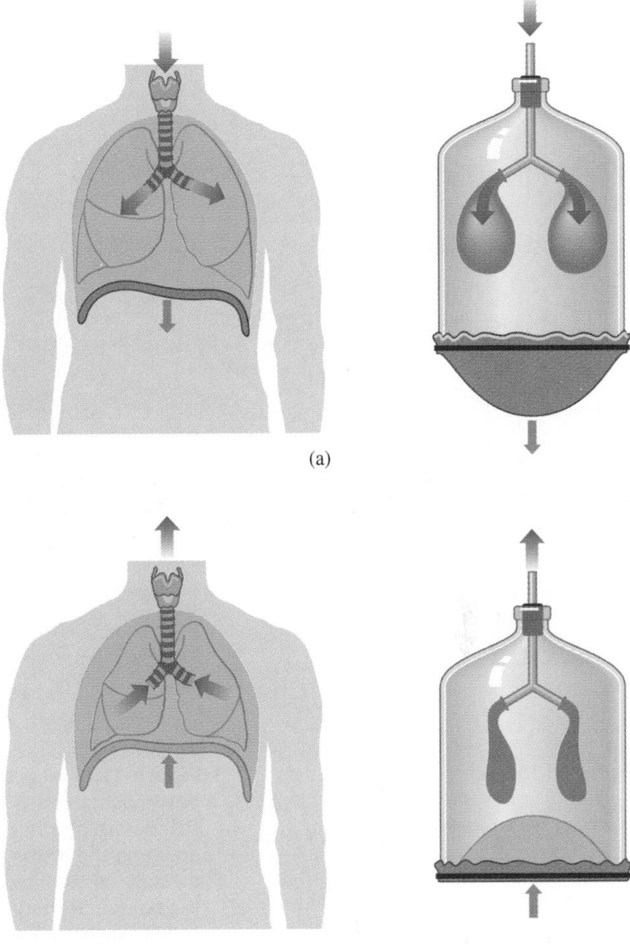

Figure 11.14 A bottomless jar surrounding two balloons and closed off with a rubber membrane can make a simple model of the lung system. (a) To inhale, the diaphragm pulls downward, enlarging the chest volume and thereby dropping the pressure (PV = constant). The chest cavity is usually at a negative gauge pressure (to keep the lungs open against their surface tension), but now the lung pressure is also negative and air flows in. (b) When the chest compresses and the diaphragm moves up, chest pressure rises, the gauge pressure in the lungs becomes positive, and air is expelled.

11.5 Pascal's Principle

Sometime around 1651, Pascal wrote a treatise entitled *On the Equilibrium of Liquids.* It contained the first precise statement* of what has come to be known as **Pascal's Principle: An external pressure applied to a fluid confined within a closed container is transmitted undiminished throughout the entire fluid.** When pressure is put on some region of a confined liquid (as, for example, when a piston pushes down on the liquid in a cylinder), the fluid compresses slightly and distributes the pressure uniformly everywhere therein. This process is quite different from the internal pressure generated by gravity, and would exist even in a

*Archimedes had come close to it 2000 years before.

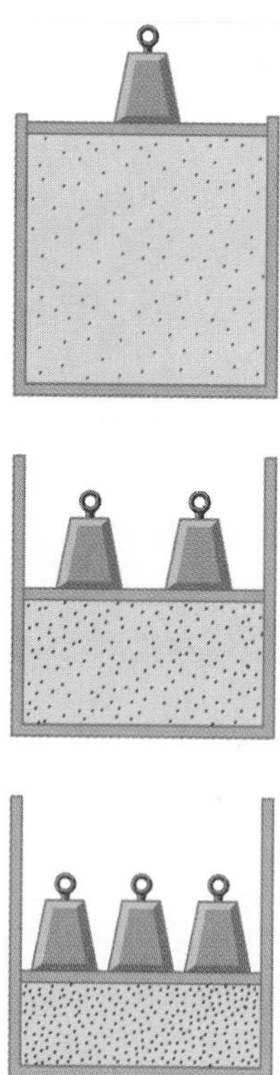

Figure 11.15 When the pressure exerted on a gas is doubled, the volume is halved; when it is tripled, the volume is reduced to one-third.

Figure 11.16 Pascal's syringe is a bottle punctured with holes and fitted with a tight piston. Pushing down on the piston increases the pressure on the fluid, which blows out the holes uniformly in all directions. This phenomenon suggests that the applied pressure is distributed equally throughout the liquid.

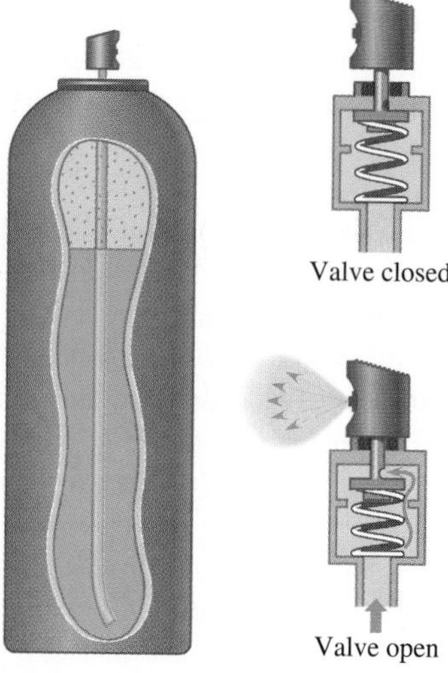

Figure 11.17 An aerosol spray can contains a gas under pressure called the propellant. It pushes down on the surface of the liquid that is to be sprayed. When the valve is opened, the top end of the long tube is at atmospheric pressure, and the bottom end is at a pressure well above that. The difference propels the liquid up and out.

weightless liquid. Pascal's syringe (Fig. 11.16) illustrates the point nicely, as does the common seltzer bottle and the aerosol spray can (Fig. 11.17).

One reason you don't let people sit on your belly when your bladder is full is Pascal's Principle. In the same way, pressure on the abdomen of a pregnant woman is transmitted to the fetus via the amniotic fluid. This mechanism is also responsible for the motion of soft-bodied animals, like the earthworm, that have hydrostatic skeletons. Using a mesh of perpendicular muscles, a worm squeezes itself into shape, becoming long and thin or short and fat as needed.

On work—long before the notion was formalized. *And it is truly admirable that there is encountered in this new [hydraulic] machine the constant rule which appears in all the older machines, such as the lever, the wheel and axle, the endless screw, etc., which is, that the path is increased in the same proportion as the force.*

BLAISE PASCAL
Traité de l'Equilibre des Liqueurs

Hydraulic Machines

The same amount of pressure can be produced within a liquid by pistons of different sizes acting with proportionately different forces (Fig. 11.18)—the larger the cross-sectional area of the piston, the larger the force needed to create a given pressure. It was at this juncture that Pascal recognized the tremendous practical significance of his principle. For the first time since antiquity, a new class of force multipliers known as *hydraulic* machines (from the Greek for *water* and *pipe*) was possible (al-

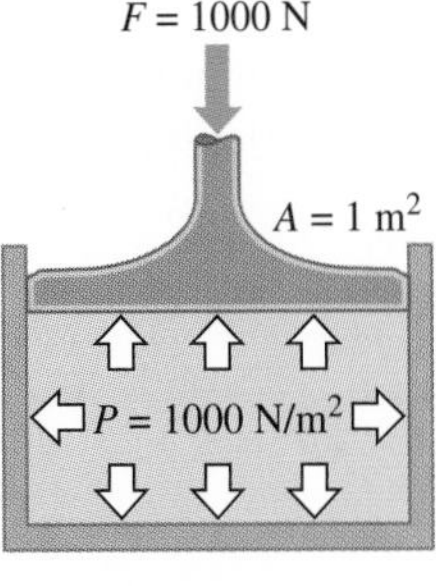

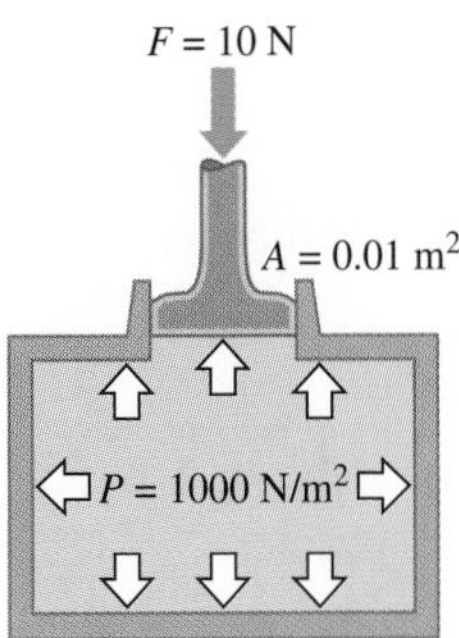

Figure 11.18 The force divided by the area of the piston determines the pressure. Different piston areas and forces here produce the same pressure.

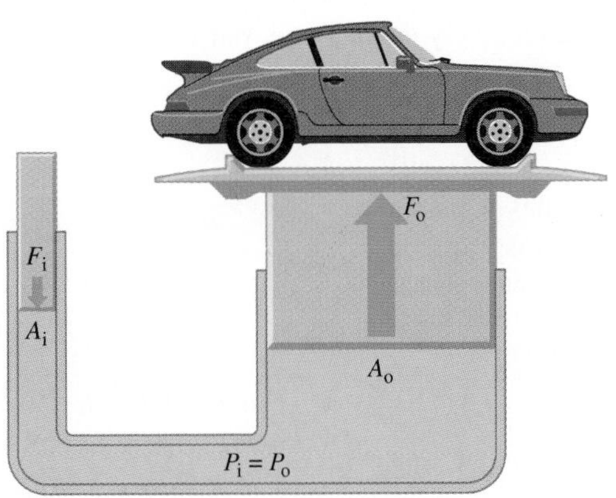

Figure 11.19 A hydraulic lift. The input force F_i creates a pressure F_i/A_i that is transmitted to the output cylinder, where that same pressure equals F_o/A_o.

though a practical device was not built until Bramah devised a functioning pressure-seal in 1796).

If two chambers fitted with different-sized pistons are connected so they share a common working fluid, the pressure generated by one will be transmitted undiminished to the other (Fig. 11.19). Years ago, gas stations dug service pits in which the mechanics stood while working on cars parked overhead. Nowadays, automobiles are unceremoniously lofted into the air on hydraulic lifts. Most are activated by compressed air pressing on oil, but the simplest arrangement is a U-tube, narrow on one side, wide on the other, with sealed movable pistons at both ends. A downward input-force F_i acting over the small input-area A_i of the narrow piston generates an input-pressure $P_i = F_i/A_i$. But this pressure is distributed uniformly and so equals the output-pressure ($P_i = P_o$), which, in turn, is given by $P_o = F_o/A_o$. Hence, $F_i/A_i = F_o/A_o$, and so

$$\frac{F_o}{F_i} = \frac{A_o}{A_i} \tag{11.6}$$

Although the pressures in and out are equal, the forces are certainly not. When the output piston under the car has 100 times the area of the input piston, it will experience an upward force 100 times greater than the input-force. We exert a force of 200 N on a hydraulic jack, and it exerts a force of 20 000 N on the car. Like all machines, this is not a *work* multiplier: at best, when energy losses (for example, those due to friction) are negligible, work-in equals work-out. In this example, the

Today, hydraulic devices that push and pull are commonplace.

small piston will have to descend a distance $y_i = 100$ cm for every 1 cm the large one rises (y_o). In other words, the volume of liquid displaced at the input side is $A_i y_i$, which is equal to the volume displaced at the output $A_o y_o$. Accordingly,

$$\frac{A_o}{A_i} = \frac{y_i}{y_o} \tag{11.7}$$

One person doing a lot of pumping over a lot of distance (y_i) can jack a car up (y_o) using a relatively small force, but *work-in* ≥ *work-out*.

Example 11.4 A barber's chair rests on a hydraulic piston 10 cm in diameter. The input side has a piston with a cross-sectional area of 10 cm², which is pumped on using a foot pedal. If the chair and the client together have a mass of 160 kg, what force must be applied to the input piston?

Solution: [Given: $A_i = 10\text{ cm}^2 = 0.0010\text{ m}^2$, $r_o = 5.0\text{ cm} = 0.050\text{ m}$, and $m_o = 160$ kg. Find: F_i.] Using Eq. (11.6)

$$F_i = \frac{F_o A_i}{A_o} = \frac{m_o g(0.0010\text{ m}^2)}{\pi(0.050\text{ m})^2} = (1568\text{ N})(0.1273)$$

and

$$\boxed{F_i = 2.0 \times 10^2\text{ N}}$$

▶ **Quick Check:** $A_i/A_o = \pi(5\text{ cm})^2/(10\text{ cm}^2) = 7.85 = F_o/F_i = 1.6\text{ kN}/F_i$ and $F_i = 0.2$ kN.

Keep an eye out for hydraulic machines. They can be found on garbage compacters, forklifts, robots, cherry pickers, plows, tractors, airplane landing gear, elevators, and even in the brake system of the family car.

11.6 Buoyant Force

It can be argued that the study of hydrostatics was begun by Archimedes in the third century B.C. The greatest physicist of ancient time, Archimedes was apparently a kinsman of Hieron II, Tyrant of Syracuse. Legend has it that the king ordered a solid gold crown to be made. But when the piece was delivered, although its weight was right, Hieron suspected that his jeweler had substituted silver for gold in the hidden interior. Archimedes was given the challenge of determining the truth without damaging the royal treasure. After pondering the problem for some time, its solution came to him while he was musing in a warm tub at the public baths. The distracted philosopher leaped from the water and ran home naked, shouting through the streets, *"Heureka! Heureka!"* I have found it! I have found it! What he found, as we shall soon see, was far more valuable than Hieron's crown.

A completely submerged body displaces a volume of liquid equal to its own volume. Experience also tells us that when an object is submerged, it appears lighter in weight; the water buoys it up, pushes upward, partially supporting it somehow. That much would be obvious to anyone who ever tried to submerge an inflated tire tube or a beach ball. Archimedes quantified the phenomenon. His **Buoyancy Principle** asserts that *an object immersed in a fluid will be lighter (that is, it will be buoyed up) by an amount equal to the weight of the fluid it displaces.* The upward force exerted by the fluid is known as the **buoyant force**. A 10-N body that displaces 2 N of water will "weigh" only 8 N while submerged.

Buoyant force is caused by gravity acting on the fluid. It has its origin in the pressure difference occurring between the top and bottom of the immersed object, a difference that always exists when pressure varies with depth (as it does for a fluid in the Earth's gravitational field). Imagine a solid cube, each face of which has an area A, somewhere below the surface of a fluid of density ρ_f, as depicted in Fig. 11.20. The gauge pressure on the bottom, $P_b = \rho_f g h_b$, is greater than the gauge pressure on the top, $P_t = \rho_f g h_t$, and that difference $\Delta P = \rho_f g(h_b - h_t) = \rho_f g h$ gives rise to the *buoyant force* F_B. The force pushing up on the bottom is greater than the force pushing down on the top by an amount

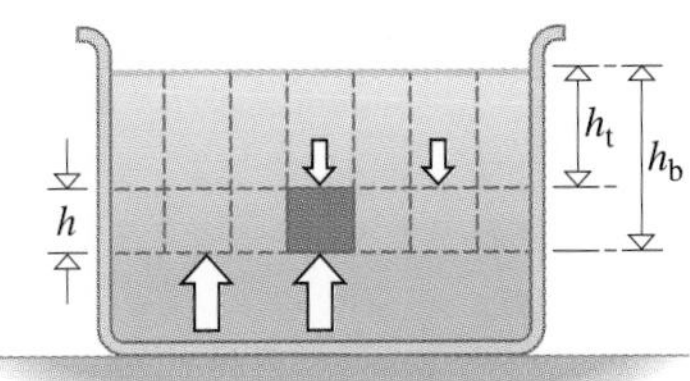

Figure 11.20 The buoyant force on a cube is the difference between the downward force on its top face and the larger upward force on its bottom face.

$$F_B = A\,\Delta P = \rho_f g A h$$

but $Ah = V$, the volume of the body or equivalently the volume of the fluid displaced. It follows, since the mass of fluid displaced is $m_f = \rho_f V$, that

$$F_B = g\rho_f V = m_f g \tag{11.8}$$

The buoyant force equals the weight of the fluid displaced.

Example 11.5 An empty spherical weather balloon with a mass of 5.00 kg has a radius of 2.879 m when fully inflated with helium. It is supposed to carry a small load of instruments having a mass of 10.0 kg. Taking air and helium to have densities of 1.16 kg/m^3 and 0.160 kg/m^3, respectively, will the balloon get off the ground?

Solution: [Given: The bag plus the load has a mass of 15.0 kg, $\rho_{He} = 0.160$ kg/m^3, $\rho_f = 1.16$ kg/m^3, and $R = 2.879$ m. Find: the buoyant force compared to the weight.] To find the weight of fluid displaced, we need the volume of the spherical body: $V = \frac{4}{3}\pi R^3 = 100$ m^3. The buoyant force is the weight of 100 m^3 of air; hence

$$F_B = \rho_f V g = (1.16\ \text{kg/m}^3)(100\ \text{m}^3)(9.81\ \text{m/s}^2)$$

and

$$\boxed{F_B = 1.14 \times 10^3\ \text{N}}$$

By comparison, the weight of the helium is

$$\rho_{He} V g = (0.160\ \text{kg/m}^3)(100\ \text{m}^3)(9.81\ \text{m/s}^2) = 156.9\ \text{N}$$

Thus, the total weight of the helium, balloon, and load is

$$F_W = (15.0\ \text{kg})(9.81\ \text{m/s}^2) + 156.9\ \text{N} = \boxed{304\ \text{N}}$$

Clearly, the buoyant force far exceeds the weight, and the balloon will accelerate upward rapidly.

▶ **Quick Check:** $F_B \approx (100\ \text{m}^3)(1\ \text{kg/m}^3)(10\ \text{m/s}^2) \approx 10^3$ N.

To see how Archimedes might have solved Hieron's problem, suppose the crown had a mass, $m = 0.982$ kg. By suspending the crown immersed in water from a balance (Fig. 11.21), he could measure $F_W - F_B = g(m - m_w)$, where m_w is the mass of water displaced. Suppose this *reduced weight* turned out to be $g(0.922$ kg$)$. This means that $g(0.922\ \text{kg}) = g(m - m_w)$ and so $m_w = 0.982\ \text{kg} - 0.922\ \text{kg} = 0.060$ kg. The mass of water displaced is 0.060 kg and its volume must be 60 cm^3, which is the volume of the crown. Is it gold? If it was, it would have a mass given by the density of gold (19.3×10^3 kg/m^3) times the volume: $(19.3 \times 10^3\ \text{kg/m}^3) \times (60 \times 10^{-6}\ \text{m}^3) = 1.158$ kg, not 0.982 kg. It is *not* pure gold. There is a shortage in the mass of $(1.158\ \text{kg} - 0.982\ \text{kg}) = 0.176$ kg, which is assumed to have arisen from the inclusion of silver. Alas, most versions of this legendary tale end with the untimely demise of the anonymous goldsmith.

11.7 Specific Gravity

The notion of density can be made more convenient by forming the ratio of the density of any material to that of water. The result is the idea of *relative density,* and it's especially useful when we float things in water. Thus, suppose an object of mass m

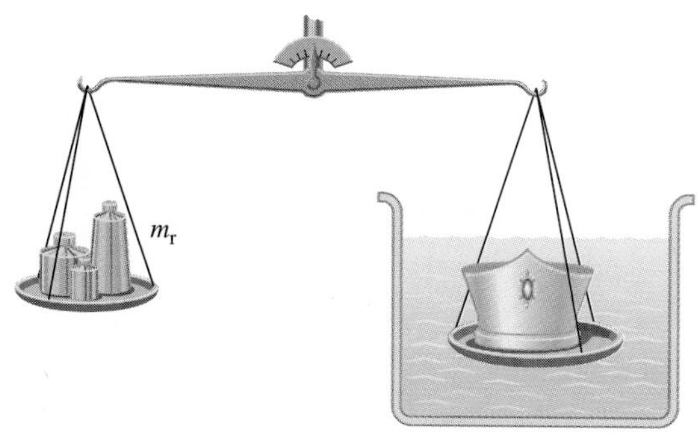

Figure 11.21 A crown of mass m, immersed in water, has a reduced mass of $m_r = m - m_w$, where m_w is the mass of water displaced. We can measure m and m_r directly and then calculate $m/(m - m_r)$, which equals the specific gravity.

is immersed in water such that the mass of fluid displaced is m_w. Using a balance, we could then directly measure the object's *reduced mass* in the water: $m_r = m - m_w$, which provides a convenient measure of the *relative density* ρ/ρ_w:

$$\frac{m}{m_w} = \frac{m}{m - m_r} = \frac{\rho}{\rho_w} \tag{11.9}$$

Equation (11.9) is equivalent to saying that *the ratio of the weight of a body to the weight of an identical volume of water is equal to the relative density,* or, as it has inappropriately been called for centuries, the **specific gravity**. Alternatively, the weight of an object in air divided by the difference between its weight in air and its weight in water is its specific gravity. The idea was introduced by the Persian al-Biruni sometime around A.D. 1025 in order to use unitless quantities and so avoid the muddle of different units existing throughout the Muslim Empire. The relative density tells us how many times more or less dense a material is than water. Equation (11.9) provides a convenient method for measuring the average density of a human body by submerging a person, sitting on a scale, in what is fondly called a "fat tank."

Example 11.6 Hieron's crown having a mass of 0.982 kg displaces 0.060 kg of water when totally submerged. Is it solid gold?

Solution: [Given: $m_g = 0.982$ kg and $m_w = 0.060$ kg. Find: the relative density of the crown.] From Eq. (11.9)

$$\frac{m_g}{m_w} = \frac{0.982 \text{ kg}}{0.060 \text{ kg}} = \frac{\rho_g}{\rho_w} = 16$$

whereas the relative density of gold (Table 10.1) is 19.3—this crown isn't solid gold!

▶ **Quick Check:** For the water $m_w/\rho_w = V$, for the gold $V\rho = m$; because the volumes are equal, $m_w\rho_g/\rho_w = m_g = (0.06 \text{ kg})(19.3 \times 10^3 \text{ kg/m}^3)/(10^3 \text{ kg/m}^3) = 1.2$ kg and not 0.982 kg.

Floating

When an object weighs more than the total volume of fluid it can displace, it sinks. If the fluid is water, this is equivalent to saying that an object whose relative density exceeds 1.0 will sink. A solid cubic foot of gold, copper, concrete, or glass weighs more than a cubic foot of water and will sink in spite of the (278-N, or 62.4-lb) buoyant force.

When an object weighs less than the total volume of fluid it can displace, it will settle down until the buoyant force equals the weight and it floats partially submerged. In water, an object whose relative density is less than 1.0 will float partially submerged. A steel ship can encompass a great deal of empty space and so have a large volume and a relatively small average density. Provided it weighs less than the maximum amount of water it can displace (that is, provided it weighs less than the maximum buoyant force), a ship—cargo and all—will float. *The weight of the water displaced by the submerged portion of a floating object equals the weight of the object* (Fig. 11.22).

Figure 11.22 A floating object displaces its own weight of liquid.

People are slightly less dense than water, especially when the lungs are filled with air, and so float partially submerged. The average specific gravity of the human body varies from person to person and breath to breath. Body fat, which is ≈18%

for a male and ≈28% for a female, has a specific gravity of ≈0.8. Muscle has a specific gravity of ≈1.0 and bone about 1.5–2.0. A lean, muscular body will tend to sink. In general, young people and women have lower average specific gravities, but 0.98 is a typical lungs-filled value. Thus, a person can float with at most 2% of the body out of the water. In seawater, the relative density $(0.98 \times 10^3\ \mathrm{kg/m^3})/(1.025 \times 10^3\ \mathrm{kg/m^3}) = 0.956$, and so about 4% of the body will be above the surface. The problem in swimming for humans is keeping the dense head above the water in such a way as to be able to breathe.

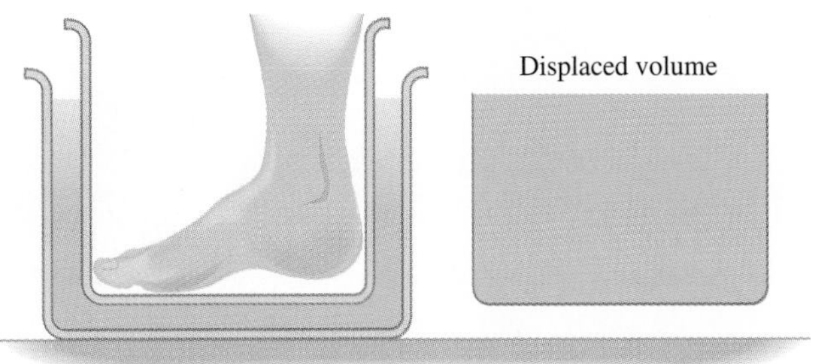

Figure 11.23 A way to displace more liquid than is actually present.

The greater the density of the fluid, the more the buoyant force. That's why it is easier to swim in the ocean than in a salt-free pool and why a penny will float on mercury. A fresh egg, one that has developed little or no gas, will be relatively dense and will sink in a glass of tap water. Dissolving a few tablespoons of salt into the water will float the egg up to the surface.

Surprisingly, *it is possible to displace more fluid than might be present,* and it is the *displaced* fluid that counts with regard to floating. However paradoxical that sounds, it really isn't, and Fig. 11.23 should make the business clear. Again, the walls of the container play a crucial role as the fluid transmits the load to them. It doesn't matter how much water is surrounding the inner beaker.

When an object's weight equals the weight of the total amount of fluid it can displace—that is, equals the maximum buoyant force—it neither sinks nor rises but hovers beneath the surface in static equilibrium. Some fish accomplish this balance by storing gas in a bladder, thereby being able to remain at a fixed depth without the need to move about. By contrast, sharks, which are rather primitive in design, lack such systems and must continually swim or they will sink. To rise, or dive, or hover even without operating its engines, a submarine takes on and discharges seawater into ballast tanks.

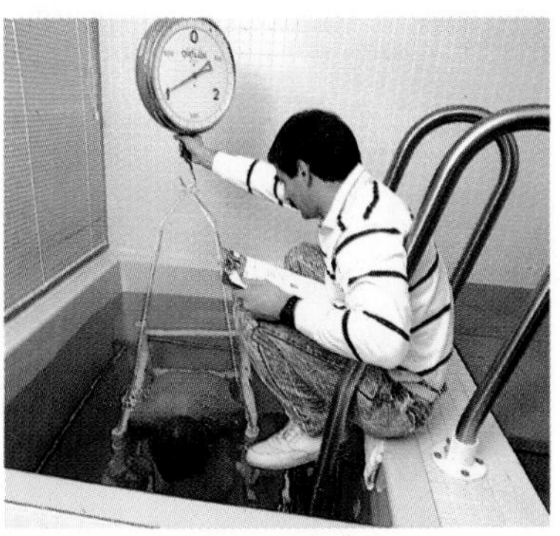

A fat tank provides a measure of a person's average density.

Example 11.7 An object of volume V floats in water with a volume V_u above the surface. (a) Write an expression for its average density. (b) Apply this solution to find the average density of a loaded barge with 20% of its volume above the waterline.

Solution: [Given: The total volume is V, the volume above water is V_u. Find: ρ.] (a) The volume of water displaced is $(V - V_u)$ and the buoyant force is

$$F_B = (V - V_u)\rho_w g$$

But this equals the weight of the object, which is also $\rho g V$, therefore

$$\rho g V = (V - V_u)\rho_w g$$

and

$$\boxed{\rho = \frac{V - V_u}{V}\rho_w}$$

(b) Thus $\rho = (1 - 0.20)\rho_w = \boxed{0.80\rho_w}$

▶ **Quick Check:** It is always wise to check the extremes. To hover below the surface, $V_u = 0$ and $\rho = \rho_w$, which makes sense. For the entire object to be above water, $V_u \to V$ and $\rho \to 0$, which also makes sense.

11.8 Surface Tension

In a liquid under the influence of gravity, each molecule descends until it can drop no more, ultimately settling into a zero net-force equilibrium. The system then has the lowest possible potential energy, which is why a liquid assumes the shape of its containers. But in the absence of gravity, or in an effectively weightless environ-

The density of the salt-rich water of Great Salt Lake is appreciably greater than that of pure water. Consequently, the swimmer can float with much of her body out of the water.

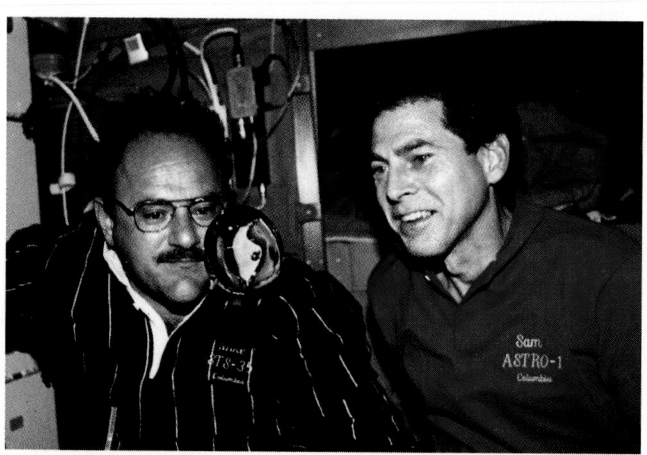

Astronauts in Earth orbit in the *Space Shuttle Columbia* examine a ball of water. In the effectively weightless environment, surface tension pulls the liquid into a minimum energy sphere.

ment, a glob of water would draw itself into a sphere. Free-falling raindrops are roughly spherical—though only roughly because of the air friction. In the same way, if we shake some mercury in an Earth-bound jar, thousands of tiny, almost spherical beads will be formed on the glass, beads quite unwilling to conform to the shape of the container's walls. These familiar liquid "tricks" are the outward manifestations of internal interactions that give rise to what we call *surface tension.*

Any molecule within the body of a liquid is surrounded by other molecules. These pull on it more or less uniformly in all directions via the cohesive interactions that bind the substance together. By contrast, a molecule on a liquid's *free surface* (that is, one bounded by gas or vacuum), will have none of its comrades above it. It will be drawn more strongly to its neighbors on the side and below. Consequently, the surface sheet of molecules reacts as if it were an elastic membrane confining the liquid, being drawn in, assuming the least area it can. About 200 years ago, Dr. Thomas Young suggested this stretchy membranous imagery—it is helpful, *but should not be taken literally.*

Most ships have markings on their hulls showing how deep they are resting in the water and therefore how heavy their cargos are.

The force exerted by a stretched elastic membrane is proportional to the amount of stretch via Hooke's Law: no displacement from equilibrium, no reaction force. By contrast, a liquid surface film is always under tension and that tension is independent of any displacement. The surface "membrane" *is* able to stretch in the usual sense, but only slightly. When the surface film is enlarged, it is accomplished by bringing up more molecules from the interior of the liquid. This increases the amount of surface, creating new exposed regions, rather than increasing the intermolecular distances within a permanent surface film, as would be the case with a true elastic membrane.

To quantify the surface tension, let's measure the force exerted by such a surface directly. The simplest practical scheme, if not the most accurate, is to attach a U-shaped wire frame (Fig. 11.24) to a balance and measure its weight with the two ends just submerged. That done, the frame is lowered straight down completely into the liquid and then raised back out, forming a film across the inverted U. The additional upward force F (beyond the original weight of the frame and the negligible weight of the film) required to hold the wire in equilibrium at its starting level is a direct measure of the downward tug F_t of the liquid. That force is parallel to the plane of the film, points in the direction of expansion, and is distributed along a

TABLE 11.5 Surface tensions of various liquids

Substances	Temperature °C	Surface tension N/m
Helium–vapor	−269	0.12×10^{-3}
Hydrogen–vapor	−258	2.9×10^{-3}
Oxygen–vapor	−183	13.2×10^{-3}
Ether–vapor	20	17.0×10^{-3}
Olive oil–water	20	18.2×10^{-3}
Ethanol–vapor	20	22.3×10^{-3}
Benzene–air	20	29×10^{-3}
Soap solution–air	20	$\approx 25 \times 10^{-3}$
Whole blood–air	37	58×10^{-3}
Water–air	100	58.9×10^{-3}
Glycerin–air	20	63×10^{-3}
Water–air	50	67.9×10^{-3}
Water–air	20	72.8×10^{-3}
Blood plasma–air	37	73×10^{-3}
Water–air	15	73.6×10^{-3}
Water–air	0	75.6×10^{-3}
Mercury–water	20	0.375
Mercury–air	20	0.465
Copper–air	1130	1.1
Tungsten–vapor	3410	2.5

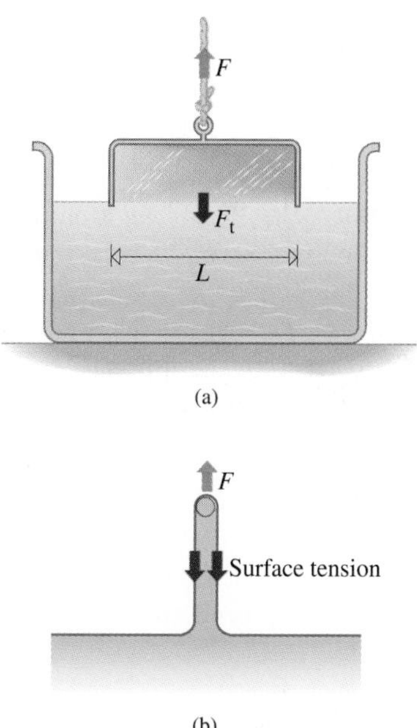

Figure 11.24 (a) An arrangement for determining surface tension. (b) The wire frame pulls up a two-sided film of length L.

narrow interface of length L. The **surface tension** γ is defined as *the force per unit length exerted by the surface.* Gamma, which has the same value everywhere on the surface, is measured in N/m. Here, where the single liquid film contained within the frame has *two* surfaces

$$F_t = 2\gamma L \tag{11.10}$$

Surface tension depends on both substances forming the interface (Table 11.5). Furthermore, as you might expect, increasing the temperature generally (there are a few exceptions) decreases the surface tension until it vanishes at a point where the substance can no longer exist in the liquid form.

Work must be done in order to bring a molecule up from the inside of a liquid, to push apart surface molecules and force this new one into the boundary layer. In the process of creating the film, in raising the U-shaped frame a distance Δy, an amount of work ΔW was done on the film, thus

$$\Delta W = F_t\Delta y = 2\gamma L\Delta y = \gamma\Delta A \tag{11.11}$$

where ΔA is the net increase in surface area (on both sides of the film). This, in turn, corresponds to an increase in potential energy stored in the surface—an idea introduced by the great astronomer and mathematician Friedrich Gauss. Alternatively, γ can be understood as the increase in stored surface potential energy per unit area. In that case, it has the units of J/m^2, which is equivalent to N/m.

Generally, a system under the influence of forces moves toward an equilibrium configuration that corresponds to a minimum PE. The sphere contains the most volume for the least surface area and therefore minimizes surface energy. Thus, *surface*-PE diminishes as a droplet approaches sphericity. That's why there are no

A water strider resting on the surface of a pond. Its "feet," or tarsi, work something like snowshoes. They are covered with fine hairs, which increase the area in contact with the liquid.

cubic raindrops. Lead shot used to be manufactured by dropping molten metal within tall temperature-controlled towers. One NASA project may lead to the commercial production of perfect ballbearings in facilities orbiting the planet.

Example 11.8 An atomizer, used for misting plants, is pumped several times, transforming 100 cm^3 of water into a fog of droplets with an average diameter of 50 μm. Determine the total amount of energy that went into creating the droplets. Ignore the surface area of the water initially.

Solution: [Given: $D = 50 \times 10^{-6}$ m, $V = 100\ \text{cm}^3 = 100 \times 10^{-6}\ \text{m}^3$, and $\gamma = 0.0728\ \text{J/m}^2$. Find: ΔW.] The energy required is the work done in producing the new area. From Eq. (11.11) that work done is $\Delta W = \gamma \Delta A$, and so we need the increase in area. Each new sphere has an area of $4\pi R^2$; hence, we need to find the number of spheres created, N. Since the volume of each droplet is $\frac{4}{3}\pi R^3$, it follows that $N = V/\frac{4}{3}\pi R^3$. The total surface area of the spheres is therefore $\Delta A = 4\pi R^2 N = 3V/R$ and

$$\Delta W = \gamma \frac{3V}{R} = \frac{(0.0728\ \text{J/m}^2)3(10^{-4}\ \text{m}^3)}{(25 \times 10^{-6}\ \text{m})} = \boxed{0.87\ \text{J}}$$

▶ **Quick Check:** For each sphere $\Delta W_s = (0.0728\ \text{J/m}^2)4\pi(25 \times 10^{-6}\ \text{m})^2 = 0.57$ nJ. $N = 3(10^{-4}\ \text{m}^3)/4\pi(25 \times 10^{-6}\ \text{m})^3 = 1.5 \times 10^9$ and $\Delta W_s N = 0.87$ J.

Plastic-coated paper clips supported by the surface tension of water. Try this demonstration yourself, but make sure the clips are flat.

Surface tension is responsible for a range of familiar occurrences from soap bubbles to teardrops. The bristles of a fine paintbrush tend to stand apart in a bushy tuft until they are wet with paint, and then the surface tension pulls the bristles together. That's why your hair, wet after a shower or swim, is a matted clump. An old-fashioned double-edged razor or a plastic-coated paper clip can be supported on top of water by surface tension even though either one is much denser and should otherwise sink.

FLUID DYNAMICS

A moment does not go by without each of us somehow interacting with fluids in motion. We walk, drive, and fly through the air, all the while breathing at least 6 quarts of it per minute. With blood pumping in our veins, we ourselves are *hydrodynamic* systems. On a larger scale, every thriving city on Earth has a lifeline of water pouring through its veins; New York City alone consumes 1.5×10^9 gallons per day. How do liquids flow through pipes, pumps, and arteries? What happens to the pressure as air blows over the wing of a plane or up the face of a skyscraper? How can we quantify fluid flow?

11.9 Fluid Flow

Experiments by O. Reynolds (1883) on the motion of fluids in pipes showed that there are two distinct flow regimes: *laminar* and *turbulent.* Gently blow air through your lips and the well-defined stream resembles the extreme of idealized smooth flow. Cough, and the burst of air is a complex swirling turbulence that represents the other extreme.

Laminar Flow

When a fluid moves such that the velocity within it at any point in space is fixed, we have steady-state motion. In the real world where fluids have internal friction, steady-state motion usually means slow flow. The velocity may be different at different points, but any speck of fluid arriving at a given spot in space will always have a fixed velocity at that location. We can inject a trickle of dye at several points within a liquid (or smoke in a gas) and trace the paths taken by the particles of fluid as they move in orderly procession, each following the route of the particle that arrived just before it. These unchanging lines of flow (Fig. 11.25) are called **streamlines**. At every point on any one of them, the tangent to the curve is in the direction of the velocity of the fluid at that point. There is no flow perpendicular to the streamlines. No two streamlines ever cross, since that would mean that at the point of intersection a particle of fluid would have two different velocities. This kind of streamline flow is also called **laminar flow** because successive layers of fluid molecules move smoothly past one another. For a normal person, about half the volume of blood is made up of cells, and so blood flow, which is smooth in most vessels, cannot be perfectly laminar, though it usually comes close.

At every point in space within the stream there is a velocity, and if this *velocity field* (Fig. 11.25) reminds you of the gravitational field, it is because the latter was derived conceptually from it. *The streamlines crowd together as the speed gets higher just as the gravity-field lines pack closer as that field gets stronger.* The boundary of a bundle of streamlines defines a *tube of flow,* wherein the fluid travels as if it were contained within an invisible pipe (Fig. 11.26).

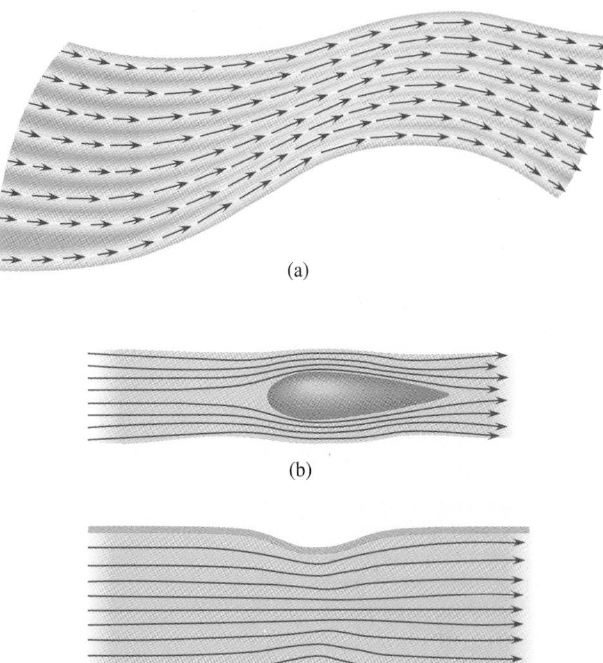

Figure 11.25 (a) The laminar flow pattern revealed by drawing the velocity vector at each point in the stream. (b) Streamlines in the laminar flow around an obstacle and (c) in a pipe. Note that the closer the lines, the higher the speed.

Turbulent Flow

Turbulent flow corresponds to nonsteady, chaotic, changing motion (Fig. 11.27). With increasing speed, the molecules of a real fluid meeting discontinuities or obstructions swirl into little shifting whirlpools, energy-carrying vortices that are spun off as curls in the flow lines (Fig. 11.28). At high enough speeds the fluid has sufficient momentum to sail past obstructions, so the flow no longer simply conforms to the shape of those obstructions. The situation is like a stream of motorcy-

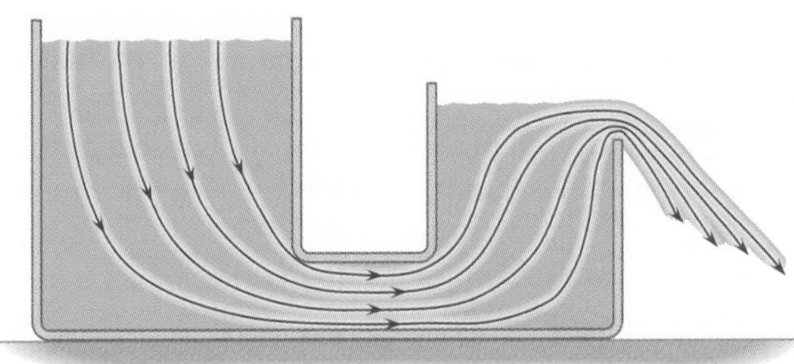

Figure 11.26 The fluid in a tube of flow travels as if in an invisible pipe, neither leaving the confines nor mixing with the contents of adjacent tubes.

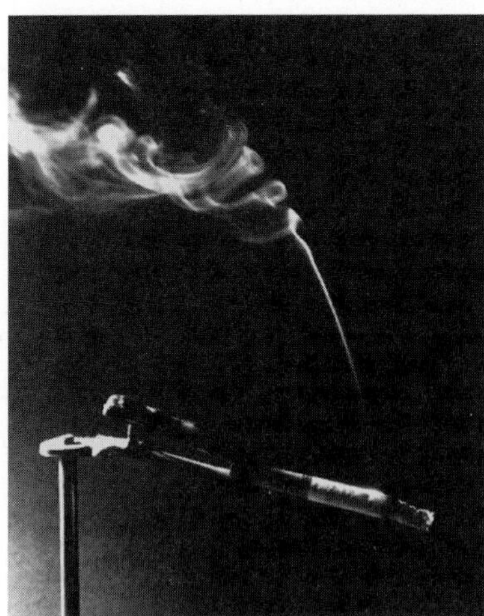

The smoke from a cigarette first rises in a laminar column and then breaks into turbulent flow.

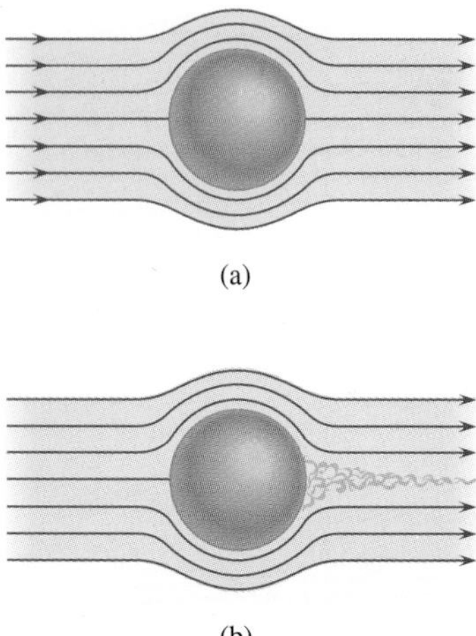

Figure 11.27 (a) An ideal fluid flows smoothly around an obstruction. (b) A real fluid cannot always follow the solid surface and forms a jumble of whirlpools that corresponds to turbulence.

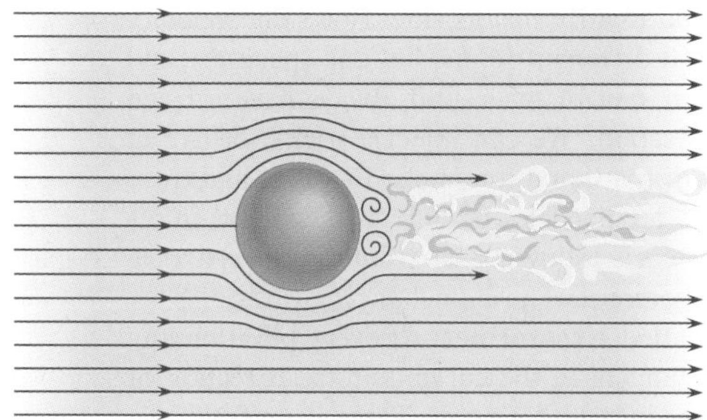

Figure 11.28 As the flow speed of a real fluid increases, its ability to follow the contours of a solid obstacle decreases. It tears away from the surface and forms a wake of turbulence that carries away energy.

To minimize air drag, the speed skater holds a low profile and keeps one arm behind the back. Smooth clothing and a hood also lessen drag.

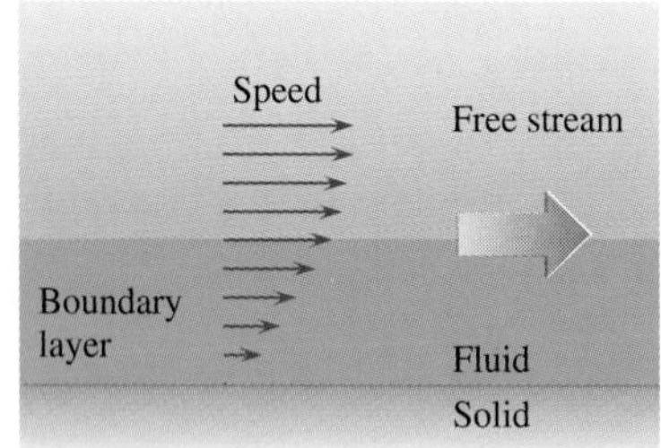

Figure 11.29 The fluid in contact with the solid surface is at rest. Its speed increases, depending on the viscosity, until it reaches the unencumbered free-stream speed, after which it's constant. The transition region is the boundary layer.

clists riding on hilly terrain—the faster they go, the more they tend to sail over the mounds and ruts rather than riding down them.

A real fluid streaming through pipes and around objects is influenced by its viscosity. Moreover, the molecules of a fluid easily come into intimate contact with those of a solid, and the short range (electromagnetic) adhesive force produces appreciable effects. *The layer of fluid in contact with a solid will adhere to that surface and remain at rest with respect to it.* The speed of flow increases from zero up to the free-stream value as the distance from the surface increases (Fig. 11.29). That takes place across a relatively thin region, depending on the medium, called the **boundary layer**.

Although we might expect fan blades running at high speeds to stay clean, they certainly don't. The dust particles they pick up low to the surface come into the static-air layer and are dragged along with the blades. Instead of the wind blowing the whirling blades clean, the dust remains at rest in the film of calm air adhering to the solid surface. Similarly, one might imagine that driving a car at 80 km/h would keep it clean, but it does not.

When the flow over an object is very rapid, the fluid may not be able to make a smooth transition from zero speed at the surface layer to some high value nearby. The result is instability and turbulence. This state is characterized by a diminished flow rate (there is a lot of backward swirling), substantial mixing (indeed, turbulence is required if efficient mixing is to occur), and noise (it is the turbulence at the heart valves that makes that familiar thumping sound). A healthy person breathes quietly, but obstructions in the air passages cause turbulence and noise that can be heard with a stethoscope.

For an object moving through a liquid or gas, the resistance offered by the medium increases with the turbulence. While the viscous drag force varies directly with speed, the turbulent drag force is proportional to the speed-squared. Since the power expended in overcoming drag is force times speed, that power varies as the speed-cubed! That is why auto manufacturers are finally streamlining their cars and why big square-front trucks are being retrofitted with wind deflectors, or fairings. The price of fuel demands that less energy be wasted making vortices. A fairing can reduce drag by 20% and save thousands of dollars a year. One wonders why it took the industry so long to figure out what kids, crouching on bicycles, have known all along.

The sloped fairing above the driver's compartment channels air up and over the body of the truck, cutting drag by about 20%.

11.10 The Continuity Equation

The constancy of the density of a flowing liquid is the basis of a fundamental relationship that allows us to understand how liquids progress in pipes and veins. Envision a tube of flow in a liquid (Fig. 11.30). There are no *sources* within the volume of the tube (no nozzles squirting in fluid), and there are no *sinks* (no drains tapping off fluid). The fluid enters at boundary 1, where area A_1 is perpendicular to the flow lines, and emerges at boundary 2, where A_2 is perpendicular to the flow. Let v_1 and v_2 be the average speeds of the fluid over A_1 and A_2. In a tiny time Δt, during which

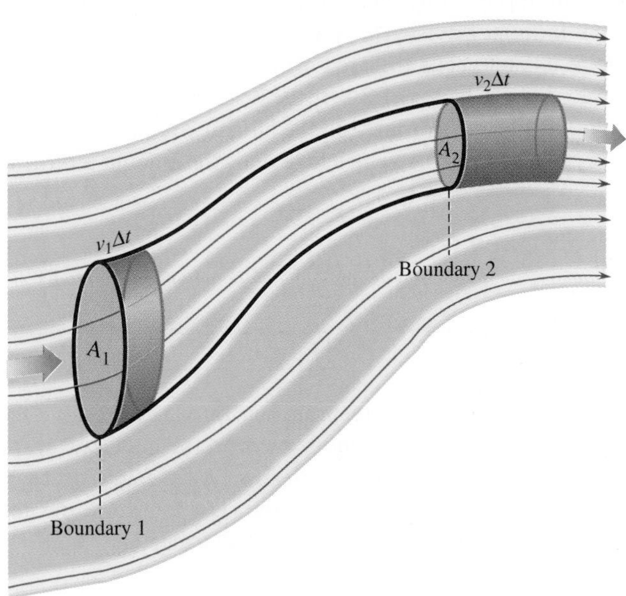

Figure 11.30 A fluid in laminar flow passes from left to right. Assuming incompressibility, the volume entering the flow tube per second at boundary 1 must equal the volume per second leaving at boundary 2, or anywhere else along its length. The flow tube, which is part of a broader stream, could also be envisioned as a pipe with a changing cross section.

American sprint cyclists in the 1992 Olympics wore suits with ribbing on the back and shoulders. The ridges reduce turbulence and cut drag by 3%.

Figure 11.31 As the water falls, its speed increases and so its cross-sectional area decreases as mandated by the Continuity Equation.

the particles of fluid entering the tube travel a distance $v_1\Delta t$, the particles leaving it travel a distance $v_2\Delta t$. Since the volume entering equals the volume leaving, we have

$$(A_1v_1\Delta t) = (A_2v_2\Delta t)$$

and

$$A_1v_1 = A_2v_2 \tag{11.12}$$

This is the **Continuity Equation**, and it points out that, as the cross-sectional area increases, the speed decreases and vice versa (Fig. 11.31). A river flowing leisurely along a broad bed speeds up as the bed narrows and all the water rushes through a smaller cross-sectional area. If instead, the river branches into a number of identical tributaries so that the *total* cross-sectional area is increased, the speed in each narrow branch will be slower than in the river. Thus, in the human body, as the blood-carrying tubing branches from the single aorta (net area $\approx 2.5 \times 10^2$ mm^2) into the many arteries (net area $\approx 2 \times 10^3$ mm^2) into the far more numerous arterioles (net area $\approx 4 \times 10^3$ mm^2) into the billions of capillaries (net area $\approx 2.5 \times 10^5$ mm^2), the speed of flow decreases proportionately even though the individual blood vessels are getting narrower.

The quantity Av, which is constant within any tube, is denoted by the symbol J and is called the *discharge rate* or **volume flux**—the latter name applies because at both boundaries $v = \Delta s/\Delta t$ and $\Delta V = A\,\Delta s$; hence

$$J = Av = \frac{\Delta V}{\Delta t} \tag{11.13}$$

This quantity is the volume of fluid flowing past a point in the tube per interval of time Δt, and it has SI units of m^3/s, though liters/min and gal/min are also common in the literature. Multiplying J by the density of the fluid provides the *mass flux,* or rate of mass flow in kg/s. Envision a system pumping water into a hose at a fixed rate. If a nozzle at its end necks the area of the opening down by a factor of 2, the exit speed must go up by a factor of 2—if the hose carries 0.000 5 m^3/s, the nozzle must discharge 0.000 5 m^3/s. Similarly, in the human circulatory system, the amount of blood pumped out of the heart's ventricles equals the amount returning to the atria.

Example 11.9 Blood is pumped out of the heart via the thick-walled (2-mm) tube known as the aorta (inside diameter about 18 mm) at an average speed, for an adult at rest, of 0.33 m/s. (a) Compute the discharge rate. The aorta branches into about 32 major arteries that are each roughly the same size—namely, 4-mm inside diameter. (b) Determine the speed of the blood through these arteries. The smallest branches of the system are the capillaries, about 8×10^{-6} m in inner diameter. (c) Given that the net cross-sectional area of capillaries is 2.5×10^5 mm^2, what is the speed of flow in a capillary?

Solution: [Given: $v_A = 0.33$ m/s, $D_A = 18$ mm, 32 arteries with $D_a = 4$ mm, $A_c = 2.5 \times 10^5$ mm^2. Find: (a) J_A, (b) v_a, (c) v_c.] (a) For the aorta, which we subscript with an A

$$J_A = Av = \pi\left(\frac{D_A}{2}\right)^2 v = \pi(9.0 \times 10^{-3}\text{ m})^2(0.33\text{ m/s})$$

and

$$\boxed{J_A = 8.4 \times 10^{-5}\text{ m}^3/\text{s}}$$

(b) The aorta branches into 32 arteries. To find the speed of the blood, we need the *net cross-sectional area of all of these arteries:* $A_a = 32\pi(D_a/2)^2 = 4.02 \times 10^{-4}$ m^2. Thus, since the flow rate from the aorta equals that in all the arteries taken together

$$J_A = J_a = A_a v_a$$

$$8.4 \times 10^{-5}\text{ m}^3/\text{s} = (4.02 \times 10^{-4}\text{ m}^2)v_a$$

and

$$\boxed{v_a = 0.21\text{ m/s}}$$

(continued)

(continued)

(c) Similarly

$$J_A = J_c = A_c v_c$$

$$8.4 \times 10^{-5}\ \text{m}^3/\text{s} = (2.5 \times 10^{-1}\ \text{m}^2)v_c$$

and

$$\boxed{v_c = 3.4 \times 10^{-4}\ \text{m/s}}$$

▶ **Quick Check:** $A_a/A_c = (4 \times 10^{-4}\ \text{m}^2)/(0.25\ \text{m}^2) = 16 \times 10^{-4}$, which should equal $v_c/v_a = 16 \times 10^{-4}$.

11.11 Bernoulli's Equation

The first modern treatment of hydrodynamics was a work completed in 1734 by the Swiss mathematician, physicist, and physician Daniel Bernoulli. He was a disciple of Leibniz and an intellectual adversary of the Newtonian school. Shunning Newton's Laws, Bernoulli used the idea of conservation of *vis viva* (p. 288) to derive the central formula of fluid dynamics. That derivation, though it came long before the concept of *energy* was formalized, turned out to be equivalent to the principle of Conservation of Energy.

The first point to consider is that a pressurized fluid must contain energy by virtue of the work done on it to establish that pressure. Pop the cap off a bottle of beer that has been well shaken, and fluid will come blasting out with plenty of KE. Clearly, potential energy is stored in the pressurized system and *a fluid that undergoes a pressure change undergoes an energy change.*

Now, imagine an ideal incompressible fluid for which there are no losses due to the conversion of mechanical energy into rotational, thermal, or any other form of energy. The pressure acting on a moving glob of such a fluid does work on it that appears as a net change in the kinetic and/or potential energy of the system; that is

$$\Delta W = \Delta \text{KE} + \Delta \text{PE}_G$$

Let's first determine ΔW for a sample of fluid.

Figure 11.32 depicts a narrow tube of flow where the specified values of pressure, speed, displacement, and height are to be thought of as averages. The pressure-force $F_1 = P_1 A_1$ acting on the glob of liquid, pushing it in the direction of motion, does an amount of work *on* it of $F_1 \Delta s_1$. In the process, molecules throughout the tube are shifted to the right with the net effect that an equal-volume glob of length Δs_2 is displaced into region 2. This time, the fluid external to the tube pushes to the left with a force $F_2 = P_2 A_2$, which is opposite in direction to the displacement. Here, the liquid in the tube is doing work pushing on the surrounding fluid, and so the work done *on* it, $F_2 \Delta s_2$, is negative. Hence, the net work done *on* the fluid is

$$\Delta W = F_1 \Delta s_1 - F_2 \Delta s_2 = P_1 A_1 \Delta s_1 - P_2 A_2 \Delta s_2$$

which can be put in a more convenient form. Recall that $\Delta s_1 = v_1 \Delta t$ and $\Delta s_2 = v_2 \Delta t$; thus

$$\Delta W = P_1 A_1 v_1 \Delta t - P_2 A_2 v_2 \Delta t$$

and from the Continuity Equation, $A_1 v_1 = A_2 v_2 = Av$; hence

$$\Delta W = vA\,\Delta t(P_1 - P_2)$$

Since the mass of the glob of liquid is $\Delta m = \rho \Delta V = \rho(vA\,\Delta t)$,

$$\Delta W = \frac{\Delta m}{\rho}(P_1 - P_2) \tag{11.14}$$

The glob of fluid, in moving from a higher to a lower pressure, has work done on it.

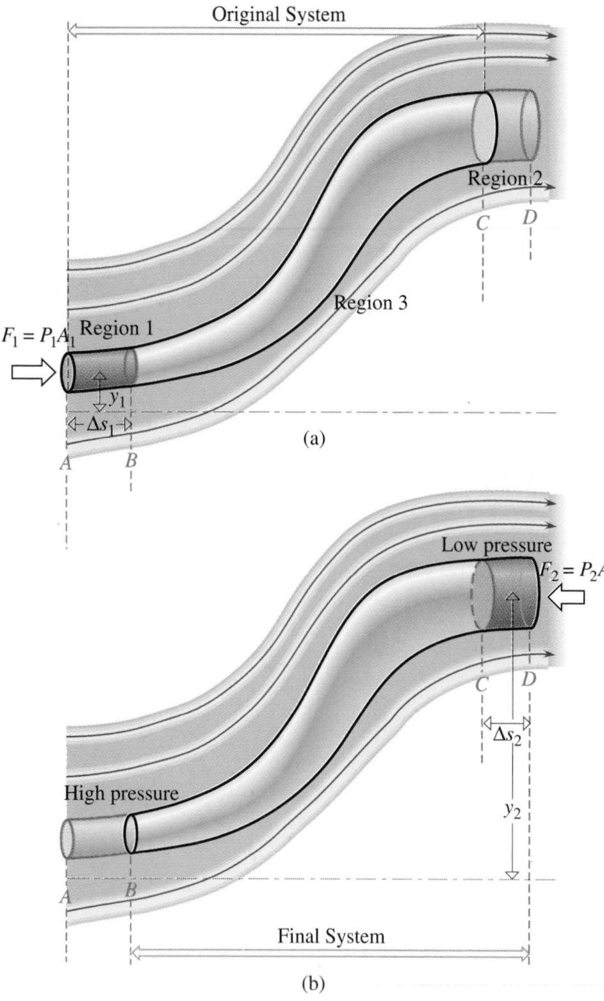

Figure 11.32 The energy analysis for an incompressible fluid in laminar flow. In a time interval Δt, the tube of liquid essentially shifts from where it was in (a) to where it is in (b). A mass of fluid Δm has been raised in the gravitational field; there's been a change in pressure and a change in speed.

Now for the accompanying change in KE. The system of interest originally is the tube of liquid (Fig. 11.32a) extending from position A to C. The glob of fluid Δm is moving at v_1 in region 1 (A to B). After a time, Δt (Fig. 11.32b), that glob has moved into region 3 and an equivalent mass of fluid that was in region 3 has moved into region 2. The system now extends from B to D. In effect, Δm has gone from region 1 at speed v_1 to region 2 at v_2. The change in kinetic energy of the fluid in the tube is

$$\Delta \text{KE} = \tfrac{1}{2}\Delta m(v_2^2 - v_1^2) \qquad (11.15)$$

The change in the *gravitational*-PE arises from the fact that the shifting of all the molecules in the tube has effectively brought a mass Δm from region 1 to region 2. Measuring the y values up from some arbitrary reference level, we can write

$$\Delta \text{PE}_G = \Delta mg(y_2 - y_1) \qquad (11.16)$$

Using the last three equations, we obtain

$$\Delta W = \Delta \text{KE} + \Delta \text{PE}_G$$

$$\frac{\Delta m}{\rho}(P_1 - P_2) = \tfrac{1}{2}\Delta m(v_2^2 - v_1^2) + \Delta mg(y_2 - y_1)$$

and $$(P_1 - P_2) = \tfrac{1}{2}\rho(v_2^2 - v_1^2) + \rho g(y_2 - y_1) \qquad (11.17)$$

Rearranging terms, we get **Bernoulli's Equation**

$$P_1 + \tfrac{1}{2}\rho v_1^2 + \rho g y_1 = P_2 + \tfrac{1}{2}\rho v_2^2 + \rho g y_2 \qquad (11.18)$$

In the steady flow of an ideal liquid, *following along a streamline*

$$P + \tfrac{1}{2}\rho v^2 + \rho g y = \text{constant} \qquad (11.19)$$

which is often referred to as *Bernoulli's Theorem.*

Each term has the dimensions of energy per unit volume, or energy density. There is a *kinetic-energy density* ($\frac{1}{2}\rho v^2$) associated with the bulk motion of the fluid, a *potential-energy density* ($\rho g y$) associated with changes in location within the gravitational field, and a *pressure-energy density* (P) arising from internal forces on the moving fluid. This latter energy is similar to that stored in a compressed spring in the sense that it's a manifestation of the interatomic electromagnetic interaction. In effect, Eq. (11.19) says that, as long as no energy enters or leaves, *the net energy density contained in the fluid is constant all along a given flow tube.* Though it was derived for an incompressible fluid, Bernoulli's Equation can be applied to gases, provided the pressure changes are only a few percent.

Example 11.10 Figure 11.33 depicts an uncovered vat of brewing beer and the narrow pipe used to take samples from it. The cross-sectional area of the vat is 1.50 m^2. At a given instant, the liquid level is falling at

(continued)

(continued)

1.0 cm/s while the beer is traveling at 50 cm/s past the gauge. Determine the absolute pressure reading at that point within the pipe, at that moment. Take the density of beer to be 1.0×10^3 kg/m^3.

Solution: [Given: gauge 2.0 m below surface, area of tank 1.50 m^2, speed of surface 1.0 cm/s, speed in pipe 0.50 m/s. Find: pressure in pipe.] Take the influx end of a tube of flow as region 1 located at the open surface of the liquid in the vessel. Take region 2 in the pipe near the pressure gauge. Since the reference level is arbitrary when dealing with potential energy, let $y_2 = 0$. The height of the liquid surface (above the gauge) is $y_1 = h = 2.0$ m. Bernoulli's Equation

$$P_1 + \tfrac{1}{2}\rho v_1^2 + \rho g y_1 = P_2 + \tfrac{1}{2}\rho v_2^2 + \rho g y_2$$

becomes

$$P_A + \tfrac{1}{2}\rho v_1^2 + \rho g h = P_2 + \tfrac{1}{2}\rho v_2^2 + 0$$

and so

$$P_2 = P_A + \tfrac{1}{2}\rho(v_1^2 - v_2^2) + \rho g h$$

$$P_2 = 1.013 \times 10^5 \text{ Pa} + \tfrac{1}{2}(1.0 \times 10^3 \text{ kg/m}^3) \times [(1.0 \times 10^{-2} \text{ m/s})^2 - (0.50 \text{ m/s})^2] + (1.0 \times 10^3 \text{ kg/m}^3)(9.8 \text{ m/s}^2)(2.0 \text{ m})$$

thus

$$P_2 = 1.013 \times 10^5 \text{ Pa} - 1.25 \times 10^2 \text{ Pa} + 1.96 \times 10^4 \text{ Pa}$$

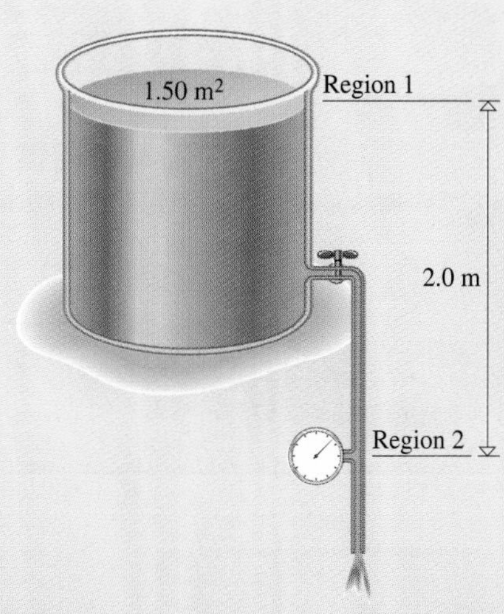

Figure 11.33 A vat of beer is sampled via a narrow tube with a pressure gauge mounted 2.0 m below the surface.

and $\boxed{P_2 = 1.2 \times 10^5 \text{ Pa}}$

▶ **Quick Check:** The pressure at a point that is 2 m down must be $P_A \approx 100$ kPa, plus the contribution due to the weight of the column of fluid $\rho g h \approx (10^3 \text{ kg/m}^3) \times (10 \text{ m/s}^2)2 \text{ m} \approx 20$ kPa, plus a drop in pressure ($\approx$0.1 kPa) due to an increase in speed or $\approx$120 kPa.

Notice that when the fluid is at rest, $v_1 = v_2 = 0$, and Eq. (11.18) becomes

$$P_1 - P_2 = \rho g(y_2 - y_1)$$

which is equivalent to the result (p. 356) for the pressure difference between any two points in a liquid.

Torricelli's Result

Consider the closed tank of liquid shown in Fig. 11.34 that has an orifice of area A_2, out of which is pouring the fluid. The rate at which the fluid leaves, the *speed of efflux* (v_2), is the quantity we want to determine. The vessel is taken to be very large so that neither the liquid level nor the pressure above it (P) changes significantly. Since the flow tube goes from region 1 out to region 2, which is open in the air, let $P_1 = P$. The emerging stream in region 2 is surrounded by air at atmospheric pressure, and nothing is there to sustain a pressure difference. *A free fluid jet must be at the same pressure as the enveloping air,* $P_2 = P_A$. Since $y_1 - y_2 = h$, Bernoulli's Equation is then

$$P + \tfrac{1}{2}\rho v_1^2 + \rho g h = P_A + \tfrac{1}{2}\rho v_2^2$$

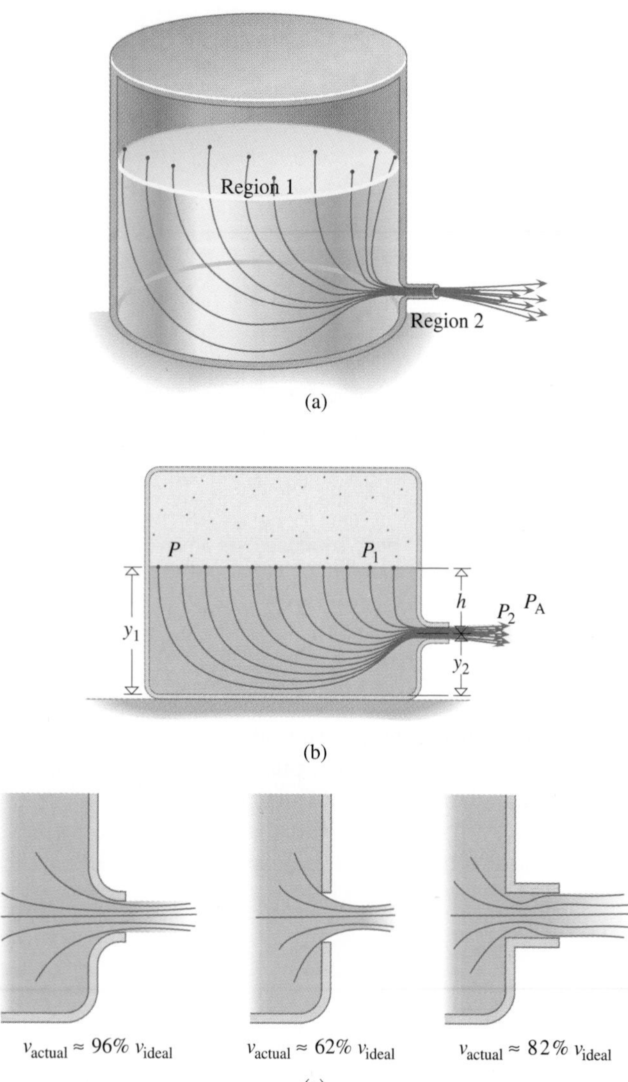

Figure 11.34 (a) The streamlines for a fluid flowing out of a closed vessel, where the pressure on the liquid's surface is *P*. (b) To apply Bernoulli's Equation, we take the top of the liquid as region 1, the outside of the nozzle as region 2, and follow a streamline from 1 to 2. (c) In actuality, the shape of the orifice affects the shape of the jet and its speed.

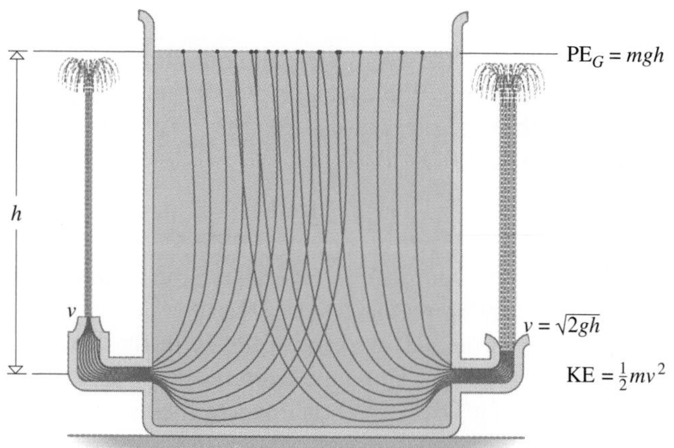

Figure 11.35 The emerging jets reach about the same heights, independent of the diameters of the nozzles. They don't have the same discharge rates, although the efflux speeds are equal. This is a different situation from that of a pump forcing *a fixed discharge rate* through two different orifices. It's also different from the garden hose fed by miles of pipe in which friction affects the flow.

and

$$v_2^2 = v_1^2 + \frac{2(P - P_A)}{\rho} + 2gh \tag{11.20}$$

From the Continuity Equation ($v_1A_1 = v_2A_2$), it follows that, when $A_1 \gg A_2$ and $v_2 \gg v_1$, then v_1^2 is negligible compared to v_2^2. In the special, though common, case where the tank is open to the air ($P = P_A$), the pressure-energy density vanishes—the fluid runs out via gravity without also being pushed out by a pressure difference, and

$$v_2 = \sqrt{2gh} \tag{11.21}$$

This relationship is *Torricelli's Result.*

Being frictionless, in theory at least, the liquid spurts from the hole with a speed equal to that which it would have gained in free-falling through the height *h*. Torricelli observed that the emerging jet, if directed straight up, would almost reach the original level (Fig. 11.35). The reason the jet doesn't quite attain the initial height is that some energy is invariably converted to thermal energy via friction.

Example 11.11 A boat on a lake crashes into a submerged rock, ripping a hole 40 cm^2 in its hull 1.00 m below the waterline. If the craft can take on 10.00 m^3 of water before getting its cargo drenched, roughly how much time does the crew have to do something clever? Assume the water flows straight in the hole and neglect any frictional effects.

Solution: [Given: area of hole 40×10^{-4} m^2, $h =$ 1.00 m. Find: time to take on 10.00 m^3 of water.] Let

(continued)

(continued)

region 1, the input end of a flow tube, be at the surface of the lake and region 2, the output end, be at the hole—water flows down and into the boat. Assuming there is atmospheric pressure in the boat at the hole, Bernoulli's Equation is

$$P_1 + \tfrac{1}{2}\rho v_1^2 + \rho g y_1 = P_2 + \tfrac{1}{2}\rho v_2^2 + \rho g y_2$$

and

$$P_1 + \tfrac{1}{2}\rho v_1^2 + \rho g(y_1 - y_2) = P_2 + \tfrac{1}{2}\rho v_2^2$$

Since the level of the lake is not going to fall very rapidly

$$P_A + 0 + \rho g h = P_A + \tfrac{1}{2}\rho v_2^2$$

which gives us Torricelli's Result

$$v_2 = \sqrt{2gh} \qquad [11.21]$$

We should have anticipated this result to begin with—the container with the leak is the whole lake pouring water into the boat, so

$$v_2 = \sqrt{2(9.81 \text{ m/s}^2)(1.00 \text{ m})} = 4.43 \text{ m/s}$$

Now

$$J = v_2 A_2 = (4.43 \text{ m/s})(40 \times 10^{-4} \text{ m}^2)$$

or

$$J = 1.77 \times 10^{-2} \text{ m}^3\text{/s}$$

and so it will take on 10.00 m^3 of water in

$$\frac{(10.00 \text{ m}^3)}{(1.77 \times 10^{-2} \text{ m}^3\text{/s})} = 565 \text{ s} = \boxed{9.4 \text{ min}}$$

This solution assumes the boat doesn't sink appreciably lower in the water in the process, thereby changing h and so J.

▶ **Quick Check:** The water effectively falls 1 m; hence, $y = \frac{1}{2}gt^2$ and $t = 0.45$ s; thus, $v = gt = 4.43$ m/s. $J \approx 160 \times 10^{-4}$ m^3/s; 10 m$^3/J \approx 10$ min.

The Venturi Effect

Some of the most practical applications of fluid dynamics arise from the interdependence of pressure and speed. There is a class of situations in which the change in *gravitational*-PE is ignorably small and Bernoulli's Equation then relates differences in pressure to differences in kinetic energy, and therefore speed. For example, envision a length of pipe (with a transverse area A_1) that necks down to a smaller area (A_2) and then opens out again (Fig. 11.36). We know that

$$A_1 v_1 = A_2 v_2 \qquad [11.12]$$

and, as we saw earlier, decreasing the area traversed by the flow of a gas or liquid results in an increase in its speed.

In any situation where the heights of the incoming stream and the constricted region are about the same ($y_1 \approx y_2$), Bernoulli's Equation can be used to express the driving pressure difference:

$$P_1 + \tfrac{1}{2}\rho v_1^2 + \rho g y_1 = P_2 + \tfrac{1}{2}\rho v_2^2 + \rho g y_2$$

becomes

$$P_1 + \tfrac{1}{2}\rho v_1^2 = P_2 + \tfrac{1}{2}\rho v_2^2$$

Using the Continuity Equation to eliminate v_1, we get

$$P_1 - P_2 = \tfrac{1}{2}\rho v_2^2 \frac{(A_1^2 - A_2^2)}{A_1^2} \qquad (11.22)$$

Inasmuch as $A_1 > A_2$, the right side of the equation is positive and $P_1 > P_2$—there is a pressure drop in the narrow region. As the fluid enters region 3, the higher pressure slows it down, and the speed returns to its original value. This decrease in pressure accompanying an increase in speed is variously called Bernoulli's Effect, or the *Venturi Effect,* after the Italian researcher who first studied it (1791).

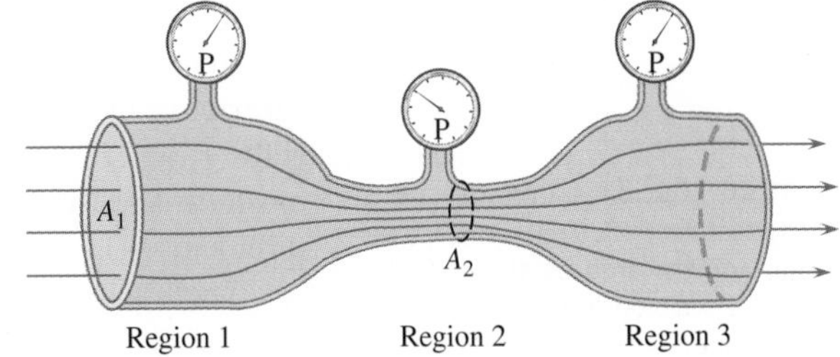

Figure 11.36 In region 1, the cross-sectional area is large, the speed is low, and the pressure is high. In region 2, the area is small, the speed is high, and the pressure is low. In region 3, the area is again large, the speed is small, and the pressure is high.

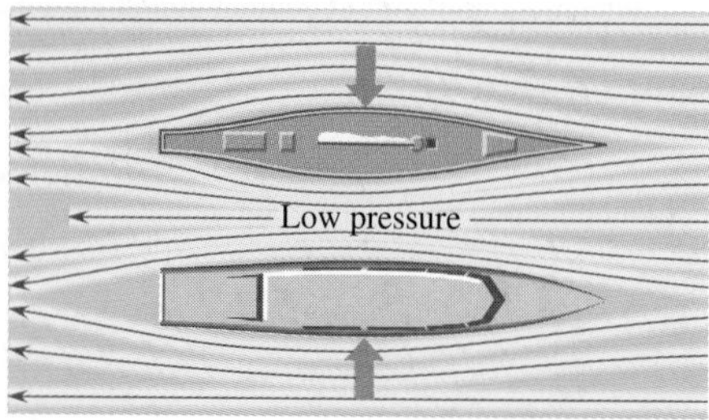

Figure 11.37 Two ships traveling side by side or moored in moving water. The deflection of the flow between the two vessels causes a Venturi drop in pressure, and the ships experience a force pushing them together.

It is possible to mechanically increase the speed of a fluid, as for example, with a pump. The KE thereby added comes from the work done on the fluid. Thus, the pressure in a high-speed free jet produced by a pump is still atmospheric. That's not to be confused with the Venturi Effect, where no additional energy enters the system and an increase in KE must come from a decrease in pressure energy, and vice versa. In that case, **a portion of a fluid made to move rapidly sustains a lower pressure than does the portion of the fluid moving more slowly**. As a large truck or bus passes alongside a car, the pressure between the two will often drop from the rushing air, giving the driver of the auto the disturbing feeling of being pulled toward the larger vehicle (Fig. 11.37).

Arteriosclerosis arises when plaque builds up on the inner walls of the arteries, restricting the flow of blood. Any such obstruction results in a Venturi pressure drop (Fig. 11.38). In an advanced stage of the disease, the pressure difference, inside and out, can be great enough to cause the artery to momentarily collapse. Backed-up blood pressure will soon force it open, only to collapse again and so on, producing a vascular flutter that can be heard with a stethoscope. The Venturi Effect in airstreams is responsible for other flutterings, such as the sound of snoring and the more melodious tones of reeded woodwinds.

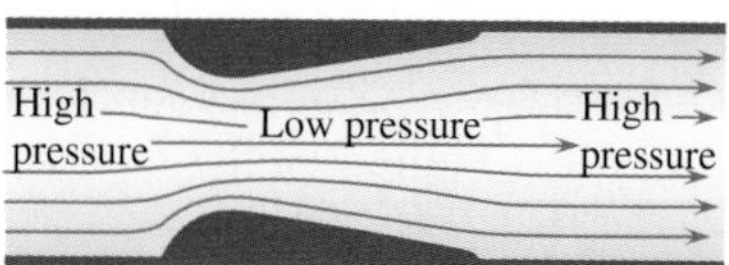

Figure 11.38 The flow of a fluid through a constricted channel.

11.12 Airfoils

An **airfoil** is an object whose shape is such that it generates a desired reaction force from the fluid through which it moves. For instance, the wing-shaped object shown in Fig. 11.39 is an asymmetrical airfoil, and so, unlike Fig. 11.38, the streamline pattern is asymmetrical. Air moves with respect to the long, curved upper surface, at a more rapid speed than the airstream beneath the wing. With an appropriately shaped wing, the corresponding drop in pressure above, as compared to below, creates a net upward force known as **lift**. Lift makes possible the level flight of powered airplanes, holds gliders in the air, and supports soaring birds.

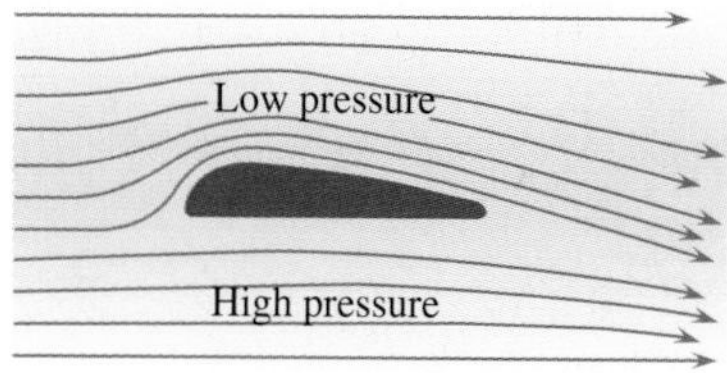

Figure 11.39 Flow past a horizontal asymmetrical airfoil. Note the high-speed, low-pressure regions where the flowlines are closer together. Compare this flow pattern with that of Fig. 11.38.

Still, it's not clear at this point why the air moves at a higher speed above the wing than below. Most popularized versions suggest that the two streams must take the same time to arrive at the rear edge and so travel at different speeds, but that explanation doesn't account for how the air finds out that it has a scheduled arrival time to meet, nor, in fact, is it even true—the traversal times need not be the same. What actually happens is affected by the viscosity of the air and its adhesion to the wing. The airfoil induces a circulation of air (Fig. 11.40) around the wing. This circulation combines with the free-air flow to produce a higher speed stream above than below the wing.

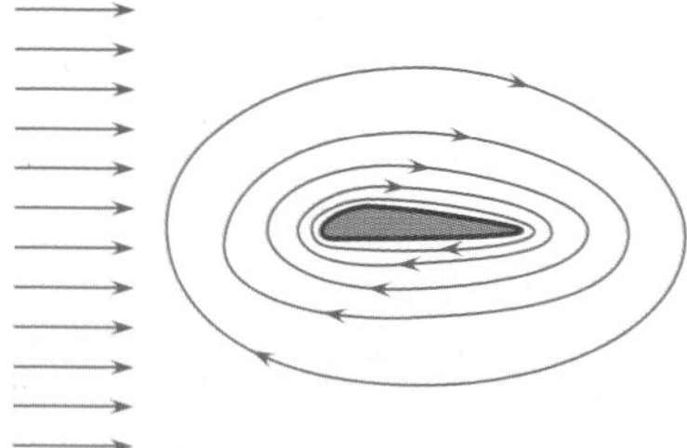

Figure 11.40 Circulation around an airfoil, in cross section. The sum of the circulation and the free-air flow produces the pattern in Fig. 11.39, where the speed is higher above the airfoil.

The amount of circulation generated depends on the shape of the airfoil, its speed, and its orientation with regard to the air flow. For example, a plane can fly horizontally, nose up, with its wings tilted at the so-called *attack angle*. The pressure above drops and the pressure below rises even more, thereby increasing the lift (Fig. 11.41). There is a practical limit to the process (typically around 15°): too much tilt and the boundary-layer air tears away from the top of the wing, becoming turbulent—the wing *stalls*, lift vanishes, and the plane falls.

There's a popular engineering adage that if a plane's engines have enough power, they can drag almost anything into the air. That being the case, the curvature of a plane's wings becomes less important. Wings that are symmetrical airfoils get their lift primarily by being pushed through the air at a nonzero attack angle—in effect, the air moving at the underside is deflected downward and transfers an upward momentum to the wing. That's how planes manage to fly upside down—if the only thing occurring was a reduced pressure due to wing curvature, the craft would be driven into the ground by the inverted "lift," but that doesn't happen.

A simple argument can be made, via Bernoulli's Equation (Problem 69), showing that *lift is proportional to the product of airspeed-squared and wing area.* This relationship turns out to be true, and it explains why big planes and big flying birds have relatively big wings. It's also why big birds have a higher minimum air speed that they must reach to take off and must maintain to stay airborne.

The skier becomes an airfoil.

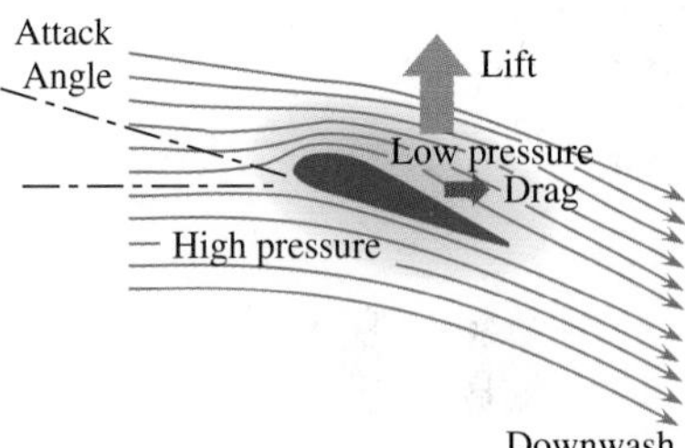

Figure 11.41 The downwash tells us that there is an upward reaction force exerted on the airfoil by the deflected air stream. The time-rate-of-change of the momentum of the air must equal the magnitude of the upward force.

11.13 Vortices

A vortex is a whirling mass of fluid bounded by a region that is irrotational. Tornadoes and hurricanes are large-scale powerful vortices; the whirling spiral of water draining from a bathtub and the great mushroom cloud of an atomic bomb are also vortex structures.

Whenever an object moves with respect to a fluid (a plane through the air or an oar through the water), it tends to spawn vortices. Small invisible whirlwinds are everywhere—behind moving ships and cars and planes and baseballs. Stirring a cup of coffee creates vortices, which is why it's done—to churn up turbulence that will disperse the ingredients.

Vortices are formed when there is a discontinuity in the flow pattern, often due to the presence of an obstruction. When two currents at different speeds come together, a transition region exists wherein the rapidly moving layer progresses beyond and around the slower layer, which essentially drags the former down. The process is like a forward moving row of skaters linked arm-in-arm right to left. If the person on the far right skates onto a bed of sand, the right side will slow dramatically while the left side of the row, still moving rapidly, swings around clockwise.

Imagine a cylindrical body immersed in a moving fluid (Fig. 11.42). Generally, there will be a more or less irregular flow pattern behind the cylinder. But if conditions are right (the speed of the fluid is high enough, its viscosity low enough), two parallel rows of equally spaced alternating eddies will be created—known as a *von Kármán vortex street.* Rows of vortices are easy to generate by moving a cylindrical object such as a pen or a finger at various speeds across the surface of a deep pan of water. The periodic shedding of vortices by an obstruction results in an oscillating pressure on the body that often tends to set it vibrating in a direction *perpendicular* to the flow. During World War II, the vibration of machine-gun barrels jutting from airplane turrets was a frequent problem; and skyscrapers are driven into oscillation by the same vortex mechanism. Similarly, vibrating (so-called singing) phone and power lines have often led to mechanical problems, sometimes ending in catastrophic failure due to fatigue.

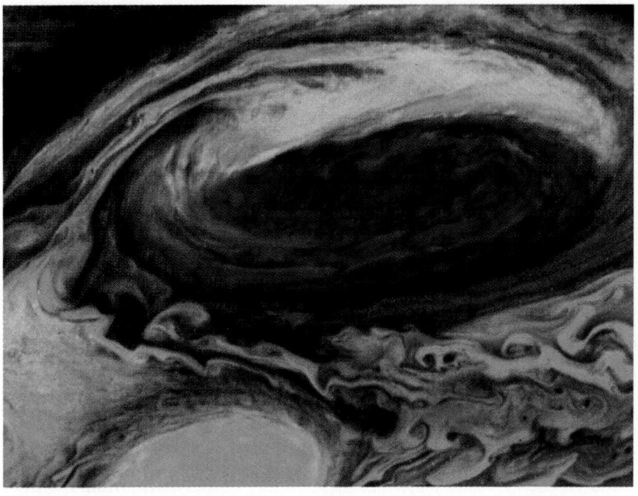

This photo of the Great Red Spot on the surface of Jupiter was taken by *Voyager 2.* The spot is a giant stable vortex that has been sustained in the deep atmosphere of that rotating planet for at least 300 years.

Surfaces that develop lift, such as airfoils, create vortices at their two ends, due to the difference in pressure at top and bottom. Air flows outward away from the fuselage, up and around the wing tips from the high-pressure region beneath to the low-pressure region above the wings (see the photo on p. 387). Figure 11.43 shows how

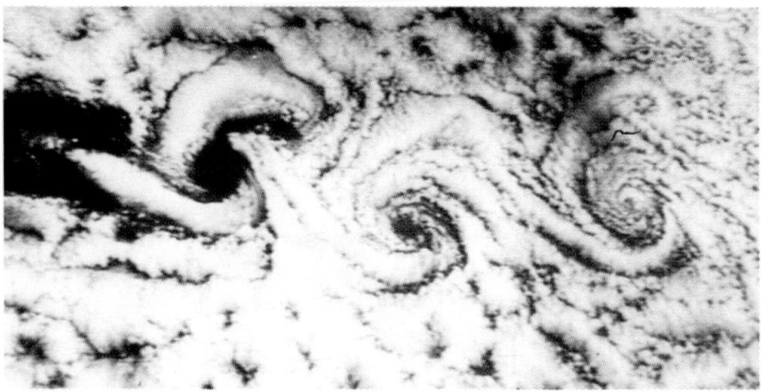

A satellite photo of vortices in the cloud cover formed in the wake behind an island in the ocean.

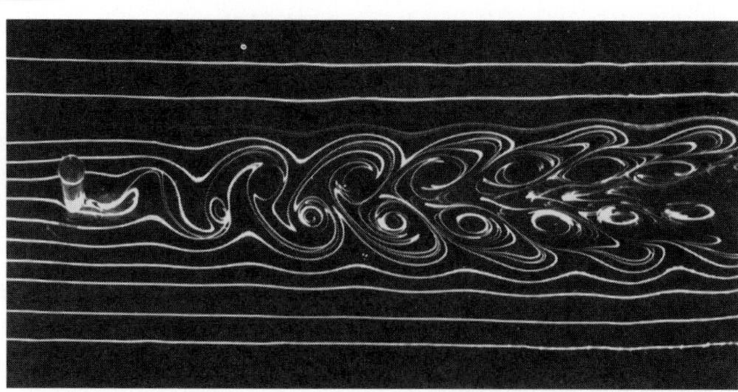

A vortex street.

different city situations can also spawn troublesome vortices. These vary from the street-level variety that steals hats to the high-altitude kind that blows out windows.

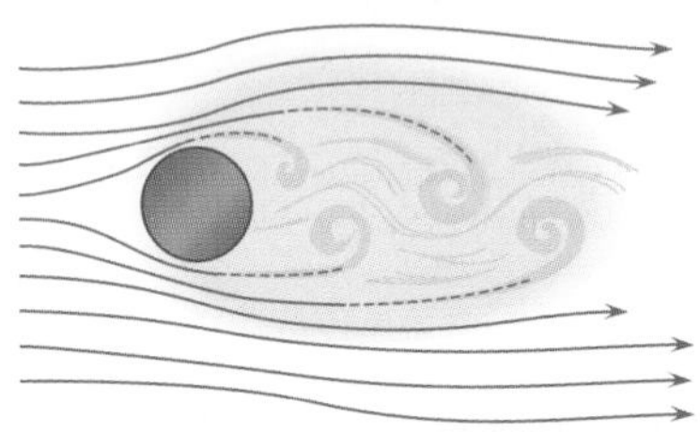

Figure 11.42 A fluid moving past a cylindrical obstruction. If conditions are right, an alternating series of vortices will be formed in the wake.

11.14 Blood Pressure

The human heart is actually two twin-chambered pumps (Fig. 11.44) mounted in a single unit: a low-pressure device (the right atrium and ventricle) and a high-pressure device (the left atrium and ventricle). The right half delivers oxygen-depleted blood to the nearby lungs. With each *systole* or rhythmic contraction of both ventricles, the right half discharges 70 to 80 cm^3 of blood at a mean gauge pressure of 2 kPa (15 mm Hg) into the pulmonary artery on the way to the lungs. An equal amount of oxygen-rich blood is returned from the lungs to the left half of the heart (at $\approx$ 8 mm Hg) for delivery to the rest of the body. The left ventricle pumps the blood out, maintaining a pulsating pressure that rises and falls 70 to 80 times per minute. The peak, or *systolic pressure,* developed during pumping (Fig. 11.45) is around 16 kPa (120 mm Hg) and the minimum or *diastolic pressure,* which occurs when the heart is essentially relaxed and refilling, is around 11 kPa (80 mm Hg).

Example 11.12 Compute the average output power of the human heart at rest. Take the average output pressure to be 1.33×10^4 Pa (100 mm Hg) and the flow rate as 5.0 liters per minute. Compare your result with the fact that the heart consumes energy at a rate of about 10 W.

Solution: [Given: $P_{av} = 1.33 \times 10^4$ Pa and $J =$ 5.0 liter/min. Find: output power.] The force doing the work equals $P_{av} \times A$, and the power developed is $\mathrm{P} = Fv = P_{av}Av = P_{av}J$. First change the units of the flow rate; since 1 liter = 1000 cm^3, then $J =$ 5.0 liter/min = (5.0 liter/min)(0.001 m^3/liter)(1/60 s/min) and $J = 8.33 \times 10^{-5}$ m^3/s. The heart takes that much blood every second from a gauge pressure of near zero to an average of about 1.33×10^4 Pa. Consequently,

$$\mathrm{P} = P_{av}J = (1.33 \times 10^4 \text{ Pa})(8.33 \times 10^{-5} \text{ m}^3/\text{s})$$

and

$$\boxed{\mathrm{P} = 1.1 \text{ W}}$$

which makes it roughly 10% efficient.

▶ **Quick Check:** 1 liter = 10^3 cm^3 = $10^3 \times 10^{-6}$ m^3; 1 liter/min = 10^{-3} m^3/60 s = 1.66×10^{-5} m^3/s. Power per 1 liter/min is P = 13 kPa(1.66×10^{-5} m^3/s) = 0.22 W.

Blood is a viscous fluid, and there will be some loss in energy and a corresponding drop in pressurization along the flow route—a real liquid drops in pressure as it negotiates even a smooth channel. The effects of viscous drag become less

Figure 11.43 City winds and vortices. Poorly designed and positioned buildings can create dangerous windstorms at street level. The average person can be blown off balance by a 65-km/h (40-mi/h) wind, and turbulent winds of 16 km/h (10 mi/h) impede walking. A gust of wind in Boston's Copley Square once blew over a half-ton mail truck. High winds can also be created by building-lined streets via the Venturi Effect.

Tip vortices made visible during crop spraying. The presence of powerful vortices behind large aircraft, such as the Boeing *747*, determines the maximum rate at which planes can follow one another down an airport runway.

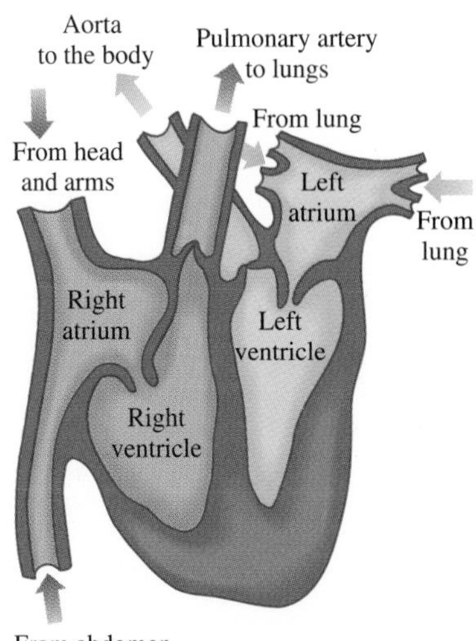

Figure 11.44 Both atria contract at the same time, forcing blood into their corresponding ventricles. Thereafter, the ventricles contract, forcing blood into the arteries and out to the body and lungs. The bluish regions carry oxygen-depleted blood, the reddish regions oxygen-rich blood.

troublesome as the pipe gets larger and more of the fluid is moving farther from the walls. Arteries have relatively large cross sections and blood progresses through them with little frictional loss and little diminution of pressure energy. That's why the pulses of blood pressure are most easily detected with arteries, especially ones close to the surface at the neck and wrist.

Insofar as viscous effects in the main arteries are not significant, Bernoulli's Equation applies:

$$P_1 + \tfrac{1}{2}\rho v_1^2 + \rho g y_1 = P_2 + \tfrac{1}{2}\rho v_2^2 + \rho g y_2 \quad [11.18]$$

Since the flow rates and cross-sectional areas in the main arteries are fairly uniform, the speeds of flow are approximately equal as well. We want to compare the pressure at the heart (P_H) and feet (P_F), so take these to be regions 1 and 2, respectively. For someone standing, $y_1 = y_H$ and $y_2 = 0$; hence

$$P_H + \rho g y_H = P_F$$

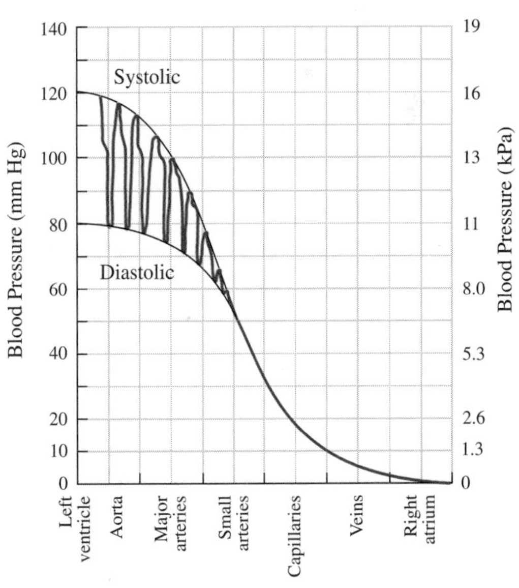

Figure 11.45 Blood pressure in the adult human male body. As the heart goes through its pumping cycle, the pressure in the main vessels fluctuates. The systolic and diastolic pressures are the maximum and minimum values developed in the arteries. Typical female pressures are slightly lower (110 mm Hg/70 mm Hg).

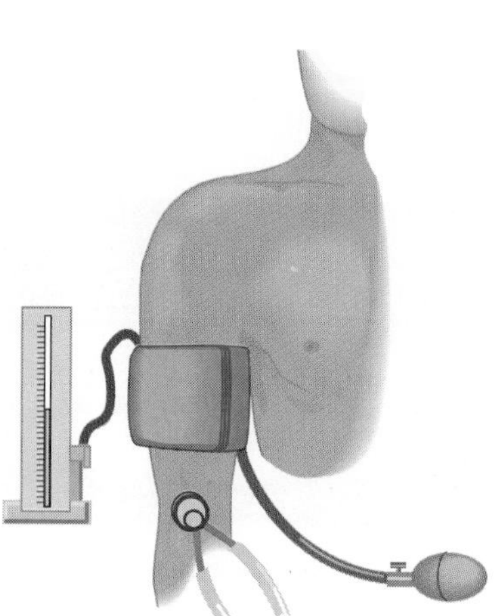

Figure 11.46 The mercury manometer reads the gauge pressure in the cuff wrapped around the arm. The small rubber bulb pumps up the cuff until it cuts off the blood flow in the brachial artery in the upper arm. The pressure is released gradually and, with the use of a stethoscope, the blood can be heard to start flowing.

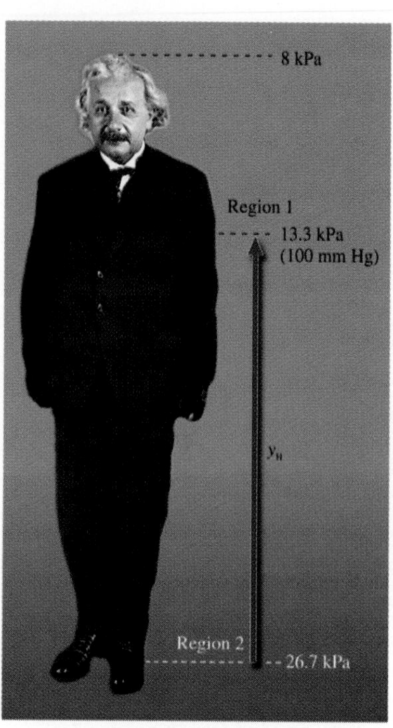

Blood pressure varies over the body, in part due to gravity. A person who jumps up from a prone position can feel momentarily dizzy because of the sudden drop in blood pressure. Similarly, standing still can cause blood to pool in the legs, and dizziness can again occur. Fainting usually lowers the head.

In effect, the weight of the blood causes an increase in pressure at the feet, just as it would in the static case. This is why it's so civilized to have a nap after lunch—the heart doesn't have to work quite as hard pumping against gravity.

The most common way to measure arterial blood pressure, though it's not very accurate, is with an inflatable bag or cuff attached to a manometer (Fig. 11.46). The cuff is tightly wrapped around the upper arm *at the level of the heart* (so there is no $\rho g y$ contribution to P) and then inflated with a rubber bulb pump. Squeezing the arm, it cuts off blood flow in the brachial artery. The overpressure in the cuff is then allowed to gradually drop while the observer listens with a stethoscope placed downstream on the artery. When the pressure in the cuff (as indicated by the manometer) equals the peak or systolic gauge pressure, blood begins to spurt past the constriction, and the turbulent, noisy flow is easily heard in the stethoscope. Further reduction of the cuff pressure is accompanied by a changing sound until at the diastolic pressure the flow is again laminar and the characteristic noise of turbulence vanishes (Table 11.6).

TABLE 11.6 The human cardiovascular system

Blood density (37°C)	1.0595×10^3 kg/m^3
Number of red cells per mm^3 of blood	5×10^6
Number of white cells per mm^3 of blood	8×10^3
Flow rate at heart	
resting	8×10^{-5} m^3/s (5 liters/min)
intense activity	40×10^{-5} m^3/s (25 liters/min)
Time for tracer to circulate	18–24 s
Time for complete circulation	54 s
Blood volume (70-kg male)	5.2×10^{-3} m^3 (5.2 liters)
Gauge pressure, right ventricle, peak	3 kPa (25 mm Hg)
Gauge pressure, left ventricle, peak	16 kPa (120 mm Hg)
Mean gauge pressure, large veins	1.07 kPa
Mean gauge pressure, large arteries	12.8 kPa
Mean gauge pressure, at the heart	13.3 kPa
Pressure drop per cm of aorta	4.3 Pa (3.2×10^{-2} mm Hg)
Power required by heart at rest	$\approx$10 W

Core Material

Pressure (P) is *the magnitude of the total force acting perpendicular to a surface divided by the area of that surface:*

$$P = \frac{F_\perp}{A} \qquad [11.1]$$

The units of pressure are N/m^2 or *pascals*. The pressure a distance h beneath the surface of a homogeneous fluid is

$$P_l = \rho g h \qquad [11.2]$$

and if there is a pressure on its surface P_s, then

$$P = P_s + P_l = P_s + \rho g h \qquad [11.3]$$

The pressure at every point at a given horizontal level in a single fluid at rest is the same.

An *atmosphere* is defined as 1 atm = 1.013×10^5 Pa, which equals 14.7 lb/in.^{2}. The absolute pressure (P) is the sum of the gauge and atmospheric pressures, or

$$P = P_A + P_G \qquad [11.4]$$

Keeping the temperature constant, the volume of any gas varies inversely with the pressure. This is **Boyle's Law**:

$$PV = \text{constant} \qquad [11.5]$$

Pascal's Principle states that an external pressure applied to a fluid that is confined within a closed container is transmitted undiminished throughout the entire fluid. Hence, for two connected pistons

$$\frac{F_o}{F_i} = \frac{A_o}{A_i} \qquad [11.6]$$

The **Buoyancy Principle** of Archimedes states that *an object immersed in a fluid will be lighter (that is, it will be buoyed up) by an amount equal to the weight of the fluid that it displaces, and the corresponding upward force exerted by the fluid is known as the buoyant force.* The *relative density*, ρ/ρ_w, is called the *specific gravity.*

Surface tension, γ, is *the force per unit length exerted by the surface.* Where the single liquid film has *two* surfaces

$$F_t = 2\gamma L \qquad [11.10]$$

The **Continuity Equation** is

$$A_1 v_1 = A_2 v_2 \qquad [11.12]$$

The quantity Av is given the symbol J and called the *discharge rate*, or *volume flux*. Applying Conservation of Energy to the ideal fluid, we get **Bernoulli's Equation**

$$P_1 + \tfrac{1}{2}\rho v_1^2 + \rho g y_1 = P_2 + \tfrac{1}{2}\rho v_2^2 + \rho g y_2 \qquad [11.18]$$

or

$$P + \tfrac{1}{2}\rho v^2 + \rho g y = \text{constant} \qquad [11.19]$$

The speed of efflux from a hole in an open container a distance h below the surface is

$$v_2 = \sqrt{2gh} \qquad [11.21]$$

which is known as **Torricelli's Result**.

Suggestions on Problem Solving

1. When doing buoyancy problems with balloons and the like, don't forget to add in, along with the load, the weight of the gas doing the lifting—it is part of the weight of the craft.

2. Some people are accustomed to working in the cgs system when it comes to density and since $\rho_w = 1.00\ \text{g/cm}^3$ and, moreover, since the specific gravity of water is 1.00, there's a tendency to lose the factor of 10^3 when substituting $\rho_w = 1.00 \times 10^3\ \text{kg/m}^3$. Remember, $1.00\ \text{g/cm}^3 = 1.00 \times 10^3\ \text{kg/m}^3$. Also remember that $1\ \text{cm}^2 = (1\ \text{cm})(1\ \text{cm}) = (1 \times 10^{-2}\ \text{m}) \times (1 \times 10^{-2}\ \text{m}) = 10^{-4}\ \text{m}^2$. Many problems deal with pressure beneath the water in an open container—*don't forget atmospheric pressure acts as well.*

3. There are a number of situations in this chapter where we have the same quantity on both sides of an equation—for example, $P_i = P_o$. In such cases, there are the same constants on both sides, and we can take some short cuts with the units. If both pressures in the expression $P_i V_i = P_f V_f$ are given, say, in mm Hg, they need not be converted to Pa before substitution because the units, whatever they are, cancel. This can save a lot of effort, but it should go without saying that you must be careful. *For the time being, use such short cuts only as checks.*

4. There are only two basic equations in this treatment of fluid dynamics, the Continuity Equation and Bernoulli's Equation. To these are added the derived expressions for Torricelli's Result and the equation for the Venturi Effect. Read the problem carefully and determine which of these concepts is involved. If the problem deals with fluid flowing through a hole or a constricted channel, one of the last two equations might be called for rather than going back to Bernoulli's Equation to rederive those results each time.

5. If you are considering a pipeline of constant diameter, the Continuity Equation requires that the speed of an ideal fluid be constant regardless of twists and turns, rises and falls.

6. When using Bernoulli's Equation, draw a diagram. If possible, locate region 1, where the fluid pressure, elevation, and speed are known. At a place where the section area is much larger than elsewhere in the flow, the fluid speed there may be negligible. The pressure in a fluid jet is the same as that of its surroundings. Locate region 2, where the unknown quantity is to be determined. Pick a zero y-level. The values of y_1 and y_2 may turn out positive or negative. If there are two unknowns, use the Continuity Equation as well. You can use either gauge or absolute pressure, but don't mix them. Bernoulli's Equation is applicable to the flow from one region to another. Accordingly, Eq. (11.18) is applicable *along a tube of flow;* in general, you cannot use the equation for *any* two points in the fluid and get exact results.

Discussion Questions

1. Imagine that you are in an elevator with a tank of water and three goldfish. A device in the gravel measures gauge pressure at the bottom of the tank. (a) Suppose it reads $4000\ \text{N/m}^2$ with the elevator car at rest. Will it increase, decrease, or stay the same when the car accelerates upward? Explain. (b) What will it read if the car drops at $9.8\ \text{m/s}^2$?

2. Consider the mercury barometer (a word coined by Boyle). What, if anything, would happen to the reading (that is, the difference in height between the two surface levels) if someone floated a coin on the open mercury? Explain.

3. Imagine a closed bottle filled with water into which is sealed a drinking straw. Could you "suck" the liquid up the straw? Why is the idea of sucking up fluids misleading? What would happen if we now punched a hole in the lid? How does the rubber suction cup on a dart stick to a wall? Why is it a common practice to wet the suction cups first? NASA is developing suction-cup shoes for use in space stations. Will they be of any use outside, for walking on the hull?

4. Figure Q4 shows a spherical (the distortion is optical, not actual) glob of olive oil ($0.9\ \text{g/cm}^3$) suspended in a solution of roughly 50% alcohol ($0.8\ \text{g/cm}^3$) and water. Explain what you see. Incidentally, by using a mix of bromobenzene ($1.497\ \text{g/cm}^3$) and xylene ($0.867\ \text{g/cm}^3$), we can determine the specific gravity of blood in this same way.

Figure Q4

5. When a helium balloon soars into the sky, it gains KE and *gravitational*-PE. Where does that energy come from?

6. In Fig. Q6, the tube is filled, pinched off, and then its ends are positioned as shown and released. The liquid is *siphoned* until the vessel on the left is emptied or the two levels become equal. What makes it function? (Hint: A siphon can work in vacuum.)

7. Picture yourself on a raft floating in a large tank of water. On board with you is a tin cup, a pine bust of Napoleon, a banana, and a chicken. What happens to the level of the water in the tank if: (a) you reach out, fill your cup with water and drink

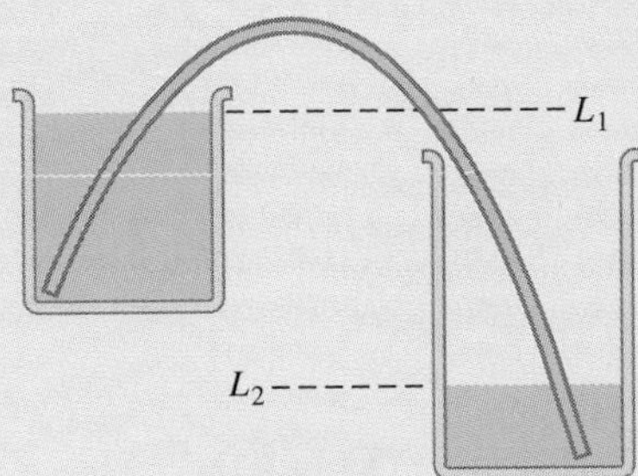

Figure Q6

it? (b) you toss the bust overboard? (c) you eat the banana? (d) the chicken flies away?

8. Imagine a metal vessel completely closed except for a small threaded hole on top. It is partially filled with water and placed on a scale that then reads 400 N. Holding on to a solid rod, you lower it down into the water. (a) Will the scale reading change? (b) If the rod were weightless, would that alter your answer? (c) What would happen to the scale reading if the rod is fixed in place by a very light collar that screws into the hole and holds the rod so that you can let go?

9. The photo in Fig. Q9 shows a glass tube open on top and bottom standing in a quantity of trichloroethylene (cleaning fluid) of density 1.44×10^3 kg/m^3. Water (with a little black ink in it for contrast) was subsequently poured into the tube. Explain what the photo shows.

Figure Q9

10. The underwater living chamber depicted in Fig. Q10 is wide open on the bottom for easy entrance. What keeps the water from rushing in? What's the pressure inside? What would happen if it were punctured at the top? Describe what happens to the volume of gas inside as the thing is hauled up to the surface.

Figure Q10

11. Figure Q11 shows an arrangement for producing negative pressures (P_T) at the top of the liquid column. Write an expression for P_T in terms of the pressure at the level of the liquid surface, the density, and the height of the column. Explain what is happening. Pressures of up to -300 atm have been achieved using water.

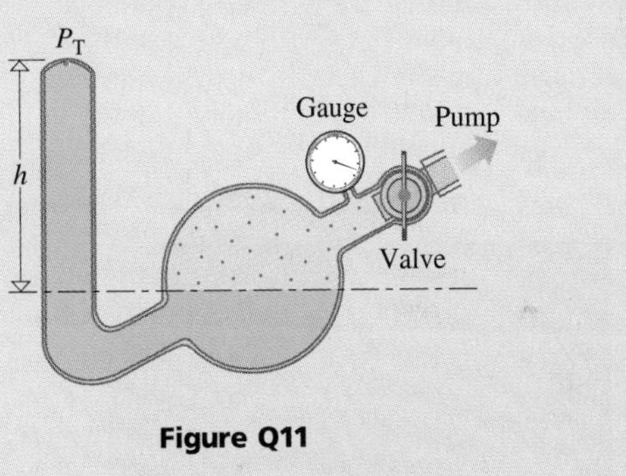

Figure Q11

Figure Q12

12. Pascal performed a rather startling demonstration: breaking a barrel using water pressure. He fit the cask with a very tall narrow tube sealed tightly into the lid (Fig. Q12). The barrel and then the tube were filled with water. Explain what happened.

13. Sprinkle some pepper on the surface of a cup of water and then touch the center of the distribution with a freshly washed and dried finger (no skin oil should be left). Now rub a little soap onto your fingertip and do it again. What happens and why? Put a little oil on your finger and do it again.

14. Rubbing alcohol and acetone both reduce the surface tension of water. With this fact in mind, float two wooden matchsticks in a bowl of water parallel to one another about 2 cm apart. What do you think will happen if you very gently introduce a drop of water between the sticks? Now do that again, this time with a drop of acetone or vodka—what happens?

15. A Ping-Pong ball can be supported in the airstream

Figure Q15

from a hair dryer even when it is tilted at an appreciable angle (Fig. Q15). Explain.

16. Concoct a bubble-making solution of 1 part water to 1 part dishwashing liquid detergent. If you have some glycerin (available at drugstores), add it in (3 parts detergent, 2 parts water, 1 part glycerin). Construct a wire ring about 2 or 3 in. in diameter and tie a small loop of thread to it. Now dip the ring in the solution so that a film stretches across the whole ring and the loop "floats" around within it. Puncture the film within the sagging thread-loop (use a hot pin if necessary). What happens? See Fig. Q16.

Figure Q16

17. The spinning baseball in Fig. Q17a carries along a boundary layer of air that revolves with it. If the spin axis is horizontal and the ball is thrown as shown in the figure, what will its flight path look like? Figure Q17b shows the flow pattern around a rotating cylinder. An experimental ship built in the 1920s crossed the Atlantic using two large rotating vertical columns and the wind as its source of power. Explain.

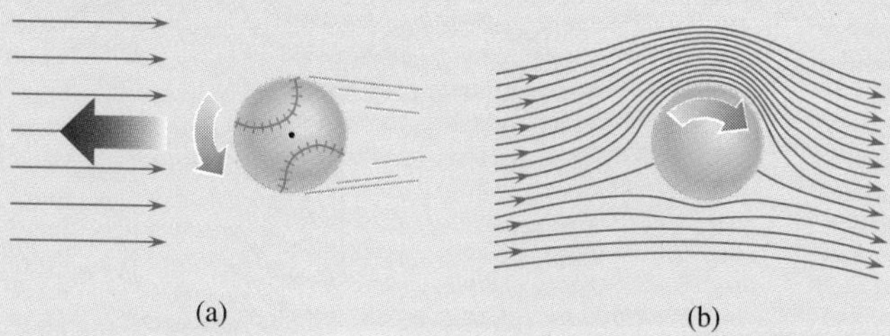

(a) (b)

Figure Q17

18. Figure Q18 shows two sets of spheres in streams of ideal fluid. Describe the forces, if any, that appear to be acting between them. An ideal fluid flows parallel to an essentially infinite smooth wall that serves as a boundary. Immersed in the fluid is a solid sphere. Describe the force, if any, that appears to exist between the sphere and the wall.

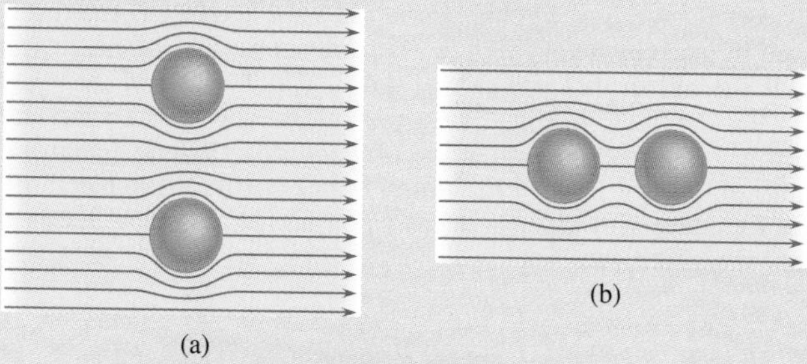

(a) (b)

Figure Q18

19. Figure Q19 shows a wing that has recently begun to move and around which has just developed a clockwise circulation of air giving rise to lift. Behind the wing we see a so-called *starting vortex,* which always accompanies the initiation of air circulation around a wing. Once formed, the starting vortex washes away downstream. Discuss why you might expect such an eddy to exist solely from momentum considerations. What do you think would happen if the wing is accelerated from rest and then rapidly stopped?

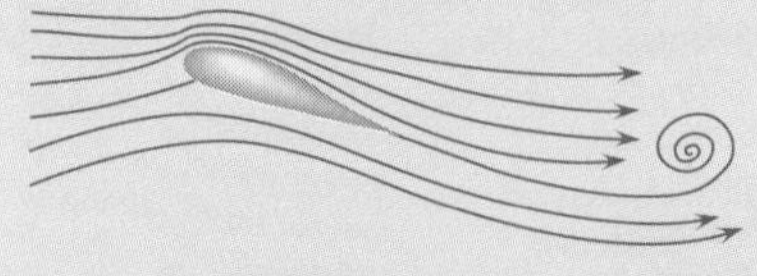

Figure Q19

20. If you cut your finger badly, why might it be wise to hold it high above your head until the bleeding stops?

Multiple Choice Questions

1. Suppose we fit a mercury barometer with a snug washer-shaped weightless disk with a hole in the middle that sits on the open surface like a piston. What would happen to the pressure reading if a bird landed on the disk?

(a) nothing (b) the pressure would exceed atmospheric and the column would rise with respect to the open surface (c) the pressure would fall slightly, and so the central column of mercury would drop a little (d) the weight of the bird would press on the mercury, increasing the pressure and decreasing the height of the column (e) none of these.

2. A sample of gas in a chamber sealed by a movable piston is kept at a constant temperature. The volume is slowly changed until it is twice its original value and the pressure therefore (a) is halved (b) is doubled (c) is unchanged (d) decreases by some unknown amount (e) none of these.
3. Boyle's Law is PV = constant, where the units of the constant are the same as those of (a) momentum (b) acceleration (c) impulse (d) work (e) none of these.
4. Is there a buoyant force acting on you at this very moment? (a) no (b) not enough information to say (c) only if you are floating around in water (d) yes, the atmosphere exerts a buoyant force (e) none of these.
5. Float an ice cube in a glass of water filled to the very brim. Will it overflow as the ice, jutting above the rim, melts? (a) no, it simply fills the displaced volume and the level is unchanged (b) yes, but only when it melts rapidly (c) no, but that is only provided not much ice is in the glass (d) yes, no matter how fast it melts (e) none of these.
6. Figure MC6 shows glass tubing, a rubber bulb, and two bottles. What has taken place in order to produce the situation depicted? (a) nothing, the arrangement is impossible (b) the tubing was filled with a single liquid, turned over, and inserted into the bottles (c) the bulb was squeezed while the tubing was in place as shown, and the two different-density liquids rose (d) the bulb was squeezed more tightly on one side than the other, so the single liquid rose as shown (e) none of these.

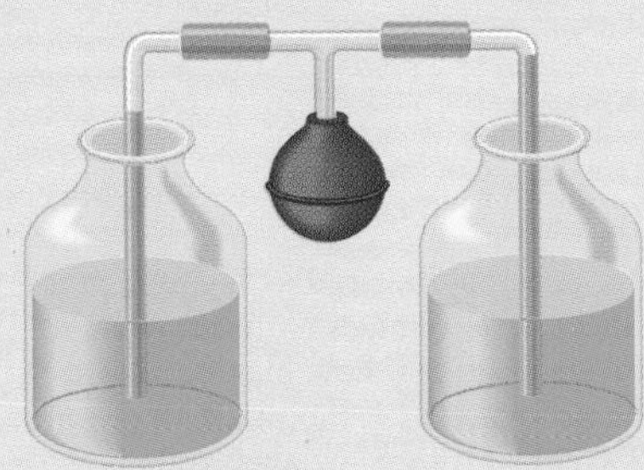

Figure MC6

7. Suppose you stick a drinking straw into a deep cup of liquid, tightly cover the upper end with a finger, and raise the straw up and out. (a) some liquid stays in the straw but its level initially drops, the gas pressure above the liquid in the straw drops, and the straw works like a poor barometer (b) the liquid simply runs out of the straw, there being nothing to keep it up (c) the liquid stays in the straw by surface tension alone (d) the liquid stays in the straw primarily because of friction with the long walls, but surface tension contributes as well (e) none of these.
8. Consider a helium-filled balloon floating around in a car. The driver suddenly stops the car. With respect to the car, (a) the driver, the air in the car, and the balloon are pushed backward (b) the driver is pushed backward, but the balloon moves forward (c) only the driver moves—and that is forward (d) the driver is pushed forward as is the air in the car, so the balloon moves backward (e) none of these.
9. A device that reads gauge pressure is suspended in a fixed position above the bottom in a large swimming pool. A helicopter with pontoons lands on the surface. Does the pressure reading change, and if so, how? (a) the water level rises and therefore the pressure rises (b) the pressure drops because the water level drops (c) the pressure remains unchanged (d) the water level remains unchanged, but the pressure drops (e) none of these.
10. The water level in a large tank supplying a small town is 37 m high. Water is pumped up for storage in the evenings and allowed to free-fall down as needed during the day. The street-level water pressure is (a) undetermined (b) 37 Pa (c) 36×10^4 Pa (d) 53×10^4 Pa (e) none of these.
11. Figure MC11 shows a hollow, narrow, open cylinder immersed in a beaker of water. Before being pushed into the fluid, a light plastic disc with a string glued to it was positioned on the bottom end of the cylinder and held in place by pulling up on the thread. Once immersed, the string could be let loose and the disc held in place by the water. That done, a liquid can now be poured into the cylinder from above until the bottom drops out. What is the main point of this experiment? (a) that it's fun to play with water (b) that liquids are more dense than air (c) that the pressure in a fluid is greater than the pressure outside (d) that the pressure in a liquid is due to the weight of the column of fluid above (e) none of these.

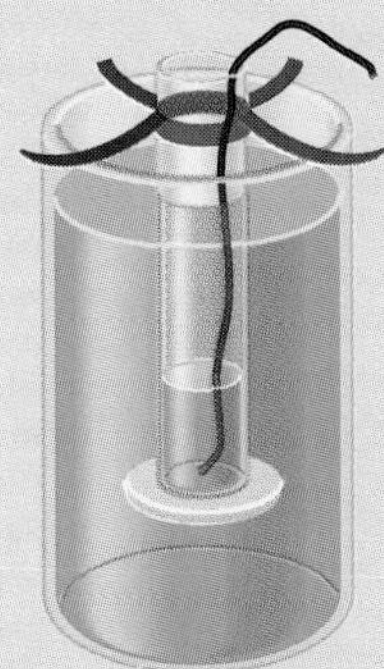

Figure MC11

12. Consider Fig. MC12, which shows a filled fish tank in the atmosphere. At which of the labeled points must the pressure always be equal? (a) A and E (b) B and C (c) A and B (d) C and D (e) none of these.
13. In the fish tank of Fig. MC12, it is possible that the pressure be equal at points (a) A and B, and B and E (b) B and C and E (c) A and C and E (d) A and C and B (e) none of these.

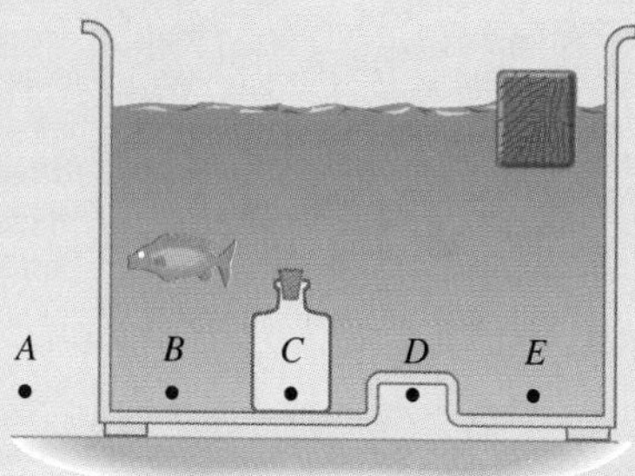

Figure MC12

14. The long glass tube of a barometer weighs 1.0 N, and the quantity of mercury almost filling it weighs 10 N (Fig. MC14). If the tube (which is very thin so the mercury-buoyant force can be ignored) is supported by a spring scale, it will read (a) 11 N (b) 9 N (c) zero (d) 1.0 N (e) none of these.

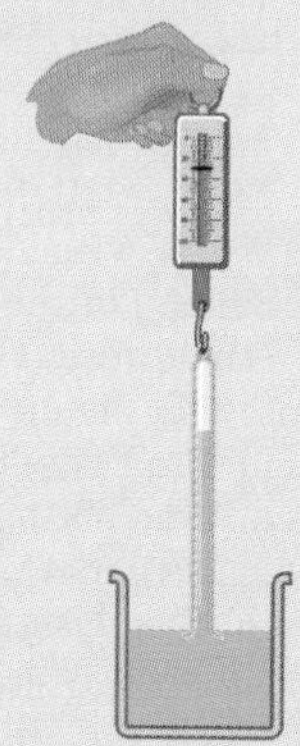

Figure MC14

15. The elderly lady in the park was running a once-in-a-lifetime sale, so you bought the 1.5-kg "solid" gold paperweight. When lowered into a fish tank back at home, it displaced 100 cm^3 of water. The known density of gold is 1.93×10^4 kg/m^3. (a) you made a great purchase (b) it may not have been a bargain, but at least it's got the right mass (c) bad luck—The density is wrong, so it's definitely not solid gold (d) the density is too high, so something must be wrong (e) none of these.
16. A stone tied to a string is supported from a spring scale that reads 10 N. A pail of water rests on a platform scale that reads 80 N. The stone is lowered into the water, and the spring scale then reads 5 N. It follows that the platform scale (a) still reads 80 N (b) now reads 70 N (c) now reads 90 N (d) now reads 75 N (e) none of these.
17. A block of wood floats in a pot of water on the floor of an elevator. Ignoring surface tension, how is the depth at which it floats changed when the elevator accelerates downward at a rate less than g? (a) the block rises because the water sinks (b) the block lowers because the water rises (c) nothing happens to it (d) the depth decreases because the water gets effectively lighter (e) none of these.
18. Water traveling at a speed of 0.1 m/s enters a pipe with a 1.0-cm radius. It emerges from a section of pipe with a radius of $\frac{1}{2}$ cm at a speed of (a) 0.4 m/s (b) 0.2 m/s (c) 1.0 m/s (d) 0.04 m/s (e) none of these.
19. A filled milk container has two holes punched in its side, one at point 1, 7.5 cm down from the liquid level, and the other at point 2, 15 cm down from that level. The two speeds of efflux are related by (a) $v_1 = v_2$ (b) $v_2 = 2v_1$ (c) $v_1 = 2v_2$ (d) $v_2 = \sqrt{2}\, v_1$ (e) none of these.
20. Two tall glass containers are connected by a narrow tube closed off with a stopcock. One container is filled with water, and when the stopcock is opened, the water passes into the other container until both levels are equal. At the end of this process, in which the "fluid seeks its own level," (a) the *gravitational*-PE has increased to a maximum (b) the KE has increased (c) the *gravitational*-PE has become a minimum (d) the *gravitational*-PE is unchanged (e) none of these.

Problems

1. [I] A swimming pool 5 m wide by 10 m long is filled to a depth of 3 m. What is the pressure on the bottom due only to the water?
2. [I] A rectangular tank 2.0 m by 2.0 m by 3.5 m high contains gasoline, with a density of 0.68×10^3 kg/m^3, to a depth of 2.5 m. What is the gauge pressure anywhere 2.0 m below the surface of the gasoline?
3. [I] An oxygen tank sitting in the corner of a laboratory has an internal gauge pressure 5.00 times atmospheric pressure. What outward force is exerted per square centimeter on the inner wall of the tank?
4. [I] At about 10 miles up in the atmosphere, air pressure drops to roughly 2 $lb/in.^2$ (down from 14.7 $lb/in.^2$ at sea level). How much is that in *atmospheres* and *pascals*?
5. [I] Prove that 1.000 $lb/in.^2$ = 6.895×10^3 N/m^2, using the fact that 1.000 lb = 4.448 23 N.
6. [I] A typical automobile tire has a gauge pressure of around 30 $lb/in.^2$. How much is that in pascals? If a car weighs 8897 N (2000 lb), how much area (in SI units) on each tire is in contact with the road?
7. [I] When Pascal's brother-in-law climbed to the top of the 3200-ft-high Mont Puy-de-Dôme, the barometer he carried showed a drop in mercury of about 3.0 in. What was the corresponding pressure drop?
8. [I] Suppose von Guericke had used two half-cubes instead of hemispheres to demonstrate atmospheric pressure. Compute the force each team of horses would have had to apply to a 1-meter-on-a-side cube. Assume zero pressure inside. How could von Guericke have been less theatrical and saved the cost of one team of horses?
9. [I] How deep must you dive in fresh water before the gauge pressure equals 1.00 atmosphere?

10. [I] A swimming pool 5.0 m wide by 10 m long is filled to a depth of 3.0 m. What is the absolute pressure on the bottom?

11. [I] Determine the gauge pressure (in both Pa and psi) at the bottom of the deep ocean trenches that reach depths of about 11 km. Assume the density of the sea is constant.

12. [I] A swimming pool 5.0 m wide by 10 m long is filled to a depth of 3.0 m. What is the total force exerted on the bottom due to the water?

13. [I] The initial pressure of the air inside a hypodermic syringe extended to 10 cm^3 is 1.013×10^5 Pa. If the cap is kept on so that the needle end is sealed, what pressure would exist when the gas is slowly compressed to 2.5 cm^3 with no change in temperature?

14. [I] A hydraulic lift consists of two interconnected pistons filled with a common working liquid. If the areas of the piston faces are 64.0 cm^2 and 3200 cm^2, and a 900-kg car rests on the latter, how much force must be exerted to raise the vehicle very slowly? If the car is to be raised 2.00 m, how far must the input piston be depressed?

15. [I] An ancient coin, which X-rays show is solid and homogeneous, has a mass of 0.0100 kg. When submerged, it displaces 0.952 g of water. What is its specific gravity and what is it probably made of?

16. [I] Envision a thin-walled jar having a mass of 10.0 g containing 0.100 g of hydrogen at atmospheric pressure. Estimate the difference we can expect to observe if we weigh the filled jar in dry air at 0°C, as opposed to weighing it in vacuum. The density of hydrogen is 90×10^{-6} g/cm^3.

17. [I] Figure P17 shows a U-shaped wire closed with a movable section of length $L = 0.10$ m. The device is then covered with a glycerin film. How much force is needed to pull the wire at a constant rate, thereby increasing the area of the film?

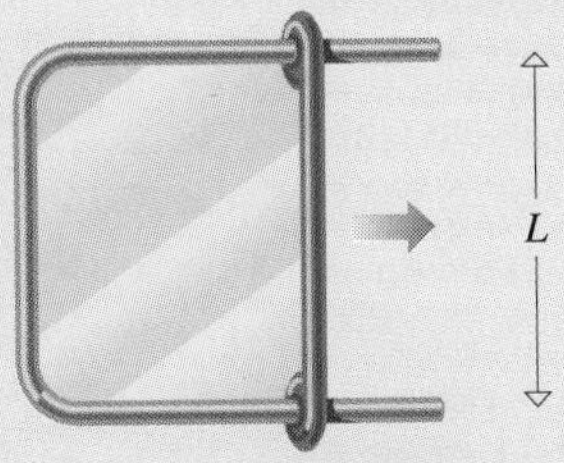

Figure P17

18. [I] A needle of length L and mass m rests on the surface of a liquid having a surface tension of γ. Show that the force needed to raise the needle is

$$F = 2L\gamma + mg$$

19. [II] The medical literature uses yet another pressure unit: the cm of water (water columns are convenient for measuring modest gauge pressures). Thus, one reads that a newborn baby is capable of developing a "momentary intrathoracic negative pressure of the order of 40 cm of water." How much is that in pascals?

20. [II] To remove unwanted fluids from the body, an aspirator such as that shown in Fig. P20 is used. A small electric pump provides the suction (typically at gauge pressure levels of −90 to −120 mm Hg), while the bottle containing water provides the control. Explain how the depth beneath the water surface of the control tube establishes the pressure in the drain tube. What happens to the suction pressure as the drain tube in the bottle becomes submerged in liquid aspirated from the patient? When will the system shut itself off? Suppose the procedure starts with the control tube 10 cm into the water—what is the suction pressure to the patient?

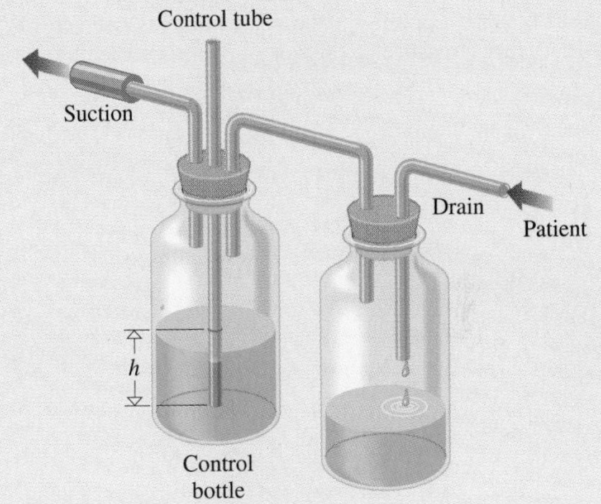

Figure P20

21. [II] Given that most people cannot "suck" water up a straw any higher than about 1.1 m, what's the lowest gauge pressure they can create in the lungs?

22. [II] The gauge pressure in a basement water pipe supplying a tall building is 3.00×10^5 Pa. In an apartment several floors up, the pressure is half this value. How high up is the apartment?

23. [II] A rubber pipe is attached to one end of an open U-tube water-filled manometer (such as the one in Fig. 11.11). A patient exhales strongly into the rubber pipe and a difference in height between the two columns of 61 cm results. What absolute pressure was developed by the lungs?

24. [II] A 227-kg block of cement, of density 2.8×10^3 kg/m^3, rests on a pedestal in front of the Al Capone Memorial Library. How much did it weigh submerged while being hauled out of the river (freshwater)?

25. [II] A ring weighs 6.327×10^{-3} N when measured in air and 6.033×10^{-3} N when submerged in water. What is its volume (to three significant figures)? What is it likely made of? What is its relative density (to three significant figures)?

26. [II] A woman weighing 500 N jumps into a swimming pool 10 m × 10 m by 5 m deep and floats around. By how much does the water level change as a result of her arrival?

27. [II] Icebergs float in the ocean with much of their huge volumes hidden below the surface. What fraction is visible above the water? What portion of an ice cube floats above the surface of a glass of tap water?

28. [II] Consider a catfish on the bottom of a freshwater lake. It opens its mouth to let loose a bubble of gas with a 1.00-mm^3 volume. Now suppose that the bubble's volume just before it bursts at the surface is 12.0 mm^3. How deep was the lake? (Assume a uniform temperature and no reabsorption.)

29. [II] A bottle of oxygen having a volume of 100 cm^3 is at a pressure of 228 cm Hg. The gas is fed into an expandable chamber maintained at atmospheric pressure and the same constant temperature. What volume will the gas occupy?

30. [II] The piston in equilibrium, shown in Fig. P30, has a face area of 0.10 m^2 and experiences a downward force of 1.00 kN. The mercury in the vessel is 10.0 cm deep. What is the gauge pressure at the very bottom of the chamber?

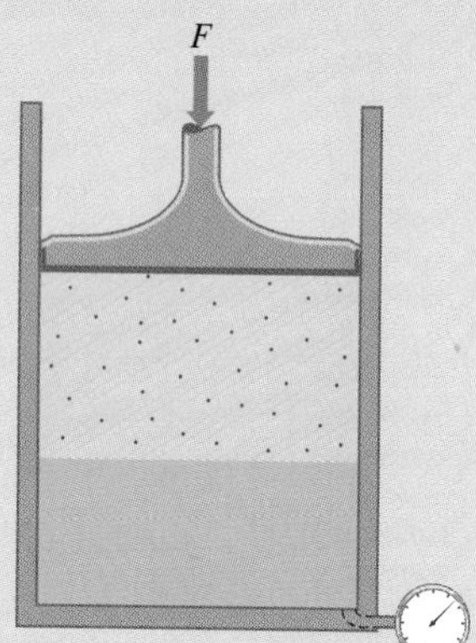

Figure P30

31. [II] A submarine rests in 20.0 m of water. How much force must a diver exert against the pressure of the sea in order to pull open a hatch 1.0 m × 0.50 m, assuming the internal pressure in the boat is 90% of atmospheric?

32. [II] A person floats on the Great Salt Lake, which has a density of 1.15×10^3 kg/m^3. Approximately how much of the swimmer's body is above the water?

33. [II] Determine the mass of helium needed to provide enough buoyancy (in dry air at 0°C) to lift a balloon and its load having a net mass of 454 kg.

34. [II] A 226.7-kg polar bear standing 6 ft tall walks onto a floating sheet of ice 0.305 m thick. How big is the ice sheet if it sinks just below the surface while supporting the bear?

35. [II] Given a pine raft, of density 0.50×10^3 kg/m^3 and dimensions 3.05 m by 6.10 m by 0.305 m, how much of a load can this raft take on for each 2.54 cm it settles into freshwater? How deep does it sink into the water when unloaded?

36. [II] The open U-tube in Fig. P36 contained some water before a less dense liquid was poured in on the right side. If the density of the unknown liquid is ρ_x, show that

$$\rho_x = \frac{\rho_w h_w}{h_x}$$

37. [II] A platinum wire ring, 6.0 cm in diameter, is suspended horizontally from a delicate balance. The ring is lowered into a liquid, and the balance is reset to zero. The

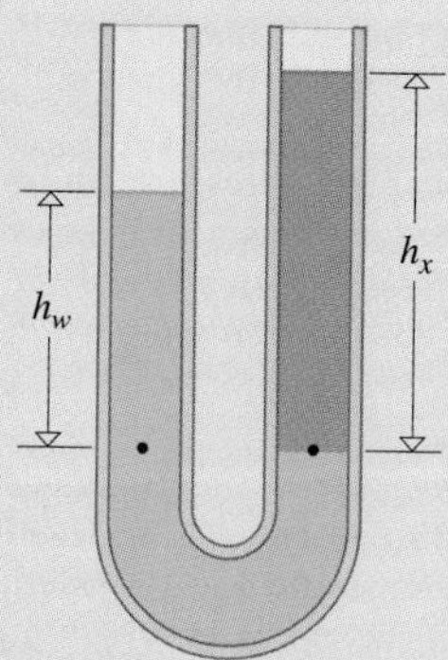

Figure P36

ring is raised again, and the force needed to pull it free of the surface is then measured. What force would be required if the liquid was water at 20°C?

38. [III] A glass plate 0.12 m × 0.20 m is mounted on the top of a submerged camera box to make an angle of 30° with the horizontal. The uppermost 0.12-m edge is 0.20 m below the surface of the water. The box is sealed, and there is air inside at atmospheric pressure. Please compute the net force acting on the plate due to the water.

39. [III] Figure P39 deals with the problem of showing that the pressure within a fluid is independent of orientation. Shown is the end of a wedge-shaped volume of fluid where the average pressure on each of its faces is P_a, P_b, and P_c. The lengths of the sides of the top face are both L, and it makes an angle of ϕ with the bottom. Find the net force on each face, and show that the fluid wedge is in equilibrium when

$$P_a L^2 \sin \phi = P_c L^2 \sin \phi$$

and $$P_b L^2 \cos \phi = P_c L^2 \cos \phi + \tfrac{1}{2}\rho g L^3 \sin \phi \cos \phi$$

Now, let the wedge shrink and show that $P_a = P_b = P_c$ in the limit as $L \to 0$.

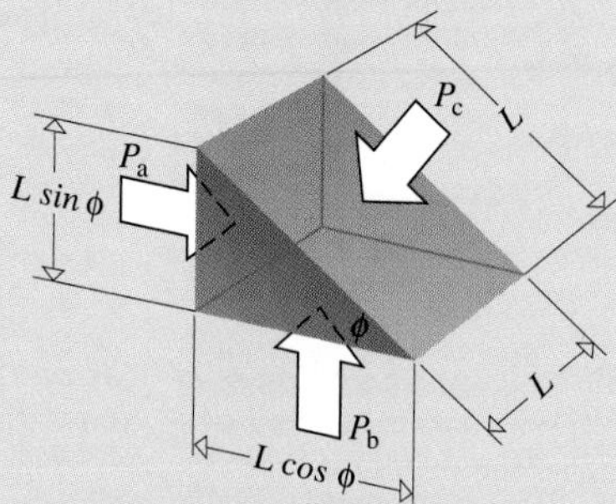

Figure P39

40. [III] The diving bell shown in Fig. P40 allows swimmers to comfortably remain under water for long intervals. Show that

$$d = \frac{P_A h}{(H - h)\rho g} + h$$

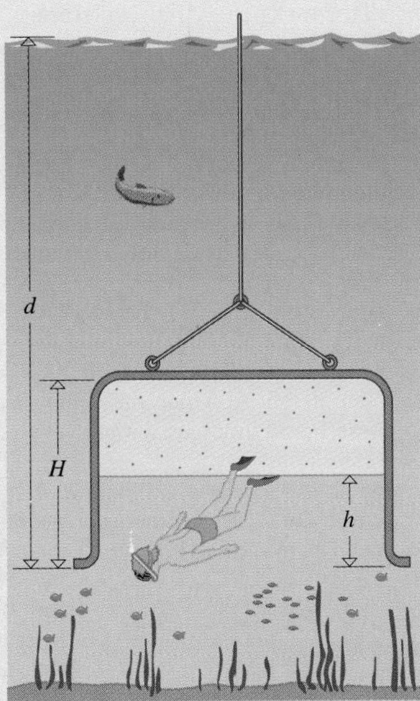

Figure P40

where P_A is the original pressure in the bell (usually atmospheric). (Hint: Don't forget atmospheric pressure on the water.)

41. [III] A swimming pool 5 m wide by 10 m long is filled to a depth of 3 m. What is the total force on either narrow wall due only to the water?

42. [III] A bronze wire, with a Young's Modulus of 120 GPa, supports a 0.000 45-m^3 block of metal. When the block is lowered completely into water, the wire decreases in length by 0.000 80 m. If the diameter of the wire is 0.10 mm, how long is it?

43. [III] An object has a weight measured in air of F_{Wa} and when immersed in water, its effective weight is F_{Ww}. It is then submerged in an unknown liquid, and its effective weight is found to be F_{Wl}. Determine the density of the unknown liquid in terms of these measured quantities and the density of water.

44. [III] Figure Q16 shows a loop of thread that will be pulled out into a circle of radius R by surface tension. Prove that the tension in the thread is given by

$$F_T = 2R\gamma$$

FLUID DYNAMICS

45. [I] Petroleum flows along a 10-cm diameter pipeline at a speed of 2.0 m/s. Determine the flow rate, assuming the liquid is ideal.

46. [I] A pipe 5.0 cm in diameter carries gasoline with a density of 0.68×10^3 kg/m^3 at a speed of 2.5 m/s. Calculate the mass-flow rate, assuming the liquid is ideal.

47. [I] Gasoline is flowing in a pipeline having a 0.50-m diameter. Taking the fluid to be ideal, what pressure change results when the pipe descends 4.0 m down an embankment?

48. [I] Determine the blood pressure at the brain, taking its height above the heart to be 40 cm.

49. [I] Someone cuts a finger and, to control the bleeding, raises the wound 85 cm above the level where the hand was being bandaged. What is the change in the pressure of the blood? Compare that to the pressure at the heart.

50. [I] A horizontal water main with a cross-sectional area of 200 cm^2 necks down to a pipe of area 50 cm^2. Meters mounted in the flow on each side of the transition coupling show a change in gauge pressure of 80 kPa. Determine the flow rate through the system, taking the fluid to be ideal.

51. [I] Derive an expression for the speed of efflux of an ideal liquid from a hole in a large tank filled with fluid in terms of the gauge pressure at the level of the orifice and the density of the liquid.

52. [I] A length of tubing is used as a siphon to empty a large upright cask of wine. At a given moment, the lower opening of the tube from which the wine emerges is 20 cm below the liquid level in the cask. Compute the speed at which the wine then emerges.

53. [I] A large vat of clear chicken soup is about to spring a leak from a small hole a distance h below the surface of the broth. Knowing that the hole is a height (y) above the floor, at what distance (x) away from the vat should you place a bowl on the floor to catch the flow, assuming it to be ideal?

54. [II] A cylindrical water tank 2.0 m in diameter has a spigot with a hole area of 5.0 cm^2 located 4.0 m below the surface. Show that for a point at the surface the $\frac{1}{2}\rho v^2$ term in Bernoulli's Equation is properly ignorable compared to the corresponding term outside of the spigot in the escaping water jet. Assume no contraction of the stream as it leaves the tank.

55. [II] A Texas pipeline carrying natural gas ($\rho = 0.90$ kg/m^3) with a mass-rate of flow of 1.0 kg/s is 35 cm in diameter. Determine the average speed at which the gas is moving along.

56. [II] Refer to the photo on p. 000. Suppose a person standing erect or sitting upright is accelerated upward at a rate a. Write an expression for the resulting blood pressure in the brain in terms of the pressure at the heart and discuss your results.

57. [II] A Pitot tube, shown in Fig. P57, is a fluid-speed measuring device that has applications in many fields, including physiology. Often used on airplanes to measure the airspeed of the craft, it can be seen mounted on the wings or fuselage. At the front opening, the streamlines part and there is a **stagnation point** there. The fluid rushes away from that point in all directions, and *its speed there is zero.* Accordingly, the pressure there and in the inner tube (P)—the stagnation pressure—is relatively high. On the other hand, the speed of the fluid over any of the small holes in the outer cylinder is the undisturbed free-stream speed (v). Derive an expression for v in terms of the pressure difference indicated by the attached manometer. By the way, the device was first used by Henri Pitot to measure water speeds in the Seine River (1730).

58. [II] A small plane, having a mass of 3000 kg, is flying in air of density 1.0 kg/m^3. Air moves over the top and bottom surfaces of the wings at 160 m/s and 130 m/s,

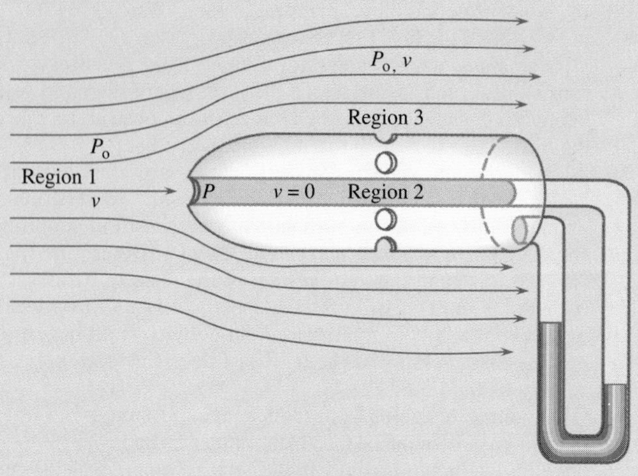

Figure P57

respectively. What is the net minimum wing area needed? Use the results of Problem 69.

59. [II] The instrument depicted in Fig. P59 is inserted in a pipeline to determine flow rates and speed. Known as a Venturi meter, it consists of a throat and two manometer tubes. Derive an expression for the speed of flow in the pipe in terms of Δy, the difference in the column heights, and the known cross-sectional areas of pipe and throat.

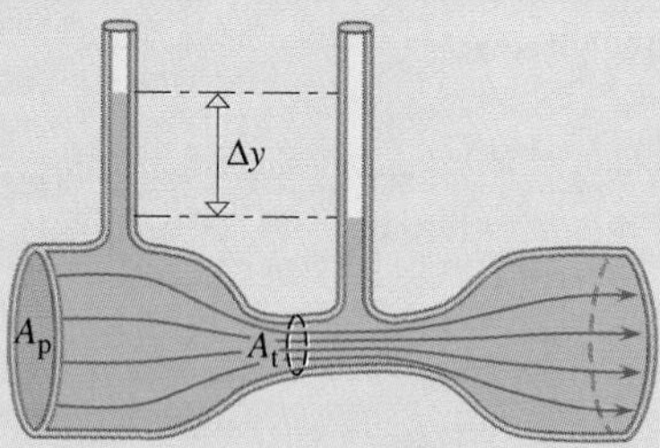

Figure P59

60. [II] The flowmeter in Fig. P59 shows a difference in height of 5.0 cm on an oil pipeline 200 cm^2 in cross section. If the throat of the device has a diameter of 10 cm, what is the pipeline flow rate? (Assume the liquid to be ideal.)

61. [II] A vaccination gun forces vaccine through a small aperture a few thousandths of an inch in diameter, at high pressures (550 psi), and so does away with the need for hypodermic needles. Compute the speed at which the fluid leaves the gun. Take the flow speed of the vaccine ($\rho = 1.1 \times 10^3$ kg/m^3) in the reservoir within the body of the gun to be negligible.

62. [II] A cardboard milk container has two holes, one above the other, punched in its side. At a given moment, the two escaping streams of milk strike the table the container is standing on, at the same point. If the heights of the upper and lower holes are, respectively, y_u and y_l, write an expression for y, the level of milk at that instant, in terms of these heights.

63. [II] A large-diameter open cylindrical storage tank stands on a high platform. It has a small horizontal spigot at its very bottom, a height Y above the ground and a depth h below the surface of the water. Use Bernoulli's Equation to follow a tube of flow from point 1 at the surface, to point 2 just outside the spigot, to point 3 at the point of impact with the ground. Write expressions for v_2 and for v_3 (the speeds at the spigot and at the point of impact with the ground) in terms of g, h, and Y. Did you expect these results?

64. [II] A large, open storage tank is being filled with water from a pipeline at a rate J. Unfortunately, it springs a leak via a hole of area A, at its base. Write an expression for y, the equilibrium height to which the water rises in the tank. Assume the liquid emerges without any contraction of the jet.

65. [II] Figure P65 shows a submerged orifice discharging a liquid from a large tank. Determine an expression for the ideal efflux speed at point 2.

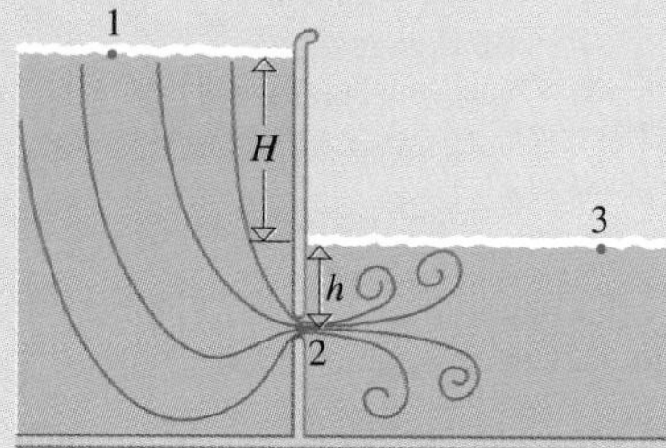

Figure P65

66. [III] During a storm, a 50-m/s (112-mi/h) wind blows horizontally across the flat level roof of a supermarket. If the roof has an area of 220 m^2, what is the most force that's likely to be exerted on it? Assume the place is locked up tight for the night. (Take the density of air to be 1.1 kg/m^3.)

67. [III] Figure P67 shows a stream of liquid emerging from a tube in the base of an open tank. Use Bernoulli's Equation applied between points 2 and 3 to get an expression for y in terms of θ and h. Can y exceed h?

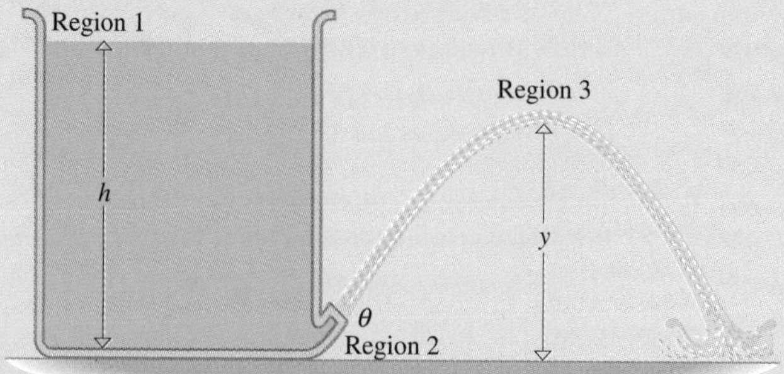

Figure P67

68. [III] A highly pressurized liquid escapes from a large, closed tank through a small nozzle with an aperture of

area A. Show that a thrust results, given by

$$F = \rho A v^2$$

which is essentially only dependent on the orifice area and the gauge pressure in the tank. There are toy rockets that you fill with water and then pressurize with compressed air via a handpump—they blast out the water and lift off quite effectively.

69. [III] Show that the lift on the wing of an airplane of area A is given by

$$F_L = \frac{1}{2}\rho(v_\alpha^2 - v_\beta^2)A$$

where v_α is the speed of the air above, v_β the speed below, and ρ is the density of air. Although it's not clear how to predict these two speeds, it is reasonable to assume that each is proportional to the airspeed of the craft, and hence $(v_\alpha^2 - v_\beta^2) \propto v^2$.

12

Oscillations and Waves

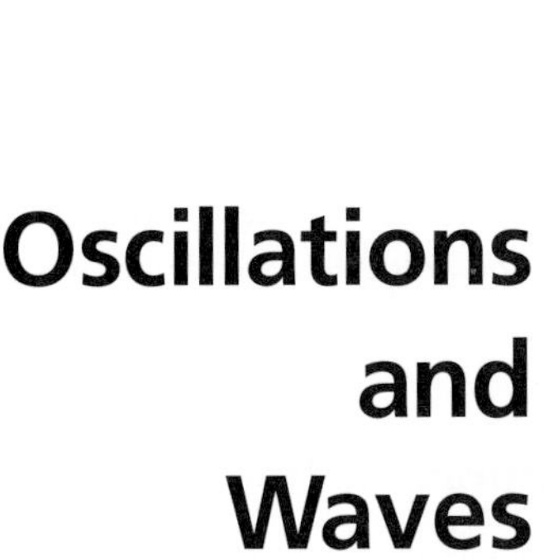

A MOTION THAT REPEATS itself over and over again, in successive constant time intervals, is said to be *periodic.* Such behavior is remarkably commonplace and seems fundamental to the physical nature of the Universe. A minute electron displays periodic characteristics that are describable in the same language we apply to analyzing a spinning neutron star. The revolving of the Earth around the Sun, the sway of a gently rocked cradle, or the rhythmic pumping of a heart are all more or less periodic (Fig. 12.1). We can distinguish two forms of repetitive motion: movement along a closed path in one angular direction (like the Earth around the Sun), and back-and-forth

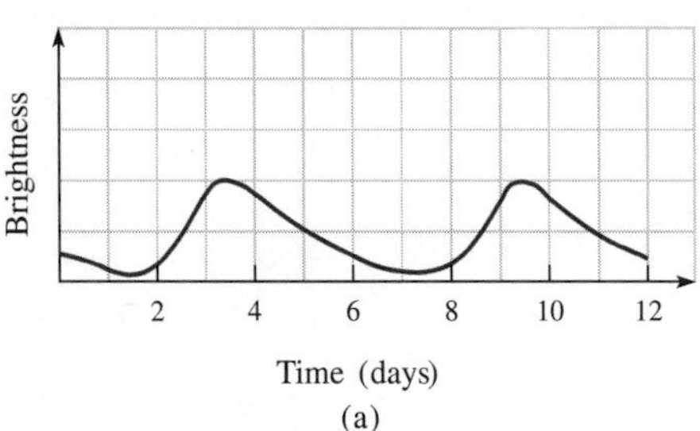

(a)

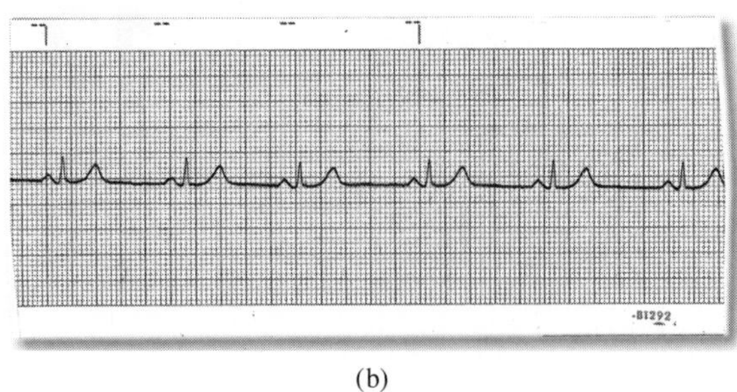
(b)

Figure 12.1 (a) The giant star Delta Cephei swells and then shrinks every 5.4 days. The plot of its brightness is therefore a regularly rising and falling curve.

(b) This graph is periodic, too, but it's just an electrocardiogram of a physicist's heart. (See Sect. 19.5.)

oscillation along the same path (like the cradle). The former was introduced in Chapter 8, and the latter, *vibrational* motion, will be considered here. This analysis leads naturally to a discussion of waves produced in vibrating mechanical systems and then, later, to sound waves (Chapter 13).

HARMONIC MOTION

The world abounds with all sorts of vibrating systems: the balance wheel in a watch oscillates back and forth; puckered lips blowing a trumpet, or a kiss, vibrate; a walker's swinging arms oscillate; so does a singing vocal cord. Vibratory periodic motion is referred to as *harmonic motion,* because of its relationship to sound and the harmony therein.

12.1 Simple Harmonic Motion

The balance wheel of a mechanical clock oscillates about its central axis.

A single sequence of moves that constitutes the repeated unit in a periodic motion is called a *cycle.* During one cycle, the system progresses in some way, returning on completion to its initial physical configuration and motion. The child on the swing leaves her father's raised hands, swoops down, rises high into the air, stops, descends backward, and comes up again to meet his hands and finish the cycle. *The time it takes for a system to complete a cycle is a* **period** (T). The period of the Earth's rotation about its spin-axis is roughly 23 h 56 min. The period is *the number of units of time per cycle;* the reciprocal of that—*the number of cycles per unit of time*—is known as the **frequency** (f), after Galileo who called it the *frequenza:*

$$f = \frac{1}{T} \tag{12.1}$$

The SI unit of frequency is the *hertz* (Hz), so named to honor Heinrich Hertz, where 1 Hz = 1 cycle/s = 1 s^{-1}. The appropriate units of *cycles per second* got distorted in common usage during the first half of this century. Radio people, especially, talked about kilocycles and megacycles, forgetting the "per second," so today, to

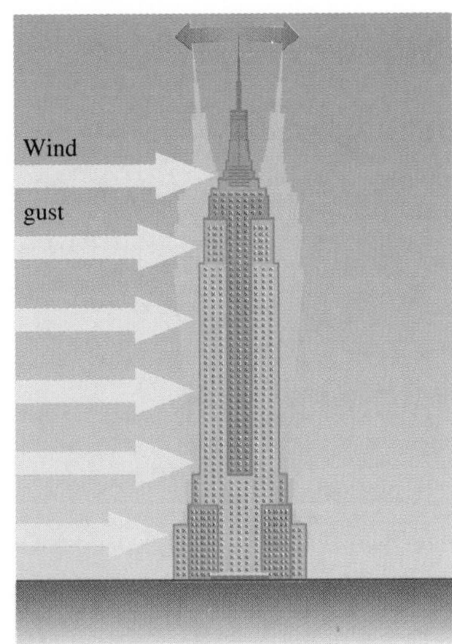

Figure 12.2 The 1250-ft-tall Empire State building set oscillating by a strong wind. Even though the displacement as drawn is exaggerated, tall buildings do sway up to several meters. In fact, on a windy day, the Twin Towers of the World Trade Center in Manhattan shift 6 or 7 ft, and the elevators have to be slowed down.

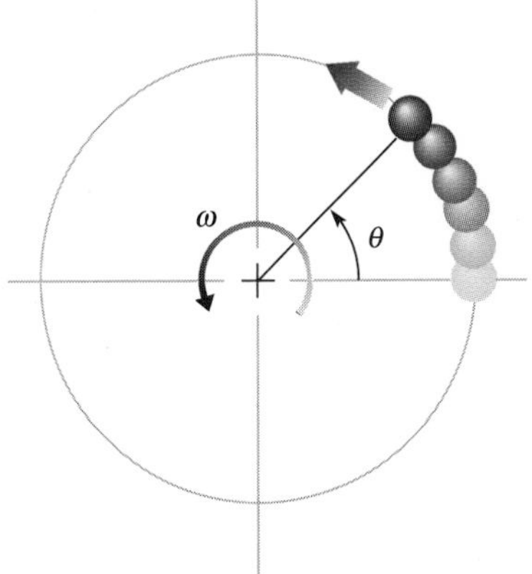

Figure 12.3 An object moving in a circle with a constant angular speed ω. Its angular position $\theta = \omega t$ changes at a constant rate.

avoid that difficulty, we have the hertz. For example, once set swaying by a wind, the Empire State building oscillates through one cycle in about 8 s, that is $T = 8$ s and $f = 1/T = 0.1\ \text{s}^{-1} = 0.1$ Hz (Fig. 12.2).

Imagine an object moving in a circular orbit (Fig. 12.3) at a constant rate: its angular speed ω is constant. Each time it makes one complete orbit, one cycle, the object sweeps through an angle of 2π rad. Since the number of cycles it makes per second is f, *the number of radians it moves through per second* is $2\pi f$, and that's exactly what angular speed ω is. Hence

$$\omega = 2\pi f = \frac{2\pi}{T} \tag{12.2}$$

and it is common practice to refer to ω as the **angular frequency**.

Most real oscillatory phenomena wiggle about in a complicated fashion with lots of different frequencies occurring at different strengths, all at once. Even so, many important systems vibrate with a single dominant frequency that far outweighs all the others. We can approximate this behavior using a single harmonic function; that is, *the periodicity has a single frequency and the motion is sinusoidal*—described by a sine or cosine function. This idealization is known as **simple harmonic motion**, or SHM for short.

Uniform rotational motion, simple harmonic motion, and periodic wave motion are all intimately related and can be analyzed with the same mathematics. We can get a sense of that interrelationship from Fig. 12.4, which depicts an arrangement

(a) $t = 0$

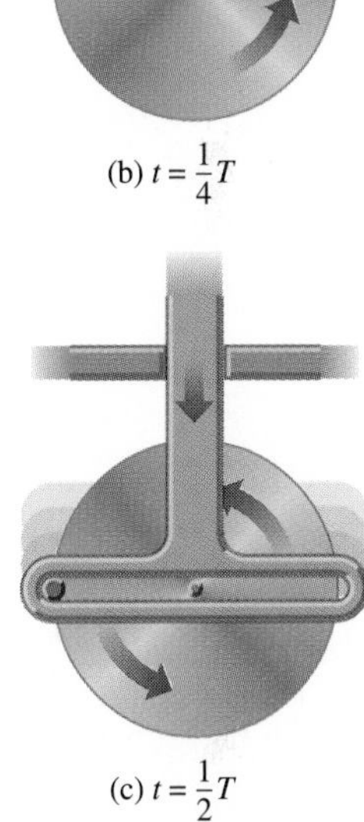

(d) $t = \frac{3}{4}T$

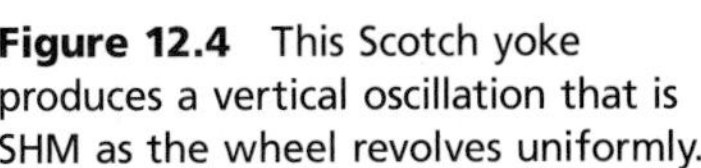

Figure 12.4 This Scotch yoke produces a vertical oscillation that is SHM as the wheel revolves uniformly.

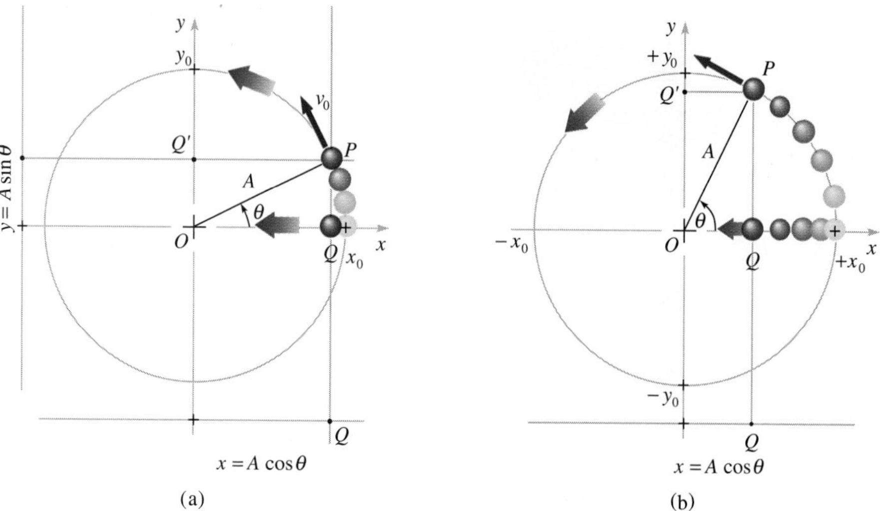

Figure 12.5 A particle P revolving around a circle at a speed v_0. Its projection on the x-axis is point Q at a distance $x = A \cos \theta$ from the nondisplaced origin O. As P revolves, Q oscillates between $+x_0$ and $-x_0$ in SHM. (a) and (b) show two locations as the motion progresses.

called a Scotch yoke. As the disk turns uniformly, the rod oscillates linearly in SHM. The frequencies and periods of the two motions are the same.

We next derive the general expressions that describe SHM for an oscillating point. These results will later be applied to specific systems, such as vibrating springs and pendulums.

Displacement in SHM

An object oscillating in SHM can be located at any instant with respect to a fixed point O called the *origin.* To do that, imagine a particle P moving at a uniform speed v_0 counterclockwise along a circle of radius A (Fig. 12.5) centered on O. At any moment, the position of the particle is given by θ. Since the linear speed is constant, the corresponding angular speed ω is also constant, $v = r\omega$. Moreover, after a time t, it follows from Eq. (8.5) that $\theta = \omega t$. The projection of P perpendicularly down onto the x-axis locates the point Q, and as P circles around with a constant speed, Q oscillates from $+x_0$ to $-x_0$ and back at a single constant frequency in SHM. The projection onto the y-axis locates Q' and it will also display SHM. The displacement of Q from 0 is given by

$$x = A \cos \theta = x_0 \cos \omega t = x_0 \cos 2\pi f t \qquad (12.3)$$

where $\omega = 2\pi f$. The displacement of Q' from 0 is given by

$$y = A \sin \theta = y_0 \sin \omega t = y_0 \sin 2\pi f t \qquad (12.4)$$

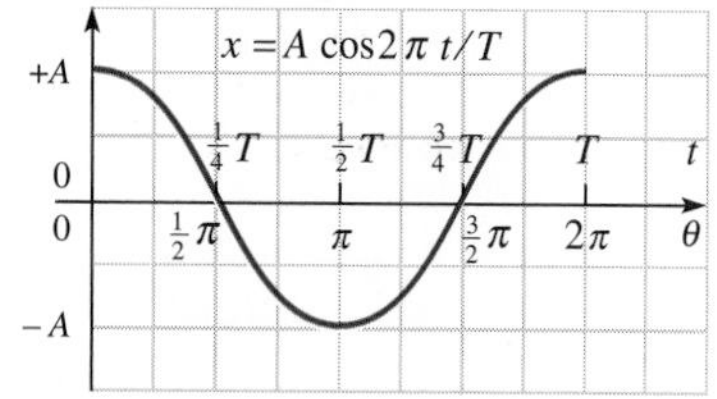

Figure 12.6 A plot of the cosine function. As t varies from 0 to T, $\cos 2\pi t/T$ varies from $\cos 0 = 1$ to $\cos 2\pi = 1$.

Because $2\pi f t = 2\pi t/T$, this factor varies from 0 when $t = 0$, to π when $t = \frac{1}{2}T$, to 2π when $t = T$, as shown in Fig. 12.6. The location of Q provided by Eq. (12.3) can be thought of as describing the top of the Empire State building wavering in the wind, or anything else oscillating in SHM about its undeflected ($x = 0$) position.

The maximum displacement of Q from 0 is the length $A = x_0$, the radius of the **reference circle**, which is called the **amplitude** of the oscillation. At any instant, the value of x is the size of the displacement from 0. The value of x varies as Eq. (12.3) and may be positive or negative; the amplitude is constant and positive.

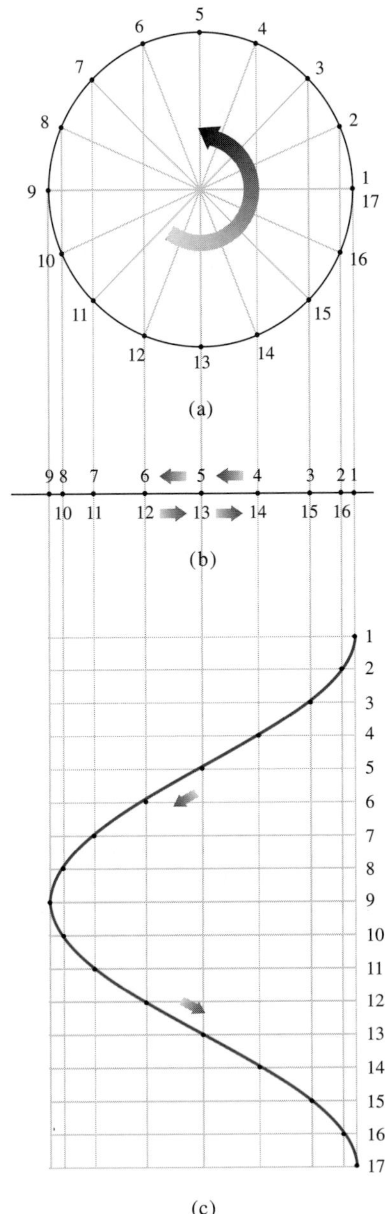

Figure 12.7 (a) Uniform circular motion projected onto a line (b), resulting in linear SHM projected on a "moving" axis (c), which results in a sinusoidal oscillation.

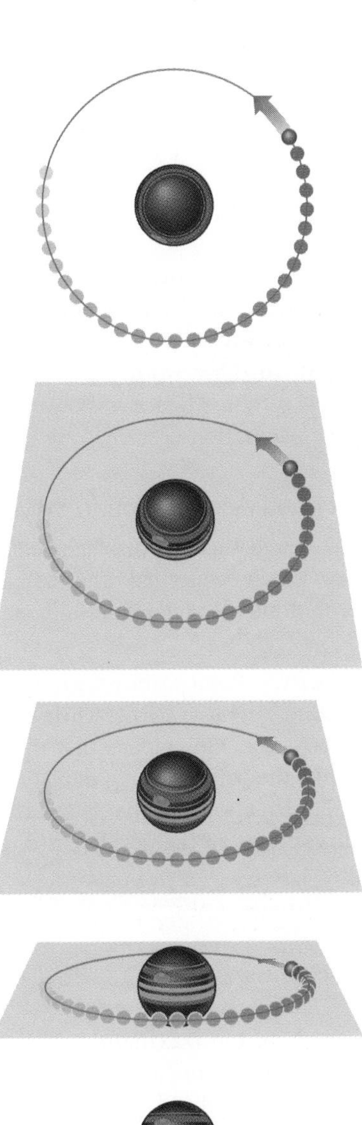

Figure 12.8 A schematic representation of a moon circling around Jupiter. As the viewing angle gets narrower, the orbit flattens, and the moon is seen to oscillate in SHM along a line in the plane of the motion.

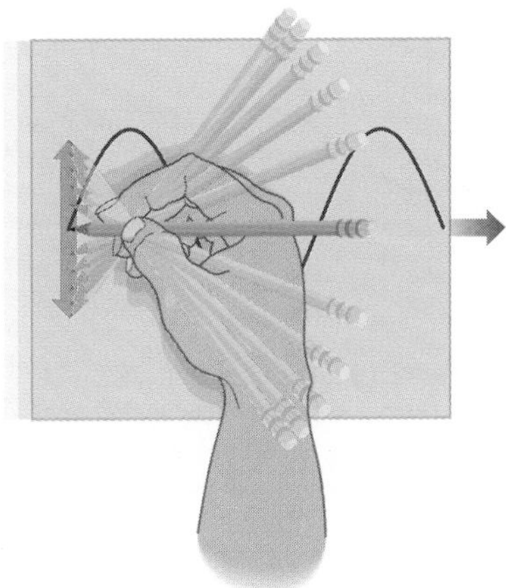

Figure 12.9 The pencil oscillates in SHM along a vertical line. Pulling the paper to the right at a constant speed results in a sinusoidal curve.

The oscillator starts out ($t = 0$) at its maximum displacement ($\cos 0 = 1$); that is, $x = A$. We now know how to describe the position mathematically as it varies in time for an object in SHM. The relationship between uniform circular motion, SHM, and the sinusoidal function is explored in Figs. 12.7, 12.8, and 12.9.

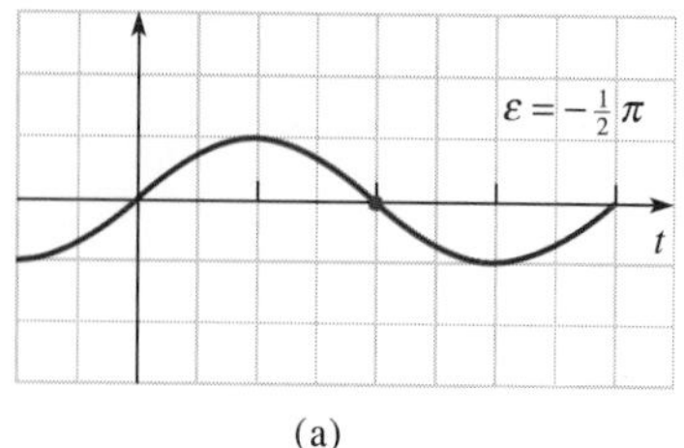

(a)

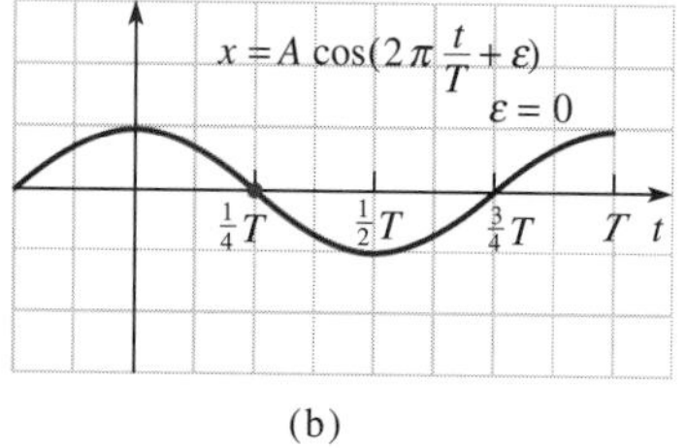

(b)

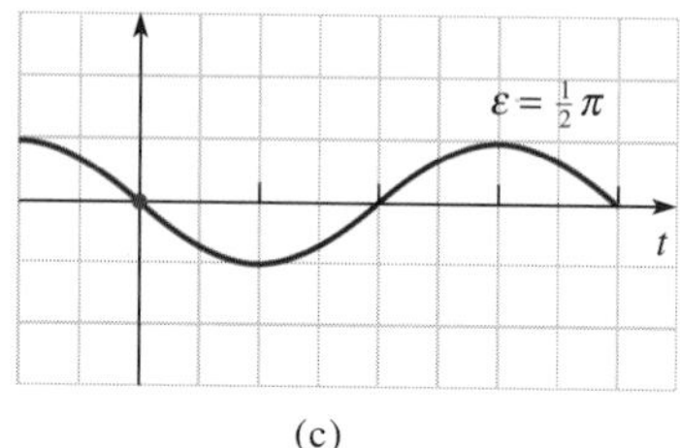

(c)

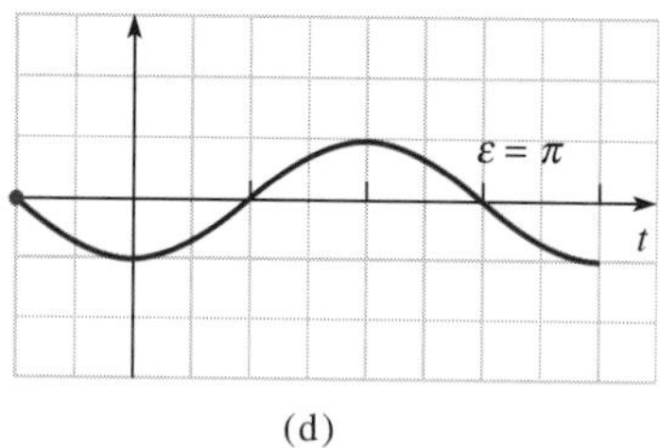

(d)

Figure 12.10 The addition of a phase term ε shifts the curve. When ε is negative, any given value occurs later in time; when ε is positive, it occurs earlier.

Notice in Eq. (12.3) that at $t = 0$, $x = +x_0$. That's fine, but what if we want our particle somewhere else at the start of the game ($t = 0$)? The argument of the sinusoidal function is the **phase**, which is often signified by the letter ϕ. In Eq. (12.3) $\phi = \theta = \omega t$, but more generally the phase will have a nonzero value at $t = 0$, a value ε, which is known as the **initial phase**. With $\phi = \omega t + \varepsilon$, we have

$$x = A \cos(\omega t + \varepsilon) = x_0 \cos(2\pi f t + \varepsilon) \tag{12.5}$$

This is the most inclusive way to express the displacement of an object undergoing SHM.

The oscillations depicted by Eqs. (12.3) and (12.5) are *out-of-phase* with respect to each other by an amount ε (Fig. 12.10). If ε is positive, the function in Eq. (12.5) will reach any desired value of x at an earlier time than the function in Eq. (12.3). In other words, $\phi = (\omega t + \varepsilon)$ will reach a given value at a smaller t when $\varepsilon > 0$ than when $\varepsilon = 0$. We say that the oscillation described by Eq. (12.5) *leads* the oscillation described by Eq. (12.3).

Velocity and Acceleration in SHM

We have found that the displacement of a system in SHM varies sinusoidally with time. As that's occurring, what happens to the system's velocity and acceleration? These, too, must change from moment to moment.

Velocity. Point P moves around its circular path ($r = A$) with a constant linear speed $v_0 = A\omega = 2\pi f A$, while its projection, point Q, always remains just below it on the x-axis. This means that the x-component of the velocity vector ($\mathbf{v}_0$) for P must equal the velocity of Q, namely, $\mathbf{v}_x$. Figure 12.11 shows the geometry. To write the scalar equation for the speed of Q, we must be careful about the signs. As shown, $\mathbf{v}_x$ points in the negative x-direction ($0 < \theta < \pi$) and so v_x is negative:

$$v_x = -v_0 \sin\theta = -A\omega \sin\theta \tag{12.6}$$

Therefore

$$v_x = -A\omega \sin\omega t = -2\pi f x_0 \sin 2\pi f t \tag{12.7}$$

For a given x_0, the maximum value of the speed ($v_0 = 2\pi f x_0$) must depend on the frequency—the point cannot oscillate over a range of $2x_0$ at a high frequency without having a high speed.

The oscillator is momentarily at rest at its two maximum displacements ($x = \pm x_0$), where $t = 0$ or $t = \frac{1}{2}T = 1/2f$ and, correspondingly, $\theta = 0$ or π, and $\sin\theta = 0$. As it begins a cycle ($t = 0$), the particle in Fig. 12.11b heads left, picking up speed that rises to a maximum at $x = 0$ when $t = \frac{1}{4}T = 1/4f$ and $\theta = \frac{1}{2}\pi$, whereupon $v_x = -v_0$. Sailing past $x = 0$, the speed of Q diminishes until it is again zero, when $t = \frac{1}{2}T$ at $x = -x_0$. Like a ball thrown straight up in Earth's gravity field, it stops for an instant at its maximum displacement.

We can also derive an expression for the speed of Q in terms of x. From Fig. 12.5 and the Pythagorean Theorem, $\overline{PQ} = y = \sqrt{A^2 - x^2}$ and so $\sin\theta = \sqrt{(A^2 - x^2)}/A$. Using $v_0 = A\omega$, we obtain

$$v_x = -A\omega \sin\omega t = \mp\omega\sqrt{A^2 - x^2} = \mp v_0\sqrt{1 - (x/A)^2} \tag{12.8}$$

which is the speed of a harmonic oscillator at any position x in terms of its maximum speed.

The trigonometric identity cos $(\alpha \pm \beta) = \cos\alpha \cos\beta \mp \sin\alpha \sin\beta$ tells us that $\cos(\omega t + \pi/2) = -\sin\omega t$, which, in turn, means that Eq. (12.7) could be written as $v_x = v_0 \cos(\omega t + \pi/2)$. Therefore, as can be seen in Fig. 12.12, the speed leads the displacement by 90°. The speed, which is a cosine function shifted one-quarter period ($t = \frac{1}{4}T$ or $\omega t = \frac{1}{2}\pi$) to the left, results in a curve that is an inverted sine function—that is, minus sine.

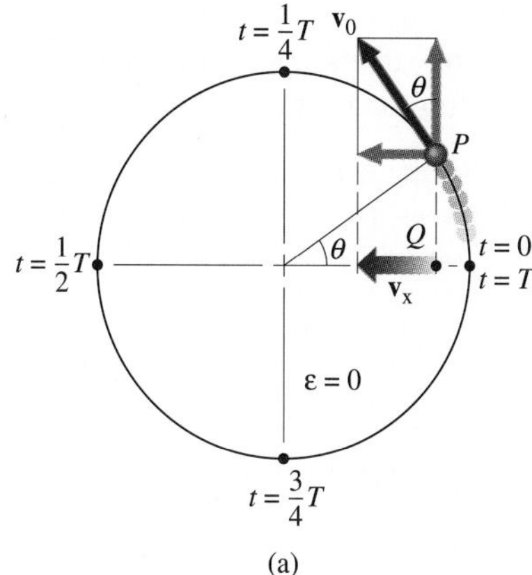

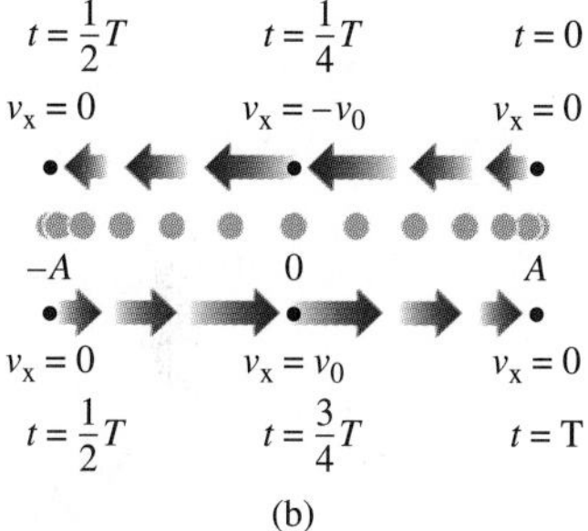

Figure 12.11 The point Q oscillates horizontally from $x = +A$ to $x = -A$, moving from a speed $v_x = 0$, to a maximum of $v_x = -v_0$ at $x = 0$, to $v_x = 0$ back at $x = -A$. (a) This reference circle shows the vector $\mathbf{v}_0$ and its component $\mathbf{v}_x$. (b) Here we see the time variation of the speed of a harmonic oscillator.

Acceleration. Particle P experiences a constant centripetal acceleration, $a_C = v^2/r = r\omega^2 = A\omega^2$, directed toward the origin (Fig. 12.13). Because Q always follows along, the horizontal component of the centripetal acceleration has to equal Q's acceleration. As before, to write $\mathbf{a}_x$ as a scalar we must include a negative sign, because it points in the negative x-direction:

$$a_x = -a_C \cos\theta = -A\omega^2 \cos\theta$$

and

$$a_x = -A\omega^2 \cos\omega t = -A(2\pi f)^2 \cos 2\pi f t \tag{12.9}$$

The oscillator has its maximum acceleration at the limits of its motion ($x = \pm A$) where the speed is zero. Prior to this chapter, we studied systems where the acceleration was either zero or constant, but that's not the case here. Moreover, because the acceleration of P is centripetal, the acceleration of Q is also center-seeking, pointing toward the equilibrium position, $x = 0$.

By comparing Eqs. (12.3) and (12.9), it is evident that

$$a_x = -\omega^2 x \tag{12.10}$$

independently of time. **The acceleration of a simple harmonic oscillator is proportional to the negative of its displacement from the midpoint of the motion.** That's the hallmark of SHM.

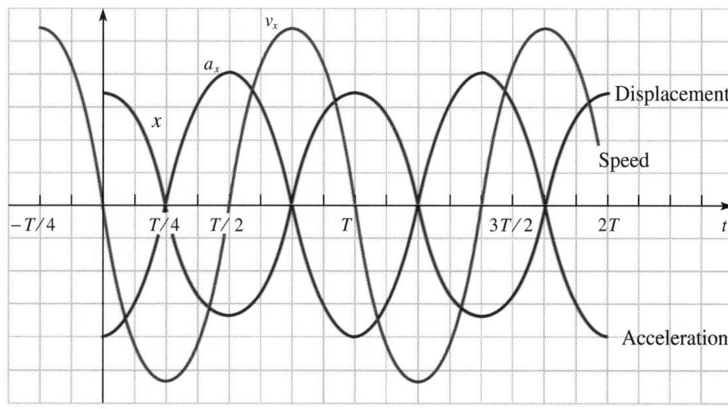

Figure 12.12 Plots of the speed, displacement, and acceleration for a simple harmonic oscillator. Note that the speed leads the displacement by $\pi/2$, while the acceleration lags the displacement by $\pi/2$.

Example 12.1 A spot of light on the screen of a computer is oscillating to and fro along a horizontal straight line in SHM with a frequency of 1.5 Hz. The total length of the line traversed is 20 cm, and the spot begins the process at the far right. Determine (a) its angular frequency, (b) its period, (c) its maximum speed,

(continued)

(continued)

and (d) its maximum acceleration. (e) Write an expression for x and find the location of the spot at $t = 0.40$ s.

Solution: [Given: $f = 1.5$ Hz and $A = 10$ cm. Find: (a) ω, (b) T, (c) $v_x(\text{max})$, (d) $a_x(\text{max})$, (e) x in general, and x at $t = 0.40$ s.]

(a) $\omega = 2\pi f = 2\pi(1.5 \text{ Hz}) = 9.4 \text{ rad/s} = \boxed{3.0\pi \text{ rad/s}}$

(b) $T = 1/f = 1/1.5 \text{ Hz} = \boxed{0.67 \text{ s}}$

(c) $v_x(\text{max}) = v_0 = A\omega = 2\pi fA = 2\pi(1.5 \text{ Hz})(0.10 \text{ m})$

and $\boxed{v_x(\text{max}) = 0.94 \text{ m/s}}$

(d) From Eq. (12.9), $a_x(\text{max}) = A\omega^2 = A(2\pi f)^2$ and

$a_x(\text{max}) = (0.10 \text{ m})(2\pi 1.5 \text{ Hz})^2 = \boxed{8.9 \text{ m/s}^2}$

(e) $x = A \cos \omega t = \boxed{(0.10 \text{ m}) \cos (9.4 \text{ rad/s})t}$

At $t = 0.40$ s

$$x = (0.10 \text{ m}) \cos (3.76 \text{ rad}) = (0.10 \text{ m})(-0.81)$$

and $\boxed{x = -8.1 \text{ cm}}$

▶ **Quick Check:** From (e) at $t = 0$, $x = +A$, as it should be. The value of x at $t = 0.40$ s; namely, -8.1 cm is reasonable since the period is 0.67 s, and so after $\frac{1}{2}T = 0.34$ s, the spot is on the negative side at $-A = -10$ cm heading right. At 0.06 s later, it's not surprising that it be at $x = -8.1$ cm.

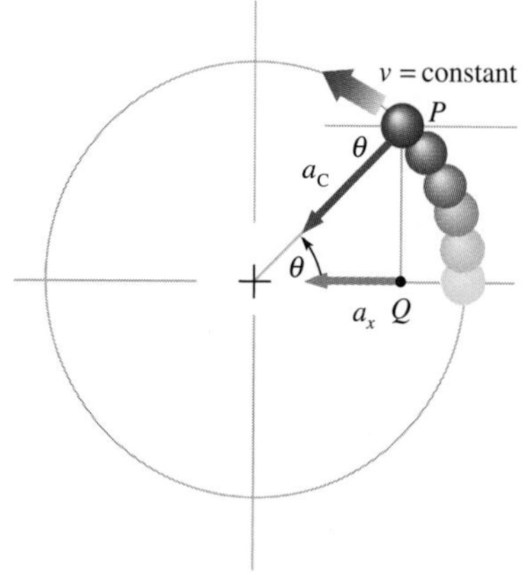

Figure 12.13 As an object moves in a circle, there is a centripetal acceleration a_C that has a horizontal component a_x. The latter is the acceleration of the oscillator.

12.2 Elastic Restoring Force

When a system oscillates naturally (without being driven by some external source of energy), it does so by moving against a force that tends to return it to its undisturbed equilibrium condition. That *restoring force* is usually gravitational or elastic (that is, electromagnetic), but other phenomena, such as surface tension, will sustain vibrations as well—raindrops oscillate as they fall. The system in equilibrium is first distorted. An additional quantity of potential energy is thereby stored in it, and then it's let loose to return toward equilibrium. Invariably, it overshoots (there being no tendency to stop at its undisplaced configuration since the system carries an excess of energy and still possesses momentum when it reaches that equilibrium point). It sails past its undistorted configuration only to be displaced in the opposite direction, to again become distorted and thereby begin the process anew. As the system vibrates, PE goes into KE and back to PE, and so on (with no losses) indefinitely. That "lossless" single-frequency ideal vibrator is known as a **simple harmonic oscillator**,

An Oscillating Spring

If a spring with a mass attached to it is slightly stretched or compressed and then let loose, it will oscillate in a way very closely resembling SHM. Many elastic systems (buildings, flagpoles, airplane wings, etc.) behave in a similar fashion, and so the spring merits a detailed examination.

Reconsider Hooke's Law [see Eq. (10.3)], $F = ks$: the force that must be exerted on an object—a spring or a wire or a tree—to make it distort *slightly* is proportional to the displacement from equilibrium that the object undergoes. Provided the distortion (stretching, compressing, twisting, bending, whatever) is not too great, $F \propto s$, and the object behaves elastically. From Newton's Third Law, the force exerted *by* an elastically stretched spring or a bent pole (the *restoring force* pulling back against that which is doing the stretching) is $F = -ks$.

If a body of mass m is attached to a helical spring of negligible mass (as in Fig. 12.14), displaced a small distance $s = x$ from equilibrium and then let go, the body will thereafter experience a force $F = -kx$, exerted on it by the spring. Yet since $F = ma$, the resulting acceleration of the body is given by $a_x = -(k/m)x$. *The acceleration of the body is proportional to the negative of its displacement*—the mass oscillates about $x = 0$ in SHM. This description assumes no energy is lost to friction (either in the spring, on the table, or through the air); otherwise the motion is *damped harmonic,* not SHM, and it will die out.

Example 12.2 Use energy considerations to derive an expression for the maximum speed of the mass oscillating in Fig. 12.14 in terms of k, m, and A. That done, confirm Eq. (12.8) for the speed as a function of displacement of a harmonic oscillator.

Solution: [Given: k, m, and A. Find: an expression for v_0.] From Eq. (10.4), we know that *elastic*-PE is given by $\Delta \text{PE}_e = \frac{1}{2}kx^2$. Let's look at the extremes: At $x = 0$, the KE is maximum and the PE is zero; hence, the total mechanical energy is $\text{E} = \frac{1}{2}mv_0^2$. At $x = \pm A$, the KE is zero ($v_x = 0$), the PE is maximum, and $\text{E} = \frac{1}{2}kA^2$. Since E is conserved

$$\tfrac{1}{2}mv_0^2 = \tfrac{1}{2}kA^2$$

and the maximum speed is

$$\boxed{v_0 = A\sqrt{\frac{k}{m}}}$$

In general

$$\text{E} = \tfrac{1}{2}mv_x^2 + \tfrac{1}{2}kx^2 = \tfrac{1}{2}kA^2$$

Solving for v_x^2, we obtain

$$v_x^2 = \frac{k}{m}(A^2 - x^2) = \frac{k}{m}A^2\left(1 - \frac{x^2}{A^2}\right) = v_0^2\left(1 - \frac{x^2}{A^2}\right)$$

which is equivalent to Eq. (12.8).

▶ **Quick Check:** We have already seen that $v_0 = A\omega$, and we will show presently that $\omega = \sqrt{k/m}$. Now let's just check that the units are right. Using $F = ks$, it follows that k has units of N/m. Thus, k/m has units of $(\text{N/m})/(\text{N}\cdot\text{s}^2/\text{m}) = 1/\text{s}^2$, and $\sqrt{k/m}$ correctly has the units of 1/s.

The restriction that the spring, diving board, or suspension bridge (whatever it is that's being stretched) must behave in a Hookean fashion is equivalent to saying that F is linear in x, which is equivalent to saying a is linear in x, which is the hallmark of SHM. Still, precise linear behavior is more an idealization than a reality. To the degree that a system behaves in a way that is approximately linear, it will oscillate in a way that is approximately SHM. To that end, the spring must only be stretched slightly and the string plucked gently—nature may love simplicity, but twist it too far and it gets very complicated.

Frequency and period. When the mass on the spring is initially displaced some distance $\pm x_0$ (that is, via compression or elongation) and *released from rest,* that displacement will remain the maximum value attainable, and the mass will oscillate between $+x_0$ and $-x_0$. Of course, this is the *amplitude* of the oscillation: $x_0 = A$. Now, compare the body's acceleration $a_x = -(k/m)x$ with Eq. (12.10) for the acceleration of an object in SHM; namely, $a_x = -\omega^2 x$. The angular frequency equals the square root of k/m, but here we will use the symbol ω_0 instead of ω to remind us that it is the *natural* angular frequency, the specific frequency at which a physical system oscillates all by itself once set in motion (as we will see, the spring can be made to vibrate at other frequencies by driving it). Thus

$$\omega_0 = \sqrt{\frac{k}{m}} \qquad (12.11)$$

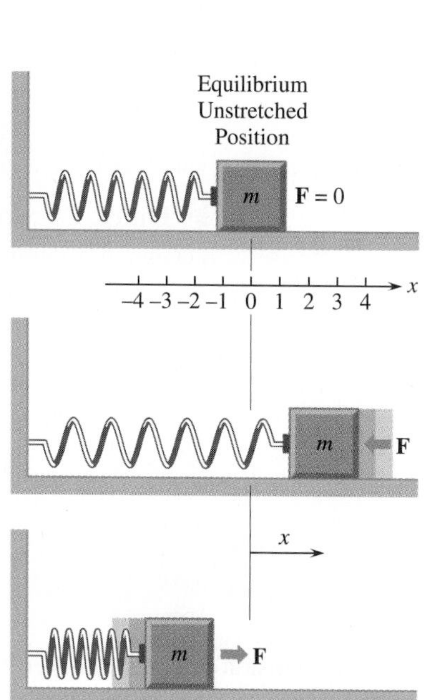

Figure 12.14 A mass on a spring vibrating horizontally in simple harmonic motion. Here, **F** is the force exerted by the spring, and there is no friction.

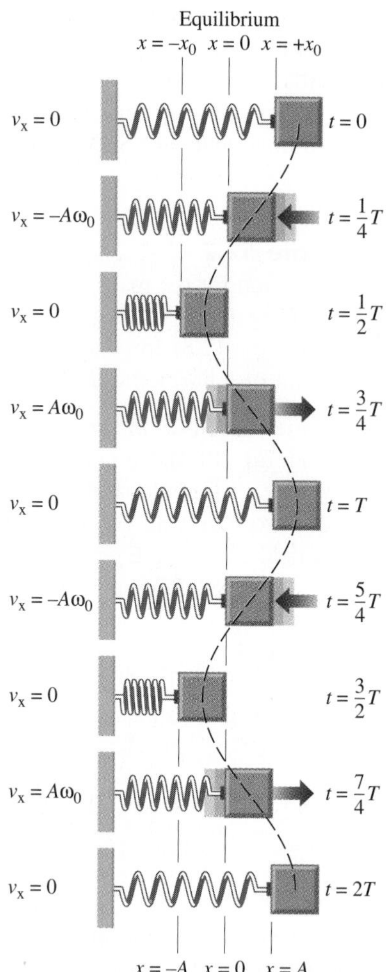

Figure 12.15 A simple harmonic oscillator shown every quarter cycle for two cycles. Indicated are the corresponding speeds and displacements.

and with $\omega_0 = 2\pi f_0$, we have

$$f_0 = \frac{1}{2\pi}\sqrt{\frac{k}{m}} \tag{12.12}$$

Since $T = 1/f_0$

$$T = 2\pi\sqrt{\frac{m}{k}} \tag{12.13}$$

Figure 12.15 shows the relationship among x, v_x, t, and T.

The stiffer the coil spring or wire or rubber band, the larger the elastic force constant k and therefore (with a given mass), the higher the vibrational frequency and the shorter the period. The springs on a car must have a large k to keep the up-and-down excursions small, but because the mass of the car is large, the frequency will nonetheless be low. For a given spring (that is, a given k) if the oscillating mass is increased, the inertia of the system will be increased, causing it to accelerate more slowly and therefore increasing the period.

A skyscraper can be made as stiff as the builder is willing to pay for, up to a point at least. It is the financial as much as the structural concerns that dictate how rigid a tall building will be. As a rule, such structures are designed so that the top never displaces more than about 1/400 of the height when acted upon by a sustained pressure of 1.4 kPa (30 lb/ft^2), which is equivalent to a 150-km/h (95-mi/h) wind. The period with which the building will rock is then determined by its mass and stiffness (k). The World Trade Center towers sway with a period of 10 s whereas the smaller Citicorp building, also in Manhattan, has a period of only 7 s.

Example 12.3 The cart in Fig. 12.16 has a mass of 1.00 kg, and someone reaches down and displaces it 5.00 cm to the right with an axial horizontal force of 10.0 N. (a) Assuming no friction, what is the period of the resulting oscillation when the cart is released? (b) Where will it be 0.200 s after release? (c) What is the elastic force constant for the system if one of the two identical springs is removed? (d) Determine the new frequency.

Figure 12.16 To displace the cart horizontally, energy must be provided to the system. That energy is conserved and ideally transforms back and forth from PE$_e$ to KE as the cart oscillates in SHM. In reality, the oscillation will continue to diminish until all the energy ends up as thermal energy via friction and internal spring losses.

Solution: [Given: $m = 1.00$ kg, $A = 0.050$ m, and $F = 10.0$ N. Find: T and x at $t = 0.200$ s, k with one spring, and f.] (a) The period can be gotten from Eq. (12.13) once we have the elastic force constant for the system. Inasmuch as the applied force produces a displacement x such that $F = kx$

$$k = \frac{F}{x} = \frac{10.0 \text{ N}}{0.050 \text{ m}} = 200 \text{ N/m}$$

This is the elastic force constant of the system—both springs acting at once, and both undergoing the displacement x as if they were attached side by side to the same end of the cart. Consequently

$$T = 2\pi\sqrt{\frac{m}{k}} = 2\pi\sqrt{\frac{1.00 \text{ kg}}{200 \text{ N/m}}} = \boxed{0.44 \text{ s}}$$

(b) To find where the cart will be at $t = 0.200$ s, we go back to the equation of x as a function of time, thus

$$x = A\cos\theta = A\cos\omega_0 t = A\cos 2\pi f_0 t \qquad [12.3]$$

where the motion began at $x = A$, at $t = 0$, so $\varepsilon = 0$. We need either ω_0 or f_0, and since we have T

$$f_0 = \frac{1}{T} = 2.25 \text{ Hz}$$

Therefore

$$x_{0.2} = A\cos 2\pi f_0 t$$

$$x_{0.2} = (0.050 \text{ m})\cos 2\pi(2.25 \text{ Hz})(0.200 \text{ s})$$

and

$$\boxed{x_{0.2} = -0.048 \text{ m}}$$

(c) If a spring is removed, displacing the body will require only half the force it did before, when one spring was being stretched as the other was compressed. It follows that now $\boxed{k = 0.10 \text{ kN/m}}$, half its previous value.

(d) The resulting oscillatory frequency, Eq. (12.12), will be $1/\sqrt{2}$ of what it was—namely, $\boxed{1.6 \text{ Hz}}$.

▶ **Quick Check:** The answer to (b) is reasonable since 0.200 s is very nearly $\frac{1}{2}T$, at which time $x = -A = -0.050$ m. (c) The new period is $T = 0.444 \text{ s} \times (1.414) = 0.63$ s; $f = 1/T = 1/0.63 \text{ s} = 1.6$ Hz.

Gravity and oscillations. An elastic band hanging freely from a hook, as in Fig. 12.17, stretches a length ΔL when it is made to support a load of mass m. At that point, the restoring force up equals the weight down: $k\,\Delta L = mg$ and the system is in equilibrium. Here, k is the net elastic force constant of the two supporting lengths of material acting together. Now, imagine that we pull down on the mass, displacing it from equilibrium an arbitrary distance $+y$ (*taking down to be positive* so that the restoring force will be negative). Newton's Second Law states that at any instant the sum-of-the-forces equals ma_y. The upward rubber band force $-k(\Delta L + y)$ plus the downward weight mg equals ma_y:

$$-k(\Delta L + y) + mg = ma_y$$

but $k\,\Delta L = mg$; hence, for any value of y

$$a_y = -\left(\frac{k}{m}\right)y$$

This is the same relationship we obtained for the horizontally oscillating mass on a spring. Gravity acting on the load establishes a new equilibrium position about which the mass will then oscillate.

Example 12.4 A 2.0-kg bag of candy is hung on a vertical, helical, steel spring that elongates 50.0 cm under the load, suspending the bag 1.00 m above the head of an expectant youngster. The candy is pulled down an additional 25.0 cm and released. How long will it take for the bag to return to a height of 1.00 m above the child?

Solution: [Given: $m = 2.0$ kg, $\Delta L = 50.0$ cm, and $A = 25.0$ cm. Find: t when $y = 0$.] The loaded system is in equilibrium when the mass is 1.00 m above the small person. It returns there from its lowest point, the starting position ($x = A$), after $\frac{1}{4}$ cycle—that is, at $t = \frac{1}{4}T$. To find T, we must first find k.

Initially $$-k(\Delta L) + mg = 0$$

and so

$$k = \frac{mg}{\Delta L} = \frac{(2.0\text{ kg})(9.8\text{ m/s}^2)}{0.50\text{ m}} = 39.2\text{ N/m}$$

We know from Eq. (12.13) that

$$T = 2\pi\sqrt{\frac{2.0\text{ kg}}{39.2\text{ N/m}}} = 1.4\text{ s}$$

and so $$\boxed{\tfrac{1}{4}T = 0.35\text{ s}}$$

▶ **Quick Check:** $F = k\,\Delta y = mg$, $m/k = \Delta y/g$, $T = 2\pi\sqrt{\Delta y/g} = 2\pi\sqrt{0.5\text{ m}/g} = 1.4$ s.

The mass of the spring. Continuing from Eq. (12.13) for the period, it follows that $T^2 \propto m$. A plot of T^2 versus m, for a number of different masses hung on a vertical coil spring, should be a straight line with a slope of $4\pi^2/k$, passing through the origin. In practice, for small values of A, the graph is straight—the motion is SHM—but it *does not* pass through the origin (Fig. 12.18). That's because the spring itself is not massless. The complete analysis is complicated by the fact that the spring's mass oscillates along with the load, but each segment of it has a different amplitude. It turns out that the effective mass of the system is actually the mass of the load plus one-third the mass of the spring.

12.3 Gravitational Restoring Force: Pendulums

Legend has it that Galileo, while an undergraduate at Pisa, was in the cathedral (1581) when an attendant pulled a candelabrum hanging from the ceiling off to one side to light it. After it was let loose, it swayed to and fro, and Galileo sensed a rhythm in the smooth, repetitive motion. At first the lamp rushed quickly along large arcs, and then as the swing diminished, the speed slowed with it so that the rhythm remained strangely constant. Being a student of medicine, Galileo naturally used the beat of his own pulse to time the swing (there were no watches then). To his astonishment, the duration of each cycle (what he named the *periodo*) was constant.

Galileo subsequently performed a series of experiments using balls of different weight hung on various lengths of string. What he soon discovered was remarkable: *the period of a pendulum is constant and determined only by the square root of its length.* The implication was clear: the string deflected the ball along a curved path, but it was otherwise free to fall, and since all objects fall at the same rate, two pendulums of the same length should, indeed, swing in step.

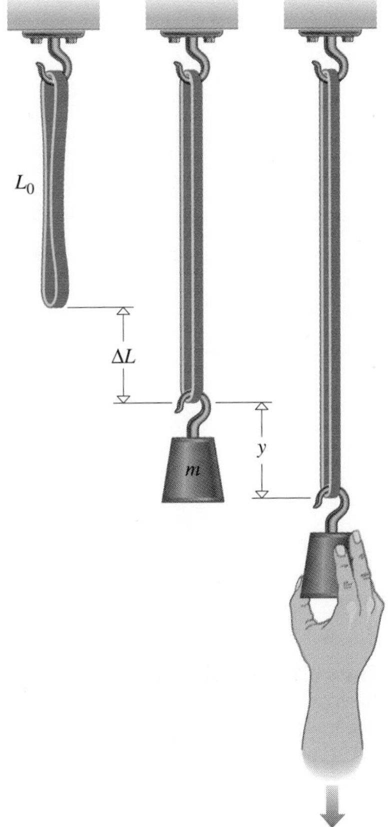

Figure 12.17 Hanging a mass m on an elastic band stretches it a length ΔL, to a new equilibrium position. Pulling down further extends it a distance y. When it is let loose, it will oscillate in SHM, provided the displacements are small and the restoring force is linear.

The Simple Pendulum

The pendulum in Fig. 12.19 is depicted in flight at some angle θ. The bob is displaced a positive distance l measured to the right along the arc of the path from the vertical zero-θ axis. Taking θ in radians, $l = L\theta$. At that moment, the weight of the bob (mg) is, as always, straight down, but it has a component tangent to the arc. It is this component, this unbalanced force

$$F = -mg \sin \theta$$

that drives the pendulum back to equilibrium ($\theta = 0$). The minus sign is needed because **F** points in the negative l-direction, toward decreasing θ. Since $F = ma_{\mathrm{T}}$, the tangential acceleration is

$$a_{\mathrm{T}} = -g \sin \theta \tag{12.14}$$

which is proportional to the *sine* of θ and not to l, and so it is *not* the condition for SHM. Still, for small angles, the value of θ in radians is very nearly the value of $\sin \theta$; they differ by less than 2% out to about 20°. Hence, $\sin \theta \approx \theta = l/L$ and

$$a_{\mathrm{T}} \approx -\left(\frac{g}{L}\right) l \tag{12.15}$$

which is SHM, provided θ is small. It follows from $a_x = -\omega_0^2 x$ and Eq. (12.15) that $\omega_0 = \sqrt{g/L}$ and so

$$f_0 \approx \frac{1}{2\pi}\sqrt{\frac{g}{L}} \tag{12.16}$$

and

$$T \approx 2\pi\sqrt{\frac{L}{g}} \tag{12.17}$$

At any location on Earth, the period of a pendulum is dependent only on the square root of its length. For small initial angular displacements, less than 23°, the actual periods vary by less than 1% from the predictions of Eq. (12.17), and the motion is approximately SHM. The agreement improves tremendously for still smaller starting angles.

To get Eq. (12.14), we set ma_{T} equal to $-mg \sin \theta$ and then canceled the mass. Well, that really was more significant than it might seem. The mass in $F = ma$ is

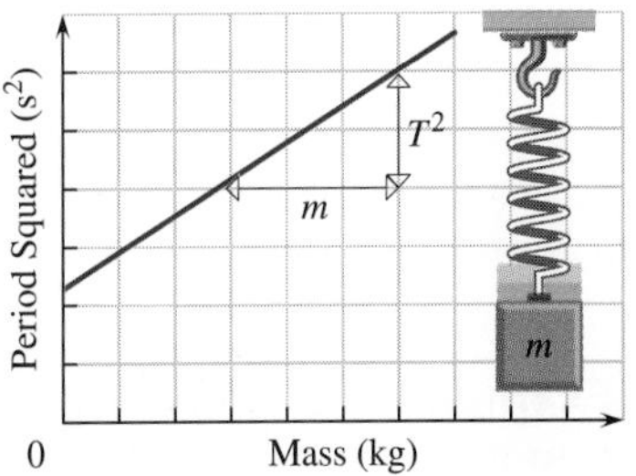

Figure 12.18 Since $T = 2\pi\sqrt{m/k}$, it follows that $T^2 = 4\pi^2 m/k$, and a plot of T^2 versus m must be a straight line with a slope of $T^2/m = 4\pi^2/k$. The reason it doesn't pass through the origin is that the spring itself has mass, which adds inertia to the system and increases the period.

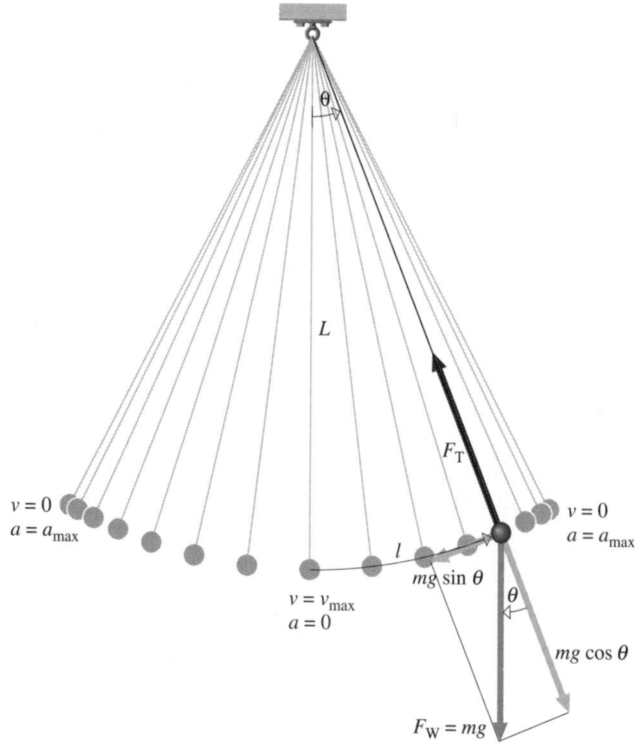

Figure 12.19 A pendulum is driven into oscillation by the nearly horizontal component of its weight $mg \sin \theta$. Provided θ is small, the mass will move in SHM.

the **inertial mass**, the mass associated with the object's tendency to resist changes in its motion. It seems to have nothing to do with gravity and perhaps might even be labeled m_i to underscore the difference in its conceptual origins. On the other hand, the weight of a body is determined by a physical property it possesses called **gravitational mass**, m_g. That property is proportional to the gravitational interaction between objects and seemingly has nothing to do with inertia. Thus, assuming these masses to be different (since they certainly seem to define different characteristics), $T \approx 2\pi\sqrt{m_i L/m_g g}$. But experiments going back to Newton himself all confirm (nowadays to within 1 part in 10^{12}) that the period is independent of the mass of the bob, which implies that $m_i = m_g$. The assertion of the equality of inertial and gravitational mass is known as the *Equivalence Principle* and it is central to the General Theory of Relativity. Perhaps gravity—the Universe acting gravitationally on each mass—*is* the cause of inertia.

It is believed that Galileo described the design of the pendulum clock to his assistant Viviani a year or so before he died.

Example 12.5 How long should a pendulum be if it is to have a period of 1.00 s at a place on Earth where the acceleration due to gravity is 9.81 m/s²?

Solution: [Given: $T = 1.00$ s and $g = 9.81$ m/s². Find: L.] From Eq. (12.17), $T^2 = 4\pi^2(L/g)$; hence

$$L = \frac{T^2 g}{4\pi^2} = \frac{(1.00\text{ s})^2(9.81\text{ m/s}^2)}{4\pi^2} = \boxed{0.248\text{ m}}$$

▶ **Quick Check:** A pendulum 1 m long is sometimes called a "seconds pendulum" because one swing takes 1 s: $T = 2\pi\sqrt{L/g} = 2.0$ s. The pendulum in this problem has half that period and therefore should have $\frac{1}{4}$ the length.

Huygens recognized that Eq. (12.17) provides a simple means of measuring g. Geologists have been using the pendulum for a long time to determine local variations in g. A fairly accurate value of g can be determined by just swinging a mass on a string and measuring T averaged over 10 or 15 cycles. Try it!

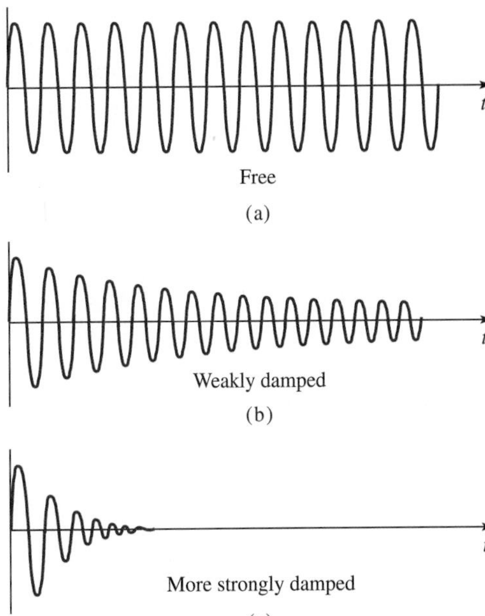

Figure 12.20 (a) An ideal harmonic oscillator, free of all forms of energy loss, will vibrate harmonically forever with no decrease in amplitude. (b) When friction is present, the oscillator will be damped, and its amplitude will decrease in time. (c) The more the damping, the more repressed is the oscillation.

12.4 Damping, Forcing, and Resonance

Thus far we have talked about freely vibrating systems, but that's an idealization. There are usually external forces acting on an oscillator in addition to the restoring force. These forces may either impede (that is, damp) the motion, or they may drive it so that its amplitude is sustained, or even grows. Sometimes this latter effect can cause catastrophic failures, as when large buildings are toppled during earthquakes.

Damping

In practice, oscillating mechanical systems lose energy in a variety of ways via friction, and the sinusoidal behavior of SHM is invariably replaced by a damped *anharmonic* motion that gradually dies away (Fig. 12.20). For example, a common friction mechanism is **viscous damping** (we are, after all, surrounded by air). At low speeds, viscous damping is proportional to the speed of the body in the damping fluid (Fig. 12.21). Any undriven system oscillating in the air must inevitably come to rest.

With weak damping, a system can continue to oscillate a relatively long time before stopping at its zero displacement position. This is what happens to an ordinary pendulum losing energy via friction. Its motion slowly decays, just as the vibrations of a tuning fork or a gong slowly decay. With still more friction, the amplitude decreases rapidly, and the mass undergoes fewer oscillations before coming to rest. Such a system is **underdamped** (Fig. 12.22). It's still periodic, although the frequency is lower than it would be without damping.

If friction is further increased, a displaced system can return to equilibrium without overshooting, in which case there will be no oscillation at all. The shock absorbers on a car should damp out any oscillations in less than a cycle. You don't want your car bobbing up and down long after it hits a bump in the road. When this nonvibratory motion occurs in the shortest amount of time, such that once let loose the system sweeps right back to its undisturbed configuration, it is said to be **critically damped**. It would be nice to have the swinging door at the kitchen of a restaurant or the entrance of a supermarket close swiftly without overshooting and slamming back to hit you from behind.

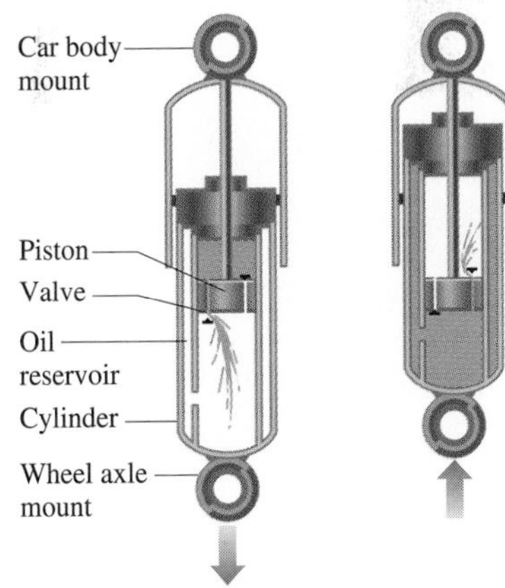

Figure 12.21 The shock absorber (or dashpot) uses viscous damping, which is proportional to the speed of the piston. A rapid jerk of the piston, as might be produced when a car hits a pothole, is met with a large damping force. On the other hand, a gradual depression of the piston is hardly resisted.

With even more damping, the system again does not oscillate, but it takes much longer to return to equilibrium. Heavy doors in public buildings almost all have hydraulic devices at their tops to keep them from slamming, and if you have ever waited interminably for one to close, you have experienced the results of **overdamping**.

A variety of damping devices are used to suppress vibrations of all kinds in physical structures and machines. Most are as simple and straightforward as a rubber collar on the shaft of a fan or inflated tires on a car, but some are more exotic. Olympic gold medals have been won using skis equipped with vibration dampers—

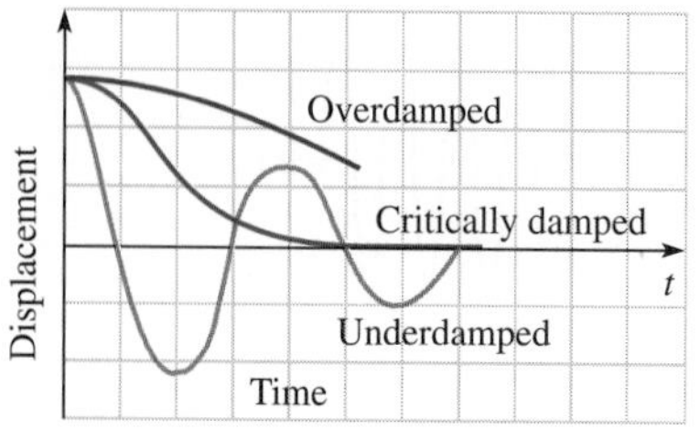

Figure 12.22 When a damped oscillating system experiences still more friction, it will stop vibrating and the motion will gradually decay. The system is then over-damped. When the motion decays as quickly as possible, the system is critically damped.

sandwiches of lead plates and spongy absorbent foam. There are space satellites that contain little carts that ride back and forth along tracks, set into motion by any inadvertent wobbling of the vehicle. Eventually the wobble is damped out and the carts come to rest—the kinetic energy of the satellite having been converted into thermal energy via friction in the carts. The World Trade Center in New York contains visco-elastic dampers: large plates coated with a sticky polymer that drag across one another whenever the towers sway in the wind. They convert troublesome shear energy into harmless thermal energy.

Forced Oscillation and Resonance

One way to sustain the oscillations of a system suffering damping is to periodically pump in energy via a force that does positive work on the system. Pushing a playground swing can keep it going indefinitely, despite friction. When mom pushes, it's best done after junior has reached the peak of the swing and is moving away from her so she does positive work on the system (**F** and Δ**s** are parallel). Energy is efficiently transferred to the system when it is pumped in in step with the natural oscillation. Accordingly, mom can push at the natural frequency, or miss a beat and push at one-half the natural frequency, or miss two and push at one-third the natural frequency, and so on.

The application of an *external alternating* driving force to a system capable of vibrating produces **forced oscillation**. The specific physical causes are many. Pulsations of pressure in lines carrying fluids will cause vibrations—some houses have "singing" plumbing, and refrigerators and air conditioners often buzz.

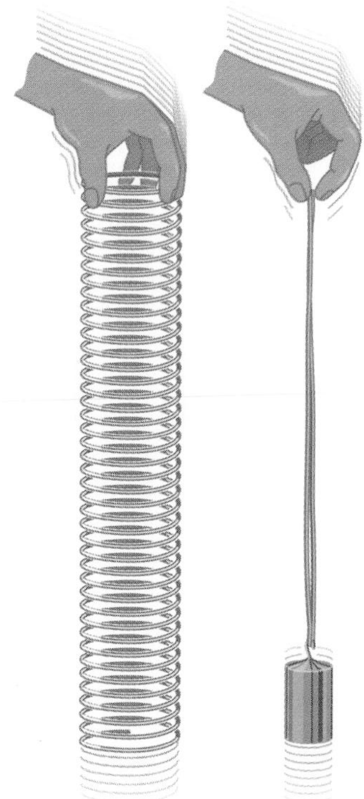

Figure 12.23 By moving up and down, the hand pumps energy into the system, forcing it to oscillate at the driving frequency.

Forced oscillation is easily seen with either a Slinky or a mass on a long elastic band (Fig. 12.23). Displace either and let it loose to oscillate at its natural frequency (f_0), which for the Slinky will be around 1 Hz. That done, stop the oscillation and now move your supporting hand up and down with a small amplitude ($\approx$2 cm) at a very low frequency ($f \approx 0.3$ Hz, $T \approx 3.0$ s), thereby driving the elastic system into motion. The Slinky will follow your hand (the driver), moving upward when your hand moves up, and downward when it moves down; that is, the Slinky will oscillate in-phase with the driver. Moreover, it will oscillate at the same frequency (f) as the driver but with a relatively small amplitude, not much different from that of your hand. Little energy is transferred from the driver to the Slinky as the process continues.

Similar behavior occurs when the driving frequency is much greater than the natural frequency ($f \gg f_0$) except now the two will be nearly 180° out-of-phase—when your hand goes down, the Slinky will move up, and vice versa. One thing will become obvious when you further vary the driving frequency—*as f approaches the natural frequency of the system, the resulting oscillation will dramatically increase in amplitude* (Fig. 12.24). The vibrational amplitude will reach a maximum when $f = f_0$, a condition known as **resonance**. At that special driving frequency, which is also known as the **resonant frequency**, energy is most efficiently transferred to the system, whether it's a hand-pumped Slinky, a windblown skyscraper, or a human internal organ set into vibration by the overamplified tumult of a rock band. It seems that every material object can oscillate somehow. As a rule, a body can bend, twist, and elongate in several different ways, displaying one or more natural frequencies at which it can resonate. Similarly, a composite object, such as a bus, in addition to vibrating as a whole, will have lots of resonant modes for its various parts—windows, doors, seats, and so on.

At resonance, most of the small amount of energy available during each cycle is stored in both the elastic medium and the moving mass; none is returned to the

An earthquake caused the collapse of the double-decked Nimitz Freeway in Oakland, California, in 1989. This portion of the highway rests on fine-grained sediments and mud that transmitted frequencies of around 2 Hz. This value is close to the natural frequency of the highway, which experienced violent resonant vibrations. Fifty-three cars were crushed when the supports for the top deck gave way.

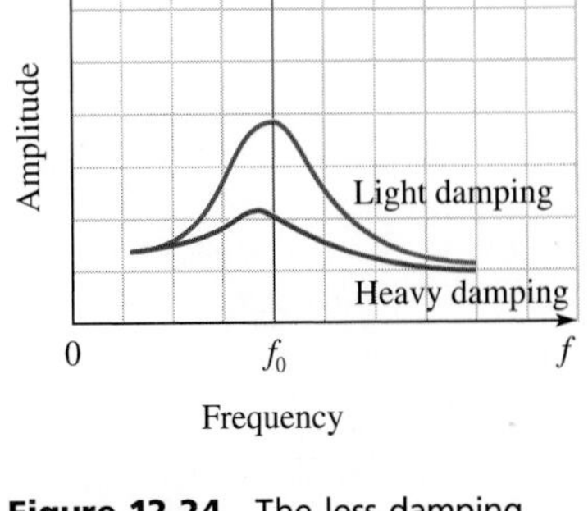

Figure 12.24 The less damping there is, the sharper and taller is the amplitude versus frequency curve. The system responds more selectively to being driven at its resonant frequency.

driver, though a bit is invariably lost to friction. The less damping there is, the sharper and taller is the curve (Fig. 12.24), the greater the resulting amplitude of the vibration and the nearer that peak is to the resonant frequency. In a lightly damped system such as a bell, gong, or turbine blade, the stresses occurring at resonance can be upwards of 300 times larger than those resulting from the same applied force acting at a much different frequency.

The phenomenon of resonance is one of the most important in all of physics. It is at the heart of a tremendous diversity of occurrences from the atomic absorption and emission of light to the tuning of a TV set, from the swaying of bridges to the rattling of china cabinets when a low-flying plane passes overhead.

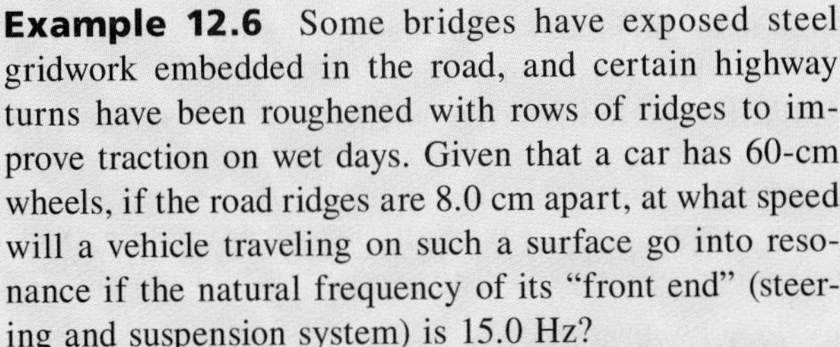

Example 12.6 Some bridges have exposed steel gridwork embedded in the road, and certain highway turns have been roughened with rows of ridges to improve traction on wet days. Given that a car has 60-cm wheels, if the road ridges are 8.0 cm apart, at what speed will a vehicle traveling on such a surface go into resonance if the natural frequency of its "front end" (steering and suspension system) is 15.0 Hz?

Solution: [Given: $D = 60$ cm, $d = 8.0$ cm, and $f_0 = 15.0$ Hz. Find: v for resonance.] There are $1/0.080 = 12.5$ ridges per meter and so at speed v (in m/s), the car is moving over $12.5v$ ridges per second, which is the driving frequency. Hence

$$12.5v = 15.0 \text{ Hz}$$

and $\boxed{v = 1.2 \text{ m/s}}$ or 3 mi/h, regardless of wheel size. That's why your car shakes so much at low speeds on that kind of surface.

▶ **Quick Check:** $v = x/T = x/(1/f_0) = (0.080 \text{ m}) \times (15.0 \text{ Hz}) = 1.2$ m/s.

Tap a good quality wineglass and it will ring, undriven, at its natural frequency, somewhere around 900 Hz. If it's now forced at that frequency, for example, by a blast of vibrating air (that is, sound) up near the rim, the glass will efficiently ab-

A man playing a tune on a street in New Orleans by resonating (via stick-slip friction) different sized and pitched glasses.

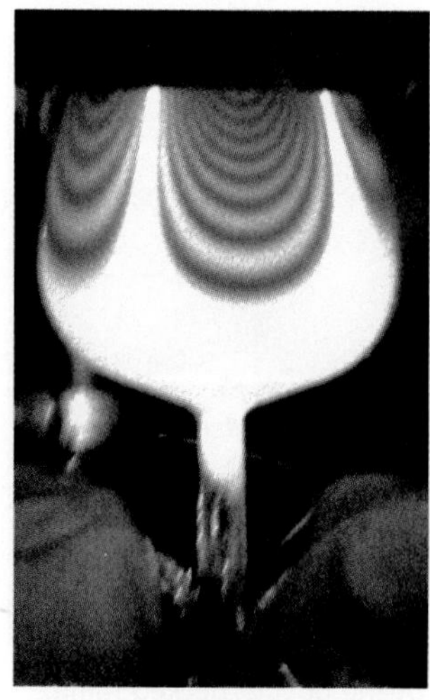

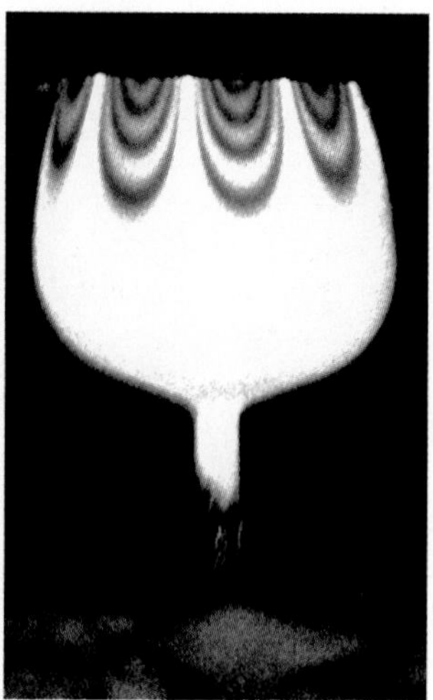

A wineglass oscillating in two of its several possible modes. The bright region is the stationary portion of the glass. The photos were made from interferometric holograms.

sorb energy and begin to vibrate. When the note is sustained and carries enough energy to overcompensate for internal damping, the rim of the glass will oscillate with increasing amplitude and ultimately will shatter because glass is brittle. The same resonance mechanism causes the ashtray in your car to clatter annoyingly to the beat of the radio, and it's the reason why troops marching over bridges are usually ordered to break step. At least two large bridges collapsed in the nineteenth century, one near Manchester, England (1831), the other over the Maine in Anjou, France (1849), both having been set into resonance by the pounding, low-frequency cadence.

Example 12.7 A room at the top of the Citicorp building houses a 400-ton block of concrete that slides "floating" on a layer of oil pumped under pressure. One end of the block is attached to pneumatic springs fixed to one side of the building while the other end is attached to a hydraulic piston that acts like a large shock absorber, mounted to the other end of the building. The device is called a "tuned-mass damper" because it's tuned to oscillate at the natural frequency of the building. As the tower begins to sway, the block goes into resonant oscillation out-of-phase with the building. The wind's energy is transferred to the tower, and then to the block, and then, by moving against the shock absorber, it's finally dissipated as thermal energy. Take the period of oscillation of the building to be 7.0 s and calculate the elastic force constant of the pneumatic spring that will produce resonance with the damping removed. (See the photo on the opposite page.)

Solution: [Given: $T = 7.0$ s and $F_W = 400$ tons $= 800 \times 10^3$ lb. Find: k.] The natural period of the spring-mass system must match that of the building—namely, 7.0 s. The period is given by Eq. (12.13), from which it follows that

$$k = \frac{4\pi^2 m}{T^2} = 4\pi^2 \frac{3.63 \times 10^5 \text{ kg}}{49 \text{ s}^2} = \boxed{2.9 \times 10^5 \text{ N/m}}$$

▶ **Quick Check:** Compute T using k; $\omega_0 = \sqrt{k/m} = \sqrt{(2.925 \times 10^5 \text{ N/m})/(3.63 \times 10^5 \text{ kg})} = 0.9$ rad/s; $f_0 = \omega_0/2\pi = 0.14$ Hz; $T = 1/f_0 = 7$ s.

12.5 Self-Excited Vibrations

Remarkably, it is possible to initiate and sustain the vibration of a particular kind of system even though the energy source is nonoscillatory. In such situations, the response of the vibrating system itself produces alternations in applied force. These **self-excited vibrations** are commonplace and often quite dangerous, though a kiss (sucking in a stream of air so that the lips vibrate) can be harmless enough.

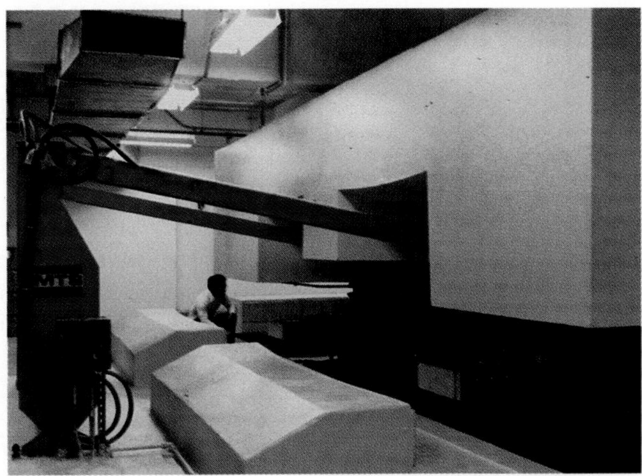

The tuned-mass damper at the top of the Citicorp tower. (See Ex. 12.7.)

Take a bow and a violin and attempt the business of making music—or at least sound. We wish to set the strings vibrating; they will set the air vibrating, and that's what a concerto is, more or less. Moving in one direction, say, to the right, the bow is made to brush over a string. Initially at rest with respect to the bow, the string is drawn to the right, and bent. The displacement continues until the growing restoring force of the string overcomes the static friction force, and it breaks loose from the bow. The string vibrates at a natural frequency, oscillating almost freely for a little while (losing energy in the form of sound). It encounters a markedly increased friction force when its relative speed with respect to the bow is low. Thus, the moving bow will recapture the string, again displacing it to the right, pumping energy in, and beginning the process over once more. The energy source is the constantly moving bow, but the motion of the system—the string—is vibratory.

This so-called **stick-slip friction** is the process responsible for the screech of a fingernail drawn across a blackboard. The high-pitched creak of an unoiled door is due to a torsional vibration of the short, stiff hinge pin. As Galileo pointed out, a wineglass can be made to ring at its resonant frequency in much the same way. Just run a moist, clean finger around the upper edge of the lip, gently pressing downward—stick-slip friction will do the rest.

There are many *self-excited vibrational* phenomena that are driven by constant streams of fluid, singing being one of them. Blowing across the mouth hole of a flute causes vortices to peel off periodically, creating a fluctuating pressure and setting the instrument vibrating. Similarly, a constant wind streaming over a banner stretched across a street can set it vibrating. The effect is easily duplicated by blowing over the broad faces of a wide rubber band.

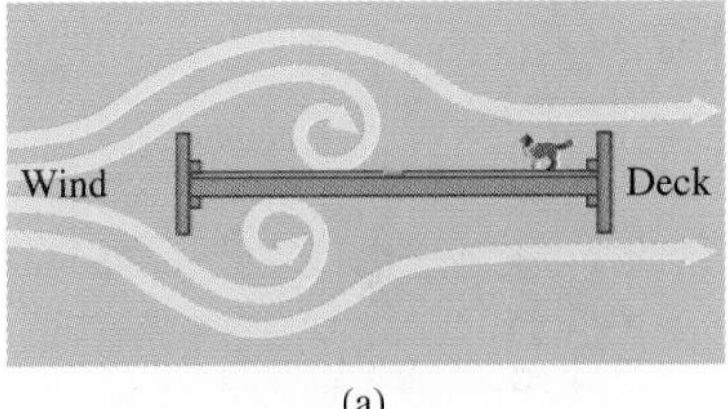

(a)

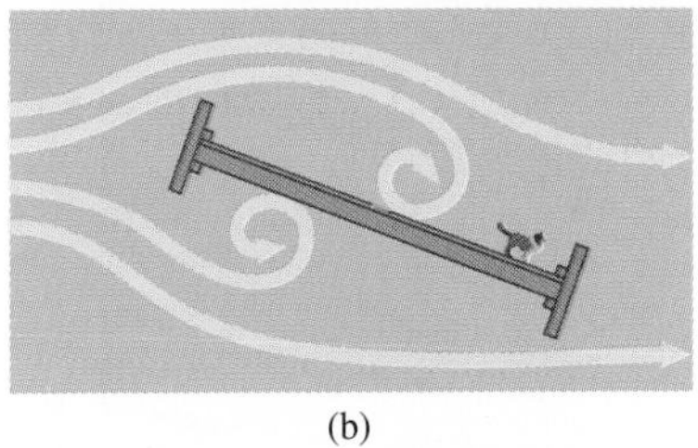

(b)

Figure 12.25 (a) Cross-sectional view of the roadway deck of the Tacoma Narrows Bridge, initially set into oscillation by alternating vortices. (b) As the deck oscillated, it caused other vortices to be induced that drove the bridge at its resonant frequency. These so-called motion-induced vortices were created in step with the twisting deck and furthered the motion.

The first bridge across the Tacoma Narrows at Puget Sound, Washington, had a main span 2800 ft long and 39 ft wide with 8-ft-tall steel stiffening girders. Opened for traffic on July 1, 1940, it soon became infamous for oscillating wildly whenever the wind was blowing. Thrill-seeking motorists were lured from miles around for the ride of a lifetime. On the morning of November 7, 1940, a wind of 40 to 45 mph set the span rippling as usual, at a frequency of 36 vibrations per minute. When the amplitude became too large, the bridge was cleared. At around 10:00 A.M., the north cable became loose in its collar, allowing the main deck to abruptly begin to vibrate in a twisting resonant mode ($f_0 = 0.2$ Hz) around the yellow centerline of the roadway.

Although there are still questions to be answered, the basic mechanism that caused the catastrophe seems to have been vortex-induced self-excited vibrations. Air rushing toward the tall windward stiffening girder broke into two streams, shedding vortices alternately above and below the deck. Once the structure began to oscillate, that motion led to the formation of other motion-induced vortices (Fig. 12.25). These were formed as the deck rose and fell, and so matched the natural vibrational frequency. The rate at which energy was absorbed from the wind soon exceeded the frictional loss, and a state existed in which the oscillations quickly grew in amplitude. Shortly after 11:00 A.M., the center span tore into shreds like so much cotton ribbon.

MECHANICAL WAVES

There are only two fundamental mechanisms for the transport of energy and momentum: a streaming of particles and a flowing of waves. And even these two seemingly opposite conceptions are subtly intertwined—there are no waves without particles and no particles without waves, but we will come back to that later.

A wave is a disturbance of a medium, and that medium can be either a field (such as the gravitational field) or a material substance (a solid or fluid). Here, the focus is on waves in material media, and these are known as **mechanical waves**. Sound is such a wave, but because of its special relationship to us, it will be treated on its own in Chapter 13.

12.6 Wave Characteristics

Consider an object—a bell, a rope, or even the Earth. Each is a vast collection of atoms forming an essentially continuous elastic medium. If the atoms or molecules in a material are pushed together, they repel and, if separated, they attract. Electrically interacting, the atoms behave as if they are connected to one another via springs. As a result of this restoring force, the medium returns to an equilibrium configuration that is its normal stable state. Now imagine that a driving force acting on the medium displaces some of its atoms; for example, the bell is struck with a hammer. Energy is imparted to these impacted atoms, which are forced off their equilibrium positions, thereby interacting more strongly with neighboring atoms, doing work on them and displacing them. These newly shifted atoms interact with still others, and so the displacement process continues, with energy transferred from atom to atom (Fig. 12.26). *The state of being displaced moves through the medium as a wave.* The disturbance of a medium under the influence of a restoring force is common to all mechanical waves.

A **progressive** *or* **traveling wave** *is a self-sustaining disturbance of a medium that propagates from one region to another, carrying energy and momentum.* There are a great variety of progressive mechanical waves, among which are waves on strings, surface waves on liquids, sound waves in the air, and compression waves in both solids and fluids. In all cases, although the energy-carrying disturbance advances through the medium, the individual participating atoms remain in the vicinity of their equilibrium positions: *the disturbance advances, not the material*

The Tacoma Narrows Bridge oscillating and ultimately collapsing. The stalled car belonged to a reporter who crawled off the bridge, leaving the car behind. The only fatality was a dog initially trapped in the car. Professor Farquharson, who was studying the bridge, risked his own life to free the dog, who then chose not to leave the scene (the dog is the small spot left of center).

medium. That's the crucial aspect of a wave that distinguishes it from a stream of particles. The wind blowing across a field sets up "waves of grain" that sweep by, even though each stalk only sways in place. Da Vinci seems to have been the first person to recognize that a wave does not transport the medium through which it travels, and it is precisely for this reason that waves are capable of propagating very rapidly.

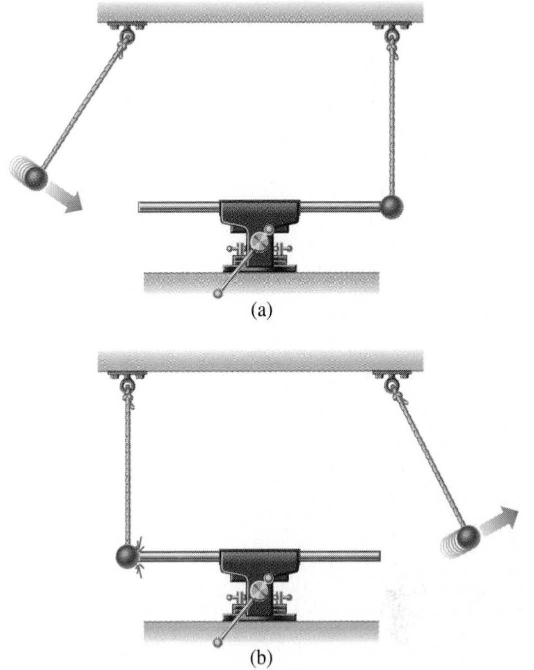

Figure 12.26 Energy is pumped into the horizontal rod by the impact of the ball on the left. A compression wave travels down the rod and slams into the ball on the right. The wave of displaced atoms transports energy and momentum from one ball to the other.

Longitudinal and Transverse Waves

There are two basic forms of waves: **longitudinal** and **transverse**. In addition to these, there are also torsion waves, which are a variation of the latter, and water waves, which are a combination of the two. Envision a long loose-coil spring, like a Slinky, held in such a way that it is straight and horizontal (Fig. 12.27). The spring itself is to be the medium that will sustain the disturbance. Now, compress several of its coils by pulling to the left and release them all at once. The region of compression, the disturbance from equilibrium, will rapidly advance to the right along the whole length of the spring. *When the sustaining medium is displaced parallel to the direction of propagation, the wave is longitudinal* (meaning "lengthwise"). The compression wave in a bell is of this variety, as are both sound waves and certain seismic waves.

If the end of the spring is displaced up and down, a hump will be produced (Fig. 12.28). A momentary disturbance will move, effectively as a pulse of energy, along the length of the spring. *When the sustaining medium is displaced perpendicu-*

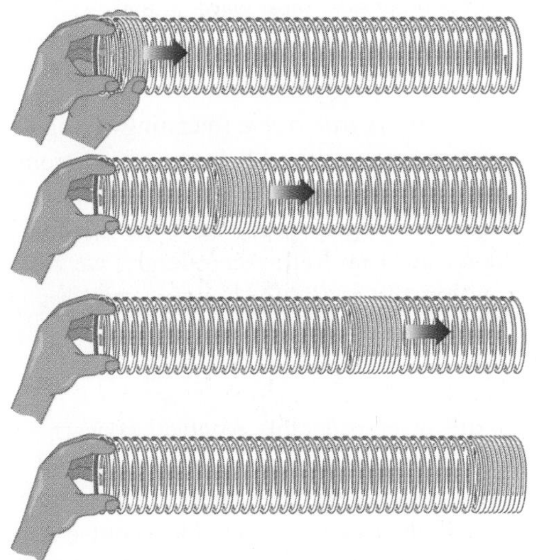

Figure 12.27 A longitudinal wave in a spring.

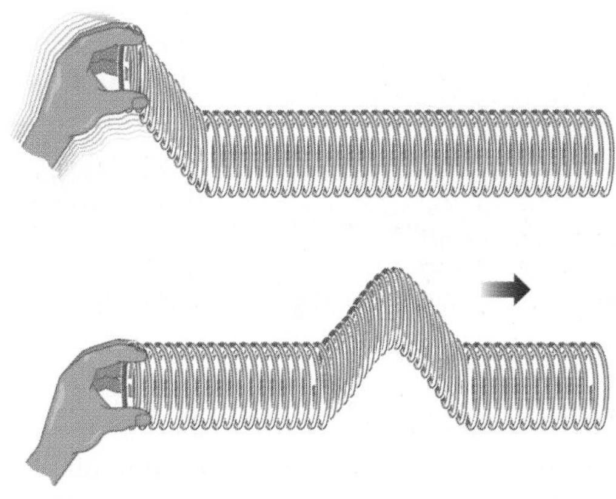

Figure 12.28 A transverse wave in a spring.

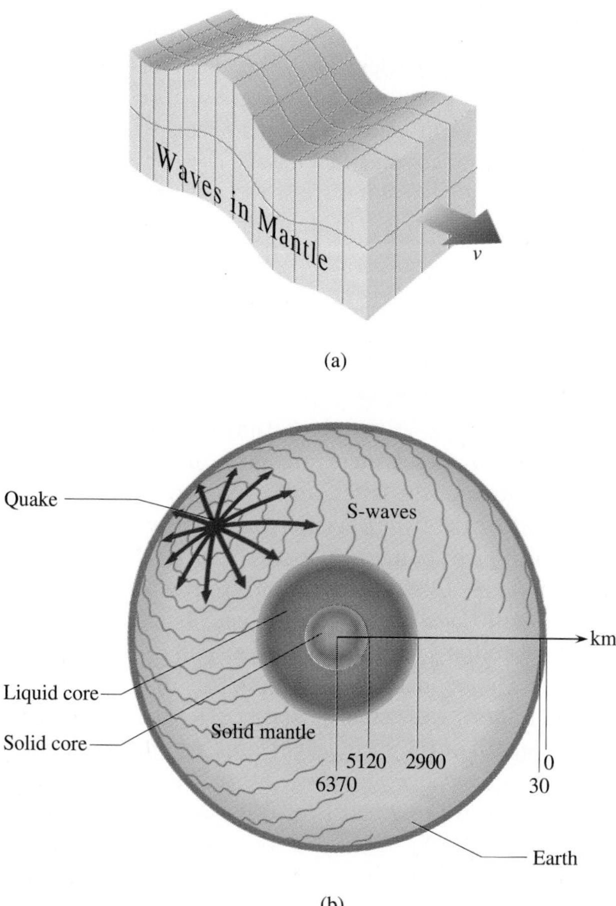

Figure 12.29 (a) Transverse seismic waves like this one are called shear waves, or S-waves. (b) Because of the Earth's molten core, which cannot support transverse waves, the S-waves from a quake on one side of the planet are picked up everywhere except diagonally across on the other side. Shear waves in the mantle travel at about 8 km/s.

lar to the direction of propagation, the wave is transverse (meaning "lying across"). Plucked guitar strings oscillate as transverse waves. All forms of electromagnetic radiation from radio waves to X-rays and light are envisioned as transverse waves.

Transverse mechanical waves can be supported provided the medium resists shear, which is something fluids don't do very well. Accordingly, we can expect these waves will not be sustained within the body of a liquid or gas (Fig. 12.29).

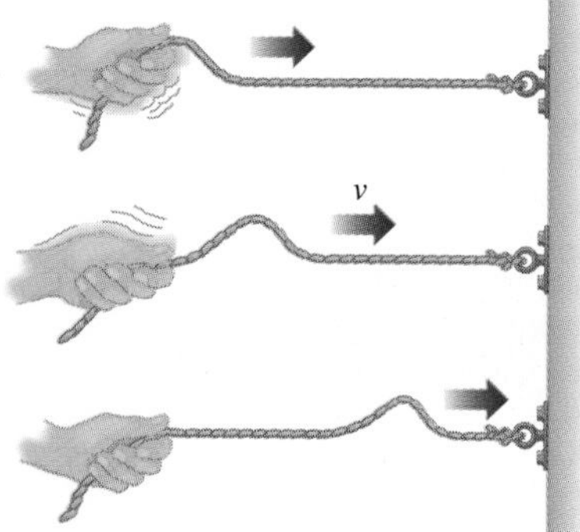

Figure 12.30 A wavepulse riding along a taut rope.

Waveforms

Waves in strings, wires, and ropes are of considerable practical concern to piano tuners and people who worry about vibrating powerlines, but our study of them is more general. For us, they provide an especially simple visual model that can be applied to all forms of waves. Figure 12.30 depicts a transverse **wavepulse** set up in a taut rope driven by a hand that smoothly moves once up and back down again. The outline, *profile* or shape of the pulse, is determined by the motion of the driver (the hand), while the physical aspects of the medium (the tension and inertia of the rope) determine the speed of the wave. A single wavepulse has its equivalent in many familiar effects, such as the sound of a shotgun blast or an ocean tidal wave.

If the hand now oscillates up and down in a regular way, it can generate a disturbance that alternately appears above and below the equilibrium zero-line like the **wavetrain** depicted in Fig. 12.31. A disturbance of this sort can be viewed as a steady sinusoidal oscillation called the **carrier**, which gradually varies in amplitude. A wavetrain, whatever the details of its shape, has a beginning and an end. A burst on a whistle will produce an acoustical wavetrain.

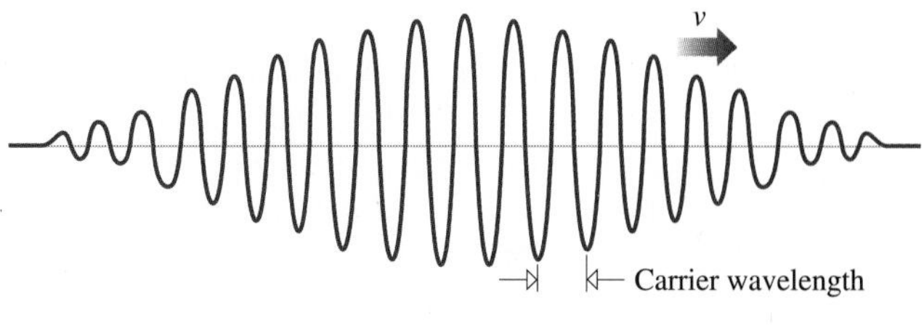

Figure 12.31 This wavetrain one can be thought of as a constant-frequency carrier whose amplitude is modulated—that is, made to vary in time.

It is possible to sustain a vibration for a long time and thus generate a continuing wave. Figure 12.32 shows the waveform produced via a microphone by a sustained note played on a saxophone. It could equally well have been created on a taut rope through some fancy jiggling. Overlooking the very slight variations, the wave can be visualized as made up of a single repeated profile-element (Fig. 12.32b). However long the overall wavetrain is, it is finite—there was a time before which the note was blown and there will be a time when it ends. That's the nature of real waves as compared with mathematical ones. In many situations, the disturbance may be extremely long with an enormous number of repetitions of the same profile-element. It then becomes much simpler mathematically to assume that the wave is infinitely long. Such an idealized disturbance composed of endless repetitions of the same profile-element is a **periodic wave**.

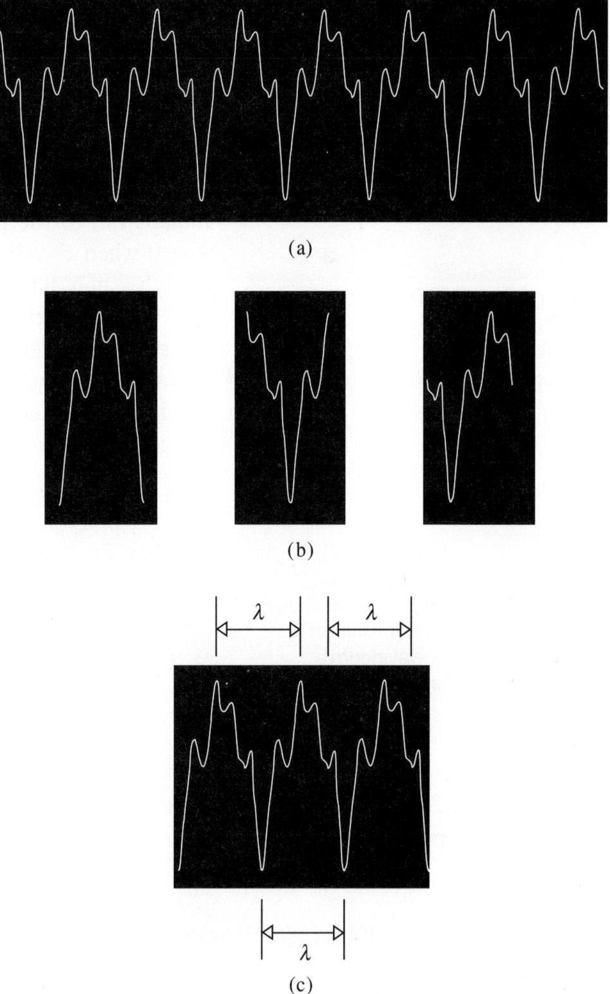

Figure 12.32 (a) The waveform generated by a saxophone. We can imagine any number of profile-elements (b) that, when repeated, create the waveform (c). The length over which the wave repeats itself is called the wavelength, λ.

To Find the Velocity of Waves

The **temporal period** (T) of a periodic progressive wave is the time it takes for one profile-element to pass a given point in space. The inverse of that ($1/T$) is the **frequency** f, the number of profile-elements passing per second. The distance in space over which the wave executes one cycle of its basic repeated form (one profile-element) is the **spatial period** or **wavelength**, λ (Greek lowercase lambda).

Imagine that you are at rest and a periodic wave on a string is progressing past you. The number of profile-elements that sweep by per second is f, and the length of each is λ. In 1 s, the overall length of the disturbance that passes you is the product $f\lambda$. If, for example, each is 2 m long and they come at a rate of 5 per second, then in 1 s, 10 m of wave flies by. This is just what we mean by the speed of the wave (v)—the rate in m/s at which it advances. Said slightly differently, since a length of wave λ passes by in a time T, its speed must equal $\lambda/T = f\lambda$. We now have an expression for the speed of any progressive *periodic* wave, be it sound, water ripple, or light:

$$v = f\lambda \tag{12.18}$$

Incidentally, Newton derived this relationship in the *Principia* (1687) in a section called "To find the velocity of waves."

Example 12.8 A youngster in a boat watches waves on a lake that seem to be an endless succession of identical crests passing with a half-second pause between each. If one takes 1.5 s to sweep straight down the length of her 4.5-m-long boat, what is the frequency, period, and wavelength of the waves?

Solution: [Given: The waves are periodic; there is $\frac{1}{2}$ s between crests; $t = 1.5$ s, $L = 4.5$ m. Find: f, T, and λ.] The time between crests is the period, so $\boxed{T = \frac{1}{2}\text{ s}}$; hence, $f = 1/T = \boxed{2.0\text{ Hz}}$. As for the speed

$$v = \frac{L}{t} = \frac{4.5\text{ m}}{1.5\text{ m}} = 3.0\text{ m/s}$$

We now know T, f, and v and must determine λ. Thus, from Eq. (12.18)

$$\lambda = \frac{v}{f} = \frac{3.0\text{ m/s}}{2.0\text{ Hz}} = \boxed{1.5\text{ m}}$$

▶ **Quick Check:** $Tv = \lambda = (\frac{1}{2}\text{ s})(3.0\text{ m/s}) = 1.5$ m. $v = f\lambda = (2.0\text{ Hz})(1.5\text{ m}) = 3.0$ m/s.

Harmonic Waves

The easiest of repeating disturbances to treat theoretically is the idealized **harmonic wave**, which rises and falls sinusoidally without end. This is the fundamental waveform because, as we will see later, all real waves can be synthesized from overlapping harmonic waves.

Remembering the treatment of SHM, examine Fig. 12.33, which is a plot of

$$y = A \sin \frac{2\pi x}{\lambda} \tag{12.19}$$

The argument of the sine function $(2\pi x/\lambda)$ is the **phase** and it's unitless, as it must always be. Equation (12.19) describes the profile of a harmonic wave frozen at $t = 0$. At any location x, the ratio x/λ is the number of wavelengths from the origin out to that point and, since there are 2π rad per wavelength, $2\pi x/\lambda$ is the number of radians out to the point at x. The profile repeats itself with a wavelenth λ so that $y = 0$ when $x = 0, \lambda, 2\lambda, 3\lambda, \ldots$. The *amplitude* is A and it is always positive even though y may be negative. The wave builds up and falls off between values of $+A$ and $-A$. **The energy associated with a wave is proportional to the amplitude of the wave squared**, and that's true for all waves. The amplitude of a sound wave determines how loud it is, and the amplitude of a light wave determines how bright it is. Figure 12.34 shows a harmonic wave advancing 1 wavelength during a time interval of 1 period (T).

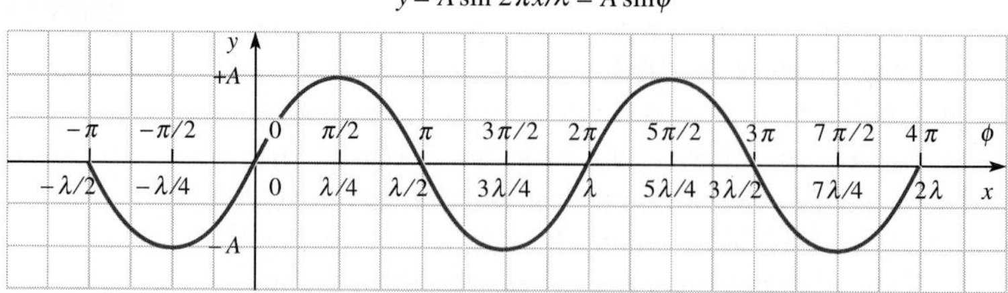

Figure 12.33 A harmonic function, which is to serve as the profile of a harmonic wave. One wavelength corresponds to a change in phase ϕ of 2π rad.

Example 12.9 The profile of a harmonic wave, traveling at 1.2 m/s on a string, is given by

$$y = (0.02\text{ m}) \sin (157\text{ m}^{-1})x$$

Determine its amplitude, wavelength, frequency, and period.

Solution: [Given: the profile, and $v = 1.2$ m/s. Find: A, λ, f, and T.] Comparing y with Eq. (12.19) tells us that $\boxed{A = 0.02\text{ m}}$. Moreover,

$$\frac{2\pi}{\lambda} = 157\text{ m}^{-1}$$

(continued)

(continued)

and so $\lambda = 2\pi/(157\ \text{m}^{-1}) = \boxed{0.0400\ \text{m}}$. The relationship between frequency and wavelength is fixed by Eq. (12.18), $v = f\lambda$, and so

$$f = \frac{v}{\lambda} = \frac{1.2\ \text{m/s}}{0.0400\ \text{m}} = \boxed{30\ \text{Hz}}$$

The period is the inverse of the frequency, and therefore $T = 1/f = \boxed{0.033\ \text{s}}$.

▶ **Quick Check:** $f/v = 1/\lambda$; hence the phase $(2\pi/\lambda)x = (2\pi f/v)x = [2\pi(30\ \text{Hz})/(1.2\ \text{m/s})]x = (157\ \text{m}^{-1})x$.

Imagine a point on the wave in Fig. 12.34 with a value of y of, say, $A/2$. Since the amplitude is constant, the point we are watching must have a particular fixed value of the phase; in this instance, the phase is $\pi/6 = 30°$. As time goes on, the wave progresses, but for the point we are watching, $y = A/2$ must always correspond to a phase of $\pi/6$. Thus, when we talk about the speed of a harmonic wave, we are referring to the rate at which a point with constant phase moves, and that's called the **phase velocity**, or phase speed.

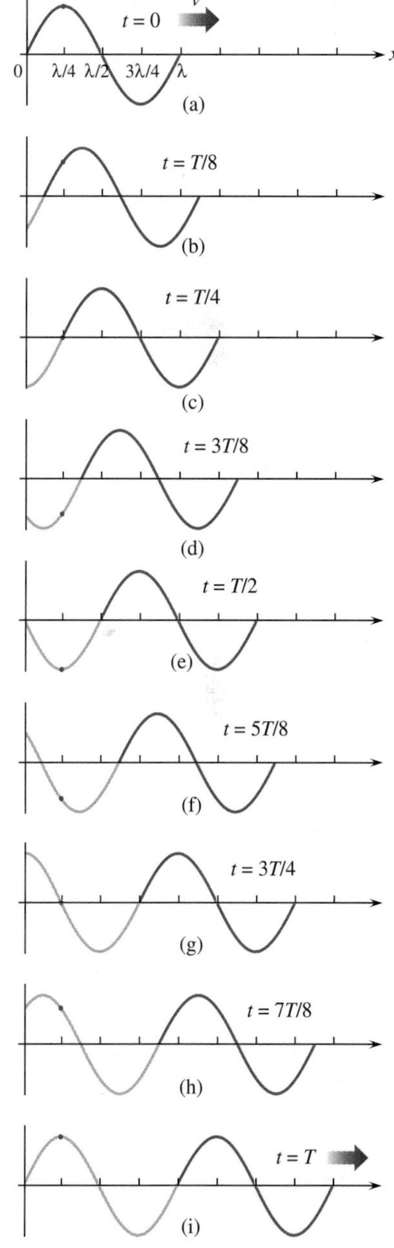

Figure 12.34 A harmonic wave moving along the x-axis during a time of one period. Note that any one point on the rope only moves vertically.

12.7 Transverse Waves: Strings

When a mechanical disturbance is created, the speed of the wave is determined by the inertial and elastic properties of the medium and not in any way by the relative motion of the source. The central concerns are how much mass is being accelerated and with how much force does the medium resist deformation. This insight can be shown rigorously for waves on strings using the calculus, but for our purposes a slightly limited though much simpler analysis (formulated by P. G. Tait in the 1800s), will do nicely.

Consider a smooth pulse traveling with a speed v to the right along a lightweight, flexible string (Fig. 12.35). The string is kept taut by the application of a constant external force that maintains a tension F_T in the system. Imagine that the wave is viewed from a reference frame moving with the pulse so it is seen to be at rest. Although the disturbance is motionless in this frame, the string is observed to move to the left at a constant speed v. The situation appears as if the string were being pulled to the left through a stationary invisible tube bent in the shape of the pulse. A tiny segment of the string that has a radius of curvature r, a length Δl, and mass Δm is essentially revolving around point C (the center of curvature of that segment) with a speed v. Since it is moving at a constant speed, the net force acting on the segment must be radially inward toward C and equal to the required centripetal force $F_C = \Delta m(v^2/r)$. From the diagram, it follows that each end of the segment experiences a force equal to F_T acting tangent to the rope. The sum of these two tensile force vectors (Fig. 12.35b) is a single radial vector of magnitude $2F_T \sin(\frac{1}{2}\Delta\theta)$: the tangential components cancel since there is no tangential acceleration.

Setting that sum equal to the centripetal force yields

$$\Delta m\left(\frac{v^2}{r}\right) = 2F_T \sin(\tfrac{1}{2}\Delta\theta)$$

but the segment is tiny compared to r; hence, $\sin(\frac{1}{2}\Delta\theta) \approx \frac{1}{2}\Delta\theta$ and

$$\Delta m\left(\frac{v^2}{r}\right) \approx F_T\,\Delta\theta \approx \frac{F_T\,\Delta l}{r}$$

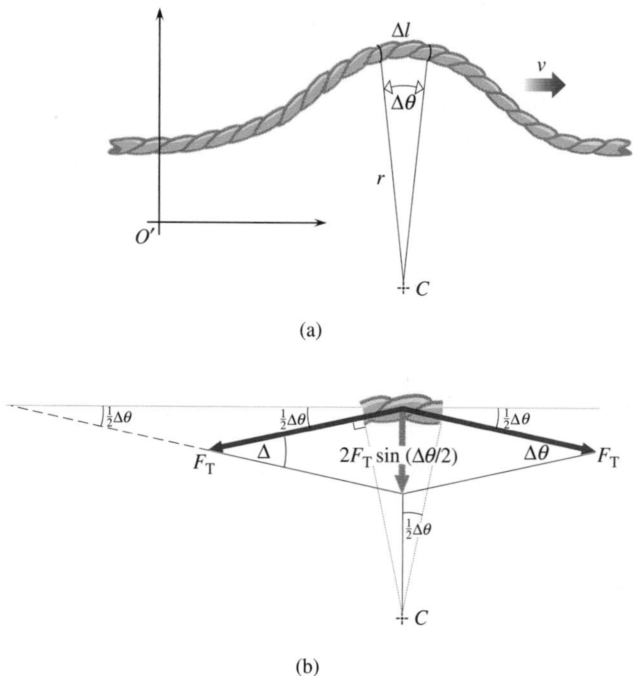

Figure 12.35 (a) A pulse on a stretched rope traveling at a speed *v*. (b) The tension in the rope, F_T, determines *v*.

where $\Delta\theta = \Delta l/r$. If we define the *linear mass-density* μ to be the mass per unit length of the string, $\mu = \Delta m/\Delta l$, and

$$v^2 \approx \frac{F_T}{\mu}$$

Provided that the pulse is not very steep, this relationship holds for all segments and

$$v = \sqrt{\frac{F_T}{\mu}} \tag{12.20}$$

The speed of the wave only depends on the force exerted and the inertial characteristics of the medium. When μ is large, there is a lot of inertia and the speed is low. When F_T is large, the string tends to spring back rapidly, and the speed is high.

Example 12.10 A 2.0-m-long horizontal string having a mass of 40 g is slung over a light frictionless pulley, and its end is attached to a hanging 2.0-kg mass. Compute the speed of the wavepulse on the string. Ignore the weight of the overhanging length of rope.

Solution: [Given: a string of $l = 2.0$ m, $m = 40$ g supporting a 2.0-kg load. Find: v.] The speed of the wave is gotten from Eq. (12.20) and so we need F_T and μ. The tension is just the load in newtons; namely, (2.0 kg) × (9.81 m/s^2) = 19.62 N. While the linear mass-density is the mass of the string divided by its total length, (0.040 kg)/(2.0 m) = 0.020 kg/m. Hence

$$v = \sqrt{\frac{F_T}{\mu}} = \sqrt{\frac{19.62\ \text{N}}{0.020\ \text{kg/m}}} = \boxed{31\ \text{m/s}}$$

which is equivalent to 70 mi/h.

▶ **Quick Check:** If the mass was 1 kg, $F_T = 9.8$ N and $v = 22$ m/s, and this quantity times 1.414 is 31 m/s.

Reflection, Absorption, and Transmission

Every variety of wave from sound to light can undergo reflection, absorption, and transmission on interacting with material media. Interesting things happen to waves at discontinuities in the media in which they travel. For the moment, we focus on waves on ropes, but the conclusions are widely applicable to other types.

In Fig. 12.36, one end of the rope is held stationary while energy is pumped in at the other end. The **reflected** wave ideally carries away all the original energy. Because it is inverted, it's said to be 180° out-of-phase with the incident wave. Alternatively, if the far end of the rope is free (Fig. 12.37), it will rise up as the pulse arrives until all the energy is stored elastically. The rope then snaps back down, producing a reflected wavepulse that is right side up. A similar effect occurs when an ocean wave slams into a breaker wall and climbs high above the crests behind it.

If we draw off energy from the oscillating end point of a rope via friction, the reflected pulse has a proportionately diminished amplitude and we call the process **absorption**. It occurs most effectively when the rope encounters friction that is speed-dependent. To understand why this is the case, realize that any small segment of the rope only moves vertically. Moreover, the vertical force exerted on any such segment due to the rope just before it is proportional to that segment's vertical speed. When a wave travels from left to right, at any point the segment on the left does work on the segment immediately to the right of it, and energy flows along with the disturbance.

Consider what would happen were we to mount a little dashpot damper to the end of the rope (Fig. 12.38). A dashpot, you recall, is a fluid-filled cylinder sealed with a movable piston. The dashpot opposes motion, via viscous damping, with a force that varies as the speed of the point to which it is fixed (that is, the speed of the piston). If the dashpot exerts just the right amount of drag on the rope, the segment immediately to the left of it experiences the same motion as if the rope continued beyond it indefinitely—the wave's energy flows via friction into the dashpot. In this ideal case, the wave will be totally absorbed. A similar absorption process happens with other waves, and that makes it of considerable interest.

When a wave passes from one medium to another having different physical characteristics, there will be a redistribution of energy. Figure 12.39 shows a wavepulse initially traveling in a low-density rope impinging on the interface with a high-density rope. Like the fixed point of Fig. 12.36, the large inertia of the second medium at the junction retards the easy motion of the boundary. There is again an oppositely directed reaction force and the reflected wave is phase-shifted by 180°. But the second medium is also displaced, and a portion of the incident energy will appear as a **transmitted** wave. The time it takes to generate the reflected and transmitted waves must be identical since they share a common source. Yet the speeds of the pulses must be different because they have the same tension and different densities. It follows that the pulses have different lengths.

When the first medium is denser than the second, the situation resembles that of Fig. 12.37 with a free end point: no phase shift results and the transmitted wave has a greater length. The process of reflection and transmission at an interface between media occurs for all waves regardless of whether they are longitudinal or transverse, though the details, such as phase shifts, differ.

Evidently, if the incident wave is periodic, the transmitted wave has the same frequency but a different speed and therefore a different wavelength: *the larger the density of the transmitting medium, the smaller the length of the wave.* The fact that the frequencies of the incident, reflected, and transmitted waves are normally the same is true for every sort of disturbance from sound to light.

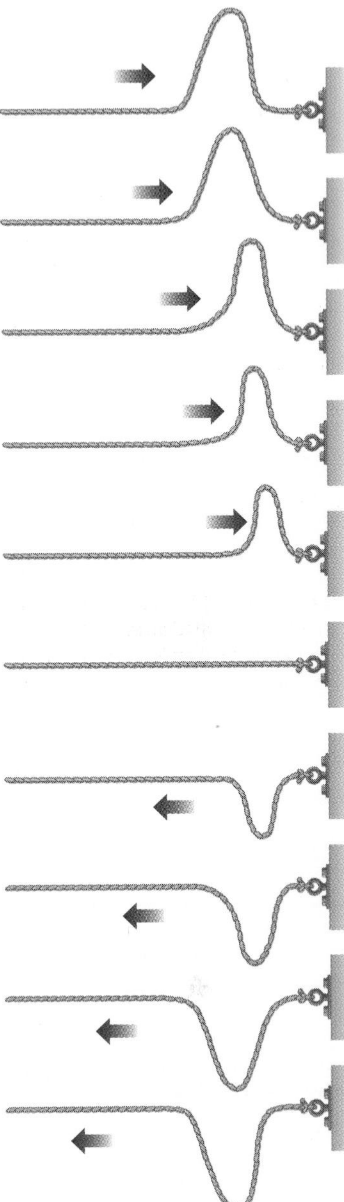

Figure 12.36 The reflection of a pulse on a rope with a fixed end point. As the pulse arrives, it exerts a vertical force on the fixed anchor point, which in turn exerts an equal and opposite force on the string. When the string tugs up, the anchor point tugs down. This downward force on the rope generates an upside-down reflected pulse traveling in the opposite direction.

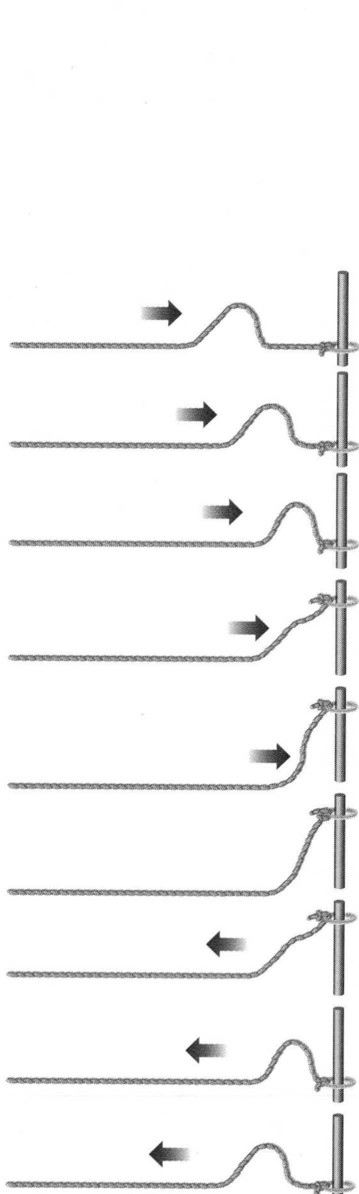

Figure 12.37 The reflection of a pulse on a rope with a free end point. That free end rises until all the energy of the end segment is stored elastically. It comes to rest at a maximum vertical displacement of twice the height of the crest. Carried up by its inertia, the end segment pulls upward on the rope, generating a reflected wavepulse that travels back toward the source, right side up and simply reversed.

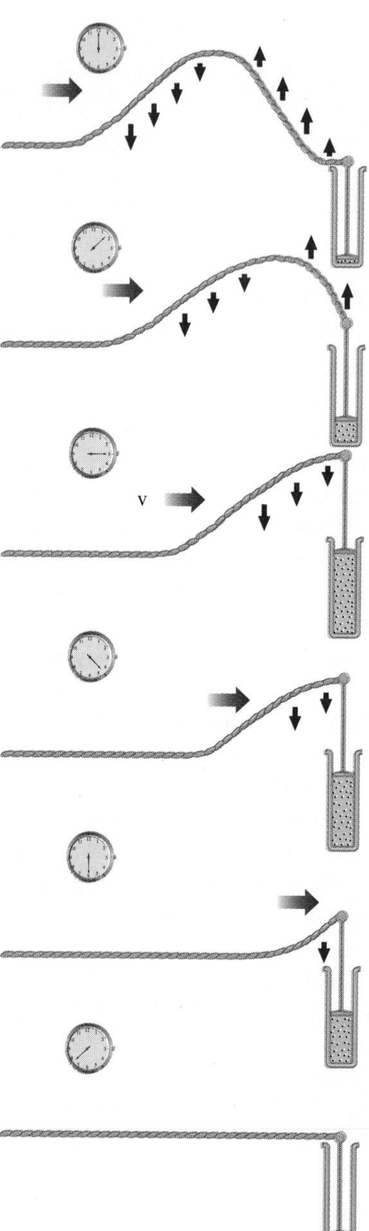

Figure 12.38 A wave in a string terminated on a dashpot. The vertical speed of the rope is proportional to the slope of the wave. The drag exerted by the dashpot is proportional to the vertical speed of the piston.

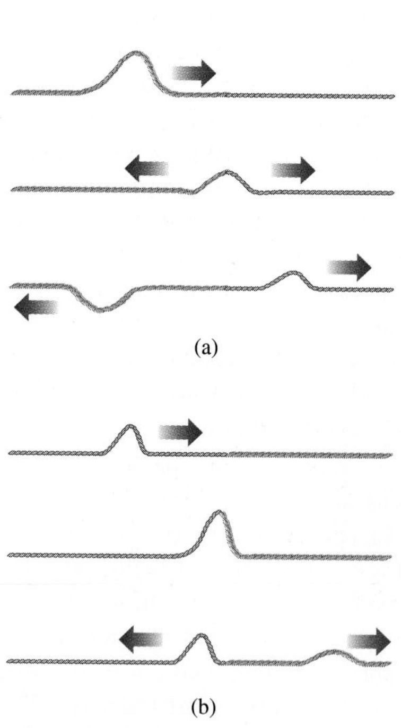

Figure 12.39 Reflection and transmission of a pulse at the junction of two media. The darker rope has a greater linear mass-density.

Example 12.11 An incident periodic disturbance of wavelength λ_i traveling at a speed v_i in one medium impinges on an interface with another medium of different density. Derive an expression relating the speeds and wavelengths of the incident and transmitted waves.

Solution: The speeds and wavelengths within the incident and transmitting media are related via $v = \lambda f$; hence

$$v_i = \lambda_i f_i \quad \text{and} \quad v_t = \lambda_t f_t$$

But we know as a rule that $f_i = f_t$ and so dividing the first equation by the second yields the desired relationship

$$\boxed{\frac{\lambda_i}{\lambda_t} = \frac{v_i}{v_t}}$$

▶ **Quick Check:** Both sides are unitless, which is fine. Recall Eq. (12.20), which tells us that when μ increases there's more inertia, and v decreases. Thus, if the speed of the wave decreases in the transmitting medium, the wavelength decreases and that makes sense.

12.8 Compression Waves

Atoms of an elastic substance can be shifted by an applied force, and the resulting state of being compressed will travel through the medium as a *compression wave.* In solids, such a disturbance is called a *longitudinal elastic wave.* In fluids, which cannot support transverse disturbances, it's called an *acoustic wave.* An earthquake or a large truck rumbling down the street can shake a building, rattling everything in it with streams of compression waves. When the frequency lies in the range from about 20 Hz to 20 kHz, compression waves are called *sound.* There is nothing special about sound waves beyond the fact that we happen to be able to *hear* them.

Let's model a succession of atoms or molecules in an elastic medium (either solid or fluid of sufficient density) using a row of spheres separated by springs (Fig. 12.40). The particle at the far left is made to vibrate in SHM along the line by some source of energy. Wherever the springs are elongated, the density of particles is diminished, and that is a *rarefaction.* Wherever the springs are compressed, the density is increased, and that is a *condensation.* **The propagation of a compression wave takes place in the direction along which the particles of the medium oscillate, and it is marked by a series of alternate condensations and rarefactions**. There are tremendous numbers of atoms in an ordinary medium, and they behave as though the distribution were continuous; rarefactions gradually blend into condensations. Figure 12.41 shows that at any instant the particles within the condensations move in the propagation direction, while those in the rarefactions move in the opposite direction.

The Speed of Compression Waves

We now derive an expression for the speed of a compression wave. It can be anticipated that any such formula involves both the restoring force generated by the sustaining medium, via some measure of its elastic properties, and its density.

Consider a compression pulse of uniform high pressure traveling to the right through a liquid at rest in a tube (Fig. 12.42, p. 431). The liquid is at a pressure P, and the pulse pressure exceeds that by ΔP. When viewed from a coordinate system moving to the right with the pulse at a speed v, the normally stationary fluid is seen to be moving left at v while the pulse is motionless. In a time Δt, a small cylinder of fluid of length $v\Delta t$ enters the high-pressure region of the pulse. The forces acting on this cylinder, the pressure changes it undergoes, and the alterations in its volume will allow us to determine the change in its speed as it becomes part of the pulse.

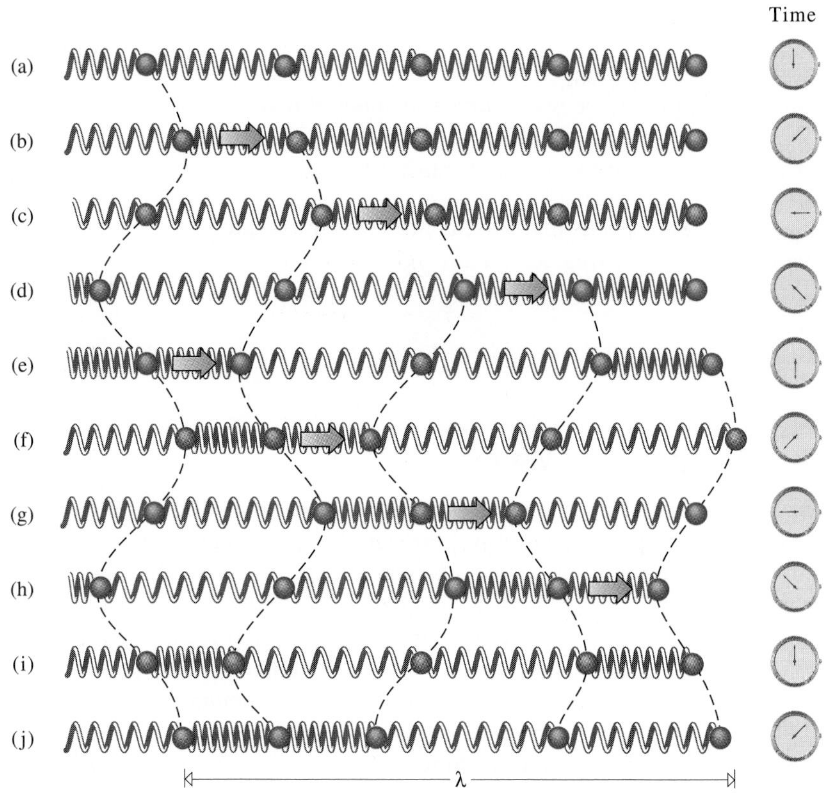

Figure 12.40 A sequence of views of a line of masses attached by springs. In (a) the system is undisturbed. The first mass on the left oscillates in SHM. The resulting disturbance propagates to the right [(b) through (j)] as a series of compressions and elongations that repeats after a distance λ. Each mass oscillates sinusoidally in SHM.

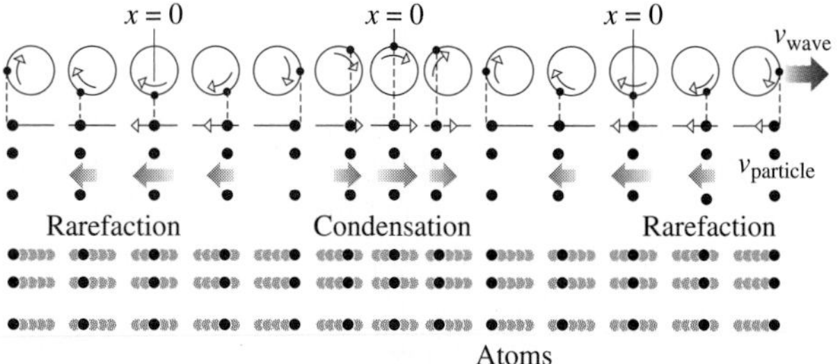

Figure 12.41 A compression wave passing to the right through a distribution of elastically interacting atoms. The peaks of the condensations and rarefactions occur at zero atomic displacements ($x = 0$). Note that the atoms in the condensations move in the direction of propagation of the wave. Start at the right and examine each reference circle in turn. Remember that each atom oscillates around its equilibrium position ($x = 0$) in SHM.

The volume of this cylinder with its cross-sectional area A is $Av\Delta t$ and its mass is therefore $\rho Av\Delta t$, where ρ is the density of the liquid. The pressure difference it experiences between its two ends is ΔP, and so it has acting on it a net force $A\,\Delta P$ to the right. By Newton's Second Law, this force produces an acceleration of $-\Delta v/\Delta t$ (the minus sign is here because a is opposite to v). In other words, $F = ma$ and

$$A\,\Delta P = (\rho Av\Delta t)\left(-\frac{\Delta v}{\Delta t}\right) \tag{12.21}$$

The elastic properties of the medium are embodied in its ability to resist compression under pressure. Recall that the Bulk Modulus B (p. 343) is defined as

$$B = -\frac{\Delta P}{\Delta V/V_0}$$

When the liquid cylinder enters the high-pressure region, it experiences a decrease in speed Δv; hence, while it originally had a volume $V_0 = Av\Delta t$, it decreases in volume by $\Delta V = A\,\Delta v\Delta t$ and

$$\frac{\Delta V}{V_0} = \frac{A\,\Delta v\Delta t}{Av\Delta t} = \frac{\Delta v}{v}$$

Equation (12.21) becomes $-\Delta P = \rho v\Delta v$, and dividing both sides by v^2 we get

$$v^2 = \frac{-\Delta P}{\rho\,\Delta v/v} = \frac{-\Delta P}{\rho\,\Delta V/V_0}$$

For compression waves in liquids

$$v = \sqrt{\frac{B}{\rho}} \tag{12.22}$$

It takes a large increase in pressure to decrease the volume of a solid as compared to a liquid, and B is generally larger for dense media—it increases as the medium becomes increasingly rigid (Table 10.7). Thus, v usually increases as ρ increases, which is not obvious from Eq. (12.22).

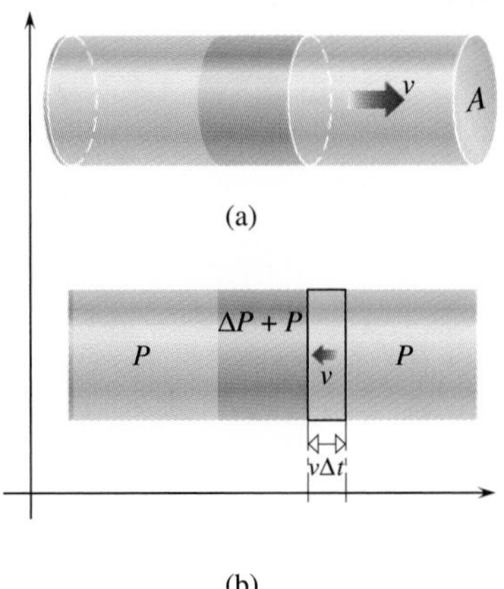

Figure 12.42 (a) A pulse of compression-wave energy traveling with a speed v through a medium with an ambient pressure P. (b) The pulse as seen in a system moving along with it at speed v. The surrounding medium enters the pulse where its pressure is changed to $P + \Delta P$.

Example 12.12 An explosion occurs not far beneath the surface of the Atlantic Ocean. Compute the speed of the resulting compression wave measured by instruments several meters below a ship.

Solution: The problem asks for the speed of a compression wave in seawater. The source is not very deep, hence the density can be taken from Table 10.1. [Given: $\rho = 1.03 \times 10^3$ kg/m^3 and from Table 10.7 $B = 2.1$ GPa. Find: v.] Substituting into Eq. (12.22), we obtain

$$v = \sqrt{\frac{B}{\rho}} = \sqrt{\frac{2.1 \times 10^9 \text{ Pa}}{1.03 \times 10^3 \text{ kg/m}^3}} = \boxed{1.4 \times 10^3 \text{ m/s}}$$

▶ **Quick Check:** $B = v^2\rho \approx (1400 \text{ m/s})^2 \times (1000 \text{ kg/m}^3) \approx 2 \times 10^9 \text{ kg/m·s}^2$.

The density and therefore the speed of compression waves in the body of the sea depends on the temperature, salinity, and depth. Which is why this example was careful not to have the waves propagating very deep. Thus, while the speed is 1.4 km/s at the surface, at a depth of 5 km it's about 1.5 km/s. Moreover, a rise in temperature of 1°C increases the speed by 3.7 m/s, while an increase of 1% in salinity adds about 1.2 m/s to it—both of which would be of great interest to a dolphin or a sonar operator in a submarine. In the rock mantle of the Earth, compression waves reach speeds of 13 km/s (Fig. 12.43).

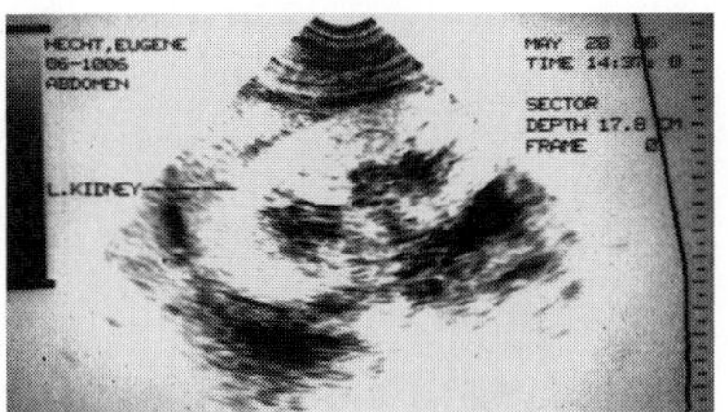

An ordinary fairly low resolution sonogram of a left kidney (the white C-shaped form is facing down toward 5 o'clock). The speed of compression waves in the body averages $\approx 1.5 \times 10^3$ m/s, and so frequencies of 1 MHz to 10 MHz have wavelengths of 1.5 mm to 0.15 mm and can reveal fine details.

Ultrasound and Infrasound

Dolphins operate their underwater tracking system by radiating compression wavetrains, high-frequency chirps with carrier wavelengths (Fig. 12.31) of about 1.4 cm. This echo ranging, which is quite similar to sonar, determines the round-trip time it takes for a pulse once emitted to be reflected back to the dolphin. *Waves reflect effectively off objects that are at least as large as about one wavelength;* here, the carrier wavelength is small enough to still "see" rather little fish. To produce emissions of that wavelength (since $v = \lambda f = 1.4 \times 10^3$ m/s), the frequency has to be up around 10^5 Hz. These waves are well beyond the human audible range (which is less

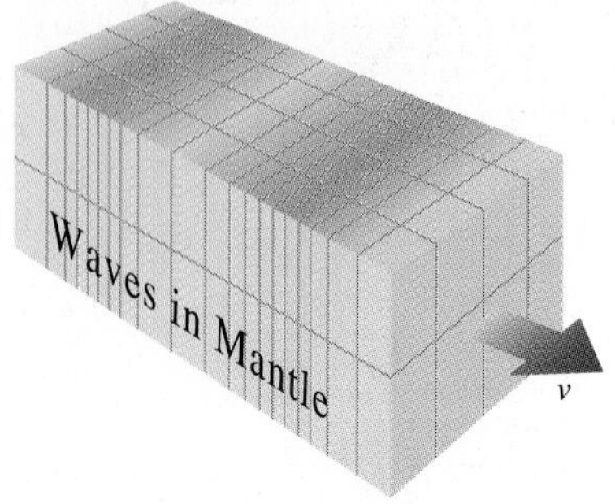

Figure 12.43 Earthquake compression waves are called primary waves, or P-waves, because they get to seismic stations before the S-waves. They predominate in the disturbances set up by underground nuclear explosions and are used to distinguish between natural quakes and weapons tests.

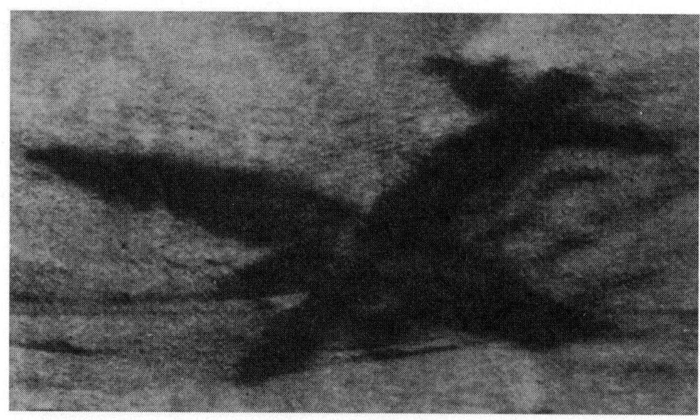

A Wellington airplane under 210 m of water at the bottom of Loch Ness, revealed by side-scanning sonar.

than 20 kHz) and are therefore called **ultrasonic**. Bats play much the same game in air emitting at frequencies of from 16 kHz to 150 kHz. The lower speed of the waves in air and the higher frequencies yield shorter wavelengths; bats hunt small insects and have to dodge narrow branches.

Example 12.13 While heading for a moth, a horseshoe bat emits a stream of ultrasonic wavetrains having a carrier frequency of 83 kHz, each lasting about 5.0 ms. Given that the speed of the emission is 340 m/s, determine the length in space of a single wavetrain. How many wavelengths of the carrier are there in one such train?

Solution: [Given: the pulse time $t = 5.0$ ms, the carrier frequency $f = 83$ kHz, and $v = 340$ m/s. Find: the pulselength L and the number of waves per pulse.] The pulselength is given by

$$L = vt = (340 \text{ m/s})(5.0 \times 10^{-3} \text{ s}) = \boxed{1.7 \text{ m}}$$

To find the number of carrier waves in such a train, we need first to compute the wavelength of the carrier. Since $v = f\lambda$

$$\lambda = \frac{v}{f} = \frac{340 \text{ m/s}}{83 \times 10^3 \text{ Hz}} = 4.1 \text{ mm}$$

Hence, the number of waves per pulse is

$$\frac{L}{\lambda} = \frac{1.7 \text{ m}}{4.1 \times 10^{-3} \text{ m}} = \boxed{4.2 \times 10^2}$$

▶ **Quick Check:** 4.1 mm is the size of a small bug, so λ is reasonable.

$$v = f\lambda \approx (83\,000 \text{ Hz})(0.004 \text{ m}) \approx 330 \text{ m/s}$$
$$t = 4.2 \times 10^2\lambda/v = 5 \text{ ms}$$

Some autofocus cameras imitate bats, sending out ultrasonic pulses and using the round-trip time to determine target range. Ultrasonic techniques are routinely used in medicine in both diagnosis and treatment. Intense concentrations of energy can be directed into the body to destroy tumors and stones. In yet another application, low-energy ultrasonic waves are used to probe the body in a way that is widely believed to be far less hazardous than X-rays.

Infrasound corresponds to compression waves of subaudible frequencies (<20 Hz). A good deal of this energy is present in the environment, especially in the vicinity of airplanes, thunderstorms, fast-moving cars, and overamplified rock bands. Really loud music booming in a relatively small bistro can impart enough compression-wave energy to set the furniture vibrating and create wave patterns on the surface of every glass of beer in the place. It can even oscillate your internal organs, which have low-frequency resonances—that's the "pleasant" buzz you may have felt in the presence of one of these bands. Human chest walls go into resonance at around 8 Hz. If the energy levels are extraordinarily high (especially around 7 Hz), dizziness, nausea, and internal bleeding can occur. It was recently discovered that elephants communicate over long distances with rumbling infrasonic calls. Similarly, the subaudible noise from the cooling systems of old Soviet nuclear submarines is transmitted so well by seawater that U.S. surveillance systems can pinpoint their locations from thousands of miles away.

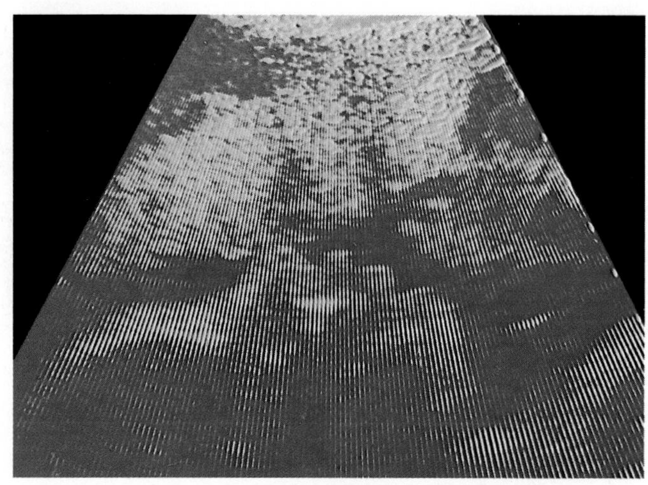

A high-resolution sonogram of the head of a fetus (facing upward) in the womb. If you can't see the baby's profile, turn the book 90° clockwise.

Core Material

The time it takes for a system to complete a cycle is its **period** (T). The reciprocal of that is the **frequency** (f), represented by

$$f = \frac{1}{T} \qquad [12.1]$$

The SI unit of frequency is the *hertz* (Hz). The **angular frequency** is given by

$$\omega = 2\pi f = \frac{2\pi}{T} \qquad [12.2]$$

Simple harmonic motion (SHM) corresponds to the situation where the periodicity has a single frequency and the motion is simply sinusoidal. In that case, the displacement is

$$x = A\cos\theta = x_0\cos\omega t = x_0\cos 2\pi ft \qquad [12.3]$$

The speed of a harmonic oscillator is

$$v_x = -A\omega\sin\omega t = -2\pi f x_0 \sin 2\pi ft \qquad [12.7]$$

Furthermore

$$v_x = - \mp\omega\sqrt{A^2 - x^2} = \mp v_0\sqrt{1 - \frac{x^2}{A^2}} \qquad [12.8]$$

which is the speed of the oscillator at any position x in terms of its maximum speed. Its acceleration is

$$a_x = -A\omega^2\cos\omega t = -x_0(2\pi f)^2\cos 2\pi ft \qquad [12.9]$$

$$a_x = -\omega^2 x \qquad [12.10]$$

The *natural frequency* of a spring carrying a load (m) is

$$f_0 = \frac{1}{2\pi}\sqrt{\frac{k}{m}} \qquad [12.12]$$

and the period is

$$T = 2\pi\sqrt{\frac{m}{k}} \qquad [12.13]$$

while the natural frequency of a *simple pendulum* is

$$f_0 \approx \frac{1}{2\pi}\sqrt{\frac{g}{L}} \qquad [12.16]$$

A progressive wave is a self-sustaining disturbance of a medium. A periodic wave propagates with a speed

$$v = f\lambda \qquad [12.18]$$

The profile of a **harmonic wave** is given by

$$y = A\sin\frac{2\pi}{\lambda}x \qquad [12.19]$$

The speed of a *transverse wave* on a stretched string is

$$v = \sqrt{\frac{F_T}{\mu}} \qquad [12.20]$$

For *compression waves* in fluids

$$v = \sqrt{\frac{B}{\rho}} \qquad [12.22]$$

Suggestions on Problem Solving

1. A system undergoing oscillation is not moving with a constant acceleration and the kinematic equations in Chapter 3 *do not apply!*

2. The basic notions to keep in mind are the definition of the spring constant ($F = ks$), the reference circle, and the fact that $\omega_0 = \sqrt{k/m}$ for a spring, and $\omega_0 = \sqrt{g/L}$ for a pendulum; most everything else can be quickly derived from these ideas, as required. If you remember an equation for ω, there's no need to also store one for $f = \omega/2\pi$ or $T = 1/f$. In general we know that $v = R\omega$, so the maximum speed, $v_0 = v_{max}$, corresponds to the maximum radius $R = A$; hence, $v_0 = A\omega$. Equations (12.3), (12.7), and (12.9) should be gotten directly from the reference circle picture without memorizing them. Physicists don't attempt to commit to memory every equation they encounter—it's much easier to *understand and retain the analysis* than the individual formulas derived from it.

3. Don't get upset by the minus signs in Eqs. (12.7), (12.9), or (12.10)—they only indicate the direction of the vectors. Remember that A is always a positive quantity.

4. Some spring problems (of middle-level difficulty) ask for f or T, thereby requiring that you have k, even though it's not explicitly given. The idea is to first determine k from the initial conditions ($F = ks$).

5. When writing the equation for the displacement, speed, or acceleration of an object in SHM, do not assume that $\varepsilon = 0$. Furthermore, *it may not be the case that A equals the initial displacement;* when the initial speed is nonzero, A is not the initial displacement.

6. The most basic equation in wave theory is $v = f\lambda$. It is likely to be encountered, in one way or another, in the solution of many of the problems. We have equations for the speeds of the different types of waves and, in conjunction with those, $v = f\lambda$ provides a means of determining either f or λ when the other is given.

7. Another fundamental relationship is $T = 1/f$. This is easy enough to remember, although the tendency at first is not to associate it with $v = f\lambda$, but, of course, $v = \lambda/T$.

8. The speed of a transverse wave on a stretched string depends on the *linear* mass-density (mass/length). Be especially careful when μ is given in units of g/m or g/cm, as it often is—convert to kg/m. All the other wave-speed equations containing density refer to the usual notion of mass/volume.

Discussion Questions

1. Draw a simple swinging pendulum and label the locations of the maxima and minima of (a) speed, (b) acceleration, (c) kinetic energy, and (d) potential energy.

2. Consider the behavior of a diving board. Compare the resulting natural frequencies if you first stood on the board at its free end and bounced up and down and then repeated the process at its middle. What would happen to its elastic constant if you shortened the board by chopping off a piece of it?

3. Imagine a hanging lightweight spring with an elastic constant k supporting a small mass m a distance L from the point of suspension. Suppose the support is free to pivot so the spring can swing in a plane. When displaced and released, the mass oscillates up and down as usual, but it soon begins to sway as the oscillation diminishes. In time, the spring oscillations vanish, and the system swings like a pendulum until that motion dies out as the oscillations again build. Explain what's happening.

4. Figure Q4 shows a spring with a mass attached to it resting on a belt. The belt is then set into constant motion counterclockwise. Assuming ordinary dry friction, describe the circumstances under which the mass will go into unstable oscillation, cycling back and forth in a way resembling a sinusoid *growing* in amplitude. Remember that friction increases as the relative speed decreases.

5. Suppose you clamp one end of a hacksaw blade to the edge of a table so it overhangs like a diving board. Now displace the free end a little and let go—twang it. What will happen to the frequency if you stick a lump of clay to the free end? Inasmuch as a tuning fork is essentially a bar bent into a U, the same thing would happen to it.

6. Why is it perfectly reasonable for the period of a pendulum to decrease as g increases? Why is it also reasonable for the period of a pendulum to increase as L increases?

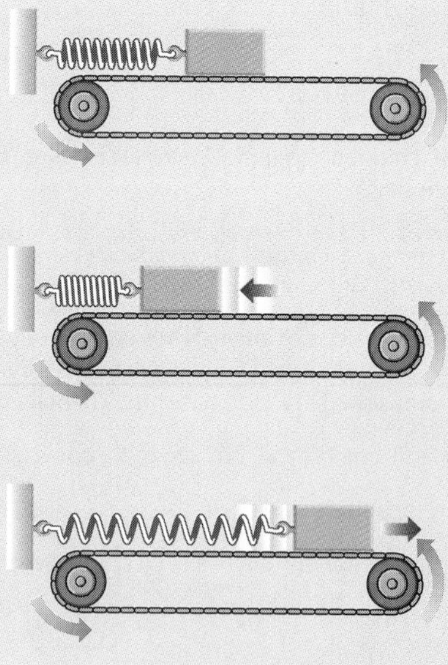

Figure Q4

7. When a person in a parachute descends, vortices are shed alternately up over one side and then the other. This phenomenon periodically changes the pressure on the chute. How might that become dangerous? (Hint: What resonance effect is of concern? Incidentally, some chutes have central holes that are there precisely to break up the vortices.)

8. Figure Q8 shows a horizontal wire supporting several simple pendulums. When the bob on the far left is displaced and

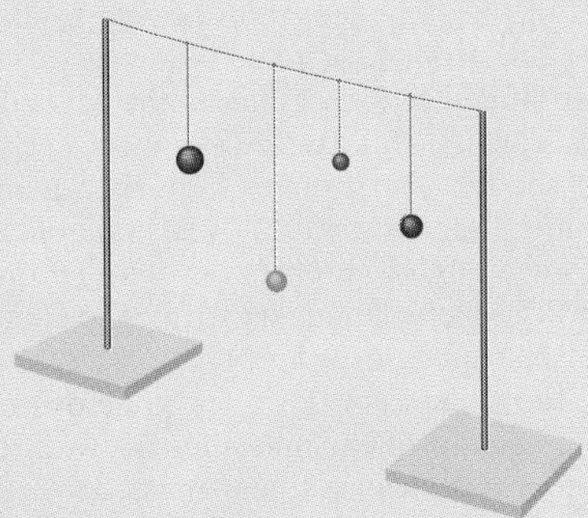

Figure Q8

set oscillating, it twists the wire a little with each swing. What do you think will happen?

9. What is the general relationship between the mass of someone jumping up and down on a trampoline (or a mattress) and the frequency of the oscillation?

10. Imagine an antiroll device for a ship: a mass on horizontal springs or a tank of water tuned to resonate at the natural rolling frequency of the ship. If waves begin to rock the ship near its resonant frequency, a very dangerous situation will develop. How does the device avert catastrophe?

11. The mythical planet Mongo is a homogeneous, nonrotating sphere whose inhabitants have drilled a hole running through the center from one side to the other. Show that a ball dropped into the hole would execute SHM.

12. Imagine a chisel held up to a piece of marble and then struck on its head with a mallet. Describe what happens in light of this chapter. Consider the question on the atomic level.

13. Imagine two symmetrical pulses that are identical in every way, except that one is inverted. These pulses are moving toward each other on a rope in opposite directions. What happens to the energy of the rope at the instant of complete cancellation?

14. In the real world, a pulse sent down a very long, taut rope diminishes in amplitude and ultimately vanishes. Describe what's happening in terms of Conservation of Energy.

15. Imagine a long string of beads hung vertically from a high ceiling with a weight fixed to its lower end. The bottom is then rapidly displaced, generating a transverse pulse. Describe the motion of the disturbance, paying particular attention to how it might differ if the string were held taut horizontally.

16. Picture a long row of standing dominos, one a little behind the other so that knocking over the first will knock over the second and so on. Now imagine an idealized tube an inch or two in diameter extending from New York to California. Suppose this tube is filled with greased, essentially frictionless marbles and you push one more in at the East Coast end only to have one pop out at the West Coast end. What's happening, and how do these arrangements of dominos and marbles relate to this chapter?

17. The accompanying list is a brief accounting of the speeds of compression waves in various materials. Considering the physical characteristics of each of the materials, discuss the relative speeds and explain why they are reasonable in light of your knowledge of the process of wave propagation.

Material	*Speed (km/s)*
Clay	1
Sandstone	2
Limestone	4
Granite	5
Salt	6

18. Extracorporeal shock wave lithotripsy (ESWL) is a technique for pulverizing kidney stones within the body. The patient is placed in a bath of water. An intense electric spark within the tank below the patient rapidly vaporizes a small amount of water, producing a shock wave that is reflected from a curved mirror that focuses it onto the stone. After a thousand or two blasts, a typical stone is destroyed. How is the shock wave formed and what kind of wave is it? What is it used for in a physical sense? Why isn't the skin destroyed as well as the stone?

19. Figure Q19 shows two different pulses heading toward each other. Draw several pictures in sequence, illustrating the result as they superimpose and separate.

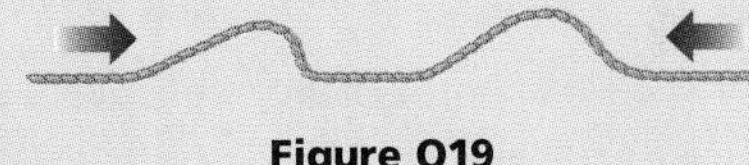

Figure Q19

Multiple Choice Questions

1. The end of a whip antenna is oscillating in SHM. Its maximum acceleration is given by the expression (a) $2\pi f_0^2 A$ (b) $4\pi^2 f_0^2 A$ (c) $4\pi^2 f_0^2 A^2$ (d) $4\pi f_0 A^2$ (e) none of these.
2. Two coiled springs are made of the same wire such that they are identical in all respects, except the first is half the length of the second. Compared to the natural frequency of the first (f), the natural frequency of the second is (a) $2f$ (b) $\frac{1}{2}f$ (c) f (d) $f/\sqrt{2}$ (e) none of these.
3. The free end of a clamped saw blade vibrates 12.8 times in 19 s. Its frequency is (a) 0.67 Hz (b) 12.8 Hz (c) 19 Hz (d) 1.48 Hz (e) none of these.
4. In SHM, the acceleration ______ the velocity by ______. (a) leads, $\frac{1}{2}\pi$ (b) lags, $\frac{1}{2}\pi$ (c) leads, $\frac{1}{4}\pi$ (d) lags, $\frac{1}{4}\pi$ (e) none of these.
5. An adult and a child are sitting on adjacent identical swings. Once they get moving, the adult, by comparison to the child, will necessarily swing with (a) a much greater period (b) a much greater frequency (c) the same period (d) the same amplitude (e) none of these.
6. A pendulum clock is set to run accurately at sea level. It is then brought to the top of a high mountain, where it is found to (a) function unchanged (b) run slow (c) run fast (d) stop running (e) none of these.
7. A tuning fork is struck and so set vibrating. Its handle is then touched to the handle of an identical fork held near the ear and this second one immediately begins to vibrate at the same frequency. This transfer of energy is an example of (a) contact potential (b) critical damping (c) resonance (d) radiation (e) none of these.
8. Sometimes a really heavy truck will go by and set the windows of the house vibrating for a moment with a low-frequency buzz. They oscillate (a) in-phase with the truck due to critical damping (b) at one of their natural

frequencies (c) at a random changing frequency (d) in an undamped mode (e) none of these.

9. An oscillator at a frequency of 1.25 Hz will make 100 vibrations in (a) 125 s (b) 12.5 s (c) 8.0 s (d) 80 s (e) none of these.
10. A pendulum clock in a free-falling elevator (a) runs normally (b) runs a little fast (c) runs a little slow (d) runs very fast (e) none of these.
11. Astronauts in space took a light, coiled spring of known elastic spring constant, attached a mass to it, and set it oscillating. Measuring the period, they could determine (a) the time of day (b) the acceleration due to gravity (c) the mass of the bob (d) the weight of the bob (e) none of these.
12. The tension in the wire of a simple pendulum is at maximum when the bob is at (a) its highest point (b) its lowest point (c) halfway down (d) any location since the tension is actually constant (e) none of these.
13. A swell of ocean waves of wavelength 1.0 m and frequency 1.25 Hz has a speed of (a) 1.25 m/s (b) 0.8 m/s (c) 125 m/s (d) 8.0 m/s (e) none of these.
14. The speed of a wave on a stretched string ______ when the tension in it is doubled. (a) doubles (b) increases by a factor of 4 (c) increases by a factor or 1.414 (d) decreases by a factor of 2 (e) none of these.
15. Tie a little piece of paper to the middle of a long horizontal taut string. Now send a transverse pulse down the string. The piece of paper will suddenly rise upward as the wave passes, proving that the disturbance transports (a) mass (b) weight (c) density (d) momentum (e) none of these.
16. Two strings of the same length are being used in a wave experiment. The first, which has twice the mass of the second, is stretched twice as taut as the second. The speed of the first's wave is (a) twice that of the second (b) the same as the second (c) greater than the second by $\sqrt{2}$ (d) less than the second by $\sqrt{2}$ (e) none of these.
17. A periodic wave passes an observer, who records that there is a time of 0.5 s between crests. (a) the frequency is 0.5 Hz (b) the speed of 0.5 m/s (c) the wavelength is 0.5 m (d) the period is 0.5 s (e) none of these.
18. A string of a certain linear mass-density is attached end-to-end to a rope whose linear mass-density is 9 times greater. The combination is stretched taut, and a pulse is sent down the string with a speed v. Its speed in the rope is (a) $9v$ (b) v (c) $3v$ (d) $v/9$ (e) none of these.
19. For a harmonic wave of a certain type in a given medium, doubling the frequency has the effect of (a) halving the speed (b) halving the wavelength (c) doubling the amplitude (d) doubling the period (e) none of these.
20. A stretched rope is dipped in water and thereby made wet. Without changing the tension, a pulse is then sent down its length. Compared to when the rope was dry, we can now expect the speed of the pulse to have (a) decreased (b) increased slightly (c) remained the same (d) doubled (e) none of these.
21. Quite generally, doubling the amplitude of a wave (a) doubles the frequency (b) halves the period (c) quadruples the energy (d) doubles the speed (e) none of these.

Problems

HARMONIC MOTION

1. [I] What is the period of a phonograph record that turns through $33\frac{1}{3}$ rotations per minute?

2. [I] The respiratory system of a medium-sized dog resonates at roughly 5 Hz so that it can pant (in order to cool off) very efficiently at that frequency. How many breaths will the dog be taking in a minute? For comparison, the dog would ordinarily breathe at around 30 breaths per minute.

3. [I] A record is turning uniformly at 78 rpm while an ant sitting at rest on its rim is being viewed by a child whose eyes are in the plane of the record. Describe the ant's motion as seen by the child. What is the frequency of the ant? What is its angular frequency?

4. [I] A hovering fair-sized insect rises a little during the downstroke of its wings and essentially free-falls slightly during the upstroke. The end result is that the creature oscillates in midair. If it typically falls about 0.20 mm per cycle, what is the wingbeat period and frequency? Compare that to the ≈10 Hz oscillation of a butterfly, which cannot hover.

5. [I] A body oscillates in SHM according to the equation

$$x = 5.0 \cos (0.40t + 0.10)$$

where each term is in SI units. What is (a) the amplitude, (b) the frequency, and (c) the initial phase? (d) What is the displacement at $t = 2.0$ s?

6. [I] A body oscillates in SHM according to the equation

$$x = 8.0 \cos (1.2t + 0.4)$$

where each term is in SI units. What is the period?

7. [I] Show that a value of $\varepsilon = +\frac{3}{2}\pi$ is equivalent to a value of $\varepsilon = -\frac{1}{2}\pi$ in Eq. (12.5). Draw a sketch of the functions.

8. [I] A point on the very end of one-half of a tuning fork vibrates approximately as if it were in SHM with an amplitude of 0.50 mm. If the point returns to its equilibrium position with a speed of 1.57 m/s, find the frequency of vibration.

9. [I] A point at the end of a spoon whose handle is clenched between someone's teeth vibrates in SHM at 50 Hz with an amplitude of 0.50 cm. Determine its acceleration at the extremes of each swing.

10. [I] A light hacksaw blade is clamped horizontally, and a 5.0-g wad of clay is stuck to its free end, 20 cm from the clamp. The clay vibrates with an amplitude of 2.0 cm at 10 Hz. (a) Find the speed of the clay as it passes through the equilibrium position. (b) Determine its acceleration at maximum displacement and compare that with g.

11. [I] A small mass is vibrating in SHM along a straight line. Its acceleration is 0.40 m/s^2 when at a point 20 cm from the zero of equilibrium. Determine its period of oscillation.
12. [I] A 1.0-N bird descends onto a branch that bends and goes into SHM with a period of 0.50 s. Determine the effective elastic force constant for the branch.
13. [I] Suppose Eq. (12.5) is to be applied to the mass in Fig. 12.14, where the cycle is to begin at $t = 0$ with m released with zero speed at -3 m. What must ε equal?
14. [I] Suppose Eq. (12.5) is to be applied to the mass in Fig. 12.14, where the timing clock is started arbitrarily and it is found that at $t = \frac{1}{4}T$, the mass is at $x = +\frac{1}{2}A$. Find ε.
15. [I] A particle is oscillating with SHM along the z-axis with an amplitude of 0.50 m and a frequency of 0.20 Hz. If it is at $z = +0.50$ m at $t = 0$, where will it be at $t = 5.00$ s, $t = 2.50$ s, and $t = 1.25$ s?
16. [I] A spot of light on a computer screen oscillates horizontally in SHM along a line 20 cm long at 50 Hz. The spot reaches the center of the line at $t = T/8$. Show that $v_x = -10\pi \sin(100\pi t + \frac{1}{4}\pi)$ is the expression for the speed of the spot as a function of time.
17. [I] A bug having a mass of 0.20 g falls into a spider's web, setting it into vibration with a dominant frequency of 18 Hz. Find the corresponding elastic spring constant.
18. [I] A 5.0-kg mass resting on a frictionless air-table is attached to a spring with an elastic constant of 50 N/m. If this mass is displaced 10 cm (compressing the spring) and is then released, find its maximum speed.
19. [I] Two kilograms of potatoes are put on a scale that is displaced 2.50 cm as a result. What is the elastic spring constant? If the scale is pushed down a little and allowed to oscillate, what will be the frequency of the motion?
20. [I] A 250-g mass is attached to a light helical spring with an elastic force constant of 1000 N/m. It is then set into SHM with an amplitude of 20 cm. Determine the total energy of the system.
21. [I] Consider the frictionless cart in Fig. 12.16. If the elastic spring constants are k_1 and k_2, respectively, determine the frequency of vibration in terms of these quantities.
22. [I] What is the speed of the mass in Problem 18 as it reaches a point 5.0 cm from its undisplaced position?
23. [I] Find the frequency of a simple pendulum of length 10.0 m.
24. [I] A small lead ball is attached to a light string so that the length from the center of the ball to the point of suspension is 1.00 m. Determine the natural period of the ball swinging through a small displacement in a vertical plane.
25. [I] How long must a simple pendulum be if it is to have a period of 10.0 s?
26. [I] One might guess that the period of a pendulum is proportional to the mass m, the length L, and the acceleration due to gravity g. Consequently, assume that

$$T = Cm^a L^b g^c$$

where C is a unitless constant. Now solve for a, b, and c using the fact that the dimensions on both sides of the equation must be the same. Of course, if this calculation produces nonsense, C wasn't unitless. Although you cannot get it from this analysis, what is the correct value of C?
27. [II] Write general expressions for both the initial displacement x_i and speed v_i of a point moving in SHM at $t = 0$. Show that

$$\tan \varepsilon = \frac{-v_i}{\omega_0 x_i}$$

What does this imply about ε if the initial speed is zero? Show that

$$A^2 = x_i^2 + \frac{v_i^2}{\omega_0^2}$$

Notice that because there is an initial speed imparted to the system at $t = 0$, $A \neq x_i$.
28. [II] A light 3.0-m-long helical spring hangs vertically from a tall stand. A mass of 1.00 kg is then suspended from the bottom of the spring, which lengthens an additional 50 cm before coming to a new equilibrium configuration well within its elastic limit. The bob is now pushed up 10 cm and released. Write an equation describing the displacement y as a function of time.
29. [II] An essentially weightless helical spring hangs vertically. When a mass m is suspended from it, it elongates an amount ΔL. When let loose, it oscillates with a measured period of T. Show how you might use this arrangement to determine g.
30. [II] A 200-g mass hung on the bottom of a light helical spring stretches it 10 cm. This mass is then replaced with a new 500-g bob that is set into SHM. Compute its period.
31. [II] An object is hung from a light vertical helical spring that subsequently stretches 2.0 cm. The body is then displaced and set into SHM. Compute the frequency at which it oscillates.
32. [II] A 5.0-kg block of wood is floating in water. It is found that a downward thrust of 10.0 N submerges the block an additional 10 cm, at which point it is let loose to oscillate up and down. Determine the frequency of the oscillation. Assume SHM.
33. [II] A uniform block with sides of length a, b, and c floats partially submerged in water. It is pushed down a little and let loose to oscillate. Given that the vertical edge has length b and the density of the block is ρ, show that the motion is SHM and determine its period. Check the units of your result.
34. [II] Derive an expression for the period of oscillation of the frictionless system shown in Fig. P34. Remember that the displacement of the mass m is the sum of the displacements of the two springs.

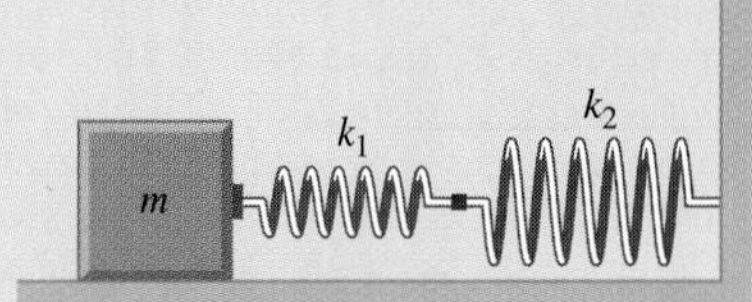

Figure P34

35. [II] A 5.0-g bullet is fired horizontally into a 0.50-kg block of wood resting on a frictionless table. The block, which is attached to a horizontal spring, retains the bullet and moves forward, compressing the spring. The block-spring system goes into SHM with a frequency of 9.0 Hz and an amplitude of 15 cm. Determine the initial speed of the bullet.

36. [II] Consider a simple pendulum of length L. Show that it gains an amount of potential energy

$$\Delta PE_G \approx \tfrac{1}{2} mgL\theta^2$$

when the bob is raised through an angle θ. (Hint: You may need to know that $1 - \cos\theta = 2\sin^2 \frac{1}{2}\theta$.)

37. [II] If the period of a simple pendulum is T, what will its new period be if its length is increased by 50%?

38. [II] Extending the ideas of Problem 36, show that the pendulum's maximum angular speed is given by $\omega_0 = \theta_0\sqrt{g/L}$.

39. [II] Using energy considerations, derive an expression for the period of oscillation of a mass m. It is fixed to a horizontal spring having an elastic constant k, and is free to move on a frictionless table.

40. [III] A vibration platform oscillates up and down with an amplitude of 10 cm at a controlled variable frequency. Suppose a small rock of mass m is placed on the platform. At what frequency will the rock just begin to leave the surface so that it starts to clatter?

41. [III] Use Fig. P41 to prove that the reciprocating motion of the piston is oscillatory but that it is not SHM.

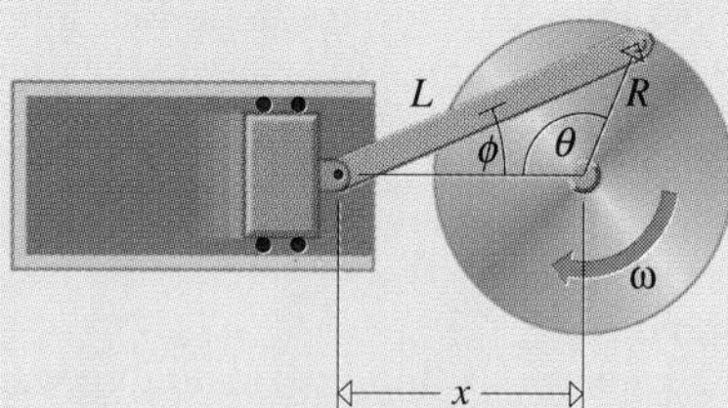

Figure P41

42. [III] A small mass is attached to the end of a long vertical string. It is displaced through an angle θ and then swung so that the mass moves in a horizontal circle. Show that if you looked at the resulting conical pendulum in the plane of motion, you would see the bob moving back and forth with a period given by

$$T = 2\pi\sqrt{\frac{L\cos\theta}{g}}$$

MECHANICAL WAVES

43. [I] Figure P43 shows two views—(a) optical and (b) acoustical—of a computer circuit. The acoustic microscope has the highly desirable ability to see several micrometers below the surface. The device focuses ultrasonic waves at a frequency of 3 GHz through a droplet of water onto the object. The speed of the waves in the water, which couples the microscope to the specimen, is 1.5 km/s. Compute the wavelength of the radiation and compare it to the wavelength of green light in air ≈500 nm.

44. [I] A long metal rod is struck by a vibrating hammer in such a way that a compression wave with a wavelength of 4.3 m travels down its length at a speed of 3.5 km/s. What was the frequency of the vibration?

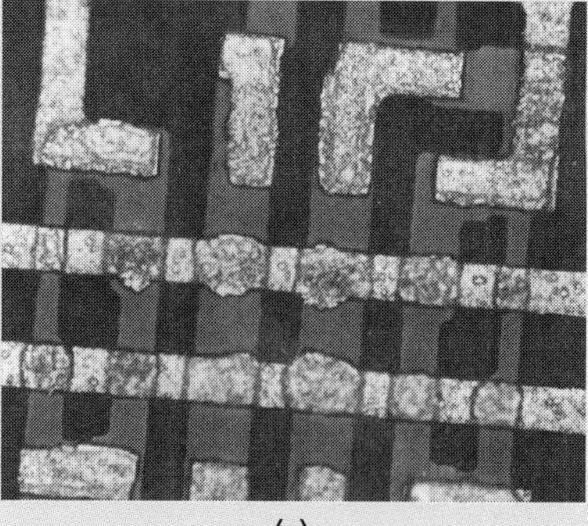

(a)

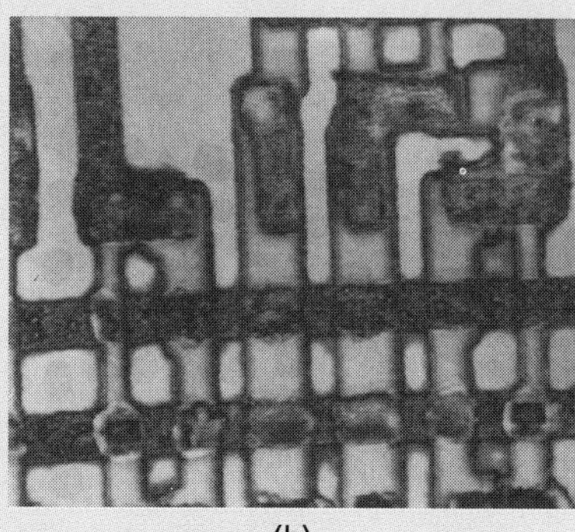

(b)

Figure P43

45. [I] An A note of 440 Hz is played on a violin submerged in a swimming pool at the wedding of two scuba divers. Given that the speed of compression waves in pure water is 1498 m/s, what is the wavelength of that tone?

46. [I] A wavepulse on a string travels the 10-m length in 2.0 s. A harmonic disturbance of wavelength 0.50 m is then generated on the string. What is its frequency?

47. [I] Show that, for a periodic wave, $\omega = (2\pi/\lambda)v$.

48. [I] A transverse harmonic wave on a beaded string has an amplitude of 2.5 cm and a wavelength of 160 cm. If at $t = 0$ a life-sized photo shows that the height of the wave at $x = 0$ is zero and at $x = 40$ cm it is +2.5 cm, then what is the string's displacement at $x = 10$ cm?

49. [I] A member of the Vespertilionidea family of bats typically emits a sequence of chirps, wavetrains lasting about 3.0 ms and having a carrier frequency that varies from 100 kHz to about 30 kHz. Generally, there is a time between individual chirps of 70 ms. Assuming the air speed of one of these wavetrains is 330 m/s, how far away can an object be and still be detected without being masked by the next outgoing chirp?

50. [I] A radar beam transmitted with a carrier frequency of 9.8 GHz is composed of a stream of nearly harmonic wavetrains each lasting 1.0 μs. Determine the number of wavelengths of the carrier that are in each pulse and show that this number is independent of v, provided the speed of the pulse equals the speed of the carrier wave.

51. [I] The A string of a violin has a linear mass-density of

0.59×10^{-3} kg/m and is stretched to a tension of 10 N. What is the speed of the transverse waves it can sustain? Incidentally, the four violin strings, when mounted and properly tuned, produce a net tension on the instrument of about 220 N (50 lb).

52. [I] A heavy rope is tied to a tree 30 m away, and it is then pulled taut with a force of 360 N. The rope is struck near its end with a stick, thus setting up a transverse wavepulse. If the rope has a linear mass-density of 0.06 kg/m, how long will it take for the pulse to make the round trip from end to tree to end?

53. [I] If the speed of a wavepulse set up on the longest string of a grand piano is 129 m/s and if the linear mass-density of the string is 64.9×10^{-3} kg/m, what is the tension? Typically, the net tension on the frame of a concert grand due to all the strings is around 2.7×10^5 N or 30 tons (60 000 lb).

54. [I] What is the speed of a compression wave in ethyl alcohol?

55. [I] Figure P55 shows a rather idealized square water wave. Show that doubling its amplitude would increase its energy by a factor of 4, thus again making the point that the energy of a wave is proportional to its amplitude squared.

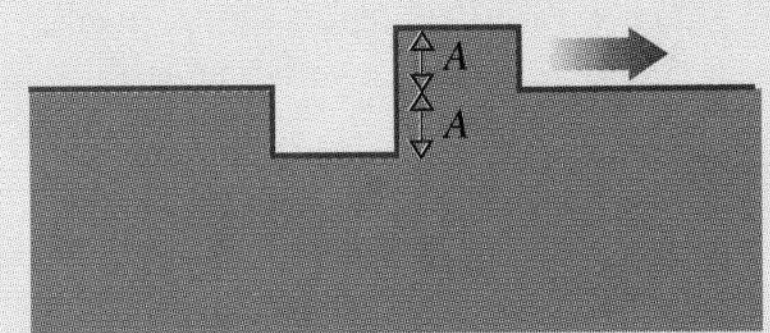

Figure P55

56. [I] Figure P56 depicts two identical symmetrical transverse wavepulses, each traveling at a speed of 5.0 m/s heading toward one another. Draw the resultant disturbance 2.0 s later and again 2.0 s after that.

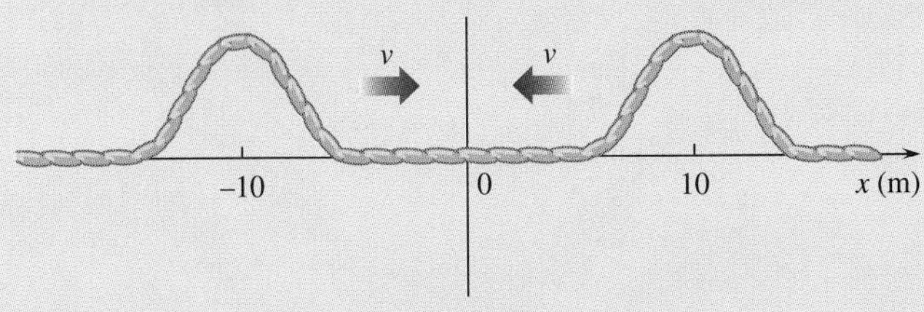

Figure P56

57. [I] Figure P57 depicts two identical symmetrical transverse wavepulses, each traveling at a speed of 5.0 m/s heading toward one another. Draw the resultant disturbance 2.0 s later and again 2.0 s after that.

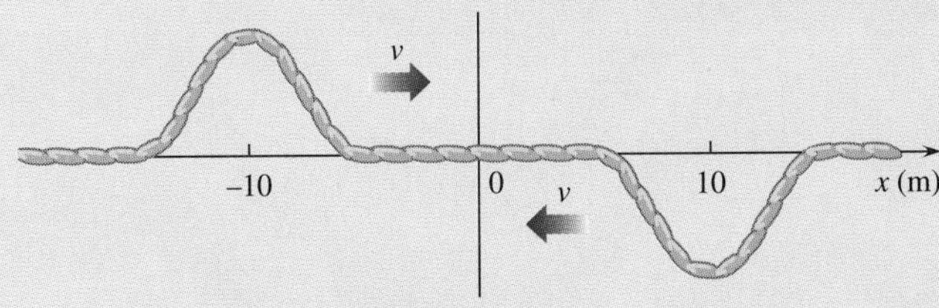

Figure P57

58. [II] Write an expression for the profile of a transverse harmonic wave of amplitude 10 cm and wavelength 1.2 m traveling along a rope in the positive z-direction and oscillating in the yz-plane. The height of the wave is zero at $t = 0$ and $z = 0$.

59. [II] Show mathematically that shifting the phase of a sinusoidal wave profile by $\frac{1}{2}\pi$ rad turns it into a cosine profile. Begin with

$$y = A \sin\left(\frac{2\pi}{\lambda}x \mp \tfrac{1}{2}\pi\right)$$

60. [II] The profile of a transverse harmonic wave on a long beaded string is described in SI units by the function

$$y = 0.020 \sin(6.28x)$$

Given that the wave has a speed of 5.0 m/s, determine the maximum *transverse* speed of any bead on the string.

61. [II] The profile of a transverse harmonic wave on a long taut nylon thread is described in SI units by the function

$$y = 0.040 \sin 2\pi x$$

Given that the wave travels at a speed of 2.0 m/s, determine the maximum *transverse* acceleration of any point on the thread.

62. [II] Use the method of dimensional analysis to "derive" an expression for the speed of a compression wave in a solid. We guess that the answer depends on Young's Modulus and the density, and write

$$v = KY^{a}\rho^{b}$$

where a and b are to be determined and K is a dimensionless constant (like 2π). Finish out the calculation by balancing the units.

63. [II] A string has a linear mass-density of 2.50 g/m and is put under a tension of 25.0 N as it is stretched taut along the z-axis. The free end is attached to a tuning fork that vibrates at 50.0 Hz, setting up a transverse wave on the string having an amplitude of 5.00 mm. Determine the speed, angular frequency, period, and wavelength of the disturbance.

64. [II] Using Problem 63, write an expression for the profile of the wave given that, at $t = 0$, the end of the rope ($x = 0$) is at $y = +5.00$ mm. The wave travels in the z-direction.

65. [II] A heavy nylon guitar string with a linear mass-density of 7.5 g/m is stretched to a tension of 80 N. What is the speed of the transverse wave that can be generated on the string?

66. [II] The speed of a transverse wave on a wire is 90 m/s when the tension in the wire is 120 N. What will be the speed if the tension is reduced to 80 N?

67. [II] A homemade telegraph system sends transverse pulses along a stretched string. It operates between two neighboring houses using 12 m of a string having a total weight of 0.20 N. What should be the tension in the string if the signals are to travel at least as fast as they would were the users simply to yell? Take the speed of sound in air to be 333 m/s.

68. [II] Determine the profile of the wave function

$$y = A \sin 2\pi\left(\frac{x}{\lambda} - \frac{t}{T}\right)$$

at $t = 0$. Find the values of y at $x = 0, \lambda/8, \lambda/4, 3\lambda/8, \lambda/2, 5\lambda/8, 3\lambda/4, 7\lambda/8$, and λ. Make a plot of the profile. Now do the same thing for the wave function

$$y' = A \sin 2\pi\left(\frac{t}{T} - \frac{x}{\lambda}\right)$$

and compare your results.

69. [III] Given the equation

$$y = A \sin 2\pi\left(\frac{t}{0.01} - \frac{x}{40}\right)$$

describing the displacement of a string of beads in SI units, write an expression for the motion of a bead at $x = 5.0$ m as a function of time. What is the value of y at $t = 0$ and $t = 0.01$ s? When is $y = 0$? Graph your results.

70. [III] A transverse harmonic wave on a long beaded string is described in SI units by the function

$$y = 0.02 \sin (6.28x - 15t)$$

If we have a detector at $x = 0$, what will be the speed of the bead at that location at a time $t = 1.2$ s?

71. [III] A uniform rope of length L and mass m hangs freely straight down from a hook in the ceiling. Taking the bottommost point as $y = 0$, write an expression for the mass of a segment of arbitrary length Δy. What is the tension in the rope at a height Δy? Derive an expression for the speed of a transverse wave set up by wiggling the lower end. What is the maximum speed? Determine an expression for the speed of the wave if a mass M is hung on the bottom of the rope.

72. [III] The very end of a wire 14.5 m long having a mass of 0.12 kg is clamped in a vise. The other end is tugged on, putting a tension of 50 N in the wire. That end is then driven transversely by a tuning fork oscillating at 440 Hz, and the wire vibrates with a maximum displacement of 0.15 mm. (a) Find the wavelength and period of the resulting disturbance. (b) Determine the maximum transverse speed of any point on the wire.

13

Sound

THE WORD **SOUND** HAS two meanings of interest to us. There is the psychological and physiological concept of sound as that which humans hear—the perception produced by the ear-brain detector. And there is the physical concept of sound as **any compression wave in a material medium that has frequency content in the range from 20 Hz to 20 kHz**. Although this chapter concentrates on sound as a longitudinal elastic wave, the primary ideas are grounded in our common human experience and, consequently, we also examine how the ear works, as well as the significance of the physiological-psychological notions of pitch, loudness, intensity-level, and timbre.

The specific *frequency bandwidth* from 20 Hz to 20 kHz is the spectrum heard by the so-called "average" person, who is a statistical fiction. Anyone in our society who can hear across that entire range is actually something of a rarity. Even so, the frequency response of the human ear-brain system diminishes with age beyond about 20 years. Moreover, the exposure to loud sounds, as is common these days, desensitizes the ear at both frequency extremes.

ACOUSTICS

The insight that sound is a wave phenomenon is remarkably ancient. It seems to have been suggested initially by the Roman architect Marcus Vitruvius Pollio, who lived about 2000 years ago. While considering the design of amphitheaters, where it's crucial to control echoing, he likened sound to water ripples moving through space "upward, wave after wave." In a sense, this wonderful image marked the beginning of the study of the physics of sound—that is, *acoustics* (from the Greek *akoustikos,* relating to hearing). The recognition that a sound wave is longitudinal follows from the fact that it propagates in air, which, like all fluids, has no stiffness, cannot resist shear, and cannot sustain a transverse wave. Because a wave does not transport the medium, its speed can be extraordinarily large as is clearly the case with sound—a bolt of lightning a mile away produces a clap of thunder heard only a moment ($\approx$5 s) later. For millennia, sound was recognized as the most rapidly propagating phenomenon (exclusive of light, which many thought was instantaneous).

Sound propagates in any medium that can respond elastically and thereby transmit vibrational energy. Air at ordinary density is certainly springy (squeeze an empty plastic soda bottle capped and uncapped, and the point will be made). The fact that the region connecting the source and detector must contain an adequate amount of matter to sustain the disturbance was established experimentally (1672) by von Guericke. He produced a partial vacuum within a chamber containing an alarm-clock mechanism. As the air was removed, the sound became weaker until it vanished altogether. Only after air was readmitted could the ringing bell again be heard. Sound does not propagate in vacuum. All the old science fiction movies that depicted roaring explosions in space have missed the point.

13.1 Sound Waves

Consider a loudspeaker being driven by a sinusoidal signal, a *pure tone.* The flexible cone vibrates harmonically, pumping the air in front of it into a series of condensations and rarefactions (Fig. 13.1). The resulting pressure variations that constitute the sound wave are really quite small. A very loud sound corresponds to a pressure change of less than about 10 Pa (10^{-5} atm), and quiet music exists at a gauge pressure of $< \pm 0.002$ Pa. The speaker cone usually moves back and forth a fraction of a centimeter, and any given air molecule moves an equally small amount (Fig. 12.41). That distance determines the amplitude of the wave and, therefore, its energy and the loudness of the sound. In the same way, a large drum head moves through a relatively large distance and makes a large sound. The time it takes the speaker cone to pass through a cycle fixes the period and frequency of the sound. The distance in air between successive condensations (or rarefactions) is the wavelength (Table 13.1). The wavelength, too, is determined by how fast the cone moves through a cycle. Energy streams into the air all the while the cone is in motion. A slow progression, even though it is through a small cone displacement (< 1 cm), can launch a wave with a λ of as much as $\approx$17 m at 20 Hz.

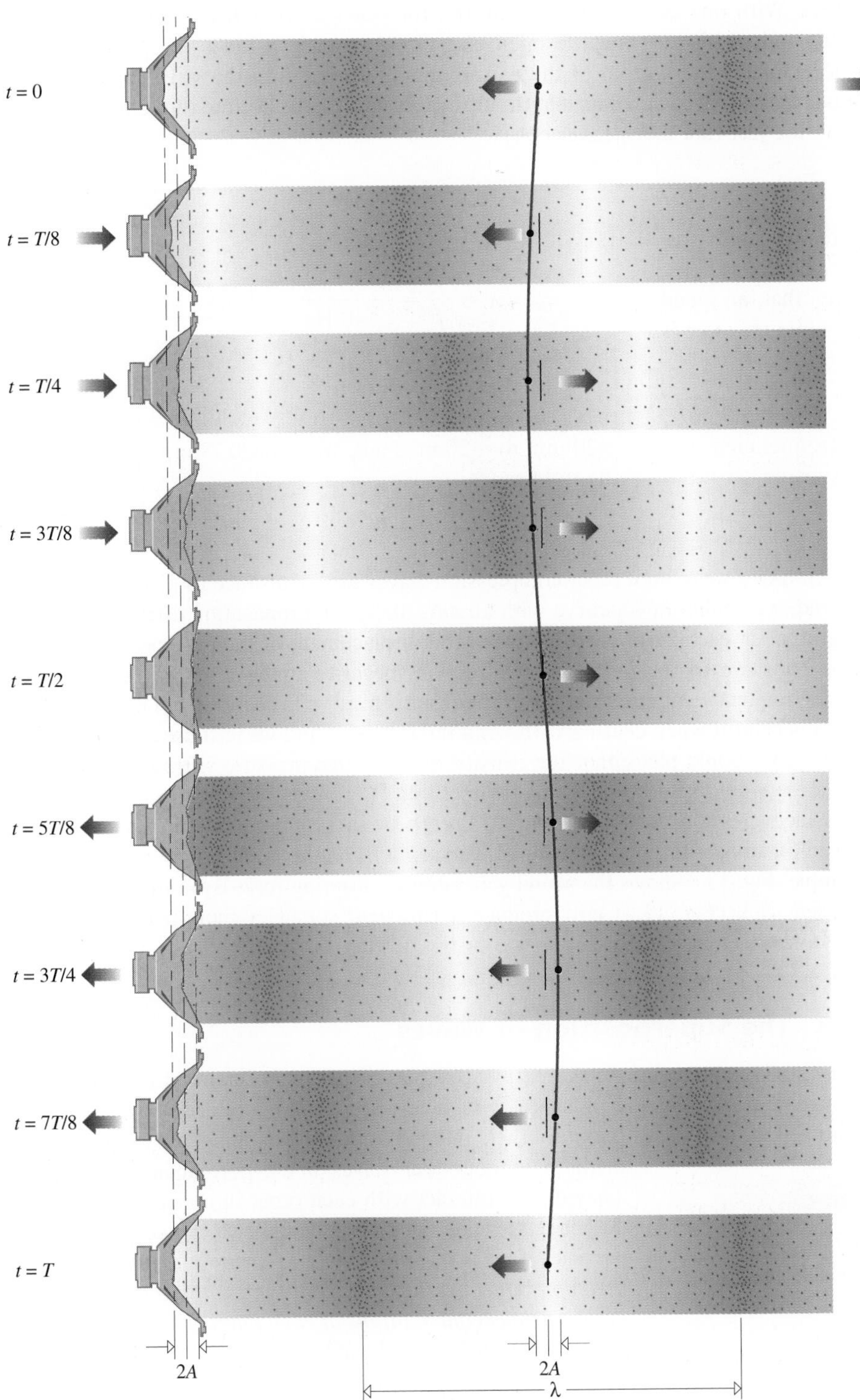

Figure 13.1 A sound wave coming from a loudspeaker. Note that the speaker cone sweeps through a distance of $2A$. Each atom oscillates through a distance of $2A$ as well. Compare that to the wavelength λ.

This same kind of thing happens in more dense media as well. For example, punch a hole in the bottom of a topless tin can, pass a string through it, and make a large knot at the end so it will not come out the hole. Then do the same thing several meters away at the other end of the string with a second can (bottoms toward each other). With one can at your mouth and the other at your listener's ear, pull the string tight and talk into the open end—you've made a kid's "telephone." Your voice vibrates the bottom of the can, pumping energy into the string and sending compression waves down its length, ultimately setting the bottom of the receiver can vibrating just like the speaker cone.

Example 13.1 By international agreement most orchestras tune to a frequency of 440 Hz, which is called A440 (the A note above middle C). Given that the speed of sound in air at room temperature is 343.9 m/s, what is the wavelength of A440?

Solution: [Given: $f = 440$ Hz and $v = 343.9$ m/s. Find: λ.] The problem deals with frequencies and wavelengths, and that should always bring to mind Eq. (12.18), $v = f\lambda$; hence

$$\lambda = \frac{v}{f} = \frac{343.9 \text{ m/s}}{440 \text{ Hz}} = \boxed{0.782 \text{ m}}$$

▶ **Quick Check:** For 100 Hz, $\lambda = 3.4$ m, which is 4.4 times smaller; hence, 4.4(0.78 m) = 3.4 m. It's a good thing to remember that sound waves extend from ≈20 mm to ≈20 m. Thus, we know 0.78 m is at least in the right range.

Unlike a wave on a rope, compression waves are a nuisance to picture directly. Instead, we could draw a curve with an axis along the propagation direction, whose magnitude at each point corresponds to the displacement of the particles of the medium (positive in the propagation direction, negative opposite to it). Thus, for a single-frequency compression wave, the resulting curve is a sinusoid. This representation is helpful when dealing with wind instruments, and we will use it later. Alternatively, we could plot either the density or the gauge pressure within the medium and these would also be sinusoidal (Fig. 13.2).

Since pressure can be measured directly with a microphone and readily displayed on an oscilloscope, plotting P_G versus t will be the preferred approach. As an example, Fig. 13.3 shows the sound generated by a tuning fork. By contrast, a struck wooden stick (Fig. 13.4) will produce a *transient* sound, a short-lived wavetrain only a few meters long.

13.2 The Superposition of Waves

A fascinating characteristic shared by all waves is that two or more of them moving through the same region of space will superimpose and produce a well-defined combined effect. Waves maintain their integrity upon overlapping; they cohabit the medium without themselves being permanently changed. They do not interact with each other in the sense that they do not scatter one another as would two crossing beams of particles. *In the region where two or more disturbances of the same kind overlap, the resultant is the algebraic sum of the various contributions at each point.* This is called the **Superposition Principle** and it was introduced by Thomas Young. As we will come to see, it's among the most far-reaching and fundamental insights in wave theory.

Figure 13.5 shows the effect of superimposing two harmonic waves of the same frequency and amplitude. At every value of x, we simply add the corresponding heights of the two sine curves, taking measures above the axis as

TABLE 13.1 Wavelengths in air at 20°C

Source	Frequency (Hz)	Wavelength (m)	Wavelength (ft)
Organ, lowest note	16	21	70
Piano, lowest note	28	12	40
Bass voice, low C	65	5.3	17
Middle C	262	1.3	4.3
Alto voice, high F	698	0.49	1.6
Soprano voice, high C	1047	0.33	1.1
Piano, highest note	4186	0.08	0.27

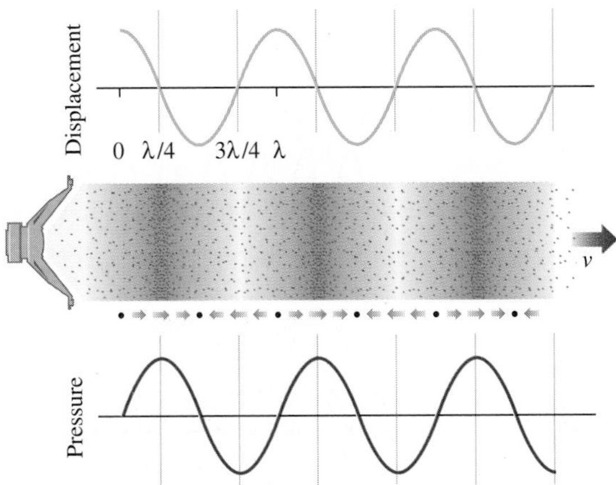

Figure 13.2 A sinusoidal sound wave. Where the pressure is maximum, above and below atmospheric, the displacement is zero. When the pressure is positive, as it is in condensations, atoms are shifted in the direction of motion of the wave.

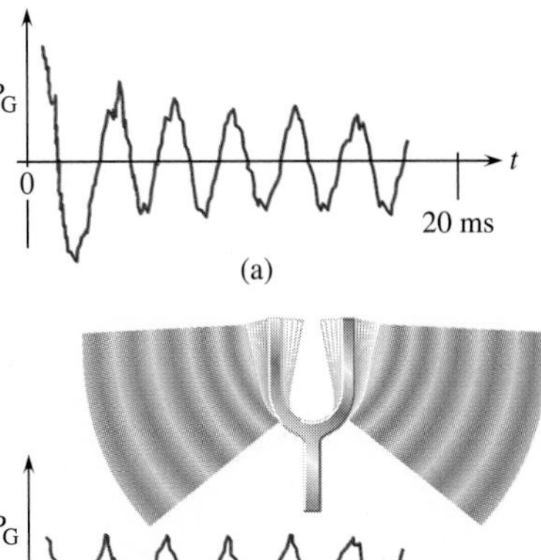

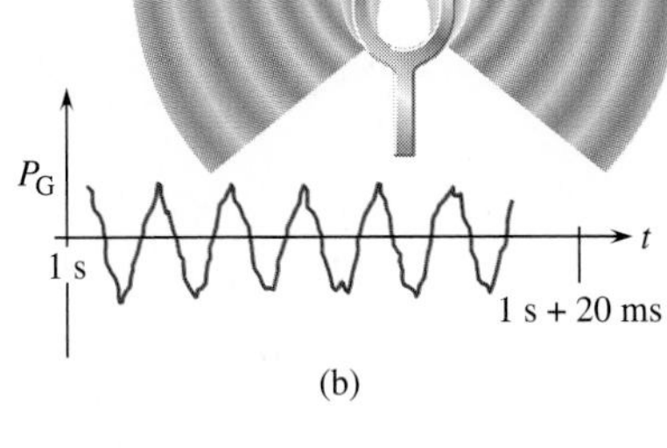

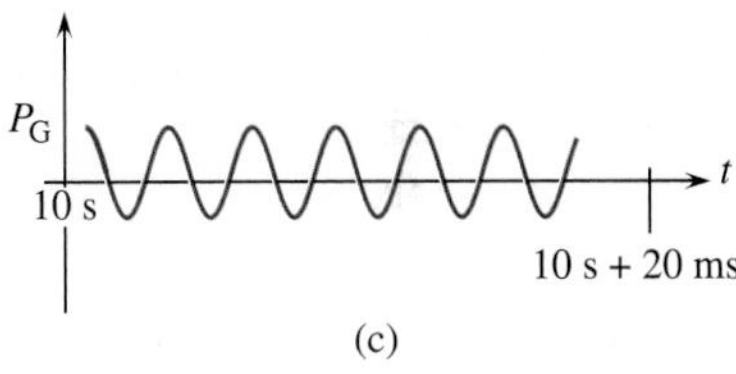

Figure 13.3 Oscilloscope traces corresponding to gauge pressure produced by a tuning fork: (a) after the fork is initially struck; (b) after oscillating about 1 s; (c) after oscillating about 10 s. As time goes on, the high-frequency, short-wavelength components vanish, leaving a smooth, nearly sinusoidal wave. Precision tuning forks can maintain their frequency with an accuracy of 1 part in 100 000.

positive and below it as negative. Interestingly, *the sum of any number of harmonic waves of the same frequency traveling in the same direction is also a harmonic wave of that frequency,* and that's true independent of the amplitudes. When sine waves with different frequencies are added to one another, the composite disturbance is not harmonic (Figs. 13.6 and 13.7).

Fourier Analysis

A beautiful mathematical technique for synthesizing waveforms was devised in 1807 by the French physicist Jean Baptiste Joseph, Baron de Fourier. He proved that *a periodic function having a wavelength λ can be synthesized by a sum of harmonic functions whose wavelengths are integral submultiples of λ* (that is, λ, λ/2, λ/4, and so forth). In fact, any wave profile encountered in nature—pulse or periodic—can be envisioned as the result of overlapping harmonic functions. Any waveform can be

Overlapping water waves pass each other and sail away unruffled.

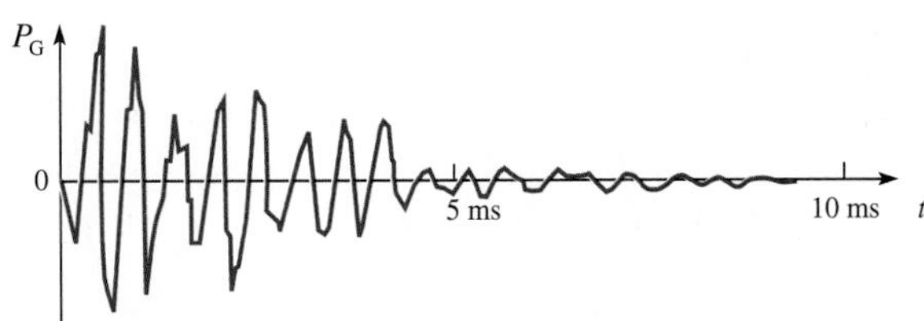

Figure 13.4 Gauge pressure produced by striking a wooden stick with a metal rod, as shown on an oscilloscope. The short-lived vibration dies out in about 10 ms.

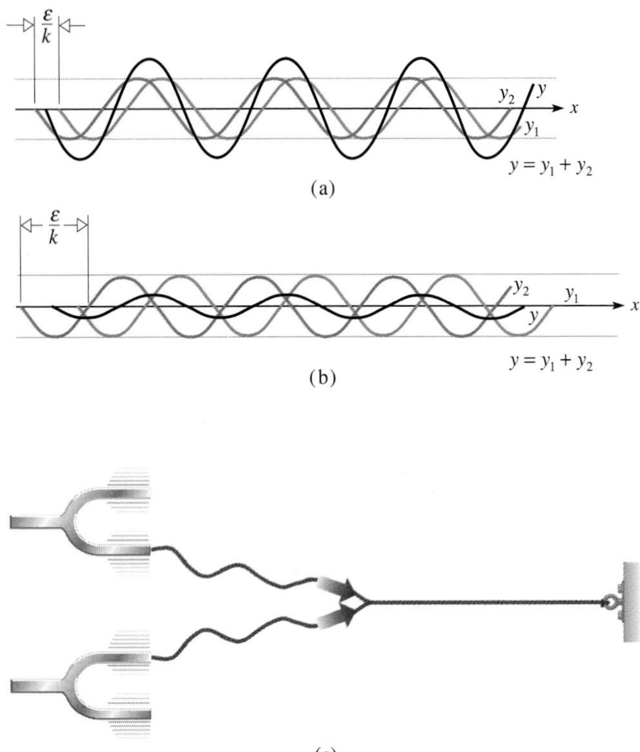

Figure 13.5 The superposition of waves y_1 and y_2, which combine to yield y. The two component waves have the same amplitude and differ in phase by ε, which corresponds to a shift in x of ε/k. In (a) the two are nearly in-phase, and the resultant is large. In (b) they are nearly out-of-phase by 180°, and the resultant is small. In (c) they are out-of-phase by exactly 180°, and the two waves cancel completely.

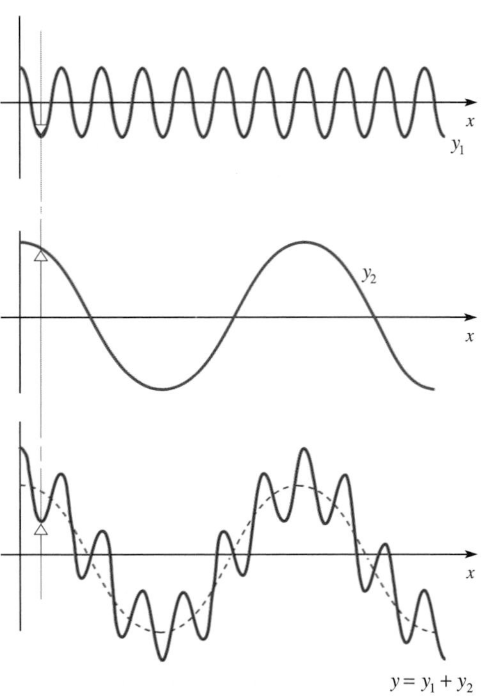

Figure 13.6 The superposition of waves y_1 and y_2. At each value of x, we add the magnitude of y_1 to that of y_2. Values above the axis are positive, those below are negative. The component waveforms have different frequencies, and the resultant is therefore not sinusoidal.

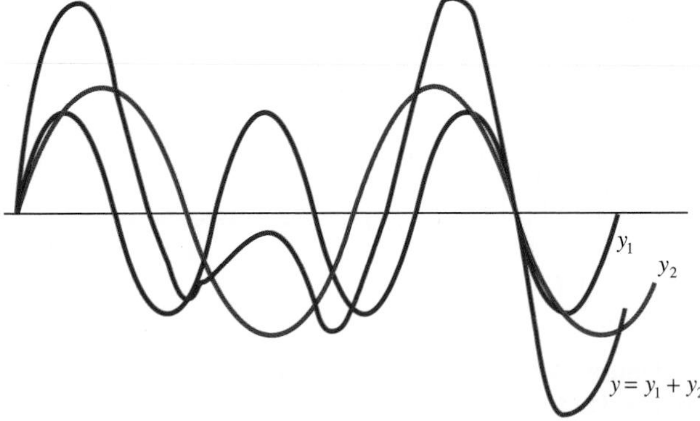

Figure 13.7 The superposition of two harmonic waves of different frequency. The resultant $y = y_1 + y_2$ is periodic, but it certainly is not harmonic. Of course, to be harmonic a disturbance must have only one frequency, and y doesn't.

decomposed into harmonic components, and this can be accomplished with sound by the human ear (or a spectrum analyzer) and with light by a prism.

According to Fourier, a periodic function of frequency f can be created out of an appropriate sum of sinusoidal terms, the first of which will also have a frequency f. This is the so-called **fundamental** or *first harmonic* ($f_1 = f$). The next term has a frequency of $f_2 = 2f$ and is called the *second harmonic,* and the next has a frequency of $f_3 = 3f$ and is called the *third harmonic,* and so on. Notice that in the time the fundamental takes to complete one cycle ($1/f_1$), every higher harmonic will complete a whole number of cycles, and the entire set will be back where it started. Thus, the frequency of the combined waveform is the frequency of the fundamental, which was made equal to f, the frequency of the periodic wave being analyzed (Fig. 13.8). If the several pure tones in Fig. 13.9a are combined, they produce the acoustic wave in Fig. 13.9b. Electronically synthesized music and talking computer chips do exactly this sort of Fourier addition (Fig. 13.10).

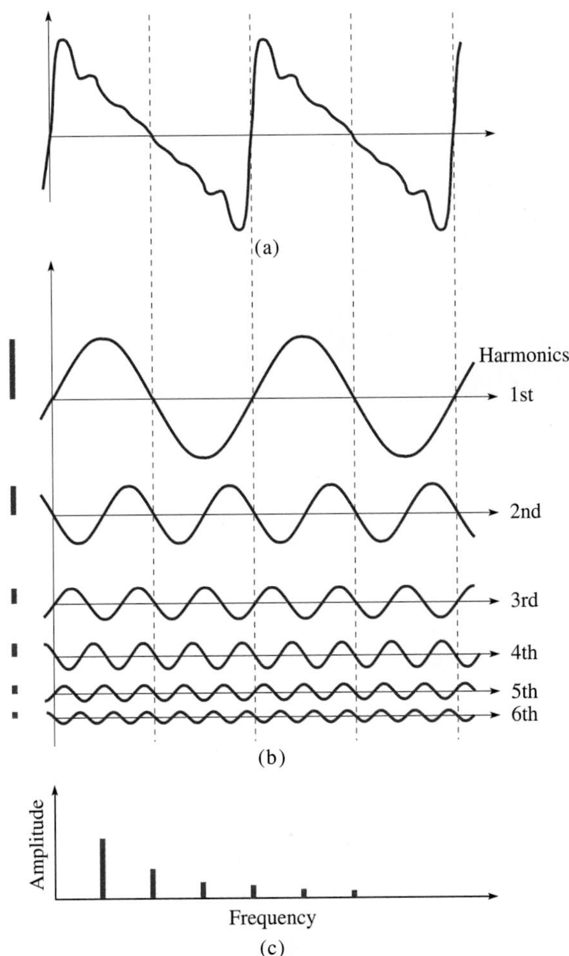

Figure 13.8 (a) The synthesis of a sawtooth wave. (b) Here, the first 6 harmonics combine to form a curve that clearly resembles a sawtooth. Another dozen terms will make it still better. (c) A plot of the amplitudes of the contributing harmonic terms plotted against frequency. This is called the *frequency spectrum.*

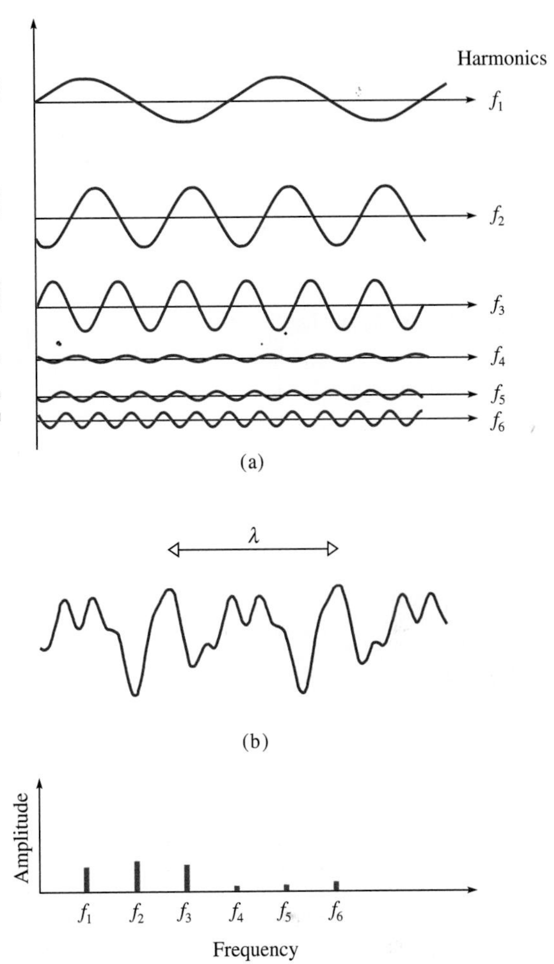

Figure 13.9 (a) Six sinusoidal waveforms. (b) The result of combining these waveforms. (c) The associated spectrum. Each vertical line represents the amplitude of the corresponding harmonic in (a). Note that $f_2 = 2f_1$, $f_3 = 3f_1$, $f_4 = 4f_1$, and so forth.

13.3 Wavefronts and Intensity

In two dimensions, the harmonic disturbance spreading out on the surface of a pool of water from a rhythmically tapping toe (Fig. 13.11) forms a pattern of expanding circular ripples. The lines connecting all the points where the water is rising or falling in-step form a series of concentric circles. At every point on any one of these circles, the ripple traveling out radially is at the same stage, or phase, of its essentially sinusoidal cycle, whether that's a crest or a trough or anything between. *The phase of the disturbance is constant at every point on a given circle.* Each curve connecting all of the neighboring points of a wave that are in-step is known as a **wavefront**.

Sound waves are three-dimensional. Imagine a *point source* radiating waves uniformly in all directions, as in Fig. 13.12. Here, **the surfaces of constant phase**, the wavefronts, are perfectly spherical (yet another mathematical idealization that does not actually exist, although it is often approximated). If we examine the com-

Figure 13.10 Each button on a Touchtone telephone produces a pair of nearly pure, widely separated tones. Push the 7-button, and you can hear the superposition of 852 Hz and 1209 Hz.

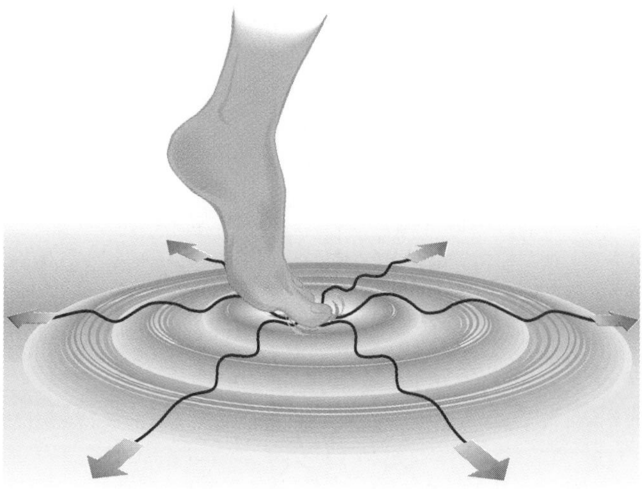

Figure 13.11 A toe wiggling in the water generates circular surface waves.

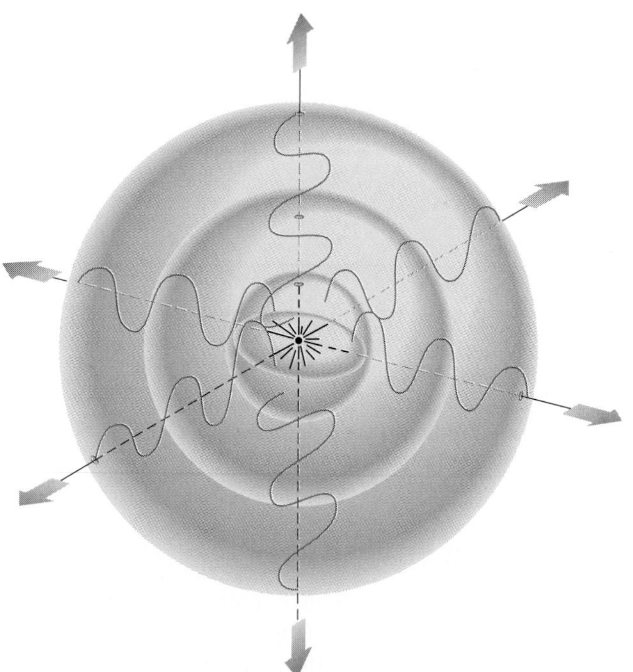

Figure 13.12 A wave from a point source spreads out uniformly in all directions. Energy flows outward in the form of a spherical wave. As the area ($4\pi R^2$) gets larger, the concentration of energy diminishes and the amplitude of the wave decreases as $1/R$.

pression wave from an exploding firecracker $\approx$10 m away, the portion of the wavefront intercepted by an ear or a microphone will be essentially spherical.

A sound wave represents a moving change in the distribution of atoms in a medium, and that change is associated with the flow of energy. The rate of transfer of energy is the *acoustic power.* A person speaking at a normal conversational level emits power at about 10^{-5} J/s and shouts at about 1 mJ/s. The energy of a mechanical wave is distributed over its wavefront and, in the case of a *homogeneous* spherical wave, that distribution is uniform in all directions. This raises a practical point: when a detector intersects a portion of a wavefront, it records an amount of energy that depends on its own receiving area and on the time during which it is receptive. Both of these depend on the particular detector, and that's inappropriate. Thus, what we must know is not the total energy arriving on the detector, but the energy per unit time, per unit area—the concentration of power. The **intensity** (I) of a wave is therefore defined as *the average power divided by the perpendicular area across which it is transported:*

A steady drip of rainwater produced these circular waves.

$$I = \frac{P_{av}}{A} \tag{13.1}$$

Intensity has the units of W/m^2. When a 2-m^2 area receives a total of 1 W of wave energy flowing perpendicularly onto it, the incident intensity is 0.5 W/m^2.

As a spherical wave expands outward, its area ($4\pi R^2$) increases, and since the same power flows through ever-increasing areas, its concentration—the intensity—diminishes *inversely with the square of the radius;* this is the *Inverse Square Law.* It is the reason why you cannot read a book by starlight or hear what the players on the field are saying from up in the stands. Remember that the amplitude of a *spherical wave* must diminish with distance because the energy it carries is spreading thinner and thinner while the wavefront gets larger.

Example 13.2 An underwater explosion is detected 100 m away, where the intensity is recorded to be 1 GW/m^2. About 1 s later, the sound wave is recorded 1.5 km away from ground-zero. What will its intensity be at that distance? Ignore absorption losses.

Solution: Assuming that the compression wave is spherical, the distances provided represent radii. [Given: $R_1 = 100$ m, $I_1 = 1$ GW/m^2, $\Delta t = 1$ s, and $R_2 =$ 1.5 km. Find: I_2.] Whenever you have energy considerations involving spherical waves of different radii, think about the Inverse Square Law. The power flowing through the first sphere will equal the power through the second; therefore

$$I_1(4\pi R_1^2) = I_2(4\pi R_2^2)$$

and

$$I_2 = \frac{I_1 R_1^2}{R_2^2} = \frac{(1 \times 10^9 \text{ W/m}^2)(100 \text{ m})^2}{(1.5 \times 10^3 \text{ m})^2}$$

Thus

$$I_2 = 4.4 \times 10^6 \text{ W/m}^2 \text{ or } \boxed{4 \text{ MW/m}^2}$$

▶ **Quick Check:** I goes as $1/R^2$ and 100 m is 15 times closer than 1.5 km; therefore, $I_1 = 15^2 I_2 \approx 225(4$ MW/m$^2) \approx 0.9$ GW/m^2.

By allowing a spherical wave to expand (Fig. 13.13) over a great distance, the wavefronts will get larger and flatter, ultimately resembling planes. The light entering a telescope from a star is usually indistinguishable from a perfectly flat wavefront.

13.4 The Speed of Sound

By the seventeenth century, Galileo, Huygens, and others were already studying the finer points of acoustics, and the wave theory was widely accepted. In Paris, Father Mersenne (ca. 1636) used a crude echo technique to first measure the speed of sound. He found a value of about 1000 ft/s, which is not far off from the present value of 331.45 m/s (1087.4 ft/s or 741 mi/h) in air at 0°C and 1 atm (Table 13.2). In

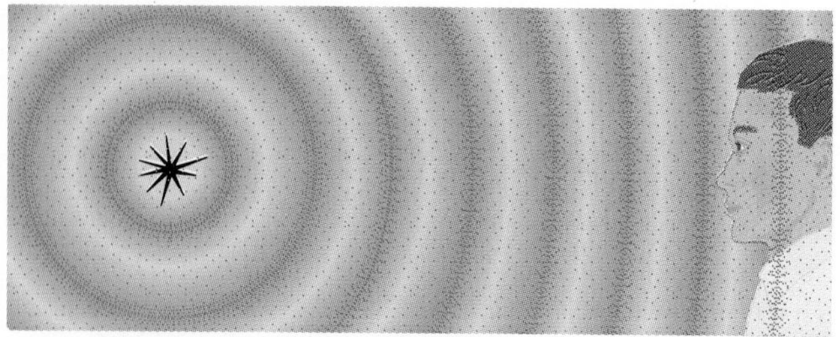

Figure 13.13 As a spherical wave expands, its radius increases and the wavefronts flatten. At very large distances, the wavefronts are quite flat and the disturbance resembles a plane wave.

a confined sample, the speed of sound increases with temperature. The springiness of a gas (for example, in an inflated ball) comes from the motion of its atoms and is measured via the pressure—the higher the temperature, the more violent the motion and the more the gas resists compression.

We have already formulated a theoretical expression for the speed of a compression wave in an elastic fluid; it is

$$v = \sqrt{\frac{B}{\rho}} \qquad [12.24]$$

which certainly ought to tell us something about the speed of sound in air. Indeed, the equation was first derived by Newton, although he called the term represented by B the "elastic force." Sir Isaac argued that this elastic force equaled the pressure of the atmosphere. That line of thought follows from Boyle's Law, Eq. (11.5), which says that *if the temperature is constant,* $PV =$ constant. Hence, if the pressure increases by a tiny amount ΔP, the volume will decrease by a tiny amount ΔV, where

$$PV = (P + \Delta P)(V + \Delta V) = PV + \Delta P\Delta V + V\Delta P + P\Delta V$$

Neglecting the very small term $\Delta P\Delta V$, this expression yields $P = -\Delta P/(\Delta V/V)$, which from Eq. (12.23) is indeed the Bulk Modulus. So Newton concluded that

$$v = \sqrt{\frac{P}{\rho}} \qquad (13.2)$$

which gives sound a speed of only 979 ft/s. Although Eq. (12.24) works fine for liquids where the compressibility is very small, this application of it to gases is clearly missing something.

The discrepancy was explained in 1816 by Pierre Simon Laplace, who recognized that Newton's derivation required the temperature in the gas to be constant, and that it wasn't constant. The regions that are significantly compressed increase in temperature, while those that are rarefied decrease in temperature (p. 495). Because the waves are long and travel quickly, there is not enough time for heat to be conducted from one region to the other, and so the effect is not neutralized. Newton had overlooked the change in elasticity due to the temperature changes induced by the wave itself. What happens can be appreciated by referring to Fig. 12.40. The springs in the rarefactions are cooled, and there k decreases; the springs in the condensations are warmed, and there k increases. The greater the difference in k on either side of an atom, the more rapidly it propagates the disturbance. There is no overall gain or loss of heat and under such conditions, Boyle's Law does not hold; instead, $PV^{\gamma} =$ constant, where $\gamma = 1.4$ for diatomic gases like air (p. 560) and

$$v = \sqrt{\frac{1.4P}{\rho}} \qquad (13.3)$$

which produces excellent agreement with observation (Table 13.2).

TABLE 13.2 The speed of sound

Material	Temperature (°C)	Speed (m/s)
Vulcanized rubber	0	54
Carbon disulfide	0	189
Air	0	331.45
Water vapor	0	401
Cork		500
Carbon tetrachloride	23	929
Helium	0	970
Chloroform	24	1001
Lead	20	1227
Hydrogen	0	1270
Mercury	≈25	1450
Fresh water	25	1493
Seawater	20	1513
Glycerin	22	1986
Platinum	20	2690
Brass		3500
Copper	20	3560
Brick		3652
Oak		3850
Aluminum		5104
Granite		6000

Example 13.3 Use Eq. (13.3) to determine the speed of sound in air at STP.

Solution: [Given: STP, from Table 10.1, $\rho = 1.29$ kg/m^3, and $P = 1$ atm $= 1.013 \times 10^5$ Pa. Find: v.] From Eq. (13.3)

$$v = \sqrt{\frac{1.4P}{\rho}} = \sqrt{\frac{1.4(1.013 \times 10^5 \text{ Pa})}{1.29 \text{ kg/m}^3}} = \boxed{332 \text{ m/s}}$$

▶ **Quick Check:** This compares well with $v =$ 331.45 m/s at STP.

One thing to observe from Eq. (13.3) is that the speed of sound does not depend on frequency. This fact is confirmed every time you listen to music in a large hall or stadium and all the sounds stay together: the high frequencies reach your ears along with the low frequencies; otherwise only the people in the front rows would hear anything recognizable.

If the pressure of some region of the atmosphere changed while the temperature remained fixed, the density would change in proportion, via Boyle's Law. That change would leave P/ρ constant and v unaltered: the speed of sound in the thin air on a mountaintop is the same as it is at sea level, provided the temperature is the same. Changing the temperature will alter the density of the air, but barometric pressure is determined by the weight of the column of air, which is independent of temperature. Since the density decreases as the temperature increases, the speed will increase with the square root of the temperature. *Over the usual range of temperatures encountered at sea level, the speed of sound changes by about* ± 0.60 m/s *per change of* ± 1.0°C.

Example 13.4 What is the speed of sound at room temperature (20°C) and normal atmospheric pressure?

Solution: [Given: a temperature of 20°C. Find: the speed of sound.] The speed increases from its 0°C value of 331 m/s by 0.60 m/s for each degree increase; hence

$$v = 331 \text{ m/s} + 20(0.60 \text{ m/s}) = \boxed{343 \text{ m/s}}$$

▶ **Quick Check:** The density of air at 20°C can be approximated from Table 10.1 as 1.2 kg/m^3; using Eq. (13.3), $v \approx 343.8$ m/s.

In deriving Newton's equation for v, it was assumed that the pressure change was small and that the $\Delta P \Delta V$ term could properly be neglected. For example, during ordinary speech, the air just in front of the mouth will experience a pressure change ($\pm\Delta P$) of no more than about one-millionth of atmospheric pressure, or about 0.1 Pa. Still, for extraordinarily loud sounds where ΔP is appreciable compared to P (which in most cases is 1 atm), $\Delta P \Delta V$ cannot be omitted. Then the above derivation leads to a Bulk Modulus of $(P + \Delta P)$, which must go into Eqs. (13.2) and (13.3). The change (ΔP) from ambient pressure in the sound wave at the mouth of a cannon can exceed atmospheric. Thus, if ΔP happens to equal atmospheric pressure, the speed of that sound wave will increase by a factor of $\sqrt{2}$ over its ordinary low-volume rate. This accounts for the peculiar fact that under the right conditions (you have to be far enough away), one can actually hear the report of a cannon *before* one hears the order to fire it. *Very loud sounds travel slightly faster than quiet ones.*

13.5 Hearing Sound

The human sound receiver, though its operation is not fully understood, is composed of three subsystems: the outer, middle, and inner ears (Fig. 13.14). Sound intercepted

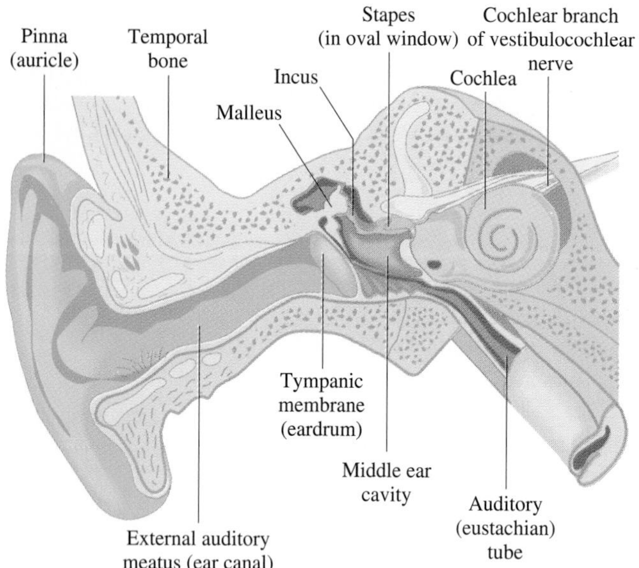

Figure 13.14 The human ear.

by the external ear is channeled via the auditory canal to the eardrum, which is driven into oscillation. This three-part structure is the *outer ear.* Besides aiding in determining the direction of sound, it provides a resonant cavity (p. 459) that amplifies (by about 100%) sounds in the range from ≈3 kHz to ≈4 kHz, making the entire system especially sensitive to this mid-frequency band. The ear is 1000 times more responsive at 1 kHz than at 100 Hz. Were the ear more sensitive to the low-frequency range, we would have to contend with a constant drone of internal body noises.

The *middle ear* links the eardrum, via three pivoted bones, to the oval window. This arrangement increases the force exerted on the flexible window into the inner ear. In addition, since the effective area of the eardrum is about 20 times larger than that of the oval window, there is another increase in pressure.

The *inner ear* is the transducer that converts a mechanical pressure input into an electrical nerve output. It begins at the oval window, which is the entrance to a fluid-filled coiled cavity known as the *cochlea.* The cochlea contains the *basilar membrane* on which is a structure of over 20 000 strands of hairlike receptor cells leading to nerve fibers. Hydraulic compression waves generated at the oval window induce vibrations in the basilar membrane, which flutters transversely along its length. At the front end of the cavity where the membrane is narrow and stiff, it vibrates at high frequencies; toward the rear where the membrane is slack, its resonant frequency is lower. As the membrane oscillates, the receptor cells in the different regions flutter and trigger impulses at a rate that depends on both the amplitude and frequency of the motion. The extent of the displacement of the basilar membrane determines the firing rate of the receptor cells, which, in turn, manifests itself in the perception of loudness.

Pitch

Pitch is a human perception resulting from the sensing of acoustic energy. It is an ear-brain response mainly to the frequency of sound. Certainly we know what we mean by the term: a soprano voice is high-pitched, a bass voice is low-pitched. The pitch of a *pure tone* (a sinusoidal wave with a single fixed wavelength, as approximated by a tuning fork) more or less corresponds to frequency—the higher the frequency, the higher the pitch (Fig. 13.15). Nonetheless, the ear-brain system is far from simple, and there is no one-to-one relationship between frequency and pitch. For example, a high-frequency tone will be heard to increase in pitch as its intensity increases, while a low-frequency tone will cause the opposite response.

Timbre

Another auditory sensation that allows us to distinguish between sounds is *timbre.* Listening to a flute, a trumpet, a saxophone, a violin, or a tuning fork, each producing the same note (same pitch) at the same loudness, there would be little difficulty telling one from the other. The catchall attribute that allows for this distinction is the timbre, which depends primarily on the waveform—that is, the frequencies present, their relative phases, and amplitudes. When the same tones made by these various instruments are picked up by a microphone and displayed on an oscilloscope

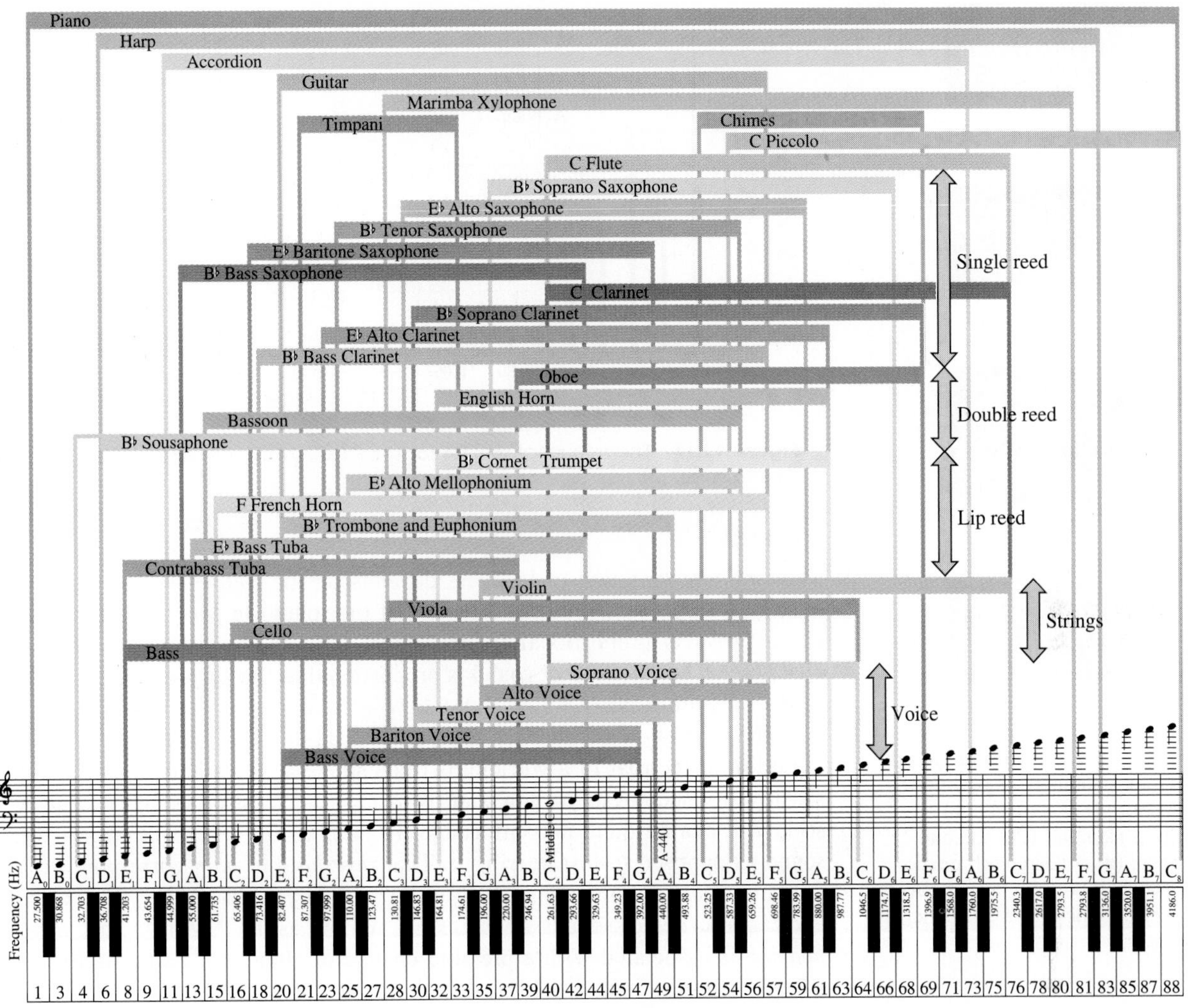

Figure 13.15 The frequency ranges of voices and instruments.

(Fig. 13.16), they are seen to be substantially different. Similarly, the same note produced by a piano and a singer have distinctly different frequency spectra (Fig. 13.17), and the ear immediately distinguishes the vocalist from the accompaniment, even when the sounds are made at the same moment.

An actual tone will contain frequencies higher than the fundamental (or first harmonic), and these are called *overtones.* An overtone need not be a whole-number multiple of the fundamental; that is, it need not be a harmonic. Thus, it is the number of overtones (whether they are harmonics or not) and their relative amplitudes that more than anything else determines timbre. Probably the most important difference between a fine musical instrument and an ordinary one is its timbre, or tone color.

An adult male generates basic voicing vibrations at $\approx$80 Hz to $\approx$240 Hz for speech and upwards of $\approx$700 Hz for song. By comparison, the adult female voice range is $\approx$140 Hz to $\approx$500 Hz in speech and up to $\approx$1100 Hz in song. Given enough control over a wide spectrum of pure-tones ($\approx$80 Hz to $\approx$1100 Hz), we presumably could artificially duplicate all human vocal sounds—a process just beginning with "talking" alarm clocks, elevators, computers, and cameras.

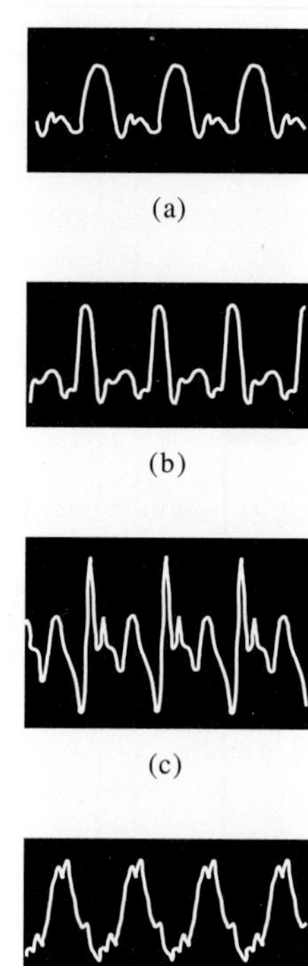

Figure 13.16 Waveforms produced by several instruments all playing the same A_4 note ($f = 440$ Hz): (a) a flute; (b) a trumpet; (c) a soprano saxophone; (d) a violin. Each portion shows roughly $3\frac{1}{2}$ cycles lasting about 8 ms, corresponding to a frequency of 0.44 kHz.

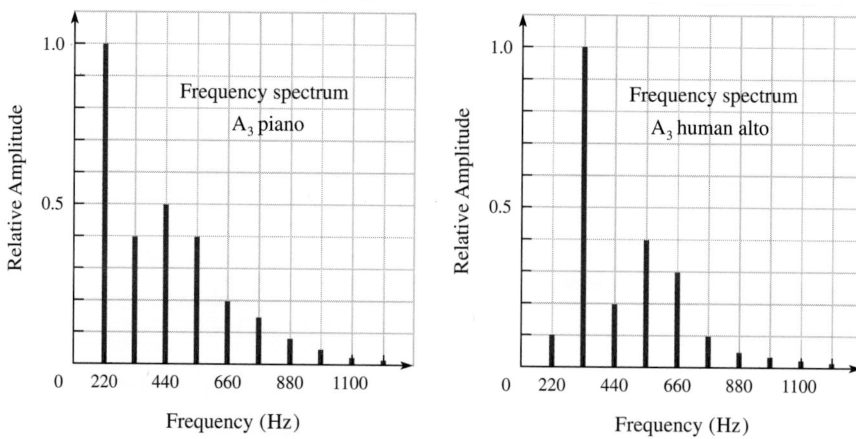

Figure 13.17 The frequency spectra for the A_3 (220 Hz) note sounded by a piano and an alto voice.

Loudness

The auditory sensation of loudness depends on the frequency spectrum, duration, and, most importantly, on the intensity of the sound. In general, the intensity of an elastic wave is proportional to the square of both the frequency and amplitude ($I \propto f^2A^2$). The intensity of sound is proportional to the square of the amplitude (P_0) of the pressure wave ($I \propto P_0^2$).

The breadth of sound intensities that occur in nature, from the faint buzz of a mosquito to the roar of a volcano, is tremendous, and the human ear is remarkably sensitive to much of it. We can bear intensities in excess of around 1 W/m^2, which is so loud it can be quite uncomfortable. In fact, at this level, one begins to feel the sound as opposed to just hearing it. That *threshold of pain* corresponds to a gauge pressure change of only about 10^{-4} atm, or roughly 10 Pa. At the other extreme, the ear can just detect sounds at intensities as low as 10^{-12} W/m^2. That *threshold of hearing* is associated with a pressure change of about 10^{-10} atm, or 10^{-5} Pa. Lord Rayleigh showed that, at 1 kHz, this change is equivalent to air molecules vibrating with an amplitude of only about 10^{-6} mm (see Problem 35).

The ear functions over a pressure range that corresponds to a factor of a million (10^{-4} atm to 10^{-10} atm), and the intensity range—the square of that—extends a factor of 10^{12}. This is an amazing breadth of operation for any sensor—it's the equivalent of using the same device to measure both the diameter of an atom and the length of a football field. Figure 13.18 depicts the limits of sensitivity as a function of frequency.

We judge the relative loudness of a sound not by the difference in intensity between it and some reference, but by their ratio. Doubling the intensity of a faint sound will produce a sensation of some increase in loudness. Doubling the intensity of an already loud sound will nevertheless produce the sensation of the same increase in loudness. In fact, **multiplying any intensity by 10 will generally be perceived as approximately doubling it in loudness**; 10^{-8} W/m^2 sounds twice as loud as 10^{-9} W/m^2, and 0.1 W/m^2 is heard to be twice as loud as 0.01 W/m^2. As we will see presently, that's equivalent to saying that the ear responds logarithmically.

13.6 Intensity-Level

The *intensity-level* (or *sound-level*) of an acoustic wave is defined as *the number of factors of 10 that its intensity is above the threshold of hearing:* $I_0 = 1.0 \times$

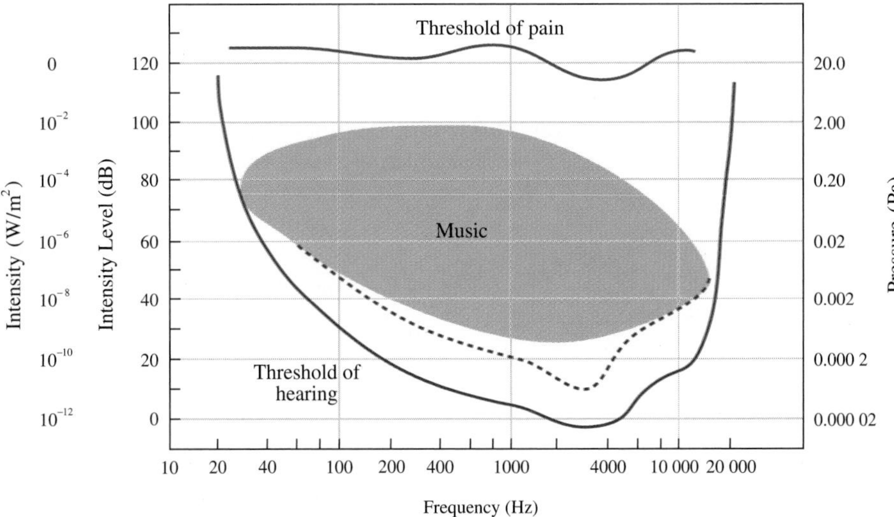

Figure 13.18 About 1% of the people in the United States can hear sound below the levels of the lower curve. Perhaps 50% can hear below the dashed curve. The shaded region corresponds to the levels of normal music.

10^{-12} W/m^2. To honor Alexander Graham Bell, who did important research in acoustics, the unit of sound intensity-level is called the *bel*. The performance of audio amplifiers, tuners, loudspeakers, and so forth, is given in terms of the smaller unit, 1/10 of a bel, which is a *decibel* (abbreviated dB). The decibel, like the radian, is not a physical unit expressible in terms of meters, kilograms, or seconds. It is simply a reminder of the meaning of the number preceding it: 10 dB means something very different from just 10 or from 10 rad.

By definition, the logarithm to the base 10 of any number X equals the power to which 10 must be raised to equal X. In other words, if $X = 10^Y$, then the $\log_{10} X = Y$. Hence, $1000 = 10^3$ and so $\log_{10} 1000 = 3$. If we form the ratio of the intensity of a wave I to the reference level I_0 (remembering how the ear responds to such ratios), and then take its logarithm, we will get the number of factors of 10 that I is above I_0: this number is the intensity-level in bels, and 10 times that gives it in decibels. Thus, the **intensity-level** β in dB of any sound is

$$\beta = 10 \log_{10} \frac{I}{I_0} \tag{13.4}$$

For example, a sound wave with an intensity of 10^{-6} W/m^2 (about the level of normal conversation with a person 1 m away) has an intensity-level of

$$\beta = 10 \log_{10} \frac{10^{-6}\ \text{W/m}^2}{10^{-12}\ \text{W/m}^2} = 10 \log_{10} 10^6 = 10(6) = 60\ \text{dB}$$

The sound has an intensity of 1 million times the threshold intensity and therefore has a sound-level of 6 bel, or 60 dB—it's 6 factors of 10 or 10^6 above threshold (see Table 13.3).

Observe that since $\log_{10} 1 = 0$, the intensity-level at the threshold of hearing is $\beta = 10 \log_{10} 1 = 0$ dB. The entire intensity range from 1.0 W/m^2 to 10^{-12} W/m^2 corresponds to 12 bels (12 multiplicative factors of 10) or spreading the scale out a bit so it can be graphed more conveniently, to 120 dB. A range of 1 to 1 million mil-

TABLE 13.3 Intensity ratios and intensity-level

Ratio of intensities	Intensity-level (dB)
100 000 000 000:1	110
1 000 000:1	60
10 000:1	40
100:1	20
10:1	10
4:1	6
2:1	3
1:1	0
1/2:1	−3
1/4:1	−6
1/10:1	−10
1/100:1	−20
1/1000:1	−30

TABLE 13.4 Approximate intensity-levels in dB

Source	Intensity-level (dB)	
Large rocket	≈180	
Jet engine	140	intolerable
Jet takeoff (30 m − 60 m)	≈125	painful
Rock concert (1.0 W/m^2)	≈120	
Yell into the ear (20 cm)	120	(immediate danger)
Pneumatic hammer	110	
Passing subway train (10^{-2} W/m^2)	100	
Car without muffler	100	(damage after 2 hours)
Shout (1.5 m)	100	very loud
Car horn, loud	95	
Heavy truck (15 m)	90	
City street	80	(damage after
Hair dryer	80	8 hours)
Loud music	80	
Freeway traffic	75	loud
Automobile interior	≈70	
Toilet flushing	≈67	
Noisy store (10^{-6} W/m^2)	60	
Conversation, average (1 m)	60	moderate
Office	50	
Living room, city	40	
Bedroom (10^{-9} W/m^2)	30	quiet
Library	30	
Broadcast studio (10^{-10} W/m^2)	20	very quiet
Whisper	20	
Cat purring	15	
Rustling leaves	≈10	barely audible
Threshold (10^{-12} W/m^2)	0	

lion is thus converted into a range of 0 dB to 120 dB (Table 13.4). *A young healthy ear can distinguish between sound-levels that differ by as little as 1 dB;* 2 dB or greater is more often the case, however.

Several logarithm identities will be useful in performing sound-level calculations. (See Appendix A-3.) These are

$$\log_{10}\frac{A}{B} = \log_{10} A - \log_{10} B \tag{13.5}$$

$$\log_{10} AB = \log_{10} A + \log_{10} B \tag{13.6}$$

$$\log_{10} A^n = n \log_{10} A \tag{13.7}$$

and reversing the process

$$10^{\log_{10} A} = A \tag{13.8}$$

For example, suppose we increase the intensity of a sound by some multiplicative factor; by how much would the intensity-level change? In general, for a change from I_1 to I_2

$$\Delta\beta = \beta_2 - \beta_1 = 10 \log_{10}\left(\frac{I_2}{I_0}\right) - 10 \log_{10}\left(\frac{I_1}{I_0}\right)$$

and so from Eq. (13.5)

$$\Delta\beta = 10 \log_{10}\left(\frac{I_2}{I_1}\right) \quad \text{in dB} \tag{13.9}$$

Increasing the intensity by a factor of 10 changes the sound-level by 1 bel or 10 dB; increasing it by 100 changes β by 2 bels, or 20 dB, and so on. A 60-dB sound compared to a 20-dB sound is 40 dB (4 bels) greater, which means—from Eq. (13.9)—that $\log_{10}(I_2/I_1) = 4$ and the intensity of one is 10^4 times the other. Since every increase of 10 dB corresponds to a doubling in loudness, an 80-dB sound is louder than a 60-dB sound by (two 10-dB steps or) a factor of 4.

Example 13.5 Two public address systems are being compared, and one is perceived to be 32 times louder than the other. What will be the difference in intensity-levels between the two when measured by a dB-meter?

Solution: [Given: a factor of 32 in *loudness*. Find: $\Delta\beta$.] We are given the change in loudness, which correlates to a change in intensity-level by way of the fact that a doubling in loudness is roughly a change of 10 dB. Hence, since $32 = 2^5$, and $\Delta\beta = 5(10 \text{ dB}) = \boxed{50 \text{ dB}}$.

▶ **Quick Check:** $\Delta\beta = 10 \log_{10}(I_2/I_1) = 50$; hence, $\log_{10}(I_2/I_1) = 5$ and $(I_2/I_1) = 10^5$, the intensities differ by 5 factors of 10, each one doubling the loudness, which increases 2^5, or 32-fold.

Example 13.6 Imagine that 10 identical violins are each about to play something at 70 dB. If they join in one at a time, what will be the sound-level as each contributes?

Solution: [Given: 10 sources at 70 dB each. Find: the corresponding sound-levels.] The intensity goes from I to $2I$ to $3I$ up to $10I$. We could solve the problem using Eq. (13.9), but let's go back to Eq. (13.4) to compute β

(continued)

(continued)

and use Eq. (13.6). For 2 violins

$$10 \log_{10} \frac{2I}{I_0} = 10 \log_{10} 2 + 10 \log_{10} \frac{I}{I_0}$$

and the sound-level becomes $\boxed{3 \text{ dB} + 70 \text{ dB}}$. *Doubling the intensity produces a 3-dB sound-level increase.* With 3 instruments playing, the sound-level rises to $\boxed{75 \text{ dB}}$; with 4, $\boxed{76 \text{ dB}}$; with 5, $\boxed{77 \text{ dB}}$, and so on up to 10, when it reaches $\boxed{80 \text{ dB}}$. Ten violins produce an intensity 10 times that of 1 and therefore 1 bel (10 dB) greater, and that essentially corresponds to a doubling of the perceived loudness.

▶ **Quick Check:** $\Delta\beta = 10 \log_{10}(I_2/I_1) = 10 \log_{10} 2 = 3.0$ and $\beta + \Delta\beta = 70 \text{ dB} + 3 \text{ dB}$.

It follows that if we have a 12-W audio system and wish to replace it by one that is equally efficient but which is capable of being twice as loud, we will need to increase its power, not twofold but by a factor of 10, raising it to 120 W.

The metal reeds in a music box mechanism are displaced by rotating pins and set vibrating. Note the different lengths of the reeds.

PRODUCTION AND PROPAGATION OF SOUND

Anything that will cause compression waves in the audible range in a material medium is a source of sound. Smack your hands together or tap a hammer and a sound will be made—it will not be sustained very long, but it certainly will be an acoustical wave (Fig. 13.4). The source itself does not have to vibrate to produce sound, even though that is most often the case. Physically, **noise** refers to an unrelated jumble of disturbances (Fig. 13.19), a *nonperiodic,* randomly changing wave with an essentially continuous frequency spectrum. When the spectrum has a *broad bandwidth encompassing the entire audible range with equal intensity everywhere,* it is called *white noise,* and we hear it as a mix of wind, pouring water, and continuous radio static.

Suppose that the tapping of a hammer occurs at a slow, regular pace. A series of separate impulses will be heard, increasing in rate as the tapping increases, until at roughly around 20 impacts per second, the ear-brain will perceive the clatter as a continuous sustained hum. Galileo noticed that he could produce a tone by drawing a knife blade over a serrated edge, generating a rapid succession of taps. The periodic nature of the process was apparent to him, and he inferred that the pitch was proportional to the rapidity of the tapping.

Between 1793 and 1801, J. Robison devised a technique for producing puffs of air at specific frequencies. A pipe (Fig. 13.20) carried a continuous stream of air to a rotating disc having a circle of evenly spaced holes in it. By adjusting the speed so that the device produced 720 puffs per second, a sound "was most smoothly uttered, equal in sweetness to a clear female voice." When de la Tour (1819) found that it would sing underwater, he gave it the name *siren* after the legendary sea nymphs. The importance of the apparatus was that it provided future researchers, notably Dr. Hermann Helmholtz, with a sound source having a precisely controlled frequency that could be operated across a wide spectrum.

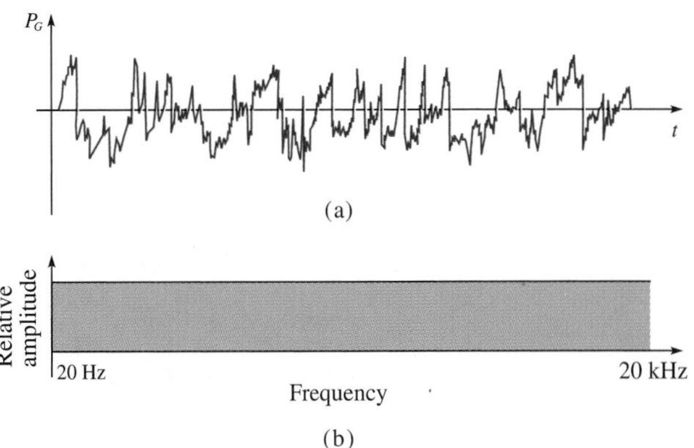

Figure 13.19 (a) Random aperiodic noise is called *white noise* when (b) its frequency spectrum has a constant amplitude all across the audible frequency range.

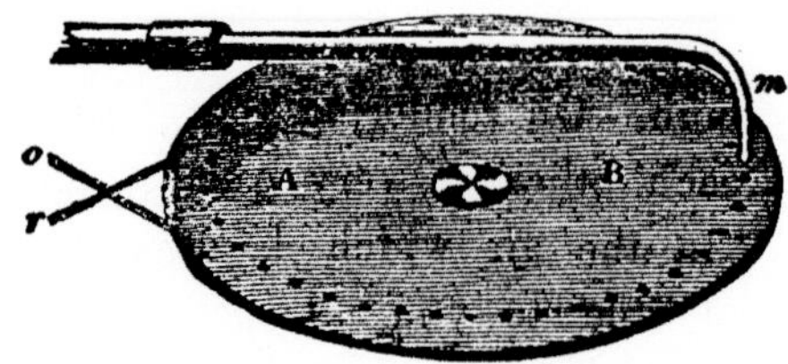

Figure 13.20 Professor Robison's siren. Puffs of air blowing through the holes in a rotating disc created a sound "equal in sweetness to a clear female voice."

13.7 Beats

With a standard source, we can determine the frequencies of other sound-generating devices, such as tuning forks, vibrating wires, and so forth. The most common technique for the purpose (still used by some traditional piano tuners) relies on the phenomenon of **beats**. Mersenne was the first to report the effect, and it was he who gave it that name. Describing the result of playing two organ pipes that differ slightly in pitch, he wrote that they "produce a throbbing or thrumming sound similar to that of beating a tambourine." The effect is a manifestation of a basic wave property called **interference**: two or more waves overlap and produce a resultant wave.

As was seen earlier (Fig. 13.5), two waves of the same frequency superimpose to form a stable resultant wave. When they are in-phase, or nearly so, crests fall on crests, troughs on troughs, and the resultant is large. When they are 180° out-of-phase, or nearly so, crests fall on troughs, and the resultant is small or even nonexistent. Figure 13.21 depicts the overlapping and interference of two pure tones having slightly different frequencies, traveling in the same direction. The two waves, because they are not quite able to stay in-step, gradually (and periodically) go in- and out-of-phase; at one instant, rising and falling together, they reinforce one another to form a large amplitude sound. Then, one rising while the other is falling, they subtract to cancel each other altogether. The two waves of frequency f_1 and f_2 combine to produce a carrier wave having the average frequency $(f_1 + f_2)/2$ that is *amplitude modulated,* rising and falling with the **beat frequency** of $(f_1 - f_2)$. What you hear is a tone—the carrier—building and fading in intensity at the beat frequency. (A mathematical analysis of the effect is explored in Problem 75.)

Example 13.7 A siren is used as a frequency reference during the fine adjustment of a middle-A tuning fork. The siren is slowly reduced in frequency and, together with the tuning fork, produces a quavering tone. This warbling also decreases in frequency until the siren reaches and holds at 440 Hz. At that point, the tone varies in loudness from maximum to minimum and back to maximum in $\frac{1}{4}$ s. What is the frequency of the fork?

Solution: The quavering tone tells us that this is a problem in beats. [Given: $f_1 = 440$ Hz and the period of the beats is $\frac{1}{4}$ s. Find: the frequency of the fork, f_2.] The beat frequency is 1 over the beat period; hence

$$\frac{1}{(\frac{1}{4}\text{ s})} = (f_1 - f_2) = 440\text{ Hz} - f_2$$

and

$$\boxed{f_2 = 436\text{ Hz}}$$

▶ **Quick Check:** $1/0.25\text{ s} = 4$ Hz and since the siren was coming down, its frequency must exceed that of the tuning fork by 4 Hz.

A variety of vibrating objects can serve as acoustical sources that will produce sustained sounds with discernable pitches. An unburdened honeybee fluttering its wings at about 440 Hz generates a buzz at what has come to be the musician's standard A_4 (middle-A) frequency. Strings and reeds, membranes, rods, bars, lips, air columns, and tongues can all be set vibrating, all making sounds, even ones we call music, whatever that is.

13.8 Standing Waves

When a wavetrain of any kind is created in some real, finite medium (whether a string, a drum, or the Earth itself), it will propagate outward until it encounters an end or boundary. There, some fraction of the wave-energy will usually be reflected backward (p. 426), and if the original disturbance is sustained, the medium will quickly fill with waves traveling back and forth over one another. These disturbances will combine or interfere to form a steady-state distribution of energy known somewhat paradoxically as a *standing* or *stationary wave.* This situation is remarkably commonplace. It happens within every sort of musical instrument from the piano to the triangle (and, for that matter, within every kind of laser as well). It occurs in the head, throat, and mouth when we speak, in the ear canal when we listen, and in the shower stall when we sing. Standing waves exist when you ring a bell, strike a tuning fork, run water from the tap into an open sink, or blow across an empty soda bottle.

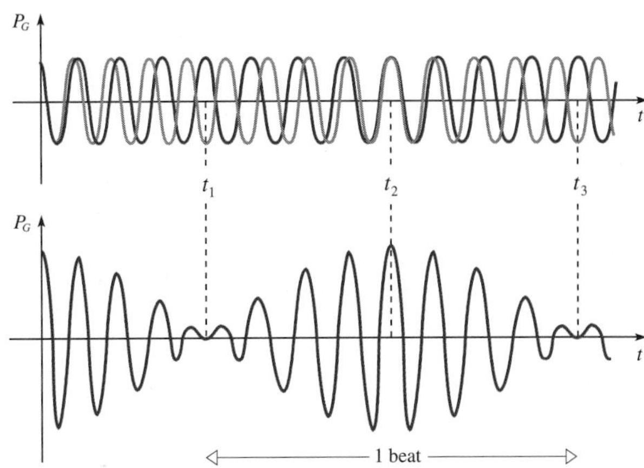

Figure 13.21 When two signals of different but nearly equal frequencies (f_1 and f_2) superimpose, they create a beat pattern, which is true of sound as well as light. The intensity rises and falls with a frequency equal to the difference ($f_1 - f_2$). Note that at t_1 and t_3 the signals are 180° out-of-phase, while at t_2 they are in-phase.

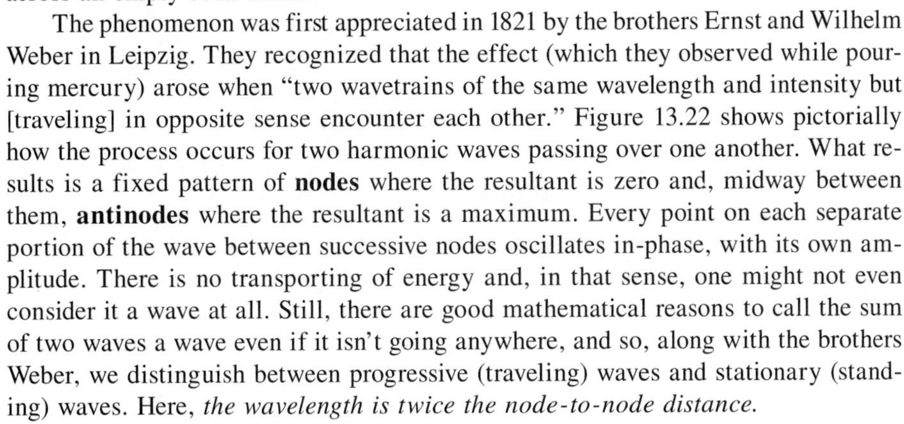

The phenomenon was first appreciated in 1821 by the brothers Ernst and Wilhelm Weber in Leipzig. They recognized that the effect (which they observed while pouring mercury) arose when "two wavetrains of the same wavelength and intensity but [traveling] in opposite sense encounter each other." Figure 13.22 shows pictorially how the process occurs for two harmonic waves passing over one another. What results is a fixed pattern of **nodes** where the resultant is zero and, midway between them, **antinodes** where the resultant is a maximum. Every point on each separate portion of the wave between successive nodes oscillates in-phase, with its own amplitude. There is no transporting of energy and, in that sense, one might not even consider it a wave at all. Still, there are good mathematical reasons to call the sum of two waves a wave even if it isn't going anywhere, and so, along with the brothers Weber, we distinguish between progressive (traveling) waves and stationary (standing) waves. Here, *the wavelength is twice the node-to-node distance.*

Standing Waves on Strings

One of the most common arrangements is the taut string with both ends held immovable, which is the basic setup for the piano and all stringed instruments (Fig. 13.23). Energy can be continuously pumped into such a system by attaching one end to a source, like a driven tuning fork or vibrator, which has such a small amplitude that it might well be considered a fixed point. Envision a harmonic wave with a tiny amplitude continuously introduced at the left, moving to the right, down the length of the string L at a speed v, as in Fig. 13.24 (p. 462). Presume that it will be reflected back to the left without any loss in amplitude but with a 180° phase shift (p. 427). This reflected wave will soon arrive back at the origin (that is, at the vibrator), there to again be phase-shifted another 180° as it is reflected (from that effectively fixed end) off to the right. What conditions have to be met if this twice-reflected wave is to move off exactly in-step with the continuously flowing original disturbance? We want both waves traveling to the right to be in-phase, meaning that the round-trip time $2L/v$ must equal the period T since the two 180° phase shifts due to reflection add up to 360° and so essentially cancel. Inasmuch as $v = f\lambda$, $2L/v = T$ yields

$$L = \tfrac{1}{2}\lambda$$

and the round trip is one wavelength long. This being the case, each trip down and back will bring the reflected wave to the fork just in time to combine in step with the newly emitted wave. Each such trip will cause the wave in the system to grow slightly larger.

A pail used to wash a floor contained a suspension of fine dirt particles in water. When placed in a curved sink, the pail gently rocked along a fixed axis, setting up standing waves and distributing the particles in ridges as they settled.

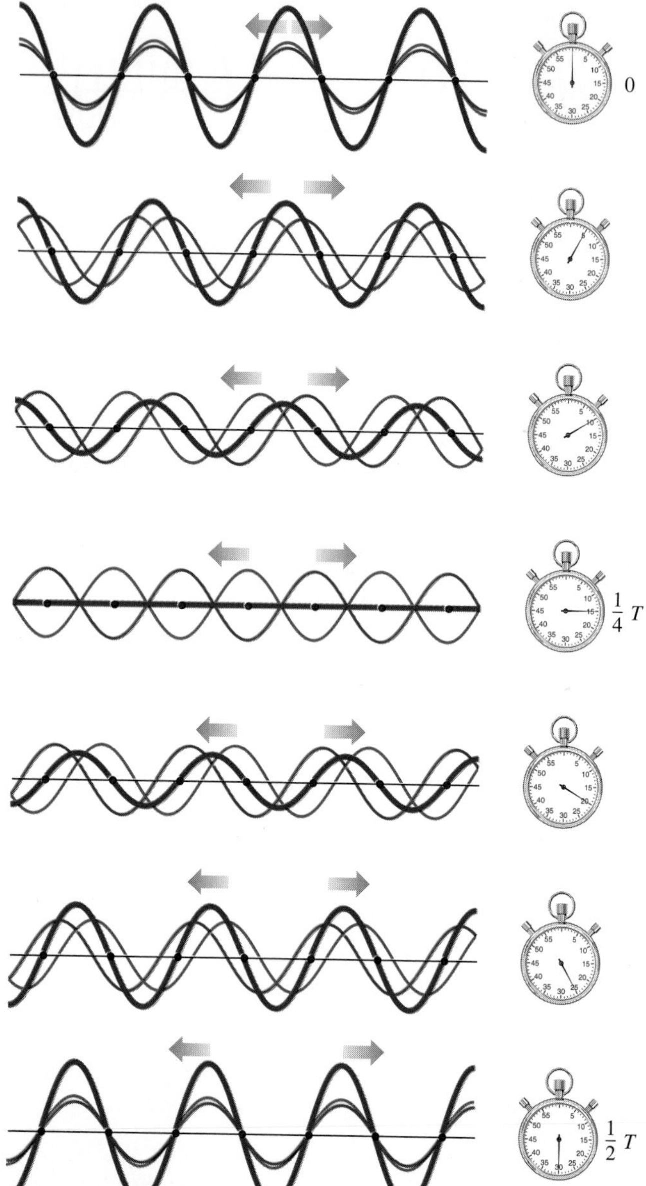

Figure 13.22 The creation of standing waves. Two waves of the same amplitude and wavelength traveling in opposite directions form a stationary disturbance that oscillates in place.

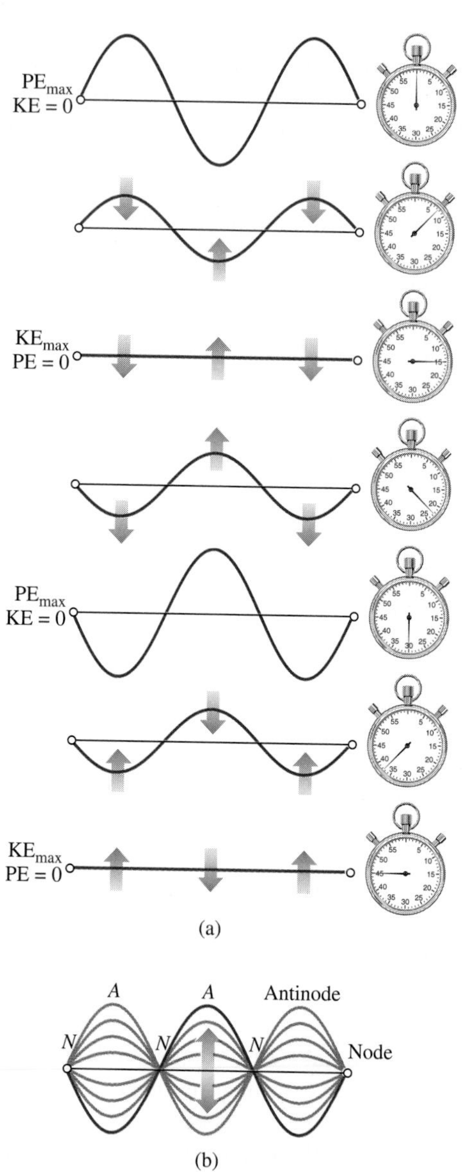

Figure 13.23 (a) A standing wave on a string that has both ends fixed. (b) represents a composite of all the configurations demonstrated in (a).

As can be seen in Fig. 13.24, if the length of the string is half a wavelength, its two ends will remain motionless as the center oscillates as an antinode. This synchronization of the waves, which depends on a matching of the vibrator's frequency with the size of the medium, is a resonance effect. Initially, the disturbance on the string will continue to build as energy is pumped in until the increasing frictional losses stabilize the system. At that point, the energy provided by the vibrator at each oscillation equals the sum of the losses, and there is no more growth of the wave, whose amplitude is nonetheless much larger than that of the source. *The ability to amplify the input is an extremely important feature of standing-wave systems.*

Similarly, if the ends are fixed and the string is displaced, it will naturally oscillate in this configuration, or *mode*. Yet, unless every point on the string is carefully displaced along a smooth sine curve, there will be higher-order standing-wave modes present as well. These occur when the round-trip time corresponds to $2T$, $3T$, $4T$, and so forth, or NT where N is a whole number (Fig. 13.25). Then

$$L = \tfrac{1}{2}N\lambda \tag{13.10}$$

and the wavelength of the Nth mode is $\lambda_N = 2L/N$. Accordingly, the pattern shown in Fig. 13.23 comes into being when $N = 3$. Stretch a long rubber band between your two hands and then pluck it. Now repeat the process holding it first horizontally and then vertically in front of the screen of a TV set or computer—these flicker rapidly and so serve as stroboscopes.

We can find the resonant frequencies f_N of the system using the fact that speed equals frequency times wavelength, along with Eq. (13.10), which becomes $L = \frac{1}{2}Nv/f_N$, and $f_N = Nv/2L$. Substituting from Eq. (12.21) for the speed of a transverse wave on a string, we get

$$f_N = \frac{N}{2L}\sqrt{\frac{F_T}{\mu}} \qquad (N = 1, 2, 3, 4, \ldots) \tag{13.11}$$

Notice that each frequency is a whole-number multiple of the *fundamental*, or lowest resonant frequency: $f_N = Nf_1$. Thus, each higher resonant frequency is a harmonic. Keep in mind that the Nth harmonic will have N antinodes, so it is easy to recognize when you see it. Usually each string of an instrument is played in its fundamental mode. In the case of a guitar, for example, the strings sound different, even though they all have the same length, because the tensions and linear mass-densities are different. By "fingering" a string; that is, by pressing it down onto a fret, thereby fixing a node at that point and effectively shortening the length of string that is able to vibrate, the player increases the fundamental frequency.

Galileo understood that an oscillating string causes "the air immediately surrounding the string to vibrate and quiver," thus producing a sound of the same frequency. A plucked or struck string actually vibrates in several modes at once and generates several tones (a fact observed by Mersenne). Strike a key on a piano, and the string will generate a strong fundamental along with a number of much quieter overtones (Fig. 13.17), which is what gives the instrument its characteristic tone color.

Example 13.8 What must be the tension in the E_5-string of a violin if it is to be tuned to 660 Hz? The length of the string from bridge to peg is 330 mm, and its mass-per-unit length is 0.38 g/m.

Solution: [Given: $f_1 = 660$ Hz, $L = 0.330$ m and $\mu =$ 0.38 g/m. Find: F_T.] The violin is tuned to a fundamental of 660 Hz; hence, $N = 1$ in Eq. (13.11), and so

$$F_T = \mu(2Lf_1)^2 = (0.38 \times 10^{-3}\text{ kg/m})4(0.330\text{ m})^2 \times (660\text{ Hz})^2$$

and

$$\boxed{F_T = 72\text{ N}}$$

or about 16 lb.

▶ **Quick Check:** $f_1 = \dfrac{1}{2L}\sqrt{\dfrac{72\ N}{0.38 \times 10^{-3}\text{ kg/m}}} =$ 660 Hz.

Because vibrating strings are not good at pushing large amounts of air, they are not, by themselves, very loud. In instruments, strings are therefore coupled to a large section of the body in such a way as to transmit vibrations to it—the sounding board in pianos or the sounding box in violins and guitars. An alternative is to couple the strings to a pick-up and amplify the sound electronically.

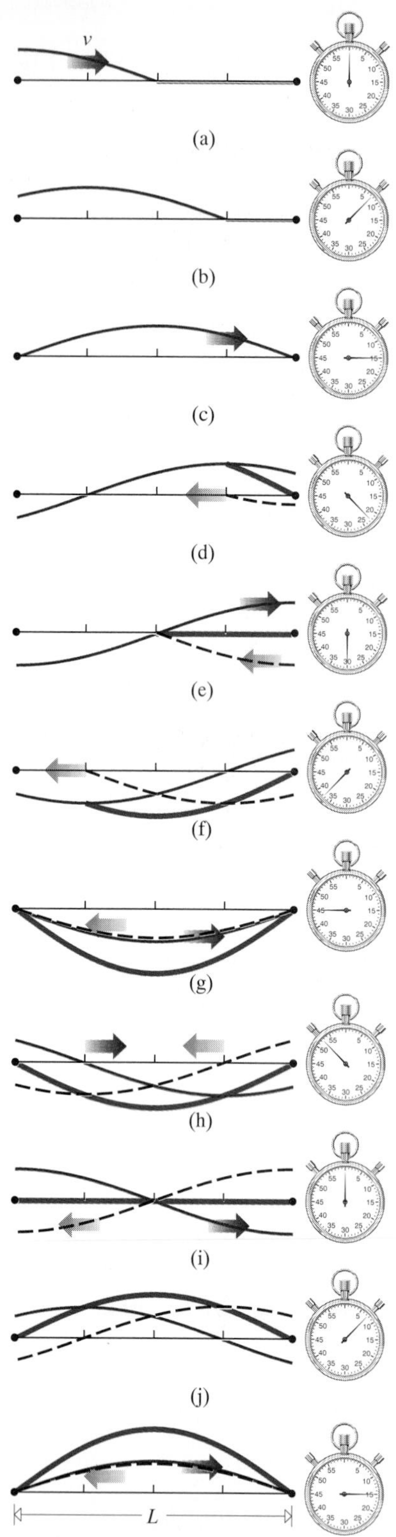

Figure 13.24 A standing wave. A wave enters from the left (a) and is reflected (dashed) from the termination point on the right (d). Now there are two overlapping waves that combine to form a standing wave (red), oscillating (g through k) in place.

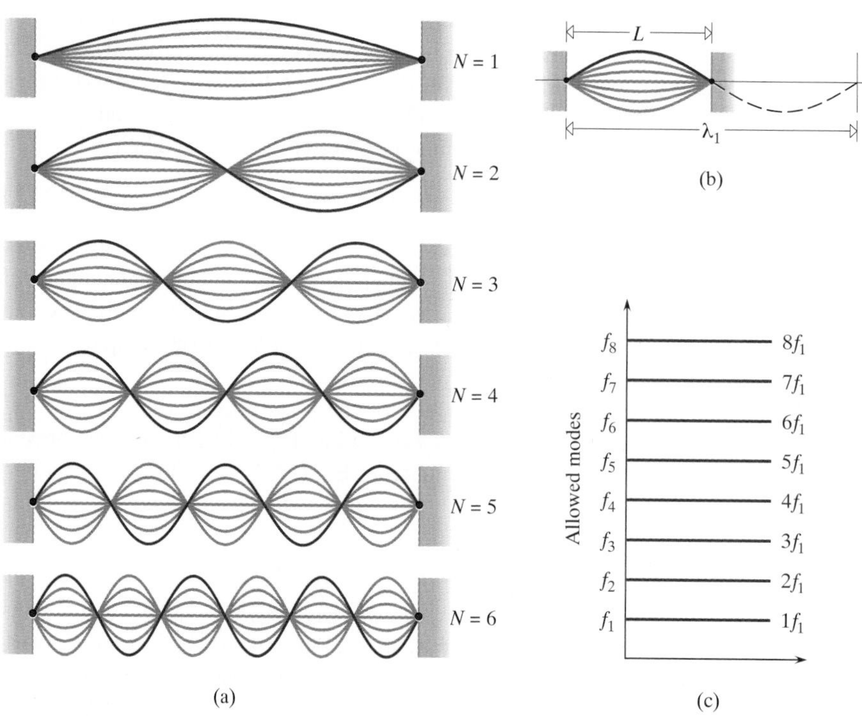

Figure 13.25 (a) Standing-wave modes with a node at both ends. (b) The wavelength of the $N = 1$ fundamental equals $2L$. (c) Allowed modes of oscillation.

Standing Waves in Air Columns

Standing waves can be sustained by the air within almost any shaped chamber from the vocal tract to the shower stall. Indeed, we speak and hear and make music via the acoustical resonance of air-filled chambers of one sort or another that amplify initial sounds. Most musical instruments driven by streams of air (organs, woodwinds, and brasses) utilize this same kind of standing-wave mechanism for increasing intensity.

There are many ways to excite a column of air into oscillation: most woodwinds use a small vibrating reed; the brasses rely on the vibration of the player's lips; and the flute and organ pipe are driven by a fluttering jet of air that pumps the column (Fig. 13.26). Similarly, blowing across the top of a soda bottle sets the air within it vibrating. Whatever the source, the air column initially oscillates with a wide range of frequencies, but only those corresponding to the standing-wave modes of the particular chamber will be sustained and amplified. There are three basic configurations for tubular enclosures: closed at both ends, closed at one end, and open at both ends.

Figure 13.27 shows a chamber closed off at the right end by a wall and sealed at the left end with a slightly movable piston or diaphragm, something like the situation that exists when you hum with your throat or mouth closed. Because the ends are essentially fixed, they will always be the locations of displacement nodes—the air molecules cannot move beyond these points (and, as we saw on p. 445, that means these are also pressure antinodes). Small oscillations of the piston will excite the air column, and a standing-wave pattern will occur. Longitudinal sound waves do not shift phase on reflection from a fixed surface as do the transverse waves on a string—*a rarefaction is reflected from a wall as a rarefaction.* Even so, because there are two 180° shifts for a string with both ends fixed, its standing-wave dis-

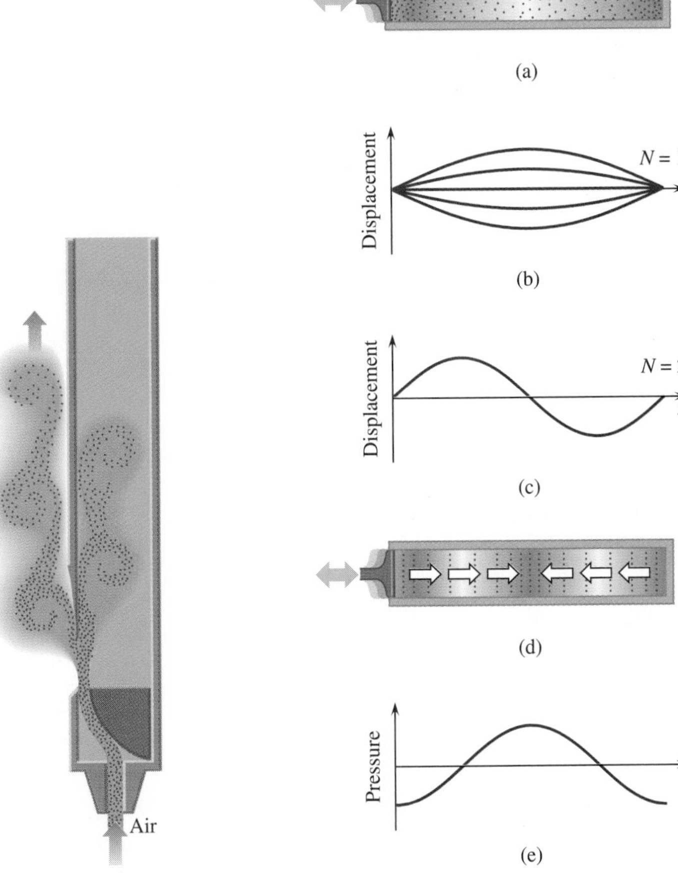

Figure 13.26 An organ pipe. The primary driving mechanism is a wavering, sheetlike jet of air from the flue-slit, which interacts with the upper lip and the air column in the pipe to maintain a steady oscillation.

Figure 13.27 (a) A sealed pipe. The dark blue regions are at density or pressure antinodes. (b) The fundamental mode. (c) The second harmonic at the instant a positive peak is at $x = \frac{1}{4}\lambda_2$. (d) The direction of motion of the molecules at that moment. (e) The associated pressure, which is 90° out-of-phase with the displacement represented in (c).

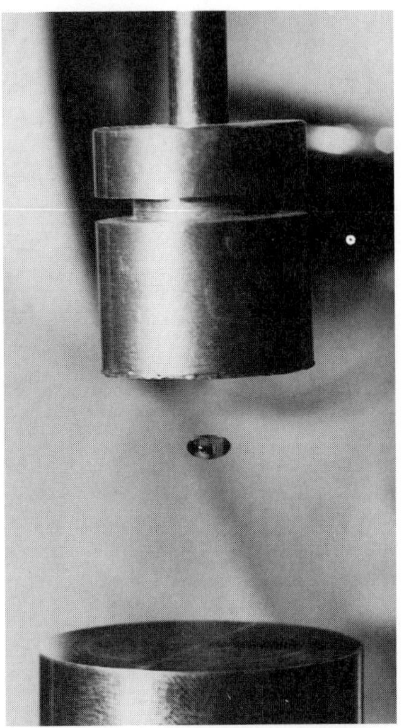

Ultrasonic levitation. A drop of water floating in a so-called energy well. Ultrasonic waves, one traveling up, the other down, create a standing-wave pattern. The suspended object is confined to a nodal region.

placement pattern will be the same as that of the sound wave in a closed chamber. Here again

$$[\textit{both ends closed}] \qquad L = \tfrac{1}{2}N\lambda \qquad [13.10]$$

and the wavelength of the Nth mode is $\lambda_N = 2L/N$. This is all fine, but if you want to really make some intense sounds, you must allow the energy to escape the instrument more efficiently: you must open your mouth and sing out, or open the end of the chamber.

The resonant chamber or cavity can be open either at one end or both, and there are organ pipes of each kind, just as there are musical instruments of each kind: the soda bottle and the trumpet versus the flute and piccolo. The organ pipe open at one end can be thought of as half the closed chamber of Fig. 13.27. At the hole where it is open to the atmosphere, the gauge pressure is zero and a pressure node (Fig. 13.28)

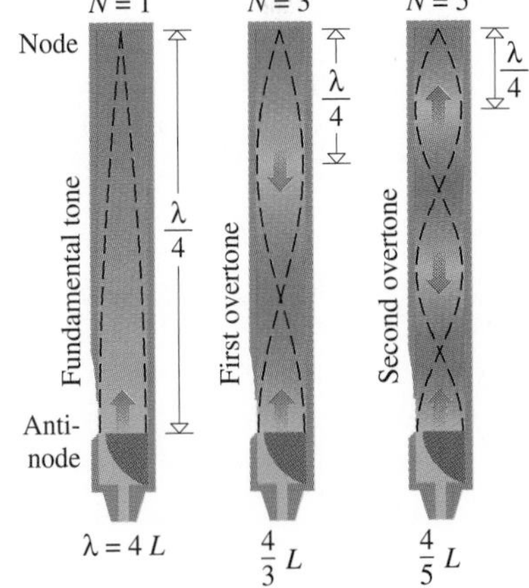

Figure 13.28 A few standing-wave modes for an organ pipe open at its lower end. The arrows show the direction in which air moves for half the cycle, whereupon the modes reverse. The dashed lines indicate the displacement nodes and antinodes. All brass instruments are closed at one end by the mouth of the player.

always exists there. Unimpeded at that point, the molecules of air can move freely and a displacement antinode exists there. This situation happens because *sound waves are phase-shifted by 180° when they reflect off an open end.* It is as if the compression wavefront travels more easily in the open air, stretching the bonds with the layers behind it and sending back a reflected rarefaction. *At an open end, a compression is reflected as a rarefaction and vice versa.* It should be mentioned that the antinodes do not actually occur precisely at the open end, but this effect is a small one for narrow pipes, and we can overlook it here. For the singly-open pipe

[*one end open*] $$L = \tfrac{1}{4}N\lambda \qquad (13.12)$$

where N is an *odd integer.* Again, since $v = f\lambda$, it follows that $L = \frac{1}{4}Nv/f_N$ and the resonant frequencies are given by

$$f_N = \frac{Nv}{4L} \qquad (N = 1, 3, 5, \ldots) \qquad (13.13)$$

Here v is the speed of sound in air. Pressing on a valve of a trumpet opens a loop of tubing into the air stream, effectively adding to the length of the pipe and increasing the wavelength and decreasing the pitch.

Example 13.9 As we saw earlier, the auditory canal, which is essentially a narrow chamber closed at one end by the eardrum, is a standing-wave cavity. Knowing that it is about 2.7 cm long determines the fundamental frequency that will be amplified at room temperature.

Solution: [Given: $L = 2.7$ cm and $v = 343$ m/s. Find: f_1.] From Eq. (13.12) with $N = 1$, $\lambda = 4L = 4(0.027 \text{ m}) = 0.108$ m and so

$$f_1 = \frac{v}{\lambda} = \frac{343 \text{ m/s}}{0.108 \text{ m}} = \boxed{3.2 \text{ kHz}}$$

▶ **Quick Check:** 3 kHz is in the mid-frequency audible region where it should be (p. 452). As sound near that frequency flows into the canal, it rapidly builds in amplitude until it is many times greater than the ambient value.

The doubly-open pipe of Fig. 13.29 must have displacement antinodes at both ends and so must have a node at its middle. Clearly, the pipe has to be half a wavelength long; that is

[*both ends open*] $$L = \tfrac{1}{2}N\lambda \qquad [13.10]$$

and $$f_N = \frac{Nv}{2L} \qquad (N = 1, 2, 3, \ldots) \qquad (13.14)$$

Except for the location of the nodes, this result is the same one we obtained for the closed pipe and the string fixed at both ends. The flute is an open pipe, open to the atmosphere both at the mouthpiece or embouchure and at the far end. Like many wind instruments, it contains a row of holes that can be covered by the fingers. Opening any one has the effect of cutting the tube off at that length and therefore shortening the wavelength and raising the pitch (Fig. 13.30). As a rule, blowing gently on a pipe (or soda bottle) primarily excites the fundamental. When air is blown much more forcefully across the opening of a flute, the column can resonate at the second harmonic, and the entire scale will be raised one octave corresponding to this doubling.

The human vocal tract—larynx, pharynx, and nasal and oral cavities—can crudely be thought of as a singly-open pipe with the vocal cords at the closed end and the mouth at the open end. It therefore sustains a series of standing-wave modes

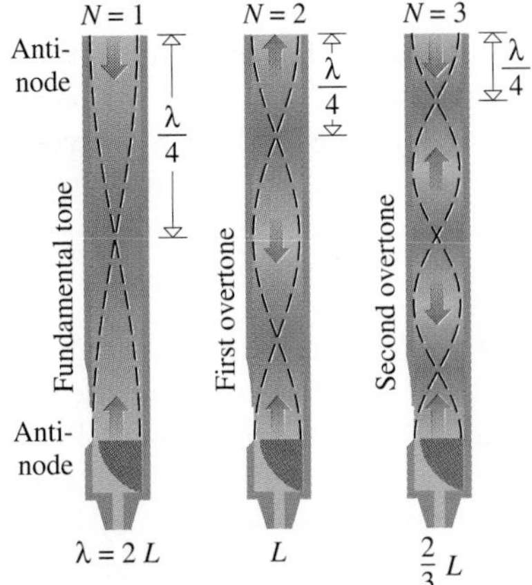

Figure 13.29 A few standing-wave modes for an organ pipe open at both ends. The arrows show the direction in which air moves for half the cycle, whereupon the modes reverse. This time, both ends of the tube are at atmospheric pressure ($P_G = 0$) and there are pressure nodes at both.

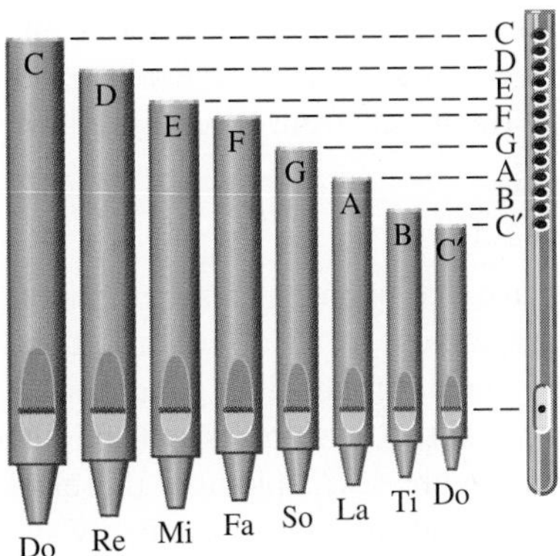

Figure 13.30 Organ pipes and a flute. The length of the resonant chamber determines the pitch.

that amplify the sounds made primarily by the vocal cords. Unlike other musical instruments, this one changes its shape. When the fundamental generated by the vocal cords is matched by a vocal tract resonance at that frequency, the amplitude of the note will be a maximum. Such is the business of a skilled singer.

The pipes of a large organ.

13.9 Sound Waves

The means of propagation of optical and acoustical energy, though totally different in physical nature, are remarkably similar in many regards. Our present emphasis will be qualitative, leaving the more quantitative treatment for the discussion of light. Here, we briefly introduce some aspects of the reflection, refraction, interference, and diffraction of sound that are applicable to all waves.

Reflection

Waves are reflected when they encounter the boundary between two physically different media. Figure 13.31 shows a succession of plane wavefronts reflecting from a smooth, flat surface. Notice how the sound reflects off at the same angle at which it impinges—that behavior has been known from observations for centuries. As da Vinci put it, "the sound of an echo will strike and rebound to the ear at equal angles"—light does the same thing.

In addition to redirecting sonic wavefronts, reflecting surfaces (sonic mirrors) can reshape them as well. When a small source such as a ticking watch is positioned at the focus of a parabolic surface (Fig. 13.32), the diverging

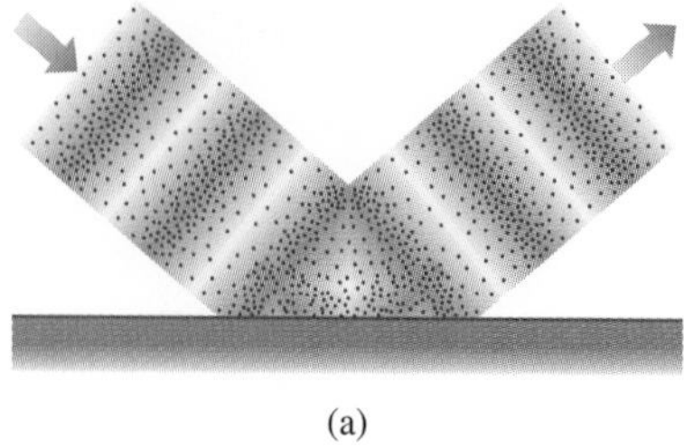

(a)

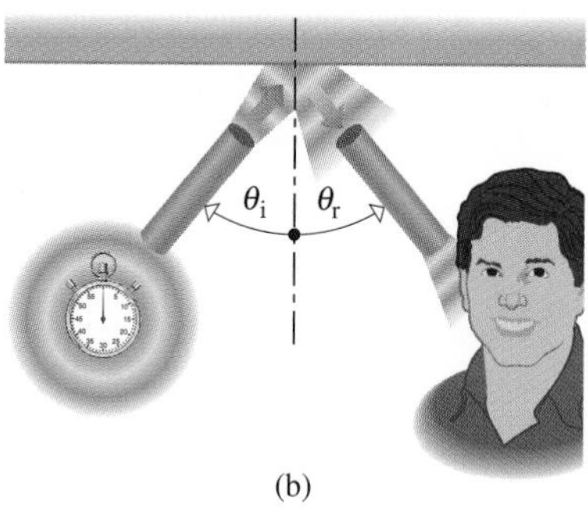

(b)

Figure 13.31 (a) Plane waves reflecting off a surface. (b) Sound reflects from a smooth surface such that the angle of incidence equals the angle of reflection.

spherical sound waves that are emitted will be converted into plane waves. This beam then travels to the other mirror, where it is detected. The arrangement makes it possible to hear the ticking at up to about 10 m. In fact, by using two ordinary soup bowls as mirrors and tilting the receiver so it points toward your ear, you can hear the watch from several feet away—try it.

Refraction

A wave that begins as a sphere in a homogeneous isotropic medium will continue to propagate out in all directions at the same speed and will stay a sphere. If, on the other hand, the medium is not uniform, the speed will not be constant in all directions and the wavefronts will be altered accordingly, a situation known as **refraction**. Figure 13.33 pictures the effects of a wind (increasing in altitude) blowing (a) *with* and (b) *against* a sound wave. Refraction occurs because the speed of sound is increased when it moves along with a wind and decreased when it moves against a wind. The same thing happens (Fig. 13.34) when the density of the medium varies as a result of temperature variations. As a consequence, it is possible to hear things out of doors from farther away on cold days (especially over a body of water, where the air temperature is more likely to increase with altitude) than on warm days. Most often, the air near the ground is warmer than that above it, and distant sounds tend to be deflected upward. That's why you can often see a flash of lightning (beyond around 23 km or so away) and never hear the associated clap of thunder. A thunderhead is typically about 4 km high, and the sound from it heading toward the ground slowly climbs upward and away, never reaching a distant observer.

Interference

As with all waves, the overlapping of two sounds of the same frequency can manifest interference. A stereo audio system (Fig. 13.35) driven by a generator at 1 kHz will fill a room with waves about 34 cm long. Where the two waves overlap, an interference pattern will exist. A listener moving around in the room (with one ear covered) would hear a dramatic variation in intensity going from almost zero, where the waves are out-of-phase by 180°, to loud maxima, where they are in-phase.

When music is playing, a room is filled with many such patterns, each corresponding to a different frequency and each with all its maxima (except for the central one) and minima located at different places. This, along with the inevitable reflections, tends to obscure the interference pattern. Still, if you want the best sound in the middle of the room, the speakers have to pump in-phase, making compressions and rarefactions in step. Reversing the leads to one speaker will cause it to be 180° out-of-phase with the other. That creates a line of nodes running down the middle of the room, along which the intensity is reduced.

Figure 13.32 The sound from the watch is reflected by the parabolic mirror. It is beamed out as a plane wave and then focused down to a microphone. The modern equivalent is used for eavesdropping and picking up the sounds of the play at football games.

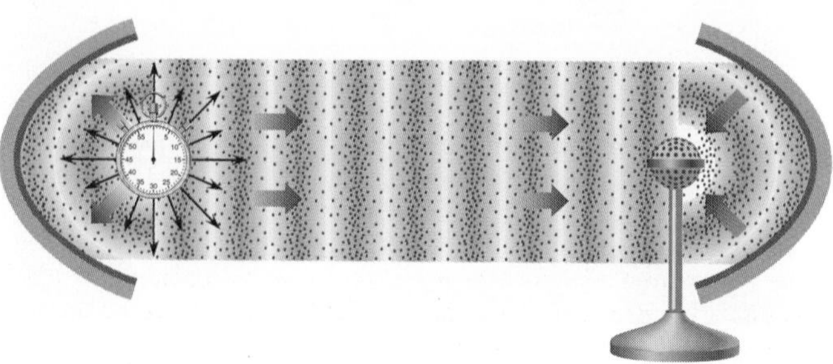

Diffraction

Two people walking through a forest can hold a normal conversation even though there might be a line of trees between them. There are never substantial "shadow" regions behind the trees where no sound reaches. Sound waves, like all waves, more or less bend around obstructions, which is in part why you can hear around corners, and why you can be heard even if you talk with a hand directly in front of your mouth. *Whenever a portion of a wavefront is obstructed in any way (such that any part of it changes its*

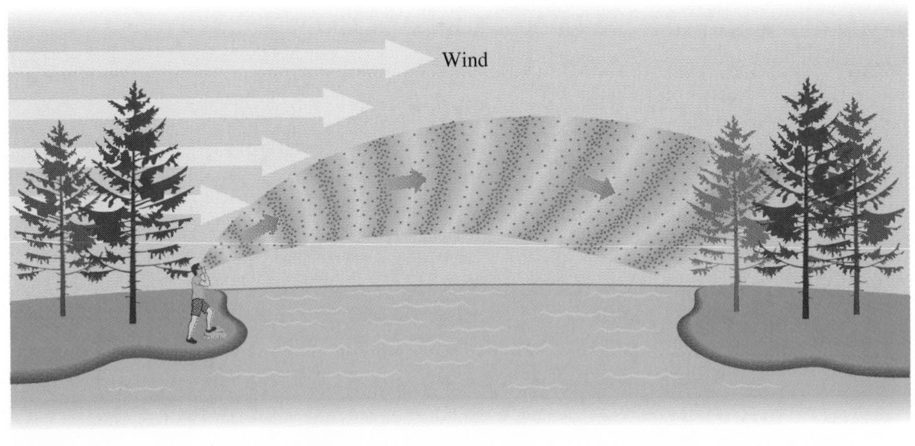

(a)

(b)

Figure 13.33 The speed of sound increases in the direction of the wind. Since the frequency of sound is unaltered, λ must change, and that bends the wavefronts and alters the direction of propagation.

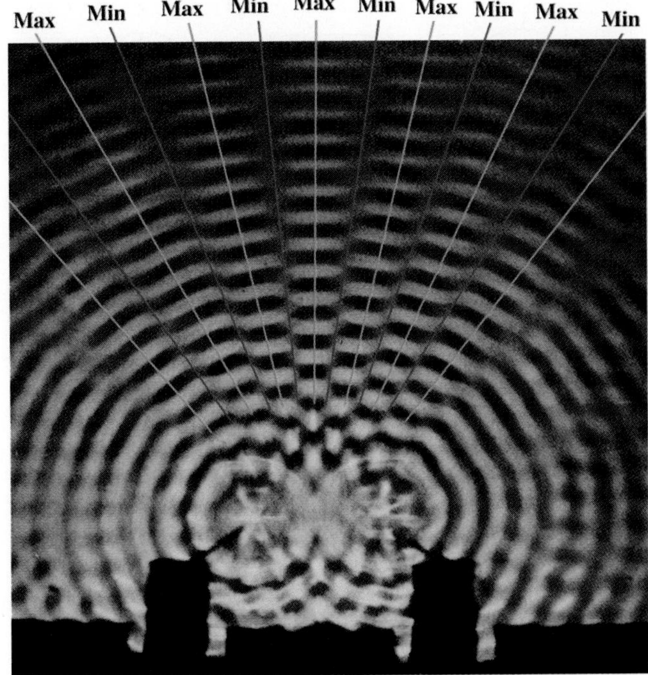

Water ripples formed by two in-phase oscillators. The waves combine to form a region up the center where the disturbance is a maximum. There, we see peaks (bright) and troughs (dark) forming a long wedge shape. On either side of this wedge is a narrow gray band along which the disturbance is canceled. These are interference minima and are highlighted with magenta lines. The gold lines lie along the maxima.

(a)

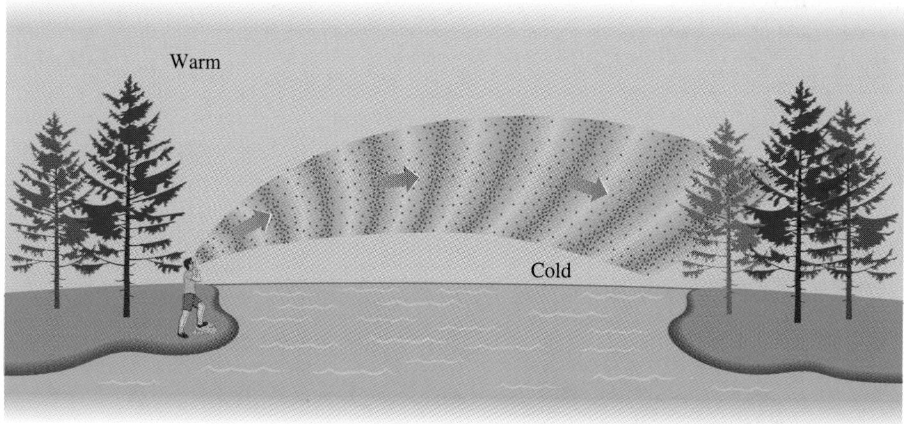

(b)

Figure 13.34 The speed of sound is greater in warm air and the wavelength is longer, which causes a bending of the wavefronts. The same idea has been used to make lenses that reconfigure sound waves. Lenses will be discussed later when light is considered.

Figure 13.35 The interference of two sound waves. Where a crest (dark region) overlaps a crest, or a trough (light region) overlaps a trough, there will be reinforcement (•). Where a trough overlaps a crest, there will be cancellation (○). Compare this with the photo on p. 467.

phase or amplitude or both), there will be some alteration in the wave's direction of propagation as it progresses beyond the obstruction. This deviation from straight-line propagation is known as **diffraction** (Ch. 27).

A general rule maintains that diffraction will be pronounced when the obstacle around which the wave is passing or the aperture through which it is traveling is small compared to the wavelength. That's why a tree will cast a substantial shadow in light (which has a tiny wavelength) and not in sound (which has a comparatively large wavelength). Sound waves quickly close up as they advance past the tree.

When most of the sound energy of a wave diffracts beyond an obstacle, little of it will remain to be reflected backward. Therefore reflection, in a sense, is the complement of diffraction. If the reflection from a body is to be enhanced, λ should be smaller than the target—a conclusion already discussed as it relates to bats, moths, and sonar (p. 432).

Imagine plane waves impinging on a hole in an opaque screen (Fig. 13.36)—the sound of distant traffic pouring indoors through an open window. If the wavelength is small compared to the size of the hole, there will be little diffraction, and the sound will propagate forward essentially along a straight line. On the other hand, if λ is comparable to or larger than the hole, the transmitted wave will "bend" well beyond the straight perpendicular projection of the aperture. You can hear outside noises way off to the side of an open window. The same thing will happen to a loudspeaker—it behaves as if it were an illuminated hole (Fig. 13.37). A speaker that is, say, 0.15 m (6 in.) in diameter projects waves of 100 Hz ($\lambda = 3.4$ m) in all forward directions almost as if it were a point source. By comparison, at 10 kHz ($\lambda = 0.03$ m), the radiation from the same speaker will be predominantly in a narrow ($< 30°$) cone in the forward direction (Fig. 13.36). That is one reason why tweeters—high-frequency speakers—are made small (and it is something to consider when finding a place to sit and listen to recorded music).

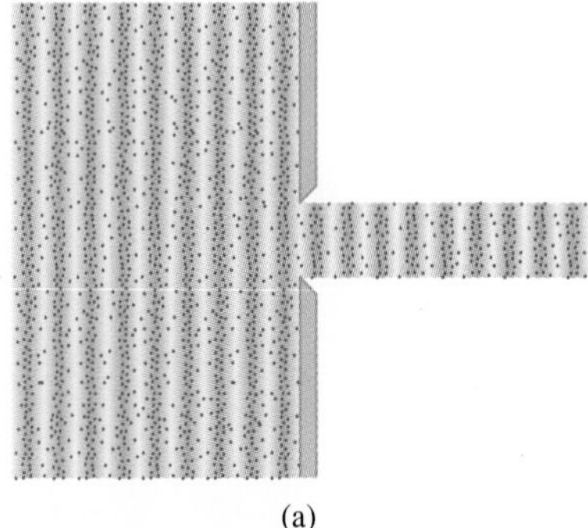

(a)

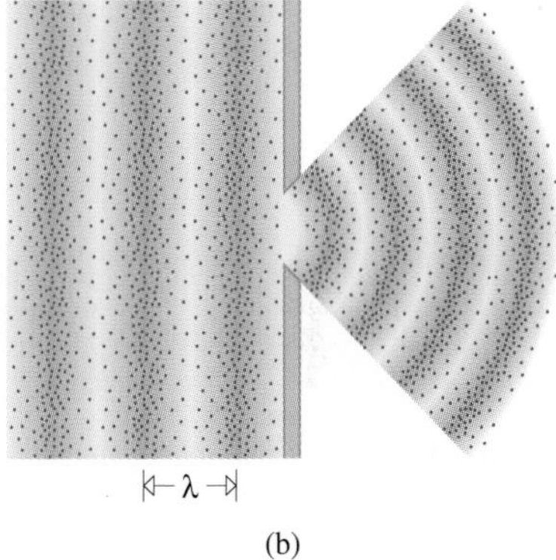

(b)

Figure 13.36 (a) When the wavelength is smaller than the aperture, the wave passes through in a fairly straight path. (b) When the wavelength is comparable to or larger than the aperture, the wavefronts bend and the disturbance propagates into regions beyond the straight-ahead path.

13.10 The Doppler Effect

Every kind of wave moves through a homogeneous medium at a constant speed that depends only on the physical properties of the medium. That's true irrespective of any motion of the source—it launches the waves and away they go. Still, the perception of a wave's frequency and wavelength can be significantly altered by a relative motion between the observer and the source. Almost everyone has heard a shift in frequency when a car blowing its horn passes by. The pitch while it is approaching is somewhat higher than when at rest, and as it passes, the pitch drops. The result is a kind of eeeeoooo sound. This phenomenon is known as the **Doppler Effect** after the Austrian physicist Johann Doppler, who first worked out the analysis for sound in 1842. He also correctly suggested that the effect would apply to light. The predictions of the theory were soon confirmed (1845) by Ballot, who arranged an acoustical extravaganza that must have been a joy to behold. For two days, a locomotive, flatcar, and band of trumpeters performed—coming and going—for a group of musician observers.

Figure 13.38 indicates how a frequency shift arises for a source moving toward a stationary observer at a rate of half the speed of sound. The motion of the source essentially closes up the gap between successive wavefronts. Thus, the observer will measure a wavelength that is halved and a frequency that is doubled. In the same way, Fig. 13.39 depicts a stationary source radiating sound to the right at speed v and an observer moving left, this time also at v. The listener rushing toward the source overtakes the wavefronts, intercepting more of them per second.

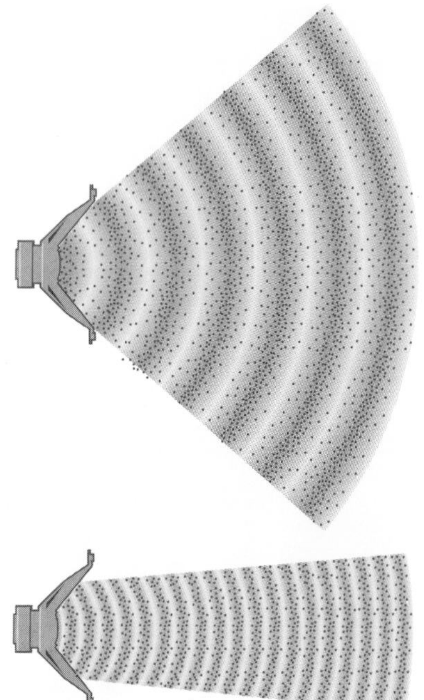

Figure 13.37 A small source behaves like an illuminated aperture. The amount of diffraction increases as the wavelength approaches the size of the aperture.

The simplest analytic situation is a medium at rest, with a point source moving at speed v_s and an observer moving at v_O, both traveling along the same line, in either direction. Figure 13.40 is a multiple exposure showing the point source's location at five consecutive moments and the associated wavefronts. Let's start time ($t = 0$) when the source was at point 0 whereupon it moved a distance $v_s t$ in a time t. Since its frequency is f_s, it has generated a number of waves equal to $f_s t$. Propagating at the speed of sound, the front emitted at $t = 0$ is now expanded out to a radius of vt, where it has encountered the observer who is moving to the right at speed v_O. Notice that the wavelength on either side of the source is different. On the left, where the observer is, λ equals the distance between the source and the observer ($vt + v_s t$) divided by the number of waves that span that space ($f_s t$), and so

$$\lambda = \frac{v + v_s}{f_s} \tag{13.15}$$

On the other side, the same number of waves is squeezed into a space of ($vt - v_s t$), and so λ in the forward direction equals $(v - v_s)/f_s$. Once we have the wavelength, we can find the frequency by dividing the relative speed by λ.

As far as the observer is concerned, she moves right at v_O (with respect to the medium) as the sound travels toward her at v (with respect to the medium), and so she sweeps through the wavefronts at a rate of $(v + v_O)$. Consequently, the observer will perceive a frequency of

$$f_O = \frac{v + v_O}{\lambda}$$

Using Eq. (13.15), this expression becomes

$$\frac{f_O}{v + v_O} = \frac{f_s}{v + v_s} \tag{13.16}$$

In this derivation, v_O and v_s are both positive in the directions shown. Changing those directions simply changes either sign. Accordingly, if the observer is at rest ($v_O = 0$) and the source is moving toward it, v_s must enter Eq. (13.16) as a negative number and $f_O = vf_s/(v + v_s)$ so that $f_O > f_s$. If instead the source moves away, v_s is positive and $f_O < f_s$, as observed. When the source is at rest, $v_s = 0$ and $f_O = f_s(v + v_O)/v$. If the observer moves toward it (as in Fig. 13.39), v_O is positive ($f_O > f_s$) and if it moves away, v_O is negative and $f_O < f_s$.

Example 13.10 An automobile traveling at 20.0 m/s (45 mi/h) blows a horn at a constant 600 Hz. Determine the frequency that will be perceived by a stationary observer both as it approaches and recedes. Take the speed of sound to be 340 m/s.

Solution: [Given: $v_s = \pm 20.0$ m/s, $v = 340$ m/s, and $f_s = 600$ Hz. Find: f_O.] Referring to Fig. 13.40, we set $v_O = 0$ and then calculate the approach from Eq. (13.16) with $v_s = -20.0$ m/s and the recession with $v_s = +20.0$ m/s. Thus, as the car approaches the observer

$$f_O = \frac{vf_s}{v + v_s} = \frac{(340\text{ m/s})(600\text{ Hz})}{(340\text{ m/s}) + (-20.0\text{ m/s})}$$

which equals $\boxed{638\text{ Hz}}$. As it recedes

$$f_O = \frac{vf_s}{v + v_s} = \frac{(340\text{ m/s})(600\text{ Hz})}{(340\text{ m/s}) + (+20.0\text{ m/s})}$$

which equals $\boxed{567\text{ Hz}}$.

▶ **Quick Check:** $v/(v + v_s) = (340\text{ m/s})/(340\text{ m/s} \pm 20\text{ m/s})$—namely, 0.944 and 1.063; 0.944(600 Hz) = 567 Hz; 1.063(600 Hz) = 638 Hz.

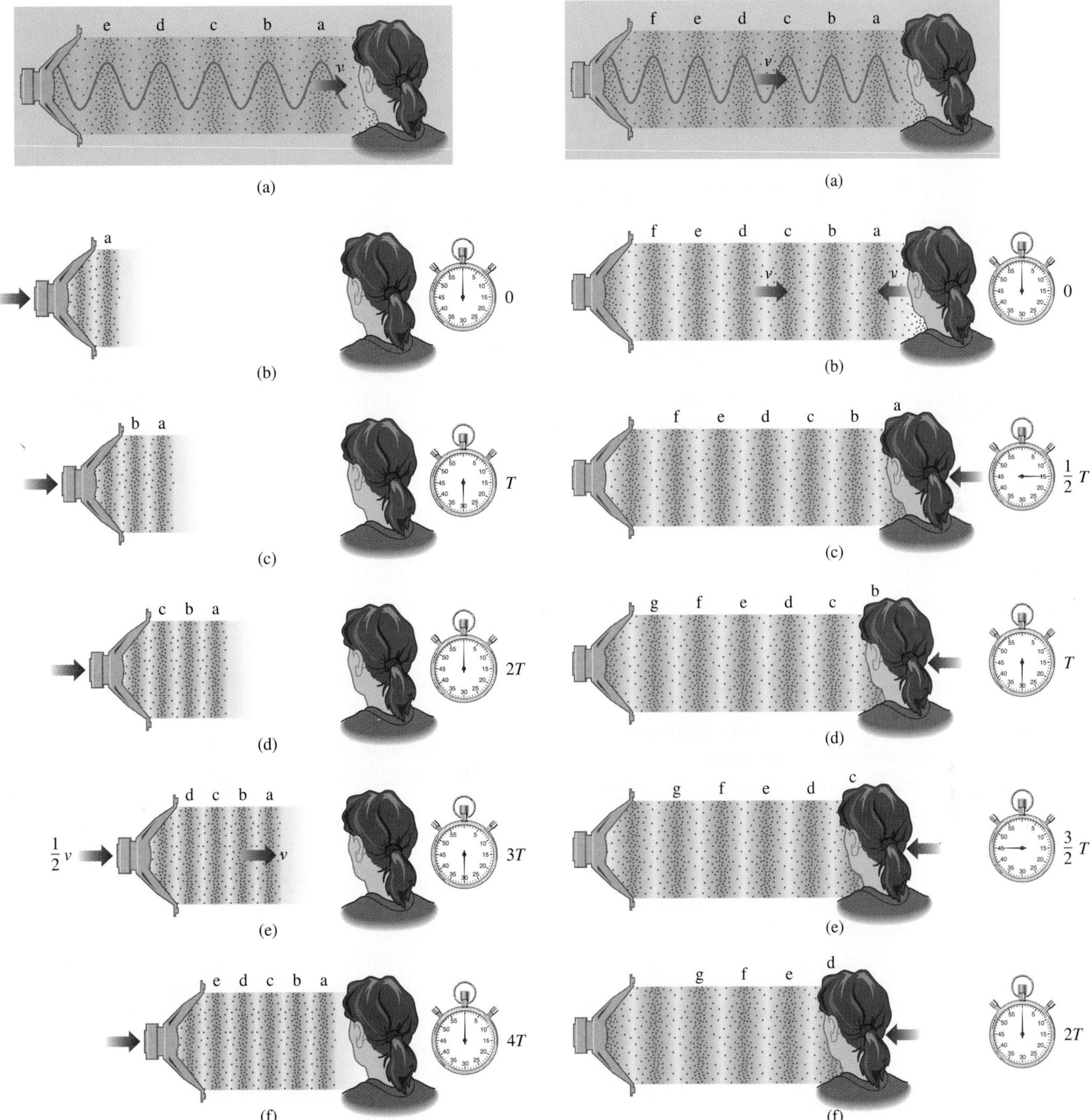

Figure 13.38 The Doppler Effect for a source moving toward a stationary observer at half the speed of sound (v). Compare the wavelength in (a), where the source is at rest, to the wavelength in (f), where it is moving at $\frac{1}{2}v$. Note how the source advances $\frac{1}{2}\lambda$ during the emission of one complete wave.

Figure 13.39 The Doppler Effect for an observer moving at speed v toward a stationary source of sound of speed v. Notice how the moving observer receives fronts *a*, *b*, *c*, and *d* in a time of two periods. Had she remained at rest as in (a), she would have received only two wavefronts *a* and *b* in that time.

All this traveling around is taking place with respect to the stationary medium, and the amount of the frequency shift actually depends on who is doing the moving—it is *not* symmetrical. The relative motion of source to observer is important

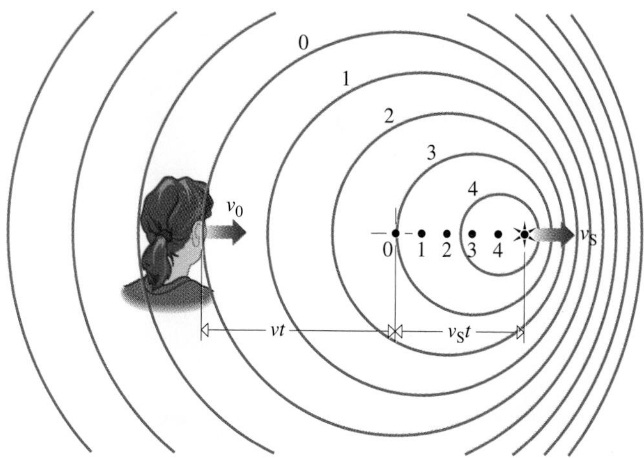

Figure 13.40 The source on the right is moving to the right at speed v_S, which is taken to be positive. The observer is also moving to the right, this time with a speed v_O, which is taken as positive.

here, but so, too, is the motion of each with respect to the medium. Whichever is moving relative to the medium will experience a wind blowing. Thus, if we recompute the first part of Example 13.10 for an observer moving toward a stationary horn at 20.0 m/s, instead of getting a frequency of 638 Hz, we get $f_O = (v + v_O)f_s/v = (340 \text{ m/s} + 20 \text{ m/s}) \times (600 \text{ Hz})/(340 \text{ m/s}) = 635$ Hz.

When waves are sent out from a stationary transmitter and reflect back off some moving target, that target functions as though it were a source of waves. The resulting Doppler shift (which is usually determined by beating the returning wave against a signal having the original frequency) provides the speed of the target. The police do this little trick with microwaves to clock passing cars on the highway, and the same approach is used to track both satellites and dangerous weather conditions. Likewise, Doppler-shifted ultrasonic waves (8 MHz) are being used to monitor the flow of blood, aiding in the diagnosis of conditions such as deep-vein thrombosis.

Example 13.11 Derive a general expression for the Doppler shift seen when a wave is bounced off an approaching target and returned to the stationary source, where it is viewed by an observer at rest.

Solution: There are two Doppler shifts involved in this process: the wave received by the target is shifted and then reemitted, and the returning wave is shifted when viewed by the final observer. Thus, the target (the observer in Fig. 13.40) receives the wave at a frequency $f_O = f_t$ where Eq. (13.16), with $v_s = 0$, yields

$$f_t = \frac{(v + v_t)f_s}{v}$$

This same frequency is next reflected, becoming the source wave to be received by the final observer. But now this reflecting source is receding (Fig. 13.40); to keep track of the signs, v_s becomes $-v_t$ in Eq. (13.16). Thus, $f_s = f_t$ in Eq. (13.16) where the observer is at rest ($v_O = 0$), and

$$f_O = \frac{vf_t}{v - v_t}$$

Notice that both of these shifts raise the frequency because the target is approaching. Substituting for f_t, we have

$$\boxed{f_O = \frac{(v + v_t)f_s}{v - v_t}} \qquad (13.17)$$

If the target were receding, the signs in front of v_t would have to be changed accordingly.

▶ **Quick Check:** First, the units are right. Second, the observed frequency (f_O) is greater than the source frequency (f_s), which is correct. Third, as $v_t \to 0, f_O \to f_s$, which is as it should be.

Example 13.12 A sound wave at 1000 Hz is sent out from a stationary source toward an approaching target. If the reflected wave is measured to have a frequency of 1200 Hz, what is the speed of the target? Take the speed of sound to be 332 m/s.

Solution: [Given: $f_s = 1000$ Hz, $f_O = 1200$ Hz, and $v = 332$ m/s. Find: v_t.] Since this is a bounce-off-the-target Doppler problem, Eq. (13.17) is immediately suggested; thus

(continued)

(continued)

$$f_O = \frac{(v + v_t)f_s}{v - v_t}$$

Hence $\quad f_O(v - v_t) = (v + v_t)f_s$

and $\quad v_t(f_s + f_O) = v(f_O - f_s)$

Thus $$v_t = v\frac{f_O - f_s}{f_O + f_s}$$

and $$v_t = \frac{(332 \text{ m/s})(200 \text{ Hz})}{2200 \text{ Hz}} = \boxed{30.2 \text{ m/s}}$$

▶ **Quick Check:** $f_O/f_s = 1.2 = (332 \text{ m/s} + 30 \text{ m/s})/(332 \text{ m/s} - 30 \text{ m/s})$.

As applied to light, the Doppler Effect provides a key insight into the creation of the Universe. The light from the vast majority of galaxies is seen to be shifted toward the low-frequency end of the spectrum. This shift implies that if the Doppler Effect is the cause (and no other possible mechanism is known), then all these thousands of millions of galaxies must be moving away from us. A notion that can only be reasonable if the Universe is expanding so that every galaxy is moving away from every other one and ours is neither at the center nor flying off on its own. If that picture is correct, then the "observed" expansion must have begun back at some moment of creation. From today's recession rates and the apparent size of the Universe, we conclude that it came into being ≈17 thousand million years ago in a cosmic fireball known as the *Big Bang.* One can almost hear the trumpeters on the flatcar celebrating it all.

Core Material

Sound is *a compression wave in any material medium oscillating within the frequency range of 20 Hz to 20 kHz.* The **intensity** (I) of a sound wave is defined as *the average power radiating divided by the perpendicular area across which it is transported:*

$$I = \frac{P_{av}}{A} \qquad [13.1]$$

The speed of sound in a diatomic gas such as air is given by

$$v = \sqrt{\frac{1.4P}{\rho}} \qquad [13.3]$$

Sound travels in air at 331.45 m/s at STP. Over the usual range of temperatures encountered at sea level, the speed of sound changes by about 0.60 m/s per change of 1.0 degree Celsius.

The **intensity-level** (or *sound-level*) of an acoustic wave is defined as *the number of factors of 10 that its intensity is above the threshold of hearing:* $I_0 = 1.0 \times 10^{-12}$ W/m^2. The intensity-level is measured in *decibels* (abbreviated dB) and is given by

$$\beta = 10 \log_{10} \frac{I}{I_0} \qquad [13.4]$$

For a change from I_1 to I_2

$$\Delta\beta = 10 \log_{10} \frac{I_2}{I_1} \quad \text{in dB} \qquad [13.9]$$

When two wavetrains of the same wavelength and amplitude traveling in opposite directions overlap, they create a **standing wave** (p. 458). For a system with zero displacement at both ends

$$L = \tfrac{1}{2}N\lambda \qquad (N = 1, 2, 3, \ldots) \qquad [13.10]$$

and the wavelength of the Nth mode is $\lambda_N = 2L/N$. The resonant frequencies of a string fixed at both ends are

$$f_N = \frac{N}{2L}\sqrt{\frac{F_T}{\mu}} \qquad (N = 1, 2, 3, \ldots) \qquad [13.11]$$

For a pipe open at one end

$$L = \tfrac{1}{4}N\lambda \qquad [13.12]$$

and

$$[\textit{antinode at one end}] \quad f_N = \frac{Nv}{4L} \qquad (N = 1, 3, 5, \ldots) \qquad [13.13]$$

For a pipe open at both ends

$$L = \tfrac{1}{2}N\lambda \qquad [13.10]$$

and

$$[\textit{antinode at both ends}] \quad f_N = \frac{Nv}{2L} \qquad (N = 1, 2, 3, \ldots) \qquad [13.14]$$

The Doppler Effect is a shift in frequency when the source and/or the observer are in motion (p. 469):

$$\frac{f_0}{v + v_0} = \frac{f_s}{v + v_s} \qquad [13.16]$$

The Doppler shift seen when a wave is bounced off an approaching target and returned to the stationary source where it is viewed by an observer at rest is given by

$$f_0 = \frac{(v + v_t)f_s}{v - v_t} \qquad [13.17]$$

Suggestions on Problem Solving

1. There are several facts associated with the notion of sound-level that are worth committing to memory. An intensity-level that is +10 dB higher or −10 dB lower than another will be heard to be twice as loud, or twice as soft, respectively. Doubling (or halving) either the acoustic power or intensity will increase (or decrease) the sound-level by 3 dB. A change of ≈1 dB is the smallest audible increment in loudness you are likely to be able to hear. Knowing this kind of thing helps to establish the physical meaningfulness of numerical answers and that should always be a primary concern.

2. When a physical entity, a string or a bar, vibrates and makes a sound, it pumps back and forth and sets the air around it vibrating at the same rate, the same *frequency.* The corresponding wavelength of the string depends on its physical conditions and the generated wavelength of the sound depends on its physical conditions, as manifested by the speed of the wave. The two wavelengths will generally *not* be the same. The natural emphasis is therefore on frequency and speed. **The energy of a wave is associated with its frequency** (p. 454).

3. Bear in mind that the standing-wave modes for all systems will have the same basic form; namely, the frequency will equal the speed over the wavelength. For a system bounded by two nodes (or two antinodes), the wavelength is some fraction of the fundamental $\lambda_1 = 2L$; that is, $2L/N$ where $(N = 1, 2, 3, \ldots)$. Hence, $f_N = v/(2L/N)$. When the system is bounded at its ends by a node and an antinode, respectively, $\lambda_1 = 4L$ and only odd fractions of that (namely, $4L/N$ where $N = 1, 3, 5, \ldots$) are possible. Each different type of wave will have its own speed, but these frequency expressions hold for all of them. Try to think in terms of $f_N = v/\lambda_N$, where the form of λ_N can be determined from a mental picture of the particular mode pattern, such as Figs. 13.25, 13.28, or 13.29—that may be easier for you than memorizing the equations. Professionals store the basics and derive the rest as needed.

4. If you are having difficulties with the signs of any of the terms when doing Doppler Effect problems, remember that if the source and observer are approaching each other there must be a shift up in pitch and, hence, down in wavelength. If they are receding from one another, the shift is down in pitch and up in wavelength.

Discussion Questions

1. The behavior of sound waves can be studied using electronic frequency generators, microphones, and oscilloscopes. Suppose, then, that the speaker of Fig. Q1 sends out a single short (5 ms) compression pulse that travels into the doubly-open tube, which is 20 m long. What signal will appear on the scope as provided by the pressure microphone at the middle of the pipe?

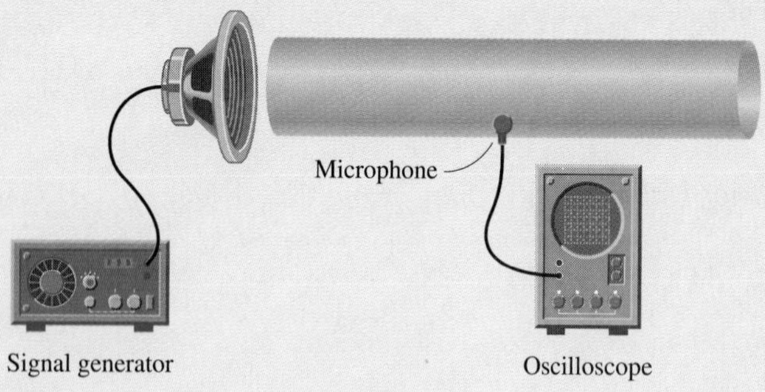

Figure Q1

2. In December 1916, a major disaster befell a contingent of the Austrian army in the Alps fighting in World War I. Almost as soon as they began a cannon barrage, an avalanche of snow came roaring down, burying thousands of soldiers. What does that suggest about sound?

3. As an orchestra literally warms up, its sound changes. What do you think happens to the pitch of the strings and the wind instruments as a result of their being played for a while? Explain.

4. According to Galileo, "a glass of water may be made to emit a tone merely by the friction of the fingertip upon the rim." What more can you now say about the ringing goblet or any bell?

5. The ear has a protective overload mechanism called the *acoustic reflex.* A sound in excess of about 85 dB causes muscles attached to the eardrum and ossicles to engage, which provides a safety margin of about 20 dB or 30 dB (the equivalent of ear plugs). This reflex takes about 30 to 40 ms to cut in, and maximum effect only occurs after about 150 ms. What is good and bad about such a system in our modern world? Why do many hunters show loss of hearing (especially in the right ear)?

6. Suppose you inhale a little helium (too much will displace oxygen and you will get into trouble even though helium is otherwise nontoxic) and speak while exhaling it. The resulting voice will have a comic high-pitched sound. Why? What would happen to an organ pipe blown with carbon dioxide?

7. When a string on, say, a violin is plucked, particularly toward one end, it initially takes on a nearly triangular shape

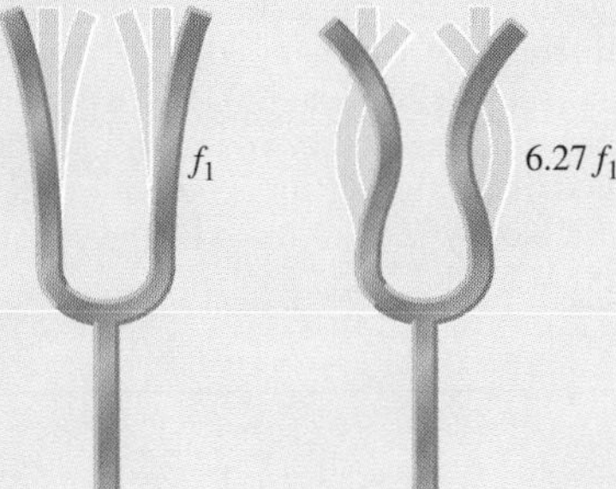

Figure Q11

rather than a sinusoidal one. Why does that result in a more "metallic" sound? Similarly, plucking the strings of a guitar with the fingers produces a more mellow sound than using a pick. Remember that the banjo is played with a pick and the harp with the fingers. Explain. Keep in mind that a signal with fine details, such as sharp corners, will require high-frequency Fourier components in order to synthesize it.

8. Suppose you sing or yell into an open piano with its dampers lifted off the strings. You are likely to hear the instrument "speak" back in response. Explain. Imagine that you have two identical tuning forks and you tap one and then place it near the other. What, if anything, will you hear if you now silence the first fork? "If one bows the bass string on a viola rather smartly," wrote Galileo, "and brings near it a goblet of fine, thin glass having the same tone as that of the string, this goblet will vibrate and audibly resound." Explain.

9. As a rule, instruments such as the piano that utilize struck strings will arrange to have the striking take place at a point 1/7th of the way down the string. Considering the fact that the 7th harmonic is generally unpleasantly dissonant, explain why the strings are struck where they are.

10. Imagine two people turning a jump rope while a third hops up and down. Describe every aspect of the process using the ideas developed in this chapter.

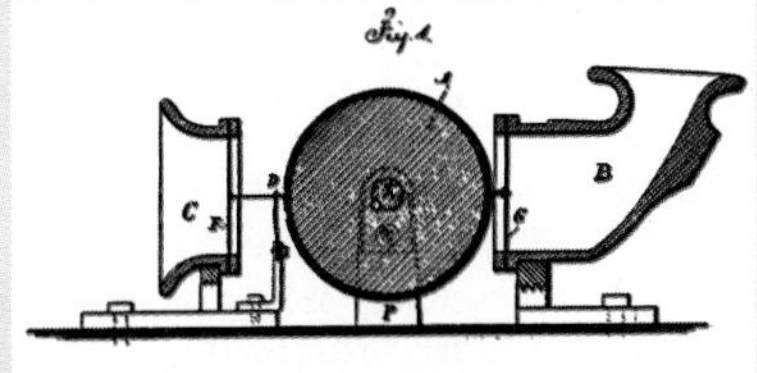

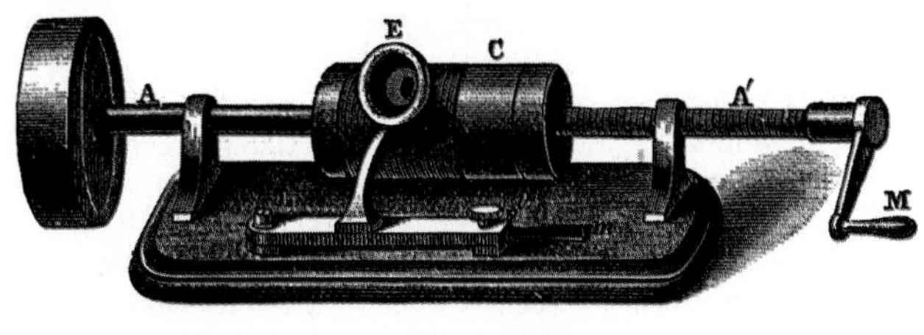

Figure Q16

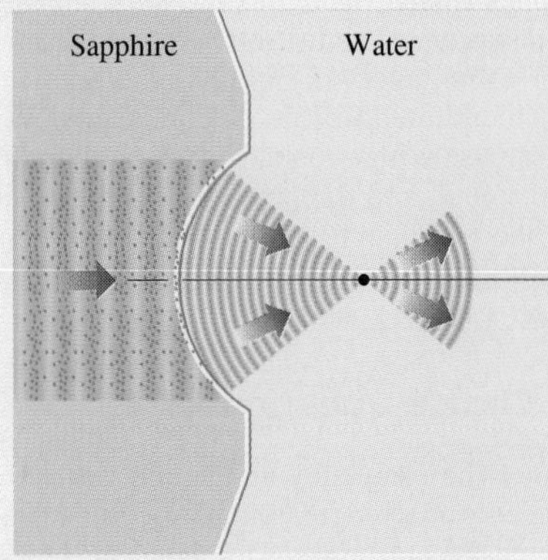

Figure Q15

11. The tuning fork, which was invented in 1712 by John Shore, Handel's trumpeter, can vibrate (Fig. Q11) briefly in a mode that produces its first overtone, or clang tone, at a frequency of 6.27 times the fundamental. If you wish to have the fork produce a nearly pure tone and so suppress the high-pitched clang tone, where should it be struck? Explain. Compare the tuning fork to the rod in the previous question.

12. Discuss two ways in which temperature changes will affect the pitch of an organ pipe.

13. A tuning fork is mounted to a hollow open-ended wooden sounding box that amplifies the sound. Explain how it manages to do that. What can you anticipate would be the difference between the sounding boxes for two such forks at, say, 1000 Hz and 50 Hz?

14. In light of the Question 13, what can you say about the reasons for the difference in size between a violin (60 cm long) and a double bass (200 cm long)?

15. Fig. Q15 shows the lens assembly for an acoustic microscope. Plane waves from a piezoelectric transducer enter from the left. Explain how the device works.

16. Fig. Q16 shows an early phonograph or "Speaking Machine" devised by Thomas Alva Edison (1877, patent number 200,521). The part labeled B is the "mouthpiece" and a clockwork mechanism advances it and the output cone C relative to the turning cylinder A. The latter has "a helical indenting groove, cut from end to end, say, ten grooves to the inch. Upon this is placed the material to be indented, preferably metallic foil. . . . It is obvious that many forms of mechanism may be used. . . . a revolving plate may have a volute spiral cut on both its upper and lower surfaces." How does the thing work?

17. The long glass tube (Kundt's tube) in Fig. Q17 is sealed at the lower end with a

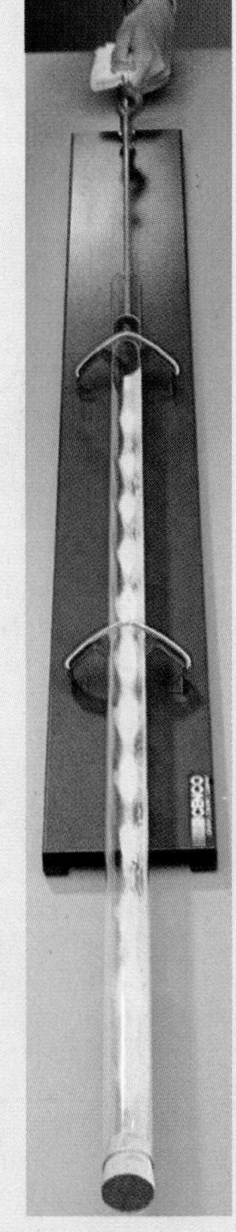

Figure Q17

movable plunger and at the upper end with a loosely fitted piston attached to a stationary rod (itself fixed at the middle). The tube contains an even layer of light, dry, very fine powder (lycopodium, precipitated silica, or cork dust). When the rod is stroked lengthwise with a rosined cloth or chamois, it gives out a high-pitched sound, and the powder flies up in violent agitation. When the tone is stopped, the powder settles back into heaps, and by adjusting the location of the plunger, these piles become well defined, uniform, and periodic. The device, which was invented by A. E. Kundt of Germany in 1876, can be used to measure the speed of sound (Problem 66). Discuss what is happening to the rod, to the air in the tube, and to the powder.

18. If you strike a tuning fork, hold it vertically near your ear, and rotate it slowly around its axis, the sound will rise and fall in intensity. Why does this happen?

Multiple Choice Questions

1. The higher the frequency of a sound wave, the (a) shorter the wavelength (b) greater the speed (c) lower the pitch (d) greater the amplitude (e) none of these.
2. When the loud music in the living room causes the doors on the china cabinet to buzz, the effect is due to (a) interference (b) diffraction (c) superposition (d) resonance (e) none of these.
3. The speed of sound in air at 40°C by comparison to the speed at 0°C is (a) much lower (b) a little lower (c) the same (d) higher (e) none of these.
4. Given: two sources of sound waves having the same amplitude, one with frequency 250 Hz and the other with frequency 500 Hz. A comparison of their relative loudness would reveal that (a) the first was twice as loud (b) the second was louder (c) the two were of equal loudness (d) the second was one-quarter as loud (e) none of these.
5. A dB-meter placed in front of an audio system reads 62 dB. The volume is turned up and the meter now reads 82 dB. Evidently, the sound got (a) twice as loud (b) half as loud (c) 20 times louder (d) 10 times louder (e) none of these.
6. A 60-dB sound-level corresponds to an intensity of (a) 10^{-10} W/m^2 (b) 10^{6} W/m^2 (c) 10^{-12} W/m^2 (d) 10^{12} W/m^2 (e) none of these.
7. A sound that has an intensity-level of 100 dB is how many times more intense than a sound of 20 dB? (a) 5 (b) 8 (c) 1000 (d) 10^8 (e) none of these.
8. On a day when you are listening to music at a sound-level of 20 dB above the threshold of audibility, an average person would say that it was (a) very loud (b) of normal listening level (c) very soft (d) slightly loud (e) none of these.
9. A sound-level meter placed in front of the loudspeaker of a 60-W audio system reads 70 dB. All else being equal, when placed in front of a 120-W system, the meter will read (a) 120 dB (b) 140 dB (c) 63 dB (d) 73 dB (e) none of these.
10. In order to cause the loudness of a sound to appear half its original value, it must be (a) increased by 50 dB (b) decreased by 50 dB (c) decreased by 10 dB (d) decreased by 3 dB (e) none of these.
11. A doubly-open organ pipe, in comparison to a singly-open pipe of the same length, has a fundamental frequency that is (a) half as great (b) twice as great (c) three times larger (d) $1/\pi$ times smaller (e) none of these.
12. The flute and piccolo are in the same family of instruments, and each functions as a pipe open at both ends. Both have the same fingering but different lengths. The flute is about 66 cm long and is an octave lower than the piccolo—that is, the piccolo plays at twice the frequency of the flute. It follows that the piccolo, in comparison to the flute, is about (a) twice as long (b) half as long (c) eight times as long (d) the same length (e) none of these.
13. Fig. MC13 shows the frequency spectrum of a flute playing a D_4 note (294 Hz). It's clear from the diagram that the flute is likely to be (a) made of metal (b) 66 cm long (c) open at both ends (d) cylindrical (e) none of these.

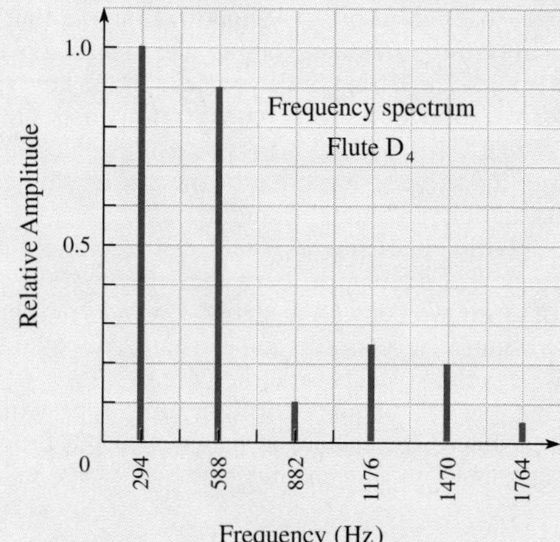

Figure MC13

14. What will happen to the fundamental frequency of an organ pipe open at both ends if a cap is placed over one end sealing it? The frequency will (a) go up 20% (b) be halved (c) stay the same (d) double (e) none of these.
15. If one stereo sounds 16 times louder than another, the difference in their sound-levels is about (a) 40 dB (b) 30 dB (c) 16 dB (d) 160 dB (e) none of these.
16. A clarinet has a single vibrating reed at one end coupled to a cylindrical resonance tube that is flared open at the other end. The coupling of reed and air column produces an interesting spectrum (Fig. MC16) even though there is a node at the reed. Typically, there is a strong first harmonic, little second, a strong third, some fourth, a strong fifth, and an appreciable sixth. It appears that the instrument functions as if it were (a) purely closed at both ends (b) purely open at one end (c) a mix of somewhat open at both ends and yet effectively closed at one (d) purely open at both ends (e) none of these.

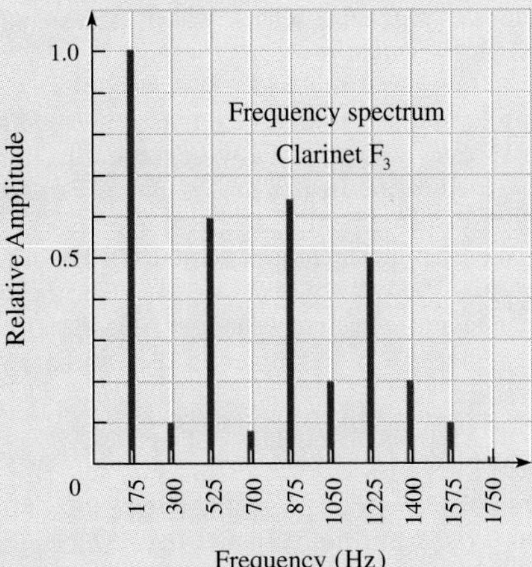

Figure MC16

17. If you stick a small lump of clay to the end of one or both tines of a tuning fork, you can expect the fundamental frequency to (a) remain unchanged (b) go down (c) go up (d) vary from moment to moment (e) none of these.
18. Two tuning forks of frequency 380 Hz and 384 Hz are excited simultaneously. The resulting sound will waver in intensity with a frequency of (a) 382 Hz (b) 0 (c) 4 Hz (d) 2 Hz (e) none of these.
19. One is likely to hear low-pitched sounds coming from behind the head better than high-pitched ones because of (a) diffraction (b) dispersion (c) interference (d) refraction (e) none of these.
20. A vehicle heading down a highway at 80 km/h is being pursued by a police car that is right behind it, also traveling at 80 km/h. If the cop's siren puts out a 500-Hz wail, what frequency will the perpetrator hear? (a) < 500 Hz (b) > 500 Hz (c) 500 Hz (d) 80 Hz (e) none of these.

Problems

ACOUSTICS
HEARING SOUND

1. [I] In music, the standard A_4 note has a frequency of 440 Hz. What is its period and wavelength at room temperature?

2. [I] A person standing at one side of a playing field on a cold winter night emits a brief yell. The short acoustical wavetrain returns 1.00 s later as an echo having "bounced off" a distant dormitory. Approximately how far away was the building?

3. [I] A groove on a monophonic phonograph record wiggles laterally such that its amplitude and frequency more or less correspond to the sound that is recorded. If at a given moment the needle is moving through the groove at 0.50 m/s, what would be the wiggle-wavelength for a 1.5-kHz tone?

4. [I] A low-frequency loudspeaker is called a woofer. At what frequency will a tone have a wavelength equal to the diameter of a 15-in. woofer? Take the speed of sound to be 344 m/s.

5. [I] A string vibrating at 1000 Hz produces a sound wave that travels at 344 m/s. How many wavelengths will correspond to 1 m?

6. [I] The intensity of the acoustic wave from an underground explosion measured 5.0 km away is 1.6×10^4 W/m^2. What intensity will be recorded at a site 50 km away? Assume no losses.

7. [I] A point source of sound waves emits a disturbance with a power of 50 W into a surrounding homogeneous medium. Determine the intensity of the radiation at a distance of 10 m from the source. How much energy arrives on a little detector with an area of 1.0 cm^2 held perpendicular to the flow each second? Assume no losses.

8. [I] At what temperature will the speed of sound in air at standard pressure equal 320 m/s?

9. [I] If a tuning fork puts out a tone at 440 Hz, what is its wavelength in air at 25°C?

10. [I] By what percentage does the speed of sound change when the air temperature rises from 0°C to 30°C?

11. [I] The contact time between the hammers and strings of a piano is one of the determining factors of the tone of the instrument; if it's too long, the higher overtones will be damped out. Though contact time varies across the keyboard, an average is about $\frac{1}{2}$ of a period. To two significant figures, what is the contact time for middle-C at 261.6 Hz?

12. [I] One of the most important acoustical characteristics of a room is its *reverberation time,* the time it takes for a sound to decrease 60 dB. What does that mean as far as the sound's intensity is concerned? For a concert hall, the reverberation time is typically 1 to 3 s.

13. [I] A radio playing quietly produces a sound intensity of about 10^{-8} W/m^2. What is the corresponding intensity-level?

14. [I] Someone playing a CD at 60 dB wants to make the music twice as loud. At what intensity-level should it be played?

15. [I] Two (otherwise identical) audio systems at a demonstration are blasting away, with one putting out 10 times the acoustic power of the other. What is the difference in their sound-levels?

16. [I] Two people having a normal conversation are 1.0 m from a sound-level meter. Approximately what will the meter read?

17. [I] Two audio systems each produce 50 W of acoustic power at the location of a microphone. What is the difference in their intensity-levels in dB at that point?

18. [I] Does a 0-dB sound-level mean there is no sound? Explain.

19. [I] The intensity of a sound is tripled. By how many decibels does it increase?

20. [I] On occasion, it turns out that we know the intensity-level of a sound and wish to find out the associated intensity. Show that

$$I = 10^{\beta/10} I_0$$

21. [I] The noise of traffic registers 77 dB on a meter. What intensity does that correspond to?

22. [I] A cannon produces a 90-dB sound-level at a certain distance from a detector. What will the device read when two such cannons at that same distance are fired?

23. [I] Twenty-five singers all singing with the same intensity produce a 65-dB sound. What is the intensity-level of the average chorus member?

24. [II] A bar of aluminum alloy 10 m long with a cross-sectional area of 1.0 cm^2 has a mass of 2.7 kg and a Young's Modulus of 7.0×10^{10} N/m^2. If the end is tapped at a rate of 100 Hz, how long will it take for the sound wave to reach the other end of the bar, given that $v = \sqrt{Y/\rho}$? What will be its wavelength?

25. [II] What is the sound-level 10 m away from a point source radiating 1.2 W of acoustic power?

26. [II] Two sounds have intensities of 10^{-2} W/m^2 and 10^{-10} W/m^2. What will be the difference in their intensity-levels as read by a sound-level meter?

27. [II] A small but noisy printer produces an acoustic intensity of 56×10^{-5} W/m^2 at a point 5.0 m away. Approximately what value will a dB-meter read at that location and, again, at 20.0 m from the printer?

28. [II] With Problem 27 in mind, what will be the *change* in intensity-level measured for any point source at 5.0 m and then at a 20.0-m distance?

29. [II] A small source emits nearly spherical waves with an acoustic power of 60 W. How far away must a sound-level meter be if it is to read 60 dB?

30. [II] The intensity-level of a large engine is measured to be 130 dB at a distance of 10.0 m. Approximately what intensity will exist at a point 100 m away?

31. [II] The sound-level 2.0 m from a pneumatic chipper is 120 dB. Assuming it radiates uniformly in all directions, how far from it must you be in order for the level to drop 40 dB down to something more comfortable?

32. [II] How much acoustic power impinges on a 10-cm^2 detector when the intensity-level is 70 dB?

33. [II] Given that the standard reference pressure for sound is $P_0 = 2 \times 10^{-5}$ Pa, write an expression for the intensity-level in terms of the pressure of the disturbance P. (Hint: p. 454.)

34. [II] Someone turns on a radio at 65.0 dB while vacuuming the floor at 80.0 dB. What will be the total intensity-level in the room?

35. [III] If the displacement of the oscillating air molecules in a sound wave is written in terms of a cosine function of amplitude A, then (with a bit of calculus and some effort) the corresponding gauge pressure can be expressed as

$$P = \frac{2\pi}{\lambda} BA \sin\left(\frac{2\pi}{\lambda}x - \omega t\right)$$

(a) Show that both sides of this expression have the same units. (b) What can you say about the displacement as compared to the pressure? (c) Show that the maximum value of the change in pressure (P_0) from the no-wave ambient value, the so-called *pressure amplitude,* is given by

$$P_0 = \frac{2\pi}{\lambda}\rho v^2 A$$

(d) Discuss, in general terms, why it is reasonable to have B and A in the equation for P.

36. [III] The pressure amplitude associated with the loudest sounds tolerable by human beings is roughly 30 Pa. Such pressure waves vary ±30 Pa with respect to atmospheric (10^5 Pa). Determine (using the results of Problem 35) the maximum displacement of the air molecules in such a disturbance having a frequency of 1000 Hz at human body temperature. Take the density of the air to be 1.22 kg/m^3.

37. [III] If one audio system sounds 6 times louder than another, what is the difference in their intensity-levels?

PRODUCTION AND PROPAGATION OF SOUND
SOUND WAVES

38. [I] Two sound waves of angular frequencies 900.0 rad/s and 896.0 rad/s overlap. What is the resulting beat frequency?

39. [I] The sound from a tuning fork of 1000 Hz is beat against the unknown emission from a vibrating wire. If beats are heard at a frequency of 4 Hz, what can be said about the frequency of the wire?

40. [I] With Problem 39 in mind, what can you say if a small piece of tape is fixed to the tuning fork and the beat frequency now increases?

41. [I] On tapping two tuning forks, an observer hears a succession of intensity maxima arriving at a rate of one every 0.99 s. What is the difference in frequency between the two forks?

42. [I] A thin wire is stretched between two posts 50 cm apart. It is then bowed and thereby set into oscillation. What are the wavelengths of the fundamental and the first overtone of the system?

43. [I] A taut string is fixed at both ends, which are 0.50 m apart. It is then set into resonance at its 6th harmonic. Determine the wavelength of the oscillation and draw the standing-wave pattern.

44. [I] A string stretched between fixed posts is 250 cm long and oscillates in its fundamental mode at 100 Hz. Determine the speed of a transverse wave on the string.

45. [I] A piece of steel piano wire is held fixed at both ends under a tension of 100 N. The free length of wire is 1.00 m, and it has a mass of 2.5 g. What is its fundamental frequency?

46. [I] A narrow tube 1.00 m long is closed rigidly at one end and with a piston at the other. Given that the speed of sound is 335 m/s, what is the frequency of the tube's fundamental oscillatory mode?

47. [I] Show theoretically that for a string fixed at both ends, its standing-wave frequencies are given by

$$f_N = \tfrac{1}{2}N\sqrt{\frac{F_T}{Lm}}$$

where L is its length and m its mass.

48. [I] A trumpet is a bent tube roughly 140 cm long and closed at one end by the player's mouth. Determine the fundamental (which will be quite difficult to blow) and the first three overtones. Take the temperature to be 20°C.

49. [I] Two loudspeakers separated by 2.0 m are fed a 688-Hz signal that drives them both in-phase (Fig. 13.35). A microphone is placed at a point 6.75 m from one and 7.00 m from the other. The magnitudes of the pressure

variations, above and below atmospheric, at that point are essentially the same when either speaker acts alone. What can you say about the intensity there when both are radiating? Take the speed of sound to be 344 m/s.

50. [I] A police car, its siren blaring at 1000 Hz, is traveling at 20.00 m/s while chasing a garbage truck moving at 15.00 m/s a block in front of it. What apparent frequency will the garbage collectors hear? The speed of sound is 330.0 m/s.

51. [I] The setup in Problem 49 is changed slightly. By reversing the wires going to one of the speakers, it is made to move forward, creating a compression, just as the other speaker is moving backward, creating a rarefaction. What, if anything, will happen to the intensity at the microphone now?

52. [I] A train whistle is blown by an engineer who hears it sound at 650 Hz. If the train is heading toward a station at 20 m/s, what will the whistle sound like to a waiting commuter? Take the speed of sound to be 340 m/s.

53. [I] A car with its horn blaring at 500 Hz passed a woman standing on the street at 25 m/s. What frequency did she hear as the car receded into the distance? The speed of sound that day was 344 m/s.

54. [I] An ultrasonic wave at 80 000 Hz is emitted into a vein where the speed of sound is about 1.5 km/s. The wave reflects off the red blood cells moving toward the stationary receiver. If the frequency of the returning signal is 80 020 Hz, what is the speed of the blood flow?

55. [I] A man running toward the stage in a theater hears an A_4 note from a stationary tuning fork to have a frequency of 441 Hz instead of its more normal 440 Hz. About how fast is he going?

56. [II] Galileo determined experimentally that the fundamental frequency of a string fixed at both ends was inversely proportional to both its diameter and to the square root of its density. Prove that this description is true theoretically.

57. [II] A B-string from a guitar is held fixed at both ends under tension with a vibrating length of 33 cm. It oscillates at its fundamental frequency of 246 Hz. What are the wavelengths on the string, and in the air at room temperature?

58. [II] Imagine a hypothetical piano with all strings made of the same material and all under the same tension. The piano extends from 27.5 Hz to 4186 Hz, which is over seven octaves—that is, seven doublings of frequency. If the highest note corresponds to a string 15 cm long, how long will the lowest string have to be? What can you conclude about this approach to piano design?

59. [II] Two identical piano strings are both tuned to 440 Hz. The tension in one is then increased by 1.00%, and both strings are activated so that they sound their fundamental frequencies. What will be the resulting beat frequency?

60. [II] A narrow glass tube 0.50 m long and sealed at its bottom end is held vertically just below a loudspeaker that is connected to an audio generator and amplifier. A tone with a gradually increasing frequency is fed into the tube, and a loud resonance is first observed at 170 Hz. What is the speed of sound in the room?

61. [II] An organ pipe that ordinarily sounds at 600 Hz at 0°C is connected to a source of helium at that temperature. At what frequency will it now operate?

62. [II] A speaker (Fig. P62) attached to a generator emits a constant 340-Hz tone, which is soon reflected off a board. The microphone is moved along a perpendicular line from source to board, and it is found that there is an intensity maximum at point P_1 and a minimum at P_2, which is 25 cm closer to the speaker. What is the speed of sound?

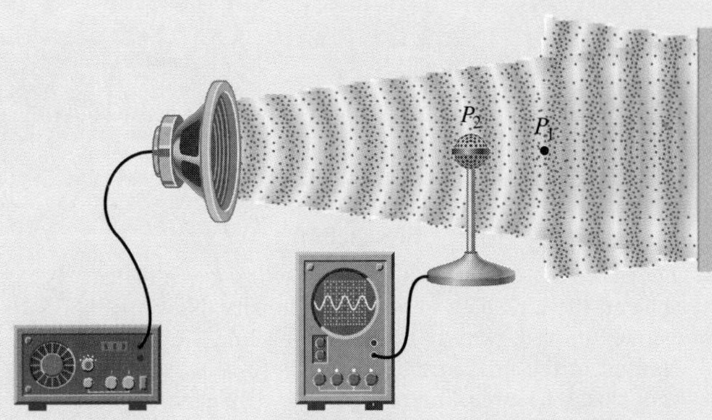

Figure P62

63. [II] A quartz tube open at both ends has a fundamental resonant frequency of 200 Hz at 0°C. Neglecting any changes in length, by how much will the fundamental change when the tube is sounded in a chamber at 40°C?

64. [II] A C-flute with all its holes covered plays a middle C (262 Hz) as its fundamental. Assuming room temperature (20°C) and overlooking end corrections, how long should the flute be from embouchure hole to end?

65. [II] A wire stretched between two posts and under a tension of 200 N oscillates at a fundamental frequency of 420 Hz. At what tension would it oscillate instead at 430 Hz?

66. [II] Referring to Kundt's tube (Fig. Q17), if v_a and v_r are the speeds of sound in the air and rod, respectively, and if L is the length of the rod and l the node-to-node distance, show that

$$v_r = \frac{v_a L}{l}$$

In other words, knowing the speed of sound in air, we can determine it in the material of the rod.

67. [II] A copper bar 1.00 m long is clamped at its middle and set vibrating. Given that $v = \sqrt{Y/\rho}$ and the bar has a density of 8.9×10^3 kg/m^3 and a Young's Modulus of 11×10^{10} N/m^2, what will its frequency be?

68. [II] Two small loudspeakers separated by a distance of 3.00 m emit a constant tone of 344 Hz (Fig. P68). A microphone is moved along a line parallel to and 4.00 m from the line connecting the speakers. It is found that intensity maxima exist on the center line and directly opposite each speaker. From this information, calculate the speed of sound. (Hint: The waves must arrive in-phase if they are to interfere to form a maximum.)

69. [II] Two small loudspeakers are set up as in Fig. P68. They emit a tone of 2000 Hz in-phase with one another. Point O is equidistant from the two and a microphone

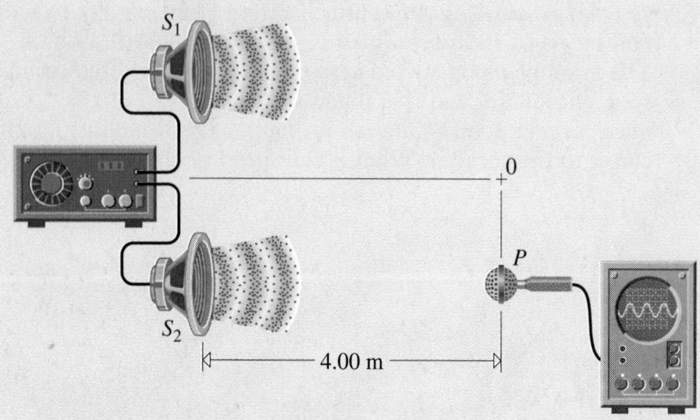

Figure P68

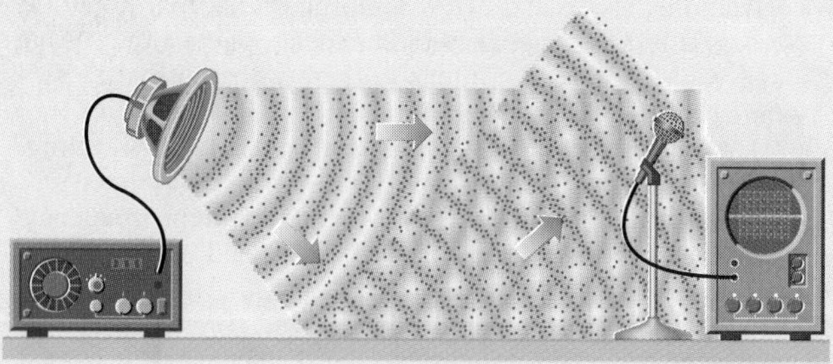

Figure P71

placed there reads a maximum intensity. Now speaker S_1 is slowly moved away to the left along line S_1O, and the output of the mike gradually drops to near zero and then goes back to a maximum when the displacement reaches 0.172 m. What is the speed of sound?

70. [II] A point source of sound on top of a police car emits a signal at 1000 Hz. If the car is traveling in a straight line at 30 m/s, what will be the wavelength perceived by people standing on the road both directly in front of and behind the car? Take the speed of sound to be 335 m/s. What is the wavelength as measured in the car?

71. [II] Figure P71 shows a loudspeaker emitting a 340-Hz sine wave toward a pressure microphone 5.00 m away. The mike is 1.146 m above the table, and its output at this location is found to be a minimum. The temperature is not known, but it's not far from room temperature. Assume there are no phase shifts on reflection and determine the speed of sound.

72. [II] A sound wave at 1000 Hz is sent out from a stationary source toward a target approaching at 20.0 m/s. What will be the frequency of the returning signal? Take the speed of sound to be 340 m/s.

73. [II] An observer with a stationary source of sound sends out a signal directly toward an approaching target. Write an expression for the frequency of the beats heard by the observer.

74. [III] Consider a long, narrow rod of length L clamped at both ends. If the rod is rubbed with a rosined cloth, a compression wave will travel its length. Show that the expression for the resulting standing-wave modes is

$$f_N = \frac{N}{2L}\sqrt{\frac{Y}{\rho}}$$

where $N = 1, 2, 3, \ldots$. A novice playing with a violin is likely to move the bow lengthwise along a string. Explain why this movement will generally produce a shrill squeaking sound.

75. [III] Consider the expressions for two waves of the same amplitude with slightly different frequencies traveling in the same direction:

$$y_1 = A \sin(\omega_1 t) \qquad \text{and} \qquad y_2 = A \sin(\omega_2 t)$$

where we simplify things a little by just looking at them at the point $x = 0$. Now, supposing the two waves were to overlap, derive the following equation for the combined waveform:

$$y = 2A \cos[\tfrac{1}{2}(\omega_1 - \omega_2)t] \sin[\tfrac{1}{2}(\omega_1 + \omega_2)t]$$

and interpret each term in reference to Fig. 13.21.

76. [III] The two loudspeakers in Fig. P68 are emitting sounds of the same frequency in-phase. Show that the separation between them (which equals a) must exceed $\frac{1}{2}\lambda$ if cancellation via interference is to be observed anywhere in the region in front of the speakers.

14

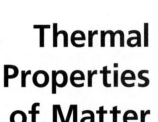

Thermal Properties of Matter

CHAPTERS 14, 15, AND 16 are about internal energy, the energy stored via the motion and interaction of the atoms of a system. The discipline is called *thermal physics* and it deals, broadly speaking, with temperature (Chapter 14), and the transfer (Chapter 15) and transformation (Chapter 16) of energy. The study begins with our rather crude bodily appreciation of hot and cold; with the what-can-you-put-in-your-mouth-without-burning-it understanding of temperature. Our sensory perception of temperature, however unreliable, must correspond to some unseen aspect of the physical world. We will find that, *even though temperature may well be a basic quantity like*

Man has no Body distinct from his Soul;
for that called Body is a portion of Soul
discern'd by the five Senses, the chief
inlets of Soul in this age.

Energy is the only life and is from the Body;
and Reason is the bound or outward
circumference of Energy.

Energy is Eternal Delight.

WILLIAM BLAKE (1793)
The Marriage of Heaven and Hell

mass and time, it can usually be associated with the concentration of thermal energy in a material system. This chapter focuses on a variety of temperature-dependent properties of bulk matter.

The firebox is at the center of our civilization. Out of sight, in power plants everywhere on the planet, there are furnaces and reactors generating thermal energy that ultimately ends up driving our TVs and electric toothbrushes. Most cars, trains, ships, and planes burn fuels in heat engines of one sort or another. The First and Second Laws of Thermodynamics (pp. 550 and 570) are about energy and change. These ideas transcend the discipline, providing us with profound insights into life itself and the very workings of the Universe.

TEMPERATURE

Like so many other physical quantities, temperature was measured long before it was understood. Galileo appears to have invented (ca. 1592) the first device for indicating "degrees of hotness." He simply placed the end (Fig. 14.1) of an inverted narrow-necked flask, warmed in his hands, into a bowl of water (or wine). As it cooled, the liquid was drawn up, partially filling the neck. Air captured in the bulb at the top either expanded or contracted when subsequently heated or cooled, and the column fell or rose proportionately. Much the same effect can be seen by putting an inflated balloon in the freezer. The gas molecules inside the balloon keep it puffed up by constantly bombarding the inner walls. Cooling the gas, removing kinetic energy from its molecules, lessens their bouncing and the balloon collapses.

The medical applications of the thermometer were recognized almost immediately, and normal body temperature became a focus of interest. In 1631, J. Rey, a French physician, inverted Galileo's device, filling the bulb with water and leaving much of the stem with air in it. In that configuration, which is more like today's thermometer, the liquid's expansion operates the device. Within 70 years of Galileo's invention, sealed pocket-sized thermometers containing either alcohol or mercury were in use (both freeze at much lower temperatures than water, and sealing the devices isolated them from atmospheric pressure changes).

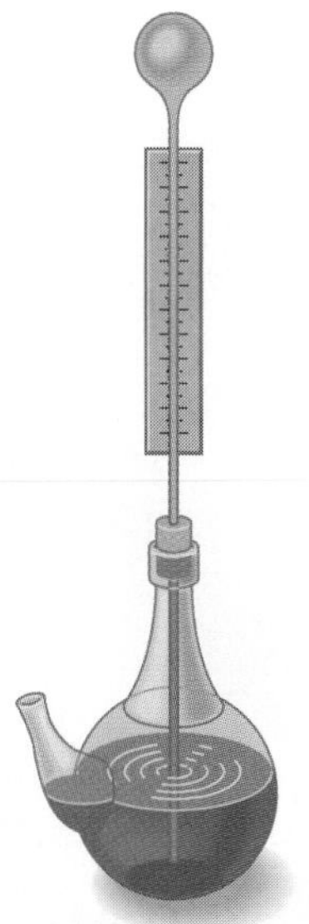

Figure 14.1 Galileo's thermoscope is not called a thermometer because the scale was arbitrary. The egg-sized globe at the top is the sensor. The gas within it expands or contracts, and the liquid level rises and falls.

By exploring the behavior of matter with the aid of the thermometer, it soon became evident that there were a number of physical occurrences that happened at fixed temperatures. Boyle, Hooke, and Huygens independently recognized (1665) that that fact could provide a reliable reference point for any thermometer. Hooke suggested using the freezing point of water, and Huygens offered its boiling point. C. Renaldini (1694) used both the freezing and boiling points of water to standardize two widely spaced points on his thermometer (something Newton would do independently seven years later).

The scale we still use in the United States is a gift of the instrument maker G. D. Fahrenheit (1717). Not wanting to deal with negative values, he set the 0 mark at the coldest temperature he could produce, that of a mixture of water, ice, and sea salt. Following Newton's lead, Fahrenheit set the upper reference at normal human body temperature, which he took as 96 (probably because that number was divisible by 12 and easily halved and quartered). Upon dividing the distance between these two fixed points into 96 equal-sized divisions, or degrees, he could then extend the scale as high or low as he wished. That done, water froze at what turned out to be 32°F (which is read *degrees Fahrenheit*) and boiled at 212°F.

The scale that is still used in modern scientific work (although it is gradually being replaced) is associated with the name of a man who neither first proposed it nor ever actually produced it. Anders Celsius (1742) used the freezing and boiling of water for the reference points and then divided the distance between them into 100 equal parts—a convenient scheme for the user. Strangely enough, Celsius set the temperature of freezing water at 100° and boiling at 0°. Some years later, with these two numbers more reasonably interchanged (Fig. 14.2) the arrangement came to be known as the *centigrade scale* (from the Latin *centum,* meaning 100, and *gradus,* degree). The Tenth International Conference on Weights and Measures (1954) changed the name to the *Celsius scale.*

Since the Fahrenheit scale goes from 32° to 212° (a total of 180 divisions), and the Celsius scale goes from 0° to 100°, each Fahrenheit degree is smaller than a Celsius degree. A change of 1 C° is equivalent to a larger change of (180/100)F°, or (9/5)F°. If we move a certain number of Celsius degrees, T_C, from freezing (0°C), that corresponds to a change from freezing (32°F) of $(9/5)T_C$ Fahrenheit degrees. In short

$$T_F = 32° + \tfrac{9}{5}T_C \qquad \text{or} \qquad T_C = \tfrac{5}{9}(T_F - 32°) \tag{14.1}$$

Rather than memorizing these equations, it might be easier to make the transformations logically. For example, human body temperature is 98.6°F, which is 66.6 Fahrenheit degrees above freezing—equivalent to (5/9)66.6 = 37.0 of the larger Celsius degrees above freezing (0°C). Hence, body temperature is 37.0°C.

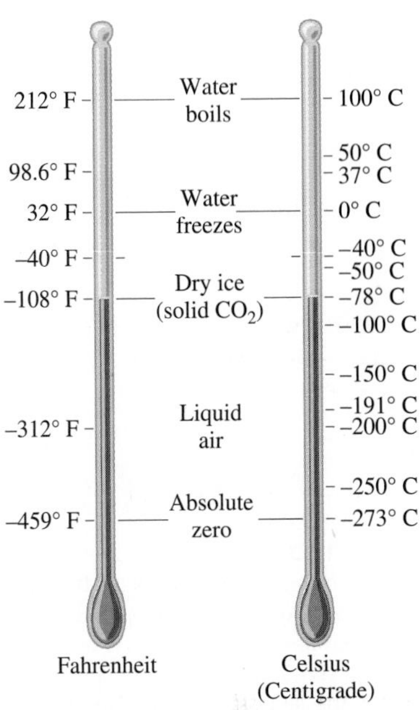

Figure 14.2 A comparison of the Fahrenheit and Celsius temperature scales.

Example 14.1 Determine the temperature at which the Fahrenheit and Celsius thermometers have the same value.

Solution: [Find the value at which $T_F = T_C$.] Using Eq. (14.1) and setting $T_F = T_C$, we have

$$T_F = 32° + \tfrac{9}{5}T_F$$

hence

$$-\tfrac{4}{5}T_F = 32°$$

and

$$\boxed{T_F = T_C = -40°}$$

The two scales are the same at −40°.

▶ **Quick Check:** $T_C = (5/9)(T_F - 32°) = (5/9) \times (-72°) = -40°$.

As we will see, the manner in which any material—mercury, alcohol, water, glass, whatever—expands on heating is characteristic of that material. What this means practically is that thermometers that differ in their physical construction and yet have the same two fixed reference points will necessarily agree exactly only at those two points.

The range of temperatures we deal with is extensive, and there is now a whole arsenal of different kinds of thermometers, each with its own virtues. The familiar mercury-in-glass instrument is only useful between the points where mercury freezes and glass melts. Moreover, it can be used reliably only when its presence doesn't affect the temperature being determined. If you want to measure the temperature of a flea or a thermonuclear fireball, the old mercury-in-glass standby will not be of much help. Accordingly, there are electrical resistance thermometers, optical thermometers, thermocouples, and constant-volume gas thermometers, to name a few. Of these, the standard for accuracy and reproducibility is still the constant-volume gas instrument (Fig. 14.3), though it is large, slow, delicate, and inconvenient.

At present we know of no purely mechanical quantity—that is, one expressible in terms of mass, length and time only—which can be used, however inconveniently, in place of temperature. We are inclined to conclude that temperature probably is itself a basic concept.

A. G. WORTHING (1940)

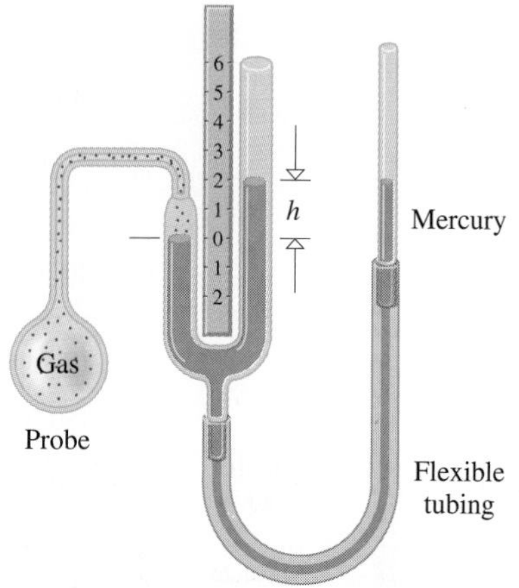

Figure 14.3 A constant-volume gas thermometer. The chamber at the left is the sensing bulb. The gas in the bulb changes volume when it experiences a temperature change. That causes the levels of mercury in the U-tube manometer to shift, and h gives us the pressure in the probe. Lifting the movable mercury column on the right shifts the other two mercury levels and allows the volume of gas to be kept constant.

14.1 Thermodynamic Temperature and Absolute Zero

The Fahrenheit and Celsius scales have little or nothing to do with the fundamental nature of the concept of temperature. After all, the freezing point of water at Earth's atmospheric pressure has no obvious relationship to any basic aspect of the Universe. Presumably, an alien scientist on a planet that has no water will devise a thermometer that measures temperature equally well. More to the point, is there a universal zero of temperature linked to the very essence of matter and energy, a zero that all scientists (human or otherwise) might discover? The answer is yes, and it is called the *absolute zero of temperature.*

Sometime around 1702, G. Amontons devised an improvement on Galileo's thermoscope that has evolved into the modern constant-volume gas thermometer (Fig. 14.3).

Figure 14.4 shows the behavior of a typical gas as its temperature is lowered. The important feature is that the graph is a straight line—the pressure falls linearly until the gas liquefies and the process abruptly ends. Remarkably, all gases behave nearly the same way. Moreover, when the straight line for any gas is extended downward, it crosses the zero-pressure axis at a temperature of −273.15°C. Similarly, if the pressure is kept constant, the volume of a gas will collapse linearly, again approaching the zero-volume axis at the same −273.15°C (−459.7°F), as in Fig. 14.5. This point is now believed to be the very limit of low temperature, **Absolute Zero.**

Amontons, studying the behavior of a number of gases, showed that each volume changed by the same percentage for a given change in temperature. Astutely, he recognized that a decrease in pressure down to zero would accompany a decrease in temperature down to some limiting value of coldness. Later, J. H. Lambert repeated these experiments with greater accuracy, concluding (ca. 1779) that there was indeed a temperature limit. "Now a degree of heat equal to zero," wrote Lambert, "is really what may be called absolute cold."

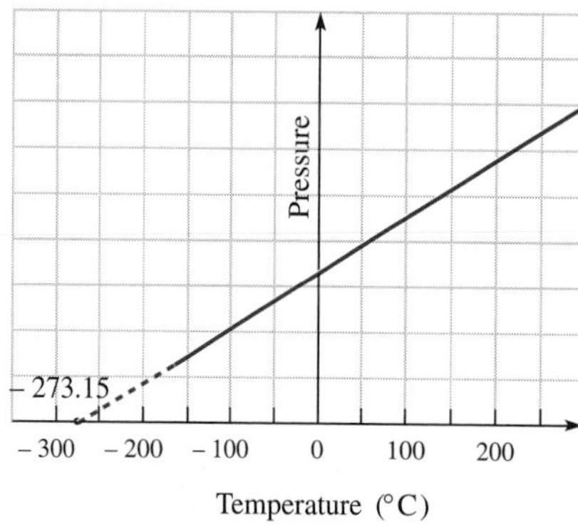

Figure 14.4 Pressure versus temperature for a gas at constant volume. As the gas is cooled, its atoms lose some of their thermal energy; they travel more slowly and collide with the chamber walls less frequently and less forcefully. As a result, the pressure in the chamber drops. All gases behave essentially the same way. Their P versus T graphs have different slopes, but they are all straight lines, and they all head toward −273.15°C.

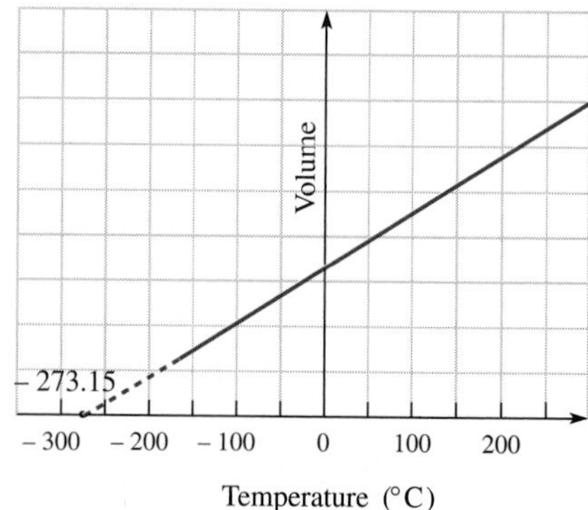

Figure 14.5 Volume versus temperature for a gas at constant pressure. As the temperature of a gas is lowered, its atoms are less energetic. To keep the pressure constant, the volume must decrease, which will make it easier for atoms to strike the chamber walls frequently, thereby sustaining the pressure. All gases shrink toward zero volume at $T = -273.15°C$.

Absolute Zero certainly seems the place to begin a temperature scale. Still, the idea received little attention for well over half a century. Then, in 1848, William Thomson (later to become Lord Kelvin), formalized the idea theoretically. On the basis of very different thermodynamic considerations (Sect. 16.5, p. 568), he proposed an **absolute temperature scale** that turned out to be in perfect agreement with the results from the constant-volume gas thermometer.

Today, the practical laboratory temperature reference is the *triple point* of water: the unique circumstance of temperature (0.01°C) and pressure (610 Pa) where solid, liquid, and vapor coexist. Absolute Zero (−273.15°C) and the triple point (0.01°C) are separated by 273.16 Celsius degrees.

Temperature is now recognized as a fundamental physical quantity like mass, length, and time, so the old degree symbol is discarded. As with all other quantities, the **thermodynamic** or **absolute temperature** is given a unit, the *kelvin* (abbreviated K). The SI value of 1 K is defined as 1/273.16 of the temperature of the triple point of water. Note that 1 kelvin is effectively the same size as 1 degree Celsius. For practical reasons (not the least of which are all the volumes of research data already in existence), the scheme is constructed to mesh with the Celsius system (Fig. 14.6). Using an arrow to mean "equivalent to," Absolute Zero is $0\text{ K} \rightarrow -273.15°\text{C}$, therefore $0°\text{C} \rightarrow 273.15\text{ K}$, and room temperature is just about 293 K (which is read 293 kelvins). The numerical equivalent is summarized symbolically by

$$T \rightarrow T_C + 273.15 \tag{14.2}$$

This expression, awkward as far as units are concerned, is more a statement of the logic of making numerical conversions than it is an equation—the 273.15 can either be degrees or kelvins.

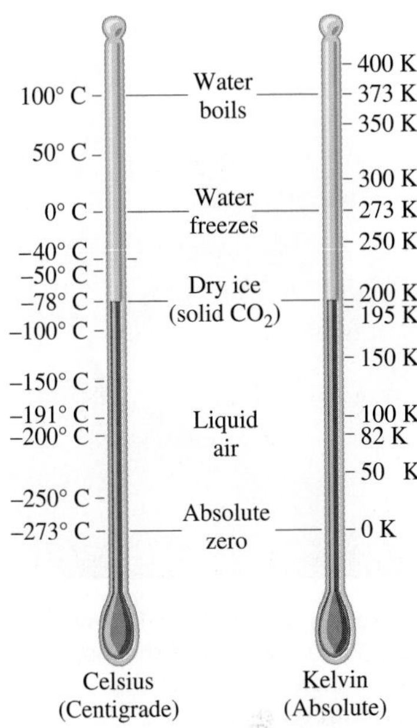

Figure 14.6 A comparison of the Celsius and Kelvin scales. All scientific work is done in these two systems, with the Kelvin scale being preferred.

Example 14.2 Gold freezes at 1337.58 K. What is the equivalent Celsius temperature?

Solution: [Given: $T = 1337.58$ K. Find: T_C.] To go from the thermodynamic temperature to Celsius, Eq. (14.2) suggests we subtract, and therefore

$$T_C = (1337.58 - 273.15)°\text{C} = \boxed{1064.43°\text{C}}$$

▶ **Quick Check:** This result is 1064.43°C above 0°C, which is ≈273°C above 0 K. Hence, it is ≈1337 K above 0 K.

14.2 The Range of Temperatures

The Universe sustains an incredible temperature range that is completely alien to our lukewarm intuition: we live out our delicate lives within a tiny band of hot and cold (see Table 14.1).

The highest temperatures likely to exist at this moment are to be found deep within certain stars: $\approx 4 \times 10^9$ K seems to be the theoretical extreme. At a temperature only about 10 times higher, matter fragments into subatomic particles. A hydrogen bomb ignites at roughly 40×10^6 K, and blazes up to nearly 10 times that, whereas the innards of the Sun churn away at a mere 15×10^6 K. Atoms can form, if only momentarily, in the seething body of a star even at 10^5 K. Such a raging mix of atoms and atomic fragments (of ion cores and electrons) is *plasma,* typical star-stuff.

At temperatures lower by a factor of 10, matter exists in clouds of atoms, ions, electrons, and an occasional molecule. There is evidence for the presence of a vari-

Lord Kelvin, born William Thomson (1824–1907), was one of the most accomplished scientists of the nineteenth century.

TABLE 14.1 Temperatures of several physical phenomena

°C		°F
−273.2	Absolute Zero	−459.7
−269	Helium boils	−436
−196	Nitrogen boils	−320
−183	Oxygen boils	−297
−79	Dry ice (CO_2) freezes	−109
−39	Mercury freezes	−38
0	Water freezes	32
3.8	Heavy water freezes	38.9
≈20	Room temperature	≈68
31	Butter melts	88
≈37	Body temperature	≈98.6
≈54	Paraffin melts	≈129
78	Alcohol boils	172
100	Water boils	212
101.4	Heavy water boils	214.6
108	Saturated salt solution boils	226
232	Tin melts	449
327	Lead melts	621
445	Sulfur boils	833
657	Aluminum melts	1215
801	Salt (NaCl) melts	1473
961	Silver melts	1762
1063	Gold melts	1945
1083	Copper melts	1981
1000–1400	Glass melts	1830–2550
1300–1400	Steel melts	2370–2550
1530	Iron melts	2786
1620	Lead boils	2948
1774	Platinum melts	3225
1870	Bunsen burner	3398
2450	Iron boils	4442
≈3410	Tungsten melts	≈6170
3500	Oxyacetylene flame	6332
5500	Carbon arc	9932
6000	Surface of the Sun	10 832
6020	Iron welding arc	10 868

ety of molecules in cool spots (<5000 K) on the surface of the Sun. If we take a cloud of steam up to ≈3000 K, about one-quarter of the water molecules rupture into separate atoms of oxygen and hydrogen. The hottest thing you are likely to find around the house is a tungsten light bulb filament, which operates at ≈2800 K (2500°C). By comparison, lava is molten at ≈2000 K. Lead melts at 327°C (621°F), paper burns at ≈230°C (450°F), and now we are down to the temperatures of baking apple pie and a hot cup of tea.

Ice melts at 273.15 K (0°C), but if salt (NaCl) is added to icy slush, the temperature of the mix drops as low as 252 K (−21°C). That's the way people traditionally cool homemade ice cream during churning. Mercury freezes at 234.29 K (−38.87°C). Adding 4 parts by weight of crystalline calcium chloride to 3 parts of water will produce a mix at 218 K (−55°C). Dry ice, solid CO_2, exists at 194 K (−79°C), and by adding ether (as was first done by M. Thilorier in 1835) or alcohol, the resulting rapid evaporation lowers the temperature further. By pumping on the mixture, carrying off the vapor, temperatures as low as ≈163 K (−110°C) can be achieved.

The mainstay of the modern low-temperature laboratory is liquid nitrogen, a clear, colorless fluid that looks like water and boils away at about 77.4 K (−196°C). It is easy to handle and relatively inexpensive (about the price of beer). Hydrogen gas can also be liquefied, but not until it is brought down to 20.4 K (−253°C). The atoms within the molecules of any substance at that penetrating cold are still capable of motion, but they do not have enough energy to rearrange themselves into new compounds; all chemical and biological activity has long ceased. Still, the atoms vibrate and a piece of frozen oxygen carries compression waves like any other solid.

The fountain effect. A tiny amount of thermal energy introduced by the spiral heater causes superfluid helium at ≈1.2 K to spurt into the air.

The most exotic of all cryogenic materials is liquid helium. Clear, colorless, and incredibly cold, it boils in an open container at 4.25 K (−268.9°C). Difficult to handle, it must be stored in a special thermos bottle surrounded by liquid nitrogen. It vaporizes easily and, since liquid helium sells at the price of a decent champagne, that's to be avoided. By pumping away the evaporating helium, the remaining liquid is further cooled, and the temperature can be lowered appreciably. At 2.18 K, helium undergoes a dramatic change, becoming not a solid, but a so-called *superfluid*. In that state, it is the best-known conductor of heat, though it remains an electrical insulator.

Dropping the temperature of a system becomes increasingly difficult and more costly in terms of energy expended, the lower we go. In order to cool a sample much below 1 K, a technique called *adiabatic demagnetization* is generally used; with it, solid specimens have been taken down to about 0.003 K. To plunge still lower, an even more subtle method of nuclear adiabatic demagnetization is employed. In the quest for Absolute Zero, no one has ever gotten bulk matter any closer than ≈0.000 000 02 K.

At these low temperatures, only a faint rustling of atoms persists. Yet, enough energy remains in the atoms to keep superfluid helium from solidifying, making it the only liquid that does not become a solid under ordinary pressure. To form crystalline helium, the atoms must be forced close together. Even below 1 K, where there is hardly any disruptive jiggling left, it still takes about 30 times atmospheric pressure to solidify helium. *Absolute zero corresponds to a state of minimum but finite atomic motion:* at 0 K, a tiny **zero-point energy** associated with the vibration of atoms remains—presumably enough to keep helium fluid, however close it approaches 0 K.

THERMAL EXPANSION

The realization that most materials expand when heated is an ancient one: centuries before the thermometer was invented, blacksmiths were putting red-hot iron rims on wooden wagon wheels (so they would cool and shrink tight). Nevertheless, it was the ability to measure temperature that made possible a quantitative study of **thermal expansion**. Early thermometers themselves operated via thermally induced changes in the volumes of gases and liquids (by 1747 there were even metal-rod thermometers), which was all the more reason to study the effect.

14.3 Linear Expansion

For the moment, let's limit ourselves to considering elongated solids so that one dimension predominates: bars, pipes, wires, girders, and so forth. Experimentally, it

(a)

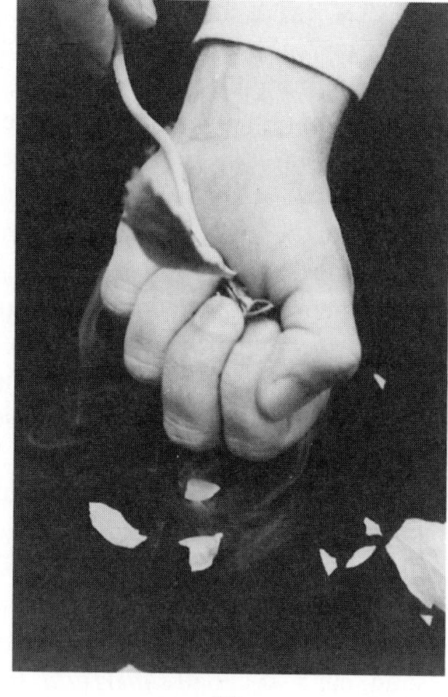
(b)

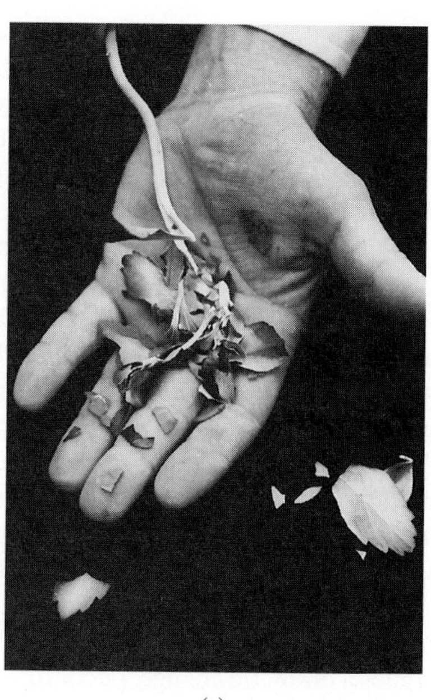
(c)

A sprig of pachysandra frozen in a bath of liquid nitrogen. Holding the plant in the liquid for a minute or so drops its temperature down to −196°C (−320°F). The leaves are then as brittle as glass and shatter into sharp fragments when squeezed. The sprig is so cold in (b) it is able to condense the water vapor in the air into streamers. The hand isn't harmed because the mass of the sprig is small, as is its specific heat capacity (p. 521).

is found that the change in length (ΔL) of a solid bar of original length L_0 depends on the change in temperature (ΔT) it experiences when heated uniformly; that is, $\Delta L \propto \Delta T$. The process is similar to the one encountered in Chapter 10 where we considered what would happen when a solid was stretched. Heating the bar causes its atoms to vibrate more actively with greater amplitudes, and *they also put more distance between each other.* At any one temperature, each atom vibrates in three dimensions about a more or less fixed equilibrium point in the solid. As thermal energy is added to the material, the equilibrium positions of the rapidly oscillating atoms separate, and the solid gradually expands.

The weaker the interatomic cohesive force is, the greater is the excursion of the atoms from their previous equilibrium separations, and the more the material expands as a whole. Since even in the liquid state the atoms are not much farther apart, it can be expected that this increase in separation will be small. A typical metal expands about 7% when its temperature rises from near 0 K to its melting point. If for some small ΔT the atoms or molecules on average separate from one another by an additional 0.001%, we can expect the rod to increase in length by that same 0.001%. Thus, a long rod will increase more than a short one, but both will change by the same percentage. As a consequence, $\Delta L \propto L_0$ and we can anticipate that $\Delta L \propto L_0 \Delta T$. The change in length depends on both the original length and the change in temperature.

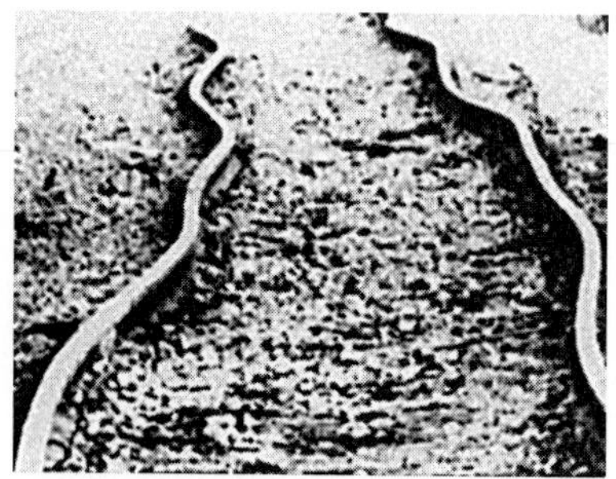

Years ago trains rattled along because fairly large spaces had to be left between the lengths of steel track. This was done so that the rails could expand in warm weather. The photo shows what would happen when that precaution wasn't taken. New low-expansion steel alloys have smoothed the ride considerably.

Once again, we have a proportionality that can be made into an equation by introducing a constant of proportionality, in this case α; thus

$$\Delta L = \alpha L_0 \Delta T \tag{14.3}$$

Here, ΔT can be in either °C or kelvins, since *the change in temperature is numerically the same for both.* At this point, we go into the laboratory and measure α (the **temperature coefficient of linear expansion**) and find that it has a specific value for every material tested (Table 14.2). The vast majority of materials have positive values of α; rubber under tension is a notable exception, and there are a number of ceramics (having strong atomic bonding) whose expansion coefficients at room temperature are nearly zero or even negative.

It is important to realize the logic of this process because it's used over and over in the development of physics. We do not yet have a theory complete enough to write out exactly what ΔL equals on an atomic level. Instead, we determine which macroscopic parameters it depends on (L_0 and ΔT) and then lump the effects of everything else into an empirical constant specific to each material, which we assume takes into account all the atomic details. If we knew more physics, we could presumably write an equation that would show exactly why lead expands 73 times more than quartz.

We can say that, because of its relatively weak interatomic bonds, lead is soft (has a low Young's Modulus), melts at a relatively low temperature (327°C), and has a high coefficient of expansion. The idea that the melting point and α are interrelated can be seen in Fig. 14.7, where a number of metals fall on the same smooth curve. The reason tin (Sn) and silicon (Si) are not closer to that curve is that they each experience a certain amount of strong *covalent* bonding. Covalent bonds, where electrons are shared, are stronger than metallic bonds (Table 10.6), making tin and silicon less able to expand for a given increase in vibrational (thermal) energy; hence they have a lower value of α than would otherwise be the case.

TABLE 14.2 Approximate values* of coefficients of linear expansion

Material	Coefficient (α) (K^{-1})
Aluminum	25×10^{-6}
Brass (yellow)	18.9×10^{-6}
Brick	10×10^{-6}
Diamond	1×10^{-6}
Cement and concrete	$10\text{–}14 \times 10^{-6}$
Copper	16.6×10^{-6}
Glass (ordinary)	$9\text{–}12 \times 10^{-6}$
Glass (Pyrex)	3×10^{-6}
Glass (Vycor)	0.08×10^{-6}
Gold	13×10^{-6}
Granite	8×10^{-6}
Hard rubber	80×10^{-6}
Invar (64% Fe, 36% Ni)	1.54×10^{-6}
Iron (soft)	$9\text{–}12 \times 10^{-6}$
Lead	29×10^{-6}
Nylon (molded)	81×10^{-6}
Paraffin	130×10^{-6}
Platinum	8.9×10^{-6}
Porcelain	4×10^{-6}
Quartz (fused)	0.55×10^{-6}
Steel (structural)	12×10^{-6}
Steel (stainless)	17.3×10^{-6}

*At temperatures around 20°C.

Example 14.3 A structural steel beam holds up the scoreboard in an open-air stadium. If the beam is 12 m long when it's put into place on a winter day at 0°C, how long will it be in the summer at 32°C? Given that it has a cross-sectional area of 100 cm^2, how much force would it take to stretch the beam that much?

Solution: [Given: $L_0 = 12$ m, $T_i = 0$°C, $T_f = 32$°C, and $A = 100$ cm^2. Find: ΔL and F.] From Table 14.2, $\alpha = 12 \times 10^{-6}\ K^{-1}$, so

$$\Delta L = \alpha L_0 \Delta T \qquad [14.3]$$

Hence

$$\Delta L = (12 \times 10^{-6}\ K^{-1})(12\ m)(32\ K) = 4.6\ mm$$

or almost $\frac{1}{2}$ cm. To find the force, remember $Y =$ stress/strain. From Eq. (10.10) and Table 10.4

$$F = YA\frac{\Delta L}{L_0}$$

$$F = \frac{(200 \times 10^9\ Pa)(100 \times 10^{-4}\ m^2)(4.6 \times 10^{-3}\ m)}{12\ m}$$

and $\boxed{F = 77 \times 10^4\ N}$ or 17×10^4 lb. This is also the force exerted by the beam on whatever might be resisting its change in length. Evidently, provision will have to be made in the design of the structure to allow for thermal expansion and contraction.

▶ **Quick Check:** Steel changes length by 0.001 2% per kelvin. Here, ΔT is 32 K, so the length change is 0.038 4% $L_0 = 4.6$ mm.

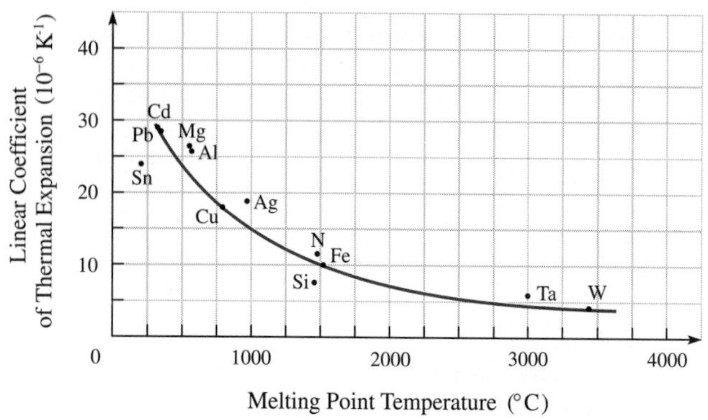

Figure 14.7 The higher the melting point of a material, the stronger the interatomic binding, and the smaller the amount of expansion.

Equation 14.3 is a useful, practical relationship, even if it is not a fundamental law of nature. Provided we are careful about the circumstances under which it is applied, it works nicely. Something to watch out for is that α really depends on T, increasing slightly as T increases (Fig. 14.8). The technical literature contains extensive tabulations of its values for a variety of substances over various ranges of temperature. Moreover, materials that have their atoms arranged differently in different directions (anisotropic materials, such as crystalline quartz and bismuth) expand and contract differently in those different directions. Indeed, when heated, graphite expands along one axis and contracts along another. But we need not worry about that, beyond realizing that none of this is as simple as it might appear.

14.4 Volumetric Expansion: Solids and Liquids

Using the same procedure as above, we can create an expression for the change in volume of a substance that has undergone a change in temperature ΔT; thus

$$\Delta V = \beta V_0 \Delta T \tag{14.4}$$

Here, V_0 is the original volume, and β is known formally as the **temperature coefficient of volume expansion** (see Table 14.3). Notice that the coefficients for liquids are about 50 times greater than for solids, which is to be expected. The weak intermolecular bonds inherent in that state account for the fact that liquids are both more compressible and more readily expanded thermally than are solids. As a rule, *β decreases as the thermodynamic temperature decreases, approaching zero as* 0 K *is approached.* For example, alcohol expands roughly 20% more per kelvin change near 373 K (100°C) than at 273 K (0°C). Fortunately, β for mercury, over the same range, varies by less than 0.01%, making it a superior thermometric substance.

Much of what was said before regarding α pertains here. The atoms again separate, but now that happens in three dimensions, not one. Indeed, a comparison of Tables 14.2 and 14.3 suggests that $\beta \approx 3\alpha$ is generally the case for isotropic solids. That relationship between α and β can be understood by imagining a rectangular solid with sides a, b, and c. If this block is changed in temperature by ΔT, we can expect each side to change its length proportionately. Accordingly, it will go from an initial volume of (abc) to a final one of $(a + \alpha a\Delta T)(b + \alpha b\Delta T)(c + \alpha c\Delta T)$. Multiplying out this expression provides the new volume, but α is a very small number and we can drop any terms containing α^2 or α^3 because they will be still smaller. The result is a final volume approximately equal to $(abc + 3\alpha abc\Delta T)$, and a change in volume $\Delta V \approx 3\alpha V_0 \Delta T$, from which it follows that $\beta \approx 3\alpha$.

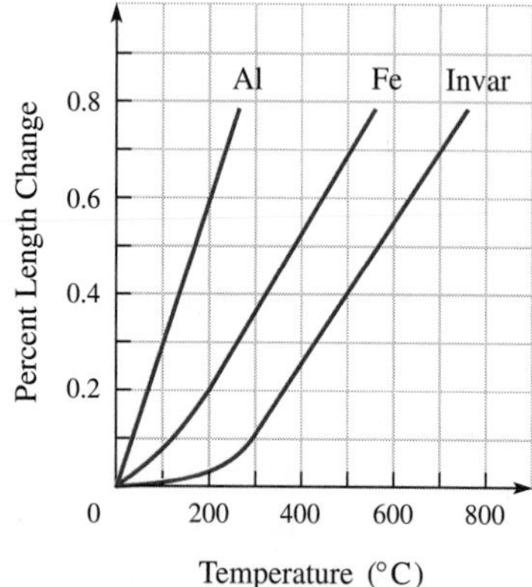

Figure 14.8 A plot of $\Delta L/L_0$ versus T. The slopes of the curves, $(\Delta L/L_0)/T$, equal α. For aluminum, over the temperature range shown, α is constant. That's not quite true for iron, and it's certainly not true for the alloy invar. The very low value of α for invar from 0°C to 200°C makes it useful for bimetallic devices and precision applications where expansion can be troublesome.

Suppose that we had a solid, homogeneous object whose shape is of no particular interest. Picture anywhere inside of it a portion of the material enclosed within a smaller imaginary sphere. When the big object is heated, it expands uniformly and, since no holes form inside of it, we can assume that the little imaginary sphere of material expands at the same rate. If the little sphere of matter is removed, the resulting cavity must expand at the same rate as the sphere itself. When an object expands or contracts, all its protuberances and cavities expand or contract proportionately.

Thus, when a bowling ball is heated the finger holes get bigger, as does the cavity within a mercury-in-glass thermometer. Similarly, a red-hot steel ring cools and contracts to fix tightly on a wagon wheel. That's why running hot water on a jar lid loosens it—the coefficient of expansion for steel is greater than for glass. A jar is larger when hot than cold and it holds more. In fact, because ordinary glass is a poor conductor of heat, putting a thick-walled glass vessel into an already hot oven is usually disastrous. The exposed faces of the vessel get hot while the region within the glass stays cool. Uneven expansion can crack the glass unless it is heated or cooled slowly. The advantage of Pyrex glass is that its expansion coefficient is 3 times smaller than that of ordinary glass.

Whenever materials are exposed to large variations in temperature, we must be concerned about expansion. If spaces are not left between slabs of ordinary concrete on streets and roads, they can buckle. Most exposed outdoor steel structures (overpasses, trestles, and so forth) have expansion joints to provide for changes in length. During a particularly cold week in 1937, the Bay Bridge in San Francisco contracted a total of 4 ft, 5 in. Often, small bridges are fastened only at one end, while the other rests on rollers to permit expansion. It has even been necessary, when finishing a large bridge in summer, to pack the final spans in ice for several hours to shrink them so they fit in place. Dental fillings and caps can experience substantial temperature variations in the mouth and should be made of materials that closely match the β for teeth.

Thermal expansion is utilized to advantage in the *bimetallic* strip, where flat lengths of two metals with different values of α are bonded together. When heated, the strip bends toward the side with the lower coefficient. These strips are used to make thermometers and circuit breakers or to open and close electrical switches in thermostats.

TABLE 14.3 Approximate values* of *coefficients of volumetric expansion*

Material	Coefficient (β) (K^{-1})
Solids	
Aluminum	72×10^{-6}
Asphalt	$\approx 600 \times 10^{-6}$
Brass (yellow)	56×10^{-6}
Cement and concrete	$\approx 36 \times 10^{-6}$
Glass (ordinary)	$\approx 26 \times 10^{-6}$
Glass (Pyrex)	9×10^{-6}
Invar	2.7×10^{-6}
Iron	36×10^{-6}
Lead	87×10^{-6}
Paraffin	590×10^{-6}
Porcelain	11×10^{-6}
Quartz (fused)	1.2×10^{-6}
Steel (structural)	36×10^{-6}
Liquids	
Acetone	1487×10^{-6}
Ethyl alcohol	1120×10^{-6}
Gasoline	950×10^{-6}
Glycerin	505×10^{-6}
Mercury	182×10^{-6}
Turpentine	973×10^{-6}
Water	207×10^{-6}

*At temperatures around 20°C.

Example 14.4 A driver pulls into a gas station and casually says "fill it up." The attendant does just that, filling the steel tank to the very brim with 56 liters of gasoline at 10°C. The trip home is a short one, and the heated garage is at 20°C. How much gasoline will overflow onto the floor?

Solution: [Given: $V_0 = 56$ liters, $\Delta T = +10$ K and that the liquid is gasoline and the tank is steel. Find: the amount of overflow.] First, convert the volume from liters to m^3: 56 liters $= 56 \times 10^3\ cm^3 = 0.056\ m^3$. The change in volume of the gasoline is

$$\Delta V_g = \beta_g V_0 \Delta T = (950 \times 10^{-6}\ K^{-1})(0.056\ m^3)(+10\ K)$$

and $$\Delta V_g = +5.32 \times 10^{-4}\ m^3$$

But the tank's capacity also increases; the tank's 0.056-m^3 cavity expands at the same rate as a solid steel mass in that shape. Hence, the tank's volume increases by

$$\Delta V_t = \beta_t V_0 \Delta T = (36 \times 10^{-6}\ K^{-1})(0.056\ m^3)(+10\ K)$$

and $$\Delta V_t = +0.202 \times 10^{-4}\ m^3$$

Therefore, about $\boxed{0.51 \times 10^{-3}\ m^3}$, or 0.51 liter overflows.

▶ **Quick Check:** $\Delta V_g/\Delta V_t = \beta_g/\beta_t = 26.4$; $(+5.32 \times 10^{-4}\ m^3)/(+0.202 \times 10^{-4}\ m^3) = 26.3$.

The Expansion of Water

Water is especially important to us, and that makes its physical behavior of great interest. Like the majority of liquids, water contracts slightly as it is cooled (Fig. 14.9),

A common device that allows an elevated roadway to expand and contract without affecting traffic. This one is part of an old wooden bridge over the Seine River in Paris.

This thick-walled glass mug cracked when it was filled with hot water.

A bimetallic strip bends when it's heated.

becoming a trifle more dense. In a unique way, however, this contraction gradually diminishes until at 277.13 K (3.98°C), it stops altogether. Between 373.15 K (100°C) and 277.13 K (3.98°C), water contracts about 4%, which is not very much, but still about 20 times more than would occur over that range for most solids. Water

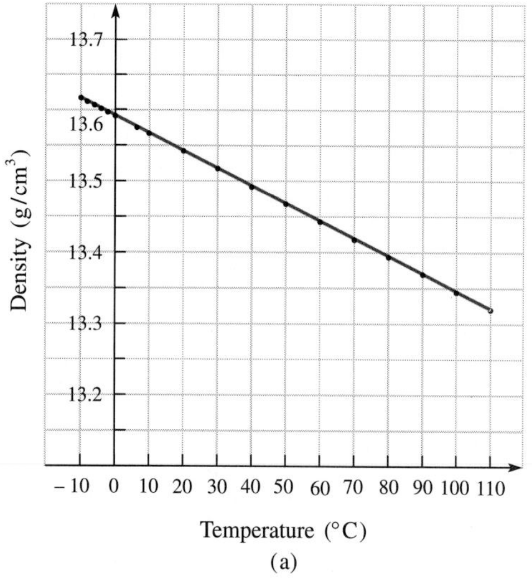

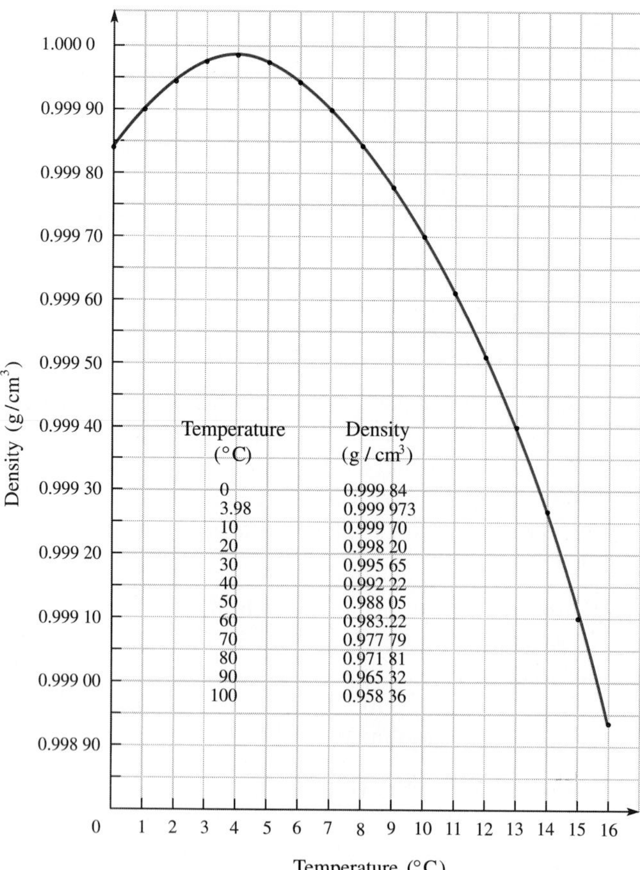

Figure 14.9 (a) The density of mercury plotted against temperature in °C. Mercury is a more representative liquid than water. (b) The density of water is a maximum at 3.98°C.

The bimetallic coil in a thermostat controlling the temperature in a room. As the coil expands, it tilts the vial of mercury, which breaks the electrical connection and shuts off the furnace.

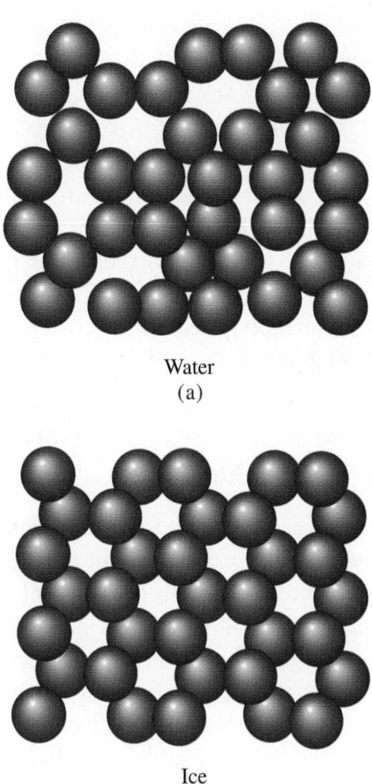

Figure 14.10 (a) The molecules of H_2O form small orderly groupings in the liquid state. (b) When the solid forms, the molecules come together in a highly open, ordered structure.

molecules can be envisioned as widely spaced, moving about fairly freely, and forming and unforming small groupings. They slowly move close together as the temperature descends. Below 277.13 K (3.98°C), the ordered clusters of water molecules rapidly get larger. Since the molecules unite with only four neighbors, these groupings contain an appreciable amount of empty space. Thus, below 3.98°C the volume of liquid begins to expand and the density decreases (Fig. 14.10).

Under controlled conditions, a pure liquid, particularly water, can be cooled below its freezing point without spontaneously solidifying; it is then said to be in a **supercooled state**. The effect was first reported in 1731 as "a somewhat strange Accident" that took place on a cold December day in Stockholm. At that time, the water in a piece of demonstration apparatus instantly froze as it was being taken down from a shelf. Working under very clean conditions, liquid water at ordinary pressure has been studied down to about 253 K (−20°C) and found to continually expand. But a mote of dust, a speck of grit, or a slight shaking will trigger the crystallization beginning at the freezing temperature 273 K (0°C).

Water crystallizes into a more orderly structure that takes up even more space. Accordingly, the volume increases abruptly—the density of water at 273 K (0°C) is 999.8 kg/m^3; the density of ice is only 917 kg/m^3. Lowering the temperature further causes the ice to contract ($\alpha = 52.7 \times 10^{-6}\ \text{K}^{-1}$ at $\approx$0°C), and it continues to do so at a diminishing rate all the way down to $\approx$75 K. By comparison to most materials, the fact that solid water floats in liquid water is a rarity. There are a few other substances, such as bismuth, antimony, and cast iron, that behave the same way, though as a rule, solids are more dense and sink in their liquids.

The implications of this for life on Earth are crucial. When, at the onset of winter, a lake or stream is gradually cooled at its surface, colder, denser water descends until the surface reaches below 3.98°C, whereupon it becomes increasingly less dense with further temperature drops and floats at the top of the lake. At 0°C, ice forms at the surface and thereafter grows downward, but since it is a much poorer conductor of heat than water, its very presence slows further ice formation. The sheet of surface ice essentially seals off the lake from the winter cold, allowing life to go on beneath it.

THE GAS LAWS

Because a gas expands to fill the container that confines it, gases do not undergo a simple temperature-dependent volume change. The change in volume of condensed matter is linked to both the amount of material present and the temperature variation. The same is true for a gas, although the relationships are more subtle. The amount of gas in a given container contributes to determining the pressure therein, and it is that pressure P, along with the temperature T, the volume V of the container, and the mass m of the gas that are the observables. We will develop a formula that ties all these variables together into an *equation of state*.

The three gas laws we will examine initially are basic relationships that were determined experimentally. They are descriptions of the behavior of gases under certain restricted conditions. Specifically, as long as the molecules of any gas do not get too close to one another, they behave *en masse* rather simply. If a single equation is to describe equally both O_2 and CO_2, which have different intermolecular cohesive forces and different molecular volumes, it will be able to do so only for conditions where there is lots of space between the particles of the gases. Thus, we are to avoid high densities, high pressures, and low temperatures (near the point of liquefaction).

A typical gas at STP (273.15 K or 0°C and 0.101 MPa or 1 atm) contains 2.7×10^{19} molecules per cubic centimeter. Each such particle, in the case of a monatomic gas such as helium or argon, has a diameter of roughly 0.2 nm. On average, they will be separated from one another by about 15 atomic-diameters. Imagining each little particle at the center of a cube 15 atomic-diameters on a side, the ratio of the empty volume to the actually occupied volume is about $15^3:1$, or 3375:1. Under such circumstances, each atom functions independently, and the gas as a whole behaves predictably.

We know from Boyle's Law (p. 364) that *if the temperature is kept constant, the volume of a given amount of any gas varies inversely with the absolute pressure*

$$[\textit{constant } T \textit{ and } m] \qquad PV = \text{constant} \qquad [11.5]$$

Given a gas-filled chamber that can expand and contract, the pressure within the gas varies inversely with V; double the volume and the pressure is halved, and vice versa.

If we keep the pressure constant and vary the temperature of a gas, we find, as did Amontons, that $\Delta V \propto \Delta T$. Indeed, the behavior of most gases agrees with Eq. (14.4), $\Delta V = \beta V_0 \Delta T$, where the expansion coefficient is almost the same for all of them, namely, $\beta \approx 0.0037\ \text{K}^{-1} = 1/(273\ \text{K})$. Thus, if we plot V versus T at constant P, as in Fig. 14.5, the slope $\Delta V/\Delta T = \beta V$ is a constant and the graph a straight line. In other words, *when the pressure is kept constant, the volume of a given amount of any gas varies directly with the thermodynamic temperature.* This is known as **Charles's Law**, after Citizen Jacques Alexandre César Charles of Paris (ca. 1787)—the man who inadvertently made the first solo balloon ascent when his assistant accidentally let go while climbing out of the gondola. Putting it another way, $V \propto T$, or

$$\frac{V}{T} = \text{constant} \qquad \text{[constant } P \text{ and } m] \qquad (14.5)$$

All gases increase in volume when the temperature increases. Doubling the temperature of a gas at constant pressure doubles its volume (Fig. 14.11).

The third basic gas relationship, called **Gay-Lussac's Law** after Joseph Louis Gay-Lussac (1802), maintains that *when the volume is kept constant, the absolute pressure of a given amount of any gas varies directly with the thermodynamic temperature;* that is, $P \propto T$, or

$$\frac{P}{T} = \text{constant} \qquad \text{[constant } V \text{ and } m] \qquad (14.6)$$

The pressure in an automobile tire (V = constant) rises in the summer as T rises, and it's dangerous to throw aerosol cans in a fire because P increases so much they explode.

Multiplying these three relationships yields

$$\frac{PV}{T} = \text{constant} \qquad \text{[constant } m] \qquad (14.7)$$

for a fixed quantity of gas. Thus, if V is decreased rapidly as in Fig. 14.12, P will rise and so too must T. Note that each of the constants in the last four equations is different.

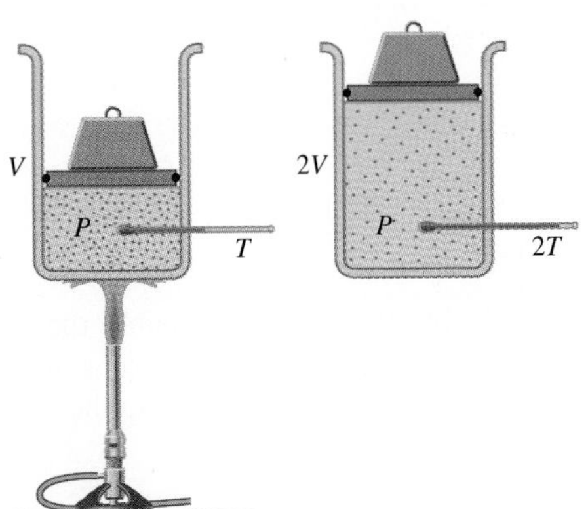

Figure 14.11 The weight on the piston keeps the pressure constant at P. Heating the gas, and thereby doubling T, must double V.

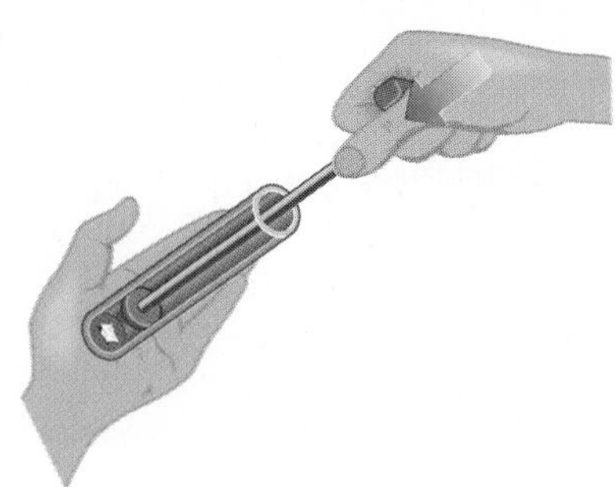

Figure 14.12 The fire syringe is a small cylinder with a tight-fitting piston. When the piston is rapidly driven in, the gas is compressed, V decreases while both P and T increase. A small piece of tissue paper will burst into flames within the syringe.

Example 14.5 A tank having a volume of 1.00 m^3 is filled with air at 0°C to 20.0 times atmospheric pressure. How much volume will that gas occupy at 1.00 atm and room temperature?

Solution: [Given: $V_i = 1.00\ m^3$, $T_i = 273.15$ K, $P_i = 20.0$ atm, $P_f = 1.00$ atm, and $T_f = 293.15$ K. Find: V_f.] Since the ratio in Eq. (14.7) is constant

$$\frac{P_i V_i}{T_i} = \frac{P_f V_f}{T_f}$$

hence

$$V_f = \frac{P_i V_i T_f}{P_f T_i}$$

(Divide pressures first so we need not convert units)

$$V_f = \frac{20(1.00\ m^3)(293.15\ K)}{273.15\ K} = \boxed{21.5\ m^3}$$

There is a slight increase in volume due to the increase in temperature, but the predominant increase is due to the 20-fold drop in pressure.

▶ **Quick Check:** The rise in temperature increases the volume by a factor of $\approx 293/273 \approx 1.07$; the pressure change increases it by a factor of 20; hence, $(1.07)20 \approx 21.5$ and $V_f \approx 21.5(1\ m^3)$.

We still have one quantity that can vary to account for, and that is the mass of the gas. Experience tells us that with all else unchanged, the volume of a gas must depend directly on the *amount* of gas present. Consider two identical bottles of gas, each with the same pressure, temperature, volume, and number of molecules. It seems reasonable that we could attach them together, remove the separating wall, and nothing significant would change on a macroscopic level—twice the number of gas molecules, twice the volume. Similarly, if V and T are fixed, increasing the number of molecules will increase the number of their collisions with the walls (p. 365) and therefore will increase the pressure. Hence $PV \propto$ m, or

[*constant T*]
$$\frac{PV}{m} = \text{constant} \tag{14.8}$$

Example 14.6 A 100-liter storage tank is slowly being filled with gas. At 5.00 times atmospheric pressure, the tank holds 0.60 kg of gas. If the temperature is kept constant, how much gas will be in the tank when the pressure is raised to 10.00 atm?

Solution: [Given: $V = 100$ liters = constant, P_i is 5.00 atm, $m_i = 0.60$ kg, and $P_f = 10.00$ atm. Find: m_f.] Since 1 liter = $10^{-3}\ m^3$, $V = 100 \times 10^{-3}\ m^3$. It follows from Eq. (14.8) that the initial and final values of the ratio PV/m must be equal; consequently

$$\frac{P_i V_i}{m_i} = \frac{P_f V_f}{m_f}$$

hence

$$m_f = \frac{P_f m_i}{P_i}$$

and

$$m_f = \frac{(1.013\ MPa)(0.60\ kg)}{0.5066\ MPa} = \boxed{1.2\ kg}$$

▶ **Quick Check:** To double the pressure, double the mass: $\Delta m = 0.60$ kg. (Notice that it would have been easier to divide the two pressures before converting their units.)

14.5 The Ideal Gas Law

The most commonly used equations of state express the amount of gas in terms of either the number of moles (n) present (p. 320) or the number of molecules present (N); that is, $PV \propto nT$ or $PV \propto NT$. The masses of the individual particles play a less central role than the total number of particles. We saw this earlier when we found that the same number of molecules of any gas at the same temperature and pressure occupies the same volume. Thus, it seems reasonable that if a single equation is to describe both H_2 and CO_2, we should deal with equal numbers of molecules, not equal masses of gas.

A mole is the mass of any substance in grams, equal numerically to the molecular mass of that substance: CO_2 has a molecular mass of 12 u + 2 × 16 u = 44 u and one mole of it corresponds to 44 g, or 0.044 kg. More to the point, a mole of any substance contains the same number of molecules. Introducing a constant of proportionality R, the equation of state becomes

$$PV = nRT \tag{14.9}$$

Remarkably, experiments reveal that R is nearly the same for all gases at low pressures; $R = 8.314\,41$ J/mol·K, and so it is called the **universal gas constant**. Still, we must limit the circumstances under which this formula is applied—it does not work well at high values of density for real gases. Accordingly, we say that a model theoretical gas for which this expression works perfectly is an *ideal gas,* and therefore Eq. (14.9) is known as the **Ideal Gas Law**. Despite its limitations, the concept has very practical applications. For example, helium at low pressure acts like an ideal gas almost to the point where it liquefies.

Example 14.7 Determine the volume of 1 mole of any gas at STP.

Solution: [Given: $T = 273.15$ K, $P = 1.013 \times 10^5$ Pa, and $n = 1.00$ mol. Find: V.] From the Ideal Gas Law

$$V = \frac{nRT}{P} = \frac{(1.00\text{ mol})(8.314\text{ J/mol·K})(273.15\text{ K})}{1.013 \times 10^5\text{ Pa}}$$

and $\boxed{V = 0.022\,4\text{ m}^3}$

▶ **Quick Check:** This result is confirmed by the discussion on p. 320.

Avogadro's number $N_A = 6.022 \times 10^{23}$ is the number of molecules per mole, and so the number of molecules N present in a sample of gas is just nN_A. Hence

$$PV = nRT = \frac{N}{N_A}RT$$

Accordingly, we define a new constant, called **Boltzmann's Constant**, $k_B = R/N_A = 1.380\,6 \times 10^{-23}$ J/K and thereby obtain the final form of the Ideal Gas Law

$$PV = Nk_BT \tag{14.10}$$

This expression may seem to be just another way of saying the same thing, but Boltzmann's Constant is actually one of the fundamental constants of nature.

Example 14.8 Estimate the number of air molecules there are in a room 4.00 m by 6.00 m by 3.00 m at standard pressure and room temperature (20°C). How many moles is that? How much volume would the gas

(continued)

(continued)
occupy if the temperature were 273 K (0°C), all else constant? Roughly how much does the air weigh? Compare that to your own weight. (See Table 14.4).

Solution: [Given: $T = 293$ K (20°C), $P = 0.101$ MPa, $V = 4.00 \times 6.00 \times 3.00$ m^3. Find: N.] From Eq. (14.10)

$$N = \frac{PV}{k_BT} = \frac{(0.1013 \times 10^6 \text{ Pa})(72.0 \text{ m}^3)}{(1.381 \times 10^{-23} \text{ J/K})(293 \text{ K})}$$

and $N = 1.803 \times 10^{27}$ molecules.

To find the number of moles, we divide by Avogadro's number, thus

$$n = \frac{N}{N_A} = \frac{1.803 \times 10^{27}}{6.022 \times 10^{23}} = \boxed{2.99 \times 10^3 \text{ moles}}$$

At STP, each mole occupies 22.414 liters; hence

$$V_{STP} = (22.414 \text{ liters/mol})(2994 \text{ mol})$$

and $$V_{STP} = 67.1 \times 10^3 \text{ liters} = \boxed{67.1 \text{ m}^3}$$

as compared to 72 m^3 at room temperature. From Table 14.4, each mole of air has a mass of 0.029 kg; hence, the total mass of air is

TABLE 14.4 The masses of one gram mole of several gases

Material	Mass (kg)
Hydrogen	0.002
Helium	0.004
Nitrogen	0.028
Air (average)	0.029
Oxygen	0.032
Argon	0.040
Carbon dioxide	0.044
Ozone	0.048

$$M = (0.029 \text{ kg/mol})(2994 \text{ mol}) = 86.8 \text{ kg}$$

which weighs $\boxed{851 \text{ N}}$ or 191 lb.

▶ **Quick Check:** Using Eq. (14.10) at STP, $V_{STP} = Nk_B(273 \text{ K})/P_A = (1.8 \times 10^{27})(1.381 \times 10^{-23} \text{ J/K}) \times (273 \text{ K})/(0.101 \times 10^6 \text{ Pa}) = 67 \text{ m}^3$.

14.6 Real Gases: Liquefaction

A given quantity of an ideal gas at a fixed temperature obeys the relationship $PV =$ constant, which is the equation of a hyperbola. Plotting P against V at any constant T produces a smooth curve (Fig. 14.13), symmetrical with the 45° line, known as an **isothermal**, or *isotherm* (from the Greek *isos*, meaning equal). When we construct the same sort of PV-diagram from data for real gases, a number of important differences appear. These were first observed by T. Andrews (ca. 1863) who used carbon dioxide under great pressure, but his results are broadly applicable (Fig. 14.14).

Each curve in the diagram is an isotherm and at high temperatures they are nearly hyperbolic. At lower temperatures, however, the isotherms drop below the corresponding hyperbolas (dashed curves) in the region of small volume and high pressure. It's here that the gas is compact enough for the intermolecular cohesive force to begin to be effective. The isotherms continue to distort as T is reduced until a special *critical temperature* is reached: the 304.3-K (31.1°C) isotherm, whose left-hand portion (beginning at the point of inflection C) marks the boundary between liquid and gas. At or below that line, there can be a change in the *phase,* or state, of the substance—it may go from liquid to gas or vice versa.

Andrews called point C the *critical point,* and the pressure there is known as the *critical pressure*. Here, the molecules are moving around, but not rapidly enough to overcome the now-potent cohesive force (Table 14.5). At C, *liquid and gas coexist at the same density* and are indistinguishable. Above the critical temperature, no amount of pressure will liquefy the gas; it can become denser and denser, but there will never be a clear gas-liquid transition. For example, above 647 K (374°C), water can exist only as a gas no matter what the pressure is. Evidently, the critical temperature is a measure of the strength of the cohesive force of the substance. The relatively high value (647 K) for water is a reflection of the sizable attraction between its polar molecules.

Let's follow along one of the isotherms below the critical temperature. That can be done by filling a cylinder with a large volume of the gas (Fig. 14.15). It is customary to call the substance *vapor,* rather than gas, when it is below the critical temperature—thus *vapors can be condensed via compression, gases cannot.* Compressing the vapor without allowing its temperature to change carries it from point 1 to point 2. Compressing it further increases the pressure and brings the vapor to point 3, where it is just about to condense. Further compression results in liquefaction at a constant pressure, bringing the system first to point 4 and then, when all is liquid, to point 5. Beyond that it takes considerably more pressure (note the slope of the curve) to compress the liquid, bringing it to point 6.

At point 4, the chamber contains both liquid and vapor, and the amount of each is stable. Some molecules that happen to be moving fast enough up near the surface will leave the liquid altogether, in the usual process of evaporation. Others in the vapor above it will crash into the liquid, becoming part of it. When the rates of evaporation and condensation are equal, the system remains in a state of *dynamic equilibrium.* If the concentration of vapor above the liquid is stable, it exerts, on average, a stable pressure on the liquid known as the **equilibrium vapor pressure**, or the *saturated vapor pressure,* or, when there is no ambiguity, just the vapor pressure. That's what the pressure gauge in the chamber (Fig. 14.15) will be reading because the piston is being moved very slowly. This pressure depends both on the nature of the liquid and on the temperature. The larger the cohesive force, the lower the vapor pressure. The higher the temperature, the higher the vapor pressure. Thus, at 293 K (20°C), strongly interacting water has a vapor pressure of 2.3 kPa, while volatile, weakly bound chloroform has a vapor pressure of 21.3 kPa—it can keep evaporating much longer before equilibrium sets in.

Systematic research on the liquefaction of gases began when Michael Faraday took up the work. In 1823, using a mixture of ice and salt as the cooling agent, Faraday was able to liquefy a number of gases, including carbon dioxide (Fig. 14.16). Still, despite all efforts, the most common gases of the air defied liquefaction, even to pressures in excess of 3000 atm. In 1869, Andrews pointed out that the failure to

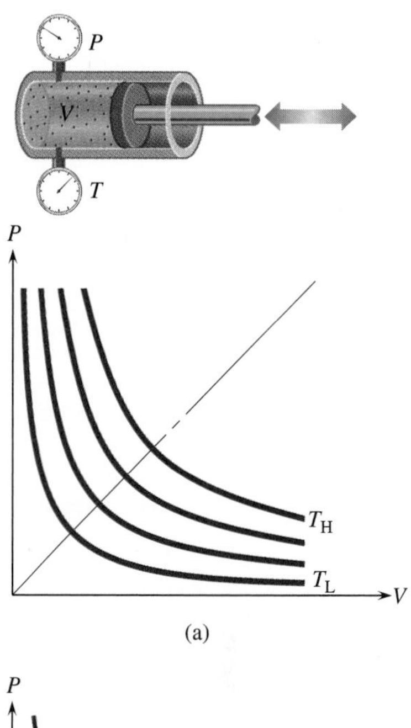

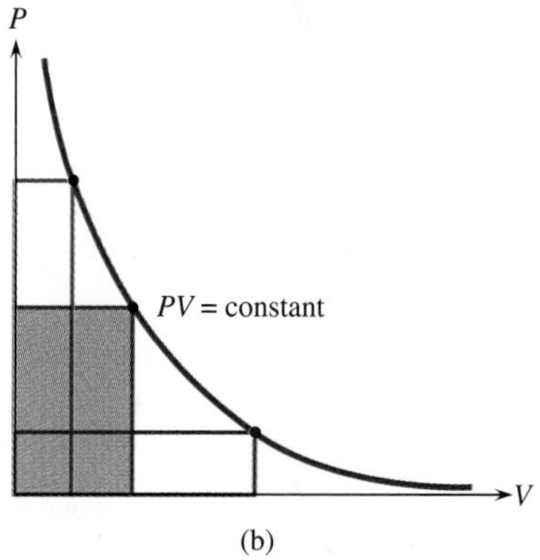

Figure 14.13 (a) A family of isothermal curves of P versus V for an ideal gas: $T_H > T_L$. (b) The area of each rectangle is $P \times V$, and that's constant.

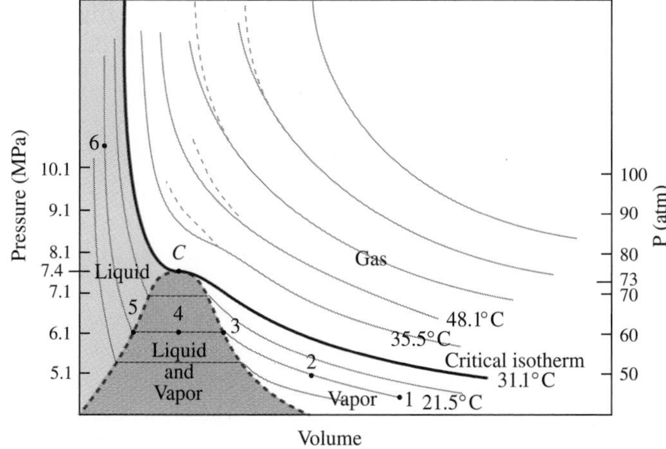

Figure 14.14 Isothermals for carbon dioxide. Below the critical isotherm and to the right, CO_2 exists as an unsaturated vapor. Liquid and saturated vapor coexist in the middle region, and only liquid exists in the region on the left.

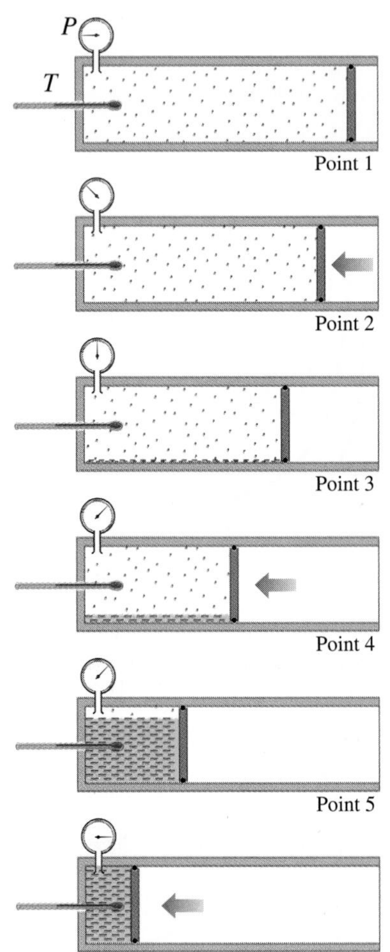

Figure 14.15 The isothermal compression of a real gas below its critical temperature. These are the six points on the isothermal shown in Fig. 14.14.

TABLE 14.5 Critical constants

Material	Critical Temperature (K)	Critical Pressure (MPa)	Critical Density (kg/m^3)
Ammonia	405.5	11.28	235
Argon	150.7	4.86	531
Bromine	584.2	10.34	1180
Carbon dioxide	304.3	7.38	468
Carbon monoxide	133.0	3.50	301
Chlorine	417.2	7.71	573
Deuterium	38.4	1.66	67
Helium	5.21	0.23	69
Hydrogen	43.24	1.30	31
Hydrogen sulfide	373.2	9.01	349
Neon	44.4	2.73	484
Nitrogen	126.3	3.39	311
Nitrous oxide	309.7	7.27	459
Oxygen	154.8	5.08	410
Water	647.4	22.12	326
Xenon	289.8	5.88	1110

liquefy these gases occurred because the early attempts used compression without cooling below the critical temperature—in short order, oxygen (1877), hydrogen, and nitrogen were liquefied.

14.7 Phase Diagrams

Figure 14.17 is a typical *PT*-diagram, also known as a *phase diagram* because it summarizes the relationships between the solid, liquid, and gaseous states of a substance, in this case, water. Two different phases coexist at temperatures and pressures lying along each of the three boundary lines. The curve from O to C corresponds to the equilibrium vapor pressure and is, as was recognized by Dalton (1804), logarithmic. It is called the *boiling-point curve.* Again, C corresponds to the critical point. If the line from O to C is crossed, there will be a clear liquid-vapor phase transition. By contrast, if the system is carried (by altering P and T) from the region on the lower right where it is a gas (up around and over C), it gradually blends into a liquid, without there being any phase transition. That's what would be seen if

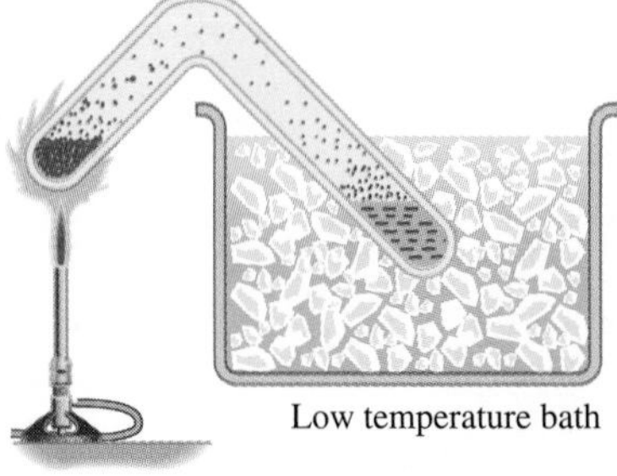

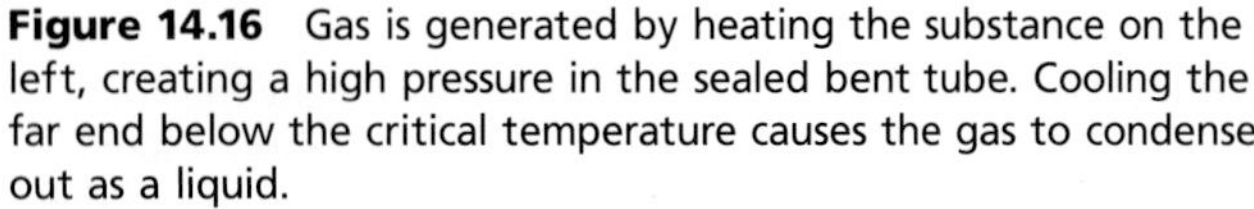
Figure 14.16 Gas is generated by heating the substance on the left, creating a high pressure in the sealed bent tube. Cooling the far end below the critical temperature causes the gas to condense out as a liquid.

we could descend into the hydrogen-helium atmosphere of the giant planet Jupiter. Under its own crushing pressure, the atmosphere gets thicker and thicker, imperceptibly merging into the liquid body of the planet.

For every pressure above that of point *O*, there is a temperature at which a pure solid will be in equilibrium with its liquid state. The locus of all such melting (or freezing) points is the line rising up from *O* on the left, which is the *melting-point curve*. Notice that this solid-liquid transition line leans to the left. Raising the pressure by 1 atm lowers the melting temperature by about 0.007 2°C (which is one reason for packing snowballs as hard as possible—it fuses the snow together). Along with friction, a rise in pressure causes the ice under the blades of a skate to momentarily melt and lubricate the surface. A person on skates might exert a pressure of 50 atm on the ice, lowering its melting point by about 0.4°C. Still, friction seems to produce most of the melting (Fig. 14.18).

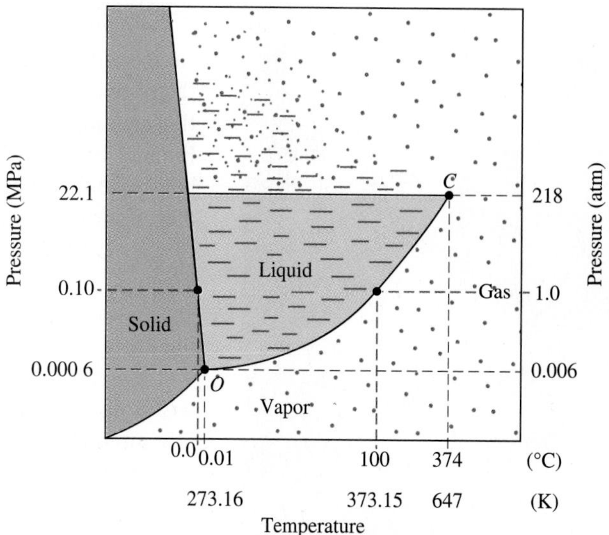

Figure 14.17 Phase diagram for water. If we take a block of ice at atmospheric pressure and slowly raise its temperature, it melts completely at 273 K (0°C) and remains liquid until 373 K (100°C), whereupon it completely vaporizes.

At point *O* in Fig. 14.17, which is the **triple point**, all three phases coexist in equilibrium (Table 14.6). Like liquids, solids also have a vapor pressure, though it may be extremely small. Thus, raising the temperature of a solid at a fixed pressure below that of the triple point will cause it to pass directly from solid to vapor, a process known as *sublimation.* The line descending from *O* with decreasing temperature corresponds to the vapor pressure and is often called the *sublimation curve.* The intermolecular bonds in dry ice are so weak that even at room temperature it will transform directly to the vapor state, as will camphor and naphthalene (mothballs). In this spirit, take a block of frozen coffee, ice cream, or even blood at a temperature below the triple point; lower the pressure and pump off the vapor as the solid is gently heated. The ice will sublime, leaving a light porous residue, which is said to be *freeze-dried.*

When the process is run the other way via a reduction in temperature, the vapor condenses out directly to the solid state. This condensation is what produces snow in the upper atmosphere and hoarfrost down at the surface of the Earth. The latter is the feathery crystalline deposit of ice that can be seen on the inside of a window pane in winter.

TABLE 14.6 Triple-point data

Material	Temperature (K)	Pressure (kPa)
Water	273.16	0.610
Carbon dioxide	216.55	517.00
Ammonia	195.4	6.07
Nitrogen	63.18	12.5
Oxygen	54.36	0.152
Neon	24.57	43.2
Deuterium	18.63	17.1
Hydrogen	13.84	7.04

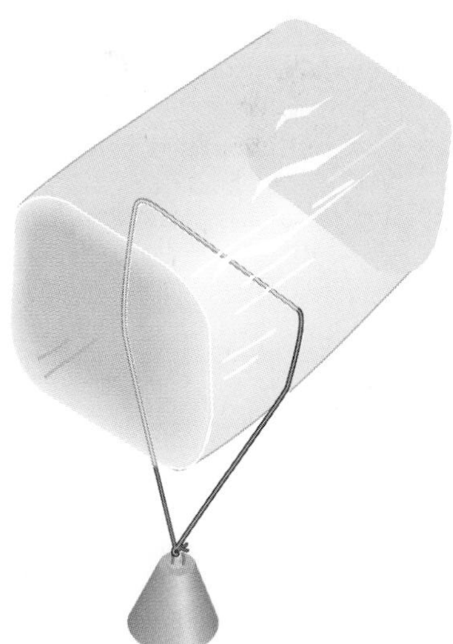

Figure 14.18 A wire that supports a load passes through a block of ice because the melting point drops due to the high pressure just under the wire. Above the wire, the water refreezes.

Hoarfrost.

14.8 Kinetic Theory

In ancient times, the Greek philosopher Plato observed that "heat and fire . . . are themselves begotten by impact and friction: but this is motion." That insight, that thermal phenomena arise from motion, was echoed almost 2000 years later by Francis Bacon, who maintained that "the very essence of heat . . . is motion and nothing else." The picture of a gas as a tumult of "very minute corpuscles which are driven hither and thither" was used as early as 1738 by Bernoulli (p. 365) to explain Boyle's Law. By the middle of the 1800s, this motion theory, now known as **Kinetic Theory**, was one of several competing formalisms concerning the nature of heat and temperature. Keep in mind that the reality of the atom was not established until the early twentieth century.

Kinetic Theory, as further developed mainly by James Clerk Maxwell and Ludwig Boltzmann, is a mathematical description of how a vast number of minute, rapidly moving particles can manifest itself macroscopically in the observed properties of bulk matter. We will examine only the basics of the theory and use it to gain some insights into the nature of gases. Following Maxwell, we model a gas as "an indefinite number of small, hard, and perfectly elastic spheres acting on one another only during impact." Clearly, we are talking about an ideal gas. To make the treatment still easier, we assume that the gas particles are essentially points having no extension, that do not collide with one another. This is certainly not the case, nor is it a necessary condition. (The inclusion of interparticle collisions would bring us into the study of Statistical Mechanics.)

Figure 14.19 depicts a piston in a cylinder of length L and cross-sectional area A. The pressure exerted on the piston originates from the countless millions of impacts made on it by the gas molecules each and every second. Inasmuch as the system is in dynamic equilibrium, this is the pressure P everywhere within the cylinder. The motion of a molecule can be resolved into its x-, y-, and z-components (Fig. 14.19a). The collisions are perfectly elastic (conserving KE) and so each time a particle bounces off the piston, its velocity's y- and z-components, parallel to the face of the piston, remain unaltered; only the x-component changes, reversing itself from $+v_x$ to $-v_x$. Thus, each impact produces a change in the momentum of a single particle of

$$\Delta p_x = 2mv_x$$

Each molecule moves along the length of the cylinder at a speed of v_x, which is constant (even though it may well be bouncing around from side to side in the process). Hence, it travels from the piston to the far face and back a distance of $2L$ in a time

$$\Delta t = \frac{2L}{v_x}$$

This is the time between collisions on the piston: the number of seconds per collision. The reciprocal of that value is the number of collisions per second, or the rate of collision: $1/\Delta t = v_x/2L$. Multiplying the number of collisions per second ($1/\Delta t$) by the change of momentum per collision (Δp) gives us the *rate of change of momentum:*

$$\frac{\Delta p_x}{\Delta t} = \frac{2mv_x}{2L/v_x} = \frac{mv_x^2}{L}$$

From Newton's Second Law, this equals the average force exerted by that one particle on the piston. Assuming, just for a moment, all N of the molecules in the cylinder behave the same way, the total average force on the piston due to the hail of

On Kinetic Theory. *So many of the properties of matter, especially when in the gaseous form, can be deduced from the hypothesis that their minute parts are in rapid motion, the velocity increasing with the temperature, that the precise nature of this motion becomes a subject of rational curiosity.*

JAMES CLERK MAXWELL (1860)

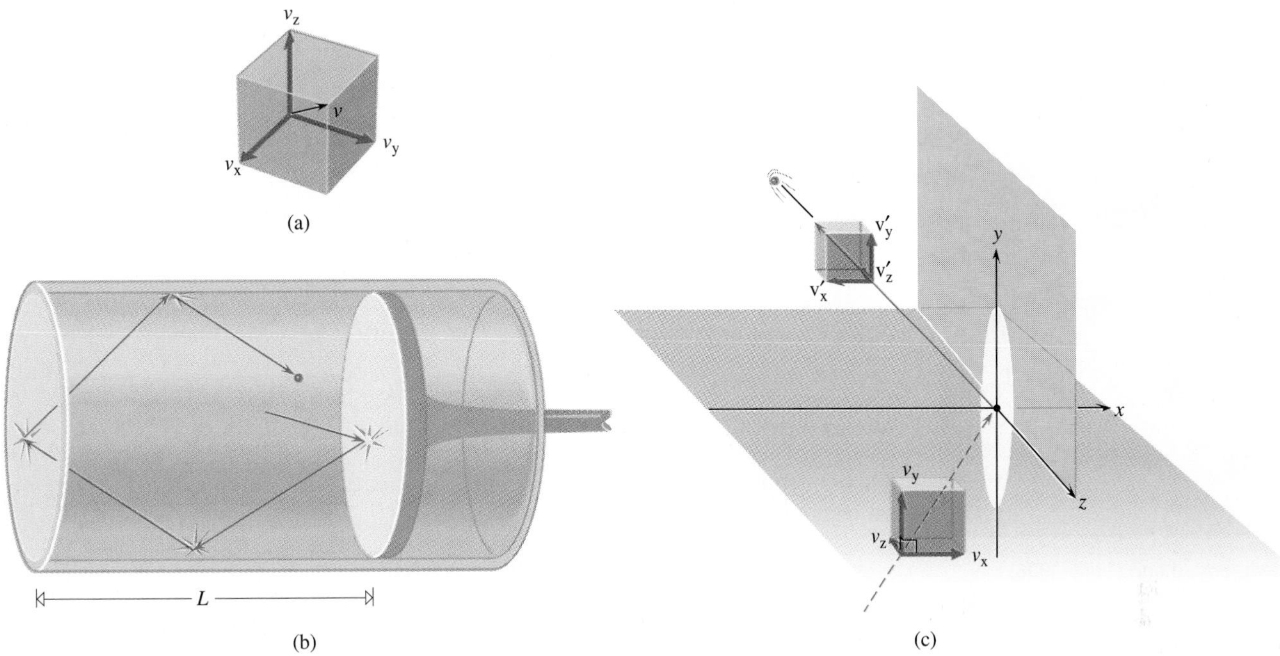

Figure 14.19 Atoms colliding with the face of a piston. Only the v_x component changes its direction on collision with the surface.

particles is

$$F_{av} = \frac{Nmv_x^2}{L}$$

By definition, the force per unit area is the pressure:

$$P = \frac{Nmv_x^2}{LA}$$

This equation can be expressed a little more conveniently using the three-dimensional equivalent of the Pythagorian Theorem, which tells us that the total speed squared equals the sum of the squares of its components; that is

$$v^2 = v_x^2 + v_y^2 + v_z^2$$

Since there are no external forces acting on the system, on average, we can expect that the gas particles will be moving in all directions with equal likelihood: $v_x^2 = v_y^2 = v_z^2$, and so

$$v_x^2 = \tfrac{1}{3}v^2$$

Accordingly

$$P = \frac{Nmv^2}{3LA} \tag{14.11}$$

This would be the pressure if each molecule had the same speed, but in fact they don't. Not surprisingly, there is a temperature-dependent distribution of speeds, something Maxwell and Boltzmann each worked out theoretically. For us, it will suffice to recognize that fact, and we will use an average speed in the pressure equation in place of v^2. The sum of the squares of the speeds of each of the N particles divided by N is the average, or *mean,* of the speed squared; consequently

Ludwig Boltzmann (1844–1906). Despondent and disturbed by the idea that his life's work was a waste, Boltzmann killed himself in 1906. Within a few decades, the atom was established as a reality, and Boltzmann's brilliant achievement was recognized as seminal.

$$(v^2)_{av} = \frac{v_1^2 + v_2^2 + \cdots + v_N^2}{N}$$

Thus

$$P = \frac{Nm(v^2)_{av}}{3LA}$$

and noticing that the volume of the cylinder is $V = LA$, we get

$$P = \frac{Nm(v^2)_{av}}{3V}$$

Since Nm is the total mass of the gas, the density is $\rho = Nm/V$ and

$$P = \tfrac{1}{3}\rho(v^2)_{av} \tag{14.12}$$

Moreover

$$PV = \tfrac{1}{3}Nm(v^2)_{av}$$

This equation is Boyle's Law. Remembering that the translational kinetic energy of an object is $\frac{1}{2}mv^2$, it follows that

$$PV = \tfrac{2}{3}N(\text{KE})_{av} \tag{14.13}$$

Pressure depends on the number of molecules per unit volume as well as on their average translational KE.

Comparing this situation with the Ideal Gas Law, Eq. (14.10), we see that

$$(\text{KE})_{av} = \tfrac{3}{2}k_B T \tag{14.14}$$

The temperature of an ideal gas is proportional to the average translational kinetic energy of its molecules. The microscopic average translational KE of the molecules manifests itself macroscopically in the form of temperature.

Still, real gases do interact, and that must be treated via Quantum Mechanics and not by Newtonian physics. Thus, although the relationship between translational kinetic energy and temperature is undeniably an important insight, one that can fruitfully guide our thinking, it has its shortcomings. Remember that a substance must have a zero-point energy even at absolute zero, something not in accord with Eq. (14.14), which therefore cannot be the whole truth at low temperatures. All of this underscores the conclusion that temperature is a fundamental quantity and, like length, time, and mass, is not expressible in terms of other quantities.

The Speeds of Molecules in a Gas

It follows from Eq. (14.14) and the fact that $(\text{KE})_{av} = \frac{1}{2}m(v^2)_{av}$ that

$$(v^2)_{av} = \frac{3k_B T}{m}$$

which is the mean of the speed squared. If we take the square root of it, we get something called the *root of the mean of the square* (*rms* for short):

$$v_{rms} = \sqrt{(v^2)_{av}} = \sqrt{\frac{3k_B T}{m}} \tag{14.15}$$

This is not quite the average speed, but a statistical analysis shows that $v_{av} = 0.92v_{rms}$, and so Eq. (14.15) gives us a measure of the typical speed of a gas particle.

Example 14.9 What is the *rms*-speed of a hydrogen molecule (H_2) at STP? Take the molar mass of hydrogen molecules to be 2.0 g/mol.

Solution: [Given: $T = 273$ K. Find: v_{rms}.] The mass of a molecule m is the molar mass divided by the number of molecules per mole, and therefore

$$m = \frac{0.0020 \text{ kg/mol}}{6.022 \times 10^{23} \text{ molecules/mol}}$$

and

$$m = 3.3 \times 10^{-27} \text{ kg}$$

Hence, from Eq. (14.15)

$$v_{rms} = \left[\frac{3(1.38 \times 10^{-23} \text{ J/K})(273 \text{ K})}{3.3 \times 10^{-27} \text{ kg}}\right]^{\frac{1}{2}}$$

and so $\boxed{v_{rms} = 1.9 \text{ km/s}}$ or about 6.3×10^3 ft/s, which is faster than a bullet.

▶ **Quick Check:** m is correctly just about the mass of two protons. From Eq. (14.12) and Table 10.1, $v_{rms}^2 = 3P/\rho = 3(1.01 \times 10^5 \text{ Pa})/(0.090 \text{ kg/m}^3)$; $v_{rms} = 1.9$ km/s.

We can expect sound waves to travel at speeds less than that of the individual molecules. The speed of sound in hydrogen (Table 13.2) is 1.27 km/s $= 0.7v_{av}$, and that's reasonable.

Our present understanding is that the molecules of a gas undergo a vast number of collisions with each other—about 5×10^9 per second for each molecule in the air around us, which is about 10^5 collisions for each centimeter of path traveled. The uninterrupted distance between collisions—the *mean free path*—is about 10^{-5} cm (Table 14.7), which is ≈100 atomic-diameters. That's why it takes time for the scent of a perfume to fill a room after the bottle is opened. The molecules, even though they move at great speeds, travel zigzag paths, only gradually diffusing outward.

The effect of the tremendous number of intermolecular collisions does *not* even out the speeds of the particles, but distributes them in a stable pattern over a broad range from zero up toward infinity. The corresponding curves represent the number of molecules in each interval of speed (Fig. 14.20). They peak at a speed that is likely to be possessed by more particles than any other, and that's roughly 13% below the average speed v_{av} and 23% below v_{rms}. Most of the particles are grouped around this peak speed; in fact, there will probably be no more than 1 out of about 10 000 molecules with speeds in excess of $3v_{av}$. Nonetheless, the immense number of particles present ensures that small but finite quantities will be traveling very slowly

TABLE 14.7 Molecular parameters at STP

Property	Hydrogen (H_2)	Oxygen (O_2)
Number of molecules per cm^3	2.7×10^{19}	2.7×10^{19}
Diameter of molecule	0.24 nm	0.32 nm
Molecules per mole	6.06×10^{23}	6.06×10^{23}
Mass per cm^3	8.99×10^{-8} kg	1.43×10^{-6} kg
Average speed	18.4×10^2 m/s	4.6×10^2 m/s
Collisions per second	10×10^9	4.6×10^9
Mass of molecule	3.3×10^{-27} kg	53×10^{-27} kg
Mean free path	1.8×10^{-7} m	1.0×10^{-7} m
Volume per gram	11.2×10^3 cm^3	700 cm^3

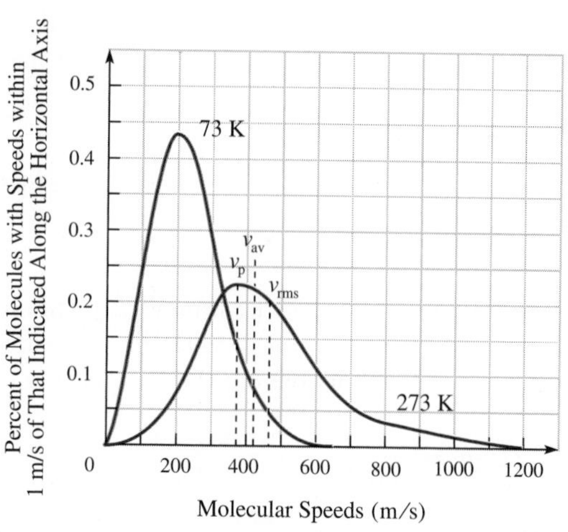

Figure 14.20 The Maxwell-Boltzmann speed distributions for molecular oxygen at $T = 73$ K and $T = 273$ K. The area under either curve between any two speeds, say, 400 m/s to 600 m/s, is the percentage of the total number of particles N in that range. The total area under either curve is 100%. Multiply the percentage in any speed range by the total number of molecules present, and you obtain the number with that range of speeds.

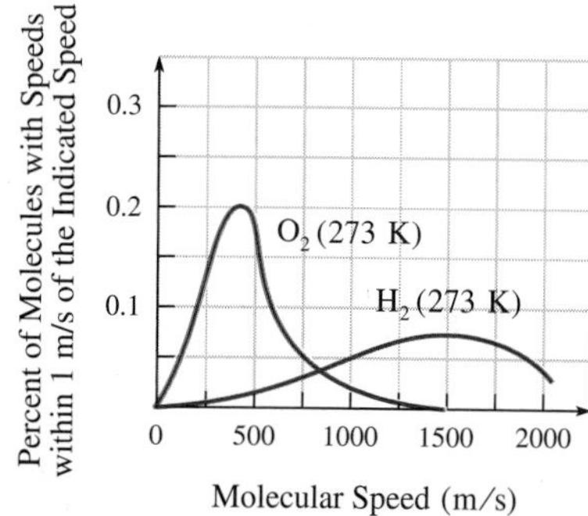

Figure 14.21 At the same temperatures, oxygen and hydrogen molecules have the same average kinetic energy; therefore, the less massive hydrogen molecules, on average, move faster.

and very rapidly. Realize, too, that the curve represents a stable pattern even though, from moment to moment, collision to collision, the speed of any one molecule will change drastically. These are the *Maxwell-Boltzmann distributions,* each displaying the most probable spectrum of molecular speeds available to the system at a particular temperature. Though purely theoretical when proposed, the distribution received direct confirmation in the late 1920s and early 1930s via molecular beam experiments.

In the derivation that led to Eq. (14.11), we assumed that the squares of the three speed components were equal. That suggests that the three possible means of sharing the total KE of the system are also equal. In other words, each mode of action, or *degree of freedom,* of the system possesses the same amount of energy. This is really a guess, but it's a guess that has been successfully generalized (and not without controversy) into the *Equipartition Theorem.* The idea was proposed by J. J. Waterston when considering a mix of two different gases held at the same temperature in a chamber. He suggested that each gas would have the same average translational KE. This same conclusion can also be seen to follow from Eq. (14.14). The implication is that massive particles will travel slower than light ones, as affirmed by Eq. (14.15) and illustrated in Fig. 14.21.

Diffusion

Presumably, if two gases of different mass are retained in a chamber that has a small hole in it, the lighter, more rapidly moving ones would escape sooner. That conclusion was confirmed experimentally (1896) by Lord Rayleigh, who partially separated two gases by allowing them to diffuse through a porous barrier into a region of vacuum.

This is the idea behind one of the processes devised during World War II to produce the atomic bomb. Bomb-stuff, fissionable uranium-235, is found in the ore mixed with heavier, but chemically identical, uranium-238. What was needed was a mechanical scheme that could separate the two kinds of atoms. Several different methods were pursued to ensure ultimate success—one of these was gaseous diffusion. Uranium was combined with fluorine, producing gaseous uranium hexafluoride. This gas was then cycled, over 4000 times, through barriers each containing thousands of millions of tiny holes smaller than one-tenth of a mean free path ($<10^{-5}$ mm). Because it was less massive, uranium-235 passed through the barriers more effectively, separating from the uranium-238. The scheme worked surprisingly well, and by the summer of 1945 we entered the age of nuclear weapons.

Core Material

Temperature is a fundamental quantity that describes an aspect of the physical state of a system. The two common temperature scales, Fahrenheit and Celsius, are related by

$$T_F = 32° + \tfrac{9}{5}T_C \quad \text{or} \quad T_C = \tfrac{5}{9}(T_F - 32°) \qquad [14.1]$$

The **thermodynamic temperature** is measured in units of *kelvins* (K):

$$T \to T_C + 273.15 \qquad [14.2]$$

and the Absolute Zero of temperature corresponds to 0 K (−273.15°C).

Most solids expand on having their temperatures raised and

$$\Delta L = \alpha L_0 \Delta T \qquad [14.3]$$

Here, ΔT can be in either °C or kelvins and α is the *temperature coefficient of linear expansion.* The change in volume of a substance that undergoes a change in temperature ΔT is

$$\Delta V = \beta V_0 \Delta T \qquad [14.4]$$

Here, V_0 is the original volume and β is the *temperature coefficient of volume expansion* (p. 490).

Boyle's Law is

$$[\textit{constant } T \textit{ and } m] \qquad PV = \text{constant} \qquad [11.5]$$

Charles's Law is

$$[\textit{constant } P \textit{ and } m] \qquad \frac{V}{T} = \text{constant} \qquad [14.5]$$

Gay-Lussac's Law is

$$[\textit{constant } V \textit{ and } m] \qquad \frac{P}{T} = \text{constant} \qquad [14.6]$$

These three relationships can be combined; thus, *for a fixed quantity of gas*

$$[\textit{constant } m] \qquad \frac{PV}{T} = \text{constant} \qquad [14.7]$$

When the mass of gas can change but the temperature is held constant, $PV \propto m$, or

$$[\textit{constant } T] \qquad \frac{PV}{m} = \text{constant} \qquad [14.8]$$

The equation of state for an ideal gas is

$$PV = nRT \qquad [14.9]$$

where $R = 8.314\,41$ J/mol·K is called the **universal gas constant**. We define a new constant, called **Boltzmann's Constant,** $k_B = R/N_A = 1.380\,6 \times 10^{-23}$ J/K and thereby obtain the final form of the Ideal Gas Law:

$$PV = Nk_BT \qquad [14.10]$$

The Kinetic Theory of gases leads to Boyle's Law, in the form

$$PV = \tfrac{2}{3}N(\text{KE})_{av} \qquad [14.13]$$

where

$$(\text{KE})_{av} = \tfrac{3}{2}k_BT \qquad [14.14]$$

The average translational KE of the molecules of an ideal gas is related to the temperature of that gas. Furthermore

$$v_{rms} = \sqrt{(v^2)_{av}} = \sqrt{\frac{3k_BT}{m}} \qquad [14.15]$$

Suggestions on Problem Solving

1. When confronted by a problem on gases, note any physical parameters that are held constant. Use the Ideal Gas Law in one of its two forms to summarize all the gas laws. Thus, if T and m (that is, n) are constant, the problem deals with variations in pressure and volume and PV = constant; if V and m (that is, n) are constant, the problem deals with variations in pressure and temperature and P/T = constant, and so on. When given or asked for N, use $PV = Nk_BT$.

2. To determine the mass of any molecule, either consult the Periodic Table or Table 14.4. Atomic masses are usually given in atomic mass units (u), where 1.000 00 u = $1.660\,54 \times 10^{-27}$ kg. The numbers in the Periodic Table are averages for naturally occurring mixes of isotopes (something we will come back to later), so there is no need to carry around all those figures.

3. In the problems concerning variations in the size of an object, the change in temperature is central. Whether you use °C or K, ΔT will be the same; however, you cannot use Fahrenheit when α or β is given in SI. **All the gas equations are formulated in terms of thermodynamic temperature, and using anything other than kelvins will produce wrong answers.**

4. The gas equations often lead to a symmetrical form like $P_iV_i/T_i = P_fV_f/T_f$, in which case the units cancel from both sides. It is then allowable to enter mixed units into the equation, but be very careful. Like it or not, there will always be a body of historical data in the literature that uses all the old, now shunned, units, and we should be able to deal with that, too. When one unit can be converted to another by a multiplicative constant, then the ratios of quantities in either set of units will be the same: P_i/P_f is the same whether it is in Pa or mm Hg. *But* when the units are converted one to the other by the addition of a numerical factor, this will not be true: T_i/T_f is *not* the same in degrees Celsius as in kelvins. (Just think about the chaos that would arise in the equations at 0°C.) So, yes you can use liters on both sides, or atmospheres, but you must use kelvins for temperature.

5. Remember that $\Delta L/L_0 = \alpha\Delta T$ and so α contains no length unit. It is the fractional change per kelvin (or per °C, which is the equivalent) and, hence, ΔL and L_0 can be entered in any unit you like.

Discussion Questions

1. List all the possible sources of errors you can think of that might limit the accuracy of a temperature reading made with a typical mercury-in-glass thermometer. Be as imaginative as you can and, as Descartes put it, "doubt of everything"; that is, take nothing for granted.

2. Plunge a sensitive mercury-in-glass thermometer in a bath of hot water. The temperature reading will first drop for a moment and then gradually rise. Explain.

3. Type metal, which is poured into molds to make printer's type, is an alloy of lead, antimony, and tin. Similarly, cast iron, which is an alloy of iron and carbon, is also used in making castings. What uncommon physical property do you think these materials might share with water? Explain.

4. A demonstration device known as a cast-iron bomb is a thick-walled sphere that is filled to the top with water and then sealed with a tightly fitting iron plug. The thing is then submerged in a pail of ice, water, and salt. What happens next? Explain. How does this same process, occurring naturally, affect rocks and mountains?

5. The two steel plates in Fig. Q5 have been attached to one another with a rivet, which was slipped into the hole running through the plates while red hot and "headed over" with a hammer. Why were the rivets put in hot? (Incidentally, the usual approach was to throw the red-hot glowing rivets, almost the size of your fist, through the air from the fire, up to the installing team, which caught them in buckets.)

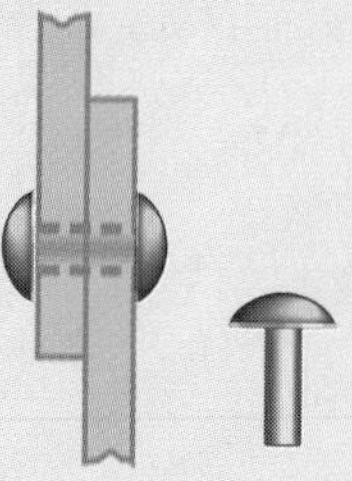

Figure Q5

6. Suppose that you are visited by an inventor who has found a liquid that has exactly the same β as glass. It is proposed that the wonder fluid be dyed red so it can be seen easily and used to make inexpensive, but incredibly accurate thermometers. Would you invest in such a scheme? Explain the physical reasons behind your decision.

7. The old-fashioned way to break boulders is to heat them up for a long time in a big fire and then quickly douse them with cold water. It is a lot easier than hitting at them with sledge hammers. Explain.

8. Anyone who has ever worked in a chemistry lab knows that such places invariably house a multitude of beakers and test tubes. Why is it that such equipment is generally made of Pyrex (or some other special glass), and why are they so thin-walled and delicate? (You never see a good, thick, sturdy test tube.)

9. In light of Question 8, why are glass incandescent bulbs so thin? What might happen to a hot light bulb if a drop of cold water fell on it? *Don't try it!* (This can be a problem to sloppy house painters.) What does that suggest about outdoor lighting? How does this relate to traditional photographic flashbulbs?

10. Though it was no doubt observed countless times before, Michael Faraday was the first person to report that two pieces of ice can be frozen together as one by simply pressing them against each other. Explain. What does this phenomenon have to do with snowballs?

11. I remember days as a kid when there was snow everywhere, but the stuff just would not pack into snowballs no matter how we tried. Although we concluded that it was just "terrible snow," what might have been the actual cause of the problem?

12. Professor James Thomson, Lord Kelvin's older brother, was the one who suggested that the three transition curves in the phase diagram (Fig. 14.17) actually meet at a single point—the triple point. Show logically that this must be the case (Fig. Q12).

13. The thick-walled sealed tube in Fig. Q13 shows liquid carbon dioxide in equilibrium with its vapor at room temperature and a pressure of about 6.2 MPa (61 atm). Describe what will happen in the tube as the temperature is slowly raised to 31.1°C.

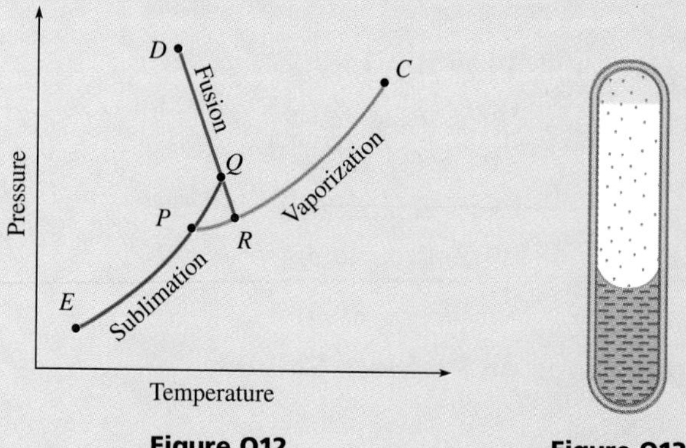

Figure Q12 **Figure Q13**

14. Under what circumstances, if any, can water freeze and boil at almost the same time? What would happen to a dish of water in an evacuated region where the vapor is efficiently drawn away?

15. Can we expect the ordinary gases of the air (oxygen, hydrogen, nitrogen, and so forth) to behave like an ideal gas? Explain.

16. Sir James H. Jeans, a physicist who made a number of important contributions to modern physics during the early part of the twentieth century, wrote the following in *The Growth of Physical Science:*

In warmer air, the molecules must, of course, move more rapidly, in cooler air more slowly. In general the energy of motion per molecule is proportional to the temperature, measured from the absolute zero at which the energy of motion is nil.

Which equations derived in the chapter confirm his comment? What, if anything, might you take exception to?

17. Given: two 1.0-m^3 sealed evacuated chambers surrounded by the same constant temperature bath. Suppose we put two moles of hydrogen in one of them and two moles of nitrogen in the other. What can be said about the pressures in the two chambers? Undeniably, the nitrogen molecules are much more massive than the hydrogen molecules. How can that difference be understood in light of your conclusion about the pressures? What, if anything, can be said about the speeds of the different molecules?

18. If we should ever be visited by some advanced extraterrestrial society, is it likely that they, too, would have arrived at the same notion of an absolute zero of temperature? How reasonable is it to expect that any thermometer they might pull out would have a scale we would be familiar with? If you answered affirmatively, which scale?

19. The air in an ordinary room heated by a stove of some sort will rise in temperature but may not increase its net amount of molecular translational energy. Explain. Be careful about the assumptions you make.

Multiple Choice Questions

1. The Celsius equivalent of 90°F is (a) 90° (b) 32° (c) 194° (d) 45° (e) none of these.
2. At the temperature known as absolute zero (a) all motion ceases (b) time ceases (c) all matter is in the solid state (d) atomic motion is at a minimum (e) none of these.
3. The surface of a very hot star has a temperature of 2×10^5 °C. Very roughly, how many times hotter is that than room temperature (293 K)? (a) 10^3 (b) 10^5 (c) 10^{-3} (d) 10^{-5} (e) none of these.
4. What is the order of magnitude of the fractional change in the length of a metal bar when it is altered in temperature by 1 K? (a) 10^{-6} (b) 10^{-4} (c) 10^{-5} (d) 10^6 (e) none of these.
5. Water has its maximum density at a temperature of about (a) 0°C (b) 32°F (c) 4°C (d) 273 K (e) none of these.
6. If the temperature of a sealed chamber full of gas is increased, the pressure will (a) remain constant (b) decrease proportionately (c) increase proportionately (d) first decrease and then increase (e) none of these.
7. Considering an ideal gas in equilibrium at a fixed temperature, volume, and pressure, we can expect that the velocity of any one of its molecules on average is (a) large (b) small (c) variable (d) zero (e) none of these.
8. Fahrenheit observed that water could be cooled below its freezing point without solidifying, and Gay-Lussac took it down to −20°C in the liquid state. This is an example of (a) superconductivity (b) supercooling (c) superfluidity (d) supersaturation (e) none of these.
9. What is the approximate mass of 11.11 liters of molecular hydrogen gas at STP? Each hydrogen atom has an average mass of 1.00797 u. (a) 1.0 g (b) 1.0 kg (c) 11.1 g (d) 22.4 liters (e) none of these.
10. The density of carbon dioxide at STP is (a) 1.96 kg/m^3 (b) 44 g/liter (c) 44 kg/m^3 (d) 6.02×10^{23} g/liter (e) none of these.
11. Given that we have 17.0 g of ammonia gas (NH_3), the volume it occupies at STP is (a) 17 liters (b) 22.4 liters (c) 22.4×10^6 m^3 (d) 17 m^3 (e) none of these.
12. According to Kinetic Theory, the molecules of a gas at a given temperature (a) all move with a speed v_{rms} (b) all move at speeds in excess of v_{rms} (c) all move at speeds less than v_{rms} (d) move at speeds above, at, and below v_{rms} (e) none of these.
13. In a mixture of oxygen and hydrogen gas, we can expect, on average, that (a) the hydrogen molecules will be moving faster than the oxygen molecules (b) both kinds of molecules will be moving at the same speed (c) the oxygen molecules will be moving more rapidly than the hydrogen molecules (d) the KE of the hydrogen will exceed that of the oxygen (e) none of these.
14. A sample of gas is held at a constant pressure in a cylinder closed by a movable piston. If the volume is halved, how will the new *rms*-speed of the molecules compare with the original *rms*-speed? It will be (a) $\sqrt{2}$ times greater (b) the same (c) 2 times greater (d) 4 times greater (e) none of these.
15. A sample of gas is held at a constant pressure in a cylinder closed by a movable piston. If the volume is doubled, how will the new *rms*-speed of the molecules compare with the original *rms*-speed? It will be (a) $\sqrt{2}$ times greater (b) the same (c) 2 times greater (d) 4 times greater (e) none of these.
16. Molecular hydrogen has a mass per mole of 2 g/mol, while the equivalent value for molecular oxygen is 32 g/mol. That means that the ratio of the *rms*-speed of a hydrogen molecule to that of an oxygen molecule will be (a) 2:1 (b) 4:1 (c) 8:1 (d) $\sqrt{2}$:1 (e) none of these.
17. An atom of argon ($^{40}_{18}A$) in the air at 20°C has an average translational kinetic energy of (a) 6.0×10^{21} J (b) 6.0×10^{-21} J (c) 6.0×10^{-21} m/s (d) 21×10^{-6} J/s (e) none of these.
18. The molecules of a gas have a certain average translational kinetic energy at 10°C. They will have twice that average value at (a) 20°C (b) 283 K (c) 566 K (d) 14.1°C (e) none of these.
19. According to Kinetic Theory, the molecules of a gas at a given temperature (a) all have the same speed (b) all

have the same direction of motion (c) all have the same kinetic energy (d) all have the same momentum (e) none of these.

20. If the temperature of an ideal gas is increased from 100 K to 400 K, the *rms*-speed of the particles will change by a multiplicative factor of (a) 2 (b) $\frac{1}{2}$ (c) $\frac{1}{4}$ (d) $\sqrt{2}$ (e) none of these.

21. A flat metal disc has a small off-center hole drilled in it. When heated, (a) the disc will expand and so will the hole (b) the disc will expand, but the hole will remain unchanged (c) the disc will expand, but the hole will shrink (d) the disc will remain unchanged, but the hole will enlarge (e) none of these.

Problems

TEMPERATURE

THERMAL EXPANSION

1. [I] What is normal body temperature (98.6°F) in degrees Celsius?

2. [I] Helium boils at −268.9°C and melts at −272°C (at 26 atm). What are the corresponding thermodynamic temperatures?

3. [I] A block of steel is heated from 0°F to 100°F. What is the change in its thermodynamic temperature?

4. [I] Hydrogen melts at 14.01 K. What is that in °C?

5. [I] If the temperature at the center of a star is 10 MK, what is the equivalent in °C and °F?

6. [I] A uranium fuel rod in a reactor operates at about 4000°F. What is that temperature in kelvins?

7. [I] The temperature of steam in a typical fossil-fuel plant is about 770 K. What would the temperature be in Fahrenheit degrees?

8. [I] For lead, what is the unitless ratio of its boiling point to its melting point computed in kelvins and in °C?

9. [I] By how much would a 1.00-m-long aluminum rod at 20°C increase if its temperature were raised 1.00°C?

10. [I] By how much would a 1.00-m-long brass rod shrink if its temperature were lowered 1.00°C?

11. [I] By how much would a 10-m-long aluminum bar change its length in the process of going from 30°C to 50°C?

12. [I] A steel girder 10 m long is installed in a structure on a day when the temperature is 5°C. By how much will it increase in length when it warms up to 35°C?

13. [I] The roadway of the Golden Gate Bridge is 1280 m long, and it is supported on a steel structure. If the temperature varies from 0°C to 35°C, how much does the length change?

14. [I] A thin sheet of copper 50.00 cm by 20.00 cm at 30.00°C is heated to 60.00°C. What will be the new area of one face? (See Problem 15.)

15. [I] Redo Problem 14 using the equation derived in Problem 18. Compare the two results.

16. [II] A steel television tower is 150.00 m tall at 10 degrees below zero Fahrenheit. What is its height at 95°F?

17. [II] A carpenter's steel measuring tape is 10.000 m long and 1.00 cm wide. Assuming that it was calibrated in the factory at room temperature (20°C), what is its length on a summer day when the temperature is 96.8°F? (For this steel, $\rho = 7860\ \text{kg/m}^3$, $\alpha = 12.15 \times 10^{-6}\ \text{K}^{-1}$, $\beta = 36.46 \times 10^{-6}\ \text{K}^{-1}$.) Comment on the practical aspects of your results.

18. [II] You have a sheet of material of area A_0 whose temperature is changed by an amount ΔT. Prove that the area will change by

$$\Delta A = 2\alpha A_0 \Delta T$$

19. [II] The mercury in a thermometer at 32°F has a volume of 0.50 cm^3. What will be its volume at 212°F?

20. [II] Imagine that we have a mercury-in-Pyrex thermometer. What is the *apparent* volume coefficient of expansion of mercury in Pyrex? That is, what is the effective coefficient taking into consideration the expansion of the body of the device?

21. [II] A soft glass beaker with an inside diameter of 10.00 cm is filled to the very top with 800.0 cm^3 of pure mercury at 95°C. How much will the level change if the temperature is lowered to 0°C?

22. [II] A thermometer has a quartz body within which is sealed a total volume of 0.400 cm^3 of mercury. The stem contains a cylindrical hole with a bore diameter of 0.10 mm. How far does the mercury column extend in the process of rising from 10°C to 90°C? Neglect any change in volume of the quartz.

23. [II] A beaker made of Pyrex is filled to the very brim with 100 cm^3 of water at 10°C. How much will overflow when it's raised to 50°C?

24. [II] A sheet of brass has a 2.000-cm-diameter hole drilled through it while at a temperature of 20°C. What will be the diameter of that hole if the sheet is surrounded by boiling water?

25. [II] A block of iron has an old label stuck to its side that gives its density as 7.85 g/cm^3 (at 20°C) and its coefficient of volume expansion as $36 \times 10^{-6}\ {}^\circ\text{C}^{-1}$. What will its density be at 0°C?

26. [II] When one cubic meter of water freezes at 0°C, it becomes how many cubic meters of ice?

27. [III] A pendulum clock is made (at 20.00°C) by supporting the *c.g.* of a bob at the very end of a 1.000-m-long aluminum rod. Determine its period. Now find its period on a winter's day when the heater goes out and the temperature drops to 0.000°C. Did the clock run fast or slow that day?

28. [III] A 1-second pendulum clock is made at 20.000°C by suspending a mass at its *c.g.* from a steel wire with a linear coefficient of expansion of $12.1135 \times 10^{-6}\ \text{K}^{-1}$. What will its period be at 40.000°C? Will it run fast or slow?

29. [III] Given that β is small, show that the final density of a solid specimen that has experienced a small temperature change of ΔT is

$$\rho \approx \rho_0(1 - \beta\Delta T)$$

30. [III] Show that the *thermal stress* (the stress due to an expansion or contraction of the object as a result of a temperature change) is given by

$$\text{Thermal stress} = Y\alpha\,\Delta T$$

31. [III] With Problem 30 in mind, suppose two concrete slabs are butted flat up against one another with their far ends fixed. Each is 10 m long and the contact area is 1000 cm^2. If the concrete was poured on a day when the temperature was 5.0°C, what will be the stress in them when the temperature is 42°C? Will they fracture? Take $\alpha = 10 \times 10^{-6}$ K^{-1} and $Y = 25$ GPa.

THE GAS LAWS

32. [I] A container of helium gas at STP is sealed and then raised to a temperature of 730 K. What will be its new pressure?

33. [I] The needle is removed from a hypodermic syringe, and that end is sealed. The air inside of it is slowly compressed (so that it remains at room temperature) to one-tenth its original volume. What is the final pressure (to two significant figures) in the syringe?

34. [I] A container of gas is kept at a constant pressure (which is close to atmospheric) by supporting a glob of mercury as in Fig. P34. If its volume is initially 500×10^{-6} m^3 at a temperature of 273 K, what will it be at 300 K?

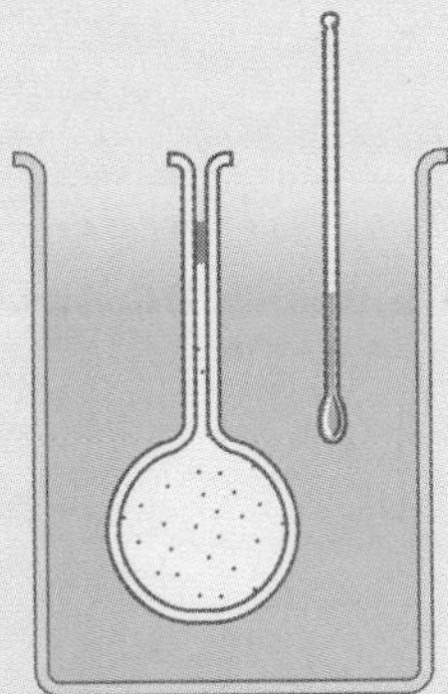

Figure P34

35. [I] Given that we have 1200 cm^3 of helium at 15°C and 99 kPa, what will be its volume at STP?

36. [I] Three moles of a gas are to be stored at a temperature of 60°C and a pressure of 93.0 kPa. How much volume will the gas occupy?

37. [I] A quantity of helium gas at atmospheric pressure occupies the 300-cm^3 volume of a cylinder fitted tightly with a piston. If the temperature is initially 25.0°C, what will be its final value when the gas occupies 200 cm^3 at atmospheric pressure?

38. [I] A chamber sealed with a movable piston contains 5.50 liters of hydrogen at 28°C and 81.3 kPa. What would its new volume be at STP?

39. [I] What phase is water in at (a) 1.0 atm and 102°C (b) 2.0 atm and 50°C (c) 0.006 atm and 5°C (d) 0.001 atm and 1.0°C (e) 0.0001 atm and 0.01°C?

40. [I] Determine the density of oxygen (O_2) at STP.

41. [I] What is the mass of a water molecule in kilograms? Assume the mass is simply the sum of the constituent masses. The masses of hydrogen and oxygen atoms are 1.008 u and 15.999 u, respectively.

42. [I] One mole of ammonia gas has a mass of 17.03 g. What is its density at STP to two significant figures?

43. [I] An expandable chamber contains 16.0 kg of molecular oxygen at STP. How many molecules are in it? How many moles?

44. [I] An expandable chamber contains 16.0 kg of molecular oxygen at STP. What's its volume?

45. [I] If 1.40 moles of ammonia gas are to be put into a 10.0-liter container at a pressure of 202.6 kPa, at what temperature should the gas be maintained?

46. [I] A very good vacuum system can pump a chamber down to about 10^{-15} atm. Assuming a temperature of 273 K in the evacuated region, how many molecules of air remain in each cubic centimeter?

47. [I] With Problem 46 in mind, how many molecules are in a cubic centimeter of air at STP?

48. [I] A spacesuit with a volume of 0.10 m^3 contains air at a pressure of 0.01 MPa at a temperature of 20°C. How many gas molecules are in the suit?

49. [I] Given five molecules with speeds of 1.0 m/s, 2.0 m/s, 3.0 m/s, 4.0 m/s, and 5.0 m/s, find their average speed.

50. [I] Given five molecules with speeds of 1.0 m/s, 2.0 m/s, 3.0 m/s, 4.0 m/s, and 5.0 m/s, find their *rms*-speed and compare it to the results of Problem 49.

51. [I] Given that the mass of an average oxygen molecule is 5.3136×10^{-26} kg, what is its *rms*-speed at 293.15 K?

52. [I] What is the average translational kinetic energy of an oxygen molecule in the air at 0°C?

53. [II] Taking the normal lung capacity to be 500 cm^3 and the pressure therein to be the equivalent of 761 mm Hg (which is still the way the medical texts are listing it), estimate the number of molecules per breath.

54. [II] An automobile tire is pumped up to an absolute pressure of 33 lb/in.^{2}, with air at a temperature of 40.0°F. After driving for several hours, the temperature in the tire reaches 120°F. Find the pressure in the tire at that point in SI units.

55. [II] A large, aluminumized-Mylar™ balloon contains 1.000 m^3 of helium when filled out of doors at 0°C on a winter's day. What will its volume be at home at 20°C?

56. [II] Hydrogen gas at 273 K fills an expandable chamber to a volume of 30.0 liters at a pressure of 2.00 atm. The gas is then compressed to 15.0 liters at a pressure of 3.00 atm. What will its new temperature be?

57. [II] At an altitude of about 12.5 km, the temperature of the Earth's atmosphere is roughly −55°C, and the pressure is around 19.4 kPa. How many kilograms of hydrogen gas (H_2) should be put in a balloon to fill it to 2000 m^3 at that altitude?

58. [II] If the density of nitrogen in a chamber at a pressure of 1.00 atm is found to be 1.245 kg/m^3, what is the temperature of the gas?

59. [II] Show that for a given amount of gas

$$\frac{P_1}{T_1\rho_1} = \frac{P_2}{T_2\rho_2}$$

60. [II] Determine the density of hydrogen gas at STP. What is its density at 1.00 atm and at a temperature of 273°C?

61. [II] What is the average amount of translational kinetic energy possessed by all the molecules of oxygen contained in 0.50 m^3 of gas at STP?

62. [II] Prove that the pressure produced by an ideal gas in a sealed chamber is proportional to both the number of molecules per unit volume and their average translational kinetic energy.

63. [II] Derive an expression for the *rms*-speed of an ideal gas in terms of its density and pressure.

64. [II] What is the ratio of the *rms*-speeds of nitrogen in the air at 0°C and 100°C?

65. [III] A gas is being collected in a beaker over mercury. This is done by inverting a graduated beaker filled with mercury and then partially submerging it in a mercury-filled bowl. Gas is introduced into the beaker from below, bubbles up through the mercury and displaces it. Suppose, at room temperature, 214 cm^3 of gas is collected, leaving the surface of the mercury 3.90 cm above the level in the bowl. A barometer on the wall reads 101.05 kPa. If the gas is removed and slowly cooled to 0.00°C at a pressure of 101.32 kPa, what will be its new volume?

66. [III] Hydrogen is being collected in an inverted beaker over water. The gas occupies a volume of 370 cm^3, and the surface of the water inside the beaker is raised 5.00 cm while a barometer reads 759 mm. The temperature of the system is 17°C. Given that water has a vapor pressure at that temperature of 1.92 kPa, find the pressure exerted by the hydrogen alone.

67. [III] Suppose 3.0 liters of oxygen at a pressure of 5.0 atm are added to 1.0 liter of nitrogen at 2.0 atm in a chamber with a volume of 4.0 liters. What is the pressure in the chamber if the temperature is kept constant throughout?

68. [III] Prove that the volume coefficient of expansion (keeping P constant) of an ideal gas equals $1/T$.

69. [III] Considering the discussion on p. 505, what is the percentage difference between the *rms*-speeds at room temperature of the two different forms of uranium hexafluoride (UF_6)—namely, those containing either ${}^{238}_{92}U$ or ${}^{235}_{92}U$?

15

Heat and Thermal Energy

THIS CHAPTER IS ABOUT internal energy, heat, and the associated behavior of bulk matter. It provides the basis for the study of Thermodynamics, which will be treated in Chapter 16. That discipline developed in the 1800s completely independently of atomic theory, and that's both a virtue and a failing. Here we will pursue an integrated approach that seeks to explain the macroscopic in terms of the microscopic.

We will attempt to answer such questions as: What is heat? What is the relationship between heat and temperature? What happens to bulk matter as thermal energy is added or removed? How is energy transferred from one

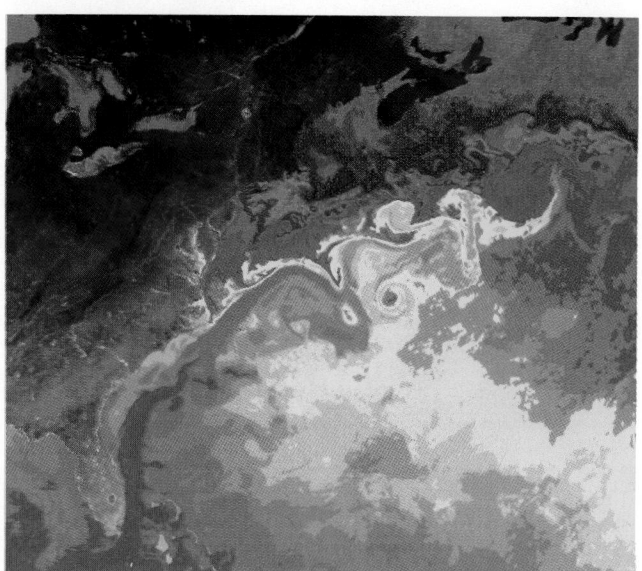

A temperature map of the surface of the east coast of the United States produced by a space vehicle orbiting the planet. Florida is outlined by warm water (deep red) at the lower left. Long Island juts into the cool blue-green coastal waters of the Atlantic Ocean. The great lakes are at the upper left. Notice the warm vortices spun off by the Gulf Stream (at the center of the picture).

place to another? The discourse is not just about boiling water and barbecues. As with all the creatures on the planet, thermal concerns are of vital personal interest at every moment of our lives. On a far grander level, these ideas lay the groundwork for a more complete understanding of the concept of energy and with that, a more complete understanding of the way change occurs throughout the Universe.

THERMAL ENERGY

The atoms of a solid, constantly colliding with one another, forever vibrate every-which-way about their fixed equilibrium positions (Fig. 15.1). In a liquid those positions shift, and much of the long-range order of the solid state vanishes, but the oscillations persist. By comparison, the gas is a tumult of impacts and recoils. Gas atoms are essentially free. There are no equilibrium positions, no oscillations; the motion is a perpetual high-speed zigzag dance punctuated by collisions. Much of this chapter treats the macroscopic manifestations of this hidden atomic motion.

Since a good deal of today's nomenclature and imagery came by way of the erroneous caloric theory of heat, we do well to examine it, if only for the roots of our present understanding.

15.1 Caloric versus Kinetic

Rarely is a hypothesis that is once decisively set aside ever resurrected, but that is exactly what happened with the Kinetic Theory. It was widely accepted, in one form or another, throughout the seventeenth century. Moreover, it was correctly believed by many that *heat* was a manifestation of this atomic motion. Galileo had said as much and Plato, long before, appreciated the equivalence of motion and heat.

The eighteenth century brought with it the gradual acceptance of a new and incorrect theory of combustion. This conception maintained that there was an invisible, weightless "matter of fire." By the 1760s, that unfortunate invisible-fluid metaphor was being extended to describe heat as well. In France and Germany especially, the consensus held that heat was an imponderable, self-repellent, indestructible fluid, which the chemist Lavoisier (ca. 1787) christened **caloric**.

The caloric theory proved to be highly useful for understanding and predicting a variety of thermal effects, much more so than Kinetic Theory, which no one at the time was able to apply practically. After all, wrong hypotheses can explain the world too, even if erroneously. We still say things like "pour on the heat" and "soak up the heat." Armed with one of the new thermometers, clever philosophers could measure the amount of caloric poured into a system—at least they thought they could.

Energy will remain in some sense the lord and giver of life, a reality transcending our mathematical descriptions. Its nature lies at the heart of the mystery of our existence as animate beings in an inanimate universe.

FREEMAN DYSON
(contemporary physicist)

The minority opinion, of heat-as-motion, would ultimately win out, but only later on, when the new concept of energy had matured to the point where heat could be recognized as a different kind of subtle "stuff"—random KE. This saga, the rise of the energy formalism and the understanding of heat as a process within that picture, was among the greatest intellectual achievements of the nineteenth century.

The Repudiation of Caloric

The first important attack on the caloric theory was delivered in 1798 by one of the most fascinating characters ever to be involved in science. The American, Benjamin

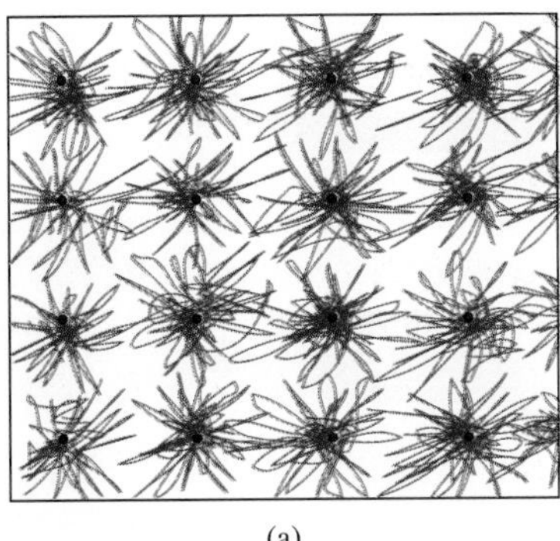
(a)

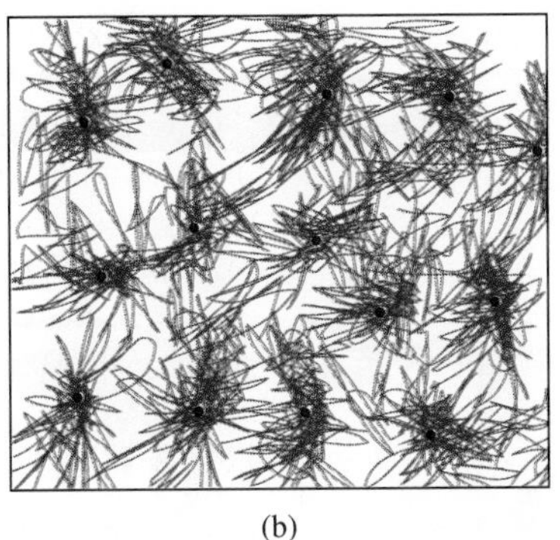
(b)

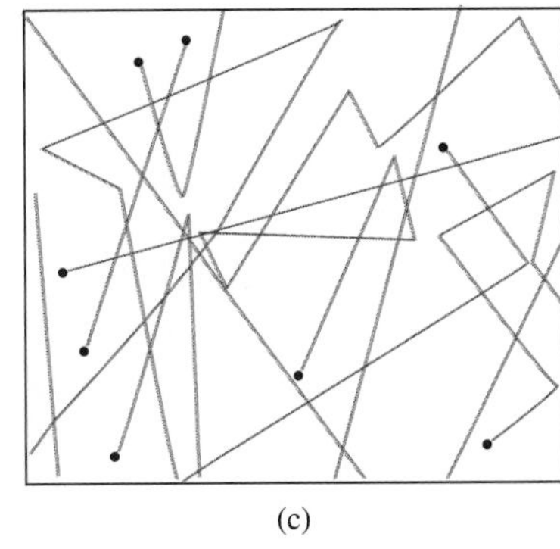
(c)

Figure 15.1 (a) In the solid, each atom oscillates about a fixed lattice site and collides (interacting electromagnetically) with its immediate neighbors. (b) In the liquid, the equilibrium positions move, and each atom vibrates over a larger region. (c) In the gas, the oscillations vanish, and the motion is essentially unconstrained until two atoms collide.

Thompson, a shrewd and elegant opportunist, was a professional soldier who saw little action yet rapidly rose to the rank of major general. Relying on an audacious blend of charm and cunning, he managed to be knighted by the King of England, became minister of war in Bavaria, and ultimately was made Count Rumford of the Holy Roman Empire. A tall, handsome, philandering, blue-eyed, likeable rogue, he fathered three children, of whom only one was legitimate. Thompson was a spy and a scoundrel, a quasi-beloved benefactor of the poor, a social reformer, a superb administrator (who even seems to have provided for a Bavarian brothel), and the founder of the British Royal Institution.

After numerous escapades our semihero, the Imperial Count Rumford, was all but running the tiny state of Bavaria. In 1787, he performed a series of very fine experiments (using his Most Serene Highness's best balance) from which he concluded that caloric, if it existed at all, had to be weightless—a fact that would become a primary objection to the theory and one that convinced him, at least, of the validity of the heat-as-motion hypothesis.

Years later, having newly outfitted the arsenal at Munich, he was watching the cannons being bored one day and realized that there was a tremendous amount of heat liberated in the process. He immediately arranged to have a brass gun-barrel blank, surrounded by several gallons of cold water, rotated against a dull steel borer (Fig. 15.2). Gradually, the temperature rose, until after $2\frac{1}{2}$ hours "the water *actually boiled!*" Heating by friction was certainly well known—aborigines and Boy Scouts still start fires that way, and people at the time even believed that forest fires began when the leaves on the trees rubbed too vigorously. The crucial point was that most calorists would have said that caloric was squeezed out of the brass by the boring tool. Moreover, the central postulate of the heat-as-matter theory was that caloric was indestructible and uncreatable. And yet Rumford showed that heat could apparently continue to be generated indefinitely as long as work was done turning the cannon barrel. It "appeared evidently to be *inexhaustible*" and could not therefore "possibly be a *material substance.*"

Though no single experiment can prove a theory, and alternative explanations were offered on behalf of caloric, Rumford's work weighed heavily on the side of the Kinetic Theory of Heat.

Count Rumford. As a young man, he had seen the Boston Massacre, but that didn't keep him from becoming a major in the King's American Dragoons. When British General Gage marched on Lexington and Concord, starting the Revolutionary War, it was on the strength of secret intelligence provided in a letter written (in invisible ink) by Benjamin Thompson.

15.2 Heat and Temperature

In previous chapters, we studied mechanical energy, the energy associated with an object that moved or interacted with other objects as a whole. In that sense, we have thus far been dealing with organized, or *ordered energy.* On the other hand, it is possible to impart uncoordinated motions to the individual constituent atoms—

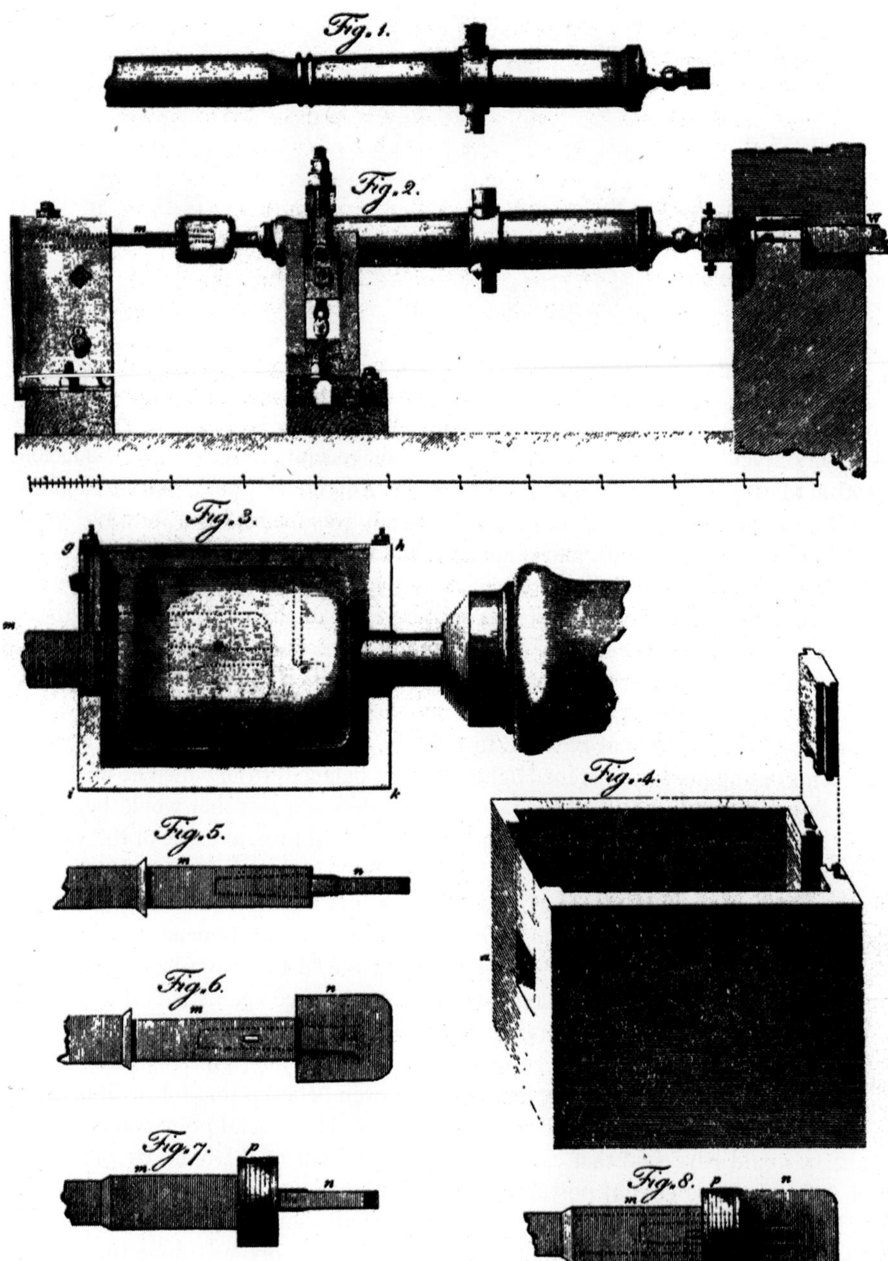

Figure 15.2 Rumford's diagrams of his experiment. A solid brass cannon blank (Fig. 1) was turned down to form a short, wide cylinder at its front end (Fig. 2), then encased (Fig. 3) in a wooden box (Fig. 4). A blunt borer (Figs. 5 and 6) was forced into the cylinder and the cannon rotated by a team of horses (not shown). The box was filled with water, which subsequently boiled.

disorganized random motion, not of the body as a whole, but within the body. **Thermal energy is the net disordered** KE **(rotational, translational, and vibrational) associated with any group of particles, usually the atoms of bulk matter**.

If we throw an apple, it will not experience an increase in thermal energy, although it will have an increased amount of KE (with respect to us). A thermometer stuck in the apple will show no change due to the motion. All the apple's atoms will move together as a whole in an ordered fashion. By contrast, if the apple crashes into a wall, it deforms and some of that organized KE will invariably be converted into an increase in thermal energy. Surprisingly, this energy perspective was essentially understood as early as 1695. Leibniz pointed out that in an inelastic collision the seeming loss of *vis viva* (which is equivalent to KE) only appeared to happen. In fact, it was "not destroyed, but dissipated among the parts. That is not to lose it, but to do like those who change large money into small." The ordered energy of the whole is shifted into the disordered energy of the parts.

In a sense, **temperature** *is a measure of the ability of randomly moving particles, usually atoms, to directly impart thermal energy to a thermometer or any other object. It reflects not the net amount of random* KE, *but its average value; the concentration of thermal energy.* The temperature of a substance is independent of the total number of atoms present. The net amount of thermal energy in the Atlantic Ocean is huge, but the average random KE of its molecules is low, and its temperature is low. By contrast, thermal energy depends on both the total number of atoms present and on their average random KE (which depends, in large measure, on temperature). The more atoms present and the more KE each possesses on average, the more thermal energy is present in the system. Because the individual atomic motions are random, altering the thermal energy of a system has no effect on its net momentum.

We now consider three basic ways in which the thermal energy of a body can be increased (or, if reversed, decreased), though much of this chapter will focus only on the last of these mechanisms.

First, we can do work on the body: rub it, compress it, or in any way *distort it;* that is, physically displace some of its atoms against an internal restoring force. Blacksmiths sometimes kindled a fire using a soft iron rod made red hot by pounding on it with rapid hammer blows. **Work** *is the organized mechanical energy transferred into or out of a system by way of a force acting through a distance.* If a force is applied to overcome interatomic forces (for example, friction) thereby displacing atoms, the energy introduced is rapidly distributed within the body through atomic collisions, and much of it will appear in the disordered form of thermal energy. That's what happens when you warm your hands by rubbing them together. Increasing the total amount of thermal energy of an object (like your hand) increases the average random KE, which means an increase in temperature.

The second means of increasing thermal energy involves shining electromagnetic radiation (light, infrared, ultraviolet, X-rays, etc.) onto the body. In other words, we can transfer radiant energy directly to atoms, which thereupon become agitated, collide with one another, and redistribute that energy into a disordered thermal form. That's what happens when you warm your face in the sunshine: radiant energy is absorbed and transformed into thermal energy, the average KE of the molecules rises, and your skin temperature goes up.

The third mechanism is perhaps the one that comes to mind first: we know from experience that the "hotness" of an object can be increased by placing it in contact with something at a higher temperature. A spoon at room temperature positioned in a flame will soon heat up. The high-temperature flame consists of atoms randomly moving about with a high average KE. This violent storm of glowing gas bombards the atoms in the surface of the spoon, which have a much lower average KE. The lower temperature spoon-atoms take up energy on impact from the more energetic flame-atoms. The former are thus increased in their random motion, and thermal energy is thereby imparted to the spoon. Energetic spoon-atoms that are touching the flame strike quiescent spoon-atoms just behind them, and the jostling continues back until the entire object becomes hot—a process we ordinarily describe by saying that "heat flows" down the spoon. More formally, **heat** (Q) *is the thermal energy transferred, via atomic collisions, from a region of high temperature* *to a region of lower temperature.* It is the energy of random motion transported by way of contact from one object to another, from one group of atoms to another—exclusively as a result of a temperature difference.

A body contains or stores thermal energy, not heat. Heat is thermal energy in transit, and once transferred it is no longer called heat. (We blow wind into a balloon, but once inflated, what is stored is air.) As before, we are again more concerned with the change in energy of a system than with the net amount of energy present at any moment. An agency of change of thermal energy is heat, just as an agency of change of mechanical energy is work. Once transmitted, the abiding mechanical energy of a system is no longer called work—what is stored is KE or PE, not work. In general, *a system contains energy when it can undergo spontaneous change.*

15.3 Quantity of Heat

Four friends dominated the intellectual life of Scotland in the latter half of the 1700s. They were Adam Smith, the political economist; David Hume, the renowned philosopher; his personal physician, Dr. Joseph Black, professor of chemistry at Glasgow; and that university's resident scientific instrument maker, James (Jamy) Watt. Both Black and Watt were reserved advocates of the caloric theory, and that was reasonable enough; caloric was a practical contrivance, and these were practical men, more concerned with using heat than with hypothetical abstractions.

Black was the great pioneer in thermal physics and probably the first to distinguish between "quantity of heat" and temperature. He saw temperature as reflecting the "intensity," or concentration, of what we would call thermal energy and he called "the matter of heat."* A stew tastes salty because of the concentration of salt, and it is hot because of the concentration of thermal energy. The bigger the potful, the more salt must be sprinkled in and the more heat must be added. The ocean, cold as it is, contains far more thermal energy than the scalding stew, and no matter how salty the dish, the ocean will contain far more salt as well. The fact that its thermal energy is spread out thin, not concentrated, accounts for the ocean's low temperature, but the large amount of random KE is there nonetheless.

James Prescott Joule (1818–1889) was a well-to-do brewer's son from Manchester, England, and a pupil of John Dalton. He was greatly influenced by Rumford's research at the cannon factory.

Sometime around 1760, Black introduced a simple scheme for quantifying heat that was operational and hence totally independent of what heat actually is—caloric or KE, it didn't really matter. One only needed to have a thermometer and a scale to weigh out some water. Following Black's lead, the **calorie** (cal) ultimately became the pre-SI metric unit of heat. It is defined as *the amount of heat that must be added to raise the temperature of 1 gram of water 1 degree Celsius*—from 14.5°C to 15.5°C. This particular temperature range is now specified because the heat required is very slightly different at other temperatures. As H. A. Rowland discovered (1878), it differs by less than 1% over the range from 0°C to 100°C which, for our purposes, is negligible. There we are then: we can say exactly what a calorie is, even without knowing what heat is, just as we can say exactly what a kilogram is, without knowing what mass is.

By definition, 1 cal will raise 1 g of water 1°C. We can assume (and experiment confirms) that 2 cal will raise the temperature of 1 g of water 2°C, and so on. In other words, the resulting temperature change is proportional to the heat-in: $\Delta T \propto Q$. Similarly, 1 cal will raise the temperature of 2 g of water only $\frac{1}{2}$°C. That is,

*Physicians were very involved in research in the early days of thermal science. Indeed, the word *temperature* appears to derive from medicine, where the goal was to temper the extremes of chill and fever.

$\Delta T \propto 1/m$; the more water, the more heat is needed to raise the temperature, $\Delta T \propto Q/m$. Introducing a constant of proportionality c (to take care of the units), we have

$$Q = cm\,\Delta T = cm(T_f - T_i) \tag{15.1}$$

In this particular case of water, $c = 1$ cal/g·C°, exactly.

Example 15.1 The equivalent of a glass of water (that is, 270 g of liquid) at 20.0°C receives 1000 cal from a heater. Assuming that all the energy goes into the water without any losses, what is the final temperature of the liquid?

Solution: [Given: $m = 270$ g, $T_i = 20.0$°C, and $Q = 1000$ cal. Find: T_f.]

$$Q = cm\,\Delta T = cm(T_f - T_i)$$

$$1000 \text{ cal} = (1.00 \text{ cal/g·C°})(270 \text{ g})(T_f - 20.0\text{°C})$$

and

$$T_f = \frac{1000 \text{ cal}}{(1.00 \text{ cal/g·C°})(270 \text{ g})} + 20.0\text{°C} = \boxed{23.7\text{°C}}$$

▶ **Quick Check:** 270 cal will raise 270 g by 1°C; 1000 cal/270 cal = 3.7, hence the new temperature is 20.0°C + 3.7°C.

The calorie turns out to be inconveniently small, especially for weight watchers. So the world of dietetics (Table 15.1) uses the **kilocalorie** (kcal), sometimes cleverly called the Calorie (1 kcal = 1 Cal = 1000 cal). More often, the fact that "calorie" is supposed to mean kilocalorie is just kept as a little secret (Table 15.2). When you

TABLE 15.1 Energy provided by various foods

Food	Kilocalories
Martini	145
Peanut butter sandwich	330
Buttered popcorn (1 cup)	55
Vanilla ice cream ($\frac{1}{3}$ cup)	145
Whiskey (1 shot)	105
Brownies (1)	140
Jelly doughnut	225
Dry red wine (1 glass)	75

TABLE 15.2 Approximate calories consumed per hour

Body weight lb Body mass kg	100 45	150 68	200 90	250 113
Sleeping it off	40	60	80	105
Sitting on the grass	65	95	130	165
Standing in line	70	100	140	170
Strolling through a park	130	195	260	320
Running from a mugger	290	440	580	730

Just 1 kilocalorie per glass. Note that the Europeans use commas in their numbers where we use periods (1 kcal = 4.18 kJ).

read on the box that your favorite breakfast cereal "contains" 110 calories per serving, that's really 110 kilocalories, and what's meant is that that amount of energy will be liberated upon the complete combustion of one serving of cereal. Some engineers and aspiring air-conditioner salesmen (mostly in the United States) hold fast to another fading anachronism, the Btu, or *British thermal unit.* One Btu is the amount of heat needed to raise 1 lb of water 1°F.

The realization that heat is energy only came gradually in the mid-nineteenth century. To the advocates of this unifying perspective, the pivotal question then became: how much heat, as defined by its effect on the temperature of water, is equal to a unit of energy, as defined by its ability to do work? If, indeed, heat is energy, then *calories* must be directly convertible into *joules.*

15.4 The Mechanical Equivalent of Heat

The German physician who took the next step was twenty-eight years old when his first essay appeared (1842). Julius Robert Mayer reinterpreted some long-forgotten results on expanding gases to determine a rough value for what came to be called the "mechanical equivalent of heat"—namely, the numerical relationship between work and heat. Meyer's bold speculative essay lacked direct experimental confirmation and was ridiculed as unprofessional.

As if the time was ripe for change, within one year of Mayer's paper there appeared in print the first report of the remarkable researches of James Prescott Joule. Unlike Mayer, Joule was a gifted experimenter who devised a whole range of apparatus using various means (for example, electric currents, friction, and compression of gases) to generate "heat." His most famous was a paddle wheel device driven by falling weights (Fig. 15.3). By equating the work done by the weights to the thermal energy imparted to the water, Joule concluded that 773 ft·lb was equivalent to 1 Btu (which is not far from the modern value of 778 ft·lb per Btu). Nowadays, we use SI units and say the **mechanical equivalent of heat** is

$$4.186 \text{ J} = 1 \text{ cal}$$

or

$$1 \text{ kcal} = 4186 \text{ J}$$

By international agreement, the calorie is no longer the recognized unit of heat. As with all forms of energy, the preferred unit is the *joule*. Thus, we can summarize much of what has been discussed so far with the statement that *the temperature of 1 kilogram of water can be increased (or decreased) by 1 kelvin through the addition (or removal) of 4186 joules of heat.*

Example 15.2 The following statement is taken from Mayer's paper of 1842: "The warming of a given weight of water from 0°C to 1°C corresponds to the fall of an equal weight from the height of about 365 metres." Compare his value of the mechanical equivalent of heat with the modern value.

Solution: [Given: $\Delta T = 1°\text{C}$ and $h = 365$ m. Find: Mayer's mechanical equivalent of heat.] By "a given weight," he means any value of mg, so let's take $m = 1.00$ kg for convenience. We must find the corresponding energy associated with a fall of 365 m in SI units. The *gravitational*-PE is given by

$$\text{PE}_\text{G} = mgh \qquad [9.13]$$

hence

$$\text{PE}_\text{G} = (1.00 \text{ kg})(9.81 \text{ m/s}^2)(365 \text{ m}) = 3.58 \text{ kJ}$$

That's the mechanical energy associated with the fall of a 1-kg mass (of water, or anything else), and it is supposed to equal the heat needed to raise the temperature of that mass (1 kg) of water 1°C, which, from Eq. (15.1), is just 1000 cal = 1 kcal. According to Mayer, then, $\boxed{1 \text{ kcal} = 3.58 \text{ kJ}}$, which is about $\boxed{14\% \text{ too small}}$.

15.5 The Zeroth Law

Joseph Black began one of his lectures with an observation so central to any complete understanding of the behavior of heat that it has since come to be called the **Zeroth Law of Thermodynamics**:

> We can perceive a tendency of heat to diffuse itself from any hotter body to the cooler ones around it, until the heat is distributed among them in such a manner that none of them is disposed to take any more from the rest. The heat is thus brought into a state of equilibrium . . . . The temperature of all of them [i.e., all the bodies] is the same.

In other words, when two bodies at different temperatures come into contact, thermal energy flows from the higher to the lower until the temperatures of the two equalize and they are said to be in **thermal equilibrium**. The idea that matter, all by itself, spontaneously tends to a uniform-temperature equilibrium state is based squarely on observation—the rest of the statement about the flow of energy is hypothesis.

Today, this equilibrium insight is often used as the starting point in a formal theoretical development that introduces the idea of thermodynamic temperature as a unique variable describing the state of the system. After all, that's really the wonderful part of the process, the fact that a cold beer left on the kitchen table will ultimately reach the same temperature as the table (and anything else sitting on it) reveals something fundamental about nature. Thus, when system 1 is placed in thermal contact with system 2 and no net energy flows from one to the other, the two exist in a special physical condition, or state, that can be identified via the concept that they have the same thermodynamic temperature.

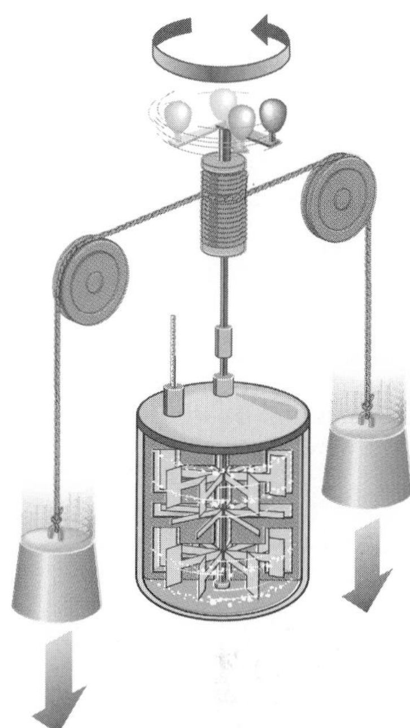

Figure 15.3 Joule's apparatus for determining the mechanical equivalent of heat. As the weights fall, the paddle turns, raising the temperature of the water. *Gravitational*-PE is converted into the KE of the paddle and the water. That KE is ultimately converted into thermal energy within the isolated chamber.

The Zeroth Law of Thermodynamics is often stated as follows: *when two systems are at the same temperature as a third, they are all three at the same temperature as each other.* It might seem at first reading that this is a trivial tautology that must be true, but that's not quite the case. Certainly, two people can be in a state of equal and mutual respect for a third person without necessarily respecting each other. The conclusion that temperature does nonetheless behave that way is an experimentally verified law. The reason for emphasizing the notion is that we tacitly assume it's true whenever we compare the temperatures of two samples using a thermometer as an intermediary third body. Thus, it is unnecessary to place our test sample in contact with a standard-temperature bath to see if they are in equilibrium. We calibrate the thermometer with the bath, and if the sample is subsequently found to be at that temperature along with the thermometer, it is necessarily at the temperature of the bath.

15.6 Specific Heat Capacity

We know that 4.2 kJ (1 kcal) will raise the temperature of 1 kg of water 1°C, but there is no reason to assume it will do the same thing to 1 kg of iron or peanut butter or pizza; still, that was the common wisdom before Dr. Black put things right. He found, quite to the contrary, that *every substance changed temperature by a characteristic amount when infused with a fixed quantity of heat.* It is informative to see how he actually managed to transport the "fixed quantity" of heat to the different substances under study.

In one scheme, he placed equal weights of different substances in identical vessels that were subsequently put on the same burner, or in front of the same fire, for the same length of time. The amount of heat transferred was then presumably proportional to the time the sample was exposed to the source (Fig. 15.4). In this way,

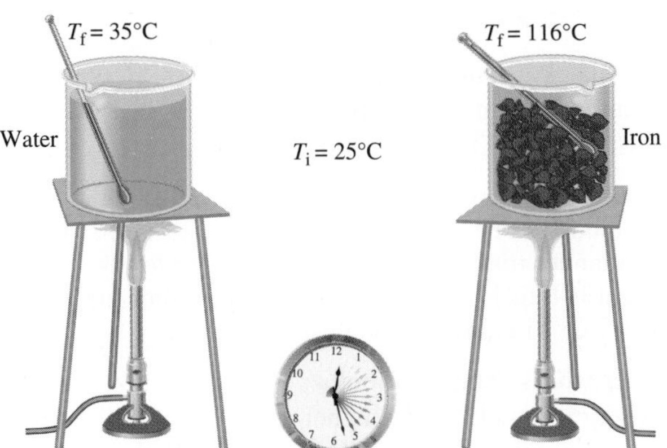

Figure 15.4 Imagine some water and an equal mass of iron at, say, 25°C. If an equal amount of heat is added to both, the temperature of the iron increases 9.1 times as much as the water. If the water rises to 35°C, the iron rises to 116°C.

it was confirmed that an equal mass of iron will get hotter than water by a factor of almost 10 in the same amount of time.

Another more controllable approach, which came to be known as the *method of mixing,* evolved into the modern technique of calorimetry. It was fairly common knowledge in the community of scientists that if a given mass of cold water at, say, 5.0°C is mixed with an equal amount of hot water at, say, 95°C, the blend will come to an equilibrium temperature of 50°C. The hot system loses a certain amount of heat to the cold system; the former drops 45°C, the latter gains 45°C—*equal changes, because equal amounts of the same material are being mixed.* We can see as much from Eq. (15.1): the *heat lost* by hot system 1 is

$$Q_1 = cm_1(T_{f1} - T_{i1})$$

where $T_{i1} > T_{f1}$. Similarly, the *heat gained* by cold system 2 is

$$Q_2 = cm_2(T_{f2} - T_{i2})$$

where $T_{f2} > T_{i2}$. But these are plainly equal; that is, $-Q_1 = Q_2$; **heat-out is negative, heat-in is positive**. Moreover $m_1 = m_2$; hence

$$-c(T_{f1} - 95°\text{C}) = c(T_{f2} - 5.0°\text{C})$$

At equilibrium, $T_{f1} = T_{f2} = T_f$ and $T_f = \frac{1}{2}(95°\text{C} + 5.0°\text{C}) = 50°\text{C}$ as expected. The crucial points here are that the final temperatures of the two samples are equal, and that the heat lost by one is gained by the other (that is, Conservation of Energy).

This approach of mixing different systems became rather important, especially after Fahrenheit performed much the same experiment using water and mercury. The hot mercury decreased in temperature more than the water rose in the process of their coming together at equilibrium. As Black interpreted the results, an equal mass of mercury required less heat to change its temperature 1 degree than did water. He concluded that "quicksilver [mercury] has less *capacity for heat*" than does water. In time, the *heat capacity* (c) of a substance was defined to be *the number of joules of heat that must be added to raise the temperature of 1 kg of the material 1 K* (that is, *1°C*). It follows from Eq. (15.1), $Q = cm\,\Delta T$, that each different substance has its own characteristic value of c (Table 15.3).

Because there were several different measures of heat in the mid-1800s, it was common to tabulate a unitless quantity called the *specific heat.* Taking a lead (as

TABLE 15.3 Specific heat capacity for some materials*

Material	Specific heat capacity kJ/kg·K	kcal/kg·K
Solids		
Aluminum	0.90	0.21
Clay (dry)	0.92	0.22
Copper	0.39	0.093
Glass	0.84	0.20
Gold	0.13	0.031
Human body (average)	3.47	0.83
Ice (water, −5°C)	2.1	0.50
Iron	0.47	0.11
Polyamides (e.g., Nylon)	1.7	0.4
Polyethylenes	2.3	0.55
Polytetrafluorethylene (e.g., Teflon)	1.0	0.25
Lead	0.13	0.031
Marble	0.86	0.21
Platinum	0.14	0.032
Protein	1.7	0.4
Silver	0.23	0.056
Stainless steel (type 304)	0.50	0.12
Wood	1.8	0.42
Liquids		
Acetone	2.2	0.53
Alcohol (ethyl)	2.4	0.57
Ammonia	4.71	1.13
Mercury	0.14	0.033
Nitrogen (−200°C)	1.98	0.474
Oxygen (−200°C)	1.65	0.394
Sulfuric acid	1.4	0.34
Water	4.186	1.000
Gases		
Air (100°C)	1.0	0.24
Argon	0.52	0.13
Carbon monoxide	1.0	0.25
Hydrogen	14.2	3.39
Methane	2.2	0.53
Steam (110°C)	2.01	0.481

*At ≈20°C.

well as its name) from the idea of specific gravity, which is a unitless ratio, specific heat was defined as the ratio of the heat capacity of the substance to the heat capacity of water. That was convenient because the heat capacity of water was defined to be unity. Today, with our standardization of units and because the SI heat capacity of water is not unity, this quantity is of less practical significance, though it persists. Regrettably, the terminology has blurred with time, and one has to be careful when consulting different sources. It's fairly common now to speak of c as the **specific heat capacity** and to give it the appropriate units. In the literature, those units are likely to be (cal/g·C°) or (kcal/kg·C°) or (J/kg·C°), though (J/kg·K) is to be preferred. Since specific heat capacity is the amount of heat (Q) that must be added or removed per kilogram to change the temperature one kelvin, it follows that

$$c = \frac{Q}{m\,\Delta T} \tag{15.2}$$

Example 15.3 A copper pot has a mass of 0.50 kg and is at 100°C. How much heat must be removed from it if its temperature is to be lowered to precisely 0°C?

Solution: [Given: Cu, $m = 0.50$ kg, $T_i = 100°\text{C}$, and $T_f = 0°\text{C}$. Find: Q.] Here Q is the heat-out of the system, which is negative since $\Delta T = -100$ K. From Eq. (15.1)

$$Q = c_{\text{Cu}} m\,\Delta T = (390\text{ J/kg·K})(0.50\text{ kg})(-100\text{ K})$$

and so $$Q = -19.5\text{ kJ}$$

or to two significant figures $\boxed{-20\text{ kJ}}$.

▶ **Quick Check:** $\Delta T = -100$ K. From Table 15.3, the specific heat capacity of copper is 9.3% that of water; if the pot were made of water, we would have to remove $\frac{1}{2}$100 kcal; for copper we need only remove 9.3% of that, or 4.65 kcal = 19.5 kJ.

Water has a high specific heat capacity, a fact that is crucial to all life on the planet. A quantity of water will change temperature only when a great deal of heat is added or removed; it heats up slowly and cools off slowly. The legendary hot-water bag is the ultimate testament. A potato or a lima bean, fished out of a hot stew, can be eaten with a little caution, but a juicy onion (mostly water) will have to be treated with respect. To lower its temperature from 100°C in the pot to something you can deal with will require removing a lot more heat first. And if you don't bother, that considerable amount of heat will be deposited in your mouth as it descends down to equilibrium with your tongue. We are mostly water ourselves, and that high value of c is one factor that helps us keep from rapid temperature changes. Cities on the coast of large bodies of water get that same benefit. It is also the reason water is used to carry thermal energy in radiators that heat houses and in cooling systems that maintain automobile engines.

Example 15.4 A quantity of what looks like lead shot, having a mass of 100 g, is poured into a test tube. The tube is then partially submerged in a beaker of boiling water and kept there until the shot reaches 100.0°C. The hot shot, as it were, is then transferred to a vessel containing 100 g of water at 20.00°C. The contents are gently stirred until they come to an equilibrium temperature of 22.41°C. What is the specific heat capacity of the shot?

Solution: [Given: $m_S = 100$ g, $m_w = 100$ g, $T_{iS} = 100.0°\text{C}$, $T_{iw} = 20.00°\text{C}$, and $T_{fS} = T_{fw} = T_f = 22.41°\text{C}$. Find: c_S.] We can do this problem in several different ways. All are equivalent, and all rely on the fact that the heat lost by the shot is gained by the water and that both are at the same final temperature. Using Eq. (15.2),

$$c_S = \frac{Q_S}{m_S\,\Delta T_S}$$

and we need Q_S and ΔT_S. The heat lost by the shot is the heat gained by the water, which we can calculate. The water has changed temperature by (22.41°C − 20.00°C) = +2.41°C. There being 100 g of water present, each 1 degree change corresponds to 100 × (4.186 J); hence, the water has gained +1008.8 J. The shot changed temperature by (22.41°C − 100°C) =

(continued)

(continued)

$-77.59°C$; thus

$$c_S = \frac{-1008.8 \text{ J}}{(0.100 \text{ kg})(-77.59 \text{ K})} = \boxed{130 \text{ J/kg·K}}$$

▶ **Quick Check:** From Eq. (15.1) and the fact that the heat-out equals the heat-in, $-Q_S = Q_w$; $-c_S m_S \Delta T_S = c_w m_w \Delta T_w$; hence, $c_S = c_w \Delta T_w / \Delta T_S = -(4186 \text{ J/kg·K}) \times (22.41°C - 20.00°C)/(22.41°C - 100°C)$, and $c_S = 130$ J/kg·K.

In all these thermal experiments, it is tacitly assumed that there is no significant heat loss. To make that a reality, we use a *calorimeter* as pictured in Fig. 15.5. It's simply a well-insulated chamber whose vacuum or styrofoam walls keep the contents from exchanging heat with the outside. A thin metal cup with a low specific heat and a small mass is used to hold a measured amount of water. The cup changes temperature easily and can store very little thermal energy in the process. Thus, it tends to remove very little from the system. That's a lot like the behavior of aluminum foil used in cooking—it can be handled, albeit carefully, right out of the oven in spite of its high temperature because the quantity of heat given off in coming to equilibrium with your fingers is necessarily small (*mc* is small). Typically, a sample whose specific heat is to be measured is raised to a known temperature and then immersed in the water. The change in temperature of the water allows the required specific heat to be computed.

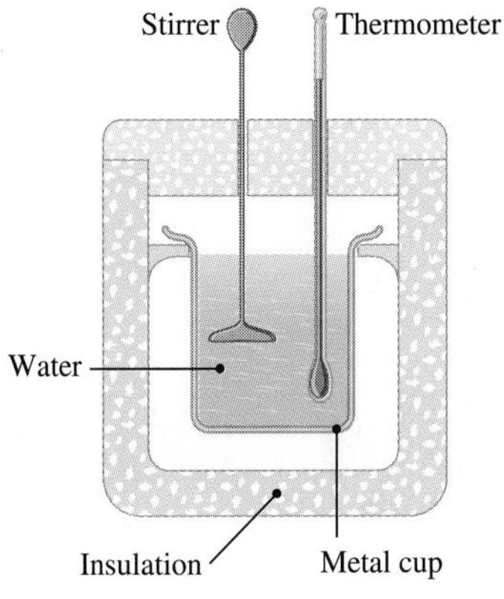

Figure 15.5 A water calorimeter.

Example 15.5 A calorimeter consisting of a thin copper cup of mass 150 g containing 500 g of water is at a temperature of 20.0°C. A 225-g sample of an unidentified material at 508°C is lowered into the bath, and the device is sealed. After a few minutes, the system reaches a constant temperature of 40.0°C. Determine the specific heat capacity of the sample. Overlook any losses due to the thermometer, stirrer, or insulation.

Solution: [Given: $m_c = 0.150$ kg, $m_w = 0.500$ kg, $T_{ic} = T_{iw} = 20.0°C$, $m_S = 0.225$ kg, $T_{iS} = 508°C$, and $T_f = 40.0°C$. Find: c_S.] The temperature change in kelvins equals that in Celsius, so there is no need to change units. The heat transferred out of the sample (which is a negative number) equals the heat transferred into the water and cup

$$-Q_S = Q_w + Q_c$$

$$-m_S c_S \Delta T_S = m_w c_w \Delta T_w + m_c c_c \Delta T_c$$

Using Table 15.3, we obtain

$$-(0.225 \text{ kg})(c_S)(40°C - 508°C) = (0.500 \text{ kg})(4186 \text{ J/kg·K})(20 \text{ K}) + (0.150 \text{ kg})(390 \text{ J/kg·K})(20 \text{ K})$$

and $105.3\ c_S = (41\,860 + 1170)$J/kg·K

hence $\boxed{c_S = 409 \text{ J/kg·K}}$

▶ **Quick Check:** $[(0.15 \text{ kg})(0.093 \text{ kcal/kg·K}) + (0.5 \text{ kg})(1.0 \text{ kcal/kg·K})](20 \text{ K}) = 10.3 \text{ kcal} = c_S(0.225 \text{ kg})(468 \text{ K})$; $c_S = 0.098$ kcal/kg·K.

The Calorie content, or better still the *energy content,* of your breakfast crunchies can be found with a modified device known as a *bomb calorimeter.* A weighed sample of dried material is placed into a sealed chamber containing a heating coil and an over-pressure of oxygen. That chamber is then positioned in the water bath of an otherwise traditional calorimeter. An electric current heats the coil

TABLE 15.4 Approximate heats of combustion

Material	Heat of combustion (MJ/kg)
Alcohol, methyl fuel	22
(denatured)	27
Anthracite (hard coal)	33
Bituminous coal	30
Bread	10
Butter	33
Carbohydrates	17
Charcoal	28
Diesel oil	45
Dung	17
Fats	38
Gasoline	48
Methane	56
Oil, furnace	44
Propane	50
Proteins	17
TNT	5
Wood	15

TABLE 15.5 Approximate energy content of various materials

Source	Energy (J)
Ton of coal	3×10^{10}
Barrel of oil	6.3×10^{9}
Ton of TNT	4.2×10^{9}
Gallon (U.S.) of oil	1.5×10^{8}
Gallon (U.S.) of gasoline	1.3×10^{8}
Kilogram of fat	3.8×10^{7}
Pound of coal	1.5×10^{7}
Pound of wood	8.6×10^{6}
Cubic foot (1 atm, 16°C) of natural gas	1.1×10^{6}

(with a known amount of energy), and the crunchies burn to dust. The same scheme can be used to determine the energy content of fuels, the so-called *heats of combustion* (see Tables 15.4 and 15.5).

An Atomic View of Specific Heat

Joseph Black could simply explain the differences in specific heat by supposing that water had more space between its atoms than iron and so could contain more caloric. The actual explanation is a good deal more subtle and depends on the structure of the molecules and on their interactions. Roughly put, the heat once entered into a sample is energy that is distributed among the molecules of the system. The more molecules there are, the less the increase in thermal energy per molecule. That means less increase in the average random KE and a smaller resulting temperature increase. A kilogram of many low-mass atoms will increase less in temperature, for a given input of heat, than will a kilogram of fewer high-mass atoms and so have a higher specific heat capacity.

That's the case for almost all the elements, provided they remain in the solid state. In fact, *the product of the specific heat capacity and the atomic mass for each of the solid elements is roughly constant*—something known as **Dulong and Petit's Law** (1819). The unified atomic mass of an atom (u) corresponds numerically to the number of grams there are of it per mole. Hence, if we multiply specific heat capacity in cal/g·K by g/mol, we get cal/mol·K, or J/mol·K, which is called the *molar specific heat capacity* (C). What Dulong and Petit found experimentally was that the molar specific heat capacity—the heat needed to raise the temperature of 1 mol, 1 K—is about 25 J/mol·K for most solid elements.

Classical theory explains this result by maintaining that the atoms of a monatomic solid behave like those of a gas, this time oscillating about a mean position with a distribution of energies similar to the Maxwell-Boltzmann distribution. Unlike the noninteracting gas, here there is an ongoing exchange between KE and PE, as there is with an oscillating spring-mass system. Remember that specific heat capacity reflects the medium's ability to manifest a change in the energy it possesses as a change in temperature. But not all forms of energy stored internally do that—PE **does not affect temperature**.

For a gas, the average KE of each atom, as given by Eq. (14.14), is $\frac{3}{2}k_B T$. Since any atom can be imagined as moving along three perpendicular axes, it is assumed that the KE is, on average, distributed equally among the three motions—that's the Equipartition Theorem. Each mode of behavior, or *degree of freedom,* is assumed to correspond to an average energy of $\frac{1}{2}k_B T$. In the case of a solid or liquid, where there is appreciable interaction among the particles, a portion of the entering energy also goes into overcoming these forces and is stored as PE. Again, there are three basic directions and three more degrees of freedom. Extending the Equipartition Theorem, it is assumed that each PE mode has the same energy as each KE mode; namely, $\frac{1}{2}k_B T$. Thus, an atom in a solid should have 6 modes of freedom, each able to take up $\frac{1}{2}k_B T$ in energy, for a total of $3k_B T$. Each mole of material, containing N_A atoms, will then have a net **internal energy** (U = KE + PE), which is the sum of the random atomic KE (the thermal energy) and the interatomic PE. At any absolute temperature T, the internal energy is given by

$$U = N_A 3 k_B T$$

where $N_A k_B$ equals R, the universal gas constant (p. 000). Hence

$$U = 3RT$$

and if an amount of heat Q is introduced, it represents a change in the internal energy ΔU such that

$$Q = \Delta U = 3R\,\Delta T$$

The specific heat capacity per mole is then

$$C = \frac{Q}{\Delta T} = 3R \tag{15.3}$$

Because the atoms interact, half the energy entering the system goes into PE, and there is that much less of a temperature change. Since R = 8.31 J/mol·K, it follows that C = 24.9 J/mol·K, which is just what Dulong and Petit found. Well, this result is wonderful, but unhappily not the whole story. What is found experimentally is that C is actually temperature dependent, generally approaching zero at 0 K, and $3R$ at temperatures around and above room temperature (Fig. 15.6).

Evidently, the interactions between atoms will be different for the various elements, and we can expect some differences in C. Substances such as diamond that are bound very strongly (are hard and have high melting points) are found to have appreciably lower room-temperature values of C than predicted—a small input of Q causes a large ΔT. It would seem that at low-to-moderate temperatures not much energy goes into PE. Why that should be cannot be understood from classical theory. It is also unclear classically what role the free electrons play in determining the specific heat of metals since they, too, should be moving around and picking up energy. A completely satisfactory theory of specific heat can only be produced using Quantum Theory, but we won't attempt that here.

Liquids allow a greater freedom of their atoms (they can both rotate and move around), which are nonetheless bound to one another. In general, liquids have higher specific heats than either solids or gases.

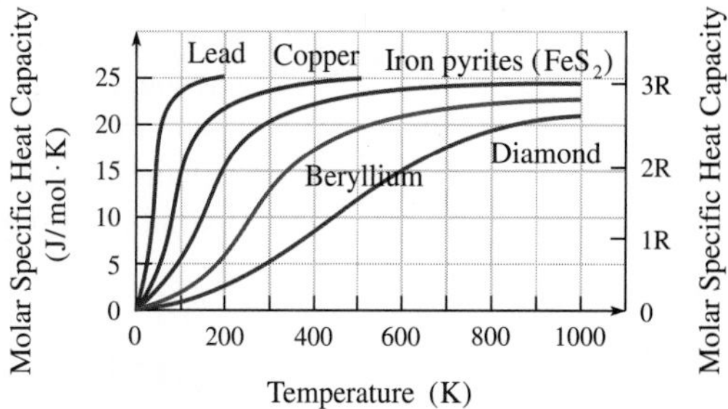

Figure 15.6 The specific heat capacity per mole for several solids. The curves approach a value of 25 J/mol·K as the temperature increases.

The Specific Heat of Water

Water stands out for its high specific heat capacity (4.2 kJ/kg·K, or 1.0 kcal/kg·K). Even in the solid state (that is, ice) water has a comparatively high specific heat capacity (2.1 kJ/kg·K, or 0.50 kcal/kg·K). The same is true for steam, which at a constant 1 atm has a specific heat capacity of roughly (2.0 kJ/kg·K, or 0.48 kcal/kg·K) that increases with pressure, and above 200°C, with temperature as well.

Water is molecularly rather lightweight; 1 kg of it contains 3 times as many molecules as there are atoms in 1 kg of iron. Additionally, those molecules effectively interact with one another, storing energy in the form of *electrical*-PE.

The hydrogen atoms in a water molecule behave like partially naked protons because their electrons tend to spend more time near the oxygen atom. As a result, there is a powerful intermolecular force binding the liquid together that forms a short-ranged, ordered structure of clusters of molecules. An individual molecule that can nonetheless slowly meander around (with relatively little energy) also oscillates about its moving equilibrium position (with a good deal more energy). Besides increasing this oscillation, which directly increases temperature, adding heat to a liquid partially breaks up the clusters, reducing their size and liberating molecules. Thus, it takes a certain amount of energy to free up the molecules and that energy does not enhance the temperature. The result of all these mechanisms is a substantial specific heat.

The Specific Heat of Gases

A similar relationship to that found by Dulong and Petit also exists among the elemental gases, though here the molar specific heat capacity is different for the different types of gas. In part, that is why hydrogen (at constant pressure and room temperature) has so large a value of c: 14.2 kJ/kg·K, or 3.4 kcal/kg·K (it's such a light molecule). In addition to this mass effect, specific heat capacity is again influenced by the various ways energy is imparted to or stored by the molecules. Monatomic gases such as helium, argon, and neon can manifest their thermal energy only as translational KE; they are not set rotating or vibrating through collisions. They are not molecular, so there can't be any internal vibration of constituent atoms. As for rotation, the moment of inertia of an atom about an internal axis is very small, and that precludes it (for quantum mechanical reasons) from transferring rotational energy via a collision. A monatomic gas easily changes temperature and has a comparatively low molar specific heat—about half that of a typical metal.

With its volume kept constant, such a simple noninteracting gas has 3 degrees of freedom due to its translational KE. Consequently, the internal energy per mole is $U = 3(\frac{1}{2}k_B T)N_A = \frac{3}{2}RT$ and $\Delta U = \frac{3}{2}R\,\Delta T$. The molar specific heat capacity at constant volume is then $C_V = \frac{3}{2}R = 12.5$ J/mol·K, or 2.98 cal/mol·K. This value is almost precisely the molar specific heat capacity of monatomic helium and argon.

Gas molecules with more than one atom can support rotations of the entire molecule about its center-of-mass, as well as internal vibrations of the atoms with respect to each other. These modes of motion are transferable via collisions and presumably can take up appreciable amounts of energy while having little effect on changing the temperature of the system. The more complicated the gas, the more the number of degrees of freedom, the greater the specific heat.

CHANGE OF STATE

The transport of heat into or out of a substance can alter it in a number of ways. For instance, it can change the temperature of the sample, and it can change the physical structure, or *phase,* of the sample.

15.7 Melting and Freezing

When a solid is transformed into a liquid by the addition of thermal energy, the process is known as *melting,* though at one time it was widely called *fusion.* The common wisdom in the mid-1700s (and a notion still afoot today) was that once at the melting point, it took only a slight addition of heat to cause a further small rise in temperature and thereafter the rapid melting of the body. Joseph Black understood that this was wrong; he knew how long it often took for the piles of winter snow to melt, even on warm days. Black was convinced that when a body melts, "a large quantity of heat enters into it . . . without making it apparently warmer. . . . This heat must be added in order to give it the form of a liquid . . . ." Imagine that we take two thin glass bottles, put a kilogram of pure water in each, and cool them both to as close to 0°C as possible, leaving one liquid and freezing the other. Now bring them both into a warm room and observe the time it takes for the water to rise 1°C (on gaining 4.2 kJ, or 1 kcal). Next, observe how long it takes the ice to completely melt. What you will find is that the melting takes about 80 times longer. Black established experimentally that *when a solid is warmed to its melting point, the slow continuing addition of heat results in the material's gradual and complete liquefaction at that fixed temperature.* Only then does the temperature again begin to rise.

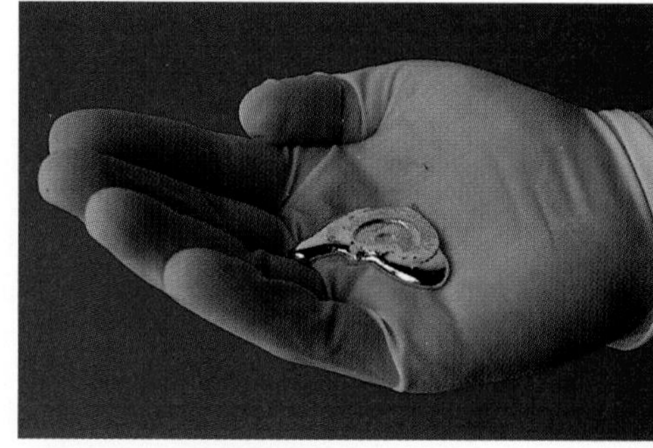

The element gallium has a heat of fusion of only 80 kJ/kg and a melting temperature of 29.8°C (85.6°F)—it melts in the hand.

The heat added to a system while it melts apparently has no effect on the temperature, and so Black called it *latent* (meaning unrevealed) *heat.* Experiments have established that 1 kg of ice at 0°C will be transformed into 1 kg of liquid water at 0°C when 334 kJ (79.7 kcal) of heat are added. The amount of heat that must be added to a kilogram of any solid at its melting temperature in order to liquefy it (or, alternatively, the same amount of heat that must be removed from a kilogram of liquid at its freezing temperature in order to solidify it) has come to be called the *latent heat of fusion,* or just the **heat of fusion**. It's an unfortunate term simply because it is energy per unit mass (J/kg) and not energy. Be that as it may, the heat of fusion (L_f) for water is

$$L_f = 334 \text{ kJ/kg}$$

which is just about 80 kcal/kg. Table 15.6 lists the heats of fusion for a representative selection of substances at a pressure of 1 atm. As we will see presently, L_f depends on pressure. It follows that to change the phase of a mass m already at its melting point requires the addition (or removal) of an amount of heat Q such that

$$Q = \pm mL_f \qquad (15.4)$$

where the positive sign corresponds to heat-in (melting) and the negative sign to heat-out (solidification).

TABLE 15.6 Approximate heats of fusion and vaporization

Material	Melting point (°C)	Heat of fusion (kJ/kg)	Boiling point (°C)	Heat of vaporization (kJ/kg)
Antimony	630.5	165	1380	561
Alcohol, ethyl	−114	104	78	854
Copper	1083	205	2336	5069
Gold	1063	66.6	2600	1578
Helium	−269.65	5.23	−268.93	21
Hydrogen	−259.31	58.6	−252.89	452
Lead	327.4	22.9	1620	871
Mercury	−38.87	11.8	356.58	296
Nitrogen	−209.86	25.5	−195.81	199
Oxygen	−218.4	13.8	−182.86	213
Silver	960.8	109	1950	2336
Water	0.0	333.7	100.0	2259

Example 15.6 A container holding 0.25 kg of water at 20°C is placed in the freezer compartment of a refrigerator. How much energy must be removed from the water to turn it into ice at 0°C?

Solution: [Given: $m_w = 0.25$ kg, $T_{iw} = 20°C$, and $T_{fw} = 0°C$. Find: Q_{out}.] The heat-out (that is, the heat that must be removed) must be enough to drop the water at 20°C down to 0°C and then transform it to ice (also at 0°C). Both processes correspond to negative amounts of heat; accordingly

$$Q_{out} = c_w m_w (T_{fw} - T_{iw}) + (-mL_f)$$

Substituting in the numerical values, we have

$$Q_{out} = (4.2 \text{ kJ/kg·K})(0.250 \text{ kg})(0°\text{C} - 20°\text{C}) - (0.250 \text{ kg})(334 \text{ kJ/kg})$$

$$Q_{out} = -21 \text{ kJ} - 83.5 \text{ kJ} = -104.5 \text{ kJ}$$

and

$$\boxed{Q_{out} = -1.0 \times 10^5 \text{ J}}$$

▶ **Quick Check:** To drop the temperature, we must remove ($\frac{1}{4}$ kg)(20 K)(1 kcal/kg·K) = 5 kcal. To freeze it, another (80 kcal/kg)($\frac{1}{4}$ kg) = 20 kcal must be removed, for a total of 25 kcal = 105 kJ.

From an atomic perspective, it is apparent that the liquid state has more internal energy than the solid state at the same temperature; after all, heat is transferred into the sample as it melts. Still, no corresponding rise occurs in temperature, and that is true for monatomic systems, like the metals, as well as molecular systems, like water. It would seem that the energy input must be going into an increase in PE. The energy associated with the heat of fusion liberates the atoms or molecules from the rigid bonding of the solid state, allowing them to move apart slightly into a less ordered, more fluid configuration.

Since heat must flow into a body when it melts, fusion is a cooling process: it removes thermal energy from the immediate environment in contact with it. A block of ice in a picnic cooler slowly melts, drawing thermal energy from the surrounding warm cans of beer. After a while, the beer is cold and the ice is melted. The same is true for ice cubes in a glass of warm soda—that's why the ice is added, and that's why it always melts. In reverse, freezing is a warming process: it exhausts thermal energy into the immediate environment. Years ago, people used to protect their fruit cellars and greenhouses from freezing by storing large vats of water in them. As the water began to freeze, it liberated thermal energy, making it that much harder to further cool the room. In part, the thermal energy exhausted into the room by a refrigerator comes from the water freezing in the ice cube trays.

15.8 Vaporization

The process of transformation of a liquid or solid into the gaseous state is known as **vaporization**. The resulting product is called a *vapor* because it deviates radically from the behavior predicted by Boyle's Law, the behavior of an ideal gas. The change from a liquid to a gas that occurs continuously at a free surface (that is, one not in contact with a solid or other liquid) is known as **evaporation**.

A liquid is composed of a large number of swarming particles, atoms, or molecules that move around with a distribution of kinetic energies much like that of a gas. Even at low temperatures, a few of those traveling in the vicinity of the surface move up toward it with speeds great enough to overcome the cohesive forces and escape the liquid altogether (Fig. 15.7). As we have already seen (p. 499), some molecules will return to the liquid as new ones leave it. As long as the vapor pressure above the liquid is less than the saturated value for that temperature, evaporation continues. And this is essentially independent of the presence of other gases (such as, air). If the vapor pressure above the liquid is made to exceed the saturated value (for example, by pressing down on the vapor with a piston), there will be condensation. Similarly, a drinking glass filled with ice cubes can cool the surrounding moist air, thereby lowering its saturated vapor pressure enough to allow water to condense onto the cold surface.

The rate of evaporation can be increased by altering the physical condition of the liquid in two ways: First, by increasing the surface area (a wet shirt dries slowly if it's balled up, and brandy evaporates from a wide snifter a lot faster than from an uncorked bottle). Second, increasing the temperature of the liquid accelerates evaporation because at greater temperatures the molecules have higher average kinetic energies and more likelihood of escaping the surface. Warming the clothes in a dryer, heating the back window of your car to drive off the condensation, or holding the bowl of the brandy snifter in a warm hand are each examples of this effect.

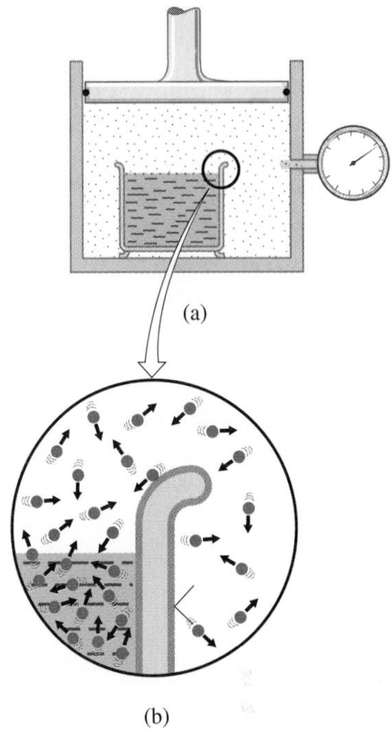

Figure 15.7 (a) A liquid in equilibrium with its vapor at a constant pressure. (b) On the average, as many molecules leave the surface from below as enter it from above.

Just as with melting, energy must be supplied to the molecules being liberated from the cohesion of the liquid state. Vaporization is an even more drastic physical change than melting: the molecules are both torn free and appreciably separated, a process requiring the input of a substantial amount of energy. The vapor contains far more energy than an equal mass of liquid at the same temperature. Accordingly, we define the **heat of vaporization** (L_v) as *the amount of thermal energy required to evaporate 1 kilogram of a liquid at a constant temperature,* or if removed from a kilogram of vapor, to condense it under the same conditions. As we will see presently, that temperature is usually the boiling point, but it certainly need not be. Moreover, the heat of vaporization generally decreases as T increases. For example, L_v for water at 0°C is 2.49×10^3 kJ/kg (595.4 kcal/kg), whereas at skin temperature, ≈33°C, it's down somewhat to 2.42×10^3 kJ/kg (578 kcal/kg). Since the density of the liquid is greater than the density of the vapor, it is evident that energy is being supplied to each escaping molecule to overcome the attraction of the other molecules. Yet this density difference decreases as the sample approaches its critical temperature (p. 498). There, the two states are indistinguishable, and the heat of vaporization is zero.

As Table 15.6 shows, each substance has a characteristic heat of vaporization. Thus, the amount of heat Q that must be supplied (or removed) in order to vaporize (or condense) a mass m of liquid (or vapor) at a specific constant temperature is given by

$$Q = \pm mL_v \tag{15.5}$$

With this equation, we could compute how much energy had to be supplied to your wet bathing suit the last time you impatiently waited around for it to dry. Clearly, one of the synthetic fabrics that does not absorb much water in the first place (so that m is small) requires the least amount of thermal energy and dries the most quickly, especially if it's a warm, dry day.

Since this rather substantial amount of thermal energy must be supplied to the system (often drawn from the immediate environment), vaporization can be considered a cooling process: the molecules that escape are the energetic ones, leaving behind their lower-average-KE brethren in the liquid, which therefore must have a reduced temperature (unless the system is being supplied heat from outside). That's in part why we perspire and why dogs pant. It's why you might lick a finger to test the wind or find your unattended bath water colder than the room even though it started out warm a few hours before.

Example 15.7 At what rate would you have to perspire to rid your body of the 75 W of thermal power you generate metabolically just sitting around resting? Assume a skin temperature of 33°C. As we will see, this is not our only cooling technique, and we generally don't have to sweat that much. Still, an athlete may lose several liters of water to perspiration during a good workout. What happens when you dry off the perspiration with a towel?

Solution: [Given: a power output of 75 W. Find: the evaporation rate.] First, 75 W = 75 J/s. From the above remarks, the heat of vaporization of water at skin temperature is 2.42×10^3 kJ/kg. Thus, you must lose energy at a rate of 75 J/s = $Q/t = mL_v/(1\text{ s})$, and $m/(1\text{ s}) = (75\text{ J/s})/L_v = (75\text{ J/s})/(2.42 \times 10^6\text{ J/kg}) = 3.1 \times 10^{-5}$ kg/s. If this were your only available cooling mechanism, you would sweat away 3.1×10^{-2} g/s, or about 0.11 kilogram per hour.

Normally we sweat about $\frac{1}{2}$ liter per day, or about the equivalent of 1.2 MJ. Toweling perspiration off the skin or letting it drip off is of no use—it has to evaporate to do any cooling.

▶ **Quick Check:** The power equals L_v times the mass evaporated per second; $(2.4 \times 10^3\text{ kJ/kg})(3.1 \times 10^{-5}\text{ kg/s}) = 74$ W.

The primary mechanism that causes hot, moist food to cool is evaporation. That's why you blow on a hot cup of soup—without air currents to carry away the vapor, it builds up above the soup and becomes saturated. That build-up decreases the evaporation rate and slows down the cooling process. To increase evaporation, you blow away the vapors every now and then. The hot food at a barbecue always seems to cool too fast because of the active air currents outdoors.

15.9 Boiling

Evaporation takes place at a liquid's free surface at any temperature, but under special circumstances it can also occur throughout the body of the liquid. Since vapor generated at any point within the liquid has a much lower density than the surrounding medium, it pushes the liquid in the immediate vicinity out of the way, creating (under hydrostatic pressure) a little gaseous sphere. The formation of such a bubble of vapor signals the beginning of the familiar process of **boiling**. To be precise, any liquid (water, olive oil, or molten gold) will boil at a specific temperature where its saturated vapor pressure equals the surrounding pressure. After all, to exist for an appreciable time, a bubble must be essentially in equilibrium with the surrounding liquid. That means the pressure inside of it—the vapor pressure—must equal the ambient hydrostatic pressure, and that's usually ≈1 atm. The vapor pressure of water, which is 0.006 atm at 0°C and rises to 0.2 atm at 60°C, equals 0.1 MPa (1 atm) at 100°C. That should be no surprise; the Celsius scale was fixed via the temperature of boiling water at atmospheric pressure, so we have just gone full circle.

When a pot of water has been heated for a while, bubbles begin to form around the walls and bottom of the vessel. The nucleation sites are centered on minute pockets of air or particles of dust. Some of the bubbles are air, driven out of solution. Once created at the hot bottom, air bubbles remain intact until they break loose. By contrast, bubbles of water vapor rapidly form, vanish, and reform. When they break loose and rise into the colder water above, they are cooled and suffer a decrease in vapor pressure, collapsing, condensing, and vanishing before reaching the surface. This compression can occur violently and is often noticeably noisy. Only when the upper regions of the water reach 100°C will true boiling take place throughout the liquid, and then bubbles reach and burst at the surface. That bursting process liberates a great deal of vapor and with it a great deal of energy—the heat of vaporization of water at 100°C is 2.259×10^3 kJ/kg (539.6 kcal/kg). If heat is continuously fed into the water, it will continue to boil, and raising the flame under the pot will only increase the rate of boiling without changing the temperature of the water. More heat-in generates more vapor, which at temperatures ≥100°C is known as **steam**.

Each kilogram of steam at 100°C can conveniently be used to carry that 2.3 MJ from one place to another, from a basement boiler to a third floor radiator. On condensing in the radiator, the steam liberates its 2.3×10^3 kJ/kg to the environment. Of course, steam can be raised to temperatures well in excess of the 1-atm boiling point, the ultimate limitation being the breaking up of the water molecules. In most modern applications, steam above 500°C is used. This *superheated steam* (as it's called above 100°C) whirls the great turbines that drive the majority of large ships and electric generators.

Example 15.8 How much heat must be added to a 1.0-kg mass of water ice at −10°C and atmospheric pressure in order to transform it into superheated steam at 110°C? Compare the energy associated with each stage of the process and confirm Fig. 15.8.

Solution: [Given: $m = 1.0$ kg, $T_i = -10$°C, and $T_f = 110$°C. Find: Q_{in}.] We have to raise the temperature of the ice to 0°C, melt it, raise the temperature of the water to 100°C, vaporize it, and raise the temperature of the steam to 110°C. Remember that each state has its own value of specific heat capacity. Using $L_f = 334$ kJ/kg, $L_v = 2.26 \times 10^3$ kJ/kg, $c_w = 4.2$ kJ/kg·K, $c_i = 2.1$ kJ/kg·K, $c_s = 2.0$ kJ/kg·K, and the relationship

$$Q_{in} = mc_i\Delta T_i + mL_f + mc_w\Delta T_w + mL_v + mc_s\Delta T_s$$

we get

$$\begin{aligned} Q_{in} = {} & (1.0\text{ kg})(2.1\text{ kJ/kg·K})[0°\text{C} - (-10°\text{C})] \\ & + (1.0\text{ kg})(334\text{ kJ/kg}) + (1.0\text{ kg}) \\ & \times (4.2\text{ kJ/kg·K})[100°\text{C} - 0°\text{C}] \\ & + (1.0\text{ kg})(2.26 \times 10^3\text{ kJ/kg}) + (1.0\text{ kg}) \\ & \times (2.0\text{ kJ/kg·K})[110°\text{C} - 100°\text{C}] \end{aligned}$$

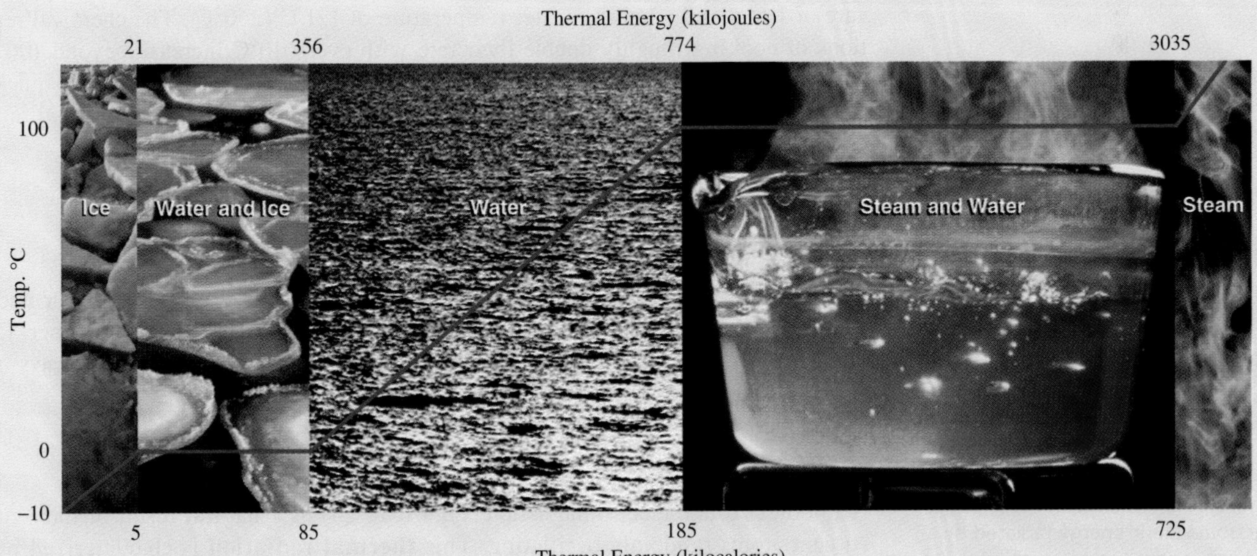

Figure 15.8 A graph of temperature versus thermal energy added to, or removed from, 1 kg of H_2O. The sample is raised from −10°C to above 100°C and passes from ice, to ice plus water, to water, to water plus steam, to steam.

(continued)

(continued)

Accordingly

$$Q_{in} = 21 \text{ kJ} + 334 \text{ kJ} + 420 \text{ kJ} + 2260 \text{ kJ} + 20 \text{ kJ}$$

thus $$Q_{in} = 3055 \text{ kJ}$$

and to two significant figures $\boxed{Q_{in} = 3.1 \times 10^3 \text{ kJ}}$. Notice that most of this energy was required to make the steam.

▶ **Quick Check:** Because the calorie was defined via the behavior of water, problems that deal exclusively with water are easier to treat in the old units:

$$Q_{in} = (1 \text{ kg})(0.50 \text{ kcal/kg·K})(+10 \text{ K}) + (1 \text{ kg}) \times (80 \text{ kcal/kg}) + (1 \text{ kg})(1.00 \text{ kcal/kg·K}) \times (+100 \text{ K}) + (1 \text{ kg})(540 \text{ kcal/kg}) + (1 \text{ kg})(0.48 \text{ kcal/kg·K})(+10 \text{ K})$$

$$Q_{in} = 5.0 \text{ kcal} + 80 \text{ kcal} + 100 \text{ kcal} + 540 \text{ kcal} + 4.8 \text{ kcal}$$

and $$Q_{in} = 729.8 \text{ kcal} = 3.1 \times 10^3 \text{ kJ}$$

The boiling point of a substance depends on the external pressure. Lower the surface pressure, and bubbles can form at a lower vapor pressure, which corresponds to a lower temperature (they are squeezed less and form easier). A pot of water on a campfire atop Mont Blanc (5 km) will boil at only 83°C, while on Mt. Everest (9 km) making a decent cup of tea on an open fire would be impossible (the pressure is ≈37 kPa; that is, 0.4 atm, and the boiling point is ≈74°C). This effect has been used commercially to boil off water from delicate foodstuffs like milk and syrups (in the manufacture of sugar) without cooking them. They're boiled in a low-pressure vessel where the vapor is continuously pumped off. If we put a small sample of water in a vacuum system and vigorously pump off the vapor, dropping the saturated pressure, the liquid will boil at room temperature. This being a cooling process, the sample will rapidly decrease in temperature and freeze into ice.

The opposite process is employed in the pressure cooker (invented by Denis Papin in the late 1600s) in which the liberated vapor builds up pressure inside a heated sealed vessel. The domestic pressure cooker usually operates at a gauge pressure of 1 atm and a boiling-water temperature of 121°C (250°F). The chemical reactions of cooking roughly double their rate with every 10°C increase beyond 100°C. Thus, the 21°C rise is significant, and the pot will cook things much more quickly than the traditional open-to-the-atmosphere culinary approach. The hospital autoclave is a large-scale pressure cooker for sterilizing instruments.

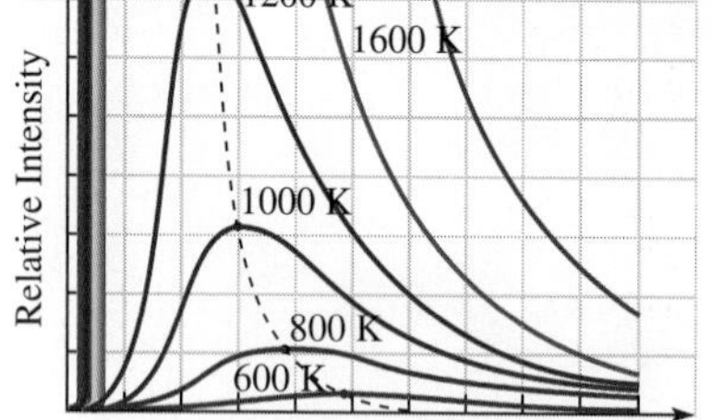

Figure 15.9 Thermal radiation: the electromagnetic energy radiated by a body at various temperatures as a function of wavelength. The narrow band from ≈350 nm to ≈700 nm is the visible region of the spectrum.

THE TRANSFER OF THERMAL ENERGY

Put a hot cup of coffee on a table in your kitchen and come back in an hour and it will certainly be cooler, just as a cold bottle of soda left out will get warmer. Objects transfer thermal energy to and from the environment in three basic ways: radiation, convection, and conduction.

15.10 Radiation

All bodies radiate electromagnetic energy (Ch. 24) as a natural result of their constituent atoms randomly oscillating. This **thermal radiation** is characterized by a broad, continuous range of frequencies (Fig. 15.9) that arises out of the electromagnetic interactions among the atoms of solids, liquids, and dense gases. At this very instant, because of the warmth of your body, you are emitting copious amounts of invisible infrared (IR) electromagnetic radiation (which can easily be observed by IR cameras and pit vipers). The reverse process of absorption becomes evident whenever you put your face in the sunshine (or near an open furnace) and feel a rush of warmth as the radiant energy, especially the IR, is converted to thermal energy in your skin.

As long ago as 1791, P. Prevost maintained that matter at a constant temperature was in dynamic equilibrium, absorbing as much radiant energy as it emitted. *The amount of thermal radiation emitted by a body depends on its surface condition (color, texture, exposed area, etc.) and on its temperature.* The higher the temperature, the more energy it radiates per second (Sect. 30.1). On the other hand, *the amount of thermal radiation absorbed by a body depends on its surface condition and on the nature of the incident radiant energy (wavelength, intensity, etc.), which, in turn, depends on the temperature of the source.*

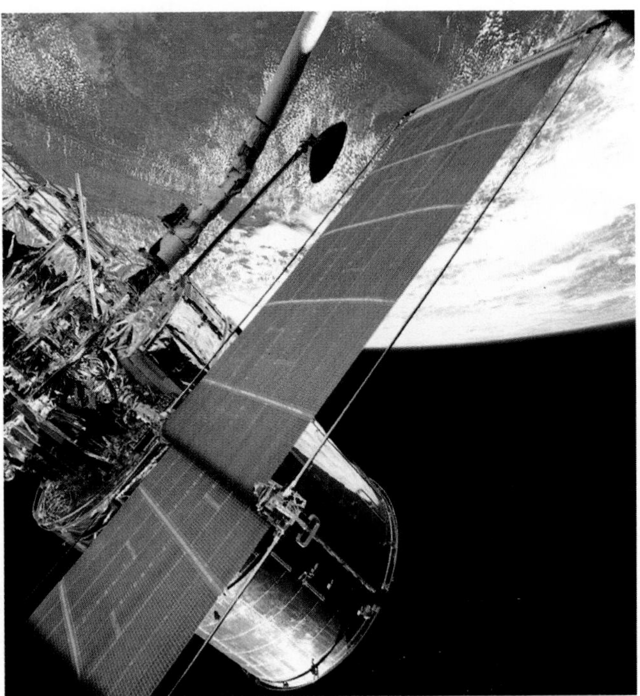

Much of the Hubble space telescope is wrapped in a highly reflecting metalized plastic covering. It reflects radiant energy from the Sun and helps maintain an appropriate range of operating temperatures.

All else equal, black rough materials emit radiant energy effusively, whereas polished metals such as silver and copper are (20 or 30 times) less given to radiate, and white surfaces are usually between the two. As a rule, *a good emitter is a good absorber,* and vice versa (it would follow that a poor absorber is a good reflector or a good transmitter—if it cannot absorb the energy that lands on it, it has to get rid of it some other way). Were it not so, we could put our great little absorber out in the sunshine, and it would get hotter and hotter indefinitely. As is, an object bathed in radiant energy from a higher temperature source (but otherwise isolated from the environment) will absorb energy and rise in temperature. It will correspondingly increase its initially (low T) low rate of emission until, at an appropriately elevated temperature, emission precisely balances absorption. At that fixed equilibrium temperature, the body will thereafter continue to radiate as much as it absorbs. That's what happens when you park your car in the sunshine: if it is black, it heats up a lot and quickly; if white, it heats up a little and slowly. In neither case does the temperature keep rising until the car melts. The more surface exposed to the radiation, the more energy will be absorbed and the more rapidly the temperature will rise—that's why we sit in the shade when it's hot and why, if there is no shade, a camel will sit facing the Sun rather than expose the length of its body to the Sun's rays.

If it were possible to put our hot cup of coffee off alone in an empty universe, it would continue to radiate away energy, slowly approaching Absolute Zero. But that process is not what happens in our far-from-empty Universe. Here, any object is actually immersed in a local sea of electromagnetic radiation, and it continuously exchanges radiant energy with that environment. The rate at which it radiates is determined by its thermodynamic temperature; the rate at which it absorbs is determined by the character of the incident radiation and therefore by the effective temperature of the surroundings. That holds true even in the distant reaches of space where the ambient radiation has a thermal spectrum consistent with a temperature of about 3 K (indicating to physicists that the radiation was once in equilibrium with matter in a denser, hotter, earlier time, presumably near the beginning of the Universe). Space is awash with randomly fluctuating electromagnetic fields that correspond to thermal radiation.

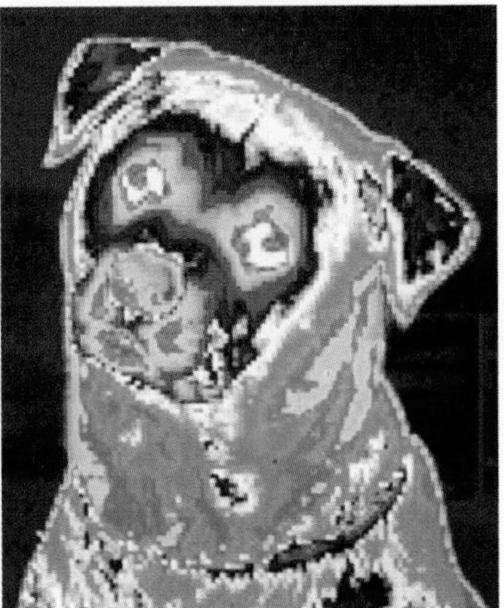

This photo of a dog was made using only the infrared radiant energy emitted by the animal. Such images are essentially temperature mappings.

Putting it rather simplistically, when an object (like our coffee) is hotter than the local radiation environment, it radiates more than it absorbs and tends to cool down to the surrounding temperature. Your kitchen is filled with a spectrum of radiant energy that is typical of a room-temperature thermal source, and the hot coffee loses energy to it. On the contrary, our cold bottle of soda still radiates, but much less than it absorbs from the radiation environment of the kitchen, and so it slowly warms up. This mechanism has nothing to do with any contact the bottle might have with the table or the air. It would not be affected if the bottle were suspended inside of a vacuum. Radiant energy passes through vacuum unimpeded, which is obvious in the case of solar radiation.

Because we humans constantly generate thermal energy internally, we must continuously exhaust it to the environment if a constant body temperature is to be maintained. Actually, we radiate rather well—light-skinned folks emit at about 60% of the maximum rate for an ideal (black) body at that temperature, while dark-skinned people approach about 80%. The difference is not unreasonable given that lighter skins probably evolved in colder climates. Sitting quietly reading a book you lose upwards of 50% of your exhausted thermal energy to the room by way of radiation.

15.11 Convection

When a region of a fluid is heated (for example, the air around a candle flame), its density decreases via thermal expansion and it rises. Displacing cooler portions in its path, the resulting fluid current constitutes a flow of thermal energy, up and away from the source. Driven by gravity, this process is known as *free* or *natural convection* (Fig. 15.10). The household radiator warms a space via convection currents that swirl up and around the room. A hot cup of coffee will produce the same effect in the air on a smaller scale. During the day on the coast, the land (heated more quickly by the Sun because of its low specific heat) heats the air above it. The warm air circulates up and around to create a cooling breeze off the sea (Fig. 15.11). At night, when the sea tends to be warmer than the land, the air current reverses. Large birds such as eagles and vultures (as well as sailplane pilots) seek out warm air updrafts. They conserve their energy by riding convection currents, hovering effortlessly or gaining altitude in the uprush of these so-called thermals. On a larger scale, the trade winds are caused by atmospheric convection currents rising up from the tropics. Ocean currents, such as the Gulf Stream, are also the result of free convection.

A reason to wear clothing (if you need one) is to prevent convection currents from carrying off too much thermal energy from your body on cold days. After all, we are tropical animals who could not survive naked very much below 20°C. Stopping convection currents is also a good reason to insulate water heaters, put up storm windows, and sleep under blankets when it's cold. The blanket, unless it's electric, does not supply heat; it just cuts down on the loss of the thermal energy you generate—covering a shut-off cold automobile engine with a blanket will not warm it up.

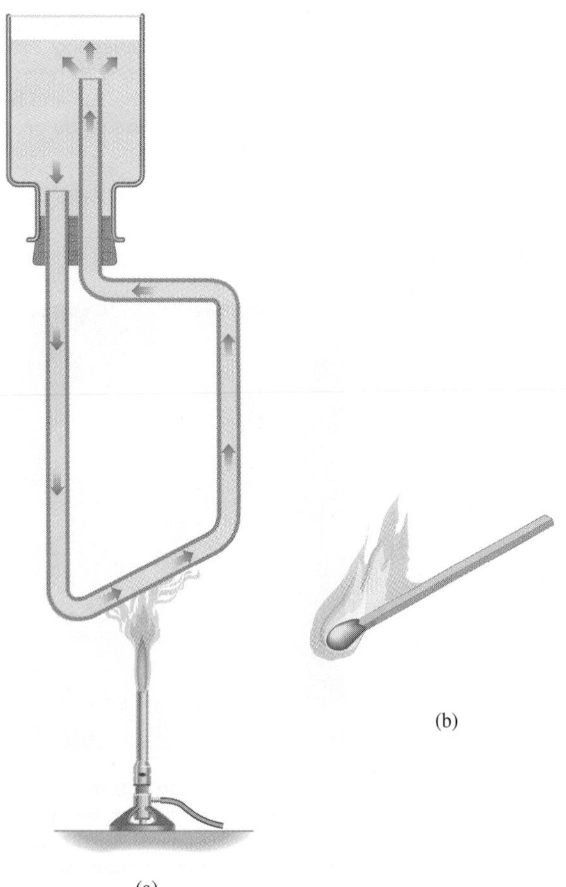

Figure 15.10 (a) Convection currents. Heated liquid rises, cooler liquid descends. The effect is due to a decrease in density as T increases and is driven by gravity—no gravity, no free convection. (b) With a tilting match, convection currents carry thermal energy up to the unburned wood, which fuels the process.

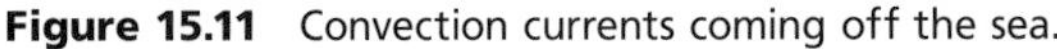

Figure 15.11 Convection currents coming off the sea.

An eagle soaring on updrafts of warm air corresponding to convection currents.

15.12 Conduction

We began this chapter with a discussion of what heat is, and now we end with a few remarks about how it is conducted—namely, by atomic collisions. At the center of the concept is the idea of a temperature difference: ΔT. Heat is only conducted from a high to a low temperature. The measure of the ability of a substance to transmit heat is known as its **thermal conductivity** (k_T), and it depends on the atomic structure of the material. Metals are roughly 400 times better conductors of heat than other solids, which are a trifle better than most liquids, which, in turn, are about 10 times better than the gases (Table 15.7).

It is usually true that a good conductor of heat is a good conductor of electricity, a parallel that is no coincidence. A metal can be envisioned as a lattice of positive ions immersed in a tenuous sea of essentially free electrons. Both heat and electricity are transported largely by the motion of these unbound electrons, which behave like a highly mobile fluid within the metal. In poor conductors of electricity, the transport of heat is primarily by the slower mode of molecular collisions. Thus, amorphous nonmetallic solids are only modest thermal conductors. But the ability to conduct energy is enhanced a bit when the substance is crystalline and the lattice of atoms can also vibrate as a continuous structure (as it does when it carries a sound wave). Whether it is via the random motion of atoms or molecules or electrons, a temperature difference propels heat much as a gravity difference propels a river.

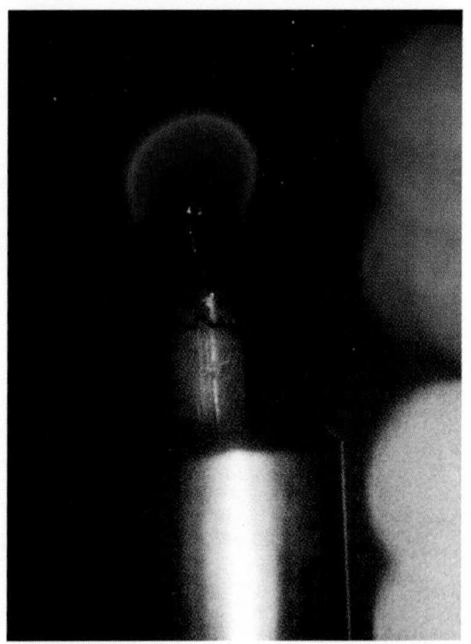

A candle burning in the Microgravity Laboratory of the *Space Shuttle*. While in Earth-orbit, convection currents are far less pronounced than they are under normal gravitational conditions. Ordinarily, hot air, being less dense than cool air, rises. Convection currents feed oxygen to the flame, which also rises.

The transfer of energy always occurs from the action of one of the fundamental Four Forces. In the case of thermal conduction, the force is electromagnetic. A group of atoms with a high average thermal energy transfers random KE to an adjacent group of atoms having a lower average thermal energy, and we observe the process as heat conduction across a temperature difference.

Body tissue without blood moving through it is a relatively poor thermal conductor (it's very slightly better than cork or felt), and the simple conduction of heat

TABLE 15.7 Approximate values* of *thermal conductivities*

Material	Thermal conductivity, k_T (W/m·K)	Material	Thermal conductivity, k_T (W/m·K)
Metals		Linen	0.088
Aluminum	210	Paper	0.13
Brass (yellow)	85	Paraffin	0.25
Copper	386	Plaster of Paris	0.29
Gold	293	Polyamides (e.g., Nylon)	0.22–0.24
Iron	73	Polyethylenes	0.3
Lead	35	Polytetrafluroethylene	
Platinum	70	(e.g., Teflon)	0.25
Silver	406	Porcelain	1.1
Steel	≈46	Rubber, soft	0.14
Other solids		Sand, dry	0.39
Asbestos	0.16	Silk	0.04
Brick, common red	0.63	Snow, compact	0.21
Cardboard	0.21	Soil, dry	0.14
Cement	0.30	Wood, fir, parallel to grain	0.13
Chalk	0.84	*Liquids*	
Concrete and cement mortar	1.8	Acetone	0.20
cinder block	0.7	Benzene	0.16
Down	0.02	Alcohol, ethyl	0.17
Earth's crust	1.7	Mercury	8.7
Felt	0.036	Oil, engine	0.15
Flannel	0.096	Vaseline	0.18
Glass	0.7–0.97	Water	0.58
fiberglass	0.04	*Gases*	
Granite	2.1	Air	0.026
Human tissue (no blood)	0.21	Carbon dioxide	0.017
fat	0.17	Nitrogen	0.026
Ice	2.2	Oxygen	0.027
Leather	0.18		

*Near room temperature.

from inside the body to the skin's surface, where it can be expelled, would be impractical. Instead, thermal energy conducted into the blood is brought near to the body's surface by the circulatory system. When it is cold and thermal energy must be retained, the capillaries near the suface contract, cutting off the flow of blood, thereby creating an insulating layer of tissue.

The molecules of a gas are widely separated, which makes them much less able to communicate thermal energy via collisions. Accordingly, air is a poor thermal conductor, and you can hold your hand beside a candle flame (as opposed to above it in the convection current) and barely feel it. Anything that will trap a layer of air will function as a thermal insulator. That's what a storm window does and why wearing many layers of clothing works nicely in the winter. Your body generates thermal energy, and the clothing, along with the air it captures, keeps it from escaping. In part, the containment of air is also what makes fibrous materials such as wool, fur, and fiberglass such effective thermal insulators. Trapped layers of motionless or "dead" air make excellent insulation. Because the thermal conductivity of air is low, our bodies don't lose much heat to it via conduction. We do exhaust heat by convection, radiation, and evaporation. A swim in a cold ocean will quickly make the point that liquids are better conductors than gases.

The steady-state amount of heat Q that flows per unit time Q/t through a sample has been studied extensively, both theoretically and in the laboratory. Because it's actually a complicated problem, we will consider only the simplest geometry and use the classic approach of compressing all the subtle atomic details into a single material constant. The amount of heat that is conducted per second through a uniform slab or rod of material (Fig. 15.12) is found to be proportional to the *fixed* temperature difference ΔT across its two faces, $Q/t \propto \Delta T$. We expect that our sample slapped against a blazing furnace wall would transmit more heat to your hand than if it were pressed against a lukewarm radiator. Moreover, it is reasonable to assume that the amount of molecular agitation that can be propagated down the sample should depend directly on its cross-sectional area A: $Q/t \propto A$.

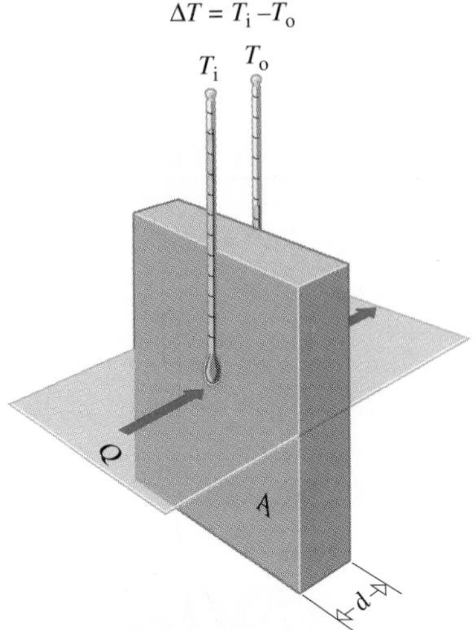

Figure 15.12 The rate (Q/t) of transport of heat is proportional to the temperature gradient ($\Delta T/d$) and the area (A).

If the sample has a thickness or length d, the amount of heat that will traverse it per second in steady-state flow is found to be inversely proportional to d. In other words, Q/t is proportional to the **temperature gradient** $\Delta T/d$ (which, if we were looking at the rate at which balls were rolling down a hill, would be the equivalent of how steep the hill was, and not just how high it was). Remember the end-face temperatures (Fig. 15.12) are held constant, and the closer those faces are, the greater the gradient and the more rapidly the collision energy is moved along. In summary, then, $Q/t \propto A(\Delta T/d)$, and if we introduce a material constant of proportionality, namely, the thermal conductivity (k_T), then

$$\frac{Q}{t} = k_T A \frac{\Delta T}{d} \tag{15.6}$$

where k_T has the units of J/s·m·K or W/m·K. This expression is usually called *Fourier's Conduction Law,* after J. B. J. Fourier (1807). It provides the rate of flow of heat across an area A in a perpendicular direction in terms of the temperature gradient in that direction. For a given ΔT, the greater d is, the lower Q/t is (the thicker your clothing is on a winter's day, the lower the rate of heat loss). When the human body is exposed to cold, the surface blood vessels contract, and that contraction goes on deeper and deeper, providing a thicker (larger d) layer of low k_T insulation as needed.

This material was developed for the heat-shield tiles of the *Space Shuttle.* It has both an extremely low thermal conductivity and small specific heat capacity; so much so, that a cube of it can be held at its edges even when red hot.

In the steady state, the amount of heat flowing into a conducting system traverses it and emerges; there is no loss or gain of energy, no storage. What that suggests (for Q/t = constant) is that regions of low conductivity correspond to large gradients and often to large temperature changes. Thus, a thin layer of insulating material generally experiences a large change in T across it—that's what a blanket does.

Example 15.9 A pane of window glass is 0.90 m wide by 1.5 m high and 4.0 mm thick. It's a cold blustery winter's day in the Midwest. The temperature of the inside face of the window is 10°C, and the outside face is at −9.0°C. How much thermal power is being transported through the windows ($k_T = 0.84$ W/m·K)?

Solution: [Given: $A = 0.90 \text{ m} \times 1.5 \text{ m}$, $d = 4.0$ mm, and $\Delta T = 10°\text{C} - (-9.0°\text{C})$. Find: Q/t.] Using

(continued)

(continued)

we obtain
$$\frac{Q}{t} = \frac{k_T A\,\Delta T}{d} \qquad [15.6]$$

$$\frac{Q}{t} = \frac{(0.84\ \text{W/m·K})(1.35\ \text{m}^2)(19\ \text{K})}{4 \times 10^{-3}\ \text{m}} = 5386.5\ \text{W}$$

and to two significant figures $\boxed{Q/t = 5.4\ \text{kW}}$. Notice that the inside face of the window is a lot colder than room temperature. That's the result of the blanketlike quality of a thin layer of stationary air that usually forms against the window. Even more power would be wasted if the inside air were moving across the inner face of the window, keeping its temperature closer to that of the room. This underscores the virtues of having drawn curtains or window shades (in addition to storm windows), especially at night when little radiation enters via the windows.

▶ **Quick Check:** $\Delta T = Qd/tk_T A = (5386.5\ \text{W}) \times (0.004\ \text{m})/(0.84\ \text{W/m·K})(1.35\ \text{m}^2) = 19\ \text{K}$.

Streaks of snow remain on the roof where the shingles stay cold, above the thick roof beams. This roof is obviously uninsulated; heat flows up from the warm room below, passing out between the beams.

When you awake on a chilly morning, miss the rug and step onto the "cold" bathroom floor, it's a lesson in conductivity. The floor and the rug must have the same temperature—they have been in equilibrium all night. But they certainly don't feel equally warm. The floor tiles have about 10 times the conductivity the rug has and will draw a great deal of heat per second from your warmer feet. That loss of heat is what produces the perception of coldness, more so than the temperature difference—we are not very good thermometers.

Core Material

Thermal energy is the disordered KE of any group of particles, usually the atoms constituting bulk matter. **Temperature** is essentially a measure of the concentration of thermal energy (p. 515). **Heat** (Q) is the thermal energy transferred, via atomic collisions, from a region of high temperature to a region of lower temperature. The amount of heat that must be added to raise the temperature of 1 gram of water 1 degree Celsius—from 14.5°C to 15.5°C— is a calorie, where

$$4.186\ \text{J} = 1\ \text{cal}$$

or

$$1\ \text{kcal} = 4186\ \text{J}$$

The **Zeroth Law of Thermodynamics** maintains that *when two systems are at the same temperature as a third, they are all three at the same temperature as each other* (p. 521). Two systems are in thermal equilibrium only when their temperatures are equal.

The **specific heat capacity** (c) of a substance is the number of joules of heat that must be added to raise the temperature of 1 kg of the material 1 K; thus

$$Q = cm\,\Delta T = cm(T_f - T_i) \qquad [15.1]$$

where the SI units of c are J/kg·K. Water stands out for its high specific heat capacity: as liquid (4.2 kJ/kg·K, or 1.0 kcal/kg·K), as ice (2.1 kJ/kg·K, or 0.50 kcal/kg·K), and as steam (2.0 kJ/kg·K, or 0.48 kcal/kg·K).

The amount of heat that must be added to a kilogram of any solid at its melting temperature in order to liquefy it is the **latent heat of fusion**. For water $L_f = 334$ kJ/kg ≈ 80 kcal/kg. Changing the phase of a mass m requires the addition (or removal) of an amount of heat Q:

$$Q = \pm mL_f \qquad [15.4]$$

The **heat of vaporization** (L_V) is the amount of thermal energy required to evaporate 1 kilogram of a liquid at a constant temperature, thus

$$Q = \pm mL_v \qquad [15.5]$$

The heat of vaporization of water at 100°C is 2.259×10^3 kJ/kg = 539.6 kcal/kg.

Objects transfer thermal energy to and from the environment in three basic ways: radiation, convection, and conduction. The amount of heat conducted per second through a uniform slab or rod of material is

$$\frac{Q}{t} = k_T A \frac{\Delta T}{d} \qquad [15.6]$$

where k_T is the **thermal conductivity**.

Suggestions on Problem Solving

1. Common practice is to represent zero degrees Celsius as 0°C with no reference to significant figures. *Take 0°C to have the same number of significant figures as the other temperatures given.*

2. Because we are generally dealing with temperature changes, we need not transform the data from C° to K. Thus, when using the equation $Q = cm\,\Delta T$, you can express the temperatures in Celsius degrees or kelvins, whichever is more convenient. Still, if you use Celsius and calculate T_f, the answer will be in Celsius.

3. Be careful with calculations when mixing various substances in *different states* (for example, water and ice, or steam and water) in a calorimeter. Whenever there is a change of state, it is entirely possible that not all of the substance will actually make the transition. The governing equation $-Q_{out} = Q_{in}$ is correct, but saying that $Q_{in} = mL_f + cm\,\Delta T$ assumes that the entire mass m has melted, and that need not be the case. We must compute the thermal energy that would be made available on descending to the transition temperature and then check to see if that is enough to accomplish the change of state for the whole mass m. If it isn't, only part of the ice melts (or part of the steam condenses, or whatever).

4. Another common error is to use L_v when you should be using L_f, and vice versa. Double check that you have made the correct substitution before carrying out the calculation.

5. The analyses of many of the mixing problems begin with $-Q_{out} = Q_{in}$. The minus sign is there because the object providing the thermal energy decreases in temperature: $\Delta T = (T_f - T_i)$ is negative.

6. In the old cgs system of units, the specific heat capacity of water is 1 cal/g·C°, which in mks is equivalent to 1 kcal/kg·C°. Watch out for the units; those are kilocalories and kilograms.

Discussion Questions

1. Discuss the physics involved in making a soft piece of steel red hot by hammering on it. Will that work as well for hard steel?

2. Suppose you leisurely stirred two identical cups of very hot coffee, one with a spoon made of steel and the other of glass, each having the same mass. If you absentmindedly put one of the spoons in your mouth, which would you rather it was? Explain.

3. The rate of energy consumed by the human body in the process of doing something, divided by the area of the body, is found to be fairly constant for different weights and body types; it's called the *metabolic rate* (usually expressed in kcal/m^2·h). Measurements show that the metabolic rates associated with bicycling and shivering are about equal at 250 kcal/m^2·h. Considering that muscles are only about 20% efficient, the rest being lost as thermal energy, compare these activities from an energy perspective. What function does shivering serve?

4. Imagine that we place a wooden rod and a brass rod, which are otherwise identical, in a freezer. Which will cool faster? Compare what they would feel like after being in the freezer for a few hours. Explain.

5. Why are steel saucepans and tea kettles sometimes covered with copper on their outside bottoms? Why are high-quality saucepans thick and tea kettles thin-walled?

6. Suppose we wrap an ice cube in aluminum foil (bright side out) and another in black cloth and place them both in the sunshine. What will probably happen and why?

7. Having evolved an expression for the rate of heat transfer via conduction, Eq. (15.6), see if you can do the same thing for convection. Explain each assumption.

8. Ordinary tungsten incandescent light bulbs are evacuated. Explain why, and discuss the effect it has on the amount of power consumed and the amount of heat wasted. (Hint: Why do light bulbs often get blacker inside the longer they are used?)

9. Is it likely that you will feel more or less comfortable in a low-humidity environment at 80°F than in a high-humidity one at 70°F? Explain. Why do many people become "flushed" when they get overheated?

10. Suppose you had to remove a large hot pot from the stove, and to protect your hands you could use either of two cloth towels that are identical, except one is dry and the other is thoroughly wet with cool water. Which would you use and why?

11. Place some water in a metal ice cube tray in a freezer (or

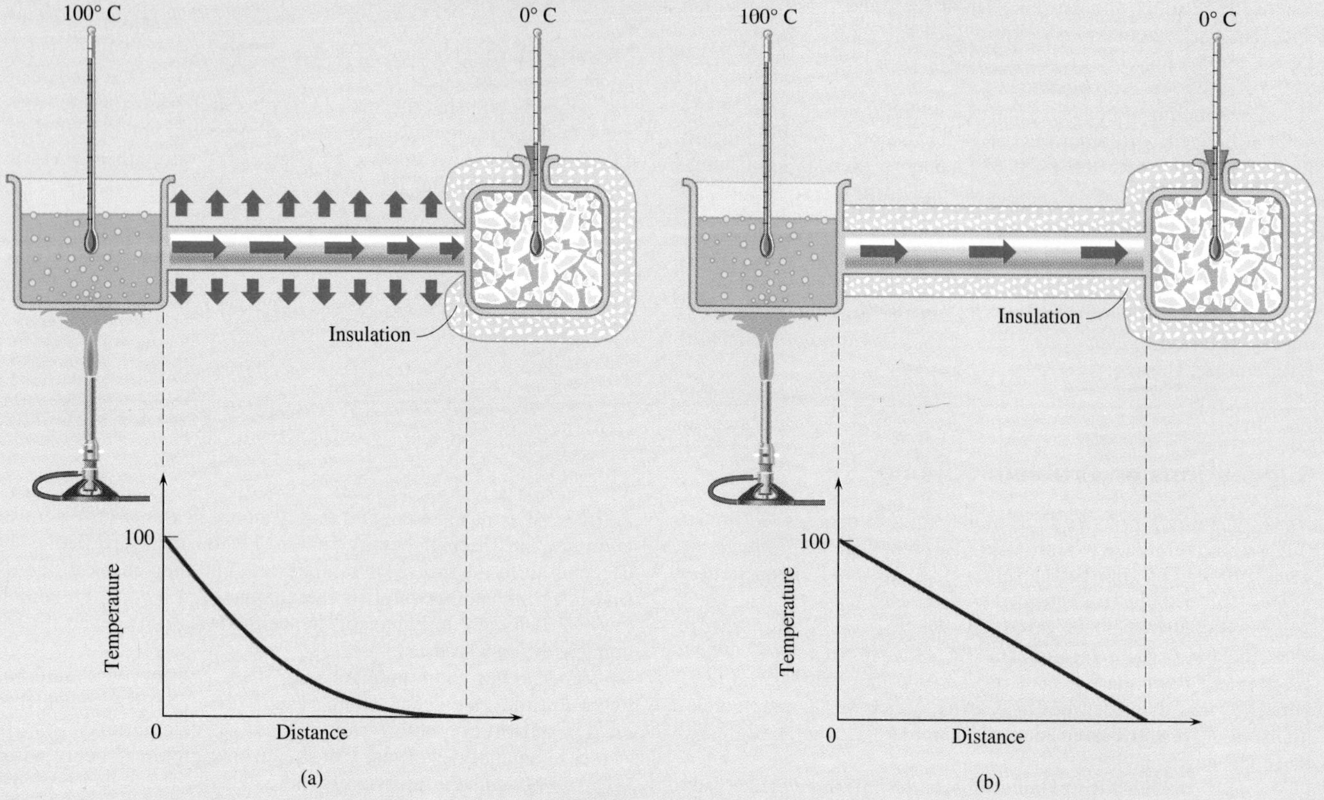

(a) (b)

Figure Q13

outside if it's below 0°C) and periodically observe what's happening. Where does the ice first begin to grow? Why is there often a region of water at the center of the cube? Now try the same thing with two trays, one with warm water and one with cool. Explain how it might be possible for the warm water to freeze first.

12. Most modern automobile cooling systems operate at a gauge pressure of 1 atm. The cooling fluid is usually a 50% mixture of water and ethylene glycol. What purpose does the ethylene glycol serve as regards the cooling process? Why is the system pressurized?

13. Figure Q13 depicts an experiment on conductivity performed by Forbes in 1864. Two metal containers at fixed temperatures of 100°C and 0°C are connected to each other by a metal rod. In the first case, the rod is exposed; in the second, it's insulated. Steady-state temperature readings are made all along the rod and the values plotted as the two *T*-versus-*d* curves in the diagram. Interpret the significance of the arrows and explain the curves. What is the meaning of the slope at each point on each curve? Does what is happening agree with the picture of heat as an incompressible liquid?

14. Given a wet shirt hanging on a clothesline, what will determine the rate at which the water evaporates from it? Why will it dry faster in the sunshine?

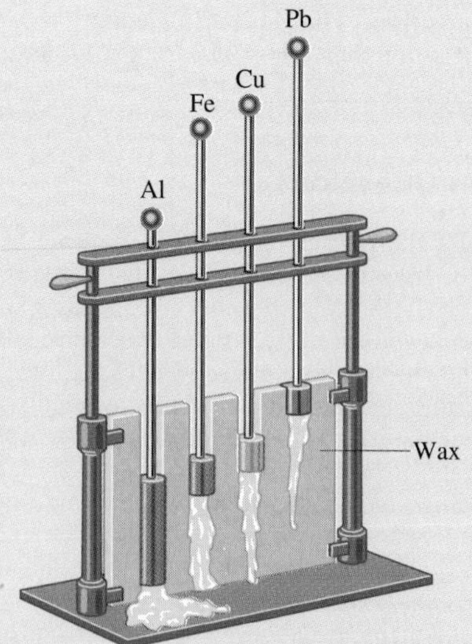

Figure Q15

15. Figure Q15 shows a demonstration devised by the English physicist John Tyndall. Four 1-kg cylinders made respectively of aluminum, iron, copper, and lead, each with the same cross-sectional area, are heated to 100°C in boiling water. They are then all placed upright upon a block of paraffin. Describe what happens and why.

16. How do storm windows work? Should they be sealed tight or left open a little? It's a hot, sunny day and you have the central air conditioning running full blast. Should you lower the blinds and draw the curtains?

17. If your goal is to keep warm in a very cold climate, should you wear a fur coat with the hair inside or out? Explain. If your goal is to keep cool in a hot environment where drinking water is plentiful, what should you wear outdoors? What if water is scarce?

18. We saw earlier that when a stream of air flows over a body, there is usually a thin layer in contact with the surface that is at rest. This same effect should exist when a flame carries hot gas over the bottom of a tea kettle. Explain how it is possible to boil water held in a paper milk container that is placed on an open flame. Why should you avoid turbulence in the hot air?

19. Imagine that you are floating around in space one or two hundred thousand miles from Earth with a thermometer. What will determine the temperature of your suit (assuming no active air conditioning), and what will the thermometer read floating nearby?

20. Leslie's cube is a hollow copper box whose outer surfaces are finished differently. One face is smooth, shiny, and gold-plated; one is roughened with sandpaper; and another is coated with black soot. The box is filled with boiling water, and the emanations from it are examined with a detector. What will be found? Should household heating system radiators be painted shiny silver, as is most often done?

Multiple Choice Questions

1. Two bodies that are not initially in thermal equilibrium are placed in intimate contact. After a while the (a) temperature of the cooler one will rise the same number of kelvins as the temperature of the hotter one drops (b) amount of thermal energy contained by both bodies will be equal (c) specific heats of both bodies will be equal (d) thermal conductivity of each body will be the same (e) none of these.
2. An open beaker of pure water is gently boiling at atmospheric pressure. A thermometer held deep in the water will likely read a temperature (a) equal to 100°C (b) a little less than 100°C (c) a little greater than 100°C (d) equal to 212°C (e) none of these.
3. Two blocks of aluminum, one with a 1.0-kg mass, the other with a 2.0-kg mass, are in thermal equilibrium with a third block of brass at 100°C. The two aluminum blocks are at temperatures, respectively, of (a) 100°C and 50°C (b) 50°C and 100°C (c) 100°C and 100°C (d) 200°C and 100°C (e) none of these.
4. Referring to Question 3, the ratio of the net thermal energy of the 1.0-kg aluminum block to that of the 2.0-kg block is (a) $\frac{1}{2}$ (b) 4 (c) 2 (d) $\frac{1}{4}$ (e) none of these.
5. An open pot of water is boiling on a gas stove when someone raises the flame. The result will be (a) a substantial increase in the temperature of the water (b) a tiny decrease in the rate of evaporation (c) an increase in the rate of boiling (d) an appreciable increase in both the rate of boiling and in the temperature of the water (e) none of these.
6. Figure MC6 shows a glass flask in which water was boiled until the steam drove out much of the air. Thereafter it was tightly corked and inverted. Cold water was then sprinkled over the spherical bottom, and the water inside (a) froze because of the evaporation (b) boiled because of the condensation of steam (c) froze because of conduction (d) boiled because of convection of thermal energy (e) none of these.

Figure MC6

7. Everyone who has ever walked barefoot on a beach in summer has noticed how fast the dry sand gets hot in the morning. That's because sand has a (a) light color (b) fairly low specific heat (c) high thermal conductivity (d) great deal of convection (e) none of these.
8. Figure MC8 shows a test tube containing water and some ice (held down by a weight). As shown, the water at the top is boiling. Is that possible and, if so, why? (a) yes, because water has a high heat of vaporization (b) no, because water has a low specific heat capacity (c) yes, because water has a low thermal conductivity (d) no, because water has a low heat of fusion (e) none of these.
9. Figure MC9 depicts a Dewar flask, or thermos bottle. The space between the two glass walls is evacuated (via a tube that is then sealed). This is done to (a) decrease conduction (b) decrease the specific heat capacity

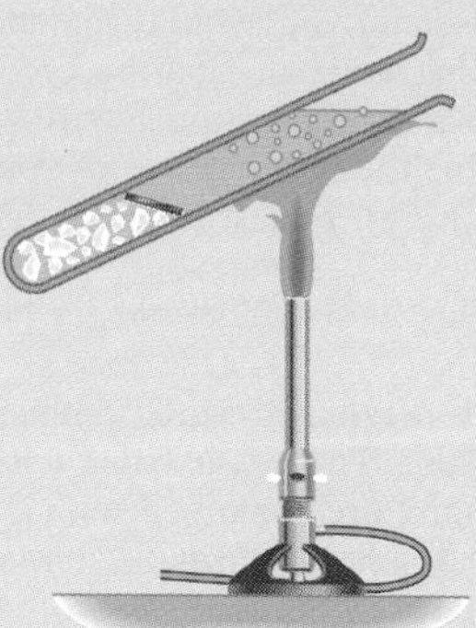

Figure MC8

(c) decrease radiation (d) decrease weight (e) none of these.

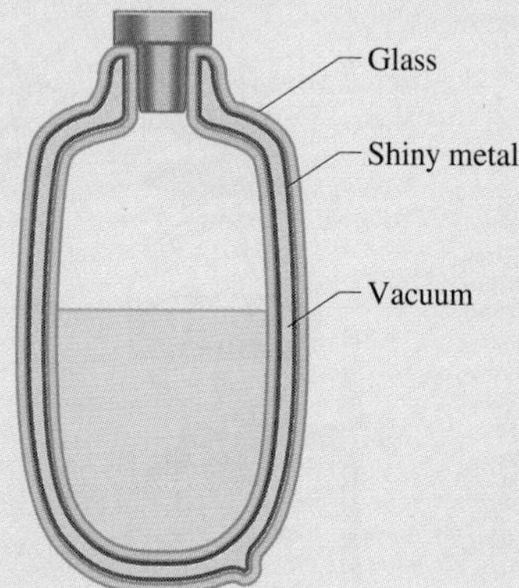

Figure MC9

10. Referring to the Dewar flask of Fig. MC9, why is the outer surface of the inner wall covered with a shiny layer of metal? (a) to decrease conductivity (b) to decrease radiation to and from the inner chamber (c) to increase convection currents in the glass (d) to increase the loss of coldness (e) none of these.
11. Given that a thermos bottle (Fig. MC9) contains a cold drink, what function does the metal film on the inner surface of the outer wall serve? (a) it keeps heat from being conducted inward beyond that point (b) it looks nice shiny (c) it reflects radiant energy incident from outside (d) it cools the inner chamber by emitting radiation (e) none of these.
12. Smoke is drawn out of a fireplace by an updraft that is initiated by wind blowing across the top of the chimney. To help matters along, it is recommended that a burning piece of newspaper be placed up the exhaust hole in the fireplace for a minute or two before lighting the fire. This is done in order to (a) increase the conductivity of the air (b) increase the updraft via convection currents (c) decrease the specific heat of the air (d) warm up the air so it burns better (e) none of these.
13. Suppose you pour a hot cup of coffee and the phone rings so you can't drink it. To keep it hot as long as possible (a) add the cool milk and sugar immediately (b) don't add the cool milk until you're ready to drink it (c) stir it once gently without the milk, but don't add the sugar (d) don't add the milk and use a black mug if possible (e) none of these.
14. Which of the following materials has the highest thermal conductivity? (a) wood (b) water (c) air (d) gold (e) none of these.
15. The purpose of a cover on a soup tureen is to (a) increase conductivity (b) decrease convection (c) increase radiation (d) decrease conduction (e) none of these.
16. The thermal effect of dissolving sugar in a cup of hot tea (a) is to lower the temperature (b) is to raise the temperature (c) is to leave the temperature unchanged (d) cannot be predicted (e) none of these.
17. Metal pots are often made shiny on the outside especially on the top and side, and that makes sense thermally because (a) this conducts heat better (b) this radiates less energy out from the pot (c) this lowers the loss to conduction (d) this appreciably decreases convection losses (e) none of these.
18. Fanning yourself on a hot day (say, 100°F, or 38°C) cools you by (a) increasing conductivity (b) decreasing the mean free path of the air (c) increasing the radiation rate of the skin (d) increasing the evaporation of perspiration (e) none of these.
19. Franklin's experiment (Fig. MC19) consists of a glass bulb connected to a wide, initially open tube. The whole thing is heated until steam drives the air from the cylinder, whereupon it is sealed. Then, by holding the bulb in the hand, the water is warmed and boils in the tube because (a) the temperature can quickly reach 100°C (b) the water is a good conductor and the heat travels to the cylinder where it is concentrated (c) the vapor pressure in the cylinder is quite low (d) radiation is trapped in the connecting tube and brought to bear on the cylinder (e) none of these.

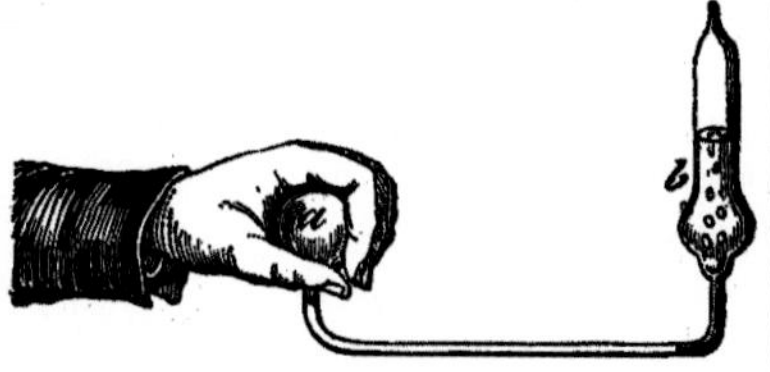

Figure MC19

20. Contemporary radiators of all sorts consist of the heated body—a pipe or transistor, for example—to which are affixed a series of spaced flat vanes, often painted black. The function of the vanes is to (a) increase the convection currents and so decrease conduction (b) increase the heated surface area to increase radiation and convection (c) increase the ability to block air currents thereby increasing conduction (d) conduct more heat to the stagnant air layer and so increase radiation (e) none of these.

Problems

THERMAL ENERGY

1. [I] If 15 kcal of heat are added to a system, how many joules of energy are added?

2. [I] A wrapped sugar cube is labeled to produce 92 kJ. How many kilocalories is that?

3. [I] A person just standing around consumes energy at a rate of about 100 kcal/h. How many watts is that?

4. [I] The human brain operates with an average power consumption of about 20 W. What is that in kcal/h?

5. [I] Alcohol has a rather high caloric value of about 7 kcal/g, which is completely converted into energy in the body. Because it contains no vitamins or minerals, people who drink to excess often have nutritional problems. How much energy is liberated by the body (as thermal energy) via one shot or drink, about 30 g?

6. [I] How much thermal energy must be added to 5.0 kg of pure water at 0°C to raise the temperature to 37°C?

7. [I] The cooling system of an engine contains 30 liters of water. If, during a brief run, the water carries off 500 kcal, by how much will its temperature rise?

8. [I] How much energy does it take to raise the temperature of 500 g of mercury from −19°C to 61°C? Give the answer in both joules and kilocalories.

9. [I] How much thermal energy must be removed from 30 g of tin with a specific heat of 0.060 to drop its temperature from 373 K to 283 K?

10. [I] How much heat does it take to raise the temperature of 0.40 kg of aluminum from 50°C to 60°C given that it has a specific heat of 0.217?

11. [I] If a bar of pure copper is found to absorb 16 kJ in the process of having its temperature raised from 293 K to 353 K, what is the mass of the metal?

12. [I] Show that the molar specific heat capacity of water is 75.3 J/mol·K.

13. [I] How much energy will be given out upon the complete combustion of 1 gallon of gasoline? One gallon equals 3785 cm^3 and the density is 0.68×10^3 kg/m^3.

14. [I] Two beakers of water, one at 15°C and the other at 95°C, each contain 1.52 kg of the liquid. If the water is combined, what will its final temperature be, assuming no heat losses from the liquid to the environment?

15. [I] It is found experimentally that by adding 100 g of iron, with a specific heat capacity of 0.113 kcal/kg·K at 80°C, to a quantity of water at 25°C, the new equilibrium temperature is 30°C. What is that amount of water?

16. [I] Suppose we combine 100 g of aluminum with a specific heat capacity of 920 J/kg·K at a temperature of 495°C with 99.9 g of water at 0.010°C. What will be the final temperature?

17. [I] A quantity of aluminum shot (c = 910 J/kg·K) at 473 K is mixed with 4.95 kg of water at room temperature, and in a little while, the whole thing comes to equilibrium at 300 K. How much aluminum was used?

18. [I] A 340-g glass mug at 20.0°C is filled with 250 milliliters of water at 96.0°C. Assuming no losses to the external environment, what is the final temperature of the mug?

19. [I] The Frenchman Gustave Hirn, in the 1800s, was the first person to make a quantitative study of the thermal energy involved in collisions. He crushed a lead ball with a steel cylinder of known mass dropping through a known height and quickly transferred it to a calorimeter. Hirn essentially found that dropping a 423-g mass a distance of 1.0 m produced 1.0 cal. Explain this result and comment on its accuracy.

20. [II] Suppose we could convert 1.00 kcal completely into work, thereby raising a 1.00-kg mass in a uniform gravitational field (g = 9.807 m/s^2). How high would it get?

21. [II] Water at 20°C flows through a 4.0-kW heater at a rate of 1.5 liter/min. Assuming no losses, at what temperature will it emerge?

22. [II] A person who has a surface area of 1.65 m^2 consumes energy at about 70 kcal/h while resting. Determine the corresponding metabolic rate (see Discussion Question 3) and compute the approximate amount of heat that will be generated in 2 h.

23. [II] According to the *Shooter's Bible,* a 357-magnum bullet has a kinetic energy of 540 ft·lb at 50 yards from the pistol. If it strikes and comes to rest inside a 1.0-kg block of wood (with a specific heat capacity of 1700 J/kg·K), by how much would the block's temperature change? The bullet is lead and weighs 158 grains (1.0 grain is equivalent to 0.0648 g).

24. [II] A 1000-kg car traveling at 60 km/h is brought to a stop by its braking system. How much thermal energy is evolved in the brakes? Friction with the ground does no work unless there is slipping.

25. [II] A 50-W immersion heater is placed in a beaker containing 1.0 kg of water at 20°C. How long will it take to raise the temperature to 100°C? Neglect any heat losses to the air or beaker.

26. [II] Suppose we have 0.30 kg of unknown material. It is heated to 371 K and placed into 0.60 kg of room temperature water. The water is in a copper cup of mass 0.14 kg surrounded by insulating material, and the final equilibrium temperature is 295 K. Compute the specific heat capacity.

27. [II] A 20-g dried sample of food is placed in an aluminum bomb calorimeter and burned completely. The device consists of an aluminum chamber (0.50 kg) containing the sample, a surrounding water bath (2.00 kg), and an aluminum cup (0.60 kg) holding the water. If the calorimeter changes temperature from 20°C to 32°C, how many kilocalories did the food make available? Neglect the mass of the ash.

28. [II] Let's model the human body as being composed of 75% water and 25% protein. Given that the specific heat capacity of protein is 0.4 cal/g·C°, approximate the specific heat capacity of the body in units of kJ/kg·K.

29. [II] Due to the presence in the body of both minerals (in the bone) and fat, its specific heat capacity is closer to 0.83 cal/g·C° than the value computed in Problem 28. Given that a 150-lb human at a temperature of 98.6°F must be dropped to a hypothermic value of 93°F, how much heat must be removed?

30. [II] A 1-kg sample of a material is surrounded by insulation material as in Fig. P30. An electric heater is

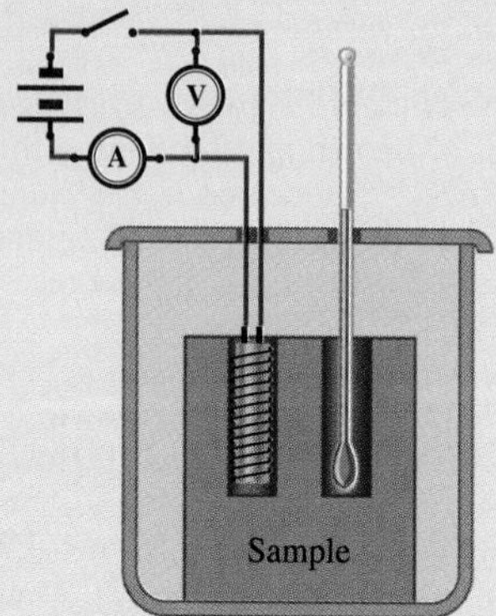

Figure P30

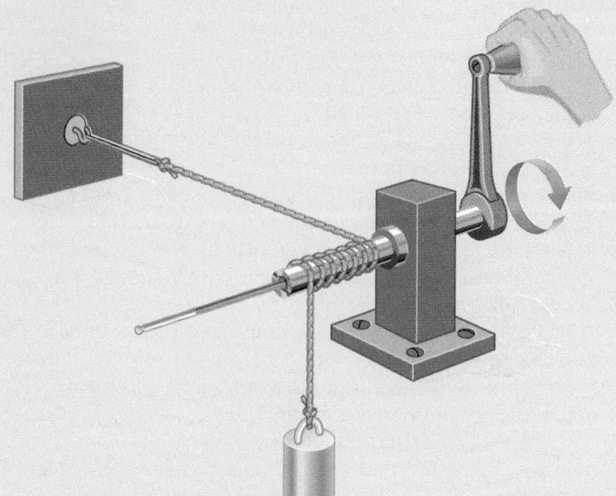

Figure P37

inserted into a hole in the sample, as is a thermometer. Given that 15 W are supplied by the heater for 16.667 min and the thermometer changes temperature by 16.67 K, what is the specific heat capacity of the sample? Ignore any losses.

31. [II] The specific heat of a 0.210-kg sample is to be determined. It's first heated to 373 K and then placed into a bath of 0.082 kg of water at 288 K, which is held in a 0.121-kg copper cup at that same temperature. If the final temperature of the system is 310 K, what is the specific heat capacity of the sample?

32. [II] A handful (405 g) of lead shot is removed from boiling water and dropped into a 99.0-g glass beaker containing 198 g of water at 20°C. If the shot and glass have specific heat capacities of 0.031 kcal/kg·K and 0.20 kcal/kg·K, respectively, find the equilibrium temperature of the system.

33. [II] A 70-kg person engaged in moderate physical activity produces thermal energy internally at a rate of 200 kcal/h. If all the cooling mechanisms become inoperable so there is no dissipation of this energy, how long would it take before the person collapses with a body temperature of 43°C (109°F)? (A body temperature of about 44°C begins to cause irreversible protein damage.)

34. [II] How much excess heat is produced when a 65-kg person with the flu experiences a rise in temperature from 98.6°F to 102°F? Take the average specific heat capacity of the human body to be 0.83 cal/g·C°.

35. [II] How much hard coal must be burned to raise the temperature of 1.00 kg of water from 0°C to 100°C?

36. [III] A house is heated electrically with a system that is 100% efficient; all the power consumed is converted into thermal energy via baseboard heaters. Suppose that it takes 2500 kW·h of energy per month to heat the place. How much wood would be needed to do the same job? Assume the wood stove, which must be exhausted to the outside, is 35% efficient. Given that a cord of wood is a stack of roughly about 1000 kg, how many cords would have to be burned each month?

37. [III] Figure P37 shows a device for measuring the specific heat of a solid rod of material by mechanical means. The rod is rotated, and the rope wrapped around it slides on its surface, producing a frictional drag. The crank is turned at a rate that takes all the tension off the rubber band holding the rope's far end. A thermometer is set in a hole in the specimen that is separated from the rest of the system by thermal insulators. Explain how the apparatus works and how it is used to find c. If the load is 6.00 kg, the mass of the sample 250 g, its diameter 3.00 cm, the number of rotations 240, and the corresponding temperature increase of the rod is 12.0 K, determine its specific heat capacity.

38. [III] Marsh gas [that is, methane (CH_4)] has a heat of combustion of 212 kcal/mol. Show that this value agrees with Table 15.4. If 10 liters of it at STP are burned completely, and

$$CH_4 + 2O_2 \rightarrow CO_2 + 2H_2O$$

how much energy is released in the process? What is the net energy released per molecule burned? Note that the formation of methane gives off an amount of energy of (−)17.9 kcal/mol, whereas the formation of carbon dioxide and water releases (−)94.0 kcal/mol and (−)68.3 kcal/mol, respectively. The heat of formation of the elements in their natural states is zero. From this fact, determine the heat of combustion of methane.

CHANGE OF STATE

THE TRANSFER OF THERMAL ENERGY

39. [I] Determine the thermal energy that must be added to a 200-g block of copper at 1356 K to melt it without raising its temperature.

40. [I] Determine the thermal energy that must be added to 200 g of molten copper at 2336°C to completely vaporize it.

41. [I] Silver vapor at atmospheric pressure condenses at a temperature of 2466 K. How much thermal energy must be removed from 1.5 kg of silver vapor if it's to form a liquid at that temperature?

42. [I] A layer of ice 2.0 mm thick covers the 0.75-m^2 windshield of a car. How much power must be supplied by a heater if the ice is to be melted in 2 min and the air temperature is 0°C? Assume no losses.

43. [I] What is the final state of the system consisting of 500.0 g of liquid mercury at its boiling temperature given that 16.244 kcal is added and there are no losses?

44. [I] Large air conditioners are often rated in tons. A 1-ton unit has a cooling capacity equivalent to changing 2000 lb

of water per day at 0°C into ice at 0°C. Determine (to two significant figures) the thermal power removed from a building by a 1-ton air conditioner.

45. [I] A 0.50-kg quantity of mercury, with an average specific heat capacity of 138 J/kg·K, at 240 K is to be brought to its boiling point and completely vaporized while being maintained at that temperature. How much thermal energy must be provided?

46. [I] A cup of water, 0.25 kg, at room temperature is set on the stove in an open pot to make some tea. The phone rings and you get distracted only to return to the pot as the last drops of water boil off completely. How much heat was transferred to the water?

47. [I] A 500-g lead mass is heated to 150°C and placed on a block of ice at 0°C. How much, if any, ice will melt?

48. [I] Compare the amount of heat it takes to raise an arbitrary amount of water from 0°C up to 100°C with the amount of heat needed to vaporize that water.

49. [I] How much ice at 0°C must be added to 0.50 kg of water at 20°C if it is all to melt, leaving only water at 0°C? (Do the problem using SI units and see Problem 50.)

50. [I] Redo Problem 49 using kilocalories rather than joules and compare your answers.

51. [I] Suppose we combine 100 g of aluminum with a specific heat capacity of 920 J/kg·K at a temperature of 495°C with 99.9 g of ice at 0.00°C. What will be the final temperature?

52. [I] Body tissue without blood flowing through it has a thermal conductivity of about 18 Cal·cm/m^2·h·C°. It's often given in those units in the life sciences because the surface area of the body is generally determined in m^2 and the thickness of the tissue in cm. Convert the conductivity to SI units.

53. [I] The following provides some idea of the inadequacy of conduction as a mode of heat transfer for the human body (see Problem 52). Suppose a person has a surface area of 1.4 m^2 and an average tissue thickness of 2.0 cm between what we might consider the inner and outer body. If the skin temperature is 33°C and the inner body temperature is 37°C, how much heat will be conducted to the surface per hour? Compare that to the fact that about 100 kcal/h are developed while just standing around staying alive.

54. [I] Determine the amount of heat that is conducted through a 4.0-m^2 portion of a brick wall 15 cm thick in the course of 1.0 h if the inside temperature is 20°C and the outside temperature is 0°C.

55. [I] A sheet of ice 1.00 cm thick with an area of 2.00 m^2 covers a piece of sidewalk. Since it's a nice clear day, we simply sprinkle a thin layer of black soot on the ice exposed to the Sun's rays. The idea is to melt the stuff rather than going out to break it up with a chopper. If the air temperature is 0°C and the solar energy per unit area per unit time impinging perpendicularly on the ice is 350 W/m^2, how long will it take before it melts? Assume 100% absorption and no losses.

56. [II] One gram of carbohydrate burned in a calorimeter liberates 4.10 Cal, and it is estimated that in the body about 98% of that energy finds its way to the cells. If a woman consumes 150 Cal/h jogging, how long must she exercise to burn off 100 g of carbohydrate?

57. [II] Knowing that the heat of vaporization of water at STP is 2492 kJ/kg, approximate the amount of energy required by one molecule to escape from the liquid.

58. [II] Compute the heats of fusion and vaporization of water in electron volts per molecule and compare those with the energy needed to tear apart a water molecule into oxygen and hydrogen (2.96 eV/molecule).

59. [II] How much steam at 100°C must be fed into 1.00 kg of water at 10°C to bring it up to 50°C?

60. [II] A lead bullet is heated up to 600 K and then fired at a stone wall, where it comes to rest in a completely molten state. Assuming no energy losses, compute its impact speed.

61. [II] A 10.0-g lead bullet at 23.0°C slams into a stone wall and, squashing, comes to rest. Assuming no loss of energy to the environment, how fast must the bullet be traveling if it is to totally melt?

62. [II] A 10-kg block of ice at 0°C falls, impacting in a way that brings it to rest with no loss of thermal energy. From what height should it drop if 0.010% of the ice is to melt?

63. [II] How many grams of steam at 110°C must be added to 1 kg of ice at 0°C to melt it without raising its temperature? Do this one in both kilocalories and joules and check your answer.

64. [II] An unheated garage in New Jersey in the winter is maintained inside at 0°C by the freezing of a tank of 1000 kg of water, despite the fact that it's colder than that outside. The garage loses heat at a rate of 10^4 cal/min. How long will it be able to stay at 0°C?

65. [II] A calorimeter contains 398 g of water in a 102-g copper cup at 5.1°C. To this is added 40.5 g of crushed ice at −7.6°C. What is the final temperature of the system?

66. [II] As little as 60 years ago, most people kept food in iceboxes that were simply well-insulated cabinets. A large block of ice, delivered roughly once a week, was placed inside above a drip pan. The remaining space was packed with food, the stuff to be colder kept nearer the ice. How does the system work? What's the equilibrium temperature inside the icebox? How much heat would a typical 100-lb block remove from the food before it melted?

67. [II] A block of ice at 0°C is placed on a balance, and an equal mass of water at 80°C is measured off. The water and ice are then combined in an isolated chamber. Determine the equilibrium temperature of the system and describe in detail what it consists of.

68. [II] How much wood would have to burn completely to provide enough energy (assuming no losses) to vaporize 1.00 kg of water boiling away at 100°C?

69. [II] What thickness of brick, with a thermal conductivity of 0.60 W/m·K, will conduct heat at the same rate as a 10-cm layer of dead air under the same conditions?

70. [II] A wall is made up of a 10-cm layer of common red brick in contact with a 20-cm layer of cinder block concrete. If the inside concrete face has a temperature of 20°C and the outside brick face is at 0°C, how much heat flows through a 1.0-m^2 area each hour? What is the power loss?

71. [III] How much thermal energy must be added to 1.00 kg of water to change its temperature from 5.0°C to 100°C? If that same quantity of energy is added to 1.20 kg of silver at 5.0°C, what will its final temperature be?

72. [III] In 1899, Callendar and Barnes devised a new method for measuring the specific heat of water as a function of temperature. It utilized a scheme (Fig. P72) in which only

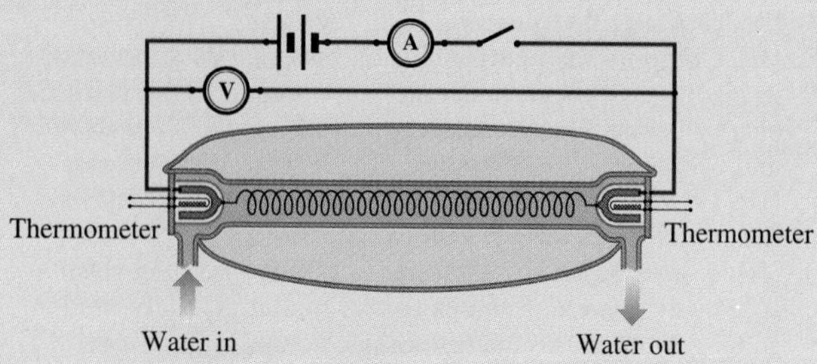

Figure P72

constant temperatures occurred that could be measured slowly and with great accuracy using platinum resistance thermometers. Water was made to flow at a constant rate through a tube that was thermally insulated by a vacuum jacket. The tube contained an electrical heating coil that generated a constant known amount of thermal power (P). In steady-state operation, the heat generated electrically is carried off by the water that enters the tube at a temperature T_i and leaves at T_o. Write an expression for the specific heat of water in terms of the experimental variables. Explain your reasoning.

73. [III] How does the system in Problem 72 lose energy and thus introduce error? Suppose the power loss is P_l. Repeat the previous experiment at a new flow rate such that a mass of water m' passes through in time t, but also change the power input to P′ so that T_o is the same. Show that the thermal losses can be eliminated from the equations, whereupon

$$c_w = \frac{P - P'}{(m - m')\Delta T}$$

Explain your reasoning.

74. [III] A double-glazed window consists of two sheets of glass (each 3.0 mm thick) separated by a 5.0-mm layer of air. Determine the ratio of the thermal conductivity per unit area to that of a single 3.0-mm pane under the same conditions.

16

Thermodynamics

NOW THAT WE KNOW a little about the basics of thermal energy, temperature, and heat, we can turn our attention to **Thermodynamics**, *the study of thermal energy, its transfer, transformation, degradation,* and *dispersal.* Although it was conceived (1824) out of narrow pragmatic concerns about steam engines, Thermodynamics has evolved into a grand discourse on energy and change. Energy is a measure of change (the change in position, speed, mass, temperature, etc.), and change informs time; at base, Thermodynamics is the study of change, the study of the unfolding of events, the progression of the Universe.

Thermodynamics treats the thermal behavior of matter, and when we consider some specific entity or group of entities, it's called a *system:* it might be a bottle of gas, a jet engine, or a yard full of chickens. Whatever else there is beyond the system (that is, the rest of the Universe) is called the *surroundings.* A system might be connected to its surroundings in any number of ways: for example, it might allow heat, or electromagnetic radiation, to cross its boundaries, or it might even emit acoustical energy. On the other hand, we can imagine a system to be completely insulated from the rest of the environment—in which case it is *isolated.* The ultimate isolated system is the entire Universe, there being no surroundings for it to interact with. Each of us can be considered a system in rather elaborate contact with our surroundings, transferring heat to and from it, exhausting and intaking gases, radiating and absorbing electromagnetic energy, ingesting and excreting material, and so forth. Living creatures are just as much the subject of Thermodynamics as steam engines.

THE FIRST LAW OF THERMODYNAMICS

The laws of Thermodynamics "control, in the last resort, the rise and fall of political systems, the freedom or bondage of nations, the movements of commerce and industry, the origins of wealth and poverty, and the general physical welfare of the race."

FREDRICK SODDY (1877–1956)
Nobel Prize-winning chemist

Perhaps the first person to grasp the nature of heat and its relationship to, and unity with, all the various forms of energy was Dr. J. Mayer (p. 520). His work contained the broad notion of Conservation of Energy, but it generated little interest beyond ridicule. Before the decade passed, Mayer was overwhelmed by a deep depression fed by derision and made more intense with the death of two of his children. Lost in despair, he leapt from a second-story window (1849) in a painfully unsuccessful attempt at suicide. Within two years, he was a straitjacketed inmate in an insane asylum.

Though Mayer was first (1842) to recognize that all the various forms of energy were one, Hermann von Helmholtz, a renowned German physiologist (1847), independently formulated the idea of Conservation of Energy, as did Joule. In 1847, Joule gave a short uninvited talk at a scientific meeting. He was cautioned to be brief, and discussion was neither expected nor encouraged, but a young man in the audience began to raise some interesting questions and soon there was a lively discussion that ultimately caused a sensation in the scientific community. That young Scotsman who was so moved by what he had heard was Professor William Thomson, the future Lord Kelvin.

16.1 Conservation of Energy

The most complete conception of the Law of Conservation of Energy (p. 297) includes *all* kinds of energy and is known as the **First Law of Thermodynamics**:

> **Energy can neither be created nor destroyed, but only transferred from one system to another and transformed from one form to another.**

Remember that, whatever energy is, it is associated with the change imparted to a system as the result of interactions. If a positive amount of work is done *on* a sample of matter (on the worked), simultaneously an equal amount of negative work will be done *by* that sample of matter (on the worker). In that sense, change is transferable and insofar as it is transferred, energy (ΔE) is transferred.

The transformation of energy from one form to another is mediated by matter and occurs in the region of that matter. Energy is a property of matter; it does not have a separate existence. Thus, it cannot spontaneously vanish at one place in the Universe and simultaneously appear at another, even though that would not explicitly violate the First Law. *The transformation of energy occurs through the action of the Four Forces.*

The First Law, fleshed out with all of the associated quantitative ideas, such as heat, work, *gravitational*-PE, *electrical*-PE, KE, and so on, allows us to build an understanding of the operation of the natural world. (Fig. 16.1). It was the contribution of the nineteenth century to add heat into the energy balance, and it was the work of Einstein in the twentieth century that showed that matter itself was part of the balance, too—but we will come back to this later, in Chapter 28, since its effect here is negligible.

When a 1-N apple falls 10 m from a tree into your hand, the original 10 J of *gravitational*-PE is transformed into 10 J of KE at the instant before it lands (minus a tiny bit transferred to the air via friction). When the apple comes to rest (neglecting a minute amount of energy lost via sound as it thumps down), the 10 J are shared as thermal energy between hand and apple, so that both increase in temperature slightly. A meteorite with much more KE may even melt on impact, but all the energy present at any moment is constant, whatever its form.

If we do 10 J of work compressing a spring, then tie it up and lower it into an insulated acid bath, what happens to the *elastic*-PE as the spring dissolves? The energy stored via the interatomic forces must be liberated when the spring vanishes. We can expect the temperature of the bath to rise slightly more when dissolving the compressed spring than when dissolving a relaxed one.

To slow down a spaceship reentering the atmosphere, some of its KE is deliberately converted into thermal energy in the air and heat shield. For that matter, every time you activate the brakes on a moving car, you are doing the same thing: converting KE—yours and the car's—to random thermal energy via friction and the First Law. As Mayer suggested, friction that occurs during the tidal action of the oceans, the back and forth rubbing against the ground, heats the waters a bit. And that ther-

Figure 16.1 The First Law of Thermodynamics established Conservation of Energy as a central principle in physics. Now thermal energy joins other forms of energy, broadening the concept of conservation. (See Figures 1.4, 4.20, 8.26, and 9.21.)

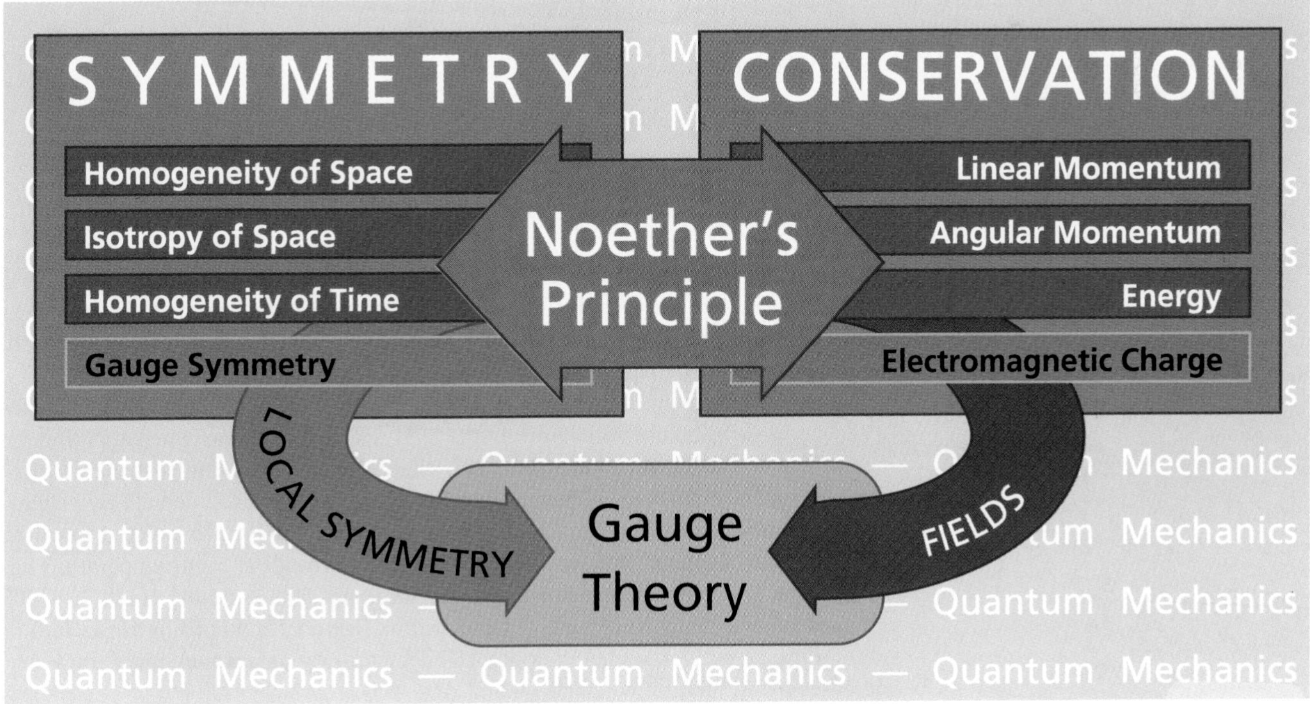

mal energy comes from the rotational KE of the planet—the Earth experiences drag and slows down.

Work, Heat, and Internal Energy

Consider the particles of an ideal gas, which are effectively noninteracting (PE = 0) point-masses. These are endowed with neither rotational nor vibrational KE, and the internal energy U is purely random translational KE. That being the case, it follows from Kinetic Theory that the internal energy of the ideal gas depends only on its temperature [Eq. (14.14)]. Because of its mathematical simplicity, we shall be using the ideal gas as a guide to the behavior of real gases, which generally respond in more or less the same way.

We already know that both heat Q and work W correspond to the transfer of energy by very specific means. The early application of these ideas was to the steam engine, where it was natural to "pour" heat in and get work out. Consequently, it made sense to define both heat-in and work-out as positive quantities, and that usage persists. Hence, work done *by* the system on the environment (work-out) is positive, while work done by the environment *on* the system (work-in) is negative. If an amount of heat Q enters the system, it could manifest itself as either an increase in internal energy, or as a resulting quantity of work performed by the system on the surroundings, or a bit of both. Consequently, the First Law of Thermodynamics becomes

$$Q = W + \Delta U \tag{16.1}$$

The definition of the internal energy of a system might then be taken as

$$\Delta U = Q - W \tag{16.2}$$

Example 16.1 A system in contact with a furnace receives 5000 J of heat while doing 2700 J of work on its surroundings. What, if any, is the change in the internal energy of the system?

Solution: [Given: $Q = +5000$ J = heat-in, $W = +2700$ J = work-out. Find: ΔU.] The relationship between Q, W, and ΔU is given by the First Law

$$\Delta U = Q - W = (+5000 \text{ J}) - (+2700 \text{ J}) = \boxed{2300 \text{ J}}$$

▶ **Quick Check:** $Q = W + \Delta U = 2700 \text{ J} + 2300 \text{ J} = 5000$ J.

Here, for the most part, we limit the discussion to the transfer of energy via work and heat. A more general treatment would simply add on terms to Eq. (16.2) as needed; for example, if the system were heated by putting it in a microwave oven, we would add in a radiation term, and so on.

Whenever there are forces acting, it is necessary to decide where the boundaries of the system will be before beginning the analysis (just as it was when applying the idea of Conservation of Momentum). The choice affects the details of the description but does not change the end results. For example, Fig. 16.2 depicts two chambers that are in contact and can exchange heat, but otherwise are isolated from the rest of the environment. An amount of heat Q flows from the high-temperature liquid into the low-temperature gas. There are three possible systems: 1 and 2, which are open, and 3, which is closed; Q flows out of 1 into 2, and nothing flows into or out of 3. Since no work is being done, ΔU_1 decreases by an amount equal to Q, ΔU_2 increases by an amount equal to Q, and $\Delta U_3 = 0$.

By definition, an isolated system is thermally insulated ($Q = 0$) and can neither do work on the surroundings nor have work done upon it ($W = 0$). It follows, then, that $\Delta U = 0$. Another equivalent statement of the First Law is then

The internal energy of an isolated system is constant (even though it may be transformed from one type of energy to another).

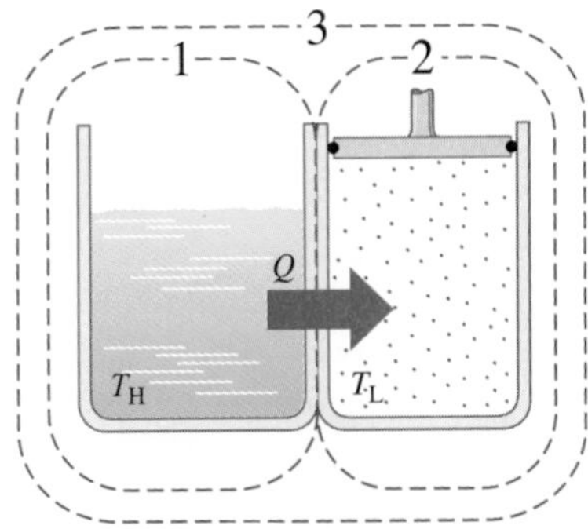

Figure 16.2 Because of the action-reaction nature of interactions, whenever forces act, it is necessary to establish the boundaries of the system being analyzed.

This applies to all isolated systems, including the Universe itself. That's a presumptuous statement to make considering our meager role in the Cosmos, but to date, the First Law has never failed, so we keep saying it. For an isolated system, the internal energy is constant regardless of what processes go on within it—be they chemical, nuclear, or biological.

The same sort of thing happens in the special case of a system that is not isolated, provided that the net amount of heat and work into and out of it turns out to be zero: $Q - W = 0$, $\Delta U = 0$, and $U =$ constant. This notion actually applies to the Earth, provided we add on a radiation term to Eq. (16.2). Take the system to be everything within a great sphere enclosing the atmosphere. Practically speaking, the amount of work done on or by the Earth is negligible ($W = 0$), and since it is thermally insulated, being surrounded by vacuum, $Q = 0$. Still, energy pours in in the form of electromagnetic radiation (mostly from the Sun and mostly visible) at a rate of about 1.7×10^{17} J/s. The Earth reflects some of this and furthermore radiates energy on its own (mostly infrared). The result is that it emits just about at a rate of -1.7×10^{17} J/s. Thus, the total internal energy of the Earth (that is, everything on it and everything in it) is more or less constant. Indeed, as long as you are not changing your weight, much the same thing can be said about you—energy goes in, changes form, and goes out.

The internal energy of a finite system is finite. Clearly, then, if work is to be extracted from it, an equivalent amount of energy must be supplied, or the internal energy will diminish and ultimately no longer be available to drive the system. A *perpetual-motion machine* is one that puts out more work than its input of energy and can presumably continue to do so forever. We call such a mythical device a *perpetual-motion machine of the first kind* because it violates the First Law. Despite the fact that people are still attempting to patent such machines, they cannot possibly work. Recently, a Los Angeles jury acquitted two men of 50 counts of fraud, theft, and conspiracy. They had raised several hundred thousand dollars from backers of a compressed air turbine that was supposedly a perpetual-motion machine. According to the deputy district attorney, "The jurors told me they believed this machine produces more energy than goes into it."

All of this is not to say that the First Law precludes the existence of isolated systems that could, for example, spin forever or bounce forever—provided they do not output work or any other form of energy, they are not forbidden by the First Law.

Processes

A system is in equilibrium when there is no tendency for it to undergo spontaneous change. Despite the fact that its atoms are moving about randomly, it manifests constant measurable macroscopic *physical properties,* such as pressure, volume, temperature, and so on. A configuration of the system, corresponding to a specific set of values of these physical properties (P, V, T, etc.), is called a *state.* As long as they are constant, the system is said to be in a particular state, even though its constituent atoms are continuously moving around. Accordingly, an ideal gas in equilibrium can be described by Eq. (14.9), $PV = nRT$, which is its *equation of state.* A *process* occurs when the system changes from one state (one set of values of its physical properties) to another state. The system returns to its original state when all of its macroscopic physical properties resume their original values.

When a process is taking place, there is some observable alteration of the system: something measurable (for example, P, V, or T) changes. A system can be varied in a variety of ways, but there are four basic modes that are especially simple to deal with:

isothermal A process is isothermal when the temperature of the system is constant. The word comes from the Greek *isos*, equal, and *therma*, heat.

isobaric A process is isobaric when the pressure of the system is constant. The word comes from the Greek *baros*, weight.

adiabatic A process is adiabatic when no heat is transferred to or from the system. The word comes from the Greek *adiabatos*, not passable.

isovolumic A process is isovolumic (or isochoric) when the volume of the system is constant.

The thermal processes involving vapors and gases are particularly important: they play a central role, for example, in the operation of the electric-power–producing steam engine, the internal combustion automobile engine, and the jet propulsion airplane engine.

Work and the First Law

Let's now express the work W in Eq. (16.1) in terms of directly measured variables. Imagine a system that expands against an external pressure, changing its volume and doing work. The primary features are the existence of an applied force and the volume change. What that system is really doesn't matter much—it could be a solid, liquid, or gas. To keep things simple, let the system consist of a cylinder of gas sealed with a weightless, frictionless piston of area A. Some influence in the surroundings pushes down on the piston with a constant force F, as shown in Fig. 16.3. The gas and piston are in equilibrium, and the downward force F is exactly opposed by an upward force produced by the gas—namely, PA. The gas is now somehow made to expand very slowly (quasi-statically) so that it always remains essentially in equilibrium with the piston (here, that means P is constant)—the process is isobaric. This could be accomplished by slowly providing heat to the gas, which will increase the average speed of the molecules thereby sustaining P, even as V increases (p. 504). If the displacement of the piston is Δs, the work done *on the surroundings by the expanding gas* is

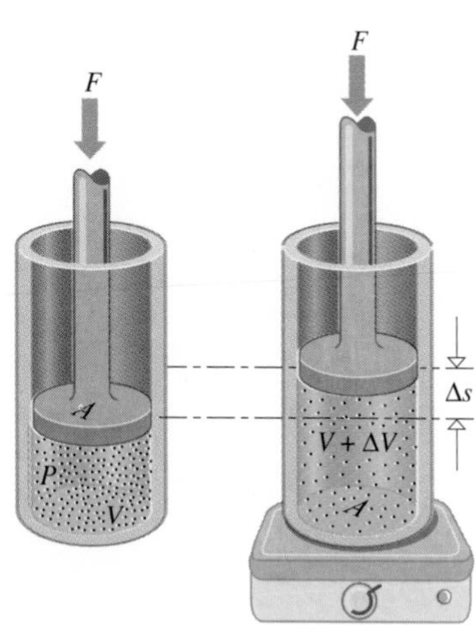

Figure 16.3 A gas sealed in a cylinder by a weightless, frictionless piston. The constant downward-applied force F equals PA, and when the piston is displaced, downward work is done on the gas.

$$W = F\Delta s = PA\,\Delta s$$

Inasmuch as the change in volume is $V_f - V_i = \Delta V = A\,\Delta s$, *the work done in an isobaric process* becomes

$$W = P\Delta V \tag{16.3}$$

When the system expands, ΔV is positive, work is done *by* the system, and W is positive. When the system contracts, ΔV is negative, work is done *on* the system, and W is negative.

Example 16.2 A cylinder closed off with a movable piston contains 10.0 g of steam at 100°C. The system is heated and its temperature rises 10.0°C as the steam expands 30.0×10^{-6} m³ at a constant pressure of 0.400 MPa. Determine (a) the work done by the steam and (b) the change in its internal energy. Take the specific heat capacity of steam to be 2.02 kJ/kg·K.

Solution: [Given: P = constant = 0.400 MPa, m = 10.0 g (of water), $\Delta T = 10.0$ K, $\Delta V = 30.0 \times 10^{-6}$ m³, and $c = 2.02$ kJ/kg·K. Find: W and ΔU.] (a) The process is isobaric, and so W is given by Eq. (16.3)

$$W = P\Delta V = (0.400 \times 10^6 \text{ Pa})(30.0 \times 10^{-6} \text{ m}^3)$$

and $$\boxed{W = 12.0 \text{ J}}$$

(b) Since $\Delta U = Q - W$, we will need Q, the heat-in. $Q = cm\,\Delta T$, and so $\Delta U = cm\,\Delta T - W$, hence

$$\Delta U = (2.02 \times 10^3 \text{ J/kg·K})(0.0100 \text{ kg})(10.0 \text{ K}) - 12.0 \text{ J}$$

and $$\Delta U = 202 \text{ J} - 12.0 \text{ J} = \boxed{190 \text{ J}}$$

▶ **Quick Check:** $Q = cm\,\Delta T = \Delta U + W = 190$ J + 12.0 J; $\Delta T = (202 \text{ J})/cm = 10$ K.

Example 16.3 An aluminum cube 20 cm on a side is heated from 50°C to 150°C in a chamber at atmospheric pressure. Determine the work done by the cube and the change in its internal energy. If the same process was carried out in vacuum, what would the change in internal energy be then?

Solution: [Given: aluminum cube, $L = 20$ cm, $T_i = 50$°C, $T_f = 150$°C. Find: ΔU when $P = 1$ atm and when $P = 0$.] This isobaric process involves ΔU and W, and that immediately suggests Eq. (16.2): the block expands against the hydrostatic pressure of the atmosphere and does work on it. We have to determine Q and W to find ΔU. To get $W = P\Delta V$, we need ΔV, and that follows from Eq. (14.4) and Table 14.3:

$$\Delta V = \beta V_0\,\Delta T = (72 \times 10^{-6} \text{ K}^{-1})(8.0 \times 10^{-3} \text{ m}^3) \times (100 \text{ K}) = 5.76 \times 10^{-5} \text{ m}^3$$

Hence

$$W = P\Delta V = (0.1013 \text{ MPa})(5.76 \times 10^{-5} \text{ m}^3) = \boxed{5.8 \text{ J}}$$

Now, to find Q, we use Eq. (15.1), $Q = cm\,\Delta T$, but we will need m first. Since $m = \rho V$ and from Table 10.1 (p. 321), it follows that

$$m = (2.7 \times 10^3 \text{ kg/m}^3)(8.0 \times 10^{-3} \text{ m}^3) = 21.6 \text{ kg}$$

Using Table 15.3:

$$Q = cm\,\Delta T = (0.90 \text{ kJ/kg·K})(21.6 \text{ kg})(100 \text{ K})$$

thus $$Q = 1.94 \text{ MJ}$$

Hence $$\Delta U = Q - P\Delta V \qquad (16.4)$$

and $$\Delta U = 1.94 \text{ MJ} - 5.8 \text{ J} = \boxed{1.9 \text{ MJ}}$$

The amount of work done is negligible by comparison to Q, so it doesn't matter if this process is carried out in vacuum or not; the internal energy change will still be 1.9 MJ.

▶ **Quick Check:** Using rounded off numbers, $\Delta V \approx (70 \times 10^{-6} \times 10 \times 10^{-3} \times 100)$ m³ $\approx 7 \times 10^{-5}$ m³. $W \approx 0.1$ MPa $\times 6 \times 10^{-5}$ m³ ≈ 6 J. $Q \approx (1 \times 10^3 \times 20 \times 100)$ J ≈ 2 MJ.

The PV-diagram for an isobaric expansion is shown in Fig. 16.4. From an initial state I, the system was transformed at a constant pressure to a different final state F. Notice that **the work done**, $P\Delta V$, **is the area under the curve**. If the pressure is now decreased (Fig. 16.5) while maintaining a constant volume $\Delta V = 0$, the system arrives at a new state F with no additional work done (the area under the curve is unchanged). In the case of an ideal gas, we can imagine the various processes of constant P, V, or T as occurring on the surface of a three-dimensional PVT-diagram (Fig. 16.6). The points A, B, and C in Fig. 16.5 correspond to those in Fig. 16.6, where the system goes from A to B to C.

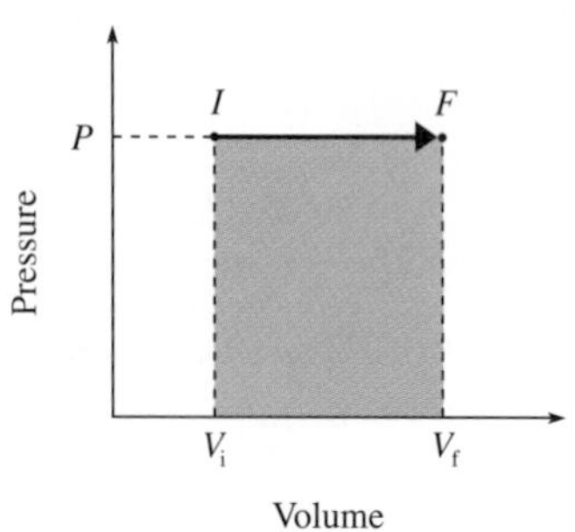

Figure 16.4 The volume of a gas is changed while the pressure is kept constant. The work done by the gas in expanding is $P\Delta V$, and that equals the area under the curve.

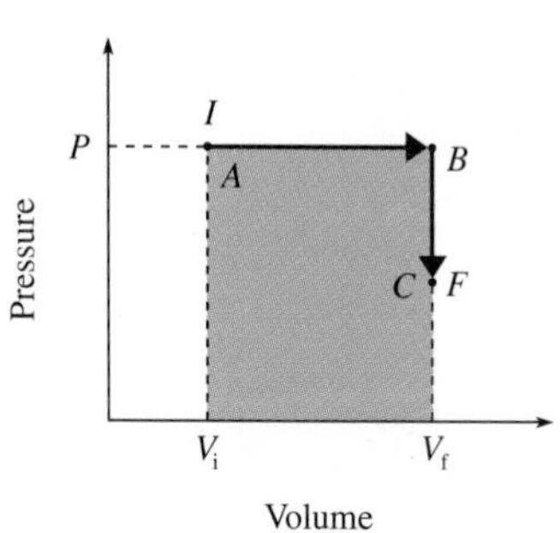

Figure 16.5 No work is done in decreasing the pressure carrying the system from *B* to *C* and a final state *F*. The work done by the gas in going from *I* to *F* is equal to the area under the curve.

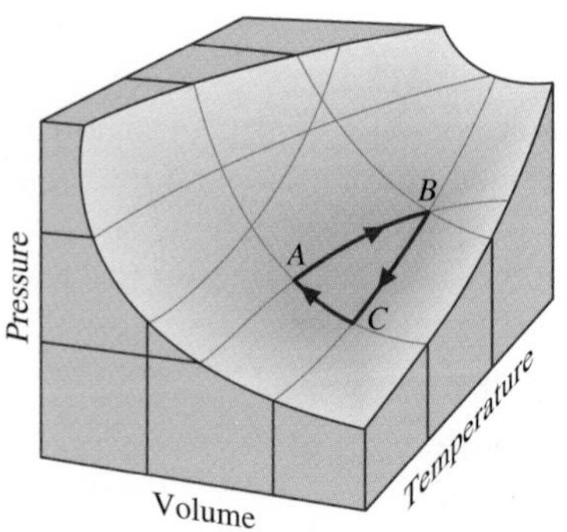

Figure 16.6 *PVT*-diagram for an ideal gas.

Example 16.4 A cylinder containing a gas is sealed with a nearly frictionless piston at a pressure of 0.20 MPa. The cylinder is placed in contact with a source of heat and the gas very slowly expands, moving the piston, whose area is 1000 cm², a distance of 5.0 cm. Assuming the process is isobaric, and given that 300 J of heat enters the system, determine its change in internal energy.

Solution: [Given: $P = 0.20$ MPa $=$ constant, $A = 1000\ \text{cm}^2$, $\Delta s = 5.0$ cm, and $Q = 300$ J. Find: ΔU.] The only formula we have for internal energy is the First Law, Eq. (16.2), $\Delta U = Q - W$, but because this process is isobaric

$$\Delta U = Q - P\Delta V \qquad [16.4]$$

Here, $\Delta V = (1000\ \text{cm}^2)(5.0\ \text{cm}) = 5000\ \text{cm}^3 = 5.0 \times 10^{-3}\ \text{m}^3$. Hence

$$\Delta U = (300\ \text{J}) - (0.20 \times 10^6\ \text{Pa})(5.0 \times 10^{-3}\ \text{m}^3)$$

and $\boxed{\Delta U = -0.70\ \text{kJ}}$

The gas does much of the work of expansion at the expense of internal energy to the extent of a 700-J reduction in U.

▶ **Quick Check:** $\Delta V = 5 \times 10^3\ \text{cm}^3 \approx 10^{-2}\ \text{m}^3$; $P\Delta V \approx 10^5\ \text{Pa} \times 10^{-2}\ \text{m}^3 \approx 10^3\ \text{J}$; $Q = \Delta U + P\Delta V \approx -700\ \text{J} + 10^3\ \text{J} \approx 300\ \text{J}$.

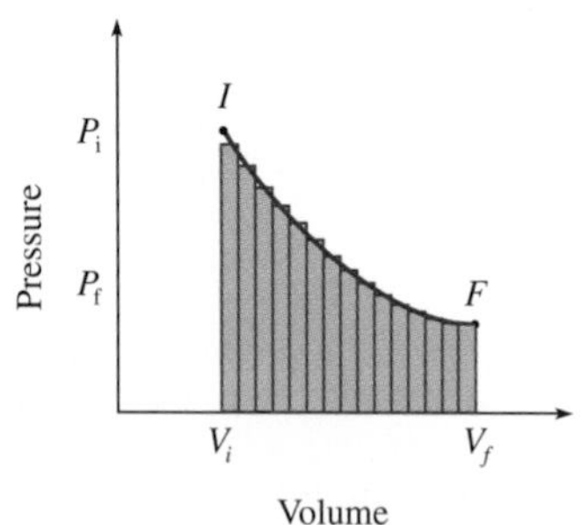

Figure 16.7 The area under a *PV*-curve equals the work done on or by the gas. Here, that area is divided into a large number of small rectangles so that it can be computed.

If the pressure varies continuously as the volume changes (Fig. 16.7), the area under the *PV*-curve would still correspond to the work done. That area can be found by dividing it up into small rectangular segments (as in Fig. 9.6, p. 284), each terminated by the average value of the pressure over that interval. The work done during the i-th constant-P segment ΔW_i is given by Eq. (16.3) as $P_i \Delta V_i$, and the sum of all such contributions, as they get narrower and their number approaches infinity, approaches the total work exactly. The only proviso is that the system be in equilibrium everywhere. If it isn't, PA will not equal F over some of those vanishingly narrow area segments, a point we will return to presently.

Depending on what we do to it, the system can be transformed from its initial condition (state I) to its final condition (state F) by various routes, each corresponding to a different amount of work done (Fig. 16.8). Clearly, first dropping P at constant V and then changing the volume encompasses less area (and so does less work) than initially keeping P constant. *The work done on or by a system depends on the manner in which it is transformed from its initial to its final state—work is not a state variable, it does not reflect the state of the system.* Remember that heat may also be entering or leaving the system, and the final state is determined by a balance of both these contributions. Thus, the path taken in the *PV*-diagram must be fully specified if the work is to be determined. Still, whatever path is taken to get there, the final state is the same—its atoms presumably should not remember the sequence of events that brought them there. Consequently, we can expect that U_f will depend on P_f and V_f but will be the same for all paths leading to that final state. Experiments bear out this conclusion: ΔU is always found to be path-independent. **The internal energy of a system is a unique measure of the state of the system.** Although U is path-independent, W is not, and it follows from the First Law that Q is not. Both W and Q depend on the history of the system and combine to sustain U so that it is independent of that history.

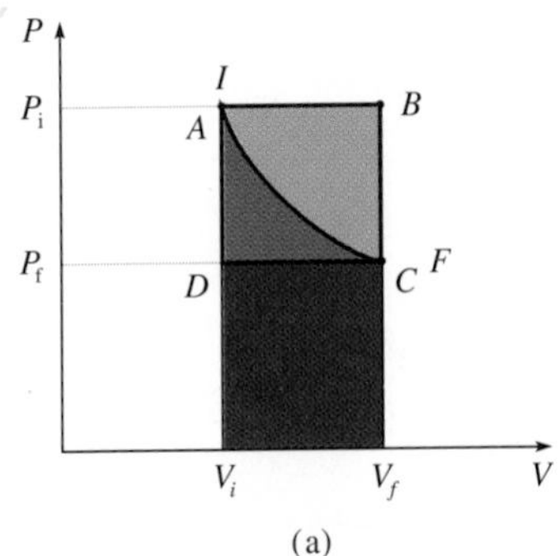

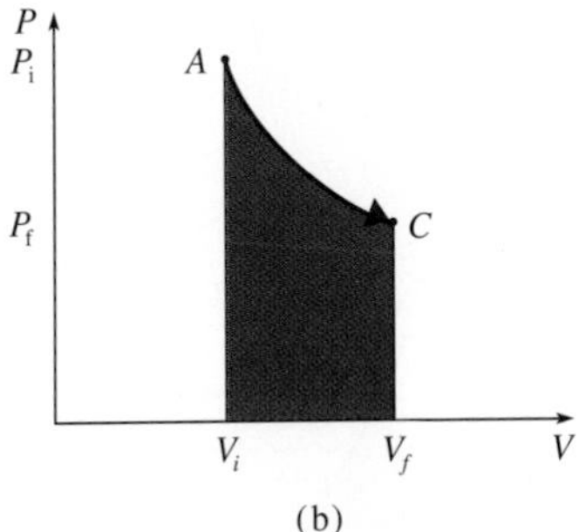

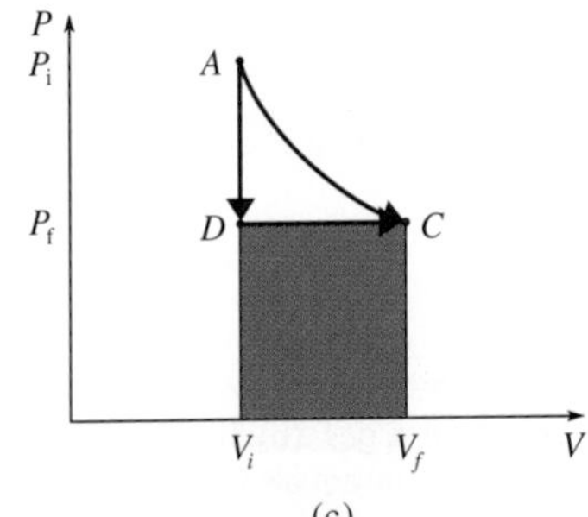

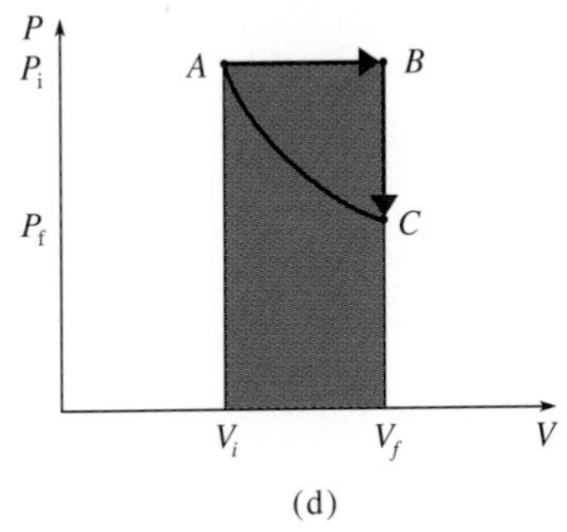

Figure 16.8 A system (a) may go from an initial state *I* to a final state *F* along several different routes, each corresponding to the doing of a different amount of work. Parts (b), (c), and (d) depict different procedures, areas, and amounts of work.

Isothermal Change

Suppose that the device in Fig. 16.9a contains an ideal gas that is to expand isothermally. Accordingly, imagine that the cylinder is made of a highly conducting material, is thin-walled, and surrounded by a large constant temperature bath (a so-called *thermal reservoir*). Now, if the gas is allowed to expand very slowly, raising the piston, heat will be able to enter the system at a rate such that the gas remains at a constant temperature. Quite generally, an expanding gas does work on the movable wall of its container, the piston. When a molecule collides with a wall that is moving away and rebounds off, it has a final speed that is less than its initial precollision speed (Fig 16.9b). The decrease in molecular KE equals the work done on the piston. That work is therefore done at the expense of internal energy. Thus, *unless heat is supplied to the gas, its temperature will drop.* Here, the cylinder is in contact with a thermal reservoir, and any tendency for the temperature to change is negated by a flow of heat across the boundary.

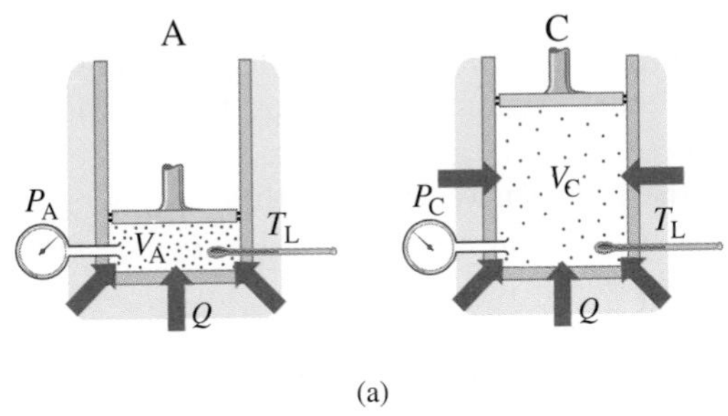

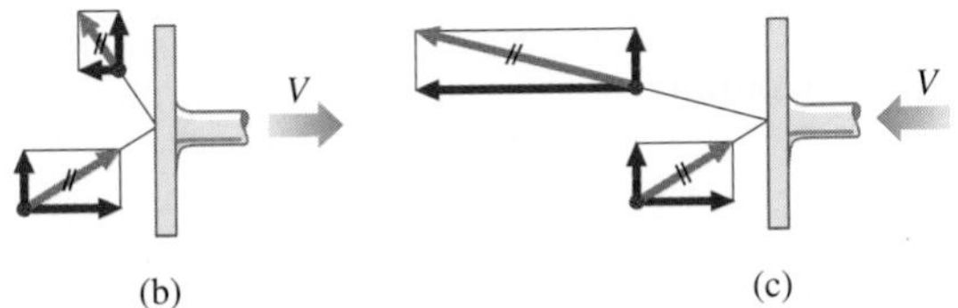

Figure 16.9 (a) A gas expanded isothermally (see Fig. 16.10) from state *A* to state *C*. As heat enters, the piston rises and pressure drops. (b) When molecules bounce off a receding piston, they lose momentum. (c) When they bounce off an advancing piston, they gain momentum.

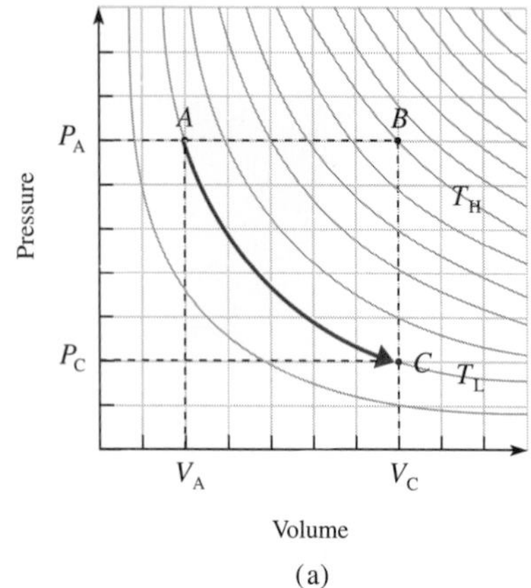

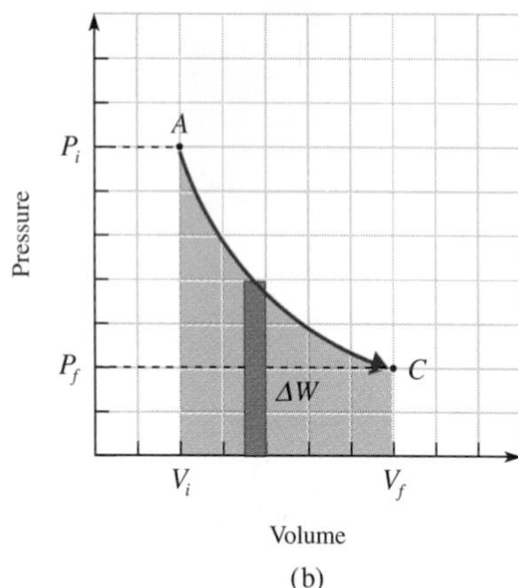

Figure 16.10 (a) An isothermal expansion of a gas. (b) The area under the *PV*-diagram equals work done.

In an isothermal expansion of an ideal gas, since U depends only on the translational KE, **which depends on T, which is constant, it follows that $\Delta U = 0$ and $Q = W$**: The work done by the gas is provided by the heat-in from the reservoir. In effect, the gas serves as a conduit that carries energy from reservoir to piston. On the other hand, when a gas is compressed, the molecules bounce off an approaching wall (Fig. 16.9c), picking up speed like a tennis ball colliding with an advancing racket. The KE of the gas increases; the piston does work on the gas, and unless heat can escape the system, its temperature will rise. Figure 16.10a depicts an ideal gas being moved along an isotherm from point A to point C, and the same path is pictured in Fig. 16.6, as well. Notice how going from A to B entails raising the gas from T_L to a higher temperature T_H, crossing several isotherms along the way. It's helpful when interpreting *PV*-diagrams to envision a grid of isothermals ranging from a low to a high temperature. Here, since $P = nRT/V = (\text{constant})/V$, the isothermals are hyperbolas.

Consider the work done by the isothermal expansion of the ideal gas. The area under the curve in Fig. 16.10b is gotten by summing up all of the tiny ΔW area elements (or, via calculus, by integration). Here, $\Delta W = P\Delta V$, but if the process is done slowly so that the system never deviates appreciably from equilibrium, $P = (\text{constant})/V$ and the pressure can be determined in terms of the volume at each element. Thus, $\Delta W = (\text{constant})\Delta V/V$; more work is done (that is, the area element is taller) initially when the volume is small and the pressure high than later, when the volume is large and the pressure low. This particular functional dependence is extremely common in the real world: ΔW is proportional to both ΔV and to $1/V$. The situation where the change in something depends both on the change in another quantity and, inversely, on how much of that other quantity exists at each step is encountered quite frequently in physics. That kind of dependence is called *logarithmic* (Appendix A-3), and we define the natural logarithm (abbreviated ln) by way of it. We encountered the $\log_{10}$ when we dealt with sound intensity. This *natural* logarithm (so called because it occurs in so many natural processes) is the log to a special base: rather than 10, it is the number known as e equal to 2.718 . . . . In other words, $\ln x = \log_e x = 2.30 \log_{10} x$, as shown in Fig. 16.11, and the corresponding values of $\ln x$ are to be found in standard tables or computed via electronic calculators.

The area under the isotherm varies logarithmically with volume—that is, W varies logarithmically with V. *The work done on or by an ideal gas during an isothermal volume change from V_i to V_f is*

$$W = nRT \ln \frac{V_f}{V_i} \tag{16.5}$$

where $W = Q$. At least for the purpose of analysis, many processes are essentially isothermal and, as a rule, they occur slowly—a very small leak in a tank of compressed gas at room temperature, an open pot of water boiling away on a low flame.

Example 16.5 Determine the amount of work needed to compress 4.0 g of oxygen at STP down to one-third its original volume, keeping the temperature constant. Assume it behaves as an ideal gas.

Solution: [Given: oxygen at STP, $m = 4.0$ g, and $V_f = V_i/3$. Find: W.] The work for an isothermal volume change follows from Eq. (16.5) and so we need n, the number of moles. The atomic mass of O_2 is $2(16) = 32$;

(continued)

(continued)

hence, there are 32 g/mol and $n = (4.0 \text{ g})/(32 \text{ g/mol}) = 0.125$ mol. Hence

$$W = nRT \ln \frac{V_f}{V_i}$$

$$W = (0.125 \text{ mol})(8.314 \text{ J/mol·K})(273 \text{ K})(\ln \tfrac{1}{3})$$

and $\boxed{W = -0.31 \text{ kJ}}$. The minus sign shows that work was done on the gas, and since the temperature did not change, it had to lose the same amount of energy as heat-out.

▶ **Quick Check:** Redoing the calculation using simplified numbers, we get 4 g $= \frac{1}{8}$ mol; $nRT \approx \frac{1}{8}(8 \text{ J/mol·K})(273 \text{ K}) \approx 273$ J/mol; $\ln \frac{1}{3} \approx -1$; $W \approx -0.3$ kJ.

Reversibility

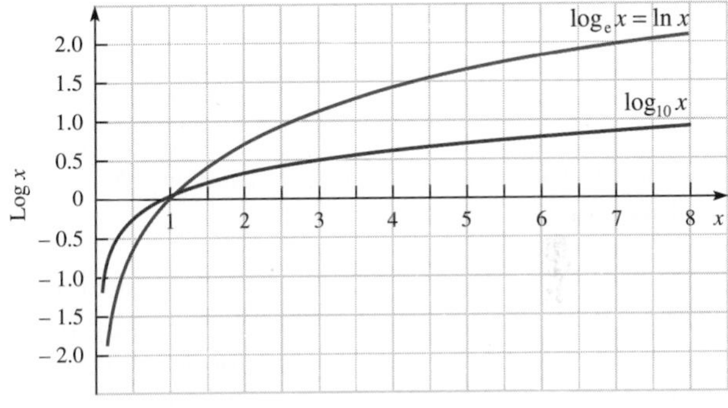

Figure 16.11 The function ln x equals $-\infty$ at $x = 0$, 0 at $x = 1$, and continues to rise slowly as x increases.

If the isothermal process that carried the gas from point A to point C in Fig. 16.10 can be retraced so that the system returns from P_f, V_f, and T to P_i, V_i, and T along the same curve, then we say that that process was **reversible**. Another way to put it is that the system must always be infinitesimally close to thermodynamic equilibrium even as it passes from one state to the next. Imagine a vertical, thermally insulated cylinder containing some gas sealed in with an upright frictionless piston on which is piled a mass of fine sand. The sand provides the downward force that is balanced by the trapped gas, leaving the piston motionless in equilibrium. If one miniscule sand grain is added to the pile, the piston will descend through a tiny displacement, in an essentially quasi-static fashion. One grain at a time, the compression can continue until we choose to stop at any moment. If sand is now slowly removed, the process will be reversed into an expansion that carries it through the same states it has already experienced. The fact that a process can be stopped anywhere along the way and turned around by a minute change in the operating conditions is the hallmark of reversibility.

In truth, reversible processes are idealizations: they can be approximated but never actually attained. For an isothermal process such as the one above to be reversible, the piston must be weightless and frictionless; moreover, the volume change would have to be done very slowly, keeping the gas in equilibrium and not generating any organized motion (such as eddies or turbulence) that would take up energy. Any process that entails friction losses is irreversible. Consequently, real processes can, at best, only resemble being reversible, but even so, such behavior is of great practical interest.

Adiabatic Change

Imagine an *adiabatic* process occurring in a system; that is, one where no heat crosses the boundary. This can be accomplished, with varying degrees of success, by thermally insulating the system. Alternatively, processes that occur suddenly tend to be adiabatic because heat takes a fair amount of time to flow. Thus, sound waves in air are adiabatic since there isn't enough time for heat to flow from the compressions to the rarefactions (see p. 450). Adiabatic processes are important in many situations, not the least of which is the operation of the internal combustion engine.

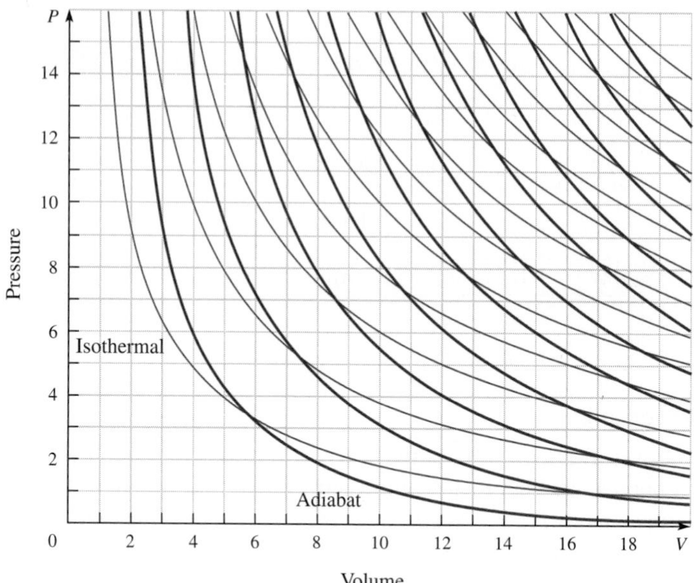

Figure 16.12 *PV*-diagram for an ideal gas; isothermals and adiabats.

Consider an ideal gas, sealed in a cylinder closed off with a piston, all of which is thermally insulated so no heat can enter or leave. The process about to be performed will be adiabatic, but it may or may not be reversible—that depends on how it is carried out. Let's stay with a reversible adiabatic process and so do everything nice and slowly to allow equilibrium to always obtain. Now, compressing the gas ($W < 0$), it follows that U must increase as the work done on the gas goes directly into internal energy ($\Delta U = -W$), and so the temperature increases. Accordingly, PV, which equals nRT, is no longer constant. The curve corresponding to this process in a *PV*-diagram is called an *adiabat* (Fig. 16.12) and is evidently everywhere steeper than an isotherm. That's the case because as the volume is decreased, the temperature rises, sustaining a higher pressure than would be if T were constant. On the other hand, as the gas expands adiabatically, it cools, sustaining a lower pressure at any value of V than would the isothermal process.

For an isothermal process, $PV =$ *constant* by way of Boyle's Law. For an adiabatic process, P drops more quickly with increasing V, and it is found experimentally that

[*reversible adiabatic*] $$PV^{\gamma} = \text{constant} \tag{16.6}$$

where γ is a constant that equals 1.67 for monatomic gases, $\approx$1.4 for diatomic gases, and $\approx$1.3 for polyatomic gases. It can be shown theoretically, though we shall not, that γ is actually the ratio of the specific heat of the gas at constant pressure to that at constant volume. Another way to write Eq. (16.6) is

$$P_i V_i^{\gamma} = P_f V_f^{\gamma} \tag{16.7}$$

and, since the gas is still ideal, the resulting temperature can be determined from

$$\frac{P_i V_i}{T_i} = \frac{P_f V_f}{T_f} \tag{16.8}$$

Example 16.6 Argon gas (which, like the other Noble Gases, is monatomic) is very slowly compressed adiabatically in a well-insulated cylinder to half its original volume of 0.100 m^3. If it was at atmospheric pressure and 27.0°C to start, what will be the final pressure and temperature of the gas?

Solution: [Given: an adiabatic process, $V_i = 0.100 \text{ m}^3$, $P_i = 1$ atm, $V_f = \frac{1}{2}V_i$, and $T_i = 27.0°\text{C}$. Find: P_f and T_f.] The process is adiabatic and involves P, V, and T, so we begin with Eq. (16.7). The gas is monatomic; hence, $\gamma = 1.67$ and

$$P_f = P_i\left(\frac{V_i}{V_f}\right)^\gamma = (0.101 \text{ MPa})(2)^{1.67}$$

and $$P_f = (0.101 \text{ MPa})(3.182) = \boxed{0.322 \text{ MPa}}$$

To find the temperature, use Eq. (16.8), and so

$$T_f = T_i\left(\frac{P_f}{P_i}\right)\left(\frac{V_f}{V_i}\right) = (300 \text{ K})(3.19)\left(\frac{1}{2}\right) = \boxed{479 \text{ K}}$$

▶ **Quick Check:** Using Eq. (16.7), $P_i V_i^\gamma = (0.1 \text{ MPa})(0.1 \text{ m}^3)^{1.67} = 2 \times 10^3$, whereas $P_f V_f^\gamma = (0.3 \text{ MPa})(0.05 \text{ m}^3)^{1.67} = 2 \times 10^3$.

A common instance of an adiabatic process is the rapid compression of air with a hand pump on the downstroke. You do (negative) work on the gas and $Q = 0$; it rises in temperature ($\Delta U = -W$), heating the pump quite noticeably. Stretch a fat rubber band, doing work on it in one quick adiabatic jerk, and its temperature will rise appreciably—you can easily sense the result with your lips.

An example of the reverse process, where the system does work, is the discharge of a carbon dioxide fire extinguisher. This happens so fast, it's again adiabatic, but now the temperature drops. Initially at room temperature, the CO_2 blasts out into a region of atmospheric pressure and does (positive) work on the surroundings. Since $\Delta U = -W$, the temperature of the gas drops, often to the point of forming dry ice crystals. Meanwhile, the canister is cooler, too, because the CO_2 remaining in it consists of the slower molecules (lower KE) left behind. The same sort of thing happens when you fill a spare tire from a can of compressed gas—the can gets cold. Hold a fat rubber band stretched for a while and then let it quickly contract, and it will cool appreciably. The elastic force does work as the band contracts that decreases the internal energy, and the temperature drops.

When the cap is removed from a beer bottle, gas in the neck expands rapidly and adiabatically. It does work on the air and also loses thermal energy, overcoming its own internal interactions. As a result, the temperature in the neck of the bottle can drop from ≈41°F to ≈−31°F. The sudden change causes water vapor in the neck to condense into a cloud of tiny droplets.

CYCLIC PROCESSES: ENGINES AND REFRIGERATORS

Let's suppose that we are working with reversible processes and, furthermore, once having carried out several of them, we wish to return the system to its original state. The PV-diagram would then represent a *cycle* and would appear as a closed figure that could be swept through without change, over and over again. The simplest program is to put an ideal gas in a cylinder closed by a piston (Fig. 16.3), surround it with a constant temperature bath, and expand it along an isothermal, as in Fig. 16.13a. An amount of heat Q_{AC} flows in as the system expands from A to C: $W_{AC} > 0$, and because $\Delta T = 0$, $\Delta U = 0$ for an ideal gas (p. 558); hence, $Q_{AC} > 0$. The work done by the gas is the area under the curve. If we return to point A via the same isothermal in reverse, as in Fig. 16.13b, work is done on the gas and heat flows out of it, where $W_{AC} = -W_{CA}$ and $Q_{AC} = -Q_{CA}$, and the system is energetically back to its starting point. Notice that if we add the negative work done on the gas during the compression leg to the positive work done during the expansion leg, the result is the net positive work done over the cycle, or in this case, zero. That result is represented by the *area enclosed by the closed curve*—zero.

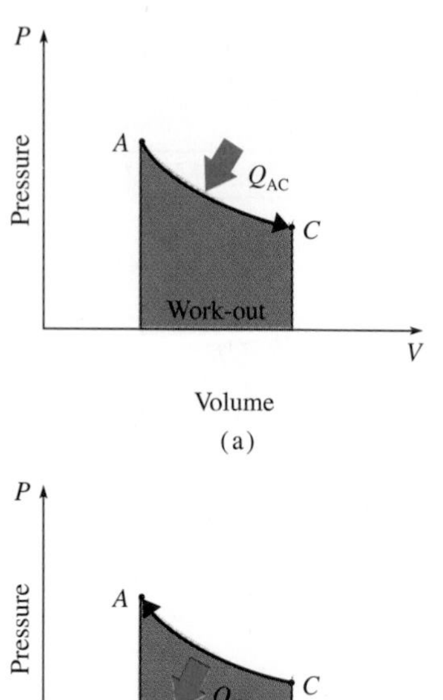

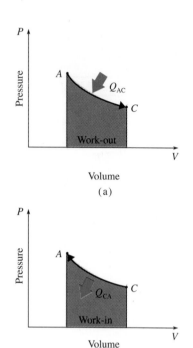

Figure 16.13 (a) The *PV*-diagram for an isothermal expansion during which a quantity of heat Q_{AC} enters the system as it goes from *A* to *C*.
(b) During the isothermal compression back to *A*, the same amount of heat Q_{CA} is ejected.

Figure 16.14 shows a cycle where the net work done by the gas (the area enclosed) is nonzero and positive. In going from *A* to *B*, work is done by the gas. Its temperature rises, so *U* increases and heat enters the system. In going from *B* to *C*, no work is done and the temperature drops, so *U* decreases and heat leaves the system. The work-out (that is, the net work done by the gas) equals the net heat-in, and $\Delta U = 0$.

This cycle suggests the operation of an engine. *An engine is a device that converts some form of energy (electrical, gravitational, chemical, thermal, etc.) into work.* In the broadest sense, both a cannon and a solid-fuel rocket motor are engines. But these are essentially one-shot affairs, and we will do better to limit the term *engine* to cyclic devices that can continue to operate for as long as we provide energy (fuel) to them. *A heat engine is a cyclic device that converts thermal energy into work-out.*

16.2 The Carnot Cycle

When Nicholas Léonard Sadi Carnot (pronounced car-no) published his masterwork *Reflections on the Motive Power of Heat* in 1824, he was twenty-eight years old. Carnot's father, Lazare, had been minister of war to Napoleon Bonaparte, and Sadi had fought in the outskirts of Paris in 1814. Recognizing that military power came with industrial mastery, Carnot turned his special ingenuity to bear on the great symbol of the age, the fire-chariot of the Industrial Revolution: the steam engine.

Carnot began his analysis with something more manageable; a simple reversible cycle that corresponds to a purely hypothetical device. Remarkably, the conclusions he drew were applicable, in one way or another, to all heat engines from the gas turbine to the power system that propels the family car. The Carnot engine (Fig. 16.15) is just a gas-filled cylinder and piston that can alternately be brought into contact with, or isolated from, either a high-temperature reservoir (steam) that serves as a source of heat or a low-temperature reservoir (cooling water) that serves as a sink into which heat is exhausted. The Carnot cycle is a four-stage reversible sequence (Fig. 16.16) consisting of: (1) an isothermal expansion (from *A* to *B*) with the reception of heat-in (Q_H) at a high temperature; followed by (2) an adiabatic expansion (from *B* to *C*); followed by (3) an isothermal compression (from *C* to *D*) with the rejection of heat-out ($-Q_L$) at a low temperature; followed by (4) an adiabatic compression (from *D* to *A*). Observe that Q_H and Q_L are traditionally positive so that the net flow of heat to the system is ($Q_H - Q_L$).

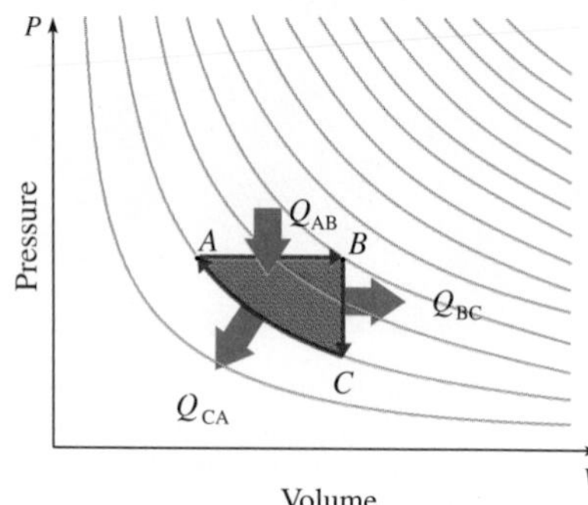

Figure 16.14 A cycle corresponding to a nonzero amount of work done by the system. The work done on the system during compression from *C* to *A* is negative.

Figure 16.17 shows the corresponding stages of the engine. The *A-B-C* portion, the expansion, is the *power stroke,* during which positive work is done. For the first part of it (*A-B*), the work is provided for by an input of heat. During the second part (*B-C*), the work is provided for by draining some of the internal energy of the gas—that's why its temperature drops. The piston may transmit work to the environment by way of a crankshaft. And it would be the inertia of the rotating crankshaft that drives the system back from *C* to *D* to *A*, doing work on the gas and returning it to its original condition.

There are several crucial points here: by using both a high-temperature source (steam) and a low-temperature sink (cooling water), the cycle rides out and back on isothermals but now encloses a nonzero area, unlike Fig. 16.13; less heat is exhausted into the sink than entered from the source (since $\Delta U = 0$, $Q_H - Q_L = W_o$). The lower the sink temperature, the more work would be done, and the less heat would be exhausted.

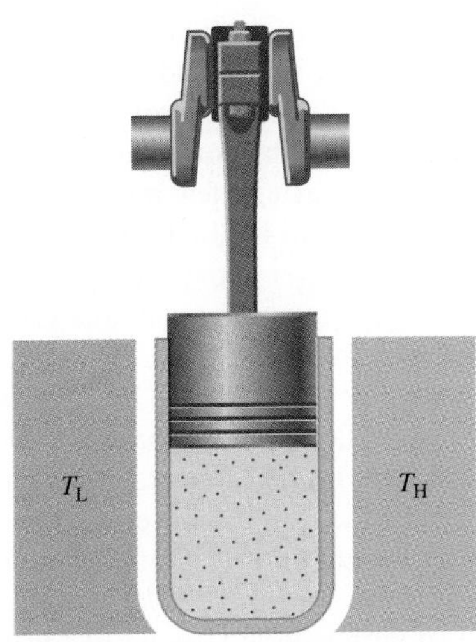

Figure 16.15 A Carnot engine consists of a cylinder and piston filled with gas. The device is able to be brought into contact with either a high-temperature or a low-temperature reservoir.

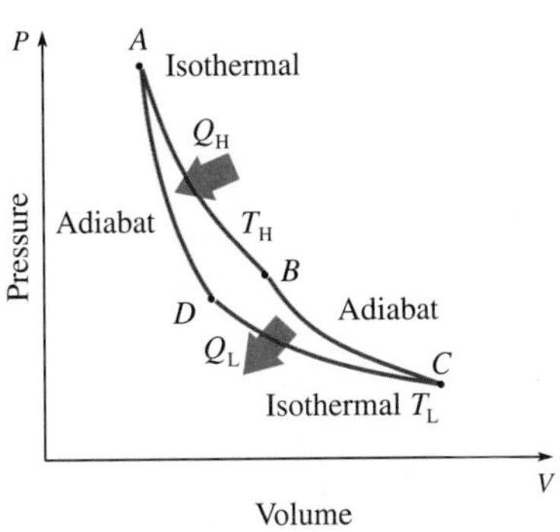

Figure 16.16 The four-stage reversible sequence representing a Carnot cycle.

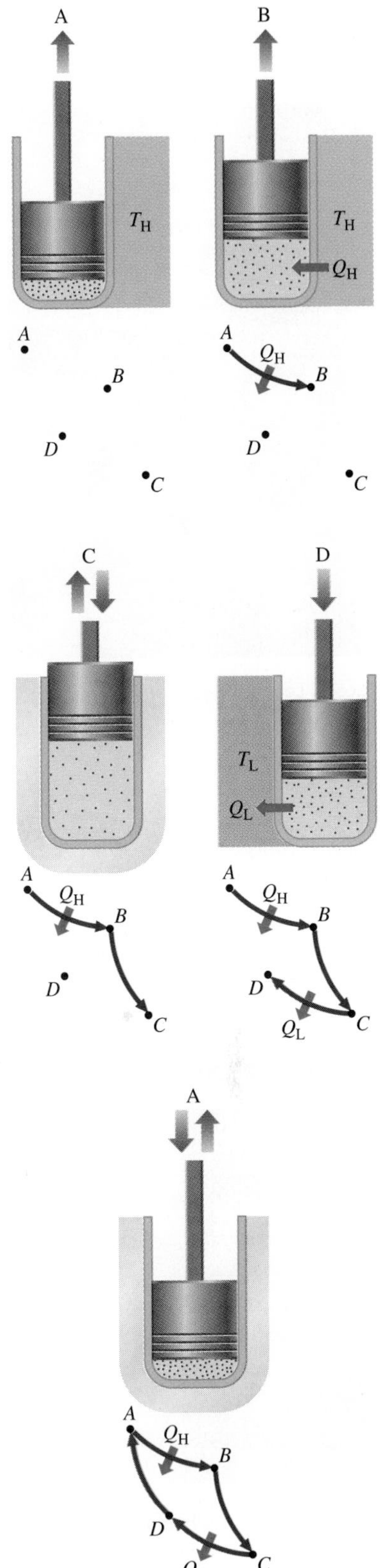

Figure 16.17 The various stages of the Carnot cycle.

Efficiency

The primary reason the Carnot engine is so important is that it represents a class of idealized devices (reversible engines) that can be shown to be the most efficient possible. Thus, it sets an upper limit on the attainable efficiency of any real heat engine, no matter how cleverly contrived, and that's a marvelous practical tool to have.

Most generally, the energy **efficiency** (e) of a process is defined as

$$\text{Efficiency} = \frac{\text{useful energy-out}}{\text{energy-in}} \tag{16.9}$$

which is the ratio of what you get out (that is of value) to what you put in (to accomplish the process). With heat engines, the useful energy output is work, and the energy input is heat; thus, over one cycle

$$\text{Efficiency} = \frac{\text{work-out}}{\text{heat-in}}$$

This expression is applicable to all heat engines, real or imagined. For an idealized cyclic heat engine (which is both frictionless and lossless), the First Law demands that $W_o = Q_H(\text{in}) - Q_L(\text{out})$, and so

$$e = \frac{W_o}{Q_i} = \frac{Q_H - Q_L}{Q_H} = 1 - \frac{Q_L}{Q_H} \tag{16.10}$$

where both Q_H and Q_L are positive numbers.

Figure 16.18 schematically depicts a heat engine operating between a source and sink at high and low temperatures, respectively. Note that the efficiency increases as

Nicolas Léonard Sadi Carnot (1796–1832).

Q_L decreases, becoming 1 (that is, 100%) only when there is no exhaust heat, a condition that has never been achieved (and one that is expressly forbidden by the Second Law of Thermodynamics, p. 570). Actual engines dissipate energy via friction and lose appreciable amounts of energy to the environment by convection, conduction, and radiation. That's why the hood of a car gets hot after a few minutes of operation, and the hotter it gets, the more wasted energy there is (Table 16.1).

The form of Eq. (16.10) in terms of heat is really not very convenient; it would be far simpler to measure if it were expressed in temperatures instead, and Carnot did just that. "According to established principles at the present time," he wrote, "we can compare with sufficient accuracy the motive power of heat to that of a waterfall." The idea was simple enough: consider a waterfall and an appropriate ideal harness-the-rushing-water-machine located somewhere in the cataract, as shown in Fig. 16.19. The maximum amount of energy available that can be converted to work by the machine is ΔPE, which is proportional to the height of the descent $(h_H - h_L)$. Thus, the fractional amount of energy available from the mill is $(h_H - h_L)/h_H$. Carnot wrongly believed heat was a liquid, and so with caloric flowing like water, he argued by analogy that the theoretical efficiency of an ideal heat engine should be $(T_H - T_L)/T_H$. Notice how the discussion has built into it a zero-temperature reference level—the denominator is the maximum change in temperature $(T_H - 0)$. Heat is naturally propelled down a temperature gradient just as water is naturally propelled down a gravity gradient. Moreover, $Q = cm\,\Delta T$ and $W = gm\,\Delta h$. Delightfully, even though there is no such heat fluid, Carnot's answer is correct. Before he died at age thirty-six in the cholera epidemic that swept through Paris in 1832, Carnot (his unpublished papers reveal) had already rejected caloric and even anticipated the First Law.

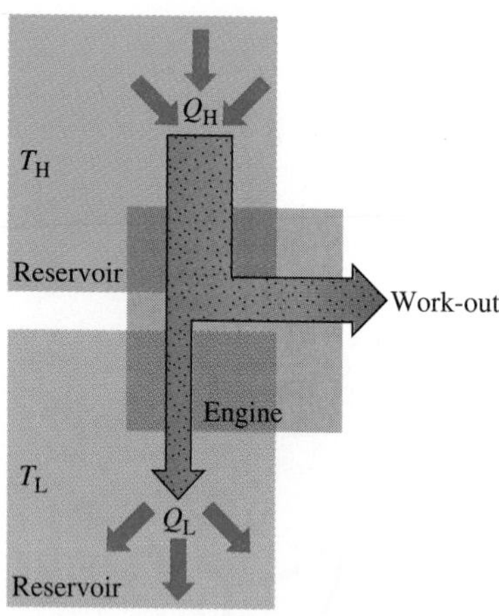

Figure 16.18 A schematic representation of a heat engine operating between a high- and a low-temperature reservoir. Q_H flows in, Q_L flows out, and the engine does work. Energy-in equals energy-out.

TABLE 16.1 Approximate efficiencies of various devices

Device	Efficiency (%)
Electric generator	70–99
Electric motor	50–93
Dry cell battery	90
Domestic gas furnace	70–85
Storage battery	72
Hydrogen-oxygen fuel cell	60
Liquid fuel rocket	47
Steam turbine	35–46
Fossil-fuel power plant	30–40
Nuclear power plant	30–35
Nuclear reactor	39
Aircraft gas turbine engine	36
Solid-state laser	30
Internal combustion gasoline engine	20–30
Gallium arsenide solar cells	>20
Fluorescent lamp	20
Silicon solar cell	12–16
Steam locomotive	8
Incandescent lamp	5
Watt's steam engine	1

The preceding temperature statement of the efficiency of a Carnot engine (e_c) can be established more rigorously, but we would gain little from the effort. Let's instead focus on understanding the concept, represented by

$$e_c = 1 - \frac{T_L}{T_H} = \frac{T_H - T_L}{T_H} \qquad (16.11)$$

Keep in mind that this expression is independent of the engine's working fluid—it could be a hot gas in an internal combustion engine or steam in a steam turbine. Remember, too, that **these are absolute temperatures**.

As we shall see presently, the **Carnot engine has the highest allowable efficiency for a heat engine operating between any two fixed temperatures**. Nonetheless, that efficiency can only equal 1 (that is, 100%) when the cold-sink temperature is 0 K. In other words, *even the Carnot engine operates at less than 100% efficiency*. One way to ensure that the limiting value of Eq. (16.11) is as large as possible is to make T_L as low as possible, but there are practical restraints on that, too. In most real situations, the cold-sink is the open air or cool water from a river, so there isn't very much latitude on the low-temperature end. Of course, you could cool the low-sink with some sort of refrigerator, but that will dissipate additional amounts of energy and negatively affect the overall efficiency of the system. On the other hand, increasing T_H to increase e_c is far more practical. That's why power plants operate with superheated steam and why modern jet engines and auto engines alike are being designed to run at increasingly higher temperatures.

Figure 16.19 A waterfall analogy for the heat engine.

Example 16.7 Determine the maximum possible efficiency for a steam engine operating between 200°C and 27.0°C.

Solution: [Given: $T_H \rightarrow 200°C$ and $T_L \rightarrow 27.0°C$. Find: e_c.] We need only substitute into Eq. (16.11), being sure to use kelvins. Hence

$$e_c = 1 - \frac{300\ \text{K}}{473\ \text{K}} = 1 - 0.634 = 0.365 = \boxed{36.5\%}$$

In practice, losses will cut that down by about one-third, but this efficiency is the very best we can hope for operating between those two temperatures.

▶ **Quick Check:** (473 K − 300 K)/(473 K) = 37%.

A modern fossil-fuel electric-generating power plant uses superheated steam to provide a high-temperature source at perhaps 500°C (≈800 K). High-pressure steam violently expands into the turbine, impacting upon and driving the spinning blades (Fig. 16.20). The steam is expelled into a cooled condenser at ≈373 K. The maximum theoretical efficiency follows from Eq. (16.11) and is 53%, although thermal losses generally reduce that to about 40% in modern facilities (much of the untransformed energy goes up the smokestack).

The automobile engine functions at approximately 3300 K, exhausting its spent fuel at about 1400 K. That gives it a theoretical efficiency of roughly 58%. Even though the exhaust gas escapes into the surrounding cool air, it leaves the engine, where it can do work, at a much higher temperature T_L. In reality, most of the fuel energy fed into the system is lost by way of thermal energy escaping to the air around the engine and out with the still hot exhaust gas.

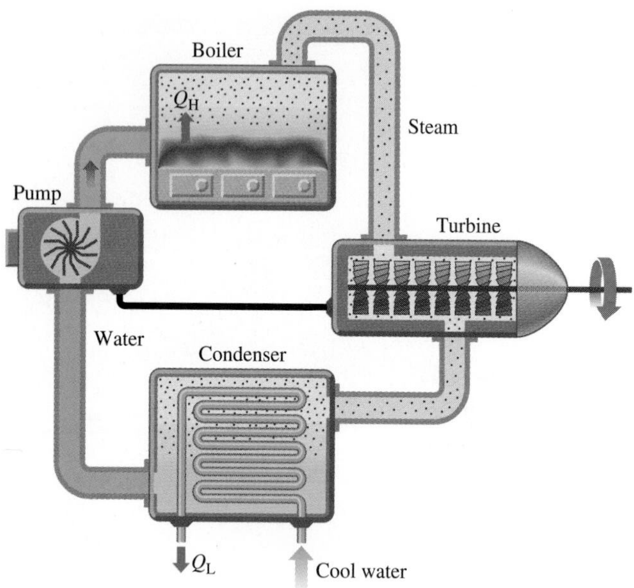

Figure 16.20 A modern fossil-fuel electric-generating power plant. Steam drives the turbine, transferring energy to it. The turbine powers a high-voltage electric generator. A large pressure difference is maintained across the turbine by condensing the steam. The steam turns back to water in the condenser (the cold-sink). Water of extreme high purity is recirculated in the sealed system.

16.3 Internal Combustion

Carnot discussed the possibilities of powering an engine by igniting an inflammable gas within a cylinder, but it was J. Lenoir (1859) who produced the first workable internal combustion engine. A traveling salesman and mechanic named Nikolaus Otto happened upon a newspaper account of Lenoir's work (1876) and within a year built and patented the first efficient four-stroke internal combustion engine. The automobile, airplane, and countless other noisy paraphernalia from the power lawnmower to the motorcycle and chainsaw all owe their existence to Otto.

Figure 16.21a shows the sequence of events occurring in each cylinder of the four-cylinder engine depicted in Fig. 16.21b. Each piston is connected by a pivoted arm to the rotating crankshaft. Starting at A, a mixture of air and gasoline (blended in a carburetor) at atmospheric pressure is drawn into a cylinder through the intake valve as the piston moves down to B. This is the *intake stroke*. The piston now rapidly (and more or less adiabatically) compresses the mix to point C with both valves closed in the *compression stroke.* The spark fires at point C, igniting the fuel, and raising the temperature and pressure up to D, while the volume stays almost constant. The piston then rushes down on what is supposed to be the adiabatic *power stroke,* carrying the system to E. The exhaust valve now opens, and the temperature and pressure drop as the volume stays constant and the system moves to F, having vented much of the waste gas. Mechanical energy stored in a rotating flywheel pushes the piston back to A, forcing the remaining waste products out of the exhaust valve and into the once fresh air. This stage is the *exhaust stroke.* That valve closes, the intake valve opens, and the cycle begins again.

16.4 Refrigeration Machines and Heat Pumps

A heat engine uses thermal energy flowing from a high-temperature reservoir to a low-temperature reservoir to produce work. A *refrigeration machine* does the reverse—it uses work to transfer thermal energy from a low-temperature reservoir to a high-temperature reservoir. When the region to be cooled is a food chest, the device, an ordinary household refrigerator, simply expels the extracted heat into the room. That's fine in the winter, but it's admittedly a little dumb in the summer. On the other hand, to cool the room itself, the refrigeration machine, known as an air conditioner, simply dumps the hot exhaust outside. If we turn the air conditioner around and cool the outdoors, regardless of its temperature, blowing the exhaust energy into the house, we have a heating system: such a reversible device is called a *heat pump.*

Figure 16.22 schematically depicts a continuously operating refrigeration machine, which is simply an engine run in reverse. If it were a Carnot refrigerator, the *PV*-diagram would be identical to Fig. 16.16, but run through counterclockwise. From the First Law it follows that (heat-out) = (heat-in) + (work-in)

or

$$Q_H(\text{out}) = Q_L(\text{in}) + W_i \tag{16.12}$$

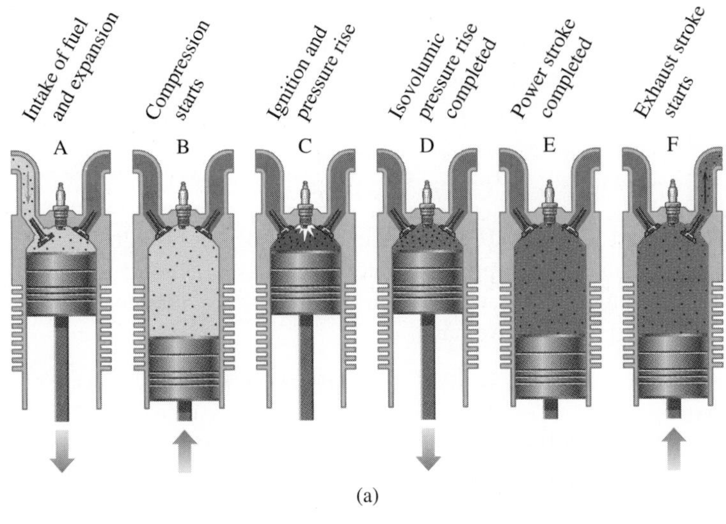

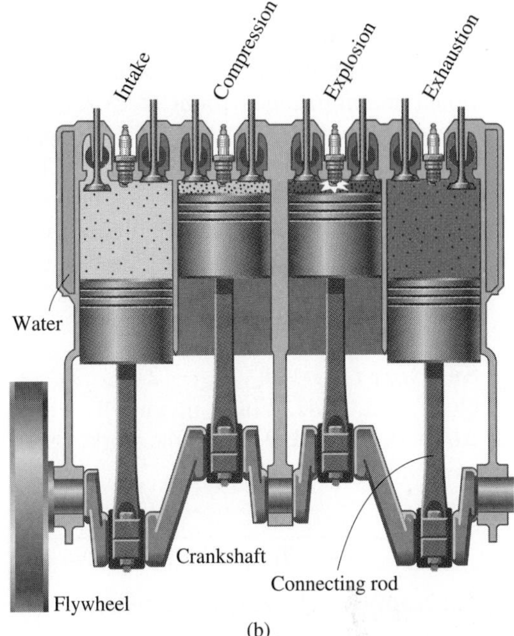

Figure 16.21 The internal compression engine. (a) The complete sequence of operations. (b) A four-cylinder engine with each cylinder undergoing a different part of the cycle.

The effectiveness of a refrigeration machine cannot be measured by way of the efficiency, Eq. (16.10), because there is no work-out. Hence, we introduce the *coefficient of performance* (η), which is defined as the ratio of the amount of heat removed from the cold reservoir (the heat-in to the system) to the amount of work done on the system, represented symbolically as

$$\eta = \frac{Q_L}{W_i} = \frac{Q_L}{Q_H - Q_L} \tag{16.13}$$

The larger is η, the more effective is the refrigeration machine, and values of about 5 are typical.

The best possible coefficient will correspond to a Carnot engine running in reverse. Accordingly, set Eq. (16.10) equal to Eq. (16.11), whereupon

$$e_c = 1 - \frac{Q_L}{Q_H} = 1 - \frac{T_L}{T_H}$$

and so

$$\frac{Q_L}{Q_H} = \frac{T_L}{T_H} \tag{16.14}$$

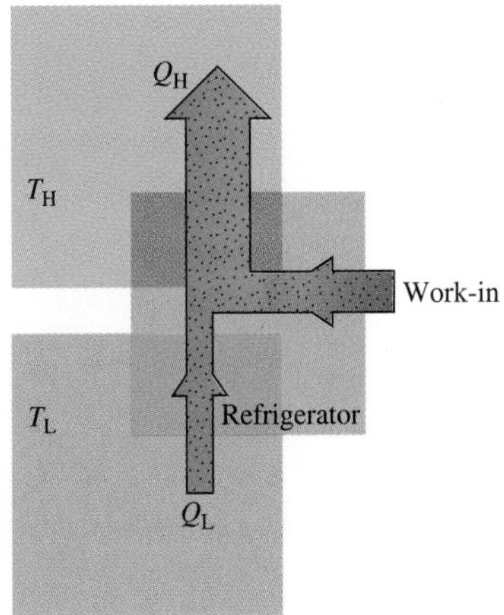

Figure 16.22 A schematic representation of a refrigeration machine. It takes heat Q_L from a low-temperature region and dumps it in a high-temperature region, at the cost of a certain amount of work-in.

The *coefficient of performance for an ideal system* in terms of the operating temperatures can now be gotten from Eq. (16.13) using Eq. (16.14); accordingly

$$\eta_c = \frac{T_L}{T_H - T_L} \tag{16.15}$$

Example 16.8 Determine the best possible coefficient of performance that can be gotten for an air conditioner maintaining a room at 21°C (70°F) when the outside temperature is 36°C (97°F). Suppose 5.0 MJ of heat leaks into the room in an hour. The machine exhausts the waste heat it generates in operating, outside the room via a cooling fan. How much work would that air conditioner have to do per hour to maintain the temperature of the room? How much heat in total is exhausted outside per hour?

Solution: [Given: $T_L \to 21°C$, $T_H \to 36°C$, and $Q(\text{in}) = 5.0$ MJ. Find: η_c and W_i.] Remember that $\Delta U = 0$, and so ΔQ equals the work. Using Eq. (16.15), we obtain

$$\eta_c = \frac{T_L}{T_H - T_L} = \frac{294\text{ K}}{(309\text{ K}) - (294\text{ K})} = 19.6$$

The hourly heat leak, $Q(\text{in})$, enters the system at the low temperature and equals Q_L. The work that must be done each hour to remove Q_L joules of thermal energy follows from Eq. (16.13); thus

$$W_i = \frac{Q_L}{\eta_c} = \frac{5.0\text{ MJ}}{19.6} = \boxed{0.26\text{ MJ}}$$

To remove 5.0 MJ, this machine does only 0.26 MJ of work. In the process, that 0.26 MJ is converted to thermal energy, which is also exhausted outside. From Eq. (16.12), we have

$$Q_H = Q_L + W_i = 5.0\text{ MJ} + 0.26\text{ MJ} = \boxed{5.3\text{ MJ}}$$

This is the net heat exhausted to the outside.

▶ **Quick Check:** For every 19.6 J it removes, the machine does 1 J of work and produces 1 J of waste thermal energy. When it does 0.26 MJ of work, it removes 5 MJ of thermal energy from the room and produces an additional 0.26 MJ itself.

Many refrigeration machines used in the home operate on what's called the vapor compression cycle (Fig. 16.23). A liquid under high pressure ($\approx$7.4 atm) passes through a very small aperture, a needle valve, into a region of lower pressure ($\approx$1.4 atm). The restriction drastically slows the speeds of the molecules and what emerges from the valve is a wet low-temperature (≈ -8°C) vapor (see Fig. 14.14). A typical *refrigerant* is an easily vaporized liquid, one with a low boiling point at atmospheric pressure, such as ammonia (-33°C) or freon (-30°C). The partially vaporized refrigerant next enters the evaporator (usually a set of coils in the freezer), where it absorbs heat from the surroundings (the food) as it isothermally and isobarically vaporizes. The low-pressure vapor (at $\approx$16°C) is then compressed roughly adiabatically via a cylinder-and-piston and emerges as a hot high-pressure vapor (Fig. 16.24). A fan blows away enough thermal energy so that the vapor condenses back to a high-pressure liquid. This is the source of the warm breeze that you may have noticed coming from the bottom or back of a refrigerator. The warm high-pressure liquid freon ($\approx$38°C) again encounters the needle valve and the process begins again. Amusingly, early electric refrigerators cleverly had their compressors mounted on top of the machine, where the hot air could rise up and away. For aesthetic reasons, the compressor was moved to the bottom, which is precisely where you would not want a constant supply of heat.

16.5 Absolute Temperature

A Carnot engine takes in a certain amount of heat (Q_H) at a high temperature (T_H) and exhausts a lesser amount (Q_L) at a lower temperature (T_L), the energy difference going into work-out. The amount of heat exhausted, and therefore essentially wasted, is directly proportional to T_L: the lower the sink temperature, the less the value of Q_L, the more work is done, and the higher the efficiency. That much has

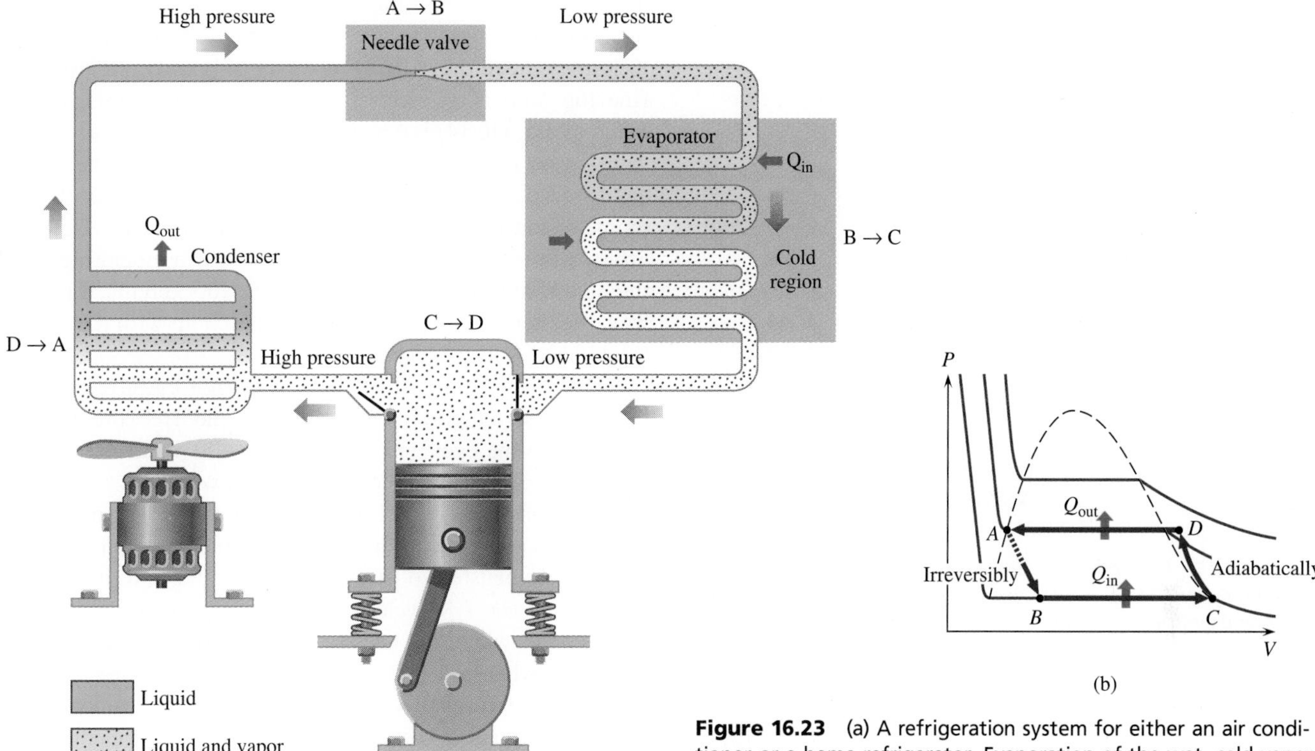

Figure 16.23 (a) A refrigeration system for either an air conditioner or a home refrigerator. Evaporation of the wet, cold vapor in the coils in the evaporator draws heat from the material to be cooled. (b) The *PV*-diagram shows the irreversible expansion through the needle valve.

been considered already:

$$\frac{Q_{\rm L}}{Q_{\rm H}} = \frac{T_{\rm L}}{T_{\rm H}} \qquad [16.14]$$

Kelvin proposed that this ratio be used to define an absolute thermodynamic temperature scale (p. 484) that utilized the Carnot engine as a thermometer. This would

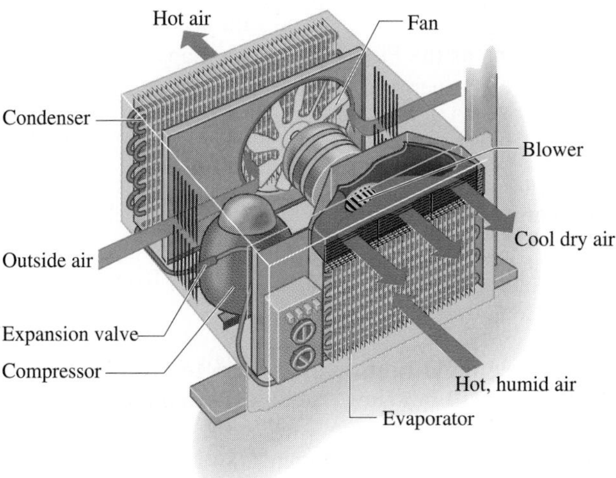

Figure 16.24 A window-mounted air conditioner. Hot, humid room air is drawn over the cold evaporation coils, where it is cooled. Water vapor condenses, runs off, and collects in the base pan, dripping away outside.

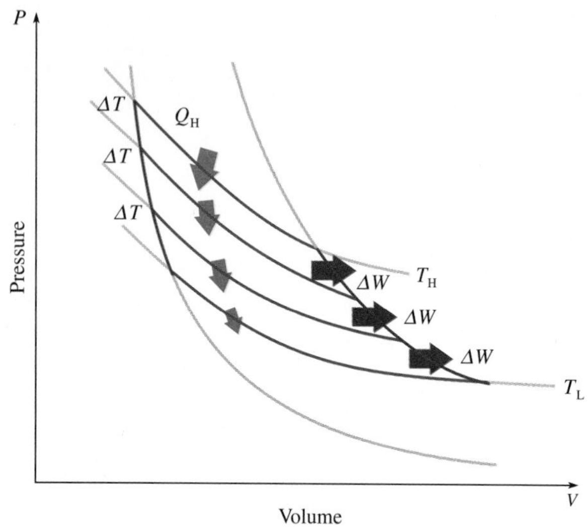

Figure 16.25 A Carnot cycle divided into a number of little Carnot cycles. Each one expels less heat than the one before.

have the virtue of eliminating the dependence on the working substance that plagues all ordinary thermometers. Kelvin assumed the validity of Eq. (16.14) and used it to define the ratio of two thermodynamic temperatures. Inasmuch as Eq. (16.14) is known to be satisfied when an ideal gas is used in a gas thermometer (since we used the ideal gas to derive the equation in the first place), the gas thermometer must agree with the Kelvin thermometer. We can arbitrarily set the difference between the temperatures of the steam and ice points of water as 100 kelvins. In that way, the thermodynamic scale will match up with the long-established Celsius scale.

Suppose, then, that we have a Carnot cycle operating between the temperatures of boiling and freezing water. Now divide the cycle (Fig. 16.25) into 100 equal-area little Carnot subcycles, each doing the same work ΔW and each residing between isotherms 1 K apart. As we progress down from one cycle to the next, more and more net work is being done, and less and less heat exhausted to the next little engine. It follows that if we continue to add more of these little 1-K cycles below the ice-point temperature, sooner or later we will reach a point where the heat-in to one of them all goes into the work-out (ΔW), and no heat is exhausted to the last low-temperature sink, which must be the Absolute Zero of temperature: 0 K. At that point the total amount of heat-in, Q_H, is transformed entirely into work-out, $\Sigma\,\Delta W$. Presumably there could be no temperature lower than 0 K, since that would imply an efficiency greater than 1, via Eq. (16.11), and more work-out than heat-in, which would violate the First Law.

As we will see later, the notion that all the heat flowing into the last little engine is converted to work violates the Second Law of Thermodynamics, and that suggests that it may be impossible to actually bring bulk matter to a temperature of Absolute Zero.

THE SECOND LAW OF THERMODYNAMICS

The Second Law of Thermodynamics was introduced into physics by way of a rather mundane statement about heat flow. But it evolved into a principle of great scope and significance, as far-reaching as the First Law, and perhaps even more revealing. Things happen via change, and energy is the measure of that change. Everything is tied together by a web of interactions, and change tends to spread out—one entity influencing another—and so energy tends to disperse. Strike a match and the smoke, sound, thermal energy, and light will all disperse. The Second Law is about the dispersal of energy, the unfolding of events. Time itself is given direction by the Second Law.

The germ of the idea was already there when scientists realized from countless observations that:

Heat flows naturally from a region at high temperature to a region at low temperature. By itself, heat will not flow from a cold to a hot body.

And this is one of several equivalent but different statements of the Second Law.

Once the concept of internal energy was formulated, this observation was enriched with the insight that the spontaneous flow of heat from one object to another is independent of the amount of internal energy in either entity. Heat is propelled by a temperature gradient—no gradient, no heat. From Carnot's work, we see that a cyclic heat engine must operate between high- and low-temperature reservoirs. If both temperatures are the same, its efficiency is zero and no heat enters the engine, across which there must be a gradient.

Furthermore, with any low-sink temperature other than zero, even for an ideal engine, there will be some unavoidable loss of heat Q_L to the reservoir: the efficiency must be less than 100%. That much follows from Eq. (16.14), if $T_L \neq 0$, $Q_L \neq 0$. There is an energy equivalence between heat and work, but there is also a *dissymmetry,* a difference on the microscopic level, that has its roots in the difference between order and disorder. We can convert all the ordered energy we wish *entirely* into thermal energy—the brakes on your car do it and Rumford's cannon borer (p. 516) did it. Nature allows the continued spontaneous transformation of work entirely into thermal energy, but it does not allow the reverse. The transformation of disordered heat into ordered work must be paid for, as it were, by an additional loss of heat Q_L (Fig. 16.17). The formal wording of the Second Law in terms of the above ideas is known as the *Kelvin statement:*

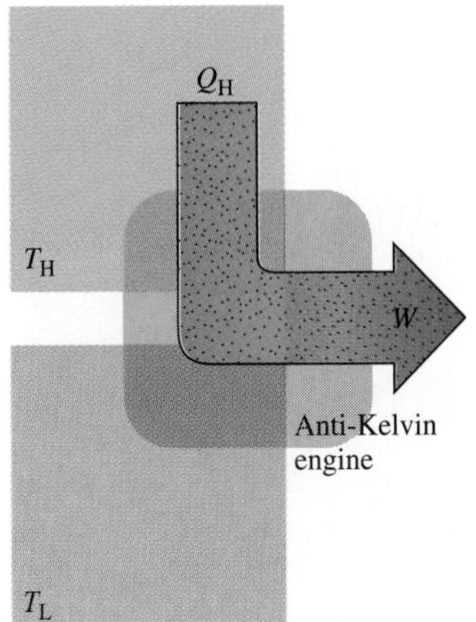

Figure 16.26 A schematic representation of an impossible anti-Kelvin device. It converts heat-in to work with no wasted heat-out.

No process is possible for which the sole result is the removal of heat from a source and its complete transformation into work.

In other words, *it is quite impossible to produce a cyclic engine that will generate work by extracting heat from a reservoir without expelling some waste heat to a lower-temperature sink* (Fig. 16.26). An anti-Kelvin device, which could continuously do work extracting thermal energy from a single reservoir with no concern about temperature gradients, is a so-called **perpetual motion machine of the second kind**. Although such a device would not violate the First Law, it would violate the Second. It could not create energy or, for that matter, run forever, but it could do the next best thing—namely, draw off energy at no cost from the vast low-temperature sources of atmosphere and ocean. No such machine capable of solely producing ordered work out of disordered heat is possible.

It is not impossible to convert thermal energy directly into work. What is impossible is to do it with no other change occurring: A pistol can transfer the thermal energy of the expanding gases into the KE of the bullet, but it will also blast out those gases into the air. The work done on the bullet will not be the sole result. Understand that that "sole result" stipulation means that the system is otherwise unchanged; the system is back in the state it was in originally, and the process can be repeated and run continuously.

16.6 The Clausius Formulation

The next major figure to appear on the thermodynamic scene was a young man in Germany, a member of the Gottlieb family. Somewhere along the way that surname received a classical rebirth, and his father was called Gottlieb Clausius. Rudolf Julius Emmanuel Clausius was a contemporary of Joule and Kelvin and, like them, was caught up in the ferment that surrounded the creation of Thermodynamics in the nineteenth century. Inspired by Carnot, Clausius put aside caloric, speculated about the relationship of heat and motion in the microworld, and recognized that

Rudolf Clausius (1822–1888).

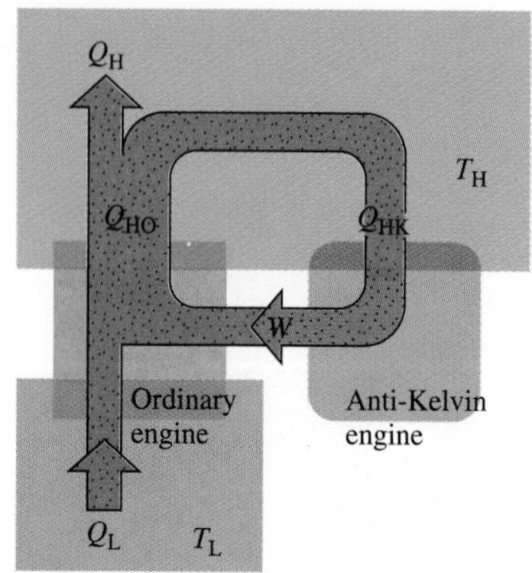

Figure 16.27 If an anti-Kelvin engine could exist, it could be attached to an ordinary engine, and the combination would violate the Clausius statement of the Second Law. It could take in Q_L and dump it as Q_H and do nothing else.

Figure 16.28 If an anti-Clausius engine could exist, it could be attached to an ordinary engine, and the combination could violate the Kelvin statement of the Second Law. It could remove heat Q_H from a source, converting it entirely into work.

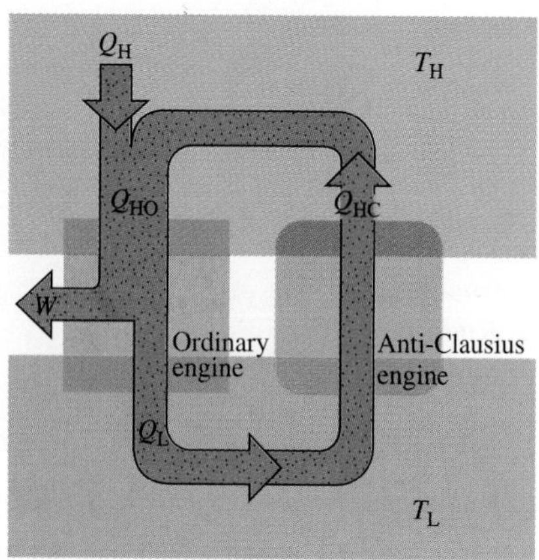

there were actually *two* energy principles that spoke, in turn, to the quantity and quality of energy.

It's commonplace for hot objects to cool spontaneously, all by themselves, losing heat to a colder environment, but we presumably never observe cold objects spontaneously getting colder by losing heat to a warmer environment. And that's true even when heat flows from a warm body with a tiny amount of internal energy (a cup of tea) into a cold body with an immense amount of internal energy (the atmosphere). In the same way, an object can start at the top of a hill and spontaneously roll down it, but never does one begin at the bottom and roll uphill unaided by any external agent. There is a specific way in which all processes (not just those involving heat) naturally unfold, and the Second Law, at its heart, is about just that: the natural direction in which energy flows as it irreversibly redistributes itself.

Clausius proposed an alternative statement of the Second Law that at first glance seems quite different from Kelvin's:

> **No process is possible in which the sole result is the transfer of a given amount of thermal energy from a body at low temperature to a body at a higher temperature.**

Refrigerators are not forbidden, provided they produce some other result in addition to the transfer of thermal energy—like the consumption of electrical power. The above statement may imply that an input of work is necessary to transport heat "uphill," but it doesn't say so explicitly. The Kelvin statement talks about the conversion of heat to work. The Clausius statement talks about the natural direction of the flow of heat. Both suggest a hidden asymmetry in the manner in which processes spontaneously unfold.

To prove that these two perspectives are in fact equivalent, refer to Fig. 16.27, which depicts an ordinary reversible engine being driven in reverse by a hypothetical anti-Kelvin engine (one that violates Kelvin's statement). The latter converts an amount of heat (Q_{HK}) directly into work, which drives the ordinary engine as a refrigerator. It, in turn, exhausts heat (Q_{HO}), some of which can be imagined circulating back to the anti-Kelvin device. The net effect is that the combined system has the sole result of taking heat from the low-temperature reservoir and dumping it into the high-temperature reservoir in direct conflict with the Clausius statement. Thus, if an anti-Kelvin device were possible, it could be used to make an anti-Clausius device. Similarly, the ordinary engine on the left in Fig. 16.28 is effectively being supplied heat by a hypothetical anti-Clausius engine that carries heat from the low to the high reservoir. The ordinary engine draws off heat (Q_{HO}), expels heat (Q_L), and performs work. The net effect of the combined system is solely to convert an amount of heat Q_H into work in direct conflict with the Kelvin statement. Thus, if an anti-Clausius engine were possible, it could be used to make an anti-Kelvin engine. We must conclude that if either statement is false, the other is false—they might be wrong, but they must be equivalent.

16.7 Entropy

In 1865 Clausius introduced a new concept, *entropy*, in order to distinguish between the ideas of *conservation* and *reversibility*. When a match burns, energy is conserved, but that's not the whole story of the change. The match will never spontaneously reform itself from the ash and smoke—the process is irreversible and the energy naturally disperses. This is very unlike the idealized occurrences of (lossless) Newtonian dynamics that are reversible, and conservation tells us all we need to know. Dynamics is independent of the direction of time: a lossless bouncing ball bounces forever and is independent of whether time runs forward or backward—the ball falls, climbs, and falls again or climbs, falls, and climbs again, either way. By contrast, heat conduction and friction are irreversible changes; indeed, without irreversible processes, life would be impossible.

When an engine operates, converting heat-in into work-out, some fraction of the input energy is irreversibly lost as heat-out. In effect, the engine takes in heat and puts out two distinct channels of energy: highly useful ordered energy (work) and not so useful disordered energy (heat). The expelled waste thermal energy is at a low temperature and will thereafter be quite difficult to extract and use again. To be sure, the more efficient the engine, the lower the sink temperature is likely to be, and the harder it will be to get at that exhausted thermal energy in the future. It's like water pouring over the falls in Fig. 16.19; once it has descended down to sea level, it is a lot less useful. In a real sense, the *quality of thermal energy (that is, its ability to do work on a macroscopic scale*) is dependent on its temperature.* When an engine operates, it does not use up energy—energy is conserved—it transforms energy, and it does that at the cost of degrading a certain amount of high-quality, high-temperature energy into low-quality, low-temperature energy. Energy does not vanish, it is dissipated. The family car is an example of how highly ordered available chemical energy (in the fuel) is converted into highly disordered, low-temperature unavailable thermal energy. Drive for 20 miles and you will convert a gallon of gasoline into a ribbon of heated air. Then try to drive back by reclaiming those 1.3×10^8 (now essentially useless) joules that are still there warming the sky.

Suppose there is an ideal gas in a chamber and it's released through a valve into an insulated evacuated vessel. The gas spontaneously expands, filling the vessel, but there is no change in internal energy because no work is done. The pressure and volume both change; the temperature remains constant. The system in its initial state transforms, all by itself, to a new final state where the energy is the same. And yet, if we want to return the gas to the original chamber, we will have to do work on it—the gas will not spontaneously squeeze back through the valve. Once the valve is opened, the gas is no longer in equilibrium, and the spontaneous rush toward equilibrium is irreversible: there is no infinitesimal change in any state variable (P, V, T) that will reverse the process. The system is directed by a hidden imperative (other than maintaining the total energy) to seek a new dispersed state from which it will not return on its own.

In the process of spontaneously expanding, the gas loses some of its ability to do work, though its energy is unchanged. The amount of work it takes to recompress the gas would seem to be equivalent to the ability to do work it lost. In any event, the final state of the gas is profoundly different from the initial state. What we want now is a new property of the system that will reflect that difference; a property that will allow us to know if spontaneous change will take place between any two states, a property that governs the direction of natural change in all systems from galaxies to embryos. Clausius proposed that there was such a measure, and he gave it the symbol S and the name **entropy**.

*The familiar definition of energy as "the ability to do work" (and that presumably means work on a macroscopic scale, an organized phenomenon) is inconsistent with the idea that energy is both conserved (First Law) and degraded (Second Law).

On Entropy. *I prefer going to the ancient languages for the names of important scientific quantities, so that they may mean the same thing in all living tongues. I propose, accordingly, to call S the* ***entropy*** *of a body, after the Greek word "transformation." I have designedly coined the word* ***entropy*** *to be similar to energy, for these two quantities are so analogous in their physical significance, that an analogy of denominations seems to me helpful.*

RUDOLF CLAUSIUS

Experience suggests that the entropy of a system be defined such that it increases as the system degrades. In other words, natural changes (that is, spontaneous ones) carry an isolated system toward disorder and are accompanied by an increase in entropy. Unnatural changes, which must be forced on a nonisolated system by an outside agency, can bring it more order and decrease *its* entropy. **Reversible changes are entropy neutral**, they leave S for an isolated system unaltered. Since entropy has something to do with degradation and disorder, it's not surprising that a reversible process carried out by an isolated system (which must be able to return to exactly the state from which it started) should have $\Delta S = 0$. The entropy equivalent of the Second Law, which is often called the *Entropy Principle,* is as follows:

When an isolated system undergoes a change, passing from one state to another, it will do so in such a way that its entropy will increase, or at best remain the same.

This hidden imperative gives direction to the natural unfolding of events. Remember that entropy describes the state of a system. It is a property, like internal energy and temperature, of engines and ice cubes and cats.

Whenever a process occurs in an isolated system, its entropy increases, or at best ideally remains the same. If the system is not isolated, then its entropy can be made to decrease by impressing order upon it from outside via external agencies. In that case (according to the Entropy Principle), the entropy of the surroundings will have to increase even more, so that the entropy of the ultimate isolated system, the entire Universe, nonetheless increases.

Every process increases (or at best ideally leaves unchanged) the entropy of the Universe.

The essential measure of the difference between reversible and irreversible processes is the change in entropy of the Universe. The essential measure of the difference between what occurs naturally and what does not occur naturally is the change in entropy of the Universe.

The Second Law and Entropy

We can get some insight into the behavior of entropy by using the Second Law as a guide. The Kelvin statement of the Second Law is subsumed into the entropy version provided that *the entropy of a system increases when it receives heat, decreases when it loses heat, and remains unchanged by work done in the absence of friction.* In that context, work is an organized mechanical change in energy and, as such, does not affect the microscopic disorder of a system. (By contrast, any mechanism—such as friction—that disperses energy into the agitation of the system's particles will increase entropy.) Consider an isolated system composed of a reservoir of thermal energy, an anti-Kelvin engine, and its immediate surroundings. Running the engine off the reservoir will have only one effect; the entropy of the reservoir will decrease, and therefore the engine is forbidden.

Similarly, the Clausius statement will be subsumed if we assume that *the higher the temperature of the heat transferred, the less entropy change results.* Imagine an isolated system consisting of two objects, one hot, the other cold, each with the same internal energy. Bring them into contact—heat spontaneously flows out of the hot one into the cold one and they both reach a new equilibrium temperature. The same amount of heat is lost by one as gained by the other. But the high-temperature heat is "high quality" and only slightly decreases the entropy of the high-temperature object. Whereas the heat delivered at a low temperature is of "low quality" and increases the entropy of that body appreciably—the net effect of this spontaneous transfer of heat is properly an increase in the entropy of the system and, hence, the Universe.

Thus, we know that entropy is directly proportional to the heat supplied and inversely proportional to the temperature at which it's supplied. Moreover, for a Carnot cycle, it follows from Eq. (16.14) that

$$\frac{Q_L}{T_L} - \frac{Q_H}{T_H} = 0 \tag{16.16}$$

We interpret this expression to represent the zero net change in entropy experienced by a Carnot engine over its reversible cycle. Entropy "flows" in from the high-temperature source, "flows" out to the low-temperature sink, and the engine returns to its original state. Accordingly, Clausius defined the change in the entropy of a system (ΔS) experienced during a *reversible* process as the ratio of the heat-in to the absolute temperature at which it enters:

[*reversible process*]

$$\Delta S = \frac{Q}{T} \tag{16.17}$$

When heat enters a system, its entropy increases; when heat leaves, its entropy decreases. Observe that the entropy of a system will increase by the same 1 unit when 1000 J enters at 1000 K or, alternatively, when 0.001 J enters at 0.001 K. To use Eq. (16.17) for finite changes (ΔS), the temperature has to be constant.

Example 16.9 By how much does its entropy change when 10 kg of ice at 0°C is completely melted into water at 0°C?

Solution: [Given: $m = 10$ kg, ice melting at 0°C. Find: ΔS.] Here T is absolute; namely, $T = 273$ K. To find the entropy for this isothermal process, we use Eq. (16.17), first determining Q for the melting. From Eq. (15.4), $Q = mL_f = (10\text{ kg})(334\text{ kJ/kg}) = 3.34$ MJ. Thus

$$\Delta S = \frac{3.34\text{ MJ}}{273\text{ K}} = \boxed{12\text{ kJ/K}}$$

▶ **Quick Check:** $Q = \Delta S T = (12\text{ kJ/K})(273\text{ K}) = 3.3$ MJ. We can expect an increase in entropy corresponding to the increase in molecular disorder as the crystalline solid gives way to the liquid.

Suppose the isolated system is a Carnot engine with the attendant source and sink (Fig. 16.16). In isothermally expanding from A to B, the source loses entropy, $\Delta S_H = (Q_H/T_H)$, and the engine gains the same amount—the net change is zero. The expansion to C is adiabatic ($Q = 0$) and therefore **isentropic** ($\Delta S = 0$). Over the isothermal compression to D, the engine loses entropy, $\Delta S_L = (Q_L/T_L)$, and the sink gains the same amount. The final adiabatic compression back to A is isentropic ($\Delta S = 0$), and it closes the cycle. Since $\Delta S_H = \Delta S_L$, from Eq. (16.16), the net change in entropy of the entire system during this reversible cycle is indeed zero. The Carnot engine is just efficient enough to exhaust the smallest allowable amount of waste heat. Had it been still more efficient, expelling less low-temperature heat, the entropy change would have been negative and that's forbidden by the Second Law. All real engines will exhaust more heat and produce a net increase in entropy.

Of course, entropy increases in irreversible processes even though we don't have a formula that spells out exactly how. Still, *we can get around that and use*

Eq. (16.17) to calculate the entropy change using a hypothetical reversible process between the same initial and final states. Since the entropy change only depends on these two states, ΔS for the reversible process will be the same as for the irreversible process. For example, imagine that an ideal gas undergoes a free expansion to a new state. Now suppose that the gas is returned to its initial state via a quasi-static isothermal compression, which decreases its entropy by an amount ΔS, given by Eq. (16.17). Since the system is back where it started, ΔS must be the increase in entropy it experienced during the irreversible expansion.

Example 16.10 A hot object at 573 K is put in contact with a cool object at 273 K, and 20.0 kJ of heat flows irreversibly from one to the other. By how much is the entropy of the Universe changed?

Solution: [Given: $T_H = 573$ K, $T_L = 273$ K, and $Q = 20.0$ kJ. Find: ΔS.] The entropy change of the Universe equals the net entropy change of the high- and low-temperature objects; thus

$$\Delta S = \Delta S_L + \Delta S_H = \left(\frac{Q}{T_L}\right) + \left(\frac{Q}{T_H}\right)$$

Remembering that heat-in is positive

$$\Delta S = \frac{+20.0 \times 10^3 \text{ J}}{273 \text{ K}} + \frac{-20.0 \times 10^3 \text{ J}}{573 \text{ K}}$$

$$\Delta S = 73.26 \text{ J/K} - 34.90 \text{ J/K}$$

and to three figures $\boxed{\Delta S = +38.4 \text{ J/K}}$.

▶ **Quick Check:** $\Delta S_H < \Delta S_L$ is proper since high-temperature heat has less entropy than low-temperature heat. Moreover, $T_H/T_L = \Delta S_L/\Delta S_H = (573 \text{ K})/(273 \text{ K}) \approx 2.1 \approx (73.3 \text{ J/K})/(34.9 \text{ J/K})$.

16.8 Order and Disorder

It was not until 1878 that Boltzmann reformulated entropy, giving it a new fundamental clarity. He recognized that entropy was a measure of the disorder of the Universe on an atomic level. Once lit, the neatly arranged molecular pattern of a match spontaneously transforms into the mayhem of gas, smoke, and flame. The disorder of the Universe increases, entropy increases; energy once concentrated is soon spread out with an evenness that is the essence of thermodynamic disorder. *Maximum disorder corresponds to a random tumult, an undifferentiated uniformity.*

We tend, at first, to think of disorder as a scene with everything tossed about, a shoe here, a sock there. But that's only crude disorder. Splendid disorder requires a fine, dispersed dust—no shoes, no socks, but a uniform cloud of agitated atoms. A bull in a china shop will spontaneously powder the pretty cups, crush the delicate vases, and increase the uncompromising entropy of the Universe. The disorder of a distribution of particles is reflected by both their separations and their speeds. If all the particles in a perfect crystal were at rest, that would correspond to its state of maximum order, only approached at 0 K.

We never measure entropy, but only entropy change, ΔS, which accompanies physical change. And a system naturally changes (energy naturally flows) so as to increase universal entropy ($\Delta S \geq 0$). **The quantity ΔS measures the degree of disorder associated with any change (ΔE) in a system.** *An isolated system in a maximum entropy configuration cannot of itself change macroscopically; it is*

OZYMANDIAS

I met a traveller from an antique land
Who said: "Two vast and trunkless legs of stone
Stand in the desert. Near them, on the sand,
Half sunk, a shattered visage lies, whose frown
And wrinkled lip, and sneer of cold command,
Tell that its sculptor well those passions read
Which yet survive, stampt on these lifeless things,
The hand that mockt them and the heart that fed;
And on the pedestal these words appear:
'My name is Ozymandias, king of kings:
Look on my works, ye Mighty, and despair!'
Nothing besides remains. Round the decay
Of that colossal wreck, boundless and bare
The lone and level sands stretch far away."

P. B. SHELLEY

Every physical and chemical process in nature occurs in such a way that the net entropy of all the matter involved increases—the disorder of the Universe inexorably moves toward a maximum.

Mixtures of any kind are more disordered than were their separate constituents. Two gases introduced into a chamber will "always" mix, even though there is no energy incentive to do so. Colored grains of sand shaken in a bag will "always" mix. Yet neither the gases nor the sands are likely to spontaneously unmix, to neatly separate, to decrease in entropy. For that matter, you can shake the bag of sand for centuries, and though it is possible, it is most improbable that the different colored specks will ever separate perfectly again. True, you could reach into the system (then no longer isolated) and sort out the sand—you could reestablish order and decrease entropy. But you would spend more order in the effort; you would disorder your breakfast more than you would order the sand—the Universe would lose again. It must lose again.

Wherever entropy is decreasing—in a factory where someone is stringing beads, among the ice cubes in a freezer, or with the cells forming a living embryo—order is rising out of disorder. But whenever that occurs locally, there will always be a hidden harvest of mayhem that overbalances the scales on the side of universal disorder. The freezer will make its ice cubes, but the system isn't isolated. The cost will be in electrical energy converted to heat and in the end, $\Delta S > 0$.

The Arrow of Time

Of all the laws we have studied, only the Second gives a direction to the unfolding of events, to the progression of processes. It points the arrow of time in the "forward" direction—toward maximum disorder. Time stops when all events stop, when nothing happens, and the Second Law provides the direction to processes and to time. Have you ever seen the gases come back, the smoke descend, the light return, the heat pour in, the flame collapse, and a match reform, unburnt and whole? Will the gray hairs spontaneously darken, the skin grow tight and supple, the sags unsag? Not likely! And yet energy would be conserved. If everything reversed and processes ran toward order, would time be going backward? All by itself the game runs downhill, and it is the Second Law that shows the way.

16.9 Entropy as Probability

Boltzmann found the mathematical fingerprint of disorder in a statistical relationship describing the likelihood that a particular molecular arrangement would occur. Were we to take all the separate molecules that make up a match and put them in a box and shake them, the chance of finding them all bound together, neatly arranged as a match, is incredibly small. It is faintly possible but, given the astounding number of possible combinations that do not lead to a match, exceedingly unlikely. Unwrap a deck of playing cards and observe the arrangement. That ordered state is just one out of 80 million, million, million, million, million, million, million, million, million, million, million possible arrangements of those 52 cards. Order is far less likely than disorder. Once the molecules in an egg are scrambled, the chance that all of them will simultaneously find their way back to so unlikely an arrangement as the original ordered ovum is slim indeed. Every process irreversibly scrambles a bit of the cosmic egg.

A system of continually agitated molecules moves from one state to the next because the new configuration is more disordered and therefore statistically more

TABLE 16.2 Possible arrangements of four coins

Number of heads	Equivalent microstates	Number of microstates
0	TTTT	1
1	HTTT, THTT, TTHT, TTTH	4
2	HHTT, THHT, TTHH, HTTH, THTH, HTHT	6
3	HHHT, THHH, HTHH, HHTH	4
4	HHHH	1
	Total number of microstates	16

probable. Throw four coins on the table and it is most likely that two will be heads and two tails. There are a total of 2^4 or 16 different ways the four coins can land (16 so-called *microstates*), as shown in Table 16.2. Of these, 6 equivalent microstates correspond to 2-heads-2-tails (that is, 1/2-heads): That's 6/16, or 38%, which is the most likely state to occur. By comparison, there is only one way (1 microstate) where all the coins will be heads, and the probability of that highly ordered state happening is only 1/16, or 6%. Clearly, the probability of finding the system in some particular state, say 2-heads-2-tails, is proportional to the number of possible equivalent microstates ($\mathcal{W}$) composing that state—in this case, 6.

A crystal is a regular structure in which, especially at low temperatures, there is very little variation in the arrangement of its atoms. Each such configuration is a microstate of the system, and there are relatively few of them possible. The crystal is highly ordered and possesses little entropy. Still, as its internal energy increases, the number of ways that energy can be distributed among the constituent particles increases, which increases the number of available microstates and, hence, the entropy. In contrast, a mass of gas has a tremendous number of possible equivalent microstates that will produce the same macrostate with the same values of temperature, volume, and pressure. The gas is highly disordered and possesses a large amount of entropy.

If there were 100 coins in our system, there would still be only 1 microstate corresponding to all-heads, but now there would be a total of 2^{100} or $\approx 10^{30}$ microstates. That makes the chance of throwing all-heads about 1 out of 10^{30}. Since there are roughly 32×10^6 seconds per year, you can throw down the coins once every second for 3×10^{22} years, and you are not likely to hit all-heads more than once, though it certainly could happen on the first shot. By comparison, there are about 10^{29} ways to get 1/2-heads and that's by far the most likely state, occurring 10% of the time. About 90% of all throws will result in between 45 and 55 heads.

Entropy.

Sequential probabilities build by multiplication. The chance of getting a head with 1 coin is 1/2 per throw (50%), while the chance of following that with another head is $1/2 \times 1/2$ and so on. Thus, the probability of throwing 4 heads in a row is $1/2 \times 1/2 \times 1/2 \times 1/2 = 1/16$, just as it was when we threw them all at once, which is reasonable since they don't talk to each other as they fall. By contrast, sequential processes would progressively *add* entropy changes. A system usually increases in entropy every time something happens to it.

The more microstates a state has, the more likely it is that that state will be observed, and the more disordered it must be. Hence, we can guess that $S \propto \mathcal{W}$, but not directly, because one is additive and the other multiplicative. This suggests using a function like the logarithm to overcome the difference because $\ln ab = \ln a + \ln b$; taking the logarithm turns products into sums.

Boltzmann (1877) proposed that the entropy could be explicitly connected to the notion of disorder by defining it in a new way, which was nonetheless consistent with Clausius's definition:

$$S = k_B \ln \mathcal{W} \tag{16.18}$$

where the k_B is Boltzmann's Constant (p. 497). Keep in mind that the logarithm (Fig. 16.11) rises rapidly at low values and levels off at high values. That would reflect the greater disordering effect of heat-in at lower temperatures. Even though fewer microstates are available at low temperatures, S is more sensitive to changes in internal energy at small T.

The probabilistic version of the Second Law does not forbid a state from spontaneously moving to a lower entropy; rather, it maintains that such a case would be most unlikely. If this statistical formulation is true to life, and it certainly seems to be, anything is possible—any process that can occur via the First Law may occur, if only remotely, via the Second. It is possible for the egg to unscramble, but it's unthinkably improbable.

To live at all is to churn through spontaneous processes, to be out of equilibrium, to pour forth entropy (while usually exhausting just about as much energy as we take in). Life feeds on order, and even as it builds, it disperses.

Core Material

Thermodynamics (or thermo, as it's affectionately called) is the study of thermal energy—its relationship to heat, mechanical energy, and work, and the conversion of one to the other.

The **First Law of Thermodynamics** is

$$Q = W + \Delta U \qquad [16.1]$$

Hence

$$\Delta U = Q - W \qquad [16.2]$$

An equivalent statement of the First Law is

> The internal energy of an isolated system is constant (even though it may be transformed from one type of energy to another).

A system is in **equilibrium** when there is no tendency for it to undergo spontaneous change (on a macroscopic level). If a system is always infinitesimally close to thermodynamic equilibrium, even as it passes from one state to the next, the process that is occurring is said to be **reversible** (p. 559).

The work done in an isobaric process is

$$W = P\Delta V \qquad [16.3]$$

The work done on or by an ideal gas during an isothermal volume change from V_i to V_f is

$$W = nRT \ln \frac{V_f}{V_i} \qquad [16.5]$$

For a reversible adiabatic process

$$PV^{\gamma} = \text{constant} \qquad [16.6]$$

where γ is a constant that equals 1.67 for monatomic gases, $\approx$1.4 for diatomic gases, and $\approx$1.3 for polyatomic gases.

A *heat engine* is a cyclic device that converts thermal energy into work-out. A *Carnot engine* is a reversible device operating in a cycle composed of two adiabats and two isothermals. The efficiency of an engine is

$$e = \frac{W_o}{Q_i} = \frac{Q_H - Q_L}{Q_H} = 1 - \frac{Q_L}{Q_H} \qquad [16.10]$$

where both Q_H and Q_L are positive numbers. The efficiency of a Carnot engine is

$$e_c = 1 - \frac{T_L}{T_H} = \frac{T_H - T_L}{T_H} \qquad [16.11]$$

A *refrigeration machine* uses work to transfer thermal energy from a low-temperature reservoir to a high-temperature reservoir.

The *coefficient of performance* (η) is defined as the ratio of the amount of heat removed from the cold reservoir (the heat-in to the system) to the amount of work done on the system; consequently

$$\eta = \frac{Q_L}{W_i} = \frac{Q_L}{Q_H - Q_L} \qquad [16.13]$$

For a **Carnot engine**

$$\frac{Q_L}{Q_H} = \frac{T_L}{T_H} \qquad [16.14]$$

The coefficient of performance for an ideal system is

$$\eta_c = \frac{T_L}{T_H - T_L} \qquad [16.15]$$

The **Kelvin statement** of the Second Law is

> No process is possible for which the *sole result* is the removal of heat from a source and its complete transformation into work.

The **Clausius statement** of the Second Law is

> No process is possible in which the *sole result* is the transfer of thermal energy from a body at low temperature to a body at a higher temperature.

A Carnot engine has the same efficiency as any other reversible engine operating between the same two temperatures. No engine can exceed the efficiency of a reversible engine (p. 565).

Clausius proposed that the measure of the disorder of a system is the **entropy** S. The Second Law in terms of entropy is

> When an isolated system undergoes a change, passing from one state to another, it will do so in such a way that its entropy will increase, or at best remain the same.

Clausius defined the change in entropy of a system (ΔS) experienced during a *reversible* process as

$$\Delta S = \frac{Q}{T} \qquad [16.17]$$

A system with maximum entropy is in equilibrium. Boltzmann proposed that entropy could be explicitly connected to the notion of disorder by defining it as

$$S = k_B \ln \mathcal{W} \qquad [16.18]$$

where $\mathcal{W}$ is the number of microstates.

Suggestions on Problem Solving

1. Keep in mind that all the calculations involving temperature in this chapter use kelvins. Look out for data given in °F; convert to degrees Celsius and then to kelvins early in the analysis.

2. Whether Q_H and Q_L are flowing into or out of the device in a particular problem can be confused, so it's a good idea to draw an energy flow diagram, like Fig. 16.18. This is especially true when dealing with refrigerators.

3. When an engine is reversible, the values of Q_L, Q_H, and W must remain the same in either mode of operation. Of course, when it's run as an engine, Q_H is the heat-in whereas, as a refrigerator, Q_H is the heat-out, but numerically the values stay the same. In other words, the energy channels in schematic diagrams like Fig. 16.28 don't change width on reversing a reversible engine; they do change if the engine is irreversible.

4. Adiabatic processes utilizing gases require values for γ, and most often air is involved, for which $\gamma = 1.4$. If not explicitly stated, you can assume that a gas behaves adiabatically if the process occurs rapidly (p. 559).

5. The heat leaving an object is negative, and so the change in entropy it experiences is also negative. Keep careful track of the signs throughout entropy calculations. If the temperature drops $\Delta T < 0$, $Q = cm\Delta T < 0$ and $\Delta S = Q/T$.

6. It often happens that a quantity you might think would be needed explicitly to solve a problem cancels out of the equations. That sort of thing happens especially with before-and-after volumes and pressures of gases. Moreover, there are situations involving the gas laws where it's useful to work on a per-unit-volume basis. Again, because the equations for gases often have before-and-after pressures and volumes (that is, P and V on both sides), you can use non-SI units such as atm or cm^3. That will save the time to convert, but it is risky if you are not careful. It's better to play it safe and convert everything to SI right away. To illustrate the point, a few of the solutions given in the back of the book use mixed units.

Discussion Questions

1. Figure Q1 is the *PV*-diagram for a diesel engine. It uses the internal combustion of a gaseous fuel in a cylinder-and-piston arrangement. Here, the fuel is sprayed into the cylinder at C. Explain each part of the cycle. Why do you think only air is compressed from B to C with the fuel injected only at C? Is work done? When does heat enter and leave the system? What kind of processes are occurring? At what point does the cycle end? Where in the cycle does the exhaust valve open to eject spent fuel? Use the gasoline engine as a guide since it's fairly similar to the diesel.

2. The gas in the combustion chamber of a rocket is in violent random motion at high temperature. It leaves the engine through a nozzle at high speeds. Describe what's happening in terms of energy. Considering that the exhaust emerges in a well-

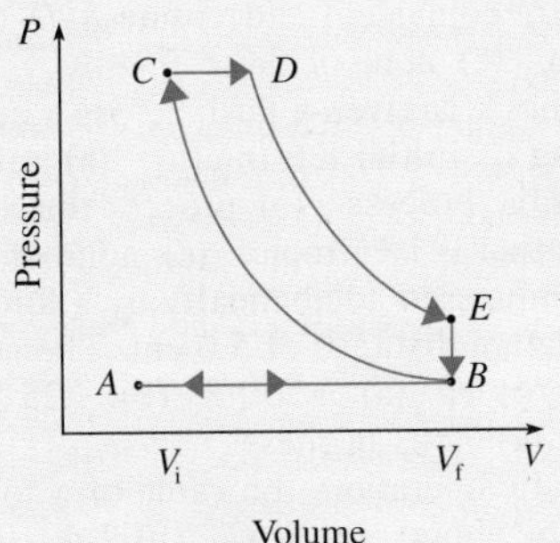

Figure Q1

defined direction, what is the temperature of the exhaust relative to that of the combustion chamber?

3. Figure Q3 depicts a reciprocating steam engine. Explain how it works.

4. The so-called *Third Law of Thermodynamics,* which to some physicists does not seem as secure as the other two, states that:

> It is impossible to bring a body to a temperature of Absolute Zero in a finite number of steps.

In other words, it is widely believed that Absolute Zero is unattainable. Discuss what this means and determine whether or not it fits in with the Second Law.

5. Discuss the thermodynamics of a possible mechanism for the formation of clouds as a result of rising columns of moist, warm air coming from near the ground.

6. You are offered the opportunity to "buy in on the ground floor" of a company selling a new machine. It is a container designed to keep drinks at a fixed temperature indefinitely. One loads it with hot liquid; the heat that inadvertently leaks off is cleverly converted into work that is fed to a stirrer that continually churns the liquid putting the thermal energy back and maintaining the temperature. Do you buy? Explain. Discuss the process from an entropy perspective.

7. What is the ultimate source of the energy stored in coal, wood, and oil? Explain. Does this arrangement violate the entropy version of the Second Law?

8. Suppose someone is placed in a totally isolated room with enough food for one day. What will inevitably happen to him? Explain your answer in terms of both the First and Second Laws.

9. What is the ultimate source of the energy extracted via a hydroelectric power plant? Explain.

10. A red-hot coal is hung on an insulating fiber inside a room-temperature metal box that is sealed, evacuated, and completely isolated from the rest of the environment with massive amounts of insulation. Discuss the situation from an entropy perspective. What happens to the coal and the box?

11. Imagine a thermally insulated cylinder-and-piston containing an ideal gas at a certain initial volume V. Suppose the gas, pushing on the piston, is allowed to expand to a volume $2V$. Discuss, in thermodynamic terms, every aspect of the process. What kind of process is it? What happens to the internal energy of the gas? What happens to P, V, and T? Is work done? By what, on what? Does the entropy change?

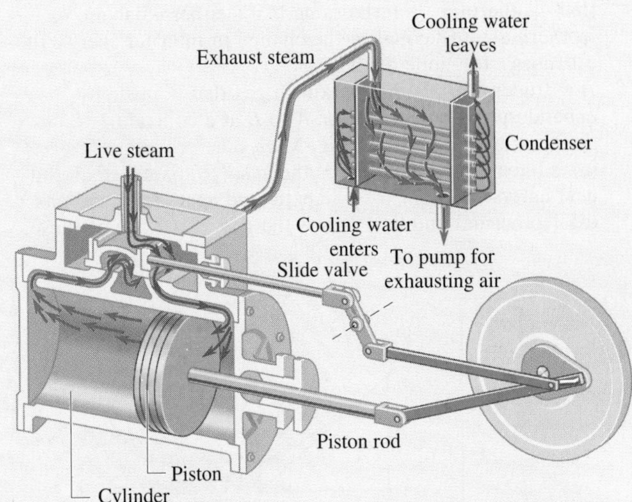

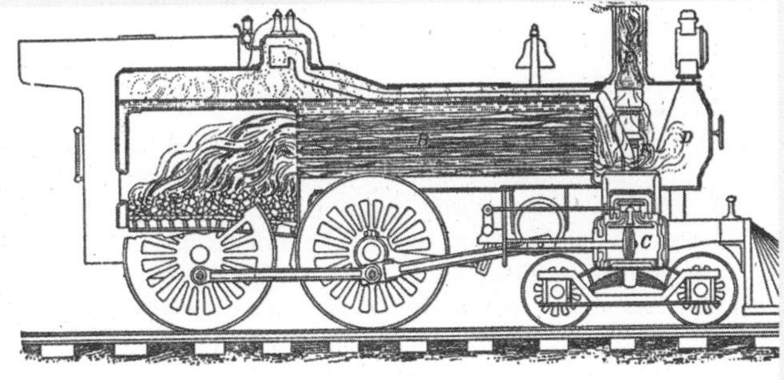

Figure Q3

12. Suppose you fly in a jet plane from New York to London. What is the net effect on the energy and entropy of the planet?

13. Suppose an automobile engine burns its hydrocarbon fuel at about 2400 K while the outside air temperature is 300 K. Explain why the actual efficiency is far less than the corresponding Carnot efficiency between those two temperatures.

14. What would you say was the overall effect of the rise of technology on the entropy of the Universe?

15. A large room may have some 10^{28} air molecules filling it. Is it possible for these molecules to suddenly all end up in one corner of the room, leaving us gasping for breath? Do we have to worry about that happening and if not, why not? How would matters change if there were only 10 molecules in the room?

Multiple Choice Questions

1. A cylinder-and-piston contains an ideal gas whose volume is to be halved. The work done on the gas will be (a) greater if the compression is isothermal rather than adiabatic (b) greater if the compression is adiabatic rather

than isothermal (c) the same if it's either adiabatic or isothermal (d) equal to the change in internal energy in all cases (e) none of these.

2. A cylinder-and-piston contains a gas that is made to expand quasi-statically from A to B at a constant temperature as shown in Fig. MC2. The process is (a) adiabatic and heat leaves the gas (b) isothermal and heat enters the gas (c) adiabatic and heat enters the gas (d) isothermal and heat leaves the gas (e) none of these.

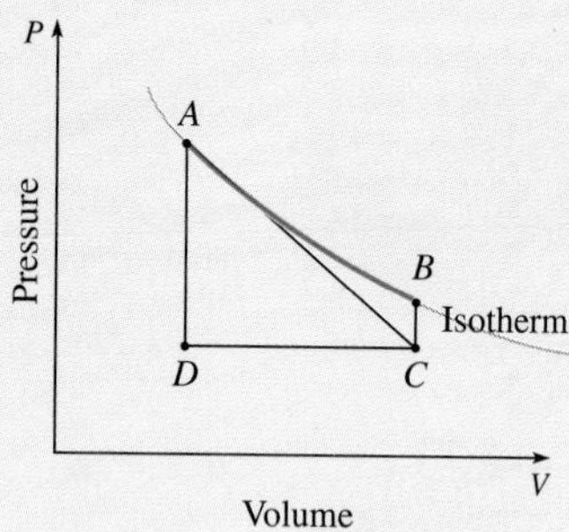

Figure MC2

3. Referring to the previous question and Fig. MC2, when the gas is carried from B to C, the process is (a) isobaric and no heat is transferred (b) isovolumic and no heat is transferred (c) adiabatic and no heat is transferred (d) isovolumic and heat leaves the gas (e) none of these.
4. Referring to the previous question and Fig. MC2, when the gas is carried from C to D, the process is (a) isothermal and no heat is transferred (b) isovolumic and no heat is transferred (c) adiabatic and heat enters the gas (d) isovolumic and heat leaves the gas (e) none of these.
5. Referring to the previous question and Fig. MC2, when the gas is carried from D to A, the process is (a) isothermal and no heat is transferred (b) isovolumic and no heat is transferred (c) adiabatic and heat enters the gas (d) isovolumic and heat enters the gas (e) none of these.
6. When Joule wrote "the grand agents of nature are, by the Creator's fiat, indestructible," (1843) he was talking about (a) the F.B.I. (b) the Second Law (c) an automobile (d) the First Law (e) none of these.
7. An ideal gas is contained in a cylinder-and-piston that is 0.5 m long, and 50 J of heat is added while the piston is held in place. The work done by the gas is (a) 50 J (b) 0 (c) 25 J (d) 250 J (e) none of these.
8. If a system does work adiabatically, (a) the temperature stays the same (b) the internal energy stays the same (c) the internal energy decreases (d) the internal energy increases (e) none of these.
9. Fig. MC2 shows the PV-diagram for a gas. In reference to the work done by the gas in going from state A to state C (a) the route A-C corresponds to a minimum (b) the route A-D-C is equivalent to route A-B-C (c) all the routes are equivalent (d) route A-D-C corresponds to a maximum (e) none of these.
10. When this book is given a push across a desktop, it comes to rest losing its initial KE in a(n) (a) reversible process (b) irreversible process (c) process that is isothermal (d) process that is isentropic (e) none of these.
11. A gas is compressed adiabatically by a force of 500 N acting through a distance of 5.0 cm. The net change in its internal energy is (a) +500 J (b) +25 J (c) −500 J (d) −25 J (e) none of these.
12. In an adiabatic operation, the value of γ for mercury vapor is likely to be about (a) 1.67 (b) 1.4 (c) 1.3 (d) 1 (e) none of these.
13. The statement that heat flows spontaneously only from a high-temperature object to a low-temperature object is (a) not always true (b) the First Law of Thermodynamics (c) the Second Law of Thermodynamics (d) only true for isothermal processes (e) none of these.
14. An ice-filled glass of soda is so cold, water from the air condenses out on it in beads. As a result, the entropy of that water must have (a) increased (b) remained unchanged (c) first increased then decreased (d) decreased (e) none of these.
15. As it regards the ordinary processes of everyday life, it can be said that the Second Law of Thermodynamics (a) distinguishes the possible from the impossible (b) is quite inapplicable (c) demands that energy be conserved, no matter what (d) requires that the sum of the heat-in and the heat-out be zero (e) none of these.
16. The addition of heat to a system (a) decreases its order and so decreases its entropy (b) decreases its order and its internal energy (c) decreases its order and increases its entropy (d) increases its disorder and decreases its entropy (e) none of these.
17. The disordering effect of heat (a) increases with increasing temperature (b) decreases with increasing entropy (c) is constant (d) increases with decreasing temperature (e) none of these.
18. The area under a process curve on a TS-diagram is the (a) heat flowing into or out of the system (b) the work done on or by the system (c) the change in pressure of the system (d) the change in entropy of the system (e) none of these.
19. The entropy associated with a particular state of a system is (a) independent of probability of any sort (b) dependent on the logarithm of the probability of the state occurring (c) inversely proportional to the number of microstates associated with that state (d) independent of the number of microstates associated with that state (e) none of these.
20. The statement that processes take place in such a way that the entropy of the Universe tends toward a maximum should be interpreted as saying (a) that the Universe is moving inextricably toward equilibrium (b) that nothing much is happening in the Universe (c) that energy is approaching a minimum (d) that order is rising out of chaos (e) none of these.

Problems

THE FIRST LAW OF THERMODYNAMICS
CYCLIC PROCESSES: ENGINES AND REFRIGERATORS
THE SECOND LAW OF THERMODYNAMICS

1. [I] An amount of heat equal to 500 J flows into a system that does 250 J of work on its surroundings. What is the change in the internal energy of the system?

2. [I] A quantity of gas in a cylinder sealed with a piston absorbs 1.00 kcal of heat as it expands by 15 liters at constant pressure. Determine that pressure if the internal energy is found to increase 286 J.

3. [I] A system has 150 J of work performed upon it while its internal energy is made to decrease by 300 J. Determine the heat transferred to or from the system.

4. [I] A brass sphere having a 15-cm diameter is heated from 0°C to 280°C. How much work does it do on the atmosphere in the process?

5. [I] A 100-g sample of gas is contained in a sealed insulated tank. Heat is slowly transferred to the gas (via an electric heater), and its internal energy increases by 500 J. Determine the amount of heat added to the gas.

6. [I] When a chunk of ice melts, it changes from a density of 920 kg/m^3 to 1000 kg/m^3, and work is done on it by the atmosphere. Compare the work done per kilogram on the ice to the heat-in per kilogram.

7. [I] An amount of heat equal to 100 cal is transferred to a system while 100.4 N·m of work is done on it. What is the net change in its internal energy?

8. [I] An ideal gas is compressed to one-half its original volume by a piston moving in a cylinder. During the process, 500 J of heat leave the gas, but its temperature remains fixed. How much work is done on or by the gas?

9. [I] An ideal gas in a cylinder-and-piston is made to follow the cycle shown in Fig. P9 from A to B to C to A. What is the work done in terms of the initial volume and pressure?

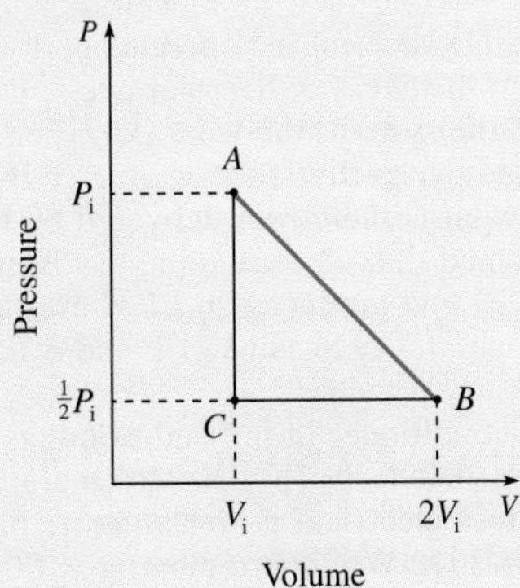

Figure P9

10. [I] A cylinder-and-piston containing an ideal gas is placed in contact with a thermal reservoir. The volume of the gas is very slowly changed from 50 liters to 10 liters as 40 J of work is done on it by an external agency. Determine the change in the internal energy of the gas and the amount of heat flowing into or out of the system.

11. [I] An ideal gas in a cylinder-and-piston is made to follow the cycle shown in Fig. P11 from A to B to C to D. What is the work done in terms of the initial volume and pressure?

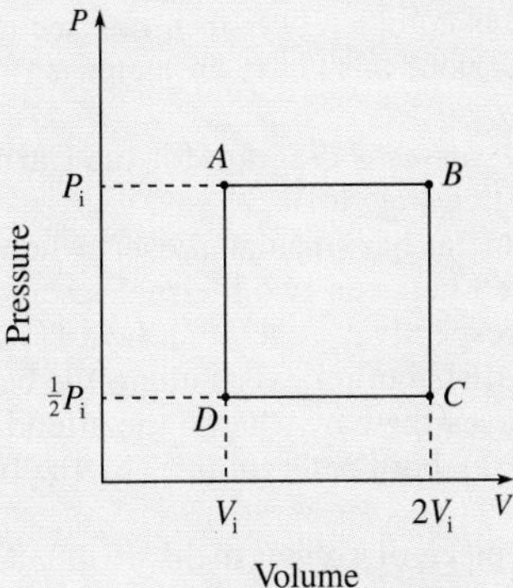

Figure P11

12. [I] An ideal gas contained by a cylinder-and-piston is made to double its volume isothermally, at 27°C. If the chamber contains 5.0 moles of gas, how much work is done in the process?

13. [I] A gas experiences a change from state I to state F as indicated in Fig. P13. Determine the work done.

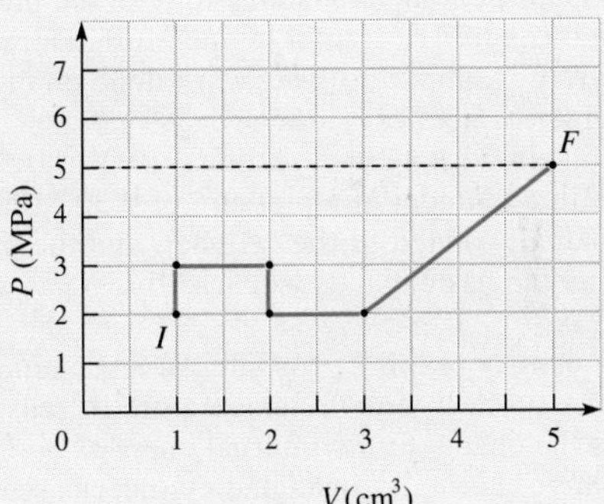

Figure P13

14. [I] An engine doing work receives 100 kJ and exhausts 75 kJ. Determine its efficiency.

15. [I] A person consumes food at 98.6°F and exhausts waste heat at 20°C. What is the maximum theoretical efficiency for a heat engine operating at the same temperatures?

16. [I] What is the maximum theoretical efficiency of a steam engine operating between 100°C and 400°C?

17. [I] To improve the theoretical efficiency of a heat engine, would it be better to lower T_L by 10 K or raise T_H by 10 K? Explain.

18. [I] In a given amount of time, a refrigerator extracts 75 kJ from a cool chamber while exhausting 100 kJ. Determine its coefficient of performance.

19. [I] A 1.0-ton air conditioner has the capacity to freeze

2000 lb of ice at 0°C in a day. What is the equivalent rate of transport of heat in watts?

20. [I] What is the coefficient of performance of a Carnot refrigerator operating between 0°C and 85°C?

21. [I] What is the change in entropy of a quantity of boiling water in a covered pot when 1.00 kJ of heat is added?

22. [I] By how much does its entropy change when 10 g of water at 0°C is completely frozen into ice at 0°C? Make it clear whether your answer is an increase or decrease in entropy.

23. [I] Draw a *T*-versus-*S* diagram for the Carnot cycle and label it to match Fig. 16.16.

24. [I] A thick metal bar, insulated everywhere but on its ends, is placed between two thermal reservoirs at 600 K and 300 K, respectively, such that heat flows and a stable temperature distribution exists along the bar. If 10.0 kcal of heat traverses the bar, what is the attendant change in entropy of (a) both reservoirs, (b) the bar, (c) the Universe?

25. [I] A 500-g piece of copper melts at 1083°C. Determine its change in entropy in the process.

26. [I] By how much does the entropy of 1.00 kg of water change when it is completely converted to steam?

27. [I] With Problem 26 in mind, compare the changes in entropy that result when 1.00 kg of water ice is melted at 0°C with that when the liquid is vaporized at 100°C. Account for the great difference in these two values.

28. [II] An ideal gas is compressed from 20 liters to 2.0 liters at a constant pressure of 3.0 atm by drawing heat from it so that its temperature drops. At that point, its volume is held constant, and heat is added to it so the pressure rises and it returns to its original temperature. Draw a *PV*-diagram. Compute the net work done on or by the gas. Determine the net heat flow.

29. [II] A piston with a cross-sectional area of 0.50 m^2 seals a gas within a cylinder at a pressure of 0.30 MPa. Heat is very gradually added to the cylinder, and the piston rises 6.5 cm as the gas expands isobarically. What work is done by the gas?

30. [II] A quantity of 5.0 kg of water in a vacuum chamber is supplied with heat, and its temperature increases by 10°C. What is the change in its internal energy?

31. [II] A large thermally insulated cylinder is sealed with a piston. Steam at 100°C enters the cylinder and pushes the piston outward against a pressure of 99 kPa, increasing the occupied volume by 4950 cm^3. How much water condenses inside the cylinder?

32. [II] A 2.0-mole sample of a gas at 0°C, which behaves like an ideal gas, is compressed to half its original volume isobarically. How much work must be done on the gas?

33. [II] A bicycle tire filled with air to a pressure of 4.5 atm is rapidly emptied into a large bag at atmospheric pressure. If the air in the tire was initially at 27°C, what will be its final temperature in the bag? Assume ideal gas behavior.

34. [II] An ideal diatomic gas at STP is adiabatically compressed to half its original volume. Determine its final temperature and pressure.

35. [II] A cylinder-and-piston with a volume of 22.4 liters contains 1.00 mole of diatomic hydrogen at a pressure of 1.00 atm. The gas is allowed to expand adiabatically to twice its original volume, whereupon it is compressed back to its initial volume isothermally. (a) What is the lowest pressure the system attains? (b) What is the lowest temperature the system attains? Draw a *PV*-diagram showing everything that's happening.

36. [II] Considering Problem 35, how much work is done on isothermally returning the system to its original volume? What is the final temperature of the system? What is the final pressure?

37. [II] An ideal gas contained in a cylinder-and-piston assembly undergoes a reversible isothermal expansion. Derive an expression for the work done by the gas in terms of the number of moles present, the temperature, the gas constant, and the initial and final pressures.

38. [II] With the results of Problem 37 in mind, compute the amount of work done by 2.00 moles of an ideal gas in an isothermal expansion at 400 K from an initial pressure of 12 atm down to 1.5 atm.

39. [II] Figure P39 shows two reversible cycles *A-B-C-A* and *C-D-E-C* carried out on an ideal gas (the curves are isothermals). Prove that the work done is the same for each cycle.

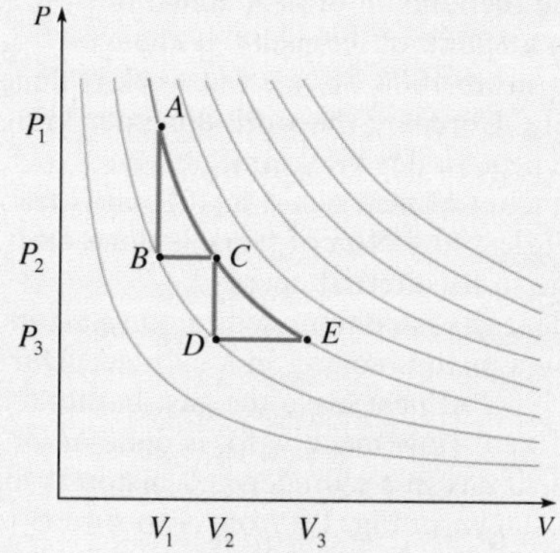

Figure P39

40. [II] A reversible heat engine operates between 325°C and 25°C taking in 610 kcal of heat per cycle. Determine the efficiency of the system, the work done per cycle, and the heat exhausted per cycle. If the engine runs at 5.0 cycles per second, what is the power delivered by it?

41. [II] It is claimed that a heat engine has been built that takes in 102 kJ and produces 26 kJ of useful work. It supposedly operates between 144°F and 5°F. Is such a device physically possible?

42. [II] The air conditioner in my bedroom has a little plaque on it that says 18 000 Btu/h, 230 V, 9.0 amps. That means that it consumes electrical power at a rate of (230 V) × (9.0 amps) = 2070 W while removing 18 000 Btu each hour from the room. Given that 1 Btu = 1.055 × 10^3 J, determine the coefficient of performance of the machine.

43. [II] An old-fashioned commercial 1-ton refrigerator removes heat from water in a chamber at 32°F at such a rate as to produce 1.0 ton (2000 lb) of ice (at 32°F) per day. The heat is exhausted to the environment at 88°F. At its very best, how much power would be needed to drive this system?

44. [II] Suppose that in the course of a day you consume one peanut butter sandwich (330 kcal), a jelly doughnut

(225 kcal), 1.0 kg of milk (650 cal/g) and two martinis (145 kcal each). You spend a quiet day averaging an output of about 60 watts of useful work for 6 hours, above and beyond the energy consumed just staying alive. What's your efficiency?

45. [II] A heat pump, which removes thermal energy from the outdoors and exhausts it inside to heat the house, has a coefficient of performance of 5.5. What is the ratio of the heat supplied to the house, to the work done? How much energy do you pay for as compared to the amount you get pumped into the house? How does this compare with heating with electricity directly?

46. [II] A refrigerator maintains food at 40°F when the room temperature is 80°F. Ideally, how much heat will be removed from the food per joule of work done by the compressor on the refrigerant?

47. [II] You are designing an electric refrigerator to remove 360 J of heat per minute from a cold chamber at −4.0°C and exhaust that energy into a room at 27°C. At the very best, how much electrical power will it require?

48. [II] We have designed a 150-W electrical refrigeration system that removes the 360 J/s of heat leaking into a test chamber and expels that thermal energy (the so-called *heat load*) into the room (which is at 23°C). Determine the lowest temperature we can expect to reach in the chamber.

49. [II] A Carnot engine operates at an efficiency of 42.2% with a high-temperature reservoir at 473 K. If the efficiency is to be raised to 50% using a new high-temperature reservoir, what will its temperature have to be?

50. [II] A Carnot refrigerator is driven by 2.00 J of mechanical work each second, from an electric motor. It, in turn, removes heat from a cold region (0°C), expelling it into the room (27°C). What is the change in entropy of the cold region?

51. [II] During each cycle of operation of a heat engine, it takes in 1200 J of heat from a hot reservoir and exhausts 450 J of heat to a cool reservoir. What is its efficiency? The engine is now run backward between the same reservoirs, and while delivering 1200 J per cycle to the high-temperature reservoir, it is driven by 1800 J of applied work. Determine whether this engine is reversible or irreversible in a thermodynamic sense.

52. [II] A 100-g chunk of ice at 0°C is placed on a large marble table, which is at 27°C. What, if any, is the change in entropy of the ice-table system when the ice melts completely?

53. [II] A 20-kg sample of pure water at 40.0°C is mixed with a 20-kg sample of pure water at 32.0°C. Using the average temperatures of each sample, determine the approximate net change in entropy of the total quantity of water.

54. [II] An ideal gas expands isothermally from V_i to V_f. Prove that the associated entropy change is given by

$$\Delta S = nR \ln \frac{V_f}{V_i}$$

55. [II] Using the data of Fig. P55, which is a *TS*-diagram for a thermal engine, determine Q_H, Q_L, the net work done by the system, and its efficiency.

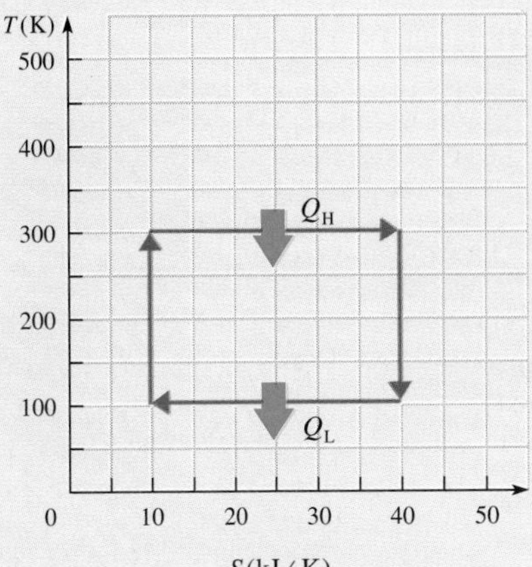

Figure P55

56. [III] A 1.0-m^2 panel of solar cells absorbs 35% of the total radiation (of intensity 0.83 kW/m^2) impinging on it from the Sun. It generates electrical energy, which is fed to a motor that raises a load, doing work at a rate of 150 J each second. Assuming no other losses, determine the rate at which the internal energy of the system (panel and motor) is changing.

57. [III] Please show that for an adiabatic volume change of an ideal gas

$$\frac{T_i}{T_f} = \left(\frac{V_f}{V_i}\right)^{\gamma-1} \qquad (16.19)$$

58. [III] A diesel engine adiabatically compresses an air-fuel mixture from its initial volume at atmospheric pressure and a temperature of 25.0°C to 1/20 that volume. What is its final temperature? (Take $\gamma = 1.35$.)

59. [III] The compression ratio (V_i/V_f) of a gasoline engine is 7.5. Determine the ratio of the temperature of the gas before and after the compression. (Take $\gamma = 1.35$.) Explain your results.

60. [III] Derive an expression for the entropy change experienced by a gas during a free expansion from a volume V_i to a volume V_f.

61. [III] Two identical thermally insulated chambers are connected by a small valve. One chamber contains 1.0 mole of an ideal diatomic gas; the other chamber is evacuated. The valve is then opened, and the gas rushes through the valve so that it occupies both chambers. What is the resulting change in entropy?

62. [III] A 200-kg boulder rolls, slides, and slips down the side of a mountain, descending 200 m vertically before coming back to rest. It's a nice cool winter's day, and the temperature of the forest is 7.0°C. What is the change in entropy of the Universe as a result of the tumble?

17

Electrostatics: Forces

THUS FAR, WE HAVE concentrated on the gravitational interaction and the associated property of matter called *mass.* Certainly, the electromagnetic interaction was always there in our considerations, but in the guise of such macroscopic notions as friction, cohesion, elasticity, contact force, and so forth. Now we focus on it and on the associated characteristic of matter called *charge.* The electromagnetic interaction binds matter in all its observable forms (Chapter 10). It is responsible for holding the subatomic particles together as atoms, for holding the atoms together as molecules, for holding the molecules together as objects, indeed, for holding your nose onto your

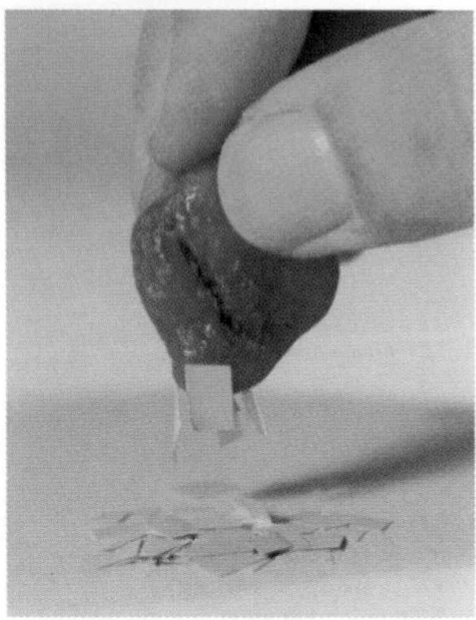

A chunk of amber rubbed on fur picks up small pieces of colored paper.

face. All biological processes are governed by the interaction of charges: seeing, feeling, moving, thinking, living—all of it.

The story begins in ancient times with amber, a yellow-brown material (fossilized pine-tree resin) used for jewelry for thousands of years. Probably while polishing amber, the Greeks noticed that it had an extraordinary quality: rubbed with cloth or fur, it attracts small bits of lightweight matter—tufts of lint, hair, etc. Even Plato wrote about the "marvels concerning the attraction of amber." The Greek word for amber is *elektron,* and by the mid-seventeenth century it was already being suggested that a substance, activated by rubbing, possessed some sort of "amber stuff," or *electricity.* And people came to refer to the *charge* of electricity imparted to an object in the same sense as an "amount" (much like a charge of gunpowder).

When rubbed with woolen cloth, a chunk of amber or a plastic comb can pick up a small piece of Styrofoam™ broken from a coffee cup. We are led by our understanding of the laws of mechanics to conclude that there is yet another force, the *electric force.* Something must be causing it, and so we call the physical attribute responsible for that interaction, electric or more precisely, **electromagnetic charge**. (Electric and magnetic phenomena are one and the same—both are manifestations of charge.)

The subject of electromagnetism is usually arranged into conceptual subdivisions, one of which is **electrostatics**, *the study of charges at rest.*

ELECTROMAGNETIC CHARGE

Charge *gives rise to electric force,* and we are only now beginning to figure out how it manages to do that. Charge is fundamental and cannot be described in terms of simpler, more basic concepts. We know it by what it does, not by what it is—if you like, it is what it does, and that's that.

If we vigorously stroke a plastic pen on a woolen glove and hold the two apart but close, fibers on the glove will stand on end, straining to reach up to the pen, making the attraction obvious. It was a French botanist, C. Dufay, who first studied the *repulsive* interactions of electricity. He found that objects of the same material electrified in the same way repelled one another. Two pieces of glass rubbed with silk will repel each other, just as two chunks of amber rubbed with fur will (Fig. 17.1). Yet the charged glass will attract both the silk and the charged amber, and vice versa. Sometime around 1734, Dufay concluded "that there are two distinct Electricities"—two kinds of electric charge—and he was quite right. Summarized in contemporary terms: **like charges repel, unlike charges attract** (Fig. 17.2).

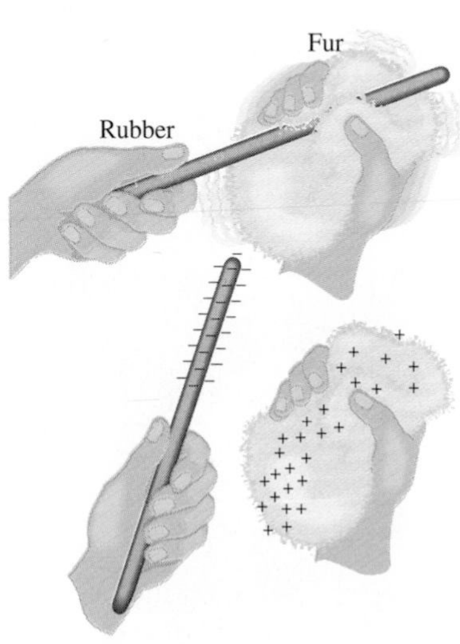

Figure 17.1 Charging by contact. Electrons are transferred from fur to rod. The hard-rubber rod becomes negatively charged, the fur becomes positively charged.

17.1 Positive and Negative Charge

Today we follow Benjamin Franklin's lead, arbitrarily calling the two kinds of charge *positive* and *negative.* In ordinary circumstances, we deal with the electrical behavior of solids (plastic combs, nylon sweaters, TV screens, etc.) and there the positive charges are locked up in the nuclei of the essentially stationary atoms, while some negative charges are more or less free to be transferred. There are two kinds of charge but, in the vast majority of cases, only one is mobile.

An object that contains the same amount of positive as negative charge in close proximity attracts and repels an external charge equally, and thereby cancels its own ability to exert a net force. It behaves as though it had no charge at all and is electrically **neutral**. That ability to combine charges to produce a null response allows us

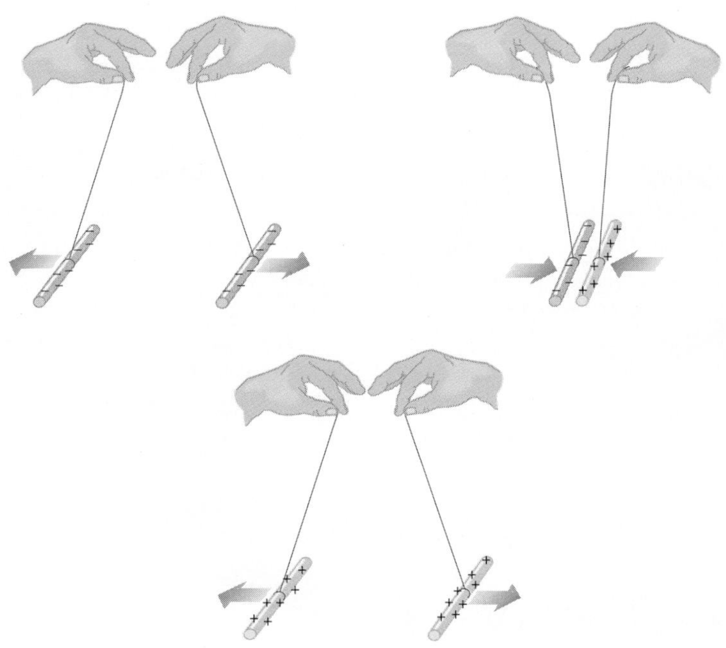

Figure 17.2 Attraction and repulsion of charges. Like charges repel; unlike charges attract.

to create a kind of electrical algebra: 10 units of positive charge and 10 units of negative charge add up to a net of zero units of charge, in the sense that together they produce no observable external electric force. Any macroscopic object, yourself included, contains a vast number of minute individual charges, but on the whole it is usually neutral. That's why you are not noticeably pulled on by the electric force as you are by the gravitational force.

Electricity is of two kinds, positive and negative. The difference is, I presume, that one comes a little more expensive, but is more durable; the other is a cheaper thing, but the moths get into it.

STEPHEN LEACOCK (1869–1944)
Canadian humorist

17.2 Charge Is Quantized and Conserved

Charge is a property of most of the subatomic entities that make up the material Universe. An electron is negatively charged; it repels other electrons and it attracts protons. The latter are positively charged. There is no way to discharge an electron, no way to peel off its charge and make it naked, neutral. If the electron is truly a fundamental particle, and most physicists believe it is, charge is an inseparable aspect of the thing in itself.

The amount of the charge of the electron, q_e, has been determined experimentally. Moreover, all electric charge comes in whole number multiples of that basic amount (the numerical value of which we will deal with shortly). Whether positive or negative, **charge is quantized**, it appears in certain specific amounts—every subatomic entity that has been observed to date has had a charge of either 0, $\pm q_e$, or $\pm 2q_e$. Since matter on the subatomic scale comes in specific lumps—only a small number of fundamental particles exist—it's not surprising that charge comes in lumps as well. Still, it is curious that the proton has a charge ($+q_e$) equal in size to that of the electron ($-q_e$), since these particles are otherwise quite different. Modern experiments have shown that the magnitudes of the charges of the electron and proton are so nearly alike that if their ratio does differ from 1 at all, it does so by less than 10^{-20}. That's fascinating, especially since the proton seems to have a complex structure. Contemporary theory maintains that most heavy subatomic particles are actually composite systems (Ch. 33) made up of several varieties of smaller fun-

This old AM radio represents so much practical physics that we'll come back to it several times throughout our study of Electromagnetic Theory.

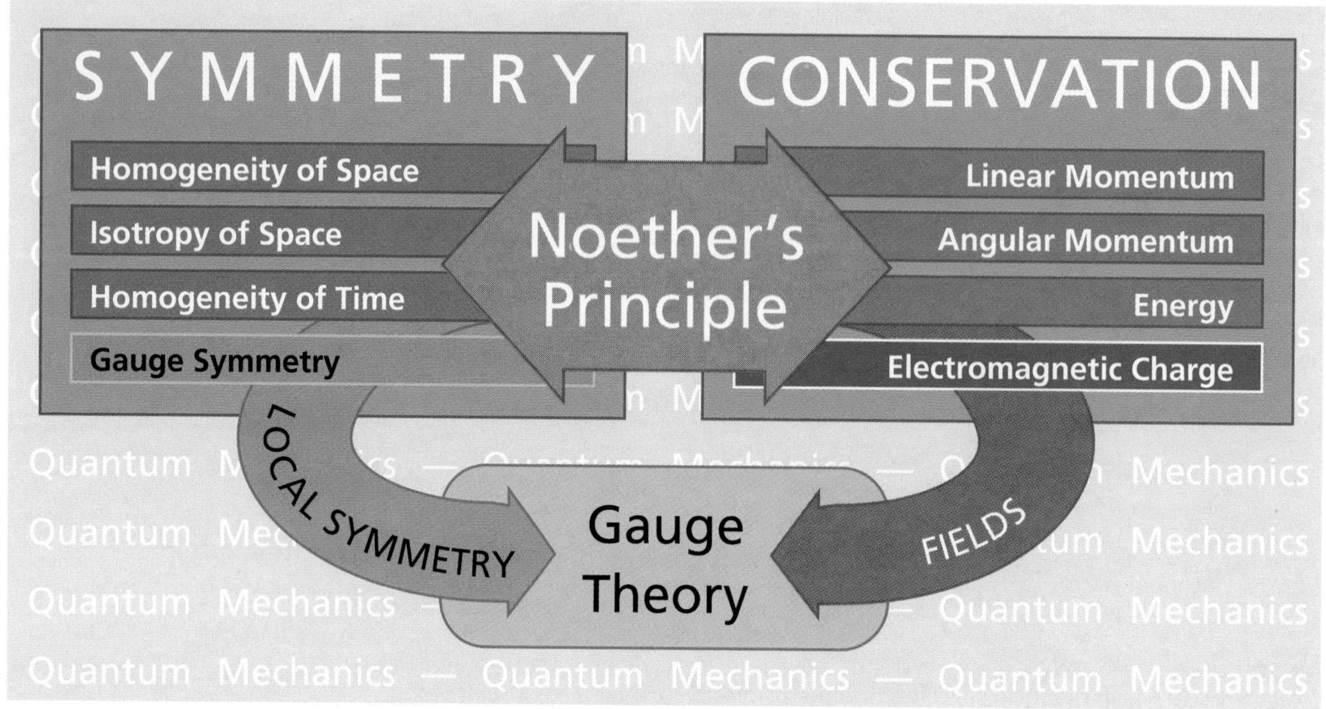

Figure 17.3 Thus far, we have recognized Conservation of Linear Momentum, Angular Momentum, and Energy. Now we add Conservation of Electromagnetic Charge. Later, on p. 618, we will relate charge conservation to symmetry.

damental entities called *quarks*. These, in turn, are supposed to have charges of $\pm\frac{1}{3}q_e$ and $\pm\frac{2}{3}q_e$. It is believed, however, that quarks cannot ordinarily exist in the free state (that is, independently), and so the observable unit of charge is indeed q_e.

A neutral "chargeless" object is electrified by either losing or gaining charge; the glass rod loses electrons, the silk used to rub it gains them. Unlike mass (which can be converted into energy, Ch. 29), electromagnetic charge is scrupulously conserved. The net charge (the difference between the amount of positive and negative charge) within an isolated system is always constant. That concept is the **Law of Conservation of Charge** (Fig. 17.3). Whenever a positive charge appears, somewhere in the vicinity there will appear an equal negative charge. As far as we can tell, the net amount of charge in the Universe is constant, which does not mean charge cannot be created; it just means that, if it is, as much positive as negative will be produced. This constant net charge of the Universe may well equal zero (that's a pretty thought), but clearly no one knows.

17.3 Charging by Rubbing

We are surrounded by charges that form a subtle electrostatic environment that we generally don't pay much attention to. For example, everyone has probably felt the sparks that often punctuate a short shuffle across a carpet on a dry winter's day. By contrast, in the potentially explosive anesthetic-rich environment of a hospital operating room, even a minute spark of this sort could be disastrous. Most of us have stroked plastic foodwrap so it adheres to some container, rubbed a balloon on a shirt to stick it on a wall, used a copy machine, or fussed with clinging clothes—the electrostatic effects of charge are everywhere.

Atoms are neutral; they possess as many electrons as protons. But the outer electrons are the least strongly bound and they are most easily shed. The process whereby these electrons pass from one object to another is not entirely understood.

TABLE 17.1 The triboelectric sequence

Asbestos Fur (rabbit) Glass Mica Wool Quartz Fur (cat) Lead Silk Human skin, aluminum Cotton Wood Amber Copper, brass Rubber Sulfur Celluloid India rubber	On contact between any two substances shown in the column, the one appearing above becomes positively charged, the one listed anywhere below it becomes negatively charged.

***Electrical action-reaction before the* Principia**. *It is commonly believed, that Amber attracts the little Bodies to itself; but the Action is indeed mutual, not more properly belonging to the Amber, than to the Bodies moved, by which it also itself is attracted; . . ."*

MAGALOTTI (1665)
Florentine Academy

Nonetheless, we do have a good overall idea of what's happening. Different materials have different affinities for electrons. When two substances are put in contact, one of them may give up some of its loose electrons while the other draws them into itself. For example, when a sheet of plastic is pressed down onto a metal plate, electrons will be transferred from the donor plastic to the grabber metal. The plastic, having lost electrons, now contains a number of immobile positive ions (atoms missing negative charge) on its surface and, as a whole, has become charged. The positive plastic attracts the negative metal and the two cling to one another.

In much the same way, when a hard-rubber rod is stroked with a piece of fur, the rod draws off electrons, becoming negatively charged, and the donor fur assumes an equal, positive charge. *The rubbing seems to do little beyond increasing the area brought into intimate contact.* Moreover, there are degrees of grabbers, and a substance that can snatch electrons away from one material may well find itself serving as donor to a still more potent grabber. Glass rubbed with asbestos will draw off electrons from that fibrous material, becoming negative, but if stroked with some persistence against silk or flannel, the glass will emerge positively charged, having lost electrons (Fig. 17.4). We can bring some order to the whole business by arbitrarily defining glass-rubbed-on-silk to be *positive,* and then any charged object that is attracted to it is *negative.*

By comparing the behavior of various materials, a listing (Table 17.1) known as the *triboelectric* (*tribo,* meaning friction) *sequence* has been formulated. The materials toward the top of the list tend to lose electrons easily, the ones near the bottom tend to gain them effectively. The farther apart on the list the two are, the more intense the resulting electrification, which is why rabbit fur and hard rubber are still the mainstays of electrostatic demonstrations (asbestos is carcinogenic and should be avoided).

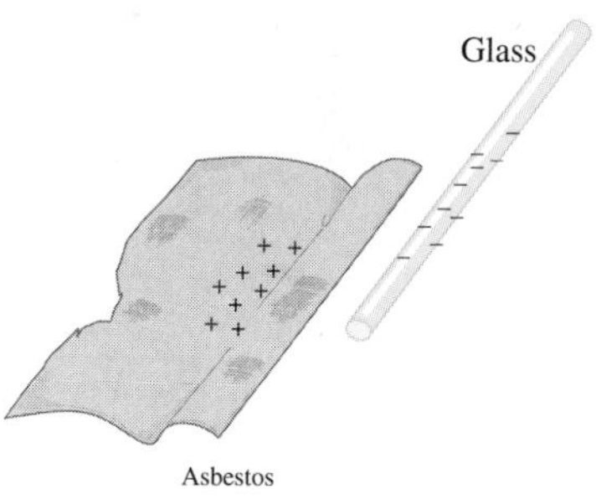

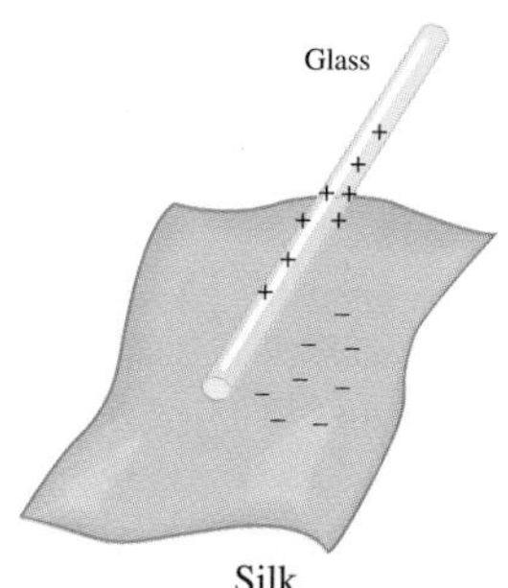

Figure 17.4 Electrons are transferred in contact from the asbestos to the glass, and from the glass to the silk.

17.4 The Transfer of Charge

A negatively charged object contains an excess of electrons that can move about, somewhat at least, and that repel each other. When such an object is placed in con-

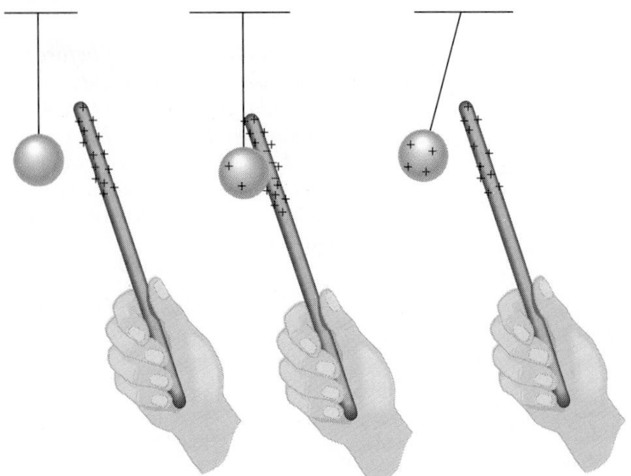

Figure 17.5 Charging a sphere of pith positively by direct transfer of electrons from ball to rod. This leaves the ball positively charged, and it is immediately repelled from the rod.

A new and extremely sensitive technique for revealing fingerprints has been developed at Los Alamos National Laboratory. It makes use of the fact that unlike charges attract. Negatively charged gold particles adhere to the positively charged proteins, which are always left behind when fingers touch a surface.

tact with a neutral body, some of those electrons are forced over onto that body, charging it negatively. Similarly (Fig. 17.5), a positively charged body has a deficiency of electrons or, equivalently, an excess of positive ions. When placed in contact with a neutral body, it attracts and draws off electrons, becoming less positive itself, while causing the now electron-diminished once-neutral body to become positive. Only electrons are transferred, but the system behaves exactly as though positive charge flowed from the charged object into the neutral one. That being the case, it's common to talk about the transfer of positive charge even though no positive charge actually moves.

The arrangement in Fig. 17.2 suggests a straightforward scheme for determining the polarity of any distribution of charge. Thus, suppose we want to find the sign of a charged elephant and don't have a modern electrometer handy. Just suspend two identical, lightweight neutral objects—two Ping-Pong balls, say, or two pith balls.* Now touch a rod of glass-rubbed-on-silk to one of the pith balls, thereby essentially transferring positive charge to it. Next, back the electrified pachyderm into the second pith ball, which will take on some of the unknown elephant-charge. If the two pith balls then attract, the elephant *may* be negative: if they repel, it surely is positive (see Discussion Question 15).

17.5 Insulators and Conductors

Any astute observer might ask, "Is it possible that our elephant got charged only at one end?" The answer is "Yes." If you rub the middle of a plastic comb with a piece of wool, that region will have the ability to pick up tissue tufts, but the comb's ends

*Pith is the pulpy, soft stuff in the center of certain dried plant stems. It was the Styrofoam of generations past, used in work with electricity for centuries and still to be found in physics storage rooms around the world.

will not—they remain neutral, even though the middle is electrified. This behavior represents a class of materials (such as wood, plastic, glass, air, hair, cloth, leather—and dry elephant), all of which are variously known as **insulators, nonconductors**, or **dielectrics**. Charges in a nonconductor have limited mobility and will only move when their mutual repulsion is great enough to overcome the tendency to be held in place by the host atoms. *When an insulator receives a charge, it retains that charge, confining it within the localized region in which it was introduced.**

A **conductor** *allows charge introduced anywhere within it to flow freely and redistribute.* Metals (copper, gold, aluminum, etc.) are among the best conductors at ordinary temperatures. Incidentally, the fact that there are both good and bad conductors of electricity was first realized in 1729 by Stephen Gray, who carried out a series of brilliant experiments, despite the severe handicap of being a pensioner in a London poorhouse.

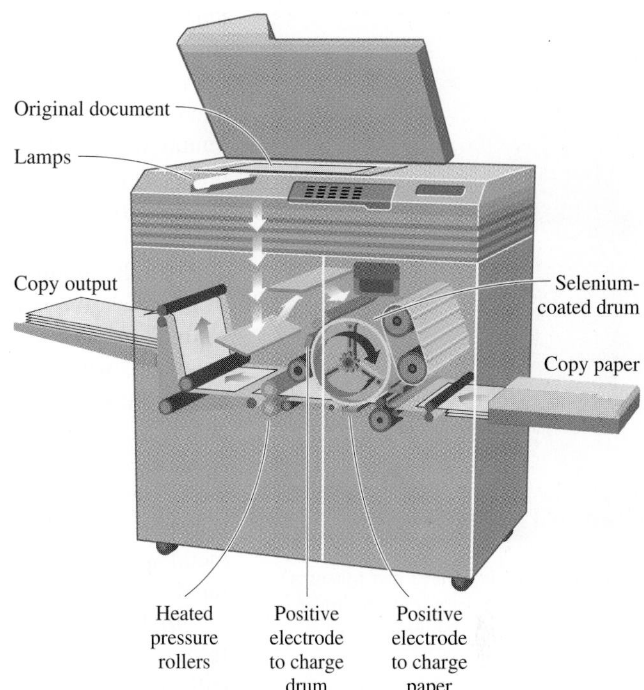

Figure 17.6 The drum is the centerpiece of both the electrostatic (photo) copier and the laser printer. It's an aluminum cylinder with a coating of photoconducting selenium, which is an insulator in the dark and a conductor in the light. The entire surface is first positively charged, then the image of the document is projected on the drum. Wherever light falls, the charge is conducted away. The dark regions remain positive and attract negatively charged toner powder. A sheet of highly positively charged paper then picks up the toner, to which it is fixed by heated rollers.

The distinction between conductors and nonconductors (and it's not always a clear one) arises from the relative mobility of charge within the material. The atoms of metals hold their own outermost electrons weakly and so a bulk sample contains a tremendous number of free electrons, roughly one per atom. Any added electrons join that sea, which moves among the unmoving positive ions. Alternatively, the atoms of a nonconductor hold fast to their own electrons and will even latch on to excess ones introduced on them (Fig. 17.6). Despite this, *no material is a perfect insulator;* all allow some redistribution of charge. Thus, human skin is a respectable nonconductor compared to copper, although it is a fairly decent conductor when compared to glass. Pure water is a modest insulator, but a pinch of some dissolved impurity such as table salt will provide enough ions to turn it into a good conductor (*ion* means *goer* or *traveler*).

Air is a good insulator, particularly dry air, even though it contains some 300 ions per cubic centimeter. Nevertheless, if enough negative charge builds up on an object, electrons under the influence of their mutual repulsion can be propelled into the surrounding gas. The air will have some of its own electrons ripped off, becoming ionized and creating a temporary conductive pathway along which the bulk of the charge then flows. Collisions with the gas increase its temperature and cause some of the atoms to emit light. The result is the familiar glowing trail known as a **spark**.

Charged Conductors

The free flow of electrons in a conductor is responsible for several interesting properties. First, unlike an insulator, where charge pretty much stays where it's put, a cluster of identical charges introduced on a conductor experiences a mutual repulsion that sends them all scurrying. Constantly pushed apart, they move until they can separate no farther and are as distant from one another as possible (Fig. 17.7). Charges tossed onto a metal sphere will very quickly stream around until they are

*The great bane of electrostatic demonstrations is dampness. Warm the objects being charged, work in a dry place, and keep metal, glass, and plastics clean by occasionally wiping them with alcohol. Airborne water molecules are electrically polarized; the hydrogen "Mickey Mouse ears" are positive and tend to attract electrons and so inevitably discharge apparatus.

Here, a bolt of lightning just misses a *Space Shuttle*. Years earlier, a similar bolt destroyed an unmanned rocket.

uniformly distributed and at rest on the outer surface. **No matter what the shape of the conductor, excess charge always resides on its outer surface**. Charges simply push each other to the very extremities of the object. With a nonspherical conductor the charge distribution will be nonuniform, bunching up somewhat in the remote regions (Fig. 17.8). Each charge is impelled to get as far away from as many others as possible, even if that means getting closer to a few charges in the process. That's why charge tends to concentrate on the sharp protrusions of a conductor, a fact well known to people who worry about sparks.

Another manifestation of the free flow of electrons in a conductor is the manner in which charge is transferred. Envision a negative conductor made to touch an uncharged metal body. Electrons are propelled onto the neutral body by their mutual repulsion, which depends on how densely packed they were to begin with. The charge flows much as a fluid flows from a filled chamber into a connecting empty chamber of arbitrary shape. That gravity-powered flow continues until the liquid levels are the same, the pressures equalize, and equilibrium is reached. As we will see in Sect. 18.7 (p. 636), a very similar balance determines the amount of charge transported. Evidently, if a total excess charge Q is placed on one of two identical metal spheres (Fig. 17.9) and those spheres are brought into contact and then separated, a charge of $\frac{1}{2}Q$ will end up on each of them.

In 1786 Rev. A. Bennet introduced a device that, until the modern era of electronics, was the premier electrostatic indicator. The **gold-leaf electroscope** is basically two extremely thin metal leaves hanging parallel to each other from a conducting wire, all surrounded by a protective glass enclosure (Fig. 17.10). A

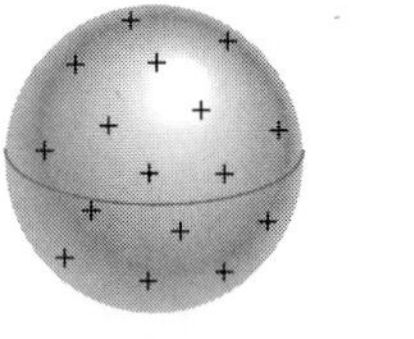

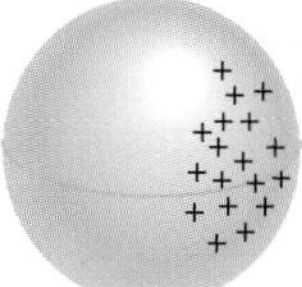

Figure 17.7 The distribution of charge placed on the surfaces of a conductor and a nonconductor.

Figure 17.8 Charge tends to bunch up on the pointed regions of a conductor.

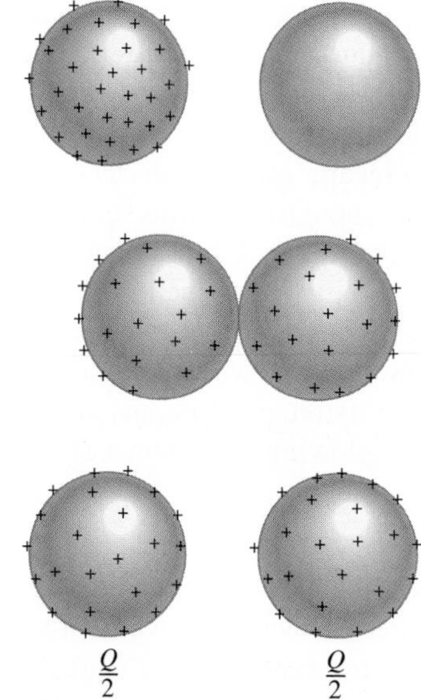

Figure 17.9 The distribution of charge on the surfaces of identical conductors. The charge divides evenly on the two spheres.

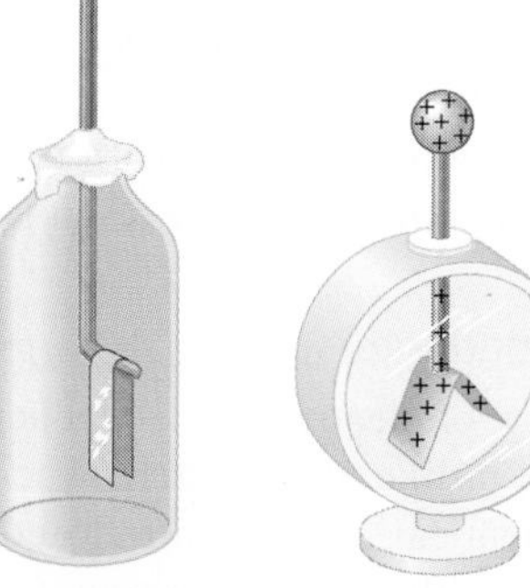

Figure 17.10 Electroscopes homemade and otherwise. No longer used in the laboratory, the electroscope is now primarily a teaching device. To make one, remove the thin aluminum foil from the wrapper on a stick of gum. Hang it on a thick wire, lower it into a bottle, and seal the bottle with wax or clay.

charge introduced onto the metal knob on top spreads out; some travels down the wire and divides equally onto each of the leaves. Repelling one another, the leaves spring apart at an angle proportional to the net charge. And there they stay, leaves apart, until the charge is deliberately removed or until it leaks off into the air.

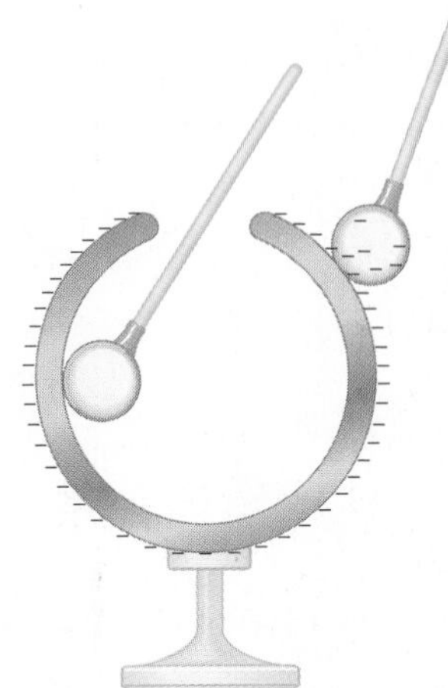

Figure 17.11 Proof-planes inside and outside a hollow charged conductor.

THE ELECTRIC FORCE

In the mid-eighteenth century, the question of the nature of the force acting between charges was a major scientific issue. Naturally, speculation had a decidedly Newtonian bent, relying heavily on discoveries regarding gravity. As early as 1760, Daniel Bernoulli confirmed that his own rather crude experiments were at least consistent with an inverse-square law for electrical attraction and repulsion.

Benjamin Franklin observed that pith balls lowered inside a metal cup appeared to be unaffected by whether the cup was charged or not. We know that excess charge repels itself to the outer extremities of the cup, leaving the inner surface chargeless. That fact is easy to confirm with a *proof-plane,* a small metal disk on an insulating handle (Fig. 17.11). If the proof-plane is touched to the outside of the electrified cup, an electroscope shows that the proof-plane is thereafter charged. Yet touching the inside wall of the cup leaves the proof-plane neutral—*there is no charge on the inside of a hollow electrified conductor.* That's clear, but it does not explain why there is no electric force inside the hollow—the pith ball is not attracted to a nearby wall even though there are charges just on the other side of that wall. Franklin communicated these findings to his English friend Joseph Priestley, the discoverer of oxygen.

Priestley made a not altogether surprising guess at the cause, proposing that the force between charges varied inversely with the square of the distance separating them. Newton had already shown (p. 216) that a particle would experience a zero gravitational force when placed inside the cavity of a uniform spherical shell of mass. And Priestley assumed that since gravity varied as $1/r^2$, the electric force did as well. Though he was correct, the analogy is really a poor one: gravitationally, the null effect is a special case where the spherical geometry works to balance the $1/r^2$ drop-off of the force. By contrast, the zero electric force inside a hollow conductor is independent of the shape of the shell, primarily because of the way the free charges distribute themselves on the outer surface.

There had been some direct experimental tests of the inverse-square law by Robison (1769) and later much more completely by Cavendish (1775), but the latter, a rather strange fellow (p. 212), never made his results public. Consequently, by the time Charles Augustin de Coulomb published his definitive study (1785), it was already widely suspected that the electrical force between two *point-charges* varied inversely with the square of their separation.

17.6 Coulomb's Law

Charles Coulomb (1736–1806).

Coulomb's apparatus (Fig. 17.12) consisted of a lightweight insulating rod hung horizontally from its middle on a long, fine silver wire. On one end of the rod was a small pith ball covered with gold leaf; on the other, counterbalancing it, was a paper disk (a damping vane). Positioned behind the suspended ball was another identical one fixed in place so that initially the two were in contact. They were then both electrified simultaneously upon being touched by a charged rod. Immediately, the suspended ball swung away from the fixed one, twisting the silver wire, and coming to rest some distance away (Fig. 17.13).

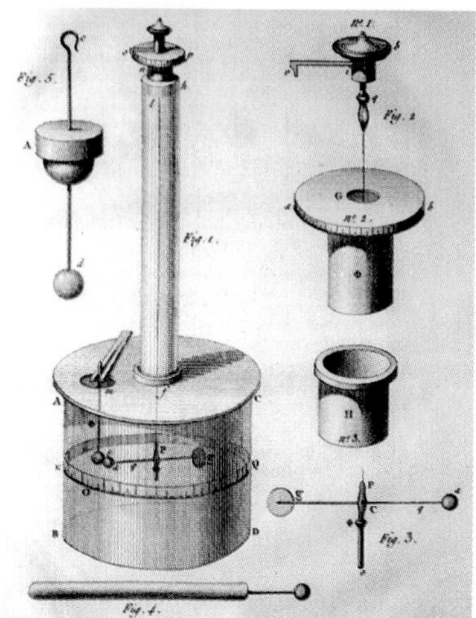

Figure 17.12 C. A. Coulomb's torsion balance (from *Histoire et Mémoires de L'Académie Royale des Sciences,* 1785). A light horizontal rod has a small sphere (a) on its left end and a disk counterweight (g) on its right end. An identical sphere (t) is inserted at the left through a hole (m). The arrangement is shown in Fig. 3.

Coulomb had already determined how much force it took to twist the wire through any angle. Now with the movable sphere pushed aside, he slowly twisted the suspension wire by hand, forcing it back closer to its original position near the fixed sphere, against the balls' mutual repulsion. While doing that, he measured the separations and determined the forces holding the two spheres at each location. His results were clear: "The repulsive force between two small spheres charged with the same type of electricity is inversely proportional to the square of the distance between the centers of the two spheres." The electric force (F_E) of interaction between the spheres (and, by Newton's Third Law, the forces are equal in magnitude and opposite in direction even if the charges are unequal) is proportional to $1/r^2$, where r is the center-to-center distance

$$F_E \propto \frac{1}{r^2}$$

just as is the gravitational interaction. Modern experiments confirm that if the power to which r is raised differs from 2, it must deviate by *less* than 3×10^{-16}.

Provided the two interacting spheres are small compared to their separations, we can expect the charge distributions on each of them to remain fairly, though not precisely, uniform. The presence of either sphere causes a shift of charge on the other sphere (that's part of the mechanism of interaction), just as a rock sags when a butterfly lands on it. When the spheres are very small, the repulsive forces between charges on either one of them will be powerful enough to keep the distribution almost uniform, especially if the other disturbing sphere is far away. With this restriction in mind, *we assume uniform charge distributions on each of the interacting spheres.*

Newton showed that a particle of mass, located outside a thin spherical shell, will be drawn gravitationally toward the sphere's center as if all the shell's mass is concentrated at that point (p. 215). In an analogous way, we can conclude that *a uniformly charged conducting sphere produces the same electric force as if all its charge is concentrated at a single point at its center.* The force law can be assumed to be precise for two point-charges separated by a distance r (and nearly so for two small charged spheres).

Coulomb ingeniously determined how the force law varied with the magnitude of the two interacting charges q_1 and q_2. By touching the charged sphere with an identical uncharged sphere (Fig. 17.9), he was able to exactly halve the charge on it, even without knowing what that charge was. In the same way, he could neatly reduce it to a quarter, an eighth, etc., of its initial value simply by repeating the touch-and-halve procedure with two spheres. He found for the two interacting spheres that halving the charge on one halved the force, while halving the charge on both quartered it. Coulomb concluded that the force of interaction was proportional to the product of the charges q_1q_2 and so

$$F_E \propto \frac{q_1q_2}{r^2}$$

The similarity with the gravity formula is obvious and striking, though quite unexplained theoretically even to this day.

The right side of this relationship could be made identical to the left side (not just proportional to it) by defining the unit of charge appropriately: one unit of charge on each sphere would produce one unit of force when the separation

Figure 17.13 A detail of Coulomb's device. Once charged, the two spheres repel and move apart. That behavior twists the wire until the force it exerts matches the electrostatic repulsion.

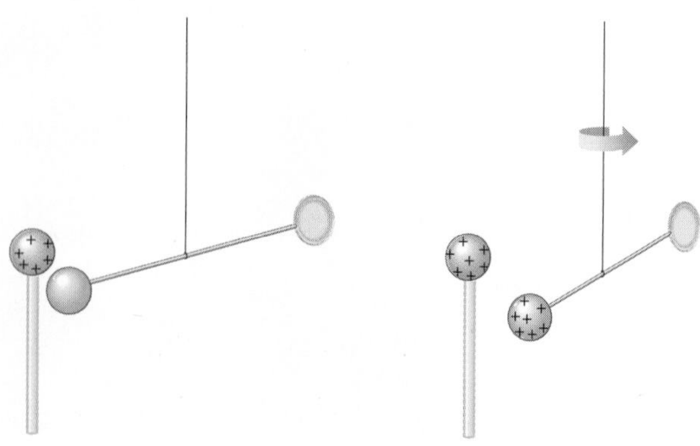

was one unit of length. That definition was formulated in the early days in the system that used *centimeters, grams,* and *seconds.* There, the unit of force was the *dyne* (1 dyne $=10^{-5}$ N) and the unit of charge was the now long-forgotten *franklin.* But the force law is really not a very good way to measure charge—the distributions on the spheres are not actually uniform. So nowadays the unit of charge is defined via the more precisely measured interactions of electrical currents (Sect. 21.10). Accordingly, there is no reason for the right side of the above proportionality to equal the left side; in fact, there is no reason to even expect that the units will be equivalent. If charge is defined in some other way, the force law must contain a constant of proportionality (k) that also has units. The electric force between two point charges q_1 and q_2 in some medium (air, vacuum, whatever), separated by a distance r is

$$F_{\mathrm{E}} = k\frac{q_1 q_2}{r^2} \tag{17.1}$$

The force acts along the line connecting the charges; like charges repel whereas unlike charges attract. This is the modern formulation of **Coulomb's Law**. The internationally accepted (SI) unit of charge is the *coulomb* (C), and k is a constant specific to the medium in which the charges are imbedded. In the special case of vacuum $k = k_0 = 8.987\,551\,79 \times 10^9\ \mathrm{N{\cdot}m^2/C^2} \approx 9.0 \times 10^9\ \mathrm{N{\cdot}m^2/C^2}$. The value of k in air is smaller than k_0 by only 0.06%, which is generally quite negligible. When the charges are imbedded in other media such as water, glass, or oil, we will have to use appropriate constants for each. For example, the value of k for water is measured to be about 1/80 that for vacuum, making the forces between charges that much smaller, but we need not worry about that here.

Until 1983, k_0 was determined by measurements, much as is the constant G. For reasons we will see later, k_0 turns out to be numerically precisely equal to 10^{-7} times the speed of light (c) in vacuum squared, and since c is now fixed by definition (p. 10), so too is k_0. Notice that the units of k are whatever it takes to cancel out all the units on the right of Eq. (17.1) and still leave N, the unit of force.

Two small spheres, each carrying 1 coulomb of charge and separated in vacuum by 1 meter, will each experience a force of about 9×10^9 N. That is a tremendous force (equal to the weight of more than 2400 jumbo jets), and it arises because a coulomb is a tremendous amount of charge. Since the best modern experimental value of the *charge of the electron* is $-1.602\,177\,33 \times 10^{-19}$ C (within an error of $\pm 0.000\,000\,49 \times 10^{-19}$ C), 1 coulomb corresponds to a charge of more than six million million million electrons.

Example 17.1 A hydrogen atom consists of an electron of mass $m_e = 9.1094 \times 10^{-31}$ kg, moving about a proton of mass $m_p = 1.6726 \times 10^{-27}$ kg at an average distance of 0.53×10^{-10} m. Determine the ratio of the electrical and gravitational forces acting between the two particles.

Solution: [Given: $\pm q_e = \pm 1.60 \times 10^{-19}$ C, $m_e = 9.1094 \times 10^{-31}$ kg, $m_p = 1.6726 \times 10^{-27}$, and $r = 0.53 \times 10^{-10}$ m. Find: F_E and F_G.] From Coulomb's Law

$$F_{\mathrm{E}} = k\frac{q_e q_p}{r^2}$$

$$F_{\mathrm{E}} = (9.0 \times 10^9\ \mathrm{N{\cdot}m^2/C^2}) \times \frac{(-1.60 \times 10^{-19}\ \mathrm{C})(+1.60 \times 10^{-19}\ \mathrm{C})}{(0.53 \times 10^{-10}\ \mathrm{m})^2}$$

and $$F_{\mathrm{E}} = -8.2 \times 10^{-8}\ \mathrm{N}$$

The minus sign here simply indicates that the force is

(continued)

(continued)

attractive, and that usage is widely adhered to in electrostatics. The sign of the force tells us only if it is attractive (−) or repulsive (+). The force on the proton due to the electron is equal in magnitude and opposite in direction to the force on the electron due to the proton, but each equals -8.2×10^{-8} N. Had we introduced this usage earlier, the Law of Universal Gravitation [Eq. (7.2)] would have had a minus sign in front of it since it's always attractive and the masses are always positive; hence

$$F_G = -G\frac{m_p m_e}{r^2} = -(6.67 \times 10^{-11}\ \text{N·m}^2/\text{kg}^2)$$

$$\times \frac{(1.67 \times 10^{-27}\ \text{kg})(9.11 \times 10^{-31}\ \text{kg})}{(0.53 \times 10^{-10}\ \text{m})^2}$$

$$F_G = -3.6 \times 10^{-47}\ \text{N}$$

And so $\boxed{F_E/F_G = 2.3 \times 10^{39}}$; the electrical force is far stronger than the gravitational force.

▶ **Quick Check:** $F_E/F_G = kq_eq_p/Gm_em_p = 2.3 \times 10^{39}$.

The electrostatic generator charges each strand of this young woman's hair; thereafter they separate as far as possible.

By human standards, the coulomb is definitely a great deal of charge. We are accustomed to taking our charge in far smaller doses, usually in the form of sparks that carry much less than a microcoulomb (1 μC = 10^{-6} C). Usually rubbing will build up charge on an ordinary-sized object with a density of up to 10 nanocoulombs per square centimeter (1 nC = 10^{-9} C), which is a practical limit beyond which there will tend to be discharging into the air.

The electrical force is a vector quantity. It has been confirmed experimentally that when several charges are present, each exerts a force given by Eq. (17.1) on every other charge. **The interaction between any two charges is independent of the presence of all other charges.** Thus, *the net force on any one charge is the vector sum of all the forces exerted on it due to each of the other charges interacting with it independently.* Only in the special case where the charges are arrayed along a straight line will the force vectors be colinear and can they be added or subtracted as scalars (provided we keep track of their directions).

Example 17.2 Figure 17.14a shows three tiny uniformly charged spheres. Determine the net force on the middle sphere due to the other two.

Solution: [Given: $q_1 = +5.0\ \mu$C, $q_2 = -4.0\ \mu$C, $q_3 = +10.0\ \mu$C, $r_{12} = 2.0$ cm, and $r_{23} = 6.0$ cm. Find: F_2.] Watch out for units—we want everything in SI. The net force on charge 2 is the vector sum of the force exerted on 2 by 1, namely, $\mathbf{F}_{21}$; and the force exerted on 2 by 3, namely $\mathbf{F}_{23}$:

$$\mathbf{F}_2 = \mathbf{F}_{21} + \mathbf{F}_{23}$$

The next step is to sketch the problem out pictorially showing the directions of the forces, as is done in Fig. 17.14b. Because the charges have opposite polarities, the forces are both attractive and act in opposite directions. Coulomb's Law provides the numerical values of those forces; accordingly

$$F_{21} = k\frac{q_1q_2}{r_{21}^2} = (9.0 \times 10^9\ \text{N·m}^2/\text{C}^2)$$

$$\times \frac{(+5.0 \times 10^{-6}\ \text{C})(-4.0 \times 10^{-6}\ \text{C})}{(2.0 \times 10^{-2}\ \text{m})^2} = -450\ \text{N}$$

(continued)

(continued)

and

$$F_{23} = k\frac{q_2 q_3}{r_{23}^2} = (9.0 \times 10^9 \text{ N}\cdot\text{m}^2/\text{C}^2)$$

$$\times \frac{(-4.0 \times 10^{-6}\text{ C})(+10.0 \times 10^{-6}\text{ C})}{(6.0 \times 10^{-2}\text{ m})^2} = -100 \text{ N}$$

These minus signs (attraction) have been integrated into the solution via the directions of the forces in the diagram and are no longer of any concern—forget about them! The two vectors in Fig. 17.14c must now be added together. As ever, when treating colinear vectors, we take the direction of the x-axis to be positive. Hence

$$F_2 = F_{21} + F_{23} = (-450 \text{ N}) + (+100 \text{ N}) = -350 \text{ N}$$

The resulting force is $\boxed{3.5 \times 10^2 \text{ N acting to the left}}$.

▶ **Quick Check:** From Coulomb's Law, $F_{23} = (2/3^2)F_{21}$; hence, $F_2 = (7/9)F_{21} = (7/9)(-450)$.

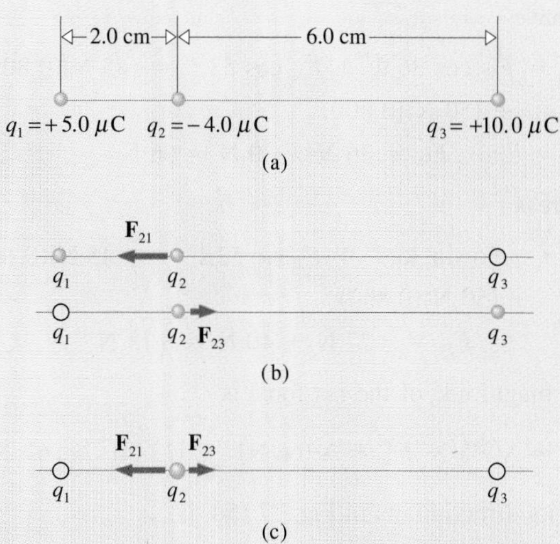

Figure 17.14 Opposite charges attract; like charges repel. Notice how strongly the distance ($1/r^2$) affects the force: $F_{23} < F_{21}$, even though $q_3 > q_1$.

In theory, if we know the details of some charge distribution, no matter how complicated, we can compute the net force it exerts on a single external charge q. Practically, the summation of all the individual Coulomb interactions can be accomplished either via calculus (approximating the situation as if the charge were distributed continuously) or, more directly, using a computer to carry out the large number of separate applications of Eq. (17.1) that would be necessary. Being more concerned with the underlying ideas, our analysis is limited to the simplest cases of just a few charges.

Example 17.3 Figure 17.15a depicts three small charged spheres at the vertices of a 3-4-5 right triangle. Calculate the force exerted on q_3 by the other two charges.

Solution: [Given: $q_1 = +50\ \mu$C, $q_2 = -80\ \mu$C, $q_3 = +10\ \mu$C, $r_{12} = 50$ cm, $r_{13} = 30$ cm, and $r_{23} = 40$ cm. Find: F_3.] To start, determine if the forces are attractive or repulsive and then draw them acting center-to-center on q_3, as in Fig. 17.15b. Since these two force vectors are not colinear, we must, as usual, resolve them into perpendicular components and sum those individually. Figure 17.15c provides the appropriate geometry. But first we need to compute the magnitudes of the two force vectors; consequently

$$F_{31} = k\frac{q_1 q_3}{r_{31}^2} = (9.0 \times 10^9 \text{ N}\cdot\text{m}^2/\text{C}^2)$$

$$\times \frac{(+50 \times 10^{-6}\text{ C})(+10 \times 10^{-6}\text{ C})}{(30 \times 10^{-2}\text{ m})^2} = +50 \text{ N}$$

which is positive because like charges repel. Furthermore

$$F_{32} = k\frac{q_2 q_3}{r_{32}^2} = (9.0 \times 10^9 \text{ N}\cdot\text{m}^2/\text{C}^2)$$

$$\times \frac{(-80 \times 10^{-6}\text{ C})(+10 \times 10^{-6}\text{ C})}{(40 \times 10^{-2}\text{ m})^2} = -45 \text{ N}$$

which is negative because unlike charges attract. Hence, from Fig. 17.15c, forgetting the attractive-minus sign,

(continued)

(continued)
we have

$$F_{3x} = F_{32}\cos 36.9^\circ + F_{31}\cos 53.1^\circ = (45\ \text{N})(0.800) + (50\ \text{N})(0.600)$$

and $$F_{3x} = 36\ \text{N} + 30\ \text{N} = 66\ \text{N}$$

whereas

$$F_{3y} = -F_{32}\sin 36.9^\circ + F_{31}\sin 53.1^\circ = -(45\ \text{N})(0.600) + (50\ \text{N})(0.800)$$

and $$F_{3y} = -27\ \text{N} + 40\ \text{N} = +13\ \text{N}$$

The magnitude of the net force is

$$F_3 = \sqrt{F_{3x}^2 + F_{3y}^2} = \sqrt{(66\ \text{N})^2 + (13\ \text{N})^2} = \boxed{67\ \text{N}}$$

and its direction, as in Fig. 17.15d, is

$$\theta = \tan^{-1}\frac{F_{3y}}{F_{3x}} = \tan^{-1}\left(\frac{13\ \text{N}}{66\ \text{N}}\right) = \boxed{11^\circ}$$

▶ **Quick Check:** There is no different easy way to arrive at the above results, but since the force between two $\pm 10\ \mu$C charges 10 cm apart is $\approx \pm 10$ N, these values are at least the right order-of-magnitude. Repelled by q_1 and attracted to q_2, the force on q_3 must be in either the first or fourth quadrants.

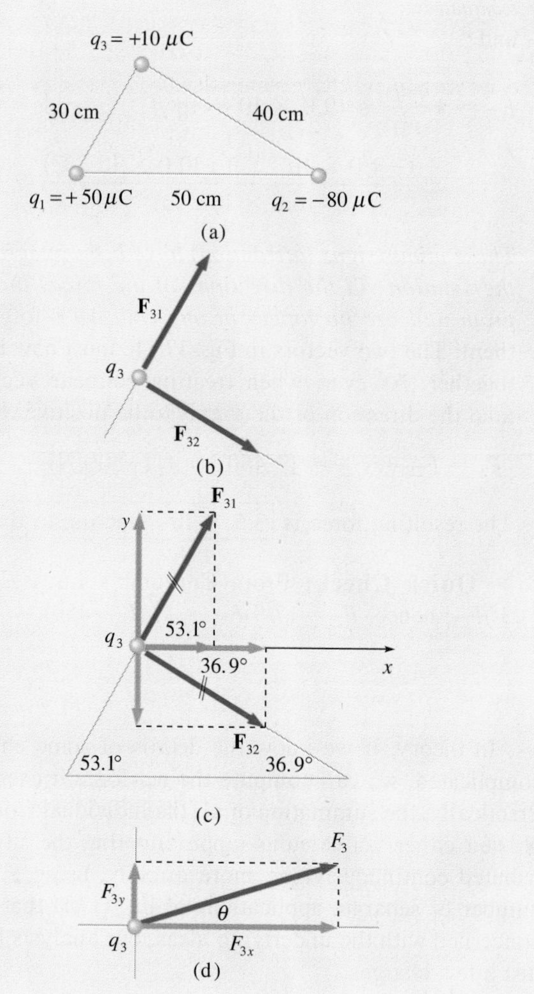

Figure 17.15 The force $\mathbf{F}_3$ on a charge q_3 due to two other charges q_1 and q_2.

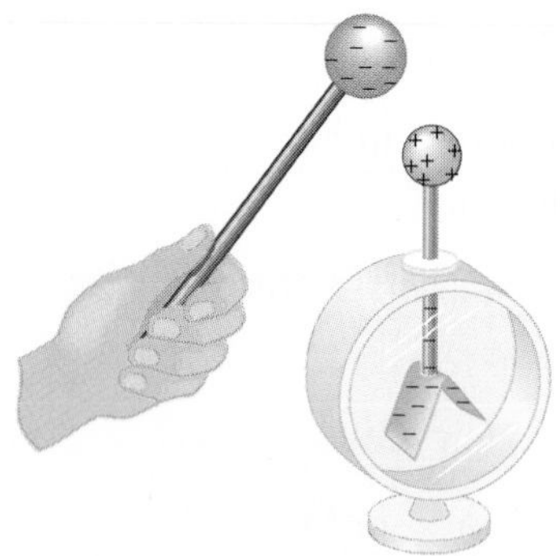

Figure 17.1C Inducing a charge on an electroscope. The negative rod repels electrons down into the leaves. The two equally charged leaves repel each other and move apart.

Storing even 1 coulomb is a formidable task, and yet the Earth appears to be carrying a horrendous charge of roughly −400 000 C. In fact, in storm-free regions, the ground leaks about 1500 C every second to the atmosphere. That flow is returned by lightning bolts (up to 20 C each). This ability of the Earth to store vast quantities of charge (because of its great size) makes it an ideal dumping ground for our excess charges. Any conductor in good contact with the Earth (like the water pipes in a building) will carry off all the charge one could possibly want to get rid of. That is precisely what is meant by the phrase to **ground** something (or in Britain, to "earth" it). Incidentally, the pictorial symbol for ground is ⏚.

17.7 Electrostatic Induction

It is not necessary for a charged object to physically touch an electroscope in order that the leaves respond to its presence (Fig. 17.16). A negatively charged object, such as a hard-rubber wand stroked with fur, located anywhere near the top of the electroscope will repel free electrons within the conducting knob and support wire.

These will be forced down into the leaves, which will spring apart and stay that way as long as the charged object remains nearby to maintain the imbalance. Figure 17.17 is a closeup of what happens in a metal under these conditions. The nearer the wand comes, the greater the Coulomb repulsion, and the more electrons enter the leaves, which makes the leaves more negative, and they spread farther apart. When the wand is removed, the displaced electrons immediately flow back, being mutually repelled by their brethren as well as attracted by the fixed positive ions of the metal (knob, wire, etc.). The leaves hang vertically, and the electroscope, which was neutral in total, reverts to its normally unsegregated charge distribution.

Instead of being transferred to the electroscope, the negative charge on the wand has **induced** a negative charge on the leaves. We could have equally well imagined that the negative wand attracted positive charges toward it, although that is not what happened. Alternatively, a positively charged wand would have attracted electrons upward and thus induced a positive charge on the leaves.

Suppose that the knob is now grounded and the game is repeated, bringing a negatively charged ball nearby (Fig. 17.18). Under the influence of the ball, mobile electrons in the metal (knob, support wire, and leaves), interacting in mutual repulsion, get a lot farther away from each other and the ball by flowing into ground. The scope becomes positively charged, and the leaves stand apart. Moving the ball away allows electrons to return up from ground, and the system relaxes back to its original neutral condition.

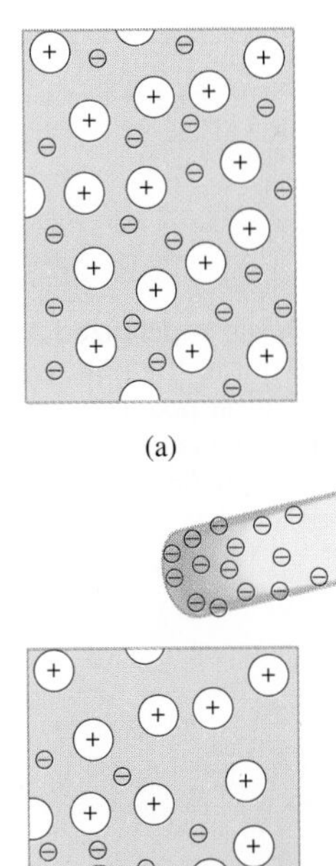

(a)

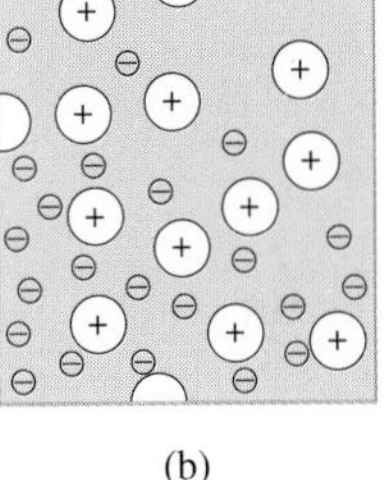

(b)

Ions (+) Electrons ⊖

Figure 17.17 (a) The conductor is neutral and its electrons are uniformly distributed. (b) A charged rod repels electrons downward. The top of the conductor becomes positive, the bottom negative.

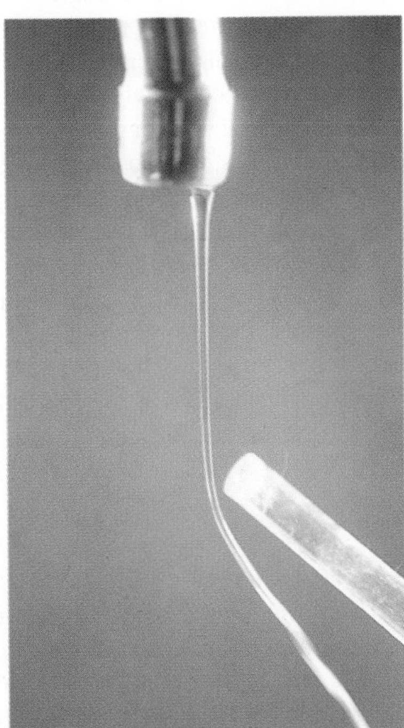

Water is composed of polarized molecules. When a charged rod is brought near the stream, these molecules align themselves so that they are drawn toward the rod. Do you think nonpolar liquids such as gasoline will show the same effect?

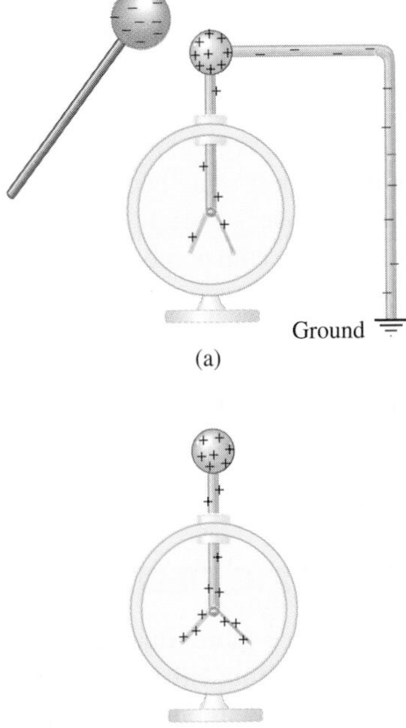

(a)

(b)

Figure 17.18 By letting electrons flow off to ground, the wire allows the electroscope to become permanently charged.

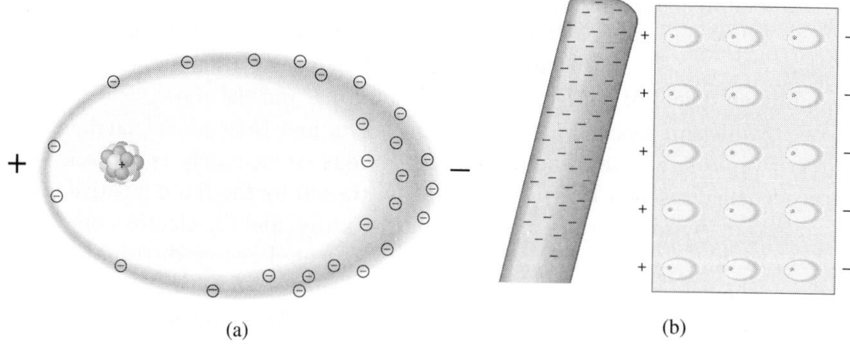

Figure 17.19 (a) A polarized atom. Electrons shifted to the right expose the nucleus. In this polarized state, the right end is negative and the left end is positive. (b) An electrically polarized dielectric.

Now bring the ball back and hold it close. The scope is again made positive, the leaves again part, but this time we go one step further and disconnect the ground lead, isolating the device. That done, the scope is shy a clutch of electrons and is more or less permanently positively charged with the leaves standing apart. Again charge was induced—the electroscope was never touched by the ball.

By contrast, electrons within a dielectric, such as a tuft of tissue paper, are far less mobile and ordinarily remain atom-bound. When a negatively charged rod is brought near such a nonconductor, it repels the electron clouds surrounding the atoms and in effect distorts or *polarizes* them (Fig. 17.19). Each nucleus, now slightly exposed, represents the positive end of the elongated atom, and the electron cloud bunched up on the opposite side constitutes the negative end. The result is that one side of the dielectric is electrified positively, the opposite side negatively. A charge has been induced, and it shows up on the surfaces of the specimen.

This is just what happens when you negatively charge a comb with a piece of wool or your clean hair and then use it to pick up bits of lint or tissue paper (Fig. 17.20). The region of the tissue closest to the comb takes on a positive charge; the region farthest away becomes negative. Since by Coulomb's Law the force drops off with distance squared, the mutual attraction of comb and tissue exceeds their slightly weaker repulsion because of the greater distance to the negative end, and the two objects experience a net tug compelling them toward one another. That's why, on a dry day, a piece of tissue will stick to the face of the cathode-ray tube (CRT) of a computer and why your TV screen is always covered with dust. It's also why a charged balloon sticks to a wall: it induces an opposite surface charge on the wall, and the two cling together until the excess electrons leak off the balloon.

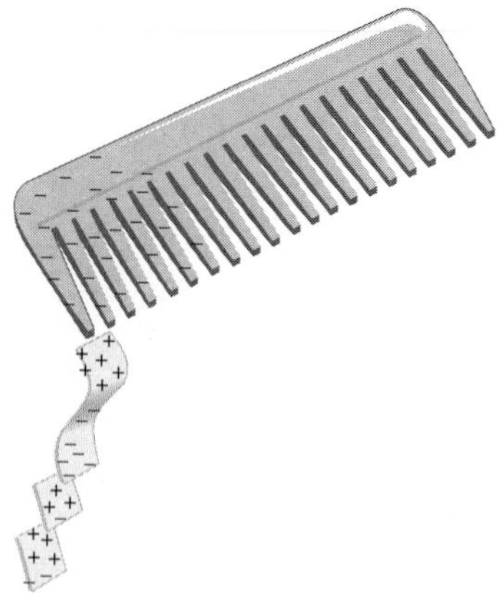

Figure 17.20 A negatively charged comb picking up bits of tissue paper.

THE ELECTRIC FIELD

Here we are once more, this time wand in hand lifting lint and pulling pith and yet not actually touching either—action-at-a-distance again, the fundamental puzzle of affecting motion without apparent contact. Physicists are still struggling to learn how one speck of matter reaches across the "void" to influence another. Today, the most prevalent theoretical picture is based on the concept of the *field,* which was introduced to help visualize the distribution of forces in the space surrounding a material object (p. 227). Inevitably, it itself became the medium for an explanation of action-at-a-distance.

Just as there is a gravitational force, there is an electromagnetic force; just as we envisioned a gravitational field, we envision an electromagnetic field. A **field of**

force exists in a region of space when an appropriate object placed at any point therein experiences a force. We can imagine a gravitational force field surrounding any object of mass m, and in the same way we can imagine an electric force field surrounding an object of charge q.

Consider a small sphere carrying a uniform positive charge. This is the *primary charge distribution,* and we want to study its influence. As a probe, we use a tiny pith ball on a thread (Fig. 17.21). By tradition, our detector (called a **test-charge**) is always positive. Accordingly, it will be repelled from the charged sphere no matter where it is positioned. At every point in the region surrounding the sphere (or primary charge distribution), the detector (of charge q_0) will sense a force of a specific strength and direction, and so at every point we assign a corresponding force vector (Fig. 17.22). It should be noted that if we actually drew a vector at *every* point, the picture would be solid black and useless; the illustration is a compromise. That, then, is the vector force field associated with this particular charged object. It happens to be radial and uniformly outward in all directions because the object is a uniformly charged sphere.

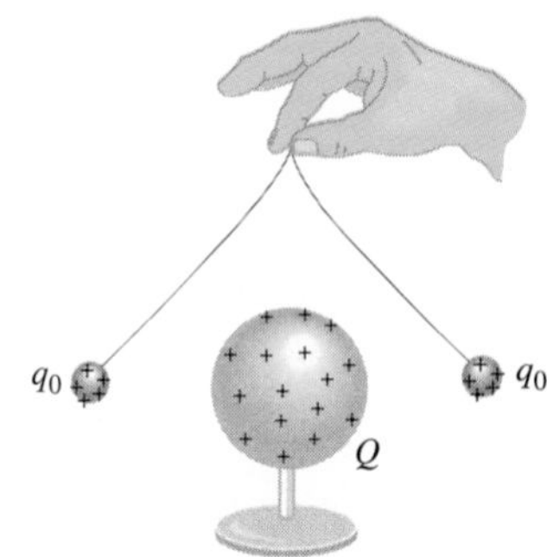

Figure 17.21 A primary charge Q affects a test-charge q_0, repelling it radially.

To appreciate the central creation in all of this—the electromagnetic field—we go back more than 100 years to the man who gave it form, the consummate experimentalist Michael Faraday. In 1812, at the age of 21, Faraday was given free tickets to the evening lectures on chemistry by Sir Humphry Davy at the Royal Institution in London. The young man was so dazzled by the experience that he sent Davy a leather-bound copy of his meticulous notes, accompanied by a request for a job as an assistant. A while later Davy fired his lab man for brawling and, remembering Faraday's flattering gesture, offered him the job of bottle washer. The bright young man accepted and quickly rose from lackey to protégé to rival.

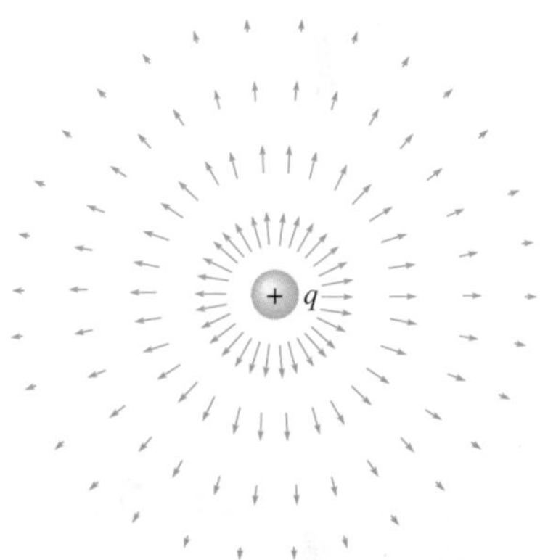

Figure 17.22 The vector force field surrounding a positive primary charge q.

Faraday was the first to introduce a visual representation of the electric force field. His scheme, using so-called *lines-of-force,* is a more convenient alternative to our field of force vectors. Faraday's version is equivalent to merging successive vectors into a continuous line that is tangent to all of them. *The force experienced by a positive test-charge at any point in space is in the direction tangent to the line of force at that point.* Notice that the lines-of-force flow radially outward from a positive point-charge, as they do for a uniformly charged sphere (Fig. 17.23).

The lines diverge, getting farther apart as they extend out, just as did the gravitational field, treated earlier. The same number of lines pass out through a small imaginary surrounding sphere centered on the charge as pass through a larger concentric sphere whose surface is more distant (Fig. 17.24). As the area ($A = 4\pi r^2$) of the encompassing sphere increases, the density of lines decreases in proportion to $1/A$ or, equivalently, to $1/r^2$. Happily, Coulomb's Law tells us that the force exerted by a point-charge q on a test-charge q_0 also drops off as $1/r^2$. It follows that the density or concentration of the lines-of-force corresponds to the strength of the force field. Since this is true for a single point-charge, it is true for the superposition of the force fields of many point-charges. *The more lines drawn in a region—that is, the denser the concentration of lines—the greater the field they represent; the farther apart the lines, the weaker the field.*

I can hardly imagine anyone who knows the agreement between observation and calculation, based on action at a distance, to hesitate an instant between this simple and precise action on the one hand and anything so vague and varying as lines of force on the other.

SIR GEORGE B. AIRY (1801–1892)
British astronomer and mathematician

To Faraday, these intricate patterns of lines came to stand for an invisible physical reality. For him, the field pervading space became an entity that reached from puller to pulled. An electrified object sends out its field into space, and the pith ball detector immersed within that web interacts with the field in proportion to its own charge. Charge-field-charge—that concept was Faraday's answer to the magic of action-at-a-distance. The field was not only an illustrative device; it was a reality

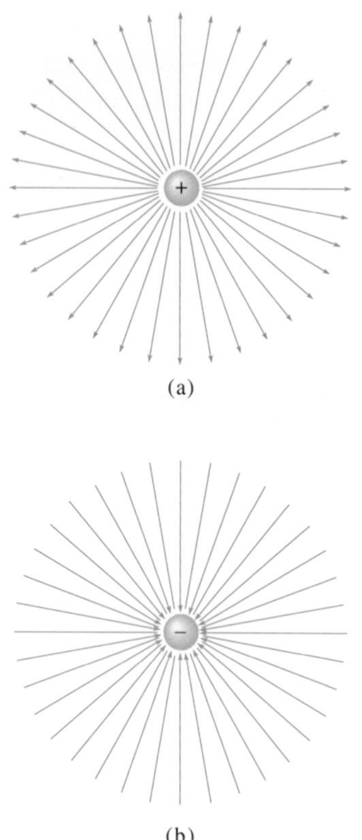

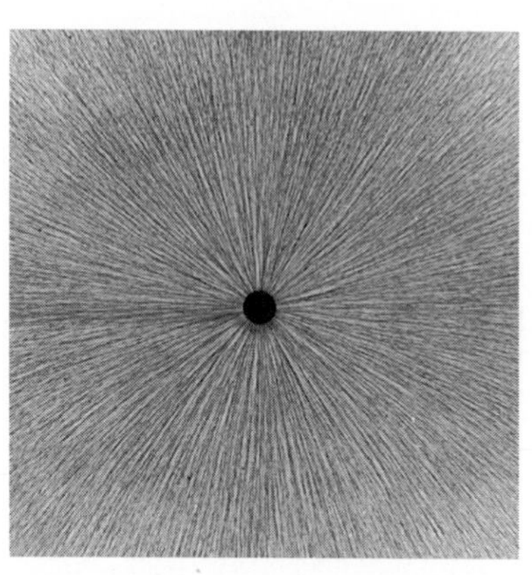

Figure 17.23 (a) Lines-of-force emanating from a positive point-charge. (b) Lines-of-force converging to a negative point-charge. (c) Fine rayon fibers suspended in oil tend to align themselves when in the vicinity of a charged object. The fiber patterns can be thought of as revealing the lines-of-force.

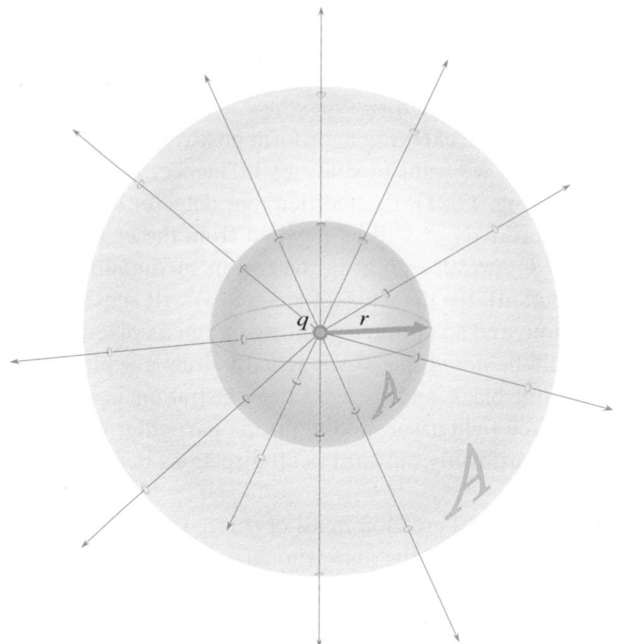

Figure 17.24 The lines-of-force stream through surrounding spheres. The area (A) of a sphere is $4\pi r^2$ and as r increases the density of lines, the number per unit area decreases as $1/r^2$, which is in accord with Coulomb's Law.

(though no less mysterious than the ghostly action-at-a-distance it now replaced). After all, what was the field made of, if anything?

One obvious disadvantage of concentrating on force is that its magnitude at every point in space depends not only on the primary charge distribution, but also on the size of the test-charge q_0. What we really want is a map showing the field of a primary body independent of the detector, a map that could be used to compute the force at every point in space when any size charge is placed there. Thus, regardless of its source, we define the **electric field** ($\mathbf{E}$) *at a point in space to be the electric force experienced by a positive test-charge at that point divided by that charge:*

$$\mathbf{E} = \frac{\mathbf{F}}{q_0} \tag{17.2}$$

Electric field has the SI units of newtons per coulomb (N/C). The tangent to a line-of-force at any point is the direction of the E-field. The custom is to draw the same pattern of lines, but to call them *electric field lines.* Only in the special case when the E-field is uniform (the field lines are parallel) is $\mathbf{E}$ a constant. More commonly $\mathbf{E}$ varies from point to point in space depending on the charge distribution.

Conversely, knowing $\mathbf{E}$ at any location in space (whatever the source) we can calculate the force $\mathbf{F}$ that would arise on any point-charge q placed at that location; accordingly

$$\mathbf{F} = q\mathbf{E} \tag{17.3}$$

Notice that $\mathbf{F}$ and $\mathbf{E}$ point in the same direction when q is positive.

Example 17.4 At a particular moment the electric field at a point 30 cm above an electric blanket is 250 N/C, straight up. Compute the force that acts on an electron at that location, at that moment.

Solution: [Given: $q_e = -1.60 \times 10^{-19}$ C, $E = 250$ N/C. Find: **F**.]

$$F = q_e E = (-1.60 \times 10^{-19}\ \text{C})(250\ \text{N/C})$$

and $$\boxed{F = -4.01 \times 10^{-17}\ \text{N}}$$

The minus sign arises from the charge and tells us that the force on the electron is in the opposite direction to the E-field; namely, downward.

▶ **Quick Check:** $E = F/q = (-4 \times 10^{-17}\ \text{N})/(-1.6 \times 10^{-19}\ \text{C}) = 2.5 \times 10^{2}$ N/C.

The electrically responsive cells in a shark allow it to detect the weak electric fields created by the operation of the muscles of its prey.

Michael Faraday (1791–1867) was one of 10 children of a blacksmith in London. As a youngster, with little formal education, Michael was apprenticed to a bookbinder. But he longed to enter "the service of Science" and ultimately became one of the greatest experimentalists of all time.

Prior to the introduction of electrical technology in the nineteenth century, the strongest electric field most humans were likely to encounter was the static atmospheric field of about 120 N/C to 150 N/C in fair weather, and up to 10 000 N/C during thunderstorms (Table 17.2).

Figure 17.23a shows the direction in which a positive test-charge q_0 would experience a force and so tend to accelerate. It is also a mapping of the electric field of the primary point-charge $+q$. Alternatively, when the primary charge is negative, the field lines converge inward toward it, as in Fig. 17.23b. The special thing about a point-charge source is that we can determine a formula for its E-field using Eq. (17.2). Thus, substituting for F from Coulomb's Law, we obtain

$$E = \frac{F}{q_0} = k\frac{qq_0}{r^2}\frac{1}{q_0}$$

and **the magnitude of the electric field of a point-charge** q is

$$E = k\frac{q}{r^2} \tag{17.4}$$

TABLE 17.2 Electric fields

Source	Field strength (N/C)
Background radiation in space	3×10^{-6}
In-house wires	10^{-2}
Radio waves	$\approx 10^{-1}$
Outside an electrified building	$\approx 10^{-1}$
Center of typical living room	≈ 3
In a fluorescent tube	10
30 cm from electric clock	15
30 cm from stereo	90
Laser beam (low power)	10^{2}
Atmosphere (fair weather)	≈ 150
30 cm from electric blanket	250
Built up by splashing water in a shower	800
Sunlight (average)	10^{3}
Atmosphere (thunderstorm)	10^{4}
Van de Graaff accelerator	2×10^{6}
Breakdown of air	3×10^{6}
X-ray tube	5×10^{6}
At cell membrane	10^{7}
Created by pulsed laser system	5.7×10^{11}
At electron in hydrogen atom	6×10^{11}
Surface of a pulsar	$\approx 10^{14}$
Surface of uranium nucleus	2×10^{21}

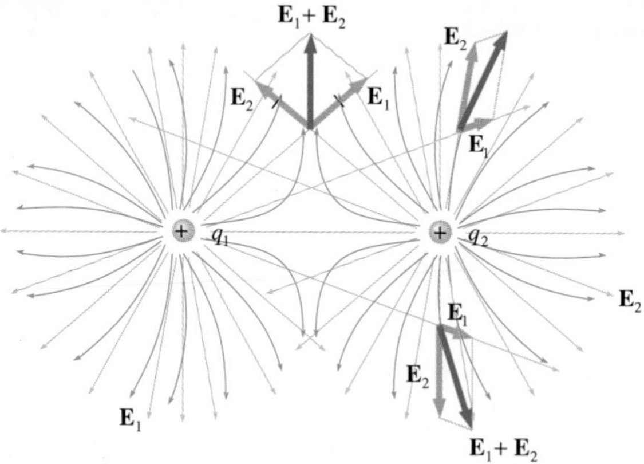

Figure 17.25 The two fields $\mathbf{E}_1$ and $\mathbf{E}_2$ superimpose, coexisting without interacting. The net field $\mathbf{E}$ is the vector sum of the two. In this diagram, at each point in space, $\mathbf{E} = \mathbf{E}_1 + \mathbf{E}_2$. Lines of $\mathbf{E}$ never intersect one another.

This result is extremely important: it allows us to calculate the **E**-field anywhere in space due to a known distribution of point-charges (electrons, protons, whatever). If there are just a few such charges, the calculation can be done directly by hand. Each charge contributes a field whose strength at any point is given by Eq. (17.4) and whose direction is radially inward (when q is negative) or outward (when q is positive), as in Fig. 17.25. These contributions add vectorially to produce the net **E**.

Example 17.5 Figure 17.26 shows two point-charges (or tiny charged spheres) each of +10 nC separated in air by a distance of 8.0 m. Compute the electric field at points A, B, and C.

Solution: [Given: $q_1 = q_2 = +10 \times 10^{-9}$ C. Find: **E** at points A, B, and C.] Starting at A, make a sketch of the layout (Fig. 17.26b) and then draw in vectors for the field $\mathbf{E}_1$ due to q_1 and $\mathbf{E}_2$ due to q_2. To do that, imagine a positive test-charge at A. The force on it due to q_1 acts along the center-to-center line, is repulsive, and so points to the right. That means the field $\mathbf{E}_1$ at A is to the right along the axis. Similarly, the force due to q_2 on our imaginary test-charge is to the left, as is $\mathbf{E}_2$. Next, calculate E_1 and E_2 via Eq. (17.4) and add them vectorially. We are spared that last effort because $E_1 = E_2$ and so the two cancel and $\boxed{\text{the field at } A \text{ is zero}}$. A test-charge exactly at A would just sit there force-free in unstable equilibrium.

At point B, the two E-fields act as drawn in Fig. 17.26c and we must find their components. First, let's calculate E_1 and E_2. Since the charges and distances happen to be the same, the magnitudes of the two contributing fields are equal; consequently

$$E_1 = E_2 = k\frac{q}{r^2} = (9.0 \times 10^9\ \text{N}\cdot\text{m}^2/\text{C}^2) \times \frac{(+10 \times 10^{-9}\ \text{C})}{(4.0/\sin 45°)^2} = 2.81\ \text{N/C}$$

Now for the vector components. From the geometry, the fields are seen to act at 45°. This time, because the charges are equal and point B is on the midline, the horizontal field components are equal and opposite, and cancel. Only the vertical components contribute

$$E_B = E_1 \sin 45° + E_2 \sin 45° = 2(2.81\ \text{N/C})(0.707)$$

and $\boxed{E_B = 4.0\ \text{N/C}}$

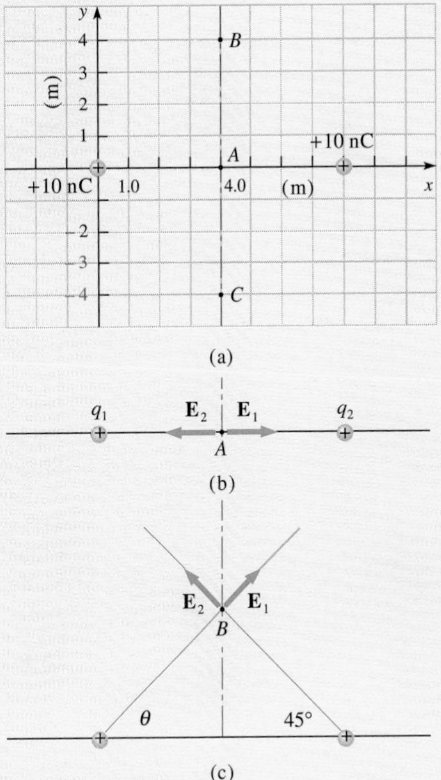

Figure 17.26 The electric fields of two charges q_1 and q_2 at points A, B, and C.

directed straight up in the positive y-direction. The same value applies to point C, where $\boxed{E_C = 4.0\ \text{N/C}}$ directed downward in the negative y-direction.

▶ **Quick Check:** As we will see presently, these results agree with the diagram (Fig. 17.27) for the field of two equal charges. The field of a 1-nC positive charge at 1 m is ≈9 N/C, so these answers are the right order-of-magnitude.

17.8 Rationalized Units

An English electrical engineer, Oliver Heaviside, pointed out sometime around 1892 that most of the equations for the E-fields of different charge distributions contained a factor of 4π. He suggested that the constant k in Coulomb's Law be written in terms of $1/4\pi$ and some new constant ε, so that

$$k = \frac{1}{4\pi\varepsilon} \tag{17.5}$$

TABLE 17.3 The permittivity (ε) and relative permittivity* ($\varepsilon/\varepsilon_0$) of some common substances

Substance	Permittivity ($C^2/N{\cdot}m^2$)	Relative permittivity ($\varepsilon/\varepsilon_0$)
Vacuum	8.85×10^{-12}	1.000 00
Air	8.85×10^{-12}	1.000 54
Body tissue	71×10^{-12}	8
Glass	44×10^{-12}–89×10^{-12}	5–10
Mica	27×10^{-12}–53×10^{-12}	3–6
Nylon	31×10^{-12}	3.5
Paper	18×10^{-12}–35×10^{-12}	2–4
Polyethylene	20×10^{-12}	2.3
Polystyrene	23×10^{-12}	2.6
Rubber	18×10^{-12}–27×10^{-12}	2–3
Silicone oil	19×10^{-12}–25×10^{-12}	2.2–2.8
Sodium chloride	50×10^{-12}	5.6
Teflon	19×10^{-12}	2.1
Ethanol (25°C)	2.2×10^{-10}	24.3
Methanol (20°C)	3.0×10^{-10}	33.6
Water (20°C)	7.1×10^{-10}	80

*Also called the *dielectric constant.*

This would have the effect of causing 4π to appear in expressions for the E-field only in situations where the charge distribution was spherically symmetric, and 2π to appear when it was axially symmetric. Since this was a purely formal device, the system of units it engendered was said to be *rationalized.* The SI system embraced the practice, and in 1960 it became official. The constant ε is called the **permittivity** (from the Latin *permittere,* to let go through). In the case of vacuum, $k_0 = 1/4\pi\varepsilon_0$, where ε_0 is referred to as the **permittivity of free space**. It follows from the value of k_0 that

$$\varepsilon_0 = 8.854\,187\,8 \times 10^{-12}\ C^2/N{\cdot}m^2$$

Table 17.3 lists the permittivities of several common materials. It also provides the corresponding **relative permittivity** ($\varepsilon/\varepsilon_0$) for each. Often called the **dielectric constant** (K_e), this ratio is unitless and applies to all systems of units. In the early days when there were several competing systems, tabulating ratios was often the only sensible way to present experimental data. The users simply multiplied K_e by the appropriate value of ε_0 in the system of units they happened to be working with.

Example 17.6 A point-charge of 10 μC is surrounded by water with a dielectric constant of 80. Calculate the magnitude of the electric field 20 cm away.

Solution: [Given: $q = 10\ \mu C$, $r = 0.20$ m and $K_e = 80$. Find: E.] The field of a point-charge is $E = kq/r^2 = q/4\pi\varepsilon r^2$, where $\varepsilon = K_e\varepsilon_0$. Hence

(continued)

(continued)

$$E = \frac{10 \times 10^{-6}\text{ C}}{4\pi(80)(8.85 \times 10^{-12}\text{ C}^2/\text{N}\cdot\text{m}^2)r^2}$$

and $\boxed{E = 28\text{ kN/C}}$

▶ **Quick Check:** In vacuum $E \approx (9 \times 10^4/0.04)$ N/C $\approx 2.3 \times 10^6$ N/C. With water, it will be a factor of 80 smaller; that is, $E \approx 28$ kN/C.

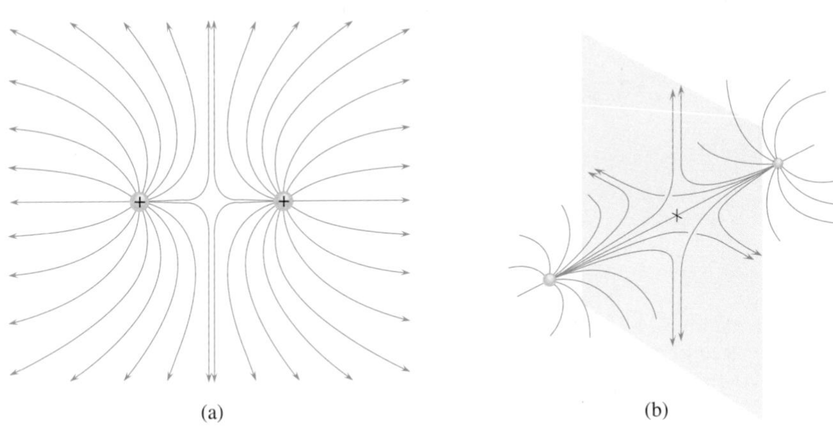

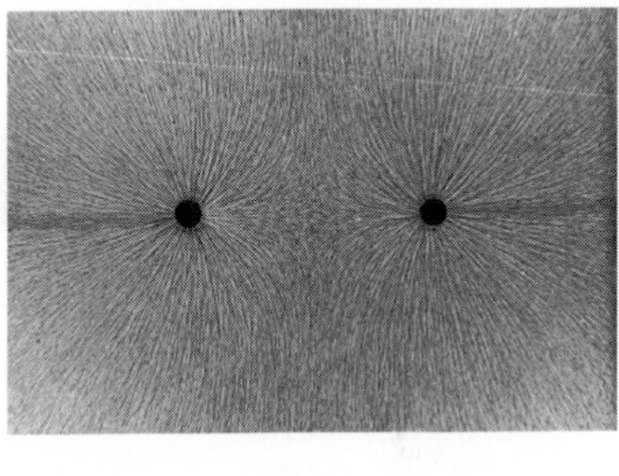

Figure 17.27 (a) The electric field lines due to two positive charges of equal value. (b) The E-field is three-dimensional. (c) Rayon fibers suspended in oil align with electric field lines.

17.9 Field Lines

Because electric field lines diverge away from a positive point-charge and converge in toward a negative point-charge, we say the former is the *source* of the field and the latter the *sink*. **In electrostatic situations, field lines always begin on positive charge and end on negative charge**. Both ends of the field lines are not always pictured, but they terminate somewhere on a charge (Fig. 17.28), even if that charge is on the walls of the room or beyond.

Figure 17.27 shows the E-field of two equal positive point-charges (each q). The pattern is three-dimensional (Fig 17.27b), and it actually corresponds to the picture we would get if we rotated the drawing (Fig. 17.27a) about the line connecting the charges. As was found analytically (Fig. 17.26), the field is zero at the very center. Very far from the pair (at a distance much greater than their separation), the field will appear as if it were due to a single positive charge ($2q$).

Lines of the net field never cross. If they did, the field would have two different values at that point, which is silly since two superimposed fields combine to yield a single net field. A positive test-charge at that point must experience a single net force, and that is the direction of **E**.

Figure 17.29 shows a positive and negative charge of the same magnitude, a configuration that is special enough to have its own name—it's called a **dipole**. Because the charges have opposite signs and equal magnitudes, the field is not zero anywhere nearby (as it was in Fig. 17.27). Still, at greater and greater distances, the lines get increasingly sparse, and the field rapidly decreases. It must, since as seen from very far away the charges will essentially coalesce and cancel. When the charges are unequal, as in Fig. 17.30, at very great distances the field should resemble the field associated with a single net charge.

Imagine two large, parallel horizontal metal plates each equally charged—the upper one positive, the lower negative (Fig. 17.31). The charge distributes itself more or less uniformly over the inner faces of the plates; the mutual attraction draws most

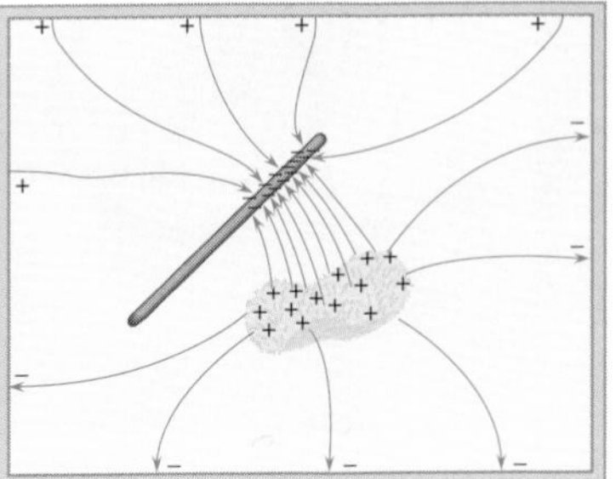

Figure 17.28 A hard-rubber rod rubbed on fur. Many of the field lines begin or end on the walls. *The net enclosed charge is zero.*

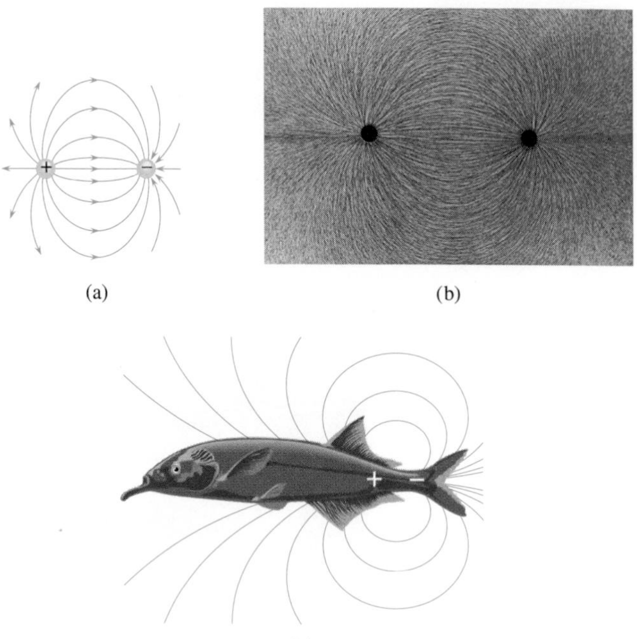

Figure 17.29 (a) A dipole field set up by two equal opposite charges. (b) Rayon fibers suspended in oil reveal the pattern of electric field lines. (c) The Elephant Gnathonemus produces a dipole electric field and detects nearby objects by their effects on that field.

+P
$-Q$
$+4Q$

Figure 17.30 A rough sketch of the electric field of two charges $+4Q$ and $-Q$. Notice that the field is zero at point P.

of the opposite charges as close together as possible (leaving just a few stray charges on the outer surfaces). A positive test-charge placed anywhere in the gap (away from the ends) would be repelled from the top plate, attracted to the bottom plate, and down the charge would go. The E-field between the plates is down and *uniform,* the field lines are parallel, and E is constant. This is the most convenient way to produce a uniform field and so is used in many devices, including cathode-ray tubes.

17.10 Conductors and Fields

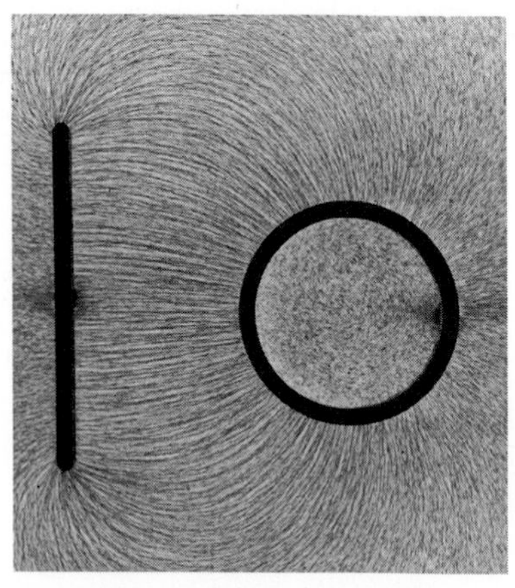

The external E-field is perpendicular to a conductor. Moreover, there is no field inside a conductor unless it surrounds a net charge.

Imagine a neutral conductor, solid or hollow, in a field-free region of space. Now suppose we add some electrons to it, either on the surface or inside. The mutual repulsion between free charges causes them to redistribute. The charges are initially bunched up, there is a transient E-field, and very quickly they move apart, coming to rest on the outer surface. The process generally only takes a fraction of a second, depending on the physical details of the conductor. Once settled, the distribution is such that each free charge experiences a zero net force—if it did not, the charge would accelerate until it could move no more, and equilibrium would ultimately be established. If none of the charges experiences a net electric force, none of them is in an electric field. Since field lines either begin or end on charge, the field could only extend inside the conductor if there were a remaining excess of free charges there and that's impossible. **The electrostatic field inside a charged conductor, anywhere beneath the surface, is zero** (provided it does not encompass a space in which there is an isolated charge). Faraday dramatically proved the point by constructing a room within a room, covering the inner enclosure with tinfoil. He sat inside this *Faraday cage,* as it has come to be called, with an electroscope at hand, while the entire structure was charged by an electrostatic generator—no field could be detected inside, even while sparks were flying outside.

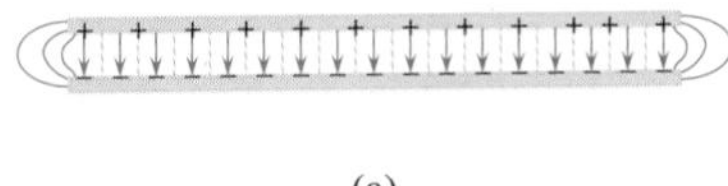

(a)

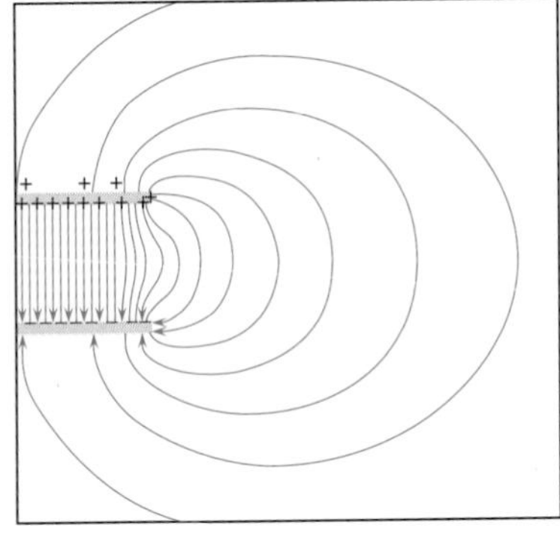

(b)

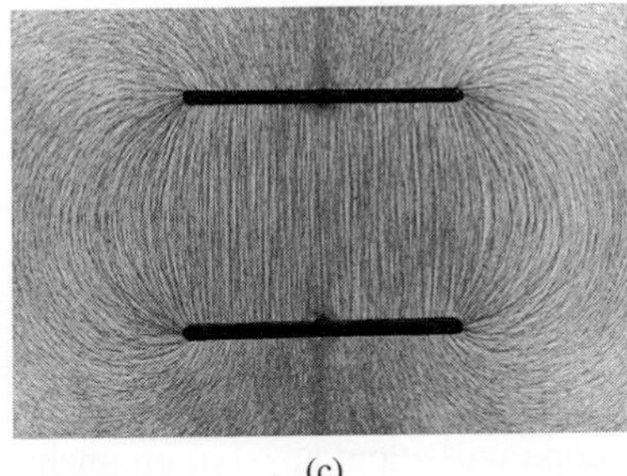

(c)

Figure 17.31 (a) The E-field of a parallel plate capacitor. (b) A closeup of edge effects. (c) The field as illustrated by rayon fibers suspended in oil.

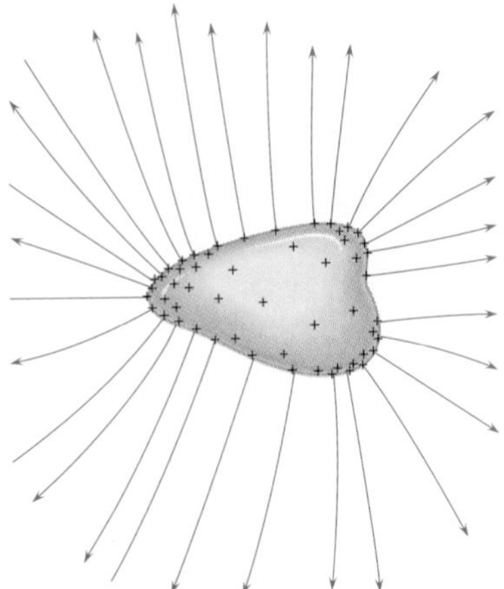

Figure 17.32 The electric field of a charged conductor is everywhere perpendicular to the surface. Charge concentrates on regions with a small radius of curvature.

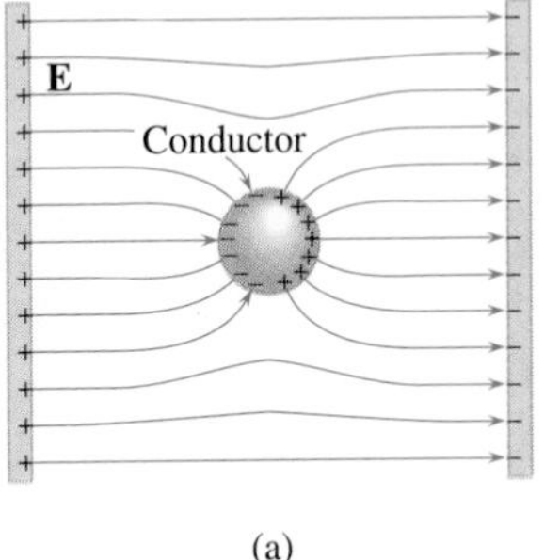

(a)

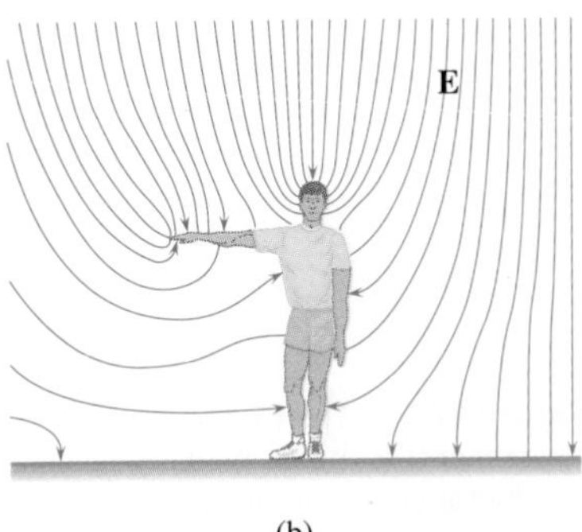

(b)

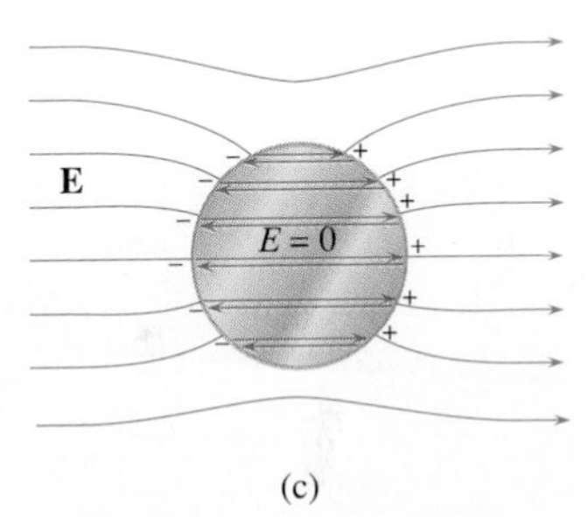

(c)

Figure 17.33 Neutral conductors immersed in an applied *E*-field. (a) A conducting sphere in the uniform field of a parallel plate capacitor. (b) A person in the *E*-field of the out-of-doors environment. The presence of a conductor distorts the field. (c) The external field polarizes the conductor, making one side positive and the other negative, which creates a self-field that cancels the applied field and leaves a zero internal electric field.

As for the electrostatic field external to the body of a conductor—*it must always be perpendicular to the surface.* It does not matter whether we are talking about a charged object (Fig. 17.32) and its own field or a neutral conductor in an external field (Fig. 17.33). If the field were not perpendicular, it would have a component parallel to the surface. There would be forces on the free surface charges, and they would move. Static equilibrium, which is ordinarily observed to set in quickly, can only be reestablished when there are no tangential field components.

Unlike gravity, we can shield against electric fields simply by surrounding the region to be isolated with a closed conductor. This is true as well in the case of nonstatic fields because the electrons can redistribute themselves very quickly. Electronic components are often encased in metal cans, and wires (for example, the leads for your stereo amplifier and cable TV) are surrounded by braided copper sheathing to keep out stray electric fields. If a portable radio is placed in a closed metal pot, the *E*-field signal will not be able to reach the antenna. That's why a car radio's reception is so poor in a metal-encased tunnel or on a steel bridge.

The electrostatic field within the material of a conductor is always zero, but it is possible to have a field in the hollow of a conductor if we put an isolated charge in-

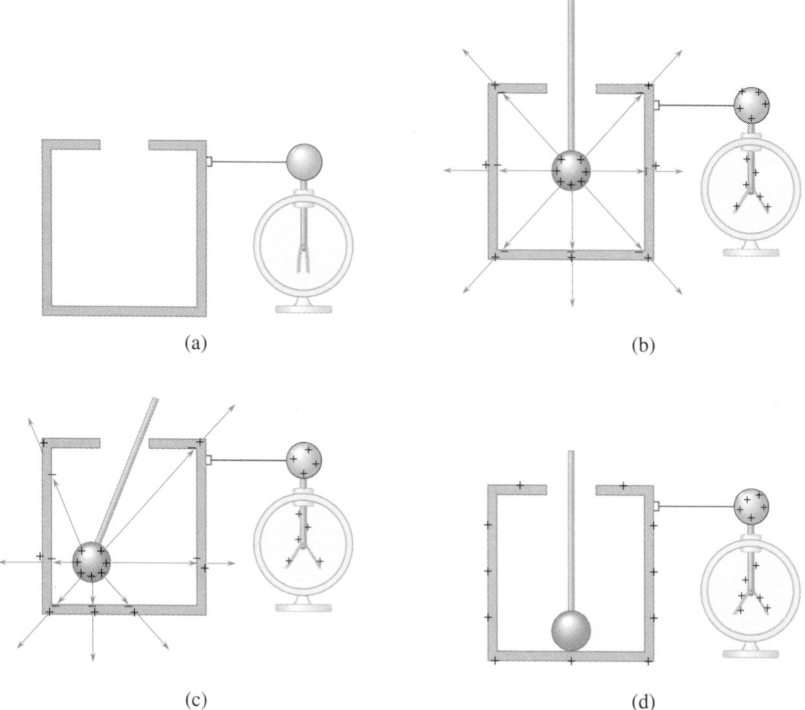

Figure 17.34 The Faraday ice-pail experiment. A charge held inside the conductor causes an equal and opposite charge to move to the outer and inner surfaces.

side it. Imagine that we introduce a positively charged ball into a nearly closed conductor attached to an electroscope (Fig. 17.34), just as Faraday did in 1843 using a small ice-pail. Once inside, it doesn't matter where the ball is, the outside will be charged positively, and the leaves of the electroscope will stand apart at an unchanging angle. If the ball touches the inside surface, it discharges, becoming neutral, but the leaves do not move at all—the same charge was induced on the outside as was carried by the ball. Notice that the net charge inside the conductor is zero both before and after the ball touched the inner surface.

During the Cold War, the CIA had the windows of the United States Embassy in Moscow screened with wire mesh and the rooms covered with metallic wallpaper. The idea was to keep out the intense electric fields (microwaves) being beamed at the building by their counterparts in the Soviet secret service. The United States had been scanning much of Moscow's microwave communications,

A tube in the old radio pictured on p. 589. It's surrounded by a metal shield that reduces interference caused by stray electric fields. Transistors and sensitive integrated circuits are commonly shielded by encasing them in metal.

The man in the Faraday cage on the left sits safely as sparks from a large Van de Graaff generator flash all around him.

using instruments in the Embassy, and now the Russians were jamming the eavesdropping. The field was not static, but the mesh and wallpaper still isolated the staff from the flood of radiation that had already made several of them ill. In effect, the Embassy was turned into a Faraday cage.

17.11 Gauss's Law

We can determine the E-field of any extended charge distribution directly using Coulomb's Law, but the process is often very difficult. There is another calculating approach that, in certain situations, is relatively simple. It's due to K. F. Gauss, the great nineteenth-century German mathematician and astronomer. Gauss's Law is about the relationship between the flux of the electric field and the sources of that flux, charge. The ideas derive from fluid dynamics, where the concept of flux was introduced, but we will have to generalize it a bit. The flow of a fluid, as represented by its velocity field, is depicted via streamlines, much as the electric field is pictured via field lines. Figure 11.30 (p. 377) shows a moving fluid within which there is a region isolated by an imaginary closed surface. The surface is a *tube of flow* bounded by flat area discs (A_1 and A_2) at either end, as shown in Fig. 17.35. The discharge rate, or *volume flux* (Av), is the volume of fluid flowing past a point in the tube per unit time as given by Eq. (11.12).

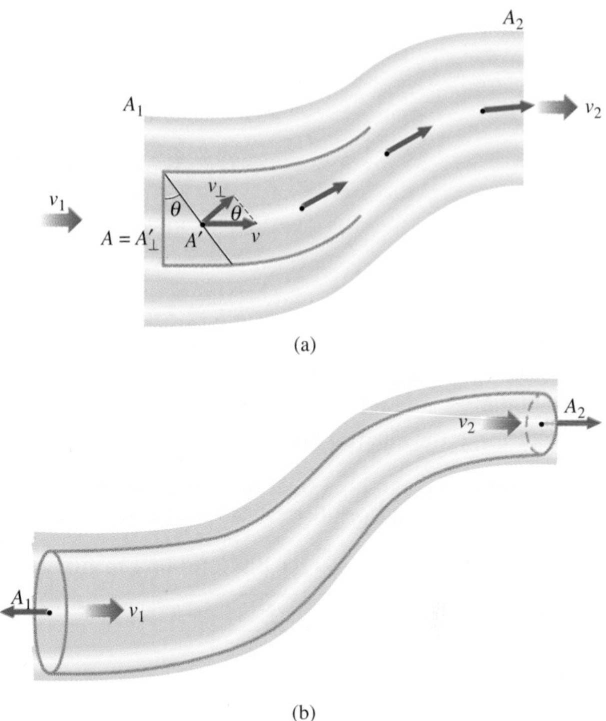

Figure 17.35 A tube of flow in a moving fluid. (a) The flux of the velocity field is defined as $v_\perp A = vA_\perp$. (b) To compute the total flux into and out of a region of space, the area vectors $\mathbf{A}_1$ and $\mathbf{A}_2$ are defined as pointing outward.

The areas through which the flow occurs in Fig. 17.35a are perpendicular to the streamlines. Still, the flux through any area, perpendicular or not, should be the same. As can be seen in Fig. 17.35b, the same amount of fluid passes per second through the different cross-sectional areas even though $Av \neq A'v$. Consequently, the flux must be defined more generally. Hence, if the imaginary surface is bounded by a tilted area A', the flux through that surface is determined by either the component of the velocity perpendicular to A' times A'; that is, $v_\perp A'$ or by v times the component of A' perpendicular to v—that is, $vA'_\perp$. Since $v = v_\perp \cos\theta$ and $A' = A'_\perp \cos\theta$, in either case, the flux is the same; namely, $vA'\cos\theta$.

Fluid flows in and out of the tube in Fig. 17.35a, and we will need to distinguish the flux-in from the flux-out. Thus, we construct area vectors ($\mathbf{A}$) normal to the surfaces and pointing *outward,* as in Fig. 17.35c. The flux is then defined as $vA\cos\theta$, where θ (as before) is the angle between $\mathbf{v}$ and $\mathbf{A}$. On the input face, the cosine of the angle between $\mathbf{v}$ and $\mathbf{A}$ is $\cos 180° = -1$, and on the output side $\cos 0° = +1$; the flux-in is $-v_1A_1$, the flux-out is $+v_2A_2$, and no flux enters or leaves through the curved side where the angle between $\mathbf{v}$ and $\mathbf{A}$ is everywhere 90°. From the Continuity Equation, this volume flux through both end surfaces is equal in magnitude—what flows in per second, flows out per second. The net flux (into and out of the closed area) summed over all the surfaces equals zero. If a small pipe is inserted into the region either sucking out fluid (a sink) or delivering fluid (a source), the net flux would then be nonzero.

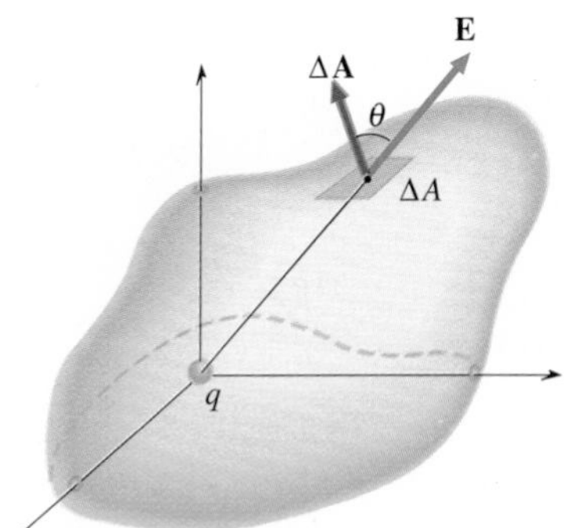

Figure 17.36 A rather free-form Gaussian surface surrounding a charge q. The electric field $\mathbf{E}$ is radially outward. The little area element ΔA has a vector associated with it that points outward ($\Delta\mathbf{A}$).

To apply these ideas to the electric field, consider the imaginary closed surface surrounding a point-charge q in Fig. 17.36. The tiny area ΔA is small enough so that the E-field can be taken as constant over its extent. In accord with the above discussion, we define the **flux of the electric field** ($\Delta\Phi_E$) through ΔA as

$$\Delta\Phi_E = E\,\Delta A\cos\theta = E_\perp\,\Delta A$$

and hence the total electric flux through the entire surface is the sum of all such contributions:

$$\Phi_E = \Sigma\,\Delta\Phi_E = \Sigma\,E_\perp\,\Delta A \tag{17.6}$$

When there are no sources or sinks of the field within the region encompassed by the closed surface, the net flux equals zero—that much is a general rule for all such fields.

Example 17.7 A uniform electric field of 2.1 kN/C passes through a rectangular area 22 cm by 28 cm. The field makes an angle of 30° with the normal to the area. Determine the electric flux through the rectangle.

Solution: [Given: area 0.22 m × 0.28 m, $\theta = 30°$, and $E = 2.1$ kN/C. Find: Φ_E.]

$$\Phi_E = EA\cos\theta = (2.1\text{ kN/C})(0.062\text{ m}^2)(0.866)$$

$$\boxed{\Phi_E = 0.11\text{ kN}\cdot\text{m}^2/\text{C}}$$

▶ **Quick Check:** Flux has the units of E-field (N/C) times area (m^2), or N·m^2/C, so the units are all right. The area is 0.22 m × 0.28 m = 0.06 m^2 and 0.866 times that is ≈0.05 m^2, so the flux should be $\approx\frac{1}{2}$ m^2 × 0.21 kN/C.

In order to find out what would happen in the presence of sources and sinks, consider a positive point-charge q. Now imagine a spherical surface of radius r centered on and surrounding that charge (Fig. 17.37). The E-field is everywhere outwardly radial and at any distance r it is entirely perpendicular to the surface: $E = E_\perp$ and $\Phi_E = \Sigma\,E\Delta A$. Moreover, since E is constant over the surface of the sphere, it can be taken out of the summation:

$$\Phi_E = E\,\Sigma\,\Delta A \tag{17.7}$$

The sum of all the area elements over a sphere of area A equals $4\pi r^2$; that is

$$\Phi_E = E4\pi r^2 \tag{17.8}$$

But we know from Eq's. (17.4) and (17.5) that the point-charge has an electric field of

$$E = \frac{1}{4\pi\varepsilon}\frac{q}{r^2}$$

and so Eq. (17.8) becomes

$$\Phi_E = \frac{q}{\varepsilon}$$

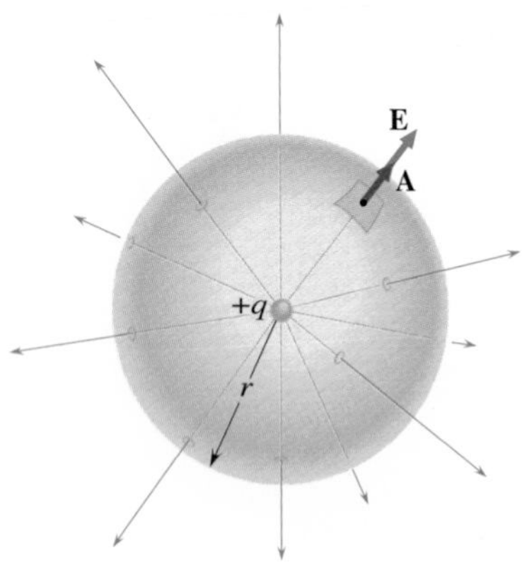

Figure 17.37 A spherical Gaussian surface surrounding a spherical primary charge distribution q at a distance r.

This is the electric flux associated with a single point-charge q within the closed surface. Since all charge distributions are made up of point-charges, it is reasonable that the net flux due to a number of **charges contained within any closed area** is

$$\Phi_E = \frac{1}{\varepsilon}\,\Sigma\,q$$

The sources provide an outwardly directed flux of the field, the sinks draw the flux inward, and the difference, in or out of the surface, is the net flux associated with the charge distribution.

If there are an equal number of positive and negative charges inside the surface ($\Sigma q = 0$), the field begins on the positive ones, ends on the negative ones, and no net electric flux emerges (Fig. 17.38). Combining this equation with Eq. (17.6) finally yields

$$\Sigma E_{\perp} \Delta A = \frac{1}{\varepsilon} \Sigma q \tag{17.9}$$

This is **Gauss's Law**, and it turns out to be even more general than Coulomb's Law, since it is applicable at any instant to both stationary and moving charge.

Because of the complexity of carrying out the calculation on the left side of Eq. (17.9), we shall limit its use to situations where the imaginary surface, the so-called *Gaussian surface*, leads to especially simple results. Thus, *we will usually construct a closed surface out of faces that are either parallel to the field (in which case, $E_{\perp} = 0$) or are perpendicular to the field (in which case, $E_{\perp} = E$), which is usually then constant over the surface selected.* With conductors, it's often useful to have part of the surface inside the material where $E = 0$. In most cases, we will stick with simple Gaussian surfaces, such as spheres and cylinders.

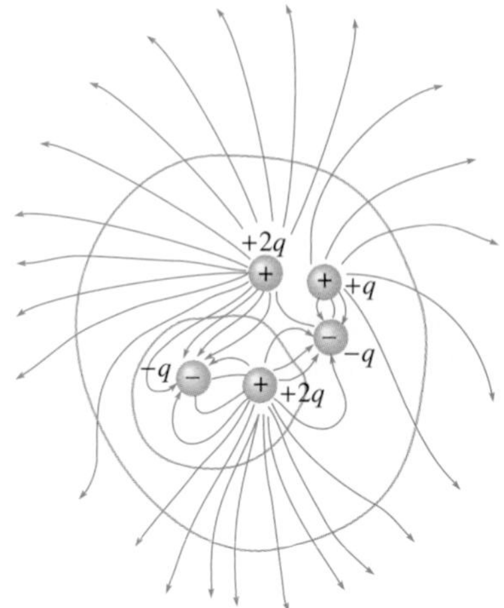

Figure 17.38 A schematic representation of the E-field of a distribution of charge. From far away it should look like the field of $+3q$. Given that there are eight lines per $\pm q$, surround any region by a closed area and subtract the number of lines in from those out.

Example 17.8 A long, straight wire of length L, in air, carries a uniform positive charge Q. Use Gauss's Law to find the electric field at a perpendicular distance R from the wire at a point that is far from the conductor's ends. To operate in a field region free of the complications of end effects, we limit the analysis to the domain where $L \gg R$.

Solution: [Given: total charge Q, distance R, and wire length L. Find: E.] First, determine the field configuration with an eye to finding its symmetries. From the facts that the charge is uniform and the field lines must begin on positive charges and be perpendicular to the conductor's surface, we can assume that *the field is radially outward* (Fig. 17.39). Because the wire is very long, the field in the middle region cannot have a component toward either end—either end is equally remote, neither is special; the field must be radial. Considerations like these are the first step in the application of Gauss's Law. Thus, given the cylindrical symmetry of the field, surround the wire with a cylindrical Gaussian surface of arbitrary radius R and length l, closed with flat end caps of area A_1 and A_3. Apply Eq. (17.9) over the closed imaginary surface made up of A_1, A_2, and A_3,

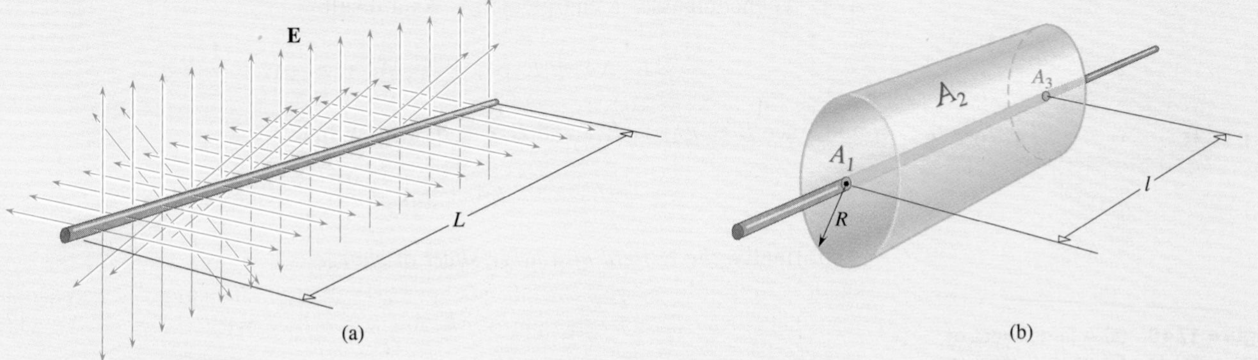

Figure 17.39 (a) The field of a line charge. (b) The corresponding Gaussian surface.

(continued)

(continued)

which will involve the field at a distance of R from the wire—just what we are after. Accordingly, processing the left side of the equation yields

$$\Sigma E_\perp \Delta A = E_{\perp 1}A_1 + E_{\perp 2}A_2 + E_{\perp 3}A_3 = \frac{1}{\varepsilon}\Sigma q$$

But $E_{\perp 1}$ and $E_{\perp 3}$ are both zero because the field is parallel to the ends. Furthermore, the entire field passes perpendicularly through the curved surface and also has a constant value over that surface; that is, $E = E_{\perp 2}$, hence

$$EA = E(2\pi Rl) = \frac{1}{\varepsilon_0}\Sigma q$$

where $A_2 = 2\pi Rl$. Now for the right side: If the total wire of length L carries a charge Q then the **charge per unit length** (λ) is Q/L and the charge inside the Gaussian surface (Σq), which we made to be of length l, is λl. Consequently

$$E(2\pi Rl) = \frac{1}{\varepsilon_0}\lambda l$$

and

$$\boxed{E = \frac{\lambda}{2\pi R\varepsilon_0}} \qquad (17.10)$$

Since R is unspecified, we really have a formula for E everywhere beyond the wire where the fringing at the ends is negligible.

▶ **Quick Check:** This result should have the units of F/q. Since λ is charge over length, it follows from Coulomb's Law that the units are all right. We can expect that at $R = 0$, $E = \infty$, and at $R = \infty$, $E = 0$, and both are the case. Moreover, the 2π is there because of the cylindrical symmetry.

The expression determined in Example 17.8 is independent of the Gaussian surface used to derive it. That's reasonable since there are an infinite number of possible surfaces that can be used (most of them with great difficulty). Thus l, which related only to the surface, cancels out of the analysis. The need sometimes to surround only a portion of the entire charge distribution within a surface will require that we introduce *charge densities*; accordingly

[*linear charge density*] $\qquad \lambda = \dfrac{Q}{L}$

[*surface charge density*] $\qquad \sigma = \dfrac{Q}{A}$

[*volume charge density*] $\qquad \rho = \dfrac{Q}{V}$

Consider the *very large flat sheet of charge* shown in Fig. 17.40. Its E-field (not far from the middle), by symmetry, must be perpendicular, outward, and *uniform*. Thus, a Gaussian surface in the shape of a cylinder with end faces of area $A = A_1 = A_2$ encompasses a charge of σA. As a result

$$\Sigma E_\perp \Delta A = E_{\perp 1}A_1 + E_{\perp 2}A_2 + E_{\perp 3}A_3 = \frac{1}{\varepsilon}\Sigma q$$

and since $E = E_{\perp 1} = E_{\perp 3}$ and $E_{\perp 2} = 0$, we say that

$$EA + EA = \frac{\sigma A}{\varepsilon}$$

And, finally, *the E-field of a large sheet of charge is*

$$E = \frac{\sigma}{2\varepsilon} \qquad (17.11)$$

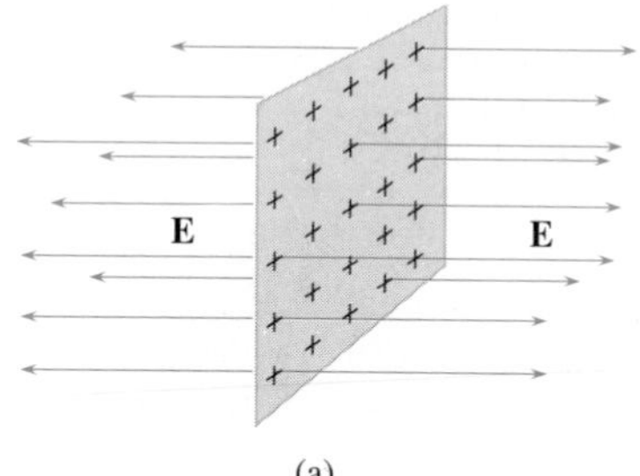

(a)

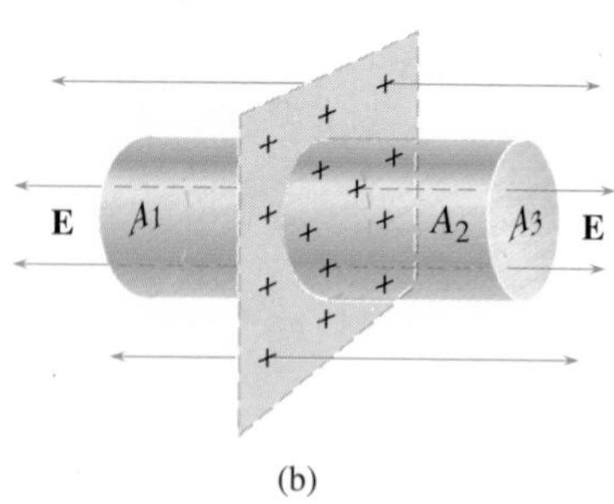

(b)

Figure 17.40 (a) A large sheet of charge. (b) The Gaussian surface (a cylinder perpendicular to the sheet) used to calculate **E**.

Figure 17.41 pictures a pair of parallel, equally and oppositely charged metal plates. The two fairly uniform sheets of opposite (excess) charge mutually attract and hold each other in place on the inner faces of the plates. Except for fringing at the edges (Fig. 17.31), the net field everywhere should be the vector sum of the two overlapping *uniform* fields set up by the two opposite sheets of charge. Outside this *parallel plate capacitor*, the two uniform fields are oppositely directed and cancel—ideally, there is no E-field outside the plates. Inside, the two fields, each given by Eq. (17.11), are in the same direction and add: $E = 2(\sigma/2\varepsilon)$ and *the field between the plates of a parallel plate capacitor is*

$$E = \frac{\sigma}{\varepsilon} \tag{17.12}$$

It's left as Problem 37 to show that this same result follows directly from Gauss's Law.

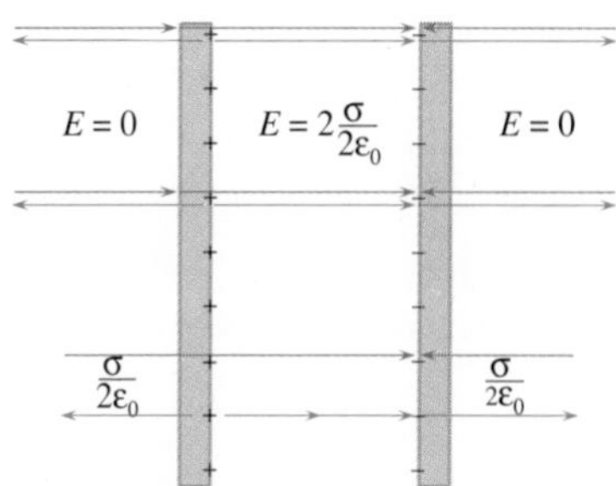

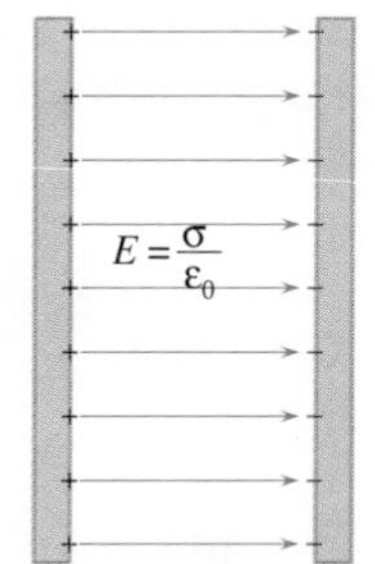

Figure 17.41 The E-field between two charged sheets. The ability of a parallel plate configuration to produce a uniform field region is used in all sorts of devices; for example, see Fig. 17.42.

17.12 Quantum Field Theory

A charged particle interacts with other charged particles. It creates a web of interaction around itself that extends out into space and corresponds to what has been called action-at-a-distance. That imagery leads to the concept of the electric field, which is a representation of the way the electromagnetic interaction reveals itself on a macroscopic level. The static electric field is, in effect, a spatial conception summarizing the interaction among charges. Classically, we say that one charge sets up an E-field in space and another charge immersed in that field interacts directly with it, and vice versa. The picture is straightforward, but many questions come to mind. Does the E-field have a physical reality? Is anything actually flowing? How does the field produce a force on a charge? Does it take time to exert its influence?

As early as 1905, Einstein already considered the classical equations of Electromagnetic Theory to be descriptions of the average values of the quantities being considered. Classical theory beautifully accounted for everything being measured, but it was oblivious to the exceedingly fine granular structure of the phenomenon. Using thermodynamic arguments, Einstein proposed that electric and magnetic fields were quantized, that they are particulate rather than continuous. After all, classical theory evolved decades before the electron was even discovered. If charge (the fundamental source of electromagnetism) is quantized, shouldn't the theory reflect that in some fundamental way?

Today, we are guided by Quantum Mechanics, a highly mathematical theory that provides tremendous computational and predictive power but is nonetheless disconcertingly abstract. Contemporary physics holds that all fields are quantized; that each of the fundamental Four Forces is mediated by a special kind of field particle. These *messenger particles* are continuously absorbed and emitted by the interacting material particles (electrons, protons, etc.). It is this ongoing exchange that *is* the interaction. The mediating particle of the electric field is the **virtual photon**. This massless messenger travels at the speed of light and transports momentum and energy. When two electrons repel one another, or an electron and proton attract, it is by emitting and absorbing virtual photons and thereby transferring momentum from one to the other, that transfer being a measure of the action of force. The messenger particles of the electromagnetic force are called *virtual* photons because

Figure 17.42 One type of ink jet printer fires charged droplets of ink at the paper. The droplets pass through a set of parallel plates, which carries a charge proportional to the control signal. The resulting E-field steers the beam of ink vertically as the paper moves by.

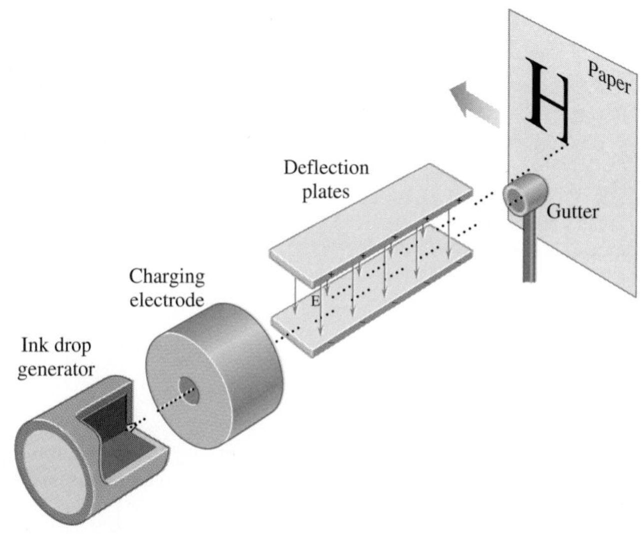

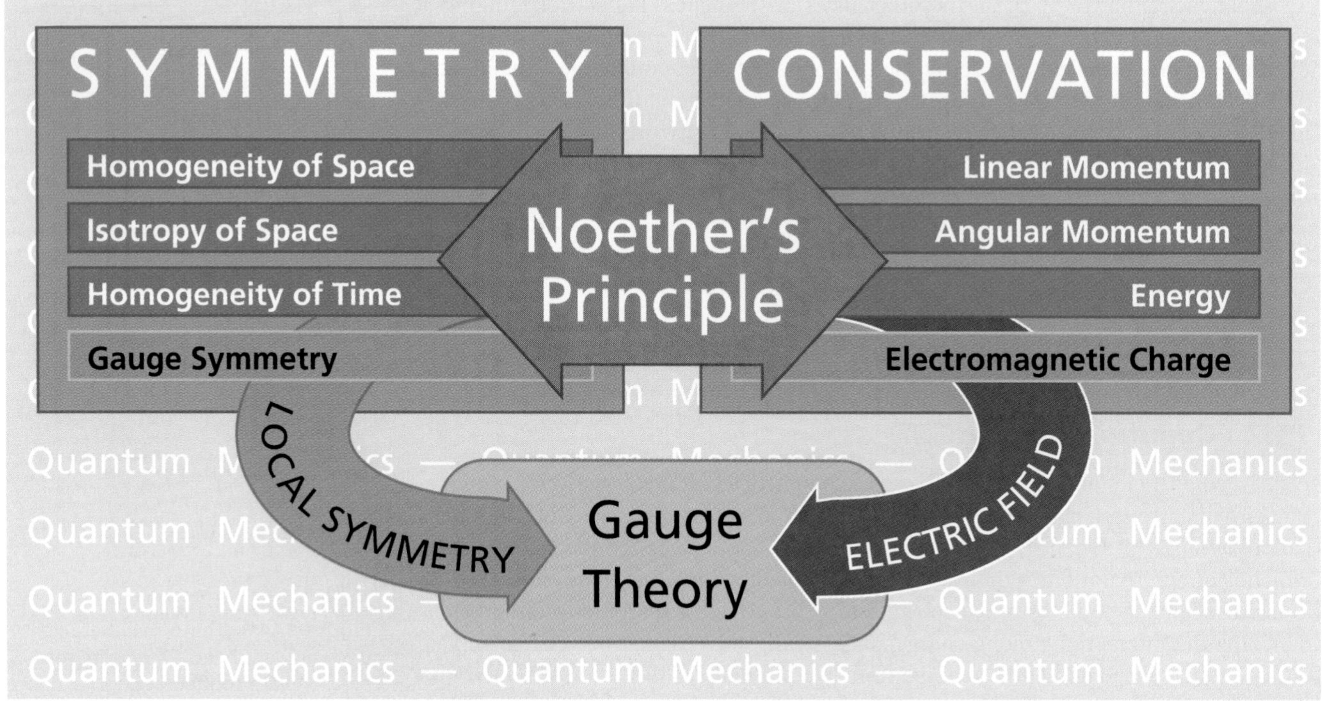

Figure 17.43 Virtual photons are the particles that mediate the electric field. They are spin-1 messengers; the interactions due to such particles are known as *gauge forces.* We will return to these concepts later when we focus on the essentials of Gauge Theory.

they are bound to the interaction. Virtual photons can never escape to be detected directly by some instrument, however unsettling that is philosophically and however hard that makes it to establish their existence. Indeed, virtual photons exist only as the means of interaction.

On a macroscopic level, messenger particles can manifest themselves as a continuous field provided they can group in very large numbers. Fundamental particles have an intrinsic angular momentum, or *spin* (p. 259), that determines their grouping characteristics. Quantum Theory tells us that the desired field behavior can only occur if forces are mediated by messenger particles having angular momenta equal to integer multiples of $h/2\pi$ (that is, $0, 1h/2\pi, 2h/2\pi, 3h/2\pi, \ldots$). The angular momentum of the virtual photon is $1(h/2\pi)$ and it's therefore referred to as a **spin-1 particle**. For reasons we shall study later, the exceedingly important class of interactions that have spin-1 messengers are known as **gauge forces** (Fig. 17.43) and the electromagnetic force is the model for all the gauge forces we shall encounter.

Today, the magic of action-at-a-distance is understood via the no-less mysterious exchange of virtual particles, but at least now there is a highly predictive mathematical theory in place that describes the phenomenon.

Core Material

Charge is the property of matter that gives rise to electric force. *Like charges repel, unlike charges attract.* Charge is conserved. The electrostatic force between two charges is

$$F_{\mathrm{E}} = k\frac{q_1q_2}{r^2} \qquad [17.1]$$

which is **Coulomb's Law**. The unit of charge is the *coulomb* (C), and k is a constant specific to the medium in which the charges are imbedded. In the special case of vacuum, $k_0 = 8.987\,551\,79 \times 10^9\ \text{N·m}^2/\text{C}^2 \approx 9.0 \times 10^9\ \text{N·m}^2/\text{C}^2$. The *charge of the electron* is $-1.602\,177\,33 \times 10^{-19}$ C. The net force on any one charge is the vector sum of all the forces exerted on it due to each of the other charges interacting with it independently (p. 598).

The **electric field** (**E**) *at any point in space is the electric force experienced by a positive test-charge at that point divided by that charge:*

$$\mathbf{E} = \frac{\mathbf{F}}{q_0} \qquad [17.2]$$

in units of newtons per coulomb (N/C). The field is represented by field lines that begin and end on charges. The denser the concentration of lines, the greater the field. The force **F** that would arise on any point-charge q placed in the field is

$$\mathbf{F} = q\mathbf{E} \qquad [17.3]$$

The *electric field of a point-charge* q is

$$E = k\frac{q}{r^2} \qquad [17.4]$$

In terms of the **permittivity** (ε)

$$k = \frac{1}{4\pi\varepsilon} \qquad [17.5]$$

The *permittivity of free space* is a constant defined as $\varepsilon_0 = 8.854\,187\,8 \times 10^{-12}\ \text{C}^2/\text{Nm}^2$.

The field inside a charged conductor in the steady state is zero. External field lines must be perpendicular to the surface of any conductor. **Gauss's Law** (p. 613) is

$$\Sigma E_\perp \Delta A = \frac{1}{\varepsilon}\Sigma q \qquad [17.9]$$

The electric field of a long, straight charged wire is

$$E = \frac{\lambda}{2\pi R\varepsilon_0} \qquad [17.10]$$

We define the *linear charge density* as $\lambda = Q/L$, the *surface charge density* as $\sigma = Q/A$, and the *volume charge density* as $\rho = Q/V$. The E-field of a large sheet of charge is

$$E = \frac{\sigma}{2\varepsilon} \qquad [17.11]$$

The field between the plates of a parallel plate capacitor is

$$E = \frac{\sigma}{\varepsilon} \qquad [17.12]$$

Suggestions on Problem Solving

1. Remember that the signs arising from the application of Coulomb's Law just relate to whether the force is attractive or repulsive. The forces between two point-charges are equal in magnitude and opposite in direction, and act along their center-to-center line. Draw a diagram, sketch in the directions of the forces, and then forget those signs. As ever, take to the right and up as the positive directions.

2. Given several point-charges, you may need to find the net force acting *on one of them.* Draw the direction of each force acting on the charge. Next, determine the magnitude of each force via Coulomb's Law. Resolve each force vector into perpendicular components and find the resultant of all the components in the usual way—that is, the net force.

3. The electric field *at a point* in space due to several point-charges is calculated much as is the force but using Eq. (17.4) instead of Coulomb's Law. Imagine a positive test-charge at that point and draw in the field vectors accordingly.

4. When applying Gauss's Law, first find the direction of the E-field so you can take advantage of its symmetries. Remember there is no field inside a conductor and embedding all or part of the Gaussian surface inside the conductor puts it in a zero-field region. The fields of all of the easy configurations have long since been figured out using Gauss's Law. Once you have learned how to make the standard dozen-or-so calculations, you have pretty much mastered the thing. Most professionals do not remember the field inside a charged coaxial cable; they just quickly derive it when needed using Gauss's Law, and that's the best reason to learn the method.

Discussion Questions

1. A common demonstration is to toss some small fragments of paper onto a highly charged conducting sphere. The paper initially sticks to the sphere and then after a little while pops off as if shot into the air. In 1676, Newton sent a note to the Royal Society describing a similar experiment where he charged a glass disk and held it horizontally above "some little fragments of paper." These he observed "sometimes leaping up to the glass, and resting there awhile; then leaping down and again resting." Explain what was happening.

2. Imagine a spherical dielectric shell—one made of glass, for example. A charge of $+Q$ is carefully distributed uniformly over the entire outer surface. Is there an electric field anywhere inside the shell? Explain.

3. Figure Q3 shows a dipole in a room (which is represented

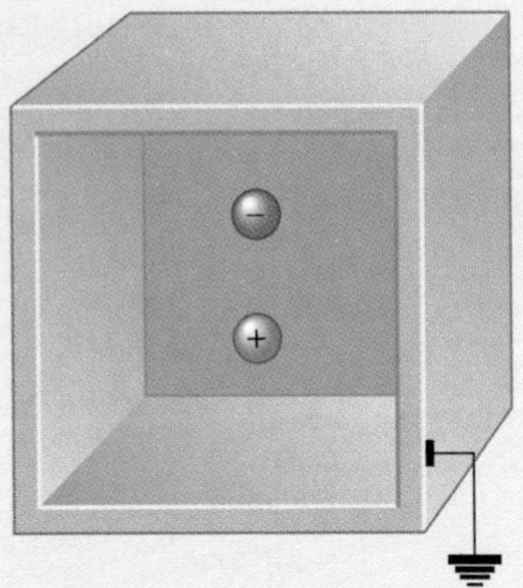

Figure Q3

by a large grounded conductor). Make a rough sketch of its E-field and explain your reasoning.

4. Static electric buildup can be troublesome as, for example, in a photographic processing laboratory where film easily becomes charged (usually positively), attracting dust, and even creating sparks. To control the effect, there are static eliminators, one variety of which uses radioactive polonium-210. This material pours out a constant stream of positively charged alpha particles. How would that help?

5. William Gilbert in the sixteenth century devised a simple instrument to detect electrification, which he called a *versorium*. To make one, just balance a length of any material "lightly pivoted on a needle." A piece of a drinking straw stuck on a tack set in clay will work nicely once it's made free to rotate (Fig. Q5). Or fold a narrow strip of paper into a long upside-down V-shape and balance it horizontally on a tack. Now rub something like a comb to electrify it. Use the versorium to see if you have succeeded—that is, wave the comb near the detector. Put your versorium near a TV tube and see if it can detect when you turn the electron beam on. What happens and why?

Figure Q5

6. Go into a dark closet with a fluorescent bulb. Rub it vigorously with a piece of cloth until it warms up a bit and starts to glow. Now stroke it with just your dry hand. In fact, simply squeeze the bulb and then quickly open your hand—the lamp will light rather brightly. Now wet it with a little water. What happens? Charge up a wand (my favorite is a clear plastic tube that a set of windshield wipers came in) and wave it around. Does its effect on the bulb change when the lamp is wet?

7. Stephen Gray attached a lead ball to the end of a glass tube with a 3-ft length of moistened "parcel string." Under the ball he placed some fragments of brass leaf (very thin foil). When he rubbed the tube, the leaf was attracted up to the ball. Explain what was happening. (By the way, Gray later redid the experiment using 80 feet of string, and "the Tube being rubbed, the Ball attracted the Leaf-Brass.")

8. Why do long strips of ordinary plastic tape freshly pulled from the roll usually crumple and stick to themselves and everything nearby? Go in a dark closet and pull off a length of transparent plastic tape—sparks will blaze brilliantly at the point of contact with the roll.

9. An electric typewriter on a desk not far from a TV set disturbs the picture, generating colored bands across the screen. The roof antenna is attached to the set via an ordinary flat twin-lead TV wire. What would you suggest as a possible solution to the problem and why?

10. Figure Q10 shows a hollow charged conductor with two proof-planes touching each other inside of it. The scheme is used to determine that no E-field is inside a charged conductor. Explain how that might be accomplished.

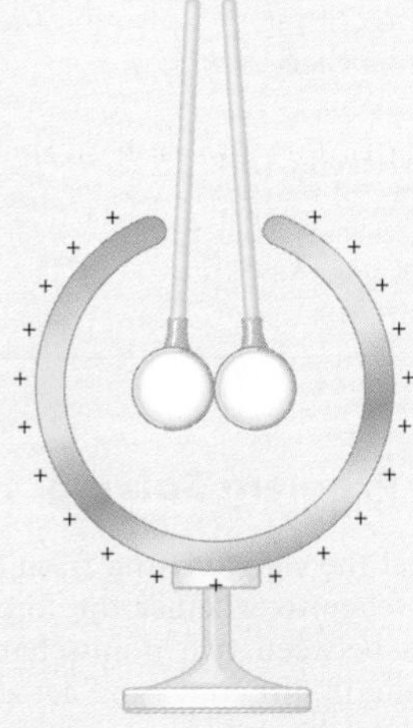

Figure Q10

11. A neutral, banana-shaped solid conductor is grounded to the plumbing via a wire. A negatively charged sphere is brought near it, the ground lead is disconnected, and the sphere removed. Describe the resulting E-field and surface charge density on the metal banana, if any.

12. Why are electrostatic phenomena, like sparks when you take off a sweater in the dark, more common in winter than in summer?

13. During the time Newton was president of the Royal Society, its curator of experiments was Francis Hauksbee. The fol-

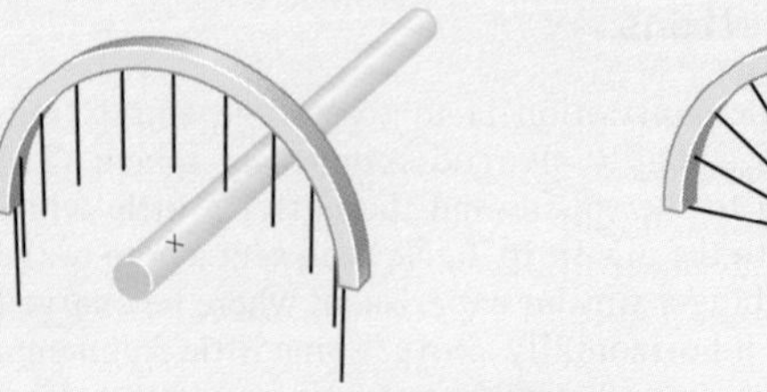

Figure Q13

lowing (Fig. Q13) is a demonstration Hauksbee performed that seems a very early hint of Faraday's lines-of-force. Onto a semicircular wire frame, Hauksbee fastened "several pieces of Woollen Thread . . . so as to hang down at pretty nearly equal distances." At the center, he placed a glass "Tube," which was subsequently charged by rubbing. The threads immediately swung about so that they pointed directly at the center of the glass. Explain what was happening and discuss the implications regarding field lines.

14. Figure Q14 shows a variation of the most widely used electrostatic generator, a device that carries the name of its inventor, Robert J. Van de Graaff (1902–1967). The two pulleys are covered with different materials so that when they are contacted by the motor-driven belt, the belt acquires a negative charge from the bottom pulley and a positive charge from the top one. In some research models, electrons are literally sprayed on the belt at the base. Figure out how the generator works and explain the crucial role of the hollow conductor dome. What, if anything, limits the amount of charge that can be built up on the dome? How would that compare to the case where the belt delivered the charge to the outside surface?

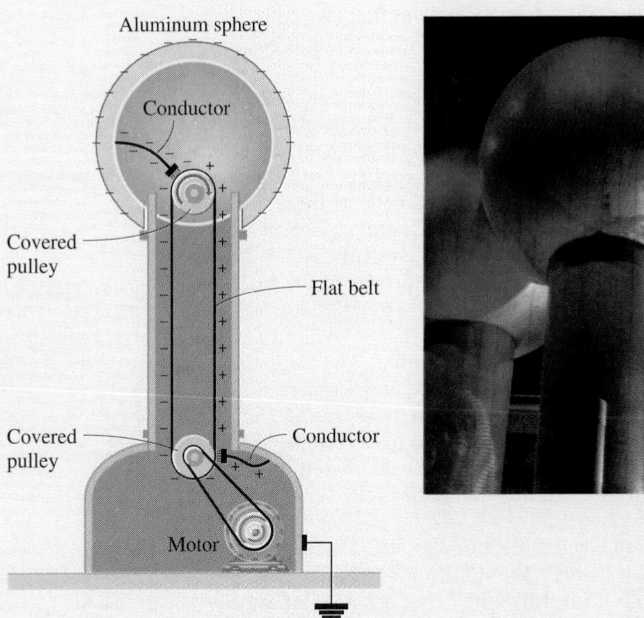

Figure Q14

15. Suppose that you approach an elephant with two known oppositely charged pith balls, in order to determine if the beast is itself charged. The first ball (+) is attracted, the second (−) repelled. What does that tell you? Which ball provides the unambiguous evidence? Explain.

16. What, if anything, might happen to the leaves of a charged electroscope if you waved a lit match around near it? Explain.

Multiple Choice Questions

1. The electrostatic force between a negative electron and a neutral neutron is (a) negative and attractive (b) positive and repulsive (c) zero (d) sometimes attractive and sometimes repulsive (e) none of these.
2. By comparison with the force of gravity, the electrical attraction between an electron and a proton (a) is just about the same size (b) is very much stronger (c) is very much weaker (d) cannot be compared (e) none of these.
3. When rubbed with a piece of wool, sulfur and glass will become charged (a) positively and negatively, respectively (b) negatively and positively, respectively (c) both positively (d) both negatively (e) none of these.
4. When the center-to-center separation between two small charged spheres is doubled, the electric force between them (a) is halved (b) doubles (c) is quartered (d) is quadrupled (e) none of these.
5. The SI units of electric flux are (a) N/C^2 (b) $N{\cdot}m/C$ (c) $N{\cdot}m^2/C$ (d) $C/N{\cdot}m$ (e) none of these.
6. Some cars and trucks (especially those carrying combustible fuels) have straplike bands attached to their bottoms that drag along the ground in a somewhat vain attempt to (a) discharge static electricity (b) get rid of moisture (c) build up negative charge (d) protect them in case they get hit by lightning (e) none of these.
7. Suppose we have three identical conducting spheres and one of them carries a charge of Q. If they are all brought into contact and then separated (a) they will each have a charge of $Q/3$ (b) they will each have a charge of Q (c) only one will be charged with Q (d) they will all be discharged (e) none of these.
8. A metal sphere is grounded through a switch, and a positively charged balloon is brought near it. The switch is opened and the balloon taken away. The sphere is now (a) neutral (b) negatively charged (c) positively charged (d) charged, but we cannot know its polarity (e) none of these.
9. A metal sphere is grounded through a switch, and a positively charged balloon is brought near it. The balloon is then taken away and the switch is opened. The sphere is now (a) neutral (b) negatively charged (c) positively charged (d) charged, but we cannot know its polarity (e) none of these.
10. If the charge on each of two identical tiny spheres is doubled while their separation is also doubled, their force of interaction will (a) double (b) become halved (c) be quartered (d) stay unchanged (e) none of these.
11. If you are stretched out in a bathtub full of water, you are likely (a) not to be grounded because the water is in a tub (b) to be grounded if the tub has metal feet (c) not to be grounded because you are inside a conductor (d) to be grounded because the water connects you electrically to the pipes (e) none of these.
12. The E-field of a point-charge 4.0 m away is measured to

be 100 N/C. The field 2.0 m away from that charge is (a) 400 N/C (b) 50 N/C (c) 100 N/C (d) 800 N/C (e) none of these.

13. A nonconductor is charged and then brought near a conductor. Consequently (a) the two electrostatically repel each other (b) the two electrostatically attract each other (c) only the nonconductor is repelled (d) there is no electrostatic interaction at all (e) none of these.
14. Three charges ($q_1 = 1$ nC, $q_2 = 2$ nC, and $q_3 = 5$ nC) are separated in space (by distances of $r_{12} = 1$ m, $r_{23} = 2$ m, and $r_{13} = 3$ m). The ratio of the magnitudes of the forces on q_3 due to q_1 and to q_2 is (a) 2/9 (b) 5/2 (c) 5/9 (d) 9/2 (e) none of these.
15. Two charges in vacuum attract each other. The same two charges, with the same separation, are now immersed in ethanol, which has a relative permittivity ($\varepsilon/\varepsilon_0$) of 25. The interaction is now (a) increased by a factor of $\sqrt{25}$ (b) decreased by a factor of $\sqrt{25}$ (c) decreased by a factor of 25 (d) increased by a factor of 25 (e) none of these.
16. An electron in an electric field of 100 N/C experiences a force of (a) 1.6×10^{-10} N (b) 1.6×10^{-21} N (c) 3.2×10^{-17} N (d) 1.6×10^{-17} N (e) none of these.
17. A charge $+q$ is placed at the origin of a coordinate system, and a charge of $+Q$ is located at $+a$ on the x-axis. The force on $+Q$ is found to be **F**. A third charge $-q$ is now placed at $+2a$ on the x-axis and the force on $+Q$ is now (a) zero (b) **F** (c) $\frac{1}{2}F$ (d) 2**F** (e) none of these.
18. In Fig. MC18, which drawing depicts the net field of a tiny positively charged conducting sphere in the gap of a charged parallel plate capacitor? (a) (b) (c) (d) (e) none of these.

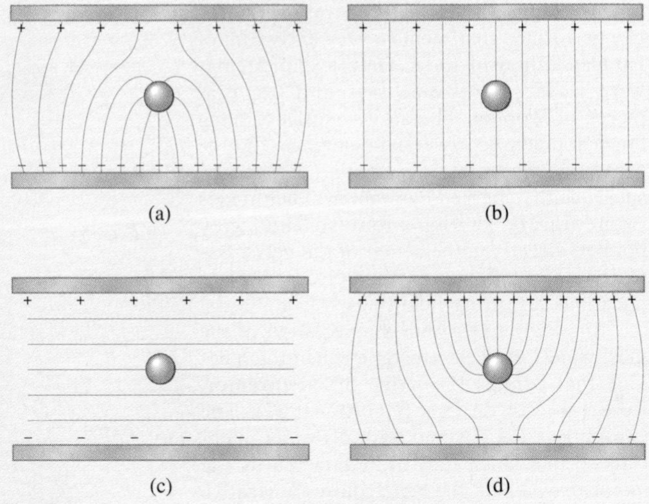

Figure MC18

19. In Fig. MC18, which drawing depicts the net field of a tiny neutral conducting sphere in the gap of a charged parallel plate capacitor? (a) (b) (c) (d) (e) none of these.

Problems

THE ELECTROMAGNETIC CHARGE
THE ELECTRIC FORCE

1. [I] By roughly how much does the mass of a copper object change when, upon being stroked with a piece of woolen cloth, it acquires an excess charge of -1.0 μC?

2. [I] How many electrons are needed to produce a charge of -1.0 C?

3. [I] Two tiny spheres carrying the same charge are 1.0 m apart in vacuum and experience an electrical repulsion of 1.0 N. What is their charge?

4. [I] A very small conducting sphere in air carries a charge of 5.0 picocoulombs and is 0.20 m from another such sphere carrying a charge Q. If each sphere experiences a mutual electrical repulsion of 2.0 μN, find Q.

5. [I] Two protons are fired directly at each other in a vacuum chamber. What is the force of repulsion at the instant they are 1.0×10^{-14} m apart?

6. [I] Two point-charges of $+0.50$ μC are 0.10 m apart. Determine the electric force they each experience in air.

7. [I] Two equally charged small spheres repel each other with an electric force of 1.0 N when 0.50 m apart, center-to-center, in air. What is the charge on each sphere?

8. [I] Compute the gravitational attraction between two electrons separated by 1.0 mm in vacuum, and compare that with the electrical repulsion they experience.

9. [I] An equilateral triangle with sides of 2.0 m is inscribed within a circle. A tiny charged sphere carrying ($+10$ μC) is then fixed at each vertex, and one of -25 μC is placed at the center of the circle. What is the net force acting on that central charge (magnitude and direction)?

10. [I] Two charged spheres each containing a quantity of excess electrons equal in number to Avogadro's number are separated by 1000 km in vacuum. Compute to two significant figures the electrical interaction between the spheres.

11. [I] Three small spheres immersed in air, each carrying a charge of $+25$ nC, are located at the vertices of a right isosceles triangle with a hypotenuse of 1.414 m, lying along the x-axis. Find the net electric force on the charge opposite the hypotenuse, located above it on the positive y-axis.

12. [I] Point-charges of $+1.0$ nC, $+3.0$ nC, and -3.0 nC are located, one each, at the three corners of an equilateral triangle of side-length 30 cm. Find the net electrostatic force exerted on the 1.0-nC charge. Assume the surrounding medium is vacuum.

13. [I] A square nonconducting framework is set up in space with a tiny metal sphere mounted at each corner. The spheres at either end of the 45° diagonal are charged equally with $+45$ nC each, while the other two spheres are both given charges of -45 nC. What is the net force on a charge of 10 nC at the very center of the square?

14. [II] How many electrons are there in a tablespoon (15 cm^3) of water? What is the net charge of all of these electrons?

15. [II] Three small negatively charged metal spheres in vacuum are fixed on a horizontal straight line, the x-axis. One ($-12.5\ \mu$C) is at the origin, another ($-5.0\ \mu$C) is at $x = 2.0$ m, and the third ($-10.0\ \mu$C) is 1.0 m beyond that at 3.0 m. Compute the net electric force on the last sphere due to the other two.

16. [II] Two charges of $+4.0$ nC and -1.0 nC are fixed to a baseline at a separation of 1.0 m. Where on the baseline should a third charge of $+2.0$ nC be placed if it is to experience zero net electric force?

17. [II] Figure P17 shows four point-charges fixed at the corners of a rectangle, in vacuum. Please compute the net electrostatic force acting on the 100-μC charge.

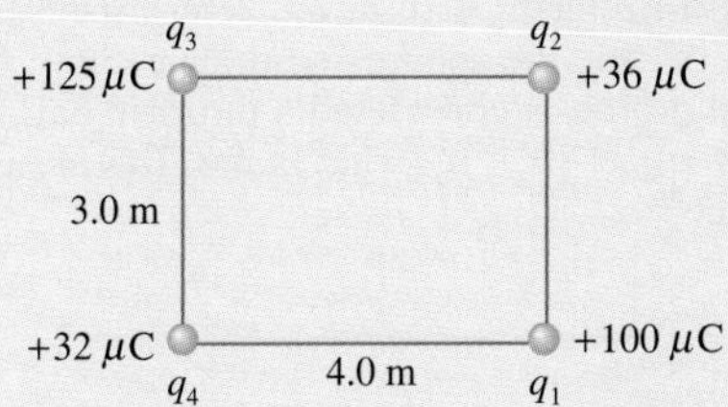

Figure P17

18. [II] Three very small charged spheres are located in a plane in air as follows: $q_1 = +15\ \mu$C at point (0, 0), $q_2 = -20\ \mu$C at point (2, 0), and $q_3 = +10\ \mu$C at point (2, 2). Find the net force acting on the last charge.

19. [II] Redo Problem 17 where now q_4 is $-32\ \mu$C.

20. [II] A cubic framework is inscribed within a sphere of 1.0-m radius. Tiny conducting spheres are then fixed at each corner and subsequently charged. There are four diagonals, and the pairs of spheres on the ends of each diagonal are equally charged with $+10$ nC, -10 nC, $+20$ nC, and -20 nC, respectively. What would be the resulting electric force on a $+100$-nC point-charge located at the center of the sphere? The surrounding medium is air.

21. [II] Two 2.0-g pith balls hang in air on cotton threads 50 cm long from a common point of support. The balls are then equally charged and they spring apart, each making an angle of 10° with the vertical. Find the magnitude of the charge on each ball.

22. [III] Two charges $+q$ and $-q$ reside in vacuum on the y-axis at locations of $-\frac{1}{2}d$ and $+\frac{1}{2}d$, respectively. Determine the force on a third charge $+Q$ located at a distance of $+x$ from the origin on the x-axis.

23. [III] Three free charges (two of which are $+Q$ and $+2Q$, separated by a distance d) are in equilibrium. Find the size, polarity, and location of the third charge.

THE ELECTRIC FIELD

24. [I] Determine the electric force acting on an electron placed in a uniform north-to-south E-field of 8.0×10^4 N/C in vacuum.

25. [I] A test-charge of $+5.0$ nC placed at the origin of a coordinate system experiences a force of 4.0×10^{-6} N in the positive y-direction. What is the electric field at that location? Assume the medium is vacuum.

26. [I] *A* $+10$-μC test-charge at some point beyond a charged sphere experiences an attractive force of 40 μN. Please compute the value of the E-field of the sphere at that point in a vacuum.

27. [I] A very small conducting sphere carrying a charge of -20 nC is attached to a force gauge and lowered into a uniform electric field. A force of 2.0 nN due east keeps the sphere in equilibrium. Describe the E-field, assuming air is the medium.

28. [I] A small positively charged object is placed, at rest, in a uniform electric field in vacuum. Write an equation giving its speed v after a time t in terms of its mass m and charge q.

29. [I] Determine the magnitude and direction of an E-field if an electron placed in it, in vacuum, is to experience a force that will exactly cancel its weight at the Earth's surface.

30. [I] What is the magnitude and direction of the electric field of an electron at a point 1.0 m away in vacuum?

31. [I] Consider a hydrogen atom to be a central proton around which an electron circulates at a distance of 5.3×10^{-11} m. Find the electric field at the electron due to the proton.

32. [I] Two point-charges of $+10$ nC and -20 nC lie on the x-axis at points $x = 0$ and $x = +10$ m, respectively. Find the electric field on the axis at point $x = +5.0$ m. Assume vacuum.

33. [I] Positive point-charges of $+20\ \mu$C are fixed at two of the vertices of an equilaterial triangle with sides of 2.0 m, located in vacuum. Determine the magnitude of the E-field at the third vertex.

34. [I] Redo Problem 33, this time with charges of $+20\ \mu$C and $-20\ \mu$C at either end of the baseline (creating the field at the remaining vertex).

35. [I] Two point-charges of $+50$ nC each are separated in air by 1.414 m. What is the value of the net electric field they produce at a point that is 1.0 m away from both of them?

36. [I] Determine a formula for the electric field of a point-charge Q in vacuum using Gauss's Law.

37. [I] Using Gauss's Law, determine the electric field in the air gap of a charged parallel plate capacitor. Use a cylindrical Gaussian surface with one endface embedded in the metal of one of the plates.

38. [I] Two flat metal plates of area 2.0 m^2 are placed parallel to each other and both are then charged, one with $+10\ \mu$C and the other with $-10\ \mu$C. Determine the electric field in the air gap anywhere far from the edges.

39. [I] A pair of flat, horizontal parallel aluminum plates 100 cm $\times$ 50 cm has a 1.0-mm vacuum gap between them. How should they be charged if there is to be a uniform upward E-field of 1000 N/C in the gap?

40. [I] Two flat metal plates of area 2.0 m^2 are placed parallel to each other and both are then charged, one with $+10\ \mu$C and the other with $-10\ \mu$C. Determine the electric field inside anywhere far from the edges when the gap is completely filled with mica (use a permittivity of 41×10^{-12} C^2/N$\cdot$m^2). How does the field now compare to the case where the gap was empty?

41. [II] A section of an advertising sign consists of a long tube filled with neon gas having electrodes inside at both ends. An electric field of 20 kN/C is set up between the electrodes, and neon ions accelerate along the length of the tube. Given that the ions each have a mass of 3.35×10^{-26} kg and are singly ionized, determine their acceleration.

42. [II] Three point-charges are placed at the corners of an isosceles triangle. At the left and right, end points of the base are $+1.0\ \mu C$ and $+1.0\ \mu C$, respectively, and at the vertex $+3.0\ \mu C$. The base of the triangle is 40 cm long, and the altitude is 30 cm high. Find the E-field at the midpoint of the baseline.
43. [II] Use Gauss's Law to determine a formula for the electric field outside of a long, uniformly charged cylinder of radius R with a positive surface charge density of σ.
44. [II] Use Gauss's Law to prove that there can be no net charge within a hollow conductor.
45. [II] Use Gauss's Law to find a formula for the electric field very close to any charged conducting surface in vacuum.
46. [II] With the previous problem in mind, compute the surface charge density on a conducting surface in the immediate vicinity of a point (in vacuum) where the electric field due to the charge is 5.0×10^5 N/C.
47. [II] An electron is placed in a uniform electric field of 1.5×10^4 N/C. Please determine its acceleration.
48. [II] Two point-charges of +10 nC and −20 nC lie on the x-axis at points $x = 0$ and $x = +10$ m, respectively. Find a point where the net electric field is zero, if such a point exists.
49. [II] Consider an idealized uniform spherical shell of charge $+Q$ surrounded at a distance by a concentric idealized shell of uniform charge $-Q$. Determine the E-fields between and beyond the shells. Assume vacuum.
50. [II] Consider the case of a small metal sphere suspended at the center of the cavity within a large hollow metal sphere. The central conductor is electrified with a charge of $+Q$. Write expressions for the E-field both inside the gap and outside the large sphere. Assume vacuum everywhere.
51. [III] A uniform ball of charge (Q) has a radius R. Determine the electric field inside the ball.
52. [III] With Problem 51 in mind, determine a formula for the field outside the sphere and draw a curve of E versus r.
53. [III] If a proton is considered a uniform ball of charge of radius 1.0×10^{-15} m, what is the E-field just beyond its surface?

18

Electrostatics: Energy

AN UNDERSTANDING OF ELECTROSTATICS is necessary before we study a number of important topics, from electric currents to atomic theory. This chapter continues the development, elaborating the primary idea of *electric potential,* the concept that corresponds in informal language to the notion of *voltage.* Of course, everybody knows what voltage is—batteries come in 1.5- and 9-volt varieties and household wall outlets deliver electricity at a voltage of 110 volts—but what does that really mean?

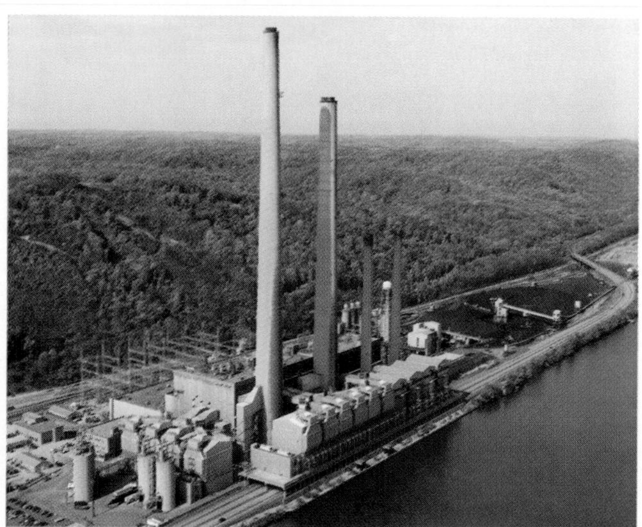

Smoke particles are removed before they leave the stacks at this plant. The gray structures at the bases of the smokestacks are the electrostatic precipitators. (See Fig. Q21.)

ELECTRIC POTENTIAL

Chapter 17 dealt with electric *force* and its more convenient equivalent, electric field. This chapter focuses on *energy* as it relates to electrostatics. Both **F** and **E** are vector quantities and therefore sometimes a little complicated to deal with. Energy is a scalar, and that will make the analysis a lot simpler.

18.1 *Electrical*-PE

It takes a force to raise a mass in the Earth's gravitational field (Fig. 18.1a). We do work on the object to overcome the downward force field. And the mass goes up from an initial value of *gravitational*-PE to a higher one. The Coulomb force has the same mathematical form as the gravitational force and so it, too, must be conservative (p. 296); potential energy is associated with each of these interactions. When a charge is made to move *against* the influence of an electric field, as for instance in Fig. 18.1b, it will experience a change in its *electrical*-PE. Such motion requires the application of an external force and the expenditure of energy by the agency providing that force (Fig. 18.2).

The opposite occurs when the field does positive work on the charge and propels it to a new lower-energy location. Thus, if an electron is released in the vicinity of a positively electrified object, it is drawn toward the plus charges; work is done *by the field* on the electron. As it "descends" in the field, the electron loses electrical potential energy (PE_E) while it gains speed, and its KE increases equivalently. The electron "falls" to the positive object (or away from a negative one), just as a proton "falls" to a negative object (or away from a positive one), just as a rock falls to the Earth.

A piano held in midair next to a pea has more *gravitational*-PE with respect to the Earth than does its little green companion because it has more mass. Similarly, a highly positively charged pith ball immersed in an E-field will experience a greater force and have a larger *electrical*-PE than will a single proton at that point in the field, because of its greater charge.

18.2 Electric Potential

Imagine a charged body of some sort. We'll refer to it as the primary charge because we want to know about *its* influence on the surrounding scene. In Chapter 17, we initially pictured the force field of a primary charge. But that was dependent on the test-charge used to make the survey, so we then eliminated that dependence by dividing out the test-charge (q_0). The result was the **E**-field, which depends only on the primary charge and which tells us all about the distribution of force (acting on *any* charge placed in the field) in the surrounding space.

What we want now is an energy measure that is again independent of the test-charge being used for the survey. Accordingly, we find the *electrical*-PE of a test-charge at all points in space within the field of the primary charge and divide each such value by q_0. This operation associates a number (positive or negative) with ev-

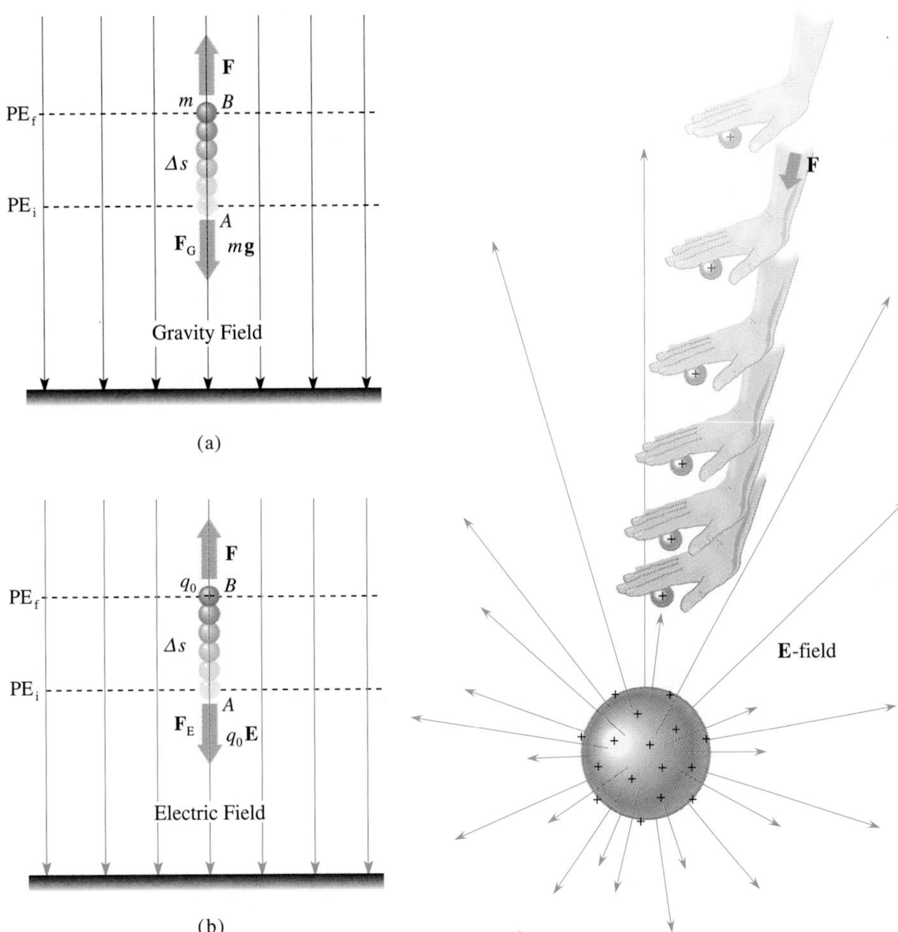

Figure 18.1 (a) An object, with a mass m, being raised in a uniform downward gravitational field by a force **F** experiences a downward gravitational force $\mathbf{F}_G = m\mathbf{g}$. (b) An object, with a charge q_0, being raised in a uniform downward electric field by a force **F** experiences a downward electric force $\mathbf{F}_E = q_0\mathbf{E}$.

Figure 18.2 Work must be done in moving a charge against an electric field. Here a positive charge is forced toward a positively charged sphere. It's being pushed against the *E*-field.

ery point in space and yields a map of the *potential energy per unit charge.* That scalar quantity is known as the **electric potential** (V) or just the *potential* for short:

$$\text{Electric potential} = \frac{\textit{electrical}\text{-PE}}{\text{charge}} \tag{18.1}$$

The units of potential are joules per coulomb (J/C) and in honor of Alessandro Volta, 1 J/C is defined to be 1 volt (1 V). It is common to speak of the electric potential as the *voltage.*

Example 18.1 The most commonly used unit of electric field is the volt per meter. Show that this is consistent with the units introduced thus far.

Solution: We want to show the equivalence between N/C and V/m. The brute force way is simply to convert everything into the most basic form possible in terms of meters, kilograms, seconds, and coulombs:

$$1\,\frac{\text{N}}{\text{C}} = 1\,\frac{\text{kg·m/s}^2}{\text{C}}$$

while

$$1\,\frac{\text{V}}{\text{m}} = 1\,\frac{\text{J/C}}{\text{m}} = 1\,\frac{\text{N·m/C}}{\text{m}} = 1\,\frac{\text{N}}{\text{C}} = 1\,\frac{\text{kg·m/s}^2}{\text{C}}$$

and the two are, indeed, identical.

▶ **Quick Check:** We could have stopped one step back and not made the first and last conversions, but we didn't know it was going to work out that way. Alternatively, multiply the top and bottom of 1 N/C by m: 1 N·m/C·m = 1 (J/C)/m = 1 V/m.

We now examine the difference in potential ΔV between two points in an arbitrary E-field. Accordingly, suppose a positive test-charge q_0 is displaced through a small distance Δs in an electrostatic field $\mathbf{E}$ and there is a corresponding change in its *electrical*-PE, ΔPE_E. For the moment, we take Δs to be so small that the E-field does not change appreciably over its extent. A force $\mathbf{F}$ is applied and the charge is brought from rest at its initial location to rest at its final location; $\Delta\text{KE} = 0$. In general, the work done against the field is $\Delta W = F_\parallel \Delta s$, where $F_\parallel$ is the component of $\mathbf{F}$ in the direction of the displacement. Because we are writing a scalar representation of vector quantities, we must assign a direction to be positive; take it as up. F and Δs are positive, E is negative; hence, $F = -q_0 E_\parallel$.

The work done on the charge by F is

$$\Delta W = F_\parallel \Delta s = -q_0 E_\parallel \Delta s$$

and that work goes into changing the potential energy:

$$\Delta W = \Delta\text{PE}_\text{E} = -q_0 E_\parallel \Delta s$$

By definition, the change in potential is

$$\Delta V = \frac{\Delta\text{PE}_\text{E}}{q_0} = -E_\parallel \Delta s \qquad (18.2)$$

When the field varies from place to place, the force varies, too, and we will have to evaluate the work (as was done before, p. 284) by taking the sum of all the contributions along each tiny displacement Δs. Accordingly

$$\Delta V = \sum_{A\to B} -E_\parallel \Delta s \qquad (18.3)$$

Potential Difference

The quantity ΔV is referred to as the **difference in potential** (or the **potential difference**) between the initial (A) and final (B) points in the field and is often written as

$$\Delta V = V_\text{B} - V_\text{A} \qquad (18.4)$$

The potential difference between two points A and B numerically equals the work done against the field in moving a unit positive charge from A to B with no acceleration:

$$V_B - V_A = \frac{W(A \rightarrow B)}{q_0} \tag{18.5}$$

The work done can be greater than zero, less than zero, or zero, and the same is true for ΔV.

Example 18.2 A small sphere carrying a positive charge of 10.0 μC is moved against an *E*-field through a potential difference of +12.0 V. How much work was done by the applied force in raising the potential of the sphere?

Solution: [Given: $q_0 = 10.0\ \mu\text{C}$ and $\Delta V = +12.0$ V. Find: *W*.] The defining relationship between ΔV, q_0, and *W* is Eq. (18.5); hence,

$$W = \Delta V\, q_0 = (+12.0\ \text{V})(10.0 \times 10^{-6}\ \text{C}) = \boxed{120\ \mu\text{J}}$$

▶ **Quick Check:** $q_0 = W/\Delta V = (120\ \mu\text{J})/(12\ \text{V}) = 10\ \mu\text{C}$.

Remember that energy is a relative quantity. There is no signpost floating in space that says, "This is the absolute zero of PE." Only changes in *electrical*-PE, and therefore *only changes in potential,* are important (Table 18.1).

A positive charge released in an electric field accelerates in the direction of the field. *It will not travel along a path that corresponds to a field line if the line is curved* (after a time Δt, $\mathbf{p}_f = \mathbf{p}_i + \Delta\mathbf{p}$, but $\mathbf{p}_i$ is not parallel to $\Delta\mathbf{p}$, which is parallel to **E**). The positive charge naturally descends from a high potential (V_A) to a low potential (V_B), losing PE_E and gaining KE. Thus, the *change in potential* ($\Delta V < 0$) *equals the negative of the work done per unit charge by the field on the positive charge.* Work done against the field always increases PE; work done by the field always decreases the PE. The change in PE equals the negative of the work done *by* a conservative force field.

TABLE 18.1 Common potential differences

Biochemical	1 mV–100 mV
Dry cell	1.5 V
Automobile battery	12 V
Household electricity	
U.S.A.	110 V–120 V
Much of Europe and Asia	240 V–250 V
Voltage induced in 600-mile Alaska pipeline by solar storm	1000 V
Power plant generator	24 000 V
Transmission lines	
Local	4 400 V
Cross country	120 000 V
Extra-high-voltage	500 kV–1 MV
Van de Graaff generator (1940)	4.5 MV
Folded tandem generator	25 MV
Lightning	10^8 V–10^9 V

The man straddling the 138 000-V power line landed there in a parasail accident. The brief surge of charge that brought him up to 138 kV burned his hands and feet, but otherwise he was fine.

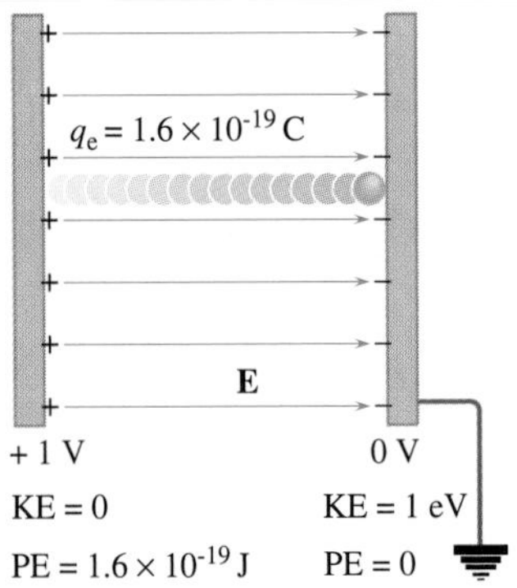

Figure 18.3 A proton falling through a potential difference of 1 volt. It follows the *E*-field being pushed along by the field and loses an amount of potential energy equal to 1 unit of charge times 1 volt, or 1 eV.

A negative charge released in an *E*-field moves spontaneously, experiencing a drop in *electrical*-PE. But it travels from a low potential to a higher potential ($\Delta V > 0$). That's evident from Eq. (18.2), where ΔPE_E is negative and q_0 is negative, so that ΔV is positive. When a negative charge moves from a point of high potential in a field to a point of low potential, it *increases* its PE_E.

The work done on a charge in moving it from point A to point B is independent of the path taken, just as the work done in climbing a mountain is independent of the path. If it were not, we could make a perpetual-motion machine, going from *A* to *B* along a small-change-in-potential path and coming back along a larger one, thereby gaining a bit of energy on each cycle.

When a particle (Fig. 18.3) with a charge equal to $+q_e$ moves from one point in a field to another, dropping 1 volt in potential, it will thereby decrease its *electrical*-PE by (p. 343) 1 electron volt (eV):

$$1 \text{ eV} = 1.6 \times 10^{-19} \text{ J}$$

Example 18.3 Figure 18.4 depicts the basic elements of a cathode-ray tube, the kind found in computer terminals and oscilloscopes. Electrons are "boiled" off a heated cathode and emerge through a pinhole, being drawn toward the first anode, which is at a relatively small positive potential above the cathode. A second anode that is 8000 V to 20 000 V above the cathode (depending on the design of the tube) accelerates the beam up to speed. Determine the change in the potential energy of each electron on traversing the gun, given that the second anode has a voltage of 20 kV above the cathode. Assuming that an electron has negligible motion at the cathode, write a general expression for its final speed in terms of ΔV. Find its specific final speed in this case.

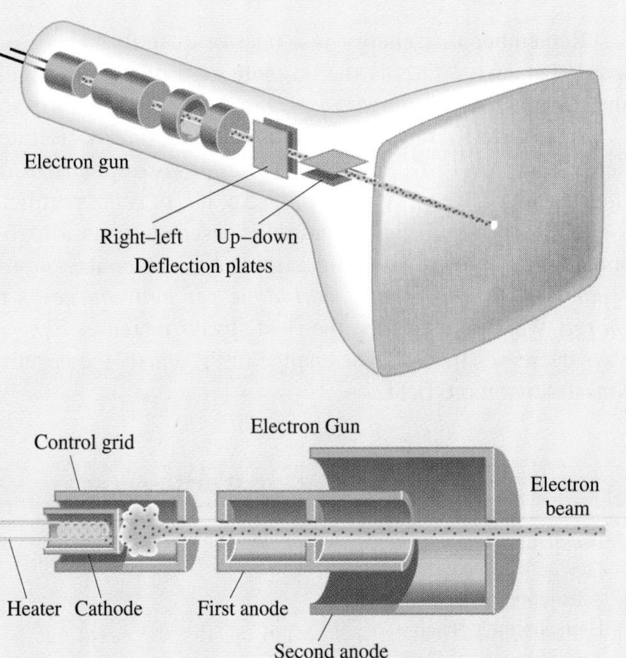

Figure 18.4 A cathode-ray tube (see Fig. Q19). A beam of electrons passes between two perpendicular sets of plates. These are appropriately charged, and the beam is deflected to any desired point on the face of the tube.

Solution: [Given: $v_i = 0$, and $\Delta V = 20$ kV. Find: v_f.] Any one electron traverses a potential difference of 20×10^3 V and so drops in potential energy by

$$\Delta PE_E = q_e \Delta V$$

This result is true regardless of the details of the field existing within the gun. Remember that a negative charge loses PE_E as it *increases* in potential (however weird that may sound).

$$\Delta PE_E = (-1.6 \times 10^{-19} \text{ C})(20\,000 \text{ V})$$

and $$\boxed{\Delta PE_E = -3.2 \times 10^{-15} \text{ J}}$$

The electron's energy is conserved and so

$$KE_i + PE_i = KE_f + PE_f$$

and since $KE_i = 0$

$$KE_f = PE_i - PE_f = -\Delta PE$$

and $$\tfrac{1}{2} m_e v_f^2 = -q_e \Delta V$$

(continued)

(continued)

Consequently

$$v_f = \left[\frac{-2q_e\Delta V}{m_e}\right]^{\frac{1}{2}} = \left[\frac{-2(-3.2 \times 10^{-15}\text{ J})}{9.1 \times 10^{-31}\text{ kg}}\right]^{\frac{1}{2}}$$

and $\boxed{v_f = 8.4 \times 10^7 \text{ m/s}}$

▶ **Quick Check:** The high voltage and tiny mass reasonably result in a tremendous speed of $\approx \frac{1}{4}$c. As we will see in Chapter 28, this rapid motion is more accurately treated using the Special Theory of Relativity. Note that the electron experiences an energy drop of 20 keV = 3.2×10^{-15} J = ΔKE.

18.3 Potential in a Uniform Field

Let's now derive an expression for the potential difference between two points in a **uniform electric field**. Consider the positive charge in Fig. 18.5. As we have already seen, the potential difference across a small displacement Δs is

$$\Delta V = -E_{\parallel}\Delta s = -E\,\Delta s\cos\theta \qquad [18.2]$$

where $E_{\parallel} = E\cos\theta$. Because the field is constant and θ is constant, the net potential difference encountered in going from A to B is the sum of the contributions given by Eq. (18.2); namely

$$\Delta V = \sum_{A\to B} -E_{\parallel}\Delta s \qquad [18.3]$$

$$V_B - V_A = -E\cos\theta \sum_{A\to B}\Delta s$$

and since

$$\sum_{A\to B}\Delta s = D$$

it follows that

$$V_B - V_A = -ED\cos\theta$$

Using $\pm D\cos\theta = d$, this becomes

$$V_B - V_A = \pm Ed \qquad (18.6)$$

The potential difference is + when the displacement has a component that is opposite to the field and − when it has a component parallel to the field. *The potential difference only depends on the displacement parallel or antiparallel to the field.*

Observe in Fig. 18.5a that

$$V_C - V_A = -Ed$$

and so $\quad V_B - V_A = V_C - V_A$

Points B and C are at the same potential. Another way to see this is to realize that if the points B and C are such that there is a path from one to the other that is everywhere

Figure 18.5 A positive charge moving diagonally across a uniform electric field. In (a) there is a component of displacement parallel to the field, and the potential drops in going from A to B. In (b) there is a component of displacement against the field, and the potential rises in going from A to B.

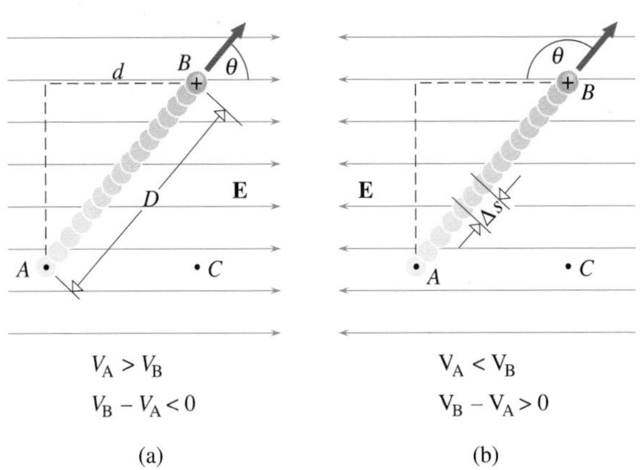

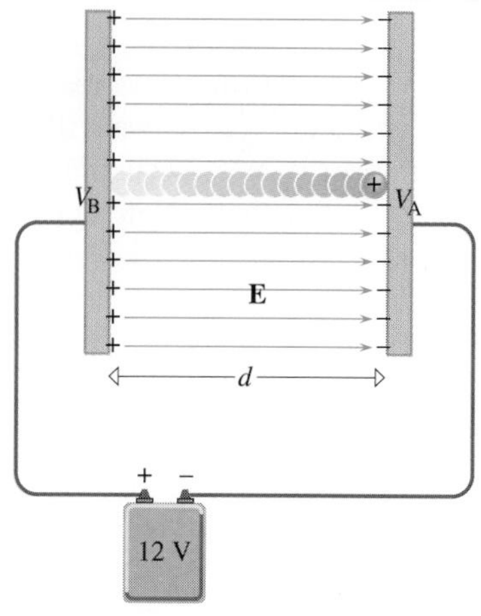

Figure 18.6 A battery connected across two parallel metal plates. There is a uniform E-field between the plates, and a positive charge experiences a drop in potential $V_A - V_B = -Ed$ upon traveling from the left plate to the right plate.

perpendicular to the field, $E_\parallel = 0$, $F_\parallel = 0$, and $\Delta W = 0$. No work is done against the field in moving from B to C, which is the case gravitationally when a mass is moved horizontally on the Earth's surface. Thus, since $\Delta W = q\Delta V$, $\Delta V = 0$, and the two points are at the same potential.

The simplest situation, that of moving a positive charge along (or opposite to) a uniform E-field is depicted in Fig. 18.6. Then $\theta = 0$ (or 180°) and $V_B - V_A = +Ed$, while $V_A - V_B = -Ed$. There are a number of devices, such as batteries, generators, fuel cells, solar cells, thermoelectric cells, etc., that can provide electrical energy and sustain a fairly constant potential difference across their two output terminals. In Fig. 18.6, **the positive terminal of the battery is 12 V higher than the negative terminal**. These terminals are connected to the parallel plates by excellent conductors, along which it is assumed there will be no appreciable loss in electrical energy and no measurable decrease in potential—that's a little unrealistic, but we will get back to that point in Chapter 19 when resistance is considered. This battery will maintain a constant 12-V difference across the parallel plates and sustain a constant field in the gap.

Example 18.4 If the separation of the plates in Fig. 18.6 is 2.0 mm, determine the magnitude of the electric field in the air gap.

Solution: [Given: $d = 2.0$ mm and $V = 12$ V. Find: E.] The plus side is at a higher potential than the minus side: $V_B > V_A$ by 12 V. We would like an expression relating E, V, and d, and so we use

$$V_B - V_A = +Ed \qquad [18.6]$$

whereupon

$$12\text{ V} = +E(2.0 \times 10^{-3}\text{ m})$$

Hence $$\boxed{E = 6.0\text{ kV/m}}$$

▶ **Quick Check:** This is a fair-sized field (Table 17.2), but that's reasonable since the spacing over which the voltage drops 12 V is only 2 mm. That corresponds to a drop of 6 V/mm, or 6 kV/m.

18.4 Potential of a Point-Charge

Earlier, we calculated the E-field of a point-charge and saw how that could be used to determine the fields for all sorts of charge distributions. The same is true for the potential; that is, once we have the potential of a point-charge we can use it to determine the potentials of other more complicated systems. To that end, return to p. 294 where we computed the change in *gravitational*-PE when a test-mass m changed position in the field of a spherical mass M. Because the gravitational field of a point-mass and the electric field of a point-charge both vary inversely with r^2, finding the corresponding change in PE is a bit complicated. The change in PE is the work done, which is the force times the displacement, but the force is now a variable. Still, the mathematical aspect of the problem has already been solved in the process of arriving at Eq. (9.14):

$$\Delta PE_G = GmM\left(\frac{1}{R} - \frac{1}{r}\right)$$

This is the change in *gravitational*-PE that occurs when a mass m is moved from R *out* to r *against* the *attractive* $1/r^2$-gravitational field produced by a sphere of mass M. It follows that

$$\Delta \mathrm{PE_E} = kq_0 Q\left(\frac{1}{r_\mathrm{B}} - \frac{1}{r_\mathrm{A}}\right)$$

This is the change in *electrical*-PE that occurs when a positive charge q_0 is moved from r_A *in* to r_B *against* the *repulsive* $1/r^2$-electric field produced by a tiny sphere of positive charge Q, as shown in Fig. 18.7. Inasmuch as

$$\Delta V = \frac{\Delta \mathrm{PE_E}}{q_0}$$

[*point-charge*]

$$V_\mathrm{B} - V_\mathrm{A} = kQ\left(\frac{1}{r_\mathrm{B}} - \frac{1}{r_\mathrm{A}}\right) \tag{18.7}$$

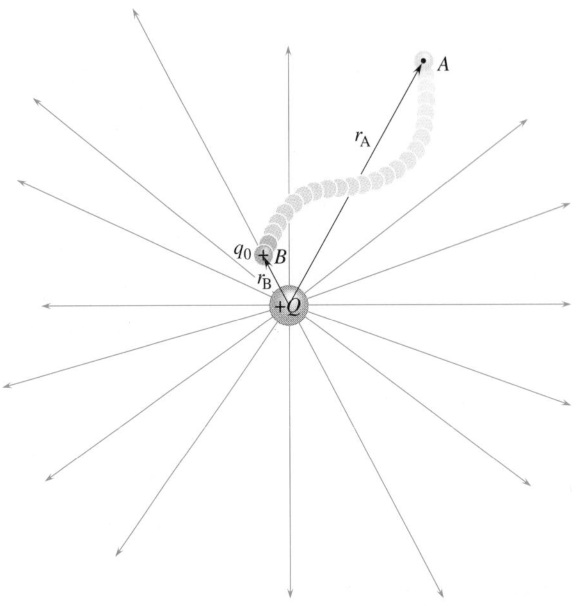

Figure 18.7 Here, a positive charge q_0 is moved inward toward Q, going from a distance r_A to a distance r_B. The potential difference ($V_\mathrm{B} - V_\mathrm{A}$) is positive because q_0 is pushed downward against the field. Its $\mathrm{PE_E}$ increases and, if released, q_0 will be pushed out and away.

This result is the sought-after potential difference encountered in moving from point A to point B in the electric field of a tiny positive sphere of charge or, equivalently, a point-charge Q.

Again, as we saw in the case of gravity, it is often useful to take the zero of potential, which is totally arbitrary, at infinity. Thus, let $r_\mathrm{A} \rightarrow \infty$, whereupon $1/r_\mathrm{A} \rightarrow 0$

and
$$V_\mathrm{B} - 0 = kQ\left(\frac{1}{r_\mathrm{B}} - 0\right)$$

represents the potential at the finite point B with respect to zero at infinity. To simplify matters, we might as well drop the subscript B since the equation applies to any point in the region beyond the sphere of positive charge. Accordingly,

$$V = kQ\left(\frac{1}{r}\right) \tag{18.8}$$

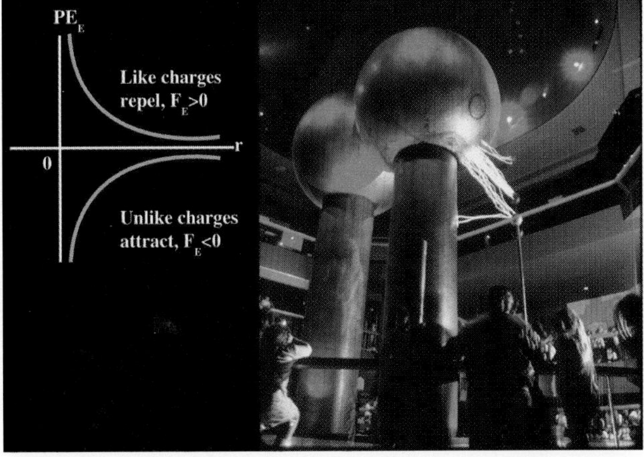

Figure 18.8 The potential energies of pairs of interacting point-charges. Here, a positive point charge is located at the origin, and the other charge is at r. Like charges repel and the $\mathrm{PE_E} > 0$; unlike charges attract and the $\mathrm{PE_E} < 0$. In each case, the force is the negative of the slope of the $\mathrm{PE_E}$ curve.

This is the potential at a distance r from an isolated positive point-charge measured with respect to the zero at ∞. Alternatively, we can say that this is *the work done per unit charge against the E-field of a positive point-charge Q in bringing a test-charge from infinity to a distance r from Q.* The potential at a point can be positive, negative, or zero depending on Q, and, of course, it is a directionless quantity, a scalar. Figure 18.8 shows the $\mathrm{PE_E}$ of a point-charge placed in the **E**-field of a point-charge; it's significant because it corresponds to the electron-proton interaction of a hydrogen atom.

Example 18.5 What is the potential difference encountered in going from point B to point A in Fig. 18.7 if the tiny sphere carries a charge of $Q = +10\ \mu\text{C}$, $r_A = 20$ cm, and $r_B = 10$ cm?

Solution: [Given: $Q = +10\ \mu\text{C}$, $r_A = 20$ cm, and $r_B = 10$ cm. Find: ΔV.] Going from point B to point A, a positive test-charge moves out along the field and so drops in potential. Using Eq. (18.7)

$$V_A - V_B = -(V_B - V_A) = -kQ\left(\frac{1}{r_B} - \frac{1}{r_A}\right)$$

$$V_A - V_B = -(9.0 \times 10^9\ \text{N}\cdot\text{m}^2/\text{C}^2)(+10 \times 10^{-6}\ \text{C}) \times (10\ \text{m}^{-1} - 5.0\ \text{m}^{-1})$$

and

$$\boxed{V_A - V_B = -0.45\ \text{MV}}$$

▶ **Quick Check:** From Eq. (18.8), the potentials at distances of r_B and r_A are +0.90 MV and +0.45 MV, respectively. Hence, going from point B to point A there is a change in potential of −0.45 MV.

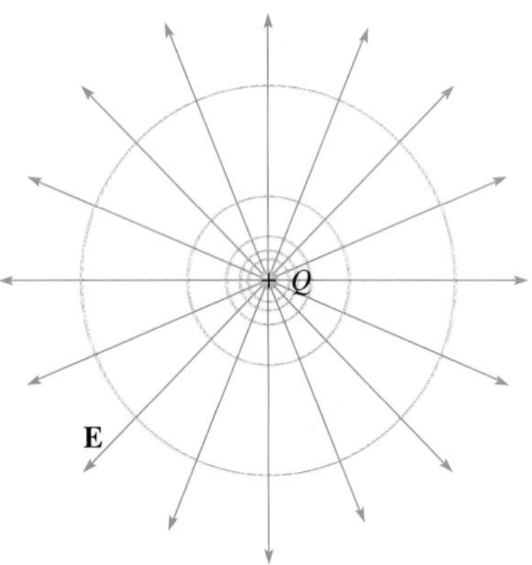

Figure 18.9 Field lines and equipotentials for a small positive charge. The equipotential surfaces are concentric spheres centered on Q.

18.5 Equipotentials

Return to Fig. 18.5 and observe that we can move a positive test-charge from point B to point C so that the displacement is everywhere perpendicular to the uniform E-field and no work is done. The potential does not change in the process, and we say that *the line from B to C, which is everywhere perpendicular to the field, is an equipotential line*—a line along which the potential energy of a test-charge remains unchanged.

Envision a positive point-charge or, equivalently, a tiny charged conducting sphere (Fig. 18.9). The potential at any fixed distance R beyond the charge is given by

$$V = kQ\left(\frac{1}{R}\right) \qquad [18.8]$$

which is constant in all directions. In other words, the **equipotential surfaces** of a point-charge are a series of concentric spheres, everywhere perpendicular to the E-field—this is a crucial part.

These ideas should bring to mind the concept of a conductor. Since we found that the electric field in the vicinity of the surface is always perpendicular to a conductor regardless of its shape, the conductor's surface must also be an equipotential. That's true whether the conductor is charged or not. Another way to appreciate this concept is to realize that there is no E-field within the substance of a conductor in electrostatic equilibrium. As a consequence, a test-charge transported from one point within a conductor to another will not have to be moved against an electric force, will not have work done on it, and will therefore experience no change in potential. By analogy, although a ball will roll down a hill by itself, if it were placed on the smooth surface of an idealized spherical planet, it would not spontaneously move at all. The ball, by itself, can only drop to a lower *gravitational*-PE, and yet everywhere on the surface it has the same PE or, if you like, all points on the surface are at the same gravitational potential.

The entire body of a conductor, devoid as it is of any E-field, must be an **equipotential volume**. Indeed, provided there are no encompassed isolated charges (as in Fig. 17.34), the total region contained within a conductor, hollow or not, is at the same potential. Because the Earth itself is a conductor, its surface is an equipotential and, by custom, it is often taken to be the zero of potential. That's just what is done in the electrical system in your home, where the "hot" terminal in each of

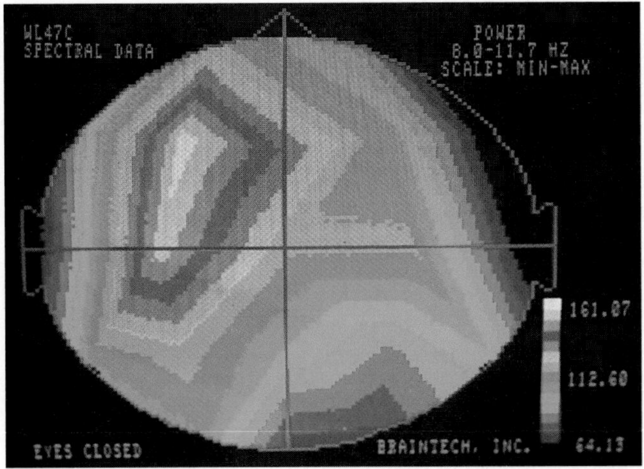

A map of the equipotentials in the brain of a person with epilepsy. This picture was made about 0.1 s after the person received a stimulus.

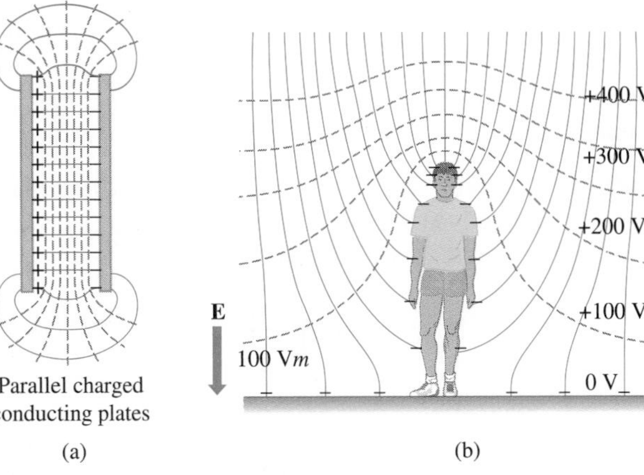

Figure 18.10 Electric fields and equipotentials. (a) A pair of charged conducting parallel plates. The person in (b) is grounded and so is at 0 potential.

the outlets oscillates between ±120 V with respect to the zero-potential terminal. The latter is physically connected to ground (p. 815).

Figure 18.10 shows both the E-field lines and the associated equipotentials for several configurations. In each case, the equipotentials are drawn successively at fixed-voltage intervals, ΔV. When the field is uniform, as in Fig. 18.11, the equipotentials are evenly spaced. The inclusion pictorially of equipotentials beautifully complements the field-line diagrams. A drawing of field lines immediately reveals the direction in which a charge will experience a force when placed anywhere in the region, and that corresponds to the direction in which it will accelerate as well. Moreover, if the field is uniform, the charge will move along a field line. A glance at a diagram containing equipotentials reveals the energy change that will occur when a charge moves from one point to another along *any* path. Together, the field lines and equipotentials provide a complete picture of the influence of the primary charge distribution everywhere in space.

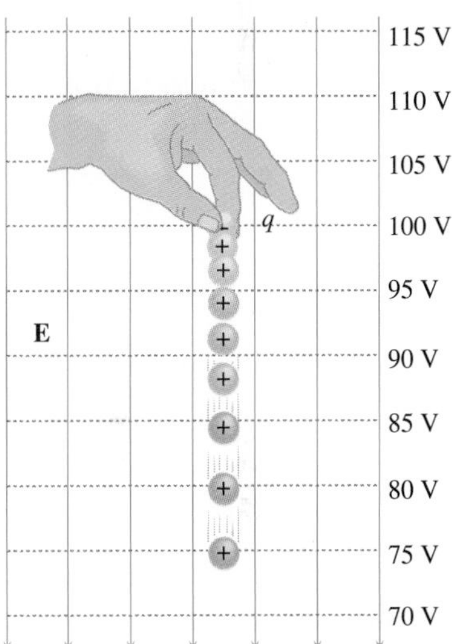

Figure 18.11 After falling through a potential difference of 25 V, the charge has lost an amount of potential energy $25q$ equal to its gain in kinetic energy.

18.6 The Potential of Several Charges

Potential is a scalar quantity. Thus, if there are several charges present, the sum of their superimposed potentials anywhere in the surrounding space is equal to the algebraic sum of the individual contributions. We can make use of this fact provided we stick with point-charges because we already have, via Eq. (18.8), the potential function for a point-charge. Therefore, *if there are two or more point-charges, the net potential at any location will be the scalar sum of all the potentials at that location due to each charge.* Consequently, it is certainly possible that at some point in space a positive potential due to one charge can be canceled by a negative potential due to another charge. If we imagine a positive test-charge brought to that zero-potential point from infinity, in the process it will be repelled by one charge as much as it is attracted by the other, and no net work will be done by or against the field. *There can be a net E-field at a point even though the potential at that point is zero and, conversely, there can be a nonzero potential at a point where the net field is zero.*

Example 18.6 Figure 17.26, p. 607, shows two charges each of +10 nC at (0, 0) and (8.0, 0). Determine the net potential at point (4.0, 0). Then let one of the charges be −10 nC and recompute the potential at that same point.

Solution: [Given: $Q_1 = Q_2 = \pm 10$ nC at (0, 0) and (8.0, 0), respectively. Find: V at point A, (4.0, 0).] For each charge, there is a contribution given by

$$V = kQ\left(\frac{1}{r}\right) \qquad [18.8]$$

and so at point A

$$V_A = kQ_1\left(\frac{1}{r_1}\right) + kQ_2\left(\frac{1}{r_2}\right)$$

Thus, with both charges positive

$$V_A = \frac{(9.0 \times 10^9\ \text{N}\cdot\text{m}^2/\text{C}^2)(+10 \times 10^{-9}\ \text{C})}{4.0\ \text{m}} + \frac{(9.0 \times 10^9\ \text{N}\cdot\text{m}^2/\text{C}^2)(+10 \times 10^{-9}\ \text{C})}{4.0\ \text{m}}$$

and $\boxed{V_A = 45\ \text{V}}$

Although the E-field is zero at that point (see Fig. 17.29), the potential is not; it is positive. Work must be done against the fields of both charges to haul a test-charge in from infinity.

With either charge negative, we have a dipole and

$$V_A = \frac{(9.0 \times 10^9\ \text{N}\cdot\text{m}^2/\text{C}^2)(+10 \times 10^{-9}\ \text{C})}{4.0\ \text{m}} + \frac{(9.0 \times 10^9\ \text{N}\cdot\text{m}^2/\text{C}^2)(-10 \times 10^{-9}\ \text{C})}{4.0\ \text{m}}$$

and $\boxed{V_A = 0\ \text{V}}$

▶ **Quick Check:** In the field of a dipole, as a test-charge is moved in from infinity, the positive charge repels it, and work must be done to overcome that repulsion. At the same time, the negative charge attracts the test-charge, doing negative work on it. The result of the push-pull is that no net work needs to be done on the test-charge to bring it to A, and the potential there is the same as it was at the start of the journey at infinity—namely, zero.

Figure 18.12 shows the fields and potentials for two equal charges. When both charges are positive (Fig. 18.12a), the field lines emerge from the system, and as we move out along them, the positive potential drops off, approaching zero at infinity. At the very center, there is a finite positive potential. Note how the equipotentials combine into a single surface that becomes increasingly more like a sphere; from very far away, the system looks like a single positive charge.

By contrast, in the case of the dipole (Fig. 18.12b), the field extends out the positive side and in the negative side. The equipotentials surrounding the positive charge are positive and those surrounding the negative charge are negative. There is a zero-potential plane down the middle to which the field is everywhere perpendicular. A test-charge can be brought from infinity by traveling within the plane and therefore along a path perpendicular to the field. No work is done, and no change from zero occurs in the potential. Of course, since the E-field is conservative, we can actually come from infinity to any point on the plane via any path, and the change in potential would still be zero.

A complex distribution of charges results in a complex pattern of field lines and equipotentials. See if you can figure out what's happening here.

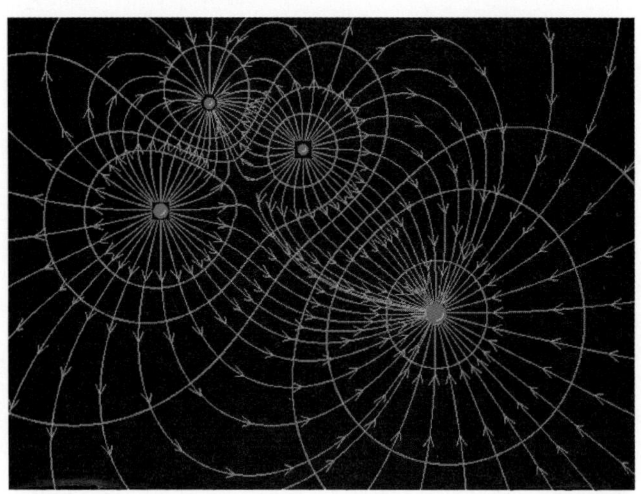

18.7 Potential and the Distribution of Charge

In 1672, Otto von Guericke devised a machine that greatly reduced the effort it took to build up charge by rubbing. He produced a large sphere of sulfur (because that material was easily electrified) and mounted it on a crank-driven mechanism that whirled it around so it could be stroked as it spun. Spinning within his cupped hand, it built up sizable quantities of "electric virtue" that sparked away impressively. Before long, all sorts of revolving rubbing machines were devised, and everything in sight from milk buckets to chickens were being charged by these *electrostatic generators.*

Imagine von Guericke's sulfur sphere. As it becomes increasingly more charged, it takes increasingly more work to further charge it. Clearly, there is a relationship between the charge, the geometry, and the work needed to bring yet another electron to the sphere. With a given charge Q distributed over a *large* sphere, the net force exerted on a new incoming electron (since the Coulomb force drops off as $1/r^2$) is smaller than if that same Q was on a tiny sphere. After all, on a large sphere, the individual charges that constitute Q are far apart from one another and from the new charge. Thus, **the potential of a charged sphere increases, both as its radius R decreases and as its net charge increases.** Indeed, the more tightly packed the surface charge, the more externally supplied work must be done against it, and the higher the potential. We already know that for a charged sphere of radius R

$$V = kQ\left(\frac{1}{R}\right) \qquad [18.8]$$

just as one might anticipate. Furthermore, since $Q = 4\pi R^2\sigma$

[*sphere*]
$$V = 4\pi kR\sigma \qquad (18.9)$$

If we have two different-size spheres, each with the same σ, the larger would hold more charge, more work would be needed to charge it, and so its potential would be higher.

An object like von Guericke's sphere cannot be charged endlessly; neither the potential nor σ keeps going up as the thing is rubbed indefinitely. The sphere, which initially strongly drew electrons from the hand, becomes highly negatively charged. The charges are increasingly crowded together until a point is reached, depending on the size of the sphere, where they strongly repel and block any further arrival of electrons. The sphere reaches a maximum potential. In contrast, the Van de Graaff generator avoids that physical limitation by having charge introduced inside a conductor, where there is no repelling field (Fig. Q14, Chapter 17). Its maximum potential can be so great that it is ultimately limited by the surrounding air. The very high electric field at the surface of the generator will ionize the air, causing it to become a conductor—a phenomenon known as *dielectric breakdown.* Air breaks down, ionizing and supporting sparks when the voltage across it is ≈30 kilovolts per centimeter. A little spark one-eighth inch long, the kind that comes from rug shuffling, corresponds to a potential difference of almost 10 kV.

Suppose we run an electrostatic generator at a sustained potential V and wish to transfer some charge from it to a neutral conducting object. Bringing the two in contact allows charge to flow from the generator to the conductor. *That flow continues until the object reaches the same potential as the generator.* Similarly, suppose a neutral conductor is brought in contact with a charged conductor. Charge will flow to the formerly neutral body *until both reach the same potential.*

When there is a potential difference between two bodies, ΔV equals the work that must be done per unit of positive charge to transfer charge from one body to the other. If the two are brought into contact, charge will spontaneously flow "downhill," and the potential will rapidly equalize at a value that may be positive, negative, or zero.

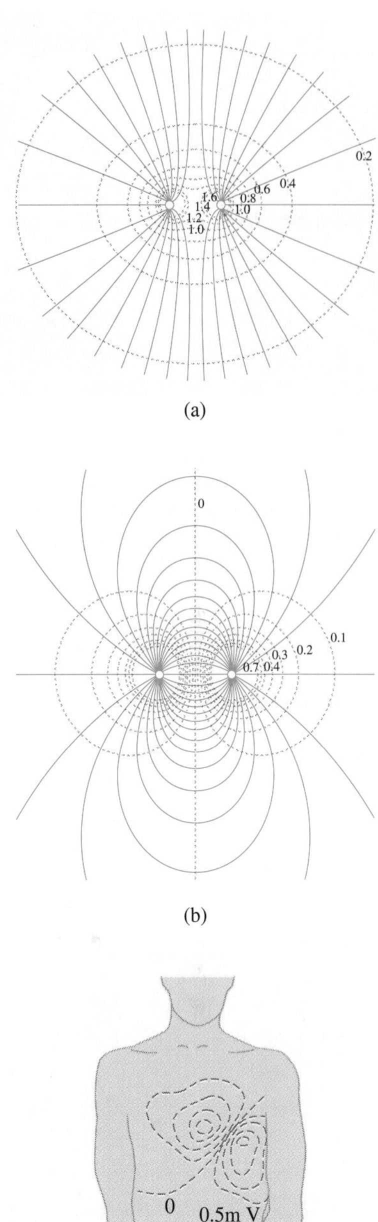

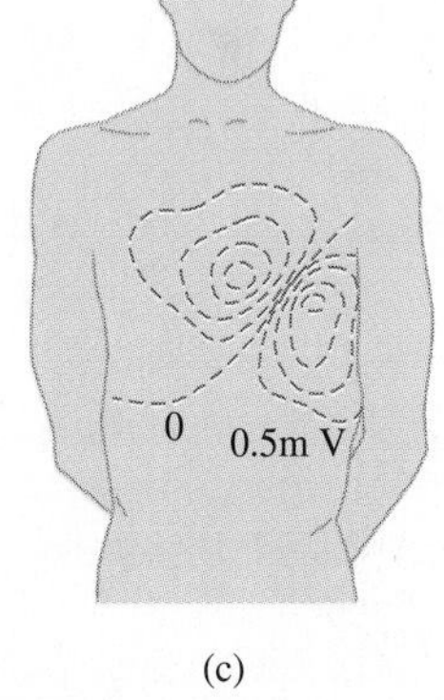

Figure 18.12 (a) Equipotentials and field lines for two equal charges. (b) Equipotentials and field lines for two opposite charges of equal magnitude (a dipole). (c) Equipotentials (in millivolts) at some instant across the chest due to heart activity.

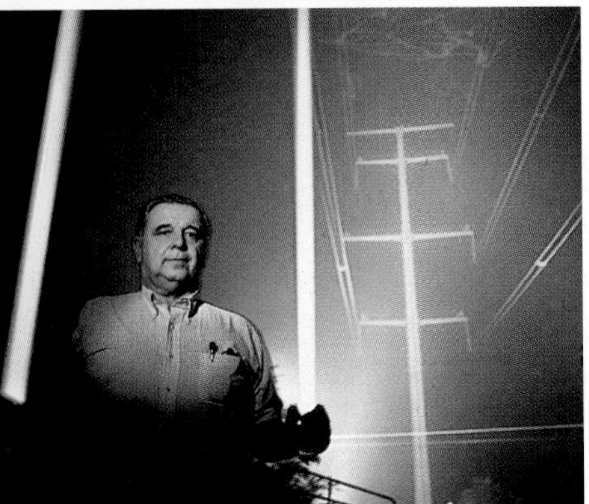

Power lines in rural areas are often operated at several hundred thousand volts. The voltage drop per meter from the line down to the zero of ground can be quite high. For this farmer, it's enough to light the fluorescent bulb he's holding in his hand.

18.8 Conservation of Charge

Conservation of Charge had been known for over a century before physicists, guided by Noether's Principle, began to search for a corresponding symmetry associated with classical Electromagnetic Theory. What they found was an invariance arising from the arbitrariness of the electric and magnetic (which we are not going to discuss) potentials. That mathematical behavior is called *gauge symmetry,* and many now believe that it is the fundamental characteristic shared by all correct theories (Fig. 18.13). Conservation of Charge is related to the fact that the proper formulation of electromagnetism is gauge symmetric. That notion sounds complicated, and we will not attempt to explain gauge symmetry until Chapter 33, but we can get a sense of the central idea very simply. Let's look at just the electric-potential part of the symmetry. There is no observable effect that depends on an absolute value of electric potential. We will show that the invariance of phenomena with respect to electric potential is consistent with Conservation of Charge.

Imagine a laboratory immersed in a uniform E-field (Fig. 18.14). We, outside the lab, fix the arbitrary zero-potential anywhere we like and then bring a positive charge up to the lab from that $V = 0$ level. We do a precise amount of work on the test-charge and it has a corresponding precise ΔPE_E, as far as we are concerned. Yet no experiment performed *inside* the laboratory can measure the PE_E of the charge *with respect to the chosen* $V = 0$ *level.* In fact, the laboratory can be translated up, down, or sidewise in the uniform field, and the experimenter inside at the new location will not observe any change—that's due to the arbitrariness of the potential.

What would happen if it were possible for a charge to vanish while all the remaining laws of physics—Conservation of Energy, Conservation of Momentum, etc.—were precisely in effect? The energy possessed by the charge would have to be liberated to the system in the lab in some way if energy is to be conserved. (In what-

Figure 18.13 We have seen this diagram five or six times before. Now note that Conservation of Electromagnetic Charge is related to gauge symmetry, and that's the first clue to the importance of Gauge Theory.

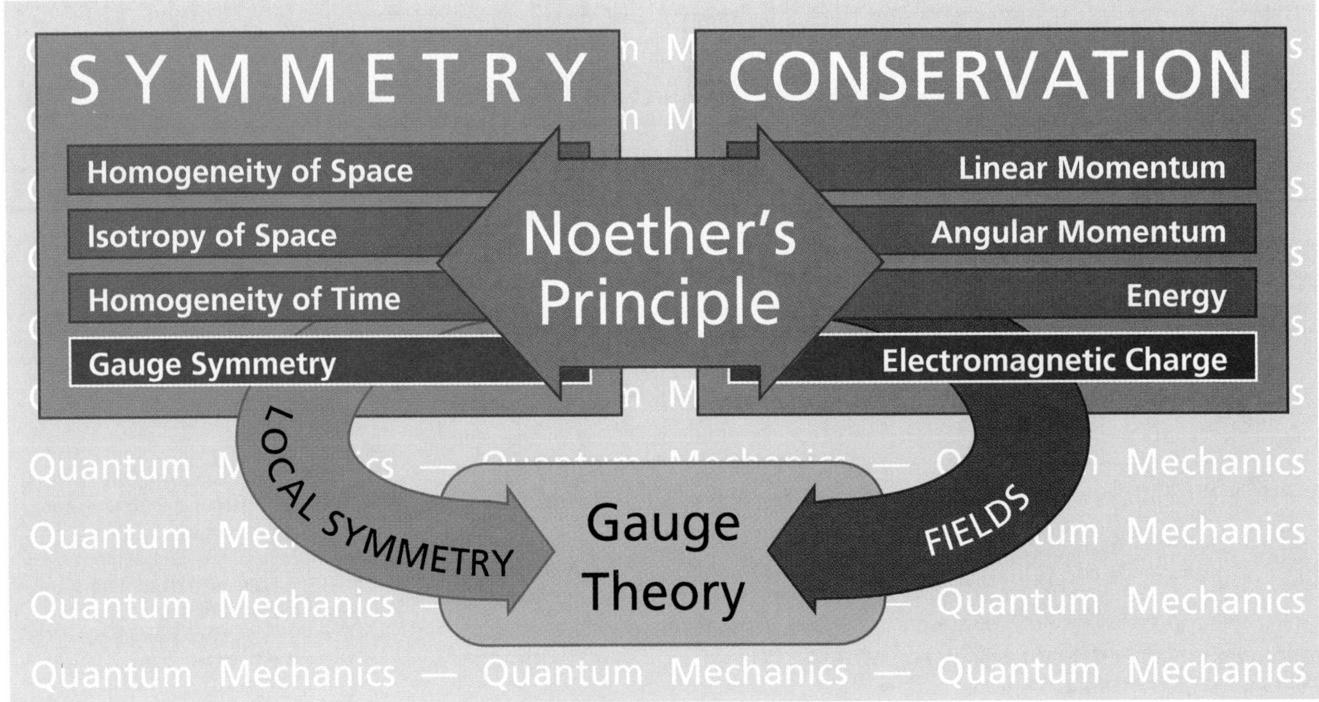

ever manner that energy is transferred, momentum must also be conserved.) The experimenter in the lab and we outside are at rest with respect to each other and must see the same amount of liberated energy. That's paradoxical—the amount of potential energy the charge possesses is arbitrary, and yet the amount of energy that would be liberated if the charge were to vanish is not. Furthermore, by measuring the energy given out, observers in the laboratory could presumably determine the charge's absolute potential, which is impossible: the $V = 0$ level can be anywhere. Hence, **a single charge cannot vanish**. On the other hand, if an equal pair of positive and negative charges was brought to the lab, *no* net work would be done against the field. They certainly could annihilate each other, or some other pair, still conserving charge, energy, and momentum. *In this Universe, energy is relative and charge must be conserved* (see the photo below).

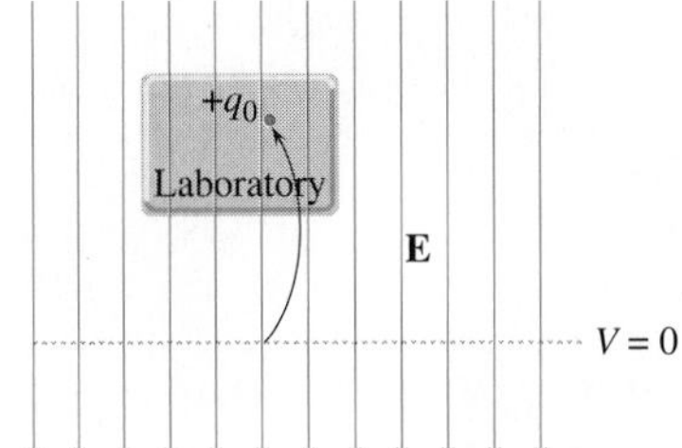

Figure 18.14 A charge in a uniform *E*-field is raised from an arbitrary zero-*V* level up to a laboratory. No experiment can be performed in the lab that will reveal the original level from which the charge was raised.

CAPACITANCE

Sometime around 1706, Hauksbee replaced von Guericke's sulfur ball with a large glass globe, thereby creating a friction electric machine that was to become popular throughout Europe. By mid-century the science of electricity was drawing eager crowds, who came to see the wonders performed by itinerant lecturers with cartloads of mysterious paraphernalia. In London, one of Hauksbee's Influence Machines was even installed at the so-called Temple of Health, that it might provide a therapeutic environment around the "magnetico-electrico" bed, ostensibly to aid in matters connubial.

In these early days, rubbing machines could develop prodigious voltages but only tiny trickles of charge—the pressing problem was to store the charge, so that it could be built up and then dumped in a single powerful blast. It was soon recognized that when a conductor was electrified, the size of the conductor determined the amount of charge it could store. In the beginning, bars of metal were used, gun barrels and the like, and sometimes even people themselves, but more ambitious practitioners suspended massive cannons, charging them to tremendous levels. Any such charge-storing device was dubbed a *condenser* by Volta, but that term has now been replaced by the word **capacitor**.

It was Volta who introduced the expression "electrical capacity" in analogy with the concept of heat capacity. At a given potential (V), the amount of charge (Q) that can be stored by a body depends on its physical characteristics, all of which we lump together under the name **capacitance** (C). The more charge, the greater the capacitance; while the less voltage that is needed to accomplish the feat, the greater the capacitance. In other words, C must vary directly with Q and inversely with V:

$$C = \frac{Q}{V} \tag{18.10}$$

The unit of capacitance is coulombs per volt, and to honor Faraday, it is called a *farad* (F): 1 farad = 1 F = 1 C/V.

One farad is a rather large capacitance; microfarad ($1\ \mu\text{F} = 10^{-6}$ F) and picofarad ($1\ \text{pF} = 10^{-12}$ F) capacitors smaller than the size of a grain of rice are commonly used in radios and TV sets. But fairly hefty capacitors (about the size of a soup can) can still be found in air conditioners where large amounts of charge are dumped into the compressor motor to get it going.

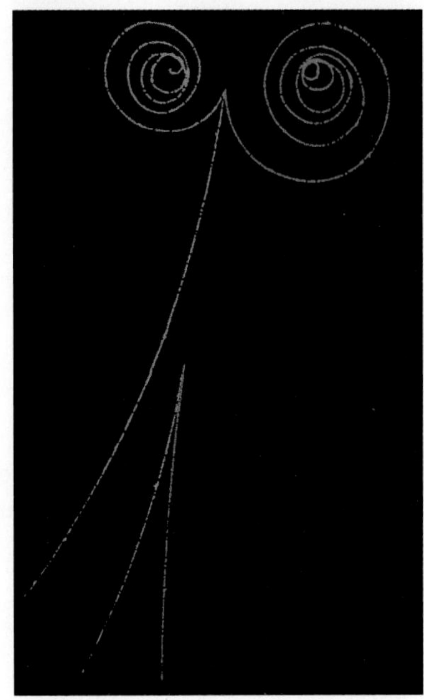

Subatomic events played out in a liquid-hydrogen bubble chamber. A gamma-ray photon (no track) descends from the top of the picture. It strikes an atom and knocks out an electron (long green track). The remainder of the photon's energy goes into creating an electron positron pair. Because of the applied magnetic field, these two low-energy particles move along tight spirals. Another invisible photon creates a second electron-positron pair further down in the picture. This time, all of the photon's energy goes into the pair. Each particle has more KE and therefore moves along a path that is nearly straight. Note how charge is conserved.

Alessandro Giuseppe Antonio Anastasio Count Volta (1745–1827).

To find the capacitance of a metal sphere of radius R, suppose it is carrying a charge Q. Its potential (p. 637) is then

$$V = \frac{kQ}{R}$$

and so

$$C = \frac{Q}{V} = \frac{Q}{\left(\frac{kQ}{R}\right)} = \frac{R}{k}$$

Hence, the capacitance of a sphere is

$$C = 4\pi\varepsilon R \tag{18.11}$$

If the sphere is surrounded by air ($k \approx k_0 = 1/4\pi\varepsilon_0$)

$$C = 4\pi\varepsilon_0 R$$

The larger the sphere, the greater the capacitance, but $(4\pi\varepsilon_0)$ is very small ($\approx 10^{-10}$), and even a large sphere will have only a modest capacitance. That's true as well for other shapes that would be much harder to calculate directly, such as your body. You are a capacitor that can store enough charge at a high enough voltage to produce observable sparks. In fact, the capacitance of a human (measured while standing on 5 cm of insulation) is roughly from 100 pF to 110 pF (for people 68 kg to 105 kg, respectively). Compare that to a cow, which is a lot larger and comes in a fairly standard 200-pF model. These values reflect a good deal of interaction with the Earth and, as we will see shortly, the presence of another conductor will appreciably increase the capacitance of any object. Thus, an isolated person, several meters above the Earth, has a capacitance of only about 50 pF.

Example 18.7 Determine the capacitance of an isolated metal sphere 50 cm in diameter and immersed in vacuum.

Solution: [Given: $R = 25$ cm. Find: C.] From Eq. (18.11)

$$C = 4\pi\varepsilon_0 R = \frac{R}{k} = \frac{0.25\text{ m}}{(9.0 \times 10^9\text{ N}\cdot\text{m}^2/\text{C}^2)} = \boxed{28\text{ pF}}$$

▶ **Quick Check:** This sphere has an area of $\approx 0.8\text{ m}^2$ as compared to the 2-m^2 area of a person whose capacitance is ≈ 50 pF.

The breakthrough in storing charge was made almost by accident in 1745 by G. von Kleist and again independently by P. van Musschenbroek. Both of these European gentlemen were experimenting with electricity and happened to insert the conductor being charged into a hand-held jar; they were probably attempting to collect "electric fluid." Van Musschenbroek dangled a brass wire, attached to a gun barrel that was being charged, into a flask "partly filled with water." It discharged, and all at once his body convulsed "as if it had been struck by lightning; . . . I thought it was all up with me," he recounted.

They had unknowingly constructed a device in which a conductor (the wire) was separated from another grounded conductor (a sweaty hand) by an insulating medium (the glass). Ordinarily, the isolated conductor being charged would rapidly reach the potential of the generator and thereafter repel any further charge. The new arrangement of conductor-insulator-conductor forestalled that cutoff. The charge put on one conductor, the central wire, induced an equal and opposite charge on the other conductor, the moist hand. That induced charge, having the opposite polarity and being relatively nearby, acted to reduce the wire's repulsion of additional charge. The result was a considerable increase in the charge stored before the device, which came to be called a *Leyden jar,* reached the potential of the generator (Fig. 18.15).

In Europe and the Colonies, anyone interested in electricity was likely to have a Leyden jar—sparks were flying everywhere (Fig. 18.16). The Abbé Nollet had 180 of Louis XV's fearless guards join hands in a circle, or *circuit.* With the first man holding the outer terminal of a charged Leyden jar, the last victim gleefully touched the central wire and shocked the whole assembly.

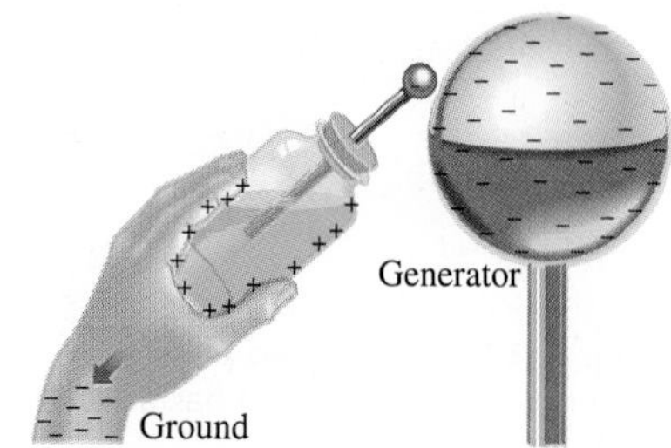

Figure 18.15 Charging a Leyden jar with the help of a generator and a grounded experimenter.

18.9 The Parallel Plate Capacitor

The renowned tamer of lightning and writer of racy prose, Ben Franklin, had his Leyden jars, too, but he went one step further. Franklin was among the first to use a new and more convenient configuration consisting of flat metal plates separated by sheets of window glass. That simple arrangement allowed for a dramatic increase in the size of the conductor-insulator-conductor sandwich. Franklin's flattened Leyden jar is a **parallel plate capacitor**.

At the end of the eighteenth century, Volta carried out a series of measurements of the potentials of large charged objects. He attached the test object via a conductor to a grounded electroscope. The leaves would then spring apart in proportion to the charge impressed on them. That, in turn, is proportional to the potential of the scope, which equals the potential of the object. In other words, the angle of the leaves of the electroscope is effectively proportional to the difference in potential between them and the grounded case. Suppose the test object is a positively charged flat plate as in Fig. 18.17. If an identical grounded plate is brought nearby, the leaves gradually descend as it approaches. In effect, the negative charge induced on the new plate will draw up some of the positive charge from the leaves, holding it fixed on the near side of the positive plate. But that means that introducing the second (now negatively charged) plate drops the potential of the first plate, which can be restored to its original value by further increasing the charge. *Thus, the second plate considerably enhances the ability of the parallel-plate capacitor to store charge at a given voltage.*

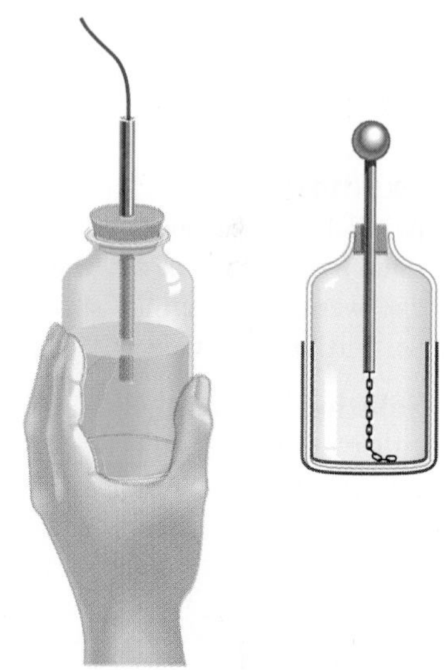

Figure 18.16 A Leyden jar. The improved version is lined with metal foil. A small chain connects the central wire with the inner metal surface.

We now determine the capacitance of the parallel plate capacitor (Fig. 18.18) as a function of its physical characteristics. If the plates carry opposite charges of $\pm Q$ and have a difference of potential ΔV between them, then $C = Q/\Delta V$. We can calculate this potential difference in terms of the E-field between the plates starting with

$$\Delta V = Ed$$

from Eq. (18.6), where d is the plate separation. Recall that

$$E = \frac{\sigma}{\varepsilon} = \frac{Q}{A\varepsilon} \qquad [17.12]$$

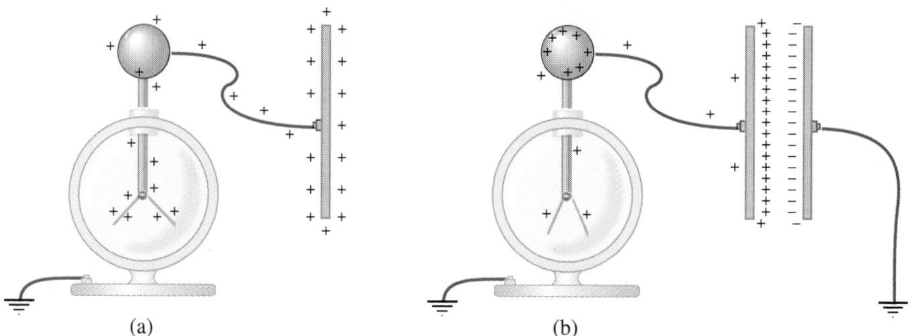

Figure 18.17 The operation of the capacitor as a charge storer. (a) The charges on a single plate experience a repulsive force, which contributes to establishing the potential. In (b), we bring a grounded neutral plate close to the positive plate. A group of negative charges are drawn onto this second plate, essentially neutralizing much of the positive charge, and reducing the potential.

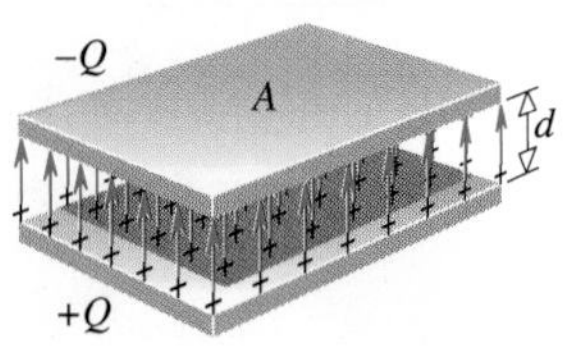

Figure 18.18 A charged parallel plate capacitor. Each plate has an area *A* and stores a charge *Q*.

wherein A is the area of *each* plate and $\sigma = Q/A$. Hence

$$\Delta V = Ed = \frac{Qd}{A\varepsilon}$$

and so

$$C = \frac{Q}{\Delta V} = \frac{Q}{Qd/A\varepsilon}$$

which simplifies to

$$C = \frac{\varepsilon A}{d} \tag{18.12}$$

To produce as large a capacitance as possible, we must make A large, d small, and use a material in the gap with a large permittivity (Table 17.3).

Example 18.8 Determine the size of a 1.00-F parallel plate capacitor if the plates are square and separated by 1.00 mm of air. How would things change if the gap were filled with a sheet of glass having a relative permittivity, or dielectric constant, of 10?

Solution: [Given: $C = 1.00$ F, $d = 1.00$ mm, $\varepsilon =$ either ε_0 or $10\varepsilon_0$, and $A = L \times L$. Find: L.] This problem involves the physical characteristics of a parallel plate capacitor and one equation should come to mind immediately:

$$C = \frac{\varepsilon A}{d} \tag{18.12}$$

Hence, with air

$$A = L^2 = \frac{dC}{\varepsilon_0} = \frac{(1.00 \times 10^{-3}\ \text{m})(1.00\ \text{F})}{(8.85 \times 10^{-12}\ \text{C}^2/\text{N}\cdot\text{m}^2)}$$

$$A = 0.113 \times 10^9\ \text{m}^2$$

and $\boxed{L = 10.6\ \text{km}}$. The plates are gigantic, about 6.6 miles on a side. When glass replaces air, ε replaces ε_0, and since $\varepsilon = 10\varepsilon_0$, the area is 10 times smaller and $L = (10.6\ \text{km})/\sqrt{10} = 3.35$ km.

▶ **Quick Check:** $C \approx (10^{-11}\ \text{C}^2/\text{N}\cdot\text{m}^2)(10^8\ \text{m}^2)/(10^{-3}\ \text{m}) \approx 1$ F.

A computer keyboard. When a key is depressed, the spacing of the plates in an air capacitor changes, thereby changing the capacitance and registering the keystroke.

Capacitors in the circuit of the AM radio shown on p. 589.

A dielectric has a larger value of permittivity than vacuum and therefore a smaller internal field ($E = \sigma/\varepsilon$), as is evident in Fig. 18.19. The dielectric becomes polarized in the external field and takes on a surface charge. As a result, a small internal *self-field* opposes the applied field within the dielectric. The effect is a weaker field in the gap and a lower voltage ($\Delta V = +Ed$) across it. Slipping a sheet of insulation into a charged capacitor (Fig. 18.20) effectively cancels some of the charge on the plates via the opposite polarity-induced charge on the surfaces of the dielectric.

Modern parallel plate capacitors come in different forms. The most common is a sandwich of metal foil (aluminum), dielectric (waxed paper, mylar, etc.), and metal foil, rolled up into a tight little cylinder and sealed (Fig. 18.21). It's represented symbolically in diagrams by two parallel lines of equal length.

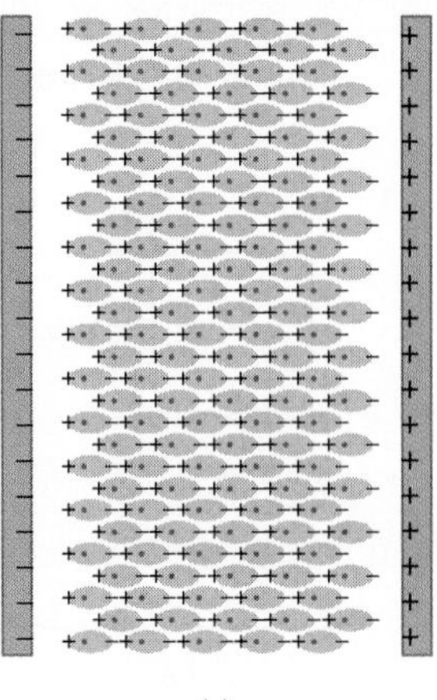

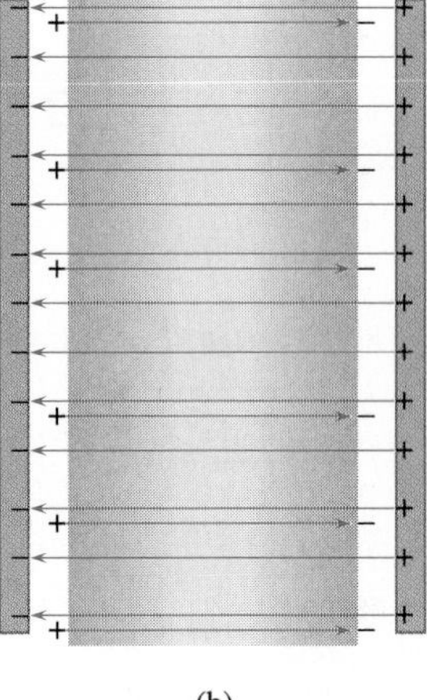

(c)

Figure 18.19 The effect of the dielectric in a parallel plate capacitor. It becomes polarized and thereby reduces the internal field. (a) Polarization of the dielectric. (b) The E-field due to the polarized dielectric is opposite the E-field due to the charge on the plates. (c) The internal E-field is reduced, and more charge can be stored at a given potential difference.

18.10 Capacitors in Combination

Inasmuch as the early interest in capacitors was purely for charge storage, it's not surprising that Leyden jars were wired to one another to increase that ability. To the same end, Franklin connected almost a dozen parallel plate capacitors together to produce great blasts of electricity.

The Parallel Circuit

Franklin, like others, used two basic wiring schemes to create two different results. Figure 18.22 shows the so-called **parallel** arrangement of three capacitors, though any number of them could be connected that way. The distinguishing characteristic

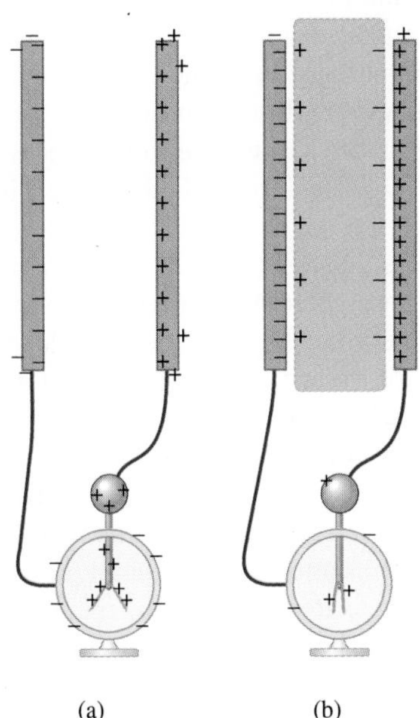

Figure 18.20 (a) A charged capacitor attached to an electroscope that indicates the potential difference. (b) Inserting a dielectric essentially neutralizes some of the charge on the plates and lowers the potential difference.

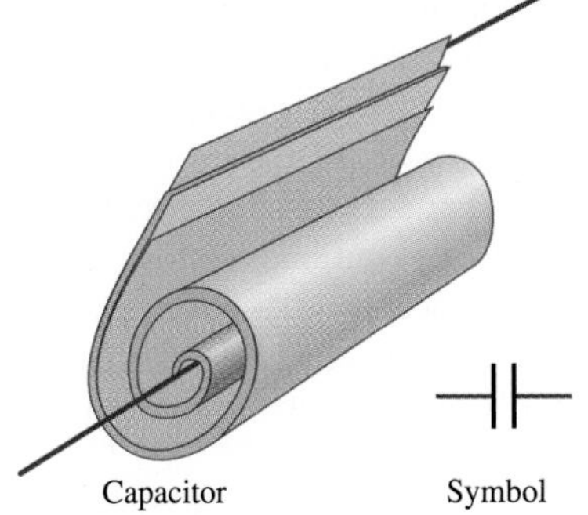

Figure 18.21 When the metal foil, dielectric, metal foil sandwich is rolled up and sealed, it forms a parallel plate capacitor with a fixed capacitance.

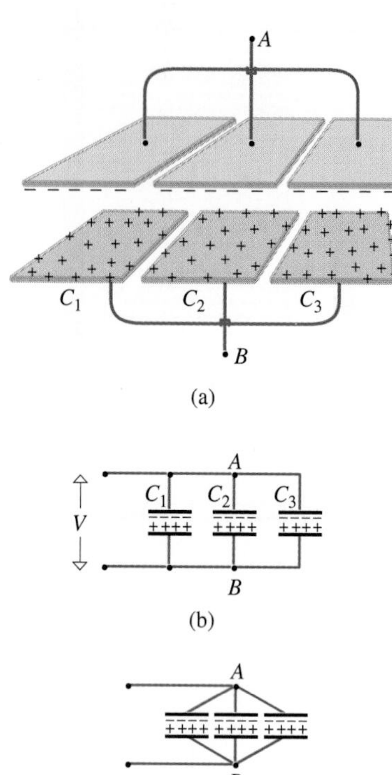

Figure 18.22 (a) Three capacitors in parallel represented in several effectively identical ways. (b) The equivalent capacitance is $C = C_1 + C_2 + C_3$. The circuits in (a), (b), and (c) are all electrically the same.

An air capacitor. This variable capacitor is from the tuning circuit of the radio shown on p. 589. Turning the knob of the radio causes a set of movable metal plates to slide between a set of stationary plates, thereby changing the capacitance and the station.

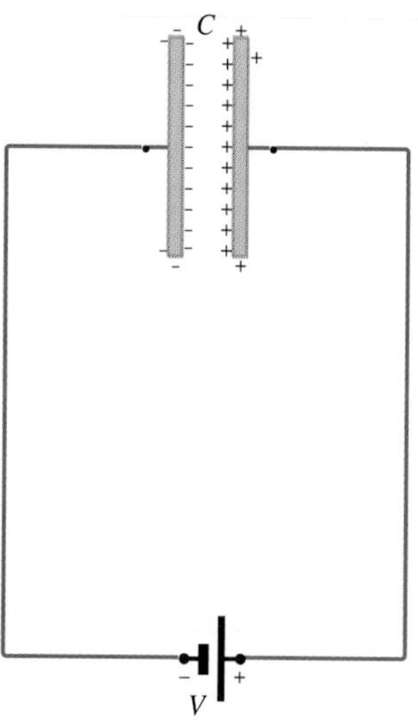

Figure 18.23 A capacitor across a battery. The capacitor's plates become charged as shown, and there is a voltage V across them.

of the parallel scheme is that one terminal (or plate) from each capacitor is connected to the same common wire, and all the remaining plates are connected to a second common wire. All the top plates are attached and must be at the same potential, and all the bottom plates are attached and must be at the same potential. **When a wire is used to connect two points together, those two points, and every point along the wire, are assumed to be at the same potential**. The result here is essentially one large capacitor whose two plates are each segmented into three connected parts.

Figure 18.23 shows a capacitor attached across a battery (p. 664), which is represented by the standard symbol, -|┣. The longer lighter line corresponds to the higher potential and is labeled +. It is V volts higher than the other terminal, labeled −. Negative charge flows out of the − terminal of the battery in the form of electrons. These charges gradually build up on the negative low-potential plate of the capacitor. They, in turn, repel an equal number of electrons off the other plate, leaving it positive, and this charge circulates back to the battery. The path is called a **circuit** and electrons continue to flow, as if around the circuit, for the brief time until the capacitor reaches the same potential difference as the battery. At that point, any further charge is repelled by the capacitor as forcefully as it is propelled by the battery. There is a field traversing the gap, and the voltage difference across the capacitor is given by Eq. (18.6), $\Delta V = Ed$. This difference increases during the charging until it finally equals the battery's voltage, at which time the charging ceases.

It is customary to write the potential difference (ΔV) across any circuit element (capacitors, batteries, resistors, etc.) simply as V. In Fig. 18.22, where the capacitors are in parallel, the voltage across the combination V must equal the voltage across each one:

$$V = V_1 = V_2 = V_3$$

The three capacitors will reach exactly the same state if they are charged to voltage V all as a single unit, or if each is charged separately and then attached together. The net charge stored (Q) is the sum of the individual amounts stored on each capacitor, and so

$$Q = Q_1 + Q_2 + Q_3$$

Hence

$$CV = C_1V_1 + C_2V_2 + C_3V_3$$

and

[*in parallel*] $$C = C_1 + C_2 + C_3 \qquad (18.13)$$

where we just add on more terms if there are more capacitors. The **equivalent capacitance of several capacitors in parallel is the sum of all the individual capacitances**. A single capacitor C, given by Eq. (18.13), would be electrically indistinguishable from the parallel array of *smaller* capacitors C_1, C_2, and C_3.

There are many kinds of circuit elements, and they can all be in parallel with each other provided both terminals of one are connected to both terminals of another. *A place where three or more leads come together is called a* **node**. The two regions in Fig. 18.22 where the three terminals are connected, top and bottom, constitute two nodes (A and B). *When elements are in parallel, any number of leads or circuit branches can converge at a node.*

Example 18.9 Figure 18.24 shows two capacitors attached to a 12-V battery. Determine the equivalent capacitance and the charge it would carry. What is the charge on each of the capacitors in the figure?

Solution: [Given: $C_1 = 20\ \mu\text{F}$, $C_2 = 30\ \mu\text{F}$, and $V = 12$ V. Find: C, Q, Q_1, and Q_2.] First, notice that the top terminals of both capacitors are connected, as are the bottom terminals—the capacitors are in parallel. You might also want to redraw the diagram, as is done in part (b). Since the potential across each capacitor is 12 V

$$V = \frac{Q_1}{C_1} = \frac{Q_2}{C_2}$$

$$Q_1 = (12\text{ V})C_1 \quad \text{while} \quad Q_2 = (12\text{ V})C_2$$

Thus

$$Q_1 = (12\text{ V})(20 \times 10^{-6}\text{ F}) = 2.4 \times 10^{-4}\text{ C}$$

and

$$Q_2 = (12\text{ V})(30 \times 10^{-6}\text{ F}) = 3.6 \times 10^{-4}\text{ C}$$

Since the capacitors are in parallel

$$C = C_1 + C_2 = (20\ \mu\text{F}) + (30\ \mu\text{F}) = \boxed{50\ \mu\text{F}}$$

The charge it would carry is gotten from $C = Q/V$; namely

$$Q = CV = (50 \times 10^{-6}\text{ F})(12\text{ V}) = \boxed{6.0 \times 10^{-4}\text{ C}}$$

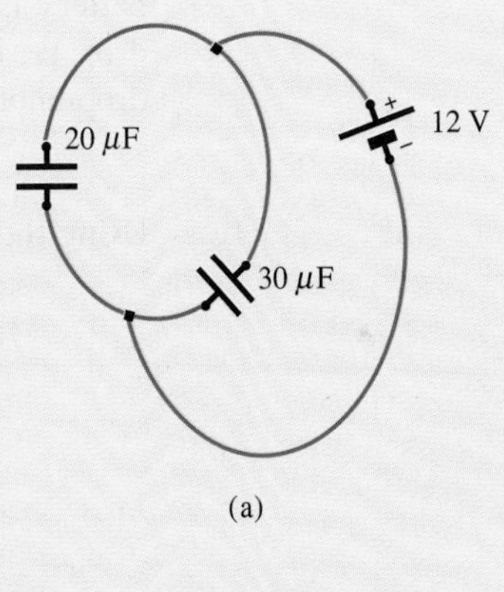

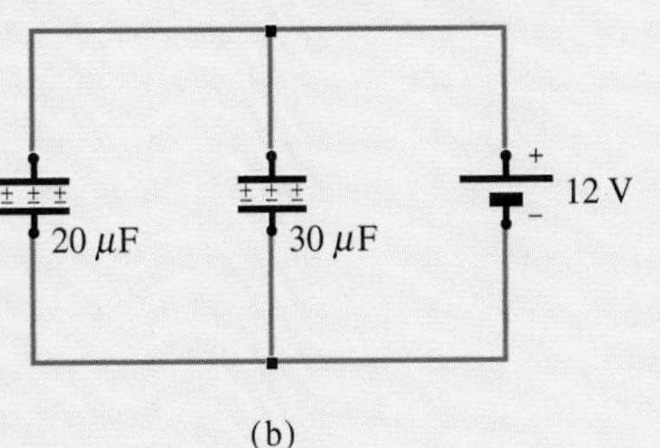

Figure 18.24 (a) A circuit of two capacitors and a battery. (b) A redrawn equivalent.

▶ **Quick Check:** For parallel capacitors we must have $Q = Q_1 + Q_2 = 2.4 \times 10^{-4}\text{ C} + 3.6 \times 10^{-4}\text{ C} = 6.0 \times 10^{-4}$ C.

The Series Circuit

Another basic way to connect circuit elements is known as **series**, and it's illustrated in Fig. 18.25. Here *one and only one terminal of a circuit element is connected to one and only one terminal of an adjacent circuit element.* **Two elements will not be in series if any other branch in the circuit connects to the point at which the two are attached**. Imagine that the three series capacitors are put across a battery of voltage V, as in Fig. 18.25c. Electrons travel from the battery to the negative plate of C_3, giving it a charge of $-Q$. These, in turn, repel an equal quantity of charge $(-Q)$ from the positive plate of C_3 to the negative plate of C_2. As the negative plate of C_2 charges up to $-Q$, it repels an equal number of electrons from its positive plate to the negative plate of C_1. An amount of electrons equivalent to $-Q$ is repelled back to the positive terminal of the battery. Hence

$$Q = Q_1 = Q_2 = Q_3$$

Notice that the net charge stored by the three capacitors is effectively $+Q$ on the left-most plate and $-Q$ on the right-most plate; the rest of the charge on the remaining plates ($-Q$ and $+Q$ and $-Q$ and $+Q$, respectively) cancels out—only Q's worth of electrons went in on the right and out on the left. Thus, the equivalent capacitor will also have a charge Q (which is *not* the sum of the charges on the individual capacitors).

The equivalent capacitor will be across the battery, just as is the series string of C_1, C_2, and C_3. Hence, the voltage across C is V. If we go from point A on the left to point D on the right, we will, by necessity, drop in potential by an amount V; the battery is connected to points A and D. Yet, in going from point A to point B, we drop V_1; in going from B to C, we drop an additional V_2; in going from C to D, we drop another amount V_3; hence

$$V = V_1 + V_2 + V_3$$

Using the definition, Eq. (18.10), $V = Q/C$, and

$$\frac{Q}{C} = \frac{Q_1}{C_1} + \frac{Q_2}{C_2} + \frac{Q_3}{C_3}$$

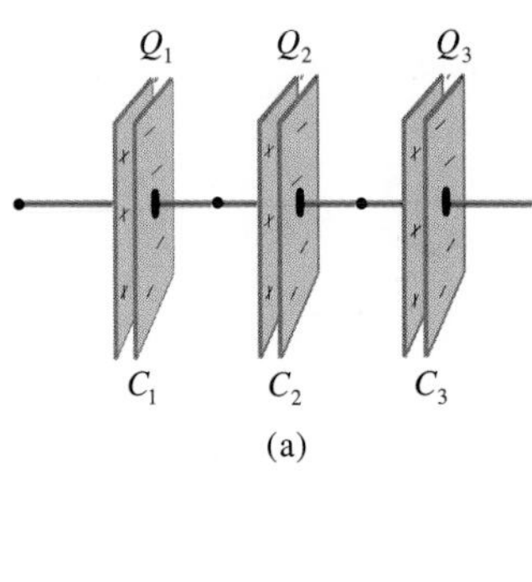

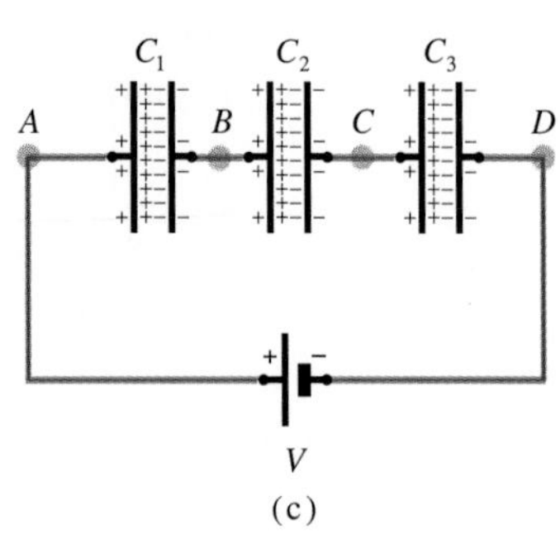

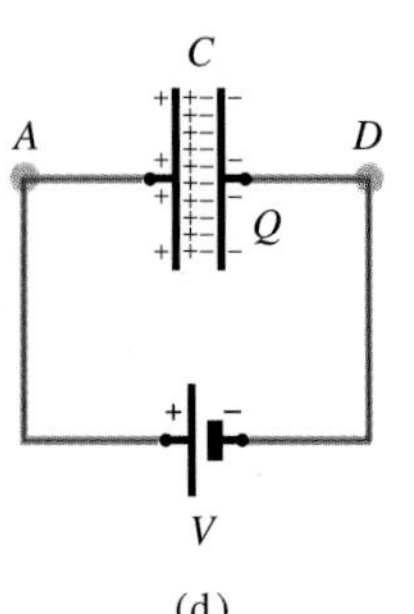

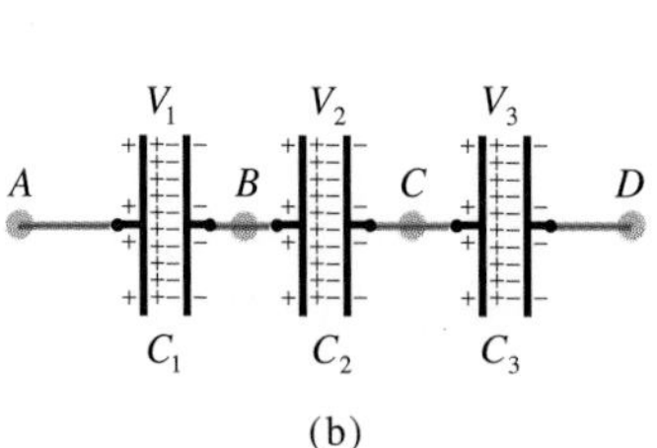

Figure 18.25 (a) Three capacitors in series and (b) their graphic representation. (c) The same series capacitors across a battery. (d) The equivalent circuit wherein $\frac{1}{C} = \frac{1}{C_1} + \frac{1}{C_2} + \frac{1}{C_3}$.

But all the charges are equal and therefore

[*in series*] $$\frac{1}{C} = \frac{1}{C_1} + \frac{1}{C_2} + \frac{1}{C_3} \tag{18.14}$$

and we just keep adding on terms if there are more capacitors.

Example 18.10 The circuit shown in Fig. 18.26a consists of a 12-V battery and three capacitors. Determine both the voltage across and charge on each capacitor after the switch S is closed and electrostatic equilibrium is established. Find the equivalent capacitance of the network.

Solution: [Given: $C_1 = 2.0\ \mu$F, $C_2 = 2.0\ \mu$F, $C_3 = 5.0\ \mu$F, and $V = 12$ V. Find: C, V_1, V_2, V_3, Q_1, Q_2, and Q_3.] First, redraw the circuit as in Fig. 18.25b to make things a bit clearer. The two 2.0-μF capacitors are in series, and their equivalent capacitance, Eq. (18.14), is

$$\frac{1}{C} = \frac{1}{C_1} + \frac{1}{C_2}$$

which is easy to compute with a calculator. However, if you would rather do it in your head, remember the equivalent form (which is left as a problem to prove)

$$C = \frac{C_1 C_2}{C_1 + C_2} = \frac{(2.0\ \mu\text{F})(2.0\ \mu\text{F})}{2.0\ \mu\text{F} + 2.0\ \mu\text{F}} = 1.0\ \mu\text{F}$$

As shown in Fig. 18.26c, this 1.0-μF equivalent of the series pair is itself in parallel with the 5.0-μF capacitor, and so they can be combined into a single capacitor via Eq. (18.13); namely,

$$C = 5.0\ \mu\text{F} + 1.0\ \mu\text{F} = \boxed{6.0\ \mu\text{F}}$$

which is the equivalent capacitance of the whole network. Now, working back from the simplified diagram of Fig. 18.26c, $V = 12$ V and so

$$Q_3 = C_3 V_3 = (5.0\ \mu\text{F})(12\ \text{V}) = \boxed{60\ \mu\text{C}}$$

There are 12 V across the combination of the two 2.0-μF capacitors and, hence, there must be a potential difference of $\boxed{6.0\ \text{V}}$ across each one. Therefore

$$Q_1 = Q_2 = (2.0\ \mu\text{F})(6.0\ \text{V}) = \boxed{12\ \mu\text{C}}$$

▶ **Quick Check:** The charge on the equivalent capacitor is $Q = CV = 72\ \mu$C. This must equal the net charge on the two series capacitors (that is, the charge on either one—namely, 12 μC) plus $Q_3 = 60\ \mu$C, and, happily, it does.

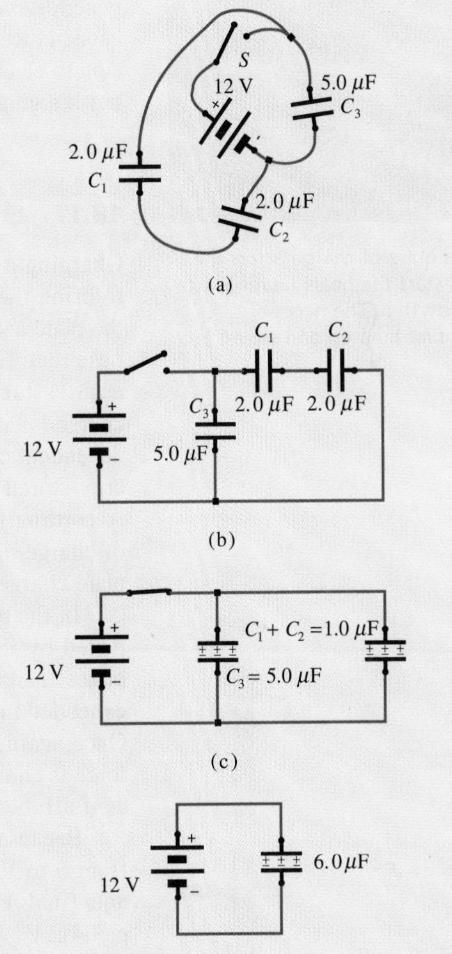

Figure 18.26 (a) A capacitive circuit and (b) a redrawn version. (c) A simplified configuration and (d) the simplest equivalent circuit.

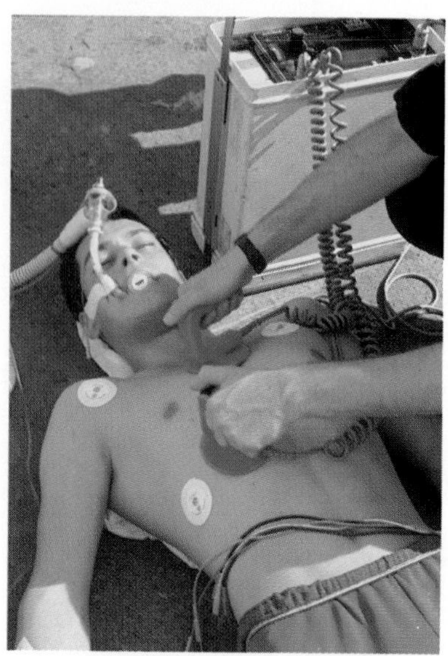
Sending a pulse of charge across the chest can start the heart beating with a steady rhythm. The necessary energy is first built up and stored in a capacitor.

Until now, we have assumed that a capacitor placed across a voltage source quickly reaches electrostatic equilibrium, which is actually something of an idealization. A real capacitor leaks charge across its dielectric spacer at a rate dependent on the geometry, the insulating material, and the voltage across it. A high-voltage mica capacitor might have over 1000 times more resistance (p. 671) to this flow of charge than does a low-voltage paper capacitor, even if they have the same capacitance. When two such capacitors are connected in series with a source of several hundred volts, some charge will continue to flow, electrostatic equilibrium will not be established, and the voltage across the mica one will be proportionately larger than that across the paper one. Thus, at substantial applied voltages, the charges are generally *not* equal on each different series capacitor even when they have the same capacitance. As a result, Eq. (18.14) does not hold in such cases. We limit our discussion to electrostatic equilibrium and will not worry about high voltages and leakage, but it's good to remember that this treatment is restricted.

18.11 Energy in Capacitors

Charging a parallel plate capacitor requires that work must be done on the charges to bring them from wherever they are to the plates. Each electron is forced over to the plate against the repulsive action of all the other electrons that preceded it. Suppose that we wish to electrify a capacitor with a total charge Q to a potential difference V, starting from a potential difference of zero. The end result is two oppositely charged plates in close proximity. The amount of energy stored in the process is independent of the details of how the capacitor got charged, just as the amount of energy stored via a boulder on a mountaintop is independent of how it got up there. Accordingly, let's work with the simplest scenario. Assume that we carry an amount of charge $+Q$ from one plate across the gap to the other, thereby leaving the first plate charged with $-Q$ (Fig. 18.27).

In the beginning, when the voltage is near zero and each new charge is only repelled by the few previous arrivers, the work done is small. As the charge on the plates increases, the voltage increases, the repulsive force increases, and the work expended increases; this process goes on until a potential difference of V is reached. Once again, we are faced with determining the work done in a process where the force is changing, but this time things are simple. We can treat this sort of problem as if all the charge were transported at once against an average potential difference.

Because $V = Q/C$, the variation in voltage as charge is built up is linear, going from 0 to V. Hence, the average potential difference is just the sum of the initial 0 and final V divided by two (remember Fig. 3.6 and the Mean-Speed Theorem, p. 64): $V_{av} = (0 + V)/2 = \frac{1}{2}V$. Hence, the work done in charging the capacitor, $W = QV_{av}$, is

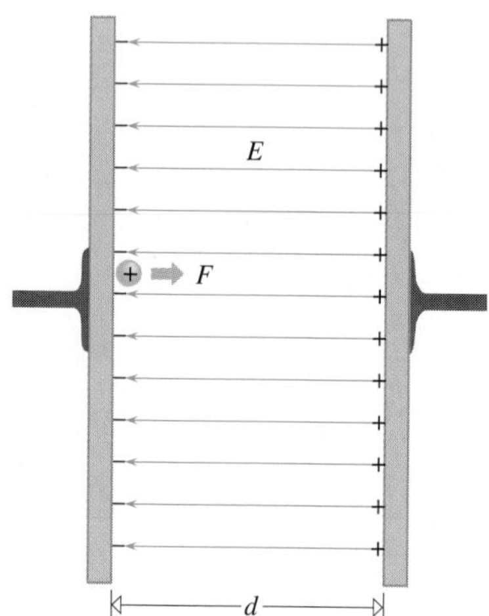

Figure 18.27 Carrying a charge through a distance *d* against a field requires an external force and the doing of work.

$$W = Q\tfrac{1}{2}V$$

and if we think of this energy as being stored as *electrical*-PE, we have

$$\text{PE}_E = \tfrac{1}{2}QV \qquad (18.15)$$

Equivalently

$$PE_E = \tfrac{1}{2}CV^2 = \tfrac{1}{2}Q^2/C \qquad (18.16)$$

If we want to know how PE_E varies with, say, Q, we must pick the expression that contains Q and no other quantity depending on Q; thus, $PE_E = \frac{1}{2}Q^2/C$ tells us that doubling Q quadruples PE_E.

Example 18.11 How much energy is stored in each of the capacitors of Fig. 18.26 in the process of charging them?

Solution: [Given: $C_1 = 2.0\ \mu F$, $C_2 = 2.0\ \mu F$, $C_3 = 5.0\ \mu F$, and $V = 12$ V. Find: the potential energy stored in each capacitor.] Using Eq. (18.15) and the results of the Example 18.10, namely, $Q_1 = Q_2 = 12\ \mu C$ and $Q_3 = 60\ \mu C$, we get

Capacitor 1 $PE_E = \frac{1}{2}QV = \frac{1}{2}(12\ \mu C)(6.0\ V)$

$$\boxed{PE_E = 36\ \mu J}$$

Capacitor 2 $PE_E = \frac{1}{2}QV = \frac{1}{2}(12\ \mu C)(6.0\ V)$

$$\boxed{PE_E = 36\ \mu J}$$

Capacitor 3 $PE_E = \frac{1}{2}QV = \frac{1}{2}(60\ \mu C)(12\ V)$

$$\boxed{PE_E = 360\ \mu J}$$

for a grand total of 432 μJ.

▶ **Quick Check:** If the equivalent capacitor is truly equivalent to all the capacitors in the system, it should store the same amount of energy. Using Eq. (18.16), it follows that

$$PE_E = \tfrac{1}{2}CV^2 = \tfrac{1}{2}(6.0\ \mu F)(12\ V)^2 = 432\ \mu J$$

Figure 18.28 shows a charged capacitor and therefore one with a potential difference across its plates. Each electron, given the opportunity, would spontaneously descend in potential energy. Thus, when a wire—a so-called *short circuit*—is connected across the capacitor, electrons immediately move from the − plate to the + plate, canceling the charge and reducing the potential difference across the capacitor to zero.

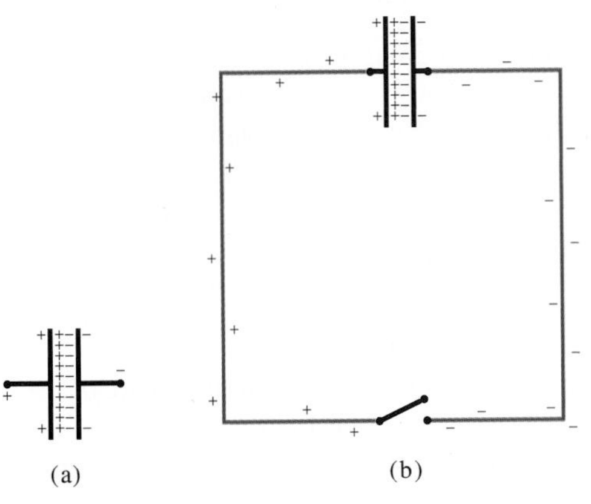

Figure 18.28 (a) A charged capacitor. (b) Closing the switch would allow charge to flow and cancel, leaving a net charge of zero.

Example 18.12 The two capacitors C_1 and C_2 in Fig. 18.29 were first put across different batteries so that they took on voltages of $V_1 = 12$ V and $V_2 = 6.0$ V, respectively. They were then attached as shown. Compute the charge and energy stored in each capacitor once the switch is closed and electrostatic equilibrium restored.

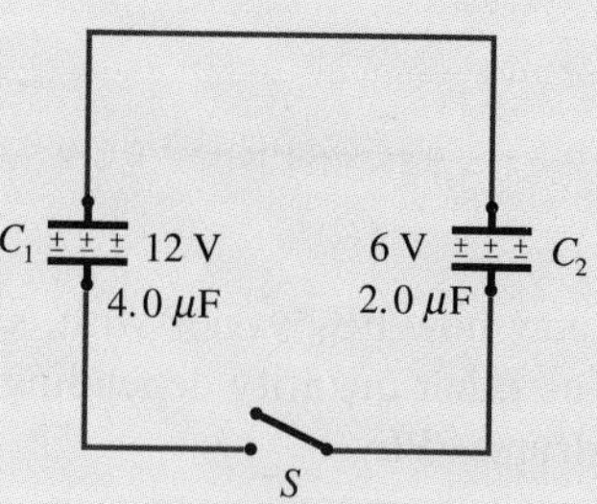

Figure 18.29 Two charged capacitors in a circuit containing a switch *S*.

Solution: [Given: $C_1 = 4.0\ \mu\text{F}$, $C_2 = 2.0\ \mu\text{F}$, $V_1 = 12$ V, and $V_2 = 6.0$ V. Find: the potential energy stored and the charge on each capacitor once the switch is closed.] A wire connects the two positive plates even before the switch is closed, but no redistribution of charge can occur because the charges are effectively bound in place on each capacitor by the oppositely charged opposing plate. Even after the switch is closed, electrons on the negative plates cannot get to the positive plates, and although the charge will redistribute itself, the net amount $Q_1 + Q_2$ is fixed. Let's find the two final charge distributions. Two unknowns will require two equations, so begin by determining the net charge. Before they were attached to one another

$$Q_1 = C_1 V_1 = (4.0\ \mu\text{F})(12\ \text{V}) = 48\ \mu\text{C}$$

and

$$Q_2 = C_2 V_2 = (2.0\ \mu\text{F})(6.0\ \text{V}) = 12\ \mu\text{C}$$

The net charge is

$$Q_1 + Q_2 = 60\ \mu\text{C}$$

Once connected, the charge redistributes itself until any potential difference between the two positive plates, and between the two negative plates, vanishes. Each capacitor has the same final voltage V across it (they're in parallel), and

$$V = \frac{Q_1}{C_1} = \frac{Q_2}{C_2}$$

whereupon

$$Q_1 = \frac{Q_2(4.0\ \mu\text{F})}{2.0\ \mu\text{F}} = 2Q_2$$

But the total charge is 60 μC; hence $\boxed{Q_1 = 40\ \mu\text{C}}$ and $\boxed{Q_2 = 20\ \mu\text{C}}$. To find the potential energy we use Eq. (18.16), and thus for C_1

$$\text{PE}_{\text{E1}} = \frac{\frac{1}{2}Q_1^2}{C_1} = \frac{\frac{1}{2}(40\ \mu\text{C})^2}{4.0\ \mu\text{F}} = \boxed{0.20\ \text{kJ}}$$

for C_2

$$\text{PE}_{\text{E2}} = \frac{\frac{1}{2}Q_2^2}{C_2} = \frac{\frac{1}{2}(20\ \mu\text{C})^2}{2.0\ \mu\text{F}} = \boxed{0.10\ \text{kJ}}$$

▶ **Quick Check:** The larger capacitor, which has twice the capacitance of the other, stores twice the charge and twice as much energy in reaching the same voltage.

The energy stored in a capacitor can be imagined as associated with the electric field that exists in the gap. We do work charging up the capacitor or, equivalently, establishing the field. Looking at the region between the plates, $C = \varepsilon_0 A/d$ and $V = Ed$, thus

$$\text{PE}_{\text{E}} = \tfrac{1}{2}CV^2 = \tfrac{1}{2}\left(\frac{\varepsilon_0 A}{d}\right)(Ed)^2 = \tfrac{1}{2}\varepsilon_0(Ad)E^2 \qquad (18.17)$$

The electrical energy stored by the E-field is proportional to the square of the strength of the field (E^2). In this sense, the electromagnetic field (as we shall see in Chapter 24) can be treated as a tenuous form of matter carrying energy and momentum.

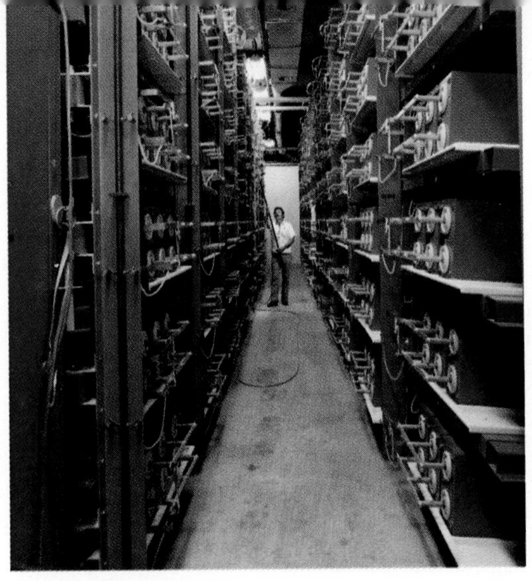

One good way to provide a tremendous blast of electrical energy is to store charge in a giant capacitor. This capacitor bank at the Lawrence Livermore National Laboratory supplies energy to the Nova laser, the most powerful in the world.

Core Material

The **electric potential** V, or just the *potential*, is defined as

$$\text{Electric potential} = \frac{\textit{electrical}\text{-PE}}{\text{charge}} \qquad [18.1]$$

where 1 J/C = 1 V. *The potential difference* ΔV *between two points A and B is the work done against the field in moving a unit positive charge from A to B*:

$$\Delta V = V_B - V_A = \frac{W(A \to B)}{q_0} \qquad [18.5]$$

In a *uniform electric field,* the potential difference is

$$V_B - V_A = \pm Ed \qquad [18.6]$$

The potential difference is + when the displacement has a component that is opposite to the field, and − when it has a component parallel to the field.

When a positive charge is moved from r_A to r_B against the $1/r^2$-electric field of a *sphere* of positive charge Q

$$V_B - V_A = kQ\left(\frac{1}{r_B} - \frac{1}{r_A}\right) \qquad [18.7]$$

The potential at a distance r from a positive point-charge, measured with respect to the zero at ∞, is given by

$$V = kQ\left(\frac{1}{r}\right) \qquad [18.8]$$

Capacitance (C) is a measure of *the capacity to store charge:*

$$C = \frac{Q}{V} \qquad [18.10]$$

the units of which are coulombs per volt, where 1 farad = 1 F = 1 C/V. For a parallel plate capacitor

$$C = \frac{\varepsilon A}{d} \qquad [18.12]$$

When two or more capacitors are connected, the equivalent capacitance for them in **parallel** is

$$C = C_1 + C_2 + C_3 + \cdots \qquad [18.13]$$

and in **series**

$$\frac{1}{C} = \frac{1}{C_1} + \frac{1}{C_2} + \frac{1}{C_3} + \cdots \qquad [18.14]$$

The energy stored in the process of charging a capacitor is

$$\text{PE}_E = \tfrac{1}{2}QV = \tfrac{1}{2}CV^2 = \tfrac{1}{2}Q^2/C \qquad [18.16]$$

Suggestions on Problem Solving

1. When calculating the PE_E of a charge, be careful with signs. An electron at a positive potential has a negative potential energy. Similarly, when a charge "falls" through a potential difference, thereby losing potential energy, the corresponding change in KE is positive. A negative charge "falls" (in the opposite direction to the E-field) to a *higher potential;* a positive charge "falls" (in the same direction as the E-field) to a lower potential.

2. If a parallel-plate capacitor has charges on its two plates of $+Q$ and $-Q$, the charge on the capacitor is Q, and that value is indicated in the defining equation $C = Q/V$. Similarly, the area A is essentially the area of overlap, generally the area of *either* plate.

3. When looking at equations, read C as capacitance and Q as charge. But C is a coulomb, which is the unit of charge. Don't confuse them.

4. Keep the following quantities handy: $\varepsilon_0 = 8.85 \times 10^{-12}\ \mathrm{C^2/N \cdot m^2}$, $1/\varepsilon_0 = 1.129 \times 10^{11}\ \mathrm{N \cdot m^2/C^2}$, $4\pi\varepsilon_0 = 1.113 \times 10^{-10}\ \mathrm{C^2/N \cdot m^2}$, and $1/4\pi\varepsilon_0 = k_0 = 9.000 \times 10^9\ \mathrm{N \cdot m^2/C^2}$.

5. Remember that Eq. (18.14) yields an equivalent capacitance for a series string of capacitors, which is always less than the smallest capacitance in the group. When applying this equation, be wary. Don't substitute into the right side, carry out the math, and present the result as the answer, forgetting that you have actually computed $1/C$ and not C. For only two capacitors you might find the following form more convenient

$$C = \frac{C_1 C_2}{C_1 + C_2}$$

In any event, you could use it for a quick check to see that the magnitude of your answer is right.

6. To remember the set of equations for a point-charge Q, memorize only Coulomb's Law: force is kqQ/r^2. By definition, divide that quantity by charge to get the field, kQ/r^2; in effect, multiply force by distance to get work and PE, kqQ/r; by definition, divide that quantity by q to get potential, kQ/r. Although simplistic and not rigorous, this scheme is helpful as a memory device.

Discussion Questions

1. A negative charge in an electric field moves from a point where the potential is zero to a point where it is -100 V. Discuss the energy change and the work done.

2. Imagine a hollow, spherical, positively charged conductor of radius R that is far away from any other bodies. Draw a graph of its potential V, as a function of r out from its center. Discuss your results.

3. With Question 2 in mind, suppose we place a neutral hollow conducting sphere in the vicinity of the charged sphere. Draw a rough graph of the potential along the center line of both spheres and compare it to the previous potential without the neutral conductor. Discuss your results.

4. With Question 3 in mind, suppose we now ground the neutral conductor. Describe the potential along the center line. Discuss your results if they are any different from those of Question 3.

5. Using equipment like that depicted in Fig. 17.34, Faraday showed that when a neutral conductor touched the inside of a hollow charged conductor, it remained neutral. Use that result to justify the conclusion that *provided there is no external field, a neutral conductor assumes the potential of the region of space in which it is introduced.* What happens to the potential of the neutral conductor (the little sphere) if the charge on the surrounding body in Fig. 17.34 is increased?

6. Describe the potential of a negative point-charge as a function of distance r. Explain your thinking.

7. A small, neutral spherical conductor is placed between the plates of a large, charged parallel plate capacitor. Describe the E-field and the equipotentials in the gap.

8. A positive point-charge is located a short distance above a large conducting horizontal plane. Describe the field lines and equipotentials. Compare your results with that of an electric dipole. Explain your conclusions.

9. Imagine a square with point-charges at each corner. At the ends of one diagonal, the charges are $+q$; at the ends of the other they are $-q$. This arrangement is called an *electric quadrupole.* Make a rough sketch of the field lines and equipotentials and explain your reasoning.

10. Figure Q10 shows a positive point-charge located at the corner of two intersecting conducting planes. Describe the E-field and the equipotentials. Compare your answer to that of the previous question. Explain your observations.

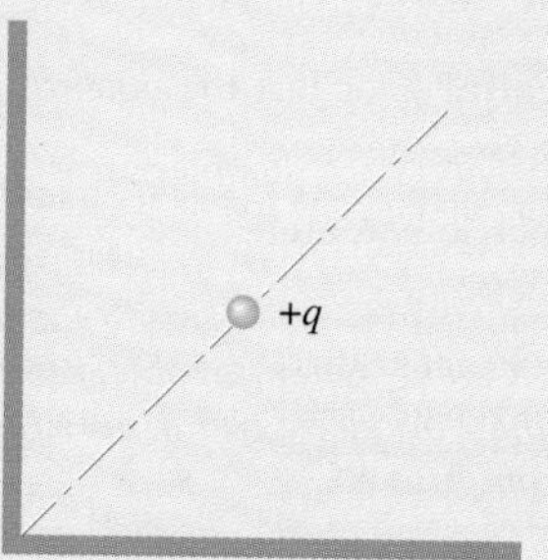

Figure Q10

11. Is it correct to maintain that *when a positive charge is deposited on a body it raises the potential of that body and, moreover, it raises the potential of the entire vicinity, including that of any other bodies in the vicinity?* Contrarily, *does the introduction of a negative charge lower the potential of every point in the neighboring region?* Explain your answer in detail. How might you prove your conclusion experimentally?

12. Is it correct to maintain that, in general, *the potential of a conductor, which certainly depends on its own charge, also depends on the distribution of charge everywhere else nearby?* Under what circumstances, if any, does the potential of a charged conductor only depend on its charge, structure, and the medium it is in?

13. We know that points infinitely far from all charge are at zero potential and yet, as a practical matter, we take the Earth, which may well be charged, to be at zero-potential. How can this inconsistency be resolved?

14. Water has a rather large dielectric constant $(\varepsilon/\varepsilon_0)$. How might that characteristic contribute to its ability to keep substances such as table salt, once dissolved, in solution?

15. Given a region that is an equipotential volume, what can you say about the possibility that there is a net charge within it?

16. Given that a particular equipotential volume (different from its surroundings) is bounded by an equipotential surface, what can you say about the charge, if any, on that surface?

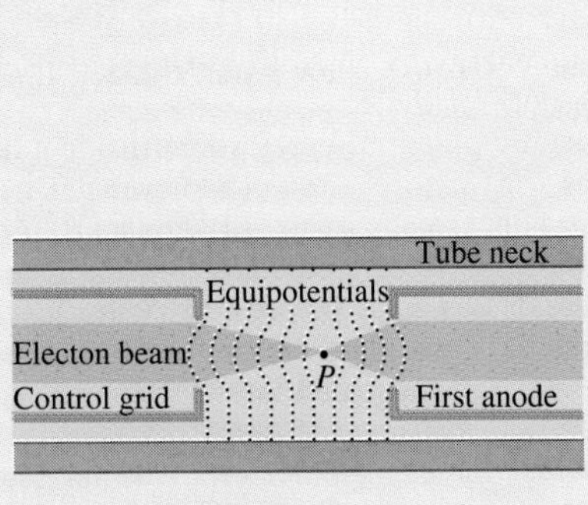

Figure Q19

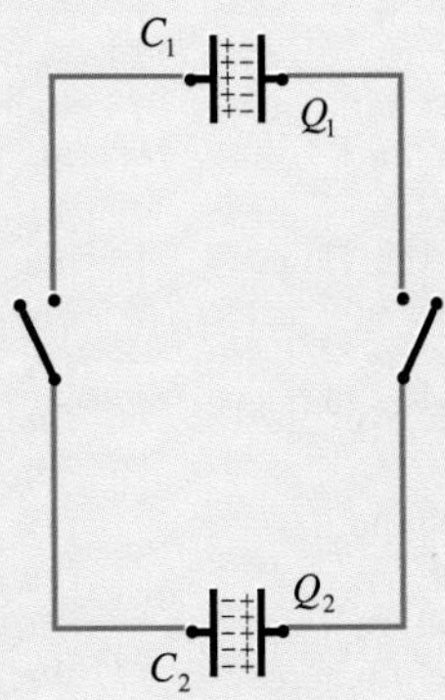

Figure Q20

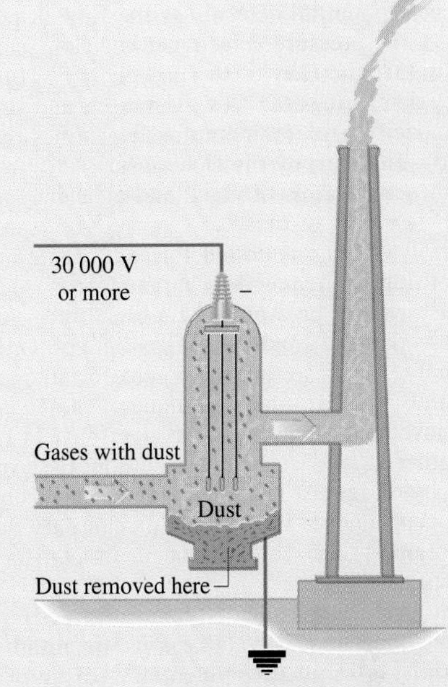

Figure Q21

17. It is sometimes desirable, especially in high-voltage applications, to replace a capacitor by an equivalent string of several capacitors in series. Discuss the possible reasons for this. How do the sizes of the several series capacitors compare to the original single one?

18. Envision a parallel plate capacitor across the terminals of a battery. When a dielectric is inserted between the plates, the capacitance increases, as does the charge, and more energy is stored by the device. Where is that additional energy stored?

19. Figure Q19 is a profile view showing the equipotentials in the space between the control grid and the first anode of an electron gun (Fig. 18.4). The holes in the ends of the metal surfaces of the grid and anode cause a distortion of the equipotentials, and the arrangement serves as an *electron lens.* The beam is brought to a focus at the "crossover" point *P*. Explain how this process happens. (Incidentally, another such lens in the second anode focuses the beam onto the screen).

20. The two identical capacitors in Fig. Q20 are charged at different voltages such that $Q_1 > Q_2$. What will be the charge on each capacitor after the two switches are closed? Explain your answer completely.

21. The Cottrell precipitator is shown mounted to a chimney in Fig. Q21. How do you think it works to remove 99% of the ash and dust that passes through it?

Multiple Choice Questions

1. The quantity 1 C·V is equivalent to (a) 1 V/m (b) 1 N·m (c) 1 C/N (d) 1 V/N (e) none of these.
2. In the case of a nonconductor (a) its surface must be at a single potential (b) its entire volume, except for the surface, is at a constant potential (c) different regions may well be at different potentials (d) the potential must be zero everywhere within it (e) none of these.
3. Electric field lines always point toward (a) ground (b) a region of higher potential (c) a region of lower potential (d) positive charge (e) none of these.
4. The potential as we get closer and closer to a point-charge (a) approaches $\pm\infty$ (b) is zero (c) is indeterminate (d) is exceedingly small, but not zero (e) none of these.
5. Given a group of nearby charges whose net value is nonzero, the equipotential surface at a very great distance is (a) nearly a plane (b) nearly a sphere (c) quite indeterminate (d) nearly a cylinder (e) none of these.
6. Any closed equipotential surface that does not surround a net charge must (a) be at zero-potential (b) be a sphere (c) enclose an equipotential volume (d) be infinitely small (e) none of these.
7. Suppose we examine a charged metal cup with a device that measures potential with respect to ground. What will happen as the probe from the device that touches the cup changes from contact with the outside to contact with the inside? (a) the reading will ascend (b) the reading will descend partway (c) the reading will go to zero (d) nothing (e) none of these.
8. A volume of space is found to have a constant potential everywhere within it. It follows that in that region (a) the *E*-field is zero (b) the potential is zero (c) the *E*-field is finite and uniform (d) the potential gradient is a nonzero constant (e) none of these.
9. When two charged metal objects are connected to each other by a conducting wire, the one that gains electrons or, equivalently, loses positive charge is said, in comparison to the other object, to have had (a) a greater electrical potential energy (b) a lower capacitance (c) a lower

dielectric constant (d) a higher potential (e) none of these.
10. Electrical potential determines the flow of positive charge just as (a) pressure determines the flow of fluid, and temperature the flow of thermal energy (b) kinetic energy determines the flow of matter, and charge the flow of electricity (c) temperature determines the flow of entropy, and entropy the flow of heat (d) power determines the flow of fluid, and efficiency the flow of work (e) none of these.
11. Generally, when any conductor is connected to ground (a) nothing happens (b) charge flows so that the conductor takes on a potential above zero (c) charge flows so that the conductor takes on a potential below zero (d) charge flows so that the conductor takes on the potential of the ground (e) none of these.
12. The movement of charge in an electric field from one point to another at a constant speed without the expenditure of work, by or against the field (a) is impossible (b) can only occur along a field line (c) can only occur along an equipotential (d) can only occur in a uniform field (e) none of these.
13. A body that when grounded takes on electrons is said to have originally had (a) a negative potential (b) zero-potential (c) a positive potential (d) an original net negative charge (e) none of these.
14. A neutron somehow picks up 10 eV. That's equivalent to it increasing its (a) charge by 10 C (b) electrical potential by 10 V (c) energy by 16×10^{-19} J (d) capacitance by 10 μF (e) none of these.
15. The electrostatic potential everywhere inside a hollow conductor is (a) always zero (b) never positive (c) always a nonzero constant (d) constant, provided there are no enclosed isolated charges (e) none of these.
16. The capacitance of a parallel plate capacitor is (a) independent of the plate separation (b) dependent on the charge (c) dependent on the voltage (d) independent of the plate area (e) none of these.
17. We can increase the capacitance of a parallel plate capacitor by (a) cooling the plates (b) bringing the plates closer together (c) decreasing the permittivity of the medium in the gap (d) increasing the voltage (e) none of these.
18. If the voltage across a capacitor is doubled, the amount of energy it can store (a) doubles (b) is halved (c) is quadrupled (d) is unaffected (e) none of these.
19. If the charge on a capacitor is halved, its stored energy (a) is halved (b) is quartered (c) is unchanged (d) is doubled (e) none of these.

Problems

ELECTRIC POTENTIAL

1. [I] A tiny sphere carrying a charge of -25.0 nC is moved 100 cm in a uniform electric field with no acceleration. It goes from a location at a potential of zero to a point where the potential is 100 V. How much work is done on it by the applied force? What is the significance of the sign of ΔW?

2. [I] A stream of singly ionized gold atoms pouring from a small oven impinges on a metal target. The target is attached to the positive terminal of a 12-V battery, and the oven is attached to the negative terminal. How much kinetic energy do the ions pick up in the process of crossing over to the target?

3. [I] What voltage should be put across a pair of parallel metal plates 10.0 cm apart if the field between them is to be 1.00 V/m?

4. [I] Two charged parallel metal plates, inside the evacuated cathode-ray tube of a radar system, are separated by 1.00 cm and have a potential difference of 25.0 V. What is the value of the electric field in the gap?

5. [I] Figure P5 shows two hollow concentric metal spheres of radii ρ and R. If the inner one is charged with $+Q$ and the outer surface is grounded, what is the potential at any point P outside the larger sphere? Explain your thinking.

6. [I] A 10.0-cm diameter metal sphere carries a charge of $+0.100$ μC. What is the potential 10.0 m away in the surrounding air?

7. [I] Two small spheres carrying charges of $+30.0$ μC and -50 μC are 100 cm apart in air. What is the potential at a point on the center-to-center line midway between them?

8. [I] Figure P8 depicts two metal objects. Sketch in some field lines and equipotentials. Discuss your answer.

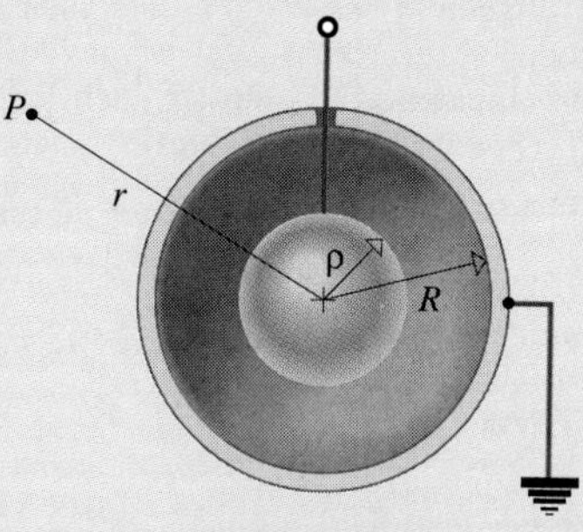

Figure P5

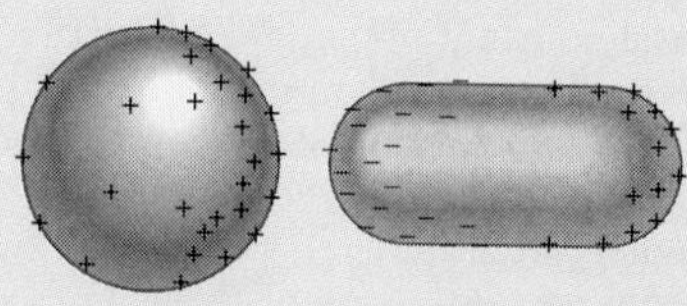
Figure P8

9. [I] An electron is to be accelerated from rest at a grounded cathode, to a metal plate at $+500$ V. Express in electron volts how much kinetic energy it will gain. How much electrical potential energy will it lose, if any?

10. [I] It is fairly easy to strip the two electrons off a helium atom leaving a bare nucleus of two neutrons and two protons. That so-called *alpha particle* is to be accelerated, essentially from rest, up to a KE of 100 keV by having it

"fall" through a potential difference. What is the necessary voltage difference?

11. [II] An electron initially at rest in an X-ray tube crosses a potential difference of 30 kV and crashes into a target that then emits radiation. Compute the KE of the electron (in joules) at impact. Determine its maximum speed.
12. [II] A proton is released from rest in a uniform electric field of 500 V/m. How fast will it be moving after traveling 40 cm in and parallel to the field?
13. [II] A metal target sphere of 20-cm diameter suspended out in space is given a charge of +1.00 nC. How much work must be done on a proton in taking it from very far away (essentially infinity) to the surface of the sphere?
14. [II] A hydrogen atom consists of a proton around which circulates an electron at an average distance of 0.053 nm. Determine the potential at that distance due to the proton and find the potential energy of the electron (in joules).
15. [II] With Problem 13 in mind, through what voltage must the proton be accelerated from rest (by some sort of space weapon) if it is to arrive at the sphere at a speed of 8.5×10^4 m/s? Neglect any gravitational effects.
16. [II] Refer to Fig. P5. If the inner sphere is charged with $+Q$ and the outer surface is no longer grounded, how does the charge distribute itself, and what is the potential in the region beyond the spheres in terms of ρ, R, and Q? Explain your answer. Write an expression for the field beyond the spheres.
17. [II] Two horizontal parallel metal plates 10.0 cm apart in a vacuum chamber are to be used to suspend an electron in "midair." What voltage must be put across the plates?
18. [II] Tiny conducting spheres carrying charges of +60 μC are fixed at each of the four corners of a square 20 cm on a side. What is the potential at the very center? What is the electric field at the center? Assume the medium is vacuum.
19. [II] Two very small metal spheres carrying charges of +10 μC and −25 μC are located at coordinates (0, 0) and (3.0 m, 0), respectively, in air. How much work would have to be done to bring a third sphere with a charge of −10 μC from very far away to the point (0, 4.0 m)?
20. [II] Point-charges of +300 nC, −700 nC, +500 nC, and −100 nC are located in sequence at the corners of a square 40 cm on a side. Determine the potential at the center.
21. [II] The fluid within a living cell is rich in potassium chloride, while the fluid outside it predominantly contains sodium chloride. The membrane of a resting cell is far more permeable to ions of potassium than sodium, and so there is a transport out of positive ions, leaving the cell interior negative. The result is a voltage of about −85 mV across the membrane, called the *resting potential.* The membrane (about 50 atom-layers) is roughly 8 nm thick. Assuming the E-field across the cell membrane is constant, determine its magnitude.
22. [III] Two very small conducting spheres surrounded by transformer oil are charged with +50.0 μC and −40.0 μC, respectively. Determine the point on the line connecting them where the potential is zero, if indeed such a point exists. The center-to-center separation is 1.00 m.
23. [III] A total of eight tiny conducting spheres, each carrying a charge −100 nC, are placed one each at the corners of a cube 1.0 m on a side. Find the potential at the center. What is the electric field at the center?
24. [III] If the inner sphere in Fig. P5 is charged with $+Q$ and the outer surface is grounded, show that

$$\Delta V = kQ\left(\frac{1}{\rho} - \frac{1}{R}\right)$$

is the expression for the potential difference across the gap.

CAPACITANCE

25. [I] A 100-pF capacitor is charged by putting it across a 1.5-V battery. What is the charge on its plates?
26. [I] A 48.0-μF capacitor, with an impregnated paper dielectric, is placed across the terminals of a 12-V battery. How much charge flows from the battery to the capacitor?
27. [I] Cathode-ray tubes used in computers and TV sets are coated inside and out with a conducting graphite paint called *Aquadag.* The inside layer is connected to the second anode (see Fig. 18.4) and helps to accelerate the beam and keep it narrow. The outer layer is connected to the chassis, which is ground. This arrangement creates an Aquadag-glass-Aquadag capacitor (which is connected across the power supply providing the high voltage for the anode, and tends to smooth out voltage variations). The capacitance is generally only about 500 pF, but all such tubes always carry a warning that even though the set has been turned off, the CRT must be discharged, otherwise you risk a dangerous shock. Explain why this warning is necessary and compute the charge for a tube voltage of 20 kV.
28. [I] Estimate the capacitance of the Earth. Its radius is 6371 km. Give the answer to two significant figures.
29. [I] What is the radius of a conducting sphere in air if it has a capacitance of 10 pF (or as it used to be called, $\mu\mu$F)?
30. [I] As a very rough estimate, assume you have about the same capacitance as a conducting sphere 3/4 m in diameter. How much charge can you store at a potential of 100 V?
31. [I] A parallel plate capacitor immersed in transformer oil carries a charge of +20 μC on one plate and −20 μC on the other, when there is a voltage of 4.0 V across it. What is its capacitance?
32. [I] What is the capacitance of two parallel metal plates each with an area of 100 cm^2 separated by 1.0 mm of air?
33. [I] What is the capacitance of two parallel metal plates each with an area of 100 cm^2 separated by 1.0 mm if the gap is filled with glass having a dielectric constant of 10?
34. [I] Thus far, we have used units for ε that relate back to Coulomb's Law. Accordingly, prove that 1 C^2/N·m^2 is equivalent to 1 F/m.
35. [I] A little flat ceramic capacitor consists of two circular plates 0.50 cm in diameter separated by a dielectric 1/3 mm thick with a relative permittivity of 4.8. Compute its capacitance.
36. [I] Show that capacitance has the dimensions of $(Q^2T^2)/(ML^2)$. (Q→charge, T→time, M→mass, and L→length.)
37. [I] Given three capacitors of values 60 pF, 30 pF, and 20 pF, determine the net capacitance when they are placed successively in series and then in parallel.
38. [I] What is the equivalent capacitance of the circuit between the terminals A and B, indicated in Fig. P38?

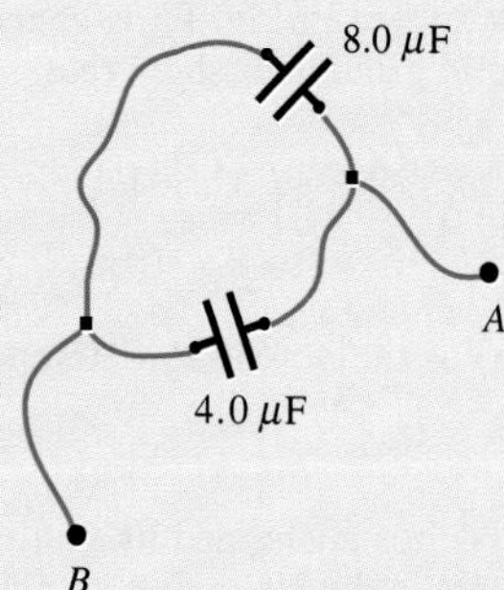

Figure P38

39. [I] Suppose we make an alternating stack of aluminum foil and paper (numbering each sheet of metal successively): 101 sheets of foil, 100 of paper, each 12 cm by 50 cm. The paper has a thickness of 0.22 mm and a relative permittivity of 4.1. If all the odd-numbered foil sheets are connected to a common wire and all the even-numbered sheets are connected to a second common wire, what will be the capacitance measured between the two wires? (Hint: There's one sheet of dielectric per capacitor.)

40. [I] What is the equivalent capacitance of the circuit between the terminals A and B, indicated in Fig. P40?

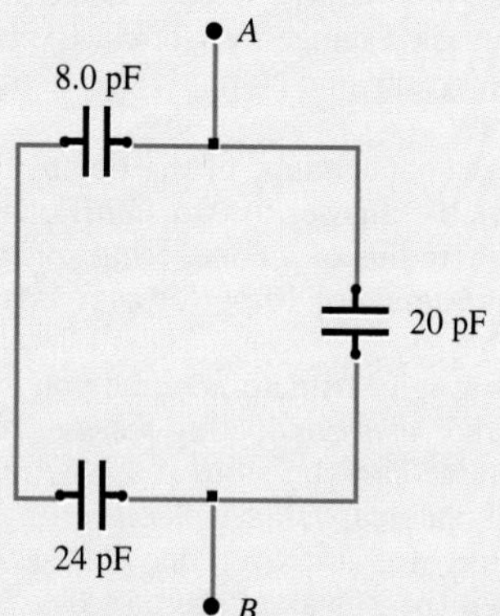

Figure P40

41. [I] What is the equivalent capacitance of the circuit between the terminals A and B, indicated in Fig. P41?

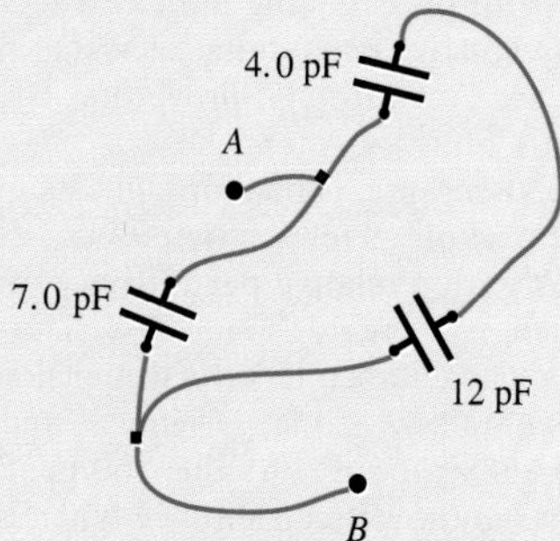

Figure P41

42. [I] What is the equivalent capacitance of the circuit between the terminals A and B, and again between A and C, indicated in Fig. P42?

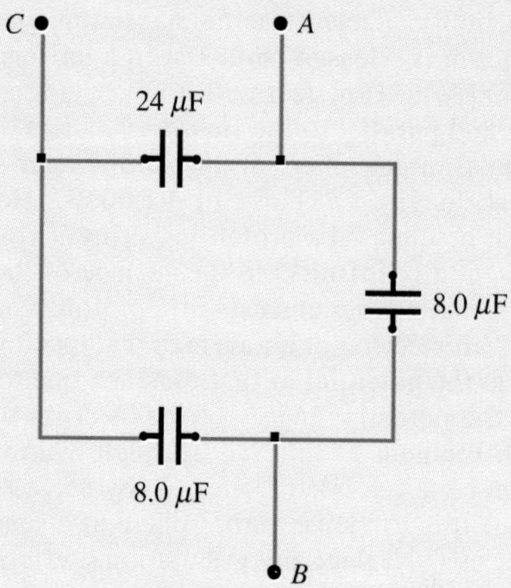

Figure P42

43. [I] A 20-pF capacitor with a mica insulating sheet $\frac{1}{2}$ mm thick is put across a 12-V battery. How much energy will it store?

44. [I] A bolt of lightning corresponding to about 20 C descends through a potential difference of upwards of 150 MV. How much energy is involved? Incidentally, at any one time there might be 2000 thunderstorms rolling over the Earth with 100 lightning bolts flashing every second!

45. [II] Figure P45 depicts a neuron, or nerve cell, and shows the long signal-carrying axon (which, from spine to fingers can be more than a meter in length). As discussed in Problem 21, the axon membrane is usually positive on the outside and negative on the inside. The dielectric constant of the membrane has been measured to be about 7. Given that the membrane wall is a mere 6.0 nm thick and that the axon radius is 5 μm, determine its capacitance per unit area. Explain any assumptions you make.

46. [II] The Leyden jar shown in Fig. 18.16 has a 22-cm-diameter base and the metal foil (inside and out) goes up to a height of 35.5 cm. The inside metal is attached to the central conductor by a wire. The glass has a thickness of 2.25 mm and a dielectric constant of 7.2. Estimate its capacitance.

47. [II] With Problem 45 in mind, determine the surface charge density of the axon. In the resting state, when there is no signal being transmitted, the potential difference across the membrane is about 70 mV.

48. [II] With the previous problem in mind, determine the capacitance per unit length of the axon.

49. [II] When a pulse is propagated down an axon, there is a shift of ions across a segment of membrane that causes a reversal of polarity and an overall change in potential of about 100 mV (see Fig. P49). This voltage spike (and the associated repolarization of the membrane) propagates along from one region to the next and constitutes the signal, just as a flame propagates down a length of fuse.

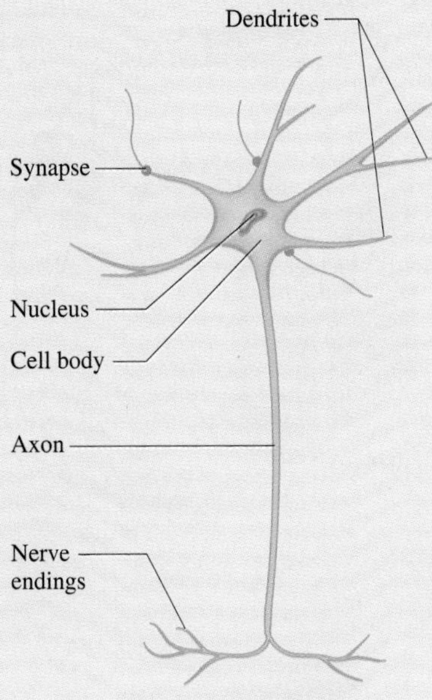

Figure P45

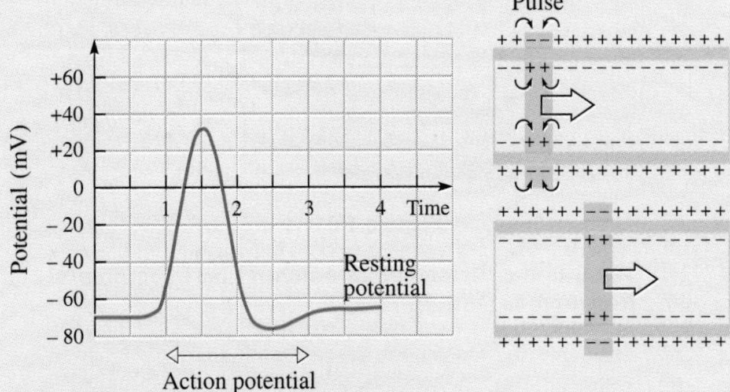

Figure P49

After a few milliseconds, this so-called *action potential pulse* passes any given point on the axon, which then returns to its resting potential (−70 mV). How much energy is required to recharge a 1-m length of axon in the wake of a pulse, so that it will be ready to transmit the next pulse?

50. [II] What is the equivalent capacitance of the grouping shown in Fig. P50?

51. [II] What is the equivalent capacitance of the circuit between terminals A and B in Fig. P51?

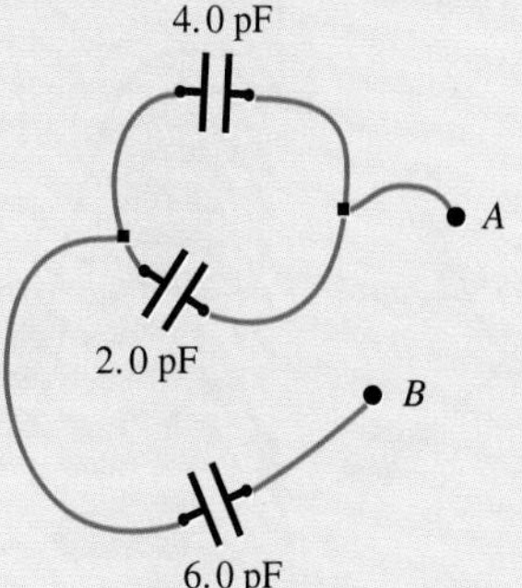

Figure P50

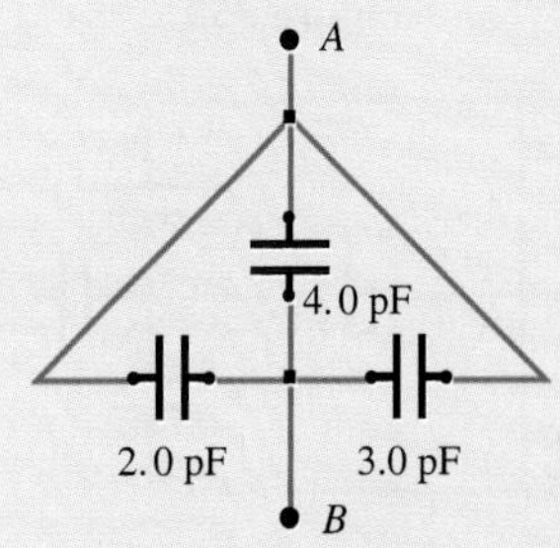

Figure P51

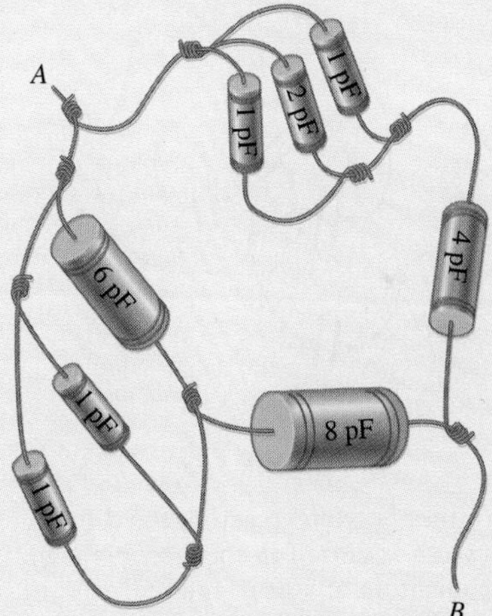

Figure P52

52. [II] What is the equivalent capacitance of the circuit between terminals A and B in Fig. P52?

53. [II] What is the equivalent capacitance of the circuit between terminals A and B in Fig. P53?

54. [II] What is the equivalent capacitance of the circuit between terminals A and B in Fig. P54, if each of the capacitors is 2.0 pF?

55. [II] What is the voltage across the 9.0-pF capacitor in the circuit shown in Fig. P55 after the switch is closed? Determine the equivalent capacitance across the battery. How much charge is drawn from the battery?

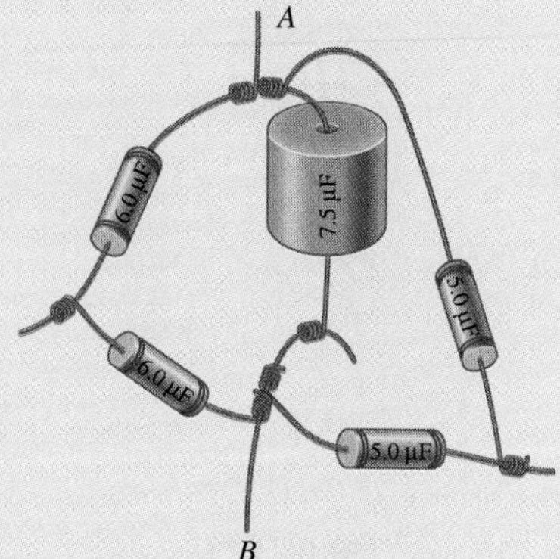

Figure P53

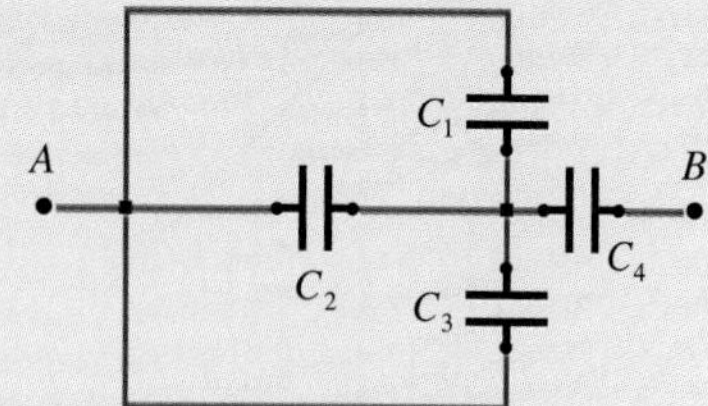

Figure P54

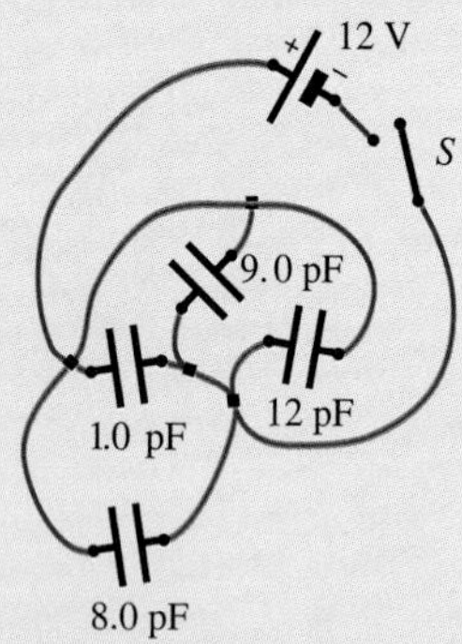

Figure P55

56. [II] Determine the voltage across the 4.0-μF capacitor after the switch is closed in the circuit shown in Fig. P56.
57. [II] Two capacitors of 100 μF and 50 μF are separately charged to 250 μC and 100 μC, respectively. They are then attached in parallel so the + plate of one goes to the − plate of the other, and vice versa. Determine the final voltage across the two.
58. [II] Figure P58 shows four metal plates with the outer pair and inner pair wired together. Given that there is air in the gaps, which are $\frac{1}{2}$ mm wide, and that the area of each plate is 0.01 m^2, what is the capacitance of the device?
59. [II] Return to Fig. P55 and, overlooking losses, find the net energy provided by the battery when the switch is closed.
60. [II] In fair weather, the constant atmospheric E-field near the surface of the Earth varies from 120 V/m to 150 V/m. Compute the corresponding range of energies stored per cubic meter in the field.

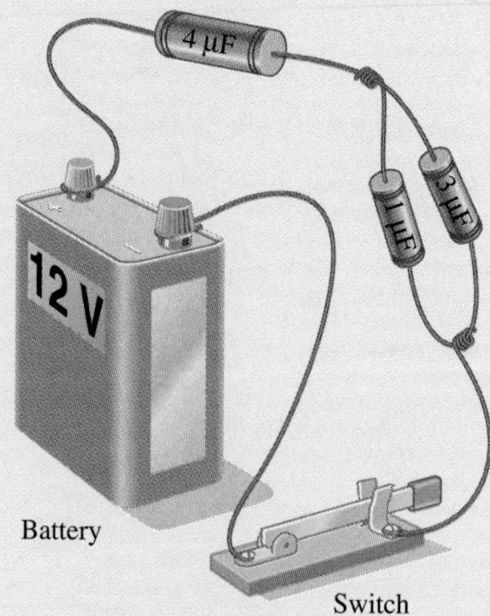

Figure P56

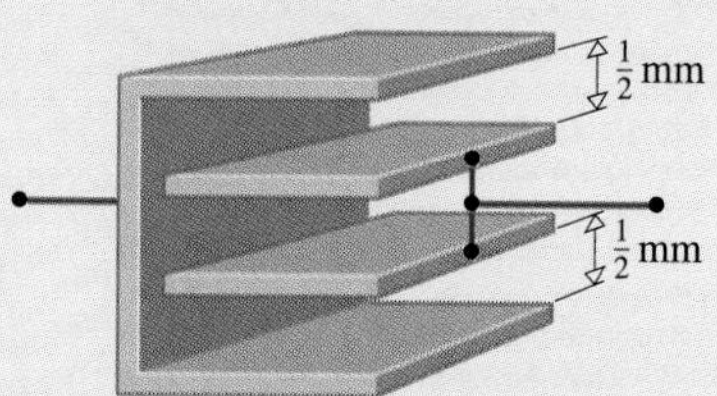

Figure P58

61. [III] Figure P61 shows three plates each 50 cm by 60 cm and separated by 1.0 mm. The plates are immersed in transformer oil with a dielectric constant of 4.5. Find the total capacitance of the system.

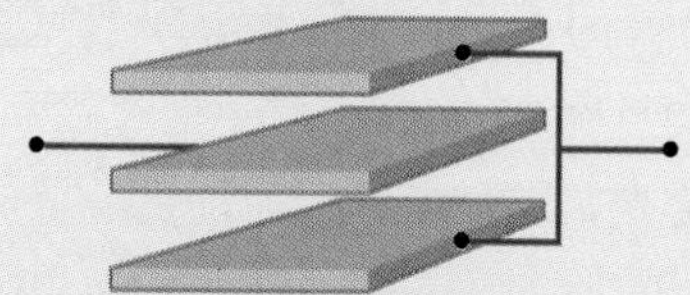

Figure P61

62. [III] What is the difference in potential between points A-D, B-D, and C-D in Fig. P62?

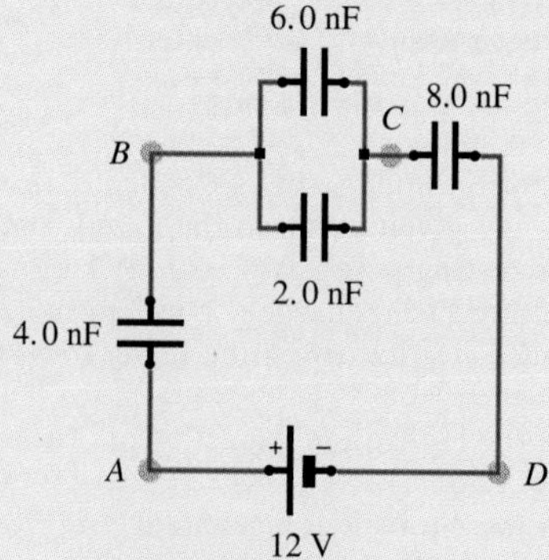

Figure P62

63. [III] If a 12-V battery is placed across terminals A and B in the circuit shown in Fig. P63, how much energy will be stored by the capacitive network? Notice that there are two places where one wire goes over another without touching it.

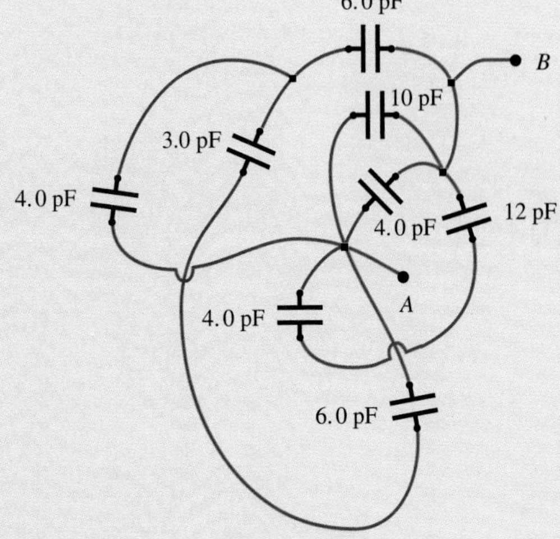

Figure P63

64. [III] Fig. P5 shows two hollow metal concentric spheres of radii ρ and R. If the inner one is charged with $+Q$ and the outer surface is grounded, show that the capacitance is

$$C = \frac{\rho R}{k(R - \rho)}$$

Moreover, if the gap (d) is very small and A is the area of either inside surface ($4\pi\rho^2 \approx 4\pi R^2$), show that

$$C \approx \frac{\varepsilon A}{d}$$

as with Eq. (18.12).

19

Direct Current

IF THERE IS A life's blood in our technology, it's most assuredly electricity—electricity coursing along wire veins, delivering power and information, trickling through a metal nervous system just as it trickles through our own. **The ordered flow of charge is called electric current,** whether we're talking about electrons propelled down a wire by a battery, or protons hurled through space by an exploding star. And currents carry energy. Much of the energy we "consume" is delivered by electricity conveniently on tap at wall outlets everywhere: refrigerators, VCRs, computers, heart-lung machines, all plug into the electric stream and draw energy from it.

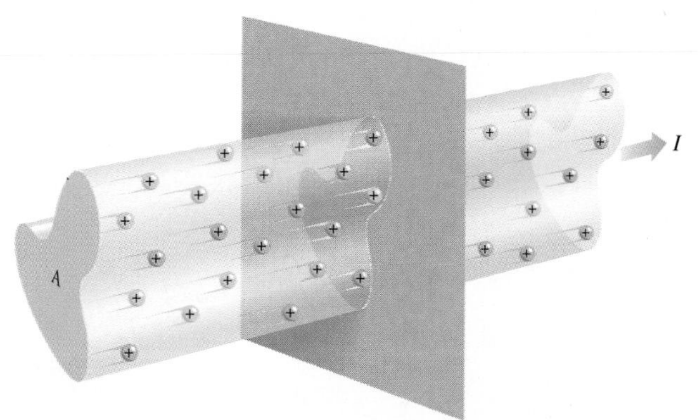

Figure 19.1 A beam of positive particles of cross-sectional area A constituting a current I. During each interval of time Δt, an amount of charge Δq passes through the plane, such that $I = \Delta q/\Delta t$.

This chapter is about electric currents, the batteries that sustain them, and the resistances that impede them.

CURRENTS

To quantify **electric current** (I), envision a stream of positive charge (Fig. 19.1). Now picture an imaginary reference plane cutting across the flow at some point of observation and determine the net charge Δq traversing the plane in a time Δt. The ratio $\Delta q/\Delta t$ is the average rate at which charge passes the point of observation during that interval, the *average current.* Often, the flow is constant, in which case

$$I = \frac{\Delta q}{\Delta t} \tag{19.1}$$

When the current changes from moment to moment, we define an *instantaneous current* as

$$I = \lim_{\Delta t \to 0} \left(\frac{\Delta q}{\Delta t} \right) \tag{19.2}$$

The SI unit of electric current is the *ampere* (A) or simply the *amp*, where 1 amp corresponds to a flow of 1 coulomb of charge per second; thus

$$1 \text{ A} = 1 \text{ C/s}$$

Minute currents in the microamp range (10^{-6} A = 1 μA) are commonplace, even within the human body. Microamp currents are generated in bone and connective tissue during exercise and seem to play a vital role in sustaining the health of these structures.

A beam of electrons (60 kiloamps, 3 MeV) deflected by both air scattering and a magnetic field. This is as much an electrical current as any that ever negotiated a toaster.

Normally, a current in a metal wire is a stream of free electrons. But, the **mobile charge carriers** constituting a current can be positive, negative, or both. The latter may be the case, for example, in a semiconductor or a plasma. Stars, streetlights, and fluorescent lamps contain plasmas wherein streams of oppositely charged carriers can be made to flow past each other in opposite directions.

The mobile charge carriers in Fig. 19.1 happen to be positive—just the sort of picture Ben Franklin had in mind for a current. Because of him, *the direction of flow of positive charge is traditionally taken to be the direction of current, regardless of the actual sign of the participating carriers.* Since electrons are the carriers in ordinary wires, this custom can be a little awkward at times, though it's easy enough to live with. **A flow of negative carriers to the left is equivalent to an equal flow of positive carriers to the right** (Fig. 19.2).

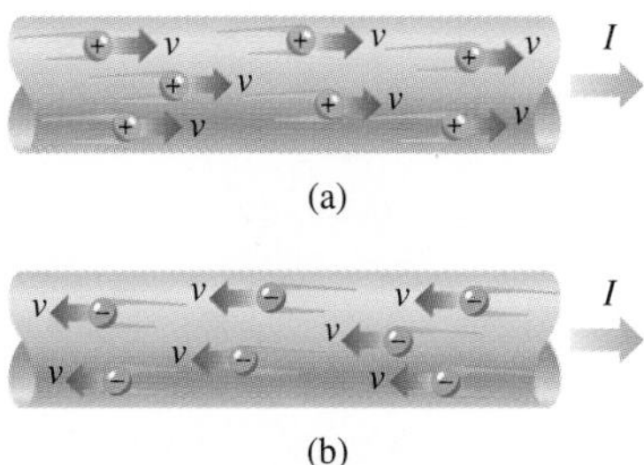

Figure 19.2 (a) The direction of an electric current is the direction of flow of positive charge. (b) Thus, it is opposite to the flow of negative charge.

Example 19.1 A constant downward electron beam transports 3.20 μC of negative charge in 200 ms across the vacuum chamber of an electron microscope. Determine the beam current and the number of electrons traversing the chamber per second.

Solution: [Given: $\Delta q = 3.20 \times 10^{-6}$ C, and $\Delta t = 200 \times 10^{-3}$ s. Find: I and the number of electrons per second.] From Eq. (19.1)

$$I = \frac{\Delta q}{\Delta t} = \frac{3.20 \times 10^{-6}\ \text{C}}{200 \times 10^{-3}\ \text{s}} = \boxed{16.0\ \mu\text{A}}$$

The current is *upward* and equal to 1.60×10^{-5} C/s. The number of electrons transported per second, each with a charge of -1.60×10^{-19} C, is

$$\frac{-1.60 \times 10^{-5}\ \text{C/s}}{-1.60 \times 10^{-19}\ \text{C}} = \boxed{1.00 \times 10^{14}\ \text{electrons/s}}$$

Again, we have to be careful with the signs since the current transfers -1.6×10^{-5} C/s.

▶ **Quick Check:** 16 μA flowing for 200 ms transports 3.2 μC. Since 1 A corresponds to (1 A)/(1.6 × 10^{-19} C) = 6.2 × 10^{18} electrons/s, 16 μA is about 10 × 10^{13} electrons/s.

Example 19.2 An electron gun fires a pulse of charge lasting 2.0 μs. The average current of the burst is 1.0 μA. How many electrons are there in the pulse?

Solution: [Given: $\Delta t = 2.0\ \mu$s, $I = 1.0\ \mu$A. Find: number of electrons.] $I = \Delta q/\Delta t$, so if we first find the amount of charge in a pulse, we can then, knowing the charge on the electron, find the number of particles in the pulse; accordingly

$$\Delta q = I\Delta t = (1.0\ \mu\text{A})(2.0\ \mu\text{s}) = 2.0\ \text{pC}$$

The number of electrons is then

$$\frac{2.0 \times 10^{-12}\ \text{C}}{1.602 \times 10^{-19}\ \text{C/electron}} = \boxed{12 \times 10^{6}\ \text{electrons}}$$

▶ **Quick Check:** $\Delta q \approx 10^{-6}\ \text{s} \times 10^{-6}\ \text{A} \approx 10^{-12}$ C: $10^{-12}\ \text{C}/10^{-19}\ \text{C/electron} \approx 10^{7}$ electrons.

A current (either a flow of electricity in a wire or water in a pipe) is usually impeded in some way by the environment through which it progresses. That *resistance* inevitably results in the expenditure of energy from the flow. To be sustained, a current must be driven by an external source of energy. In the case of electricity, a nonelectrostatic source must continuously supply energy to the mobile charge carriers, pushing them along. It's not obvious how to perform such a process; indeed, sustained currents were unknown up until the end of the 1700s. It was only then that a strange series of events led to the development of the electric battery, thereby plunging the world into the Age of Electricity.

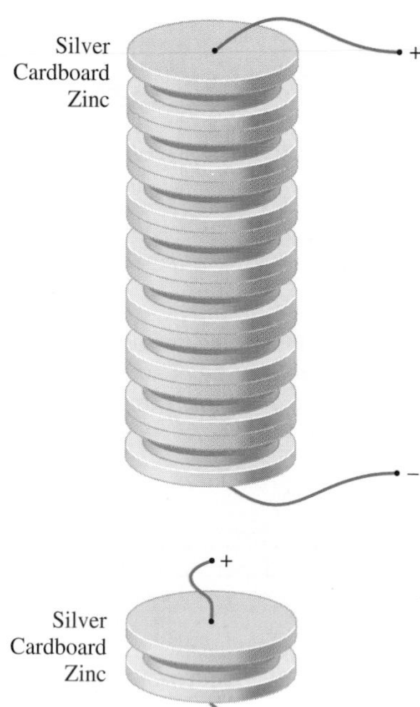

Figure 19.3 The voltaic pile, in which the elementary cell was a zinc-cardboard-silver sandwich. Volta wrongly thought that it was the zinc-silver contact that provided the effect.

19.1 The Battery

By the late 1700s, many researchers working with Leyden jars had experienced muscle spasms due to accidental shocks. Fascinated by the relationship between electricity and life, they began to study the electrical excitation of muscles in animals. The frog, whose muscular legs were a popular delicacy, became a logical martyr to the cause. In a short time, dissected frog's legs were twitching and convulsing on command from one end of Europe to the other. Even Faraday kept a froggery in the basement of the Royal Institution.

In 1780, Luigi Galvani, a noted Italian anatomist, made the first of several accidental discoveries that would arouse a storm of controversy. Each of his little green participants was humanely terminated, being impaled through the spinal cord on a sharp bronze hook. Thus prepared, the frogs waited their turns hanging on an iron trellis in the garden. To his surprise, Galvani noticed that every now and then the disembodied legs would spastically convulse for no apparent reason. He concluded that he had discovered "in the animal itself" a natural source of electricity.

Alessandro Giuseppe Antonio Anastasio Volta, already renowned as an "electrician," soon became Galvani's staunchest critic. It seemed to Volta that the source of the electricity was not the animal, but the two different metals brought in contact. The frog was simply a current meter. Recognizing that his tongue was a highly sensitive and convenient muscle, Volta became his own guinea pig (or, in this case, frog). He placed different pairs of metals on his tongue and brought them into contact. Expecting to feel a contractive spasm, he was surprised when the arrangement produced a metallic taste that lasted as long as the two metals were in contact.* This implied the incredible notion that "the flow of electricity from one place to another is continuing without interruption." He had created a source of sustained current.

Volta soon produced two devices that were the forerunners of the modern electric battery. He replaced the hook, frog, and trellis with something much more convenient. The **voltaic pile** was a stack of small disks of zinc, brine-soaked cardboard, and silver, layered in order: zinc-cardboard-silver, and so on (Fig. 19.3). Each zinc-cardboard-silver sandwich formed a unit (much like the aluminum, wet tongue, and silver) called a *cell.* These were repeated about 20 times (with the cells effectively in series) to build up the voltage. The result was a device that produced an appreciable continuous current. The other variation on the theme was the *voltaic wet cell.* It consisted of a drinking glass filled with brine or a dilute acid into which two metal strips were immersed, one of zinc, the other of copper. By putting several of these cells in series (Fig. 19.4), Volta created the world's first *electric battery.* **A battery is two or more cells electrically attached to one another**. By 1838, the Great Western Railway already had a voltaic pile powering a telegraph link. The granddaddy of the cheap flashlight battery, the *carbon-zinc cell* was conceived by Bunsen in 1842. In

Volta demonstrating his "pile" to Napoleon in a painting by a contemporary, A. E. Fragonard. Some years later, Volta gave Faraday a battery of this sort.

*Volta was not the first to perform this little experiment, nor did he ever fully understand it. He used tin and silver or gold, but aluminum foil and sterling silver work well. Try it.

1859, Planté gave the world the lead-plate-dipped-in-sulfuric-acid storage device that has come to be known as the automobile battery.

How Batteries Work

Almost any two different solid conductors immersed in a variety of active solutions, known as *electrolytes,* function more or less as a battery. Chemical energy stored in the interatomic bonds (typically at less than about 3 eV) is converted into *electrical*-PE as the solution and one or both of the conducting plates, the *electrodes,* become involved in the chemical reaction. In the voltaic wet cell, the acid attacks the copper (Cu), and some of its positive Cu^{++} ions go into solution, leaving behind a negatively charged plate. Similarly, zinc (Zn) ions go into solution, too, but zinc is more soluble than copper and the zinc plate becomes even more negative—with even more excess electrons. The copper is slightly lower in potential than the acid; the zinc is a lot lower than the acid. The result is a difference in potential between the electrodes, with the copper higher and so positive, and the zinc lower and negative (Fig. 19.5).

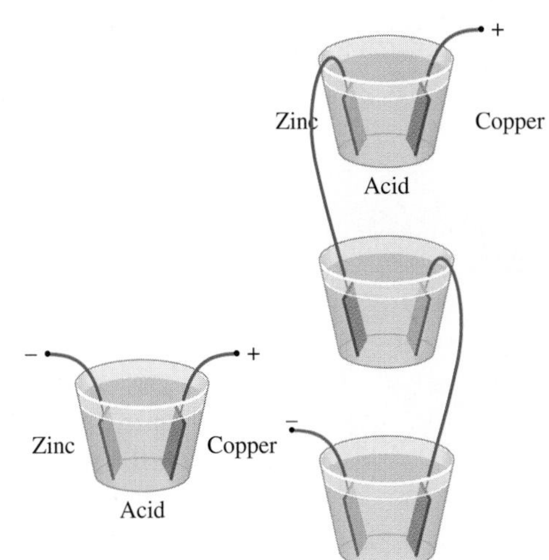

Figure 19.4 A single voltaic wet cell and three wired in series to form a battery.

The *electromotive series* (Table 19.1) is a list of metals in decreasing order of the tendency of each to become ionized by losing an electron. If we construct a simple cell with two metal electrodes immersed in a uniform electrolyte, the cell's voltage can be approximated from the values in the table. Accordingly, the potential of copper is +0.34 V, whereas zinc is −0.76 V (each with respect to hydrogen as the reference). Thus, copper is (0.34 V) − (−0.76 V) = +1.1 V higher than zinc, and this value is the largest voltage we can expect to produce across the terminals of a simple copper-zinc cell.

Figure 19.5 A voltaic wet cell composed of a copper and zinc plate immersed in a solution of dilute sulfuric acid (H_2SO_4). This device is the kind the young poet Percy Shelley was playing with at Oxford, burning holes in the carpet of his apartment with the splashed acid.

TABLE 19.1 Electromotive series for several metals*

Substance	Electrode potential (V)
Lithium	−3.0
Potassium	−2.9
Sodium	−2.7
Aluminum	−1.7
Zinc	−0.76
Iron	−0.44
Tin	−0.14
Lead	−0.13
Copper	+0.34
Mercury	+0.80
Silver	+0.80
Platinum	+1.2
Gold	≈+1.3

*These values correspond to the potentials required to ionize the corresponding metal atoms. The signs are a matter of convention referenced to hydrogen (and measured at 25°C).

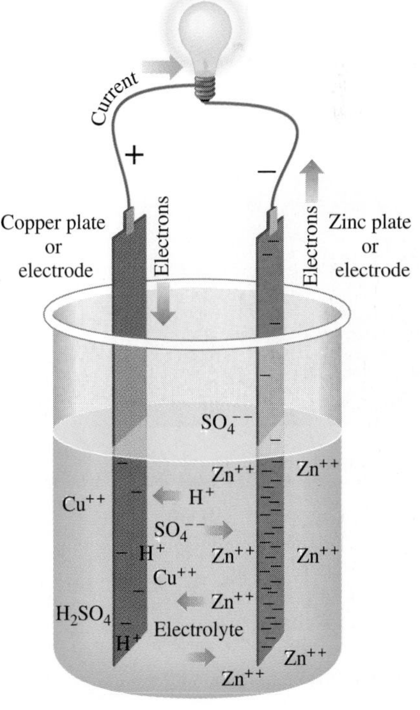

A little dry cell torn open to show the central carbon rod. The zinc casing is corroding in spots.

A potential difference that can be used to supply energy and thereby sustain a current in an external circuit is an **electromotive force**, or *emf* (pronounced *ee em ef*), although that's a misnomer—it's not force at all. Practically, **the emf is the voltage measured across the terminals of a source when no current is being drawn from or delivered to it.**

A strip of copper (some wire or a penny) and a strip of zinc (a few galvanized nails) immersed in salt water easily puts out $\approx$0.7 mA at $\approx$0.6 V. Add some vinegar and the emf will go up to perhaps 0.9 V. A small speaker from an old radio makes a fine current detector. Attach one wire, or *lead,* from the cell to the speaker and then tap the other lead from the cell on the remaining terminal. The speaker will clatter with each touch. Stick a carbon rod (a pencil lead will do) into a lemon and surround it with a bunch of galvanized nails. The resulting cell will deliver $\approx$0.5 mA at a voltage of roughly 0.7 V. Replace the carbon with something as ordinary as a paper clip, and the lemon cell will still generate an emf, though down now to $\approx$0.2 V. Try a grapefruit or even some sauerkraut.

A given kind of cell will generate a voltage difference that is determined by its chemical makeup, independent of size. The size determines the total current a cell can deliver, not the voltage; the greater the quantity of each substance chemically reacting, the more charge is liberated. An ordinary flashlight "battery," a *dry cell* (Fig. 19.6a), has an emf of 1.5 V. A *mercury cell,* one of those little button-sized

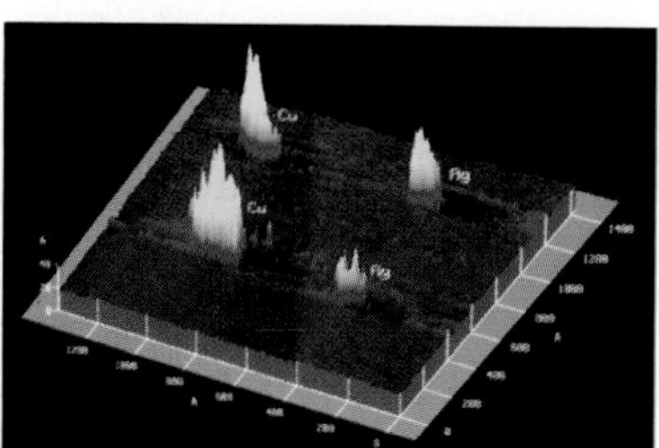

The world's smallest battery. Measuring 70 nm long and made of fewer than 5×10^5 atoms, it is 1/100 the diameter of a red blood cell. It consists of tiny pillars of copper and silver deposited on a graphite surface. When immersed in a copper sulfate solution, it produces an emf of 20 mV for about 45 minutes.

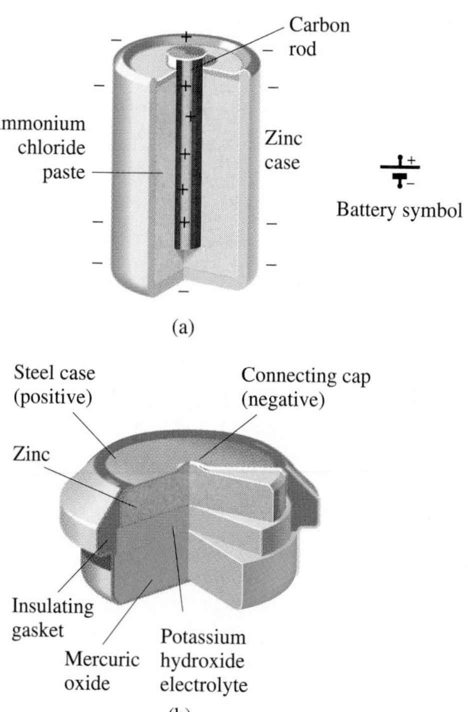

Figure 19.6 (a) The basic dry cell, essentially as Leclanché developed it in 1868. The modern version has a so-called *depolarizing* region of manganese dioxide around the carbon rod. That region controls the buildup of hydrogen ions that would otherwise reduce the emf. (b) A mercury cell used in calculators and watches.

calculator, watch, and hearing-aid "batteries," has an emf of about 1.4 V, while the *lead storage cell* of auto fame is a 2-V device. One of its great virtues is that it can be recharged by the generator in the car. The *nickel-cadmium cell* used in rechargeable computer battery packs has an emf of 1.2 V.

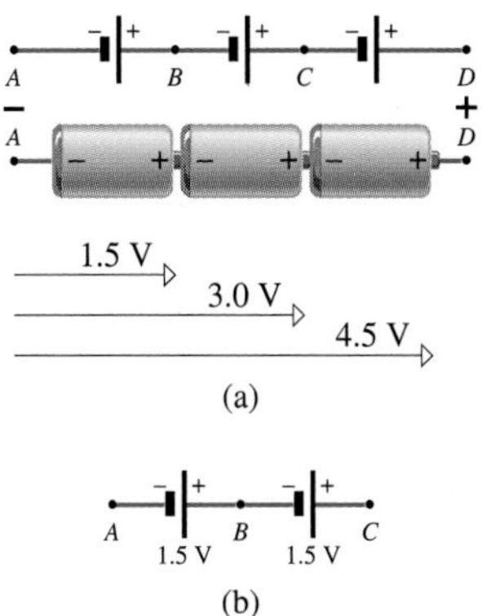

Figure 19.7 Cells in series. As connected in (a), the voltages add up, and point D is 4.5 V above point A. As connected in (b), the voltages subtract and A and C are at the same potential.

Cells in Series and Parallel

To boost the potential difference provided, cells are connected in series. The crucial point is that *the voltage across the series-connected battery is the sum of the voltages across each constituent cell.* Refer to Fig. 19.7a. Point B is 1.5 V higher than A, and point D is 4.5 V higher than A. This sort of series stacking is just what you are doing when you load two, three, or four D-cells, top (+) to bottom (−), into a flashlight or a portable radio in order to get the 3.0 V, 4.5 V, or 6.0 V needed to operate the device. It's also the way the cells are wired in a car battery to yield 12 V.

If a wire is connected across the terminals of a cell (as in Fig. 19.8a), a current will move around the closed conducting path. In practice, this is a bad idea since a wire having little resistance corresponds to a *short circuit* that will drain the cell. It could also be dangerous if enough current flows to melt the wire. As a rule, the battery provides power to a *load*—for example, a motor, radio, or light bulb (Fig. 19.8b). **A steady-state current can only exist in a closed circuit** and the same current flows in and out of the load. **Current is never used up by a circuit element**. Because the polarity of the cell is fixed, the direction of the flow of charge is constant, and we say that I is a **direct current** (dc). If the same thing is done with a battery comprising several cells in series (Fig. 19.9), the same amount of current will pass through each cell. In series the voltages add, while the current remains unaltered as it passes in and out of each element (reminiscent of the way series capacitors behave where the voltages add and the net charge is the charge on any one). By contrast, cells attached in parallel (Fig. 19.10a) form a battery whose voltage is the same as the individual voltages but whose current capacity is the sum of the individual current outputs (again matching the behavior of capacitors). If you want a lot of current at a low voltage, put the cells in parallel; if you want both a large current and a large voltage, stack the cells in parallel and then put the stacks in series.

But when I took the frog into a closed room, laid it on an iron plate and began to press the hook that was fixed in its spinal cord against the plate, lo and behold, the same contractions and the same kicks!

GALVANI

Battery manufacturers provide (though not often right on the battery where it ought to be) a crude measure of the current capacity in the form of an **amp-hour rating**. A little 1.5-V AA-cell, the common penlight finger-sized "battery," is rated at about 0.6 amp-hour, whereas the larger 1.5-V D-cell can deliver as much as 3 amp-hours. Presumably, one can draw a steady 0.3 amps for 10 hours or 0.1 amp for

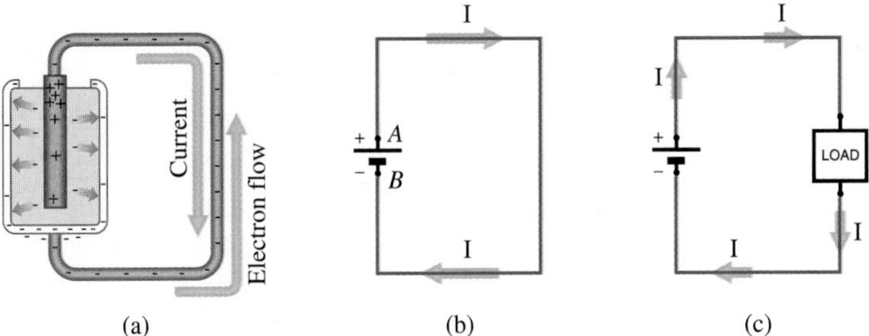

Figure 19.8 (a) (b) The flow of charge through a dry cell. This diagram represents a short circuit, which will quickly destroy a cell. (c) A more practical arrangement has the battery send current through a load, such as a motor or a light bulb.

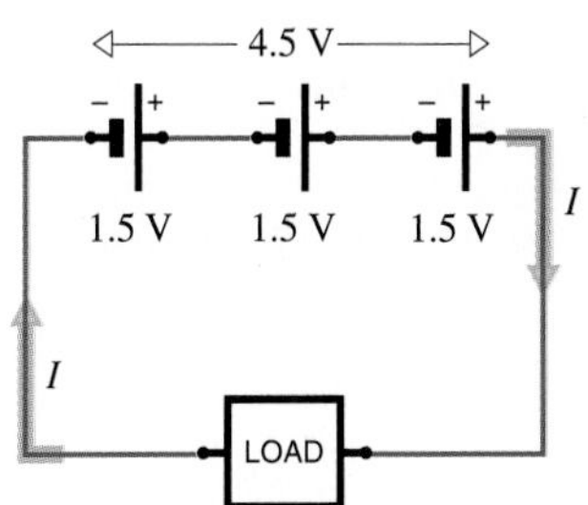

Figure 19.9 When cells in series form a battery, the same current I passes through each cell: I circulates around the circuit and is undiminished as it enters and leaves each element.

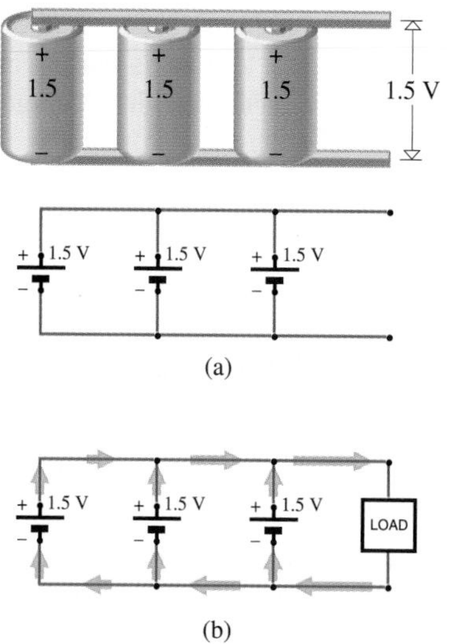

Figure 19.10 (a) Three 1.5-V cells in parallel. The voltage across the battery is the voltage across each cell. (b) Currents in the various segments of a parallel array of cells with a load across the terminals. The current through the load is the sum of the currents provided by each cell.

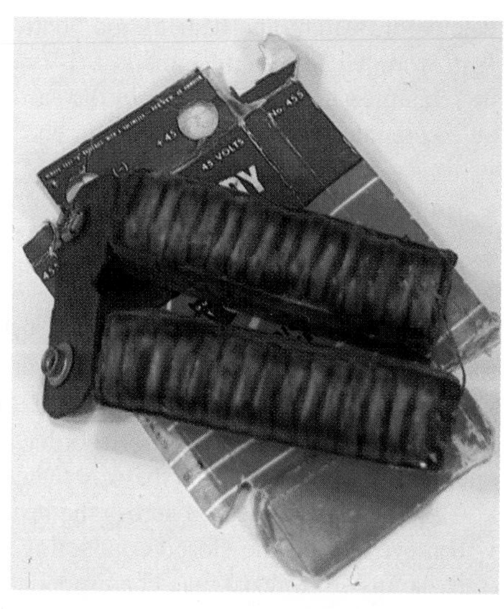

An old 45-V battery made up of thirty 1.5-V dry cells in series.

30 hours before discharging a D-cell. It will never be able to put out 300 amps for 0.01 hour, but you get the point. Most operating batteries tend to produce positive hydrogen ions which, if unchecked, build up and rapidly degrade the operation of the cell—a process known as *polarization.* Modern dry cells contain a region of manganese dioxide around the anode that eliminates the hydrogen, gradually *depolarizing* the cell. The discharge rate is limited by the need to have the depolarization keep pace with the liberation of H^+ ions. Dry cells are designed to function intermittently so they have a chance to recuperate as the depolarizer works. Have you ever noticed how a dim flashlight will come on strong again after a rest? Mercury cells are self-depolarizing, removing the need for rest periods. They are especially well suited for the continuous operation of low-current solid-state commercial, scientific, and medical devices—cameras, hearing-aids, alarm systems, and so on.

On the voltaic pile. *It may be conjectured that we have carried the power of the instrument to the utmost extent of which it admits; and it does not appear that we are at present in the way of making any important additions to our knowledge of its effects, or of obtaining any new light upon the theory of the action.*

JOHN BOSTOCK
Account of the History and Present State of Galvanism (1818)

What you want from a car battery is a lot of current to power the starter motor that turns the engine until it gets going on its own. A heavy-duty 12-V truck battery with an amp-hour rating of near 160 can provide a tremendous current; even 10 amps is a great deal—but of course it's only at a meager 12 V. Plenty of charge is available to flow from such a battery, but there's little push to propel it, so it's fairly safe. That's like having all the water of the Atlantic Ocean behind a dam one foot deep; lots of water stored, but little pressure. Typically a lead-acid automobile battery stores about 0.5 kW·h (that is, 0.5×10^3 W·h $\times$ 60 min/h $\times$ 60 s/min = 1.8 MJ) of energy. By comparison, the batteries that power diesel submarines while submerged weigh as much as several hundred tons and consist of row upon row of interconnected rechargeable cells that can retain 10 000 times the energy of a car battery.

Example 19.3 Design a battery to be constructed of D-cells (each rated at 3 amp-hours) that will provide a maximum operating current of 5.0 A with an emf of 4.5 V. One restriction is that no cell be required to supply in excess of 1.0 A. What can you expect will be the lifetime of your battery while delivering maximum current?

Solution: The 1.0-A restriction per cell demands we put five cells in parallel to form a 5.0-A unit, as in Fig. 19.11a. The maximum current through each cell would then be 1.0 A as the unit provides the required 5.0 A with an emf of 1.5 V. To bring the voltage up to 4.5 V, we put three such units in series, as in Fig. 19.11b. With a 3.0-amp-hour rating, operating at a maximum of 1.0 A, we anticipate a lifetime of $\boxed{3\text{ h}}$ for the battery.

▶ **Quick Check:** The 5.0-A current entering an adjacent unit splits up into five separate 1.0-A currents, one of which passes through each cell. These reunite as a 5.0-A stream on leaving the unit.

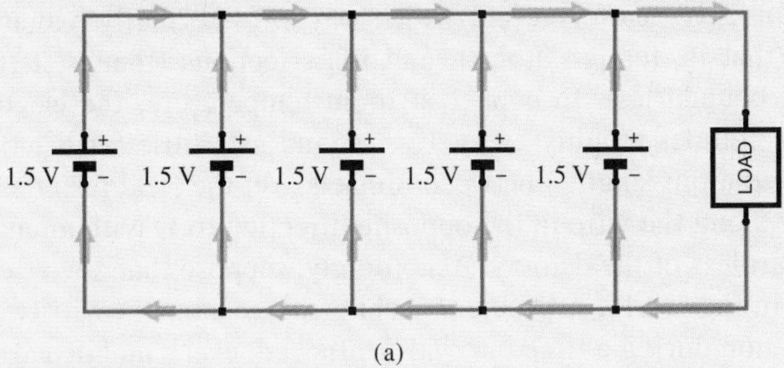

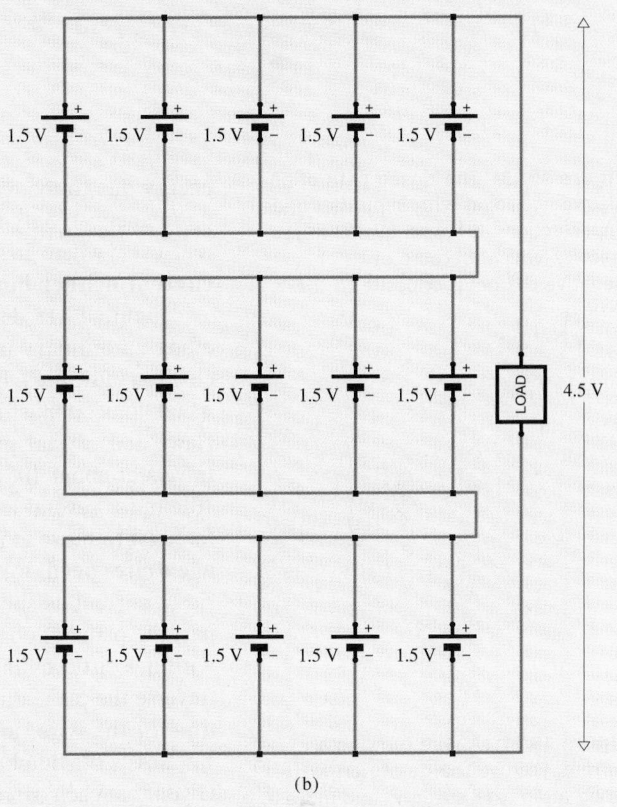

Figure 19.11 A battery made up of D-cells each rated at 3 amp-hours. (a) The voltage across this battery is the voltage across each of its cells—namely, 1.5 V. Thus, the voltage across the load is 1.5 V. (b) By putting three strings of cells in series, the voltage across the load is now 4.5 V.

19.2 Electric Fields and the Drift Velocity

The electric field produced by an isolated battery rises out of the positive terminal (*anode*) and returns to the negative terminal (*cathode*), looking more or less like the field of a dipole (Fig. 19.12). A negative mobile charge carrier in the vicinity of the anode would be drawn toward it, and one near the cathode would be repelled. Now suppose we attach a few meters of ordinary copper hookup wire across the terminals of the battery. Electrons, forced away from the cathode, stream into the wire, repelling each other and spreading out to the wire's surface as they progress along its length. The surface becomes nonuniformly charged, although there is no net charge on the wire as a whole. The copper resists the flow of charge (p. 671), and so there must be a sustained driving force (that is, an E-field within the wire maintained by a battery, generator, or power supply). Once the surface charges accumulate in a region, they naturally inhibit any further lateral flow of charge and the bulk of the current progresses down the length of the wire. Thus, a surface charge quickly builds up that creates a uniform axial field along the length of the wire (and a field outside it as well). Open the circuit, the surface charges redistribute, the field van-

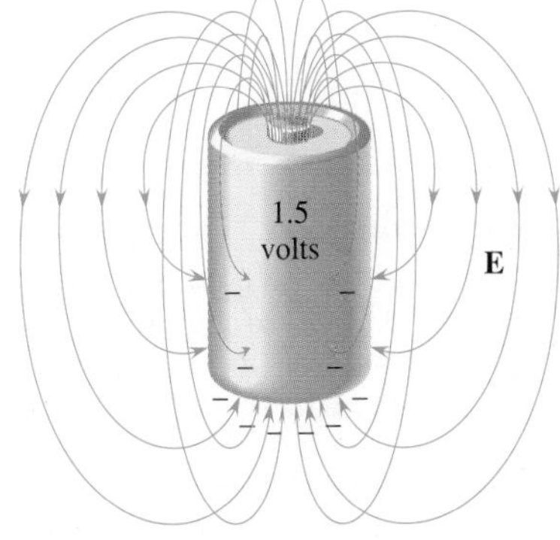

Figure 19.12 The electric field of a dry cell. The field lines are distributed symmetrically around the central axis.

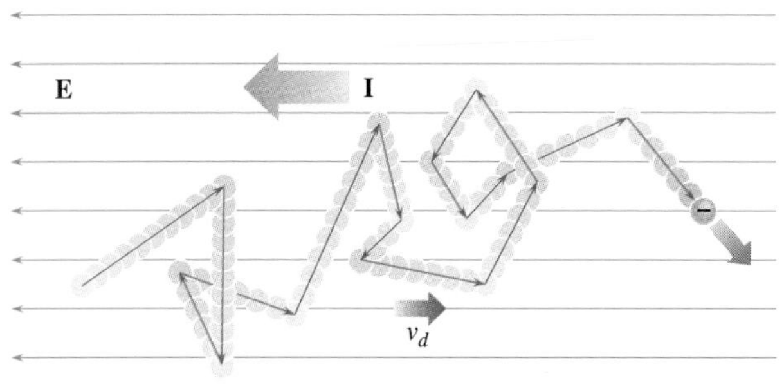

Figure 19.13 The zigzag path of an electron colliding with impurities and imperfections. It has an overall drift velocity, which (because of its negative charge) is opposite to the *E*-field.

ishes, and the current ceases; an *E*-field cannot be maintained inside a conductor under electrostatic conditions.

If there are any kinks or turns in the wire, a few additional electrons initially bunch up on the surface at those locations until the resulting internal steady-state *E*-field they contribute to is completely axial. After a very short time, a stable distribution of surface charges is in place, a uniform axial *E*-field exists, and a constant current progresses around the closed circuit. That current may be imagined as continuously emerging from the positive terminal of the battery, moving along the wire (where it loses energy due to the resistance), and coming back into the battery (where it's pumped up in energy), only to be sent out and around again. In the steady state there is current everywhere in the closed circuit and no further buildup of charge anywhere; **current neither bunches up nor gets used up**.

Figure 19.13 depicts the zigzag path of a typical electron as it makes its way along an ordinary metal conductor. Pushed by an externally sustained *E*-field, the electron advances but quickly scatters every-which-way off thermally vibrating metal ions, impurities (that is, foreign atoms), and imperfections (that is, lattice flaws and crystal grain boundaries). In copper at room temperature, the electron travels at about 10^6 m/s, collides roughly every 10^{-14} s, and gets little farther than 10^{-8} m (or several hundred atom lengths) between impacts. Yet the field relentlessly forces it to move axially along the wire in the opposite direction to **E** with an **average drift speed**, v_d, regardless of the detours. Accordingly, suppose that a wire carries a current as shown in Fig. 19.14. A group of mobile charge carriers will sweep past the plane of observation during a time interval Δt. Indeed, every mobile carrier within a little cylindrical volume of length $v_d \Delta t$ and cross-sectional area A will traverse the plane during that interval. If η is the **number of carriers per unit volume** in the wire, and $v_d \Delta t A$ is the volume of carriers swept past the plane, then $\eta v_d \Delta t A$ is the number of mobile charges passing a point of observation in a time Δt. Given that each carrier has a charge q_e, the net amount of charge (Δq) transported in a time Δt is

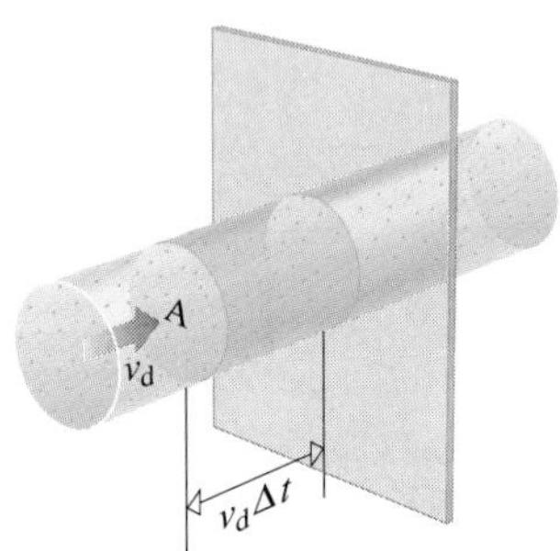

Figure 19.14 A wire carrying a current. Positive charge carriers drift along at an axial speed v_d and move a distance $v_d \Delta t$ in a time Δt.

$$\Delta q = (\eta v_d \Delta t A) q_e$$

It follows from Eq. (19.1) that the current is

$$I = \frac{\Delta q}{\Delta t} = \eta v_d A q_e \tag{19.3}$$

Example 19.4 Determine the order-of-magnitude of v_d in copper, assuming a current of 1 A, a cross-sectional area of $1.0 \text{ mm}^2 = 1.0 \times 10^{-6} \text{ m}^2$, and the availability of one conduction electron per atom. Discuss the notion that this is only an order-of-magnitude calculation.

Solution: [Given: $I = 1$ A, $A = 1.0 \times 10^{-6}$ m^2, and 1 electron per Cu atom. Find: v_d.] Equation (19.3) relates I, A, and v_d and immediately suggests that we find η. It follows from Chapter 10 that η equals the number of moles of copper per cm^3 multiplied by Avogadro's number of atoms per mole—that is, the number of atoms per cubic centimeter. Copper has an atomic mass of 63.5 u and a density of 8.9 g/cm^3, and so the number of moles per cm^3 is (8.9 g/cm^3)/(63.5 g/mol). Thus η equals

$$\frac{(1 \text{ electron/atom})(6.0 \times 10^{23} \text{ atom/mol})(8.9 \text{ g/cm}^3)}{63.5 \text{ g/mol}}$$

(continued)

(continued)

or $\eta = 8.4 \times 10^{22}$ electrons/cm^3

and $\eta = 8.4 \times 10^{28}$ electrons/m^3

From Eq. (19.3)

$$v_d = \frac{I}{\eta A q_e}$$

$$v_d = \frac{1\ \text{A}}{(8.4 \times 10^{28}\ \text{electrons/m}^3)(1.0 \times 10^{-6}\ \text{m}^2)(1.6 \times 10^{-19}\ \text{C})}$$

and so

$$\boxed{v_d = 0.7 \times 10^{-4}\ \text{m/s}} \approx 0.1\ \text{mm/s}$$

Notice that had we used a current of 10 A, the speed would have been 1 mm/s, so this should be considered only an order-of-magnitude result.

▶ **Quick Check:** The size of our computed η compares well with those given in Table 10.2, p. 324. This relatively low-drift speed is much like a wind powered by a pressure difference: the molecules are randomly zigzagging around at about 1 km/s, while the organized breeze is much slower.

Electrons at room temperature make progress down a wire very slowly, typically at a mere 1 mm/s or so. How, then, can the telephone manage to transmit electrical signals along its clutter of wires at close to the speed of light? The answer is simple: the electron we push on in New York is not the one that tickles the phone in San Francisco. The starting electron might take 16 minutes to travel the first meter of the journey; it may not even be out the door before the message is over! The situation is like a long pipe filled with water. You push inward at this end, and a pulse of water carrying energy spurts out the other end. A disturbance of the water rapidly propagates down the pipe even though any given sample of liquid hardly moves. It's the electric field that can be thought of as traveling down the wire at near the speed of light, carrying the signal, setting the electrons in motion before it. When we buy electrical energy, we don't buy electrons, there are plenty of those in the wires we already have—we buy additional amounts of electron motion. *The power company pushes around our electrons and sends us a bill for how much work they did in the process.*

Georg Simon Ohm (1787–1854).

RESISTANCE

There is a great diversity in the ability of materials to conduct electricity, and Georg Simon Ohm set himself the task of finding order in the seeming chaos. His experiments had to be modest in scale. Ohm was a high school teacher in Cologne and never far from poverty. He had been inspired to the particular challenge by the recent publications of Fourier, whose work established that the rate of flow of heat along a conducting rod was proportional to the temperature difference between its ends. Ohm wondered whether the rate of flow of charge along a conducting rod was likewise proportional to the voltage difference between its ends. Remember that in the steady state there is a uniform E-field along the current-carrying rod and so $V_B - V_A = \pm Ed$. As we move a distance d from point A to point B in the direction of $\mathbf{E}$, that is, in the direction of I in Fig. 19.15, there will be a voltage drop, $V = -Ed$. If we attach a length L of almost any kind of hookup wire across a battery, there will be a total voltage drop of $-EL$ along the wire and that, in turn, is determined by the voltage the battery sustains across its terminals.

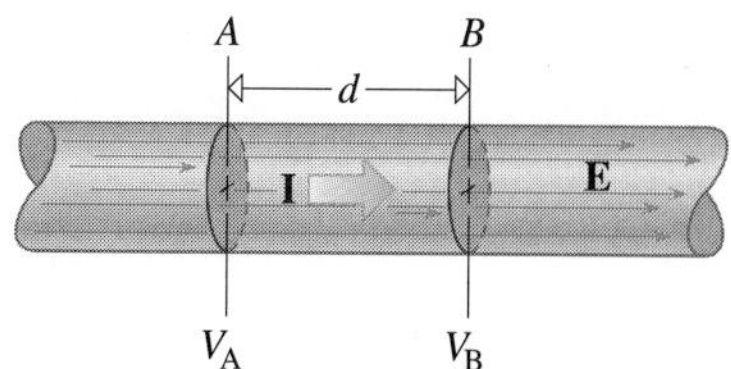

Figure 19.15 Moving along the uniform E-field from A to B a distance d, there is a voltage drop $V = -Ed$.

Suppose we take a sample of metal wire and attach it successively to the terminals of different batteries, thereby applying different known voltages across it. In each case, we could measure the resulting current passing through the sample with a device known as an *ammeter* (Ohm used a kind of magnetic torsion balance to accomplish the same thing). What we would find is that the current the battery could force through a specimen depends linearly on the applied voltage, $I \propto V$; doubling the voltage doubles the current. And this relationship is true for a variety of different conducting materials—gold, copper, brass, etc.—though each behaves in a characteristic way. Ohm suggested that every sample manifested a **resistance** (R) to the flow of charge. The greater the resistance (symbolized diagrammatically by -\/\/\/-), the less current any battery could push through it. The current in the circuit, the current through the sample, varies directly with the applied V and inversely with R. In 1826, Ohm published his results:

$$I = \frac{V}{R} \tag{19.4}$$

or

$$V = IR \tag{19.5}$$

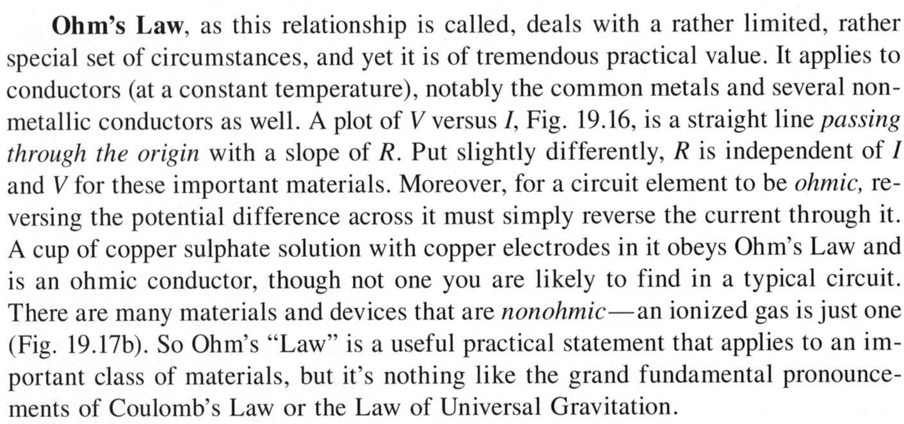

Ohm's Law, as this relationship is called, deals with a rather limited, rather special set of circumstances, and yet it is of tremendous practical value. It applies to conductors (at a constant temperature), notably the common metals and several nonmetallic conductors as well. A plot of V versus I, Fig. 19.16, is a straight line *passing through the origin* with a slope of R. Put slightly differently, R is independent of I and V for these important materials. Moreover, for a circuit element to be *ohmic,* reversing the potential difference across it must simply reverse the current through it. A cup of copper sulphate solution with copper electrodes in it obeys Ohm's Law and is an ohmic conductor, though not one you are likely to find in a typical circuit. There are many materials and devices that are *nonohmic*—an ionized gas is just one (Fig. 19.17b). So Ohm's "Law" is a useful practical statement that applies to an important class of materials, but it's nothing like the grand fundamental pronouncements of Coulomb's Law or the Law of Universal Gravitation.

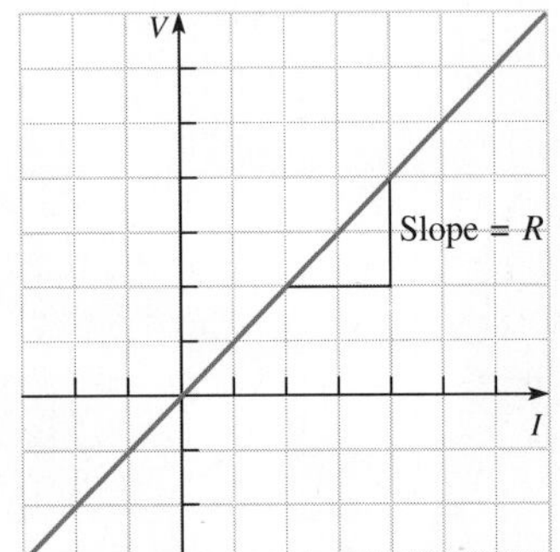

Figure 19.16 An ohmic circuit element has a linear V versus I curve. The slope of the line equals R.

To honor Ohm—recognition for his work was depressingly late in coming—the unit of resistance was named the *ohm* and symbolized by the Greek capital letter omega, Ω. Ohm's Law serves to define resistance as $R = V/I$, and so 1 ohm = 1 volt per ampere.

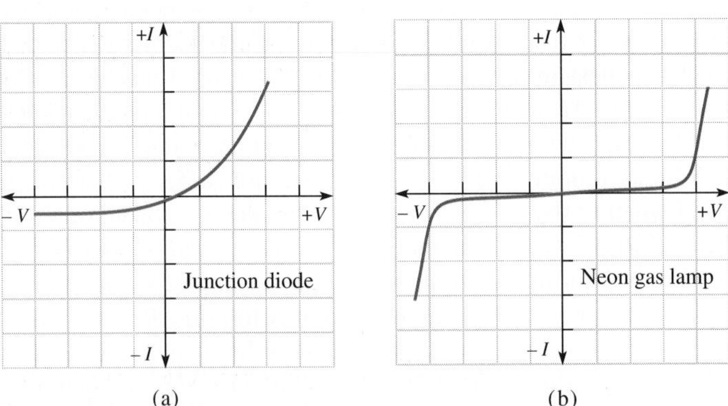

Figure 19.17 Many electrical devices, like the two shown here, are nonohmic.

Example 19.5 A small ohmic light bulb is placed in series with two D-cells, as shown in Fig. 19.18. The ammeter in series with the bulb reads the current in the circuit (0.50 A) without introducing any appreciable voltage drop across its own terminals. The voltmeter attached to the terminals of the bulb reads the voltage across it (3.0 V) without introducing any appreciable change in the current through the bulb. What is the resistance of the bulb?

Solution: [Given: at the bulb, $V = 3.0$ V, and $I = 0.50$ A. Find: R.] Note that the total voltage across the bulb is the net voltage produced by the batteries, 1.5 V + 1.5 V. From Ohm's Law

$$R = \frac{V}{I} = \frac{3.0\text{ V}}{0.50\text{ A}} = \boxed{6.0\ \Omega}$$

▶ **Quick Check:** $V = IR = (0.50\text{ A})(6.0\ \Omega) = 3.0$ V. The low resistance draws a sizable current even at a low voltage.

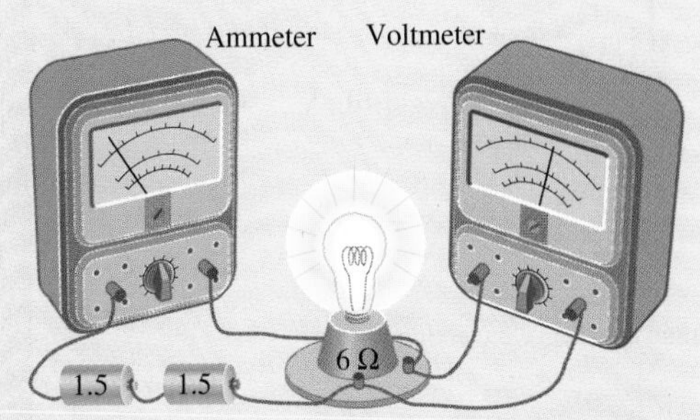

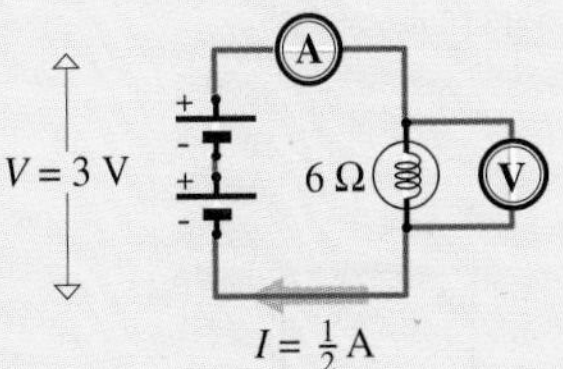

Figure 19.18 A circuit consisting of two 1.5-V cells and a lamp, all in series. The ammeter -Ⓐ- is placed in the arm of the circuit through which the desired current passes. The voltmeter -Ⓥ- is placed across the two points where a potential difference is to be determined.

All kinds of electrical devices, including the very wires that connect them, have resistance. Even so, there are specific circuit elements called **resistors** whose primary function is to introduce a certain known resistance and in so doing, control the currents and voltages in a circuit. They range from fractions of an ohm to millions of ohms (megohms, MΩ). Used in almost all electronic devices from radios to computers, they regulate the flow of charge (Fig. 19.19).

Example 19.6 Suppose someone falling out of a tree grabs an overhead power line. The wire has a resistance of 60 microhms per meter and is carrying a dc current of 1000 amps. With hands a meter apart, what is the voltage across him? Will the unfortunate soul get much of a shock?

Solution: [Given: $R/L = 60\ \mu\Omega/\text{m}$, $I = 1000$ A, and $d = 1.0$ m. Find: V.] The voltage drop across 1.0 m of the line is

$$V = IR = (1000\text{ A})(60\ \mu\Omega/\text{m} \times 1.0\text{ m})$$

hence

$$V = 0.060\text{ V} = \boxed{60\text{ mV}}$$

You know you can handle the terminals of a 1.5-V dry cell without feeling any electricity, so 60 mV is much too small a voltage to push a detectable amount of current through a human body (see Sect. 23.6).

▶ **Quick Check:** $V = IR = (10^3\text{ A})(6 \times 10^{-5}\ \Omega) = 6 \times 10^{-2}$ V.

19.3 Resistivity

The resistance of a piece of wire, or anything else for that matter, is specific to that sample—it tells us nothing about the material that the wire is made of. Alternatively, it would be nice to have some measure of how each kind of material behaves independent of geometry. We did the same sort of thing when we expressed the

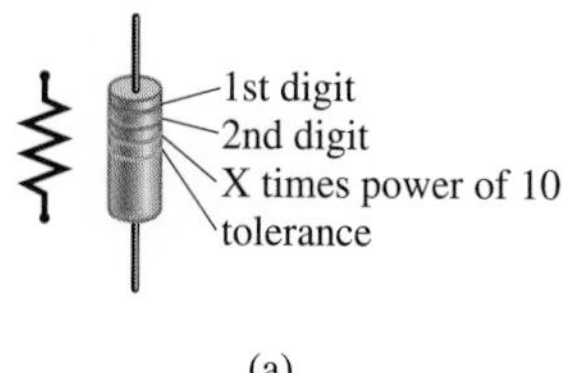

(a)

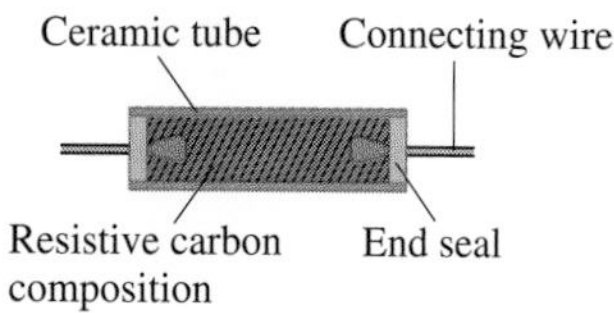

(b)

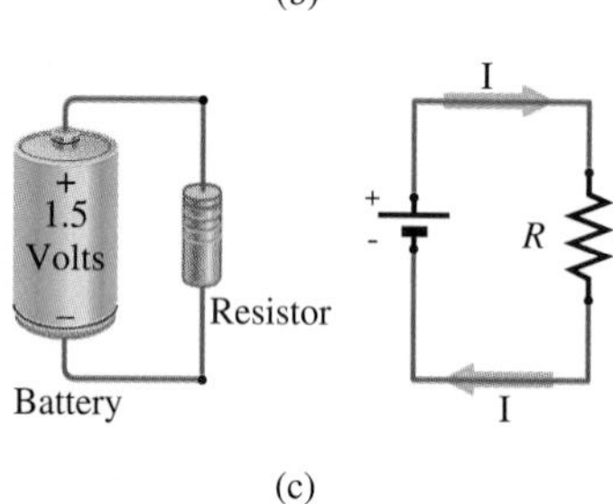

(c)

Color Code	
Black	0
Brown	1
Red	2
Orange	3
Yellow	4
Green	5
Blue	6
Violet	7
Gray	8
White	9
Gold	± 5%
Silver	± 10%

Figure 19.19 (a) Although resistors come in many forms, the most common is the little striped brown cylinder. These are carbon composition resistors, as shown in (b). The stripes are a color code indicating the resistance. (c) shows the schematic representation.

A close-up of the circuitry in the radio pictured on p. 589. The gray cylinder is a 1.5-kΩ wire-wound resistor. Such resistors are made by wrapping wire of some exotic alloy such as nichrome, manganin, or constantan around an insulating core. They are more precise and have better temperature stability than the less expensive carbon resistors.

spring constant (which is specific to each spring) in terms of Young's Modulus (p. 343), which is characteristic of the material making up the spring. Thus, we follow Ohm's lead, and for the very simplest geometry, we can guess at what he found experimentally. How does the resistance of a rod vary with its shape and composition? The analogy with water flowing through pipes suggests that the resistance is directly proportional to the length of the conductor (L)—the longer the rod, the more scattering the electrons will experience in traversing it. Similarly, the narrower the pipe, the more the resistance, and so we can anticipate that electrical resistance will also vary inversely with the cross-sectional area, $R \propto 1/A$. From Eq. (19.3), $I \propto Av_{\mathrm{d}}$, where the drift velocity should depend on E, and E in turn depends on V. Hence, $I \propto AV$, and so from Ohm's Law, $R \propto 1/A$. In any event, Ohm found experimentally that $R \propto L/A$ and introduced a material-dependent constant of proportionality ρ, the **resistivity**. To make the statement an equation:

$$R = \rho \frac{L}{A} \tag{19.6}$$

A long extension cord (large L) should have heavy-gauge wire (large A) to keep R small. This result is similar to Eq. (15.6) which describes the heat current (Q/t) driven by a temperature difference (ΔT): the ratio of driving influence to resulting thermal current is proportional to the sample's length over its cross-sectional area.

Table 19.2 lists the resistivities, in units of ohm-meters (Ω·m), for a number of important materials. Substances with resistivities of less than about 10^{-5} Ω·m, such as silver and copper, are called **conductors**. **Insulators** such as glass, rubber, and

TABLE 19.2 Resistivities*

Substance	Resistivity (ρ) (in $\Omega\cdot$m)
Aluminum	2.8×10^{-8}
Brass	$\approx 8 \times 10^{-8}$
Constantan (60% Cu, 40% Ni)	$\approx 44 \times 10^{-8}$
Copper	1.7×10^{-8}
Iron	$\approx 10 \times 10^{-8}$
Manganin ($\approx$84% Cu, $\approx$12% Mn, $\approx$4% Ni)	44×10^{-8}
Mercury	96×10^{-8}
Nichrome ($\approx$59% Ni, $\approx$23% Cu, $\approx$16% Cr)	100×10^{-8}
Platinum	10×10^{-8}
Silver	1.6×10^{-8}
Tungsten	5.5×10^{-8}
Carbon	3.5×10^{-5}
Germanium	0.46
Silicon	100–1000
Glass	10^{10}–10^{14}
Neoprene	10^{9}
Polyethylene	10^{8}–10^{9}
Polystyrene	10^{7}–10^{11}
Porcelain	10^{10}–10^{12}
Teflon	10^{14}
Sodium chloride (saturated solution)	0.044
Blood	1.5
Fat	25

*Values determined at or near 20°C.

teflon typically have resistivities greater than about $10^5\ \Omega\cdot$m. Between 10^{-5} and $10^5\ \Omega\cdot$m are the so-called **semiconductors**, like silicon and germanium.

The plastic polyacetylene is ordinarily a semiconductor, but when doped with iodine, it becomes a conductor. (A material is said to be *doped* when small amounts of a foreign substance are introduced into it.) An iodine atom removes an electron from a carbon atom in the polymer chain, leaving behind a "hole" that behaves like a positively charged particle. The holes advance in the direction of the E-field as if they were a flow of positive charge, a current. This metallic-looking plastic is, ounce-for-ounce, twice as conductive as copper and, though still in the developmental stage, it promises a new age of inexpensive plastic electronic devices.

Example 19.7 A length of nichrome ribbon with a rectangular cross section of 0.25 mm × 1.0 mm is to be used as the heating element in a toaster. How long should it be if it's to have a total resistance of 1.5 Ω at room temperature?

Solution: [Given: cross section 0.25 mm × 1.0 mm, nichrome ribbon; $R = 1.5\ \Omega$. Find: L.] The resistance, cross section, length, and type of wire are related via Eq. (19.6). The cross-sectional area is

$$A = (0.25 \times 10^{-3})(1.0 \times 10^{-3}) = 0.25 \times 10^{-6}\ \text{m}^2$$

Using Table 19.2, we have

$$L = \frac{RA}{\rho} = \frac{(1.5\ \Omega)(0.25 \times 10^{-6}\ \text{m}^2)}{100 \times 10^{-8}\ \Omega\cdot\text{m}} = \boxed{0.38\ \text{m}}$$

▶ **Quick Check:** The resistance of this wire per meter ($L = 1$ m) is $\rho l/A = 4\ \Omega$/m. Hence, $1.5/4 = L/1$, and $L = 0.38$ m.

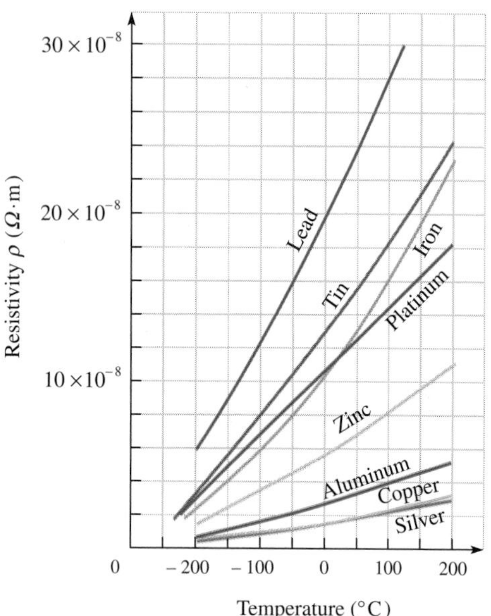

Figure 19.20 The resistivities of several metals over a range of temperatures. All rise with T and all are nearly linear.

The Temperature Dependence of Resistivity

When the temperature of a conductor increases, the corresponding increase in the random vibrations of its atoms and ions increases the scattering of electrons, impeding their progress and elevating the resistivity of the material. For example, during the fraction of a second while the tungsten filament in a light bulb rises roughly 2000°C as it becomes incandescent, its resistance increases by a factor of about 10. Experiments show that ρ usually varies *almost* linearly with modest changes in temperature (ΔT), as shown in Fig. 19.20. Accordingly, we write an expression for ρ at any temperature relating it to a known value (ρ_0) at some reference temperature:

$$\rho \approx \rho_0(1 + \alpha_0 \Delta T) \qquad (19.7)$$

where values of ρ_0 appear in Table 19.2. Values of α_0, the *temperature coefficient of resistivity,* are given in Table 19.3, and ΔT is the difference in temperature (either in °C or K) from the reference value. Because α_0 also varies somewhat with temperature, it has been subscripted to indicate that it, too, must be correlated to the reference temperature. For precise work, there are handbooks of physical data that supply values of both ρ_0 and α_0 at different temperatures. The approximation sign in Eq. (19.7) is a reminder that the formula holds best in the vicinity of the reference temperature at which α_0 and ρ_0 are given. Notice that for most pure metals $\alpha_0 \approx 1/273\ \mathrm{K}^{-1}$ (see Problem 69).

The temperature coefficients of the semiconductors in Table 19.3 are negative; they become less resistive when the temperature increases. As with conductors, in-

TABLE 19.3 Temperature coefficients of resistivity*

Substance	α_0 (K^{-1})
Aluminum	0.003 9
Brass	0.002
Constantan (60% Cu, 40% Ni)	0.000 002
Copper	0.003 93
Iron	0.005 0
Manganin (≈84% Cu, ≈12% Mn, ≈4% Ni)	0.000 000
Mercury	0.000 89
Nichrome (≈59% Ni, ≈23% Cu, ≈16% Cr)	0.000 4
Platinum	0.003 927
Silver	0.003 8
Tin	0.004 2
Tungsten	0.004 5
Carbon	−0.000 5
Germanium	−0.05
Silicon	−0.075
Sodium chloride (saturated solution)	−0.005

*Values determined at or near 20°C.

The tungsten helical filament of an ordinary incandescent light bulb. It has a high resistance and becomes white hot at 110 V.

creasing T increases the scattering of charge carriers. The density of charge carriers in a semiconductor increases strongly with T—carriers that were initially not free to move can be set loose after absorbing thermal energy. The resulting increase in η has the effect of increasing I for a given V and, hence, decreasing R, via Equations (19.3) and (19.4).

Example 19.8 One of the most useful devices for measuring temperature is the platinum-resistance thermometer. Typically, about 2.0 m of pure platinum wire 0.1 mm in diameter is formed into a coil with a resistance at 0°C of 25.5 Ω. Taking the temperature coefficient of resistivity to be 0.003 927 K^{-1}, determine the change in resistance corresponding to a 1.00°C change in temperature. At what temperature is the thermometer if its resistance is 35.5 Ω?

Solution: [Given: $R_0 = 25.5\ \Omega$, and $\alpha_0 = 0.003\,927\ K^{-1}$. Find: ΔR corresponding to $\Delta T = 1.00°C$, and T corresponding to $R = 35.5\ \Omega$.] Since

$$\rho \approx \rho_0(1 + \alpha_0 \Delta T) \qquad [19.7]$$

it follows from Eq. (19.6) that

$$R \approx R_0(1 + \alpha_0 \Delta T) \qquad (19.8)$$

We want the change in resistance $(R - R_0)$ that occurs when $\Delta T = \pm 1$, which can either be in °C or K; thus

$$R - R_0 = R_0\alpha_0 \Delta T = (25.5\ \Omega)(0.003\,927\ K^{-1})(1.00\ K)$$

and

$$R - R_0 = 0.100\ \Omega$$

The thermometer changes resistance by $\boxed{0.100\ \Omega \text{ per } 1\ K}$ or 1°C; hence, a change of +10.0 Ω from its 0°C reading means a temperature of $\boxed{+100°C}$.

▶ **Quick Check:** For $\Delta T = 100$ K, $R - R_0 = (25.5\ \Omega)(0.003\,927\ K^{-1})(100\ K) = 10.0\ \Omega$.

19.4 Superconductivity

Resistance, although helpful in controlling currents, is the bane of electrical technology; it limits the operation of almost everything. If we could get rid of it, we could put nuclear power plants safely away from populated areas, cut the cost of electricity, float cars frictionlessly on magnetic fields, make tiny powerful motors, and improve computers. We could revolutionize the entire technology. Today, the goal of eliminating resistance seems just over the horizon in the realm of superconductivity.

Soon after H. Kamerlingh Onnes succeeded in liquifying helium in 1908, he began a study of the temperature dependence of the dc resistance of metals. The use of liquid helium as a refrigerant allowed him to routinely operate down to about 1 K (−458°F). Onnes, working first with platinum, found that its resistance dropped as the temperature descended, though it leveled off at a fairly constant value below roughly 4 K. He next selected mercury to examine because it could be obtained at ultrahigh purity and he wrongly believed that the resistance of any pure metal would vanish as it approached 0 K—something certainly suggested by Fig. 19.20. The resistance of mercury did slowly decrease with T, but at 4.2 K it inexplicably plunged to an unmeasurably small value. "Mercury has passed into a new state," wrote Onnes, "which on account of its extraordinary electrical properties may be called the superconducting state."

This total absence of dc resistivity below a **critical temperature** (T_c) is known as **superconductivity**. Onnes was lucky in selecting mercury: only 27 elements become superconducting under ordinary pressure, and many of those do so at critical temperatures well below 4.2 K (see Table 19.4). Incidentally, platinum does not make the transition to superconductivity, nor do the other good conductors, copper, silver, and gold. Still, well over a thousand alloys and compounds undergo this remarkable transformation.

TABLE 19.4 Critical temperature of some superconducting elements

Element	T_c (K)
Aluminum	1.175
Beryllium	0.026
Cadmium	0.52
Gallium	1.083 3
Indium	3.405
Lead	7.23
Mercury (α)	4.154
Molybdenum	0.916
Niobium	9.25
Osmium	0.655
Protactinium	1.4
Tantalum	4.47
Tin	3.721
Titanium	0.39
Tungsten	0.015 4

A computer model of a high-temperature, superconducting compound. The atoms are yttrium (gray), barium (green), copper (blue), and oxygen (red).

Although a substance in its normal state has resistance and so must have an emf across it and an E-field within it to sustain a current, no such electric field exists in a superconductor. A current once initiated will continue on its own in a closed superconducting loop, perhaps indefinitely. One such experiment ran for over $2\frac{1}{2}$ years with no observed diminution in the circulating supercurrent. By contrast, a current circulating in a normal resistive material without any driving force would last for less than a second. Careful measurements indicate that the decay time for a supercurrent is at least 10^5 years, implying that if there is any resistance at all in the superconducting state, it's at least 10^{-12} times that of the normal state.

The basic understanding of low-temperature superconductivity is known as the BCS Theory after its Nobel-laureate (1972) creators John Bardeen, Leon Cooper, and Robert Schrieffer, who proposed it in 1957. Theirs is a quantum mechanical formulation maintaining that, unlike a conductor in the normal state where electrons behave independently, the electrons in a superconductor are paired. There is a long-range attractive interaction between paired electrons that is mediated by the electrons' interaction with the vibrating lattice. In effect, the densely packed electrons are all linked together in a kind of overlapping web of two-by-two interactions.

The situation is like a dance hall strewn with tables and chairs and yet crowded with people dancing independently—that's the normal state of a conductor. To make the transition to the superconducting state, partners are designated and each couple is provided a long scarf. The two people in every pair go on dancing far apart, but each holds an opposite end of their scarf. While bound together in pairs with an attractive scarf-interaction, every dancer repels all the other people on the floor. If the group is now impelled to drift in one direction—each person being pushed somehow (for example, by tilting the floor)—no single dancer can collide with an obstacle in the room and bounce away from the collective flow. The group acts as a coherent unit restraining all the participants to move together. Likewise, no single electron can be scattered by some imperfection from the collective motion of a supercurrent—no scattering, no resistance.

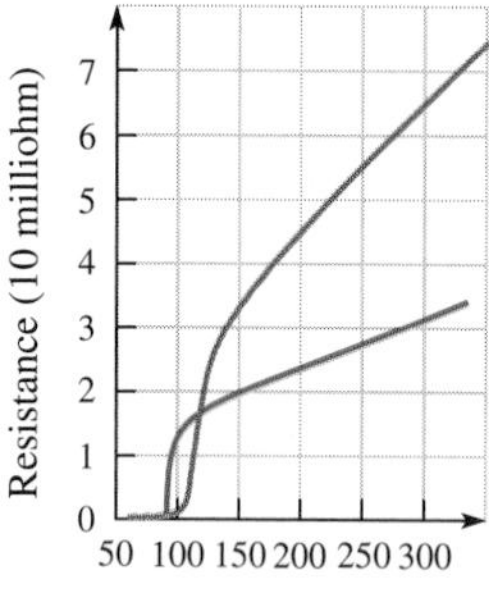

Figure 19.21 Resistance versus temperature of a thallium compound (blue) and a europium compound (red). Both become superconducting below around 100 K. These are typical of high-temperature superconductors. (Adapted from *Physics Today*, April 1988.)

A breakthrough occurred in 1986, when it was discovered that an exotic ceramic compound of barium, lanthanum, copper, and oxygen had an unprecedentedly high critical temperature of 35 K. The race was on. By the beginning of 1987, physicists had prepared a ceramic (substituting yttrium for lanthanum) for which T_c was 98 K (-283°F), well above the temperature of liquid nitrogen (77 K). And by early 1988, a thallium compound (Tl-Ca-Ba-Cu-O) with a rather balmy critical temperature of 125 K had been produced (Fig. 19.21). Though still far from room temperature, values of T_c continue to creep upward and a variety of new high-temperature superconducting devices now exists.

19.5 Voltage Drops and Rises

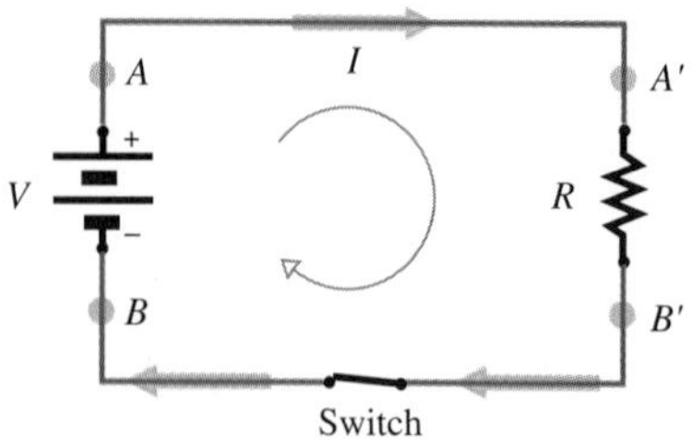

Figure 19.22 Points A and A', and B and B', are at the same potential.

Figure 19.22 depicts an idealized battery (having no internal resistance of its own) connected to a resistor by two ideal zero-resistance leads (generally, heavy-gauge copper wire will do nicely, but you can imagine these to be superconductors). The voltage across the battery (V) is its emf, and since the hookup wires have no resistance, there is no drop in potential along them—points A and A' are at the same voltage, as are points B and B'. In this simple circuit, the voltage across the terminals of the battery equals its emf, equals the voltage across the resistor.

In the steady state, charges introduced from the battery into that idealized wire coast along freely ($E = 0$) until they encounter the resistor. If we suppose that the mobile charge carriers are positive, they effortlessly traverse the ideal wire and arrive at the resistor (at point A'). Impeded in its progress, positive charge accumulates in the vicinity of that point, repelling positive charges out and away from the other end of the resistor, which becomes negatively charged. An E-field now exists within the resistor that has an axial component along its length. With the potential across R equal to its maximum value, namely that of the battery (V), the E-field drives a steady-state current I through the resistor. From point B' the carriers coast at a constant potential back to the battery. Thus, there are surface charges distributed around the circuit with concentrations $(+)$ at A and A', and $(-)$ at B and B'.

Note that I leaves the high potential (positive) side of the battery and enters the high potential (positive) side of the resistor. In passing through the resistor, the mobile charge carriers are scattered somewhat, imparting random KE (that is, thermal energy) to the atoms of the resistive medium. As a result, energy is transferred from the current to the resistor—each positive charge q drops in potential V and thus in potential energy qV, on traversing the resistor. The carriers return to the negative terminal of the battery, which raises their potential by V, thus pumping them up in energy qV, and sending them out again. Clearly, the battery supplies electrical energy, and the resistor dissipates that exact amount as thermal energy (in accord with the Law of Conservation of Energy). Current will circulate until the battery supplies all the energy it can and inevitably runs down. **The current is the transporter of energy**.

No point in the circuit is grounded, and so no point has a potential referenced to the zero of ground. The circuit is said to *float*, and what is important is the voltage of one point with respect to another: **voltage differences are the practical quantities**.

Figure 19.23 depicts a variable resistor, or *potentiometer*, a pot for short. This device is usually a long coil of metal wire (its end points being A and C) with a central slide pressing against it, making contact with the wire (at a variable point B). When placed across an ideal battery, the emf (V) equals V_{AC}, and V_{AB} is any desired fraction thereof. This sort of *voltage divider* is very useful. When you turn the knob on the volume control of a radio or stereo, you're moving the slide on just such a pot.

(a)

(b)

Figure 19.23 A potentiometer is a variable resistor. It is commonly found in radios, stereos, amplifiers, and tape decks.

19.6 Energy and Power

A current in a circuit is like a moving fluid capable of transporting energy from the source of emf (a battery, generator, solar cell, etc.) to some device (a toaster or a TV), where it can subsequently be utilized. The transport of energy by a current is one of the fundamental features of electricity, one that we must quantify.

Suppose that a small amount of charge Δq traverses a circuit element and moves through a constant potential difference of V. It will, in doing so, change its electrical potential energy (ΔPE_E) by an amount $\Delta q\, V$. The rate at which this happens, the rate

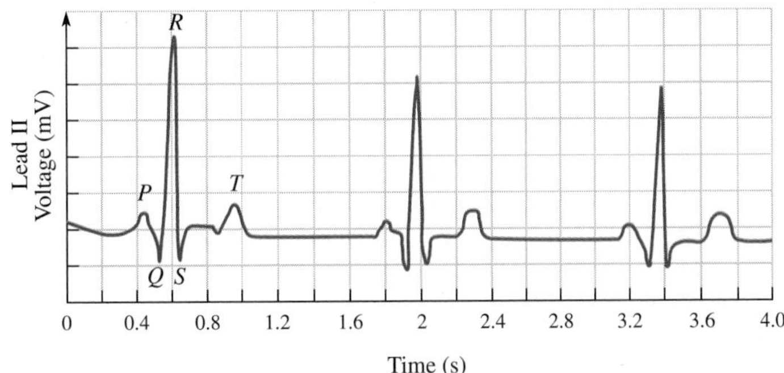

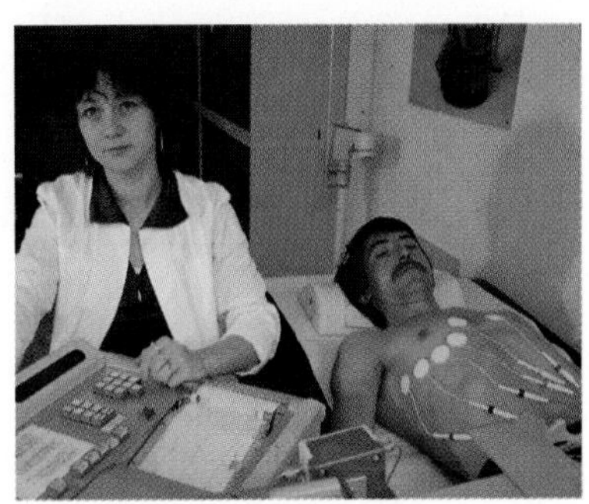

The electrocardiograph records potential differences at the body's surface due to the electrical activity of the heart. The voltage shown here is between the left leg and the right arm. The region labeled P arises when the atria contract. The contraction of the ventricles corresponds to QRS, and T marks the preparation of the ventricles for the next contraction.

of transfer of energy, is

$$\frac{\Delta \mathrm{PE_E}}{\Delta t} = \frac{\Delta q}{\Delta t} V$$

which, by definition (p. 286) is the **power**, P (either delivered or consumed), over a time Δt. Since $\Delta q/\Delta t$ is the current, the power at any instant is

$$\mathrm{P} = IV \tag{19.9}$$

The unit of electical power is an ampere·volt where

$$1 \text{ A·V} = (1 \text{ C/s})(1 \text{ J/C}) = 1 \text{ J/s} = 1 \text{ W}$$

In Fig. 19.22, the power delivered by the battery to the current I in raising its potential V is IV. Similarly, the power delivered thermally to the resistor by the current I while dropping in potential by V is also IV. If several batteries are connected in a circuit, it is possible that a current may *enter* one of them at its positive terminal. Such a current would emerge after a drop in potential—that is, after depositing energy in that battery. It is by supplying energy to a battery in this fashion that it can be recharged.

Using Ohm's Law, $V = IR$, **the power dissipated in a resistance** R can be expressed as

$$\mathrm{P} = I(IR) = I^2R \tag{19.10}$$

or

$$\mathrm{P} = \left(\frac{V}{R}\right)V = \frac{V^2}{R} \tag{19.11}$$

J. P. Joule (1841) first showed experimentally that the "heating-power" of an electric current through a resistance had the form of Eq. (19.9). To recognize his accomplishment, we speak of the thermal energy produced in this way as **joule heat**.

Example 19.9 The circuit in Fig. 19.24, containing a battery of unknown voltage, carries a current of 5.0 A. Determine the power either dissipated or provided by each circuit element.

Solution: [Given: a 12-V battery, an unknown battery, $I = 5.0$ A, and $R = 10\ \Omega$. Find: power for each.] The same 5.0-A current passes through both elements in going from A to B. Hence, the power *delivered to* (since I enters on the + side) the 12-V battery is

$$\mathrm{P} = IV = (5.0 \text{ A})(12 \text{ V}) = \boxed{60 \text{ W}}$$

It's receiving energy from the current. Resistors only dissipate power, and here that is in the amount of

$$\mathrm{P} = I^2R = (5.0 \text{ A})^2(10\ \Omega) = \boxed{0.25 \text{ kW}}$$

The net power dissipated is therefore 0.060 kW + 0.25 kW = 0.31 kW and that's provided by the second battery—current leaves its + terminal.

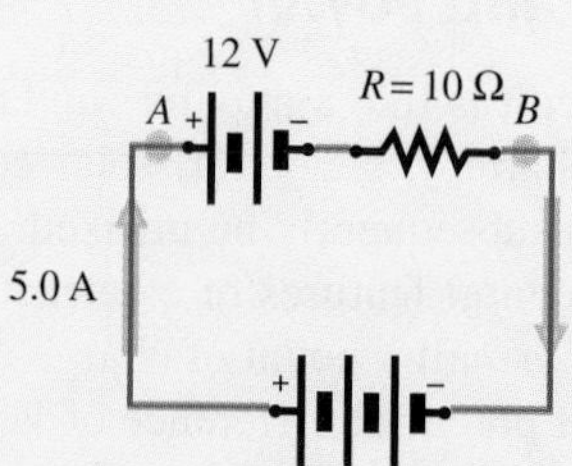

Figure 19.24 Each circuit element either provides or dissipates power.

▶ **Quick Check:** The voltage of the second battery is $V_{AB} = 12\text{ V} + IR = 12\text{ V} + 50\text{ V} = 62$ V. The power it delivers is therefore $\mathrm{P} = IV = (5.0\text{ A})(62\text{ V}) = 0.31$ kW, which is the power dissipated.

Example 19.10 Calculate the price of electrical energy in dollars per kilowatt-hour as supplied by a 50¢ D-cell having a 3.0 amp-hour rating. Compare that with the ≈10¢ per kW·h for electricity on tap from a wall outlet.

Solution: [Given: 50¢ for 3.0 amp-hours. Find: price per kW·h.] First, 1.0 kW equals 1000 J/s; hence, 1.0 kW·h = (1000 J/s)(3600 s/h), or 3.6 MJ. The cell delivers 3.0 amp-hours at 1.5 V; $IV = \mathrm{P}$, which is the power delivered, and that multiplied by the time during which it is delivered is the energy; consequently

$$\mathrm{P}t = (3.0 \text{ amp-hours})(1.5 \text{ V}) = 4.5 \text{ W·h}$$

or 4.5×10^{-3} kW·h. At 50¢ per cell, that's (50¢)/(4.5×10^{-3} kW·h) or $\boxed{\$111 \text{ per kW·h}}$.

▶ **Quick Check:** The energy of the cell is 4.5 W·h divided by the operating voltage, which yields the number of amp-hours: 3.

As a rule, for a given type of cell, the bigger it is, the higher the amp-hour rating, the more energy it stores. That's why portable devices that need a lot of energy, such as heavy-duty flashlights or large speakers, operate on D-cells rather than A-cells, even though both have the same emf of 1.5 V.

Core Material

The ordered flow of charge is called **electric current** (I):

$$I = \frac{\Delta q}{\Delta t} \qquad [19.1]$$

The SI unit of current is the *ampere:* 1 A = 1 C/s. A nonelectrostatic potential difference that sustains a current in an external circuit is called an **electromotive force**, or emf (p. 666).

The current in a wire of cross-sectional area A, in terms of the **average drift speed**, v_d, the number of carriers per unit volume, η, and the charge per carrier, q_e, is

$$I = \frac{\Delta q}{\Delta t} = \eta v_d A q_e \qquad [19.3]$$

The current passing through a resistor varies directly with the applied V and inversely with R:

$$V = IR \qquad [19.5]$$

which is **Ohm's Law**, and it applies to a limited range of materials. The unit of resistance is the *ohm* (Ω). The resistance of a wire can be expressed in terms of its length L, cross-sectional area A, and a material-dependent constant of proportionality ρ, the **resistivity**, as

$$R = \rho \frac{L}{A} \qquad [19.6]$$

Resistivity is temperature-dependent (p. 676); accordingly

$$\rho \approx \rho_0(1 + \alpha_0 \Delta T) \qquad [19.7]$$

α_0 is called the *temperature coefficient of resistivity.* The total absence of dc resistivity below some **critical temperature** (T_c) is known as **superconductivity** (p. 677).

The power either provided or dissipated by a circuit element is

$$\mathrm{P} = IV \qquad [19.9]$$

The unit of electrical power is the *watt.* The power *dissipated* in a resistance R is

$$\mathrm{P} = I(IR) = I^2R \qquad [19.10]$$

or

$$\mathrm{P} = \left(\frac{V}{R}\right)V = \frac{V^2}{R} \qquad [19.11]$$

Figure 19.25 summarizes the diagrammatic representations of the circuit elements introduced thus far.

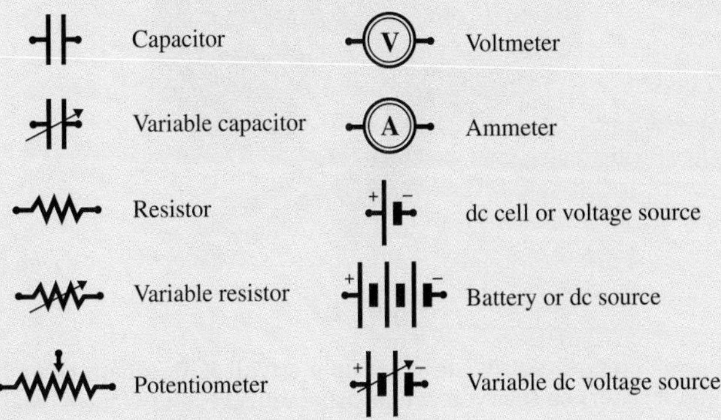

Figure 19.25

Suggestions on Problem Solving

1. An important type of derivation involves determining the current, given a flow of charge or vice versa. For example, a detector measures N particles, each carrying a charge q, arriving *per second per unit of its surface area*—what current strikes the instrument if its area is A? The basic definition is $I = \Delta q/\Delta t$, but to use this equation you must first find Δq, the amount of charge that strikes the detector during the time Δt. We are given N and so Nq is the amount of charge arriving *per second* per unit of surface area. Consequently, NqA is the total amount of charge arriving on the detector per second and is equal to I. We obtained I without finding Δq explicitly because N had the time built into it to begin with. This kind of analysis of flow is very important in physics—we've seen it before and we'll see it again.

2. Remember that all parts of an ideal hookup wire in a circuit are at the same potential. Drops or rises in potential occur only across circuit elements (resistors, batteries, etc.), and in the case of a resistor, that means only when a current is passing through it (take a look at Multiple Choice Question 9). **A steady-state current cannot exist in a length of conductor unless it forms part of a closed path**.

3. Whenever you have a circuit diagram with a battery in it, place + and − signs at the appropriate terminals. Current, being imagined as composed of positive charge, always emerges from the positive terminal (provided there are no other batteries bucking it) and re-enters at the negative one. In a simple circuit where you know the direction of the current, follow it around, labeling all the resistors with + signs where current enters and − signs where it leaves. This will show the voltage drops and rises, something that will be explored further in the Chapter 20.

4. The voltage of any point in a circuit is only known relative to some other point in that circuit. Thus, one side of a battery might be 12 V higher than the other side (though it could be 10 000 V higher than your nose). For example, the entire chassis of an automobile serves as a common conductor called "ground" in the trade, though it isn't usually grounded and so actually floats—the + terminal of the battery is 12 V higher than the engine block. Often a point in a circuit is grounded, as is done with stereo systems and computers. Its potential is then taken as zero, and all voltage drops or rises are referenced with respect to it.

Discussion Questions

1. Ohm once remarked that the laws of electricity "are so similar to those given for the propagation of heat . . . that even if there existed no other reasons, we might with perfect justice draw the conclusion that there exists an intimate connection between these natural phenomena." What was he alluding to? Discuss the nature of electrical and thermal conductivity for metals.

2. Figure Q2 shows several 9-V batteries opened up so we can look inside. Explain what you see.

Figure Q2

3. Suppose you wish to set up a circuit with a battery and some resistors so that you can measure currents and voltages and confirm the ideas of this chapter. How should you select the hookup wire, or doesn't it matter?

4. What happens when you turn the key in the ignition of your automobile? Should you start your car on a cold rainy night with the wipers, lights, heater, and defroster all on? Explain.

5. Figure Q5 shows a 1.5-V dry cell attached to a 1.5-V flashlight bulb via two single-pole double-throw knife switches. Each switch attaches the central post (or pole) to either the right or left terminal on the device—the blade is only thrown to a horizontal position. You probably have an arrangement of this kind in your house, especially if it has a long flight of steps that are lighted. What's it for? Make a simplified drawing of the circuit and discuss its operation. As shown, is the light on? What happens when either switch A or switch B is thrown?

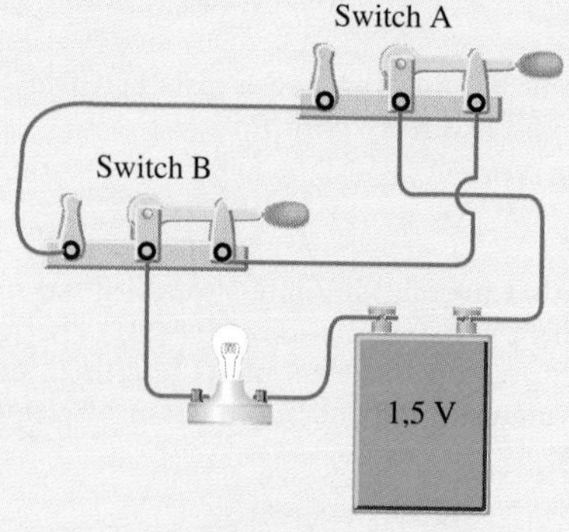

Figure Q5

6. Suppose you parked your car and left the lights on all night. Why would you have trouble starting it the next day? Would it be more bothersome to start on a really cold winter's day? After several tries, the engine just makes some dishearten-

ing clicking noises and you've had it—what has happened? A friend comes to your aid and "jump starts" your car. How is that done and what does it accomplish?

7. With Question 6 in mind, it's assumed that once the engine is running, all will return to normal. Explain. In other words, how does your engine keep running once the jumper cables are removed, given that the battery is still exhausted? Is it then a good idea to let the engine run or drive for a while without turning on any unnecessary electrical devices? Why? What's happening in the electrical system while you drive the car with the radio playing?

8. Figure Q8 shows an ordinary metal-body flashlight. Explain how it works and discuss the significance of the arrangement of the dry cells.

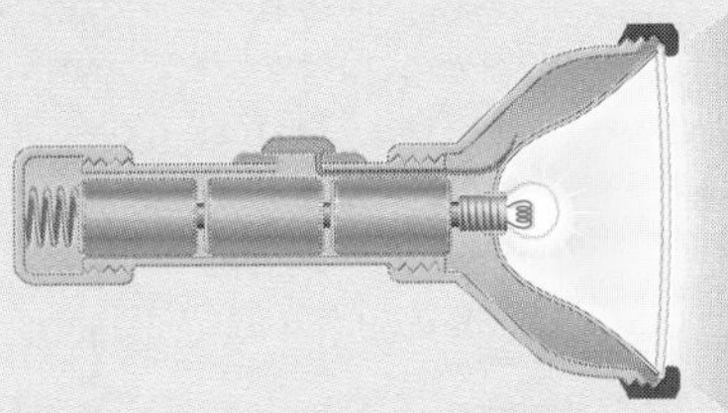

Figure Q8

9. Describe Volta's tin-tongue-silver cell. Why don't you taste anything until the two metals touch? Volta thought that the current of "electric fluid" in his pile arose from a "contact force" between the two metals. It was not until about 25 years later that the chemical action taking place was recognized (by C. A. Becquerel, A. De La Rive, G. F. Parrot, and others).

10. Figure Q10 shows a vertical arrangement of pipes with water flowing around the circuit at a rate that keeps the levels constant. This system is a liquid analogy to an electrical circuit. Draw the corresponding electrical diagram and discuss the relationship between the various parameters. Consider the system from an energy perspective. What does the pump "pump" besides water? No wonder nineteenth-century scientists liked the notion of "electric fluid." Have you ever heard a modern electrician mutter things like "turn on the juice"?

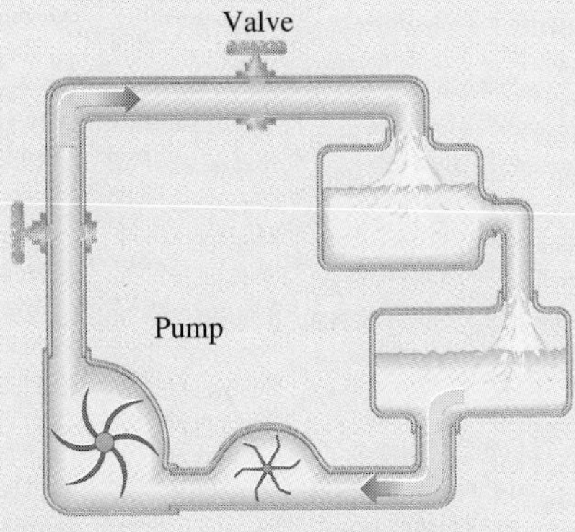

Figure Q10

11. Discuss everything of significance you know about the dry cell—that is, the ordinary flashlight "battery." Would it still work if it was actually dry? Ask a few elderly people if they remember how leaky old-fashioned D-cells were when they were kids—explain. Why do expensive electronic devices recommend that they not be stored with the batteries in them?

12. Question 10 deals with a fluid system using gravity. Figure Q12 represents a horizontal variation that is independent of gravity. Relate the liquid and electrical parameters. Considering fluid friction, talk about the influence of the narrow-bore tube.

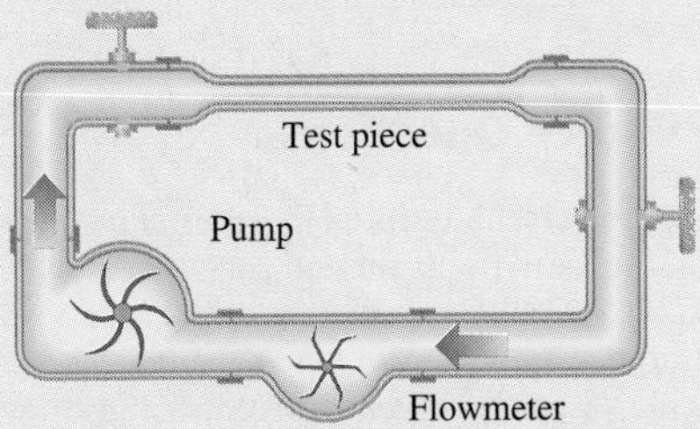

Figure Q12

13. When an ordinary tungsten light bulb burns out, it's likely to flash bright blue-white for an instant. Explain what is happening. Suggest a possible reason these lamps usually burn out soon after being turned on.

14. Suppose that you are watching a demonstration performed by your instructor using a small battery (anywhere from 4.5 V to 9 V) and a conductor with a high resistance (a small-diameter strand of steel wool will serve nicely). A single filament of steel is stretched across the terminals, and within seconds

NEWSDAY, WEDNESDAY, JULY 29, 1987

Lightning Kills 3

Three men whose bodies were found yesterday at a hilltop campsite in Darien, N.Y., about 30 miles east of Buffalo, were killed when lightning struck a metal tent pole, ripped scars in the ground and electrocuted the trio, officials said.

The three men, whose names were not released, arrived Saturday in a car registered to an owner in Vernon in Oneida Country. Two men were in one tent, supported by a 5-foot long, three-piece aluminum pole. Investigators believe lightning struck the pole and electrocuted the two men sleeping with their heads beside it. The bolt ripped a 3-inch deep path 6 feet away to the second tent, where it killed the third man.

Figure Q15

a region of it glows red hot and bursts into a shower of molten beads. (*Don't try this—it's dangerous!*) Why did the wire melt? I once saw someone accidentally drop a wrench across the terminals of a 12-V car battery. There was a tremendous sparking flash, and molten beads of metal flew all over the place. Explain. Why did the wrench only melt where it touched the battery?

15. When lightning hits the ground, it often spreads out radially as it penetrates downward. Thus, it is possible for a large four-legged animal standing with its bodyline radial with the bolt to be electrocuted. Explain. Considering ground currents, what is the best posture for a human out in a thunderstorm—lying down, squatting, or standing up? Read the newspaper article on p. 683 (Fig. Q15).

16. High-voltage cables are generally uninsulated, and a bird standing on one of them may have its feathers puffed up. Why? Why aren't our little feathered friends electrocuted on their perches? Could a chicken walk safely on the third rail of an electrified railroad?

Multiple Choice Questions

1. If 10 amperes circulate in a closed circuit, how much charge passes any point therein in 2 s? (a) 10 C (b) 5 C (c) 20 C (d) 200 C (e) none of these.
2. The difference in potential between the electrodes of a voltaic cell when there is no current being drawn is (a) zero (b) 1.5 V (c) its emf (d) the power (e) none of these.
3. The emf of a voltaic cell (a) is independent of the chemical interactions taking place within it (b) is dependent on the size of the cell (c) is dependent on its amp-hour rating (d) is independent of the plate size (e) none of these.
4. Usually, the larger a voltaic cell is, the more (a) voltage it can supply (b) current it can supply (c) potential it can develop (d) emf it can sustain (e) none of these.
5. The moving nonconducting belt of a Van de Graaff generator carries 10 μC of charge up to the metal sphere (Fig. Q14, Chapter 17) each second. The steady-state potential difference between the sphere and the grounded source of charge is 3×10^6 V. What, if any, is the emf of the generator? (a) 10 μV (b) 3 μV (c) 3 MV (d) 30 μV (e) none of these.
6. A statue of a chicken made of pure gold has a resistance of 0.10 mΩ between the beak and the tail. An exact duplicate of the piece is made in an alloy that has a resistivity 10 times greater than gold. The resistance between the same two points will now be (a) 0.10 mΩ (b) 1.00 mΩ (c) 0.01 mΩ (d) 10.0 mΩ (e) none of these.
7. A metal wire has a resistance of 1.0 Ω. What will be the resistance of a wire made of the same material but twice as long and with half the cross-sectional area? (a) 0.40 Ω (b) 2.00 Ω (c) 0.02 Ω (d) 40.0 Ω (e) none of these.
8. The resistance of a superconducting material drops, essentially to zero, rather suddenly when (a) the sample is heated above T_c (b) the sample is exposed to a magnetic field (c) the sample is cooled below T_c (d) the sample experiences a current I_c (e) none of these.
9. Referring to Fig. MC9, the voltage at point B is (a) +12 V (b) −12 V (c) 0 V (d) −6 V (e) none of these.
10. Referring to Fig. MC9, the voltage at point D is (a) +12 V (b) −12 V (c) 0 V (d) −6 V (e) none of these.
11. Referring to Fig. MC9, the voltage at point C is (a) +12 V (b) −12 V (c) 0 V (d) −6 V (e) none of these.

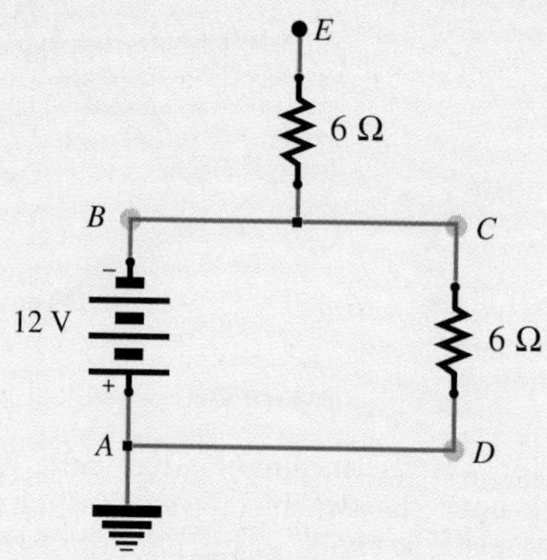

Figure MC9

12. Referring to Fig. MC9, the voltage at point E is (a) +12 V (b) −12 V (c) 0 V (d) −6 V (e) none of these.
13. An 8-Ω speaker is connected across the output terminals of a power amplifier that delivers 64 W to it. The current supplied to the speaker is (a) 2.83 A (b) 8.00 A (c) 2.00 A (d) 64.0 A (e) none of these.
14. Figure MC14 shows the voltage across a tungsten light bulb filament plotted against the current through it. The curve bends upward because (a) the filament is getting hot and running out of electricity (b) it takes more voltage to

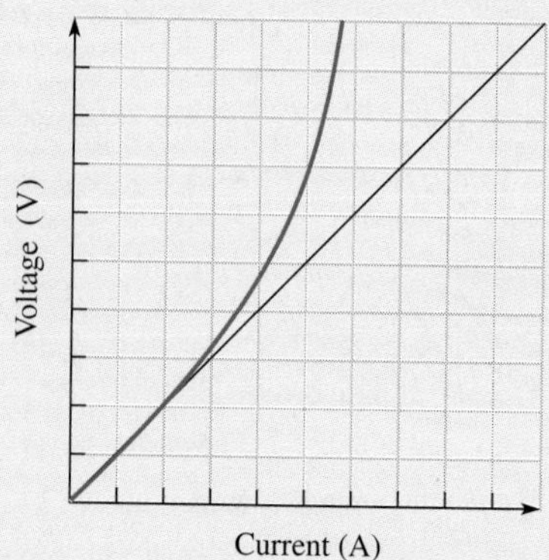

Figure MC14

propel the same current than it would if the device were ohmic because the resistance increases (c) the resistance goes down, so we get more voltage for a given current (d) the resistance stays constant, but the current decreases (e) none of these.

15. A copper wire has a resistance of 10 Ω. What will be its new resistance if the wire is shortened by cutting it in half? (a) 20 Ω (b) 10 Ω (c) 5 Ω (d) 1 Ω (e) none of these.
16. When the temperature of a length of aluminum wire is lowered, its resistance (a) increases slightly (b) decreases correspondingly (c) stays the same (d) increases as ΔT (e) none of these.
17. If a potential difference of 12 V across a resistor results in a current of $\frac{1}{2}$ A, how much power is dissipated? (a) 48 W (b) 12 W (c) 4 W (d) 6 W (e) none of these.
18. Imagine a length of ordinary insulated hookup wire. The wire's resistance is not dependent on (a) the conductor's length (b) the conductor's radius (c) the material making up the insulator (d) the material making up the conductor (e) none of these.
19. The units of $\Omega \cdot A^2$ correspond to (a) current (b) energy (c) power (d) voltage (e) none of these.
20. One microvolt (1 μV) corresponds to (a) 10^6 V (b) 10^{-6} V (c) 1000 V (d) 1/1000 V (e) none of these.
21. A resistor operated at 100 V generates joule heat at a rate of 20 W. When placed across a 50-V source, it will draw (a) 0.1 A (b) 5 A (c) 20 A (d) 4 A (e) none of these.

Problems

CURRENTS

1. [I] The moving nonconducting belt of a Van de Graaff generator carries 10 μC of charge up to the metal sphere (Fig. Q14, Chapter 17) each second. Determine the corresponding current.

2. [I] A beam of positrons carries 1.4 C past a point in space in 2.0 s. What current does that correspond to?

3. [I] If a quantity of singly ionized sodium ions (Na^+) equal to Avogadro's number streams past a point in 1000 s, what is the current?

4. [I] The photo on p. 662 shows an electron beam representing a current of 60 kA. How many electrons flow out of the device per second?

5. [I] The starter motor in an automobile is a small but powerful electrical device that "turns over" the main gasoline engine, moving it through a cycle to get it started. Typically, it will draw about 180 A from the battery for perhaps 2.0 s. How much charge flows through the circuit?

6. [I] A particle accelerator contains two beams flowing side-by-side in opposite directions; the beams will ultimately be made to collide. One is a stream of protons (each with a charge of $+1.60 \times 10^{-19}$ C), the other a stream of antiprotons (each with a charge of -1.60×10^{-19} C). Given that either beam can deliver 1.0×10^{14} particles per second to the colliding region, what is the net current in the machine as the two streams race past each other?

7. [I] A synchrotron accelerates protons up to nearly the speed of light, imparting energies to them of 500 MeV. If the beam current is 1.0 mA, how many protons will hit a target in 0.10 s?

8. [I] With the previous problem in mind, how many protons are there in each 1.0-cm-long segment of the beam? Assume a uniform particle density throughout the beam.

9. [I] A wire is connected across the terminals of a battery for precisely 60.0 s during which time a constant current of 2.00 A circulates around the loop. What was the net charge that flowed past any point on the wire?

10. [I] A portable tape recorder is powered by six 1.5-V AA-cells in series. What is its operating voltage?

11. [I] A torpedo, a giant saltwater ray that can develop a voltage of 220 V, is covered with cells known as electroplaques, each of which produces a potential difference of about 0.15 V. Several thousand rows (each made up of a series-connected array of cells) are then connected in parallel to build up a sizeable current. How many cells would you guess form each row? Why do freshwater electric fish in general develop higher voltages than saltwater ones?

12. [I] A NiCd cell has a voltage of 1.2 V and an amp-hour rating of 34. It is a sealed storage cell with a nickel anode and a cadmium cathode immersed in an alkaline electrolyte. How long can it operate when providing 2.0 A to some load?

13. [I] A battery is rated at 10 amp-hours. How much charge does that correspond to?

14. [I] The positive terminal of a 1.5-V dry cell is attached to ground via a length of heavy hookup wire. The negative terminal is then connected, via the same kind of wire, to a neutral brass ball. What is the potential of the ball? Describe its state of charge. What is the potential of the central carbon electrode?

15. [II] A pure gold wire with a 1.00 mm × 1.00 mm cross section carries a flow of electrons having a current density (I/A) equal to 1.0 MA/m^2. How long will it take for an amount of electrons equal to Avogadro's number to pass a point on the wire?

16. [II] An ion generator used to clean room air puts out a stream of negatively charged molecules that attach themselves to airborne pollutants, which are then collected electrostatically. Oxygen molecules tend to pick up electrons, thereby becoming negative oxygen ions. According to one company's literature, at 1.0 m from a particular generator, a detector would record the arrival of "168 million ions/sec./cm^2." Assuming that each ion is singly charged, what current impinges on a 10.0-cm^2 target at 1.0 m from the device?

17. [II] Figure P17 is a diagram of a portion of the electrical system of an automobile. List which switches—*A*, *B*, *C*, *D*, *E*, and *F*—must be closed in order to (a) blow the horn;

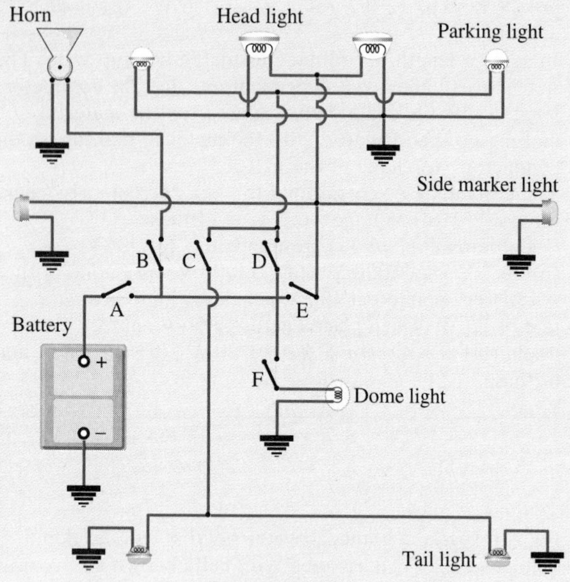

Figure P17

(b) turn on the headlights; (c) turn on the tail lights; (d) turn on only the parking lights; (e) activate the inside dome light. When do the side marker lights go on? *Note that the whole body of the car, including the engine block, is generally wired together as a common "ground."*

18. [II] A particular model of automobile battery, when fully charged, can deliver roughly 4.0×10^5 C before becoming completely run down. What is the amp-hour rating of such a battery?

19. [II] We wish to make up a panel of silicon solar cells that will provide at least 440 mA at 9.0 V when placed in an appropriately illuminated region. Given that each cell provides 22 mA at 0.45 V, design the panel.

20. [II] A rectangular aluminum wire with a cross section 1.0 mm × 2.0 mm carries an electron current of 0.10 A. Given that aluminum has 6.0×10^{22} atoms per cm^3 and each contributes roughly one free electron, determine the electron's drift speed.

21. [II] If in Problem 20 the current is kept constant while the cross-sectional area of the wire is halved, how long will it take a typical electron to travel down a 1.0-m length of the aluminum?

22. [II] The drift velocity of electrons in a pure copper wire carrying a current of 1 A and having a cross-sectional area of 1.0 mm^2 was found in Example 19.4 to be ≈0.1 mm/s. How would that change if the current was increased to 20 A, all else kept as is? What would happen if the current was doubled and the cross-sectional area was halved?

23. [III] In the process of recharging a rundown automobile battery, it's attached to an electronic 12-V charger. As soon as the device is turned on, an ammeter shows that the battery draws 7.0 A, but as it revives, the current slowly drops until after 6.0 hours it's down to 3.0 A. Assuming the current decreased linearly with time, how much charge passed through the battery?

RESISTANCE

24. [I] A small high-torque variable-speed dc motor requires an input of 10 mA at 3.0 V. Determine the electrical resistance of the motor.

25. [I] A 100-V electric heater draws 10 A. What is its resistance?

26. [I] If electric contacts are placed on the scalp, time-varying differences in potential will be observed. These can be recorded by an electroencephalograph. Voltage differences of ≈0.5 mV will appear across resistances of ≈10 kΩ. What size currents are involved?

27. [I] A wooden stick in contact with the metal sphere of a 100-kV Van de Graaff generator carries a current of 2.0 μA down to ground. Calculate the stick's resistance.

28. [I] If the current in a 10-Ω resistor is 500 mA, what is the voltage across its terminals?

29. [I] Given a solid cube of metal, under what circumstances, if any, will $R = \rho$?

30. [I] A 1.0-m-long wire of pure silver at 20°C is to have a resistance of 0.10 Ω. What should be its diameter?

31. [I] Considering a length of metal wire, show that Ohm's Law can be written as

$$I = \frac{E}{\rho/A}$$

32. [I] A nichrome wire with a cross-sectional area of 1.5×10^{-6} m^2 is to be used in a heater. If the design calls for a 3.0-Ω coil, what length of wire will be needed?

33. [I] A 5.0-m-long wire has a cross-sectional area of 2.0 mm^2 and a resistance of 40 mΩ. What is the resistivity of the material constituting the wire?

34. [I] According to the American Wire Gauge system, No. 0000 wire, which is the heaviest, has a diameter of 11.7 mm. What would be the resistance of 100 m of copper AWG No. 0000 at 20°C?

35. [I] Copper telegraph wire has a resistance of about 10 Ω per mile. What's its diameter in millimeters?

36. [I] A wire of pure gold is drawn through a die so that it is stretched out to twice its original length. Given that its volume is unchanged in the process and its new cross-sectional area is constant, compare the new with the original resistance.

37. [I] A narrow rod of pure iron has a resistance of 0.10 Ω at 20°C. (a) What is its resistance at 50°C? (b) A narrow rod of manganin has a resistance of 0.10 Ω at 20°C. What is its resistance at 50°C?

38. [I] A carbon rod used to generate the bright light in a movie-theater projector has a resistance of 110 Ω at 20.0°C. What will be its resistance at 520°C? (Incidentally, the hottest point on a functioning carbon arc, and the point of greatest luminosity, is typically at about 3500°C.)

39. [I] A portable tape recorder has a plate on its underside indicating that it uses 1 W at 9 V dc. What net current does it draw from its battery?

40. [I] Referring to Problem 39, what is the net resistance of the tape recorder?

41. [I] A high-torque dc motor designed to drive cassette decks has a no-load speed of 7400 rpm at 9 V with a torque of 9.6 in.·oz and a no-load current of 16 mA. How much power will it draw from a 9-V battery when turning freely?

42. [I] What is the maximum current that should be passed through a 100-Ω, 10-W resistor?
43. [I] The little speaker in a portable radio is labeled 8 Ω, 0.2 W. What current does that correspond to?
44. [I] A windmill with 6-ft propellers generates dc at 12 V. In a strong wind it will produce up to 200 W of electrical power, which is fed to a 230-amp-hour battery. What's the maximum current the windmill will deliver?
45. [I] An automobile starter motor will draw about 180 A from the 12-V battery of a car for perhaps 2.0 s in the process of starting the gasoline engine. How much power does it use?
46. [II] Figure P46 shows a variable resistance (total 100 Ω) across which is a voltage drop of 12 V supplied by a battery. What must be the resistance of that portion of the resistor between A and B (namely, R_{AB}) if the voltage V_{AB} is to be (a) 12 V; (b) 6.0 V; (c) 3.0 V?

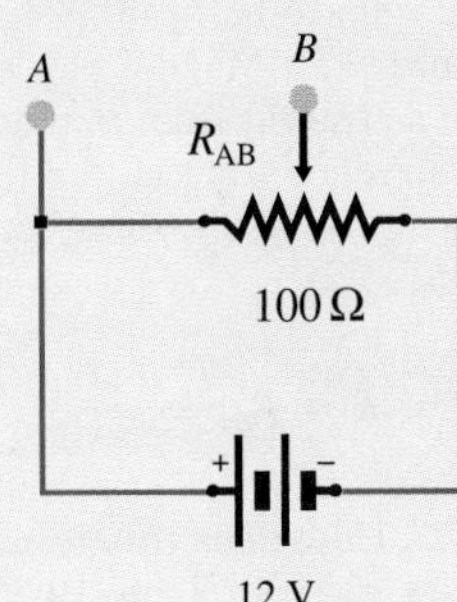

Figure P46

47. [II] A length of copper wire 1.0 m long with a cross-sectional area of 1.0 mm^2 is part of a circuit carrying 1.0 A. What is the value of the steady-state electric field within the wire?
48. [II] A variable slidewire resistor with a total of 50 Ω is made by wrapping a single layer of 200 turns of varnished copper wire around an insulating cylindrical core. When it carries a current of 3.0 A, what is the voltage drop across each turn of wire?
49. [II] It is said that the carbon filaments in the early incandescent light bulbs that Edison produced lost more than two-thirds of their resistance soon after they were turned on. Explain this occurrence and compute their approximate operating temperature. The latter was actually about 1900°C.
50. [II] Imagine that you are going to connect a remote speaker to your stereo system. The speaker has a resistance of 4.0 Ω, and so you want to use hookup wire whose total resistance is small by comparison, say, 0.25 Ω. If the speaker is to be 15 m away, what diameter copper wire should be used? (Hint: You'll need two leads.)
51. [II] A platinum resistance thermometer made up of a coil of wire with a resistance of 10 Ω at 20°C is placed in a chamber at 420°C. What will be its new resistance if α_0 is fairly constant at 0.0039?
52. [II] Show that if a conductor has its temperature changed, it will experience a fractional change in its resistivity given by

$$\frac{\Delta\rho}{\rho_0} = \alpha_0 \Delta T$$

What assumption must be made here?
53. [II] An iron wire at 20°C is heated until its resistance doubles. At what temperature will that occur? Assume the temperature coefficient is constant over that temperature range.
54. [II] When currents are transported at very high voltages, it's desirable to keep the electric fields surrounding the conductors down to levels that will not break down the surrounding air, thereby controlling sparking. Accordingly, conductors at power plants and high-voltage laboratories are often large-diameter pipes. What is the resistance per meter of a copper pipe 2.5 cm thick with an inside diameter of 15.0 cm?
55. [II] A powerhouse near a waterfall has a large dc generator that produces electricity for a factory 0.50 mile away. The energy is transmitted over two cables each with a resistance of 0.25 Ω/mile. Given that the factory requires 45 kW at a voltage of 110 V to run its equipment, what must be the output of the powerhouse?
56. [II] A 0.8-in.-square silicon solar cell delivers 90 mA at 0.45 V when illuminated by sunlight (at 100 mW/cm^2). Suppose 10 such cells are connected in series; how much power could the panel deliver?
57. [II] For greater flexibility, electrical wires often consist of several fine strands twisted together rather than being constructed of one thick lead. A copper wire is made up of 10 fine fibers, each with a resistance of 2.0 mΩ. When placed across a voltage difference, a total current of 0.12 A traverses the wire. (a) What is the net resistance of the length of wire? (b) What voltage exists across it? (c) How much power is dissipated by each strand? (Hint: How much current does each strand carry?)
58. [II] A credit-card–size calculator uses two tiny 1.5-V cells in series that provide a normal operating power of 0.00018 W. Determine the current passing through each cell when the device is in use.
59. [II] An old-fashioned trolley car draws 12 A from an overhead wire at +500 V (the rails are grounded). What power is delivered to the motor? If the motor is 86% efficient, what power does it develop in propelling the car?
60. [II] When put across the terminals of two D-cells in series, a small flashlight bulb draws 330 mA. How much power does it consume? How much energy does it take from the cells in 1.0 minute of operation?
61. [II] The moving nonconducting belt of a Van de Graaff generator carries 10 μC of charge up to the metal sphere (Fig. Q14, Chapter 17) each second. The steady-state potential difference between the sphere and the grounded source of charge is 3 MV. What minimum power must be supplied to the generator to sustain its operation?
62. [II] A stereo tuner-amplifier with a maximum power output of 50 W per channel (that is, 50 W to each of two speakers) has its right channel connected as shown in Fig. P62. Since the speaker will be destroyed if it receives more than 36 W, a fuse is installed to limit the current

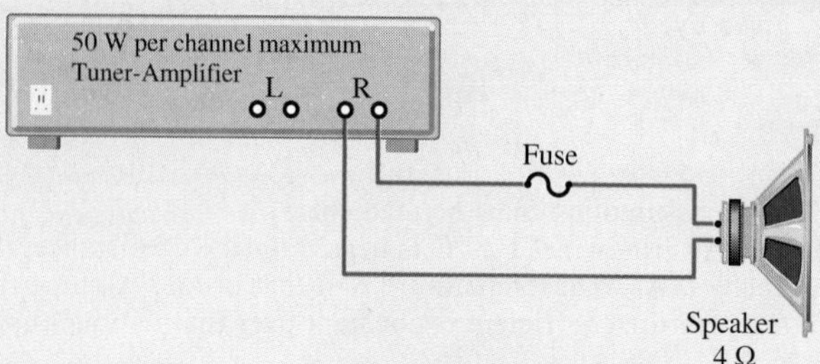

Figure P62

entering it. What should be the rated maximum current of the fuse?

63. [II] When turned on, a flashlight bulb draws 1/3 A at 3.9 V. How much power does it require? Incidentally, more than 95% of that power appears in the form of thermal energy and not light. What is the resistance of the filament in the lamp?

64. [II] How much current does an ideal 1.00-hp dc electric motor draw when operating off a portable 100-V generator?

65. [II] A NiCd battery has an amp-hour rating of 10 and consists of five cells connected in series. What is the maximum power the battery will deliver if it is to operate for 5.0 h?

66. [III] A silicon solar cell 5 mm × 4 mm produces a current of 5 mA at 0.45 V under sunlight illumination of 100 mW/cm^2. Determine its efficiency.

67. [III] A modern incandescent lamp has a tungsten filament with a melting point of 3400°C. Ordinarily, with the bulb evacuated, it's operated at about 2200°C. The efficiency can be improved almost threefold by raising the temperature to 2800°C, but that causes the tungsten to evaporate and shortens the life appreciably. The alternative (an idea introduced by Langmuir) is to fill the bulb with nitrogen or argon to suppress evaporation of the tungsten. What is the fractional change in the resistance of the filament when raised from 2200°C to 2800°C? Use the value of α_0 found in Table 19.3. Given the following values of α_0 for tungsten (from the *Handbook of Chemistry and Physics*)—0.004 5 at 18°C, 0.005 7 at 500°C, 0.008 9 a 1000°C— how good was the calculation you just made? Discuss your answer and make a very rough estimate of α_0 at 2500°C. Now estimate the fractional change in resistance using this new value.

68. [III] Determine the power rating in kilowatts of an electric heater that will raise the temperature of 10 liters of water from 25°C to 85°C in 15 minutes, assuming no loss of thermal energy. Given that the heater coil has a resistance of 10 Ω, how much current does it draw?

69. [III] Figure 19.20 shows how the resistivities of various pure metals each seem to be heading toward zero at some Celsius temperature $-T'$. Make a plot of R versus T (in °C), assuming it to be a straight line. Is that reasonable? Taking $T = 0$°C as the reference (that is, the line crosses the R-axis at $R = R_0$ and $T = 0$), show that for any point (T, R) on the line

$$R = R_0 \frac{T' + T}{T'}$$

and that $\alpha_0 = 1/T'$. How does this result compare with the values of resistivity given in Table 19.3? Notice how most values approximate $1/273 = 3.7 \times 10^{-3}$.

20

Circuits

NOW THAT WE ARE familiar with the basics of dc—batteries, resistors, capacitors, and Ohm's Law—we can apply this knowledge to the treatment of circuits, where several elements are attached together. For example, we can study the electrical system of an automobile or design an experimental setup to measure the response of pigeons to the sight of popcorn. Today, virtually every field of science uses electric circuits in its research.

A variety of sensors, or *input transducers,* convert nonelectrical signals into electrical ones—the stereo cartridge on a record player and the microphone are familiar examples. Any up-to-date hospital intensive care unit can

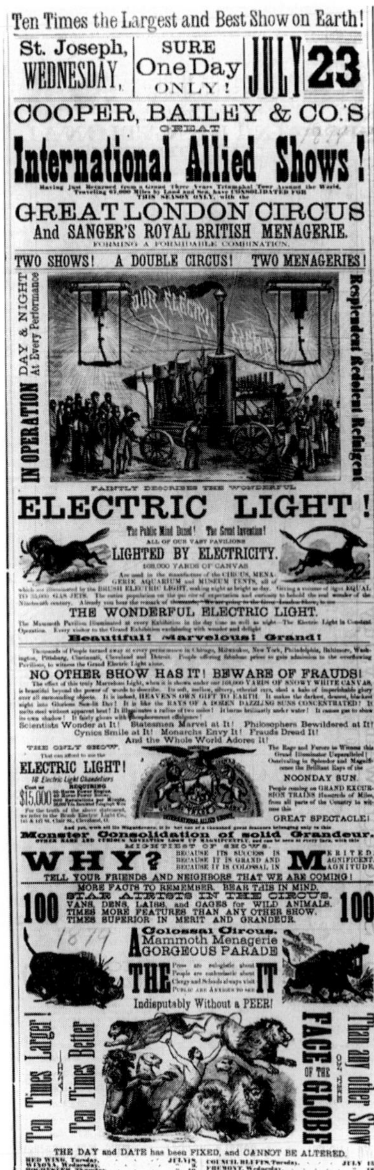

A circus poster from 1879. A steam engine powered a generator that supplied current to the arc lamps that lit this circus "day and night."

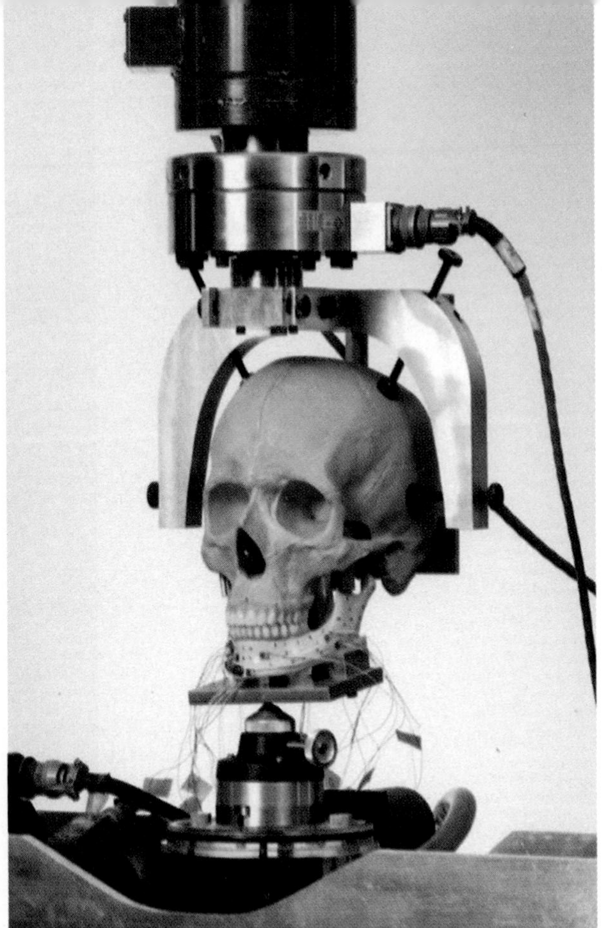

A testing device applies axial loading to the jaw. Here the mandible was fractured and repaired with a surgical plate assembly. Strain gauges around the jaw convert the resulting strain distribution into electrical signals.

electronically monitor a patient's body temperature, blood pressure, and respiration, measuring infusions and drainage while recording electrocardiograms, all with transducers. One kind of strain gauge is made from fine alloy wire that changes its resistance as it's stretched and compressed. Such gauges are widely used to monitor mechanical apparatus and in medical studies of bones, joints, and the behavior of muscles. Clearly, there's more to resistive devices and sources than we have considered thus far. Learning to deal with simple circuits lays the groundwork for understanding a wide range of practical electrical systems.

CIRCUIT PRINCIPLES

A circuit, in simple form, is a few elements (for example, a resistor and a battery) joined together to make at least one closed current path (usually with some purpose in mind beyond merely confounding students). Often a system converts electrical energy into light via an incandescent lamp or into thermal energy via a toaster. There might be an input transducer, a microphone, at one end and an output transducer, a speaker, at the other end. In any event, we are generally interested in determining a variety of circuit parameters, such as voltage drops, currents, power dissipated, and so on—you can't send a 100-W signal out to a 20-W speaker, at least not twice. The only limitation in this chapter is that the currents and voltages are

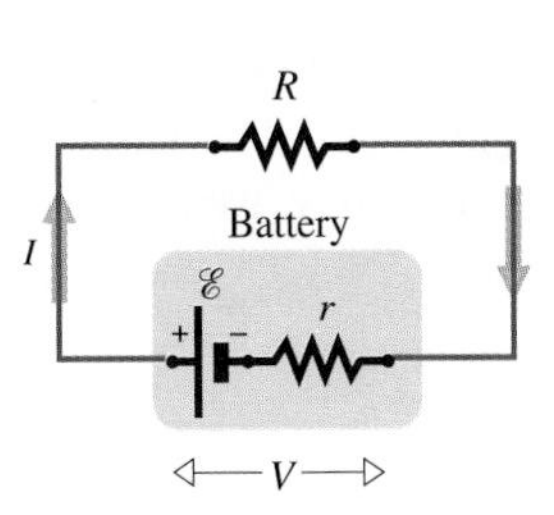

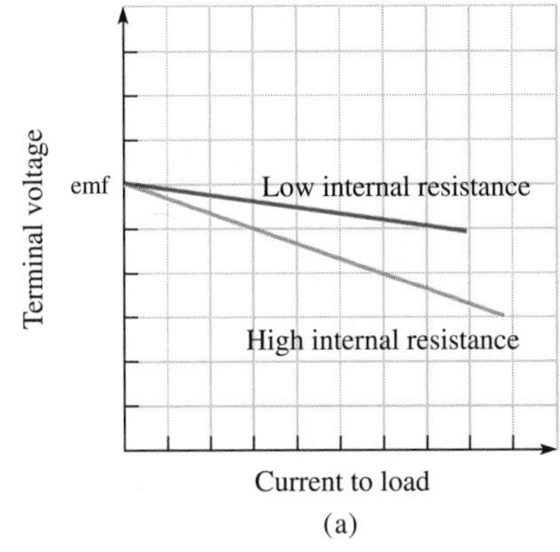

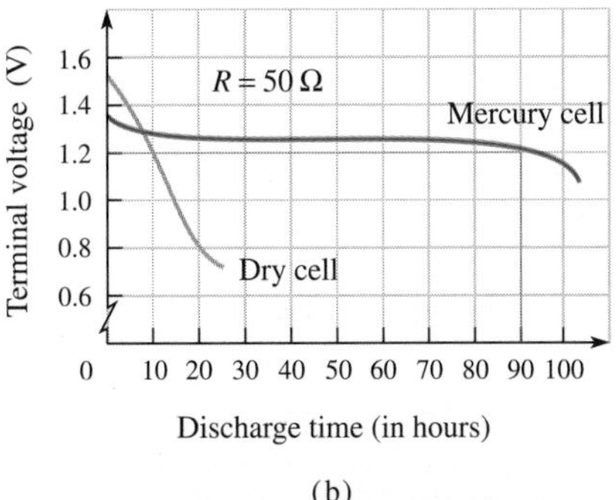

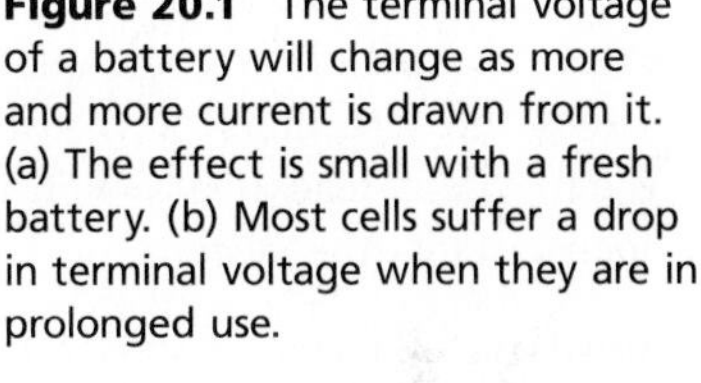
Figure 20.1 The terminal voltage of a battery will change as more and more current is drawn from it. (a) The effect is small with a fresh battery. (b) Most cells suffer a drop in terminal voltage when they are in prolonged use.

dc—they may rise and fall, but they do not fall below zero; they do not reverse direction, the sources do not change polarity.

20.1 Sources and Internal Resistance

So far, the battery has been taken to be an ideal constant-voltage source. That assumes that its terminal voltage remains fixed regardless of the resistive load across it or the length of time it provides power. Alas, that assumption is generally not the case. Although mercury cells approximate this behavior, they are small and usually supply very little current. Most other batteries show a marked decrease in terminal voltage as the current supplied by them increases (Fig. 20.1). Start your car with the headlights on. As the starter motor draws hundreds of amps, the battery's terminal voltage drops and the lights become noticeably dimmer.

When a voltaic cell provides current to an external circuit, there is a transport of charge from one electrode to the other across the electrolyte, and that does not happen in an unimpeded way. The cell resists current traversing it in either direction. This **internal resistance** (r) is an inseparable aspect of any real battery, cell, or other power source. Insofar as the plot in Fig. 20.1a is linear, r is ohmic, and the battery can be represented most simply by a resistor in series with an ideal emf ($\mathscr{E}$), as in Fig. 20.2. The + and − terminals of the source are the points A and B, respectively, and it doesn't matter on which side the internal resistor is drawn as long as it is between those terminals.

When the battery supplies current to an external circuit, as in Fig. 20.3a, I leaves the positive terminal A, traverses the circuit, returns to the battery at terminal B and passes internally to A. (Label the resistor with + and − signs, indicating the ends at which current enters and leaves, respectively.) If we trace from B to A, there will be a voltage drop (−) across the internal resistor of $-Ir$ followed by a rise (+) equal to the emf, and so with $V = V_A - V_B$, we have

$$V = \mathscr{E} - Ir \tag{20.1}$$

The terminal voltage V, *the potential difference measured*

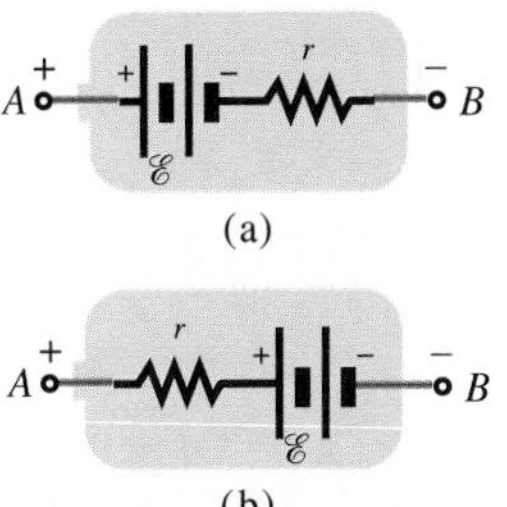

Figure 20.2 A real battery can be represented by an emf in series with an internal resistance r. The order doesn't matter since (a) and (b) are equivalent. The voltage difference between A and B is the terminal voltage of the battery.

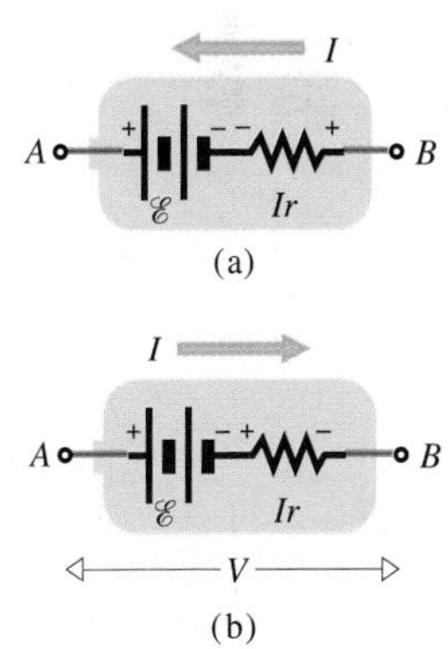

Figure 20.3 (a) When the battery supplies current to an external circuit (not shown here), there is a voltage drop across r, and $V < \mathscr{E}$. (b) When the battery receives current from an external source (not shown here), $V > \mathscr{E}$.

by an ideal voltmeter across the battery, is less than the emf when the current goes from B to A and the battery provides power.

When the external circuit supplies current to the battery, as in Fig. 20.3b, I enters at the positive terminal A and passes internally to terminal B. The battery is having power supplied to it—it's being charged. Tracing from B to A, we move first from $-$ to $+$ across the internal resistor, which corresponds to a voltage rise $(+)$ of $+Ir$, followed by another rise $(+)$ equal to the emf. Hence

$$V = \mathscr{E} + Ir \tag{20.2}$$

and the terminal voltage exceeds the emf.

Example 20.1 A voltage source with an emf of 12 V and an internal resistance of 0.40 Ω supplies 1.5 A to an external load. If a high-quality voltmeter is placed across the terminals of the source, what will it read? What will it read if a switch in series with the source open-circuits the system?

Solution: [Given: $\mathscr{E} = 12$ V, $r = 0.40$ Ω, and $I = 1.5$ A. Find: V when $I = 1.5$ A and when $I = 0$.] First, draw a circuit diagram with the source functioning as in Fig. 20.3a. The terminal voltage when current exists is given by Eq. (20.1),

$$V = \mathscr{E} - Ir = 12\text{ V} - (1.5\text{ A})(0.40\ \Omega)$$

and $V = 11.4$ V or, to two significant figures, $\boxed{V = 11\text{ V}}$. Opening the switch causes I to become zero, whereupon $\boxed{V = \mathscr{E} = 12\text{ V}}$.

▶ **Quick Check:** The drop across r is $Ir = (1.5\text{ A}) \times (0.40\ \Omega) \approx 1$ V. $V \approx 12\text{ V} - 1\text{ V} \approx 11$ V.

The internal resistance of a fresh battery is small: ≈0.05 Ω or so for a D-cell and only ≈2 mΩ for a 12-V lead storage battery. As a battery ages, its internal resistance increases while its emf gradually decreases. Still, it's possible to have an old 1.5-V dry cell with an internal resistance as high as several ohms and an emf only a few percent less than normal. As a consequence, using a meter to measure the terminal voltage of a worn D-cell with no load on it ($I = 0$) may yield a value near its normal emf. Clearly, that's an unreliable indication of the condition of the cell. Measuring V while there is a load on the cell (for example, while there is a flashlight bulb across it) is a better test. For most applications, if V does not exceed at least 1.1 V, the cell should be replaced.

Example 20.2 A 5.9-Ω load resistor is placed across the terminals of a battery that has an emf of 12 V and an internal resistance of 0.10 Ω. Write a general expression for the current in the circuit and then compute its specific value. What's the terminal voltage of the battery?

Solution: [Given: $R = 5.9$ Ω, $r = 0.10$ Ω, and $\mathscr{E} = 12$ V. Find: I in general and in particular, and V.] First draw a circuit diagram (Fig. 20.4). Current leaves the + side of the battery. We can find I using Ohm's Law if we know the voltage across R; namely, the terminal voltage. So, the first thing is to find V. From Eq. (20.1), $V = \mathscr{E} - Ir$, which in turn equals IR, so

$$V = IR = \mathscr{E} - Ir$$

and

$$I(R + r) = \mathscr{E}$$

so that

$$\boxed{I = \frac{\mathscr{E}}{R + r}} \tag{20.3}$$

(continued)

(continued)

Specifically

$$I = \frac{12\ \text{V}}{5.9\ \Omega + 0.10\ \Omega} = \boxed{2.0\ \text{A}}$$

As for the terminal voltage

$$V = \mathscr{E} - Ir = 12\ \text{V} - (2.0\ \text{A})(0.10\ \Omega) = 11.8\ \text{V}$$

or to two significant figures $\boxed{V = 12\ \text{V}}$.

▶ **Quick Check:** $V = IR = (2.0\ \text{A})(5.9\ \Omega) = 11.8\ \text{V}$.

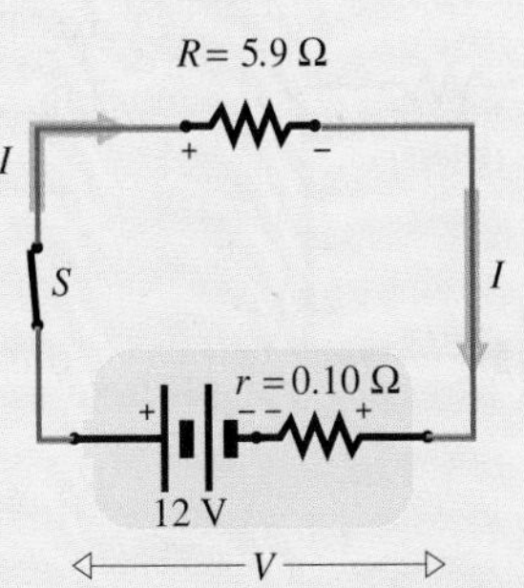

Figure 20.4 A battery with an emf of 12 V and internal resistance r supplies a current I to the resistor R when the switch S is closed. The current (a flow of positive charge) leaves the positive terminal of the battery and returns, with less energy, to the negative terminal.

If an ammeter and voltmeter are added to the circuit of Fig. 20.4, as indicated in Fig. 20.5, it becomes an easy matter to determine r. Note that *the ammeter is always placed in the branch of the circuit through which the required current passes,* whereas *the voltmeter is always placed across the two points whose potential difference is to be measured.*

Before we move on, let's standardize some of the terminology: a **branch** is one or more circuit elements (in series) carrying a single current. A **node** is a place (often a single point) where three or more branches come together. Nodes are interconnected by branches, and branches begin and end on nodes. A circuit that has no nodes can be thought of as a single branch that closes on itself. Any closed current path is a **loop**. The circuit in Fig. 20.4 is a single loop, and the ammeter is positioned in that loop. The two points where the voltmeter attaches to the circuit (Fig. 20.5) are nodes, and the voltmeter constitutes one of three branches.

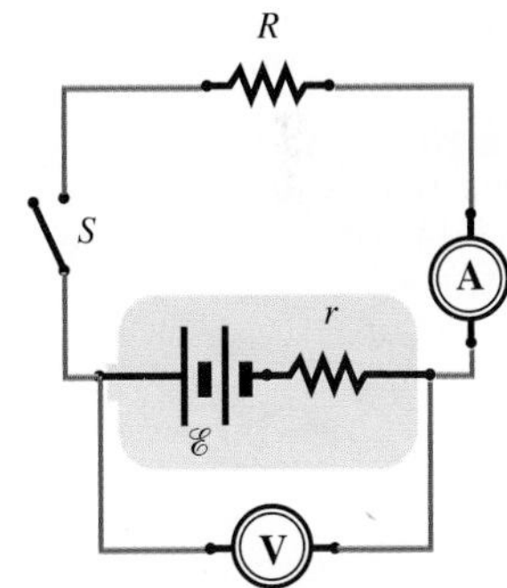

Figure 20.5 Using high-quality meters, we can accurately determine r. Here, an ammeter Ⓐ is placed in series with the battery, and a voltmeter Ⓥ is placed across it.

20.2 Resistors in Series and Parallel

Suppose we have a circuit with several branches and lots of resistors and sources and we want to compute the voltage across, and the current through, each element. The computation is likely to be difficult, but two special configurations are particularly easy to deal with. When the resistors are in *series* or in *parallel* or a combination of both, they can be replaced by a single *equivalent resistor* (as was done earlier with capacitors, p. 643). Then Ohm's Law can be applied and the problem solved step-by-step. How do we arrive at this **equivalent resistance** (R_e)?

Figure 20.6a shows an ideal voltage source in series with two resistors R_1 and R_2. *The same current I enters and leaves each element in series.* The difference in potential between points A and B is the voltage of the source V. If we go from B to A across the battery, we encounter a voltage rise of $+V$. Similarly, tracing from B to A across the resistors, we encounter a rise of V_2 followed by another rise of V_1, which must also equal V; thus

$$V = V_1 + V_2$$

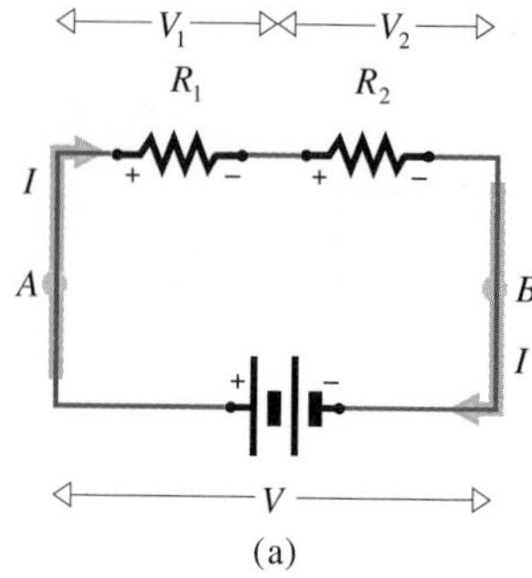

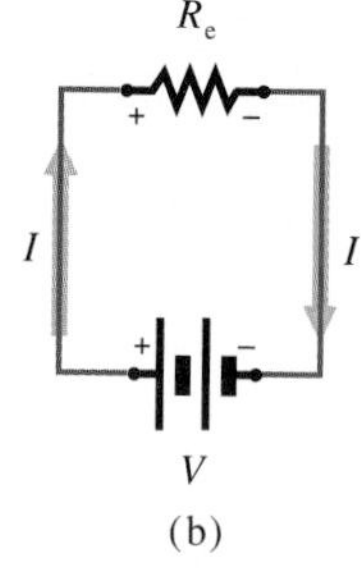

Figure 20.6 (a) An ideal voltage source in series with two resistors. (b) The equivalent circuit where $R_e = R_1 + R_2$.

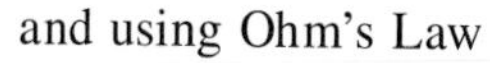

and using Ohm's Law $$V = IR_1 + IR_2$$

We want an equivalent resistance that will replace the load, as indicated in Fig. 20.6b, so that the same current I circulates. Hence

$$V = IR_e$$

and so

$$IR_e = IR_1 + IR_2$$

Consequently

[*in series*] $$R_e = R_1 + R_2 \tag{20.4}$$

The equivalent resistance is simply the sum of the series resistances, no matter how many there happen to be. To see how this works, suppose $R_1 = 5.0\ \Omega$, $R_2 = 1.0\ \Omega$, and $V = 12$ V. From Eq. (20.4), $R_e = 6.0\ \Omega$, and using Ohm's Law for Fig. 20.6b, we find the current to be $I = V/R_e = 2.0$ A. Returning to the original circuit of Fig. 20.6a, we see that $V_1 = IR_1 = 10$ V and $V_2 = IR_2 = 2.0$ V and, as expected, $V_1 + V_2 = 12$ V. The analysis of the circuit in Fig. 20.4 can now be reinterpreted in light of the above discussion. Accordingly, $R_e = R + r$, and since $I = V/R_e$, $I = V/(R + r)$, which is Eq. (20.3).

Figure 20.7a shows two resistors in parallel with a constant voltage dc source, though there could equally well have been ten of them. The (as yet undetermined) current I supplied by the source comes to node A and splits into two branch currents I_1 and I_2, which are (as we will find presently) inversely proportional to the resistances of the two branches. These currents recombine into I at node B:

$$I = I_1 + I_2$$

The voltage across each resistor is V, and therefore Ohm's Law gives

$$I = \frac{V}{R_1} + \frac{V}{R_2}$$

But in the equivalent circuit of Fig. 20.7b $I = V/R_e$, and so

$$\frac{V}{R_e} = \frac{V}{R_1} + \frac{V}{R_2}$$

Canceling V yields

[*in parallel*] $$\frac{1}{R_e} = \frac{1}{R_1} + \frac{1}{R_2} \tag{20.5}$$

and the summation would just continue on were there additional resistors in the circuit.

Equation (20.5) for two resistors can be rewritten as

$$R_e = \frac{R_1 R_2}{R_1 + R_2} \tag{20.6}$$

which makes it apparent that when either resistor is much larger than the other, say, $R_1 \gg R_2$, then $R_1 + R_2 \approx R_1$ and $R_e \approx R_2$; the equivalent resistance approximates the *smaller* of the two. **The equivalent parallel resistance is always less than either of the contributing resistances.** All of the circuits in Fig. 20.8 are electrically identical, and all have the same equivalent resistance between A and B.

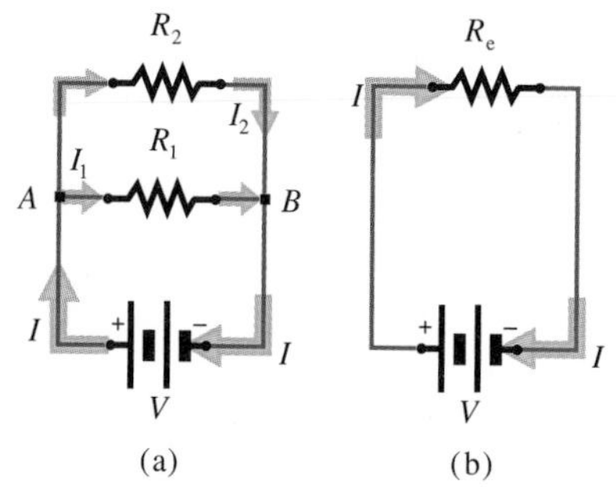

Figure 20.7 (a) An ideal voltage source across two resistors in parallel. Notice how the current I splits at node A and recombines at B. (b) The equivalent circuit where $1/R_e = 1/R_1 + 1/R_2$.

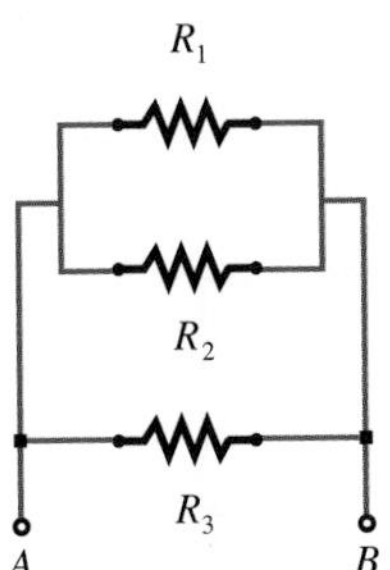

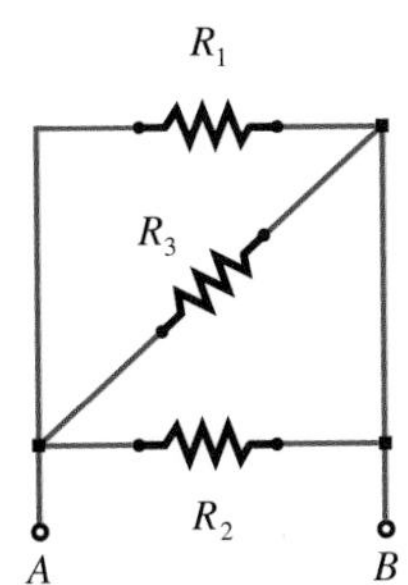

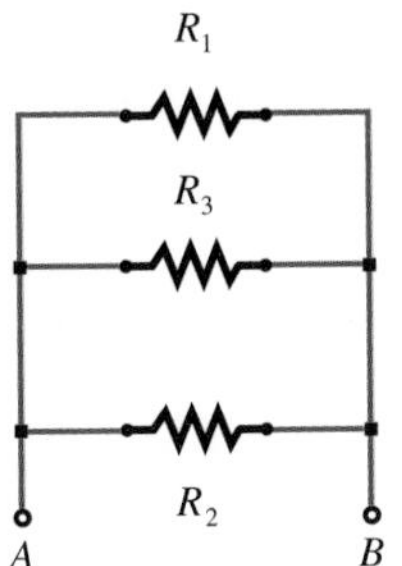

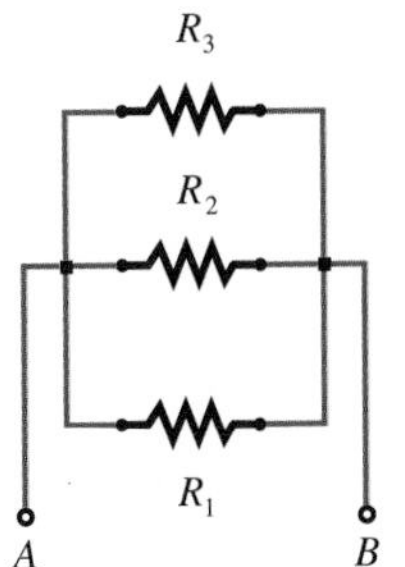

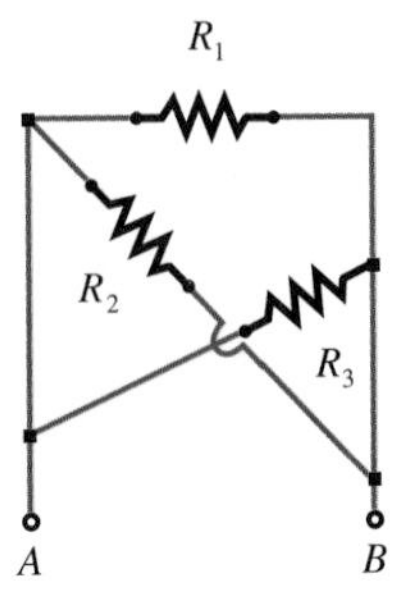

Figure 20.8 Each of these circuits has the same equivalent resistance between A and B. In each version, R_1, R_2, and R_3 are in parallel.

Example 20.3 Figure 20.9a represents three light bulbs with resistances of 2.0 Ω, 4.0 Ω, and 8.0 Ω attached across a source with an emf of 6.0 V and an internal resistance of 1.0 Ω. Find the current through each bulb.

Solution: [Given: $R_1 = 2.0\ \Omega$, $R_2 = 4.0\ \Omega$, $R_3 = 8.0\ \Omega$, $r = 1.0\ \Omega$, and $\mathscr{E} = 6.0$ V. Find: I_1, I_2, and I_3.] Draw a diagram putting in all the required quantities. The three bulbs are in parallel so their equivalent (R) is gotten from Eq. (20.5):

$$\frac{1}{R} = \frac{1}{2.0\ \Omega} + \frac{1}{4.0\ \Omega} + \frac{1}{8.0\ \Omega} = 0.875\ \Omega^{-1}$$

and $R = 1.14\ \Omega$, which is the equivalent resistance of the load (Fig. 20.9b). R and r are in series, and so using Eq. (20.4), the equivalent resistance of the entire circuit (R_e) is $(R + r) = 2.14\ \Omega$. With R_e across the emf in Fig. 20.9c, the current provided by the source (via Ohm's Law) is

$$I = \frac{\mathscr{E}}{R_e} = \frac{6.0\ \text{V}}{2.14\ \Omega} = 2.8\ \text{A}$$

Going back to Fig. 20.9b, we can find the terminal voltage across R and then each branch current. The terminal voltage follows from Eq. (20.1); namely, a drop of $-Ir = -2.8$ V followed by a rise of $\mathscr{E} = +6.0$ for a total of $V = +3.2$ V. The branch currents are then

$$I_1 = \frac{V}{R_1} = \frac{3.2\ \text{V}}{2.0\ \Omega} = \boxed{1.6\ \text{A}}$$

$$I_2 = \frac{V}{R_2} = \frac{3.2\ \text{V}}{4.0\ \Omega} = \boxed{0.80\ \text{A}}$$

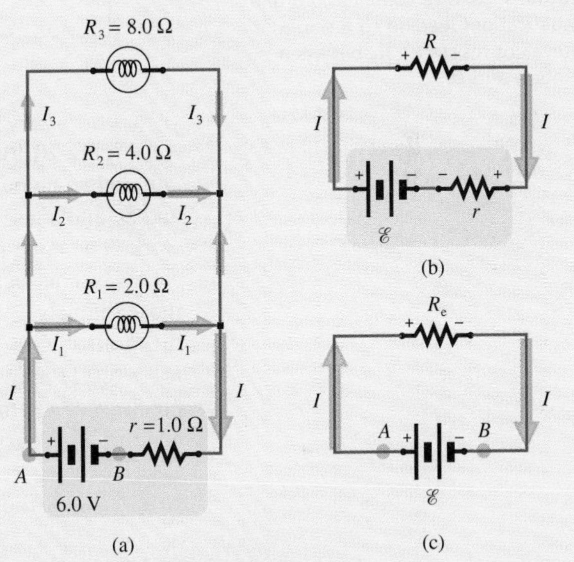

Figure 20.9 (a) Three light bulbs in parallel with a real battery. (b) The bulbs have resistances that are in parallel such that $1/R = 1/R_1 + 1/R_2 + 1/R_3$. (c) The equivalent resistance "seen" by the emf between points A and B is R_e.

and $$I_3 = \frac{V}{R_3} = \frac{3.2\ \text{V}}{8.0\ \Omega} = \boxed{0.40\ \text{A}}$$

▶ **Quick Check:** Note that R is less than any of the constituent resistors, which is as it should be. $I = I_1 + I_2 + I_3 = (1.6\ \text{A}) + (0.80\ \text{A}) + (0.40\ \text{A}) = 2.8$ A, which is appropriate. The smaller the resistance of the branch, the more current it carries.

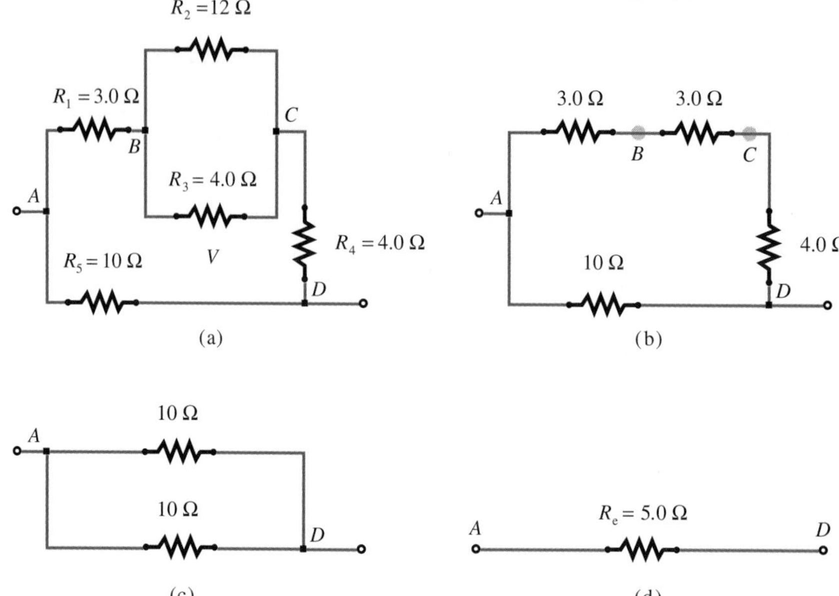

Figure 20.10 A cluster of resistors and the successive combinations and simplifications leading to a single equivalent resistance R_e between points *A* and *D*.

Figure 20.10a depicts a cluster of attached resistors. There is no source in the circuit and so no current, but we can still think about adding one, say, across points *A* and *D*, and consequently still be concerned with the equivalent resistance of the network. The node points are labeled to keep track of the process as the circuit is successively contracted. Resistors R_2 and R_3 are in parallel, and the result of adding those (namely, 3 Ω) is in series with R_1 and R_4 for a total of 10 Ω. This resistance is in parallel with R_5, since both resistors are attached at *A* and *D*, thus yielding $R_e = 5.0\ \Omega$ between points *A* and *D*. This means that the whole cluster behaves in precisely the same way as would a single 5-Ω load. Put a battery across *A* and *D*, and the same current will be delivered to either configuration. This value is *not* necessarily the equivalent resistance between *A* and *C*, or *A* and *B* (Problems 14, 15, and 16)—each of those is quite distinct.

Example 20.4 Imagine a wire square and label its corners *A*, *B*, *C*, and *D* going clockwise. Now place a larger square surrounding (and with sides parallel to) the first, and label its corners *E*, *F*, *G*, *H* clockwise, with *E* closest to *A*. Insert a 1.0-kΩ resistor in each arm of each square (eight in total) and attach another between *A* and *E*. Place a 12-V DC source between *C* and *G*. (a) If possible, simplify the circuit and determine an equivalent resistance between *C* and *G*. (b) What current is provided by the source? (c) What is the voltage across points *G* and *E*? (Perhaps it should be said that this is about as complicated a circuit as you are likely to see.)

Solution: [Given: nine resistors, $R = 1.0\ \text{k}\Omega$ each, and $V = 12$ V. Find: R_e, *I*, and *V* between *G* and *E*.] Figure 20.11a is the circuit diagram. We can simplify it as is or, alternatively, imagine the wires to be flexible and lift up the inside square, with the resistor and source attached, and place it outside *E–F–G–H*, as in Fig. 20.11b. Branches *A–B–C* and *A–D–C* are in parallel, as are *E–F–G* and *E–H–G*, and each has a resistance of $1.0\ \text{k}\Omega + 1.0\ \text{k}\Omega = 2.0\ \text{k}\Omega$ (Fig. 20.11c). Thus, the resistance of each square *E–F–G–H* and *A–B–C–D* reduces to

$$\frac{1}{2}\ \text{k}\Omega + \frac{1}{2}\ \text{k}\Omega = \frac{1}{R}$$

and $R = 1.0\ \text{k}\Omega$. The three 1.0-kΩ resistors in Fig. 21.11d are in series with the source. (a) The equiva-

(continued)

(continued)

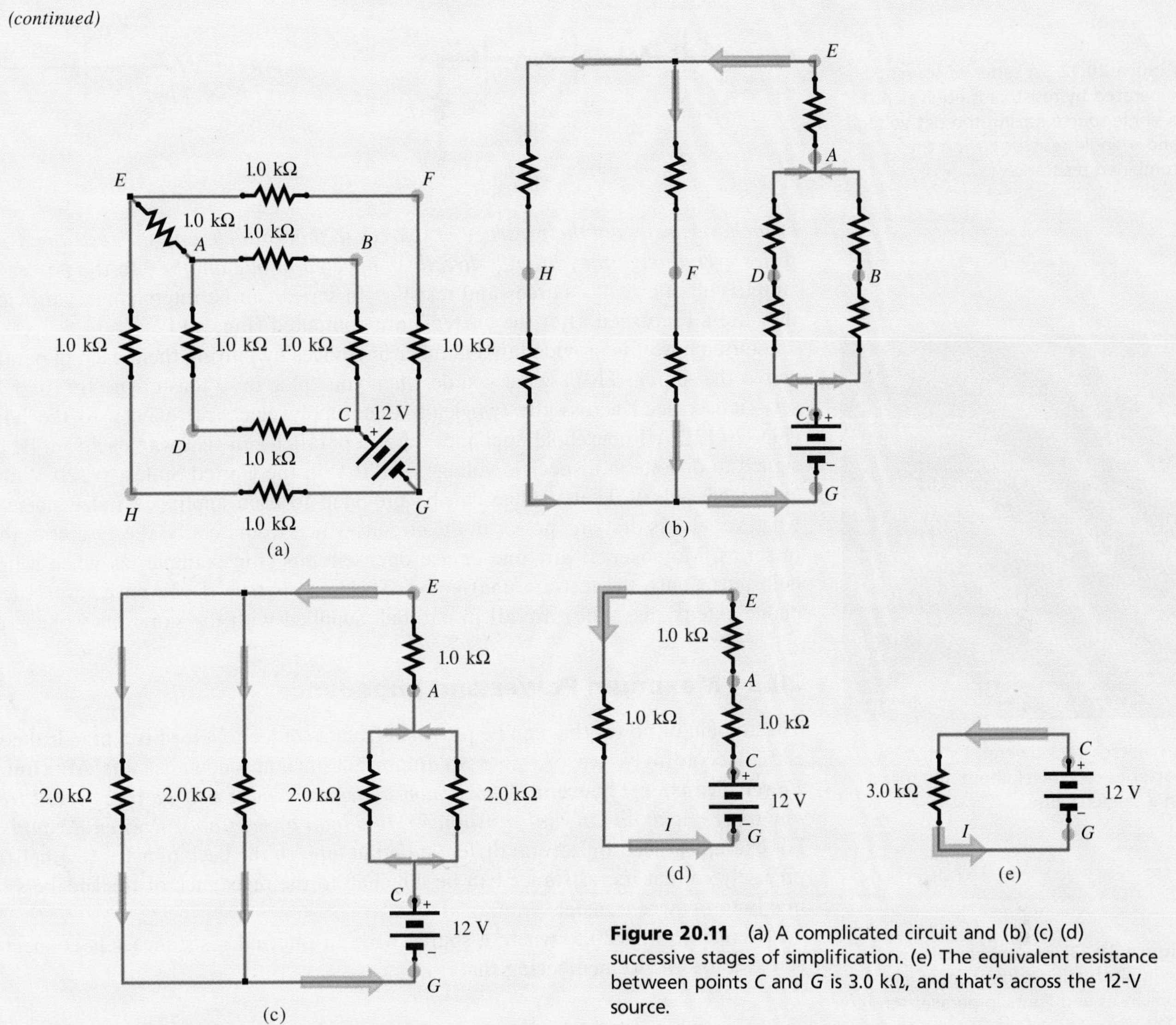

Figure 20.11 (a) A complicated circuit and (b) (c) (d) successive stages of simplification. (e) The equivalent resistance between points *C* and *G* is 3.0 kΩ, and that's across the 12-V source.

lent resistance in Fig. 20.11e is $\boxed{3.0\ \text{k}\Omega}$. (b) Since $V = IR_e$, $\boxed{I = 4.0\ \text{mA}}$. (c) A current of 4.0 mA leaves the battery and splits at *C*. Because the two branches *C–D–A* and *C–B–A* have the same resistance, the current divides into two equal streams of 2.0 mA each. The voltage drop in going from *C* to *A* is given by $V_{AC} = IR = (2.0\ \text{mA})(1.0\ \text{k}\Omega + 1.0\ \text{k}\Omega) = 4.0\ \text{V}$. In going from *A* to *E*, there is another drop of $V_{EA} = IR = (4.0\ \text{mA})(1.0\ \text{k}\Omega) = 4.0\ \text{V}$. *C* is 12 V above *G*, *A* is 8.0 V above *G*, and $\boxed{E \text{ is } 4.0\ \text{V above } G}$.

▶ **Quick Check:** 4.0 mA leaves *A*, enters *E*, and splits so that 2.0 mA goes through both branches *E–F–G* and *E–H–G*. The voltage drop from *E* along either branch to *G* is (2.0 mA)(2.0 kΩ) = 4.0 V.

The resistors in Fig. 20.6a can be slid around so that the battery is between them and nothing changes. *The same current passes through every resistor in a given*

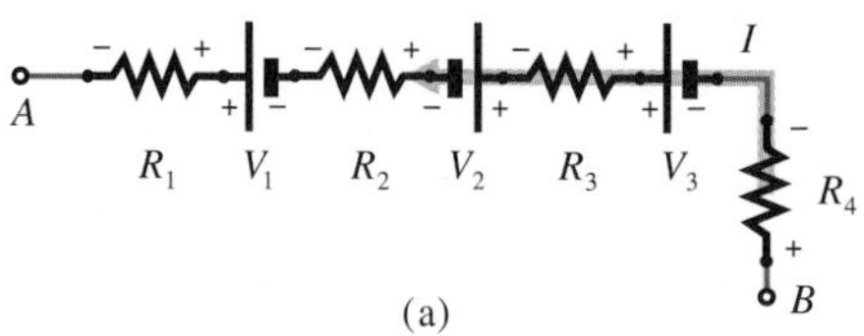

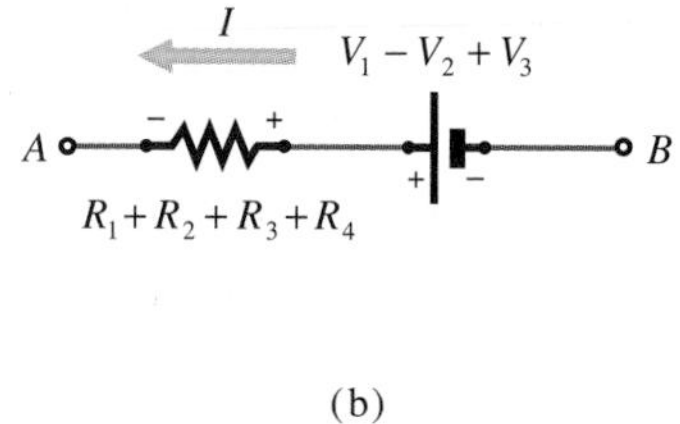

Figure 20.12 A series of sources separated by resistors is equivalent to a single source having the net voltage and a single resistor having the combined resistance.

branch, regardless of the presence of sources in that branch, and the resistors are in series even though they are not directly connected to one another. In the process of simplifying a circuit, sources and resistors in series can be interchanged, provided they are all returned after the currents are computed (Fig. 20.12).

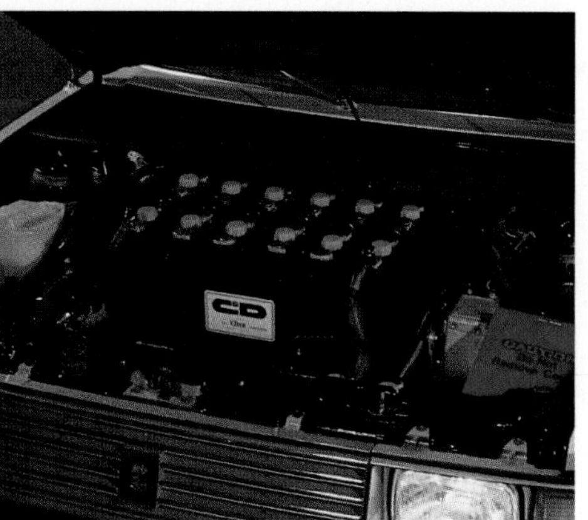

An electric car powered by storage batteries can travel about 100 miles on a single charge.

Often power is provided to a number of devices by putting them each in parallel across the source. That's what you do when you "plug in" a toaster (p. 815) or a TV set—it is placed across the two leads that supply electrical energy to the home (Fig. 20.13). All household appliances are in parallel—in that way each can be designed to operate at a specific voltage (≈120 V in the United States, ≈220 V most everywhere else). That voltage can be provided to each appliance independent of whatever else is drawing power in the circuit, which would not be the case were they in series. Moreover, if any one device open-circuits (for example, as when a light bulb burns out), the rest are unaffected. The same is true if you have several telephone extensions—they are all in parallel, supplied with the same low-voltage dc.

20.3 Maximum Power and Impedance

The amount of power that can be provided from a source to a load is a practical concern. Accordingly, we examine an important insight known as the **Maximum Power Transfer Theorem**: *A maximum amount of power will be transferred from any source, with an internal resistance r, to a load of resistance R when R equals r.* For example, there are several different output taps on the back of a stereo amplifier precisely so that its resistance can be matched to the resistance of the speakers for maximum power transfer.

Return to Fig. 20.5, where a source with an internal resistance r is connected across a load R. Remembering that

$$I = \frac{\mathscr{E}}{R + r} \qquad [20.3]$$

the power delivered to the load is

$$\text{P} = I^2 R = \frac{\mathscr{E}^2}{(R + r)^2} R \qquad (20.7)$$

Notice that when $R \gg r$, the denominator is essentially R^2, and the power drops as $1/R$ to a low value. When $R \ll r$, the denominator is essentially r^2, but the power, which varies with R, must also be small. Somewhere between these extremes, P has a maximum. A plot of P versus R, Fig. 20.14, makes the same point.

The process of transferring energy from one system to another is fundamental, and the above ideas apply to a wide range of physical phenomena beyond electricity (for example, light, sound, mechanical vibrations, etc.). It is found

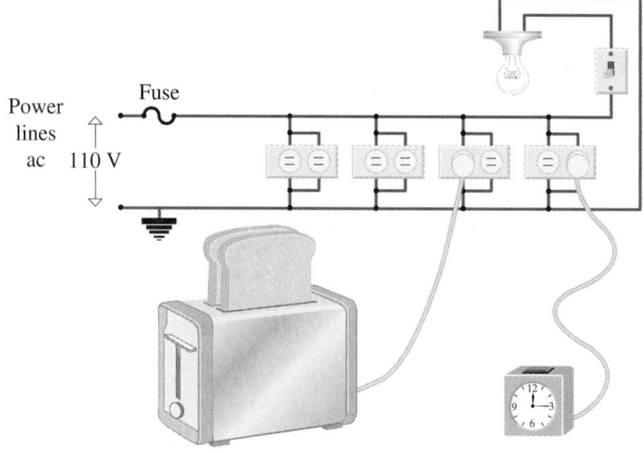

Figure 20.13 Household wiring, using alternating current, puts all appliances and lights in parallel across ≈110 V.

that by matching a certain characteristic of the provider-of-energy to that of the receiver-of-energy, a maximum transfer will take place. That characteristic is a generalized notion that corresponds to the resistance offered to the transport of energy, and it's known as the *impedance* (p. 808). Figure 12.39 shows this effect for mechanical waves on discontinuous ropes—maximum transmission occurs when the two ropes have equal densities (that is, equal impedances). Similarly, the kinetic energy transferred from a moving object to a stationary one (as the result of an elastic collision) has the same form where the impedance corresponds to the measure of the resistance to the change in motion—namely, the mass. When the masses are equal (that is, when the impedances are matched), the energy transferred is a maximum. One can contemplate the impedance of the ear, the impedance of a musical instrument, the impedance of an antenna, and so on, wherever energy is transferred.

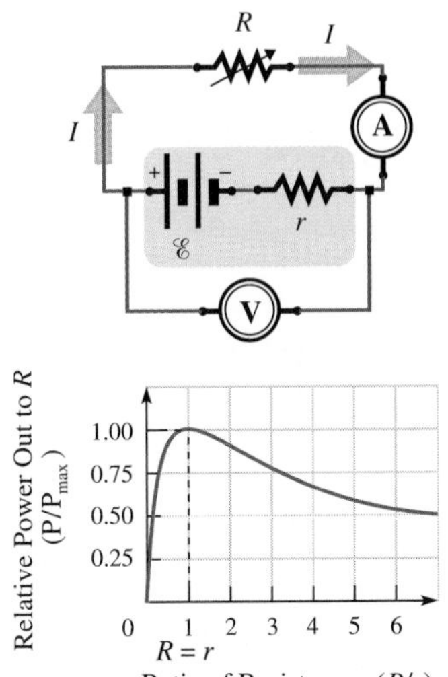

Figure 20.14 The ratio of the power out to a load resistor R, divided by the maximum power supplied to it, plotted against the ratio of R to the internal resistance of the source r. When $R = r$, the power P equals P_{max} and $P/P_{max} = 1$.

20.4 Ammeters and Voltmeters

Electrical meters of all kinds come in two varieties, analog and digital. An analog meter usually has a pointer that moves across a scale, whereas a digital meter provides direct numerical values of whatever is being measured. An analog meter is constructed around a moving coil galvanometer (p. 749). The latter is a coil of wire with a resistance r of perhaps 200 Ω or less, having a pointer fixed to it, placed between the poles of a magnet. Passing a current through the coil causes it to twist proportionately and the pointer swings around. The digital meter is an electronic device that takes a reading from the circuit, which it amplifies, digitizes, and displays. Digital meters tend to be more accurate, sturdier, and more expensive than analog meters, but they are both used in the same way.

Figure 20.15a represents the basic analog detector, a galvanometer comprising a resistor in series with an ideal rotating coil. The input that causes a maximum deflection of the pointer is called the *full-scale current,* and for an inexpensive device, that's about 1 mA. To use the galvanometer as an **ammeter**, a current meter, we must direct a small fixed fraction of the main stream of current through the galvanometer. That's accomplished by placing a small *shunt* resistor ($R_s \ll r$) across the galvanometer (Fig. 20.15b). The resulting instrument is an ammeter, and it is always positioned within the branch where the current I is to be measured. The branch current I entering the ammeter splits with part I_g going through the galvanometer and most of it, I_s, being shunted through the small resistor R_s. The specific value of R_s depends on the full-scale current sensitivity of the galvanometer. Thus, if the coil has a full-scale current of $I_g = 1.0$ mA and we want the ammeter to read $I = 1.0$ A at full scale, then, since $I = I_g + I_s$, $I_s = 0.999$ A. Because the voltage drops across R_s, and $r = 200\ \Omega$ are equal

$$I_g r = I_s R_s$$

hence $$R_s = \frac{I_g r}{I_s} = \frac{(1.0\text{ mA})(200\ \Omega)}{0.999\text{ A}} = 0.20\ \Omega$$

Notice that the analog dc ammeter has a specific polarity; the current must enter at the + terminal or the meter may be damaged. One nice feature of the digital ammeter is that we usually needn't worry about polarity.

A **voltmeter**, Fig. 20.15c, is a galvanometer in series with a resistance R. The voltmeter is always placed *across* the two points whose potential difference is to be measured. To ensure that little current passes through the coil, R is very large (usu-

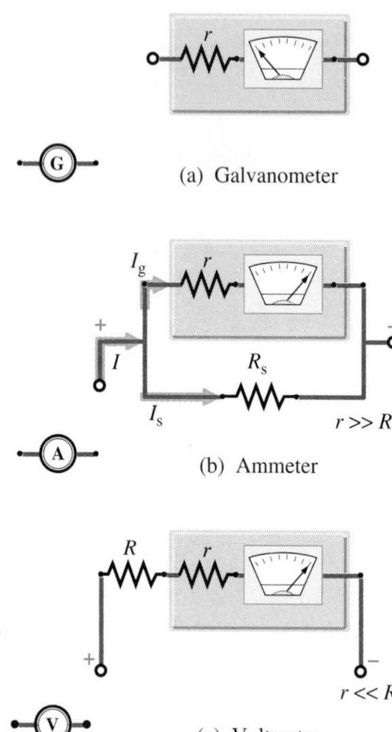

Figure 20.15 (a) A galvanometer having an internal resistance *r*. (b) When the galvanometer is placed in parallel with a low-resistance shunt R_s, we have an ammeter. (c) When the galvanometer is in series with a high-resistance *R*, we have a voltmeter.

ally in the high kilohm to megohm range). For example, suppose the voltmeter is to measure from 0 to 100 V, using our cheap galvanometer with a full-scale current of $I_g = 1.0$ mA and $r = 200\ \Omega$. That current must exist when the voltage is a maximum of 100 V; thus

$$V = I_g(R + r) = 100 \text{ V} = (1.0 \text{ mA})(R + 200\ \Omega)$$

and $R = 10 \times 10^4\ \Omega$. This meter is not a very good one because *R* is not really large enough. If it were placed across a several-kilohm resistor in a circuit, it would appreciably affect the current and produce an erroneous reading. By contrast, a fine electronic voltmeter might have a resistance of $10^8\ \Omega$ or more. Notice that the dc analog voltmeter also has a specific polarity and must be connected with its + terminal at the higher potential.

20.5 *R-C* Circuits

Any circuit containing a capacitor also has some resistance, even if it's only due to the wiring. Such ***R-C* circuits** are both commonplace and important—especially if you happen to be wearing a cardiac pacemaker. And that points up one of the most interesting features of *R-C* circuits. The transient states of the resistive circuits considered thus far are extremely brief, and so voltages and currents are taken to be constant at their steady-state values. By contrast, *R-C* circuits have time-dependent voltages and currents. That makes them very useful for creating a variety of circuits that produce time-varying signals (in a pacemaker, for example).

Imagine a capacitor with its plates charged at $\pm Q_i$ as shown in Fig. 20.16. There is an initial voltage $V_i = Q_i/C$ across the capacitor, but no charge flows while the switch *S* is open. At $t = 0$, the switch is closed, the voltage across the resistor is immediately V_i, and the charge on the capacitor begins to redistribute itself under the influence of the Coulomb force. For an instant there is an initial current through the circuit of

$$I_i = \frac{V_i}{R} = \frac{Q_i}{RC}$$

But as the charge on the capacitor decreases ($\Delta Q < 0$), the voltage decreases, and that decreases the current: $I = -\Delta Q/\Delta t$. At any moment, the voltage across the resistor *IR* equals the voltage across the capacitor Q/C, such that

$$IR = -\frac{\Delta Q}{\Delta t}R = \frac{Q}{C}$$

and

$$\frac{\Delta Q}{\Delta t} = -\frac{1}{RC}Q \tag{20.8}$$

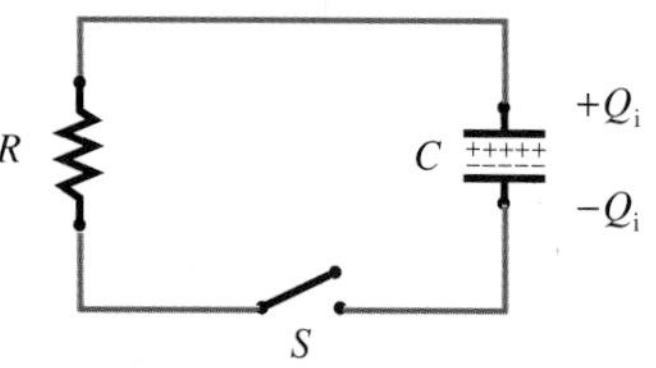

Figure 20.16 A charged capacitor *C* in series with a resistor and an open switch *S*. The charge on each plate is $\pm Q_i$.

which is a rather special and yet wonderfully common situation. **Whenever a physical quantity changes ($\Delta Q/\Delta t$) at a rate that depends on the quantity itself (Q), the quantity varies exponentially** (Appendix A-3). If we plot *Q* versus *t* (Fig. 20.17a), *the slope of the curve is proportional to the value of the curve:* when the charge is large, the slope ($\Delta Q/\Delta t$) is negative and large; when *Q* is small, the slope is small. We will examine the exact shape of the e^x curve later, in Chapter 32. For the moment, it will suffice just to see what the solution to Eq. (20.8) looks like; namely

$$Q = Q_i e^{-t/RC} \tag{20.9}$$

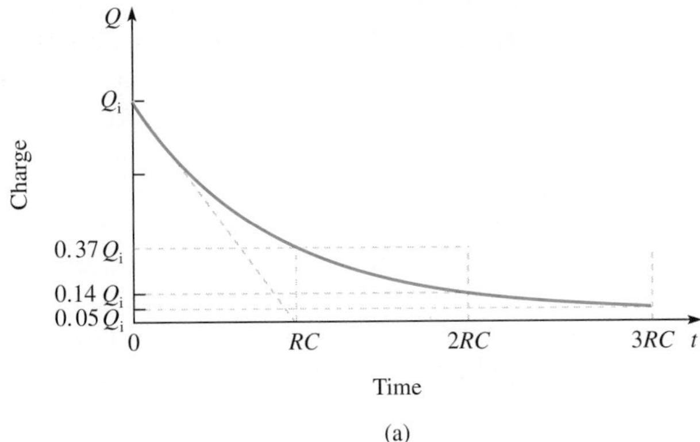

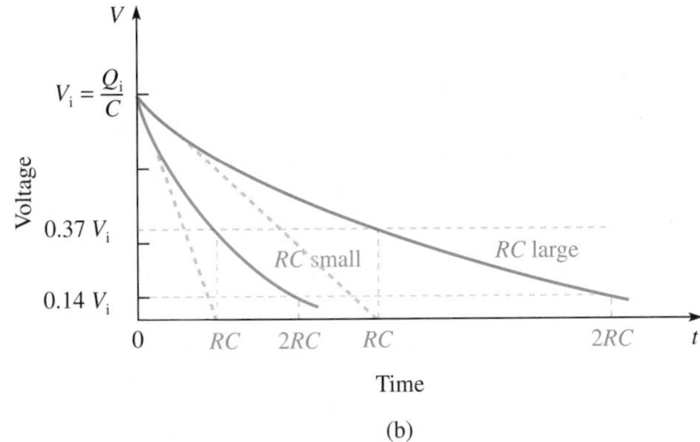

Figure 20.17 Curves for the charge and voltage of the capacitor in Fig. 20.16 once *S* is closed. (a) After each interval of time equal to *RC*, the curve of *Q* versus *t* drops to 36.787 9% of the previous value. (b) The voltage also decays exponentially.

Since $e^0 = 1$, at $t = 0$, $Q = Q_i$, which makes sense. At $t = \infty$, $e^{-\infty} = 0$ and $Q = 0$, which also makes sense; the capacitor is discharged. Notice from Eq. (20.8) that at $t = 0$, the slope is $-Q_i/RC$. The initial slope intersects the time axis at $t = RC$. At that time, $Q = Q_i e^{-1} = 0.37Q_i$. The quantity *RC* is called the **time constant** because it influences the temporal behavior of the circuit. That's evident in the plots of the voltage across the capacitor in Fig. 20.17b for two different values of *RC*. The voltage drops 63% of the way down toward zero in one time constant, and the smaller *RC* is, the faster that drop happens. The smaller is *R*, the less the resistor retards the flow of charge, and the smaller is *C*, the faster the capacitor discharges.

The circuit in Fig. 20.18 shows an uncharged capacitor in series with a resistor and an ideal battery. When the switch is closed, charge flows and a time-dependent current begins to circulate. At that moment, the charge on the capacitor *Q* is essentially zero as is the voltage drop across it; hence, $\mathscr{E} = I_i R$. At any instant thereafter

$$\mathscr{E} = IR + Q/C$$

Since $\mathscr{E}$ is constant, as the charge builds up (Fig. 20.19a), the current correspond-

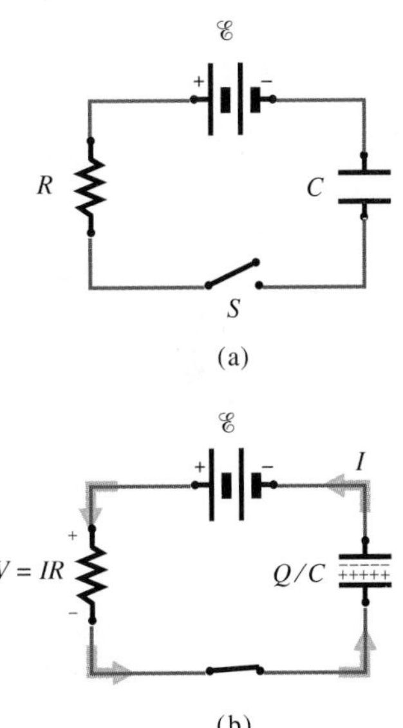

Figure 20.18 (a) When the switch in the circuit is closed, a transient current appears. (b) Positive charge goes out of the battery, across the resistor, and builds up on one plate of the capacitor. An equal amount of charge is repelled off the other plate and returns to the negative terminal of the battery.

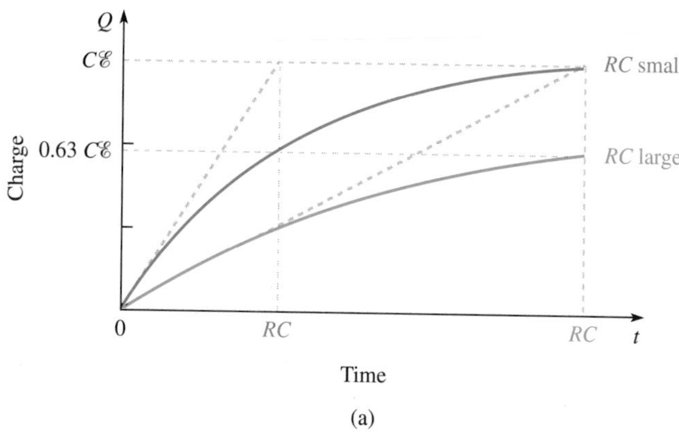

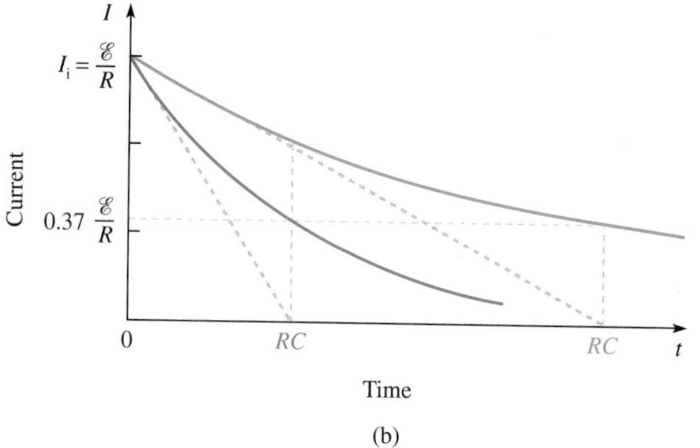

Figure 20.19 (a) As the charge on the capacitor in Fig. 20.18 builds up, (b) the current in the circuit dies away.

ingly dies off (Fig. 20.19b). As the current approaches zero, $\mathscr{E}$ approaches Q/C, and the maximum charge on the capacitor approaches $C\mathscr{E}$. The current never actually reaches zero; after 20 time constants ($t = 20RC$), it's down to $I = I_i e^{-20} = 2 \times 10^{-9} I_i$, which isn't much.

R-C circuits are often used as timers to control the periodic activation of other devices. For example, the intermittent action of an automobile windshield wiper is controlled by an R-C circuit. A cardiac pacemaker circuit with a time constant RC of 0.8 s can be used to reach a certain triggering voltage at a frequency of $1/(0.8 \text{ s}) = 1.25$ times per second, thereupon activating a signal to the heart at a rate of 75 pulses per minute.

NETWORK ANALYSIS

The first approach to analyzing any circuit is to reduce it to a manageable equivalent scheme, but there are many configurations that can't be simplified. Figure 20.20 shows a representative selection wherein the elements are neither in series nor in parallel. Complicated circuits are often called *networks*—Fig. 20.21 is an example of the genre. But before we get into the systematic approach to the solution of such circuits, it's appropriate to look at some special cases that are both important and remarkably simple. For example, although the circuit in Fig. 20.21 appears to be a small horror, there are aspects of it that are easily solved. What is the current through, and the voltage drop across, the 6.0-Ω resistor (R_1)? Happily, the voltage drop across the whole uppermost branch is apparent—it's 12 V because the two resistors (6.0 Ω and 4.0 Ω in series) are directly across the terminals of the 12-V source. All the complex circuitry in the middle has no effect on the voltage across the upper (10-Ω) branch. Thus, whatever the current may be, the drop across R_1 is 6/10 of 12 V, or 7.2 V. The current is then $I = V_1/R_1 = (7.2 \text{ V})/(6.0\ \Omega) = 1.2$ A or, alternatively, $I = V/(R_1 + R_2) = 1.2$ A.

Figure 20.22 depicts another rather formidable network, but there's something special here, too. What is the voltage drop across, and the current through, the 6.0-Ω resistor? Notice that when the current reaches node A, it "sees" two identical paths and therefore splits into equal parts. The voltage drop across R_1 is equal to the drop across R_3, and so points C and D are at the same potential. There is no voltage difference across R_5 and no current through it; R_5 has no effect on this so-called *bridge circuit!* It's as if the branch from C to D were removed.

20.6 Kirchhoff's Rules

The systematic analysis of networks proceeds from two basic rules set forth by the German physicist Gustav Robert Kirchhoff in the late 1800s. There's really nothing new about these ideas; what *is* new is the application. Kirchhoff's first rule is

> *The algebraic sum of the voltage rises and drops encountered in going around any closed path formed by any portion of a circuit must be zero.*

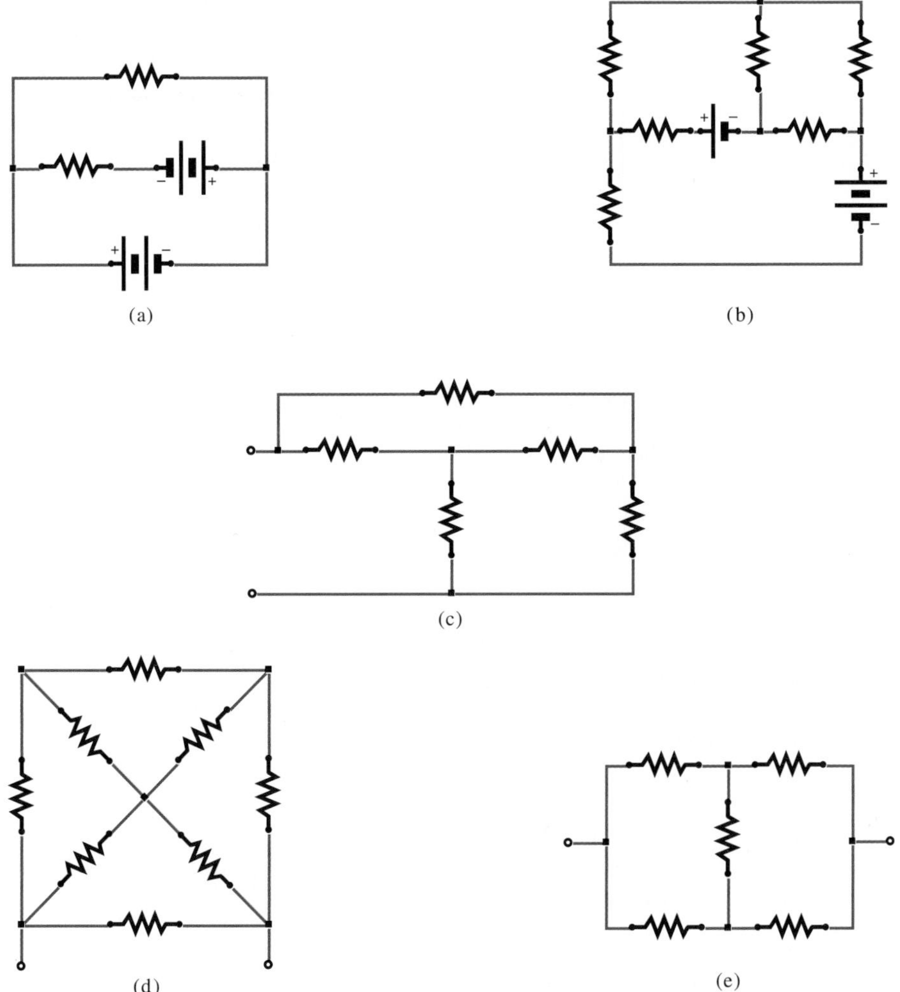

Figure 20.20 Several circuits in which the elements are neither in series nor in parallel.

An electronic photographic flash unit. The rechargeable battery is a stack of four cells at the upper left. The large cylinder is the capacitor that stores charge and energy. To fire the flash, the capacitor dumps its charge, via the red and blue leads, into the lamp. After a short wait, determined by *RC*, it's recharged and ready to go again.

Figure 20.21 We can easily calculate the current through R_1 since the voltage across the upper branch is 12 V.

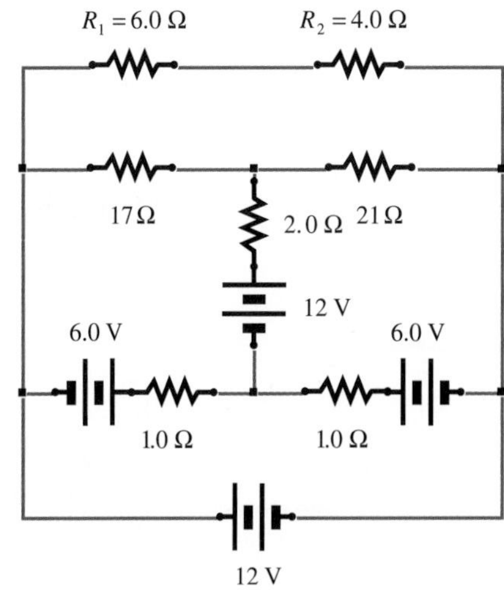

This **Loop Rule** maintains that if we start at some arbitrary point in a circuit at a given potential and follow any path that returns to that same point, there can be no net change in potential.

In exactly the same way, there is no change in *gravitational*-PE if you return to the same spot after meandering up and down a mountainside. Thus, the sum of the voltage rises (taken as positive) and the voltage drops (taken as negative) must equal zero. For a charge traversing a loop and returning to its starting point, the net energy gained must equal the net energy lost. The Loop Rule is a restatement of the Law of Conservation of Energy.

To apply the Loop Rule, let's determine the potentials at points *A*, *B*, *C*, *D*, *E*, and *F* in Fig. 20.23. This circuit has only one loop, and therefore only one branch, and one un-

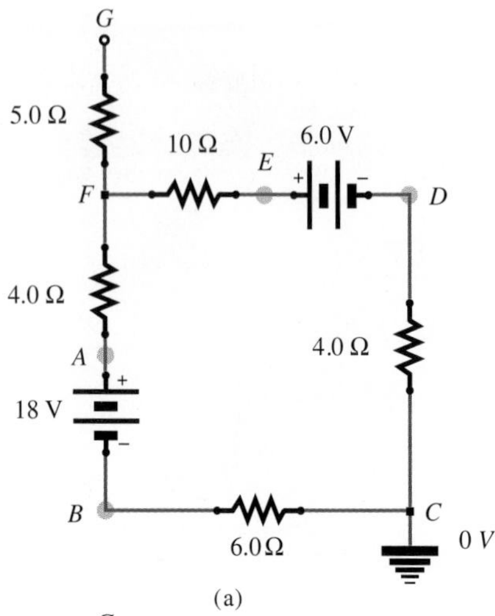

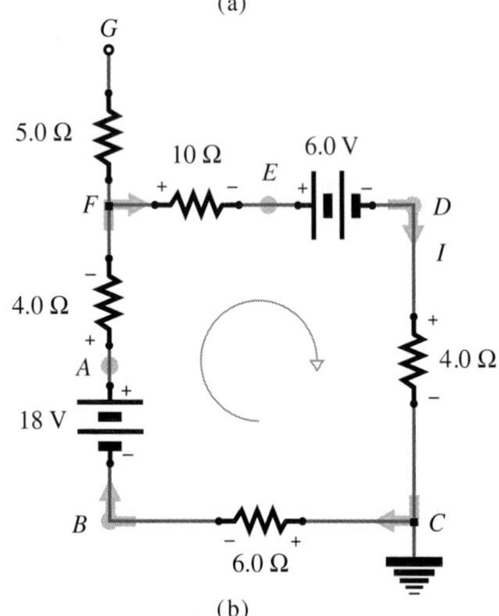

Figure 20.23 The analysis of a circuit using the Loop Rule.

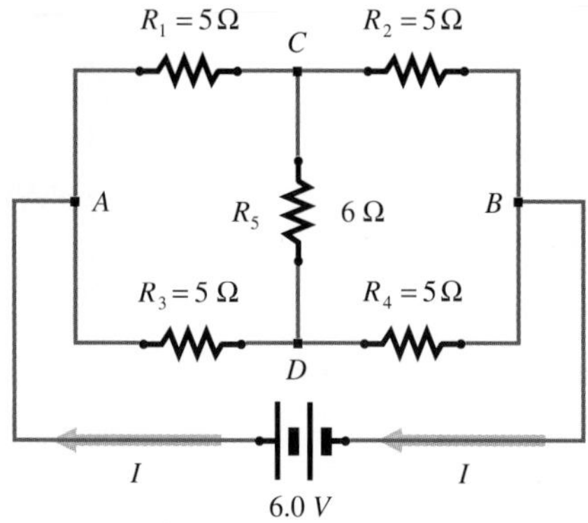

Figure 20.22 By symmetry, the voltage at C equals the voltage at D, so $V_{CD} = 0$. Therefore, no current passes through R_5, and R_1 and R_2 are in series, as are R_3 and R_4.

known branch current, I. Kirchhoff's Rules must be applied (thereby producing an equation) as many times as there are unknowns in the system.

The next step is to guess at the directions of the branch currents and draw them in. Assume that the larger battery dominates and the current emerges from its positive terminal. But not to worry, if ever the guess is wrong, the solution for I will simply come out negative, indicating that it's really in the opposite direction. Now label all the rises and drops with + and − signs depending on the direction of the current. *The signs of the emfs are independent of the current.* At this point, we are ready to apply the Loop Rule. Start at any point, say A, and go around the loop in either direction, writing the loop equation as you go. Let's do it clockwise: from A to F, there's a drop of $-4.0I$; from F to E, a drop of $-10I$; from E to D, a drop of -6.0 V; from D to C, a drop of $-4.0I$; from C to B, a drop of $-6.0I$; and from B to A, a rise of $+18$ V. Thus

$$-4.0I - 10I - 6.0 - 4.0I - 6.0I + 18 = 0$$

and

$$-24I + 12 = 0$$

hence

$$I = \tfrac{1}{2}\,\text{A}$$

The current is positive, and therefore all the drops are correct as indicated. Since C is grounded and at a potential of zero, B is at -3.0 V. That is, to go from B to C, we rise $(\frac{1}{2}\,\text{A})(6.0\,\Omega)$: A is at $+15$ V, F is at $+13$ V, E is $+8.0$ V, and D is $+2.0$ V. Notice that the potential at any point—for example, E—is indeed the same whether we go from C to D to E or from C to B to A to F to E.

Kirchhoff's second rule is known as the **Node Rule**:

The sum of all the currents entering any node in a circuit must equal the sum of all the currents leaving that node.

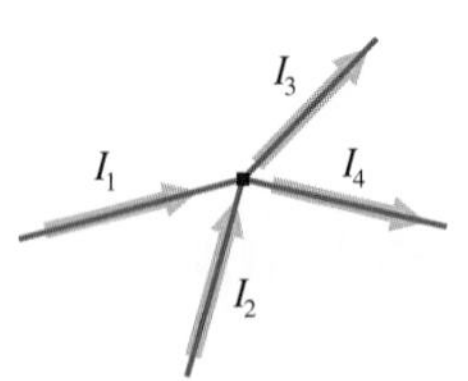

Figure 20.24 The Node Rule states that the sum of the currents entering a node must equal the sum of the currents leaving the node.

In the steady state, there is neither an accumulation nor a diminution of charge anywhere in the circuit. Hence, at a node, the net amount of charge flowing in must equal the net amount flowing out—that's the Principle of Conservation of Charge. Figure 20.24 illustrates the point: $I_1 + I_2 = I_3 + I_4$.

To see how all of this comes together in practice, let's apply Kirchhoff's Rules to the network depicted in Fig. 20.25a, determining the current through every element. First, notice that the circuit has only two nodes—C and E—and therefore there are three branches C–B–A–E, C–D–E, and C–E. Assign a current arbitrarily to each branch, as shown in Fig. 20.25b and label with signs all the voltage drops and rises.

Three branches means three unknown currents, which requires three equations. But when it comes to picking those three equations, we have to be cautious. There are three loops: the lower one A–B–C–E–A, the upper one A–B–C–D–E–A, and the little one C–D–E–C, as well as two nodes C and E. Thus, there are five possible equations, *but not any three will do*—we need three independent equations. To that end, *use a mix of loop and node equations and always avoid using all the equations of either kind.* For instance, here the two node equations are

$$I_1 + I_2 = I_3 \qquad \text{at } C$$

and

$$I_3 = I_1 + I_2 \qquad \text{at } E$$

These are identical and not independent—only one of them can be used. *If two nodes involve the same set of currents, they will not produce independent equations* (and you should not waste time playing with them both). The three loop equations are also not independent. Consequently, we utilize any two loop equations and either node equation.

Using loop A–B–C–E–A and going around clockwise starting at A, we have

$$-10I_1 - 20I_1 - 30I_3 + 40 = 0$$

or

$$-30I_1 - 30I_3 + 40 = 0 \tag{i}$$

Similarly, going clockwise around loop C–D–E–C starting at C, yields

$$-20 + 10I_2 + 30I_3 = 0 \tag{ii}$$

These two equations can be solved simultaneously if we use the node equation to get rid of one of the variables, say, $I_2 = I_3 - I_1$. Hence, Eq. (ii) becomes

$$-20 + 10(I_3 - I_1) + 30I_3 = 0$$

and

$$-10I_1 + 40I_3 - 20 = 0 \tag{iii}$$

Multiplying this by 3 and subtracting it from Eq. (i) yields

$$-150I_3 + 100 = 0$$

and $I_3 = 2/3$ A. It follows from Eq. (i) that $I_1 = 2/3$ A. And, finally, something of a surprise: $I_2 = 0$. As a *quick check,* we see that the voltage difference in going from C to E is -20 V via either branch and, moreover, the sum of the drops and rises is indeed zero around each loop.

As a last point, observe that loop A–B–C–D–E–A produces (going clockwise from A) the equation

$$-10I_1 - 20I_1 - 20 + 10I_2 + 40 = 0$$

which is

$$-30I_1 + 10I_2 + 20 = 0 \tag{iv}$$

If we subtract this result from Eq. (i), we get Eq. (ii) and vice versa, proving that we could not solve the problem with just those three—they're not independent.

If there are N nodes, use $N-1$ node equations along with as many loop equations as is necessary to form a total number of equations equal to the number of unknowns. Every element of the circuit must appear in at least one loop equation.

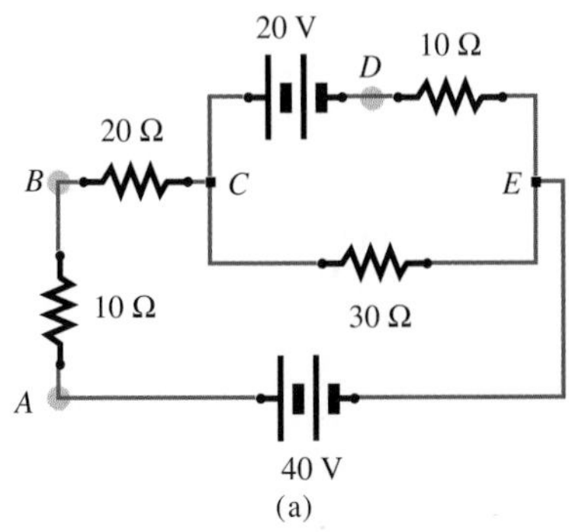

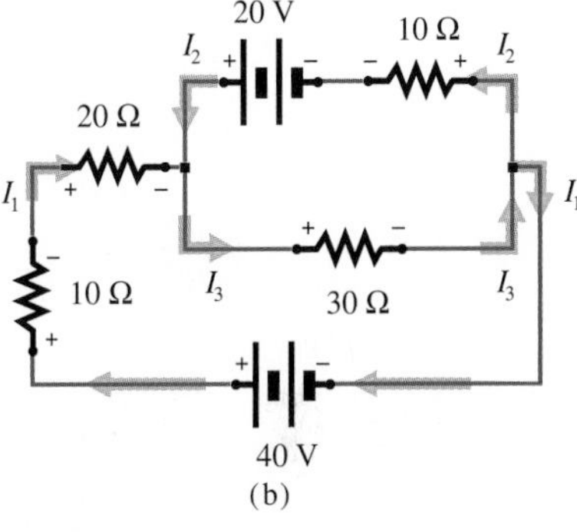

Figure 20.25 A network to be analyzed using Kirchhoff's Rules. Each branch has a current assigned to it.

Example 20.5 Determine the power delivered to the circuit by the 12-V source in Fig. 20.26.

Solution: [Given: circuit. Find: P delivered by 12-V source.] What's needed is the current through the 12-V source: $P = IV$. Let's first simplify the circuit as much as possible—Fig. 20.26b, c, and d show the series and parallel combinations. There are two nodes in the final circuit configuration A and B and three branches B–F–A,

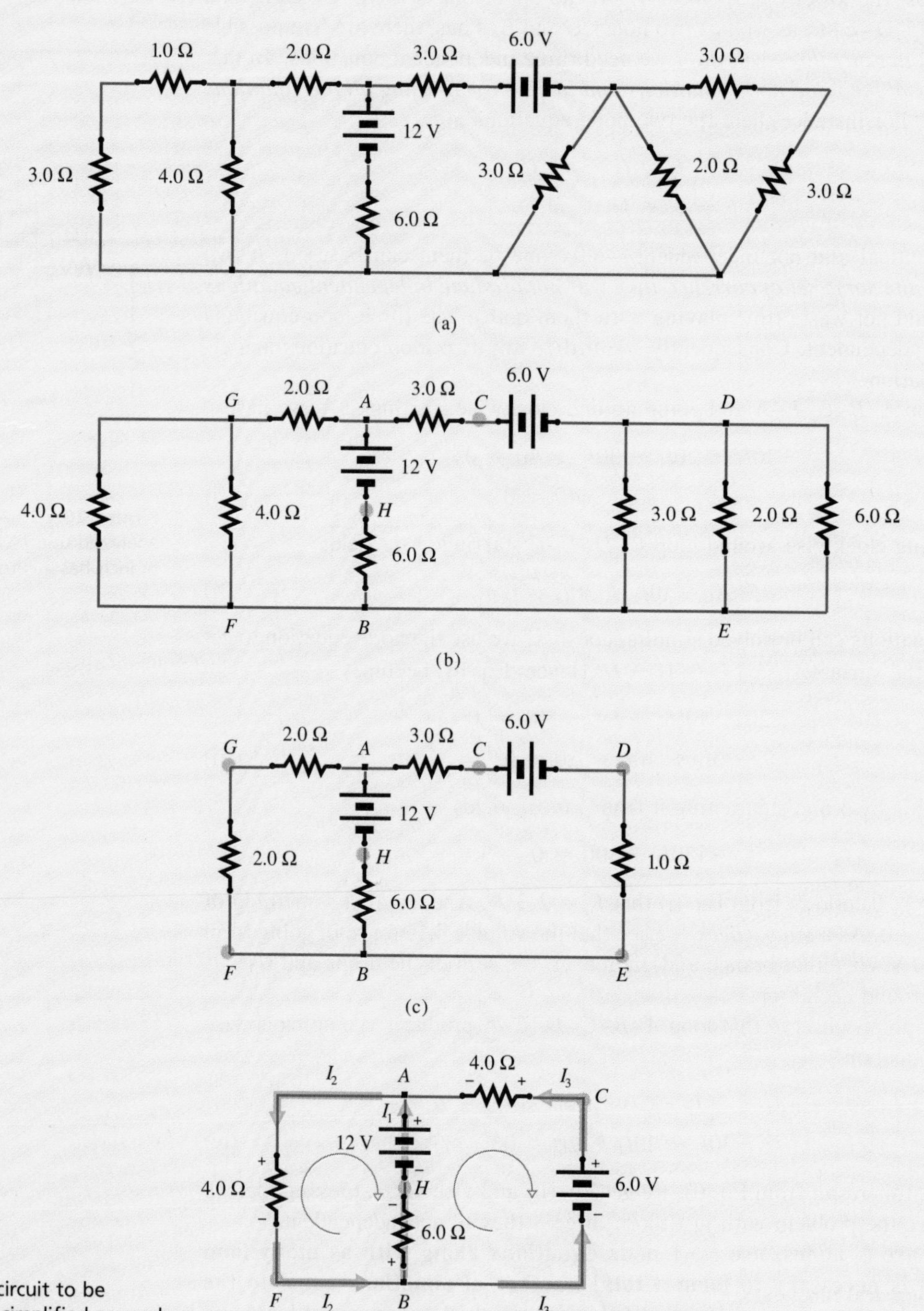

Figure 20.26 A circuit to be analyzed. It is first simplified as much as possible.

(continued)

(continued)
B–H–A, and B–C–A. Hence, we assign three branch currents. The same currents are involved at A and B so only one node equation will be used; accordingly

$$I_1 + I_3 = I_2 \qquad \text{(i)}$$

Starting at B and going clockwise around loop B–F–A–H–B, the Loop Rule yields

$$+\,4.0I_2 - 12 + 6.0I_1 = 0 \qquad \text{(ii)}$$

Starting at B and going clockwise around loop B–H–A–C–B, the Loop Rule yields

$$-\,6.0I_1 + 12 + 4.0I_3 - 6.0 = 0$$

or

$$-\,6.0I_1 + 4.0I_3 + 6.0 = 0 \qquad \text{(iii)}$$

We just want I_1, so use Eq. (i) to get rid of I_2 in Eq. (ii), which then becomes

$$10I_1 + 4.0I_3 - 12 = 0 \qquad \text{(iv)}$$

Subtracting Eq. (iv) from Eq. (iii) results in

$$-16I_1 + 18 = 0$$

and $I_1 = +1.125$ A. Current passes through the source from − to +; hence, power *is* delivered by the source. Consequently

$$P = IV = (1.125 \text{ A})(12 \text{ V}) = 13.5 \text{ W}$$

or to two figures $\boxed{P = 14 \text{ W}}$.

▶ **Quick Check:** From Eq. (ii), $I_2 = 1.31$ A and the potential difference from B to A via F is +5.24 V, which must equal the rise from B to A via H; namely, $-(6.0\ \Omega) \times (1.125\text{ A}) + 12 = +5.25$ V.

Core Material

The voltage (V) across the terminals of a source of emf ($\mathscr{E}$) having an internal resistance (r) is

$$V = \mathscr{E} - Ir \qquad [20.1]$$

when the source provides power, and

$$V = \mathscr{E} + Ir \qquad [20.2]$$

when power is provided to it (p. 691).

The **equivalent resistance** of two or more resistors in series is

$$R_e = R_1 + R_2 + \cdots \qquad [20.4]$$

The equivalent resistance of two or more resistors in parallel is

$$\frac{1}{R_e} = \frac{1}{R_1} + \frac{1}{R_2} + \cdots \qquad [20.5]$$

For two resistors, this equation can be expressed as

$$R_e = \frac{R_1 R_2}{R_1 + R_2} \qquad [20.6]$$

The **Maximum Power Transfer Theorem** states that *a maximum amount of power will be transferred from any source with an internal resistance r to a load of resistance R when r equals R* (p. 698).

Kirchhoff's **Loop Rule** is *the algebraic sum of the voltage rises and drops encountered in going around any closed path formed by any portion of a circuit must be zero.* Kirchhoff's **Node Rule** is *the sum of all the currents entering any node in a circuit must equal the sum of all the currents leaving that node.*

Suggestions on Problem Solving

1. A common error is to use Eq. (20.5) for the equivalent resistance of a parallel string of resistors, computing and adding $1/R_1$ and $1/R_2$ and then presenting that as R_e rather than $1/R_e$; look out for this one. Note, too, that the reciprocal of the sum is *not* equal to the sum of the reciprocals: $1/(R_1 + R_2) \neq 1/R_1 + 1/R_2$. Equation (20.6), $R_e = (R_1R_2)/(R_1 + R_2)$, is very useful for getting a quick sense of parallel combinations without a calculator. The reciprocal function on the calculator is wonderfully helpful in computing Eq. (20.5) directly. For example, to compute R where $1/R = 1/4 + 1/3$, hit the following sequence of keys on your calculator: [4] [1/x] [+] [3] [1/x] [=] [1/x] to get 1.7. The expression $R_1R_2R_3/(R_1 + R_2 + R_3)$ is *not* the correct formula for three resistors in parallel!

2. When analyzing a circuit, study it carefully. Work with a drawing of the network. If the original version is not neatly diagrammed, redraw it before proceeding. Is there anything extraordinary about the circuit (as, for example, in Fig. Q3)? Are there any obvious shortcuts to finding the required quantities (as, for example, in Fig. 20.21)? Is the circuit symmetrical in such a way that some sort of bridge is present so that elements might be redundant (as, for example, in Fig. 20.22)? Remember, if no current traverses a particular resistor, it can be removed from the

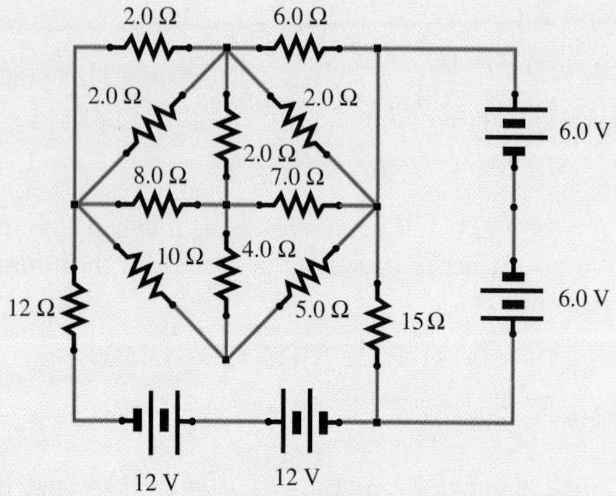

Figure 20.27

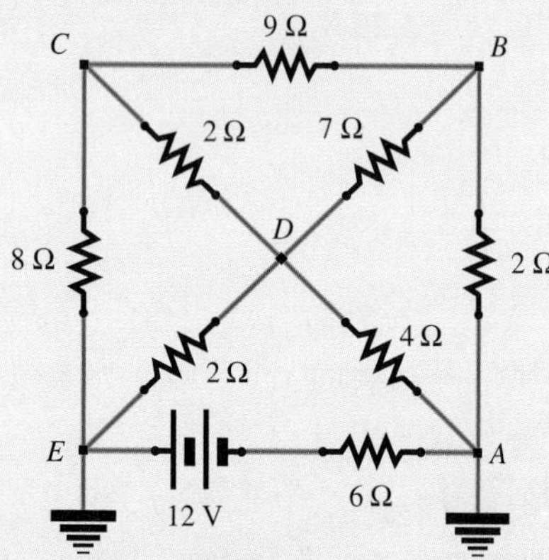

Figure 20.28

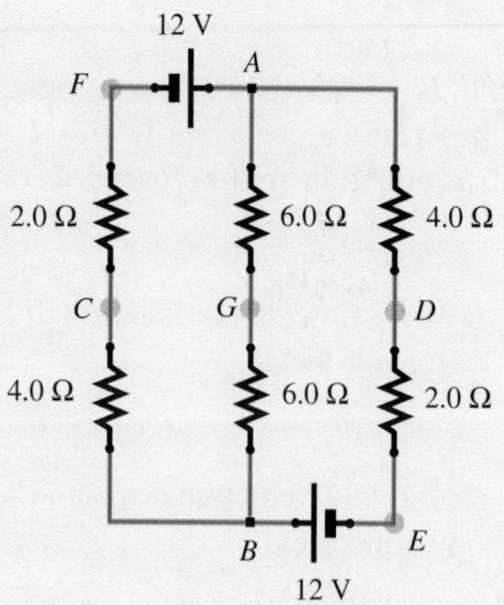

Figure 20.29

circuit. All right, then, what current passes through the 8-Ω resistor in Fig. 20.27? Zero! All the sources buck each other—there's no current anywhere. Try Fig. 20.28. What's the current in the 6-Ω resistor? By inspection, it must be 2 A to the left. Both ends of the bottom branch are grounded, so the net potential difference between E and A is zero. Going from A to E, there must be a drop of 12 V across the resistor since there is a rise of 12 V across the battery. In Fig. 20.28, what is the current in the 9-Ω resistor?

Again, it's zero because there is no voltage across E–A. All the current from the battery ($I = 2$ A) goes from E into the zero-resistance path back to A via ground. When ground connections are indicated, it should be assumed that all such leads are wired to a common line even if not shown.

Referring to Fig. 20.29, symmetry demands that current only circulate in the outer elements. Points A and B are at the same potential because the circuit is symmetrical. Therefore, no current can go across the resistors in that branch. The current through both batteries is 2 A.

3. The first step in an analysis is to simplify the circuit by combining series and parallel groupings wherever possible. Redraw the circuit whenever needed, keeping track of what you have done so that you can work your way back to the original. (a) If the system reduces to one source across an equivalent resistance, use Ohm's Law to find what is unknown and work backwards reconstructing the original circuit step-by-step. (b) If the circuit cannot be further simplified, apply Kirchhoff's Rules to what you have. (c) Line up the several variables, essentially making columns of them so you can see what needs to be done at a glance. For example

$$-3I_2 - 2I_3 \qquad + 12 = 0 \qquad \text{(i)}$$

$$+ 2I_3 - 3I_4 - 6 = 0 \qquad \text{(ii)}$$

$$-3I_2 \qquad - 3I_4 + 6 = 0 \qquad \text{(iii)}$$

Discussion Questions

1. Figure 20.5 depicts a battery with an internal resistance of r in series with a load R. What experimental procedure might be used to determine r?

2. What is meant by a complete or closed circuit and what is the relationship between such a circuit and a steady-state direct current? When the switch in Fig. 20.18 is closed, a transient current circulates in the circuit. Explain how this could happen in spite of the gap in the capacitor that stops charge from literally crossing it.

3. Figure Q3 is a reproduction of a circuit taken from a simulated MCAT exam created by a well-known company that prepares students for that test. The question asks for the current I and the answer given is "9 amps toward G." What's wrong with that and why do you think they give 9 A as the answer?

4. In light of Question 2, discuss the statement: *In the steady state, a capacitor acts as an open circuit to dc.*

5. The circuit of Fig. Q5 contains a gas-filled lamp—the sort of flashing light often used at construction sites. At low voltages, the lamp has nearly infinite resistance, but at a certain breakdown voltage, which is less than the terminal voltage of the

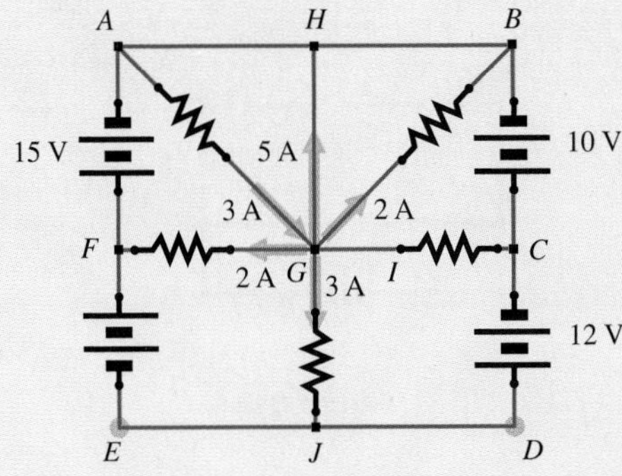

Figure Q3

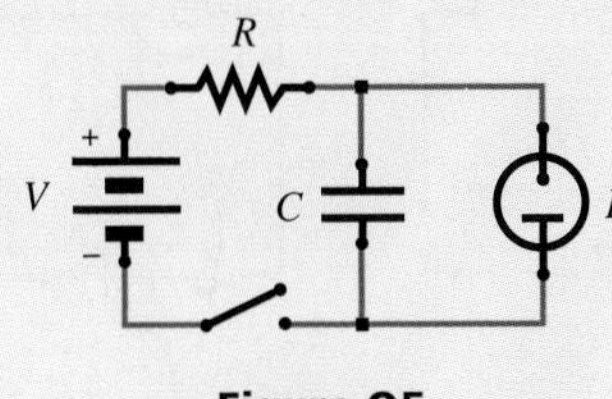

Figure Q5

battery, it becomes a very good conductor. A blast of charge through the lamp causes it to flash brightly. What happens once the switch is closed?

6. Suppose we have a fluid voltaic cell across which we place a load resistor. Next we measure the current from, and voltage across, the source with an ammeter and a voltmeter, respectively. It can be observed that when the plates are moved apart, the current diminishes. Moreover, when the plates are partly lifted out of the electrolyte, the current again drops, although in both cases the voltage remains unchanged. What's happening to the internal resistance of the cell? Relate your conclusions to the fluid-flow analogy of current.

7. *R-C* circuits are often used to change the shape of a signal. The arrangement of Fig. Q7 shows a network being fed a square-wave signal comprising a series of rectangular dc voltage pulses rising to $+V$ and falling to 0. Part (b) shows the corresponding voltage across the capacitor. Explain its shape. What can you say about the value of RC as compared to the width of each pulse? Draw a curve of the output signal as seen across the resistor and explain its features. Write a general expression relating V, V_R, and V_C.

8. Explain how a flashlight bulb works. Draw a picture showing the relationship between a bulb's filament and its two terminals. Why does the brightness depend on the current through the bulb? What effect would increasing the voltage across a bulb have on its brightness? Incidentally, light bulbs are not ohmic.

9. Figure Q9 depicts several voltage-versus-time curves. These might be the outputs of a generator, transducer, etc. Which of them corresponds to dc? Explain.

10. Draw a flashlight bulb and show where the current usually enters and leaves. Draw a D-cell and label its two terminals. Now draw a diagram showing how to connect the bulb and cell with two lengths of wire. How might you connect them with one wire? Show the path of the current in each case.

11. Figure Q11 shows three arrangements of identical flashlight bulbs along with an ideal 6-V source. Compare the brightness of all the bulbs, listing them in descending order. Explain.

12. Compare the brightness of all the bulbs in Fig. Q12, listing them in descending order. The battery is ideal. Explain.

13. Compare the brightness of each identical bulb in Fig. Q13. Explain.

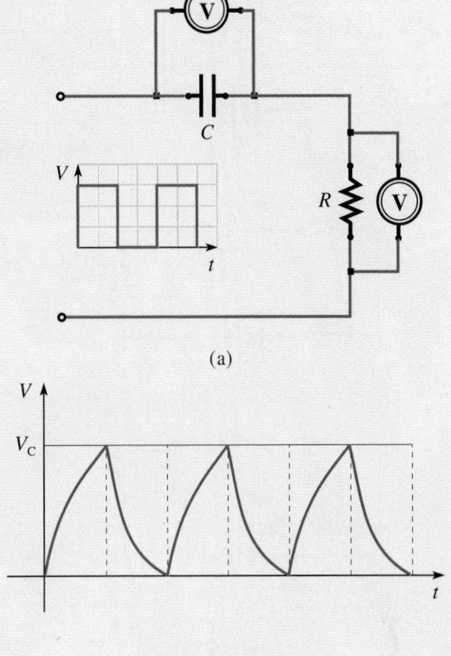

Figure Q7

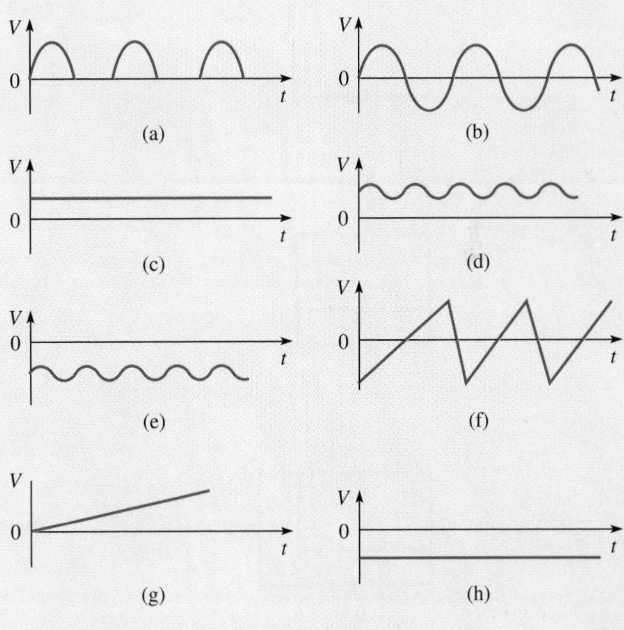

Figure Q9

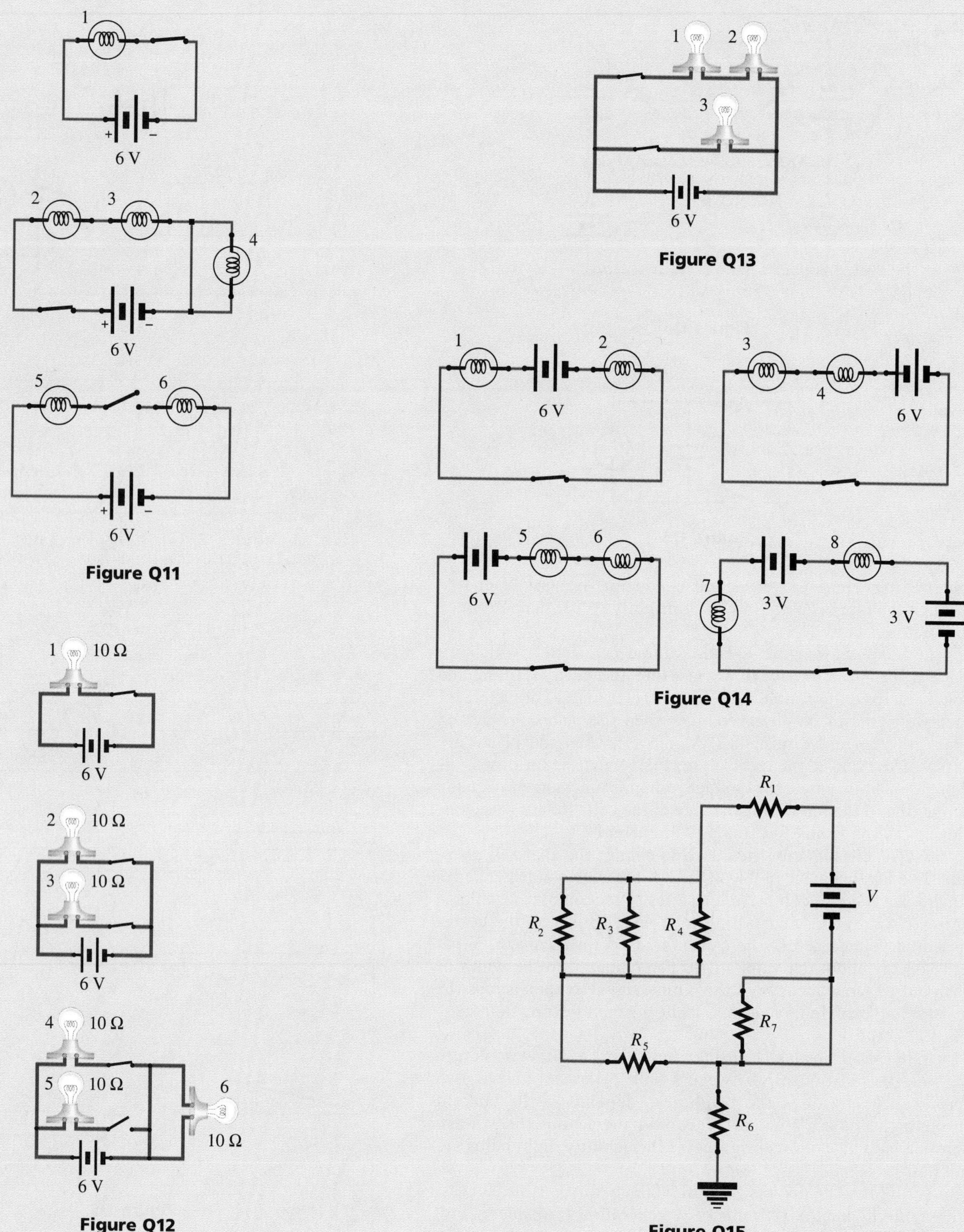

Figure Q11

Figure Q12

Figure Q13

Figure Q14

Figure Q15

14. All the bulbs in Fig. Q14 are identical. Discuss the relative brightness of each and explain your answer.

15. If all the resistors in Fig. Q15 are identical, rate them in order of the amount of current passing through each one. Explain.

Multiple Choice Questions

1. The equivalent resistance of a 4-Ω resistor in parallel with a 12-Ω resistor is (a) 48 Ω (b) 8 Ω (c) 16 Ω (d) 3 Ω (e) none of these.
2. An ideal source of emf is connected in series with a resistor in a closed loop, and a current of 0.50 A circulates. An additional 6.0-Ω resistor is added in series, and the current drops to 0.30 A. The original resistance was (a) 9.0 Ω (b) 6.0 Ω (c) 15 Ω (d) 3.0 Ω (e) none of these.
3. In the previous question, what was the emf? (a) 9.0 V (b) 12 V (c) 15 V (d) 4.5 V (e) none of these.
4. A 24-W, 12-V ohmic lamp is placed in a circuit with 6 V across it, at which time it draws a current of (a) 2 A (b) 1 A (c) 0 (d) $\frac{1}{2}$ A (e) none of these.
5. The device in Fig. MC5 consists of a rotating switch, five different resistors, and a galvanometer. What is the apparatus? (a) an ohmeter (b) a multirange voltmeter (c) a multirange ammeter (d) a multirange wattmeter (e) none of these.

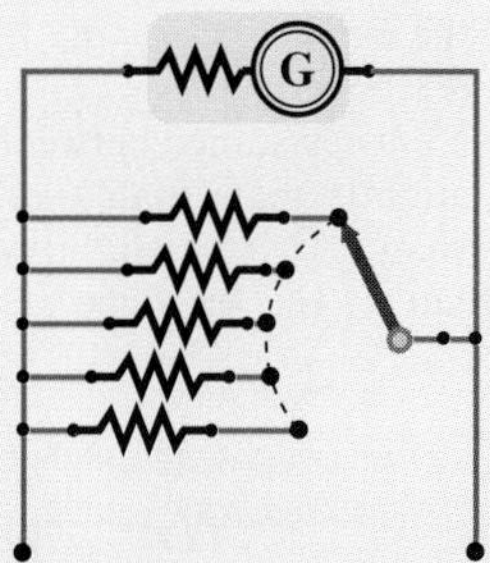

Figure MC5

6. The circuit in Fig. MC6 contains four identical light bulbs in series with an ideal battery. A wire is then clipped to points B and E and (a) lamps 3 and 4 become brighter (b) lamp 1 gets dimmer (c) lamps 2, 3, and 4 remain unaffected (d) lamp 1 brightens (e) none of these.

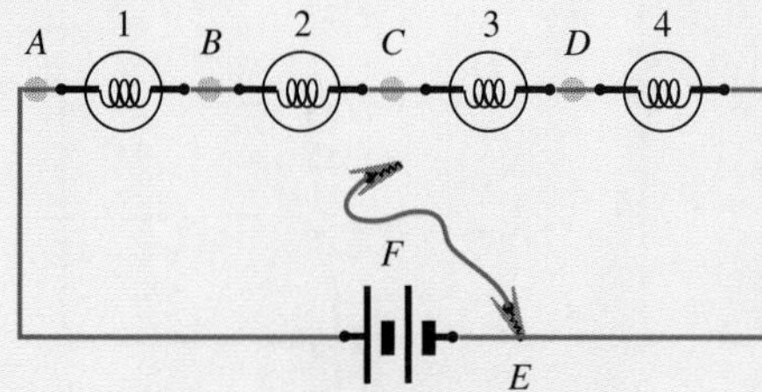

Figure MC6

7. In Fig. MC6, a wire is clipped to points B and E and (a) lamps 2, 3, and 4 become brighter (b) lamps 1 and 2 get dimmer (c) lamps 2, 3, and 4 go out altogether (d) lamp 1 goes out (e) none of these.
8. In Fig. MC6, a wire is clipped to points F and E and (a) lamps 1, 2, 3, and 4 become brighter (b) lamp 1 gets slightly dimmer (c) lamps 2, 3, and 4 remain unaffected (d) lamps 1, 2, 3, and 4 get slightly dimmer (e) none of these.
9. In Fig. MC6, a wire is clipped to points C and E and (a) more current is drawn from the battery (b) less current is drawn from the battery (c) the same current is drawn from the battery (d) not enough information is given to tell what will happen to the current (e) none of these.
10. If Fig. MC6, a wire is clipped to points C and E and (a) the voltage across lamps 1 and 2 increases (b) the voltage across lamps 3 and 4 increases (c) the voltage across only lamp 1 increases (d) the voltage across lamp 2 decreases (e) none of these.
11. In Fig. MC6, a wire is clipped to points C and E and (a) the power delivered by the battery remains unchanged (b) the power delivered by the battery increases (c) the power dissipated by the circuit decreases (d) the power dissipated by the circuit is halved (e) none of these.
12. Suppose the four lamps in Fig. MC6 are rewired so they are in parallel across the battery: (a) each lamp will put out the same amount of light as before (b) each lamp will put out more light than before (c) each lamp will put out less light than before (d) there is not enough information to tell what will happen (e) none of these.
13. The two circuits (i) and (ii) shown in Fig. MC13 are each composed of identical resistors connected to a battery. The resistors in (i) and the resistors in (ii), respectively, are hooked up in (a) series and series (b) parallel and

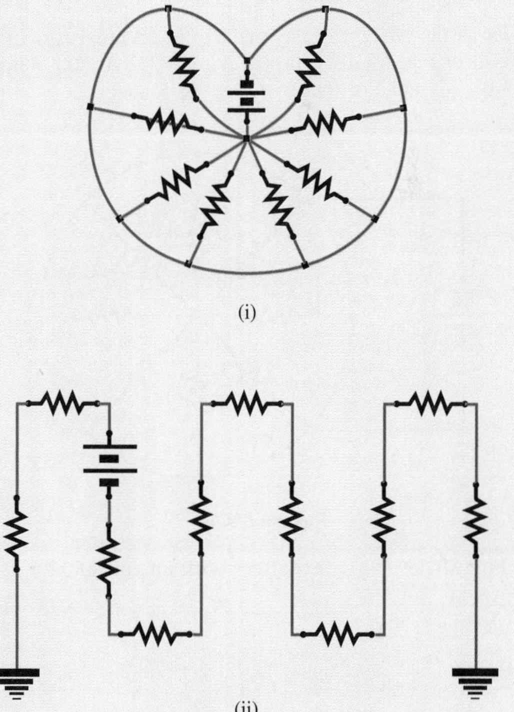

Figure MC13

parallel (c) series and parallel (d) parallel and series (e) none of these.

14. A battery of emf 12 V and internal resistance 0.2 Ω is placed across a variable resistor. The resistor is adjusted until it dissipates a maximum amount of power, at which point the current through it is (a) 12 A (b) 0.4 A (c) 30 A (d) 60 Ω (e) none of these.

15. A battery attached to a load supplies 2 A with a terminal voltage of 12 V. If the battery dissipates 0.4 W, its emf is (a) 11.8 V (b) 12 V (c) 12.2 V (d) 12.4 V (e) none of these.

16. Two resistors of 5 Ω and 20 Ω are connected in parallel across an ideal source of 20 V. The current supplied by the source is (a) 4 A (b) 5 A (c) 20 A (d) 1 A (e) none of these.

17. Figure MC17 shows a portion of a network. The current I is (a) +2 A (b) +4 A (c) not enough information to tell (d) −4A (e) none of these.

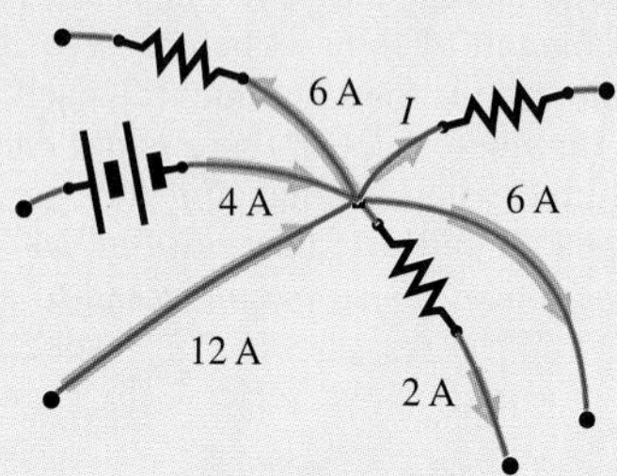

Figure MC17

18. What is the current in any one of the 4-Ω resistors in the circuit of Fig. MC18? (a) 12 A (b) 1.2 A (c) 3 mA (d) 0 (e) none of these.

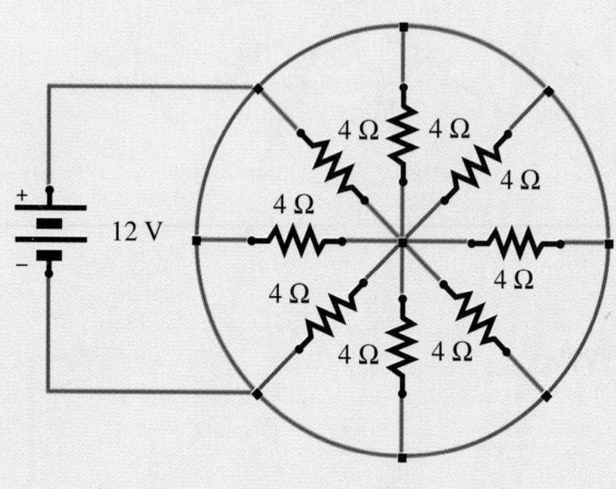

Figure MC18

19. In Fig. MC19, (a) the battery with an emf of 12 V is being charged (b) the battery with an emf of 6 V is being charged (c) the battery with an emf of 6 V is being discharged (d) neither battery is being charged (e) none of these.

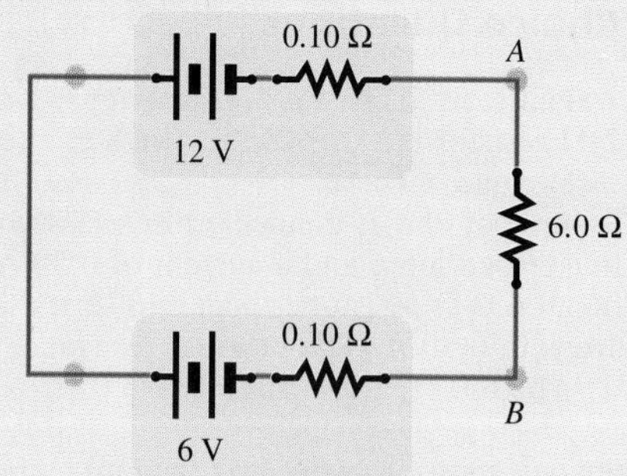

Figure MC19

20. In Fig. MC19, the higher-voltage battery (a) internally dissipates more power via joule heating than the lower-voltage battery (b) internally dissipates less power via joule heating than the lower-voltage battery (c) dissipates more power than the 6-Ω resistor (d) has a terminal voltage of 14 V (e) none of these.

21. What is the equivalent resistance between A and B in Fig. MC21? (a) slightly more than 1 kΩ (b) 4 kΩ (b) slightly more than 1 Ω (c) slightly more than 17 kΩ (d) slightly less than 1 kΩ (e) none of these.

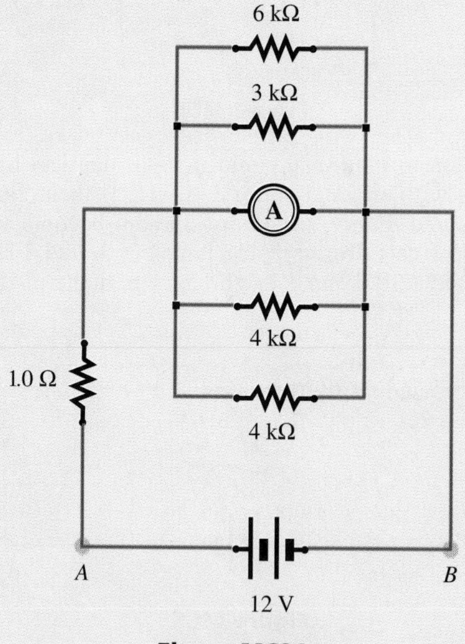

Figure MC21

Problems

CIRCUIT PRINCIPLES

1. [I] What is the current in the circuit of Fig. P1 before and after the wire is clipped on at C and B? What happens to the 12-Ω bulb after the wire is attached? The battery has a negligible internal resistance.

2. [I] A dc source with an internal resistance of 0.10 Ω is

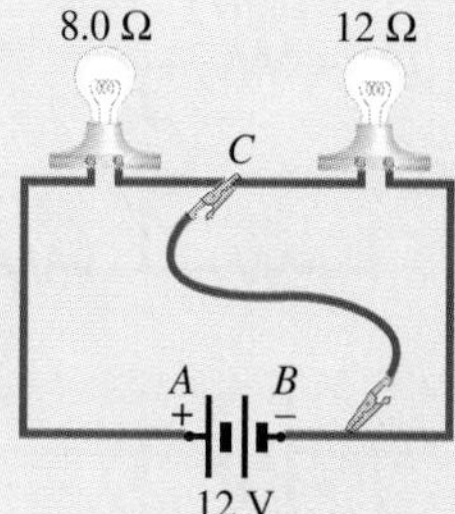

Figure P1

connected across a length of nichrome wire having a resistance of 20 Ω. If a voltmeter across the nichrome indicates a drop of 10 V, what is the emf of the source?

3. [I] Each lamp in Fig. P3 has a resistance of 20 Ω. How much current is drawn from the battery before and after the switch is closed? How does the power supplied to the circuit change when the switch is closed? The battery has a negligible internal resistance.

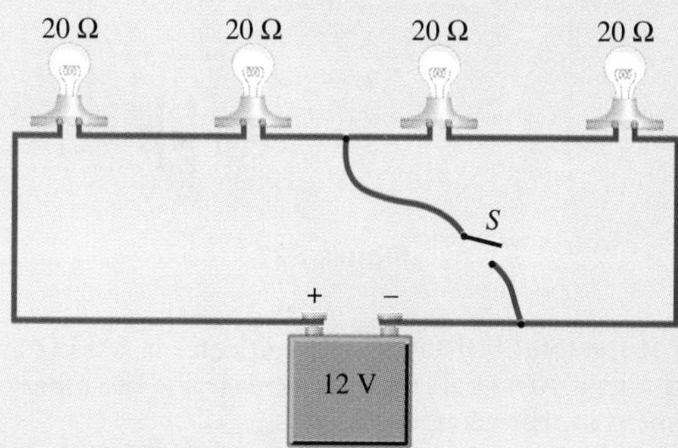

Figure P3

4. [I] A dc source with a terminal voltage of 100 V internally dissipates 40 W as it delivers 2.0 A. What is its emf?
5. [I] Two 2.0-Ω resistors are in parallel. What is their equivalent resistance?
6. [I] Three resistors with values of 2.0 Ω, 3.0 Ω, and 6.0 Ω are connected in parallel. What is the equivalent resistance?

7. [I] A portable generator in a field hospital produces dc at 100 V. Five 100-W lamps are attached in parallel and the string placed across the terminals of the generator. How much current must the source provide for the lamps to operate as designed?

8. [I] Three resistors with values of 1/2 Ω, 1/3 Ω, and 1/6 Ω are connected in parallel. What is the equivalent resistance?
9. [I] Show that if there are a number N of identical resistors in parallel, each with a value R, then $R_e = R/N$.
10. [I] If in Fig. P10 the ideal ammeter reads 3.20 A, what will the ideal voltmeter read?

11. [I] What is the equivalent resistance of the circuit in Fig. P11 between terminals A and B? Note that the wires cross but do not make contact at the center.

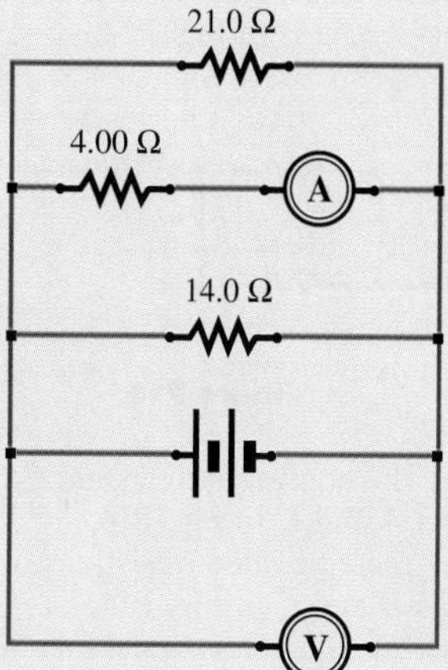

Figure P10

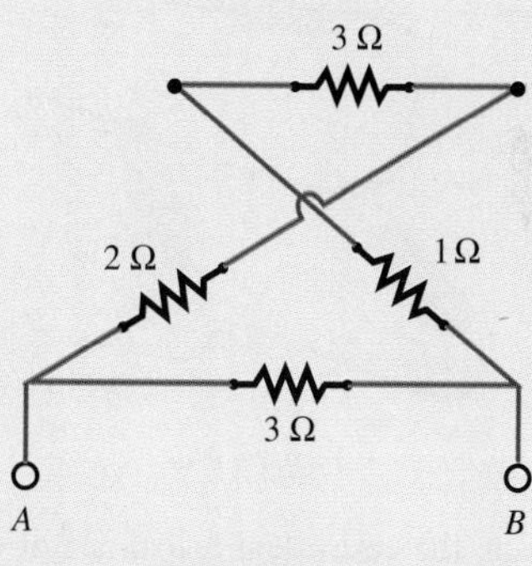

Figure P11

12. [I] Determine the equivalent resistance of the circuit between points A and B in Fig. P12.

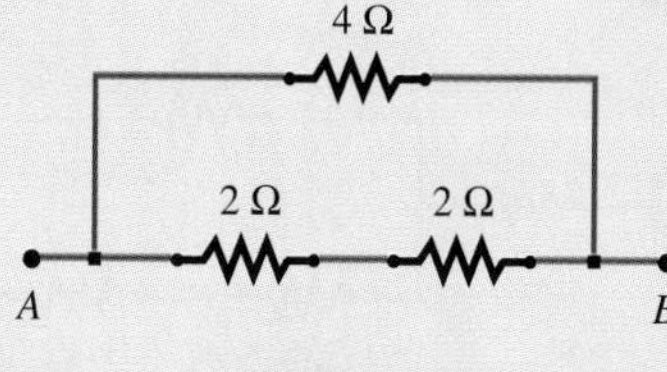

Figure P12

13. [I] Determine the equivalent resistance of the circuit between points A and B in Fig. P13.

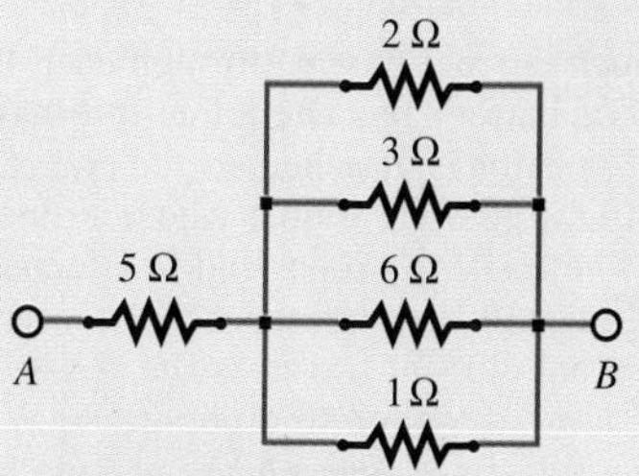

Figure P13

14. [I] Determine the equivalent resistance of the circuit between points A and B in Fig. P14.

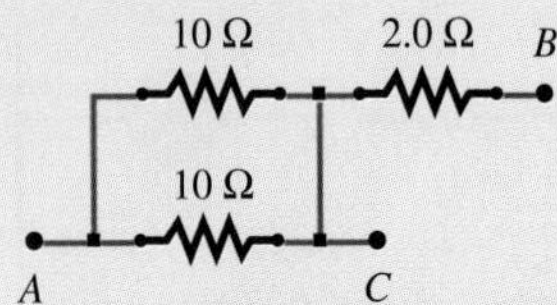

Figure P14

15. [I] Determine the equivalent resistance of the circuit between points A and C in Fig. P14.
16. [I] Determine the equivalent resistance of the circuit between points B and C in Fig. P14.
17. [I] Determine the equivalent resistance of the circuit between points A and B in Fig. P17.

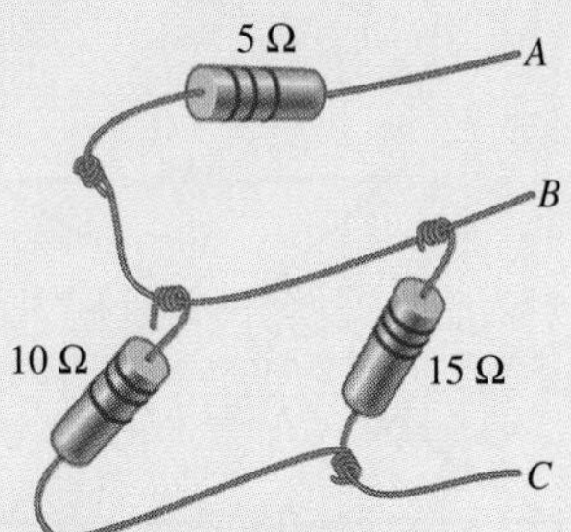

Figure P17

18. [I] Determine the equivalent resistance of the circuit between points C and B in Fig. P17.
19. [I] Figure P19 shows a portion of a circuit, the rest of which contains both sources and resistors. If the ammeter reads 9 A, what current passes through each resistor shown?

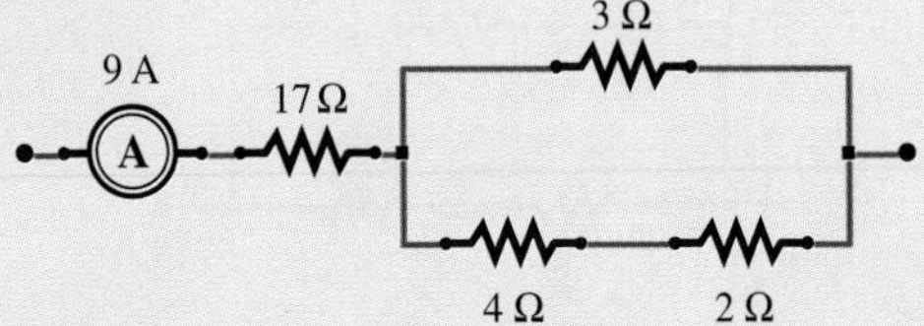

Figure P19

20. [I] Given that the ammeter in Fig. P20 reads 1.0 A, what is the emf of the ideal dc source as indicated by the voltmeter?
21. [I] How much current passes through each lamp in Fig. P21? The battery has negligible internal resistance.
22. [I] A 10-Ω resistor is attached at one end to the + terminal of a 20-V dc source whose − terminal is grounded. The resistor's other end is attached to the + terminal of a 40-V dc source whose − terminal is grounded. What current traverses the resistor?
23. [I] A 20-V lamp designed to dissipate 80 W is placed in series with a resistor R and a 60-V dc source. What value should R be for the lamp to operate properly?

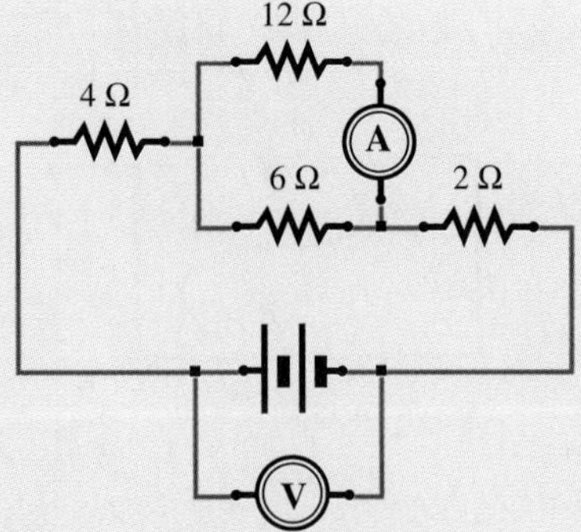

Figure P20

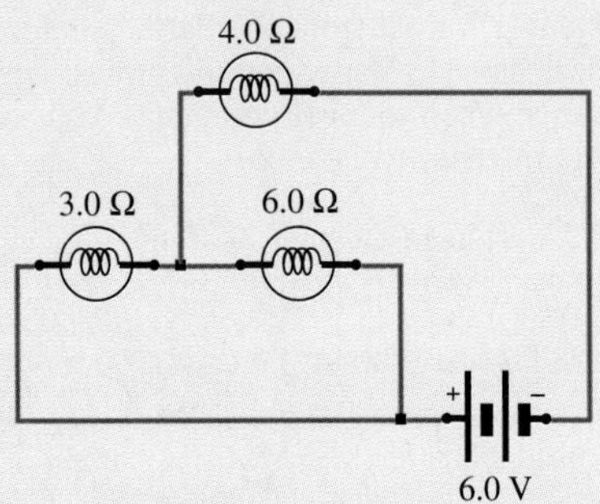

Figure P21

24. [I] If the total resistance of a voltmeter is 6.0 kΩ and it contains a 50-Ω coil movement, describe the other circuit element in the meter.
25. [I] An ammeter movement consists of a 50.0-Ω coil. If 0.10% of the current entering the meter is to pass through the coil, how big must the shunt resistor be?
26. [I] Show that the time constant of an *R-C circuit* has the correct units.
27. [I] An electronic flash fires a blast of energy from a 800-μF capacitor into a xenon lamp. It recharges through a series resistor of 5.0 kΩ. How long will it take to recharge 63% of its maximum charge?
28. [I] A charged 600-μF capacitor is in series with an open switch and a 4.0-kΩ resistor. If the switch is closed, how long will it take for the charge on the capacitor to decay down to 37% of its original value?
29. [I] An uncharged 1.0-μF capacitor is in series, through a switch, with a 2.0-MΩ resistor and a 12.0-V battery (with negligible internal resistance). The switch is closed at $t = 0$ and a current I_i immediately appears. Determine I_i. How long will it take for the current in the circuit to drop to $0.37I_i$?
30. [I] A 3.0-μF capacitor is put across a 12-V ideal battery. After an hour, it is disconnected and put in series, through a switch, with a 200-Ω resistor. (a) What is the initial charge on the capacitor? (b) What is the initial current when the switch is closed?
31. [II] Determine the equivalent resistance of the circuit between points A and B in Fig. P31.

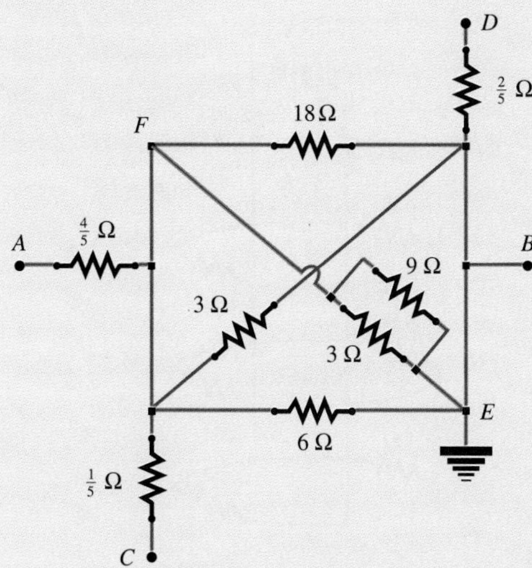

Figure P31

32. [II] Two resistors R_1 and R_2 are in parallel with each other and with an ideal source ($r = 0$) having a terminal voltage V. Show that the branch currents are given by

$$I_1 = I\left(\frac{R_2}{R_1 + R_2}\right) \quad \text{and} \quad I_2 = I\left(\frac{R_1}{R_1 + R_2}\right)$$

where the larger current goes through the smaller resistor. (These are good relationships to remember.)

33. [II] Figure P33 shows a portion of a circuit, the rest of which contains both sources and resistors. If the ammeter reads 9 A, what current passes through each resistor in the diagram?

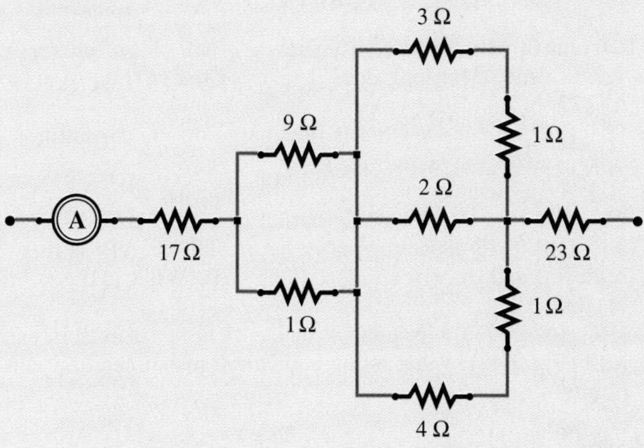

Figure P33

34. [II] Imagine that N is the number of identical cells in series, each having an internal resistance r and emf $\mathscr{E}$. Derive an expression for the current through a load resistor R placed across the terminals of the battery.

35. [II] What is the equivalent resistance between points A and B of the circuit shown in Fig. P35? How much power would be dissipated by this circuit if a constant 20-V dc source with a 0.10-Ω internal resistance was placed across A and B?

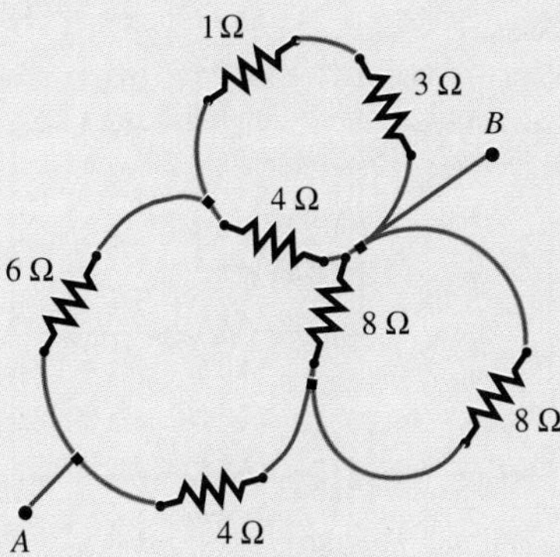

Figure P35

36. [II] How much current passes through each of the 5.0-Ω resistors in Fig. P36? How much power is delivered by the dc source?

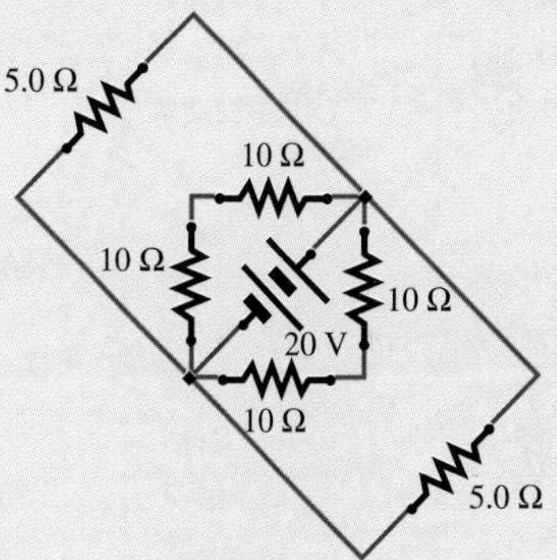

Figure P36

37. [II] Given the circuit in Fig. P37, calculate the current in each resistor. What power is delivered by the battery? What is the potential difference between A and C?

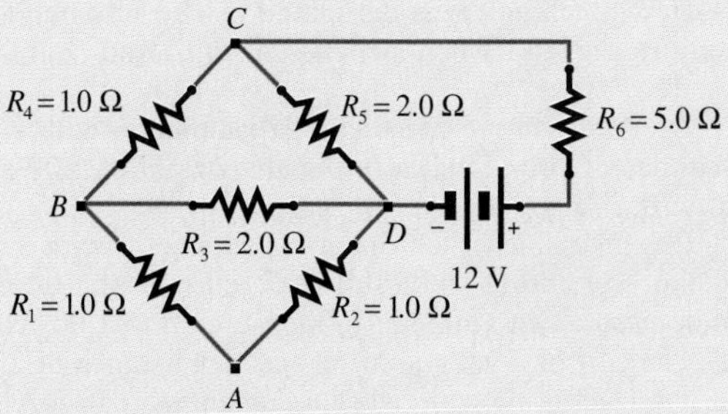

Figure P37

38. [II] The battery in Fig. P38 has an emf of 9.0 V and an internal resistance of 0.50 Ω. (a) What current does it supply? (b) What is the total power dissipated by the

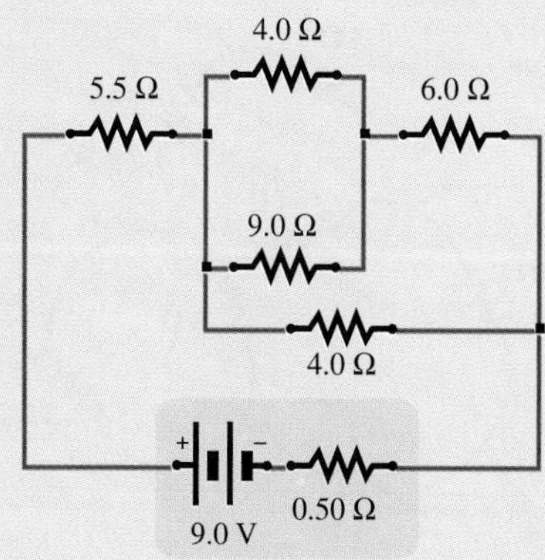

Figure P38

entire circuit? (c) What is the terminal voltage of the battery?

39. [II] What is the current provided by the battery in Fig. P39, given that its internal resistance is 0.50 Ω? What is its terminal voltage?

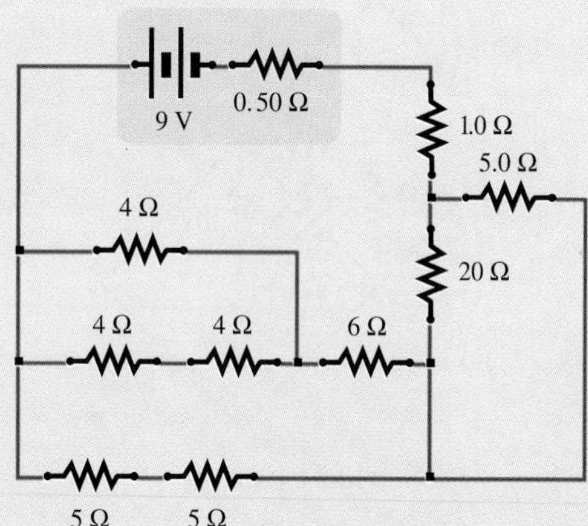

Figure P39

40. [II] Find the current supplied by the source in Fig. P40. The resistors are mounted around a cylindrical form.
41. [II] How much power is dissipated by the automobile circuit in Fig. P41 when switches *A*, *B*, *C*, and *D* are all closed?
42. [II] Design a voltmeter using a galvanometer with a coil resistance of 100 Ω and a full-scale current of 1.00 mA that will measure 100 V full scale.

43. [II] Design a shunt such that a galvanometer with a coil resistance of 100 Ω and a full-scale current of 1.00 mA can be used as an ammeter to measure up to 1.00 A.

44. [II] The coil of a galvanometer has a resistance of 20 Ω, and it deflects full scale when a current of 0.50 mA passes through it. By shunting the coil with a 2.0-mΩ resistor, it becomes an ammeter. What full-scale current will it now read?

45. [II] A 6.0-μF capacitor is charged up to 12 V and subsequently connected through a switch to a 100-Ω resistor. At $t = 0$, the switch is closed. What is the initial current through the circuit? Draw a rough plot of current versus time. How long does it take for the current to drop to 37% of its initial value?

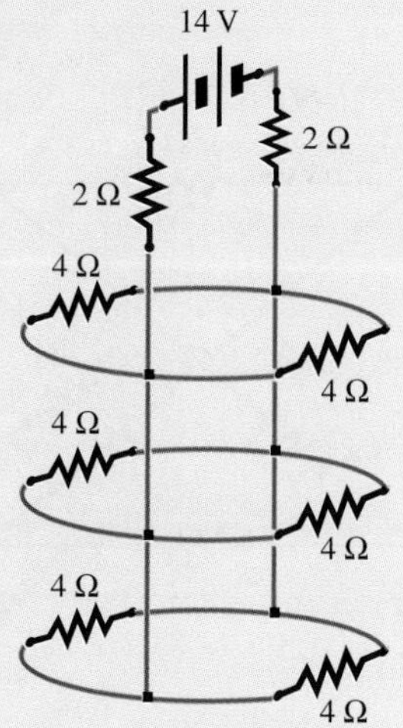

Figure P40

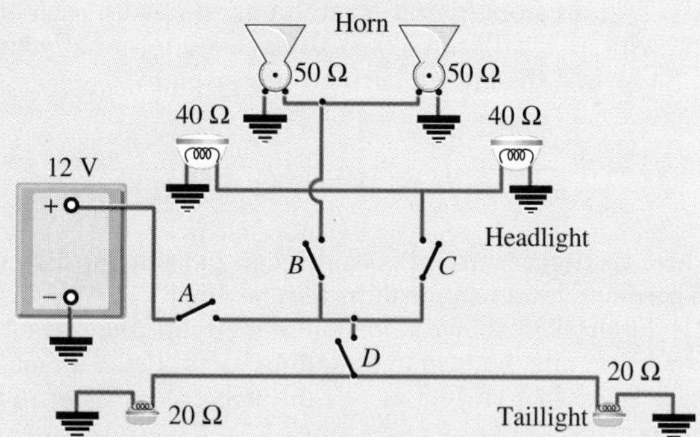

Figure P41

46. [II] In Problem 45, what is the charge on the capacitor 6.0 ms after the switch is closed?
47. [II] A 12-V battery with negligible internal resistance is placed across a series combination of a 1.0-MΩ resistor and a 12.0-μF capacitor for 10 hours. The battery is then removed and the circuit closed at $t = 0$. What is the current through the resistor at $t = 24$ s?
48. [III] Imagine a wire cube with identical resistors *R* in each arm (Fig. P48). What is the equivalent resistance between points *A* and *B*?

NETWORK ANALYSIS

49. [I] Use Kirchhoff's Loop Rule to solve for the currents in branches *A–D–C* and *A–B–C* of the circuit in Fig. P49. Then use the Node Rule to find the current in branch *A–C*.

50. [I] Find the current through each element of the circuit in Fig. P50.

51. [I] Apply Kirchhoff's Rules to the circuit of Fig. P51, and solve for the three branch currents. Next, simplify the network, determine all the currents, and check your answers.

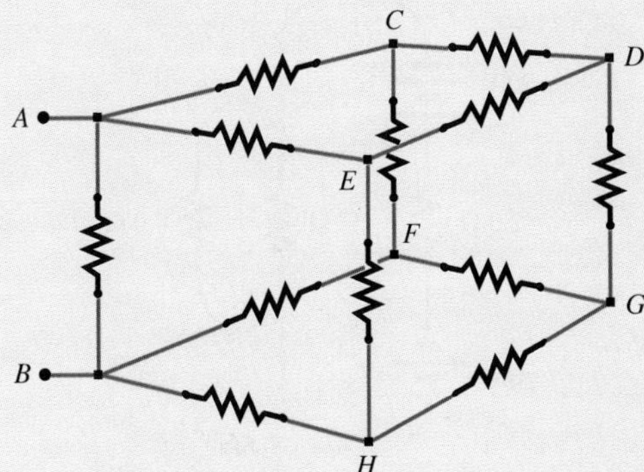

Figure P48

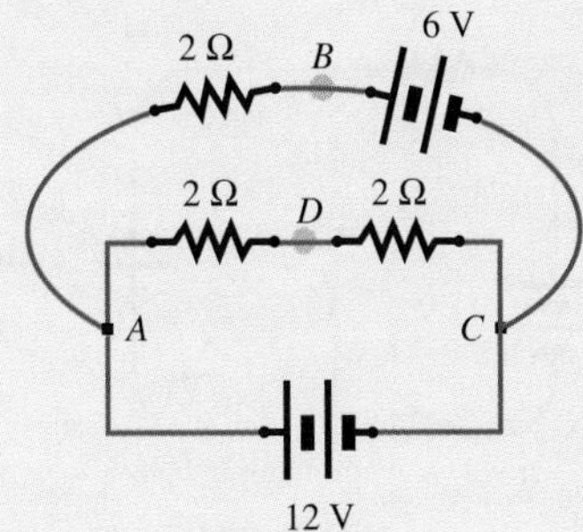

Figure P49

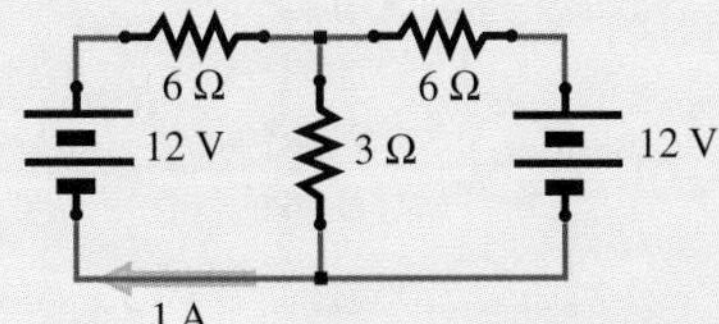

Figure P50

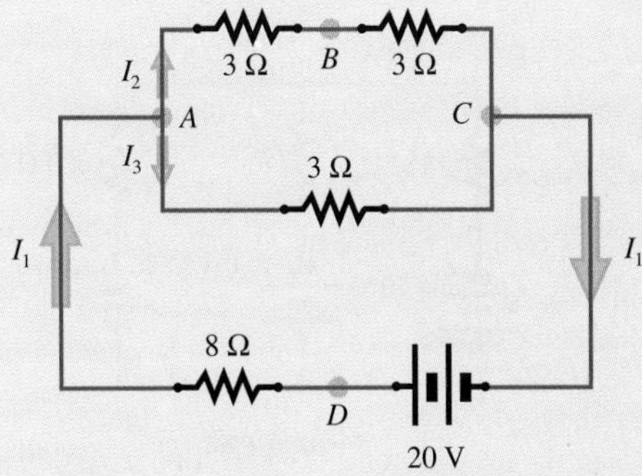

Figure P51

52. [I] Solve for the unknown source voltage and the power delivered by the 12-V battery in Fig. P52.

53. [I] The transistor circuit of Fig. P53 is to be checked for proper operation. According to specifications, the collector voltage (point C) should be at a constant $+6$ V with respect to ground. If that's the case, what should be the voltage measured at point D? (Hint: Read Discussion Question 5 to learn that a capacitor is an open circuit to dc.)

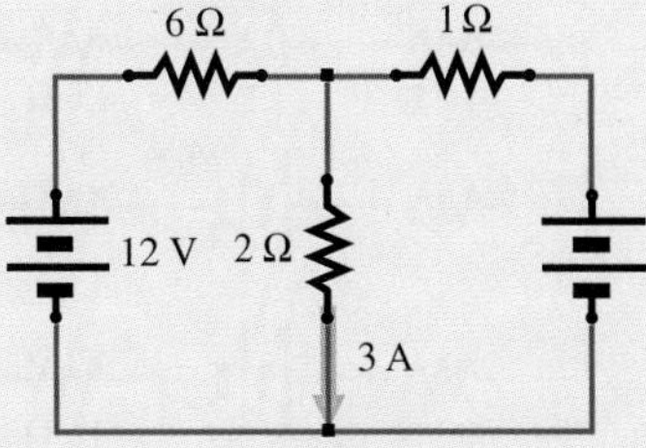

Figure P52

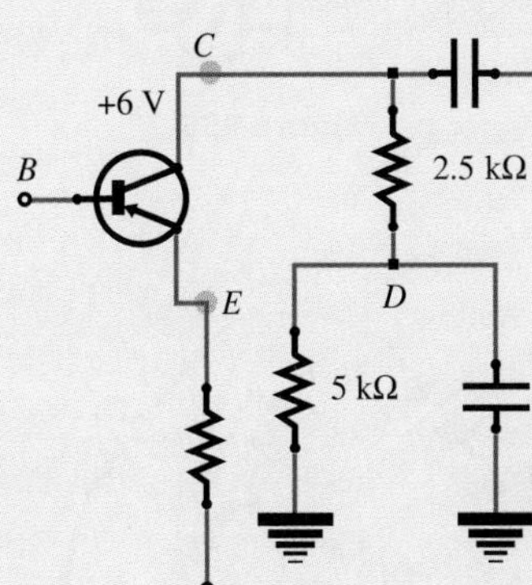

Figure P53

54. [I] If the ammeter and voltmeter in Fig. P54 read 2.0 A and 10 V, respectively, what current passes through the 1.0-Ω resistor?

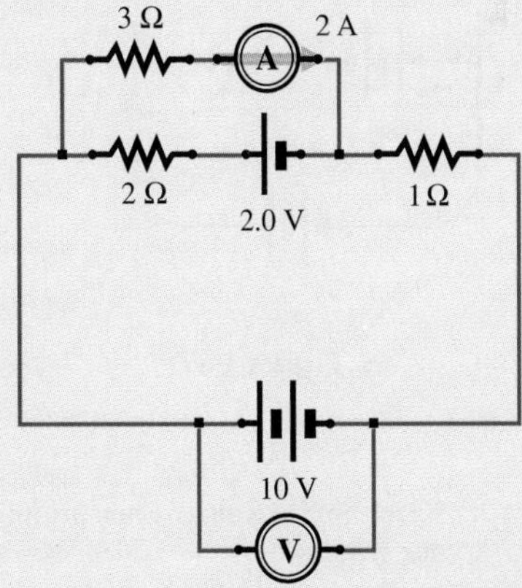

Figure P54

55. [I] The ammeter in Fig. P55 reads 2.0 A; what will the voltmeter read?

56. [I] Referring to the bridge circuit of Fig. P56, if $I = 6$ A, $I_2 = 4$ A, and $I_3 = 0$, find I_1, I_4, and I_5.

57. [I] Determine the current in each branch of the circuit of Fig. P57.

58. [I] Find the current in each resistor of the circuit in

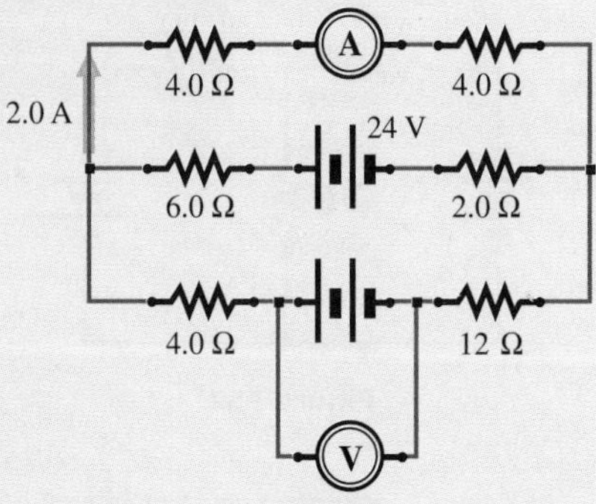

Figure P55

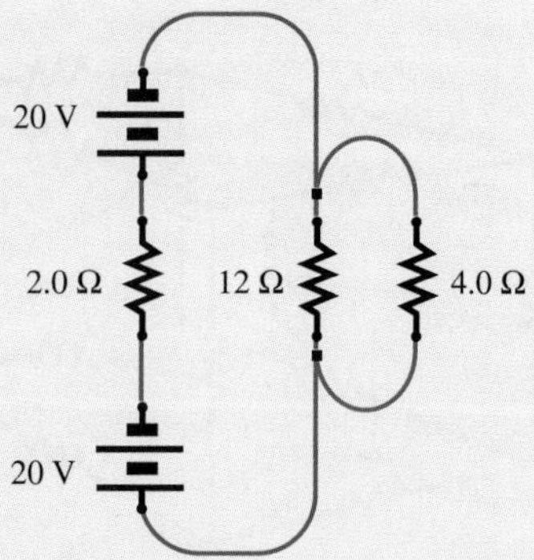

Figure P58

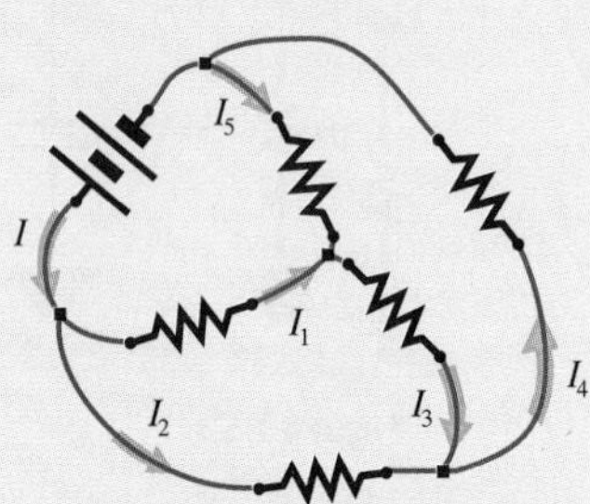

Figure P56

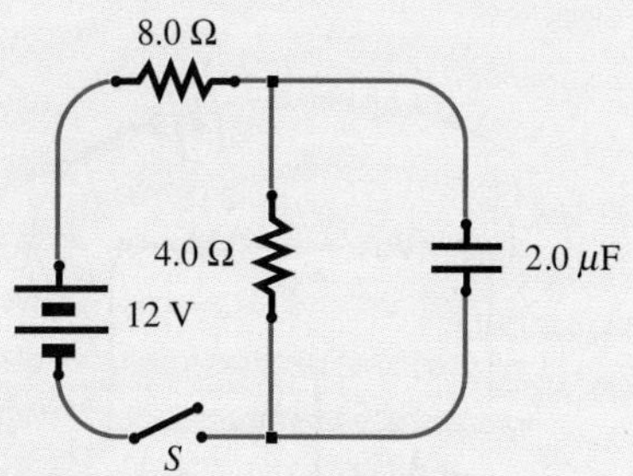

Figure P59

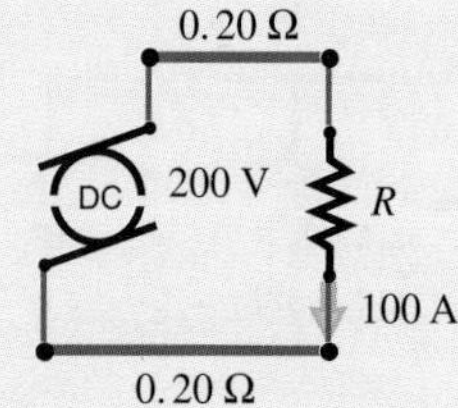

Figure P60

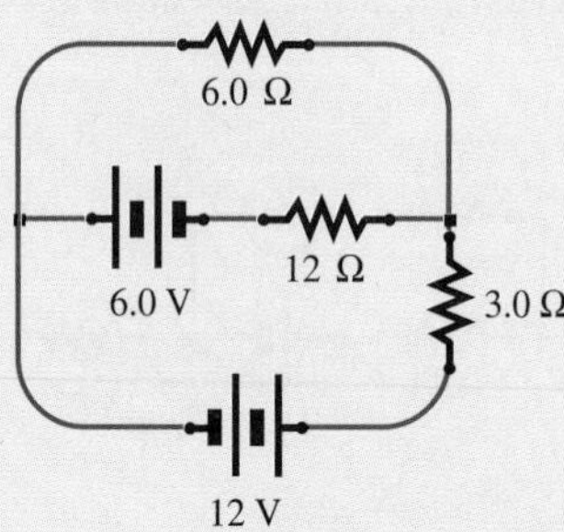

Figure P57

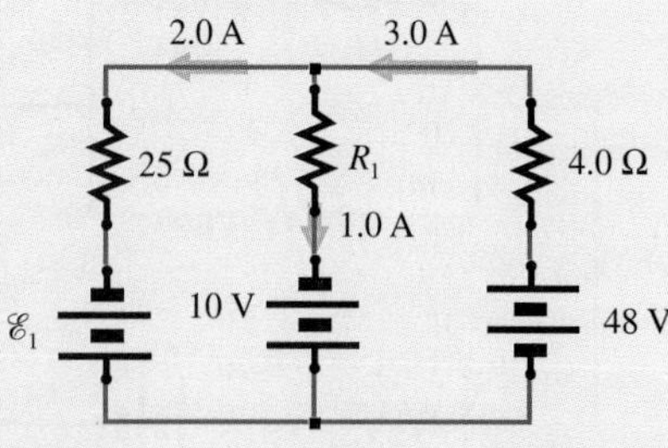

Figure P61

Fig. P58 using Kirchhoff's Rules. Then simplify the circuit and compare your results.

59. [I] The switch in the circuit of Fig. P59 is closed, and a steady state is established. What is the charge on the capacitor?

60. [I] Figure P60 shows a 200-V dc generator (the pictorial symbol used is another fairly common one) supplying 100 A to a load via a two-lead cable having a resistance of 0.20 Ω per length of conductor. What is the voltage across the load?

61. [II] Find the values of R_1 and $\mathscr{E}_1$ in Fig. P61.

62. [II] The variable resistor in Fig. P62 is adjusted to 20 Ω, whereupon the ammeter (which has a negligible internal resistance) reads zero. Use Kirchhoff's Rules to determine the power provided by the sources. What is the voltage at points B, C, and D?

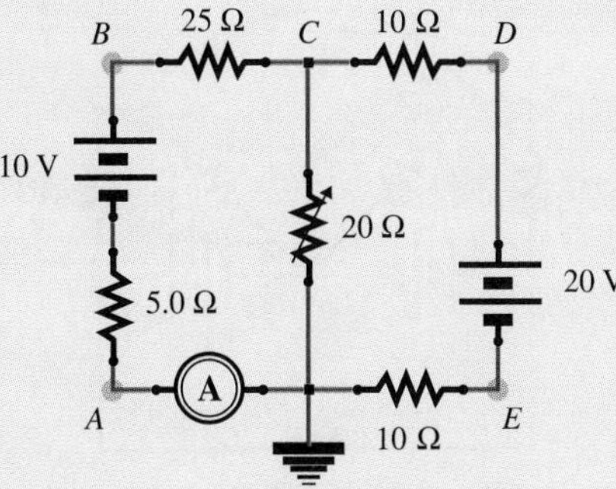

Figure P62

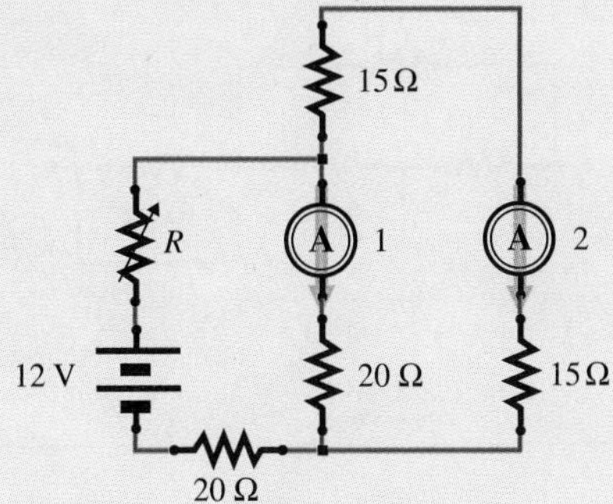

Figure P63

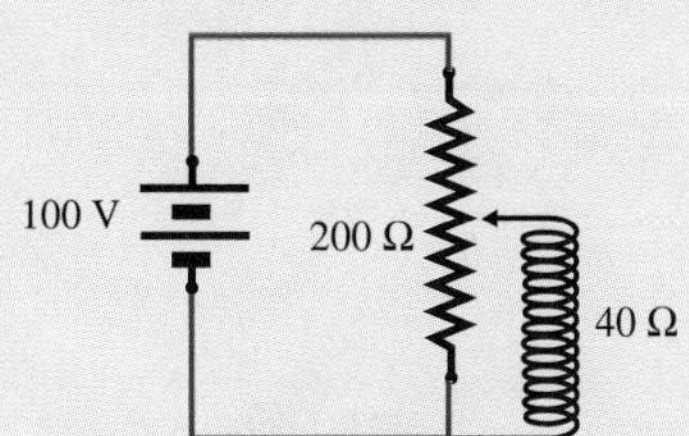

Figure P64

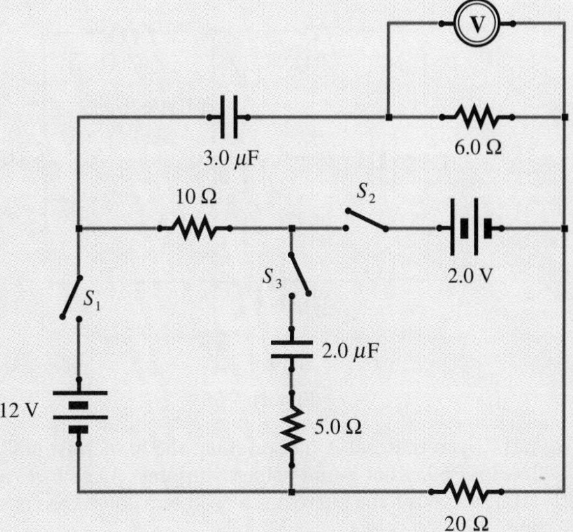

Figure P65

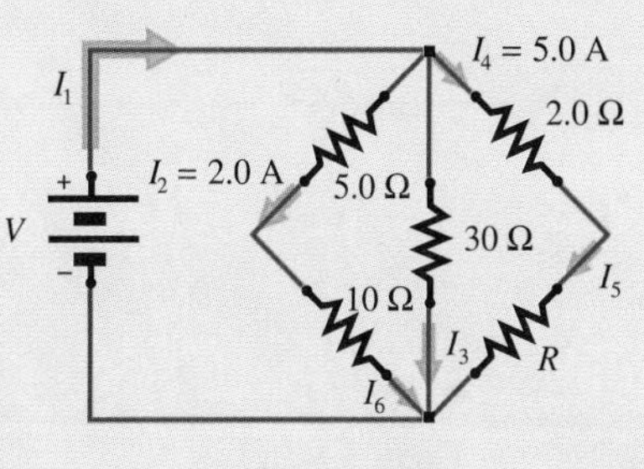

Figure P67

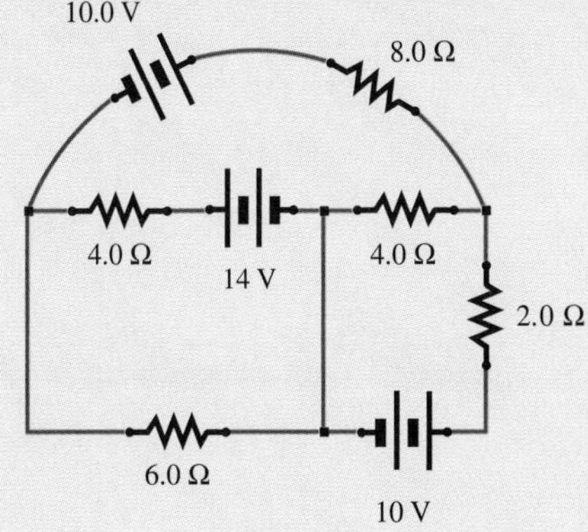

Figure P68

63. [II] The variable resistor in Fig. P63 is adjusted until the ammeters read (#1) 120 mA and (#2) 80 mA, with the directions of the currents as shown. Find the value of R.

64. [II] The potentiometer in Fig. P64 has a total resistance of 200 Ω. At what position must the slider be placed so that the 40.0-Ω coil of wire receives 1.00 A? Check your results using Kirchhoff's Loop Rule.

65. [II] With only switch S_1 closed in Fig. P65, (a) what is the steady-state reading of the voltmeter? (b) What is the charge on the 3.0-μF capacitor?

66. [II] With Problem 65 in mind, how much power does the 12-V battery supply in the steady state a few minutes after all the switches are closed? What's the charge on the 2.0-μF capacitor?

67. [II] Find the values of R, V, and all the unknown branch currents in the network of Fig. P67, given that $I_3 = 1.0$ A.

68. [II] Solve for the currents in each branch of the circuit in Fig. P68.

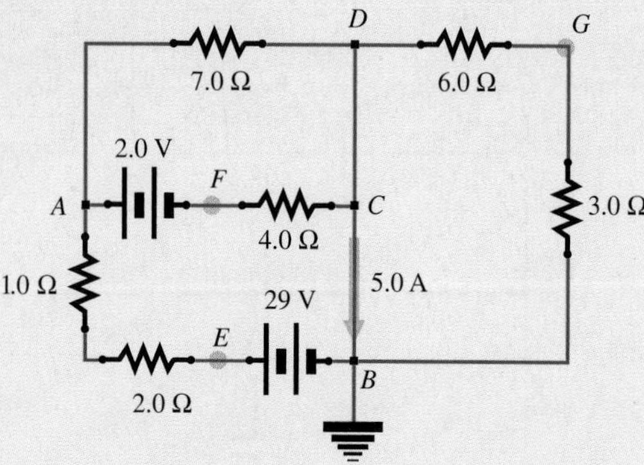

Figure P69

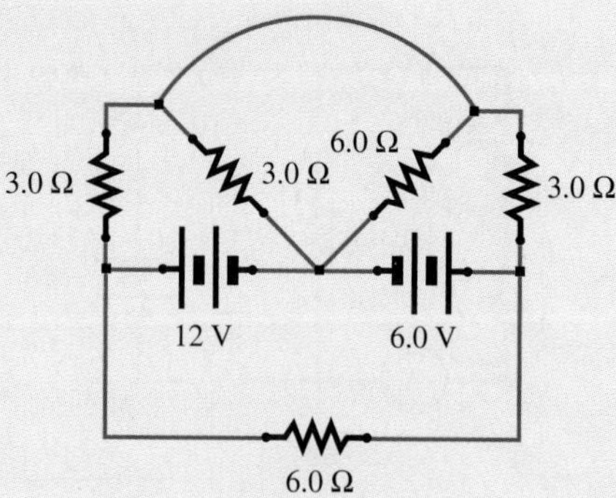

Figure P70

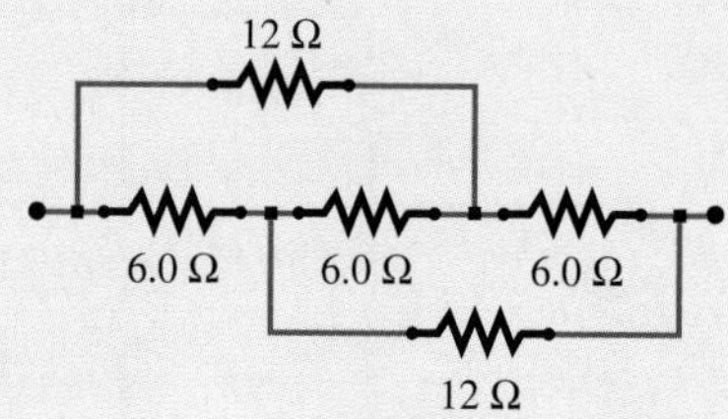

Figure P71

69. [II] Given that 5.0 A passes along the branch from *C* to *B* in Fig. P69, what is the voltage of points *A*, *D*, *E*, *F*, and *G*?
70. [III] Solve for the currents in each branch of the circuit in Fig. P70.
71. [III] Determine the equivalent resistance between the terminals of the group of resistors shown in Fig. P71 using Kirchhoff's Rules.

21

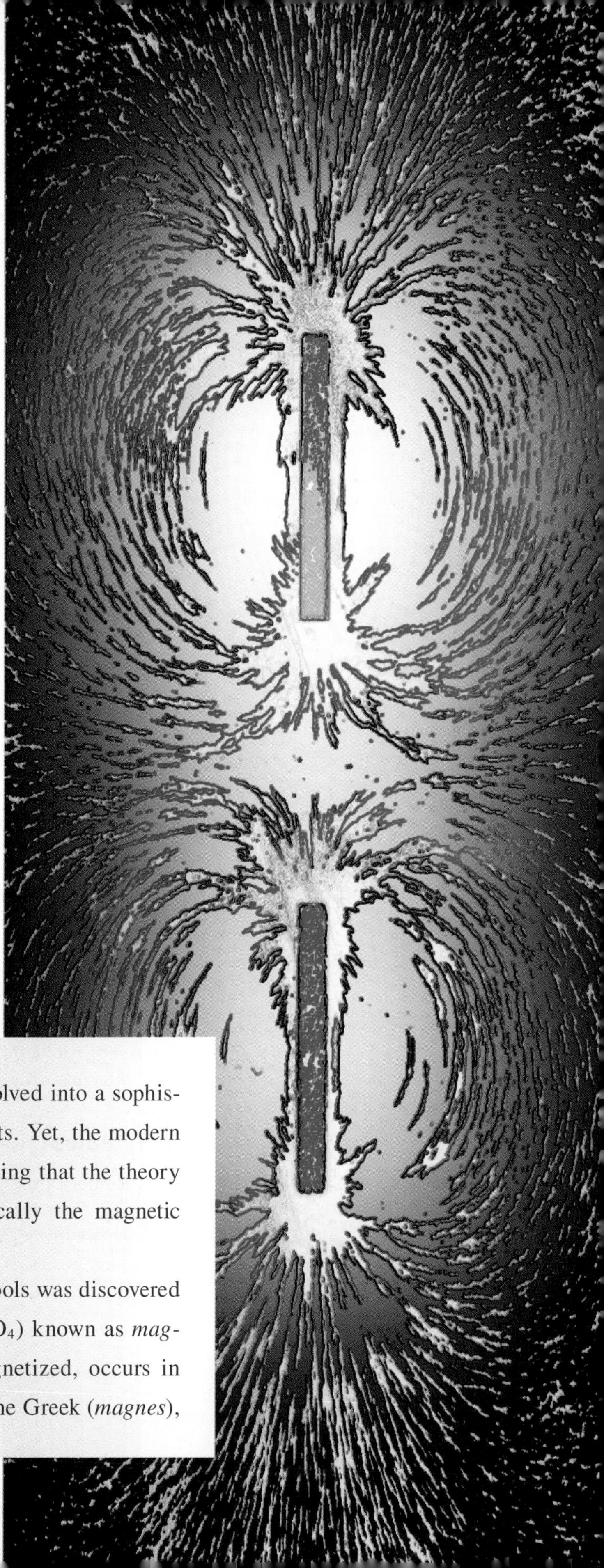

Magnetism

OUR UNDERSTANDING OF MAGNETISM has evolved into a sophisticated picture of whirling electrons, of fields and currents. Yet, the modern era of magnetism only began in 1820, and it's not surprising that the theory remains incomplete—we still can't determine theoretically the magnetic characteristics of the proton or the neutron.

The strange power of the **lodestone** to cling to iron tools was discovered in ancient times. The lodestone is an oxide of iron (Fe_3O_4) known as *magnetite.* That rich iron ore, some of it permanently magnetized, occurs in many parts of the world. The word *magnet* comes from the Greek (*magnes*),

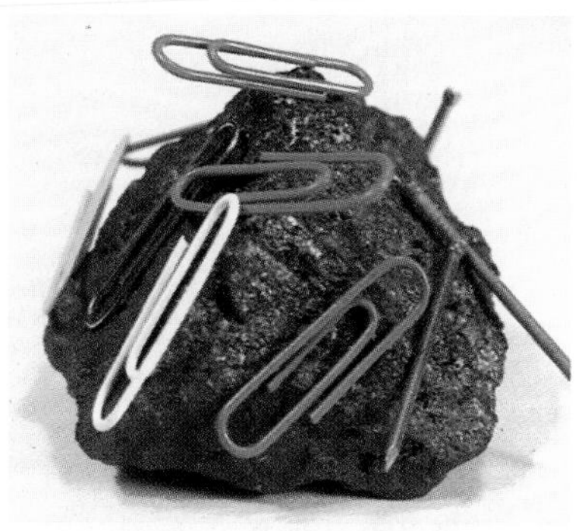

A lodestone with a few steel nails clinging to it. It is a chunk of magnetite, a common iron ore, magnetized by the Earth's magnetic field.

which probably derives from the ancient colony of Magnesia, where the ore was mined 2500 years ago.

Chinese legend has it that Emperor Hwang-ti (ca. 2600 B.C.) was guided in battle through dense fog by a small pivoting figure that always pointed south, a lodestone embedded in its outstretched arm. By about 1100 A.D., the magnetic compass had come to the West, and the lodestone's ability to align itself with the north-south axis of the "Universe" was awesome.

For all the centuries from Hwang-ti to the 1800s, the only practical source of magnetism (apart from the gradual realization that the Earth itself was the mother of all magnets) was the dark lodestone. The surge of activity and innovation that began in the nineteenth century followed the discovery by Oersted (p. 731) that electric currents give rise to magnetic force. For centuries, electricity and magnetism had been taken as two distinct powers, and now they were connected. *Charges generate electric fields. Charges in motion, in addition, generate magnetic fields.* The two fields are different manifestations of a single phenomenon—**electromagnetism**.

MAGNETS AND THE MAGNETIC FIELD

The earliest scholar to study the lodestone was probably Thales (ca. 590 B.C.). Almost 150 years later, the philosopher Socrates dangled soft iron rings clinging to one another beneath a lodestone. That's the same game most of us have played with a magnet and a box of paper clips. Just as a charged comb can induce charge on scraps of paper, a magnet can magnetize nearby pieces of iron.

There are many variations on what passes for a chunk of iron, and they all behave differently magnetically. Cast iron is a hard, brittle material rich in carbon (from 2% to about 7%), whereas pure iron is rather soft. Between these two extremes is the carbon-iron alloy *steel.* Soft iron retains its magnetized condition only so long as the inducer (the lodestone) is kept nearby. When removed, the specimen quickly demagnetizes, just as a charged piece of paper quickly depolarizes. Low-carbon soft steel, the stuff of paper clips and common nails, demagnetizes a bit more gradually. In contrast, a piece of hard steel, once magnetized, retains much of its power, and we speak of it as a **permanent magnet**, although that's an exaggeration.

The Chinese were versed in the art of making permanent magnets by the beginning of the second century A.D. A manuscript from the period suggests stroking an iron rod or needle from end to end along a lodestone, repeatedly, and *always in the same direction.* Try it with a magnet and a screwdriver made of hard tool steel.

21.1 Poles

On a summer's day in 1269 a French engineer, Peter de Maricourt—alias Peter Peregrine, Peter the Pilgrim—sat down to write a long letter to a friend. (Peter was whiling away time in the trenches with an army that was laying siege to a city in Italy.) In the letter, he described his researches on magnetism: military engineers were concerned about the compass, often being required to construct long tunnels leading under fortress walls.

Peregrine was the first to introduce the concept of the **magnetic pole**. He had several lodestones ground into the shape of a sphere to resemble the Earth and called them *terrellas.* A steel needle placed anywhere on the surface of

ON THE LODESTONE

The stone not only attracts iron rings but also imparts to them a similar power whereby they attract other rings.

Plato recounting the words of Socrates (ca. 400 B.C.)

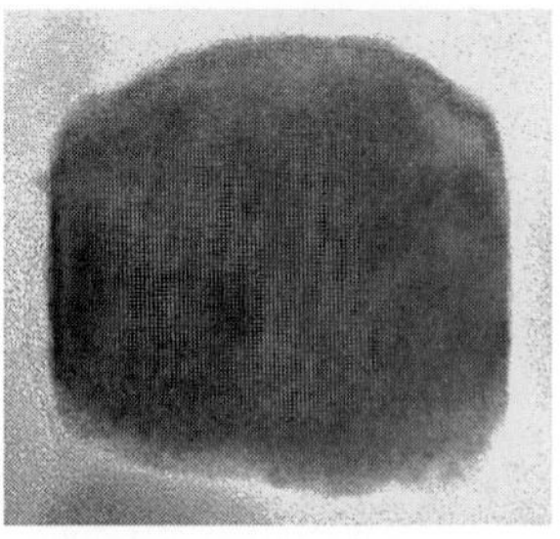

A magnetite crystal found in a bacterium.

Similar crystals, this one about a millionth of an inch long, have been found in the human brain.

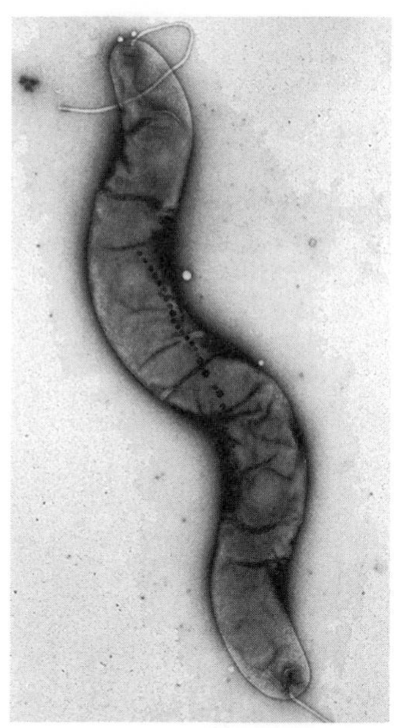

A magnetotactic bacterium. The dark line of dots is a chain of magnetite crystals that functions as a compass needle.

a terrella aligned itself in a particular way (Fig. 21.1). By drawing lines on the stone in the directions assumed by the needle, he determined that they all crossed at two opposing points, just as "all the meridian circles of the Earth meet in the two opposite poles of the world." If a piece of a needle was placed in contact with the lodestone, it would stand straight upright at, and only at, the poles.

Most magnets have two poles, where the force is clearly strongest. A straight bar magnet is the simplest two-pole configuration (known as a **dipole**), and Peregrine made iron bar magnets as well. Beyond that, it is possible for a magnet to have any number of poles, odd or even, provided it's two or more. Some modern flexible magnets are made in long strips with hundreds of poles.

Peregrine next put a terrella in a wooden bowl and set it afloat in a large vessel of water. As soon as it was released, it spun around, bowl and all, the *north-seeking* or **north pole** always pointing northward, the *south-seeking* or **south pole** always pointing southward. Of course, this was a compass, not all that different from the crude floating-needle instruments already in widespread use. Holding another terrella, whose poles were determined and marked, he approached the floating stone. When the north pole of one was brought near the south pole of the other, the little boat lunged toward the hand-held stone, but when either two north poles or two south poles were positioned near each other, the boat was pushed away. Peregrine had discovered the basic mutual interaction of all magnets: **like magnetic poles repel and unlike magnetic poles attract** (Fig. 21.2).

Naturally, Peregrine tried to isolate a single **monopole**, a piece of magnet that was simply and only north polar or south polar. And what more obvious way to do that than to split a magnet in two (Fig. 21.3)? Surprise! No matter how we break a magnet, the fragments are always bipolar—*the monopole cannot be isolated.* It's as if a magnet were composed of a succession of microscopic bar magnets with opposite

Figure 21.1 Several small steel needles attracted to the surface of a spherical magnet. Only at the poles do the needles stand straight up. A compass pointer pivoted so that it can swing vertically shows the same behavior in the Earth's magnetic field.

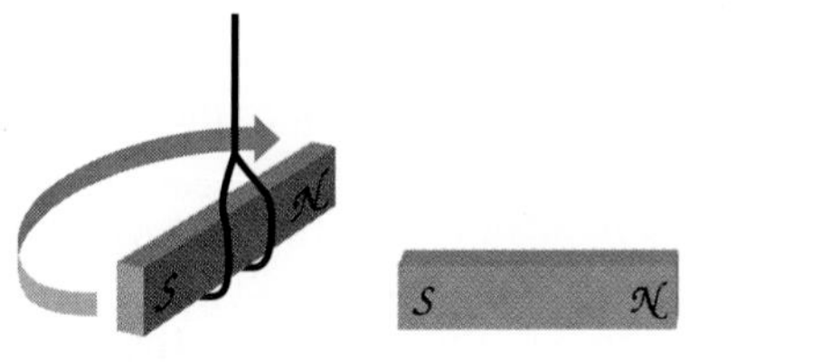

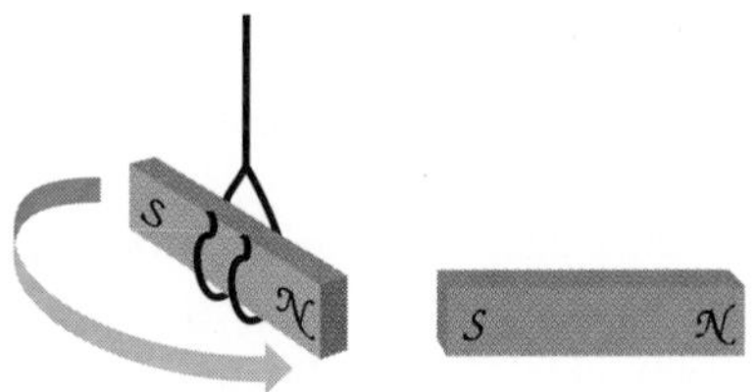

Figure 21.2 The attraction of unlike poles and the repulsion of like poles can be observed easily by suspending one bar magnet on a thread. The suspended magnet will swing away from the like pole of a second, stationary magnet and swing toward the stationary magnet's unlike pole.

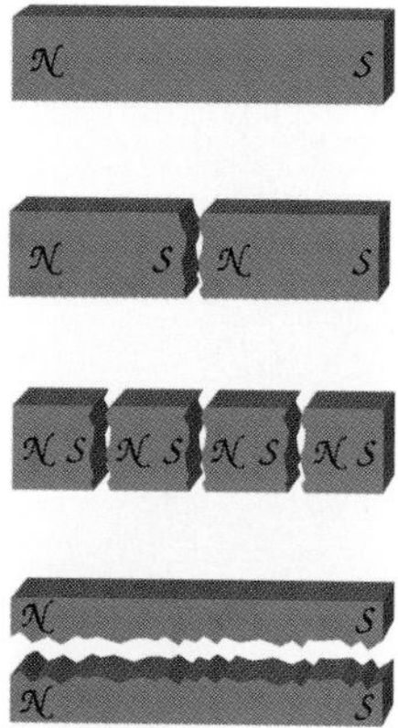

Figure 21.3 The fragments of a bar magnet always have two poles.

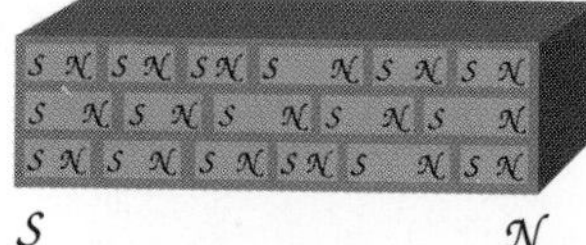

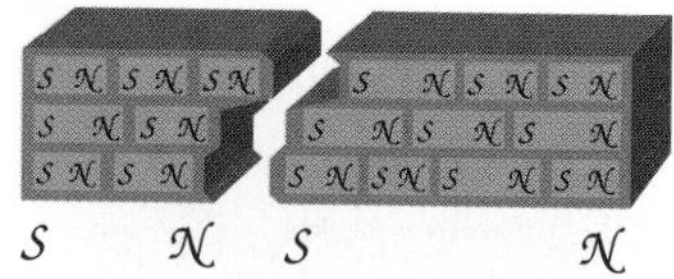

Figure 21.4 A magnet behaves as if it were composed of tiny bipolar units, tiny bar magnets, or dipoles.

poles touching and neutralizing each other everywhere but at the ends. When the magnet is broken, the appropriate poles appear (Fig. 21.4). That behavior will be understood only after we learn that **the electron itself is the fundamental dipole magnet**.

21.2 The Magnetic Field

A high point in the study of magnetism was provided in 1600 by William Gilbert (p. 208), then physician to Queen Elizabeth I. Gilbert compared the Earth to a large spherical lodestone, maintaining that the compass needle was drawn to the planet's magnetic pole, not to the heavens as everyone else had thought. It was simply a matter of one magnet pulling on another.

Dr. Gilbert probed the region surrounding a magnet with a little compass and concluded that "Rays of magnetick virtue spread out in every direction in an orbe." The statement seems almost identical in spirit to the nineteenth-century vision of Faraday's *lines-of-force of the magnetic field*. One need only connect the little compass arrows with smooth arcs to transform the imagery from an "orbe of virtue" into a "magnetic field." Less than half a century after Gilbert, Descartes carried the mapping process one step further. He sprinkled iron filings around a magnet, and they aligned themselves like minute compass needles to form curved continuous filaments, which even more potently *suggest* lines-of-force, lines of *magnetic field* (Fig. 21.5).

Again we are faced with the mystery of action-at-a-distance, and again our answer is to assume that every magnet establishes, in the space surrounding it, a **magnetic field** (B). As before, we say a field exists in a region of space when an appropriate object placed at any point therein experiences a force. The fields span the space and communicate the interaction—the fields mediate the Third Law's action-reaction. Remember that we probed the gravity field with a test-mass, just as we probed the electric field with a test-charge. Now we will map the magnetic field, not with a monopole, but with the next best thing, a dipole, a tiny test-compass.

A compass needle, able to turn freely, placed near the south pole of a bar magnet simultaneously has its own south pole repelled and its north pole attracted (Fig. 21.6a). It experiences a net torque and twists around into a new equilibrium orientation such that the torque vanishes (Fig. 21.6d). That's why a compass needle aligns itself with the local field. The strength of a magnetic field (B) at every point in space can be determined from the torque tending to realign the test-compass. For the moment this approach will suffice, although a more practical one will be forthcoming. The SI unit for B is the *tesla* (after Nikola Tesla), abbreviated T (Table 21.1). The compass is a tiny bar magnet that will settle tangent to the field, and *we arbitrarily take the arrow from its south to its north pole as the direction of the field in which it is immersed.*

Once again the concentration of field lines, the number per unit cross-sectional area, is proportional to the strength of the field. The field lines in Fig. 21.7 are more densely concentrated near the poles, where B is largest. By convention **the field**

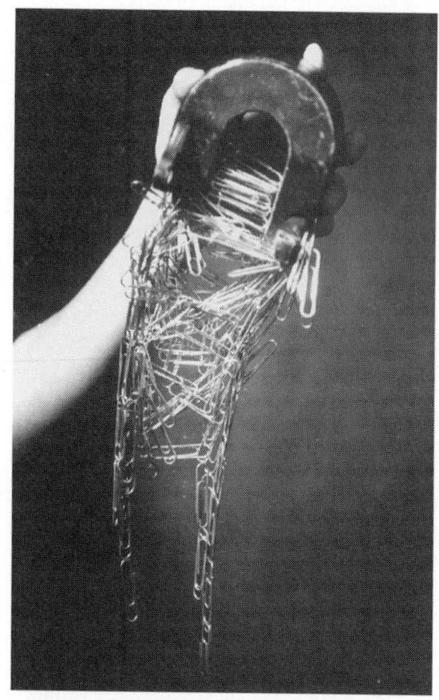

A magnet and a box of paper clips. Each clip is magnetized, becoming a small temporary magnet. The clips pack in densely where the field is strong. They form bridges that arch from pole to pole, crudely suggesting a pattern of field lines.

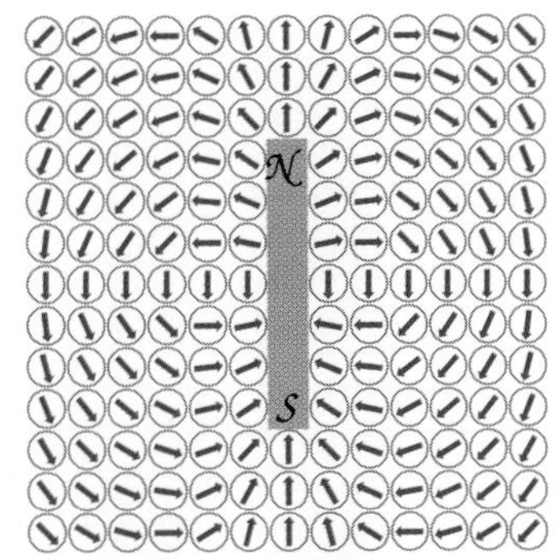

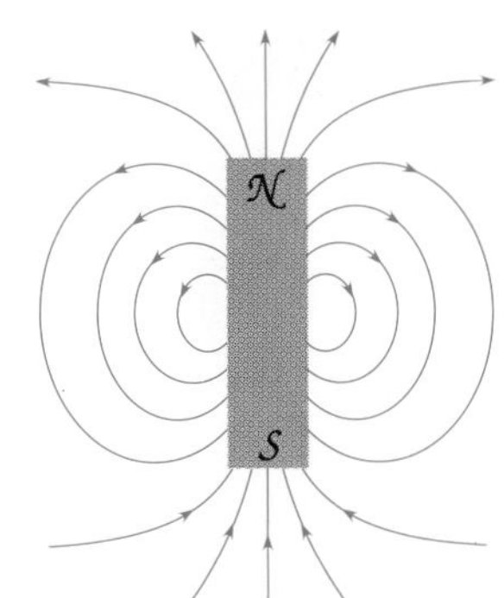

Figure 21.5 The force field around a bar magnet, as revealed by an array of small compasses. This drawing shows what is happening in only one plane. The field is three-dimensional. The photo shows iron filings lining up in the vicinity of a small bar magnet. By tradition, we say the field lines emerge from the north pole, curve around, and enter the south pole. But by just looking at this photo, we can't tell which pole is which.

TABLE 21.1 Magnetic fields

Source	Field (T)
Nucleus (at surface)	10^{12}
Neutron star (at surface)	$\approx 10^{8}$
Highest yet attained in laboratory	
explosive compression ($\approx 10^{-6}$ s)	1.5×10^{3}
pulsed coils ($\approx 10^{-3}$ s)	100
constant (dc, superconducting, 1993, MIT)	37.2
constant (dc, room temperature)	23.5
No acute effects on bacteria, mice, or fruit flies	14
Large laboratory electromagnet	5
Sunspot (within)	0.3
Human exposure limit (full body, dc, for minutes)	≈ 0.2
Small ceramic magnet (nearby)	≈ 0.02
Small bar magnet (near pole)	10^{-2}
Sun (at surface)	10^{-2}
Hair dryer (60 Hz, nearby)	1×10^{-3}–2.5×10^{-3}
Can opener (60 Hz, nearby)	0.5×10^{-3}–1×10^{-3}
Jupiter (at poles)	8×10^{-4}
Blender (60 Hz, nearby)	10^{-4}–0.5×10^{-3}
Earth (dc, at surface)	0.5×10^{-4}
Color TV (60 Hz, nearby)	10^{-4}
Transmission line (maximum under, 765 kV, 4 kA)	$\approx 0.5 \times 10^{-4}$
Toaster (60 Hz, nearby)	0.1×10^{-4}–1×10^{-4}
Sunlight (rms)	3×10^{-6}
Refrigerator (60 Hz, nearby)	10^{-6}
Mercury (at surface of planet)	2×10^{-7}
Human body (produced by)	$\approx 310^{-10}$
Interstellar space	$\approx 10^{-10}$
Earth (50–60 Hz, at surface)	10^{-12}
Shielded region (smallest value measured)	1.6×10^{-14}

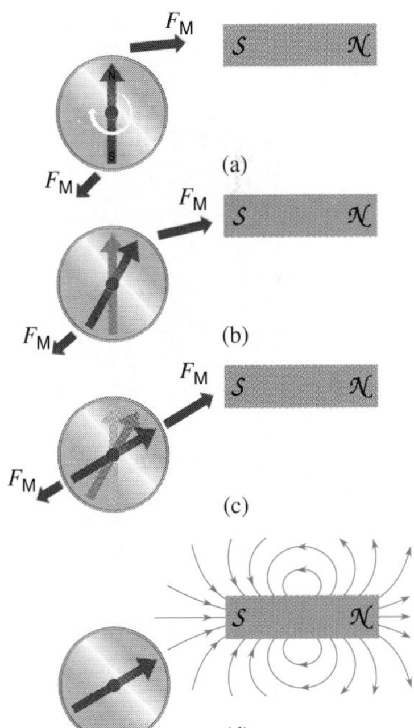

Figure 21.6 A compass needle mounted so that it can turn freely experiences a torque in the B-field of a magnet and swings around until that torque vanishes.

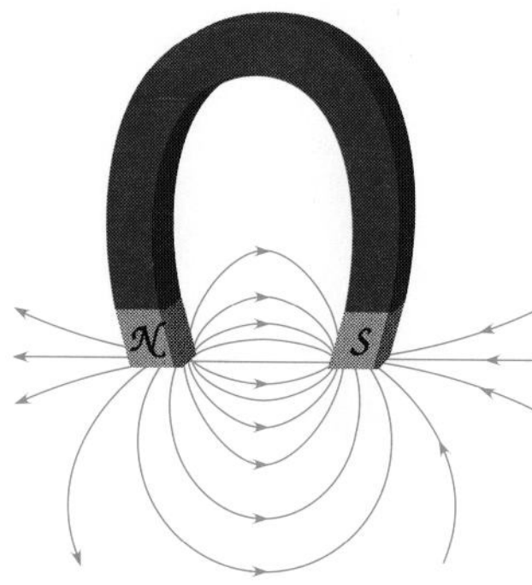

Figure 21.7 The horseshoe magnet concentrates its field in the immediate region between its poles.

(a)

(b)

If we suppose that each iron filing is a compass needle, then the pattern they form reveals the magnet's B-field lines. In (a), the two magnets have like poles facing each other. In (b), unlike poles face each other. Of course, these views are in only one plane; the field is three-dimensional.

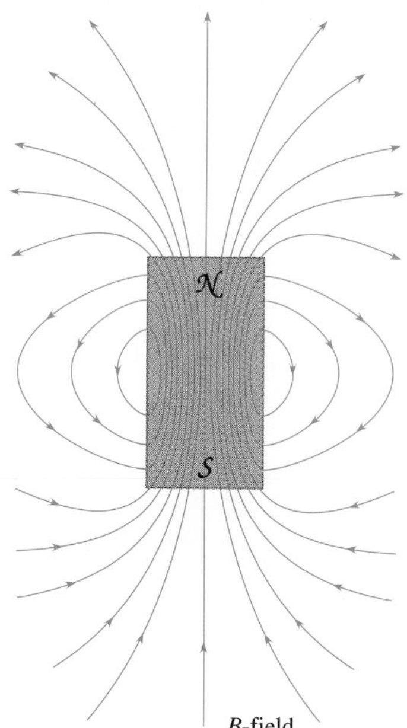

Figure 21.8 The magnetic field inside and outside of a permanent magnet. Electric field lines begin on positive charges and end on negative charges. Here there are no magnetic charges (monopoles), so magnetic field lines are closed. They must turn back on themselves. The field in three dimensions is like a system of concentric misshapen doughnuts.

extends out from the north pole and into the south pole. The north end of the test-compass is attracted to the south pole of the magnet responsible for B (and the south end is attracted to its north pole). As usual, *field lines never intersect, nor should they begin or end on anything but sources and sinks:* if there were monopoles clustered at both ends of a bar magnet, the field lines would begin and end there. But there are not, and **the lines of B form closed loops** (Fig. 21.8).

21.3 The Earth's Magnetism

The Chinese of the eleventh century were already aware that the compass needle does not align itself everywhere along a true north-south direction. That knowledge came to Europe at least as early as 1436, but it's still a surprise to some to learn that a magnetic compass rarely points true north. While a compass in Chicago does align itself almost due north, one in Los Angeles tilts about 15° to the east, and one in New York leans almost 15° west of north.

Gilbert was right, at least to the extent that the Earth behaves as if it contained, as its main field source, a dipole. It acts as if a relatively short bar magnet were embedded at its center (Fig. 21.9), tilted over at about 11.5° along the 70° W meridian. But matters are much more complicated than that.

The core of the planet is too hot to be totally solid and even too hot ($\approx$2500 K) for magnetic materials to remain magnetic—we are not dealing with a buried bar magnet. Besides that, the field changes in time, both in size and direction. There is also compelling evidence that the field has reversed itself some 300 times in the past 170 million years and may have done so just 30,000 years ago.

As a further complication, the *dip poles*, where the field points straight up and down, not only wander, but are 500 miles from the *geomagnetic poles* (the poles of Gilbert's magnet toward which a compass points). Moreover, neither the geomagnetic nor the dip poles are at the geographic poles, as fixed by the planet's spin axis. The final irony is linguistic: the Earth's northern magnetic pole is actually a *south* pole, and the southern magnetic pole is—you guessed it—a *north* pole. That's why the north-seeking pole of a compass needle points north—it's attracted toward the Earth's south magnetic pole.

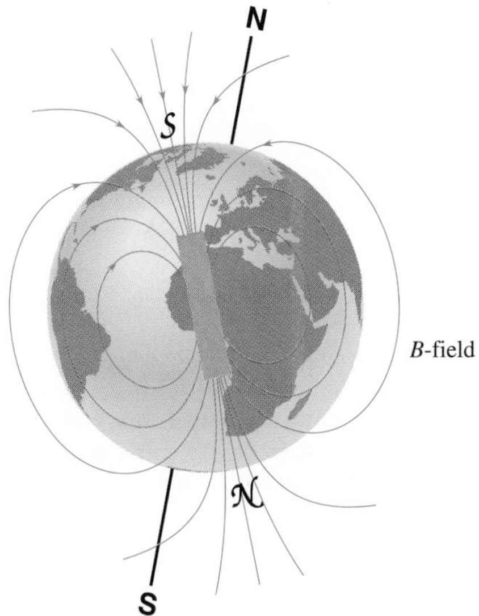

Figure 21.9 The Earth's magnetic influence resembles that of a tilted bar magnet. A compass needle aligns itself with the field and points roughly toward the north geographic pole, which is not far from the Earth's south magnetic pole. The field extends thousands of kilometers out into space and is rotationally symmetrical around the axis of the hypothetical bar magnet.

21.4 Monopoles

Unfortunately, in 1785, Coulomb showed that a workable magnetic force law could be stated in an identical form to the electrostatic force law, provided one introduced the notion of magnetic charges, or *monopoles*. That theory was so reasonable it took over 100 years before it was discredited. *Magnetic charges are not the source of magnetism.*

Even so, in 1931, P. A. M. Dirac presented a theoretical argument resurrecting the magnetic monopole. Besides establishing a symmetry between electric (+, −) and magnetic (N, S) particles, the existence of the monopole would also provide a ready explanation for the quantization of charge. Today, there are competing versions of the so-called Grand Unification Theory (Ch. 33), which attempts to unify all the forces of nature. These formulations maintain that monopoles were created in high-energy collisions during the Big Bang that birthed the Universe. If they still survive at all, it's not likely there are many around and what few there are, are flying through space (certainly they're not clustered at the ends of magnets). The slowest, and by far the most massive, of all the elementary particles (weighing about as much as a bacterium), monopoles would be fascinating little creatures, each much smaller in volume than a proton. For one thing, they would exert an attraction between their opposite numbers almost 5000 times greater than the attraction that exists between the electron and proton.

At present, there are at least 35 major monopole hunts in progress worldwide. To date, not a single monopole has been spotted. But of course that incredible shyness may spring from the fact that monopoles are exceedingly scarce, or they just don't exist anymore, or perhaps they never did.

21.5 Magnetism on an Atomic Level

As we shall see, charge in motion produces magnetic force, and in particular a current moving in a circular path is magnetically identical to a dipole (p. 735). Moreover, electrons behave in ways that suggest they are perpetually spinning.* It follows that *the electron itself corresponds to a circulating charge and is the ultimate subatomic dipole magnet* (Fig. 21.10). The observed magnetic response of electrons is consistent with each having a purely dipole field down to a radius of 10^{-12} m. The orbitlike motion of an electron about the nucleus of an atom also constitutes a current and produces an additional magnetic field. Atomic nuclei can generate dipole fields too, but these are typically a thousand times weaker. Together, the two elec-

Having made many and divers compasses . . . I found continuallie that after I had touched the yrons with the stone, that presentlie the north point thereof wouilde bend or decline downwards under the horizon in some quantitie.

ROBERT NORMAN
The Newe Attractive (1581)

*Although physicists commonly talk about spinning subatomic particles in the same way they talk about spinning basketballs, in their heart of hearts they know better. Quantum Mechanics has made it clear that the old notion of spin needs a modern interpretation. By spin, we mean a fundamental quality (like mass and charge, whatever they are) that is associated with the manifestation of angular momentum, but not necessarily with the existence of spinning, with turning round and around. The latter picture produces relativistic inconsistencies and must be rejected. Thus, an electron has spin though it is not spinning in the usual sense of the word. Whatever it's doing, we are confident that an electron has an intrinsic angular momentum and an intrinsic magnetic dipole field.

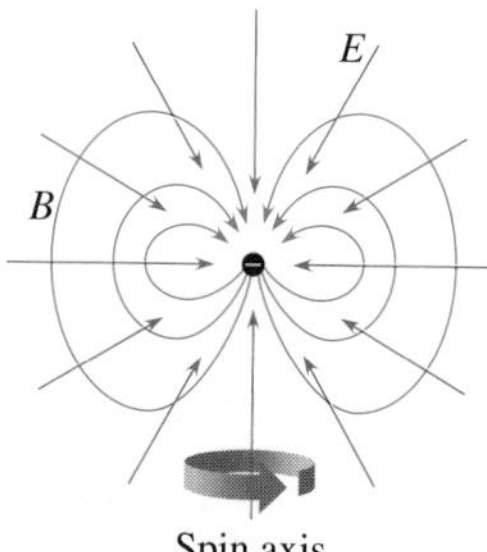

Figure 21.10 The electric and magnetic fields of an electron. The E-field is radially inward. The B-field forms a system of concentric doughnut shapes about the spin axis. The magnetic field is that of a minute dipole.

I have seen Samothracian iron rings even leap up and at the same time iron filings move in a frenzy inside brass bowls, when this Magnesian mineral was placed beneath.

LUCRETIUS (First century B.C.)

tron mechanisms (spin and orbit) account for the magnetic behavior of all the various forms of bulk matter.

Usually, electrons in atoms come in oppositely moving pairs and their magnetic fields cancel—the vast majority of atoms and molecules have no net magnetic field. Still, no material exists that will not respond magnetically to an applied B-field. Most substances (water, glass, copper, lead, salt, rubber, diamond, wood, and so on), the majority of gases (such as nitrogen, carbon dioxide, and hydrogen), and millions of organic compounds (for example, plastics) are very weakly magnetic in what might at first seem a startling way. They are *repelled* by the pole of a strong magnet. Faraday was the first to observe this peculiar phenomenon (1845), and he called it **diamagnetism**. It's still amusing to swing a small piece of glass dangling on a thread into a field almost 100 000 times stronger than the Earth's, only to have the glass pop right out and stay out at some gravity-defying angle. Humans are mostly water and should be able to feel this repulsion. But that requires a tremendously strong B-field (you feel nothing even at 4 T, which is quite large), and very few people have actually experienced the repulsion.

Diamagnetism is associated with the orbital motions of atomic electrons. Turning on a B-field changes their angular momenta, and that added motion produces a field that opposes the applied field (p. 768). Accordingly, diamagnetism is present in all substances, although it is observable only when not swamped out by other much stronger effects.

When an atom has an odd number of electrons or a structure in which not all its electrons are paired, the atom will have a net magnetic dipole field of its own. In a substance made up of countless such atoms, these dipoles *en masse* produce only a feeble response because the ordinary thermal agitation of the atoms keeps them disoriented. Substances of this kind are called **paramagnetic** and they include the elements aluminum, oxygen, sodium, platinum, and uranium, among others. If placed near a powerful magnet, they will be drawn in toward a pole, but only weakly.

The last of the three major magnetic classes of substances (there are others that are less important) is known as **ferromagnetic**. This group, to which newly concocted materials are always being added, includes magnetite, a troop of alloys such as steel and Alnico, and a number of elements. Iron, cobalt, and nickel have been known for centuries to be "magnetic" at room temperature, and there are a half-dozen other elements that become ferromagnetic at low temperatures. On average, ferromagnetic materials have a greater number of unpaired, spin dipoles per atom. But more importantly, *these dipoles enter into large-scale cooperative alignments.* In other words, the uncompensated spin dipoles of each atom interact strongly with the dipoles of adjacent atoms, locking together in a parallel orientation that tends to persist even at room temperature.* Such substances are strongly attracted to the poles of a magnet and are themselves easily magnetized.

An interesting application arises when we place a thin layer of a ferromagnetic material on a plastic substrate. Typically, that's done with extremely fine particles of $\gamma\text{-}Fe_2O_3$. These needlelike grains can conveniently be magnetized in microscopic patterns, thereby "permanently" recording information. Of course, what you have then is magnetic recording tape, computer memory disks, etc. (See Discussion Question 8.)

The magnetic-domain structure of a polycrystalline sample of cobalt samarium. Each little domain has its own random magnetic field orientation. The sample as a whole is said to be unmagnetized.

Magnetic Domains

Ferromagnetic substances are composed of very many microscopic **domains**—islands of order—throughout each of which tremendous numbers of atomic spin dipoles

*That powerful coupling is a quantum-mechanical effect, first explained by Werner Heisenberg in 1928.

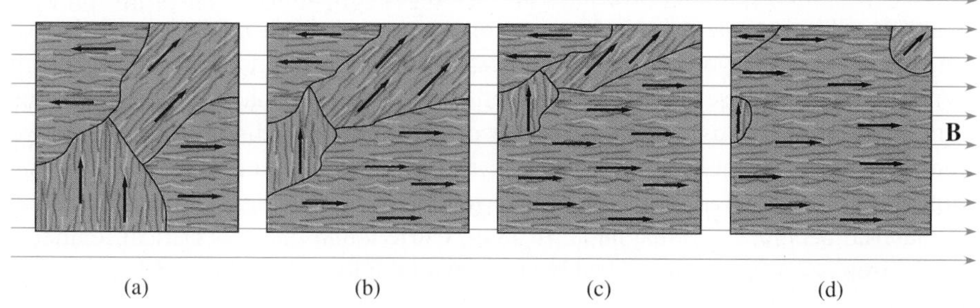

Figure 21.11 The growth of domains aligned in the direction of an applied magnetic field. As the applied *B*-field increases, the pattern goes from (a) to (b) to (c) to (d). Regions parallel to the field grow at the expense of those not parallel.

are aligned parallel to one another. Each domain is a tiny (roughly 5×10^{-5} m across) magnet that can be viewed under a microscope. In an unmagnetized specimen, the orientations of these domains are random, and their fields cancel (Fig. 21.11). That configuration is a compromise between total order of the spin dipoles and total disorder, and exists because it is energetically the most economical.

When a ferromagnetic sample is brought into a magnetic field, its domain structure can be drastically altered in two basic ways. The low-energy process (which occurs even with fairly weak fields) is one in which *the many domains that happen to be already aligned with the applied field grow at the expense of the domains that are misaligned at the start and subsequently shrink.* This is what occurs when soft iron is placed in a field and becomes magnetized (Fig. 21.12). This state with aligned domains and a resulting induced magnetic field (in which there is a good deal of energy) is unstable. Without support from outside, the induced field collapses and the iron spontaneously demagnetizes, rearranging domains.

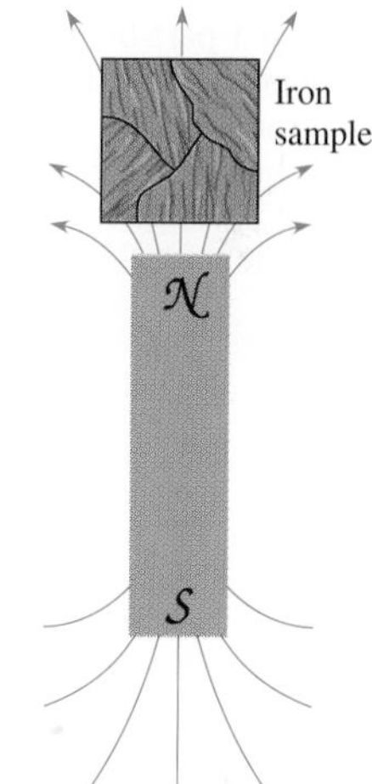

Figure 21.12 An iron sample placed in the magnetic field of a bar magnet. The domains of the sample are influenced, and they align with the applied field. The sample becomes magnetized and is drawn toward the bar magnet.

The other magnetizing process, which requires a higher applied *B*-field, results in the *irreversible reorientation of the domains* (Fig. 21.13). All the electron dipoles coupled together within each domain can literally be rotated into alignment with the applied field, just like a compass needle. This mechanism prevails in substances that become permanently magnetized. Domains in materials with irregular internal structures, such as steel, cannot easily change shape. The domains rotate instead, and having once been forced to do so against a kind of internal friction, they tend to stay put. When a steel knife blade (some stainless steel is "nonmagnetic") is stroked with a strong magnet, the effect is to rotate the domains into that direction, to "comb" them into alignment.

Permanent magnets and ferromagnetic materials are used in VCRs, TVs, stereo headsets, automobiles, speakers, tape decks, motors, and telephones. They float in space in a thousand different satellites, are on the backs of millions of credit cards, and in the inks on countless dollar bills and personal checks. (Try holding a dollar bill near a powerful magnet.)

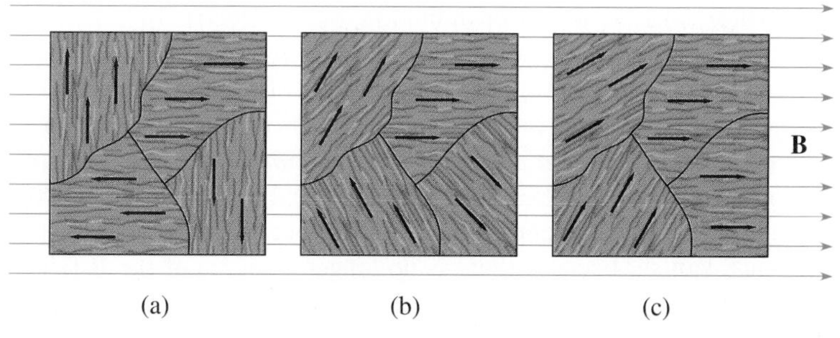

Figure 21.13 The reorientation of domains, bringing about an alignment with the applied field. As the applied magnetic field increases from (a) to (b) to (c), the electrons in the domains rotate so that their fields align with the applied field. The domains do not change shape as they did in Fig. 21.12. This realignment mechanism occurs in materials that become permanently magnetized.

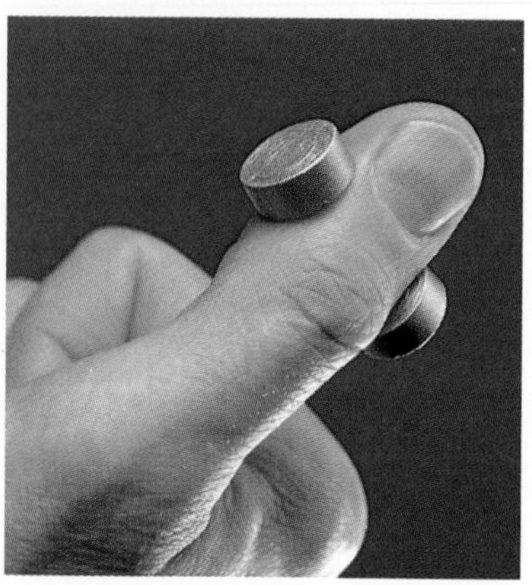

These two extremely powerful magnets are made of a new (1985) alloy of 14 parts iron, 2 parts neodymium, and 1 part boron. The alloy is usually powdered, the grains aligned in a strong B-field, and then the material is heated until it fuses.

Anything that reorients the domains of a piece of magnetized steel diminishes its supposedly permanent field. Banging on a magnet with a hammer will do just that, by knocking regions out of alignment—providing the domains with enough energy to unpin themselves. If the magnet's temperature is made high enough, the vibrating atoms will jostle themselves out of alignment and disrupt the long-range order. Pierre Curie realized that there is a limiting temperature, now called the **Curie temperature**, for each substance beyond which ferromagnetism vanishes and the material becomes paramagnetic. In 1894, Curie found that this critical temperature for iron is 770°C, though it had been known for a long time that red-hot iron has no magnetic power. The Curie temperature for magnetite is roughly 575°C, and for the other important iron ore, hematite, it's about 675°C. Interestingly, in all three cases, the Curie temperature is well below the melting point.

Permeability

The presence of a dielectric in an applied electric field has the effect of producing a weaker net internal E-field (p. 643). Similarly, when a diamagnetic material is placed in an applied magnetic field, it decreases the net B-field within the medium, but the effect is very small. On the other hand, the presence of a paramagnetic material will very slightly enhance the field within the medium. By contrast, on being immersed in a B-field, a ferromagnetic material becomes strongly magnetized. That, in turn, contributes an additional field component that adds to the original field and tends to cause the new net field to follow the contours of the metal (Fig. 21.14a). Lord Kelvin called this property **permeability**, observing that iron is hundreds of times more permeable than air. Special alloys that are millions of times more permeable are used to magnetically shield things such as delicate wristwatches and color TV picture tubes.

Just the opposite effect occurs in superconductors, which are both perfectly conducting and perfectly diamagnetic. Below a certain applied field strength, a superconductor will maintain itself in a state where, internally, $\mathbf{B} = 0$. When the sample (immersed in a magnetic field) is cooled below its critical temperature, it becomes superconducting. As in Fig. 21.14c, it then almost completely expels the B-field from its interior and is therefore said to be *perfectly diamagnetic.* To be precise, an external field does penetrate a superconductor, but only in a very thin surface layer 10^{-5} cm to 10^{-6} cm thick. Only if the applied field is made to exceed the so-called *critical field* will it again penetrate the body of the specimen. The diamagnetic behavior of superconductors was discovered experimentally in 1933 and is known as the **Meissner Effect**.

ELECTRODYNAMICS

A charged particle, whether at rest or in motion, has an electric field **E**, which is not much more than saying that charges interact electrically. Now, suppose that the particular charged particle has no intrinsic magnetism of its own. Nonetheless, when such a charge moves in space, it exerts magnetic forces and possesses a magnetic field. This *magnetism arises out of motion, and motion is relative.* A beam of protons is a current I and as such exerts both electric and magnetic forces. And yet, if we run along with the flow, essentially causing I to become zero, the magnetic force vanishes. It must vanish, because there is no longer a source of the B-field.

The Special Theory of Relativity provides the realization that current-generated magnetism is a facet of electricity. A modification of the electrical interaction arising from relative motion appears as a force, the magnetic interaction. On a theoretical level well beyond our needs, Coulomb's Law can be modified to include the effects of relative motion. Though that modification is typically quite small, its consequences are far-reaching, and we call them *magnetism.* Indeed, if the speed of light were infinite, currents would not generate magnetic fields. Electricity and magnetism are the two sides of a single phenomenon—*electromagnetism*—that looks different to observers in relative motion.

The study of the electromagnetic interaction in all its manifestations is known as **electrodynamics**, a word prophetically coined by Ampère, who initiated the unification over 150 years ago when he linked the source of magnetism to currents.

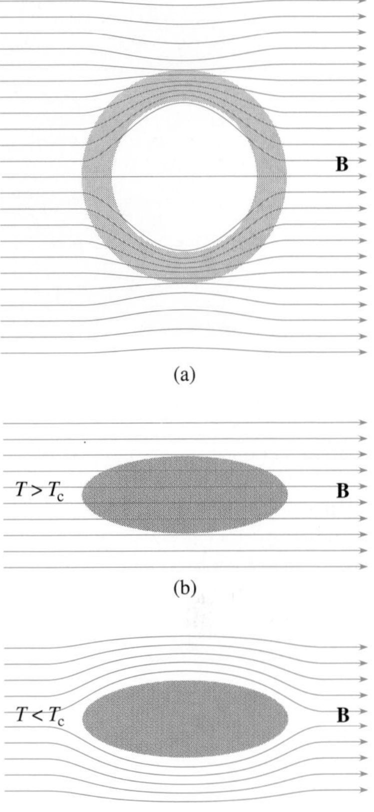

Figure 21.14 (a) When a material with a high permeability is placed in a *B*-field, most of the field lines pass through the material, effectively shielding the region it surrounds. (b) By contrast, a substance (such as lead) in its nonsuperconducting or normal state allows the *B*-field to pass through it. (c) When the temperature drops and the material becomes superconducting, it completely expels the applied *B*-field via the Meissner Effect.

21.6 Currents and Fields

Soon after Volta's invention of the battery, there were renewed attempts to find some relationship between electricity and magnetism. Evidence that a connection existed had been at hand for decades. The 1735 volume of the *Philosophical Transactions of the Royal Society of London* carried a paper entitled "Of an Extraordinary Effect of Lightning in Communicating Magnetism." It was the report of a bolt of lightning that struck a tradesman's house, blasting apart a box full of knives and forks, hurling them "all over the room . . . but what was most remarkable" was that they were all strongly magnetized afterwards!

On July 21, 1820, Hans Oersted, professor of physics at Copenhagen University, delivered a lecture on electricity to some advanced students. By chance, a wire leading to a voltaic pile was nearly parallel to and above a compass that happened to be on the table along with other paraphernalia. When the circuit was closed, the needle swing around almost perpendicular to the current-carrying wire as if gripped by a powerful magnet (see photo below).

The news of Oersted's discovery reached Paris on September 4, when Dominique F. J. Arago reported it to a skeptical gathering of the Paris Academy of Sciences. A young professor, André Marie Ampère attended the talk, and within two weeks he completed a series of experiments of his own. Ampère showed that the magnetic force experienced by a compass needle in the vicinity of a current-

(a)

(b)

Oersted's demonstration. (a) With no current in the wire, the compass needle points north. (b) When a current exists, the needle swings so that it almost aligns with the new field created by the current. The Earth's field causes a small northerly deflection of the needle.

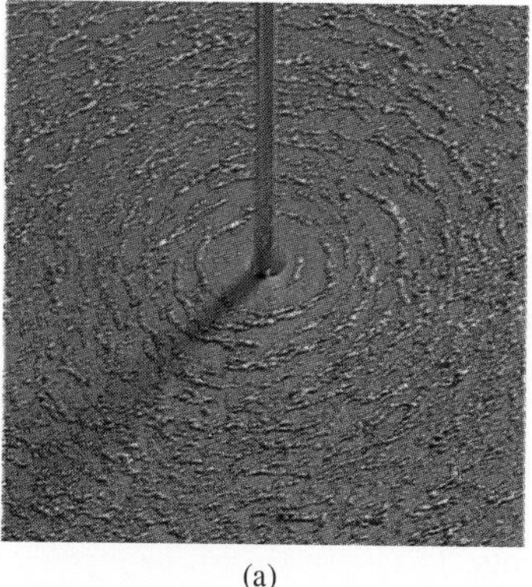

(a)

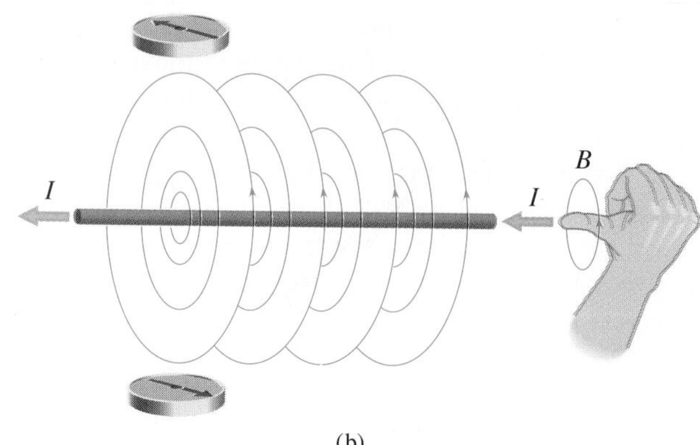

(b)

Figure 21.15 (a) The circular magnetic field surrounding a current-carrying wire, as revealed in an iron-filing pattern. (b) When the thumb on the right hand points in the direction of the current (I), the fingers curl in the direction of the circular magnetic field (B).

carrying straight wire acted at right angles to the current along a series of concentric circles (Fig. 21.15). *Point the thumb of the right hand in the conventional direction of the current and the direction in which a compass will point, the direction of the B-field, anywhere in the surroundings, is given by the direction of the fingers curling around the wire at that location.* (Call this the **Right-Hand-Current Rule**.) That rather surprising perpendicular character of the force explains why it took so long to discover.

The Current-Carrying Wire

A straight current-carrying wire generates a circular, or more accurately, a cylindrical magnetic field in the space surrounding it. *That field is constant at any given perpendicular distance from the wire and gets weaker as that distance increases.* Experiments, principally by Jean Biot and Félix Savart (1820), established that the B-field near a long straight wire in air is directly proportional to the current I and inversely proportional to the perpendicular distance r from the wire: $B \propto I/r$. The next logical step is to introduce a constant of proportionality that balances the units, yielding teslas on both sides, and thereby produce an equality. Since the SI system is "rationalized," the constant is defined so that $1/2\pi$ shows up in the equations for fields when there is axial symmetry, as there is here (that is, a cross section of the field over a plane perpendicular to the symmetry axis is the same wherever the plane is located). Furthermore, it was found that the B-field depends on the magnetic behavior of the medium in which the wire is immersed. Hence, we follow tradition and introduce the constant, $\mu/2\pi$. The Greek letter mu (μ) represents the medium-dependent constant that, naturally enough, is called the **permeability**.*

Hans Christian Oersted (1777–1851). Besides his work in electricity and magnetism, Oersted was the first to prepare pure metallic aluminum (1825).

The magnetic field at any point outside a long straight current-carrying wire (but not so far from it that distortions due to the ends of the wire show up) is then

$$B = \frac{\mu}{2\pi}\frac{I}{r} \tag{21.1}$$

In vacuum, the value of the permeability is, by definition

$$\mu_0 = 4\pi \times 10^{-7}\ \mathrm{T{\cdot}m/A}$$

*The constant $\mu/2\pi$ happens to have the μ on top because the permeability was originally defined so that Coulombs's Law for magnetic charge (no longer of any interest) would have a factor of $1/4\pi\mu$ in front to match the $1/4\pi\varepsilon$ in front of Coulomb's Law for electrical charge.

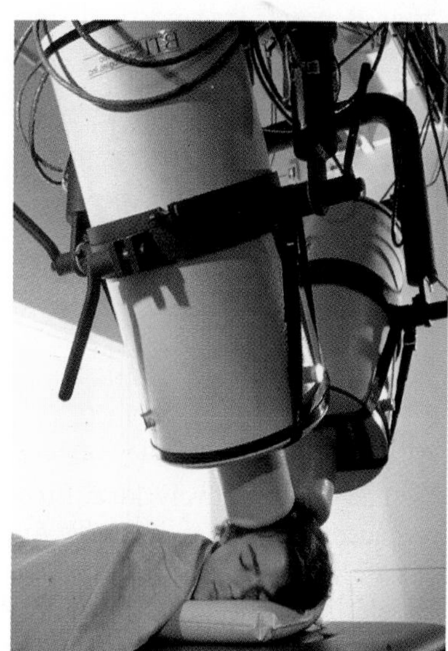

Today, superconducting quantum interference devices (SQUIDS) are being used to measure the minute ($\approx 10^{-13}$ T) magnetic fields generated by currents in the brain and heart. The magnetoencephalograph can locate the source of nerve signals within the brain to within a few milimeters.

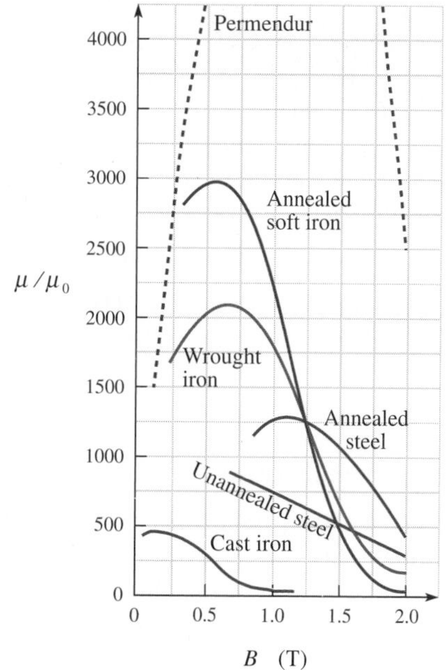

Figure 21.16 The permeability of magnetic materials is not constant, but instead changes as the field affects the material. Notice how the permeability initially rises, peaks, and then falls off as the material reaches saturation. Permendur, which is half iron and half cobalt, keeps a high μ even at large values of B.

An expression related to Eq. (21.1) will be used later to define the unit of current, and that's where this value of μ_0 comes from. It might be helpful to think of μ_0 as a scale factor for the units rather than a reflection of some physical characteristic of free space. Equation (21.1) works fine provided μ is constant, which it is for diamagnetic and paramagnetic media. It is not constant for ferromagnetic media, where μ changes as the B-field changes (Fig. 21.16), and Eq. (21.1) should then be used with care. For ferromagnetic media, μ is usually much larger than μ_0. Typically, in media that are likely to be of concern to us such as air or water, $\mu \approx \mu_0$. For diamagnetic media, $\mu/\mu_0 < 1$, whereas for paramagnetic media $\mu/\mu_0 > 1$, but in both cases $\mu/\mu_0 \approx 1$. The differences are small. In air $\mu_{\text{air}}/\mu_0 = 1 + 3.6 \times 10^{-7}$; whereas $\mu_{\text{water}}/\mu_0 = 1 - 0.88 \times 10^{-5}$. Thus, we can rewrite Eq. (21.1), **the magnetic field of a long straight wire in vacuum**, as

$$B = \frac{\mu_0}{2\pi}\frac{I}{r} \tag{21.2}$$

This equation provides the strength of the field (Fig. 21.17) of any line of current—a straight beam of protons transporting I coulombs per second through space will produce a B-field given by Eq. (21.2).

A. M. Ampère (1775–1836). Legend has it that Ampère was the classic absent-minded professor, who once even forgot to attend a dinner with the Emperor Napoleon.

Example 21.1 The overhead power cable for a street trolley is strung horizontally 10 m above the street. A long straight section of it carries 100 amps dc due east. Describe the magnetic field produced by the current and determine its value at ground level just under the wire. Compare that to the strength of the Earth's field.

Solution: [Given: $r = 10$ m and $I = 100$ A. Find: B.] First, draw a diagram. Fig. 21.15 will suffice here with the current assumed heading east. At ground level (at a point beneath the easterly current), the Right-Hand-Current Rule tells us that **B** points due north. Using Eq. (21.2)

$$B = \frac{(4\pi \times 10^{-7}\ \text{T·m/A})(100\ \text{A})}{2\pi(10\ \text{m})} = \boxed{2.0 \times 10^{-6}\ \text{T}}$$

which, from Table 21.1, is only 4% of the Earth's field.

▶ **Quick Check:** $\mu_0 = 1.2566 \times 10^{-6}$ T·m/A, $B \approx (10^{-6}\ \text{T·m/A})(10^2\ \text{A})/60\ \text{m} \approx 2 \times 10^{-6}$ T.

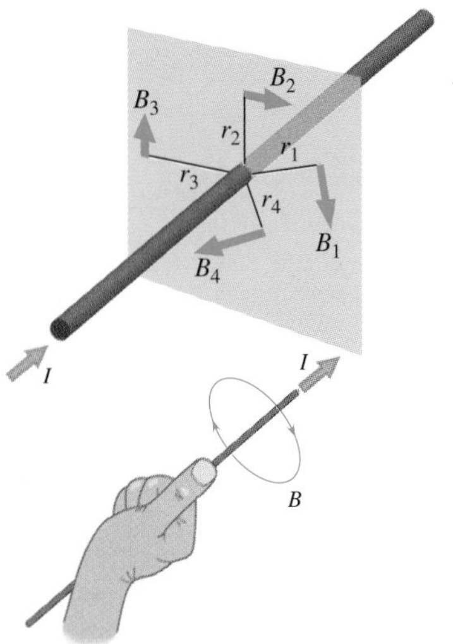

Figure 21.17 The B-field surrounding a current-carrying straight wire lies in a plane perpendicular to the wire. At any point on that plane, the B-field is perpendicular to the line from the wire to the point, and decreases inversely with the distance from the wire to the point.

The Current Loop

During those first weeks of excitement, Ampère had another lovely idea. Since a straight current-carrying wire is surrounded by concentric rings of magnetic force, bending the conductor into a loop should concentrate that force (as does bending a bar magnet into a horseshoe magnet). Using the field imagery, we can imagine the lines of B that extend far out into space on one side of a straight wire, being carried around and crowded within the small region then encompassed by the loop (Fig. 21.18). The field inside the loop is much stronger than the field outside. Again the Right-Hand-Current Rule provides the direction of B. The field will effectively vanish if the loop is squashed such that the current doubles back on itself—oppositely directed fields then overlap and cancel. The current-circle brings to mind the picture of a dipole field. Recognizing this similarity, Ampère boldly suggested that currents were the basic cause of magnetism (and to the degree that the intrinsic magnetism of an electron is due to the motion of its charge, he was right).

Biot and Savart determined experimentally that the field at the very center of a current loop points axially outward (Fig. 21.19) along the z-axis. They found B to be directly proportional to I and inversely proportional to the radius R of the loop:

[circular loop, center] $$B_z = \frac{\mu_0 I}{2R} \qquad (21.3)$$

In this case, the field's cross section over a plane perpendicular to the central z-axis is different at different values of z, and so the factor of 2π is not present.

Stacking several loops in parallel results in the overlapping of their individual fields, providing a proportionately increased net effect. Each loop simply adds (vectorially) its field to the fields of all the others. Thus, a tight short coil (Fig. 21.19) composed of N closely wrapped turns of wire each carrying a current I has a field at its center of

[circular coil, center] $$B_z = N\frac{\mu_0 I}{2R} \qquad (21.4)$$

This sort of coil, with a negligible length compared to its diameter, resembles a stubby disk magnet. If the coil is delicately pivoted so it can rotate freely about a

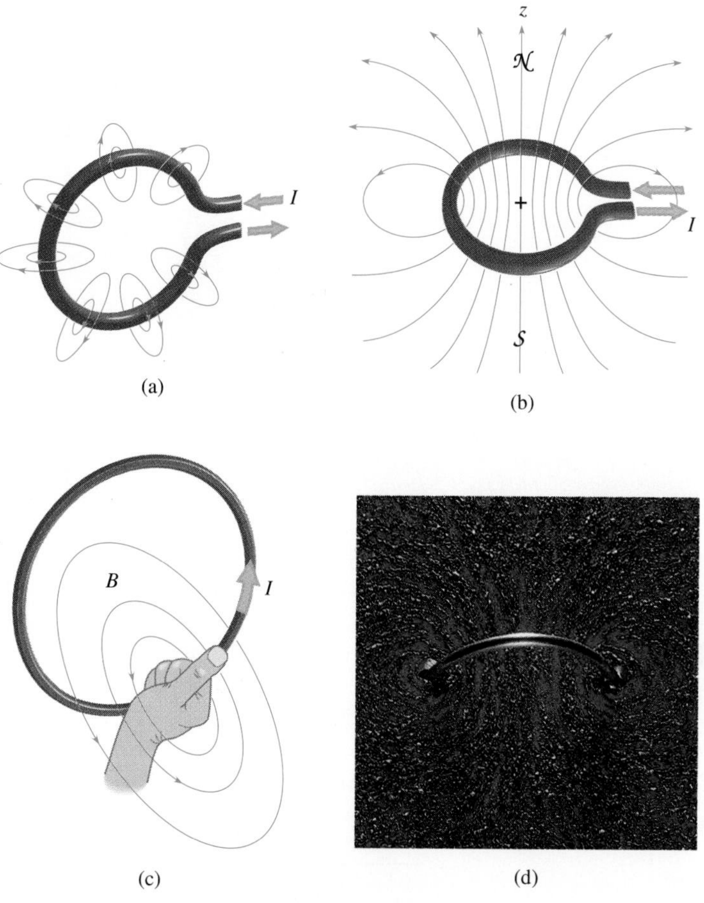

Figure 21.18 (a) Each segment of a current-carrying loop is surrounded by a circular B-field. (b) These combine to produce a dipole field very much like the field of a bar magnet. Remember that the field is 3-dimensional and more or less axially symmetrical around the central z-axis. (c) The Right-Hand-Current Rule gives the direction of $\mathbf{B}$. (d) The field pattern, as revealed with iron filings, in a plane perpendicular to the loop.

diameter, it will swing into alignment with an applied field just as a compass would. Ampère observed as much in 1820. This twisting behavior is the basis of both the moving coil galvanometer and the electric motor (p. 749).

The Solenoid

Carrying the coil concept one step further, Ampère wound wire into a long helix (Fig. 21.20) or *solenoid* (from the Greek, *solen* meaning "tube") and found that, with a current passing through it, it acted like a bar magnet. Within the space encompassed by a long, narrow (at least 10 times longer than it is wide), tightly wound solenoid, the B-field is strong and quite uniform, especially in the middle and around the central z-axis. The solenoid is one of the most useful magnetic devices—a typical home has dozens of them operating bells, chimes, and speakers. The solenoid is the central component of the relay that mechanically controls equipment such as washing machines, dishwashers, clothes driers, and furnaces. As a circuit element, the solenoid is in radios, TVs, and computers.

A solenoid is helical and not quite the same as stacking a bunch of separate flat loops; here, the current progresses from one end of the coil to the other and that adds a small additional contribution to B (p. 740). On the other hand, if the solenoid is wound with overlapping turns, an even number of layers will bring the end wire back to the beginning and cancel the longitudinal current.

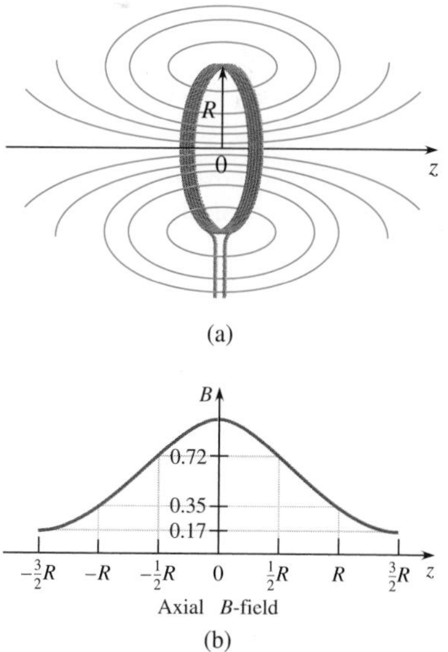

Figure 21.19 (a) A narrow circular current-carrying coil. (b) The B-field measured along the central z-axis. The curve of B is fairly straight around $z = R/2$, where there is a turning point (see Discussion Question 18).

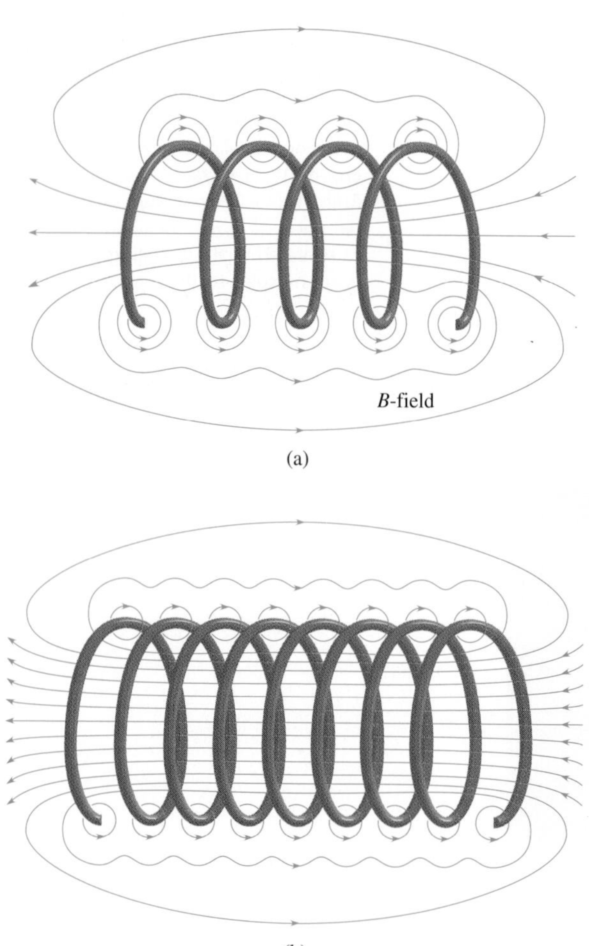

Figure 21.20 The solenoid. (a) The magnetic field of a loosely wound current-carrying coil. (b) When the coil is wound tighter and there are more loops, the field inside becomes larger and more uniform. The Right-Hand-Current Rule provides the direction of **B**.

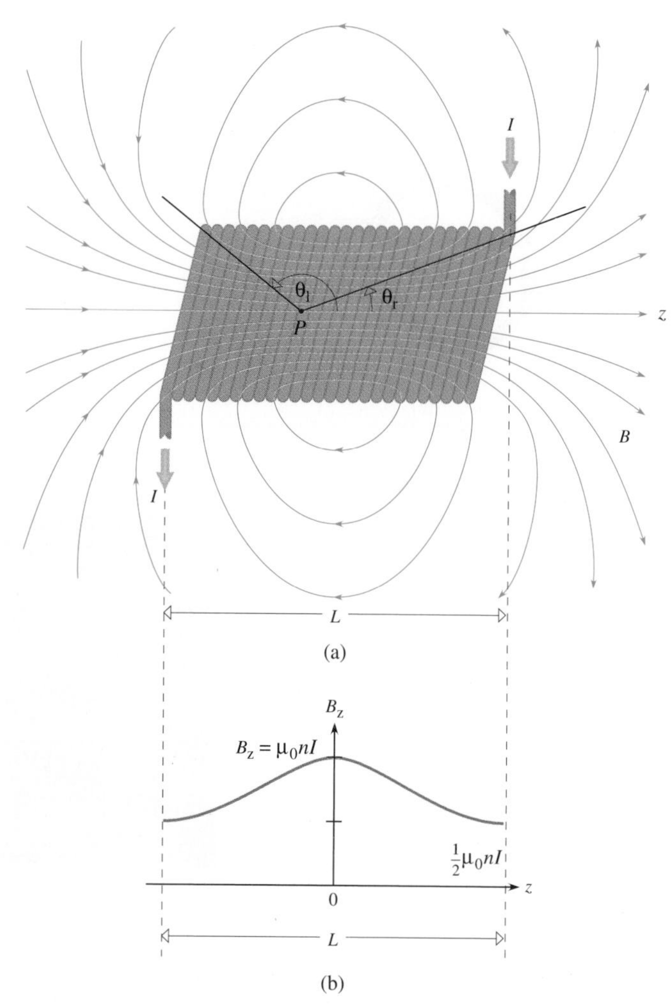

Figure 21.21 (a) The *B*-field of a finite solenoid carrying a current *I*. (b) The value of the field at any point *P* on the central *z*-axis is called B_z, and it varies with θ_l and θ_r. At the very center of the coil ($z = 0$), the field is a maximum equal to $\mu_0 nI$.

It's reasonable to assume that the field inside a solenoid increases directly with *I*. Moreover, the field should increase with the number of current loops contributing, but here there's the additional concern of packing the loops as close together as possible. Carrying the same current, 100 turns spread over a length of a meter will produce a much weaker inside field than 100 turns distributed over a centimeter. What's needed is a measure of the density of the winding, namely the *number of turns per unit length* (*n*) of solenoid. Hence, if *L* is the length of the coil and *N* the total number of turns, $n = N/L$. Then $B \propto nI$, and it's found experimentally (Fig. 21.21) that in air

[*solenoid, center, axial field*]

$$B_z \approx \mu_0 nI \tag{21.5}$$

The field at any axial point P is actually given by

$$B_z = \tfrac{1}{2}\,\mu_0 nI(\cos\theta_r - \cos\theta_l)$$

where θ_r and θ_l are the angles made with the right and left edges of the coil. As $\theta_r \to 0$ and $\theta_l \to 180°$, as they would for an infinitely long coil, the field approaches that of Eq. (21.5). For a coil 10 times longer than its diameter, Eq. (21.5) yields results that are only about 0.5% too large, which is good enough for our purposes.

If such a coil is imagined cut across the middle into two equal-length solenoids carrying the same current as the original, we can expect that the fields at their ends (which add up to $\mu_0 nI$ in the middle of the uncut coil) should be

[*solenoid, ends*] $$B_z \approx \tfrac{1}{2}\mu_0 nI$$

About half the field "leaks out" from between the windings of a real solenoid.

The magnetic field lines produced by a current-carrying coil. The pattern is again formed using iron filings.

Example 21.2 A 20-cm-long solenoid with a 2.0-cm inside diameter is tightly wound on a hollow quartz cylinder. There are several layers with a total of 20×10^3 turns per meter of a niobium-tin wire. The device is cooled below its critical temperature and becomes superconducting. Since the wire is then without resistance, it can easily carry 30 A and not develop any I^2R losses. Compute the approximate field inside the solenoid near the middle. What is its value at either end?

Solution: [Given: a solenoid where $n = 20 \times 10^3\ \text{m}^{-1}$, and $I = 30$ A. Find: B_z.] The solenoid is long and narrow and will obey the approximations that lead to Eq. (21.5), thus

$$B_z \approx \mu_0 nI = (1.257 \times 10^{-6}\ \text{T·m/A}) \times (20 \times 10^3\ \text{m}^{-1})(30\ \text{A})$$

and $$\boxed{B_z \approx 0.75\ \text{T}}$$

which is a formidable field, over 10^4 times that of the Earth. The field at either end is about half this, $\boxed{0.38\ \text{T}}$.

▶ **Quick Check:** $B_z \approx \mu_0 nI \approx (10^{-6}\ \text{T·m/A}) \times (2 \times 10^4\ \text{m}^{-1})(3 \times 10^1\ \text{A}) \approx 0.6$ T.

Sometime around 1825, W. Sturgeon wrapped 18 turns of bare wire around a varnished iron bar and sent a current through the coil; in so doing, he created the first powerful *electromagnet*. The field set up by the current aligned the domains within the iron to produce a combined magnetic field of unprecedented strength (Fig. 21.22). The American physicist Joseph Henry heard about the feat and set about to better it. Legend has it that he tore apart his wife's petticoats so that he could insulate his wires with the silk stripping. By using many turns of insulated wire, Henry enhanced the field while keeping I relatively low. In 1831, he produced a modest-sized device powered by an ordinary battery that could lift more than a ton of iron. When the current was interrupted, the soft iron core almost completely demagnetized spontaneously, and the load was released.

__Talking about Ampère.__ He further deduced from this analogy the consequence that the attractive and repulsive properties of magnets depend on electric currents which circulate about the molecules of iron and steel.

D. F. J. ARAGO (1820)
French physicist

21.7 Confirming Ampère's Hypothesis

By the beginning of the twentieth century, Ampère's hypothesis—that all magnetism is due to currents—was widely held, though it certainly had not been con-

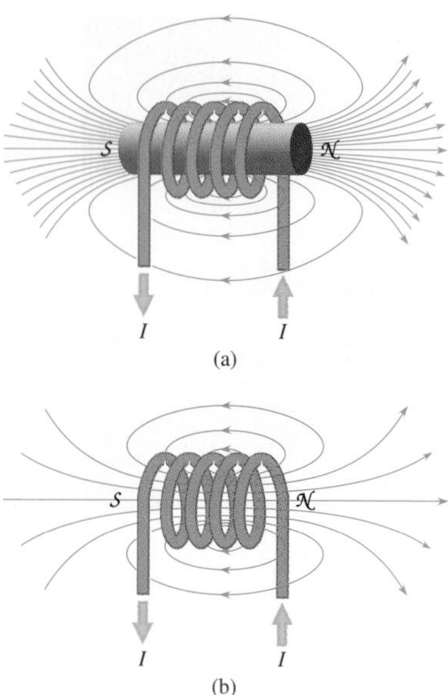

Figure 21.22 The field of a current-carrying coil (a) with and (b) without a ferromagnetic core. The B-field set up by the current in the coil magnetizes the core, aligning its domains, and that increases the field even more.

Steel nails align themselves to reveal the B-field configuration surrounding the opposite poles of two powerful superconducting magnets. Frost around the necks of the stainless steel Dewar flasks is a clue to the fact that they're operating at liquid helium temperature (around 4.2 K).

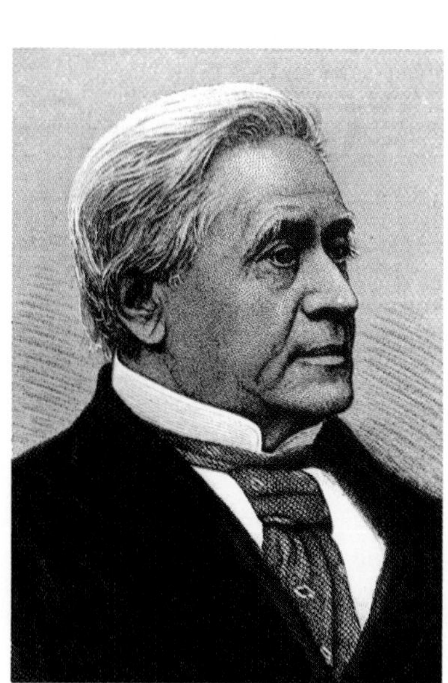

Joseph Henry (1797–1878). To honor him for his many original contributions, the International Electrical Congress of 1893 named the unit of inductance the henry.

firmed. And that was despite the fact that decades earlier, the great physicist James Clerk Maxwell had proposed several experimental ways of resolving the issue. One of his suggestions was to see whether or not a magnet behaved like a gyroscope. If the magnetism of a bar magnet is due to hidden circulating currents, these whirling charges should have angular momentum and display gyroscopic behavior.

In 1909, S. J. Barnett demonstrated the inverse effect: simply *rotating a ferromagnetic specimen magnetizes it* precisely as if it had been placed in a uniform external B-field directed along the axis of rotation. Barnett assumed there were microscopic current whirls initially at random orientations throughout the sample. (This was decades before the introduction of the notion of the intrinsic spin of the electron, and ferromagnetism was understood in terms of atomic current loops.)

To model the arrangement, imagine several small gyroscopes (mounted so they can swivel in any direction) attached to the horizontal platform of a centrifuge. With the little gyros spinning in random directions, the centrifuge is turned on and rotates about a vertical axis. All of the gyros now revolving with the platform will align themselves vertically, having the same sense of rotation as the centrifuge. This is quite similar to the way a gyrocompass aligns itself with the planet's spin axis. In Barnett's demonstration, the rotation of the sample as a whole caused the tiny current loops to swing around into alignment with the bulk spin axis. That alignment was subsequently revealed by the magnetization of the specimen. The argument applies equally well to spinning electrons, which is the way it is interpreted today.

Maxwell's idea was finally confirmed in 1915, by no less a personage than Albert Einstein, who worked in conjunction with W. J. de Haas. Their experiment verified that the magnetization of a sample causes it to rotate. Since this work was done before the appreciation of electron spin, their theoretical arguments were somewhat in error, and their numerical result turned out to be off by a factor of 2, but that didn't seem to bother Einstein. In the experiment, a ferromagnetic cylinder was hung vertically inside a solenoid whose *B*-field was periodically reversed. Einstein and de Haas explained that magnetizing the sample aligns current whirls, which, in turn, must align their angular momentum vectors, thus increasing the overall internal angular momentum of the sample. But the Law of Conservation of Angular Momentum demands that the net change in **L** must be zero; the cylinder as a whole must therefore rotate in the opposite direction. And rotate it did. "We have given firm proof," wrote Einstein, "of Ampère's molecular currents . . . ."

The speaker of the radio shown on p. 589. You can see the permanent magnetic housing at the rear of the speaker. The paper cone is attached to a current-carrying coil that causes the cone to vibrate. (See Fig. Q9, p. 754.)

21.8 Calculating *B*-Fields

There are two equivalent schemes for calculating magnetic fields due to currents, one devised by Laplace based on the data of Biot and Savart, the other by Ampère. We will study *Ampère's Law,* which is simpler to use, provided the geometry is simple. We cannot *derive* the law from basics (that is, from the properties of the electron)—physicists simply don't know enough yet to do that. It can, however, be *deduced* from experimental observations.

Ampère's Law

Ampère's Law is a little obscure physically—it will take a bit of doing to justify it, but it's worth it, as we'll see in future chapters. Accordingly, imagine a straight current-carrying wire and the circular *B*-field surrounding it. We know from experiments that $B = \mu_0 I/2\pi r$, which is Eq. (21.2). Now, suppose we put ourselves back in time to the nineteenth century when it was common to think of magnetic charge (q_{m}). Let's define this monopole charge so that it experiences a force when placed in a magnetic field B equal to $q_{\mathrm{m}}B$ in the direction of B, just as an electric charge q_{e} experiences a force $q_{\mathrm{e}}E$. Suppose we carry this north-seeking monopole around a closed circular path perpendicular to and centered on a current-carrying wire and determine the work done in the process. Since the direction of the force changes, because **B** changes direction, we will have to divide the circular path into tiny segments (Δl) and sum up the work done over each. Work is the component of the force parallel to the displacement times the displacement: $\Delta W = q_{\mathrm{m}}B_{\parallel}\Delta l$, and the total work done by the field is $\Sigma q_{\mathrm{m}}B_{\parallel}\Delta l$. In this case, **B** is everywhere tangent to the path, so that $B_{\parallel} = B = \mu_0 I/2\pi r$, which is constant around the circle. With both q_{m} and B constant, the summation becomes

$$q_{\mathrm{m}}\Sigma B_{\parallel}\Delta l = q_{\mathrm{m}}B\Sigma\Delta l = q_{\mathrm{m}}B2\pi r$$

where $\Sigma\Delta l = 2\pi r$ is the circumference of the circular path.

If we substitute for B the equivalent current expression, which varies inversely with r, the radius cancels—the work is independent of the circular path taken. Since no work is done in traveling perpendicular to **B**, the work must be the same if we move along a radius from one circular segment to another as we go around. Indeed, W is independent of path altogether—the work will be the same for *any closed path* encompassing the current. Putting in the current expression for B and canceling the "charge" we get the rather remarkable expression

$$\Sigma B_{\parallel}\Delta l = \mu_0 I$$

which is to be summed over any closed path surrounding I. The magnetic charge has disappeared, which is nice, since we no longer expect to be able to perform this

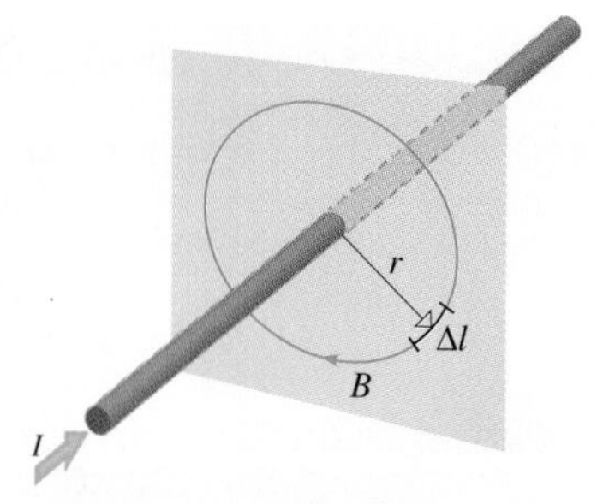

Figure 21.23 Using Ampère's Law. The circle of radius r surrounding the current-carrying wire is the Ampèrian path we have chosen. The B-field is everywhere parallel to the path elements Δl.

little thought experiment with a monopole. Still, the physics was consistent, and the equation should hold, monopoles or no. Moreover, if there are several current-carrying wires encompassed by the closed path, their fields will superimpose and add, yielding a net field. The equation is true for the separate fields and must be true as well for the net field. Hence

$$\Sigma B_{\parallel} \Delta l = \mu_0 \Sigma I \tag{21.6}$$

Today this equation is known as **Ampère's Law**, though at one time it was commonly referred to as the "work rule."

Ampère's Law can be used to compute B generated by some pattern of currents (without having to rely on calculus) when the configuration of the field is known and is not very complicated. For example, consider the simplest case of a straight wire carrying a current I. *Wherever possible, we want* **B** *either perpendicular* ($B_{\parallel} = 0$) *or parallel* ($B_{\parallel} = B$) *to the encompassing path.* The field here (Fig. 21.23) is known to be circular, so we choose a circular Ampèrian path, around which $B_{\parallel} = B$. Hence, $B \Sigma \Delta l = B 2\pi r$ and because the net current is I

$$B = \frac{\mu_0 I}{2\pi r} \qquad [21.2]$$

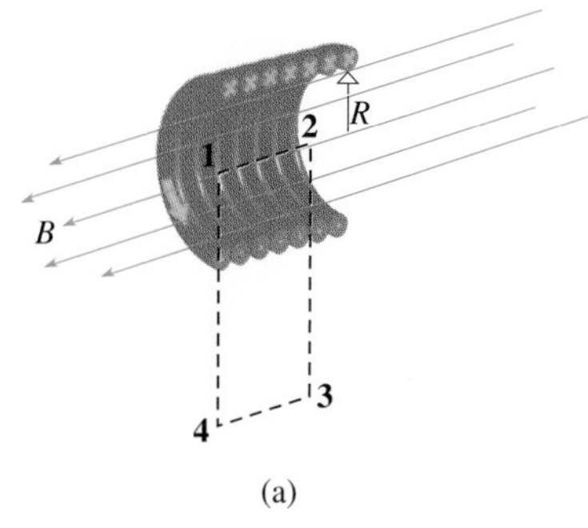

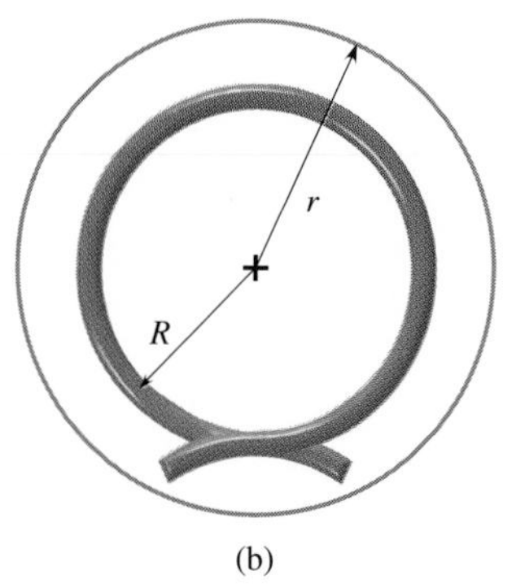

Figure 21.24 Using Ampère's Law. A helical coil carrying a current (out of the page and to the right at the bottom and into it and to the left at the top). (a) To find the field inside the solenoid, use path 1–2–3–4. (b) To find the field outside, use a circle of radius r.

As another and last example, let's find the field inside a very long solenoid (Fig. 21.24). Its asymmetry requires a bit of tricky maneuvering. As an Ampèrian loop that follows the B-field, with segments either parallel or perpendicular to it, we construct the path 1–2–3–4–1. The summation $\Sigma B_{\parallel} \Delta l$ is carried out over the four straight segments. Since the field lines close on themselves and since the system is symmetrical, there cannot be a radial component of the field. Thus B must be perpendicular to segments 2–3 and 4–1, and they make no contribution. Even if there were a radial field, in going around the loop, the contribution along 2–3 would be equal and opposite to that from 4–1, and they would cancel. Now for segment 3–4: the field outside an infinitely long solenoid is zero, and the field outside a finite but long solenoid must be small, axial, and drop off with r. In any event, we take 3–4 so far from the solenoid that the field there is negligible and that segment makes no appreciable contribution. What remains is segment 1–2, over whose length L the field is parallel (that is, $B_{\parallel} = B_z$). Thus, if the number of turns of wire encompassed by the path is N, the net current is NI, and Ampère's Law yields

$$B_z L = \mu_0 N I$$

but N/L is the number of turns per unit length n and so

$$B_z = \mu_0 n I \qquad [21.5]$$

Note that the field inside is independent of its distance from the central z-axis and must be uniform. It's also independent of the cross-sectional shape of the coil.

If the solenoid is wound in a single helical layer, charge is transported from one end to the other, and there is a net current I in the z-direction that produces a circular external field. Applying Ampère's Law using the path shown in Fig. 21.24b, we find that the field has a component B_ϕ around the solenoid, given by

[*solenoid, outside*] $$B_\phi = \mu_0 I / 2\pi r$$

This is the field of a straight wire, and it's very small compared to the axial field inside the solenoid, where n might be 10^4 turns per meter. (See Discussion Question 11.)

MAGNETIC FORCE

Oersted demonstrated that a current exerted a force on a compass needle. To establish that this was a purely magnetic interaction, Ampère did away with the iron needle altogether. He passed a current through two parallel wires, one of which was suspended so that it could swing in response to the *B*-field of the other, and swing it did. Inasmuch as currents exert forces on magnets, it follows from Newton's Third Law that magnets ought to exert forces on currents. As we will see, Ampère's two-wire experiment proves the point, as do a number of other elegant arrangements, including the electric motor (p. 750) and generator (p. 774).

We know now that these interactions, which are describable on a macroscopic level in terms of currents, are fundamentally due to a magnetic force experienced by mobile charge carriers. Nowadays it's easy enough to send a beam of charged particles (for example, in a cathode-ray tube) through a known field and observe the effects firsthand. Several conclusions are forthcoming: a charge q moving through a magnetic field **B** with a velocity **v** experiences a force $\mathbf{F}_M$, which, reasonably enough, is proportional to q, v, and B; that is, $F_M \propto qvB$. Thus, no relative motion ($v = 0$), no magnetic force. Further, the two vectors **v** and **B** determine a plane, and the force is perpendicular to that plane, as shown in Fig. 21.25. *If a particle with an opposite charge is introduced, the force reverses*—the sign of q affects the sign of F_M.

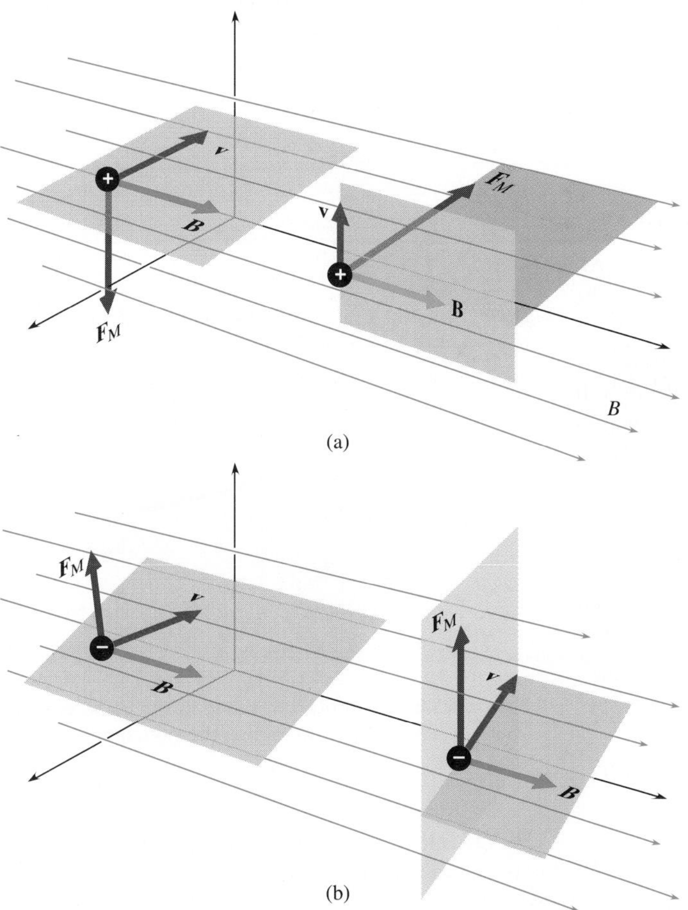

Figure 21.25 Charged particles, (a) positive and (b) negative, moving in a magnetic field. In each case, the resulting magnetic force $\mathbf{F}_M$ is perpendicular to the plane of **v** and **B**.

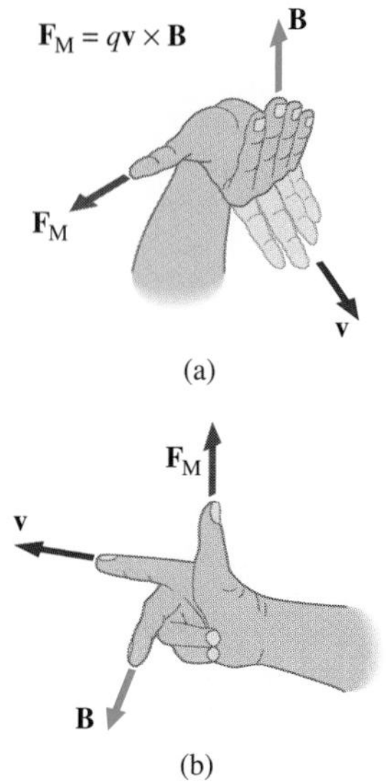

Figure 21.26 (a) Point the fingers of the right hand in the direction of **v**. Close them (through the smallest angle) toward **B**, and the extended thumb points in the direction of $\mathbf{F}_M$. (b) A positive charge moving with a velocity perpendicular to a magnetic field experiences a force perpendicular to both.

The magnitude of the force depends on the angle θ between **v** and **B**: in particular, *when* $\theta = 0$ *or* $180°$ *and the particle is moving along or opposite to the field, the force is zero. When* $\theta = 90°$ *or* $270°$ *and the particle is moving perpendicular to the field, the force is a maximum* (qvB). In other words, $F_M \propto \sin\theta$, and putting it all together

$$F_M = qvB \sin\theta \tag{21.7}$$

There is no constant of proportionality here because the equation is basically the one that will be used to define B. The zero-force line can be defined as the direction of the magnetic field. From the formula, a 1-N force will be exerted on a 1-C charge moving at 1 m/s at 90° to a 1-T magnetic field.

Not surprisingly, the force, the field, and the velocity are related by a right-hand rule. If you put the fingers of your *right* hand in the direction of **v** and curl them through the smallest angle to **B**, your thumb will point in the direction of $\mathbf{F}_M$, as in Fig. 21.26. All of this information can be represented in a single vector equation using the cross product (p. 177)

$$\mathbf{F}_M = q\mathbf{v} \times \mathbf{B} \tag{21.8}$$

To keep the Earth's B-field from exerting forces on the electron beam in a color TV set, the large end of the picture tube is shrouded in a casing of a high-permeability material that shields the internal region.

Example 21.3 A conventional water-cooled electromagnet produces a 3.0-T uniform magnetic field in the 4-inch gap between its flat pole pieces. The field is aligned horizontally pointing due north. A proton is fired into the field region at a speed of 5.0×10^6 m/s. It enters traveling in a vertical north-south plane, heading north and downward at 30° below the horizontal. Compute the force vector acting on the proton at the moment it enters the field.

Solution: [Given: a proton with $v = 5.0 \times 10^6$ m/s, at 30° below the horizontal in the northerly direction, $B = 3.0$ T, north. Find: F_M.] First, make a drawing—Fig. 21.27. The proton has a *positive* charge of $+1.60 \times 10^{-19}$ C and so $\mathbf{v} \times \mathbf{B}$ is due east, $\mathbf{F}_M$ is due east. From Eq. (21.7)

$$F_M = q_e vB \sin\theta = (+1.6 \times 10^{-19}\ \text{C})(5.0 \times 10^6\ \text{m/s}) \times (3.0\ \text{T})(\sin 30°)$$

and $\boxed{F_M = 1.2 \times 10^{-12}\ \text{N}}$

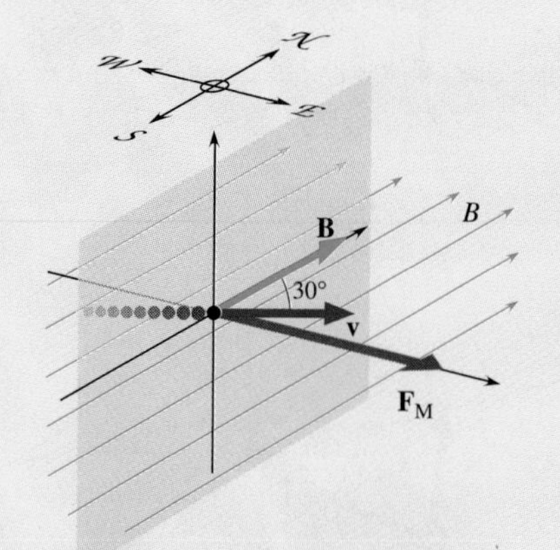

Figure 21.27 A proton traveling in the plane enters the field region at an angle of 30° below the horizontal.

▶ **Quick Check:** $F_M = q_e vB \sin\theta \approx (10^{-19}\ \text{C}) \times (10^7\ \text{m/s})(1\ \text{T})(1) \approx 10^{-12}$ N.

A TV special—"Monty Hall meets the magnet." The *B*-field of this fairly strong horseshoe magnet exerts forces on the electron beam and distorts the TV picture. This demonstration can be done with a black-and-white set with no risk of harm to it, but it's not advisable to try it on a color picture tube.

Because $\mathbf{F}_{\mathrm{M}}$ is perpendicular to **v**, it's purely a *deflecting* force; it changes the direction of **v** without altering v. Since there can never be a component of magnetic force along the motion, there will be no tangential acceleration. Thus, **no work will be done on a moving charge by a B-field, and no change in its energy can occur in the process.**

21.9 The Trajectory of a Free Particle

Imagine a positively charged particle q entering perpendicularly into a uniform magnetic field (Fig. 21.28). Because the magnetic force is always perpendicular to the velocity, an otherwise free particle will experience a centripetal acceleration that is also always perpendicular to the motion; that is, radial. The particle will be forced to move along a circular arc. If the field is strong enough and the particle stays in it long enough, and moreover doesn't lose any energy, it will swing out a complete circle. In reality, free accelerating charges radiate electromagnetic energy and spiral inward.

Figure 21.28 (a) A positive particle traveling through a uniform *B*-field experiences a force that causes it to move in a curved path. (b) The force is always perpendicular to the velocity, so the particle will ideally move in a circle. (It actually radiates, loses energy, and spirals inward.)

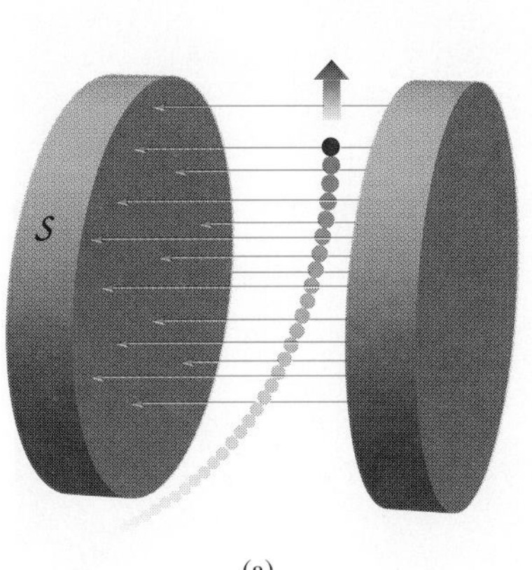

(a)

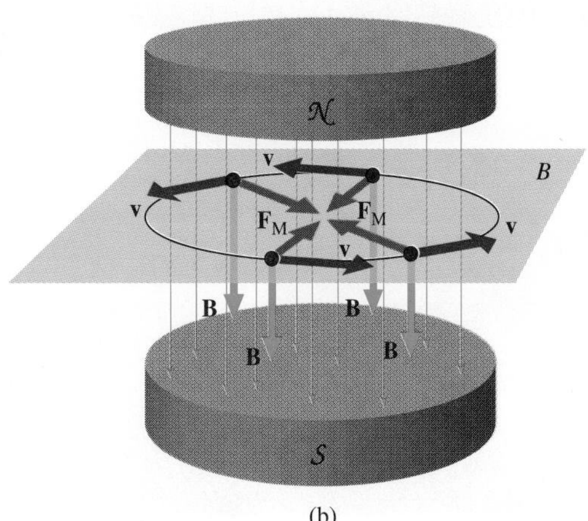

(b)

The B-field coils, or yoke, which surround the neck of a TV picture tube. They steer the electron beam so that it sweeps across the screen.

Here, a positive particle is carried into a counterclockwise trajectory; had the particle been negative, it would have simply swung the other way into a clockwise circle, as in the photo on p. 639. We know that an object will move in a circle of radius R provided there is a centripetal force on it whose magnitude is

$$F_C = \frac{mv^2}{R} \quad [5.11]$$

Here, with $\theta = 90°$

$$qvB = \frac{mv^2}{R}$$

and

$$R = \frac{mv}{qB} \quad (21.9)$$

For a given q, the radius R of the path is determined by the momentum mv and the field. If we somehow impart energy to the particle (for example, via an applied E-field), thereby increasing its momentum, and if at the same time we increase B proportionately, the radius of the orbit can be kept fixed. This is the central feature of all modern ring-shaped particle accelerators. To get the largest possible $mv = qBR$, we need both the largest available field and orbital radius. Today, the approach is to construct a great doughnut-shaped hollow chamber miles in diameter, pump out all the air within, and surround it with powerful superconducting electromagnets, so B can be as large as possible. A beam of, say, protons is fired in tangentially and, by adjusting mv and B for the fixed R of the chamber, the particles are brought up to tremendous energies (Ch. 33).

The *deflection yoke* on the neck of a TV picture tube consists of a pair of coils that create a set of crossed magnetic fields, one for vertical deflection and one for horizontal deflection. The current through each coil creates a variable magnetic field that steers the beam across the face of the tube.

(a) A beam of electrons in an old cathode-ray tube. (b) The beam is bent using a horseshoe magnet. (c) A beam of electrons bent into a circular orbit by the magnetic field of a set of large Helmholtz coils. The small amount of gas in the tube glows when it's ionized by collisions with the electrons, making the trajectories of the electrons visible.

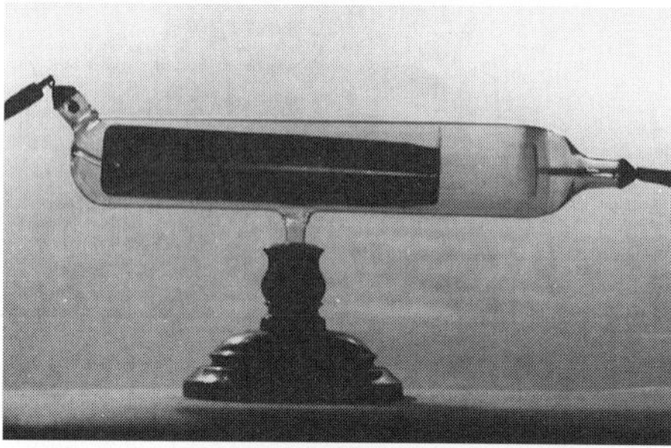

(a)

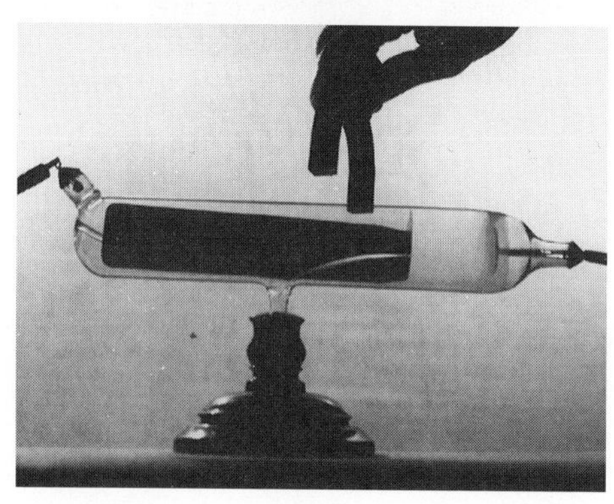

(b)

(c)

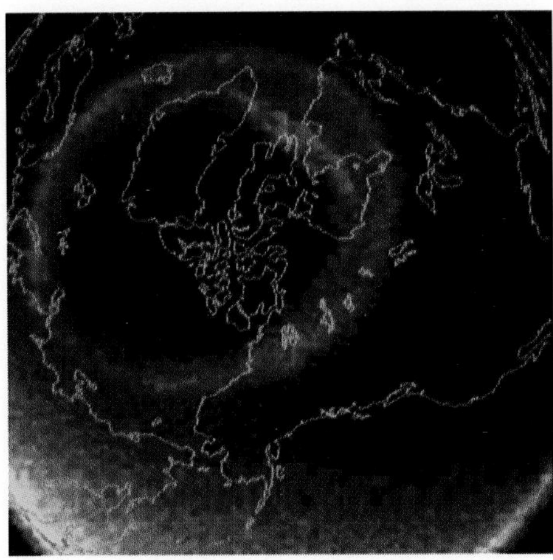

A high-altitude view of the aurora taken by a satellite about three Earth-radii away from the planet. A computer graphic shows the location of land masses. The U.S.A. is at the right of center.

The aurora borealis seen from the surface of the Earth.

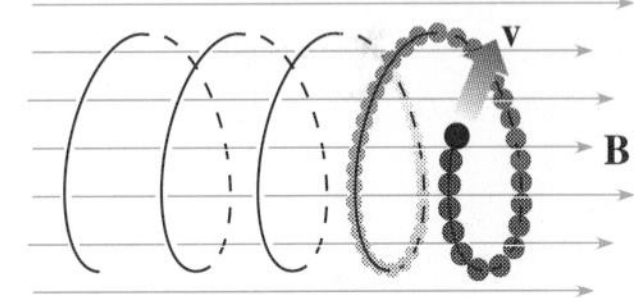

Figure 21.29 A particle, which has an initial component of **v** in the direction of **B**, moving along a spiral.

Auroras and Radiation Belts

On a grand scale, the magnetic field acts on charged particles as a kind of cosmic accelerator provoking and guiding a range of violent occurrences from X-ray emissions and stellar flares to the blazing northern and southern lights—the auroras—in the Earth's atmosphere.

When a charged particle enters a uniform magnetic field with an initial component of velocity parallel to **B**, it progresses as it spirals, following a helical path (Fig. 21.29). Things get much more complicated in a nonuniform B-field. Figure 21.30 depicts a little dipole initially traveling to the right in a field that gets stronger in that direction. The magnetic forces on each end are different. The dipole is pushed to the left, decelerates, and may ultimately reverse direction if the field is strong enough. Something similar happens to a spiraling charge, which is a tiny current (Fig. 21.30b). The charge, too, behaves like a dipole and will be decelerated and reversed as if reflected from a so-called *magnetic mirror.* If two such regions exist, the particle may be trapped, bouncing back and forth in what is called a *magnetic bottle,* Fig. 21.30c. In the laboratory such arrangements are used, for example, in the study of controlled thermonuclear fusion, to retain immensely hot plasmas that would otherwise be destroyed by contact with the walls of an ordinary container.

In space, particles known as *cosmic rays* come streaming toward the Earth (mostly from the Sun) and encounter the planet's B-field. Some are deflected away,

Figure 21.30 (a) A magnetic dipole moving in the direction of a nonuniform B-field. That field may be thought of as arising from an unseen south pole at the right. The south end of the dipole is pushed back more strongly than the north end is pulled forward. The net force is F_M. (b) A spiraling positive charge is the equivalent of a dipole, and it, too, experiences a backward force. (c) When the field is properly shaped, it behaves like a magnetic bottle and can capture charges that spiral back and forth within it.

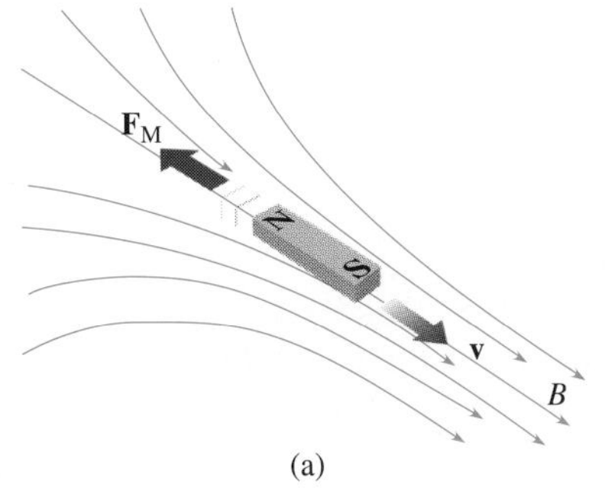

(a)

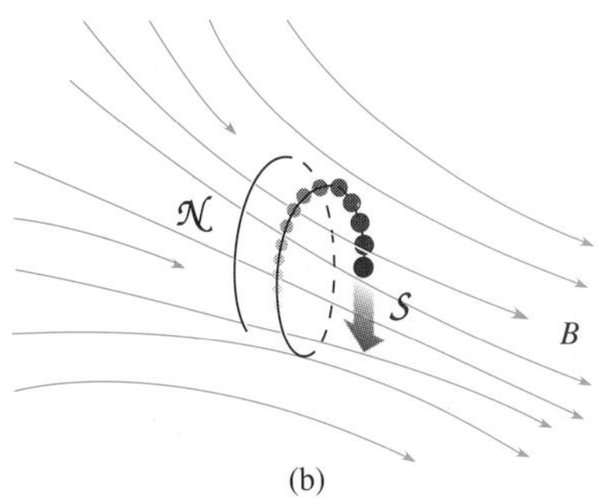

(b)

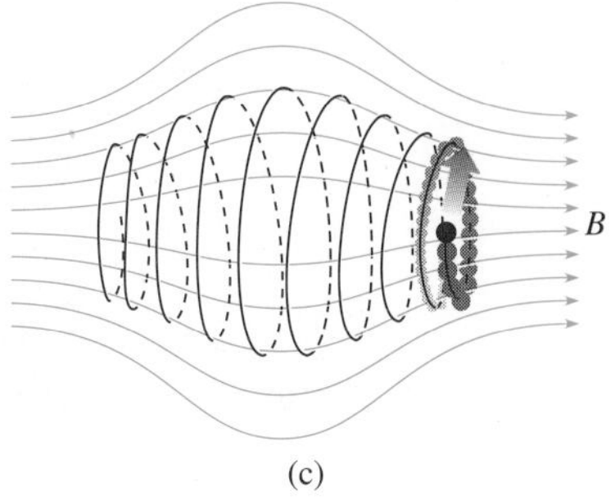

(c)

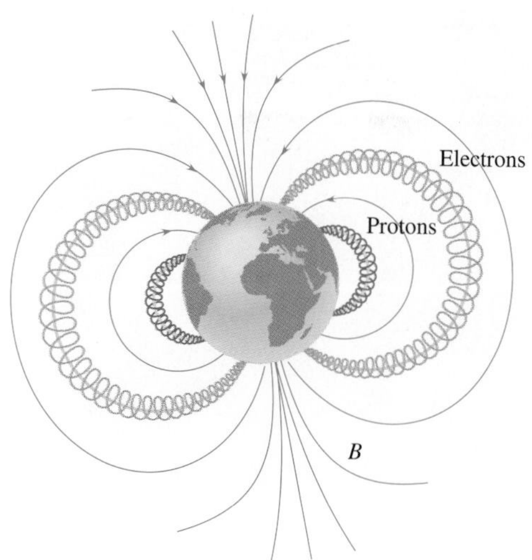

Figure 21.31 The Van Allen belts and the charged particles spiraling within them. Shown here is a cross-sectional view; the belts surround the planet.

others spiral down along the field (Fig. 21.31). Some of these particles become trapped in a form of magnetic bottle pinched off at its ends near the poles where the field increases rapidly. These form the Van Allen radiation belts that surround the planet. When some of those fast-moving charges collide with air molecules, the latter emit light in a wonderful display of color known as the aurora borealis. When a nuclear bomb was detonated high above the Johnston Islands in the Pacific (1960s), particles spiraling along the Earth's magnetic field caused an auroral display that was seen as far south as Hawaii.

21.10 Forces on Wires

Free charges experience forces when they traverse a magnetic field, something that would happen whether they were in a beam in vacuum or traveling as an ordinary current down a copper wire. When constrained to move within a conductor, the charges impart an average force to the conductor.

Return for a moment to Fig. 19.14 where a current traveling along a wire is viewed in terms of the drift of mobile charge carriers. A little segment of the wire of length l contains a total number of carriers equal to its volume (Al) times the number of mobile charges per unit volume (η): $Al\eta$. Moving at a speed equal to the drift speed (v_d) through a B-field, all these charges experience a net force

$$F_M = (q_e v_d B \sin\theta)(Al\eta)$$

But the current is

$$I = \eta v_d A q_e \qquad [19.3]$$

Hence, for a wire segment of length l

$$F_M = IlB \sin\theta \qquad (21.10)$$

and the direction of the force on the wire is the same as the direction of the force on the individual mobile carriers (always taken as positive charges). In other words, the

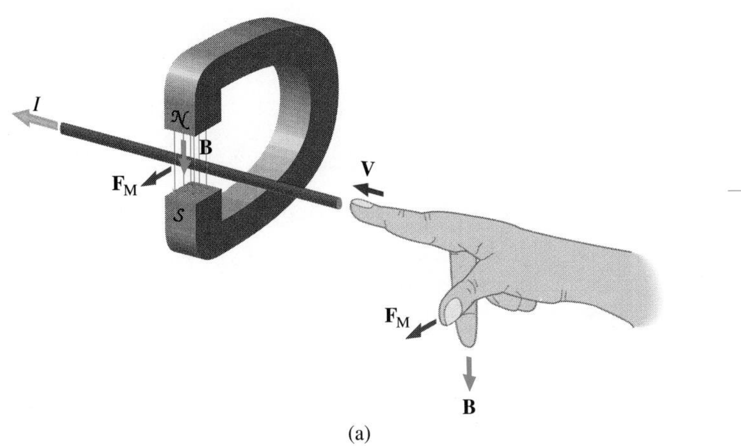

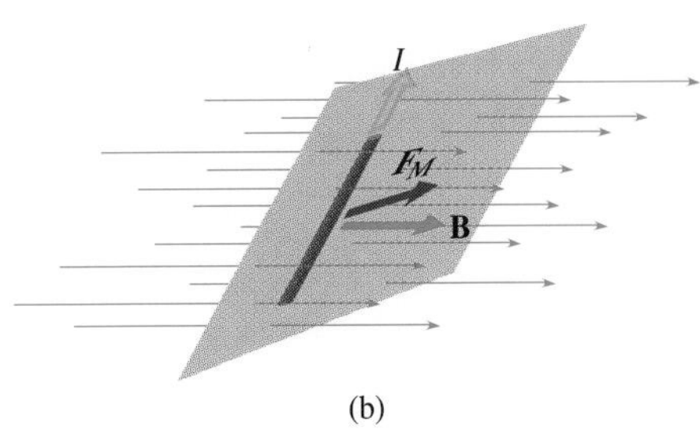

Figure 21.32 (a) The force on a current-carrying wire in a magnetic field. (b) Positive charges moving in the direction of I. $\mathbf{F}_M$ is then in the direction of $\mathbf{v} \times \mathbf{B}$.

direction of the force is given by the direction of $\mathbf{v} \times \mathbf{B}$, as shown in Fig. 21.32 (or if you like, index finger in the direction of **v**, middle finger in the direction of **B**, and thumb in the direction of **F**). Ampère used the arrangement of Fig. 21.33 to establish that no matter how he applied a field, the force on the current-carrying wire segment was always perpendicular to it.

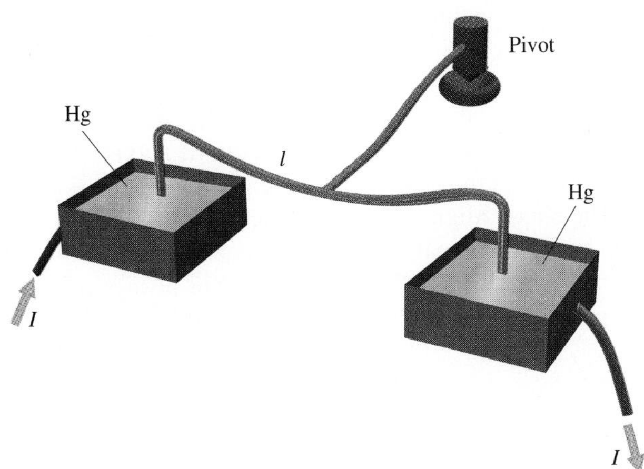

Figure 21.33 An arrangement used by Ampère to prove that F_M is always perpendicular to I. A segment of wire was set so as to freely rotate about a pivot. No matter what the direction of the applied magnetic field was, the wire never rotated because there was never a force tangent to the wire.

Example 21.4 A flat, horizontal rectangular loop of wire is positioned, as shown in Fig. 21.34, in a hypothetical 0.10-T uniform vertical magnetic field. The sides of the rectangle are F–C equal to 30 cm and C–D equal to 20 cm. Determine the total force acting on the loop when it carries a current of 1.0 A.

Solution: [Given: $B = 0.10$ T, F–$C = 30$ cm, C–$D = 20$ cm, and $I = 1.0$ A. Find: F_M.] Current travels from the positive terminal of the battery clockwise around the circuit. The directions of the forces on each segment are arrived at via $\mathbf{v} \times \mathbf{B}$ and are indicated in the diagram. Because the forces on segments F–C and D–E are equal and opposite, they cancel. The total force is that acting on segment C–D. Using Eq. (21.10), we have

$$F_M = IlB \sin\theta = (1.0 \text{ A})(0.20 \text{ m})(0.10 \text{ T})(\sin 90°)$$

and

$$\boxed{F_M = 0.020 \text{ N}}$$

▶ **Quick Check:** A 1-A current in a 1-m wire at 90° to a 1-T field experiences a force of 1 N. Here, the length is 0.2 of that value and the field is 0.1 of it, so we can expect a force 0.02 times smaller.

(continued)

(continued)

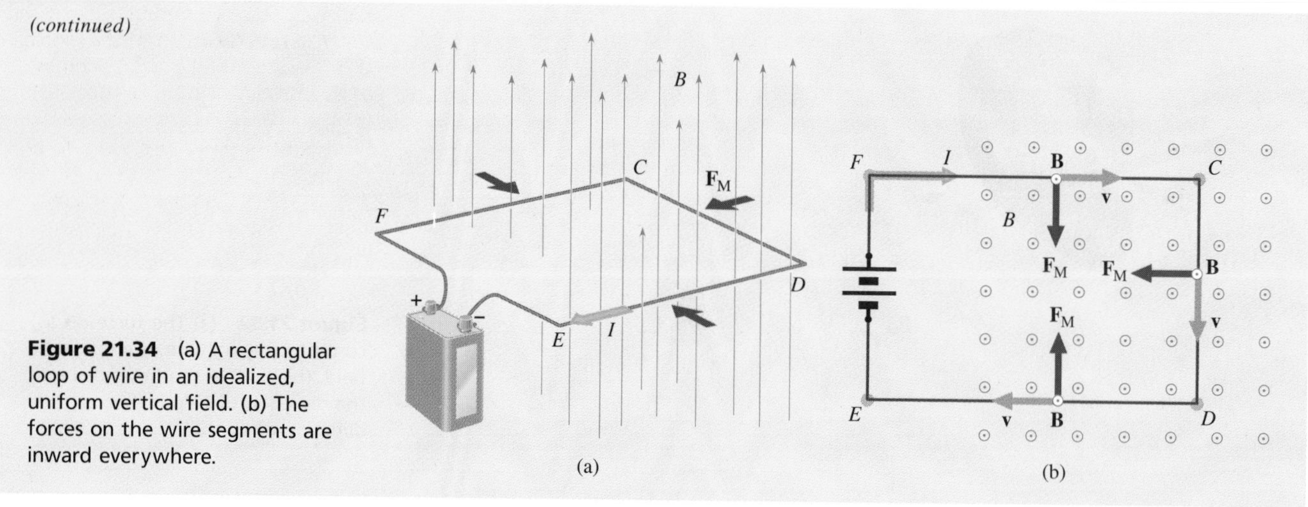

Figure 21.34 (a) A rectangular loop of wire in an idealized, uniform vertical field. (b) The forces on the wire segments are inward everywhere.

The Torque on a Current Loop

Imagine a lightweight current-carrying coil in the shape of a rectangle (Fig. 21.35) supported in a vertical plane so it turns easily about a vertical axis. The lengths of its horizontal and vertical sides are l_h and l_v, respectively. When placed in a uniform horizontal B-field, the forces on the top and bottom wires are oppositely directed and act parallel to the axis of rotation and so have no effect on the allowed motion of the coil. But the forces on the vertical segments act at a distance from that axis and produce a torque that tends to twist the coil so that it becomes perpendicular to the field. The coil is equivalent to a little dipole with the z-axis (Right-Hand-Current Rule) pointing from the south to the north pole. Like a compass, it tends to swing around such that the z-axis aligns with the B-field and $\phi = 0$.

The forces on the two vertical segments are equal and constant, independent of the angle between the z-axis and **B** because they are always perpendicular to the field. For each segment formed of N wires

$$F_M = NIl_vB$$

Moving coil meters are used in a tremendous variety of gauges. Transducers provide electrical signals that are proportional to the physical quantities being measured, and these signals are displayed by galvanometer gauges.

On the other hand, the torque (τ) changes because the moment-arm (the perpendicular distance from the line-of-action of the force to the axis of rotation) for each force varies with ϕ as $\frac{1}{2}l_h \sin\phi$. At any orientation, the torque due to either vertical segment is $F_M\frac{1}{2}l_h \sin\phi$, as shown in Fig. 21.35d. Hence, the total torque on the coil is twice this or

$$\tau = F_Ml_h \sin\phi = NIl_vl_hB \sin\phi$$

and since l_vl_h is the area A

$$\tau = NIAB \sin\phi \tag{21.11}$$

The fact that this formula depends on A rather than on the details of the geometry suggests that it will be the same for any shaped coil, and that is confirmed by a more general treatment. If the coil is hung on a fine wire (Fig. 21.36), it becomes the movement of a galvanometer.

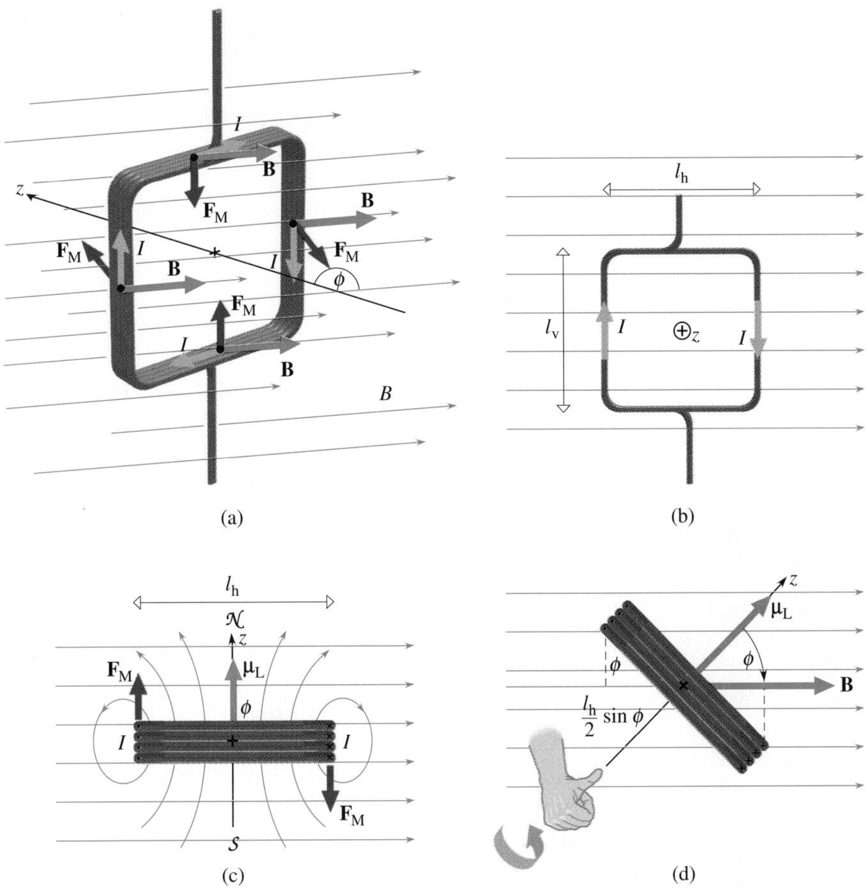

Figure 21.35 (a) Here the z-axis was chosen as the direction of the dipole field. (You can find the direction by using the Right-Hand-Current Rule.) (b) The coil swings around to align its dipole field with the applied external field. (c), (d) Views looking down from the top. The vector **u** is defined below.

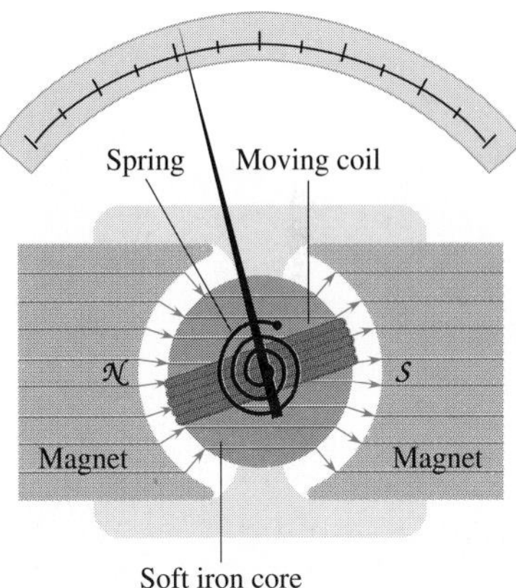

Figure 21.36 A current-carrying coil hung so that it can turn in a permanent B-field. The variation of torque with ϕ complicates matters, but it can be removed. If a stationary cylindrical ferromagnetic core is placed within the coil, the field lines can be concentrated and made almost radial, thereby rendering ϕ effectively equal to 90° regardless of the position of the coil.

The Magnetic Dipole Moment

Equation (21.11) looks very much like the torque formula of Eq. (6.3), suggesting a vector cross product. Accordingly, we define IA to be the magnitude of the **magnetic dipole moment** ($\boldsymbol{\mu}_1$) of a single current loop. For a coil of N turns, the magnitude of the dipole moment is NIA, and it has the units A·m^2. Using the Right-Hand-Current Rule, we assign a direction to $\boldsymbol{\mu}_1$ such that it points along the extended thumb of the right hand when the fingers curl in the direction of I (Fig. 21.35d). The torque on any current loop [Eq. (21.11)] then becomes

$$\boldsymbol{\tau} = \boldsymbol{\mu}_1 \times \mathbf{B} \tag{21.12}$$

The magnetic moment tends to align with the B-field just like a compass needle; indeed, the compass needle is a little magnetic dipole.

Example 21.5 The Bohr model depicts the hydrogen atom as an electron circulating around a proton in an orbit with a radius of 0.0529 nm at a speed of 2.2×10^6 m/s. Compute the orbital magnetic moment of the electron μ_B, also called the *Bohr magneton.*

Solution: [Given: $r = 0.0529$ nm and $v = 2.2 \times 10^6$ m/s. Find: μ_B.] By definition $\mu = IA$, so we need the current and the area of the orbit. The current is due to an electron of charge q_e moving past a point once in an interval of time equal to one period T: $I = q_e/T$. The

(continued)

(continued)

circumference is $2\pi r$ and therefore $T = 2\pi r/v$, whereas $A = \pi r^2$ and so

$$\mu_B = \left(\frac{q_e v}{2\pi r}\right)(\pi r^2) = \tfrac{1}{2} q_e v r$$

$$\mu_B = \tfrac{1}{2}(1.60 \times 10^{-19}\ \text{C})(2.2 \times 10^6\ \text{m/s})(0.0529\ \text{nm})$$

and

$$\boxed{\mu_B = 9.3 \times 10^{-24}\ \text{A·m}^2}$$

▶ **Quick Check:** The units of $\frac{1}{2} q_e v r$ are C(m/s)m = (C/s)m^2 = A·m^2.

The intrinsic magnetic moment of the electron μ_e has a magnitude of very nearly 1 Bohr magneton. The electron is the archetypal dipole magnet. Because electrons in an atom tend to pair up with opposing spins in closed shells, the magnetic moments of atoms are due to unpaired electrons and are typically equal to just a few Bohr magnetons.

The DC Motor

Now suppose we again suspend a pivoted electromagnet, or *armature,* in a *B*-field and send a current through it, as in Fig. 21.37a. That coil swings around into alignment with the fixed magnetic field; north and south poles come together (c). The coil has inertia and will slightly overshoot the mark. If at the instant (d) it sails past alignment, the polarity of the armature is inverted (by reversing the current through it), and the coil will be violently rotated all the way around until it is once more horizontal (f). But it again overshoots, and if the current is reversed for a second time (g) as it passes the pole, the armature will swing around back to the original position, and so on, round and around.

The timed current reversals are easy to accomplish automatically whenever the armature rotates through 180°. Current is fed into the coil via a split-ring arrangement known as a *commutator.* In the diagram, two metal strips rest on either side of the commutator (each making contact with one-half of the split-ring), but in commercial motors that's done with spring-loaded carbon blocks known as *brushes.* Whenever they pass over the split, the brushes essentially change places and the current through the armature reverses. This continuous whirling machine is a rudimentary version of the ever popular dc motor—the great mover that has powered a range of devices from the trolley car and golf cart to automobile starters, toy trains, electric cars, and so on.

(a)

(b)

(c)

(d)

(e)

(f)

(g)

(h)

(i)

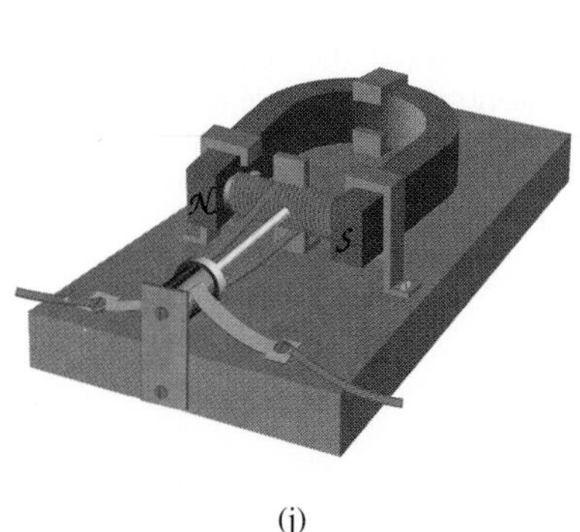

(j)

Figure 21.37 A simple electric motor. (a) The coil swings around with like poles repelling each other. (b) Then unlike poles attract and the rotation continues. (c) Just as the magnetic poles are about to align, the direction of the current reverses and the process continues.

Two Parallel Wires—the Ampère

Ampère passed a current through two long parallel wires and found that they experienced equal and opposite forces. The idea was to demonstrate that a current-carrying wire

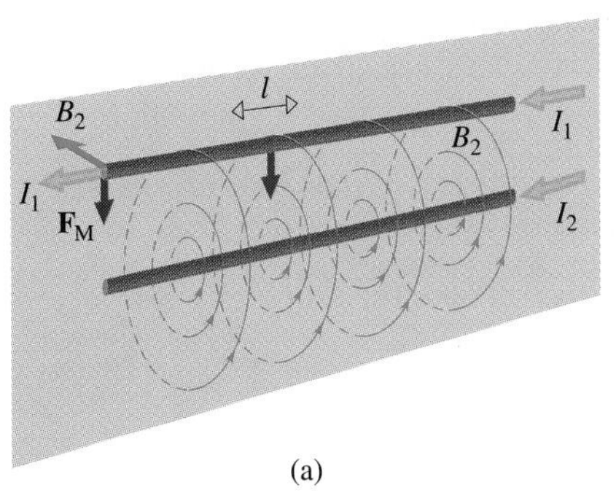

(a)

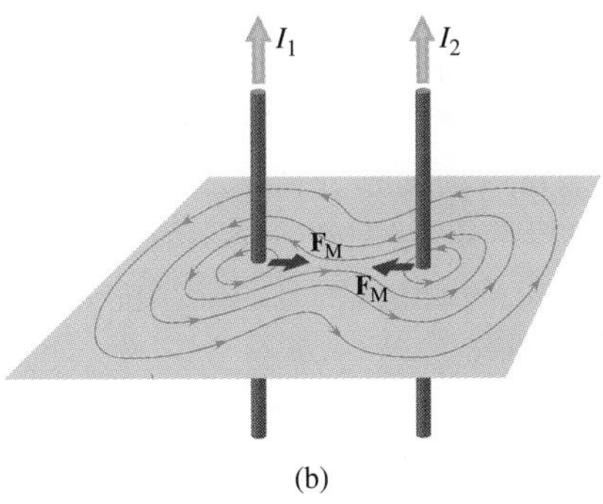

(b)

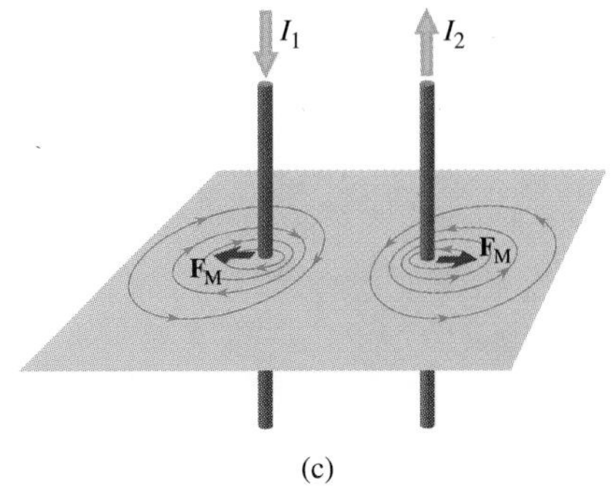

(c)

Figure 21.38 (a) Two parallel wires carrying current. Each wire is in the B-field of the other. (b) When the currents are parallel, they are forced toward one another. (c) When the currents are antiparallel, they are forced apart.

interacts with the B-field it is immersed in, regardless of whether that field is due to a permanent magnet or another current.

Figure 21.38a shows how the field of I_2 extends over to I_1, a distance d away, where it has a value given by Eq. (21.2) of

$$B_2 = \frac{\mu_0 I_2}{2\pi d}$$

and points directly to the left. If l is some length of the upper wire, Eq. (21.10) provides the force on it where $\theta = 90°$. Thus, the force per unit length on wire 1 due to the field of wire 2 is

$$\frac{F_M}{l} = I_1 B_2 = \frac{\mu_0 I_1 I_2}{2\pi d} \qquad (21.13)$$

The same force acts on wire 2 due to the field of wire 1 (the equation is symmetrical in the currents), except that B_1 is to the right and the force (in the direction of $\mathbf{v} \times \mathbf{B}$) is upward in the plane. Newton's Third Law makes the same prediction. The wires attract each other. Reversing either one of the currents reverses its field and the forces both reverse, becoming outwardly directed and therefore repulsive.

Example 21.6 What is the force per unit length experienced by each of two extremely long parallel wires carrying equal 1.0-A currents in opposite directions while separated by a distance of 1 m in vacuum?

Solution: [Given: $I_1 = I_2 = 1.0$ A and $d = 1$ m. Find: F_M.] From Eq. (21.13)

$$\frac{F_M}{l} = \frac{\mu_0 I_1 I_2}{2\pi d} = \frac{(4\pi \times 10^{-7}\ \text{T·m/A})(1.0\ \text{A})(1.0\ \text{A})}{2\pi(1\ \text{m})}$$

and so

$$\boxed{\frac{F_M}{l} = 2 \times 10^{-7}\ \text{N/m, repulsion}}$$

▶ **Quick Check:** $B_2 = \mu_0 I_2/2\pi r = (2 \times 10^{-7}\ \text{T})$; $F_M/l = I_1 B_2 = 2 \times 10^{-7}$ N/m.

The ideas in Example 21.6 form the basis for the SI definition of the ampere:

One ampere is the constant current that, if present in two infinitely long parallel straight wires one meter apart in vacuum, produces a force of exactly 2×10^{-7} newtons per meter of length.

Practically, we can't use infinitely long wires, but very precise measurements can be made with a delicate instrument called a *current balance.* It turns out to be much more reasonable to define the ampere first and use it to define the coulomb:

One coulomb is that amount of charge transported in one second through any cross section of a wire carrying a current of one ampere.

Along with the kilogram, the meter, and the second, the ampere is considered a basic unit.

Core Material

Like magnetic poles repel and unlike magnetic poles attract. The SI unit for B, the magnetic field, is the *tesla* (T). The field of a magnet extends out from the north pole and into the south pole. The field lines form closed loops (p. 726).

A straight current-carrying wire generates a circular magnetic field in the space surrounding it given by

[*straight long wire*] $$B = \frac{\mu_0}{2\pi}\frac{I}{r} \qquad [21.2]$$

where μ is the permeability and for vacuum $\mu_0 = 4\pi \times 10^{-7}$ T·m/A. For a current loop of radius R

[*circular loop, center*] $$B_z = \frac{\mu_0 I}{2R} \qquad [21.3]$$

If there are N turns of wire

[*circular coil, center*] $$B_z = N\frac{\mu_0 I}{2R} \qquad [21.4]$$

For a long coil with n turns per unit length

[*solenoid, center*] $$B_z \approx \mu_0 nI \qquad [21.5]$$

[*solenoid, ends*] $$B_z \approx \tfrac{1}{2}\mu_0 nI$$

Ampère's Law

$$\Sigma B_{\parallel}\Delta l = \mu_0 \Sigma I \qquad [21.6]$$

provides a means of computing magnetic fields (p. 740).

The magnitude of the force on a charge moving in a B-field depends on the angle (θ) between $\mathbf{v}$ and $\mathbf{B}$:

$$F_M = qvB \sin\theta \qquad [21.7]$$

and $$\mathbf{F}_M = q\mathbf{v} \times \mathbf{B} \qquad [21.8]$$

The force on a current-carrying wire of length l is

$$F_M = IlB \sin\theta \qquad [21.10]$$

The torque on a coil of N turns with an area A, making an angle ϕ with the field is

$$\tau = NIAB \sin\phi \qquad [21.11]$$

With the orbital **magnetic moment** defined as NIA, the torque becomes

$$\boldsymbol{\tau} = \boldsymbol{\mu}_l \times \mathbf{B} \qquad [21.12]$$

The force per unit length exerted on each other by two current-carrying straight wires is

$$\frac{F_M}{l} = \frac{\mu_0 I_1 I_2}{2\pi d} \qquad [21.13]$$

One ampere is the constant current that, if present in two infinitely long parallel straight wires one meter apart in vacuum, will produce a force of exactly 2×10^{-7} *newtons per meter of length.*

Suggestions on Problem Solving

1. This chapter has several equations that depend directly on q and that produce reversed results when the sign of the charge changes. As soon as you see you are dealing with negative charges, *be extra careful of directions.*

2. It's helpful to remember that $\mu_0/4\pi = 10^{-7}$ N/A^2 and so $\mu_0/2\pi = 2 \times 10^{-7}$ N/A^2, where $\mu_0 = 1.257 \times 10^{-6}$ N/A^2.

3. Physicists don't usually memorize all the equations for the fields arising from the various current configurations. In-

stead, many just remember Ampère's Law and derive what they need. When applying Ampère's Law to a new situation, first establish the direction of the B-field. That's usually done via symmetry arguments and the realization that the field lines are closed. Since there are no monopoles in magnets, B-field lines are not going to be radial, as they were with the E-fields of a small spherical charge or a line charge. Select an Ampèrian loop that is everywhere perpendicular or parallel to **B**. Any lengths or distances that are specific to your Ampèrian path and not arbitrary (as is the radius of the loop in Fig. 21.23) must cancel out of the final equation for B.

Discussion Questions

1. Assuming that space is isotropic, discuss the symmetry of the electric field of a point charge, both at rest and in uniform motion. What can you expect for the symmetry of the magnetic field of a charge moving with a constant velocity (that is, a current)?

2. Why is a chunk of iron attracted to either pole of a magnet? What does this tell you about the reliability of attraction as a test of whether something is permanently magnetized or not? A classic puzzle involves two seemingly identical rods, one steel and magnetized (with poles at its ends), the other soft iron and not magnetized. How can you tell which rod is which, using nothing else and not bending or breaking either rod?

3. James Clerk Maxwell confirmed experimentally that the B-field of a long straight current-carrying wire drops off inversely with distance. His apparatus is pictured in Fig. Q3 and it consists of a lightweight disk, free to rotate, on which rest four bar magnets. No matter how large the current through the central wire, there was never any rotation of the disk. Given that the poles are at distances of R_S and R_N from the wire, use the idea of fictitious magnetic charges and the torques they would experience to explain the experiment.

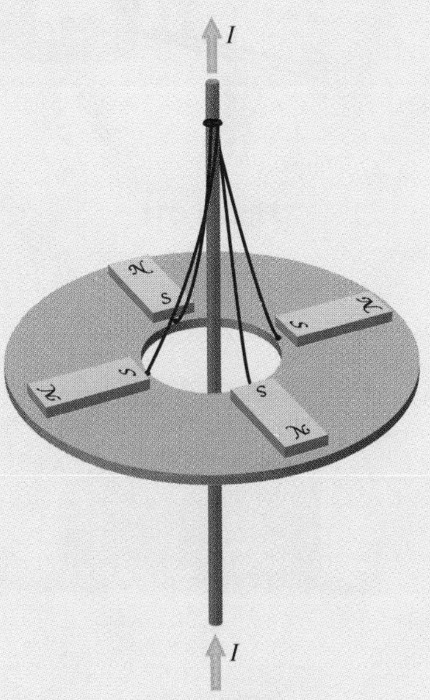

Figure Q3

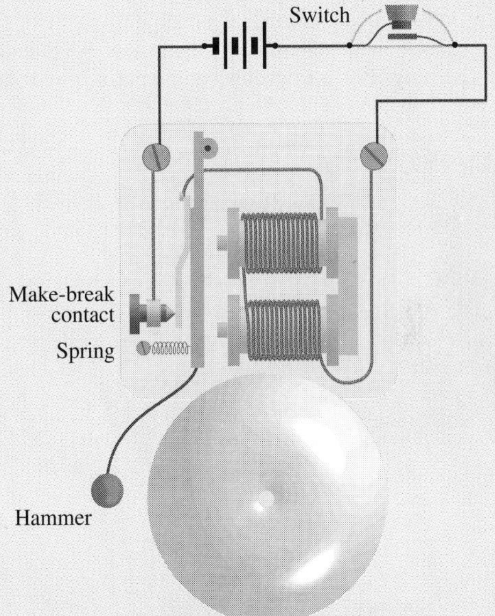

Figure Q4

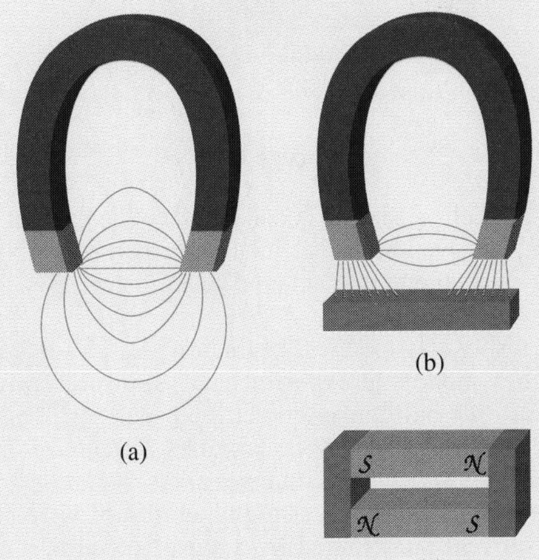

Figure Q5

4. An electric door bell is basically an automatically interrupted electromagnet. Referring to Fig. Q4, describe how it works.

5. Figure Q5 shows a horseshoe magnet with and without a piece of iron, called a *keeper,* across its poles. Explain what happens to the field. To understand the virtues of a keeper, realize that an ordinary bar magnet produces a field that extends externally from N to S and then continues through the magnet to form closed loops. But we can think of the poles at the bar's ends setting up an opposing internal field (N to S), which tends to

demagnetize the magnet. The keeper essentially cancels these bare poles via induction. Figure Q5c shows the arrangement (with two soft-steel keepers) usually used to package a pair of bar magnets—explain how it works.

6. Suppose you placed a candle flame between the poles of a powerful electromagnet so that it was in a nonuniform high-field region. What, if anything, would happen to the flame? Why? What would happen to a soap bubble filled with smoke?

7. A good way to demagnetize the heads of a tape recorder (or anything else) is to send an alternating current—one that reverses direction periodically—through a nearby coil, which is then slowly moved away from the heads, decreasing B. Discuss how this process will demagnetize an object.

8. Figure Q8 is a schematic diagram of a setup for tape-recording the output of a microphone. Explain how it works.

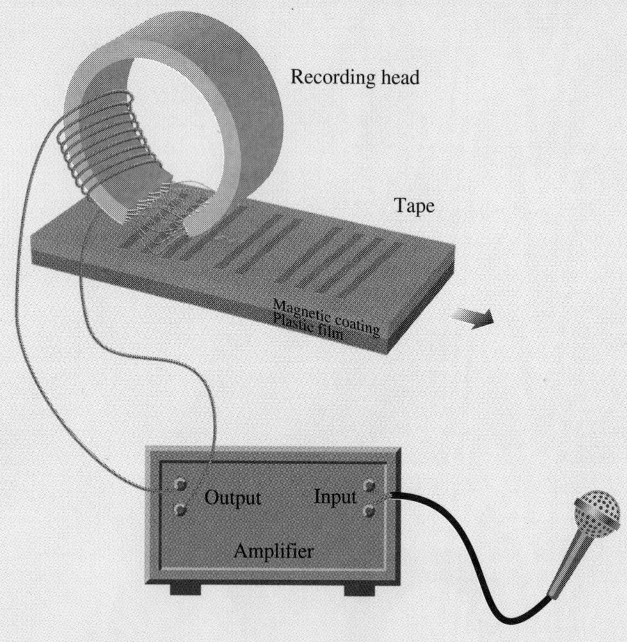

Figure Q8

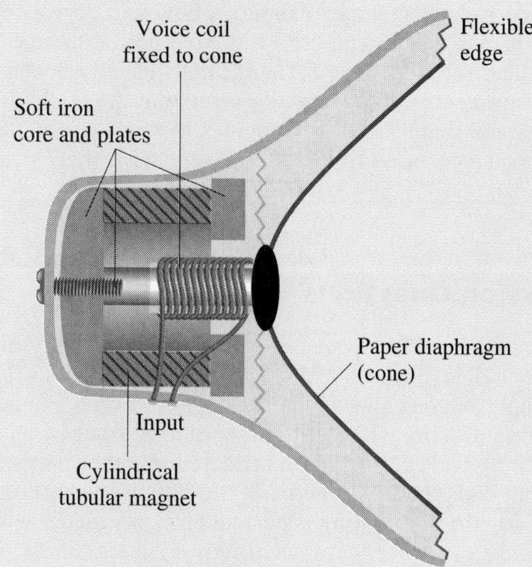

Figure Q9

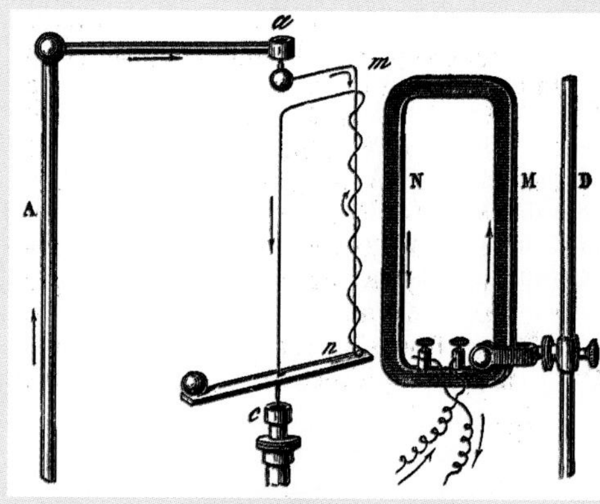

Figure Q11

9. A speaker consists of a coil fixed to the back of a flexible cone, as in Fig. Q9. The coil is mounted in an assembly that is a cylindrical permanent magnet and a soft steel core. Describe how the device works.

10. A lovely way to shield against the Earth's magnetic field was used in a monopole experiment at Stanford University. A deflated lead-foil balloon is cooled below its critical temperature. Inflating the balloon then provides a field-free region inside of its hollow. How does that work?

11. Figure Q11 shows a straight section of wire carrying a downward current surrounded by a spiral of wire carrying the same current but upward. What will happen to these wires (they are pivoted and free to rotate as a unit), when a current-carrying coil or a bar magnet is brought nearby? Discuss the situation from the perspective of Ampère's Law. How does it relate to the field outside of a solenoid as considered in Fig. 21.24b?

12. It's been known since the late nineteenth century that a short coil can act like a lens, deflecting charged particles toward the central symmetry axis (Fig. Q12). Accordingly, explain how the electron initially traveling parallel to the z-axis, but displaced from it, ends up moving in toward that axis.

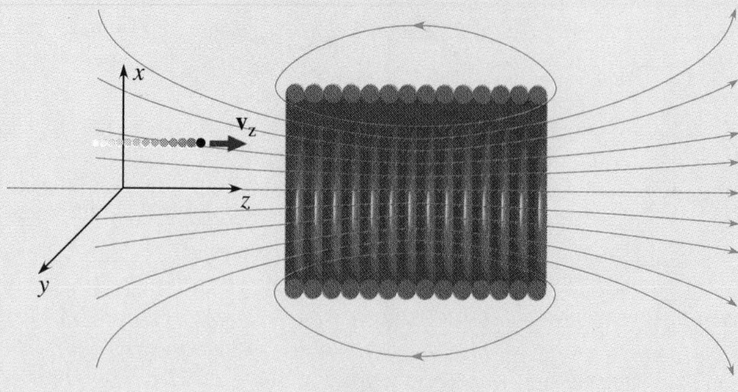

Figure Q12

13. A very flexible helical coil is suspended (Fig. Q13) so that its lower end just dips into a cup of mercury. What will happen when a current is sent through the coil? Incidentally, this

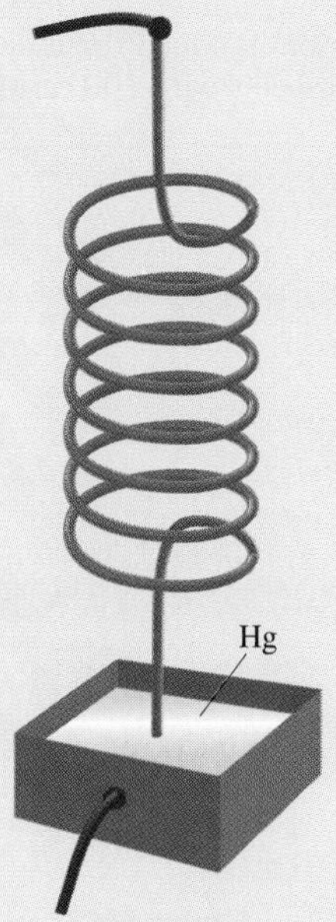

Figure Q13

is called *Roget's Spiral*. What would be the effect of putting an iron rod up the middle of the spiral?

14. Most radiators, steel wastepaper baskets, and metal garbage cans in the world are magnetized. Explain how this might happen naturally, and figure out the likely polarity. Steel ships are also magnetized, which prompted someone to wrap ac current-carrying coils around ships during World War II in order to foil magnetic mines. W. Gilbert makes the following observation (in his book of 1600) concerning a "glowing mass of iron": "Let the smith be standing with his face to the north, his back to the south. Let him always, while he is striking the iron direct the same point of it toward the north and let him lay down that end toward the north [during cooling]." What will happen to the iron?

15. In 1821, Faraday devised a primitive motor he called a *rotator.* Actually, he produced two versions of it in a single unit (Fig. Q15). One had a pivoted wire carrying a sizable current swinging around a vertical, fixed bar magnet immersed in mercury. The other had a pivoted magnet rotating around a fixed current-carrying wire, also in mercury. Explain how they worked.

16. Describe what you think will happen to a long glass tube filled with very fine iron filings when it is placed in a strong magnetic field, shaken, and then gently removed. What effect will subsequently shaking it have? Compare this with a solid bar of iron from the perspective of domains.

17. Figure Q17 shows a small cylindrical permanent magnet floating above a superconducting tin disk bathed in liquid helium at ≈1.2 K. The magnet was placed on the disk, and the latter was cooled below its transition temperature, at which point the magnet spontaneously jumped into the air. Explain what happened. This same kind of magnetic levitation is being applied via high-temperature superconductors to produce frictionless magnetic bearings for gyroscopes, computer disk drives, and the like.

18. Since the field of a narrow coil changes linearly with z at a distance of around $R/2$, as shown in Fig. 21.19b, Helmholtz put two such coils (each with N turns) parallel to one another a distance R apart. The result is a fairly uniform B-field over the large central region given by

$$B_z \approx \frac{0.72\mu_0 NI}{R}$$

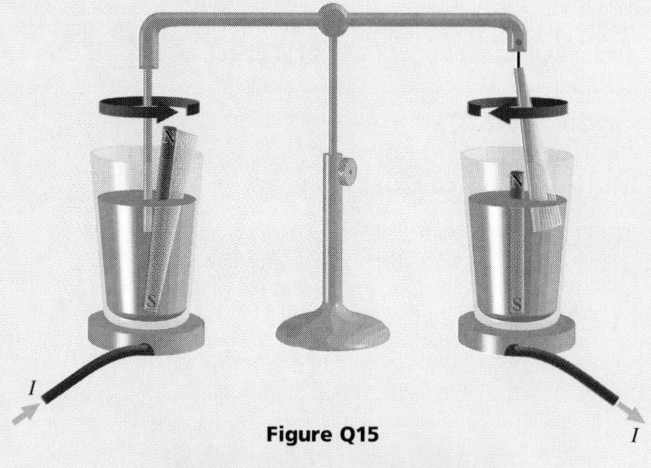

Figure Q15

Figure Q17

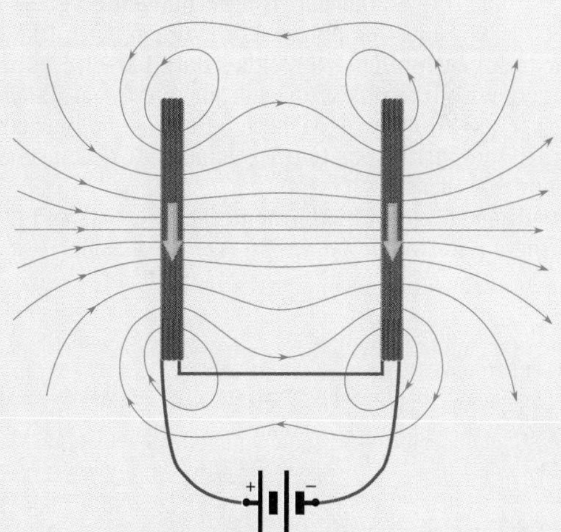

Figure Q18

and depicted in Fig. Q18. Discuss the advantages of these Helmholtz coils as a provider of uniform field (see the photo on the bottom right of p. 744). In what direction does the current progress in the coils in the diagram? How much current would it take to provide a region in which the Earth's field was canceled for coils of 1.0-m diameter with 200 turns each?

Multiple Choice Questions

1. The field-line pattern around the two bars in Fig. MC1 shows that (a) neither bar is a permanent magnet (b) both bars must be permanent magnets with like poles adjacent to each other (c) either both bars are permanent magnets with like poles adjacent, or one is permanent and one is a soft iron bar (d) both must be identical permanently magnetized bars with opposite poles adjacent to each other (e) none of these.

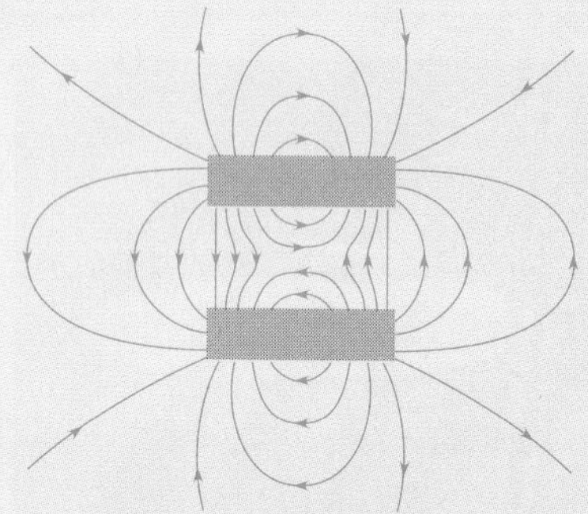

Figure MC1

2. The field-line pattern around the two bars in Fig. MC2 shows that (a) neither bar is a permanent magnet (b) both bars must be permanent magnets with like poles adjacent to each other (c) either both bars are permanent magnets with like poles adjacent, or one is permanent and one is a soft iron bar (d) both bars must be permanently magnetized with opposite poles adjacent to each other (e) none of these.
3. Iron filings are sprinkled around the two bars in Fig. MC3 making a pattern showing that (a) neither bar is a permanent magnet (b) both bars must be permanent magnets with like poles adjacent to each other (c) the top bar is soft iron, the bottom is a permanent magnet (d) both must be identical permanent bar magnets with opposite poles adjacent to each other (e) none of these.

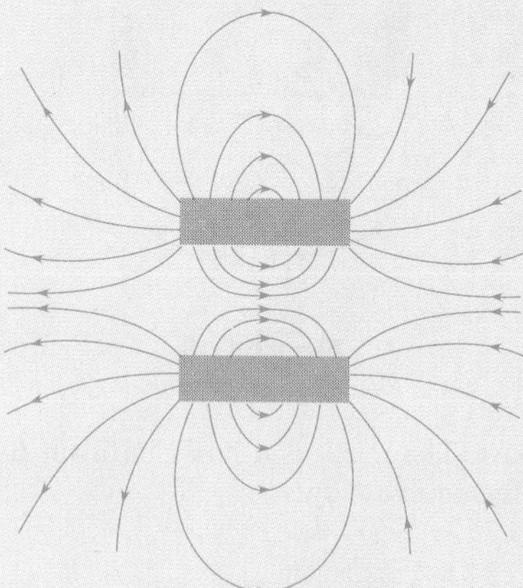

Figure MC2

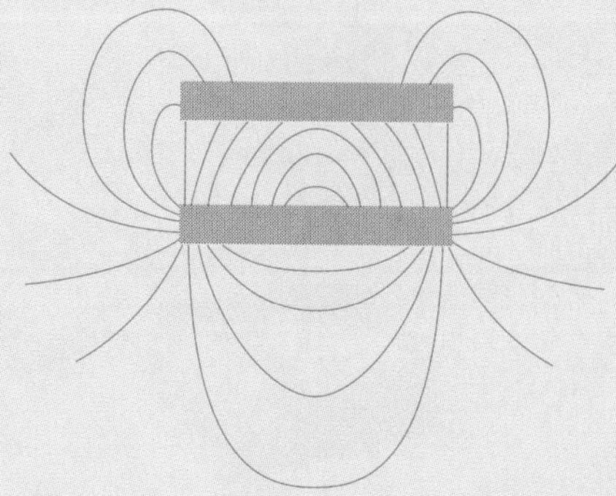

Figure MC3

4. A long, magnetized needle is floated vertically with its north pole up. It is held in place, and a bar magnet is brought near, as shown in Fig. MC4. When released, the floating needle will (a) stay exactly where it is (b) rush toward the north pole of the magnet in a straight line (c) swing in an arc away from the north pole and over to the south pole (d) move in a straight line to the south pole (e) none of these.

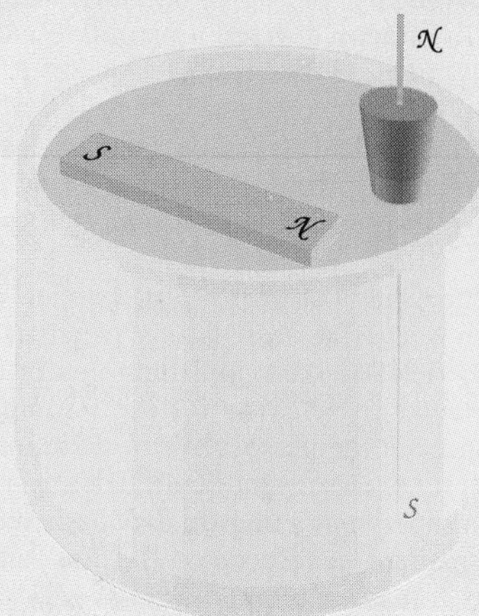

Figure MC4

5. The net force on a magnetic dipole in a uniform magnetic field is (a) toward the north pole (b) toward the south pole (c) zero (d) not enough information given (e) none of these.

6. At the instant shown in Fig. MC6 (assuming no interaction between them), which of the little magnets experiences a net downward magnetic force? (a) 1, 2, and 5 (b) 1 and 5 (c) 2 and 4 (d) 1, 3, 5, and 6 (e) none of these.

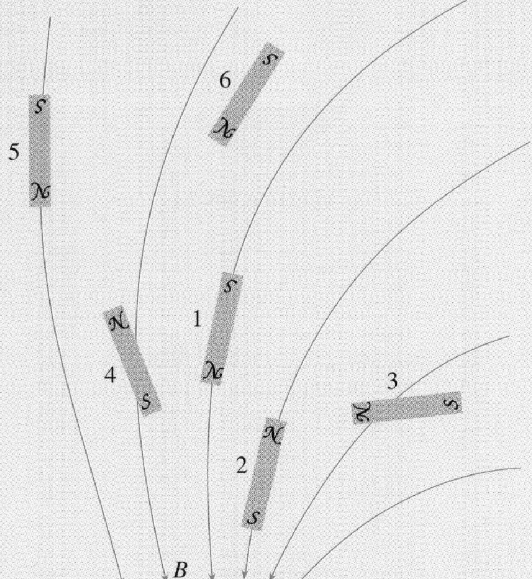

Figure MC6

7. Figure MC7a shows the *B*-field of a long current-carrying wire immersed in a uniform magnetic field. Figure MC7b depicts the resultant field. The wire experiences (a) zero force (b) a force to the right (c) a downward force (d) a force to the left (e) none of these.

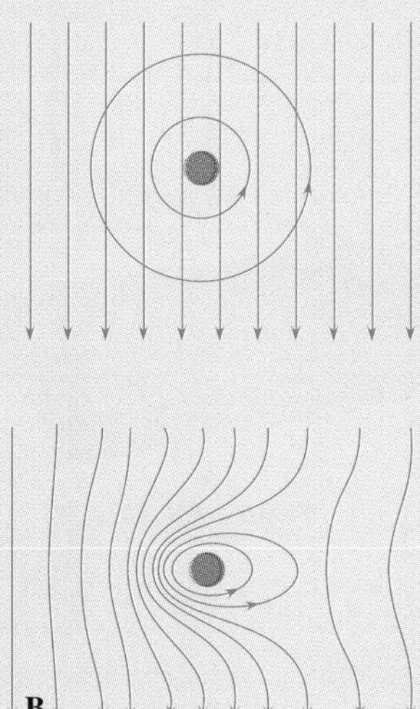

Figure MC7

8. Referring to Fig. MC8 (and taking "up" as out of the plane) (a) particle 1 experiences an upward force while particle 3 experiences a downward force (b) particle 1 experiences an upward force while particle 4 experiences a force due north (c) particle 3 experiences an upward force while particle 2 experiences no force at all (d) particle 4

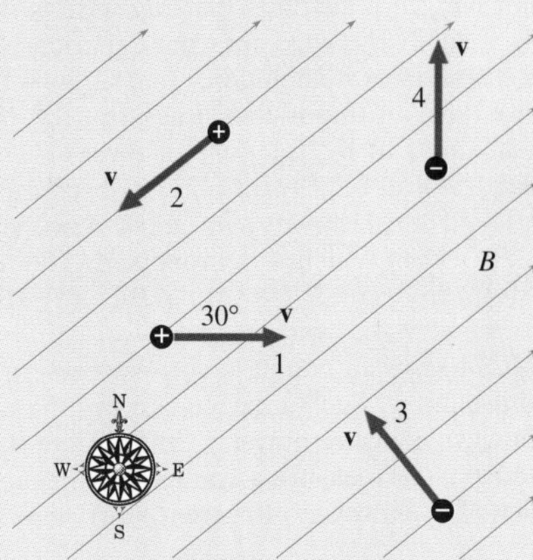

Figure MC8

experiences an upward force while particle 2 experiences a downward force (e) none of these.

9. A charge can move through a magnetic field and not experience a force by (a) moving quickly (b) traveling parallel to **B** (c) moving perpendicular to **B** (d) traveling very slowly (e) none of these.

10. The diagram of Fig. MC10 shows a device for measuring the vertical force on a sample in a magnetic field. If we hang a chunk of fresh apple from its end (a) it will be drawn downward (b) it will be pushed upward (c) it will swing to the left (d) nothing will happen (e) none of these.

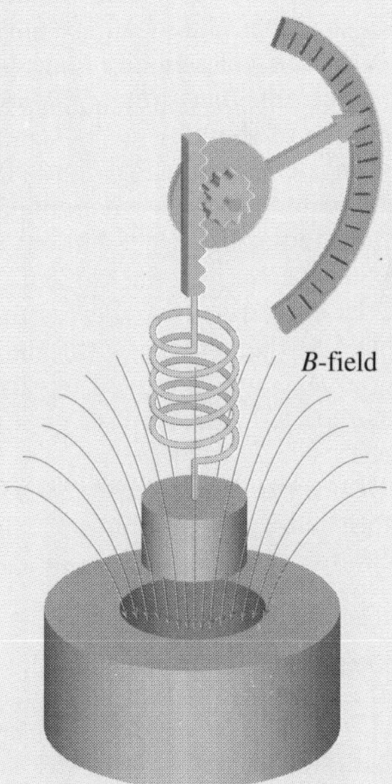

Figure MC10

11. If the force on a current element is imagined as arising from a flow of positive charge carriers, how will it change if the same current is considered to be due to an oppositely directed flow of negative carriers? (a) it will be unchanged (b) it will reverse direction (c) it will be perpendicular (d) it will be at 45° in the first quadrant (e) none of these.

12. A straight length of current-carrying wire is in a uniform magnetic field. If the wire does not experience a force, (a) everything is as it should be (b) it must be parallel to **B** (c) we have an impossible situation (d) the wire must be perpendicular to **B** (e) none of these.
13. A circular flat coil of N turns and enclosed area A, carrying a current I, has its symmetry z-axis parallel to a uniform B-field in which it is immersed. The torque on the coil is (a) zero (b) $NIBA$ (c) NBA (d) IBA (e) none of these.
14. Figure MC14 shows a compass with several turns of wire wrapped around it. Such a device can best be used in a circuit to (a) measure power (b) indicate the presence of resistance (c) measure a voltage difference (d) indicate the presence of current (e) none of these.

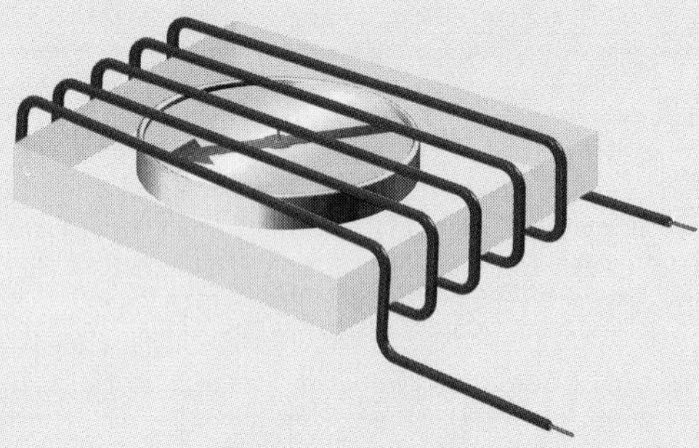

Figure MC14

15. Just wrap 150 turns of a heavy (#22) insulated wire around an iron rod (for example, a door hinge pin), attach the leads to a 1.5-V D-cell via a switch and you have (a) a radio (b) an electromagnet (c) a galvanometer (d) an ammeter (e) none of these.
16. The American physicist Henry Rowland (1876) placed some charges (in a fixed location) on a nonconducting disk, which he then rotated at high speed about its central axis near a delicate compass. (a) the current created a B-field that deflected the compass (b) there was no E-field, so nothing happened to the compass (c) the charges attracted the compass, which moved toward the disk (d) the charges repelled the compass via Coulomb's Law (e) none of these.
17. Which way will the bar move in Fig. MC17 once the current is established? (a) straight up (b) to the left (c) to the right (d) it will not move at all (e) none of these.

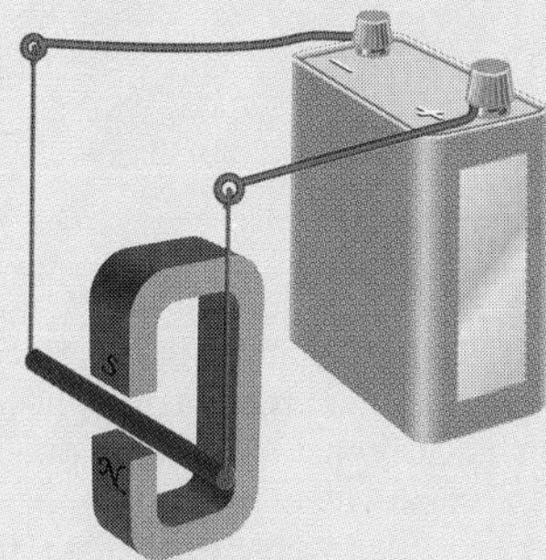

Figure MC17

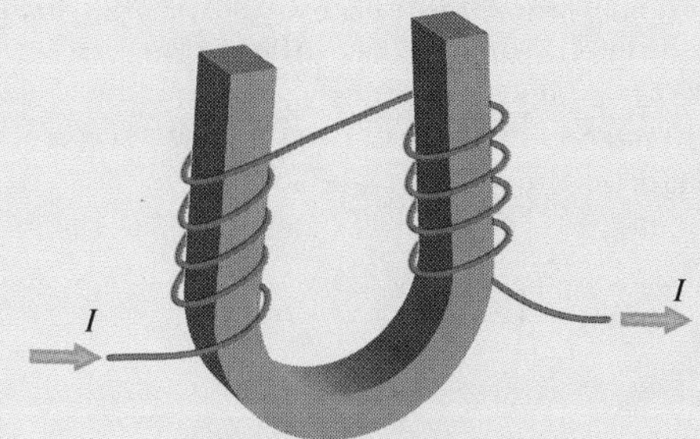

Figure MC18

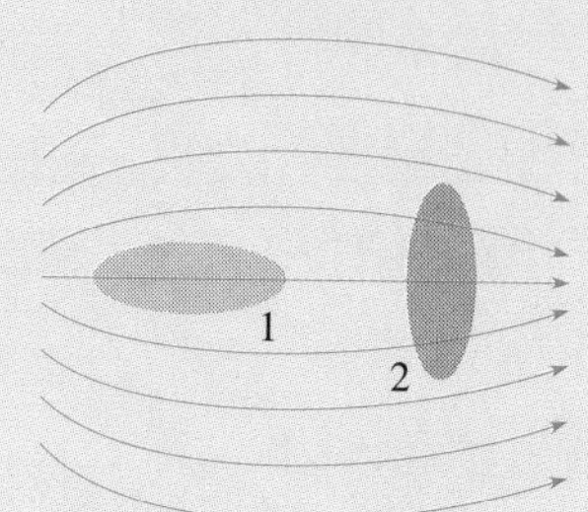

Figure MC20

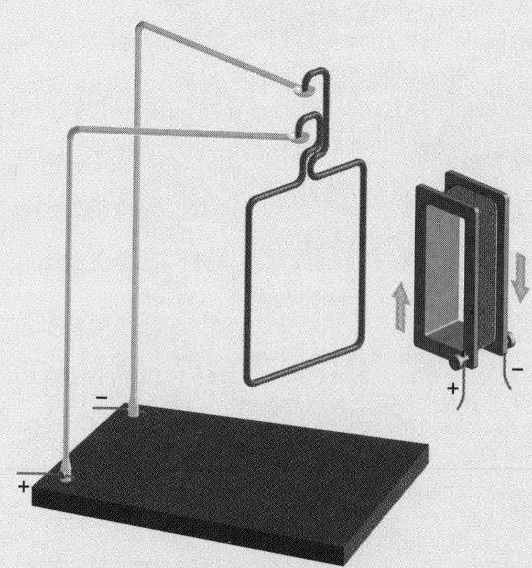

Figure MC21

18. The horseshoe-shaped iron bar in Fig. MC18 is wrapped with a wire, and once the dc current is turned on (a) it becomes demagnetized (b) the right end becomes a south pole and the left a north pole (c) the left end becomes a south pole and the right a north pole (d) both ends become north poles and the bottom becomes a south pole (e) none of these.
19. If the current through a long solenoid is doubled while the coil's length is also doubled, keeping the total number of turns constant, the magnetic field at a point inside near the axis is (a) four times larger (b) half the original size (c) unchanged (d) one-quarter the size (e) none of these.

20. The two (noninteracting) samples (1 and 2), which have swung into alignment in the B-field, as shown in Fig. MC20 (p. 758), may be, respectively, (a) ferromagnetic and paramagnetic (b) diamagnetic and ferromagnetic (c) paramagnetic and diamagnetic (d) paramagnetic and ferromagnetic (e) none of these.

21. The movable loop in Fig. MC21 (p. 758) and the coil carry currents in the directions indicated. The loop will (a) looking down from above, rotate counterclockwise (b) experience no force (c) looking down from above, rotate clockwise (d) experience an upward force (e) none of these.

Problems

MAGNETS AND THE MAGNETIC FIELD
ELECTRODYNAMICS

1. [I] Show that the units T·m/A and N/A^2 for μ are equivalent.

2. [I] The most common pre-SI unit of magnetic field, one still to be found in use, is the *gauss* (1 tesla = 10^4 gauss). The magnetic field in intergalactic space is around 10^{-6} gauss, as compared with the 8-kilogauss field of a powerful samarium-cobalt permanent magnet. Express these quantities in teslas.

3. [I] Determine the B-field 50 cm from a long narrow wire carrying a current of 10 A immersed in air.

4. [I] A magnetic field of 6.0 μT is to be produced 10 cm from a single long straight wire in air. How much current must it carry?

5. [I] The return stroke of a bolt of lightning typically carries a peak current of 20 kA up from the ground. What is the maximum magnetic field associated with the bolt 1.0 m away?

6. [I] A long straight wire carrying a current of 4.00 A is suspended in air and the field around it is measured with a magnetometer. It is found that at some point P the field is 0.660×10^{-5} T. How far from the wire is P?

7. [I] A straight wire 1.00 m long having a resistance of 1.2 Ω is attached to a 12-V battery. What is the magnitude of its B-field 2.0 cm away in air?

8. [I] A long, somewhat wiggly copper wire carries a current of 20 A. What is the B-field at a distance of 50 cm away, given that this dimension is much greater than those of any of the bends in the wire? (See Discussion Question 11.)

9. [I] A single flat loop of superconducting wire 50 cm in diameter carries a current of 25 A. What is the magnitude of the magnetic field at its center?

10. [I] A narrow flat circular coil with a diameter of 20 cm consists of 100 turns of wire. What is the magnetic field at its center when it carries a current of 5.0 A?

11. [I] A beam of protons is made to travel in a nearly circular orbit by a perpendicularly applied external B-field. Determine the magnetic field at the center of a 20-cm-diameter orbit produced by a 0.10-mA proton beam.

12. [I] A flat circular coil having a diameter of 25 cm is to produce a B-field at its center of 1.00 mT. If it has 100 turns, how much current must be provided to it?

13. [I] We wish to make a hollow solenoid 10 cm long and 1.5 cm in diameter having 200 turns of wire. How much current must we send through it to produce a field of roughly 0.50 mT inside the coil?

14. [I] An air-filled solenoid is to be 80 cm long and carry a current of 20 A. How many turns of a superconducting wire should it have if the field inside it near the middle is to be roughly 2.0 T?

15. [I] An air-core solenoid has 100 turns per centimeter and a resistance of 60 Ω. Determine the magnetic field inside it near its middle when it is connected across a 12-V battery.

16. [I] A 5.0-cm-diameter solenoid 50 cm long is made by wrapping four layers each of 1000 turns of wire on a thin hollow plastic core. Calculate the approximate B-field generated near the coil's ends when it carries a current of 1.5 A.

17. [II] Suppose we remove the magnetic deflection yoke from the neck of a TV picture tube so the electron beam travels straight down the central axis. For the brightness level given, there are 6.0×10^{12} electrons arriving at the screen per second. Determine the magnetic field (magnitude and direction) caused by the beam at a radial distance of 1.5 cm from it.

18. [II] Two long horizontal straight parallel wires are 28.28 cm apart and each carries a current of 2.0 A in the same direction, namely due south. What is the **B**-field at a point that is a perpendicular distance of 20 cm from both wires?

19. [II] A superconducting niobium wire will return to the normal state when the magnetic field at its surface exceeds 0.100 T. If the wire has a diameter of 2.00 mm, what is the *critical current;* that is, what maximum current can it carry without quenching the superconducting state?

20. [II] Two long current-carrying wires are depicted in Fig. P20. What is the value of the B-field at point P, 1.0 m away from the crossing point?

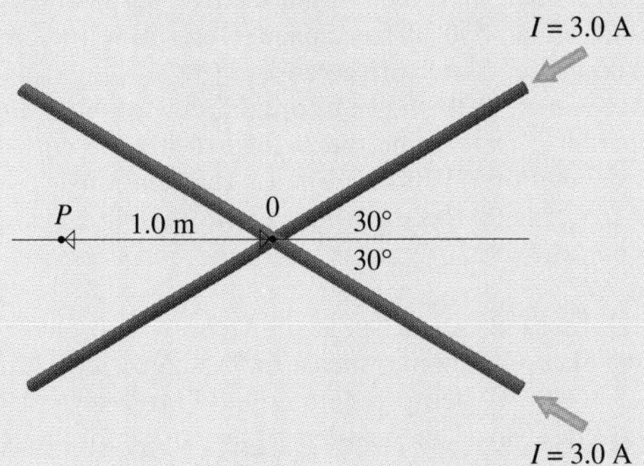

Figure P20

21. [II] Considering Problem 20, what would be the magnitude of B at P if one of the currents was reversed?

22. [II] Show that the dimensions of $\varepsilon_0\mu_0$ correspond to those of 1 over speed squared: $[T^2]/[L^2]$. This point will come up again later when we see that $\varepsilon_0\mu_0 = 1/c^2$. Indeed, since the speed of light is now defined to be exactly 299 792 458 m/s, both ε_0 and μ_0 are also exact.

23. [II] A long straight vertical wire carries a vertically upward current of 15 A. A small horizontal compass is placed 10 cm away due north of the wire. At that point, the Earth's B-field has a horizontal component of 0.50×10^{-4} T directed due north. Determine the equilibrium direction of the compass needle.

24. [II] A rather simplistic model of the hydrogen atom has a single electron revolving around a nuclear proton with an orbital radius of 0.53×10^{-10} m at a speed of 2.19×10^6 m/s. Determine the magnetic field at the proton due to the electron.

25. [II] The field along the central z-axis of a current-carrying loop is

$$B_z = \frac{\mu_0 I R^2}{2(R^2 + z^2)^{\frac{3}{2}}}$$

where $z = 0$ is at the center of the circular loop of radius R. Show that this expression is consistent with Eq. (21.3). What is the approximate form of the equation at great distances away along z?

26. [II] A solenoid 30 cm long and 2.0 cm in diameter is wrapped around an aluminum core. The coil contains 500 turns of fine insulated wire carrying 5.0 A. If the permeability of aluminum is 1.257×10^{-6} T·m/A, what is the value of B produced at the center of the coil?

27. [II] Consider the Earth (radius 6.4×10^3 km) from the perspective of the Dynamo Theory, which maintains that the B-field is due to currents in the spinning planet. Given that its liquid iron outer core has a mean radius of about 2.3×10^3 km and that the field at either pole is about 0.6×10^{-4} T, approximate the net current that would produce the dipole magnetic field. (Make your model as simple as possible, but discuss its shortcomings.) Draw a rough sketch of the planet and the currents. In what direction does I progress? (Take a look at Problem 25 and Fig. 21.19.)

28. [II] Consider the solenoid of Problem 26, this time with a silicon-iron core having a relative permeability of 6000 times that of free space. Determine the approximate field at the center of the coil.

29. [II] Figure P29 shows a toroidal coil of N turns carrying a current I. Determine the B-field inside the coil. How does it depend on position? What is the relationship between the B-field inside the torus and that of a single straight wire along the axis of symmetry of the coil carrying a current of NI?

30. [II] In 1962, scientists at the Geophysical Observatory in Tulsa, Oklahoma, measured a magnetic field of 1.5×10^{-8} T being generated by a large tornado about 9.0 km away. Use Ampère's Law to approximate the current swirling downward within the funnel.

31. [II] With Problem 29 in mind, use Ampère's Law to prove that the B-field lies entirely within the confines of the torus.

32. [II] Why is it that a long solenoid can be regarded as a toroidal coil with an infinite radius?

33. [II] Figure P33 shows a portion of an infinite conducting sheet carrying a current (traveling in the z-direction) per unit of width (along the x-axis) given by i amps per meter. Imagine the sheet as if it were made up of an infinite number of adjacent parallel wires. Use symmetry arguments and the Right-Hand-Current Rule to establish

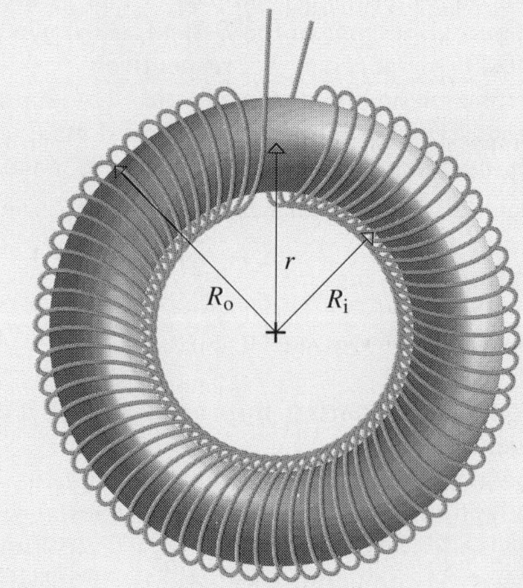

Figure P29

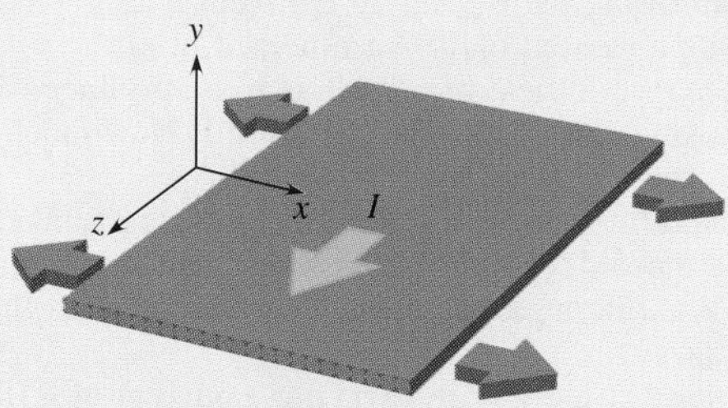

Figure P33

that **B** is a constant everywhere (independent of distance) above and below the sheet. Further, show that B points in the negative x-direction above the sheet and in the positive x-direction below the sheet. Using Ampère's Law, show that

$$B = \tfrac{1}{2}\mu_0 i$$

34. [III] A long straight hollow cylindrical conductor with an inner radius of R_i and an outer radius of R_o carries a current I, which is distributed uniformly over its cross section and which propagates along its very long length. Given that it is immersed in air, determine the magnetic fields in the inner and outer air-filled regions, as well as in the conductor. Draw a graph of B against radial distance from the central axis.

35. [III] Imagine two horizontal parallel infinite planes, one above the other. If these are each thin conductors carrying equal currents, the top one traveling west to east, the bottom one east to west, find the B-field between them if the current per unit width is i. Take a look at Problem 33. What is the field outside the sheets?

MAGNETIC FORCE

36. [I] A small ball of plastic carries a positive charge and is shot at high speed along the x-axis in the horizontal x–y plane. It passes through a uniform horizontal B-field, which makes an angle with the x-axis of $+23.5°$. In what direction is the magnetic force it will experience?

37. [I] A negatively charged projectile is fired upward in the vertical z–y plane at an angle of $+20°$ to the y-axis. It travels through a uniform magnetic field directed downward in the negative z-direction. What is the direction of the resulting magnetic force?

38. [I] A proton travels downward in the vertical x–z plane at an angle of $-40°$ with respect to the x-axis. It passes through a uniform horizontal B-field directed in the $+y$-direction. In what direction is the resulting magnetic force?

39. [I] A proton in a research facility is propelled at 2.0×10^6 m/s perpendicular to a uniform magnetic field. If it experiences a force of 4.0×10^{-13} N, what is the strength of the field?

40. [I] Show that 1 T = 1 kg/s·C.

41. [I] Show that 1 T = 1 V·s/m^2.

42. [I] Show that 1 T = 1 N/A·m.

43. [I] An electron of mass 9.11×10^{-31} kg travels in a circular orbit within a large evacuated chamber. The orbit has a 2.0-mm radius and is perpendicular to a B-field of 0.050 T. What's the electron's speed?

44. [I] An electron moving at 2.5×10^6 m/s within a cathode-ray tube (CRT) makes an angle of 45° with a uniform 1.2-T magnetic field. What is the magnitude of the force it will experience?

45. [I] A beam of protons of various speeds enters a region where it encounters perpendicular electric and magnetic fields both at right angles to **v**. Show that only particles for which $v = E/B$ will pass through undeflected. Accordingly, such an arrangement is called a *velocity selector.*

46. [I] A proton moving at 3.00×10^6 m/s through a uniform magnetic field experiences a maximum force of 5.2×10^{-12} N, straight upward when it's traveling due east. Determine **B**.

47. [I] A straight wire carrying an upwardly directed current is in the x–z plane making an angle of $+20°$ with the horizontal x-axis. The wire is immersed in a uniform B-field in the positive z-direction. What is the direction of the force on the wire?

48. [I] A 10-m-long wire is strung more or less horizontally in an east-west direction, and a current of 10 A is sent from west to east along it. Assuming the Earth's B-field at that location has a magnitude of 0.5×10^{-4} T and is horizontal and due north, find the magnitude of the force on the wire.

49. [I] A 1.50-m-long straight wire experiences a maximum force of 2.00 N when in a uniform 1.333-T B-field. What current must be passing through it?

50. [I] A wire is positioned in a uniform magnetic field so that the force on it—4.00 N—is a maximum. If the wire is 20 cm long, and carries 10.0 A, what is the magnitude of the B-field?

51. [I] A straight current-carrying ($I = 6.0$ A) wire makes an angle of 31.2° with a 0.01-T uniform B-field. What is the magnitude of the force exerted on a 1.0-cm length of the wire?

52. [I] A narrow flat coil wound on a square frame has 200 turns and sides of 20 cm. It carries a current of 1.25 A and is positioned in a 0.50-T magnetic field. What is the maximum torque that can be exerted by the field on the coil?

53. [I] A circular flat coil of wire encompassing an area of 1.3×10^{-3} m^2 has 20 turns and carries a current of 1.5 A. If its rotational symmetry axis makes an angle of 32° with a B-field of 0.90 T, what is the torque acting on it?

54. [I] Two parallel wires 50.0 m long are 25.0 cm apart and each carries a current of 10 A in the same direction. What is the force (magnitude and direction) between them?

55. [II] An electron having a velocity of 5.0×10^6 m/s along the x-axis enters a region where there is a uniform 5.0-T B-field making an angle of 60° to the x-axis. Determine the magnitude of its acceleration.

56. [II] A cosmic ray proton traverses a uniform B-field perpendicularly at a speed of 2.0×10^7 m/s in the positive x-direction, experiencing an acceleration of 4.0×10^{12} m/s^2 in the positive y-direction. Determine **B**.

57. [II] A doubly charged positive helium ion, one missing two orbital electrons, has a mass of 6.7×10^{-27} kg and is accelerated through a potential difference of 10 kV. It then enters a region in which there is a uniform perpendicular B-field of 1.50 T. What is its subsequent path in the field?

58. [II] A tiny plastic sphere of mass 0.10 mg is shot horizontally at a speed of 200 m/s in the x–y plane. Given that it carries a charge of $-10\ \mu$C and travels along the positive y-axis, determine the nature of the smallest uniform magnetic field that will keep the particle moving horizontally despite the Earth's gravity.

59. [II] Write an expression for the momentum of a particle of charge q and mass m moving in a circular orbit of radius R in a uniform magnetic field B.

60. [II] A proton is sailing through the outer region of the Sun at a speed of 0.15c. It traverses a locally uniform magnetic field of 0.12 T at an angle of 25°. What is the radius of its helical orbit? (Hint: $v_{\parallel}$ and $v_{\perp}$ can be considered separately.)

61. [II] The American physicist E. H. Hall discovered (1879) that when a current travels along a conducting plate of width l, which is perpendicular to a magnetic field, a potential difference V appears across the plate as shown in Fig. P61. Prove that

$$V = vBl$$

The Hall probe makes a very convenient magnetometer. Discuss the difference you might expect if the probe is

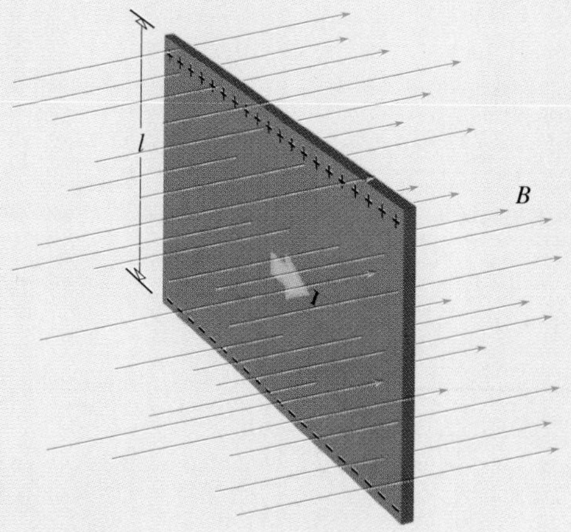

Figure P61

made of copper in one case, where the charge carriers are negative, and germanium, where the charge carriers are positive "holes." The Hall effect reveals a difference between positive charge moving to the right and negative charge moving left. Explain.

62. [II] A single power line 50.0 m long is stretched more or less horizontally at an angle of 20° east of north at a location where the Earth's field is 0.50×10^{-4} T at 5.0° west of north. What is the total force on the wire when it carries 1.0 kA?

63. [II] With Problem 61 in mind, suppose we want to use the Hall effect to measure the blood's flow rate (v) in an artery. If we apply a transverse magnetic field, it will cause the positive and negative ions in the blood to separate, thereby producing a voltage across the artery. Given that the artery has an inner diameter of 4 mm, if a uniform 0.50-mT field produces a voltage difference of 1.0 μV, what is the flow rate?

64. [II] A flat rectangular coil of 10 turns is suspended vertically from one arm of a beam balance that is brought into horizontal alignment with a few weights on the other pan. The bottom section of the coil, which is 10 cm long, is next exposed to a uniform perpendicular magnetic field of 0.60 T. A current of 300 mA is sent through the coil, which is pulled downward. How much weight must now be added to the other pan to restore balance?

65. [II] In a moving-coil galvanometer, the coil, which has N turns, surrounds a steel core so that the B-field always makes an angle of $\phi \approx 90°$ with the symmetry axis (the z-axis). If the current through the coil is reduced by 25%, how does the torque on it change?

66. [II] A single circular loop of wire 10 cm in radius is hung on a fine silver ribbon so that the plane of the loop is parallel to the magnetic field in a uniform region between the pole pieces of a large electromagnet. If the torque on the coil is 0.100 N·m when 5.0 A passes through the coil, what is the magnitude of the B-field?

67. [II] A long wire is stretched out horizontally to a 1.0-m-diameter tree, around it and back. Both ends are connected to the terminals of a battery and 10 A is drawn by the wire. If both lengths are parallel and the Earth's field is negligible, what force exists between each meter of the wires?

68. [III] A positive particle of charge q and mass m is accelerated through a potential V and is subsequently bent into a circular orbit of radius R by a uniform perpendicular magnetic field B. Write an expression for R^2 in terms of V, B, and m/q; the latter is called the charge-to-mass ratio. A modern mass spectrometer allows R to be measured directly, and from that m is determined with a precision of about 1 part in 10^7.

69. [III] A positively charged particle is moving in a circle under the influence of a perpendicular magnetic field. Write an expression for the frequency of its orbital motion and show that it is independent of both radius and speed. Overlook the fact that it radiates.

70. [III] Suppose we have two tiny spheres 1.00 m apart, each carrying 1.00 C. If they now move along parallel straight paths (1.00 m apart) at 1.00 m/s, compare the electric force to the magnetic force they experience.

22

Electromagnetic Induction

ELECTRICITY PRODUCES MAGNETISM; THAT much had been established in the early 1820s, and so the converse—magnetism produces electricity—seemed a reasonable thing to expect. And yet, the best researchers of the day could only come up with results that were ambiguous and unconvincing. Still, the agenda was obvious enough: a charge can electrify a nearby object by induction, and a magnet can magnetize a nearby piece of iron by induction; it was only reasonable to expect that a current should induce a current in a nearby conductor.

Since a steady current generates a steady magnetic field, should not a steady magnetic field generate a steady current? However logical that was, it was wrong—currents *were* being induced, right before their eyes, but these were transient currents, something no one expected and no one was ready to "see." A steady magnetic field does not impart energy to free charges, it does no work on them, and yet for a current to exist it must get energy from somewhere. A constant current in a wire sitting next to and at rest with respect to another wire will not induce a current in that second wire. But a changing magnetic field is something very different. It can impart energy to charges and it can produce currents.

ELECTROMAGNETICALLY INDUCED EMF

In 1821, Ampère conducted an experiment during which he observed the momentary effects of what has come to be known as **electromagnetic induction**, but that was not the main thrust of his research and he did not appreciate what was at hand. "Convert magnetism into electricity" was the brief remark Faraday jotted in his notebook in 1822, a challenge he set himself with an easy confidence that made it seem so attainable. Two years later, Arago by chance observed that the needle of a fine compass tended, after it was jolted, to oscillate for a shorter-than-normal time when it was in a housing with a copper bottom (p. 778). Arago then performed the inverse experiment and rotated a copper disk beneath a compass needle and found, inexplicably, that the needle followed the disk around.

After several years doing other research, Faraday returned to the problem of electromagnetic induction in 1831. His first apparatus made use of two coils mounted on a wooden spool. One, called the *primary,* was attached to a battery and a switch; the other, the *secondary,* was attached to a galvanometer (Fig. 22.1a). Initially, the results were totally negative until he installed a much more powerful battery composed of a hundred cells. Even then, the effects were faint and transient, and yet their mere existence was enough for him. Faraday found that the galvanometer deflected in one direction just for a moment whenever the switch was closed, returning to zero almost immediately, despite the constant current still in the primary. Whenever the switch was opened, interrupting the primary current, the galvanometer in the secondary circuit momentarily swung in the opposite direction and then promptly returned to zero.

Using a ferromagnetic core to concentrate the "magnetic force," Faraday wound two coils around opposing sections of a soft iron ring (Fig. 22.1b). Now the effect was unmistakable—a changing magnetic field generated a current. Indeed, as he

Figure 22.1 (a) A current in one coil produces a time-varying magnetic field that couples to the other coil, inducing a transient current in it. (b) An iron core becomes magnetized by the field of the primary current, increasing the field and improving the coupling to the secondary current.

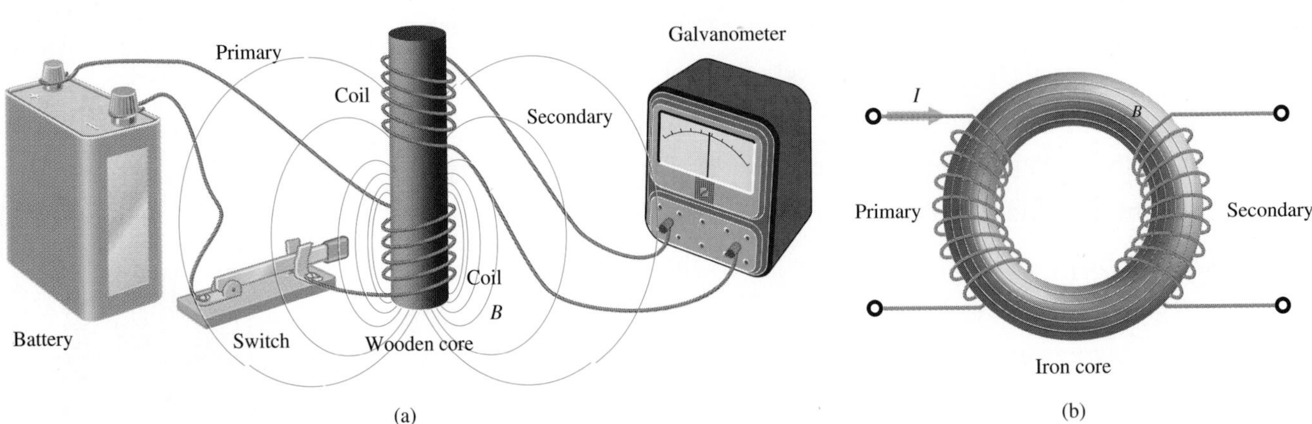

would continue to discover, *change* was the essential aspect of electromagnetic induction. Within a few weeks, the modest, untutored genius had explored the phenomenon and was ready to present his first paper on the subject.

Joseph Henry in New York didn't hear of Faraday's achievement until April, and when he did, the news was a bitter disappointment. He had performed much the same experiment as Faraday well over a year before (1830) though the results went unpublished. Like Faraday, Henry appreciated the implications of the discovery even if he felt no urgency to publish them. When he finally dashed off an article in response to the news from Europe, the experiment he described was somewhat more sophisticated than that performed in London, and it even contained an important feature Faraday had missed (p. 779).

Mr. Faraday (he declined knighthood) went on to use his new knowledge to build an electric generator (p. 774), one of the single most far-reaching accomplishments of the era. Later, when asked by Prime Minister Gladstone if his research had any practical value, Faraday quipped, "Why, sir, you will soon be able to tax it." Electromagnetic induction is now at the very heart of the generation, delivery, and utilization of electrical energy—and they do indeed tax it.

22.1 Faraday's Induction Law

Figure 22.2 depicts a modern version of an experiment performed by Faraday to explore the dependence of electromagnetic induction on variations in B. The oscilliscope displays, as a function of time, the voltage—otherwise known as the **induced emf** ($\mathscr{E}$)—across the coil's terminals. The exact shape of the resulting curve is not important; it depends on the details of the motion of the magnet. Observe that thrusting a south pole toward the coil produces a positive emf and yanking it away produces a negative emf (for this particular winding). Approaching the coil with a north pole reverses the polarity of the induced emf. Futhermore, the amplitude of the emf depends on how rapidly the magnet is moved; when $v = 0$, the emf $= 0$. In this arrangement, *the induced emf depends on the rate of change of B* through the coil and not on B itself. A weak magnet moved rapidly can induce a greater emf than a strong magnet moved slowly.

When the same changing B-field passes through two different wire loops, as in Fig. 22.3, the induced emf is larger across the terminals of the larger loop. In other

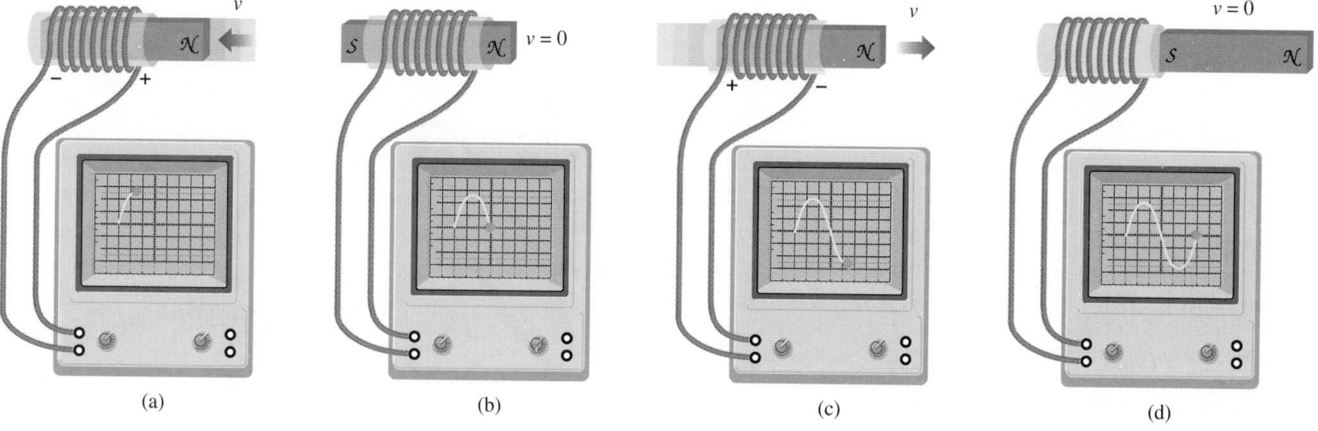

Figure 22.2 (a) As the S-pole approaches the right end of the coil, a current circulates from left to right through the coil, and the right end of the coil becomes an S-pole. The right side of the coil becomes positive, the left side negative, and it acts like a battery, sending current clockwise through the external circuit. (b) When the magnet is stationary, the emf is zero. (c) Removing the magnet induces a north pole at the right end of the coil, which attracts the receding S-pole of the magnet. (d) No motion, no flux change, and $\mathscr{E} = 0$.

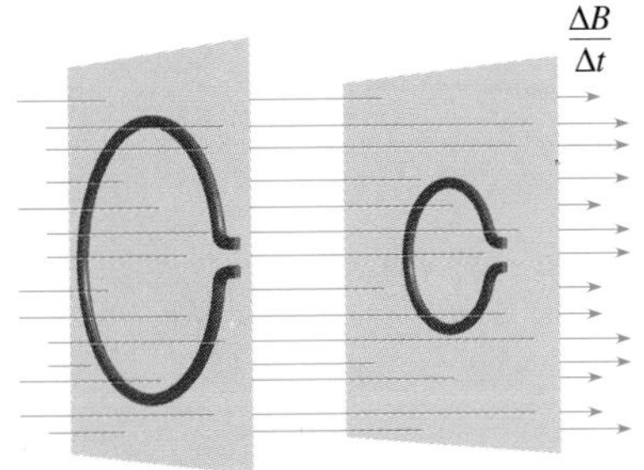

Figure 22.3 A time-varying magnetic field passes through both loops. It induces a larger emf in the larger loop, which embraces a greater perpendicular area.

words, here where the B-field is changing, *the induced emf is proportional to the area A of the loop penetrated perpendicularly by the field.* If the loop is successively tilted over, as in Fig. 22.4, the area presented perpendicularly to the field ($A_\perp$) varies as $A \cos \theta$ and, when $\theta = 90°$, the induced emf is zero because no amount of B-field then penetrates the loop: when $\Delta B/\Delta t \neq 0$, emf $\propto A_\perp$. The converse also holds: *when the field is constant, the induced emf is proportional to the rate-of-change of the perpendicular area penetrated.* Thus, if a coil is twisted or rotated or even squashed (Fig. 22.5) while in a constant B-field so that the perpendicular area initially penetrated is altered, there will be an induced emf $\propto \Delta A_\perp/\Delta t$ and it will also be proportional to B. In summary, then, when $A_\perp$ = constant, emf $\propto A_\perp \Delta B/\Delta t$ and, when B = constant, emf $\propto B\Delta A_\perp/\Delta t$ (see Appendix A-6).

All of this suggests that the emf depends on the rate-of-change of both $A_\perp$ and B; that is, on the rate of change of their product. This brings to mind the notion of the flux of the field (p. 613)—the product of field and area where the penetration is perpendicular. Accordingly, we define the **flux of the magnetic field** as

$$\Phi_M = B_\perp A = BA_\perp = BA \cos \theta \tag{22.1}$$

The units of magnetic flux are tesla-meter^2, which is called a **weber** (Wb) to honor one of the early workers in magnetism, Wilhelm Weber: 1 Wb = 1 T·m^2. Clearly,

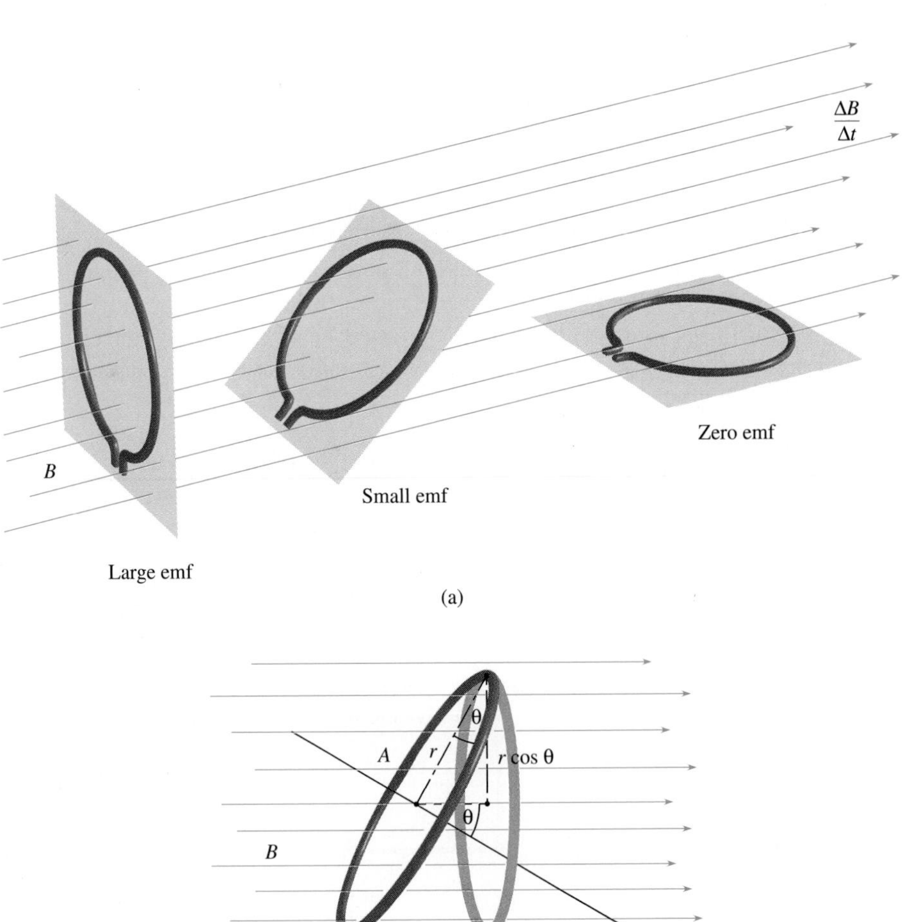

Figure 22.4 (a) The induced emf is proportional to the perpendicular area intercepted by the field. (b) The area of the ring ($A = \pi r^2$) projects as an ellipse. Vertically r projects to $r \cos \theta$; horizontally r projects to r. The area of the ellipse is $A_\perp = \pi(r \cos \theta)r = A \cos \theta$.

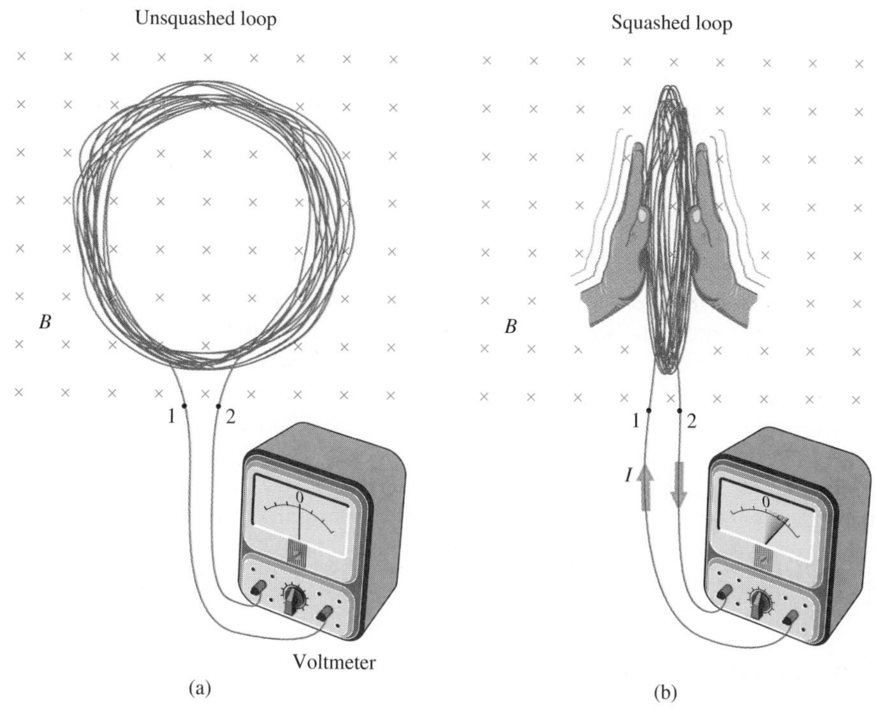

Figure 22.5 (a) A downward B-field perpendicular to a wire loop. The coil starts at point 1, winds clockwise through a number of turns, and ends at point 2. (b) Squashing the loop reduces $A_{\perp}$, changes the flux, and, therefore, induces an emf. The voltmeter will show a momentary jump in voltage, after which it will return to zero.

B must be measured in Wb/m^2, which is the old unit that was replaced by the tesla: 1 Wb/m^2 = 1 T. It's common in the literature to find B referred to as **flux density** (flux per unit area).

The results of the above observations for a one-turn loop of wire can be stated as

$$\mathscr{E} = -\frac{\Delta\Phi_{\mathrm{M}}}{\Delta t} \tag{22.2}$$

where the significance of the minus sign will be considered presently. A single loop of wire will experience an induced voltage (in volts) that equals (in SI units the constant of proportionality is unity) the time-rate-of-change of magnetic flux through it at any given instant. If there are N turns of wire in a coil, each has an induced voltage that is in series with the others so that the net *average induced emf* is

$$\mathscr{E} = -N\frac{\Delta\Phi_{\mathrm{M}}}{\Delta t} \tag{22.3}$$

which is one of the *fundamental equations of electromagnetism*. It is a summary of observations and not derivable from any previous formulas. The equation is generally known as **Faraday's Induction Law**, though Faraday himself never wrote it.

Example 22.1 A circular flat coil of 200 turns of wire encloses an area of 100 cm^2. The coil is immersed in a uniform perpendicular magnetic field of 0.50 T that penetrates the entire area. If the field is shut off so that it drops to zero in 200 ms, what is the average induced emf? Given that the coil has a resistance of 25 Ω, what current will be induced in it?

(continued)

(continued)

Solution: [Given: $N = 200$, $A = 100\text{ cm}^2$, $B_i = 0.50\text{ T}$, $B_f = 0$, $\Delta t = 200\text{ ms}$, and $R = 25\ \Omega$. Find: emf and I.] The B-field links the entire area perpendicularly; hence, the initial flux is

$$\Phi_M = BA = (0.50\ T)(0.0100\text{ m}^2) = 0.0050\text{ T·m}^2$$

and so $\Delta\Phi_M = -0.0050\text{ T·m}^2$ since the final flux is zero. It follows from Eq. (22.3) that

$$\mathscr{E} = -N\frac{\Delta\Phi_M}{\Delta t} = -200\frac{(-0.0050\text{ T·m}^2)}{0.200\text{ s}} = \boxed{5.0\text{ V}}$$

(Don't worry about the signs here; we'll deal with that part presently.) From Ohm's Law

$$I = \frac{V}{R} = \frac{(5.0\text{ V})}{(25\ \Omega)} = \boxed{0.20\text{ A}}$$

▶ **Quick Check:** Remember that a change of flux of 1 T·m^2 through 1 turn in 1 s produces 1 V. Here, $N\Delta\Phi_M$ (which is the change in the total flux linking the coil) equals one, so 5.0 V is reasonable.

Equation (22.3) makes the point that the emf depends on the rate-of-change of flux linking all the turns of wire, and it's useful to formalize the notion by speaking about the **flux linkage** in the coil—namely, $N\Phi_M$. Thus, **the induced emf is the negative of the time-rate-of-change of the flux linkage**.

22.2 Lenz's Law

The negative sign in Eq. (22.3) relates the polarity of the induced emf to the flux change, which can occur in a variety of ways: the field can be increased, decreased, or moved; the loop area can be turned, squashed, or yanked out of the field, and yet there is always a consistent, reproducible sense in which the emf appears. It was a physicist working in Russia—Heinrich Friedrich Emil Lenz (pronounced *lents*)—who first (1834) published an elegant statement of the phenomenon that has come to be known as **Lenz's Law**:

The induced emf will produce a current that always acts to oppose the change that originally caused it.

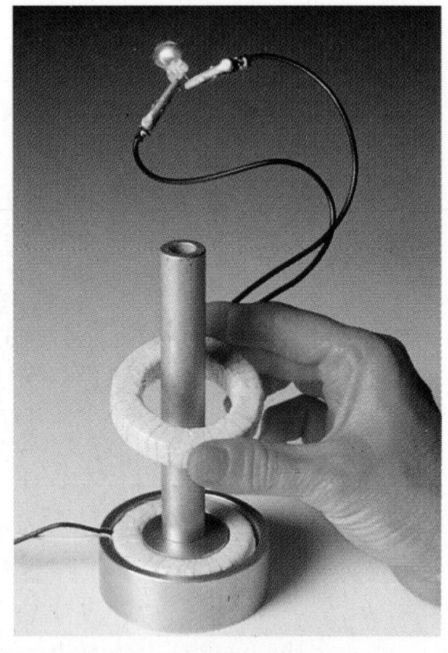

A time-varying current in the lower coil produces a time-varying magnetic flux that passes through the upper coil. That, in turn, induces a current in the upper coil, and the bulb lights.

In Fig. 22.6a, a south pole approaches a coil, inducing an emf across its terminals. The emf will be such as to cause an induced current, which in turn creates an induced magnetic field (B_I). This field opposes the *change* of flux linking the coil. The instigating effect is an *increasing B-field directed to the right* (toward the approaching magnet's south pole). The opposing induced field, tending to cancel an increasing field to the right, will be to the left, and the current in the coil circulates appropriately. **The polarity of the emf indicates the direction in which it drives current in the external circuit** and would be measured by a voltmeter across the terminals of the coil. In other words, the approach of a south pole induces a current that causes the coil to manifest a south pole at its near end, which opposes the advance of the magnet and thereby the change of the field. If the magnet is withdrawn, as in Fig. 22.6b, the change—*a decreasing field to the right*—is opposed by an induced field to the right. The current direction is now reversed, as is the emf. A north pole is induced at the coil's right end, which attracts the magnet tending to oppose its motion away.

Change is at the heart of the phenomenon of induction, just as it underlies the concept of energy—change is a manifestation of energy. Not surprisingly, Lenz's Law can be appreciated as a result of the Law of Conservation of Energy. For

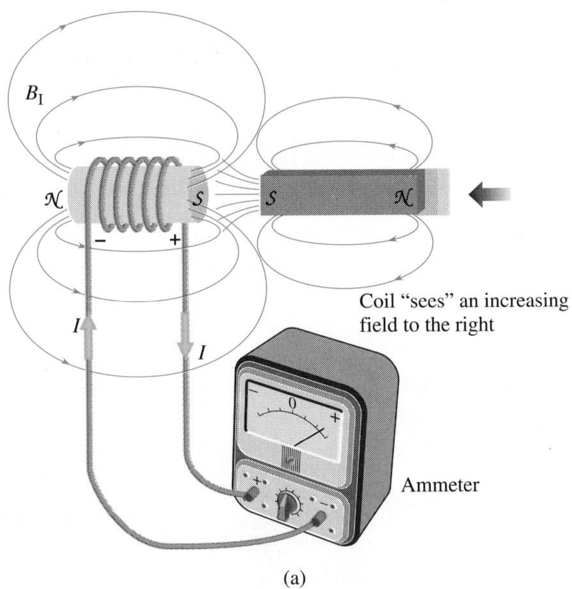

(a)

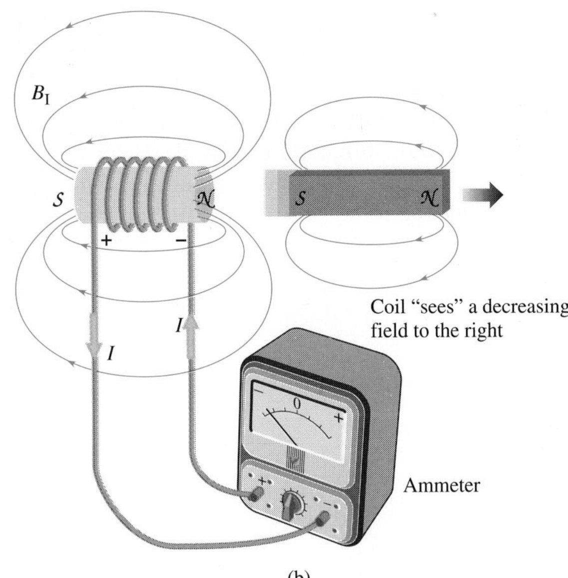

(b)

Figure 22.6 A coil with an air core, an ammeter, and a bar magnet. (a) The approach of an S-pole induces an emf (+ on the right, − on the left) that causes a current to circulate clockwise in the external circuit. (b) Removing the S-pole induces a counterclockwise current and the reversal of the emf, driving it. Note the polarity of the connections on each meter and the sign of its reading.

example, when an external agent moves the magnet in Fig. 22.6a, it must overcome the counterforce immediately exerted by the coil. The work done in the process provides the electrical energy needed to build up and sustain the induced current. The magnetic field is the intermediary between the external mover pushing the magnet and the resulting current. Were that not the case, the induced current would be created at no cost of mechanical work, and energy would not be conserved. Thus, no matter how the flux changes, work must be done on the system and the induced current, which is a manifestation of that work, must oppose the change—it doesn't necessarily stop it, but it always opposes it. *The work done by the magnet on the current* (W_{mc}) *must equal the work done by the current on the magnet* (W_{cm}), and that work is negative: $W_{mc} = -W_{cm}$.

Example 22.2 The loop in Fig. 22.5a has an area of 0.25 m^2 and is immersed in a uniform perpendicular 0.40-T downward magnetic field. If the coil has 200 turns, a resistance of 5.0 Ω, and is squashed to a zero area in 100 ms, what average current will be induced in it during the collapse? Discuss the direction of the current and the polarity of the emf. What happens to the current?

Solution: [Given: $A_i = 0.25$ m^2, $A_f = 0$, $B = 0.40$ T, $N = 200$, $R = 5.0\ \Omega$, and $\Delta t = 0.100$ s. Find: I.] The initial flux is

$$\Phi_M = BA = (0.40\text{ T})(0.25\text{ m}^2) = 0.100\text{ T·m}^2$$

and so $\Delta\Phi_M = -0.100$ T·m^2 since the final flux linking the loop is zero. It follows from Eq. (22.3) that the average emf is

$$\mathscr{E} = -N\frac{\Delta\Phi_M}{\Delta t} = -200\,\frac{(-0.100\text{ T·m}^2)}{0.100\text{ s}}$$

and
$$\mathscr{E} = 2.0 \times 10^2\text{ V}$$

Hence, from Ohm's Law, the average current is

$$I = \frac{V}{R} = \frac{(200\text{ V})}{(5.0\ \Omega)} = \boxed{40\text{ A}}$$

The current circulates clockwise around the loop so as to induce a B-field within its confines, which will tend to counter the decrease in downward flux. Thus, terminal 2 is positive with respect to terminal 1.

(continued)

(continued)

▶ **Quick Check:** The sides of the coil as it is crushed will experience a force via Eq. (21.10)—a force on a current-carrying wire in a B-field. That force will everywhere be in the direction of $\mathbf{v} \times \mathbf{B}$, where $\mathbf{v}$ is in the direction of motion of the positive current. For a clockwise current, the force is to the left on the left side and to the right on the right side. The induced force will oppose the external force that is crushing the coil. Because of resistance, the energy associated with the current is dissipated as joule heat, and the current stops as the emf $\rightarrow 0$; it would continue to circulate if the coil was superconducting.

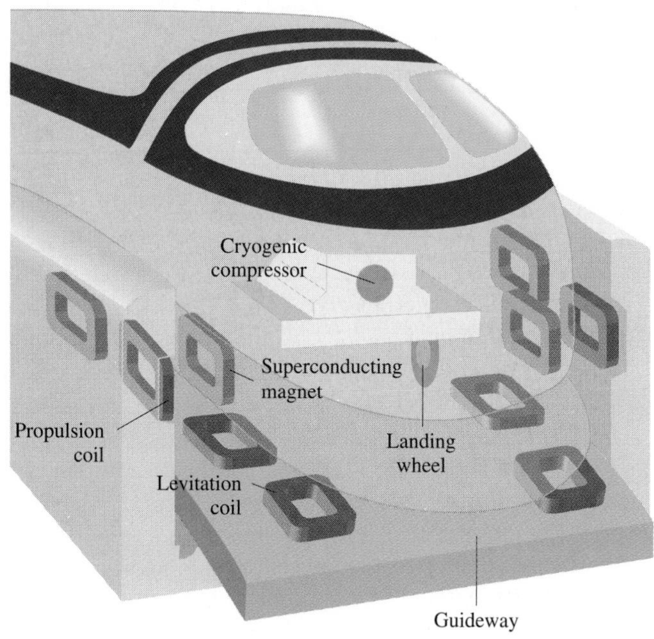

Figure 22.7 A high-speed maglev. Superconducting coils on the bottom of the car induce opposing B-fields in aluminum coils in the guideway that lift the car.

In one version of the magnetically levitated train (Fig. 22.7), there are superconducting electromagnets in the bottom of the car. These pass above two rows of closed coils mounted in the guideway, and as they do, currents are induced in the coils creating an opposing field (via Lenz's Law) that lifts the vehicle six inches into the air. It is propelled by successively energizing the coils in the sides of the guideway.

22.3 Motional Emf

Suppose a length of ordinary wire is moved perpendicularly across a uniform magnetic field at a speed v (Fig. 22.8). The mobile charge carriers within it are dragged along to the right with the wire at the same speed. Moving through the field, they experience a force

$$\mathbf{F}_{\mathrm{M}} = q\mathbf{v} \times \mathbf{B} \qquad [21.8]$$

parallel to the wire. In this case, the carriers are electrons and they flow downward. Accordingly, there is an induced emf across the wire—the top is at a higher potential than the bottom. As *within a battery,* negative charge is propelled from the + terminal to the − terminal.

Once a charge begins to move along the wire with some drift speed v_d, it will experience yet another force ($F_\perp = qv_d B$) due to this new motion in the B-field. The charge is then also propelled perpendicularly across the wire, but that lateral motion is restrained by the boundaries of the conductor. In accord with Lenz's Law, the net transverse force on all the mobile charges is opposite to the velocity of the wire and must be overcome by an externally applied force if the motion of the wire is to be sustained. *The external agency that moves the wire supplies energy to the system* (by doing work on the wire against the net transverse force). That energy is imparted to the induced current.

The motional emf (the change in potential) equals the work done on a positive test-charge in bringing it from one end of the rod to the other (that is, the change in its potential energy) per unit charge (p. 627). The force qvB acting parallel to the wire's length l does work on a charge in the amount of $qvBl$, and so the induced emf is

$$\mathscr{E} = vBl \tag{22.4}$$

Here the wire moves perpendicular to the B-field, the angle θ between $\mathbf{v}$ and $\mathbf{B}$ is 90°, and $\sin 90° = 1$ in Eq. (21.7). It need not be so, in which case $v_\perp = v \sin\theta$ instead of v must appear in Eq. (22.4).

The downward transit of electrons makes the bottom end of the wire negative, leaving behind a positive upper end. This charge buildup continues until the repulsion thus produced on any subsequent approaching charge matches the driving force qvB, and the current stops. Simply put, the bottom end gets so negatively charged that the induced **motional emf** cannot push any more electrons down to it. If the motion of the wire stopped, the emf would go to zero and electrons would flow back up or, if you like, the hypothetical mobile positive charges would descend from the higher potential and move downward.

Because there is no closed circuit, there cannot be a steady current produced by this wire generator. Instead, at equilibrium, the separated charges will have created an electric field E that opposes any further motion of charge—the induced current is transient. As we saw earlier (p. 631), since emf $= El$, it follows that the electric field in the wire, which exactly counters the motional emf, is

$$E = vB \tag{22.5}$$

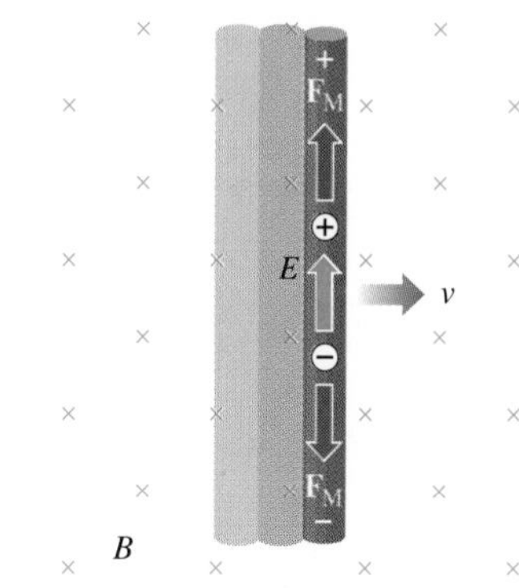

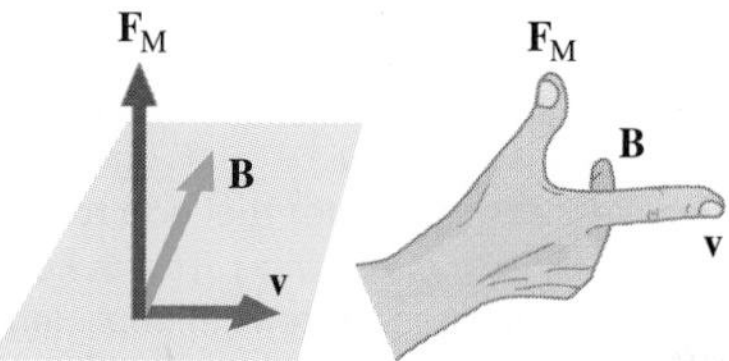

Figure 22.8 A positive charge experiences an upward force $\mathbf{F}_M = q\mathbf{v} \times \mathbf{B}$ as the wire cuts across the B-field.

Example 22.3 A 1.0-meter-long wire held in a horizontal east-west orientation is dropped at a place where the Earth's magnetic field is 2.0×10^{-5} T, due north. Determine the induced emf 4.0 s after release.

Solution: [Given: $l = 1.0$ m, $t = 4.0$ s, and $B = 2.0 \times 10^{-5}$ T. Find: emf.] From Eq. (22.4), we will need the speed, thus

$$v_f = v_i + gt = 0 + (9.8 \text{ m/s}^2)(4.0 \text{ s}) = 39.2 \text{ m/s}$$

hence

$$\mathscr{E} = vBl = (39.2 \text{ m/s})(2.0 \times 10^{-5} \text{ T})(1.0 \text{ m})$$

and

$$\boxed{\mathscr{E} = 0.78 \text{ mV}}$$

▶ **Quick Check:** $v_f \approx (10 \text{ m/s}^2)(4 \text{ s}^2) \approx 40$ m/s; $\mathscr{E} \approx 80 \times 10^{-5}$ V.

Cutting Field Lines

With the moving wire in Fig. 22.8, there is no closed loop and so no flux change; the induced emf arises from the motion of the charges across the field. We might expect

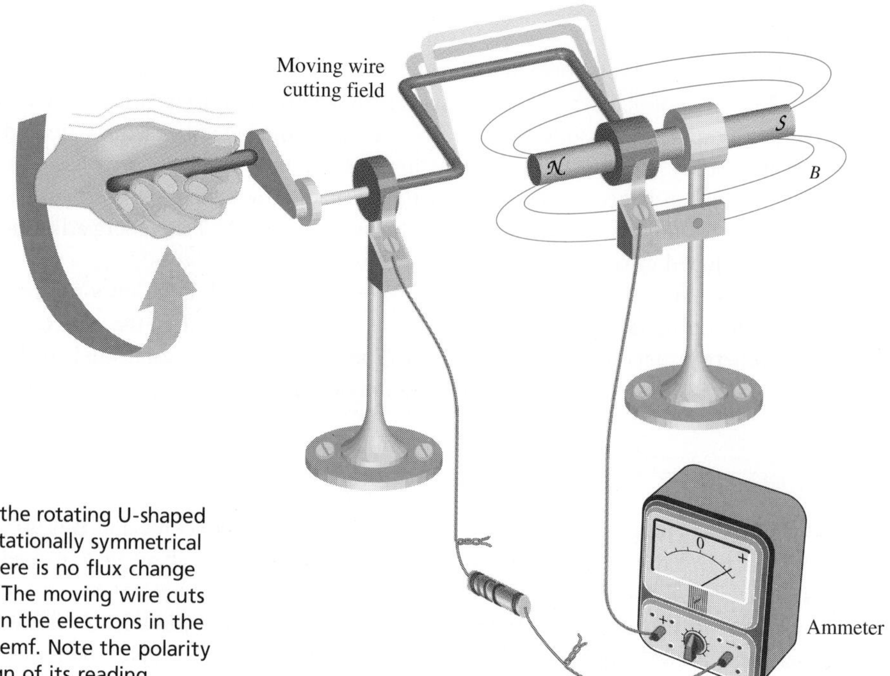

Figure 22.9 Consider the loop formed by the rotating U-shaped rod and the hookup wires. The B-field is rotationally symmetrical and does not penetrate the loop. Hence, there is no flux change as the rod rotates, and yet there is an emf. The moving wire cuts the field. There is a relative motion between the electrons in the wire and the field, and there is an induced emf. Note the polarity of the connections to the meter and the sign of its reading.

that if the wire was kept stationary and the field moved to the left, the same relative motion would exist and the same emf would be induced and, indeed, that's what happens experimentally. With this in mind, Faraday provided a useful alternative model in the spirit of Eq. (22.4). He envisioned the emf arising when the wire cut across magnetic field lines. Remember that the strength of a field B can be related to the number of field lines per unit area. Thus, through any perpendicular area A, the number of field lines is defined to be BA. When the wire in Fig. 22.8 moves with speed v, it sweeps across an area equal to vl per second. The number of field lines "cut" per second is therefore vlB, which, according to Eq. (22.4), equals the emf: *the induced emf equals the number of field lines cut per second by a conductor.*

Figure 22.9 shows the field of a permanent magnet being cut by a revolving wire. The field is rotationally symmetrical, and there is no flux through the loop bounded by the device, and yet if the crank is turned at a steady rate, a steady emf will be induced and a steady current will circulate. Both the emf and the current will vary linearly with the rotation rate.

Now suppose the moving wire of Fig. 22.8 is made to be part of a closed loop by placing it on a conducting U-shaped track (Fig. 22.10) having an appreciable resistance. Closing the loop will allow a current I to circulate, thereby extracting energy from the arrangement, if only as joule heat. The moving wire again produces an emf $= vBl$ much like a battery, and the resulting current provides energy at a rate of $\mathrm{P} = I\mathscr{E}$; we have a dc generator. Notice that if the rod moves at a constant speed v in a time Δt it travels a distance $v\Delta t$ and sweeps out an area $\Delta A = v\Delta tl$. It therefore cuts a number of field lines equal to $B\Delta A = Bv\Delta tl$, and it does so at a rate of $B\Delta A/\Delta t = Bvl$, which indeed equals the emf! From Lenz's Law, the induced current progresses counterclockwise and thereby generates a force to the right opposing the applied force that is moving the wire.

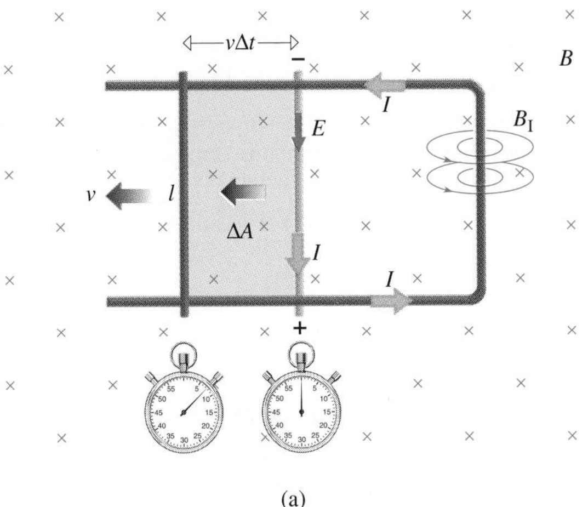

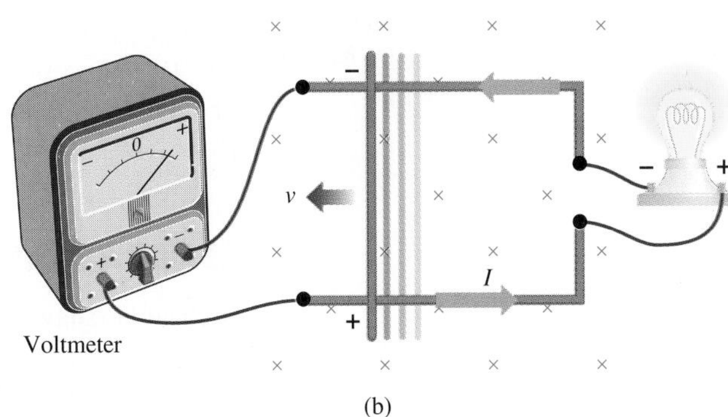

Figure 22.10 A straight wire sliding along, and in contact with, a stationary U-shaped conductor. (a) As the wire moves to the left, there is a time-rate-of-change of the flux equal to $B\Delta A/\Delta t = B(lv\Delta t)/\Delta t = Bvl$. Notice how the induced field B_I *within* the loop opposes the increase of flux. (b) Current circulates counterclockwise as if the load were placed across a battery with + and − terminals.

Alternatively, we can say that the flux through the loop increases at a rate $B\Delta A/\Delta t = Bv\Delta tl/\Delta t = Bvl = emf$, which is the same result. Again, Lenz's Law maintains that the induced current will generate a B-field (out of the plane, inside the loop) that tends to decrease the downward flux increase. Inasmuch as the flux through a perpendicular area is BA and the number of field lines through that area is BA, it's not surprising that these two interpretations of the experimental results are *nearly* equivalent. What is remarkable is that we need two distinct notions (Eq's. 22.3 and 22.4) to deal with the whole range of physical possibilities associated with induced emfs. One might wonder whether or not we are missing something.

22.4 Induced Magnetic and Electric Fields

A stationary charge has a constant E-field at every surrounding point in space, whereas a moving charge—a current—produces a time-varying E-field. But a current is also surrounded by a B-field. This suggests a fresh way of looking at things from a field perspective; namely, **a time-varying E-field produces a B-field**. Let's now examine the reciprocal idea that a time-varying B-field is always accompanied by an E-field.

Picture a long solenoid of radius R carrying a current (Fig. 22.11). To probe for an electric field, we place within the solenoid a small coaxial coil of radius r having one or more turns, a so-called *search coil* (reminiscent of the test-charge). If the primary current is now increased, the flux through the search coil will increase, an emf will be induced across its terminals, and thus an induced E-field will exist along it. Alternatively, we can envision the increasing current creating additional B-field lines that arise out of the wires and move inward toward the center of the primary (decreasing in density away from the wire). These lines cut the search coil, causing a positive charge therein to have an outward relative motion with respect to the field and thus be moved (in the direction of $\mathbf{v} \times \mathbf{B}$) to flow as a clockwise current, as required by Lenz's Law (Fig. 22.11b). At any point in the conductor of the search coil, a positive charge would experience a tangential force and a tangential electric field.

Since the wire loop of the search coil doesn't physically contribute to the E-field, it's useful to reconsider the above situation without the search coil in place. Experi-

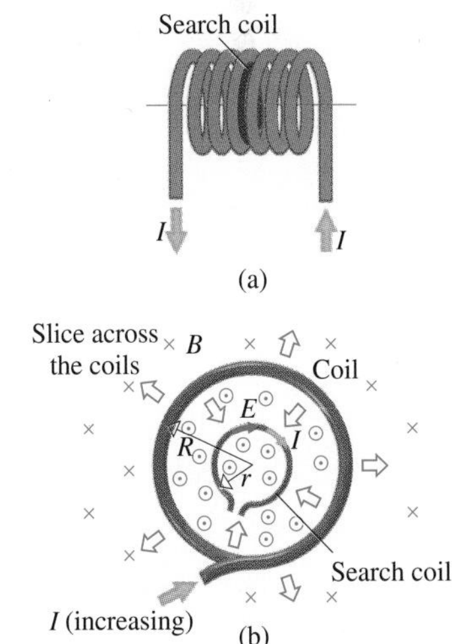

Figure 22.11 (a) A current-carrying coil. As I increases, B increases, and inside the coil, field lines move in toward the center. (b) A search coil within the solenoid experiences an E-field.

What led me more or less directly to the special theory of relativity was the conviction that the electromotive force acting on a body in motion in a magnetic field was nothing else but an electric field.

ALBERT EINSTEIN

ments show that a charged particle will be accelerated in a uniformly varying B-field precisely as if there was a circular E-field present. In general, **an induced electric field accompanies a time-varying magnetic field** (p. 836). There not being any sources or sinks in the form of charges, *the induced E-field lines must close on themselves.* The search coil simply allowed us to detect the presence of an independent E-field associated with the time-varying B-field. This is exactly what the circular UHF loop antenna you might have on top of your TV set is supposed to do.

GENERATORS

Michael Faraday devised the world's first electric generator in 1831 (Fig. 22.12). It is a seemingly simple thing: a copper disk hand-cranked to rotate between the poles of a permanent magnet. Charges in the metal, set in motion across the field as the disk turns, experience a radial force acting along the conductor. A conventional current of positive charges moves outward in the direction of $\mathbf{v} \times \mathbf{B}$. Work done on the mobile charges raises them in potential such that the rim terminal is positive and the axis terminal is negative. As long as the rotation rate is kept constant (which is easy to do), the emf is quite steady, and a continuous current is supplied to a load by this dc generator, or **dynamo**. The emf produced was small and the device impractical (modern versions have been used in electroplating where precise dc is desirable); still, its effect on the world was probably as far-reaching as any other single invention in all of history.

22.5 The AC Generator

Currently, the people of the world use electrical energy at a tremendous rate of about 10^{13} W. Almost all the electric current supplied commercially is generated with induction machines that produce a relative rotation between coils of wire and magnetic fields. The external power to rotate either the coils or the field is usually supplied by steam turbines (steam blasting against fanlike blades that are forced into rotation).

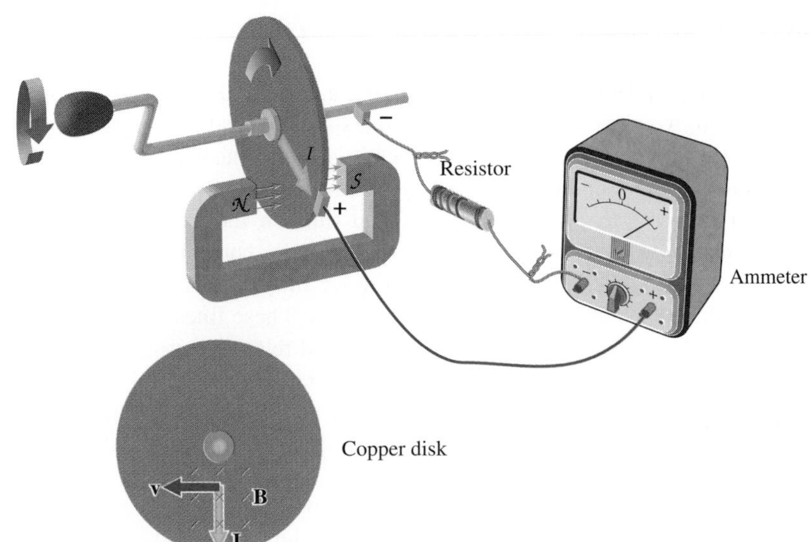

Figure 22.12 The first generator. Positive charge carriers moving in the direction of **v** experience a force $q\mathbf{v} \times \mathbf{B}$ that is radially outward along the disk. As long as the disk is turned in one direction, this device is a dc generator, also called a *dynamo.*

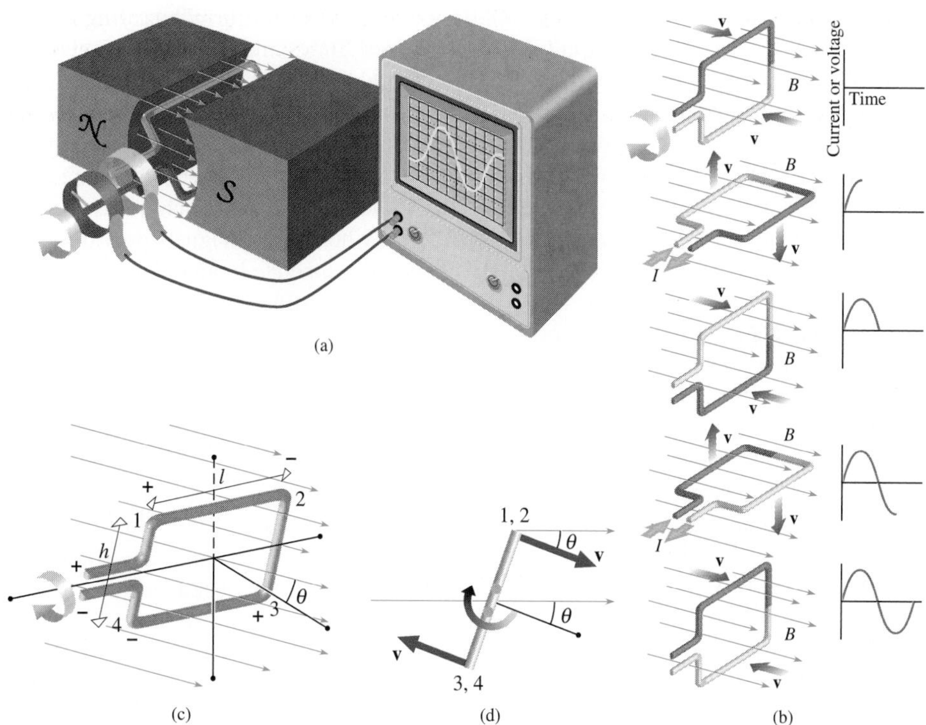

Figure 22.13 (a) A simple ac generator consisting of a loop rotating in a magnetic field. (b) Various stages in the rotation cycle, along with graphs of the corresponding generator outputs. (c) Detail of the conducting loop tilted at an angle θ. (d) The same loop shown in profile.

The steam generally comes either from burning fossil fuels or from nuclear reactors. A simplified version of an alternating current generator is shown in Fig. 22.13. Usually a coil, of several turns of wire wound on an iron armature, is rotated in the constant field of a magnet. Brushes rubbing against two slip rings attached to the ends of the coil carry off the induced current. As the coil rotates, its two long parallel sides (1–2 and 3–4), each of length l, move through the field in opposite directions. The emf induced across side 1–2, moving at a speed $v_\perp = v \sin\theta$ with respect to B, follows from Eq. (22.4):

$$\mathscr{E} = v_\perp Bl = Blv \sin\theta$$

At the moment pictured in Fig. 22.13c, $\mathbf{v} \times \mathbf{B}$ points from 2 to 1, so point 2 is at a lower potential than point 1 (work is done in bringing positive charges over to point 1). The same emf appears across the length 4–3 with point 4 lower than point 3. Since no emf is induced along the lengths of 1–4 and 2–3, the net emf around the single loop is $\mathscr{E} = 2(Blv \sin\theta)$ with point 4 negative (lower) and point 1 positive (higher): a conventional current will circulate from 4 to 1 driven by induction and then out (from the + terminal, near 1) into the external circuit, around and back into the loop (at the − terminal, near 4). When there are N turns of wire forming a coil, the emf of each loop is in series with the next, yielding a net emf of

$$\mathscr{E} = 2NBlv \sin\theta \tag{22.6}$$

When θ exceeds 180°, the polarity reverses. Since both θ and v can be written in terms of ω, we use the fact that when the angular speed ω is constant, $\theta = \omega t$. Moreover, $v = r\omega$, where here $r = \frac{1}{2}h$. Keeping in mind that lh is the area A, $lv = l \times (\frac{1}{2}h\omega) = \frac{1}{2}A\omega$ and the emf becomes

$$\mathscr{E} = NAB\omega \sin\omega t \tag{22.7}$$

Regardless of the actual shape of the coil, the emf will be alternating with a frequency $f = \omega/2\pi$, via Eq. (12.2). In the United States and Canada, f is typically 60 Hz, whereas in much of the rest of the world f is 50 Hz. AC voltages can be changed quite easily and so you can plug an American hairdryer (using a converter) into a British wall outlet—the dryer isn't noticeably affected by the frequency difference. On the other hand, you will have a lot more trouble with something requiring timing signals, like playing an American video tape on a French VCR.

The simple generator of Fig. 22.13, with a few horseshoe magnets providing the field, was once commonplace (you've probably seen someone in an old movie cranking a telephone or perhaps a detonator just prior to setting off explosives). In any event, when large voltages (several kilovolts) and currents (of 50 amps and more) are involved, the slip rings and brushes, which tend to spark and deteriorate, become too troublesome. Today, this sort of *revolving-armature* machine (the armature is the component in which the emf is induced) is far less important commercially than it once was. To avoid the difficulties associated with voltage exceeding about 600 V, the coil assembly carrying the induced current is made stationary, and electromagnets are mounted on the turning shaft to produce a *revolving-field* machine. The current creating the revolving B-field is comparatively small, and slip rings and brushes can handle the task well. Meanwhile, the large induced current passes through a motionless, heavy, continuous wire. This kind of *alternator* powers the electrical system in most automobiles, where it's turned by the engine via belts and pulleys. The ac output must subsequently be converted to dc in order to charge the battery.

22.6 The DC Generator

There are many applications where dc electrical power is required, even though it's somewhat more difficult to produce. For example, although small motors (for things like hand drills and vacuum cleaners) run nicely on ac, large electrical motors generally don't do quite so well. The really big motors used on electric railway systems (trolley cars, subway trains, etc.) usually operate on dc.

The polarity reversals of the voltage from an ac generator can be eliminated using a split-ring *commutator* (Fig. 22.14), just like that on a simple dc motor. Thus, the negative half of the ac signal is reversed and made positive. The result is a bumpy direct current that rises and falls but never goes negative (Fig. 22.15). The

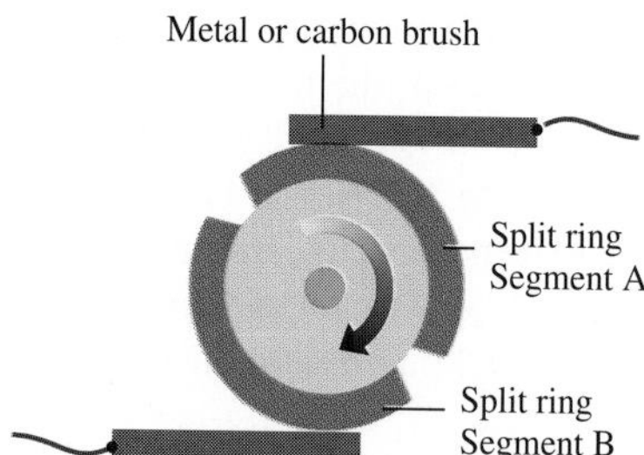

Figure 22.14 A two-segment commutator. At the moment shown, the top brush is touching segment *A* and the bottom one, segment *B*. As the ring turns, segment *A* moves into contact with the bottom brush. Reversing the connection reverses the signal.

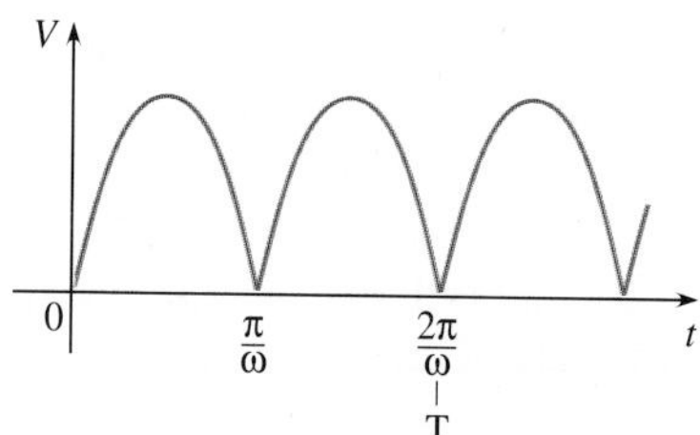

Figure 22.15 The pulsating dc from a single-coil generator with a two-segment commutator. Just as the sinusoidal voltage is about to go negative, the commutator flips it.

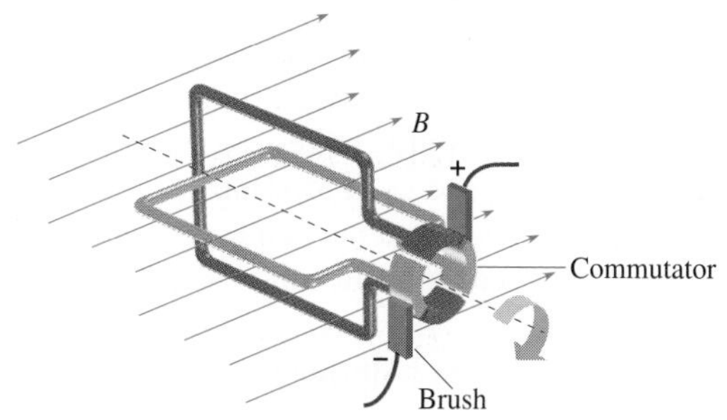

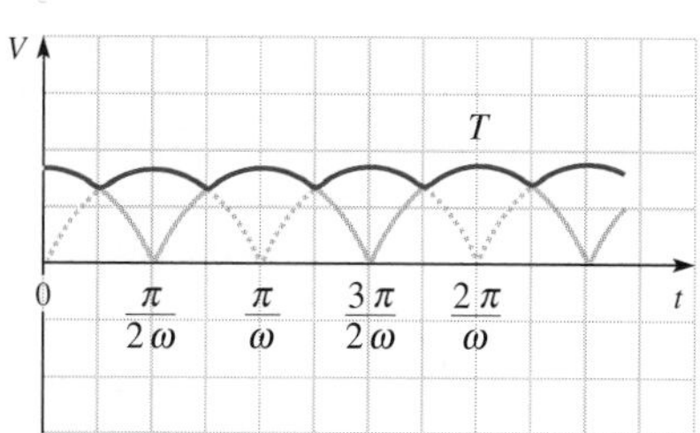

Figure 22.16 (a) A two-coil dc generator. Compare this device with the one depicted in Fig. 22.13 and note how the brushes are in contact with one of the loops at maximum voltage. (b) Emf of the two-coil dc generator. Notice how this emf is equivalent to two voltages like the one shown in Fig. 22.15, shifted by $\pi/2\omega = \frac{1}{4}\tau = 90°$ with respect to one another. A single coil contributes for $\frac{1}{4}$ of a cycle until the brushes shift and the other coil kicks in. It then contributes for the next $\frac{1}{4}$ cycle, and so on. At any instant only one coil is providing the voltage.

single-coil arrangement produces an emf that is zero twice during each revolution. Moreover, the voltage is relatively small for much of the cycle. By contrast, if the armature comprises two perpendicular coils (Fig. 22.16) and the commutator has four segments, so that the brushes are always in contact with the coil having the greatest emf, the output voltage across the terminals will be appreciably smoother, and the zeros will no longer occur. With 90° between coils, there will be a relative shift in the induced voltages of 90°, which is equivalent to a shift in time of $\pi/2\omega$, as shown in Fig. 22.16b. With three coils 60° apart, three voltages each shifted in time by $\pi/3\omega$ will be induced, yielding a final voltage equal to the envelope of the three curves, as in Fig. 22.17. Still more coils can be added, and the output voltage can thereby be made almost constant with only a slight ripple.

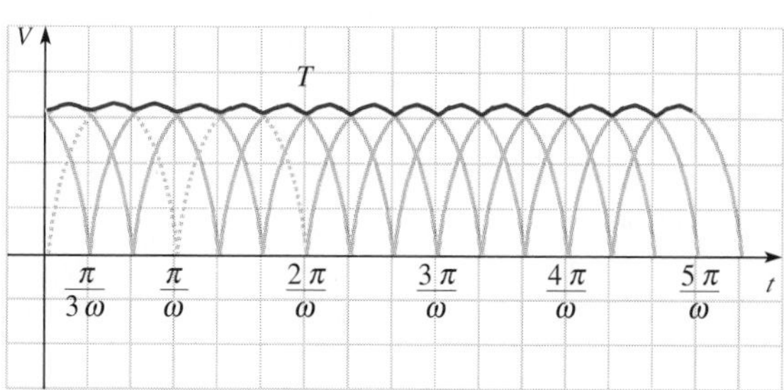

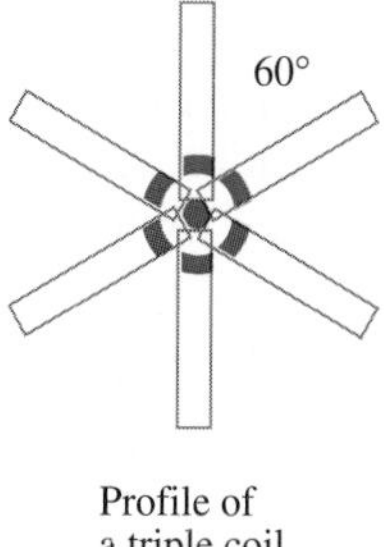

Profile of a triple coil

Figure 22.17 The emf of a three-coil dc generator. Each component signal is shifted by $\pi/3\omega = \frac{1}{6}\tau = 60°$.

Example 22.4 A simple single-coil dc generator rotates at a constant frequency of 60 Hz in a 0.40-T magnetic field. Given that the coil has 10 turns and encompasses an area of 1200 cm^2, what will be its maximum emf?

Solution: [Given: $f = 60$ Hz, $A = 1200$ cm^2, $B = 0.40$ T, and $N = 10$. Find: the maximum emf.] The maximum emf ($\mathscr{E}_m$) is the amplitude of the oscillating voltage given by

$$\mathscr{E} = NAB\omega \sin \omega t \qquad [22.7]$$

namely

$$\mathscr{E}_m = NAB\omega$$

Here $\omega = 2\pi f = 2\pi(60 \text{ Hz}) = 376.99$ rad/s and

$$\boxed{\mathscr{E}_m = 0.18 \text{ kV}}$$

▶ **Quick Check:** $\omega \approx 6(60) \approx 360$ rad/s; $\mathscr{E}_m \approx 10 \times (0.12 \text{ m}^2)(0.4 \text{ T})(360 \text{ rad/s}) \approx 0.17$ kV.

It's curious that the similarity between the dynamo and the dc motor was not recognized until the mid-nineteenth century and not really appreciated until a remarkable accident occurred in 1873. Not only do they look alike, they are actually identical: the action of the dynamo is just the reverse of the action of the dc motor. The dynamo converts mechanical energy into electrical energy—the armature is turned mechanically from the outside, and out comes a current. The motor converts electrical energy into mechanical energy—in comes a current from the outside, and around goes the armature.

The large hall at the 1873 Vienna Exposition was filled with modern gadgets. One of the Gramme dynamos driven by a steam engine was pouring forth electrical power when a workman unwittingly connected the output leads from another dynamo to the energized circuit. In an instant, the second device began to spark and whine and come alive, whirling around at great speed. The dynamo had become a motor. That impromptu mating of machines gave birth to the future as we have come to know it.

22.7 Eddy Currents

When Arago rotated a copper disk beneath a compass and found that the needle soon began revolving around with the disk, he was actually observing the effects of electromagnetic induction. The rotating disk "sees" a nonuniform time-varying magnetic field. An emf is induced and loops of current are set up in the disk. These currents generate their own counter B-field that opposes, via Lenz's Law, the cause of the induction—the needle is thereby made to rotate such that its relative motion with respect to the disk tends to vanish.

If an extended conductor translates with respect to a B-field, which is not uniform over the entire conductor, or if different portions of the conductor move at different velocities (that is, it rotates) with respect to the field, currents will be induced within it that circulate in closed paths. Instead of the transient current in the uniform-field situation of Fig. 22.8, a closed current loop can now form. For example, with Faraday's disk (open circuited, as shown in Fig. 22.18) the induced current loops, known as **eddy currents**, set up two dipole field regions, each north on one face of the disk and south on the other. The leading dipole, emerging from the external B-field on the left, attracts the poles of the magnet, thus tending to oppose the rotation. The trailing eddy-current dipole, moving into the B-field of the magnet, presents its north to the magnet's north and its south to the magnet's south, thus repelling the magnet and again resisting the motion. Consequently, the eddy currents produce a Lenz's Law drag on the rotating disk.

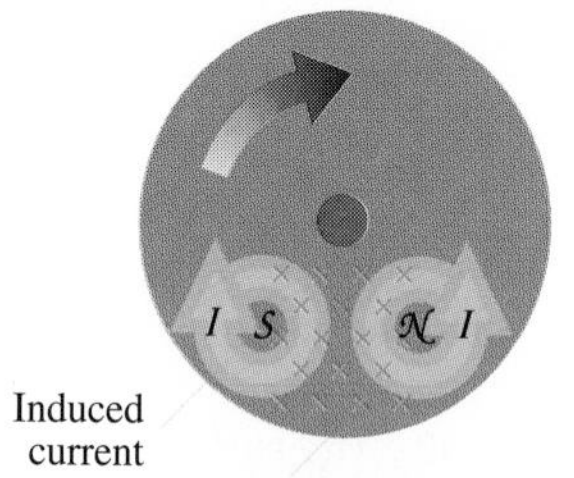

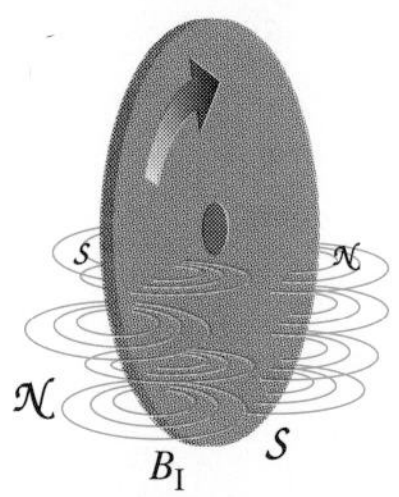

Figure 22.18 A copper disk rotating across a perpendicular magnetic field. Swirls of eddy currents circulate, opposing the motion. The induced magnetic field B_I opposes the turning of the disk by interacting with the applied field.

Similarly, the iron armature of a motor or generator (or even the core of an electromagnet) will harbor eddy currents that waste energy in the form of joule heating. To reduce this effect, most cores are built up of stacks of separate insulated sheets of iron.

Eddy currents caused by the geomagnetic field are commonplace, even if they are rarely of much consequence—they're induced in a metal spoon whenever you stir a cup of coffee, and in a coin when you flip it. In the early days of the Space Age, engineers were surprised to find that many of their low-altitude spin-stabilized satellites were gradually "despinning." The cause was traced to eddy currents induced in the walls of spacecraft, which were rotating in the planet's magnetic field. Not all such effects are detrimental: eddy currents are used to great advantage in heating induction furnaces, in damping the oscillations of delicate devices, and in the operation of a variety of instruments, including the metal detector now familiar to air travelers and beachcombers alike. In the latter, a time-varying magnetic field induces currents in a metal object, and the resulting induced B-field reveals its presence to a pickup coil.

Faraday delivering one of his famous Christmas lectures (1855). On the left in the front row is the Prince Consort between his two princely charges. This room at the Royal Institution in London still exists, and they still give demonstrations there.

SELF-INDUCTION

Whenever a voltage from some external source is placed across the terminals of a coil, the resulting current will produce a magnetic field. But this phenomenon raises an interesting point first considered by Joseph Henry. A coil laced with an increasing field must experience an induced emf—what does it matter that the coil itself is involved in creating the field? Field lines stream out from the center of the wire of the coil "cutting" it, just as they would cut a search coil immediately adjacent to it. Mobile charges within the wire experience a time-varying B-field, a perpendicular force, and a resulting induced emf along the length of the conductor. Or, if you prefer, the flux linking the coil will be changing in time, and Eq. (22.3) therefore requires that there be an induced emf. With either interpretation, it follows from Lenz's Law that the induced emf must oppose the cause of itself and so it is called a *back-emf.* Because of this opposition, the steady state is not reached instantaneously; the current builds gradually, as does its associated B-field. This process of **self-induction** retards the increase or decrease of current in a coil (and to a lesser extent in other circuit elements, hookup wires, transmission lines, and so on). As such, it is an extremely important aspect of almost all real ac devices from stereos to satellites.

22.8 Inductance

To quantify the self-induction of a current-carrying coil, we realize that the flux linkage $N\Phi_M$ is proportional to the current I producing it:

$$N\Phi_M = LI \tag{22.8}$$

The constant of proportionality L is called the **self-inductance**, or just the *inductance,* for short. In a sense, it is the electrical equivalent of inertia, a resistance to change. The SI unit of inductance is the **henry** (H); a flux linkage of 1 weber is established in a coil by a current of 1 ampere circulating therein when the coil's inductance is 1 henry. As we have seen so many times before, the constant of proportionality L is a composite of several of the system's physical characteristics. Here, it depends on the size and shape of the coil and on the surrounding medium. The flux linkage does not vary linearly with I if the permeability of the medium is not constant—that is, if it depends on B and therefore on I. Consequently, if there is a ferromagnetic material within the coil, or even nearby, L becomes a function of I and though Eq. (22.8) still holds, it is inherently more complicated than it would be were L constant.

In practice, when the precise value of an inductance is needed, it's usually measured, but if the geometry is simple enough, it can be approximated theoretically. As

The antenna of the old radio pictured on p. 589. The antenna consists of a coil wrapped around a ferrite core (the gray, cylindrical bar). The time-varying magnetic field component of an incoming electromagnetic wave induces a signal, an emf, in the coil.

an example, to compute the inductance of a long hollow coil of length l with n turns per unit length ($n = N/l$) and a cross-sectional area A, we recall that

$$B_z \approx \mu_0 n I \qquad [21.5]$$

Assuming B to be constant across the region encompassed by the solenoid (although we know the field actually decreases in toward the center), we should get a useful expression, even if it's likely to be slightly too large. Using Eq. (22.8) the inductance of a long air-core solenoid becomes

$$L = \frac{N\Phi_M}{I} = \frac{NBA}{I} \approx \frac{\mu_0 N^2 A}{l} \qquad (22.9)$$

If some material with a permeability μ is made to fill the hollow, we simply replace μ_0 by μ. When that material is ferromagnetic, the equation still works, provided we know μ at the particular current being used. There are commercial inductors that allow their inductances to be adjusted by sliding a ferrite slug partway into the coil. (See the photo on p. 781). Ferrites (oxides of magnesium, manganese, zinc, or nickel) are especially useful in surpressing core eddy currents at the high operating frequencies of radio and television circuits.

Example 22.5 A solenoid 3.0 cm long with a cross-sectional area of 0.50 cm^2 and comprising 300 turns of fine copper wire in a single layer is to be used as the antenna for a radio. The magnetic field component of the incoming electromagnetic signal will oscillate within the coil and induce an emf that will then be processed by the rest of the radio, and out will come music. (a) Determine the coil's inductance when the core is air-filled. (b) Approximate the inductance when a ferrite core is used instead, given that its relative permeability at the anticipated current level is 400.

Solution: [Given: $l = 0.030$ m, $A = 0.50 \times 10^{-4}$ m^2, $N = 300$, and $\mu/\mu_0 = 400$. Find: L.] (a) Using Eq. (22.9)

$$L \approx \frac{\mu_0 N^2 A}{l}$$

$$L \approx \frac{(1.26 \times 10^{-6}\ \text{T·m/A})(300)^2(0.50 \times 10^{-4}\ \text{m}^2)}{(0.030\ \text{m})}$$

and to two significant figures $L \approx 1.88 \times 10^{-4}$ H $\approx$ $\boxed{0.19 \text{ mH}}$. (b) Given $\mu = 400\mu_0$, L is increased by a factor of 400 and $\boxed{L \approx 75 \text{ mH}}$.

▶ **Quick Check:** $N^2A \approx 9 \times 10^4(0.5 \times 10^{-4}) \approx 4.5$; $N^2A/l \approx 150$; $L \approx 1\frac{1}{4}(150)\ \mu\text{H} \approx 190\ \mu\text{H}$.

22.9 The Back-Emf

Faraday's Induction Law can be reformed in terms of the inductance, thereby providing an expression for the back-emf. From Eq's. (22.3) and (22.9) the average self-induced emf is

$$\mathscr{E} = -N\frac{\Delta\Phi_M}{\Delta t} = -\frac{\Delta(LI)}{\Delta t}$$

Provided that the inductance is constant (see Appendix A-6), $\Delta(LI) = L\Delta I$ and

$$\mathscr{E} = -L\frac{\Delta I}{\Delta t} \qquad (22.10)$$

The average induced emf is proportional to the time-rate-of-change of the current in the coil. *The positive direction is that of I.* An inductance of 1 H will induce a back-emf of 1 V when the current through it changes at a rate of 1 A/s. This expression also tells us that 1 V = 1 H·A/s and so 1 H = 1 V·s/A, or 1 H = 1 Ω·s. *If the coil itself has negligible resistance, there will not be an appreciable voltage drop across it due to the current traversing it, and the back-emf will be the voltage measured across its terminals.*

Suppose an externally supplied current is made to *increase* through a coil passing from its terminals A to B (Fig. 22.19a). In response to the changing flux, an induced current (I_I) then moves from B to A, opposing ΔI and reducing I in the circuit, which in effect raises the potential of point A making it positive with respect to B. This back-emf is in the opposite direction to I and is negative—it's a voltage drop. The coil's inductance impedes the original *changing* current, thereby producing a voltage drop across its terminals (as if it now had some kind of "resistance" even though $IR = 0$). While the current is increasing, there will be a voltage across the terminals as shown. If the current begins decreasing, an induced current will pass from A to B to oppose the decrease. Point B will be positive with respect to point A, and the back-emf will reverse, becoming positive (Fig. 22.19b).

This adjustable ferrite-core inductor is part of the tuning circuit of the AM radio on p. 589. We'll come back to what it does when we consider resonant circuits (p. 811).

Example 22.6 The current in a 50-μH coil (for which R is negligible) goes from 0 to 2.0 A in 0.10 s. Determine the average self-induced emf measured across its terminals.

Solution: [Given: $L = 50\ \mu\text{H}$, $I_i = 0$, $I_f = 2.0$ A, and $\Delta t = 0.10$ s. Find: $\mathscr{E}$.] From Eq. (22.10) the numerical value of the back-emf is

$$\mathscr{E} = -L\frac{\Delta I}{\Delta t} = -(50 \times 10^{-6}\ \text{H})\frac{2.0\ \text{A}}{0.10\ \text{s}} = \boxed{-1.0\ \text{mV}}$$

The negative sign tells us what we presumably know—namely, that the back-emf opposes the increase in current. The polarity shown in Fig. 22.19 already took the sign into consideration.

▶ **Quick Check:** The current is changing at a moderate rate of 20 A/s, but the inductance is only 50 microhenries, so we can expect an emf in the millivolt range.

An **inductor** is a device designed expressly to introduce inductance into a circuit, something that is done for a variety of reasons. For instance, an inductor (or *choke*) opposes a changing current and will therefore impede the progress of alternating currents while passing, with very little opposition, steady currents. Not surprisingly, this impedance (p. 808) to ac increases with both the frequency of the current and the inductance of the inductor. Accordingly, inductors are often used to separate or filter out ac from dc. Typically, an inductor consists of a coil of wire wound on a hollow cylinder that may contain air or some kind of ferromagnetic core. Values usually range from microhenries (μH), used at high frequencies as in radio and television, to several henries, used, for instance, in low-frequency (60 Hz) choke-filtered power supplies. The circuit symbols for an inductor with and without a ferromagnetic core are [inductor symbol with core] and [inductor symbol], respectively.

Wire-wound resistors are also coils, but in that case having an appreciable inductance is generally not desirable. The solution in practice is to double the wire over on itself before winding the coil. Current then circulates in opposite directions simultaneously, and the flux in these *noninductive* coils is zero.

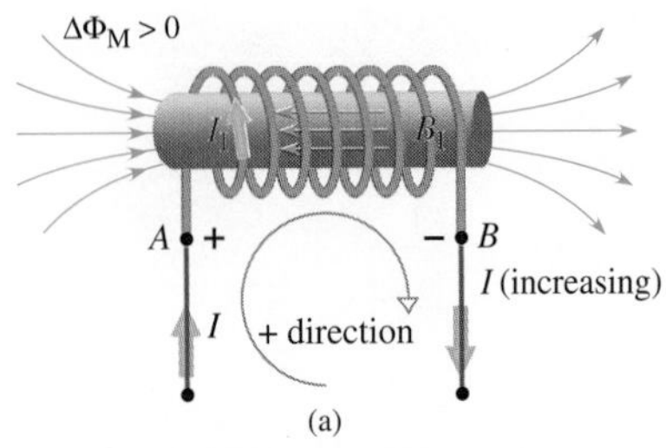

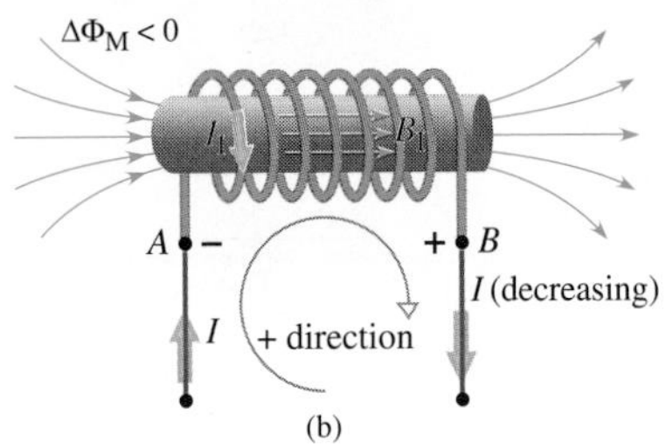

Figure 22.19 (a) I is increasing, an induced current I_I opposes that increase, and a negative back-emf (a voltage drop) appears across the terminals. (b) I is decreasing, I_I opposes that decrease, and a positive back-emf (a voltage rise) appears across the terminals.

22.10 The *R-L* Circuit: Transients

Figure 22.20 shows a battery in series with an inductor L and a resistor R. The inductor is imagined to have zero resistance, although we could have just as well lumped its more realistic nonzero resistance into R. As soon as the switch is closed, an increasing current I just begins to circulate, but it's opposed by the back-emf, so it builds only gradually. From Kirchhoff's Loop Rule (p. 702), at any instant

$$V = L\frac{\Delta I}{\Delta t} + IR \tag{22.11}$$

Remember that V, R, and L are fixed but I varies from moment to moment. Inasmuch as the battery voltage V is constant, it follows from the equation that as the current increases, it must increase ever more slowly. This occurs because as IR increases, the back-emf decreases, and $\Delta I/\Delta t$ decreases. That's shown in the exponential I-versus-t curves of Fig. 22.21a. The slopes do indeed decrease with time. The inductor causes the current to rise slowly, reaching a maximum value of V/R as $\Delta I/\Delta t \to 0$, as $t \to \infty$.

Since $$L\frac{\Delta I}{\Delta t} = \text{emf} = V - IR$$

at the instant the switch is closed ($t = 0$), the current begins to rise from its value of zero and, hence, the emf initially equals V; that is, $L(\Delta I/\Delta t) = V$. Thus, the initial slope of the curve at $t = 0$ is

$$\frac{\Delta I}{\Delta t} = \frac{V}{L}$$

Had the current continued to rise at its initial rate of increase (that is, with the slope it had immediately after the switch was closed—namely, V/L) it would have reached its maximum value (V/R) at a time L/R (such that V/R divided by L/R equals V/L), which is the **time constant**. After an interval of 1 time constant, the current reaches an amount of 0.632 of its final value (V/R). After an interval 5 time constants long, the current is within 1% of its steady-state value.

Notice that when R is very large, most of the voltage drop will be across the resistor, the back-emf will be small, and the delay it causes in the current buildup will be small. On the other hand, the larger the inductance, the greater the back-emf, the less tilted the I-t curve is initially, the larger the time constant, and the longer it takes for the current to reach maximum [which brings to mind the R-C circuit (p. 700), where the time constant is RC].

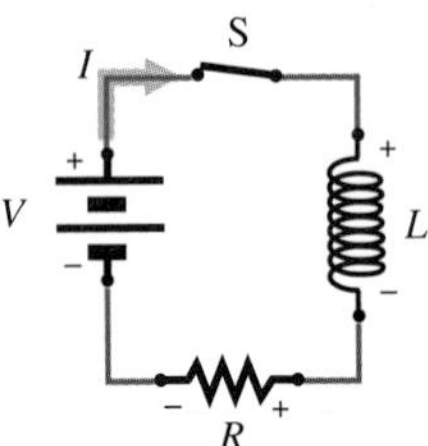

Figure 22.20 An R-L circuit with the switch just closed.

Example 22.7 Determine the time constants for a series R-L circuit consisting in one case of a 100-Ω resistor and a 10-H inductor, and in the other case of the same resistor with a 1.0-H inductor. Assuming the circuit of Fig. 22.20, compare the two resulting currents.

Solution: [Given: $L = 10$ H and 1.0 H, and $R = 100\ \Omega$. Find: the time constants.] The time constant is L/R; hence

$$\frac{L}{R} = \frac{10\text{ H}}{100\ \Omega} = \boxed{0.10\text{ s}}$$

as compared to

$$\frac{L}{R} = \frac{1.0\text{ H}}{100\ \Omega} = \boxed{10\text{ ms}}$$

With the larger inductor, we will have to wait 0.10 s for the current to reach 63% of its maximum value, as compared to only 10 ms for the smaller one.

▶ **Quick Check:** The inductances differ by a factor of 10, so the time constants L/R must also: (0.10 s)/(10 ms) = 10.

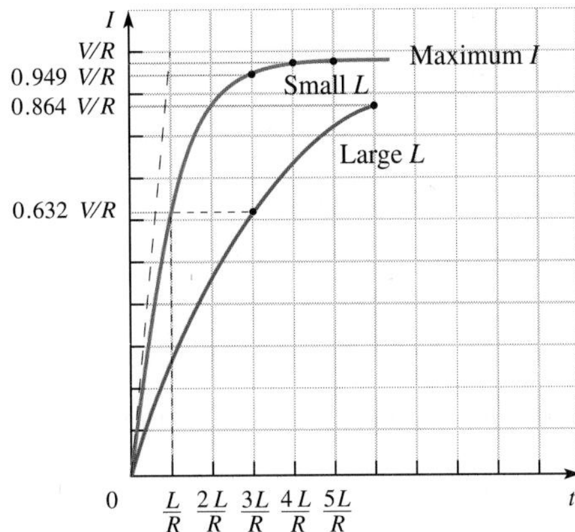

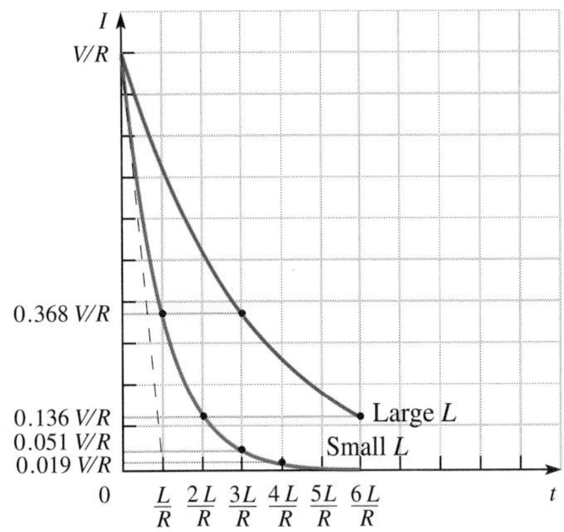

Figure 22.21 (a) The graph shows I versus t for an R-L circuit with a time constant of L/R. For comparison, another curve is included where the time constant is 3 times as large. Note that the curves rise to 63.2% of maximum value after L/R, and 63.2% of the remaining 36.8% after $2L/R$, and 63.2% of the remaining 13.6% after $3L/R$, and so on. (b) When the switch in Fig. 22.22 is opened and the current decays, the current drops 63.2% of the remaining amount during each successive one-time-constant interval. Thus the current falls to 36.8% V/R after L/R, and then 63.2% of 36.8% V/R to 13.6% V/R at $2L/R$, then 63.2% of 13.6% V/R to 5.1% V/R at $3L/R$, and so on.

Opening the switch once the steady state has been established will cause the current to gradually decay, being sustained by the back-emf (Fig. 22.21b). Have you ever pulled the plug on an appliance that was still turned on and seen a spark at the ends of the prongs? If a lamp with a large resistance, compared to that of the coil, is positioned as in Fig. 22.22 and the switch is closed, it will only glow faintly in the steady state: most of the current will then be passing through the inductor since the back-emf is zero and the resistance low. On opening the switch, the lamp will not immediately go out, but instead will burn even more brightly for a moment and then gradually diminish in intensity. The back-emf at the instant S is opened can be greater than the steady-state voltage across the lamp— that condition will not last long, but it is what makes the lamp flare up initially. The larger the L, the larger the time constant, and the longer the lamp glows on.

One might well ask at this point, "Where is the energy coming from that powers the lamp when the switch is opened?" Clearly, it's not from the battery and that only leaves the inductor itself. Classical theory maintains that the B-field must possess energy that it imparts to the coil in the form of an induced current when the steady-state current, which sustained the field, is disrupted. The model suggests that when the switch is opened the field returns its energy to the coil. The realization that the field carries energy (and momentum), which are properties of matter, raises a number of interesting questions that will have to be dealt with later. For example, how exactly is energy transferred from the field to a charged particle?

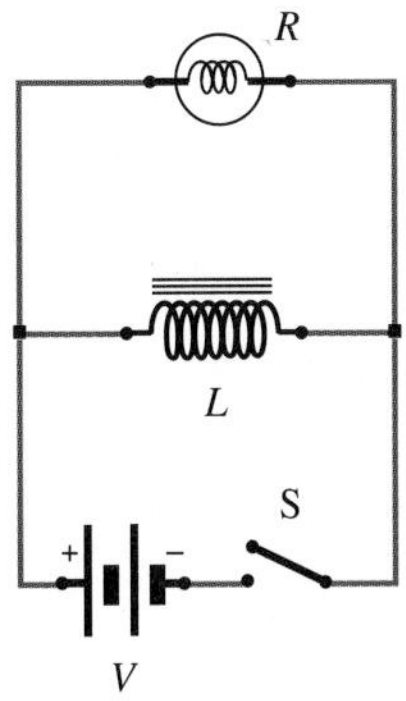

Figure 22.22 An inductor, a battery, and a lamp. The resistance of the lamp (R) is much greater than that of the coil, and the lamp gets little current when S is closed. Remarkably, it glows brightly for a moment after the switch is opened.

22.11 Energy in the Magnetic Field

In Chapter 18, we talked about the energy stored in a charged capacitor in the process of building up the electric field that spans the gap between its plates:

$$\text{PE}_\text{E} = \tfrac{1}{2}CV^2 = \tfrac{1}{2}\varepsilon_0(Ad)E^2 \qquad [18.17]$$

The field in the gap can be imagined as retaining this energy uniformly within the space of volume Ad, which it occupies. Thus, if we introduce the concept of *energy per unit volume of the electric field* (u_E), the above expression becomes

$$u_\text{E} = \tfrac{1}{2}\varepsilon_0 E^2 \qquad (22.12)$$

This is the **energy density** of the electric field; wherever there is an E-field in space, this will be the energy per unit volume associated with it.

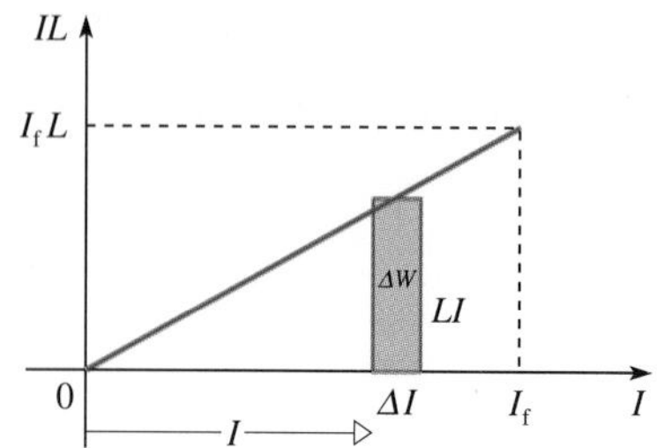

Figure 22.23 A plot of IL versus I. The area under the curve is the work done in building up a current of I_f through an inductor L.

The analogous situation exists when we establish a current through an inductor, thereby building up a magnetic field. In Fig. 22.20, the battery must do work W against the back-emf if it is to send a current through the inductor. During a tiny interval of time Δt, the battery, which provides power to the inductor at a rate P, does a small amount of work

$$\Delta W = \text{P}\Delta t = I\mathscr{E}\,\Delta t = IL\frac{\Delta I}{\Delta t}\Delta t$$

and so

$$\Delta W = IL\,\Delta I$$

where I is the current at any instant. We want the total amount of work done in increasing the current from $I = 0$ to $I = I_f$:

$$\Sigma\,\Delta W = \Sigma\,IL\,\Delta I$$

The right side is the area under a plot of IL versus I, as shown in Fig. 22.23. Since L is constant, this plot is a straight line and the total area (W) under it between 0 and I_f is the triangular area of height $I_f L$ and base I_f; namely, $W = \frac{1}{2}LI_f^2$. This is the work done on the inductor and it's also the magnetic potential energy PE_M stored. Remembering that we are considering the final current, the subscript f can be dropped for simplicity. The energy stored in an inductor carrying a current I is then

$$\text{PE}_\text{M} = \tfrac{1}{2}LI^2 \tag{22.13}$$

where we *assume* it's associated with the B-field. Thus, with a long solenoid of length l, where from Eq. (22.9) $L = NBA/I$, the field existing inside is $B \approx \mu_0 NI/l$, and so

$$\text{PE}_\text{M} = \tfrac{1}{2}LI^2 = \tfrac{1}{2}NBAI = \tfrac{1}{2}(Al)\frac{B^2}{\mu_0} \tag{22.14}$$

Representing the *energy per unit volume of the magnetic field* as u_M and realizing the Al is the volume occupied by the B-field, we have

$$u_\text{M} = \tfrac{1}{2}\frac{B^2}{\mu_0} \tag{22.15}$$

Although this formula was derived for a long solenoid, it turns out to be generally valid. Apparently, it is possible to store electromagnetic energy in electric and magnetic fields in vacuum. As we shall see in Chapter 24, it is even possible to transport this energy across free space—indeed, that's what light is.

So far, we have dealt with a variety of subtle electromagnetic phenomena in terms of classical field theory. It allows us to understand everything we have looked at in terms of electric and magnetic fields using rather picturesque nineteenth-century images of flux and field lines. Indeed, these fields seem to become more substantial with each extension of the analysis. **The classical field appears as a continuous something that can store, transfer, and transport energy** (although it's not obvious exactly how it does that). This picture will have to be profoundly modified when we confront the modern concept that fields are quantized and radiant energy exists only in minute discrete bursts. Usually however, when observed on a macroscopic scale, quantum fields average out to, and become indistinguishable from, classical fields. Though the classical electromagnetic field is an indispensable conceptual tool, as we'll see, the model can no longer be taken quite so literally.

Core Material

The **flux of the magnetic field** is

$$\Phi_M = B_\perp A = BA_\perp = BA \cos\theta \qquad [22.1]$$

with units of **webers**: 1 Wb = 1 T·m^2. As a result of flux changes, the average induced emf is

$$\mathscr{E} = -N\frac{\Delta\Phi_M}{\Delta t} \qquad [22.3]$$

which is **Faraday's Induction Law** (p. 767). **Lenz's Law** is

The induced emf will produce a current that always acts to oppose the change that originally caused it.

When a wire of length l moves at a speed v perpendicularly through a B-field, the induced **motional emf** (p. 770) is

$$\mathscr{E} = vBl \qquad [22.4]$$

An ac generator revolving at a rate ω produces an emf given by

$$\mathscr{E} = NAB\omega \sin \omega t \qquad [22.7]$$

The net flux through a current-carrying coil—the flux linkage $N\Phi_M$—is proportional to the current I producing it:

$$N\Phi_M = LI \qquad [22.8]$$

The constant of proportionality, L, is the **self-inductance**, or just the *inductance*. The self-inductance of a long air-core solenoid is

$$L \approx \frac{\mu_0 N^2 A}{l} \qquad [22.9]$$

The induced, or **back-emf**, in a coil is

$$\mathscr{E} = -L\frac{\Delta I}{\Delta t} \qquad [22.10]$$

When such a coil is placed in series with a battery of voltage V and a resistor R, we have an R-C circuit where

$$V = L\frac{\Delta I}{\Delta t} + IR \qquad [22.11]$$

The *energy per unit volume of the magnetic field* is

$$u_M = \frac{1}{2}\frac{B^2}{\mu_0} \qquad [22.15]$$

Suggestions on Problem Solving

1. When determining the polarity of an emf induced in a coil, it's useful to imagine a resistor placed across the terminals, forming an external circuit. The current enters the resistor on the (+) high side and exits on the (−) low side, as in Figs. 22.10b, 22.13c, and 22.19. The induced current *inside* the coil travels from (−) to (+) just as it does in a battery, which often seems to be a sticky point for students. The induced emf is the work done, in accord with Lenz's Law, on the charge, per unit charge.

2. Be aware that $\Delta\mathbf{B}$ may point in a different direction from the $\mathbf{B}$-field. Thus, if the field is directed to the east and decreasing, $\Delta\mathbf{B}$ points west. If there is an induced current, its induced field will oppose this change and so must point east.

3. Imagine an electrostatic E-field and suppose that we carry a charge from point 1 to point 2. Work will be done and an emf will exist between 1 and 2, an emf that is independent of the path taken. The electrostatic E-field is conservative, but the E-field arising from a changing magnetic flux is *not* conservative. The emf does depend on the path taken between 1 and 2 since different amounts of work may well be needed. Given a time-varying B-field in some localized region of space, it's often possible to go around a closed loop so the path either encompasses the varying flux totally or in part, or even avoids it, thereby leading to different path-dependent values of the emf. The work done in moving a charge around a closed path in an electrostatic E-field is zero—you come back to the same energy you started with. By contrast, in the case of an induced E-field, if we move a positive charge around a loop, work may be done continuously. *The concept of potential is ambiguous with induced E-fields* (see Discussion Question 15).

4. The sign in Eq. (22.10) for the back-emf is there to remind us that the emf opposes that which causes it. If the current is building at a rate of, say, 2.0 A/s, then $\Delta I/\Delta t$ is positive and the emf is negative (meaning it opposes the buildup). However, without a clear picture of the system, the minus sign by itself doesn't provide a complete picture. In such cases, it's often easier to put aside the sign, remembering that the emf opposes its cause. For example, given that a certain back-emf is measured to be 10 V while the current buildup is 2.0 A/s, find the inductance. Since this situation represents an increase of current, the emf must be put into Eq. (22.10) as −10 V or you will get a negative L, which is impossible. Had the current been decreasing, it would be entered as −2.0 A/s and the emf would then be positive, again yielding a positive L. It's common to just give the numerical value of the emf in a problem (without talking about which terminal is higher or lower), so you must be careful not to come up with a negative L.

Discussion Questions

1. In 1825, long before the successful work by Henry and Faraday on induction, Jean Daniel Colladon attempted to observe the effect by attaching a helical coil to a sensitive galvanometer. To shield the delicate instrument from the direct

influence of the moving magnet, which he planned to wave around near the coil, Colladon wired the galvanometer to the rest of the circuit via two long leads so it could be safely located in an adjacent room. Because he had no assistant, whenever he moved the magnet he had to walk over to the galvanometer to observe its response. Since the name of Colladon is never included with that of Faraday and Henry, what do you think he saw and what did he not see? Explain.

2. During a major period of solar activity, magnetic storms of such severity occur that voltages of upwards of 1 kV can appear across the length of the Alaska pipeline from Prudhoe Bay to Valdez. Explain how this happens. (In 1956, a particularly strong magnetic storm severely affected the first transatlantic voice cable. Such effects were a troublesome problem for early telegraph operators.)

3. Using rechargeable batteries, it is possible to power small appliances such as electric toothbrushes with watertight sealed units. Rather than having the customary two exposed terminals that plug directly into a dc power supply, these devices are simply positioned, handle down, in a well within a holder that is continuously attached to ac. During the long periods of nonuse, both the holder well and the handle become quite warm. As the toothbrush is slightly lifted from the base, you can feel a vibration (at what seems to be 60 Hz). How might such a system work? Does any "electricity" actually pass from the base to the batteries? A similar arrangement could be used to power an artificial heart by passing electromagnetic energy into the chest without the need for wires through the skin. Discuss how this might be done.

4. Imagine a superconductor in its normal state in the shape of a ring immersed in a magnetic field parallel to the central axis (that is, perpendicular to the plane of the ring). Suppose the ring is cooled and becomes superconducting. (a) Describe what happens to the field. (b) If the ring is pulled perpendicularly out of the field, what will happen to the flux in the hole in the "doughnut"? (c) Account for the energy associated with the work done on the ring (if any) in yanking it from the field region.

5. Imagine a superconducting ring supported horizontally so that it can be approached from below by a bar magnet. Describe and explain what will happen as the magnet (north pole upward) is brought near the ring. Since a superconductor has zero resistance, an induced E-field would result in an infinite current. How, then, must a superconductor behave in the presence of a changing B-field in order that this impossibility not occur?

6. A promising future source of energy is the controlled fusion reaction, the same phenomenon that powers the stars. Deuterium and tritium atoms are ionized. The resulting plasma of positive nuclei and electrons is confined to a ring-shaped region within a vacuum chamber (Fig. Q6) by magnetic fields. The plasma is raised up to temperatures in excess of 100 million K, whereupon the nuclei undergo fusion and liberate great amounts of energy. The confinement field, which actually spirals around the toroid, is provided in part by external current-carrying coils wrapped around the chamber. A crucial component of this confinement field is generated by toroidal currents circulating in the plasma itself. Explain how such a toroidally directed current (along the axis of the chamber) could be induced in the plasma. Describe the kind of current required and trace the transfer of energy into the plasma.

7. Imagine a hand-cranked generator in parallel with a 50-W light bulb and a 100-W light bulb. Suppose each bulb can be switched out of the circuit. Compare the amount of effort it would take to light each bulb steadily with the effort needed to crank the generator in a sustained fashion without a load. Explain what causes the difference if there is one.

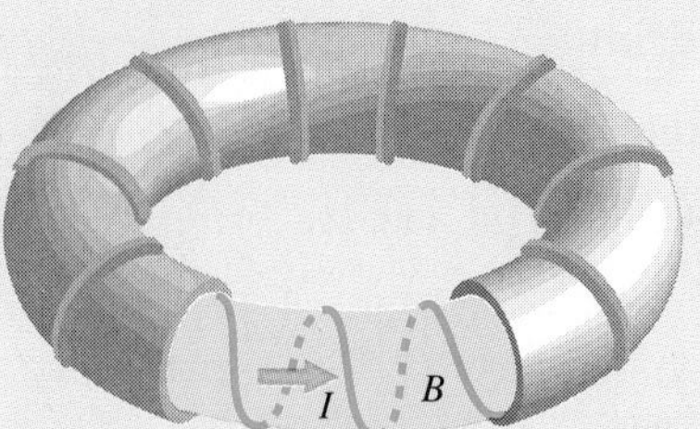

Figure Q6

8. The switch in the circuit shown in Fig. Q8 is closed, and after a long wait, the resistor is set so the two lamps are equally bright, at which point the switch is again opened. After another long interlude, the switch is closed. Describe what subsequently happens to the two lamps.

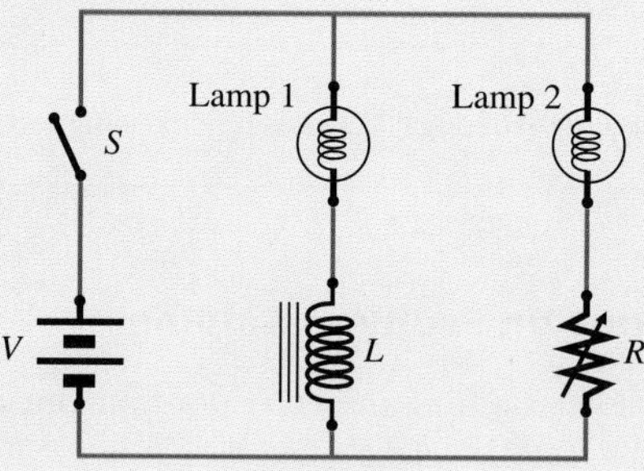

Figure Q8

9. Figure Q9 shows a 70-kV, 60-Hz power line in a remote area of the countryside. A shifty local resident has erected a large open loop just below the line with the intention of drawing off power. Is this possible and, if so, how would it be transferred? Where would the energy stolen come from? Would anyone be able to detect the loss? Might it be possible to bug a telephone using the same approach? Explain.

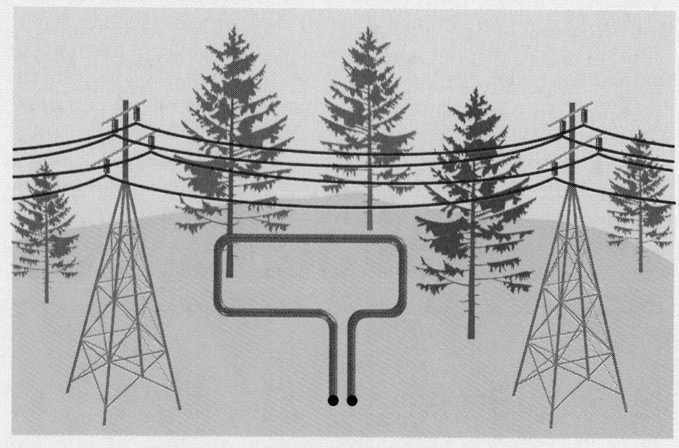

Figure Q9

10. Consider several inductors (L_1, L_2, and L_3) connected alternatively in series and then in parallel. What do you think will be the equivalent inductance in each case? Explain your reasoning.

11. A circuit contains an air-core coil of inductance L and resistance R in series with a power supply. Discuss and compare the amounts of energy stored with and without an iron core in place. Compare the final currents established. Where does the difference in energy come from?

12. Figure Q12 illustrates the construction of a so-called variable reluctance microphone. A diaphragm is attached to a light flexible rod made of ferromagnetic material that, in turn, is fixed via a permanent magnet to a C-shaped structure also made of magnetic material. How does it work? Comment on the way the two coils are wound.

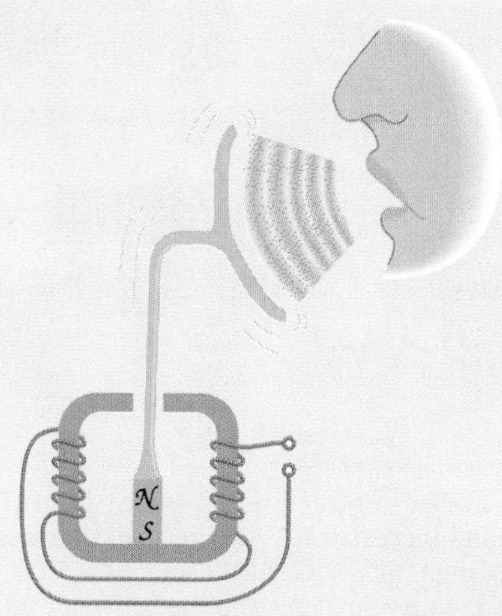

Figure Q12

13. Motors and generators are the same device, the distinction appearing in the operation rather than in the construction—mechanical power in and electrical out and you have a generator; electrical in and mechanical out, and you have a motor. It's reasonable, therefore, to ask, "Is a motor in some way a generator while it's turning?" Explain. If there is a generated back-current, how might it depend on the load? What do you think limits the speed of a free-turning motor? Describe the operation of the motor with a load attached. Why do many motors have open slots on their sides and little internal fan blades attached to their shafts? If you jam a motor—bind a drill or a blender so it's receiving current but not turning—it won't be long before you smell burning insulation. What's happening, and why?

14. Figure Q14 depicts a strip of recorded magnetic tape (which is like a succession of little magnets) passing under a playback head. The latter is a ferromagnetic C-shaped structure with a coil wrapped around it. How does the playback head read the tape? Comment on the relationship between the fineness of the read and write heads, the speed of the tape, and the density of information.

15. The circuits in Fig. Q15 are adapted from an article by R. H. Romer entitled "What do 'voltmeters' measure?: Faraday's law in a multiply connected region" (*Am. J. Phys.*, **50** no. 12, Dec. 1982, 1089). Part (a) illustrates the nonconservative nature of the induced E-field, in this instance surrounding a long sole-

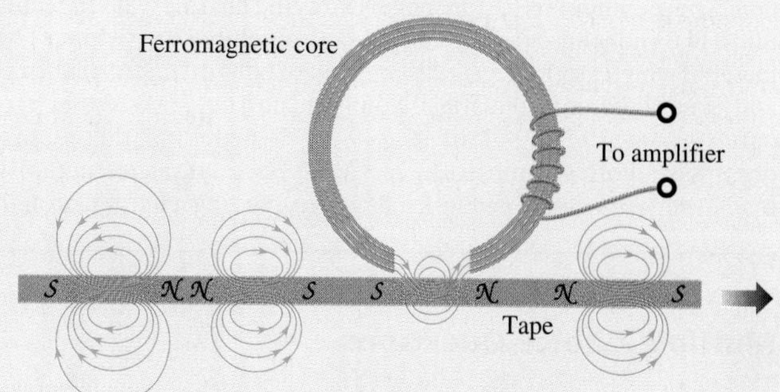

Figure Q14

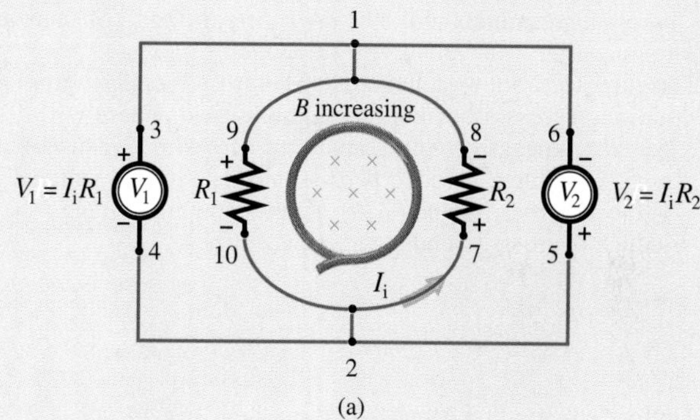

(a)

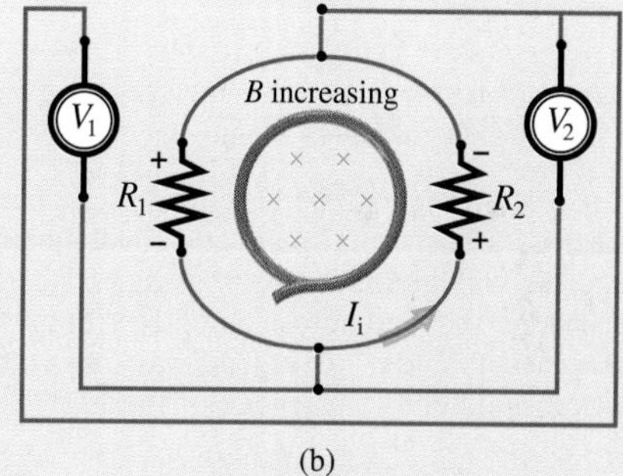

(b)

Figure Q15

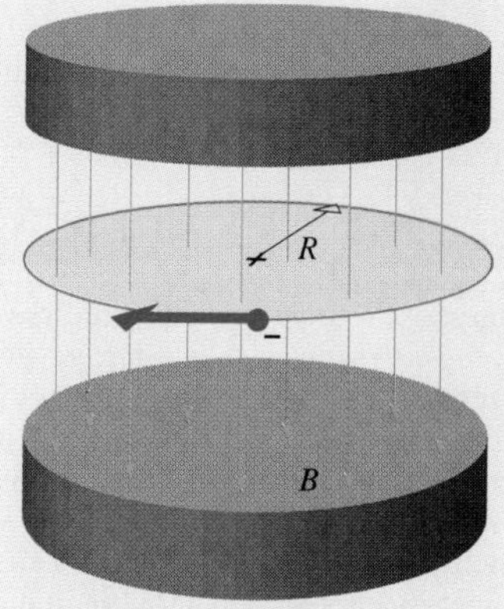

Figure Q16

noid (perpendicular to the page) carrying a time-varying current. The induced current, I_i, passes through both resistors. The meters (which draw negligible current) show different readings and even different polarities. Note that no flux links either circuit 1–3–4–2–10–9–1 or 1–8–7–2–5–6–1, and these must obey Kirchhoff's Loop Rule (p. 702). Each voltmeter actually reads the work done per unit charge in moving charge through the meter itself. Discuss what's happening. What does the meter on the left in part (b) read and why?

16. Figure Q16 (p. 787) shows an electron orbiting between the poles of an electromagnet in a device called a betatron. The field is gradually being increased. How does the machine accelerate the electron? What keeps it in orbit?

Multiple Choice Questions

1. A single horizontal loop of wire is moving in a horizontal plane at a constant speed *across* a uniform vertical magnetic field that "fills" and surrounds the loop. The emf induced across its terminals will be (a) time varying (b) constant (c) negative (d) positive (e) none of these.
2. Figure MC2 shows a horizontal length of copper wire moving across a horizontal magnetic field. There will be (a) a negative voltage induced across its ends (b) a positive voltage induced across its ends (c) no voltage induced across its ends (d) a time-varying voltage induced across its ends (e) none of these.

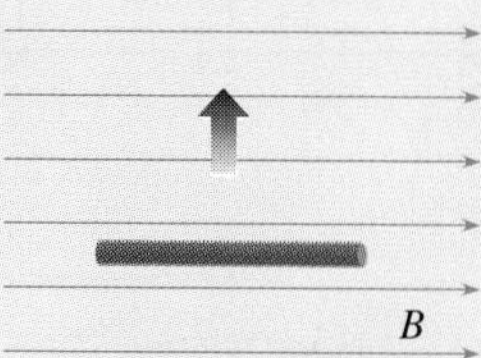

Figure MC2

3. Figure MC3 shows a horizontal length of copper wire moving across a horizontal magnetic field. There will be (a) a negative voltage induced across its ends (b) a positive voltage induced across its ends (c) no voltage induced across its ends (d) a time-varying voltage induced across its ends (e) none of these.

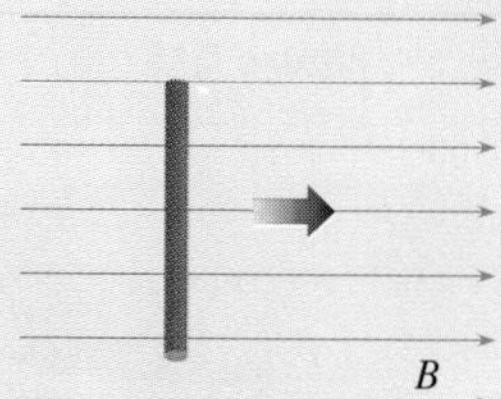

Figure MC3

4. A closed loop moves at a constant speed parallel to a long straight current-carrying wire, as in Fig. MC4. (a) the induced current in the loop will progress clockwise (b) there will be no induced current in the loop (c) the induced current in the loop will progress counterclockwise (d) the induced current in the loop will vary with the speed at which the loop moves (e) none of these.
5. The bar magnet in Fig. MC5 is moving at a constant speed toward the coil. The voltage measured across points A and B is (a) higher at B than A and increasing (b) higher at A than B and increasing (c) zero (d) higher at A than B and decreasing (e) none of these.

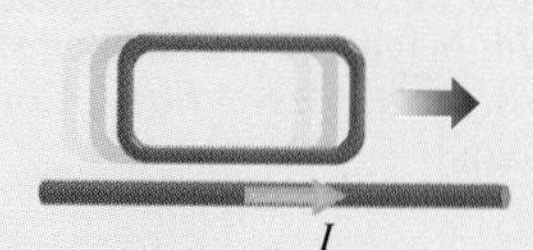

Figure MC4

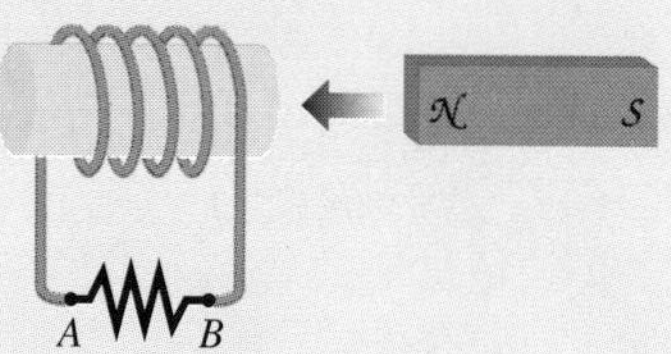

Figure MC5

6. The more rapidly a magnet approaches a coil (as in Fig. MC5), the (a) lower the current in the coil (b) greater the resistance of the coil (c) greater the induced voltage across the coil (d) more it is attracted (e) none of these.
7. The wire loop of area 0.050 m^2 in Fig. MC7 is in a uniform downward B-field, which is increasing at 0.010 mT/s. The induced emf is such that the potential of (a) C is higher than A by 0.50 μV (b) A is higher than C by 0.50 μV (c) A is the same as that of C (d) C is higher than A by 0.50 mV (e) none of these.

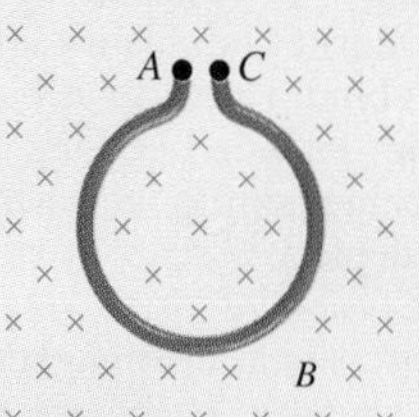

Figure MC7

8. The two coils in Fig. MC8 are wrapped on an iron bar. When the switch is closed (a) a current momentarily passes through R from right to left (b) a constant current circulates through R from right to left (c) a current momentarily passes through R from left to right (d) a

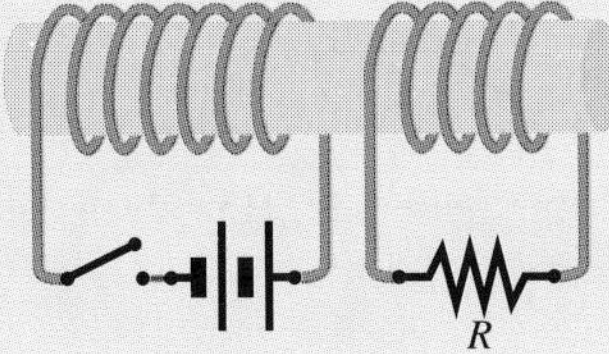

Figure MC8

constant current circulates through R from left to right (e) none of these.

9. The coil in Fig. MC9 has a core made up of a stack of insulated iron wires. The reason for using such a configuration is to (a) generate as much thermal energy as possible (b) be able to make it inexpensively (c) reduce eddy current losses (d) make it strong to resist magnetic bending (e) none of these.

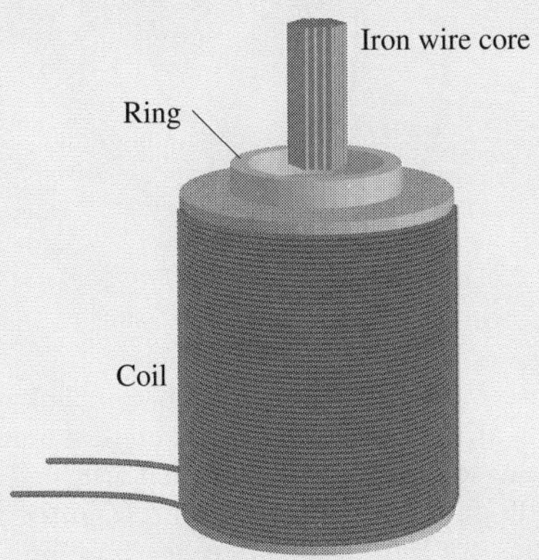

Figure MC9

10. When a loose metal ring is placed on the coil in Fig. MC9 and the latter is suddenly fed a large current, the ring will (a) pop into the air (b) initially vibrate and then stop (c) initially remain stationary but get very hot (d) have nothing happen to it (e) none of these.
11. The solenoid and battery of Fig. MC11 are moving at a constant speed toward the coil on the left. The voltage measured across points 1 and 2 is (a) zero (b) higher at 1 than 2 and increasing (c) higher at 2 than 1 and increasing (d) higher at 1 than 2 and decreasing (e) none of these.

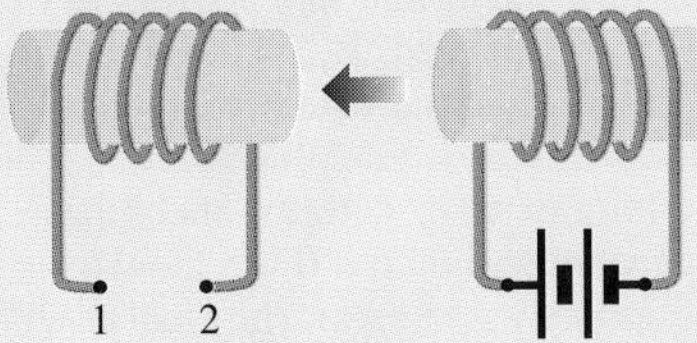

Figure MC11

12. We wish to produce a clockwise (looking down) current in the loop on the right in Fig. MC12. The variable resistor should be (a) left as is (b) decreased (c) increased (d) reversed (e) none of these.

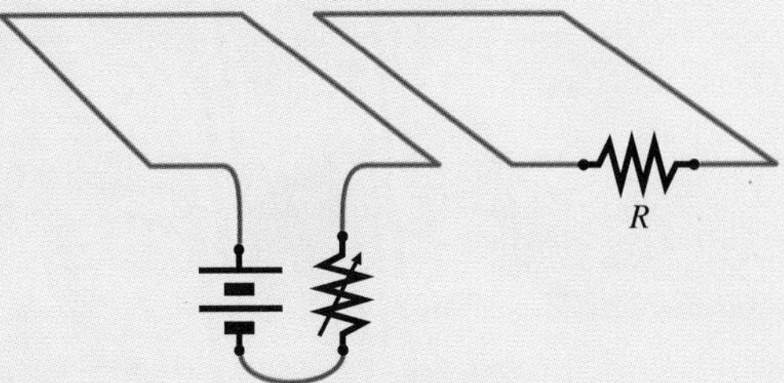

Figure MC12

13. The copper ring in Fig. MC13 is in a uniformly increasing magnetic field. The induced electric field within it is (a) clockwise and constant (b) counterclockwise and constant (c) clockwise and changing (d) counterclockwise and changing (e) none of these.

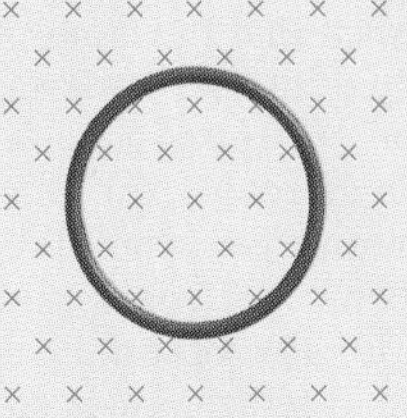

Figure MC13

14. Figure MC14 shows an aluminum ring and the current induced in it by the nearby magnet that is free to move along its central axis. (a) the magnet must be stationary (b) the magnet must be moving to the right (c) the magnet must be moving to the left (d) not enough information to say anything about the magnet (e) none of these.

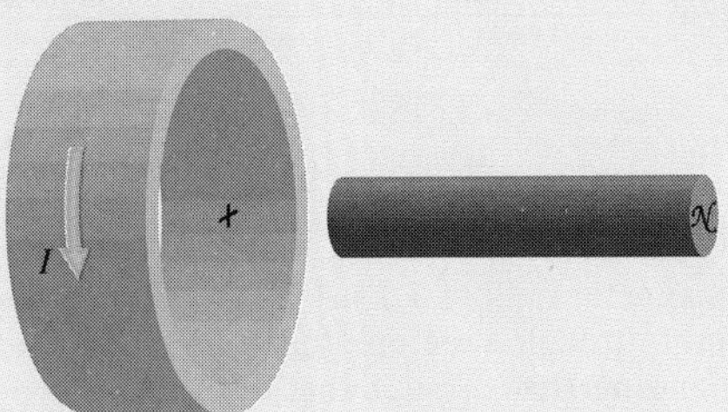

Figure MC14

15. A solenoid is physically altered by doubling the number of turns it has while halving the current through it, leaving everything else unchanged. (a) its self-inductance stays the same (b) its self-inductance doubles (c) its self-inductance is halved (d) its self-inductance is four times greater (e) none of these.
16. The coil in Fig. MC16 is rotating at a constant rate about an axis perpendicular to the field. The induced voltage across its terminals will (a) always be zero when $\theta = 0$ (b) always be zero when $\theta = 90°$ (c) sometimes be zero when $\theta = 90°$ (d) never be zero (e) none of these.

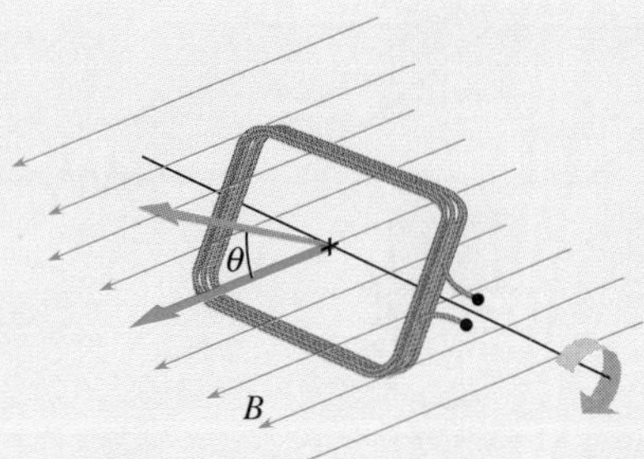

Figure MC16

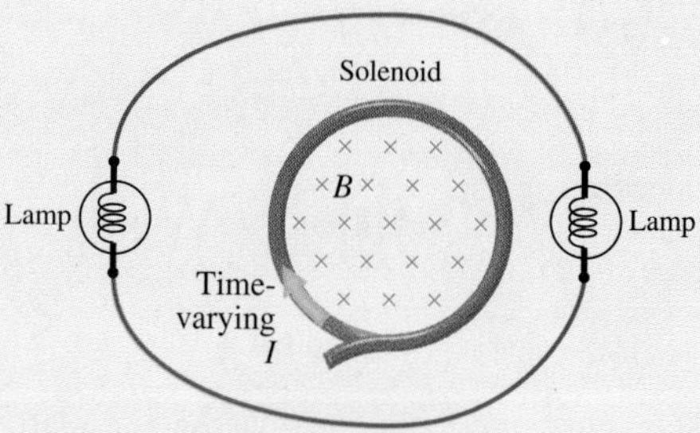

Figure MC20

17. A small light bulb is in series with an air-core coil and a dc power supply such that the lamp glows brightly. An iron core is then inserted into the coil, and an hour later the lamp (a) is brighter (b) goes out completely (c) grows dimmer (d) is unaffected (e) none of these.
18. A small light bulb is in series with an air-core coil and an ac power supply such that the lamp glows brightly. An iron core is then inserted into the coil, and the lamp (a) grows brighter (b) goes out completely (c) grows dimmer (d) is unaffected (e) none of these.
19. An electric locomotive going uphill draws power from the feeder lines. In its most efficient mode of operation, when it goes downhill, it (a) must draw even more power (b) can generate and return power to the lines (c) will neither draw nor produce power (d) draws the same amount of power (e) none of these.
20. Figure MC20 shows an end-view looking down onto a long narrow solenoid carrying an increasing clockwise current. Two identical light bulbs are connected in a circuit that encircles the solenoid. (a) the bulb on the right glows, the one on the left does not (b) the bulb on the left glows, the one on the right does not (c) both bulbs glow equally (d) neither bulb glows (e) none of these.
21. Figure MC21 shows an end-view looking down onto a long narrow solenoid carrying an increasing clockwise current. Two identical light bulbs are connected in a circuit that encircles the solenoid. A wire is then attached as shown. (a) the bulb on the right glows more brightly, the one on the left does not light at all (b) the bulb on the left glows more brightly, the one on the right does not light at all (c) both bulbs glow equally (d) neither bulb glows (e) none of these.

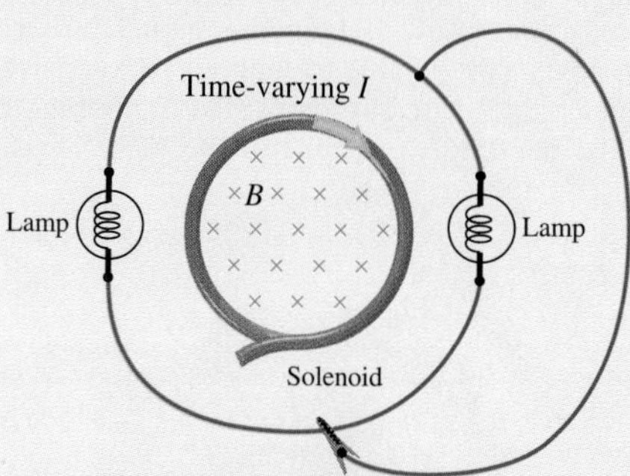

Figure MC21

Problems

ELECTROMAGNETICALLY INDUCED EMF

1. [I] A magnetic field of 1.2 mT passes perpendicularly and uniformly through a region of area 25 cm^2. What is the corresponding flux through that region?

2. [I] A uniform magnetic field of 100 mT passes through a loop of wire having an area of 0.020 m^2. The field makes an angle of 30° with the perpendicular to the plane of the loop. What is the magnetic flux through the loop?

3. [I] A flux of 6.0 mWb passes uniformly along an iron bar of cross-sectional area 50 cm^2. Compute the flux density in the bar.

4. [I] A coil wrapped around a portion of an iron ring generates a flux that passes through another coil also wrapped around the ring. This secondary has 100 turns and is penetrated by a flux of 0.016 Wb. Given that the ring has a cross-sectional area of 8.0 cm^2, what is the magnitude of the magnetic field in the toroid?

5. [I] A changing magnetic field has an initial value $\mathbf{B}_i$, which points due south and has a strength of 0.5 T. If the final value of the field, $\mathbf{B}_f$, is 0.6 T pointing due south, what is the magnitude and direction of the change in the field, $\Delta\mathbf{B}$?

6. [I] A changing magnetic field has an initial value $\mathbf{B}_i$, which points downward and has a strength of 0.6 T. If the final value of the field, $\mathbf{B}_f$, is 0.5 T pointing upward, what is the magnitude and direction of the change in the field, $\Delta\mathbf{B}$?

7. [I] A single loop of wire with an encompassed area of 0.25 m^2 is perpendicular to a 0.40-T uniform magnetic field. The loop is yanked from the field in 200 ms; what is the induced emf?

8. [I] A 100-turn coil of wire is removed laterally from an axial B-field in 20 ms, and an emf of 1.0 V is induced across it in the process. How big was the field if the cross-sectional area of the coil is 4.0 cm^2?

9. [I] A magnet is moved close to a coil of 150 turns in which it introduces a flux change of 30 mWb. If a voltage of 60 V appears across its terminals and if the resistance of the coil is negligible, how much time elapsed during the flux change?

10. [I] A coil of 100 turns is penetrated by a flux of 0.050 Wb in 0.020 s. Determine the numerical value of the average induced emf.
11. [I] The flux passing through a coil of 200 turns changes uniformly from a constant 0.010 Wb to 0.070 Wb in a time of 1.5 s. Find the numerical value of the emf induced during that interval.
12. [I] A 10-turn coil of wire of area 0.25 m^2 is in a perpendicular magnetic field that changes in time as illustrated in Fig. P12. Determine the induced voltage across the loop as a function of time.

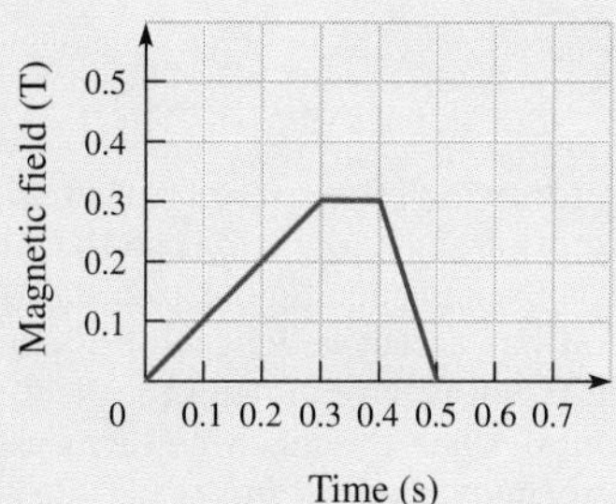

Figure P12

13. [I] A Boeing *747* jumbo jet with a wingspan of 60 m is flying in an area where the Earth's *B*-field is 0.050 mT. Given that the plane traverses the field perpendicularly at a speed of 200 m/s, what is the potential difference induced between its wing tips?

14. [I] A straight 30-cm length of copper wire is moved at a constant speed of 0.50 m/s perpendicularly across a uniform magnetic field of 1.5 T. Determine the emf appearing across its ends.

15. [I] How could you measure the potential difference in Problem 13? If we attached a small light bulb across the wing tips, would it light?

16. [I] An emf of 0.45 V is induced in a straight conductor having a length of 20 cm moving at right angles to a magnetic field at a constant speed of 600 cm/s. Calculate the value of the field.

17. [I] A straight copper rod 100 cm long is initially held horizontally pointing due east and then released. It falls across the Earth's field of 0.5×10^{-4} T, which is pointing north and 50° below the horizontal. What is the induced emf in the rod when it reaches a speed of 2.8 m/s?

18. [II] A specimen is to be exposed to a controllable magnetic field and is therefore positioned inside a 0.55-m-long narrow air-core solenoid of cross-sectional area 2.0×10^{-4} m^2 comprising 10 turns per cm. To monitor the field, a small search coil (connected to a voltmeter) is wrapped around the outside of the solenoid at its middle. When the current in the solenoid is increased from zero to 4.9 A in 5.00 ms, what will be the emf across the search coil given that it consists of 240 turns of fine wire?

19. [II] The rectangular loop in Fig. P19 has a resistance R and is moving left with a constant speed v into a region of uniform magnetic field. The field is a little unrealistically assumed to be constant, dropping immediately to zero beyond its rectangular boundaries. Describe the induced current as a function of time.

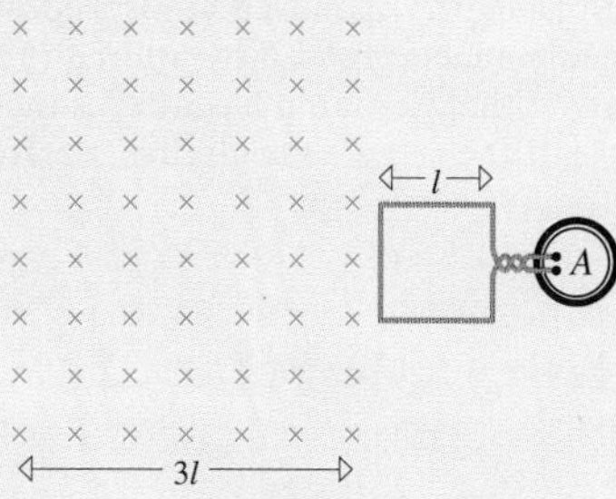

Figure P19

20. [II] A flux of 8.0 mWb generated by a large electromagnet links a coil of 100 turns placed between its poles. Suppose the current in the electromagnet is gradually and continuously reduced to zero and then reversed so that a flux of 8.0 mWb is uniformly reestablished in the opposite direction in a time of 200 ms? What is the emf induced in the coil?

21. [II] Imagine a vertical magnetic field decreasing at a constant rate of 20 mT/s. A flat circular coil of 220 turns with a 20-cm diameter has a 30-μF capacitor across its terminals. If the coil is placed perpendicular to the field, what is the steady-state charge on the capacitor?

22. [II] A flat circular coil of 10 turns and diameter 10 cm (shown in Fig. P22) is located in a uniform magnetic field of 0.20 T that passes through the coil area at 45°. What will be the voltage, on average, across the terminals if the plane of the coil is smoothly rotated (through 45°) so that it's parallel to the field in a time of 0.10 s? Indicate which terminal has the higher potential. What will the voltage be when the coil comes to a stop?

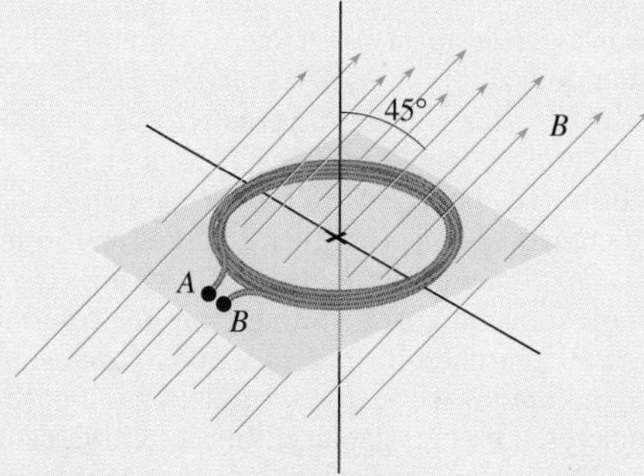

Figure P22

23. [II] A single loop of copper wire in the shape of a square 4.0 cm on each side is lying flat on a horizontal table. A large electromagnet is positioned with its north pole above and to the left a little so that the uniform magnetic field is downward onto the loop, making an angle of 30° with the vertical. Compute the average induced emf across the loop as the field varies linearly from 0 to its final value of 0.500 T in 200 ms. What is the direction of the induced current?

24. [II] A wire of length l is moved at a speed v across a perpendicular magnetic field B by a force F, as shown in Fig. P24. Given that the lamp has a resistance R while the resistance of all the wires is negligible, derive an expression for F.

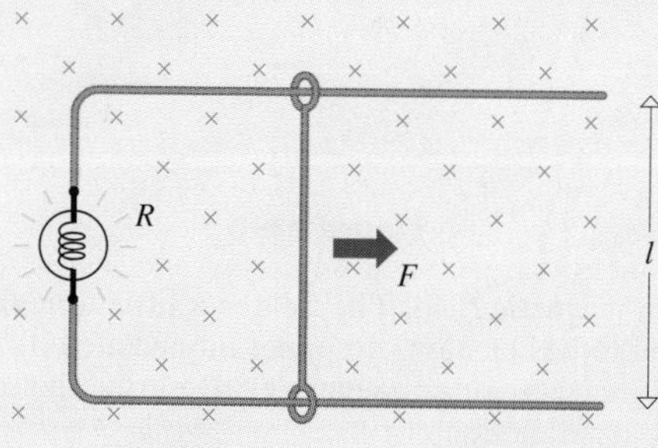

Figure P24

25. [II] With Problem 24 in mind, what is the power supplied to the moving wire by the external agency? Show that this power is equal to I^2R dissipated in the lamp.
26. [II] Imagine a flat circular coil of N turns and resistance R connected to a ballistic galvanometer (that is, one with negligible damping). The flux through the coil (Φ_M) is changed and an induced charge ΔQ flows through the galvanometer, whose pointer is initially displaced by an amount $\theta \propto \Delta Q$, known as the *throw*. Show that

$$\Delta Q = \frac{N\Delta\Phi_M}{R}$$

27. [II] With Problem 26 in mind, suppose a small search coil is attached to a ballistic galvanometer for which $\Delta Q = K\theta$, where K is a measured constant and θ is the initial deflection. The coil is first placed perpendicularly in a uniform B-field that is to be measured. It is then rapidly yanked out, causing the galvanometer to swing. Write an expression for B in terms of θ, R, K, N, and A, the area of the search coil.
28. [II] A flat circular coil of 20 turns with an area of 5.0×10^{-2} m^2 is located in a uniform magnetic field of 10 mT. Initially, the magnetic flux passes through the coil perpendicularly. The coil is then rotated, in 150 ms, so that its central axis makes an angle of 50° with **B**. Determine the average induced emf resulting from the rotation.
29. [II] A small search coil (of area 0.50 cm^2, resistance 0.50 Ω, and having 12 turns) is attached to a ballistic galvanometer. The coil is placed inside, at the center of, and coaxial with, a large solenoid. When the current in the solenoid is rapidly reversed, the galvanometer deflects, indicating a pulse of 2.0 μC of charge. Determine the initial B-field of the solenoid. (Hint: Remember Problems 26 and 27.)
30. [III] Imagine two stationary (zero-resistance) vertical conductors with a horizontal wire crossbar (of length l) capable of moving downward freely while staying in contact (Fig. P30). The crossbar (of mass m) is dropped and it falls perpendicular to the uniform field B. Describe its motion completely, writing an equation for the acceleration and the maximum speed, if there is one.

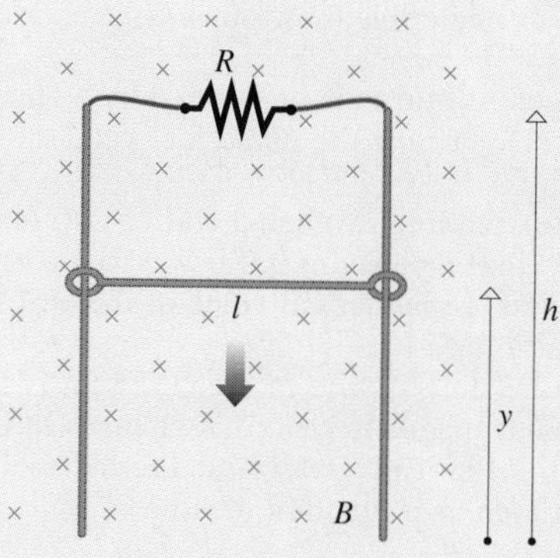

Figure P30

GENERATORS

SELF-INDUCTION

31. [I] A generator, which consists of a flat coil, is rotated about its central axis (Fig. 22.13a) so that the longer portion of it moves perpendicular to a 0.80-T magnetic field. The coil has an active length of 50 cm (that is, there is a total of 50 cm of wire cutting the field normally). What is the emf when the machine turns at a rate such that each conductor moves at 4.0 m/s?
32. [I] We wish to design an ac generator using an armature on which is wound a flat tight rectangular coil (8.00 cm × 20.0 cm) of 150 turns. The generator output is to be sinusoidal with a maximum voltage of 20.0 V and a frequency of 50.0 Hz. How strong should the B-field be and at what angular speed should the coil turn?
33. [I] A 25-turn coil is made to rotate in a uniform 0.35-T magnetic field. Each turn has two 10-cm wire lengths that cut the field at varying angles as the coil rotates. What is the emf when the wire is moving with a speed of 22 m/s at 90° to the field?
34. [I] In the United States, the national power grid supplies ac to ordinary consumers with a maximum voltage of 170 V at 60 Hz (the standard 120-V "average" household service). Given that a generator has a 100-turn coil of area 0.50 m^2, what strength magnetic field should be used?
35. [I] An ac generator produces an emf that (in SI units) has the form $\mathscr{E} = 100 \sin 376.99t$. What is the maximum voltage? What is the frequency of the output?
36. [I] If a current of 2.0 A in a coil produces a flux linkage of 6.0 mWb, what is the self-inductance?
37. [I] A long solenoid of 500 turns carrying a current of 3.8 A produces within itself a uniform magnetic flux of 2.0 mWb. Compute the self-inductance of the coil.
38. [I] An air-core coil having a self-inductance of 3.0 mH carries a current that creates a magnetic field within it of magnitude B. The coil is slid onto a ferromagnetic rod that has a relative permeability of 2000 when in a field of magnitude B. What will be the new inductance with the core in place?
39. [I] An air-core coil with 500 turns has an inductance of 200 mH. If a current of 2.0 A passes through the coil, what will be the flux within it at that moment?
40. [I] Show that the unit of μ_0 is H/m.
41. [I] What is the number of turns of a 2.5-H coil if, when 1.80 A is being carried, the flux linking it is 1.80 mWb?

42. [I] Determine the inductance of a 0.50-m-long solenoid having 200 turns wound on a hollow core 4.0 cm in diameter.
43. [I] A 40-cm-long air-core solenoid with a cross-sectional area of 4.5 cm^2 is to be constructed for a radio circuit so that it has an inductance of about 1.0 mH. How many turns of wire should it have?
44. [I] A coil with an inductance of 10 H has a current within it that is increasing at a rate of 4.0 A/s. What is the back-emf induced across the coil?
45. [I] If an emf of 10 V is induced across a coil when the current within it is changing at a constant rate of 2.0 A/s, what is its inductance?
46. [I] A current in a 5.0-H coil drops from 2.5 A to zero uniformly in a time of 10 ms. What is the induced emf?
47. [I] A current in a coil rises from zero to 4.0 A in 0.020 s. If the back-emf is measured to be 500 V, what is the inductance of the coil?
48. [I] A 10.0-H inductor is in series with a 12-V battery, a 6.0-Ω resistor, and a switch. What is the steady-state current attained after the switch is closed?
49. [I] In Fig. 22.21 the current reaches a value of $0.63V/R$ in 1 time constant. Show that $1 - (1/e) = 0.632$, where $e = 2.7183$.
50. [I] How much energy is stored in a 200-mH inductor when a current of 10 A circulates through it?
51. [I] What is the inductance of a coil if it stores 15 J of energy when the current through it is 5.0 A?
52. [I] What is the total energy density in a region where the Earth's surface magnetic field is 0.50×10^{-4} T and its surface electric field is 100 V/m? Notice the difference in the energies stored in the two fields.
53. [II] We wish to make a dc generator that puts out a constant current. Accordingly, a flat copper disk with a radius of 25 cm is placed in a uniform magnetic field so that its axis of rotation is parallel to B. A 20-Ω resistor is connected between the axis and the rim of the disk. The disk is rotated at a rate of 360 rpm and a dc current of 1.25 mA passes through the resistor. Find B. (Hint: Consider a radial strip of conductor and show that the rate at which it sweeps out area is $\frac{1}{2}r^2\omega$. Also, look at Problem 55.)
54. [II] A simple generator, which consists of a flat rectangular coil (Fig. 22.13) with a turning radius of 10 cm, is rotated about its central axis so that it moves perpendicularly across a 0.60-T magnetic field. The coil has an active length of 30 cm (that is, there is a total of 30 cm of wire cutting the field). What is the emf when the machine turns at 10 rev/s?
55. [II] A helicopter hovering in the air has its 5.0-m rotor blades revolving at 240 rpm in a plane that intersects the local B-field of the Earth (0.050 mT) at an angle (with the normal to the plane) of 40°. Determine the induced emf between the hub and the tip of each rotor blade. (Hint: Look at the hint for Problem 53.)
56. [II] A 25-turn flat coil is made to rotate in a uniform 0.35-T magnetic field. Each turn has two 10-cm wire lengths that cut the field at varying angles as the coil rotates. What is the emf when the wire is moving with a speed of 22 m/s at 30° with respect to the field?
57. [II] A simple ac generator consists of a coil, each turn of which has two 20-cm lengths of wire that cut the uniform 0.50-T B-field as they revolve at a radius of 5.0 cm from the central axis. Write an expression for the emf as a function of time given that the coil has 200 turns and revolves at 6000 rpm.
58. [II] An inductor in the motor circuit of a robot consists of an air-core solenoid of 100 turns with an inductance of 200 mH. If the coil carries a current of 5.0 A, what is the flux linkage? How much energy does the coil store?
59. [II] A 5.0-A current through a 1.5-H coil is reversed, and in the process an average emf of 100 V is induced. Compute the time it took to reverse the current.
60. [II] We wish to make an inductor that will be used in an experiment and must be formed around an iron rod for which $\mu = 1.5 \times 10^3\mu_0$. The rod has a diameter of 2.0 cm and is 27 cm long. Given that the inductance is to be 300 mH, approximately, how many turns of wire will be needed? Why should we stress the word *approximately* here?
61. [II] A long narrow solenoid having a 2.0-cm radius is made of copper wire wrapped with two turns per millimeter on a plastic core. How much energy is stored per unit length of coil when 5.0 A circulates through it?
62. [II] Write an expression for the self-inductance of an air-core toroidal coil of N turns, with a mean radius of b and a cross-sectional area A.
63. [II] An experimental setup is comprised of an air-core solenoid (having an inductance of 300 mH and negligible resistance) connected in series to a sample of bone (of 25 Ω), an ammeter, and a 150-V dc source via a switch. The intent is to provide a gradually building current to the sample. (a) What will be the steady-state current? (b) Determine the time constant of the system. (c) Determine the current after such a time has elapsed.
64. [II] An inductor ($L = 2.00$ H and $R = 4.0$ Ω) is connected in series with an ammeter (of negligible resistance) and a varying voltage source. Figure P64 is a plot of the current in the circuit versus time. Draw a corresponding plot of the voltage across the inductor versus time.

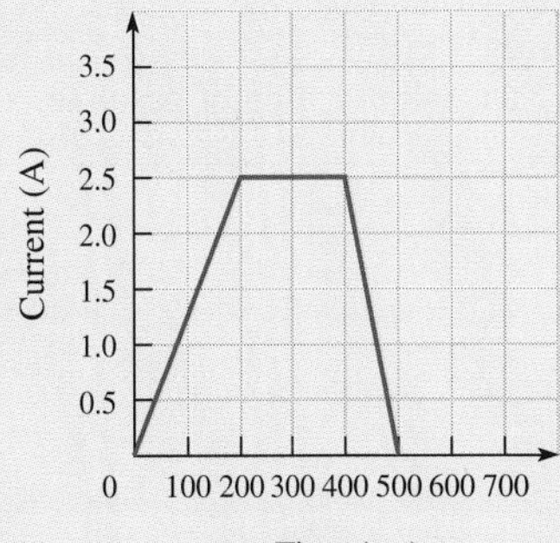

Figure P64

65. [II] A 24-H iron-core inductor of negligible resistance is placed in series with a 12-V battery, a 6.0-Ω resistor, and a switch. Determine: (a) the steady-state current; (b) the time constant; and (c) roughly how long it would take for the current to reach within 1% of its maximum value.
66. [II] A 24-H iron-core inductor of negligible resistance is placed in series with a 12-V battery, a 6.0-Ω resistor,

and a switch. Once the switch is closed, the current in the circuit is

$$I = \frac{V}{R}\left\{1 - \exp\left[\frac{-t}{(L/R)}\right]\right\}$$

Compute the current at a time of 2.0 s after the switch is closed.

67. [II] Inductance transducers, mounted in catheters (narrow tubes inserted into the body), have been used to measure blood pressure. Blood presses against a diaphragm (a few millimeters in diameter), which bends a proportionate distance backward. On the back side of the diaphragm is a small ferromagnetic shaft (of permeability μ) that is thereby displaced, entering into an air-core coil of length l (a distance d). The resulting change in inductance ΔL is measured electrically and calibrated to correspond to the blood pressure. Show that

$$\Delta L \approx \frac{d(\mu - \mu_0)N^2A}{l^2}$$

68. [II] The so-called "solenoid" in your car is mounted on the starter motor. When the key is turned, 12 V are applied to the solenoid, which then carries a current that generates a magnetic field. The ferromagnetic plunger (Fig. P68) is drawn into the coil, and that moves a rod that does two things: it closes a heavy-duty switch sending about 100 A to the starter motor and it engages a gear so that the motor can "turn over" the engine. If the solenoid (which is 200 turns wrapped on a hollow core 5.0 cm in diameter and 9.5 cm long) has a resistance of 1.20 Ω, determine the approximate energy density in the coil, given that the plunger fills the space and has a relative permeability of 500 at that field.

69. [II] Determine the approximate amount of energy stored in the Earth's magnetic field in the first 200 km above the planet's surface. Even though the dipole field drops off as $1/r^3$, this distance is comparatively small, so assume $B = 0.4 \times 10^{-4}$ T throughout. (That value averages-in the variation of B between the equator and pole.) The mean radius of the Earth is 6371.23 km. Compare your answer to the energy in a gallon of gasoline, $\approx 10^8$ J.

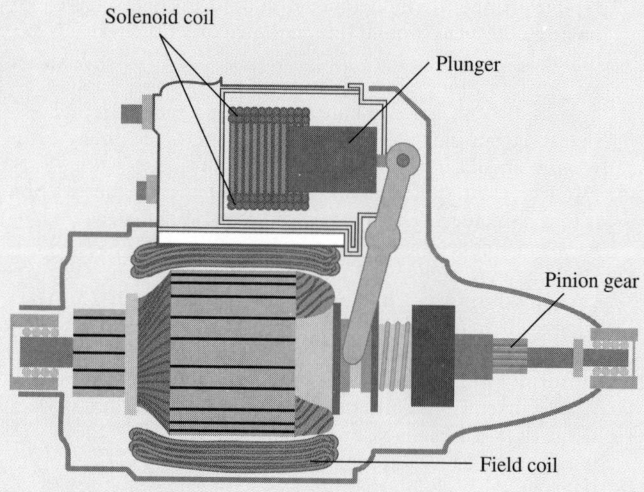

Figure P68

70. [III] Derive the expression for the emf of a rotating coil

$$\mathscr{E} = NAB\omega \sin \omega t \qquad [22.7]$$

by considering the work done by a torque acting through an angle.

71. [III] Derive an expression for the time-rate-of-change of the current in an R-L circuit. Interpret your result in light of Fig. 22.21. Suppose a 200-Ω resistor is placed in series with a 50-mH inductor, a switch, and a 120-V dc power supply. Compute the rate-of-change of the current when $t = 0$, $t = L/R$, and $I = 1.0$ A.

72. [III] An air-core solenoid is attached to a 12-V battery via a switch. The switch is closed, and after a while the current reaches a constant level of 2.0 A. If the current changes at a rate of 12 A/s when $I = 1.0$ A, what is the resistance and inductance of the coil?

73. [III] The current in a long solenoid of 210 turns generates a flux within it of 10 mWb. If that current is gradually reversed in a time of 200 ms, what will be the induced emf?

23

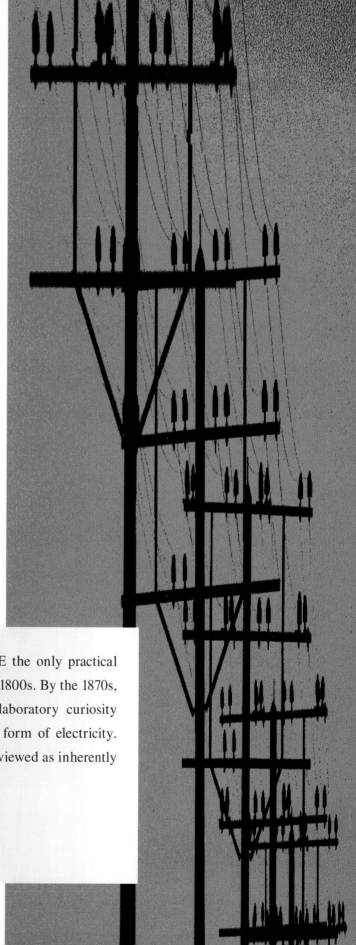

AC and Electronics

HOWEVER RUDIMENTARY, BATTERIES WERE the only practical source of sustained electric current throughout the early 1800s. By the 1870s, the dynamo (p. 776) had made the transition from laboratory curiosity to practical workhorse, and dc was still the preferred form of electricity. As yet undeveloped, alternating current (ac) was widely viewed as inherently inferior.

Nikola Tesla (1856–1943) was one of the leading figures in the development of alternating current. The spectacular display crackling around Tesla was produced by high-frequency (20,000 Hz) high-voltage ac. The handwritten inscription reads "To my illustrious friend Sir William Crookes of whom I always think and whose kind letters I never answer! June 17, 1901."

ALTERNATING CURRENT

Instead of maintaining a fixed polarity, each terminal of an ac generator, though always opposite to the other, alternates between + and −. The electrons that constitute a typical alternating current move first forward then backward, oscillating essentially in place at some given number of cycles per second corresponding to the generator frequency. Remember that the electrons in a dc current drift quite slowly; what moves at nearly the speed of light is the disturbance of the electrons, the front behind which the electrons are moving and beyond which they are not yet affected by the driving voltage source. An alternating current transports energy in the same way as does a direct current—namely, in the form of the organized kinetic energy of mobile-charge carriers. And joule heat arises with ac, just as it does with dc.

When the practical high-resistance, low-current incandescent lamp was introduced by Edison around 1880, each installation came with its own 110-volt generator, and it was dc. Most other lamps of the era were low-resistance devices that had to be put in series so that a high current could pass through each one (P = I^2R and for a required P, a small R means a large I). No one lamp could be shut off without open-circuiting the system. Edison's carbon-filament lamps had a high resistance and could be put in parallel, where the feeder current would be divided into many small branch currents, each powering one bulb that could be turned on and off at will (Fig. 20.13, p. 698)—which, of course, is the way your home is wired.

Meanwhile, ac was quietly being transformed by a brilliant, wildly eccentric, young engineer named Nikola Tesla, who had briefly been associated with (and subsequently came to despise) Mr. Edison. With Tesla's invention of a practical ac induction motor, alternating current became far more appealing, particularly to shrewd visionaries like George Westinghouse who hired Tesla and bought the rights to his motor. Slowly, as the market expanded, there developed a titanic struggle for control of the industry between the hustlers of *high-voltage ac* (primarily Westinghouse and General Electric, the J. P. Morgan combine to which Edison would ultimately sell out) and those of *low-voltage dc* (led rather unscrupulously by Edison).

When central generating stations came into being and especially when they were located at remote energy sources like Niagara Falls (1895), the electrical power produced had to be transmitted over long distances. But the very wires that carry electricity have some resistance, and that poses a major problem. A medium-sized city might easily require ≈10 MW of power (P = IV). If that amount is to be provided at a modest 100 V or so, then 100 000 A will have to be supplied. There's the difficulty: the joule heating (P = I^2R) in the delivery wires varies as I^2, not just as I. A two-wire line of one-quarter-inch-diameter copper wire has a resistance of ≈1.7 Ω/mile. Carrying 10^5 A, the joule heating losses are ≈1.7×10^{10} W/mile. For each mile, that's ≈1.7×10^7 kW and every hour the line loses 1.7×10^7 kW·h of energy. At a cost of roughly 10¢ per kW·h, sending 10^5 A down the line wastes 1.7 million dollars per hour per mile!

There was no economically feasible way out but to lower the current. Clearly, if the voltage was raised to 100 000 V, the same power could be efficiently delivered by 100 A! Thus, raising the voltage by a factor of 10^3 allows for the lowering of the cur-

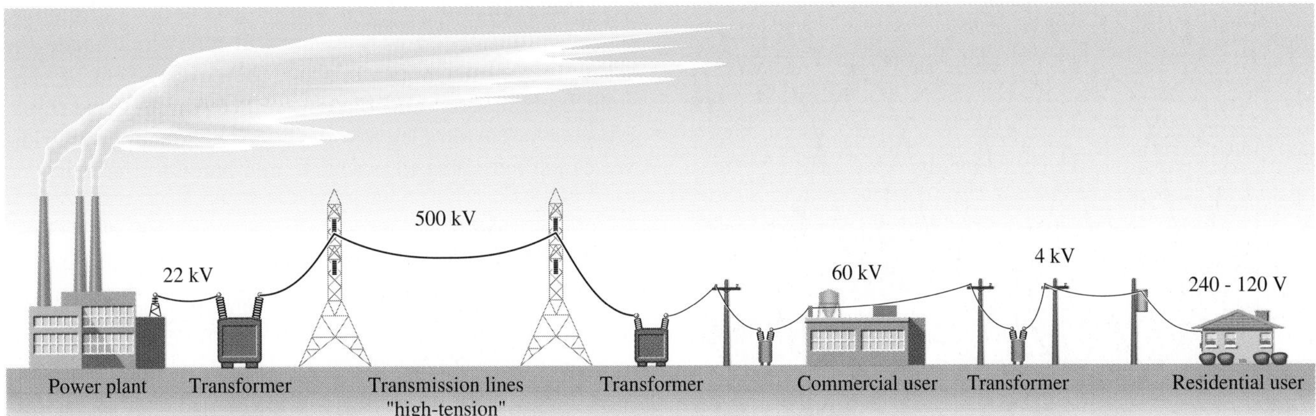

Figure 23.1 The transmission of ac power. Transformers are used to increase the voltage so power can be efficiently sent long distances via "high-tension" wires. The voltage is then stepped down for commercial and residential use. For simplicity, the ground and other "hot" leads are not shown.

rent by 10^3, and that drops the associated power lost over the same lines by a factor of 10^6. Since there already existed a very simple way to raise and lower the voltage of ac (via transformers, p. 812) but no comparable means for dc (until recently), the contest ultimately went to the high-voltage-ac people.

Electrical power is usually generated at around 22 kV and boosted by step-up transformer to about 500 kV, which is then transmitted long distances via high-tension lines (Fig. 23.1). Using a transformer, the voltage is then reduced to perhaps 60 kV for heavy industrial consumers—usually the more power required, the higher the voltage at which it's delivered. Substations in each area step the voltage down further to about 4 kV for distribution to local communities. This voltage is what's carried by those endless ugly overhead wires that crisscross most suburban towns. Thereafter, the voltage is finally reduced by small pole-mounted transformers, each of which feeds a cluster of buildings. **Line voltage** in the United States and Canada is ordinarily 110 V, 115 V, or 120 V and can be anywhere in that range, varying from time to time.

In the early days, there were a number of different frequencies supplied (at first 125 Hz and 133 Hz were common and later 25 Hz, 35 Hz, 50 Hz, and 60 Hz became popular), depending on the producer. Lower frequencies were more suitable for use with the induction motor, but they still had to be high enough to keep incandescent lamps from flickering. Today, most of Europe and Asia uses 50 Hz (220 V), whereas the United States and Canada have adopted 60 Hz (110 V) as standard.

Alternating current was initially supplied (as was dc) using conventional two-wire lines in what is called a *single-phase* system—the voltage rises and falls sinusoidally (p. 774), as does the current. Under such circumstances, power is inefficiently available in pulses. Tesla was among the few to realize that by simultaneously generating several sinusoidal voltages, each shifted in phase with respect to the other, a transmission line could carry more power and actually do so at a constant, continuous rate. Using three wires instead of two and sending out three (120°-shifted) sinusoidal voltages between them (Fig. 23.2), Tesla was able to triple the average amount of power delivered by the line. Equally as important, this kind of *polyphase* ac was necessary to drive the big new induction motors that required no brushes or commutators and were extremely reliable. Under the irresistible pressure of these benefits, the entire ac power system was overhauled, and the transmission of *three-phase* ac became almost universal. Today, although industrial users are often supplied directly with three-phase ac, the typical domestic consumer is usually provided with two single-phase 120-V ac lines (p. 815).

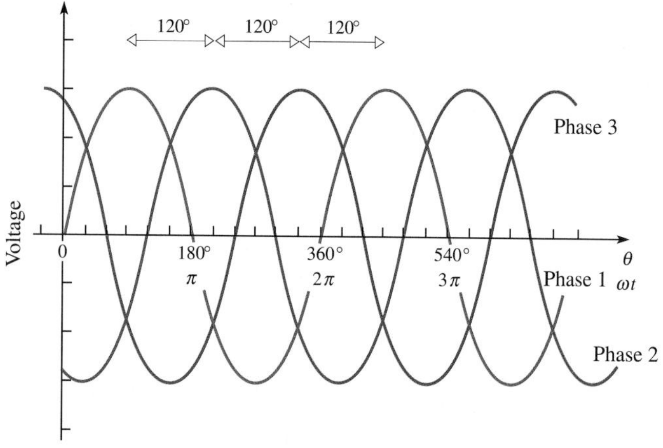

Figure 23.2 The emf produced by a three-phase ac generator. Each signal is shifted by 120°.

The availability of efficient ac generators and motors, the invention of the ac watt-hour meter (which made the selling of ac energy practicable), the development of efficient transformers, and the economical three-phase transmission system have led to today's vast national power grids. We can simply plug in, sit back, and listen to the music.

23.1 AC and Resistance

As we saw earlier [Eq. (22.7)], the emf produced by an ac generator can be represented by a sine function of angular frequency $\omega = 2\pi f = 2\pi/T$. In addition to the ubiquitous wall outlet, there are electronic devices known as *oscillators* that also provide a harmonically varying potential difference. They're often used in the laboratory, though there are many gadgets that have oscillators built into them—the modern radio receiver is one. Whatever the source, the terminal voltage then has the form

$$v = V_m \sin \omega t = V_m \sin 2\pi f t \qquad (23.1)$$

where v is the **instantaneous voltage**. The lowercase letter v is used here to distinguish it from the several different possible practical measures of ac voltage that will be considered presently. A plot of v (Fig. 23.3) rises and falls between two peak or **maximum voltage** values, $+V_m$ and $-V_m$, which is what would be seen if we plugged the source into an oscilloscope. Then we could read the maximum value directly from the scale on the screen just as we could read the so-called **peak-to-peak voltage** (V_{pp}), which is simply $2V_m$.

With a purely resistive load R across the terminals of an ac source, a current will circulate that increases and decreases, reversing direction over and over again, precisely in step with the oscillating voltage—the applied emf directly drives the **instantaneous current** i. Ohm's Law takes the form

$$i = \frac{v}{R} = \frac{V_m}{R} \sin \omega t$$

and with the **maximum current** given by $I_m = V_m/R$, it becomes

$$i = I_m \sin \omega t = I_m \sin 2\pi f t \qquad (23.2)$$

An ac shaded-pole induction motor of the kind that powers devices such as blenders and can openers. The field coil sets up an oscillating B-field in the stator. Currents induced in the copper wire "shades" retard the changing B-field in those regions, producing a primitive rotating field that induces currents in the rotor. These currents result in a torque that sets the rotor turning.

Both the voltage and the current are zero when $2\pi f t$ equals either 0, π, 2π, etc., since $\sin 0 = \sin \pi = \sin 2\pi = 0$. This occurs when $t = 0$, $t = 1/2f = \pi/\omega$, or $t = 1/f = 2\pi/\omega$, and so on, as shown in Fig. 23.3. For 60-Hz ac, the zeros happen at $t = 0$, 1/120 s, 1/60 s, etc. The current reverses itself every 1/120 s, and the period is $T = 1/f = 1/60$ s.

The early workers with electricity were faced with a practical problem when it came to measuring ac; they couldn't have the pointer on an ammeter or a voltmeter flutter back and forth at 60 Hz. Indeed, what does one even mean by the statement "the ac voltage* is 120 V?" Engineers were accustomed to dealing with dc, where

*Despite the obvious awkwardness, such terms as ac current and ac voltage are now part of the jargon. In such cases, the ac should be read as *ay cee* and not alternating current.

it makes sense to speak of a 10-A current, but how does one describe the strength of an alternating current? It was decided to use the *ampere* as the unit of alternating current and define it in such a way as to be, insofar as possible, equivalent to the direct-current ampere. There were three possible methodologies that could be used: chemical, magnetic, or thermal. But only the thermal effect of a current varies as i^2 and is independent of the direction, or sign, of the current. Since both ac and dc produce joule heating, *the alternating-current ampere was defined as the quantity of current that produces the same amount of heating in a resistor as does a direct-current ampere during the same interval of time.*

Nowadays, ac ammeters are calibrated to read this **effective current** (I_{eff}): 10 amps ac effective will generate the same amount of thermal energy as 10 amps dc. So common is the usage, it is automatically assumed that a device rated at simply 10 amps ac refers to 10 A effective. And the same is true for ac voltmeters, which are, as a rule, calibrated in **effective voltage** (V_{eff}). Moreover, since we are likely to deal only in effective values, the subscripts are often dropped, and it is understood that the I and V mean effective current and voltage, respectively.

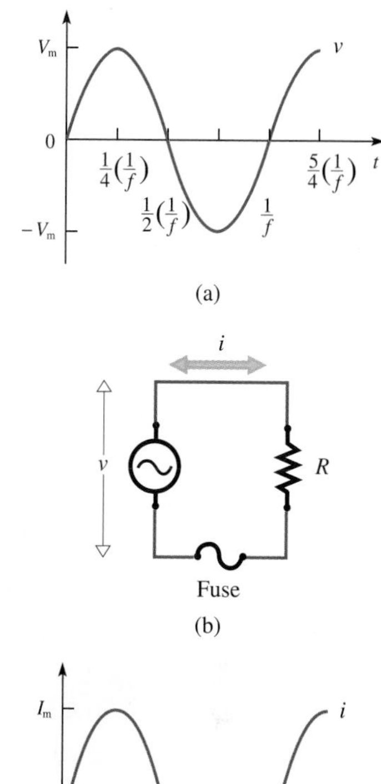

Figure 23.3 A sinusoidal voltage (a) applied to a resistive load (b) results in a sinusoidal in-phase current (c).

The effective values can be related to the maximum values by looking at the power; I_{eff} is defined by its ability to cause heating in a resistor, which is equivalent to its ability to transfer an average level of power. The **instantaneous power** p dissipated in a resistor is

$$\mathrm{p} = i^2R$$

What we actually measure via a calorimeter is the effect of such power averaged over many cycles. Inasmuch as the R is constant

$$\mathrm{P}_{\text{av}} = [i^2]_{\text{av}}R \tag{23.3}$$

and this result by definition is to equal I^2R or, if you like, $I^2_{\text{eff}}R$. The average rate at which thermal energy is developed in a resistor by an alternating current is

$$\mathrm{P}_{\text{av}} = I^2_{\text{eff}}R \tag{23.4}$$

Comparing these two equations, it follows that

$$I = I_{\text{eff}} = \sqrt{[i^2]_{\text{av}}}$$

The effective current equals the root of the mean of the square of the instantaneous current or, for short, the **rms current**. It's not uncommon in the literature to find the symbol I_{rms} used instead of I or I_{eff}. Had we started with $\mathrm{p} = v^2/R$, we would have come upon a precisely analogous definition for the **rms voltage**.

Returning to Eq. (23.3) and substituting Eq. (23.2) into it yields

$$\mathrm{P}_{\text{av}} = [(I_{\text{m}} \sin \omega t)^2]_{\text{av}}R = [I^2_{\text{m}} \sin^2 \omega t]_{\text{av}}R$$

Because I^2_{m} is constant

$$\mathrm{P}_{\text{av}} = I^2_{\text{m}}[\sin^2 \omega t]_{\text{av}}R \tag{23.5}$$

The average of any time-varying function taken over some interval is equal to the area under the curve divided by the length of the interval. In Fig. 23.4, the area under the curve equals the area under the constant $\frac{1}{2}$-line since the tops of the peaks can be imagined cut off and moved so they fill the troughs. For any interval ωt, of several periods, the area ($\frac{1}{2} \times \omega t$) divided by the duration (ωt) is just $\frac{1}{2}$: $[\sin^2 \omega t]_{\text{av}} = \frac{1}{2}$. Another way to see this result is to start with the identity $\sin^2 \omega t + \cos^2 \omega t = 1$. Realizing that $\sin^2 \omega t$ and $\cos^2 \omega t$ are identical except for a 90° phase shift, they must both have the same average value over an interval greater than one period. But

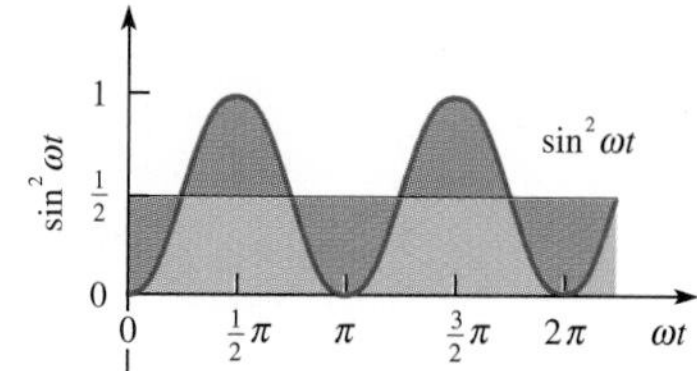

Figure 23.4 The area under the curve $\sin^2 \omega t$ divided by the extent of the curve (ωt) equals the average value; $[\sin^2 \omega t]_{\text{av}} = 1/2$.

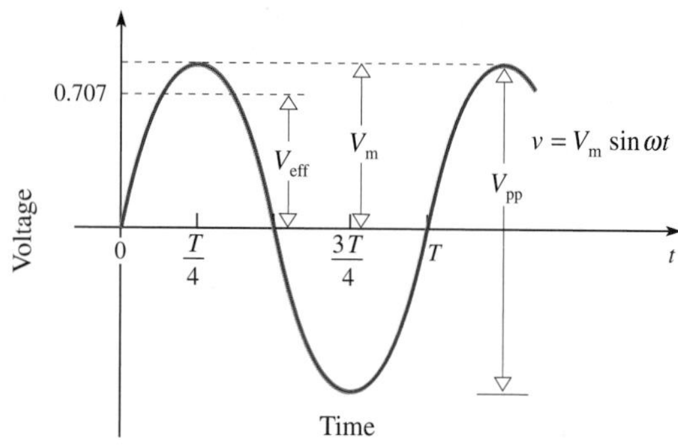

Figure 23.5 A sinusoidal voltage and the corresponding effective, maximum, and peak-to-peak values.

if we take the average of both sides of the identity, it follows that $[\sin^2 \omega t]_{av} = [\cos^2 \omega t]_{av} = \frac{1}{2}$. Thus, from Eq. (23.5)

$$P_{av} = \tfrac{1}{2} I_m^2 R \tag{23.6}$$

and since $P_{av} = I_{eff}^2 R$, we have the sought-after expression for the effective current in terms of the maximum current, namely

$$I = I_{eff} = \frac{I_m}{\sqrt{2}} \tag{23.7}$$

And, similarly

$$V = V_{eff} = \frac{V_m}{\sqrt{2}} \tag{23.8}$$

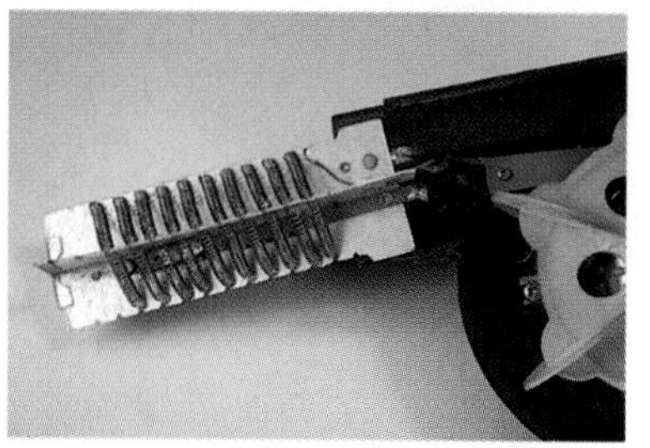

The heating coil in a hair dryer. Current passing through the coil causes it to become red hot ($P = I^2R$) via joule heating. The white plastic star-shaped fan blows air over the coil and out the nozzle.

The effective current or voltage is just 0.707 times the corresponding maximum value (Fig. 23.5). A wall outlet providing 120 volts effective is actually supplying a sinusoidal emf with a maximum terminal voltage of (120 V)/(0.707) = 170 V, which is a good reason for you to be even more careful with 120-V ac than 120-V dc.

From Ohm's Law, which is valid at any instant, $V_m = I_m R$ and

$$V_{eff} = I_{eff} R \qquad \text{or} \qquad V = IR \tag{23.9}$$

Moreover

[*resistive load*]
$$P_{av} = I_{eff}^2 R = I_{eff} V_{eff} \tag{23.10}$$

which is nice and simple.

Example 23.1 According to the little metal plate on a hair dryer, it is rated at 120 V 1200 W. Assuming the load to be purely resistive (it really isn't because of the motor), how much current does it draw and what is its resistance? What is the maximum value of the current in the dryer?

Solution: [Given: $V = 120$ V and $P_{av} = 1200$ W. Find: I, I_m, and R.] From Eq. (23.10), the effective current is

$$I = \frac{P_{av}}{V} = \frac{1200 \text{ W}}{120 \text{ V}} = \boxed{10.0 \text{ A}}$$

The maximum current then follows from Eq. (23.7), as

$$I_m = 1.414 I = \boxed{14.1 \text{ A}}$$

From Ohm's Law

$$R = \frac{V}{I} = \frac{120 \text{ V}}{10.0 \text{ A}} = \boxed{12\ \Omega}$$

▶ **Quick Check:** $P_{av} = I^2R = (10.0 \text{ A})^2(12\ \Omega) = 1200$ W.

23.2 AC and Inductance

The distinctive aspects of ac become evident when there are inductors or capacitors in the circuit. It's then that the currents through, and the voltages across, these elements are *not in-phase.* As a result of these phase differences (which are different for capacitors and inductors), all sorts of interesting things happen. For example,

Ohm's Law doesn't apply in the same simple way it did with dc: in general, the current in an ac circuit (which is not purely resistive) is not equal to the applied voltage divided by the resistance of the circuit. Moreover, the algebraic sum of the ac voltage drops across each element of a series circuit (as measured by individual voltmeters) may not equal the applied voltage (as it does in a dc circuit). Energy is still conserved and Kirchhoff's Rules apply, but for sinusoidal voltages (and currents) that are out-of-phase, we can't simply add their effective values algebraically.

To study the ac behavior of an inductor, consider the circuit shown in Fig. 23.6a. It consists of an inductor L, having negligible resistance, placed across the terminals of an ac source. At the instant shown in Fig. 23.6b, an increasing clockwise current exists, creating an increasing flux in the coil. Remember (p. 780) that in response, an induced current i_I will appear in the windings, and it will oppose the change in the flux via Lenz's Law. The induced current at that moment emerges from the inductor at point A, and there is a potential difference across the terminals of the inductor, which is $+$ at point A and $-$ at point B. The inductor then behaves much like a battery being charged. Because the current from the source is sinusoidal, there will be a continuously changing flux and a sustained harmonic back-emf. In the steady state, the back-emf is 180° out-of-phase with the emf of the source. In Fig. 23.6, trace around the circuit and when the back emf is a drop, the source voltage will be a rise. In the unattainable case where $R = 0$, the two are numerically equal. At any instant, the sum of the voltage rises and drops around the loop is then zero.

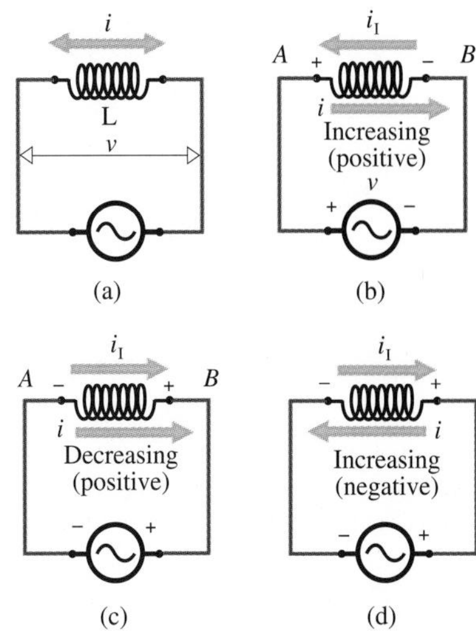

Figure 23.6 (a) An inductor placed across the terminals of an ac source. (b) Note that when i is positive (clockwise) and increasing, the terminal voltage v is positive. (c) When i is positive and decreasing, v is negative, and (d) when i is negative and increasing, v is negative.

Ideally, the coil has no resistance so that the source emf need not be larger than what it takes to overcome the back-emf in order to sustain current in the circuit. On average there cannot be any energy dissipated by an ideal inductor. However, a real inductor has resistance; consequently, the difference between the source voltage and the back-emf will equal iR [Eq. (22.11)], and some small amount of energy provided by the source will be dissipated as i^2R.

The voltage measured across the terminals of the inductor ($v_L = L\Delta i/\Delta t$) equals the voltage v of the source. To write an expression for the *instantaneous* voltage, we take the limit of $L\Delta i/\Delta t$ as $\Delta t \to 0$; thus

$$v = L \lim_{\Delta t \to 0} \left(\frac{\Delta i}{\Delta t}\right)$$

Here, $v = V_m \sin \omega t$. At this point, we would like to deduce an expression for the instantaneous current in the circuit, and that could be done nicely using calculus. Apart from that, we can determine the form of i, a bit less elegantly, from the discussion of the oscillator in Chapter 12 and the fact that position and instantaneous speed are related by an identical limiting expression (p. 30) to that above. It was shown (p. 406) that when the position is given by $x = A \cos \omega t$, the corresponding speed is given by $v_x = -A\omega \sin \omega t$. Now, position x and speed v_x are related in the same way as are current i and voltage v: actually, as i and v/L. Comparing voltage/inductance with speed, we see that v_m/L corresponds to $-A\omega$ and so A corresponds to $-V_m/L\omega$. The expression for x therefore suggests that

$$i = -\frac{V_m}{\omega L} \cos \omega t = -I_m \cos \omega t \qquad (23.11)$$

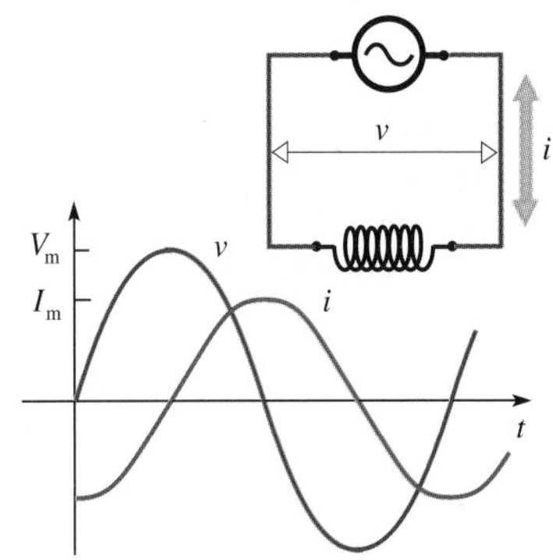

Figure 23.7 An inductor effectively holds back the current so that i (the current through it) lags v (the voltage applied by the source). (See Fig. 23.6.)

Figure 23.7 is a plot of this instantaneous current through, and the voltage across, the inductor. Notice how the current peaks at a later time than the voltage: **the current lags the voltage** (or the voltage leads the current) *by one-quarter cycle* (90°). The instantaneous voltage depends on the product of the inductance and the *rate-of-*

change of the current. The voltage is zero when the current curve levels off (that is, when its slope is zero) at its maximum values. The voltage is maximum when the current curve is steepest (that is, when its slope is maximum) at points where it crosses the time axis. *The higher the frequency, the tighter the curves are, and the greater the slopes at these points.*

A real inductor inhibits current both because of its resistance and because of its inductance (via the associated back-emf). We can speak of the total effect as representing an *impedance* (p. 808), part of which, the *reactance*, is due purely to the inductive behavior exclusive of any resistance. To elaborate, observe that it follows from Eq. (23.11) that

$$I_m = \frac{V_m}{\omega L} \tag{23.12}$$

and since $I = 0.707 I_m$ and $V = 0.707 V_m$

$$V = \omega L I \tag{23.13}$$

This result is so much like Ohm's Law, with the ωL term impeding the current, that we introduce a new quantity called the **inductive reactance** X_L, defining it as

$$X_L = \omega L \tag{23.14}$$

whereupon

$$V = I X_L \tag{23.15}$$

Consequently, 1 volt per 1 amp corresponds to an inductive reactance of 1 ohm.

In this ideal situation, there is no resistance and the reactance equals the impedance. The greater the inductance, the greater the back-emf, and the more the inductor inhibits or "chokes" the current. Note that in the case of dc ($\omega = 0$), there is no back-emf—the reactance is zero and there is no inhibition of current. An inductor with a large L and small R is usually called a *choke coil* because of its ability to control ac currents without wasting nearly as much power as would a resistor. Because the inductive reactance increases with frequency, a choke can be used to suppress the high end of a mix of frequencies. For example, in a loudspeaker system there are usually at least two drivers, a small high-frequency tweeter and a large low-frequency woofer. By putting an inductor in series with the woofer (Fig. 23.8), high-frequency currents (representing high-pitched sound) are suppressed, allowing the woofer to respond to the signal range it was designed to handle best.

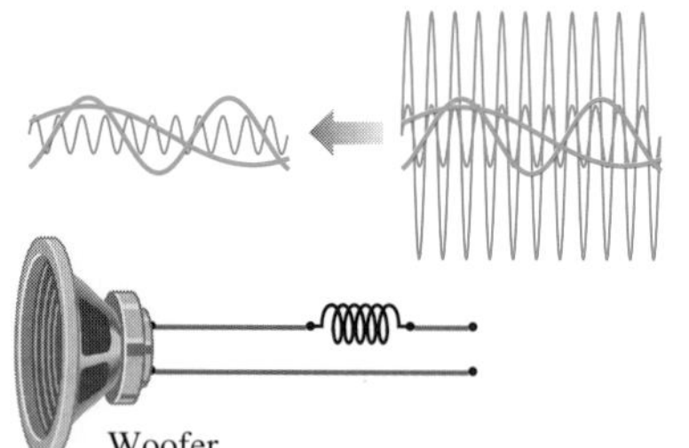

Figure 23.8 The inductor's impedance increases with frequency, so it passes low-frequency signals and blocks high-frequency signals.

Example 23.2 A radio circuit contains a 400-mH inductor with a resistance of 0.50 Ω. This is supplied (across its terminals) with an 80-V-effective ac signal of 100 Hz. Determine both the reactance of the coil and its rms current.

Solution: [Given: $L = 0.400$ H, $R = 0.50\ \Omega$, $V =$ 80 V, and $f = 100$ Hz. Find: X_L and I.] Let's first find the inductive reactance:

$$X_L = \omega L = 2\pi f L = 2\pi(100\text{ Hz})(0.400\text{ H}) = \boxed{251\ \Omega}$$

By comparison, R is negligible and we can treat the coil as if it were a pure inductance. Hence

$$I = \frac{V}{X_L} = \frac{80\text{ V}}{251\ \Omega} = \boxed{0.32\text{ A}}$$

▶ **Quick Check:** $X_L \approx (6)10^2(4 \times 10^{-1})\Omega \approx 0.24\text{ k}\Omega$; $V = IX_L \approx (0.32\text{ A})(\frac{1}{4}\text{ k}\Omega) \approx 0.08\text{ kV}$.

Figure 23.9 is a plot of the instantaneous power ($p = iv$) associated with the ideal inductor. Where i and v are both positive or both negative, p is positive, and where either one (i or v) is negative, p is negative. Thus, the area under the curve (subtracting what is below the axis from what is above it) per cycle is zero. Energy is stored in the alternating magnetic field, but *the average power supplied per cycle in a purely inductive circuit is zero.* Power is alternately absorbed from the generator (positive portion of the cycle) and then returned to it (negative portion of the cycle).

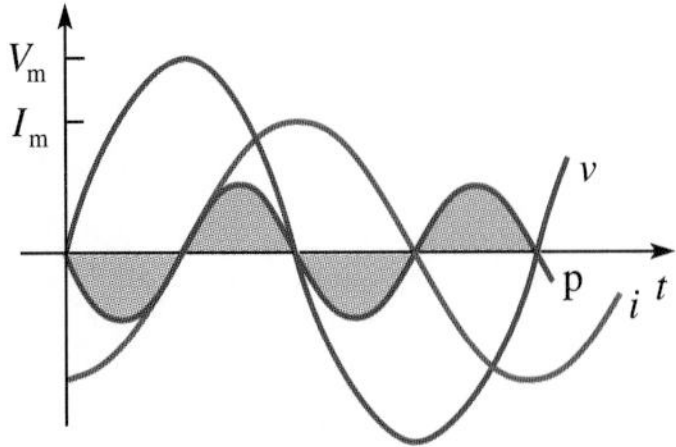

Figure 23.9 The power drawn by a pure inductance averages to zero. Note how p = 0 whenever either v or i is zero. The relative heights of the curves are unimportant.

23.3 AC and Capacitance

Imagine a capacitor C placed across the terminals of a battery. We are already familiar with what happens in the transient state: charge piles up on the plates until the mutual repulsion between the charges stops any more from arriving. A *reverse potential* appears (via $v = q/C$) that increases until it reaches the battery potential, at which point the current stops. Because C is constant, $\Delta v = \Delta q/C$; a change in the charge is associated with a change in the voltage. By definition, the current is the time-rate-of-change of q, which corresponds to a time-rate-of-change of v: no $\Delta v/\Delta t$, no steady state i.

Now suppose the battery is replaced by an ac source (Fig. 23.10). We want to compare v across the capacitor to i passing through it. As the sinusoidal voltage of the generator rises positively (0° to 90° in the ac cycle), we can imagine positive charges traveling to the upper plate and an equal number repelled away from the lower plate. A large positive current immediately circulates, because at $t = 0$ there is no charge on the plates and no reverse potential to inhibit it. As the impressed voltage rises, the charge on the plates increases in step with it, and the reverse potential increases too, making it harder for more charge to be deposited—the clockwise current dies off.

At the moment the voltage peaks (90°) so that $\Delta v/\Delta t = 0$, the current is zero. But then the voltage begins to decrease (90° to 180° in the ac cycle), whereupon the capacitor starts to discharge: $\Delta v/\Delta t \neq 0$ and $i \neq 0$. Current appears in the negative counterclockwise direction. After the voltage passes through zero, it reverses and increases (180° to 270° in the ac cycle), increasing the charge on the plates in the reverse direction. That again decreases the counterclockwise current until it becomes zero, when the voltage peaks in the negative direction (270°). The negative applied voltage then decreases, and positive charge leaves the capacitor flowing clockwise and constituting a positive current (270° to 360° in the ac cycle). Apparently, **the instantaneous current in a capacitor leads the instantaneous voltage across it by one-quarter of a cycle** (90°).

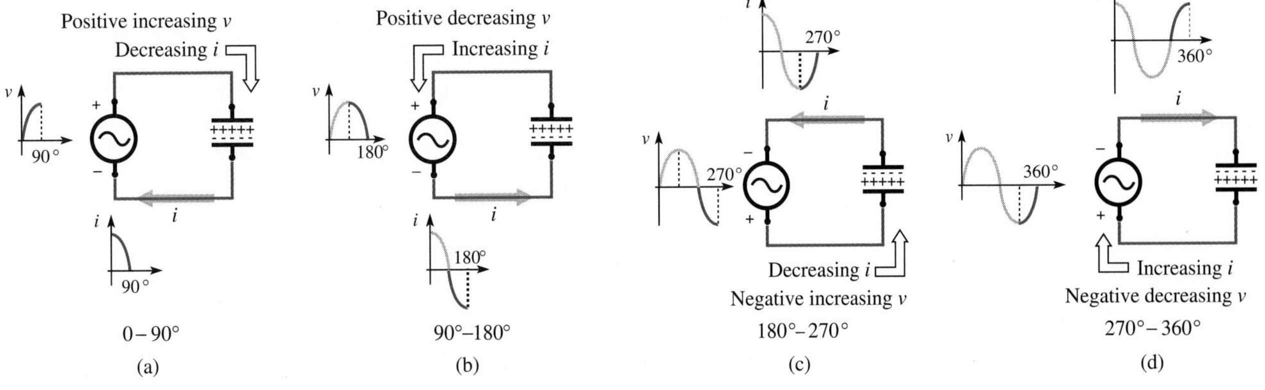

Figure 23.10 The voltage and current in an ac capacitive circuit. The positive direction is taken to be clockwise.

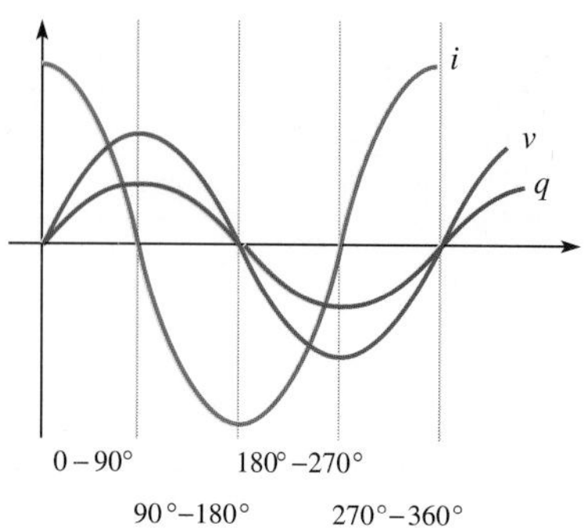

Figure 23.11 The current, charge, and voltage for an ac capacitive circuit.

We can describe this behavior analytically in the following way. Since $v = q/C$ and $v = V_m \sin \omega t$, it follows that

$$q = CV_m \sin \omega t \tag{23.16}$$

The charge and voltage are in-phase as shown in Fig. 23.11. Now to find the instantaneous current i. The average current is $\Delta q/\Delta t$, and in the limit as $\Delta t \to 0$, that ratio becomes i: the time-rate-of-change of the charge is the current. Again, using Chapter 12, Eq. (23.16) suggests that we begin with the sinusoidal speed of the oscillator $v_x = -A\omega \sin \omega t$ and examine its time-rate-of-change; namely (p. 407), the instantaneous acceleration $a_x = -A\omega^2 \cos \omega t$. Comparing speed and charge, we see that $-A\omega$ corresponds to CV_m. Hence, with acceleration having the form of the current, where $-A\omega^2$ corresponds to $CV_m\omega$, we get

$$i = \omega CV_m \cos \omega t = I_m \cos \omega t \tag{23.17}$$

This time, $\omega CV_m = I_m$ and so the *rms* values share the same relationship—namely, $\omega CV = I$. Hence

$$V = \frac{1}{\omega C} I \tag{23.18}$$

and the form of Ohm's Law suggests that we introduce a **capacitive reactance** X_C, defined as

$$X_C = \frac{1}{\omega C} \tag{23.19}$$

whereupon $V = X_C I$

The reactance, in ohms (Problem 28), is a measure of the capacitor's opposition to the passage of alternating current.

When $\omega = 0$, the reactance is infinite, and no finite dc voltage will produce a current through an ideal capacitor—**capacitors block dc**. As the frequency goes up, the capacitive reactance goes down, in contrast to the inductive reactance; a given source voltage results in a higher and higher current as ω is increased. What stops the current is the peaking of the voltage arising from the buildup of charge on the plates. At high frequency, the charge has little opportunity to pile up and thereby decrease the current before the applied voltage is reversed. Alternatively, increasing ω increases the slope of the v curve, which means a larger value for the time-rate-of-change of q, and so a larger current. If the capacitor is getting charged and discharged more rapidly, the current must increase. Raising the capacitance lowers the reactance; a large capacitor can store more charge before the voltage across its plates rises to appreciably diminish the current. Alternatively, since $\Delta q/\Delta t = C\Delta v/\Delta t$, the larger is C, the greater the current.

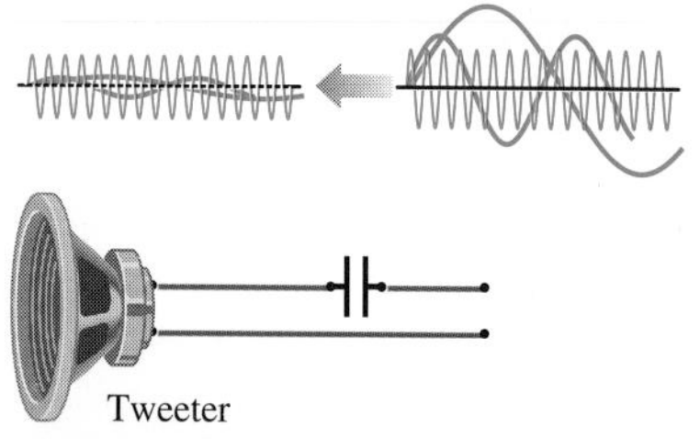

Figure 23.12 The capacitor represses low frequencies and passes high frequencies.

Figure 23.12 indicates how a capacitor in series with a tweeter in a speaker system will suppress the low frequencies in the input signal; ($X_C \propto 1/\omega$). It will pass on, with little attenuation, the high-frequency components that the tweeter was designed to convert into sound.

Example 23.3 A 50-μF capacitor is connected across the terminals of an oscillator set to have a sinusoidal output at 50 Hz with a maximum voltage of 100 V. Determine the effective current in the circuit.

(continued)

(continued)
How would this change if the frequency were raised to 5 kHz?

Solution: [Given: $C = 50\ \mu\text{F}$, $V_m = 100$ V, and $f = 50$ Hz. Find: I.] Knowing the maximum voltage, we can find the effective voltage and, from that and the reactance, we can find I. So, figure out X_C:

$$X_C = \frac{1}{\omega C} = \frac{1}{2\pi f C} = \frac{1}{2}\pi(50\text{ Hz})(50\ \mu\text{F}) = 63.7\ \Omega$$

Now $V = 0.707V_M = 70.7$ V and therefore

$$I = \frac{V}{X_C} = \frac{70.7\text{ V}}{63.7\ \Omega} = \boxed{1.1\text{ A}}$$

At 5 kHz, the reactance would be lower by a factor of 100 and the current raised by a factor of 100.

▶ **Quick Check:** $V = IX_C = I/\omega C = I/2\pi f C \approx (1\text{ A})/6(50\text{ Hz})(50 \times 10^{-6}\text{ F}) \approx 1/0.015\text{ V} \approx 0.07\text{ kV}$.

As with the inductor, because the instantaneous current and voltage are out-of-phase by 90°, no average power will be dissipated by an ideal capacitor. Energy stored in the electric field between the plates is returned to the source. Thus, *only resistance will dissipate power in an ac circuit, converting electrical energy into thermal energy.*

L-C-R AC NETWORKS

Electronic devices such as radios, televisions, stereos, and so on, utilize inductors, capacitors, and resistors to process electrical signals—that is, currents and voltages. Arrangements of these elements can be used to reshape a signal, to filter out or perhaps accentuate certain frequencies, or to remove any dc that might be present, and so on.

23.4 Series Circuits

Suppose we put an inductor, a capacitor, and a resistor in series across the terminals of an oscillator with an instantaneous voltage v, as shown in Figure 23.13. Now represent the current by either a sine or cosine function. Let it be

$$i = I_m \sin \omega t \qquad (23.20)$$

This same current exists in each element of the circuit. And since the instantaneous voltage across the resistor (v_R) is in-phase with the current, we have

$$v_R = I_m R \sin \omega t \qquad (23.21)$$

By comparison, the instantaneous voltage across the capacitor (v_C) lags the current by 90°, or $\pi/2$ radians; the sine function describing the voltage reaches the same value as $\sin \omega t$ at a later time. Using Eq. (23.18), we obtain

$$v_C = \frac{I_m}{\omega C} \sin\left(\omega t - \frac{\pi}{2}\right) \qquad (23.22)$$

Thus, $\sin \omega t = 0$ at $t = 0$, whereas $\sin(\omega t - \pi/2) = 0$ *later,* when $t = \pi/2\omega$. On the other hand, the instantaneous voltage across the inductor leads the current by 90° or

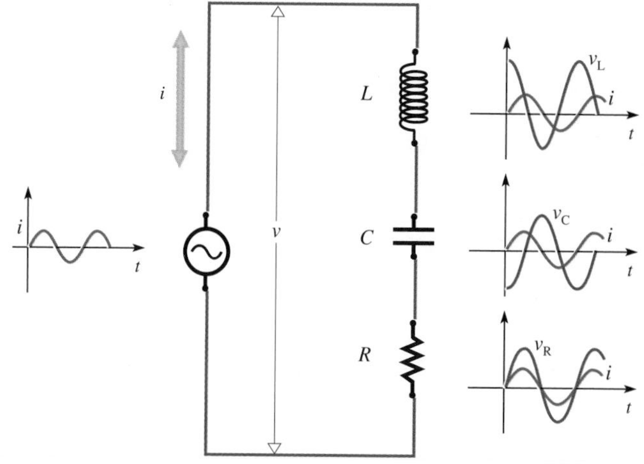

Figure 23.13 The same current passes through each element in a series circuit. The voltages across each are usually not in-phase and $v = v_L + v_C + v_R$.

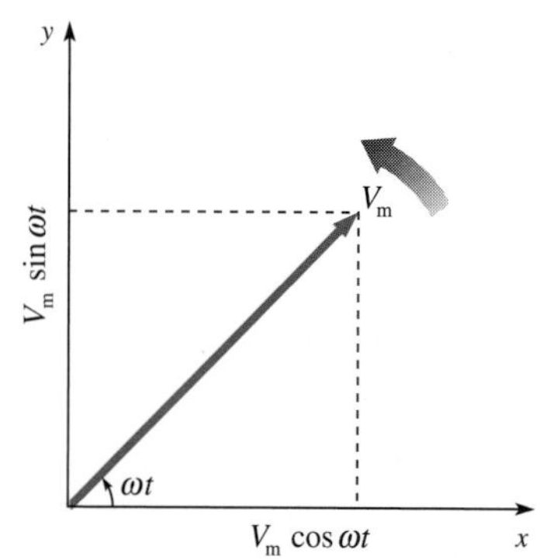

Figure 23.14 As the arrow rotates, the x- and y-components oscillate harmonically.

$\pi/2$ radians, as

$$v_L = I_m \omega L \sin\left(\omega t + \frac{\pi}{2}\right) \tag{23.23}$$

Thus, $\sin \omega t = 0$ at $t = 0$, whereas $\sin(\omega t + \pi/2) = 0$ *earlier*, when $t = -\pi/2\omega$.

In order for the applied voltage v to drive the current i through the circuit, it must equal the sum of the instantaneous voltages

$$v = v_R + v_C + v_L \tag{23.24}$$

If we added the three sinusoids via algebra and trigonometry (and the mathematics is laborious), we would end up with a resultant sinusoid

$$v = V_m \sin(\omega t + \theta) \tag{23.25}$$

Interestingly, the sum of any number of sine functions of the same frequency is a sine function of that same frequency (Fig. 13.5, p. 446). The complete analysis yields expressions for V_m and θ:

$$V_m = \sqrt{(I_m R)^2 + \left(I_m \omega L - \frac{I_m}{\omega C}\right)^2} \tag{23.26}$$

and

$$\theta = \tan^{-1} \frac{\omega L - 1/\omega C}{R} \tag{23.27}$$

which, when computed and substituted into Eq. (23.25), gives the potential difference across all three elements. We have not provided the details of that calculation because there is a much simpler practical scheme for arriving at the same results, a method called **phasor addition**.

Figure 12.5 (p. 404) illustrated how a line rotating at a rate of ωt can be projected onto either axis in order to generate harmonic functions. Suppose, then, that we draw an arrow of length V_m and have it rotate counterclockwise at a rate ω, where at any instant it makes an angle ωt. The corresponding sine and cosine components are shown in Fig. 23.14. This revolving "arrow," which looks like a vector and has some of the properties of a vector, is formally a different beast called a *phasor* (designated by boldfaced type). The important thing is that *phasors add like vectors* and, in so doing, we in effect add their components—namely, the sinusoids. Combine the phasors and we combine the sinusoids, which is what we want, and that is easily done graphically.

Since the current and the voltage across the resistor are in-phase, we draw these two phasors one on top of the other in Fig. 23.15. Their lengths are I_m and $V_{Rm} = I_m R$, respectively, and their y-components are the sine functions of Eqs. (23.20) and (23.21). Because all of the phasors will be referenced to the current phasor, it's customary to simplify things a little by working at $t = 0$ (and remembering to put ωt in the expression for the phase when needed), which has the effect of placing the current phasor on the x-axis. Henceforth, all phasors will be drawn with respect to the x-axis. If its phase is $(\omega t + \alpha)$, it is tilted α radians above the x-axis; if its phase is $(\omega t - \beta)$, it is tilted β radians below the x-axis. Figure 23.16 shows how two phasors **A** and **B** are added tip-to-tail like vectors. The projection on the x-axis is then the sum of the individual cosine functions, and the projection on the y-axis is the sum of the sine functions. *Our phase shifts will only be* $\pm 90°$.

Figure 23.15 The phasors **i** and $\mathbf{v_R}$ are in-phase; that is, the ac current is in-phase with the voltage across the resistor.

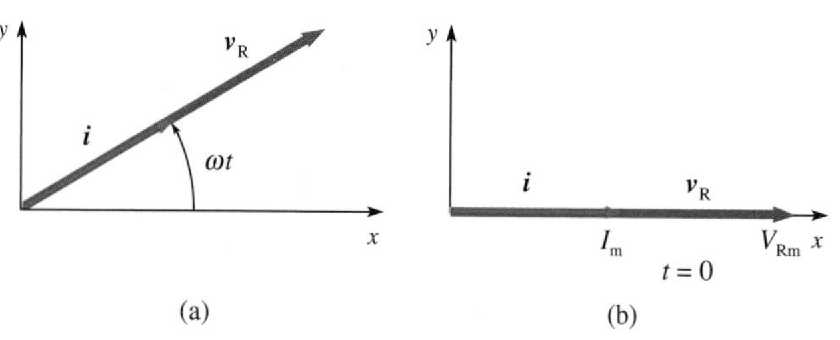

Consider a circuit with just an inductor and a resistor (Fig. 23.17). Because the voltage across the inductor leads the current by 90°, $\mathbf{v}_L$ leads $\mathbf{v}_R$ by 90°, and the two phasors (with lengths $V_{Lm} = I_m\omega L$ and V_{Rm}) are drawn as shown in Fig. 23.17b. For simplicity, this configuration is redrawn at $t = 0$ in Fig. 23.17c. If we put these instantaneous voltages across the terminals of a dual-beam oscilloscope, we'll see two sinusoids shifted 90° with respect to each other. The two phasors are next added "vectorally" to produce a resultant phasor $\mathbf{v}$ in Fig. 23.27d. The length of $\mathbf{v}$ corresponds to the maximum voltage (V_m), which will be measured across the combination of R and L. The angle θ (positive if it's a lead, negative if a lag) is the phase angle of the resultant voltage sinusoid with respect to the current through the circuit. And that's it; Eq. (23.25) provides the expression for v once we compute V_m and θ from the diagram. Figure 23.17e shows the individual instantaneous voltages that can be added point-by-point to get v. Notice how v is indeed shifted by some angle (namely, $\theta < 90°$) with respect to v_R. The larger is L, the larger is θ. Similarly, if R is made larger, θ becomes smaller.

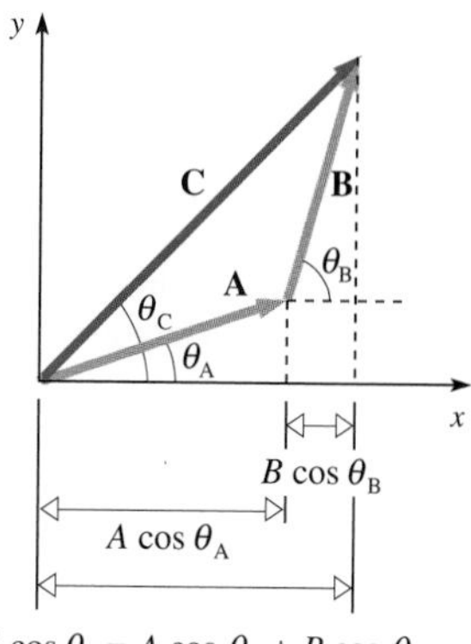

Figure 23.16 The addition of phasors **A** and **B**. Phasors add like vectors, and their projections onto the x-axis add like cosines.

Return to the L-C-R circuit of Fig. 23.13. The voltage across the capacitor lags v_R by 90° and its phasor, of length $V_{Cm} = I_m/\omega C$, is added in along with $\mathbf{v}_R$ and $\mathbf{v}_L$ in Fig. 23.18. The two opposing phasors on the y-axis add to yield a single phasor of length $|V_{Lm} - V_{Cm}|$, which may or may not point in the positive y-direction, depending on which is larger in a particular situation. Finally, using the Pythagorean Theorem to find the length of the resultant phasor (Fig. 23.18c), we obtain

$$V_m = \sqrt{(V_{Rm})^2 + (V_{Lm} - V_{Cm})^2} \tag{23.28}$$

and this is identical to Eq. (23.26). Moreover

$$\tan \theta = \frac{(V_{Lm} - V_{Cm})}{V_{Rm}} \tag{23.29}$$

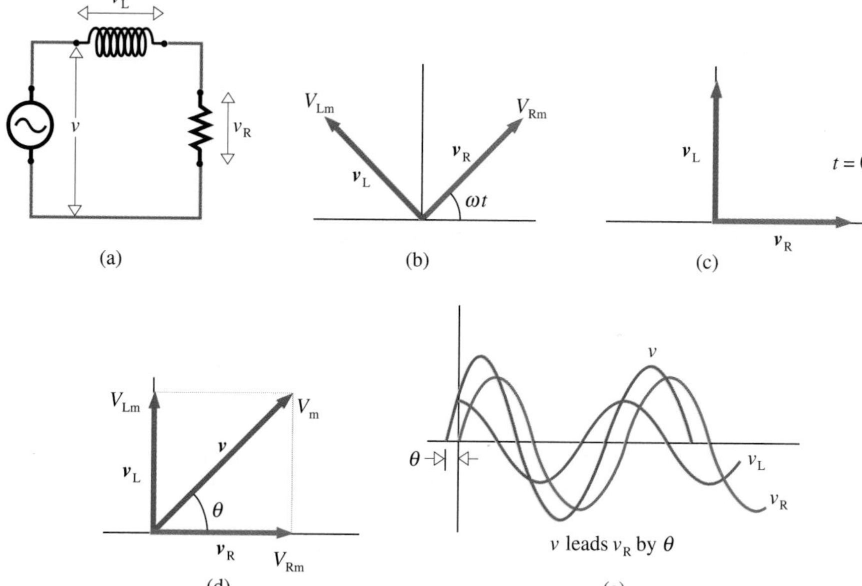

Figure 23.17 (a) An ac circuit containing inductance and resistance. (b) The voltage phasors actually rotate at a rate ω. (c) Rotation is eliminated when $t = 0$. (d) The amplitude of $\mathbf{v}$ is V_m. (e) v_L and v_R are 90° out-of-phase.

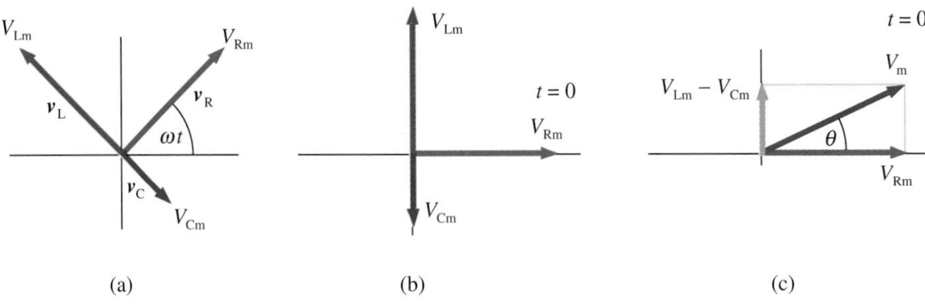

Figure 23.18 (a) The phasor diagram for a circuit containing resistance, inductance, and capacitance. (b) The magnitudes of the voltage phasors at $t = 0$. (c) The magnitude of the voltage across all the elements R, L, and C is V_m.

which is equivalent to Eq. (23.27). Inasmuch as the same expression holds for the effective voltages, Eq. (23.28) tells us that the sum of the separate voltage readings made with a meter across each element will generally exceed the voltage across the source.

Example 23.4 A series circuit contains a 240-Ω resistor, a 3.80-μF capacitor, and a 550-mH inductor. It's placed across the terminals of an oscillator set to 100 Hz. If an ammeter in the circuit reads 250 mA effective, what is the maximum voltage of the oscillator?

Solution: [Given: $R = 240\ \Omega$, $C = 3.80\ \mu\text{F}$, $L = 550$ mH, $f = 100$ Hz, and $I = 250$ mA. Find: V_m.] We have an expression for the maximum voltage, namely Eq. (23.26), and so must first determine I_m; thus

$$I_m = 1.414I = 353.5 \text{ mA}$$

Hence, $I_m R = 84.84$ V, $I_m \omega L = 122.2$ V, and $I_m/\omega C = 148.1$ V. And so

$$V_m = \sqrt{(I_m R)^2 + (I_m \omega L - I_m/\omega C)^2} \quad [23.26]$$

hence
$$V_m = \sqrt{(84.84 \text{ V})^2 + (-25.9 \text{ V})^2}$$

and
$$\boxed{V_m = 88.7 \text{ V}}$$

Notice that *the voltages across the capacitor* (148 V) *and inductor* (122 V) *are higher than the voltage across all three components taken together* (88.7 V).

▶ **Quick Check:** The reactances are $X_L = \omega L = 346\ \Omega$ and $X_C = 1/\omega C = 419\ \Omega$, which are comparable. We can therefore expect the voltage drops V_{Lm} and V_{Cm} to be roughly equal, yielding only a small vertical phasor $V_{Cm} - V_{Lm}$. The net voltage should be a little larger than $V_{Rm} = 84.8$ V, and it is.

Going back to Eq. (23.26), factor out the current and divide both sides by $\sqrt{2}$ to get effective values, whereupon

$$V = I\sqrt{R^2 + \left(\omega L - \frac{1}{\omega C}\right)^2} = I\sqrt{R^2 + (X_L - X_C)^2}$$

The quantity $(X_L - X_C)$ is the **reactance** of the circuit; it's a measure of the net non-resistive influence impeding the current, and it's denoted by X:

$$X = (X_L - X_C) \quad (23.30)$$

It follows that
$$V = I\sqrt{R^2 + X^2}$$

Ohm's Law again suggests that the measure of a circuit's entire ability to restrain ac current (inductively, capacitively, and resistively) can be defined by its **impedance** (Z), where

$$Z = \sqrt{R^2 + X^2} \quad (23.31)$$

given in ohms. And so Ohm's Law survives into ac provided that it's written as

$$V = IZ \tag{23.32}$$

Notice how the impedance can also be thought of as a kind of vector quantity (one that isn't time-varying) in the sense that Z is equal to the magnitude of the resultant of adding X and R as if they were vectors. The associated diagram (Fig. 23.19) is often called the *impedance triangle,* and we see immediately that $\tan\theta = X/R$, which is equivalent to Eq's. (23.27) and (23.29).

Figure 23.20 summarizes some of the results for two-element series circuits. Of course, a circuit may contain several resistors, capacitors, or inductors. We already know how to add any number of resistors (p. 693) and capacitors (p. 646) in series; suffice it to say without proof that inductors add as do resistors (see Discussion Question 10 in Chapter 22). Thus, to analyze an ac series circuit with many components, we first combine all of the same kinds of elements, whereupon all the above equations apply, with R, L, and C being the resultant values.

On average, the power drawn by an L-C-R circuit is dissipated by the resistor: $P_{av} = I^2R$. This process can be expressed in terms of the voltage by noting from Fig. 23.18c that

$$\cos\theta = \frac{V_{Rm}}{V_m} = \frac{I_m R}{I_m Z} = \frac{R}{Z}$$

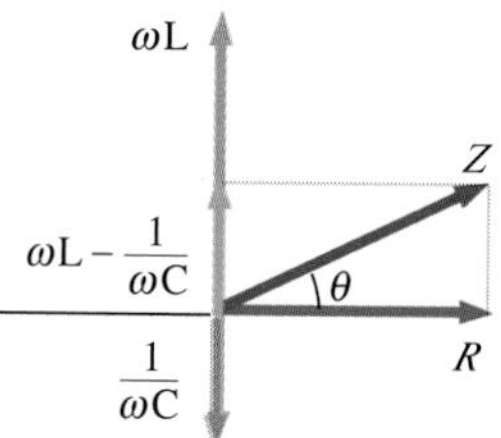

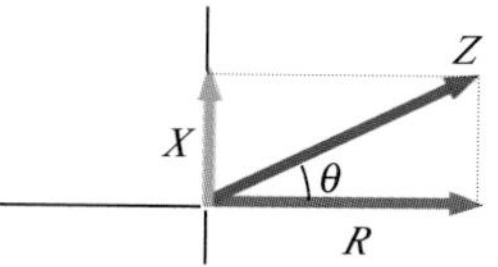

Figure 23.19 The impedance triangle. Reactances and resistance can be added as if they were vectors. The resultant is the impedance (Z) of the circuit. The "vector" $\mathbf{X}_C$ always points down, whereas $\mathbf{X}_L$ always points up. The "vector" $\mathbf{X}$ is the sum of the two, and so $X = X_L - X_C$. When $X_L > X_C$, θ is above the x-axis and positive. When $X_C > X_L$, X is negative, $\mathbf{X}$ points down, and θ is negative; it is beneath the "vector" $\mathbf{R}$, which always points in the positive x-direction.

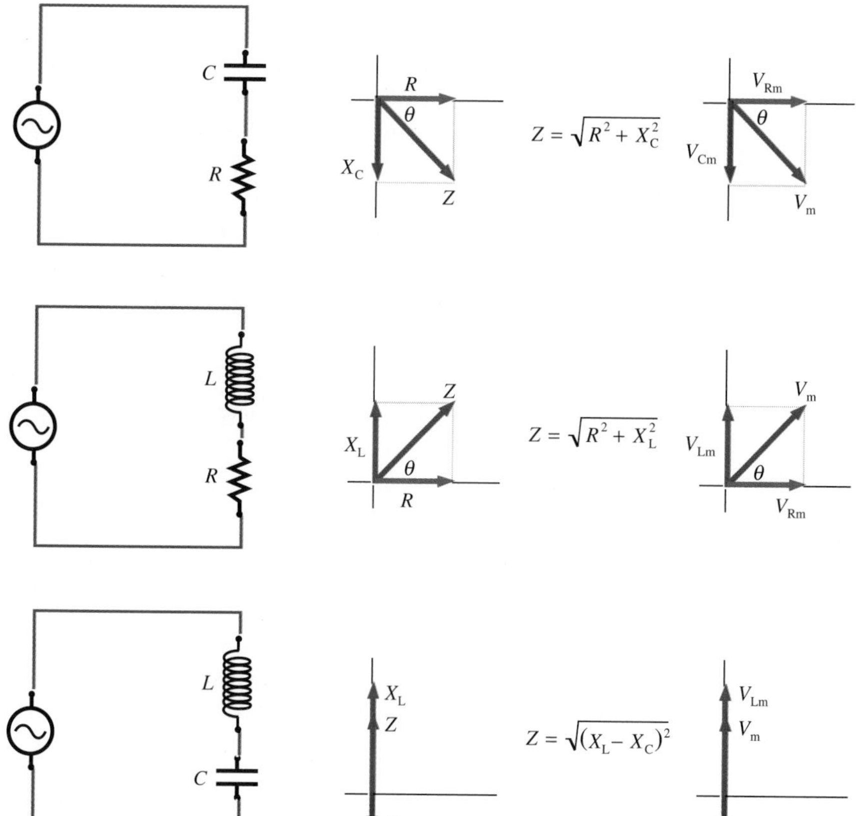

Figure 23.20 A summary of the behavior of various ac series circuits.

Accordingly $$P_{av} = I^2 Z \cos\theta$$

and so $$P_{av} = IV \cos\theta \tag{23.33}$$

This expression is known as the **real**, or *average,* or **dissipated power**. It differs from the corresponding dc equation by the term $\cos\theta$, which is called the **power factor** of the circuit. The power factor is a measure of the relative influence of the resistance in dissipating power. With a purely resistive circuit, $Z = R$, $\cos\theta = 1$, and the real power is IV as expected. For either a purely inductive or capacitive circuit, the phase angle is either $\pm 90°$, and no real power is transferred from the source. A typical circuit has a power factor of less than 1 (or, since the power factor is often given in percent, less than 100%). For example, in a series ac circuit containing a resistor and an inductor for which the resistance equals the inductive reactance, it follows from the impedance triangle that $\theta = 45°$ and $\cos\theta = 0.707$. Small single-phase motors have power factors of around 0.6 to 0.8.

The product IV is called the **apparent power** and is measured in volt-amps (V·A) to distinguish it from real power. Practically, an amount of power equal to IV must be supplied even if a portion of that, $(1 - \cos\theta)IV$, is stored in the fields and returned to the source. A piece of electrical equipment that has an appreciable reactance, such as a transformer, fluorescent lamp, motor, etc., has a power factor much less than 100% and must, in order to operate properly, be supplied more power than it consumes. With a power factor of 80%, a motor that consumes 800 W must be supplied with 1000 V·A in order to operate. A large commercial user pays for the power that has to be provided, even if some of it is returned.

Example 23.5 An oscillator set for 500 Hz puts out a sinusoidal voltage of 100 V effective. A 24.0-Ω resistor, a 10.0-μF capacitor, and a 50.0-mH inductor in series are wired across the terminals of the oscillator. (a) What will an ammeter in the circuit read? (b) What will a voltmeter read across each element? (c) What is the real power dissipated in the circuit?

Solution: [Given: $R = 24.0\ \Omega$, $C = 10.0\ \mu\text{F}$, $L = 50.0$ mH, $f = 500$ Hz, and $V = 100$ V. Find: I, V_L, V_C, V_R, and P_{av}.] (a) Having V, to find the current we'll need the impedance; accordingly

$$X_L = \omega L = 2\pi(500\text{ Hz})(50.0 \times 10^{-3}\text{ H}) = 157.1\ \Omega$$

$$X_C = \frac{1}{\omega C} = \frac{1}{2\pi}(500\text{ Hz})(10.0 \times 10^{-6}\text{ F}) = 31.8\ \Omega$$

and therefore

$$Z = \sqrt{(24.0\ \Omega)^2 + (125.3\ \Omega)^2} = 127.5\ \Omega$$

Thus $$I = \frac{V}{Z} = \frac{100\text{ V}}{127.5\ \Omega} = \boxed{784\text{ mA}}$$

(b) Across each element, a voltmeter will read

$$V_R = IR = (784\text{ mA})(24.0\ \Omega) = \boxed{18.8\text{ V}}$$

$$V_L = IX_L = (784\text{ mA})(157.1\ \Omega) = \boxed{123\text{ V}}$$

$$V_C = IX_C = (784\text{ mA})(31.8\ \Omega) = \boxed{24.9\text{ V}}$$

(c) To determine the power, we must first compute the power factor:

$$\cos\theta = \frac{R}{Z} = \frac{24.0\ \Omega}{127.5\ \Omega} = 0.188$$

$$P_{av} = IV\cos\theta = (0.784\text{ A})(100\text{ V})(0.188) = \boxed{14.7\text{ W}}$$

▶ **Quick Check:** From the fact that $\cos\theta = 0.188$, $\theta = 79.2°$ and, using Eq. (23.29), $\tan\theta = (123\text{ V} - 24.9\text{ V})/(18.8\text{ V}) = 5.22$ and $\theta = 79.2°$.

Series Resonance

An ac series circuit can function in a remarkable way at a specific frequency, $\omega_0 = 2\pi f_0$, known as its **resonant frequency**. The phenomenon is the electrical equivalent of the mechanical concept of resonance considered earlier (p. 416). Figure 23.21 depicts the frequency-dependent behavior of R, X_L, and X_C, and what we see is that, at the resonant frequency, the capacitive and inductive reactances are equal. Consequently, inasmuch as

$$Z = \sqrt{R^2 + (X_L - X_C)^2} \tag{23.34}$$

at resonance

$$Z = R$$

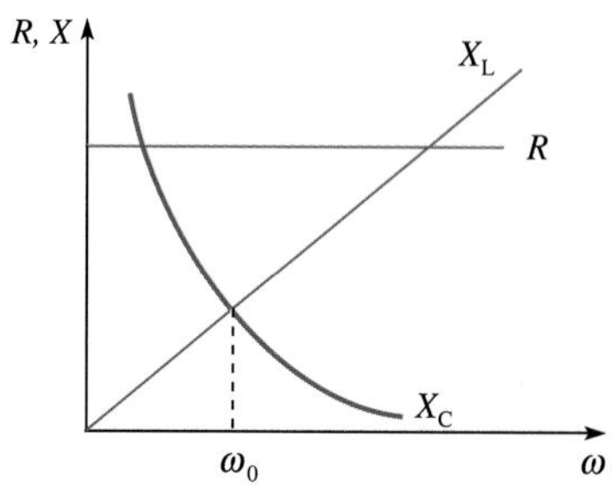

Figure 23.21 At the resonant frequency (ω_0), the inductive and capacitive reactances are equal.

The condition for resonance exists when

$$\omega L = \frac{1}{\omega C}$$

and that occurs at $\omega_0 = 2\pi f_0$, whereupon

[*resonance*]
$$f_0 = \frac{1}{2\pi\sqrt{LC}} \tag{23.35}$$

Thus, at resonance $\theta = 0$, $Z = R$, and $P_{av} = IV$. Because the impedance is then a minimum, for a given voltage V, the current I is a maximum: $V = IR$. In other words, suppose that a wide range of frequencies is fed into an L-C-R circuit. If we adjust or *tune* L or C or both so that Eq. (23.35) holds at a particular frequency, say 1 kHz, then the circuit will have a peak current at 1 kHz—all other currents at other frequencies will be considerably less. Figure 23.22 indicates how the current curves are affected by the resistance of the circuit.

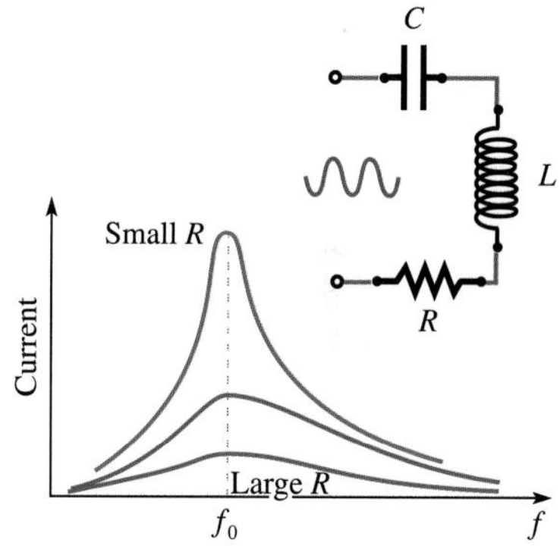

Figure 23.22 At resonance, the current in a series L-C-R circuit reaches a maximum. The smaller the resistance, the sharper the peak, and the narrower the range of frequencies selected by the circuit. Compare this to Fig. 12.24 (p. 417).

Example 23.6 A series circuit contains a 50.0-Ω resistor adjacent to a 200-mH inductor attached to a 0.050-μF capacitor, all connected across an oscillator with a terminal sinusoidal voltage of 150 V effective. (a) What is the resonant frequency? (b) What voltages will be measured by voltmeters across each element at resonance? (c) What is the voltage across the series combination of the inductor and capacitor?

Solution: [Given: $R = 50.0\ \Omega$, $C = 0.050\ \mu\text{F}$, $L = 200$ mH, and $V = 150$ V. Find: f_0, V_L, V_R, and V_C.] From Eq. (23.35)

$$f_0 = \frac{1}{2\pi\sqrt{LC}} = \boxed{1.59\text{ kHz}} \quad \text{and} \quad \omega_0 = 10.0\text{ krad/s}$$

To find the voltages, we need the reactances and the current:

$$X_L = \omega L = (10.0\text{ krad/s})(0.200\text{ H}) = 2.00\text{ k}\Omega$$

$$X_C = \frac{1}{\omega C} = \frac{1}{(10.0\text{ krad/s})(0.050 \times 10^{-6}\text{ F})} = 2.00\text{ k}\Omega$$

and $Z = R$. Consequently

$$I = V/R = (150\text{ V})/(50.0\ \Omega) = 3.00\text{ A}$$

and therefore

$$V_R = IR = (3.00\text{ A})(50.0\ \Omega) = \boxed{150\text{ V}}$$

$$V_L = IX_L = (3.00\text{ A})(2.00\text{ k}\Omega) = \boxed{6.00\text{ kV}}$$

$$V_C = IX_C = (3.00\text{ A})(2.00\text{ k}\Omega) = \boxed{6.00\text{ kV}}$$

Notice that although there is 6.00 kV across the inductor

(continued)

(continued)

and 6.00 kV across the capacitor, the corresponding instantaneous voltages are 180° out-of-phase—the voltage across the combination is zero!

▶ **Quick Check:** The fact that the reactances are equal at resonance is a good indication that we haven't messed up the numbers.

AM Radio

A radio broadcasting system essentially converts sound (20 Hz–20 kHz) into electromagnetic waves (p. 834) that travel a lot faster and farther. We could transmit such waves at the same frequencies as the information, the sound, but that would require an antenna of tremendous size and is totally impractical. The solution is to use a convenient high-frequency radiowave (the electromagnetic **carrier wave**) and impress the information on it. In AM, or *amplitude modulation,* the carrier's amplitude is made to vary with the information. Thus, the high-frequency signal of Fig. 23.23 carries all the music or talk in the form of relatively low-frequency changes in height, or strength, of the signal. The valuable information is the envelope of the signal and the carrier itself will ultimately be discarded by the receiver.

Figure 23.24 is a rudimentary AM radio receiver. The antenna picks up a tumult of signals composed of the transmissions from all of the stations reaching it. That hodgepodge is available to the tuning circuit by way of the coupling between L_1 and L_2. When you turn the tuning knob on a radio, you are adjusting the capacitor C_1 to resonate the input circuit at the frequency of, say, WCBS. Only that frequency and its immediate surroundings will then be passed to the next stage, the crystal diode (p. 819). This is a one-way gate that chops off the negative portion of the signal. The resulting positive voltage is applied to a filter formed by C_2 and R. When the diode drops the current to zero, the capacitor discharges through R, but the time constant of this RC circuit is large compared to the period of the carrier; the discharge is slow, and voltage across the filter hardly decreases before the diode passes current again and C_2 recharges. The result is a slightly wiggly, low-frequency voltage across R that otherwise corresponds in shape to the envelope of the carrier. The original information (Fig. 23.23b) oscillated above and below the axis, whereas this signal is only positive—it's been displaced by a constant positive voltage. That dc component is eliminated by a blocking capacitor (C_3). The resulting slowly oscillating voltage corresponds almost exactly to the oscillating sound wave. When it's fed into headphones, out comes a sound wave identical to the one that was heard in the studio at WCBS.

Figure 23.23 An amplitude modulated (AM) signal. (a) A constant amplitude, constant frequency carrier is made to carry information, as in (b), by modulating its amplitude, as in (c).

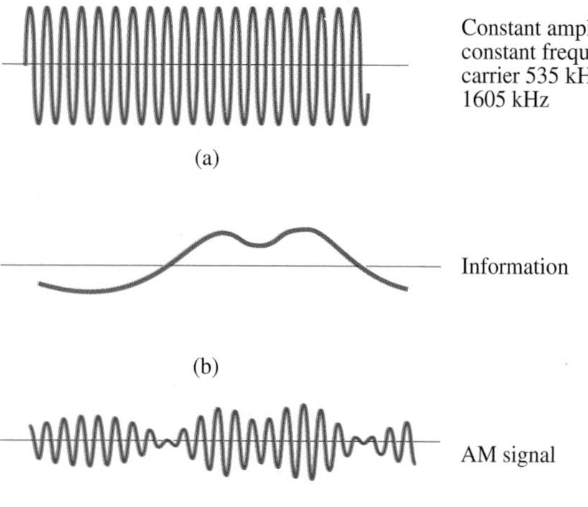

23.5 The Transformer

In 1831, Faraday discovered the principle of electromagnetic induction that underlies the transformer, but it took about 50 years before the latter became a practical instrument. In broad terms, the **transformer** *is an induction device used to convert energy in the form of a large time-varying current at a low voltage into nearly the same amount of energy in the form of a small time-varying current at a high voltage (or vice versa).*

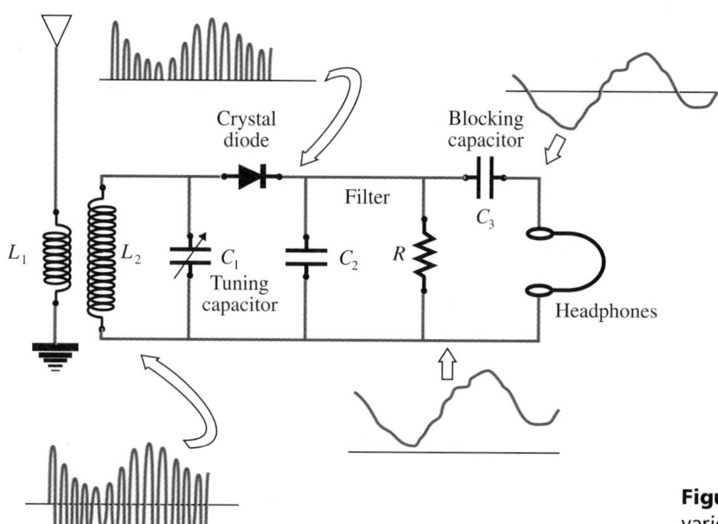

Figure 23.24 An early radio receiver showing a signal at various stages of being processed. The two coupled coils L_1 and L_2 constitute a transformer, a device we'll deal with next.

Imagine two coils wrapped around an iron core, as shown in Fig. 23.25. An ac power source is applied to the *primary* winding, and the switch is initially left open in the circuit of the *secondary* winding. The time-varying primary current I_p creates a time-varying magnetic flux that circulates through the secondary coil. Because of its high permeability, the iron core enhances the flux generated by the primary current by a factor of about 10 000. It provides an easy path for the field, which is very nearly totally constrained to pass within its volume, effectively coupling the two coils.

It might seem that closing the switch S_1 would essentially short-circuit the source, causing a tremendous input current; after all, the primary winding has a low resistance. Indeed, that's exactly what would happen if we put a constant dc source across the primary. With an ac source, the sizable self-inductance L, due largely to the core, will result in a substantial induced back-emf via Eq. (22.10). This back-emf will oppose the applied voltage, keeping the primary current I_p very small. Since the load on the source is essentially purely inductive, almost no energy is drawn from the source, and with the secondary open-circuited, no energy is transferred.

Ideally, the same time-varying flux passes through all the turns of both coils and so the induced emf on any one turn is the same as that on any other. Hence, the total induced emf on the primary is proportional to the total number of its turns (N_p) just as the total induced emf on the secondary is proportional to its number of turns (N_s). Assuming the windings have negligible resistance and therefore sustain no IR voltage drops, the induced emf across the primary will be numerically equal to the terminal voltage across it (V_p). Similarly, the emf induced across the secondary will equal its terminal voltage (V_s). It follows, then, that the ratio of the effective voltages will equal the ratio of the numbers of turns:

$$\frac{V_p}{V_s} = \frac{N_p}{N_s} \tag{23.36}$$

The coil with the higher number of turns corresponds to the higher voltage, which could be either the primary or the secondary, depending on how we wire it. *The ratio of the number of turns in the higher-voltage winding to the number of turns in*

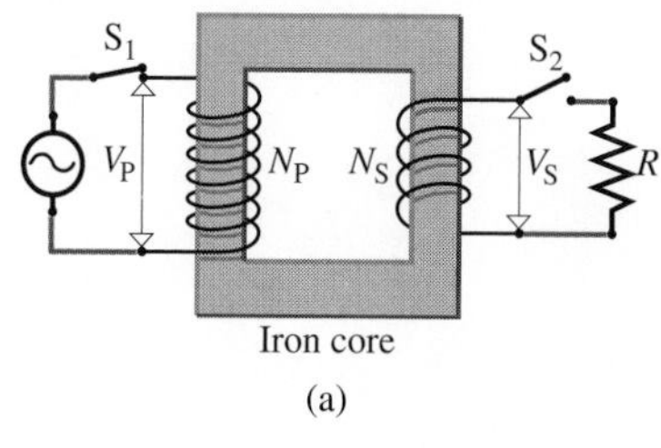

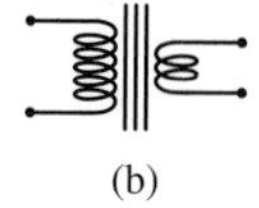

Figure 23.25 An iron-core transformer. (a) The input side is the primary; the output side is the secondary. (b) The symbol for an iron-core transformer.

the lower-voltage winding is called the **turn ratio**. For example, a transformer with a 10:1 turn ratio has 10 times the number of turns on one coil as on the other.

For any given primary voltage, we need only select an appropriate turn ratio in order to produce any desired secondary voltage. When $V_p > V_s$, it's called a **step-down transformer**, and when $V_p < V_s$, it's a **step-up transformer**. The spark coil in an automobile is a commonplace example of the step-up variety. The 12-V dc from the battery is chopped into pulses by a switching device. The "coil" is a transformer that boosts this pulsating 12-V input up to 20 kV, which activates the spark plugs that ignite the gasoline in each of the cylinders (p. 567).

Example 23.7 The ac adapter for a pocket calculator contains a transformer, two diodes, and a capacitor (p. 822). The transformer takes the 120-V ac at the wall down to 15.0-V ac, which the rest of the circuit converts to dc. If the transformer's secondary consists of exactly 50 turns, how many turns must the primary have? What is the turn ratio?

Solution: [Given: $V_p = 120$ V, $V_s = 15.0$ V, $N_s = 50$. Find: N_p and N_p/N_s.] We have the primary and secondary voltages as well as the secondary number of turns, which suggests Eq. (23.36), and consequently

$$N_p = \frac{N_s V_p}{V_s} = \frac{50(120\text{ V})}{15.0\text{ V}} = \boxed{400\text{ turns}}$$

This is a step-down transformer, and so the turn ratio is

$$\frac{N_p}{N_s} = \frac{400}{50} = \boxed{8}$$

▶ **Quick Check:** The transformer is of the step-down variety, and we expect $N_p > N_s$. Further, the ratio of the voltages (120 V)/(15 V) does equal the turn ratio.

Transformers lose energy because of the resistance of the coils and because of eddy currents and residual magnetization of the core. To control resistance losses, the low-voltage, high-current coil is usually made of relatively thick wire. A typical core is a laminate of insulated sheets, thereby removing the possibility of large-scale eddy currents flowing in cross-sectional loops. A well-built modern transformer will have a remarkable efficiency of better than 98%.

A speaker and output transformer from the AM radio shown on p. 589. The transformer allows the output amplifier of the radio to efficiently supply power to the speaker. (See Problem 73.)

With the switch S_2 in Fig. 23.25 closed, a secondary current progresses around the circuit. If we then make the reasonable approximation that energy losses are negligible, the average power-in equals the average power-out, and so from Eq. (23.33)

$$I_p V_p \cos\theta_p = I_s V_s \cos\theta_s \tag{23.37}$$

It's not at all obvious that the two power factors are equal and will cancel, but that's the way it will turn out. The presence of a secondary current means there will be an induced flux in the secondary coil that is proportional to $N_s I_s$. By Lenz's Law, this induced flux opposes the cause of the secondary current—namely, the alternating core flux due to the small initial ac primary current. In other words, the secondary flux momentarily decreases the primary flux and therefore the induced primary back-emf is decreased, which allows more current to come from the generator that, in turn, increases the back-emf until it again equals the generator voltage. The primary current I_p, which is then no longer negligible, provides the energy supplied to the secondary. The process stabilizes when the increase in the primary flux (which is proportional to $N_p I_p$) equals the induced secondary flux; thus

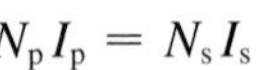

$$N_p I_p = N_s I_s$$

Combining this equation with Eq. (23.36) yields

$$I_p V_p = I_s V_s \tag{23.38}$$

and comparing this with Eq. (23.37), it follows that *the power factors are equal.* When the load on the secondary is large and resistive, the current and emf are essentially in-phase. Thus, the secondary power factor is 100% (as is the primary power factor), and Eq. (23.38) corresponds to the equality of the average powers in and out.

Example 23.8 Assume that the little transformer in the ac adapter in Example 23.7 is lossless (which it certainly isn't). (a) If it supplies 450 mA at 15.0-V ac, what is the current in the primary for a line voltage of 120 V effective? (b) If the secondary circuit has a power factor of 80%, what is the average power supplied by the wall outlet?

Solution: [Given: $V_p = 120$ V, $V_s = 15.0$ V, $I_s = 450$ mA, $\cos\theta = 0.80$. Find: I_p and P_p.] (a) From Eq. (23.38)

$$I_p = \frac{V_s I_s}{V_p} = \frac{(15.0\text{ V})(0.450\text{ A})}{120\text{ V}} = \boxed{56.3\text{ mA}}$$

(b) The wall supplies the same average power as the secondary provides, and so from Eq. (23.33)

$$P_p = I_s V_s \cos\theta_s = (0.450\text{ A})(15.0\text{ V})(0.80) = \boxed{5.4\text{ W}}$$

▶ **Quick Check:** Since the power factors are equal, $P_p = I_p V_p \cos\theta_p = (0.0563\text{ A})(120\text{ V})(0.80) = 5.4\text{ W}$.

Nowadays, transformers can be found around the home in a variety of appliances: fluorescent lamps, television sets, toy-train power supplies, high-intensity desk lights, battery chargers, stereos, door chimes, radios, and so on.

23.6 Domestic Circuits and Hazards

Although three-phase ac is supplied to large commercial users, the ordinary consumer is provided—via three wires down from the pole (Fig. 23.26)—with a pair of single-phase 120-V lines. Two of these wires are "hot" and the third is neutral, grounded back at the transformer. Inside the building, the neutral is usually attached to the entering water pipe and all electrical metal conduits, receptacle boxes, wall brackets, etc. are connected to it. The two hot leads (one red and one black), along with the neutral, go through a meter to a breaker or fuse box. The instantaneous voltages between red and ground (120 V) and black and ground (120 V) are 180° out-of-phase and can be combined to yield 240 V effective (between red and black), for use with heavy-duty air conditioners, water heaters, etc. Emerging from the breaker box are several parallel pairs of leads (black and white, hot and neutral, respectively) that carry power to the rest of the house. This arrangement is what's known as 240-V ac service, and it's fairly standard in North America. When you plug your TV into the wall, you are attaching one of the leads in the power cord to a black (hot) wire in the outlet and the other lead to a white (neutral) wire, across the two of which there is 120-V-effective, single-phase ac.

A pole-mounted transformer used to drop the voltage from the supply cables. Three leads provide two separate 110-V lines for domestic service. One of the three wires from the transformer is common to both lines. It attaches to a horizontal, uninsulated support cable, around which the other two wires are wrapped. These two wires go off on the right and left to individual houses.

A typical home is wired internally with several separate lines leading to the main, each attached in series with a fuse or circuit breaker of its own. The idea is to limit the amount of current that can be drawn by any single line—too much current and the wires in the wall can get dangerously hot (via I^2R). In fact, that's how the fuse

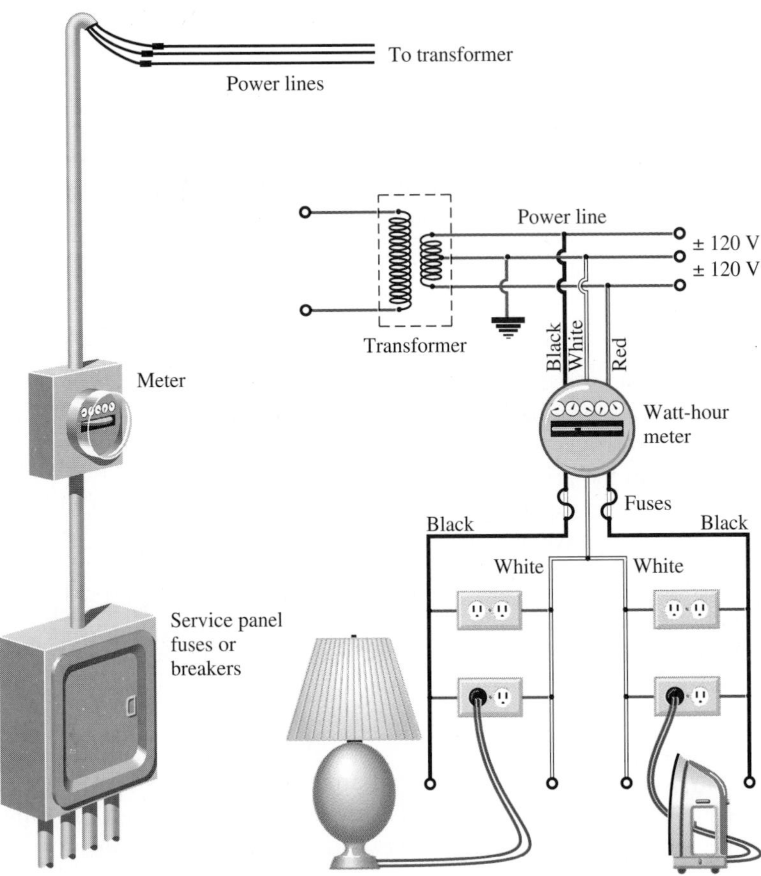

Figure 23.26 Domestic ac service.

works: current passes through a thin metal element that will melt and open-circuit when I exceeds the designated value. Household wiring is typically rated at 15 or 20 amps effective and fused accordingly. If a line contains too large a fuse and is made to carry too large a current by plugging in too many appliances, there will be an appreciable voltage drop in the wiring itself. The temperature of the line will rise, and its resistance will therefore also rise. The terminal voltages at the outlets may then be appreciably less than 120 V, and lights will dim, TV pictures will shrink, and you will be risking burning the place down. If a line is rated at 20 A, it will provide (120 V)(20 A) = 2400 W, which is just enough power to run a dishwasher, simultaneously make a slice of toast, and allow you to watch it all happen under a 100-W light bulb (Table 23.1).

The newest three-wire system includes an additional ground lead connected to the metal housing of the appliance. Thus, the whole steel body of a washing machine or refrigerator is attached to ground via the round prong on the three-prong plug (Fig. 23.27). Experience has shown that occasionally the hot wire in an electric device can be exposed and the whole appliance become "hot." For example, suppose you move your washing machine to do some cleaning and then while pushing it back against the wall, a sharp corner of the cabinet cuts into the power cord and touches the hot lead. In the old two-wire system, the entire machine would sit there at 120 V waiting for someone (standing on a damp basement floor) to touch it and get a nasty shock. In the new system, as soon as the hot lead makes contact with the grounded case, the current is shorted out and the fuse blows.

The electrical lines enter a house through an energy meter, which records the number of kilowatt hours delivered.

TABLE 23.1 Electrical power consumption

Appliance	Typical wattage (W)
Range	12 000–16 000
Clothes drier	5000–8000
Oven	4000–8000
Hair dryer	1000–1300
Dishwasher	1200–1500
Furnace blower	1200
Iron	1100
Toaster	1100
Waste disposal	1000
Oil burner	800
Refrigerator (big, double-door)	800
Vacuum cleaner	600
Washing machine	550
Blender	400
Fan	200
TV	100
Typewriter	90
Humidifier	40
Clock	4
Motors:	
1 hp	1500
1/2 hp	1000
1/4 hp	700
1/6 hp	450

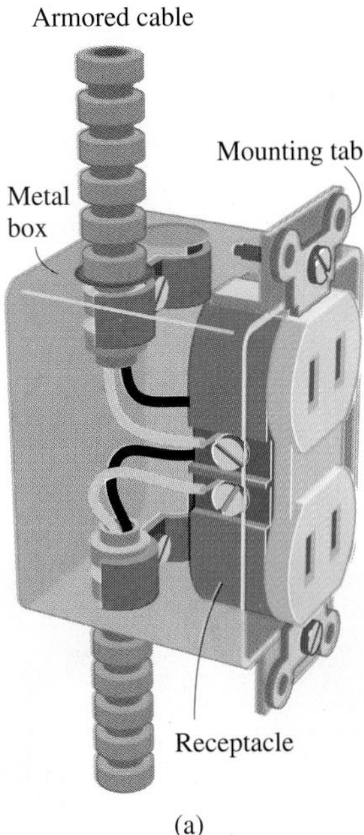

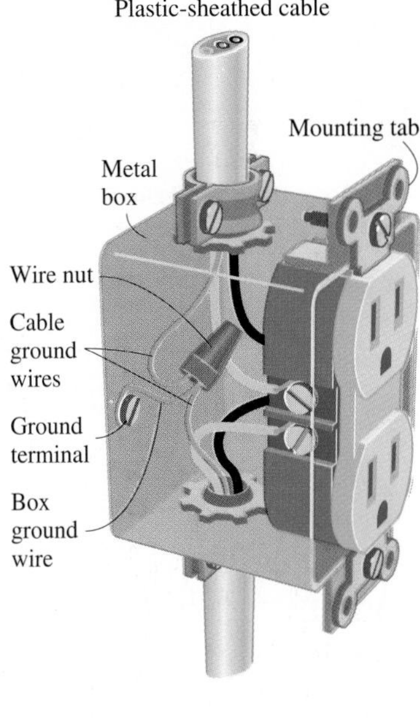

Figure 23.27 (a) An old-style two-prong outlet. (b) A modern three-prong outlet. The round third prong opening is internally connected to the metal mounting tab.

This raises the interesting question of the effect of electricity on the human body. Over 1000 people a year are accidentally electrocuted in the United States alone. Included are the foolhardy who balance plugged-in appliances on bathroom shelves within reach of the tub, as well as the poor souls who invariably try to retrieve kites from high-voltage lines, usually with long damp sticks (that are fair conductors at high enough voltages). Of course, anyone in a tub of water is grounded via the pipes. In the end "it's the current that kills ya"; the voltage is relevant only to the extent that it must be sufficiently high to force the current through the body. If very little current is available, as in a small electrostatic generator, even tremendous voltages, in excess of 100 kV, will be harmless for short durations.

Usually, we can feel a 1-mA current, and up to about 5 mA, however unpleasant, is as a rule harmless. When a substantial amount of charge flows (>10 mA) through muscles, it causes wrenching spasms. If those muscles are in the hand, the effect may be little more than tiny burns and a deep ache. At above 15 mA or so, one loses voluntary muscle control—rather awkward if you are holding on to a "hot" wire and can't let go. Up to roughly 50 mA, currents will cause considerable pain, but probably no massive malfunction of any crucial body process. More sizable currents across the torso can paralyze the respiratory system and disrupt the steady pumping of the heart. A current of approximately 100 mA sustained for a second or more through the heart will cause it to go into a lethal condition of ventricular fibrillation (irregular beating). The idea is to avoid letting current pass through the body in general and certainly to keep it away from the heart. Electricians working with high voltages (120 V can be lethal, 240 V demands great care), will often position one arm well away from the circuit; they risk blasting the fingers of one hand but avoid creating a hand-to-hand pathway across the chest.

The resistance of the body is determined to a large extent by the contact resistance with the outer layer of the skin. The wet human stuff within each of us is rich in ions and is a fairly good conductor. Therefore, depending on the condition of the skin, the area of contact, and the intimacy of that contact, the resistance (hand-to-hand or head-to-foot) may vary from perhaps 100 kΩ to over 1.5 MΩ dry, and possibly 100 times less, wet. If we assume a body resistance of 200 kΩ and an outlet voltage of $V = 120$ V $= IR$, it follows that $I = 1.2$ mA—far from problematic. If you are dripping wet and $R = 1$ kΩ, $I = 120$ mA and you are in big trouble. Be exceedingly careful with 120-V ac and don't even consider puttering around with 240-V ac. One is not likely to be able to let go of a high-voltage line, and the resulting burns will quickly lower the skin resistance, along with the chance of survival.

ELECTRONICS

Over the past three decades, electronics has become dominated by solid-state components, principally semiconductor diodes and transistors. Each of us is likely to use dozens of intregrated circuits (IC) in the course of an ordinary day; they are in computers, calculators, telephones, pacemakers, TVs, answering machines, radios, VCRs, elevators, wristwatches, cameras, cars, microwave ovens, and so on. An IC microprocessor "chip" (perhaps $\frac{1}{4}$ in. by $\frac{1}{4}$ in.) in the realm of what's called "very large-scale integration" might contain 450 000 minute transistors, along with a multitude of diodes, resistors, and capacitors. A digital watch requires about 5000 transistors; a little pocket calculator utilizes roughly 20 000 transistors; a computer, whose equivalent might once have filled a room with vacuum tubes, now contains a tiny 100 000-transistor chip and comfortably fits on a person's lap. By the end of the century, the 1 000 000-transistor chip will probably be commonplace.

A 4-megabit integrated circuit (IC) memory chip.

23.7 Semiconductors

An isolated atom can exist in any one of a number of distinct energy levels; its electron cloud can only have certain configurations. The atom typically has many electrons (for example, silicon has 14), distributed in closed shells about the nucleus. Only the outer group—the *valence electrons*—are involved in the behavior we are concerned with here. When these outer electrons (in silicon, there are 4) are in their lowest-energy configuration, the atom as a whole is in its lowest-energy state (the one it usually occupies), called the *ground state.* When two atoms are near each other, their interaction causes a slight shift in the allowed levels. When there are a tremendous number of interacting atoms, as in a solid, the shifted levels are so numerous and so close together that they form energy bands. The ground state is called the **valence band**, and the electrons therein are usually held to their respective atoms.

Given enough energy, one of the outer electrons can be ripped from its atom, to move relatively freely through the lattice of atoms. That electron is then said to be in the more energetic **conduction band**. Figure 23.28 shows how, in an insulator, the almost totally empty conduction band is separated from the occupied valence band by a sizable gap (10 eV or greater) from which electrons are prohibited. Very few electrons can pick up enough energy thermally to be propelled across the gap. Thus, very few move freely in the insulator; hence, the high resistivity. By comparison, in a conductor the two bands overlap, and no clear distinction between the valence and conduction bands exists. Valence electrons are free to wander from atom to atom, and the resistance is low. An *elemental* (or *intrinsic*) semiconductor such as silicon or germanium has a small forbidden energy gap (in silicon it's only about 1.1 eV). Still, at room temperature (300 K), the average KE of the electrons ($kT \approx 4.1 \times 10^{-21}$ J) is about 0.026 eV so that very few high-energy electrons can jump the gap—the conduction band will be nearly empty.

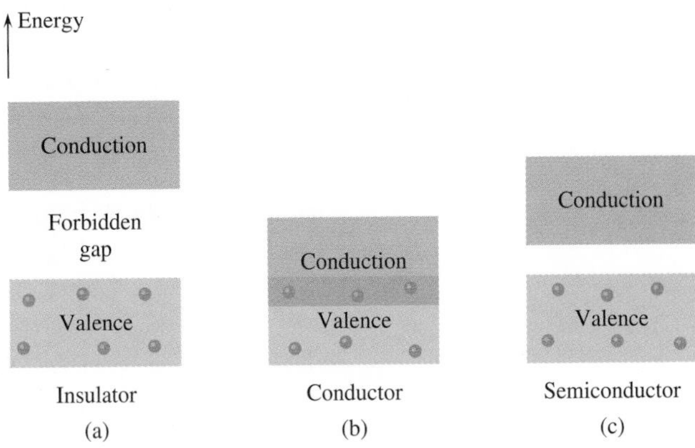

Figure 23.28 A representation of the band structure of solids. (a) The valence band in an insulator is occupied with charge carriers (electrons), but it's separated from the conduction band by a gap, so it is a poor conductor. (b) In the conductor, the two bands overlap and electrons flow easily. (c) A semiconductor has a gap, but it's small.

Doping

Devices such as transistors and diodes are fabricated using *impurity* semiconductors prepared by adding minute quantities of foreign atoms (just a few parts per million) to an intrinsic semiconductor. The process is known as **doping**, and it affects the availability of mobile-charge carriers, producing two distinct kinds of systems. Figure 23.29 shows how the electrons of a silicon single crystal are shared (in so-called covalent bonds) between the four nearest neighbor atoms. When such a crystal is doped with a five-valence-electron atom (Fig. 23.30) like arsenic, the fifth electron is not locked in place—it does not fit and can move around freely within the crystal. Such an electron resides in an energy level just below the conduction band, into which it can easily be made to jump. Because these mobile-charge carriers are negative, the system is referred to as an ***n*-type** semiconductor.

If the silicon is doped with a three-valence-electron atom like gallium (Fig. 23.31), there will be a deficiency of one electron; in effect, there will be a **hole** in the negative distribution of electrons. An outer electron from a nearby silicon atom can drop out of its cloud and fill the hole, but this will leave a new hole in the place where the electron originally was. The hole moves about like a bubble in a cup of water—it is the absence of negativity and so behaves as if it were a positive mobile-charge carrier. Because the carriers are positive, this system is known as a ***p*-type** semiconductor. The presence of the gallium gives rise to a number of empty levels just above the valence band. Electrons can jump into these levels, leaving behind holes in the valence band that can then move around in response to an applied electric field, thus constituting a current.

23.8 The *pn*-Junction and Diodes

One of the most practical arrangements of semiconductors is the ***pn*-junction**, the interface formed by joining a *p*-type to an *n*-type semiconductor. In practice, a polished slice of single crystalline *p*-type silicon is heated to around 1000°C and exposed to a vapor of arsenic or phosphorus, which diffuses into the surface. The uppermost layer is transformed into *n*-type silicon, which is then coated with a protective insulating layer of silicon dioxide. Because the single crystal structure is continuous across the junction, electrons from the *n*-type region can diffuse across to the *p*-type region, where they fill an equivalent number of holes. The *n*-type region is left positive (because of a deficiency of electrons) and the *p*-type region is negative (because of an excess of electrons). The resulting internal potential difference cuts off further transfer of charge, leaving a central region depleted of carriers. This **depletion layer** is essentially an insulator and the junction then resembles a charged capacitor (Fig. 23.32).

To see how such a *pn*-junction can function as a **diode** (that is, as a one-way gate passing current in one direction and blocking it in the other), examine Figure 23.33.

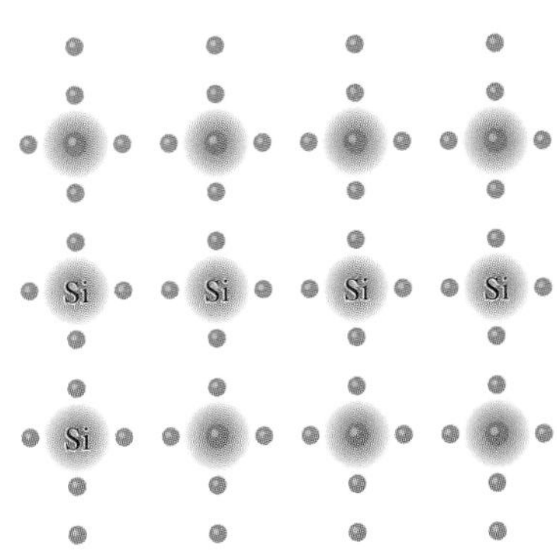

Figure 23.29 Silicon atoms sharing electrons in a crystal.

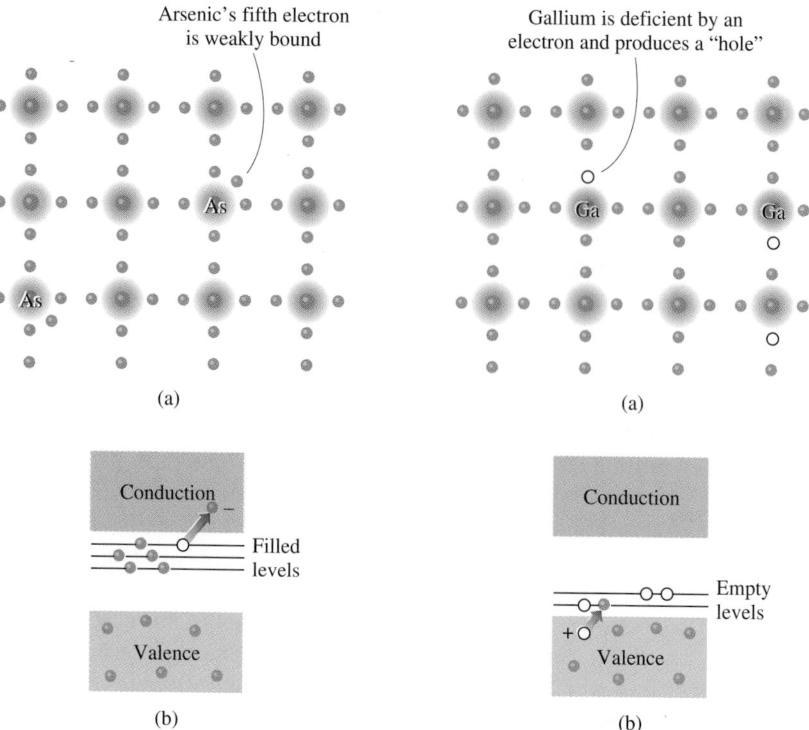

Figure 23.30 Silicon doped with arsenic. The impurity atoms result in filled levels just below the conduction band. Electrons from these levels can reach the conduction band relatively easily. These are the mobile-charge carriers.

Figure 23.31 Silicon doped with gallium. The impurity atoms result in empty levels just above the valence band. Electrons from the valence band can reach these levels relatively easily. The holes left behind in the valence band are the mobile-charge carriers.

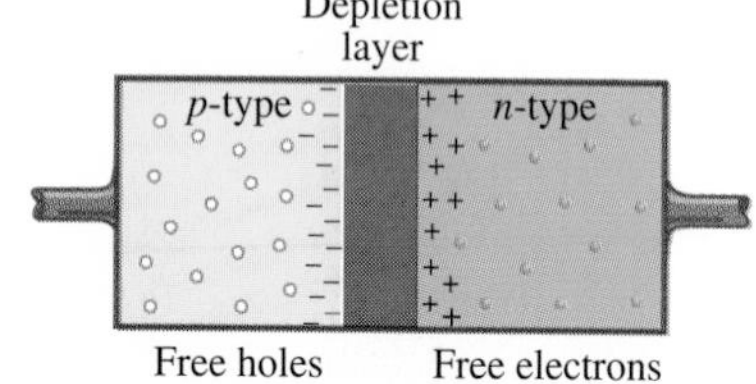

Figure 23.32 Schematic representation of a junction diode.

Here, an external potential difference is applied across the junction diode such that the positive terminal is attached to the p-side and the negative to the n-side. This opposes the internal potential difference, and the diode is **forward biased**. The positive terminal repels holes into the junction, where they are met by electrons repelled by the negative terminal—at first a tiny current traverses the diode. The depletion layer shrinks as the voltage is raised to about 650 mV (or about 300 mV for germanium), at which point the layer vanishes and the amount of current increases abruptly as carriers flow freely across the diode.

By contrast, when the diode is *reverse biased,* electrons are attracted away from the junction toward the positive terminal just as holes are attracted away toward the negative terminal. The depletion layer broadens, and only an exceedingly small current can traverse the diode.

Although we can take the opportunity to study only the diode, there are a number of other important applications of the pn-junction. These include the photovoltaic (solar) cell, the light-emitting diode (LEDs are the bright little red lights in VCR, camera, and stereo displays), and the *diode laser.*

Rectification

The process of converting ac to dc is called **rectification**, and the junction diode is a popular rectifier. The vast majority of electronic devices require some dc, and if

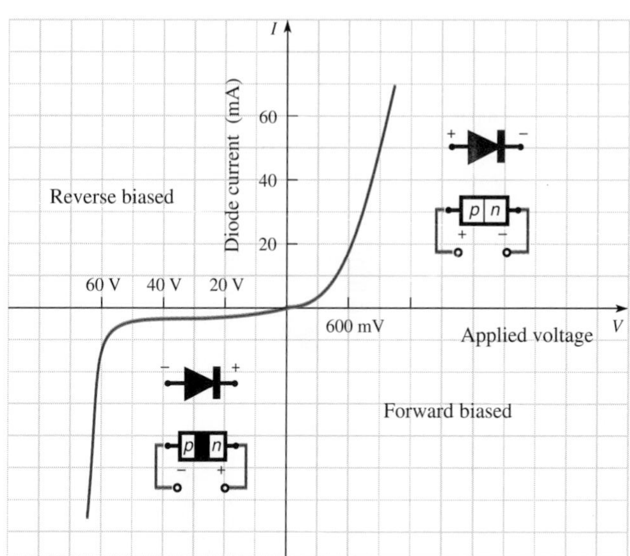

Figure 23.33 Current versus voltage for a junction diode. Observe the different voltage scales. A forward-biased junction diode conducts. A reverse-biased diode does not conduct. The symbol for the diode is an arrow in the direction it conducts a conventional current.

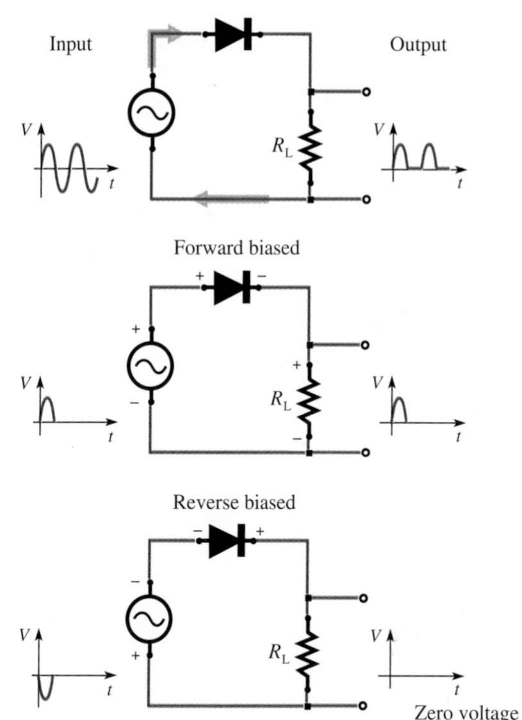

Figure 23.34 Rectification using a diode. Only when it's forward biased does the diode conduct. Think of it as an active switch that opens and closes depending on the polarity of the signal.

the power source is 60 Hz ac, a rectifier will be needed. As arranged in Fig. 23.34, just the positive portion of the signal is passed by the diode, which will allow only a clockwise current. The output voltage across the *load resistor* R_L is dc, but hardly constant. To smooth out the voltage, we add a capacitor to make an *RC*-filter (Fig. 23.35). The voltage across *C* rises positively, and it becomes charged. As the voltage starts to drop from its maximum, the capacitor discharges into the load resistor (it can't send a current backwards through the diode). With a long-time constant (*RC*), that discharge will be so slow that the next positive voltage peak will interrupt it and recharge the capacitor. The output, which still isn't flat, is said to have a *ripple* and, for obvious reasons, the circuit is known as a **half-wave rectifier**. A **full-wave rectifier** makes use of both the positive and negative peaks, producing a much smoother output with less ripple. One version, which you're likely to find if you cut open the ac adapter for your calculator or portable CD player (p. 822), uses a center-tapped transformer and two diodes (Fig. 23.36).

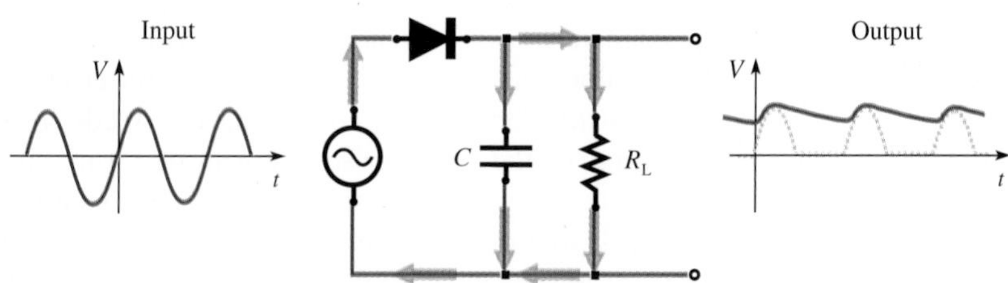

Figure 23.35 Rectification and filtering. The diode passes on only the positive peaks, and the *RC*-filter smooths them out.

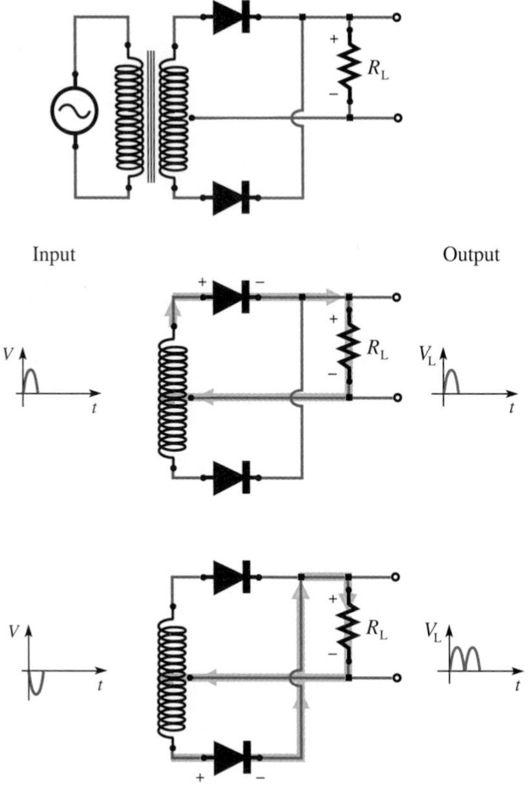

Figure 23.36 A full-wave rectifier. Note that the direction of the current through the load resistor is the same regardless of the polarity of the input signal. The output is unfiltered, but a capacitor across the output would smooth the signal appreciably.

The full-wave rectifier in an ac-to-dc adapter. You can see the transformer (attached to the plug prongs via two yellow wires), two little (black) diodes, and a (blue) filter capacitor. This is one of those black-box plugs that supplies power to calculators, rechargeable flashlights, portable tape recorders, CD players, etc.

23.9 Transistors

The transistor is effectively two diodes, back-to-back. Usually made of silicon, it is formed in a three-tiered sandwich of either *pnp* or *npn* doped regions. These *epitaxial* layers (from the Greek *epi* meaning *upon* and *taxis* meaning *arranged*) are grown in place so as to preserve the single-crystal structure (Fig. 23.37). One layer, called the *emitter* (labeled E), is highly doped. It therefore has a low resistance and is rich in mobile-charge carriers. The middle, very thin layer is the lightly doped *base* (labeled B). The remaining layer is the *collector* (labeled C) and it, too, is lightly doped. Reasonably enough, **the emitter emits mobile-charge carriers to the collector**. In a *pnp* transistor, the principal carriers are positive holes, and the arrow in the symbolic representation of the transistor is in the direction of traditional current. In an *npn* transistor, the principal carriers are negative electrons, and a positive current is imagined to pass from collector to emitter while the actual electron flow goes from emitter to collector. In either case, within the transistor symbol the emitter-base arrow always points from a p-type to an n-type region.

The transistor is usually placed in series with a dc source able to provide an appreciable current. It then serves as an electrical control valve, opening fully or partially, and allowing a proportionately large current to pass, or closing and cutting off the current altogether. This it does, in effect, by varying the emitter-collector resistance (indeed, the name *transistor* comes from combining the words *transfer* and *resistor*). The electrical control of the valve arises by way of a tiny current through the base—the base or signal current. Variations in this very small input swing the valve open proportionately and thus control the very much larger emitter-collector current. In other words, the tiny input signal current is *amplified* in the form of an identically shaped, but much larger, output current.

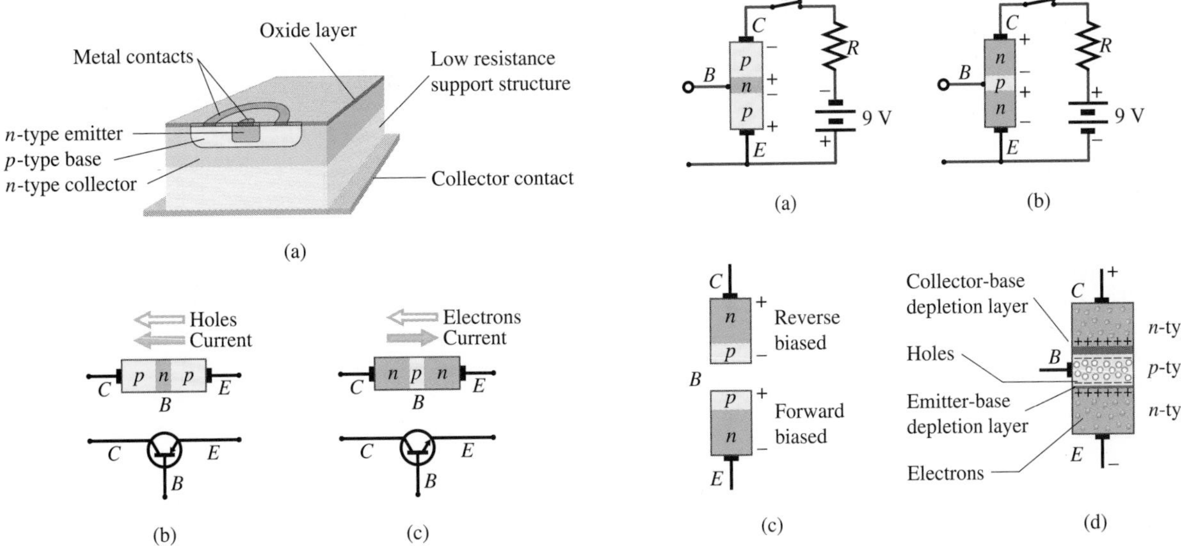

Figure 23.37 (a) The transistor is formed of doped layers of *p*- and *n*-type semiconductors. (b) The mobile-charge carriers in the *pnp* device are holes. (c) Electrons are the mobile-charge carriers in the *npn* transistor.

Figure 23.38 (a) A *pnp* transistor and (b) an *npn* transistor, each with a voltage between collector and emitter. (c) The device functions as if it were two *np* diodes oppositely biased. (d) In the *npn* transistor, electrons flow from the emitter (*n*) to the collector. In the *pnp* transistor, holes flow from the emitter (*p*) to the collector.

Figure 23.38 shows both a *pnp* and an *npn* transistor, each with a battery supplying a voltage between collector and emitter. Consider either transistor—say, the *npn*. It can be imagined as two *np* diodes back-to-back (*np*-*pn*)—two junctions formed, one on each surface of the base and separated by the layer's small ($\approx$10-μm) thickness. Recall that when the *p*-region is positive with respect to the *n*-region, the junction is forward biased, as in Fig. 23.33. Thus, while the *E*-*B* junction is forward biased by the battery, the *C*-*B* junction is reverse biased (and the same is true for the *pnp* transistor). Immediately after the switch is closed, if the forward bias exceeds about 650 mV for silicon (and $\approx$300 mV for germanium), electrons will easily flow from the emitter into the base. The majority of these will continue on, cross the thin base, move into the collector and out into the circuit, driven by the battery. But this current will not continue long. Holes in the base will soon be depleted, and that will change things drastically. The holes migrate across the *E*-*B* junction toward the negative terminal of the battery, and they can also be lost by recombining with electrons flowing toward the base from the emitter. The total effect is to cause the appearance of a net negative charge on the base that, opposing the electron flow from the emitter, will soon stop that current almost entirely (Fig. 23.38d). A relatively small charge inhibits the progress of a good deal of current that could be provided by the battery. The transistor is then like an opened switch with a nearly infinite resistance, which is identically the case with the *pnp* transistor, except there the principal carriers flowing from *E* to *C* are holes.

This blocked situation can be reduced or eliminated by injecting holes into the base of the *npn* transistor or, equivalently, drawing electrons out of it. By sending a small positive current into the base, a proportionately large positive current will pass from *C* to *E* and around the external circuit (Fig. 23.39). Again, the same thing will happen with a *pnp* transistor, provided electrons are injected into (or holes removed from) its base. *The presence of a small base current controls the flow of charge from emitter to collector.*

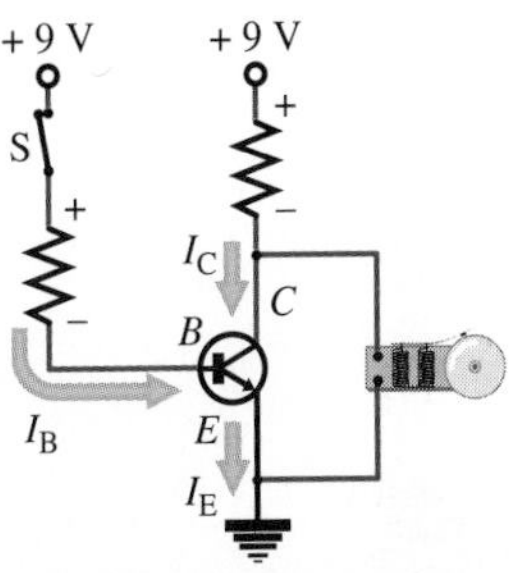

Figure 23.39 An *npn* transistor used as a switch, perhaps in a burglar alarm. Opening the switch *S* rings the bell. Here $I_E = I_B + I_C$. As long as current enters the base, the collector-emitter current is uninhibited. Interrupting I_B will cause the transistor to essentially open the circuit. With the path to ground blocked, current I_C goes through the bell instead, and it rings.

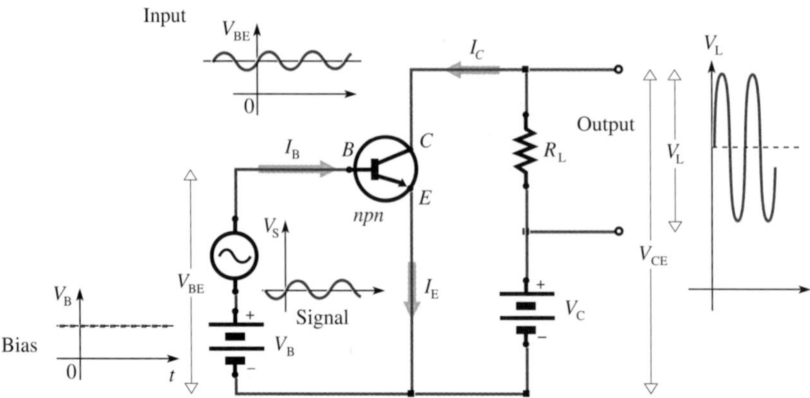

Figure 23.40 The common-emitter amplifier. The signal V_S is superimposed on a bias voltage V_B so that the input V_{BE} is always positive. The amplified output is taken off the load resistor. This kind of amplifier has a large current gain (50–250), a high voltage gain, and a medium input resistance ($\approx$2 kΩ).

Amplifiers

Most transducers, used either in the home or laboratory, rarely generate enough electrical output to directly power an appropriate indicating device. Some high-quality microphones produce voltages of only a few tenths of a microvolt or less. The ac signals from a radio antenna or a tape recorder pick-up head are all a few millivolts and much too weak to be fed directly to a loudspeaker, especially if you want to hear a thundering performance. Such signals are usually amplified several times in successive stages before sending them out to the speakers or some other *load*. The design of an amplifier (often with one or two transistors) depends on the type of input and load, and there is an endless variety of possibilities.

Figure 23.40 depicts one of the most useful basic amplifier configurations, the *common emitter*. Its name comes from the fact that the emitter is common to both the base and collector circuits. An alternating signal is fed to the base on top of a constant dc bias voltage of the battery (perhaps 1.5 V). That way, the input into the base is never negative, and the base always receives the necessary influx of positive charge. Without the bias, the transistor would clip off the negative portion of the signal.

A small change in base or *input current* ΔI_B (typically several microamps) results in a large change in the collector or *output current* ΔI_C (perhaps a few milliamps), and we define the ratio $\Delta I_C/\Delta I_B$ as the *current gain* of the amplifier. And there is a corresponding definition for the *voltage gain*. According to the manufacturer, a general purpose *npn* transistor like the 2N3904 has a maximum current gain of 400.

An integrated circuit.

Though individual transistors are still being used to fabricate amplifiers and countless other devices, especially in small production runs, the integrated circuit is dominant in many areas of electronics. A *monolithic IC* is formed on a single crystal wafer, usually silicon, but there are other competing semiconductors as well. Using diffusion techniques (the vapor deposition of layers of materials and etching), hundreds of thousands of components can be formed and "wired" together. A microscopic monolithic resistor is just a tiny sheet of isolated semiconductor. A monolithic capacitor is a little reverse-biased *pn*-junction. Monolithic transistors are usually *npn*, formed by successive diffusions. Although several hundred thousand transistors and all their accompanying resistors and capacitors can be made to fit on a dime, the basic operation of such a technological wonder is not any different from the kinds of circuits we have discussed.

Core Material

A sinusoidal ac source has a terminal voltage of the form

$$v = V_m \sin \omega t = V_m \sin 2\pi ft \quad [23.1]$$

and with a load resistor R across its terminals

$$i = I_m \sin \omega t = I_m \sin 2\pi ft \quad [23.2]$$

The **effective** (or *rms*) **current** is given by (p. 802)

$$I = I_{eff} = \frac{I_m}{\sqrt{2}} \quad [23.7]$$

and the **effective** (or *rms*) **voltage** is

$$V = V_{eff} = \frac{V_m}{\sqrt{2}} \quad [23.8]$$

Thus the average power dissipated is

[*resistive load*] $$P_{av} = I_{eff}^2 R = I_{eff} V_{eff} \quad [23.10]$$

With a purely inductive load, *the current lags the voltage* (or the voltage leads the current) *by one-quarter cycle* (90°). The **inductive reactance** X_L is defined as

$$X_L = \omega L \quad [23.14]$$

whereupon $$V = IX_L \quad [23.15]$$

The average power supplied per cycle in a purely inductive circuit is zero (p. 803).

The instantaneous current in a capacitor leads the instantaneous voltage across it by one-quarter of a cycle (90°). The **capacitive reactance** X_C is defined as

$$X_C = \frac{1}{\omega C} \quad [23.19]$$

whereupon $$V = IX_C$$

Only resistance will dissipate power in an ac circuit converting electrical energy into thermal energy (p. 805).

In a series L-C-R circuit

$$V_m = \sqrt{(I_m R)^2 + \left(I_m \omega L - \frac{I_m}{\omega C}\right)^2} \quad [23.26]$$

and $$\theta = \tan^{-1} \frac{\omega L - 1/\omega C}{R} \quad [23.27]$$

The quantity $(X_L - X_C)$ is called the **reactance** of the circuit:

$$X = X_L - X_C \quad [23.30]$$

It follows that $$V = I\sqrt{R^2 + X^2}$$

and the **impedance** is

$$Z = \sqrt{R^2 + X^2} \quad [23.31]$$

in ohms. Ohm's Law for ac is then

$$V = IZ \quad [23.32]$$

The average power dissipated is

$$P_{av} = IV \cos \theta \quad [23.33]$$

where the product IV is called the *apparent power* (p. 809).

At **resonance**, where the capacitive and inductive reactances are equal (p. 811), we have

[*resonance*] $$f_0 = \frac{1}{2\pi\sqrt{LC}} \quad [23.35]$$

$\theta = 0$, $Z = R$, and $P_{av} = IV$.

For a transformer, the ratio of the effective voltages will equal the ratio of the numbers of turns, expressed as

$$\frac{V_p}{V_s} = \frac{N_p}{N_s} \quad [23.36]$$

Moreover $$V_p I_p = V_s I_s \quad [23.38]$$

A *pn-junction* is the interface formed by joining a p-type to an n-type semiconductor. A *pn*-junction can function as a **diode**. The process of converting ac to dc is called *rectification* and the junction diode is a popular rectifier (p. 820). The junction transistor, usually made of silicon, is formed in a three-tiered sandwich of either *pnp* or *npn* doped regions. It's a kind of electrical switch that can be used to control a large current and, in that sense, to amplify a signal.

Suggestions on Problem Solving

1. When converting from effective values to maximum values, or vice versa ($V = 0.707V_m$ or $V_m = 1.414V$), keep in mind that you are simply multiplying by either $1/\sqrt{2}$ or $\sqrt{2}$. Which of these to use is easy to remember: if you are computing V_m, it's larger than V, and you multiply by 1.414 to get a larger number.

2. Remember that ω is in radians per second. Don't mix up degrees and radians when dealing with the phase angle. Accordingly, $\sin \omega t$ must be computed using the radian setting on your calculator. For example, when $f = 60$ Hz and $t = 0.01$ s, $\sin 2\pi ft = \sin 3.77 \text{ rad} = -0.59 = \sin 216°$.

3. In effect, an inductor holds back, or chokes, ac current, allowing voltage to develop, and a capacitor holds back ac voltage, allowing current to develop. In series circuits, the phasor diagrams are for voltages (the current is the same for each element). In series, *whenever the phase angle is positive, voltage leads current;* the applied voltage leads the circulating current. *Whenever the phase angle is negative, voltage lags current.*

4. Many of the problems you will confront in this chapter can be checked by recomputing the results using a different approach. That is especially true in series circuits where the impedance triangle usually provides a complementary perspective.

Discussion Questions

1. Heavy wire connects a light bulb to a large air-core coil of several hundred turns (Fig. Q1). The coil goes to a pair of fuses, which connect to a double-pole single-throw switch, and finally the leads plug into the power line via a wall outlet. Both leads contain fuses just to make sure the hot line, which is often unknown, is fused. When the switch is closed, will the lamp light? What, if anything, will happen to the light level after the iron core is inserted?

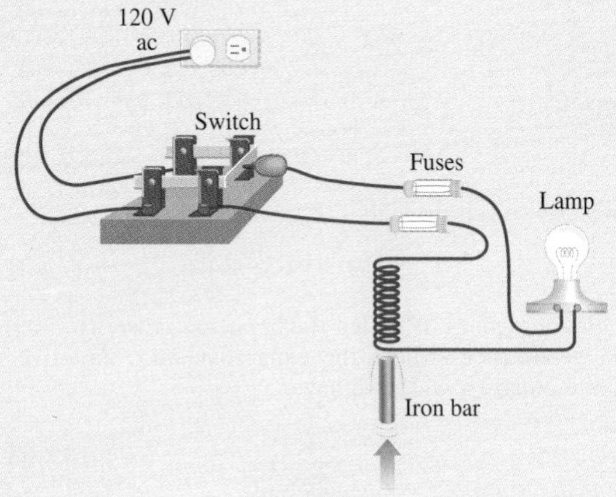

Figure Q1

2. Figure Q2 shows a galvanometer in a circuit with four diodes. Explain how this arrangement works as an ac voltmeter.

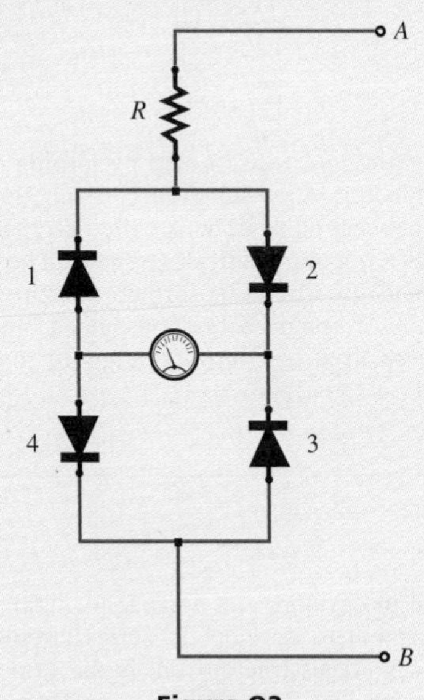

Figure Q2

3. The transmission of electrical power via high-voltage dc is becoming increasingly attractive. List a few reasons why this might be the case.

4. Does the input to a transformer have to be dc for there to be an output?

5. The laboratory setup in Fig. Q5 is used for testing diodes. The oscilloscope is essentially a voltage-measuring instrument, and that determines how it is used in the circuit. Voltages fed across the horizontal and vertical terminals cause the beam to be deflected proportionately along those directions. Explain what's happening here. What is being measured on the scope?

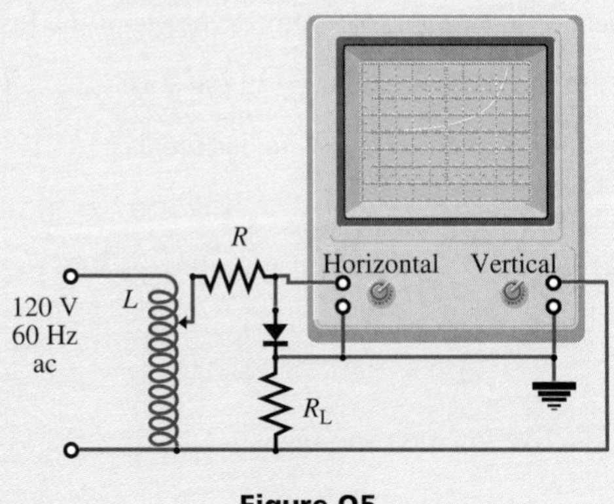

Figure Q5

6. The voltage tester in Fig. Q6 is a little neon discharge lamp that glows brightly when 110 V is across its terminals (with no concern for polarity). Explain what should be observed in each part of the diagram if all is functioning normally. The shorter slot (here on the right) is supposed to be wired "hot" in modern receptacles. What would happen if one lead of the tester was inserted in the right-hand slot and the other touched the faucet on the kitchen sink?

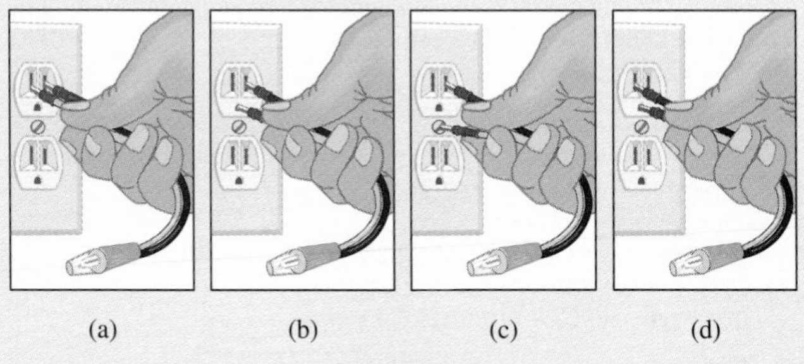

Figure Q6

7. The circuit in Fig. Q7 is being studied, using an oscilloscope. The scope is essentially a voltage meter. Here, it's set to scan horizontally and at the same time be deflected vertically by the voltage signal coming in across its two terminals (see Problem 22). Sketch the curves that would result with the scope's probes across points A and B, B and C, and A and C. Discuss.

8. Imagine that you have a faulty wall outlet and wish to replace it. Why would it be wise to shut off the voltage to the outlet? You go to the fusebox and unscrew the fuse and return to the receptacle. Need you be concerned about touching the cover plate, which you want to unscrew and remove? What can you conclude if the tester held as in Fig. Q8 lights up? What do you do next?

9. Imagine that a fuse blows (or a breaker pops) and the lights in a room go out. You remove the toaster and shut off the hair dryer and then screw in a new fuse, but the lights don't go

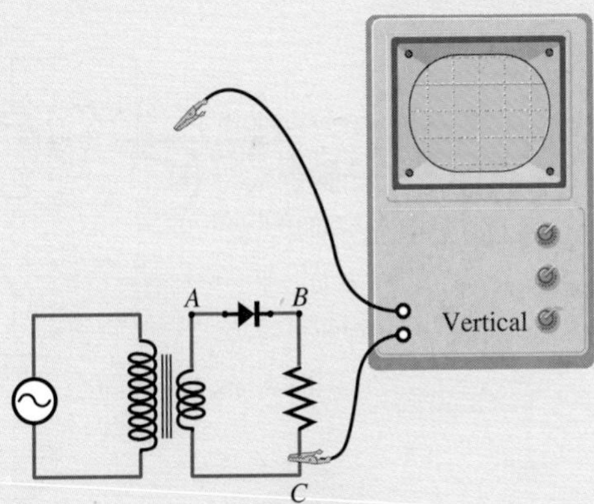

Figure Q7

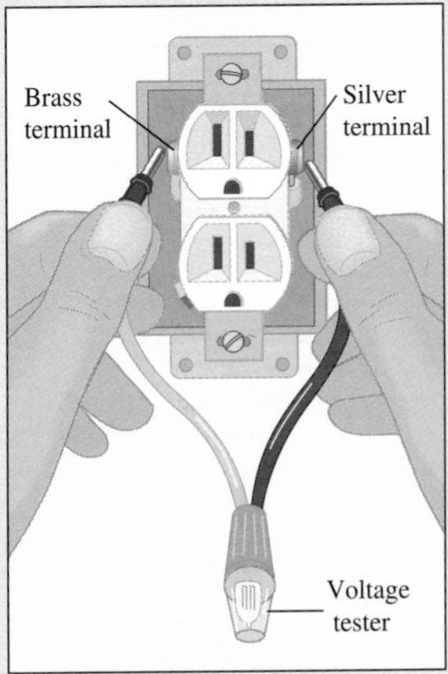

Figure Q8

back on. How can you check the installation of the fuse with a neon tester?

10. Suppose a 240-V air conditioner doesn't seem to be working (at least it doesn't do anything when the on-switch is turned). A check of the outlet (Fig. Q10) reveals that a neon tester lights brightly between terminals B and C, but not between A and B or A and C. What's wrong?

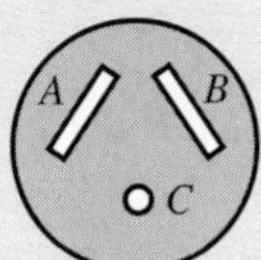

Figure Q10

11. Figure Q11 shows an R-C combination being fed by an ac signal, with the voltage across the capacitor as the output. It is left for Problem 67 to show that the ratio of the output to the input voltage is

$$\frac{V_o}{V_i} = \frac{1}{\sqrt{1 + (2\pi f RC)^2}}$$

What happens to this ratio when $f \to 0$? When $f \to \infty$? Why is it called a low-pass filter?

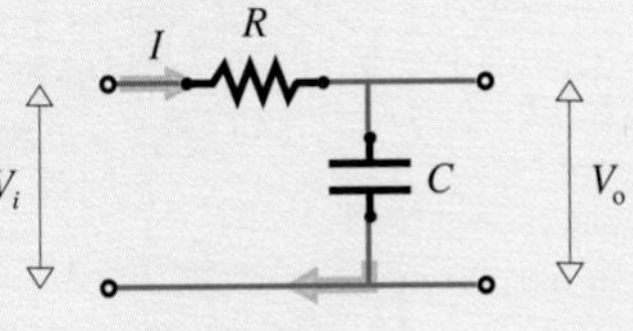

Figure Q11

12. Figure Q12 shows a three-prong adapter. What purpose does it serve and how? What will happen if you just plug in the adapter and don't bother attaching the plug's ground contact to anything?

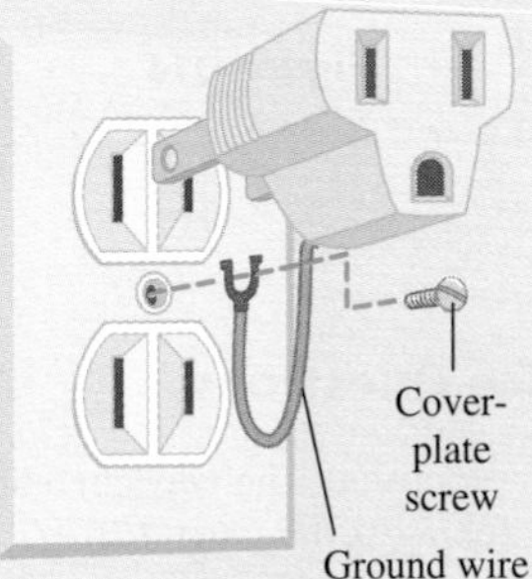

Figure Q12

13. Figure Q13 shows an R-C combination being fed by an ac signal, with the voltage across the resistor as the output. It is left for Problem 66 to show that the ratio of the output to the input voltage is

$$\frac{V_o}{V_i} = \frac{1}{\sqrt{1 + \dfrac{1}{(2\pi f RC)^2}}}$$

What happens to this ratio when $f \to 0$? When $f \to \infty$? Why is it called a high-pass filter?

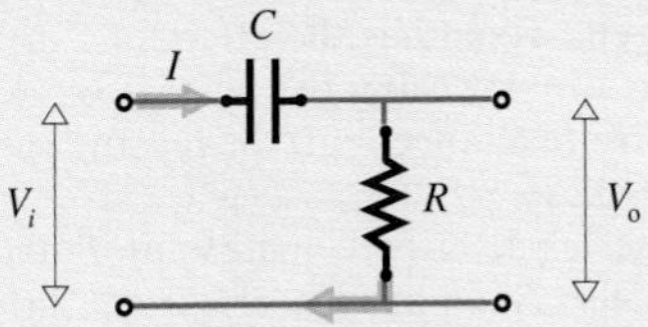

Figure Q13

14. The circuit shown in Fig. Q14 represents the wiring arrangement for a switch and a light fixture. As is customary in such installations, two or more wires are joined (that is, twisted together) by "wire nuts"—those are the small plastic caps wherever there is a splice. The cable is plastic-sheathed and contains two insulated leads and a third bare ground wire. Describe the circuit and draw a simplified diagram. Where does power enter? What does the switch do? What do the gray wires do? Which is the hot lead?

15. What is the device pictured in the circuit diagram in Fig. Q15? How does it work—that is, what is the function of each portion of the circuit?

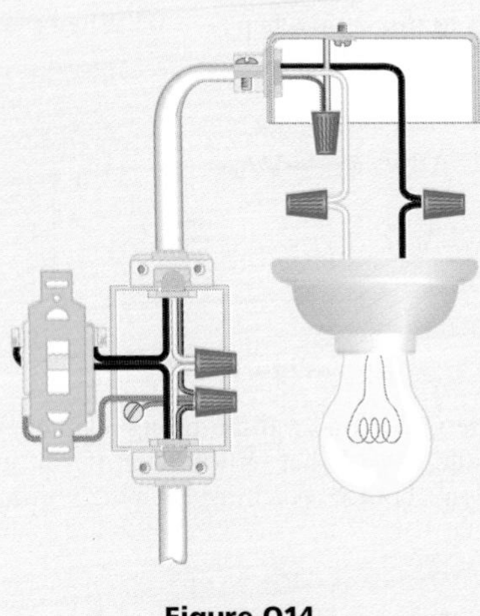

Figure Q14

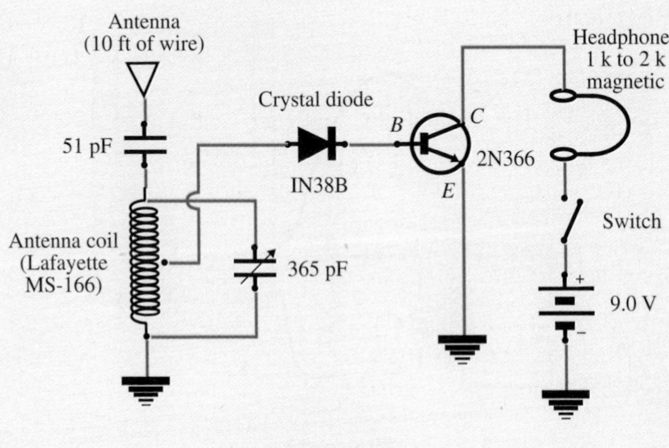

Figure Q15

Multiple Choice Questions

1. In a sinusoidal-ac circuit, the *rms* current is (a) $1.414I_m$ (b) $I_m/1.414$ (c) $I_m/0.707$ (d) $I/0.707$ (e) none of these.
2. In a sinusoidal-ac circuit, the peak-to-peak voltage equals (a) 2 V (b) 2(0.707) V (c) 2(1.414) V (d) 1.414 V/2 (e) none of these.
3. If we double the frequency in a sinusoidal-ac circuit, the effect on a capacitor in that circuit is to (a) double its reactance (b) increase its reactance by a factor of four (c) leave its reactance unchanged (d) halve its reactance (e) none of these.
4. When the instantaneous voltage and current in an ac circuit are in-phase, we know that (a) the total reactance is zero (b) the capacitive reactance is zero (c) the inductive reactance is zero (d) the resistance is zero (e) none of these.
5. In an *R*-*L* series ac circuit, increasing *L* and leaving everything else fixed has the effect of (a) lowering the impedance (b) lowering the reactance (c) increasing the capacitive reactance (d) increasing the power factor (e) none of these.
6. In an *R*-*L* series ac circuit, increasing *R* and leaving everything else fixed has the effect of (a) lowering the impedance (b) lowering the reactance (c) increasing the capacitive reactance (d) increasing the power factor (e) none of these.
7. In an *L*-*C*-*R* circuit, when the inductive and capacitive reactances are equal, (a) the resistance is zero (b) the phase factor is zero (c) the phase angle is zero (d) the current is zero (e) none of these.
8. In an *R*-*L* ac series circuit (a) the instantaneous current and voltage are in-phase everywhere (b) the instantaneous current leads the voltage across *L* (c) the instantaneous voltage across *L* leads the current (d) the instantaneous voltage across *L* lags the current by 90° (e) none of these.
9. A variable capacitor is wired across the terminals of an ac source. Increasing the capacitance (a) increases the reactance and decreases the current (b) decreases the reactance and increases the current (c) decreases the reactance, leaving the current unchanged (d) decreases the reactance and decreases the current (e) none of these.
10. In an ac series *L*-*C* circuit where the resistance is zero, we can expect that the phase angle will (a) always be zero (b) never be zero (c) be + or −90° (d) be 180° (e) none of these.
11. Measuring across the two terminals that span an *R*-*C* ac series circuit, we find that (a) the instantaneous current leads the voltage by 90° (b) the instantaneous current lags the voltage by 90° (c) the instantaneous current and voltage are in-phase (d) the instantaneous power leads the resistance (e) none of these.
12. A capacitor with a reactance of 100 Ω is in series with a 100-Ω resistor. The ac voltage across the two (a) lags the current by 90° (b) lags the current by 45° (c) leads the current by 90° (d) leads the current by 45° (e) none of these.
13. A pure 1.0-H inductor is in series with a 0.20-μF capacitor across a 110-V, 60-Hz wall outlet. The resulting power factor is (a) zero (b) infinite (c) 1.414 (d) 0.707 (e) none of these.
14. In an ac circuit, the power factor equals the (a) apparent power (b) real power (c) real divided by the apparent power (d) apparent divided by the real power (e) none of these.
15. True or average power equals apparent power when (a) the voltage is equal to the current (b) the voltage is in-phase with the current (c) the phase factor equals 0.707 (d) the current is very large (e) none of these.

Problems

ALTERNATING CURRENT

1. [I] The maximum output voltage of an ac generator is +120 V at a time equal to $\frac{1}{4}$ cycle. What will be the instantaneous terminal voltage (a) $\frac{1}{4}$ cycle later, and (b) $\frac{1}{8}$ cycle later (at $\frac{3}{8}$ of a cycle)?
2. [I] A sinusoidal oscillator puts out an instantaneous voltage represented by $v = (75\text{ V})\sin(376.99\text{ rad/s})t$. What is its frequency and maximum voltage?
3. [I] Write an expression for the instantaneous current delivered by an ac generator supplying 10 A effective, at 50 Hz.
4. [I] The maximum potential difference across the terminals of a 20.0-Hz sinusoidal-ac source is +50.0 V (occurring at $t = \frac{1}{4}$ cycle). If at $t = 0$, $v = 0$, find v at $t = 2.00$ ms.
5. [I] If the *rms* voltage read across a resistor by a voltmeter in a sinusoidal-ac circuit is 100 V, what is the maximum voltage?
6. [I] The effective current passing through a resistor R in a sinusoidal-ac circuit is 2.00 A. What is the maximum voltage drop across the resistor if $R = 100\ \Omega$?
7. [I] Determine the current drawn by a lit 100-W light bulb plugged into a 120-V wall outlet.
8. [I] The instantaneous voltage measured across an ac source is $v = 200 \sin 2\pi 70t$. What is its effective voltage and frequency?
9. [I] A resistor is in series with an ac generator. An ammeter in series with the resistor reads 1.50 A, and a voltmeter across the resistor reads 75.0 V. What average power is being supplied by the source?
10. [I] If the generator in Problem 8 is placed across a 200-Ω resistor, what current will be measured by an ac ammeter in series with the resistor?
11. [I] A 1.0-Ω resistor is attached across an ac generator. An oscilloscope shows that the sinusoidal current in the circuit has a maximum value of 0.50 A. What average power does the resistor dissipate?
12. [I] An 80-μF capacitor C is placed across the terminals of a 60-Hz sinusoidal-ac oscillator. What is the capacitive reactance of C?
13. [I] If a capacitor in series with a sinusoidal oscillator has a reactance of 200 Ω when $f = 50.0$ Hz, what reactance will it have when the frequency is raised to 5.00 kHz?
14. [I] Determine the current that will be drawn by a 45.0-μF capacitor connected across a 240-V, 50.0-Hz source of sinusoidal-ac.
15. [I] A 60-Hz ac generator has a terminal voltage of 120 V. What size capacitor should be placed in series with it so that a current of 1.00 A circulates?
16. [I] Figure P16 depicts a simple circuit for studying the ac behavior of capacitors. What current will the milliammeter read? If either capacitor is disconnected, what will happen to the current?
17. [I] A 0.15-H coil with a negligible resistance has a reactance of 10 Ω when wired into a sinusoidal-ac circuit. Determine the frequency of the source.
18. [I] Determine the inductive reactance of a 10.0-mH coil connected to a 100-Hz ac generator.
19. [I] A high-quality 250-mH inductor (with negligible resistance) is attached across the terminals of a 60-Hz generator having an *rms* output of 125 V. Determine the reactance of the coil.

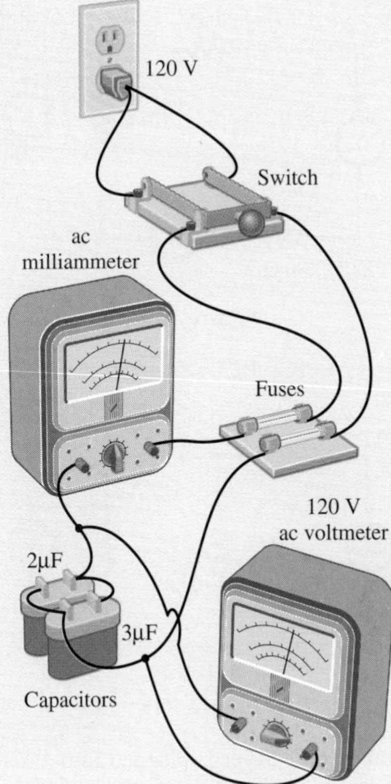

Figure P16

20. [I] What is the inductance of a coil having negligible resistance if at an angular frequency of 628.3 rad/s, its reactance is 200 Ω?
21. [II] If at a time of 2.00 ms after the start of a cycle, a sinusoidal oscillator has an output of 95.1% of its maximum voltage, at what frequency is it operating?
22. [II] An ac generator and a load resistor are in series. An oscilloscope is connected across the resistor. The vertical deflection on the scope is caused by the voltage input and corresponds to a setting of 150 mV/cm. The horizontal axis is set for scan and is swept internally by the scope, producing a sinusoidal image on the screen. If the peak-to-peak displacement of the signal is 4.0 cm, what is the *rms* voltage across the resistor?
23. [II] The circuit in Fig. P23 needs your help. Every time the switch is closed the fuse blows. What could be wrong? (Hint: One of the components is malfunctioning.)

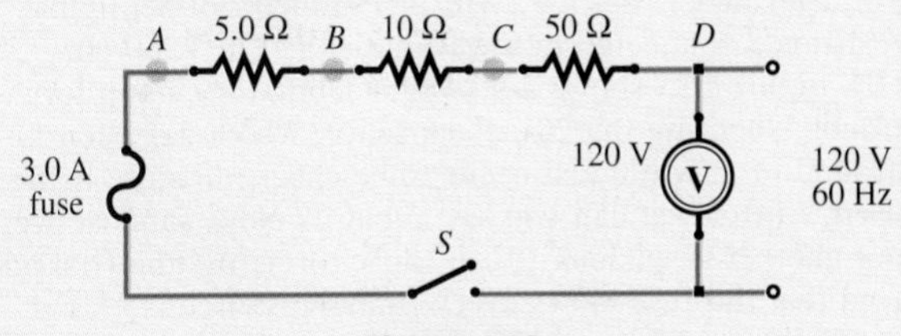

Figure P23

24. [II] Referring to Fig. P24, what will the two meters in the circuit read?

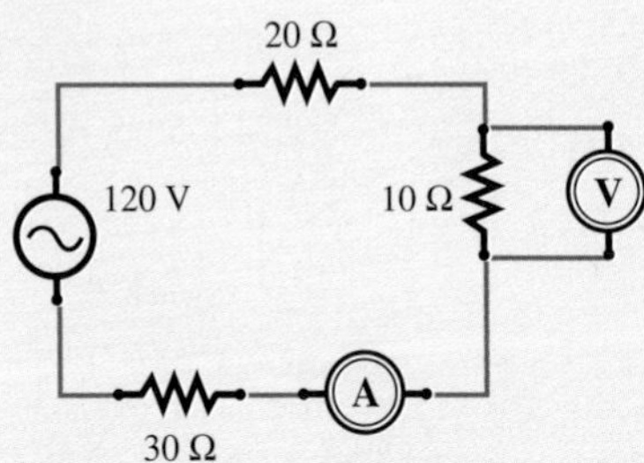

Figure P24

25. [II] Referring to Fig. P25, what will the two meters in the circuit read?

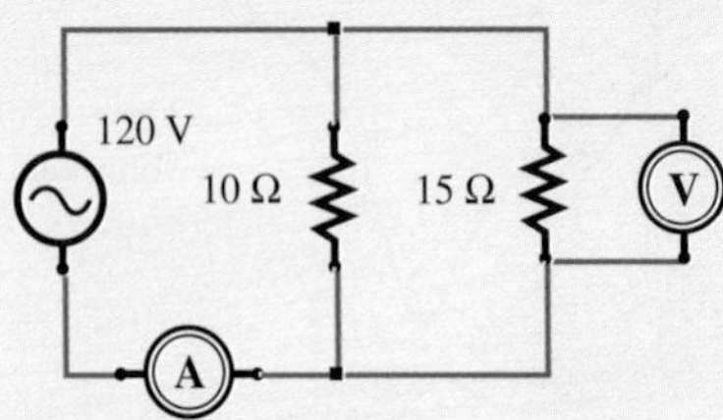

Figure P25

26. [II] A 1200-W hair dryer is plugged into a 120-V line. What's its resistance and how much current does it draw?
27. [II] Figure P27 shows a laboratory setup for studying the behavior of a resistive load (R) in an ac circuit. Here, a transformer inputs the 120 V from the wall outlet and supplies 30 V at 60 Hz, and the milliammeter reads 500 mA effective. Assuming the losses in the wiring to be negligible, what average power does the resistor dissipate? How would that change if the transformer was removed and the circuit plugged directly into the wall outlet?
28. [II] Prove that the ohm is the SI unit of capacitive reactance.

29. [II] Prove that the ohm is the SI unit of inductive reactance.

30. [II] A sinusoidal oscillator with a terminal voltage of 125 V at a frequency of 55 Hz is put in series with a capacitor of 100 μF. What effective current will be provided by the oscillator?
31. [II] A high-quality coil with an inductive reactance of 100 Ω and negligible resistance is placed across the terminals of a 150-V ac generator. Determine the maximum current through the coil.
32. [II] Determine the current that would be read by an ammeter in series with a 410-mH inductor of negligible resistance connected to a wall outlet at 120 V, 60 Hz.

33. [II] Figure P33 shows a 60-Hz ac laboratory setup for trouble-shooting the 20-μF capacitor, which has been around for a while and needs to be checked before being used. (a) What can you say about it? Now, suppose we put another suspicious 10-μF capacitor across the first and find that the fuse blows every time the switch is closed. (b) What would you then conclude?

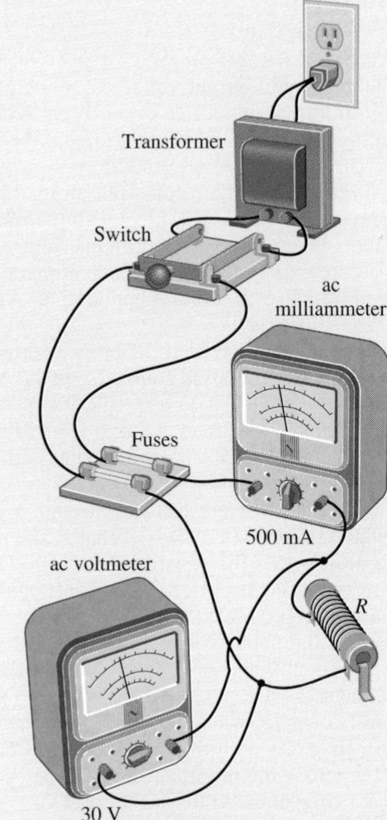

Figure P27

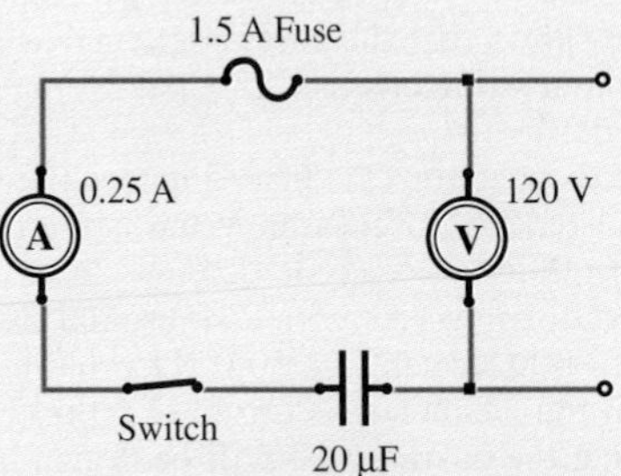

Figure P33

34. [III] A 100-Ω resistor is wired across the terminals of an oscillator, which sends a current through it given by

$$i = 2.40 \cos 180t$$

in SI units. (a) Determine the period of the current oscillation. (b) Find the effective voltage of the source. (c) Write an expression for the instantaneous voltage across the load.

35. [III] Write an expression for the maximum instantaneous power dissipated in a sinusoidal-ac circuit consisting of a source and a resistor. How does this expression compare with the average power?

L-C-R AC NETWORKS

36. [I] A 6.0-μF capacitor in series with a 500-Ω resistor are both placed across a 120-V, 60-Hz ac source. Draw the impedance triangle for the circuit. Check your answer by computing Z.

37. [I] A coil has a resistance of 6.10 Ω and a reactance of 8.05 Ω. What is its impedance?

38. [I] An inductor has a reactance of 8.0 Ω and a resistance of 6.0 Ω. When connected across a 120-V ac outlet, how much current will it draw?

39. [I] A capacitor with a reactance of 160 Ω is in series with a 150-Ω resistor and an ac source. What is the net impedance of the load?

40. [I] A coil has an inductance of 32.0 mH and a resistance of 30.5 Ω. What is its impedance when placed across a 240-V, 200-Hz sinusoidal oscillator?

41. [I] If the total reactance of an *L-C-R* circuit is 1200 Ω and the total resistance is 500 Ω, what is the impedance?

42. [I] In a series *L-C-R* circuit, the inductive reactance is 1800 Ω, the capacitive reactance is 2600 Ω, and the resistance is 1000 Ω. What is the impedance?

43. [I] An *L-C-R* ac series circuit has a value of $(X_L - X_C) = +200\ \Omega$ and a resistance of 400 Ω; what is the phase angle and does the current lead or lag the applied voltage?

44. [I] A 2.0-kΩ resistor and a 2.0-μF capacitor are connected in series across a 60-Hz ac generator. What is the impedance of the circuit?

45. [I] A series circuit contains a capacitor, a resistor, and a 40-V, 240-Hz sinusoidal-ac source. If the net impedance of the load (resistor and capacitor) is 160 Ω, what value of current will be measured by an ammeter in the circuit?

46. [I] A coil with a reactance of 150 Ω and a resistance of 25 Ω is placed across a 120-V ac harmonic source. Determine the phase angle between the instantaneous voltage across the coil and the current.

47. [I] A capacitor with a reactance of 0.20 kΩ is placed in series with a 100-Ω resistor and the combination is attached to the terminals of an ac source. What is the phase angle between the instantaneous current and the instantaneous voltage across the combination?

48. [I] In the previous problem, what is the rms current if the source has a terminal voltage of 0.12 kV?

49. [I] A coil has a reactance of 1.55 kΩ and draws 26 mA from an ac generator with a terminal voltage of 60 V effective. What is its resistance?

50. [I] In an *L-C* series circuit with negligible resistance, the voltage drops across the inductor and capacitor are 100 V and 120 V, respectively. What is the phase angle?

51. [I] A circuit consists of a capacitor with a reactance of 2500 Ω in series with an inductor of negligible resistance and a reactance of 2000 Ω. What is the impedance of the circuit?

52. [I] A choke coil (of negligible resistance) with a reactance of 2.5 kΩ and a 2.5-kΩ resistor are in series across a 120-V ac generator. What average power is dissipated in the circuit?

53. [I] An *L-C-R* series circuit contains a 500-Ω resistor, a 5.0-H choke coil, and a capacitor. What value of capacitance will cause the circuit to resonate at 1000 Hz? What can you say about the practicality of your result?

54. [I] An *L-C-R* series circuit contains a 1.00-μF capacitor, a 5.00-mH coil, and a 100-Ω resistor. What is its resonant frequency?

55. [I] A little radio has a tuning circuit consisting of a variable capacitor and an antenna coil. If the circuit has a maximum current when $C = 300$ pF and $f = 40.0$ kHz, what is the inductance of the coil?

56. [I] A transformer has a 500-turn primary and a 2500-turn secondary. If the primary voltage is 120-V ac, what is the secondary voltage?

57. [I] A typical neon sign operates at a voltage of about 11.5 kV. If a transformer is to raise the 115-V line voltage up to that level, what should be its turn ratio?

58. [I] The primary of a transformer has 600 turns and a voltage across it of 240 V. How many turns should the secondary have if the output voltage is to be 60.0 V?

59. [I] A load resistor is placed across the secondary terminals of a step-down transformer with a turn ratio of 4:1. If the primary voltage and current are 120 V and 1.0 A, respectively, what will be the secondary voltage and current? Assume 100% efficiency.

60. [II] Suppose we want to find the inductance of a coil experimentally. We use an ohmmeter to determine that its resistance is 50 Ω. Next, we attach the coil to an oscillator, which supplies 300 mA to it at 20.0 V and 1.20 kHz. Determine L.

61. [II] A 30-mH coil with a resistance of 30 Ω is supplied by a 240-V, 200-Hz sinusoidal-ac source. Calculate the effective current it draws and the phase angle between the instantaneous current and the supply voltage.

62. [II] A 5.2-H choke coil with a resistance of 20 Ω is in series with a 1200-Ω resistor. The circuit is plugged into a transformer that puts out 60-V ac at 60 Hz. What voltage drop will appear across the resistor? Across the inductor?

63. [II] A 400-mH inductor with negligible resistance and a 185-Ω resistor are in series across a 60.0-Hz sinusoidal-ac source. If the voltage drop across the resistor is 370 V, what is the voltage across the inductor? What is the voltage of the source?

64. [II] Imagine yourself in a lab. On the table is an inductor rated at 1.5 H, supposedly with a resistance of 50 Ω. It's in series with a 0.5-A fuse. Every time the circuit is plugged into the 110-V, 60-Hz wall outlet the fuse blows. (a) Determine the current that should be in the circuit if all were well. (b) What would you guess was wrong? How might you confirm that?

65. [II] In an experimental setup, a capacitor and a 20-Ω resistor are in series across a 120-V, 60-Hz sinusoidal-ac source. An ammeter indicates that the circuit carries an *rms* current of 1.20 A. Determine the capacitance.

66. [II] Figure Q13 shows an *R-C* combination known as a high-pass filter (see Discussion Question 13). It's being fed an ac signal composed of a broad range of different frequency information. The voltage across the resistor is the output that is passed on to the next circuit for further processing. Show that the ratio of the output to the input voltage is

$$\frac{V_o}{V_i} = \frac{1}{\sqrt{1 + \dfrac{1}{(2\pi f RC)^2}}}$$

67. [II] Figure Q11 shows an R-C combination known as a low-pass filter (see Discussion Question 11). It's being fed an ac signal composed of a broad range of different frequency information. The voltage across the capacitor is the output that is passed on to the next circuit for further processing. Show that the ratio of the output to the input voltage is

$$\frac{V_o}{V_i} = \frac{1}{\sqrt{1 + (2\pi fRC)^2}}$$

68. [II] A series L-C-R circuit is connected across a 10.0-kHz source. The 1.2-H inductor and 1000-Ω resistor are fixed whereas the capacitor is variable. (a) At what capacitance will the current in the resistor be a maximum? (b) What will be the voltage drop across the resistor if the source voltage is 50 V? (c) What average power will be dissipated?

69. [II] A transformer is used to power a 6.0-V door chime. The 240-turn primary, which has an inductance of 3.0 H and negligible resistance, is connected across the standard 120-V, 60-Hz ac. (a) What turn-ratio should the transformer have? (b) With the secondary open-circuited, determine the primary current.

70. [II] Draw the impedance triangle for the circuit in Fig. P70 and determine the phase angle and impedance. Check your answers by direct computation. How much current is circulating? What is the resonant frequency of the circuit, and if it were driven at that frequency, how much current would be present?

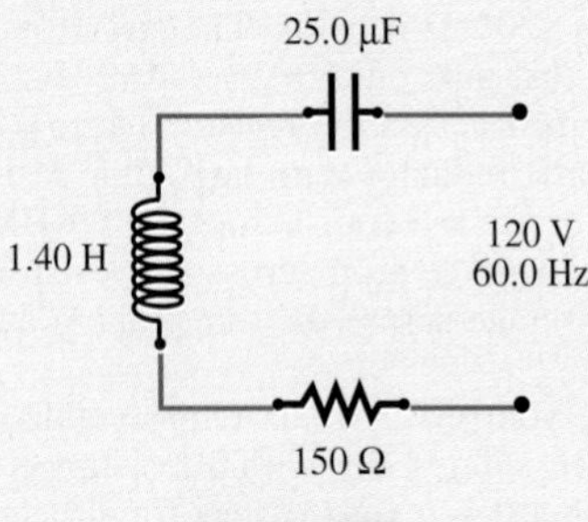

Figure P70

71. [II] A transformer delivers power at a rate of 45 kW. If it loses 300 W in hysteresis and eddy current effects (the so-called iron loss) and 500 W in joule heat (the so-called copper loss), what is its efficiency?

72. [III] A transformer is supplied 8.20 kW by an ac source across its primary. The terminal voltage at the secondary is 220 V and the load is such that there is a power factor of 82%. Determine the current in the secondary circuit.

73. [III] To transfer a maximum amount of power from a source to a load (p. 698), the two should have the same impedance but often don't (audio amplifiers have a high impedance, speakers have a low impedance). A transformer can be used to match impedances (see photo on p. 814). Let Z_s be the load attached to the secondary and V_p and I_p be the voltage and current in the primary when the transformer and load are connected to the source. The source then sees an effective load Z_p; that is, it behaves as though Z_p were attached across its terminals (rather than the transformer and Z_s). Show that

$$Z_p = Z_s\left(\frac{N_p}{N_s}\right)^2$$

What turn-ratio should a transformer have if it is to match a 3.2-kΩ amplifier to an 8.0-Ω speaker?

24

Radiant Energy: Light

"WHAT *IS* LIGHT?" IS one of those superbly simple questions we humans have pondered, and struggled with, for well over 2000 years. We are not without answers, marvelous answers that reflect the highest accomplishment of human ingenuity and creativity, but our understanding even now is far from complete, far from satisfactory. This chapter, and the next three as well, are about electromagnetic radiant energy, of which light is only that small part we happen to see. Thus, we begin the study of *Optics,* the study of the behavior of radiant energy.

Light. Caloric is the name assigned by the new nomenclature to the element of heat. It is the most subtile of all bodies, and exists in nature in very great quantities. . . . Light is occasioned by the transmission of caloric, or the matter of heat, which travels through the air, in one second of time, about 170,000 miles. . . . It is heat projected. The identity is the same; and consists of infinitely small particles thrown off in every direction from a luminous body.

H. G. SPAFFORD
General Geography and Rudiments of Useful Knowledge (1809)

Sources of Light Waves. Just as sound waves are disturbances set up in the air by the vibrations of bodies of ordinary dimensions, so light waves are disturbances set up in the ether probably by the vibrations of the minute corpuscles, or electrons, *of which the atoms of ordinary matter are supposed to be built up.*

ROBERT A. MILLIKAN AND HENRY G. GALE
A First Course in Physics (1906)

Augustin Jean Fresnel (1788–1827).

THE NATURE OF LIGHT

Among the jumble of ideas that came from the ancient Greeks were the seeds of two significant lines of thought. The Atomists evolved the *emission theory,* which pictured light as a torrent of exceedingly minute, high-speed particles. The other influential conception was advanced by Aristotle. To the four elements of nature (fire, air, earth, and water), he added a fifth—the *aether.** There was no such thing as empty space. All the void was filled with ether and so Aristotle could propose that human vision "arises from a movement, produced by the body we perceive, in the interposed medium. . . ." Light was ethereal motion.

24.1 Waves and Particles

In the 1660s, Robert Hooke studied the color patterns associated with thin transparent films, such as soap bubbles. The repetitive form of those patterns guided him to propose a rudimentary *wave theory* in which light consisted of very rapid vibrations of the ether propagating at tremendous, though finite, speed. The fastest naturally occurring thing known at the time was sound, and light was even faster—the lag between the perception of thunder and lightning proved that. So it seemed reasonable to suppose that only a wave, which is a *self-sustaining disturbance of a medium without the transport of matter,* could travel at such tremendous speed. Furthermore, two beams of light can cross (as they must, for example, when you look through a small hole at two separate sources), and yet they never scatter each other as streams of particles surely would.

Although Newton seems earnestly to have tried to stay out of the wave-versus-particle debate, he nonetheless ultimately favored the corpuscular picture. His rejection of the wave theory came from his inability to explain the observed straight-line propagation of light. To Newton's mind, a wave should spread out markedly as it passes through an aperture (Fig. 24.1), filling almost the whole region beyond. It couldn't possibly produce a narrow beam, and yet such beams could easily be made. Unhappily, neither he nor anyone else of that era knew very much about waves. Waves do spread out, just as Newton had maintained, but will only do so in the way he envisioned when the wavelength is roughly the same size as the hole through which they pass and the wavelength of light is minute (Fig. 13.36, p. 469, makes the point nicely).

Aware of Hooke's work with thin films, Newton proposed a remarkably prophetic solution. Light has a dual nature: it is a stream of material particles, but these are capable of setting up vibrations in the ether. The resulting pattern in the distribution of ether channels the stream of corpuscles—waves of ether guide particles of light.

Although the wave theory did have a few important backers in the eighteenth century, such as Benjamin Franklin and Leonhard Euler, the age was dominated by the works of Newton. His doting disciples soon forgot the master's own doubts and, in his name, light *became* a stream of particles.

In the first few decades of the 1800s, the wave theory was reborn with a new analytic vigor. Resurrected principally and independently by English physician Thomas Young and French civil engineer Augustin Fresnel, the wave conception

*Though no longer accepted, the notion of a material aether (the modern spelling is *ether*) has played a tremendously important role in the development of physics. This divine, invisible, elastic "goop" would ultimately become the transmitter of all sorts of influences from gravity and electricity to heat and "life force." The word itself was first used by the poet Hesiod (700 B.C.) to refer to the uppermost tenuous reaches of the Earth's atmosphere.

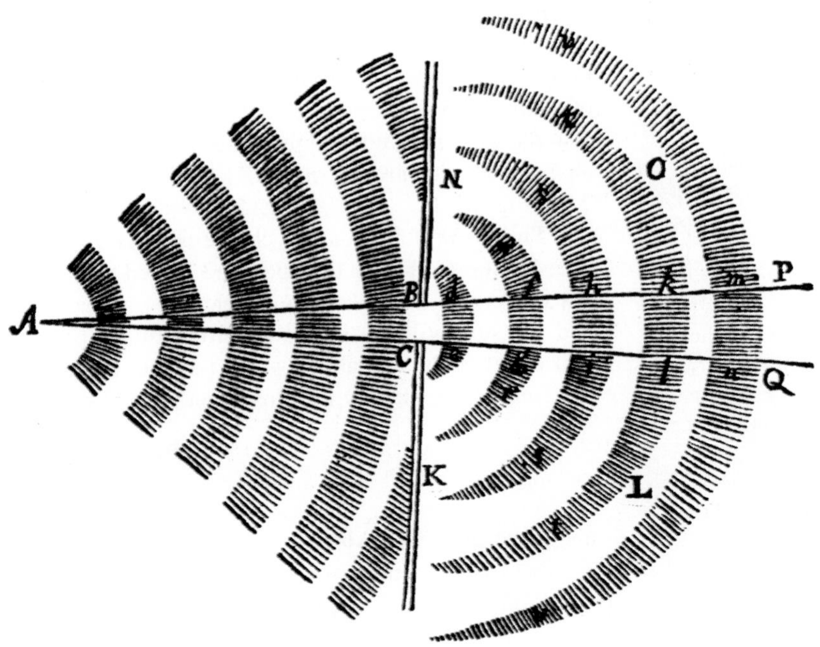

Figure 24.1 Taken from the *Principia,* this is Newton's picture of the diffraction of a wave as it passes through a hole. It's a fairly good representation, provided the hole is smaller than the wavelength.

slowly gained acceptance. By supposing that light was a *transverse* disturbance (p. 421), they could understand everything that had been observed over the centuries. Light was once again a wave—presumably an elastic wave in the ethereal sea.

24.2 Electromagnetic Waves

While this saga was being played out, and quite independent of it, the study of electricity and magnetism was reaching a high point of its own. The masterwork of the age was created by a good-humored, bright young man, one James Clerk Maxwell of Kirkcudbrightshire in the Lowlands of Scotland.

William Thomson (later to become Lord Kelvin) was, in the 1840s, the first to begin to mathematize the notion of *lines-of-force.* Faraday himself was ill at ease with mathematics, never having had any formal training in the subject. One day in 1846, Faraday was asked to step in at the last moment and give the evening lecture at the Royal Institution in place of Charles Wheatstone who had had stagefright and run off minutes before. During the impromptu remarks that followed, Faraday conjectured that light might be a wave of some sort propagated along the lines-of-force. So it was that Faraday's visions and Thomson's lovely though disconnected thoughts became the stimulus for Maxwell, who began to pursue his own researches within months of graduating from Cambridge in 1854. He drew together everything fundamental that was known about electricity and magnetism in a set of four equations. Two of these are credited to Gauss, the third is Faraday's Induction Law, and the fourth, Ampère's Circuital Law—from these Maxwell built his theory.

James Clerk Maxwell (1831–1879).

We saw earlier that *in the presence of a varying magnetic field, an emf will be induced in any closed path;* that is,

$$\mathscr{E} = -\frac{\Delta\Phi_M}{\Delta t} \qquad [22.2]$$

A Letter From Maxwell to Kelvin

Trin. Coll., Feb. 20, 1854

Dear Thomson

Now that I have entered the unholy estate of bachelorhood I have begun to think of reading. This is very pleasant for some time among books of acknowledged merit wh one has not read but ought to. But we have a strong tendency to return to Physical Subjects and several of us here wish to attack Electricity.

Suppose a man to have a popular knowledge of electrical show experiments and a little antipathy to Murphy's Electricity, how ought he to proceed in reading & working so as to get a little insight into the subject wh may be of use in further reading?

If he wished to read Ampère Faraday &c how should they be arranged, and at what stage & in what order might he read your articles in the Cambridge Journal?

If you have in your mind any answer to the above questions, three of us here would be content to look upon an embodiment of it in writing as advice. . . .

where $\Phi_M = B_\perp A$ is the flux. Because there is an emf, any charges in the presence of this varying B-field would experience forces and that, in turn, is equivalent to the existence of a nonelectrostatic electric field. By definition, the emf is also the work per unit charge performed by that E-field as a charge is carried around the path (a little length of which is Δl in Fig. 24.2).

This work done per unit charge is given in general by $\Sigma E_\parallel \Delta l$, summed over the specified closed path (p. 682). Since this is a nonelectrostatic field, it is not conservative and the sum will not be zero. Setting this expression for emf equal to the one above where the flux passes through the area A, bounded by the closed path C, yields

$$\Sigma_C E_\parallel \Delta l = \mathscr{E} = -\frac{\Delta \Phi_M}{\Delta t} \tag{24.1}$$

and since $\Delta\Phi_M = A \Delta B_\perp$

$$\Sigma_C E_\parallel \Delta l = -\frac{\Delta B_\perp}{\Delta t} A \tag{24.2}$$

This variation of Faraday's Law is today known as one of **Maxwell's Equations**. In essence, it says that **a time-varying B-field generates an E-field**: when the right side is nonzero, the left side is nonzero.

Propagation of Light by the Ether. As the atoms of matter vibrate in the ether in which they are immersed, they communicate their vibration to it. The vibrations thus started in the ether are propagated through it in every direction in minute waves and with an inconceivable velocity.

J. A. GILLET AND W. J. ROLFE
Natural Philosophy for the Use of Schools and Academies (1882)

Figure 24.3 depicts a localized upwardly directed increasing magnetic field **B** surrounded by an induced electric field **E**. Three points are crucial here: **E** and **B** are everywhere perpendicular to each other; the lines of **E**, which do not now begin and end on charges, must therefore close on themselves, and **E** is not restricted to the region actually containing the flux—it extends beyond that.

The second equation that will be of special interest is Ampère's Circuital Law (p. 740), relating the amount of **B** parallel to a closed path C with the total current due to flowing charges, ΣI, passing within the confines of C; that is, passing through *any* area bounded by C:

$$\Sigma_C B_\parallel \Delta l = \mu_0 \Sigma I \tag{21.6}$$

This expression says that moving charges are the source of the magnetic field, and although that's true, it's not the whole truth. That much is evident by the fact that, while charging or discharging a capacitor, one can measure a magnetic field in the region between the plates, even though no actual current traverses the device (Fig. 24.4). Moreover, Ampère's Law is not very particular about the area used, provided it's bounded by the curve C, which makes for an obvious problem when charging a capacitor, as shown in Fig. 24.5. If flat area A_1 is used, a net current of I flows through it and there is a B-field along curve C—the right side of the equation is nonzero, so the left side is nonzero. But if area A_2 is used instead to encompass C, no net current passes through it and the field must now be zero, even though nothing physical has actually changed. Something is obviously wrong! Notice that if Q is the charge on either plate and A the area thereof, it follows from Eq. (17.12) that

$$E = \frac{Q}{\varepsilon_0 A}$$

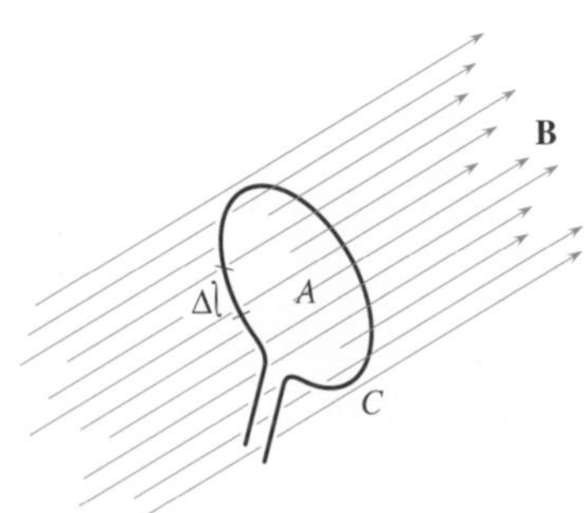

Figure 24.2 A loop enclosing an area A in a magnetic field **B**. Here, the area is flat and perpendicular to the field.

is the electric field spanning the gap. As the charge builds on the plates, the field changes as well, $\Delta E = \Delta Q/\varepsilon_0 A$. In a time Δt,

$$\varepsilon_0 A \frac{\Delta E}{\Delta t} = \frac{\Delta Q}{\Delta t}$$

where the expression on the right is the current actually flowing into, and out of, the capacitor. What this result suggests is that charge building up on the plates creates a changing electric field that crosses the gap and is in some way equivalent to a current. Maxwell hypothesized the existence of just such a mechanism, which he called the *displacement current,* $I_D = \varepsilon_0 \Delta\Phi_E/\Delta t$, where $\Phi_E = E_\perp A$ is the flux of the E-field. In this case, where $E_\perp = E$ and $\Delta\Phi_E = A\,\Delta E$, it follows that $I_D = \varepsilon_0 A\,\Delta E/\Delta t$. In other words, a time-varying E-field functions as an effective current I_D, and he added this term into the sum in Eq. (21.6) along with the actual currents: ΣI henceforth would be $\Sigma(I + I_D)$. So modified, Ampère's Law becomes another of Maxwell's Equations; namely

$$\Sigma_C B_\parallel \Delta l = \mu_0 \Sigma\left(I + \varepsilon_0 \frac{\Delta\Phi_E}{\Delta t}\right) \tag{24.3}$$

In free space, where there are no real currents ($I = 0$), this expression simplifies to

$$\Sigma_C B_\parallel \Delta l = \mu_0 \varepsilon_0 \frac{\Delta E_\perp}{\Delta t} A \tag{24.4}$$

and it points out that **a time-varying E-field generates a B-field** (Fig. 24.6). This idea is totally new—it had never been proposed nor observed prior to Maxwell. It's the crucial creation that allows us to understand electromagnetic waves.

Each form of radiant energy (radiowaves, microwaves, infrared, light, ultraviolet, X-rays, and γ-rays) is a web of oscillating electric and magnetic fields inducing one another. A fluctuating E-field creates a B-field perpendicular to itself, surrounding and extending beyond it. That magnetic field sweeping off to a point further on in space is varying there, and so it generates a perpendicular electric field that moves out and beyond that point as well, and so on: $\Delta E/\Delta t$ creating a perpendicular B, $\Delta B/\Delta t$ creating a perpendicular E, extending out indefinitely in an independent undulation of interlacing fields, an **electromagnetic wave**.

When an electromagnetic wave interacts with bulk matter, the E-field component has a much greater effect on charges than does the B-field. Accordingly, experiments have revealed that it is principally the electric field of the wave that effectuates vision, photochemistry, fluorescence, and so on. We shall usually explicitly consider only the E-field component assuming that the B-field tags along—the two are inseparable.

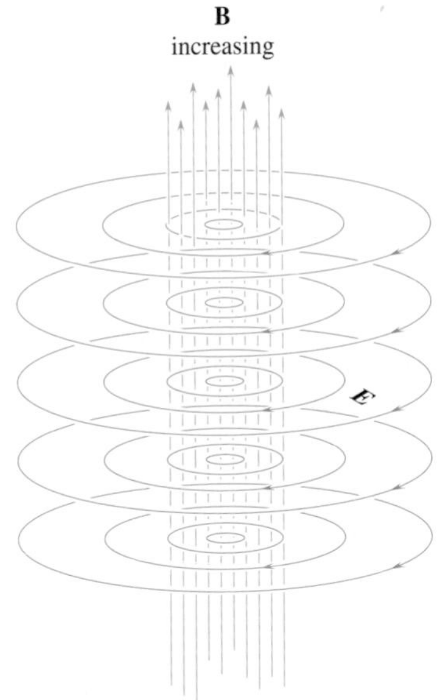

Figure 24.3 Imagine a wire loop surrounding the increasing B-field. A current would be induced such that the induced magnetic field $\mathbf{B}_I$ is downward, telling us the direction of the current and, therefore, the direction of $\mathbf{E}$.

Figure 24.4 As a capacitor gets charged, a transient current exists, producing a B-field. In the gap where no actual current exists there is nonetheless a B-field. It's due to the time-varying E-field.

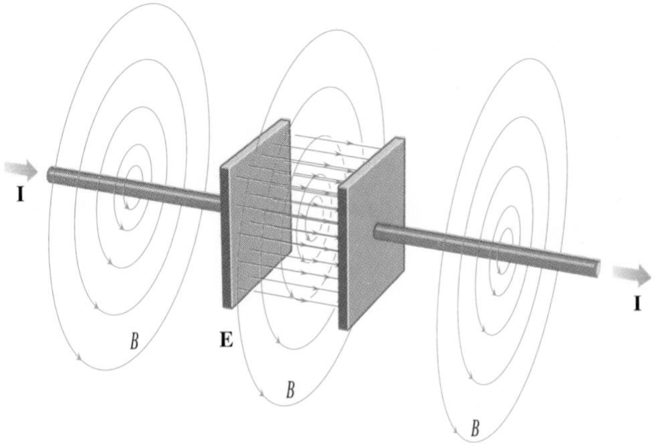

24.3 Waveforms and Wavefronts

A *progressive* electromagnetic wave is a self-supporting, energy-carrying disturbance that travels free of its source (the light from the Sun sails through space, unleashed and on its own, for 8.3 minutes before arriving at Earth).

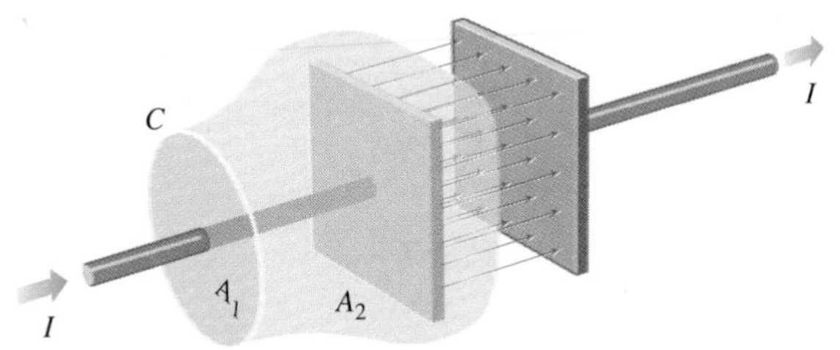

Figure 24.5 Ampère's Law is oblivious to which area A_1 or A_2 is bounded by the path C. But a current passes through A_1 and not through A_2, and that means something is very wrong.

Armed with an E-field meter,* we could presumably measure the variations in electric field as a wave swept by. Figure 24.7 attempts to show pictorially what might be seen as the meter-pointer wiggles, first positively then negatively, while the size of the E-field at that point in space fluctuates with the passing wave. The sizes of the drawn arrows correspond to the instantaneous values of the strength of the E-field—*nothing is actually displaced in space.* This phenomenon is not a wave rising and falling so many centimeters in the air like a rope. What's being pictured is a succession of readings in time, on a single meter, at one point in space. Alternatively, Fig. 24.7b can also be thought of as showing the many readings from a whole row of meters at different points along a line in the direction of propagation of the wave, all at the same instant. The construct is of a disturbance whose "outline," or **profile**, is irregular.

Figure 24.8 depicts an electromagnetic wave whose profile is sinusoidal. (Take another look at Chapter 12 for a discussion of waveforms and the definitions of period T, wavelength λ, frequency f, and phase speed v.) Here, as with all periodic waves,

$$v = f\lambda \qquad [12.18]$$

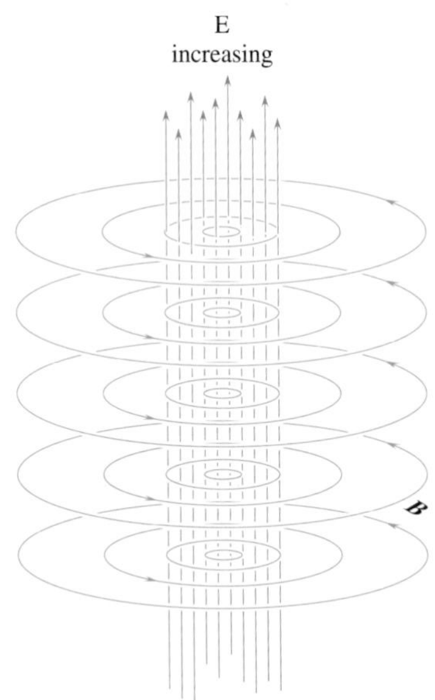

Figure 24.6 The increasing E-field shown here has the effect of an upward current. The Right-Hand-Current Rule tells us the direction of the associated B-field.

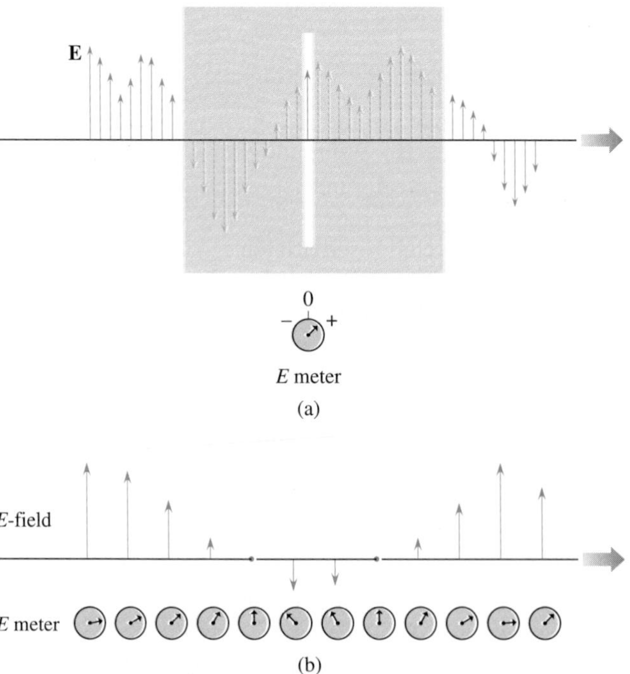

Figure 24.7 The electric field of an electromagnetic wave. (a) If we observe the wave at a single, stationary location, we see an oscillating field as the wave passes. (b) If we observe the wave at many different locations all at the same time, we see the profile of the wave.

*Though these do exist for relatively low-frequency signals, a field meter that would function fast enough to follow the exceedingly rapid swings of the electric field associated with light is still beyond the capabilities of present technology.

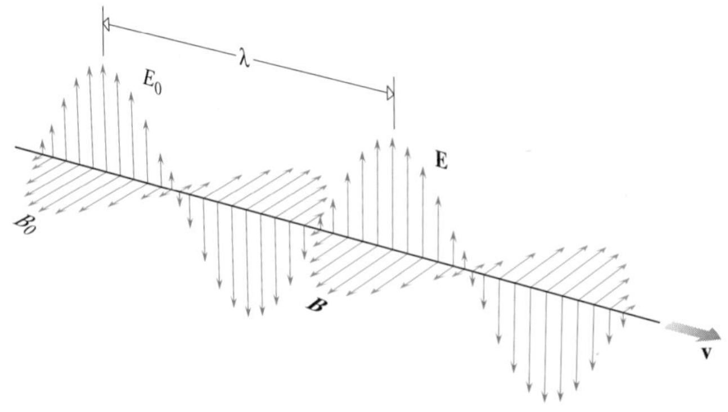

Figure 24.8 A harmonic electromagnetic wave. The coupled *E*- and *B*-fields are perpendicular to each other, and the propagation direction is given by **E** × **B**.

In vacuum, $v = c$, whereas in all other material media, v is usually less than c, though it can be greater under certain circumstances (p. 874).

Example 24.1 In New York City, WNYC broadcasts FM radio signals at 93.9 MHz. Assuming that the speed of propagation of these electromagnetic waves in air is negligibly different from c, please determine the corresponding wavelength.

Solution: [Given: $v = c$ and $f = 93.9$ MHz. Find: λ.] Rewriting Eq. (12.18), we have

$$\lambda = \frac{v}{f} = \frac{2.998 \times 10^8 \text{ m/s}}{93.9 \times 10^6 \text{ Hz}} = \boxed{3.19 \text{ m}}$$

▶ **Quick Check:** $v = f\lambda \approx 300 \times 10^6$ m/s. Incidentally, TV sound and FM radio are VHF (very high frequency) signals ranging in wavelength from 1 m to 10 m.

In Fig. 24.8, along with the vertical *E*-field is included the ever-present accompanying *B*-field. Every point on the disturbance experiences a sinusoidal oscillation in time, and the whole wave extends along the line-of-travel as an advancing sinusoid moving with a speed v. The *E*-field has an *amplitude* of E_0. It oscillates between $\pm E_0$, and so the profile of the wave can be represented mathematically as

$$E = E_0 \sin \frac{2\pi}{\lambda} x \tag{24.5}$$

very much like that of the mechanical wave of Eq. (12.19). To get a sense of the size of E_0, realize that bright sunlight on Earth might have an *E*-field amplitude of about 10 V/cm as compared with the tremendous value of 10^{10} V/cm that can occur in the focused beam of a high-power laser. The quantity $2\pi/\lambda$ arises so often in the analysis of waves that it's given its own name, the **propagation number**, and is symbolized by the letter k.

***An Anticipation of the Wave Theory**. Just as a stone thrown into water becomes the cause and center of various circles, sound spreads in circles in the air. Thus, every body placed in the luminous air spreads out in circles and fills the surrounding space with infinite likenesses of itself and appears all in all and in every part.*

LEONARDO DA VINCI (ca. 1508)

Example 24.2 Write an expression for the profile of the *E*-field of a harmonic electromagnetic wave propagating in vacuum in the positive *x*-direction if its frequency is 600 THz (green light) and its amplitude is 8.00 V/cm.

Solution: [Given: $v = c$, $E_0 = 800$ V/m, and $f = 600$ THz. Find: the profile.] Equation (24.5) is what's needed, but we will first have to find $k = 2\pi/\lambda$, and that means finding λ:

(continued)

(continued)

$$\lambda = \frac{v}{f} = \frac{2.998 \times 10^8 \text{ m/s}}{600 \times 10^{12} \text{ Hz}} = 500 \times 10^9 \text{ m} = 500 \text{ nm}$$

Hence, $k = 12.6 \times 10^6 \text{ m}^{-1}$ and so

$$\boxed{E = (800 \text{ V/m}) \sin(12.6 \times 10^6 \text{ m}^{-1})x}$$

▶ **Quick Check:** $k = 2\pi/\lambda$ and $\lambda = 2\pi/k$; hence, $2\pi f/k = f\lambda = 2\pi(600 \text{ THz})/(12.6 \times 10^6 \text{ m}^{-1})$ should equal c, and it does.

Refer back to the harmonic electromagnetic wave traveling in vacuum (shown in Fig. 24.8) and notice that the direction of propagation corresponds to the direction of $\mathbf{E} \times \mathbf{B}$. The wave is *transverse;* **E** and **B** are perpendicular to each other and to the direction of propagation as well. Such **transverse electromagnetic waves**—so-called **TEM waves**—exist in vacuum, but things usually get much more complicated for waves traversing material media, where the fields may not be totally transverse.

Electromagnetic waves, like sound waves, are three-dimensional. An ideal point source would radiate sinusoidal waves uniformly in all directions, as in Fig. 13.12. Here, *the surfaces of constant phase*—the **wavefronts**—are spherical. More practically, if we examine the light from a street lamp even a few hundred meters away, the portion of the wavefront intercepted by an eye or a small telescope will be essentially spherical. By allowing a spherical wave to expand out (Fig. 13.13) over a great distance, the wavefronts will get larger and flatter, ultimately resembling planes.

It might seem easier to just say that a wavefront corresponds to a surface over which the disturbance has some constant strength or magnitude. Indeed, quite often waves are *homogeneous;* they have constant magnitudes over their wavefronts—but not always! A homogeneous harmonic plane wave traveling in free space can be envisioned as a moving stack of flat surfaces, like a deck of cards, over each of which E and B are constant, arranged so that as we go from one plane to the next, the fields change in a sinusoidal way (Fig. 24.9). The most popular form of laser light resembles plane waves that have a far greater electric field strength at the center (and are much brighter there) than at the edge of the beam—such a wave is *inhomogeneous.*

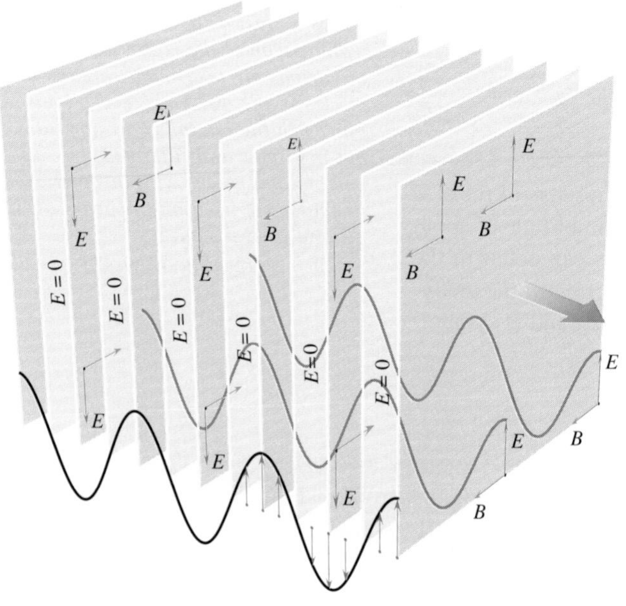

Figure 24.9 A plane harmonic electromagnetic wave. Notice how the fields change from one plane to the next. Over each plane both **E** and **B** are constant.

24.4 The Speed of Propagation: c

Let's now return to Maxwell's Equations and make sure that our wave picture is consistent with them. Accordingly, imagine a sheet or flat pulse of uniform magnetic field **B** pointing in the z-direction laced with a uniform electric field **E** pointing in the y-direction. The combination, resembling a rudimentary plane wave, is traveling through empty space in the x-direction at speed c (Fig. 24.10). To find out what kind of E-field is accompanying the B-field, we need only apply Faraday's Law. Envision an imaginary closed rectangular loop C of width L and indefinite length suspended in the xy-plane. In a time interval Δt, the pulse advances a distance $(c\,\Delta t)$ and sweeps out an area of loop equal to $(c\,\Delta t)L$. Hence, the change of flux through the loop is $\Delta\Phi_M = B(cL\,\Delta t)$ and so from Eq. (24.2), we obtain

$$\Sigma_C E_\parallel \Delta l = -\frac{\Delta\Phi_M}{\Delta t} = -BcL$$

The minus sign just tells us that the induced emf opposes the flux change. Consequently, we will take the reminder, make sure that everything is in its proper direction, and drop the minus sign, which is unnecessary clutter at this point. Now focus on the left side of the equation. Since **E** is upward, it is perpendicular to the top and bottom sides of the loop and therefore the only nonzero contribution to the sum is from the single vertical side, along which $E_\parallel = E$. The entire sum equals EL and, as a result

$$E = cB \tag{24.6}$$

The pulse we fashioned is indeed in accord with Faraday's Law, provided that the strengths of the electric and magnetic fields are related by Eq. (24.6). That relationship turns out to be true in general for electromagnetic waves in space.

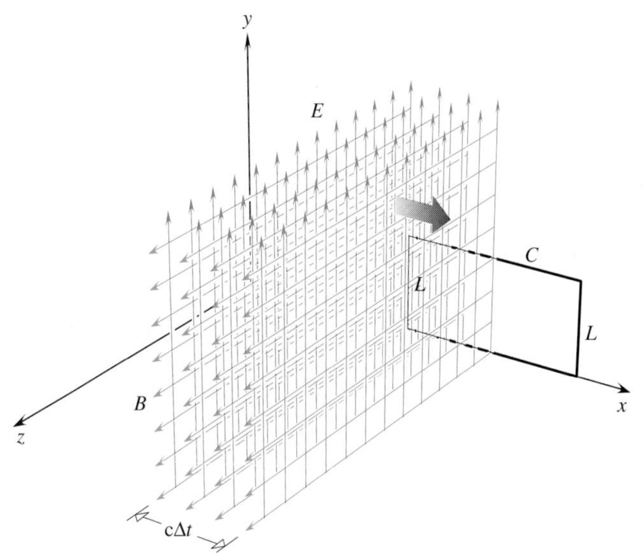

Figure 24.10 An electromagnetic pulse propagating past a loop. The change in magnetic flux through the loop induces a current in the loop and, therefore, an *E*-field as well.

It's worth considering what is happening from a slightly different perspective. As the *B*-field sweeps by, the magnetic flux in the region of space of the loop increases in the *z*-direction. That, in turn, induces an *E*-field in that region of space. The direction of this field is the direction of the induced current, which must be upward, so its induced *B*-field will oppose the increasing flux in the loop. In short, a traveling *B*-field generates an accompanying *E*-field, and only such coupled perpendicular fields propagate through space as a wave.

We could now put the loop in the *zx*-plane and do the whole analysis over for the *E*-field, using Maxwell's version of Ampère's Law. Instead, let's simply compare the two of Maxwell's Equations for free space, using the result that $E = cB$; thus

$$\Sigma_C E_\parallel \Delta l = -\frac{\Delta B_\perp}{\Delta t} A \qquad [24.2]$$

and

$$\Sigma_C B_\parallel \Delta l = \varepsilon_0 \mu_0 \frac{\Delta E_\perp}{\Delta t} A \qquad [24.4]$$

Forgetting about the minus sign, if we substitute cB for E in both formulas and then multiply the second one by c, the two will turn out to be identical, provided that $\varepsilon_0\mu_0 c^2 = 1$; that is

$$c = \frac{1}{\sqrt{\varepsilon_0 \mu_0}} \tag{24.7}$$

and that must be the case—both equations must be satisfied simultaneously. Substituting in the values of the constants, we get

$$c = \frac{1}{\sqrt{(8.85 \times 10^{-12}\ \mathrm{C^2/N{\cdot}m^2})(4\pi \times 10^{-7}\ \mathrm{N{\cdot}s^2/C^2})}} = 3.00 \times 10^8\ \mathrm{m/s}$$

Amazing! This is the speed of light in vacuum.

Maxwell's calculation was far more elegant than the one we just went over. He had begun with the several laws of Electromagnetic Theory and combined them to form an expression that was well known to describe all wave phenomena. And where there should have been a variable associated with the medium, there alternatively was E or B. And where there should have been the speed of the wave, there was $1/\sqrt{\varepsilon_0\mu_0}$. He had in hand these two constants (they had been determined in 1856 by Weber and Kohlrausch), and he had the speed of light (3.15×10^8 m/s) measured by Fizeau with a rotating toothed wheel in 1849. The conclusion was inescapable: *light was "an electromagnetic disturbance in the form of waves" propagated in the ether.* And this description was true, as well, for "radiant heat, and other radiations if any" (light and "radiant heat"—that is, infrared—were known at the time, as was something called chemical rays, now referred to as ultraviolet). Maxwell's analysis, which forms the basis of classical Electromagnetic Theory, stands as one of the greatest theoretical achievements in the history of physics.

All electromagnetic waves propagate in vacuum at exactly

$$\mathrm{c} = 2.997\,924\,58 \times 10^8 \text{ m/s}$$

which is a tremendous speed. Light travels 1 ft in $\approx 10^{-9}$ s. As we'll see, this wave speed is a macroscopic manifestation of the fact that photons exist only at the speed c.

24.5 Energy and Irradiance

If we are to deal with light quantitatively, we have to measure the amount of radiant energy arriving on a surface—a photographic plate, a retina, whatever. It would be nice to determine the strength of the E- and B-fields directly, but that's technically impractical at the high frequencies ($\approx 10^{15}$ Hz) of light. The next best thing to find is the average amount of light energy flowing during a convenient interval of time. A surface is illuminated and we measure the amount of energy arriving per second on each square unit of area. Earlier (p. 449), we called that quantity *intensity,* but in modern optics the *average amount of energy-per-unit-area-per-unit-time* is known as the **irradiance** (I) and it's specified in J/s·m^2 or W/m^2.

Figure 24.11 shows a beam of light of cross-sectional area A impinging on a plane in vacuum. It travels at c and in a time Δt, all the light in the cylindrical section of length $(\mathrm{c}\,\Delta t)$ and volume $V = (\mathrm{c}\,\Delta t)A$ will sweep down onto the plane. If we knew the energy per unit volume u in the beam, then the amount of energy striking the surface in the time Δt would just be $uV = u(\mathrm{c}\,\Delta t)A$. But we do know u from Eqs. (22.12) and (22.15)—it's the energy density in the E-field, $u_\mathrm{E} = \frac{1}{2}\varepsilon_0 E^2$, plus the energy density in the B-field, $u_\mathrm{M} = \frac{1}{2}B^2/\mu_0$. Before we press on, notice that for an electromagnetic wave $E = \mathrm{c}B$ and $\mathrm{c} = 1/\sqrt{\varepsilon_0\mu_0}$, and so $u_\mathrm{M} = \frac{1}{2}E^2/\mathrm{c}^2\mu_0 = \frac{1}{2}\varepsilon_0 E^2 = u_\mathrm{E}$. Not surprisingly, the energy stored in the B-field of an electromagnetic wave equals the energy stored in the E-field. Thus, $u = \varepsilon_0 E^2$. Since the amount of energy arriving in a time Δt is $u(\mathrm{c}\,\Delta t)A$, the energy-per-unit-area-per-unit-time is then $u(\mathrm{c}\,\Delta t)A/A\,\Delta t = u\mathrm{c} = \mathrm{c}\varepsilon_0 E^2$.

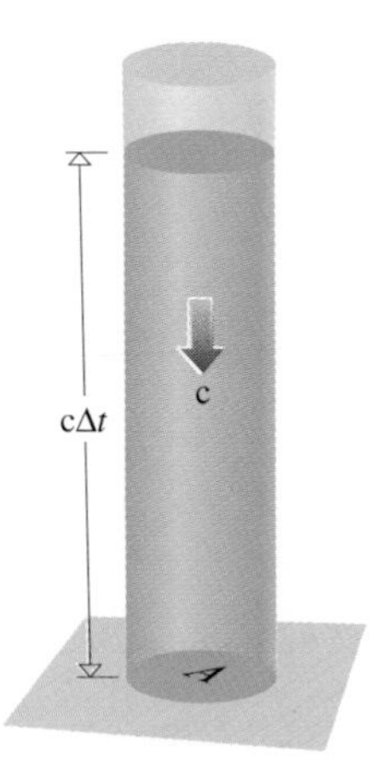

Figure 24.11 Traveling at a speed c, the column of light cΔt high will impinge on the plane in a time Δt. Thus, a volume of the beam equal to $A\mathrm{c}\,\Delta t$ will sweep down onto the plane in a time Δt.

Because E varies very rapidly, this quantity has to be averaged in time to obtain I:

$$I = [\mathrm{c}\varepsilon_0 E^2]_\mathrm{av}$$

Here E is a time-varying sinusoidal function that we will write as $E = E_0 \sin\phi$, keeping in mind that ϕ varies in time. Thus

$$I = [\mathrm{c}\varepsilon_0 E_0^2 \sin^2\phi]_\mathrm{av} = \mathrm{c}\varepsilon_0 E_0^2[\sin^2\phi]_\mathrm{av}$$

since $c\varepsilon_0 E_0^2$ is constant. We saw in Fig. 23.4 that $[\sin^2 \phi]_{av} = \frac{1}{2}$ and so

$$I = \tfrac{1}{2}c\varepsilon_0 E_0^2 \tag{24.8}$$

The irradiance (or intensity), which we measure with a meter, is proportional to the square of the amplitude of the electric field of the wave.

Example 24.3 A 1.0-mW laser beam with a frequency of 4.74×10^{14} Hz has a cross-sectional area of 3.14×10^{-6} m^2. Determine (a) the energy arriving in 1.00 second on a screen intercepting the beam perpendicularly; (b) the *radiant flux density,* or irradiance; and (c) the amplitude of the electric field. Take the medium to be vacuum.

Solution: [Given: beam-power $\text{P} = 1.0 \times 10^{-3}$ W, cross-sectional area $A = 3.14 \times 10^{-6}$ m^2, and $f = 4.74 \times 10^{14}$ Hz. Find: (a) energy arriving in 1.00 s, (b) irradiance I, and (c) E_0.] (a) The beam-power or *radiant flux* is P; hence, in $\Delta t = 1.00$ s, the energy that arrives is

$$\text{P}\Delta t = (1.0 \times 10^{-3}\ \text{W})(1.00\ \text{s}) = \boxed{1.0 \times 10^{-3}\ \text{J}}$$

(b) The irradiance or energy-per-unit-area-per-unit-time is

$$I = \frac{\text{P}}{A} = \frac{1.0 \times 10^{-3}\ \text{W}}{3.14 \times 10^{-6}\ \text{m}^2} = \boxed{3.2 \times 10^2\ \text{W/m}^2}$$

(c) Once we have I, we can get the amplitude from Eq. (24.8); thus

$$E_0^2 = \frac{2I}{c\varepsilon_0} = \frac{2(3.2 \times 10^2\ \text{W/m}^2)}{(3.00 \times 10^8\ \text{m/s})(8.85 \times 10^{-12}\ \text{C}^2/\text{N}\cdot\text{m}^2)}$$

and

$$\boxed{E_0 = 0.49\ \text{kV/m}}$$

▶ **Quick Check:** $I \approx (1.0\ \text{mW})/(3 \times 10^{-6}\ \text{m}^2) \approx 0.3\ \text{kW/m}^2$; $\frac{1}{2}c\varepsilon_0 \approx 1.3 \times 10^{-3}\ \text{C}^2/\text{N}\cdot\text{m}\cdot\text{s}$; $E_0^2 = I/\frac{1}{2}c\varepsilon_0 \approx (0.3 \times 10^3)/(1.3 \times 10^{-3})(\text{V/m})^2 \approx 0.2 \times 10^6\ (\text{V/m})^2$ and $E_0 \approx 0.5 \times 10^3$ V/m.

24.6 The Origins of EM Radiation

Though all forms of electromagnetic (EM) radiation share a single vacuum speed c, they do differ in frequency and wavelength. Still, there really is only one entity, one essence of electromagnetic "stuff." Maxwell's Equations, which are independent of frequency, do not suggest any fundamental differences in kind. Thus, it is natural enough to look for a common basic source-mechanism. What we find is that all the various types of radiant energy seem to have a common origin in that they are all associated somehow with *nonuniformly moving charges.* Of course, classically, we are dealing with waves in the electromagnetic field, and charge *is* that which gives rise to the field.

Free Charge

A stationary charge will have a constant E-field, no motional B-field and, hence, produce no radiation. (Where would the energy come from if it did radiate, and what would turn it on or off?) A uniformly moving charge has both an E- and a B-field, but these do not abruptly disengage, the motion is continuous, and, again, there is no radiation. If you moved along with the charge, the current would thereupon vanish; hence, B would vanish, and we would be back to the previous case of $v = 0$. That's reasonable since it would make no sense at all if the charge stopped radiating just because you started walking along next to it. Besides, if uniformly moving charges

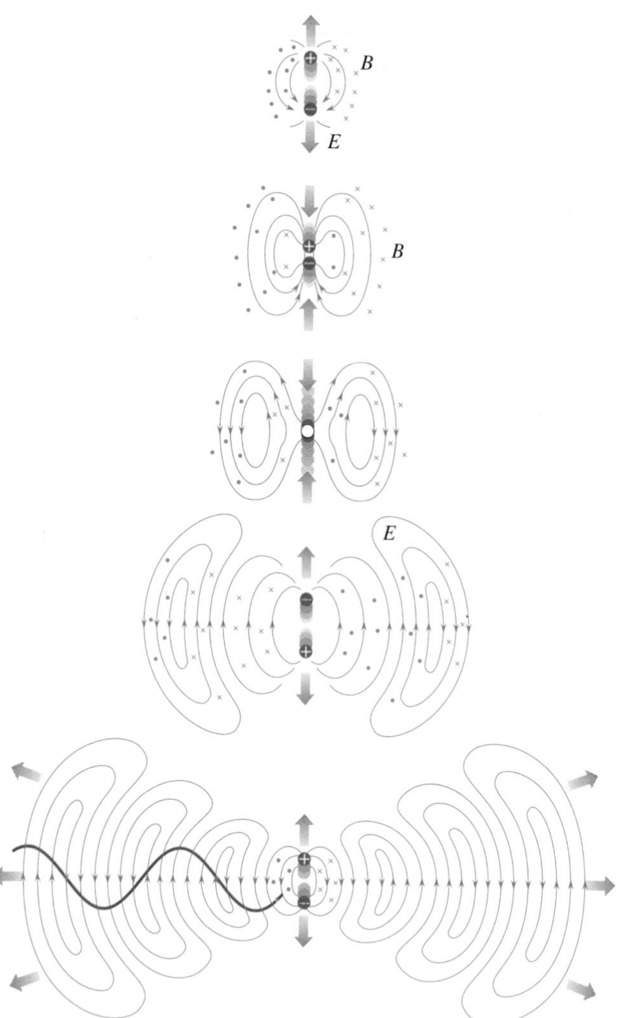

Figure 24.12 A cross-sectional view of the three-dimensional pattern of electromagnetic waves radiated by an oscillating dipole. The loops are electric field lines. The dots and crosses depict the emergence and exit of the circles of magnetic field, which are in planes perpendicular to the page and surround the dipole.

radiated, they would not move uniformly very long (not without an external agent supplying energy) and what then would happen to the Law of Inertia? That leaves *nonuniformly moving charges,* which, assuredly, will be accompanied by time-varying E- and B-fields.

We know, in general, that *free charges* (those not bound within an atom) emit electromagnetic radiation when accelerated. This much is true for charges sailing around in circles within a synchrotron, moving at a changing speed on a straight line in a linear accelerator, or simply oscillating back and forth in a radio antenna—**if charge accelerates, it radiates**. A portion of its kinetic energy is converted into radiant energy.

The Dipole

Perhaps the simplest electromagnetic wave-producing mechanism to visualize is the oscillating dipole: two charges, one plus and one minus, vibrating to and fro along a straight line. The electric field lines, which begin and end on charges, close on themselves whenever the two charges overlap and thereby effectively vanish (Fig. 24.12). Moving charge constitutes a current (in this instance, an oscillating current), which is accompanied by an oscillating magnetic field whose closed circular lines-of-force are in planes at right angles to the motion. The E- and B-fields, which are in-phase, rising and falling together, are perpendicular to each other and to the propagation direction as well. The frequency of the radiation is determined by the oscillation of the current. In brief, far from the dipole there will be developed an outgoing TEM wave. Keep in mind that this pattern is actually three-dimensional, roughly resembling a series of almost spherical concentric tires expanding radially outward.

It's an easy matter to attach an ac generator between two conducting rods and send currents oscillating up and down that transmitting "antenna." The result, Fig. 24.13, is a fairly standard AM radio tower. The antenna will function most efficiently if its length corresponds to the wavelength being transmitted or, more conveniently, to half that wavelength. The radiated wave is then formed at the dipole in sync with the oscillating current producing it. Unhappily, AM radiowaves are typically several hundred meters long. Accordingly, the antenna shown uses the conducting Earth so that it functions as if half of the $\frac{1}{2}\lambda$-dipole were buried in the ground, which allows us to build the thing only $\frac{1}{4}\lambda$ tall.

To receive such a signal, one with a vertically oscillating E-field, we need only stick a straight length of wire up into the air, more or less parallel to E. The electric field strength of the incoming signal might be anywhere from microvolts per meter to millivolts per meter depending on where you are and what you want to receive. A signal with an amplitude of 1.0 mV/m reaching an antenna 2.0 m tall will induce a voltage having an amplitude of 2.0 mV. An old wire coat hanger will work adequately well as a replacement antenna for your car when you are driving around in the city, where the signals could be as strong as 10 mV/m.

Photons

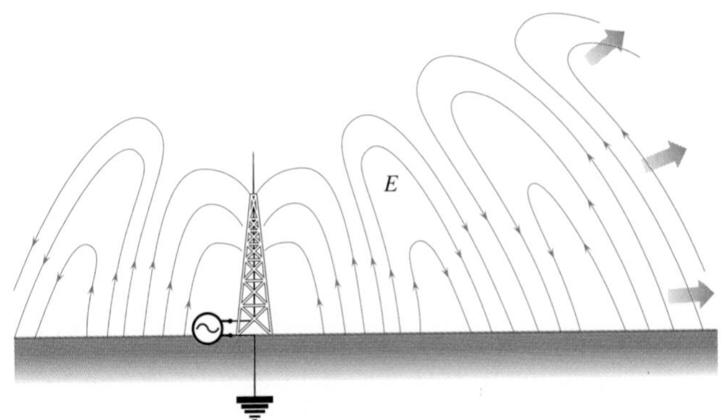

Figure 24.13 The *ground wave* generated by an AM antenna. The wave tends to hug the planet's surface, where most of the people with the radios and TVs are likely to be. Typically, this gives a commercial radio station a range of somewhere between 25 and 100 miles.

Einstein's Special Theory of Relativity revolutionized Classical Mechanics, completely removing the theoretical need for a stationary ether. Deprived of the necessity for an all-pervading elastic fiction, physicists simply had to get used to the idea that electromagnetic waves could propagate through an etherless-free space. The conceptual emphasis passed from ether to field—*light became an electromagnetic wave propagating in the electromagnetic field.* Ether vanished and field appeared as an entity in itself.

In 1905 Einstein shocked the scientific world with another brilliant insight. He boldly introduced a novel form of corpuscular theory that immediately explained several experimental problems that had developed since the late 1800s (Sect. 30.2). The energy carried by light (and all other forms of electromagnetic radiation) is not smoothly spread out across the wave, but is somehow concentrated at points within it. Einstein asserted that light consists of massless quanta, "particles" of electromagnetic radiation.

24.7 Energy Quanta

Every physicist thinks he knows what a photon is. I spent my life to find out what a photon is and I still don't know it.

ALBERT EINSTEIN

Each *quantum of electromagnetic radiation,* or **photon**, as it came to be called in the 1920s, has an energy proportional to its frequency. The constant of proportionality, $h = 6.626 \times 10^{-34}$ J/Hz (or 4.136×10^{-15} eV/Hz), is known as Planck's Constant. The energy of a photon of frequency f is

$$\mathrm{E} = hf \tag{24.9}$$

Example 24.4 A helium-neon laser puts out a beam of red light containing a very narrow range of frequencies centered at 4.74×10^{14} Hz. Determine the energy of each photon at that frequency.

Solution: [Given: $f = 4.74 \times 10^{14}$ Hz. Find: E.] The energy per photon follows from Eq. (24.9), where

$$\mathrm{E} = hf = (6.626 \times 10^{-34}\ \mathrm{J/Hz})(4.74 \times 10^{14}\ \mathrm{Hz})$$

and

$$\boxed{\mathrm{E} = 3.14 \times 10^{-19}\ \mathrm{J}}$$

▶ **Quick Check:** Using h in eV/Hz, we get E = 1.96 eV, and since 1 eV = 1.602×10^{-19} J, E = 3.14×10^{-19} J. Photons of light have an energy of roughly 2 or 3 eV.

The emission and absorption of light (that is, the transfer of electromagnetic energy and momentum from, or to, a material object) takes place in a corpuscular way. Because photons have zero mass, each quantum of light carries very little energy. Thus, even an ordinary flashlight beam must be thought of as a torrent of perhaps 10^{17} photons per second. In general, when we "see" light, what is recorded with eye, detector, or film is the average energy-per-unit-area-per-unit-time arriving at some surface.

Example 24.5 Reconsider the laser beam in Example 24.3. Taking the frequency of the light to be 4.74×10^{14} Hz, determine (a) the *photon flux*—the average number of photons per second impinging on a perpendicular flat target; and (b) the *photon flux density*—the number of photons per second per unit area hitting that target.

Solution: [Given: beam-power $P = 1.0 \times 10^{-3}$ W, cross-sectional area $A = 3.14 \times 10^{-6}$ m^2, and $f = 4.74 \times 10^{14}$ Hz. Find: (a) photon flux and (b) photon flux density.] (a) We have the power or energy per second, 1.0×10^{-3} W; hence, if we divide that quantity by the energy of each photon (from Example 24.4), we will get the photon flux

$$\frac{P}{hf} = \frac{1.0 \times 10^{-3}\text{ W}}{3.14 \times 10^{-19}\text{ J}} = \boxed{3.2 \times 10^{15}\text{ photons/s}}$$

(b) The photon flux density is just the flux per unit area, or

$$\frac{3.18 \times 10^{15}\text{ photons/s}}{3.14 \times 10^{-6}\text{ m}^2} = \boxed{1.0 \times 10^{21}\text{ photons/s·m}^2}$$

▶ **Quick Check:** The photon flux density (1.013×10^{21}) times the energy per photon equals 0.32 kW/m^2, the irradiance.

Light, and all other forms of electromagnetic radiation, interacting with matter in the processes of emission and absorption, behave like streams of particlelike concentrations of energy that exist only at the speed c, *and which otherwise propagate in a wavelike fashion.* We cannot say whether light *is* particle or wave; it seems to be both and so is likely neither. The "wee beasties" of the microworld, whether they be protons or photons, don't play our game of waves *or* grains—of "either/or." The electromagnetic wave (like the water wave made up of countless molecules) is an illusion of continuity. On the most fundamental level, there is no such thing as a continuous electromagnetic wave. Yet torrents of photons behave exactly as if they were dissolved into a smooth classical TEM wave.

So here we have the latest picture of what seems to defy picturing: powerful in its ability to explain, wonderful in its subtlety—it is the latest theory of light, but probably not the last.

24.8 Atoms and Light

By far the most important mechanism for the emission and absorption of radiant energy—especially of light—is the *bound charge,* electrons confined within atoms. Much of the chemical and optical behavior of a substance is determined by only its outer electrons; the remainder of the cloud is formed into closed, essentially unresponsive shells around and tightly bound to the nucleus. Although it's not clear what exactly happens when an atom radiates, we do know with some certainty that light is emitted during changes in the outer charge distribution of the electron cloud (Sect. 30.5). It is the absorption and emission of light via these electrons that determines almost all optical phenomena in nature.

Each electron is usually in the lowest possible energy state available to it, and the atom as a whole is in its **ground state**. There it will remain, indefinitely, if left undisturbed. Any mechanism that can pump energy into an atom, such as a collision with another atom, with a photon or an electron, will affect this situation. In addition to the ground state, there are specific well-defined higher energy levels, so-called **excited states**.

At low temperatures, atoms (and molecules) tend to be in their ground states but, as the temperature rises, more and more of them become excited through collisions. This process is indicative of a class of relatively gentle excitations (glow discharges,

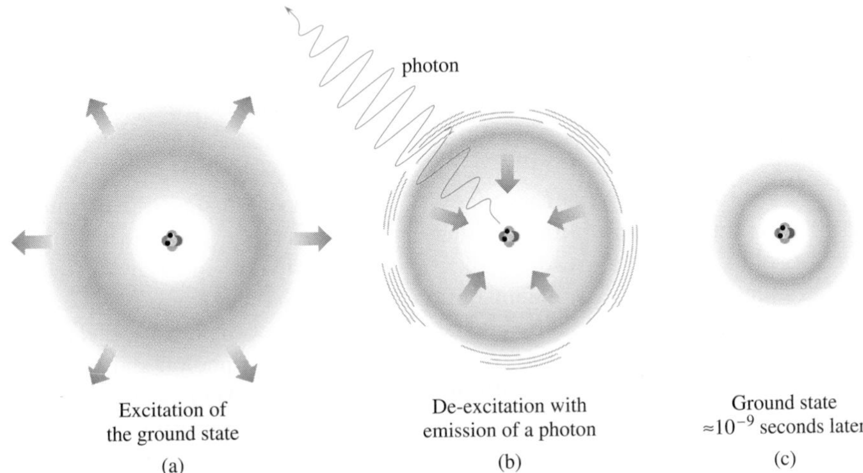

Figure 24.14 A schematic representation of an atomic emission. (a) An atom is raised to an excited state. (b) With the emission of a photon, it (c) spontaneously drops back to the ground state.

flames, sparks, etc.), which energize only the outermost unpaired valence electrons rather than the far more strongly bound inner ones. For the moment, we will concentrate on these outer-electron transitions, which give rise to the emission of light, infrared, and ultraviolet.

When just the right amount of energy (ΔE) is imparted to an atom (that is, to a valence electron), it can respond by suddenly jumping from a lower to a higher energy level. Thus, the amount of energy that can be absorbed by an atom is quantized (that is, limited to specific, well-defined amounts). This excitation of the atom is a short-lived resonance phenomenon. Usually, after a time of about 10^{-8} s or 10^{-9} s, the excited atom spontaneously relaxes back to a lower state, most often the ground state, losing the excitation energy along the way (Fig. 24.14). This transition can occur by the emission of light or (especially in dense materials) by conversion to thermal energy via interatomic collisions within the medium.

When the downward atomic transition is accompanied by the emission of light, the energy of the photon (hf) exactly matches the quantized energy decrease of the atom (ΔE). Thus, $\Delta E = hf$ and there is a specific frequency associated with both the emitted photon and the atomic transition between the two states. The latter is effectively a *resonance frequency of the atom, one of several at which it very efficiently absorbs and emits energy.*

Even though we don't know exactly what is going on during that 10^{-8} s, it is helpful to imagine the orbital electron somehow making the downward energy transition via a diminishing, oscillatory motion at the specific resonance frequency. Light can therefore be imagined as emitted in a short oscillatory pulse, or wavetrain, that is somehow representative of the radiated photon. Keep in mind that there are two intertwined models that have evolved to represent radiant energy: the particlelike and the wavelike—and light is both. Without the more complete development of Quantum Theory (which even then leaves something to be desired), we will have to irreverently talk about particles at one moment and waves at another.

24.9 Scattering and Absorption

The process whereby an atom absorbs a photon and emits another photon is known as **scattering**. The transmission of light through a windowpane, the reflection from a mirror or a face, the coloration of the sunset sky, are all governed by scattering.

We see most things not because they are self-luminous but because they absorb part of the incident ambient light and redirect some of the remainder toward our eyes. An atom (one on your cheek, for instance) can react to incoming light in essentially two different ways depending on the incident frequency (color) or, equivalently, on the incoming photon energy ($E = hf$). The atom can absorb the light or, alternatively, its ground state can scatter the light.

If the photon's energy matches that of one of the excited states, the atom will absorb it, making a quantum jump to that higher energy level. In the dense atomic clutter of solids and liquids, it's very likely that the excitation energy will be transferred, via collisions, to the random KE of the atoms rather than being reemitted when the atom returns to the ground state—the photon vanishes, its energy converted into thermal energy. This process (the taking up of a photon and its conversion into thermal energy) is called **dissipative absorption**. Your skin dissipatively absorbs certain frequencies and scatters others, which is what gives it the color it has under white light illumination. A red apple appears red because it has a resonance in the blue and absorbs out the yellow-BLUE-green band, reflecting mostly red.

In contrast to this resonant process, *ground-state,* or **nonresonant elastic scattering**, occurs for incoming light of other frequencies; that is, other than resonance frequencies. Even now, the light passing through your eyes is progressing via elastic scattering. Envision an atom in its lowest state and suppose that it interacts with a photon of frequency f, whose energy is too small to cause an excitation up into any of the higher states. Nonetheless, the electromagnetic field of the light can drive the electron cloud into oscillation. There can be no resulting atomic transitions; the atom will remain in its ground state while the cloud vibrates ever so slightly at the frequency of the incident light. The electron, once it begins to oscillate, is, of course, an accelerating charge and so will immediately begin to reemit light of that same frequency, f. This scattered light consists of a photon that sails off in some direction carrying the same amount of energy as did the incident photon. In effect, we are imagining the atom to resemble a little dipole oscillator. When a material (like a piece of glass) is bathed in light, this almost-omnidirectional scattering gives each atom the appearance of being a tiny source of spherical wavelets. As we'll see in the next chapter, it is precisely this nonresonant scattering that accounts for the transmission of light through all transparent material media and for the reflection of light from most surfaces.

THE ELECTROMAGNETIC-PHOTON SPECTRUM

The whole spread of radiant energy, which conceptually ranges in wavelength between zero and infinity, is referred to as the **electromagnetic spectrum**. It is usually subdivided into seven more or less distinct regions. These were delineated originally as much by historical circumstance as by physical necessity, and so there tends to be a good bit of overlap in the categories. Needless to say, light was discovered first, then infrared (1800), ultraviolet (1801), radiowaves (1888), X-rays (1895), gamma rays (1900), and, finally, it was just a technical matter of filling in the microwaves, which was done in the 1930s, primarily with an eye toward radar (Fig. 24.15).

24.10 Radiowaves

Great, slowly rising and falling electromagnetic waves (more than 18 million miles long) have been measured impinging on the Earth, streaming in from the depths of

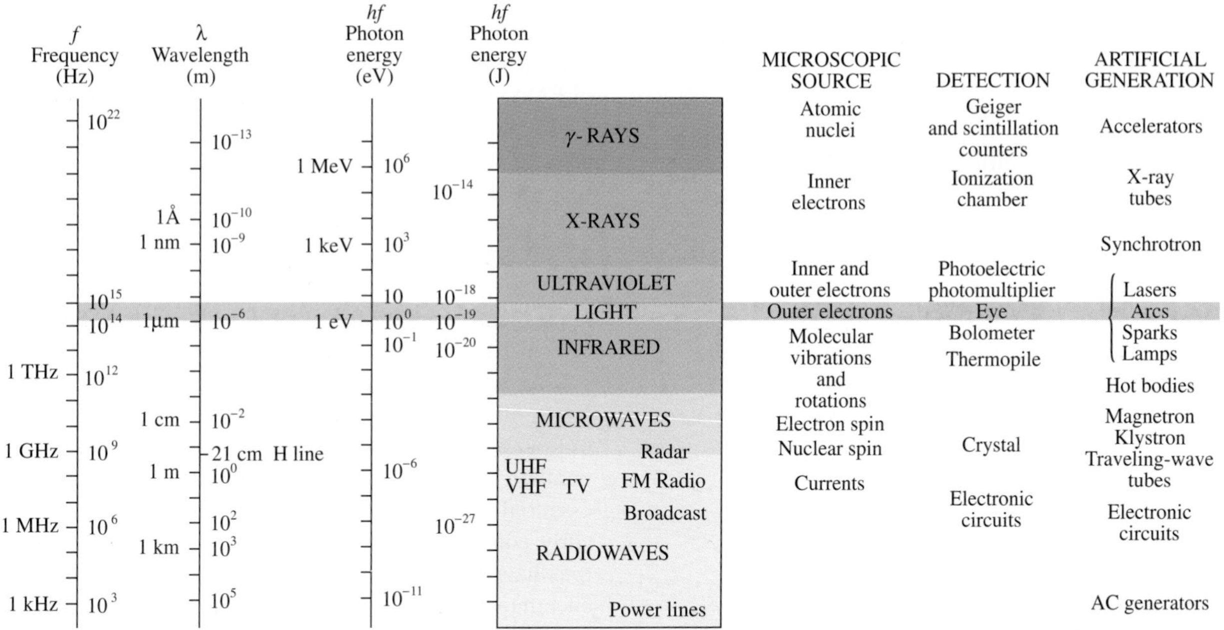

Figure 24.15 The electromagnetic spectrum.

space. These faint cosmic flutters are one extreme of the radiant energy grouping known as **radiowaves**, which extends in wavelength down to about 0.3 meter.

AM radio and television transmissions in the United States are confined to the frequency band from 535 kHz to 1605 kHz (p. 812). The major failing of the AM technique is that it picks up extraneous noise sent out by everything from lightning and electric motors to car ignitions. In fact, electronic calculators transmit radiowaves—not very powerfully and therefore not very far, but far enough to be picked up by a nearby AM radio. Turn the radio's dial to the low-frequency end between stations and punch in a beeping tune on your calculator. You can even watch the effects of lightning—flashing white dashes running horizontally across your TV picture—during a storm. The picture is transmitted AM, the sound, FM. Or hold a telephone receiver next to your ear near a computer and listen to the radiated hum.

Example 24.6 The AM receiver of Fig. 23.24 is tuned to accept a carrier frequency of 1.0 MHz. If the inductance in the tuning circuit has a value of $L = 300\ \mu\text{H}$, (a) determine the value of the tuning capacitor. (b) What is the wavelength of the carrier wave? (c) What is the energy of a radio frequency carrier photon in electron volts?

Solution: [Given: $f_0 = 1.0$ MHz and $L = 300\ \mu$H. Find: C, λ, and E.] (a) Using Eq. (23.35), $f_0 = 1/(2\pi\sqrt{LC})$

$$C = \frac{1}{4\pi^2 f_0^2 L} = \frac{1}{4(3.14)^2(1.0 \times 10^6\ \text{Hz})^2(300 \times 10^{-6}\ \text{H})}$$

and $$C = 84 \times 10^{-12}\ \text{F} = \boxed{84\ \text{pF}}$$

(b) The wavelength follows from $c = f\lambda$; namely

$$\lambda = \frac{c}{f} = \frac{3.0 \times 10^8\ \text{m/s}}{1.0 \times 10^6\ \text{Hz}} = \boxed{3.0 \times 10^2\ \text{m}}$$

(c) $E = hf = (6.63 \times 10^{-34}\ \text{J/Hz})(1.0 \times 10^6\ \text{Hz}) = 6.63 \times 10^{-28}$ J. Since $1.00\ \text{J} = 6.24 \times 10^{18}$ eV;

$$\boxed{E = 4.1 \times 10^{-9}\ \text{eV}}$$

▶ **Quick Check:** Light has a frequency of about 500 THz and an energy of around 1 eV. These radio photons are roughly 0.5×10^9 lower in frequency and should be that much lower in energy, thus $E = (1\ \text{eV})/(0.5 \times 10^9) = 2 \times 10^{-9}$ eV, which is the same order of magnitude as above.

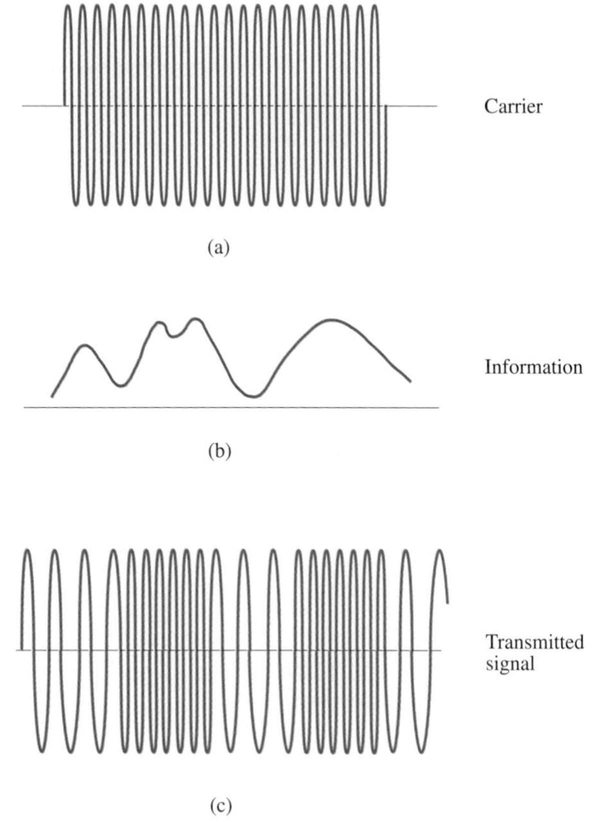

Figure 24.16 Frequency modulation (FM). (a) A constant-frequency, constant-amplitude carrier is modulated by (b) an information signal such that the amplitude of the information determines the frequency of the FM signal, as shown in (c).

Frequency modulation (FM) was introduced primarily to decrease the amount of noise and to extend the frequency content of the information transmitted. The range of FM is from about 88 MHz to 108 MHz, corresponding to wavelengths of from 3.4 m to 2.8 m. The carrier is made to vary in frequency (Fig. 24.16) in proportion to the amplitude of the audio message. Noise, which only affects the amplitude of the signal, is simply cropped off by the receiver since all the useful information is in the frequency distribution. FM signals, and therefore TV as well, are often transmitted with the *E*-field horizontal, which is why a typical TV receiving antenna has several $\frac{1}{2}\lambda$ horizontal bars.

At 1.0 MHz, a radio frequency photon has an energy of 6.6×10^{-28} J, a very small quantity by any measure. In an ordinary flow of radio emission, there is an incredible number of photons, each individually so weak as to be essentially undetectable on its own. The result is an apparently continuous wavelike transport of energy. Radio photons have such low energies that they are not very likely to encounter atomic resonances, meaning that we can expect nonconductors (such as glass, bricks, and concrete) to be fairly transparent to radiowaves, while metals (with their free electrons) certainly are not. Roll yourself in a ball around a portable radio and see what happens to the reception. Next, put some metal foil around the radio. Note the difference. Incidentally, the human body most effectively serves as a conducting antenna in the range from about 30 MHz to 300 MHz, which is in the FM and VHF regions. Perhaps you have noticed as much while playing with an indoor TV antenna.

24.11 Microwaves

The frequency region extending from about 10^9 Hz (1 GHz) up to roughly 3×10^{11} Hz is the domain of the **microwaves**. In wavelength, that corresponds to the range from approximately 30 cm to 1.0 mm. Electromagnetic radiation in the region from less than 1 cm to about 30 m can penetrate the Earth's atmosphere, and that makes microwaves especially useful for space-vehicle communications and radio astronomy. The ground state of the cesium atom consists of two very closely spaced levels, separated by only 4.14×10^{-5} eV. When the atom drops from one level to the other, the resulting microwave emission has a splendidly precise frequency of $9.192\,631\,77 \times 10^9$ Hz. This is the basis for the cesium clock, the present-day laboratory standard of frequency and time (p. 13).

To understand how the microwave oven works, recall that molecules can absorb and emit energy by altering the state of motion of their constituent atoms. The molecule can be made to vibrate and/or rotate; again, the energy associated with either such motion is quantized, and molecules therefore possess a number of rotational and vibrational energy levels in addition to those due to their electrons. Only when a molecule is polar will it experience forces via an electromagnetic wave that cause it to rotate into alignment with the changing *E*-field. Because molecules are massive and not able to swing around easily, we can anticipate low-frequency rotational resonances (infrared of 0.1 mm to microwave of 1 cm).

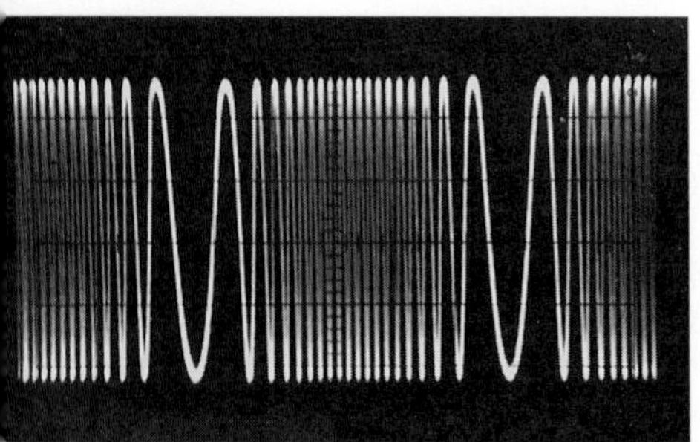

A frequency-modulated signal.

For example, the water molecule is polar—the hydrogen end is positive, the oxygen end is negative. When exposed to an alternating electromagnetic wave, it will swing around, trying to stay lined up with the *E*-field. Water molecules will very efficiently absorb microwave radiation at or near a resonant frequency, thereupon exhibiting large-amplitude oscillations. The oscillatory KE of these excited molecules is rapidly converted into thermal energy via collisions with other molecules. The microwave oven (12.2 cm, 2.45 GHz) is an obvious application. Clearly, the thing to be heated has to contain water—a dry paper plate will remain quite cool. The diathermy machine, used to warm muscles and joints in order to relieve soreness, works on the same principle, at the same frequency.

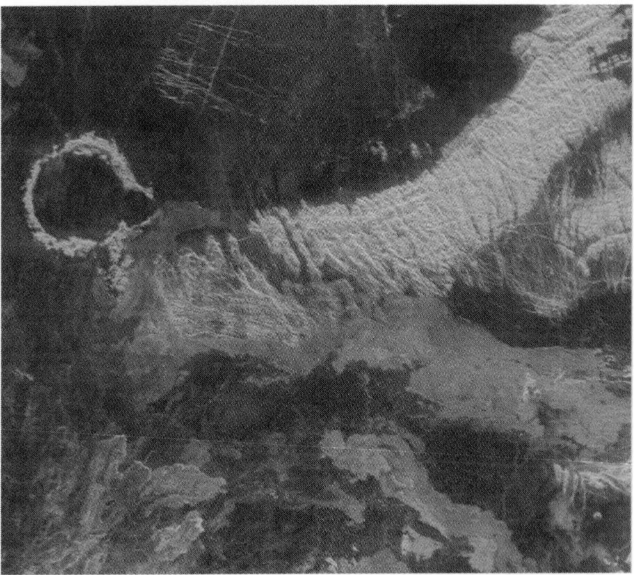

A microwave (radar) image of the surface of Venus taken by the *Magellan* spacecraft in 1992.

Microwaves are now used for everything from carrying phone conversations and interstation TV to guiding planes and catching speeders (via radar) to studying the origins of the Universe and chatting with astronauts. Even though individual photon energies are small (in general, you want to avoid absorbing large numbers of them), people have been killed by massive exposure. In the United States, the supposedly safe limit to environmental microwave exposure is assumed to be 10 mW/cm^2, which some maintain is far too high.

24.12 Infrared

In 1800, the renowned astronomer (musician and deserter from the Hanoverian Foot Guards) Sir William Herschel made a surprising discovery. Using a prism, he had been studying the different amounts of "heat" conveyed by the various colors of sunlight. To his astonishment, he found that his thermometer registered its greatest increase just "beneath" the red region. Herschel rightly concluded that he was observing the effects of an invisible radiation, now called **infrared** (beneath the red).

The infrared (or IR) band merges with microwaves at around 300 GHz (1.0 mm) and extends to about 385 THz (780×10^{-9} m; that is, 780 nm), where the photons have enough energy (1.6 eV) to break apart certain molecules. Most any material will radiate IR via thermal agitation of its molecules—just heat it up and it will pour forth IR. Infrared is copiously emitted from glowing coals and home radiators—roughly half the radiant energy from the Sun is IR, and an ordinary light bulb puts out far more IR than light. Like all warm-blooded creatures, we ourselves are IR emitters. The human body radiates very weakly, starting practically around 3000 nm; it peaks in the vicinity of 10 000 nm and trails off from there. This fact is exploited by some rather nasty "heat"-sensitive snakes (Crotalidae pit vipers and Boidae constrictors) that are active at night (p. 534).

Microwave antennae on the top of the Eiffel Tower in Paris.

In addition to rotating, a molecule can vibrate in several different modes. The corresponding vibrational emission and absorption spectra are, generally, in the infrared (1000 nm to 0.1 mm). Many molecules have vibrational and rotational resonances in the IR and are good absorbers, converting radiant energy into thermal energy, which is one reason IR is often misleadingly called "heat waves" (just put your face in the sunshine and you'll warm to the notion).

Heat lamps (1000 nm–2000 nm) used in physical therapy (and in bad restaurants) offer a more penetrating radiation than light. Typically, near-IR (that is, near

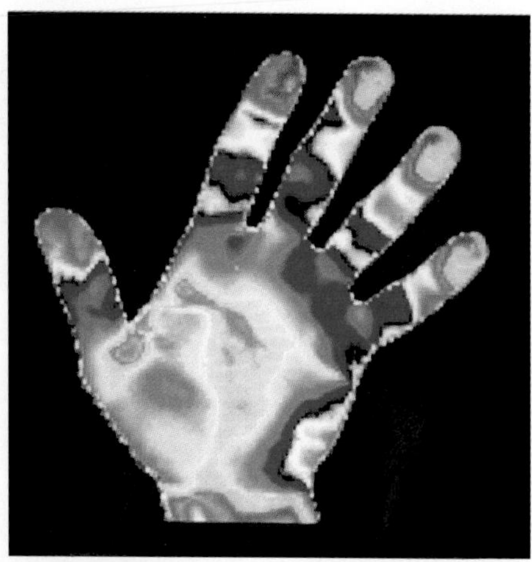

An infrared picture of a normal hand showing a typical temperature distribution. Abnormal blood flow becomes immediately obvious in IR images of this kind.

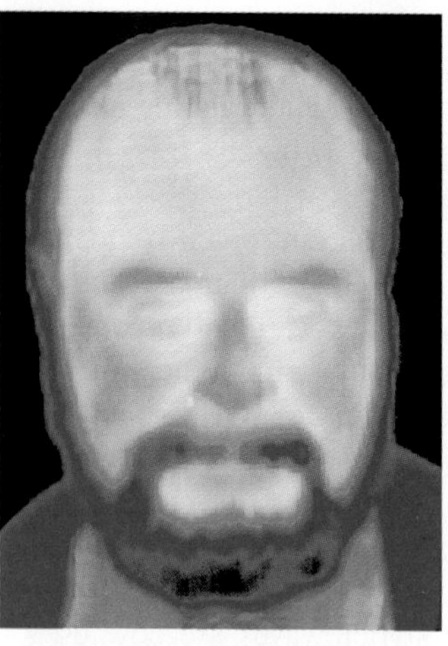

A portrait of the author taken using self-radiated IR. The beard is cool, and there seem to be no abnormal hot spots. The top of the mustache is warm from exhaled air.

the visible) enters the human body to a depth of not much more than about 3 mm below the skin, regardless of its color. If you've ever warmed yourself in sunlight streaming through a window, you already know that ordinary glass passes a good fraction of the incident near-IR; so does the cornea and lens of the human eye, and that's a good reason not to stare at the Sun, especially through cheap sunglasses that deceptively often pass much more IR than light.

There are photographic films sensitive to near-IR (<1300 nm); TV systems that produce continuous infrared pictures known as *thermographs;* there are IR spy satellites that look out for rocket launches, IR satellites that look out for crop diseases, and IR satellites that look out into space; there are "heat-seeking" missiles that are guided by infrared; IR lasers and IR astronomical telescopes peering at the sky. You probably change stations on your TV with an IR-emitting remote control unit. Wherever subtle variations in temperature are of concern, from detecting brain tumors and breast cancer to simply spotting a lurking burglar, IR systems have found practical use.

24.13 Light

The narrow band of the spectrum that we humans "see" is often referred to as *light.* That's a rather inaccurate specification because we can "see" X-ray shadow patterns cast directly on the retina. Furthermore, many of us can see—if only poorly—into both the IR (up to roughly 1050 nm) and ultraviolet (down to about 312 nm). At those extremes, the sensitivity of the eye has dropped by a factor of about a thousand. Accordingly, let's fix the meaning of the word *light* to stand for that tiny range of the electromagnetic spectrum from 780 nm to 390 nm, one octave.

Newton was the first to realize that **white light** is actually a mixture of all the colors of the visible spectrum; that the prism does not create color by changing white light to different degrees, as had been thought for centuries, but simply fans out the light, separating it into its constituent colors. Not surprisingly, the very concept of white light seems dependent on our perception of the Earth's daylight spectrum. The phenomenon of "daylight" changes from moment to moment and place to place, but it is nonetheless recognizable as a broad, slightly wiggly frequency distribution that falls off more rapidly in the violet than in the red (Fig. 24.17).

What we perceive as whiteness is a wide mix of frequencies roughly in the same amounts as that of daylight, and that's what we mean when we talk about "white light." Be that as it may, lots of different distributions will still appear more or less "white." We recognize a piece of paper to be white whether it's seen indoors in incandescent light or outside in skylight, even though those whites are quite different. Indeed, many pairs of colored light beams (for example, 656 nm red and 492 nm cyan) will produce the sensation of whiteness, and the human eye cannot always distinguish one white from another—it cannot frequency analyze light into its harmonic components the way the ear can analyze sound (p. 445).

Colors themselves are the subjective human physiological and psychological responses primarily to the various frequency regions extending from about 384 THz for red, through orange, yellow, green, and blue, to violet at about 769 THz (Table 24.1). Thus, a harmonic wave of the single frequency 508 THz will be perceived as yellow, as will 507 THz and 509 THz, etc. Each such single-frequency, single-color light is spoken of as *monochromatic*. But to be monochromatic, the wavetrain must be perfectly sinusoidal, extending from $-\infty$ to $+\infty$. It can have no beginning and no end, and that's impossible. In reality, the best we can hope for is light having a narrow band of frequencies—**quasimonochromatic light**.

Color corresponds to the human perception of photon energy or frequency. It is not a property of the light itself but a manifestation of the electrochemical sensing system: eye, nerves, brain. To be precise, we should not say "yellow light," but instead refer to light that is "seen to be yellow." Actually, a variety of different frequency mixtures can evoke the same color response from the eye-brain sensor. For example, a beam of red light (peaking at, say, 690 THz) overlapping a beam of green light (peaking at, say, 540 THz) will result, believe it or not, in the perception of *yellow* light, even though there are no frequencies present in the yellow band. The eye-brain averages the input and "sees" yellow! Furthermore, the nice reddish-blue or purple known as *magenta* does not even exist as a single frequency—it's not in the white-light spectrum.

The human eye, under daytime illumination levels, is most responsive to yellow-green (one reason that sodium-yellow street lights are so common and why yellow-tinted eyeglasses can be useful). Sensitivity gradually diminishes at both higher (blue, violet, ultraviolet) and lower (orange, red, IR) frequencies. It should be no surprise to learn that the solar spectrum also peaks at around 560 nm (2.2 eV) in the yellow-green so that there are usually plenty of the "right" photons flying around.

The wavelength of even red light at 0.000 000 780 m is rather small—780 nm is roughly 1/100 the thickness of this page. On an atomic scale the wavelengths of light are immense, several thousand times the size of an atom. If a uranium atom were enlarged to the size of a pea, a single wavelength of red light would then be about 54 feet long. This disparity is one of the crucial factors in determining the way light reflects off material objects. Remember that most of what we see is via reflected light. We make it a point not to look directly at bright, self-luminous objects even when they are around (apart from TV sets and campfires, which aren't very bright). Seeing the world almost exclusively in reflected light gives rise to a richly colored, shadow-filled, high-contrast, detailed picture—there's essentially no background noise.

In bright sunlight, where more than 10^{17} photons arrive each second on every square centimeter of a surface, the quantum nature of the process can easily be overlooked. Even so, light quanta are energetic enough ($hf \approx 1.6$ eV up to 3.2 eV) to pro-

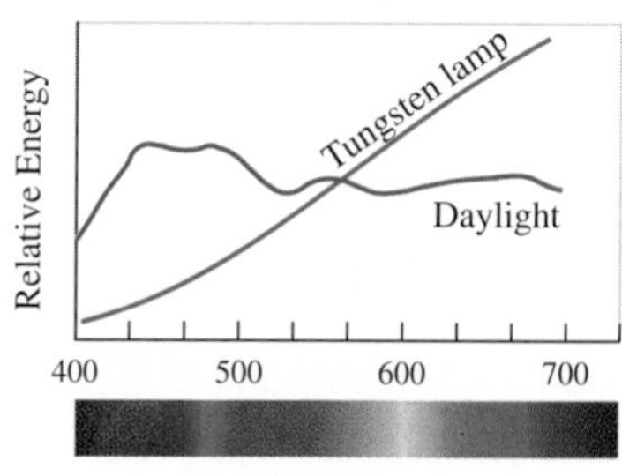

Figure 24.17 The distribution of the various frequencies in the light from a tungsten lamp and in sunlight.

TABLE 24.1 ***Approximate frequency and vacuum wavelength ranges for the various colors***

Color	λ_0(nm)	f(THz)*
Red	780–622	384–482
Orange	622–597	482–503
Yellow	597–577	503–520
Green	577–492	520–610
Blue	492–455	610–659
Violet	455–390	659–769

*1 terahertz (THz) = 10^{12} Hz, 1 nanometer (nm) = 10^{-9} m.

duce effects on a distinctly individual basis. For instance, the human eye can detect as few as 10 photons impinging on it and perhaps as few as 1 arriving at the retina. Quanta of light can break up delicate chemical bonds, and so substances such as aspirin and wine are protected by dark bottles, and light-sensitive photographic films are commonplace. Premature infants sometimes develop jaundice, due to an excess of bilirubin in the blood, a condition that is successfully treated by exposing them to light. Blue-light photons have enough energy to dissociate the bilirubin molecule. Light is a major agency for the Sun → Earth transport of energy. Powered by sunlight, the process of photosynthesis results in the removal of upwards of 200 thousand million tons of carbon yearly from atmospheric carbon dioxide and the subsequent generation of complex organic molecules to the greater good of life on the planet.

24.14 Ultraviolet

A year after IR was discovered, J. Ritter (1801) found yet another invisible radiation. It was well known that white silver chloride would turn black, liberating metallic silver in the presence of light, especially blue light (this reaction was the precursor of photography). Ritter found that silver chloride did its little trick even more efficiently when exposed to the spectral region "beyond" the violet, where there was no visible radiation. This was the discovery of what came to be known as **ultraviolet** (UV) radiation, which corresponds to the range from about 8×10^{14} Hz to 2.4×10^{16} Hz. This is the so-called "black light," which is neither black nor light. Ultraviolet controls certain annoying dermatological conditions, tans the skin, and activates the synthesis of vitamin D within it.

At wavelengths of ≈300 nm and below, at the edge of the solar spectrum, UV can cause sunburn as well as tanning. Interestingly, this is roughly the energy needed (4 eV) to break a carbon-carbon bond. Passage through the atmosphere filters out much of this radiation, especially in the higher latitudes and at the low sun-angles that occur in winter and in the early and late parts of the day, even in summer. Someone in Chicago has to be a lot more determined to get tanned than someone in Florida. But then again, solar UV is the major cause of skin cancer in human beings. Our continued concern for the ozone (O_3) layer stems from the fact that this gaseous envelope absorbs (<320 nm) what would otherwise be a lethal stream of solar UV photons.

An ultraviolet photo of Venus taken by *Mariner 10.*

Ultraviolet in the wavelength range less than 300 nm will depolymerize nucleic acids and destroy proteins, both of which are strong absorbers, making UV quite incompatible with life on this planet. Extended exposure to UV, in time, causes wrinkles, liver spots, actinic keratosis (precancerous dark blotches), and finally cancer (80% of which is the curable form of basal-cell carcinoma). UV also inhibits the body's immune system, which may explain why some viral diseases, such as fever blisters and chicken pox, get more severe when exposed to sunshine.

Some materials reflect UV much as they reflect light, so that long exposure while playing around on snow or water can be deceptively hazardous. The same is true for lying about on a beach on a totally overcast summer day, since water vapor passes a good deal of UV ($\approx$50% in this case). In contrast, ordinary window glass invariably contains iron oxide contaminants, which make it quite opaque to near-UV. It's a waste of time to attempt to tan yourself behind such a window no matter how warm it gets.

Humans don't see ultraviolet very well because the cornea absorbs it, especially at the shorter wavelengths, while the eye's lens absorbs most effectively beyond 300 nm. Someone who has had a lens removed because of cataracts can see UV ($\lambda > 300$ nm). It now seems that in addition to insects such as honeybees, a fair number of creatures can visually respond to UV as well. Pigeons, for one, are quite capable of recognizing patterns illuminated by only UV and likely employ that ability to navigate by the Sun, even on overcast days.

An atom emits a UV photon when the electron makes a long jump down from a highly excited state. For example, the outermost electron of a sodium atom can be raised into higher and higher energy levels until it's simply torn loose altogether at 5.1 eV. The atom is then said to be ionized. Should it subsequently recombine with a free electron, the latter will quickly descend to the ground state, most likely in a series of jumps to ever lower levels, each resulting in the emission of a photon. If, however, the electron makes one long plunge to the ground state, a single 5.1-eV ultraviolet photon will result. Still more energetic UV photons can be generated when the inner electrons of an atom are excited.

Example 24.7 Suppose that a singly ionized copper atom recombines with an electron. The ionization energy of copper is 7.72 eV. What is the shortest possible wavelength that can be emitted via the recombination?

Solution: [Given: E = 7.72 eV. Find: the minimum λ.] The maximum amount of energy available, corresponding to a transition directly to the ground state, equals 7.72 eV. This value, in turn, is associated with a maximum frequency given by $E = hf$ and therefore a minimum wavelength where $E = hc/\lambda$; hence

$$\lambda = \frac{hc}{E} = \frac{(4.136 \times 10^{-15} \text{ eV/Hz})(3.00 \times 10^{8} \text{ m/s})}{7.72 \text{ eV}}$$

and

$$\boxed{\lambda = 161 \text{ nm}}$$

This, the most energetic radiation, is in the UV.

▶ **Quick Check:** 161 nm is very roughly $\frac{1}{4}$ the wavelength of light, and so the energy of the photon should be approximately 4 times the energy of a light photon, which is roughly 1.5 eV. That value yields about 6 eV, which is close to the given value of 7.72 eV. Thus, 161 nm must be at least the right order-of-magnitude.

The unpaired valence electrons of isolated atoms are the source of much colored light. But when these atoms combine to form molecules or solids, those valence electrons can get paired up in the very process of forming the chemical bonds that hold the thing together. As a direct consequence, the electrons are often more tightly bound and their molecular-excited states are higher up; that is, in the UV. This gives rise to selective molecular UV absorption. Molecules in the atmosphere, such as N_2, O_2, CO_2, and H_2O, have such electronic resonances in the UV (which contributes to making the sky blue, p. 864).

Each UV quantum can carry enough energy (from 3.2 eV to 100 eV) to independently ionize an atom or rip apart a chemical bond. The particlelike aspects of radiant energy begin to become increasingly more evident as the frequency increases. At wavelengths less than around 290 nm, UV is germicidal; that is, it kills microorganisms.

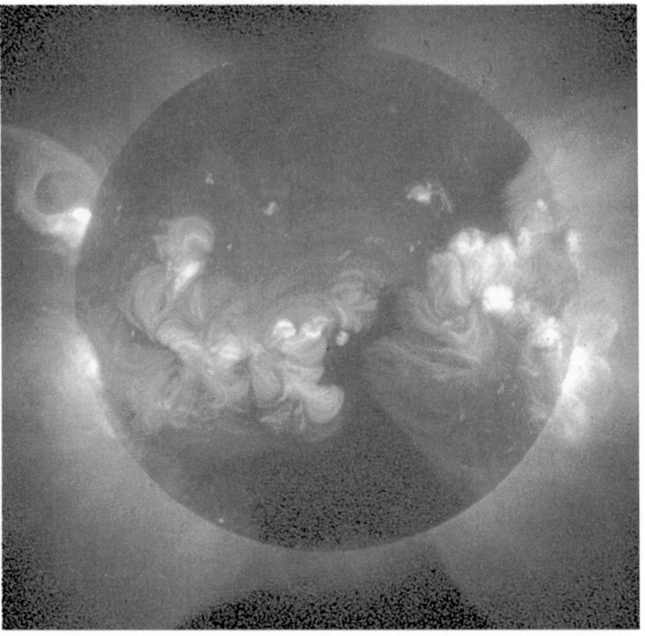

An X-ray photograph of the Sun. Modern focusing methods are producing detailed images of distant celestial sources. Orbiting X-ray telescopes have given us an exciting new view of the Universe.

Wilhelm Konrad Röntgen (1845–1923). As a result of his discovery of X-rays in 1895, Röntgen became the instant hero of his age and the first winner of the Nobel Prize in Physics.

24.15 X-rays

When Wilhelm Röntgen discovered X-rays on November 5, 1895, it was quite by accident (p. 1022). Within months, the marvelous rays were at work everywhere. Without the slightest inkling of their inherently dangerous nature, X-rays were used for everything conceivable, from removing facial hair to examining luggage. Too often, the results of that cavalier attitude were horribly tragic. (And, of course, the Victorian ladies were well warned of the only danger anyone seems to have been concerned with: lurking X-ray peeping Toms.) Incredibly, X-rays were even being used as late as the 1950s to treat acne, though it's only recently that the victims began developing cancer, particularly of the thyroid.

Extending in frequency from roughly 2.4×10^{16} Hz to 5×10^{19} Hz, X-rays have exceedingly short wavelengths; most are smaller than the size of an atom. The individual photon energies (100 eV to 0.2 MeV) are so large that X-ray quanta can interact with matter one at a time in a clearly granular fashion, almost like bullets of energy. The primary mechanism for the production of this radiation is the rapid deceleration of high-speed charged particles.

Example 24.8 An electron flies across an X-ray tube. Traversing a voltage difference of 1.00×10^4 V, it crashes into a metal target. Assuming it makes a rare head-on collision and comes to rest with the emission of a single photon, what is the wavelength of the radiation?

Solution: [Given: $V = 1.00 \times 10^4$ V and $q_e = 1.60 \times 10^{-19}$ C. Find: λ.] An electron that falls through a potential difference of V picks up an amount of energy Vq_e, which then appears as the photon energy $E = hf = hc/\lambda$; hence

$$\lambda = \frac{hc}{E} = \frac{hc}{Vq_e}$$

$$\lambda = \frac{(6.626 \times 10^{-34}\ \text{J/Hz})(3.00 \times 10^8\ \text{m/s})}{(1.00 \times 10^4\ \text{V})(1.60 \times 10^{-19}\ \text{C})}$$

and $\boxed{\lambda = 1.24 \times 10^{-10}\ \text{m}}$

(continued)

(continued)

Notice that the calculation could have been done a bit more easily using electron volts (Problem 35).

▶ **Quick Check:** This value of λ is 0.12 nm, which is just about the size of an atom, and we know that's the correct wavelength for X-rays.

Diagnostic X-rays used in medicine have energies from 20 keV to 100 keV. Traditional medical film-radiography of this simple variety produces shadow castings. What arrives at the film is a crude mapping of the absorption that took place as the beam crossed all the interposed tissue.

In the 1970s, the marriage of the X-ray machine and the computer gave rise to a marvelous advance in X-ray technique known as *computed tomography.* Tomography (from the Greek *tomos,* or slice) is the process of creating an image of a cross-sectional region of a three-dimensional object—in this case, a person. The CT (or CAT) scan is done by rotating an X-ray source 360° around the patient and recording the transmission across the body at hundreds of different viewing angles. From that data, the computer then constructs a picture of the traversed region (see photo), producing a splendidly detailed representation showing all the soft tissue in that "slice" of the body.

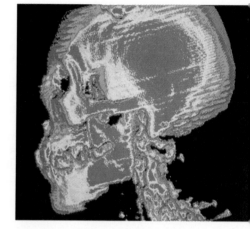

A modern X-ray CAT scan can reveal, slice-by-slice, structural details never before visible in ordinary X-ray photos.

24.16 Gamma Rays

Just as the atom is an interacting system of charged particles that exists only in well-defined configurations with well-defined energies, so, too, is the nucleus. Accordingly, the nucleus has a lowest-energy configuration, a ground state to which it will return once excited. A nucleus in one of its several well-defined excited states drops back to its ground state with the emission of a **gamma-ray** photon. Because the nuclear energy states are tightly bound, γ-ray energies range from keV to MeV.

There is no substantive difference between γ-rays and X-rays, other than the superficial historical one that the former were first seen to come from nuclei and the latter from atomic electrons—high-energy photons from a synchrotron come from neither. Thus, the distinction between X-rays and gamma rays vanished with the introduction of modern high-energy machines. Hundreds of modern hospitals are equipped with sophisticated devices such as linear accelerators and betatrons that provide multi-MeV electron beams that generate energetic photons for cancer treatment.

Core Material

Light (and all other forms of radiant energy) is electromagnetic. A *progressive* wave is a self-supporting, energy-carrying disturbance that travels free of its source (p. 836). The most important form of disturbance, from a theoretical perspective, is the **harmonic**, or sinusoidal, wave, which has a profile given by

$$E = E_0 \sin kx \qquad [24.5]$$

$k = 2\pi/\lambda$ is the **propagation number**. As with all waves

$$v = f\lambda \qquad [12.18]$$

In particular, for electromagnetic waves in vacuum

$$E = cB \qquad [24.6]$$

and

$$c = \frac{1}{\sqrt{\mu_0 \varepsilon_0}} \qquad [24.7]$$

where

$$c = 2.997\,924\,58 \times 10^8 \text{ m/s}$$

The energy-per-unit-area-per-unit-time conveyed by a lightwave is its **irradiance** I, where

$$I = \frac{1}{2}c\varepsilon_0 E_0^2 \quad [24.8]$$

Light is wavelike, though unlike traditional waves, electromagnetic energy is not smoothly spread out across the wavefront but somehow concentrated within it. These energy concentrations are known as **photons**. The energy of a photon is

$$E = hf \quad [24.9]$$

where $h = 6.626 \times 10^{-34}$ J/Hz (or 4.136×10^{-15} eV/Hz) is Planck's Constant.

The **electromagnetic spectrum** ranges from radiowaves to microwaves, infrared, light, ultraviolet, X-rays, and gamma rays (p. 848).

Suggestions on Problem Solving

1. Planck's Constant is given as either 6.6×10^{-34} J/Hz or 4.1×10^{-15} eV/Hz and, depending on how the data of the problem is provided, one or the other is the more convenient. Note that 1 Hz is 1 cycle per second, so the units of h are also J·s.
2. Many of the introductory wave problems make use of two equations ($v = \lambda f$ and $f = 1/T$) in endless variations. That was the case with sound, and it's true here as well.
3. Keep in mind that although $\lambda = c/f$, if we have a wavelength range $\Delta\lambda$, there will be a corresponding frequency range, but $\Delta\lambda \neq c/\Delta f$. We have to calculate $\lambda_1 = c/f_1$ and $\lambda_2 = c/f_2$ and then determine $\Delta\lambda = \lambda_2 - \lambda_1$ and $\Delta f = f_2 - f_1$.
4. When a charge q falls through a potential difference V, the energy change, qV, is in joules, *not* electron volts. That's a common point of confusion.

Discussion Questions

1. Why are fire engines sometimes painted yellow?
2. The Stanford Linear Accelerator imparts energies of up to 20 000 MeV to electrons, accelerating them along its straight two-mile course. Discuss the behavior of these particles as it pertains to electromagnetic radiation.
3. Does your hair dryer radiate radiowaves? How might you determine as much?
4. When NASA sends space probes to bodies beyond Earth, they routinely bathe the vehicles on the launch pad in ultraviolet radiation for several hours. Why? What's the significance of the so-called hole in the ozone layer?
5. People such as Ben Franklin rejected the corpuscular theory of light on the basis of momentum considerations. Devise an argument he might have used.
6. Is the region of space you are in right now crisscrossed by electromagnetic waves? In fact, are you permeated by them even as you read this question? Explain.
7. How does Newton's corpuscular theory of light compare with the modern quantum picture of light?
8. Central to Maxwell's theory of electromagnetic waves is the insight that the electric and magnetic fields generate one another. Explain. In vacuum, electromagnetic waves (including light) are said to be *transverse*. What does that mean?
9. How is it possible to cook a piece of meat on a paper plate in a microwave oven? What would happen to a dry cube of ice immediately after being placed in an operating microwave oven?
10. In his experiments (1780) on animal electricity, Luigi Galvani at one point tied a long wire to a dissected frog via a nerve, and another wire from the feet was grounded down a well. Waiting for a thunderstorm, he observed the legs twitching in rhythm with the lightning. In what sense was this the first frog-radio? What was happening? Incidentally, H. Hertz (1880s) vainly attempted to use a frog to detect the TEM waves he was generating in his laboratory.

Figure Q15

11. What is the physical meaning of the idea of whiteness? That is, when do we see a thing as being white? In what way is whiteness a property of the detector (that is, the human eyeball)?
12. One often hears terms such as "ultraviolet light" and "infrared light" (although X-ray light and radio light are never spoken of). Why are such terms at best misleading?
13. Is a pulse a wave, or must a wave be something that "rises and falls" over and over again? What waves in a lightwave?

14. In what regard is it better to say "light that is perceived to be yellow" rather than "yellow light"?

15. Figure Q15 shows the phenomenon of optical levitation. A tiny glass sphere one-thousandth of an inch in diameter is suspended in an upward laser beam. What can you say about the ability of radiant energy to transfer momentum? Might it be possible to travel around in space not far from the Sun using a sail craft?

16. A few decades ago, some doctors were treating tonsillitis with X-rays. What would you expect would be a possible consequence of receiving such therapy?

Multiple Choice Questions

1. The source of electromagnetic waves is (a) a constant current (b) a charge moving only in circles (c) any accelerating charge (d) any accelerating particle (e) none of these.
2. The radio antenna for an AM station is a 75-m-high tower that is equivalent to $\frac{1}{4}\lambda$; another $\frac{1}{4}\lambda$ corresponds to the ground reflection. At what frequency does the station transmit? (a) 10 kHz (b) 75 kHz (c) 1 MHz (d) 300 kHz (e) none of these.
3. When sodium atoms are excited (for example, by heating salt in a flame) they emit bright yellow light at two wavelengths of 589.0 nm and 589.6 nm. That emission implies that the sodium atom has two nearby energy levels separated by only (a) 2 eV (b) 2×10^{-3} eV (c) 0.6 eV (d) 0.6 MeV (e) none of these.
4. In comparison to UV, light has (a) wavelengths that are shorter (b) frequencies that are higher (c) wavelengths that are equal (d) frequencies that are equal (e) none of these.
5. Dental X-ray photos are usually taken while operating the machine at about 50 000 V. The minimum wavelength of such radiation is (a) 0.025 nm (b) 0.25 nm (c) 2.5 nm (d) 25 nm (e) none of these.
6. Which statement is not true for a photon? (a) it is electromagnetic in nature (b) it always travels at the speed c independent of frequency (c) it possesses energy independent of frequency (d) it has wavelike properties (e) none of these.
7. An X-ray tube produces a beam of photons by bombarding a dense metal target with high-energy electrons. The resulting beam possesses a wide range of frequencies and is terminated at a maximum energy that (a) depends only on the target material and the temperature (b) depends on the voltage across the tube and is constant provided the voltage is constant (c) is not constant and changes even when the voltage is kept constant (d) depends on the tube length and the shielding (e) none of these.
8. Which of the following requires a physical medium to travel in? (a) lightwaves (b) radiowaves (c) sound waves (d) gamma rays (e) none of these.
9. The propagation number of a wave (a) varies inversely with wavelength (b) has to do with the number of pulses in a burst (c) varies inversely with frequency (d) depends on the speed of the wave (e) none of these.
10. Nowadays it is possible to directly measure, using electronic techniques, the frequencies of electromagnetic oscillations ranging up to 500 MHz. That corresponds to a wavelength of (a) 0.6 m (b) 0.6 cm (c) 6.0 m (d) 6.0 cm (e) none of these.
11. What is the frequency of 1-GeV gamma rays? (a) 2.4×10^{20} Hz (b) 1.5×10^{42} Hz (c) 3×10^{8} Hz (d) 2.4×10^{23} Hz (e) not enough information given.
12. For all practical purposes, the light from the following source can best be considered a plane wave (a) a street light overhead (b) a nearby desk lamp (c) a match held in the hand (d) a star in the constellation Orion (e) none of these.
13. An oscillating dipole is best described as (a) two poles moving in a circle (b) a single pole moving in two directions (c) two equal and opposite charges moving to and fro along a line (d) two like charges oscillating (e) none of these.
14. Light travels 1.000 m in vacuum in (a) 1.000 s (b) 3.336×10^{-19} s (c) 0.334 s (d) 3.336×10^{-9} s (e) none of these.
15. A typical AM radiowave is (a) 1.0 m long (b) 1.0 cm long (c) millions of meters long (d) hundreds of meters long (e) none of these.
16. Is there a relationship between the electric and magnetic fields of an electromagnetic wave in vacuum? (a) no (b) yes, $B = cE$ (c) yes, $E = cB$ (d) yes, $E = B/v$ (e) none of these.
17. An AM radio station transmits a signal whose electric field is received with a strength of 1.5 mV/m. If the antenna on a portable radio at that location is 0.75 m long and straight up in the air, the input voltage will be (a) 1.5 mV (b) 1.1 mV (c) 7.5 mV (d) 1.5 V (e) none of these.
18. A typical atomic transition lasts about (a) 10^{-5} s to 10^{-6} s (b) 1.0 s to 10.0 s (c) 0.01 s to 0.001 s (d) 10^{8} s to 10^{9} s (e) none of these.
19. In order of decreasing frequency, the entire electromagnetic spectrum is made up of (a) radiowaves, microwaves, IR, light, UV, and γ-rays (b) γ-rays, X-rays, UV, IR, microwaves and radiowaves (c) radiowaves, microwaves, IR, light, X-rays, and γ-rays (d) light, IR, and UV (e) none of these.
20. The ground state of an atom is the state (a) with the highest energy (b) nearest the ground (c) with no energy (d) with the least energy (e) none of these.
21. The molecular rotational and vibrational energy states are primarily responsible for the emission of (a) radiowaves (b) light and UV (c) IR and microwaves (d) X-rays and gamma rays (e) none of these.

Problems

THE NATURE OF LIGHT
PHOTONS
THE ELECTROMAGNETIC SPECTRUM

1. [I] A lightwave has a wavelength of 500 nm. What is its propagation number?

2. [I] An infrared electromagnetic wave has a propagation number of 2000π m^{-1}. What is its wavelength?

3. [I] The propagation number of a harmonic electromagnetic wave is 6.283×10^{-4} m^{-1}. What is its wavelength?

4. [I] An electromagnetic wave has a profile given by

$$E = E_0 \sin kx$$

If the amplitude of the wave is 20.0 V/m, what is the size of the field at $x = 0$?

5. [I] An electromagnetic wave at $t = 0$ has the profile

$$E = E_0 \sin kx$$

Draw a plot of E versus x showing E at points $x = 0$, $\lambda/4$, $\lambda/2$, $3\lambda/4$, and λ. Remember that $k = 2\pi/\lambda$.

6. [I] Draw a plot of the function

$$E = E_0 \sin (kx - \pi/2)$$

and compare it to your result from Problem 5. How far has the profile advanced?

7. [I] The electric field of a microwave is given by

$$E = (20 \text{ V/m}) \cos \frac{2\pi}{1.00 \text{ mm}}[x - (3.00 \times 10^8 \text{ m/s})t]$$

What is its value at $x = 0$ and $t = 0$?

8. [I] The electric field of a TEM wave has the form

$$E = (5.0 \text{ V/m}) \sin k(x - vt)$$

What is the value of the E-field at $x = \lambda/4$ and $t = 0$?

9. [I] Make a sketch of the profile or shape of the wave $E = (10 \text{ V/m}) \sin [k(x - vt) + \varepsilon]$ when $\varepsilon = 0$ and again when $\varepsilon = \pi/2$. Set $t = 0$ and plot the curve for various values of x (namely, $x = 0, \lambda/4, \lambda/2, 3\lambda/4, \lambda$), remembering that $k = 2\pi/\lambda$. What does the phase-shifted wave look like?

10. [I] Make a sketch of the wave

$$E = E_0 \sin k(x - vt)$$

at $t = 0$ given that $\lambda = 10$ m and $E_0 = 2.0$ V/m.

11. [I] In the article "The Longest Electromagnetic Waves" (*Sci. Am.*, March 1962), J. R. Heirtzler described the detection of waves 18.6×10^6 miles in wavelength. What type of waves were they and what was their period?

12. [I] How far does light travel in vacuum in 10^{-9} s?

13. [I] What is the vacuum wavelength of a 20.0-Hz electromagnetic wave?

14. [I] If light is emitted from an atom in little wavetrains (Fig. P14) lasting up to 10^{-8} s, how long, at most, is such a disturbance in space? If we approximate the wavelength as 500 nm, roughly how many waves long is the train?

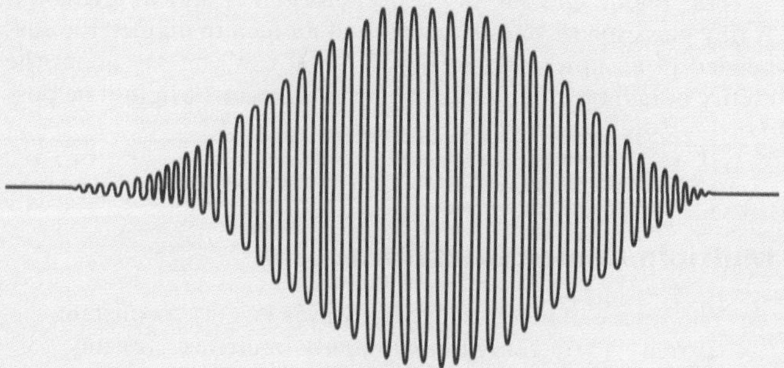

Figure P14

15. [I] A tuning circuit in an FM radio is designed to pick up a station at 100 MHz. If the capacitance of the input circuit is 0.5 pF, how much is the inductance?

16. [I] In 1887, Heinrich Rudolf Hertz succeeded in generating and detecting long-wavelength electromagnetic waves. His transmitter was an induction coil (a device that converted low-voltage dc into high-voltage ac) attached to a loop of wire ending in a spark gap. Across the room, a wire loop with its own open gap served as the receiver. When the circuit discharged, an oscillatory spark flashed across the transmitter's gap, and the event was almost immediately reported by a fainter spark at the distant receiver. If the frequency of the waves was 75 MHz, what was their wavelength?

17. [I] Determine the frequency of a 200-m-long AM radiowave, assuming the conditions to be that of vacuum.

18. [I] On December 12, 1901, Marconi, using a 20-kW transmitter attached to a 200-ft antenna, sent electromagnetic signals (with a 1-km wavelength) across the Atlantic for the first time. Compute the frequency of the emission. Assume the speed in air is the same as that in vacuum.

19. [I] The energy arriving per second on a 2.00-m^2 detector held perpendicular to the light from a bright star is about 2.4×10^{-9} J. Determine the irradiance of the radiation.

20. [I] The irradiance 1 m from a candle flame is just about 1.5×10^{-3} W/m^2. How much energy will arrive in 2.00 s on a disk having a 1.00-cm^2 area held as close to perpendicular as possible 1.00 m from the flame?

21. [I] The maximum irradiance I of solar radiation arriving on the lawn of the White House is 1.05×10^3 W/m^2. What maximum energy will impinge each minute on a flat collector with a 1.00-square-meter area?

22. [I] What is the energy of a tangerine-colored photon where λ_0, the vacuum wavelength, equals 616 nm?

23. [I] Determine the vacuum wavelength, frequency, and energy in joules of a 2.00-eV photon. What type of radiation is it?

24. [II] Show that $\varepsilon_0(\Delta\Phi_E/\Delta t)$ has the units of current.

25. [II] Show that the wave equation in Problem 10 can be written as

$$E = E_0 \sin (kx - \omega t)$$

26. [II] Show that the electric field of a progressive harmonic electromagnetic wave, as given in Problem 10, can be written as

$$E = E_0 \sin 2\pi f\left[\left(\frac{x}{v}\right) - t\right]$$

27. [II] What is the magnitude of the electric field given in SI units by $E = 20 \cos(kx - \omega t + \pi)$ at $x = 0$, when $t = 0$, $t = T/4$, and $t = T$?

28. [II] The electric field of an electromagnetic wave is given by

$$E = 2.0 \times 10^2 \sin[3.0 \times 10^6 \pi(x - 3.0 \times 10^8 t)]$$

where everything is in SI units. What is the wave's frequency? (Hint: Compare this expression with the one in Problem 10.)

29. [II] What is the amplitude of the magnetic field associated with the TEM wave in the previous problem?

30. [II] The antenna shown in Fig. P30 is called a folded dipole. How long should it be if it's to receive FM signals at 90 MHz?

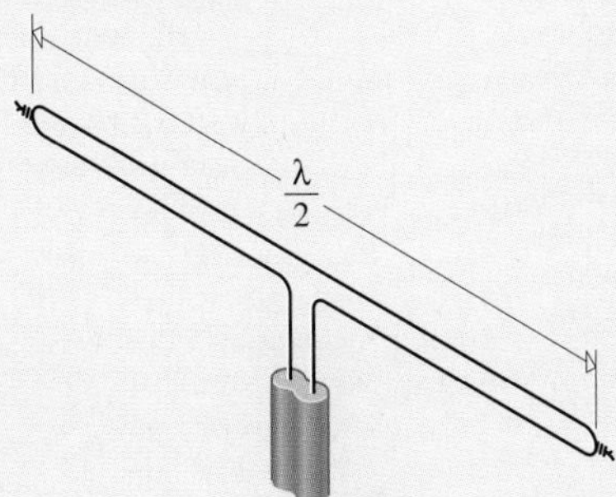

Figure P30

31. [II] In 1982, workers at Bell Labs produced optical pulses lasting 30.0 femtoseconds. How many wavelengths of the 620-nm red light correspond to one of these little wavetrains?

32. [II] Light with a frequency of 6.50×10^{14} Hz is traveling through vacuum. How many of these waves (end-to-end) are there per centimeter?

33. [II] A TEM wave given by

$$E = 2.0 \times 10^2 \sin[3 \times 10^6 \pi(x - 3.0 \times 10^8 t)]$$

where everything is in SI units, impinges on a perpendicular surface in space. What is the irradiance?

34. [II] A beam of TEM harmonic waves has an irradiance of 13.3 W/m^2. What is the amplitude of the electric field?

35. [II] An electron falls through a potential difference of 1.00×10^4 V. What is the maximum energy of a resulting X-ray photon? What is the minimum corresponding wavelength? Do this calculation in electron volts and then compare the results with that of Example 24.8.

36. [III] It takes an energy of 33 keV to remove the innermost electron from an atom of iodine. Show that iodine will be a powerful absorber of X-rays at a frequency of 8.0×10^{18} Hz.

37. [III] If the electric field of an electromagnetic wave traveling in vacuum, pointing in the y-direction, is given in SI units by

$$E_y = 200 \sin(10^7 x - \omega t)$$

find (a) the vacuum wavelength and (b) the frequency. (c) In what direction does the wave progress?

38. [III] Using the results of Problem 37, write an expression for the accompanying B-field.

39. [III] A laser that emits pulses of UV lasting 2.00 ns has a beam diameter of 2.5 mm. If each burst contains an energy of 3.0 J, (a) what is the length in space of each pulse? (b) what is the average energy per unit volume (J/m^3), the energy density, in one of these pulses?

40. [III] Given the wave function for an electromagnetic harmonic disturbance expressed in SI units

$$E = 10^2 \sin \pi(3 \times 10^6 x - 9 \times 10^{14} t)$$

find (a) the amplitude, (b) the speed, (c) the frequency, (d) the wavelength, (e) the period, and (f) the direction of propagation.

25

The Propagation of Light: Scattering

OUR PRESENT CONCERN IS with the two basic laws that describe the *reflection* (p. 869) and *refraction* (p. 876) of light. But the underlying questions are: How does light move through bulk matter? And, what happens to it as it does? Each such encounter is a stream of photons sailing through an array of atoms suspended in the void. The details of that marvelous journey determine why the sky is blue and blood is red, why your cornea is transparent and your hand opaque, why snow is white and rain is not.

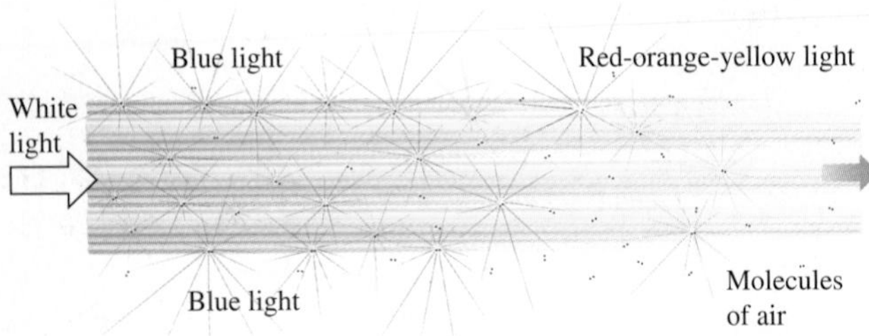

Figure 25.1 A beam of light traversing a region of widely spaced molecules. The light laterally scattered is mostly blue, and that's why the sky is blue. The unscattered light, which is rich in red, is viewed only when the Sun is low in the sky at sunrise and sunset.

SCATTERING

At its core, this chapter is about **scattering**, the absorption and prompt re-emission of electromagnetic radiant energy by atoms and molecules. The processes of reflection and refraction are macroscopic manifestations of scattering occurring on a submicroscopic level. Let's first consider the propagation of radiant energy through various homogeneous media.

25.1 Rayleigh Scattering: Blue Skies

Imagine a narrow beam of sunlight, a torrent of photons having a broad range of frequencies, advancing through empty space. As it progresses, the beam spreads out very slightly, but apart from that, all the energy continues forward at c. There is no scattering, and the beam cannot be seen from the side (the photons move straight ahead). Nor does the light tire or diminish in any way. When a star in a nearby galaxy 1.7×10^5 ly away was seen to explode in 1987, the flash of light that reached Earth had been sailing through space for 170,000 years. Photons are timeless.

Now, suppose we mix a wisp of air into the void—some molecules of nitrogen, oxygen, and so forth. Since these molecules have no resonances in the visible (pp. 848 and 855), no one of them can be raised into an excited state by absorbing a quantum of light, and the gas is transparent. Instead, each molecule behaves as a little oscillator whose electron cloud can be driven into a ground-state vibration by an incoming photon. Immediately upon being set oscillating, the molecule initiates the re-emission of light. A photon is absorbed, another photon of the same frequency is emitted; the light is *elastically scattered.* The molecules are randomly oriented, and photons scatter out every which way (Fig. 25.1). Even when the light is fairly dim, the number of photons is immense, and it looks as if the molecules are scattering little classical spherical wavelets (Fig. 25.2)—energy streams out in every direction. Still, the scattering process is quite weak and the gas tenuous, so the beam is very little attenuated unless it passes through a tremendous volume of air.

The amplitudes of these ground-state vibrations, and therefore the amplitudes of the scattered light, increase with frequency because the molecules all have electronic resonances in the UV. The closer the driving frequency is to a resonance, the more vigorously the oscillator responds (p. 416). So, violet light is strongly scattered laterally out of the beam, as is blue to a slightly lesser degree, as is green to a considerably lesser degree, as is yellow to a still lesser degree, and so on. The beam that traverses the gas will thus be richer in the red end of the spectrum, while the light scattered out (sunlight not having much violet in it in the first place) will be richer in blue. That, in part, is why the sky is blue.

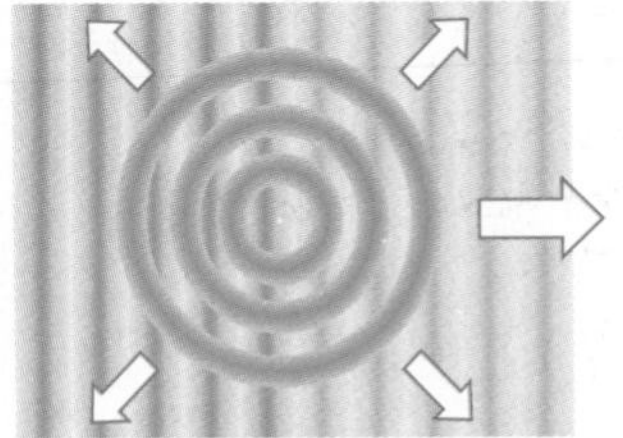

Figure 25.2 Scattering of some of the energy out from a plane wave in the form of a spherical wavelet. The process is continuous and hundreds of millions of photons per second stream out of the scattering atom in all directions.

Long before Quantum Mechanics, Lord Rayleigh (1871) analyzed scattered sunlight in terms of molecular oscillators and correctly concluded that the intensity of the scattered light was proportional to $1/\lambda^4$, which increases with f^4. Before this work, it was widely believed that the sky was blue because of scattering from minute dust particles. Since that time, *scattering involving particles smaller than a wavelength* has been referred to as **Rayleigh scattering**. A human's blue eyes, a blue jay's feathers, the blue-tailed skinks's blue tail, and the baboon's blue buttocks are all colored via Rayleigh scattering.

Example 25.1 A beam of white light traverses a medium composed of randomly distributed particles that are each much smaller than a typical wavelength. Compare the amount of scattering occurring for the red (710 nm) component with that of the violet (400 nm) component.

Solution: [Given: $\lambda_r = 710$ nm and $\lambda_v = 400$ nm, compare their scattered intensities.] The degree of Rayleigh scattering is proportional to $1/\lambda^4$. But $\lambda_r = 1.775\lambda_v$ and so $1/\lambda_r^4 = (1/1.775\lambda_v)^4$; hence, violet is scattered $(1.775)^4 = \boxed{9.93}$ times more intensely than red.

▶ **Quick Check:** We can check the order-of-magnitude of the answer by noting that red is very roughly twice the wavelength of violet and the answer must be between $1^4 = 1$ and $2^4 = 16$; at least we pushed the right keys on the calculator.

As we will see in a moment, a dense uniform substance will not appreciably scatter laterally, and that applies to much of the lower atmosphere. Something else beyond Rayleigh scattering must be contributing to making the lower dense regions of the sky blue. What happens in the atmosphere is that thermal motion of the air results in rapidly changing *density fluctuations* on a local scale. And these momentary, fairly random fluctuations cause more molecules to be in one place than another and to radiate more in one direction than another. A theory of scattering from these fluctuations gives very much the same results as Rayleigh obtained for a tenuous gas.

Sunlight streaming into the atmosphere from one direction is scattered in all directions. Without an atmosphere, the daytime sky would be as black as the void of space, as black as is the Moon sky. When the Sun is low over the horizon, its light passes through a great thickness of air (far more so than it does at noon). With the blue-end appreciably attenuated, the reds and yellows propagate along the line-of-sight from the Sun to produce Earth's familiar fiery sunsets.

25.2 Scattering and Interference

In dense media, there are a tremendous number of close-together atoms or molecules contributing an equally tremendous number of scattered electromagnetic wavelets. These wavelets overlap and interfere in a way that does not occur in a tenuous medium. As a rule, the denser the substance through which light advances, the less the lateral scattering.

The phenomenon of interference has already been discussed (p. 466) and will be treated in further detail in Chapter 27; here, the basics suffice. Interference is *the superposition of two or more waves producing a resultant disturbance that is the sum of the overlapping wave contributions.* Figure 25.3 shows two harmonic waves

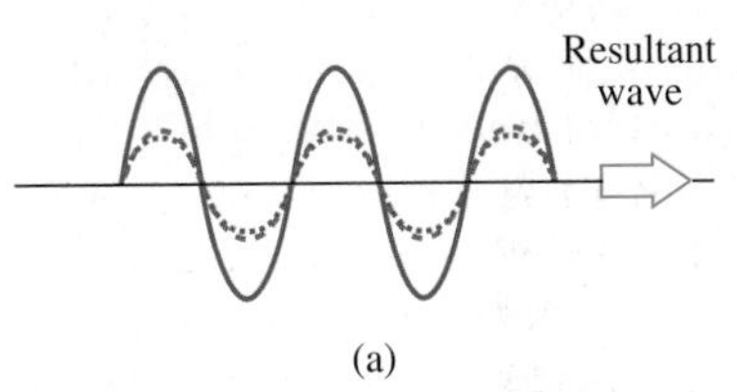

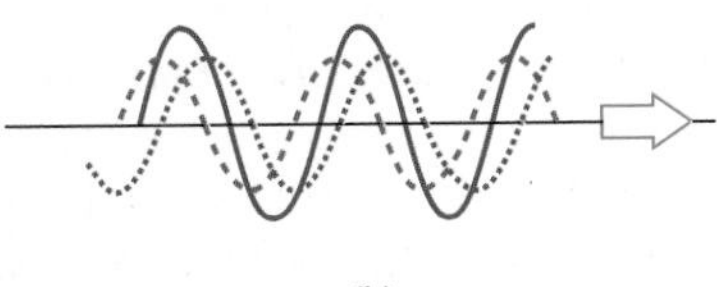
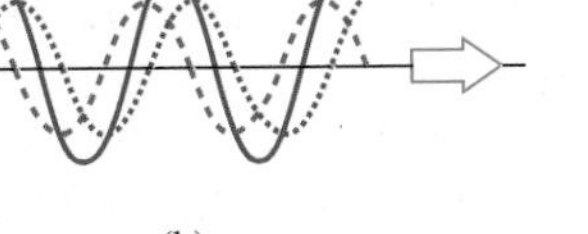

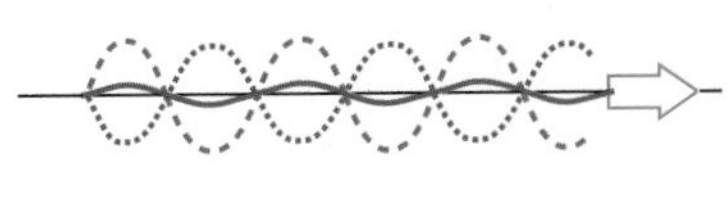

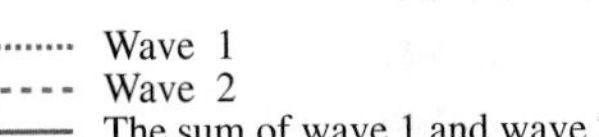

Figure 25.3 The superposition of two waves. These combine to form a resultant wave. (a) When the component waves are in-phase, we have *constructive interference,* and the resultant is large. (b) As the phase difference increases, the resultant decreases. (c) When the component waves are 180° out-of-phase, the resultant has its smallest value, and we have *destructive interference.*

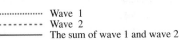

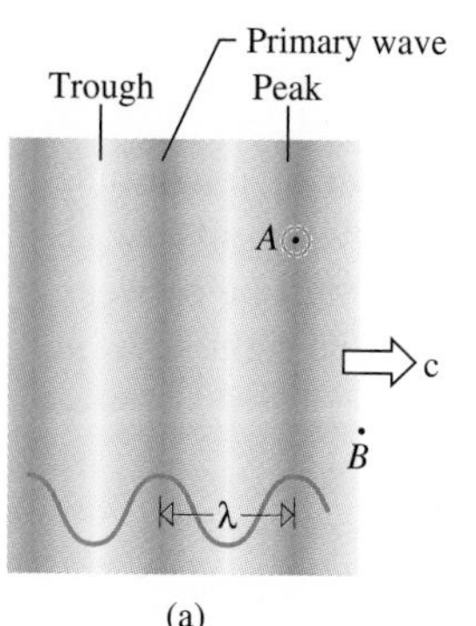

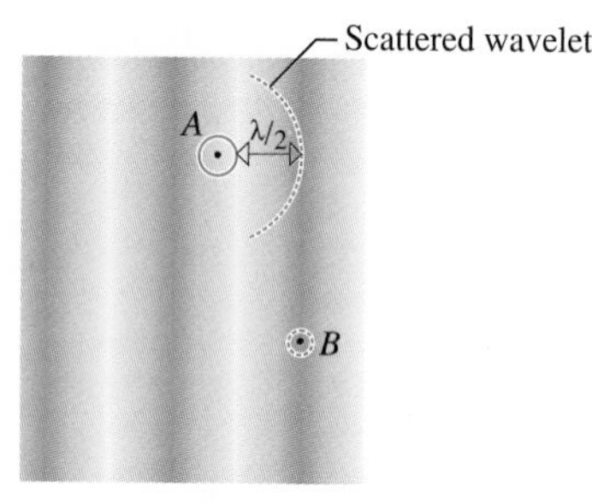

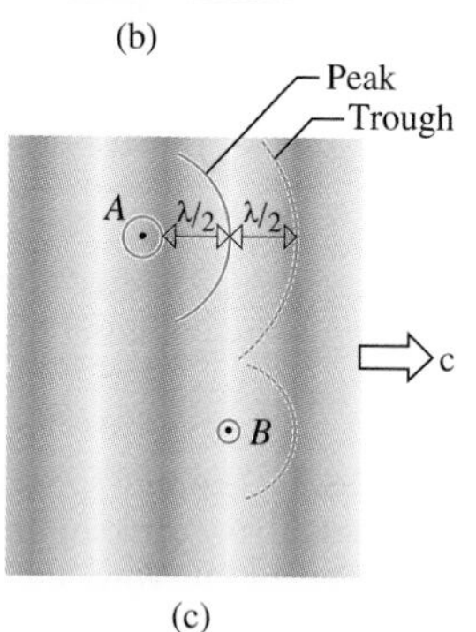

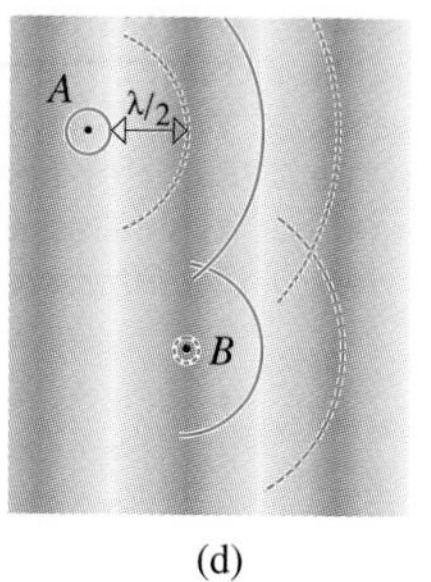

Figure 25.4 Plane waves of a beam of light moving to the right, passing over molecules *A* and *B*. These scatter wavelets that travel in-phase in the forward direction. In actuality, billions of such wavelets combine constructively, scattering light in the forward direction.

of the same frequency traveling in the same direction. When they are in-phase, the resultant at every point is the sum of the two wave-height values (Fig. 25.3a). This extreme case is called *total constructive interference.* When the phase difference reaches 180°, the waves tend to cancel, and we have the other extreme, called *total destructive interference* (Fig. 25.3c).

The simple theory of Rayleigh scattering has the molecules randomly arrayed in space so that the phases of the wavelets scattered off to the side have no particular relationship to one another and there can be no sustained interference between them. That situation will occur when the separation between the molecular scatterers is roughly a wavelength or more, as it is in a tenuous gas. When there are many scatterers, the random hodgepodge of overlapping waves effectively averages away the interference. *Random, widely spaced scatterers driven by an incident wave emit wavelets that are essentially independent of one another in all directions except forward.* Laterally scattered light, unimpeded by interference, streams out of the beam. And this is approximately the situation existing about 100 miles up in the Earth's tenuous high-altitude atmosphere, where a good deal of blue-light scattering takes place.

To see why the forward direction is special, why the wave advances in any medium, consider Fig. 25.4. It depicts a sequence in time showing two molecules *A* and *B*, fairly far apart, interacting with an incoming plane wave—a solid line represents a wave peak (a positive maximum); a dashed line corresponds to a trough (a negative maximum). In (a), the incident wavefront impinges on molecule *A*, which begins to scatter a spherical wavelet. For the moment, suppose the wavelet is 180° out-of-phase with the incident wave. Thus, *A* begins to radiate a trough (a negative *E*-field) in response to being driven by a peak (a positive *E*-field). Part (b) shows the spherical wavelet and the plane wave overlapping, marching out-of-step but marching together. The incident wavefront impinges on *B* and it, in turn, begins to reradiate a wavelet, which must also be out-of-phase by 180°. In (c) and (d), we see the point of all of this, namely, that both wavelets are moving forward with the incident wave—they are in-phase with each other, but out-of-phase with the incident wave. And that condition would be true for all such wavelets regardless of both how many molecules there were and how they were distributed. Because of the asymmetry introduced by the beam itself, **all the scattered wavelets add constructively with each other in the forward direction**.

25.3 The Transmission of Light through Dense Media

Now, suppose the amount of air in the region under consideration is increased. In fact, imagine that each little cube of air, one wavelength on a side, contains a great many molecules, whereupon it is said to have an appreciable *optical density.* At the wavelengths of light, the Earth's atmosphere at STP has about three million molecules in such a λ^3-cube. The wavelets ($\lambda \approx 500$ nm) radiated by sources so close together ($\approx$3 nm) cannot properly be treated as random. The light beam effectively encounters a fairly uniform medium with no discontinuities to destroy the symmetry. Again the scattered wavelets interfere constructively in the forward direction (that much is independent of the arrangement of the molecules), but now destructive interference predominates in all other directions. *No light ends up scattered laterally or backwards.*

For example, Fig. 25.5 shows the beam moving through an ordered array of close-together scatterers. Some molecule *A* radiates spherically out of the beam, but because of the ordered close arrangement, there will be a molecule *B*, a distance

$\approx\lambda/2$ away, such that both wavelets cancel in that transverse direction. Here, where λ is thousands of times larger than the scatterers and their spacing, it is very likely that there will always be pairs of molecules that tend to negate each other's wavelets in any given direction. Even if the medium were not perfectly ordered, the net electric field at a point in any direction will be the sum of a great many tiny scattered fields, each somewhat out-of-phase with the next (Fig. 25.6), so that the sum will always be negligibly small. The more dense, uniform, and ordered the medium is, the more complete will be the destructive interference and the less the nonforward scattering. The beam advances undiminished.

The scattering phenomenon on a per-molecule basis is extremely weak. In order to have half its energy scattered, a beam of green light will have to traverse $\approx$150 km of atmosphere. Since about 1000 times more molecules are in a given volume of liquid than in the same volume of vapor (at atmospheric pressure), we can expect to see an increase in scattering. Still, the liquid is a far more ordered state with much less pronounced density fluctuations, and that should suppress the nonforward scattering appreciably. Accordingly, an increased scattering, per unit volume, is observed in liquids, but it's more like 5 to 50 times as much rather than 1000 times. *Molecule for molecule,* liquids scatter substantially less than gases. Put a few drops of milk in a tank of water and illuminate it with a bright flashlight beam. A faint but unmistakable blue haze will scatter out laterally, and the direct beam will emerge decidedly reddened.

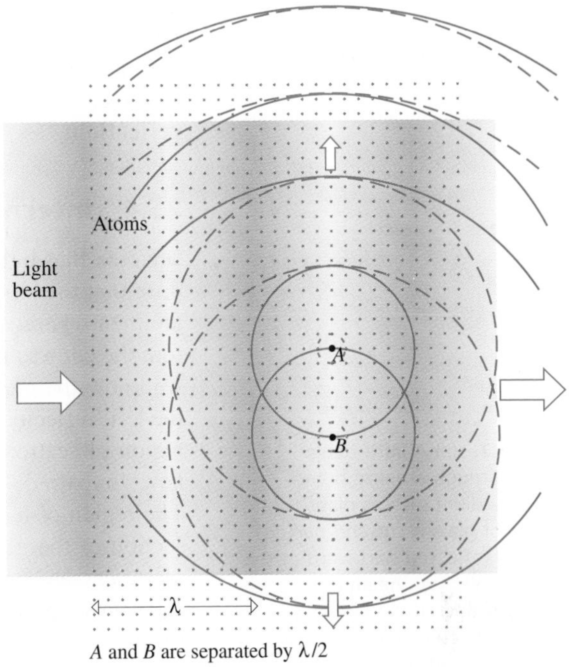

Transparent amorphous solids, such as glass and plastic, will also scatter light, but very weakly. Good crystals, like quartz and mica, with their almost perfectly ordered structures, scatter even more faintly. Of course, imperfections of all sorts (dust and bubbles in the liquids; flaws and impurities in the solids) will serve as scatterers, and when these are small, as in the gem moonstone, the emerging light will be bluish. In the same way, some inexpensive plastic food containers and white garbage-bag plastic look pale blue-white in scattered light and are distinctly orange in transmitted light.

Light beam

Atoms

A and B are separated by $\lambda/2$

Figure 25.5 A plane wave impinging from the left onto a material composed of many closely spaced atoms. Among countless others, a wavefront stimulates two atoms, *A* and *B*, that are very nearly one-half wavelength apart. The wavelets they emit interfere destructively. Trough overlaps crest and they completely cancel each other in the direction perpendicular to the beam. That process happens over and over again and little or no light is scattered laterally.

REFLECTION

When a beam of light impinges on the surface of a transparent material, such as a sheet of glass, the wave "sees" a vast array of very closely spaced atoms that will somehow scatter it. Remember that the wave may be $\approx$500 nm long while the atoms and their separations ($\approx$0.2 nm) are thousands of times smaller. In the case of transmission through a dense medium, the scattered wavelets cancel each other in all but the forward direction and just the ongoing beam is sustained. But that can only hap-

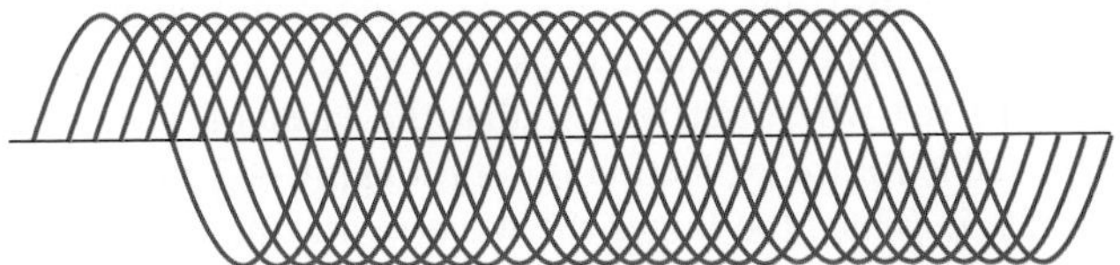

Figure 25.6 When a great many slightly shifted waves arrive at a point in space, there is generally as much positive *E*-field as negative, and the resultant disturbance is very nearly zero.

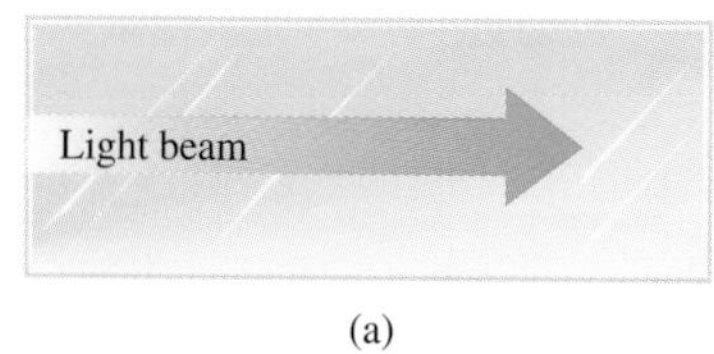

(a)

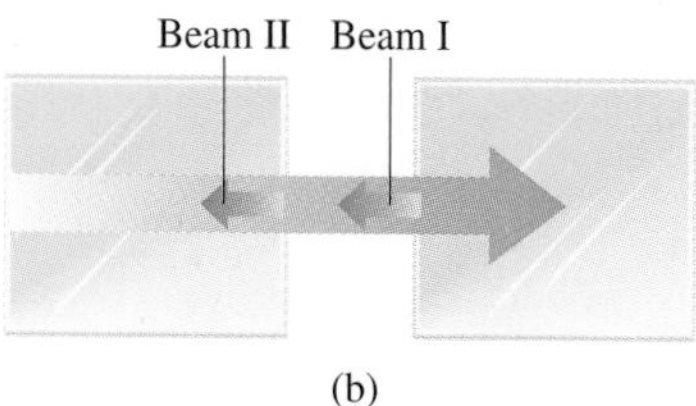

(b)

Figure 25.7 (a) A light beam traveling through a dense homogeneous medium such as glass. (b) If the block of glass is cut and parted, the light is reflected backward at the two new interfaces. Beam I is externally reflected, and beam II is internally reflected. When the two pieces are pressed back together, the two reflected beams cancel one another.

pen if there are no discontinuities, whereupon the scattering is uniform. This is not the case at an interface between two different transparent media (such as air and glass), which is a jolting discontinuity. When a beam of light strikes such an interface, some light is always scattered backward, and we call this phenomenon **reflection**.

25.4 Internal and External Reflection

Imagine that light is traveling across a large block of glass (Fig. 25.7a). Now, suppose that the block is sheared in half perpendicular to the beam. The two segments are then separated, exposing the smooth flat surfaces depicted in Fig. 25.7b. Just before the cut was made, there was no lightwave traveling to the left inside the glass—we know the beam only advances. Now there must be a wave (beam I) moving to the left, reflected from the surface of the right-hand block. The implication is that a region of scatterers on and beneath the exposed surface of the right-hand block is now "unpaired," and the backward radiation they emit can no longer be canceled. The region of oscillators that was adjacent to these, prior to the cut, is now on the section of the glass that is to the left. When the two sections were together, these scatterers presumably also emitted wavelets in the backward direction that were 180° out-of-phase with, and canceled, beam I. Now they produce reflected beam II. Each molecule scatters light in the backward direction and, in principle, each and every molecule contributes to the reflected wave. Nonetheless, in practice, it is a thin layer ($\approx\lambda/2$ deep) of unpaired atomic oscillators near the surface that is effectively responsible for the reflection. For an air-glass interface, about 4% of the energy of an incident beam falling perpendicularly *in* air *on* glass will be reflected straight back out by this layer of unpaired scatterers. And that's true whether the glass is 1.0 mm thick or 1.00 m thick.

Beam I reflects off the right-hand block, and because light was initially traveling from a less to a more optically dense medium, this is called **external reflection**. Since the same thing happens to the unpaired layer on the section that was moved to the left, it, too, reflects backwards. With the beam incident perpendicularly *in* glass *on* air, 4% must again be reflected, this time as beam II. And this process is referred to as **internal reflection**. If the two glass regions are made to approach one another increasingly closely (so that we can imagine the gap to be a thin film of, say, air—p. 956), the reflected light will diminish until it ultimately vanishes as the two faces merge and disappear and the block becomes continuous again.

Remember this 180° *relative phase shift between internally and externally reflected light*—we will need it later on. This kind of reflection is of practical importance when you have a microscope with lots of compound lenses and perhaps a dozen or two interfaces each kicking back $\approx$4% of the incident light (just try looking through 20 or 30 layers of plastic food wrap). To overcome that difficulty, modern high-quality lenses are covered with antireflection coatings (p. 958).

We know from experience with the common mirror that white light is reflected as white—it certainly isn't blue. To see why, first realize that the layer of scatterers responsible for the reflection is very roughly $\lambda/2$ thick (per Fig. 25.5). Thus, the larger the wavelength, the deeper the

The cruiser *Aurora,* which played a key role in the Communist Revolution (1917), docked in St. Petersburg. Where the water is still, the reflection is specular. The image blurs where the water is rough.

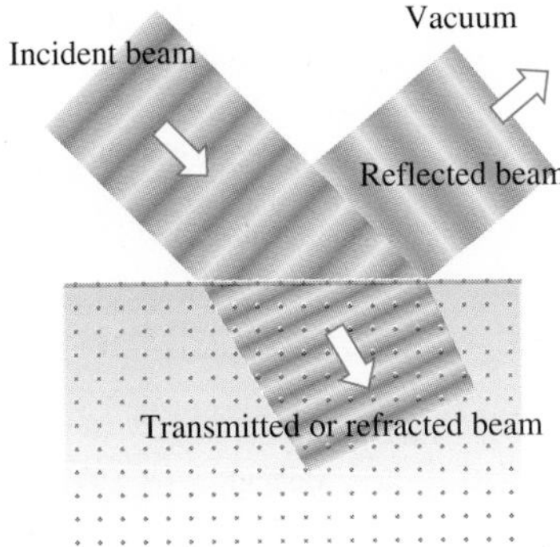

Figure 25.8 A beam of plane waves incident on a distribution of molecules constituting a piece of clear glass or plastic. Part of the incident light is reflected and part refracted.

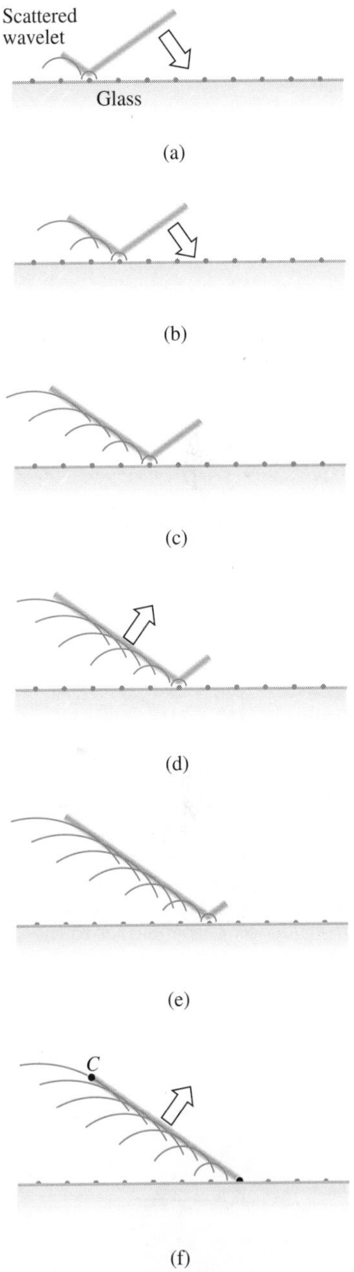

Figure 25.9 An incoming plane wave scattering off a layer of molecules. Because the wavelength is thousands of times larger than the molecular spacings, the reflection can be treated as occurring at the surface.

layer contributing, and the more scatterers there are acting together. This tends to balance out the fact that each scatterer is less efficient as λ increases (remember $1/\lambda^4$). The combined result is that *the surface of a transparent medium reflects all wavelengths about equally and doesn't appear colored in any way.* That, as we will see, is why this page looks white under white-light illumination.

25.5 The Law of Reflection

Figure 25.8 shows a beam composed of plane wavefronts impinging at some angle on the smooth flat surface of an optically dense medium (let it be glass). Assume that the surrounding medium is vacuum. Follow one wavefront as it sweeps in and across the molecules on the surface. For the sake of simplicity, in Fig. 25.9 we have omitted everything but one molecular layer at the interface. As the wavefront descends, it excites one scatterer after another, each of which reradiates a stream of photons that can be thought of as a hemispherical wavelet in the incident medium. Because the wavelength is so much greater than the separation between the molecules, the wavelets advance together and add constructively in only one direction, and there is one well-defined *reflected* beam. (That would not be true if the incident radiation was short-wavelength X-rays, in which circumstance there would be several reflected beams. And it would not be true if the scatterers were far apart compared to λ, as they are for a diffraction grating, in which case there would also be several reflected beams.)

In Fig. 25.10, the line $\overline{AB}$ lies along an incoming wavefront while $\overline{CD}$ lies on an outgoing wavefront—in effect, $\overline{AB}$ transforms on reflection into $\overline{CD}$. With Fig. 25.9 in mind, we see that the wavelet emitted from A will arrive at C in-phase with the wavelet just being emitted from D (as it is stimulated by B), so long as the distances $\overline{AC}$ and $\overline{BD}$ are equal. In other words, if all the wavelets emitted from all the surface scatterers are to overlap in-phase and form a single reflected plane wave, it must be that $\overline{AC} = \overline{BD}$. Then, since the two triangles have a common hypotenuse

$$\frac{\sin\theta_{\mathrm{i}}}{\overline{BD}} = \frac{\sin\theta_{\mathrm{r}}}{\overline{AC}}$$

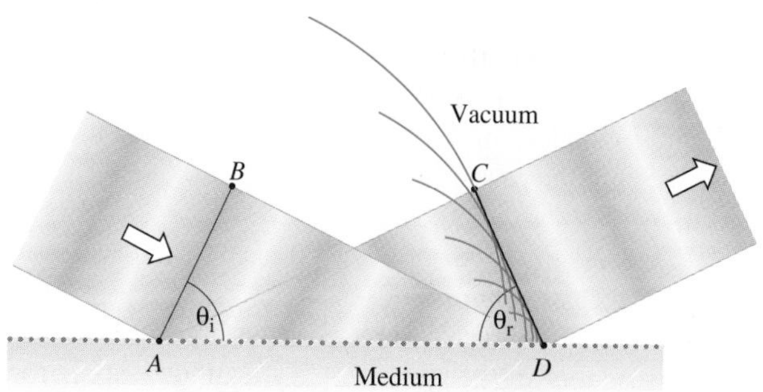

Figure 25.10 Wavefront geometry for reflection. The reflected wavefront $\overline{CD}$ is formed of waves scattered by the atoms on the surface from A to D. Just as the first wavelet arrives at C from A, the atom at D emits, and the wavefront along $\overline{CD}$ is completed.

All the waves travel in the incident medium with the same speed v_i. It follows that in the time Δt it takes for point B on the wavefront to reach point D on the surface, the wavelet emitted from A reaches point C. In other words, $\overline{BD} = v_i \Delta t = \overline{AC}$, and so from the above equation, $\sin \theta_i = \sin \theta_r$, which means that

$$\theta_i = \theta_r \tag{25.1}$$

The angle of incidence equals the angle of reflection. This equation is the first part of the **Law of Reflection**.

Drawing wavefronts can get things a bit cluttered, so we now introduce another convenient scheme for visualizing the progression of light. The imagery of antiquity was in terms of straight-line streams of light, a notion that got into Latin as "radii" and reached English as "rays." **A ray is a line drawn in space corresponding to the direction of flow of radiant energy**. It is a mathematical device and not a physical entity. In a medium that is uniform (homogeneous), rays are straight. If the medium behaves in the same manner in every direction (isotropic), **the rays are perpendicular to the wavefronts**. Thus, for a point source emitting spherical waves (Fig. 13.12, p. 448), the rays, which are perpendicular to them, point radially outward from the source. Similarly, the rays associated with plane waves are all parallel (Fig. 25.11a). Rather than sketching bundles of rays, we can simply draw one incident ray and one reflected ray (Fig. 25.11b). *All the angles are measured from the perpendicular (or normal) to the surface,* and thus θ_i and θ_r have the same numerical values as before (Fig. 25.10).

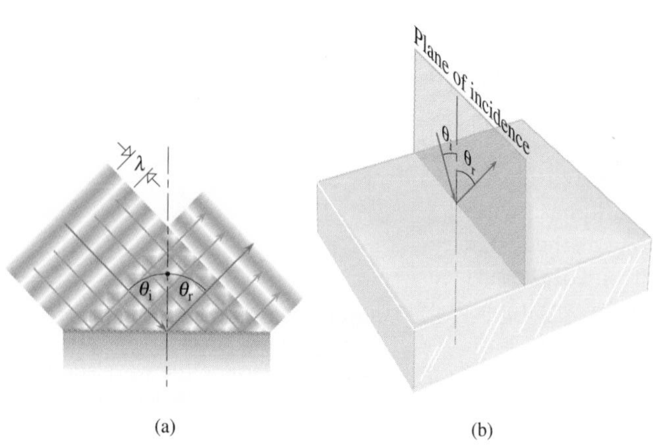

Figure 25.11 (a) We select one ray to represent the beam of plane waves. Now both the angle of incidence θ_i and the angle of reflection θ_r are measured from a perpendicular drawn to the reflecting surface. (b) The incident ray and the reflected ray define a plane, known as the *plane-of-incidence,* perpendicular to the reflecting surface.

The ancient Greeks knew the Law of Reflection—it can be deduced by observing the behavior of a flat mirror, and nowadays that observation can be done most simply with a flashlight or, even better, a laser. The second part of the Law of Reflection maintains that **the incident ray, the perpendicular to the surface, and the reflected ray all lie in a plane** called the **plane-of-incidence** (Fig. 25.11b)—this is a three-dimensional business. Try to hit some target in a room with a flashlight beam by reflecting it off a stationary mirror and the importance of this second part of the law becomes obvious!

When a beam is incident upon a reflecting surface that is smooth (one for which any irregularities are small compared to a wavelength), the light re-emitted by millions upon millions of atoms will combine to form a single well-defined beam in a process called **specular reflection**. If, on the other hand, the surface is rough, although the angle of incidence will equal the angle of reflection for each ray, the whole lot of rays will emerge every which way, constituting what is called **diffuse reflection** (Fig. 25.12). Both of these conditions are extremes—the reflecting behavior of most surfaces lies somewhere between them (see the photo below).

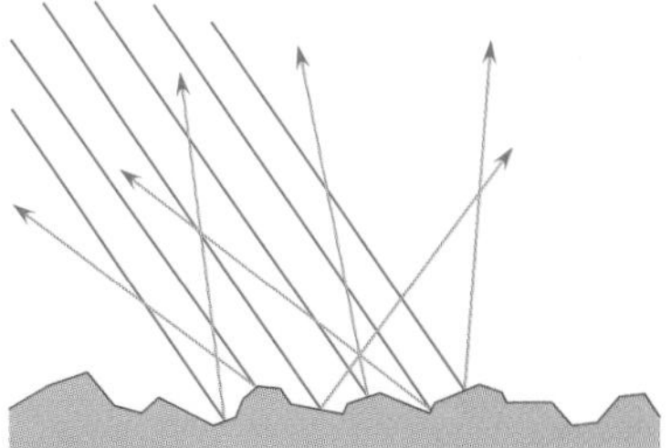

Figure 25.12 Diffuse reflection. When the surface roughness is large compared to λ, the scattering molecules are far apart, and there is no longer a direction in which all the wavelets will add constructively to produce a single reflected wave. Energy goes off in a broad range of directions as shown in the photo below.

25.6 The Plane Mirror

While standing in front of a plane (flat) mirror* (that is, any polished smooth surface), every point on your body that is illuminated by some external source scatters light, some of which heads toward the mirror (Fig. 25.13). Every molecule on your face that is driven by the incident E-field reradiates a more or less spherical wavelet or, if you prefer, sends out a cone of rays. Some portion of these wavelets will reflect from the mirror and subsequently reach your eye. But the eye-brain receiver, accustomed to straight-line propagation, will see the source as if it were behind the mirror. *The rays received by the observer diverge from the image point* and, as such, the image is said to be **virtual**—*it appears behind the mirror* and *cannot be projected onto a screen.* Every point on your hand, for instance, will have its corresponding image point behind the mirror, recreating the appearance of that hand (Fig. 25.14).

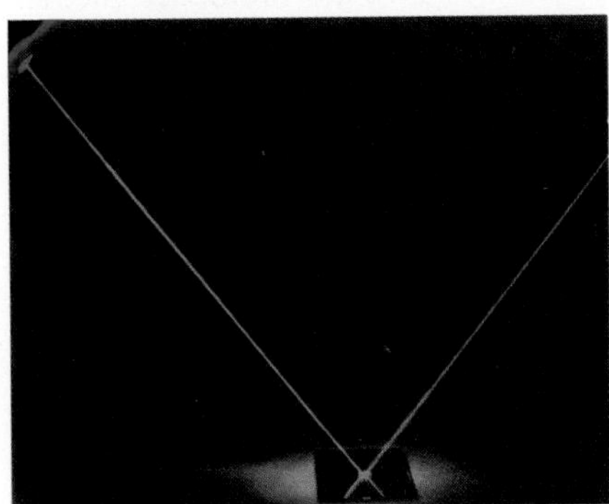

Specular reflection. A laser beam reflected from a mirror in a well-defined beam. Note that the angle of incidence equals the angle of reflection. The laser is at the upper left in both photos.

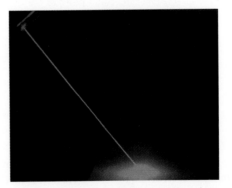

Diffuse reflection. Here the surface is rough compared to λ, and the light (unable to interfere effectively) scatters in every direction. There is no well-defined reflected beam.

*Most mirrors in common use outside of the laboratory are back-silvered—they reflect light off a layer of metal *behind* the glass; that's done in order to protect the delicate reflecting coating. For simplicity, we'll draw only front-silvered mirrors of the type used in optics labs.

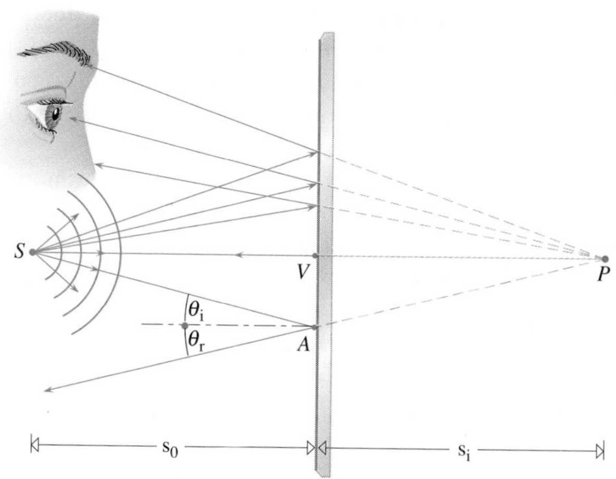

Figure 25.13 The light from an atom (S) on the face of someone looking into a mirror. Rays from S strike the mirror, $\theta_i = \theta_r$, and some reflect back to the viewer's eye. The eye-brain system apprehends the rays as if they were straight and being emitted from P, the image of S behind the mirror.

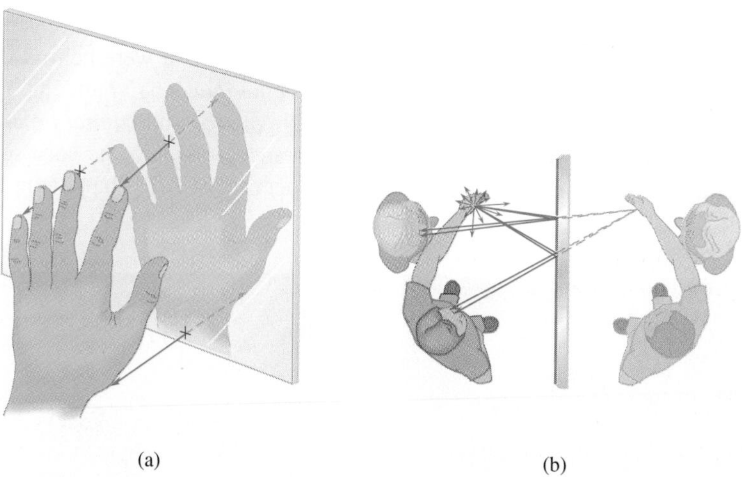

Figure 25.14 (a) The image of the left hand can be easily traced by using rays from it perpendicular to the mirror ($\theta_i = 0 = \theta_r$). They simply reflect back on themselves. A left hand becomes a life-size right hand. (b) The light from the young man's hand reflects both to his eye and the eye of the woman standing next to him. Both can therefore see the image of his hand behind the mirror.

Example 25.2 Using Fig. 25.13, prove that the image is located the same distance behind the mirror as the object is in front of it. That is, the **object distance** s_0 equals the **image distance** s_i.

Solution: The exterior angle of triangle SAP equals $\theta_i + \theta_r$, and that equals the sum of the alternate interior angles of that triangle; namely, $\angle VSA + \angle VPA$. But $\angle VSA = \theta_i = \theta_r$ and therefore $\angle VSA = \angle VPA$, which means that triangles VAS and VAP are congruent by angle-angle-side. It follows that $s_0 = s_i$, **the image of the source is the same perpendicular distance behind the mirror as the object is in front of it.**

The fact that the mirror is flat and that the angle of incidence equals the angle of reflection, combine to produce an undistorted, right-side up, life-size image standing just as far behind the surface as the object is in front of it. Realize that the image of a left hand (the outline of which can be determined using rays falling perpendicularly on the mirror as in Fig. 25.14a) is a right hand. Smack your left hand up against a mirror (the reflected fingers must be directly against the actual fingers)—the image must be a right hand. *A single reflection changes a right-handed system into a left-handed one and vice versa.*

Example 25.3 What is the length of the smallest vertical plane mirror in which you can see your entire standing body all at once, and how should it be positioned?

Solution: [Given: that $s_o = s_i$, determine the minimum height of the mirror.] Refer to Fig. 25.15. A ray from your toe will enter your eye, striking point H somewhere such that $\angle DHC = \angle BHC$. Triangles CHD and CHB are congruent and so $\overline{GH} = \overline{HI} = \frac{1}{2}\overline{BD}$. Similarly, if you are to see the top of your head, $\overline{EF} = \overline{FG} = \frac{1}{2}\overline{AB}$. Thus, a mirror of length $\overline{FH}$ will do the job, where

$$\overline{FH} = \overline{FG} + \overline{GH} = \tfrac{1}{2}\overline{AB} + \tfrac{1}{2}\overline{BD} = \tfrac{1}{2}\overline{AD}$$

A mirror half your height with its upper edge lowered by half the distance between your eye and the top of your head serves nicely.

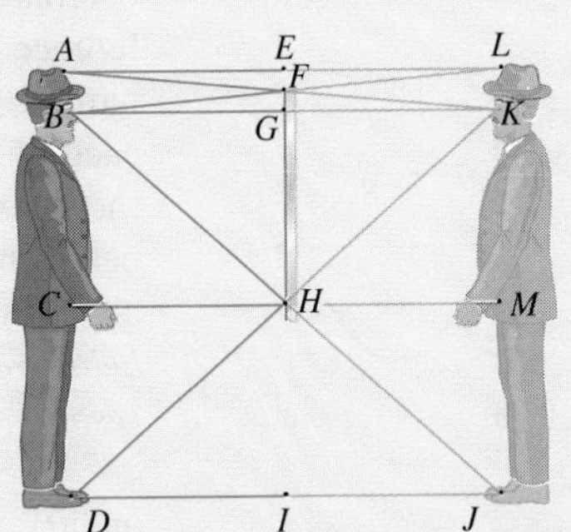

Figure 25.15 A determination of the smallest mirror ($\overline{FH}$) in which a person can see one's entire body.

Generally, the amount of light reflected from a surface increases as the angle of incidence increases. The amount reflected actually approaches 100% at glancing incidence where $\theta_i \approx 90°$. Furthermore, even the ability of a rough surface to reflect specularly improves as the light skims across it at glancing incidence. Hold this book horizontally, right at eye level so that the light from a lamp grazes off a page into your eye—you'll see a bright image of the bulb reflected in the paper.

REFRACTION

We saw earlier that as light propagates through a transparent homogeneous medium, there will be elastic scattering from the atoms. The initial beam, let's call it the *primary wave,* results in the emission of scattered wavelets. These, in turn, cancel each other in all but the forward direction, wherein they combine to form what we shall call the *secondary wave* (Fig. 25.4). The result is two waves (the primary and secondary), not necessarily in-phase with each other, overlapping and propagating together as one net disturbance—the **refracted wave**.

Before we press on, realize that *both waves travel at* c: **photons do not exist at any speed other than** c. *When light traverses any material, it travels in the interatomic void.* And yet, if we were to measure the speed of a macroscopic beam of light in a material, we would generally find some value other than c! All that atomic absorbing and re-emitting has the effect of producing a net speed, which can be either greater than, less than, or equal to c, depending on the medium and the frequency of the radiant energy. This process is marvelously subtle, and not every aspect of it can be fully developed here; still, it's so fundamental that it must be examined, if only superficially.

25.7 The Index of Refraction

The crucial feature here is that the secondary wave is generally somewhat out-of-phase with the primary wave. What happens is that the primary wave drives the molecular oscillators into a ground-state vibration. But this vibration will not usually follow in step with the driver; there will be a phase difference, as we saw earlier when we studied Slinkies (p. 416). When the driving frequency is much below the oscillator's resonant frequency, there is only a slight phase difference. But that difference increases as the frequency of the incident beam increases toward the resonance. The result is the emission by every driven oscillator, of a scattered wavelet, each of which is out-of-phase by the same amount (from 0° to 180°) with respect to the primary wave. To this must be added another phase shift of 90°, which arises between the oscillators and the reconstituted secondary wave. We needn't worry about all the details; what's of interest is that *for a primary wave with a frequency below resonance, the secondary wave lags behind it, and for a primary wave above resonance, the secondary wave leads it* (Fig. 25.16).

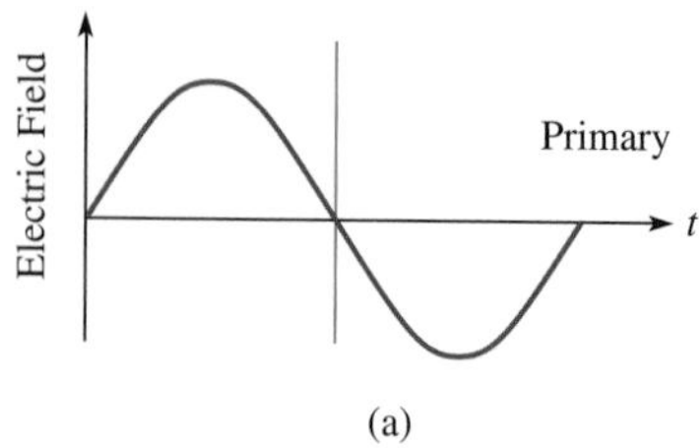

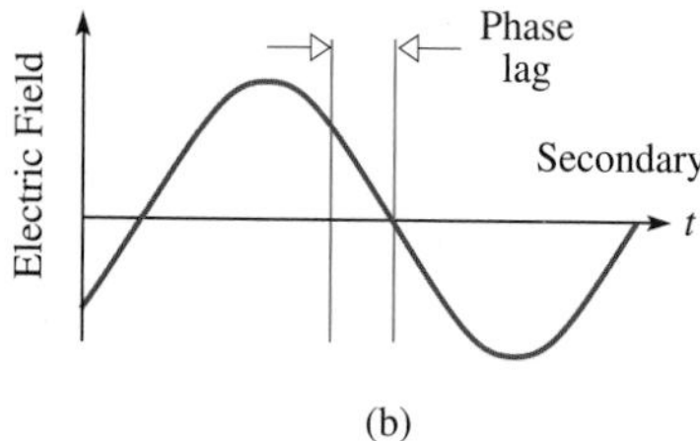

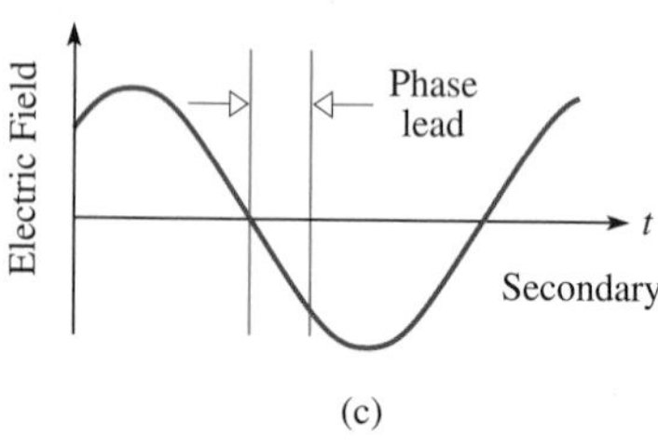

Figure 25.16 (a) A primary wave and two possible secondary waves. In (b), the secondary wave lags behind the primary—it takes longer to reach any given value. In (c), the secondary wave reaches any given value at an earlier time than the primary; that is, it leads the primary.

The refracted wave is the sum of the primary wave and this phase-shifted secondary wave. Remember that the phase speed of any wave is the rate at which any point of constant phase, like a crest, moves. But adding the secondary wave to the primary has the effect of changing the phase. Figure 25.17 shows how the phase of the refracted wave can be advanced or retarded, and that is exactly equivalent to increasing or decreasing its average speed by some small amount. Notice that a small secondary wave produces only a small phase shift. When the phase is retarded, a crest will get to some point beyond it in the medium just a little later (at a larger value of t) than it would have otherwise. A phase lag means a time lag (the crest will take longer to get anywhere in the forward direction), and that means a speed $v < \mathrm{c}$. Keep in mind that an electromagnetic wave is the averaged macroscopic manifestation of a torrent of photons. Such a wave effectively slows down as it enters the medium, even though the individual photons, without which there is no wave, only travel at c.

X-rays have frequencies in excess of the electron resonances and so generally $v > \mathrm{c}$. On the other hand, in transparent media, light frequencies are most often lower than the electron resonances of the medium, and $v < \mathrm{c}$. This situation ($v < \mathrm{c}$) is usually the case, but it most assuredly is not always the case, especially for colored materials that have resonances in the visible range. *Because of the ongoing process of absorption and re-emission, electromagnetic waves propagate through material media at speeds other than* c. The dependence of the speed of propagation on the frequency of the radiant energy is known as **dispersion**; it's responsible for the separation of white light into its constituent colors via a prism.* The ratio of the speed of an electromagnetic wave in vacuum to that in a medium is defined as the **index of refraction** n; accordingly

$$n = \frac{\mathrm{c}}{v} \tag{25.2}$$

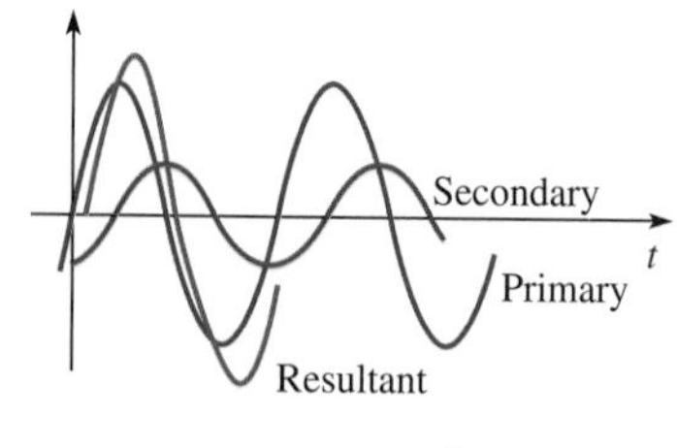

Figure 25.17 If the secondary lags the primary, the resultant will also lag it and vice versa.

*One need not fret about violating Relativity Theory with speeds in excess of c—these are phase speeds, there is no modulation and the waves carry no information. If they were modulated somehow, the signal would travel at a reduced rate known as the *group velocity*.

Keep in mind that n always varies somewhat with the frequency of the illumination. For most transparent materials, this variation, although significant, is not very large across the visible spectrum, and so it's common to use a single value of n (see Table 25.1) for such a substance illuminated with light of any frequency (Fig. 25.18). However, as f approaches that of an atomic resonance, n changes drastically, and there is a large increase in dissipative absorption as the amplitudes of the atomic oscillations increase. That's what happens to ordinary glass in the UV and IR (and it's why you can't get a tan behind a window that you can easily see through).

Example 25.4 What is the apparent speed of light (589 nm) in diamond?

Solution: [Given: from Table 25.1, $n_d = 2.42$. Find: v.] From the definition of the index of refraction, Eq. (25.2), we have

$$v = \frac{c}{n_d} = \frac{3.00 \times 10^8 \text{ m/s}}{2.42} = \boxed{1.24 \times 10^8 \text{ m/s}}$$

▶ **Quick Check:** The index of vacuum is 1, and since diamond has an index of around 2.4, the speed in diamond is 2.4 times slower than $c \approx 3 \times 10^8$ m/s, or roughly 1×10^8 m/s.

TABLE 25.1 Approximate indices of refraction of various substances*

Substance	n
Air	1.000 29
Ice	1.31
Water	1.333
Ethyl alcohol (C_2H_5OH)	1.36
Fused quartz (SiO_2)	1.458 4
Carbon tetrachloride (CCl_4)	1.46
Turpentine	1.472
Benzene (C_6H_6)	1.501
Plexiglass	1.51
Crown glass	1.52
Sodium chloride (NaCl)	1.544
Light flint glass	1.58
Polystyrene	1.59
Carbon disulfide (CS_2)	1.628
Dense flint glass	1.66
Lanthanum flint glass	1.80
Zircon ($ZrO_2 \cdot SiO_2$)	1.923
Fabulite ($SrTiO_3$)	2.409
Diamond (C)	2.417
Rutile (TiO_2)	2.907
Gallium phosphide	3.50

*Values vary with physical conditions—purity, pressure, etc. These correspond to a wavelength of 589 nm.

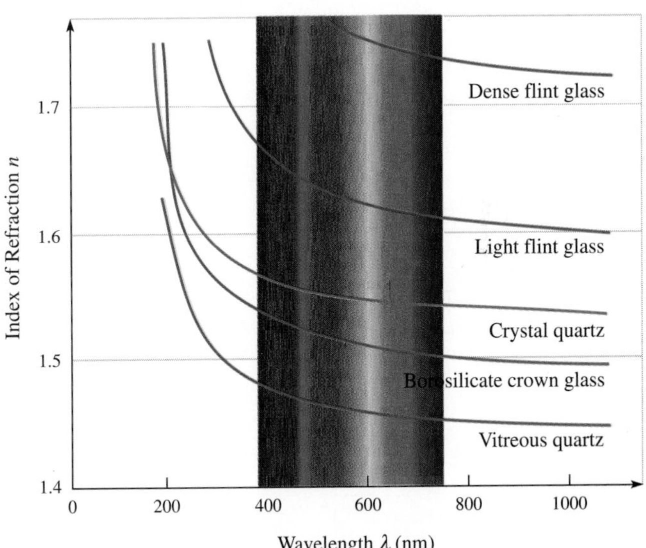

Figure 25.18 The wavelength dependence of the index of refraction for various materials. Transparent substances such as these usually have resonances in the UV (the region below 380 nm) and in the IR (far off to the right), which is why all of the curves here rapidly rise in the UV. That means that glass will dissipatively absorb UV. Can you guess why UV lamps are made of vitreous quartz rather than glass?

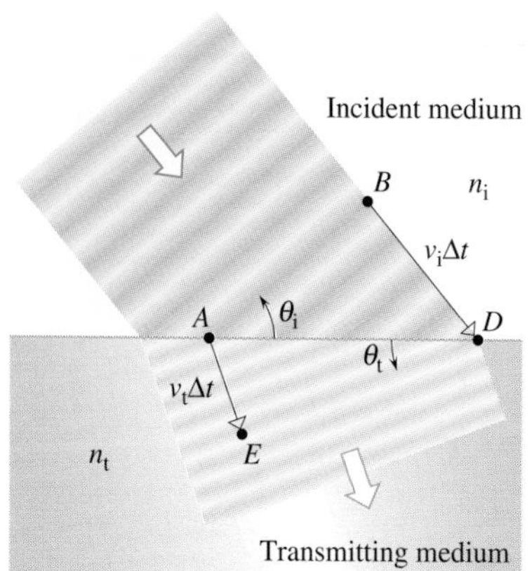

Figure 25.19 The wave picture of refraction. The atoms in the region of the surface of the transmitting medium reradiate wavelets that combine constructively to form a refracted beam.

25.8 Snell's Law

Let's examine what is happening to the transmitted beam in Fig. 25.8. Each molecule radiates wavelets into the glass that expand out at the speed c. These can be imagined as combining into a secondary wave that then recombines with the unscattered remainder of the incoming beam, the primary wave, to finally form the refracted wave. However we visualize it, immediately on entering the transmitting medium, there is a single net field, a single net wave—the refracted, or transmitted, light. As we have seen, this transmitted wave usually propagates with an effective speed $v_t < c$.

Fig. 25.19 picks up where we left off with Fig's. 25.8 and 25.10. The diagram depicts several wavefronts, all shown at a single instant in time. Remember that each wavefront is a surface of constant phase and, to the degree that the phase of the net field is retarded by the transmitting medium, each wavefront is held back, as it were. The wavefronts bend as they cross the boundary because of the speed change. Alternatively, we can envision Fig. 25.19 as a multiple-exposure picture of a single wavefront showing it after successive equal intervals of time. Notice that in the time Δt, which it takes for point B on a wavefront (traveling at speed v_i) to reach point D, the transmitted portion of that same wavefront (traveling at speed v_t) has reached point E. If the glass ($n_t = 1.5$) is immersed in an incident medium that is vacuum ($n_i = 1$) or air ($n_i = 1.0003$) or anything else where $n_t > n_i$, $v_t < v_i$ and $\overline{AE} < \overline{BD}$, the wavefront bends. The refracted wavefront extends from E to D, making an angle with the interface of θ_t. As before, the two triangles ABD and AED in Fig. 25.19 share a common hypotenuse ($\overline{AD}$), and so

$$\frac{\sin\theta_i}{\overline{BD}} = \frac{\sin\theta_t}{\overline{AE}}$$

where $\overline{BD} = v_i\,\Delta t$ and $\overline{AE} = v_t\,\Delta t$. Hence

$$\frac{\sin\theta_i}{v_i} = \frac{\sin\theta_t}{v_t}$$

Multiply both sides by c, and since $n_i = c/v_i$ and $n_t = c/v_t$

$$n_i \sin\theta_i = n_t \sin\theta_t \tag{25.3}$$

This equation is the first portion of the **Law of Refraction**, also known as **Snell's Law**. At first, the indices of refraction were simply experimentally determined constants of the physical media. Later on, Newton was actually able to derive Snell's Law using his own corpuscular theory. By then, the significance of n as a measure of the speed of light was evident. Still later, Snell's Law was shown to be a natural consequence of Maxwell's Electromagnetic Theory.

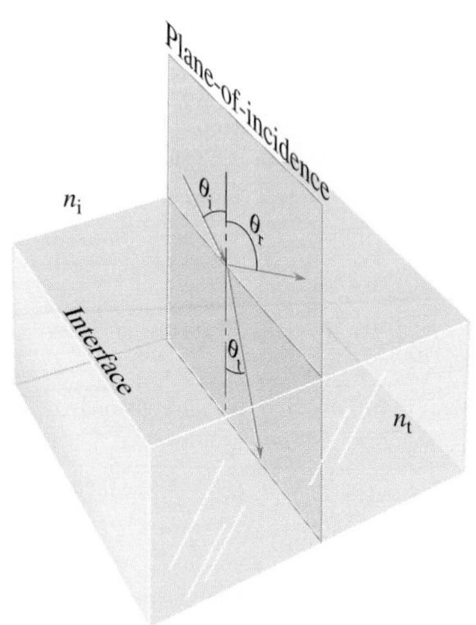

Figure 25.20 The incident, reflected, and transmitted beams all lie in the plane-of-incidence. The incident medium, containing the incoming beam, has an index of refraction n_i. The transmitting medium has an index n_t.

It is again convenient to transform the diagram into a ray representation (Fig. 25.20) wherein all the angles are measured from the perpendicular. Along with Eq. (25.3), there goes the understanding that **the incident, reflected, and refracted rays all lie in the plane-of-incidence**. When $n_i < n_t$ (that is, when the light is initially traveling within the lower-index medium), it follows from Snell's Law that $\sin\theta_i > \sin\theta_t$, and since the sine function is everywhere positive between 0° and 90°, then $\theta_i > \theta_t$. Rather than going straight through, **the ray entering a higher-index medium bends toward the normal** (Fig. 25.21a). The reverse is also true; that is, on entering a medium having a lower index, the ray, rather than going straight through, will bend *away* from the normal (Fig. 25.21b). Notice that this im-

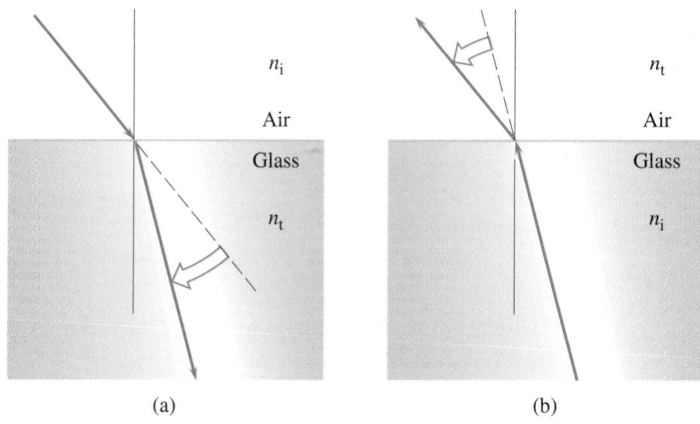

Figure 25.21 (a) When a beam of light enters a more optically dense medium, one with a greater index of refraction ($n_i < n_t$), it bends toward the perpendicular. (b) When a beam goes from a more dense to a less dense medium ($n_i > n_t$), it bends away from the perpendicular.

plies that the rays will traverse the same path going either way, into or out of either medium—the arrows can be reversed and the resulting picture is still true.

Example 25.5 Suppose that the beam of light in Fig. 25.22 travels in air before it's incident on the glass plate ($n_g = 1.5$) at 60°. (a) At what angle will it be transmitted into the block? (b) Show that the beam emerges from the far side parallel to the original incoming light.

Solution: [Given: an air-glass interface and $\theta_i = 60°$. Find: θ_{t1} and θ_{t2}, using Fig. 25.22.] (a) Applying Snell's Law at the first interface

$$n_a \sin \theta_{i1} = n_g \sin \theta_{t1}$$

and

$$\sin \theta_{t1} = \left(\frac{n_a}{n_g}\right) \sin \theta_{i1} = \left(\frac{1.0}{1.5}\right) \sin 60° = 0.577$$

Hence $\theta_{t1} = 35.3°$, that is, $\boxed{35°}$. (b) Applying Snell's Law to the second interface where $\theta_{t1} = \theta_{i2}$, we obtain

$$n_g \sin \theta_{i2} = n_a \sin \theta_{t2}$$

and

$$\sin \theta_{t2} = \frac{n_g}{n_a} \sin \theta_{i2} = \frac{1.5}{1.0} \sin 35.3° = 0.866$$

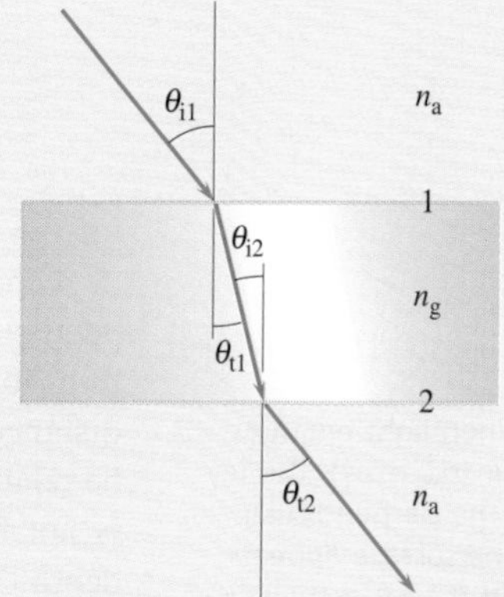

Figure 25.22 Light incident on a parallel plate of glass of index n_g immersed in air of index n_a.

Hence $\theta_{t2} = 60°$, and the light emerges parallel to the incident beam.

▶ **Quick Check:** On crossing the first interface, the ray bent toward the normal, which is what it should have done. It emerged parallel to the incident ray, which it must do if the interfaces are parallel, and that's a check of the calculation in itself.

Fig. 25.19 illustrates the three important changes that occur in the beam traversing the interface. (1) It changes direction. Because the leading portion of the wavefront in the glass slows down, the part still in the air advances more rapidly, sweeping past and bending the wave toward the normal. (2) The beam in the glass has a broader cross section than the beam in the air; hence, the energy is spread thinner. (3) The wavelength decreases because the frequency is unchanged while the speed decreases, $\lambda = v/f$. This latter notion suggests that the color aspect of light is better thought of as its frequency (or energy, $E = hf$) than its wavelength, which changes with the medium through which the light moves. When we do talk about wavelengths and colors, we should always be referring to *vacuum wavelengths* (henceforth to be given as λ_0).

Example 25.6 Someone is wearing a scarlet-red bathing suit that reflects primarily $\lambda_a = 629$ nm in air ($n_a \approx 1.00$). What is the corresponding wavelength in water ($n_w = 1.33$)? Do bathing suits change color underwater? Explain.

Solution: [Given: $\lambda_a = 629$ nm for $n_a = 1.00$. Find: λ_w when $n_w = 1.33$.] $n_a = c/v_a = c/f\lambda_a$; hence, $f = c/n_a\lambda_a = 4.77 \times 10^{14}$ Hz. Using the same reasoning, in water we have

$$\lambda_w = \frac{c}{n_w f} = \frac{3.00 \times 10^8 \text{ m/s}}{1.33(4.77 \times 10^{14} \text{ s}^{-1})} = \boxed{473 \text{ nm}}$$

which, were it in vacuum, would be seen as blue. Of course, the suit still appears red in water.

▶ **Quick Check:** Since wavelength decreases as index increases, $\lambda_w/\lambda_a = n_a/n_w$, and $\lambda_w = \lambda_a(1/1.33) = 473$ nm.

Notice that had we substituted $(\lambda_0 f)$ for c in Example 25.6, we would have gotten

$$\lambda = \frac{\lambda_0}{n} \tag{25.4}$$

for the wavelength in any medium of index n.

The fact that rays leaving a medium of higher index bend away from the perpendicular gives rise to the familiar effect in which distances within liquids appear foreshortened when viewed from above. Figure 25.23 shows a fish underwater a real distance down d_R, which seems to be at an apparent depth d_A. These distances can be related in a particularly simple way if we limit the problem to one where the fish is not far away horizontally from the viewer; that is, x is small compared to the depth. Then θ_i and θ_t are both small. For small angles, the cosine is approximately 1.0 and the tangent then equals the sine:

$$\sin\theta_i \approx \tan\theta_i = \frac{x}{d_R} \quad \text{and} \quad \sin\theta_t \approx \tan\theta_t = \frac{x}{d_A}$$

It follows from Snell's Law that

$$\frac{n_i x}{d_R} = \frac{n_t x}{d_A}$$

and

$$\frac{n_i}{n_t} = \frac{d_R}{d_A}$$

which is why a pencil seems to bend when it's dipped into water—the immersed end appears higher than it should in comparison to the portion remaining in the air.

In all the situations thus far treated, it was assumed that the reflected and refracted beams always had the same frequency as the incident beam, and ordinary experience tells us that that is a very reasonable assumption. Light of frequency f impinges on a medium and presumably drives

Figure 25.23 When light emerges from water into air ($n_w > n_a$), the ray bends away from the perpendicular. Anything below the surface appears higher in the liquid than it actually is, which makes spear fishing tougher than it might otherwise seem.

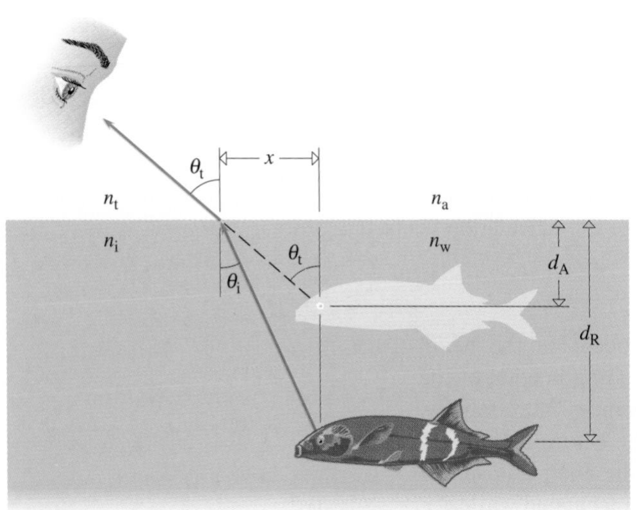

the molecules into simple harmonic motion. That's certainly the case when the amplitude of the vibration is fairly small, as it is when the electric field driving the molecules is small. The E-field for bright sunlight is only about 1000 V/m (while the B-field is less than a tenth of the Earth's surface field). This isn't very large compared to the fields keeping a crystal together, which are of the order of 10^{11} V/m, just about the same magnitude as the cohesive field holding the electron in an atom. We can usually expect the oscillators to vibrate in simple harmonic motion and so the frequency will remain constant—the medium will ordinarily respond linearly. That will not be true, however, if the incident beam has an exceedingly large-amplitude E-field, as can be the case with a high-power laser. So driven, at some frequency f the medium can behave in a nonlinear fashion, resulting in reflection and refraction of harmonics ($2f$, $3f$, etc.) in addition to f. Nowadays, second-harmonic generators are available commercially; you shine red light (694.3 nm) into an appropriately oriented transparent nonlinear crystal (of, for example, potassium dihydrogen phosphate, KDP, or ammonium dihydrogen phosphate, ADP) and out will come a beam of UV (347.15 nm).

The rays of light from the submerged portion of the pencil bend on leaving the water as they rise toward the viewer.

25.9 Total Internal Reflection

Often a beam of light originates in air (a low-index medium) and impinges on glass or water (a high-index medium). The reverse is also possible. When we look at a swimming fish, the reflected light from the fish originates in the water (a high-index medium) and impinges on the air (a low-index medium). When the light is incident on an interface where $n_i > n_t$, a curious and rather important thing happens. As the angle of incidence is made larger and larger (Fig. 25.24), the transmitted beam bends more and more away from the normal and toward the interface. As that occurs, the transmitted beam grows weaker, and the reflected beam—the beam traveling back into the higher-index medium—becomes stronger. When a particular incident angle is reached, known as the **critical angle** (θ_c), all the light striking the interface will be reflected back; no light will be transmitted. This **total internal reflection** continues for all incident angles greater than the critical angle. If, for example, $n_i = 2$ while $n_t = 1$, Snell's Law maintains that $2 \sin \theta_i = \sin \theta_t$ and, pro-

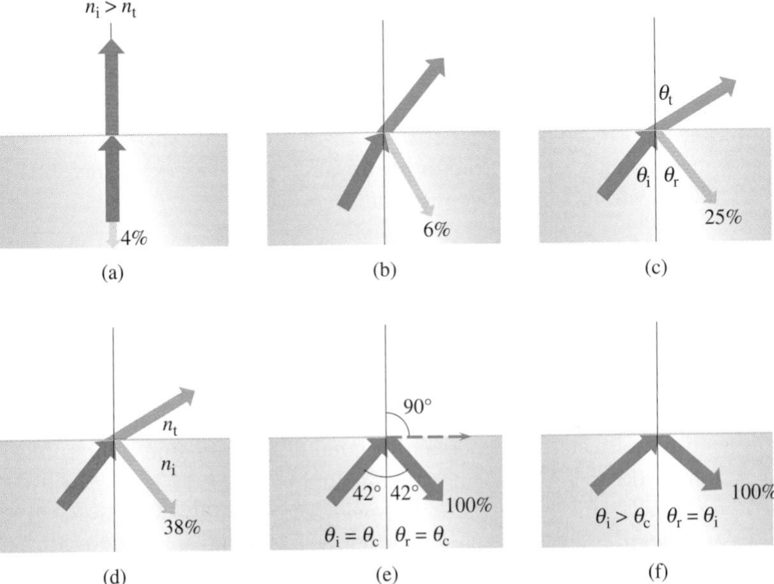

Figure 25.24 Internal reflection ($n_i > n_t$). As the incident angle increases, more and more light is reflected. (a) When $\theta_i = 0$, only about 4% of the light is reflected, while ≈96% is transmitted. (b) (c) (d) As θ_i increases, more light is reflected and less transmitted. (e) At the critical angle ($\theta_i = \theta_c$), all the light is internally reflected. (f) At values of θ_i greater than θ_c, the incident beam continues to be totally internally reflected.

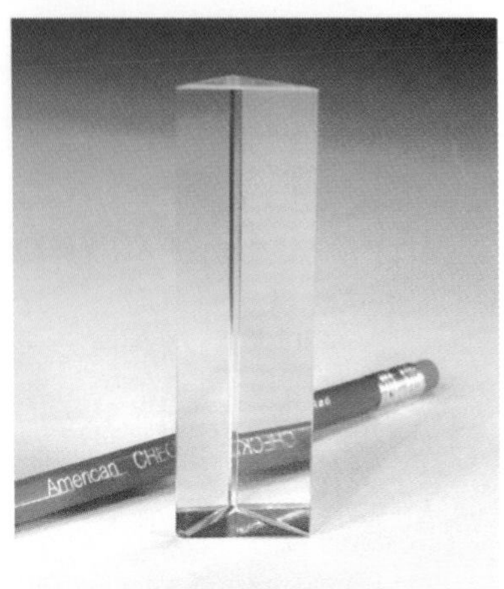

The prism behaves like a mirror and reflects a portion of the pencil (reversing the lettering on it). The operating process is total internal reflection.

vided $\sin\theta_i \leq \frac{1}{2}$, the equation can be satisfied. But for incident angles greater than this critical value of $\theta_i = 30°$, the $\sin\theta_t$ would have to be greater than 1, and that cannot be—there is no transmitted beam.

Snell's Law yields

$$\sin\theta_i = \frac{n_t}{n_i}\sin\theta_t$$

but when $\theta_i = \theta_c$, we know that $\theta_t = 90°$, then $\sin 90° = 1$, and we get the defining equation

$$\sin\theta_c = \frac{n_t}{n_i} \tag{25.5}$$

For incident angles equal to or greater than the critical angle, 100% of the energy in the incoming beam is reflected back into the incident medium, and no energy is transmitted. Recall that at the critical angle the transmission angle becomes 90°; the "transmitted" light propagates along the interface. Because that wave is limited to the boundary, its energy flows back and forth across the interface with no average transmission into the second medium.

Example 25.7 Imagine a beam of light traveling in a block of glass for which $n_g = 1.56$. Now suppose the light comes to the end of the block and impinges on an air-glass interface. What is the minimum incident angle that will result in all the light being reflected back into the glass?

Solution: [Given: $n_i = n_g = 1.56$, $n_t = n_a = 1.00$. Find: θ_c.] From Eq. (25.5), we have

$$\sin\theta_c = \frac{1.00}{1.56} = 0.641$$

and

$$\boxed{\theta_c = 39.9°}$$

▶ **Quick Check:** For $n_t = 1.5 = 3/2$, $\sin\theta_c = 2/3$ and $\theta_c = 42°$, so the above result is certainly reasonable.

The fact that the critical angle for air-glass typically ranges from 36° to 43° (depending on the kind of glass) is utilized in the reflecting prisms of Fig. 25.25. This is a convenient way to redirect a beam, and it's used in cameras, binoculars, telescopes, and a variety of other optical devices.

Fiberoptics

Easily the most important application of total internal reflection is **fiberoptics**. Techniques have evolved in recent times for efficiently conducting light and near-IR (1300 nm) energy along thin, transparent, dielectric fibers. Figure 25.26 depicts a glass (or plastic) fiber having a diameter of, say, 50 μm, just about the thickness of a human head-hair. Because of the small diameter, most of the light entering one face goes on to strike the cylindrical wall at a large angle and is subsequently totally internally reflected. This occurs over and over again, typically thousands of times per foot, as the beam propagates along the fiber. The smooth surface of a single filament must be kept clean of contamination, or the boundary conditions change and

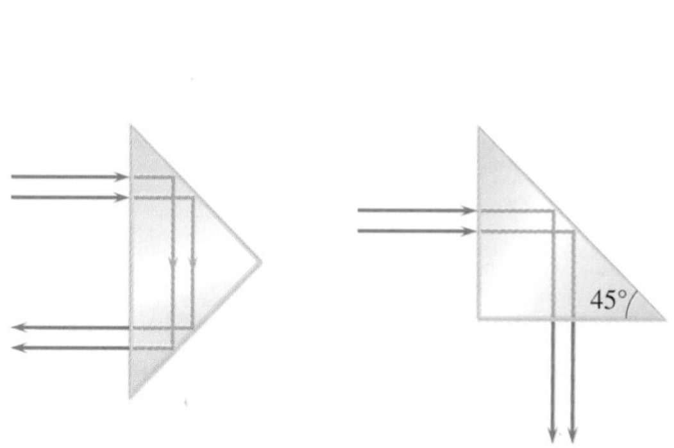

Figure 25.25 The rays strike the walls of these prisms at 45°, which is in excess of the critical angle ($36° \lesssim \theta_c \lesssim 43°$). The rays are then totally internally reflected.

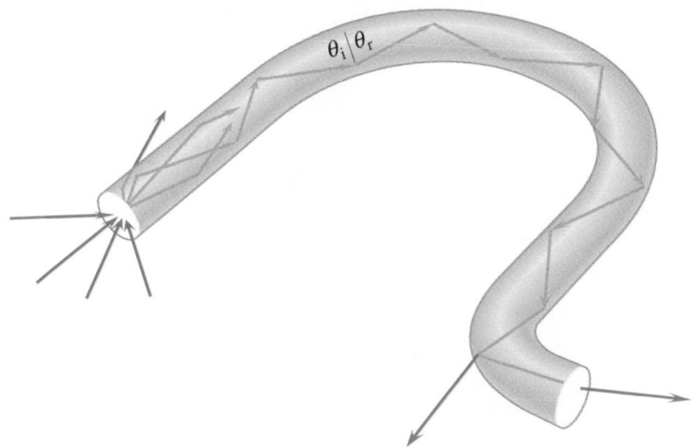

Figure 25.26 Light captured within a thin transparent fiber. Note that if a bend is too tight $\theta_i < \theta_c$, and some light leaks out. Similarly, if a ray enters at too large an angle, it will leak out.

light leaks out at those spots. Accordingly, each fiber core is usually enshrouded in a transparent sheath of a lower-index material called a *cladding.*

All such thin filaments are quite flexible and, if not bent too tightly, will transmit light through twists and turns with relatively little loss. Bundles of free fibers whose ends are bound together and then polished form flexible light guides. If no attempt is made to keep the fibers in an ordered array so that the two end-faces are each a different jumble, the thing is called an *incoherent bundle.* These are light carriers. Easy to produce and inexpensive, they are used for remote light sensing and illumination. They conduct light, for example, from a conveniently positioned bulb to some otherwise inaccessible place such as an instrument panel in an airplane or deep into a human body.

Conversely, when the fibers are arranged so that their terminations on both end-faces are exactly the same, the bundle is said to be *coherent,* and we have a flexible image carrier. Frequently, these bundles are tipped off with a small lens so that they need not be in contact with the surface being viewed. Today, it is commonplace to use fiberoptic apparatus to poke into all sorts of unlikely places from nuclear reactor cores and jet engines to stomachs and reproductive organs. When a device is used to examine internal body cavities, it's called an *endoscope,* and there are bronchoscopes, colonoscopes, gastroscopes, and so on. An additional incoherent bundle incorporated into the device usually supplies the illumination.

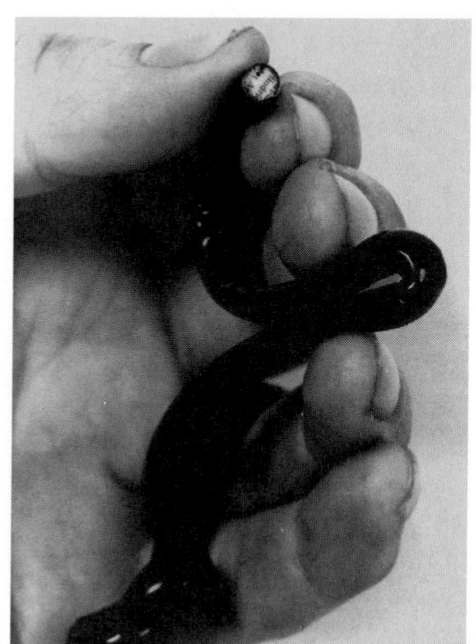

A bundle of thousands of carefully arranged thin glass fibers transmitting an image. One end of the bundle rests on a page of print, and the other end reveals the words below, despite the knots and bends.

The world is now in the first stages of a new era of optical communications, of light flashing along fibers, replacing electricity moving in metal wires—not for transmitting power, but information. The much higher frequencies of light allow for an incredible increase in data-handling capacity. For example, using some sophisticated transmitting techniques, a pair of copper telephone wires can be made to carry up to about two dozen simultaneous conversations. To get a feel for how much information that is, consider the fact that a single ordinary TV transmission is equivalent to about 1300 simultaneous phone conversations, which, in turn, is roughly the equal of sending some 2500 typewritten pages each and every second! So, at present, it's quite impractical to attempt to send television over copper phone lines. By comparison, it's already possible to transmit in excess of 12 000 simultaneous conversations over a single pair of fibers, which is more than nine TV channels—and this is only the beginning; the technology is in its infancy. Achieved capacities to date don't even begin to approach the theoretical limit. A pair of fibers will someday connect your home to a vast network of communications and computer facilities that will make the era of the copper wire seem charmingly primitive.

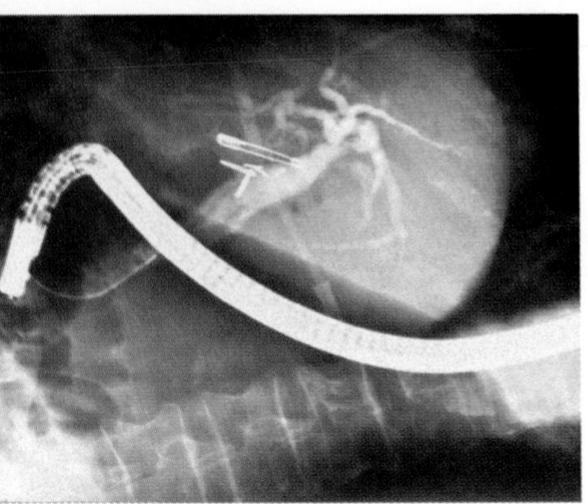

A colonoscope being used to examine a colon for cancer.

THE WORLD OF COLOR

We are now in a position to explore why the objects in our environment look the way they do. What makes paper white and coal black? Why is silver gray and grass green?

25.10 White, Black, and Gray

A reflecting surface is white when it diffusely scatters a broad range of frequencies under white-light illumination. This page is white. So are salt and snow. Clouds are white, as are powders and pills and cloth. Even the foam on a glass of beer is white, steam and soap bubbles and chalk and sugar and gray hair are white—and all this diversity shares a common structural similarity. All of these objects are composed of transparent grains or particles or fibers or bubbles that are large in size compared with the wavelength of light, but otherwise quite small. Each is much too large to produce Rayleigh scattering, too large to act as a single oscillator preferentially scattering at the higher frequencies. Despite the common opinion that things white are opaque, each grain or fiber is actually transparent. There is no such thing as a single particle of white pigment that by itself is opaque white. Materials that have no atomic resonances in the visible are transparent to light of all frequencies—they do not appear colored because they do not absorb any range of colors. Sprinkle a single layer of salt or sugar on this page, and the print will still be legible through the transparent crystals. Crush a piece of clear glass and the grains, like grains of sand, will appear white. Water is clear, but snow and steam and clouds are white.

Whiteness arises when the incident white light is reflected back out of the medium in all directions as if from countless point sources, from scatterers that show no preferential absorption. All colors of light come in, and all colors scatter out. When the medium is composed of many small transparent bodies, like the matted fibers of this page or the random jumble of sugar crystals on a spoon, each reflects some light, transmits some light, and reflects again (Fig. 25.27). And that process happens over and over, layer upon layer, with much of the light eventually reflected back into the general direction from which it came. Slightly lift a few pages of this book and peep under them—lots of light will be transmitted through the pages but, like a cloud, if the mass is thick enough, very little light will emerge and it will appear gray or black.

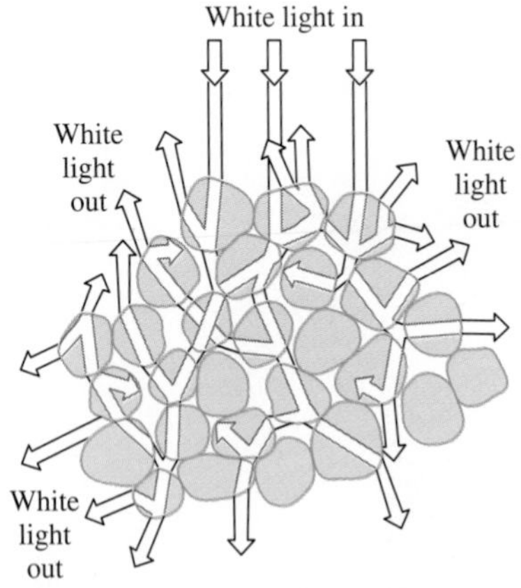

Figure 25.27 A number of transparent particles, which are large compared to the wavelengths of light. These might be cloth fibers, sugar crystals, talcum powder, or snowflakes. White light enters, is reflected and refracted out.

The fraction of light energy reflected at each surface depends on the difference between the indices of the two media—if there is no difference, the boundary essentially vanishes. *The closer the indices are to each other, the less light will be reflected,* and the less luminous will be the surface, Accordingly, to make white paint, we need only mix a powdered transparent material, the pigment, with a clear vehicle (such as oil or acrylic). If the refractive indices are equal, the pigment simply vanishes, and the paint is transparently useless. If the indices differ appreciably, there will be strong reflections at the countless surfaces and the paint will be a brilliant white. A really splendid white surface can reflect up to 98% of the incident light.

Because of the low index of air, small particles such as salt, talc, and sugar, as well as all kinds of undyed fibers, surrounded by air appear bright white. That brightness will change tremendously if these fibers are immersed in a substance having an index between that of fiber and air, a substance like water. What happens to the bright whiteness of a tissue or a piece of cloth when it becomes wet?

A diffusely reflecting surface that absorbs somewhat uniformly right across the spectrum will reflect a bit less than a white surface and so appear matt gray. The less it reflects, the darker the gray, until it absorbs almost all the light and appears black. A surface that reflects perhaps 70% or 80%, but reflects specularly, will appear the familiar shiny gray of a typical metal. Metals possess tremendous numbers of free electrons that scatter light very effectively independent of frequency—the free electrons are not bound to the atoms and have no associated resonances. The amplitudes of the vibrations are an order-of-magnitude larger than they are for the bound electrons. The incident light cannot penetrate into the metal any more than a fraction of a wavelength or so before it's canceled completely. There is little or no refracted light—most of the energy is reflected out; the small remainder is absorbed.

If the metal is thin enough (only a few atom layers thick), it will transmit light and one can see through it. Look at a bright lamp through a CD. Nowadays, "two-way" mirrors that can be seen through are a common security device. A number of snack food products are packaged in shiny metal-coated plastic films. Hold one up to your eyes and look through it at a bright light. Note that the primary difference between a gray surface and a mirrored surface is one of diffuse versus specular reflection.

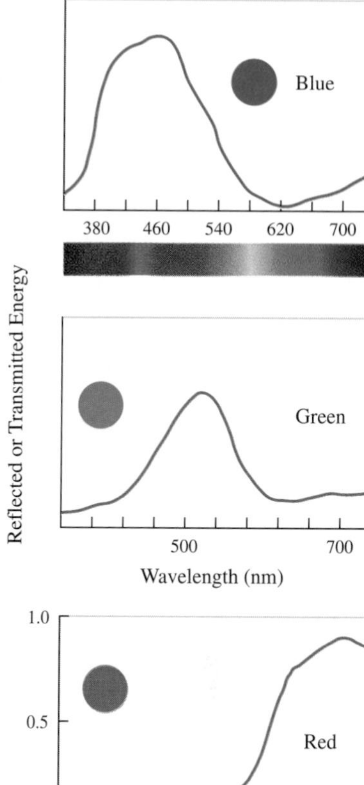

Figure 25.28 Reflection curves for blue, green, and red pigments are typical, but there is a great deal of variation within each color.

25.11 Colors

When the distribution of energy in a beam of light is fairly uniform across the spectrum, the light appears white; when it is not, the light usually appears colored. Figure 25.28 depicts typical frequency distributions for what would be perceived as red, green, and blue light. These curves show the dominant frequency regions, but beyond that there can be a great deal of variation in the distributions, and they will still provoke the eye-brain responses of red, green, and blue. Thomas Young, in the early 1800s, showed that a broad range of colors can be generated by mixing three beams of light, provided their frequencies were widely separated. When three such beams combine to produce white light, they are called **primary colors**. There is no single unique set of these primaries, nor do they have to be monochromatic. Since the widest range of colors can be created by mixing *light beams* of red (R), green (G), and blue (B), these tend to be the most commonly used. They are the three components (emitted by three phosphors) that generate the whole gamut of hues seen on a color TV set (p. 884). Keep in mind that at the moment we are talking about mixing light beams and not paint pigments.

A white screen illuminated with red, blue, and green light appears white. Anyone standing between a lamp and the screen casts a shadow that has the complementary color.

Figure 25.29 summarizes the result of overlapping these three primaries in a number of different combinations: red plus blue is *magenta* (M), a reddish blue; blue plus green is *cyan* (C), a bluish green; and, most surprising, red plus green is *yellow* (Y). And the sum of all the three

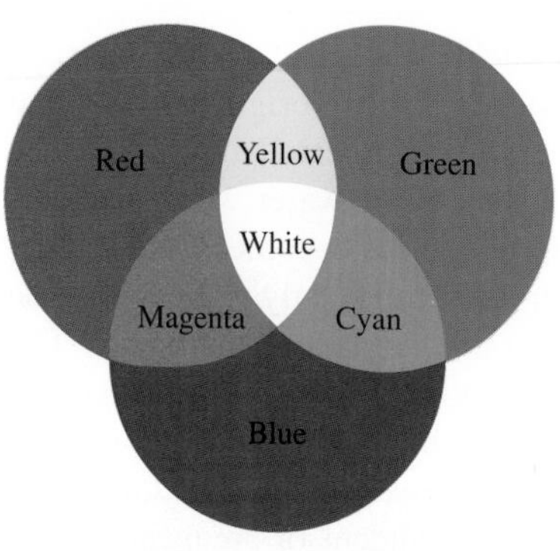

Figure 25.29 Three overlapping beams of colored light. A color TV set uses the same three primary light sources—red, green, and blue.

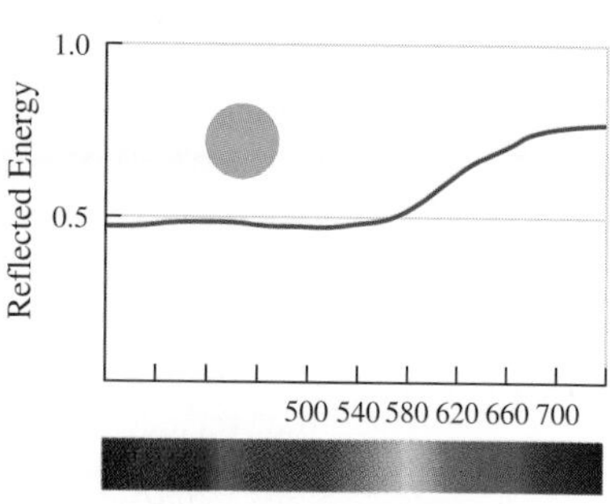

Figure 25.30 Spectral reflection of a pink pigment. The broad range of wavelengths produces a white background; adding the red peak yields pink. A shirt that reflects about half the light incident on it over most of the spectrum, but reflects even more in the red, looks pink.

primaries is white:

$$\mathrm{R + B + G = W}$$

$$\mathrm{M + G = W}, \quad \text{since} \quad \mathrm{R + B = M}$$

$$\mathrm{C + R = W}, \quad \text{since} \quad \mathrm{B + G = C}$$

$$\mathrm{Y + B = W}, \quad \text{since} \quad \mathrm{R + G = Y}$$

Any two colored light beams that together produce white are said to be **complementary**, and the last three symbolic statements exemplify that situation. Now suppose we overlap a beam of magenta and a beam of yellow, represented symbolically by

$$\mathrm{M + Y = (R + B) + (R + G) = W + R}$$

The result is pink, a combination of white and red. And that raises another point: we say that a color is **saturated**, that it is deep and intense, when it does not contain any white light. As can be seen in Fig. 25.30, pink is unsaturated red; red superimposed on a background of white.

Imagine a piece of yellow stained glass; that is, glass having a resonance in the blue, which it strongly absorbs. Looking through it at a white-light source composed of red, green, and blue, the glass would absorb blue, passing red and green, which is yellow (Fig. 25.31). Yellow cloth, paper, dye, paint, and ink all selectively absorb blue and reflect what remains—yellow—and that's why they appear yellow. And if you peer at something that is a pure blue through a yellow filter, the object will appear black. Here, the filter or the paint colors the light yellow by removing blue, and we speak of the process as *subtractive* coloration. On the other hand, *additive* coloration can only result from the overlapping of light beams. Red light and green light makes yellow; red paint and green paint makes brown.

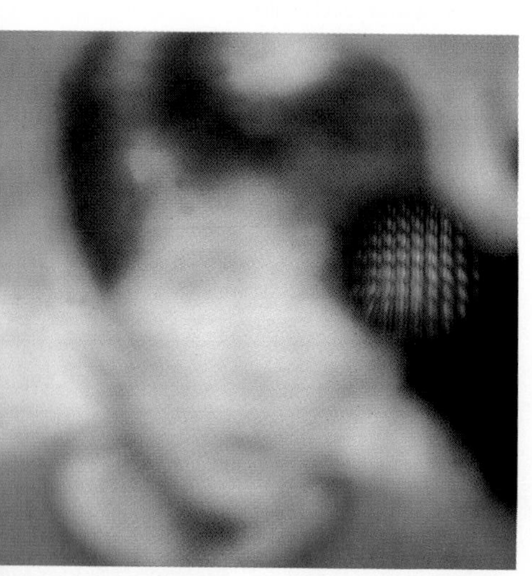

A color TV screen is made up of red, blue, and green regions that can be made visible with a magnifying glass.

A wide range of colors (including red, green, and blue) can be produced by passing light through various combinations of magenta, cyan, and yellow filters (Fig. 25.32). Magenta, cyan, and yellow are the primary colors of subtractive mix-

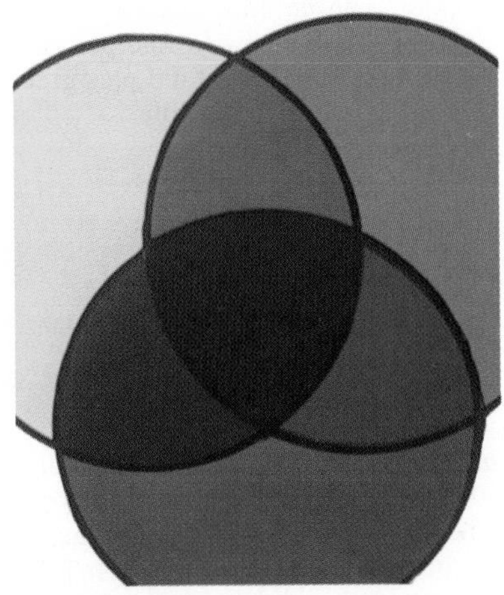

Filters of magenta, yellow, and cyan. These absorb certain frequencies and the process is called subtractive coloration. Where all three overlap, no light is passed.

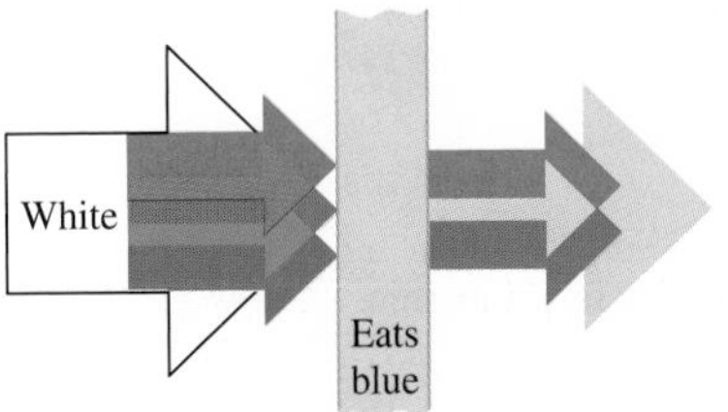

Figure 25.31 Yellow stained glass absorbs blue light, passing yellow. The incoming white beam corresponds to R + B + G. Subtracting blue yields R + G, which is yellow.

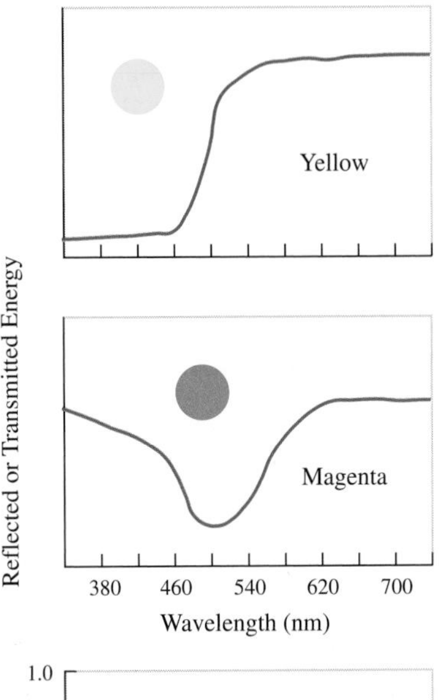

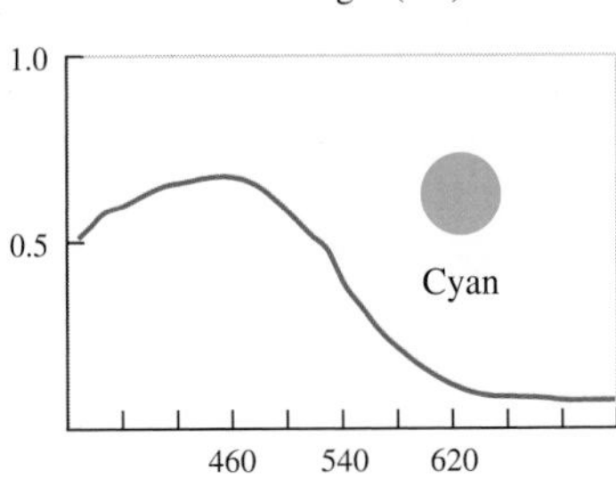

Figure 25.32 Transmission curves for colored filters. Or reflection curves for colored paints.

ing. (They are the primaries of the paint box, although they are very often mistakenly spoken of as red, blue, and yellow.) They are the basic colors of the dyes used to make photographs and the inks that printers use to print them. To color paint or cloth or anything else, we need only surround the transparent grains or fibers with a colored filter, a dye layer. White light passes through the film of dye, loses a select portion of the spectrum due to dissipative absorption, and emerges with the complementary color. Ideally, if you mix all the subtractive primaries together (either by combining paints or by stacking filters), you get no color, no light, just blackness. Each removes a region of the spectrum, and together they absorb it all. Selective absorption is the principal mechanism that colors the world from green grass to red roses to pink lips.

Core Material

The fundamental mechanism underlying the propagation of light through matter is **scattering**—the absorption and re-emission of radiant energy. *A ray is a line drawn in space corresponding to the direction of flow of radiant energy* (p. 870).

When a beam of light reflects from a smooth surface

$$\theta_i = \theta_r \qquad [25.1]$$

The angle of incidence equals the angle of reflection. This equation is the first part of the **Law of Reflection**. The second part maintains that *the incident ray, the perpendicular to the surface, and the reflected ray all lie in a plane* called the **plane-of-incidence**.

The ratio of the speed of an electromagnetic wave in vacuum to that in matter is the *index of refraction n:*

$$n = \frac{c}{v} \qquad [25.2]$$

The **Law of Refraction**, or **Snell's Law**, is

$$n_i \sin \theta_i = n_t \sin \theta_t \qquad [25.3]$$

where *the incident, reflected, and refracted rays all lie in the plane-of-incidence.* When light enters a refracting medium, its wavelength changes, and

$$\lambda = \frac{\lambda_0}{n} \qquad [25.4]$$

provides the wavelength in any medium of index n.

When the light impinges on an interface where $n_i > n_t$, there will be a special incident angle, the **critical angle** (θ_c), at which all the incoming light will be reflected back and none will be transmitted. This **total internal reflection** continues for all incident angles greater than the critical angle defined by

$$\sin \theta_c = \frac{n_t}{n_i} \qquad [25.5]$$

The widest range of colors can be created using a mix of light beams of red (R), green (G), and blue (B). Red plus blue is *magenta* (M); blue plus green is *cyan* (C); and red plus green is *yellow* (Y). And the sum of all three (884) is white.

Suggestions on Problem Solving

1. The equation for the apparent depth of an object in a liquid (Fig. 25.23) can be remembered by keeping in mind that the ratio of indices equals the ratio of depths: smaller-to-larger, (or possibly, larger-to-smaller). Thus, for example, when air is above water, it's the index of air to the index of water as the apparent depth is to the real depth.

2. An error commonly made when dealing with total internal reflection is to substitute the wrong values in for the two indices—most often they are simply interchanged. Remember here $\sin \theta_c = n_t/n_i$ and $n_i > n_t$. The index ratio must be less than one.

3. When dealing with problems involving plane mirrors, always draw a diagram. The central notion is that the angle of incidence equals the angle of reflection—all the rest is geometry. Often it is useful to construct rays that touch the very edges of the mirror and thereby define the limits of the system.

4. When treating a problem that pertains to wavelengths, remember to distinguish between the vacuum wavelength and the wavelength in the medium ($\lambda_0 > \lambda$). Keep in mind that $n = \lambda_0/\lambda$.

Discussion Questions

1. (a) Describe the kind of reflection that is taking place off this page. (b) Why is it that glossy paints can be deeper and richer in color than flat paints? (c) Why do the colors on a wet watercolor painting look so much more vivid than when they dry?

2. In what regard can we say that both reflection and refraction are manifestations of scattering?

3. Shaving lather is often substituted for meringue in pie-throwing fests. Why do they look so similar?

4. Suppose you have the bad luck to own a car whose exhaust has a bluish tint to it. What can you say about the effluvium?

5. (a) What is the angle of reflection for a ray incident normally on a smooth surface? (b) What is the angle of refraction for a beam striking an air-glass interface perpendicularly?

6. Modern picture frames sometimes come with "nonreflecting" glass (that is, glass whose surface has been deliberately made rough). What does it actually accomplish?

7. Is Venus in Velasquez's painting *The Toilet of Venus* (Fig. Q7) looking at her own face? Draw a ray diagram and in it drop a perpendicular from her nose to the plane of the mirror.

Figure Q7 *The Toilet of Venus* by Diego Rodriguez de Silva y Velasquez—National Gallery, London.

8. How is it that you can see your image in the surface of a polished *black* automobile? Explain what's happening. What allows you to see the gin in a glass, or the glass itself, for that matter?

9. The girl in Manet's painting, *The Bar at the Folies-Bergères* (Fig. Q9), is standing in front of a large plane mirror. We see reflected in it her back and the face of a man she seems to be talking to. From the Law of Reflection what, if anything, is amiss?

10. A properly cut diamond has a brilliant, lustrous appearance arising from the fact that most of the light entering the stone re-emerges. Explain how this happens and why diamond is especially well suited for the business.

11. Glass is often colored by adding to the silica, ions of the transition elements (Cr, Mn, Fe, Co, Ni, and Cu), usually in the form of oxides—all have a partially filled electron shell. Ordinary window glass contains, as contaminants, traces of iron oxides (in both ferrous and ferric forms) that have weak absorptions in the red and blue and a strong resonance in the near-UV. What radiation can you expect will traverse a sheet of glass? What color do you expect the glass to have if seen edge-on? Explain your answers.

12. Can a magenta light beam always be separated by a prism into red and blue? Similarly, can a cyan light beam always be separated into green and blue?

13. (a) What "color" will a red apple appear under red-light illumination? (b) What "color" will a cyan ink appear under the

Figure Q9 *The Bar at the Folies-Bergères* by Edouard Manet—Courtauld Institute Galleries. London.

same conditions? (c) Why is a yellow filter often used when taking black-and-white photos of the clouds in the sky?

14. Which colored light appears the most saturated—the blue of the sky, the red of a traffic signal, or the pink of a rose?

15. (a) What "color" will a sheet of white paper appear under 650-nm illumination? (b) Given that you have a beam of cyan light shining on a white wall, what color beam should be added so that the reflected light is white?

16. Figure Q16 is a plot of the amount of light transmitted across the spectrum by a piece of stained glass. (a) What color does it appear? (b) Two beams of light are projected onto a white sheet of paper. If one passes through a cyan filter and the other through a yellow filter, what color results?

17. (a) A beam of cyan light passes into a yellow filter. What color emerges? (b) A beam of yellow light passes into a magenta filter. What color emerges? (c) What color results when two beams of light, one cyan and one magenta, are made to overlap on a white screen? (c) White light passes through a cyan filter followed by a magenta filter. What color emerges?

18. Figure Q18 shows the transmission spectrum of a sheet of colored plastic. What color is it? Redraw the curve so that it corresponds to the frequency distribution reflected from an orange.

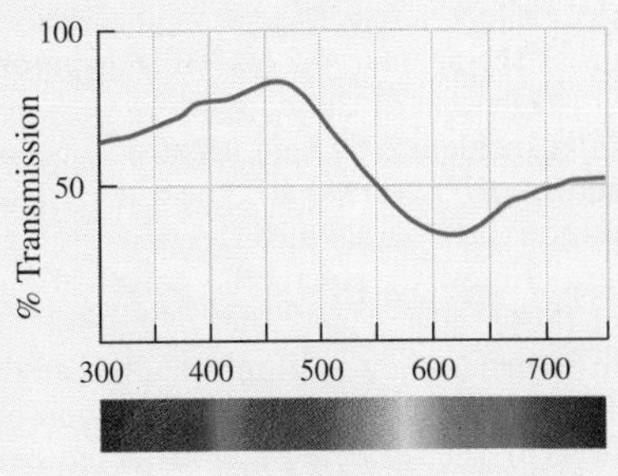

Figure Q16

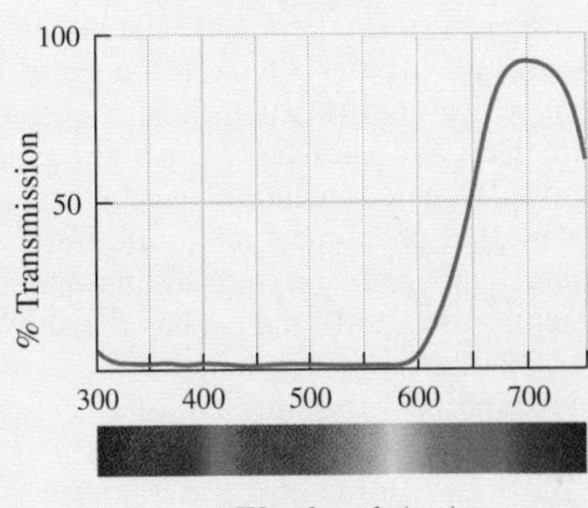

Figure Q18

Multiple Choice Questions

1. Standing on its surface looking upward, the Moon's sky is black because, unlike the Earth, the Moon (a) has no street lights (b) is very cold (c) has no atmosphere (d) has no oceans to reflect sunlight (e) none of these.
2. Cigarette smoke rising from the burning end has a bluish cast, while smoke exhaled is whitish. This difference happens because (a) the smoke contains a blue material that is left behind in the lungs (b) the body changes the chemistry so that the exhaled smoke is white (c) the smoke cools in the process and as a result turns white (d) water droplets from the body surround the inhaled smoke particles and, because they are large, they subsequently scatter white light (e) none of these.
3. Take a piece of ordinary window glass and crush it into a fine powder; holding the mound in hand, it will (a) appear black (b) become green (c) be as transparent as ever (d) appear white (e) none of these.
4. If you have ever scratched a piece of clear plastic or the varnished finish on a table top, you are familiar with the white marks that result. They have that characteristic appearance due to (a) changes in the chemistry of the surface (b) melting of the surface so it emits white light (c) electron bombardment of the surface via static charge that causes fluorescence (d) leaving behind a streak of powdered material that scatters a broad frequency band in the visible (e) none of these.
5. A material is transparent to a certain band of frequencies of electromagnetic radiation. Accordingly, we can say that the atoms in the material (a) are very small (b) have resonances in that band (c) are not interacting with each other (d) are not close together (e) none of these.
6. If the atmosphere were very much deeper than it is now, the Sun might appear (a) blue at sunset (b) red at noon (c) green at sunrise (d) violet at noon (e) none of these.
7. From our knowledge of the atom, it seems reasonable that all types of matter must have resonances somewhere in the electromagnetic spectrum and therefore must, if the matter is dense enough, (a) dissipatively absorb some radiant

energy (b) reflect across the spectrum (c) absorb broadly across the spectrum (d) be transparent at those resonances (e) none of these.

8. When light is incident on a smooth surface, the angle of incidence ordinarily (a) equals the angle of scattering (b) doesn't equal the angle of reflection (c) equals the angle of refraction (d) equals the polarization angle (e) none of these.
9. When light is incident on an interface, the angle(s) of (a) the reflected and refracted beams depend on the wavelength (b) the reflected and refracted beams are independent of frequency (c) the refracted beam is independent of wavelength (d) the reflected beam is independent of frequency (e) none of these.
10. Compared to an object in front of it, the image in a plane mirror is always (a) smaller (b) virtual (c) three times as far away (d) distorted (e) none of these.
11. A handful of white sandlike material is poured into a beaker of clear oil and vanishes from sight as it passes into the liquid. We can conclude that (a) the material dissolved (b) this phenomenon is quite impossible and could not have happened (c) the oil and material have mutually decomposed (d) the oil has the same index of refraction as the material (e) none of these.
12. A chicken is standing 1.0 m in front of a vertical plane mirror. A woman is standing 5.0 m from the mirror, behind and in line with the bird. How far from her will she see the image of the chicken? (a) 5.0 m (b) 1.0 m (c) 6.0 m (d) 4.0 m (e) none of these.
13. When a beam of light traveling in air enters a glass block, it ordinarily undergoes a change in (a) speed only (b) frequency only (c) wavelength only (d) speed and wavelength (e) none of these.
14. The bending or refraction of a beam of light as it enters a medium that has a greater index of refraction than the incident medium is due to a change in its (a) amplitude (b) effective speed (c) frequency (d) period (e) none of these.
15. What "color" will a yellow shirt appear under blue-light illumination? (a) blue (b) yellow (c) white (d) black (e) none of these.

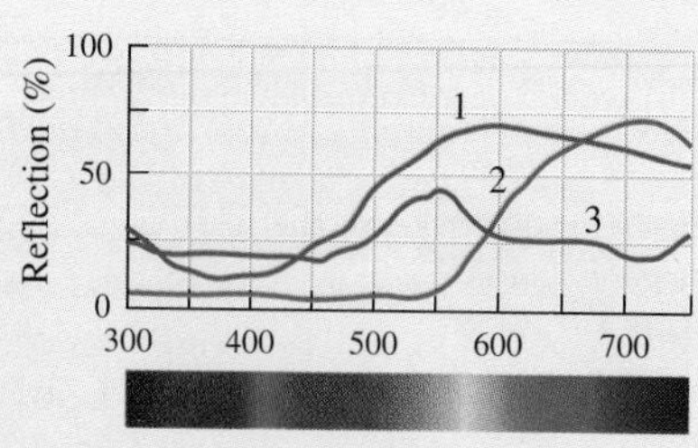

Figure MC20

16. What combination of colored beams of light when overlapped on a white screen will produce black? (a) R + B + G (b) M + C + Y (c) M + R + C + Y + G (d) M + C + Y + R + B + G (e) none of these.
17. A beam of yellow light passes through a cyan filter. The color that emerges is (a) yellow (b) blue (c) green (d) red (e) none of these.
18. The color that results from the mixing of magenta and yellow paint is (a) red (b) blue (c) green (d) black (e) brown.
19. What color emerges when white light passes through a magenta filter followed by a green filter? (a) white (b) yellow (c) green (d) blue (e) none of these.
20. Figure MC20 shows three plots of the amount of light reflected over the visible region of the spectrum from a lemon, a ripe tomato, and a lettuce. These correspond, in turn, to curves (a) 1, 3, and 2 (b) 3, 2, and 1 (c) 1, 2, and 3 (d) 2, 3, and 1 (e) 3, 1, and 2.
21. A yellow surface is illuminated with magenta light. What color will it appear in reflected light? (a) blue (b) green (c) yellow (d) red (e) none of these.

Problems

SCATTERING
REFLECTION

1. [I] Two beams of light, one red ($\lambda_r = 780$ nm) and one violet ($\lambda_v = 390$ nm), pass through several hundred meters of air. What is the ratio of the amount of scattering of red to violet?
2. [I] A beam of white light crosses a large volume occupied by a tenuous molecular gas mixture of mostly oxygen and nitrogen. Compare the amount of scattering occurring for the yellow (580 nm) component with that of the violet (400 nm) component.
3. [I] A laser beam strikes a front-silvered mirror at an angle of 30° to the perpendicular. At what angle will it be reflected?
4. [I] A beam of parallel light strikes a flat polished metal surface perpendicularly. At what angle will light be reflected?
5. [I] Rays of light impinge on a smooth flat mirror at glancing incidence, making a tiny angle with respect to the surface. Approximately, what are the angles of incidence and reflection?
6. [I] A narrow beam of light impinges on a smooth glass plate at 25° measured from the perpendicular to the surface. What is the angle between the incident and reflected beams?
7. [I] Two rays leave a point source and strike a front-silvered mirror, making angles with the normal of 30° and 40°, respectively. What is the angle between the two reflected rays?
8. [I] A very narrow beam of light from a laser is incident on a horizontal mirror at an angle of 58°. The reflected beam strikes a wall at a spot 5.0 m away from the point of incidence where the beam hit the mirror. How far horizontally is the wall from that point of incidence?

9. [I] The tomb of FRED the Hero of Nod is a dark closed chamber with a small hole in a wall 3.0 m up from the floor. Once a year, on FRED's birthday, a beam of sunlight enters via the hole, strikes a small polished gold disk on the floor 4.0 m from the wall and reflects off it, lighting up a great diamond imbedded in the forehead of a glorious statue of FRED, 20 m from the wall. Roughly how tall is the statue?

10. [I] There are a number of practical devices that use rotating plane mirrors. Show that if a mirror rotates through an angle α, the reflected beam will move through an angle of 2α.

11. [I] A woman can see an object clearly when it's held at a distance of 25 cm from her eyes. How far should she hold a flat mirror to see her own image clearly?

12. [I] A man is walking at 1.0 m/s directly toward a flat mirror. At what speed is his image approaching him?

13. [I] Two upright 1.00-m-tall plane mirrors are placed parallel to each other 2.8 m apart. The top of the mirror on the right is then moved back a little so that its surface tilts away from the other mirror at an angle of 10.0° off the vertical. A narrow laser beam passes perpendicularly through a small hole in the very bottom of the mirror on the left. It subsequently strikes the tilted mirror from which it reflects. How many times will it reflect off the upright mirror on the left?

14. [I] Suppose both mirrors in Problem 13 were, say, 10 m high. What then would be the angle of reflection off the tilted mirror when the beam encounters it for the second time?

15. [I] A woman stands between a vertical mirror $\frac{1}{2}$ m tall and a distant tree whose height is H. She is 1.0 m from the mirror, and the tree is 11.0 m from the mirror. If she sees the tree just fill the mirror, how tall is the tree?

16. [II] If 4.00% of the light energy incident normally on a microscope slide is reflected from the top surface, how much will be transmitted through two such slides?

17. [II] Suppose you are standing in front of a 4.0-ft-tall flat vertical mirror in which you can see some fraction of your body. What will happen to that fraction when you step farther from the mirror? Draw a ray diagram that proves your contention.

18. [II] A man 6 ft tall standing 10 ft from a 3-ft-high vertical mirror can see his entire image in it. If his eyes are 4 in. below the very top of his head, how high above the floor is the bottom edge of the mirror?

19. [II] We wish to set up an eye test in a rather modest-sized office and so must use a plane mirror to get the full required distance to the chart (C_1C_2) as shown in Fig. P19. If the observer is a distance d from the front-silvered mirror (which is H meters tall) and the chart is h meters in height, write an expression for the minimum value of H that will suffice, in terms of h, d, and the object distance s_0.

20. [II] Figure P20 shows a ray striking one of two perpendicular mirrors at angle θ_{i1} and then striking the other at θ_{i2}. Find a relationship between these two angles and describe the paths taken by the incoming and outgoing rays.

21. [II] Return to Fig. P20, but this time imagine there is a frog sitting between the mirrors looking at himself. Draw a ray diagram showing the locations of all the possible images.

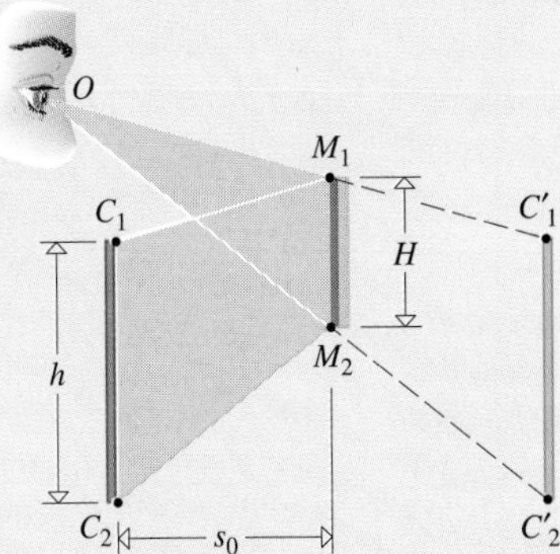

Figure P19

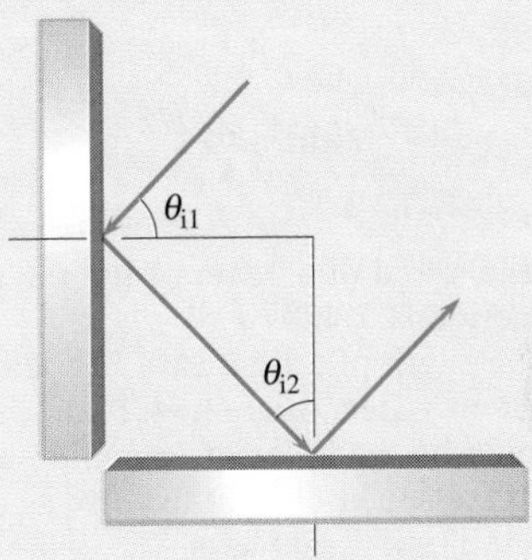

Figure P20

You should actually try this one with two ordinary mirrors held together. Look directly into the corner and hold up your right hand. Which hand will the image hold up? Notice where the seam is. Most people will see it run down the middle of their stronger eye. (A few with equal-strength eyes will see it down the middle of their faces.)

22. [II] Two vertical plane mirrors are brought together so that they make a wedge-angle of 35°, and a point source of light (S) is placed midway between them on a line bisecting the wedge-angle. Draw a simplified ray diagram locating all of the images of S by just using normal rays and making the object distance equal the image distance.

23. [III] Figure P23 shows an arrangement of mirrors forming a rangefinder for a camera. Mirror M_1 is partially silvered and fixed in position while M_2 is totally silvered and rotates about a vertical axis. Looking into M_1, you would see two images, and by rotating M_2, they are made to exactly overlap. Since d is known accurately and the angle through which the second mirror is turned can be easily measured, a scale can be computed that displays the distance L directly on the camera. Write an expression for L in terms of ϕ and d, taking the tilt of the fixed mirror to be 45°.

24. [III] Most mirrors in common use are back-silvered. Suppose you stand 1.000 0 m from such a mirror, which is made of 5.00-mm-thick plate glass ($n = 1.50 = 3/2$). How far behind the front surface will your image appear? Draw a ray diagram.

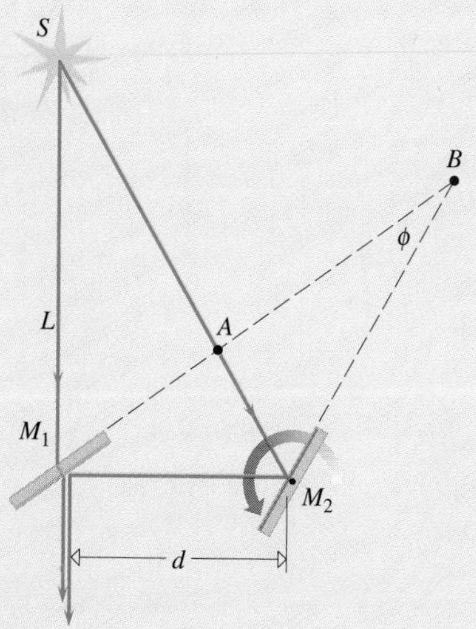

Figure P23

REFRACTION

25. [I] What is the speed of a beam of light in diamond if the index of refraction is 2.42?
26. [I] If the wavelength of a lightwave in vacuum is 540 nm, what will it be in water, where $n = 1.33$?
27. [I] What should be the index of refraction of a medium if it is to reduce the speed of light by 10% as compared to its speed in vacuum?
28. [I] If the speed of light (that is, the phase speed) in Fabulite ($SrTiO_3$) is 1.245×10^8 m/s, what is its index of refraction, to three significant figures?
29. [I] How far does yellow light travel in water in 1.00 s? What is the speed of yellow light in ethyl alcohol?
30. [I] If a narrow beam of light is incident on the surface of a tank of turpentine at 30°, at what angle will it be transmitted?
31. [I] A beam of light impinges on an air-liquid interface at an angle of 55°. The refracted ray is observed to be transmitted at 40°. What is the refractive index of the liquid?
32. [I] A swimmer shines a beam of light in water up toward the surface. It strikes the air-water interface at 35°. At what angle will it emerge into the air?
33. [I] Prove that to someone looking straight down into a swimming pool, the water will appear to be 3/4 of its true depth.
34. [I] A pool of water is 3.00 m deep and 4.00 m wide. Someone lying with his face close to the water is looking for a coin that is directly across the pool on the bottom far edge. Will he be able to see it?
35. [I] Radiant energy can be conveniently converted from IR (1.06 μm) to UV using nonlinear crystals. First, a portion of the IR is converted into its second harmonic on passing through an appropriate crystal. Determine the resulting frequency. Then, these two frequencies are mixed in a second crystal to generate the third harmonic. Find the wavelength of the third harmonic.
36. [I] What is the critical angle for diamond? Give your answer to two significant figures. What, if anything, does the critical angle have to do with the luster of a well-cut diamond?
37. [I] What is the critical angle for an air-ice interface?
38. [I] Using a block of a transparent, unknown material, it is found that a beam of light inside the material is totally internally reflected at the air-block interface at an angle of 48.0°. What is its index of refraction?
39. [I] What is the minimum incident angle for internal reflection at the interface between two materials of indices 1.33 and 2.40, respectively?
40. [II] A 500-nm lightwave propagating in vacuum impinges normally on a glass sheet of index 1.60. How many waves span the glass if it's 1.00 cm thick?
41. [II] The photo in Fig. P41 shows a pulse of green light (530 nm, the second harmonic of 1.06 μm from a neodymium-doped glass laser) lasting about 10 picoseconds. The cell contains water ($n \approx 1.36$) and the scale is in millimeters. The exposure time was also about 10 ps. Explain what is shown in the picture. How far did the pulse travel during the exposure? How many wavelengths long (of green light) is the pulse?

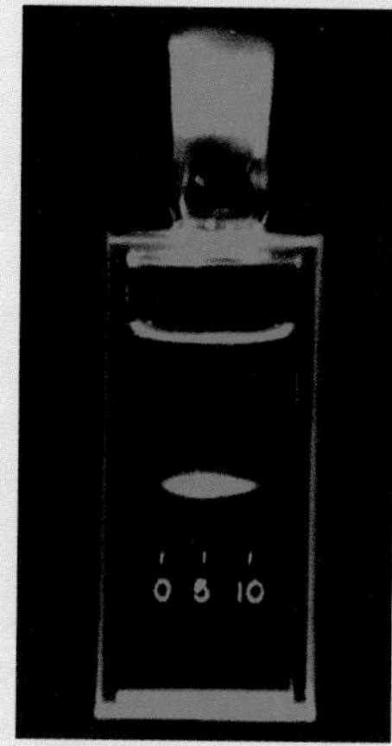

Figure P41

42. [II] The American physicist R.W. Wood (1868–1955) pointed out the following: a pulse of red light entering a block of glass (with a measured value of $n = 1.52$) 12.00 miles thick will emerge 1.80 miles ahead of a blue pulse that entered at the same moment. (a) How long will it take the red light to traverse the glass? (b) How fast does the blue pulse travel? (c) How much time will elapse between the emergence of the red and then the blue? Incidentally, Albert Michelson actually observed the effect in carbon disulfide.
43. [II] We wish to determine the index of refraction of a liquid filling a small tank as shown in Fig. P43. By moving up and down, it is found that the back edge of the container is just visible at an angle of 20.0° above the horizontal. Find the index.
44. [II] Yellow light from a sodium lamp ($\lambda_0 = 589$ nm) traverses a tank of glycerin (of index 1.47), which is 20.0 m long, in a time t_1. Now, if it takes a time t_2 for the light to pass through the same tank when filled

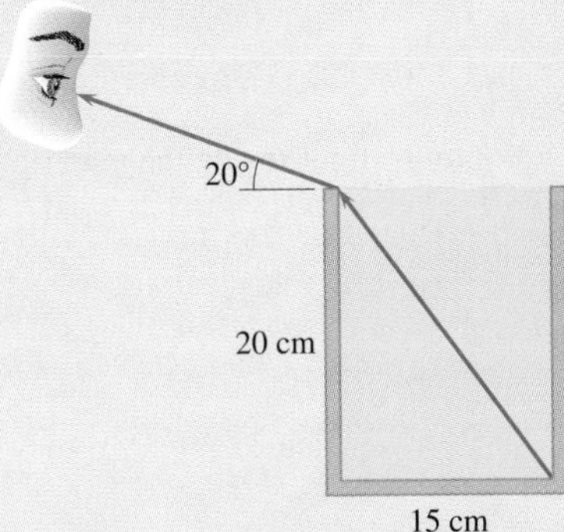

Figure P43

with carbon disulfide (of index 1.63), determine the value of $t_2 - t_1$.

45. [II] A beam of light impinges on the top surface of a 2.00-cm-thick parallel glass ($n = 1.50$) plate at an angle of 35°. How long is the actual path through the glass?

46. [II] Consider an air-glass interface with light incident in the air. If the index of refraction of the glass is 1.70, find the incident angle such that the transmission angle is to equal $\frac{1}{2}\theta_i$?

47. [II] Imagine that you focus a camera with a close-up bellows attachment on a letter printed on this page. Now, suppose the letter is covered with a 1.00-mm-thick microscope slide ($n = 1.55$). How high must the camera be raised in order to keep the letter in focus?

48. [II] A prism, ABC, is cut such that $\angle BCA = 90°$ and $\angle CBA = 45°$. What is the minimum value of its index of refraction if, while immersed in air, a beam traversing face AC is to be totally internally reflected from face BC?

49. [II] A fish looking straight upward toward the surface receives a cone of rays and sees a circle of light filled with the images of sky and ships and whatever else is up there (Fig. P49). This bright circular field is surrounded by darkness. Explain what is happening and compute the cone-angle.

50. [II] A block of glass with an index of 1.55 is covered with a layer of water of index 1.33. For light traveling in the glass, what is the critical angle at the interface?

Figure P49

51. [III] A lightwave propagates from point A to point B in vacuum. Suppose we introduce into its path a flat glass plate ($n_g = 1.50$) of thickness $L = 1.00$ mm. If the vacuum wavelength is 500 nm, how many waves span the space from A to B with and without the glass in place? What phase shift is introduced with the insertion of the plate?

52. [III] Refer to Fig. 25.22. If the plate has a thickness τ, show that the emerging beam is laterally displaced by a perpendicular distance d from the incident beam where

$$d = \frac{\tau \sin(\theta_{i1} - \theta_{t1})}{\cos \theta_{t1}}$$

53. [III] A coin rests on the bottom of a tank of water ($n_w = 1.33$) 1.00 m deep. On top of the water floats a layer of benzene ($n_b = 1.50$), which is 20.0 cm thick. Looking down nearly perpendicularly, how far beneath the topmost surface does the coin appear? Draw a ray diagram.

54. [III] With the results of Problem 52 in mind, by how much would a beam, incident at 30°, be displaced on traversing a plate of glass 5.00 cm thick having an index of 1.60?

26

Geometrical Optics and Instruments

AN OBJECT THAT IS either self-luminous or externally illuminated can be imagined to have a surface covered with radiating point sources. Your face in the sunshine is a distribution of countless atomic scatterers each sending out little spherical wavelets. The associated rays emanate radially outward in the direction of the energy flow (Fig. 26.1); they *diverge* from each point source S. Now suppose the object stands before some arrangement of reflecting and/or refracting surfaces constituting an *optical system* whose function it is to collect the light and cause it to *converge*. The energy in the diverging cone arrives at P, which is called the *image* of S. A

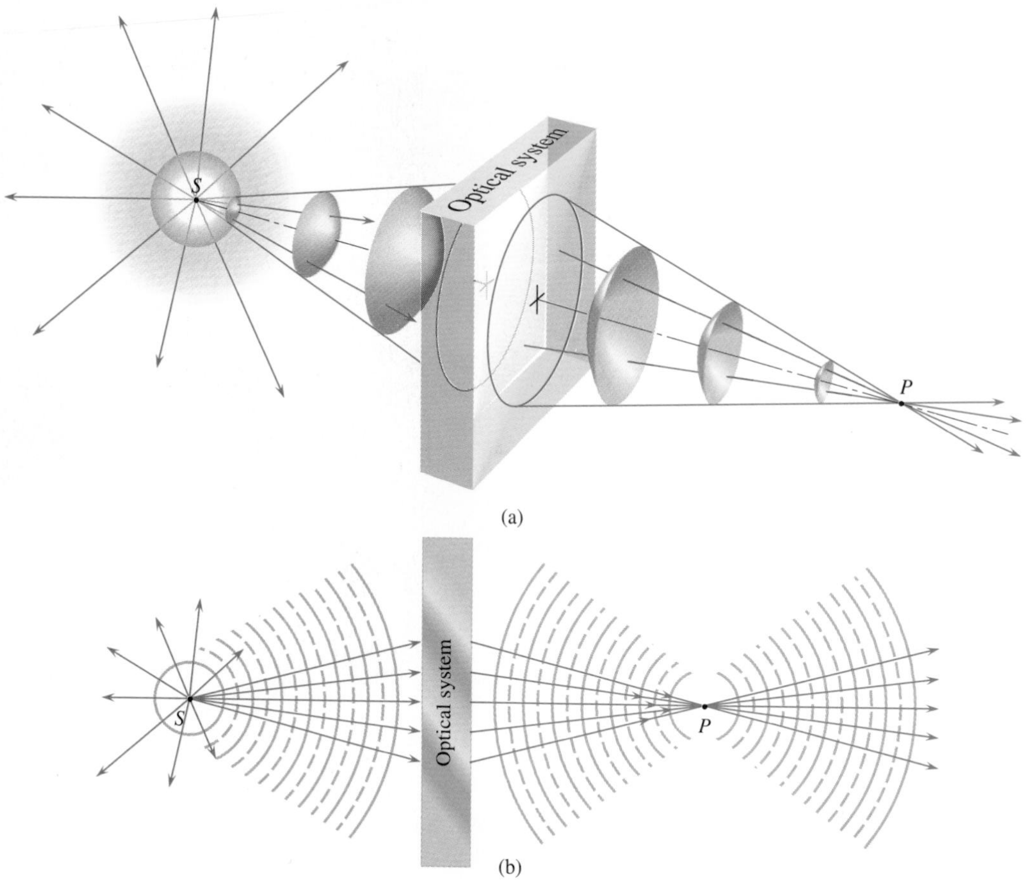

Figure 26.1 (a) A point source S sends out spherical waves. A cone of rays enters an optical system that inverts the wavefronts, causing them to converge on point P. (b) Rays diverge from S and a portion of them converge to P. If nothing stops the light at P, it continues on.

billion points on a face contribute light to a billion points making up the image of that face.

Real systems—cameras, telescopes, and eyeballs—cannot possibly collect *all* the light emitted from an object. When only a portion of any incident wavefront enters the system, there will inevitably be some diffraction. No matter how well made the optical system is, an object point will invariably be imaged as a somewhat enlarged *blur spot.* An otherwise perfect imaging system is thus said to be *diffraction limited.*

As the dimensions of the optical system become increasingly large in comparison to the wavelength, the effects of diffraction diminish. Accordingly, it is often reasonable to simply neglect diffraction and assume that all the rays travel in straight lines and go where they go via the Laws of Reflection and Refraction. As the wavelength becomes vanishingly small, diffraction disappears, rectilinear propagation obtains, and we have the idealized domain of **geometrical optics**. *That discipline deals with the manipulation of wavefronts (or rays) by means of interposing reflecting and/or refracting objects, neglecting any diffraction effects, and assuming rectilinear propagation.*

LENSES

The lens is no doubt the most widely used optical device, and that's not even considering the fact that we see the world through a pair of them. Human-made lenses date back at least to the *burning-glasses* of antiquity, which, as the name implies, were used to start fires long before the advent of matches—there's a reference to one of these in Aristophanes' play *The Clouds* (424 B.C.). From our perspective, *a lens is a refracting device (or discontinuity in the transmitting media) that reconfigures an incoming energy distribution.* And that much is true whether we are dealing with UV, lightwaves, IR, microwaves, radiowaves, or even sound waves (Fig. Q15, p. 475).

The configuration of a lens is determined by the reshaping of the wavefront it is to perform. Point sources are basic, and so it is often desirable to convert diverging spherical waves into a beam of plane waves; flashlights, projectors, and searchlights all do this in order to keep the beam from spreading out and weakening as it progresses. In just the reverse, it is frequently necessary to collect incoming parallel rays and bring them together at a point, thereby focusing the energy, as is done with a burning-glass or a telescope lens.

A lens for short-wavelength radio waves. The disks serve to refract these waves much as rows of atoms refract light.

26.1 Aspherical Surfaces

We can begin to understand how a lens works by interposing in the path of the wave a transparent substance in which the wave's speed is different than it was initially. Fig. 26.2a shows a diverging spherical wave traveling in an incident medium of index n_i, impinging on the curved interface of a transmitting medium of index n_t. When n_t is greater than n_i, the wave slows as it enters the new substance. The central area of the wavefront travels more slowly than its outer extremities, which are still moving through the incident medium. These extremities overtake the mid-region, continuously straightening out the wavefront. If the interface is properly configured, the spherical wavefront bends into a plane wave.

To find the required shape of the interface so we can make such a device, refer to Fig. 26.2c, wherein point A can lie anywhere on the boundary. *One wavefront is transformed into another provided the paths along which the energy propagates are all equal, thereby maintaining the phase of the wavefront.* A little spherical surface of constant phase emitted from S must evolve into a flat surface of constant phase at $\overline{DD'}$. That means that whatever path the light takes from S to $\overline{DD'}$, it must always be the same number of wavelengths long, so that the disturbance begins and ends in-phase. Radiant energy leaving S as a single wavefront must arrive at the plane $\overline{DD'}$, *having traveled for the same amount of time* to get there, no matter what the actual route taken by any particular ray. In other words, $\overline{F_1A}/\lambda_i$ (the number of wavelengths along the arbitrary ray from F_1 to A) plus $\overline{AD}/\lambda_t$ (the number of wavelengths along the ray from A to D) must be constant regardless of where on the interface A happens to be. Now, if we add these and then multiply by λ_0, we get

$$n_i(\overline{F_1A}) + n_t(\overline{AD}) = \text{constant} \tag{26.1}$$

Each term on the left is the length traveled in a medium multiplied by the index of that medium, and each represents what is called the **optical path length** (O.P.L.) traversed. The O.P.L. is the equivalent length in vacuum—if it is divided by c, we get the time it takes the light to travel the actual distance it did in the medium. If Eq. (26.1) is divided by c, the first term becomes the time it takes to travel from S to A and the second term, the time from A to D—the right side remains constant (not

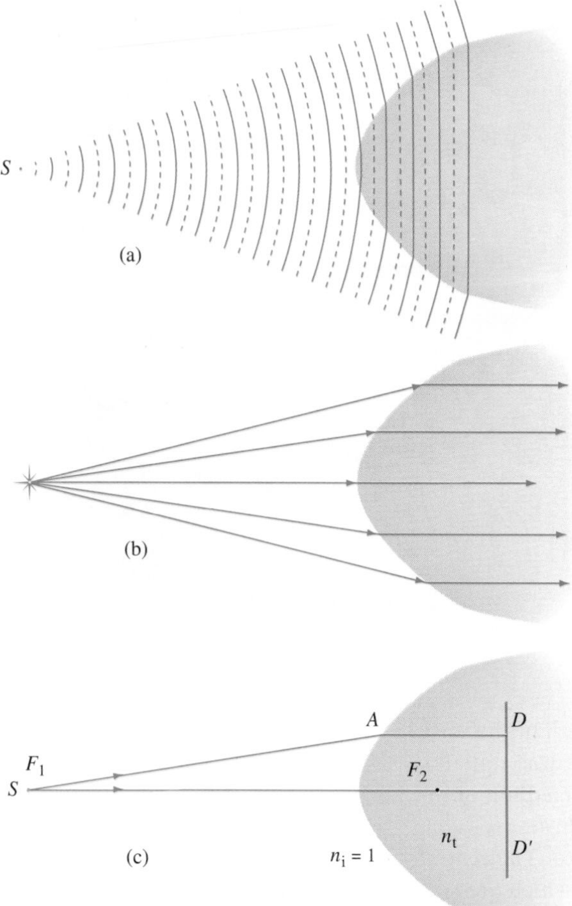

Figure 26.2 Light striking a hyperbolic interface between air and glass. (a) The wavefronts bend and straighten out. (b) The rays become parallel. (c) The hyperbola is such that the optical path from S to A to D is the same no matter where A is.

the same constant, but constant). Equation (26.1) is equivalent to saying that all paths from S to $\overline{DD'}$ must take the same amount of time to traverse.

Now, back to finding the shape of the interface. Dividing Eq. (26.1) by n_i, it becomes

$$\overline{F_1A} + \left(\frac{n_t}{n_i}\right)\overline{AD} = \text{constant} \tag{26.2}$$

This is the equation of a *hyperbola* where the eccentricity, which measures the bending of the curve, is given by $(n_t/n_i) > 1$; the greater the eccentricity, the flatter the hyperbola. When a point source is located at the focus F_1 and the interface between the two media is hyperbolic, plane waves will be transmitted into the higher-index material. Ellipsoidal and paraboloidal surfaces are also useful, but we'll limit our discussion here to hyperboloidal ones only (Fig. 26.3).

It's an easy matter now to construct lenses such that both the object and image points (or the incident and emerging light) will be outside of the medium of the lens. In Fig. 26.4a diverging incident spherical waves are made into plane waves at the first interface via the mechanism of Fig. 26.3b. These plane waves within the lens strike the back face perpendicularly and emerge unaltered; $\theta_i = 0$ and $\theta_t = 0$. And because the rays are reversible, *plane waves incoming from the right will converge to point F,* which is known as the **focal point** of the lens. Exposed to the parallel rays from the Sun, our rather sophisticated lens would serve nicely as a burning-glass.

In Fig. 26.4b, the plane waves within the lens are made to converge toward the axis by bending the second interface. Both of these lenses are thicker at their midpoints than at their edges and are therefore said to be **convex** (from the Latin *convexus,* meaning arched). Each lens causes the incoming beam to converge somewhat, to bend a bit more toward the central axis, and so they are referred to as **converging lenses**.

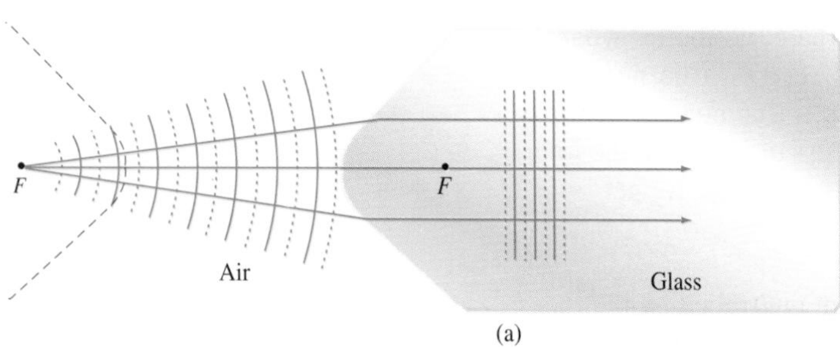

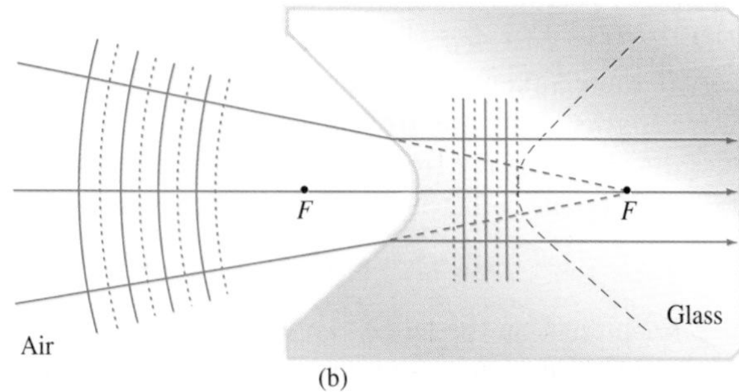

Figure 26.3 Hyperbolic surfaces have two foci, F. In (a), light originating at F passes into the glass in the form of plane waves. In (b), light converging toward F passes into the glass as plane waves. In both cases, the rays can be reversed so that they emerge into the air as either converging or diverging cones.

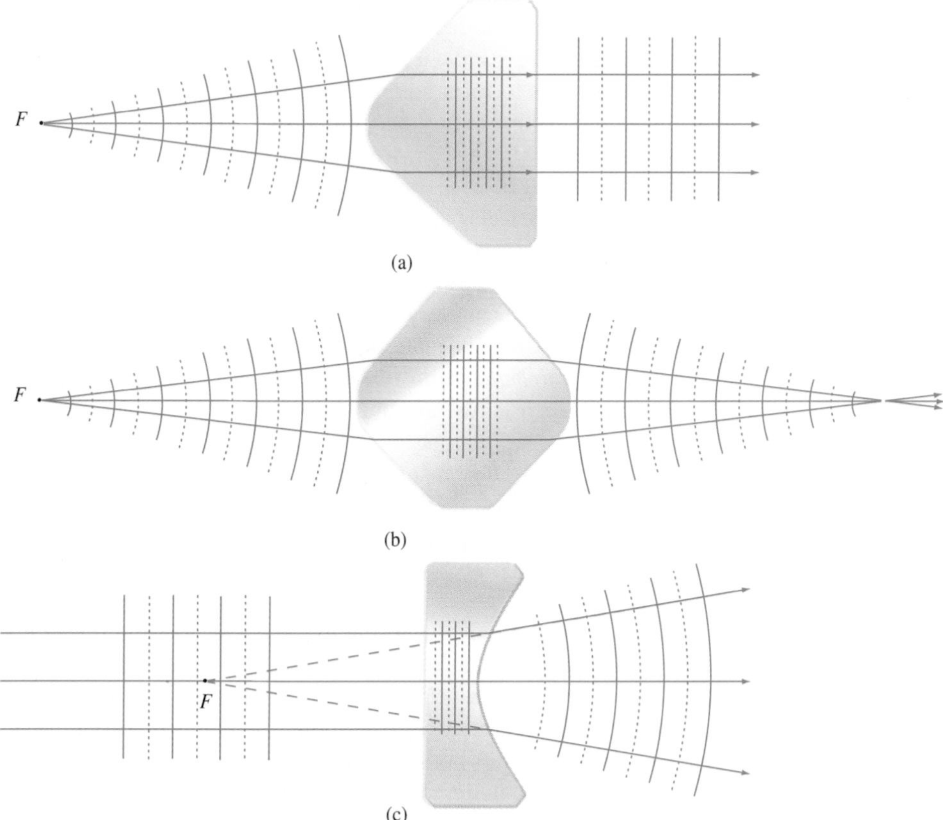

Figure 26.4 Several hyperbolic lenses.

In contrast, a **concave** lens (from the Latin *concavus,* meaning hollow—and most easily remembered because it contains the word *cave*) is thinner in the middle than at the edges, as is evident in Fig. 26.4c. It causes the rays that enter as a parallel bundle to diverge. All such devices that turn rays outward away from the central axis (that in so doing add divergence to the beam) are called **diverging lenses**. In Fig. 26.4c, parallel rays enter from the left and, on emerging, seem to diverge from F; still, that point is taken as a focal point. *When a parallel bundle of rays passes through a converging lens, the point to which it converges (or when passing through a diverging lens, the point from which it diverges) is a focal point of the lens.*

Optical elements (mirrors and lenses) having at least one curved surface that is not spherical are referred to as *aspherical.* Although they are easy to understand and they perform certain tasks exceedingly well, such elements are difficult to manufacture with great accuracy, Even so, many millions of aspherics have been made, and they can be found in instruments across the whole range of quality: in telescopes, projectors, cameras, and spy satellites.

26.2 Spherical Thin Lenses

In comparison to aspherics, spherical surfaces (actually, segments of spheres) are easy to fabricate. Unlike the hyperboloid, ellipsoid, or paraboloid, the sphere has no single central symmetry axis. All its diameters are alike, and the sphere can be generated by simply randomly rocking a rough-cut glass disk against an approximately spherical grinding tool. The two will wear each other away until both are perfectly spherical. The vast majority of lenses in use today have surfaces that are segments of spheres—despite the fact that those spherical surfaces are not ideal and will result in imaging errors known as *aberrations*. By using several components made of different materials to form compound lenses, these errors can be controlled so well that image quality need only be limited by diffraction.

As has already been seen, we cannot expect perfect imagery from spherical surfaces; hence, it will be necessary to place limitations on the way a spherical lens can be used so that it behaves appropriately. Thus, we only allow the lens to receive rays that strike it *not far from the central axis and enter only at shallow angles*—rays of this kind are said to be **paraxial**. No matter how the rays are drawn (and they will often be depicted making large angles for the sake of clarity), the rays are nonetheless paraxial. To simplify matters further, we will only deal with **thin lenses**; that is, *lenses for which the radii of curvature of the surfaces are large compared to the thickness.* Such lenses are quite common—most telescope and eyeglass lenses are thin.

The distances of objects and images are usually measured from the lens and can be on either side of it. It is important to associate a specific sign with each such distance so that it can be properly manipulated algebraically. *As a rule, light enters from the left,* and Fig. 26.5 shows how a typical ray is twice bent toward the central axis as it traverses a convex spherical lens. In Fig. 26.6a, we see two rays leaving the axial point source S and converging to the corresponding image point P. This is the basic geometry for which several possible sign conventions exist. We will take an object distance s_o to the *left* of the lens as positive and an image distance s_i to the *right* of the lens as positive. Furthermore, the radius of curvature of a lens surface is positive when its center point C is to the right of the surface. Here R_1, the radius of the first surface encountered, is positive while R_2, whose center C_2 is to the left, is negative. *All interfaces that bulge toward the left have positive radii, and all interfaces that bulge right have negative radii.*

To derive an equation for the operation of a spherical lens, note that the optical path lengths traversed by all rays from S to P are equal. All portions of the diverging wavefront take the same time to reach P, and all arrive in-phase. Let's examine two such paths, one along the central axis and one higher up, some arbitrary distance y. The O.P.L. from S to A to G to P must be the same as that from S to H to J to P. The wavefronts just in front of and behind the lens, Σ and Σ', have radii $\overline{SH} = \overline{SA}$ and $\overline{GP} = \overline{JP}$, respectively, and so if the O.P.L. from A to G equals that from H to J, the overall optical path lengths along the two routes will be equal. The paraxial limitation requires that y be small and since the lens is thin, we can approximate the path from A to G as a straight line (Fig. 26.6b). For simplicity, assume the lens has an index n_l and it is immersed in air with an index of one. Thus, S will be imaged at P provided that

$$n_l\overline{HI} + n_l\overline{IJ} = \overline{AB} + \overline{BC} + n_l\overline{CD} + n_l\overline{DE} + \overline{EF} + \overline{FG} \tag{26.3}$$

for all allowed values of y.

Since all the surfaces are spheres, the curves in the figure are circles, and there is a nice way to represent each of these little distances. Figure 26.6c shows a chord cutting a diameter ($2R$) of a circle perpendicularly. The piece σ is called the *saggita* (from the Latin for *arrow*) because it looks like an arrow resting on a bow. There is a theorem (proved in Problem 34) stating that for two such intersecting lines, the product of the two segments of one equals the

Figure 26.5 The radius drawn from C_1 is normal to the first surface, and as the ray enters the lens, it bends down *toward* that normal. The radius from C_2 is normal to the second surface, and as the ray emerges, since $n_l > n_a$, the ray bends down *away* from that normal.

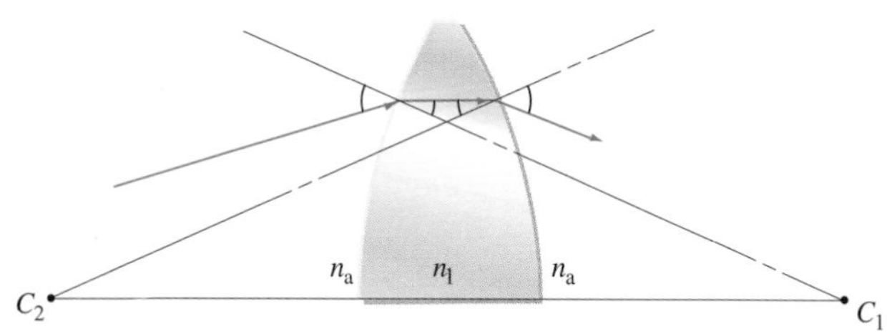

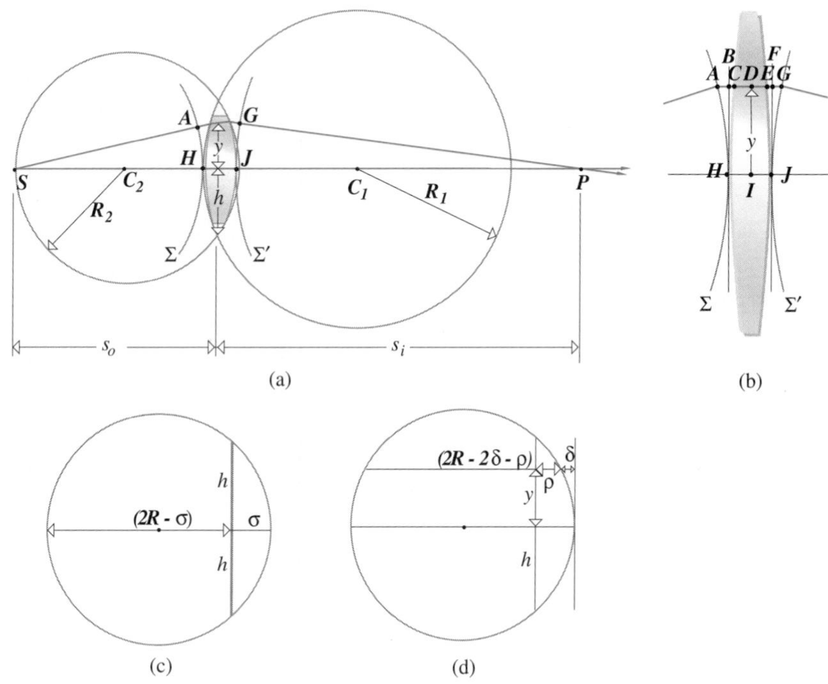

Figure 26.6 (a) The geometry of the thin lens. (b), (c), and (d) are associated with the approximations needed to determine the optical path lengths for rays from *S* to *P*.

product of the segments of the other:

$$\sigma(2R - \sigma) = (h)(h)$$

Hence
$$2R\sigma - \sigma^2 = h^2$$

But we are only interested in cases where $R \gg \sigma$ and there σ^2 is negligible compared to $2R\sigma$; hence, $2R\sigma \approx$ h^2 and

$$\sigma \approx \frac{h^2}{2R} \tag{26.4}$$

Figure 26.6d shows another chord, the segment ρ resembling a saggita, and the length δ extending beyond the circle. In the same way as above, it follows (Problem 35) that

$$\rho \approx \frac{(h^2 - y^2)}{2R} \quad \text{and} \quad \delta \approx \frac{y^2}{2R} \tag{26.5}$$

These three approximations correspond to the segments in Fig. 26.6b where the surfaces of the lens form circles of radii R_1 and $-R_2$ and the radii of the wavefronts Σ and Σ' are represented by s_o and s_i, respectively. This last approximation is equivalent to saying that the lens is so thin we can measure the object and image distances from *either* its center or from its faces. Pressing on, we have

$$\begin{aligned}
\overline{AB} &= \delta_0 \approx y^2/2s_0 & \overline{DE} &= \rho_2 \approx (h^2 - y^2)/-2R_2 \\
\overline{BC} &= \delta_1 \approx y^2/2R_1 & \overline{EF} &= \delta_2 \approx y^2/-2R_2 \\
\overline{CD} &= \rho_1 \approx (h^2 - y^2)/2R_1 & \overline{FG} &= \delta_i \approx y^2/2s_i \\
\overline{HI} &= \sigma_1 \approx h^2/2R_1 & \overline{IJ} &= \sigma_2 \approx h^2/-2R_2
\end{aligned}$$

Now, substituting all of these into Eq. (26.3) and shifting things around, we see that

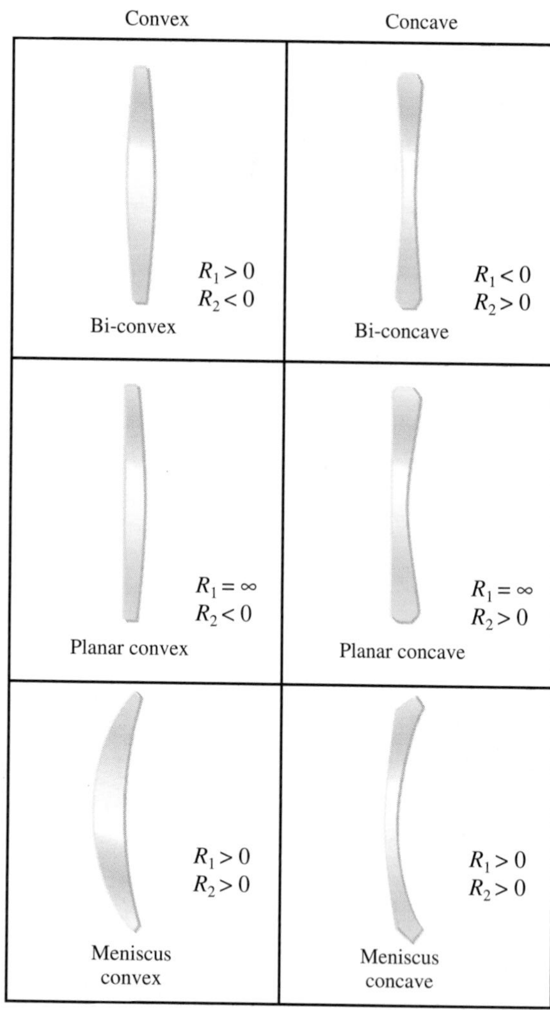

Figure 26.7 Cross sections of various centered spherical simple lenses. Take the surface on the left as number one, since it's encountered first by light coming from the left.

every term with h^2 vanishes; the $y^2/2$ factors out and cancels, leaving

$$\frac{1}{s_o} + \frac{1}{s_i} = (n_l - 1)\left(\frac{1}{R_1} - \frac{1}{R_2}\right) \tag{26.6}$$

This is the **Thin-Lens Equation**, often referred to as the **Lensmaker's Formula** because the left side (which treats what is going on external to the lens) is given in terms of the physical variables that would have to be selected to fabricate the lens (Fig. 26.7).

It can be shown (Problem 37) that, had the lens been immersed in a medium of index n_m rather than air, the Lensmaker's Formula would again result, but instead of the first term on the right being n_l, or equivalently $n_l/1$, it would now be n_l/n_m.

Example 26.1 A small point of light lies on the central axis 120 cm to the left of a thin bi-convex lens having radii of 60 cm and 30 cm. Given that the index of refraction of the lens is 1.50, find the location of the resulting image point. As far as the image distance is concerned, does it matter which way the lens is facing;

(continued)

(continued)

that is, which surface is toward the object? Unless otherwise informed, always assume the surrounding medium is air.

Solution: [Given: $s_o = 1.20$ m, $R_1 = 0.60$ m, $R_2 = -0.30$ m and $n_l = 1.50$. Find: s_i.] Because the lens is bi-convex, the second radius encountered by incoming rays from the left is taken as a negative number. It follows from Eq. (26.6) that

$$\frac{1}{1.20\text{ m}} + \frac{1}{s_i} = (1.50 - 1)\left(\frac{1}{0.60\text{ m}} - \frac{1}{-0.30\text{ m}}\right)$$

$$\frac{1}{s_i} = (0.50)\left(\frac{1}{0.20\text{ m}}\right) - \frac{1}{1.20\text{ m}} = \frac{2}{1.20\text{ m}}$$

and $\boxed{s_i = 0.60\text{ m}}$. The image distance is positive, and so the image lies to the right of the lens on the axis. Had we let $R_1 = 0.30$ m and $R_2 = -0.60$ m, nothing would have changed—*it doesn't matter which way a thin lens faces!*

▶ **Quick Check:** The right side of the Lensmaker's Formula equals (0.50)/(0.20 m) and that quantity should equal the left side of the formula. Hence, 1/(1.20 m) + 1/(0.60 m) should equal 1/(0.40 m), and it does.

26.3 Focal Points and Planes

Suppose that the object point S in Fig. 26.6a is moved far to the left. As $s_o \to \infty$, the rays enter the lens as a parallel bundle and are brought together at a specific image point known as the **image focal point**, F_i. The distance from the lens to this point is called the **image focal length**, f_i, where as $s_o \to \infty$, $s_i \to f_i$ as depicted in Fig. 26.8a. Similarly, the rays will emerge from the lens as a parallel bundle as $s_i \to \infty$ and the special object point for which this occurs is called the **object focal point**, F_o. The distance from the lens to this point is called the **object focal length**, f_o, where as $s_i \to \infty$, $s_o \to f_o$ as depicted in Fig. 26.8b. A thin lens surrounded by the same medium on both sides is a special case for which the image and object focal lengths are the same and the subscripts can be dropped altogether. Accordingly, go back to the Lensmaker's Formula and let $s_o \to \infty$, whereupon $(1/s_o) \to 0$ while $s_i \to f$, and we get

$$\frac{1}{f} = (n_l - 1)\left(\frac{1}{R_1} - \frac{1}{R_2}\right) \tag{26.7}$$

and the same thing results as $s_i \to \infty$. The focal length of a lens is determined by its physical makeup and can be positive or negative. In Figs. 26.8a and b, $R_1 > 0$ and $R_2 < 0$; hence, each focal length is positive. In Fig's. 26.9a and b, $R_1 < 0$ and $R_2 > 0$,

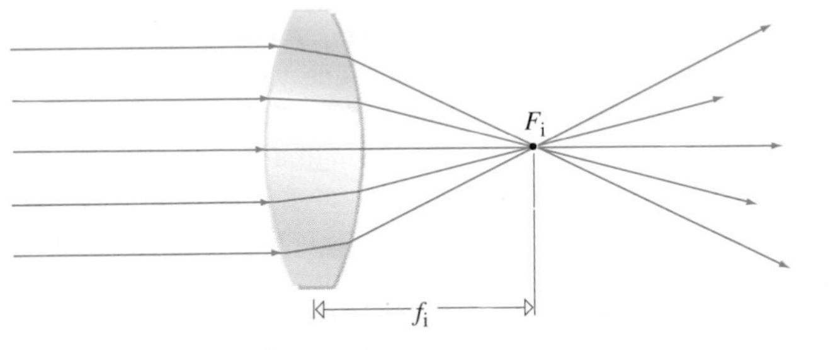

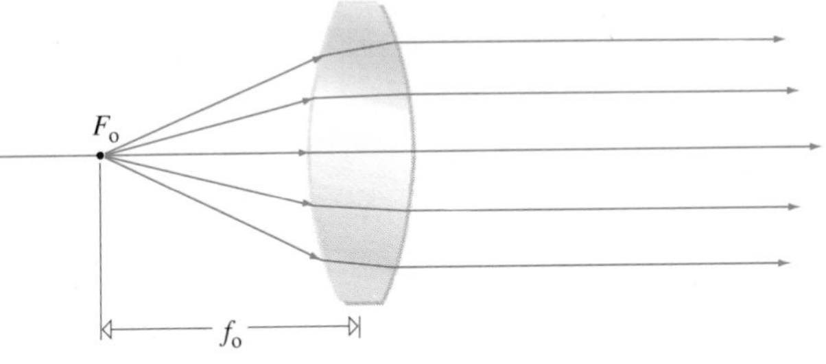

Figure 26.8 Focal points for a converging lens. (a) A parallel bundle of rays passing through a thin lens is brought to convergence at the image focal point F_i. (b) A point of light at the object focal point F_o emits light that emerges from the lens as a parallel beam.

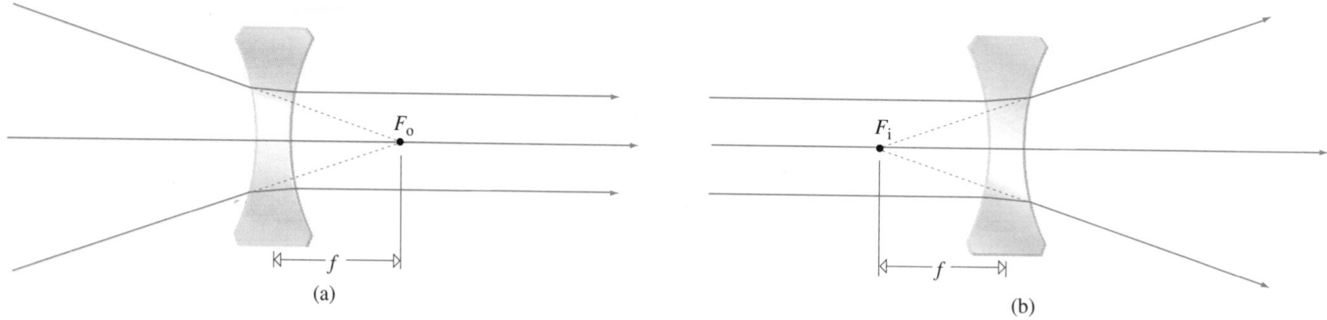

Figure 26.9 Focal points for a diverging lens. (a) Rays heading for F_o emerge parallel. (b) Rays entering parallel emerge as if from F_i.

and each focal length is negative. Going by the sign of its focal length, we commonly refer to a converging lens as a **positive lens** and a diverging one as a **negative lens.**

If for a bi-convex lens the radii are small (the lens has bulging faces), then the right side of the equation will be large, and the left side must also be large, meaning f must be small (the lens has a short focal length). The more "bent" the surfaces are, the more they bend the rays. On the contrary, when the surfaces are fairly flat (the radii are large), the focal length will be large. In fact, if both radii are equal and the lens is made of glass ($n_l \approx \frac{3}{2}$), it follows from Eq. (26.7) that $f \approx R$. For bi-convex and bi-concave lenses, f is ordinarily of the order-of-magnitude of the smaller radius, which is a good thing to keep in mind. This need not apply with meniscus lenses, where the magnitude of the focal length can be much greater than the radii.

Example 26.2 Determine the focal length in air of a thin spherical planar-convex lens having a radius of curvature of 50 mm and an index of 1.50. What, if anything, would happen to the focal length if the lens is placed in a tank of water?

Solution: [Given: the fact that the first surface is flat means that it has an infinite radius of curvature, $R_1 = \infty$; $R_2 = -0.050$ m and $n_l = 1.50$. Find: f when $n_m = 1.00$ and 1.33.] From Eq. (26.7),

$$\frac{1}{f} = (1.50 - 1)\left(\frac{1}{\infty} - \frac{1}{-0.050\ \text{m}}\right)$$

$1/\infty = 0$ and $\boxed{f = +0.10\ \text{m}}$. When the lens is surrounded by a medium of index n_m rather than air, n_l must be replaced by $n_l/n_m = 1.50/1.33$. The effect is to reduce the lens's ability to bring the rays into convergence and to increase the focal length to $+0.39$ m.

▶ **Quick Check:** f is positive as it should be for a convex lens. Moreover, $f = 2|R_2|$, which is also the right order-of-magnitude.

It is especially convenient to draw a ray along the central axis of a lens because it strikes each surface perpendicularly and passes straight through undeviated (Fig. 26.10). We now examine a tilted off-axis ray that enters the lens and emerges parallel to the incident direction. Such a ray must pass through a fixed point on the axis known as the **optical center** of the lens, O. It actually enters, bends a little, and emerges parallel to its incident direction. But because the lens is thin, the lateral displacement of the emerging ray is negligible. Thus, we can assume that **any paraxial ray heading toward the center of any thin lens will pass through O undeviated and may be drawn as a straight line.** It is customary when treating a thin lens to simply place the point O at the geometric center of the lens (Table 26.1).

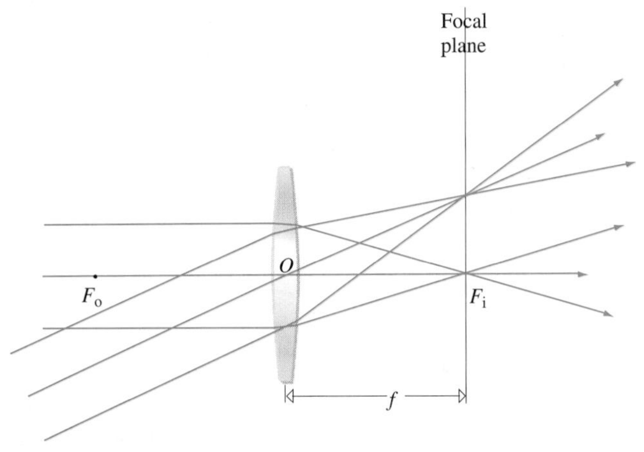

Figure 26.10 A ray headed for the center O of a thin lens passes straight through without bending. Ideally, all parallel bundles of rays focus on a plane, but actually only paraxial rays focus on that focal plane.

TABLE 26.1 Sign convention for spherical refracting surfaces and thin lenses*

s_o, f_o	+ left of O
s_i, f_i	+ right of O
R	+ if C is right of O
y_o, y_i	+ above optical axis

*Light enters from the left.

A bundle of parallel rays impinging on the lens at some angle with the central axis will be focused at a point on the central ray, the ray passing through O (Fig. 26.10). All such convergence points really lie on a curved surface, but if we limit the discussion to paraxial rays, that surface closely resembles a plane perpendicular to the central axis at the focal point. This is the **focal plane**, located a distance f from the lens, given by Eq. (26.7).

26.4 Finite Imagery: Lenses

Any point on an extended object sends out light in a great many directions, and if some of that light enters a positive lens, it can be made to converge to an image point. In that way, point-by-point, the image is formed somewhere in what is called the *image space.* Wherever two rays from an object point are made to converge, all rays from that point, passing through the lens, ideally converge. *Find where any two rays from any object point cross, and you have found the corresponding image point.* And that is particularly easy to do using any two of the three special rays whose behavior we already know (Fig. 26.11). **Ray 1** heading for the center of any type of thin lens goes straight through. **Ray 2** entering a positive lens parallel to the central axis emerges passing through the image focus; a similar ray entering a negative lens emerges so that it can be extended back to pass through the image focus. **Ray 3** passing through the object focus of a positive lens emerges from the lens parallel to the central axis; a ray heading for the object focus of a negative lens emerges parallel to the central axis.

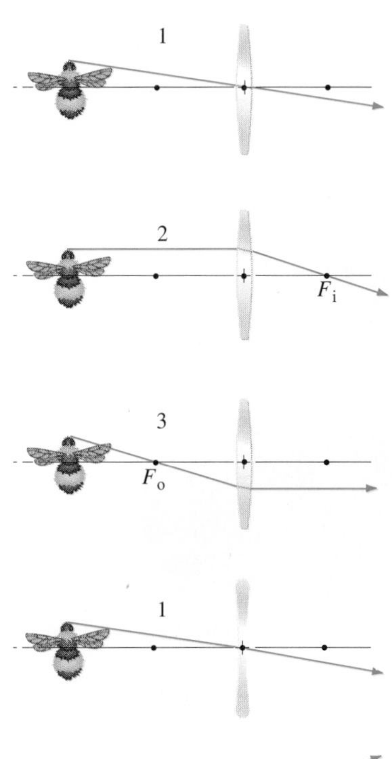

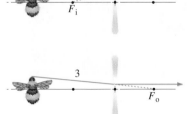

Figure 26.11 Tracing a few key rays through a positive and negative lens.

Since we know where these rays are going, we can simplify the drawings by constructing ray-paths with a single refraction taking place on a vertical line through the center of the lens. There really are two refractions, one at each face, but this way of proceeding will save us a lot of effort. Begin a ray diagram by setting down a horizontal axis and then sketching in a centered lens of arbitrary size and roughly the right shape. *The most important construction feature of the diagram is the location of the object and image focal points, which are at equal distances (f) on each side of the lens.* The vertical center line, the focal length, and the location of the object determine the entire geometry. As a rule, when not drawing everything to scale, make the lens about the same size as the focal length. Should you inadver-

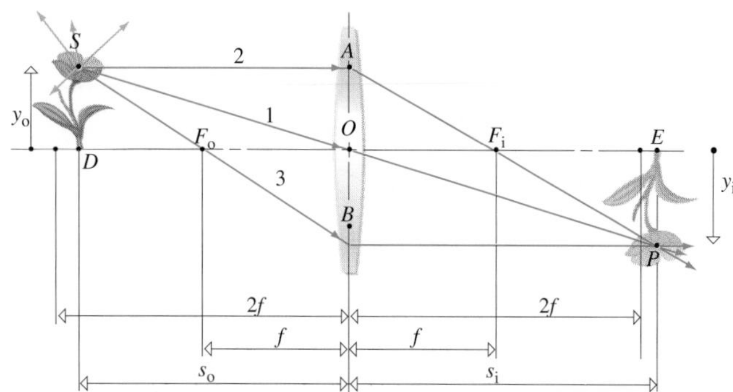

Figure 26.12 The geometry of image formation via a thin convex lens. For simplicity, refractions take place at the *A-O-B* plane.

tantly draw a lens too small to accept a particular ray, it can still properly be drawn refracted from the vertical line as if the lens was pictured appropriately. At a distance of f on each side of the lens, put a mark on the axis to represent the focal points and then draw in another pair of points at distances of $2f$. These four points divide the object and image spaces into four distinctive regions.

Figure 26.12 depicts an object at some distance between f and $2f$ in front of a positive lens. To locate the resulting image, trace through the lens any two rays from each and every point on the flower—practically, a point on the very top and one on the bottom will do. From the top most object point S, draw ray 1 straight through the center O of the lens. Now, we can either draw ray 2 or ray 3. If you like, draw all three; they, and indeed *all* rays entering the lens from that same object point, converge at the same conjugate image point P—that locates the top of the flower's image. The bottom of the image E can be found by doing the same thing over again for the bottom-most object point.

A real image projected on the viewing screen of a 35-mm camera, much as the eye projects its image on the retina. Here a prism has been removed so you can see the image directly.

Here, **the rays converge toward the image**, and we say that the image is **real**. The light actually arrives at each such point and if there is a viewing screen there (as in a movie theater), the image would appear on it, which is why it's called a *real* image. By comparison, when an image is formed of light diverging from it (as was the case with the plane mirror, p. 871), we say that the image is **virtual**—it cannot be projected onto a screen, but it certainly can be viewed directly. In much the same way, we refer to an object point as *real* when rays diverge from it and *virtual* when rays appear to converge toward it; the former situation is by far the most common, though the latter can occur (Fig. 26.9a).

The rays in Fig. 26.12 don't just end on the image, they go on past it to diverge again. Moving the viewing screen toward or away from the lens intercepts a more or less blurred version of the flower, whose sharpest image exists at the one location in space where the rays converge to points. Unless the rays are intercepted (for example, by a film-plate), nothing of the image can be seen from the side. There is no transverse propagation—the image is not visible laterally as the flower itself *is*. The lens forms an image of the flower, but it does not recreate the original lightwave coming from the flower; something is lost in the transformation (p. 969).

The ray diagram can provide us with an analytic relationship between the object and image distances and the focal length. In Fig. 26.12, triangles F_iEP and F_iAO are similar, all their angles are equal, and so

$$\frac{\overline{PE}}{\overline{AO}} = \frac{s_i - f}{f}$$

wherein $\overline{AO} = \overline{SD}$. Triangles SOD and POE are also similar and therefore

$$\frac{\overline{PE}}{\overline{SD}} = \frac{s_i}{s_o} \tag{26.8}$$

Setting these two equations equal and rearranging terms, we get

$$\frac{1}{s_o} + \frac{1}{s_i} = \frac{1}{f} \tag{26.9}$$

which is the famous **Gaussian Lens Equation** first derived by C. F. Gauss in 1841.

Now that we see how Eq. (26.9) follows from the geometry of the lens and rays, it should be noted that the same expression results on comparing Eqs. (26.6) and (26.7).

Example 26.3 We wish to place an object 45 cm in front of a lens and have its image appear on a screen 90 cm behind the lens. What must be the focal length of the appropriate positive lens?

Solution: [Given: $s_o = 0.45$ m and $s_i = 0.90$ m. Find: f.] The problem does not mention the n or the radii of the lens—all of that is summarized in f, and so the analysis proceeds from Eq. (26.9), as

$$\frac{1}{0.45\text{ m}} + \frac{1}{0.90\text{ m}} = \frac{1}{f}$$

Using a calculator, this can easily be evaluated term by term with the $(1/x)$ function

$$2.222 + 1.111 = \frac{1}{f}$$

and $\boxed{f = +0.30\text{ m}}$

▶ **Quick Check:** Working without a calculator, recognize that Eq. (26.9) has the same form as that for two resistors adding in parallel; hence

$$f = \frac{s_i s_o}{s_i + s_o} = \frac{(0.45\text{ m})(0.90\text{ m})}{1.35\text{ m}} \tag{26.10}$$

and $f = +0.30$ m.

26.5 Magnification

The ratio of any transverse dimension of the image formed by an optical system to the corresponding dimension of the object is defined as the **transverse magnification** or, more often, simply the *magnification* M_T. In Fig. 26.12, the magnification is the height of the image divided by the height of the object:

$$M_T = \frac{y_i}{y_o} \tag{26.11}$$

Here y_i is below the central axis and is traditionally taken to be a negative number. The image is upside down, meaning that **the magnification is negative whenever the image is inverted and positive when the image is right-side-up**. Bear in mind that the magnification refers to the ratio of image size to object size and need not only correspond to enlargement ($|M_T| > 1$). The image can certainly be minified ($|M_T| < 1$) or life-size ($|M_T| = 1$) as well (Table 26.2).

TABLE 26.2 Meanings associated with the signs of various thin lens parameters

Quantity	Sign +	Sign −
s_o	Real object	Virtual object
s_i	Real image	Virtual image
f	Converging lens	Diverging lens
y_o	Right-side-up object	Inverted object
y_i	Right-side-up image	Inverted image
M_T	Right-side-up image	Inverted image

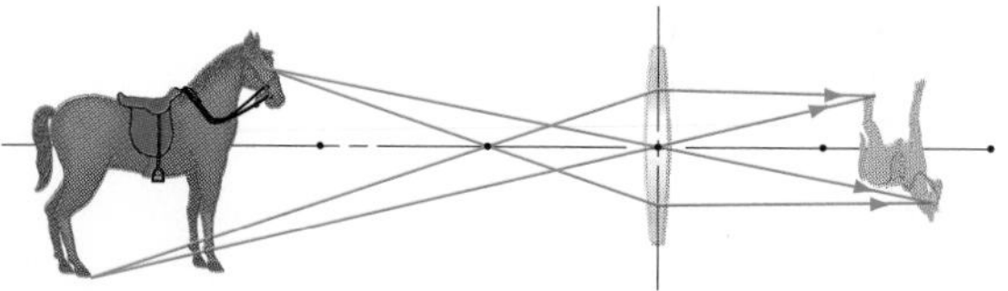

Figure 26.13 The image of a 3-dimensional object is itself 3-dimensional. The image exists in space and any portion of it can be viewed on a screen even though it cannot be seen from the side as drawn here.

From the similar triangles in Fig. 26.12, we got Eq. (26.8), which provides a convenient alternative statement for the magnification:

$$M_T = -\frac{s_i}{s_o} \tag{26.12}$$

The minus sign here is necessary because both the object and image distances are positive and yet the image is inverted; that is, the magnification is negative.

It's easy to forget that the image exists in 3-dimensional space. As a reminder, Fig. 26.13 depicts a horse standing in front of a very big lens (a small one would do, but the ray diagram would be awkward to draw). The resulting minified image is real, and it exists in an extended region of space (although we generally view it in transverse slices on a flat screen).

Example 26.4 The horse in Fig. 26.13 is 2.25 m tall, and it stands with its face 15.0 m from the plane of the thin lens, whose focal length is 3.00 m. (a) Determine the location of the image of the equine nose. (b) What is the magnification? (c) How tall is the image? (d) If the horse's tail is 17.5 m from the lens, how long—nose-to-tail—is the image of the beast?

Solution: [Given: (a) $s_o = 15.0$ m, $f = +3.00$ m, $y_o = +2.25$ m; (d) $s_o = 17.5$ m. Find: s_i, front and rear; M_T; image height; and describe the image.] (a) From the Gaussian Lens Equation (26.9), we obtain

$$\frac{1}{15.0 \text{ m}} + \frac{1}{s_i} = \frac{1}{3.00 \text{ m}}$$

and $\boxed{s_i = +3.75 \text{ m}}$. (b) Computing the magnification from Eq. (26.12), we have

$$M_T = -\frac{s_i}{s_o} = -\frac{3.75 \text{ m}}{15.0 \text{ m}} = \boxed{-0.25}$$

(c) From the definition of magnification, Eq. (26.11), it follows that

$$y_i = M_T y_o = (-0.25)(2.25 \text{ m}) = \boxed{-0.563 \text{ m}}$$

where the minus sign tells us that the image is inverted. (d) Again from the Gaussian Equation, for the tail

$$\frac{1}{17.5 \text{ m}} + \frac{1}{s_i} = \frac{1}{3.00 \text{ m}}$$

and $s_i = +3.62$ m. The entire equine image is only $\boxed{0.13 \text{ m}}$ long.

▶ **Quick Check:** Because the image-distance is positive, the image is *real.* Because the magnification is negative, the image is *inverted,* and because the absolute value of the magnification is less than one the image is *minified.* Moreover, the image is between 1 and 2 focal lengths from the lens. All of which matches Fig. 26.13, where the object distance exceeds $2f$.

There are two important things to observe about the situation in Example 26.4 that apply to all real images formed by a positive lens. First, the image is evidently distorted, in that its length is reduced more than its height—the magnification transverse to the axis (M_T) is greater than the *longitudinal magnification* along the axial direction. This should not be surprising; the image of everything behind the horse, as far as the eye can see—out to infinity—must be compressed into the small space between the image-horse's rump and the lens (in fact, as we will see shortly, it occupies even less than that, ending at F_i).

The second important feature is that aside from being inverted, the image is oriented in an interesting way quite differently from that of the plane mirror. *The horse's nose, which is closer to the lens, is imaged farther away.* Were we to place a transverse observing screen far from the lens on the right and gradually move it to the left, closer in, we would first encounter the horse's face (with the saddle blurred). And then with the screen moved closer, the horse's head would be blurred, and the saddle would appear sharply "in focus." Nearest the lens, with horse and saddle fuzzy, only the tail might be clear.

26.6 A Single Lens

We are now in a position to understand the entire range of behavior of a single convex or concave lens. To that end, suppose that a distant point source sends out a cone of light that is intercepted by a positive lens (Fig. 26.14). If the source is at infinity (that is, so far away that it might just as well be infinity), rays coming from it entering the lens are essentially parallel and will be brought together at the focal point F_i. If the source point S_1 is closer, but still fairly far away, the cone of rays entering the lens is narrow, and the rays come in at shallow angles to the surface of the lens. Because the rays do not diverge greatly, the lens bends each one into convergence, and they arrive at point P_1. As the source moves closer, the entering rays diverge more and the resulting image point moves farther to the right. Finally, when the source point is at F_o, the rays are diverging so strongly that the lens can no longer bring them into convergence, and they emerge parallel to the central axis. Moving the source point closer results in rays that diverge so much on entering the lens that they still diverge on leaving. The image point is now virtual—*there are no real images of objects closer in than f.*

A positive lens operates with three distinct regions of object space. Suppose a man with an umbrella is standing near a tree somewhere in the most distant region, extending from ∞ to $2f$, as indicated in Fig. 26.15a. His real, inverted image will be formed on the right of the lens between f and $2f$; the farther away he is, the closer is his image to F_i (if he were at ∞, it would be at F_i). This situation corresponds to the way an eye or a camera works. The image on the retina is minified so that a panorama fits on the small screen; thus, we can see a whole oak tree at once rather than one acorn at a time. *As the man walks toward the lens, his image grows in size and slowly moves away from the lens.* This domain, from ∞ to $2f$, is the first region of object space. It ends for the man at a distance of two focal lengths. When he stands at $2f$, he is at the symmetry point of the system, where right and left sides (object and image) are identical. Figure 26.15b shows the situation: the triangles are now not just similar, they are congruent, and the real image is life-sized. This optical setup is that of a photocopy machine.

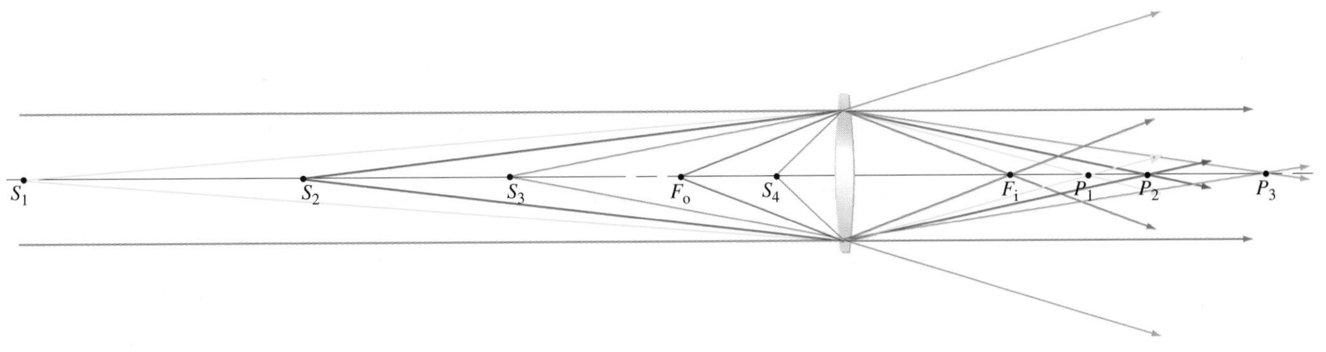

Figure 26.14 As the source moves closer, the rays diverge more and the image point moves out away from the lens. The emerging rays no longer converge once the object reaches the focal point; nearer in still, they diverge.

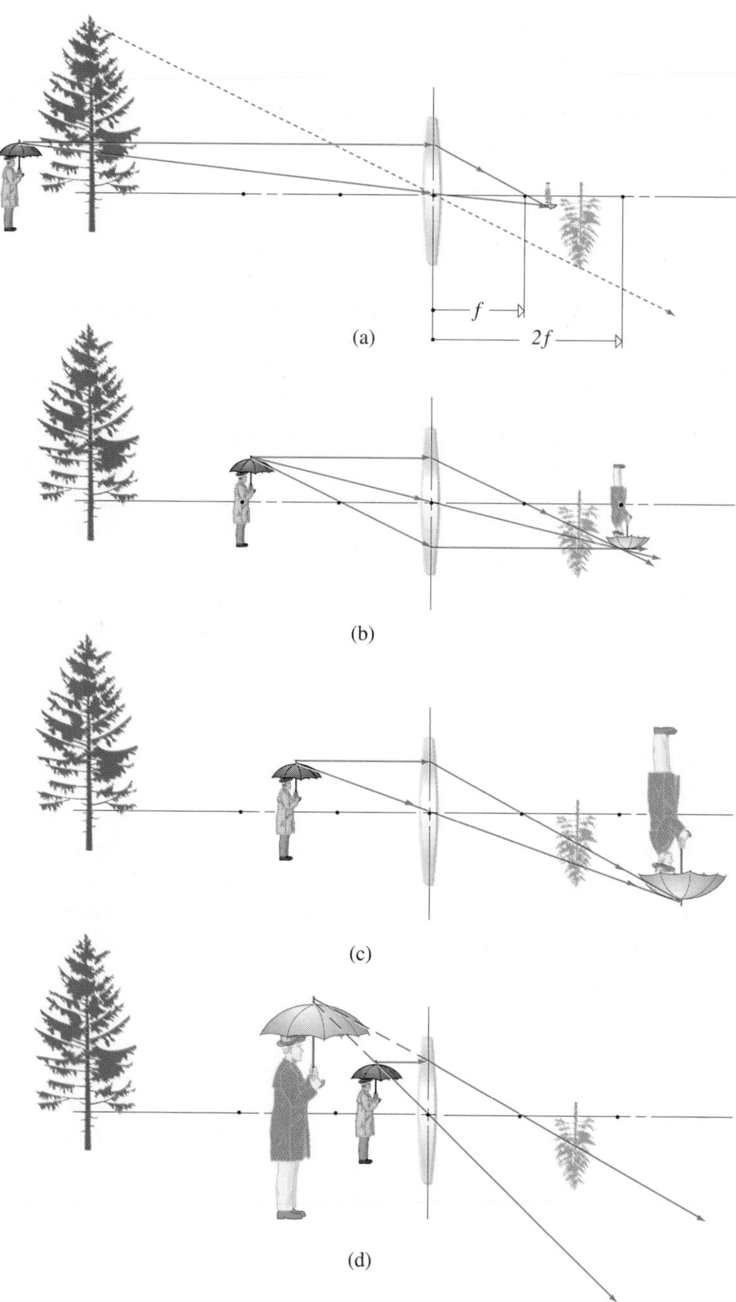

Figure 26.15 The operation of a thin positive lens.

As the man continues to walk toward the lens, moving from $2f$ toward f, his initially life-sized image gradually grows while it moves away from the lens, advancing to the right from $2f$ toward ∞. The second region of object space, from $2f$ to f, transforms into the image space from $2f$ to ∞. Notice that everything thus far has happened in a smooth, continuous way. As he walks in from very far away, his real image walks out; starting small, it reaches life-size at $2f$ and continues to grow, becoming immense as it moves away. This arrangement is that of a projector. The slide or film is the object located between $2f$ and f, each point of which will have a corresponding image point on the picture screen. Thereon will appear a real, enlarged, upside-down image of the illuminated portion of the slide. The projector is focused by just changing the slide-to-lens distance. And the inverted image is corrected by simply putting the slide or movie film in upside down.

It follows from Eq. (26.12) that the magnification approaches ∞ as the object nears f and s_i approaches ∞. When the object is exactly at the focal point, exactly at the boundary of the second region of object space, rays from each object point emerge from the lens parallel to one another. The image is no longer clearly observable on a screen, no matter how far away the screen is. Only a very large blur will be seen as if a real image were in focus somewhere far beyond the observing screen (namely, at ∞).

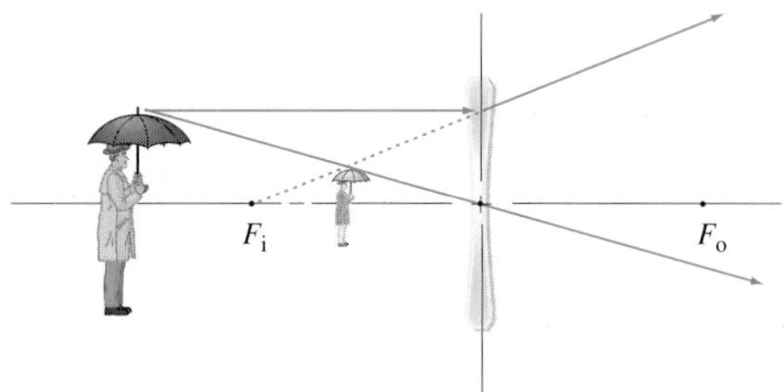

Figure 26.16 A concave lens forms a virtual, minified, right-side-up image.

When the little man crosses beyond f into the third region of object space nearest the lens, he is so close and the rays so strongly diverging, that no real image is possible. An eye or a camera looking into the lens could bring the emerging diverging rays together. What one would see is a virtual, right-side-up image much like the image formed by a plane mirror (Fig. 26.15d), except enlarged. With the object at f, the image is an immense blur, but if the little man leans back a bit, his image forms—large, real and inverted; if he leans forward, his image blurs, vanishes, and then reappears—large, virtual, and right-side-up. It flips over, but that discontinuity happens when he is at f and, reasonably enough, cannot be observed because the image is gone. As he walks toward the lens, his image (as seen through the lens) will diminish ($M_T > 1$) until his face is flat up against the lens and he appears life-size. A positive lens operating on the light from an object located in this third region is known as a **magnifying glass**.

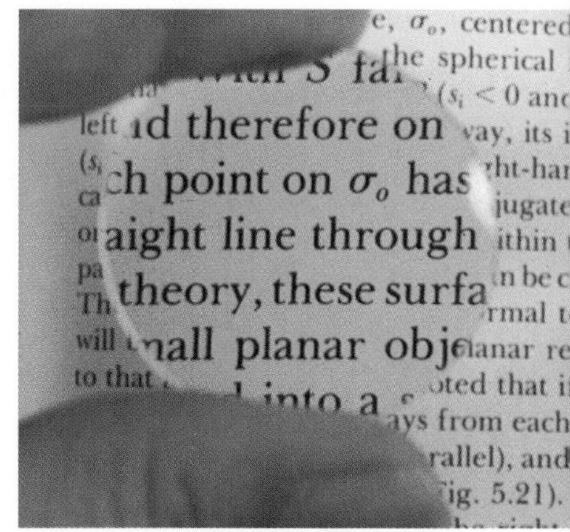

A positive lens serves as a magnifying glass when the object is closer to the lens than one focal length ($s_o < f$).

The concave lens (Fig. 26.16) operates in only one way and so is much easier to keep track of. It produces *only* virtual, right-side-up, minified images no matter where the object is located. Rays diverging from any object point are made to diverge even more. An object pressed up against the lens appears very nearly life-size, but one more distant is minified correspondingly. In all cases, the image distance is negative, and the image always appears on the left side of the lens—on the same side the light enters from. Table 26.3 summarizes these conclusions for both types of lens.

TABLE 26.3 Images of real objects formed by thin lenses

Convex

Object	Image			
Location	**Type**	**Location**	**Orientation**	**Relative size**
$\infty > s_o > 2f$	Real	$f < s_i < 2f$	Inverted	Minified
$s_o = 2f$	Real	$s_i = 2f$	Inverted	Same size
$f < s_o < 2f$	Real	$\infty > s_i > 2f$	Inverted	Magnified
$s_o = f$		$\pm\infty$		
$s_o < f$	Virtual	$\lvert s_i \rvert > s_o$	Right-side-up	Magnified

Concave

Object	Image			
Location	**Type**	**Location**	**Orientation**	**Relative size**
Anywhere	Virtual	$\lvert s_i \rvert < \lvert f \rvert$, $s_o > \lvert s_i \rvert$	Right-side-up	Minified

Example 26.5 A 5.00-cm-tall matchstick is standing 10 cm from a thin concave lens whose focal length is −30 cm. Determine the location and size of the image and describe it. Draw an appropriate ray diagram.

Solution: [Given: $y_o = 0.0500$ m, $s_o = 0.10$ m, and $f = -0.30$ m. Find: s_i and y_i.] From the Gaussian Lens Formula, Eq. (26.9),

$$\frac{1}{0.10\text{ m}} + \frac{1}{s_i} = \frac{1}{-0.30\text{ m}}$$

and $s_i = -1/13.3 = -0.075\text{ m} = \boxed{-7.5\text{ cm}}$. The image distance is negative, meaning it is to the left of the lens, and therefore the image is virtual. Using Eq. (26.12), we can compute the magnification and from that the size of the image via Eq. (26.11):

$$M_T = -\frac{s_i}{s_o} = -\frac{-0.075\text{ m}}{0.10\text{ m}} = +0.75$$

and so

$$y_i = M_T y_o = 0.75(0.05\text{ m}) = \boxed{0.038\text{ m}}$$

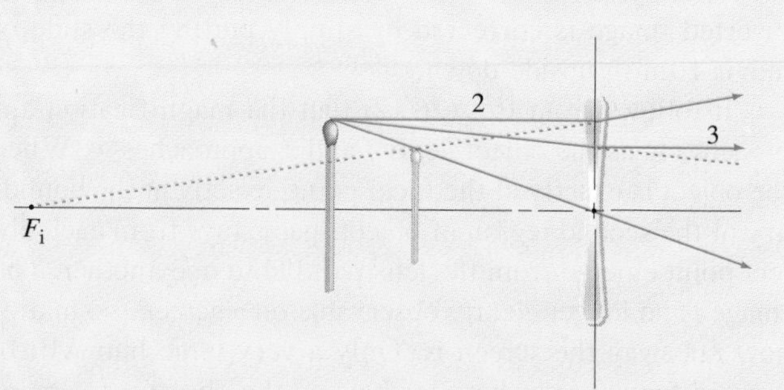

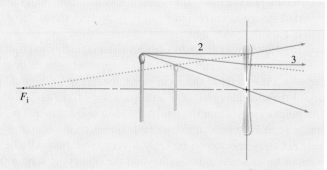

Figure 26.17 The image of a matchstick formed by a concave lens.

Figure 26.17 is the corresponding ray diagram. Notice how ray 3 heading for the object focus off to the right emerges parallel to the axis, as in Fig. 26.9a, while ray 2 entering parallel to the axis appears, on emerging, to be coming from the image focus.

▶ **Quick Check:** The image is right-side-up ($M_T > 0$), minified ($|M_T| < 1$), virtual ($s_i < 0$), and $s_o > |s_i|$, as Table 26.3 says it should be.

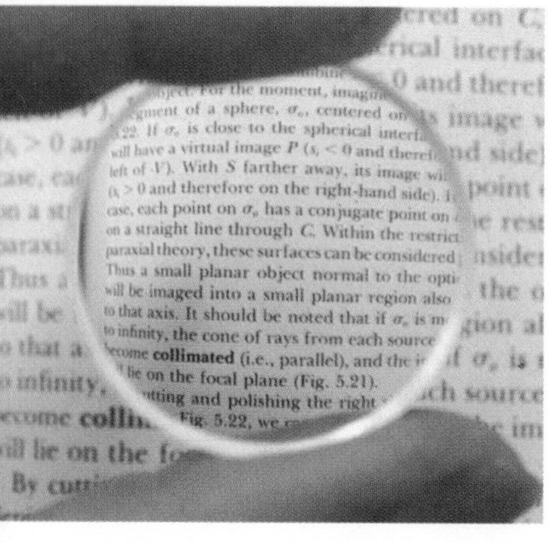

The minified, right-side-up, virtual image formed by a negative lens.

The Human Eye

The eye (Fig. 26.18) is an almost spherical (24 mm long by about 22 mm across) jellylike mass contained within a tough, flexible shell, the *sclera*. With the exception of the front portion or *cornea*, which is transparent, the sclera is white and opaque. The cornea is the first and strongest converging element of the eye system. Most of the bending impressed on a bundle of rays by the eye takes place at the air-cornea interface. The reason you can't see very well underwater ($n_w \approx 1.333$) is that the liquid's index is so close to that of the cornea ($n_C \approx 1.376$) that adequate refraction can no longer occur.

Having passed into the cornea, the light is only made slightly more convergent on emerging because it enters a chamber filled with a watery fluid known as the *aqueous humor* ($n_{ah} \approx 1.336$). Immersed in the aqueous is a variable diaphragm called the *iris*, which controls the amount of light entering the remaining portion of the eye by way of an aperature or *pupil*.

Just behind the iris is the *crystalline lens*. The lens (9 mm in diameter and 4 mm thick) is a complex, layered fibrous mass surrounded by an elastic membrane. In structure, it is somewhat like a small transparent onion, formed of roughly 22 000 very fine layers. As a whole, the lens is quite pliable, albeit less so with age. Its index of refraction varies from about 1.406 at the inner core to roughly 1.386 at the less dense cortex. The crystalline lens provides the needed fine-focusing mechanism via changes in its shape.

The cornea and crystalline lens can be treated as forming a double-element lens whose object focus is about 15.6 mm in front of the outer surface of the cornea and whose image focus is about 24.3 mm behind it on the retina. The combined lens has an optical center 17.1 mm in front of the retina, just at the rear edge of the crystalline lens. As a rule, $s_o > 2f$.

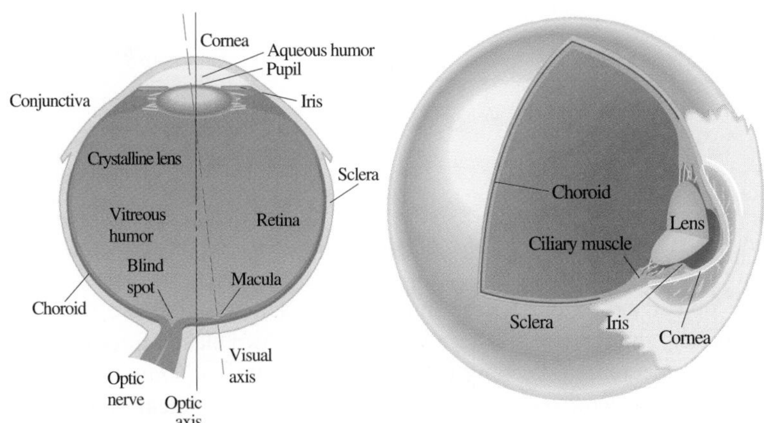

Figure 26.18 The human eye. The cornea is slightly flattened to control spherical aberration; the fact that rays from the edges of a lens usually do not focus at the same point as do rays near the center. Moreover, the spherical shape of the eyeball and retina eliminates problems associated with the fact that the image "plane" is not really flat. The retina contains 125 to 130 million photoreceptor cells.

Example 26.6 As we saw earlier (p. 241), the Moon subtends an angle of about 0.009 rad as seen from Earth. How big is its image formed by the human eye?

Solution: [Given: see Fig. 26.19. Find: y_i.] Since the Moon subtends an angle of 0.009 rad, it follows from Eq. (8.1) and the geometry that the image size is just $r\theta$ or, equivalently

$$|y_i| = s_i\theta = (17.1 \text{ mm})(0.009 \text{ rad}) = 0.15 \text{ mm}$$

and

$$\boxed{|y_i| = 0.2 \text{ mm}}$$

The absolute value is used because the image is inverted, and so y_i is negative by convention. The fact that the diameter is 0.15 mm means that the retinal image of the face of the Moon is a dot smaller than the size of a period on this page!

▶ **Quick Check:** $|y_i|/s_i \approx (15 \times 10^{-5} \text{ m})/(17 \times 10^{-3} \text{ m}) \approx 1 \times 10^{-2}$ rad.

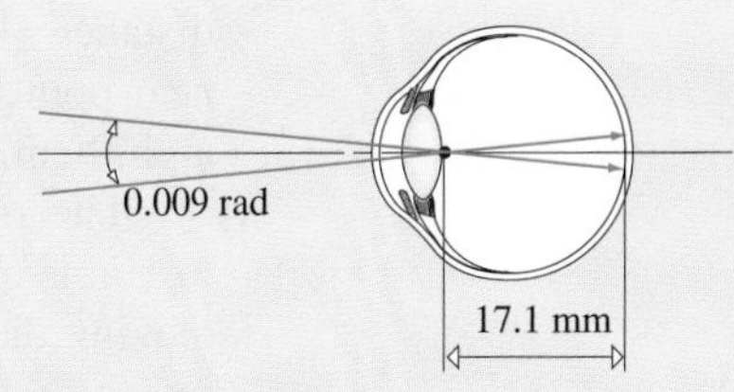

Figure 26.19 The angle subtended by the Moon on the retina.

Behind the lens is another chamber filled with a transparent gelatinous substance, the *vitreous humor* ($n_{vh} \approx 1.337$). A thin, delicate, transparent multilayer of cells (from 0.5 mm to 0.1 mm thick) covers about 65% of the interior surface of that chamber. This structure is the light-sensitive *retina* (from the Latin *rete*, meaning net). The retina is the transducer that converts electromagnetic energy impinging on it into electrical nerve impulses that can be processed by the brain.

Accommodation. The fine focusing or **accommodation** performed by the human eye is carried out by the crystalline lens. Since the image distance for the eye is fixed, the only way we can see things clearly at different object distances is if the focal length is changed. The lens is suspended by ligaments that are connected to a circular yoke of muscles. Ordinarily, these are relaxed and elongated, the aperture they encompass is large, and in that state they pull back on the network of fine fibers holding the rim of the lens. This draws the pliable lens into a fairly flat configuration, increasing its radii of curvature (especially of the anterior surface), which increases its focal length. With the muscles completely relaxed, the light from an object at infinity (which is practically speaking anywhere beyond about 5 m) is focused on the retina (Fig. 26.20). Not all eyes will do that well and so the **far-point**,

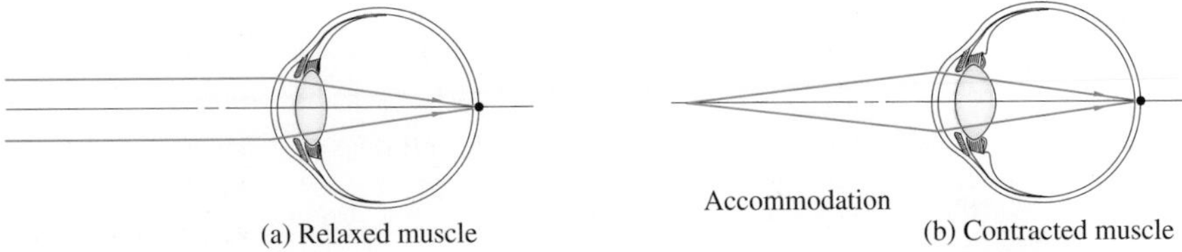

Figure 26.20 The human eye focuses on an object by contracting the ciliary muscles around the edge of the crystalline lens, which relaxes the tension on the lens so it contracts and bulges.

the point that is seen clearly by the unaccommodated eye, is frequently closer in than infinity (or even 5 m).

As the object moves closer, the muscles contract, the aperture encircling the lens gets smaller, and the lens bulges a little under its own elastic forces. In so doing, the focal length decreases, keeping the image on the retina. *The closest point that can be clearly seen with maximum accommodation is the* **near-point**. The lens doesn't change thickness by much more than about 0.5 mm over the whole range. A 10-year-old with a flexible lens may have a near-point as close as 7 cm, but that will typically move out to 12 cm by age 25, and 28 cm at around age 45—still a workable distance (11 inches) for reading without any inconvenience. But by 50 years, the near-point jumps to about 40 cm; by 60 years, it's out to 100 cm and by 70, it's at 400 cm.

The amount of accommodation required when an object is moved from infinity to 1 m is quite small; just enough to shift the image forward 0.06 mm. By contrast, moving the object from 1 m to 1/8 m requires enough accommodation to shift the image an additional 3.51 mm in order to keep it on the retina. Reading this page demands a disproportionate effort. Long study of things too close is one reason for eyestrain; we were apparently designed to look at, and out for, things afar.

The Camera

The prototype of the modern photographic camera is the *camera obscura,* the earliest form of which was simply a dark room containing a small hole in one wall. Light entering the aperture formed an inverted image of the sunlit outside scene on an inside screen (Fig. 26.21). By replacing the viewing screen with a photosensitive sur-

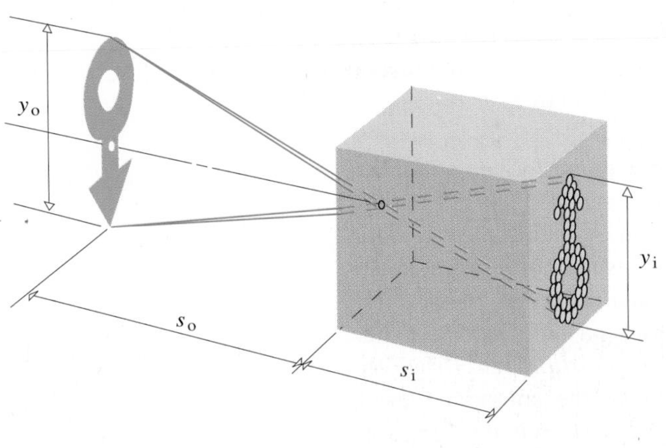

Figure 26.21 (a) Ideally, light entering the pinhole travels in a straight line, forming an image built up by countless little spots. (b) A pinhole camera photo (Science Building, Adelphi University). Hole diameter 0.5 mm, film plane distance 25 cm, A.S.A. 3000 shutter speed 0.25 s. Note depth of field.

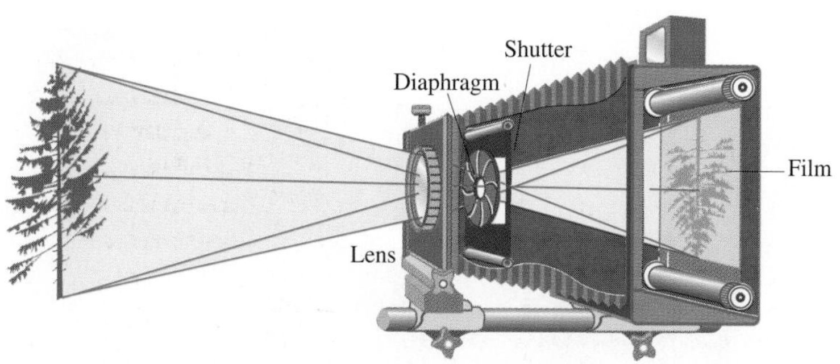

Figure 26.22 A large-format camera consists of a lens, followed by an adjustable diaphragm, a shutter that can open and close, and a sheet of film on which the image is formed.

face and filling the much-enlarged hole (which will let in a lot more light) with a lens, we have a photographic camera (Fig. 26.22), a TV camera, or a human eye, depending on the details.

Figure 26.23 depicts a typical camera lens with a focusing or distance scale that runs from ∞ to a fraction of a meter. When the camera is focused on infinity, the image distance takes on its smallest value; namely, f. At any lesser object distance, the image distance is greater than it was, and to keep the image on the film plane, the lens must be moved forward an amount δ such that $s_i - f = \delta$. From the lens equation, $s_i = s_o f/(s_o - f)$, and so $\delta = s_o f/(s_o - f) - f(s_o - f)/(s_o - f) = f^2/(s_o - f)$. For a camera lens where $s_o \gg f$, we have $\delta \approx f^2/s_o$.

The lens in the diagram has a focal length of 52 mm and to refocus it from infinity down to an object at, say, 10 m, the above equation tells us that the lens must be advanced a distance $\delta = 0.27$ mm. If it is moved another 0.27 mm from the film plane, it is now focused on an object at (10 m)/2 = 5 m; yet another movement of 0.27 mm sets it at (10 m)/3 = 3.3 m away, and so on. Rounded off, these are the numbers on the distance scale marked on the barrel of the lens.

Light enters the camera by way of a hole, and the size of that aperture and the duration over which it stays open together determine the amount of light allowed in. The time interval during which light is admitted is controlled with a shutter. The shutter is traditionally marked with a selection of settings designated 1000,

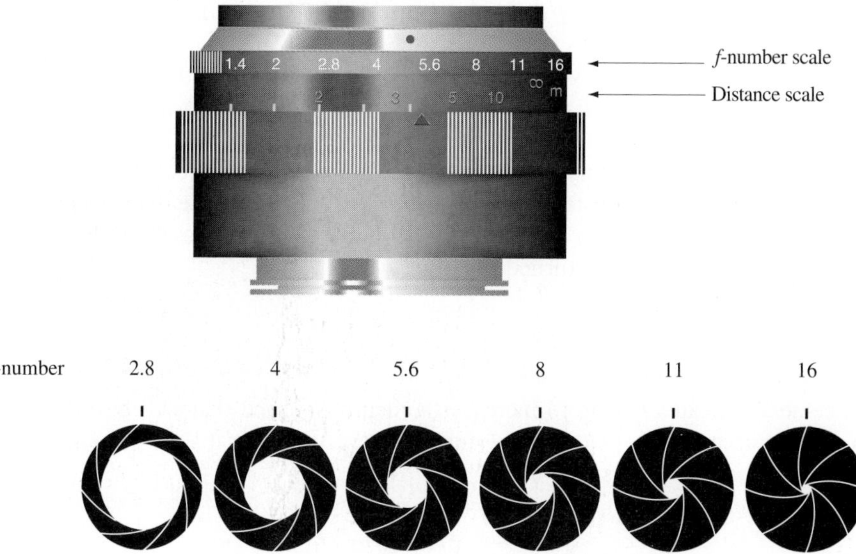

Figure 26.23 A camera lens showing possible settings of the variable diaphragm that is usually located within the lens.

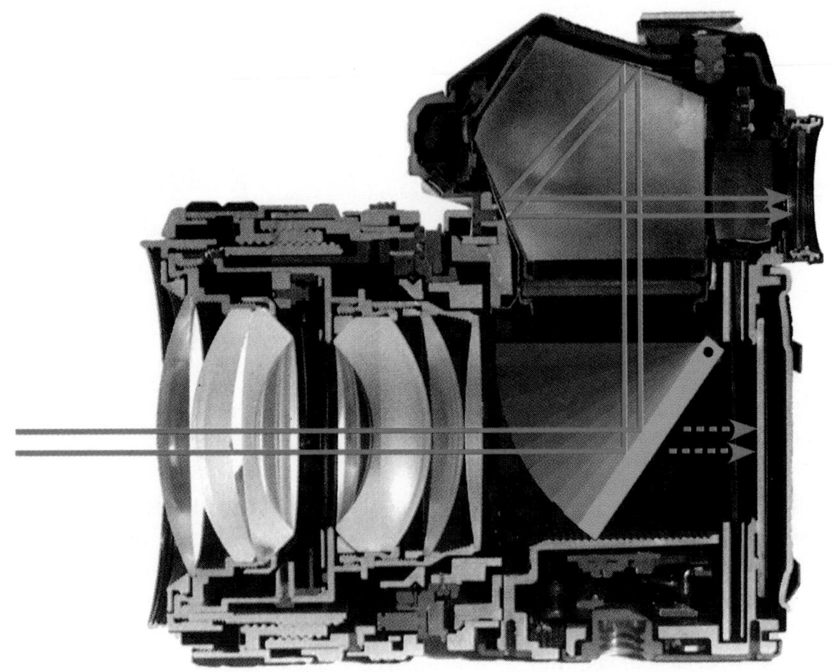

A single-lens reflex camera. Light from the lens hits the mirror and goes up to the prism and out to the eye. When the shutter is released the mirror pops up, the light goes directly to the film, and then the mirror pops back down.

500, 250, 125, 60, 30, 15, 8, 4, 2, 1. These are fractional open-time intervals to be understood as 1/1000 s, 1/500 s, etc. up to 1 s, *each being twice as long as the one before it.*

The second light-control mechanism is a variable diaphragm, like the pupil of the eye (Fig. 26.23). The amount of light entering the camera from a broad source is proportional to the area of the open aperture, and that is proportional to its diameter squared (D^2). The concentration of light reaching the back of the camera, the *energy per unit area,* depends inversely on the area of the image (each side of which is dependent on the focal length) and, hence, is inversely proportional to f^2. The energy density on the image plane therefore goes as $(D/f)^2$. The ratio D/f is the *relative aperture,* whereas its inverse, f/D, is the ***f*-number**. *The smaller the f-number, the more light reaches the film.*

Since the energy density depends inversely on the square of the *f*-number, multiplying the *f*-number by $\sqrt{2} = 1.4$ *halves the amount of light reaching the film.* The seemingly strange values marked on the lens in Fig. 26.23 (1.4, 2, 2.8, 4, 5.6, 8, 11, 16, etc.) are consecutive *f*-numbers where each *stop,* as they are called, is a multiple of $\sqrt{2}$. As the lens is adjusted from $f/1.4$ to $f/2$ to $f/2.8$, etc., the diaphragm within it is set to close down by an amount that will halve the light passed by the previous opening (or double it if it's turned the other way).

Example 26.7 The lightmeter on a camera indicates that the proper amount of energy to expose a certain film will be delivered when the aperture is set at $f/2.8$ with a shutter speed of 1/120 s. The photographer wishing to increase the depth in space that will be in focus sets the aperture at $f/8$ instead. What should the shutter speed be?

(continued)

(continued)

Solution: [Given: *f*/2.8 and 1/120 s to pass same energy as *f*/8. Find: corresponding shutter speed.] *f*/8 is three stops down from *f*/2.8; that is, in going to *f*/8, the energy density will be halved three times. Accordingly, the energy must be doubled three times via the shutter if the total amount reaching the film is to be maintained. Rather than taking the photo at 1/120 s, the shutter time should be doubled three times—it should be left open eight times longer, that is, for $\boxed{1/15\ \text{s}}$.

The Magnifying Glass

To examine an object in detail, you simply bring it nearer the eye so that the retinal image increases in size. That procedure can continue until the object is at the near-point, beyond which the eye can no longer provide adequate accommodation because the rays diverge too much (Fig. 26.14). To enlarge the object further, a single positive lens can be used to add convergence to the visual system, allowing the object to be brought still closer. A lens so used is a *magnifying glass.* Its function is *to provide an image of a nearby object that is larger than that seen by the unaided eye.* It would be nice to have a right-side-up, magnified image where the rays entering the eye are not converging, and that's satisfied by placing the object within one focal length of the positive lens.

To deal with how large an object appears in some optical device, we must consider the size of its retinal image. Consequently, the *magnifying power,* or **angular magnification** M_A, of an instrument is defined as *the ratio of the size of the retinal image formed by the device to the size of the retinal image formed by the unaided eye at normal viewing distance.* The latter is taken as the distance to the near-point d_n. In Fig. 26.24, M_A is equivalent to the ratio of the angles α_a (aided) and α_u (unaided):

$$M_A = \frac{\alpha_a}{\alpha_u} \tag{26.13}$$

Being restricted to the paraxial region, $\tan \alpha_a = y_i/L$ is very small and therefore approximately equal to α_a while $\tan \alpha_u = y_o/d_n \approx \alpha_u$. Since $-s_i/s_o = y_i/y_o$

$$M_A = \frac{y_i d_n}{y_o L} = -\frac{s_i d_n}{s_o L}$$

wherein both y_i and y_o are above the axis and positive. Taking all the previously unspecified distances in the diagram such as l, d_n, and L to be positive as indicated makes M_A positive as well. In the most commonly encountered application of the

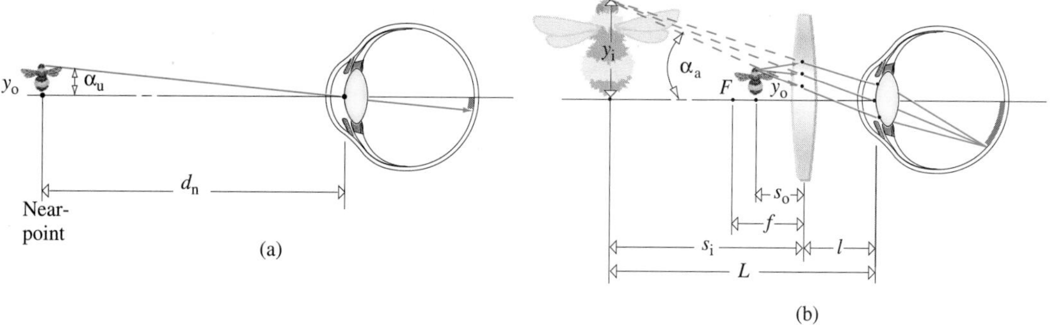

Figure 26.24 A magnifying glass. (a) An object is examined directly by placing it at the near-point. The retinal image is then as large as possible. (b) With a magnifying glass, the same object results in a much larger retinal image.

magnifying glass, the object is located at the focal point of the lens. In that case, the image is at infinity ($-s_i \approx L = \infty$; $s_o = f$) and

$$M_A = \frac{d_n}{f} \tag{26.14}$$

for all practical values of l. Rays emerging from the lens are parallel and can be viewed with a relaxed eye, which is a very important practical consideration.

Single lens magnifiers are usually limited by their aberrations to powers of about 2× or 3×. The famous Sherlock Holmes reading glass is an example of the type. More complicated multi-element magnifiers can be made in the range from 10× to 20×, and these are also used in microscopes and telescopes.

26.7 Thin-Lens Combinations

Optical systems are usually composed of several lenses and so we now examine a procedure for treating such arrangements. Figure 26.25 depicts two positive thin lenses L_1 and L_2 separated by a distance d, which here happens to be smaller than the focal length of either lens. The resulting image can be determined graphically by ray-tracing, using the following procedure. Imagine that L_2 is no longer there and

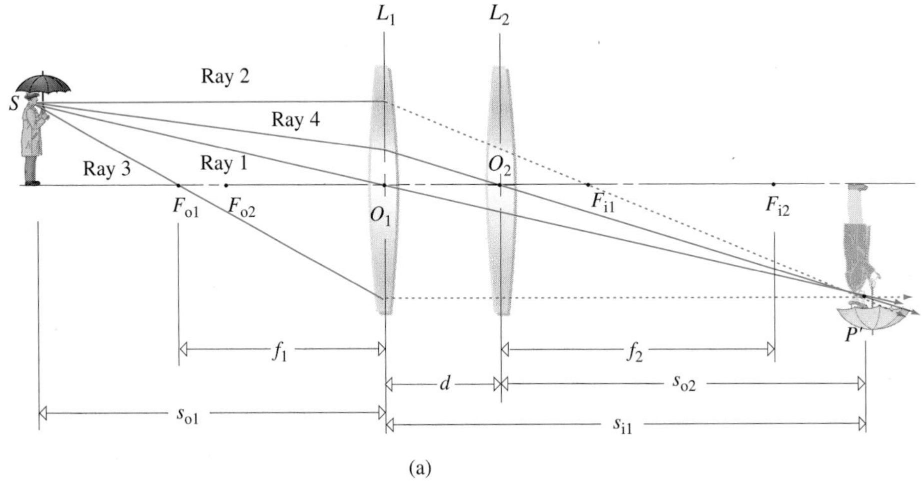

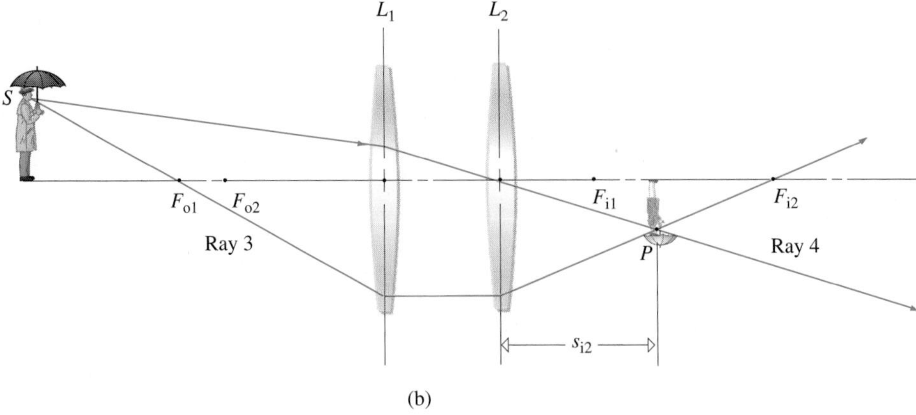

Figure 26.25 The image formed by two thin lenses. In (a) we imagine that lens L_2 is not there and we locate the image formed by L_1. Ray 4 is found passing through point O_2. It will be unchanged when lens L_2 is put back (b) and, along with ray 3, it locates the final image.

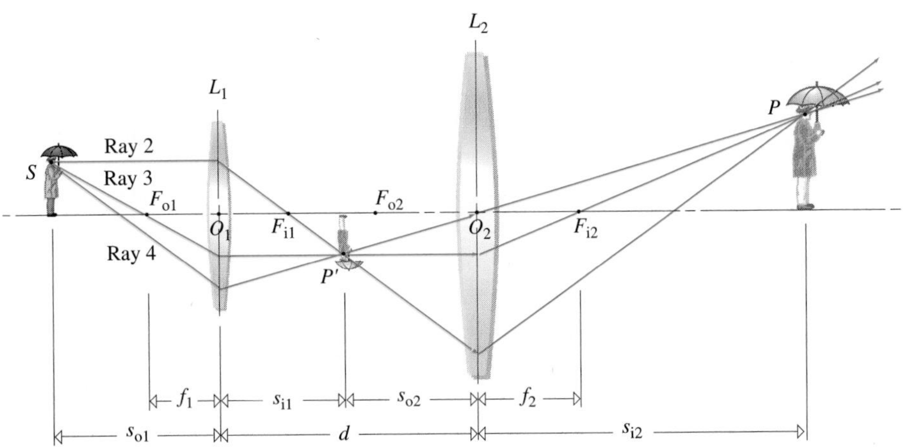

Figure 26.26 Two thin lenses separated by a distance $d > (f_1 + f_2)$. Now a real intermediate image is formed at P', and rays from it can be traced through L_2 to produce the final image.

find the image formed exclusively by L_1. This is easily done with ray 1 drawn through the center of L_1 and ray 2 entering parallel to the axis. These two rays fix point P' and the *intermediate image.* If L_2 was in place, a ray through its center would be undeviated; hence, construct ray 4 running backwards from P' through O_2 to L_1 and back to S. This ray is one of the two crucial rays that will locate the final image, since it passes properly through both lenses. The second necessary ray is just ray 3 through F_{o1}, which emerges from L_1 parallel to the central axis and therefore passes through F_{i2} on leaving the second lens. The point of intersection of these two rays determines the final image P of S. Notice that the intermediate image created by L_1 falls to the right of L_2 and can be thought of as a virtual object for L_2, which then goes on to create the final image.

Another example of the method is shown in Fig. 26.26 wherein the lenses are farther apart; in fact, $d > (f_1 + f_2)$. Again ray 1 and ray 2 fix the intermediate image at P'. And again ray 4 passing through P' and O_2 is drawn backwards through L_1 to S. This ray and ray 3 once more locate the final image P. Here, the intermediate image is real, and we could have proceeded by simply treating it as the real object for the second lens, forgetting about the first lens. Any two convenient rays from P' through the second lens will suffice to locate the final image.

The system can be treated analytically in much the same way—that is, by finding the intermediate image formed by the first lens and letting that function as the object for the second lens.

Example 26.8 Two positive lenses with focal lengths of 0.30 m and 0.50 m are separated by 0.20 m, as in Fig. 26.25. A red jellybean rests on the central axis 0.50 m in front of the first lens. Locate the resulting image with respect to the second lens.

Solution: [Given: $f_1 = +0.30$ m, $f_2 = +0.50$ m, $d = 0.20$ m, and $s_{o1} = 0.50$ m. Find: s_{i2}.] Applying the lens formula to L_1 only yields

$$\frac{1}{s_{i1}} = \frac{1}{f_1} - \frac{1}{s_{o1}} = \frac{1}{0.30\text{ m}} - \frac{1}{0.50\text{ m}} \qquad (26.15)$$

and $s_{i1} = 0.75$ m. The intermediate image falls $(0.75\text{ m} - 0.20\text{ m}) = 0.55$ m to the right of the second lens. This value, then, is the virtual object distance for L_2, and because it is to the right, it is negative ($s_{o2} = -0.55$ m). Again, the lens formula for the second lens with this object yields

(continued)

(continued)

$$\frac{1}{s_{i2}} = \frac{1}{f_2} - \frac{1}{s_{o2}} = \frac{1}{0.50\text{ m}} - \frac{1}{-0.55\text{ m}} \qquad (26.16)$$

and $\boxed{s_{i2} = +0.26\text{ m}}$. The image is real and to the right of the last lens.

▶ **Quick Check:** The second lens adds more convergence, pulling the image in closer to the lens as if the object was farther away than it is. The image is real, inverted, and minified. See Example 26.9.

Example 26.9 Considering two thin lenses separated by a distance d, derive an expression for s_{i2} in terms of s_{o1}, d, and the focal lengths.

Solution: [Given: Fig. 26.25. Find: s_{i2}.] For L_1, using Eq. (26.15)

$$s_{i1} = \frac{s_{o1}f_1}{s_{o1} - f_1} \qquad (26.17)$$

From Eq. (26.16), for L_2

$$s_{i2} = \frac{s_{o2}f_2}{s_{o2} - f_2}$$

Substituting in the fact that $s_{o2} = d - s_{i1}$, s_{i2} becomes

$$s_{i2} = \frac{(d - s_{i1})f_2}{(d - s_{i1} - f_2)}$$

Getting rid of s_{i1} using Eq. (26.17), we have

$$\boxed{s_{i2} = \frac{f_2 d - [f_2 s_{o1} f_1/(s_{o1} - f_1)]}{d - f_2 - [s_{o1} f_1/(s_{o1} - f_1)]}} \qquad (26.18)$$

▶ **Quick Check:** Using the numbers from Example 26.8

$$s_{i2} = \frac{(0.50\text{ m})(0.20\text{ m}) - (0.50\text{ m})(0.50\text{ m})(0.30\text{ m})/(0.50\text{ m} - 0.30\text{ m})}{0.20\text{ m} - 0.50\text{ m} - (0.50\text{ m})(0.30\text{ m})/(0.50\text{ m} - 0.30\text{ m})}$$

and $s_{i2} = (-0.275\text{ m}^2)/(-1.05\text{ m}) = +0.26\text{ m}$.

Suppose that the individual lenses are now brought close enough to touch one another, as is often done in compound systems. When $d = 0$, in Eq. (26.18) we can find the focal length of the combination by letting $s_{o1} \to \infty$, whereupon $s_{i2} = f$. The terms with d vanish, $(s_{o1} - f) \to s_{o1}$ and

$$f = \frac{f_1 f_2}{f_1 + f_2}$$

or

$$\frac{1}{f} = \frac{1}{f_1} + \frac{1}{f_2} \qquad (26.19)$$

The focal lengths add like resistors in parallel.

Eyeglasses

Spectacles were probably invented some time in the late thirteenth century, possibly in Italy or China. A Florentine manuscript (1299), which no longer exists, spoke of "spectacles recently invented for the convenience of old men whose sight has begun to fail." In 1804, Wollaston, recognizing that traditional (fairly flat bi-convex and concave) eyeglasses provided good vision only while looking through their centers, patented a new deeply curved lens. These were the forerunners of modern meniscus lenses that allow the turning eyeball to see through them from center to margin without significant distortion.

It is customary in physiological optics to speak about the **dioptric power** $\mathcal{D}$ of a lens, which is simply the *reciprocal of the focal length.* A lens possesses great power when it strongly bends rays, which happens when it has a *short* focal length. Power has the units of inverse meters, or *diopters* (D): 1 m^{-1} = 1 D. For instance, a converging lens with a focal length of +10 m has a power of 0.10 D, while a diverging lens with a focal length of −2 m has a power of $-\frac{1}{2}$ D. It follows from Eq. (26.7) that

$$\mathcal{D} = (n_1 - 1)\left(\frac{1}{R_1} - \frac{1}{R_2}\right) \tag{26.20}$$

The combined focal length of two lenses in contact is given by Eq. (26.19), and so their total power is the sum of the individual powers

$$\mathcal{D} = \mathcal{D}_1 + \mathcal{D}_2 \tag{26.21}$$

Example 26.10 Two lenses with focal lengths of +0.100 m and −0.333 m are held close together on a common centerline. Compute both the focal length and the power of the combination.

Solution: [Given: $f_1 = +0.100$ m and $f_2 = -0.333$ m. Find: f and $\mathcal{D}$.] The focal length is obtained from Eq. (26.19), as

$$\frac{1}{f} = \frac{1}{f_1} + \frac{1}{f_2} = \frac{1}{+0.100\text{ m}} + \frac{1}{-0.333\text{ m}}$$

and $\boxed{f = 0.143\text{ m}}$. Equation (26.21) provides the power; thus

$$\mathcal{D} = \mathcal{D}_1 + \mathcal{D}_2 = (10.0\text{ D}) + (-3.0\text{ D}) = \boxed{7.0\text{ D}}$$

▶ **Quick Check:** $1/f = 1/(0.143\text{ m}) = 7.0$ D. A negative lens combined with a stronger positive lens yields a positive lens.

The human eye has a total power of roughly +59 D for the unaccommodated state (of which the cornea provides about +43 D). In the normal eye, that's just the refractive power needed to focus a parallel bundle of rays onto the retina. All too commonly, however, the image focus does not lie on the retina. This condition can arise either because of abnormal changes in the refracting mechanism (cornea, lens, and so on) or because of alterations in the length of the eyeball that upset the lens-retina distance. The latter is by far the more common cause. About 25% of the young adult population falls in the class of requiring as little as about ±0.5 D or less of eyeglass correction, and perhaps as many as 65% need only ±1.0 D or less.

Farsightedness, or *hyperopia,* is the defect that causes the image focus of the unaccommodated eye to fall behind the retina (Fig. 26.27). It is most often (perhaps 90% of the time) due to a shortening of the anteroposterior axis of the eye—the lens is too close to the retina. As a result, the image on the photoreceptors is formed of overlapping blotches of light and the picture is somewhat blurred. The *relaxed* hyperopic eye cannot bend the rays enough because it lacks the needed convergence and cannot see anything, near or far, clearly. But it can accommodate, thereby increasing its power and bringing into focus light from far away, which isn't very divergent to begin with (Fig. 26.27b). By accommodating, the farsighted eye can see clearly everything from infinity inward to some **near-point**. This near-point will be a lot farther away than it is in the normal eye. Any closer and the rays diverge too much; strain as the eye may, the image will be blurred.

To increase the power of the hyperopic visual system, a positive spectacle lens can be placed in front of the eye. This lens will allow the unaccommodated eye to see very distant objects clearly (Fig. 26.27d) while effectively pulling in the near-point so that it is at some close, convenient distance. Another way to appreciate this

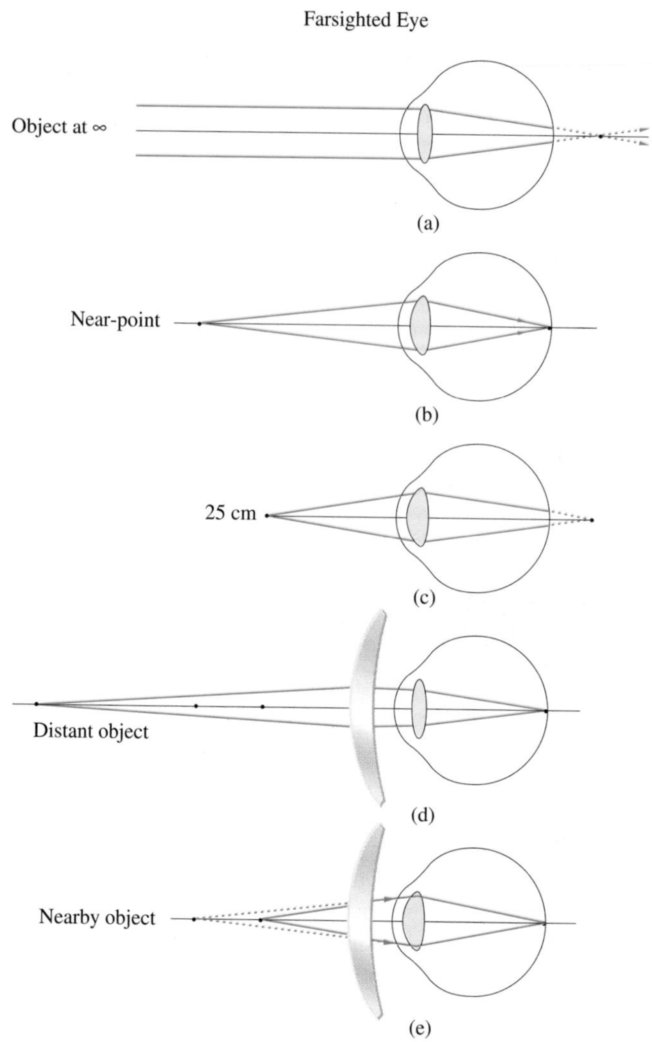

Figure 26.27 Correction of the farsighted eye (a), which focuses parallel light beyond the retina. (b) The near-point is now farther away than 25 cm, which is normal (c). By adding more convergence to the eye, via a positive eyeglass lens, distant objects can be viewed in a relaxed state (d). Moreover, objects at 25 cm, which were blurred, are now clearly seen at the near-point (e).

is to realize that a nearby object (closer in than the focal length of the corrective lens) results in a distant right-side-up, virtual image. That image is located farther out than is the eye's unaided near-point and so can be seen clearly. These spectacles will cast real images—try it if you happen to be hyperopic.

Example 26.11 It is determined that a patient has a near-point at 50 cm. Approximate the eye to be 2.0 cm long. (a) Roughly how much power does the refracting system have when focused on an object at infinity? (b) When focused at 50 cm? (c) How much accommodation is required to see an object at 50-cm distance? (d) What power must the eye have to see clearly an object at the standard near-point distance of 25 cm? (e) How much power should be added to the patient's vision system via reading glasses?

Solution: [Given: actual near-point is 50 cm, $s_i =$ 2.0 cm, standard near-point at 25 cm. Find: (a) power focused at ∞; (b) power focused at 50 cm; (c) accommodation for 50-cm vision; (d) power focused at 25 cm; (e) prescription for correction.] (a) When focused at in-

(continued)

(continued)

finity with the image falling on the retina 0.020 m beyond, the eye has a power of

$$\mathcal{D} = \frac{1}{f} = \frac{1}{\infty} + \frac{1}{0.020 \text{ m}} = \boxed{50 \text{ D}}$$

(b) Similarly, with $s_o = 0.50$ m and $s_i = 0.020$ m, the power is

$$\mathcal{D} = \frac{1}{f} = \frac{1}{0.50 \text{ m}} + \frac{1}{0.020 \text{ m}} = \boxed{52 \text{ D}}$$

(c) Thus, the eye adds 2 D of power via accommodation when changing focus from infinity to 0.50 m. This particular patient cannot provide any more than $\boxed{+2 \text{ D}}$ of accommodation. (d) When $s_o = 0.25$ m, the eye must bring to bear a power of

$$\mathcal{D} = \frac{1}{f} = \frac{1}{0.25 \text{ m}} + \frac{1}{0.020 \text{ m}} = \boxed{54 \text{ D}}$$

(e) Accordingly, this person is lacking $\boxed{+2 \text{ D}}$ of power that can be provided by correction lenses.

▶ **Quick Check:** (a) $f = \infty(0.02 \text{ m})/(\infty + 0.02 \text{ m}) = 0.02$ m. (b) $f = (0.50 \text{ m})(0.02 \text{ m})/0.52 \text{ m} = 1.9$ cm. (d) $f = (0.25 \text{ m})(0.02 \text{ m})/0.27 \text{ m} = 1.85 \text{ cm} = 1/(54 \text{ D})$.

Example 26.12 An optometrist finds that a far-sighted person has a near-point at 125 cm. What power contact lenses will be required if they are to effectively move that point inward to a more workable distance of 25 cm? Use the fact that if the object is imaged at the near-point, it can be seen clearly.

Solution: [Given: $s_o = 0.25$ m and $d_n = 1.25$ m. Find: $\mathcal{D}_c$.] The eye can see the near-point clearly so we want 1.25 m to be the image distance of the correction lens. This s_i must be on the left of the lens and so it is negative, and the lens is being used as a magnifying glass. Accordingly, $s_i = -1.25$ m and $s_o = 0.25$ m. Both the focal length and power of the contact lens can be found via Eq. (26.9); accordingly

$$\frac{1}{s_o} + \frac{1}{s_i} = \frac{1}{f_c} = \mathcal{D}_c = \frac{1}{0.25 \text{ m}} + \frac{1}{-1.25 \text{ m}} = \boxed{+3.2 \text{ D}}$$

The contact lens will form a virtual image of the book that appears at the near-point of the eye. By adding 3.2 D of power to the eye, the near-point of the corrected system becomes 0.25 m instead of 1.25 m.

▶ **Quick Check:** s_o is less than $f = 0.31$ m. $1/(1/4) - 1/(5/4) = 4.0 \text{ D} - 0.8 \text{ D} = 3.2$ D.

Nearsightedness or *myopia* is the condition where parallel rays are brought to a focus in front of the retina; the power of the eye's refractive system is too large for the anterior-posterior axial length (Fig. 26.28). The problem occurs primarily because the eye elongates or the cornea changes shape. It is a situation that usually becomes noticeable in the teens and then levels off in severity at about age 25 or so. Interestingly, myopia hardly exists in "primitive" populations, whereas it is exceedingly common in so-called "advanced" civilizations (there are perhaps 40 million myopes in the United States).

With the myopic eye, images of faraway objects fall in front of the retina. And that's true for object distances from infinity inward to the so-called **far-point**, where the rays diverge enough so that the image is finally right on the retina and clearly visible. It is the farthest point that can be seen sharply by the unaided myopic eye. Depending on the degree of the problem, the far-point can be very much closer in than infinity, and all objects beyond it in space appear blurred. Moreover, the near-point is also closer than normal, which is a convenience for doing detailed work because it provides a bit more magnification.

In effect, the myopic eye has too much convergence; its positive power is too great. To correct the symptoms, we need only place a negative lens in front of the

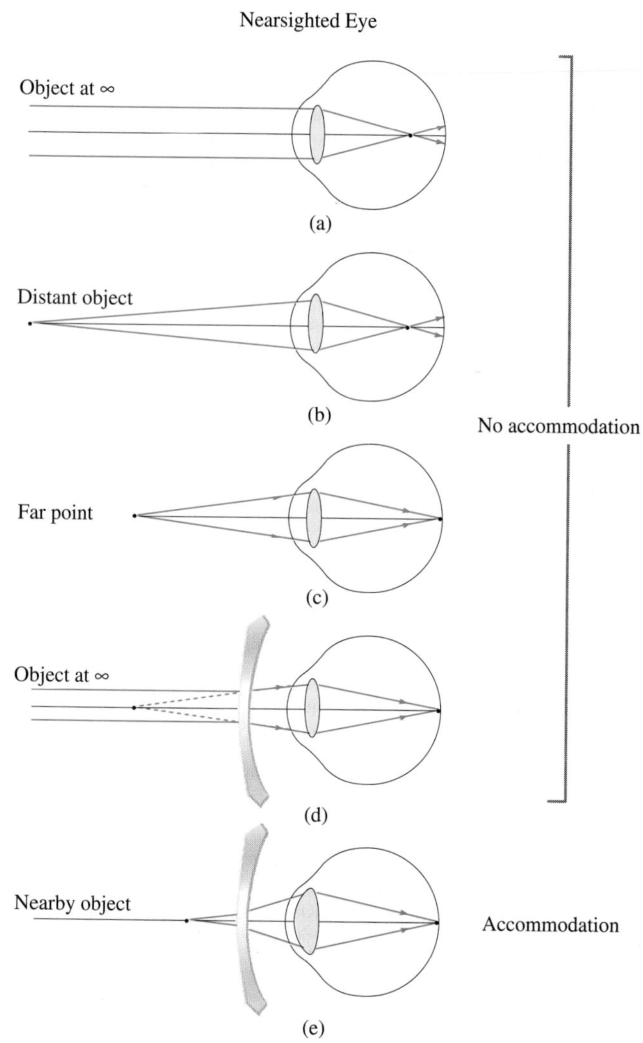

Figure 26.28 Correction of the nearsighted eye (a), which focuses parallel light in front of the retina. By adding some divergence to the eye, via a negative eyeglass lens, distant objects can be viewed (d) in a relaxed state. The far-point (c) is brought in closer (e).

eye. The image focus of the combined lens-eye system must fall on the retina. In other words, *parallel light is made to diverge just enough so that it appears to come from the far-point,* which can then be seen clearly by the unaccommodated eye. The intermediate image formed by the spectacle lens is at the far-point, which is the location of the object for the eye. *The far-point distance from the correction lens equals its focal length.* The eye views the right-side-up virtual images of all objects formed by the correction lens, and those images are located between its far- and near-points. The near-point also moves away a little, which is why myopes will often prefer to remove their spectacles when reading small print; they can then bring the material closer to the eye, increasing the magnification. If you are wearing glasses to correct myopia, try casting a real image with them—it can't be done.

Example 26.13 A person sees objects beyond a distance of 2.0 m to be blurred, but otherwise everything seems fine. What kind of contact lenses, with what power, should they wear?

(continued)

(continued)

Solution: [Given: a far-point of 2.0 m. Find: $\mathcal{D}$.] We want the image of an object at infinity to appear at the far-point of the eye ($s_o = \infty$, $s_i = -2.0$ m): Again, the far-point is on the left side of the lens and therefore the image distance is *negative*. Every distant object will then be imaged closer to the eye than the far-point and will be seen clearly. Hence, a lens must be added to the eye that has a power given by

$$\frac{1}{s_o} + \frac{1}{s_i} = \frac{1}{f} = \mathcal{D} = \frac{1}{\infty} + \frac{1}{-2.0 \text{ m}} = \boxed{-\tfrac{1}{2}\text{ D}}$$

▶ **Quick Check:** $f = \infty(-2.0 \text{ m})/(\infty - 2.0 \text{ m}) = -2.0$ m.

The Compound Microscope

The compound microscope goes the next step beyond the simple magnifier, providing still higher angular magnification (from 15× to around 1200×), achieved this time with a two-step arrangement. A simple version is illustrated in Fig. 26.29. The lens system closest to the object is the **objective**. It forms a real, inverted, magnified image of the object that is then viewed by the **eyepiece**. The latter is essentially a magnifying glass that looks at and enlarges the image created by the objective. Rays diverging from this intermediate image emerge from the eyepiece as a parallel bundle that can comfortably be viewed by a relaxed eye.

The problem is to examine an object that is close at hand, and since $M_T = -s_i/s_o$, the objective should have as small an s_o and as large an s_i as possible, which means, first, that the objective must be close to the object. Rearranging the lens equation yields $s_i = fs_o/(s_o - f)$, suggesting that the image distance will be appropriately large when $s_o \approx f$: *the objective must have a short focal length,* and the object must be positioned just beyond it so that the intermediate image is real. This image is then viewed by the eyepiece, which also must have a short focal length (f_E) since its magnification is $M_{AE} = d_n/f_E$. The intermediate image falls near the focal plane of the eyepiece so that the rays emerge parallel or almost so.

The eyepiece magnifies the intermediate image, which is a magnified version of the object; in other words, the total angular magnification of the system is the product of the magnifications of the objective (M_{TO}) and the eyepiece:

$$M_A = M_{TO} M_{AE}$$

Return to Fig. 26.12 and notice that since triangles AOF_i and PEF_i are similar, $y_i/y_o = -(s_i - f)/f = M_T$. Applied to the objective in Fig. 26.29, this expression becomes $M_{TO} = -L/f_O$, where the image distance minus the focal length of the objective is symbolized by L and is known as the **tube length**. Many manufacturers design their microscopes such that L is standardized at a length of about 160 mm. Using $M_{AE} = d_n/f_E$ and the fact that *it is customary to take the near-point d_n at 254 mm (10 in.),* we have

$$M_A = \frac{-L}{f_O}\frac{d_n}{f_E} = \left(-\frac{160 \text{ mm}}{f_O}\right)\left(\frac{254 \text{ mm}}{f_E}\right) \tag{26.22}$$

where the focal lengths on the right are in *millimeters.*

The barrel of an objective with a focal length of, say, 32 mm, is engraved with the mark 5×, indicating a *magnification* of 5 = 160 mm/32 mm. Combined with a 10× eyepiece (f_E = 25.4 mm), the microscope then has a magnification of 50×—the apparent size is 50 times the actual size.

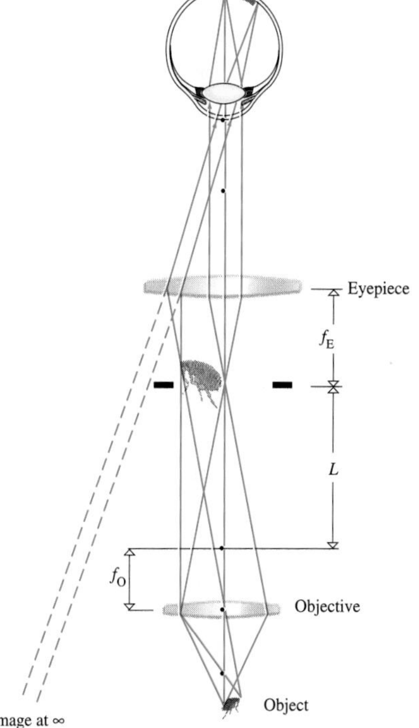

Figure 26.29 A rudimentary compound microscope. The objective forms a real magnified image of the object. That image is further magnified by the eyepiece. Because the intermediate image is at the focal point of the eyepiece, rays enter the eye in parallel bundles, and the eye is relaxed. Notice the large retinal image, which is, of course, what we are after.

Example 26.14 Suppose we wish to make a microscope (which can be used with a relaxed eye) out of two positive lenses both of focal length 25 mm. Assuming the object is positioned 27 mm from the objective, (a) how far apart should the lenses be, and (b) what magnification can we expect?

Solution: [Given: $f_O = f_E = 25$ mm and $s_o = 27$ mm. Find: (a) lens separation; (b) M_A.] (a) The intermediate image distance is obtained from the lens formula applied to the objective; thus

$$\frac{1}{27\text{ mm}} + \frac{1}{s_i} = \frac{1}{25\text{ mm}}$$

and $s_i = 3.38 \times 10^2$ mm. This value is the distance from the objective to the intermediate image, to which must be added the focal length of the eyepiece to get the lens separation, and so

$$3.38 \times 10^2\text{ mm} + 25\text{ mm} = \boxed{3.6 \times 10^2\text{ mm}}$$

(b) $M_{TO} = -s_i/s_o = -(3.38 \times 10^2\text{ mm})/(27\text{ mm}) = -12.5\times$, while the eyepiece has a magnification of $d_n \mathscr{D}_E = (254\text{ mm})(1/25\text{ mm}) = 10.2\times$. Since $M_A = M_{TO}M_{AE}$, the total magnification is $M_A = (-12.5) \times (10.2) = \boxed{-1.3 \times 10^2}$; the minus sign just means the image is inverted.

▶ **Quick Check:** $L = s_i - f_O = 338\text{ mm} - 25\text{ mm} = 313\text{ mm}$; $M_{TO} = -L/f_O = -(313\text{ mm})/(25\text{ mm}) = -12.5\times$. $M_{AE} = (254\text{ mm})/f_E = 10.2\times$.

The Refracting Telescope

The primary function of the telescope is to enlarge the image of a *distant* object. The device shown in Fig. 26.30 has an objective and an eyepiece just like a microscope but because the job to be done is different, the structure is also different. The object is at a finite far distance from the device so that the intermediate image is located beyond the image focus of the objective. As with the microscope, this real, inverted image serves as the object for the eyepiece, which functions as a magnifier. Hence, the intermediate image is made to fall within one focal length (f_E) of the eyepiece so that the resulting final image it creates is virtual, enlarged, and remains inverted. In practice, *the position of the intermediate image is fixed and only the eyepiece is moved in order to focus the instrument.*

The central piece of design information is that the object is far away; that is, the object distance for the objective is very large in comparison to all the other distances in the system. The Gaussian Lens Equation then tells us that since $1/s_o \approx 0$, it follows that $s_i \approx f_O$. Unlike the microscope, where the intermediate image was magnified, here it must be *minified.* That might seem strange at first, but realize how

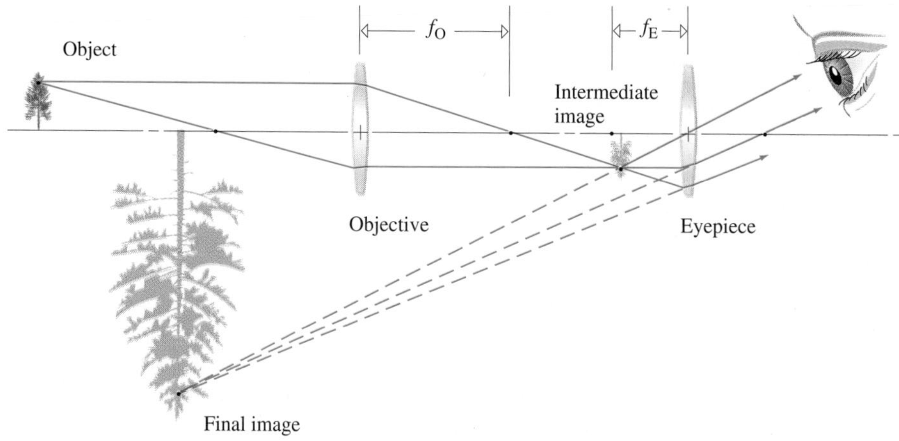

Figure 26.30 Kepler's astronomical telescope. It's used in astronomy where it doesn't matter if the image of the Moon or a star is upside down.

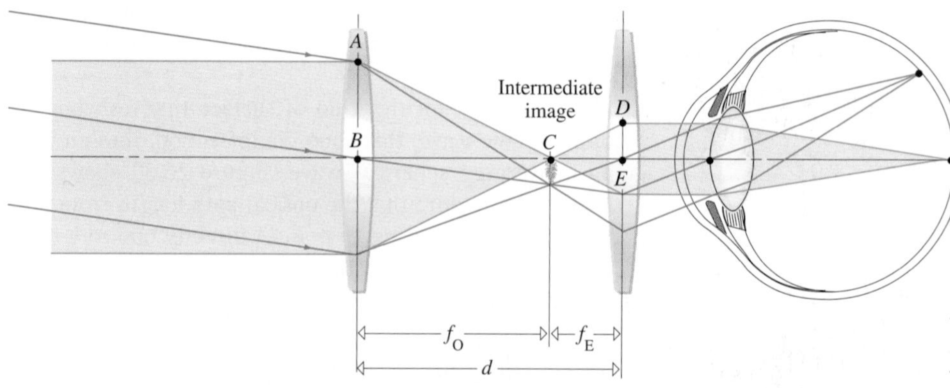

Figure 26.31 An astronomical telescope operating so that parallel light enters the objective and leaves the eyepiece.

awkward it would be to try to fit a larger-than-life image of, say, the Moon inside the tube of a telescope. The important feature of the real image is that it is now so close by that it can easily be examined with a magnifier.

Suppose the object focus of the eyepiece, which lies in front of it, overlaps the image focus of the objective, which lies behind it, as it does in Fig. 26.31. Then the separation between the two lenses equals the sum of their focal lengths. Parallel rays entering the scope from a very distant object exit in parallel bundles that can be viewed comfortably by the relaxed eye. That concern is an important one, and this arrangement is most often used.

In general, we want a minified intermediate image that is as large as is practical. Since $M_{TO} = -s_i/s_o \approx -f_O/s_o$, we need an objective lens with *as long a focal length as possible.* Again, the eyepiece views the intermediate image and magnifies it. Since the magnification varies inversely with f_E, the eyepiece should have a *short focal length.* In fact, the magnifying power of a telescope, adjusted so parallel rays emerge, is

$$M_A = -\frac{f_O}{f_E}$$

This equation is the reason why high-power refracting telescopes usually have long tubes into which one inserts a short focal-length eyepiece. When you look through the back end of a telescope, everything appears minified. The roles of eyepiece and objective are reversed, and because their focal lengths are quite different, the effect is striking. Reversing the telescope reduces the image size by the same factor it was previously increased by.

MIRRORS

Mirror systems are finding increasingly more extensive and important applications, particularly in the infrared, ultraviolet, and X-ray regions of the spectrum. It is relatively easy to construct a reflecting device that performs satisfactorily across a broad range of frequencies; the same cannot be said for refracting systems. Today, mirrors play a significant role in all sorts of devices, from spy satellites and copy machines to cameras, microscopes, and lasers.

26.8 Aspherical Mirrors

Curved mirrors that form images very much like those of lenses have been known since the ancient Greeks. Fortunately, we have already developed much of the con-

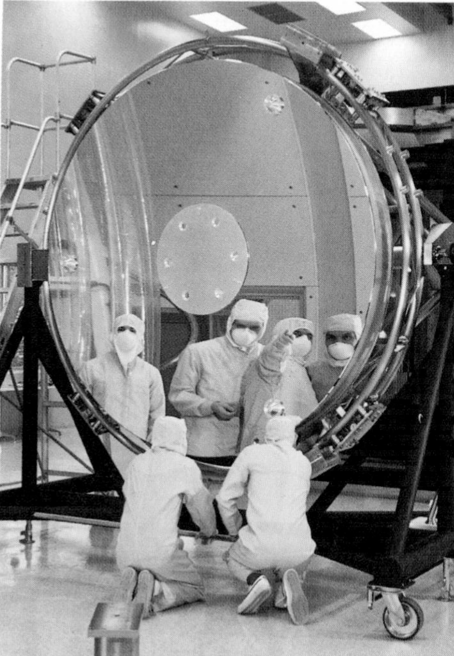

The 2.4-m-diameter primary mirror of the Hubble Space Telescope.

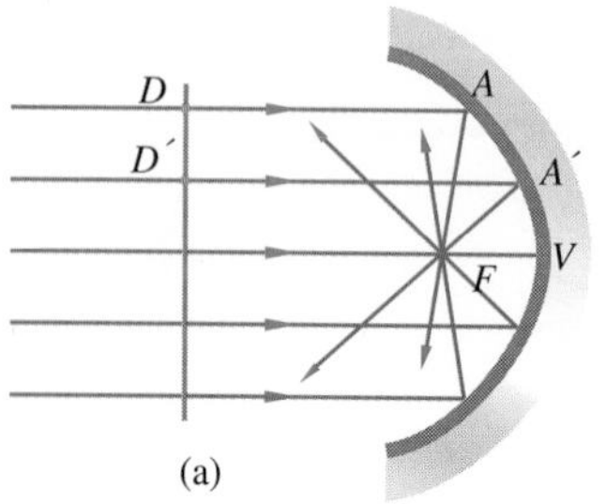

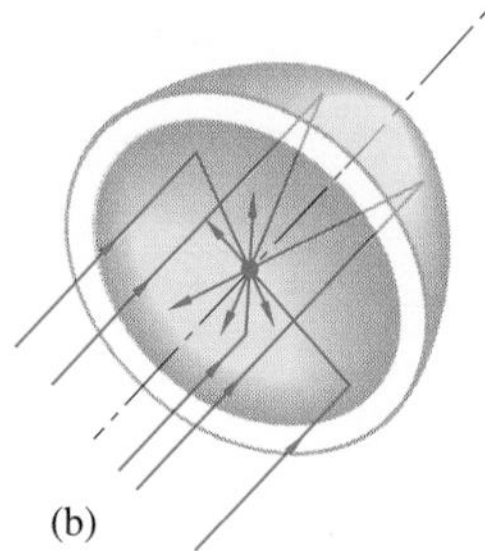

Figure 26.32 A paraboloidal mirror. (a) Parallel axial rays are brought to a focus at point *F*. (b) The configuration is 3-dimensional and symmetric about the central axis.

ceptual basis for analyzing curved mirrors and will be able to evolve the subject quickly without introducing many new ideas.

We again ask for the kind of surface that reshapes an incoming plane wave, this time via reflection, into an outgoing converging spherical wave. Figure 26.32 shows the geometry. We require that the optical path length from any point *D* on the plane wave to point *A* directly opposite it on the surface, and thence to the fixed point *F*, must be constant. Since there is now only one medium involved

$$\overline{DA} + \overline{AF} = \text{constant} = \overline{D'A'} + \overline{A'F'}$$

In two dimensions, this expression corresponds to the equation of a parabola (of eccentricity $n_t/n_i = 1$, p. 896) with its focus at *F*. A bundle of parallel rays reflecting off a paraboloidal mirror will be brought to a focus at *F*, a distance *f* from the **vertex** *V* of the mirror. Most of the world's older astronomical reflecting telescopes have energy-gathering mirrors that are paraboloidal.

Figure 26.33 depicts the behavior of several other aspherics. In recent years, these have been used to form images, via reflected X-rays, of a variety of phenomena from solar emission to laser-induced fusion. Today, the hyperboloid is the overwhelming choice for large telescopes, including the Hubble Space Telescope.

26.9 Spherical Mirrors

A spherical mirror has no one particular symmetry axis, and that can be a great advantage, especially when the device is not movable. In many applications restricted to paraxial optics, spherical mirrors perform quite well. Problem 76 deals with the proof that a sphere and a paraboloid coincide in the region close to the symmetry axis provided that $|f| = |R|/2$, as shown in Fig. 26.34. Thus, insofar as the rays are paraxial, they will encounter a region where sphere and paraboloid are nearly identical and the sphere will behave like the mirror in Fig. 26.32.

Absolute values, $|f|$ and $|R|$, are used because we have not yet agreed upon signs for mirror quantities. Using the previous convention, *R* in Fig. 26.35 is negative be-

Figure 26.33 Two aspherical mirrors.

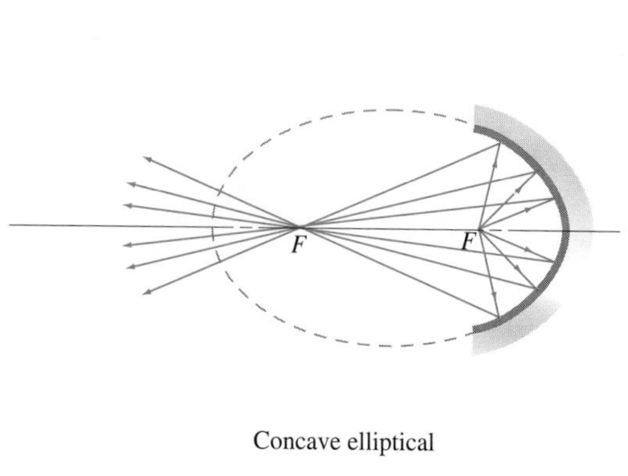

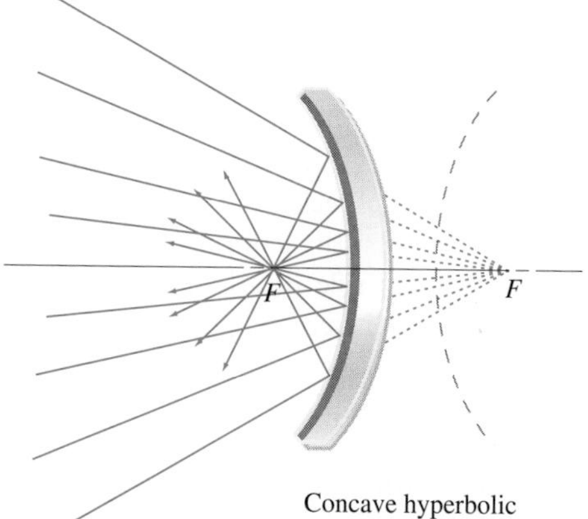

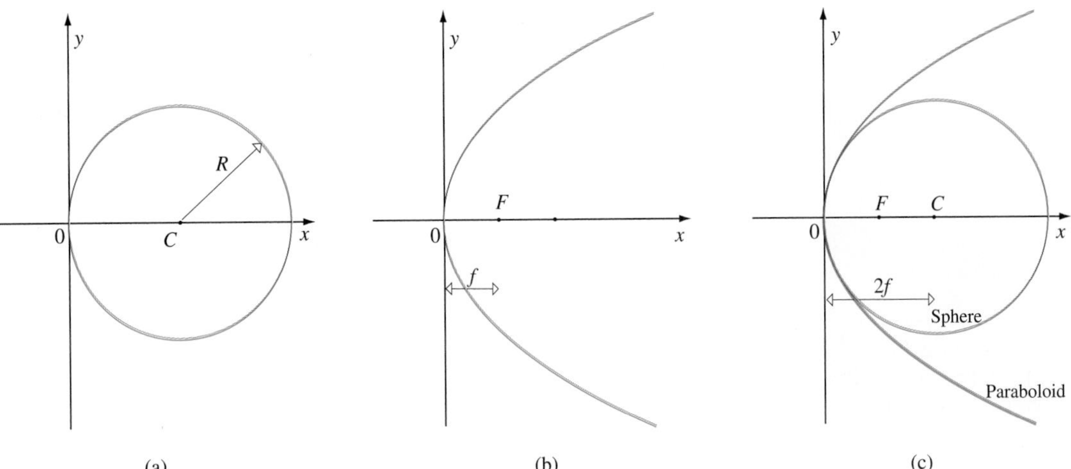

Figure 26.34 A sphere (a) and a paraboloid (b) coincide (c) in the region near the axis, such that $|R| = 2|f|$.

cause C is to the left of V. A parallel bundle of axial rays will converge to F, and so we take f to be positive, despite the fact that it is now measured to the left of the mirror (as compared with to the right of the lens). That being the case

$$f = -\frac{R}{2} \tag{26.23}$$

The focal length of a spherical mirror equals one-half its radius.

The image P of S in Fig. 26.35 is real, but it's *on the left of the mirror,* unlike the situation with the positive lens. If we continue to require that real images have positive image distances, then we must henceforth take s_i to be *positive when left of the vertex,* exactly the same way that we took the focal length to be positive. These differences in the signs of f and s_i will be the only deviations from the convention for lenses, and with them all the appropriate equations will turn out the same as before. The image-formation geometry for the concave mirror is identical to that of the convex lens, and we need not rederive all the equations. Suffice it to say that

The 1000-ft radiotelescope at Arecibo, Puerto Rico, operates at 21 cm. The spherical bowl reflects radiant energy up to the focal-point detector suspended above it on cables attached to three towers.

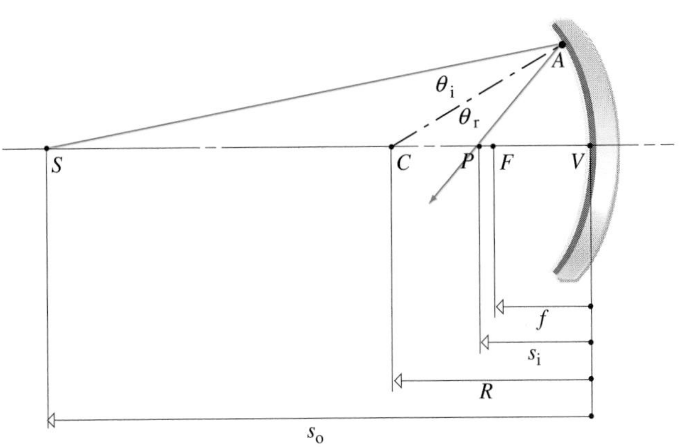

Figure 26.35 The geometry of a spherical mirror in the paraxial region.

Just like a convex lens, a concave mirror can form a real image. Here the mirror in the foreground casts a real, minified, inverted image of the candle flame.

paraxial rays in Fig. 26.35 obey the relationship

$$\frac{1}{s_o} + \frac{1}{s_i} = -\frac{2}{R} \tag{26.24}$$

which is called the **mirror formula**. If $s_o \to \infty$, $s_i \to f_i = -R/2$ while if $s_i \to \infty$, $s_o \to f_o = -R/2$. In other words, both the image and object focal lengths equal f and

$$\frac{1}{s_o} + \frac{1}{s_i} = \frac{1}{f} \tag{26.25}$$

Observe that f is positive for concave mirrors ($R < 0$) and negative for convex mirrors ($R > 0$). In the latter case, the image is formed behind the mirror in diverging light and cannot be projected upon a screen—it is virtual (Fig. 26.36a).

Example 26.15 A point source lies on the central axis 1.00 m in front of a concave spherical mirror having a radius of 20 cm. Locate and describe the resulting image.

Solution: [Given: $s_o = 1.00$ m and $R = -0.20$ m. Find: s_i.] From Eq. (26.24)

$$\frac{1}{s_i} = -\frac{2}{R} - \frac{1}{s_o} = -\frac{2}{-0.20 \text{ m}} - \frac{1}{1.00 \text{ m}}$$

and $\boxed{s_i = 0.11 \text{ m}}$. The image distance is positive and so the image itself must be real.

▶ **Quick Check:** $f = -R/2 = 0.10$ m $= (1.00$ m$) \times (0.11$ m$)/(1.00$ m $+ 0.11$ m$) = 0.10$ m.

Figure 26.36 The focal point, F, of a convex spherical mirror.

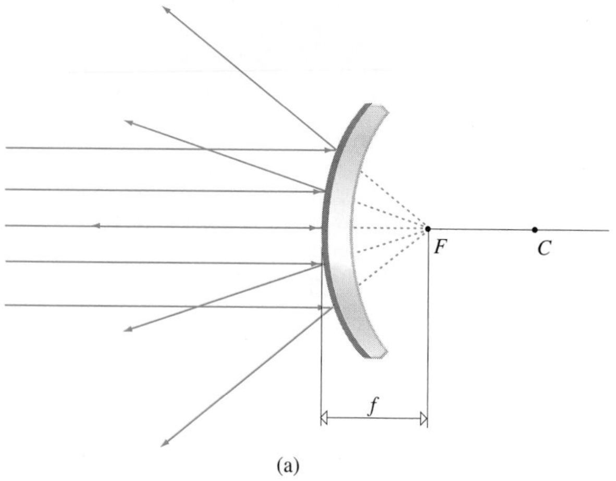

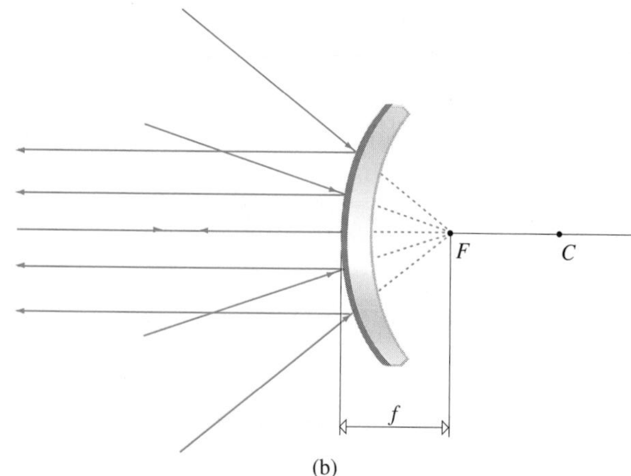

TABLE 26.4 Sign convention for spherical mirrors

Quantity	Sign +	Sign −
s_o	Left of *V*, real object	Right of *V*, virtual object
s_i	Left of *V*, real image	Right of *V*, virtual image
f	Concave mirror	Convex mirror
R	*C* right of *V*, convex	*C* left of *V*, concave
y_o	Above axis, right-side-up object	Below axis, inverted object
y_i	Above axis, right-side-up image	Below axis, inverted image

26.10 Finite Imagery: Mirrors

The remaining mirror properties are so similar to those of lenses that we need only mention them briefly without repeating the development. Accordingly, within the restrictions of paraxial theory, *any off-axis parallel bundle of rays will be focused to a point on the focal plane* a distance *f* from *V*. The image of any point on an object can again be located using any two easily drawn rays (Fig. 26.37). Unlike the lens, which is transparent and has separate object and image foci (one on each side), the mirror has only one focus. Table 26.4 summarizes the sign convention.

The ray diagram (Fig. 26.38) shows how the image of an extended object is created by a concave mirror. Because triangles *SDV* and *PEV* are similar (look at Fig. 26.12), their sides are proportional. Hence, taking distance measured down from the axis to be negative, $y_i/y_o = -s_i/s_o$, which, of course, is the transverse magnification M_T as defined earlier (p. 905). The striking similarity between the behavior of a concave mirror and a convex lens on one hand, and a convex mirror and a concave lens on the other, is evident on comparing Tables 26.3 and 26.5. Figure 26.39 graphically illustrates the entire range of responses of the concave mirror. Except for the fact that the mirror folds the rays over (so that object and image space overlap when the image is real and do not when it's virtual), the diagram is the same as Fig. 26.15.

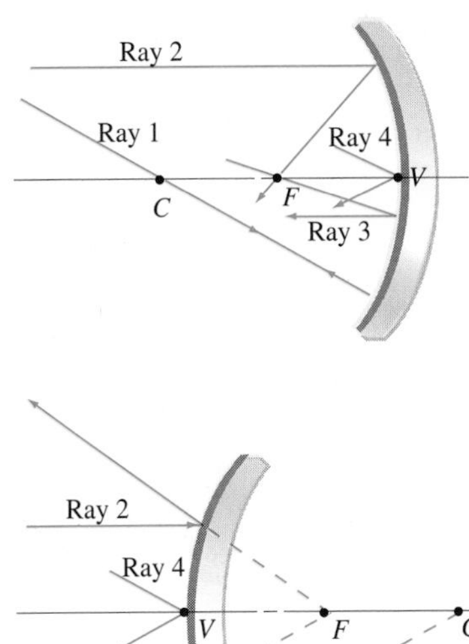

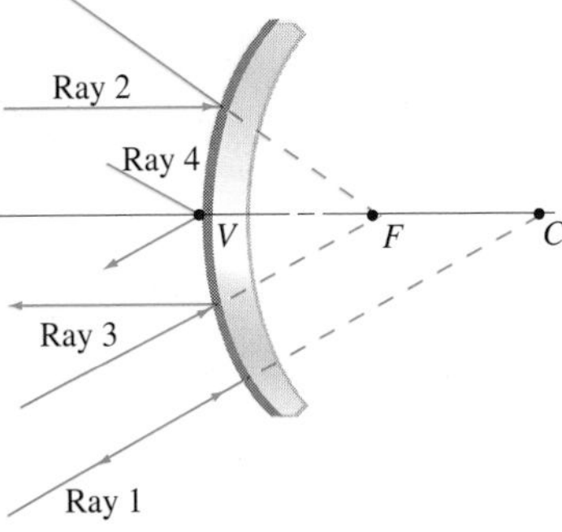

Figure 26.37 Four easily drawn rays. Ray 1 heads toward *C* and reflects back along itself. Ray 2 comes in parallel to the central axis and reflects toward (or away from) *F*. Ray 3 passes through (or heads toward) *F* and reflects off parallel to the axis. Ray 4 strikes point *V* and reflects such that $\theta_i = \theta_r$.

A convex spherical mirror forming a virtual, right-side-up, minified image.

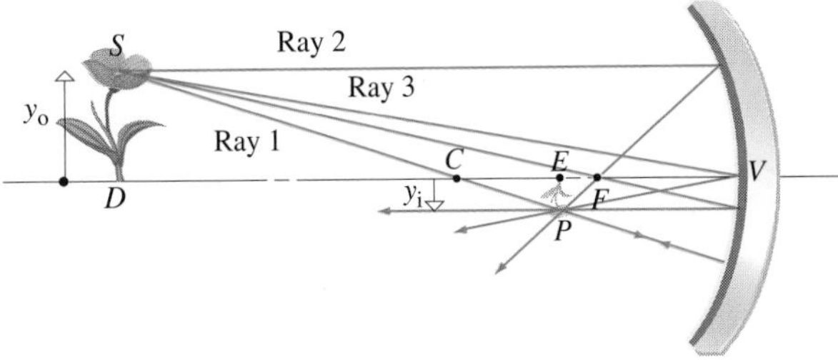

Figure 26.38 An extended image formed by a spherical concave mirror.

TABLE 26.5 Images of real objects formed by spherical mirrors

Concave

Object	**Image**			
Location	**Type**	**Location**	**Orientation**	**Relative size**
$\infty > s_o > 2f$	Real	$f < s_i < 2f$	Inverted	Minified
$s_o = 2f$	Real	$s_i = 2f$	Inverted	Same size
$f < s_o < 2f$	Real	$\infty > s_i > 2f$	Inverted	Magnified
$s_o = f$		$\pm\infty$		
$s_o < f$	Virtual	$\lvert s_i \rvert > s_o$	Right-side-up	Magnified

Convex

Object	**Image**			
Location	**Type**	**Location**	**Orientation**	**Relative size**
Anywhere	Virtual	$\lvert s_i \rvert < \lvert f \rvert$, $s_o > \lvert s_i \rvert$	Right-side-up	Minified

Example 26.16 A youngster looking into the shiny convex back of a spoon with a spherical bowl sees herself reflected therein. The bowl has a radius of 3.00 cm, and her nose is 25.0 cm from its surface. Where will the image of her nose appear? Describe the image completely.

Solution: [Given: $R = +0.030$ m and $s_o = 0.250$ m. Find: s_i and M_T.] To locate the image, use the mirror formula

$$\frac{1}{0.250\text{ m}} + \frac{1}{s_i} = -\frac{2}{0.030\text{ m}}$$

and $\boxed{s_i = -0.014\text{ m}}$. The image distance is negative and so the virtual image is located to the right, behind the mirror. The magnification is

$$M_T = -\frac{s_i}{s_o} = -\frac{-0.014\text{ m}}{0.250\text{ m}} = \boxed{+0.056}$$

The image is minified and right-side up.

▶ **Quick Check:** $f = -R/2 = -0.015$ m $= (0.250\text{ m})(-0.014\text{ m})/(0.250\text{ m} - 0.014\text{ m}) = -0.015$ m.

The Reflecting Telescope

In addition to providing magnification, serious astronomical telescopes must meet another demand: they must gather in as much light as possible because the objects being viewed are generally extremely faint. As with the camera, the energy entering the system is proportional to the diameter of the objective—the bigger, the better. But there is a real difficulty in making big lenses. The largest such instrument in the world is the 40-inch diameter Yerkes refracting telescope in Wisconsin as compared to the 200-inch Palomar reflector in California. The problems are evident: a lens has to be perfectly transparent and free of internal flaws. A front-silvered mirror need not even be transparent. A lens can only be supported by its rim and may sag under its own weight; a mirror can be supported over its entire back. For these and other reasons (better frequency response, better aberration control, and so on), reflectors predominate in the domain of large telescopes, from observatories to spy satellites.

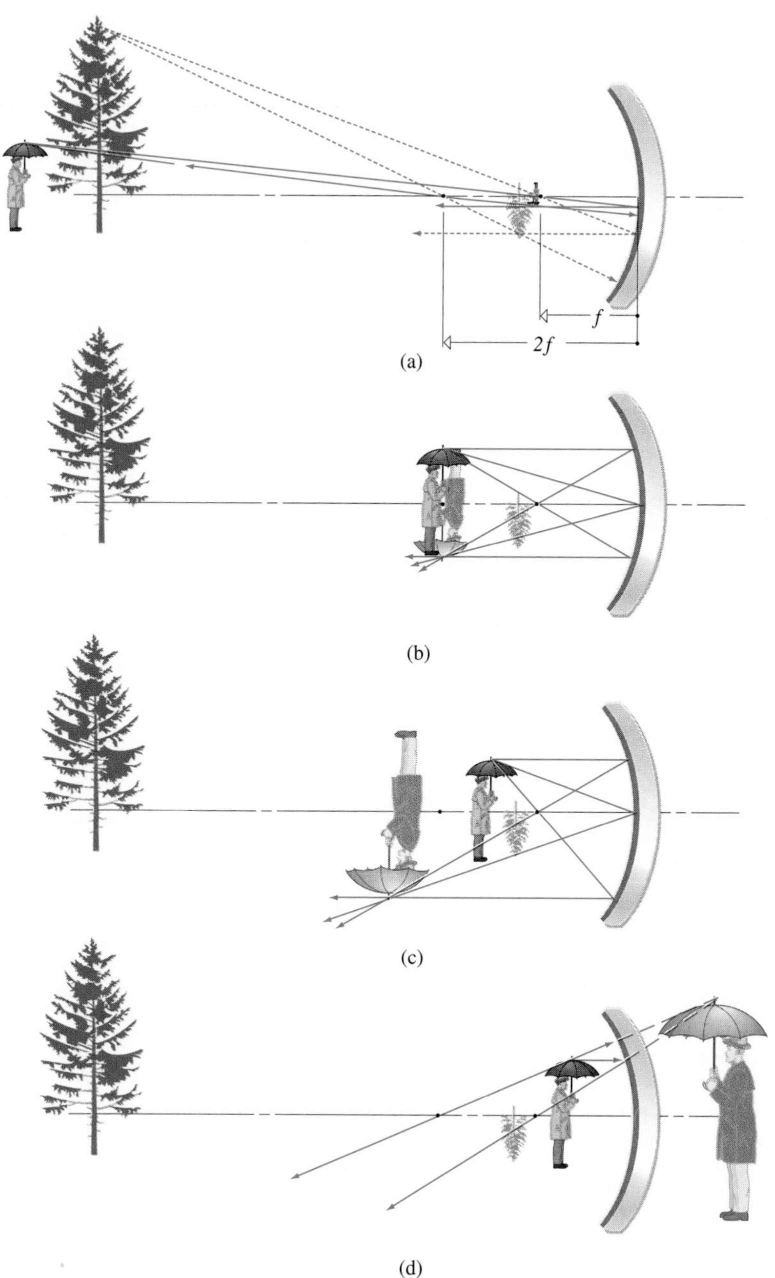

Figure 26.39 The images formed by a concave spherical mirror.

Invented by the Scotsman James Gregory in 1661, the reflecting telescope was first successfully constructed by Newton in 1668 (primarily because he had mistakenly concluded that the aberrations suffered by lenses were unavoidable). Two common reflectors are shown in Fig. 26.40. A plane mirror or prism brings the beam out to the side in the Newtonian version. The traditional Cassegrainian arrangement uses a convex hyperboloidal secondary mirror to increase the effective focal length of a paraboloidal primary. A modernized Cassegrainian with a hyperboloidal primary is now the leading telescope configuration.

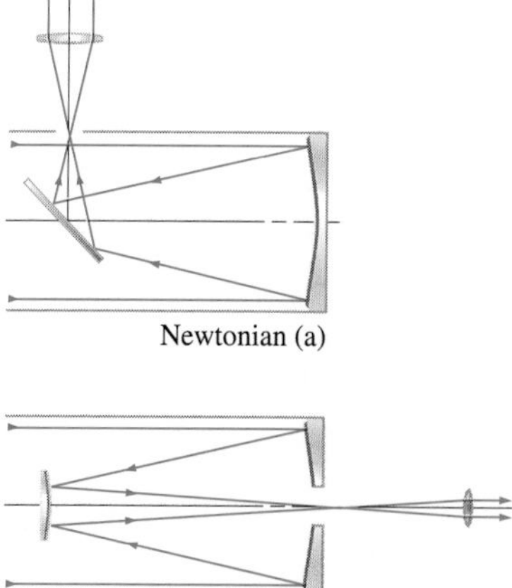

Figure 26.40 Reflecting telescopes.

Core Material

As the wavelength of the radiant energy being processed becomes vanishingly small, diffraction disappears, rectilinear propagation obtains, and we have the idealized domain of **geometrical optics**. For thin spherical lenses, there is the **Lensmaker's Formula**

$$\frac{1}{s_o} + \frac{1}{s_i} = (n_l - 1)\left(\frac{1}{R_1} - \frac{1}{R_2}\right) \qquad [26.6]$$

where the **focal length** f is given as

$$\frac{1}{f} = (n_l - 1)\left(\frac{1}{R_1} - \frac{1}{R_2}\right) \qquad [26.7]$$

Comparing these two expressions yields the **Gaussian Lens Equation**

$$\frac{1}{s_o} + \frac{1}{s_i} = \frac{1}{f} \qquad [26.9]$$

The ratio of any transverse dimension of the image to the corresponding dimension of the object is the **transverse magnification** or, more often, simply the **magnification**, M_T:

$$M_T = \frac{y_i}{y_o} \qquad [26.11]$$

Alternatively

$$M_T = -\frac{s_i}{s_o} \qquad [26.12]$$

If the object is at the focal point of a **magnifying glass**, its image is at infinity and

$$M_A = \frac{d_n}{f} \qquad [26.14]$$

When two lenses are being used at once, the image of the first is the object of the second, and the analysis can be carried out one lens at a time (p. 916). The focal length of two thin lenses in contact is given by

$$\frac{1}{f} = \frac{1}{f_1} + \frac{1}{f_2} \qquad [26.19]$$

The **dioptric power** $\mathcal{D}$ of a lens is the reciprocal of the focal length, and its units are inverse meters or *diopters* (D). The total power of two lenses in contact is

$$\mathcal{D} = \mathcal{D}_1 + \mathcal{D}_2 \qquad [26.21]$$

A spherical mirror behaves very much like a thin lens. It has a focal length given by

$$f = -\frac{R}{2} \qquad [26.23]$$

But we take f to be positive to the left of the mirror. The relationship between object distance, image distance, and focal length is represented by

$$\frac{1}{s_o} + \frac{1}{s_i} = -\frac{2}{R} \qquad [26.24]$$

and this is the **mirror formula**.

Suggestions on Problem Solving

1. Read the problem carefully for *given* information that is subtly stated. If the image created by a single thin lens appears on a wall, the lens is positive, its focal length is positive, the image is real, the image distance is positive, the image is inverted, and the magnification and image height are both negative. All of that detail may *not* be explicitly spelled out. Make a sketch of the lens or mirror, the object and image, putting everything in roughly where it belongs, with all the numbers included so the setup can be visualized as a whole. Worry about the ray diagram later.

2. When working problems, keep in mind the sign convention as you substitute in the numbers. Read your diagram to confirm the signs. If you know the image formed by the lens is inverted, y_i must go in as a negative number, just as M_T must be negative. If that image is 3 m tall, $y_i = -3$ m. Watch out for the signs of the radii of curvature of lenses and mirrors—these are often messed up. The sign convention requires that the *light enter from the left.* If this is not the case in a particular situation, redraw the figure.

3. Several of the equations in this chapter, such as the Lensmaker's Formula and the Gaussian Lens Equation, are written in terms of the reciprocals of quantities that are of interest. Among the most common errors made is to neglect to invert the final result. For example, computing the numerical value for $1/f$ and then giving that number as your answer for f. Furthermore, remember that you *cannot* take an equation like $1/s_o + 1/s_i = 1/f$ and invert both sides!

4. In most cases, especially with only one lens or mirror involved, you should already have a good idea of what the image will be from Table 26.3 and/or Table 26.5. *Whenever possible, check the results of your calculations with reality; that is, with those tables.* If the object is just a little bit farther from the positive lens than one focal length and the image is computed to appear roughly one focal length beyond the lens, then you should know something is terribly wrong!

5. When drawing ray diagrams, remember that objects nearer to a positive lens produce real images that are farther away. Also keep in mind that all images formed by a convex mirror lie between the vertex and the focal point behind the mirror. If you have a rough idea of what to expect, your ray diagrams will not go wild.

6. Notice that the angular magnification of both the compound microscope and the astronomical telescope are negative because the images are inverted. That means that you will have to be a little careful about signs when doing problems dealing with these devices. A 10× telescope means $M_A = -f_O/f_E = -10$; both focal lengths are positive.

Discussion Questions

1. What happens to the focal length of a glass lens when it's taken from the air and placed in water? Explain your answer.

2. If a lens of glass surrounded by air is negative, what can be said about the identically shaped lens made of air surrounded by glass? A bubble in a glass of beer is a tiny lens. What kind?

3. Explain why it is that the focal length of a lens actually depends on the color of the light being transmitted.

4. How do goggles work to allow an underwater swimmer to see clearly?

5. What is a quick physical means of determining the approximate focal length of a converging lens? How might you use a known strong positive lens in order to find the focal length of a negative lens?

6. If a horse stands facing a positive lens, which part of the beast will be closest the lens in the real image? In the virtual image? Draw the appropriate ray diagrams.

7. Imagine a converging glass lens in a chamber filled with a gas under a few atmospheres of pressure. Suppose the lens is illuminated by parallel light. What, if anything, will happen to the point at which the beam converges if the gas is gradually pumped out?

8. If a real image is formed by a large plane mirror, what can be said about the incoming rays? Incidentally, a tiny flat mirror can form a real image just as a pinhole does.

9. What is the focal length of a plane mirror? What does Eq. (26.24), the mirror formula, say about the image distance? What then is the magnification of a flat mirror according to the equations?

10. Figure Q10 depicts a hyperboloidal mirror and its accompanying geometry. Explain what is happening in the diagram. Describe the wavefronts before and after reflection from the hyperboloidal mirror.

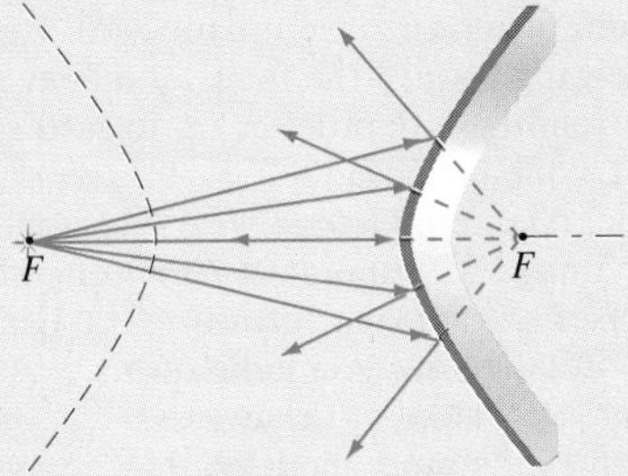

Figure Q10

11. Legend has it that Archimedes (ca. 287 B.C.–212 B.C.) burned the invading Roman fleet by focusing sunlight onto the sails. One version of the story has him on a mountainside lining up soldiers holding brightly polished shields. How could he have arranged the soldiers to accomplish this task?

12. Suppose we take two positive lenses and place them in contact with one another. In what sense is the combination more powerful than either lens separately?

13. Figure Q13 is a diagram of an X-ray camera used for diagnosing the implosion of tiny laser-fusion targets (Sect. 32.10) at the Lawrence Livermore Laboratory in California. Explain how it works to form an image of the target. (You should be able to figure out what's happening even though some aspects were not discussed in the text explicitly.)

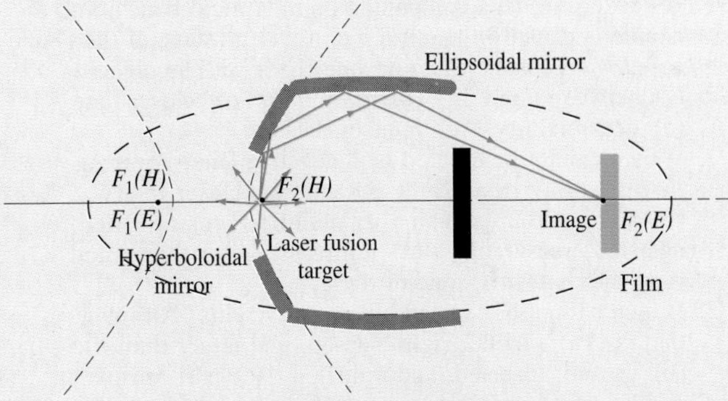

Figure Q13

14. A person with normal vision puts on a pair of eyeglasses made for a hyperope. What will things look like near and far with the eye relaxed? Not relaxed?

15. Microscopes operating in the visible are limited in magnification to about 1200×. How big an object would produce an image seen in the eyepiece to be $\frac{1}{2}$ mm across? What does that suggest about the limitations of the instrument? Will a 12 000× light microscope resolve any finer details of the object?

16. The spotlight in Fig. Q16 is a rather common, very simple one used in theaters in the United States. Explain what it does and how it works. Where is the filament located with respect to both the lens and the mirror? Why?

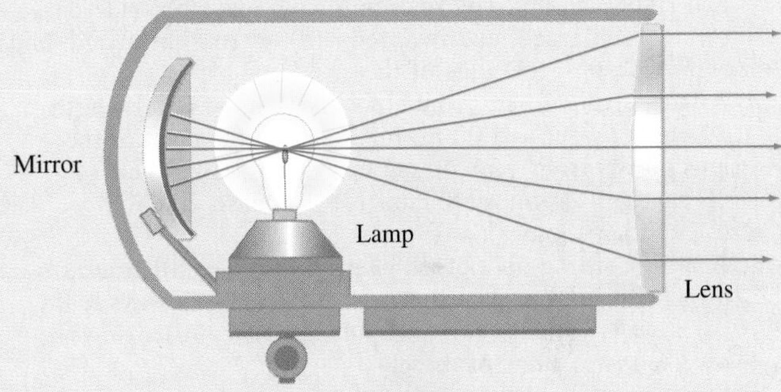

Figure Q16

17. Figure Q17 shows an ellipsoidal reflector spotlight, which is probably the most widely used form of the device. How does it work?

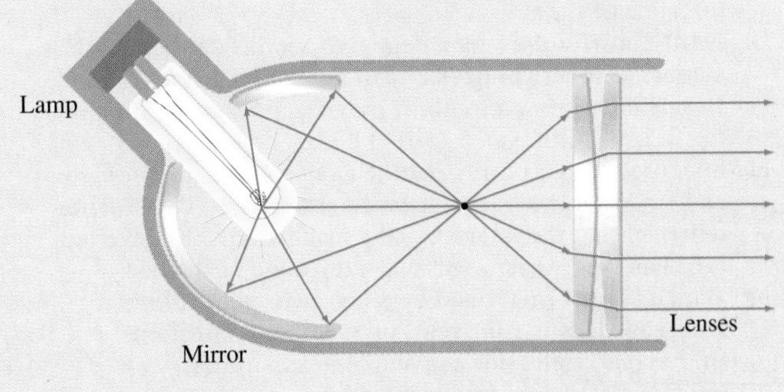

Figure Q17

Multiple Choice Questions

1. A converging lens is mounted on an optical bench, and a candle is placed in front of it so that an image of the flame appears on a screen $1.5f$ beyond the lens. The object is (a) between f and $2f$ (b) beyond $2f$ (c) closer than f (d) near infinity (e) none of these.
2. Which cannot be created by a negative lens? An image that is (a) virtual, right-side-up, and closer than f (b) right-side-up and smaller than life (c) minified, inverted, and virtual (d) minified, closer than f, and right-side-up (e) none of these.
3. Which cannot be formed by a positive lens? An image that is (a) virtual, right-side-up, and larger than life (b) virtual, inverted, and minified (c) real, inverted, and minified (d) real, inverted, and magnified (e) none of these.
4. A convex lens with a very small focal length is placed in contact with a convex lens with a very large focal length. The combined focal length will be (a) much larger than either (b) much smaller than either (c) approximately equal to the smaller (d) approximately equal to the larger (e) none of these.
5. A lens made of glass ($n = 1.50$) has a convex first surface with a radius of curvature of 2.0 m and a second concave surface with a radius of curvature of 1.0 m. Its focal length is (a) 3.0 m (b) $\frac{1}{2}$ m (c) $-\frac{1}{4}$ m (d) -4.0 m (e) none of these.
6. When an object is placed between the vertex and focal point of a negative lens, its image is (a) virtual, enlarged, and right-side-up (b) real, minified, and right-side-up (c) real, enlarged, and inverted (d) virtual, minified, and right-side-up (e) none of these.
7. A frog sitting 40 cm from a positive lens of focal length 20 cm hops out to 1.00 m. Its image on the right of the lens moves from (a) 40 cm to 1.00 m (b) 40 cm to 25 cm (c) 20 cm to 40 cm (d) 40 cm to 20 cm (e) none of these.
8. A bird is sitting on a perch peering into a small convex mirror in its cage. The image it sees is (a) always real (b) always virtual (c) always magnified (d) always inverted (e) none of these.
9. Suppose you make a calculation of the magnification of the image of a real object located somewhere in front of a convex spherical mirror and it turns out to be -1.5. You can then conclude that (a) the calculation was wrong (b) the image is smaller than life (c) the object was upside down (d) the image is larger than the object (e) none of these.
10. A little mirror used by a dentist to see inside the mouth is a concave spherical device with a focal length of 25 mm. Held 2.0 cm from a tooth, it provides a magnification of (a) 0.5 (b) 10 (c) 5 (d) 0.1 (e) none of these.
11. Suppose that you carry out the analysis of a mirror system and find that the image distance is negative. This finding tells you that the image is (a) magnified (b) inverted (c) black and white (d) real (e) none of these.
12. Which cannot be formed by a concave spherical mirror? An image that is (a) real, inverted, and minified (b) virtual, right-side-up, and larger than life (c) magnified, upside down, and beyond $2f$ (d) life-size, real, and inverted (e) none of these.
13. A concave makeup mirror magnifies, by a factor of 2, the face of anyone looking into it from a distance of 25 cm. (The image is, of course, right-side-up.) The mirror's focal length must therefore be (a) 5 cm (b) 50 cm (c) 500 cm (d) 5000 cm (e) none of these.
14. The off-axis mirror forming a perfect point image depicted in Fig. MC14 is (a) a segment of a paraboloid (b) planar (c) a segment of an ellipsoid (d) a segment of a hyperboloid (e) none of these.

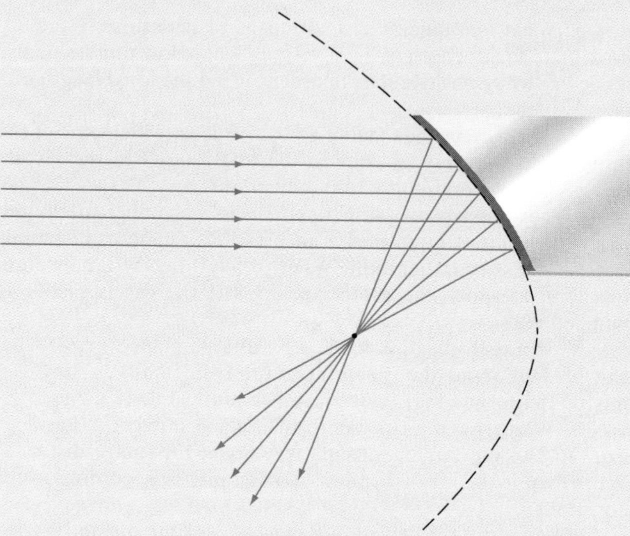

Figure MC14

15. Can a paraboloidal mirror be used to create a reflected diverging spherical wave? (a) no (b) yes, by sending a parallel beam into a concave mirror (c) yes, by sending a beam converging toward the focus of a convex mirror (d) yes, by sending a parallel beam toward a convex mirror (e) none of these.
16. George is nearsighted. George wears negative corrective lenses. The image focal point of either one of George's eyeglass lenses (a) lies at infinity (b) lies at his near-point (c) lies in the middle of his eye (d) lies at his far-point (e) none of these.
17. If the human eye, simply modeled, has a power of about 59 D, its focal length is (a) +17 mm (b) −5.9 mm (c) +5.9 mm (d) −17 mm (e) none of these.
18. A human eye is about 2 cm long and at night has a maximum pupil diameter of 8 mm. The f-number of the system is (a) 0.25 (b) 4 (c) 16 (d) 2.5 (e) none of these.
19. Reducing the radius of the diaphragm on a camera by a factor of 2 (a) decreases the f-number by a factor of 2 (b) increases the f-number by 2 (c) increases the f-number by a factor of 2 (d) decreases the f-number by 2 (e) none of these.
20. Suppose you slip out the eyepiece from a microscope and replace it with one having twice the focal length. The magnification of the instrument will (a) double (b) halve (c) remain unchanged (d) quadruple (e) none of these.

21. Suppose we come upon a bag of lenses that are marked +1.0 D, −4.0 D, +10 D, +0.10 D, +95 D, and −0.50 D. Which would you use to build a microscope with the largest possible magnification? (a) +0.10 D as objective and +95 D as eyepiece (b) −4.0 D as objective and −0.50 D as eyepiece (c) +95 D as objective and +0.10 D as eyepiece (d) +10 D as objective and +0.10 D as eyepiece (e) none of these.

22. Suppose we come upon a bag of lenses that are marked +1.0 D, −4.0 D, +10 D, +0.10 D, +95 D, and −0.50 D. Which would you use to build an astronomical telescope with the largest possible magnification? (a) +0.10 D as objective and +95 D as eyepiece (b) −4.0 D as objective and −0.50 D as eyepiece (c) +95 D as objective and +0.10 D as eyepiece (d) +10 D as objective and +0.10 D as eyepiece (e) none of these.

Problems

LENSES

1. [I] What is the radius of curvature of the spherical surface of a planar-convex lens that has a diameter of 20 cm, is $\frac{1}{2}$ cm thick at its center, and curves gradually to a sharp edge all around its periphery?

2. [I] A thin glass lens ($n = 1.50$), which is fatter in the middle than at its edges, has one flat face and one face with a radius of curvature of 1.00 m. How far from the lens will it focus sunlight?

3. [I] A bi-convex lens made of plastic ($n = 1.58$) has radii of curvature of +1.00 m and −1.00 m. What is its focal length?

4. [I] A glass ($n = 1.5$) bi-convex thin lens has radii of curvature having magnitudes of 2.0 m and 5.0 m. Determine its focal length in air.

5. [I] We measure the radius of curvature of the face of a planar-convex lens to be 2.00 m and find its focal length to be 1.60 m. What is its index of refraction?

6. [I] A cat is 30.0 cm in front of a positive lens that has a 10.0-cm focal length. Locate and describe the image.

7. [I] A convex lens having a 60.0-cm focal length is placed 100.0 cm from a frog. Where will the image of the frog be located?

8. [I] A window on a spaceship in the Mongoian Royal Fleet is flat on one side and slightly concave on the other. As a result, it has a focal length of −10 m. What does the Universe look like through that porthole? A beacon light is 50 m from the window. How far away will it appear to someone looking out?

9. [I] A light bulb is 0.75 m in front of a thin positive lens that has a focal length of 0.25 m. (a) Completely describe the image. (b) Draw a ray diagram.

10. [I] A sharp image of a bus appears on a piece of paper held 2.5 m behind a positive eyeglass lens that has a focal length of 2.0 m. How far away is the vehicle?

11. [I] A light bulb is 80.0 mm in front of a positive lens with a focal length of 120 mm. Determine the location and type of image formed.

12. [I] The lens in a camera has a focal length of 60.0 mm and is 100 mm from the film plane. How far away should the bug be that is having its picture taken?

13. [I] A camera lens has a focal length of 50 mm and an aperture with a radius of 12.5 mm when its diaphragm is wide open. What is the speed of the lens—that is, what is its minimum *f*-number?

14. [I] A 52.0-mm focal-length camera lens is focused on a very distant motorcycle heading toward the photographer. How far must the lens be advanced from the film plane if the driver is to be in focus when 5.00 m away?

15. [I] A 35-mm camera lens has a focal length of 52 mm and is marked *f*/1.4. What is the maximum diameter of the lens aperture?

16. [I] A photo of a racehorse is perfectly exposed but somewhat blurry when taken at 1/30 s and *f*/16. To "stop" the motion, the shutter speed is raised to 1/500 s. What must now be the new *f*-number?

17. [I] A lens used as a magnifying glass is found to provide an angular magnification of 2.0× when viewed by a normal eye in the relaxed state. What is its focal length?

18. [I] Holding the magnifying glass close to his eye, Dr. Watson brought the bloodstained scrap of cloth up toward it until his eye was relaxed and the image clear. If the glass has a focal length of 10 cm and the good doctor has a near-point of 25 cm, what was the magnification?

19. [I] A little magnifier marked 8× is for sale in a camera store. It is designed for examining photographs and allows light to enter from the sides through a clear cone-shaped stand that keeps the lens one focal length above whatever flat surface the thing sits on. How high above the surface should the lens be?

20. [I] The largest refracting telescope in the world at the Yerkes Observatory in Wisconsin has a 40-in.-diameter lens with a focal length of 63 ft. The reflector on Mount Palomar in California has a 200-in.-diameter parabolic mirror with a focal length of 55.5 ft. Determine the *f*-number for both. Which is the faster? (That is, all else being equal, which would require the shorter exposure time?) Roughly how many times faster is one than the other?

21. [I] The 20-cm-diameter image of the face of a clock appears on a screen 1.0 m from a positive lens. A negative lens is then placed in the path 90 cm behind the first lens, and the image moves out an additional 10 cm beyond where it was. Determine the focal length of the second lens. How big is the final image?

22. [I] A negative lens with a focal length of magnitude 15.0 cm is in contact with a positive lens having a focal length of 24.0 cm. What is the focal length of the combination?

23. [I] We wish to spread a narrow parallel laser beam out into a diverging cone of light so that it makes a large blotch on a nearby screen. At our disposal are two positive lenses, each with a focal length of 20 cm. If one lens is placed in the beam and the resulting blotch is still too

small, will it help any to put the other lens in right up against the first? What is the combined focal length?

24. [I] An optical system consists of three lenses with focal lengths of +10 cm, +20 cm, and +5.0 cm. The first and second of these are separated by 30 cm, the third and fourth by 5.0 cm. If parallel light is shone into the first lens, how far from the third will it be brought to a focus by the system?

25. [I] A rather expensive well-corrected (for aberrations) lens consists of three simple lenses of focal lengths +10 cm, −20 cm, and +5.0 cm all glued together. What is the combined focal length? Can it form real images?

26. [I] A lens positioned 1.00 m from a light bulb produces a sharply focused image of the filament on a screen 33 cm beyond the lens. What is the refractive power of the lens?

27. [I] The near-point of a person's eye is 100 cm away rather than a more desirable 25.4 cm. What contact lens should be prescribed?

28. [I] The far-point of a person's eye is 100 cm away rather than a more desirable ∞. What contact lens should be prescribed?

29. [I] Someone wearing corrective contact lenses having a focal length of −5.0 m can see quite normally. Determine this person's unaided far-point.

30. [I] A compound microscope is made with a 20× objective and a 5× eyepiece. What is the total magnification of the device?

31. [I] Two lenses used to form a homemade microscope are mounted in a tube with a separation of 10.0 cm. If the objective has a focal length of 10 mm and the eyepiece has a focal length of 30 mm, what is the so-called tube length?

32. [I] Typically, a good eyepiece will have a focal length of around an inch. Suppose we have one with a focal length of 2.5 cm and we wish to make a 10× astronomical telescope with it. What objective shall we use and how long will the scope end up?

33. [I] Suppose we wanted to make a telescope to look at the stars. At our disposal is an old eyeglass lens (+1.00 D) donated by hyperopic Aunt Jane and a little magnifier having a focal length of 3.0 cm that came with a stamp collection we got from someone. How long should the tube be for relaxed viewing? What magnification can be achieved?

34. [II] Prove that the products of the segments of two intersecting chords of a circle are equal. (See p. 899.)

35. [II] Referring to Fig. 26.6, prove that

$$\rho \approx \frac{(h^2 - y^2)}{2R} \quad \text{and} \quad \delta \approx \frac{y^2}{2R} \qquad [26.5]$$

36. [II] The image of a face is to be projected life-sized onto a screen via a bi-convex lens whose both radii equal 0.60 m. The lens is made of glass ($n = 1.5$), and the whole thing is taking place in air. (a) Compute the necessary location of the object. (b) How far must the lens be from the screen? (c) Draw a ray diagram.

37. [II] Starting with Eq. (26.3), show that, had the lens been immersed in a medium of index n_m rather than air, the Lensmaker's Formula would result, but instead of the first term on the right being n_l, there would now be an n_l/n_m term.

38. [II] Design the lens for a simple 35-mm slide projector that will cast an image on a screen 10 m away that is enlarged 100 times.

39. [II] A grasshopper sitting 10 cm to the left of a convex lens sees its image on a screen 30 cm to the right of the lens. It then jumps 7.5 cm toward the lens. (a) Where will its image be now? (b) Describe the image in both instances. (c) Draw a ray diagram of the situation after the jump.

40. [II] It's a clear cold winter's day in Nebraska, and a youngster playing outside decides to warm herself with an ice-burning glass. She takes a flat sheet of ice 5.0 cm thick, cuts it into a 50-cm-diameter disk, and then shapes one surface so that it's spherical, gradually thinning out to a sharp edge. If it's 5.0 cm thick at its center, determine the focal length of her lens.

41. [II] A photographer wishes to take a picture of his pet chicken Fred, who happens to have a fine face 5.0 cm high. While standing 2.0 m away, he selects a lens that will fill the film (24 mm top-to-bottom) with Fred's poultry physiognomy. What lens should be used?

42. [II] Suppose you wanted to take a picture of the Moon (which has a diameter of 0.273 times that of Earth and is 3.84×10^8 m away) using a 35-mm camera with a normal lens having a 50-mm focal length. How large will the image be on the film? The diameter of the Earth is about 1.27×10^7 m.

43. [II] A person who is nearsighted has a far-point at 1.00 m and a near-point at 25.0 cm. (a) What corrective lens should this individual wear if it's to be mounted 15.0 mm from the eye? (b) Where is the near-point when glasses are worn?

44. [II] George has been nearsighted since he was eighteen and now that he's fifty-five, his far-point has moved in to 3.00 m. But even more annoying, his near-point has recently migrated out to 45 cm. Assuming that both his eyes are the same, prescribe bifocals with the tops for distance and the bottoms for reading to be worn at 2.0 cm from the cornea. Incidentally, it was Ben Franklin who made the first pair of bifocals (by sawing two lenses in half).

45. [II] Using Fig. 26.21, show that the equation for the transverse magnification produced by a pinhole is the same as for a lens. What does this similarity tell us about the pinhole-film plane distance and the magnification?

46. [II] A combination lens consists of a positive lens with a focal length of +0.30 m located 10 cm in front of a negative lens with a focal length of −0.20 m. Determine the location and magnification of the image of an object 30 cm in front of the first lens by finding the effect of each lens in turn.

47. [II] A compound microscope formed of two lenses, a 20× objective and a 10× eyepiece, is adjusted for viewing by a relaxed eye. It has a standard tube length of 160 mm. (a) What is the total magnification of the device? (b) What is the focal length of each lens? (c) Compute the object distance.

48. [II] Suppose we have two lenses with focal lengths of 2.0 cm and 2.0 mm. How should they be arranged to make a microscope to view, with a relaxed eye, an object 2.5 mm from the front lens? What is the separation between lenses? What is the magnification of the device?

49. [III] The image projected by an equiconvex lens ($n = 1.50$) of a 5.0-cm-tall frog standing 0.60 m from a screen is to be 25 cm high. Compute the necessary radii of the lens.
50. [III] A thin double convex glass lens (with an index of 1.56) surrounded by air has a 10-cm focal length. If this lens is placed underwater (having an index of 1.33) 100 cm beyond a small fish, where will the guppy's image be formed?
51. [III] Write an expression for the focal length (f_w) of a thin lens that is immersed in water ($n_w = \frac{4}{3}$) in terms of the focal length it had when it was surrounded by air (f_a). Take the index of the lens to be 1.5.
52. [III] An equiconvex thin lens L_1 is placed in close contact with a thin negative lens L_2, whereupon the combination has a focal length of 0.50 m in air. If the lenses are made of glass with indices of 1.50 and 1.55, respectively, and if the focal length of L_2 is -0.50 m, compute the radius of each surface. Note that a convex surface of L_1 fits precisely against a concave surface of L_2.
53. [III] A homemade TV projection system uses a large positive lens to cast the image of the screen onto a wall. The final picture is enlarged 3 times and although rather dim, it's nice and clear. If the lens has a focal length of 60 cm, what should be the distance between the screen and the wall? Why use a large lens? How should we mount the set with respect to the lens?
54. [III] A nearsighted person has a far-point of 200 cm. What power eyeglass lens should be worn 2.0 cm from the cornea? What contact lens is equivalent to this? Show that both have the same far-point. Verify that Eq. (26.3) agrees with your results.
55. [III] Mary Lou got her first pair of reading glasses (+2.0 D) when she was forty-eight, in 1979. Now, in order to peruse her mail (still wearing those spectacles 2.0 cm down on her nose), she finds she must hold it 80 cm away from her eyes. Prescribe new glasses for her.

MIRRORS

56. [I] A convex mirror has a radius of curvature whose magnitude is 0.50 m. What is its focal length?
57. [I] A polished 50-cm-diameter steel ball in the hands of a statue reflects the scene around it. Determine the ball's focal length.
58. [I] We wish to convert light from a candle flame into a parallel beam using an inexpensive mirror 30 cm away from it. Design the appropriate arrangement and draw a ray diagram.
59. [I] If an object 200 cm from the vertex of a spherical concave mirror is imaged 400 cm in front of the mirror, what is the latter's focal length?
60. [I] A 15-cm-tall Teddy bear stands 60 cm from the vertex of a concave mirror having a radius of curvature of 60 cm. Determine the resulting image and describe it in detail. Draw a ray diagram illustrating the phenomenon.
61. [I] A display lamp having a bright 5.0-cm-long vertical filament is positioned 30 cm from a concave mirror that projects the bulb's image onto a wall 9.0 m from the vertex. (a) What is the radius of curvature of the mirror? (b) How big is the image?
62. [I] Suppose you had a spherical mirror with a 0.20-m radius. If you wished to project the image of a candle flame onto a piece of paper 1.10 m away, where should the candle be located? Describe the image, giving the magnification as well. Draw a ray diagram.
63. [I] If your nose is 20 cm from a convex spherical mirror having a radius of 100 cm, what will the image look like? Draw a ray diagram.
64. [I] The cornea of the human eye behaves like a little convex spherical mirror when seen from nearby. Suppose you look at yourself reflected in someone's eye a distance of 20 cm away. Taking the radius of curvature of the cornea to be 8.0 mm, what will your image look like? Where will it be located?
65. [II] An Earth satellite in the shape of a 2.0-m ball is orbiting at an altitude of 500 km. It is being tracked by a telescope having a spherical concave mirror with a 1.0-m radius of curvature. How big is the image of the craft formed by the mirror?
66. [II] We wish to design an eye for a robot using a concave spherical mirror such that the image of a 1.0-m-tall object 10 m away fills its 1.0-cm-square photosensitive detector (which is movable for focusing purposes). Where should this detector be located with respect to the mirror? What should be the focal length of the mirror? Draw a ray diagram.
67. [II] You are herewith requested to design a little dentist's mirror to be fixed at the end of a shaft for use in someone's mouth. The requirements are (1) that the image be right-side-up as seen by the dentist, and (2) that when held 1.5 cm from a tooth, the mirror should produce an image twice life-size.
68. [II] Prove that with a spherical mirror of radius R, an object at a distance s_o will result in an image that is magnified by an amount

$$M_T = \frac{R}{2s_o + R}$$

69. [II] With the results of Problem 68 in mind, suppose that we have a concave mirror with a radius of curvature of 60 cm, which is forming an image of an object 2.4 m away. What will be the resulting magnification? Describe the image. Draw a ray diagram.
70. [II] A keratometer is a device used to measure the radius of curvature of the cornea of the eye, which is useful information when fitting contact lenses. In effect, an illuminated object is placed a known distance from the eye, and the reflected image off the cornea is observed. The instrument allows the operator to measure the size of that virtual image. Suppose, then, that the magnification is found to be 0.037× when the object distance is set at 100 mm. What is the radius of curvature?
71. [II] Consider a spherical mirror. Show that the locations of the object and image are given by

$$s_o = \frac{f(M_T - 1)}{M_T} \quad \text{and} \quad s_i = -f(M_T - 1)$$

72. [II] Looking into the bowl of a spherical soup spoon, someone 25 cm away sees their image reflected with a

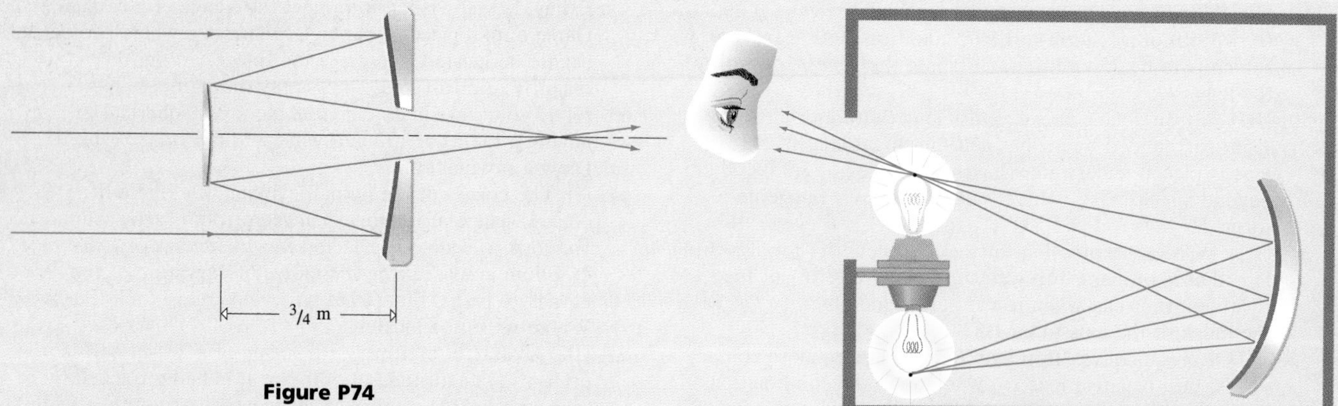

Figure P74

Figure P75

magnification of −0.064. Determine the radius of curvature of the spoon.

73. [III] A large, upright, convex spherical mirror in an amusement park is facing a plane mirror 10.0 m away. A youngster 1.0 m tall standing midway between the two sees herself twice as tall in the plane mirror as in the spherical one. In other words, the angle subtended at the observer by the image in the plane mirror is twice the angle subtended by the image in the spherical mirror. What is the focal length of the latter?

74. [III] The telescope depicted in Fig. P74 consists of two spherical mirrors. The larger (which has a hole through its center) has a radius of curvature of 2.0 m and the smaller of 60 cm. How far from the smaller mirror should the film plane be located if the object is a star? What is the effective focal length of the system?

75. [III] Figure P75 shows the arrangement of a classic illusion that works surprisingly well. A person looks through a window into the box and sees a glowing bulb in a socket. The bulb is turned off and vanishes, leaving only the socket seen in roomlight entering the window. Explain what is happening. If the bulb-mirror distance is 1.0 m, compute the mirror's radius of curvature.

76. [III] Starting with the equation $y^2 + (x - R)^2 = R^2$ for a circle whose center is displaced by a distance R from the coordinate origin, solve for x and expand the appropriate term in a binomial series. Compare the first term with the equation of a parabola having its vertex at the origin, $y^2 = 4fx$.

27

Physical Optics

LIGHT IS ELECTROMAGNETIC, AND it is wavelike, and (if we are careful to remember that what we usually observe are the macroscopic manifestations of torrents of photons) we can say that *light is an electromagnetic wave.* Because the wavelengths associated with the visible region of the spectrum are quite small, the wave nature of light can sometimes be neglected. Geometrical optics can be developed using only rays to indicate the direction light travels, without saying anything about the nature of what's traveling. Still, there are a variety of phenomena—*polarization, interfer-*

Crystals of potassium chloride, calcium carbonate (calcite), and sodium chloride (table salt). Only the calcite produces a double image.

ence, and *diffraction*—that reveal the underlying wave characteristics of radiant energy, and these form the study of **physical optics**.

POLARIZATION

A curious discovery made in 1669 led to one of the great insights into the nature of light. A newly found crystal, now called calcite, was observed to have the remarkable ability to produce double images of everything seen through it. That strange separation, or **polarization** of light, was first studied in depth by Huygens, who explained some aspects of it via his wave theory. In the early 1800s, Augustin Fresnel, joined by Arago, conducted a series of experiments to see if polarization had any effect on the process of interference. Their positive results were utterly bewildering because they, like Huygens, believed light was a longitudinal wave in the ether. For the next few years, Fresnel in France, allied with Thomas Young in England, wrestled with that stubborn problem until finally Young (in 1817) had the crucial revelation: **light is a transverse wave** (something Robert Hooke had suggested a century before).

We can wiggle a length of rope up and down in a vertical plane while the wave progresses horizontally (Fig. 27.1). The disturbance remains in that fixed plane, and we say that the wave is plane-polarized. When viewed end-on, the vibration appears to be along a line perpendicular to the propagation direction, and the situation is also referred to as *linear polarization.* The plane in which such a transverse wave oscillates is the *plane of polarization,* and it certainly need not be vertical—any orientation is equally possible. When the **E**-field oscillates in a plane, the TEM wave is **plane-polarized**.

It's even possible for the **E**-field to swing around in a circle, in which case the wave is said to be **circularly polarized**. If the advancing wave revolves clockwise (looking toward the source), then it's said to be *right-circularly polarized*; if counterclockwise, it's *left-circularly polarized.* The ordinary mode in which a jump rope is operated is a familiar example of the kind. With circular light, the electric-field vector remains constant in magnitude while it revolves once around with every advance of one wavelength (Fig. 27.2). Each sustained configuration is known as a *state of polarization,* and so there are $\mathcal{P}$-state (plane-polarized), $\mathcal{R}$-state (right-circular), and $\mathcal{L}$-state (left-circular) lightwaves.

We needn't concern ourselves with circular light other than to realize that photons can exist in two states, $\mathcal{R}$- and $\mathcal{L}$-, which carry oppositely directed angular momenta. All forms of light, all observed states, are a mix of these two photon states.

Figure 27.1 Plane-polarized waves. The rope oscillates in a plane. Notice how the end point vibrates along a line as the wave advances. As a result, the wave is also said to be linearly polarized.

27.1 Natural Light

The light coming from an ordinary (nonlaser) source is a polychromatic jumble of overlapping emissions from a tremendous number of more or less independent atoms. Each atom radiates a tiny polarized photon-wavetrain lasting less than $\approx 10^{-8}$ s, and these all arrive with no particular phase relationship or orientation, one to the other. The resultant field is the sum of all the separate wavetrain contributions at that location and is constantly changing as new photons arrive and others sweep by. Thus, the net optical field has some polarization at every point, at every instant, but that polarization varies exceedingly rapidly in a totally random fashion. Because the state of polarization of the light is sustained for less than $\approx 10^{-8}$ s, that state is essentially undetectable, and we refer to the light as **unpolarized**, or **natural**. This is the case even though, strictly speaking, there is no such thing as unpolarized light. Often, the light we experience in the environment is a blend of polarized and unpolarized known as *partially polarized.* Ordinary sky light, the light reflected from objects, and the light transmitted through windows are all partially polarized.

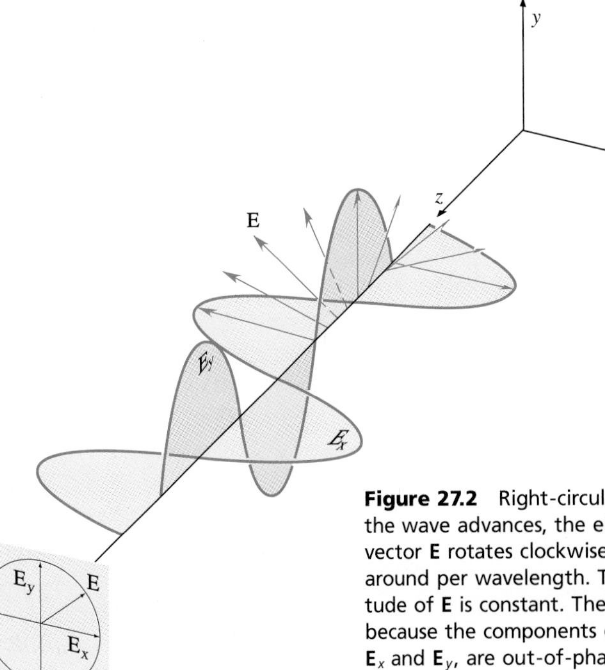

Figure 27.2 Right-circular light. As the wave advances, the electric field vector **E** rotates clockwise once around per wavelength. The magnitude of **E** is constant. The effect arises because the components of the field, $\mathbf{E}_x$ and $\mathbf{E}_y$, are out-of-phase by $\lambda/4$.

There are several ways to conceptually visualize unpolarized light. One of the easiest is to think of it as a superposition of many differently oriented plane-polarized waves of different amplitudes, all changing rapidly and randomly (Fig. 27.3a). In time, the resultant E-field points every which way, swiftly jumping from one orientation and amplitude to another.

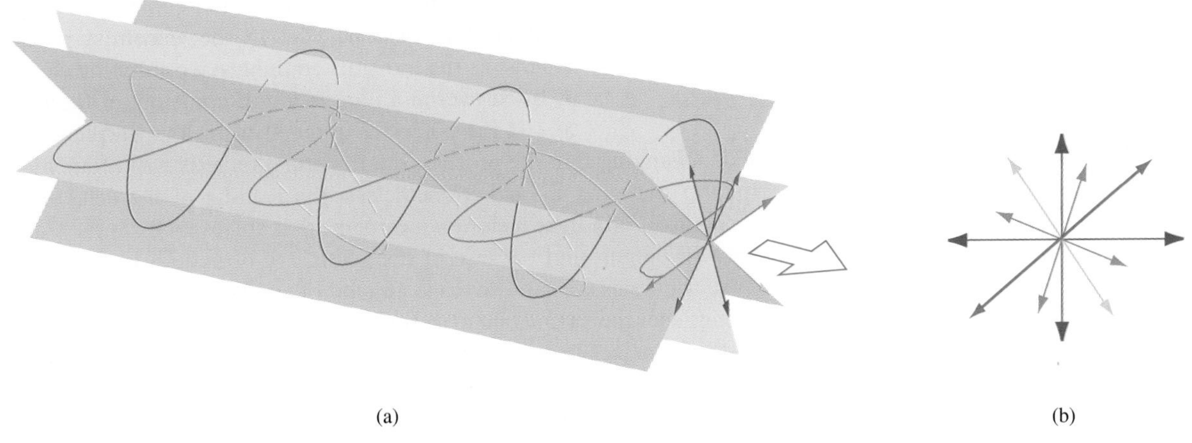

Figure 27.3 (a) Natural light is a jumble of random, rapidly changing fields. (b) Looking toward the source, one sees a resultant field oscillating, first in one direction, then another, each lasting for a fraction of a period before **E** jumps abruptly to a new random orientation.

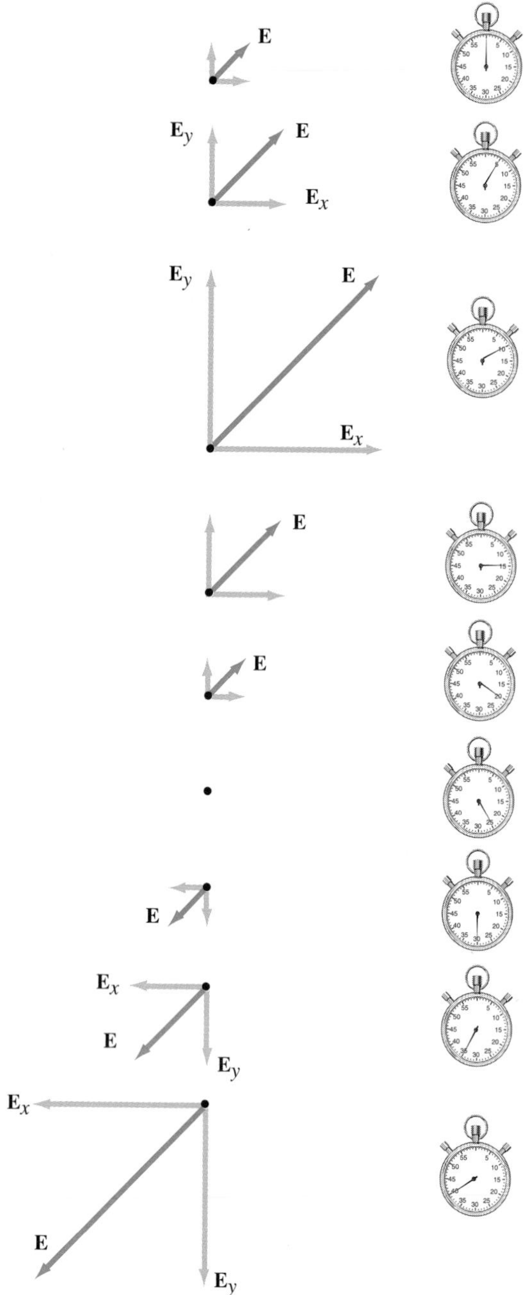

Figure 27.4 A sequence in time showing the oscillation of a linearly polarized light field. **E** is always along a 45° line in the first and third quadrants. The two component fields $\mathbf{E}_x$ and $\mathbf{E}_y$ are in-phase—when one is positive, the other is positive. As the wave sweeps past a fixed point in space, the field oscillates in time, as shown.

Any electric field vector can be resolved into two mutually perpendicular components (which we take for convenience as horizontal and vertical). For example, if at some instant the electric field vector points up and to the right, it can be thought of as the resultant of two smaller constituent fields, one to the right and one up (Fig. 27.4). As the tilted field oscillates, it can be resolved into—and replaced by—the two perpendicular fields that oscillate at the same rate in-step with each other (Fig. 27.5).

The natural-light field vector depicted in Fig. 27.3b can similarly be resolved into horizontal and vertical components that change from moment to moment as the vector changes. The result at any instant is a net horizontal field and a net vertical field. *We can think of the unpolarized jumble as equivalent to two independent, equal-amplitude $\mathcal{P}$-states varying rapidly and randomly along any two perpendicular axes.* Each of these carries $\frac{1}{2}$ the total irradiance (I_0) of the light.

27.2 Polarizers

A device that takes an input of natural light and transforms it into an output of polarized light is a **polarizer**. Natural light is a superposition of two independent, perpendicular $\mathcal{P}$-states. Anything that separates these two equal-irradiance components, discarding one and passing on the other, will produce a beam of light that has its **E**-field fixed in a plane, a beam that is plane-polarized. Such a device is a *linear polarizer.* **Ideally, if natural light of irradiance I_0 is incident on a linear polarizer, $\mathcal{P}$-state light of irradiance $\frac{1}{2}I_0$ will be transmitted.**

Before we go on, we must establish a means of experimentally verifying that a device is in fact a linear polarizer. When natural light is incident on the ideal linear polarizer of Fig. 27.6, only $\mathcal{P}$-state light emerges. It has an orientation parallel to a specific direction called the **transmission axis** of the polarizer. This is not literally a single axis but a direction—any line on the polarizer parallel to that axis is also a transmission axis. Only the component of the incident electric field parallel to the transmission axis will emerge as the transmitted light. An oscillating electric field **E** tilted over at an angle θ will come out of the polarizer.

If the polarizer is revolved in its own plane about the central axis, changing θ in the process, the transmission axis will rotate along with it, as will the emerging **E**-field. Nonetheless, the reading of the detector (for example, a photocell or a light meter on a camera) that corresponds to the transmitted irradiance will be constant because of the symmetry of natural light—each transmitted $\mathcal{P}$-state carries the same average amount of energy. Similarly, the human eye cannot distinguish the various polarization states and will see only a constant irradiance as the polarizer is revolved.

Now, suppose we introduce a second identical polarizer called an **analyzer**, whose transmission axis is vertical (Fig. 27.7). The first polarizer transmits a wave of amplitude E_{01} tilted over at an angle θ. Only the vertical component of that wave,

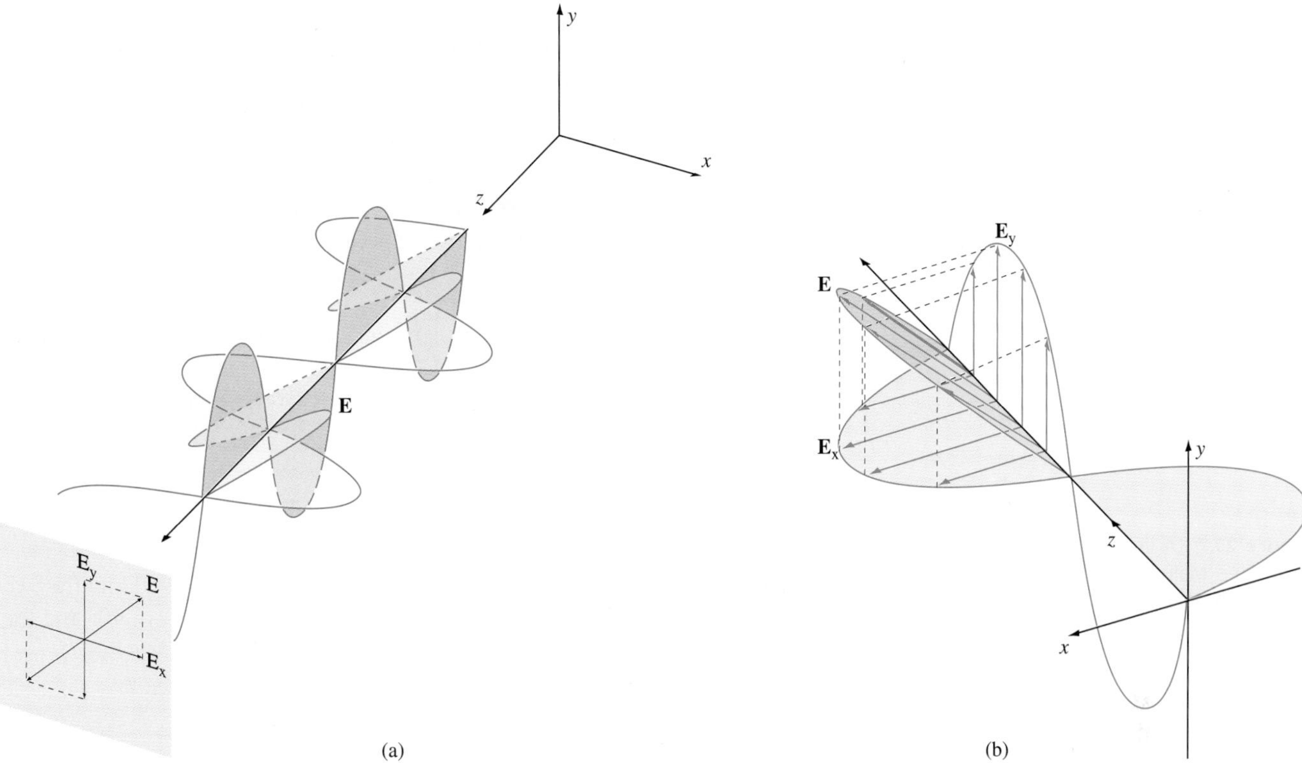

Figure 27.5 A linearly polarized wave. In Fig. 27.4, we examined the disturbance at a fixed point in space as time progressed. (a) Here, we examine the disturbance spread out in space at a given instant in time. (b) The resultant field at any instant is **E**, the sum of $\mathbf{E}_x$ and $\mathbf{E}_y$.

namely $E_{01}\cos\theta$, will be passed by the analyzer because its transmission axis is vertical. (Note that θ is the angle between the transmission axes of the polarizer and analyzer.) According to Eq. (24.8) the irradiance reaching the detector is proportional to the square of the amplitude ($E_{02} = E_{01}\cos\theta$) of the wave coming out of the analyzer; thus

$$I = \tfrac{1}{2}c\varepsilon_0(E_{01}\cos\theta)^2 = (\tfrac{1}{2}c\varepsilon_0 E_{01}^2)\cos^2\theta = I_1\cos^2\theta$$

where the irradiance leaving the first polarizer is I_1. Slowly revolving the analyzer about the central axis varies θ, and the transmitted irradiance changes as the cosine-

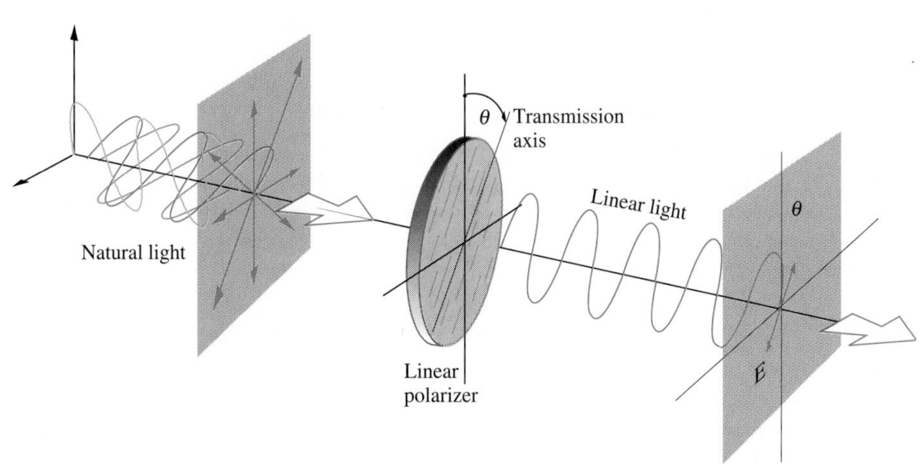

Figure 27.6 Natural light incident on a linear polarizer. The transmission axis of the polarizer is tilted at an angle θ and only the **E**-field parallel to θ is passed. In this case, the polarizer absorbs half the light that enters it.

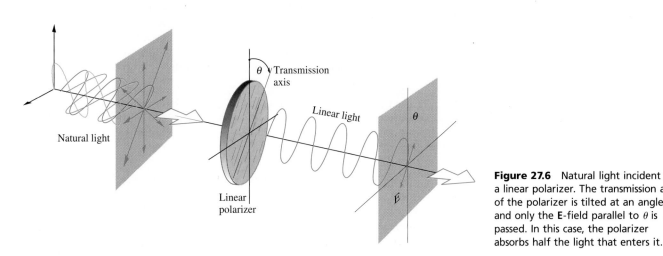

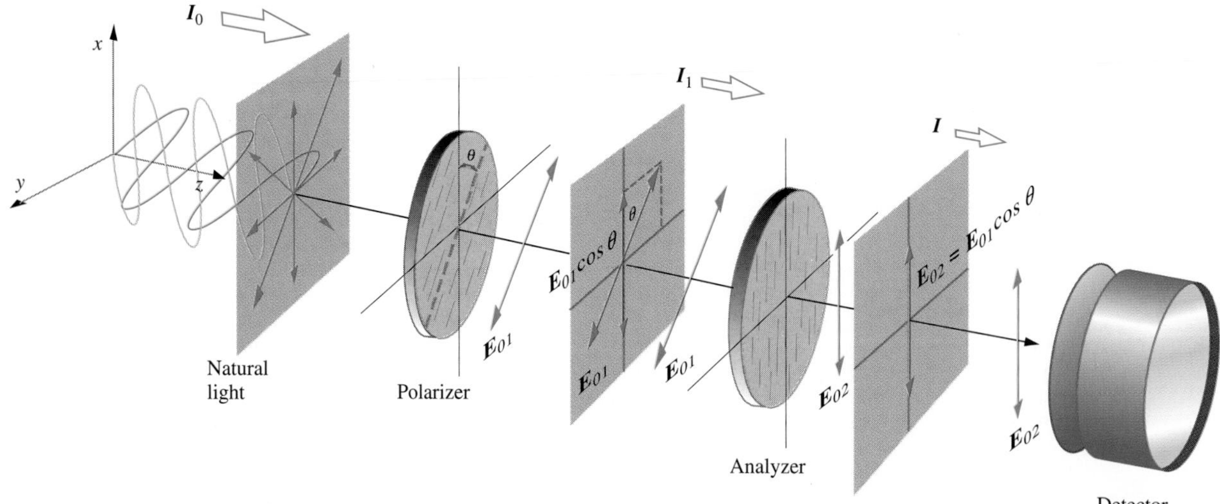

Figure 27.7 A linear polarizer and analyzer—Malus's Law. Natural light enters the polarizer and emerges in a $\mathcal{P}$-state at an angle θ. The analyzer passes a fraction of that light, such that $E_{02} = E_{01} \cos \theta$. The irradiance of the natural light incident on the polarizer is I_0, and it, in turn, transmits an irradiance $I_1 = \frac{1}{2} I_0$. This is incident on the analyzer, and it transmits an irradiance $I = I_1 \cos^2 \theta$.

squared. That is, it varies from its maximum value ($I = I_1$ when $\theta = 0$ and the two transmission axes are parallel) to its minimum value ($I = 0$ when $\theta = 90°$ and the axes are said to be **crossed**). The expression

$$I = I_1 \cos^2 \theta \tag{27.1}$$

is known as **Malus's Law**, having first been published in 1809 by Étienne Malus, military engineer and captain in the army of Napoleon. Using the setup of Fig. 27.7 and Malus's Law, we can test polarizers to see if they are linear by simply revolving one in front of the other. In fact, if we analyze a beam of light through a revolving polarizer and find that it varies in accord with Eq. (27.1), going completely dark and then bright every 90°, the beam must be $\mathcal{P}$-state light.

Example 27.1 A beam of unpolarized light with an irradiance of 1000 W/m^2 impinges on an ideal linear polarizer whose transmission axis is vertical. The light is studied using a second ideal linear polarizer, and it is found that the final emerging beam has an irradiance of 250 W/m^2. (a) How much light leaves the first polarizer? (b) What is the orientation of the second polarizer?

Solution: [Given: $I_0 = 1000$ W/m^2 incident and $I = I_2 = 250$ W/m^2. Find: (a) I_1 and (b) θ.] (a) If natural light is incident with an irradiance I_0, half of that will be "eaten" by the first polarizer, which must discard half the light if it is to pass only a $\mathcal{P}$-state. Hence, $I_1 = \frac{1}{2} I_0 = \boxed{500 \text{ W/m}^2}$. (b) Since the second polarizer passes $I_2 = 250$ W/m$^2 = \frac{1}{2} I_1$, it follows from Malus's Law that $\cos^2 \theta = \frac{1}{2}$. Consequently, $\cos \theta = \sqrt{\frac{1}{2}}$ and $\boxed{\theta = 45°}$; the second polarizer is 45° from the vertical.

▶ **Quick Check:** Substituting back into Eq. (27.1), $I_2 = (500 \text{ W/m}^2) \cos^2 45° = 250$ W/m^2.

There are many kinds of linear polarizers, among them several that operate by the dissipative absorption of one of the two perpendicular $\mathcal{P}$-state components of the incident light, a process broadly known as *dichroism*. Of these, we will examine two types—wire grids and Polaroids—the first only because it explains the second.

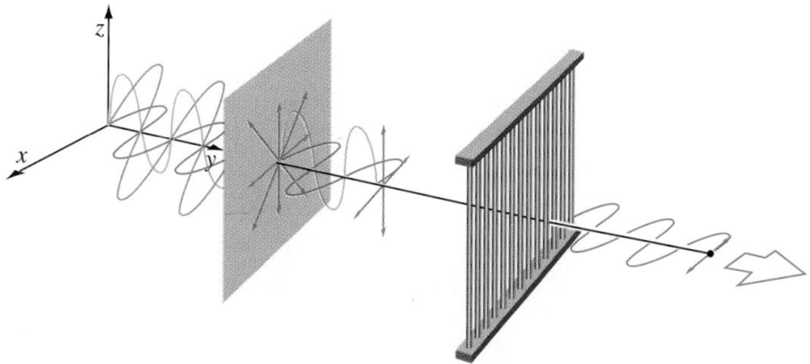

Figure 27.8 Natural light incident on a wire-grid polarizer. The unpolarized light can be represented as two uncorrelated $\mathcal{P}$-states, one horizontal and one vertical. The grid absorbs the vertical E-field and passes the horizontal $\mathcal{P}$-state.

The Wire-Grid Polarizer

The **wire-grid polarizer** consists of a grid of parallel conducting wires each spaced less than a wavelength apart. The electric field of an incident unpolarized electromagnetic wave can again be imagined resolved into two independent, perpendicular components—one parallel to the wires and the other transverse to them (Fig. 27.8). The vertical component of the field effectively drives the free electrons within the wires. Oscillating, up-and-down currents result along the length of each conductor. These currents dissipate some of the wave's energy via joule heating, thereby diminishing the vertical field. At the same time, the accelerating charges reradiate like tiny dipoles. But these emissions are out-of-phase with the incoming wave, thus canceling most of the remaining vertical field in the forward direction. The reradiation in the backward direction constitutes the reflected wave and carries off the remainder of the energy. By contrast, the currents oscillating across the wires are very restricted, and there is hardly any effect on the transverse electric field, which passes through the grid almost the same as it arrived.

The transmission axis of the grid is perpendicular to the wires. It is an extremely common error to assume that the field somehow slips between the wires and is vertically polarized—it doesn't and it isn't! Such grids have been made to operate in the visible (one had 2160 wires per mm) and infrared, though they are much easier to fabricate for microwaves.

Polaroids

In 1928, Edwin H. Land, then a nineteen-year-old undergraduate at Harvard College, invented the first dichroic *sheet polarizer.* That was the humble start of the now-famous Polaroid Corporation. In 1938, Land devised the much improved *H-sheet Polaroid,* which is now easily the most widely used linear polarizer. A sheet of clear polyvinyl alcohol is heated and stretched, thereby aligning—in nearly parallel rows—its long-chain hydrocarbon molecules. The sheet is then dipped into a dye solution rich in iodine. The iodine atoms impregnate the plastic and attach to each polymeric molecule, coating it and effectively forming a conducting chain along it. The conduction electrons associated with the iodine can now move down each chain, from one atom to the next, as if the coated molecule was a long, thin microscopic wire. The result is essentially a wire-grid polarizer with its transmission axis transverse to the aligned molecules. In natural light, each sheet looks gray because it absorbs roughly half the incident light.

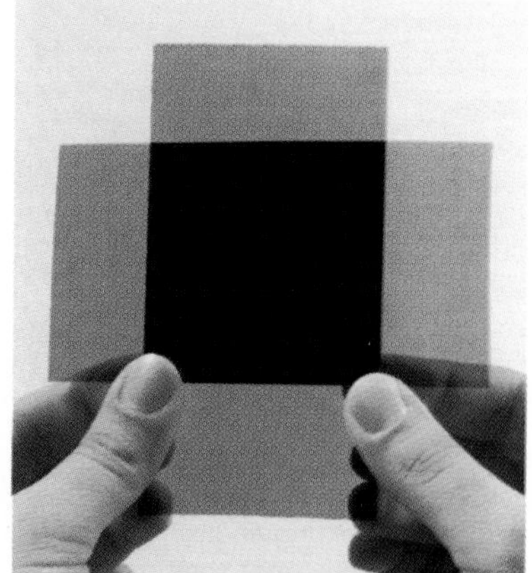

A pair of crossed Polaroids. The electric field passed by the first polarizer is perpendicular to the transmission axis of the second polarizer.

Example 27.2 An unpolarized light beam of 800 W/m^2 is incident on an ideal pair of crossed linear polarizers. Now a third such polarizer is inserted between the other two with its transmission axis at 45° to that of each of the others. Determine the emerging irradiance before and after the insertion of the third polarizer and explain what's happening.

Solution: [Given: polarizers at $\theta = 0°$, $\theta = 90°$, and $\theta = 45°$, and $I_0 = 800$ W/m^2. Find: I_2 and I_3.] Before the third polarizer is inserted, no light passes through the crossed pair. It might seem that inserting yet another polarizer could have no effect, but that's wrong! If the first filter has a vertical transmission axis, then $I_1 = \frac{1}{2}I_0 = 400$ W/m^2 of vertical $\mathcal{P}$-state light will emerge from it. The second polarizer makes an angle of 45° with the first; hence, by Malus's Law

$$I_2 = I_1 \cos^2 45° = \boxed{200 \text{ W/m}^2}$$

The transmitted electric field is now tilted at 45°—*the presence of the second polarizer has changed the field reaching the third polarizer.* The field is at 45° with respect to the last polarizer whose transmission axis is horizontal. Again from Malus's Law

$$I_3 = I_2 \cos^2 45° = \boxed{100 \text{ W/m}^2}$$

which is the amount of horizontal $\mathcal{P}$-state light finally emerging through all three filters.

▶ **Quick Check:** $E_{01} \propto \sqrt{I_1}$; $E_{02} = E_{01} \cos 45°$; $I_2 \propto E_{02}^2 = E_{01}^2 \cos^2 45°$; $I_2 = I_1(1/\sqrt{2})^2 = \frac{1}{2}I_0(1/\sqrt{2})^2 = \frac{1}{4}\,800$ W/m^2.

27.3 Polarizing Processes

There are several important natural processes that produce polarized light. One of the most commonplace is the reflection from dielectric media. The glare spread across a window pane, a sheet of paper, or a classroom desk; the sheen on a balding head or a shiny nose are all partially polarized.

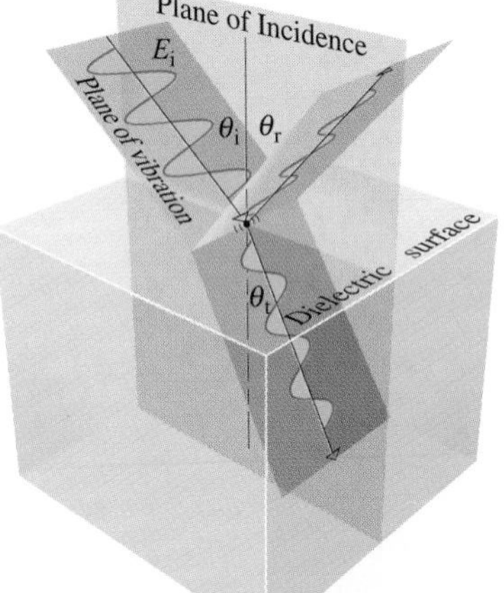

Figure 27.9 A wave with its E-field perpendicular to the plane-of-incidence reflecting and refracting at an interface. Electrons then oscillate perpendicular to the plane-of-incidence and reradiate light perpendicular to that plane.

Polarization by Reflection

We can understand what's happening in a simple way by again considering the electron-oscillator model of the atom, although the description is a little simplistic. Figure 27.9 shows an incident wave plane polarized *perpendicular to the plane-of-incidence.* The atoms in the surface are driven by the E-field parallel to the interface and they reradiate, producing the usual reflected and refracted waves, both of which are similarly polarized. On the other hand, if the incident E-field is linearly polarized *in the plane-of-incidence,* something rather different happens to the reflected wave. The atoms at and near the surface are again driven into oscillation by the E-field of the transmitted wave (Fig. 27.10), but this time that's not parallel to the interface. The oscillations are all perpendicular to the refracted ray, making a small angle θ with the reflected ray. Because these dipoles do not radiate along their oscillatory axes, the amount of light reflected is relatively small. In fact, if we could arrange things so that $\theta = 0$, whereupon the reflected and transmitted rays would be perpendicular ($\theta_r + \theta_t = 90°$), no light would be reflected at all.

The special angle of incidence for which this occurs is designated by θ_p and is called the **polarization angle**, or *Brewster's angle,* where $\theta_p + \theta_t = 90°$. It follows from Snell's Law and the fact that $\theta_t = 90° - \theta_p$ that

$$n_i \sin \theta_p = n_t \sin \theta_t = n_t \sin(90° - \theta_p) = n_t \cos \theta_p \tag{27.2}$$

where shifting the sine by 90° makes it a cosine. Dividing both sides by cos θ_p and again by n_i yields

$$\tan \theta_p = \frac{n_t}{n_i} \tag{27.3}$$

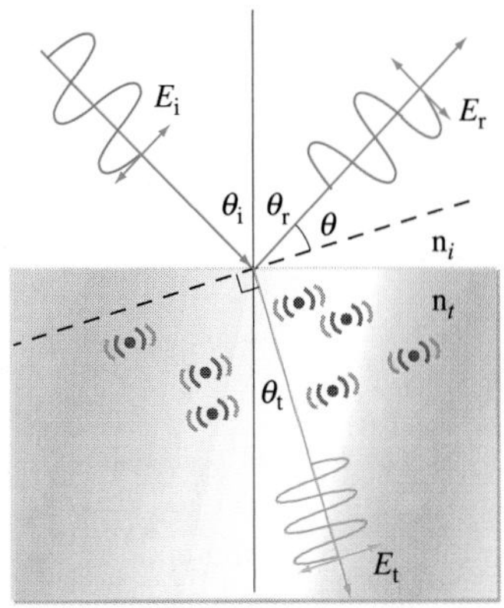

Figure 27.10 Electron-oscillators and Brewster's Law. Here, the incident E-field is in the plane-of-incidence. When $\theta = 0$, the electrons near the surface will oscillate parallel to what would ordinarily be the direction of the reflected beam. But they do not radiate along their axis of vibration, so there will be no reflected wave; $E_r = 0$.

This equation is known as **Brewster's Law** after the man who discovered it empirically, Sir David Brewster, the inventor of the kaleidoscope.

When an incoming unpolarized beam is incident on a dielectric surface at an angle θ_p, *only the component polarized normal to the plane-of-incidence will be reflected*: **the reflected light will be totally plane-polarized parallel to the interface** (Fig. 27.11). At any other angle, the reflected light will be partially polarized. This provides a handy way to find the transmission axis of a linear polarizer. Locate the reflected glare of some light source on a flat, horizontal, nonconducting surface (at an angle around 50°, which is a typical value for θ_p). Now, slowly revolving the polarizer, view the glare through it and when the reflected light essentially vanishes, the transmission axis is vertical (see the photo on p. 948).

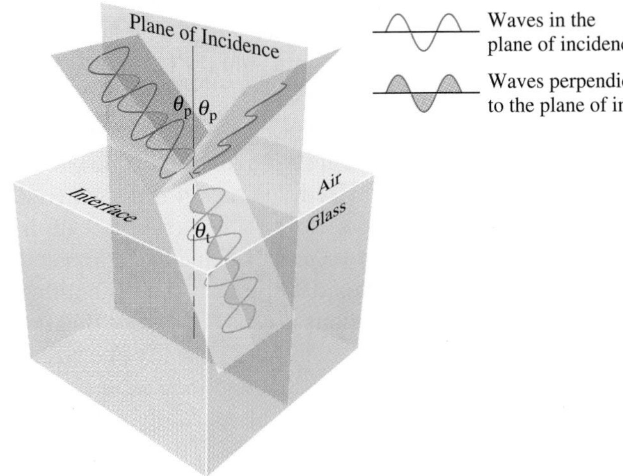

Figure 27.11 The polarization of light that occurs on reflection from a dielectric, such as glass, water, or plastic. At θ_p, the reflected beam is a $\mathcal{P}$-state perpendicular to the plane-of-incidence. The transmitted beam is strong in $\mathcal{P}$-state light parallel to the plane-of-incidence and weak in $\mathcal{P}$-state light perpendicular to the plane-of-incidence—it's partially polarized.

Example 27.3 The image of a child reflects off a wet city street. At what angle should the reflection be viewed if it is to be seen in totally polarized, linear light? Give your answer to three significant figures.

Solution: [Given: $n_i = 1.00$ and $n_t = 1.33$. Find: θ_p.] The relationship that governs polarized reflections is Brewster's Law:

$$\tan \theta_p = \frac{n_t}{n_i} = \frac{1.33}{1.00} = 1.33$$

and so $\boxed{\theta_p = 53.1°}$.

▶ **Quick Check:** tan 53.1° = 1.33.

(a)

(b)

Light reflecting off the puddle is partially polarized. When viewed through a Polaroid filter whose transmission axis is parallel to the ground, the glare is passed and visible (a). When the Polaroid's transmission axis is perpendicular to the water's surface, most of the glare vanishes (b).

Polarization by Scattering

When light is scattered by molecular-sized particles, as in Rayleigh scattering (p. 864), the particles behave like tiny dipole oscillators. Because the dipole radiation pattern is not the same in all directions, the scattered light will be polarized. Figure 27.12a depicts an incident vertically polarized plane wave driving a molecule, just as sunlight drives the molecules of the atmosphere. Light is reradiated in all directions, except along the axis of vibration—straight up and straight down. In Fig. 27.12b, the incident beam is horizontally polarized, and again it is scattered into all directions, except along the axis of vibration. Now, if the incident beam is unpolarized, it can be envisioned as equivalent to two perpendicular, uncorrelated $\mathcal{P}$-states. Figure 27.12c shows the scattering of such a disturbance, and it's nothing more than the superposition of the two previous situations (a) and (b). Notice that the light coming off perpendicular to the original beam is completely linearly polarized.

If you happen to have a piece of Polaroid, locate the Sun and then examine the sky roughly at right angles to the solar rays. It will appear only partially polarized, mainly because of the depolarizing effects of multiple scattering. The light is scattered and re-scattered several times before reaching you. That mechanism can be illustrated by putting a piece of waxed paper between crossed Polaroids. The multiple reflections within the paper jumble up the E-field orientations and depolarize the transmitted light. As a last example, return to the "put a few drops of milk in a tank of water" experiment of Chapter 25 (p. 867) where suspended, minute fat particles do the scattering. This time examine the scattered light transverse to the beam through a Polaroid—it will be polarized, just as is the sky light. Apparently, bees and pigeons can distinguish polarized sunlight and navigate using it even when the Sun cannot be seen during overcast days.

A piece of waxed paper between crossed polarizers. Light, having passed through the first filter, is plane-polarized, but the hodgepodge of fibers depolarizes the beam. Try it with an ordinary piece of wet paper.

Birefringence

Many crystalline substances, such as ice, quartz, and calcite, have optical properties that are not the same in all directions. Because the atoms of these substances are not

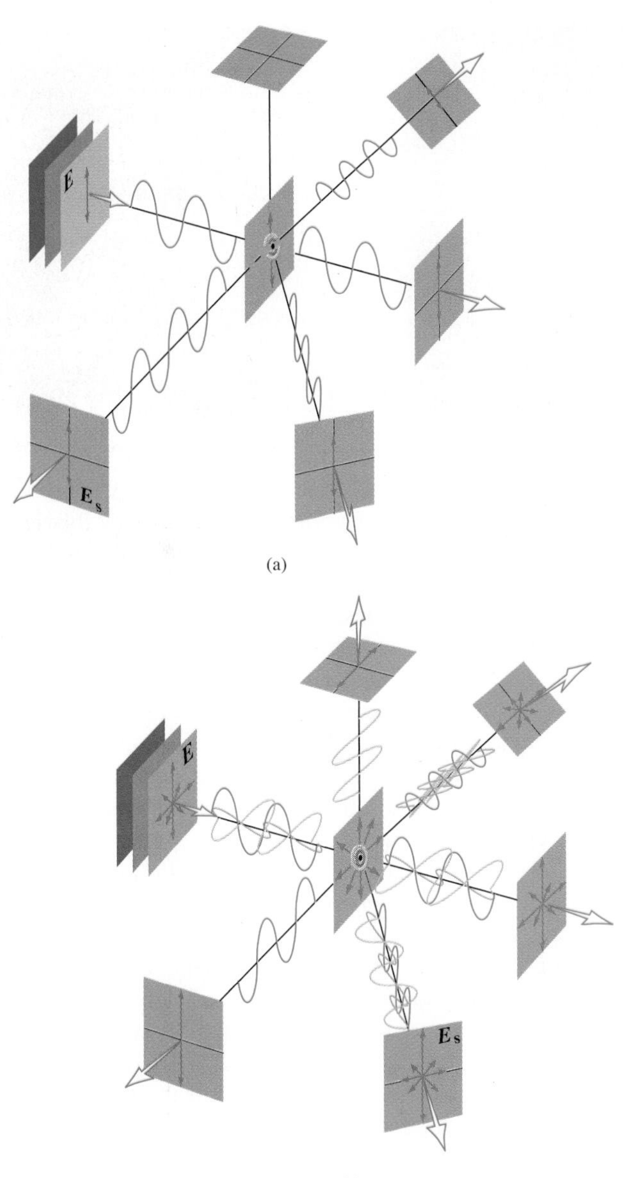

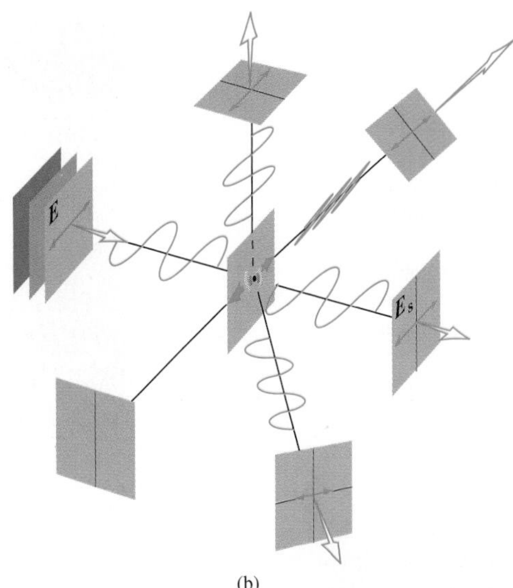

Figure 27.12 Scattering of polarized light by a molecule. Vertically polarized light scatters, as shown in (a); horizontally polarized light scatters as in (b). Since natural light (c) can be imagined as the sum of two perpendicular 𝒫-states, it scatters as the superposition of (a) and (b). Notice how the light scattered perpendicular to the propagation direction in (c) is linearly polarized.

A pair of crossed polarizers. The lower one is noticeably darker than the upper one, indicating that scattered sky light is partially polarized.

uniformly arrayed, the forces on their electron clouds are different in different directions. Thus, the response of the atomic oscillators depends on the direction of the incoming E-field. As a result, such crystals display two different indices of refraction. The word *refringence* used to be used instead of our more modern term *refraction,* hence the name **birefringence**, meaning doubly refractive.

When unpolarized light enters a calcite crystal, each of the two constituent, independent, perpendicular 𝒫-states encounters an essentially different medium, and each travels with a different speed. The two beams refract at different angles (ac-

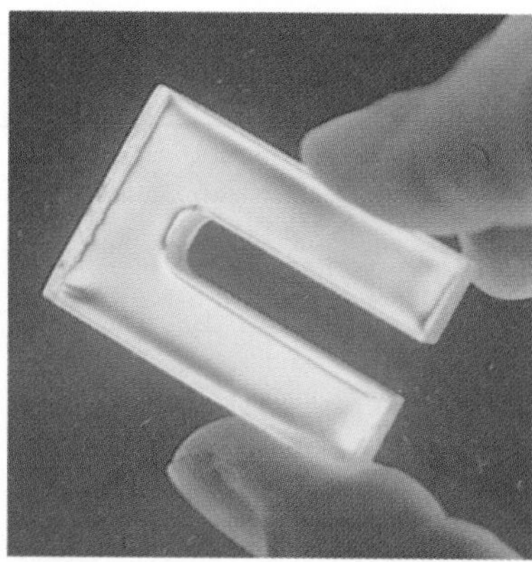

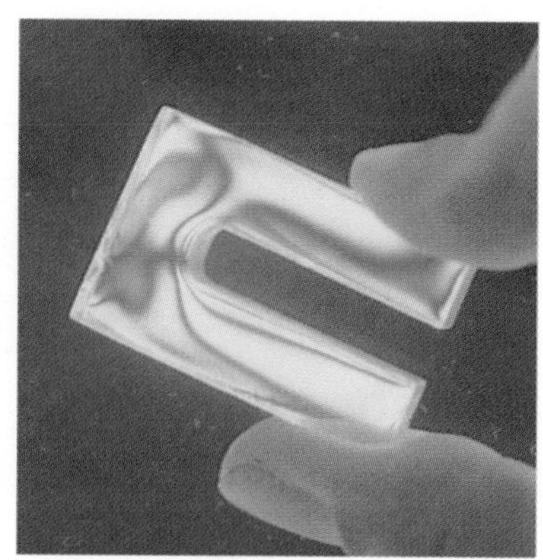

Stress within a material can cause shifts among its molecules, which then affect the transmission of polarized light. Here a piece of clear plastic between crossed Polaroids is being squeezed to increase that stress. Notice how dust, deliberately left between the crossed Polaroids scatters and depolarizes light, thereby becoming quite visible.

A piece of crumpled cellophane placed between crossed Polaroids. Depending on its thickness and the frequency of the light, the cellophane rotates the **E**-field by different amounts. Rotating one of the Polaroids will shift the colors to their complements.

cording to Snell's Law) and therefore separate, producing the double image discussed earlier (see the photo on p. 940). By viewing both images through a linear polarizer, we can easily see that the two are plane-polarized in perpendicular directions. Many human-made substances, such as certain transparent food wraps, clear plastic tapes, and plastic microscope slides, are birefringent because of the alignment of their long-chain molecules—the best of all of these is cellophane. Crumple up a sheet of cellophane (still to be found wrapped around things like imported cigars) and insert it between crossed Polaroids. A profusion of multicolored regions will appear, resulting from the variations in thickness of the birefringent material—the indices of refraction of all substances are frequency dependent.

Some transparent isotropic materials that ordinarily show no polarization effects can have their molecules shifted by an applied force, producing an anisotropy and what is known as *stress birefringence.* Most clear plastic implements—rulers, spoons, cups, etc.—have this stress birefringence frozen in as they solidify. Between crossed polarizers, they reveal the internal stress pattern. To study the stresses within a machine part or an architectural element, we need only make a model of the piece in plastic and examine it in polarized light. Glass lenses when stressed in manufacture or by improper mounting will also show birefringence. Federal regulations require that eyeglasses be heat-treated to remove internal stresses that make them more likely to break. A simple test with Polaroids is used to evaluate the annealing. The back windows of automobiles are often deliberately stressed so that they will shatter into small harmless pieces on impact, and that induced birefringence is often visible, even without polarizers, at Brewster's angle.

INTERFERENCE

The basic idea of interference was introduced earlier (p. 466): in the region of overlap, two waves of the same frequency can combine constructively or destructively depending on their relative phase (Fig. 25.3, p. 865). Thus, if two identical point sources S_1 and S_2 emit spherical waves with the same wavelength, we can expect that the space surrounding them will contain some pattern of interference. Assuming the two sources continue to pump out light in-step with each other, the waves reaching an arbitrary point P in Fig. 27.13 will interfere in a stable observable fashion. Their relative phase (δ) at P will determine exactly how they interfere, and δ

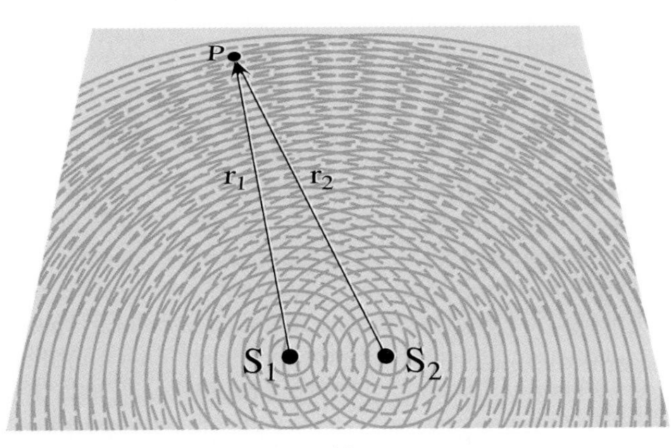

(a)

(b)

Figure 27.13 (a) Two point sources S_1 and S_2 sending out waves in-phase. Here we examine what's happening in a plane and draw the waves as circular. Where peak overlaps peak (two solid lines meet) and trough overlaps trough (two dashed lines meet), there is constructive interference and a maximum. Where peak overlaps trough (a solid line and a dashed line meet), there is destructive interference and a minimum. (b) Water waves in a ripple tank. (See the photo on p. 467.)

will depend on the difference between the two optical path lengths (that is, the index of the medium times the actual path length) traveled by the waves. Here, we will assume the index of refraction is one. Thus, if one route (r_1) is greater than the other (r_2) by half a wavelength ($r_1 - r_2 = \frac{1}{2}\lambda$), then $\delta = 180°$ and destructive interference occurs—the waves arrive out-of-step; trough overlays peak, there is cancellation, and P appears as a dark spot. The same thing would happen if the path difference was $1\frac{1}{2}\lambda$ or $2\frac{1}{2}\lambda$ or $3\frac{1}{2}\lambda$ and so, quite generally, a **minimum** in the irradiance occurs when

$$(r_1 - r_2) = m'\tfrac{1}{2}\lambda \tag{27.4}$$

where $m' = \pm1, \pm3, \pm5, \ldots$. Similarly, if the path difference corresponds to λ or 2λ or 3λ, the overlapping waves will arrive in-phase, $\delta = 0$ or, equivalently, 2π or 4π or 6π, etc. In general then, a **maximum** occurs when

$$(r_1 - r_2) = m\lambda \tag{27.5}$$

where $m = 0, \pm1, \pm2, \ldots$.

Suppose that the right side of either of these two expressions is held fixed (that is, m' = constant or m = constant). The left side then represents all the possible locations of P for which $(r_1 - r_2)$ = constant. But if the difference in the distances from any point P to two fixed points S_1 and S_2 is a constant, as it is here, then the locations of P, which satisfy Eqs. (27.4) and (27.5), form a family of hyperbolas; that's the definition of a hyperbola. In three dimensions, a vertical viewing screen placed at P, perpendicular to the plane of S_1 and S_2, shown in Fig. 27.13, will be covered with vertical bright and dark bands known as **interference fringes**. The energy, instead of being uniformly distributed, is redirected out of certain areas and into others—what is missing at minima appears at maxima: energy is conserved.

27.4 Coherence

It should be noted that the two sources producing interference need not really be in-phase with each other. A somewhat shifted but otherwise identical pattern will oc-

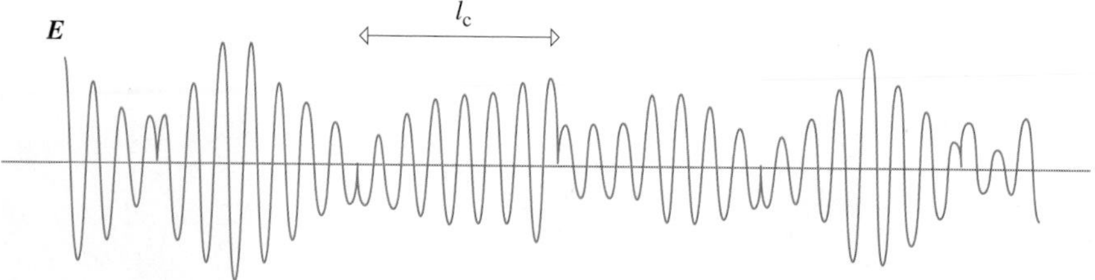

Figure 27.14 A light beam will be nicely sinusoidal for a time t_c, extending in space a length l_c, before it changes randomly. The more nearly monochromatic the light is, the longer is l_c.

cur even if there is some *initial phase difference* between the sources, so long as it remains constant. Such sources (which may or may not be in-step, but are always marching together) are said to be **coherent**. Remember that because of the granular nature of emission from atoms, a conventional quasimonochromatic source produces light that is a mix of photon wavetrains. At each illuminated point in space, there is a net field that oscillates nearly sinusoidally through fewer than a million cycles or so for less than 10^{-8} s before it randomly changes phase (Fig. 27.14). This interval (over which the wave resembles a sinusoid) is a measure of its **temporal coherence**. *The average time interval during which the lightwave oscillates in a predictable way* is known as the **coherence time** (t_c). The longer the coherence time, the greater the temporal coherence of the light.

The corresponding spatial extent over which the lightwave oscillates in a regular, predictable way is called the **coherence length**, l_c, where in vacuum $l_c = ct_c$. It is often convenient to picture the light beam as a progression of well-defined, more or less sinusoidal wavegroups of average length l_c whose phases are quite uncorrelated one to the other. Note that *temporal coherence is a manifestation of spectral purity*. If the light were ideally monochromatic, the wave would be a perfect sinusoid with an infinite coherence length. By comparison, a good laboratory discharge lamp has a coherence length of only several millimeters, whereas certain special lasers provide coherence lengths of tens of kilometers.

Two ordinary sources, two light bulbs or candle flames, can maintain a constant relative phase for a time no greater than t_c (which is $<10^{-8}$ s), and so the interference pattern they produce will randomly shift around in space at an exceedingly rapid rate, averaging out and making it quite impossible to observe. Until the advent of the laser, no two different sources could produce an observable interference pattern. Several decades ago interference was electronically detected using independent lasers, and recently (1993), fringe patterns produced by two separate lasers have even been photographed directly.

27.5 Young's Experiment

The main problem in producing interference is the sources: they must be *coherent*. And yet separate, independent, adequately coherent sources, other than the modern stabilized laser, don't exist! That dilemma was first solved two hundred years ago by Thomas Young in his classic double-slit experiment. He brilliantly took a single wavefront, split off from it two coherent portions, and had them interfere. In effect, he produced two coherent sources using a single wavefront. The technique is straightforward (Fig. 27.15a)—an opaque screen σ_a containing two identical small apertures is illuminated by a symmetrical wave that impinges on the two holes in the same way. That wave can be planar, spherical, or cylindrical provided that the phases of the two oscillating E-fields across the holes are identical. The light streaming from the apertures S_1 and S_2 pours out (diffracts) as if from two coherent identical sources.

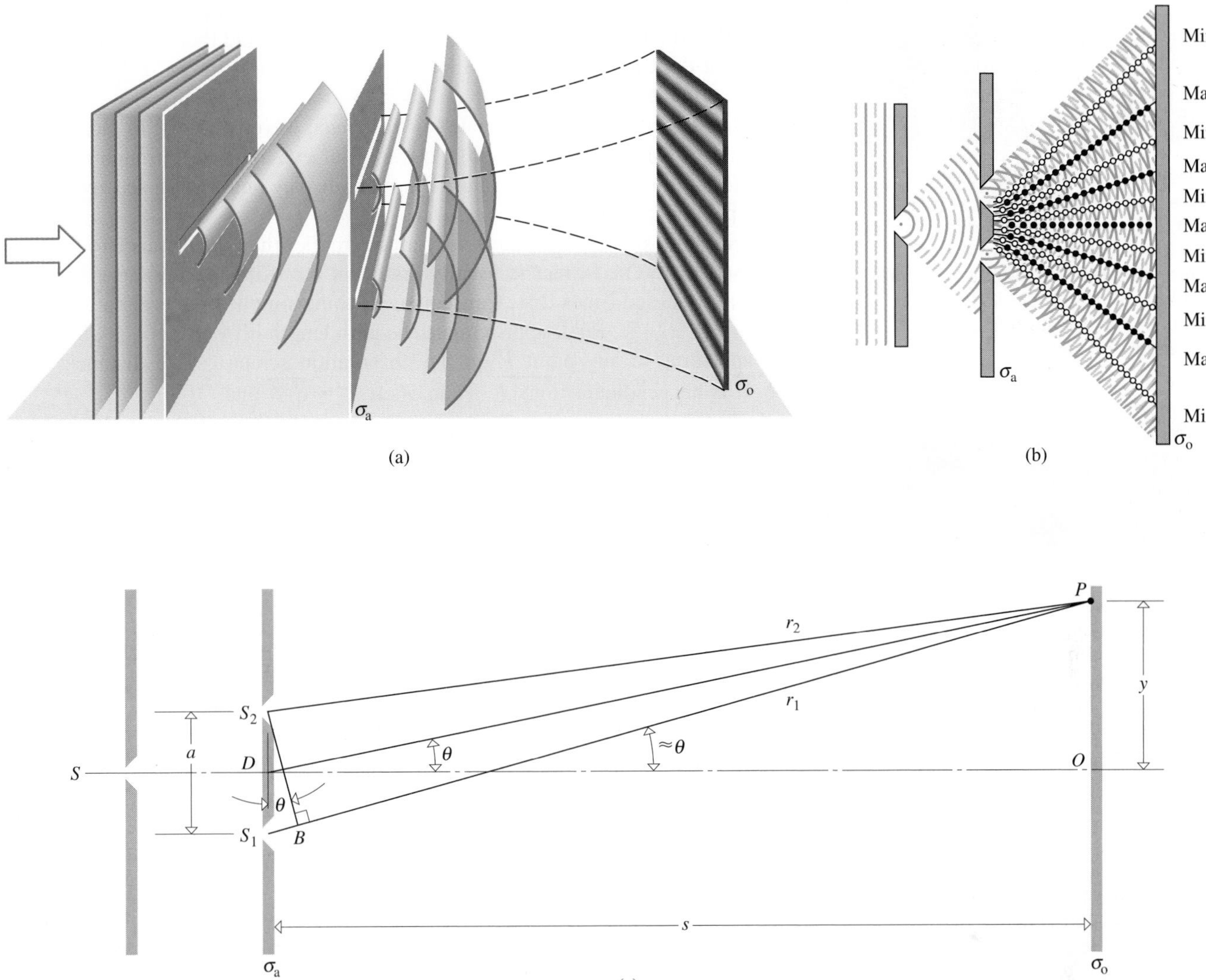

Figure 27.15 Young's Experiment. S_1 and S_2 are two coherent sources—they can be slits or holes (slits provide more light and a brighter pattern). The waves from them overlap, as in Fig. 27.13, and fill the space with fringes. The fringes that appear on a screen σ_o are horizontal in (a), (b), and (c). (b) is an edge-view of (a). With a laser beam, σ_a can be illuminated directly without the need for the single slit at far left. (c) Usually $s \gg a$ and $s \gg y$, in which case angle $S_1S_2B \approx \theta$.

The apertures could be pinholes, in which case the setup closely resembles Fig. 27.13. In Fig. 27.15a, at every point where trough lies on trough (dashed line crosses dashed line) or peak overlaps peak (solid line crosses solid line), the waves are in-phase and a maximum or bright spot appears. The lines connecting all of these maxima are rather straight hyperbolas. Similarly, where trough overlaps peak (dashed line crosses solid line), the waves are out-of-phase and a minimum or dark spot appears. These, too, lie on fairly straight hyperbolas. As the wavefronts sweep out and away from the apertures, the points of overlap—the maxima and minima—travel out along the respective hyperbolas.

Put two pinholes (about the size of a period) one diameter apart in an opaque piece of paper. Hold it very close to your eye and look through them at a distant streetlamp at night. The characteristic fringe pattern will appear on your retina.

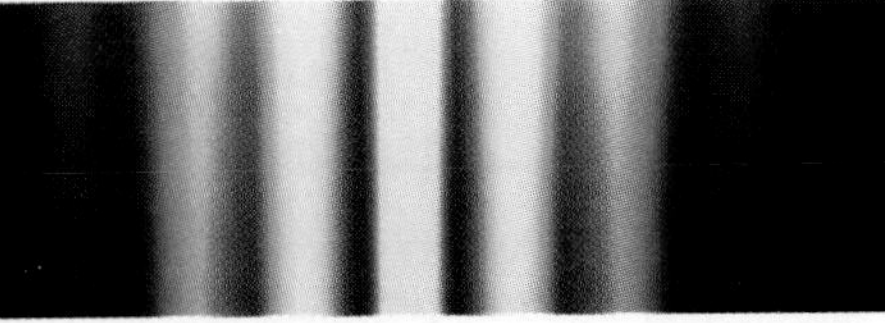

A double-slit fringe pattern. When white light is used in Young's Experiment, each wavelength produces its own fringe pattern slightly shifted from the others. The result is a central white band ($m = 0$ for all λ) surrounded by fringes that are increasingly colored.

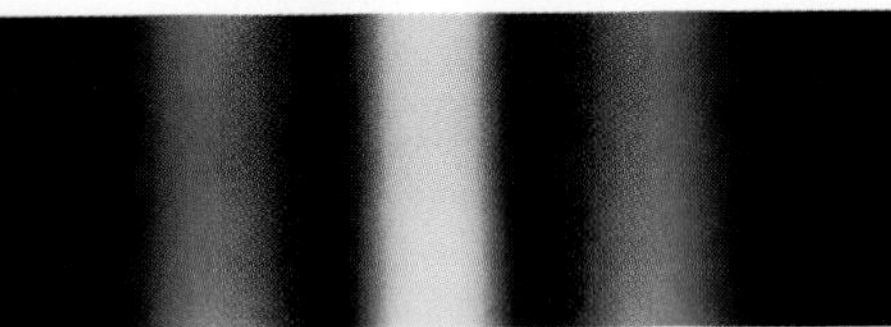

A double-slit fringe pattern. When the separation between slits, a, is decreased, the distances the fringes are from the central axis, $y_m = sm\lambda/a$, increases. Similarly, the fringes broaden since $\Delta y = s\lambda/a$. All of this is easily seen with an ordinary long-filament display light bulb.

In order to pass a greater amount of light and get a brighter pattern, narrow slits are used rather than pinholes. An observation screen σ_o would then be covered with fairly straight dark and light bands running parallel to the slits, as in Fig. 27.15b. There we see plane waves impinging on a slit to produce a cylindrical wave, but a laser beam could just as well illuminate σ_a with plane waves directly. In fact, if you place σ_a (a 3 × 5 index card will do) close to your eye and look through the slits (cut with a razor) at an ordinary straight-filament display bulb, you'll see a wonderful fringe pattern cast on your retina.

Generally, the slit separation a is very much smaller than the distance to the observation screen s. The former is usually less than a millimeter, the latter is typically several thousand times that. Consequently, some simplifying assumptions can be made in Fig. 27.15c. First, approximate the path length difference $(r_1 - r_2)$ from the two apertures to any point P on the observation screen as the distance $\overline{S_1 B}$ where $\overline{S_2 B}$ is perpendicular to $\overline{S_1 P}$. Then $a(\sin\theta) = \overline{S_1 B}$ and

$$(r_1 - r_2) = a\sin\theta \tag{27.6}$$

But θ is very small and so $\sin\theta \approx \theta$, in which case

$$(r_1 - r_2) \approx a\theta$$

Inasmuch as $\angle PDO = \theta$, $\tan \angle PDO = y/s \approx \theta$ and

$$(r_1 - r_2) \approx \frac{ay}{s}$$

Yet, for maxima

$$(r_1 - r_2) = m\lambda \qquad [27.5]$$

which means that the height of the mth bright band (counting the central one as the zeroth) above the axis is

$$y_m \approx \frac{s}{a} m\lambda \tag{27.7}$$

Figure 27.16 The plane waves from a laser illuminate the double-slit experiment. The $m = \pm 1$ maxima occur where the optical path length difference $(r_1 - r_2)$ equals $\pm\lambda$.

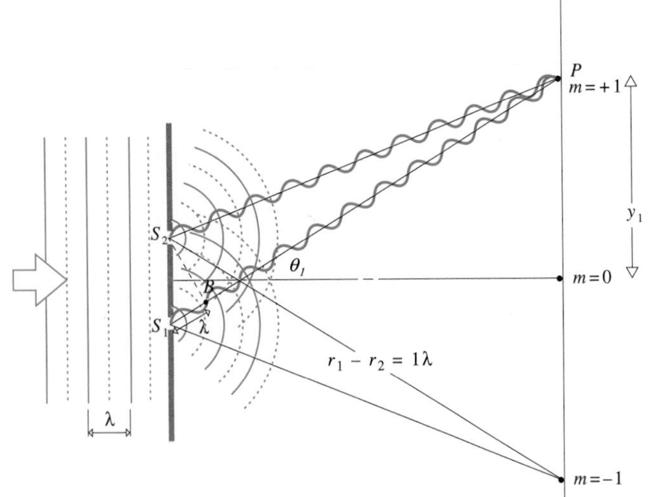

These maxima correspond to locations of P where, as shown in Fig. 27.16, the optical path difference is either 0 (where $m = 0$), $\pm\lambda$ (where $m = \pm 1$), $\pm 2\lambda$ (where $m = \pm 2$), and so forth. The height of the fringes is λ-dependent, which means that with incident white light we can expect to see a white central band ($m = 0$) where all wavelengths overlap but that all other fringes will show some coloration.

The angular height of the mth maximum is obtained by comparing Eqs. (27.5) and (27.6); thus

$$a\sin\theta_m = m\lambda \tag{27.8}$$

or, since the angles are small, $\theta_m \approx m\lambda/a$.

The spacing between fringes on the screen, Δy, is just the difference between the locations of consecutive maxima. From Eq. (27.7) it follows that $\Delta y = y_{m+1} - y_m$ is

$$\Delta y \approx \frac{s}{a}\lambda \tag{27.9}$$

For a particular wavelength of light, the spacing of the fringes is constant and inversely proportional to the slit separation: the wider apart the slits, the finer the fringes. Ideally, the irradiance would vary as a constant-amplitude cosine-squared, but because of diffraction, that's only approximated in reality.

As P gets farther from the axis, $\overline{S_1 B}$ (which is less than or equal to $\overline{S_1 S_2}$) increases. If the primary source has a short coherence length, as the optical path difference increases, identically paired wavegroups will no longer be able to arrive at P exactly together—there will be an increasing amount of overlap of portions of uncorrelated wavegroups, and the contrast of the fringes will degrade (Fig. 27.17). It is possible for $l_c < \overline{S_1 B}$. Then, instead of two correlated portions of the same wave-

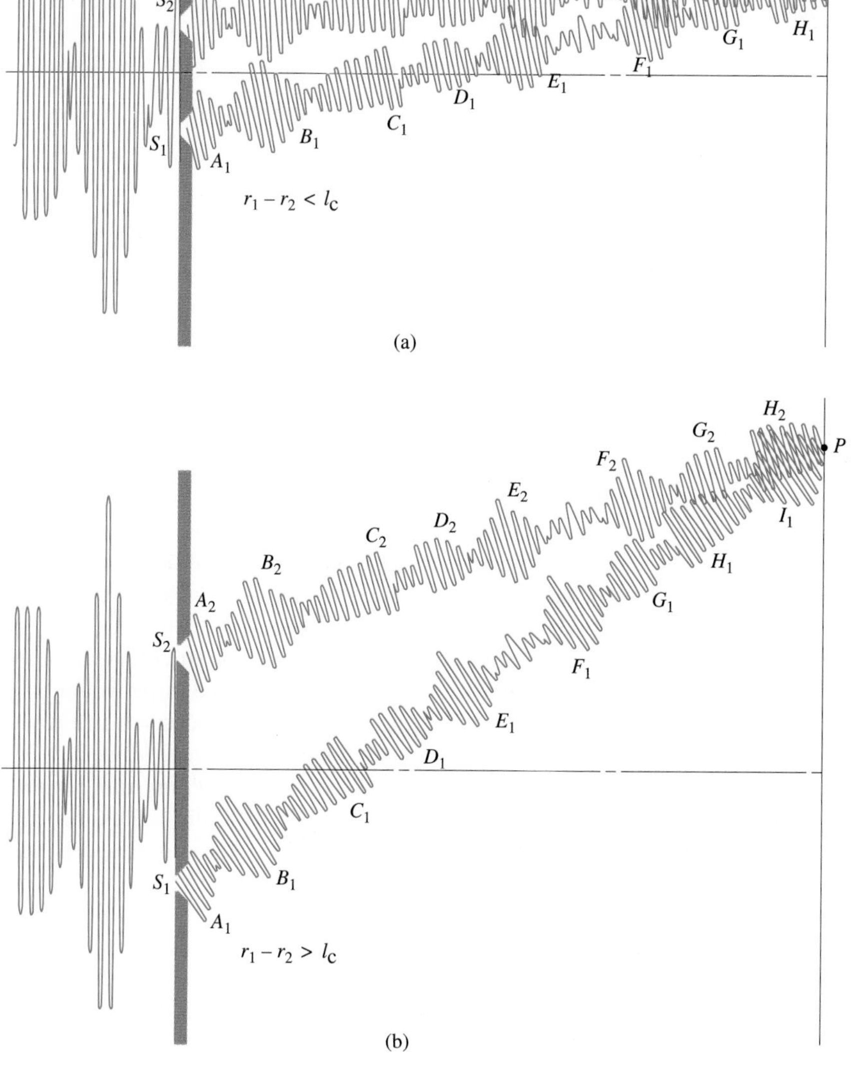

Figure 27.17 A schematic representation of interfering lightwaves. (a) Where $l_c > (r_1 - r_2)$, the interference will be sustained and observable. In (b), wavegroup E_1 from source S_1 arrives at P with wavegroup D_2 from S_2, and there is interference, but it lasts only for a tiny time before the pattern shifts when wavegroup D_1 begins to overlap wavegroup C_2, since the relative phases are different. Had either the l_c been larger or $(r_1 - r_2)$ smaller, wavegroup D_1 would more or less interact with its clone wavegroup D_2, and so on for each pair. The phases would then be correlated and the interference pattern stable, as in (a).

group arriving at P, only segments of different wavegroups will overlap, and the fringes will vanish.

Since a white-light source has a coherence length of only about 1 wavelength or so, it follows from Eq. (27.5) that very few fringes will be seen on either side of the central maximum; the more nearly monochromatic the light, the more fringes.

Example 27.4 Red light from a He-Ne laser (λ = 632.8 nm) is incident on a screen containing two very narrow horizontal slits separated by 0.200 mm. A fringe pattern appears on a white piece of paper held 1.00 m away. (a) Approximately how far (in radians and mm) above and below the central axis are the first zeros of irradiance? (b) How far (in mm) from the axis is the fifth bright band?

Solution: [Given: $\lambda = 632.8$ nm, $a = 0.200$ mm, and $s = 1.00$ m. Find: the angle for the first minimum, and y_5, the height of the fifth maximum.] (a) According to Eq. (27.4), the first minimum occurs when $m' = \pm 1$ and

$$(r_1 - r_2) = \pm\tfrac{1}{2}\lambda$$

Hence

$$a \sin\theta_1 = \pm\tfrac{1}{2}\lambda$$

and

$$\theta_1 \approx \frac{\pm\frac{1}{2}\lambda}{a} \approx \pm\tfrac{1}{2}\frac{(632.8 \times 10^{-9}\text{ m})}{(0.200 \times 10^{-3}\text{ m})}$$

$$\boxed{\theta_1 \approx \pm 1.58 \times 10^{-3}\text{ rad}}$$

or $y = s\theta_1 = (1.00\text{ m})(\pm 1.58 \times 10^{-3}\text{ rad}) = \boxed{\pm 1.58\text{ mm}}$. (b) From Eq. (27.7)

$$y_5 \approx \frac{s5\lambda}{a} \approx \frac{(1.00\text{ m})5(632.8 \times 10^{-9}\text{ m})}{(0.200 \times 10^{-3}\text{ m})}$$

$$\boxed{y_5 \approx 1.58 \times 10^{-2}\text{ m}}$$

▶ **Quick Check:** The fringe irradiance varies as cosine-squared and the answer to (a) is half a fringe width. The answer to (b) is a distance of 5 full fringes and therefore it is 10 times larger.

27.6 Thin Film Interference

The vivid iridescence of a peacock feather is due to interference of the light reflected from its complex layered surface.

In addition to the triumph of the double-slit experiment, Young was able to explain the colors arising from *thin films*: the colors of soap bubbles and oil slicks on a wet pavement; the rainbow of hues that appears on oxidized metal surfaces; the iridescence of peacock feathers and mother-of-pearl.

When light impinges on the first surface of a transparent film, a portion of the incident energy re-emerges as a reflected wave while the remainder is transmitted. The refracted energy subsequently encounters the second surface where, again, a portion is reflected and the remainder transmitted out of the film (Fig. 27.18). The film has the effect of dividing the wave into three segments: one reflected from the top surface, one reflected from the bottom, and one transmitted. Rather than utilizing two separate regions of the incident wavefront (each with the same amplitude as the incoming wave) as was done in Young's Experiment, the film shears the entire wavefront. These two reflected waves come off in the same direction and can be made to overlap at some point on the viewer's retina. The same thing would happen if this situation involved a thick block of glass, but there is a crucial difference. *If the film is thin in comparison to the coherence length of the light traversing it, the two waves will be correlated; that is, they will be fairly coherent and capable of interfering in a sustained fashion.* Because they travel different routes, depending on the film's thickness, they will ultimately interfere in some way that depends on that thickness.

Consider a nonuniform oil film floating on a puddle so that it thins out to a thickness of only a few wavelengths (and the dark surface beneath absorbs any transmitted light, keeping it from scattering back to the viewer). Wherever the film is exactly the right thickness for the two waves of emerging red light to undergo constructive interference, the film will appear to reflect a spot of red light (and so on across the spectrum and across the film). These so-called *fringes of equal thickness* create a kind of colored topographical map of the film.

Let's analyze the situation for the simplified case of nearly perpendicularly incident light. Figure 27.19 shows a thin film of index n_f between media with indices n_1 and n_2. Ray 2 travels through the film, down and back up, crossing it twice. If the thickness at that point is d, then ray 2 traverses an additional optical path length of $2dn_f$ before rejoining ray 1, which means that ray 2 traveled an additional number of waves $2dn_f/\lambda_0$ farther. To see this, notice that the extra distance $\approx 2d$ in the film corresponds to an extra number of wavelengths equal to $2d/\lambda_f$, which, because $\lambda_f = \lambda_0/n_f$, equals $2dn_f/\lambda_0$. Since each wavelength is equivalent to a phase change of 2π rad, the two waves will have a relative phase difference δ of

$$\delta = \frac{4\pi d n_f}{\lambda_0} \tag{27.10}$$

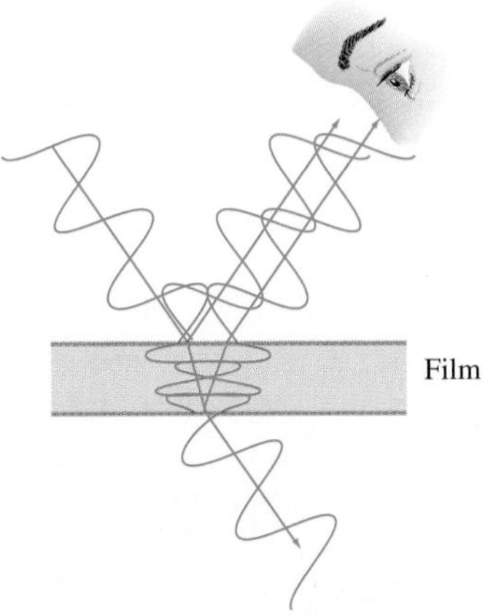

Figure 27.18 Thin film interference. Light reflected from the top and bottom of the film interferes to create a fringe pattern.

When this quantity equals a whole number (m) multiple of 2π, the two waves will be back in-phase, and that particular wavelength of light will undergo constructive interference:

$$\delta = 2\pi m = \frac{4\pi d n_f}{\lambda_0}$$

and therefore (CASE 1)

$$[\textit{maxima}] \qquad d = \frac{m\lambda_0}{2n_f} = \frac{m\lambda_f}{2} \qquad m = 1, 2, 3, \ldots \tag{27.11}$$

Maxima in normally reflected light occur when the film thickness is a whole number multiple of half the wavelength. And in the same way, *minima occur when the thickness is an odd multiple of one-quarter of the wavelength.*

A wedge-shaped film made of liquid dishwashing soap, showing interference fringes. The top part is drained thin.

As the film becomes thicker, exceeding several wavelengths, it becomes possible for two different colors to have maxima at the same spot. For example, a film 1000 nm thick will produce maxima for light with wavelengths within it of both 400 nm and 500 nm. At still greater thicknesses, where many wavelengths can interfere constructively at the same time, the reflected color becomes increasingly unsaturated, the fringe contrast decreases, and the pattern ultimately vanishes. Monochromatic illumination with its infinite coherence length will encounter no such limitations, and a sheet of window glass is still effectively a thin film for laser light.

Actually, Eq. (27.11) provides the thickness of the film for an interference maximum *only* when $n_1 > n_f > n_2$ or $n_1 < n_f < n_2$. In the first case, the reflections are both *internal* and, in the second, they're both *external.* As we saw earlier (p. 868), nearly normally incident light will undergo a relative phase shift of π rad between its internally and externally reflected components. That's not relevant in either of the above cases, which is why Eq. (27.11) applies. Notice that as the film gets vanishingly thin, all wavelengths will interfere more or less constructively everywhere, to gradually produce the uniform reflection from the interface (between the two surrounding media) that must exist when the film disappears altogether.

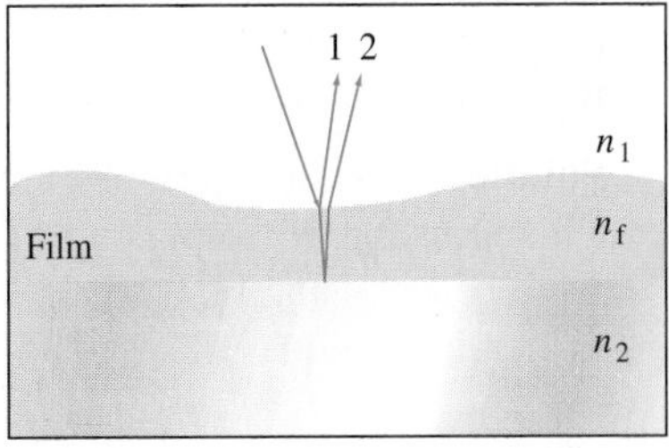

Figure 27.19 The reflection of light from the top and bottom of a thin film of index n_f.

An even more commonly occurring situation has $n_1 < n_f > n_2$ or $n_1 > n_f < n_2$, as with a soap film in air in the first case, and an air film between two sheets of glass in the second. Now, one reflection is internal and the other is external, and there will be an additional $\pm\pi$ phase shift. Which sign we select is irrelevant, accordingly, using the minus

$$\delta = \frac{4\pi d n_f}{\lambda_0} - \pi$$

Interference from the thin air film between a convex lens and the flat sheet of glass it rests on. The illumination was quasimonocromatic, which is why the fringes aren't more colorful. These fringes were first studied in depth by Newton and are known as Newton's rings. See Problem 42, p. 976.

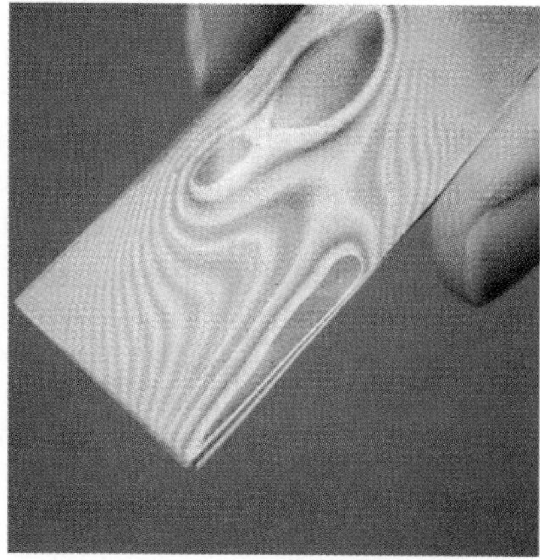

The interference pattern produced by a thin film of air between two microscope slides.

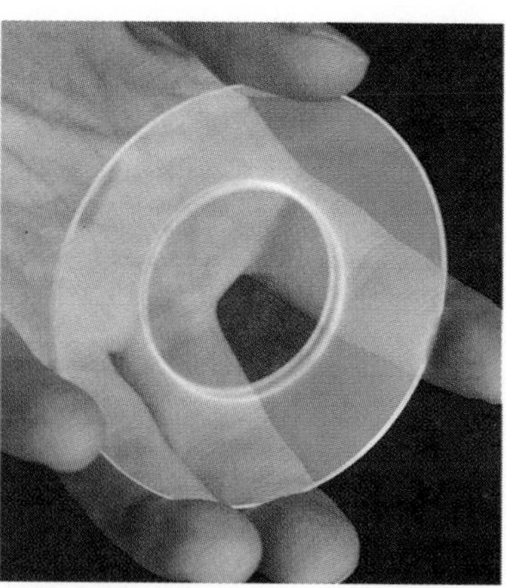

An antireflection coating in the shape of a circle, applied at the center of each side of a glass disk.

Hence, constructive interference in reflected light now occurs when

$$\delta = 2\pi m = \frac{4\pi d n_f}{\lambda_0} - \pi$$

and therefore (CASE 2)

$$[maxima] \qquad d = \frac{(m + \frac{1}{2})\lambda_0}{2n_f} = \frac{(m + \frac{1}{2})\lambda_f}{2} \qquad m = 0, 1, 2, \ldots \tag{27.12}$$

As the film becomes vanishingly thin ($d \to 0$, $d \ll \lambda_f$), the total phase difference, δ, approaches $-\pi$, there is total destructive interference, and the film appears uniformly black in reflected light and uniformly transparent in transmitted light. That's reasonable enough—since it must disappear altogether when $d = 0$, it should reflect less and less and transmit more and more as it gets there. Biological bimolecular lipid films (5 nm to 10 nm thick) appear black in reflected light, as does an ordinary liquid-soap film when it drains and becomes a small fraction of a wavelength thick (see the photo on p. 957).

Example 27.5 A soap film in air has an index of refraction of 1.34. If a region of the film appears bright red ($\lambda_0 = 633$ nm) in normally reflected light, what is its minimum thickness there?

Solution: [Given: $\lambda_0 = 633$ nm and $n_f = 1.34$. Find: d.] This situation corresponds to CASE 2, $1.00 < 1.34 > 1.00$; hence, using Eq. (27.12) with $m = 0$, which corresponds to the minimum thickness, we have

$$d = \frac{(0 + \frac{1}{2})(633 \text{ nm})}{2(1.34)} = \boxed{118 \text{ nm}}$$

▶ **Quick Check:** O.P.L. = 2(1.34)(118 nm) = 3.16×10^{-7} nm; O.P.L./$\lambda_0 = 0.5$, which, correctly, is an odd whole number multiple of $\frac{1}{2}$.

One of the most important practical applications of these ideas is the *antireflection coating*. Whenever light reflects off an interface separating two dielectrics (such as glass and air), some fraction is reflected, and that can become very problematic in complex instruments where there might easily be a dozen or more such surfaces. As a solution, each optical element is coated with a thin, transparent, solid, dielectric film such that it does not reflect a specific range of wavelengths. Often that range is chosen to be in the yellow-green region of the spectrum, where the eye is most sensitive. Lenses coated for the yellow-green reflect in the blues and reds, giving the surface a familiar purple color.

Example 27.6 A glass microscope lens with an index of 1.55 is to be coated with a magnesium fluoride ($n = 1.38$) film to increase the transmission of normally incident yellow light ($\lambda_0 = 550$ nm). What minimum thickness should be deposited on the lens?

Solution: [Given: $n_f = 1.38$, $n_g = 1.55$, and $\lambda_0 = 550$ nm. Find: d for a minimum.] The refracted wave will traverse the film twice, and there will be no relative phase shift on reflection. Hence, we want a film one-quarter wavelength thick if the two waves are to be out-of-phase and interfere destructively; consequently

$$d = \frac{\lambda_0}{4n_f} = \frac{(550 \text{ nm})}{4(1.38)} = \boxed{99.6 \text{ nm}}$$

▶ **Quick Check:** O.P.L. $= n_f d = (1.38)(99.6$ nm); O.P.L./$\lambda_0 = 0.25$, as it should; this result equals the number of waves, or $\frac{1}{4}$.

27.7 The Michelson Interferometer

There are a number of practical devices known as **interferometers** that produce fringe patterns very much in the nature of the thin film effects just considered. The most important of these, both historically and practically, is the *Michelson Interferometer,* shown in Fig. 27.20. Here, an extended source emits a wave that enters from the left. This could be an expanded laser beam, the light from a discharge lamp, or an ordinary tungsten bulb. The wave is then sheared into two equal-amplitude parts by a beamsplitter (M_S), which is usually a half-silvered mirror. The two waves are subsequently reflected by mirrors M_1 and M_2 and return to the beamsplitter. The

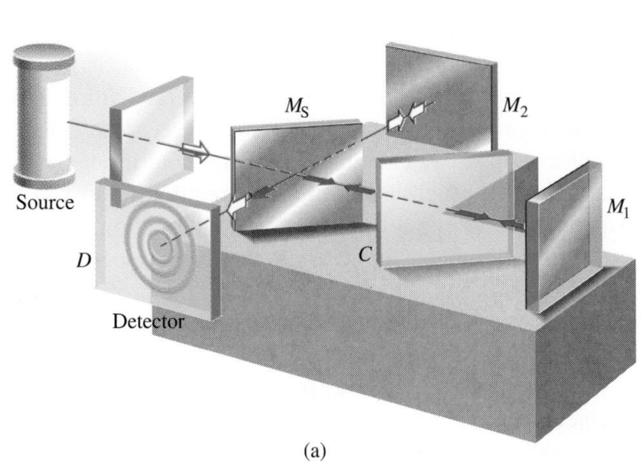

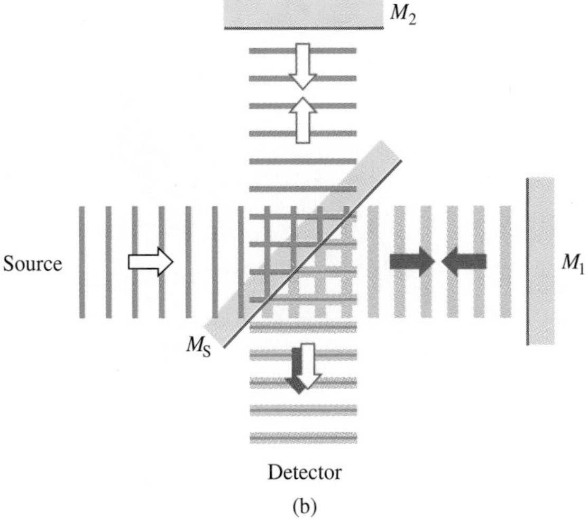

Figure 27.20 The Michelson Interferometer. Light enters at the left and is split into two equal beams by the beamsplitter M_S. Part goes on to mirror M_1 and part to mirror M_2. These beams are reflected back to M_S, and both pass on to the detector. When M_1 and M_2 are perpendicular, the fringes are circular and centered on the axis of the observer's eye. (b) is a simplified schematic representation of the wavefronts ignoring refraction.

wave from M_1 reflects toward the detector, and the wave from M_2 traverses M_S, passing on to the detector. The waves are reunited, and we can expect to see an interference pattern that depends on the path length differences traversed as well as whatever phase shifts are introduced via reflections from the beamsplitter.

Since one beam passes through M_S three times and the other only once, there will be a good deal of difference in the optical path lengths, even when the two arms are of equal actual length. Moreover, the optical path length will depend on λ because of dispersion (p. 874) in the glass. The inclusion of a compensator plate, C, which is a duplicate of the beamsplitter (without the silvering), equals out the number of thicknesses of glass traversed and negates the effects of dispersion. With the compensator in place even white light can be used; without it, the illumination must be quasimonochromatic.

What we see when we look into the device at the position of the detector is the image of both mirrors superimposed. We see the beamsplitter and, at the same time, mirrored in it M_1 and, through it, M_2. In effect, we are looking at M_1 and M_2, one behind the other, with a layer of air between them whose thickness corresponds to the difference in the two path lengths.

If M_2 and the image of M_1 are not parallel, fringes of equal thickness appear. Thus, when the mirrors are set to form a triangular-shaped air film, a system of parallel, equal-spaced straight fringes aligned with the edge of the wedge (corresponding to lines of equal film thickness) will be seen, as in the photo. And here we see another fundamental advantage of this kind of interferometer: there is space within the arms that we have access to. A hot soldering iron inserted into one of the arms will change the index of refraction of the air, shift the fringes, and reveal an otherwise invisible phenomenon. There are several variations on Michelson's instrument that exploit this marvelous ability to study the structure of transparent systems (such as gases, plasmas, lenses, and so forth).

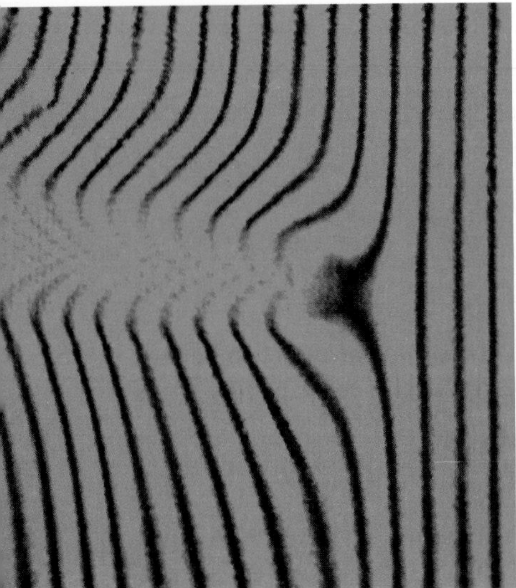

Wedge fringes in a Michelson Interferometer. The distortions are caused by a hot soldering-iron placed in one arm.

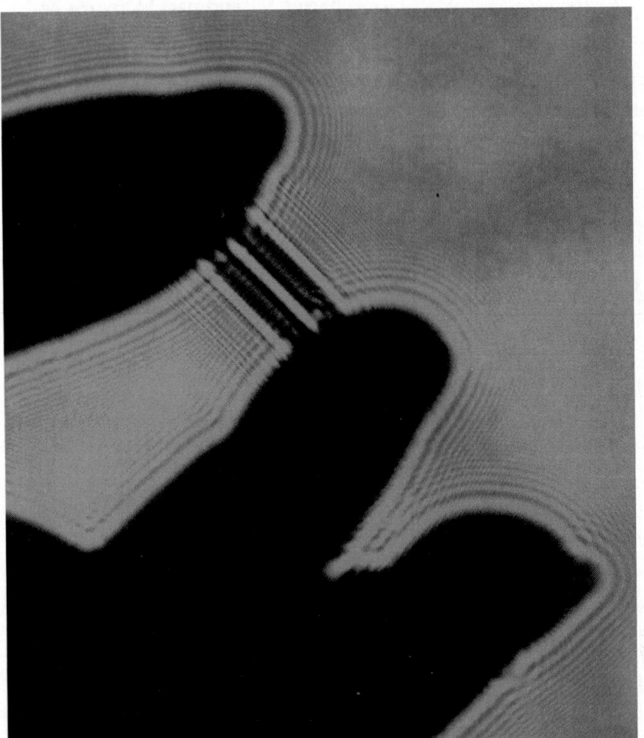

Diffraction pattern of a hand-held paper clip. This is simply the shadow cast on a wall using a He-Ne laser beam (a distant point source would do as well).

DIFFRACTION

An object placed between a point source and a screen casts an intricate shadow made up of bright and dark regions quite unlike anything one might expect from the tenets of geometrical optics. This deviation from rectilinear propagation, known as **diffraction**, is *a general characteristic of all wave phenomena occurring whenever a portion of a wavefront is obstructed in any way.* When in the course of encountering an obstacle, either transparent or opaque, a region of the wavefront is altered in amplitude or phase, diffraction occurs. Anyone who has ever walked at night in the rain wearing eyeglasses has probably seen the effect. Just put a drop of water on a glass plate, hold it near your eye, and look through it at a distant streetlight, and you'll see a complex system of bright and dark diffraction fringes. Similarly, the amoeba-like floaters that can be seen within your own eye when you squint at a bright broad source are diffraction patterns of drifting cellular debris cast on your retina.

Nowadays (because these fringes are characteristic of the objects that give rise to them, and because they can be analyzed by computer) all sorts of automatic processors

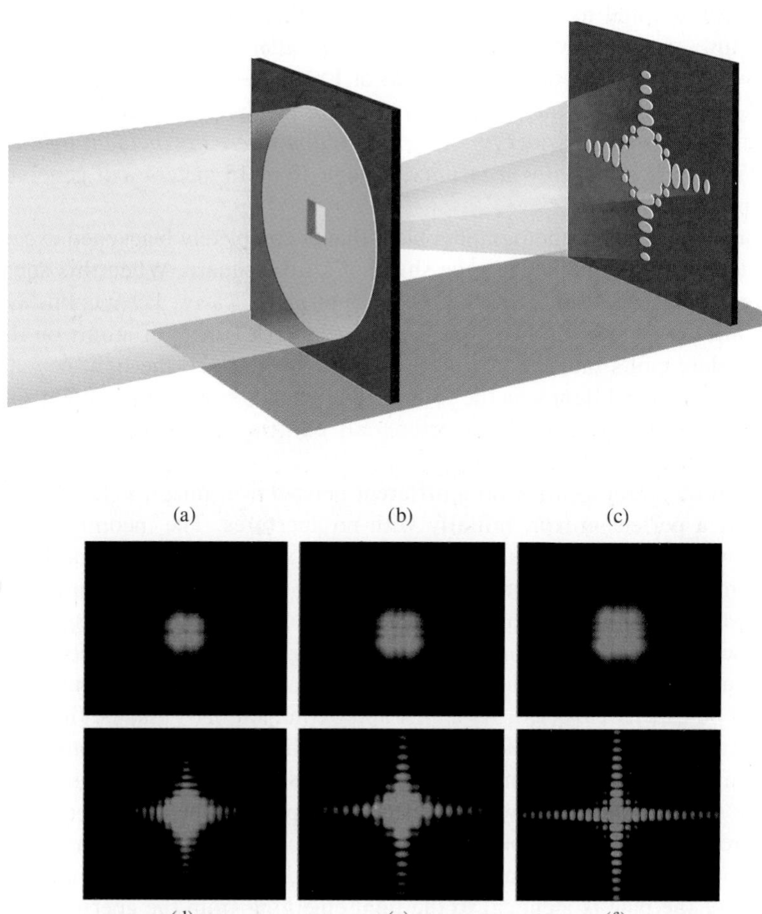

Figure 27.21 The shadow or diffraction patterns arising from a single square aperture. The hole was sequentially decreased, producing Fresnel diffraction (a), which gradually transformed (b), (c), (d), (e), into Fraunhofer diffraction (f). All apertures were illuminated by the plane waves from a He-Ne laser and the viewing screen was fixed in place.

utilize diffraction, for things as diverse as spotting tanks in aerial photos and checking fingerprints to measuring blood cell sizes.

The various unobstructed segments of a wavefront that propagate beyond an obstacle (be it sound or light) interfere to produce the particular energy-density distribution referred to as the diffraction pattern. There is no profound physical difference between interference and diffraction—in fact, the two are inseparable, though we tend to talk about them individually because that's the way the notions developed historically.

As far as the analysis is concerned, we recognize two distinct kinds of diffraction, though that, too, is more for practical reasons than fundamental ones. Figure 27.21 will help make the point. It shows a succession of diffraction patterns produced by a square hole illuminated by a laser. When the hole is large and the film very close to it by comparison, the pattern is complicated internally but the overall shape is recognizable—that's what is seen in (a), (b), and (c). But as the hole is made still smaller, *the pattern increasingly spreads out in directions perpendicular to the hole's edges* until it becomes quite unrecognizable. Finally, as in (f), a point will be reached where making the hole smaller simply makes the pattern grow larger without changing shape.

Since the effect depends on the relative size of the hole compared to its distance from the film, the same set of photos could just as well have been gotten by leaving the aperture fixed and moving the film plane farther and farther away. The first several pictures, from (a) to (e), represent what is called *near-field* or **Fresnel diffraction**, while (f) corresponds to *far-field* or **Fraunhofer diffraction**. The latter is a special case of the former, one that is comparatively easy to deal with mathematically. We will only consider *far-field diffraction where the incident light is planar and the film is essentially at infinity* (though 10 or 15 meters will generally do nicely in practice).

Consider a thin glass photographic plate that is completely blackened except for a small, clear aperture, perhaps in the shape of a little square. When this square is illuminated by a succession of normally incident plane waves, layer upon layer of atoms within the glass will scatter the light until the last sheet of atoms on the far side of the plate emits into the air in all directions. In effect, the light beyond the screen—the diffracted light—is the result of a tremendous number of point sources distributed uniformly across the aperture, all emitting secondary, more or less spherical, wavelets.

To make the point again from a different perspective, imagine the obstructing screen to be a perfect mirror, initially with no apertures. The incoming light reflects backward and none is transmitted. Whatever radiation comes from the atoms of the mirror cancels the forward-moving primary wave and simply produces the reflected wave. Now, cut out the same hole as before and remove the square plug. The back-scattered radiation from the screen no longer completely cancels the wave, and light emerges from the region of the hole. Still, if the plug were reinserted, the light would be exactly canceled. As a first approximation, let's assume that the mutual interaction of all the atomic oscillators is negligible; that is, the atoms in the screen are unaffected by the removal of the atoms in the plug. The field in the region beyond the screen will be that which existed there prior to the removal of the plug, namely zero, minus the contribution of the plug alone. All of which suggests that, except for being 180° out-of-phase, the light emitted by all the atoms spread across the surface of the plug is identical to the light emerging from the aperture.

The diffraction field, in this approximation, can be pictured as arising from a set of fictitious (noninteracting) oscillators distributed uniformly over the unobstructed area of the aperture. These imagined secondary point sources emit wavelets beyond the obstruction that mutually interfere to create the diffraction pattern.

Figure 27.22 illustrates how the notion is applied to the diffraction of a wave by an aperture. In (a), we envision the unobstructed wavefront replaced by a very large number of point sources emitting essentially spherical waves in the forward direction. This is equivalent (b) to rays traveling out from each secondary source-point in all directions. And this, in turn, is equivalent (c) to plane waves being diffracted in all directions. Since we are limiting the discussion to Fraunhofer diffraction, it is

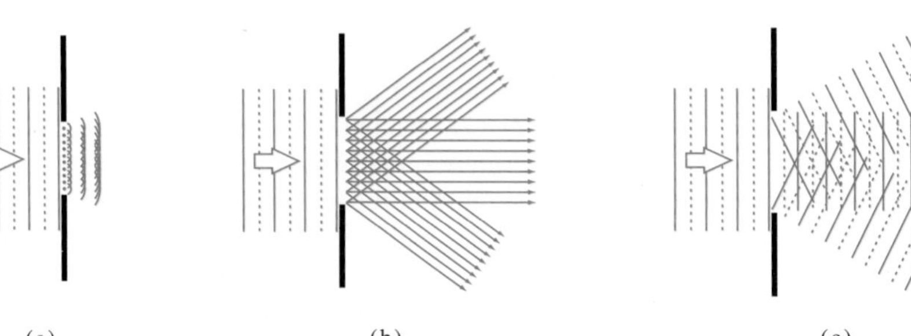

Figure 27.22 The diffraction of light through a narrow slit, represented by (a) spherical wavelets, (b) rays, and (c) plane waves.

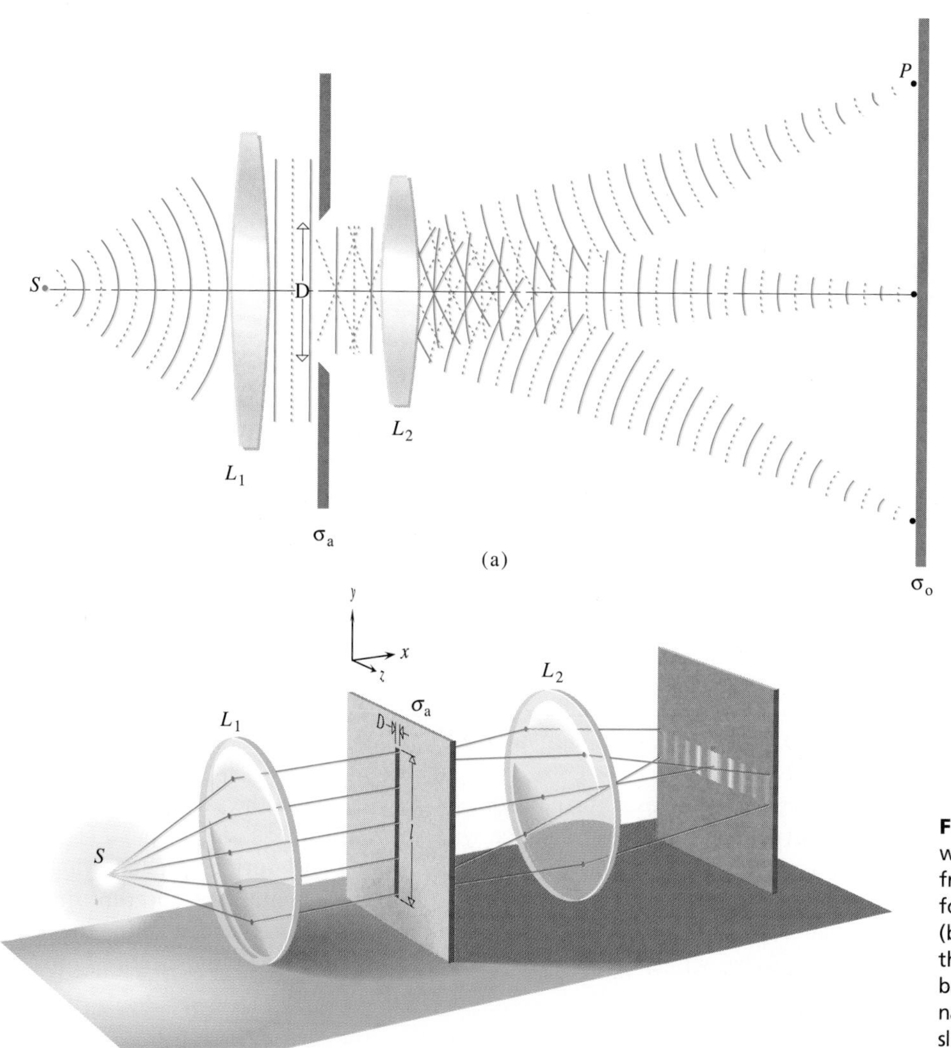

Figure 27.23 (a) The diffracted waves (seen in a horizontal plane) from the aperture are brought to a focus on a nearby screen by lens L_2. (b) When the aperture is a single slit, the diffraction pattern is a series of bands. With a point source S or a narrow laser beam illuminating the slit, the fringe pattern is narrow as well.

more convenient to use a few lenses to compress the setup rather than having to observe the pattern at infinity (Fig. 27.23).

27.8 Single-Slit Diffraction

Suppose that the aperture in Fig. 27.23a is a long narrow slit of width D, perpendicular to the plane of the diagram. Under monochromatic plane-wave illumination, we envision every point in the aperture emitting rays in all directions. The light that continues to propagate directly forward (Fig. 27.24a) is the undiffracted beam, all the waves arrive on the viewing screen in-phase, and a central bright region is formed. Figure 27.24b shows the specific bundle of rays coming off at an angle θ_1 where the path length difference between the rays from the very top and bottom, $D \sin \theta_1$, is made equal to one wavelength. A ray from the middle of the slit will then lag $\frac{1}{2}\lambda$ behind a ray from the top and exactly cancel it. Similarly, a ray from just below center will cancel a ray from just below the top and so on; all across the aper-

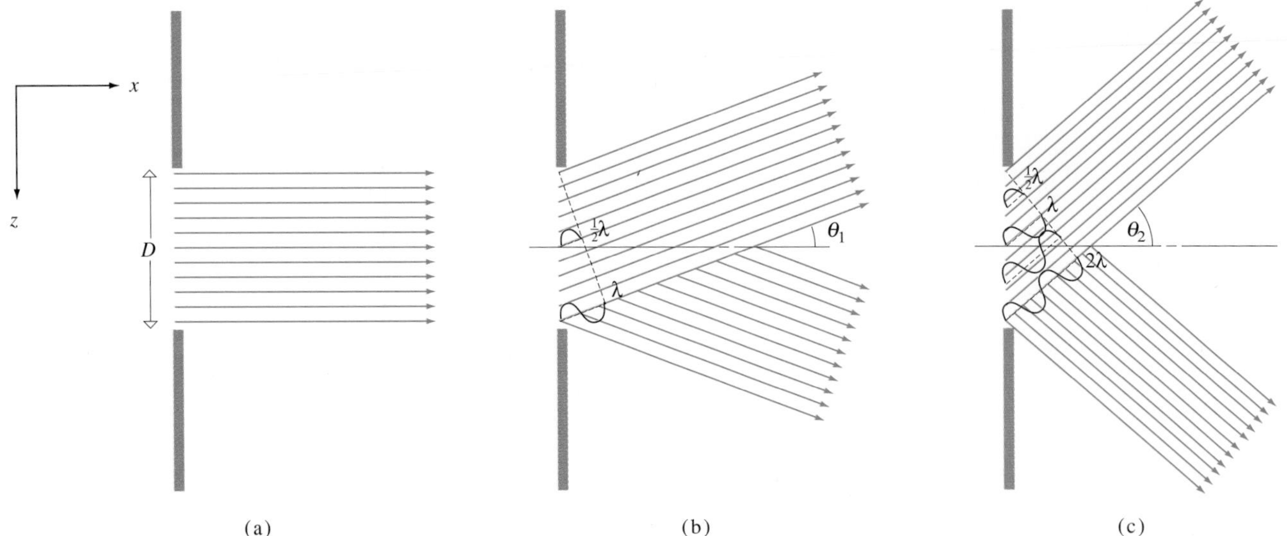

Figure 27.24 The zeros in the irradiance distribution produced by a single slit. (a) The central maximum is produced by undiffracted light. (b) The first pair of minima ($m' = \pm 1$) occur when the waves from the edge and center of the slit are out-of-phase by $\lambda/2$. (c) The second pair of minima ($m' = \pm 2$) occur when waves from each quarter of the slit cancel.

ture ray-pairs will cancel, yielding a black minimum. The irradiance has dropped from its high central maximum to the first zero on either side (Fig. 27.25) at $\sin\theta_1 = \pm\lambda/D$.

As the angle increases further, some small fraction of the rays will again interfere constructively, and the irradiance will rise to form a subsidiary peak less than 5% of the central maximum. Increasing the angle still further produces another minimum, Fig. 27.24c, when $D \sin\theta_2 = 2\lambda$. Now, imagine the aperture divided into quarters. Ray by ray, the top quarter will cancel the one beneath it and the next, the third quarter, will cancel the last quarter. Ray-pairs at the same locations in adja-

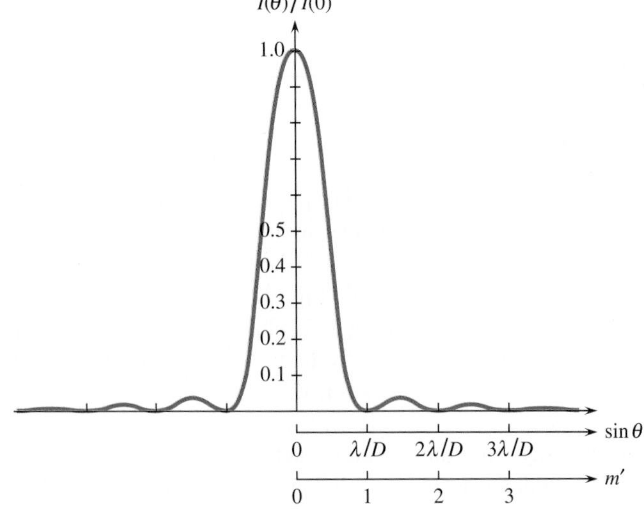

Figure 27.25 The Fraunhofer diffraction pattern of a single slit. This is a plot of the relative irradiance. Notice that the central maximum is twice the width of the little maxima.

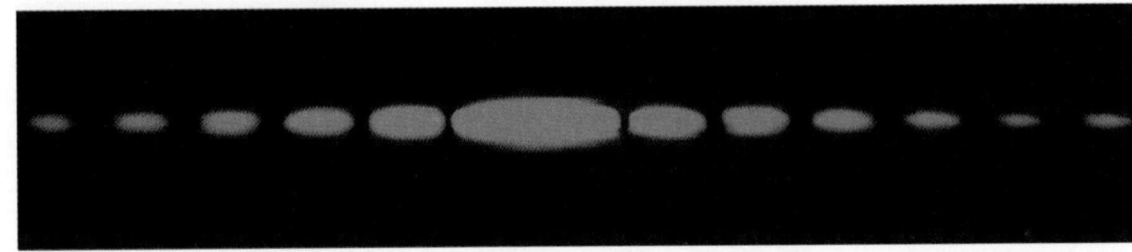

Diffraction pattern of a single vertical slit under narrow laser beam illumination.

cent segments are $\lambda/2$ out-of-phase and destructively interfere. In general, then, zeros of irradiance will occur when

$$D \sin \theta_{m'} = m'\lambda \tag{27.13}$$

where $m' = \pm1, \pm2, \pm3, \ldots$.

Example 27.7 A single slit 0.10 mm wide is illuminated by plane waves from a helium-neon laser (λ = 632.8 nm). If the observing screen is 10 m away, determine the **width of the central maximum** (*defined as the distance between the centers of the adjoining minima*).

Solution: [Given: $D = 1.0 \times 10^{-4}$ m, $\lambda = 632.8 \times 10^{-9}$ m, and $s = 10$ m. Find: the width of the central maximum.] The two minima bounding the central maximum correspond to $m' = \pm1$ in Eq. (27.13); hence

$$\sin \theta_1 = \frac{\lambda}{D} = \frac{632.8 \times 10^{-9}\text{ m}}{1.0 \times 10^{-4}\text{ m}} = 632.8 \times 10^{-5}$$

and $\theta_1 = 0.36° = 6.3 \times 10^{-3}$ rad. The central band is $2(6.3 \times 10^{-3}$ rad) wide. Since the linear half-width is $s(\tan \theta_1)$, which in radians is approximately $s\theta_1$, the overall width of the central fringe is

$$s2\theta_1 = (10\text{ m})(0.0126\text{ rad}) = \boxed{0.13\text{ m}}$$

▶ **Quick Check:** Using the above results, $2\theta_1 = (0.13\text{ m})/s = 0.013 \approx 2 \sin \theta_1 = 2(632.8 \times 10^{-5}) = 0.013$. Alternatively, $\tan \theta_1 = y_1/s$; $2y_1 = 2s \tan \theta_1 = 2(10\text{ m}) \tan 0.36° = 0.13$ m.

*As the slit narrows, and D gets smaller, the diffraction pattern spreads out and the central peak gets wider.** Again, the light fans out perpendicularly against the edges of the aperture just as it did in Young's Experiment. In fact, during that experiment, very narrow slits are used and each produces a single-slit diffraction pattern with a very wide central bright band. These two diffraction patterns overlap to create the familiar double-slit cosine-squared fringes (Fig. 27.26a and b).

27.9 The Diffraction Grating

Something rather interesting develops as the number of parallel slits (each spaced a distance a) is increased beyond two (Fig. 27.26). Identical single-slit diffraction patterns again overlap to produce maxima at the same locations as in Young's Experiment, but now these principal peaks are narrower, and faint subsidiary maxima appear between them. Three slits will produce one subsidiary maximum; four slits, two subsidiary maxima; and N slits, $(N - 2)$ subsidiary peaks between successive principal bright bands. As N increases, the secondary peaks become more numerous

*This interrelationship is very important for wave phenomena. It will appear over and over again in optics and later as well, when we consider the Heisenberg Uncertainty Principle (Sect. 31.9). The finer the details of the aperture, the more broadly is the light diffracted and vice versa.

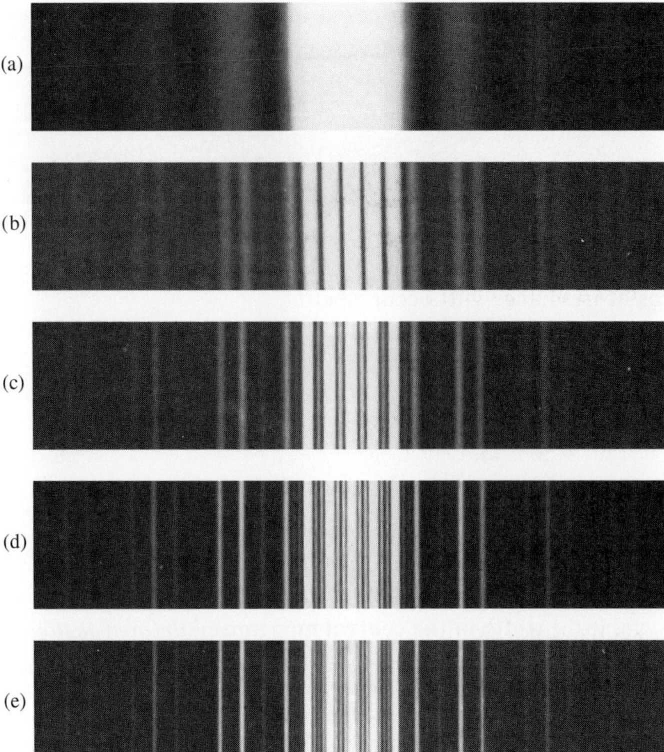

Figure 27.26 Diffraction patterns for slit systems composed of (a) 1, (b) 2, (c) 3, (d) 4, and (e) 5 identical, equally spaced, vertical slits. Notice that the maxima for several slits are where they were in the two-slit case. The more apertures, the finer the maxima become and the more faint the fringes appear between them. When there are thousands of identical slits, the various order maxima are quite fine and they appear separated by black regions.

and even fainter, with the effect that more of the diffracted light appears in the sharpened, widely spaced, principal bands. When there are thousands of slits, the bright bands are, for all practical purposes, separated by black regions where essentially no light arrives. It follows from Eq. (27.8) for Young's fringes that the principal maxima are to be found where

[*principal maxima*] $$a \sin \theta_m = m\lambda \qquad [27.8]$$

and $m = 0, \pm 1, \pm 2, \ldots$. A repetitive array of apertures or obstacles that alters the amplitude or phase of a wave is a **diffraction grating**, and Eq. (27.8) is known as

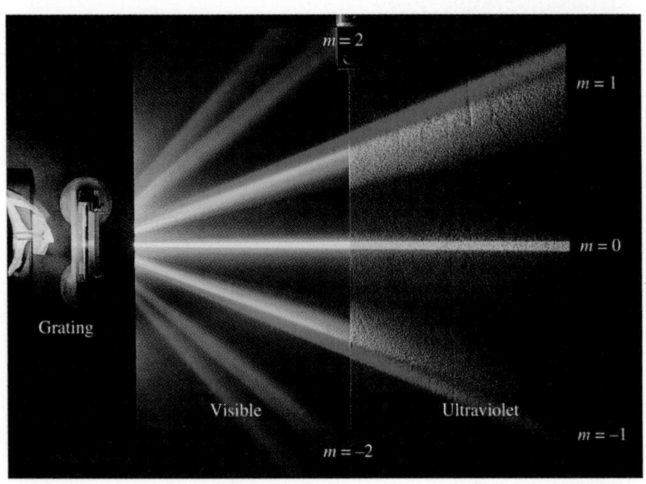

Light and UV passing through a transmission grating. The region on the left shows the visible spectrum, that on the right reveals the ultraviolet.

the *grating equation.* Master gratings with over 10 000 lines inscribed per centimeter are either ruled using a fine diamond point or generated holographically. Most gratings are plastic copies of master gratings, and they can be transparent *transmission gratings* or metal-coated *reflection gratings.*

Example 27.8 Light from a distant star enters a telescope and then passes through a diffraction grating. Each component of the emerging dispersed light is focused onto a curved strip of film. The grating is located on the central axis of the setup. It is found that a red beam (known as the hydrogen-α line) appears off axis, in the first order ($m = \pm 1$) spectrum, at an angle of 25.93°. If the lines of the grating are separated by 1.50×10^{-6} m, determine the wavelength of that light.

Solution: [Given: $\theta_1 = 25.93°$, $a = 1.50 \times 10^{-6}$ m, and $m = 1$. Find: λ.] From the grating equation, Eq. (27.8), we have

$$\lambda = \frac{a(\sin \theta_1)}{m} = \frac{(1.50 \times 10^{-6}\ \text{m})(0.4373)}{1} = \boxed{656\ \text{nm}}$$

▶ **Quick Check:** $a \sin \theta_m = m\lambda$, $\theta_m = \sin^{-1} m\lambda/a = \sin^{-1}(0.437) = 25.9°$.

Gratings are used across much of the electromagnetic spectrum in research applications. Moreover, they are finding increasing commercial use as well. For example, one compact disk (CD) system uses a grating to split the laser beam that reads the disk into three ($m = 0$ and $m = \pm 1$) tracking beams.

Notice in Eq. (27.8) that it is only because a is greater than λ that more than the single $m = 0$ wave exists. The smaller a is, the fewer are the values of m that will satisfy the formula. Moreover, if $\lambda > a$, since the sine cannot exceed one, only $m = 0$ is possible. This is equally true for both transmission and reflection gratings, and it makes an important general point. When light scatters specularly from a smooth surface (p. 871), it does so such that one beam is reflected and the angle of incidence equals the angle of reflection. Because $\lambda \gg a$, where a is the spacing between atomic scatterers, only the $m = 0$ order can be sustained. By comparison, when the wavelength is reduced to the X-ray region where λ is comparable to a, the higher-order reflections emerge.

A multi-aperture diffraction pattern. This is a picture of a white-light point source shot through a piece of tightly woven cloth.

A CD has its information stored in the form of tiny raised areas that lie along a spiral path. The resulting $\frac{1}{4}\lambda$-high ridges cause the CD to behave like a reflection grating, as does an ordinary phonograph record held at glancing incidence. There are also a number of natural systems that possess the appropriate degree of order to display grating behavior. Certain beetles, wasps, and butterflies are decorated with diffraction-grating colors.

Look at a bright source through the regular array of ridges forming a bird's feather, and you'll see transmission-grating colors. Similarly, the fabric of an umbrella, a regular mesh nylon curtain, or an undergarment can serve as a 2-dimensional transmission grating. And the gemstone opal gets its shifting internal coloration from an ordered array of tiny silica (SiO_2) spheres that make it a 3-dimensional diffraction grating. As we'll see, for X-rays all crystals are 3-dimensional gratings (p. 1023).

27.10 Circular Holes and Obstacles

When light from a distant point source (like a star or, more mundanely, an atom on a star's nose) is focused by a lens, the image formed is a small blotch rather than a perfect point. It must be so because the lens captures only a portion of the wavefront, and diffraction must occur—the blotch is the far-field diffraction pattern of the aperture of the lens. In the end, every image formed by a circular lens, whether in your

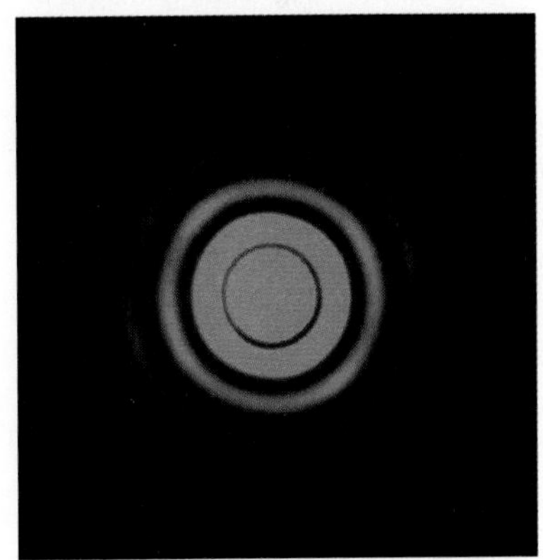

Airy rings (1.0-mm-hole diameter). The center disk was allowed to get overexposed in order to make a few rings visible.

eye or a TV camera, is composed of countless overlapping circular-aperture Fraunhofer diffraction patterns. And that alone makes this particular configuration of the greatest importance. Unfortunately, even though a complete analysis was provided in the nineteenth century by Sir George B. Airy, Astronomer Royal of England, it's far too complicated for our purposes.

All Fraunhofer patterns, no matter how weirdly shaped the aperture, are symmetrical about the center point. Draw a line through that center, and the distribution of light outward along one-half of the line is the same as along the other. A circular hole must generate a circularly symmetric pattern. In fact, a circular aperture produces a circular disk of light, the so-called **Airy disk**, surrounded by a system of increasingly faint concentric rings. Actually, 84% of all the energy lies in the Airy disk. By the way, because there is so little energy in the higher-order fringes, circular lenses are better image-formers than rectangular ones.

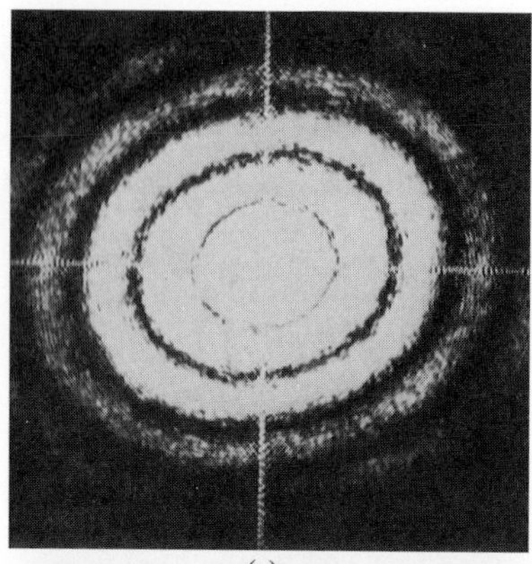

(a)

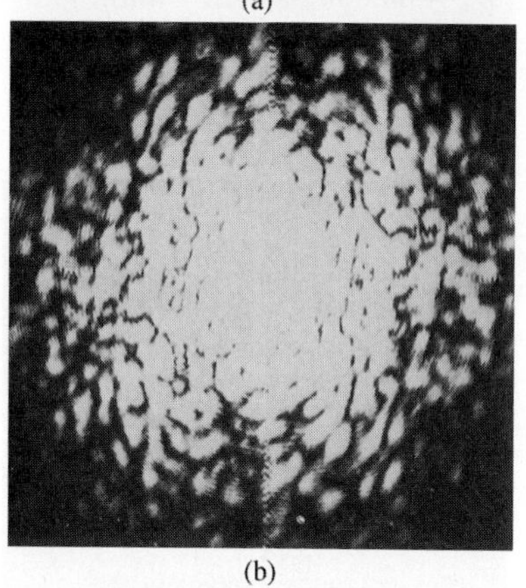

(b)

(a) The Fraunhofer diffraction pattern of a normal cervical cell. (b) The diffraction pattern of a malignant cervical cell is very different. Diffraction is being studied as a possible means of rapid automatic analysis of Pap tests for cancer.

A hole with a diameter D followed by a lens of focal length f will generate a diffraction pattern on a screen (in the lens's focal plane) having an Airy disk of radius r_a that can be shown to be

$$r_a \approx \frac{1.22 f\lambda}{D} \tag{27.14}$$

Alternatively, using the fact that $\theta_a \approx \sin\theta_a = r_a/f$, we can express the size of the central bright circle in terms of its angular half-width θ_a as

$$\theta_a \approx \frac{1.22\lambda}{D} \tag{27.15}$$

in radians.

Suppose we wish to examine two equal-irradiance, incoherent, point sources—two close-together stars seen through a telescope, or two points on a virus viewed in a microscope. Because of diffraction, the images will spread out, and only if their angular separation exceeds this smearing will they appear distinct. As the two sources come closer together, their images come together, commingling into a single blend of fringes. Lord Rayleigh suggested that the criterion for *just being able to resolve the separate images* be that their center-to-center distance equal the radius of their Airy disks (Fig. 27.27). We can actually do a little better, but Rayleigh's criterion is both simple and standard. Equation (27.15) therefore specifies the **angular limit of resolution**, and the *resolving power* of an image-forming system is taken as the reciprocal of θ_a.

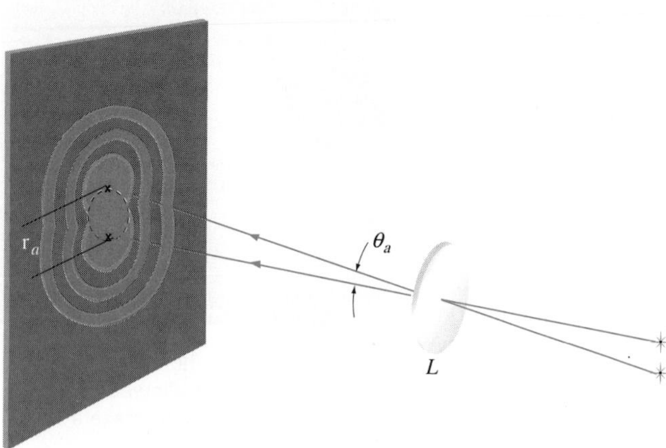

Figure 27.27 Overlapping images of two point sources (separated by θ_a) that are just resolvable.

By decreasing the wavelength, we can increase the resolving power of an instrument. Ultraviolet microscopes can "see" finer details than light microscopes (which at best can distinguish two points as distinct that are no closer than about 0.12×10^{-6} m). Similarly, electron microscopes with wavelengths equivalent to roughly 10^{-4} to 10^{-5} times that of light have a limit of resolution of about 0.5 nm. By increasing the diameter of the objective lens or mirror of a telescope, it can collect more light, but the images it forms will be sharper as well. That's why spy satellites have large-diameter cameras.

Example 27.9 Compute the angular limit of resolution of the eye, assuming that it's determined only by diffraction. Take the pupil diameter to be 2.0 mm and $\lambda = 550$ nm. How far apart will two points be if they are just able to be distinguished at a distance of 25 cm from the eye?

Solution: [Given: $D = 2.0$ mm, $\lambda = 550$ nm, and $d_o = 25$ cm. Find: θ_a and y_o.] The limit of resolution follows from Eq. (27.15); thus

$$\theta_a \approx \frac{1.22\lambda}{D} = \frac{1.22(550 \times 10^{-9}\text{ m})}{2.0 \times 10^{-3}}$$

and $$\boxed{\theta_a \approx 3.4 \times 10^{-4}\text{ rad}}$$

which is 1.9×10^{-2} degrees. At a distance of 25 cm, this angle corresponds to a linear distance of $y = d_o\theta_a$ or

$$y_o = (0.25\text{ m})\theta_a = \boxed{8.4 \times 10^{-2}\text{ mm}}$$

which, at roughly 1/10 mm, is just about right for a normal eye.

▶ **Quick Check:** $\theta_a D/\lambda$ should equal 1.22, therefore $(0.34\text{ mrad})D/\lambda = (0.34\text{ mrad})(2.0\text{ mm})/(550\text{ nm}) \approx 680/550 \approx 1.2$.

In 1818, the young Fresnel entered his new wave theory in a competition sponsored by the French Academy. The judging committee consisted of such luminaries as Pierre Laplace, Jean Biot, Siméon Poisson, Dominique Arago, and Joseph Gay-Lussac. An ardent antagonist of the wave hypothesis, Poisson managed to deduce a remarkable and seemingly untenable conclusion from Fresnel's theory. He showed that the treatment predicted that a bright spot should appear at the very center of the shadow of a circular opaque obstacle; such an absurdity must surely disprove the entire theory! Arago, who was not one to accept anything on face value, went to the laboratory to learn the truth of Poisson's death blow. And there, wonder of wonders, at the center of the shadow he saw the "absurd"—a bright spot of light. Fresnel was right!

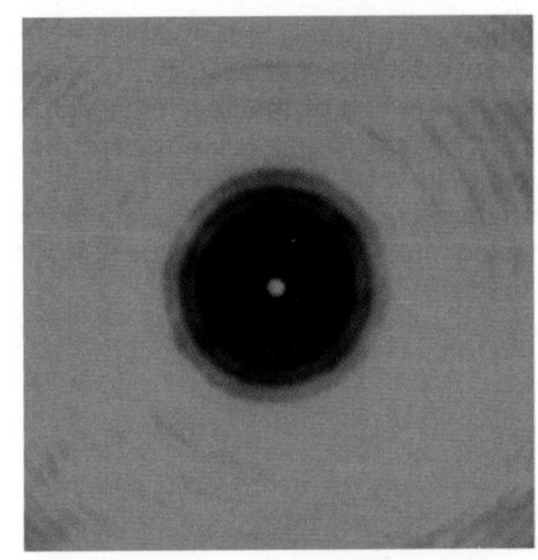

The shadow of an $\frac{1}{8}$-in.-diameter ball bearing illuminated with He-Ne laser light. To hold it in place, the ball bearing was glued to a microscope slide. The effect can easily be seen with a halogen lamp as the point source.

27.11 Holography

The technology of photography has been with us for a long time, and we're all accustomed to seeing the 3-dimensional world flattened onto a scrapbook page. A photograph is a record of the energy per unit area per unit time that once impinged on each of its surface points, and the light it scatters (the light we see when we look at it) is nothing more than a reflection of that frozen irradiance distribution. It is not an accurate reproduction of the original light field that came from the subject, but only a record of the square of the field's amplitude averaged over the exposure time. It reveals nothing about the phases of the waves that formed the image. On the other hand, if we could somehow reconstruct both the amplitude and phase of the wave coming from the object, the resulting light field would be indistinguishable from the original. One would then see the reformed image in perfect 3-dimensionality, exactly as if the object were there before us, generating the wave.

The technique of *image reconstruction* was invented by Dennis Gabor sometime around 1948. His research, which won him the 1971 Nobel Prize in physics, led to a practical means of recording and playing back the complete wave emanating from an object. The process is now called **holography** (from the Greek *holos* meaning whole). The crucial insight is that both the amplitude and phase information can be captured in a coded form via interference and preserved photographically. When two monochromatic waves generate an interference pattern, the shape,

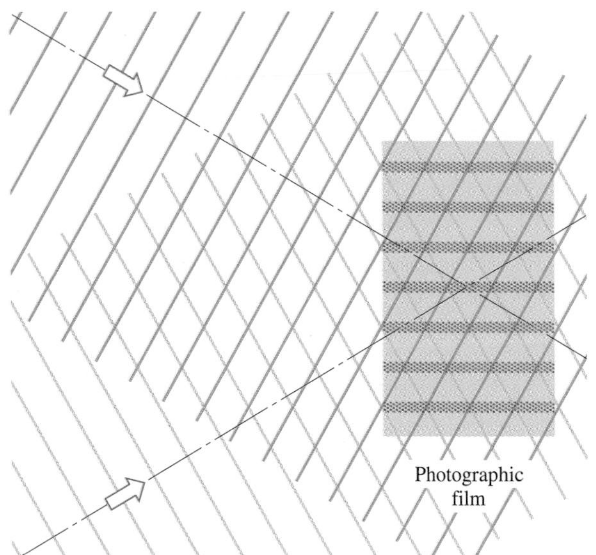

Figure 27.28 A schematic representation of the interference of two plane waves traveling toward the same side of a photographic film, thereby creating a transmission hologram. (For simplicity, refraction has been ignored.)

contrast, and spacing of the fringes is a record of the physical characteristics of the overlapping waves themselves. If we wish to capture the vital features of an **object-wave**, we can cause it to interfere with a known, simply configured, **reference-wave**. The two should have roughly the same amplitude and, of course, be coherent. This last requirement kept Gabor's work in semi-unnoticed oblivion for about 15 years, until the laser was invented.

Consider the simple case of two plane waves traveling in the same general direction (Fig. 27.28) with a photographic plate inserted in the region of overlap. The wedge-shaped film viewed in a Michelson Interferometer produces the same configuration of two tilted plane waves, and so we can anticipate a fringe pattern of parallel bands across the emulsion. To appreciate this more clearly, just lay a pencil along each wavefront and move them both in their respective propagation directions at the same speed. The point of overlap, the maximum, will travel to the right along a straight line—along the fringe. As the shape of the object-wave deviates increasingly from a simple plane, the fringe pattern becomes more modulated and complex. Thus, if the object-wave is the light reflected from some one's face, the fringe system will be an unrecognizable tumult of fine bright and

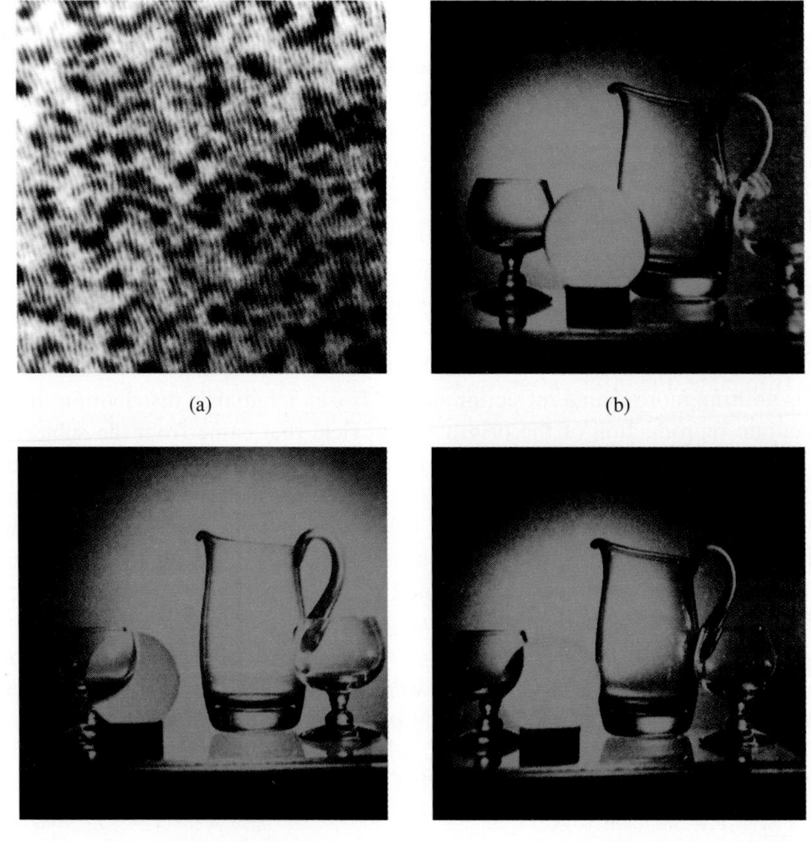

Figure 27.29 (a) A hologram. (b), (c), (d) Three different views photographed from the same holographic image generated by the hologram in (a). By moving your head (or in this case, a camera), you can see different regions of the scene. The lens in the hologram will magnify different objects depending on how you look through it.

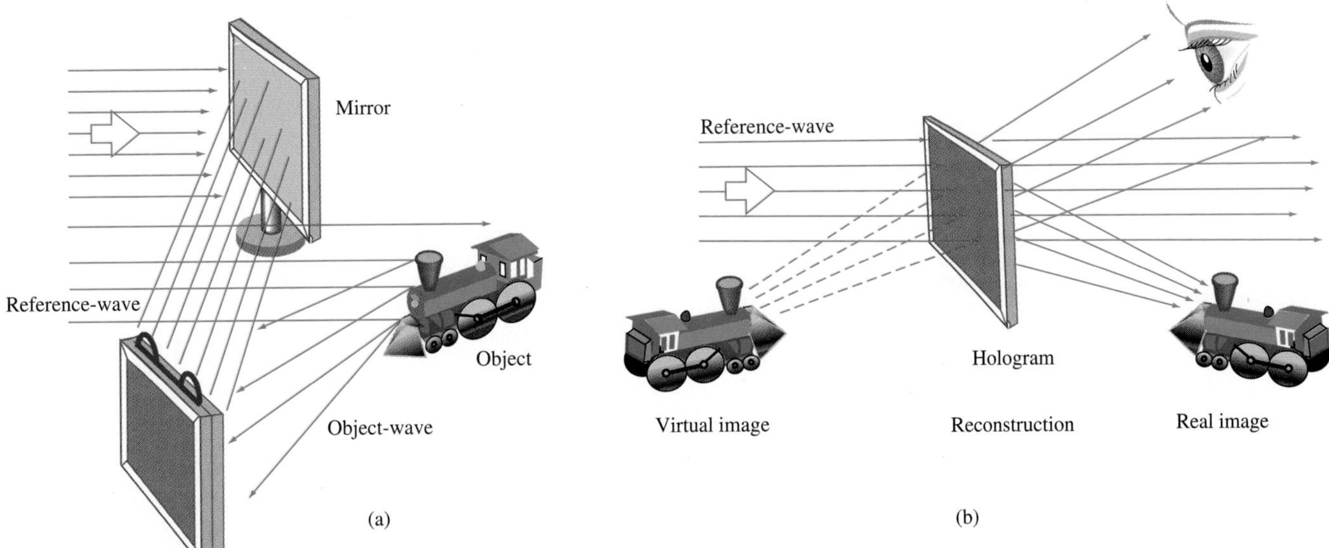

Figure 27.30 Holographic recording and reconstruction of an image. (a) The photo plate receives light from the object and a reference beam directly from the mirror. (b) To view the hologram, it is illuminated with a wave identical to the reference-wave.

dark regions. When developed, the silver atoms in the emulsion form a kind of 3-dimensional diffraction grating, which is the hologram (Fig. 27.29).

All we need do is illuminate a fine-grained film with both a scattered object-wave and a coherent reference-wave (Fig. 27.30). The developed film is a *transmission hologram,* and when it is illuminated from the back by a wave identical to the reference-wave (Fig. 27.30b), the original light field is reconstructed. The hologram during playback functions much like a diffraction grating that scatters the incoming plane waves into two off-axis beams. One of these gives rise to an extraneous real image, while the other corresponds to a duplicate of the original lightwave diverging from the object.

The light emerging from a hologram is identical to the light originally coming from the object (it's as if you were simply looking through a window). In the case of an extended scene, your eyes would have to refocus as you viewed different regions at various distances. And if you chose to photograph the vision, you would only have to point the camera through the "window," focus on the desired object, and shoot. Furthermore, since the light from any point on the subject spreads across much of the film plate, the information about that point is redundantly recorded all across the hologram. Consequently, just as you can peer out of a small opening in a window shade and see the entire scene beyond, you can peep through a small piece of the hologram and see everything as well. It's been suggested that the human brain functions in a similarly redundant way and that memory is, in that sense, holographic.

Core Material

When the electric field of a lightwave remains in a fixed plane the wave is **plane-polarized**, or *linearly polarized.* With *circular light,* the electric-field vector remains constant in magnitude while it revolves once around with every advance of one wavelength (p. 940). Because the state of polarization of light from an ordinary source is sustained for less than 10^{-8} s, that state is essentially undetectable, and the light is said to be **unpolarized**, or **natural** (p. 941).

Light that passes through two consecutive linear polarizers whose transmission axes make an angle θ, emerges with an irradiance given by

$$I = I_1 \cos^2 \theta \qquad [27.1]$$

where I_1 leaves the first filter. This equation is **Malus's Law** (p. 944).

When an unpolarized beam is incident on a surface at an angle θ_p, known as **Brewster's angle**, the reflected light will be totally plane-polarized parallel to the interface. The **polarization angle** is given by

$$\tan \theta_p = \frac{n_t}{n_i} \qquad [27.3]$$

Light can also be linearly polarized via scattering (p. 948).

In a region of overlap, two waves of the same frequency can combine constructively or destructively, depending on their relative phase, to produce a redistribution of energy in that area—this is the essence of **interference**. *The average time interval during which a lightwave oscillates in a predictable way* is known as the **coherence time** (t_c) of the radiation (p. 951). *The corresponding spatial extent over which the lightwave oscillates in a regular, predictable way* is called the **coherence length** (l_c), where in vacuum $l_c = ct_c$.

The double-slit setup, known as Young's Experiment, is the premier demonstration of interference. Bright maxima appear (p. 952) symmetrically about the centerline at distances of

$$y_m \approx \frac{s}{a} m\lambda \qquad [27.7]$$

The maxima are displaced from the central axis by the angles θ_m such that

$$a \sin \theta_m = m\lambda \qquad [27.8]$$

or since the angles are small, $\theta_m \approx m\lambda/a$. The spacing between fringes on the screen is

$$\Delta y \approx \frac{s}{a}\lambda \qquad [27.9]$$

Thin transparent films generate fringe patterns when the two waves reflected from the front and back surfaces interfere (p. 956). Where the film is nonuniform, the pattern of so-called *fringes of equal thickness* creates a topographical map. When $n_1 > n_f > n_2$ or $n_1 < n_f < n_2$, reflected maxima occur where the thickness is

$$d = \frac{m\lambda_0}{2n_f} = \frac{m\lambda_f}{2} \qquad m = 1, 2, 3, \ldots \qquad [27.11]$$

An even more commonly occurring situation has $n_1 < n_f > n_2$ or $n_1 > n_f < n_2$. Now maxima occur where

$$d = \frac{(m + \frac{1}{2})\lambda_0}{2n_f} = \frac{(m + \frac{1}{2})\lambda_f}{2} \qquad m = 0, 1, 2, \ldots \qquad [27.12]$$

The deviation of light from rectilinear propagation is **diffraction**. We concentrate on *Fraunhofer* diffraction, which effectively obtains *where the incident light is planar and the observation screen is at infinity.* In the case of a single slit of width D, zeros of irradiance will occur on both sides of the broad central maximum when

$$D \sin \theta_{m'} = m'\lambda \qquad [27.13]$$

where $m' = \pm1, \pm2, \pm3, \ldots$. Similarly, when many parallel slits, each separated by a distance a, are present, narrow principal maxima will appear at the same locations as Young's fringes.

The angular half-width of the Airy disk is given, in radians, by

$$\theta_a \approx \frac{1.22\lambda}{D} \qquad [27.15]$$

Suggestions on Problem Solving

1. A common error associated with calculations of the polarization angle is to invert the n_t/n_i—watch out for that. Another frequent oversight is to forget that a linear polarizer reduces the irradiance of incident *unpolarized* light by 50%. Don't fail to square the cosine in Malus's Law.

2. When dealing with interference in thin films, be sure to determine whether or not there is a relative phase shift, due to internal and external reflection, before making any numerical calculations.

3. Many of the equations from the analysis of Young's Experiment carry over directly to the treatment of the diffraction grating. That means that Eqs. (27.7) through (27.9) apply to gratings, as does the expression $\theta_m = m\lambda/a$. Keep in mind that this last equation is always in radians!

4. The number density of lines on a grating is often given in lines per centimeter, *not* per meter, so on inverting this factor to compute a, the line spacing, you must first convert to lines/m if a is to be in meters.

Discussion Questions

1. Is it correct to say that ideally monochromatic light must be polarized? Explain.

2. Suppose that you had a beam of monochromatic light that you knew to be the in-phase superposition of two equal-amplitude, plane-polarized waves having their electric fields respectively horizontal and vertical. How might you verify this situation experimentally?

3. Can polychromatic light be polarized? In what sense is there no such thing as unpolarized light? Explain your answers.

4. Natural light is incident on a flat pane of glass at an angle of 45°. Describe the polarization of both the reflected and transmitted beams. How would that change if the light came in at Brewster's angle instead?

5. Suppose that you are looking through a linear polarizer, with its transmission axis vertical, at the surface of a pot of water, and at some angle of reflection find that the glare on the surface vanishes. What is happening? Would the vision change if you now replace the water with benzene, all else held constant? Explain.

6. What was the central problem that Young solved via his double-slit arrangement? How will the fringe pattern in his experiment change if the light source is gradually altered so that

its coherence time diminishes? Describe what will happen to the fringe system in some region where the optical path length difference from the two sources ultimately exceeds the coherence length.

7. An antireflection coating on glass has a thickness that corresponds to one-quarter of a wavelength of red light in the medium of the film. What color light will be reflected and what color transmitted when white light is incident normally? Assume $n_a > n_f > n_g$.

8. If you put a few drops of liquid soap in a cup with a little water and shake it around, you can make a great froth of bubbles. After several seconds, the bubbles will begin to show bright rainbow color patterns. Why must you wait before the colors appear?

9. Take a very close look at a worn surface under direct sunlight. A fingernail, an old coin, or the roof of an old car will all show a very fine granular pattern of tiny colored dots. This phenomenon is the so-called *speckle effect.* What might be its cause?

10. Fog up a piece of glass with your breath and look through the fogged plate in a darkened room at a white-light point source. You will see a system of concentric colored rings. What does this remind you of, and what do you think is causing the effect? Figure Q10 shows a point source seen through a glass plate covered with a layer of transparent, spherical lycopodium spores.

Figure Q10

11. Consider Young's Experiment where the apertures are two small circular holes, one next to the other, separated by a distance a, which is similarly small but still much larger than any of the incoming wavelengths. Assuming normally incident monochromatic plane-wave illumination, describe—qualitatively but in detail—the resulting fringe pattern on a very distant screen parallel to the plane of the apertures. Make a sketch of it.

12. Discuss the interrelationship between aperture size, illuminating wavelength, and *finite* distance to the observation screen (no lens after the hole) for near- and far-field diffraction. In other words, what will happen to the pattern on a screen as each of the above is varied?

13. How could you use a Michelson Interferometer to measure the coherence length of a discharge lamp?

14. Why do bright stars appear to the eye to be larger than faint ones?

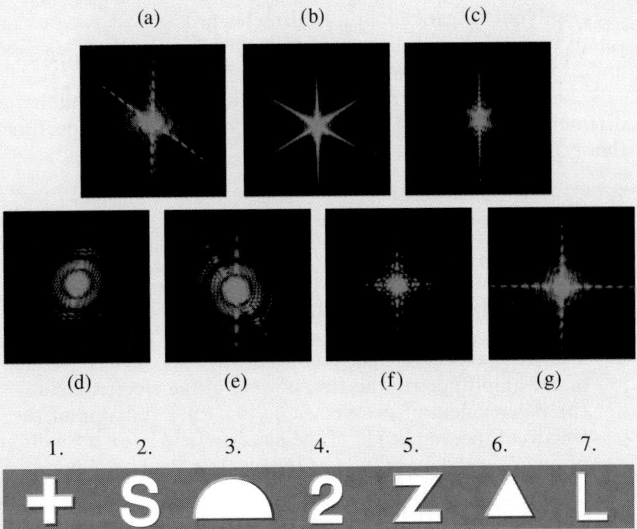

Figure Q17

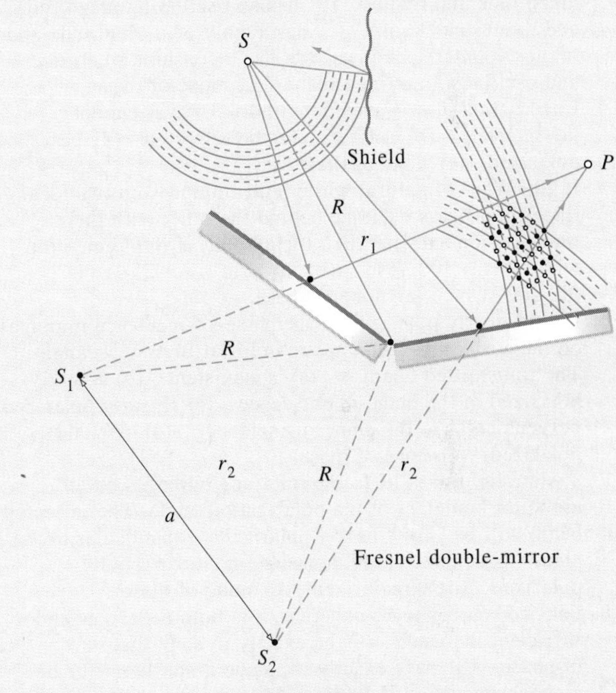

Figure Q18

15. A nearsighted person looking through an adjustable circular aperture at a distant object can improve his vision by making the hole a bit smaller than his pupil—people squint to improve acuity. Still, if the aperture is made less than about 1.0 mm in diameter, the image will again blur. What is happening?

16. Using a TV or computer screen as a convenient broad source, stand a few feet from it and peer at it through a narrow slit (for example, the space between two straight fingers). Hold your hand about five or six inches from your eye and form a slit

between your fingers about a millimeter or less wide—you can rotate your hand slightly to effectively narrow the slit. Describe and explain the distribution of light within the slit.

17. Figure Q17 shows seven apertures and seven Fraunhofer diffraction patterns. Match up the patterns with the apertures that produced them.

18. Explain how the Fresnel double-mirror works to produce interference fringes (Fig. Q18). What will the pattern look like?

19. Imagine that you have a long-filament light bulb held vertically. If you look at it from several meters away through a vertical slit cut in a 3 × 5 card, you will see a lovely fringe system. Explain what's happening.

Multiple Choice Questions

1. The irradiance of a polarized lightwave is (a) a function of the amplitude (to the first power) of the electric field (b) independent of the electric field (c) a function of the square of the amplitude of the electric field (d) a function of one over the amplitude of the electric field (e) none of these.
2. Natural light is light that is (a) only found in nature (b) linearly polarized (c) completely unpolarized (d) polarized but rapidly changing (e) none of these.
3. The electric field of $\mathcal{R}$-state light (a) has a constant magnitude and rotates (b) has an oscillating magnitude and rotates clockwise (c) has a sinusoidal amplitude and rotates counterclockwise (d) has a constant amplitude and oscillates in a fixed plane (e) none of these.
4. Unlike transverse waves, longitudinal waves cannot (a) interfere (b) diffract (c) be reflected (d) be polarized (e) none of these.
5. If 100 W/m^2 of natural white light impinges on two ideal linear polarizers held one behind the other with their transmission axes parallel, the amount of light emerging will be (a) 100 W/m^2 (b) 50 W/m^2 (c) 25 W/m^2 (d) 200 W/m^2 (e) none of these.
6. Light linearly polarized in the plane-of-incidence impinges on the surface of a glass plate in air at Brewster's angle. The transmitted beam is (a) nonexistent (b) linearly polarized in the plane-of-incidence (c) linearly polarized perpendicular to the plane-of-incidence (d) partially polarized (e) none of these.
7. $\mathcal{P}$-state light parallel to the interface impinges on an air-water boundary at the polarization angle. The reflected beam will be (a) linearly polarized perpendicular to the plane-of-incidence (b) nonexistent (c) partially polarized (d) unpolarized (e) none of these.
8. For two sources to be coherent, it is both necessary and sufficient that each (a) be exactly in step, that is, in-phase (b) have exactly the same amplitude (c) be monochromatic (d) be linearly polarized (e) none of these.
9. Coherent lightwaves never arise from (a) two lasers (b) two pinholes (c) two candles (d) two slits (e) any of these.
10. Increasing the coherence time of a source is equivalent to (a) decreasing the coherence length (b) increasing the speed (c) making it more nearly monochromatic (d) making it less monochromatic (e) none of these.
11. Imagine that we cover each slit in Young's Experiment with one of two identical linear polarizers such that one transmission axis is parallel to the slits while the other is transverse to them. On the distant screen, we will observe (a) no light at all (b) a fairly uniform illumination (c) cosine-squared fringes shifted $\frac{1}{2}\lambda$ from normal (d) the usual fringe pattern (e) none of these.
12. The middle of the first-order maximum, adjacent to the central bright fringe in the double-slit experiment, corresponds to a point where the optical path length difference from the two apertures is equal to (a) λ (b) 0 (c) $\frac{1}{2}\lambda$ (d) $\frac{1}{4}\lambda$ (e) none of these.
13. Suppose you take a ring-shaped wire and dip it in liquid soap so that a circular film forms. Now, if you place it horizontally on a platform, which is then spun about a central vertical axis, and look at it under white light, (a) a series of straight colored fringes will appear (b) you will see a uniformly bright surface (c) a series of concentric colored circular fringes surrounding a black central disk will appear (d) a set of concentric circular colored bands surrounded by a black fringe will appear (e) none of these.
14. A narrow slit of width D is illuminated by light coming from a monochromator so that the wavelength can be varied all across the visible spectrum. As λ decreases, the Fraunhofer diffraction pattern, viewed in the focal plane of a lens, (a) shrinks with all the fringes getting narrower (b) spreads out with all the fringes getting wider (c) remains unchanged (d) alters such that only the central maximum broadens (e) none of these.
15. A small aperture is located some distance from an observation screen. As the wavelength of the illumination is gradually decreased, the far-field diffraction pattern seen on the screen (a) remains unchanged (b) changes to a near-field pattern (c) changes to a Fraunhofer pattern (d) stays as a far-field pattern but gradually expands (e) none of these.
16. The effect of increasing the number of lines per centimeter of a grating is to (a) increase the number of orders that can be seen (b) allow for the use of longer wavelengths (c) increase the spread of each spectral order (d) produce no change in the diffracted light (e) none of these.
17. A grating diffracts red light through an angle that is (a) greater than the angle for blue light (b) independent of frequency (c) less than the angle for blue light (d) the same as the angle for blue light (e) none of these.

Problems

POLARIZATION

1. [I] What is the irradiance of a beam of linearly polarized electromagnetic radiation traveling through vacuum if the maximum value attained by its electric field is 1000 V/m?

2. [I] Plane-polarized light, oscillating in a vertical plane and having an energy flux density of 2000 W/m^2, impinges normally on a linear polarizer whose transmission axis is horizontal. What is the transmitted irradiance?

3. [I] A 100-W/m^2 beam of linearly polarized light with its electric field vertical impinges perpendicularly on an ideal linear polarizer with a vertical transmission axis. What is the irradiance of the transmitted beam?

4. [I] Light from an ordinary tungsten bulb arrives at an ideal linear polarizer with a radiant flux density of 100 W/m^2. What is its corresponding value on emerging?

5. [I] A beam of natural light with an irradiance of 500 W/m^2 impinges on the first of two consecutive ideal linear polarizers whose transmission axes are 30.0° apart. How much light emerges from the two?

6. [I] Many substances such as sugar and insulin are *optically active;* that is, they rotate the plane of polarization in proportion to both the path length and the concentration of the solution. A glass vessel is placed between crossed linear polarizers, and 50% of the natural light incident on the first polarizer is transmitted through the second polarizer. By how much did the sugar solution in the cell rotate the light passed by the first polarizer? This sort of arrangement can be used to determine such things as the amount of sugar in urine.

7. [I] A beam of white light linearly polarized with its electric field vertical and having an irradiance of 160 W/m^2 is incident normally on a linear polarizer whose transmission axis is 30° above the horizontal. How much light is transmitted?

8. [I] $\mathcal{P}$-state light aligned with its electric field vector at +40° from the vertical impinges on an ideal sheet polarizer whose transmission axis is at +10° from the vertical. What fraction of the incoming light emerges?

9. [I] The reflection of the sky coming off the surface of a pond ($n = 1.33$) is found to completely vanish when seen through a Polaroid filter. At what angle is the surface being examined? (Give the answer to two significant figures.)

10. [I] What is the polarization angle for reflection of light from the surface of a piece of glass ($n_g = 1.55$) immersed in water ($n_w = 1.33$)?

11. [II] Two ideal linear sheet polarizers are arranged with respect to the vertical with their transmission axes at 10° and 70°, respectively. If $\mathcal{P}$-state light at 40° enters the first polarizer, what fraction of its irradiance will emerge?

12. [II] Four ideal linear polarizers are stacked one behind the other with the transmission axes of the first vertical, the second at 30°, the third at 60°, and the fourth at 90°. What fraction of the incident unpolarized light emerges?

13. [II] A beam of light is reflected off the surface of a cup of benzene, and the light is examined with a linear sheet polarizer. It is found that when the central axis of the polarizer (that is, the perpendicular to the plane of the sheet) is tilted down from the vertical at an angle of 56.33°, the reflected light is completely passed, provided the transmission axis is parallel to the plane of the interface. From this information, compute the index of refraction of the liquid.

14. [II] Light reflected from a glass ($n_g = 1.65$) plate immersed in ethyl alcohol ($n_e = 1.36$) is found to be completely linearly polarized. At what angle will the partially polarized beam be transmitted into the plate?

15. [III] Two ideal linear polarizers are in place one behind the other. What angle should their transmission axes make if the incident unpolarized beam is to be reduced to 30% of its original irradiance?

INTERFERENCE

16. [I] A fine mercury arc lamp, emitting light at 546.1 nm, has a coherence length in vacuum of 0.60 m. Determine its coherence time.

17. [I] Red light from a He-Ne laser ($\lambda = 632.8$ nm) is incident in air on a screen containing two very narrow horizontal slits separated by 0.100 mm. A fringe pattern appears on a screen 2.00 m away. How far (in mm) above and below the central axis are the first zeros of irradiance?

18. [I] Red plane waves from a ruby laser ($\lambda = 694.3$ nm) in air impinge on two parallel slits in an opaque screen. A fringe pattern forms on a distant wall, and we see the fourth bright band 1.0° above the central axis. Calculate the separation between the slits.

19. [I] Two parallel slits 0.100 mm apart are illuminated by plane waves of quasimonochromatic light, and it is found that the fifth bright fringe is at an angle of 1.20°. Determine the wavelength of the light.

20. [I] A 3 × 5 card containing two pinholes, 0.08 mm in diameter separated center-to-center by 0.10 mm, is illuminated by red light from a He-Ne laser ($\lambda = 632.82$ nm). If the fringes on an observing screen are to be 10 mm apart, how far away should the screen be?

21. [I] Parallel rays of blue light from an argon ion laser ($\lambda = 487.99$ nm) are incident on a screen containing two very narrow slits separated by 0.200 mm. A fringe pattern appears on a sheet of film held 1.00 m away. How far (in mm) from the central axis is the fourth irradiance maximum?

22. [I] A collimated beam of light ($\lambda = 550$ nm) falls on a screen containing a pair of long, narrow slits separated by 0.10 mm. Determine the separation between the two third-order maxima on a screen 2.00 m from the apertures.

23. [I] A thin film of ethyl alcohol ($n_e = 1.36$) spread on a flat glass plate and illuminated with white light shows a lovely color pattern in reflection. If a region of the film reflects only green light (500 nm) strongly, how thick is it?

24. [I] A glass camera lens with an index of 1.55 is to be coated with a cryolite film ($n \approx 1.30$) to decrease the reflection of normally incident green light ($\lambda_0 = 500$ nm). What thickness should be deposited on the lens?

25. [I] A soap film of index 1.35 appears yellow (580 nm) when viewed from directly above. Compute several possible values of its thickness.

26. [II] White light falling on two long, narrow slits emerges and is observed on a distant screen. If red light ($\lambda_1 = 780$ nm) in the first-order fringe overlaps violet in the second-order fringe, what is the latter's wavelength?
27. [II] Considering the double-slit experiment, derive an equation for the distance $y_{m'}$ from the central axis to each irradiance *minimum*. (Hint: Consider Eq. (27.4). The first dark bands on either side of the central maximum correspond to $m' = \pm 1$.)
28. [II] A strip of photographic film is completely black except for two horizontal parallel slits, each 1.0 cm long by 0.10 mm wide, separated center-to-center by 0.50 mm. When illuminated by sunlight, all but the zeroth-order fringe show a spread of colors. If violet light appears on a screen 3.0 m away at a distance of 2.40 mm above and below the central axis in the first colored bands, what's its wavelength?
29. [II] Two 1.0-MHz radio antennas emitting in-phase are separated by 600 m along a north-south line. A radio placed 20 km east is equidistant from both transmitting antennas and picks up a fairly strong signal. How far north should that receiver be moved if it is again to detect a signal nearly as strong?
30. [II] Sunlight incident on a screen containing two long, narrow slits 0.20 mm apart casts a pattern on a white sheet of paper 2.0 m beyond. What is the distance separating the violet ($\lambda = 400$ nm) in the first-order band from the red ($\lambda = 600$ nm) in the second order?
31. [II] Two stereo speakers are 5.00 m apart, up against the east-west wall of someone's living room. They have been wired carelessly so that when one cone is displaced forward by the driving signal, the other is displaced backward. What's the main disadvantage of this arrangement to someone sitting on the midway line equidistant from the speakers? How far due east should the listener move if she is sitting up against the opposite wall 10.0 m away and she wishes to hear a peak in the power level of a 1000-Hz sound? Take the speed of sound to be 346 m/s.
32. [II] It's an easy matter to see fringes of equal thickness in a wedge-shaped film. Take two flat sheets of glass, one on top of the other, and separate them with a piece of paper (Fig. P32). If the wedge angle is α and the illumination is near normal at a wavelength in the film of λ_f, derive an expression for the distance (x_m) measured from the apex A to the successive maxima.

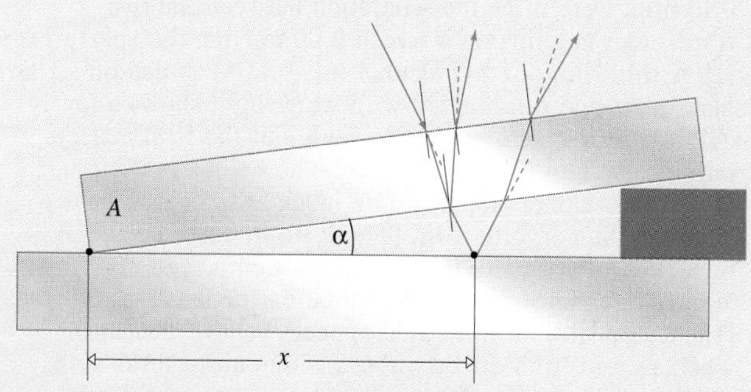

Figure P32

33. [II] Refer to the wedge-shaped film in Problem 32. What is the increase in film thickness as we move along from one maximum to the next? Derive an expression for the spacing Δx between consecutive bright fringes.
34. [II] With Fig. P32 in mind, suppose a wedge-shaped air film is made between two sheets of glass using a piece of paper 7.618×10^{-5} m thick as the spacer. If light of wavelength 550 nm comes down from directly above, determine the number of bright fringes that will be seen across the wedge.
35. [III] A Michelson Interferometer is illuminated with monochromatic light. One of its mirrors is then moved 2.53×10^{-5} m, and it is observed that 92 fringe-pairs, bright and dark, pass by in the process. Determine the wavelength of the incident beam.
36. [III] One of the mirrors of a Michelson Interferometer is moved, and 1000 fringe-pairs shift past the hairline in a viewing telescope during the process. If the device is illuminated with 500 nm light, how far was the mirror moved?
37. [III] Suppose we place a chamber 10.0 cm long (measured inside) with flat parallel windows in one arm of a Michelson Interferometer that is being illuminated by 600-nm light. If the refractive index of air is 1.000 29 and all the air is pumped out of the cell, how many fringe-pairs will shift by in the process?
38. [III] With regard to Young's Experiment, derive a general expression for the shift in the vertical position of the mth *maximum* as a result of placing a thin parallel sheet of glass of index n and thickness d directly over one of the slits. Identify your assumptions.
39. [III] Plane waves of monochromatic light impinge at an angle θ_i on a screen containing two narrow slits separated by a distance a. Derive an equation for the angle measured from the central axis that locates the mth maximum.
40. [III] Every source of light is composed of a range of frequencies distributed around some mean value. As a measure of the spectral purity of a source, we define the *ratio of the mean period to the coherence time* as the *frequency stability*. Accordingly, a typical He-Ne laser with a mean wavelength of 632.82 nm might have a frequency stability of 10^{-6}. Determine the coherence length of such a laser.
41. [III] A soap film of index 1.340 has a region where it is 550.0 nm thick. Determine the vacuum wavelengths of the radiation that is not reflected when the film is illuminated from above with sunlight.
42. [III] Figure P42 shows a setup for examining the shape of a lens. The lens is placed on an optical flat and illuminated at normal incidence by quasimonochromatic light. The gap between the lens and optical flat constitutes a circularly symmetric, wedge-shaped air film. The amount of uniformity in the resulting concentric system of circular fringes, known as Newton's rings, is a measure of the degree of perfection (see photo on p. 958). Derive the expression

$$x_m = [(m + \tfrac{1}{2})\lambda_f R]^{\frac{1}{2}}$$

for the radius of the mth bright ring. Use the fact that $R \gg d$.

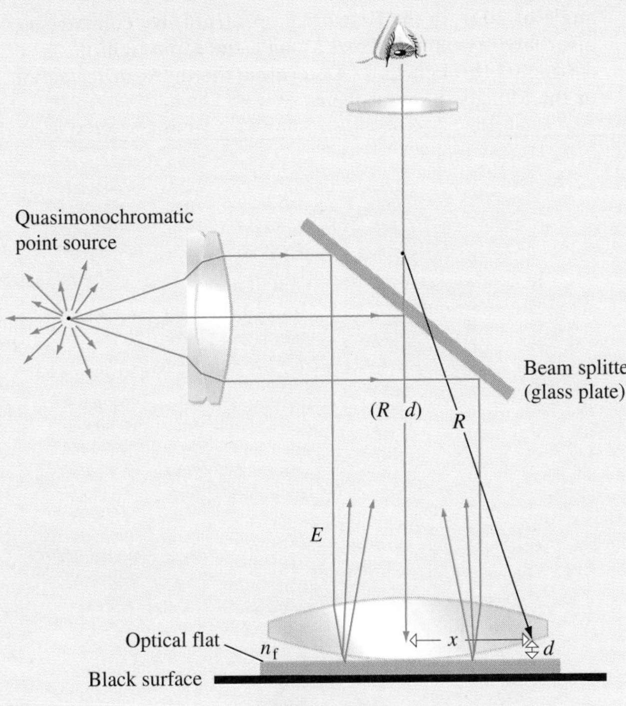

Figure P42

DIFFRACTION

43. [I] A single slit 0.10 mm wide is illuminated (in air) by plane waves from a krypton ion laser ($\lambda = 461.9$ nm). If the observing screen is 1.0 m away, determine the angular width of the central maximum as defined on p. 965.

44. [I] A narrow single slit (in air) is illuminated by IR from a He-Ne laser at 1152.2 nm, and it is found that the center of the tenth dark band lies at an angle of 6.2° off the central axis. Please determine the width of the slit.

45. [I] A parallel beam of microwaves impinges on a metal screen that contains a 20-cm-wide, long horizontal slit. A detector moving parallel to the screen locates the first minimum at an angle of 36.87° above the central axis. Determine the wavelength of the radiation.

46. [I] A transmission grating whose lines are separated by 3.0×10^{-6} m is illuminated by a narrow beam of red light ($\lambda = 694.3$ nm) from a ruby laser. Spots of light, on both sides of the undeflected beam, appear on a screen 2.0 m away. How far from the central axis is either of the two nearest spots?

47. [I] We wish to study several different gratings by holding them each near one eye and looking through each at an ordinary light bulb, blocked off by two pieces of cardboard so that it forms a slit source. What will happen to the locations of the various-order spectra as we go from 200 lines/cm to 400 lines/cm to 800 lines/cm?

48. [I] A diffraction grating with slits 0.60×10^{-3} cm apart is illuminated by light with a wavelength of 500 nm. At what angle will the third-order maximum appear?

49. [I] A diffraction grating produces a second-order spectrum of yellow light ($\lambda = 550$ nm) at 25°. Determine the spacing between lines.

50. [I] If you peeped through a 0.75-mm-diameter hole at an eye chart, you might notice a decrease in visual acuity. Compute the angular limit of resolution, assuming that it's determined only by diffraction; take $\lambda = 550$ nm. Compare your results with the value of 1.7×10^{-4} rad that corresponds to a 4.0-mm pupil.

51. [I] The Mount Palomar telescope has a 508-cm-diameter objective mirror. Determine its angular limit of resolution at a wavelength of 550 nm, in radians, degrees, and seconds of arc.

52. [I] With Problem 51 in mind, how far apart must two objects be on the surface of the Moon if they are to be resolvable by the Palomar telescope? The Earth-Moon distance is 3.844×10^8 m; take $\lambda = 550$ nm.

53. [I] Compare the results of Problem 52 with the human eye; that is, how far apart must two objects be on the Moon if they are to be distinguished by eye? Assume a pupil diameter of 4.00 mm.

54. [II] Plane waves with a wavelength of 500 nm are incident on a long, narrow aperture 0.10 mm wide. How wide will the central maximum be on a screen 3.00 m from the slits? (Use the fringe-width defined in Example 27.7, p. 965.)

55. [II] A beam of parallel rays of red light from a ruby laser ($\lambda = 694.3$ nm in air) incident perpendicularly on a single slit produces a central bright band that is 40 mm wide on an observing screen 3.50 m beyond. How wide is the slit? (See Problem 54.)

56. [II] Consider Problem 44. At what angle would the tenth minimum appear if the entire arrangement were immersed in water ($n_w = 1.33$) rather than air ($n_a = 1.000\,29$)?

57. [II] Bats use ultrahigh-frequency sound to reflect off targets. Estimate the typical frequency of this rodent sonar, presuming that bats eat moths. Take the speed of sound to be 330 m/s and a moth to be about 3.0 mm across.

58. [II] White light falls normally on a transmission grating that contains 1000 lines per centimeter. At what angle will we find red light ($\lambda = 650$ nm) in the first-order spectrum?

59. [II] Light from a sodium lamp has two strong yellow components at 589.592 3 nm and 588.995 3 nm. How far apart in the first-order spectrum will these two be on a screen 1.00 m from a grating having 10 000 lines/cm? (a) Assume Eq. (27.7) is accurate enough for our purposes. (b) Show that a more exact solution, arrived at using Eq. (27.8), yields a separation of 1.13 mm.

60. [II] Sunlight impinges on a transmission grating having 5000 lines per centimeter. Does the third-order spectrum overlap the second-order spectrum? Take red to be 780 nm and violet to be 390 nm.

61. [III] Two long slits 0.10 mm wide, separated (center-to-center) by 0.20 mm, in an opaque screen are illuminated by light of wavelength 500 nm. How many Young's fringes will be seen within the central bright band on a screen 2.0 m away? Determine a general rule for this situation in terms of the number N where $a = ND$. (Hint: Count fractions of fringes too.)

62. [III] Light with a frequency of 4.0×10^{14} Hz is incident on a grating having 10 000 lines per centimeter. What is the highest-order spectrum that can be seen with this device? Explain.

63. [III] A device used to measure the diffracted angle, and therefore the wavelength, of light passing through a grating is called a *spectrometer.* Suppose that such an instrument while in vacuum on Earth sends 500-nm light off at an angle of 20.0° in the first-order spectrum. By comparison, after landing on the planet Mongo, the same light is diffracted through 18.0°. Determine the index of refraction of the Mongoian atmosphere.

28

Special Relativity

THE SPECIAL THEORY OF Relativity represents the culmination of Classical Physics. It arises from a rethinking of the essence of space and time and the very nature of interaction. So profound are the ideas that they led to a flood of new conclusions, many of which were no less than shocking: time is relative and flows differently in systems that are in motion with respect to each other; the length of an object is not absolute, but depends on its relative motion with regard to each particular observer; the speed of light, c, is invariant and, moreover, it is the upper limit of all speeds in the Universe; the gravitational and electromagnetic interactions propagate at c rather than infinitely swiftly; except for Newton's Second Law (as he gave it, p. 103), most of the equations of classical dynamics are only approximations; mass and energy are equivalent in a way never conceived of before. These ideas and more follow from two simple postulates that Einstein introduced and then elegantly developed.

BEFORE THE SPECIAL THEORY

The research that laid the groundwork for the Special Theory was concerned with the behavior of light. Ironically, Relativity has nothing to do with light *per se,* although it has a lot to do with the constancy of the speed of light.

Prior to Einstein's work in 1905, it was widely believed that space was filled with ether and light was an electromagnetically induced vibration of that invisible elastic "goop." The common wisdom was that ether was a transparent solid that could support transverse lightwaves and at the same time a perfect fluid that could be parted effortlessly by the wandering planets.

28.1 The Michelson-Morley Experiment

The ether, even though it doesn't exist, played a crucial part in the development of the Special Theory. If we are to appreciate the context in which Relativity was conceived, it would be helpful to know a little about the ether as envisioned at the turn of the twentieth century. At that time, the most important single observation was the Michelson-Morley experiment, which had failed to detect any evidence of the ether and which seemed to suggest the invariance of the speed of light. It is arguably one of the most important experiments ever performed, and at the same time, it was a disappointing failure for the man who performed it.

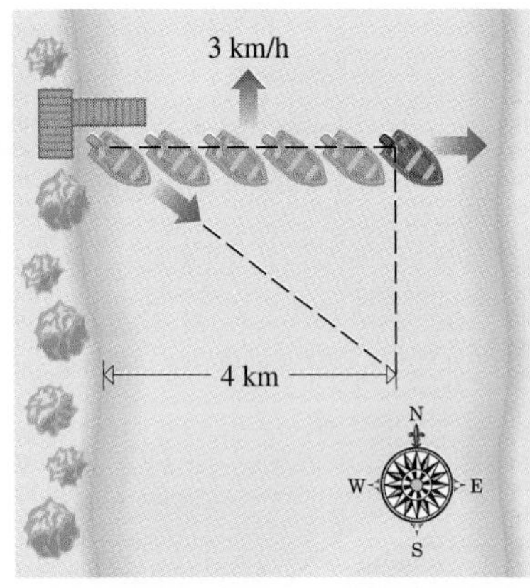

(a)

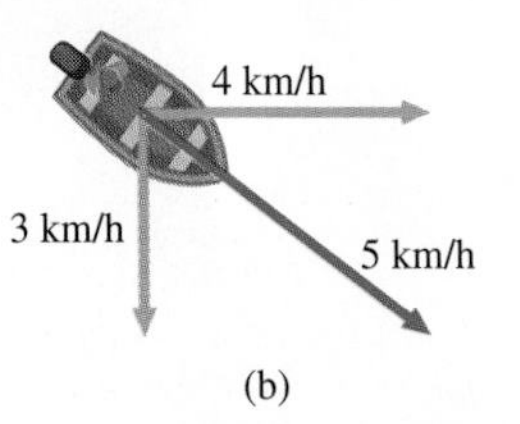

(b)

Figure 28.1 A boat traveling due east across a river flowing north at 3 km/h. It accomplishes this by having a southerly velocity component of 3 km/h in addition to its easterly component of 4 km/h.

James Clerk Maxwell, in 1879, the last year of his life (and the year Einstein was born), wrote a letter that was ultimately published. In it, he discussed a scheme for measuring the speed (v) at which the orbiting Earth plows through the ether. As the Earth moves, the ordinarily motionless ether presumably streams over the planet. The back-and-forth time of transit of a light beam in this ether wind should be different along or across the flow. To see this, imagine a wide river streaming north at 3 km/h and a motorboat capable of a single speed with respect to the water of 5 km/h (Fig. 28.1). Compare the transit times for two trips; 4 km upstream and back, versus 4 km across stream and back. Going upstream, bucking the current, the boat makes 5 km/h − 3 km/h = 2 km/h and reaches its destination in 2 h, whereupon it turns around and returns at 5 km/h + 3 km/h = 8 km/h, arriving back after an additional $\frac{1}{2}$ h; the total trip taking $2\frac{1}{2}$ h. By contrast, to cross the river directly, the boat must be headed south a bit so that it has a southerly velocity component equal to the river's northerly 3 km/h (p. 45). The two north-south velocities cancel and the boat travels due east. The 3-4-5 right-triangle geometry gives the boat a net easterly speed of 4 km/h. It travels out for 1 h and then returns (again heading somewhat south) after another 1 h—total transit time: 2 h. The time difference between the two journeys (here, $\frac{1}{2}$ h) reveals the presence of the moving medium. If v (the river's speed) was zero, the time difference would vanish.

Unfortunately, as Maxwell pointed out, in the case of an ether wind (instead of a river) flowing at v and a light beam (instead of a boat), the effect depends on $(v/c)^2$, and that "is quite too small to be observed." Even with the ether streaming past the planet at the Earth's orbital speed of $v = 3 \times 10^4$ m/s (that is, ≈66 000 mi/h), the quantity $(v/c)^2 \approx 10^{-8}$ and the transit-time difference for a 4 km-out-and-4 km-back journey is a mere tenth of a millionth of a millionth of a second. Not very promising.

Maxwell's letter came to the attention of a young instructor at the United States Naval Academy, Albert Abraham Michelson. Michelson took a leave of absence in 1880 to study in Europe, and while there, accepted Maxwell's challenge. In the past, several experimenters had used interferometers to examine the effects of motion on the transmission of light through the ether. But all of these efforts were limited in their precision, and though they failed to detect the ether, the results were unconvincing. Now Michelson designed an entirely new instrument that would allow him to make the decisive measurement precisely enough to reveal $(v/c)^2$ variations. His interferometer (funded by none other than Alexander Graham Bell) would surely detect an ether wind carrying along the lightwaves.

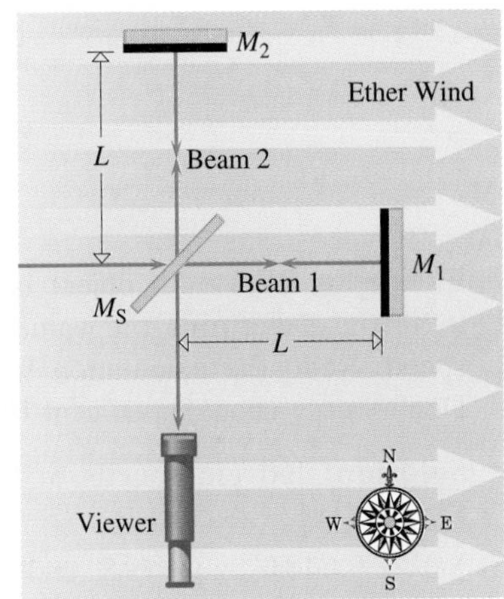

Figure 28.2 A simplified version of the Michelson Interferometer placed in the supposed ether wind.

Figure 28.2 is a schematic drawing of the Michelson Interferometer; it's the same device we studied earlier (p. 959). The difference between the along-the-ether and the across-the-ether transit times, Δt, is

$$\Delta t \approx \frac{L}{c}\beta^2 \tag{28.1}$$

where $\beta = v/c \approx 10^{-4}$. One wave travels a little longer than the other, and they arrive a bit out-of-phase. Hence, the fringes are shifted from where they would have been were the Earth at rest (Fig. 28.3). Since we can't stop the Earth, this shift is not immediately apparent, but that can be rectified by rotating the entire interferometer while observing the fringes. This will gradually interchange the two arms so that the one that was along the ether wind is now across it and vice versa. The cross-stream wave will go from leading by Δt to lagging by Δt as it becomes an upstream-downstream wave. That change of $2\,\Delta t$ in time is equivalent to an introduced path-length difference of $c2\,\Delta t$, which causes the fringe pattern to move. A shift of one wavelength results in a displacement of one bright-dark fringe-pair, and the observer (following around as the device is rotated through 90°) should see $N = c2\,\Delta t/\lambda \approx 2L\beta^2/\lambda$ fringe-pairs move past a cross hair in a viewing telescope.

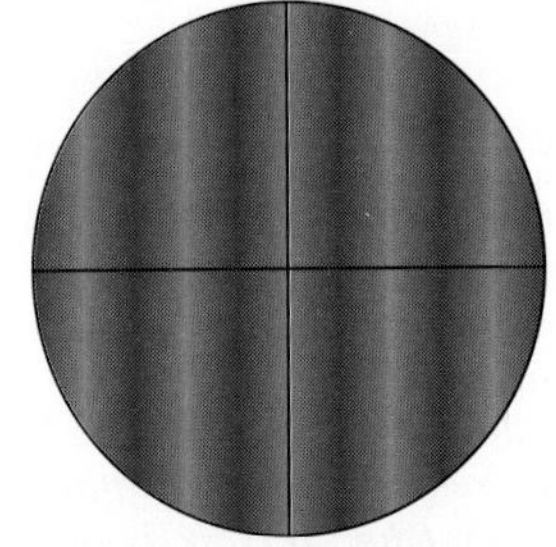

Figure 28.3 With the interferometer set up to produce straight fringes, we would see a pattern like this one in the viewing telescope. It occurs when M_1 and M_2 are perpendicular to the plane of the instrument but not to each other.

When Michelson first performed the experiment in 1881, N should have been about four one-hundredths of a fringe, but he saw no shift at all. Confident of his measurements, he published the startling result that "there is no displacement of the interference bands." Still, his so-called *null result* was not very persuasive. With the urging of Lord Rayleigh and Lord Kelvin, Michelson decided to redo the experiment in an improved fashion and settle the matter. He was joined by E.W. Morley and they enlarged the apparatus to the point where the expected shift was now four-tenths of a fringe. Still no appreciable shift was observed—and this time the results could not be neglected—there was no detectable ether wind. *The speed of light is not influenced by the motion of the Earth.*

A modern version of the Michelson-Morley experiment performed in 1979 used stable lasers to improve the precision by a factor of 4000; it, too, found no sign of an ether wind.

The Lorentz-FitzGerald Contraction

In 1892, G. FitzGerald proposed a rather imaginative hypothesis to get around the Michelson-Morley result and still keep the ether wind blowing. The idea was elaborated by Lorentz and is today known as the **Lorentz-FitzGerald Contraction**. However *ad hoc* the notion was, it would re-emerge over a decade later as a natural consequence of Einstein's Special Theory of Relativity. What FitzGerald proposed was that the ether wind exerts a pressure on a body moving through it and the body

The luminiferous ether, that is the only substance we are confident of in dynamics. . . . One thing we are sure of, and that is the reality and substantiality of the luminiferous ether.

LORD KELVIN
Popular Lectures and Addresses (1891)

compresses slightly: *every object moving at a speed v contracts along the direction of motion by a factor equal to* $\sqrt{1-\beta^2}$. If the $M_S M_1$ arm of the interferometer shrinks so that L becomes $L\sqrt{1-\beta^2}$, $\Delta t = 0$ (see Problem 1). There is no such ether pressure, but strange as it may seem, there is a Lorentz-FitzGerald Contraction.

Example 28.1 An object is moving at a speed of 0.200 0c. Determine the value of $1/\sqrt{1-\beta^2}$ at that speed. Redo the calculation for a speed of 0.002 0c. Don't worry about significant figures.

Solution: [Given: speeds of 0.200 0c and 0.002 0c. Find: $1/\sqrt{1-\beta^2}$.] $\beta = v/c = 0.2000$, $\sqrt{1-\beta^2} = \sqrt{1-0.0400} = 0.9798$, and $1/\sqrt{1-\beta^2} = \boxed{1.021}$. Now for $v = 0.0020c$, $\beta = 0.0020$, $\beta^2 = 4.0 \times 10^{-6}$, $(1-\beta^2) = (1-0.0000040) = 0.9999960$ and $\sqrt{1-\beta^2} = 0.9999980$. Thus $1/\sqrt{1-\beta^2} = \boxed{1.000002}$.

▶ **Quick Check:** Knowing that we'll need it again (p. 1003), we now derive a useful approximation. From the binomial expansion with $x = -\beta^2$ and $n = -\frac{1}{2}$, we get

$$(1 + x)^n = 1 + nx + \frac{n(n-1)x^2}{2} + \cdots$$

All the terms beyond the second one can be dropped since β^2 is tiny, and therefore $x^2 = (-\beta^2)^2$ is miniscule. Hence $1/\sqrt{1-\beta^2} \approx (1 + nx) \approx 1 + \beta^2/2 = 1 + 0.0000040/2 = 1.000002$.

FitzGerald maintained that there was no fringe shift because the arm of the interferometer along the wind shrunk. And there's no easy way to measure that shrinkage because all rulers will likewise shrink. It was not until 1932, long after Einstein had put things right, that the notion of an ether-wind-induced contraction was laid to rest experimentally by R. Kennedy and E. Thorndike. In 1990, a modern laser version of their experiment accurate to within 70 parts per million confirmed their result.

How Comfortable the Old Ideas. *The beginner will find it best to accept the ether theory, at least as a working hypothesis. . . . Even if future developments prove that the extreme relativists are right and that there is no ether, it is likely that the change will involve no serious readjustments so far as explanations of the ordinary phenomena are concerned.*

A. A. KNOWLTON
Physics for College Students (1928)

Stimulated by Michelson's observation, physicists subsequently performed many elaborate optical, electrical, and magnetic experiments, all of which failed to detect the motion of the Earth with respect to the ether. In 1900, the brilliant French mathematician Jules Henri Poincaré wrote:

> Our ether, does it really exist? I do not believe that more precise observations could ever reveal anything more than *relative* displacements.

THE SPECIAL THEORY OF RELATIVITY

Long after his billowing shock of hair had turned gray, Albert Einstein recalled how, when he was sixteen, he began to struggle with a troubling paradox. Light was understood to be electromagnetic, an intricate oscillatory web of linked time-varying electric and magnetic fields rippling through space. If we could travel out at speed c alongside a pulse of light, what would we see? Moving in imagination next to a wave peak, the disturbance would appear unchanging, motionless; yet electromagnetic theory does not allow such a situation. A stationary, nonvarying field-pulse cannot exist. In that regard, light is totally unlike a sound wave or even a stream of bullets, each of which can be followed and examined as if frozen in flight. Light is self-sustained by *change*. It is a thing of interwoven fields that, by alternation, generate each other—no variation in time, no existence. A stationary observer must see the pulse, even as the observer moving at c sees nothing—how strange. At sixteen,

There was a young fencer named
Fisk,
Whose thrust was exceedingly brisk.
So fast was his action,
The Lorentz-FitzGerald contraction
Reduced his rapier to a disk.

ANONYMOUS

Einstein had recognized a profound dilemma, a conflict between Newton's mechanics, which allowed travel at lightspeed and Maxwell's electrodynamics, which couldn't abide with it. One of these two formalisms was wrong.

Albert Einstein (1879–1955).

28.2 The Two Postulates

In 1905, Einstein was an unknown clerk in the Bern Patent Office, Switzerland. Newly married, shy but friendly, the young man had plenty of time to think and create (reviewing patent applications wasn't very taxing). He often worked on his own ideas, hiding the calculations in a drawer whenever footsteps approached. That was the year he published five papers in the prestigious *Annalen der Physik,* representing three extraordinary new developments in physics. One of these was his first paper on Relativity, "*Zur Elektrodynamik bewegter Körper,*" ("On the Electrodynamics of Moving Bodies").

At twenty-six, Einstein had created the *special* or *restricted* theory—restricted in the sense of specialized, since it pertained only to *uniform motion.* The theory assumed the validity of two *postulates* that he believed were correct but could not otherwise prove. He then set out to derive the physical implications of those postulates, and in the process completely recast our understanding of space and time.

The Principle of Relativity

The first of the two postulates is called the *Principle of Relativity.* It's a generalization of work done by Galileo and Newton. Both men recognized that uniform motion had no perceivable effect on mechanical systems. One can play pool or Ping Pong aboard a ship and never know the vessel is moving, regardless of the velocity, so long as it is constant. The woman juggling oranges in the lounge of a *747* can't tell from the behavior of the fruit in the air whether she is cruising at 600 mi/h or sitting at rest on the runway. This, then, is the *Classical Principle of Relativity: the laws of mechanics are the same for all observers in uniform motion.*

Newton had struggled with the distinction between absolute and relative motion. Was there something somewhere in the vast Universe that was totally stationary, something absolutely at rest from which all motion could be reckoned absolutely? "I hold space to be at rest" wrote Newton—space, unchanging and immovable, was his fixed reference frame. By the end of the 1880s, a motionless ether filled all space and provided a material backdrop for absolute rest throughout the Universe. A thing moves absolutely when it moves with respect to the ether. The ether had two theoretical functions that justified its existence in the face of untold contradictions: it provided the medium for lightwaves, and it was the signpost of absolute rest.

We speak of a uniformly moving observer as an **inertial observer**—someone standing still in a train, plane, or rocket ship that is itself moving at a constant velocity (p. 39). *A system that is moving at a constant velocity is an* **inertial system**. The name comes from the fact that the Law of Inertia holds in all inertial systems. A body at rest does not tend to stay at rest, nor does a body in motion tend to stay in uniform motion in a straight line if it's traveling in a system that is accelerating.

While in an inertial system, jet-plane commuters expect no novel experiences. Our battery-operated toothbrushes still hum along, computers compute, popcorn pops, cola tastes the same, and pizza smells the same. Experience suggests that not just the laws of mechanics are the same, but all the laws of physics are the same for inertial observers. All aspects of the physical environment are unaffected by uniform motion, and life goes on quite normally at 2 km/h or 2000 km/h. This conjecture Einstein raised to the status of his **First Postulate**, his **Principle of Relativity**:

All the laws of physics are the same for uniformly moving observers.

It is impossible, using experiments performed within inertial systems, to observe results that can distinguish between such systems, and no experiment whatsoever can establish whether a particular inertial system is uniformly moving or at "absolute rest." The concept of absolute rest thereby loses all significance; if it can never be determined, motion is relative, not absolute. The Principle of Relativity logically abolishes the concept of absolute rest and along with it the concept of absolute motion—**motion is relative**.

All the failed experiments that had vainly tried to measure the absolute motion of the Earth with respect to the ether had led Poincaré to the Principle of Relativity, and now Einstein embraced the same conclusion. He dismissed the ether wind, boldly dispensing with the concept of ether entirely:

> The introduction of a "luminiferous ether" will prove to be superfluous, inasmuch as the view here to be developed will not require an "absolutely stationary space."

The Constancy of the Speed of Light

Einstein's **Second Postulate** is the **Principle of the Constancy of the Speed of Light**:

Light propagates in free space with a speed c that is independent of the motion of the source.

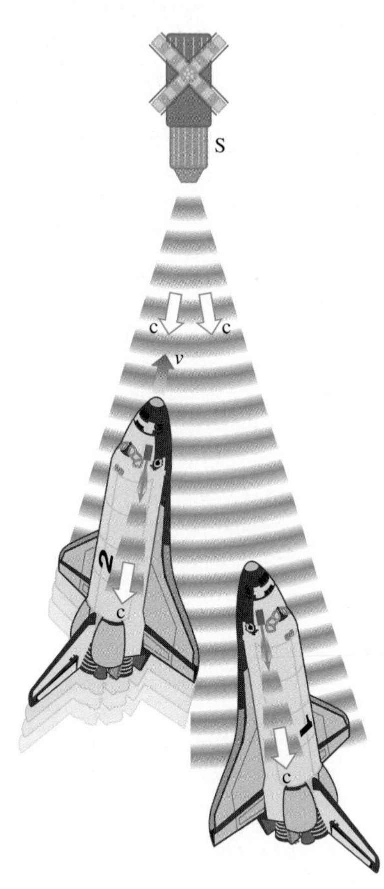

Figure 28.4 Two spaceships, one at rest, the other moving at speed v with respect to a source S. Both have on-board light beams, which they measure to travel at c. Both must measure the light from S to travel at c as well.

Now this much in itself is both orthodox and reasonable; after all, the speed of sound is independent of the motion of the source. A sound wave is launched into a medium, and the speed of the disturbance is only determined by the physical characteristics of the medium. The speed of the source is irrelevant. If light is a wave in the ether, this statement makes obvious sense. Still, there's more here than meets the eye; just three months before this paper, Einstein had submitted an article wherein he maintained that light was a stream of particles, in which case the Second Postulate is not so obvious.

Remember that the equations of Electromagnetic Theory—Maxwell's Equations—led to a wave equation that provided the speed of light in vacuum (p. 842). Assuming Maxwell's Equations are right and given the First Postulate, it must be that this same wave equation is applicable in all inertial systems—the vacuum speed of light measured on Earth or inside a rocket ship must be the same, independent of any uniform relative motion. In Maxwell's theory, the speed of light is a constant, not a motion-dependent variable. In other words, *the speed of light measured with respect to an inertial system must be the same for all such systems.*

The fact that the speed of light is independent of the motion of the source is not at all troublesome, but is it also independent of the motion of the detector (that is, the observer)? Certainly, the speed of sound is not; if the detector rushes toward the source, moving with respect to the air, the measured speed of sound increases. Just imagine two identical ships headed toward a motionless sound-emitting buoy, one steaming along at full speed and the other dead in the water. On both ships, the time it takes a blast of sound to sweep from bow to stern is measured, and the speed of the wave is computed in that inertial system. Clearly, for the ship moving toward the buoy, that time will be shorter (during the interval it takes the sound to traverse the ship, the stern will advance somewhat toward the pulse, shortening the effective length) and the wave speed will be determined to be faster.

Figure 28.4 depicts the same arrangement, this time with spaceships and lightwaves. Both ships receive light from the outside source S, and both compare the speed of that light to the speed of the light from their own sources. Ship 1 at rest with respect to the beacon must measure both its beam and the beacon's beam to have the same speed c. Ship 2 (moving at v with respect to S) must measure the beam from its own source to travel at c because of the Second Postulate. But it must also measure the light from the beacon to travel at c, as well. If it didn't, that would mean that the two beams could be used together as a motion detector to establish that ship 2 was actually moving, which is not allowed by the First Postulate. The conclusion is an astonishing one: *no matter how fast a light source moves toward or away from an inertial observer, and no matter how fast the observer moves toward or away from the source, the speed of the light passing from one to the other in vacuum will always be* c—**the speed of light is constant**; it *is* absolute. The Michelson-Morley experiment ends with a null result because the speed of light along each arm is identically the same—there is no fringe shift because there is no transit-time difference, $\Delta t = 0$.

The two postulates separately are innocent enough. It's when we mix them together, when we demand that they both apply at once, that things seem to become fantastic. The paradox of Einstein's youth is now no longer a paradox—one simply cannot travel next to a light beam and catch up to it. The spaceship in Fig. 28.5 could be rushing either toward or away from the source at any speed you like, say, 99% of c, and still the inertial observer aboard it will measure the speed of the beam to be c! This is one of the premier conclusions of the analysis; its disturbingly "illogical" nature suggests that our familiar, comfortable understanding of space and time requires revision.

To our best knowledge, neutrinos and gravitons both also travel at c, and there must equally well be a *Principle of the Constancy of the Speed of Neutrinos.* The important thing is the speed c, not the photons doing the traveling—they just happen to be the most convenient, easily detected, c-speed probe.

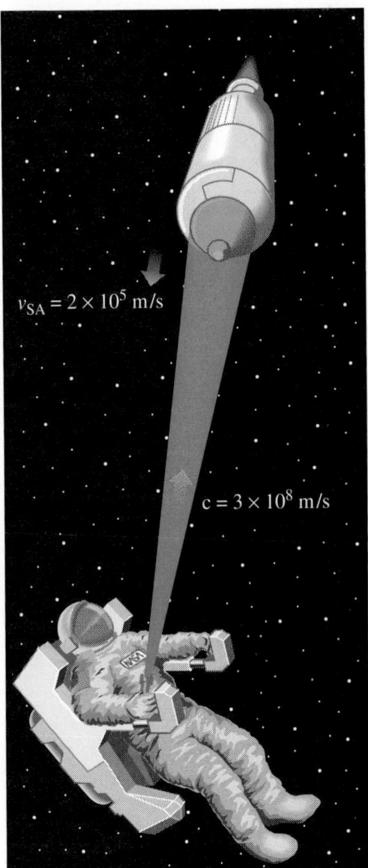

Figure 28.5 The speed of light in vacuum will be measured to be constant, regardless of the motion of either the source or the receiver. Here, a ship receives a light beam from an astronaut toward whom it is rushing at a speed v_{SA}. Regardless of the relative motion between the ship and the astronaut, a person on board the vessel will measure the light to arrive at c.

Experimental Confirmations of the Second Postulate

At the time Einstein proposed the Second Postulate, there had been no direct experimental evidence to confirm it—it was at birth a purely logical conjecture. That situation has changed considerably over the intervening decades, and the extensive confirmation that now exists has transformed the Second Postulate into what might better be called the *Law of the Constancy of* c.

In a most convincing contemporary experiment, subatomic particles known as neutral pions (p. 1138) were produced at a tremendous speed of 99.98% of c. Free neutral pions have a mean life of only 8.7×10^{-17} s before naturally decaying—vanishing into gamma rays. So here we have a pulse of pions traveling at 0.999 8c, and in a short while each of them transforms into two photons. Detectors in the forward direction record bursts of photons, and their speeds can be determined over the 31-m flight path. Instead of finding a speed of $0.9998c + c \approx 2c$, which would have been expected from Newtonian kinematics, researchers found a speed of $2.9977 \pm 0.0004 \times 10^8$ m/s, in excellent agreement with the standard value of c measured with a stationary source.

Ether Persisted. *We have learned too that radiant heat energy is believed to be transmitted by a medium called the ether. At the present time, some scientists believe that other ether waves produce various other effects. . . . It is possible, then, that light waves are ether waves.*

CHARLES E. DULL
Modern Physics (1939), a high school text

28.3 Simultaneity and Time

Einstein was troubled by the fact that, on one hand, the speed of light must be invariant and, on the other, such invariance violates the customary addition rules for velocities. He spent most of a year struggling with the problem. Then it came to him: "Time cannot be absolutely defined, and there is an inseparable relation between time and signal velocity. With this new concept, I could resolve all the difficulties . . . . Within five weeks the Special Theory of Relativity was completed."

Light is the swiftest instrumentality at our disposal for communication and for the perception of events (which are themselves the basis of time). Yet its speed is finite, and that shapes our understanding. If c were infinite, the distinction between Special Relativity and Newtonian theory would vanish. In fact, wherever c is so large in comparison with the relevant motions that it can be considered effectively infinite, Newtonian theory works very well. This is precisely why the discrepancies went unnoticed so long—we live, for the most part, in a comparatively slow-moving world. Running, driving, even spaceship-flying at thousands of miles per hour are all mere crawling compared to 2.998×10^8 m/s (Table 28.1).

The link between space and time and c hardly intrudes in our everyday lives—it doesn't matter that the people you are talking to 10 feet away are also ten-thousandths of a millionth of a second away. Light takes about 10^{-9} s to travel 1 foot, and the people are seen not as they are now, but as they were 10×10^{-9} s ago. When you look out the window, the scene on your retina—the scene you see—is not all happening at the same moment; the view of the Sun in the sky shows it as it was 8.3 minutes before Uncle George smiled in the foreground. The stars in a photo of the night sky were not all there looking as they do at the same moment, even though the light from them arrived on the film at the same instant. For that matter, the stars you see might no longer exist even while you are "looking" at them: a star 100 light-years away appears as it was 100 years ago when the light you see now first left it. And if any of this seems strange, it is because you have always thought of the world as if you saw things the instant they happened, as if c = ∞, but it doesn't.

TABLE 28.1 Various speeds and the corresponding β and γ values*

Object	Speed v	β v/c	γ $1/\sqrt{1-(v/c)^2}$
Human walking	8 km/h	0.000 000 007	1.000 000 000
100-yard dash (max.)	10.0 m/s	0.000 000 033	1.000 000 000
Commercial automobile (max.)	62 m/s	0.000 000 21	1.000 000 000
Sound	333 m/s	0.000 001 11	1.000 000 000
SR-71 reconnaissance jet	980 m/s	0.000 003 27	1.000 000 000
Moon around Earth	1000 m/s	0.000 003 33	1.000 000 000
Apollo 10 (re-entry)	11.1 km/s	0.000 037	1.000 000 001
Escape speed (Earth)	11.2 km/s	0.000 037	1.000 000 001
Pioneer 10	14.4 km/s	0.000 048	1.000 000 001
Earth around Sun	29.6 km/s	0.000 099	1.000 000 005
Mercury orbital speed	47.9 km/s	0.000 16	1.000 000 013
Helios B solar probe	66.7 km/s	0.000 22	1.000 000 025
Earth-Sun around galaxy	2.1×10^5 m/s	0.000 70	1.000 000 245
Electrons in a TV tube	9×10^7 m/s	0.3	1.05
Muons at CERN	2.996×10^8 m/s	0.999 4	28.87
Electrons in Stanford Linear Accelerator (SLAC)	$2.997 9 \times 10^8$ m/s	0.999 999 999 7	4×10^4

*To better show the behavior of β and γ, little concern is given to significant figures.

The National Institute of Standards and Technology (NIST) sends out radio signals with which we can set our clocks—"When you hear the tone it will be . . . BEEP." Yet the farther you are away from the transmitter, the later the BEEP will arrive. An observer in the Andromeda galaxy will have to wait 2,200,000 years for the BEEP to arrive marking *now.* That's presumably no problem since we can make the necessary corrections for communication-time lags, knowing distance and c. Still, what does *now* mean for us in regard to our friend on Andromeda? That is, what does it mean to say that two events occur simultaneously, one here and one there, now? To be sure, if two events occur simultaneously at the same location, there's no problem—if two comets are seen to blow up immediately on crashing, we know the two events occurred absolutely simultaneously. Every other inertial observer in the Universe will see the same thing. Difficulties arise when the events occur at separate locations, and the greater their separation in space, the longer the period of inconclusiveness in time.

Figure 28.6 illustrates the problem quite simply. Part (a) is the view of a pulse of light as seen sequentially by an observer at rest inside the ship. She sees the expanding lightwave strike the front and back walls simultaneously. Note that *once the light is launched into space, it's on its own and it must travel at* c *in all directions as seen by any and all inertial observers.* Part (b) is the view as seen by an observer outside with respect to whom the ship is uniformly moving. He sees the lightwave, traveling at c in all directions, but now the rear wall advances on the source and is struck first while the front wall recedes and is struck second—the

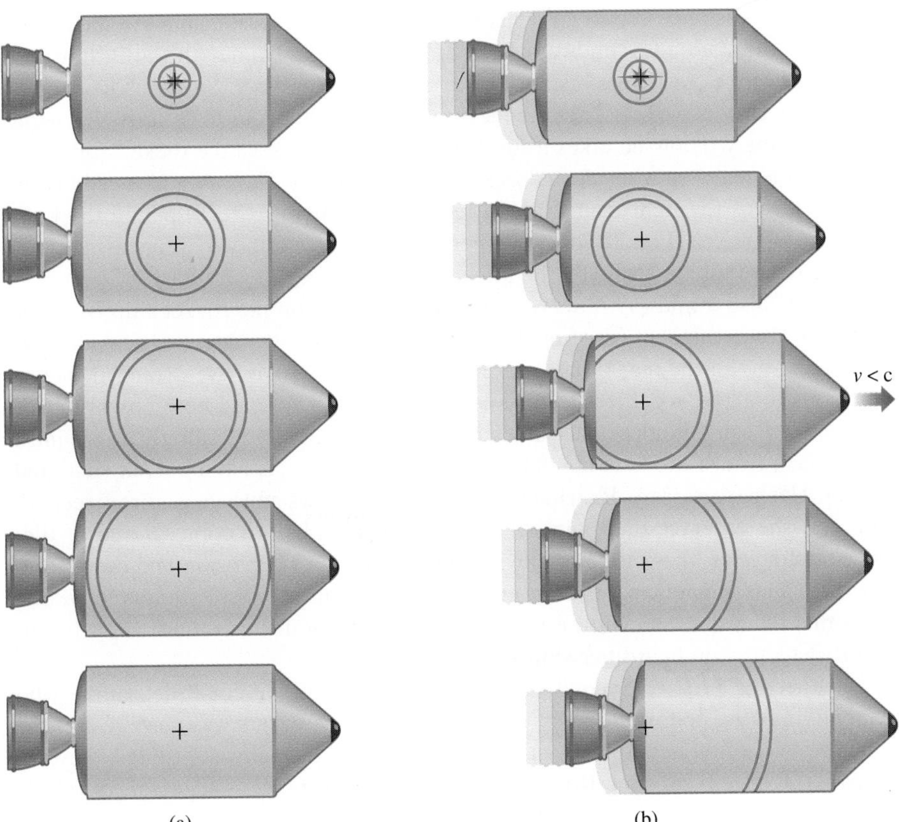

Figure 28.6 A flash of light in (a) a "stationary" and in (b) a "moving" ship as seen from outside. Or, equivalently, a flash in a ship seen (a) inside and (b) by an observer outside with respect to whom the ship is moving at a constant speed *v*.

events are *not simultaneous.* Notice that if the walls in the craft are mirrored, the waves will be reflected and meet in the center of the chamber in Fig. 28.6a. There's no ambiguity then; the same thing must happen in (a) and (b)—events that occur simultaneously at one place in space must be simultaneous for any inertial observer. What all of this suggests is that, like rest and motion, simultaneity is not absolute. You might say "who cares?" but if simultaneity, which is at the heart of the measurement of when events occur, is not absolute, time itself is not absolute.

Since that's really a big step to take, let's be a bit more rigorous about it. *Two spatially separated events are simultaneous if they are seen to occur at the same time by an observer located midway between the sites where the events happen.* This way, we will not have to introduce any corrections for different distances as was the case with the BEEPs from NIST. The question now is, do two events occurring simultaneously for one **midway inertial observer** occur simultaneously for all midway inertial observers regardless of their motion?

Imagine a rocket ship and a space station gliding past each other at any constant relative speed (Fig. 28.7). *We select the station commander,* a stubborn but honest fellow named Stan, *to be the one who will see the two events we are arranging as simultaneous.* Playing out his part, he switches on two explosive flares that just happen to be floating a few meters apart in space between him and the rocket. At the very moment they explode (as determined by him later), he finds himself lined up nose-to-nose with the rocket's pilot, a woman named Rosie no less willful than himself. Both are in the middle of their respective crafts. (It doesn't matter how these flares are themselves moving, or how they got there. We are only interested in them as light sources that will conveniently leave dent marks on each hull.) According to Stan, points A and A' are adjacent to each other at the same moment that B and B' are adjacent—the distances between dents on the station and on the ship are the same.

Stan in the station is a midway observer, and the light from both flares, fore and aft, reaches him at the same instant (we arranged it that way). So far as he is concerned, the two explosions are unquestionably simultaneous. Naturally enough, looking out the window he takes himself to be at rest and sees the rocket ship moving off to his rear. As a result, he sees Rosie advance on flare A and recede from flare B. He sees the light from flare A reach her first (Fig. 28.7b) before the light from flare B arrives (Fig. 28.7d). He expects Rosie to see the two events as not having occurred simultaneously, but he knows that he's right.

Rosie is also a midway observer, and the two dents on the rocket ship from the explosions are permanent proof of that. She looks out her window and sees the station moving off to the rear of her rocket ship (Fig. 28.8). We have arranged for the two beams to arrive together at Stan's location (Fig. 28.8c), and both observers see that happen because it takes place at a single point in space. Accordingly, Rosie must see the flash from flare A (Fig. 28.8b), which she records as having fired first, and then the flash from flare B, which she therefore knows must have fired second. Moreover, she observes that points A and A' coincided before points B and B' did. Understanding that the station is moving toward the later explosion and away from the earlier one, she anticipates Stan giving an erroneous report of simultaneity. She knows that the events she witnessed did not occur simultaneously, and all of the automatic detectors on board her ship will verify it.

If either the First or Second Postulates were wrong, we could determine who was really moving and would know which of the two observers was correct. But the discrepancies cannot be resolved—they are inherent in the nature of things; they are the reality. **There is no such thing as absolute simultaneity**—*events separated in*

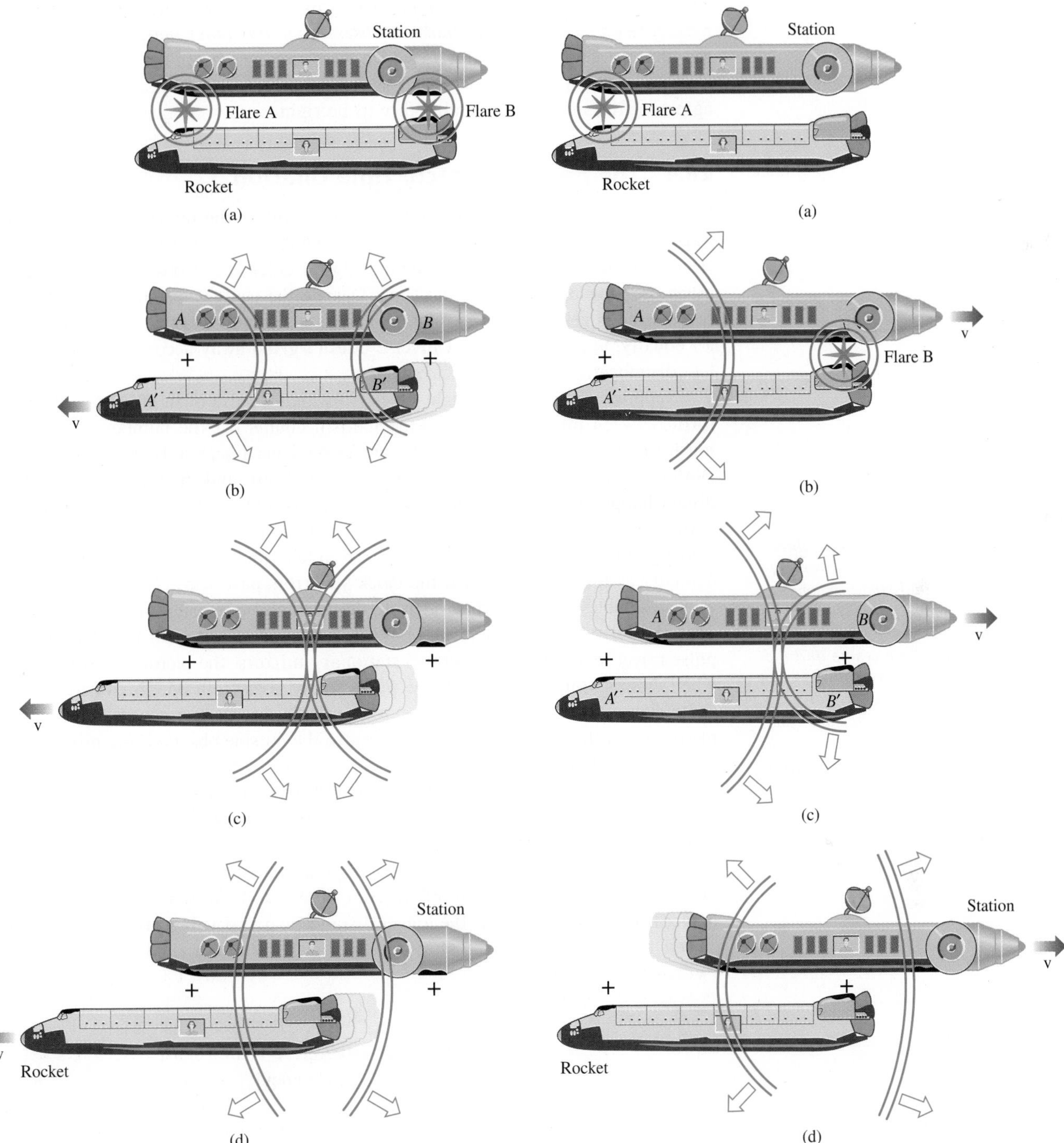

Figure 28.7 From the perspective of Stan, the Station Commander, the flares lit simultaneously and the rocket ship sailed off to the rear of his station. It follows from the Principle of Relativity that either inertial observer can assume that he or she is at rest and that it is the other person who is moving.

Figure 28.8 From the perspective of Rosie, the rocket's pilot, flare A fires first and the station sails off to the rear of her ship. The drawing neglects the fact that the station must appear to Rosie to be somewhat shrunken, a point we'll come back to later.

space that are simultaneous for midway observers in one inertial system are not necessarily simultaneous for midway observers in any other inertial system. Notice, too, that the lengths $\overline{AB}$ and $\overline{A'B'}$, which were seen by Stan in the station to be equal (Fig. 28.7), are not seen to be equal by Rosie in the rocket (Fig. 28.8); they will not agree on times or lengths, and both will be right.

There was a young lady named Bright
Who traveled much faster than light;
She started out one day,
In the relative way,
And returned on the previous night.

A. H. REGINALD BULLER

28.4 The Hatter's Watch: Time Dilation

We reckon time by comparing simultaneous events—the moment the horse's nose touches the wire, and the placement of the hands on a stopwatch. Evidently, if the front flare in Fig. 28.7 was a flashing digital clock, each observer would see the other flare explode at a different time. Despite the centuries of thought to the contrary, **time is relative**, not absolute. There is no universal time that dances over us all equally, unalterably—the Universe doesn't grind away according to a single silent drumbeat.

Suppose we construct a light-clock, in which a pulse of light bounces back and forth between mirrors and some mechanism counts off the number of traversals, much like the familiar tick-tock. Figure 28.9a depicts such a device in a spaceship. With respect to the pilot on board, the clock is at rest, and she sees nothing extraordinary happening—everything aboard ship is normal. Quite the contrary is true according to an observer on the ground as the ship whisks by with a uniform relative velocity (Fig. 28.9b). He sees a light pulse in her clock behave as if it advanced diagonally as the vehicle carried the clock's mirrors past him. (Notice how the situation resembles the transverse arm of the Michelson Interferometer—indeed, the analysis will yield similar results.) While someone at rest with the clock sees the pulse travel to and fro between two stationary mirrors, the picture is very different when viewed from the ground. The pulse surely goes from one mirror to the other (both observers see the pulse strike either mirror—at such moments, the pulse and mirror are at the same point in space), but to the outside observer, the mirrors move along the flight path of the ship.

There is one single and invariable time, which flows in two movements in an identical and simultaneous manner. . . . Thus, in regard to movements which take place simultaneously, there is one and the same time, whether or no the movements are equal in rapidity. . . . The time is absolutely the same for both.

ARISTOTLE

As seen by the pilot who is stationary with respect to the clock, the pulse travels a distance given by $c\,\Delta t_S$, where Δt_S is the elapsed interval on her clock (one tick's worth). To the observer on the ground with respect to whom the ship is moving at speed v, the pulse travels a longer diagonal path. Bound by the Second Postulate, the pulse (which must be seen by all inertial observers to progress at c) must take a longer time, Δt_M, to traverse the longer distance, $c\,\Delta t_M$. During that interval, as perceived from the ground, the ship advances a distance $v\Delta t_M$. It follows from the Pythagorean Theorem that

$$(c\,\Delta t_M)^2 = (v\Delta t_M)^2 + (c\,\Delta t_S)^2$$

Hence

$$(c\,\Delta t_M)^2 - (v\Delta t_M)^2 = (c\,\Delta t_S)^2$$

and

$$(\Delta t_M)^2 = \frac{c^2(\Delta t_S)^2}{c^2 - v^2} = \frac{(\Delta t_S)^2}{1 - \dfrac{v^2}{c^2}}$$

$$\Delta t_M = \frac{\Delta t_S}{\sqrt{1 - v^2/c^2}} \tag{28.2}$$

Time travels in divers paces with divers persons. I will tell you who time ambles withal, who time trots withal, who time gallops withal, and who he stands still withal.

WILLIAM SHAKESPEARE
As You Like It (Act 3, Scene 2)

Inasmuch as $\sqrt{1 - \beta^2}$ must be less than 1, the time interval seen by the outside observer with respect to whom the clock is moving (Δt_M) must be greater than the cor-

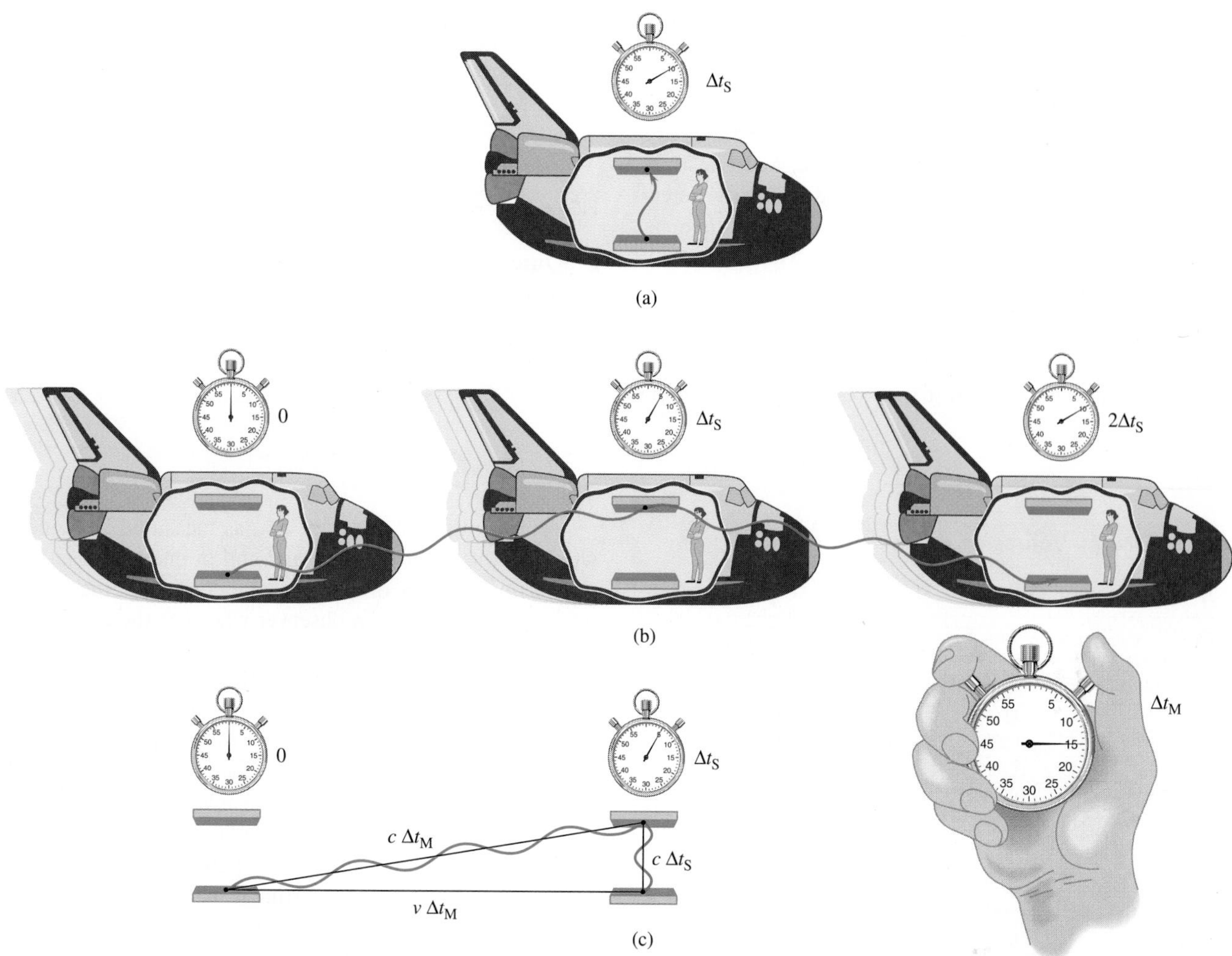

Figure 28.9 (a) A person at rest relative to a light-clock sees what we might call an interval of stationary time Δt_S go by. (b) When that same clock is viewed by an observer who sees it moving, (c) that observer sees an interval Δt_M go by. He measures the moving clock to be running slow compared to his own stationary time. Only when an observer times two events occurring at the same place will he measure Δt_S.

responding time interval (Δt_S) seen by the inside observer with respect to whom the clock is stationary. The interval between ticks and tocks is longer for the observer who sees the clock moving than for the one who doesn't. Let $\gamma = 1/\sqrt{1 - \beta^2}$, whereupon we can write

$$\Delta t_M = \gamma \Delta t_S \tag{28.3}$$

with $\gamma > 1$. This slowing down of time is known as **time dilation**, and it's a very small effect. A clock aboard a commercial plane flying at top speed for $\approx$70,000 years would lose about 1 s compared to a clock on the ground.

Time on a clock that is moving with respect to an observer is seen to run slower than time on a clock that is stationary with respect to that observer. And this is true for any clock (a wristwatch, a pendulum, a beating heart, a fertility cycle, or a dividing cell); all must slow down, all must match the light-clock. Otherwise we could easily learn from the difference who was actually moving, which is nonsense; no one is *actually* (or absolutely) moving—absolute motion violates the First Postulate.

Example 28.2 A college physics laboratory is under observation by aliens traveling on an asteroid. An undergrad seen measuring the period of a mass oscillating on a spring gets a value of 2.00 s. Given that the aliens are cruising by at a constant speed of 0.50c, and that they have nothing better to do, what period will they determine? Because they are moving rapidly, there can be substantial communication-time lags, so we assume they make any necessary corrections.

Solution: [Given: $v = 0.50c$ and $\Delta t_S = 2.00$ s. Find: the period determined by the aliens.] The period measured by an observer at rest with respect to the pendulum is 2.00 s. From Eq. (28.3), the aliens record an interval given by

$$\Delta t_M = \gamma \Delta t_S = \frac{1}{\sqrt{1-\beta^2}} \Delta t_S$$

Here, $\sqrt{1-\beta^2} = 0.866$ and $\boxed{\Delta t_M = 2.3 \text{ s}}$.

▶ **Quick Check:** The time interval is properly dilated; the aliens see the oscillations taking longer than the student does. Moreover, when $\beta = \frac{1}{2}$, Table 28.2 agrees with this value of γ.

TABLE 28.2 Values of β, $1/\gamma$, and γ

β	$1/\gamma$	γ
v/c	$\sqrt{1-(v/c)^2}$	$1/\sqrt{1-(v/c)^2}$
0.000 000	1.000 00	1.000 000
0.100 000	0.994 987	1.005 038
0.200 000	0.979 796	1.020 621
0.300 000	0.953 939	1.048 285
0.400 000	0.916 515	1.091 089
0.500 000	0.866 025	1.154 701
0.600 000	0.800 000	1.250 000
0.700 000	0.714 143	1.400 280
0.800 000	0.600 000	1.666 667
0.900 000	0.435 890	2.294 157
0.990 000	0.141 067	7.088 812
0.999 000	0.044 710	22.366 27
0.999 900	0.014 142	70.712 45
0.999 990	0.004 472	223.607
0.999 999	0.001 414	707.107

To keep things neat, the proper number of significant figures has not been kept consistently.

The duration of an event Δt_S, as measured by someone who sees the event to begin and end in one place is always shorter, by a factor of γ,$^{-1}$ than the corresponding interval Δt_M, as measured by an observer who sees the event to occur in a moving system.

Everyone at rest on the Earth experiences the same Earth-time, and we call the time measured by an observer at rest with respect to the clock, the **proper time**. Since none of us really rushes around very quickly, $c \gg v$, $\gamma \approx 1$, and quite generally $\Delta t_M \approx \Delta t_S$ for *everyone* on the planet, which is just what we expect. At its extremes, Relativity must yield the same results as our well-tested Classical Mechanics. As the Earth sails through space, it essentially has its own proper time. Anyone riding an asteroid past the planet will see our time running slower than their own time as seen on their "stationary" clocks. Someone else flashing by at a greater speed will see our "moving" clocks running still slower compared to their time. And the inverse is true: we will see their "moving" clocks run slow, as our own "stationary" clocks run at their normal rate. This reciprocity doesn't make the process less meaningful—time dilation has been measured. Certain nuclei vibrate and emit gamma-rays with very precise frequencies. When a sample of such a substance is heated, the gamma-ray frequency is reduced. The atoms move around more rapidly, and with respect to an observer at rest in the laboratory, their nuclear clocks run slower.

There have been many other experimental confirmations of time dilation over the years—the following is among the more interesting. Muons are subnuclear particles that are like heavy electrons. They are unstable, decaying into electrons and neutrinos. A muon at rest in the laboratory has a mean life of 2.2 μs, and this provides us with a convenient natural clock having a 2.2-μs tick-tock interval. In 1976, experimenters at the European Council for Nuclear Research (CERN) created a beam of muons traveling at 0.999 4c. These were injected into a large doughnut-shaped storage ring, where they were confined by powerful magnets and circulated until they decayed. Although a typical muon might, on the basis of Newtonian theory, be expected to survive 14 or 15 trips around the ring, most muons actually made in excess of 400 orbits. Electron detectors surrounding the ring established that the rapidly moving muons had a mean life about 30 times longer than when they were at rest ($\gamma = 28.87$). Equation (28.2) was confirmed to an accuracy of 0.2%.

More recently (1985), fast-moving neon atoms (excited by a laser so that they emitted light of a precise frequency) were used to confirm the time dilation to within an accuracy of 40 parts per million.

What then is time? If no one asks me, I know: if I wish to explain it to one that asketh, I know not.

ST. AUGUSTINE

28.5 Shrinking Alice

Neither time nor space is absolute in this Universe where the speed of light is constant. The fall of absolute simultaneity takes with it both absolute time and absolute length (or distance). *Only if both ends of a moving rod can be located at exactly the same instant can its length be measured accurately, and that cannot be done absolutely.* It's not much good in finding the length of a rod to say that the front end lined up with the 3-m mark of a ruler at 1:00 P.M., and 2 seconds later the rear end was next to the 2-m mark. We must watch the alignment with the ruler at both ends (separated in space) simultaneously, and different observers will not agree about that. If the flares in Fig. 28.7 are replaced by clocks, they could be synchronized to the same time by someone at rest with respect to them, but they will be seen to be unsynchronized by anyone moving relative to their inertial system. What is the distance between the exploding flares in Figs. 28.7 and 28.8? Remember there were two different distances seen by the two observers just because they couldn't agree on whether the flares fired off simultaneously or not. A rod has one **proper length** measured by any observer at rest with respect to it, but it can also have different shorter lengths measured by people who are in uniform motion with respect to it.

Imagine a meter stick in a rocket ship flying past you, and you also have a meter stick. *The sticks are aligned parallel to the relative velocity*, and along with appropriate clocks and sources, they are used to measure the speed of light. As seen by Rosie in the rocket, a pulse of light travels the length of her ruler (L_S) in a certain proper time at speed c. As seen by you, with respect to whom that experiment is moving, the pulse in the rocket traverses a length L_M (which may or may not equal L_S; that is, 1 m). You, looking into the rocket, see the pulse travel for fewer seconds on the rocket's clock, which you see running slow. You watch the pulse traverse the apparatus, in a room where time runs "slow," just as Rosie does, in a room where time runs "normally," and both of you must determine the speed to be c. For this outcome to be the case, the pulse in the rocket ship must travel along a path that to you is apparently shorter than it "ought to be" (shorter than L_S, shorter than 1 m) by a factor of γ; so that

$$L_M = \gamma^{-1} L_S \tag{28.4}$$

or

$$L_M = L_S\sqrt{1 - v^2/c^2} \tag{28.5}$$

Pilot Rosie moving along with her apparatus sees everything in the ship as properly normal and measures c. You, the outside viewer, see the rocket ship *shrunk along the line of motion*, the experiment on board it shrunk, and everything in the cabin happening slowly. And you understand why Rosie found the pulse to travel at c (her time was running slow by a factor of γ, but her experiment was shrunk by a factor of $1/\gamma$). With your proper length stick and proper time, your light pulses will also travel at c. And when Rosie views you on Earth, she sees everything of yours appropriately shrunk and your time running slow compared to her clock—the contractions and dilations are symmetrical via the First Postulate.

A moving observer measures an object to have a length (along the direction of motion) that is shorter than the length measured by an observer at rest with respect to the object (that is, shorter than the proper length). This is the **length contraction**, and it applies only to the direction of motion; *transverse distances are unaltered.* Equation (28.5) is mathematically identical to the Lorentz-FitzGerald Contraction (p. 981), and it's often called by that name.

Example 28.3 A flying saucer descending straight down toward the Earth at 0.400 0c is first observed by an astronomer on the planet when it passes a satellite at an altitude of 3000 km. At that instant, what will be the ship's altitude as determined by its navigator?

Solution: [Given: $L_S = 3000$ km and $v = 0.400\,0c$. Find: the altitude.] The height measured by an observer with respect to whom the distance is stationary is $L_S =$ 3000 km. Using Eq. (28.5), we have

$$L_M = L_S\sqrt{1 - v^2/c^2} = (3000 \text{ km})\sqrt{1 - 0.160\,0}$$

and $\boxed{L_M = 2750 \text{ km}}$

▶ **Quick Check:** The speed is not very high so one expects a relatively small contraction. A glance at Table 28.2 confirms that $\sqrt{1 - v^2/c^2} = 0.92$.

There is no absolute distance between New York and Chicago; an atlas provides the proper distance as measured by people at rest with respect to the planet. But every traveler moving with regard to the surface sees his or her own version of that spatial interval, depending on his or her relative speed. Go fast enough, and London and San Francisco can be a meter or two apart. As for the reality of all of this (that is, is the spaceship actually squashed?), the business is similar to what happens with the Doppler Effect. Run toward a source and the sound pitch increases. The wave itself doesn't physically change, but the perception of it certainly does change. What you hear and measure—the reality of the experience—most assuredly depends on how you move with respect to the source.

You might find it satisfying to know that the electromagnetic field seen by differently moving inertial observers will be different. And that a detailed theory of the electromagnetic forces between the atoms within an object shows that the object must thereby contract by a factor of $1/\gamma$.

Example 28.4 A starship (some time in the very distant future) is headed for a galaxy that, according to human astronomy texts, is 200 light-years away from Earth. Flying a direct course, the ship reaches a cruising speed of 0.999c. What will be the Earth-galaxy distance as then determined by the navigator?

Solution: [Given: $L_S = 200$ ly and $v = 0.999c$. Find: the distance.] The first thing to settle is which distances are which in Eq. (28.5). L_M is the length as seen by someone moving with respect to the physical system in which the proper length is L_S. Thus

$$L_M = L_S\sqrt{1 - v^2/c^2} = (200 \text{ ly})\sqrt{1 - (0.999)^2}$$

and $\boxed{L_M = 8.94 \text{ ly}}$

▶ **Quick Check:** The speed is very high so one expects a considerable contraction. A glance at Table 28.2 confirms that $\sqrt{1 - v^2/c^2} = 0.044\,7$. At these great speeds, the length is still a few percent of the proper length. (Get a feeling for the numbers you can expect by looking over the table.)

(a)

(b)

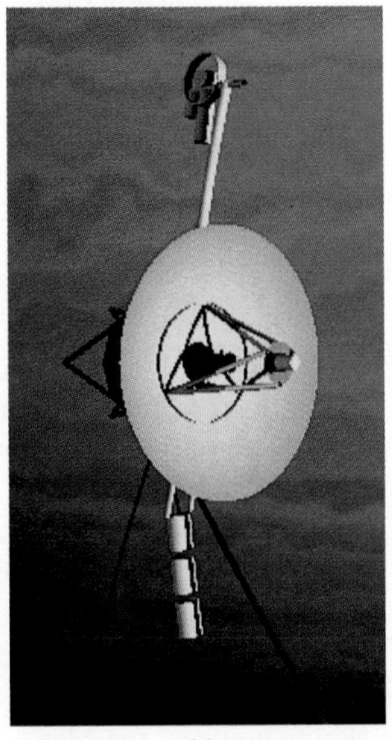
(c)

(d)

Computer images of a space probe as seen by an observer traveling with respect to it at four different speeds: (a) 0, (b) 0.25c, (c) 0.5c, and (d) 0.75c. Moving away at high speed, the light is red-shifted via the Doppler Effect (p. 469).

The question of what objects look like to a rapidly moving observer because of the length contraction is a complicated one. The semi-enlightened comic book renditions of things simply shrunk along the direction of motion—the long-skinny-people vision—is erroneous. Complications arise because light arriving at the retina (or on a piece of film) at some instant must have left different regions of the object at different times, depending how far away they are. Three-dimensional objects will therefore seem to be twisted and distorted, the more so the greater v (see photo above).

28.6 The Twin Effect

Someday, we may have rocket ships that can attain speeds high enough to experience significant time dilations and length contractions. This feat requires engines that can continue to exert thrust for very long periods so that the craft can accelerate at a humanly bearable rate (1 g would be nice) for years. Such an achievement is far beyond our present capabilities and may be for centuries. Practicalities aside, suppose we have such a starship. You and I meet at the launch pad, engage our identical stopwatches, shake hands, and you fly off to some star a distance $L_S = 50.00$ light-years away. The mission is to arrive at the star, plant the flag as it were, and promptly come home. To make the calculations simple, you quickly get the craft up to a wonderfully unrealistic cruising speed v of 0.999 8c and settle in for the journey. A glance out the window reveals to the passengers that the Earth-star distance is now $L_S\sqrt{1 - v^2/c^2}$. Since we both agree on *our* relative speed (and neglecting the comparatively short times it takes to negotiate the several accelerations), the trip out takes a proper on-board time of $(L_S\sqrt{1 - v^2/c^2})/v$. Thus, if T_S is the total round-trip flight time recorded by you, then

$$T_S = \frac{2L_S}{v}\sqrt{1 - v^2/c^2}$$

Similarly, if T_M is the total flight time I record while watching you through my telescope, then $vT_M = 2L_S$ and

$$T_M = \frac{2L_S}{v}$$

If we substitute this expression into the equation for T_S we're right back to the time dilation equation.

For this particular trip, $\gamma = 50.00$, $T_M = 2(50.00 \text{ ly})/(0.999\,8c) = 100.0$ y while $T_S = T_M/\gamma = 2.000$ y. (Note that in the expression for T_M, we could have entered L_S in meters and v in meters per second, but putting L_S in in light-years is equivalent to replacing it with the number-of-years-traveled-at-c multiplied by c; that is

$$1 \text{ light-year} = c(1 \text{ year})$$

in which case, the two c factors, regardless of their units, cancel yielding T_M in years directly). I watch you travel 100.00 ly total (out and back) at a speed nearly that of light, thus taking about 100 y. You see a contracted journey and travel only 1.000 ly out and 1.000 ly back taking about 2.000 y on your clock. I will greet you on the big day of your return in a wheelchair being around 100 y older than when you left, and you will saunter out having aged a mere 2 y.

Example 28.5 The nearest galaxy to ours in all the Universe is the shapeless star-island known as the Magellanic Cloud (located about 1.70×10^5 ly from midtown Manhattan). Assuming you could get up to a speed of 0.999 99c in a negligible amount of time (which is sheer nonsense), how long would you say the trip to that galaxy will take? Incidentally, the fastest anyone has ever gone is only around 0.000 037c.

Solution: [Given: $v = 0.999\,99c$ and $L_S = 1.70 \times 10^5$ ly. Find: the flight time T_S on the traveler's clock.] You, the traveler, see a contracted distance $L_S\sqrt{1 - v^2/c^2}$, which is to be traversed at a speed $v = 0.999\,99c$. Hence, your proper time is

$$T_S = \frac{L_S\sqrt{1 - v^2/c^2}}{v}$$

and

$$T_S = \frac{(1.70 \times 10^5 \text{ ly})(4.472\,12 \times 10^{-3})}{0.999\,99c}$$

Thus

$$T_S = \frac{760 \text{ ly}}{c} = \boxed{760 \text{ y}}$$

It would seem we're not likely to travel to other galaxies using the technology we have at hand.

▶ **Quick Check:** Table 28.2 confirms the value of $\sqrt{1 - v^2/c^2}$. Multiplying 760 y by γ yields 1.7×10^5 y, which is the time in which we on Earth would see you make the 1.70×10^5-ly journey traveling at 0.999 99c.

This **twin effect** has often been called the *twin paradox* (usually enunciated with one twin staying and one traveling), but it's no paradox at all. It may be a startling result, but it's quite understandable within the context of the two postulates. Even so, Einstein pointed out that because of the accelerations on the part of the traveler, the analysis should more properly be done using General rather than Special Relativity. Until now, the situations we have treated have been symmetrical; inertial observers in relative motion see each other's clocks run slow. Here, however, there are accelerations, and we know from the inertial forces that the traveler is moving and not the stay-at-home. That determination can be made in an isolated laboratory aboard the vessel. It is those accelerations that change the path of the traveler through space and time, impressing on him the burden of being out of step with all the friends he left behind. In that sense, the time elapsed between two moments in a journey is route-dependent, just as is the distance traveled. As we will see (p. 998), accelerations put bends in the space-time path and alter the time of the journey.

One of the most compelling confirmations of the reality of these conclusions was made in October of 1971. Then, four exceedingly accurate cesium-beam atomic clocks were flown around the world twice, on regularly scheduled commercial jet flights. The idea was to compare the clocks with those at the U. S. Naval Observatory before and after they had circumnavigated the Earth. Because of the planet's spin, there were two trips around, first eastward and then westward. Things were complicated by the presence of gravity, which affects what's happening via the General Theory of Relativity. (The problem is that time speeds up as the gravitational potential decreases with altitude. This was in addition to the speed-dependent slowdown of time expected from the Special Theory.) In any event, the eastward clocks should have lost 40 ± 23 ns in the journey, and the westward-flying clocks should have gained 275 ± 21 ns. After $7600 was provided for airfare, it was found that, with respect to terrestrial reference standards, the eastward clocks lost 59 ± 10 ns, and the westward clocks had gained 273 ± 7 ns—in breathtaking agreement with theory!

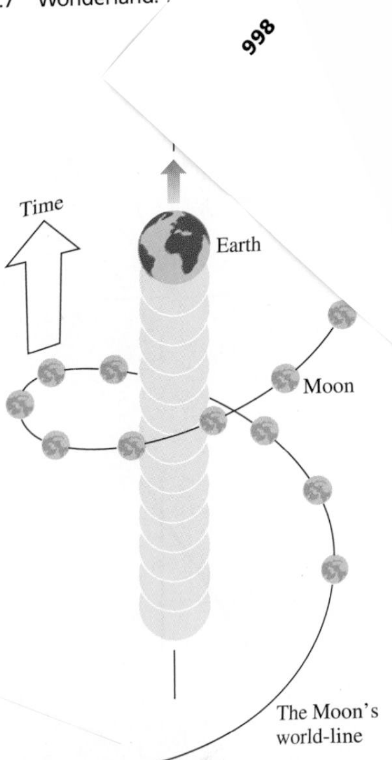

Figure 28.10 The Moon's unfolding world-line as it revolves about the Earth.

28.7 Wonderland: Space-Time

> Gentlemen! The views of space and time which I wish to develop before you have sprung from the soil of experimental physics, and therein lies their strength.

Standing at the podium, Herman Minkowski delivered his lecture to the Eightieth Assembly of German Natural Scientists and Physicians, 1908. At that time, he was a renowned professor at Göttingen University. (Minkowski had once taught young Albert but was unimpressed—"in his student days Einstein had been a lazy dog.") Ironically, here he stood years later presenting a paper entitled "Space and Time" that represented an elegant reformulation of the Special Theory. Minkowski read on:

> Henceforth space by itself, and time by itself, are doomed to fade away into mere shadows, and only a kind of union of the two will preserve an independent reality.

He had recast Relativity in a 4-dimensional geometrical framework that would later prove invaluable to Einstein.

"Let's meet for lunch on the 30th floor of the hotel at the corner of 43rd and Park at 1:30 P.M." You know how to play the game—it takes three coordinates to locate anything in space. But in a real sense the event, our meeting, is not specified until we provide a fourth coordinate, 1:30 P.M. We can envision our own lives as a sequence of events in the 4-dimensional realm of **space-time**. We are swimming, as it were, through a space-time continuum.

The history of our lives, of any object in the Universe, can be imagined as a sequence of points sweeping out a smooth curve in 4-dimensional space-time known as a **world-line** (Fig. 28.10). Figure 28.11 is a simplified graphing of time versus a single spatial coordinate, where any motion is limited to running either way along the x-axis. The diagram is being drawn by you and me who are observers sitting together watching the passing scene. As with the more usual plots of x versus t, a straight line corresponds to a constant speed. Here, the slope $\Delta t/\Delta x$ equals $1/v$ so the faster an object moves, the less the tilt of its world-line. Indeed, an object at rest at $x = 0$ (namely, you or me) has a world-line corresponding to the t-axis. The coordinates are chosen so that the world-line of any pulse of light in the positive x-direction is a diagonal at 45°. That is, if t is plotted in seconds, x is plotted in light-distance light travels in one second). Since, as we'll see presently, c speed, all world-lines must have greater slopes (lower

***Dimension**. I have already said that it is impossible to conceive more than three dimensions. A learned man of my acquaintance, however, believes that one might regard duration as a fourth dimension. . . . The idea may not be admitted, but it seems to be not without merit, if it be only the merit of originality.*

DIDEROT
Encyclopédie (1777)

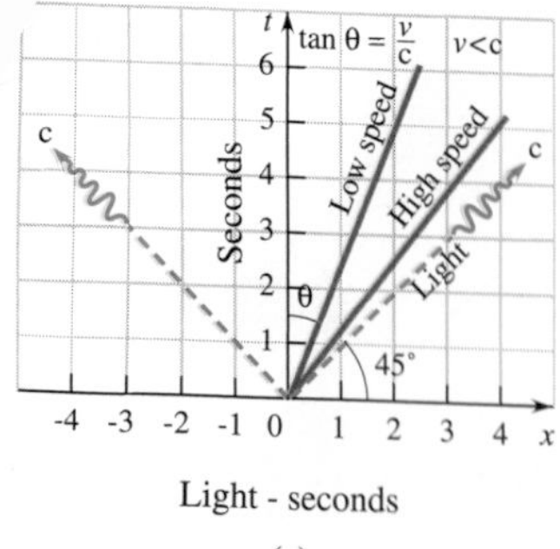

(a)

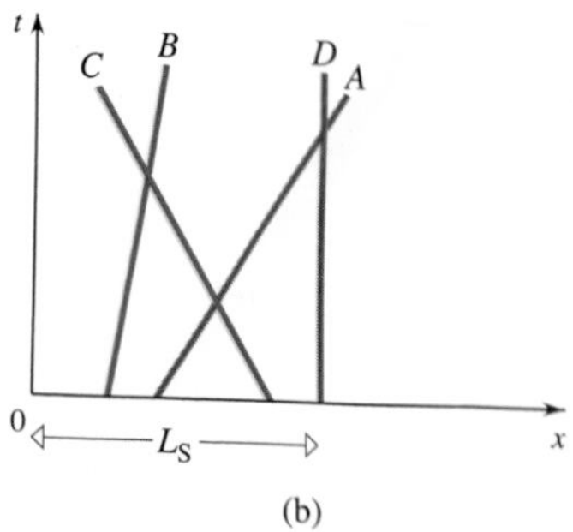

(b)

Figure 28.11 World-lines for objects moving at different speeds. (a) A photon has the least-tilted world-line, traveling the greatest value of x in a given time interval. (b) With you and me at rest at $x = 0$, inertial observer A moves away from us, whereas inertial observer C moves toward us.

speeds) than this. In Fig. 28.11b, we see four inertial travelers: A and B moving away from you and me (who are still sitting at $x = 0$) at different speeds, traveler C coming toward us, and D just standing there a proper distance L_S away from us.

As something of a *tour de force,* Fig. 28.12 depicts a traveler who first accelerated away, then coasted out at a constant speed, slowed and stopped at a distance L_S, rested a while, and then came back to meet us after a time we measure to be T_M. This journey is essentially that of the traveler in the twin effect. The proper time measured by the traveler is marked out along his curved world-line. Note how the time intervals depend on the relative speed, matching those of the stay-at-home (played by you or me) only when $v = 0$. The larger is the relative speed between observer and observed, the more nearly the curve approaches the slope of the photon line, and therefore the larger is the difference between an interval of your proper time and an interval of the traveler's proper time. The total time elapsed from the start to the end of a journey in space-time is always less along a curved world-line than along a direct one. The slower the traveler moves out and back, the more nearly the route approaches the direct one, and the less difference there will be in the two proper times.

As the curve in Fig. 28.12 approaches the photon line, the interval of time stretches out. *For the photon, time does not pass.* In the frame traveling at c, time is stopped (p. 1003); photons don't age, on their clocks they cross the Universe in no time at all.

Billy Pilgrim says that the Universe does not look like a lot of bright little dots to the creatures from Tralfamadore. The creatures can see where each star has been and where it is going, so that the heavens are filled with rarefied, luminous spaghetti. And Tralfamadorians don't see human beings as two-legged creatures, either. They see them as great millipedes—"with babies' legs at one end and old people's legs at the other," says Billy Pilgrim.

KURT VONNEGUT, JR.
Slaughterhouse-Five (1968)

28.8 Addition of Velocities

Classical theory maintains that velocities add vectorially (p. 41), and at everyday speeds that certainly seems to be true. Accordingly, consider Fig. 28.13, which shows a coordinate system S' moving with respect to another such system S. The speed of O' relative to O is $v_{O'O}$. Now consider some object at point P moving in the space of both systems. For simplicity, we limit its motion to be in the $\pm x$-directions. Suppose that P travels at a speed relative to O of v_{PO}, and at a speed relative to O' of $v_{PO'}$. Classically, we learned that

$$v_{PO} = v_{PO'} + v_{O'O} \qquad (28.6)$$

for all objects in uniform motion. But this certainly couldn't be true for a photon, for which it follows from the Second Postulate that $v_{PO} = v_{PO'} = c$ regardless of $v_{O'O}$. This dilemma can be sorted out by carefully applying the two postulates to the broader question of how coordinates in one system relate to those in another. We will skip the derivation—which isn't nearly as difficult as it is long—and simply state the *1-dimensional relativistic formula for the addition of velocities:*

$$v_{PO} = \frac{v_{PO'} + v_{O'O}}{1 + \dfrac{v_{PO'}v_{O'O}}{c^2}} \qquad (28.7)$$

Note that the individual speeds can be either positive or negative. F-
differs from Eq. (28.6) only because of the term $v_{PO'}v_{O'O}/c^2$. Thus
moves slowly with respect to S' (that is, $v_{PO'} \ll c$), or S' mo-
to S (that is, $v_{O'O} \ll c$), it follows that $v_{PO'}v_{O'O}/c^2 \ll 1$. In
pression for v_{PO} becomes identical to the classical
used for about two centuries before there was

Now for the true test: let's find the s-
Observer O', carrying a flashlight, i-

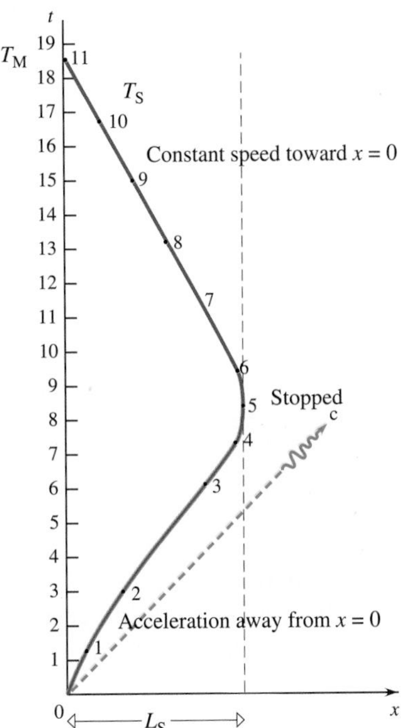

Figure 28.12 The world-line of a traveler who accelerates in the positive x-direction, slows to a stop, and then returns to $x = 0$ at a constant speed. Note that the journey took 11 h on the traveler's clock and $18\frac{1}{2}$ h on our stationary clock.

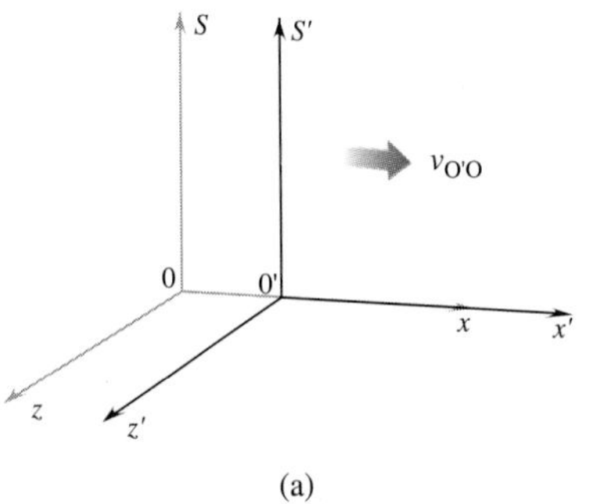

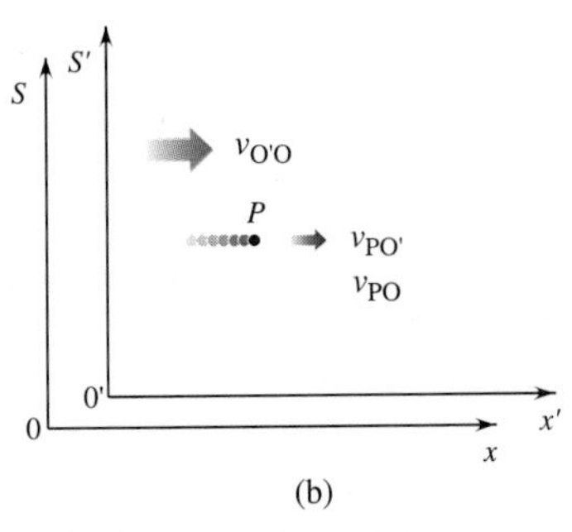

Figure 28.13 Two inertial systems S and S' moving with respect to each other at speed $v_{O'O}$. (a) Here, S is at rest and S' advances in the x-direction. (b) A particle P moves at a speed v_{PO} relative to S, and $v_{PO'}$ relative to S'.

"Clearly," the Time Traveller proceeded, "any real body must have extension in four *directions: it must have Length, Breadth, Thickness and—Duration. But through a natural infirmity of the flesh. . . we incline to overlook this fact. There are really four dimensions, three which we call the three planes of Space, and a fourth, Time. . . . Some philosophical people have been asking why* three *dimensions particularly—why not another direction at right angles to the other three?—and have even tried to construct the Four-Dimension geometry."*

H. G. WELLS
The Time Machine (1895)

relative to observer O. A stream of photons is sent out in the positive x-direction at a speed $v_{PO'} = c$, and we want to find its speed v_{PO} with respect to O, namely,

$$v_{PO} = \frac{v_{PO'} + v_{O'O}}{1 + \dfrac{v_{PO'}v_{O'O}}{c^2}} = \frac{c + v_{O'O}}{1 + \dfrac{cv_{O'O}}{c^2}} = \frac{c(c + v_{O'O})}{c + v_{O'O}} = c$$

Wonderful! No matter what value $v_{O'O}$ has, each observer sees the speed of light to be c.

As another example, consider the rocket ship (S') in Fig. 28.14. It is flying away from us (S) at speed $v_{O'O} = \frac{1}{2}c$ when it emits a pulse of light. A traveler aboard the ship sees the pulse moving at $v_{PO'} = -c$; the minus sign shows that it's advancing in the negative x-direction. The pulse's speed with respect to us is

$$v_{PO} = \frac{-c + \frac{1}{2}c}{1 + \dfrac{(-c)(\frac{1}{2}c)}{c^2}} = -c$$

The light comes toward us at c even though it was emitted from a platform receding at a speed of $\frac{1}{2}c$.

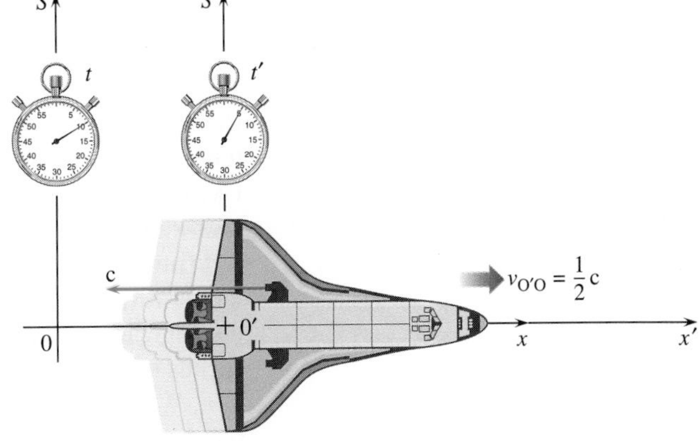

Figure 28.14 A rocket ship in system S' moving at speed $v_{O'O} = \frac{1}{2}c$ emits a pulse of light toward the axis of system S.

Example 28.6 Two galaxies are speeding away from the Earth along a line in opposite directions, each with a speed of 0.75c with respect to the planet. At what speed are they moving apart with respect to each other?

Solution: [Given: galactic speeds of 0.75c. Find: relative speed.] We have three moving bodies, and in Fig. 28.15, we let one galaxy be S, the Earth be S', and the other galaxy be P. Thus, S' moves forward with respect to S at $v_{O'O} = 3c/4$; P moves forward with respect to S' at $v_{PO'} = 3c/4$, and we want to find v_{PO}, the speed of P with respect to S. From Eq. (28.7)

$$v_{PO} = \frac{v_{PO'} + v_{O'O}}{1 + \dfrac{v_{PO'}v_{O'O}}{c^2}} = \frac{0.75c + 0.75c}{1 + \dfrac{(0.75c)(0.75c)}{c^2}} = \boxed{0.96c}$$

The galaxies separate at 96% of c.

▶ **Quick Check:** This result is at least reasonable because the galaxies could not separate from each other at a speed equal to or greater than c since Eq. (28.7) says that even photons traveling in opposite directions ($v_{PO'} = c$ and $v_{O'O} = c$) only separate at c.

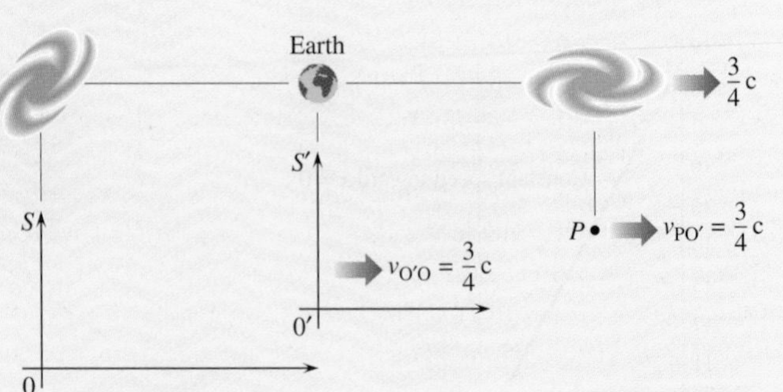

Figure 28.15 The galaxy on the right at point P is moving away from O' in S' at $v_{PO'} = \frac{3}{4}c$. With respect to the galaxy on the left (at rest in S), the Earth (at rest in S') is moving to the right at $v_{O'O} = \frac{3}{4}c$.

The velocity transformations are such that **the speed of one entity with respect to another is always less than or equal to** c. As far as the two galaxies in Example 28.6 are concerned, they separate from each other at 0.96c and can intercommunicate directly (via light beams traveling at c). An important point here is that *any two entities, regardless of their motions, can interact;* they can transport energy (at speed c) from one to the other. Thus, even though the gravitational and electromagnetic interactions propagate at the finite speed c, two objects cannot outrun their mutual interactions.

RELATIVISTIC DYNAMICS

Classical dynamics developed along two conceptual paths illuminated by the fundamental principles of Conservation of Momentum and Conservation of Energy. And so, faced with the necessity to reformulate dynamics, Einstein was guided by these same principles. Rather than attempt a complete derivation of relativistic dynamics, we limit our study to a few main results, their origins, and implications.

28.9 Relativistic Momentum

Both the energy and momentum of an object depend on the inertial reference system in which they are determined: the passenger holding a bowling ball on her lap flying to Paris "sees" it to have neither momentum nor kinetic energy. Since we now know, via Eq. (28.7), that the speed of an object with respect to different inertial frames has to be treated relativistically, it should be no surprise that both momentum (p) and energy (E) will have to be reformulated. Here, we are guided by our knowledge that the classical definition of momentum $\mathbf{p} = m\mathbf{v}$ works well at low speeds ($v \ll c$) and must be an approximation of a more precise *relativistic momentum.* We can get a sense of how the momentum might be modified by examining the details of a simple collision.

Let's first establish that if momentum is to be conserved relativistically, it cannot be the classical notion given by $\mathbf{p} = m\mathbf{v}$. To that end, Fig. 28.16 depicts an elastic collision between two identical balls. There is a "stationary" frame S and a person therein, observer 1, holding ball 1. Passing nearby is a uniformly "moving" frame S' (traveling at speed $v_{O'O}$), in which there is a second person, observer 2, holding ball 2. Just at the right moment, both observers throw their balls vertically toward one another at identical speeds equal to u. Classically, the balls have momenta of $p_y = mu$ that are equal in magnitude and opposite in direction, and the initial net vertical momentum equals zero. After the head-on collision, the balls are both reversed; the net vertical momentum is still zero, and classical momentum is conserved.

Now, let's run through this situation again, taking into account that the relative speed of S' with respect to S is appreciable. Observer 2 sees his ball travel a vertical round-trip distance of $2d$ at a speed u in a time $2d/u$. The same is true of observer 1, who also sees her ball travel a vertical distance of $2d$ at a speed u in a time $2d/u$. But observer 1 sees ball 2 to be moving and therefore to be experiencing a time dilation (by a factor of γ). Thus, observer 1 must see ball 2 take a longer time to make the round trip from, and back to, the hand of observer 2. Observer 1 sees ball 2 travel a vertical distance of $2d$ in a longer time, and therefore at a slower speed, u/γ, than his own ball moves. And yet the collision reverses the motion of his ball—classical momentum is not conserved. So we need a new statement of what is conserved and, although we might guess that it look like γmv, we certainly haven't proven it.

Nonetheless, the **relativistic linear momentum**, which is conserved, is indeed given by

$$\mathbf{p} = \gamma m\mathbf{v} = \frac{1}{\sqrt{1 - v^2/c^2}} m\mathbf{v} \qquad (28.8)$$

As v/c becomes negligible, $\beta \rightarrow 0$, $\gamma \rightarrow 1$, and $\mathbf{p} \rightarrow m\mathbf{v}$, which is the classical value. Fig. 28.17 is a plot of p/mc, which classically equals β and relativistically equals $\gamma\beta$. As the speed of a body increases beyond $\beta \approx \frac{1}{2}$, the relativistic momentum climbs away from the classical value, approaching infinity as $v/c \rightarrow 1$.

Example 28.7 Electrons have a mass of 9.1094×10^{-31} kg. If one is traveling at 0.99c, what is its momentum?

Solution: [Given: $m = 9.1094 \times 10^{-31}$ kg and $v = 0.99c$. Find: p.] The momentum follows from Eq. (28.8), where we first compute γ to be 7.0888. Hence

$$p = \gamma mv$$

$$p = (7.0888)(9.1094 \times 10^{-31}\text{ kg})(2.9679 \times 10^8\text{ m/s})$$

and $p = 1.9165 \times 10^{-21}$ kg·m/s $= \boxed{1.9 \times 10^{-21}\text{ kg·m/s}}$.

▶ **Quick Check:** From Table 28.2, γ looks okay at 7. The classical momentum is roughly $(9.1 \times 10^{-31})(3 \times 10^8) = 27 \times 10^{-23}$ kg·m/s, so our answer should be about $7(27 \times 10^{-23})$, which it is.

Though experiments have verified the validity of Eq. (28.8), the interpretation of $p = \gamma mv$ is unsettled at this moment. We can associate the γ with mass and suppose that there is a *relativistic mass* $m_R = \gamma m$ that is a function of speed (p. 1003). Then $p = m_R v$, which is nice. Einstein's early work seems ambivalent about the idea of a speed-dependent mass. Nonetheless, in 1948, he made it very clear that introducing the concept of relativistic mass was "not good." Experimentally, what we

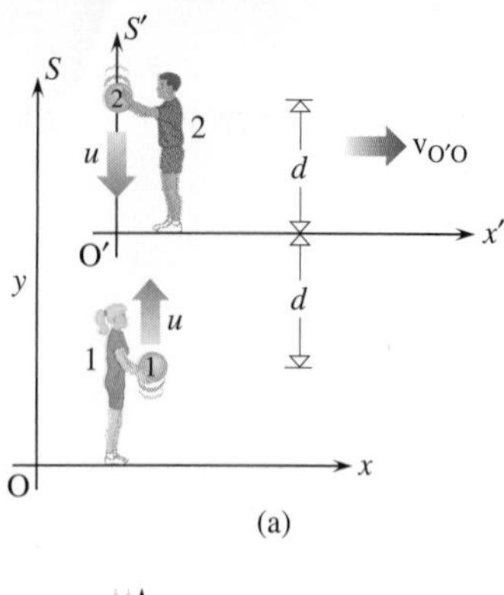

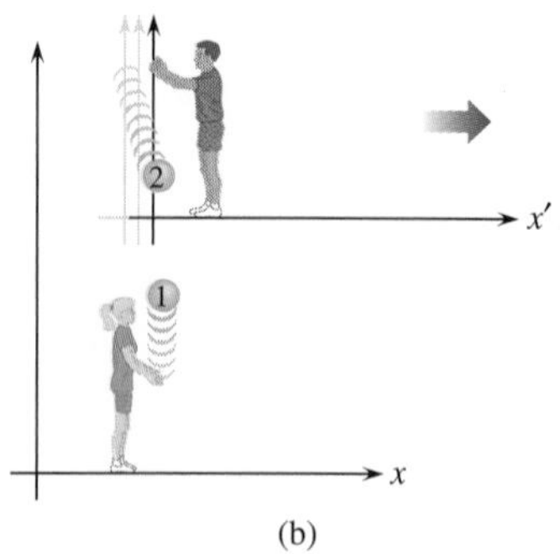

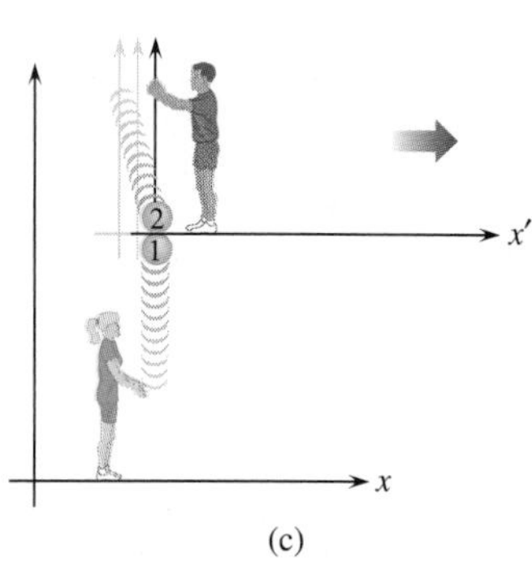

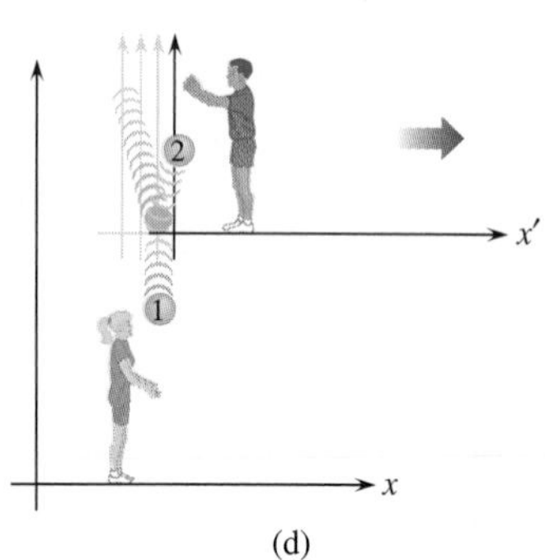

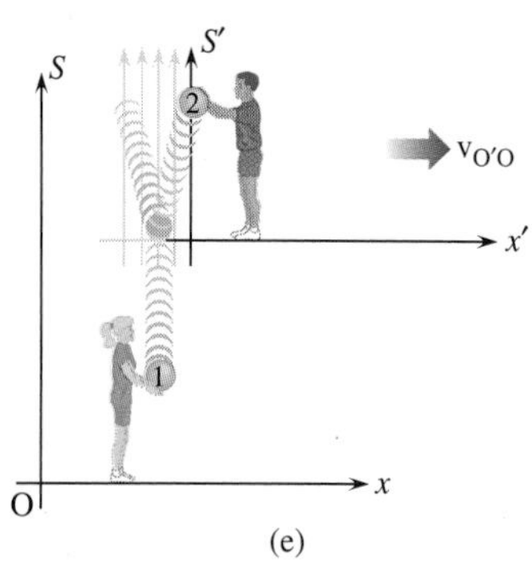

Figure 28.16 An elastic collision as viewed by observer 1 at rest in inertial system *S*. Observer 2 is at rest in system *S*′, which moves relative to *S* at a speed $v_{O'O}$. Observer 1 sees time running slow for ball 2 by a factor of γ, which depends on its net relative speed.

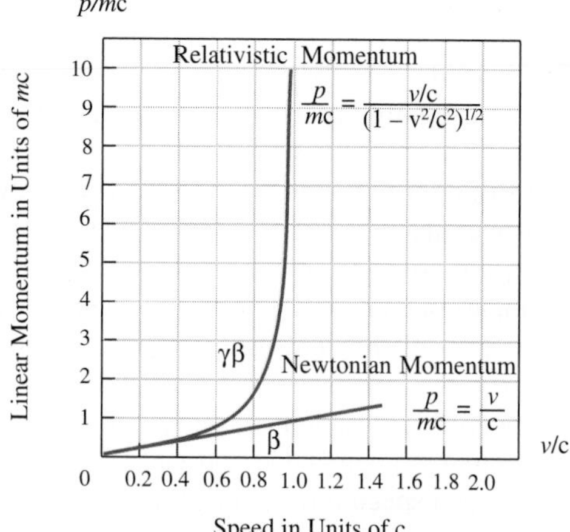

Figure 28.17 Linear momentum in units of *mc* plotted against speed in units of c; that is, *p*/*mc* versus *v*/c. Here, we compare relativistic momentum against Newtonian momentum. Observe that although they both increase with *v*, the former approaches ∞ as *v* approaches c.

measure is a change in momentum with speed; there is no way to measure the mass of a moving body directly. Many physicists prefer to think of mass as speed dependent. A growing number of others find it more appealing to hold that mass is an invariant property of matter like charge. Both views lead to the same measured results; the difference is a matter of interpretation. **We shall take *m* to be the invariant mass**—the speed independent, reference-frame-independent, almost Newtonian mass (m)—though that idea will need some more discussion (p. 1005).

Newton's Second Law (for an average force) is, as ever

$$\mathbf{F}_{\text{av}} = \frac{\Delta \mathbf{p}}{\Delta t} \qquad [4.2]$$

wherein **p** is the relativistic momentum given by Eq. (28.8). Now, suppose a constant force is applied to a body that is initially at rest. What happens to v? In that case, $F = p/t$, where p is the momentum after the elapse of a time t. Solving Eq. (28.8) for v^2, we obtain

$$v^2 = \frac{p^2/m^2}{1 + p^2/m^2c^2}$$

Dividing by c^2 and taking the square root gives

$$\beta = \frac{v}{c} = \frac{Ft/mc}{\sqrt{1 + (Ft/mc)^2}} \qquad (28.9)$$

Now consider the possible motion of an object. If the force and time are such that Ft

is small and Ft/mc is therefore much less than 1, Ft/mc is negligible in the denominator and $\beta \approx Ft/mc$, whereupon $mv = Ft$, and we have the classical change-in-momentum-equals-impulse expression, Eq. (4.3). On the other hand, if Ft is large (as, for instance, if a large force is applied for a very long time), $(Ft/mc)^2 \gg 1$ and $\sqrt{1 + (Ft/mc)^2} \rightarrow Ft/mc$, and so as $Ft \rightarrow \infty$, $\beta \rightarrow 1$. What this means is that **regardless of how great the force or how long it is applied, the body will neither reach nor exceed lightspeed. As a body moves faster and faster under the influence of a force, it takes longer and longer (because of the time dilation) for the speed to further increase.** In effect, an inertial observer will see the acceleration decrease (much as if the body's mass had actually increased).

As v approaches c, time slows [Eq. (28.3)] until at $v = c$, $\beta = 1$, $\sqrt{1 - \beta^2} = 0$, $\gamma = \infty$, and a second stretches out into infinity. Nor is the Lorentz-FitzGerald Contraction any more comprehensible when $v = c$ and everything shrinks to nothingness. For speeds beyond c, all this becomes unreal, literally and mathematically. Only one conclusion is evident: **the speed of light is an upper limit on the rate of propagation of objects that have mass.** Nothing with mass can be accelerated to c (Question 11). To be precise, this conclusion does not preclude the existence of particles that have somehow been created with speeds in excess of c. These fantastic hypothetical entities, known as *tachyons,* have never been observed and may well never be. They are a product of the theoretical school that maintains that whatever is not explicitly forbidden must exist. But since we are not possessed of all the theories we are ever going to have, such a notion seems a trifle premature.

28.10 Relativistic Energy

We want now to arrive at an expression for the relativistic kinetic energy (KE) of a particle. The rigorous, though rather mathematically involved, way to proceed is to go back to the idea of the work done by a force in the process of changing the body's speed (in a zero-PE situation). Instead, let's take a less formal and easier approach. As we saw earlier (p. 982), when $\beta \ll 1$, $\gamma = 1/\sqrt{1 - \beta^2} \approx 1 + \beta^2/2$, and so $\gamma m \approx m(1 + \beta^2/2)$. Multiplying this by c^2 gives us

$$\gamma mc^2 \approx mc^2 + \tfrac{1}{2}mv^2$$

The term on the far right is the familiar low-speed $\text{KE} = \frac{1}{2}mv^2$; thus

$$\text{KE} \approx \gamma mc^2 - mc^2$$

The rigorous analysis leads to the same expression for all speeds:

$$\text{KE} = \gamma mc^2 - mc^2 \tag{28.10}$$

This is the **relativistic kinetic energy**, and it's no longer simply $\frac{1}{2}mv^2$. (See Fig. 28.18 for a lovely experimental confirmation.) Consequently

$$\gamma mc^2 = \text{KE} + mc^2 \tag{28.11}$$

and each term has the units of energy. The second quantity on the right is independent of speed and is known as the **rest energy** (E_0). All of this suggests that we interpret the

Figure 28.18 Here, electrons were given various values of kinetic energy by accelerating them across appropriate potential differences. Their speeds were then measured by finding the flight times over a fixed path. The dots represent the measured values, for which the vertical bars indicate the expected range of experimental error. Clearly, the results at high speeds confirm the validity of the relativistic expression for KE (the purple curve). The classical formula for KE corresponds to the straight dashed line, which is accurate at low speed. [Adopted from W. Bertozzi, *Am. J. Phys.* 32, 551 (1964).]

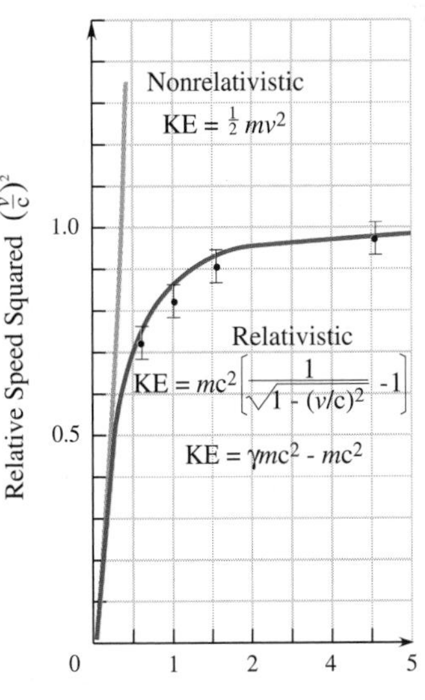

term on the left as the **total energy** of the body ($E = \gamma mc^2$):

Total energy = kinetic energy + rest energy

$$E = KE + E_0 \tag{28.12}$$

Example 28.8 Every electron has a rest energy of 0.511 MeV. If one in particular is traveling at a speed of 0.900c, determine its total energy and its kinetic energy. Give your answers in MeV.

Solution: [Given: $E_0 = 0.511$ MeV and $v = 0.900c$. Find: E and KE.] Since $E = \gamma mc^2$, let's first determine γ by using

$$\gamma = \frac{1}{\sqrt{1-\beta^2}} = \frac{1}{\sqrt{1-(0.900)^2}} = 2.2942$$

Then realize that $mc^2 = 0.511$ MeV and so

$$E = 2.2942(0.511 \text{ MeV}) = \boxed{1.17 \text{ MeV}}$$

From Eq. (28.12)

$$KE = E - E_0 = 1.172 \text{ MeV} - 0.511 \text{ MeV}$$

and $\boxed{KE = 0.661 \text{ MeV}}$

▶ **Quick Check:** γ is greater than 1, which is a good sign, and it also checks with Table 28.2. As we will see in Problem 61, rest energy for a particle equals kinetic energy when $\beta = 0.866$, so this result where $\beta = 0.900$ is certainly the right magnitude.

When the body is at rest ($\gamma = 1$), $KE = 0$, and the total energy ($E = \gamma mc^2$) equals the rest energy, and so

$$E_0 = mc^2 \tag{28.13}$$

This is perhaps the most famous outcome of the Special Theory. Confirmed experimentally, it stands beyond doubt. If there can be such a thing as the premier equation of the twentieth century, this is it. It has revolutionized Modern Physics and hurled us all into the age of nuclear weapons. Even so, the precise meaning of the relationship is still being argued in the scientific literature, primarily because we have not yet satisfactorily defined matter, mass, and energy. For example, is mass a congealed form of energy, or are mass and energy very different concepts that are simply proportional to each other and only interrelated via Eq. (28.13), just as F and a are interrelated by $F = ma$? Still, it has been established experimentally that matter possessing mass (for example, electrons and positrons) can be transformed into electromagnetic radiation and vice versa. At the level of our discussion, it's probably best to ignore some of these subtleties and follow the most common usage.

Mass can be transformed into energy, and energy can be transformed into mass; hence

$$1 \text{ kg} \leftrightarrow 8.987 \times 10^{16} \text{ J}$$

$$1 \text{ kg} \leftrightarrow 5.609 \times 10^{29} \text{ MeV}$$

There is one unified concept: *mass-energy.* Had we known that a few centuries ago, we would not now have joules, Btus, calories, and kilowatt-hours to fuss with: the kilogram would do for all forms of mass-energy. The constant c^2 is a scale factor that numerically relates mass to energy. It's an immense number, $\approx 9 \times 10^{16}$ m^2/s^2, so even a tiny change in mass corresponds to an enormous change in energy. In

TABLE 28.3 Masses and rest energies for some particles and atoms

Object	Mass (kg)	Rest energy (MeV)
Photon	0	0
Neutrino	0	0
Electron (or positron)	$9.109\,389\,7 \times 10^{-31}$	0.510 999
Proton	$1.672\,623\,1 \times 10^{-27}$	938.272
Neutron	$1.674\,929 \times 10^{-27}$	939.566
Muon	$1.883\,54 \times 10^{-28}$	105.659
Pion (+)	$2.416\,5 \times 10^{-28}$	135.56
Deuteron	$3.343\,584 \times 10^{-27}$	1875.612
Triton	$5.007\,357 \times 10^{-27}$	2808.920
Alpha	$6.644\,72 \times 10^{-27}$	3727.41
Hydrogen atom ($^1_1$H)	$1.673\,534 \times 10^{-27}$	938.783
Deuterium atom ($^2_1$H)	$3.344\,497 \times 10^{-27}$	1876.12
Tritium atom ($^3_1$H)	$5.008\,270 \times 10^{-27}$	2809.43
Helium-3 atom ($^3_2$He)	$5.008\,237 \times 10^{-27}$	2809.41
Helium atom ($^4_2$He)	$6.646\,482 \times 10^{-27}$	3728.40

Modern Physics, mass is often given in units of MeV/c^2; thus

$$1 \text{ MeV/c} = 5.344\,29 \times 10^{-22} \text{ kg·m/s}$$

$$1 \text{ MeV/c}^2 = 1.782\,663 \times 10^{-30} \text{ kg}$$

whereupon, for example, the mass of an electron is 0.510 999 MeV/c^2 (Table 28.3).

Example 28.9 A 1.00-kg chicken is placed on the transporter of a fictitious starship whereupon it is converted directly into electromagnetic energy—a process that is theoretically possible, though technologically quite beyond our poor powers—so that it could be "beamed" down to the galley. What is the equivalent energy of the chicken? How much is that in kilowatt-hours?

Solution: [Given: m = 1.00 kg. Find: E_0.] From Eq. (28.13)

$$E_0 = mc^2 = (1.00 \text{ kg})(2.998 \times 10^8 \text{ m/s})^2$$

and

$$\boxed{E_0 = 8.99 \times 10^{16} \text{ J}}$$

1 kilowatt-hour = 1000 J/s × 60 s/min × 60 min = 3.60 MJ; hence, dividing this result into the energy yields $\boxed{E_0 = 2.50 \times 10^{10} \text{ kW·h}}$. That's the equivalent of running ten 100-W bulbs for 2.5×10^{10} hours, about 3 million years.

▶ **Quick Check:** Since 1 kg → 5.6×10^{29} MeV and 1 MeV = 1.6×10^{-13} J, 1 kg → 9×10^{16} J.

A single material particle at rest, by virtue of its very existence, has a rest energy. Similarly, a body composed of several particles possessing internal energy (thermal, potential, whatever), when taken as a whole, also has a net rest energy and a net mass. As Einstein put it, "the mass of a body is a measure of its energy-content." A hot apple pie has more rest energy and more mass than an otherwise identical cold apple pie. The greater mass is due purely to its greater energy content—**energy possesses inertia**. Thus, a flashlight emitting energy (ΔE) decreases in mass (by $\Delta E/c^2$), just as a plant absorbing that light gains in mass. A spring must weigh more after *elastic*-PE is stored in it, than before. No one has ever measured these miniscule variations in mass, but there is ample confirmation elsewhere. In the final analysis, all the familiar occurrences that liberate energy—from burning marshmallows to exploding dynamite—transform a minute amount of mass into energy. Ultimately, that is the source of the reaction energy, and this is as true for chemical energy as it is for energy liberated by a nuclear weapon.

In Chapter 9 we talked about Conservation of Mechanical Energy, and in Chapter 16 we generalized that concept to include thermal energy. The result was the First Law of Thermodynamics. Now we come to the final generalization: there is one all-encompassing law of **Conservation of Energy**:

> **The total energy of an isolated system always remains constant although any portion of it can be converted from one form to another, including rest energy.**

Example 28.10 There are several fusion reactions that convert mass directly into energy and can power stars and drive hydrogen bombs. One such process fuses two nuclei of heavy hydrogen (deuterium) together, resulting in a still heavier hydrogen (tritium) nucleus, an ordinary hydrogen nucleus (proton), and the KE they fly off with. It's customary to write such a reaction in terms of the neutral atoms involved (neglecting the tiny amounts of energy holding the electrons to each atom, $\approx$10 eV). Thus

$$ {}^2_1\text{H} + {}^2_1\text{H} \rightarrow {}^3_1\text{H} + {}^1_1\text{H} + \text{energy} $$

Determine the energy liberated per fusion.

Solution: [Given: the reaction. Find: energy liberated.] Energy is conserved; the total energy on the right in the reaction equals the total energy on the left. In other words, the difference in the rest energy before and after is the liberated KE. From Table 28.3, 1876.12 MeV + 1876.12 MeV = 2809.43 MeV + 938.783 MeV + energy

$$ \text{energy} = 4.03 \text{ MeV} = \boxed{6.45 \times 10^{-13} \text{ J}} $$

where 1 MeV = 1.602×10^{-13} J.

▶ **Quick Check:** Nuclear reactions typically involve several MeV, so this result is the right order-of-magnitude.

It is not always possible practically to convert a quantity of mass completely into energy, or vice versa, although both effects are now commonplace. Still, one single gram of mass is equivalent to 9×10^{13} J, which is enough energy to raise 200 000 000 kg of water from 0°C to 100°C. That corresponds to the peak power output of Boulder Dam operating for 19 hours; namely, 25 million kilowatt-hours. An exploding kilogram of TNT liberates about 5 million joules, which is certainly formidable, though it represents a mass loss of only about 6×10^{-11} kg—far too little to measure. Chemical reactions, which are at heart relatively weak electrical interactions, release energies of the order of a few eV. By comparison, nuclear transmutations involving the far more powerful *strong force* (p. 1107) correspond to mass changes of about 0.1%. This is roughly a million times more than the mass change in a chemical reaction (see Table 28.4) and, hence, directly observable.

The total energy can be written in terms of p without any explicit reference to v, and that provides some insights about photons. Starting with $\text{E} = \gamma mc^2$, square both sides and write it as

$$ \text{E}^2 = \gamma^2 m^2 c^2 (c^2 + v^2 - v^2) $$

Now

$$ \text{E}^2 = \gamma^2 m^2 c^2 (c^2 - v^2) + \gamma^2 m^2 c^2 v^2 $$

TABLE 28.4 The fractional change of mass for various kinds of processes

Process releasing energy $\Delta E = \Delta(mc^2)$	$\Delta m/m$
Chemical	$\approx 1.5 \times 10^{-8}\%$ to $\approx 10^{-7}\%$
Nuclear fission	$\approx 0.1\%$
Nuclear fusion	$\approx 0.6\%$
Electron-positron annihilation	100%
Neutral pion decay into two photons	100%

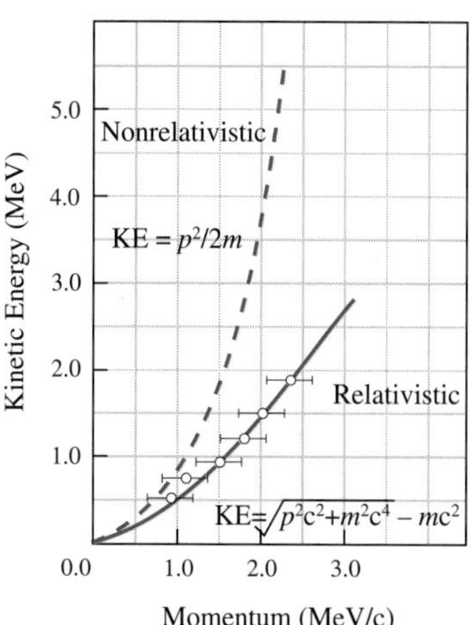

Figure 28.19 A comparison of the classical and relativistic relationships between KE and *p*. Momentum and energy are measured using electrons emitted via radioactive decay. Note how nicely the data fit the predicted relativistic curve. [Adopted from R. Kollarits, *Am. J. Phys* 40, 1125(1972).]

Putting in the expression for γ, we have

$$E^2 = \frac{m^2c^2(c^2 - v^2)}{(1 - v^2/c^2)} + \gamma^2m^2c^2v^2$$

and using $p = \gamma mv$, substitute in $p^2 = \gamma^2m^2v^2$ to get

$$E^2 = m^2c^4 + p^2c^2 \tag{28.14}$$

or

$$E^2 = E_0^2 + (pc)^2 \tag{28.15}$$

Figure 28.19 shows the kind of striking confirmation of these conclusions that, today, is available experimentally.

Notice that inasmuch as $E_0^2 = (mc^2)^2$ is a constant independent of any reference frame, $E^2 - (pc)^2$ must also be invariant, even though E and p separately depend on v and are thus relative quantities. We shall not explore the implications of this, other than to point out that the invariance of $E^2 - (pc)^2$ implies that **in space-time there is a single notion that ties together what until now were the two separate ideas of energy and momentum**.

Example 28.11 A proton (of mass 938.3 MeV/c²) is accelerated across a potential difference of 202.0 MV so that it has a kinetic energy of 202.0 MeV. Determine its total energy (in MeV) and momentum (in MeV/c). What is the speed of the particle?

Solution: [Given: $m = 938.3\ \text{MeV/c}^2$, KE = 202.0 MeV. Find: E and p.] $E = KE + E_0 = KE + mc^2 = 202.0\ \text{MeV} + (938.3\ \text{MeV/c}^2)c^2 = \boxed{1140\ \text{MeV}}$. From Eq. (28.15)

$$p = \frac{\sqrt{E^2 - E_0^2}}{c} = \boxed{647.5\ \text{MeV/c}}$$

To find the speed, we use the fact that $E = \gamma E_0$; $E = E_0/\sqrt{1 - v^2/c^2}$ and so

$$v = c\sqrt{1 - \frac{E_0^2}{E^2}} = c\sqrt{1 - \left(\frac{938.3}{1140}\right)^2} = \boxed{0.5683c}$$

▶ **Quick Check:** $p = \gamma mv$, $v = p/\gamma m = p/(E/c^2) = (647.5\ \text{MeV/c})/(1140\ \text{MeV/c}^2) = 0.5680c$.

In addition to the photon, physicists have found it necessary to deal with other zero-mass particles (p. 1139) such as the several different neutrinos and the graviton. The existence of neutrinos is well confirmed experimentally, although the graviton is still a matter of speculation. When $m = 0$ in Eq. (28.14)

[*zero-mass*]

$$E = pc \tag{28.16}$$

Moreover, because $E = \gamma mc^2$, $E/\gamma = mc^2$ and when $m = 0$, $E/\gamma = 0$. Since E is nonzero, $1/\gamma = 0 = \sqrt{1 - v^2/c^2}$, and it follows that for particles of zero mass $v = c$. **Particles of zero mass exist only at the speed** c.

Example 28.12 A neutral pion has a mass 264 times larger than the electron mass and a rest energy of 135 MeV. It is unstable and decays into two oppositely directed γ-ray photons. Assume the pion to be at rest and determine the energy and momentum of the photons.

Solution: [Given: $E_0 = 135$ MeV. Find: E and p for photons.] Since the pion is at rest when it decays, initially $p = 0$, and the momenta of the photons are equal and opposite. The total energy of the two photons (2E) must equal the energy of the pion: 135 MeV = 2E, and for each photon $\boxed{E = 67.5\text{ MeV}}$. From Eq. (28.16) for massless particles,

$$p = \frac{E}{c} = \boxed{67.5\text{ MeV/c}}$$

and since 1 MeV/c $= 5.3443 \times 10^{-22}$ kg·m/s, $p = 3.61 \times 10^{-20}$ kg·m/s.

▶ **Quick Check:** $m = 264(9.11 \times 10^{-31}\text{ kg}) = 2.4 \times 10^{-28}$ kg; the photon energy is half the pion energy $E = \frac{1}{2}mc^2 = 1.08 \times 10^{-11}\text{ J} = 67.5$ MeV.

Though we shall not elaborate the argument, Relativity establishes that if an interaction is to be *causal* (in the sense that the act of initiation of an event must precede the event in time for all inertial observers), then the speed of propagation of the interaction (that is, the signal) cannot exceed c. The uncovering of a radioactive source sets a distant counter clattering; the emitted γ-rays travel from source to detector at c, and the sequence of events is causal. The transport of information (matter-energy in a discriminated form) in a cause-and-effect sequence cannot occur at a speed in excess of c. That does not mean that there cannot be apparent motions that are greater than c. Shine a laser beam at observer *A* leaning on a very distant screen. Now quickly jerk the laser so the light spot moves across the screen to observer *B*. Clearly, if you shift the laser fast enough and if the target is far enough away, the light spot travels from *A* to *B* in excess of c (oscilloscopes actually perform this trick all the time). Still, the photons that get to *B* were never at *A*—new photons a little behind in the stream arrive at each successive point on the screen. Light never literally travels from *A* to *B*.

Relativity does not forbid noncausal interactions. In fact, the concept is at the center of a great deal of modern experimentation and interpretation as it relates to Quantum Mechanics. There are some wonderfully mind-boggling contemporary experiments that are challenging our familiar notions of reality.

Core Material

The Michelson-Morley experiment established that the motion of the Earth through the supposed surrounding ether could not be detected (p. 980). The **Principle of Relativity**, the *First Postulate* of the Special Theory, states that *all the laws of physics are the same for nonaccelerating observers.* Einstein's *Second Postulate,* the **Principle of the Constancy of the Speed of Light**, states that *light propagates in free space at the invariant speed* c (p. 983).

There is no such thing as **absolute simultaneity**—events separated in space that are simultaneous for midway observers in one inertial system are not necessarily simultaneous for midway observers in any other inertial system (p. 985). Time on a clock that is moving with respect to an observer runs slower than time on a clock that is stationary with respect to that observer; thus

$$\Delta t_M = \frac{\Delta t_S}{\sqrt{1 - v^2/c^2}} \qquad [28.2]$$

A moving observer sees an object to have a length (along the direction of motion) that is shorter than the length seen by an observer at rest with respect to the object; thus

$$L_M = L_S\sqrt{1 - v^2/c^2} \qquad [28.5]$$

The 1-dimensional relativistic velocity transformation is

$$v_{PO} = \frac{v_{PO'} + v_{O'O}}{1 + \dfrac{v_{PO'}v_{O'O}}{c^2}} \quad [28.7]$$

The relativistic momentum of a body of mass m is

$$\mathbf{p} = \gamma m\mathbf{v} = \frac{1}{\sqrt{1 - v^2/c^2}}\, m\mathbf{v} \quad [28.8]$$

Here we take the approach that *m is not a function of v*. The relativistic energy of a body is given by

$$E = KE + E_0 \quad [28.12]$$

where

$$E = \gamma mc^2$$

and the rest energy is

$$E_0 = mc^2 \quad [28.13]$$

The mass of a composite body changes if the rest energy of its component parts changes (p. 1003). In terms of momentum

$$E^2 = E_0^2 + (pc)^2 \quad [28.15]$$

For zero-mass particles

$$E = pc \quad [28.16]$$

Suggestions on Problem Solving

1. The factor $\gamma = 1/\sqrt{1 - \beta^2}$ is so ubiquitous in the work of this chapter that it merits a few words. It's common to forget either to square the β or to take the square root once having computed $1 - \beta^2$. Remember that $\gamma > 1$. A nice way to compute γ on an electronic calculator if v is given as a number (for example, $v = 2.43 \times 10^7$ m/s) with no reference to c is

v [÷] c [=] [x^2] [+/−] [+] 1 [=] [$\sqrt{\ }$] [$1/x$]

If you have v in terms of c (for example, $v = 0.9c$), then the calculation is more straightforward because you can enter v/c as a number (in this case, $v/c = 0.9$)

1 [−] v/c [x^2] [=] [$\sqrt{\ }$] [$1/x$]

Many calculators have constants stored in memory, and c is usually one of them.

2. When doing problems on time dilation and length contraction, the "hard part" is just figuring out which piece of information to call L_M and which L_S. For example: A kid on an asteroid holding a meter stick is going to measure the length of a space probe as it flies by at 0.9c parallel to the stick. As seen by the pilot, how long is the meter stick? "As seen by the pilot"—the stick moves past the probe at 0.9c, the stick is moving with respect to the frame we are interested in, the pilot's frame. Hence, the *proper length* of the stick measured by the kid is $L_S = 1$ m. We place ourselves in the frame of the pilot and see a moving stick of length L_M. The rest of the problem is just plug-in using Eq. (28.5).

Another useful approach is to *figure out who sees what shrunk or dilated before getting into the problem numerically.* In the above example, "As seen by the pilot" tells us whose perspective we are to determine. The pilot sees the ruler moving and *shrunk.* The answer must be less than 1 m. Keeping in mind that $\sqrt{1 - v^2/c^2} < 1$,

$$(\text{Length seen by pilot}) = (\text{length seen by kid})\sqrt{1 - v^2/c^2}$$

The stick has only one proper length (measured at rest), and that length is the longest possible measure.

3. There are clever things that can be done regarding units in momentum and energy problems. For example, given rest energy in MeV, if you are asked to find momentum, determine it in terms of MeV/c. You needn't convert MeV to joules. Thus, if $v = \frac{1}{2}c$, $p = \gamma mv = \gamma m\frac{1}{2}c$; *now multiply top and bottom by* c; $p = \gamma\frac{1}{2}mc^2/c = \frac{1}{2}\gamma E_0/c$ and you enter E_0 in MeV and get p in MeV/c.

Discussion Questions

1. In terms of Classical Physics only, suppose you are sitting in a chair resting. (a) What will you see if you jump up, rush away at, say, a speed of 100c, spin around, and stop? (b) In a similar vein, how might you observe Lincoln's Gettysburg address? (c) Is all of this actually possible?

2. Prior to the Special Theory, Lorentz and others were faced with an interesting dilemma. Figure Q2 shows a beam of light and two observers moving along with it, one at a speed $v < c$ and the other at $v > c$. From a classical perspective, how would each see the light to move? What about the notion that the beam propagates in the direction of $\mathbf{E} \times \mathbf{B}$?

3. How might the history of physics been different if c were 3×10^2 m/s rather than 3×10^8 m/s?

4. Could we connect two distant spacecraft by a long rigid rod and communicate faster than we might with light by just wiggling the ends of the rod? Explain.

5. Imagine an ideally rigid long pair of scissors. As the two blades are closed, the cutting point moves out away from the pivot. The more nearly parallel the blades become, the more rapidly that point travels away. If the scissors are long enough and closed rapidly enough, the contact point will travel out faster than c. An ax slammed almost horizontally onto an equally

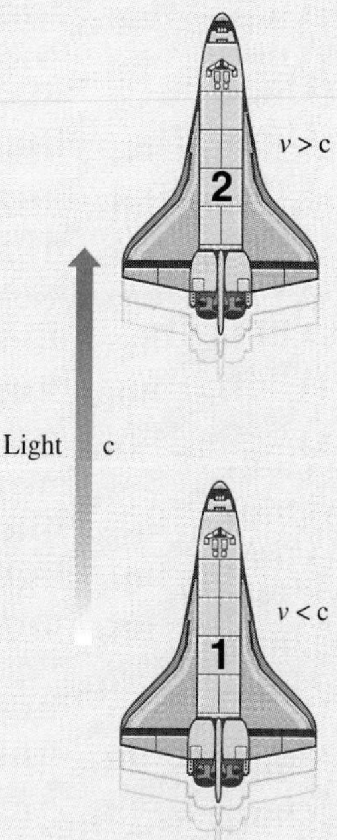

Figure Q2

large horizontal sheet of wood suggests the same effect. Think of the point of overlap of two crossed laser beams in the same plane that are moved toward being parallel. Does any of this violate Relativity? Explain. Can you think of another "thing" that moves faster than c?

6. Primary cosmic rays impacting on the upper atmosphere at altitudes of around 15 km create high-speed ($v = 0.999c$) muons. Despite their 1.5-μs half-life, about 200 muons are detected at ground level per square meter per second. That's surprising because half of them should disappear after 1.5 μs (or 450 m) and after 33.3 half-lives (15 km) a miniscule 10^{-8}% of them should survive (much fewer than do). Explain what is happening from the perspective of the muon.

7. Suppose we are receiving a stream of photons of frequency f in vacuum and then decide to fly directly away from the source. Discuss what we will observe to happen to the photons' speed, momentum, total energy, and rest energy.

8. Astronaut Suzy inside her spaceship shines a flashlight beam on the front wall just as the craft passes the Earth. If the ship is traveling at or in excess of c, what would Suzy see? What would you on Earth see? Using the two postulates, what can you conclude?

9. As we have seen, when two parallel wires carry currents in the same direction, there is a "magnetic" attraction between them. Amazingly, it turns out that this effect is actually an electrostatic one and, more generally, that *magnetism is a relativistic manifestation of electrostatics.* There really is only one force—the electromagnetic force. Thus, envision two current-carrying wires and their moving electrons and stationary positive ions. What do things look like within wire 1 as seen by a moving electron in wire 2? Is there a Lorentz-FitzGerald Contraction and, if so, what is its apparent effect on the positive and negative charge densities?

10. For warm-ups, draw part of the world-line of a book at rest on your desk. Figure Q10 is a space-time diagram showing several world-lines. Describe the motion associated with each as completely as possible.

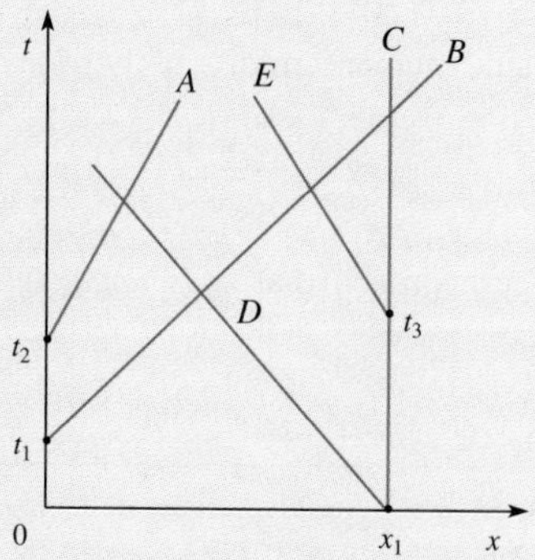

Figure Q10

11. Given that $E = \gamma mc^2$, what happens to a material object as it approaches the speed of light? What does this tell us about v compared to c?

12. Figure Q12 is a space-time diagram illustrating how light pulses can be used as time markers. Explain what's happening in detail. Why are the intervals seen by observer *B* longer than those seen by observer *A*? What does observer *C* see? Consider the flashes arriving at *C* to have emanated either from *A* or *B*. Each observer is in a different inertial reference frame.

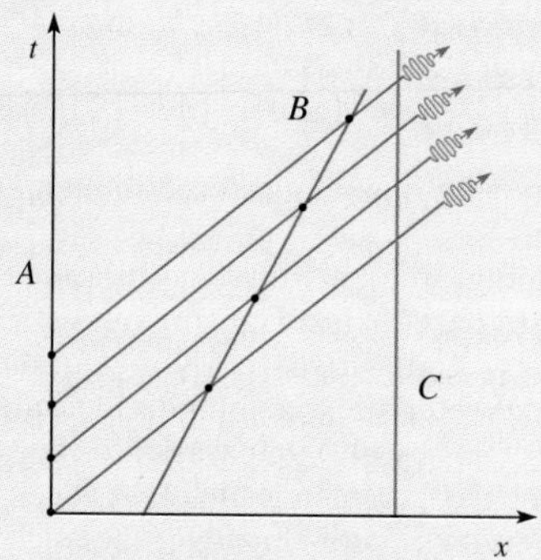

Figure Q12

13. Newton said "Absolute, true, and mathematical time, of itself, and from its own nature, flows equally without relation to anything external." Discuss the significance and validity of this statement.

14. The energy of a photon is entirely kinetic. Discuss the relationships between E, p, f, and λ for photons. How does the Doppler Effect influence these considerations? Can a light beam exert pressure on a target?

15. Starting with Eq. (28.7), show that either $v_{PO'}/c = 1$ or $v_{O'O}/c = 1$ implies $v_{PO}/c = 1$. When is v_{PO} less than the classical value? When is it greater?

16. You may have learned in chemistry class somewhere along the line that if you "weighed" all the reactants before and after a chemical reaction, the sum would be unchanged. That's

the so-called Law of Conservation of Mass and it's confirmed to an accuracy of one part per ten million ($1/10^7$). Inasmuch as Relativity shows us that most chemical reactions involve mass changes ($\Delta m/m$) of a few parts per ten thousand million ($\approx 2/10^{10}$), what can be said about the Law of Conservation of Mass?

17. Poincaré challenged the idea of immediate action-at-a-distance, although Newton had accepted it. From Poincaré's perspective, for example, a change in the gravitational field of the Sun would take a finite time to reach the Earth. Generalizing this assumption, show that nothing can travel at an infinite speed. Assuming there is an upper-speed limit, discuss why it must be invariant with respect to all inertial reference frames. What does this suggest about the Michelson-Morley experiment and c?

18. You've no doubt heard in school the adage "Matter can neither be created nor destroyed" or its haunting corollary "Energy can neither be created nor destroyed." What can you say about these statements?

19. Here are two postulates: (1) All interactions propagate at a finite speed. (2) An object cannot interact with itself. Discuss how it follows that there must be an upper-speed limit.

Multiple Choice Questions

1. The Michelson-Morley experiment established (a) that the Earth does not move with respect to the Sun (b) that the ether moves at c as the Earth travels in its orbit (c) that the ether is an elastic solid that streams over the Earth (d) that there is no observable ether wind at the surface of the Earth (e) none of these.
2. Simultaneity is (a) dilated (b) absolute (c) invariant (d) relative (e) none of these.
3. A clock is moving at a uniform velocity with respect to an observer. The latter, comparing things to her clock, reports that the time on the moving clock is (a) perfectly accurate (b) fast (c) slow (d) running backwards (e) none of these.
4. According to the Lorentz-FitzGerald Contraction (a) a body contracts only when it accelerates (b) a body contracts along the direction of motion (c) the time it takes for a light clock to tick contracts (d) a body contracts transverse to the direction of its motion (e) none of these.
5. How long will a vertical meter stick appear to someone moving horizontally with respect to it at a speed of 0.99c? (a) 99 cm (b) 100 cm (c) 0 cm (d) 0.01 cm (e) none of these.
6. Two inertial observers separating from each other at a speed of 99.99% of c each shine a light beam at one another. (a) each will see the light arrive at a speed of c (b) each will see the light arrive at a speed of 2c (c) the light will not reach either of them (d) the light will reach one at c and the other at 2c (e) none of these.
7. A spaceship observes an asteroid coming directly toward it at a speed v. The ship fires its engines and travels away along the line of approach. It measures the asteroid's (a) v and p to decrease, but E_0 is unaltered (b) v, p, and E to be unchanged (c) v only to decrease, p and E being unaltered (d) p and E to increase, leaving v and E_0 unaltered (e) none of these.
8. A long thin rocket takes off straight up from the ground. It is observed by three people: a man selling ice cream on the beach, a woman riding up an adjacent gantry elevator, and the pilot looking out a window. Referring to the rocket, ideally (a) all three see the same length (b) of the three, the pilot sees the shortest length (c) of the three, the woman sees the shortest length (d) of the three, the man with the ice cream sees the longest length (e) none of these.
9. An astronaut heading out toward a star at a constant high speed can determine that he is in motion by (a) the contraction of on-board meter sticks (b) the slowing down of time on his clocks (c) the increase of his mass (d) the speeding up of his heart (e) none of these.
10. An arrow 30 cm long is fired at very high speed directly toward a hollow cylinder 30 cm long. The arrow passes axially inside the cylinder. As seen by an observer at rest with respect to the cylinder, (a) the arrow is shrunken and fits inside for an instant (b) the arrow and cylinder are unchanged and stay exactly the same size (c) the cylinder is shrunken and the arrow never fits completely inside (d) the arrow and the cylinder are both shrunken and remain the same size (e) none of these.
11. An arrow 30 cm long is fired at very high speed directly toward a hollow cylinder 30 cm long. The arrow passes axially inside the cylinder. As seen by an observer at rest with respect to the arrow, (a) the arrow is shrunken and fits inside for an instant (b) the arrow and cylinder are unchanged and stay exactly the same size (c) the cylinder is shrunken and the arrow never fits completely inside (d) the arrow and the cylinder are both shrunken and remain the same size (e) none of these.
12. An observer in the laboratory section of a spacecraft performs a series of experiments to determine whether the ship is at rest or in uniform motion. He (a) can succeed by making very precise time measurements (b) can succeed by making very precise mass measurements (c) can succeed by making very precise length and time measurements (d) cannot succeed no matter what he does (e) none of these.
13. When a battery is drained in ordinary use, (a) its mass decreases because its internal rest energy decreases (b) its mass decreases, but its rest energy remains unchanged (c) its mass remains unchanged, but its internal rest energy decreases (d) its mass decreases because its internal rest energy increases (e) none of these.
14. When a photon is red-shifted via the Doppler Effect by having an observer moving away from the source, (a) its energy is unchanged (b) its momentum increases (c) its energy increases (d) both its momentum and energy decrease (e) none of these.
15. An arrow 30 cm long is fired at very high speed v directly toward a hollow cylinder 30 cm long. The arrow passes

axially inside the cylinder. An observer moving parallel to the cylinder in the direction of the arrow at a speed $\frac{1}{2}v$ sees the arrow moving forward at the same speed he sees the cylinder moving backward. He observes that (a) the arrow is shrunken and fits inside for an instant (b) the arrow and cylinder are unchanged and exactly the same size (c) the cylinder is shrunken and the arrow never fits completely inside (d) the arrow and the cylinder are both shrunken and are the same size (e) none of these.

16. The proper time of a system is (a) the time measured on clocks at rest in that system (b) the time measured on clocks at rest in an inertial system moving properly with respect to the first system (c) the time measured on clocks moving uniformly in that system (d) the time that agrees with the National Institute of Standards and Technology (e) none of these.
17. A photon's proper time (a) is a function of its speed (b) changes slowly with respect to a moving observer (c) never changes (d) changes faster with respect to systems at rest (e) none of these.
18. Given that there are about 2.56×10^9 heartbeats in a statistically average lifetime of 70 years, people who are born and die on a spaceship moving at a constant speed of 0.600c can expect their hearts to beat a total of (a) $(0.600)(2.56 \times 10^9)$ times (b) 2.56×10^9 times (c) $(0.800)(2.56 \times 10^9)$ times (d) $(1.25)(2.56 \times 10^9)$ times (e) none of these.

Problems

BEFORE THE SPECIAL THEORY
THE SPECIAL THEORY OF RELATIVITY

1. [I] In the Michelson-Morley experiment, the time it takes beam 1 to travel out and back, with and against the ether wind, is $t_\parallel$, and the time for beam 2 to go from M_S to M_2 to M_S is $t_\perp$, where

$$t_\parallel = \frac{2L}{c}\frac{1}{(1-\beta^2)} \quad \text{and} \quad t_\perp = \frac{2L}{c}\frac{1}{\sqrt{1-\beta^2}}$$

Suppose the $M_S M_1$ arm of the Michelson Interferometer shrinks as suggested by FitzGerald, so that L becomes $L\sqrt{1-\beta^2}$. Show that $\Delta t = t_\parallel - t_\perp$ then equals zero.

2. [I] A light-minute is often symbolically written as 1 c·min. Define the concept and give its value in SI units.
3. [I] Show that $\beta^2 = (\gamma^2 - 1)/\gamma^2$.
4. [I] A computer in a spacecraft takes 2.00 μs to make a calculation, as measured by a co-moving observer on board. Someone traveling at a relative speed of 0.998c looks in the window. How long will that person determine the calculation to take?

5. [I] According to a biology text, it takes 45 s for blood to circulate around the human body. How long on his clocks will it take blood to circulate around the body of an astronaut traveling such that with respect to Earth $\gamma = 2$? How long will a ground-based observer see the astronaut's circulation take?

6. [I] A stopwatch in the hands of a terrestrial observer is seen by a passer-by in a flying saucer traveling relative to the planet with a γ of 2.00. If the earthling sees two events separated by a proper time interval of 60.0 s, how long will the interval be as seen by the traveler?
7. [I] The vaudeville team of Shultz and O'Hara, playing to a full house at Minsky's Burlesque, takes 60.0 s to tell their worst joke. How long will an observer in a rocket ship flying by at 0.995c have to wait to lip-read the punch line? Will the observer start laughing after seeing the audience respond? Assume that the observer is equidistant from the stage when he perceives the start and finish of the joke so we needn't worry about communication-time lags.
8. [I] An astronaut on the way to Mars at 0.600c wishes to take a one-hour nap. She contacts Ground Control requesting a wake-up call. Neglecting any complications due to the time it takes the signal to reach the ship, how long should the flight controller let her sleep as measured on his clock?

9. [I] Suppose you look at the inhabitants of a passing asteroid and see all their clocks running such that 52.0 s goes by on them while 60.0 s elapse on your planetary clock. What is the relative speed of the asteroid and Earth?

10. [I] An astronaut is on the way directly to a star system along a straight line from Earth. Her instruments indicate that at the speed relative to the Earth at which she is traveling, her γ is 5/3 and the total distance she has to traverse is (6.0c)(1.0 y). What is the corresponding proper distance shown on Earth-based instruments?
11. [I] How large would a 1.00-m-long Italian bread look to a hungry observer (with very sharp eyes) if he sees the bread to have a speed of 0.900c in a direction parallel to the loaf?
12. [I] A long pepperoni is fired out of a new secret weapon developed by the Mediterranean arm of NATO. As it flashes past the reviewing stand, its length is measured to be 44.0% of what it was before launch. How fast is it moving?

13. [I] If an electron is hurtling down the two-mile long Stanford Linear Accelerator at 99.98% of c, how many meters long is the trip as seen by the electron?

14. [I] A 1.00-m-long rocket is to be fired past an observer who will locate both its ends simultaneously and determine its length. If that length is found to be 0.500 m, how fast was the rocket moving?

15. [I] A spacecraft passes a platform traveling parallel to it (at 0.600c) along its full length. The pilot of the spacecraft determines the platform to be 400.0 m long. You witness this event from the platform lounge where you happen to be having a drink. What is the proper length of the platform?

16. [I] A spaceship moving at a speed of 0.70c with respect to you fires a rocket in the forward direction at 0.40c relative to the ship. What is the speed of the rocket with respect to you? What would be expected classically?

17. [I] Two pulses of light are sent out in vacuum, one to the right and another colinearly to the left. What is the speed of either one with respect to the other? Show your work, please. What would be expected classically?

18. [I] A spaceship moving at a speed of 0.70c with respect to you fires a rocket in the backward direction at 0.40c relative to the ship. What is the speed of the rocket with respect to you? What would be expected classically?

19. [I] An electron traveling at 0.992c collides head-on with a positron traveling at 0.981c. Both speeds are measured with respect to the laboratory in which the experiment occurs. At what speed do they approach each other?

20. [II] Calculate the approximate time dilation that would be observed by someone on the ground watching a clock in a supersonic jet flying at 1800 mi/h. Use the binomial expansion (which is applicable at speeds less than around 0.3c). Try doing this calculation exactly—you'll see why the approximation is so useful.

21. [II] A bright yellow line painted along the length of a rocket is seen to be 1.000 yd long by an observer who is standing on the Earth as the vehicle sails by. Knowing that the proper length of the line is actually 1.000 m, how fast is the rocket traveling?

22. [II] Calculate the percentage length contraction for a high-speed jet plane traveling at 3600 mi/h. Try doing it exactly and then use the binomial expansion (which is okay for speeds less than around 0.3c).

23. [II] Two inertial observers pass each other at a relative speed v, and each notices a 10.0% length contraction of the other. Determine v.

24. [II] The navigator on a Federation starship traveling toward Earth along a path parallel to the Earth-Sun line sees the separation between these bodies to be 6.11 c·min. Consulting charts, the navigator finds the proper distance to be 8.33 c·min. How fast is the ship going?

25. [II] Two friends, Harvey (age 20) and Lisa (age 24), say their goodbyes. She boards a space shuttle for a round-trip flyby of the planet Mongo. Rapidly reaching cruising speed, she settles in for a long journey. Neglecting the short acceleration times (which is quite impractical), Lisa returns home 24 months later (as seen on her calendar). She is met by Harvey, who is 26 years old. At what speed was Lisa cruising?

26. [II] Mars is 80.00×10^6 km from Earth when a settler sends a laser beam message back home. As fate would have it, just as the flash of light goes out, a spacecraft flies by on the way toward Earth at 0.500 0c. How long will someone on Earth see the ship take to make the journey? How long will the message take to arrive on Earth? How long will the crew of the ship say the trip took? How long will they say the message took to get to Earth?

27. [II] A 1.000-m bar of steel travels at a constant speed of 0.600c with respect to an observer. The bar moves lengthwise along a line perpendicular to the observer's telescope. How long does it take the bar to fly past the cross hairs in the telescope?

28. [II] A starship passes a space station at a uniform 0.600c. The pilot looking into the gym area sees a runner traveling perpendicular to the flight path. The runner covers the straight 100-m gym track in 10.0 s by his watch. What was the runner's average speed as seen by the pilot? (Hint: How long is the track as seen by the pilot?)

29. [II] A laser beam fired from Earth is received by a spacecraft heading directly toward the source at a speed v. Show that a technician aboard the ship will measure the speed of the incoming beam to be c regardless of the value of v.

30. [II] Show that if the speed of an object in any one inertial frame is less than c, its speed in all other inertial frames (each moving with respect to the first at a speed less than c) will be less than c. (Hint: Let $v_{PO'} = ac$ and $v_{O'O} = bc$, where a and b are < 1.)

31. [III] Consider the passengers aboard a jet plane cruising at 1000 km/h with respect to an observer on the ground. Neglecting all other effects (for example, gravity), how long would the flight have to last before the clock aboard the plane was 1.00 s behind a clock on the ground with which it had been synchronized?

32. [III] With Problem 31 in mind, consider a satellite in orbit or a plane flying around the Earth at a uniform speed. We wish to determine how long the flight must last for there to be a discrepancy of 1.00 s between clocks on the ground and in the air. Show that this will indeed be the case when the proper time on the moving craft is

$$\Delta t_S = \frac{1.00\ \text{s}}{\gamma - 1}$$

33. [III] The TV mast on a spaceship passing Earth is fixed at 21.0° to the line of flight by the communications officer. If the ship zips by at 0.852c, at what angle will someone on Earth see the mast?

RELATIVISTIC DYNAMICS

34. [I] An electron ($m = 9.110 \times 10^{-31}$ kg) is traveling at 0.866c with respect to the face of a TV picture tube toward which it is heading. How much momentum does it have with respect to the tube?

35. [I] A neutron ($m = 1.675 \times 10^{-27}$ kg) is traveling at 0.50c with respect to a "stationary" target. How much momentum does it have with respect to the target?

36. [I] A proton has a mass of $1.672\,6 \times 10^{-27}$ kg. Determine its rest energy in joules and in MeV and compare your results with Table 28.3.

37. [I] A hypothetical particle is traveling with respect to an observer such that its total energy is seen to be twice its rest energy (which is 1000 MeV). How fast is it moving? What is its kinetic energy?

38. [I] An electron with a rest energy of 0.511 MeV has a kinetic energy of 0.089 MeV. What is its total energy?

39. [I] The total energy of a muon is 106.7 MeV. If its rest energy is 105.7 MeV, what is its kinetic energy?

40. [I] A proton ($m = 1.672\,6 \times 10^{-27}$ kg) moves with respect to the laboratory frame such that it has a $\gamma = 1.50$. What is its total energy in joules and electron-volts?

41. [I] A particle of mass 1.000×10^{-6} kg is fired from a device such that it has a muzzle "velocity" of 0.200 0c with respect to an inertial observer. Determine the rest energy, total energy, and kinetic energy of the projectile with respect to the observer.

42. [I] An electron with a rest energy of 0.511 MeV travels with a γ of 5/3 with respect to a laboratory. What is its total energy?

43. [I] A deuteron is the heavy hydrogen nucleus ${}^2_1$H composed of a proton and a neutron. Its rest energy is 1875.6 MeV. How much energy must be supplied to rip it apart? That is, how much energy is liberated (as KE and as a γ-ray) when a deuteron is formed of a separate proton and neutron?

44. [I] A hypothetical particle has a rest energy of 1.500 MeV and a total energy of 3.000 MeV. What is its momentum?

45. [I] A clock, in the course of being wound, has 89.88 J of work done on its mainspring. By how much does its mass change?

46. [I] A proton and a neutron together form the nucleus of a deuterium atom that is completed with a single orbiting electron. The mass of such an atom is 3.3445×10^{-27} kg. When the particles come together, mass is transformed, and the composite atom has less mass than the separate constituents; the equivalent energy difference is called the **binding energy**. Find the binding energy of deuterium in units of J and MeV to three figures.

47. [I] Compute the amount of energy liberated in the fusion reaction

$$^2\text{H} + {}^2\text{H} \rightarrow {}^3\text{He} + {}^1\text{n} + \text{energy}$$

Here, the neutron flies off with about 3 times the KE of the helium.

48. [I] Another important fusion reaction involving deuterium and tritium is

$$^2\text{H} + {}^3\text{H} \rightarrow {}^4\text{He} + {}^1\text{n} + \text{energy}$$

where the neutron flies off with about 4 times the KE of the helium. Determine the total amount of energy liberated.

49. [I] A photon is to have the same energy as a 1.000-MeV electron (that is, an electron with a KE of 1.000 MeV). What must be its frequency? (Hint: That's total energy.)

50. [II] Show that as the momentum of a body of mass m becomes very large, its speed approaches c, but does not exceed it.

51. [II] Suppose that an object traveling uniformly has a value of γ relative to an inertial observer. Derive an expression in terms of γ for the percent error introduced by using the classical momentum (p_c) instead of the relativistic momentum (p); that is, $(p - p_c)/p$. What is this error for a speed of 0.600c?

52. [II] An electron ($m = 9.1095 \times 10^{-31}$ kg) has a momentum of 8.603×10^{-23} N·s. What's its speed?

53. [II] If the ratio of the momentum to the mass of a proton is 0.101, determine its speed. Compare this ratio to its classically determined speed. When will the relativistic speed approximate the classical speed?

54. [II] An electron with rest energy 0.511 MeV travels at 0.80c with respect to the laboratory. What is its momentum in the lab frame in units of MeV/c?

55. [II] An object has a total energy equal to twice its rest energy. Show that $p = \sqrt{3}\, mc$.

56. [II] An electron has a kinetic energy of 0.2000 MeV. Find its speed. Compare that with its classical speed.

57. [II] An electron at rest accelerates through a potential difference of 1.50 MV. Determine its speed on emerging and compare it with the erroneous classical value.

58. [II] We wish to accelerate an electron from rest up to 0.990c. Through what potential difference must it pass?

59. [II] A proton with a rest energy of 938.3 MeV has a momentum of 100.0 MeV/c. How fast is it moving?

60. [II] Show that for an object moving at speed v, the ratio of the relativistic KE to the classical KE, call it Γ, is

$$\Gamma = 2(\gamma - 1)/\beta^2$$

61. [II] Given a particle whose rest energy equals its kinetic energy, determine its β value.

62. [II] With Problem 60 in mind, show that

$$\gamma = \tfrac{1}{4}(\Gamma + \sqrt{\Gamma^2 + 8\Gamma})$$

63. [II] A photon is to have the same momentum as a 1.000-MeV electron (that is, an electron with a KE of 1.000 MeV). What must be its frequency?

64. [III] An object has an initial speed v and a corresponding momentum p. Show that if the momentum is to be doubled, the final speed must become

$$v_f = \frac{2vc}{\sqrt{c^2 + 3v^2}}$$

29

The Origins of Modern Physics

WE NOW BEGIN A sequence of five chapters dealing with Modern Physics. That discipline is a study of the submicroscopic world of atoms, and the particles that compose them, and the particles that compose those particles. It is essentially a creation of the twentieth century, but most of the work had its foundations in the late 1800s, a period when even the existence of the atom was not totally accepted. This chapter deals with the establishment of the atom as a real entity, and, beyond that, with the realization that it is a complex structure capable of coming apart into its constituent subatomic particles.

Here we study a series of important experiments that laid the groundwork for all that was to follow. We trace the discoveries of the electron, X-rays, radioactivity, the atomic nucleus, protons, and neutrons. The understanding of the atomic domain that evolved from these fundamental insights now guides much of the work done in contemporary physics, chemistry, and biology.

SUBATOMIC PARTICLES

The first subatomic particle to reveal itself was the electron, and that discovery was predicated on the study of electric currents in solids, liquids, and gases. We know now that charge is a property of substantial matter and that it is quantized; it comes in whole number multiples of a minimum amount corresponding to the magnitude of the charge on the electron. A little over a hundred years ago, physicists had no idea of what charge was; whether it was continuous or particulate, whether it was material or ethereal, like light.

29.1 The Quantum of Charge

Soon after the invention of the battery, Nicholson and Carlisle used one to decompose water via a process known as *electrolysis.* Their discovery was accidental; to improve the contact between a wire and their battery, they put some water on the connection and the liquid filled with gas bubbles. Thereafter, they inserted two wires into a beaker of water and passed a current through it. Bubbles of oxygen were released at the *anode,* and hydrogen was formed at the *cathode.* Humphry Davy and his assistant Michael Faraday used electrolysis to explore chemical interactions. When the conducting solution, the electrolyte, contains a dissolved salt of some metal like silver, that metal will be liberated at the cathode just as hydrogen is (Fig. 29.1). To account for the passage of current across the electrolyte, Faraday supposed that there was a flow of charged particles, which he called *ions.* Although he did not speculate as to their nature, we know that ions are atoms that have lost or gained one or more orbital electrons and therefore carry a net charge.

Faraday measured the mass of monovalent silver deposited on the cathode and the net charge provided (that is, current multiplied by the time it was applied) in the process. To liberate one *mole* of a single-valence element (singly charged $\pm$ ion), he found that a large quantity of charge F, now called a **faraday** and equal to 96 485 C, had to be provided. Remember that a mole of any element comprises $N_A = 6.022\,136\,7 \times 10^{23}$ atoms. Thus, for example, silver has an atomic mass of 107.87 units, and 1 faraday of charge liberates 6.02×10^{23} silver atoms with a net mass of 107.87 grams. Faraday teetered on the edge of electron theory, hinting that electricity was composed of particles of charge e. Given that there are N_A atoms per mole, and assuming each monovalent ion has a charge e, then a faraday of charge must correspond to an amount

$$\mathrm{F} = N_A e \tag{29.1}$$

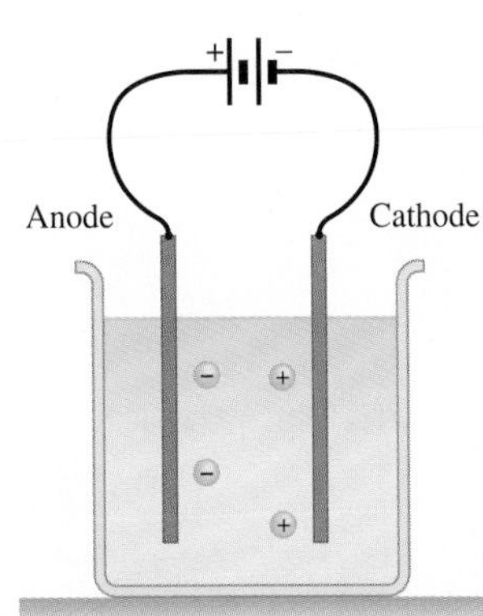

Figure 29.1 The passage of a current through an electrolyte. Negative ions move to the anode, positive ions move to the cathode.

The striking thing is that the same quantity F is found for all monovalent elements regardless of their chemical properties; $e = \mathrm{F}/N_A$ is a constant. For bivalent ions like copper, a charge of 2F must be supplied to liberate a mole of Cu, and each ion carries a charge of $2e$. Though F was directly measurable, Avogadro's number was not determined for several decades and that left e unknown. Still, 1 faraday of charge is transported by 1 mole of monovalent substance. It follows that $\frac{1}{2}$ faraday is transported by $\frac{1}{2}$ mole, and so on down, *presumably* to the smallest unit of charge e associated with the smallest mass m; that is, the mass of a single atom of the material liberated. For example, we find from electrolysis that the **charge-to-mass ratio** for singly ionized hydrogen (H^+) is

$$\left(\frac{e}{m}\right)_{H^+} = 9.58 \times 10^7 \text{ C/kg}$$

Hydrogen is the lightest of the elements, and this value turns out to be the largest charge-to-mass ratio of any element. Although neither e nor m was known at the time, the corresponding ratios could be determined experimentally.

Example 29.1 During electrolysis, how much silver will be deposited by a current of 1.00 A applied for 1.00 s?

Solution: [Given: element silver, $I = 1.00$ A, and $t = 1.00$ s. Find: m.] We have the current and the time during which it flowed, so the net charge passed is

$$\Delta q = I\Delta t = (1.00 \text{ A})(1.00 \text{ s}) = 1.00 \text{ C}$$

Since 1 faraday of charge liberates 1 mole of silver (that is, 107.9 g), 1.00 C of charge liberates an amount m where

$$\frac{96\,485 \text{ C}}{107.9 \text{ g}} = \frac{1.00 \text{ C}}{m}$$

and

$$\boxed{m = 1.12 \text{ mg}}$$

▶ **Quick Check:** We are passing about 10^{-5} faraday, and that quantity should deposit about 10^{-5} of a mole of Ag, which equals 1.1 mg.

Remarkably, in 1874, the Irish physicist George Stoney used Eq. (29.1) along with the best available values of F and Avogadro's number to determine the basic unit of charge. His estimate of e was too small by a factor of about 20, but even that was still quite good, all things considered.

By 1881, Helmholtz asserted that, "If we accept the hypothesis that the elementary substances are composed of atoms, we cannot avoid concluding that electricity also, positive as well as negative, is divided into definite elementary portions, which behave like atoms of electricity." For a while, the German literature referred to e as "*das Helmholtzsche Elementarquantum*" (the Helmholtz elementary quantity, or **quantum**). In 1891, Stoney christened that fundamental unit of charge the *electron.*

29.2 Cathode Rays: Particles of Charge

A totally different route to the electron began one dark night in 1675 when the French astronomer J. Picard noticed in amazement that the barometer he was swinging as he walked began to give off an eerie flickering light. The effect was duplicated in the laboratory by Hauksbee using an electrostatic generator to provoke a rarefied gas in a bottle into the mysterious luminosity. Thus began the study of electrical discharges in low-pressure gases.

Outstanding among the many scientists in the field during the 1870s was Sir William Crookes, a rather unorthodox fellow who believed he could communicate with the dead. A "Crookes tube" with two sealed electrodes is illustrated in Fig. 29.2. (It's the forerunner of all the blazing bar and motel signs—the neon uglies, p. 1015.) When the tube was connected to the terminals of a high-voltage source, a glowing beam spread down its length. Different gases glowed with differ-

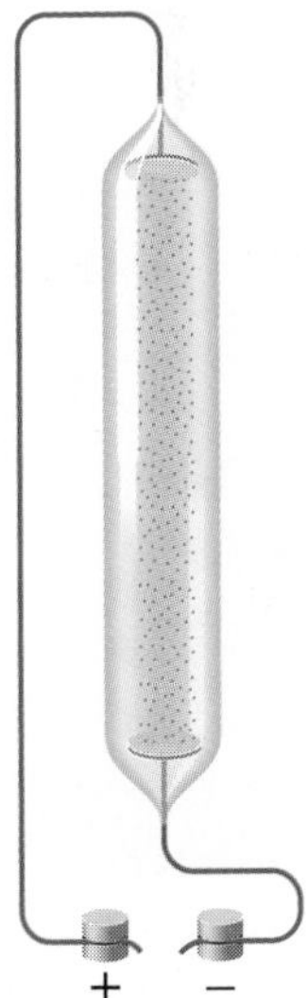

Figure 29.2 A cathode-ray or Crookes tube.

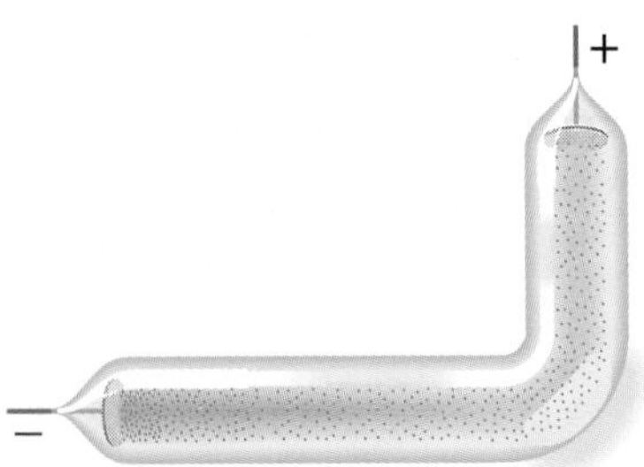

Figure 29.3 When the tube is bent, it becomes clear that the rays emanate from the negative terminal, or cathode.

ent colors: mercury emitted greenish blue, while neon gave off a bright orange-red. By displacing the anode to the side (Fig. 29.3), Crookes verified that the emanations were actually streaming from the negative plate, not always making it to the anode. When these **cathode rays** struck the walls of the tube, the glass gave off a pale green fluorescent glow that betrayed the impact of the beam.

Objects that were inserted into the stream cast sharp shadows in the glow at the far end of the tube. This, too, suggested straight-line propagation. Crookes even put a little paddle wheel inside a tube and spun it around with the bombarding beam, which more and more appeared to be a stream of particles carrying energy and momentum. Later, he was able to deflect the beam with a magnet (p. 741) in the same way a current of negative charges would be deflected. What emerged from this experimentation was the "British" view that cathode rays were streams of negatively charged submicroscopic particles.

The opposing "German" view was most vigorously pressed by P. Lenard. Heinrich Hertz had discovered that cathode rays could pass through thin sheets of metal without leaving any holes, and that was enough to convince his assistant Lenard that the rays must be nonmaterial. Today, we know that bulk matter is mostly empty space and that some subatomic particles can sail through yards of concrete and steel as if they weren't there. To Lenard, cathode rays were without substance; they were waves like light—"phenomena in the ether."

A so-called neon sign is just a long tube filled with an appropriate gas, across which there is a sustained high-voltage discharge.

There the matter stood, befogged by nationalism—corpuscle versus wave—a controversy that would last at least twenty years.

Discovering the Electron

When Joseph John Thomson began his research into the nature of cathode rays, he was the director of the famed Cavendish Laboratory at Cambridge University. In 1894, Thomson showed that cathode rays traveled much slower than light (another blow to the notion that they were "phenomena in the ether"). And then J. Perrin (1895) found that a metal obstacle located in the path of the beam acquired a negative charge. It remained for Thomson (his friends called him J. J.) to demonstrate that the cathode rays and the charge were one and the same.

To that end, Thomson constructed the device shown in Fig. 29.4, which is similar in many respects to a TV picture tube, a so-called CRT (cathode-ray tube). A high voltage between cathode (C) and anode (A) ionized the trace of residual gas, creating a cloud of positively charged atoms and negative electrons; the latter were the cathode rays. These rays were accelerated toward the positive anode reaching tremendous speeds (of as much as 60 000 mi/h). A small hole in the front of the anode allowed a narrow unobstructed stream to pass into the region beyond A. The speed v of the beam, which could be controlled via the cathode-anode voltage, was fairly uniform. Further along the tube, the beam traveled between two metal plates (P) that were attached to a variable dc voltage, creating a vertical electric field **E**. Coils straddling the tube produced a controllable horizontal magnetic field **B** crossing the same region.

A cathode-ray tube. The metal cross intercepts the beam and casts a shadow on the face of the tube.

The idea was to measure the charge-to-mass ratio in a two-step procedure. The beam was assumed to be composed of a stream of discrete particles of charge e and mass m_e that could be deflected in the vertical plane by either field. The first step was to determine v, which was done by adjusting E and B so that the forces produced were equal and opposite and the beam sailed through undeflected (the effect of gravity was negligible). This arrangement is the "velocity selector" we considered

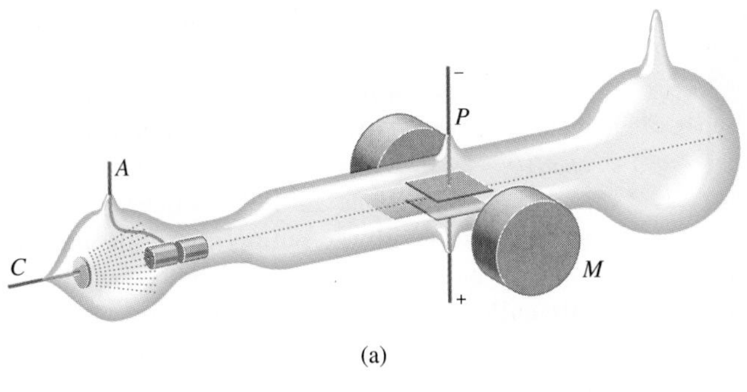

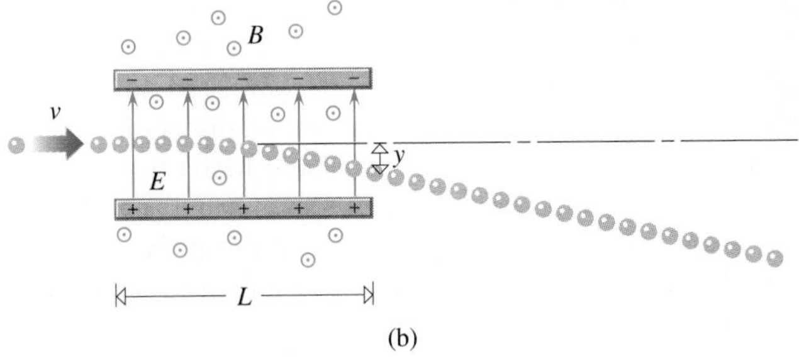

Figure 29.4 (a) J. J. Thomson's electron-beam device showing the horizontal electrical deflecting plates and the vertical magnetic coils. (b) Electrons experience a downward electric force and (with the B-field coming out of the page) an upward magnetic force.

J. J. Thomson's original tube.

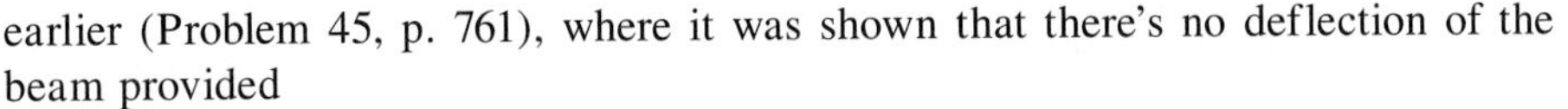

earlier (Problem 45, p. 761), where it was shown that there's no deflection of the beam provided

$$v = \frac{E}{B}$$

Next, the B-field was shut off, and the beam was deflected by the remaining E-field. The cathode-ray particles experienced a downward Coulomb force (eE) and so accelerated down uniformly, via Newton's Second Law (eE/m_e). They each "fell," much as a stone would fall in a gravity field, for all the time (t) that they were between the plates. Since the horizontal speed v is constant, the traversal time is $L/v = L/(E/B)$. We want the distance each particle drops (y) in the course of passing through the plates. Thus, from Eq. (3.9), $y = \frac{1}{2}at^2$, or

$$y = \frac{1}{2}\left(\frac{eE}{m_e}\right)\left(\frac{LB}{E}\right)^2$$

Inasmuch as y can be determined directly from the point where the beam hits the screen (using the geometry of the tube, p. 1040), and since we know E from the plate voltage and B from the coil current, the only unknowns are e and m_e. Thus

$$\frac{e}{m_e} = \frac{2yE}{L^2B^2} \tag{29.2}$$

The present-day accepted value of the ratio is

$$\frac{e}{m_e} = (1.758\,819\,62 \pm 0.000\,000\,53) \times 10^{11}\ \text{C/kg}$$

It is clear that, in some sense, there are electrons and protons, and we cannot well doubt the substantial accuracy of their estimated masses and electric charge. That is to say, these constants evidently represent something of importance in the physical world, though it would be rash to say that they represent exactly what is at present supposed.

BERTRAND RUSSELL
The Analysis of Matter (1954)

The electrical matter consists of particles extremely subtle since it can permeate common matter, even the densest, with such freedom and ease as not to receive any appreciable resistance.

BENJAMIN FRANKLIN (ca. 1750)

J. J. Thomson at the Cavendish Laboratory.

Thomson suggested that the reason this ratio was so much larger (namely, 1836 times larger) than the corresponding one for hydrogen was that the mass of the electron was that much smaller than the mass of the H^+ ion. *The electron was a tiny fragment detached from a complex atom.* Not everyone in 1897 believed in atoms, and few were pleased with the notion of **subatomic particles**.

J. J. put the finishing touches on what today is generally called his "discovery of the electron" with two further experiments in 1899. First, he repeated the above measurement using a totally different source of electrons, namely the *photoelectric effect* (p. 1049), and got the same results for e/m_e. His second experiment was to measure e, again using a totally new procedure. His student C. T. R. Wilson (the inventor of the cloud chamber) had been working on the formation of clouds and found that droplets could form around charged particles. We shall not describe the methods used by Thomson and his colleagues other than to say that by 1901 he was able to arrive at a value for e that was only about 30% off. The work inspired a classic measurement by the American Robert Millikan, which we will examine.

The Oil-Drop Experiment

Millikan's famous oil-drop experiment is pictured in Fig. 29.5. A fine mist of oil is squirted from an atomizer above a small hole in the top plate of what can be thought of as a large parallel-plate capacitor. A few droplets descend through the hole into the region occupied by a variable E-field. Lit from the side, each minute droplet shines like a tiny star when seen through a viewing telescope. Most of the droplets

On Electrons. *At first there were very few who believed in the existence of these bodies smaller than atoms. I was even told long afterwards by a distinguished physicist who had been present at my lecture at the Royal Institution that he thought I had been "pulling their legs."*

J. J. THOMSON
English physicist

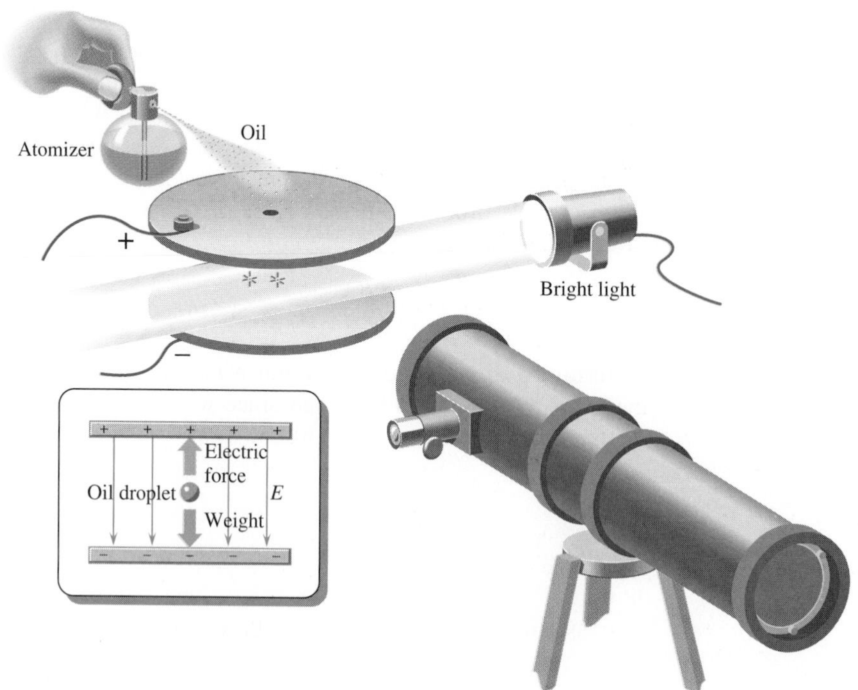

Figure 29.5 The Millikan oil-drop experiment. A mist of oil causes a few drops to enter between the plates of a capacitor. There, they experience a force due to the applied electric field.

get negatively charged on passing through the nozzle, picking up some small but unknown number of electrons. Once a drop is caught sight of, the voltage on the plates is carefully varied, thereby controlling a downward E and slowing the fall until that sphere of oil is suspended motionless midair in a balance between gravitational (mg) and electrical (qE) forces. At that moment $qE = mg$, and if the mass of the droplet (m) were known, the charge that it carried (q) would be known. The next procedure determines m.

The E-field is shut off and the same droplet watched as it again falls, soon reaching a constant terminal speed. By timing the motion of the droplet as it passes from one hairline down to another in the telescope, the terminal speed is measured. The theory of air resistance tells us that the terminal speed depends on the radius of the droplet, as well as on a number of other factors, all of which are known. Thus, from the terminal speed we have the radius, and from the radius we have the mass of the sphere. Millikan actually used a somewhat more complicated, though less arduous approach, but the idea is the same. His graduate student H. Fletcher methodically found the net charge on thousands of droplets, one by one. Millikan and Fletcher were then able to show that the droplets carried whole number multiples of a basic charge $q_e = e$, which was ascribed to the electron itself. They arrived at an average value of e of 1.592×10^{-19} C. Today, the best value of this fundamental quantum of charge (Table 29.1) is

$$e = (1.602\,177\,33 \pm 0.000\,000\,49) \times 10^{-19}\ \text{C}$$

All charged subatomic matter observed to date, whether positive or negative, carries a net charge that is an integer multiple of *e*.

Combining e with the charge-to-mass ratio provides another remarkable number—the mass of the electron:

$$m_e = (9.109\,389\,7 \pm 0.000\,005\,4) \times 10^{-31}\ \text{kg}$$

Using Eq. (29.1) and the measured value of the faraday, Millikan computed Avogadro's number to be 6.062×10^{23} molecules per mole (as compared to the present-day value of 6.022×10^{23}—not bad at all).

Given a charge-to-mass ratio for a hydrogen ion of 9.58×10^7 C/kg and the value of e, we can make a rough estimate of the size of an atom—silver, for instance. The relative atomic mass of hydrogen is 1.008, whereas that of silver is 107.9. The mass of a hydrogen ion is e divided by the charge-to-mass ratio: $(1.602 \times 10^{-19}\ \text{C})/(9.58 \times 10^7\ \text{C/kg}) = 1.67 \times 10^{-27}$ kg. Thus, a silver atom has a mass of $(107.9/1.008)(1.67 \times 10^{-27}\ \text{kg}) = 1.79 \times 10^{-25}$ kg. The density of silver is $10.4 \times 10^3\ \text{kg/m}^3$, and so there must be $(10.4 \times 10^3\ \text{kg/m}^3)/(1.79 \times 10^{-25}\ \text{kg/atom}) = 5.8 \times 10^{28}\ \text{atoms/m}^3$. The reciprocal of that is the number of cubic meters occupied per silver atom; namely, $1.72 \times 10^{-29}\ \text{m}^3\text{/atom}$. Each atom sits in a little imaginary cube of this volume. Assuming the silver atoms are tightly packed

TABLE 29.1 Some physical characteristics of the electron*

Mass m_e	$9.109\,389\,7(54) \times 10^{-31}$ kg
Charge e	$1.602\,177\,33(49) \times 10^{-19}$ C
Rest energy	0.510 999 06(15) MeV
Charge-to-mass ratio e/m_e	$1.758\,819\,62(53) \times 10^{11}$ C/kg

*The parentheses show the uncertainty in the last two figures.

spheres, the cube root of 1.72×10^{-29} m^3 should be roughly the diameter of one such atom; that is, 0.26 nm.

29.3 X-Rays

Wilhelm Conrad Röntgen was fifty-five in 1895, a well-respected professor of physics at Würzburg. He was a private person who conducted his research quietly, almost secretively, so much so that his students nicknamed him *der Unzugängliche,* the unapproachable one. He had become interested in the work done on cathode rays by Hertz and Lenard, especially the latter's study of the passage of the rays though an aluminum-foil window and thence outside beyond the vacuum tube into the air. Lenard had surrounded his apparatus with lead and iron sheet to shield it from light and electric fields, and Röntgen also covered the tube, but luckily he used only thin black cardboard. In the darkened room, he noticed that one of his luminescent screens (a piece of paper coated with a barium salt) some distance away glowed brightly for no apparent reason. Astonished, he explored the effect until he was convinced that it was real. After a while, the cause seemed clear: some new invisible penetrating radiation was being emitted from the discharge tube. He called it **X-rays**, because x in mathematics represents an unknown.

Like light, X-rays expose photographic film, and because of their tremendous penetrating power, he was able to produce "photographs . . . of the shadows of the bones of the hand." Though the rays could ionize gases, they were similar to light in that they were not deflected by either electric or magnetic fields. To prove that X-rays were waves and not particles, Röntgen tried but failed to observe refraction, specular reflection, and polarization. Still, J. J. Thomson, among others, favored what we know to be the proper interpretation; namely, that X-rays are very short-wavelength electromagnetic radiation ($\lambda \approx 0.1$ nm). The correct wave picture was only slowly gaining acceptance when C. Barkla, in 1906, managed to observe the partial polarization of X-rays, thus establishing their transverse nature (see Discussion Question 7). Still, the wavelength of the radiation needed to be found.

X-Ray Diffraction

The question of the wavelength of X-rays was settled in 1912 by the German physicist Max von Laue (pronounced *fun lowùh*). He reasoned that X-rays could not be

Van Dyck's painting of *Saint Rosalie Interceding for the Plague-Stricken of Palermo,* as viewed in reflected light and in transmitted X-rays. Because the X-rays pass through the painting to reach the photographic film placed behind it, they reveal information about what is beneath the surface. (See photos on p. 1123.)

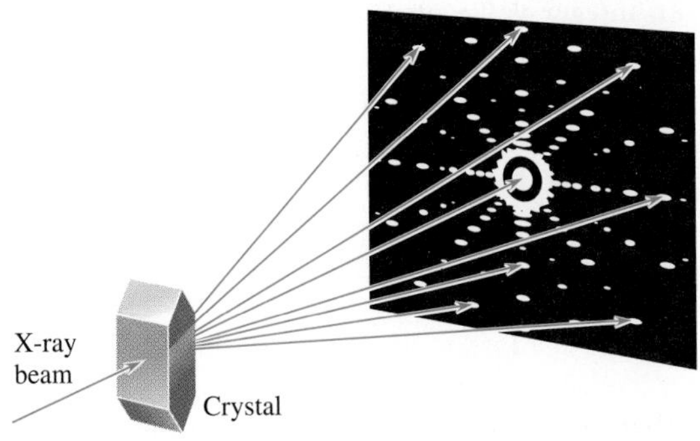

Figure 29.6 X-ray diffraction. A narrow X-ray beam enters a crystal, which diffracts it into many exiting beams. The process is akin to the operation of a 3-dimensional transmission grating.

diffracted by ordinary gratings because their wavelengths were much too small. As a rule, if an array of scatterers is to produce an observable diffraction pattern, their spacing must be of the order of the wavelength of the wave. Von Laue proposed that the ordered array of atoms in a crystal (which at the time were correctly believed to be separated by about 0.1 nm to 0.2 nm) might serve to produce diffraction. The atoms would scatter X-rays in all directions, but because of the regularity of the array, the waves traveling in certain directions would interfere constructively.

A crystal was placed in a narrow X-ray beam in front of a photographic plate (Fig. 29.6). If all went according to plan, the crystal would produce a number of off-axis diffracted beams, and an ordered pattern of dots would appear in the photo. The very first attempt was a resounding success, "proving" both the transverse nature of X-ray waves and the periodicity of the atoms in a crystal. The published accounts of this work caught the attention of W. H. Bragg, professor of physics at the University of Leeds, and his son, W. L. Bragg, a student at Cambridge. They devised a simpler diffraction process now known as Bragg scattering.

Figure 29.7 shows a regular array of atoms. We can imagine various sets of planes passing through the array in lots of different directions and containing different groups of atoms. Each such plane scatters rays in specific directions. For simplicity, let's deal with the set of horizontal planes. A parallel beam strikes the atoms and is scattered every which way. We have already seen that, for each plane, the reflected radiation will reinforce only in the direction such that it leaves the plane at the same angle at which it entered. That was what gave rise to the Law of Reflection (and it's true separately for each plane in the diagram). Thus, in Fig. 29.8 (p. 1024), ray 1 and ray 2 correspond to waves that will exactly reinforce each other. The question now is, What is the condition for rays from the second plane (and all others as well) to reinforce those from the first plane?

The waves associated with ray 1 and ray 3 will be in-phase, provided that their path-length difference is a whole number of wavelengths. Ray 3 travels farther than ray 1 by a distance equal to $2d \sin \theta$. For special values of θ; namely, θ_m, $2d \sin \theta_m$ equals $m\lambda$, where m is a whole number (1, 2, 3, . . .). It follows that constructive interference (Fig. 29.8b) occurs between waves from the various planes only when

$$2d \sin \theta_m = m\lambda \qquad (29.3)$$

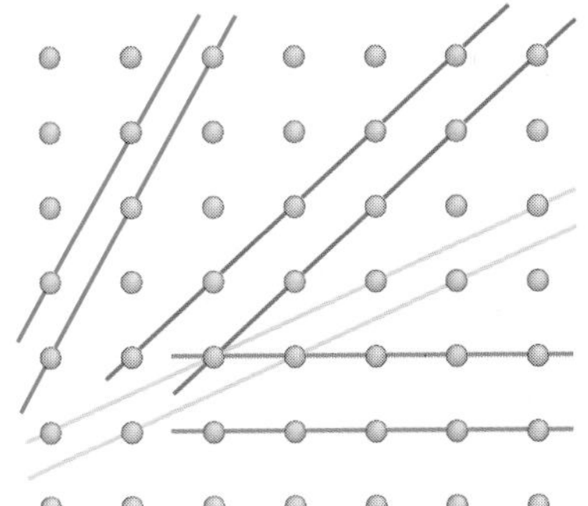

Figure 29.7 Several of the atomic planes off which X-rays may scatter. Notice that d is different in each case.

This formula has come to be known as the **Bragg equation**. Consistent with the equation (for a fixed d and λ), there may be several angles at which diffraction can occur, each corresponding to an integer value of m known as the *order.* Although the angle of incidence equals the angle of reflection as with light, *only those incident angles satisfying the Bragg equation will result in reflection (or diffraction),* and even then the exiting beam will be quite weak.

Since $\sin\theta$ must be equal to or less than 1

$$\frac{m\lambda}{2d} = \sin\theta_m \leq 1$$

hence with $m = 1$

$$\lambda \leq 2d$$

Generally d is at most 0.3 nm, which sets a practical limit of 0.6 nm on λ.

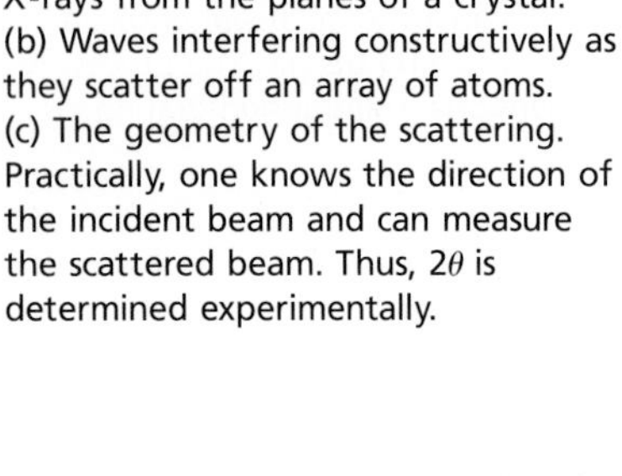

Figure 29.8 (a) The scattering of X-rays from the planes of a crystal. (b) Waves interfering constructively as they scatter off an array of atoms. (c) The geometry of the scattering. Practically, one knows the direction of the incident beam and can measure the scattered beam. Thus, 2θ is determined experimentally.

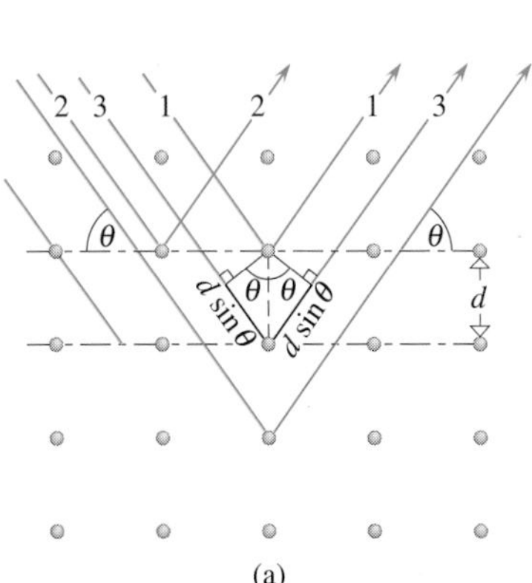

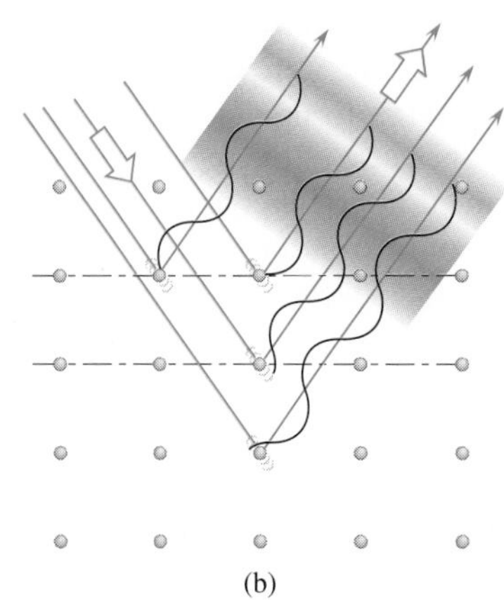

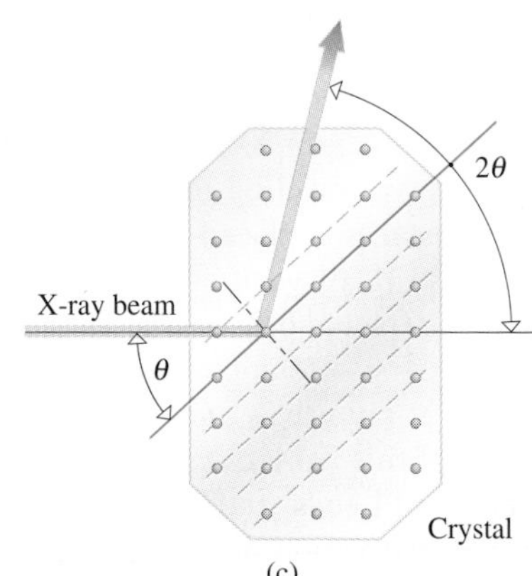

Example 29.2 A beam of quasimonochromatic X-rays with a wavelength of 0.15 nm is incident on a crystal. What is the space between atomic planes if the first-order diffraction maximum occurs at $\theta_1 = 16.0°$?

Solution: [Given: $\lambda = 0.15$ nm, $m = 1$, and $\theta_1 = 16.0°$. Find: d.] From the Bragg equation

$$d = \frac{m\lambda}{2\sin\theta_m} = \frac{(1)(0.15\text{ nm})}{2\sin 16.0°} = \boxed{0.27\text{ nm}}$$

▶ **Quick Check:** The answer is the right order-of-magnitude (that is, a few tenths of a nanometer). $2d\sin\theta_m = 1.49 \times 10^{-10}\text{ m} = m\lambda$.

X-rays are generated when high-speed electrons crash into a material target and rapidly decelerate, thereupon emitting radiant energy. A modern X-ray tube (Fig. 29.9) boils electrons out of a hot metal filament in a process called *thermionic emission* (discovered by Edison while he was working with light bulbs). These electrons are then accelerated across a potential difference of from 10^4 V to 10^6 V, in vacuum, whereupon they impact on a metal anode, decelerate, and radiate.

X-ray diffraction has become a routine technique for determining crystal structure. Equally as important, it has become a powerful tool for studying molecules, especially in the case of large repeating configurations such as proteins and DNA. Indeed, it was with the guidance of X-ray data that J. D. Watson and F. H. C. Crick were able (1953) to figure out the double-helix structure of DNA.

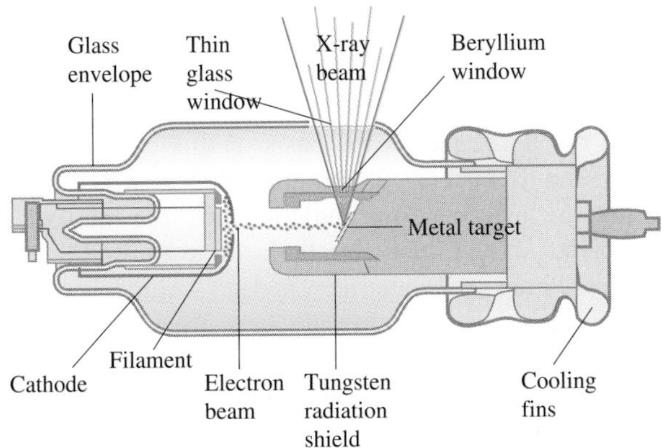

Figure 29.9 A modern X-ray tube. A beam of electrons crashes into a metal target, and X-rays are emitted.

29.4 The Discovery of Radioactivity

Röntgen's chance discovery of X-rays sent much of the scientific community into a flurry of activity. Henri Becquerel, in France, returned to his father's earlier work on fluorescent substances. These absorb and re-emit light, as does the pale green paint on the hands of a watch. The emission of X-rays had been related to the glowing region of the Crookes tube, and Becquerel wondered whether his glowing materials might also send out X-rays. As fate would have it, he began to work with one of his father's compounds, potassium uranylsulfate—a *uranium* salt. He exposed the salt to sunlight, and it actually did emit radiation that could penetrate the heavy paper wrapping on a photographic plate and fog the film.

On a sunless day in the winter of 1896, Becquerel placed the fluorescent compound atop a wrapped film plate as usual but, this time, he put it aside in the darkness of a drawer to wait for clearer skies. A few days later, almost as an afterthought, he developed the plate; there, amazingly, was the blackened outline of the piece of uranium salt! Prior exposure to light was apparently quite unnecessary—the radiation was emitted without it, without the fluorescence. By accident, he had discovered what Madame Curie would later name **radioactivity**, the property of certain substances to give out, all by themselves, penetrating radiation.

Entitled "Radium," this lithograph appeared in the December 22, 1904, issue of *Vanity Fair.* It depicts the Curies, Pierre holding a sample of the glowing element and Marie standing properly behind him.

Becquerel studied the radiation and reported several of its properties (a few of them erroneously), including that it could discharge a charged electroscope, as could X-rays. By 1897, he turned to other things. Unlike X-rays, these new emanations couldn't produce those exciting pictures of bones, and they were generally greeted with indifference. This limbo of neglect lasted for about a year and a half until the problem was taken up by a promising young student at the Sorbonne in search of a Ph.D. problem. Marie (Manya) Sklodovska Curie accepted the challenge and began her life's work.

Mme. Curie determined that thorium, like uranium, was radioactive. She and her husband, Pierre, discovered an element about 400 times more active than uranium and called it *polonium* (the name deriving from Marie's homeland, Poland). No one knew at the time that the radiation was essentially debris flung out from the nucleus of an unstable atom as it spontaneously readjusted itself. Even so, Marie was the first (1898) to report "that radioactivity is an atomic property." Working for 45 months under the most dreadful conditions in an abandoned, leaking wooden shed, the Curies finally isolated yet another radioactive element: *radium.* After four years of incredible labor, they had distilled from over a ton of the uranium ore a mere 1/30 ounce of the pure radium salt. But this new element was two million

Marie Sklodovska Curie (1867–1934). She is the only person to win Nobel prizes in physics and chemistry: in 1903 for her work on radioactivity, and in 1911 for her discovery of two new elements.

times more radioactive than uranium. Over the years, a single ounce of radium would spew out an amount of energy equal to that of burning ten tons of coal. That quietly blazing precious salt stays warm from its own internal heat and glows, self-luminous, like a firefly. The term *atomic energy* entered the vocabulary of science.

Marie received her Ph.D. at the Sorbonne on the basis of this epochal work. Within six months she, Pierre, and Becquerel shared the 1903 Nobel Prize in physics. Pierre was probably already suffering from radiation sickness (he used to keep the vials of radium in his pockets), and they were both too exhausted to even attend the ceremony in Stockholm. Three years later he was dead—run down in the street by a horsecart. In 1934, after a long illness, Marie Curie (the only person to ever win Nobel Prizes in both physics and chemistry) died of leukemia, a victim of many years of overexposure to radiation—too close to the firefly. Even the pages of her lab notebook were later found to be contaminated with radioactive fingerprints.

Alpha, Beta, and Gamma

Ernest Rutherford was twenty-four when he arrived at the Cavendish Laboratory in 1895. There, he began to study "Becquerel rays" and found that they were "complex, and that there are present at least two distinct types of radiation—one that is very readily absorbed, which will be termed for convenience α [**alpha**] radiation, and the other of a more penetrative character, which will be termed the β [**beta**] radiation." Only a short time before, Sir William Ramsay had discovered helium on Earth in a uranium-bearing mineral (and, along with F. Soddy, determined that it was liberated from radium). Thus, there was an early connection between α-rays and helium. (An alpha particle is a helium nucleus, two protons and two neutrons, but that would not be known for years.)

In 1903, Rutherford succeeded in deflecting α-rays using strong electric and magnetic fields, thus "proving" that they were positively charged particles (Fig. 29.10). Measurements of the charge-to-mass ratio of α-particles produced a value several thousand times smaller than that of the electron, again suggesting a large mass. Working with a young Ph.D., Hans Geiger (of Geiger counter fame, p. 1034), Rutherford concluded that the α-particle carried a charge of $+2e$, that is, twice the positive fundamental charge (twice the charge of the hydrogen ion).

In a beautiful experiment (1908), Rutherford and Royds captured alpha particles (Fig. 29.11). Alpha-emitting radon gas was collected above mercury in a thin-walled tube. After a week, the alpha particles that had passed into the surrounding vacuum region were pushed up into a capillary tube by raising the mercury level. On electrically exciting the accumulated gas, they found it to produce the characteristic emission spectrum of helium. It was not yet clear what atoms were, but "the α-particle, after it has lost its positive charge [by gaining two orbital electrons], is a helium atom." **An alpha particle is the nucleus of the helium atom.**

Ernest Rutherford (1871–1937) came to the Cavendish in 1895. Fresh from New Zealand, this unpolished colonial was the first of the new research students and the first to work with J. J. Thomson himself. Rutherford received the Nobel Prize in chemistry, strangely enough, in 1908.

The work on β-rays progressed quickly in a number of European laboratories: they were deflected by magnetic fields (1899), found to have a negative charge (1900), revealed to possess a charge-to-mass ratio very near that of cathode rays (1900), and finally were determined to have a mass equal to that of the electron (1902). **Beta rays are electrons.**

P. Villard, in Paris, also studied the emissions from radium and, in 1900, reported the presence of highly penetrating rays. They were not even slightly deflected by strong magnetic fields, thus suggesting a kinship with light. Gradually, the evidence grew that this γ (**gamma**) radiation was electromagnetic; somewhat more energetic and shorter in wavelength, but otherwise identical to X-rays. The is-

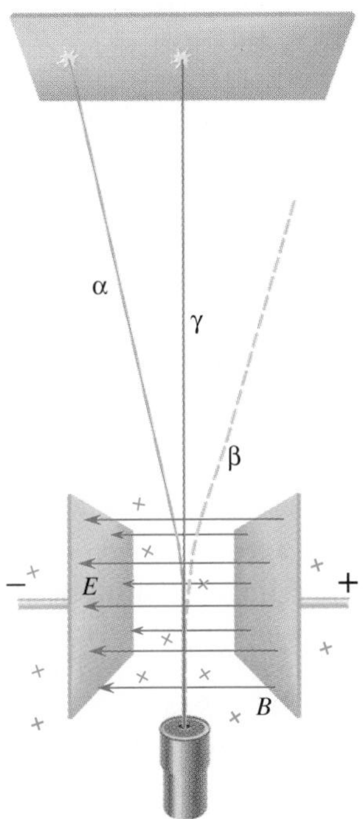

Figure 29.10 Alpha and beta rays, as charged particles, can be bent off course by both electric and magnetic means. Gamma rays, being electromagnetic radiant energy, are unaffected by these fields.

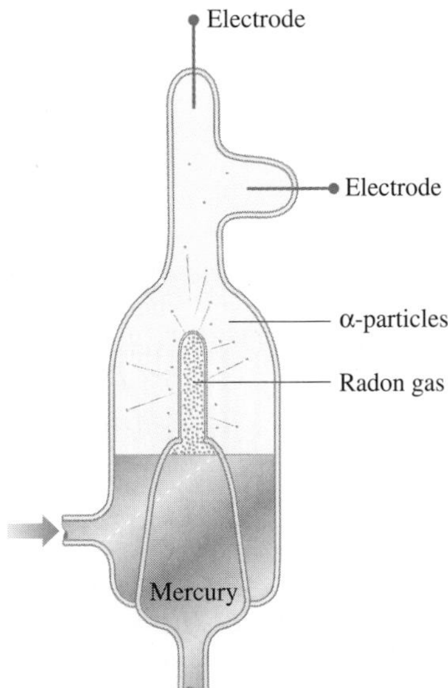

Figure 29.11 Radioactive radon in the thin-walled chamber emits α-particles, which are captured within the thick outer vessel. When the mercury level is raised, the alphas are forced into the discharge tube at the top. There, a helium spectrum is produced.

sue was settled in 1914 when Rutherford and Andrade succeeded in reflecting γ-rays off the surface of a crystal.

Gamma photons (typically 0.01 MeV to 10 MeV) are highly penetrating. They can be completely absorbed only after passing through several feet of concrete or about one to five centimeters of lead. More potent (and more dangerous) than X-rays, gamma rays have no trouble whisking through a human body and destroying molecules along the way. Beta particles (0.025 MeV to 3.2 MeV), traveling at roughly 25%c to 99%c, can sail through upwards of 15 meters of air. A millimeter or so of aluminum will block them totally, and they will not burrow very deeply into people ($\approx$1 mm to 2 cm). They are about 100 times more penetrating than alphas, though far less so than gammas.

Alpha particles (4 MeV to 10 MeV) have a mass of 6.642×10^{-27} kg, roughly 7300 times that of an electron. They're ejected from atoms at formidable speeds (14×10^3 km/s to 22×10^3 km/s)—the alpha particles from radium are emitted at $\approx$15 190 km/s. Easily stopped, alphas will barely make it through a single sheet of paper, or 0.3 cm to 8.6 cm of air. Even so, breathing in radioactive dust can put alpha emitters inside the lungs, where they can be quite lethal. They are very highly ionizing (1000 times more so than β-rays). A 5-MeV alpha can create 40 000 ion pairs in the process of traversing 1 cm of dry air. As a result, alphas lose their energy rapidly and are especially hazardous to biological organisms (perhaps 20 times more damaging than beta or gamma radiation). Only very energetic alphas

A smoke alarm. A small amount of a radioactive α-emitter ionizes the air between two parallel metal plates. A voltage across the plates causes the ions to drift, creating a current. The presence of smoke cuts off the current and triggers the alarm.

TABLE 29.2 Ionizing radiation sources in the United States

Sources	Percent
Human activity	Total 18%
Medical and dental X-rays	11
Nuclear medicine	4
Consumer products	3
Occupational	0.3
Fallout	<0.3
Nuclear fuel cycle	0.1
Miscellaneous	0.1
Natural sources	Total 82%
Radon	55
Internal body emission	11
Terrestrial minerals	8
Cosmic rays	8

(>7.5 MeV) can penetrate human skin, and external radiation is usually not a problem (Table 29.2).

THE NUCLEAR ATOM

Today, we know beyond any doubt that an atom is composed of a central dense core of positive charge—the nucleus—surrounded by a fast-moving cloud of electrons. This nuclear model of the atom was advanced by Rutherford, but others before him paved the way. Earnshaw (1831) had argued that no system of charges interacting via an inverse-square-law force could be in a stable static equilibrium—if they're bound by a Coulomb force, *the particles that constitute the atom must be in motion.* It followed from the cathode-ray experiments that those particles "must" be electrons. J. Larmor (1900) proposed a vague scheme with *electrons attracted to a central positive charge* and, soon after that, J. Perrin (1901) put forth a Solar System model with *orbiting electrons.* But as we'll see, there were serious problems with the stability of all such systems—*orbiting electrons are accelerating and so should radiate,* lose energy, and collapse into the positive center.

Lord Kelvin in 1902 proposed that the atom might be imagined as a jellylike sphere of positive charge, embedded throughout with an equal negative charge in the form of electrons, like the raisins in a glob of pudding. The following year J. J. made a thorough investigation of the stability of such a scheme, and as a result, it came to be known as the "Thomson atom." This *raisin-pudding atom* was widely considered, for want of anything better.

29.5 Rutherford Scattering

One day in early 1909, Geiger went to Rutherford to ask if his student, an undergraduate named Marsden, might be allowed "to begin a small research." Being of a similar mind, Rutherford suggested a simple scattering experiment, whose results he was sure he knew beforehand. Rutherford had been the first to scatter α-particles from matter, and Geiger had done some work in the area as well. The idea was to study the way alpha particles were deflected as they traversed a thin foil, and from that, possibly learn something about the hidden structure of the atoms of the target.

A few milligrams of a radium compound in a hollow lead tube served as the gun, shooting α-particles in a well-defined beam. The target, an exceedingly delicate gold foil about 0.000 06 cm (about 20 millionths of an inch) thick, corresponded to only $\approx$1000 layers of atoms. Alphas that easily penetrated the foil would slam into a distant zinc sulphide screen (Fig. 29.12). With each impact, the screen would give off a minute flash of light that could be seen in a totally dark room and counted with great effort. (The experiment is nerve-wracking, to say the least.) Positioning the detector straight down in the forward direction, the experimenters had found that very few alphas were even slightly deflected away from the original beam, which for the most part went right through undeviated. That finding was reasonable enough; after all, the alpha particles were very massive and moving exceedingly fast. They would presumably sail right through the tenuous positive pudding and could hardly be pulled aside appreciably by the minute electron raisins

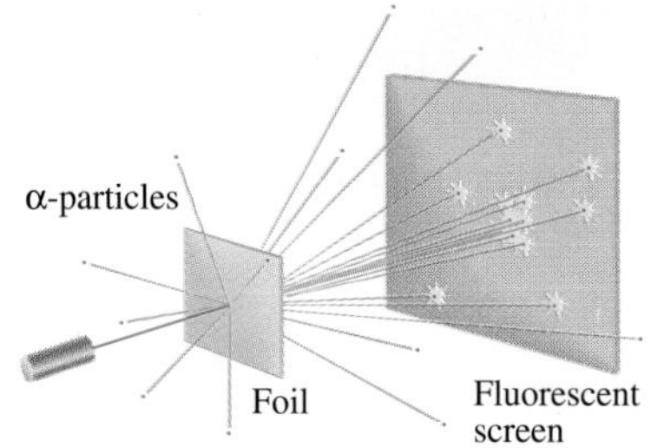

Figure 29.12 Rutherford's alpha particle scattering experiment.

(Fig. 29.13). So, when Rutherford suggested that Marsden look for alphas scattered through large angles (greater than 90°), he knew the eager undergraduate simply wouldn't find any.

Two or three days later, Geiger rather excitedly rushed back to "Papa," as they called the great man in those days, to tell him the unbelievable news: "We have been able to get some of the alpha particles coming backwards." Rutherford recalled, "It was almost as incredible as if you fired a 15-inch shell [that is, in diameter] at a piece of tissue paper and it came back and hit you." It took about two years for Rutherford to work out the details of a theory that would explain those extraordinary observations (Fig. 29.14).

It was clear from the start that each alpha particle could be blasted backward toward the source by a head-on collision with a concentrated, highly charged, positive, massive object—the **atomic nucleus**. We know from the study of elastic collisions (p. 306) that a projectile (the α-particle) will bounce back off a target (a nucleus) only when its mass is exceeded by that of the target (Fig. 9.20b). This kind of collision must be a very rare event (the nucleus must be exceedingly small, or else the undeviated penetration of the bulk of the beam would never occur). To be sure, the back-scattering happened only about once in every 10 000 or 20 000 impacts. Nonetheless, there were many millions of alphas being fired (a gram of radium undergoes about 4×10^{10} atomic disintegrations every second).

"Papa" saw the connection with the arching flight of a comet in the Sun's attractive gravitational field. In time, he was able to derive an equation for the scattering that would occur in a repulsive interaction between a target and projectile when both were positively charged. The α-particles sail off in hyperbolic orbits (Fig. 29.14). One morning in 1911, Rutherford happily sauntered into Geiger's lab to share one of the great secrets of the Universe: *Each atom consists of a tiny massive concentration of positive charge, the nucleus* (a term he introduced in 1912), *surrounded by a distribution of electrons.* Geiger immediately began to test the theoretical predictions—the specific dependence of the scattering on foil thickness, nuclear charge, alpha velocity, etc. Within a year, his measurements would convincingly bear out the power of the image, however fuzzy: the nuclear atom had arrived.

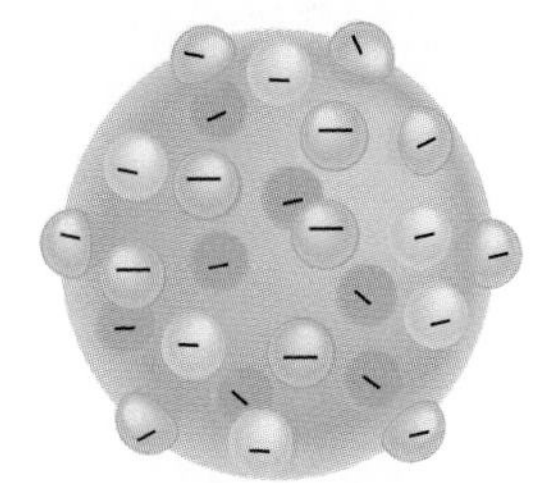

(a)

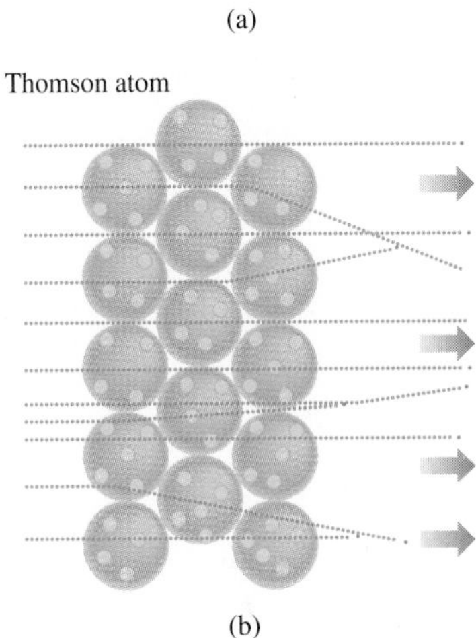

(b)

Figure 29.13 (a) The raisin-pudding atom—negative electrons embedded within and throughout a large positive fluff. In its simplest form, the electrons hover in circular patterns within fixed planes. (b) Bombarding α-rays should pass right through the raisin-pudding atom with very little deflection.

The Size of the Nucleus

Atoms were known to be about 10^{-10} m across (p. 322). Now Rutherford's scattering experiments provided a means of approximating the size of the nucleus. Consider a head-on collision between an α-particle and a nucleus, which we suppose to contain a number of positive charges Z, each of magnitude e. The α-particle initially sails in with a kinetic energy of

$$\mathrm{KE}_\alpha = \tfrac{1}{2} m_\alpha v_\alpha^2$$

and it gradually slows down as it approaches closer and closer to the repelling nucleus. At any distance r from the nucleus, the electric potential is given by Eq. (18.8); namely

$$V = \frac{kQ}{r}$$

where $Q = Ze$, the nuclear charge (assumed to be a point, which it isn't). Thus, at any distance r, the alpha has an electric potential energy (gravity is negligible) equal

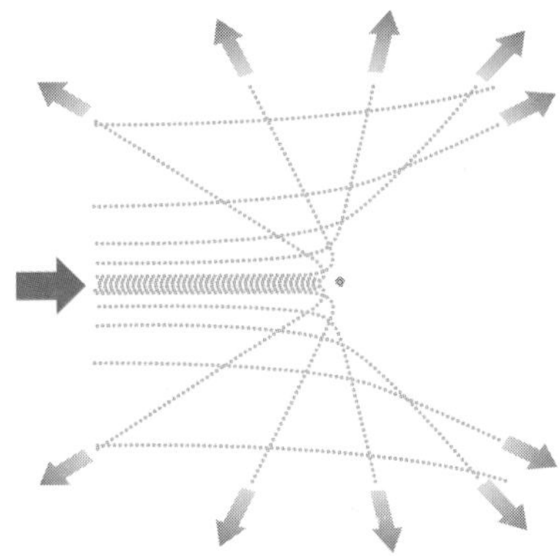

Figure 29.14 The scattering of a beam of α-particles by a positive massive nucleus.

to its charge $2e$ times V; that is

$$\mathrm{PE}_\alpha = \frac{k(Ze)(2e)}{r}$$

This is only approximate since it neglects the cloud of orbital electrons (which will soon be penetrated anyway). The alpha will reach a point of closest approach to the nucleus at a distance $r = R$, where it momentarily stops (as if it had compressed a spring and is about to pop backward). At that point, its initial KE has been transformed into PE:

$$\mathrm{KE}_\alpha = \mathrm{PE}_\alpha$$

$$\tfrac{1}{2}m_\alpha v_\alpha^2 = \frac{k2Ze^2}{R}$$

and

$$R = \frac{4kZe^2}{m_\alpha v_\alpha^2} \tag{29.4}$$

This distance represents an upper limit on the nuclear radius. Thus, given $v_\alpha = 1.5 \times 10^7$ m/s and $m_\alpha = 6.6 \times 10^{-27}$ kg, we have to come up with a value of Z. At the time when Rutherford was doing these experiments, it was reasonable to guess that (as with the α-particles themselves) a nucleus would contain a number of charges equal to half the atomic weight; that is, $Z = \frac{1}{2}197 \approx 99$. Actually, for gold, $Z = 79$, but that knowledge would have to wait until the work of Moseley (p. 1065). We can use either number; the choice will not affect things much. Hence

$$R = \frac{4(9.0 \times 10^9\ \mathrm{N{\cdot}m^2/C^2})(79)(1.60 \times 10^{-19}\ \mathrm{C})^2}{(6.62 \times 10^{-27}\ \mathrm{kg})(1.5 \times 10^7\ \mathrm{m/s})^2} = 4.9 \times 10^{-14}\ \mathrm{m}$$

The nucleus is about ten thousand times smaller than the atom. Thus, if in imagination the nucleus of a typical atom is enlarged to the size of an apple (10 cm), the atom as a whole would be about 1 km across, most of it a vast emptiness.

In several regards Rutherford was quite lucky. He happened to be using both low-energy alphas ($\approx$5 MeV) and targets with large nuclear charges. As a result, the alphas had enough kinetic energy to easily pierce the surrounding atomic electron clouds (which could then be neglected) rather than being appreciably scattered by them. Still, the alphas were not energetic enough to penetrate deeply into the so-called *nuclear Coulomb barrier*. In other words, working against the Coulomb repulsion, they never got much closer than about 5×10^{-14} m to the center of the nucleus. Had the α-particles been traveling with much more initial energy (say, around 25 MeV or so) Rutherford's results would have been very different; at shorter and shorter distances, the scattering deviates more and more from a purely Coulomb interaction (p. 1108). What he could not have known was that there is an additional nuclear force lurking at very short ranges and that its influence would have messed up his neat, simple understanding of what was happening; that challenge would come, but happily only later.

Another bit of good fortune for Rutherford was that the alphas were moving slowly enough to be in the classical rather than the relativistic regime. Given this case, the correct quantum-mechanical behavior for a Coulomb potential is identical to the predicted classical scattering. In short, Rutherford was lucky, and we, rather fortuitously, were given the nuclear atom.

Rutherford never satisfactorily addressed three outstanding issues concerning his model: the question of the stability of the atom, the necessity to explain atomic spectra, and the need to understand the Periodic Table within the context of atomic structure. That too would come later.

29.6 Atomic Spectra

In the late 1800s, there developed a body of research called *spectroscopy* that provided an accurate means of distinguishing between atomic species. The formulas describing the patterns of spectroscopic phenomena are of little interest to us in themselves. But they reflect, with extraordinary precision, the hidden structure of the atom. Ultimately, those formulas will serve as one of the most powerful tests of atomic theory (p. 1057).

Spectroscopy began with Newton and his experiments using white light and prisms, but it didn't become an analytic tool for exploring matter until the 1800s. A. J. Ångström, in 1853, used a discharge tube filled with various gases to study their spectra. Light from a specimen is made to pass through a slit in a screen and then through a prism or grating that separates the narrow beam into its constituent color bands, the **spectral lines** (Fig. 29.15). When a gas is excited, it emits specific wavelengths, and we see colored lines on a black background. This is known as the **emission spectrum** (Fig. 29.16). Inversely, when white light passes through the same gas, the atoms absorb those same specific wavelengths. This is the **absorption spectrum**, and we see black lines on a smoothly varying, bright-colored background. Ångström first measured the wavelengths of the four bright visible emission lines of hydrogen. Because hydrogen is the simplest of all the elements, these data would play a special role in the later development of theoretical atomic physics. J. Plücker gave the lines the names appearing in Table 29.3 and, by 1858, he had rightly suggested that the spectrum of any substance was a fingerprint that unambiguously specified its identity.

The very existence of spectral lines (and the knowledge that light was an oscillatory wave of some sort) suggested that atoms had a complex structure that could sustain many different internal vibrations. Maxwell pointed *that* fact out in 1875 but, even before then, George Stoney noticed that the hydrogen spectral wavelengths formed simple ratios:

$$H_\alpha : H_\beta : H_\delta = \frac{1}{20} : \frac{1}{27} : \frac{1}{32}$$

Stoney talked about, "the 32nd, 27th, and 20th harmonics of a fundamental vibration" of the atom. In 1885, Johann Balmer, having investigated the notion of atomic

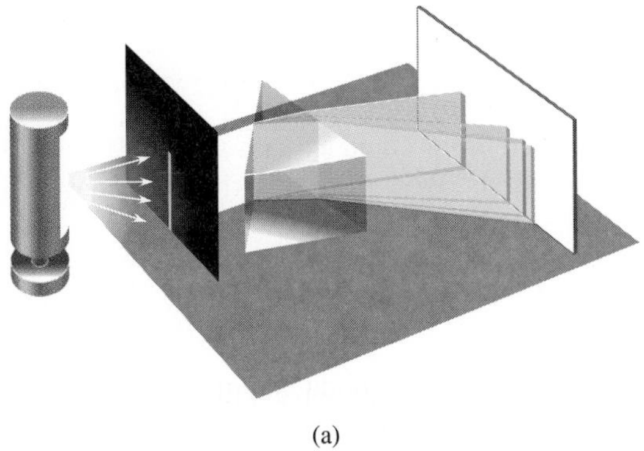

(a)

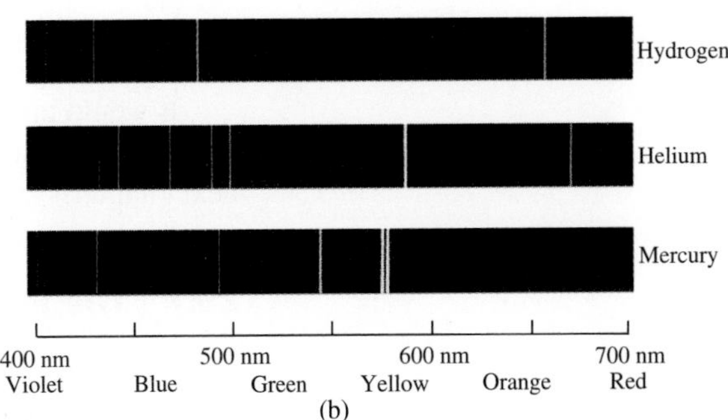

(b)

Figure 29.15 (a) The formation of spectral lines by a prism. (b) The emission spectra of, from top to bottom, hydrogen, helium, and mercury.

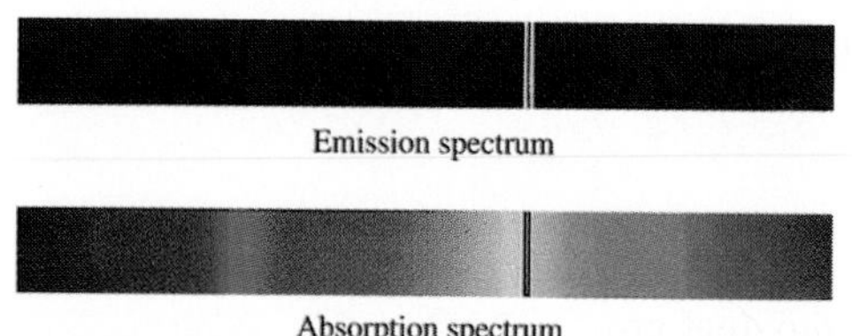

Figure 29.16 Emission and absorption spectra of sodium.

TABLE 29.3 Hydrogen visible spectra

Line	Wavelength (nm)	Color
H_α	656.28	red
H_β	486.13	blue-green
H_γ	434.05	violet
H_δ	410.12	violet

harmonics, published a simple formula that rather remarkably yielded the observed hydrogen wavelengths. In its modern form, it is

$$\frac{1}{\lambda} = R\left[\frac{1}{2^2} - \frac{1}{n^2}\right], \qquad n = 3, 4, 5, \ldots \tag{29.5}$$

The equation generates the wavelengths of the various *visible* lines—of what has come to be known as the **Balmer series**—when we substitute in turn, $n = 3$ or 4 or 5, and so on. When λ is expressed in meters, R, which is called the *Rydberg constant,* is

$$R = 1.097\,373\,15 \times 10^7\ \text{m}^{-1}$$

which is quite close to the number Balmer came up with. Note that the series continues with the wavelengths getting shorter and shorter as $n \to \infty$ at the *series limit* (Problem 19); in time, this too, was confirmed experimentally.

Example 29.3 Use Balmer's formula to compute the wavelength of the red line in the hydrogen spectrum.

Solution: [Given: Balmer's formula. Find: λ for red line.] The red line has the longest wavelength and the lowest n; namely, $n = 3$. Thus

$$\frac{1}{\lambda} = (1.097 \times 10^7\ \text{m}^{-1})\left[\frac{1}{4} - \frac{1}{9}\right]$$

and

$$\boxed{\lambda_\alpha = 656.3\ \text{nm}}$$

▶ **Quick Check:** This λ is appropriate for red light and, of course, it agrees with Table 29.3.

The first term in the brackets of Eq. (29.5) contains a denominator of 2^2, and Balmer was so sure of himself that he proposed (off the top of his head, as it were) that there might well be other sets of lines for 1^2, 3^2, 4^2, and so on. As luck would have it, he was essentially right, and in time the *Lyman series*

$$\frac{1}{\lambda} = R\left[\frac{1}{1^2} - \frac{1}{n^2}\right], \qquad n = 2, 3, 4, \ldots \tag{29.6}$$

in the ultraviolet was observed. As were the *Paschen series*

$$\frac{1}{\lambda} = R\left[\frac{1}{3^2} - \frac{1}{n^2}\right], \qquad n = 4, 5, 6, \ldots \tag{29.7}$$

and two others (with $1/4^2$ and $1/5^2$) in the infrared (p. 1061).

It would take until 1913 before Niels Bohr could explain spectral lines in terms of atomic transitions between energy levels. We will deal with that discovery in the next chapter.

29.7 The Proton

Recall the cathode-ray experiments in which a cathode and anode were placed in a tube containing a low-pressure gas. A high potential across the electrodes caused the gas to become ionized, and a beam of electrons accelerated away from the cathode

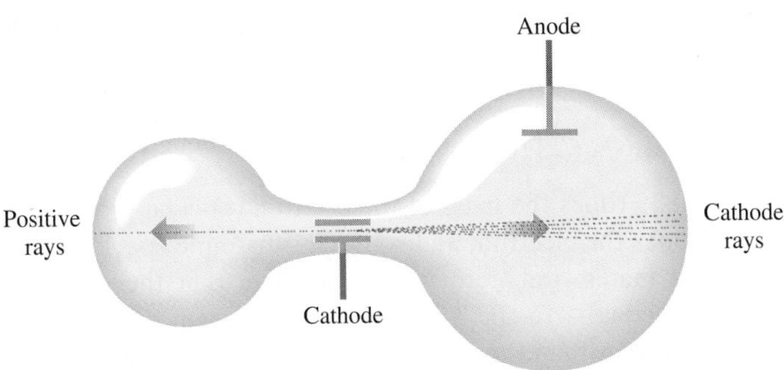

Figure 29.17 Cathode rays stream out toward the general location of the anode. Positive rays sail toward, through, and beyond the cathode.

out in the general direction of the anode (Fig. 29.17). It was noticed in 1886 that the glowing trail of a beam could also be projected in the opposite direction. The backward streaming of that emanation, along with the observation that it could be deflected by a magnetic field, was convincing evidence that these rays were positive particles; J. J. Thomson christened them *positive rays.* When their charge-to-mass ratio was measured, several rather low values were found (indicative of particles with large masses), on the order of those of the atoms themselves.

After the Rutherford model had been put forward, it was reasonable to suppose that positive rays were simply ionized atoms. Naturally enough, the e/m ratios depended on the particular trace gas in the tube. When Thomson performed the experiment with hydrogen (1907), he found two values of e/m, the larger for the light H^+ ion and the smaller for the ionized molecule H_2^+. The former compared nicely with the value arrived at via electrolysis. More and more it seemed that the positive nuclear particle (the thing many people called the H-particle) was the hydrogen atom stripped of its single orbital electron.

The clincher came when Rutherford, pounding the nuclei of various elements with alpha particles, discovered H-particles among the flying fragments. The apparatus (Fig. 29.18) consisted of an evacuated chamber containing an alpha source. The chamber was sealed with a piece of foil just thick enough to stop all the alphas. Nitrogen gas was then introduced into the chamber (later on, fluorine and other elements were used; in all cases, the results were the same). Scintillations were observed, indicating the presence of a more penetrating radiation that proved to have the same range and charge as the H-particle. The range in air was easily determined by moving the screen, and it was indicative of the kind of radiation being dealt with. Apparently, this positive particle must be a basic component of many if not all the elements. It wasn't until 1920 that Rutherford formally proposed the name **proton**

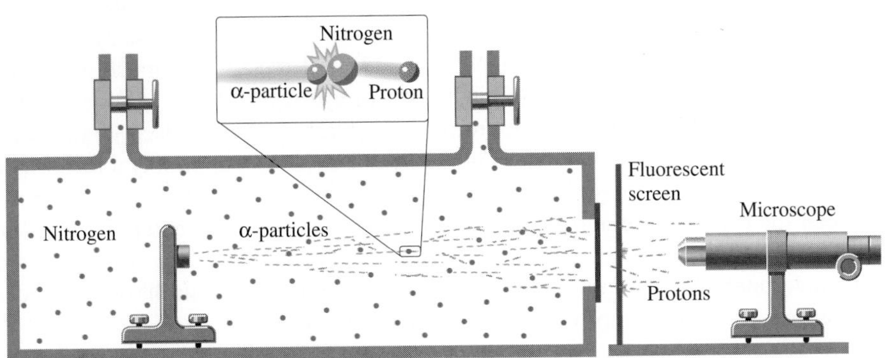

Figure 29.18 Alpha particles crashing into nitrogen atoms liberate protons, which strike the fluorescent screen, causing flashes that can be seen with a microscope. The chamber is sealed with a thin foil that will pass protons but not alpha particles.

(from the Greek *protos* for "first"), a word that had already quietly been in use for a dozen years.

With a mass of

$$m_p = 1.672\,623\,1(10) \times 10^{-27}\ \text{kg} = 938.272\,31(28)\ \text{MeV}$$

the proton (p) is 1836 times the mass of the electron (e). The proton carries a quantum of charge as does the electron, and so these subatomic specks logically seem, if not twins, at least mates. Laboratory tests confirm that the proton and electron charges are indeed equal—in the most precise measurement ever carried out, the hydrogen atom has been found to be neutral to an accuracy of twenty-two decimal places. This equality of charge becomes even more remarkable when we learn that the proton is a complex structured thing; in fact, it remains one of the fundamental puzzles of Modern Physics (p. 1148).

29.8 The Neutron

We picture the hydrogen atom as a single nuclear proton coupled by an attractive Coulomb force to one orbital electron. The next element in the Periodic Table, helium, might then consist of two protons and two surrounding electrons; lithium might have three and three; and so on, all the way up to uranium with 92 protons and 92 electrons. Yet, the α-particle has a charge of $+2e$ and an e/m ratio of half that of the hydrogen nucleus; it has a mass of 4. And this is *not* quite right; something else must be happening in the nucleus. Such differences between **atomic number** (Z) and **atomic mass** (A) occur throughout the Table all the way up to uranium, which has 92 nuclear protons and a mass of 238. There is a good deal of mass unaccounted for when the uranium nucleus is considered to be a cluster of 92 protons. If everything is made up of the only two then-known particles—electrons and protons—we have a little problem here.

Rutherford, in 1920, proposed that all of these difficulties could be resolved by assuming the existence of tightly bound *proton-electron pairs,* neutral units that would add mass but not charge. The idea was wonderfully simple, but quite erroneous. Rutherford even suggested the name *neutron* for the pair-particle. And he promptly set a group of his "boys" to work to track down the neutral object; thus, James Chadwick began a 12-year search.

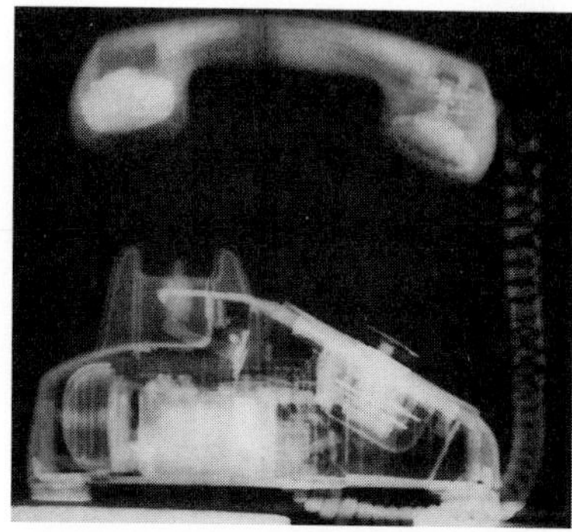

A radiograph of a telephone created with a beam of neutrons. These neutrophotos usually show considerably more of the subtle variations and details than do X-ray pictures, which they otherwise closely resemble.

In 1930, it was discovered that when beryllium (Be) was bombarded with alpha particles, out streamed a flux of very energetic radiation that was assumed to be γ-emission, but which seemed a bit odd in some respects. It was soon determined that, as with γ-rays, the new radiation was both very penetrating and not deflected by a magnetic field. However, unlike γ-rays, it was not ionizing and could not discharge an electroscope. The Joliet-Curies, Irène (Mme Curie's daughter) and her husband, Frédéric Joliot-Curie, observed that the presence of the mystery emission (which they thought was electromagnetic) could be detected in a much enhanced way by making it impinge on a substance rich in hydrogen, that is, protons. When the beam struck a sheet of paraffin, protons were blasted out, and these could easily be picked up with a Geiger counter (Fig. 29.19). (The latter was simply a metal tube containing a low-pressure gas such as argon. A voltage of about 1000 V was applied between the central-wire anode and the tube. Ionizing radiation entered via a thin mica window and knocked electrons out of a few gas atoms. These rushed toward the anode, ionizing more atoms in the process. The resulting discharge created a current pulse that was amplified and sent to a counter.)

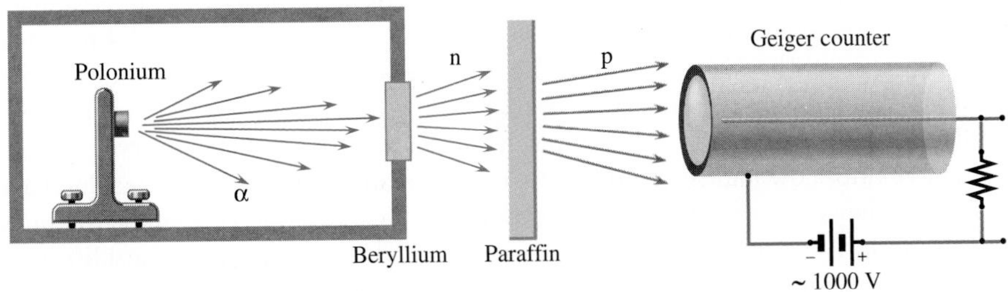

Figure 29.19 Polonium emits alphas, which knock neutrons out of the beryllium. These knock protons out of the paraffin.

Chadwick (1932) argued that a nonionizing particle capable of ejecting protons had to be the long-sought neutron (n). He could determine the speed of the protons from their range and was able to show from collision theory that all of his observations could be understood quantitatively by hypothesizing the presence of a neutral particle having about the mass of the proton. When the alpha bombardment of beryllium was finally understood, Chadwick wrote it out as

$$^{4}_{2}\alpha + {}^{9}_{4}\text{Be} \rightarrow {}^{12}_{6}\text{C} + {}^{1}_{0}\text{n}$$

(mass as the superscript, charge as the subscript). Beryllium absorbs an alpha and, with the emission of a neutron, is transformed into carbon.

Rutherford had been close (but no cigar): electrons don't live in the nucleus, and the neutron is as much a single particle as is the proton, though it is slightly more massive:

$$m_{\text{n}} = 1.674\,928\,6(10) \times 10^{-27}\ \text{kg} = 939.565\,63(28)\ \text{MeV}$$

Very soon after the announcement of the discovery of the neutron, Werner Heisenberg proposed the now accepted neutron-proton model of the atomic nucleus. **All nuclei are composed exclusively of neutrons and protons**, the total number of which (A) determines the atomic mass. Thus, uranium has 92 protons (an atomic number Z of 92) and 146 neutrons, yielding an atomic mass (A) of 238—everything works out neatly, at last.

Once removed from the nucleus, the neutron is unstable, decaying into an electron, a proton, and a neutrino (actually an antineutrino, p. 1116) at a rate such that half of the neutrons present at any moment will decay within roughly 10.4 minutes. This process is an example of β-decay, and it explains how electrons can be ejected from a nucleus wherein they do not exist—they are a byproduct of the transformation, created in the process. While nestled with the protons in the nucleus, the neutrons interact with (and seem to be more or less stabilized by) the others, so they don't decay in all nuclei. Moreover, neutrons and protons provide a glue (by way of the strong force) that helps keep the nucleus together. To underscore this intimate relationship (and because they are the exclusive components of all nuclei), the neutron and proton are called **nucleons**.

Neutrons are about 1.3 MeV/c^2 more massive than protons, and with this excess rest energy, it is the free neutron that decays "down" to the proton. Had the reverse been true, we could expect free protons to decay into neutrons. Happily for us, it's not the case, and hydrogen, water, and human beings are stable. (A proton can decay into a neutron but only in a fairly large nucleus that can provide the needed mass-energy.)

When nuclei decay, as when neutrons decay, the process is described statistically. We know that half the free neutrons in a box will decay after ≈10.4 minutes, but we have no idea which ones will be left. Each survivor seems identical, each seems unaffected by the passage of ≈10.4 min. The data suggests that the likelihood of a particular neutron decaying during any ≈10.4-minute interval is constant, no matter how many such intervals it has already survived. Neutrons don't age and, when they do decay, it is for no apparent reason. This strange result is the same one Rutherford observed for radioactive atoms, and it suggests that causality (and the accompanying predictability we are so fond of) has given way on some hidden level to the roulette wheel and the rule of probabilities. Today, it is widely believed that our inability to predict the individual behavior of a neutron or a nucleus is not the result of a correctable ignorance but is a fundamental manifestation of the quantum-mechanical nature of the Universe.

Core Material

A faraday of charge (96 485 C) corresponds to

$$\mathrm{F} = N_A e \qquad [29.1]$$

For single-valence elements, 9.65×10^4 C will liberate 1 mol, or equivalently, 9.65×10^7 C will liberate 1 kmol (p. 1016). The **charge-to-mass ratio** for singly ionized hydrogen (H^+) is

$$\left(\frac{e}{m}\right)_{H^+} = 9.58 \times 10^7 \text{ C/kg}$$

The charge-to-mass ratio of the electron is

$$\frac{e}{m_e} = \frac{2yE}{L^2B^2} \qquad [29.2]$$

The accepted value of that ratio for electrons is

$$\frac{e}{m_e} = 1.7588 \times 10^{11} \text{ C/kg}$$

where $e = 1.602\,177\,33 \times 10^{-19}$ C and $m_e = 9.109\,389\,7 \times 10^{-31}$ kg. The **Bragg equation** (p. 1023) for X-ray diffraction is

$$2d \sin \theta_m = m\lambda \qquad [29.3]$$

Each atom consists of a tiny massive concentration of positive charge, the nucleus, surrounded by a distribution of electrons (p. 1028). The nucleus is at least ten thousand times smaller than the atom. The **Balmer series** for the spectrum of hydrogen is

$$\frac{1}{\lambda} = R\left[\frac{1}{2^2} - \frac{1}{n^2}\right] \qquad n = 3, 4, 5, \ldots \qquad [29.5]$$

wherein $R = 1.097\,373\,15 \times 10^7 \text{ m}^{-1}$

The masses of the proton and neutron are

$$m_p = 1.672\,623\,1(10) \times 10^{-27} \text{ kg} = 938.272\,31(28) \text{ MeV}$$

and

$$m_n = 1.674\,928\,6(10) \times 10^{-27} \text{ kg} = 939.565\,63(28) \text{ MeV}$$

All nuclei are composed exclusively of neutrons and protons.

Suggestions on Problem Solving

1. Return to Fig. 29.8c and notice that the angle between the transmitted and diffracted beams is 2θ. It is this angle that is usually measured experimentally, and it is called the *diffraction angle.* Among the first things to determine in a given X-ray diffraction problem is which angle is being provided or asked for.

2. When dealing with the spectral series for hydrogen, the order is Lyman (1), Balmer (2), Paschen (3), Brackett (4), Pfund (5); these are the numbers that are squared in the first term of the denominators for $1/\lambda$ [that is, Eq's. (29.5–29.7)]. The first spectral line in any particular series is the next number up; for example, substituting $n = 3$ into Eq. (29.5) generates the first Balmer line. Remember to square those integers. Once again, we have one-over the quantity of interest (that is, $1/\lambda$), and it's common to forget to invert the result to get the final answer—be alert to this.

3. Remember that it's commonplace to specify the uniform electric field between parallel plates via $E = V/d$; that is, V and d are given, and it is assumed you therefore know E. Because d is small, it is often provided in centimeters; make sure to convert it into meters—failure to convert is a common error.

Discussion Questions

1. A molten bath of NaCl has a current passed through it. Chlorine gas bubbles off at the anode, and sodium is deposited at the cathode. Explain what's happening. Why did we start with molten NaCl rather than a water solution?

2. The chemical cell in Fig. Q2 depicts a silver nitrate solution, a silver anode, and a copper cathode. Explain what will happen when the switch is closed. How is it that the concentration of the solution remains unchanged? Describe the nature of the current both in the solution and in the external circuit. What, if anything, happens to the silver of the anode?

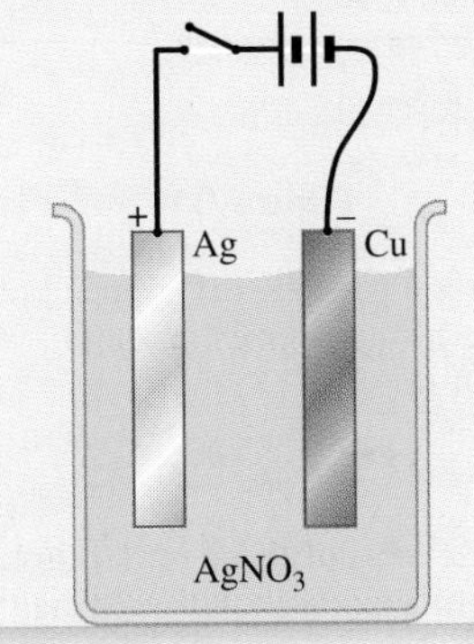

Figure Q2

3. Let's free our imaginations and speculate about the nature of the electron. Experiments to date have shown that if the electron has a "hard core," it's smaller than 10^{-16} cm. What are your feelings about what the electron "looks like"; that is, size, shape, charge distribution, etc? What questions immediately come to mind? Incidentally, no one knows what configuration it has or even if it has one. Nor do we know if it's finite or simply a point (whatever that means). We do believe that it is an elementary particle (that is, it has no smaller constituent parts). How does mass enter in your picture?

4. An equation for the mass M of material liberated at either electrode when a net charge of Q is passed across an electrolyte is

$$M = \frac{(Q)(\text{mass per mole})}{(\text{F})(\text{valence})}$$

Explain how this expression comes about.

5. Beyond their dynamical parameters (that is, velocity, momentum, etc.), is it reasonable to assume that all electrons are identical?

6. Compare X-ray diffraction from a crystal with the reflection of light from a smooth surface. How do the two processes differ? Although we speak of the "reflection" of X-rays, that description is somewhat misleading. Why?

7. Using detectors that were not themselves sensitive to polarization, Barkla (1906) performed a crucial experiment that showed that X-rays could be polarized. His arrangement consisted of two blocks of carbon as depicted in Fig. Q7 and an

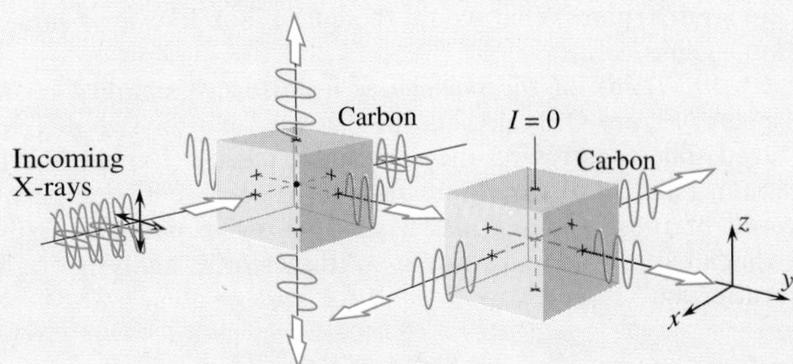

Figure Q7

incident unpolarized beam (entering at the left along the x-axis). From what you learned via Fig. 27.12, explain why the absence of scattered radiation from the second block outside the xy-plane proved the wave nature of the process.

8. Figure Q8 depicts a monochromatic beam of X-rays incident on a single crystal. Explain what is happening.

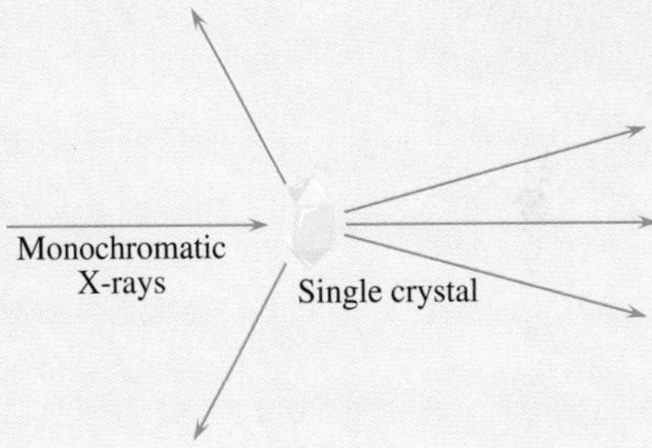

Figure Q8

9. Consider X-ray diffraction using both a single crystal and a chamber holding a monatomic gas. Figure Q9 shows the

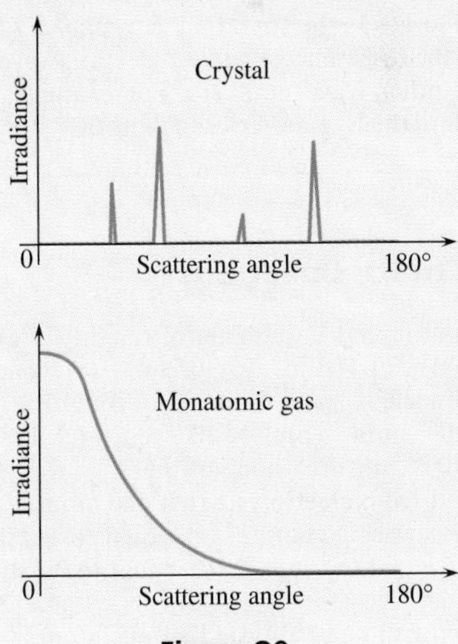

Figure Q9

irradiance reaching a detector as a function of the angle from the beam axis. Explain the features of the two curves and why they are so different. What would the curve look like for a liquid or amorphous solid?

10. Think of the nucleus as a uniformly charged positive sphere. Figure Q10a depicts the electric field for two different-sized spheres carrying the same total charge. Explain the diagram. Part (b) of the figure shows a head-on collision between each of these spheres and a small positive particle. Explain what's happening and relate it to Rutherford's analysis of alpha scattering.

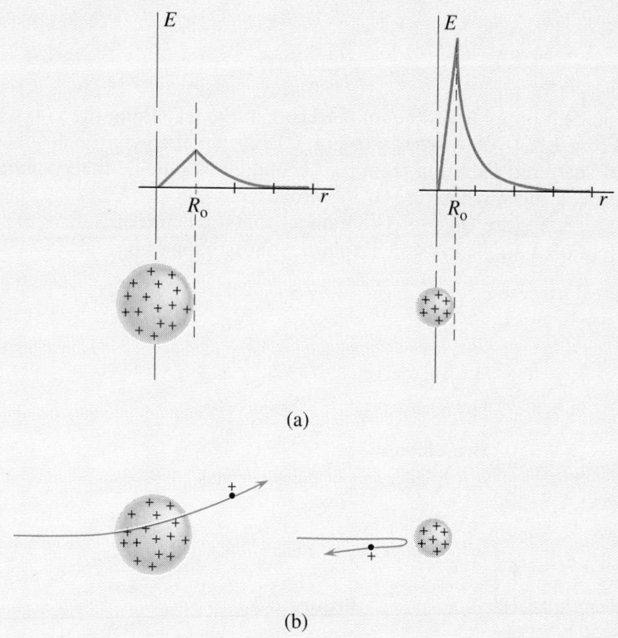

Figure Q10

11. Because of the form of Eq's. (29.5) through (29.7), it's useful to define $1/\lambda$ as the so-called *wave number.* It was noticed early on that the differences between certain wave numbers corresponded to other wave numbers. For example, the first four wave numbers in the Lyman series in units of $\times 10^3\ \text{cm}^{-1}$ are 82.3,

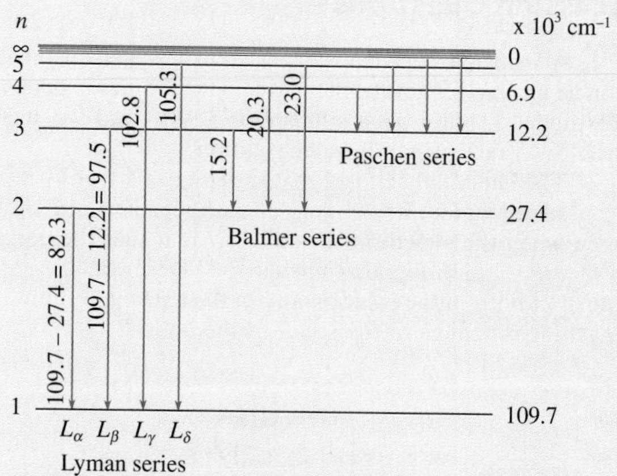

Figure Q11

97.5, 102.8, and 105.3. Thus, the difference between the first two wave numbers in the Lyman series is the H_α wave number in the Balmer series, which is

$$97.5 - 82.3 = 15.2$$

Figure Q11 displays several of these remarkable relationships and we see that the H_α line corresponds to what might be called a *transition* from the *3rd* to the *2nd* "level." **All the spectral lines can thus be seen as transitions from one level to another,** which is essentially the *Ritz Combination Principle,* first proposed in 1908. What is the significance of the lowest level (109 677), of the next lowest level (27 419), and so on? What does all of this suggest about what might be happening in the atoms giving off the light? (The Bohr atom, treated in Sect. 30.5, p. 1057, is predicated on similar reasoning.)

12. The electron, proton, and neutron all have magnetic properties; they each have a magnetic moment (p. 749) and each behaves like a tiny bar magnet. The electron's magnetic moment is roughly 666 times stronger than the proton's, which is about 1.5 times stronger than the neutron's. Experiments show that the magnetic moments of ordinary nuclei are comparable to those of the proton. Does this relationship suggest anything about electrons residing in the nucleus of an atom? Explain.

Multiple Choice Questions

1. A typical atom has a diameter of roughly (a) 0.2 mm (b) 0.2 pm (c) 0.2 m (d) 0.2nm (e) none of these.
2. A typical nucleus has a diameter of roughly (a) 1×10^{-14} mm (b) 1×10^{-14} cm (c) 1×10^{-14} m (d) 1×10^{-14} nm (e) none of these.
3. We learned from electrolysis that one atom of a univalent substance carries a charge (a) equal to F (b) equal to F/N_A (c) equal to $2F/N_A$ (d) equal to N_A/F (e) none of these.
4. We learned from electrolysis that one atom of a bivalent substance carries a charge (a) equal to F (b) equal to F/N_A (c) equal to $2F/N_A$ (d) equal to N_A/F (e) none of these.
5. The magnitude of the charge of the electron is (a) equal to F (b) equal to F/N_A (c) equal to $2F/N_A$ (d) equal to N_A/F (e) none of these.
6. When a monochromatic X-ray beam impinges on a crystal (λ and d are fixed), there may be several values of the (a) angle θ at which diffraction will occur, and these correspond to different values of m (b) crystal spacing at which the reflected beam angle will exceed the incident angle (c) energy at which the reflected beams emerges at an angle in excess of 90° (d) reflection angle for each incident angle and these correspond to different values of f (e) none of these.
7. The largest Bragg angle through which a beam of X-rays

can be bent is (a) 0° (b) 45° (c) 90° (d) 180° (e) none of these.

8. The light emitted via the excitation of low-pressure hydrogen gas is (a) a continuous spectrum (b) made up of a discrete spectrum of bright lines (c) a mix of bright spectral lines superimposed on a continuous bright background (d) composed exclusively of X-rays (e) none of these.
9. Naturally occurring radioactive atoms can spontaneously emit (a) δ-, γ-, and ξ-rays (b) α-, β-, and γ-rays (c) N-rays (d) X-rays (e) none of these.
10. Neutrons are highly penetrating because they (a) are very thin and can slide between atoms (b) never strike an atom because they twist as they advance (c) are neutral and do not lose energy via Coulomb interactions with the atoms (d) always travel at extremely high speeds in comparison to protons or electrons (e) none of these.
11. Gamma-ray emissions can (a) be distinguished from β-ray emissions using a magnetic field (b) not be distinguished from β-ray emissions (c) not be distinguished from α-ray emissions using a magnetic field (d) be distinguished from a neutron beam by bending the latter's path via a B-field (e) none of these.
12. An element emits spectral lines (a) that are exactly the same as all other elements (b) that are exactly the same as all other elements in its column of the Periodic Table (c) that are characteristic of that element (d) that are evenly spaced (e) none of these.
13. In reference to the Balmer series, the longest wavelength line (a) is associated with the smallest n, namely 1 (b) is associated with the smallest n, namely 3 (c) is associated with the largest n^2, namely 9 (d) is associated with the largest n namely ∞ (e) none of these.
14. In reference to the Balmer series, the shortest wavelength line (a) is associated with the smallest n, namely 1 (b) is associated with the smallest n, namely 3 (c) is associated with the largest n^2, namely 9 (d) is associated with the largest n, namely ∞ (e) none of these.
15. The magnitude of the charge-to-mass ratio of the electron (a) is zero (b) is less than that for the proton (c) is greater than that for the proton (d) equals that of the neutron (e) none of these.
16. The proton and the electron (a) have the same size charge and mass (b) have the same mass but differ in charge by a factor of about 2000 (c) have the same size charge and differ in mass by a factor of about 2000 (d) differ in both charge and mass by a factor of about 2000 (e) none of these.
17. The neutron (a) has a mass slightly greater than that of the proton (b) has a charge slightly greater than that of the proton (c) has a mass slightly greater than that of the electron (d) has a mass slightly less than that of the proton (e) none of these.

Problems

SUBATOMIC PARTICLES
THE NUCLEAR ATOM

1. [I] How many electrons would you get if you bought a gram of them? How does that compare with the total number of stars in the entire Universe ($\approx 10^{22}$)?
2. [I] Determine the total mass of all the electrons in 2.00 g of hydrogen (H_2) gas at STP.
3. [I] A current of 965 A passes through a molten solution of NaCl for 100.0 s. How much chlorine gas will be liberated?
4. [I] For single-valence elements, verify that 1 faraday is equivalent to 9.65×10^7 C/kmole.
5. [I] What mass of metallic sodium will be deposited at the cathode if 50.0 A passes through a molten bath of NaCl for 5.00 minutes?
6. [I] How much charge must pass through a molten bath of NaCl in order to liberate 11.2 liters of chlorine gas (Cl_2) at STP?
7. [I] A current of 2.00 A is passed through molten $BaCl_2$ for 5.00 h. How much barium and chlorine will be liberated?
8. [I] In the Millikan oil-drop experiment, a droplet of mass 1.111×10^{-15} kg is held motionless by an electric field of 34.0 kV/m. How many extra electrons is it carrying?
9. [I] A tiny droplet of oil carries an extra charge of e and is held motionless between parallel plates 2.00 cm apart across which there is a potential difference of 60.0 kV. What is the mass of the droplet?
10. [I] A beam of electrons in a 45-kV X-ray tube bombards the target, producing 750 W of thermal energy per second. Nonetheless, only 1.00% of the beam's energy ends up as X-rays. Compute the average rate at which electrons strike the target. (Hint: Use the relationship between power, current, and voltage.)
11. [I] A beam of X-rays having a wavelength of 0.090 nm impinges on an unknown crystal. The strongest reflection maximum is observed to occur at an angle of 25°. What is the spacing between the atomic planes doing the scattering?
12. [I] A crystal such as that in Fig. 29.8c can be used as a monochromator. It can select a single frequency (at a particular angle) out of a polychromatic incident beam. If the Bragg planes are known to be separated by 1.50 Å (that is, 0.150 nm), what wavelength will have a first-order peak at 30° from the beam axis?
13. [I] Determine the angle at which a narrow beam of monochromatic X-rays of wavelength 0.090 nm has a first-order reflection off a calcite crystal ($d = 0.303$ nm).
14. [I] A silver bromide crystal has atomic planes spaced by 0.288 nm. Given that an X-ray beam diffracts at an angle $2\theta = 35.0°$, producing a 1st-order maximum at a detector, what is the wavelength?
15. [I] X-rays reflect off a salt crystal whose atomic planes are 0.28 nm apart. If the first-order maximum is at 22° to the planes, what is the wavelength of the radiation?
16. [I] Confirm Stoney's notion that the hydrogen spectral wavelengths formed simple ratios, such as $H_\alpha : H_\beta : = \frac{1}{20} : \frac{1}{27}$.
17. [I] Compute (to 4 significant figures) the wavelength of the second line of the hydrogen Balmer series.
18. [I] What is the frequency (to 4 significant figures) of the third line in the Balmer series?

19. [I] Compute (to 4 significant figures) the wavelength that corresponds to the series limit of the Balmer series.

20. [I] Compute (to 4 significant figures) the wavelength of the first line of the Paschen series.

21. [I] Determine (to 4 significant figures) the longest wavelength in the Balmer series for hydrogen.

22. [I] Compute (to 4 significant figures) the wavelength of the second line of the Paschen series.

23. [II] Imagine a block of pure metallic copper with a mass of 63.55 g and a density of 8.96 g/cm^3. Approximate the size of a copper atom.

24. [II] If 0.754 5 g of metallic silver are deposited onto an electrode using a current of 0.500 A for 22.5 minutes, what is the atomic mass of silver? The ions are singly charged. Refer to Discussion Question 2.

25. [II] Use Fig. P25 depicting Thomson's electron-beam apparatus to show that

$$\frac{e}{m_e} \approx \frac{E\theta}{B^2L}$$

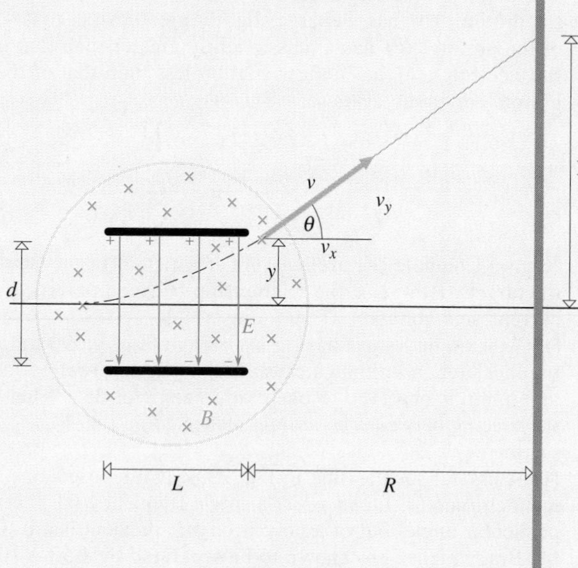

Figure P25

26. [II] If r is the radius of the path taken by an electron in the B-field region of Fig. P25, show that

$$\frac{e}{m_e} = \frac{E}{B^2r}$$

and prove that this equation is equivalent to the result gotten in Problem 25.

27. [II] An electron in Thomson's apparatus moves under the influence of a B-field along a path with a radius of 15.00 cm. If an E-field of 20.0 kV/m makes the path straight and horizontal, find B.

28. [II] With the previous few problems in mind, knowing e/m for the electron, determine the strength of the B-field that will result in a deflection of 0.25 rad when $V = 180$ V, and $L = 4.5$ cm, and the plates are separated by 1.66 cm.

29. [II] An electron in a cathode-ray tube is emitted from a hot filament at a negligible speed, whereupon it accelerates across a potential difference of 1.00×10^6 V. Determine its final speed. (Hint: The correct answer is about half the classical value.)

30. [II] The continuous X-ray emission from a copper target impinges in a narrow beam on a calcite crystal with an atomic spacing of 0.303 nm. A detector picks up the first strong (first-order) maximum at 24.0° to the beam axis. What is the shortest wavelength present in the radiation?

31. [II] A narrow beam of X-rays of wavelength 0.200 nm impinges on a crystal as in Fig. 29.8c. The detector picks up the first-order maximum at an angle of 50.0° from the beam axis. What is the spacing and orientation of the atomic planes causing the reflection?

32. [II] X-rays of unknown wavelength are incident on a nickel single crystal having an interatomic separation of 0.215 nm, as shown in Fig. P32. The diffracted beam has a first-order maximum at an angle $\phi = 45°$. Determine the wavelength. (Hint: First compute d. Note that the planes *do not* run diagonally across each little square of atoms.)

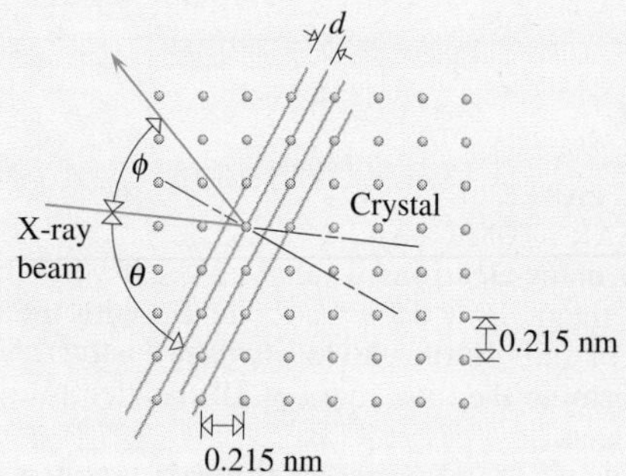

Figure P32

33. [II] A beam of 0.020 0-nm X-rays impinges on a crystal having an array of Bragg planes separated by 1.22×10^{-10} m. What is the highest-order reflected maximum?

34. [II] A polychromatic narrow beam of X-rays impinges on a crystal at 30° to a set of atomic planes. What wavelengths will be reflected at 30° if the atomic spacing is 0.050 0 nm?

35. [II] An alpha particle of mass 6.6×10^{-27} kg is fired at a speed of c/20 directly at an iron nucleus. About how close will it come to the center of the nucleus?

36. [II] What is the shortest wavelength of the Lyman series of hydrogen? Give your answer to 5 significant figures.

37. [II] If light is defined to correspond to the wavelength range from 390 nm to 780 nm, what is the shortest wavelength of the Balmer series that falls within that range?

38. [II] What is the longest wavelength of the Lyman series of hydrogen? Give your answer to 5 significant figures.

39. [III] Use Fig. P25 depicting Thomson's electron-beam apparatus to show that

$$Y = \left(\frac{e}{m_e}\right)\frac{B^2L}{2E}(L + 2R)$$

40. [III] An oil droplet carrying a net charge of Q and having a mass m falls in air at a steady vertical terminal speed between two vertical parallel plates separated by a distance d. When a potential difference V is applied across the plates, the droplet moves uniformly at an angle θ with the vertical. Show that

$$\tan\theta = \frac{VQ}{mgd}$$

41. [III] An imperfect crystal has Bragg "planes" spaced on average by 0.100 nm, which, instead of being fixed everywhere, vary by ±0.001 nm from place to place. A polychromatic beam of X-rays diffracts from the crystal into a first-order peak at an angle of 30°. Determine the spread in the wavelengths of the emerging first-order beam at 30°.

42. [III] Rock salt (NaCl) is a cubic crystal with a molecular mass of 58.5 and a density of 2.16×10^3 kg/m^3. Approximate the space between its atoms.

30

The Evolution of Quantum Theory

QUANTUM MECHANICS DEVELOPED IN two fairly distinct phases. First came the shocking basics: the appreciation of Planck's Constant, the quantum of action; the discovery of the quantization of energy and then, later, momentum; the conception of the photon; a crude but effective atomic model; and the wave-particle duality of matter. All these ideas form the body of knowledge known as the Old Quantum Theory. Building on that foundation, from 1925 to the present, physics moved into the contemporary, more theoretically abstract and mathematically driven stage of Quantum Mechanics, which will be discussed in Chapter 31.

A red-hot furnace emits a spectrum of radiation that is independent of the furnace wall materials and is contingent only on the temperature.

THE OLD QUANTUM THEORY

It shouldn't be surprising that if physics was to be turned upside down, it would be done while trying to figure out what *light* (that is, radiant energy) was about. Quantum Mechanics had its earliest tentative beginnings in the theory of blackbody radiation, which itself began back in 1859. That year, Darwin published *The Origin of Species,* and it was the year that Gustav Robert Kirchhoff proffered an intellectual challenge that would lead to a revolution in physics.

30.1 Blackbody Radiation

Kirchhoff analyzed the way bodies in thermal equilibrium behave in the process of exchanging radiant energy. This *thermal radiation* (p. 534) is electromagnetic energy radiated by all objects, the source of which is their thermal energy. Suppose that the abilities of a body to emit and absorb electromagnetic energy are characterized by an **emission coefficient** ε_λ and an **absorption coefficient** α_λ. Epsilon is the energy per unit area per unit time emitted in a tiny wavelength range around λ (in units of $W/m^2/m$); any energy-measuring device admits a range of wavelengths. Alpha is the fraction of the incident energy absorbed per unit area per unit time in that wavelength range; it's unitless. These coefficients depend on both the nature of the surface of the body (color, texture, etc.) and the wavelength—a body that emits or absorbs well at one wavelength may emit or absorb poorly at another.

Now, imagine an isolated chamber of some sort in thermal equilibrium at a fixed temperature T. Clearly, it would be filled with radiant energy at lots of different wavelengths—just think of a glowing electric furnace. Assume that there is some formula, or **distribution function** I_λ, which depends on T and which tells us the intensity or amount of energy present at each wavelength. Put in numbers for T and λ, and the formula tells us the quantity of energy in the radiation within the cavity. Apparently, the *total* amount of energy at all wavelengths being absorbed by the walls versus the amount emitted by them must be the same, or else T will change. Kirchhoff argued that if the walls were made of different materials (which behave differently with T) that same balance would have to apply for *each* wavelength range individually. Thus, the energy absorbed at λ, namely, $\alpha_\lambda I_\lambda$, must equal the energy radiated, ε_λ, *and this is true for all materials no matter how different.* **Kirchhoff's Radiation Law** is thus

$$\frac{\varepsilon_\lambda}{\alpha_\lambda} = I_\lambda$$

wherein the distribution I_λ is a universal function the same for every type of cavity wall regardless of material, color, size, and shape and is only dependent on T and λ. That's extraordinary! Still, the famous British ceramist Thomas Wedgwood (1792) had long before noted that the objects in a fired kiln all turned glowing red together with the furnace walls regardless of their size, shape, or material constitution.

Although Kirchhoff did not provide the energy distribution function, he did point out that a perfectly absorbing body, one for which $\alpha_\lambda = 1$, will appear black and, in that special case, $I_\lambda = \varepsilon_\lambda$. The distribution function for a perfectly black object is the same as for an isolated chamber at the same temperature. This means that the radiant energy at equilibrium inside an isolated cavity is in every regard the same, "as if it came from a completely black body of the same temperature." The energy leaking from a small hole in the chamber should be identical to the radiation coming from a perfectly black object at the same temperature and that, as we will see, has important practical consequences.

Even though the scientific community accepted the challenge of determining I_λ, technical difficulties caused progress to be very slow. A simplified experimental setup is shown in Fig. 30.1. Radiant energy from a source passes through a slit and into a nonabsorbing prism. The wavelengths present are spread out into a continuous band that is sampled by a detector. Data must be extracted that is independent of the specifics of the detector. Thus, the best thing to plot is the radiant energy per unit time, which enters the detector per unit area (of the entrance window) per unit wavelength range (admitted by the detector). Figure 30.2 shows the kind of curves that were ultimately recorded (see also Fig. 15.9, p. 534), and each is a plot of I_λ at a specific temperature.

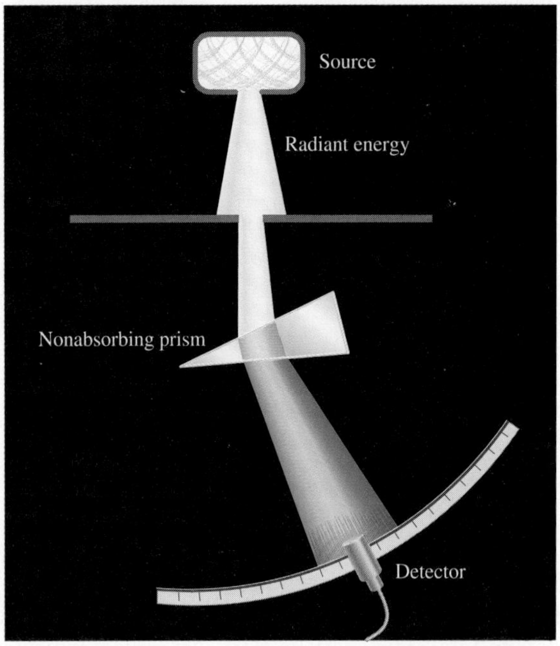

Figure 30.1 A schematic setup for measuring blackbody radiation.

The Stefan-Boltzmann Law

In 1865, J. Tyndall published the result that the total energy emission of a heated platinum wire was 11.7 times as much operating at 1200°C (1473 K) as it was at 525°C (798 K). Amazingly, Josef Stefan (1879) noticed that the ratio of $(1473\ \text{K})^4$ to $(798\ \text{K})^4$ was 11.6, nearly 11.7, and he inferred that the rate at which energy is radiated is proportional to T^4. In this he was quite right (and quite lucky), because Tyndall's results were actually far from those of a blackbody. In any event, the conclusion was subsequently proven via a theoretical argument carried out by L. Boltzmann (1884). This was a traditional analysis of the radiation pressure exerted on a piston in a cylinder using the Laws of Thermodynamics and Kirchhoff's Law. The discussion progressed in much the same way one would treat a gas in a cylinder, but instead of atoms, the active agency was electromagnetic waves. The resulting **Stefan-Boltzmann Law** for blackbodies is

$$\text{P} = \sigma A T^4 \tag{30.1}$$

where P is the total power radiated at all wavelengths, A is the area of the radiating surface, T is the absolute temperature in kelvins, and σ is a universal constant now given as

$$\sigma = 5.670\,3 \times 10^{-8}\ \text{W/m}^2\text{·K}^4$$

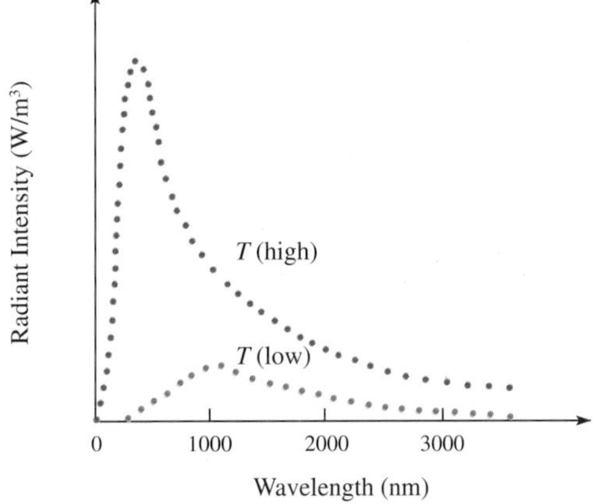

Figure 30.2 A plot of energy (J) per unit time (s), per unit area (m^2), per unit wavelength interval (m), entering a detector successively centered at a number of different wavelengths. The source is a blackbody. Each curve represents I_λ at a given temperature.

TABLE 30.1 Representative values of total emissivity*

Material	ε
Aluminum foil	0.02
Copper, polished	0.03
Copper, oxidized	0.5
Carbon	0.8
White paint, flat	0.87
Red brick	0.9
Concrete	0.94
Black paint, flat	0.94
Soot	0.95

*$T = 300$ K, room temperature.

The total area under any one of the blackbody-radiation curves of Fig. 30.2 for a specific T is the power per unit area, and from Eq. (30.1) that's just $\text{P}/A = \sigma T^4$.

No real object is a perfect blackbody; carbon black has an absorptivity of nearly one, but only at certain frequencies (obviously including the visible). Its absorptivity is much lower than that in the far infrared. Still, most objects resemble a blackbody (at least at certain temperatures and wavelengths)—you, for instance, are nearly a blackbody for infrared. Thus, it's useful to write a similar expression for ordinary objects. To that end, we introduce a factor called the **total emissivity** (ε), which relates the radiated power to that of a blackbody for which $\varepsilon = 1$, at the same temperature, thus

$$\text{P} = \varepsilon\sigma A T^4 \tag{30.2}$$

Table 30.1 provides a few values of ε (at room temperature), where $0 < \varepsilon < 1$. Note that emissivity is unitless.

Example 30.1 A cube of rough steel 10 cm on a side is heated in a furnace to a temperature of 400°C. Given that its total emissivity is 0.97, determine the rate at which it radiates energy from each face.

Solution: [Given: $L = 10$ cm, $T = 400$°C, and $\varepsilon = 0.97$. Find: P.] From the Stefan-Boltzmann Law, with $T = 673$ K and $A = 0.01$ m^2

$$\text{P} = \varepsilon\sigma A T^4$$

$$\text{P} = (0.97)(5.67 \times 10^{-8}\ \text{W/m}^2\cdot\text{K}^4)(0.01\ \text{m}^2)(673\ \text{K})^4$$

and $\boxed{\text{P} = 1.1 \times 10^2\ \text{W}}$

▶ **Quick Check:** $\text{P} \approx (1)(6 \times 10^{-8}\ \text{W/m}^2\cdot\text{K}^4) \times (10^{-2}\ \text{m}^2)(7 \times 10^2)^4 \approx 1.4 \times 10^2\ \text{W}$.

Suppose a body with a **total absorptivity** of α is placed in an enclosure such as a cavity or a room having an emissivity ε_e and a temperature T_e. The body will radiate at a rate $\varepsilon\sigma A T^4$ and absorb energy inside the enclosure at a rate $\alpha(\varepsilon_e\sigma A T_e^4)$. But at any equilibrium temperature between body and enclosure (that is, $T = T_e$), these rates must be equal; hence, $\alpha\varepsilon_e = \varepsilon$ and that must be true for all temperatures. We can write an expression for the net power radiated (when $T > T_e$) or absorbed (when $T < T_e$) by the body:

$$\text{P} = \varepsilon\sigma A(T^4 - T_e^4) \tag{30.3}$$

All bodies not at zero kelvin radiate, and the fact that T is raised to the fourth power makes the radiation sensitive to temperature increases. If a body at 0°C (273 K) is brought up to 100°C (373 K), it radiates about 3.5 times the previous power. Raising the temperature raises the net power radiated; that's why it gets harder and harder to increase the temperature of an object. (Just try heating a steel spoon to 1300°C.) Raising the temperature also shifts the distribution of energy among the various wavelengths present. When the filament of a light bulb "blows," the resistance, current, and temperature momentarily rise, and it goes from its normal operating reddish-white color to a bright flash of blue-white.

The Wien Displacement Law

In 1893, Wilhelm "Willy" Wien derived what has come to be known as the **Displacement Law**. Each blackbody curve reaches a maximum height at a value of

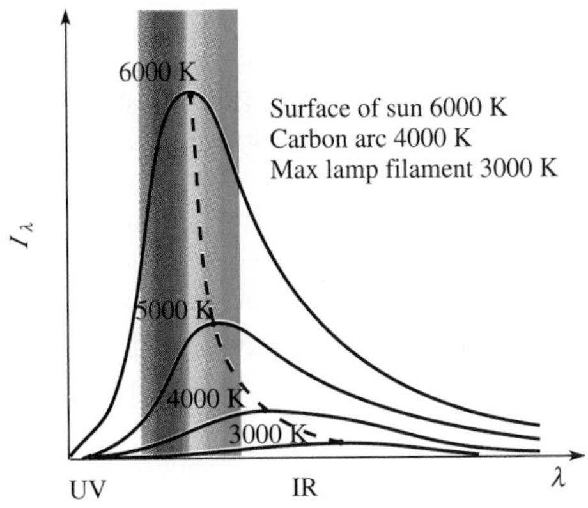

Figure 30.3 The amount of radiant energy emitted by a hot object at various wavelengths. Each curve peaks at a point where $\lambda_{max} T$ = constant; that is the Wien Displacement Law.

wavelength (λ_{max}), which is particular to it and therefore to the absolute temperature T. At that wavelength, the blackbody radiates the most energy. Wien correctly arrived at the fact that

$$\lambda_{max} T = \text{constant} \tag{30.4}$$

where the constant was found experimentally to be 0.002 898 m·K. The peak wavelength is inversely proportional to the temperature. *Raise the temperature, and the bulk of the radiation moves to shorter wavelengths and higher frequencies* (see the dashed curve in Fig. 30.3). As a glowing coal or a blazing star gets hotter, it goes from IR warm, to red-hot, to blue-white. A person or a piece of wood, both only approximating blackbodies, radiates mostly in the infrared and would only begin to glow faintly in the visible at around 600°C or 700°C, long after either had decomposed. The bright cherry red of a chunk of hot iron (p. 481) sets in at around 1300°C.

A hot filament and the spectra it emits. As the temperature rises from (a) to (b) to (c), the corresponding emission curves shift, as shown in Fig. 30.3. The peaks of the curves move toward the yellow, and the blue end of the spectrum increases in intensity as well. The result is that the filament shifts from cherry red to white-hot.

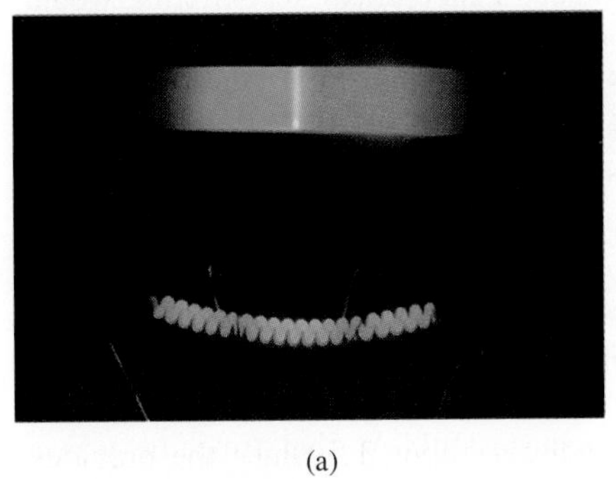

(a)

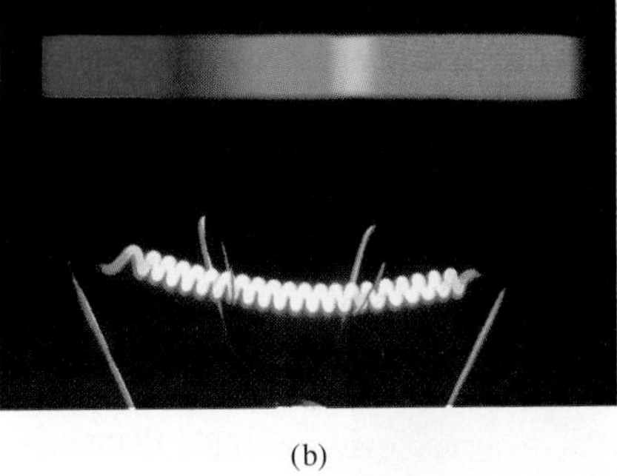

(b)

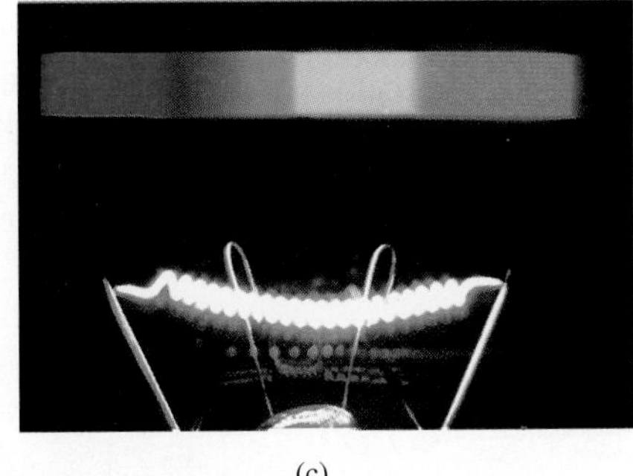

(c)

Example 30.2 On the average, your skin temperature is about 33°C. Assuming you radiate as does a blackbody at that temperature, at what wavelength do you emit the most energy?

Solution: [Given: T = 33°C. Find: λ_{max}.] Using Eq. (30.4), we obtain

$$\lambda_{max} = \frac{0.002\,898 \text{ m·K}}{306 \text{ K}} = \boxed{9.5\ \mu\text{m}}$$

or 9.5×10^3 nm.

▶ **Quick Check:** This result is well into the infrared and therefore reasonable. It's also independent of skin color.

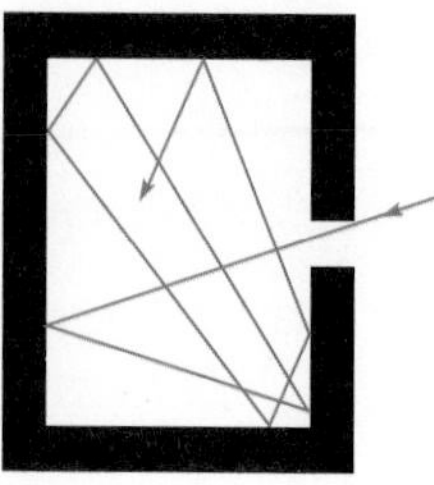

Figure 30.4 Energy entering a small hole in a chamber rattles around until it's absorbed. In reverse, the aperture in a heated enclosure appears as a blackbody source.

In 1899, researchers tremendously advanced the state of experimentation by using, as a source of blackbody radiation, a small hole in a heated cavity (Fig. 30.4). Energy entering such an aperture reflects around inside until it's absorbed (the pupil of the eye appears black for precisely this reason). A near-perfect absorber is a near-perfect emitter, and the region of a small hole in an oven is a wonderful source of *blackbody radiation.*

Planck and His Energy Elements

Max Karl Ernst Ludwig Planck at forty-two was the reluctant father of Quantum Theory. Like so many other theoreticians at the turn of the century, he, too, was working on blackbody radiation. But Planck would succeed not only in producing Kirchhoff's distribution function, but he would turn physics upside-down in the process. We cannot follow the details of his derivation—they are far too complicated and, besides, the original derivation is wrong (Einstein corrected it years later). Still, it had such a powerful impact that it's worth looking at some of the features that are right.

Planck knew that if an arbitrary distribution of energetic molecules was injected into a constant-temperature chamber, it would ultimately rearrange itself into the Maxwell-Boltzmann distribution of speeds as it inevitably reached equilibrium. Presumably, if an arbitrary distribution of radiant energy is injected into a constant-temperature cavity, it, too, will ultimately rearrange itself into the Kirchhoff distribution of energies as it inevitably reaches equilibrium.

In October of 1900, Planck produced a distribution formula that was based on the latest experimental results. This mathematical contrivance, concocted "by happy guesswork," fit all the data available. It contained two fundamental constants, one of which (h) would come to be known as **Planck's Constant** (p. 845). That much by itself was quite a success, even if it didn't explain anything. Although Planck had no idea of it at the time, he was about to take a step that would inadvertently revolutionize our perception of the physical Universe.

Naturally enough, Planck set out to construct a theoretical scheme that would logically lead to the equation he had already devised. He assumed that the radiation in a chamber interacted with simple microscopic oscillators of some unspecified type. These vibrated on the surfaces of the cavity walls, absorbing and re-emitting radiant energy independent of the material. (In fact, the atoms of the walls do exactly that. Because of their tightly packed configuration in the solid walls, the atoms interact with a huge number of their neighbors. That completely blurs their usual characteristic sharp resonance vibrations, allowing them to oscillate over a broad range of frequencies and emit a continuous spectrum.) Try as he might, Planck was unsuccessful. At that time, he was a devotee of E. Mach, who had little regard for the reality of atoms, and yet the obstinate insolubility of the problem ultimately led Planck to "an act of desperation." He hesitantly turned to Boltzmann's "distasteful" statistical method, which had been designed to deal with the clouds of atoms that constitute a gas.

Most of the participants in the first Solvay Conference of 1911. Among the standees are Einstein (second from right), Planck (second from the left), and Rutherford (fourth from right). The young man next to him in the casual suit is Jeans. The lone woman is of course Madame Curie. J. J. Thomson missed the picture altogether.

Boltzmann, the great proponent of the atom, and Planck were intellectual adversaries for a while. And now Planck was forced to use his rival's probabilistic formulation of entropy (p. 577), which—ironically—he would actually misapply. If Boltzmann's scheme for counting atoms was to be applied to something continuous, such as energy, some adjustments would have to be made in the procedure. Thus, according to Planck, the total energy of the oscillators had to be thought of, at least temporarily, as apportioned into "energy elements" so that they could be counted. These energy elements were given a value proportional to the frequency f of the resonators. Remember that he already had the formula he was after and in it there appeared the term hf. Planck's constant

$$6.626\,075\,5 \times 10^{-34}\ \text{J}\cdot\text{s} \qquad \text{or} \qquad 4.135\,669\,2 \times 10^{-15}\ \text{eV}\cdot\text{s}$$

is a very small number and so hf, which has the units of energy, is itself a very small quantity. Accordingly, he set the value of the energy elements equal to it:

$$\mathrm{E} = hf \tag{30.5}$$

This was a statistical analysis, and counting was central. Still, when the method was applied as Boltzmann intended, it naturally smoothed out energy, making it continuous as usual. Again, we needn't worry about the details; the amazing thing was that Planck had stumbled on a hidden mystery of nature: **energy is quantized**—it actually comes in tiny bursts as given by Eq. (30.5), but he certainly didn't realize it then.

Max Karl Ernst Ludwig Planck (1858–1947). In 1889 he became Professor of Physics at the University of Berlin. There, he soon took up the problem of blackbody radiation first raised by his old teacher Kirchhoff, who had recently died (1887), and whose former position Planck now held.

The equation that finally resulted is presented here only so that you can see the answer to Kirchhoff's challenge, namely

$$I_\lambda = \frac{2\pi hc^2}{\lambda^5}\left[\frac{1}{e^{\frac{hc}{\lambda k_B T}} - 1}\right]$$

This is **Planck's Radiation Law**, and it fit blackbody data splendidly. Notice how the expression contains the speed of light, Boltzmann's Constant (p. 497), and Planck's Constant (h). It bridges Electromagnetic Theory to the domain of the atom. In 1901, Planck used blackbody data to arrive at a numerical value of k_B. With that and the universal gas constant R (p. 497), he determined N_A. And using Avogadro's number, the faraday, and Eq. (29.1), he calculated the charge on the electron! Remarkably, his answer was only 2% too small (which was much better than the result J. J. Thomson had gotten), and it was arrived at eight years before Millikan's measurements.

Although Eq. (30.5) represents a great departure from previous ideas, Planck did not mean to break with classical theory. It would have been unthinkable for him to even suggest that radiant energy was anything but continuous. "That energy is forced, at the outset, to remain together in certain quanta . . . ," Planck later remarked, "was purely a formal assumption and I really did not give it much thought." It was only around 1905, at the hands of a much bolder thinker, Albert Einstein, that we learned that the atomic oscillators were real, and that their energies were quantized. Each oscillator could only exist with an energy that was a whole number (n) multiple of hf (a little like the *gravitational*-PE of someone walking up a flight of stairs). Moreover, **radiant energy itself is quantized**, existing in localized blasts of an amount $\mathrm{E} = hf$.

30.2 Quantization of Energy: The Photoelectric Effect

The concept of the energy quantum went unnoticed for almost five years. It languished until Einstein completely recast blackbody theory. He had published three brilliant papers between 1902 and 1904 that laid out the principles of the discipline

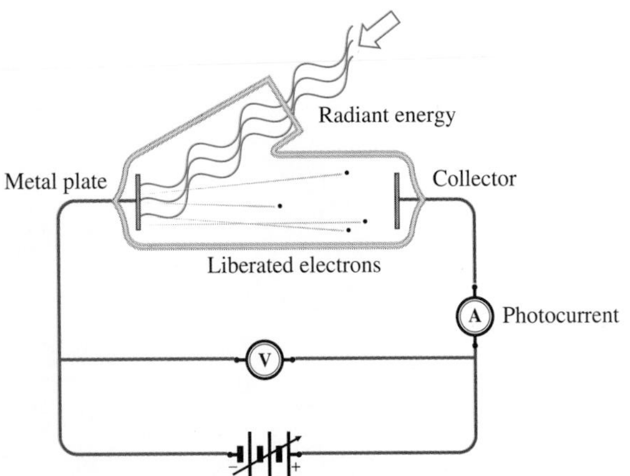

Figure 30.5 The Photoelectric Effect—radiant energy liberating electrons from a metal.

known as *Statistical Mechanics*. Building on this work, Einstein in 1905 published a paper in which he showed that radiant energy behaved as if it was a collection of particles, light-quanta, each of energy hf. That same paper is most remembered for its analysis of the **Photoelectric Effect**.

The Photoelectric Effect was discovered by Heinrich Hertz, the man who had experimentally established the existence of electromagnetic waves. Ironically, the Photoelectric Effect, which he considered a *minor* observation, would ultimately lead to the overthrow of the classical understanding of electromagnetic waves. The Photoelectric Effect corresponds to the fact that *radiant energy (in the form of X-rays, ultraviolet, or light) impinging on various metals ejects electrons from their surfaces* (Fig. 30.5).

Classical theory suggests that the incident "light" arrives as an electromagnetic wave. If we use a uniform beam, its energy is presumably evenly spread across the entire wavefront (as it is with a water wave). The brighter the light, the greater its intensity, the larger the amplitudes of the E- and B-fields everywhere on a wavefront, and the more energy the wave delivers per second. These fields exert forces on the electrons in the metal and can even liberate some of them from the surface.

When the collector plate is made positive with respect to the emitter plate, *photoelectrons* easily traverse the tube, and that constitutes a *photocurrent* that is measurable with a microammeter. Classical theory predicts that increasing the brightness of the light beam provides more energy, thus liberating more electrons, and increasing the photocurrent (Fig. 30.6). This proportionality between photocurrent and irradiance (brightness) was shown to hold across a broad range from a low, where the eye cannot even see the light, to a high, where it cannot bear it.

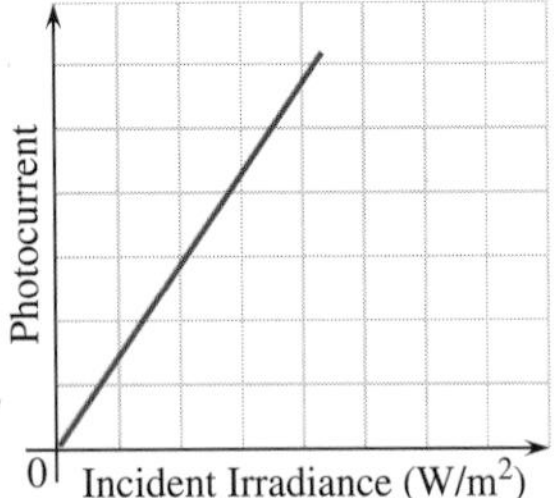

Figure 30.6 Photocurrent is directly proportional to the incident irradiance (that is, the energy per unit area per unit time).

If the collector's positive potential is gradually decreased, we can expect the photocurrent to slowly decrease as well. Figure 30.7 illustrates the effect for monochromatic light. Some electrons make it to the collector even at zero potential difference; and the brighter the light, the more current there will be. Evidently, if the collector is made negative, it will repel photoelectrons, decreasing the current even further. At a negative potential—specific to each metal—known as the **stopping potential** (V_S), all the electrons hurled out of the metal are turned back, and the photocurrent goes to zero. The existence of a stopping potential characteristic of the metal, but *independent of intensity,* was troubling. It would seem that V_S ought to

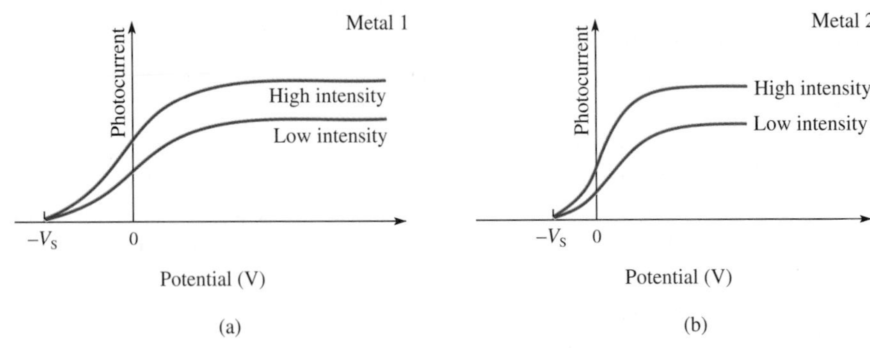

Figure 30.7 The variation of photocurrent with collecting potential and light intensity. Different metals have different stopping potentials.

depend on how bright, and therefore energetic, the light is. An electron has to traverse the region where the E-field is pushing against it and thus overcome a potential energy barrier in order to arrive at the collector. Reaching the maximum negative potential finally turns back even those electrons with a maximum KE, where

$$\text{KE}_{\text{max}} = eV_S \tag{30.6}$$

The surprise is that KE_{max} is independent of the light intensity. With a low light level and very little energy coming in, the electron must somehow store up what it needs to pop out of the metal, and yet at high light levels, it must refrain from absorbing any more energy than KE_{max}—very strange.

However disconcerting these experimental results proved, when attention was turned to the frequency of the radiation, things became utterly bizzare. For any given metal, there is a specific threshold frequency below which photoemission does not occur, no matter how intense the incident radiant energy. That simply should not be; provided enough energy (via a bright beam), electrons should come pouring out. Moreover, the astounding facts were established (1903) that *the photoelectron's maximum kinetic energy was independent of beam intensity, and yet it was proportional to the frequency of the illumination.* Kinetic energy should only depend on the incident energy; it shouldn't have anything to do with frequency. How could frequency affect energy?

The time delay between the arrival of radiant energy and the emission of photoelectrons is now known to be less than 3×10^{-9} s; switching on even the dimmest source can "instantly" release electrons. That behavior is simply impossible within orthodox wave theory. Each incoming wave should spread uniformly over the surface of the illuminated metal, imparting only a minute amount of energy to each of the millions upon millions of atoms in its path. We can calculate, at least crudely, how long it would take an electron to absorb enough energy to attain the observed speeds (Problem 48). There is some ambiguity depending on how much absorbing "area" the electron presents to the beam, but all such calculations conclude that it will take from seconds to months of irradiation before an electron can be ejected—and yet there they were, flying off almost instantaneously. Something fundamental was obviously wrong with the accepted understanding of the way electromagnetic energy interacted with matter.

Remarkably, J. J. Thomson (1903) suggested that electromagnetic waves might well be radically different from other waves; perhaps a sort of concentration of radiant energy actually existed. After all, if one shone X-rays on a gas, only certain of the atoms, here and there, would be ionized, as if the beam had "hotspots" rather than being uniform. The notion was prophetic.

Einstein and the Photon

Planck had earlier introduced the concept of energy elements, and regardless of the inconsistencies, he had supposed that radiant energy was continuous and it behaved like a classical wave. Now Einstein, in 1905, stepped into the muddle. Audaciously breaking with the long-standing traditional view, he proposed that **light itself is granular**, that it is actually composed of discrete bursts, particles of energy, or **photons**, as they came to be called.* The interaction between electromagnetic radiation

*The word *photon* was coined by G. N. Lewis in 1926.

and substantial matter occurs in jolts. Radiant energy is absorbed and emitted discontinuously because it itself is discontinuous!

Einstein's first paper on light-quanta, "On a Heuristic Point of View Concerning the Generation and Conversion of Light," was published in 1905 (before Relativity). And *heuristic* means something that serves as a guide in the solution of a problem but is otherwise itself unproved. In that spirit, he postulated that *every electromagnetic wave of frequency f is actually a stream of energy quanta, each with an energy*

$$\mathrm{E} = hf = \frac{hc}{\lambda} \tag{30.7}$$

Example 30.3 In the atomic domain, energy is often measured in electron-volts. Accordingly, arrive at an expression for the energy of a light-quantum in eV when the wavelength is in nanometers. What is the energy of a quantum of 500-nm light?

Solution: [Given: λ in nm and particularly $\lambda = 500$ nm. Find: E in eV.] Using $\mathrm{E} = hc/\lambda$ with $h = 4.135\,67 \times 10^{-15}$ eV·s (p. 1049), we get

$$\mathrm{E} = \frac{1239.8 \text{ eV·nm}}{\lambda}$$

For $\lambda = 500$ nm, $\boxed{\mathrm{E} = 2.48 \text{ eV}}$.

▶ **Quick Check:** The numerator here can be remembered as 1234.5, which is accurate to about 0.4%. The energy of a 500-nm quantum is in keeping with the energies associated with valence-electron processes, a few eV.

Nobody knows what a photon "looks like," but it is both localized and wavy (it has a frequency). The higher the frequency, the greater the energy of the individual photons. The irradiance of a monochromatic beam, the energy per unit area per unit time, is determined by the number of photons in the stream. The brighter the beam, the more photons.

A photon colliding with an electron in a metal can vanish, imparting essentially all of its energy to the electron. *The electron cannot be totally free*: because of the demands of momentum conservation, momentum must be transferred to the metal atoms, which nonetheless only pick up a negligible amount of energy. Even in the faintest beam, a single adequately energetic photon can kick loose an electron, and this can happen as soon as the illumination is turned on—there is no time delay. Raising the irradiance (intensity) of the beam increases the number of photons and therefore proportionately increases the current (Fig. 30.6). It takes energy to bring an electron up to the surface of the metal and subsequently liberate it. If electrons were not bound to the metal, they would be escaping all the time, leaving it positively charged, which doesn't happen. For an electron already up near the surface, this liberation energy is a minimum called the **work function** ϕ (Table 30.2). A photon's energy goes into freeing the electron, and whatever is left appears as KE. When the electron is at the surface, the liberating energy is a minimum, and the electron takes on a maximum KE given by

$$hf = \mathrm{KE}_{\max} + \phi \tag{30.8}$$

TABLE 30.2 Representative work function values

Metal	Work function (ϕ in eV)
Na	2.28
Co	3.90
Al	4.08
Pb	4.14
Zn	4.31
Fe	4.50
Cu	4.70
Ag	4.73
Pt	6.35

This wonderfully simple expression is known as **Einstein's Photoelectric Equation**, and it explains every aspect of the effect. Since $KE_{max} = hf - \phi$, increasing the intensity of the light leaves the maximum kinetic energy unchanged. Only by changing f is the KE_{max}, or equivalently, the stopping potential, changed for a given metal. The threshold frequency f_0 corresponds to the initiation of emission where $KE_{max} = 0$. Hence, $hf_0 = \phi$ and below a frequency of $f_0 = \phi/h$, the photocurrent will be zero.

It follows from Eq's. (30.8) and (30.6) that

$$eV_S = hf - \phi$$

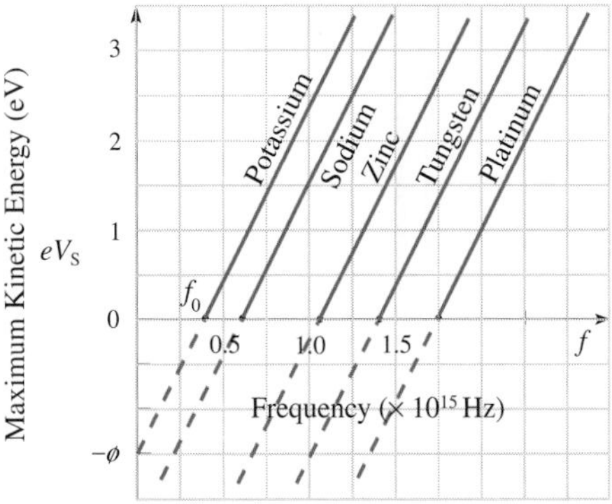

Figure 30.8 Each line follows the equation $eV_S = hf - \phi$. Thus, each intercepts the f-axis at f_0 and the energy axis at $-\phi$.

which has the familiar form of the equation of a straight line ($y = mx + b$). The theory predicted that for any and all metals, a plot of stopping-potential-multiplied-by-electron-charge (y) against frequency (x) will be a straight line of slope (m) equal to Planck's Constant, and y-intercept (b) equal to the negative of the work function. No such relationship was known prior to Einstein's paper of 1905 and, amazingly, exactly that relationship was subsequently observed in every respect! Not until 1914–1915 would Robert Millikan finally and conclusively determine that the Photoelectric Equation was in complete agreement with experiment. He had spent 10 years of meticulous labor intent on showing that Einstein was totally wrong, and in the end he had established "the exact validity" of the theory. And yet he was unwilling to accept the reality of the photon even as he accepted the Nobel Prize for his work. Figure 30.8 shows the kind of results Millikan got. Each metal produces a straight line that intersects the horizontal axis (where the voltage is zero) at its threshold frequency. And each intersects the vertical axis (where $f = 0$) at $-\phi$. Moreover, each line has a slope equal to Planck's Constant!

The above analysis is increasingly being called the *Single Photon Photoelectric Effect* because, nowadays, lasers can put out such a torrent of nearly monochromatic photons that it's possible to blast an electron with two or more light-quanta before it leaves the metal. The result is the *Multiple Photon Photoelectric Effect* (see Discussion Question 5).

Light propagates from one place to another as if it were a wave. Several centuries of work had established *that* with convincing credibility, and yet now it became equally as clear that *light interacts with matter in the processes of absorption and emission as if it were a stream of particles.* This, then, is the so-called **wave-particle duality**, the schizophrenia of light (and, as we will see, of matter in general). Radiant energy appears and disappears in minute localized blasts and is seemingly transported via spreading waves.

A few frames of a motion picture showing the optical sound track at the right. Light passing through the track illuminates a photoelectric detector that converts the signal into a varying voltage.

The idea of light-quanta was not received well at all, and it had very few early advocates. Paul Ehrenfest (who seems to have been independently thinking along similar lines in 1905), Max von Laue, and Johannes Stark (who in 1909 anticipated the Compton Effect, p. 1055) were the leading exceptions. Despite its clearly demonstrated theoretical power, it was so disconcerting to people educated in classical wave theory that it was especially slow to be accepted. When Planck recommended Einstein for membership in the Prussian Academy in 1913, he felt that, notwithstanding Einstein's demonstrated genius, he still had to apologize for him because "he may sometimes have missed the target in his speculations, as, for example, in his hypothesis of light-quanta."

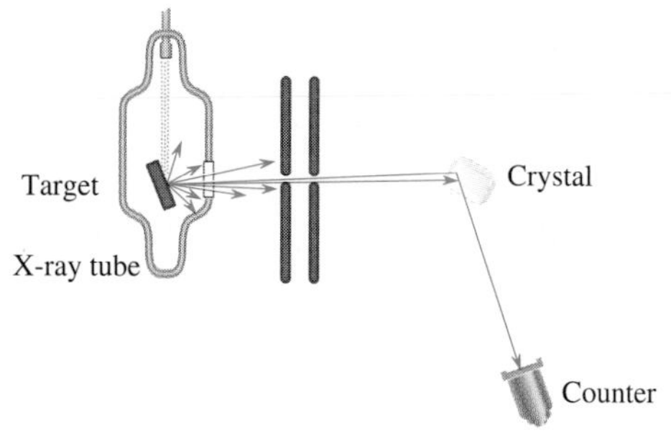

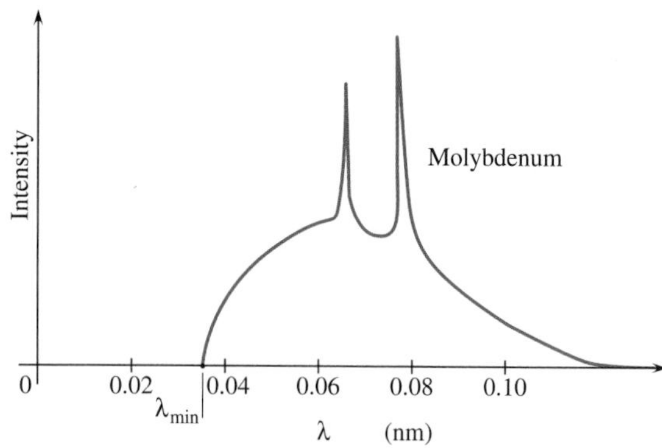

Figure 30.9 X-ray spectrum for a molybdenum target. The broad peak in intensity is due to Bremsstrahlung.

Despite the general indifference, Einstein continued to use his heuristic principle to make a number of splendid predictions, one of which (1911) concerned the generation of X-ray quanta. It would be another 10 years before the strange and wonderful notion that light was no longer a simple TEM wave became so potent that it had to be accepted, even if not quite "understood." Einstein received the 1921 Nobel Prize "for his services to Theoretical Physics, and especially for his discovery of the law of the photoelectric effect."

30.3 Bremsstrahlung

X-rays are generated when high-speed electrons impacting on a dense target rapidly decelerate and thus radiate. The resulting broad continuous X-ray spectrum, illustrated in Fig. 30.9, is known as Bremsstrahlung (p. 1024). The spikes in the curve arise from the atomic structure of the specific target, and we'll deal with them later in this chapter. The process of X-ray production is something of an inverted Photoelectric Effect; in come electrons, out goes radiant energy. And both can only occur in the presence of heavy atoms that can take up some momentum. Using the notion of light-quanta, Einstein predicted there would be a high-frequency (low-wavelength) limit on the radiation.

Imagine an incident electron arriving with an energy eV after being accelerated across a potential of V. As it slows down within the target, it radiates one or more photons. The maximum photon frequency (f_{max}) occurs when all the electron's kinetic energy is radiated as a single photon, whereupon

$$\mathrm{E} = hf_{max} = eV$$

Accordingly, there will be a cut-off frequency of

$$f_{max} = \frac{eV}{h}$$

or since $c = f\lambda$, a minimum wavelength of

$$\lambda_{min} = \frac{ch}{eV} \tag{30.9}$$

In other words, the electron cannot emit more energy than it has, and that maximum amount corresponds to the maximum frequency (minimum wavelength) end of the Bremsstrahlung curve.

Example 30.4 Determine the smallest wavelength X-rays that can be emitted by an electron as it crashes into the metal mask on the front face of a color TV tube operating with an accelerating voltage of 20.0 kilovolts.

Solution: [Given: $V = 20.0$ kV. Find: λ_{min}.] From Eq. (30.9)

$$\lambda_{min} = \frac{ch}{eV}$$

(continued)

(continued)

$$\lambda_{\min} = \frac{(2.9979 \times 10^8 \text{ m/s})(4.135\,67 \times 10^{-15} \text{ eV·s})}{20.0 \times 10^3 \text{ eV}}$$

Notice how entering h in electron-volts allows us to enter the denominator in electron-volts. Finally

$$\boxed{\lambda_{\min} = 0.062\,0 \text{ nm}}$$

▶ **Quick Check:** An atom is about 0.1 nm across, and X-rays range in wavelength from there down to roughly 0.006 nm.

The existence of a maximum frequency was confirmed experimentally in 1915 by W. Duane and F. Hunt at Harvard. Using Eq. (30.8), they determined h to an accuracy of better than 4%.

30.4 The Compton Effect

Evidence of the reality of photons and the fact that they behave like particles with a well-defined energy and *momentum* was provided by the American Arthur H. Compton. He shone X-rays onto targets of low atomic number, like carbon. These have many loosely bound electrons that are essentially "free," and they scatter the radiant energy in a characteristic fashion. Classical theory is complicated by several factors, one of which is that the electrons recoil from the collision, and that introduces a Doppler Shift in the radiation they emit. Be that as it may, Compton reported in 1922 that the evidence at hand was in severe conflict with classical theory.

In 1923, Compton and Debye independently applied relativistic kinematics to the problem of a photon colliding with a "free" electron and thereby explained all the bewildering observations. Recall Eq. (28.16) for a massless particle, $E = pc$; hence, for photons for which $E = hf$

$$p = \frac{E}{c} = \frac{hf}{c}$$

and

$$p = \frac{h}{\lambda} \tag{30.10}$$

Thus, both the momentum and the energy of a high-frequency (short-wavelength) photon, such as an X-ray quantum, exceed that of a low-frequency (long-wavelength) photon, such as a microwave quantum.

Imagine a beam of wavelength λ_i incident on a target. When photons collide with strongly bound electrons that stay put (and do not absorb energy), the photons are elastically scattered in all directions, and one sees this as radiation coming out with the incident wavelength unaltered. By contrast, picture an incident photon of momentum $\mathbf{p}_i$ (Fig. 30.10). If it strikes a free, essentially motionless, electron and imparts some of its momentum to that electron ($\mathbf{p}_e$), a new scattered photon with a longer wavelength and less momentum ($\mathbf{p}_s$) will come flying out at some angle θ.

This is an elastic-collision problem, and we analyze it just as we did earlier (p. 306). Applying Conservation of Momentum yields

$$\mathbf{p}_i = \mathbf{p}_s + \mathbf{p}_e \tag{30.11}$$

Similarly, the initial energy of the system is associated with the incident photon

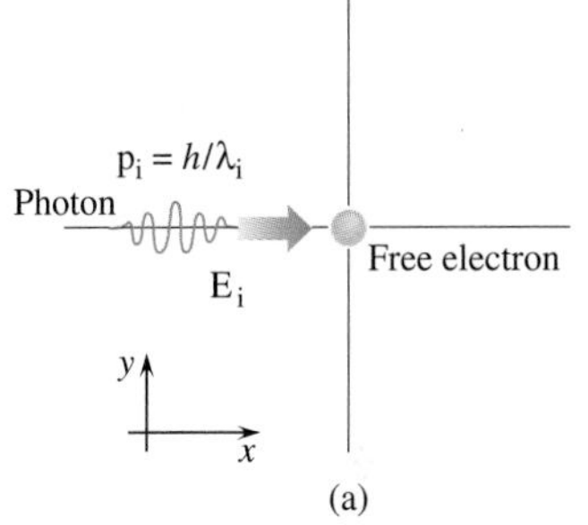

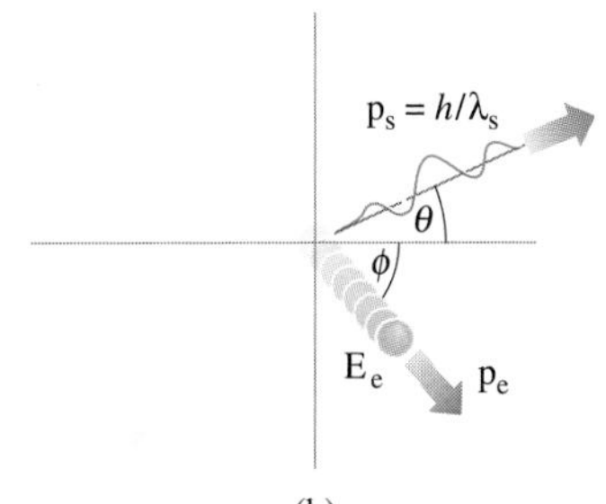

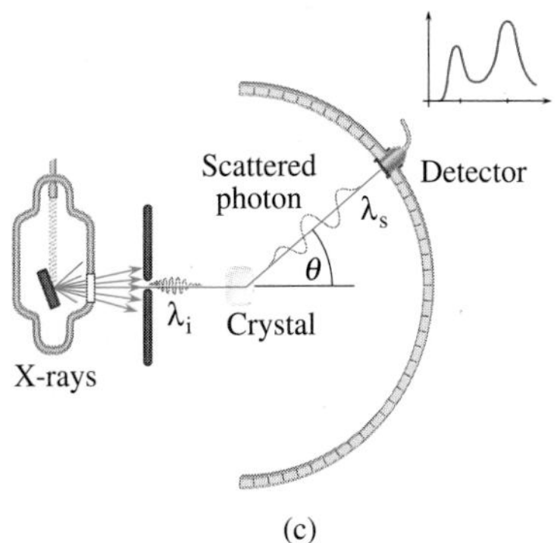

Figure 30.10 Compton scattering of X-rays. (a) A photon collides with a free electron and (b) both are scattered. (c) The scattered X-rays come off at an angle θ.

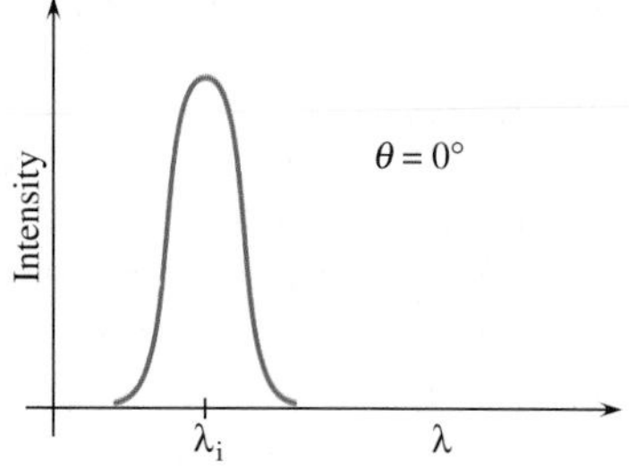

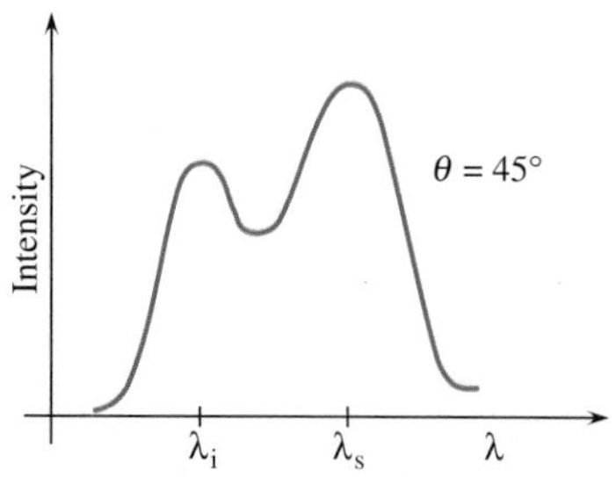

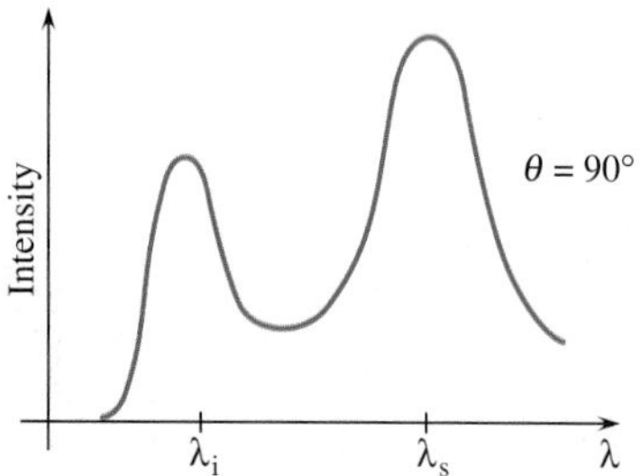

Figure 30.11 Scattered intensity at several angles.

($E_i = hf_i$) and the rest energy of the electron ($m_e c^2$). After the collision, the energy is that of the scattered photon ($E_s = hf_s$) and the moving electron (E_e). Conservation of Energy states that

$$E_i + m_e c^2 = E_s + E_e \tag{30.12}$$

Equations (30.11) and (30.12) can be combined with the relativistic expression [Eq. (28.15), p. 1007] describing the energy of the electron

$$E_e^2 = c^2 p_e^2 + m_e^2 c^4$$

to yield (see Problem 47) a formula for the increase in wavelength

$$\Delta\lambda = \lambda_s - \lambda_i = \frac{h}{m_e c}(1 - \cos\theta) \tag{30.13}$$

Figure 30.11 shows the sort of results that are observed and that agree very nicely with Eq. (30.13). Note that the wavelength shift $\Delta\lambda$ only depends on the photon-scattering angle and is independent of the initial wavelength. It's also independent of the scattering material. The accuracy of the experiment was found to be within 1% by determining the multiplicative constant ($h/m_e c$), which is called the **Compton Wavelength** and which should theoretically equal 0.002 426 nm. In 1925, Bothe and Geiger, using counters, showed that the scattered electron and the scattered X-ray photon actually do fly off at the same time. More recent experiments (1950) verify that simultaneity to within 0.5 ns.

In the 1920s, the Compton Effect had the impact of eliminating the last vestiges of doubt: **whatever light is, it is a stream of particles capable of transferring energy and momentum that can also somehow behave as a wave**. Interestingly, though the contemporary consensus is that the Compton Effect is strong "proof" of the photon, people have nonetheless come up with alternative derivations in terms of the wave model—the business is far from settled.

Example 30.5 A 50.0-keV photon is Compton-scattered by a quasi-free electron. If the scattered photon comes off at 45.0°, what is its wavelength?

Solution: [Given: $E_i = 50.0$ keV and $\theta = 45.0°$. Find: λ_s.] To use Eq. (30.13), first compute λ_i using the results in Example 30.3, p. 1052:

$$\lambda_i = \frac{1240 \text{ eV·nm}}{50.0 \times 10^3 \text{ eV}} = 0.0248 \text{ nm}$$

Since

$$\lambda_s - \lambda_i = \frac{h}{m_e c}(1 - \cos\theta)$$

we have

$$\lambda_s = (0.0248 \text{ nm}) + (0.002\,426 \text{ nm})(1 - 0.707)$$

and

$$\boxed{\lambda_s = 0.0255 \text{ nm}}$$

▶ **Quick Check:** $\Delta\lambda = \lambda_s - \lambda_i = 0.0255$ nm $-$ 0.0248 nm $= 7 \times 10^{-4}$ nm as compared to Eq. (30.13), where $(0.002\,426 \text{ nm})(1 - 0.707) = 7 \times 10^{-4}$ nm.

ATOMIC THEORY

Niels Henrik David Bohr was twenty-six when he arrived at Cambridge from Denmark in the autumn of 1911. He had come to work with J. J. Thomson, but they didn't hit it off well. As fate would have it, the guest speaker at the annual Cavendish dinner was Ernest Rutherford, who, only a few months before, had published his theory of the nuclear atom. Bohr met with Rutherford and, by the spring of 1912, he was happily immersed in the whirlwind of atomic research that churned at Rutherford's lab. "This young Dane," Rutherford once remarked, "is the most intelligent chap I've ever met." Bohr stayed only for four months before returning to Copenhagen, but he had already become convinced of the validity of the nuclear atom.

The atomic theory Bohr later created was a brilliant accomplishment at the time. It has, however, been surpassed by the far more powerful formulations of Quantum Mechanics. Still, we will study the Bohr Theory because it provides the basic conceptual vocabulary for modern atomic physics and because many of its conclusions are essentially valid.

30.5 The Bohr Atom

Bohr began his creation by guessing about the nature of the simplest atomic configuration, the single-electron atom; his first postulate assumed that *the electron sails around the nucleus in a circular orbit.* He then postulated that, unlike a planet, which can revolve permanently at any distance from the Sun, *atomic electrons exist in only certain stable, lasting orbits about the nucleus.* These are now called **stationary states**, and the one with the lowest energy is the **ground state** (p. 846). *While in such a stationary state, the atom does not radiate.* Everyone knew that something was wrong with the planetary model of the atom; electrons orbiting the nucleus, accelerating, should continuously radiate. Losing energy, they should wind inward, finally crashing into the nucleus. And it's easy to calculate that this death spiral will take about a hundred-millionth of a second. Our very existence is therefore an embarrassment to such a theory, which ridiculously insists that all the atoms in the Universe should have long ago collapsed. Why doesn't Bohr's orbiting electron radiate? Here, Bohr was refreshingly outrageous. He simply insisted it didn't, and that was that. Of course, his stance implied that Maxwell's Electromagnetic Theory somehow reaches profound limitations on the atomic level, but he didn't bother with that.

Bohr was not the only one groping toward an orbital model. J. W. Nicholson, an English astrophysicist, was perhaps his chief rival. Like Maxwell long before him, Nicholson raised the question of there being a relationship between the structure of an atom and the kind of spectrum it emits. Bohr believed that any such interdependence must be exceedingly complicated, and at first he avoided the notion altogether. But when a friend, H. M. Hansen, returned to Copenhagen after studying spectroscopy and the two began to discuss Bohr's work, the issue of spectral colors immediately came up. Hansen suggested that spectra (p. 1031) were not really so complex. "As soon as I saw Balmer's formula," Bohr recalled, "the whole thing was immediately clear to me."

Niels Henrik David Bohr (1885–1962). Before Bohr left Denmark in 1943 with the Nazis on his trail, he dissolved in acid the gold Nobel Prize medals that Von Laue and Franck had given him for safekeeping. The bottle containing the gold solution was left on a shelf in Bohr's lab throughout the war. When he returned to Copenhagen, Bohr precipitated the gold and had the medals recast.

Consider the simplest of all atoms, the hydrogen atom. It has a nuclear proton orbited by a single electron in its ground state. When the atom is appropriately stimulated (perhaps thermally via collisions, electrically, or even by absorbing light), the electron is excited into a higher energy orbit that is more distant from the nucleus. There it resides, ordinarily for about a nanosecond, before spontaneously descending to some inner orbit, ultimately dropping back to the ground state. During each drop (a moment when the theory goes impotent), the electron emits its excess energy as a burst of electromagnetic radiation—a photon. This is the now-famous **quantum jump**.

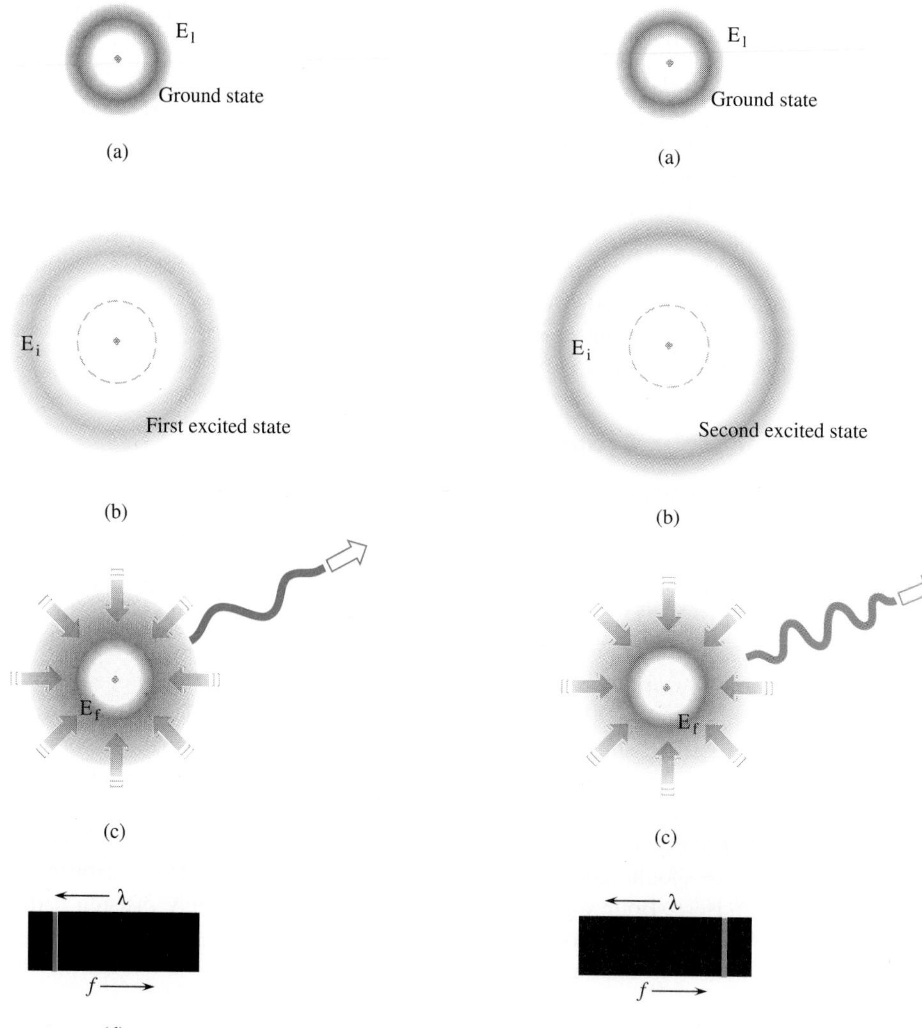

Figure 30.12 An atom in the first excited state drops back to the ground state with the emission of a long-wavelength, low-energy photon. The emission shows up as a long-wavelength, low-frequency spectral line.

Figure 30.13 An atom in the second excited state drops back to the ground state with the emission of a short-wavelength, high-energy photon.

The picture is a little like a Greek amphitheater, in which a ball revolving aroung the lowest tier is boosted up to some higher ring, where it orbits for a while, only to drop back to ground level in either one large or many small descents. The sequence of Balmer's frequencies became a ladder of orbits, of **energy levels**. When the electron in an excited atom drops from an initial (E_i) to a final (E_f) energy level, it emits the difference as a quantum hf; that is

$$E_i - E_f = hf \tag{30.14}$$

The larger the drop, the greater the photon frequency (Fig's. 30.12 and 30.13).

By the time Bohr began his work, the notion that an oscillator could have quantized energy levels that were whole number multiples (n) of hf was widely accepted;

$E = nhf$ was an essential feature of the new blackbody theory. Planck recognized that h had the units of energy multiplied by time (J·s), or equivalently of momentum multiplied by distance. Since that product, known as *action,* was important in classical mechanics, he called h the **quantum of action**. The idea that action was quantized would be generalized into a guiding principle. For example, for an electron in the nth circular orbit of radius r_n, the distance traveled per revolution is $2\pi r_n$ and, since $p_n = m_e v_n$, we might guess that the action (momentum times distance) for each orbit would equal a whole number multiple of h: $(m_e v_n)(2\pi r_n) = nh$. This immediately leads to $m_e v_n r_n = nh/2\pi$, and that we recognize (p. 836) as the angular momentum (L) of the orbiting electron. In any event, Nicholson (1912) suggested that "the angular momentum of an atom can only rise and fall by discrete amounts." A year later, Bohr adopted the same idea as his second postulate; that is

$$L_n = m_e v_n r_n = n\frac{h}{2\pi}, \qquad n = 1, 2, 3, \ldots \tag{30.15}$$

Here, each successive value of n corresponds to a higher orbit of larger radius and lower speed. The term $h/2\pi$ comes up so frequently that P. A. M. Dirac gave it its own symbol, $\hbar$, and we refer to it as "h-bar":

$$\hbar = 1.054 \times 10^{-34} \text{ J·s}$$

Angular momentum is quantized in whole number multiples of $\hbar$.

If we continue with this simple pictorial model (which we *will* have to abandon soon enough), we can derive an expression for the radii of the allowed orbits (Fig. 30.14) of a one-electron atom. Of course, hydrogen is the simplest such atom, but the theory works for such systems as singly ionized helium as well. The electron is attracted to a nucleus (which in general has Z protons and a net charge of $+Ze$) via a Coulomb interaction. Equating that force, $F_E = k(Ze)(e)/r^2$ with the centripetal force, $F_C = mv^2/r$, leads to

$$k\frac{Ze^2}{r_n^2} = \frac{m_e v_n^2}{r_n} \tag{30.16}$$

which can be solved for

$$r_n = \frac{kZe^2}{m_e v_n^2}$$

The quantity v_n, which we do not know, can be eliminated from this expression by solving Eq. (30.15) for it. Whereupon $v_n = nh/2\pi m_e r_n$, and substituting this equation in the above and simplifying (canceling a factor of r_n) gives

$$r_n = n^2\frac{\hbar^2}{m_e kZe^2} = n^2 r_1 \tag{30.17}$$

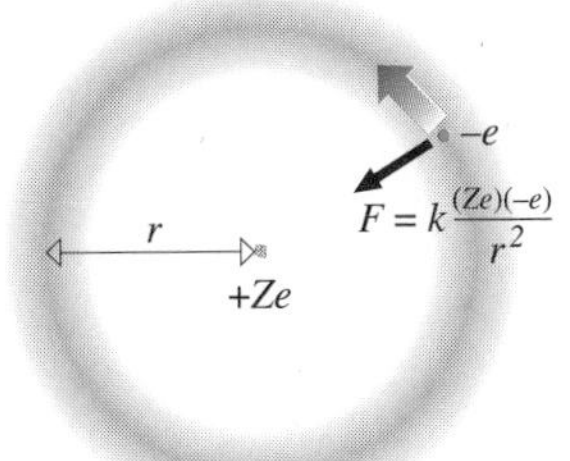

Figure 30.14 A single-electron atom showing the Coulomb interaction between electron and nucleus.

For hydrogen, $Z = 1$, and putting the appropriate numbers in Eq. (30.17) leads to $r_1 = 0.052\,917\,7$ nm. This is the radius of the ground state of the hydrogen atom,

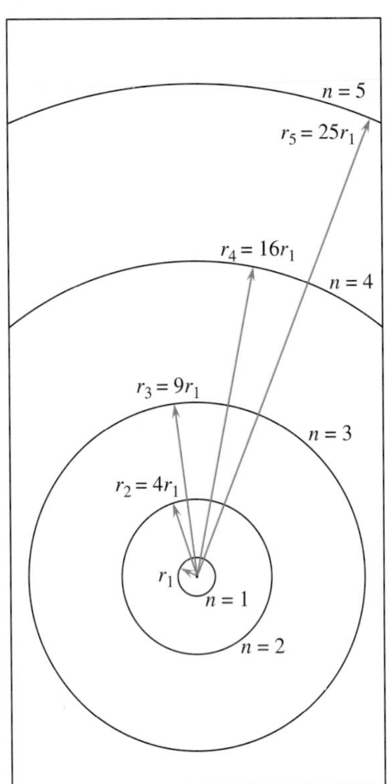

Figure 30.15 The first five Bohr orbits.

and it's called the **Bohr radius**. We know that an atom is about 0.1 nm across so this result is very encouraging (Fig. 30.15). The theory suggests (it only deals directly with single-electron systems) that for heavier atoms (larger Z), the increased nuclear charge draws more strongly on the surrounding electrons, shrinking their orbits. Thus, atomic diameters are all close to the same size. Uranium, which is 238 times more massive than hydrogen, has a diameter only about 3 times as great.

The model Bohr devised pictured the discrete spectral lines resulting from transitions between discrete energy levels. Clearly, energy is the next thing to consider. The total energy of an orbital electron (E_n) is its KE plus its *electrical*-PE $=$ $qV = -eV$. Here, V is due to the nucleus and so $V = kq/r = k(Ze)/r$. Thus

$$E_n = \tfrac{1}{2}m_e v_n^2 - \frac{kZe^2}{r_n}$$

which is the energy of the atom as a whole. It follows from Eq. (30.16) that $\tfrac{1}{2}m_e v_n^2 = kZe^2/2r_n$ and so

$$E_n = -\frac{kZe^2}{2r_n}$$

Notice that the total energy of the atom is negative, only becoming zero when the electron is removed to infinity, or the practical equivalent thereof (see Discussion Question 10). The dependence on r_n can be removed using Eq. (30.17), yielding

$$E_n = -\frac{2\pi^2 k^2 e^4 m_e}{h^2}\frac{Z^2}{n^2} \tag{30.18}$$

The integer n, which enumerates both the orbital radii and the energies, is known as the **principal quantum number**. Notice that as n gets larger, the radii increase as n^2, and the orbits get farther apart. On the other hand, the energy levels become less negative and approach zero as $n \rightarrow \infty$; they get closer together, ultimately becoming continuous when the energy is positive and the electron is unbound.

Equation (30.18) provides the energy of each orbit, and it can be simplified considerably by noting that the ground-state energy ($n = 1$) for hydrogen ($Z = 1$) turns out to equal -13.6 eV. This value is the hydrogen atom's largest negative energy and its most tightly bound configuration. In that case

[*for hydrogen*]
$$E_n = -\frac{13.6 \text{ eV}}{n^2} \tag{30.19}$$

The *first excited state* ($n = 2$) of the hydrogen atom has an energy of $-(13.6/2^2)$ eV, or -3.40 eV. The minus sign tells us that that much energy must be added to raise the electron to the zero level. We can think of it as a kind of energy well with the electron usually down -13.6 eV, at the bottom. Accordingly, the **ionization energy** needed to remove the electron from the atom setting it free (to raise it up and out of the potential well) is 13.6 eV, and this agrees precisely with the measured value! Being able to derive that from basic principles is an amazing accomplishment.

Example 30.6 How much energy must a hydrogen atom absorb if it is to be raised from the ground state to the first excited state? If that excitation energy is to come in the form of a photon, what must be its frequency?

(continued)

(continued)

Solution: [Given: H atom in ground state. Find: $E_2 - E_1$ and f.] We already know that $E_1 = -13.6$ eV and that $E_2 = -3.40$ eV; hence, the atom must receive an energy of

$$(-3.4 \text{ eV}) - (-13.6 \text{ eV}) = \boxed{10.2 \text{ eV}}$$

if it's to be raised into its first excited state. Hence, a photon for which

$$\Delta E = hf = 10.2 \text{ eV}$$

should have a frequency of

$$f = \frac{(10.2 \text{ eV})}{(4.136 \times 10^{-15} \text{ eV·s})} = \boxed{2.47 \times 10^{15} \text{ Hz}}$$

▶ **Quick Check:** $\lambda = 120$ nm, which is in the UV where it should be (see Fig. 30.16).

If the transition from an upper-excited state down to a lower state is accompanied by the emission of a photon as per Eq. (30.14), we should now be able to write an explicit expression for the resulting wavelength, or better yet, for $1/\lambda$. Since $hf = hc/\lambda$, rewrite Eq. (30.14) as

$$\frac{1}{\lambda} = \frac{1}{hc}(E_i - E_f)$$

or using Eq. (30.18)

$$\frac{1}{\lambda} = \frac{2\pi^2 k^2 e^4 m_e Z^2}{h^3 c}\left[\frac{1}{n_f^2} - \frac{1}{n_i^2}\right] \tag{30.20}$$

where $n_f < n_i$, the atom is initially in a higher state than the one it drops down to. Now compare this equation to the Balmer series ($Z = 1$), for which $n_i = n$ and $n_f = 2$; that is

$$\frac{1}{\lambda} = R\left[\frac{1}{2^2} - \frac{1}{n^2}\right], \qquad n = 3, 4, 5, \ldots \qquad [29.5]$$

where all those constants in front of the bracketed term in Eq. (30.20) must equal RZ^2. When Bohr carried out the calculation in 1913 using the best values of the day, he got the measured R to within 1%—spectacular!

Each line in the Balmer series arises when the hydrogen atom in an excited state relaxes back in a single quantum jump to the first excited state ($n_f = 2$). Similarly, when the transition is down to the ground state ($n_f = 1$), the energies are greater and the resulting Lyman series, which wasn't discovered until 1916, is in the far-UV. The Paschen series (first observed in 1908) corresponds to $n_f = 3$. Brackett found a new series in 1922 in the IR for which $n_f = 4$, and Pfund, in 1924, located yet another in the IR for $n_f = 5$ (Fig. 30.16); all precisely as predicted by the Bohr Theory.

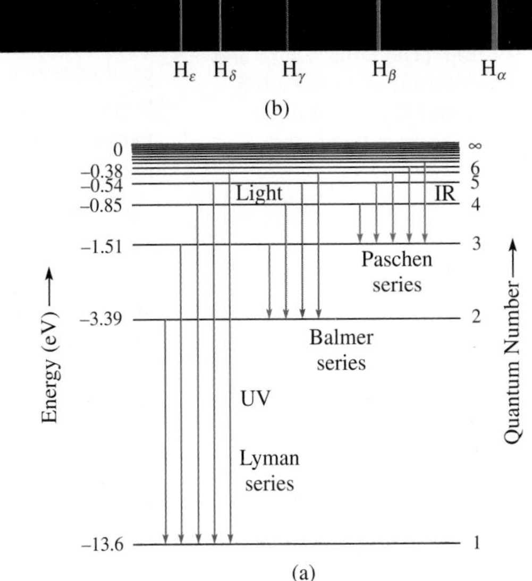

Figure 30.16 (a) Energy levels of the hydrogen atom and the transitions between them corresponding to the emission of radiation. Shown are the Lyman, Balmer, and Paschen series. (b) The spectral lines corresponding to the Balmer series. (Take a look at Fig. Q11, p. 1038.)

Energy Levels

An atom can be raised into any excited state by absorbing an amount of energy given by Eq. (30.14). This can happen, for instance, via collisions with other atoms or from bombardment by projectiles such as electrons (via an electric current) or photons. If an incoming photon does not have enough energy to raise the atom into its first excited state, the atom remains in its ground state, immediately

The first photograph of a solitary atom. The tiny blue-green dot at the center of this photo is a single barium ion. It was slowed or "cooled" by laser beams and is suspended in a radio frequency trap. (The large, white and red structure surrounding the ion is part of the trap.) The ion absorbs energy from the laser beam and re-emits at 493 nm.

elastically scattering away the energy (p. 848). Thus, a visible photon with an energy of around 3 eV cannot raise hydrogen into its first excited state and so hydrogen, normally in its ground state, is transparent to light; photons enter and leave the gas without any energy losses. Only if the photon (or any other impacting particle) has energy equal to, or in excess of, that needed for excitation will the atom "rise up" to a higher level. For example, consider a monatomic gas at room temperature. The atoms collide, but they don't give off light. Why not? According to Eq. (14.14), the average KE of such an atom is $\frac{3}{2}k_B T = \frac{3}{2}(1.38 \times 10^{-23}\ \text{J/K})(300\ \text{K}) = 6.2 \times 10^{-21}\ \text{J} = 0.04\ \text{eV}$. That's much too small to excite any of the atoms, and the collisions remain perfectly elastic (which is why the air doesn't glow on a warm day).

In recent times, it has become possible to examine the transitions made by single atoms from one specific energy level to another. For example, in 1986, an individual barium ion held, almost at rest, in a electromagnetic-field trap was stimulated using laser beams of precise frequency. The experiments (see Discussion Question 13) directly confirmed the existence of quantum jumps between energy levels.

30.6 Stimulated Emission: The Laser

In a conventional light source, such as a tungsten lamp or a neon sign, energy is usually *pumped* into the reacting atoms via collisions with an electrical current. An excited atom drops back to its ground state *spontaneously,* without any external inducement, emitting a randomly directed photon. The atoms are all essentially independent, and each photon in the emitted stream bears no particular phase relationship to any other.

Now, imagine a bunch of atoms somehow pumped up into an excited state. What happens next is not obvious. In 1917, Einstein pointed out that an excited atom can relax to a lower state via photon emission in two distinct ways. In one, the atom emits energy spontaneously, while in the other, it is triggered into emission by the presence of a photon of the proper frequency. The former process is known as **spontaneous emission**, the latter as **stimulated emission**. Suppose that the energy difference between an upper- and lower-energy state is $\Delta E = E_u - E_l$ such that if the transition down occurs, a photon hf_{ul} will be emitted. Let the atom be in the E_u-state with no concern as to how it got there. If the excited atom is exposed to a photon of frequency f_{ul}, it will immediately be stimulated into dropping down to the E_l state. A remarkable feature of the process is that *the emitted photon is in-phase with, has the polarization of, and propagates in the same direction as, the stimulating radiation.* This is a manifestation of the basic nature of photons to cluster in the same state (p. 1084). The incident lightwave is thereby increased in irradiance (Fig. 30.17).

Since most atoms are usually in their ground states, any incoming light is far more likely to be absorbed than to produce stimulated emission. This raises an interesting point: what happens if a substantial number of atoms could be excited into an upper state, leaving the lower state almost empty? Such a condition is called a

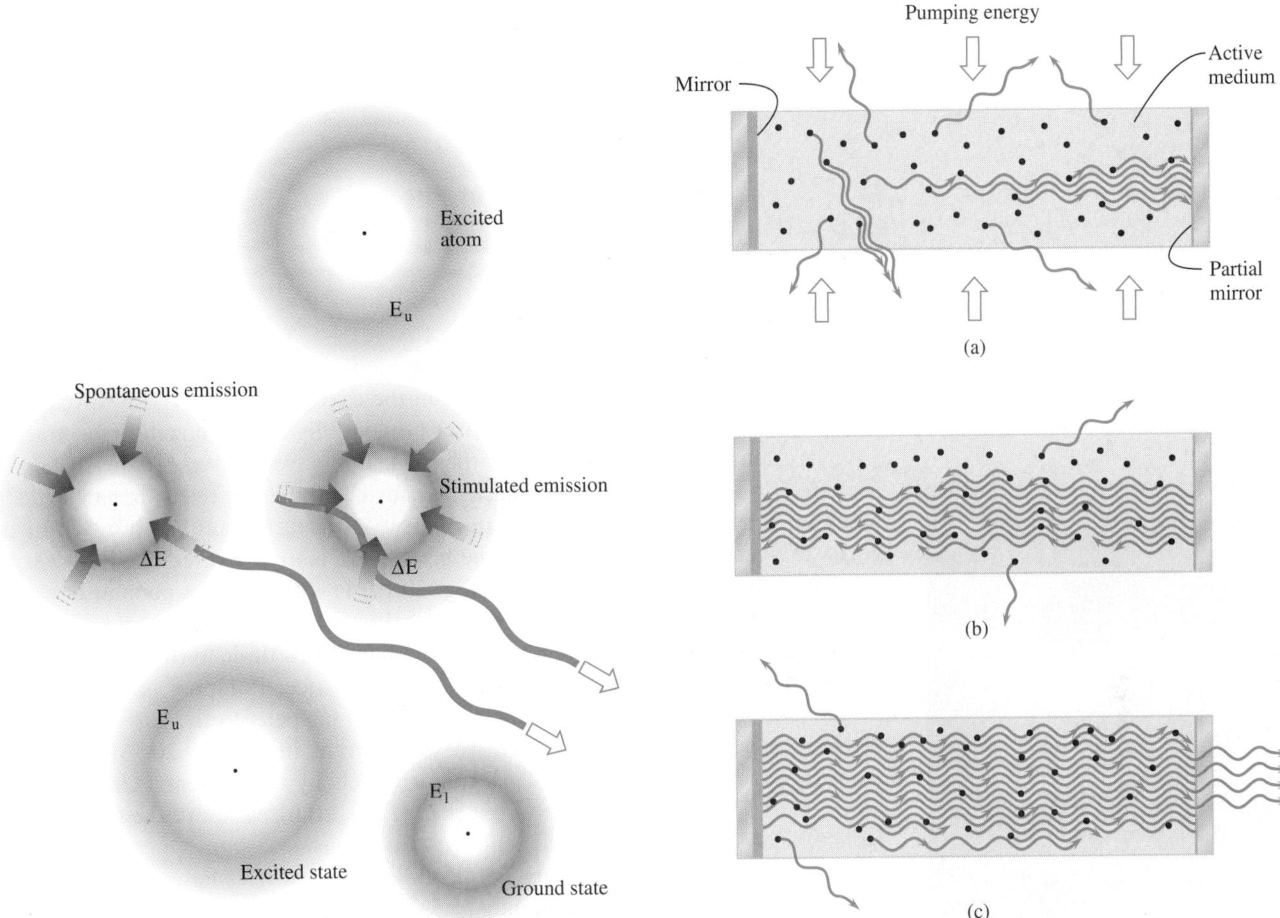

Figure 30.17 A population inversion: five atoms in a group where all but one are in an excited state. One atom (on the left) spontaneously returns to the ground state, emitting a photon, which causes a nearby atom to be stimulated into emitting an in-phase photon of its own.

Figure 30.18 A schematic representation of a laser. An active medium with most of its atoms in an excited state is located between two mirrors, one totally reflecting, the other 99% reflecting. Atoms spontaneously emit. Only photons along the axis stay within the cavity and stimulate additional emission. As the wave sweeps back and forth across the active medium, the wave builds until a portion leaks out as the laser beam. Energy can be pumped in continuously, in which case the beam can be continuous.

population inversion. An incident photon of the right frequency could then easily trigger an avalanche of stimulated photons—all in-phase, all perfectly in step. The initial triggering wave would continue to build as it swept across the active medium, as long as there were no dominant competitive processes (such as scattering) and provided the population inversion could be maintained. Moreover, the process could be enhanced by placing the active medium between two mirrors so that the lightwave passed back and forth through it many times. To allow some light to escape in the form of a beam, one of the mirrors could be made to be less than 100% reflecting; it would leak the beam (Fig. 30.18). In effect, energy (electrical, optical, chemical, whatever) would be pumped in to sustain the inversion, and a beam of light would be extracted. Such a device for **l**ight **a**mplification by **s**timulated **e**mission of **r**adiation is known as a **laser**.

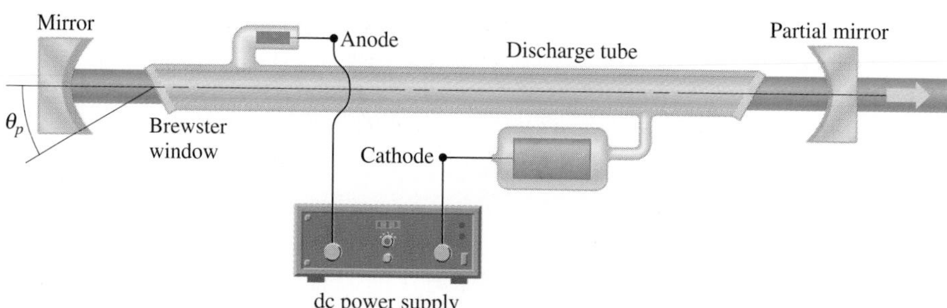

Figure 30.19 An early version of a helium-neon (He-Ne) continuous laser. The Brewster windows serve to polarize the beam.

The two mirrors constitute a **resonant cavity**. Only certain wavelengths that "fit" (p. 462) can set up standing waves in the cavity, and only those wavelengths will be sustained. Consequently, the laser produces light of unmatched spectral purity; it is quasimonochromatic and therefore has a very long coherence length. Moreover, only waves sweeping back and forth parallel to the axis can undergo many reflections and thereby build in intensity. Thus, the rays emerging from the cavity are very nearly parallel, and the laser beam is highly directional.

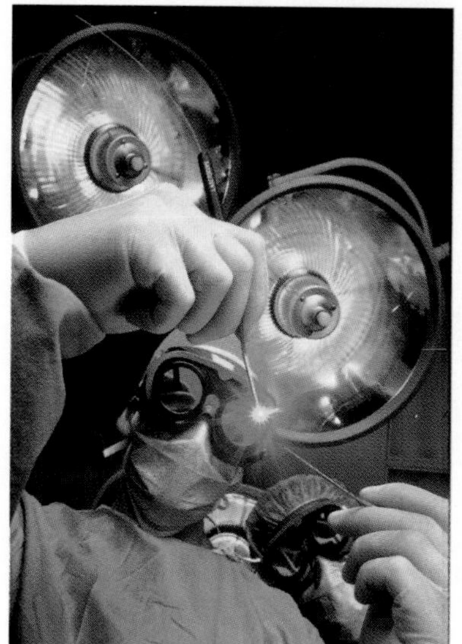

Surgery using a laser.

The helium-neon (He-Ne) laser (Fig. 30.19) is still the most popular device of its kind. It puts out a continuous beam, usually of a few milliwatts of red light at 632.8 nm. The active medium, a mixture of about 10 Pa of neon (the active centers) to 100 Pa of helium, is placed in a gas-discharge tube. Pumping is accomplished through a high-voltage electrical discharge. What happens to the gases can be understood using the simplified energy-level diagram in Fig. 30.20. Many helium atoms are kicked up into a number of upper levels by the current. After dropping down, most accumulate in a long-lived state (E_{H2}), 20.61 eV above the ground level, from which there are no allowed radiative transitions.

The excited He atoms inelastically collide with and transfer energy to ground-state neon atoms, raising them into a long-lived energy state E_{N4}. This level is 20.66 eV above the ground level, the difference (0.05 eV) having been supplied from the KE of the colliding atoms. There then exists a population inversion, among the neon atoms, with respect to the lower E_{N3} level. Spontaneous photons initiate stimulated emission, and a chain reaction begins down from E_{N4} to E_{N3}. The result is the emission of bright red light at 632.8 nm. The E_{N3} level readily drains off to the E_{N2}

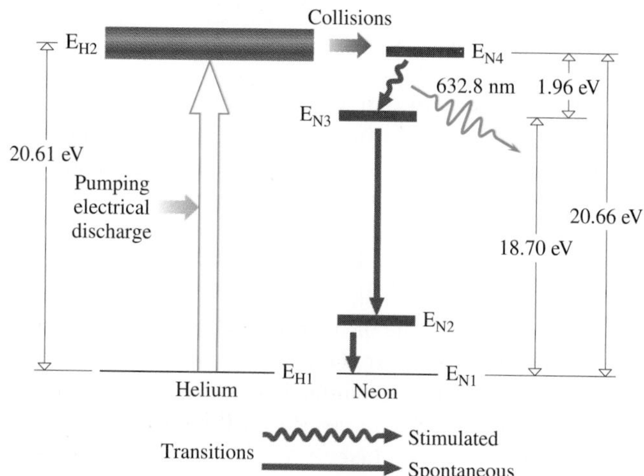

Figure 30.20 Simplified He-Ne laser energy levels. Here the lasing transition $E_{N4} \rightarrow E_{N3}$ is not down to the ground state. E_{N3} quickly dumps to E_{N2}, so the population inversion between E_{N3} and E_{N2} is constantly sustained.

level, sustaining the inversion. The latter is above the ground level, but that isn't significant—the important point is that lasing takes place between two upper levels. As a result, since the E_{N3} level is only sparsely occupied, the inversion is easy to maintain continuously, without having to half-empty the ground state.

Example 30.7 The helium-neon laser puts out a bright red beam at a wavelength of 632.8 nm. Please determine the difference in energy between the two states defining the transition.

Solution: [Given: $\lambda = 632.8$ nm. Find: ΔE.] There is a photon and an atomic energy transition, and that brings us back to Eq. (30.14): $E_i - E_f = hf$. Hence

$$\Delta E = \frac{hc}{\lambda} = \frac{(6.6262 \times 10^{-34}\ \text{J·s})(2.9979 \times 10^{8}\ \text{m/s})}{632.8 \times 10^{-9}\ \text{m}}$$

and $\Delta E = \boxed{3.139 \times 10^{-19}\ \text{J}} = 1.959$ eV.

▶ **Quick Check:** The order-of-magnitude of the answer, a few eV, is appropriate for outer-electron transitions. The 1.96-eV transition corresponds to an energy in joules of $(1.96\ \text{eV})(1.6018 \times 10^{-19}\ \text{J}) = 3.14 \times 10^{-19}$ J, and since that value equals hf, $f = 4.738 \times 10^{14} = c/\lambda$ and $\lambda = 632.8$ nm.

30.7 Atomic Number

The **characteristic X-rays** that appear as sharp spikes superimposed upon the broad Bremsstrahlung spectrum (Fig. 30.9) come in groups much like the Balmer and Paschen series. Barkla (1911) named them the *K*, *L*, *M*, . . . and so forth series and, today, the spikes in each group are labeled (recall the hydrogen lines) K_α, K_β, K_γ, and so on. Just as the Balmer lines are an indication of the hidden outer electron structure of the atom, these series are equally as significant for the inner electrons. Bohr discussed this matter with his friend Henry Moseley, and Moseley, in 1913, brilliantly found the pattern. After a tremendous solitary effort, he was able to devise an empirical equation describing the frequencies of the K_α lines in terms of the particular atomic structure of the target; that is, in terms of something he called the *atomic number.*

Each element has an integer atomic number, and it increases, one unit at a time, from hydrogen (1), to helium (2), to lithium (3) and beryllium (4), up to uranium (92), just like a place number marking consecutive positions in the Periodic Table (inside back cover). Moseley astutely concluded: "This quantity [atomic number] can only be the charge on the central positive nucleus." *The* **atomic number** *is the number of units of nuclear charge, the number of protons in the nucleus* (*Z*). The formula Moseley deduced was $f = C(Z - 1)^2$, where the constant *C* was subsequently shown to have the value $3cR/4$; hence

$$[\textit{for } K_\alpha \textit{ lines}] \qquad f = \left(\frac{3cR}{4}\right)(Z - 1)^2 \qquad (30.21)$$

Bohr conjectured that for all atoms, as the nuclear charge increased, the number of orbital electrons would also increase. One by one, they would build across the Periodic Table. Unhappily, the heavier atoms were beyond treating analytically with his simple theory. Despite that, W. Kossel (1914) suggested that if an inner tightly bound electron of one such atom was somehow removed from its orbit (for example, by bombarding it with cathode rays), all the other whirling electrons would cascade downward until that vacancy was filled. He called the innermost electron energy

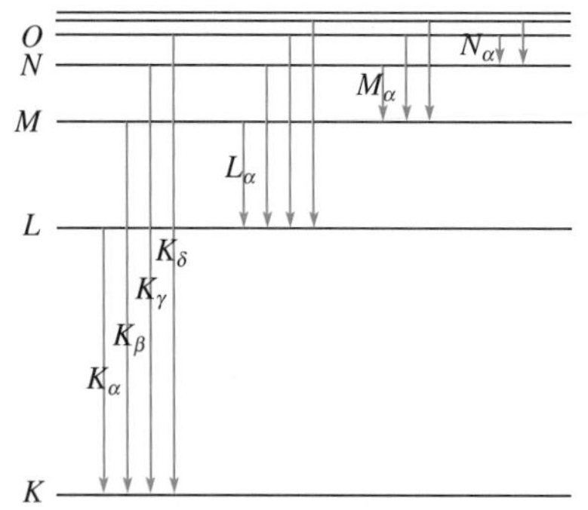

Figure 30.21 Inner-electron energy levels and the transitions that give rise to X-ray emission.

level *K*, the next *L*, and so on (Fig. 30.21), and that scheme led to a picture of electron transitions (much like the ones corresponding to the outer electrons) and the associated spectral lines. By comparison, these jumps were far greater in energy, and so emitted X-rays. For example, while 7.4 eV will remove the outermost electron from a lead atom, it takes 88 keV to remove either one of the *K* electrons.

Equation (30.21) can be rewritten for wavelengths, and it then takes on a familiar form:

$$\frac{1}{\lambda} = R(Z - 1)^2\left[\frac{1}{1^2} - \frac{1}{2^2}\right] \tag{30.22}$$

This equation looks just like Eq. (30.20) except for the $(Z - 1)^2$ instead of Z^2. Ordinarily, the $n = 1$, *K*-level has two electrons in it, but now one has been removed, leaving one remaining. As an electron from an upper level plunges toward the hole, it sees an effective nuclear charge reduced (via Gauss's Law) by one negative unit because of the single nearby orbital electron; it sees $+e(Z - 1)$, not $+eZ$.

Example 30.8 A target emits K_α radiation of wavelength λ. Show that

$$Z = 1 + \sqrt{\frac{4}{3\lambda R}}$$

Then suppose that $\lambda = 0.2510$ nm. What metal is being used?

Solution: [Given: $\lambda = 0.2510$ nm. Prove above expression and find *Z*.] Starting with Eq. (30.22)

$$\frac{1}{\lambda} = R(Z - 1)^2\left[\frac{1}{1^2} - \frac{1}{2^2}\right] = \frac{R(Z - 1)^2 3}{4}$$

and $$(Z - 1)^2 = \frac{4}{3\lambda R}$$

Hence $$Z = 1 + \sqrt{\frac{4}{3\lambda R}}$$

Thus

$$Z = 1 + \sqrt{\frac{4}{3(0.2510 \times 10^{-9}\text{ m})(1.097\,373 \times 10^7\text{ m}^{-1})}}$$

and $Z = 23.0$, therefore the metal is **vanadium**.

▶ **Quick Check:** Using the expression on the left, $1/\lambda = R(22)^2 3/4$; $\lambda = 0.251$ nm.

A Letter from the Army. Let it suffice to say that your son died the death of a hero, sticking to his post to the last. He was shot through the head, and death must have been instantaneous. In him the brigade has lost a remarkably capable signalling officer and a good friend.

REGARDING H. MOSELEY
Signalling Officer, 38th Brigade

The idea that the elements of the Periodic Table were to be ordered not by mass but by nuclear charge had been around for several years already. However, Moseley could directly observe the emitted X-rays and from them immediately tell whether or not the material under study was elemental and, if so, where in the Table it belonged! For instance, there was some controversy about argon (atomic weight 39.9) and potassium (atomic weight 39.1), since the former is an inert gas and should be located with the other inert gases (inside back cover), which would have it strangely preceding the lighter alkali metal. Moseley's X-rays revealed argon to have an Atomic Number of 18, whereas that of potassium was 19. Argon clearly belonged before potassium.

Unfortunately, Harry Moseley would solve no more of nature's mysteries; in 1915, at the age of twenty-eight, he was killed at Gallipoli in one of the most useless campaigns of World War I ("the war to end all wars").

Core Material

The total power radiated at all wavelengths by a blackbody is given by the **Stefan-Boltzmann Law**, namely

$$P = \sigma A T^4 \tag{30.1}$$

where *A* is the area, *T* is the absolute temperature in kelvins, and

$$\sigma = 5.6703 \times 10^{-8}\text{ W/m}^2\cdot\text{K}^4$$

Given the **total emissivity** (ε), the power radiated by a surface is

$$P = \varepsilon\sigma A T^4 \qquad [30.2]$$

For a body in an enclosure, the net power radiated (when $T > T_e$) or absorbed (when $T < T_e$) is

$$P = \varepsilon\sigma A(T^4 - T_e^4) \qquad [30.3]$$

For blackbodies, **Wien's Displacement Law** is

$$\lambda_{max} T = \text{constant} \qquad [30.4]$$

where the constant is 0.002 898 m·K. **Planck's Radiation Law** is

$$I_\lambda = \frac{2\pi hc^2}{\lambda^5}\left[\frac{1}{e^{hc/\lambda k_B T} - 1}\right]$$

where $h = 6.626\,075\,5 \times 10^{-34}$ J·s (p. 1048).

In the Photoelectric Effect, the stopping potential is given by

$$KE_{max} = eV_s \qquad [30.6]$$

According to Einstein, **light is a stream of energy quanta**, each with an energy

$$E = hf = \frac{hc}{\lambda} \qquad [30.7]$$

The **Photoelectric Equation** is then

$$hf = KE_{max} + \phi \qquad [30.8]$$

The minimum wavelength of Bremsstrahlung (p. 1054) is given by

$$\lambda_{min} = \frac{ch}{eV} \qquad [30.9]$$

In the **Compton Effect**, a photon can scatter off a free electron with an accompanying wavelength shift given by

$$\Delta\lambda = \lambda_s - \lambda_i = \frac{h}{m_e c}(1 - \cos\theta) \qquad [30.13]$$

According to Bohr, for a hydrogen atom, a photon is emitted when the atom makes a transition from one energy level to another; that is

$$E_i - E_f = hf \qquad [30.14]$$

The resulting orbits have radii given by

$$r_n = n^2\frac{\hbar^2}{m_e k Z e^2} = n^2 r_1 \qquad [30.17]$$

and energies given by

$$E_n = -\frac{2\pi^2 k^2 e^4 m_e}{h^2}\frac{Z^2}{n^2} \qquad [30.18]$$

Moseley showed (p. 1065) that

$$[\textit{for } K_\alpha \textit{ lines}] \qquad f = \left(\frac{3cR}{4}\right)(Z - 1)^2 \qquad [30.21]$$

Suggestions on Problem Solving

1. Some useful constants to keep in mind are

$$1\text{ eV} = 1.602\,177 \times 10^{-19}\text{ J}$$

$$hc = 1.239\,842 \times 10^{-6}\text{ eV·m} \approx 1240\text{ eV·nm}$$

$$h = 6.626\,08 \times 10^{-34}\text{ J·s} = 4.135\,67 \times 10^{-15}\text{ eV·s}$$

$$\hbar = 1.054\,573 \times 10^{-34}\text{ J·s} = 6.582\,12 \times 10^{-16}\text{ eV·s}$$

2. In problems involving electrons, we might be given the energy in eV or, equivalently, told the potential across which an electron moves. It's then convenient to use $m_e = 0.511$ MeV/c^2. The 1/c^2 tells us we are dealing with mass, not energy (MeV). To get the electron's mass in kg, *multiply* 0.511 MeV/c^2 by 1.602×10^{-13} J/MeV and then *divide* by c^2 in (m/s)^2. The 1/c^2 is not to be treated as if it were a unit like s in m/s, which we get rid of by multiplying by s. The quantities 1/c^2 or 1/c are literal divisions; they're not carried out numerically because we anticipate that they will soon cancel and it's convenient to leave them in this form, provided it doesn't confuse the issue.

3. When calculating things such as the minimum X-ray wavelength $\lambda_{min} = hc/eV$, the energy will often be encountered in electron-volts, whereupon it's helpful to use $hc = 1.239\,842 \times 10^{-6}$ eV·m.

4. Computations concerning the Compton Effect can be simplified a bit by utilizing the Compton wavelength: $(h/m_e c) = 0.002\,426$ nm.

5. Many modern calculators have built-in constants like c, h, e, k_B, m_e, etc. It will often be easier to use these built-in numbers and let them determine the form of the units (that is, eV versus J, etc.).

Discussion Questions

1. A candle or match flame usually has a distinct yellow color to it. One might guess that the color comes from sodium, but this is generally *not* the case. It's actually due to blackbody radiation from minute hot (≈2000 K) particles of soot. How might you confirm this fact? (Incidentally, the bluish light at the base of the flame is from CH, C_2, and CO_2 molecular emission.)

2. Not very many solid materials can be heated to incandescence, and even fewer liquids can. Discuss and explain this observation. How do the following substances behave regarding incandescence: iron, water, glass, carbon, plastic, gasoline, banana, soap, wood, ceramics?

3. In order to test the photon hypothesis, Walther Bothe

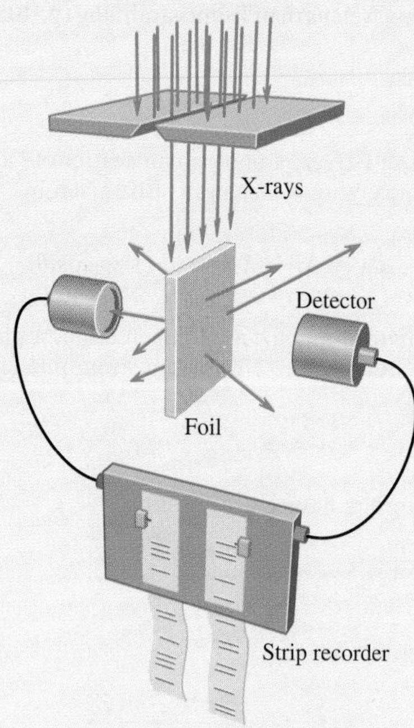

Figure Q3

conducted the following experiment (Fig. Q3). A low-intensity beam of X-rays was shone on a thin foil, which subsequently emitted X-rays (via X-ray fluorescence) toward two counters. The counts from the two detectors were recorded as they occurred and no correlation in time was observed. What does that "prove"? Explain.

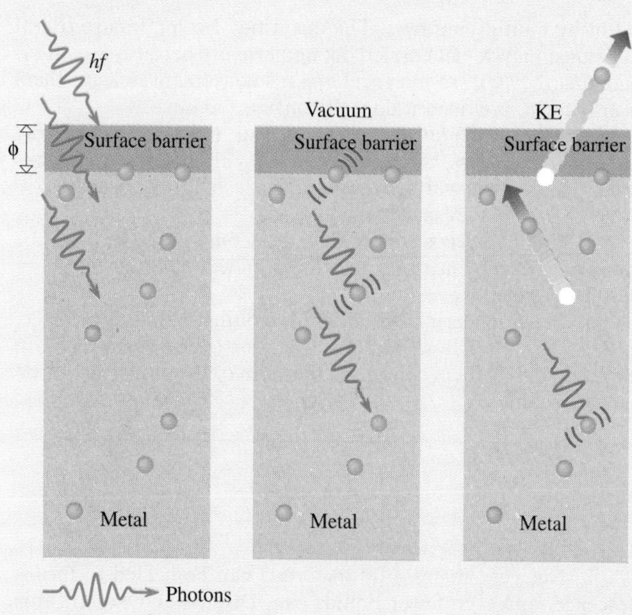

Figure Q4

4. Figure Q4 shows a sequence of drawings of three photons interacting with a metal. Explain each portion of the figure. Why does the liberated electron have a maximum kinetic energy?

5. With a bright source, such as a high-powered laser, it is possible to have a photoelectron absorb two or more photons before it leaves the metal (p. 1053). Given that an electron absorbs N identical photons of frequency f, how should Eq. (30.8) be modified, if at all? What will happen to the stopping potential as compared to the single photon effect? Does the work function change? Does the threshold frequency change? Explain your answers in detail.

6. Figure Q6 shows a current-versus-voltage curve for the Photoelectric Effect using polychromatic light. Explain every detail of the curve. Which frequency is higher?

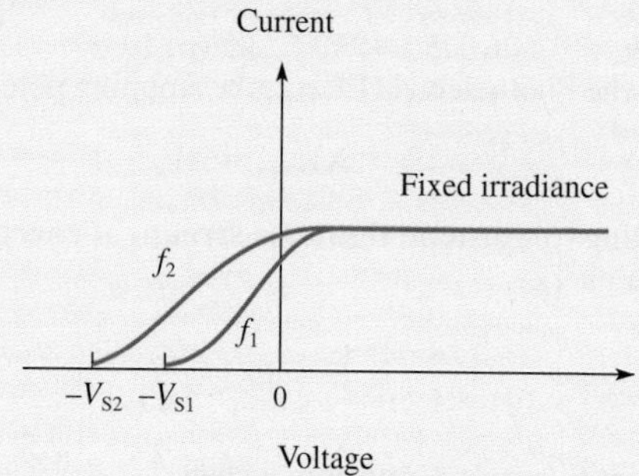

Figure Q6

7. Light absorbed by a substance may result in a chemical change of that substance via a process known as a *photochemical reaction* (which is why wine and aspirin are kept in dark-colored bottles). For each such reaction, there is a threshold frequency below which it will not occur. Assuming that the molecule must absorb an amount of energy that will trigger the process E_a, known as the *activation energy*, explain (as Einstein first did) the existence of a threshold frequency.

8. Figure Q8 shows some data for gamma-ray Compton scattering. Even though the gamma-ray wavelengths are much smaller than those of X-rays, the slope was found to be 2.4×10^{-12} m, in fine agreement with theory. Explain the curve in every regard.

9. A photon cannot be absorbed by a free electron—prove that statement by using Conservation of Energy and Momentum. Assume the electron is initially at rest. First, draw a diagram of the problem as seen from a coordinate system fixed with the center-of-mass—show it before and after the collision. Note that by definition the initial momentum in the center-of-mass system is zero. What is the final momentum? Write expressions for the initial and final energy. Now write the statement of Conservation of Energy (as seen from the center-of-mass). The fact that this implies that $m > \gamma m$ makes the initial assumption impossible.

10. Discuss the relationships between E, KE, and PE for any Bohr orbit of hydrogen. In the process explain Fig. Q10 and offer a definition of the binding energy of the electron. Which of these quantities is positive and which negative?

11. Figure Q11 shows a hypothetical 4-level laser. How does it work and what are the advantages of the arrangement?

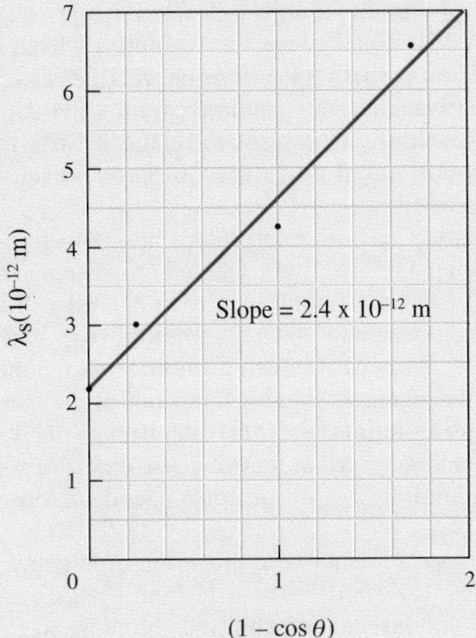

Figure Q8

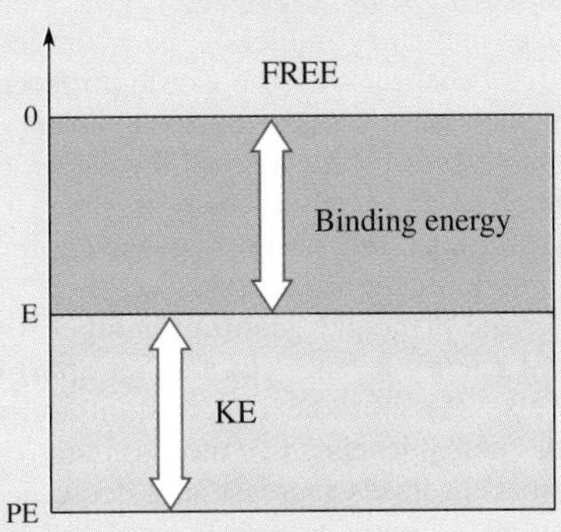

Figure Q10

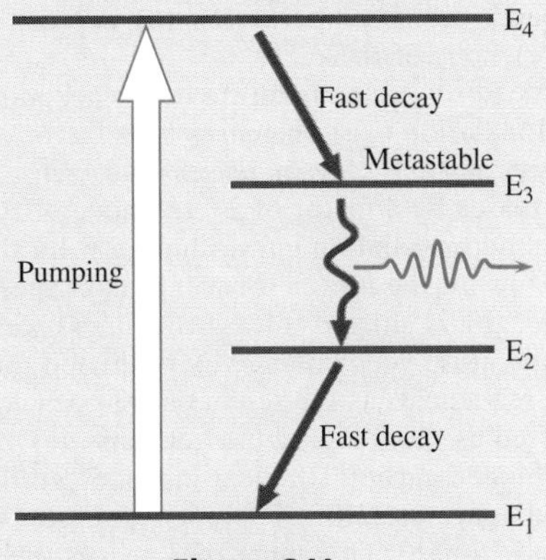

Figure Q11

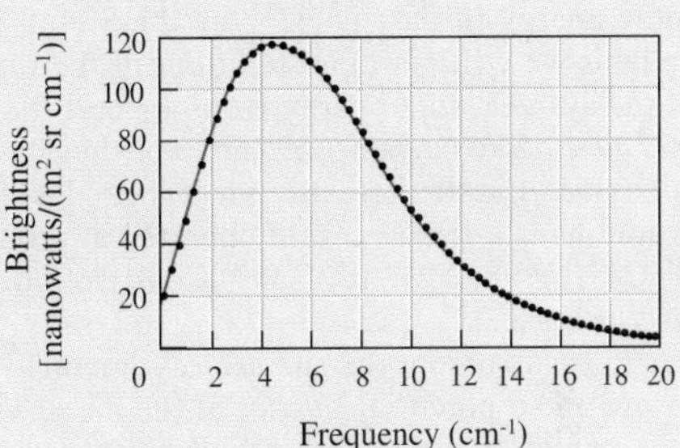

Figure Q12

12. It is theorized that the Universe was created roughly 17 thousand million years ago and is still expanding from that explosive moment. If that's the case, the original radiant fireball should have cooled considerably (via the Doppler Effect) as the Universe spread out to its present size. Less than a year after the so-called Big Bang, the radiation and matter should have reached equilibrium. By 300,000 years later, the radiation was red-shifted down to $\approx$4000 K and neutral atoms formed. Since then, the Universe has expanded about a thousandfold, and we can now expect isotropic cosmic background radiation with a blackbody temperature of about 3 K. Figure Q12 is a plot of brightness versus wavelength of the radiation detected by the *Cosmic Background Explorer* (*COBE*) satellite. The measured microwave spectrum corresponds to the black dots, the solid line is the Planck blackbody curve for a temperature of 2.735 ± 0.06 K. What can you conclude from all of this?

13. In 1986, a single barium ion (and in a different experiment, a mercury ion) was held in a "trap" and excited by laser light (p. 1062). The ion has two energy levels above the ground state that were utilized. The first was an ordinary short-lived state, and when bathed in light of the proper frequency, the atom absorbed and emitted 100 million times per second, putting out a seemingly continuous stream of light that was easily observed in a microscope. The other higher state was metastable and, once excited, the atom took several seconds to make the transition back down to the ground state via the emission of a lone photon (which went undetected). When stimulated into the first state, the atom shone with a "continuous" light, but whenever it was twinked into the metastable state, the shining light blinked off only to blink back on seconds later. Explain what happened.

Multiple Choice Questions

1. Pour hot water into a box whose outer walls are painted either black or white. Place thermal detectors at each face. Even though the entire surface of the box is at one temperature, (a) the black faces will be hotter (b) the white faces will be hotter (c) the black faces will radiate more energy (d) the white faces will radiate more energy (e) none of these.
2. It follows from the Stefan-Boltzmann Law that the energy per unit area per unit time radiated by a blackbody (a) depends on its total mass (b) depends on its color

(c) depends on its temperature (d) depends on its total area (e) none of these.

3. When the absolute temperature of a blackbody is doubled, its total radiation (a) increases by a factor of 16 (b) stays constant (c) increases by a factor of 4 (d) decreases by a factor of 2 (e) none of these.
4. All blackbody radiation curves have a peak that (a) is shifted toward the longer wavelengths as the temperature goes up (b) is shifted toward the longer wavelengths as the temperature goes down (c) is shifted toward the higher frequencies as the temperature goes down (d) is not shifted as the temperature goes up (e) none of these.
5. Old-fashioned carbon filament incandescent light bulbs operated at around 2100 K. Modern tungsten lamps operate at around 2500 K. This more than triples the efficiency because (a) it's always better to have the strength of metal (b) at the higher temperature the blackbody curve puts more energy in the visible (c) the Stefan-Boltzmann Law tells us we will get more light out of the tungsten (d) we always get more energy from a metal than from a semiconductor like carbon (e) none of these.
6. For us humans the theoretically ideal incandescent light bulb would have a filament that operated at around (a) 20° C (b) 98.6° F (c) 2500 K (d) 6000 K (e) none of these.
7. In order to escape from the surface of a metal, an electron must overcome a potential barrier at the metal-vacuum interface, which requires (a) a high temperature $T = hf/k_B$ (b) an energy equal to ϕ (c) a stopping potential (d) a photon of energy $h\lambda$ (e) none of these.
8. The work function of a metal depends on (a) the frequency of the incident radiation (b) the voltage applied (c) the maximum kinetic energy of the electrons (d) the stopping potential (e) none of these.
9. In the photoelectric effect, if $f > f_0$ and the irradiance of the incident beam is doubled, the photocurrent (a) is unchanged (b) decreases by a factor of 4 (c) doubles (d) is halved (e) none of these.
10. When dealing with photoelectrons, a plot of (a) e times the stopping potential against photon frequency is a straight line for each metal (b) photocurrent against potential is a straight line of slope h for each metal (c) photocurrent against incident intensity is a straight line of slope h for each metal (d) incident intensity against photon frequency is a straight line of slope ϕ for each metal (e) none of these.
11. When a photon is Compton-scattered, the (a) minimum shift $\Delta\lambda$ will equal twice the Compton wavelength (b) minimum shift $\Delta\lambda$ will equal h times twice the Compton wavelength (c) maximum shift $\Delta\lambda$ will equal h times twice the Compton wavelength (d) maximum shift $\Delta\lambda$ will equal twice the Compton wavelength (e) none of these.
12. Conservation of both energy and momentum lead to the conclusion that (a) only free electrons can absorb a photon (b) only bound electrons can absorb a photon (c) no electrons can absorb a photon (d) both free and bound electrons can absorb a photon (e) none of these.
13. As the hydrogen atom goes from one Bohr orbit to another, increasing in radius, the (a) electron's speed increases (b) electron's speed decreases (c) electron's speed remains unaltered (d) proton's speed increases (e) none of these.
14. The higher the principal quantum number of a Bohr orbit, the (a) smaller the orbit and the less negative the energy (b) larger the orbit and the more negative the energy (c) larger the orbit and the less positive the energy (d) larger the orbit and the more positive the energy (e) none of these.
15. The classical planetary model of the hydrogen atom is unacceptable because (a) the Coulomb force is too weak to hold the electron (b) the nucleus is too powerful to allow the electron to orbit it (c) the electron is accelerating and must radiate away its energy (d) the electron's angular momentum is all wrong to be in such an atom (e) none of these.
16. The greater the principal quantum number, the (a) closer are adjacent energy levels (b) farther apart are adjacent energy levels (c) more nearly constant are the separations of adjacent energy levels (d) more rapidly the separations between adjacent levels increase and decrease (e) none of these.
17. A population inversion means that (a) there are more atoms in one gas than in another (b) there are more atoms in some excited state than in a lower state (c) there are more states populated than unpopulated (d) the lower states are filled rather than the higher ones (e) none of these.
18. To pump a laser means to (a) lower all its electron states (b) remove some gas from it (c) put pressure on the active medium (d) excite its atoms (e) none of these.

Problems

THE OLD QUANTUM THEORY

ATOMIC THEORY

1. [I] Suppose that a person has an average skin temperature of 33° and a total naked area of 1.4 m^2. If the person's total emissivity is 97%, find the net power radiated per unit area, the irradiance, when the environment is room temperature. How much energy does that person radiate per s?

2. [I] An object resembling a blackbody is raised from a temperature of 100 K to 1000 K. By how much does the amount of energy it radiates increase?

3. [I] Assume you are a blackbody at 33°C (external temperature). What is the wavelength at which you radiate most energy? (See Problem 2).

4. [I] An object resembling a blackbody at room temperature (20°C) radiates energy into the environment. What is the wavelength that carries away the most energy?

5. [I] A class O blue-white star has a surface temperature of around 40×10^3 K. At what frequency will it radiate most of its energy?

6. [I] The Sun radiates the most energy at 470 nm. Taking it to be a blackbody, determine the temperature of its outer layer, the photosphere.
7. [I] Show that the units of action defined as either energy multiplied by time, or momentum multiplied by position, are equivalent.
8. [I] What is the energy in joules of a 0.100-nm photon?
9. [I] What is the wavelength of a 2.50-eV photon?
10. [I] What is the energy in keV of a 0.20-nm photon?
11. [I] A typical light level for good reading corresponds to about 2×10^{13} photons per second per square centimeter. If these have an average wavelength of 550 nm, what is the corresponding irradiance?
12. [I] What is the threshold frequency that will result in photoemission from zinc?
13. [I] Photons of wavelength 650 nm liberate photoelectrons with negligible kinetic energy from a metal target. What is the threshold frequency of the metal?
14. [I] Determine the cut-off wavelength for the emission of photoelectrons from tungsten, which has a work function of 4.52 eV.
15. [I] Ultraviolet radiation of wavelength 200 nm is shone on a zinc target. What's the maximum kinetic energy of the emitted electrons?
16. [I] In the photoelectric effect, what frequency of radiant energy should be shone on iron to produce electrons with a maximum kinetic energy of 1.50 eV?
17. [I] X-rays with a wavelength of 110 pm are scattered off free electrons at an angle of 20.0°. Find the change in λ.
18. [I] What is the shortest-wavelength X-ray emanation that can be expected from a tube operating at a voltage of 30.0 kV?
19. [I] At what scattering angle does the photon come off in the Compton Effect when the electron has a maximum kinetic energy?
20. [I] Compton shone 71.0-pm photons at a target containing loosely bound electrons and examined the scattered photon beam at 90.0°. What was the wavelength of the scattered photon?
21. [I] A 60-keV photon is Compton-scattered off an electron at rest. In the process, a 25-keV photon comes sailing off at 35°. What is the kinetic energy of the scattered electron?
22. [I] Determine (to 3 significant figures) the energy in electron-volts and in joules of the first excited state of the hydrogen atom.
23. [I] Compute the energy in eV of the second excited state of a hydrogen atom (to 3 significant figures).
24. [I] With Problems 22 and 23 in mind, what is the energy (in joules) of the quantum of light given out when a hydrogen atom drops from its second to its first excited state?
25. [I] With the last problem in mind, what is the frequency and wavelength of the light given out when a hydrogen atom drops from its second to its first excited state?
26. [I] The carbon dioxide laser usually emits at 10.6 μm in the IR. Typically such a device puts out a continuous emission of from a few watts to several kilowatts. Determine the energy difference in eV between the two laser levels for this wavelength.
27. [I] What is the radius of the second excited state of the hydrogen atom according to the Bohr Theory?
28. [I] A certain variation of He-Ne laser operates between two levels that are 3.655×10^{-19} J apart. What color is the beam? Compute the wavelength.
29. [I] Compute the principal quantum number corresponding to a hydrogen atom with a 1-m orbital radius.
30. [I] The anode of an X-ray tube emits K_α radiation of wavelength 0.075 9 nm. What metal is the target made of?
31. [II] At what temperature will an object resembling a blackbody emit a maximum amount of energy per unit wavelength in the red end of the visible region of the spectrum ($\lambda = 650$ nm)?
32. [II] A blackbody is at a temperature of 6000 K. At what wavelength will it radiate the most energy per unit wavelength?
33. [II] The current in a Photoelectric Effect experiment decreases to zero when the retarding voltage is raised to 1.25 V. What is the maximum speed of the electrons?
34. [II] Suppose that 60-nm radiant energy is incident on a metal whose work function is negligibly small. Compute the maximum speed of the liberated photoelectrons.
35. [II] Photons of wavelength 220 nm impact on a metal target and liberate electrons with kinetic energies ranging from 0 to 61×10^{-20} J. Determine the threshold frequency and wavelength.
36. [II] In the Photoelectric Effect, the target has a stopping potential of 4.00 V when the incident energy has a momentum of 3.50×10^{-27} kg·m/s. What is the threshold frequency?
37. [II] An electron traveling at 1.00×10^8 m/s is fired at a metal target. On impact, it rapidly decelerates to half that speed emitting a photon in the process. What is the wavelength of the photon?
38. [II] In the Compton experiment, derive an expression for the kinetic energy of the scattered electron in terms of the incident and scattered photon wavelengths.
39. [II] On scattering via the Compton Effect, a photon undergoes a *fractional wavelength change* $(\lambda_s - \lambda_i)/\lambda_i$ equal to 6.00%. If the incident photon has a wavelength of 0.020 0 nm, at what angle is the detector to the incident beam?
40. [II] X-ray photons of wavelength 0.220 0 nm are Compton-scattered at 45°. What is the energy of the scattered photons (in eV)?
41. [II] What is the speed of the electron in the ground state of the hydrogen atom according to Bohr?
42. [II] Determine the force holding the electron in orbit in the ground state of the hydrogen atom.
43. [II] Compute the ground-state energy of singly ionized helium using the Bohr Theory.
44. [II] Use the Bohr Theory to prove that for hydrogen-like atoms

$$v_n = \frac{kZe^2}{n\hbar}$$

45. [II] Derive an expression for the frequency (f_n) of the electron revolving in the nth Bohr orbit of a hydrogen-like atom.
46. [II] With Problem 45 in mind, use the Bohr Theory to prove that

$$|E_n| = \tfrac{1}{2}nhf_n$$

for the various orbits (and not $E_n = nhf_n$).

47. [III] The Compton equation (30.13) is a formula in terms of photon parameters and the constant m_e. To derive it, start with Eq. (30.11) and write expressions for conservation of the x- and y-components of momentum. Show that

$$p_i^2 + p_s^2 - 2p_i p_s \cos\theta = p_e^2$$

Then, using Conservation of Energy, establish that

$$E_e = E_i + m_e c^2 - E_f$$

Substitute both of these formulas into the equation for the electron's energy

$$E_e^2 = c^2 p_e^2 + m_e^2 c^4$$

thereby getting an expression containing the electron's constant rest-energy and the energies and momenta of only the photons. Using $E = pc$ for photons, simplify, and Eq. (30.13) follows in short order.

48. [III] Approximate the time it takes for an electron in the Photoelectric Effect to absorb enough energy ($\approx$1 eV or 2 eV) from a classical electromagnetic wave to be liberated. First assume a He-Ne laser source of 10 W/m^2. Compare this result with the time corresponding to an irradiance of about 1 μW/m^2, which is detectable by sodium. Assume the electron in a sodium atom absorbs over an area equal in radius (0.10 nm) to its orbit (which should be giving it the benefit of the doubt).

49. [III] When a photon is absorbed by an atom in the process of excitation, the atom must recoil if it is to conserve momentum. Consider a photon able to raise a 939-MeV/c^2 hydrogen atom into its first excited state. Determine the recoil energy of the atom and notice that it is negligible.

50. [III] Compute the energy of the third excited state of the helium atom in the Bohr Theory and compare it to the first excited state of hydrogen.

31

Quantum Mechanics

THE PACE OF THINGS intellectual picked up after the First World War (caution seemed less fitting in a battered world). The Jazz Era was largely a period of revolt and irreverence, and that selfsame turbulent mood gave form to the new physics that rose like a flash out of the Roaring Twenties. The hodgepodge of *ad hoc* ideas that constituted the Old Quantum Theory had gone about as far as it could. Quantum Mechanics emerged around 1925 as a complete theoretical structure that subsumed, and went far beyond, Classical Physics and the Old Quantum Theory.

Quantum Mechanics is far too mathematically complex for us to attempt

Prince Louis de Broglie (1892–1987) spent much of his life teaching physics at the Sorbonne and at the Institut Henri Poincaré, both in Paris.

to understand it here. Indeed, as Richard Feynman (1967) penetratingly remarked, "nobody understands quantum mechanics." But we can get a sense of it, see where it came from, learn a little about its strengths and weaknesses, and appreciate some of its accomplishments.

THE CONCEPTUAL BASIS OF QUANTUM MECHANICS

Amazingly, when physicists set about reformulating classical theory in the 1920s, they seemed at first to have come upon an overabundance of riches. Heisenberg created Matrix Mechanics, Dirac produced Transformation Theory, and Schrödinger devised Wave Mechanics, all almost simultaneously. Though their approaches were distinctive and their resulting formalisms seemed equally unique, it was soon shown that all three theories were essentially equivalent. These formulations are the basis of what is broadly called Quantum Mechanics. Our study will be cursory at best, and limited to only Wave Mechanics.

31.1 De Broglie Waves

In the summer of 1923, the French aristocrat Prince Louis Victor Pierre Raymond de Broglie (pronounced to rhyme with *Troy*) proposed that the wave-particle duality (p. 1053) manifested by radiant energy might be a fundamental characteristic of *all* entities (for example, electrons, protons, and neutrons). Einstein had already shown that substantial matter (matter possessing mass) and radiant energy are interconvertible. Why should they not display similar properties? In particular, might not substantial matter have some sort of wave aspect associated with it? Suppose that Eq. (30.10), $p = h/\lambda$, which describes photons, applies to all matter. In that eventuality, with $p = mv$ for material entities, we have

$$\lambda = \frac{h}{mv} \tag{31.1}$$

Electron micrograph of a neuron (see Fig. Q1). With an accelerating voltage of around 10^5 V, the electrons have a de Broglie wavelength of about 0.004 nm. The photo of an insect shown on p. 1073 was also made using electrons.

Matter in motion has a wavelength. Not that de Broglie knew how to visualize such waves or even what was doing the waving. A particle moving with a constant momentum is associated with a monochromatic wave ($\lambda = h/p$) having a single frequency determined by its *total energy* ($f = E/h$) via Eq. (24.9). As with all waves, it moves with a phase speed given by $v_p = f\lambda$, which is usually different from, though related to, the speed of the particle (Problem 24).

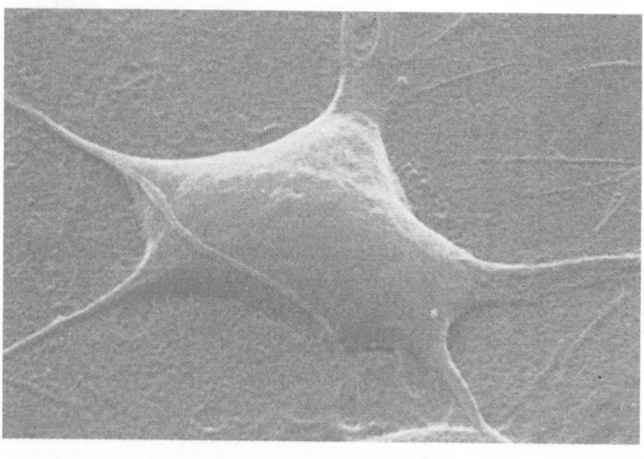

However theoretical, the idea that electrons and protons were that much more like photons brought a pleasing symmetry to nature. This fascinating speculation was the basis of a Ph.D. thesis that de Broglie submitted to his physics professors in Paris. Still, the concept was considered outlandish, and a copy of the work was sent to Herr Professor Einstein for his opinion. The great man was enthusiastically supportive, and the degree was promptly granted. Six years later (1929), de Broglie received the Nobel Prize for the idea that particles are waves—but a remarkable accident happened first.

In April of 1925, the American C. J. Davisson was busy scattering electrons off a polycrystalline nickel target.

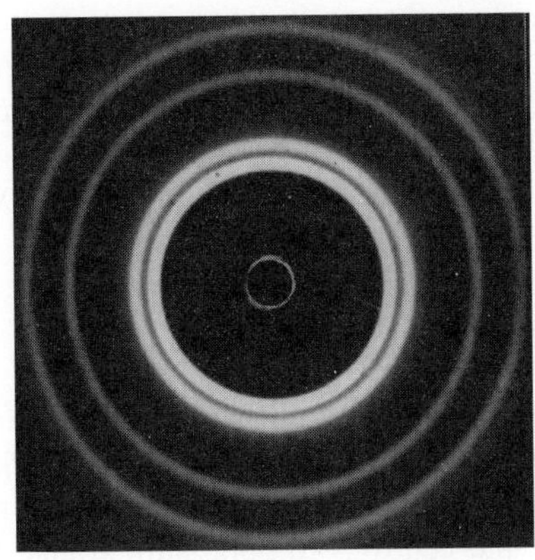
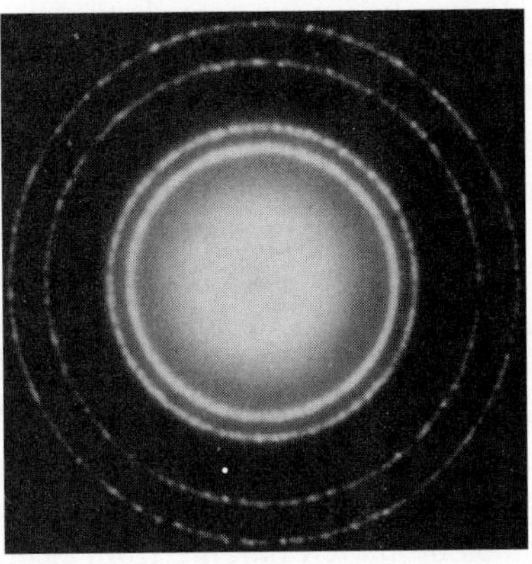

The concentric ring patterns produced when a beam of X-rays (left) and a beam of electrons (right) passes through the same thin polycrystalline aluminum foil. The two diffraction patterns are almost identical.

While collaborating with L. Germer, an explosion rocked the lab, but after putting the experiment back together, it was clear that something very strange had happened: the data "completely changed." Unbeknownst to them, while cleaning up the target through prolonged heating, the nickel sample had reformed into a few large crystals. When they subsequently directed an electron beam at the target, they saw a diffraction pattern identical to the one produced by X-rays (p. 1023), even though it's impossible for a stream of particles to do that. Another year passed before anyone realized that Davisson and Germer had verified de Broglie's hypothesis; Eq. (31.1) matched the data in every detail. *Electrons with comparable momenta to those of X-ray quanta are diffracted as if they were waves of the same wavelength.*

Example 31.1 Calculate the wavelength of an electron once it has been accelerated through a potential of 110 V and compare that to X-rays. Assume the speed is nonrelativistic.

Solution: [Given: $V = 110$ V for an electron. Find: λ.] To get λ, we will apply Eq. (31.1), and so we first need v. Since for the electron $KE_f = PE_i$, $\frac{1}{2}m_e v^2 = eV$

$$v = \sqrt{\frac{2eV}{m_e}} = 6.22 \times 10^6 \text{ m/s}$$

it follows that

$$\lambda = \frac{h}{m_e v} = 1.17 \times 10^{-10} \text{ m} = \boxed{0.117 \text{ nm}}$$

This is in the wavelength range of X-rays, just about the size of an atom.

▶ **Quick Check:** We'll first derive a useful expression for λ: $KE = p^2/2m$, $\lambda = h/p = h/\sqrt{2m(KE)}$, multiplying top and bottom by c, and using the fact that $hc = 1239.8$ eV·nm (p. 1067), we have $\lambda = (1239.8 \text{ eV·nm})/\sqrt{2(mc^2)(KE)}$. For an electron, $mc^2 = 0.511$ MeV, and using the kinetic energy in eV, we have

$$[\textit{for electrons}] \qquad \lambda = \frac{1.226}{\sqrt{KE}} \text{ nm} \qquad (31.2)$$

Here, where $KE = 110$ eV, $\lambda = 0.117$ nm.

Davisson shared the 1937 Nobel Prize with G. P. Thomson, J. J. Thomson's son, and therein lies another irony. The younger Thomson started out deliberately to test the idea of matter waves. By passing electrons through a thin metal foil, he confirmed that they behaved exactly as do X-rays of equal wavelength. And so J. J. Thomson "proved" that electrons were *particles,* and G. P. Thomson "proved" they were *waves.*

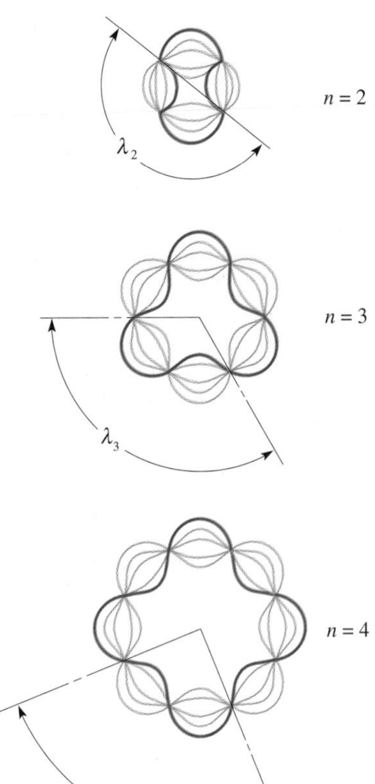

Figure 31.1 De Broglie standing waves for three excited states ($n = 2$, 3, 4) of hydrogen. Each Bohr orbit corresponds to a standing-wave configuration for the electron.

De Broglie's wave-particle theory offered another provocative interpretation of the Bohr atom. The quantization of angular momentum, $L = n\hbar$, means that $mvr = nh/2\pi$ and $mv = nh/2\pi r$. But if $mv = h/\lambda$, $n\lambda = 2\pi r$; in other words, the circumference of each of Bohr's allowed circular orbits ($2\pi r$) exactly equals an integer multiple of the electron wavelength ($n\lambda$). The orbits are electron standing-wave configurations (Fig. 31.1).

31.2 The Principle of Complementarity

When light propagates beyond an obstruction or through a hole, it casts a complex pattern in the process of diffracting, a pattern that can be analyzed in all its detail only if the light is taken as a wave. It is remarkable that researchers using all sorts of beams of matter, ranging from electrons and neutrons to potassium atoms, have now observed exactly the same complex interference and diffraction patterns. Classically, there is just no way to understand how beams of atoms can come together on a screen in clumps that exactly match optical interference patterns.

If the "logic" of Eq. (31.1) holds, and no one knows if it does, it applies equally well to oranges, baseballs, and fire engines as it does to electrons. The momentum of an ordinary macroscopic object in motion is so gigantic that λ is much too small to ever measure, and that even applies to something as small as a tiny dust mote gently falling through the air. Bullets may well have de Broglie wavelengths, though we'll never see them bending around obstacles. Unless you can pass a camel through the eye of a needle, observing diffraction (and thereby λ) is out of the question for structures made up of millions of atoms.

Microparticles (electrons, protons, photons, atoms, and so forth) propagate as if they were waves and exchange energy as if they were particles—that's the *wave-particle duality*. When we measure the arrival of one of them at a detector, we always measure the energy it delivers to a single point, even though that point may be part of a pattern that could only be created by a wave. In any experimental event, such entities will do one or the other—they cannot simultaneously manifest the properties of wave and particle. The imagery of particle and wave are two aspects of the complete picture of the photon, and they do not compete; rather, they complement each other. That notion (which does not immediately make things any less disconcerting) was first mentioned by Bohr in 1925 and came to be known as the **Principle of Complementarity**. Energy multiplied by temporal period ($T = 1/f$) equals momentum multiplied by spatial period (λ):

$$\mathrm{E}T = p\lambda = h$$

The particle notions of E and p complement the wave notions of T and λ.

The rabbi spoke three times; the first talk was brilliant: clear and simple. I understood every word. The second was even better: deep and subtle. I didn't understand much, but the rabbi understood all of it. The third talk was a great and unforgettable experience. I understood nothing and the rabbi himself didn't understand much either.

A story told by Bohr about a young man reporting on his visit to a great rabbi as related by VICTOR WEISSKOPF

31.3 The Schrödinger Wave Equation

In 1657, Pierre Fermat proposed the intriguing idea that light, in going from one place to another, traveled along the route that took the least time. In a single simple statement, Fermat seemed to have gotten an overview of optics. His successes (for example, the derivation of Snell's Law) stimulated a great deal of effort to supersede Newton's Laws with a similar formulation. The resulting dynamical *Principle of Least Action* was introduced, albeit in a confused fashion, by Maupertuis (1747) and formalized by William Rowan Hamilton about a hundred years later. The principle asserts that a dynamical system moves in such a way as to keep its action ($\Sigma p\,\Delta q$) a minimum (here p is the momentum and q is the corresponding, or *conjugate,* space variable, for example, p_x and x or L and θ). The kinship between the propagation of light and the motion of a material particle is already there in this early work.

Hamilton found that there are mathematical surfaces over which the action of a system is constant, and that these *surfaces of constant action* propagate in a completely analogous fashion to the way surfaces of constant phase (that is, wavefronts) propagate in optics. The same equations apply to both! The angular temporal frequency ($\omega = 2\pi f$) of the light corresponds to the energy (E) of the particle; the angular spatial frequency ($k = 2\pi/\lambda$) of the light corresponds to the momentum (p) of the particle. Hamilton maintained that Newtonian Mechanics corresponded to the ray representation of Geometrical Optics. But the latter is only an approximation of the complete description of Wave Optics. What it lacks is interference, and what Hamilton's ray-dynamics for material entities lacked was also interference. Of course, this all happened almost a hundred years before de Broglie introduced the notion of the interference of electrons (the electron wasn't even discovered yet). For us, the question almost leaps from the page: *is Newtonian Mechanics an approximation of some as yet undiscovered Wave Mechanics?*

Erwin Schrödinger (1887–1961). In 1925, he became interested in de Broglie waves, and by the end of that year he derived a correct relativistic version of Wave Mechanics. Thinking he was in error (because, without the concept of spin, the theory does not predict fine structure), he abandoned it and published the more limited nonrelativistic wave equation.

The Viennese physicist Erwin Schrödinger learned about de Broglie's wave hypothesis from a footnote in one of Einstein's papers—he was already quite familiar with Hamilton's work. Just before Christmas, 1925, he went off to the Swiss Alps for vacation, leaving his wife at home. He took along a copy of de Broglie's thesis and an old girlfriend (Erwin was a rather blatant philanderer). When he and the lady returned from their holiday two and a half weeks later, the great breakthrough had already been made. Starting essentially where Hamilton had left off, he produced an extraordinary formula now universally known as **Schrödinger's Wave Equation**:

$$\frac{1}{2m}(i\hbar)^2\nabla^2\Psi + \mathrm{U}\Psi = i\hbar\frac{\partial\Psi}{\partial t}$$

It was literally created by Schrödinger and presented as a postulate because it works. The equation cannot be derived directly from Classical Theory, although it has its roots there. Mathematically, it's well beyond what we can deal with here, but because it's one of the most important intellectual accomplishments of the twentieth century, it's worth examining. During the derivation, the expression had a constant K in it. Upon applying the analysis to the orbital electron of the hydrogen atom, Schrödinger immediately got the Bohr energy levels, provided K was set equal to $\hbar$. Interestingly, the theory naturally produced the quantum number n, but (like both Heisenberg's and Dirac's previous analyses), $\hbar$ had to be inserted into the formalism. The above equation is the full, time-dependent three-dimensional formulation. The symbol $\nabla^2\Psi$ stands for a set of derivatives of Ψ in terms of the three space variables and the $\partial\Psi/\partial t$ is the time derivative of Ψ. Thus, the equation talks about the "motion" in space and time of a function Ψ (Greek letter capital *psi*) referred to as the **wavefunction**.

Schrödinger's Equation is the equation of motion of de Broglie waves, Ψ-waves, whatever they are. The symbol U represents the *potential energy* of the system under consideration. When U is constant in time, as it often is, the time-dependent part of the wave function can be canceled, leaving only the space-dependent part ψ. The one-dimensional, time-independent Schrödinger equation is then

$$\frac{1}{2m}(i\hbar)^2\frac{d^2\psi}{dx^2} + \mathrm{U}\psi = \mathrm{E}\psi$$

When we treat a particle, such as an electron in an atom, m is its mass, U is its potential energy, and E is its total energy. Recalling that $p^2/2m$ = KE, we see that the Schrödinger Equation is suggestive of the classical expression for the energy (KE + U = E) as Hamilton first wrote it, namely

$$\frac{1}{2m}p^2 + \text{U} = \text{E}$$

A particle constrained to a region by an interaction forms a *bound system*—for example, an electron in an atom. *The quantization of the energy of a bound system is a consequence of the wave equation and certain restrictions placed on the wavefunction.* Those "reasonable" restrictions simply require that ψ have only one value at each point, that it be continuous, and that it be finite everywhere. The quantum numbers that were introduced into the Bohr Theory in an *ad hoc* way now emerge, as Schrödinger put it, "in the same way as the integers specifying the number of modes in a vibrating string."

In general, Ψ is complex; it contains $i = \sqrt{-1}$ and thus is a so-called imaginary quantity. Accordingly, Ψ is not directly measurable; it has no physical existence. In other words, Ψ does not carry energy, as do all physical waves, and cannot be detected first-hand via that energy. Still, it has all the usual mathematical attributes of a wave: it has a frequency, amplitude, and phase; it obeys the superposition principle and so mathematically undergoes interference and diffraction. When two subatomic particles have identical physical characteristics (mass, charge, velocity, and so on), we say they are in the same *quantum state,* wherein they have identical wavefunctions. It is assumed that such particles are physically indistinguishable from one another.

New theories are rarely complete at the moment of presentation, and Schrödinger's was no exception. Although it proved to be remarkably effective, several features were quite troubling from the outset. In particular, what was the physical significance of Ψ? In the early days, Schrödinger assumed that the negative charge of the atom was actually spread out in space around the nucleus and that Ψ was related to the density of that charge-cloud. Curiously, although Schrödinger's theory became an immense success, the spread-out-charge interpretation of Ψ was a disappointing failure.

Probability Waves

Max Born, who was a professor of physics at Göttingen, was troubled by the idea that Ψ could undergo diffraction and be split up into separate portions—how could it characterize an electron that doesn't fragment? One can detect individual electrons with a Geiger counter and see their separate tracks in a cloud chamber. These are real manifestations of a highly localized entity.* In the spring of 1926, guided by a suggestion Einstein had made with regard to photons and the electric field, Born proposed what has now come to be the orthodox interpretation of Ψ. He suggested that the Ψ-wave associated with a particle was a probability function, an information wave that would tell us not where the particle is, but only where it is likely to be. Today, the first postulate of the orthodox interpretation of Quantum Mechanics is that **the state of a system is completely specified by its wavefunction**.

If we shine the image of some scene on a photon detector and then drastically lower the light level, we can watch the flash of each light-quantum arriving, one at a time. The recorded spots of light seem unpredictably random, but the image

*Electrons interact with each other over distances through their electric fields. Thus, two colliding beams scatter electrons via this field interaction. If the electrons have some size, some extent in space, as the beams come closer and closer, there should be a deviation from pure field scatter. As of this writing, distances down to 10^{-16} cm have been probed with no sign of a hard core—if the electron has a size, it's extremely small.

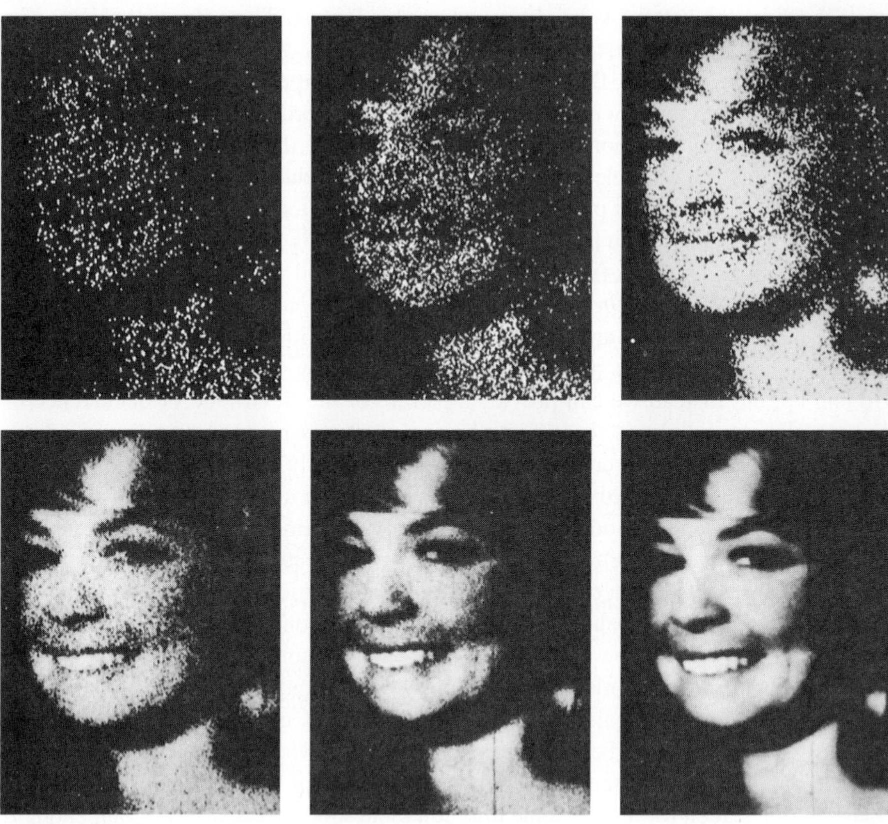

Figure 31.2 Under exceedingly faint illumination, the pattern (each spot corresponding to one photon) seems random, but as the light level increases, the quantal character of the process gradually becomes obscured. The torrent of photons blends into a seemingly continuous flow of radiant energy. (See *Advances in Biological and Medical Physics* V, 1957, 211–242.)

gradually builds until it is recognizable (Fig. 31.2). Under ordinary conditions, a great torrent of photons produces the picture almost all at once, obscuring the fundamental grainy nature of the process. Similarly, when light impinges on the double-slit arrangement of Young's Experiment (Fig. 31.3), it produces a flutter of flashes on the screen that blend into the familiar irradiance distribution of bright and dark bands. The irradiance (I) is proportional to the square of the amplitude of the electric-field wave (E_0^2) as per Eq. (24.8). The most likely place for a photon to

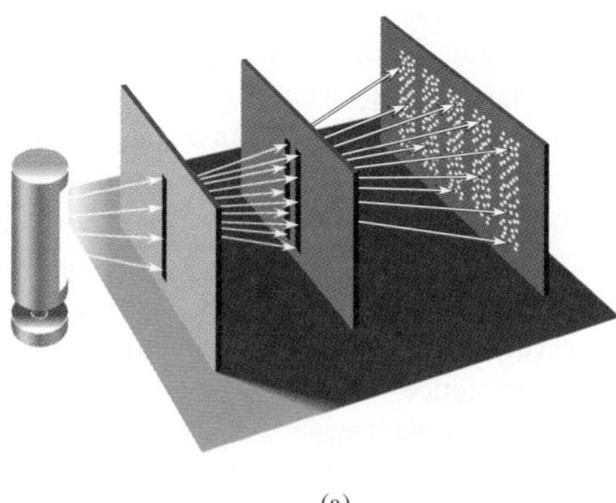

(a)

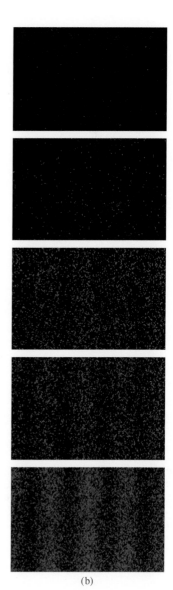

(b)

Figure 31.3 (a) Young's double-slit experiment. Flash by flash, photon by photon, the pattern gradually develops. (b) The double-slit interference pattern, but this time it's generated by a beam of electrons rather than photons. The number of electrons in each frame runs from 10 to 100 to 3000 to 20 000 and, finally, to 70 000.

arrive is where the pattern is brightest, where I is largest. Thus, the probability of a photon arriving at any particular location is given by the amplitude of the field squared. Naturally enough, Born suggested that if the experiment is carried out with electrons, or neutrons, or with any microparticle, the probability of finding such a particle at a specific location is given by the square of the amplitude of the wavefunction. Just as waves of electric fields interfere to produce complicated fringe patterns, Ψ-waves interfere to produce much the same patterns.

Schrödinger's Equation is quite as deterministic as is Newton's; that is, once the wavefunction is known, its future form can be computed precisely. Nonetheless, *the Ψ-wave can only tell us the probability of finding a particle at a given point in space.* Determinism as it applies to measurable quantities vanishes in the microworld; what is observable is probabilistic. We never know where a particular electron will end up, but instead we can provide a picture of the likelihood of occurrence of all its various possibilities. Large numbers of particles respond as a group whose overall behavior can be accurately predicted statistically. Even if the individual is lost in the crowd, the crowd *en masse* does its thing in a well-defined fashion that can be measured, understood, and anticipated. If we shine a beam of electrons at a pair of tiny slits, we can calculate the resulting fringe pattern precisely even though we know nothing of the specific motion of any one particle. Similarly, an actuarial table can tell you the probability of dying for any person in your age group. But it obviously cannot reveal how long you, a particular person, will live.

Still, Quantum Mechanics as we currently understand it challenges us at almost every turn. Suppose we again shine a beam of photons, electrons, or neutrons on a double slit. Now lower the beam intensity to the point where only one microparticle enters the apparatus at a time. It will pass through one of the slits (or not) and cause a flash on the screen (or not) at some unpredictable spot. But if we wait long enough, perhaps a few months, flash by flash, the traditional fringe system will appear. It seems that *each photon, electron, or neutron interferes with itself* (whether it interferes *only* with itself is another question). And it follows that the wavefunction, which undergoes the interference, must extend over both apertures and pass through both apertures. In that sense, the individual photon passes through both slits.

Envision a laser beam comprising a stream of photons. For what it's worth, one might picture the wavefunction of one such light-quantum as a thin, flat disk of progressing undulations—the photon has a probability of being anywhere within the confines of the beam and so the wavefunction is defined by the beam diameter. Some would say that the photon was a flying disk, while others keep their wavefunctions purely mathematical and think of the photon as a localized concentration of energy. Still others don't worry about what a photon looks like since it can't be seen in transit anyway. It's even possible in the case of light with a broad classical wavefront (starlight, for example) to separate the slits in Young's Experiment by 10 or 20 meters and still observe interference. If we buy the Ψ-wave picture, we must buy the idea that the photon's wavefunction then also extends over tens of meters.

The photon itself must somehow "see" both slits, however far apart they are, and yet arrive at the observation screen as a tiny localized blast of energy. If we cover either aperture, the interference vanishes. Indeed, if we place a particle detector behind either aperture so as to determine if the photon (or electron) has passed through it alone (as a particle might), the interference pattern also vanishes (see Discussion Question 5). And that is presumably true even if the beam is not obstructed as it is detected. Once the particle aspect is observed—once we know it has passed through one slit or the other—the wave aspect is obliterated. It's safe to say that many physicists are still not totally happy with the orthodox (or any other) explanation of these strange phenomena.

On the concept of quantum jumps. *The whole idea of quantum jumps necessarily leads to nonsense. . . . If we are still going to have to put up with these damn quantum jumps, I am sorry that I ever had anything to do with quantum theory.*

ERWIN SCHRÖDINGER

Just as surprising, if we set up several million identical double-slit experiments, shine a single identical photon (or electron or neutron) at each one at the same moment, and then superimpose the results from each, the total pattern should be the same fringe system as always. Each photon addresses the diffracting apertures individually and it progresses, guided by its wavefunction pattern, in a totally random way. Where on the theoretically computable, observably verifiable distribution it will land is indeterminant, but that it will land on the distribution is certain. This state of affairs is very peculiar! *It appears that identical particles in identical situations need not behave identically*—that's the essence of *quantum randomness,* which must be postulated in order to account for the observations. And again, not every physicist is happy wth these weird goings on, though most are pragmatic enough to keep on cranking out amazing solutions to real physical problems.

If we determine Ψ from Schrödinger's Equation for an electron moving in the Coulomb field of a proton, we get a whole range of potential electron locations, a cloud of probabilities. The most likely place to find the electron in the ground state is out at a distance identically equal to the Bohr radius, 0.052 9 nm (Fig. 31.4). There is no longer any reason to think of atomic electrons flying around in little orbits, though what they are doing from one moment to the next is a mystery. Figure 31.5 is a representation of the first few excited states of hydrogen. Keeping in mind that we are dealing with one rapidly moving electron, imagine a cloudlike afterimage around the nucleus, revealing where the electron has been and where it is likely to be.

The indeterministic substructure of the Universe was hinted at decades before Quantum Mechanics, when it was discovered that the decay of radioactive atoms was only predictable statistically (p. 1036). Generally, that realization was rationalized as simple classical ignorance, which we could hope to overcome, rather than profound quantum-mechanical indeterminism, which could never be eliminated. Even though it is now the consensus, the idea of a probabilistic world (wherein things like particle decay happen spontaneously, without apparent cause) does not sit well with all physicists. It certainly didn't sit well with Einstein, who complained that God does not play dice with the Universe.

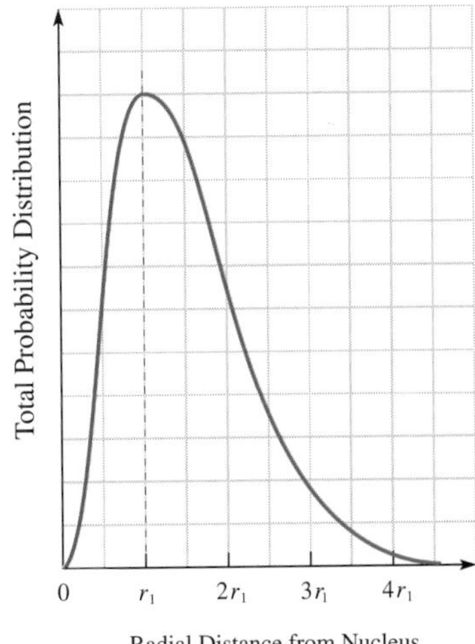

Figure 31.4 A plot of the probability of finding the electron at a distance from the nucleus of the hydrogen atom in its ground state ($n = 1$, $l = 0$). The most likely location is at the Bohr radius.

QUANTUM PHYSICS

Guided by challenging experimental observations and powered by a potent theoretical machinery, Quantum Mechanics developed rapidly in the 1920s and 30s. Attempts to understand the Zeeman Effect stimulated the introduction of the fundamental concept of *spin*. Another revelation, the *Pauli Exclusion Principle,* helped to uncover the detailed structure of the atomic electron cloud. The *Uncertainty Principle,* a profound overview of the microworld, was formulated by Heisenberg. A relativistic quantum theory, *Quantum Electrodynamics,* was introduced by Dirac, and it predicted the existence of *antimatter.* All of these milestones are explored in the remainder of this chapter.

31.4 Quantum Numbers

Modern Quantum Mechanics generates a set of fundamental quantum numbers in a mathematical way that flows naturally from the theory of differential equations. These numbers describe the state of a quantum-mechanical system and therefore specify the wavefunction. The

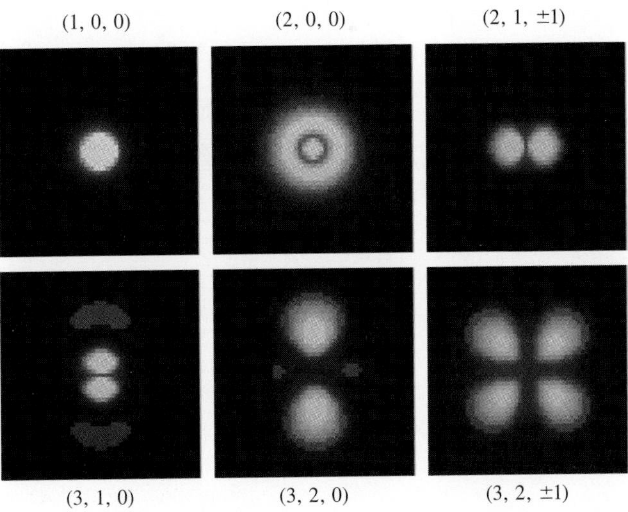

Figure 31.5 These are plots of the probability density, $|\Psi|^2$, for the electron in the ground state and in several excited states of the hydrogen atom. Each picture is a slice of the 3-dimensional electron cloud with the nucleus at the center. The brighter the region, the higher the probability of finding the electron. Each pattern corresponds to a specific set (p. 1085) of quantum numbers (n, l, m_l). For example, the ground state is specified by (1, 0, 0).

Schrödinger Equation has solutions that are energy values and, in one dimension, these are each associated with a value of the **principal quantum number** n. For a three-dimensional atomic system, there must be at least three quantum numbers (n, l, and m_l), and the rules for specifying these numbers come directly from the mathematics. These same rules for quantizing dynamical systems were more or less anticipated by the Old Quantum Theory, though in an *ad hoc* fashion.

Ever since Michelson's work (1891), it was known that each of the hydrogen spectral lines actually comprised several lines, very close together. With that in mind, Arnold Sommerfeld (1915) generalized the Bohr Theory to encompass this **fine structure**. What he needed was a very small additional energy term that would produce the observed multiplet structure. Accordingly, he assumed the orbits were elliptical and would precess about the nucleus. Surprisingly enough, when this variation on Bohr's theme was framed relativistically, it produced the desired effect in full agreement with observation. Thus, was the fine structure of hydrogen "explained." Sommerfeld's theory, which is now totally obsolete, was so accurate and visually appealing that the picture of interweaving ellipses persists even today. (Just ask any youngster to draw an atom).

Each possible motion of a system is known as a *degree of freedom* and, in effect, Sommerfeld had added another degree of freedom to the one circular orbital motion of the Bohr electron; elliptical orbits could swing around the nucleus. To the principal quantum number n, which essentially determined the orbital energy, he added a new quantum number—an important insight. To put things in a modern context, we call it the **orbital angular momentum quantum number**, symbolized by l and having integer values from 0 to $n - 1$ (that is, $n \neq l$). This quantum number determines the magnitude of the total orbital angular momentum (**L**), fixing it in steps of multiples of $\hbar$. Here the modern theory departs dramatically from Bohr's treatment in that it allows for an $l = 0$, zero angular momentum state. That possibility does great mischief to any thoughts of *orbiting* electrons!

31.5 The Zeeman Effect

The next problem to demand another degree of freedom (and a new quantum number) was remotely necessitated by Faraday, who was convinced that magnetism should affect light. Following his lead, Pieter Zeeman (1896), who had the advantages of a powerful electromagnet and a fine diffraction grating, placed a sodium flame in a B-field. Whenever "the current was put on, the two D lines were distinctly widened"—the magnetic field influences the way atoms emit light. The **Zeeman Effect** was quickly explained classically: the orbiting electron is a tiny current loop that has a magnetic moment and experiences a torque. As a result, the angular momentum vector revolves around **B**, and that alters the energy slightly, producing a splitting of the spectral lines (Fig. 31.6).

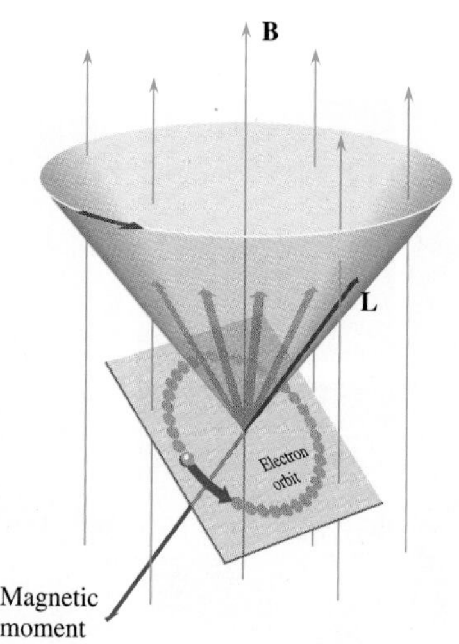

Figure 31.6 An orbiting electron creates a magnetic moment that revolves around **B**.

Debye and Sommerfeld were able to interpret the Zeeman Effect within the Old Quantum Theory by associating a new quantum number with the magnetic moment of the orbital electron. Today, this number is written as m_l and is known as the **orbital magnetic quantum number**. The energy of a bar magnet, a dipole equivalent to a current loop, in a B-field depends on its orientation with respect to the field. Accordingly, consider the magnetic moment of an orbital electron pointing in the opposite direction of its angular momentum. When the electron is placed in a magnetic field, **L** can only assume certain orientations with respect to **B**. Sommerfeld called this *space quantization* ("spatial quantization" might be better). In other words, if **B** is in, say, the z-direction, L_z is quantized such that

$$L_z = m_l \hbar$$

wherein m_l goes from $-l$ to 0 to $+l$ in integer steps. For example, when $l = 2$, m_l can be -2, -1, 0, $+1$, or $+2$. And if $m_l = 0$, $L_z = 0$ and **L** will be perpendicular to **B**. For states other than the $n = 1$ ground state (for which $l = 0$ and $m_l = 0$), an applied magnetic field will split the energy levels into 3. 5, etc., sublevels whose separation depends on the strength of the field (Fig. 31.7).

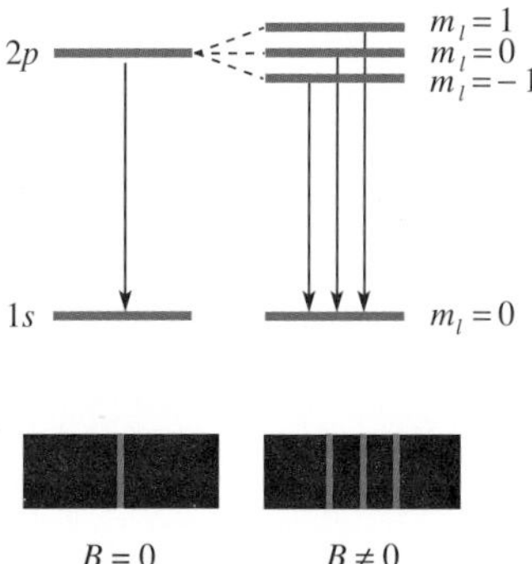

Figure 31.7 The splitting of energy levels in a magnetic field; the Zeeman Effect. The application of a magnetic field causes the atom to have three close-together excited states. The result is three emission lines where there ordinarily is only one.

31.6 Spin

Almost as soon as the problem of the Zeeman Effect was solved, it was complicated again. Although there were lines that behaved according to theory (the so-called Normal Zeeman Effect), it was discovered in 1897 that others, like those of sodium, actually split into a bewildering array of lines in a magnetic field (the Anomolous Zeeman Effect).

R. Kronig, from Columbia University, conceived the idea (1925) that the electron might be spinning about its own axis and therefore have a magnetic moment. Although Compton (1921) had already suggested that "the electron itself, spinning like a tiny gyroscope, is probably the ultimate magnetic particle," he never applied the idea. Kronig did, but there were relativistic problems with the notion of a finite spinning charged particle, and when several colleagues (among them, Pauli and Heisenberg) rejected the hypothesis, he decided to let it go unpublished. Ironically, S. A. Goudsmit and G. E. Uhlenbeck together independently introduced the idea of electron spin after reading a paper by Pauli in which he showed the necessity for associating four quantum numbers with the electron. They, too, hesitated, but fortunately they were working under Ehrenfest who told them "that it was either highly important or nonsense" and, in any event, he had already sent their paper off to be published.

Classically, an object such as the Earth can have any value of spin angular momentum—the quantity has always been assumed to be continuous. But that's not possible with particles in the atomic domain. To match experimental results, the **spin angular momentum quantum number** (s) for electrons must have only one value, $s = \frac{1}{2}$; because of this, electrons are referred to as *spin-$\frac{1}{2}$ particles*. Spatial quantization again applies because the spin magnetic moment orients itself with regard to an applied B-field. Thus, the component of the spin angular momentum vector in the direction of **B** is quantized and equals $m_s \hbar$ where m_s is the **spin magnetic quantum number**. Moreover, m_s can only change by steps of 1 going from $+s$ to $-s$; in other words, $m_s = \pm\frac{1}{2}$. The component of spin angular momentum of an electron in the direction of B can only equal $\pm\frac{1}{2}\hbar$. We speak of **spin-up** ($+\frac{1}{2}$) and **spin-down** ($-\frac{1}{2}$) electrons. The electron's energy in the magnetic field depends on its spin orientation, and so all of the levels are split by the applied B-field, thereby producing the Anomalous Zeeman Effect.

A sunspot. These dark blotches on the face of the Sun are associated with powerful localized magnetic fields. The thin black vertical line in (a) corresponds to the location of the slit opening in a spectrograph that analyzes the light within that narrow region. In (b) we see three spectral lines running vertically. Where the slit opening passes into the strong magnetic field of the sunspot, the middle line clearly splits (b) into three lines as a result of the Zeeman Effect. By measuring the separations of these lines, the magnetic field (≈4 T) can be determined.

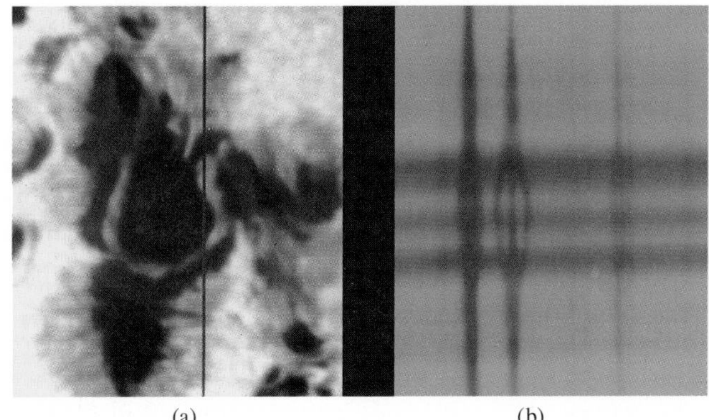

TABLE 31.1 Atomic quantum numbers

Name	Symbol	Values
Principal quantum number	n	1, 2, 3, . . .
Orbital angular momentum quantum number	l	0, 1, 2, . . . , $(n - 1)$
Orbital magnetic quantum number	m_l	0, ±1, ±2, . . . , $\pm l$
Spin angular momentum quantum number	m_s	$\pm\frac{1}{2}$

The notion of spin is not supposed to be taken literally. Particles do have angular momentum—a beam of light can twist a torsion pendulum—yet there are serious relativistic problems associated with picturing particles as tiny revolving entities. The idea of the electron being a minute spinning charged sphere is appealing but very problematic. What is said instead is that it has an inherent spin angular momentum (without the need for revolving), but that concept doesn't completely satisfy those who like their angular momentum to involve a clear angular motion. Still, in 1928, Dirac demonstrated that a relativistic quantum theory of the electron naturally provides intrinsic spin via an additional quantum number (Table 31.1).

31.7 The Pauli Exclusion Principle

The quantum numbers specify the state of a system and are therefore extraordinarily important. As Pauli showed, the state of any atomic electron can be determined via four quantum numbers: n, l, m_l, and m_s. Furthermore, *no two atomic electrons can occupy the same state;* that is, no two electrons in the same atom can have the same four quantum numbers. In a generalized form, this idea, known as the **Pauli Exclusion Principle**, is a basic precept of nature applicable to a wide range of situations.

We learn from Quantum Mechanics that the Exclusion Principle applies to particles that are known as **fermions**. These are particles (whether elementary or compound) that have spins that are odd integer multiples of $\frac{1}{2}$; particles such as the proton, neutron, and neutrino. It does not apply to the group of particles called **bosons**, which have zero or integer spins. Alpha particles, which are composed of even numbers of fermions, are bosons. Photons are spin-1 particles and therefore bosons.

The state of a photon is specified by its momentum vector and polarization. Thus, a planar monochromatic lightwave can be thought of as a stream of photons all in the same state; a circumstance that is possible because they are bosons. Indeed, bosons tend to accumulate in the lowest possible energy state. If that were the case for fermions, all atomic electrons would collect in the same lowest-energy level and the atom would compress. Moreover, when atoms are brought very close to one another (so close that their wavefunctions overlap), the Exclusion Principle obtains, which is why two hydrogen atoms can form a molecule only when the atoms have spin-up and spin-down electrons, respectively.

To date, tests of the Exclusion Principle have never found it wanting; researchers have seen no violations of it down to an accuracy of less than 2 parts in 10^{26}.

31.8 Electron Shells

During the 1920s, Bohr, Stoner, and others constructed a model of the electron structure of all the elements in the Periodic Table (inside back cover). Moseley's

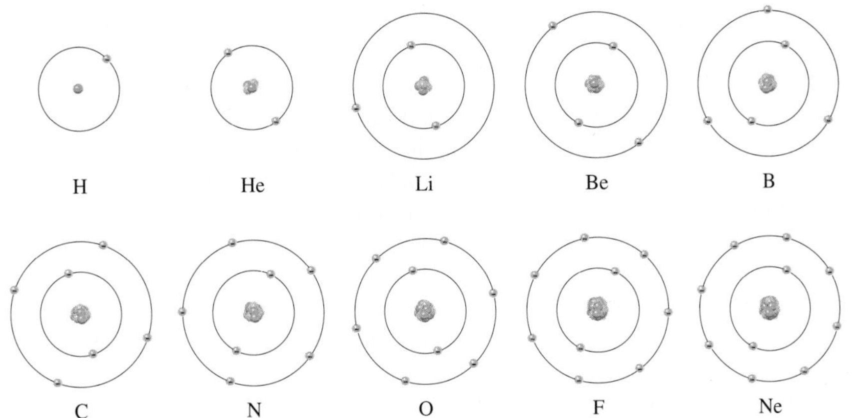

Figure 31.8 A schematic representation of the electronic structure of the first 10 atoms in the Periodic Table. (The sizes of the nuclei are tremendously exaggerated, as are the electrons.)

work provided a knowledge of the number of nuclear protons and, since the atom is neutral, that's the number of orbital electrons. It's no simple matter to sort out the electron structure (Fig. 31.8); for that analysis, chemical behavior and atomic spectra were the guides.

Electrons are ordered in shells and subshells about the various nuclei according to rules associated with their quantum numbers. Each shell corresponds to a specific value of n, and they are traditionally given letter names, where

$$\text{Shell} = K\ L\ M\ N\ O\ \ldots$$
$$n = 1\ \ 2\ \ 3\ \ 4\ \ 5\ \ldots$$

Remember that l ranges from $(n-1)$ to 0; that m_l goes from $-l$ to $+l$; and that m_s is either $+\frac{1}{2}$ or $-\frac{1}{2}$. Again, by tradition, the states of an electron are given letter names, where

$$\text{State} = s\ p\ d\ f\ g\ h\ \ldots$$
$$l = 0\ \ 1\ \ 2\ \ 3\ 4\ \ 5\ \ldots$$

A **shell** is a group of states that have the same principal quantum number. A **subshell** is a smaller group of states that has both the same value of n and l (Fig. 31.9). An **orbital** is specified by the three quantum numbers n, l, and m_l, and it can contain two electrons; one spin-up, one spin-down. (Each short horizontal line in Fig's. 31.9 and 31.10 represents an orbital.) And a **state** is specified by all four quantum numbers and contains one electron, as per the Exclusion Principle.

Figure 31.9 Ordering of atomic energy levels in terms of quantum numbers n, l, m_l. For example, the $3d$ subshell corresponds to $n = 3$ and $l = 2$. It comprises five orbitals corresponding to $m_l = -2, -1, 0, +1,$ and $+2$.

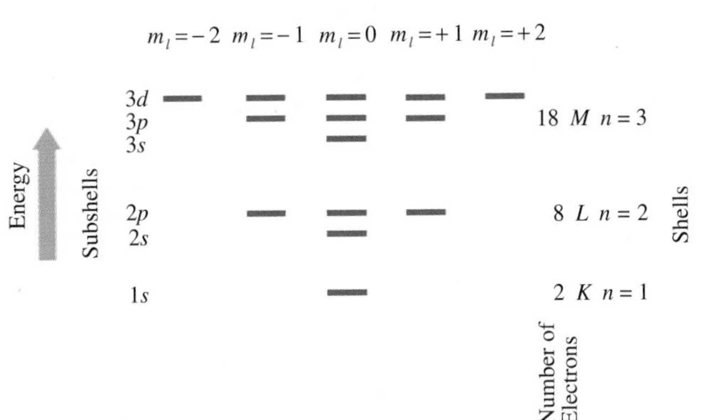

Hydrogen ($Z = 1$) has one electron and is chemically active. Its one unpaired electron gives it a valence of one and requires that it enter into covalent bonds in which the electron is shared with another atom. In the ground state, $n = 1$, $l = 0$, $m_l = 0$, and $m_s = \pm\frac{1}{2}$ (Fig. 31.11). The electron configuration is said to be a $1s$ orbital ($n = 1$, $l = 0$) of the K-shell (Fig. 31.9). Once two hydrogen atoms have combined (H_2) by sharing their spin-up, spin-down electrons, there are no longer any unpaired charges, and a third hydrogen cannot join the group (Table 31.2, p. 1087).

Helium ($Z = 2$) has two electrons, and because it's a stable *Noble Gas* (in the right-most column of the Periodic Table), we can conclude that two electrons must correspond to a completed system. Thus (even before the Exclusion Principle), it was presumed that the first (innermost) two electrons of any atom form a closed K-shell. In the ground state, the quantum numbers (n, l, m_l, m_s) for one electron are $(1, 0, 0, -\frac{1}{2})$ and for the other are $(1, 0, 0, +\frac{1}{2})$. The K-shell has only two states and is filled. The 2-electron configuration is $1s^2$, where the superscript is the number of electrons (Table 31.3, p. 1088). Both electrons are close to the nucleus, and the atom has a high ionization energy of 24.6 eV (Fig. 31.12); it holds tightly to its electrons and does not interact effectively with other atoms.

Lithium ($Z = 3$) is next. It has three electrons, two in the first closed K-shell and the third rather far off in the second, or L-shell. With no room in the $n = 1$ shell, the third electron goes into the $l = 0$, s-subshell of the $n = 2$ shell; (n, l, m_l, m_s) for the third electron corresponds to $(2, 0, 0, -\frac{1}{2})$. The overall configuration is $1s^2{\cdot}2s^1$.

Beryllium ($Z = 4$) has four electrons, the last of which fills the $l = 0$, s-subshell and has quantum numbers $(2, 0, 0, +\frac{1}{2})$. The 4-electron ground-state configuration is $1s^2{\cdot}2s^2$; there are no unpaired electrons and the valence is zero. As it happens, the energy difference between the $(2, 0, 0, \frac{1}{2})$ state and the $(2, 1, 0, \frac{1}{2})$ state is very small and the beryllium atom can easily go into the $1s^2{\cdot}2s^1 2p^1$ configuration; that is, one electron can be raised up into the p-subshell. Accordingly, it can have a valence of either 0 or 2, but nothing else (Fig. 31.10).

Boron ($Z = 5$) has five electrons, the last of which goes into the next open level in the p-subshell; $(2, 1, +1, -\frac{1}{2})$. The 5-electron ground-state configuration is $1s^2{\cdot}2s^2 2p^1$, and a valence of 1 is to be expected. Here again, the configuration can be altered with the input of a small amount of energy, which is available to the atom via collisions even at room temperature. It then takes the form $1s^2{\cdot}2s^1 2p^2$.

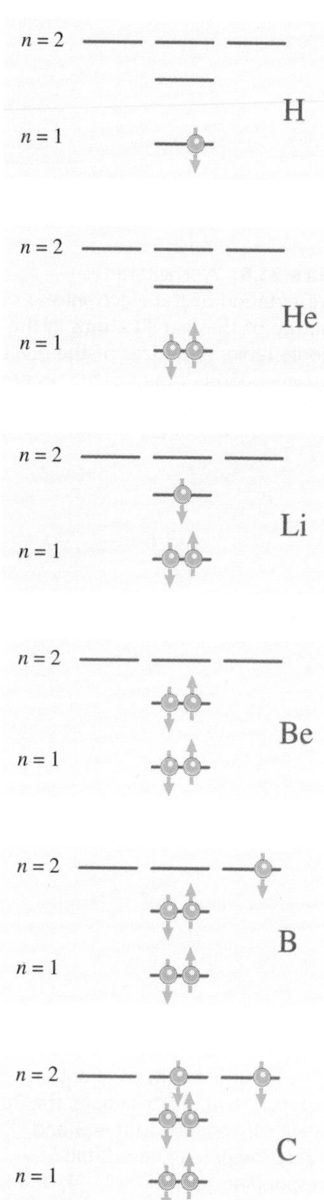

Figure 31.10 Distributions of electrons in the energy levels for the first six elements: hydrogen (H), helium (He), lithium (L), beryllium (Be), boron (B), and carbon (C).

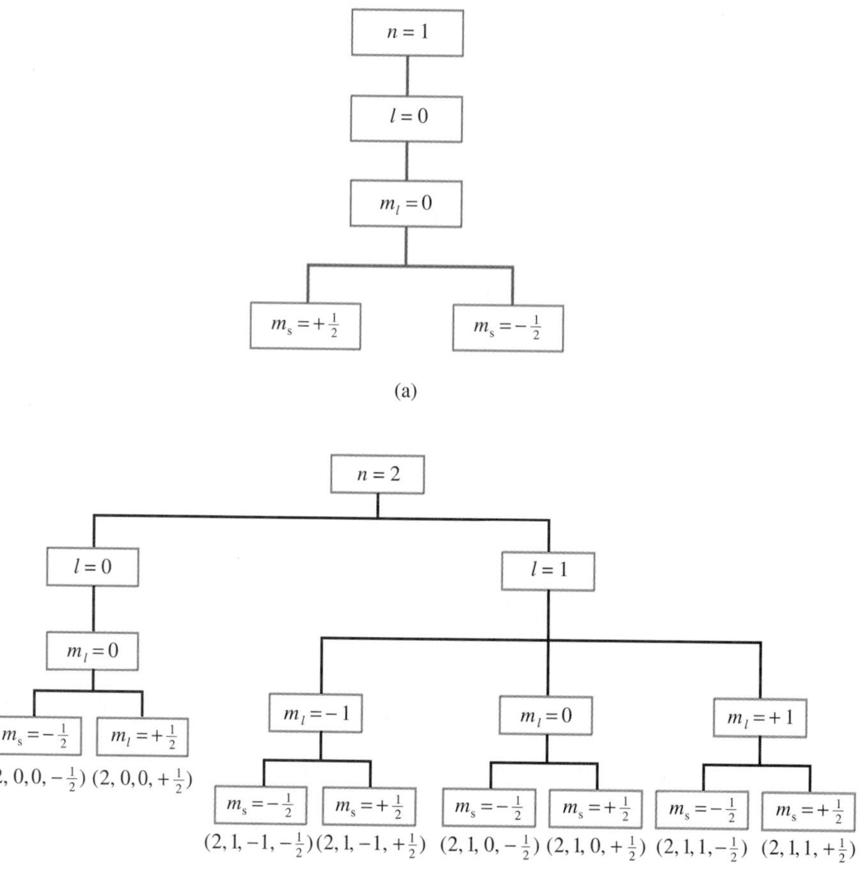

Figure 31.11 The quantum numbers and corresponding states (n, l, m_l, m_s) for the principal quantum numbers (a) $n = 1$ and (b) $n = 2$ for hydrogen.

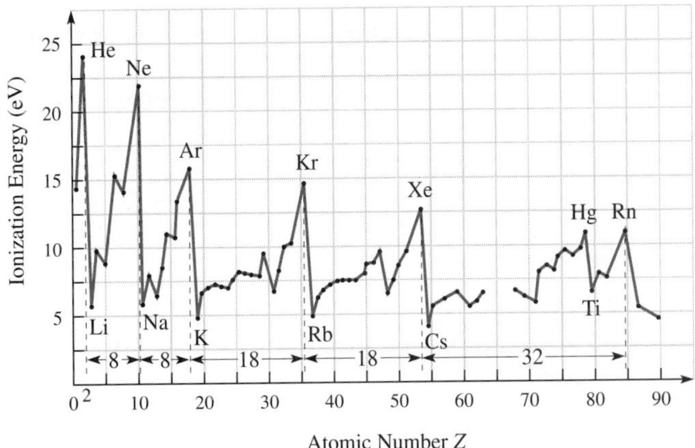

Figure 31.12 Ionization energy versus Atomic Number. Closed shells occur at $Z = 2$, 10, 18, 36, and 54. An atom such as sodium has one electron beyond a closed shell ($Z = 11$), and it's shielded from the nucleus, so its binding energy is small.

TABLE 31.2 Ground-state configurations of several elements

Element	Symbol	Atomic number, Z	Electronic configuration
Hydrogen	H	1	$1s$
Helium	He	2	$1s^2$.
Lithium	Li	3	$1s^2{\cdot}2s$
Beryllium	Be	4	$1s^2{\cdot}2s^2$
Boron	B	5	$1s^2{\cdot}2s^22p$
Carbon	C	6	$1s^2{\cdot}2s^22p^2$
Nitrogen	N	7	$1s^2{\cdot}2s^22p^3$
Oxygen	O	8	$1s^2{\cdot}2s^22p^4$
Fluorine	F	9	$1s^2{\cdot}2s^22p^5$
Neon	Ne	10	$1s^2{\cdot}2s^22p^6$.
Sodium	Na	11	$1s^2{\cdot}2s^22p^6{\cdot}3s$
Magnesium	Mg	12	$1s^2{\cdot}2s^22p^6{\cdot}3s^2$
Aluminum	Al	13	$1s^2{\cdot}2s^22p^6{\cdot}3s^23p$
Silicon	Si	14	$1s^2{\cdot}2s^22p^6{\cdot}3s^23p^2$
Phosphorus	P	15	$1s^2{\cdot}2s^22p^6{\cdot}3s^23p^3$
Sulfur	S	16	$1s^2{\cdot}2s^22p^6{\cdot}3s^23p^4$
Chlorine	Cl	17	$1s^2{\cdot}2s^22p^6{\cdot}3s^23p^5$
Argon	Ar	18	$1s^2{\cdot}2s^22p^6{\cdot}3s^23p^6$.

Carbon ($Z = 6$) has six electrons, two in the $n = 1$ shell and four in the $n = 2$ shell. The last electron goes into the p-subshell; $(2, 1, 0, -\frac{1}{2})$. As a rule, *electrons filling a subshell do not double up in an orbital until each orbital has one.* Because of their mutual repulsion, the electrons get as far apart as possible by going into different orbitals. The 6-electron ground-state configuration is $1s^2{\cdot}2s^22p^2$. Thus, we would expect carbon (Fig. 31.10) to participate in covalent bonds and have a valence of 2. And yet it almost always displays a valence of 4. This occurs because it takes a mere $\approx$2 eV to break the $2s^2$ pair, raising one of the electrons into the empty orbital of the p-subshell. There are then four unpaired electrons, yielding a valence of 4. Figure 31.13 shows how carbon can have valences of 0, 2, or 4, but nothing else. It's this range of valences that allows carbon to share electrons, forming single, double, and triple bonds (for example, C—O, C═C, C═O, C═N, C≡C, C≡N), and

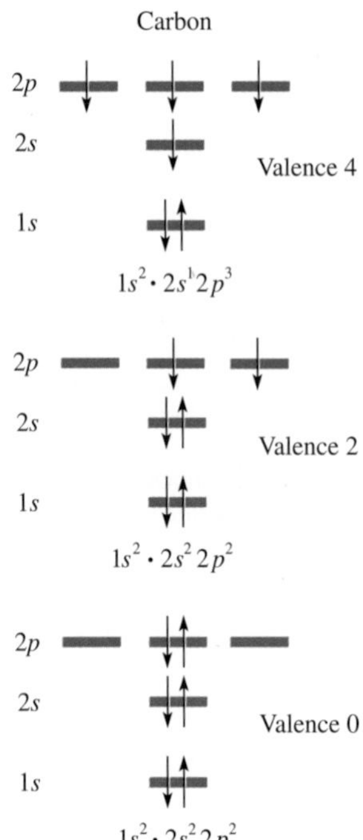

Figure 31.13 The distribution of electrons in carbon corresponding to valences of 0, 2, and 4. Each orbital fills up only when it contains two electrons. Accordingly, the valence corresponds to the number of single electrons that can be shared.

TABLE 31.3 Electron configurations of some atoms in their ground states

Z	Element		K (n: 1) s	L (n: 2) s	L p	M (n: 3) s	M p	M d	N (n: 4) s	N p
1	Hydrogen	H	1							
2	Helium	He	2							
3	Lithium	Li	2	1						
4	Beryllium	Be	2	2						
5	Boron	B	2	2	1					
6	Carbon	C	2	2	2					
7	Nitrogen	N	2	2	3					
8	Oxygen	O	2	2	4					
9	Fluorine	F	2	2	5					
10	Neon	Ne	2	2	6					
11	Sodium	Na	2	2	6	1				
12	Magnesium	Mg	2	2	6	2				
13	Aluminum	Al	2	2	6	2	1			
14	Silicon	Si	2	2	6	2	2			
15	Phosphorus	P	2	2	6	2	3			
16	Sulfur	S	2	2	6	2	4			
17	Chlorine	Cl	2	2	6	2	5			
18	Argon	Ar	2	2	6	2	6			
19	Potassium	K	2	2	6	2	6	·	1	
20	Calcium	Ca	2	2	6	2	6	·	2	
21	Scandium	Sc	2	2	6	2	6	1	2	
22	Titanium	Ti	2	2	6	2	6	2	2	
23	Vanadium	V	2	2	6	2	6	3	2	
24	Chromium	Cr	2	2	6	2	6	5	1	
25	Manganese	Mn	2	2	6	2	6	5	2	
26	Iron	Fe	2	2	6	2	6	6	2	
27	Cobalt	Co	2	2	6	2	6	7	2	
28	Nickel	Ni	2	2	6	2	6	8	2	
29	Copper	Cu	2	2	6	2	6	10	1	
30	Zinc	Zn	2	2	6	2	6	10	2	
31	Gallium	Ga	2	2	6	2	6	10	2	1
32	Germanium	Ge	2	2	6	2	6	10	2	2
33	Arsenic	As	2	2	6	2	6	10	2	3

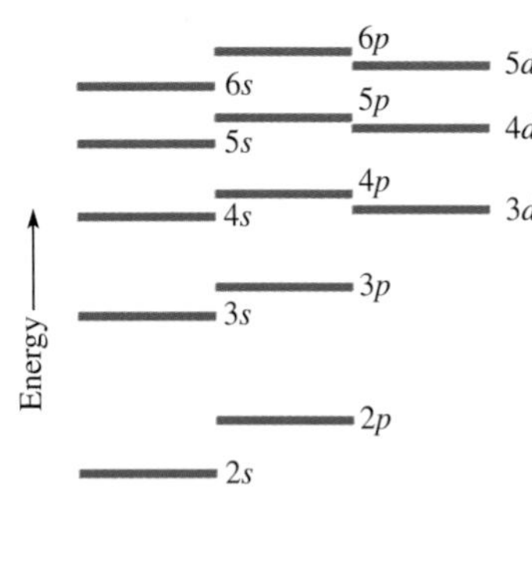

Figure 31.14 Relative positions of atomic orbitals.

thereby producing a wide range of complex molecules, including the ones that are the basis of all known life.

At the point where stable neon ($Z = 10$) is reached (Table 31.4), its ten electrons complete the first two shells, with two and eight electrons, respectively, ($1s^2 \cdot 2s^2 2p^6$). Argon ($Z = 18$) is the next Noble Gas, and it fills the $3p$-subshell in the M-shell with a configuration of $1s^2 \cdot 2s^2 2p^6 \cdot 3s^2 3p^6$. After argon, the order of several subshells changes because, for example, the $4s$ level is a little less energetic than the $3d$, so it fills first (Fig. 31.14). Krypton ($Z = 36$) fills the M-shell and has a configuration of $1s^2 \cdot 2s^2 2p^6 \cdot 3s^2 3p^6 \cdot 3d^{10} 4s^2 4p^6$. And so on up the Periodic Table.

With few exceptions, *all the atoms in any one column of the Periodic Table have the same number of unpaired electrons.* The very active Alkali Metals all have only one such electron, which is easily given up or shared; the less active Alkaline

Earths (such as magnesium and calcium) have two; the boron, carbon, nitrogen, and oxygen families have three, four, five, and six, respectively; and the highly reactive Halogens each contain an almost complete shell of seven electrons.

Example 31.2 Determine the ground-state electron configuration for $_{12}$Mg. Start with neon, for which $(1s^2{\cdot}2s^22p^6)$.

Solution: [Given: $_{12}$Mg. Find: its electron configuration.] Neon has 10 electrons $(1s^2{\cdot}2s^22p^6)$ and a closed p-subshell. The $n = 2$ shell allows $l = 1$ or 0, so there is an s- and a p-subshell. The eleventh electron $(n = 3)$ corresponds to $(1s^2{\cdot}2s^22p^6{\cdot}3s^1)$ and the twelfth to $\boxed{(1s^2{\cdot}2s^22p^6{\cdot}3s^2)}$.

▶ **Quick Check:** Look at Table 31.2, p. 1087.

31.9 The Uncertainty Principle

Born's statistical interpretation suggests that we can compute only the likelihood of individual events. This "blurriness" of vision, which is inherent in the formulation of Wave Mechanics, might seem peculiar. After all, if we know the precise location of a proton to begin with, as well as how fast and in what direction it's moving, why can't we follow it *exactly*? Perhaps that's not the right question, however; perhaps we should be bolder and ask whether or not we really can simultaneously measure both the position and velocity precisely in the first place. The product of position and momentum, or time and energy, brings us back to *action* and the quantum-of-action (h), which is at the heart of things quantum-mechanical. Questions about the measurability and definability of quantum concepts had occupied Werner Heisenberg in 1927 before he came up with the crucial insight known as the **Uncertainty Principle**.

Ordinarily, we observe the position of something by scattering something else off it: radar waves off a car; sound waves off a whale; lightwaves off your face; or the end of a cane off the curb. A stream of sunshine photons scattering from a flying baseball hardly affects its path, but one photon slamming into an electron drastically and unpredictably alters the electron's motion. Evidently, to measure anything in Atomland we will need a probe as well, but even the most delicate one possible will obtrude on the measurement. In order to observe the position of a microparticle precisely, we must—to cut down on diffraction—use a short-wavelength (and, therefore, a high-energy) probe. And that, in turn, will blast the particle away, making knowledge of its velocity and momentum even less precise. To approximate the effect, we use a photon of wavelength λ to locate a tiny object along the x-axis. Reasonably enough, assume the position can be measured to an accuracy of $\Delta x \approx \lambda$. At most, the photon can transfer all its momentum (h/λ) to the object, whose own momentum will then be uncertain by an amount $\Delta p_x \approx h/\lambda$. Hence

$$\Delta p_x \Delta x \approx h \tag{31.3}$$

and this is a crude form of the Heisenberg Uncertainty Principle. It quantifies the "giveth and taketh away" interrelationship between so-called conjugate variables, such as position and momentum, and energy and time (p. 1076). Even though it's clear that the very act of measuring unavoidably obtrudes on the atomic level, most physicists now believe that the Uncertainty Principle has a yet more fundamental basis.

Consider a particle moving along the x-axis with a speed v and a momentum p_x. Momentum, according to de Broglie $(p = h/\lambda)$, depends on wavelength, which, in

TABLE 31.4 Groups of elements

Halogens	Number of electrons
Fluorine	9
Chlorine	17
Bromine	35
Iodine	53
Astatine	85
Noble Gases	**Number of electrons**
Helium	2
Neon	10
Argon	18
Krypton	36
Xenon	54
Radon	86
Alkali Metals	**Number of electrons**
Lithium	3
Sodium	11
Potassium	19
Rubidium	37
Cesium	55
Francium	87

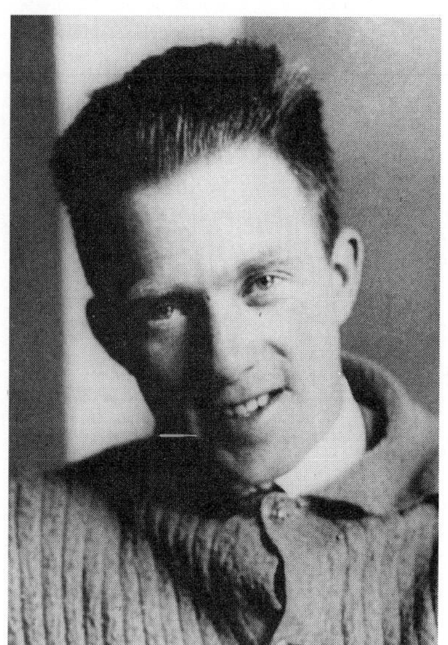

Werner Karl Heisenberg (1901–1976) was awarded the Nobel Prize for physics in 1932 for his discovery of the Uncertainty Principle.

turn, corresponds to an extension in space. To specify λ, we must think of observing a cycle of the wave in space. The wave-mechanical concept of a precisely determined momentum is thus in direct conceptual conflict with the notion of a precisely defined conjugate position (x). Similarly, consider a particle of energy E moving at a time t. Energy ($E = hf$) depends on frequency, which, in turn, corresponds to an extension in time. To specify f, we must think of observing a cycle of the wave in time. The wave-mechanical concept of a precisely determined energy is thus incompatible with the notion of a precisely defined conjugate time (t).

Suppose we set out to simultaneously determine a pair of conjugate variables, say, p_x and x. Having made a number of measurements of any physical quantity, we would find a spread in the data about an average value. Accordingly, we record a momentum spread of Δp_x and a position spread of Δx. Classically, we would expect to be able to simultaneously sharpen up these values, making them both more and more precise, with no inherent conceptual limitations keeping us from improving things. We need only measure v over a vanishingly small track (Δx) centered at x. But the wave nature of the particle makes that process much less straightforward. To fix the momentum, we must fix the wavelength with precision, and to do so will necessitate allowing it some extension in space, thereby blurring x.

This linkage between conjugate variables is at the center of Heisenberg's discovery of the Uncertainty Principle, which is

$$\Delta p_x \Delta x \geq \tfrac{1}{2}\hbar \tag{31.4}$$

The product of the simultaneous uncertainties in position and momentum is at best equal to $\frac{1}{2}\hbar$. And the conceptually related expression

$$\Delta E \, \Delta t \geq \tfrac{1}{2}\hbar \tag{31.5}$$

holds as well. We cannot simultaneously know both the momentum and the position to any accuracy we wish. Homing in on one conjugate variable decreases its uncertainty and increases the uncertainty in the other variable.

Consider the implications of the above conclusions as they relate to complementarity; that is, to the wave-particle duality. The Uncertainty Principle sees to it that we will never be able to experimentally resolve the either/or issue of whether a photon is a particle or a wave. Similarly, precisely measuring the position of an electron fixes it with a sharp spatial localization, whereupon it may correctly be considered a particle. Yet when $\Delta x = 0$, it follows that $\Delta p = \infty$, and the electron has zero wavelength ($p = h/\lambda$); it has no wavelike attributes. The basis of the wave-particle duality thus seems to be the conceptual conflict between position and momentum as reflected in the Uncertainty Principle, and that sets the agenda for much of Quantum Mechanics. If we cannot know precisely an electron's present, we will have to rely on statistical means to tell its future. The very existence of the quantum of action smears certainty into probability and blurs cause and effect so that reality is not quite as sharp as we once thought it was. *Just as the unique character of Relativity arises from the fact that* c *is finite and not infinite, the unique character of Quantum Mechanics arises from the fact that* h *is finite and not zero.* The classical world behaves as if $c = \infty$ and $h = 0$. The real world behaves as if $c = 3 \times 10^8$ m/s and $h = 7 \times 10^{-34}$ J·s. It's only on the macroscopic scale, where h is relatively negligible, that energy and momentum seem continuous and the world appears to be classical.

The more important fundamental laws and facts of physical science have all been discovered, and these are now so firmly established that the possibility of their ever being supplanted in consequence of new discoveries is exceedingly remote. Our future discoveries must be looked for in the sixth place of decimals.

ALBERT MICHELSON (1899)
American physicist

Δp and Diffraction

Not surprisingly, a similar interrelationship to that which exists between momentum and position appears with lightwaves: closing down a single slit (p. 963) increases the spread in the diffraction pattern. Recall the equation for single-slit diffraction of light of wavelength λ; namely,

$$D \sin \theta_{m'} = m'\lambda \qquad [27.13]$$

Making D smaller increases θ. To see how the Uncertainty Principle can be used to provide more than just a numerical limit, consider the beam of electrons or photons impinging on a slit of width D in Fig. 31.15. The hole confines the beam, restricting it in the y-direction to a width D and making the uncertainty in the sidewise position of any electron $\Delta y = D$. The electron must acquire a sidewise y-component of momentum Δp_y on passing through the slit such that at best

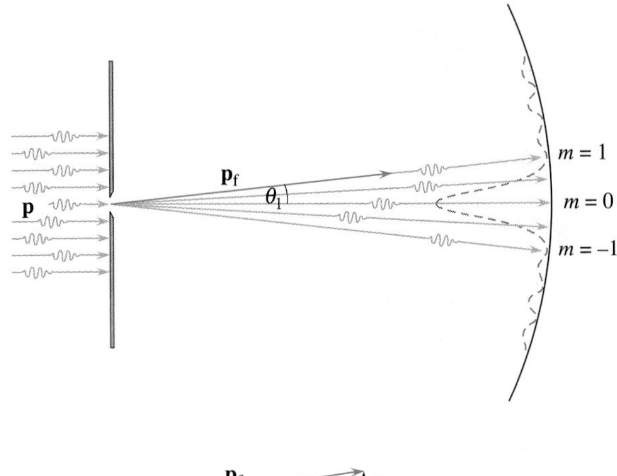

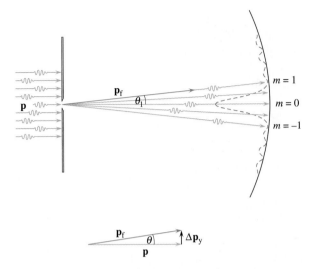

Figure 31.15 The Uncertainty Principle as applied to a stream of photons passing through a narrow slit of width *D*.

$$\Delta p_y \Delta y \approx h$$

That's what is traditionally called diffraction; by passing through the slit, the electron was effectively localized in y, but its y-momentum component simultaneously spread out. From the diagram $\Delta p_y = p \tan \theta \approx p \sin \theta = (h/\lambda) \sin \theta$, and since $\Delta p_y \Delta y \approx h$

$$\left(\frac{h}{\lambda} \sin \theta\right) D \approx h$$

At best, the electron flies off at an angle somewhere between 0 and θ, the latter given by

$$\sin \theta \approx \frac{\lambda}{D}$$

From Eq. (27.13), that's the location of the first minimum ($m' = 1$). It effectively defines the central diffraction maximum, where most of the electrons will end up. If we knew nothing about diffraction, the Uncertainty Principle would tell us to expect the electrons to be spread out over the central region and beyond, rather than to cast a sharp image of the slit.

Uncertainty and Nuclear Electrons

It was pointed out earlier (p. 1034) that electrons do not reside in the nuclei of atoms, and a persuasive argument to that effect can be made using the Uncertainty Principle. An electron imprisoned in a nucleus must have a position uncertainty not greater than the size of the nucleus. The Uncertainty Principle then demands a corresponding indeterminancy in momentum, allowing values of p and therefore of energy quite inconsistent with observation.

Example 31.3 Assume that an atomic nucleus is 1.0×10^{-14} m in diameter and calculate the kinetic energy of an electron confined to such a space. Compare that to the energy range for beta rays (0.025 MeV to 3.2 MeV). Recall that an electron has a rest energy of 0.511 MeV.

(continued)

(continued)

Solution: [Given: nuclear diameter of 1.0×10^{-14} m. Find: KE.] The electron can be anywhere in the nucleus, and so $\Delta x \approx 10^{-14}$ m. Hence, the smallest uncertainty in momentum is

$$\Delta p_x \approx \frac{\frac{1}{2}\hbar}{\Delta x} = \frac{5.27 \times 10^{-35}\ \text{J·s}}{1.0 \times 10^{-14}\ \text{m}} = 5.27 \times 10^{-21}\ \text{J·s/m}$$

The electron's momentum must be at least as large as this result, and so its total energy is

$$E^2 = E_0^2 + (pc)^2 \qquad [28.15]$$

For an electron, $E_0 = 8.19 \times 10^{-14}$ J and

$$E^2 = 6.70 \times 10^{-27}\ \text{J}^2 + (5.27 \times 10^{-21}\ \text{J·s/m})^2 \times (2.99 \times 10^8\ \text{m/s})^2$$

Thus $\quad E = 1.58 \times 10^{-12}\ \text{J} = 9.85$ MeV

and with $E_0 = 0.511$ MeV we have

$$KE = E - E_0 \approx \boxed{9.3\ \text{MeV}}$$

This is the smallest energy, and it's appreciably larger than the observed β-ray energies. That difference strongly suggests that there are ordinarily no electrons confined to the nucleus.

▶ **Quick Check:** To redo the calculation in MeV, use $h = 4.14 \times 10^{-15}$ eV·s in the expression $\Delta p \approx \frac{1}{2}\hbar/\Delta x$. Hence, $p = 0.0329$ eV·s/m and $p^2c^2 = 9.73 \times 10^{13}$ eV^2. Since $E_0 = 0.511$ MeV, $E^2 = E_0^2 + p^2c^2$ and $E \approx$ 9.9 MeV.

Electrons can only exist inside the nucleus if there is a tremendous force restraining them there. A large collapsing star can have such immense pressures at its core that orbital electrons are crushed into the nuclei to create a giant ball of neutrons—a neutron star.

Determinism

An intelligence which at a given moment knew all the forces that animate nature, and the respective positions of the beings that compose it, could condense into a single formula the movement of the greatest bodies of the Universe and that of the least atom: for such an intelligence nothing could be uncertain, the past and future would be before its eyes.

PIERRE SIMON, MARQUIS DE LAPLACE (1749–1827)

Ultimately, we do not even know if a particle simultaneously has a perfectly defined position and momentum. There are many physicists who believe that it does not. In any case, we could not measure it if it did; not because we're momentarily without the proper method, but because no such method is theoretically possible. Heisenberg's original German word for his concept was *Unbestimmtheitsprinzip,* which better translates as "Indeterminacy Principle." "Uncertainty" suggests that a particle has definite simultaneous values of conjugate variables whose determination is uncertain, which was not Heisenberg's belief, though not everyone agrees.

There was a time when scientists like famed Laplace idly fancied that if they knew the initial location and momentum of all the particles in the Universe, they could (theoretically at least) foretell the future and retell the past. (This, despite free will and passion, as if your next warm kiss had been set in motion some 17 thousand million years ago when the Universe began.) That's the ultimate in *determinism,* and however beyond our actual computational abilities, the logic still disturbs. Though Newtonian theory weaves a deterministic plot where every sigh results from the inexorable confluence of physical law, Quantum Mechanics happily puts a touch of chance in the Hatter's clockwork. The future of even one lone electron is quite beyond prediction; the Uncertainty Principle makes it impossible to precisely know its initial conditions. So kiss whomever you like, change your mind however you will, and rest assured that physics, at least, knows that *it* cannot hope to calculate the future.

I think I can safely say that nobody understands quantum mechanics.

RICHARD P. FEYNMAN
The Character of Physical Law (1967)

31.10 QED and Antimatter

In 1928, things were still simple, ominously simple: atoms were composed only of electrons and protons; the neutron had not yet been discovered; and there were no other microparticles in the zoo except for the photon. That was the year P. A. M. Dirac published "The Relativistic Theory of the Electron," for which he would share the 1933 Nobel Prize with Schrödinger. It was Dirac's intention to reformulate Quantum Mechanics so that it was consistent with Special Relativity. A relativistic quantum theory would naturally mate with Maxwell's Equations and, as such, is now known as **Quantum Electrodynamics** and affectionately called QED. The resulting *Dirac equation* was a triumph: among other things, electron spin came out naturally, Sommerfeld's fine structure formula was derived, and the magnetic moment of the electron was calculated. However splendid, the theory seemed to be flawed by one small aspect, one enigmatic mathematical quirk: it had twice as many energy states as were presumably called for.

Paul Adrien Maurice Dirac (1902–1984). He shared the 1933 Nobel Prize in physics with Schrödinger.

Dirac traced the problem back to the relationship $E^2 = c^2p^2 + m^2c^4$; as with all square roots, it has two solutions; that is

$$E = \pm c\sqrt{p^2 + m^2c^2}$$

Yet the energy of a free electron cannot be negative. It made no sense to keep the negative solution—its implications were too bizarre to be real. As is customary in such cases, Dirac simply put aside the ugly twin as a mathematical aberration. We do that all the time in Classical Physics. It soon became clear, however, that the negative portion could not be overlooked without seriously weakening the whole theory. Now there was a dilemma—a richly potent theory embarrassed by what seemed a bit of technical minutia.

It took Dirac until the end of 1929 to come up with a bold solution. The negative-energy states are real: many electrons have presumably radiated their energy and descended down into these lowest-possible states. Since the $\approx 10^{80}$ ordinary electrons in this Universe of ours are not observed to tumble into the oblivion of the negative Wonderland, it must be that all the negative states are occupied, in accordance with the Exclusion Principle. Wonderland has no vacancies. The situation was like a large hotel with half its floors below ground and half above. With most of the upper rooms empty, there could be lots of shifting around by the guests, but below ground level all the rooms were occupied. No one new could move down and in until someone moved up and out.

Instead of being empty, vacuum was now to be imagined filled with an invisible multitude of unseen electrons sailing around with negative energies. If one of them was somehow removed, pulled up into the ordinary positive world, it would leave behind an empty hole, a bubble in the negative sea of electrons. Dirac recognized that this hole would appear to us as a positive charge and proposed that it was the only known positive particle, the proton. J. Robert Oppenheimer quickly showed that this proposal was untenable. The hole had to behave as if it were the mirror image of the electron. "A hole, if there is one would be a new kind of particle, unknown to experimental physics, having the same mass and opposite charge to the electron." So wrote Dirac in 1931. "We may call such a particle an anti-electron."

This flight of creative imagination might have remained little more than a mathematical fantasy if not for a cosmic-ray study taking place at the California Institute of Technology. Carl Anderson, working with Millikan, was examining these particles (predominantly protons, possibly emitted from supernovas), which stream in on Earth from space. They were being observed using a cloud chamber placed in a uniform magnetic field so that positive and negative particles would follow oppositely curved paths. Among the thousands of photographs taken, Anderson noticed one that seemed curious. It clearly showed two oppositely curved tracks, one of which was certainly made by an electron; the other suggested an antielectron, but

All Nature is but Art, unknown to thee;
All Chance, Direction, which thou canst not see;
All Discord, Harmony not understood.

ALEXANDER POPE (1688–1744)

> *The Quantum Mechanics is very imposing. But an inner voice tells me that it is still not the final truth. The theory yields much, but it hardly brings us nearer to the secret of the Old One. In any case, I am convinced that He does not play dice.*
>
> ALBERT EINSTEIN (1926)

that conclusion was still "very radical at the time." Others had seen these positive tracks and either ignored them or dismissed them as "dirt." But Dirac's paper changed the world view, and the facts in the cloud chamber changed accordingly. By the summer of 1932, Anderson had clear evidence of the existence of the antielectron! It was Anderson who christened it the **positron**.

Today, it is commonplace to create electron-positron pairs—to have substantial matter materialize out of radiant energy (see the photo on p. 639). A blast of electromagnetic energy, a gamma ray with zero mass, can disappear and in its place an electron and a positron pop into existence. This ultimate alchemy occurs provided that the gamma-ray photon carries an energy equal to at least the total rest energy of the two particles (and provided there is a heavy object nearby to conserve momentum). Just as pair creation is possible, so is pair annihilation. A positron and *any* electron it approaches can come together and totally obliterate each other, vanishing in a puff of gamma rays (as if the electron had dropped back into the hole and disappeared).

Example 31.4 Determine the minimum energy (in joules and MeV) that a gamma photon must have to create an electron-positron pair.

Solution: [Given: an electron-positron pair. Find: minimum energy to create.] The minimum energy, E, will equal the rest energy of the pair; namely

$$\mathrm{E}_0 = 2m_e c^2$$

$$\mathrm{E}_0 = 2(9.11 \times 10^{-31}\ \mathrm{kg})(3.00 \times 10^8\ \mathrm{m/s})^2$$

and so

$$\boxed{\mathrm{E} = 1.64 \times 10^{-13}\ \mathrm{J}}$$

or

$$\boxed{\mathrm{E} = 1.02\ \mathrm{MeV}}$$

▶ **Quick Check:** Since $\mathrm{E} = hf$

$$f = \frac{\mathrm{E}}{h} = \frac{1.64 \times 10^{-13}\ \mathrm{J}}{6.63 \times 10^{-34}\ \mathrm{J \cdot s}} = 2.48 \times 10^{20}\ \mathrm{Hz}$$

This result corresponds to 0.001 2 nm, which is a gamma ray.

Richard Phillips Feynman (1918–1988).

The proton is 1836 times more massive than the electron. Thus the creation of the **antiproton**, which would require that much more energy, could not be attempted until a new generation of accelerators was available. In 1955, the bevatron at the University of California, Berkeley, produced the first human-made antiproton—a negative speck with the same mass as the proton. Highly energetic beams of protons (p) and antiprotons ($\bar{\mathrm{p}}$) can slam together, creating all sorts of new subnuclear particles (p. 1142). On rare occasions, there is a cancellation of the equal and opposite charges and the creation of a neutron and an **antineutron** (1956). The similarity of neutrons (n) and protons is underscored by the fact that protons can similarly interact with antineutrons ($\bar{\mathrm{n}}$), as can antiprotons (often called p-bars) and neutrons.

Nowadays, antimatter is familiar in the laboratory; there are even artificial radioactive isotopes, such as $^{22}\mathrm{Na}$, that emit positrons on decaying and so serve as convenient sources. For example, by injecting someone with glucose laced with such a radioactive tracer, we can get a picture of the metabolism of various regions of the brain (Fig. 31.16). Antinucleons have been joined together to make compound structures such as antideuterons and antialpha particles. Researchers have made *positronium,* a hydrogen-like atom composed of a positron and an electron bound together in a somewhat stable form. And an exotic atom called *antiprotonic hydrogen,* a proton and antiproton orbiting each other, was created in 1978. Protons, electrons, neu-

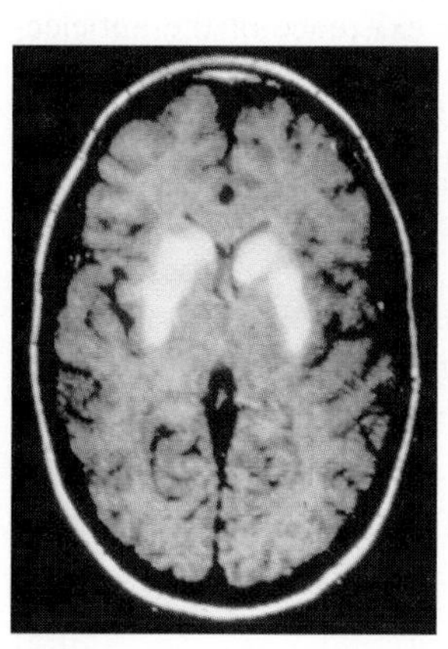

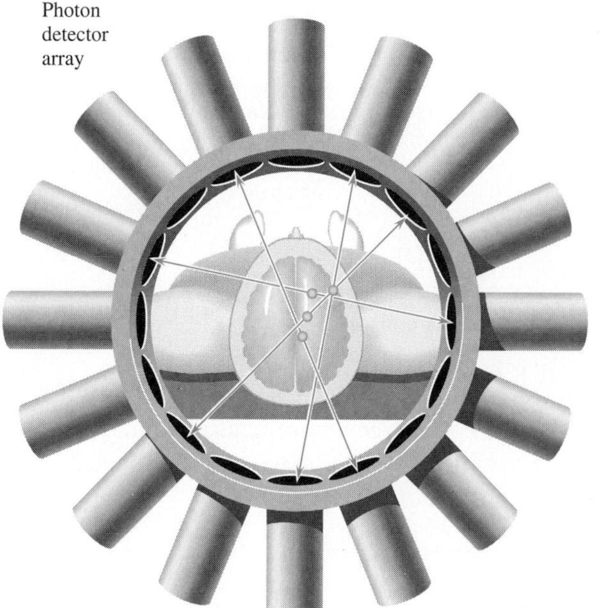

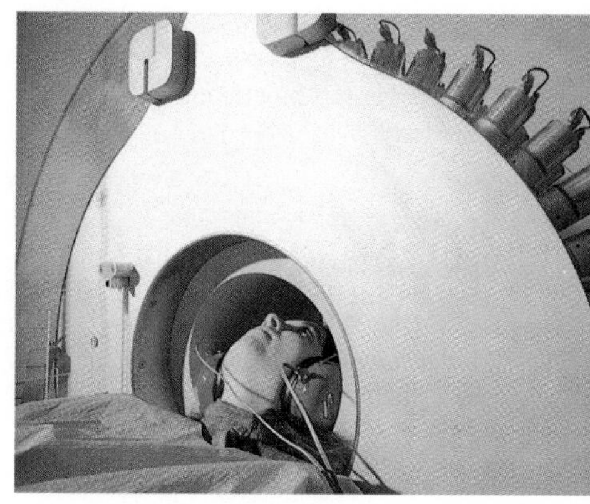

Figure 31.16 Positron Emission Tomography (PET). A patient receives an injection of some substance such as glucose that has been tagged with a radioactive element. The glucose soon goes to the brain, where it emits positrons. The positrons annihilate with electrons, producing pairs of gamma photons that are readily measured by a surrounding array of detectors. The images that result can reveal Alzheimer's disease, schizophrenia, and a wide range of other disorders of the brain.

trons, neutrinos (indeed, all the material subnuclear particles) have their antimatter twins, their mirror-image annihilators.

Dirac's theory worked well; in its first approximation, it agreed nicely with the known experiments of the day. His prediction of the electron's magnetic moment was accurate to two decimal places. But there gradually developed some very troublesome features: an atomic electron should interact both with the field of the nucleus and, to some extent, with its own field. When attempts were made to add in appropriate contributions, the theory went wild, producing totally weird results—things took on infinite values. In 1947, Willis Lamb at Columbia University used microwave techniques to discover that two quantum states of hydrogen actually had slightly different energies even though the Dirac theory predicted that they had to be *exactly* identical. This *Lamb shift* stimulated a modern-day reformulation of Quantum Electrodynamics, principally by Julian Schwinger and Richard Feynman of the United States and Sinitiro Tomonaga of Japan. They devised a mathematical scheme for making the theory workable known as *renormalization.* It got around the infinites, producing "normal" results.

Despite the fact that QED, the theory of the electron and photon, is certainly limited, it nonetheless is among the premier theoretical formalisms of physics. In the several instances where QED can be carried through, it provides an amazing degree of accuracy. For example, the magnetic moment of the electron as determined theoretically via QED has the numerical value 1.001 159 652 46, with an error of about 20 in the last two digits. The accepted measured value is 1.001 159 652 193, with a possible error of about 10 in the last two digits. A theory that can do that will engender a great deal of faith in its practitioners.

As we'll see (p. 1142), the structure of QED is very special in a way that few people were concerned about before 1954; it's a *gauge theory,* derivable from a symmetry principle. In time, QED would become a model for all the theoretical advances in High Energy Physics, the great work of the second half of the twentieth century.

Despite its enormous practical success, quantum theory is so contrary to intuition that, even after 45 years, the experts themselves still do not agree what to make of it.

BRYCE DEWITT
Physics Today (September 1970)

Core Material

According to de Broglie, the wavelength of a particle is

$$\lambda = \frac{h}{mv} \qquad [31.1]$$

and using the kinetic energy in eV, for nonrelativistic situations

$$\lambda = \frac{1.2264}{\sqrt{\mathrm{KE}}}\ \mathrm{nm} \qquad [31.2]$$

Material particles that are moving have momentum and therefore a wavelength.

Heisenberg, Schrödinger, and Dirac each formulated a quantum-mechanical theory, all of which are closely related. **Schrödinger's Equation** in simple form is

$$-\frac{\hbar^2}{2m}\frac{d^2\psi}{dx^2} + \mathrm{U}\psi = \mathrm{E}\psi$$

and it is the basis of Wave Mechanics (p. 1076). The wavefunction, ψ, determines the probability of occurrences.

No two fermions in any system can occupy the same state; that is, have the same quantum numbers. This concept is the **Pauli Exclusion Principle** (p. 1084). The state of an atomic electron is fixed by four *quantum numbers: n, l, m_l,* and m_s.

One form of the **Heisenberg Uncertainty Principle** (p. 1089) is

$$\Delta p_x \Delta x \approx h \qquad [31.3]$$

or, more precisely

$$\Delta p_x \Delta x \geq \tfrac{1}{2}\hbar \qquad [31.4]$$

The position and linear momentum of an object cannot both be measured with unlimited precision at the same time. Similarly

$$\Delta \mathrm{E}\, \Delta t \geq \tfrac{1}{2}\hbar \qquad [31.5]$$

Suggestions on Problem Solving

1. Some useful constants to keep in mind are

$$1\ \mathrm{eV} = 1.602\,177 \times 10^{-19}\ \mathrm{J}$$

$$hc = 1.239\,842 \times 10^{-6}\ \mathrm{eV\cdot m} \approx 1240\ \mathrm{eV\cdot nm}$$

$$h = 6.626\,08 \times 10^{-34}\ \mathrm{J\cdot s} = 4.135\,67 \times 10^{-15}\ \mathrm{eV\cdot s}$$

$$\hbar = 1.054\,573 \times 10^{-34}\ \mathrm{J\cdot s} = 6.582\,12 \times 10^{-16}\ \mathrm{eV\cdot s}$$

$$\tfrac{1}{2}\hbar = 3.29 \times 10^{-16}\ \mathrm{eV\cdot s}$$

2. We often have to find such things as the wavelength of, say, an electron, and so need its momentum ($p = h/\lambda$). And here a decision has to be made as to whether to compute the classical value $p = \sqrt{(\mathrm{KE})2m}$ or the relativistic value $p = \sqrt{(\mathrm{E}^2 - \mathrm{E}_0^2)}/c$. When $\mathrm{E}_0 \gg \mathrm{KE}$, we can use the classical formulation. For example, for an electron with a KE of 0.1 MeV (as compared to its rest energy of 0.5 MeV), using the classical momentum in calculating wavelength introduces a 5% error.

3. When calculating the de Broglie wavelength ($\lambda = h/p$) and frequency ($f = \mathrm{E}/h$) for anything other than a photon, remember that λf is the phase speed of the wave, *not* the speed of the particle and *not* c.

4. The following summary of levels and quantum numbers is handy to have:

Principal	$n = 1, 2, 3, 4, \ldots$
Orbital	$l = 0, 1, 2, 3, \ldots, n-1$
Magnetic	$m_l = l, l-1, \ldots, 0, \ldots, -l+1, -l$
Spin	$m_s = -\frac{1}{2}, +\frac{1}{2}$

Discussion Questions

1. Given that the resolving power of a microscope is proportional to the wavelength of the illumination, why might an electron microscope be appealing? Figure Q1 shows a transmission electron microscope invented by Ruska in 1935. How does it work? Incidentally, electrons traveling within the axially symmetric magnetic field inside the gap of a current-carrying coil tend to be brought inward to a focus. Compare this tendency with the behavior of X-rays. While an optical instrument might have a resolution of 200 nm or so, a modern electron microscope can resolve objects as small as about 0.1 nm. (The limit on resolution is actually set by the spherical aberration of the lenses and not λ.)

2. A narrow, very sparse beam of monoenergetic protons is shone on an even narrower slit in an opaque screen. Make a set of drawings showing the pattern you would expect to observe after several increasing intervals of time. Explain your conclusions. What if the beam were not monoenergetic, what would happen?

3. Draw a crude pictorial representation of the electron shell structure of sodium fluoride (NaF) and explain how the molecule is held together. Do the same for hydrogen chloride (HCl).

4. Considering the shell model of the atom, derive an expression for the number of electrons in the nth shell and also the number in the lth subshell. Explain your reasoning.

5. Imagine a screen with two narrow slits in it and suppose there is an incident beam of monochromatic electrons that encompasses both apertures (Fig. Q5). An interference pattern typical of Young's Experiment will be observed on a distant surface. Now suppose that each slit is surrounded by a coil of wire such that, as an electron passes through the opening and then the

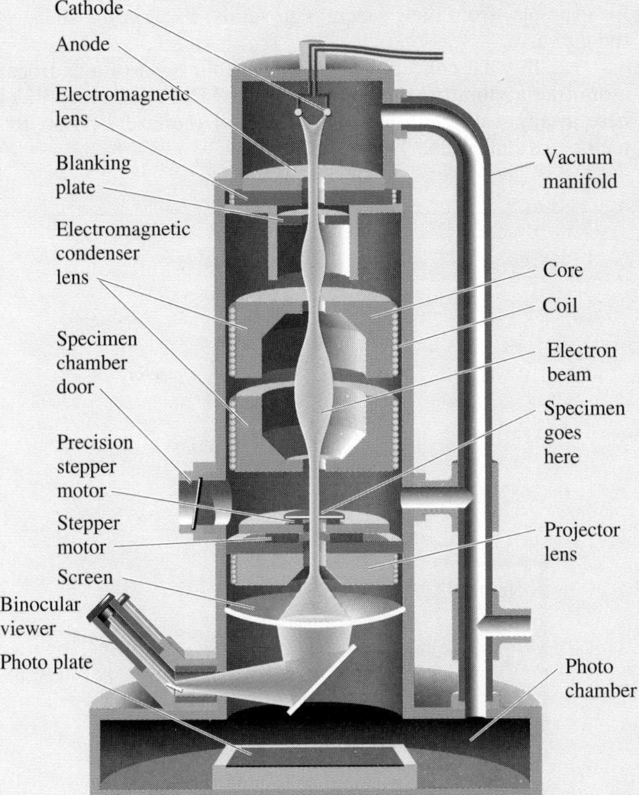

Figure Q1

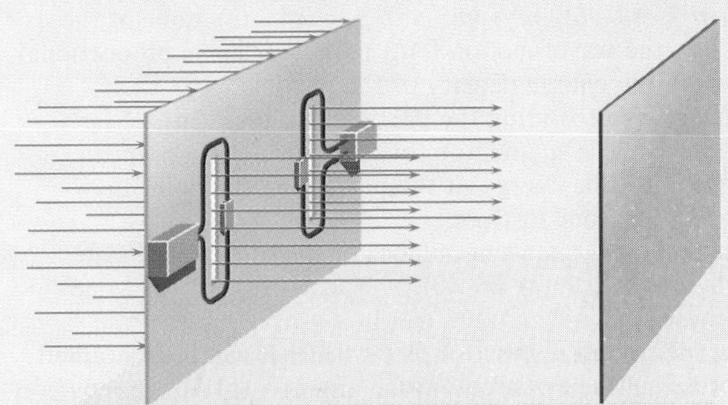

Figure Q5

coil, a current will be induced and the passage of the particle appropriately recorded. Use the arguments of Quantum Mechanics to describe what, if anything, will happen to the interference pattern once the switches on the coils are closed. If only one coil was used, would that change things?

6. Like photons, electrons have a wavelength, frequency, spin, and phase. Might it be possible to build a laserlike device that will produce tremendously intense beams of coherent electrons? Explain.

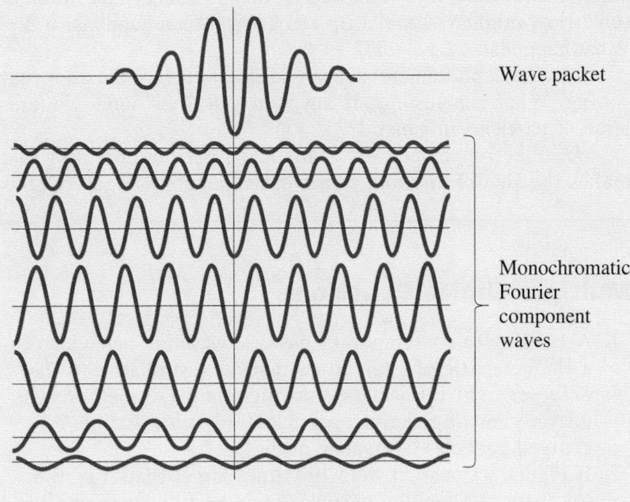

Figure Q9

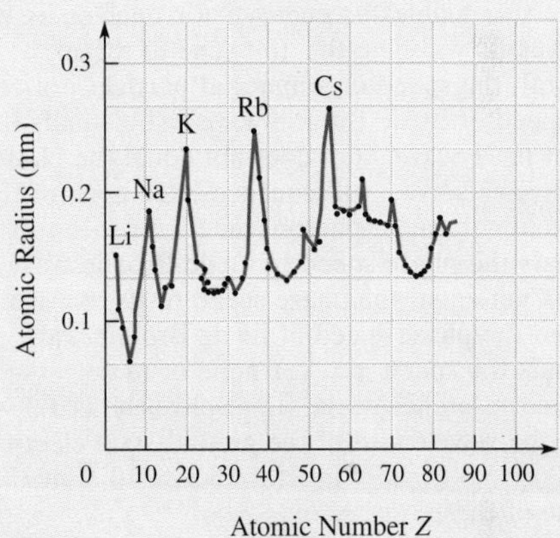

Figure Q11

7. A far-reaching consequence of the Uncertainty Principle is that a particle confined to a region of space cannot have zero kinetic energy. The energy it does have is called the **zero-point energy**. Explain this statement and discuss how it might apply to the energy of a material at a temperature of absolute zero.

8. In 1929, Otto Stern shone a beam of neutral helium atoms (with a de Broglie wavelength of about 0.13 nm) at an angle to the surface of a crystal. What do you think he saw when he examined the pattern of the reflected helium atoms? Explain.

9. A reasonable way to represent a particle mathematically is to use a localized wavefunction like that of Fig. Q9. Such a pulse is known as a *wave packet,* and it's equivalent to the superposition of a number of monochromatic waves, as indicated in the figure. The tighter (shorter in space) the packet, the more numerous the needed monochromatic contributions (remember Fourier's analysis). In what sense does this description of the

particle have built into it an uncertainty in energy and momentum? How might you make Δp zero? What then happens to Δx? What happens as $\Delta x \to 0$?

10. Can a photon create a single electron? Explain your reasoning. What conclusions, if any, can you draw about the creation of particles in general?

11. Electron clouds are statistical distributions, and that makes the size of an atom somewhat ambiguous. Figure Q11 is nonetheless a summary of several atomic radii as determined, for example, from their spacings in solids. Explain the shape of the curve.

12. In what conceptual sense is a Bohr orbit of a hydrogen atom like a vibrating circular steel hoop? Does the central idea here in any way relate to the structure of the laser? (Hint: Remember Kundt's tube?)

Multiple Choice Questions

1. According to contemporary physics, wavelike behavior is a characteristic of (a) all particles at rest relative to the observer (b) all particles moving relative to the observer (c) only moving charged particles (d) only stationary charged particles (e) none of these.
2. If Planck's Constant were 100 times larger than it is, the (a) mass of a moving particle would be 100 times smaller (b) momentum of a moving particle would be 100 times larger (c) wavelength of a moving particle would be 100 times smaller (d) wavelength of a moving particle would be unchanged (e) none of these.
3. Doubling the momentum of a neutron (a) decreases its energy (b) doubles its energy (c) doubles its wavelength (d) halves its wavelength (e) none of these.
4. In general, the speed of a material particle (a) equals the phase speed of its de Broglie wave (b) equals the phase speed of its Ψ-wave (c) does not equal the phase speed of its de Broglie wave (d) equals c (e) none of these.
5. For a monochromatic photon, the phase speed in vacuum (a) equals the phase speed of its de Broglie wave (b) does not equal the phase speed of its Ψ-wave (c) does not equal the phase speed of its de Broglie wave (d) equals the speed γc (e) none of these.
6. If the circumference of the first Bohr orbit is 3.3×10^{-10} m, what is the wavelength of the ground-state electron? (a) 0 (b) ∞ (c) 3.3×10^{-10} m (d) $(3.3 \times 10^{-10}\text{ m})/h$ (e) none of these.
7. Doubling the total energy of a meson has the effect of (a) doubling its momentum (b) doubling its wavelength (c) quartering its frequency (d) doubling its frequency (e) none of these.
8. If a moving particle's energy is increased by a factor of 10, (a) its frequency increases by a factor of 10 (b) its frequency decreases by a factor of 10 (c) its frequency remains unchanged (d) its wavelength increases by a factor of 10 (e) none of these.
9. Neutrons from a reactor are slowed down by passing them through graphite (Fig. MC9) so that they have a de Broglie wavelength of about 0.1 nm. The beam is directed at a crystal, and the neutrons (a) because they are uncharged pass right through the crystal, totally unaffected by it (b) reflect off at an angle equal to the incident angle, just like a stream of baseballs (c) are totally absorbed (d) reflect off in several beams, as determined by the Bragg equation (e) none of these.
10. It follows from the de Broglie hypothesis that (a) $E = \omega\hbar$ (b) $E = \omega h$ (c) $E = f\hbar$ (d) $E = \lambda\hbar$ (e) none of these.

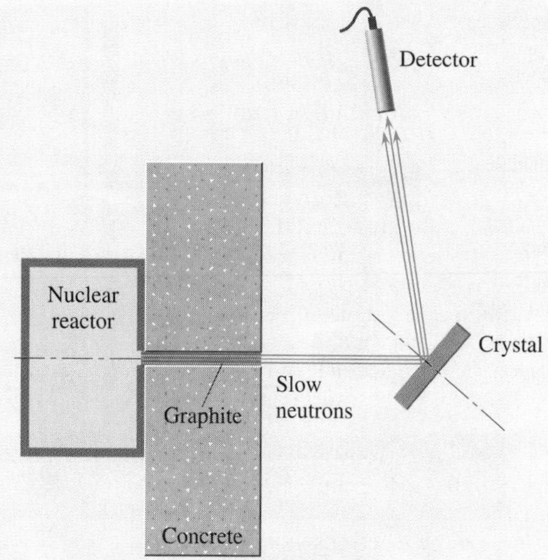

Figure MC9

11. It follows from the de Broglie hypothesis that (a) $p = \omega\hbar$ (b) $p = \lambda h$ (c) $p = k\hbar$ (d) $p = \lambda\hbar$ (e) none of these.
12. If Ψ is the wavefunction for a particle, $|\Psi|^2$ is proportional to (a) the charge density of the particle (b) the probability of finding the particle at a point in space (c) the momentum of the particle at a point in space (d) the energy of the particle at the point in space (e) none of these.
13. The principal quantum number of the 6th excited state of hydrogen is (a) 6 (b) 5 (c) 4 (d) 7 (e) none of these.
14. The maximum number of electrons that can be contained in the *f*-subshell of an atom is (a) 14 (b) 10 (c) 6 (d) 12 (e) none of these.
15. The size of the space in which an electron is confined determines the uncertainty in its (a) linear momentum (b) maximum angular momentum (c) spin angular momentum (d) its mean lifetime (e) none of these.
16. We might say that an electron in the ground state cannot radiate because (a) if it did, its wavelength would increase and it could not fit in any smaller lower-energy orbit (b) it has no energy to radiate in the ground state (c) it would decrease in wavelength and lose momentum (d) it would not conserve both energy and momentum (e) none of these.

Problems

THE CONCEPTUAL BASIS OF QUANTUM MECHANICS
QUANTUM PHYSICS

1. [I] Calculate the wavelength of a 60-kg person jogging along at 2.0 m/s.

2. [I] What is the de Broglie wavelength of a 10.00-g bullet traveling at 331 m/s?

3. [I] Determine the wavelength of an electron traveling at a speed of c/10.

4. [I] Determine the frequency of an electron traveling at a speed of c/10. Use relativistic considerations.

5. [I] A *thermal neutron* is one that is traveling at a speed comparable to that of a gas molecule at room temperature. In other words, one for which $(3/2)k_B T = \text{KE}$, and that (at 293 K) turns out to be 6.068×10^{-21} J. What is the wavelength of a thermal neutron?

6. [I] Use the results of Problems 3 and 4 to determine the phase speed of the de Broglie wave of an electron traveling at c/10. Note that v_p is greater than c.

7. [I] Suppose we are to build an electron microscope and we want it to operate at a wavelength of 0.10 nm. What accelerating voltage should be used?

8. [I] What are the values of the quantum numbers n and l for a 3d electron state?

9. [I] Consider the second excited state of the hydrogen atom. What are the values of the appropriate quantum numbers n, l, and m_l?

10. [I] Can two electrons in an atom have quantum number sets of $(2, 0, 0, +\frac{1}{2})$ and $(2, 0, 0, -\frac{1}{2})$, respectively? Explain.

11. [I] How many electrons can reside in the K, L, M, and N shells of an atom?

12. [I] How many electrons can reside in each subshell of the M-shell of an atom?

13. [I] Imagine a particle flying along the x-axis in a box of length L. What is the uncertainty in its momentum along the x-axis?

14. [I] A 100-g ball is confined to move in a 1.00-m-long frictionless tube lying along the x-axis. What is the minimum uncertainty in its speed? And how far will it move in a year at that speed?

15. [I] The position of a 0.001 00-kg particle along the x-axis is measured to be 1.2437 ± 0.0005 cm from the tip of a probe. What is the minimum uncertainty in its speed? (Hint: That's $\pm 0.0005 \times 10^{-2}$ m.)

16. [I] Since a charged pi meson at rest exists on average for only 26 ns, its energy will not be able to be measured with unlimited precision. Determine the minimum uncertainty in the meson's rest energy.

17. [I] A typical excited atomic state has a lifetime of 10^{-8} s. Determine the uncertainty in the energy of such a state in joules and electron-volts. (This is the unavoidable "blurriness" of the energy level and, rather than a sharp line, it produces an emission band known as the *natural linewidth.*)

18. [I] With Problem 17 in mind, determine the minimum uncertainty in the frequency of the emitted photon when there is a transition down to the ground state.

19. [I] A rho meson at rest has a mean lifetime of 4.4×10^{-24} s and an energy of 765 MeV. Determine the minimum uncertainty in the energy and write it as a fraction of the rest energy.

20. [II] What is the wavelength of an electron whose KE is 4.00 MeV? Since this energy is comparable with the rest energy, use relativistic arguments.

21. [II] What is the wavelength of an electron with a KE of 20 eV?

22. [II] Determine the kinetic energy of an electron that has a wavelength of 1.00 m.

23. [II] Consider a particle that has a large KE compared to its rest energy. Show that its wavelength is approximately equal to the wavelength of an equal-energy photon.

24. [II] If v_p is the phase speed of the de Broglie wave of a particle traveling at a speed v, prove that $v_p v = c^2$. Use Special Relativity.

25. [II] For a lightwave, the phase speed is given by $c = E/p$. Assume the same form for a de Broglie wave and find a relationship between its phase speed (v_p) and the particle's speed v. Assume classical conditions; that is, let E = KE. Your result neglects the rest energy and will not compare very well with the relativistic expression $vv_p = c^2$.

26. [II] With Problem 25 in mind and assuming an electron in a hydrogen atom has an energy given by $E_n = -hf_n$, where f_n is the frequency of the electron in the nth "orbit," show that de Broglie's hypothesis leads to the same orbital energies as the Bohr Theory; namely,

$$E_n = -\frac{n\hbar v_n}{2r_n}$$

provided E = KE.

27. [II] Using Bohr Theory, determine the energy of a photon that would excite the electron in a hydrogen atom from the K-shell to the M-shell.

28. [II] Make a table of all of the allowed four quantum numbers for the first three shells of the hydrogen atom. How many electrons can each shell accommodate?

29. [II] An excited hydrogen atom with its electron in the O-shell drops down to the L-shell, emitting a photon in the process. Determine the energy of that photon using Bohr Theory.

30. [II] What are the possible values of n, l, and m_l for a 5f atomic state? (First, check to see if it's allowed.)

31. [II] What are the possible values of n, l, and m_l for a 3f atomic state, and what can be said about that state?

32. [II] A 10.0-μg particle is traveling at 2.00 cm/s. Given that there is a 1.00% uncertainty in its speed, what is the least uncertainty in its position?

33. [II] Determine the ground-state electron configuration for $_{17}$Cl.

34. [II] Prove that a 10-g beetle whose position is known to within 1.00 nm can move with a speed uncertainty of 1.00 nm per year and still not violate the Uncertainty Principle.

35. [II] A 10.0-g particle is traveling at 20.0 cm/s. If the uncertainty in its momentum is 1 part in 1000, determine the uncertainty in its position.

36. [II] Approximate the minimum kinetic energy of an

electron confined to a region the size of an atom (0.10 nm).

37. [II] The position of an electron is measured within an uncertainty of 0.100 nm. What will be its minimum position uncertainty 2.00 s later?

38. [II] Angular momentum (L) and angle (θ), in radians, are conjugate variables. Accordingly, beginning with $\Delta p_x \Delta x \geq \frac{1}{2}\hbar$, apply it to circular motion and derive the uncertainty relationship for angular momentum.

39. [III] Gold is placed in an oven, melted, vaporized, and kept at a temperature of 1600 K. A parallel beam of atoms (each of mass 3.271×10^{-25} kg) emerges directly toward a circular hole of radius r in a screen (Fig. P39). Taking all the atoms to have the same speed, use the Uncertainty Principle to approximate the spot size formed by the gold on a screen that is 1.000 m away.

40. [III] In the Davisson-Germer experiment, 54-V electrons were scattered from a nickel crystal (in the first order) at 65°. The spacing of the crystal's atomic planes was found, using X-rays, to be 0.091 nm. Show that these results conform to de Broglie's hypothesis.

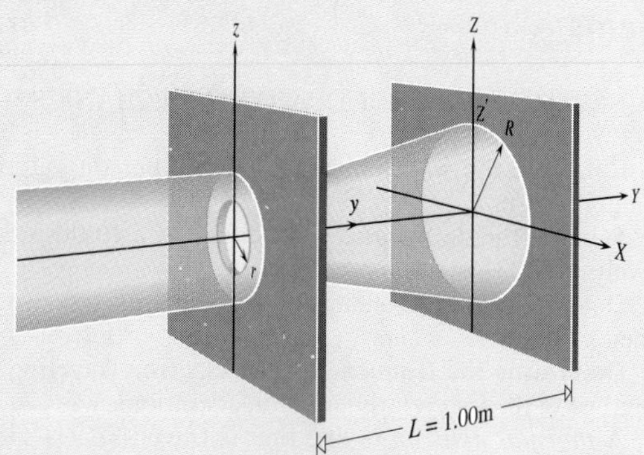

Figure P39

32

Nuclear Physics

AN ATOM, WITH A radius of about 10^{-10} m, consists of a cloud of electrons moving at great speeds around a positively charged nucleus comprised of protons and neutrons. That miniscule core, containing over 99.9% of the total mass, is roughly 10^4 times smaller than the atom as a whole. Like the atom, the nucleus is a bound system and can exist in a number of quantum states beyond its lowest-energy ground state.

We now turn our attention to the atomic nucleus. Though crucial to the structure of the atom, it plays an unobtrusive role hidden deep below the swirling cloud of electrons. On Earth, the atomic nucleus reveals itself in

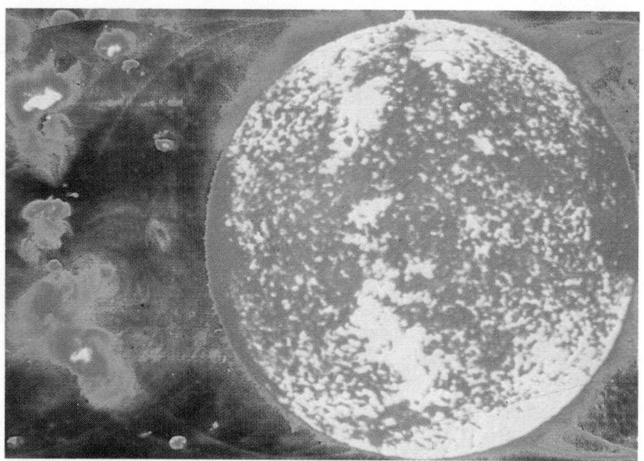

The Sun is a swirling sphere of plasma. At its center is a thermonuclear furnace driven by a gigantic fusion reaction.

the phenomenon of radioactivity and in nuclear weapons that unleash vast amounts of energy. On a much grander scale, nuclear fusion powers the stars (p. 1129).

NUCLEAR STRUCTURE

During the past several decades, experiments have provided reliable information about the size, shape, the distribution of charge within, and the magnetism of, the nucleus. Three principal techniques are used: (1) the nucleus can be probed with high-energy ($\approx$10 GeV), short-wavelength beams, usually of electrons; (2) orbital electrons interact with the nucleus, and the electromagnetic energy they emit provides information about nuclear structure; (3) a beam of positive particles can be scattered inelastically, transferring some of its energy to the nucleus, which becomes excited. The gamma radiation emitted as the nucleus returns to the ground state provides data on the nuclear configuration. It has been found that the nucleus is a complex shifting structure made of rapidly moving parts. The first clues to its composition came from studying isotopes.

32.1 Isotopes: Birds of a Feather

All atoms of a particular element were supposed to be identical. That assertion had been central to atomic theory since before Dalton's time, but it was wrong. There *are* different kinds of every one of the elements, often six or seven distinct variations. Rutherford and F. Soddy traced the decay of several heavy, naturally occurring radioactive elements, such as uranium and thorium. These decayed, spewing out particles and continuously transforming into different elements until they ended up as lead. Along the way, elements appeared that were chemically identical to others but that had different radioactive characteristics. Although ordinary lead has an atomic mass of 207.20 (we'll straighten out the units in a moment), lead present in uranium ore had a mass of only 206.05. Soddy named these variations of a given element **isotopes** (from the Greek *isos* meaning "same" and *topos* for "place"—having the same place in the Periodic Table).

In 1913, J. J. Thomson and F.W. Aston were the first to separate the isotopes of an element. They had put neon gas (atomic mass 20.2) into a positive ion tube and deflected the beam with charged plates and magnets to measure the atomic mass. At the face of the tube *two* distinct spots appeared. The conclusion was unmistakable: "Neon is not a single gas, but a mixture of two gases, one of which has an atomic weight [sic] of about 20, and the other of about 22."

A particular species of nucleus specified by a characteristic value of both **atomic number** (Z) and mass number, or **nucleon number** (A), is called a **nuclide**, and there are about 1500 nuclides. Every distinctly different nucleus is a specific nuclide. **Isotopes** of a given element are nuclides having the same Z (that is, the same nuclear charge), but having a different A; the total number of nucleons is different. Since $A - Z$ is the number of neutrons present (N), it follows that *isotopes of an element differ from one another only in their value of N.* Letting X be the chemical symbol for any element, we can specify any nuclide by writing it as

$$^{A}_{Z}\mathrm{X}_{N}$$

The name of an element is associated with a specific value of Z, a specific place in the Periodic Table. Any atom containing 10 protons is neon no matter what its nucleon number. Moreover, 10 protons and 10 neutrons make neon ($^{20}_{10}Ne_{10}$), just as 10 protons and 12 neutrons make a different form of neon ($^{22}_{10}Ne_{12}$). Apparently, this notation is redundant: $Z = 10$ means neon, and vice versa. The distinction is also often made by referring to neon-20, or Ne-20.

When found in our atmosphere, 90% of natural neon is of the first and lighter variety while the other 10% is the heavier kind. The *chemical atomic mass* of an element referenced in the Periodic Table reflects the natural abundances of the isotopes in a typical sample. These values were arrived at before the knowledge of the existence of isotopes and correspond to a weighted average of all the isotopes naturally present in the environment. For neon that average turns out to be 20.179 7.

In the early part of the nineteenth century, it made sense to determine relative atomic masses. Thus, hydrogen was set at 1, and all the other elements were measured with respect to it. Today, masses in the atomic domain are often specified in **unified atomic mass units** (u) where a neutral carbon atom ($^{12}_{6}C$) is defined to have a mass of precisely 12.000 000 u (see Table 32.1). Because of the practical link to the laboratory, with its accelerating electric fields, the unit MeV/c^2 is also widely used:

$$1\text{ u} = 1.660\,540 \times 10^{-27}\text{ kg} = 931.494\text{ MeV/c}^2$$

As a rule, A for each isotope differs slightly, but significantly, from its mass in units of u. (The exception is ^{12}C.)

TABLE 32.1 Atomic mass of some representative nuclides

Element	Symbol	Mass (u)
Hydrogen	$^{1}_{1}H$	1.007 825
Deuterium	$^{2}_{1}H$ (D)	2.014 102
Tritium*	$^{3}_{1}H$ (T)	3.016 049
Helium	$^{3}_{2}He$	3.016 029
Helium	$^{4}_{2}He$	4.002 603
Lithium*	$^{5}_{3}Li$	5.012 54
Lithium	$^{6}_{3}Li$	6.015 121
Beryllium	$^{9}_{4}Be$	9.012 182
Nitrogen	$^{14}_{7}N$	14.003 074
Nitrogen	$^{15}_{7}N$	15.000 109
Nitrogen*	$^{16}_{7}N$	16.006 100
Oxygen	$^{16}_{8}O$	15.994 915
Oxygen	$^{17}_{8}O$	16.999 131
Oxygen	$^{18}_{8}O$	17.999 160
Lead	$^{204}_{82}Pb$	203.973 020
Lead*	$^{205}_{82}Pb$	204.974 458
Lead	$^{207}_{82}Pb$	206.975 872
Uranium*	$^{233}_{92}U$	233.039 628
Uranium*	$^{235}_{92}U$	235.043 924
Uranium*	$^{238}_{92}U$	238.050 785

*Radioactive. (Remember Table 29.3.)

Example 32.1 Show that the chemical atomic mass of neon should be about 20.18 u, given that ^{20}Ne and ^{22}Ne have natural abundances of about 90.51% and 9.22%, and masses of 19.99 u and 21.99 u, respectively. (Your error will be due to the fact that we have overlooked trace amounts of ^{21}Ne.)

Solution: [Given: isotopes with masses of 19.99 u and 21.99 u, and abundances of 90.51% and 9.22%. Find: the chemical atomic mass.] The lighter neon is almost 10 times more abundant and therefore 10 times more influential in determining the chemical atomic mass of any natural sample than is the heavier isotope. The weighted average mass is thus

$$90.51\%\,(19.99\text{ u}) + 9.22\%\,(21.99\text{ u})$$

or $18.09\text{ u} + 2.03\text{ u} = \boxed{20.12\text{ u}}$.

▶ **Quick Check:** As already given, the actual value is 20.179 7 u.

Some 280 isotopes of the naturally occurring elements are stable and presumably will last for all time (Fig. 32.1). Around 1200 others (human-made and natural) are radioactive and transient. All the elements beyond uranium, from $Z = 93$ to ≈ 110, are produced in the laboratory and are radioactive; if they ever existed in abundance in nature, most were short-lived enough to have decayed to unobservably low levels. Some elements, such as xenon and iodine, have more than a dozen known

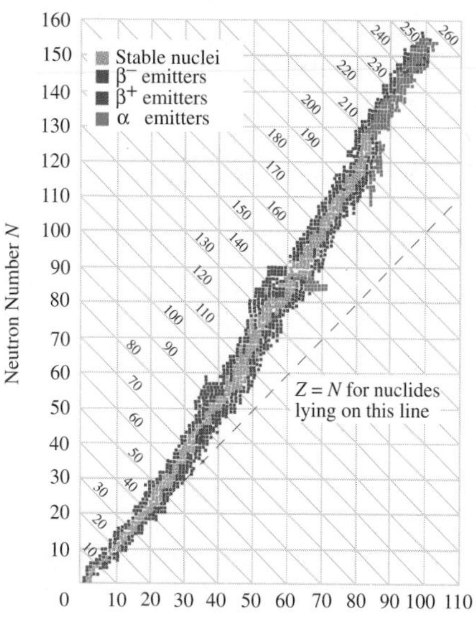

Figure 32.1 A plot of *N* versus *Z* for stable and unstable nuclides. The so-called *line of stability* runs up through the gold-colored band, which represents the stable nuclei. Each integer value of *N* and *Z* lying on a colored region corresponds to a nuclide.

isotopes each. Nine of the xenon forms are stable, but iodine has only one stable isotope, as do aluminum and gold.

The three isotopes of hydrogen (${}^{1}_{1}\mathrm{H}$, ${}^{2}_{1}\mathrm{H}$, ${}^{3}_{1}\mathrm{H}$) are so different from each other and so important that, unlike all others, they have their own names (Fig. 32.2). Ordinary *hydrogen* with a single proton ($Z = 1$, $A = 1$) is the lightest and most common (99.985%). **Deuterium** (also written ${}^{2}_{1}\mathrm{D}$) with an added neutron ($Z = 1$, $A = 2$) is quite rare (0.015%). For every 6500 ordinary hydrogen atoms in a natural sample, there is only 1 deuterium atom. Radioactive **tritium** (${}^{3}_{1}\mathrm{T}$) with still another neutron ($Z = 1$, $A = 3$) is by far the least abundant (for every 10^{18} atoms of ${}^{1}\mathrm{H}$ there is 1 of ${}^{3}\mathrm{T}$). Since these isotopes differ enormously in mass, and since the nuclei must interact somewhat differently with the single orbital electron, it's not surprising that they are distinctive chemically. For example, living organisms respond differently to water formed of oxygen and deuterium, known as **heavy water**, than they do to the ordinary brew, which differs as well in its freezing and boiling points. Heavy-water ice cubes, though they look and taste ordinary enough, sink to the bottom of a glass of tap water.

Insofar as the nucleus has little influence on the outer electrons, isotopes behave identically chemically except for some minor effects. Different isotopes of the same element can be separated, but as a rule only by mechanical means. The fact that a uranium-based atomic bomb requires large amounts of ${}^{235}\mathrm{U}$, which must be separated from ${}^{238}\mathrm{U}$, makes their production quite difficult for all but the most technologically advanced nations (p. 1123).

32.2 Nuclear Size, Shape, and Spin

Experiments pioneered by R. Hofstadter in the 1950s have revealed that nuclei are fairly spherical, with most being slightly ellipsoidal, though they come elongated, flattened a little, and pear-shaped as well. Figure 32.3 is a plot of nuclear charge density measured out along a radial distance in femtometers. The same results hold for the mass-density, with neutrons and protons being distributed in much the same way. Recall that $1\ \mathrm{fm} = 10^{-15}\ \mathrm{m}$; it's a distance that's convenient in the nuclear domain just as the nanometer was convenient in the atomic domain. (Nowadays, 1 fm is often called a *fermi* in honor of the Italian-American physicist Enrico Fermi.) The density drops off gradually over an outer thickness of roughly 2.5 fm—the nucleus thins out across this one-nucleon-thick surface region.

The nuclear radius R is often taken as the distance from the center to the half-density point (Fig. 32.3b). Independent of A, the density of a nucleus is constant for much of its radius. That means that the number of nucleons contained in a nucleus (assumed to be spherical) simply depends on its volume $\frac{4}{3}\pi R^3$. Hence, $A \propto R^3$ and $R \propto A^{1/3}$. Using a proportionality constant R_0 to make this relationship into

The 60-inch cyclotron at the Argonne National Laboratory. A beam of deuterons (the nuclei of deuterium) is streaming out at a speed of ≈28 000 miles per second.

an equality, we have

$$R = R_0 A^{1/3} \tag{32.1}$$

R_0 turns out to be ≈ 1.2 fm (Fig. 32.4). This situation should remind us of a drop of water, which also has a volume proportionate to the number of molecules it contains.

Example 32.2 Determine the radius of the carbon nucleus ($A = 12.0$).

Solution: [Given: $A = 12.0$ u. Find: R.] Using Eq. (32.1), we have

$$R = (1.2\ \text{fm})A^{1/3} = (1.2\ \text{fm})(2.29) = \boxed{2.7\ \text{fm}}$$

▶ **Quick Check:** Compare this result with Fig. 32.3.

The density of *nuclear matter,* as it's called, is the mass over the volume, and since A is equivalent to the mass in units of u, where $1\ \text{u} = 1.66 \times 10^{-27}$ kg

$$\rho = \frac{m}{\frac{4}{3}\pi R^3} = \frac{A(1.66 \times 10^{-27}\ \text{kg})}{\frac{4}{3}\pi R_0^3 A} = \frac{1.66 \times 10^{-27}\ \text{kg}}{7.24 \times 10^{-45}\ \text{m}^3} = 2.3 \times 10^{17}\ \text{kg/m}^3$$

This value is tremendously large; by comparison, the density of water is a mere $10^3\ \text{kg/m}^3$. Nuclear matter in bulk does not exist on Earth. You would know if it did because a cubic inch of it would weigh about 4×10^9 tons. Still, it is likely that certain celestial objects, such as neutron stars, consist of nuclear matter. Each such object is essentially a gigantic nucleus with gravity forcing it together.

Neutrons and protons are fermions with spins of $\frac{1}{2}$ and angular momenta of $\frac{1}{2}\hbar$. The *Shell Model* (p. 1110) assumes that the nucleus has energy levels very like the atom. The Exclusion Principle requires that only two protons (one spin-up, one spin-down) and two neutrons (one spin-up, one spin-down) can occupy any level. The total spin of a nucleus is the sum of the spins of its parts—*odd-A nuclides are fermions, even-A nuclides are bosons.* Thus, for an even-even nucleus (even Z, even N), such as ${}^4_2\text{He}$, ${}^{12}_6\text{C}$, or ${}^{16}_8\text{O}$, the total spin is zero. The nucleus of deuterium, the **deuteron** (${}^2_1\text{H}$), is typical of odd-odd nuclides; it has unfilled sublevels (one unpaired neutron and one unpaired proton) and a spin of 1. Even-odd and odd-even nuclides have one unpaired nucleon and therefore total spins that are odd-integer multiples of

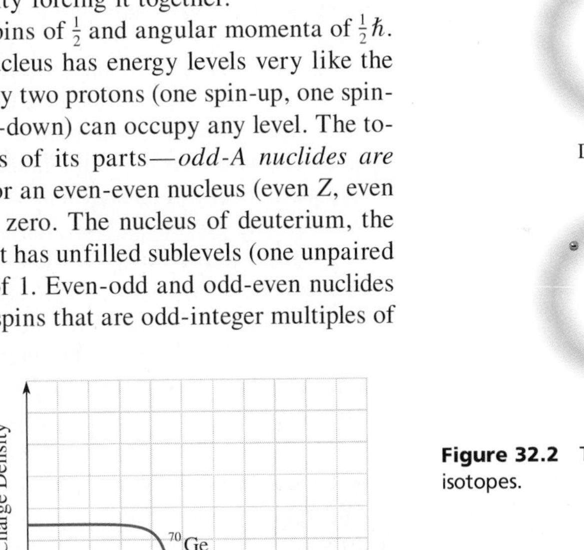

Hydrogen

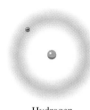

Deuterium

Tritium

Figure 32.2 The three hydrogen isotopes.

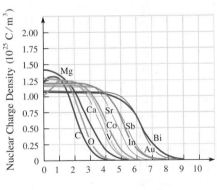

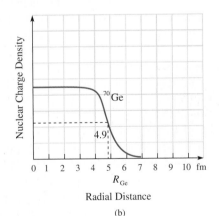

Figure 32.3 (a) The charge density versus distance from the center of the nucleus for several nuclei. [Data from R. Hofstadter, *Annual Reviews of Nuclear Science* **7**, 231 (1957).] (b) The radius of the germanium nucleus is about 4.9 fm.

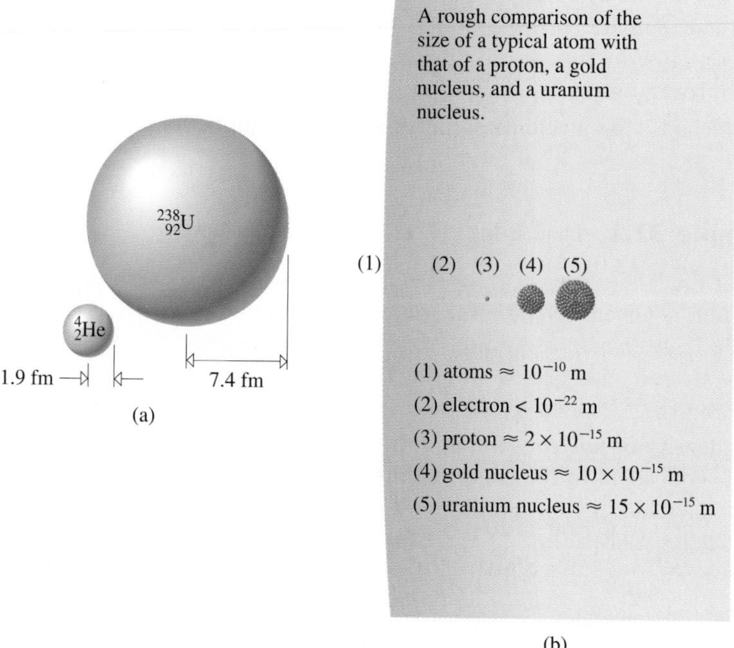

Figure 32.4 (a) shows the relative sizes of helium and uranium nuclei. (b) gives a sense of the size difference between a nucleus and an atom.

$\frac{1}{2}$. Thus, the spin of $^{7}_{3}Li$ (3 protons, 4 neutrons) arises from a group of two protons and two neutrons with spin zero plus the remaining 3 spin-$\frac{1}{2}$ nucleons, yielding a total of $\frac{3}{2}$.

The presence of spin suggests the possibility of a magnetic moment, which is certainly the case for a charged particle or an entity composed of charged particles. The magnetic moment of the proton is about 0.15% that of the electron and is in the same direction as its spin. This much is determined experimentally; as yet there is no complete theory of nuclear magnetism that accounts for the details. The magnetic moment of the proton ($+1.41 \times 10^{-26}$ J/T) is about 3 times larger than expected from its mass and the way the electron behaves. We can already anticipate that the proton will be a far more complicated entity than was first thought. The neutron, too, has a magnetic moment; it's smaller than the proton's and in the opposite direction to its own spin. The nonzero magnetic moment implies that the neutron, though it is neutral, has some internal nonuniform charge distribution. We now know that the proton and neutron are each composed of three still smaller charged quarks. A typical nucleus will have a net magnetic moment that depends on the number and arrangement of its nucleons.

By applying magnetic fields to an atom, it's possible to alter the spin state of the nucleus in a process discovered in 1946 called **nuclear magnetic resonance** (NMR). In the case of hydrogen, a proton can align its magnetic moment either parallel or antiparallel to an applied static *B*-field. The spin-up state has a slightly lower energy (ΔE) than the spin-down state. If a photon with an energy (hf) equal to the difference in energy between these two spin states (namely, ΔE) impinges on the proton, it can be absorbed. Thus, a pulse of radiofrequency electromagnetic energy of exactly the correct frequency can resonate the protons, exciting them into the higher-energy state. The same is true of more complex nuclei that have their characteristic resonant frequencies. When the nuclei relax back to their ground states, they each emit a photon of the same resonant frequency, which is easily detected. Either the absorption or re-emission of photons at different points in the specimen can be used to produce images of its internal structure. The human body is about 75% water and contains an abundance of hydrogen and therefore of protons. **Magnetic resonance imaging** (MRI) usually uses the NMR of protons to noninvasively form images of body tissue of living patients (Fig. 32.5).

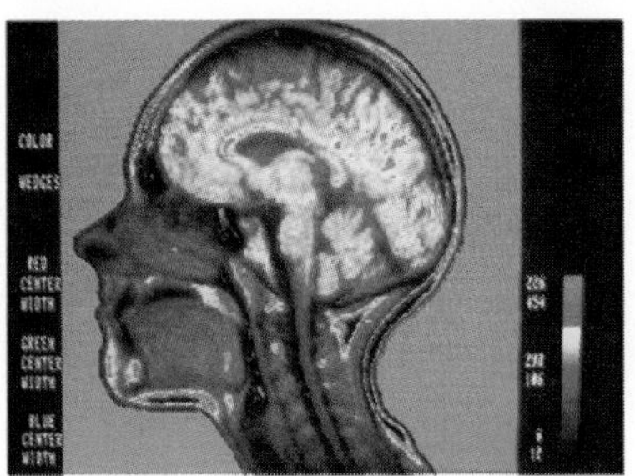

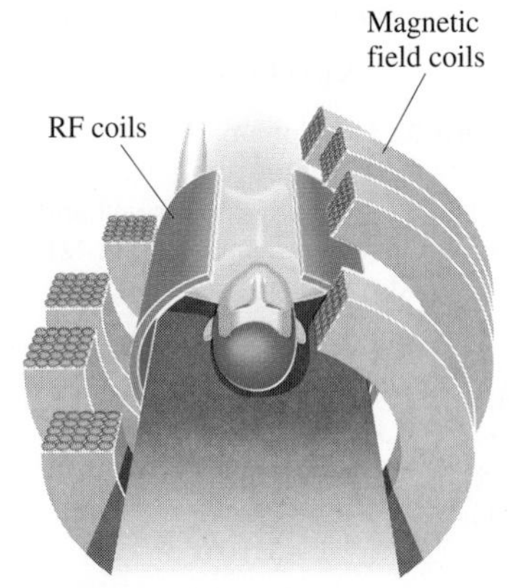

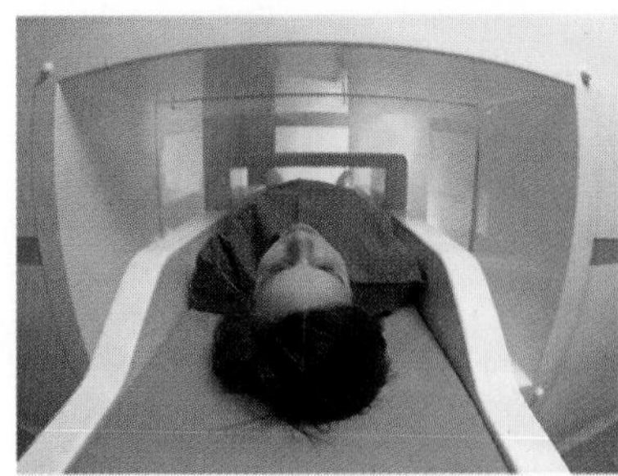

Figure 32.5 Magnetic resonance imaging (MRI). A human body is subjected to a powerful constant *nonuniform* magnetic field. The energy levels available to the hydrogen nuclei split in two (much like the Zeeman Effect), separating by $\Delta E \propto B$. A broad band radiofrequency (RF) pulse of electromagnetic energy then excites some of the nuclei ($hf = \Delta E \propto B$). When the nuclei subsequently relax, they emit RF photons ($f \propto B$), which are used to construct an image of the distribution of nuclei based on a knowledge of the spatial configuration of B.

32.3 The Nuclear Force

No sooner had the nucleus been revealed, than there arose the obvious question: what holds it together? By 1925, there was a recognition of the need for a new type of force. A cluster of positively charged particles must repel one another with an electrostatic force (like charges repel). Moreover, we know from scattering experiments that the $1/r^2$ Coulomb force works right down to nuclear distances. A simple calculation of the repulsion between two protons separated by a distance that puts them just about in contact within a nucleus yields a value of around 50 N (about 11 lb). That interaction is enormous when considered in light of the tiny masses of the protons; no nuclei other than hydrogen should ever have formed, and those that exist should explode immediately. Evidently, there is another kind of force, the **nuclear force**, operating within the nucleus. **The nuclear force binds neutrons and protons together to form nuclei**. It had its conceptual origins in a proposed short-range, neutron-proton force first suggested (1932) by Heisenberg. Now known to have an effective range of only about 1 fm, the nuclear force is powerfully attractive, imparting potential energies to nucleons of as much as 100 MeV. It is repulsive at distances less than about 0.5 fm (two nucleons cannot occupy the same space), and it depends on the spins of the interacting particles.

The nuclear force is a manifestation of the more fundamental and less restricted *strong force,* which affects a whole class of particles (hadrons), not just nucleons. The most potent of all known interactions, the strong force (at a proton-proton separation of 2 fm) is about 100 times stronger than the electromagnetic force and roughly 10^{34} times stronger than the gravitational force (Table 32.2). While it might take as much as 8 MeV to remove a nucleon from a nucleus, the electron in a hydrogen atom can be ionized with a mere 13.6 eV. For that reason, pound for pound, a nuclear reaction can liberate millions of times more energy than a chemical reaction. We will discuss the strong force at greater length in Chapter 33.

From experiments starting in 1936, we have found that the nuclear force exists between any two nucleons. The evidence for the $n \leftrightarrow n$ attraction is inferential, but the $p \leftrightarrow p$ and $p \leftrightarrow n$ interactions can be measured, although indirectly. This is done using beams of neutrons or protons scattered from a target consisting mostly of hydrogen (that is, protons). Electrons are completely immune to the nuclear force, which is why they were so effective at probing the nuclear charge distribution.

Rutherford's scattering experiments of 1913 established, that down to distances of roughly 10^{-14} m, α-particles interacted with nuclei via the Coulomb force. In 1919, he fired 5-MeV α-particles at low-Z nuclei, thereby minimizing the Coulomb repulsion. What he found (using hydrogen nuclei as targets) was that at distances of $\approx$3.5 fm the resulting scattering markedly deviated from that predicted by electrodynamics (Fig. 32.6).

At relatively large distances, protons repel one another via the Coulomb force, but there is a change at roughly 3 fm, where the interaction becomes increasingly attractive. A monoenergetic neutron beam colliding with protons shows little or no interaction to about 2 fm, whereupon there is again an increasingly strong attractive

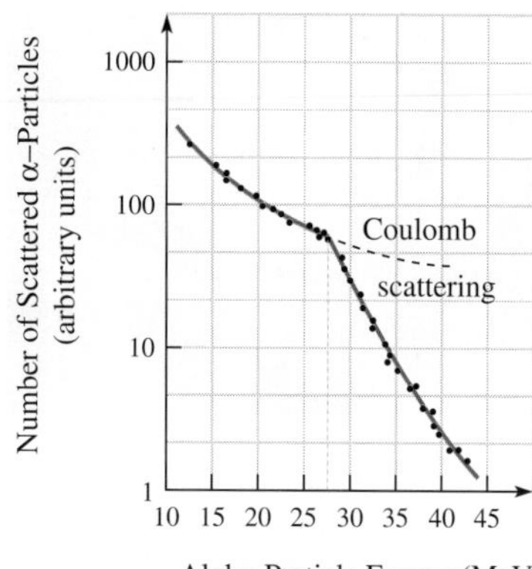

Figure 32.6 Alpha particles elastically scattered from a lead target. [Data from G.W. Farwell and H. E. Wegner, *Phys. Rev.* Vol. 95: 1212 (1954).] At around 27.5 MeV, the curve abruptly deviates from purely Coulomb scattering. The nuclear force now comes into play as the α-particles get close enough to the nucleus for that short-range force to be influential.

TABLE 32.2 The Four Forces of nature*

Force	Interacts between	Strength[1]	Effective range
Gravitational	All mass-energy[2]	10^{-34}	Unlimited
Weak	All material particles (quarks and leptons)	10^{-2}	$\approx 10^{-17}$ m
Electromagnetic	Electromagnetic charges	10^{2}	Unlimited
Strong	Many subnuclear particles (quarks and gluons)	10^{4}	$\approx 10^{-15}$ m

*At the temperatures that exist today, we see four distinct interactions. At much higher energies, these blend together (p. 1158).
[1]The strengths (in newtons) are for two protons separated, center-to-center, by 2 fm.
[2]Gravity and the strong force both act on their own field quanta.

force. At very small separations, the interaction quickly becomes repulsive. The nuclear force is so powerfully repulsive at very small distances that nucleons rarely get closer to one another than about 0.4 fm. Estimates of the nucleon radius range from about 0.3 fm to 1 fm.

Figure 32.7a depicts a crude potential-energy curve for a proton interacting with a nucleus of radius R. At large distances, the curve is positive and Coulombic—an approaching proton experiences a repulsion. The positive *electrical*-PE increases as $1/r$. By contrast, a neutron does not "feel" any electrical force—it approaches the nucleus along the zero-PE axis in Fig. 32.7b. At the surface of the nucleus, a nucleon is tremendously influenced by the attractive nuclear force. The resulting negative potential energy confines the nucleon to the tiny region of the nucleus. Bound neutrons and protons rattle around inside the nucleus like submicroscopic bees flying in a well as much as 50 MeV deep.

The nuclear force is *saturable:* each bound nucleon interacts with only a few of its nearest neighbors. If a nucleon is added to a nucleus, it will not interact with all the other particles via the nuclear force. That means that the more massive nuclei should have nearly the same density as small nuclei, which is borne out by experiment (Fig. 32.3). Remember that atoms, held together by the $1/r^2$ Coulomb force, are nearly all the same size regardless of A. Adding more charge to both the nucleus and the electron cloud increases the long-range interaction, and that essentially

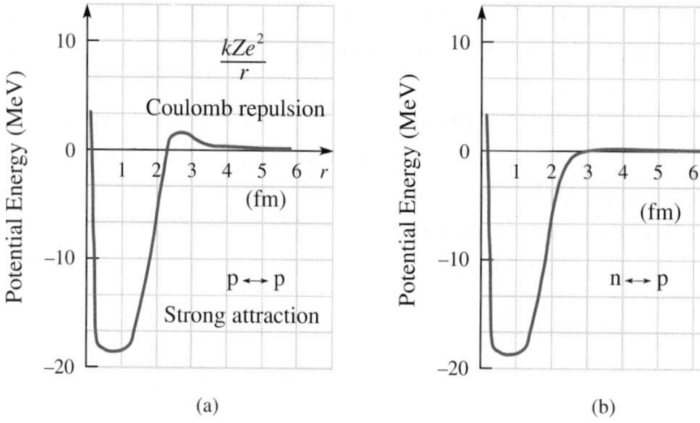

Figure 32.7 Approximate potential wells for (a) proton-proton and (b) neutron-proton interactions.

compresses the atom. By contrast, nuclei get larger with A via Eq. (32.1). The nucleus will prove to be more similar in behavior to a droplet of water than it is to an atom.

The nuclear force performs as do the forces between molecules—both fall off rapidly with separation. Moreover, two water molecules also repel one another when they come very close together and their electron clouds start to overlap. Nucleons and water molecules, while shifting around, will tend to stay separated by a distance such that their mutual attraction is maximized. The separation between nucleons is comparable to the range of the nuclear force, and the surface of the nucleus is about that thick as well. As with a liquid, the nucleus also experiences surface tension; nucleons at the surface will be pulled inward, which is why nuclei are essentially spherical. In a sense, the nucleus is a droplet of nuclear matter.

When a neutron and a proton come together to form a deuteron, they do it only if the two spins are parallel, suggesting yet another complication in the way the nuclear force works: it has a substantial spin dependence. When nucleon spins are antiparallel, the nuclear force between the two particles is about a factor of 2 weaker.

32.4 Nuclear Stability

Figure 32.1 reveals that stable light nuclei up to about $A = 20$ are almost all composed of equal numbers of neutrons and protons ($Z = N$). The hydrogen nucleus ($A = 1$) is a single proton and naturally stable (though, according to the Grand Unified Theories, protons should decay, no evidence has yet been found to support that conclusion—see p. 1158). Protons repel one another with the Coulomb force, and it takes some doing to get two of them close enough together to stick via the nuclear force. There is evidence that the di-proton has been created, but it lasts for less than 10^{-18} s. On the other hand, neutrons experience the nuclear force while being immune to electrostatic repulsion; they therefore serve as a source of nuclear glue, but that's a bit simplistic, since the di-neutron is unstable. The next heavier nucleus ($A = 2$) results when a neutron clings to a proton with parallel spin, forming a stable deuteron ($Z = 1$, $N = 1$). Two protons can be held together with the inclusion of a neutron, thereby making stable ${}^3_2\text{He}$. Two deuterons can combine to create helium ($Z = 2$, $N = 2$).

Both the spin and the magnetic moment of the α-particle are zero. Since the magnetic moments of the neutron and proton are different, that tells us that the two neutrons (spin-up and spin-down) pair together, as do the two protons (spin-up and spin-down). The corresponding closed stable system (the α-particle) plays an important role in the scheme of the nuclides (p. 1112). That's borne out by the observation that, while there are only four stable odd-odd nuclei (${}^2_1\text{H}$, ${}^6_3\text{Li}$, ${}^{10}_5\text{B}$, and ${}^{14}_7\text{N}$, for which $Z = N$), there are 160 stable even-even nuclei. Furthermore, when a nuclide is blasted with a nucleon, it's much more likely that an α-particle will be emitted than a deuteron.

Most of the radioactive nuclides in Fig. 32.1 are of the *artificially induced* variety, made in the laboratory by bombarding other nuclides. There are only about 20 naturally occurring radioactive isotopes in the range up to lead ($Z = 82$). Beyond that, all nuclides are radioactive, though bismuth decays so slowly it might as well be considered stable.

The Shell Model

Nuclides with certain numbers of neutrons or protons are especially stable and abundant. These so-called **magic numbers** with N or Z equaling

$$2, 8, 20, 28, 50, 82, 126$$

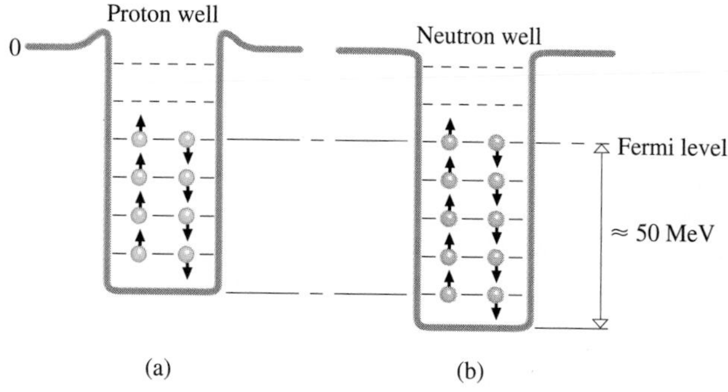

Figure 32.8 Potential-energy wells for (a) protons and (b) neutrons in the same nucleus. To bring the Fermi levels in line, protons can convert to neutrons by emitting positron-neutrino pairs. Note the unfilled higher-energy levels available when the nucleus is excited.

suggested a closed shell structure reminiscent of the configurations of the Noble Gases in atomic theory (p. 1084). A nucleus with either N or Z magic is tightly bound. One with both N and Z magic is very tightly bound, highly stable, and abundant. Such is the case with ${}^{4}_{2}\text{He}_2$, ${}^{16}_{8}\text{O}_8$, ${}^{40}_{20}\text{Ca}_{20}$, ${}^{48}_{20}\text{Ca}_{28}$, and ${}^{208}_{82}\text{Pb}_{126}$, the five known nuclei with double magic (see Discussion Question 1).

A nucleon within the nucleus is surrounded by other nucleons, and the nuclear forces they exert on it in opposing directions should tend to cancel; an interior nucleon will experience little local influence via the nuclear force. Don't picture the nucleus as a static cluster of tightly packed cannon balls! Instead, imagine it to be more like a cloud of flying bees, each of which sees and attracts only its nearest neighbors. Although there is a complex tumult of motion, the swarm hangs together as a single entity. In fact, if the bees fly around in pairs, the analogy would be still better. In any event, we can assume that a nucleon "sees" an average force that retains it within the nucleus. The Uncertainty Principle requires that particles confined to a small region of space have a correspondingly large uncertainty in their motion and so at least a minimum kinetic energy.

Suppose that each nucleon moves around in a complex independent way within the nucleus. Each is restrained by the surface of the nucleus, as if in a potential-energy well. Figure 32.8 depicts a crude representation of the average potential-energy wells for a proton and neutron in a medium- to large-sized nucleus. The difference in shape and depth is due to the Coulomb force. The proton's well is not as deep because of the repulsion.

Each nucleon is accompanied by a de Broglie wave, a standing wave, fitting into the nucleus, just as the electron's wave fit into the atom. Thus, each nucleon should correspond to a particular standing wave pattern or, equivalently, should occupy a particular quantum orbital. A group of energy levels that are close to one another is called a shell, and this scheme is the **shell model** of the nucleus. In the same way that an atomic level can contain at most two different electrons, each nuclear energy level can harbor two protons and two neutrons. No exclusion acts between neutrons and protons, and they can enter the same states. The levels fill up from the bottom. The highest occupied level corresponding to the greatest kinetic energy is called the **Fermi level**. Although we might expect a moving nucleon to undergo frequent collisions and be scattered every which way, the Exclusion Principle and the demand for filled levels will keep that from happening. There are very few available energy levels for a scattered nucleon to enter, and so they cannot easily be knocked off course.

Once in identical states, a proton and neutron experience a maximum attractive interaction, and so the simplest closed subshell system, that of the helium nucleus, is energetically favorable. When the energy levels were computed quantum-mechanically, it was found that the nucleus had a system of closely spaced subshells. These come in groups constituting major shells that are themselves much farther apart energetically. M. Goeppert Mayer and J. H. D. Jensen further showed (1947) that the magic numbers corresponded to the number of states in the major shells. When the major shells are filled, the corresponding nuclei were especially stable. For example, a nucleus with 50 neutrons or 50 protons has a filled shell. Thus, tin ($Z = 50$), which is relatively abundant in nature, has 10 stable isotopes.

Maria Goeppert Mayer (1906–1972). German-born American physicist shared the 1963 Nobel Prize in physics with J. H. D. Jensen.

Nuclides off the **line of stability** that runs through the stable nuclei spontaneously decay, often changing neutrons to protons or vice versa, and becoming more tightly bound and energetically stable. Observe how the line of stability in Fig. 32.1 increasingly bends toward the N-axis, indicating a larger percentage of neutrons. Because of the long range of the Coulomb force, the greater the number of protons, the more the nucleus tends to decay, and the more neutrons are needed to stabilize it. The progression of stable nuclei ends at lead; further increasing the percentage of neutrons will not keep things together. There are no pea-sized nuclei, and not until the mass becomes immense can gravity help to hold nuclear matter together on a large scale as in a neutron star.

Binding Energy

If we fire a neutron directly at a proton so that they come very close, the two can grab hold of each other via the strong force. Rammed together, the system will emit a burst of electromagnetic energy, a 2.224-MeV gamma-ray photon. Thus formed, the deuteron (2.013 553 u) has shed mass; its constituents have drawn tightly together and lost some of the plumpness they had apart. This transformed mass is known as the **mass defect** (Δm), and it indicates how tightly a nuclide is bound. In this case

$$m_p + m_n = (1.007\,276 \text{ u}) + (1.008\,665 \text{ u}) = 2.015\,941 \text{ u}$$

whereas the deuteron mass (m_d) is only 2.013 553 u. The difference between the mass of the separate components and the combined nucleus is the mass defect; namely

$$\Delta m = 0.002\,388 \text{ u}$$

Since 1 u is equal to 931.494 MeV/c^2, the mass defect corresponds to a **binding energy** ($E_B = \Delta m\, c^2$) of 2.224 MeV—exactly the energy ejected as a photon. In reverse, to split a deuteron into a neutron and a proton, 2.224 MeV (or 3.56×10^{-13} J) has to be supplied, for example, via a photon or a collision. Of course, 2.224 MeV is an immense amount of energy on an atomic scale. It takes only 2×10^{-18} J, one hundred thousand times less energy, to pull the electron out of a deuterium atom, and even that is 10 times the energy released when the "burning" of a carbon atom forms CO_2.

Example 32.3 When a neutron is removed from a $^{43}_{20}$Ca atom (of mass 42.958 766 u), it will be transformed into a $^{42}_{20}$Ca atom (of mass 41.958 618 u). What minimum energy must be provided to accomplish the removal?

Solution: [Given: atomic masses of 42.958 766 u and 41.958 618 u. Find: the energy to remove a neutron.] Let's first find the difference between the masses of the two nuclei. If that difference is less than the mass of a neutron, the deficiency (Δm) has to be made up by pumping energy in. We were given atomic masses because they are what's generally tabulated rather than nuclear masses. That fact doesn't matter here because taking the difference cancels the mass of the atomic electrons; thus

$$(\text{mass of } ^{43}\text{Ca}) - (\text{mass of } ^{42}\text{Ca}) = 1.000\,148 \text{ u}$$

Not surprisingly, this is less than the neutron mass (1.008 665 u); hence, the difference

$$\Delta m = (1.008\,665 \text{ u}) - (1.000\,148 \text{ u}) = 0.008\,517 \text{ u}$$

(at the least) will have to be supplied as energy. Consequently

$$E = (0.008\,517 \text{ u})(931.494 \text{ MeV/u}) = \boxed{7.934 \text{ MeV}}$$

▶ **Quick Check:** As we'll see presently, 8 MeV is typical for the binding-energy-per-nucleon (which is an average value); for heavy nuclei, our answer is reasonable. (See Problem 29.)

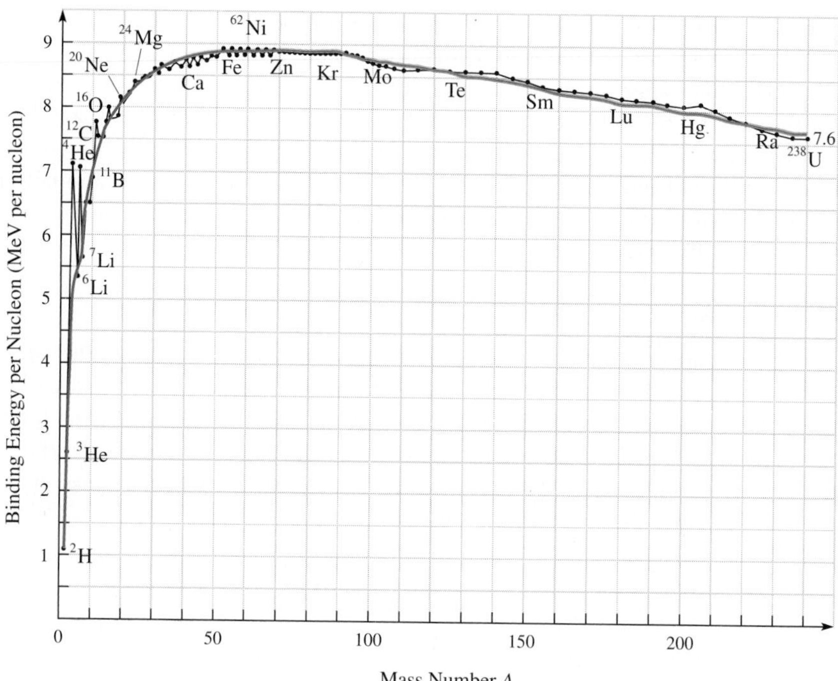

Figure 32.9 Average binding-energy-per-nucleon for the nuclides that occur naturally (including short-lived ^{8}Be, the second spike near He) versus mass number.

Planck first realized that **any bound system should have less mass than the sum of its constituents** and, in 1913, Langevin applied that insight to the nucleus. *The coming together of nucleons to form a nuclide is accompanied by a conversion of mass to energy.* The mass defect is equivalent to the total binding energy of the nuclide. Dividing that energy by the total number of nucleons yields an average measure of how strongly each nucleon is bound to the composite system; namely (E_B/A), the **binding-energy-per-nucleon**. (In a way, it's like our swarm of bees flying around in a well where we ask how much energy it will take on average to lift each of them up and out.)

A plot of binding-energy-per-nucleon for all the elements from hydrogen to uranium is given in Fig. 32.9. Hydrogen, with one proton, has no binding energy at all; from there, the curve rises to 1.1 MeV for the deuteron, 2.8 MeV for ^{3}H, 2.6 MeV for ^{3}He, then on up to a 7.074-MeV spike for tightly bound helium, falling back to 5.3 MeV for ^{6}Li, then rising to 5.6 MeV for ^{7}Li, and gradually continuing upward. The significance of the tightly bound α-particle structure is manifest by nuclides that can be thought of as comprising whole number multiples of ^{4}He. Thus, ^{8}Be (2 alphas), which decays almost immediately, nonetheless has a substantial binding energy of 7.06 MeV. Stable ^{12}C (3 alphas) and ^{16}O (4 alphas) have relatively high binding energies compared to their neighbors. Moreover, those nuclides, whose Z and A values correspond to whole number groupings of α-particles, also have zero spin and zero magnetic moment.

The binding-energy-per-nucleon curve reaches a maximum of 8.795 MeV at nickel-62, which is the most stable and most tightly bound of all nuclides. That fairly flat peak in the curve (around $Z = 60$) contributes to the abundance of iron in the Universe. From there on, the graph gradually drops, because of the Coulomb repulsion of the protons, all the way to uranium. Except for the lightest nuclei, E_B/A is fairly constant at about 8 MeV per nucleon. That this value is independent of A

again suggests that the nucleus is held together by a short-range force. With a long-ranged force such as gravity, the size of the composite body is crucial; it's much more difficult to remove a 10-kg stone from the surface of the Earth than from the surface of the Moon. To the contrary, where the force is short-ranged, as with the intermolecular force holding water together, there is the same independence of quantity; it takes the same amount of energy to evaporate 10 kg of water from a kiddy pool as from an ocean.

The binding-energy-per-nucleon is the energy above and beyond its kinetic energy (the Fermi level) that must be added to a nucleon to remove it from the nucleus (Fig. 32.8). For small nuclei, the well depth is small because there are fewer nucleons acting, and therefore the binding energy is also small. For large nuclei, the electrostatic potential energy raises the bottom of the well and decreases the binding energy at the top.

Notice that if we select a nuclide way off on either side of the peak in the binding-energy-pér-nucleon curve and alter its structure so as to move up the curve toward Ni, a very large amount of energy could be liberated. Thus, if two light nuclei (say, of hydrogen) could be joined together (that is, fused), the resulting nuclide would reside further up the curve and would have a greater binding-energy-per-nucleon. Each of its nucleons would be more tightly squeezed and individually less massive than it was prior to the union. The resulting mass defect would appear as liberated energy in the well-known process of *nuclear fusion.* On the other hand, splitting a large nucleus (from the right side of the curve) into small fragments also transforms mass. The binding-energy-per-nucleon of the fragments is higher than it was for the original nucleus that broke up. This process of *nuclear fission* liberates copious amounts of energy. It's no accident that the key elements of the nuclear age, hydrogen and uranium, are the end points of the binding energy curve.

NUCLEAR TRANSFORMATION

The transformation of one nuclide into another can take place either spontaneously or as a result of an external stimulus. The remainder of this chapter deals with a variety of nuclear transformations. We examine the spontaneous processes of alpha, beta, and gamma decay; introduce the weak force; study the mathematical description of the rate of radioactive decay; and explore the phenomenon of induced radioactivity. The discussion ends with an introduction to fission and fusion as applied to bombs and stars.

32.5 Radioactive Decay

As we saw earlier, nuclei spontaneously transform themselves into more energetically favorable configurations via alpha, beta, and gamma emission in a process called **radioactive decay**. Usually, when a nuclide resides above the line of stability in Fig. 32.1, it decays by emitting an electron, thereby transforming a neutron into a proton and moving downward and to the right, coming closer to the line. Similarly, when a nuclide resides beneath the line of stability, it emits a positron or, occasionally, an alpha, transforming to the left toward the line.

A given radioactive decay can be a single step in a long sequence of transformations from one unstable nuclide to another, ultimately ending in a stable form. There are three *naturally occurring* radioactive series beginning with ^{238}U, ^{235}U, and ^{232}Th (Table 32.3), all ending in different isotopes of lead. The series shown in Fig. 32.10

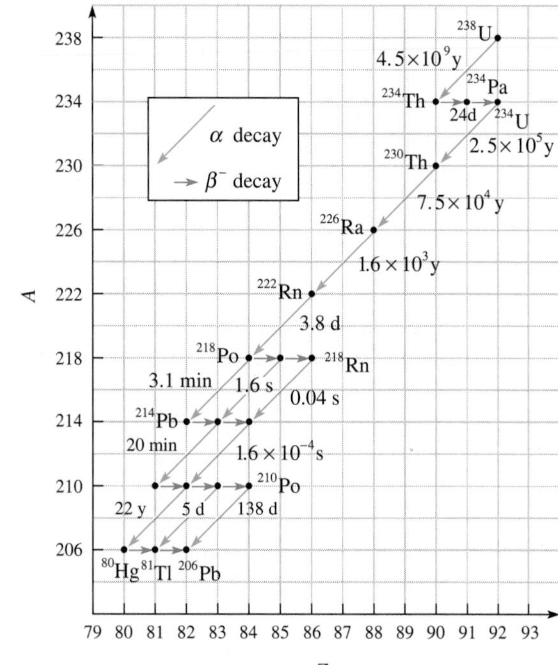

Figure 32.10 The naturally occurring decay series beginning with uranium-238.

TABLE 32.3 The three natural radioactive series

Series	Starting nuclide	Half-life (years)	Stable end nuclide
^{235}U-Actinium	$^{235}_{92}U$	7.04×10^8	$^{207}_{82}Pb$
^{232}Thorium	$^{232}_{90}Th$	1.41×10^{10}	$^{208}_{82}Pb$
^{238}U-Radium	$^{238}_{92}U$	4.47×10^9	$^{206}_{82}Pb$

begins with the slowly decaying nuclide uranium-238. A little less than half of all the U-238 present at the formation of the Solar System some 5×10^9 years ago still remains. By contrast, radium-226 decays away fairly rapidly (half of it is transformed every 1600 years), and all of Earth's original radium atoms have long ago disappeared. Yet, radium is still plentiful because, like so many other radioactive substances, it is continuously replenished (Fig. 32.10).

Alpha Decay

Alpha emission is rare in the light nuclides, although at around $Z = 60$, there is a small cluster of α-emitters. It is beyond $Z = 82$, where there are no stable nuclides, that alpha emission predominates. There, it drives the nuclides down in both Z and N, toward the line of stability. The emission of an α-particle ($^4_2$He, or $^4_2\alpha$) decreases N by 2, decreases Z by 2, and decreases A by 4. In general, the decay can be written as

$$^A_ZX \rightarrow {}^{A-4}_{Z-2}Y + {}^4_2He + Q \qquad (32.2)$$

where X is called the **parent nucleus**, Y is the **daughter nucleus**, and Q is the **disintegration energy**. The rules of the game require that the process *conserve* *charge, nucleon number (A), angular and linear momentum,* and *mass-energy.* Alpha decay is a rearranging: it doesn't change the total number of nucleons present in the formula, nor does it alter the net charge, but it does alter the net mass. Whenever such a rearrangement happens by itself, energy is liberated in the amount Q such that

$$Q = (m_X - m_Y - m_\alpha)c^2 \tag{32.3}$$

where m_X, m_Y, and m_α are the masses of the parent, daughter, and alpha, respectively. The disintegration energy (here given in joules) appears as the total kinetic energy of the daughter nuclide and the α-particle. A positive Q means that the process takes place spontaneously; a negative Q means it cannot. Using the binding energy data, it can be shown that for heavy nuclei, the average value of Q is around 5 MeV. Typical of the process is the *spontaneous decay* (Fig. 32.11) of uranium-238 into thorium-234:

$$^{238}_{92}\text{U} \rightarrow {}^{234}_{90}\text{Th} + {}^{4}_{2}\alpha + 4.3 \text{ MeV}$$

A self-luminous 1000°C radioactive sphere. At its core are six ounces of plutonium (^{238}Pu) dioxide, clad in an iridium shell, surrounded by a graphite casing. The α-particles emitted in the decay are absorbed in the surrounding shell, where they impart thermal energy at a rate of 100 W and will continue to do so for decades.

Example 32.4 A too-common hazard in many homes is the radioactive gas radon-222. It's produced in the ground by the alpha decay of radium-226 and often leaks into basements. Write out the transformation equation and determine (to 3 figures) the total kinetic energy of the decay products in MeV. The masses of the radium, radon, and helium *atoms* are 226.025 406 u, 222.017 574 u, and 4.002 603 u, respectively. As a rule, most of the KE is carried off by the light particle; here, the alpha. Very little (≈0.1 MeV) goes to the massive recoiling daughter.

Solution: [Given: $m_X = 226.025\,406$ u, $m_Y = 222.017\,574$ u, and $m_\alpha = 4.002\,603$ u. Find: KE.] Using

$$^{A}_{Z}\text{X} \rightarrow {}^{A-4}_{Z-2}\text{Y} + {}^{4}_{2}\text{He} + Q \tag{32.2}$$

we obtain

$$^{226}_{88}\text{Ra} \rightarrow {}^{222}_{86}\text{Rn} + {}^{4}_{2}\text{He} + Q$$

The KE equals Q, where

$$Q = (m_X - m_Y - m_\alpha)c^2 \tag{32.3}$$

The atomic masses can be used because the electrons cancel. In MeV

$$Q = (226.025\,406 \text{ u} - 222.017\,574 \text{ u} - 4.002\,603 \text{ u})(931.494 \text{ MeV/u})$$

and

$$Q = (0.005\,229 \text{ u})(931.494 \text{ MeV/u}) = 4.870\,8 \text{ MeV}$$

The net KE of the decay products is $\boxed{4.87 \text{ MeV}}$.

▶ **Quick Check:** Our answer is roughly 5 MeV, as we know it should be.

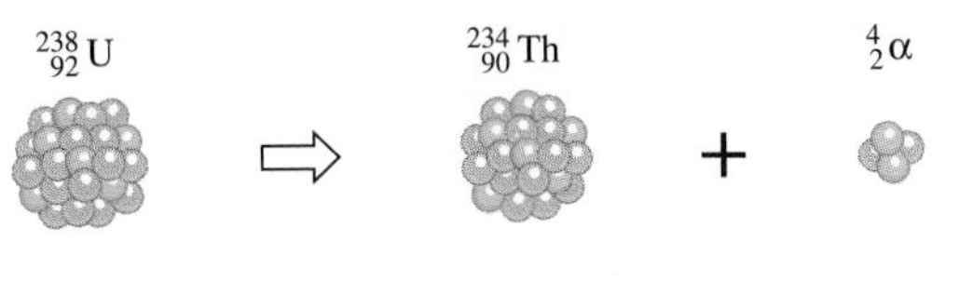

Figure 32.11 An unstable parent nuclide (U-238) decaying, via α emission, into a daughter nuclide (Th-234).

Beta Decay and the Neutrino

Three distinct processes are each a form of *beta decay*. In **β^- decay**, an electron ($_{-1}^{0}\text{e}$) is emitted from a nucleus as a neutron transforms into a proton. In **β^+ decay**, a positron ($_{+1}^{0}\text{e}$) is emitted from a nucleus as a proton transforms into a neutron. In **electron capture**, one of the orbital electrons in an inner shell of the cloud is drawn into the nucleus, transforming a proton into a neutron.

By the end of the 1920s, experiments on electron beta decay had revealed some extremely puzzling aspects. A parent nucleus was assumed to decay into a daughter after the creation and immediate emission of an electron. One problem was an apparent violation of Conservation of Linear Momentum. When a parent nucleus (the best results are obtained with a monatomic gas) more or less at rest decays, the electron and recoiling daughter nucleus should move in opposite directions. That's the only way the net linear momentum before and after can be zero, and surprisingly, that didn't always happen. The basic process occurring was assumed to be the spontaneous transformation of a neutron into a proton and an electron, but each of these particles is a spin-$\frac{1}{2}$ fermion. The original neutron spin is $\frac{1}{2}$, whereas the resulting total proton-electron spin could be either 0 or 1. Clearly, this situation violated Conservation of Angular Momentum.

An even more confounding observation was made by Chadwick in 1914. Figure 32.12 shows the broad energy spectrum measured for emitted electrons from a typical beta emitter. Each parent atom is identical, as is each daughter, and so every escaping electron should be identical and have the same energy, but they do not. If the basic process corresponds to $\text{n} \rightarrow \text{p} + \text{e}^-$, we can expect a disintegration energy of

$$Q = (m_\text{n} - m_\text{p} - m_\text{e})c^2$$

which equals 0.783 MeV (see Problem 38). Except for a very small amount of energy appearing as the recoil KE of the proton, Q ought to be exactly the KE of every emitted electron. In the case of a heavy nucleus, even less energy is given to the recoil. In other words, all electrons should appear with the maximum kinetic energy (KE_max) terminating the curve in Fig. 32.12 for each emitter. That energy (Q) is equivalent to the parent-daughter-electron mass difference. Experimentally, we find that all electrons are ejected with less than this energy. To explain that evident violation of Conservation of Energy, it was suggested that the β-particles lost energy via collisions while emerging from the atom. That suggestion was rejected by C. D. Ellis (who, incidentally, became a physicist only after being interned in a German prison camp with Chadwick during World War I) and W. A. Wooster. Using a thick lead-walled calorimeter, they proved that the total amount of energy deposited corresponded to the average energy of the spectrum, which is about 40% of what it ought to be (either energy was not conserved, or 60% of it mysteriously escaped from the calorimeter).

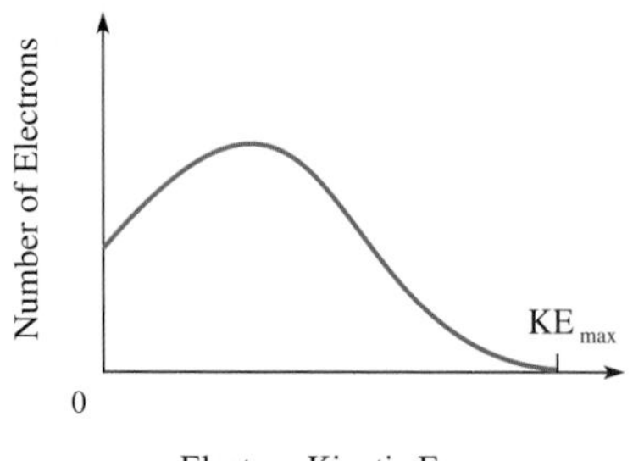

Figure 32.12 Energy spectrum of the electrons in a typical beta decay.

Some physicists felt compelled to reject Conservation of Energy—Madame Curie and Niels Bohr were among those skeptics. Rather than accept the obvious facts that were there for all to see, Wolfgang Pauli remained steadfast. In 1930, he postulated that an invisible particle emitted along with each electron carried off precisely the amount of energy needed to sustain conservation. Sometime later, Enrico Fermi, while answering a question, almost jokingly referred to this phantom speck as a **neutrino** (which in Italian means "little neutral one"), and the name stuck. Today (for reasons that will become clear) we call it an *electron-antineutrino* ($\overline{\nu}_\text{e}$). Not surprisingly, Pauli proposed that it was uncharged and so immune from the electromagnetic force. Because this neutral particle could pass through the thick lead walls of the Ellis-Wooster calorimeter, it had to be unaffected by the strong force. To conserve angular momentum, Pauli argued that his phantom particle had a spin of $\frac{1}{2}$. Today, it's known that neutrinos and antineutrinos do have spin-$\frac{1}{2}$. When an electron and an antineutrino are created as a pair, their spins are oppositely directed.

The kinetic energies of some emitted electrons approach KE_{max}, meaning that, in the extreme, the accompanying antineutrino can have a total energy that is vanishingly small. This total energy equals its rest energy (mc^2) plus its kinetic energy. The only way that sum can be vanishingly small is if m is zero or nearly zero, while $KE \approx 0$ as well. By 1933, Perrin had argued from the β-spectrum "that the neutrino has *zero intrinsic mass.*" Like a photon, it would then travel at c, and like a photon with zero rest energy, it could carry ample kinetic energy to balance out the demands of conservation. Intrigued by these ideas, de Broglie (1934) suggested that, as with the electron, the neutrino should have a mirror image, the antineutrino.

Electron-beta decay arises when a neutron decays into a proton, an electron, and an electron-antineutrino,

$$\mathrm{n} \rightarrow \mathrm{p} + {}^{0}_{-1}\mathrm{e} + \bar{\nu}_{\mathrm{e}} \tag{32.4}$$

In the case of a radioactive nuclide, the decay has the form

$${}^{A}_{Z}\mathrm{X} \rightarrow {}_{Z+1}^{A}\mathrm{Y} + {}^{0}_{-1}\mathrm{e} + \bar{\nu}_{\mathrm{e}} + Q \tag{32.5}$$

where Q is the kinetic energy of all three particles emerging from the transformation. Similarly, positron-beta decay arises when a proton decays into a neutron, a positron, and an electron-neutrino:

$$\mathrm{p} \rightarrow \mathrm{n} + {}^{0}_{+1}\mathrm{e} + \nu_{\mathrm{e}} \tag{32.6}$$

In the case of a radioactive nuclide, the decay has the form

$${}^{A}_{Z}\mathrm{X} \rightarrow {}_{Z-1}^{A}\mathrm{Y} + {}^{0}_{+1}\mathrm{e} + \nu_{\mathrm{e}} + Q \tag{32.7}$$

Electron-capture, the process wherein a parent nucleus absorbs one of its own orbital electrons, competes with β^+ decay; that is

$${}^{A}_{Z}\mathrm{X} + {}^{0}_{-1}\mathrm{e} \rightarrow {}_{Z-1}^{A}\mathrm{Y} + \nu_{\mathrm{e}} + Q \tag{32.8}$$

For example, beryllium-7 captures an electron and transforms into lithium-7:

$${}^{7}_{4}\mathrm{Be} + {}^{0}_{-1}\mathrm{e} \rightarrow {}^{7}_{3}\mathrm{Li} + \nu_{\mathrm{e}}$$

The distinction between the electron-neutrino (ν_e) and the electron-antineutrino ($\bar{\nu}_e$) is that the former is left-handed (its spin angular momentum vector is antiparallel to its linear momentum vector) and the latter is right-handed (its angular and linear momentum vectors are parallel). *The emission of a neutrino is equivalent in many regards to the absorption of an antineutrino, and vice versa.*

Because it doesn't "feel" either the strong or the electromagnetic force, a neutrino can sail through a thickness of over a light-year of solid lead before undergoing a single collision. In the 1930s, there was little hope of ever confirming their existence. Twenty-five years later, the feat was accomplished. Using a nuclear reactor (which is a copious source of antineutrinos, most arising from the decay of free neutrons), C. L. Cowan and F. Reines bombarded the protons in a huge quantity of water. If antineutrinos existed, the inverse process to that of Eq. (32.6)—add an antineutrino to both sides—should produce observable effects. Roughly 1 in 10^{12} antineutrinos struck a proton head-on, converting it into a neutron and a positron, both of which were then detected. The electron-antineutrino exists. More recently (1986), it has been determined that its mass (if it has mass) is less than a mere 27 eV/c^2 (which is about 19 000 times less than the electron). The issue of a nonzero neutrino mass is crucial in astrophysics. Even if the neutrino corresponded to just a few eV, they are so numerous that much of the mass of the Universe would be invisible neutrinos. That hidden mass might well determine the future fate of the entire Universe.

At 7:35 A.M. on February 23, 1987, counters on two huge particle detectors deep below the surface of the Earth, one near Cleveland and the other near Tokyo, began flashing for several seconds (p. 1158). The cause of that unprecedented event was revealed several hours later when the light from an exploding star in a nearby galaxy (170 000 ly away) arrived at Earth. The cataclysmic eruption of Supernova 1987A emitted a blast of neutrinos that immediately flew out into space at or very near the speed of light. Several hours later, the shock wave reached the surface of the star, and it began emitting a blaze of light a million times brighter than the Sun. The pulse of neutrinos and flood of photons had flown toward Earth for 170,000 years, arriving about two hours apart.

32.6 The Weak Force

In 1934, Fermi introduced a highly successful theory of beta decay predicated on the existence of a new interaction in nature, the **weak force**. The very fact that neutrinos were so penetrating suggested that they were uninfluenced by both the strong and the electromagnetic forces, and yet something (other than gravity) both blasted them into existence and promoted their occasional interaction with matter.

In its broadest meaning, a force is an agent of change; it may manifest itself as a push or a pull (a changer of motion), or it may be a changer of some aspect of the state of a system, a transformer. The electromagnetic force can transform mass-energy in an atomic system, thereby creating a photon (or it can transform a stick of dynamite into a puff of smoke). In the case of beta decay, it was necessary to conceive of a new force that could transmute a neutron into a proton, or vice versa.

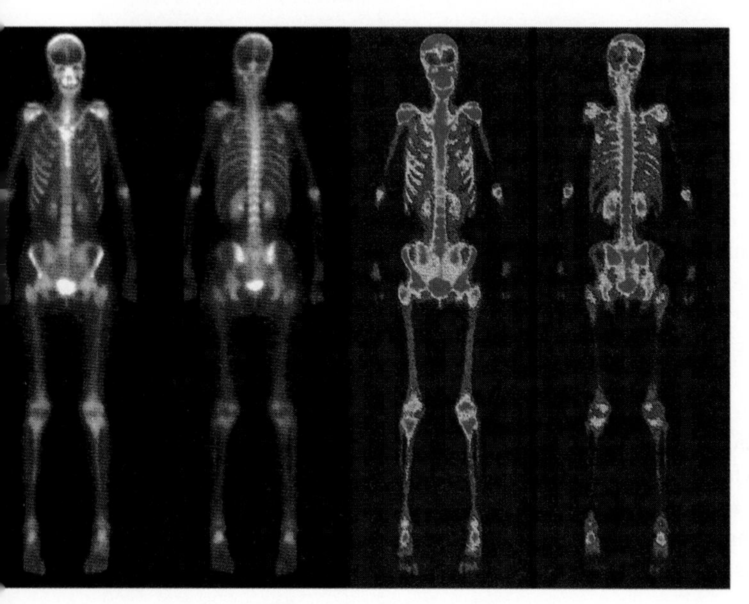

Full-body bone scans. Gamma-ray images of a patient three hours after being injected with 25 millicuries (see Problem 40) of technetium-99, a γ-ray emitter with a half-life of 6 h.

A free neutron has a half-life of 10.4 minutes (a very long time in Atomland), and that implies that the operative force is very weak—it's argued that the stronger the force, the faster the processes it drives. As we'll see in Chapter 33, there are hundreds of unstable exotic subnuclear particles. Some of these decay via the strong force (producing strongly interacting particles) and have lifetimes of only $\approx 10^{-23}$ s or so. By contrast, a decay resulting in a photon and so driven by the electromagnetic force might take $\approx 10^{-20}$ s. Weak decays involving neutrinos happen much more slowly, with lifetimes of perhaps $\approx 10^{-8}$ s.

The weak force is a million times more feeble than the strong force (Table 32.2), though it is immensely ($\approx 10^{32}$ times) more powerful than gravity. It reveals its presence in the macroscopic world primarily through radioactive transformation (there being hundreds of β-decaying nuclides) and also via the results of fission and fusion occasionally rather spectacularly in a supernova. Most subnuclear particles (the vast majority of them are unstable) interact via the weak force. The spontaneous transmutation of many subnuclear particles occurs through the weak force. It is extremely short-ranged, so much so that it was only in the 1980s that accurate determinations of its extent could be attempted. Beyond about 10^{-17} m (10^{-2} fm), the weak force becomes exceedingly small. It is for the most part an inside-the-particles force rather than a between-the-particles force. Still, as we become more sophisticated experimentally, we are able to detect the minute influence of the weak force reaching from the nucleus to the electron cloud of an atom.

32.7 Gamma Decay

After an alpha or beta decay, a daughter nucleus may momentarily end up in an excited state; that is, with a nucleon in a higher energy level than the ground state (Fig. 32.8). The nucleus then quickly relaxes, inevitably reaching the lowest available energy configuration. The energy difference ($\approx$1 keV to $\approx$1 MeV) is emitted as one or more gamma photons (*hf*). As with an atom, it is possible to excite a nucleus by pumping in energy. This might be accomplished via the absorption of a gamma photon of the proper frequency, or it might happen as the result of a collision with a massive particle. For example, a slow-moving neutron can be absorbed by a ^{238}U nucleus, whereupon it becomes an excited ^{239}U* nucleus (the asterisk indicates an excited state). Returning to its ground state, it emits a gamma photon; that is

$$^{1}_{0}\text{n} + {}^{238}\text{U} \rightarrow {}^{239}\text{U}^* \rightarrow {}^{239}\text{U} + \gamma$$

When observed independent of its nuclear source, a gamma photon is identical in all respects to a photon of the same energy arising from any other means. It's only because most nuclear energy levels are widely separated that gamma rays are usually more energetic than photons emitted from atomic electron transitions.

Example 32.5 Nitrogen-12 beta decays into carbon-12. The latter, ^{12}C*, then emits a 4.43-MeV gamma photon during de-excitation. Compute the mass of the ^{12}C* atom.

Solution: [Given: gamma energy of 4.43 MeV and ^{12}C system. Find: ^{12}C* atomic mass.] The mass of the ^{12}C atom is 12.000 000 u. The equivalent mass of the photon is gotten from $E = mc^2$, or the fact that 1.000 00 u = 931.494 MeV/c^2:

$$m_\gamma = \frac{4.43 \text{ MeV}}{931.494 \text{ MeV/c}^2} = 0.004\,76 \text{ u}$$

Hence, the total mass of the excited carbon-12 atom is

$$12.000\,00 \text{ u} + 0.004\,76 \text{ u} = \boxed{12.004\,76 \text{ u}}$$

▶ **Quick Check:** Let's redo the calculation using kilograms. (4.43 MeV)($1.782\,663 \times 10^{-30}$ kg/MeV) = $7.897\,197 \times 10^{-30}$ kg; (12.000 00 u) × ($1.660\,540 \times 10^{-27}$ kg/u) = $1.992\,648 \times 10^{-26}$ kg; hence, the total mass is $1.993\,438 \times 10^{-26}$ kg; or 12.004 76 u.

32.8 Half-Life

While working with highly radioactive ^{220}Rn, Rutherford observed that the intensity of the emissions decreased with time in a precise and predictable way. The amount of radiation emerging from a sample of a radioactive element is virtually independent of the surrounding environment (that is, the chemical compound it's in, the temperature, pressure, etc.). The quantitative measure of radioactive intensity is the *number of disintegrations per second,* known also as the **decay rate**, or the **activity**.

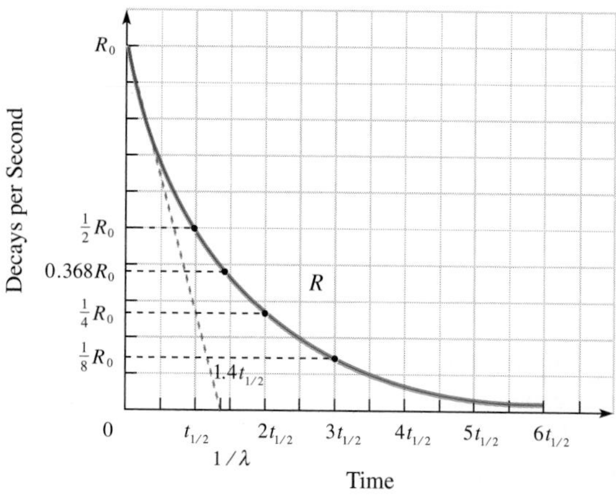

Figure 32.13 The exponential decay of the activity R of a radioactive nuclide.

The SI unit of decay rate is the **becquerel** (Bq), which is *one disintegration per second.* Activity seems to be dependent only on the type and amount of radioactive material present at that moment, which is both a common and a very special behavior. Many processes unfold at a rate that depends on the amount of participating ingredient present at that instant (for example, the decay of the charge on a capacitor, Fig. 20.19).

Letting N stand for the number of radioactive atoms present at a given instant, $|\Delta N/\Delta t|$ represents the decay rate (R). This rate can be made independent of the size of the particular sample by dividing it by N. Thus, $|\Delta N/\Delta t|/N$ is the fraction of the atoms of a given species disintegrating per unit time, regardless of the sample size. This quantity has been found experimentally to be constant over a wide range of sources and for long periods of time. It's called the **decay constant** and is usually symbolized, rather unfortunately, by λ, which is *not* to be confused with wavelength (sorry about that):

$$\lambda = \frac{|\Delta N/\Delta t|}{N} = \text{constant} \tag{32.9}$$

The unit of λ is s^{-1}, and it should be thought of as the fractional disintegration rate.

Suppose a detector is set up next to the sample, and the counting rate per convenient interval of time is recorded. Start at $t = 0$ with a rate R_0. From Eq. (32.9), $R_0 = \lambda N_0$ where N_0 is the number of atoms present initially. Figure 32.13 is a typical plot of R against time. The measured decay rate drops to $\frac{1}{2}R_0$ after a time interval Rutherford called the **half-life** ($t_{1/2}$). The known radioactive nuclides have half-lives ranging from roughly 10^{-22} s to about 10^{21} y. Upon waiting two half-lives, the decay rate will be one-quarter its original value; after three half-lives, it will be half of that or one-eighth its original value, and so on, with R approaching, but never reaching, zero. Thus, after n half-lives, the decay rate is

$$R = (\tfrac{1}{2})^n R_0 \tag{32.10}$$

Each radioactive element has its own decay curve, some declining more steeply than others (just like the two curves in Fig. 22.21), but all will have the same basic shape. A straight line having the initial slope of each curve will intersect the time axis at the *time constant,* or **mean lifetime** (τ), at which moment $R = 0.368\ R_0$. The greater the activity ($R = |\Delta N/\Delta t|$), the greater λ is, the steeper the slope, and the smaller is τ. In fact, $\tau = 1/\lambda$. The red curve in Fig. 32.14 is a replot of Fig. 32.13, with R_0 set equal to 1 and τ set equal to 1; the fundamental decay curve. Let's try to fit this nice smooth graph with a simple mathematical function: 2^{-t} or $1/2^t$ is too slow descending, and 3^{-t} is a little too fast. After some routine calculation, it can be determined that $(2.718\,281\,8\ldots)^{-t}$ fits the decay exceedingly well. It's traditional to call 2.718 281 8 . . . simply e (after the mathematician Euler), whereupon the basic decay curve is e^{-t} (Appendix A-3).

Figure 32.14 A comparison of 2^{-t}, e^{-t}, and 3^{-t}.

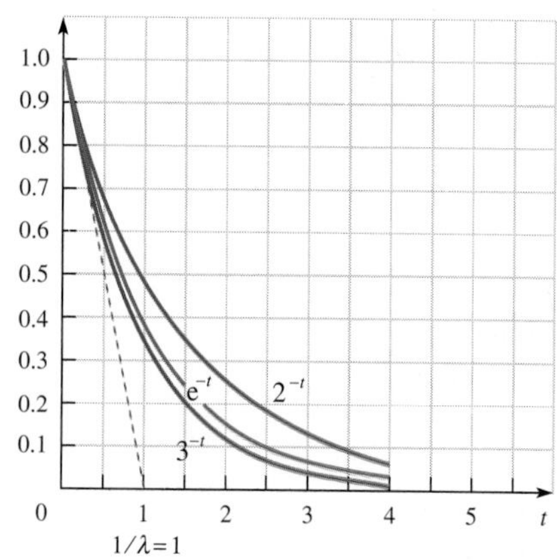

All decay curves drop to $1/e = 0.368$ of their initial values at $t = \tau = 1/\lambda$, and so the basic curve can be particularized to a specific decay rate by writing it as $1e^{-\lambda t}$. Moreover, letting the initial activity be some value R_0, the decay rate for any radioactive element becomes

$$R = R_0 e^{-\lambda t} \tag{32.11}$$

At $t = 0$, $e^0 = 1$, and $R = R_0$, as it should be. Now, let's consider the half-life. When $t = t_{1/2}$, $R = R_0 e^{-\lambda t_{1/2}} = \frac{1}{2}R_0$, thus $e^{-\lambda t_{1/2}} = \frac{1}{2}$; yet as an electronic calculator can show $e^{-0.693} = \frac{1}{2}$, hence

$$\lambda t_{1/2} = 0.693 \tag{32.12}$$

or $t_{1/2} = 0.693\tau$. The average value of the exponential is reached at $t = \tau$, which is why it's called the *mean lifetime.* If we know the half-life or the lifetime, we can calculate the decay constant or vice versa. A large decay constant means a short half-life.

It follows from Eq's. (32.9) and (32.11) that

$$R = \lambda N_0 e^{-\lambda t} = \lambda N \tag{32.13}$$

hence the number of atoms remaining after a time t is

$$N = N_0 e^{-\lambda t} \tag{32.14}$$

Notice that both sides of this equation could be multiplied by the mass of the nuclide, thereby yielding an expression for the total mass of the remaining undecayed element.

Example 32.6 Radium-226 is found to have a decay constant of 1.36×10^{-11} Bq. Determine its half-life in years. Given that the Curies had roughly 200 g of radium in 1898, how much of it will remain 100 years later?

Solution: [Given: $\lambda = 1.36 \times 10^{-11}$ decays/s, $m_0 = 200$ g, and $t = 100$ y. Find: $t_{1/2}$ and m.] Using Eq. (32.12), we have

$$t_{1/2} = \frac{0.693}{1.36 \times 10^{-11}\ \text{s}^{-1}} = 5.096 \times 10^{10}\ \text{s}$$

$$\boxed{t_{1/2} = 1.62 \times 10^3\ \text{y}}$$

Since $t = 100$ y $= 3.15 \times 10^9$ s, and $\lambda = 0.693/t_{1/2} = 1.36 \times 10^{-11}\ \text{s}^{-1}$

$$m = m_0 e^{-(1.36 \times 10^{-11}\ \text{s}^{-1})(3.15 \times 10^9\ \text{s})}$$

and $$m = (200\ \text{g})(0.958) = \boxed{192\ \text{g}}$$

▶ **Quick Check:** The factor $(-\lambda t)$ to which the exponential is raised must be a pure number; its units must cancel as they do here. With a half-life of 1620 y, $\lambda = 0.693/(1620\ \text{y}) = 4.28 \times 10^{-4}\ \text{y}^{-1}$ and $e^{-\lambda t} = e^{-0.0428} = 0.958$.

The half-life of radioactive carbon-14 is 5730 years (see Table 32.4), and it serves as an archeological clock. The ratio of the amount of ^{14}C to stable ^{12}C in the atmosphere is fairly constant at about 1.3×10^{-12}. Thus, any living organism takes in both isotopes in that fixed proportion as it assimilates carbon. Once dead, the organism takes in no new carbon, the ^{14}C decays, and the ratio changes in a way that marks the passage of time. For instance, if a piece of charcoal from an ancient fire-

TABLE 32.4 Half-lives of some radioisotopes

Isotope	Decay mode	Half-life
Rubidium-87	e^-	4.7×10^{10} y
Uranium-238	α	4.5×10^9 y
Plutonium-239	α	2.4×10^4 y
Carbon-14	e^-	5730 y
Radium-226	α	1600 y
Strontium-90	e^-	28 y
Tritium-3	e^-	12.26 y
Cobalt-60	e^-	5.24 y
Iodine-131	e^-	8 d
Radon-222	α	3.82 d
Technetium-104	e^-	18 min
Fluorine-17	e^+	66 s
Polonium-213	α	4×10^{-6} s
Beryllium-8	α	1×10^{-16} s

place contains one-quarter of the original ratio of ^{14}C to ^{12}C, then a time of $2t_{1/2} = 11\,460$ years has passed since the wood died.

Radioactivity is a quantum-mechanical phenomenon and Einstein (1916) first realized that Eq. (32.13) was a manifestation of the inherent statistical nature of the process. It is now the common wisdom that an unstable nucleus decays spontaneously (just as an atom in an excited state emits a photon spontaneously); that there is no knowable external trigger that fires them off in a way that can be anticipated; that identical atoms behave, all by themselves, in nonidentical ways that conform *en masse* to the rules of probability. It would seem that radioactivity provides a glimpse of the quantum-mechanical microworld playing havoc with classical cause and effect. An atom that is ten thousand years old is supposedly identical to an atom of the same species that is ten seconds old; from this moment on, one of these may live for ten thousand years and the other for ten seconds, and yet we don't know which will do what. Whether that ignorance is a profound one, forced on us by the probabilistic nature of the Universe, or just a practical one inflicted by our own limitations remains to be seen.

32.9 Induced Radioactivity

Irène and Jean Frédéric Joliot-Curie were not having much luck in the early 1930s. Chadwick's announcement of his discovery of the neutron in 1932 was a shock. They had often produced these new particles in their own laboratory but had thought they were gamma rays. Hot on the trail of the positron, they lost that glory to Anderson in 1933. Still, their triumph (and the Nobel Prize in chemistry) would soon come (1934) out of a series of experiments bombarding light elements with α-particles. In particular, they transmuted aluminum into phosphorus:

$$^{4}_{2}\text{He} + ^{27}_{13}\text{Al} \rightarrow ^{31}_{15}\text{P}^* \rightarrow ^{30}_{15}\text{P} + ^{1}_{0}\text{n}$$

The α-particle was captured into the Al-27 nucleus, forming a highly unstable *compound nucleus,* P-31, that immediately decayed into P-30. The surprising thing was that the phosphorus went on emitting positrons even after the alpha bombardment ceased, as

$$^{30}_{15}\text{P} \rightarrow ^{30}_{14}\text{Si} + ^{0}_{+1}\text{e}$$

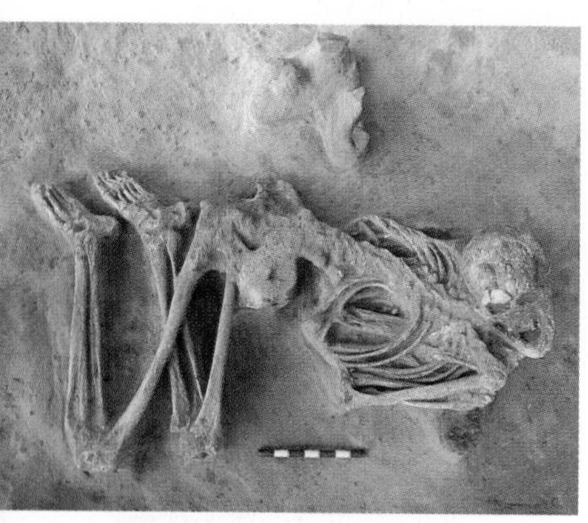

Archeological remains, like this skeleton, can be dated using carbon-14.

They had **artificially induced radioactivity** and, within a year, created a whole group of **radioisotopes**. The stage was set for our present-day medical, biological, and industrial use of these materials—even then the implications were enormous.

Enrico Fermi in Rome was particularly fascinated by the work of the Joliot-Curies, and he set out to test the novel idea he had of using neutrons to create radioisotopes. Since these neutral bullets experienced no electrical repulsion, he rightly reasoned they would easily approach and enter a target nucleus. Within weeks, Fermi published his first positive results. Systematically, he bombarded every element he could get his hands on, all the way up to uranium. Along the way, he realized that neutrons that had been slowed down in their passage through certain materials were even more potent at instigating transmutations than were ordinary fast neutrons. These **thermal neutrons** (moving at the speeds of room temperature air molecules) spent more time in the vicinity of a nucleus as they traveled by and could be captured more effectively. "The Pope" (that was Fermi's nickname because of his strong advocacy of the "new faith" of Quantum Mechanics) concluded that the light hydrogen atoms (protons) in a sheet of paraffin would be very efficient at slowing the neutrons. Colliding objects that are about the same mass transfer the most momentum (p. 306). To everyone's amazement, inserting a sheet of paraffin into the neutron beam increased its ability to produce transmutations a hundredfold.

Uranium bombarded by thermal neutrons exhibited some surprising behavior in that it subsequently emitted beta rays. The natural conclusion was that a nucleus had swallowed up a neutron, emitted an electron, and ended up with an additional proton. Fermi believed that uranium had thus been transformed into a new element one box higher in the Periodic Table, the first *transuranic* element. Although that transformation does occasionally take place, the predominant effect was something far more spectacular: nuclear fission.

Enrico Fermi (1901–1954), Italian-American physicist, was one of the leading figures in the development of the atomic bomb. After years of working with radiation, he died from cancer at the age of 53.

32.10 Fission and Fusion

Shortly after Fermi's announcement, Ida Noddack, a German chemist, published an alternative scenario that at the time, though it was correct, seemed absolutely crazy. She proposed that the incoming neutron had ruptured the uranium nucleus "into several big fragments which are really isotopes of already known elements." Everyone, including the world's foremost radiochemist, Professor Otto Hahn at the Kaiser Wilhelm Institute in Berlin, completely dismissed the idea as "impossible." After all, how could a small, low-energy neutron burst the largest known nucleus?

(a)

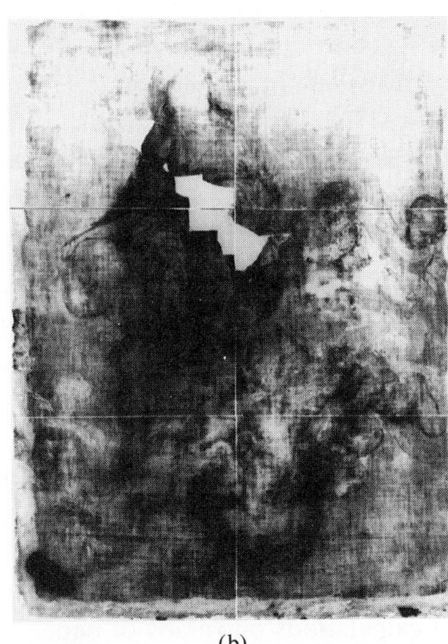

(b)

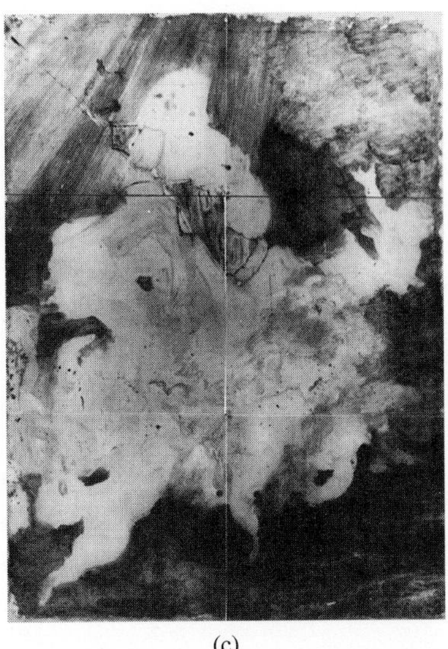

(c)

By bombarding a painting (a), in this case one by Van Dyck, with neutrons, different elements in the paint and varnish can be made radioactive. For example, manganese, once common in the dark brown pigment umber, absorbs neutrons with the subsequent emission of electrons. (b) A special film sensitive to electrons reveals the original layers of umber. The white region free of manganese is a modern repair. (c) Four days later, after the emission from the umber faded away, radiation from phosphorus in charcoal revealed a whole new aspect to the painting. Notice the inverted charcoal sketch of the head of a person, now clearly visible. (See p. 1022.)

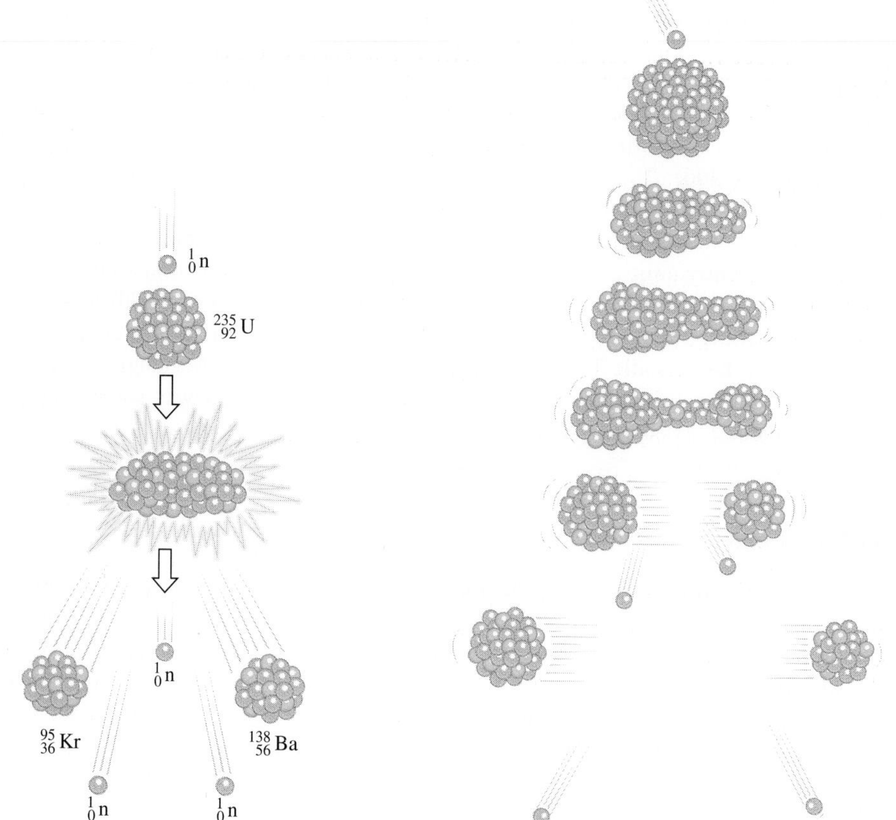

Figure 32.15 The fission of U-235.

Figure 32.16 Nuclear fission.

Hahn and his colleague Lise Meitner, joined by F. Strassmann, repeated the Italian experiments with uranium. The chemical analysis was extremely difficult, and the results remained inconclusive, but that wasn't their only worry. Dr. Meitner was an Austrian Jew, and despite Planck's personal appeal to Hitler, she was dismissed from the Institute where she had worked for 30 years. That summer (1938), she fled to Stockholm. Just before Christmas, Hahn and Strassmann finally realized that they were dealing with isotopes of the light element barium. It was inescapable: the uranium nucleus had indeed split in two (Fig. 32.15). They dashed off a letter to the editor of the journal *Naturwissenschaften* and sent a copy to their exiled friend Meitner.

Just then Meitner was being visited by her nephew, a young physicist named Otto Frisch who had fled the Nazis and was working at Bohr's Institute in Copenhagen. With Hahn's letter in hand, the two went for a walk in the woods. Frisch was well versed in Bohr's new model, which maintained that the nucleus resembled a droplet of liquid bound by surface tension. They sat down on a tree stump and began making calculations on scraps of paper. A very large nucleus such as uranium with its many protons would experience an internal electrostatic repulsion that would all but cancel the cohesive surface tension. Even the slight nudge of an incoming neutron might cause it to oscillate and elongate (Fig. 32.16), whereupon the Coulomb repulsion would tend to increase the distortion until, swamped by the short-range strong force, the two pieces would snap into a separate existence and fly apart (all in $\approx 10^{-22}$ s).

Remembering the binding energy curve (p. 1112), Meitner could assume that the fragment nuclei might have a binding-energy-per-nucleon of about 8.5 MeV compared to about 7.6 MeV for a heavy nucleus. The energy released would be the difference (8.5 − 7.6) MeV for each of the roughly 240 nucleons, or about 216 MeV. That's a lot of energy for a single atomic event (roughly a million times the amount liberated by the combustion of one molecule of gasoline). About 1% of the original nuclear mass would be converted into energy, most of it (168 MeV) going into the KE of the two fragments.

Lise Meitner (1878–1968).

A few days later, Frisch caught up with Bohr as he was about to sail off to America to visit Einstein. "Oh, what fools we have been!" Bohr blurted out, "We ought to have seen that before." Hurriedly, Frisch and Meitner wrote a paper announcing their results and, borrowing a term from the biology of cell division, Frisch named the process **nuclear fission**.

Naturally occurring uranium is composed of three isotopes, U-234, U-235, and U-238. By far the most abundant at 99.27% is U-238, followed by U-235 at only 0.72%, and a mere trace of U-234. On the basis of theoretical considerations, Bohr and his former student J. A. Wheeler concluded that it was the rare U-235 that had undergone slow-neutron fission. In fact, it's the only naturally occurring element that can do that little trick. (The common, and far more stable, isotope U-238 can be split, but only by fast-neutron bombardment.) A slow neutron captured by U-235 results in the highly unstable compound nucleus U-236, which immediately ruptures into two large pieces. One of the common fragment pairs is barium-138 and krypton-95:

$$^{1}_{0}\text{n} + {}^{235}_{92}\text{U} \rightarrow {}^{236}_{92}\text{U}^* \rightarrow {}^{141}_{56}\text{Ba} + {}^{92}_{36}\text{Kr} + 3\,{}^{1}_{0}\text{n}$$

The Chain Reaction: On the Way to the Bomb

In the 1930s, most people (including Einstein) shared Rutherford's view: "Anyone who expects a source of power from the transformation of these atoms is talking moonshine." Leo Szilard, a brash young Hungarian physicist, was incensed by Rutherford's remark, and in short order he conceived the concept of the **chain reaction**. Szilard speculated that if a nuclide could be found that, when struck by a neutron, would emit *two* or more neutrons, the resulting fission process could continue on its own, building up into a horrendous cascade (Fig. 32.17). Sensing that Europe would soon be engulfed in war, Szilard accepted a post at Columbia University and arrived in America in January, 1939. Fermi came to Columbia that same month. His mother's antecedents were Jews, as was his wife, and so after stopping in Stockholm to pick up the 1938 Nobel Prize, the family continued on to New York.

Only weeks after Fermi's arrival, Bohr reached New York with the news of the discovery of fission. "The Pope" immediately appreciated the possibilities of making a nuclear bomb, but there were troubling questions that had to be resolved. Only if each fission event emitted an average of at least two neutrons could an explosive device be developed. In March, Szilard and Zinn determined experimentally that, on average, between two and three fast neutrons were indeed emitted per fission. A single such event liberates about 3.2×10^{-11} J, which is a great deal of energy in Atomland, but not very much on the human scale of things. That was why the chain reaction—and, in particular, a chain reaction that fanned out and grew—was so crucial if a vast amount of energy was to be liberated. If the first nucleus to split fired out two neutrons, those two could split two other nuclei in the second generation, and so on. Ideally, after, say, 80 generations, an incredible 1.2×10^{24} atoms

SEPTEMBER 12, 1933

Lord Rutherford

Atom-Powered World Absurd, Scientists Told

Lord Rutherford Scoffs at Theory of Harnessing Energy in Laboratories

By The Associated Press

LEICESTER, England, Sept. 11.—Lord Rutherford, at whose Cambridge laboratories atoms have been bombarded and split into fragments, told an audience of scientists today that the idea of releasing tremendous power from within the atom was absurd.

He addressed the British Association for the Advancement of Science in the same hall where the late Lord Kelvin asserted twenty-six years ago that the atom was indestructible.

Describing the shattering of atoms by use of 5,000,000 volts of electricity, Lord Rutherford discounted hopes advanced by some scientists that profitable power could be thus extracted.

"The energy produced by the breaking down of the atom is a very poor kind of thing," he said. "Any one who expects a source of power from the transformation of these atoms is talking moonshine. . . . We hope in the next few years to get some idea of what these atoms are, how they are made and the way they are worked."

would have shattered in a fraction of a millisecond. About 0.5 kg of uranium would vanish, releasing 3.8×10^{13} J, or the equivalent of roughly 10 kilotons of TNT.

When Hitler embargoed the export of uranium from Czechoslovakian mines, Szilard was convinced the Germans were developing nuclear weapons.* In August of 1939, he went to Einstein with a letter to President Roosevelt, which the highly respected scientist signed. It advised "that extremely powerful bombs of a new type may thus be constructed." The all-out push to develop the atomic bomb began on December 6, 1941, just one day before Pearl Harbor was attacked by the Japanese.

Nuclear Reactors

A chunk of natural uranium is mostly U-238, and these nuclei don't easily split. Moreover, fast neutrons from a fissioning nucleus can be captured equally well by U-238 and U-235. Thus, pure natural uranium will not support a chain reaction. But U-238 has much less appetite for slow neutrons than does U-235, and that difference was exploited in the design of the first **fission reactor**, the controlled chain-reaction machine. If uranium was interspersed within a *moderator* (some light substance that slowed neutrons down), it would have almost the same effect as removing the U-238 altogether, leaving only the U-235 to interact. The most promising candidates for moderators were beryllium, heavy water (regular hydrogen absorbs too many neutrons), and carbon. Beryllium was crossed off the list because it is rare and toxic. Heavy water would have been perfect, but there were only a few quarts of it in the United States, and there wasn't time to prepare the several tons needed. The choice fell to carbon in the form of pure graphite.

Fermi and his group moved the top-secret operation to the University of Chicago. They set up shop in the squash courts (where the ceilings were more than 25 feet high) under the abandoned football stands of Stagg Field. There, they stacked 40 000 graphite bricks, into which were inserted chunks of uranium oxide. Several *control rods* of cadmium, which is a voracious absorber of neutrons, were built into the reactor to keep a leash on the chain reaction. On December 2, 1942, the cadmium rods were carefully withdrawn, and the world's first self-sustained controlled nuclear chain reaction occurred.

The absorption by U-238 of a neutron (Fig. 32.18) within the core of the reactor produced a new radioactive transuranic element called neptunium. With a half-life of $2\frac{1}{2}$ days, neptunium decays into yet another new element called *plutonium*. It was predicted that Pu-239 would undergo slow-neutron fission even easier than U-235,

The time will come when atomic energy will take the place of coal as a source of power. . . . I hope that the human race will not discover how to use this energy until it has brains enough to use it properly.

SIR OLIVER LODGE (1920)

*Szilard was right. Germany had an active nuclear weapons program led by Heisenberg. Because he seems never to have actually believed in the possibility of making a bomb, Heisenberg's efforts were rather half-hearted. In 1943, the Japanese joined the race to build an atomic bomb. A lack of uranium made things difficult for them, and when a submarine from Hitler carrying two tons of it was sunk by the Allies, their project faltered.

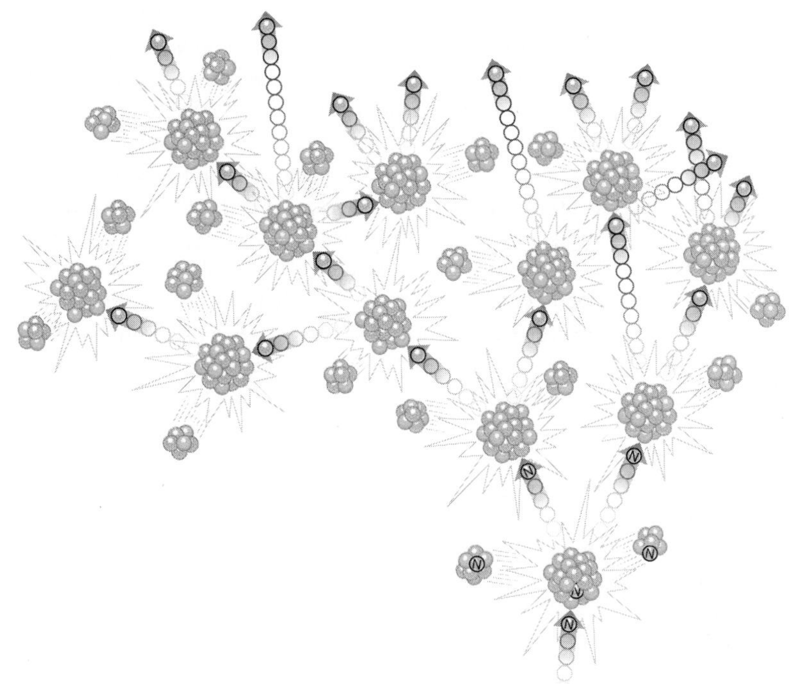

Figure 32.17 A branching chain reaction arising from uranium fission.

SECOND NEWS SECTION

For News, Subscriptions or Advertising Call ATlantic 6100.

SATURDAY MORNING,

Pittsburgh Post-Gazette

DECEMBER 29, 1934.

SPORTS, FINANCIAL, CLASSIFIED SECTION

This WORLD of OURS

Constantly Expressive—Einstein Warms Up to His Subject

ATOM ENERGY HOPE IS SPIKED BY EINSTEIN

Efforts at Loosing Vast Force Is Called Fruitless.

SAVANT TALKS HERE

Now Indicates Doubt Of Relativity Theory He Made Famous.

Energy of Atom

The article in the right column reports Einstein's opinion (1934) that attempts at loosing the "energy of the atom" would be fruitless—an opinion he soon changed.

and that made it a prime candidate for use in a bomb. With that as the driving force, the multimillion-dollar Plutonium Project got underway almost immediately. Three giant water-cooled production reactors were constructed in 1943 on the Columbia River near Hanford, Washington. After cooking for a few months, intensely radioactive slugs of U-238 were taken out of the reactor and the ${}^{239}_{94}\text{Pu}$ extracted. Each reactor could produce about half a pound of plutonium a day from otherwise plentiful and comparatively worthless U-238.

The modern power reactor is very similar in principle to these early machines. It usually uses natural uranium *enriched* to contain a few percent U-235, with ordi-

We are justified in reflecting that scientists who can construct and demolish elements at will may also be capable of causing nuclear transformations of an explosive character.

FRITZ HOUTERMANS
Berlin (1932)

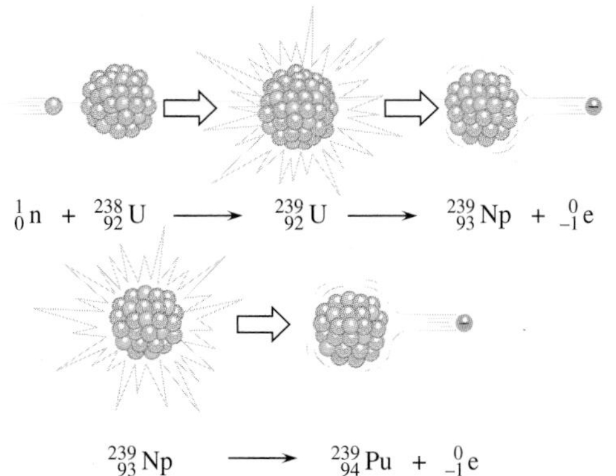

Figure 32.18 Plutonium-239 produced from uranium-238 by neutron bombardment.

nary water as the moderator (Fig. 32.19). The reactor is used essentially as a furnace to supply "heat" (via steam) that drives a traditional turbine and ultimately generates electricity. The kinetic energy of fission fragments and neutrons absorbed in the reactor transforms into copious amounts of thermal energy. A liquid coolant circulating around the fuel rods carries this energy out of the core. A modern 1000-million-watt reactor is charged with about 90 000 kilograms of fuel, a third of which is replenished each year. It's because a rather innocently appearing power reactor can generate plutonium that it becomes very difficult to control the spread of nuclear weapons (which is why the United States, in 1991, bombed Iraq's purported power reactors in the first few days of the Gulf War).

The Atomic Bomb

The actual design of the bomb was carried out at Los Alamos, New Mexico, under the direction of J. Robert Oppenheimer ("Oppie" to almost everyone). The design team included, among many others, Bohr, Chadwick, Frisch, Fermi, and Feynman. General Groves, who headed the entire bomb program (code-named the Manhattan Project), once remarked "At great expense, we have gathered here the largest collection of crackpots ever seen."

The principle underlying an A-bomb is simple. A small spherical chunk of fissionable material under bombardment by neutrons can support only a limited chain reaction; too many neutrons released from the fissioning nuclei escape from the sample through its surface. If a sphere of U-235 or Pu-239 is made larger, its volume increases faster than its surface area, more neutrons are produced within it, and proportionately fewer escape from its surface. There is therefore a *critical size* at which point the average number of neutrons from each fission that will cause another fission (rather than being lost somehow) equals one. Under such circumstances, the chain reaction is just initiated. Any increase in the amount of fissionable material

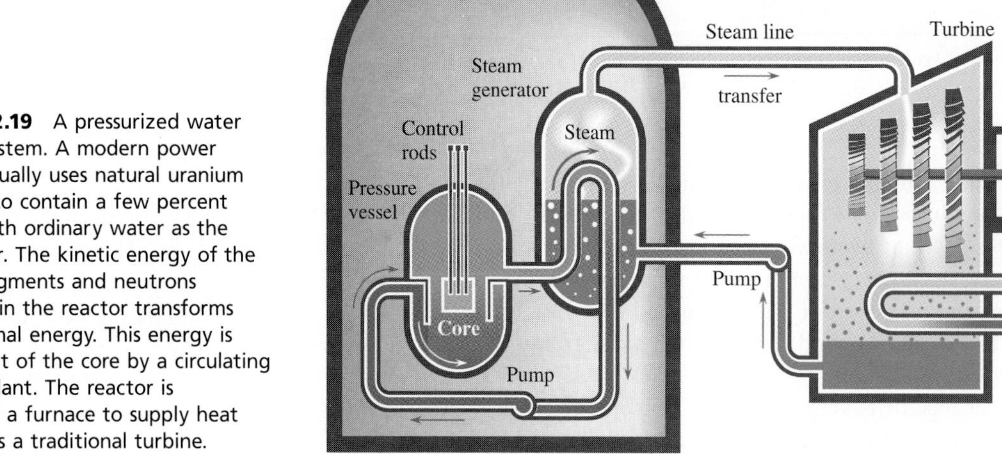

Figure 32.19 A pressurized water reactor system. A modern power reactor usually uses natural uranium enriched to contain a few percent U-235, with ordinary water as the moderator. The kinetic energy of the fission fragments and neutrons absorbed in the reactor transforms into thermal energy. This energy is carried out of the core by a circulating liquid coolant. The reactor is essentially a furnace to supply heat that drives a traditional turbine.

beyond that produces a branching, rapidly growing chain reaction—such a mass is *supercritical*. A solid sphere of U-235 about the size of a softball, struck by a stray neutron, will roar into a furious, multimillion-degree plasma in a millionth of a second. That's all it takes: one chunk in excess of the *critical mass* (a few kilograms) and off it goes. Of course, the U-235 or Pu-239 has to be kept in a subcritical configuration so it can be safely transported. The hard part is to assemble a supercritical mass fast enough so that it doesn't just melt and fizzle. That was especially difficult to do with plutonium, which is very unstable.

A depiction of the historic start-up of the world's first nuclear reactor under the direction of Fermi.

The first bomb to be designed was a brute force device so simple it was used in war without having been previously tested. Two pieces of U-235, each too small to ignite on its own, were rapidly brought together into a *supercritical* mass. The bomb was little more than a short cannon that fired a subcritical projectile of U-235 into another subcritical target screwed into the end of the muzzle. On August 6, 1945, at 8:15$\frac{1}{2}$ in the morning, it dropped from the belly of a B-29 over Hiroshima, Japan. Seconds later, 1500 feet above the ground, it burst into a 15-kiloton fireball.

Starfire: Thermonuclear Fusion

Technicians loading the fuel into a modern reactor at a nuclear power plant.

Stars and hydrogen bombs are both powered by a process called **thermonuclear fusion**. Figure 32.9 makes the point that fusing two light nuclei together produces a heavier nucleus higher up the binding-energy-per-nucleon curve. On bonding more tightly, a fraction of each nucleon's mass is converted to energy in accord with Einstein's $E_0 = mc^2$.

Stars are born, live, and die—some explode and disperse; others, latecomers, rise out of their ashes. First-generation stars are composed of primordial hydrogen laced with helium. They are formed when a great interstellar cloud of gas slowly collapses under its own gravitational force, heating up and growing more dense as the material plunges inward. At densities 100 times that of water and temperatures of 15 to 20 million K, an almost-star is a seething plasma ball where bare nuclei rush about at tremendous speeds. Two protons, each with a charge ^+e, can overcome their Coulomb repulsion and come close enough ($r < 10^{-14}$ m) for the strong force to lock them together. That happens provided they approach one another with a kinetic energy that can override the repulsion; a net KE equal to their electrical potential energy at a center-to-center separation r:

$$\text{PE}_\text{E} = k\frac{e^2}{r} = (9 \times 10^9\ \text{N}\cdot\text{m}^2/\text{C}^2)\frac{(1.6 \times 10^{-19}\ \text{C})^2}{10^{-14}\ \text{m}}$$

or

$$\text{PE}_\text{E} = 2.3 \times 10^{-14}\ \text{J} = 0.14\ \text{MeV}$$

which follows from Eq. (18.8). A gas of protons with an average KE of 0.14 MeV has a temperature given by Eq. (14.14); accordingly

$$(\text{KE})_\text{av} = \tfrac{3}{2}k_\text{B}T$$

and so

$$T = \frac{(0.66)(2.3 \times 10^{-14}\ \text{J})}{1.38 \times 10^{-23}\ \text{J/K}} = 1 \times 10^9\ \text{K}$$

On the Atomic Bomb. *That is the biggest fool thing we have ever done. The bomb will never go off, and I speak as an expert in explosives.*

ADM. WILLIAM LEAHY (1945)

There is another form of temptation even more fraught with danger. This is the disease of curiosity. . . . It is this which drives us on to try to discover the secrets of nature, those secrets which are beyond our understanding, which can avail us nothing, and which men should not wish to learn. . . . In this immense forest, full of pitfalls and perils, I have drawn myself back, and pulled myself away from these thorns. In the midst of all these things which float unceasingly around me in my everyday life, I am never surprised at any of them, and never captivated by my genuine desire to study them. . . . I no longer dream of the stars.

ST. AUGUSTINE
Confessions

A nuclear explosion.

J. R. Oppenheimer and General Groves in front of the remains of the vaporized 100-foot steel tower that had held the first atomic bomb (a plutonium device) ever exploded. The desert sand was seared into a glassy jade green crust. One wonders at the wisdom of visiting such a place so soon after an explosion.

Fusion can take place at temperatures roughly 50 times lower than this for two reasons. First, this value is an average KE, and many particles have much higher energies. Second, there are quantum-mechanical effects that allow the particles to penetrate the Coulomb barrier. In any event, high speeds and high temperatures are called for, which is why the process is known as *thermo*nuclear fusion. Once fusion begins, the furnace at the center of a star puffs it up with an outrushing of released energy, and the gravitational collapse stops; the plasma ball stabilizes and a star is born.

Many thermonuclear fusion processes can drive a star, depending on its physical circumstances (temperature, mass, available fuel, and so on). Each process generates energy out of matter, cooking up heavier elements along the way. Except for the hydrogen in your body (which goes back roughly 17 thousand million years to the first moments of creation), you are star dust, the ash created in the belly of some supernova long before our Sun was born. We believe the Sun (a second-generation star) is predominantly powered by a process known as the *proton-proton chain.* It begins with two protons coming together to form a deuteron ($^2_1$H), a positron, and a neutrino:

$$^1_1\text{H} + {}^1_1\text{H} \rightarrow {}^2_1\text{H} + {}^{\;0}_{+1}\text{e} + \nu_e$$

The deuteron then captures another proton, producing helium:

$$^2_1\text{H} + {}^1_1\text{H} \rightarrow {}^3_2\text{He} + \gamma$$

Finally, two helium-3 nuclei fuse in the transformation

$$^3_2\text{He} + ^3_2\text{He} \rightarrow ^4_2\text{He} + ^1_1\text{H} + ^1_1\text{H}$$

forming He-4 ash, two protons, and a great deal of energy. In effect, the total process is

$$4\text{p} \rightarrow \alpha + 2\,^{\;0}_{+1}\text{e} + 2\nu_e + \gamma$$

The liberated energy (6.2 MeV per proton) is carried off mostly by gamma rays, positrons, and neutrinos. The gamma photons are absorbed by the plasma. The positrons are annihilated when they come in contact with electrons generating more photons (which are also absorbed and add more energy). About 2% of the energy liberated comes off with the escaping neutrinos. The Sun unleashes 3.8×10^{26} J each second; each second it pours into space the energy equivalent of burning three million million million gallons of gasoline. Every second roughly 660 000 000 tons of hydrogen are transmuted into helium. In the process, about 4 600 000 tons (4.2×10^9 kg) of matter transform into radiant energy. And so it has raged for at least five thousand million years.

Twinkle, twinkle, little star!
How I wonder what you are,
Up above the world so high,
Like a diamond in the sky!

ANN TAYLOR
Rhymes for the Nursery

Core Material

Masses in the atomic domain are often specified in *unified atomic mass units* (u) where a neutral $^{12}_{6}\text{C}$ atom is defined to be precisely 12.000 000 u:

$$1\ \text{u} = 1.660\,540 \times 10^{-27}\ \text{kg} = 931.494\ \text{MeV/c}^2$$

The nuclear radius R is

$$R = R_0 A^{1/3} \qquad [32.1]$$

R_0 turns out to be ≈ 1.2 fm (p. 1105).

When nucleons combine to form a nucleus, an amount of mass Δm known as the **mass defect** is converted into **binding energy** $E_B = \Delta mc^2$. Alpha decay has the form

$$^A_Z\text{X} \rightarrow {}^{A-4}_{Z-2}\text{Y} + ^4_2\text{He} + Q \qquad [32.2]$$

where X is called the **parent nucleus**, Y is the **daughter nucleus**, and Q is the **disintegration energy** (p. 1114),

$$Q = (m_X - m_Y - m_\alpha)c^2 \qquad [32.3]$$

Here m_X, m_Y, and m_α are the masses of the parent, daughter, and alpha, respectively. Electron-beta decay arises when (p. 1116)

$$\text{n} \rightarrow \text{p} + \text{e}^- + \overline{\nu}_e \qquad [32.4]$$

In the case of a radioactive nuclide, the decay has the form

$$^A_Z\text{X} \rightarrow {}_{Z+1}^{\;\;A}\text{Y} + {}^{\;\,0}_{-1}\text{e} + \overline{\nu}_e + Q \qquad [32.5]$$

Positron-beta decay arises when

$$\text{p} \rightarrow \text{n} + \text{e}^+ + \nu_e \qquad [32.6]$$

In the case of a radioactive nuclide, the decay has the form

$$^A_Z\text{X} \rightarrow {}_{Z-1}^{\;\;A}\text{Y} + {}^{\;\,0}_{+1}\text{e} + \nu_e + Q \qquad [32.7]$$

Electron-capture, the process wherein a parent nucleus absorbs one of its own orbital electrons, competes with β^+ decay (p. 1117):

$$^A_Z\text{X} + {}^{\;\,0}_{-1}\text{e} \rightarrow {}_{Z-1}^{\;\;A}\text{Y} + \nu_e + Q \qquad [32.8]$$

The SI unit of decay rate is the *becquerel* (Bq), which is equivalent to one disintegration per second. After n half-lives, the decay rate R will be

$$R = (\tfrac{1}{2})^n R_0 \qquad [32.10]$$

and (p. 1120)

$$R = R_0 e^{-\lambda t} \qquad [32.11]$$

where λ is the decay constant. The half-life is related to λ by

$$\lambda t_{1/2} = 0.693 \qquad [32.12]$$

Since

$$R = \lambda N \qquad [32.13]$$

where N is the number of atoms remaining, it follows that

$$N = N_0 e^{-\lambda t} \qquad [32.14]$$

Suggestions on Problem Solving

1. Some useful numerical quantities for this chapter are

$$1\ \text{u} = 1.660\,540 \times 10^{-27}\ \text{kg} = 931.494\ \text{MeV/c}^2$$
$$1\ \text{eV/c}^2 = 1.782\,663 \times 10^{-36}\ \text{kg}$$
$$1\ \text{eV} = 1.602\,177 \times 10^{-19}\ \text{J}$$
$$m_e = 9.109\,389\,7(54) \times 10^{-31}\ \text{kg}$$
$$= 0.510\,999\,06(15)\ \text{MeV/c}^2 = 5.485\,80 \times 10^{-4}\ \text{u}$$
$$m_p = 1.672\,623\,1(10) \times 10^{-27}\ \text{kg}$$
$$= 938.272\,31(28)\ \text{MeV/c}^2 = 1.007\,276\ \text{u}$$
$$m_n = 1.674\,928\,6(10) \times 10^{-27}\ \text{kg}$$
$$= 939.565\,63(28)\ \text{MeV/c}^2 = 1.008\,665\ \text{u}$$
$$m_\alpha = 4.001\,506\ \text{u}$$

2. Most tabulations of masses for the various isotopes are *atomic* mass listings, not nuclear mass. When you are given that the atomic mass for carbon-10 is 10.016 856 u, it is for the neutral atom, and it includes the six electrons. One way to handle this situation is to use atomic masses throughout any calculations (for example, of mass defect). Just *keep track of the electrons.* There will be a difference in the electron binding energies, but that's generally totally negligible. Accordingly, it may often be more convenient to use the mass of the hydrogen atom (1.007 825 u) in place of the proton mass—the difference is the electron.

3. A typical problem concerning radioactive decay might involve one or two of Eq's. (32.10) through (32.13). If you need the activity (or decay rate) R, that's Eq. (32.11). Any question about the mass, or fractional amount, number of atoms, or percentage of a substance remaining is addressed to Eq. (32.13). You may be given τ, λ, or $t_{1/2}$ and be required to find one of the others. Similarly, the equations for R and N are written in terms of λ, though you may be provided τ or $t_{1/2}$. *Don't forget the minus signs* in those two exponential expressions. Review Appendix A-6 on exponentials and logarithms. You can be asked to solve for λ in either Eq. (32.11) or (32.13), and that will require undoing the exponential by taking the natural log. Watch the time units in the exponential equations: you can use λ in decays/year provided you enter t in years—just don't mix years with days or seconds.

4. The advent of the electronic calculator has made it just as easy to use Eq. (32.10) $R = (\frac{1}{2})^n R_0$ directly as to use Eq. (32.11), $R = R_0 e^{-\lambda t}$. Accordingly, you'll be able to check your work and are advised to always do so—everyone makes numerical errors, we just have to catch them.

5. To calculate $R = R_0 A^{1/3}$ use the $x^{1/y}$ key on your electronic calculator. Enter the number for A, which is x, hit the $x^{1/y}$ key, enter 3, and hit the = key.

Discussion Questions

1. Figure Q1 is a plot of the excitation energy necessary to raise a nuclide with a given number of neutrons up to the next highest energy level. What's the significance of the data in light of the magic numbers?

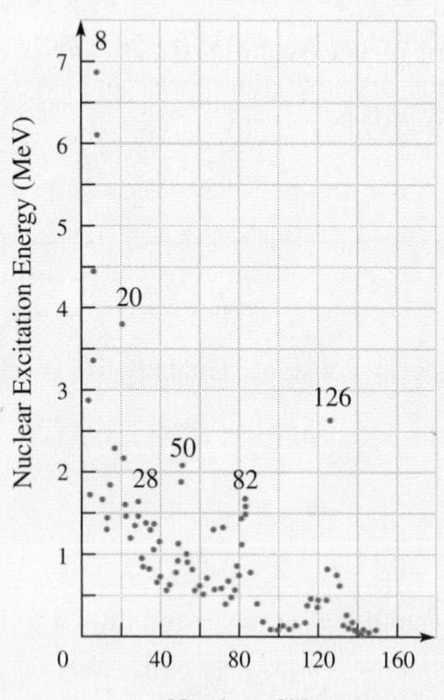

Figure Q1

2. The two nuclides ^{235}U and ^{239}Pu undergo fission via bombardment with neutrons of very little KE (even less than 0.01 eV). By contrast, ^{238}U and ^{232}Th will fission, provided the neutrons have kinetic energies in excess of 1 MeV. How is this fact used in a fission reactor? What does it suggest about fuels for bombs? Incidentally, it's been proposed that small atomic explosions could be used to propel a spaceship up to tremendous speeds. (Perhaps some good will come of this yet.)

3. Your standard H-bomb (what a thought) is essentially an ellipsoidal chunk of lithium-6 deuteride, an opalescent white solid, surrounded by an atom-bomb trigger. The whole thing is often encased in a U-238 shell, which initially reflects energy back onto the ^{6}LiD. The cheap U-238 also undergoes fission, allowing one to increase the output even more. (The Russians, with their big missiles and poor guidance systems, prefer large bombs.) The United States tested the first lithium-deuteride fission-fusion-fission device in 1953, ushering in a new era of human folly with a 15-megaton roar. Given the reactions

$${}^2_1\text{D} + {}^3_1\text{T} \rightarrow {}^4_2\text{He} + \text{n} + 17.6\ \text{MeV}$$

and

$${}^6_3\text{Li} + \text{n} \rightarrow {}^4_2\text{He} + {}^3_1\text{T} + \text{energy}$$

describe as much of what happens in an H-bomb as you can. Comment on the limitations, if any, on the mass of the U-238 shell. How might tritium be produced in a factory situation?

4. During World War II, the only large producer of heavy water in the world was the Norsk Hydro plant at Vemork, Norway. Toward the end of 1942, a group of Anglo-Norwegian commandos carried out two fairly successful raids on the facility.

Figure Q5

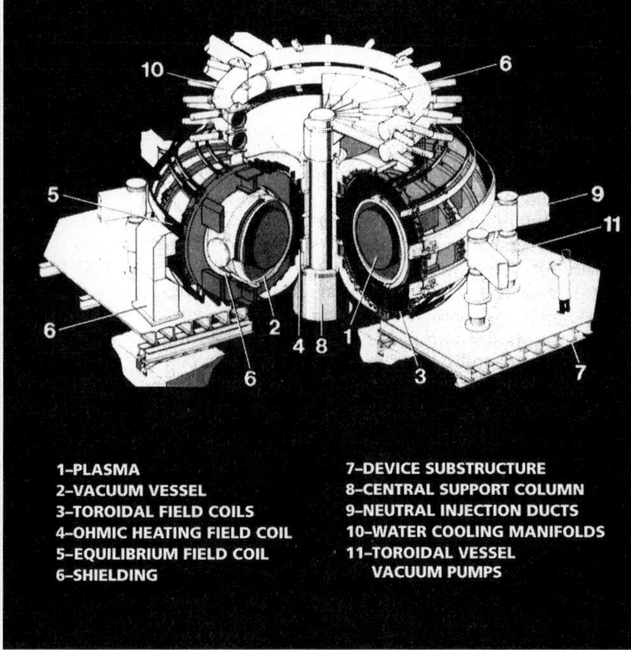

Figure Q6

The plant was in Nazi hands, and the Nazis were shipping all of its output to Germany. In the fall of 1943, after the plant had been repaired, American planes attacked it again. At that point, it should have been clear that we knew and that they knew that we knew. Why did we attack the plant? In the United States, we often use enriched U-238 and ordinary water in reactors. By contrast, the Canadians generally use D_2O and natural uranium. Explain.

5. Figure Q5 depicts a vacuum chamber into which are pointed a number of powerful laser beams. A tiny pellet of deuterium and tritium is dropped inside and blasted from all sides by the beams. This is a so-called *inertial confinement* reactor. What does it do and why? Knowing that neutrons are a likely by-product, see if you can figure out the formula for a possible nuclear reaction that might take place.

6. Figure Q6 shows a Tokamak Fusion Reactor. A high-temperature deuterium-tritium plasma is confined by powerful magnets so that it stays suspended within the vacuum chamber. By pumping exceedingly large currents through the plasma, temperatures of tens of millions of degrees can be attained. What do you think this thing is supposed to do? What are the high temperatures for? Why is fusion power appealing?

7. Discuss the difference between the half-life and the mean life of a radioactive substance. Relate these ideas to the life of a human being. Given a typical sample of Americans, 90% will make it to age 30, 80% to age 50, 50% to age 68, 25% to age 77, 10% to age 84, 1% to age 92, 0.1% to age 97, 0.01% to age 100, and 0% to age 200.

8. Hans Bethe at Cornell proposed a high-temperature fusion process known as the *carbon-nitrogen cycle* that may well power some stars:

	MeV
$^{12}C + p \rightarrow {}^{13}N + \gamma$	2.0
$^{13}N \rightarrow {}^{13}C + e^+ + \nu_e$	2.2
$^{13}C + p \rightarrow {}^{14}N + \gamma$	7.5
$^{14}N + p \rightarrow {}^{15}O + \gamma$	7.3
$^{15}O \rightarrow {}^{15}N + e^+ + \nu_e$	2.7
$^{15}N + p \rightarrow {}^{12}C + {}^4He$	5.0
	26.7

Explain what's happening. What form does the energy produced take? What is the net result of running through the cycle?

9. Return to Fig. 32.13, the exponential curve, which is a rather special function. For example, consider the slope of the curve. To do that qualitatively, imagine short line segments tangent to the curve at every point. What is the significance of the slope? Now plot the slope versus time. What is the shape of the curve?

10. Let's have a little fun and do a bit of speculating. What would happen to the Universe if the nuclear force vanished, *now*?

11. What would the Universe look like if the electromagnetic force vanished, *now*?

12. Suppose we produce the molecules of ${}^{12}_{6}C\,{}^{16}_{8}O$ and $({}^{14}_{7}N)_2$. Then we singly ionize them (that is, remove one electron) and accelerate them inside a vacuum chamber so we can determine their masses. Each molecule has a total of 14 protons and 14 neutrons. Explain in general terms what we would observe.

Multiple Choice Questions

1. A nuclide is specified completely by (a) its atomic number (b) its proton number (c) the number of nucleons it has (d) giving both Z and A (e) none of these.

2. The quantity A is the (a) proton number (b) nucleon number (c) neutron number (d) atomic number (e) none of these.

3. Isotopes are nuclides having (a) the same A and Z

values (b) the same Z, but different values of A (c) the same A, but different values of Z (d) different values of both A and Z (e) none of these.

4. When a nucleus undergoes alpha decay, the daughter as compared to the parent has (a) the same Z and an N reduced by 4 (b) Z reduced by 4 and A reduced by 2 (c) A reduced by 4 and Z reduced by 2 (d) Z increased by 2 and N reduced by 2 (e) none of these.
5. Beta particles are (a) protons emitted from a nucleus (b) electrons emitted from a nucleus (c) neutrons emitted from the nucleus (d) electrons emitted from the atomic cloud (e) none of these.
6. The mass of a nucleus composed of several nucleons is (a) always less than the sum of masses of its constituents (b) sometimes less than the sum of masses of its constituents (c) always more than the sum of masses of its constituents (d) always equal to the sum of masses of its constituents (e) none of these.
7. When a nucleus undergoes electron-beta decay, the daughter as compared to the parent has (a) the same Z and an N reduced by 1 (b) Z reduced by 1 and A reduced by 1 (c) A reduced by 1 and Z increased by 1 (d) Z increased by 1 and N reduced by 1 (e) none of these.
8. The nucleus $^{234}_{90}\text{Th}$ β decays into (a) $^{234}_{89}\text{Ac}$ (b) $^{235}_{91}\text{Pa}$ (c) $^{234}_{91}\text{Th}$ (d) $^{234}_{91}\text{Pa}$ (e) none of these.
9. The missing particle in the reaction

$$^{2}_{1}\text{D} + ^{199}_{80}\text{Hg} \rightarrow ^{197}_{79}\text{Au} + (?)$$

is a(n) (a) gamma (b) alpha (c) neutron (d) proton (e) electron.

10. The missing particle in the reaction

$$^{2}_{1}\text{D} + ^{196}_{78}\text{Pt} \rightarrow ^{197}_{79}\text{Au} + (?)$$

is a(n) (a) deuteron (b) alpha (c) neutron (d) proton (e) gamma.

11. The missing particle in the reaction

$$^{1}_{0}\text{n} + ^{198}_{80}\text{Hg} \rightarrow ^{197}_{79}\text{Au} + (?)$$

is a(n) (a) deuteron (b) alpha (c) neutron (d) proton (e) electron.

12. The missing particle in the reaction

$$^{1}_{0}\text{n} + ^{196}_{78}\text{Pt} \rightarrow ^{197}_{78}\text{Pt} + (?)$$

is a(n) (a) deuteron (b) alpha (c) neutron (d) proton (e) gamma.

13. The missing particle in the reaction

$$\gamma + ^{198}_{80}\text{Pt} \rightarrow ^{197}_{79}\text{Au} + (?)$$

is a(n) (a) electron (b) alpha (c) neutron (d) proton (e) gamma.

14. An antineutrino always accompanies (a) α decay (b) β^- decay (c) gamma emission (d) neutron emission (e) none of these.
15. In a nuclear reaction, the value of Q (a) is always negative (b) is positive in a spontaneous decay (c) is positive only for gamma emission (d) is always negative for proton emission (e) none of these.
16. The process whereby a nucleus splits into two very roughly equal parts is known as (a) fusion (b) beta decay (c) fission (d) nucleation (e) none of these.
17. A nuclide with a half-life of 10 days has a decay constant of (a) 0.069 decays/day (b) 6.9 decays/day (c) 1/10 decays/day (d) 10 decays/day (e) none of these.
18. Given that a freshly prepared nuclide has a half-life of 10 days, the percentage of it remaining after 30 days is (a) 30% (b) 10% (c) 12.5% (d) 72.5% (e) none of these.
19. As a naturally occurring radionuclide decays, (a) its λ increases (b) its τ decreases (c) its $t_{1/2}$ decreases (d) its R remains constant (e) none of these.
20. The Sun is powered by (a) nuclear beta decay (b) nuclear fission (c) thermonuclear fusion (d) nuclear gamma decay (e) none of these.

Problems

NUCLEAR STRUCTURE

1. [I] How many nucleons does the nuclide $^{111}_{50}\text{Sn}$ possess?

2. [I] What is the proper symbol for a nucleus having 14 protons and 15 neutrons?

3. [I] How many neutrons and how many protons does the nuclide $^{15}_{7}\text{N}$ possess?

4. [I] How many neutrons are in a ^{15}O nucleus?

5. [I] Identify each of the following nuclides: $^{211}_{87}\text{X}$, $^{202}_{82}\text{X}$, $^{105}_{47}\text{X}$, and $^{142}_{59}\text{X}$.

6. [I] What is the difference structurally between $^{183}_{76}\text{Os}$ and $^{193}_{76}\text{Os}$?

7. [I] The stable isotope of nitrogen, $^{14}_{7}\text{N}$, has an atomic mass of 14.003 074 u. How much is that in MeV/c^2 and GeV/c^2? Give the answers to six significant figures.

8. [I] Inasmuch as nuclear matter has a density of 2.3×10^{17} kg/m^3, how many cubic meters of water would have to be compressed into a cubic centimeter to match that density?

9. [I] The mass of the lithium-6 isotope is 5.603 051 GeV/c^2. How much is that in kg? Give your answer to seven significant figures.

10. [I] There are two stable isotopes of chlorine, Cl-35 (with an atomic mass of 34.968 853 u) and Cl-37 (with an atomic mass of 36.965 903 u). They are found with relative abundances of 75.77% and 24.23%, respectively. Determine the atomic mass of chlorine as it would be listed in the Periodic Table.

11. [I] Niobium has several isotopes, but only Nb-93 is stable. Its atomic mass is $1.542 748 \times 10^{-25}$ kg. Determine its mass in unified atomic mass units and compare it with the atomic mass listed for niobium in the Periodic Table. Explain your results.

12. [I] What is the radius of a gold nucleus? Gold has a nucleon number of 197.
13. [I] Determine the diameter of the uranium isotope ^{235}U.
14. [I] The radius of a nucleus doubles whenever the number of nucleons increases by a multiplicative factor of how much?
15. [I] If a nucleus is determined to be 7.2 fm in diameter, what mass number does it correspond to?
16. [I] A boron atom ($^{10}_{5}$B) has a mass of 10.012 937 u. What is the mass defect of its nucleus?
17. [I] Using the results of Problem 16, determine the binding energy of the boron nucleus.
18. [I] Use Fig. 32.9 to approximate the binding-energy-per-nucleon for boron-10. What is the total binding energy of this nuclide? How does that compare with the computed value?
19. [I] Using the results of Problem 17, determine the average binding-energy-per-nucleon for boron-10. Compare your answer with that of Problem 18.
20. [I] Use Fig. 32.9 to determine an approximate value for the total binding energy of a U-238 nucleus.
21. [I] Find the binding energy of the last neutron in the nucleus of oxygen-16. The masses of the atoms ^{16}O and ^{15}O are 15.994 915 u and 15.003 065 u, respectively.
22. [I] Calculate the binding energy of the last neutron in $^{13}_{6}$C. The atomic mass of carbon-13 is 13.003 355 u.
23. [II] Bromine has at least two stable isotopes: Br-79 (with an atomic mass of 78.918 336 u) and Br-81 (with an atomic mass of 80.916 289 u). Given that the relative abundance of Br-79 is 50.7% and the Periodic Table lists its atomic mass as 79.909 u, is it likely to have any other long-lived isotopes? Explain via numerical analysis.
24. [II] Approximate the ratio of the density of an atom to the density of its nucleus. Take the radius of the atom to be 0.05 nm and that of the nucleus to be 1.2 fm.
25. [II] How much bigger is the radius of the nucleus of U-238 than that of He-4?
26. [II] What is the atomic number of the nuclide that has a diameter one-quarter that of tellurium-128?
27. [II] Use Avogadro's Number and the definition of the atomic mass unit to show that $1 \text{ u} = 1.66 \times 10^{-27}$ kg.
28. [II] Determine the binding-energy-per-nucleon of helium to four figures. (Hint: Watch out for the electrons.)
29. [II] Iron-54 has an atomic mass of 53.939 613 u. Determine its nuclear mass defect in atomic mass units. Find the binding-energy-per-nucleon in MeV (four figures will do).
30. [II] With Problem 29 in mind, find the binding-energy-per-nucleon for iron-55 (atomic mass 54.938 296 u) and compare results.
31. [II] The neutrons in an isotope tend to pair and thereby bind strongly. For example, compute the minimum energy needed to remove a neutron from calcium-41 (atomic mass 40.962 278 u) as compared to calcium-42 (atomic mass 41.958 618 u). The former has 21 neutrons, the latter 22. Calcium-40 has an atomic mass of 39.962 591 u.
32. [II] What is the minimum amount of energy necessary to remove a proton from the nucleus of a $^{42}_{20}$Ca atom, thereby converting it into a $^{41}_{19}$K atom? The former has a mass of 41.958 618 u, the latter 40.961 825 u, and a hydrogen atom has a mass of 1.007 825 u.
33. [II] Uranium-232 (with an atomic mass of 232.037 13 u) is radioactive. It emits an α-particle and transforms into thorium-228 (with an atomic mass of 228.028 73 u). Determine the KE available to the decay products. Assuming the U-232 was at rest, this KE must be shared by the Th-228 and the alpha so as to conserve momentum. (Hint: The above are the masses of the atoms.)

NUCLEAR TRANSFORMATION

34. [I] When boron is struck by an alpha, it is transmuted into nitrogen-13, the reaction being

$$^{10}_{5}\text{B} + ^{4}_{2}\text{He} \rightarrow ^{13}_{7}\text{N} + ^{1}_{0}\text{n}$$

which then decays (with a half-life of $\approx$10 minutes) via positron emission. Write out that decay reaction.
35. [I] Irène and Pierre Joliot-Curie bombarded a foil of aluminum with α-particles, transmuting some of the nuclei and producing neutrons in the process. Write out the transformation formula.
36. [I] Rubidium-87 undergoes electron-β decay. What is the daughter nuclide? Write out that decay reaction.
37. [I] Samarium-147 decays via alpha emission. What is the resulting daughter nuclide? Write out that decay reaction.
38. [I] Calculate the maximum kinetic energy available (in MeV) to the electron and electron-antineutrino created by the decay of a neutron.
39. [I] Consider the reaction in which lithium-7 is struck by a proton; that is

$$\text{p} + ^{7}_{3}\text{Li} \rightarrow \alpha + \alpha$$

Compute the difference in mass before and after the collision. How much KE will the alphas have (in excess of the KE the proton delivered)? The mass of the ^{1}H, ^{7}Li, and ^{4}He *atoms* are 1.007 825 u, 7.016 003 u, 4.002 603 u, respectively.
40. [I] An old unit of activity that approximated the decay rate of radium is the *curie* (Ci) where $1 \text{ Ci} = 3.7 \times 10^{10}$ decays/s. What is the equivalent activity of one microcurie (μCi) in becquerels?
41. [I] A cobalt atom is at the core of the vitamin B_{12} molecule. Accordingly, radioactive Co-60 has been used as a tracer to study B_{12} absorption defect in pernicious anemia. Cobalt-60 has a half-life of 5.3 years. What is its decay constant in disintegrations per second?
42. [I] With Problem 40 in mind, given that the level of radioactive activity in the human body is naturally 10 nCi, how many nuclear disintegrations occur per second inside you?
43. [I] A radioactive sample is studied for a period of 36 hours, during which time its decay rate decreases to one-eighth its original value. What is its half-life?
44. [I] In humans, iodine is readily taken up by the thyroid, which requires it in the making of the hormone thyroxine. To study metabolism and treat thyroid disease, the isotope $^{131}_{53}$I is often introduced into the body. Given that it has a half-life of 8 days, what fraction of the original activity remains after 8 weeks? Assume none is excreted. Take a look at Problem 46.

45. [I] Radium-226 emits an α-particle and decays into radon-222 with a half-life of 1620 years. What is its decay constant in decays/year?

46. [I] Radionuclides that have entered the human body are sometimes biologically excreted in an approximately exponential way. The effective half-life is then

$$1/t_{1/2}(\text{eff}) = 1/t_{1/2}(\text{bio}) + 1/t_{1/2}$$

Given that iodine-131 has a half-life of 8.0 d and a biological half-life of 138 d, what is its effective half-life? How much of it will remain after 8.0 weeks? (See Problem 44.)

47. [I] Given that radon has a half-life of 3.8 days, what is its mean life?

48. [I] Radon undergoes α decay with a half-life of 3.8 days. Given some original amount, how much of it will remain after 11.4 days?

49. [I] A radioactive sample has a decay constant of 7.69×10^{-3} decays/s. What is its average lifetime and its half-life?

50. [I] A sample of radioactive oxygen-15 has a half-life of 2.1 minutes and a decay rate of 5.5×10^{-3} decays/s. By how much will the amount of this isotope be diminished after 4.0 s?

51. [I] Protactinium-234 has a half-life of 1.18 minutes. If 1.00 mg of it is freshly created, what fraction will be left in 1.00 h?

52. [II] Polonium-210 is radioactive. Confirm that it is energetically possible for it to decay via alpha emission in the following way

$$^{210}_{84}\text{Po} \rightarrow {}^{206}_{82}\text{Pb} + \alpha$$

What is the net kinetic energy of the decay particles, assuming the polonium nucleus to be at rest? The masses of the polonium and lead atoms are 209.982 848 u and 205.974 440 u, respectively. (Hint: Watch out for the electrons.)

53. [II] By bombarding beryllium with alphas, we get the following reaction

$$^{9}_{4}\text{Be} + {}^{4}_{2}\alpha \rightarrow {}^{12}_{6}\text{C} + {}^{1}_{0}\text{n} + Q$$

The mass of the Be atom is 9.012 182 u, the mass of the He atom is 4.002 603 u. Please compute the value of Q in excess of the alpha's incoming KE.

54. [II] Given the reaction

$$^{6}_{3}\text{Li} + \text{p} \rightarrow \alpha + {}^{3}_{2}\text{He} + Q$$

If $Q = 4.0185$ MeV in excess of the proton's KE and the lithium was at rest, find the mass of the helium-3 atom.

55. [II] Determine the value of Q, which is required in order to initiate the following reaction

$$\alpha + {}^{14}\text{N} \rightarrow {}^{17}\text{O} + \text{p}$$

This energy must come by way of the KE of the alpha, which actually has to be a bit higher than Q because some KE must be given to the other particles if momentum is to be conserved.

56. [II] Can lead-206 (of atomic mass 205.974 440 u) spontaneously decay via alpha emission? Explain your answer completely. The atomic mass of mercury-202 is 201.970 617 u.

57. [II] Calculate the activity of one milligram of radon-222, which has a decay constant of 2.1×10^{-6} decays/s.

58. [II] Uranium-238 has a half-life of 4.5×10^9 years. The Earth was formed about 5×10^9 years ago. What fraction of the ^{238}U present then is still around today?

59. [II] Radioactive phosphorus-32 has been used to study bone metabolism and for the treatment of blood diseases. What is the activity of a sample of 5×10^{16} atoms if it has a half-life of 14.3 days?

60. [II] Carbon-11 is radioactive with a half-life of 20.4 minutes. If a sample initially has 1.0×10^{17} carbon-11 atoms in it, what is its activity after 10 minutes?

61. [II] Oxygen-15 has a half-life of 2.1 min. How many ^{15}O atoms are present in a source with an activity of 4.1 mCi? (See Problem 40.)

62. [II] Consider a sample of freshly cut wood that has been reduced to pure carbon. What is the activity per gram resulting from the decay of ^{14}C (half-life of 5730 y) in this material?

63. [II] Determine the maximum permissible concentration in the air of tritium in the workplace, given that the maximum permissible annual intake of tritium is 444 MBq. A typical person inhales $\approx$10 m^3 of air per working day. Use a 50 work-week year.

64. [II] A 50-g chunk of charcoal is found in the buried remains of an ancient city destroyed by invaders. The carbon-14 activity of the sample is 200 decays/min. Roughly when was the city destroyed (more accurately, when was the tree felled from which the charcoal came)? Refer back to Problem 62.

65. [II] Determine the amount of energy liberated in the fusion reaction

$$4\text{p} \rightarrow \alpha + 2\,{}^{0}_{+1}\text{e} + 2\nu_e + \gamma$$

The masses of the proton, electron, and alpha particle are 1.007 276 u, 0.000 548 580 u, and 4.001 506 u, respectively.

66. [II] Radioisotopes are sometimes used to produce electricity to power such things as interplanetary probes and pacemakers. Consider the radioactive nuclide Po-210, which has a half-life of $\approx$140 d, emitting 5.30-MeV alphas. If all the KE of the alphas is absorbed in a radioisotope thermal generator (known as an RTG, in the trade), what's the average thermal power (in watts) developed per gram during the first 140 days of operation?

67. [III] Show that $t_{1/2} = (\ln 2)/\lambda$.

68. [III] A radioactive sample shows a decrease by a factor of 10 in activity over a period of 5.0 minutes. What is its decay constant?

33

High-Energy Physics

As THE TWENTIETH CENTURY unfolded, the number of observed subnuclear particles gradually increased. The electron, proton, and neutron were joined by the muon, pion, positron, and so forth, until by the 1970s, several hundred distinct particles were identified. The obvious problem was to determine which of these tiny specks of matter was *elementary* in the sense of being a single homogeneous entity having no internal structure. It now seems certain that most of these subnuclear particles are clusters of two or more fundamental entities called *quarks* (p. 1148). These are the primary building blocks that fuse together via the strong force to create neutrons,

TABLE 33.1 Some fairly long-lived elementary particles

Category	Name	Particle	Antiparticle	Mass (MeV/c^2)	B	L_e	L_μ	L_τ	S	Lifetime (s)
Leptons	Electron	e^-	e^+	0.51100	0	±1	0	0	0	Stable
	Neutrino (e)	ν_e	ν_e	0 (<14 × 10^{-6})	0	±1	0	0	0	Stable
	Muon	μ^-	μ^+	105.659	0	0	±1	0	0	2.197 × 10^{-6}
	Neutrino (μ)	ν_μ	ν_μ	0 (<0.25)	0	0	±1	0	0	Stable
	Tau	τ^-	τ^+	1784.2 ± 3.2	0	0	0	±1	0	(4.6 ± 1.9) × 10^{-13}
	Neutrino (τ)	ν_τ	ν_τ	0 (<35)	0	0	0	±1	0	Stable
Hadrons										
Mesons	Pion	π^+	π^-	139.567	0	0	0	0	0	2.60 × 10^{-8}
		π^0	Self	134.96	0	0	0	0	0	0.87 × 10^{-16}
	Kaon	K^+	K^-	493.7	0	0	0	0	±1	1.24 × 10^{-8}
		K^0	K^0	497.7	0	0	0	0	±1	0.9 × 10^{-10}
Baryons	Proton	p	p	938.3	±1	0	0	0	0	Stable
	Neutron	n	n	939.6	±1	0	0	0	0	898
	Lambda	Λ^0	Λ^0	1115.6	±1	0	0	0	∓1	2.6 × 10^{-10}
	Sigma	Σ^+	Σ^-	1189.4	±1	0	0	0	∓1	0.80 × 10^{-10}
		Σ^0	Σ^0	1192.5	±1	0	0	0	∓1	5.8 × 10^{-20}
		Σ^-	Σ^+	1197.3	±1	0	0	0	∓1	1.5 × 10^{-10}
	Xi	Ξ^0	Ξ^0	1315	±1	0	0	0	∓2	2.9 × 10^{-10}
		Ξ^-	Ξ^+	1321	±1	0	0	0	∓2	1.64 × 10^{-10}
	Omega	Ω^-	Ω^+	1672	±1	0	0	0	∓3	0.82 × 10^{-10}

protons, pions, and so forth. The complementary group of elementary entities, the *leptons,* do not experience the strong force and do not combine to form composite subnuclear particles. They exist only individually and, to within the limits of our most powerful techniques, each (the electron, neutrino, and so on) appears to be a distinct structureless object.

Table 33.1 provides a representative sample of the more long-lived subnuclear particles. They are listed primarily according to their modes of interaction. All of them experience or *couple to* the gravitational interaction, which nonetheless plays a negligible role in particle physics because it's so feeble. The more dominant influences arise from the electromagnetic, strong, and weak interactions, and these are transmitted by a set of elementary virtual particles known as the *gauge bosons* (p. 1152).

Our low-energy world of trees and rocks and politicians is made up almost entirely of three fundamental material particle types: the electron (a lepton), the u-quark, and the d-quark. Along with the neutrino, these constitute the *first generation* of matter, the ordinary stuff still remaining in the cool Universe as it presently exists. At higher temperatures and greater energies, we can produce more exotic forms of matter—the *second* and *third generations* (p. 1151). These highly unstable particles once existed in abundance in the blazing early moments of the primordial Universe (p. 1156). Today, we can just begin to recreate some of those incredibly violent conditions. In studying the subnuclear domain, we gain a view, however darkly, of the first moments of Creation.

ELEMENTARY PARTICLES

The hundreds of subnuclear particles discovered in the second half of the twentieth century offered physicists an opportunity to begin to sort out the fundamental nature of matter and its interactions. The process was similar to finding the electron-cloud structure of the atom beginning with the elements of the Periodic Table. First, they (elements or particles) were grouped by way of shared behavior. Then they were arranged according to measurable characteristics (such as mass and charge) underlying that behavior. And, finally, the hidden structure that gave rise to all the patterns of experimental data was hypothesized and essentially confirmed.

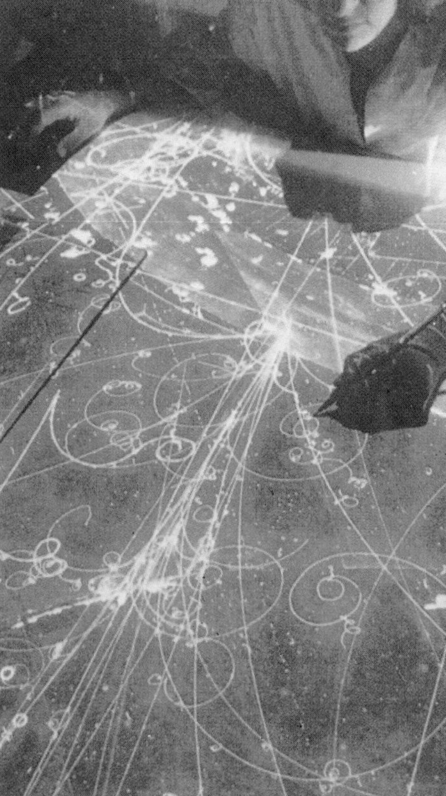

Subnuclear particle tracks left in a bubble chamber. (See Discussion Question 7.)

33.1 Leptons

The first group of twelve particles in Table 33.1 constitutes the **leptons** and **antileptons** (from the Greek *leptos,* meaning "slight"). These are the particles that couple to the weak force, and if they are electrically charged, they react to the electromagnetic force as well. *All are immune to the strong force.* The most familiar and the lightest of the charged leptons is the electron. The muon (μ^-), found among the products of cosmic-ray bombardment (1936), was the first unstable subnuclear particle to be discovered. Muons are commonplace, rushing toward the surface of the planet in a flow of roughly 1 per cm^2 per minute. Even as you read these lines, muons are more or less harmlessly streaming downward through your body like a gentle penetrating cosmic rain. The muon decays (in about 2.2 μs via the weak interaction) into an electron and two neutrinos:

$$\mu^- \rightarrow e^- + \nu_\mu + \bar{\nu}_e$$

Otherwise, it behaves in every way like an overweight electron. The same can be said about the tau (τ), although, since it was only discovered in the 1970s, we have less experience with it. The tau is almost twice as massive as the proton; nonetheless, it is believed to be truly elementary. Both muons and taus play no known significant role in the scheme of things, and their very existence is puzzling in a Universe that seems to have a preference for economy.

At our current limits of observation, leptons appear to be structureless, pointlike entities. All are spin-$\frac{1}{2}$ fermions. Each lepton having substantial mass is created in conjunction with its own variety of essentially massless neutrino. Thus, there are electron-neutrinos, muon-neutrinos, and tau-neutrinos. In total, there are probably more neutrinos than anything else—they outnumber electrons and protons by a factor of perhaps a thousand million. The Universe is awash with neutrinos, some of which are harmlessly passing through your body even now. Each lepton comes in both particle and antiparticle varieties, making a total of six electrically charged and six neutral leptons. These form three—electron, muon, and tau—particle-neutrino couplets: e and ν_e; μ and ν_μ; and τ and ν_τ.

$$\begin{bmatrix} e \\ \nu_e \end{bmatrix} \quad \begin{bmatrix} \mu \\ \nu_\mu \end{bmatrix} \quad \begin{bmatrix} \tau \\ \nu_\tau \end{bmatrix}$$

The systematic patterns of production and decay of leptons suggested that an underlying conservation law might be at work. Remember the special feature that leptons are created in couplets. It is believed that *lepton number* (L) is conserved in all presently attainable processes involving any members of this family. Thus, each of the three subsets of leptons (electron-type, muon-type, and tau-type) conserves its own lepton number (L_e, L_μ, L_τ). In each subset, the particle (for example, e^-) has a lepton number (L_e) of $+1$, whereas the antiparticle (e^+) has a lepton number of -1, and the same is true of the associated neutrino and antineutrino. Thus the decay reaction

$$\mu^- \rightarrow e^- + \nu_\mu + \bar{\nu}_e$$

for which $\quad L_e$: $\quad 0 \rightarrow 1 + 0 - 1$

and $\quad L_\mu$: $\quad 1 \rightarrow 0 + 1 + 0$

separately conserves both muon- and electron-lepton numbers.

33.2 Hadrons

Hadrons are the strongly interacting composite particles. They are all fairly massive; ergo, the name, which derives from the Greek *hadros,* meaning "bulky." A succession of hadrons was discovered between 1947 and 1954 in cosmic-ray studies and at the Brookhaven Cosmotron in New York. That machine, a 3-GeV accelerator, made it possible for the first time to bring to bear the needed amounts of energy to create heavy particles via $E_0 = mc^2$. These conjured hadrons were christened the kaon (K: $\approx$500 MeV/c^2), the lambda (Λ: $\approx$1100 MeV/c^2), the sigma (Σ: $\approx$1200 MeV/c^2), and the xi (Ξ: $\approx$1300 MeV/c^2). All were observed via photographs of the several-centimeter-long tracks they left in visual detectors. These trails, marking their passage between creation and decay, corresponded to lifetimes of around 0.1 ns to 10 ns. In contrast to the twelve leptons, there are hundreds of hadrons, which increasingly suggested that hadrons might not be elementary. Every hadron couples to the strong, weak, and gravitational interactions, and some also couple to the electromagnetic interaction. All of them, with the possible exception of the proton, decay—*some rapidly, via the strong interaction; some less rapidly, via the electromagnetic interaction*; *and others still more slowly, via the weak interaction.* Hadrons form two distinct subgroups (defined by their spins): *mesons* and *baryons.*

Mesons

The Greek root *meso* means "middle," and it was applied in 1939 to particles whose mass was between that of the electron and proton. Today, the term **meson** refers to hadrons that are bosons (they have spins of 0, 1, 2, . . .). Table 33.1 lists only the several pions (or π-mesons) and kaons (or K-mesons). Although there are dozens of other known mesons, these serve our purposes for the time being.

The three members of the *pion* family (π^+, π^-, π^0) are the lightest of the mesons. Proton-proton collisions carrying enough energy create pions:

$$\mathrm{p} + \mathrm{p} \rightarrow \mathrm{p} + \mathrm{p} + \pi^+ + \pi^-$$
$$\mathrm{p} + \mathrm{p} \rightarrow \mathrm{p} + \mathrm{n} + \pi^+$$

and pion-proton collisions produce more of them, for example

$$\mathrm{p} + \pi^- \rightarrow \mathrm{p} + \pi^- + \pi^0$$
$$\mathrm{p} + \pi^- \rightarrow \mathrm{n} + \pi^0$$

Each process conserves angular momentum (that is, spin; **the pion has zero spin**), electromagnetic charge, and nucleon number (that is, the number of nucleons). Whereas π^+ and π^- can be thought of as particle and antiparticle, neutral π^0 is its own antiparticle. The π^0 has a fleeting lifetime (0.87×10^{-16} s) usually decaying (via the electromagnetic interaction) into two photons

$$\pi^0 \rightarrow \gamma + \gamma$$

Notice that there is *no* conservation of meson number. The most common decay route for the charged pions is fixed by the fact that there are no hadrons lighter than themselves; they must decay into leptons (muons) via the weak interaction; thus

$$\pi^+ \rightarrow \mu^+ + \nu_\mu \qquad \pi^- \rightarrow \mu^- + \overline{\nu}_\mu$$

Pions play a very special role in nature: they make a major contribution to mediating the nuclear force between nucleons (p. 1107).

The kaon (K-meson) is produced in collisions via the strong interaction. There are four zero-spin kaons: K^+, K^-, K^0, and its antiparticle $\overline{K}^0$. They are unstable and undergo a number of different decay reactions. For example, each can decay into a pair of pions:

$$K^0 \rightarrow \pi^+ + \pi^-$$
$$\overline{K}^0 \rightarrow \pi^0 + \pi^0$$
$$K^+ \rightarrow \pi^+ + \pi^0$$

Two surprising things were observed here and with a group of other newly discovered particles as well. First, the reaction products were always created in twos, even though there was no known reason for it. For example, every time a K^0 was produced, a Λ^0 was also produced. Second, instead of a decay time typical of the strong interaction ($\approx 10^{-23}$ s), these transformations sauntered along with lifetimes of from 0.1 ns to 10 ns. It was as if the particles, born out of the strong force, died by the weak force.

That mysterious behavior was denoted by calling these objects **strange particles**. To explain what was happening, Murray Gell-Mann, in the 1950s, introduced a new quantum number that reflects a new conserved quantity. Like electromagnetic charge or spin, he proposed that these particles uniquely possess a quality of *strangeness* (as if they carried a strangeness charge), and he assigned a numerical value of it (S) to each hadron, according to its observed behavior. *The strong and electromagnetic interactions both conserve strangeness. The weak interaction does not, and the strangeness may change, but by no more than one unit.* Pions and nucleons, which are not at all strange, have a strangeness of 0. The kaon (K^+ and K^0) has a strangeness of +1, whereas the strange baryon known as lambda has a strangeness of −1.

The strong interaction could create a pair of strange particles out of more ordinary matter (with zero strangeness), provided the resulting pair had canceling values of strangeness. Thus, a pion and a proton (net strangeness, 0) can interact strongly to produce a K-zero and a lambda-zero of strangeness +1 and −1, respectively: $\pi^- + p \rightarrow K^0 + \Lambda^0$. The logic forbids strange particles (like K^0), once created, from decaying to lighter nonstrange particles ($K^0 \rightarrow \pi^+ + \pi^-$) by the rapid route of the strong interaction. Strange particles can accomplish such decay only via the weak interaction and only slowly. This kind of *ad hoc* conceptual fine-tuning reveals the way physics develops; in the absence of a proven theoretical formalism, phenomenological relationships are derived, or guessed at, from observed patterns. The idea, however well it worked then, really makes sense only when viewed from the perspective of the quark theory (p. 1148). Strangeness simply depends on the number of strange and antistrange quarks composing the particles.

Baryons

Baryons derive their name from the Greek *barys,* meaning "heavy." They are the heavy hadrons but, more importantly, they are all fermions. Neutrons and protons, the nucleons, are the most well-known baryons. Table 33.1 displays a representative selection of baryons. Generalizing from the concept of nucleon number, it was proposed that the number of baryons and antibaryons in a reaction is conserved, and so they are assigned a *baryon number* (B). For baryons $B = +1$, whereas antibaryons have $B = -1$, and all nonbaryons have $B = 0$ (see Discussion Question 11). This system has the effect of "explaining" why the proton does not decay into still lighter particles, even though such a decay is not otherwise in violation of any known principle. In all particle reactions (at the energies we can attain), the total baryon number must, it is assumed, be conserved. Thus, the strong reaction

$$\pi^- + p \rightarrow K^0 + \Lambda^0$$

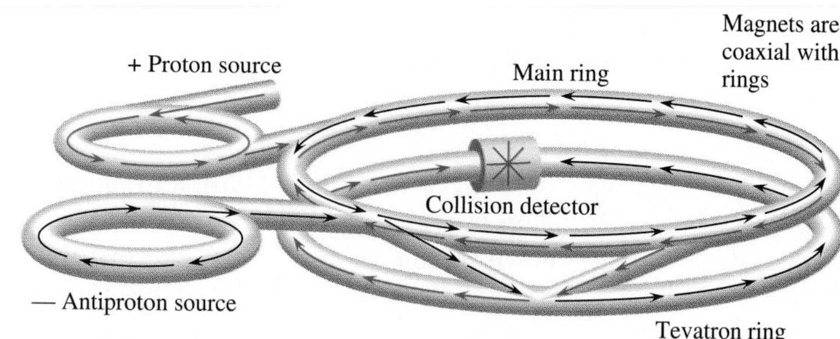

The world's most powerful particle accelerator at the Fermi National Accelerator Laboratory. When it actually has the funds to operate, it accelerates protons in one ring and antiprotons in the other. These beams are then made to collide with a total energy of ≈2000 GeV. Superconducting magnets force the particles to stay within the rings.

conserves electromagnetic charge (zero on both sides), strangeness (zero on both sides), angular momentum ($\frac{1}{2}$ on both sides), and baryon number (+1 on both sides), and therefore can and does take place. The heavier strange baryons (for example, Λ, Σ, Ξ, Ω) decay to lighter baryons and ultimately to the proton.

Whereas there are conservation laws for leptons and baryons, there are no restrictions on the number of photons or mesons entering and leaving an encounter. These particles can appear and vanish at will. In fact, the photon and the pion have $L = S = B = 0$. In the case of the photon and the neutral pion, each is its own antiparticle. With all of these quantum numbers equaling zero, including the electromagnetic charge, it is impossible to distinguish particle from antiparticle. As we'll see presently, photons and pions play a very special role in Quantum Field Theory.

QUANTUM FIELD THEORY

Prior to the introduction of Quantum Mechanics, particles and fields were considered interrelated though distinct entities; particles possessed intrinsic features (for example, mass and/or electromagnetic charge) that gave rise to external fields (for example, gravitational and/or electromagnetic). Force fields emanated from particles and filled the surrounding space. They carried energy and were, in a sense, real continuous media that interconnected all interacting particles and mediated their interactions. Particles were composed of matter, fields were composed of energy. The force field was the nineteenth century's answer to the age-old mystery of action-at-a-distance. Still, particles that do not react to any force fields are unobservable and physically meaningless. Force fields that do not act upon any particles are equally without significance. The ideas of particle and field take meaning from their interrelationship.

The concept of field began to change drastically with the introduction of Einstein's photon. The electromagnetic field does not, after all, have its energy continuously spread out in space. **The photon is the quantum of the electromagnetic field, and it carries the energy and momentum of the field**. The interaction of two charged particles corresponding to the electromagnetic force, transmitting energy and momentum from one to the other, must take place through the exchange of electromagnetic energy quanta—photons. Quantum Electrodynamics (QED), the theory of such interactions, was the first successful application of these ideas (p. 1093).

Figure 33.1 is a representation of two electrons as they undergo an elastic collision. It is called a **Feynman diagram**. Suppose each electron is initially traveling at the same speed. The electrons first approach and then recede from one another along a line in space that is projected upward in the increasing time direction in the diagram. The electron on the left emits a photon (the wiggly line), and for a moment (Δt) there are two electrons and one photon. The electron on the right absorbs the photon, and the interaction is momentarily over; other photons will subsequently go back and forth between the electrons. The average force is proportional to the rate of transfer of momentum mediated by the exchange of the photons. The measure of the probability of both emission and absorption of photons is the charge. Hence, the force must be proportional to both interacting charges (recall Coulomb's Law). Think of the repulsive interaction between two astronauts floating in space, throwing a ball back and forth (p. 115).

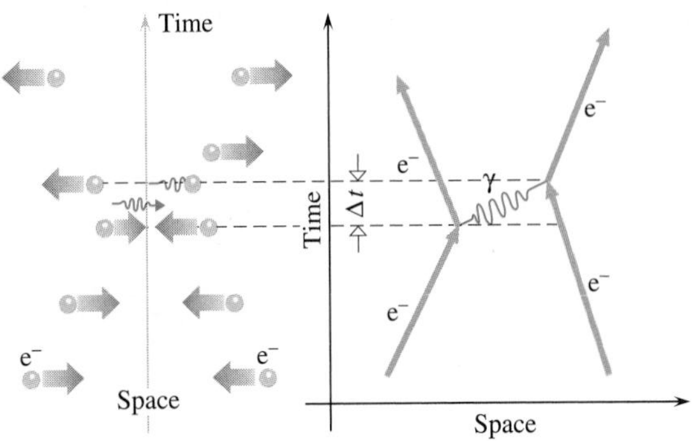

Figure 33.1 A Feynman diagram showing the scattering of two electrons via the exchange of a virtual photon.

The electrons' exchange interaction is a quantum effect and cannot be visualized in classical terms. Still, repulsion by way of an exchange force can be thought of via the astronaut-ball analogy. However, attraction between an electron and a proton through exchange is unvisualizable, unless you resort to nonsense like the astronauts facing away from each other catching boomerangs backwards so that they're pushed together. Figure 33.2 depicts the attraction, but in a way that does not attempt to be faithful to the kinematics. Feynman diagrams are symbolic—they're computational devices in QED and are not concerned with accurately picturing the particle trajectories. Thus, the horizontal distances are of little significance, and the arrangement of Fig. 33.1 is often used for both attraction and repulsion. The important part is the interaction.

The collision in Fig. 33.1 is elastic; the energy of either electron is unchanged throughout, and yet during the time Δt, the system contains an additional amount of energy hf corresponding to the photon. For a time Δt, Conservation of Energy is seemingly violated! Can this situation be tolerated? One answer offered by modern physics is *yes,* provided it can never be observed. In other words, there is always some uncertainty (ΔE) in the measured value of the energy of a system. The Heisenberg Uncertainty Principle (p. 1089) tells us that

$$\Delta \mathrm{E}\, \Delta t \geq \tfrac{1}{2}\hbar$$

Nonconservation of energy up to an amount ΔE will be hidden by the ever-present energy uncertainty, provided the time available to make the observation (Δt) is restrictively small; namely

$$\Delta t \leq \tfrac{1}{2}\frac{\hbar}{\Delta \mathrm{E}}$$

(If a moment of nonconservation is totally unobservable, is Conservation of Energy actually violated?) The energy uncertainty will exceed the photon energy hf if the photon exists for a time less than

$$\Delta t = \tfrac{1}{2}\frac{\hbar}{hf} = \frac{1}{4\pi f}$$

This unobservable photon can travel a maximum distance of

$$R = \mathrm{c}\, \Delta t = \frac{\mathrm{c}}{4\pi f} \tag{33.1}$$

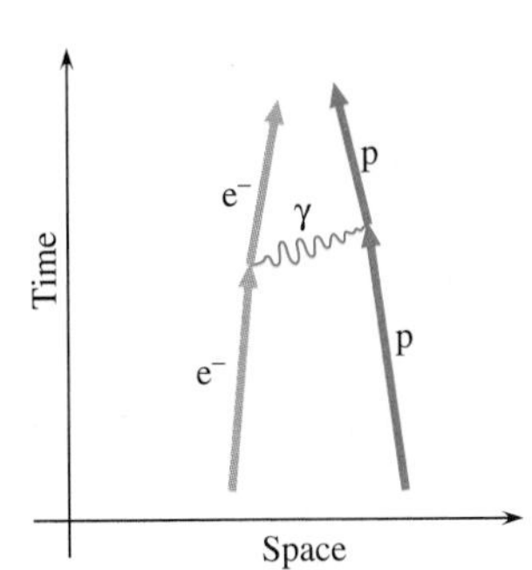

Figure 33.2 A Feynman space-time diagram of the electron-proton interaction.

and since its frequency can be arbitrarily small, the range of the force transmitted

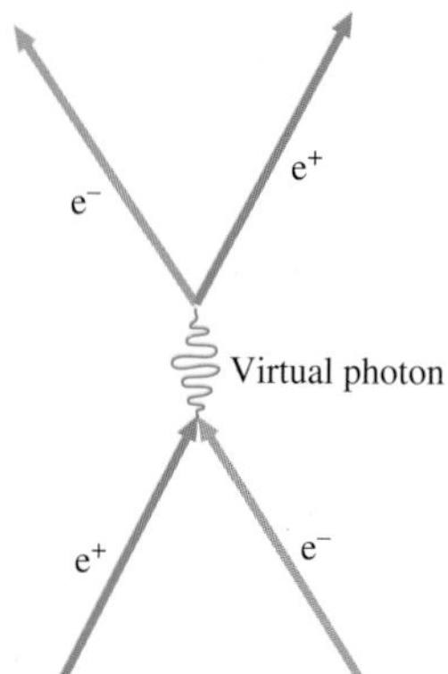

Figure 33.3 The annihilation and creation of electron-positron pairs.

by the massless photon is unlimited. Such unobservable exchange quanta are called **virtual photons**.

In contrast to *real* quanta, these *virtual* quanta are the messengers of the interaction. In the Feynman diagrams, they are the internal segments that begin and end within the figure. They effectively "tell" the material particles what's happening. A photon that is observable in the sense that it is detected by an eyeball or a Geiger counter is real enough. A photon that never leaves the region of interaction between charges (Fig. 33.3) and vanishes in the process of communicating the electromagnetic force—a photon that cannot therefore be seen by a detector (so that we never observe any violation of the basic conservation laws)—is a virtual photon. It need not obey the equation for the total energy of a real particle: $E^2 = m^2c^4 + p^2c^2$.

The distinction between real and virtual photons is not always obvious. At the extreme of short range and short existence, messenger photons can have tremendous energies (modern accelerators can generate virtual quanta of upwards of 100 GeV). At the other extreme, because the photon is massless and moves at the ultimate speed, there is no limitation on the time it can travel with its message (its clocks don't appreciate time anyway, p. 998). The range of virtual photons is limitless, and the range of the electromagnetic interaction is limitless. If Δt is tiny, as it usually is in particle physics, there's no issue; if Δt is large, as it might be in astronomy, the imagery can get a little murky. The question of whether a particular photon that has traveled through space for a million years is real or virtual loses significance: ΔE is vanishingly small.

By the late 1920s, it was recognized that the known material particles (protons and electrons) could each be considered as the quantum of a specific particle field. From this perspective, there are electron fields, proton fields, and so on; the Universe is a set of quantized fields. Reality (a cup of coffee and a hamburger) is the totality of observable manifestations of field quanta. And from that perspective, the problem of the wave-particle duality no longer exists: matter is field.

When Fermi formulated his theory of the weak interaction in 1932, he founded it on the principles of QED. Not long after that, the Japanese physicist Hideki Yukawa proposed (1934) that the strong interaction was mediated by the exchange of a massive virtual boson. At the time, the only known particles were the electron, proton, and neutron. As we'll see, he was able to predict that the mass of this new messenger would be between m_e and m_n and thus it came to be called the meson. It has since been identified specifically as the π-meson, or pion. Strong (hadronic) interactions occur with equal strength between electrically positive, negative, and neutral particles, and so the idea was extended to include three exchange particles: one positive (π^+), one negative (π^-), and one neutral (π^0). It was proposed that virtual particles, emitted and absorbed, constantly fly back and forth between nucleons which, in turn, are transformed—the proton and neutron are two alternative states of the nucleon (Fig. 33.4). The *virtual meson* of mass m_π and rest energy $m_\pi c^2$ traveling at nearly the speed of light has a maximum range provided by Eq. (33.1), of

$$R \approx c\,\Delta t \approx c\left(\frac{\frac{1}{2}\hbar}{\Delta E}\right) \approx c\left(\frac{\frac{1}{2}\hbar}{m_\pi c^2}\right) \approx \frac{h}{4\pi m_\pi c}$$

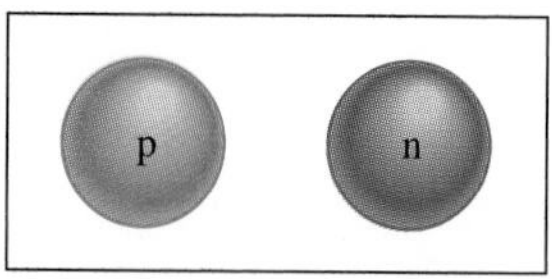

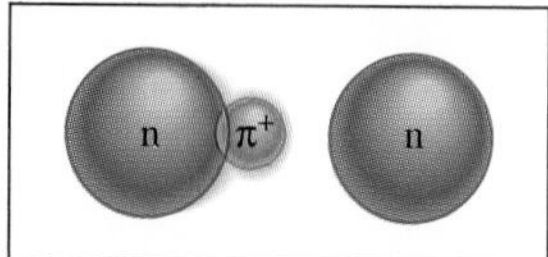

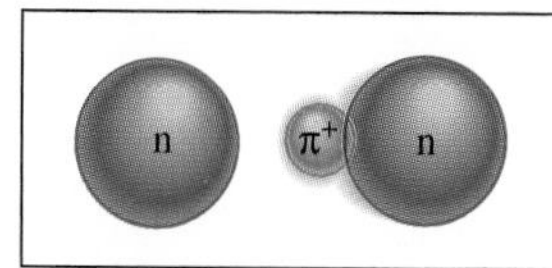

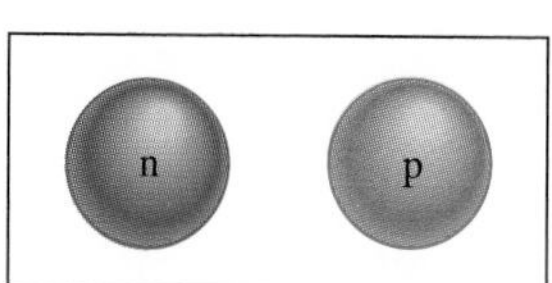

Figure 33.4 The attractive interaction between a neutron and proton through the exchange of a positive pion.

Given the known range of the nuclear force ($\approx$1 fm), the predicted particle mass comes out about $200m_e$ or $\approx$100 MeV/c^2. As a rule, **when enough energy is present such that $E_0 = mc^2$, a real particle corresponding to the virtual one can be created**. In 1934, the only source of sufficient energy was cosmic radiation. But not until 1947 was it possible to study high-energy cosmic-ray collisions using photographic emulsions. And only then was the Yukawa particle finally found: the **pion** was discovered (Fig. 33.5).

Contemporary Quantum Field Theory operates under several assumptions: (1) the essential reality is the set of quantum fields—nothing else exists; (2) these fields obey the rules of Special Relativity and Quantum Mechanics; (3) the intensity of a field at some location is a measure of the likelihood of finding an associated particle at that location; (4) the field quanta interact as the fields interact. **All particle interactions (all forces) are mediated by field quanta**. We now believe that pion exchange between hadrons is actually a low-strength residual manifestation of a still more basic and more powerful interaction involving the exchange of gluons among quarks (p. 1148).

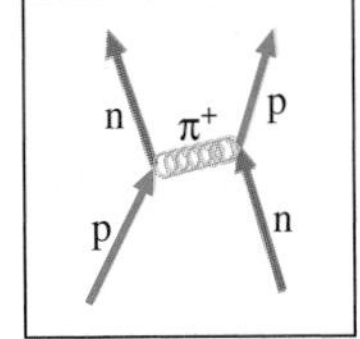

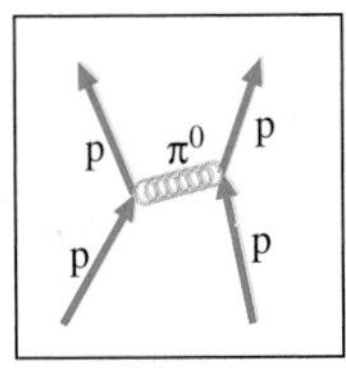

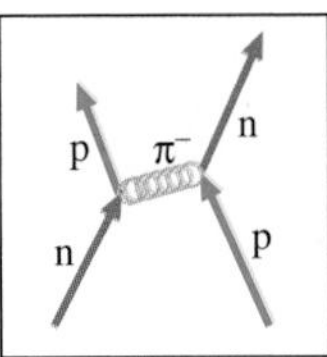

Figure 33.5 Nucleon-nucleon interactions through pion exchange.

33.3 Gauge Theory

It is now widely accepted that all field theories that accurately portray nature must possess a particular type of mathematical structure known as **gauge symmetry**. The meaning of the term is subtle and depends on several subsidiary ideas, and so we will develop it slowly. The discovery of the significance of gauge symmetry, one of the greatest of the century, was prompted by the realization that the General Theory of Relativity was gauge symmetric. Einstein's analysis of mass (or gravitational charge) in terms of the curvature of space and time held the secret to treating all forms of charge, all theories of the very essence of the cosmos. That insight was underscored by the awareness that renormalized QED (p. 1095) happened also to be gauge symmetric. Today, it is believed that we have a mathematical test of the legitimacy of any new field theory; even better, we have a potent guide to formulating such a theory—that's an amazing thing to be able to say! Physicists maintain that the hidden structure of the Universe, whatever detailed form it takes, is gauge invariant.

The philosophical insight that sprang from Relativity Theory was that all the laws of physics are the same for all observers—anyone anywhere must experience nature to behave in the same way. And yet a thousand scientists on a thousand planets across the Universe will surely have their own definitions and formulations, their own arbitrary sets of units, arbitrary base levels for things such as zero speed, zero voltage, and zero potential energy. If one body of physical law rules the cosmos, different local constructs must be equivalent—the mindset of the scientist cannot alter the underlying physical reality. A correct theory must be expressible in different local languages in terms of local conceptions and still carry the same truth. All such theories should be translatable from one foreign mathematical language to any other.

The question of units isn't a problem: we can *transform* units from one system to another—an inch is as good as 2.54 centimeters. There are no completely natural units; all are dreamed up in the minds of scientists and cannot possibly affect physical law. Nor can the arbitrarily assigned base levels of relative concepts such as potential energy have any effect on law. There is no natural zero of voltage, and however it may be assigned, we can again transform from one level to another with no problem (that's why there are zero reset knobs on meters). This freedom to assign base levels of certain important physical quantities is a manifestation of an underlying symmetry.

In the broadest terms, a physical system possesses symmetry if something happens (whereupon there is a change of some kind) and no observable change in the

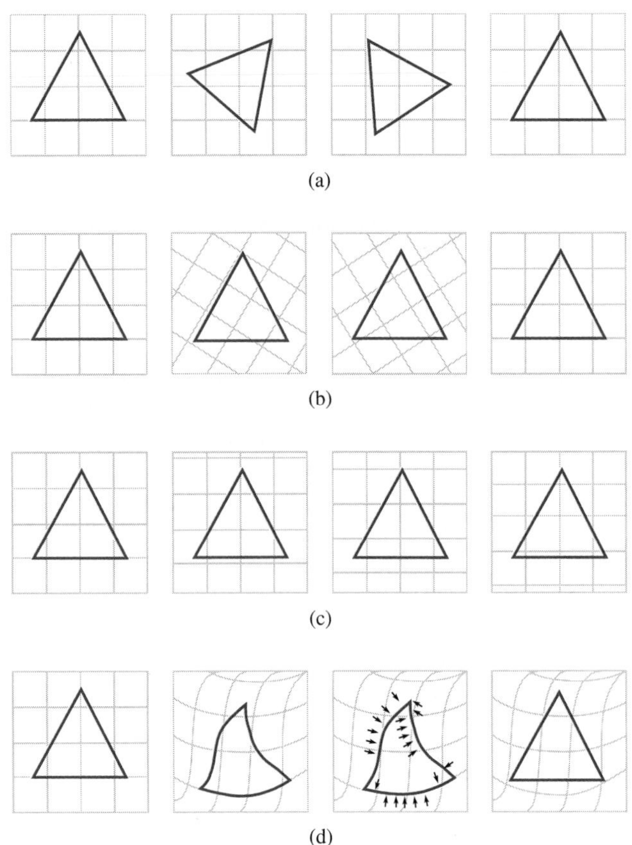

Figure 33.6 (a) The rotational symmetry of an equilateral triangle. (One side is made distinguishable so you can recognize the rotation.) (b) Rotating the entire coordinate system has no effect on the shape of the triangle. (c) Translating the entire coordinate system downward has no effect on the shape of the triangle. (d) Altering the coordinate system in a point-to-point, local way distorts the triangle. Even so, overlaying a force field, a gauge field, will reestablish the original configuration of the triangle.

final state of the system results. We've already talked about the geometrical symmetry of a snowflake (p. 7), which can be rotated through 60° and come up unaltered. Here, however, we are concerned with *abstract,* or *internal, symmetry* such that the alteration of some aspect of a system described by a set of equations does not change the solution of those equations. An equilateral triangle looks the same if we rotate it through 120° (Fig. 33.6a). The angle is continuously variable, and the symmetry operation is said to be *continuous.* By comparison, the triangle is also symmetric via a reflection through any one of its altitudes, but this is a jump, a discontinuous symmetry operation. Recall (p. 6) the discussion of **Noether's Principle**: *for every continuous symmetry, there is a corresponding conservation law and vice versa* (Fig. 33.7). Accordingly, we will only be concerned with continuous symmetries and indeed only with ones of a very restricted kind; namely, *local* rather than *global* symmetries. It is this last stricture that is the defining feature of gauge invariance.

Think of an object, a wire triangle, and imagine a mathematical coordinate grid extending in all directions against which the shape of the figure is to be described mathematically. Instead of an object, we might transform a physical quantity represented by a mathematical function (for example, the potential), but the triangle is easier to visualize. Every point on the wire is transformed to a point on the coordinate grid. Rotating the grid (Fig. 33.6b) will not alter the lengths of the sides or the angles of the projected triangle, nor will translating the grid (Fig. 33.6c). These are global transformations; they happen everywhere identically and do not affect our system (the shape of the triangle). When a feature of the system is invariant under such transformation, we say the system has a *global symmetry.*

To give the idea a more subtle application, consider yourself coasting around on a bicycle on top of a hill. It's a nice smooth mound, so the equipotential lines (the lines of equal height) are smooth concentric closed curves. If the entire hill is now transformed upward 100 m, you'll not notice a thing as you roll up and down its trails; assuming gravity is constant, the potential energy is independent of absolute height, and the forces on the bike only depend on the slopes of the surfaces. Similarly, the gravitational potential contour map, the potential field, is unchanged—this is a global symmetry.

By contrast, if the grid of Fig. 33.6d is altered independently from point to point so that it's twisted every which way, the shape of the triangle as formulated in this distorted space changes, and the symmetry is lost. But suppose we bend the wire here and there applying a force (call it a gauge force) at every point to counter the distortion and re-symmetrize this local transformation. We then have imparted **local symmetry** in the presence of an added compensating **gauge force field**. Go back up the hill on that bike. With a local height transformation that changes the potential from point to point, the hillside will be distorted, covered with humps and potholes; the contour lines will be full of wiggles. But suppose we somehow introduce a re-symmetrizing force field at every point that miraculously readjusts the net force field acting on you to its original configuration; we make a transformation of

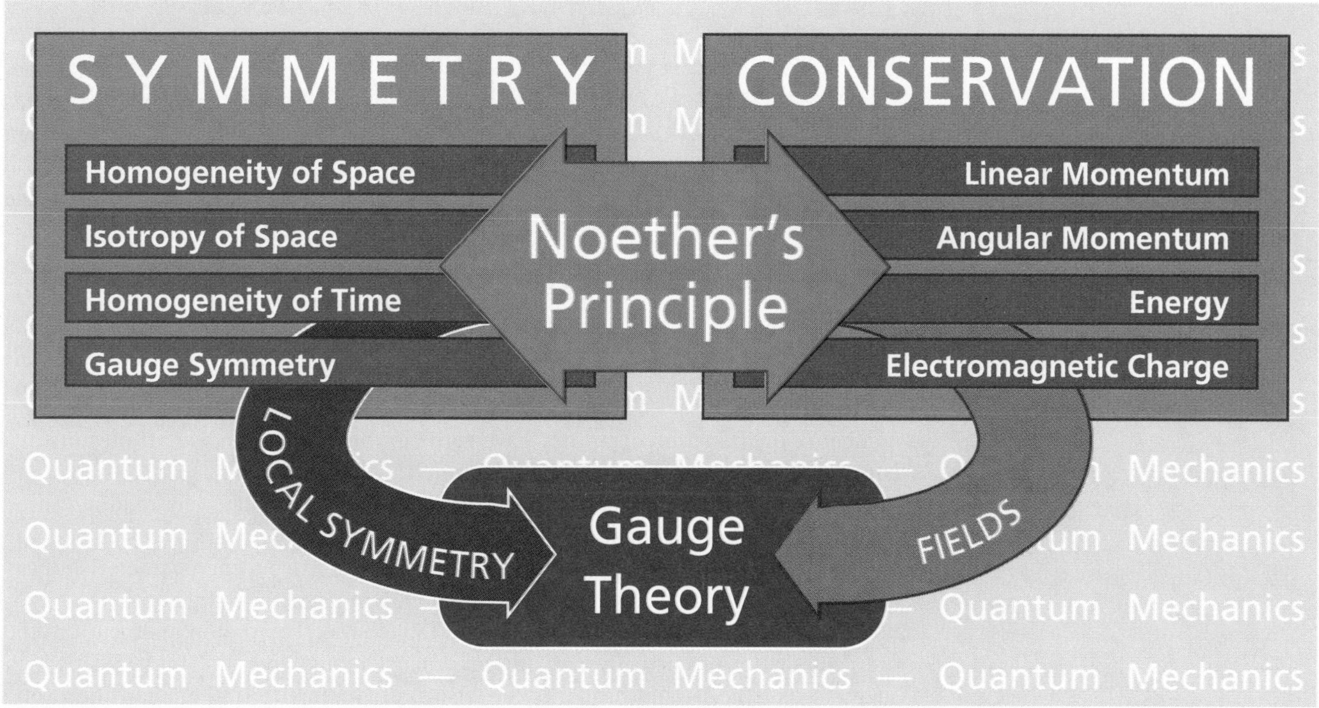

Figure 33.7 Gauge Theory is predicated on the notion of local symmetry and the concept of field.

the potential field. You would then happily bike around as if nothing had changed. The compensating gauge field establishes a local symmetry, a gauge symmetry.

Alternatively, if we place a mapmaker at "every" point on the hill and ask that they all independently establish a potential for just the one location at which each resides, we'll again get contours that are a wiggly mess—the cartographers choose the zero altitudes wherever they like, making the height of each point quite unrelated to the next. Now, let each person make up a complete map (leaving out the numerical values of the contours): they'll all be identical—the hill is the hill. If we add the proper field, the gauge field, to the wiggly-mess field, the mess will be transformed into a correct contour map of the hill. *A correct theory must be independent of all the arbitrary definitions that went into it; it must be gauge invariant.* It must be transformable from one measuring framework, or gauge, to another without affecting the conclusions of the formalism.

Classical electromagnetism is gauge invariant; Conservation of Charge is associated with symmetry in the electromagnetic potentials (p. 638), and for that symmetry to be local, one must introduce a compensating gauge field that turns out to be Maxwell's familiar electromagnetic field. The conservation of electromagnetic charge inexorably leads to classical Electromagnetic Theory. This gauge invariance carries over into the quantized domain of QED. Here, the phase of the de Broglie wave of a charged particle is made locally symmetric. To accomplish that, a new compensating field interacting with the charged particles is required, and it turns out to be the photon field. **The photon is the quantum of the gauge field associated with electromagnetic charge**.

Both the gravitational and Coulomb force fields are long-range gauge fields. These are mediated by massless gauge quanta traveling at the speed of light (gravitons and photons). Moreover, the interaction is proportional to the source's quantum number; that is, its "charge" (mass, electromagnetic charge, etc.).

33.4 Quarks

The theoretical quandary produced by the proliferation of hadrons was addressed independently in 1963 by Murray Gell-Mann and George Zweig. They proposed that all known hadrons were composite particles, each made up of a cluster of a two or more truly elementary particles. These, Gell-Mann lightheartedly called **quarks**, from the phrase "Three quarks for Muster Mark" in James Joyce's novel *Finnegans Wake*. How many quarks make up a baryon? Baryons are fermions, and if they have parts, the parts must be fermions—any cluster of bosons is a boson. Thus, all baryons must be made up of an odd number of quarks. The simplest thing is to assume quarks are spin-$\frac{1}{2}$ fermions. Accordingly, Gell-Mann and Zweig proposed that **all baryons are composed of three quarks and all antibaryons are composed of three antiquarks**. Each quark has a baryon number of $+1/3$ and each antiquark a baryon number of $-1/3$. Similarly, mesons are bosons and must have an even number of quarks. Moreover, they have zero baryon number, and so **all mesons are composed of one quark and one antiquark**.

To account for every then-known hadron, Gell-Mann and Zweig required three varieties, or **flavors**, of quark. With a whimsical flare, these quark types came to be called **up** (u), **down** (d), and **strange** (s). Their most off-putting characteristic was the requirement that quarks have fractional electromagnetic charge, either $\pm 1/3$ or $\pm 2/3$ of the fundamental electron charge (Table 33.2). The proton and neutron, the ordinary matter of our Universe, are quark clusters uud and udd (Fig. 33.8), which suggests that the u- and d-quarks are also ordinary matter rather than the exotic matter created in high-energy collisions. Table 33.3 shows the quark combinations forming a number of hadrons. Note that if a meson is composed of a quark and antiquark of the same type or flavor (as is π^0), the constituents can rapidly annihilate one another and the lifetime is very brief. **The strong interaction cannot**

TABLE 33.2 Characteristics of quarks and antiquarks

Quarks								
Flavor	**Symbol**	**Charge**	**Spin**	**Baryon number**	**Strangeness**	**Charm**	**Bottomness**	**Topness**
Up	u	$+\frac{2}{3}e$	$\frac{1}{2}\hbar$	$\frac{1}{3}$	0	0	0	0
Down	d	$-\frac{1}{3}e$	$\frac{1}{2}\hbar$	$\frac{1}{3}$	0	0	0	0
Strange	s	$-\frac{1}{3}e$	$\frac{1}{2}\hbar$	$\frac{1}{3}$	−1	0	0	0
Charmed	c	$+\frac{2}{3}e$	$\frac{1}{2}\hbar$	$\frac{1}{3}$	0	+1	0	0
Bottom	b	$-\frac{1}{3}e$	$\frac{1}{2}\hbar$	$\frac{1}{3}$	0	0	+1	0
Top	t	$+\frac{2}{3}e$	$\frac{1}{2}\hbar$	$\frac{1}{3}$	0	0	0	+1

Antiquarks								
Flavor	**Symbol**	**Charge**	**Spin**	**Baryon number**	**Strangeness**	**Charm**	**Bottomness**	**Topness**
Up	$\bar{u}$	$-\frac{2}{3}e$	$\frac{1}{2}\hbar$	$-\frac{1}{3}$	0	0	0	0
Down	$\bar{d}$	$+\frac{1}{3}e$	$\frac{1}{2}\hbar$	$-\frac{1}{3}$	0	0	0	0
Strange	$\bar{s}$	$+\frac{1}{3}e$	$\frac{1}{2}\hbar$	$-\frac{1}{3}$	+1	0	0	0
Charmed	$\bar{c}$	$-\frac{2}{3}e$	$\frac{1}{2}\hbar$	$-\frac{1}{3}$	0	−1	0	0
Bottom	$\bar{b}$	$+\frac{1}{3}e$	$\frac{1}{2}\hbar$	$-\frac{1}{3}$	0	0	−1	0
Top	$\bar{t}$	$-\frac{2}{3}e$	$\frac{1}{2}\hbar$	$-\frac{1}{3}$	0	0	0	−1

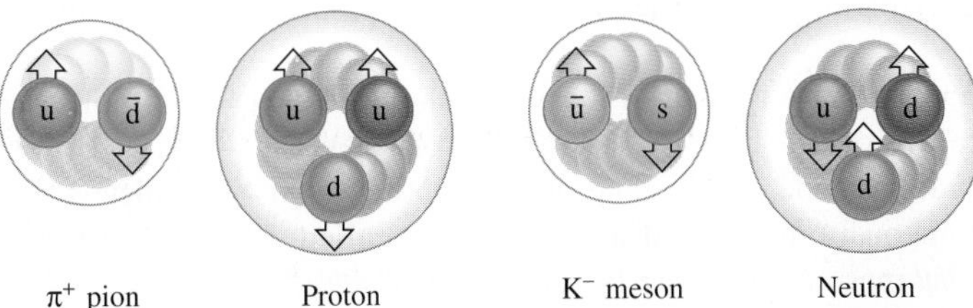

Figure 33.8 The quark composition of the pion, proton, meson, and neutron. As we'll see, each hadron also contains a swarm of gluons (p. 1150) that bind the quarks together. These gluons contribute significantly to the hadron's physical characteristics. For example, much of a neutron's or proton's spin is carried by its gluons.

transform flavor. If the flavors are different, the quark and antiquark will eventually annihilate and the meson decay, but only via the weak interaction and only relatively slowly. **The weak interaction can transform flavor**.

One might expect that blasting two protons together would produce a shower of quarks—they ought to be easy to generate and easy to identify, and yet their shyness proved a continuing embarrassment and impediment to the acceptance of the theory. No free quark has ever been observed. Accordingly, theoreticians have made a case for **quark confinement**, the notion that free quarks cannot exist, but that contention is not entirely convincing (it certainly wasn't in the early 1970s). Still, a series of experiments begun in 1969 at the Stanford Linear Accelerator Center (SLAC) in California and repeated using neutrinos at CERN gave the theory a needed boost. Probing with high-energy electrons (20 GeV), the SLAC group established that the proton and neutron were made up of three small, hard lumps of charge. The nucleon is composed of three pointlike charges that move around quite freely, like three bees inside a small balloon. But even that picture is now known to be an oversimplification—nucleons seem to be much more complicated.

A theoretical objection to the quark model was raised in the early 1960s: there are several hadrons composed of the same flavor quarks, and their existence violates the Pauli Exclusion Principle. For example, the Δ^{++} baryon corresponds to uuu, three u-quarks (three fermions in the same state is a no no no). A way out of that quark quandary developed that has since proven to be remarkably fruitful for a number of other reasons. It was proposed that each flavor of quark (at the time u, d, and s) actually came in three varieties, or **colors**: red, green, and blue. Each quark carries one of three types of *color charge*. The u-quarks in Δ^{++}—namely, $(u_R u_G u_B)$—were not really identical, and therefore there is no problem with the Exclusion Principle (Fig. 33.9).

Of course, nothing is colored in the ordinary sense of the word, but the analogy with light is convenient and makes the details easy to remember. Quarks carry color charges—they possess redness, blueness, or greenness. Antiquarks have anticolor charges of antired (think of it as cyan), antiblue (think of it as yellow), and antigreen (think of it as magenta). Red, blue, and green color are like positive electromagnetic charge; antired, antiblue, and antigreen anticolor are like negative electromagnetic charge—*it is these qualities that give rise to the strong force fields.*

Since no hadron displays an observable property that can be associated with color, it follows that color is an internal characteristic—**all hadrons are color neutral** (just as the neutron is electromagnetic-charge neutral). This means that either the total amount of each color is zero (as with mesons that are color-anticolor pairs, for example, $q_R\bar{q}_{\bar{R}}$), or that all the colors are present in equal amounts (as with baryons, $q_R q_B q_G$). The latter is analogous to the fact that red, blue, and green light beams add to make white. This scheme demands that no observable particle be composed of either two (qq) or four (qqqq) quarks—nor can free quarks exist. We might say that an atom is electromagnetically white and that an ion is colored. And because it attracts opposite charges, an ion has a higher energy and tends to be-

TABLE 33.3 The quark composition of several hadrons*

Particle	Quarks
Mesons	
π^0	$u\bar{u}$, $d\bar{d}$ mix
π^+	$u\bar{d}$
π^-	$\bar{u}d$
η	$d\bar{d}$, $u\bar{u}$ mix
η'	$s\bar{s}$
K^0	$d\bar{s}$
$\bar{K}^0$	$\bar{d}s$
K^+	$u\bar{s}$
K^-	$\bar{u}s$
J/ψ	$c\bar{c}$
Υ	$b\bar{b}$
Baryons	
p	uud
n	udd
Δ^0	udd
Δ^{++}	uuu
Δ^+	uud
Δ^-	ddd
Σ^+	uus
Σ^-	dds
Σ^0	uds
Ξ^0	uss
Ξ^-	dss
Λ^0	uds
Ω^-	sss

*Where the quark compositions are the same, their spin alignments are different.

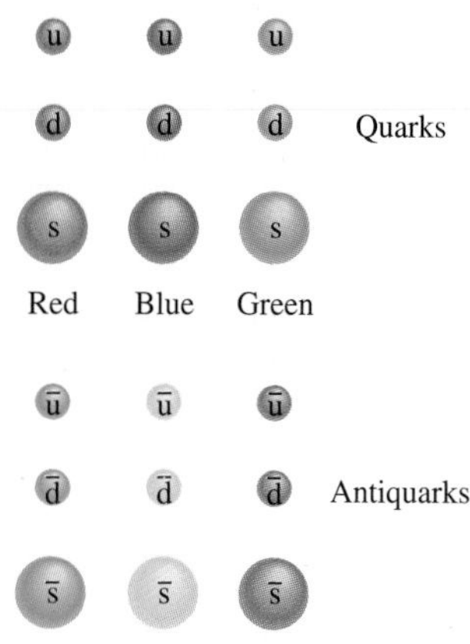

Figure 33.9 A few quarks and antiquarks, each coming in three colors or anticolors.

come white by picking up an electron. Similarly, quarks tend to form white neutral composites.

As fate would have it, quark theory, which was none too popular, was revivified by a chance discovery. In 1974, two teams of scientists—one at Brookhaven National Laboratory under C. C. Ting, and the other at SLAC under B. Richter—almost simultaneously discovered a new hadron, called J by one and ψ (psi) by the other. The J/ψ meson was three times more massive than the proton and, remarkably, lived for 10^{-20} s before decaying. That's 10^3 times longer than is normal for a hadron of that mass—there was some fundamentally new physics at work here. Moreover, the quark theory as it stood was no help; it was all filled up, there were no more particles that could be accounted for with three quarks.

The difficulty was soon settled by appealing to an idea suggested by Sheldon L. Glashow that had been around for some time but had found little support. Because there were then four known leptons, it had been proposed (on the grounds of natural symmetry among elementary particles) that there ought to be four quarks. The new addition had already been named **charm** (c), though up until the discovery of J-psi, there was no reason to take it seriously. It is now accepted that J/ψ corresponds to the two-quark bound state ($c\bar{c}$). A whole clutch of very massive charmed mesons and baryons, particles containing the heavy charm quark, have since been discovered.

This story played itself out again in 1975 when the tau (τ) lepton was discovered. Assuming there was also a τ-neutrino to be found, there were then six leptons and four quarks. By 1977, a new meson, the upsilon, was observed. It is 75 times more massive than the pion. With little hesitation, a fifth heavy quark flavor—**bottom**, or *beauty* (b)—was conjured up, and the upsilon was recognized to be ($b\bar{b}$). The mesons ($b\bar{d}$) and ($b\bar{u}$) were found in 1983. We are now patiently waiting for the discovery of the sixth heavy quark flavor, already named **top**, or *truth* (t). Experiments (1990) strongly suggest that nature probably cannot accommodate more than three kinds of neutrinos, and therefore there will be a total of six leptons and presumably no more than six quark flavors forming three generations of elementary particles (Fig. 33.10).

One might wonder what kind of interquark force is at work. Could some theoretical Coulomb-like interaction be devised that is proportional to color charge, one that would match all the diversity of strong reactions? In 1954, the first powerful step in that direction was taken by C. N. ("Frank") Yang and R. L. Mills. They produced the mathematical framework on which would ultimately rest modern Quantum Field Theory. Their work would make possible a satisfactory description in terms of the quantum of the gauge field of color: the gluon.

33.5 Quantum Chromodynamics

The interaction between two point-like quarks is mediated by the exchange of a boson messenger amusingly called the **gluon**. This is the basis of the fundamental *strong interaction,* or **color force** (Fig. 33.11). The quantum of the color field, the gluon, is a spin-1, electromagnetically neutral, massless particle referred to as a *vector boson.* Spin-1 particles are bosons that have wavefunctions in the form of a 4-dimensional vector; hence, the name. Because the quarks come in three colors and the absorption or emission of a gluon can change the quark's color (though not its flavor), it turns out that there are eight possible different couplings and Color Gauge Theory postulates eight massless gluons (Fig. 33.12). These differ significantly from the photon, which is chargeless, in that six of them carry color charge.

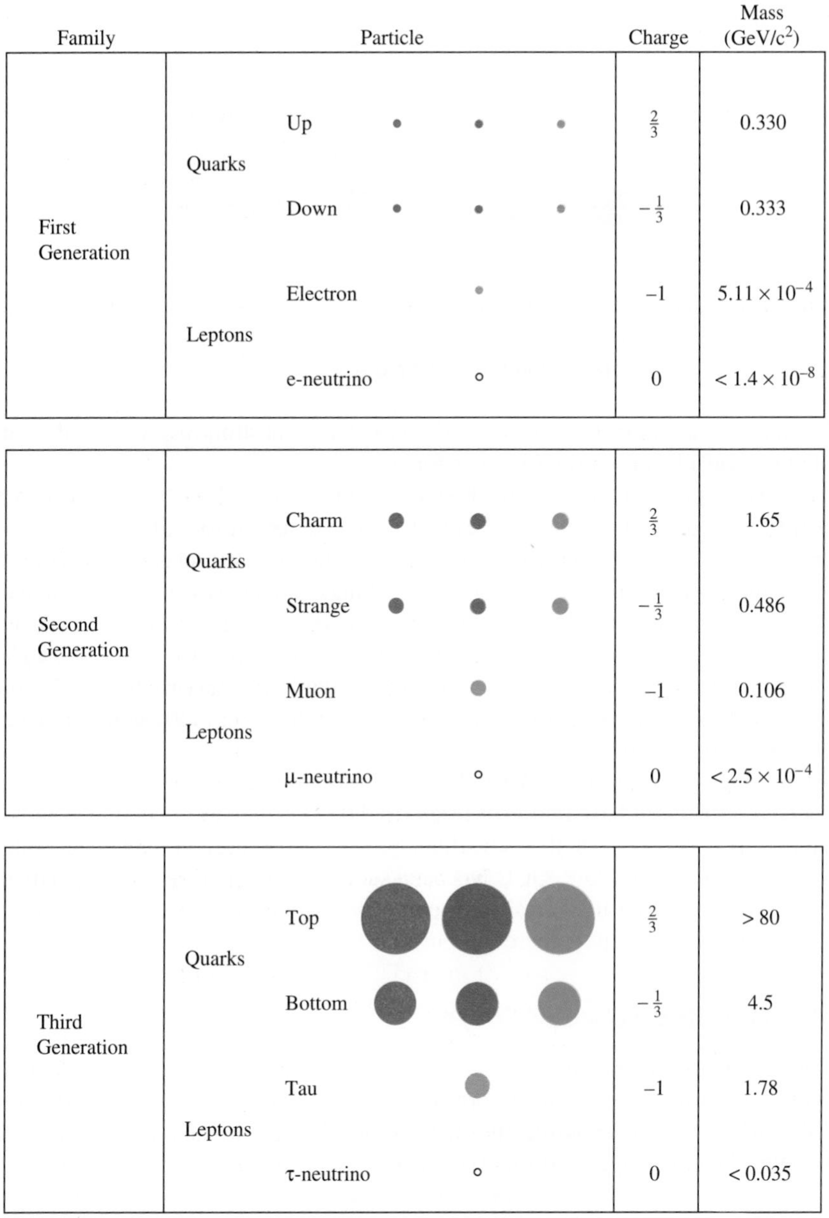

Family		Particle	Charge	Mass (GeV/c^2)
First Generation	Quarks	Up	$\frac{2}{3}$	0.330
		Down	$-\frac{1}{3}$	0.333
	Leptons	Electron	−1	5.11×10^{-4}
		e-neutrino	0	$< 1.4 \times 10^{-8}$
Second Generation	Quarks	Charm	$\frac{2}{3}$	1.65
		Strange	$-\frac{1}{3}$	0.486
	Leptons	Muon	−1	0.106
		μ-neutrino	0	$< 2.5 \times 10^{-4}$
Third Generation	Quarks	Top	$\frac{2}{3}$	> 80
		Bottom	$-\frac{1}{3}$	4.5
	Leptons	Tau	−1	1.78
		τ-neutrino	0	< 0.035

Figure 33.10 The three generations of elementary fermion (the associated antiparticles are not shown). Both the top quark and the τ-neutrino are presumed to exist, although they remain experimentally unconfirmed.

Each of these transports color and anticolor. Figure 33.13 shows how a red quark radiating a red-antigreen gluon loses red and has left behind green—it becomes charged green.

The strong interaction acts via color (for example, as with the meson depicted in Fig. 33.14). Gluon exchange also holds the hadrons together (Fig. 33.15) as composite entities (in twos and threes), but the transfer of individual gluons between separate hadrons is essentially precluded. Hadrons interacting with other hadrons experience the effects of the strong force mostly via the exchange of quark-antiquark composite particles (Yukawa's mesons). Figure 33.16 depicts the quark picture of the strong force proton-proton interaction. The creation and annihilation of quark pairs allows the transfer of a mediating pion, but the basic interaction is between quarks

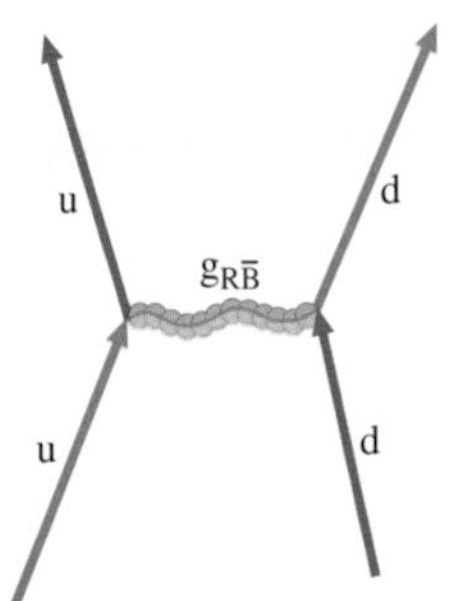

Figure 33.11 The interaction of an up-quark and a down-quark mediated by the exchange of a red-antiblue gluon.

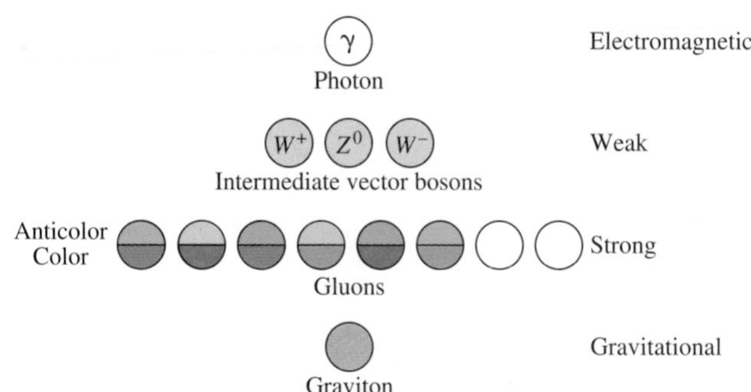

Figure 33.12 Boson force mediators.

and gluons. The strong force holding the nucleus of an atom together is the rather feeble remnant of the interquark color force.

In summary then, quarks are bound to one another (via the strong or color force) to form hadrons through the exchange of gluons. Unlike the electromagnetic force, which is communicated by chargeless photons, most gluons carry color charge, and their absorption or emission changes the color of the participating quark. The strong interaction leaves the flavor (u, d, s, c, t, b) of a quark unaffected, although its color is generally altered. Only quark-antiquark pairs of the same flavor can annihilate each other in a strong interaction (that's not the case for a weak interaction). Moreover, quarks cannot decay via the strong force, although they can via the weak force.

The gauge theory of the color field, the strong interaction, was rather colorfully christened **Quantum Chromodynamics** (QCD)—*chroma* in Greek means color—by Gell-Mann. This is a mathematically sophisticated, renormalizable gauge theory modeled after QED. To date, QCD has been successful in dealing with experimental findings, and although it's certainly incomplete, it may even be the right and true theory (or more likely, part thereof), but that remains to be seen.

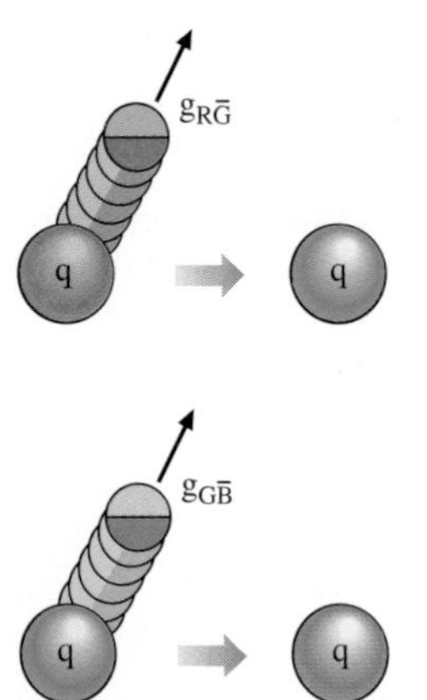

Figure 33.13 The transformation of a quark from one color to another. A $g_{R\bar{G}}$ gluon carries away red and antigreen (or magenta), leaving behind only green, and the quark becomes charged green. The emission of a $g_{G\bar{B}}$ gluon carries away green and antiblue (or yellow), leaving behind a blue quark.

33.6 The Electroweak Force

Fermi's nonrenormalizable theory of the weak interaction (1932) pictured the process of neutron decay at a single point in space-time (Fig. 33.17a, p. 1154). He had oversimplified things, avoiding the question of the carrier of his new force. Two years later, Yukawa suggested that the weak force (like the electromagnetic interaction of QED) was mediated by a massive messenger particle. After decades of neglect, the idea was revived by Julian Schwinger (1956), who attempted to describe the weak force in terms of Gauge Theory. He called the intermediary the W-particle (for *weak*). Like the photon, it had to have a spin of 1 if angular momentum was to be conserved; as a result, it came to be known as the *intermediate vector boson.* The extremely short range ($\approx$0.01 fm) of the weak force required that its mediator be quite massive and quite small.

Schwinger postulated the existence of two charged vector bosons. The neutron decays into a proton and a virtual W-particle, which must therefore carry a negative electromagnetic charge (Fig. 33.17b). The W^- then decays into an electron and an electron-antineutrino. Similarly, he argued that the creation of a positron and a neutrino must be the result of the decay of a positively electromagnetically charged W-particle, the W^+ (Fig. 33.18a). The emission or absorption of a charged intermediate vector boson by a fundamental particle results in the prompt transformation of

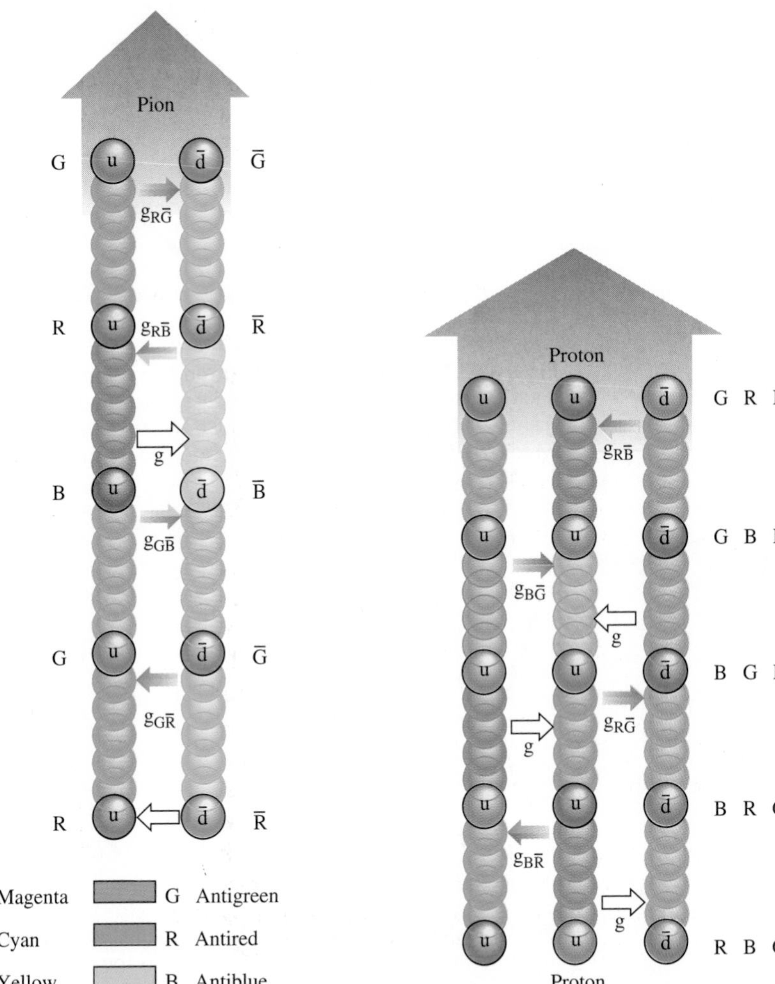

Figure 33.14 The color force. A schematic representation of the interaction between the two quarks $(u, \bar{d})$ comprising a pion. The attraction is mediated by the exchange of color-charged and uncharged gluons.

Figure 33.15 The color force. A schematic representation of the interaction between the three quarks $(u, u, \bar{d})$ comprising a proton. The attraction is mediated by the exchange of color-charged and uncharged gluons.

that particle. That's something new among the forces we have studied. The emission or absorption of a photon alters the phase of the de Broglie wave, but it doesn't transform the emitting particle. Similarly, gluon exchange alters color, not flavor, so there's no change of particle type there either.

The weak force acts differently, depending on the handedness of the participants. Thus, the effect of the weak force on an electron changes depending on the particle's motion; that is, on the alignment or antialignment of its linear and spin angular momenta (a property we shall not discuss further). The weak force operates between fermions, and so leptons and quarks (that is, left-handed particles and right-handed antiparticles) couple to it; they possess weak charge. The Ws carry both electromagnetic charge and weak charge.

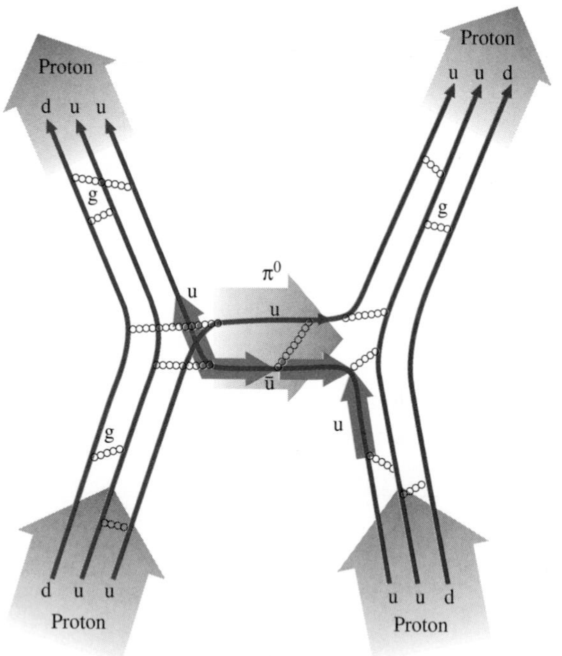

Figure 33.16 The strong interaction between protons. Two protons exchanging a neutral pion as understood via the quark model. The creation of one u-$\bar{u}$ pair and the annihilation of another allows the transfer of a π^0. Gluons exchanged between quarks hold the hadrons together.

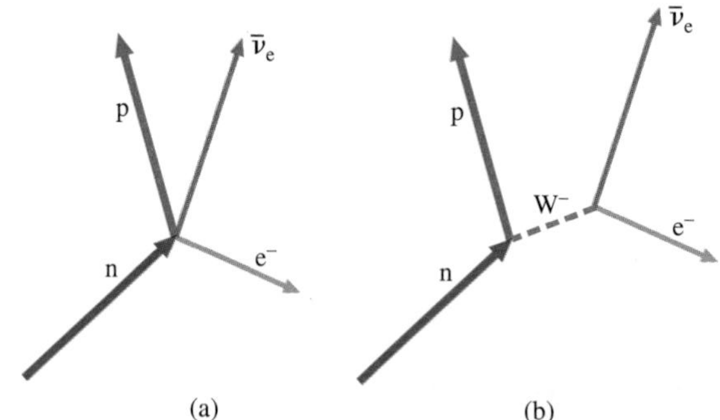

Figure 33.17 The decay of a neutron into a proton, electron, and electron-antineutrino. (a) The decay occurring at a single point. (b) The decay as mediated by a negative vector boson, which itself transforms into an electron and neutrino.

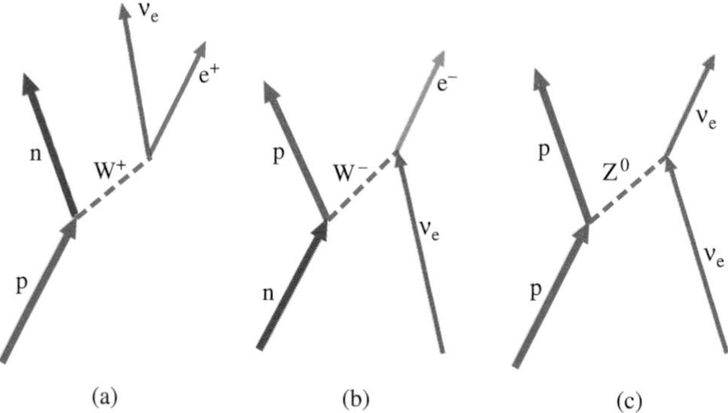

Figure 33.18 (a) Weak interactions as mediated by the three intermediate vector bosons. (b) A change in the charge of any of the particles creates a *charged current*. When, as in (c), there is no change in charge, we have a so-called *neutral current*. Note that if we slide the incoming neutrinos up so that they are created by the W^- or Z^0, they become outgoing antineutrinos.

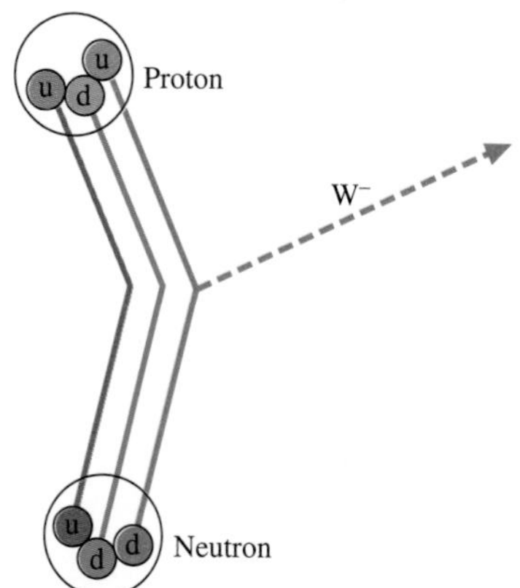

Figure 33.19 The transformation of a down-quark into an up-quark via the emission of a W^- intermediate vector boson. The result is the transformation of a neutron into a proton.

If, as a result of emitting a W^-, a neutron (udd) is transformed into a proton (udu), it must be that a d-quark is transformed into a u-quark (Fig. 33.19). The weak interaction can alter the flavor of a quark without altering its color. In fact, the emission or absorption of a $W^\pm$ results in the prompt transformation of one type of lepton (for example, e^-) into another (for example, $\bar{\nu}_e$), as shown in Fig. 33.20. *All the various nuclear decays that take place via the weak force involve quark flavor transformations.*

The fact that the W-particles are electromagnetically charged suggested to Schwinger that there is an intimate relationship between the weak and electromagnetic interactions; the W-particles carry both weak charge and electromagnetic charge. In the late 1950s, Schwinger turned to other concerns, but before he did, he asked one of his Ph.D. students, "Shelly" Glashow, to think about a possible connection between the weak and electromagnetic interactions. Years later (1961), Glashow published a seminal paper introducing a third messenger, the neutral Z boson, but few people paid it much attention. Like the $W^\pm$, these messengers were quite massive and yet gauge bosons presumably had to be massless—that, along with the fact that Glashow's model was nonrenormalizable, left much to be desired. Amazingly, Steven Weinberg, an old high school rival of Glashow, was independently working on the same problem at the Massachusetts Institute of Technology. He devised a scheme (1967) by which the vector bosons could take on mass and still fit into a gauge theory but again, no one, not even Glashow, seemed to notice it. A year later, Abdus Salam published a very similar unifying treatment that met with the same neglect. In 1969, a young graduate student in Holland, Gerhard 't Hooft, demonstrated how Yang-Mills gauge fields (p. 1150) could be made renormalizable, and the whole area of study came alive. The Glashow-Weinberg-Salam synthesis, the **Electroweak Theory**, won these three physicists the Nobel Prize in 1979. They had devised a quantum gauge theory that naturally resulted in four gauge quanta: three massive vector bosons, W^-, W^+, and a new neutral one, Z^0 (zee-zero), along with a fourth massless force-carrier, the familiar, though no less mysterious, photon (Table 33.4).

Gauge quanta are supposed to be massless. How did the theory produce massive gauge bosons? How did the symmetry demanded by Gauge Theory get suspended, or "broken"? The idea of a symmetry spontaneously being disrupted or broken is not new. Think of a sample of ferromagnetic material (p. 728) above its Curie temperature; its atoms have spins pointing every which way. A microscopic observer would encounter magnetic uniformity in all directions. Indeed, Maxwell's Equations are symmetric in space; there are no preferred directions for the theory. Yet, if the sample is cooled, that symmetry is spontaneously broken as the atoms align themselves arbitrarily into tiny domains—the energy of an aligned set of interacting magnets is less than the energy of a random grouping. The system (without any external asymmetrical influences) spontaneously breaks the symmetry of the laws of electromagnetism without violating those laws. In a quantum-mechanical Universe, where subatomic particles shift around constantly, this sort of energy-driven descent to the ground state is quite reasonable.

The W- and Z-particles "should" be massless like the photon. According to Electroweak Theory, they acquire their mass as a result of the *ad hoc* presence of an additional field known (after the person who first studied it) as a *Higgs field.* The latter is a spinless, directionless scalar field (as yet undetected experimentally). Space, at the low temperature at which it exists today, is permeated by the Higgs

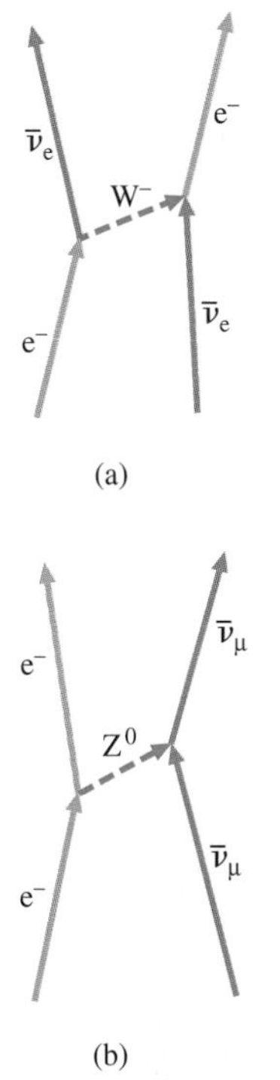

Figure 33.20 (a) An electron scattering with an electron-antineutrino via the weak interaction mediated by a W^- boson. The W carries charge, and we see an electron transformed into a neutrino. (b) An e^- and $\bar{\nu}_\mu$ come in and go out, but now the scattering is mediated by a Z^0 boson. The Z^0 is neutral and there is no change in electromagnetic charge. Note how the interaction resembles that mediated by a photon.

TABLE 33.4 The Four Forces of nature*

Force	Applicable to	Strength[1]	Effective range	Mediator
Gravity	All leptons and hadrons[2]	6×10^{-39}	Unlimited	Graviton
Weak	All leptons and hadrons	10^{-5}	$\approx 10^{-17}$ m	$W^\pm$ and Z^0
Electromagnetic	Charged leptons and hadrons	1/137	Unlimited	Photon γ
Strong	All hadrons	1	$\approx 10^{-15}$ m	Gluons g

*At the temperatures that exist today, we see four distinct interactions. At much higher energies, these blend together, presumably becoming one.

[1] Here is a slightly different scheme from the one we saw before for comparing the strengths relative to the strong force as 1.

[2] Gravity and the strong force both act on their own field quanta.

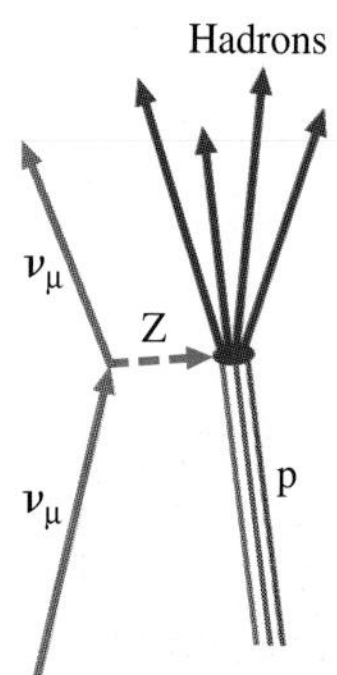

Figure 33.21 Inelastic scattering corresponding to a neutral current.

field, which drags on the W- and Z-particles like cold, thick honey. In other words, vast numbers of hypothetical spin-0 Higgs bosons condense in the ground state, producing a classical field that acts on the Ws and Zs. No longer able to travel through space at lightspeed, the W and Z bosons behave like massive particles compared to the photon, which doesn't interact with the Higgs "honey" field. At high temperatures (that is to say, at high energies), the Higgs "honey" thins out and does not affect the W- and Z-particles. They supposedly become massless (at about 1000 GeV), thereby unifying in a blazing bliss with the ever-constant photon. The weak and the electromagnetic interactions are one and the same force, mediated by the exchange of these four field quanta.

All of this cannot help but seem bizarre, and yet the Electroweak Theory made several striking predictions that were subsequently confirmed in every regard. First was the existence of the *neutral-current process,* which is a weak interaction without any resulting change in electromagnetic charge. For example, the theory proposed that a very energetic neutrino might collide with a proton. Much of the neutrino's energy ($\approx$100 GeV) could be given up to the creation of a neutral Z that could pass over to a quark in the proton (Fig. 33.21). The neutrino could scatter off physically unaltered while the energy transformed into a burst of quark-antiquark pairs that would form a spray of hadrons with a net electromagnetic charge of plus one. That bit of high-energy alchemy was confirmed at CERN and later at Fermilab in 1973.

Second, the theory predicted that the masses of the intermediate vector bosons would be 80 GeV/c^2 for $W^\pm$ and 90 GeV/c^2 for Z. These values were way beyond the energy of any existing accelerators, and the direct observation of the real incarnations of these bosons had to wait for the construction of the proton-antiproton collider at CERN. Finally, in 1983, a team of 130 physicists led by Carlo Rubbia and Simon Van der Meer (both of whom shared the 1984 Nobel Prize for their work) succeeded in producing and detecting the intermediate vector bosons. One out of roughly five million p-$\bar{\text{p}}$ collisions fused together a quark from a proton and an antiquark from an antiproton, creating a vector boson that then disintegrated in less than 10^{-24} s. From the tracks made by the debris, the researchers could unambiguously identify the bosons. The masses determined for $W^\pm$ (81 GeV/c^2) and Z^0 (91 GeV/c^2) were in splendid accord with the predictions.

The Higgs boson is like a toilet. Everyone has one, but it's not necessarily an object of pride and beauty.

SHELDON LEE GLASHOW (1984)

A recent test of the Electroweak Theory used laser beams to detect tiny distortions of heavy atoms such as cesium. The weak force between the nucleus and the orbital electrons of an atom has a minute but measurable effect. To date, all the results are in excellent agreement with theory. Still, the whole idea of the Higgs mechanism (which seems inelegant, but necessary) remains to be resolved, and until it is, few physicists will be completely content with the existing formalism.

33.7 GUTs and Beyond: The Creation of the Universe

Just as the electric and magnetic forces were unified by Maxwell's Theory into a single electromagnetic field, contemporary theory has produced a two-fold unification, a merger of the weak and the electromagnetic fields into a single **electroweak field**. What we see as the separate weak and electromagnetic interactions is a result of the cool environment of today's Universe—the unity is hidden. Only in the largest machines ever made can we even begin to simulate the inferno of primordial creation. Our Universe, space and time, began some seventeen thousand million

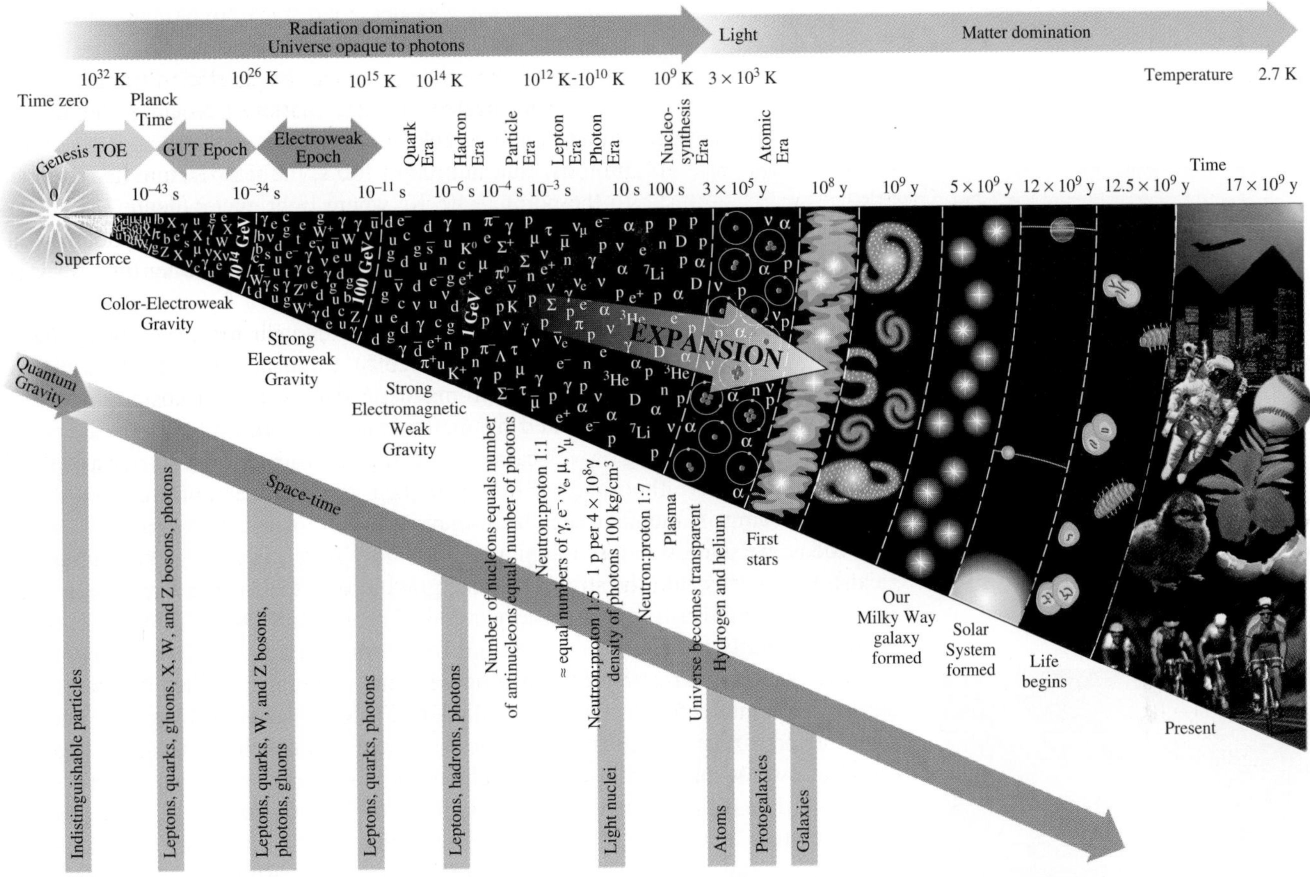

Figure 33.22 The creation of the Universe.

years ago. The interval from $\approx 10^{-34}$ s to $\approx 10^{-11}$ s after the initiation of Genesis (Time Zero) can be called the *electroweak epoch* (Fig. 33.22). Ending at an incredible temperature of around 3×10^{15} K, all there was was a dense seething soup of leptons, quarks, and massless gluons, photons, Ws and Zs. During the epoch (at temperatures above 3×10^{15} K), the weak force and the electromagnetic force shared the same inverse-square behavior and were of equal strength (Fig. 33.23). As the Universe cooled further, the electroweak symmetry shattered, Ws and Zs took on mass and vanished, and the cosmos became ruled by the Four Forces we know.

As the Universe continued to expand, it cooled to around 10^{14} K in less than its first microsecond. Free quarks and gluons vanished, coming together to form all the hadrons physicists have worked so hard to recreate. Then the heavy hadrons also annihilated, leaving (at $\approx 10^{-4}$ s) a dense quantum brew of leptons and light hadrons (protons, neutrons, pions, and so on). Neutrons and protons in almost equal number were transmuted into one another via weak interactions: $p + e^- \leftrightarrow n + \nu$ and $n + e^+ \leftrightarrow p + \bar{\nu}$. Because neutrons are heavier and require more energy to create, more protons resulted as the Universe cooled. After 1 s, the temperature was $\approx 10^{10}$ K. The unstable heavy leptons, the muons and taus, disappeared leaving neutrons, protons, electrons, positrons, and neutrinos—by now, there were five protons to every neutron.

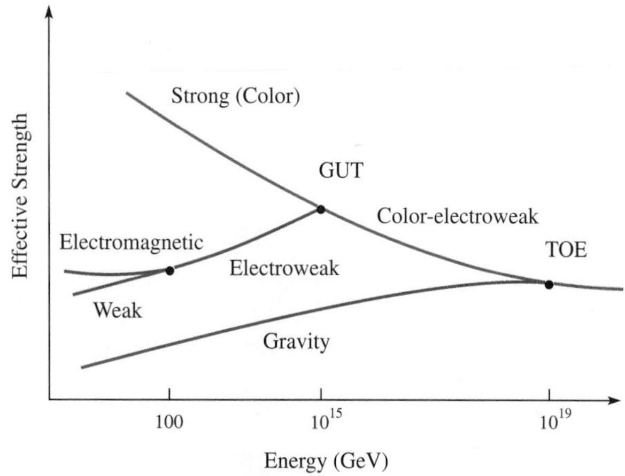

Figure 33.23 The Four Forces coalesce as the energy (or temperature) of the interacting particles increases and the closer together they can approach.

The world was pervaded by a fireball of radiant energy; photons ruled the cosmos for the next $\approx$100 s. The Universe continued to expand, the particles separated; collisions and nucleon transformations ceased. After several minutes, appreciable numbers of free neutrons decayed into protons. But, at around 100 s, the temperature had dropped to the point ($\approx 10^9$ K) where light nuclei (deuterons and α-particles) could form via fusion and remain bound—all was plasma. Alpha production stopped when the available neutrons were locked away. After a few hundred thousand years, the Universe cooled enough for atoms of hydrogen (75%) and helium (25%) to form without being blasted apart by electromagnetic radiation. The cosmos was then dominated by matter, and in a billion years or so, protogalaxies came into being, first-generation stars flashed on, lived, and died. Five thousand million years ago, the Sun was born, and the Earth came into being. A moment later life began, and you turned the page.

Today, we understand all the marvels of nature in terms of three primary gauge fields: the electroweak, the strong, and the gravitational fields. The first two, treated quantum mechanically via the Electroweak Theory and QCD, form the basis of what is generally referred to as the **Standard Model**. Driven by the aesthetic of *unification,* gauge theories have been proposed that attempt to unite the strong and electroweak interactions. These so-called **GUT**, or Grand Unification Theories, seek to find a relationship that will allow a threefold unification bringing together the electroweak gauge quanta and the gluons. The six quark flavors and the six lepton types then dissolve into a single lepton-quark fermion field.

One such renormalizable GUT supposes that quarks and leptons brought together at a distance of around 10^{-31} m will transform one into the other. The gauge bosons mediating such transformations, as well as transformations among quarks, are known as X particles. They would have to be incredibly massive, perhaps 10^{15} times the mass of the proton ($\approx 10^{15}$ GeV/c^2)—well beyond our laboratory capabilities for the foreseeable future.

A 60 × 70 × 80-ft tank containing 8000 tons of ultra-pure water (a rich supply of protons) forms the body of a giant particle detector. Here a diver inspects the photomultiplier tubes. Although designed to detect proton-decay debris predicted by GUTs, in February of 1987 it recorded a flux of neutrinos from supernova 1987A.

A prominent prediction of the simplest form of GUT is the ultimate decay of the proton. In the unlikely event that two quarks inside a proton happen to come closer than 10^{-31} m, they might exchange an X gauge boson, converting the proton into a positron and a pion. Rather than being timeless, the theoretical proton lifespan is $\approx 10^{31}$ years (which, albeit finite, exceeds the age of the Universe by a factor of 10^{21}). To date, despite considerable experimental effort, no sign of proton decay has been observed. The proton's minimum life expectancy is now set at 10^{32} years, and climbing.

The GUT epoch of the Universe lasted over the miniscule interval from $\approx 10^{-43}$ s to $\approx 10^{-34}$ s after Creation. Within the continuum of space-time, a dense mass of leptons, quarks, and gauge bosons interacted, paying no attention to color, flavor, or electromagnetic charge. Only two forces operated: the color-electroweak and gravity. Unlike the Standard Model, GUTs predict that neutrinos have mass, and here experiments are as yet inconclusive. Studies of supernova 1987A (p. 91) suggest that if the electron-neutrino has any mass at all, it's less than a mere 16 eV. Though certainly unconfirmed, GUTs nonetheless remain compelling.

The Universe came into existence as simple as it could be, becoming increasingly diverse only as it cooled enough for complexity to crystallize—for symmetries to spontaneously break. Already by $\approx 10^{-43}$ s beyond Time Zero, gravity had settled out from the once single *superforce* that had ruled from the beginning. Even more ambitious than GUTs are the attempts to bring a quantum theory of gravity (with its mediator, the graviton) into a fourfold unification. That ultimate goal is amusingly called the *Theory of Everything* (TOE). It is estimated that the effects of quantum gravity will become significant at a particle separation of 10^{-35} m, which corresponds to a phenomenal energy of 10^{19} GeV. This value is totally beyond direct experimental observation—our most powerful machines will not likely exceed 10^5 GeV for decades. Since gravity acts on both fermions and bosons, it would seem that the TOE should account for the transformation of these two particle types into one another.

During the TOE epoch prior to $\approx 10^{-43}$ s, the entire Universe was subnuclear in size and surely quantum-mechanical in disposition. The ambient temperature ($\approx 10^{32}$ K) and density ($\approx 10^{92}$ times greater than water) were unimaginable. Every point in the inferno could spontaneously become a Black Hole dropping through the fabric of the cosmos and then evaporating back. That disconnected seething foam of space and time was quantized. It is here in the primal whirlwind, even before spacetime congealed into a gravitational continuum, that our physics is most challenged. It is here, in the distant reaches of imagination, that we are most bold. Yet who could doubt that the whirlwind was a gauge whirlwind?

All there is today, stars, wind, baseballs, lovers, all there is, was there in the foam of beginning. Shattered symmetries, epoch upon epoch, fanned out diversity. Quarkstuff, you and I, for a moment contemplating Creation—how naive, how wonderful.

Core Material

Leptons are the particles that couple to the weak force and, if they are electrically charged, to the electromagnetic force, but leptons *are immune to the strong force.* **Hadrons** are the strongly interacting composite particles that form two distinct subgroups: *mesons* and *baryons.* Mesons are bosons; baryons are fermions (p. 1140). On a nonfundamental level, the nuclear force arises from the exchange of mesons (p. 1144). With a mass m, these have a maximum range of

$$R \approx c\Delta t \approx c\left(\frac{\frac{1}{2}\hbar}{m_\pi c^2}\right) \approx \frac{h}{4\pi m_\pi c}$$

All field theories must possess a particular type of mathematical structure known as gauge symmetry (p. 1145).

Baryons are composed of three quarks; antibaryons are composed of three antiquarks. Mesons are composed of one quark and one antiquark. There are six flavors of quark: *up* (u), *down* (d), *strange* (s), *charm* (c), *bottom* (b), and *top* (t) (p. 1148). Each flavor comes in three varieties or *colors*—red, blue, or green. Antiquarks have anticolor charges of antired, antiblue, and antigreen. Hadrons are color neutral (p. 1149). The interaction between quarks is mediated by the exchange of eight massless gluons (p. 1150). The gauge theory of the color field is called Quantum Chromodynamics (QCD).

The weak interaction is mediated by three *intermediate vector bosons:* W^+, W^-, and Z^0 (p. 1152). The weak interaction can alter the flavor of a quark without altering its color. *Electroweak Theory* combines the electromagnetic and weak forces (p. 1156). GUTs, or Grand Unification Theories, seek a threefold unification, bringing together the electroweak gauge quanta and the gluons.

Suggestions on Problem Solving

1. Some useful numerical quantities for this chapter:

$$1 \text{ eV} = 1.602\,177 \times 10^{-19} \text{ J}$$

$$hc = 1.239\,842 \times 10^{-6} \text{ eV}\cdot\text{m}$$

$$k_B = 1.380\,66 \times 10^{-23} \text{ J/K} = 8.6174 \times 10^{-5} \text{ eV/K}$$

2. How would you go about determining if the reaction

$$\mu^+ + \nu_\mu \rightarrow \pi^+$$

is possible? First, you might check the electromagnetic charge; $+1 + 0 \rightarrow +1$ (it's okay). Then check the spin; $\pm\frac{1}{2} + \pm\frac{1}{2} \rightarrow 0$

(it's okay). Then check the baryon number; $0 \rightarrow 0$ (it's okay). Then check the lepton number. There are only μ-type leptons; hence, on the left, we have -1 for the μ^+ antimuon and $+1$ for the μ-neutrino, yielding a total of zero on both sides. These are not strange particles, so we needn't worry about S. Finally, checking the energy, the muon has less mass than the pion, and the neutrino can balance the formula by supplying the energy difference. Conclusion: the reaction is allowed. Moreover, the inverse reaction

$$\pi^+ \rightarrow \mu^+ + \nu_\mu$$

must also be allowed.

3. Notice that adding the same particle or antiparticle to both sides doesn't change any of the requirements and must result in allowed processes as well. Thus, adding an antiparticle to both sides, we see that the reaction

$$\pi^+ + \bar{\nu}_\mu \rightarrow \mu^+ + \nu_\mu + \bar{\nu}_\mu$$

results in

$$\pi^+ + \bar{\nu}_\mu \rightarrow \mu^+$$

which is fine, provided the μ has the right amount of KE. This is equivalent to the notion that *carrying a particle from one side of the formula to the other changes it to an antiparticle,* and vice versa. And *that's equivalent in a Feynman diagram to changing an incoming particle to an outgoing antiparticle,* or vice versa.

4. Remember that *the production of strange particles via the strong force conserves strangeness;* only weak-force decays do not conserve strangeness.

Discussion Questions

1. Figure Q1a is a drawing of the Stanford Linear Collider (SLC) showing its three-kilometer straightaway. The cathode fires two successive bunches of electrons. The damping rings condense the beams, which are focused down to a diameter of a few millionths of a meter later on. Discuss, in general terms, what this 100-GeV machine does and how it does it. What might this process have to do with the Z^0 intermediate vector boson? These are the heaviest known real particles and are of great interest because, among other reasons, they have many possible decay modes. Figure Q1b depicts the simplest linac, the drift-tube accelerator. How does it work? (Hint: The particles spend the same amount of time drifting along inside each cylinder.)

2. In many respects, the Z^0 behaves like a heavy photon. Discuss this notion, using the diagrams of Fig. Q2. Here, the generic label *quark* or *lepton* means that any such particle coming in also sails out. Explain what's happening in each illustration. An interaction between quarks can take place via Zs or Ws. How do these differ?

3. The μ^- and μ^+ are antiparticles of each other, and because the μ^- decays to an e^-, we take it as the matter and the other as the antimatter. By contrast, π mesons are not so easy to categorize; π^0 seems to be its own antiparticle, but the π^+ and π^- always decay into a lepton and an antilepton. Discuss this situation as it relates to the quark picture. Is the classification of matter and antimatter unambiguous? Does it matter which pion we call the antiparticle?

4. It is known that a Feynman diagram can be turned on its side to yield an entirely new physical insight, provided that particles moving backward in time are interpreted as antiparticles. Is this attribute true of Fig. Q2? Explain.

5. Every hadron interaction must be able to be analyzed in terms of the quark model. Accordingly, explain in detail how the reaction

$$\pi^- + p \rightarrow \Lambda^0 + K^0$$

takes place.

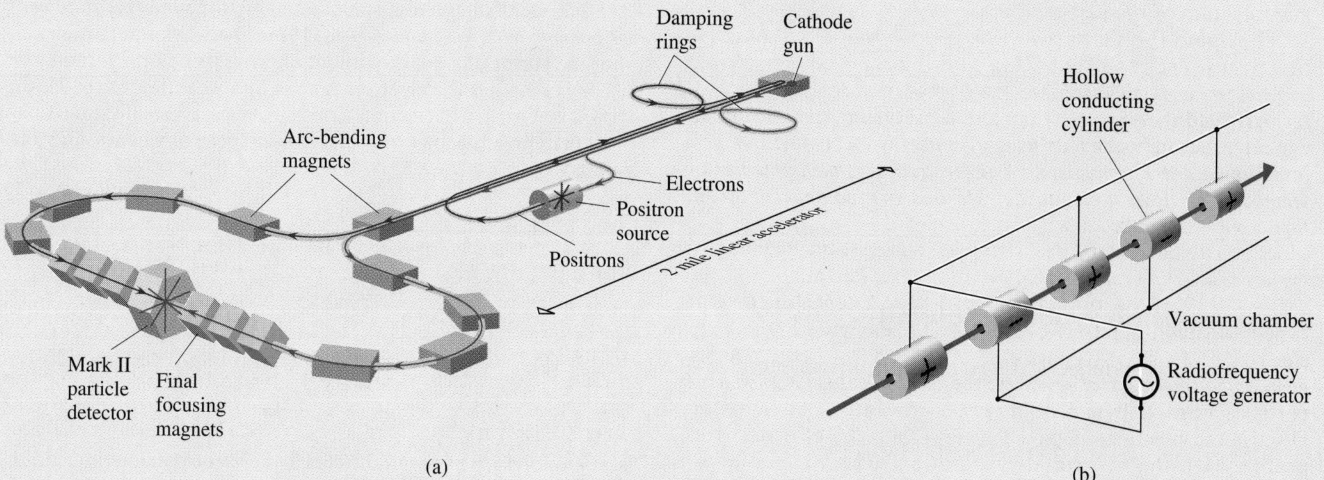

Figure Q1

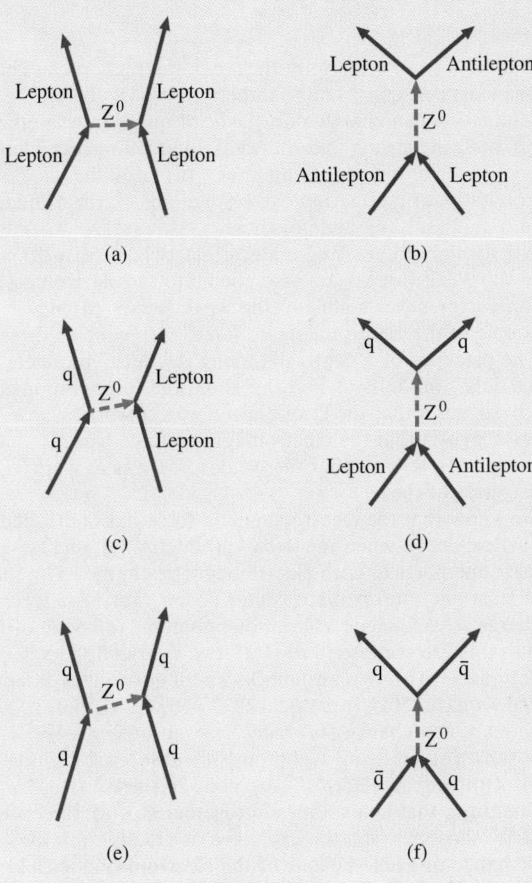

Figure Q2

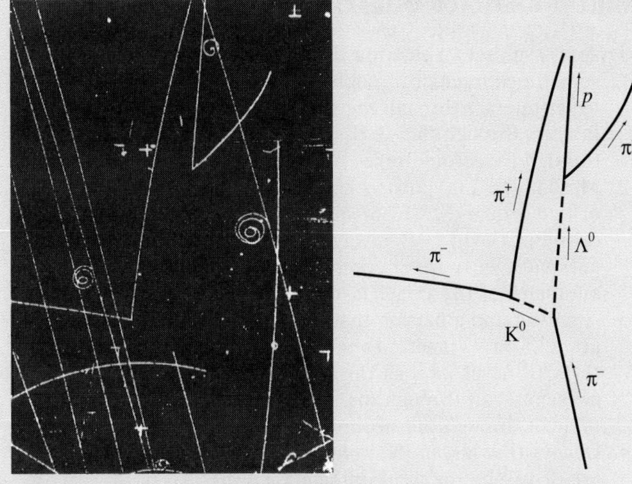

Figure Q7

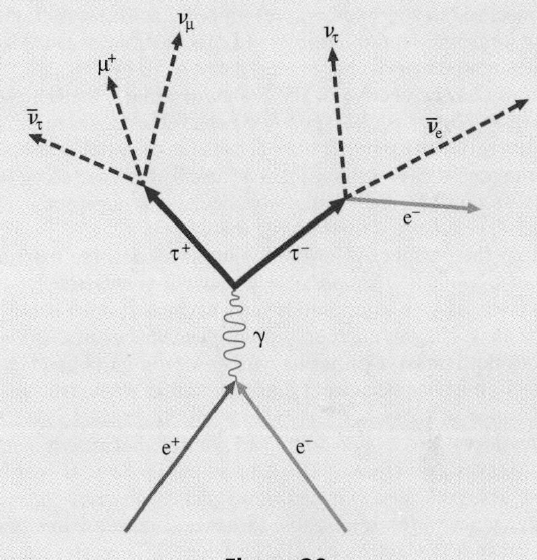

Figure Q9

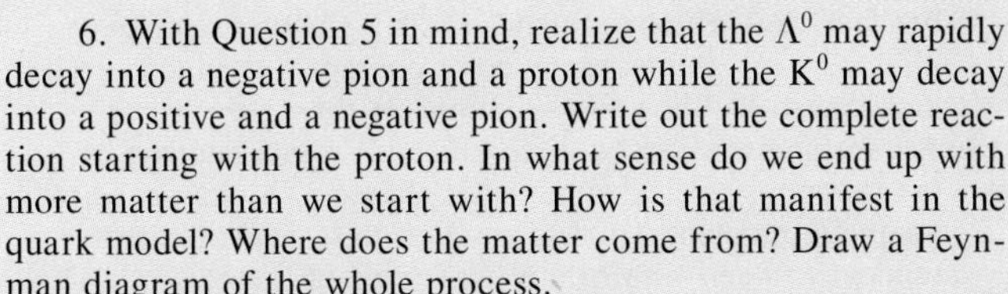

6. With Question 5 in mind, realize that the Λ^0 may rapidly decay into a negative pion and a proton while the K^0 may decay into a positive and a negative pion. Write out the complete reaction starting with the proton. In what sense do we end up with more matter than we start with? How is that manifest in the quark model? Where does the matter come from? Draw a Feynman diagram of the whole process.

7. Figure Q7 shows a bubble chamber photo of some particle tracks. The chamber is filled with supercooled liquid hydrogen. When a particle ionizes the hydrogen atoms, tiny bubbles form, revealing the path. Compare the photo and the sketch, and explain what happened. Why did a few of the participants remain invisible? The tracks are bent by a known magnetic field—what can we tell from the radius of curvature of each track? Write out a formula for what took place.

8. Is it correct to say that *every observable subnuclear event occurs via creation and annihilation?* Explain.

9. In 1975, Martin Perl reported the existence of some unexpected muon-electron events associated with high-energy electron-positron collisions: $e^+ + e^- \rightarrow e^- + \mu^+ + ?$. He suggested that the observations were the result of the decay of a new particle-antiparticle pair of very massive leptons he called *tau*. Discuss the reactions depicted in Fig. Q9. The effect first appeared when the two colliding particles had a total combined energy of about 3.6 GeV. Why is that significant?

10. Neutrons decay when they are free—why might they be stable when bound to protons in a nucleus? (Hint: Consider the situation in terms of energy.) In light of GUTs, might they be less than totally stable, even in the nucleus?

11. The idea of baryon number as a conserved quantity was introduced by Weyl and Stückelberg (1930) to supposedly explain the proton's stability. Let's reconsider the notion. First, does it *explain* anything? Discuss the nature of the "law"—is it arbitrary? Is it applicable beyond particle physics, as is, for example, Conservation of Angular Momentum? What does the fact that the Universe contains far more protons than antiprotons suggest? How does the probable electrical neutrality of the Universe fit into this discussion? What about lepton number? What do GUTs say about all of this?

Multiple Choice Questions

1. A proton and an electron that are close together (a) will sometimes interact through the strong force (b) will never interact through the gravity force (c) will never interact through the weak force (d) will never interact through the strong force (e) none of these.
2. Mesons are (a) always composed of either two u-quarks or two d-quarks (b) always composed of two quarks of different flavor (c) always composed of a quark and an antiquark (d) never composed of a quark and an antiquark of the same flavor (e) none of these.
3. A meson and a baryon that come very close together (a) will only sometimes interact through the gravity force (b) will never interact through the gravity force (c) will never interact through the weak force (d) will never interact through the strong force (e) none of these.
4. Decays that last in the range from 10^{-20} s to 10^{-23} s are (a) driven by the weak interaction (b) driven by gravity (c) driven by the strong force (d) driven by the electromagnetic force (e) none of these.
5. The reaction $e^- + p \rightarrow n + \bar{\nu}_e$ is (a) possible because it conserves everything (b) impossible because it does not conserve baryon number (c) impossible because it does not conserve lepton number (d) impossible because it does not conserve charge (e) none of these.
6. From the perspective of the Standard Model, the reaction $p \rightarrow \pi^+ + \pi^0$ is (a) possible because it conserves everything (b) impossible because it does not conserve strangeness (c) impossible because it does not conserve lepton number (d) impossible because it does not conserve charge (e) none of these.
7. From the perspective of the Standard Model, the reaction $p \rightarrow e^+ + \gamma$ is (a) possible because it conserves everything (b) impossible only because it does not conserve baryon number (c) impossible because it does not conserve either baryon or lepton number (d) impossible because it does not conserve charge (e) none of these.
8. The decay $K^+ \rightarrow \gamma + \gamma$ is (a) possible because it conserves everything (b) impossible for several reasons, one being because it is electromagnetic and must conserve strangeness (c) impossible for several reasons, one being because it is electromagnetic and does not conserve meson number (d) impossible for several reasons, one being because it does not conserve baryon number (e) none of these.
9. The reaction $\nu_e + n \rightarrow e^- + p$ is (a) possible because it conserves everything (b) impossible because it does not conserve baryon number (c) impossible because it does not conserve lepton number (d) impossible because it does not conserve charge (e) none of these.
10. Which is true? (a) only the weak interaction can change one flavor of quark into another (b) only the strong interaction can change one flavor of quark into another (c) both the strong and the weak interactions can change one flavor of quark into another (d) only the gravitational interaction can change one flavor of quark into another (e) none of these.
11. All strongly interacting material particles (a) also couple to the weak force (b) also couple to the electromagnetic force (c) never couple to the weak force (d) never couple to the electromagnetic force (e) none of these.
12. The presence of a neutrino among the decay products of a particle (a) tells us nothing about the force responsible (b) suggests that the strong force was at work (c) suggests that the electromagnetic force was at work (d) tells us that the weak force was at work (e) none of these.
13. We know that the electromagnetic force has controlled a particle decay when the decay products (a) include at least one particle with electromagnetic charge (b) include at least one intermediate vector boson (c) have a zero net charge (d) include at least one photon (e) none of these.
14. Baryons are composed of (a) two quarks and an antiquark (b) two antiquarks and a quark (c) a lepton and a quark (d) three quarks (e) none of these.
15. A quark can change flavor by (a) absorbing a W boson (b) emitting a photon (c) absorbing a gluon (d) emitting an electron (e) none of these.
16. The thing that holds a meson together is (a) the exchange of W^- bosons (b) the exchange of gluons (c) the exchange of gauge bosons of the electromagnetic field (d) the absorption and emission of quarks (e) none of these.
17. Gluons can attract each other to form so-called *glueballs* because (a) gluons are sticky (b) gluons exchange photons (c) gluons carry color charge (d) gluons emit and absorb $W^\pm$ bosons (e) none of these.
18. An antineutron can be distinguished from a neutron by (a) its opposite electromagnetic charge (b) its opposite magnetic moment (c) its smaller mass (d) its greater tendency to self-annihilate (e) none of these.
19. Prior to 10^{-43} s after Time Zero, we believe that the Universe (a) was the size of the Moon (b) was ruled by a single superforce (c) was in the so-called GUT era (d) did not exist (e) none of these.
20. We believe that a correct fundamental field theory must (a) be simple (b) predict that the proton is unstable (c) have global symmetry (d) be gauge invariant (e) none of these.

Problems

ELEMENTARY PARTICLES
QUANTUM FIELD THEORY

1. [I] Knowing that Δ^* is a baryon with spin $\frac{3}{2}$ and 0 strangeness and that η and η' are mesons with 0 spin, 0 electromagnetic charge, and 0 strangeness, indicate which forces determine the following reactions: (a) $\eta \rightarrow \gamma + \gamma$ (b) $\Delta^* \rightarrow p + \pi$ (c) $\Lambda^0 \rightarrow \pi^- + p$ (d) $\eta' \rightarrow \eta + \pi + \pi$. Explain your answers fully.

2. [I] Determine the forces that are responsible for each of the following reactions: (a) $\pi^+ \rightarrow \mu^+ + \nu_\mu$

(b) $\pi^0 \rightarrow \gamma + \gamma$ (c) $\pi^- + p \rightarrow K^0 + \Lambda^0$. Explain your answers fully.

3. [I] State some of the conservation laws that are violated in each of the following reactions:
(a) $\Lambda^0 \rightarrow \pi^+ + \pi^- + e^- + e^+$
(b) $\pi^+ \rightarrow \mu^+ + \gamma$ (c) $\Lambda^0 \rightarrow \pi^- + \pi^+ + p$
(d) $p + p \rightarrow \pi^+ + \Sigma^+$

4. [I] Give at least one conservation law violated in each of the following reactions? (a) $n + \nu_e \rightarrow e^+ + \pi^-$
(b) $p + \bar{\nu}_e + e^- \rightarrow \pi^0 + \gamma$
(c) $p + p \rightarrow \pi^+ + K^+ + n + n$
(d) $p + n \rightarrow K^- + K^+ + \pi^0$

5. [I] Does the reaction

$$p + n \rightarrow p + \bar{p} + p$$

occur and if not, why not?

6. [I] Is the reaction

$$\mu^- \rightarrow e^- + \bar{\nu}_e$$

possible and if not, why not?

7. [I] Is the reaction

$$\mu^- \rightarrow e^- + \nu_e + \nu_\mu$$

possible and if not, why not?

8. [I] Is the reaction

$$p + \pi^- \rightarrow p + \pi^- + \pi^0$$

possible and if not, why not? By what interaction could it proceed?

9. [I] Is the reaction

$$K^0 \rightarrow \pi^+ + \pi^- + \pi^0$$

possible and if not, why not? By which interaction will it proceed?

10. [I] Is the reaction

$$\pi^- + p \rightarrow K^0 + n$$

likely to be observed?

11. [I] Is the reaction

$$K^- \rightarrow \mu^- + \bar{\nu}_\mu$$

forbidden by any conservation law? How does it decay?

12. [I] Is the reaction

$$\pi^- + p \rightarrow K^+ + K^-$$

forbidden by any conservation law?

13. [I] Is the reaction

$$\pi^+ \rightarrow \mu^+ + \nu_\mu$$

forbidden by any conservation law? If not, how much energy will be released as KE as a result of the decay? (Assume the neutrino to have negligible mass and the pion to be at rest initially.)

14. [I] Determine the particle that is missing in each of the following reactions: (a) $p + () \rightarrow n + \mu^+$
(b) $p + p \rightarrow p + K^+ + ()$
(c) $p + p \rightarrow p + n + \pi^+ + ()$.

15. [I] Consider the decay

$$\Lambda^0 \rightarrow \pi^- + p$$

which takes a rather long 10^{-10} s. Assuming the lambda to be at rest, what is the net KE of the resulting particles? What conservation law does the reaction not have to obey?

16. [I] If the reaction

$$\mu^- \rightarrow e^- + \bar{\nu}_e + \nu_\mu$$

can occur, what is the maximum energy carried by the two essentially massless neutrinos assuming the muon to be at rest?

17. [I] If a proton and an antiproton, both essentially at rest, were to completely annihilate each other, how much energy could be liberated?

18. [I] What is the maximum total kinetic energy shared by the pion and electron in the reaction

$$K^+ \rightarrow \pi^0 + e^+ + \nu_e$$

assuming the kaon to be at rest?

19. [I] Approximately how much energy must be provided in order to create a separated u-quark and $\bar{u}$-antiquark pair?

20. [I] What is the quark composition of an antiproton?

21. [I] If the mean lifetime of a proton were a mere 10^{30} y, how many protons would we have to put in a tank in order to observe an average of one decay per year?

22. [II] Is the decay $\Xi^0 \rightarrow n + \pi^0$ forbidden and if so, why?

23. [II] Consider a mediating particle of mass m; derive an expression for its rest energy in terms of the range of the associated force.

24. [II] Two protons moving head-on toward each other at the same speed collide, causing the reaction

$$p + p \rightarrow p + p + \pi^0$$

Determine the minimum kinetic energy of each proton.

25. [II] Two protons moving toward one another at the same speed collide to produce the reaction

$$p + p \rightarrow n + p + \pi^+$$

What is the minimum amount of KE each must have in order for the reaction to take place?

26. [II] Consider the reaction

$$p + p \rightarrow p + \Lambda^0 + K^0 + \pi^+$$

Determine the minimum KE the protons must have if this event is to occur. For simplicity, we take the protons to be moving toward one another at the same speed.

27. [II] Determine the Q value of the reaction

$$\pi^- + p \rightarrow K^0 + \Lambda^0$$

Could this reaction take place if the pion and proton were moving slowly toward each other just before impact?

28. [II] How much energy would appear in the form of gamma rays if a positron with a KE of 10.0 MeV was to annihilate an electron with a KE of 20.0 MeV?

29. [II] A neutrino has an energy of 100 MeV. Assuming it has zero mass, what is its frequency, wavelength, and momentum? Use SI units.

30. [II] A D^+ meson has quantum numbers $B = 0$, $S = 0$, $C = +1$, and $Q = +1$ (and no topness or bottomness). What is its quark configuration?

31. [II] A D^0 meson has quantum numbers $B = 0$, $S = 0$, $C = +1$, and $Q = 0$ (and no topness or bottomness). What is its quark configuration?

32. [II] A K^- meson has quantum numbers $B = 0$, $S = -1$, $C = 0$, and $Q = -1$ (and no topness or bottomness). What is its quark configuration? What is the configuration of K^+? Explain your answer in detail.

33. [II] A Λ^0 baryon has quantum numbers $B = +1$, $S = -1$, $C = 0$, $Q = 0$, spin $= \frac{1}{2}$ (and no topness or bottomness). What is its quark configuration? Explain your answer in detail.

34. [II] Explain how, according to the quark model, this reaction takes place

$$\pi^0 \rightarrow \gamma + \gamma$$

35. [II] Explain how, according to the quark model, this reaction takes place

$$K^0 \rightarrow \pi^+ + \pi^-$$

(First write it out in terms of quarks and then discuss what happens.)

36. [II] Explain how, according to the quark model, this reaction takes place

$$\Omega^- \rightarrow \Lambda^0 + K^-$$

(First write it out in terms of quarks and then discuss what happens.)

37. [II] Suppose an electron and a positron, each very nearly at rest, annihilate forming a single *virtual* photon. What would be its frequency?

38. [II] The energy at which the weak and electromagnetic forces unify into the electroweak force is around 100 GeV (Fig. 33.23). What temperature does that correspond to? (Check your answer against Fig. 33.22.)

39. [II] One GUT formulation supposes that quarks and leptons brought together at a distance of around 10^{-31} m will transform one into the other. Determine the mass of the gauge boson that would have such a range. Give your answer in GeV/c^2.

40. [II] Assume the body of a typical human being contains roughly 10^{28} protons. If the proton had a half-life of only 10^{10} years, what would be its disintegration rate in decays per second?

41. [II] With Problem 39 in mind, at what temperature will this unification occur? (Check your answer against Fig. 33.22.)

42. [II] GUTs predict that both neutrons and protons can decay in the nucleus via routes that do not conserve baryon number. Roughly how many nucleons are in a liter of water? If the mean lifetime of a nucleon was 10^{20} y, how many decays would occur in the liter of water per year? What does this result suggest about the lifetime of a proton?

Appendixes: A Mathematical Review

APPENDIX A ALGEBRA

Algebra is essentially a body of rules and procedures for logically exploring the relationships between concepts using symbols to provide a concise and easily read format. Our concern is usually with equations describing interdependencies among physical quantities. Thus, the length traveled (l) in a time (t) by a bird moving at a speed (v) is given by $l = vt$. *When we have one equation we can solve for only one unknown quantity.* Thus, suppose the speed is known to be 2 m/s and the length traveled is 500 m (ignoring units for the moment), the equation becomes $500 = 2t$; one equation, one unknown (namely, t). How long did it take the bird to make the trip? That is, solve for t—*get the unknown all by itself on one side of the equation.* To do that, remember the following three logical rules:

1. The same quantity (constant or variable) can be added to or subtracted from both sides of an equation without changing the equality: $100 = 100$; $100 - 2 = 100 - 2$.
2. Both sides of an equation can be multiplied or divided by the same quantity (constant or variable) without changing the equality: $100 = 100$; $100/2 = 100/2$.
3. Both sides of an equation can be raised to the same power (squared or square rooted, cubed or cube rooted) without changing the equality: $100 = 100$; $\sqrt{100} = \sqrt{100}$.

Wherever possible, apply rule 1 first followed by rule 2. Isolate the variable to whatever power it is raised to and if that power is other than one, use rule 3 to make it one.

Back to the bird: $2t = 500$. To get the t alone, remove the 2 (using rule 2) by dividing both sides by 2; hence, $t = 250$ seconds.

Let's solve $8x + 2 = 42$ for x. To get the x alone, first remove the 2 via rule 1, $8x = 42 - 2 = 40$. Next, use rule 2 to remove the 8; $x = 40/8 = 5$.

Now solve $5x^2 + 12 = 57$ for x. Apply rule 1: $5x^2 = 57 - 12 = 45$. Next isolate x^2 using rule 2 to divide both sides by 5: $x^2 = 9$. Now use rule 3, taking the square root of both sides: $x = \pm 3$. Note that there are two solutions: $+3$ and -3.

Solve $\frac{3}{4}t^2 - 6 = 0$ for t. Use rule 1 to move the 6, yielding $\frac{3}{4}t^2 = 6$. Now, using rule 2, multiple both sides by 4 to get $3t^2 = 24$. Use it again, dividing both sides by 3 to get $t^2 = 8$. Now, using rule 3, take the square root of both sides: $t = \sqrt{8}$.

Some Trouble Spots

1. Dividing by fractions can be troublesome. $8/\frac{1}{4}$ is *not* 2. A *fraction is not changed when the top and bottom are multiplied by the same quantity.* Here, multiply top and bottom by 4 to get 32/1. Similarly, given $\frac{1}{4}/\frac{1}{2}$ multiply top and bottom by 2 to get $\frac{1}{2}/1 = \frac{1}{2}$.
2. $(a + b)^2$ is *not* equal to $a^2 + b^2$! $(a + b)^2 = (a + b)(a + b) = aa + ab + ba + bb = a^2 + 2ab + b^2$.

3. $\dfrac{1}{a} + \dfrac{1}{b}$ is *not* equal to $\dfrac{1}{a+b}$. We can only add terms with the same denominators. To get a common denominator (ab), multiply top and bottom by whatever is needed: $\dfrac{b}{ba} + \dfrac{a}{ab}$ and add. Thus

$$\frac{1}{a} + \frac{1}{b} = \frac{a+b}{ab}$$

Similarly $$\frac{1}{2} + \frac{2}{5} = \frac{5}{10} + \frac{4}{10} = \frac{9}{10}$$

4. Remember that $\sqrt{a}\sqrt{b} = \sqrt{ab}$; thus, $\sqrt{4}\sqrt{9} = \sqrt{36}$, or $2 \times 3 = 6$. Similarly, $\sqrt{8} = \sqrt{4}\sqrt{2} = 2\sqrt{2}$; hence, $\sqrt{25gt^2} = 5t\sqrt{g}$.
5. Given the relationship $F = GmM/R^2 = 2GmM \neq R^2$, how does F change if m is doubled? The new F is $G(2m)M/R^2$, which is twice the original F. How does F change when R is doubled? The new F is $GmM/(2R)^2 = GmM/4R^2$, which is one-quarter the original F.

A-1 Exponents

A square with sides of length a has an area of $a \times a$, which is more concisely written as a^2 (a raised to the second power). Because of that, a^2 is read "a squared." In the same way, a cube with sides of a has a volume of $a \times a \times a = a^3$ or "a cubed." In general, a raised to the nth *power* is a^n, where the *exponent* n can be any number, fractional or whole, positive or negative. Quantities raised to various powers are often multiplied and divided. Thus, $a^2 \times a^3 = (a \times a) \times (a \times a \times a) = a^2a^3 = a^5$ and, in general

$$(a^n)(a^m) = a^{n+m} \tag{A.1}$$

The base numbers (a) being raised to the powers n and m are the same. Alternatively, if the base numbers (a and b) are different and the powers are the same, then

$$(a^n)(b^n) = (ab)^n \tag{A.2}$$

Now, for division: $2^3/2^2 = 8/4 = 2 = 2^1$; the two 2s on the bottom cancel two 2s on the top or, equivalently, the exponent on the bottom subtracts from that on the top. More generally

$$\frac{a^3}{a^2} = \frac{a \times a \times a}{a \times a} = a^{3-2} = a$$

As a rule $$\frac{a^n}{a^m} = a^{n-m} \tag{A.3}$$

which suggests that we define

$$\frac{1}{a^m} = a^{-m} \tag{A.4}$$

Hence, $1/5 = 5^{-1}$, $1/3^2 = 3^{-2}$, and $1/4^{-2} = 4^2$. Inasmuch as $a/a = 1$, it follows from Eq. (A.3) that $a^{1-1} = a^0 = 1$. *Any quantity raised to the zero power is* 1.

Fractional exponents correspond to roots, thus

$$a^{1/n} = \sqrt[n]{a} \tag{A.5}$$

Thus, using Eq. (A.1), we have

$$\sqrt{a}\sqrt{a} = a^{\frac{1}{2}}a^{\frac{1}{2}} = a^1 = a$$

Since powers undo roots and vice versa

$$(a^{1/n})^n = 1$$

For example, $(25^{\frac{1}{2}})^2 = 5^2 = 25$. In general

$$(a^n)^m = a^{nm} \tag{A.6}$$

where n and m can be either whole or fractional numbers.

A-2 Powers of Ten: Scientific Notation

We deal with numbers that are extremely small (like the mass of an electron, 0.000 000 000 000 000 000 000 000 000 000 911 kg) and extremely large (like the number of stars in the Universe, ≈10 000 000 000 000 000 000 000). The *scientific notation* is a shorthand way of writing numbers in terms of powers of ten:

$10^0 = 1$	$10^0 = 1$
$10^1 = 10$	$10^{-1} = 1/10 = 0.1$
$10^2 = (10 \times 10) = 100$	$10^{-2} = 1/(10 \times 10) = 0.01$
$10^3 = (10 \times 10 \times 10) = 1000$	$10^{-3} = 1/(10 \times 10 \times 10) = 0.001$
$10^4 = (10 \times 10 \times 10 \times 10) = 10\,000$	$10^{-4} = 1/(10 \times 10 \times 10 \times 10) = 0.0001$

and so forth. For positive exponents, *the number of zeros corresponds to the power of ten.* Thus, the number of stars in the Universe is a 1 with 22 zeros, or 1×10^{22}, or just 10^{22}.

Suppose we want to express a number greater than 1.0 given in ordinary form (such as the number of seconds in a year—31 560 000) in scientific notation. Start by indicating the decimal at its reference position (31 560 000.0). Then decide where you would like to move it to (for example, 31**560 000*.0). Here the decimal is to be relocated six places to the left yielding 31.56, and multiplying by 10^6 will move it back six places to the right, where it started. Thus, $31\,560\,000 = 31.560 \times 1\,000\,000 = 31.56 \times 10^6$, or 3.156×10^7, or 0.3156×10^8. To write a number less than 1.0 in scientific notation (for example, 0.000 000 000 000 000 000 000 000 000 000 911 kg), again locate where you would like the decimal to appear (0.*000 000 000 000 000 000 000 000 000 000 9**11 kg). That's 31 places to the right, so multiplying 9.11 by 10^{-31} will shift the decimal back 31 places to the left where it started. The mass of an electron is 9.11×10^{-31} kg.

When multiplying or dividing numbers in scientific notation, process the numerical terms separately from the exponents; as

$$(1.1 \times 10^{12})(5.0 \times 10^{17}) = (1.1 \times 5.0)(10^{12} \times 10^{17}) = 5.5 \times 10^{29}$$

$$\frac{(1.1 \times 10^{12})}{(5.0 \times 10^{17})} = \frac{(1.1)}{(5.0)} \times \frac{(10^{12})}{(10^{17})} = 0.22 \times 10^{-5}$$

Scientific calculators have an "exp" key that allows you to enter exponents, and they will keep track of the decimal automatically.

A-3 Logarithms

Suppose we have a positive number y expressed as a power of b where $b > 0$ and $b \neq 1$; accordingly

$$y = b^x$$

Let's define *x to be the logarithm of y, to the base b;* thus

$$x = \log_b y$$

The logarithm x is the exponent, and the two equations are just different ways of saying the same thing. For example, if $2^3 = 8$, then $3 = \log_2 8$; if $8^{-4/3} = 1/16$, then $-4/3 = \log_8 1/16$. Because logarithms are exponents, the laws governing the two are very similar. Thus, independent of the base

$$\log(ab) = \log a + \log b$$

$$\log \frac{a}{b} = \log a - \log b$$

$$\log a^n = n \log a$$

where

$$\log_b b = 1$$

$$\log 1 = 0$$

and

$$b^{\log_b a} = a$$

There are two widely used bases, 10 and $e = 2.718\,281\,828 \cdots$. The *common logarithms* arise when $y = 10^x$, whereupon $x = \log_{10} y$. Before the advent of the electronic calculator, common logs were used to carry out complicated calculations. Our interest in them arises out of the idea of intensity-level (p. 455) and the dB. The *natural logarithms* arise when $y = e^x$, whereupon $x = \ln y$ (it's customary to write $\ln y$ rather than $\log_e y$). The exponential function e^x (Fig. A1) and the natural log (Fig. A2) occur in a great many physical situations in which the rate-of-change of some quantity depends on that quantity. We can see this dependence in Fig. A1: the rate-of-change of the curve $y = e^x$ equals e^x. When the curve has small values, the slope is small; when the curve has large values, the slope is large. This property is very special and it means that if some quantity N changes in time exponentially, then $\Delta N/\Delta t \propto N$.

It follows from the above that

$$\ln e^x = x \qquad \text{and} \qquad e^{\ln x} = x$$

Notice that with $y = e^x$ when $y = 1$, $x = 0$; hence, $y = 2$ when $2 = e^x$, whereupon $x = \ln 2 = 0.693$. Doubling y yields $y = 4$ when $x = \ln 4 = 1.386 = (0.693 +$

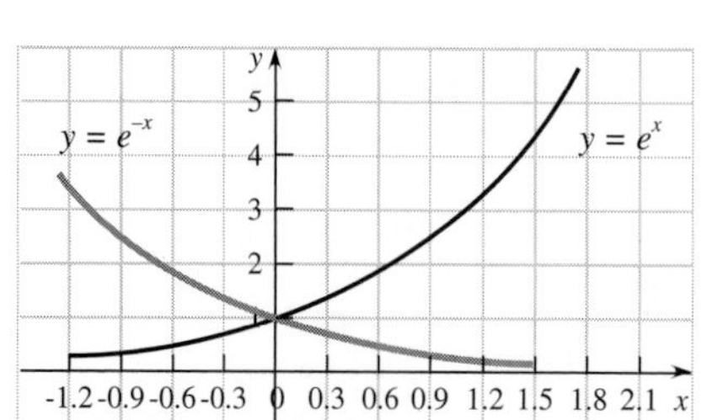

Figure A1

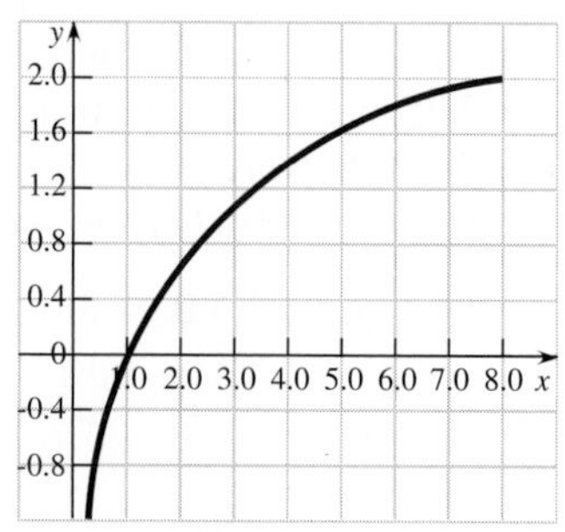

Figure A2

0.693), and again $y = 8$ when $x = \ln 8 = 2.079 = (0.693 + 1.386)$. Every time x increases by 0.693, y doubles. Indeed, any time x increases by a fixed amount, y will increase by a fixed multiplicative factor—y increases by the same factor in equal intervals of x. When x increases by $\ln 10 = 2.302\,6$, y increases by a factor of 10.

A-4 Proportionalities and Equations

We frequently find that one physical quantity depends on another: the force exerted by a spring depends on the amount it's stretched; the current through a resistor depends on the voltage across it. Suppose the quantities A and B depend on one another such that doubling either doubles the other. They are said to be **directly proportional** to each other or, symbolically, $A \propto B$. On the other hand, it's possible that doubling one could halve the other, in which case $A \propto 1/B$, and they are **inversely proportional**. Table A1 gives a list of values of A and B. At first, there may seem no relationship between them, but form the ratio B/A and we see that there is a consistent pattern: $B/A = 3$. Thus, $B = 3A$ and 3 is known as a **constant-of-proportionality**. When two quantities are proportional, we can always turn the proportionality into an equality using an appropriate constant-of-proportionality. The circumference of a circle (C) is directly proportional to the diameter (D), where the constant of proportionality is π: $C = \pi D$. The area of a sphere (A) is directly proportional to the square of the radius (r), where the constant-of-proportionality is 4π: $A = 4\pi r^2$.

TABLE A1

A	B
3.1	9.3
18.0	54
5.2	15.6
71.0	213

When the variables in an equation occur only to the first power, it's called a **linear equation**; for example, $y = mx$, where m is a constant. Here, when $x = 0$, $y = 0$, and the curve (Fig. A3) is a straight line passing through the origin. For any two points on the line, $\Delta y/\Delta x$ is its **slope**, or tilt, and it equals m. If a constant (b) is added to y such that $y = mx + b$, the line shifts parallel to itself; b is the place where the line crosses the y-axis, the *y-intercept* (Fig. A4).

The Quadratic Equation

When an object is uniformly accelerating, it travels a distance s given by Eq. (3.9); namely, $s = v_i t + \frac{1}{2}at^2$. This expression is typical of a class of equations that contains a variable (t) raised to the second power, and it's called a **quadratic equation**. The standard form has everything on the left set equal to zero on the right:

$$ax^2 + bx + c = 0$$

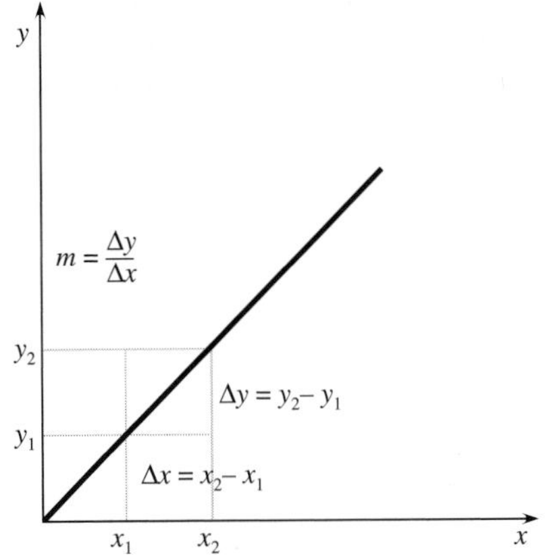

Figure A3

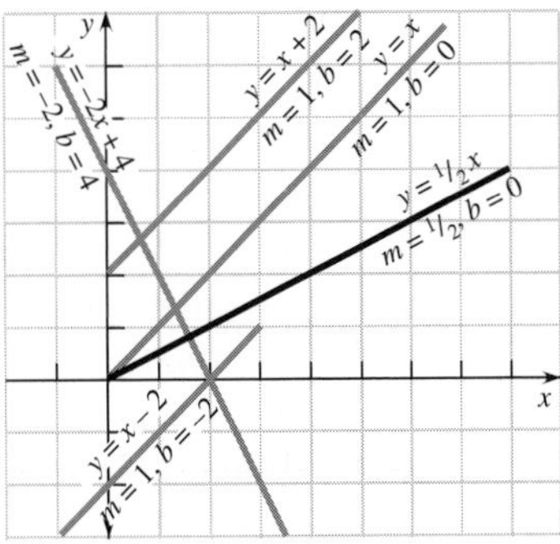

Figure A4

where a, b, and c are constants. This equation can be solved directly, but it's common practice to memorize the solution in the form

$$x_{\pm} = \frac{-b \pm \sqrt{b^2 - 4ac}}{2a}$$

where, provided $b^2 > 4ac$, there are two real solutions. For example, $2x^2 + 6x - 20 = 0$ corresponds to $a = 2$, $b = 6$, and $c = -20$; therefore

$$x_{+} = \frac{-6 + \sqrt{36 - 4(2)(-20)}}{2(2)} = \frac{-6 + \sqrt{196}}{4} = +2$$

$$x_{-} = \frac{-6 + \sqrt{36 - 4(2)(-20)}}{2(2)} = \frac{-6 - \sqrt{196}}{4} = -5$$

It may happen in a physical analysis that only one of the two solutions corresponds to a possible situation; we then simply ignore the other. For example, we might solve Eq. (3.9) for the time and find one of the solutions to be negative—it works mathematically, but not physically.

Simultaneous Equations

For every unknown, we must have an independent equation (one that isn't equivalent to any other equation in the set). Situations with one and two unknowns are common throughout the discipline. Indeed, we will encounter problems in circuit analysis where there are as many unknowns as there are loops in the circuit, but three will be our limit. Suppose there are two unknowns, x and y, and two equations

$$4x - 2y = 16$$
$$3x + 4y = 23$$

Lots of numbers satisfy each of these equations (for example, in the first one, $x = 4$, $y = 0$ or $x = 0$, $y = -8$). We want to solve these *simultaneously* so that the solutions satisfy both equations at the same time. There are two ways to accomplish this goal. (1) *Solve either equation for either unknown in terms of the other unknown, and substitute that into the remaining equation.* From the first equation, $4x = 16 + 2y$, $x = 4 + \frac{1}{2}y$. Now put this result into the second equation so it has only one unknown $3(4 + \frac{1}{2}y) + 4y = 23$ and so $12 + 3y/2 + 4y = 23$, $3y/2 + 8y/2 = 23 - 12$, $11y/2 = 11$, $y/2 = 1$, $y = 2$. Put that back into the first equation: $4x - 2(2) = 16$, $4x = 20$, and $x = 5$. (2) Alternatively, *multiply either or both of the equations by whatever numbers it takes to make the same coefficient appear for either one of the unknowns. Then add or subtract the two, thereby removing one unknown.* Here, to get rid of y, multiply the first equation $4x - 2y = 16$ by 2

getting $$8x - 4y = 32$$
and add $$3x + 4y = 23$$
$$11x \qquad = 55$$

or $x = 5$. Substitute this result into either equation and solve for y.

A-5 Approximations

It's often desirable to approximate the solution to a problem, either as a quick check or because the exact solution is too elaborate to deal with. Newton's *Binomial Theo-*

rem, expressed as

$$(a + b)^n = a^n + na^{n-1}b + \frac{n(n-1)}{2 \times 1}a^{n-3}b^2 + \frac{n(n-1)(n-2)}{3 \times 2 \times 1}a^{n-3}b^3 + \cdots \quad \text{(A.7)}$$

produces some very helpful approximations. In general, if n is a positive integer, the series ends in a finite number of terms, otherwise it has an infinite number. Expressions of the form $(1 + x)^n$ arise frequently and can be approximated using the Binomial Theorem ($a = 1$, $b = x$); thus

$$(1 + x)^n = 1 + nx + \frac{n(n-1)}{2}x^2 + \cdots \quad \text{(A.8)}$$

When x is very small ($x \ll 1$), the x^2 and higher terms will be negligibly small and

$[x \ll 1]$ $\qquad (1 + x)^n \approx 1 + nx$

with $n = 2$ $\qquad (1 + x)^2 \approx 1 + 2x$

with $n = 3$ $\qquad (1 + x)^3 \approx 1 + 3x$

with $n = \frac{1}{2}$ $\qquad (1 + x)^{\frac{1}{2}} = \sqrt{1 + x} \approx 1 + \frac{1}{2}x$

with $n = -\frac{1}{2}$ $\qquad (1 + x)^{-\frac{1}{2}} = \dfrac{1}{\sqrt{1 + x}} \approx 1 - \frac{1}{2}x$

with $n = -1$ $\qquad (1 + x)^{-1} = \dfrac{1}{(1 + n)} \approx 1 - x$

with $n = \frac{1}{2}$ $\qquad (1 - x)^{\frac{1}{2}} = \sqrt{(1 - x)} \approx 1 - \frac{1}{2}x$

Try taking the square root of 1.000 000 001 0 on your calculator. Now, let $x = 0.000\,000\,001\,0$ and the square root of $(1 + x)$ is $\approx 1.000\,000\,000\,5$. For $a \gg b$

$$(a + b)^n = a^n\left[1 + \frac{b}{a}\right] \approx a^n\left[1 + n\frac{b}{a}\right] \quad \text{(A.9)}$$

What is the value of $a/(a + b)^2$ when $a \gg b$? Since b is negligible compared to a, $(a + b)^2 \approx (a)^2$ and $a/(a + b)^2 \approx 1/a$. Check this expression with Eq. (A.9), $n = 2$ and $(a + b)^2 \approx a^2(1 + 2b/a) \approx a^2$. Suppose $b \gg a$, $(a + b)^2 \approx (b)^2$ and $a/(a + b)^2 \approx a/b^2$. We frequently know how a system behaves at its extremes and can confirm the analysis at those extremes when a key quantity is very large or very small. For example, any equation for the force between two magnets must go to zero as the separation between them gets very large.

A useful means of improving the likelihood that an analysis is correct is to simplify the numbers by rounding them off and then run quickly through the calculation to get a crude answer. When a number is rounded off to the nearest power of 10, we say it's an **order-of-magnitude** figure. Thus, the order-of-magnitude of the acceleration due to gravity on Earth (9.81 m/s^2) is 10 m/s^2; the great mountains have heights of the order-of-magnitude of 10 km; the electron's mass ($9.109\,389\,7 \times 10^{-31}$ kg) is of the order-of-magnitude of 10^{-30} kg; the number of atoms per cubic centimeter of a solid is of the order-of-magnitude of 10^{23}.

A-6 The Change in a Quantity

We will often encounter quantities that are formed of products such as $z = xy$. The question arises, what happens to z when either x or y or both change? The initial value of z is $z_i = xy$. Its final value, after a change in x of Δx, and a change in y of Δy, is

$$z_f = (x + \Delta x)(y + \Delta y)$$

$$z_f = xy + x\Delta y + y\Delta x + \Delta x\Delta y$$

Hence $$\Delta z = z_f - z_i = x\Delta y + y\Delta x + \Delta x\Delta y$$

When Δx and Δy are small, $\Delta x\Delta y$ is negligibly small and

$$\Delta z \approx x\Delta y + y\Delta x$$

This is an important relationship that we will make use of often.

APPENDIX B GEOMETRY

Figure B1 provides a number of useful relationships for the angles formed by intersecting lines and triangles. Of special interest is the Pythagorean Theorem, which relates the sides of any right triangle; consequently

$$C = \sqrt{A^2 + B^2}$$

as in Fig. B2. There are a number of right triangles with whole number sides—the 3–4–5 and 5–12–13 are the most commonly encountered (Fig. B3).

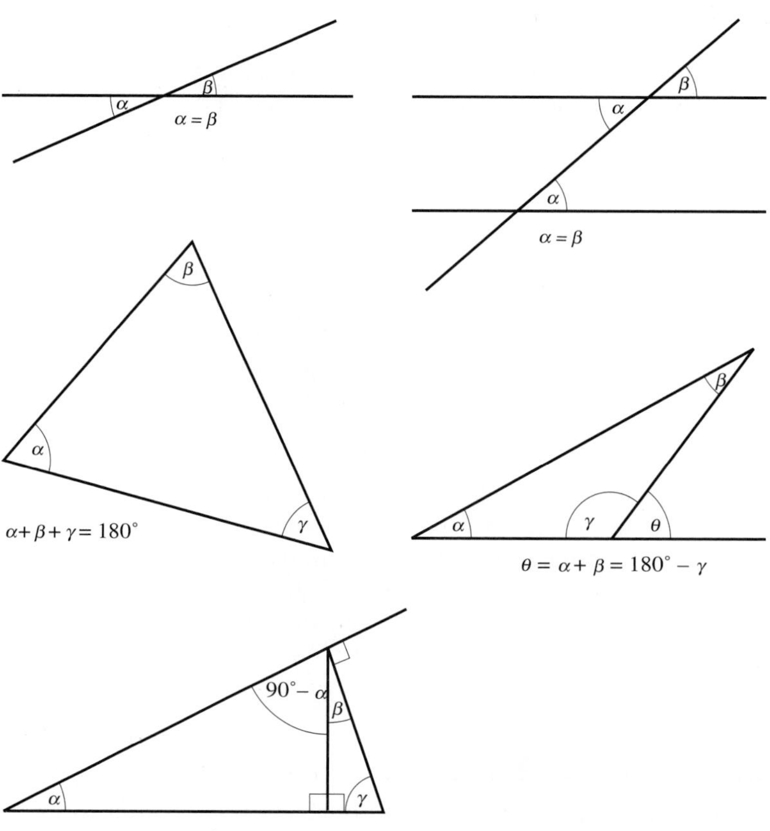

Figure B1

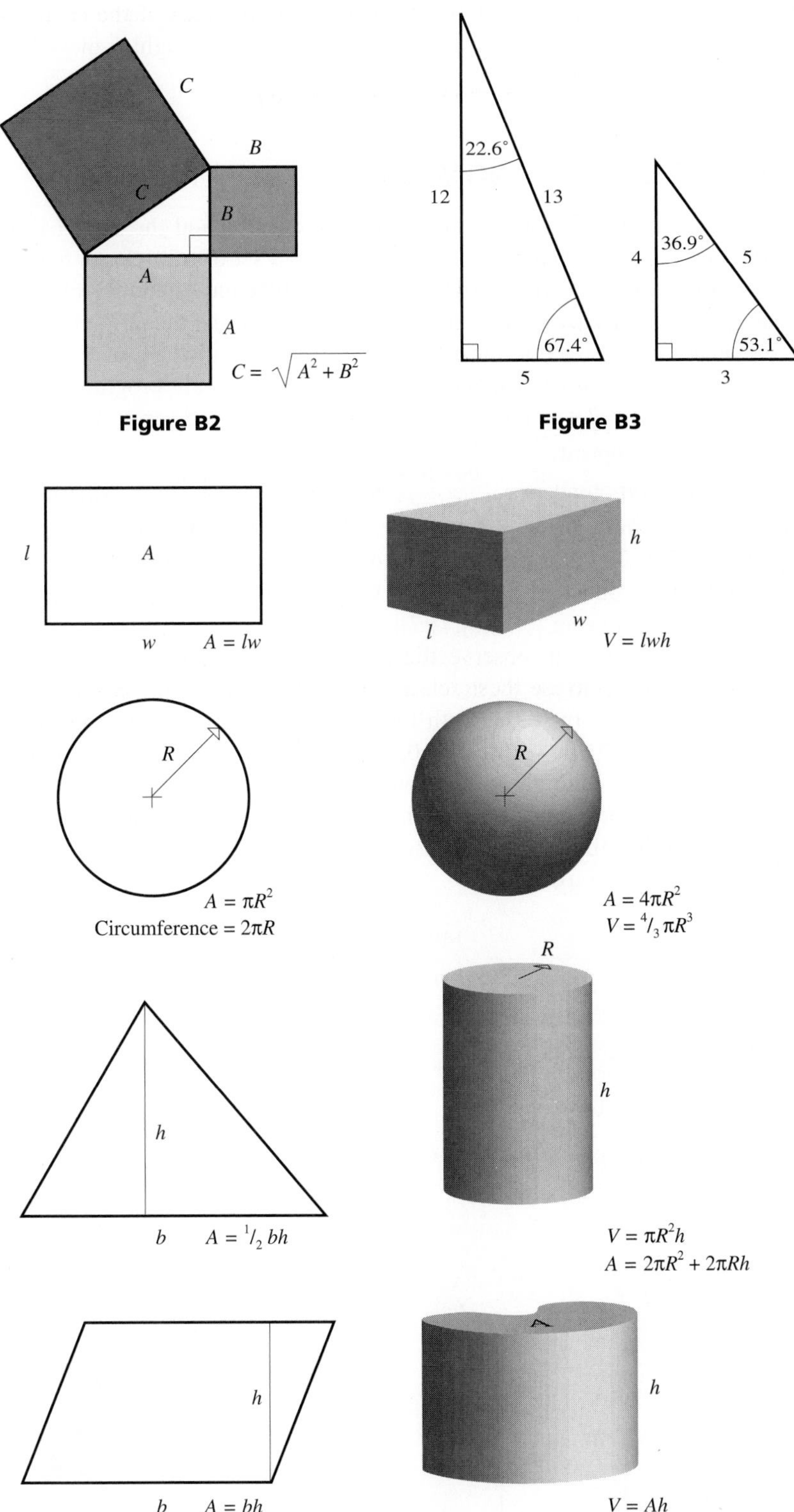

Figure B2

Figure B3

Figure B4

Figure B4 displays the areas and volumes of shapes that are frequently used in this text. Notice that the area of the curved surface of a circular cylinder ($2\pi Rh$) is that swept out by a line of height h moved around the circumference of the base

$(2\pi R)$. In a similar way, the volume of a cylinder is the area of the base times the height (that is, the volume swept out by a vertical line of height h moved to every point on the base area).

APPENDIX C TRIGONOMETRY

Figure C1a shows a right triangle with an acute angle θ and the sides (A) adjacent and (B) opposite to it. A particular value of θ fixes a set of triangles, a few of which are shown in Fig. C1b. Although the triangles are different, the ratios of the sides in any one of them is the same for all of them. In other words, the ratio of the opposite over the adjacent is the same for each triangle and is indicative of θ. The various possible ratios are given names, and electronic calculators are programmed to compute them. Knowing θ we can determine any ratio and vice versa. Thus

$$\sin\theta = \frac{\text{opposite}}{\text{hypotenuse}} \qquad \cos\theta = \frac{\text{adjacent}}{\text{hypotenuse}} \qquad \tan\theta = \frac{\text{opposite}}{\text{adjacent}}$$

For example, for an angle of 30°, the ratio of the opposite to the adjacent side will always be 0.5774. To determine as much on a calculator, first make sure it's in the *degree* mode, then punch in [3][0][tan]. Alternatively, knowing the ratio, you can find the angle by entering [0][·][5][7][7][4][$\tan^{-1}$].

Referring to Fig. C1a, observe that $\sin\theta = B/C = \cos\phi$ and that $\cos\theta = A/C = \sin\phi$; don't confuse these relationships. In Fig. C1c, increasing θ, with the adjacent side constant, increases both the opposite side and the hypotenuse. When $\theta \approx 0$, $B \approx 0$, $A \approx C$ and so $\sin\theta = 0$, $\cos\theta = 1$ and $\tan\theta = 0$. Similarly, when $\theta \approx 90°$, $C \approx B$ and both are tremendously large. Thus, $\sin 90° = 1$, $\cos 90° = 0$, and $\tan 90° = 1$. These results are summarized in Fig. C2, which provides $\sin\theta$, $\cos\theta$, and $\tan\theta$ for all values of θ between 0 and 360°.

Some useful relationships are

$$\tan\theta = \frac{\sin\theta}{\cos\theta}$$
$$\sin^2\theta + \cos^2\theta = 1$$
$$\sin 2\theta = 2\sin\theta\cos\theta$$

For all triangles (Fig. C3)

$$\text{Law of Sines} \qquad \frac{\sin\alpha}{A} = \frac{\sin\beta}{B} = \frac{\sin\gamma}{C}$$
$$\text{Law of Cosines} \qquad C^2 = A^2 + B^2 - 2AB\cos\gamma\,.$$

Figure C1

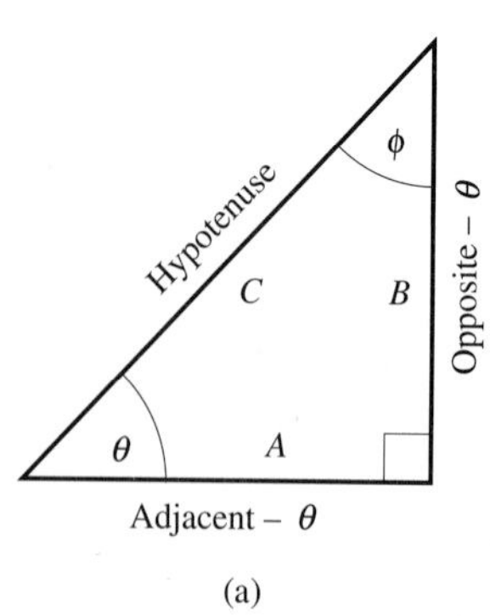

(a)

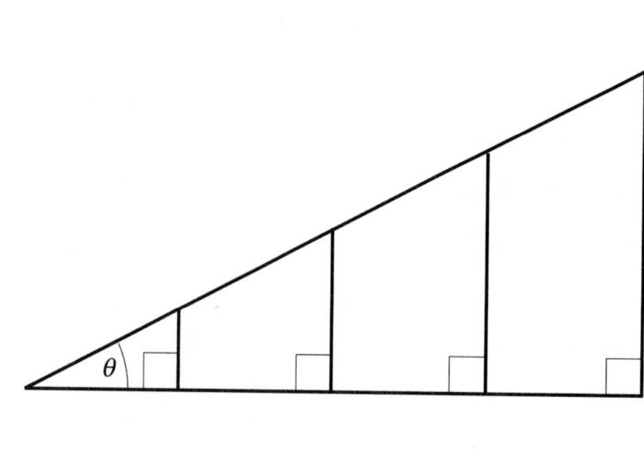

(b)

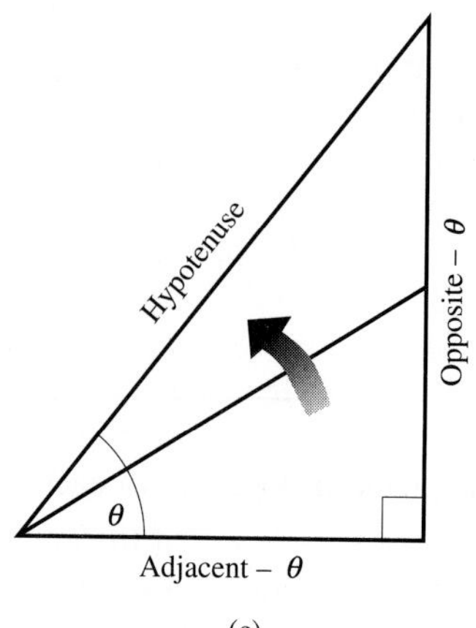

(c)

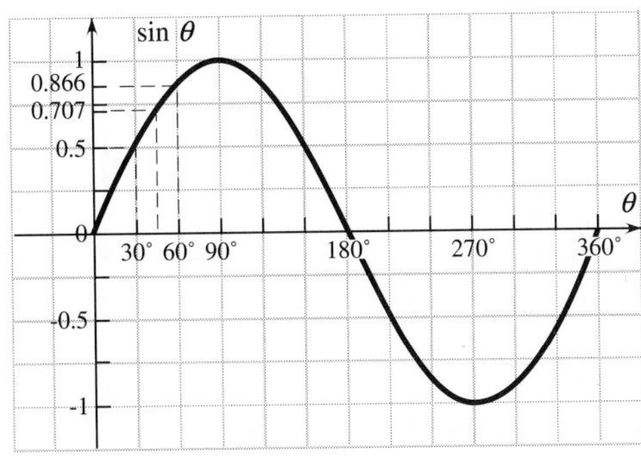

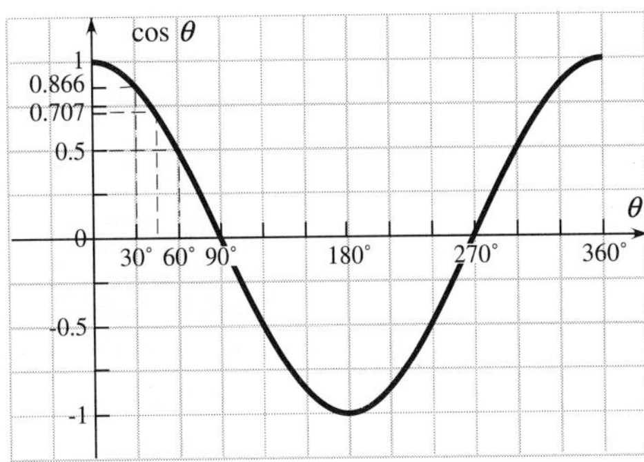

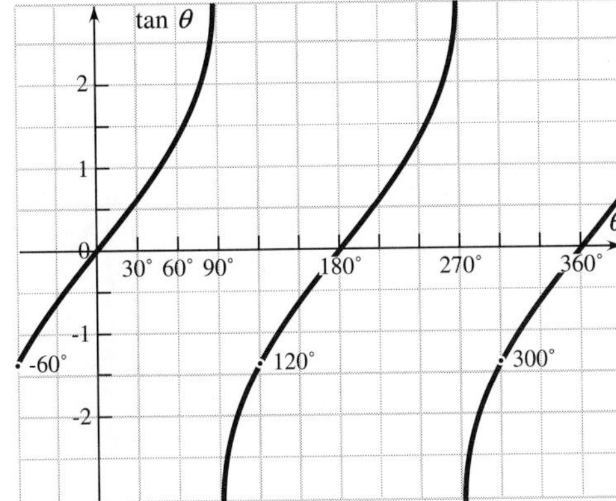

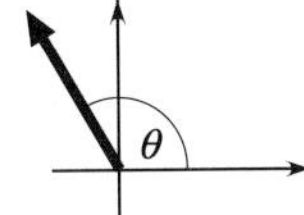

Figure C2

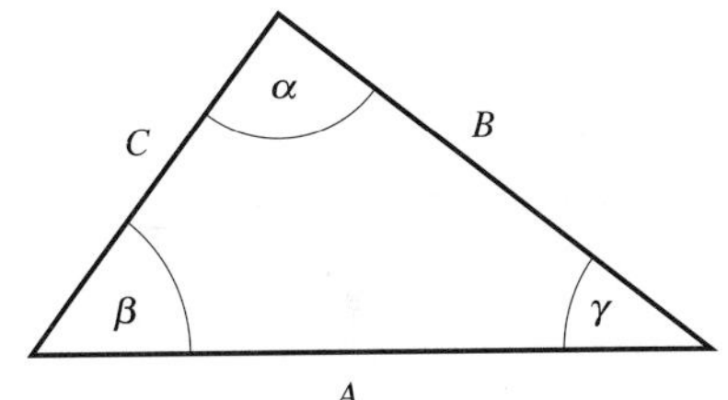

Figure C3

APPENDIX D VECTORS

Vectors are discussed at length in Chapter 2. Here, we'll go over some of the trouble spots often encountered when dealing with them. Figure D1 shows a vector **C** in each of the four quadrants and the corresponding x and y components. In each case

$$C_x = C \cos \theta \qquad \text{and} \qquad C_y = C \sin \theta$$

where θ is measured either up or down from the x-axis and is always acute. Now, suppose we add several vectors along the x-axis and get $\mathbf{C}_x$ and add several along the y-axis and get $\mathbf{C}_y$. The next step in determining **C** is to use $\tan \theta = C_y/C_x$ to find θ. It's best to write the tangent as $\tan \theta = |C_y|/|C_x|$, using the absolute values of the components so that $\tan \theta$ is always positive and θ is always between 0° and 90°. Figure C2c is a plot of $\tan \theta$ against θ from 0° to 360°, where θ is always measured from the positive x-axis. Thus, the inverse tangent of -1.732 is both 120° and 300°, lying in either the second or fourth quadrants, as expected from the figure. A calculator provides another form of the answer—namely, $-60°$ (equivalent to 300°). It neglects 120° altogether, which means that if we take the ratio of the components when finding the tangent and one or both of them are negative, there will be some ambiguity in the resulting angle. Use the ratio of absolute values; determine which quadrant the resultant is in by drawing C_x and C_y. Then measure $\theta < 90°$ up or down from the horizontal axis, as in Fig. D1.

We now wish to analytically add the two vectors **A** and **B** in the four situations depicted in Fig. D2. In each case, find $C_x = A_x + B_x$ and $C_y = A_y + B_y$. Since the angles are between 0 and 90°, the sines and cosines will be positive. Thus, oppositely directed components must be assigned plus and minus signs. Enter any component as plus (+) if it is in the positive x- or y-direction, and minus (−) if it is in

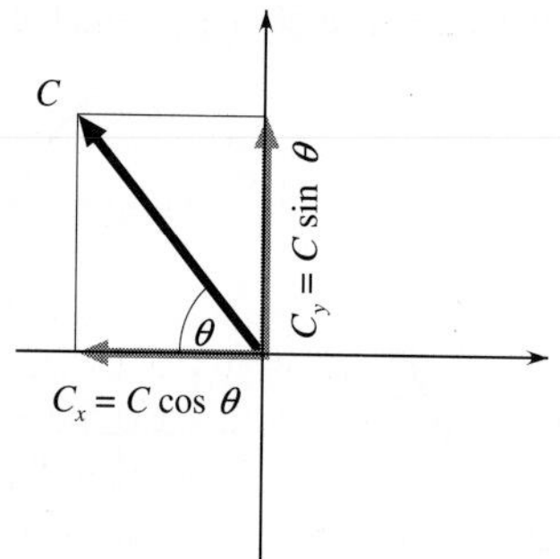
C
$C_y = C \sin \theta$
θ
$C_x = C \cos \theta$

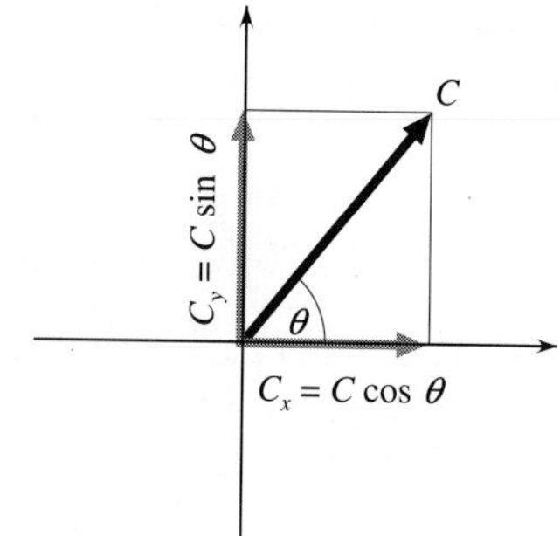
C
$C_y = C \sin \theta$
θ
$C_x = C \cos \theta$

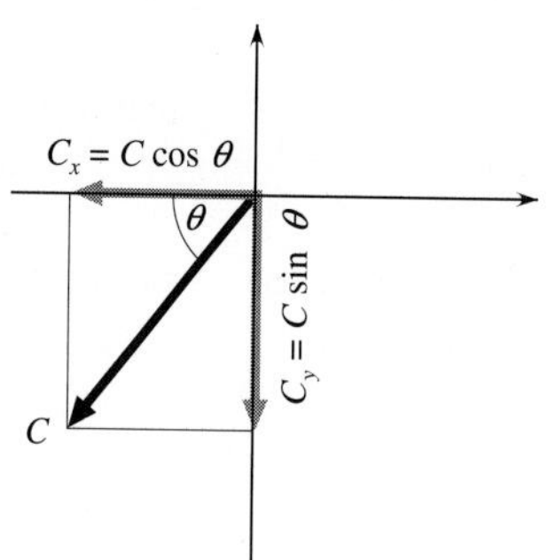
$C_x = C \cos \theta$
θ
$C_y = C \sin \theta$
C

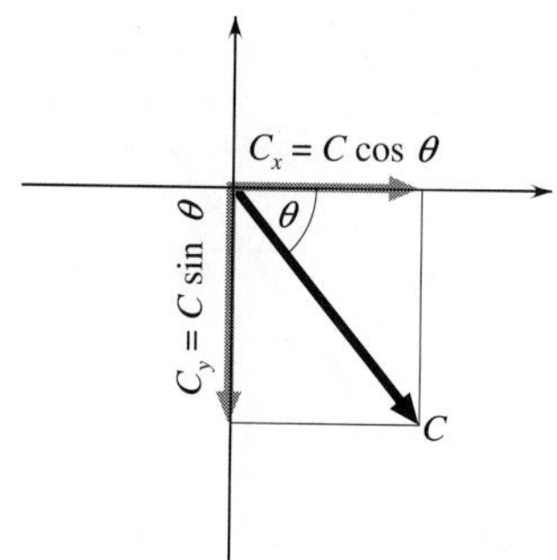
$C_x = C \cos \theta$
θ
$C_y = C \sin \theta$
C

Figure D1

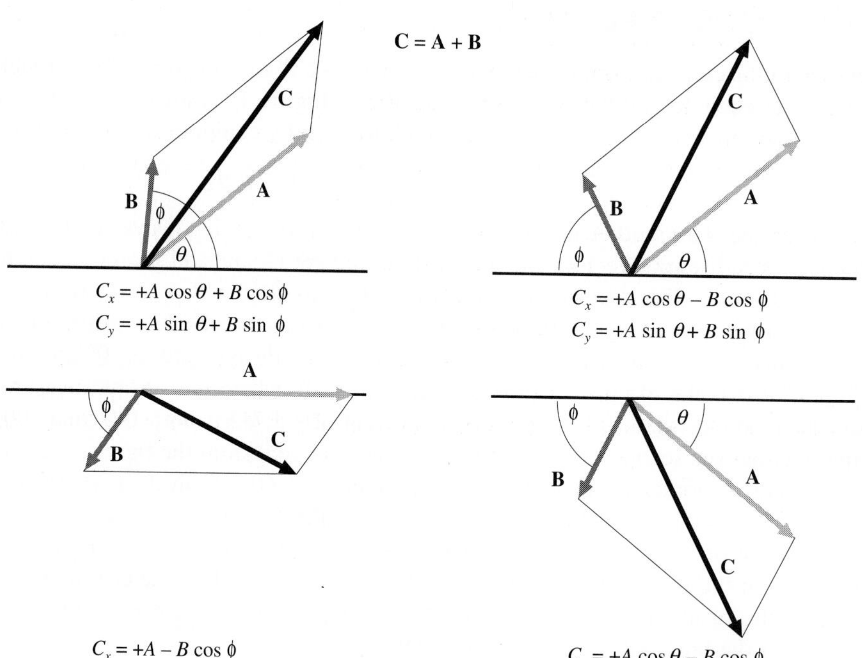
C = A + B
C
B
A
ϕ
θ
$C_x = +A \cos \theta + B \cos \phi$
$C_y = +A \sin \theta + B \sin \phi$
$C_x = +A \cos \theta - B \cos \phi$
$C_y = +A \sin \theta + B \sin \phi$
$C_x = +A - B \cos \phi$
$C_y = - B \sin \phi$
$C_x = +A \cos \theta - B \cos \phi$
$C_y = -A \sin \theta - B \sin \phi$

Figure D2

the negative x- or y-direction. The appropriate expressions are shown in each case in Fig. D2.

APPENDIX E DIMENSIONS

To determine a distance, we must specify the units in which the measurement will be made:—feet, yards, light-years, meters, and so forth. But all such measurements have something in common—they are all lengths, regardless of the specific unit. The type of unit is the *dimension* of the quantity; in this case, the dimension of length indicated as $[L]$. Area is a distance times another distance and so has the dimensions of $[L][L] = [L^2]$. We can assign dimensions to all the quantities in mechanics using the basic dimensions of *length* $[L]$, *time* $[T]$, and *mass* $[M]$. Thus, speed, which is measured in units of m/s, ft/s, mi/h, km/h, and so on, has the dimensions of length/time or $[L]/[T]$.

The volume of a sphere of radius R is $\frac{4}{3}\pi R^3$, whereas that of a cube of side a is a^3. Each has units of m^3 or ft^3 and each has the dimensions of volume $[L^3]$, suggesting that a technique of dimensional analysis, with appropriate logical rules, can be used to check our equations. To that end, consider the equation representing three physical quantities A, B, and C, as

$$C = A + B$$

There are three rules. The first is: *dimensions can be treated as algebraic quantities*—added, subtracted, multiplied, and divided accordingly. Furthermore, A and B must have the same units (we can't add seconds to meters). Our second rule is: *only quantities having the same dimensions can be added or subtracted.* Whatever it might be, it's clear that C must have the same units as $A + B$. More generally, our third rule is: *the dimensions on either side of an equal sign must be the same.* All physical equations must obey these rules.

If you come up with an expression that obeys the rules, it *may* be correct; if it doesn't obey them, it's *definitely* incorrect. For example, in Chapter 3 we find that the distance traveled s by an object that was initially moving at speed v_i after it accelerates at a constant rate a for a time t is given by $s = v_i t + \frac{1}{2}at^2$. Knowing that the dimensions of speed and acceleration are $[L]/[T]$ and $[L]/[T^2]$ respectively, apply dimensional analysis. The expression in dimensional form is

$$[L] = \frac{[L]}{[T]}[T] + \frac{[L]}{[T^2]}[T^2]$$

Cancelling time on the right, the equation obeys the second rule since each term being added has the dimension $[L]$. And it obeys the first rule as well. Conclusion: the equation may be correct.

Answer Section

CHAPTER 1

Answers to Odd-Numbered Multiple Choice Questions

1. a **3.** a **5.** c **7.** d **9.** d **11.** d **13.** d **15.** b **17.** a **19.** c

Answers to Odd-Numbered Problems

1. 10 000 000 000 = 10^{10} **3.** (a) 0.010 s (b) 0.001 s (c) 10 000 s (d) 100 000 000 s (e) 0.000 001 s **5.** 1.00×10^7 mm **7.** 76.2 mm **9.** 5×10^2 nm **11.** 2×10^2 **13.** 86 400 s **15.** 9.46×10^{15} m **17.** 8.4×10^{10} bacteria **19.** 13 times greater **21.** (a) 0.000 001 00 g (b) 0.000 000 000 001 g (c) 0.100 0 g (d) 0.010 000 g (e) 10 000 g **23.** 3×10^8 m/s, 3.00×10^8 m/s, 2.998×10^8 m/s, $2.997\,924\,6 \times 10^8$ m/s **25.** $(1.602\,177\,33 \pm 0.000\,000\,48) \times 10^{-19}$ C **27.** 0.092 90 m^2 **29.** $V = 2.2 \times 10^{19}$ m^3 **31.** 0.393 7 in./cm **33.** 1.6×10^3 m **35.** 0.092 90 m^2 **37.** 237×10^{-6} m^3 **39.** 1×10^{-7} g; 0.3×10^2 kg **41.** 1806.8 m **43.** 1.00 kg **45.** 6.1×10^{-3} g^3

Solutions to Selected Problems

3. (a) 0.010 s; (b) 0.001 s; (c) 10 000 s; (d) 100 000 000 s; (e) 0.000 001 s. **7.** 1.00 in. = 2.54 cm = 25.4 mm; 3.00 in. = 76.2 mm. **11.** 1 kg = 1000 g; (1000 g)/(5 g) = 200 = 2×10^2. **15.** $(5.88 \times 10^{12}$ mi)(1.609 km/mi) = 9.46×10^{15} m. **21.** (a) 0.000 001 00 g; (b) 0.000 000 000 001 g; (c) 0.100 0 g; (d) 0.010 000 g; (e) 10 000 g. **25.** $(1.602\,177\,33 \pm 0.000\,000\,48) \times 10^{-19}$ C. **29.** $V = (4/3)\pi R_☾^3 = 2.199\,1 \times 10^{19}$ $\text{m}^3 = 2.2 \times 10^{19}$ m^3. **33.** (45 hairs/d) × (365 1/4 d/y) = 16 436 hairs/y; (16 436) × (0.10 m) = 1.6×10^3 m. **37.** 1 liter = 1000 $\text{cm}^3 = 10^{-3}$ m^3; hence, 1.00 cup = 237×10^{-6} m^3. **41.** These add up to 1806.818 m, and the least number of decimal places is one, so the answer is 1806.8 m. **45.** (0.002 1 g)(655.1 × 10^3 g)(4.41 × 10^{-6} g) = 6.1×10^{-3} g^3.

CHAPTER 2

Answers to Odd-Numbered Multiple Choice Questions

1. d **3.** b **5.** c **7.** c **9.** c **11.** d **13.** b **15.** a **17.** c

Answers to Odd-Numbered Problems

1. 29.5 km/s **3.** (a) 1.0×10^{-9} s (b) 3.336×10^{-6} s **5.** 1.000 m/s = 3.600 km/h; 1.000 m/s = 2.237 mi/h; 1.000 m/s = 3.281 ft/s **7.** 2.0 m; v_{max} = 4.0 m/s, v_{min} = 0; from 1.5 s to 3.3 s and from 4.0 s to 5.0 s **9.** (a) 20 km/h (b) 30 km/h **11.** 15 m **13.** 0.33 m/s **15.** 1.7 km **17.** 120 km/h **19.** 6.7 km/h **23.** 1.4 m **25.** 15 km **27.** D = 10 m **29.** zero, 36 km **31.** **s** = 24 m–straight up **33.** **C** = 5 units–due N **35.** v_{av} = 4.3 m/s; $\mathbf{v}_{av}$ = 0.30 m/s–south **37.** **s** = 63 m–72° up from the horizontal **39.** (a) zero (b) **v** = 0.4 m/s–south **41.** v_{av} = 150 m/s; $\mathbf{v}_{av}$ = 150 m/s–upward: round trip v_{av} = 150 m/s; $\mathbf{v}_{av}$ = 0 **43.** $\mathbf{v}_{av}$ = 2.1 m/s–southeast **45.** (a) 2×10^2 m/s–north (b) 2×10^2 m/s–north **47.** 100 s **49.** 10×10^2 m **51.** θ_1 = 67.38°; θ_2 = 22.62° **53.** 10 m **55.** 12 m **57.** 74 km/h **59.** (a) v_H = 254 m/s (b) v_V = 159 m/s (c) v = 300 m/s **61.** $v_{av} = 1.3 \times 10^3$ km/h **63.** θ = 11°; v = 10 m/s **65.** (a) no (b) θ = 42° (c) 35 km/h **67.** 22 km/h–3.0° W of N **69.** (a) v_{RS} = 1470 km/h; v_{RC} = 0 (b) v_{RC} = 2940 km/h (c) eastward (d) Florida is closer to the equator.

Solutions to Selected Problems

1. $v_{av} = l/t$ = (885 km)/(30.0 s) = 29.5 km/s. **5.** 1.000 m/s =

$$\frac{(1.000\text{ m})/(1000\text{ m/km})}{(1.000\text{ s})/(60.00\text{ s/min})(60.00\text{ min/h})}$$

= 3.600 km/h: 1.000 m/s =

$$\frac{(1.000\text{ m})(3.281\text{ ft/m})/(5280\text{ ft/mi})}{(1.000\text{ s})/(60.00\text{ s/min})(60.00\text{ min/h})}$$

= 2.237 mi/h: 1.000 m/s =

$$\frac{(1.000\text{ m})(3.281\text{ ft/m})}{(1.000\text{ s})} = 3.281\text{ ft/s}.$$

9. (a) $v_{av} = l/t$ = (15 km)/(0.25 h) = 60 km/h; $v_{av} = l/t$ = (15 km)/(0.75 h) = 20 km/h; (b) $v_{av} = l/t$ = (30 km)/(1.0 h) = 30 km/h. **15.** Assuming the time for the light to reach you is negligible, $l = (3.3 \times 10^2$ m/s)(5.0 s) = 1.7 km. The error incurred by not including the time for light to travel is a few millimeters. **19.** L is the one-way distance; $v_{av} = l/t = 2L/[L/(10\text{ km/h}) + L/(5.0\text{ km/h})] = 2/[1/(10\text{ km/h}) + 1/(5.0\text{ km/h})] = 20/(3.0\text{ km/h}) = 6.7$ km/h. **21.** $l = Ct^2$; $(l + \Delta l) = C(t + \Delta t)^2 = C[t^2 + 2t\Delta t + (\Delta t)^2]$; $\Delta l = Ct^2 + 2Ct\Delta t + C(\Delta t)^2 - 1$; $v_{av} = \Delta l/\Delta t = [2Ct\Delta t + C(\Delta t)^2]/\Delta t = 2Ct + C\Delta t$. Letting $\Delta t \to 0$, in $v_{av} = 2Ct + C\Delta t$, the second term vanishes and $v = 2Ct$. **25.** This is a 3-4-5 right triangle, so the displacement is 15 km. **29.** Zero, 36 km. **35.** $v_{av} = l/t$ = (43 m)/(10 s) = 4.3 m/s; $\mathbf{v}_{av} = \mathbf{s}/t$ = (3.0 m–south)/(10 s) = 0.30 m/s–south. **39.** (a) The slopes are zero, as are both velocities. (b) The slope is −2.0 m in 5.0 s or v = (−2.0 m)/(5.0 s) = −0.4 m/s and so **v** = 0.4 m/s–south. **43.** From Eq. (2.6), $\mathbf{v}_{av} = (\mathbf{s}_f - \mathbf{s}_i)/\Delta t$; $\Delta\mathbf{s} = \sqrt{(3.0\text{ m})^2 + (3.0\text{ m})^2}$–southeast = 4.24 m–southeast; $\mathbf{v}_{av}$ = 2.1 m/s–southeast. **45.** (a) $\mathbf{v}_{av}$ = (600 m)/(3 s)–north = 2×10^2 m/s–north; (b) also 2×10^2 m/s–north. **47.** Each runs 500 m in time t at speed 5.00 m/s; (500 m)/(5.00 m/s) = 100 s. Alternatively, together they cover the track at a speed of 5.00 m/s + 5.00 m/s = 10.0 m/s; hence, (1000 m)/(10.0 m/s) = 100 s. **51.** tan θ_1 = 12/5 = 2.400, θ_1 = 67.38°; tan θ_2 = 5/12 = 0.416 7, θ_2 = 22.62°. **55.** tan 25° = h/(25 m) = 0.466; h = 11.66 m, or 12 m. **61.** cos $\theta = R_\theta/R$; $R_\theta = R\cos\theta$ = (6400 km) cos 41° = 4830 km; the circumference is $2\pi R_\theta = 3.0 \times 10^4$ km; $v_{av} = (3.0 \times 10^4$ km)/(24 h) = 1.3×10^3 km/h, or 7.9×10^2 mi/h. **65.** (a) No. Those velocity vectors cannot form a right triangle because the hypotenuse ($\mathbf{v}_{BW}$) must be larger than either of the other sides, and here $v_{BW} = v_{WE}$. (b) $\mathbf{v}_{BE}$ cuts directly across the river. Head at $\mathbf{v}_{BW}$ somewhat upstream at an angle θ such that $\mathbf{v}_{BE} = \mathbf{v}_{BW} + \mathbf{v}_{WE}$, sin θ = (20 mi/h)/(30 mi/h) = 0.667, θ = 42°. (c) $v_{BE} = [(30\text{ mi/h})^2 - (20\text{ mi/h})^2]^{1/2}$ = 22 mi/h = 35 km/h. **69.** (a) v_{RS} = 1470 km/h; $\mathbf{v}_{RC} = \mathbf{v}_{RS} + \mathbf{v}_{SC}$ = 1470 km/h–west + 1470 km/h–east = 0. (b) $\mathbf{v}_{RC}$ = 1470 km/h–east + 1470 km/h–east = 2940 km/h–east. (c) Eastward. (d) Florida is closer to the equator and is moving faster than Maine.

CHAPTER 3

Answers to Odd-Numbered Multiple Choice Questions

1. a **3.** c **5.** b **7.** b **9.** e **11.** a **13.** d **15.** c **17.** d

Answers to Odd-Numbered Problems

1. 10 m/s^2 **3.** 0.17 m/s^2–north **5.** 0.23 m/s^2 **7.** 18 m/s^2 **9.** 5.0 m/s^2 **11.** 20 m/s^2 **13.** 3.5 m/s^2 **15.** (a) a_{av} = 0 (b) a = −1.1 m/s^2 **17.** −17 km/s^2 **19.** $a = A + 2Bt$ **21.** 1×10^2 m/s^2 **23.** v_f = 4.7 m/s **25.** v_i = 48 m/s **27.** $a = -0.9 \times 10^6$ m/s^2 **29.** 0.2 s **31.** $s_3 - s_2$ = 1.6 m **33.** 244.5 m and they don't collide **35.** (a) 70 m (b) 90 m **37.** 0.1×10^2 m **39.** 10 m/s^2 **41.** 74 m **43.** 0.12 km **45.** 3.1 m/s; 10 ft/s; 6.9 mi/h **47.** −9.8 m/s, −20 m/s, −49 m/s, −98 m/s; −4.9 m,

−20 m, −0.12 km, −0.49 km **49.** $v_f =$ 423 m/s = 947 mi/h **51.** s versus t graph is a parabola. s versus t^2 graph is a straight line **53.** proof **55.** bullet falls 49.0 cm below target center **57.** 20 m **59.** the equation must be wrong **61.** $v_i = 4.9$ m/s **63.** 127.5 m **65.** $a = -gs_{max}/s_a$ **69.** 11 m/s **71.** 24.6 m, $v_f = 50.8$ m/s **73.** proof **75.** $s_B = 6.1$ m **77.** $24t^4 - 255t^3 + 901t^2 - 400 = 0$

Solutions to Selected Problems

3. Taking north as positive, $a_{av} = \Delta v/\Delta t = [(+10 \text{ m/s}) - (-10 \text{ m/s})]/(120 \text{ s}) = 0.17 \text{ m/s}^2$–north. **7.** $a_{av} = \Delta v/\Delta t = (244 \times 0.4470 \text{ m/s})/(6.2 \text{ s}) = 18 \text{ m/s}^2$. **11.** Draw a line on the curve from $t = 0$ to $t = 3$ s: the rise is 6(10 m/s); the run is 3.0 s; the slope is (60 m/s)/(3.0 s) = 20 m/s². **17.** $a_{av} = \Delta v/\Delta t = (-60 \text{ km/h} \times 1000 \text{ m/km} \div 3600 \text{ s/h})(1 \text{ s}/1000) = -17 \text{ km/s}^2$. **19.** $v = At + Bt^2$; $(v + \Delta v) = A(t + \Delta t) + B(t + \Delta t)^2 = At + A\,\Delta t + Bt^2 + B(\Delta t)^2 + B2t\,\Delta t$; $\Delta v = A\,\Delta t + B(\Delta t)^2 + 2Bt\,\Delta t$; $a_{av} = \Delta v/\Delta t = A + 2Bt + B\,\Delta t$; as $\Delta t \to 0$, $a = A + 2Bt$. **23.** $v_f^2 = v_i^2 + 2as = (1.5 \text{ m/s})^2 + 2(1.0 \text{ m/s}^2)(10 \text{ m}) = 22 \text{ m}^2/s^2$; $v_f = 4.7$ m/s. **29.** From Eq. (3.12), $s = v_i t_R - v_i^2/2a$; $t_R = (23.3 \text{ m})/(16.7 \text{ m/s}) + (16.7 \text{ m/s})/2(-7 \text{ m/s}^2) = 0.2$ s. **33.** To stop, the Express needs $s_E = -v_i^2/2a = -(26.67 \text{ m/s})^2/2(-4 \text{ m/s}^2) = 88.9$ m; the Flyer needs $s_F = -v_i^2/2a = -(30.56 \text{ m/s})^2/2(-3 \text{ m/s}^2) = 155.6$ m; for a total of 244.5 m and they don't collide. **39.** It covered 20 m in 2.0 s; $s = v_i t + \frac{1}{2}at^2$; the initial relative speed was zero; $a = 2s/t^2 = 10 \text{ m/s}^2$. **43.** His initial speed with respect to the train is 50 km/h, and he passes it as if it were standing still. Hence $s = v_i t + \frac{1}{2}at^2$, $500 \text{ m} = (13.89 \text{ m/s})t + \frac{1}{2}(10 \text{ m/s}^2)t^2$, and using the quadratic formula, $t = [-(13.89 \text{ m/s}) \pm \sqrt{(13.89 \text{ m/s})^2 - 4(5.0 \text{ m/s}^2)(-500 \text{ m})}]/2(5.0 \text{ m/s}^2) = (-13.89 \text{ m/s} \pm 100.96 \text{ m/s})/(10 \text{ m/s}^2) = 8.7$ s; the train travels $s = (13.89 \text{ m/s})(8.7 \text{ s}) = 0.12$ km. **49.** Taking down as positive, $v_f^2 = v_i^2 + 2as = 0 + 2(9.81 \text{ m/s}^2)(9144 \text{ m})$; $v_f = 423$ m/s = 947 mi/h. **53.** $v_{iH} = v_i \cos\theta$, $v_{iV} = v_i \sin\theta$, hence Eq. (3.16) yields the desired expression. **57.** (Taking down as positive) $s_V = \frac{1}{2}gt^2 = \frac{1}{2}(9.81 \text{ m/s}^2)(2.0 \text{ s})^2 = 19.6$ m, or to two figures, 20 m. **61.** $s = v_i t + \frac{1}{2}at^2$; $0 = v_i(1.0 \text{ s}) + \frac{1}{2}(-9.81 \text{ m/s}^2)(1.0 \text{ s})^2$; $v_i = 4.9$ m/s. **65.** The lift-off speed is v_l, where $v_l^2 = 2as_a$. During the deceleration, $0 = v_l^2 + 2gs_{max}$ or $v_l^2 = -2gs_{max} = 2as_a$; $a = -gs_{max}/s_a$. **71.** To avoid the quadratic, use Eq's. (3.10) and (3.8) applied to the vertical motion, where $v_{iV} = (40.0 \text{ m/s}) \times \sin 60.0^\circ = 34.64$ m/s. Taking down as positive, $v_{fV}^2 = (34.64 \text{ m/s})^2 + 2(9.81 \text{ m/s}^2)(50.0 \text{ m})$; and $v_{fV} = 46.70$ m/s. $s_V = 50.0 \text{ m} = \frac{1}{2}(v_{iV} + v_{fV})t = (40.67 \text{ m/s})t$ and $t = 1.23$ s. It strikes the water a distance downrange of $v_H t = (40.0 \text{ m/s})(\cos 60.0^\circ) \times (1.23 \text{ s}) = 24.6$ m. It hits at a speed of $v_f = (v_{fH}^2 + v_{fV}^2)^{1/2} = 50.8$ m/s. **75.** Down is positive; over his height $s = v_i t + \frac{1}{2}at^2$; $2 \text{ m} = v_i(0.20 \text{ s}) + \frac{1}{2}(9.81 \text{ m/s}^2)(0.20 \text{ s})^2$; $v_i = 9.02$ m/s at top of his head; at ground $v_f = v_i + gt = 9.02 \text{ m/s} + (9.81 \text{ m/s}^2) \times (0.20 \text{ s}) = 10.98$ m/s; for total fall, $v_f^2 = v_i^2 + 2as_B$; $(10.98 \text{ m/s})^2 = 0 + 2(9.81 \text{ m/s}^2)s_B$; $s_B = 6.1$ m. **77.** Using the Pythagorean Theorem, we obtain $(20 \text{ m})^2 = (v_H t)^2 + (v_{iV}t + \frac{1}{2}gt^2)^2$, $v_{iH} = v_i \cos\theta = (30 \text{ m/s}) \times \cos 60^\circ$ and $v_{iV} = v_i \sin\theta = (30 \text{ m/s}) \sin 60^\circ$. $24t^4 - 255t^3 + 901t^2 - 400 = 0$.

CHAPTER 4

Answers to Odd-Numbered Multiple Choice Questions

1. a **3.** a **5.** c **7.** c **9.** e **11.** c **13.** b **15.** c **17.** a **19.** c

Answers to Odd-Numbered Problems

1. $\mathbf{v}_M = 10$ m/s–horizontal **3.** 1.0000 m **5.** 8.36 km compared to 110 m **7.** 16.7 km **9.** 100 N **11.** 100 N on each person **13.** $F = 298$ N at $\theta = +35^\circ$ **15.** 0.12 kN at $\theta = 54^\circ$ **17.** (a) 24.0 m (b) 96.0 m **21.** 500 N at 76.9° up from x-axis **23.** one due east; the other two, ±60° **25.** 400 N–straight up **27.** 37 m **29.** −10 N·s **31.** unchanged, 10 kg·m/s **33.** −2 MN **35.** 0.012 s **37.** 3.3 m/s–south **39.** −0.12 m/s **41.** 1.53 kN **43.** 0.3 N **45.** 203 N **49.** $s_h = 1.7$ m **51.** 42 MN **53.** 1.0 kN **55.** −1.00 kN **57.** 18 m/s **59.** 0.0202 m/s opposite to snowball **61.** 83.9 kN **63.** −0.12 MN **65.** $v_{Nf} = -0.100$ m/s, $v'_{Sf} = 0.198$ m/s, $v''_{Sf} = 0.400$ m/s

Solutions to Selected Problems

1. At maximum altitude, $v_V = 0$, hence $\mathbf{v}_M = 10$ m/s–horizontal. **5.** Using Eq. (3.15), $s_p = -v^2/2g = 8.36$ km, which is tremendously different from 110 m! The difference is due to drag, which is large because of the high speed and effective because of the small mass. **11.** The force is 100 N on each person. **13.** $F_{1x} = (100 \text{ N}) \times \cos 45^\circ = 70.7$ N; $F_{2x} = (200 \text{ N}) \cos 30^\circ = 173.2$ N; $F_{1y} = (100 \text{ N}) \sin 45^\circ = 70.7$ N; $F_{2y} = (200 \text{ N}) \sin 30^\circ = 100$ N; $F^2 = (243.9 \text{ N})^2 + (170.7 \text{ N})^2$; $F = 298$ N; at $\theta = \tan^{-1} 170.7/243.9 = +35^\circ$. **17.** Just to use something other than Eq. (3.15), $v_{iV} = (30.7 \text{ m/s}) \sin 45^\circ = 21.71$ m/s, take up as positive; $v_{fV} = v_{iV} + gt$, $0 = 21.71 \text{ m/s} + (-9.81 \text{ m/s}^2)t_p$; $t_p = 2.21$ s. $s_{Hp} = v_H t_p = (21.7 \text{ m/s})(2.21 \text{ s}) = 48.0$ m. (a) $s_p = \frac{1}{2}(v_{iV} + v_{fV})t_p = \frac{1}{2}(21.7 \text{ m/s})(2.21 \text{ s}) = 24.0$ m above the launch height. (b) $s_R = 2s_{Hp} = 96.0$ m. **21.** There is 90° between the two vectors, which form a 3-4-5 right triangle, so the resultant is 500 N at an angle of 36.9° + 40° up from the x-axis. **25.** By symmetry, the horizontal components of the forces cancel, leaving only 4(200 N) × cos 60° = 400 N acting straight up. **29.** Taking the direction of motion as positive, $\Delta p = m\,\Delta v = (1.0 \text{ kg}) \times (-10 \text{ m/s}) = -10$ N·s. **33.** To find F_{av}, the force exerted on the gas, take the direction of the exhaust as positive and use $F_{av}\,\Delta t = m\,\Delta v$; $F_{av} = (1000 \text{ kg})(2 \text{ km/s})/(1.000 \text{ s}) = 2$ MN. Thrust is −2 MN. **37.** The initial momentum of the boat-person system is zero, $\mathbf{p}_i = \mathbf{p}_f = 0$, and $m_p v_{pf} + m_b v_{bf} = 0$; taking motion north as positive, $(50 \text{ kg})(+10 \text{ m/s}) = -(150 \text{ kg})v_{bf}$; $v_{bf} = -3.3$ m/s; or 3.3 m/s–south. **43.** 2 oz = 2/16 lb; $m = (2/16 \text{ lb}) \times (0.4536 \text{ kg/lb}) = 0.0567$ kg; $F_{av} = m\Delta v/\Delta t = (0.0567 \text{ kg})(0.50 \text{ m/s})/(0.1 \text{ s}) = 0.284$ N, or 0.3 N. **47.** $F_{av} = m\,\Delta v/\Delta t$; impact speed from Eq. (3.10), $v_I = \sqrt{2gs_h}$. The duration of the collision follows from Eq. (3.8), $s_c = \frac{1}{2}v_I\,\Delta t$, and so $\Delta t = 2s_c/v_I$; hence, $F_{av} = m\,\Delta v/\Delta t = mv_I/(2s_c/v_I) = mv_I^2/2s_c = mgs_h/s_c$. **53.** Taking the initial direction as positive, $F_{av} = m\,\Delta v/\Delta t$; from Eq. (3.8), $s = \frac{1}{2}v_i\,\Delta t$, $\Delta t = 2(0.20 \text{ m})/(200 \text{ m/s}) = 2.0$ ms; $F_{av} = (0.010 \text{ kg})(200 \text{ m/s})/(0.0020 \text{ s}) = 1.0$ kN. **57.** $p_i = (10\,000 \text{ kg})(20 \text{ m/s}) = 2.0 \times 10^5 \text{ kg·m/s} = p_f = (10\,000 \text{ kg} + 900 \text{ kg})v_f$; $v_f = 18.3$ m/s, or 18 m/s. **61.** Let's see what a is first, then t, and then F. $v_f^2 = v_i^2 + 2as$; $a = (69.5 \text{ m/s})^2/2(914.4 \text{ m}) = 2.64 \text{ m/s}^2$; $v_f = v_i + at$; $t = (69.5 \text{ m/s})/(2.64 \text{ m/s}^2) = 26.3$ s; m = 31 751 kg; $F = m\,\Delta v/\Delta t = (31\,751 \text{ kg})(69.5 \text{ m/s})/(26.3 \text{ s}) = 83.9$ kN. We could have started with Eq. (3.8). **63.** 120 mi/h = 53.64 m/s; if v_T is his terminal speed and s_D the depth of the crater, Eq. (3.8) yields $\Delta t = 2s_D/v_T = 0.0398$ s. $F_{av} = m\,\Delta v/\Delta t = (90 \text{ kg})(-53.64 \text{ m/s})/(0.0398 \text{ s}) = -0.12$ MN. **65.** $p_{Ni} + p_{ai} = 0 = p_{Nf} + p_{af} = (100 \text{ kg})v_{Nf} + (0.500 \text{ kg}) \times (20.0 \text{ m/s})$; $v_{Nf} = -0.100$ m/s—he moves in the opposite direction to the asteroid, whose direction is taken as positive; she catches it and so $p'_{ai} = p'_{af} + p'_{Sf}$; $(0.500 \text{ kg}) \times (20.0 \text{ m/s}) = v'_{Sf}(m_a + m_S) = v'_{Sf}(50.5 \text{ kg})$; $v'_{Sf} = 0.198$ m/s. Now she throws it back, $p''_{af} + p''_{Sf} = p''_{Si} + p''_{ai}$; $(50.5 \text{ kg}) \times (0.198 \text{ m/s}) = (0.500 \text{ kg})(-20.0 \text{ m/s}) + (50.0 \text{ kg})v''_{Sf}$; $v''_{Sf} = 0.400$ m/s.

CHAPTER 5

Answers to Odd-Numbered Multiple Choice Questions

1. c **3.** d **5.** e **7.** c **9.** b **11.** c **13.** d **15.** d **17.** a **19.** c

Answers to Odd-Numbered Problems

1. −71 N **3.** >4.9 N **5.** 1.5 F_w **7.** 592 N, 392 N **9.** 0.586 g **11.** 508 N, upward **13.** 4.81 m/s² **15.** $\theta = 22^\circ$ **17.** $F_{T1} = 1.00$ kN, $F_{T2} = 475$ N, $F_f = 125$ N **19.** $v_f = 12$ m/s **21.** $a = 1.7 \text{ km/s}^2$, 7.8×10^{-4} N **23.** $F_{T1} = 0.03$ kN, $F_{T3} = 0.02$ kN, $a = \frac{1}{4}g$, the tensions are internal to the 3-body system, the weights are external **25.** $g_M = 3.4 \text{ m/s}^2$ **27.** 17 m/s **29.** $F_C = 6.6 \times 10^2$ N, friction **31.** $F_C = 2.1$ kN **33.** $R = 9.5 \times 10^2$ m **35.** $a_C = 1.8 \times 10^6 \text{ m/s}^2$ **37.** 0.12 kN **39.** $F_N = m(v^2/r + g)$ **41.** 19.2 m/s **43.** (a) time to go once around, $T = 55$ s, or 0.018 rotations per second (b) $a_C = 4\pi^2 r/T^2$ **45.** 99.7 N **47.** $\mu_s = 0.53$ **49.** $a = \mu_s g$ **51.** 0.31 **53.** crate slides toward back of truck **55.** 0.10 **57.** 20 m **59.** 3 s **61.** $\mu_k = 0.070$ **63.** $\mu_k = 0.30$ **65.** (a) $a = 0$ (b) $a = 1$ m/s^2 **67.** $\theta = 0.2 \times 10^2$ degrees **69.** $\mu_k = 0.11$

Solutions to Selected Problems

1. $F = ma = 0$; $F_f = -F \cos 45° = -(100 \text{ N})(0.707) = -71$ N. **7.** $+\uparrow\Sigma F_V = ma = F_N - F_W$; $F_N = ma + F_W = m(a + g) = (40.0 \text{ kg})(14.81 \text{ m/s}^2) = 592$ N. $F_W = mg = 392$ N. **11.** $+\downarrow\Sigma F_V = ma = F_W + F_f$; $m(2.00 \text{ m/s}^2) = mg + F_f$; $F_f = -(7.81 \text{ m/s}^2)(65.0 \text{ kg}) = -508$ N, or 508 N, upward. **17.** The tension in the front rope is 1.00 kN. For the first barge (taking the pull of the tug to be positive and to the right), $\xrightarrow{+}\Sigma F_{H1} = m_1 a = 1.00 \text{ kN} - F_{T2} - F_f = (4.00 \times 10^3 \text{ kg})(0.100 \text{ m/s}^2)$; for the second barge, $\xrightarrow{+}\Sigma F_{H2} = m_2 a = F_{T2} - F_f = (3.50 \times 10^3 \text{ kg})(0.100 \text{ m/s}^2)$. Adding the two equations, $-2F_f = (7.50 \times 10^3 \text{ kg}) \times (0.100 \text{ m/s}^2) - 1.00$ kN; $F_f = +125$ N, opposing the motion; $F_{T2} = 475$ N. **21.** $s_R = 12$ in. $= 0.3048$ m $= [2v_i^2/(9.8 \text{ m/s}^2)] \cos\theta \sin\theta$; $v_i = 1.729$ m/s; (a) $a = \Delta v/\Delta t = (1.729 \text{ m/s})/(1.0 \times 10^{-3} \text{ s}) = 1.7$ km/s^2; (b) $F = ma = (4.5 \times 10^{-7} \text{ kg}) \times (1.73 \text{ km/s}^2) = 7.8 \times 10^{-4}$ N. **23.** For mass-1, $+\downarrow\Sigma F_{V1} = m_1 a = F_{W1} - F_{T1}$; for mass-2, $\xrightarrow{+}\Sigma F_{H2} = m_2 a = F_{T1} - F_{T2}$; for mass-3, $+\uparrow\Sigma F_{V3} = m_3 a = F_{T2} - F_{W3}$; adding all three of these, $a(m_1 + m_2 + m_3) = F_{W1} - F_{W3} = g(m_1 - m_3)$; (b) $a = g(2 \text{ kg})/(8 \text{ kg}) = \frac{1}{4}g$; (c) the tensions are internal to the 3-body system, the weights are external; (a) $m_1 a = m_1 g - F_{T1}$; $F_{T1} = m_1(g - a) = 29.4 \text{ N} = 0.03$ kN. $m_3 a = F_{T3} - F_{W3}$, $F_{T3} = m_3(a + g) = 24.5 \text{ N} = 0.02$ kN. **27.** $F_C = mv^2/r$, $v = \sqrt{(20 \text{ N})(0.3556 \text{ m})/(0.025 \text{ kg})} = 17$ m/s. **31.** The distance traveled once around is $2\pi r$, hence $v = 2(2\pi r)/(1.0 \text{ s})$; $r = 1.828$ m; $v = 22.98$ m/s; $F_C = mv^2/r = (7.257 \text{ kg})v^2/r = 2096.8$ N, or 2.1 kN. **35.** $a_C = v^2/r$; $r = (0.100 \text{ m}) \sin 30° + 0.050 \text{ m} = 0.10$ m; $v = 2\pi r(40\,000)/60 = 418.88$ m/s; $a_C = 1.8 \times 10^6$ m/s^2. **39.** Using Fig. 5.24 with F_T replaced by F_N, we get $F_N = m(v^2/r + g\cos\theta)$; at the bottom, $\theta = 0$, $\cos\theta = 1$; $F_N = m(v^2/r + g)$. **43.** (a) $a_C = v^2/r = 1.0g$; $v^2 = \frac{1}{2}(1500 \text{ m})g$; $v = 85.76$ m/s; the cylinder must go once around a distance $(2\pi r)$ in a time $(2\pi r)/v = T = 54.9$ s, or 55 s; (b) $a_C = v^2/r = (2\pi r/T)^2/r = 4\pi^2 r/T^2$, where the time to go once around is the same throughout the station. Hence, the simulated gravity varies linearly with r, decreasing with distance "up" from the floor. **45.** $+\downarrow\Sigma F_V = ma_C = F_G - F_N$; F_N is equal and opposite to the effective weight; F_G is the actual gravitation force downward, namely, 100 N. $F_N = F_G - ma_C = F_G - (F_G/g)v^2/R_\oplus = 100 \text{ N} - 0.345 \text{ N} = 99.7$ N. **49.** $F = ma$; $a = F_f/m = mg\mu_s/m = \mu_s g$. **53.** For no slipping, $+\swarrow\Sigma F_\parallel = 0 = F_W \sin\theta - F_f(\text{max})$; and $\mu_s = \tan\theta$, $0.3 = \tan\theta$; $\theta = 16.7°$; hence, the crate slides toward the back of the truck. **57.** $v_i = 50.0 \text{ km/h} = 13.89$ m/s; to find the stopping distance, $v_f^2 = v_i^2 + 2as$; $s = -v_i^2/2a$; where $a < 0$ is determined from $F_f(\text{max}) = \mu_s mg = ma$; hence, $a = \mu_s g$ and so $s = -v_i^2/2(-\mu_s g) = 20$ m. **61.** $+\swarrow\Sigma F_\parallel = 0 = F_W \sin\theta - F_f$; $\tan\theta = \mu_k = 0.070$. At -10°C, the water film that usually forms via melting under the skis would not. **65.** (a) $F_H = F\cos\theta = (300 \text{ N})(0.866) = 259.8$ N, which must exceed $F_f(\text{max}) = 0.5[(50 \text{ kg})g + (300 \text{ N}) \times \sin 30°] = 320$ N, but it does not, so there is no acceleration; (b) $F_f(\text{max}) = 256$ N—it moves! $\xrightarrow{+}\Sigma F_H = ma = 259.8 \text{ N} - 0.3(640.3 \text{ N}) = 67.7$ N and $a = 1.4$ m/s^2, or because the friction coefficients are known to only one figure, $a = 1$ m/s^2. **69.** $v_f^2 = 2as$, $a = (8.33 \text{ m/s})^2/2(30.0 \text{ m}) = 1.158$ m/s^2; $v_f = at$, $t = (8.33 \text{ m/s})/(1.158 \text{ m/s}^2) = 7.196$ s; the box accelerates with respect to the truck at a_b, where $s_b = \frac{1}{2}a_b t^2 = 1.00$ m, and $a_b = 0.0386$ m/s^2. If there were no friction, it would have accelerated back at 1.158 m/s^2; hence, $(1.158 \text{ m/s}^2 - 0.0386 \text{ m/s}^2)m = F_f = \mu_k mg$, $\mu_k = 0.11$.

CHAPTER 6

Answers to Odd-Numbered Multiple Choice Questions

1. d **3.** d **5.** c **7.** e **9.** b **11.** a **13.** e **15.** d **17.** d **19.** c

Answers to Odd-Numbered Problems

1. (a) 250 N (b) 200 N **3.** (a) 30 N (b) 38 N, 2.0 N **5.** they slide equally **7.** $F_{Tb} = 0$, $F_{T1} = 98$ N **9.** $\theta = 38°$, $F_T = 16$ N **11.** 39.7 N **13.** $F_T = 50.0$ N, 100 N **15.** $F_{BC} = 0$, $F_{DE} = 0$, $F_{FG} = 0$ **17.** 6.7 N **19.** $F_T = 3.1$ kN **21.** 75°, 193 N **23.** $\mu_s = 0.93$ **25.** $F_{RB} = 2.00$ kN, $F_{RA} = 1.73$ kN $= F_{RC}$, $F_{RD} = 3.00$ kN **27.** 100 N·m **29.** 140 N **31.** $F_R = 355.9$ N, $x = 0.7317$ m **33.** $F_{RB} = 6000$ N, $F_{RA} = 2000$ N **35.** $F_{N1} = F_{N2} = 69.3$ N **37.** 1.0 kg **39.** $F_{RD} = 360$ N, $F_{RC} = 80.0$ N, $F_{RA} = 40$ N **41.** $F_n = 0.83$ kN **43.** $F_f = 11$ N **45.** $F_{AH} = -1.5$ kN (to the left); $F_{AV} = -0.50$ kN (down) **47.** midface in the glue 8 cm from the end **49.** $x_{cg} = 0.063$ m, $y_{cg} = 0.063$ m **51.** $F_{T2} = 100$ N, $F_W = 173$ N **53.** yes **55.** yes **59.** $F_{RG} = 0.31$ kN at 74° **61.** $F_{Nh} = 0.21$ kN, $F_{Nt} = 0.11$ kN **63.** $x_{cg} = F_{N2}h/(F_{N1} + F_{N2})$ **65.** $F_T = 90.4$ N

Solutions to Selected Problems

3. (a) $(2.0 \text{ N/m})(20 \text{ m})(3/4) = 30$ N. (b) 1.0 m from the top, the tension is 38 N; 1.0 m from the bottom, it's 2.0 N. **5.** For each one, $+\swarrow\Sigma F_\parallel = F_W \sin\theta - \mu_s F_W \cos\theta = 0$; and $\mu_s = \tan\theta$ independent of weight, so they tend to slide equally. **9.** $+\uparrow\Sigma F_V = (20 \text{ N})\cos 30° + F_T\cos\theta - 30 \text{ N} = 0$; $\xrightarrow{+}\Sigma F_H = F_T\sin\theta - (20 \text{ N})\sin 30° = 0$; hence, $F_T\cos\theta = 12.679$ N and $F_T\sin\theta = 10$ N. We could divide these and get $\sin\theta/\cos\theta = 0.7887 = \tan\theta$ and $\theta = 38°$, or square and add them, since $\sin^2\theta + \cos^2\theta = 1$, and get $F_T^2 = (12.697 \text{ N})^2 + (10 \text{ N})^2$; $F_T = 16$ N. **15.** Since BC is perpendicular to AC and CE at joint C, there can be no vertical force in BC; hence, force in BC is zero. The same is true at joint D (so $F_{DE} = 0$) and at joint G (so $F_{FG} = 0$). **19.** $+\uparrow\Sigma F_V = 2F_T \sin 5.0° - 533.8 \text{ N} = 0$; $F_T = 3.1$ kN. **25.** On the small ball, $\xrightarrow{+}\Sigma F_H = F_{RA} - F_{RB}\cos 30° = 0$ and $+\uparrow\Sigma F_V = F_{RB}\sin 30.0° - 1.00 \text{ kN} = 0$; $F_{RB} = 2.00$ kN. $F_{RA} = 1.73 \text{ kN} = F_{RC}$; $F_{RD} =$ the net weight $= 3.00$ kN. **31.** $+\uparrow\Sigma F_V = 533.8 \text{ N} - 889.66 \text{ N} + F_R = 0$; $F_R = 355.9$ N; $\circlearrowleft{+}\Sigma\tau_S = (355.9 \text{ N}) \times (1.829 \text{ m}) - (889.66 \text{ N})x = 0$; $x = 0.7317$ m from Selma. **35.** The reaction forces are each normal to the surfaces because there is no friction. $F_{N1} = F_W \sin 45.0° = F_W \cos 45.0° = F_{N2} = 69.3$ N. **41.** $\circlearrowleft{+}\Sigma\tau = (300 \text{ N})(0.25 \text{ m}) - F_n(0.09 \text{ m}) = 0$; $F_n = 0.83$ kN. **45.** Take the horizontal bar out and draw a free-body diagram; $\circlearrowleft{+}\Sigma\tau_A = (1000 \text{ N})(3.0 \text{ m}) - (2.0 \text{ m})F_C \sin 45° = 0$; $F_C = 2121.6$ N. At A, the horizontal force is $F_{AH} = -F_{CH} = -1.5$ kN (force is to the left). $+\uparrow\Sigma F_V = F_C \sin 45° - 1000 \text{ N} - F_{AV} = 0$; $F_{AV} = -0.50$ kN (force is down). **49.** Taking the upright side as $y = 0$, $\Sigma mx = (0.50 \text{ kg/m})(0.25 \text{ m})(0.125 \text{ m}) = 0.0156$ kg·m and $x_{cg} = (0.0156 \text{ kg·m})/(0.25 \text{ kg}) = 0.063$ m, and, by symmetry, $y_{cg} = 0.063$ m. **55.** If each is of length L, the *c.g.* of the topmost block is at the edge of the one beneath it. Measured from the left end, for the top two blocks, $x_{cg} = (mL + mL/2)/2m = 3L/4$, and that is located just above the edge of the third block. For the top three blocks, $x_{cg} = (2mL + mL/2)/3m = 5L/6$. For the top four blocks, $x_{cg} = (3mL + mL/2)/4m = 7L/8$. The overhangs are $L/8$, $L/6$, and $L/4$, which is more than $L/2$. **59.** $+\uparrow\Sigma F_V = F_N - g(30 \text{ kg}) = 0$; $F_N = 0.29$ kN. $\circlearrowleft{+}\Sigma\tau = 0 = F_N(8.0 \text{ m})\cos 60° - F_f(8.0 \text{ m})\sin 60° - g(30 \text{ kg})(4.0 \text{ m})\cos 60°$; $F_f = 85$ N, $F_{RG} = 0.31$ kN at 74°. **63.** $\circlearrowleft{+}\Sigma\tau_1 = F_W x_{cg} - F_{N2}h = 0$; $\circlearrowleft{+}\Sigma\tau_2 = F_{N1}h - F_W(h - x_{cg}) = 0$; $F_W = F_{N2}h/x_{cg}$; $F_{N1}h - F_{N2}h(h - x_{cg})/x_{cg} = 0$; $x_{cg} = F_{N2}h/(F_{N1} + F_{N2})$. **65.** Taking O at the base of the wall, $\circlearrowleft{+}\Sigma\tau_o = (4.00 \text{ m})F_{N2} - (3.00 \text{ m})F_{N1} - F_W(2.00 \text{ m}) = 0$; and $\xrightarrow{+}\Sigma F_H = F_T\cos\theta - F_{N1} = 0$, where θ is the angle the rope makes with the ground. From the geometry (we are dealing here with a 3-4-5 right triangle), $\theta = 10.62°$. $+\uparrow\Sigma F_V = F_{N2} - F_W - F_T\sin\theta = 0$. Eliminating F_{N1} from the first and second equations, $-(3.00 \text{ m}) \times F_T\cos\theta - F_W(2.00 \text{ m}) + (4.00 \text{ m})F_{N2} = 0$; and substituting F_{N2} into the third equation, $(0.75\,F_T\cos\theta + 0.50\,F_W) - F_W - F_T\sin\theta = 0$. Hence, $F_T(0.75\cos\theta - \sin\theta) = 0.50\,F_W$; $F_T = 90.4$ N.

CHAPTER 7

Answers to Odd-Numbered Multiple Choice Questions

1. c **3.** d **5.** c **7.** c **9.** b **11.** d **13.** c **15.** b

Answers to Odd-Numbered Problems

1. would be halved **3.** m = 1.22×10^5 kg **5.** 4 times larger **7.** 2.5×10^{16} N **9.** $s_{♀} = 1.1$ m **11.** $F_E = 2 \times 10^{39} F_G$ **13.** 0.11 $M_\oplus$ **15.** 4.4 m **19.** 0.379 g_0 **21.** $g_n = 1 \times 10^{12}$ m/s^2 **23.** $F_G = GmMr/R^3$ **25.** $T = 1 \times 10^{-3}$ s **29.** 1.62 m/s^2 **31.** 1.65×10^3 m/s **35.** $T_2 = 2.8\,T_1$ **37.** 3.32×10^{-5} m/s^2 **39.** 5.9×10^{-3} m/s^2 **41.** 8.000 Earth-years **45.** 1.078 AU, 1.12 Earth-years **47.** $GM/2R^2$ **49.** 1.05×10^3 kg/m^3 **51.** 9.7 d, 1.440 km/s

Solutions to Selected Problems

3. $F_G = Gm^2/r^2$; 1.00 N = $(6.67 \times 10^{-11}$ N·m^2/kg$^2)m^2/(1.00\text{ m})^2$; $m = 1.22 \times 10^5$ kg. **9.** $v_f^2 = 0 = v_i^2 - 2gs$; $s = v_i^2/2g$, $s_\oplus/s_{♀} = g_{♀}/g_\oplus$, $(1.00\text{ m})/s_{♀} = 0.88$, $s_{♀} = 1.1$ m. **13.** $g_{♂} = GM_{♂}/R_{♂}^2$, $M_{♂} = (3.7\text{ m/s}^2)(\frac{1}{2}6.8 \times 10^6\text{ m})^2/(6.67 \times 10^{-11}\text{ N·m}^2/\text{kg}^2) = 6.4 \times 10^{23}\text{ kg} = 0.11\,M_\oplus$. **15.** To find $g_{♀}$, we need the mass; $M_{♀} = \frac{4}{3}\pi r^3\rho = 4.82 \times 10^{24}$ kg; $g_{♀} = GM_{♀}/R_{♀}^2 = 8.79$ m/s^2; $s = \frac{1}{2}g_{♀}t^2 = 4.4$ m. **19.** $g_{♂} = GM_{♂}R_{♂}^2 = G(0.108\,M_\oplus)/R_\oplus^2(0.534)^2 = 0.379\,g_0$. **23.** $F_G = GmM_r/r^2$, where $M_r = \frac{4}{3}\pi r^3\rho$, and $\rho = M/(4\pi R^3/3)$; hence, $M_r = Mr^3/R^3$ and $F_G = GmMr/R^3$. **27.** $g_p = GM/(R + h)^2 = GM/R^2(1 + h/R)^2 = (GM/R^2)(1 + h/R)^{-2}$. From the binomial expansion, the bracketed term becomes $1 + (-2)(1)h/R + \frac{1}{2}(-2)(-3) \times (1)(h/R)^2 + \ldots$, which is $\approx(1 - 2h/R)$ since the $(h/R)^2$ term is very small. **29.** $g_☾ \approx GM_☾(1 - 2h/R)/R^2 \approx GM_☾[1 - (200\text{ m})/(1.74 \times 10^6\text{ m})]/R_☾^2 \approx 0.999\,885\,GM_☾/R_☾^2 \approx 1.62$ m/s^2. The difference is negligible. **33.** From Eq. (7.7), $T^2 = 4\pi^2 r_\oplus^3/M_\oplus G = 9.90 \times 10^{-14} r_\oplus^3$; $T = 3.15 \times 10^{-7} r_\oplus^{3/2}$. **37.** $g_☾ = GM_☾/r_{☾\oplus}^2 = 3.32 \times 10^{-5}$ m/s^2. **41.** From Eq. (7.6), $T^2 = r_{♂}^3/C_\odot = (4.000\text{ AU})^3/(1\text{ AU}^3/\text{Earth-year}^2) = 64.00$ Earth-years^2; $T = 8.000$ Earth-years. **45.** $r_{\odot I} = \frac{1}{2}(0.186\text{ AU} + 1.97\text{ AU}) = 1.078$ AU. $r_{\odot I}^3/T_I^2 = 1 = (1.078\text{ AU})^3/T_I^2$; $T_I = 1.12$ Earth-years. **51.** From Eq's. (7.6) and (7.7), you find that $C_\oplus = GM_\oplus/4\pi^2 = 1.010 \times 10^{13}\text{ m}^3/\text{s}^2 = r_{\oplus s}^3/T_s^2 = (\frac{1}{2}3.844 \times 10^8\text{ m})^3/T_s^2$ and $T_s = 0.838\,5 \times 10^6$ s = 9.7 days; $v = 2\pi r_{\oplus s}/T_s = 1.440$ km/s. Check this answer using Eq. (7.8).

CHAPTER 8

Answers to Odd-Numbered Multiple Choice Questions

1. b **3.** e **5.** b **7.** c **9.** b **11.** b **13.** d **15.** a **17.** d **19.** b

Answers to Odd-Numbered Problems

1. 0.017 453 rad **3.** 1.6 rad, 0.02 rad, 0.02 rad, 2.0 rad **5.** $D = 3 \times 10^5$ m **7.** 2.0 m **9.** 69 rev **11.** 12.2 rad/s **13.** 2.3 m **15.** 105 rad/s^2 **17.** 4.2 s **19.** 2.97×10^4 m/s, 1.99×10^{-7} rad/s **21.** 0.2 m **23.** 0.029 rad/s, 0.26 m/s^2 **25.** 2.7 m/s **27.** $v_E = 0.1$ m/s, upward **29.** $\alpha = 9.82$ rad/s^2, 16.0 s **31.** 4.84×10^3 km, $\theta = 40.6°$ **33.** $\omega_4 = 1200$ rpm, $\omega_5 = 400$ rpm, $\omega_n = \omega_1 N_1/N_n$; if n is odd, the direction is unchanged **35.** 1.4 m **37.** 0.75 m/s–upward **39.** 25 rev **41.** -2.4×10^{-22} rad/s^2, 6.51×10^{-5} rad/s, 27 h **43.** MR^2 **45.** 450% decrease **47.** $\frac{1}{3}ml^2$ **49.** 3×10^{-3} kg·m^2/s **51.** 0.20 rad/s^2 **53.** 426 rad/s^2 **55.** $a = mg/(m + I/R^2)$ **57.** $a = 1.5$ m/s^2 **59.** (a) 1.6×10^{-4} kg·m^2 (b) 8.0×10^{-2} kg·m^2 (c) ≈ 0.080 kg·m^2 (d) 0.010 kg·m/s (e) 4.0×10^{-3} kg·m^2/s (f) 5.0×10^{-2} rad/s **61.** $a = \frac{1}{2}g \sin\theta = g/4$ **65.** 1.2 m/s, 1.8×10^4 kg·m^2/s, 24 m/s, 17×10^3 N **69.** $I \approx \frac{1}{3}Ml^2$

Solutions to Selected Problems

3. (ii) 90° = 1.6 rad; $\tan\alpha = 1/100$, $\alpha = 5.7 \times 10^{-1}$, $2\alpha = 1.15° = 0.020\,1$ rad; $\theta = l/r = (2.00\text{ m})/(100\text{ m}) = 0.020\,0$ rad. (i) $\theta = l/r = 2.00$ rad. **5.** $l = r\theta = (3.8 \times 10^8\text{ m})(8 \times 10^{-4}) = 3 \times 10^5$ m, which is the diameter. **9.** $2\pi r = 2\pi(0.15\text{ m}) = 0.942$ m, $(65\text{ m})/2\pi r = 69$ rev. **13.** $\omega = 2\pi(1800\text{ rpm})/(60\text{ s/m}) = 188.5$ rad/s, $v = R\omega = (0.12\text{ m}) \times (188.5\text{ rad/s}) = 22.62$ m/s; $l = vt = (22.62\text{ m/s})(0.10\text{ s}) = 2.3$ m. **17.** $\omega_i = 200(2\pi)/(60\text{ rad/s}) = 20.94$ rad/s; $0 = \omega_i + (-5\text{ rad/s}^2)t$; $t = 4.2$ s. **21.** $\omega^2 = 0 + 2\alpha\theta$, $(20 \times 2\pi/60)^2 = 2(5\text{ rad/s}^2)\theta$, $\theta = 0.4$ rad, or 0.07 turns. (0.4 rad)(0.5 m) = 0.2 m. **25.** $v = 2\pi R(100\text{ rpm})/(60\text{ s/min}) = 8.7$ ft/s = 2.7 m/s. **29.** $\omega_f^2 = \omega_i^2 + 2\alpha\theta$, $(1500 \times 2\pi/60)^2 = 2\alpha(200 \times 2\pi)$, $\alpha = 9.82$ rad/s^2. $\omega_f = \omega_i + \alpha t$, $t = (1500 \times 2\pi/60)/(9.82\text{ rad/s}^2) = 16.0$ s. **33.** 1500 rpm × 20 = $\omega_4 25 = \omega_5 75$; $\omega_4 = 1200$ rpm; $\omega_5 = 400$ rpm. $\omega_n = \omega_1 N_1/N_n$, where N is the number of teeth and n is the number of gears. If n is odd, the direction is unchanged. **35.** $a_C = R\omega^2$, $a_T = R(4.0\text{ rad/s}^2)$, $a = \sqrt{a_C^2 + a_T^2}$, $(8.0\text{ m/s}^2)^2 = R^2\omega^4 + R^2(4.0\text{ rad/s}^2)^2$; $R^2 = 2.0$ m^2, $R = 1.4$ m. **37.** $l_3 = (0.24\text{ m})\theta_1$, $l_5 = (0.12\text{ m})\,\theta_1 = (0.20\text{ m})\theta_2$, $l_4 = (0.15\text{ m})\theta_2$, so $\theta_2 = 0.60\theta_1$ and $l_4 = (0.15\text{ m})(0.60\,\theta_1) = (0.90 \times 10^{-1}) \times \theta_1 = 0.90 \times 10^{-1}(l_3/0.24) = 0.375l_3$, $v_4 = 0.375v_3 = 0.375(2.0\text{ m/s}) = 0.75$ m/s, upward. **45.** 450% decrease; $L_i = L_f$. **49.** $L = I\omega$; $I \approx mr^2 = (2 \times 10^{-3}\text{ kg}) \times (0.5\text{ m})^2 = 5 \times 10^{-4}$ kg·m^2; $L \approx (5 \times 10^{-4}\text{ kg·m}^2)(2\pi\text{ rad/s}) = 3 \times 10^{-3}$ kg·m^2/s. **53.** $\Sigma\tau = 50.0$ ft·lb $= I\alpha$, $I = mR^2 = (F_W/g)R^2$, 50.0 ft·lb = $[(3.22\text{ lb})/(32.2\text{ ft/s}^2)] \times [(13\text{ in.})/(12\text{ in./ft})]^2\alpha$; $\alpha = 426$ rad/s^2. **57.** $+\!\downarrow\Sigma F_y = ma = F_W - F_T$; $\circlearrowright\!+\Sigma\tau = I\alpha = F_T R - 0.060$ N·m; $a = R\alpha$; $F_T = (I\alpha + 0.060\text{ N·m})/R = Ia/R^2 + (0.060\text{ N·m})/R$; $F_W - Ia/R^2 - (0.060\text{ N·m})/R = ma$, $a(m + I/R^2) = Mg - (0.060\text{ N·m})/R$, $a = 9.2/6.0 = 1.5$ m/s^2, as compared to 1.6 m/s^2 previously. **61.** $+\!\searrow\Sigma F_\parallel = F_W \sin\theta - F_f = ma$; $\Sigma\tau_{cm} = F_f R = I_{cm}\alpha = mR^2\alpha$; $a = R\alpha$; $F_f R = mRa$; $F_W = mg$; $a = \frac{1}{2}g \sin\theta = g/4$. **65.** $v = a_T t = 1.0 \times 10^{-3}g(120\text{ s}) = 1.2$ m/s; $L = rmv = 1.8 \times 10^4$ kg·m^2/s. $L = (5.0\text{ m})(150\text{ kg})v = 1.8 \times 10^4$ kg·m^2/s, $v = 24$ m/s, or 54 mi/h. $F_C = mv^2/r = 17 \times 10^3$ N, or 4×10^3 lb! **67.** $+\!\searrow\Sigma F_\parallel = mg \sin\theta - F_f = ma$; $\circlearrowright\!+\Sigma\tau_{cm} = F_f R = I\alpha$, $\alpha = a/R$, $I_{cm} = \frac{1}{2}mR^2$, $F_f = \frac{1}{2}ma$; $mg \sin\theta = (3/2)ma$, $a = (2/3)g \sin\theta$; $s = h/\sin\theta$, $v_f^2 = 2as = 2(2/3)(g \sin\theta)h/\sin\theta = (4/3)gh$. $v_f = 2\sqrt{gh/3}$.

CHAPTER 9

Answers to Odd-Numbered Multiple Choice Questions

1. a **3.** a **5.** a **7.** c **9.** c **11.** a **13.** a **15.** c **17.** e **19.** c **21.** d

Answers to Odd-Numbered Problems

1. 23 J **3.** 2.0 J **5.** 20 MJ **7.** 4.00 N **9.** 5.0 W **11.** 10 m/s **13.** 5.0 m **15.** 0.50 kN·m **17.** 1.2×10^2 J **19.** (a) 0.20 kN·m (b) 2.5 N·m/cm^2 (c) 0.20 kN·m **21.** 5.3 s **23.** 60 W/m^2 **25.** 74 W/m^2 **27.** 90 J **29.** 10^{15} W **31.** 0.98 m/s, $KE_G = 0.95$ J, $KE_B = 0.29$ kJ **33.** 6×10^2 m **35.** 1.2×10^6 J **37.** (a) 2.9×10^5 J (b) 7.2×10^4 J/s **39.** 8 min **41.** (a) 100 J (b) no rope (c) +100 J **43.** (a) 0.16 $g_\oplus$ (b) 1.6×10^5 J **45.** 1.7×10^3 m **47.** 5.2 m **49.** both **51.** 4.85×10^3 J, $h = 9.00$ m or 10.0 m above ground **53.** 0.46 km/s **55.** -6.2×10^7 J, -3.1×10^7 J, -2.1×10^7 J, -1.6×10^7 J, -1.2×10^7 J, -0.62×10^7 J, E = -1.0×10^7 J **57.** 2.4 km/s **59.** $v_{esc}^2(\text{Moon})/v_{esc}^2(\text{Earth}) = 0.046$ **61.** 2121 m/s **63.** 0.07 m/s **65.** $v_{2f} = 15$ m/s, $v_{1f} = -10$ m/s **67.** $v = \sqrt{2(m_2 - m_1)gy/(m_1 + m_2)}$ **69.** $v_B = [(m_B + m_C)/m_B]\sqrt{2gh}$

Solutions to Selected Problems

3. $W = Fs = (20\text{ N})(0.10\text{ m}) = 2.0$ J. **5.** The thrust is constant at 10×10^3 N; hence, $W = (2.0 \times 10^3\text{ m})(10^4\text{ N}) = 20$ MJ. **11.** $P = Fv = 1.0 \times 10^4\text{ J/s} = (1.0 \times 10^3\text{ N})v$; $v = 10$ m/s, or 22 mi/h. **15.** $\sin\theta = 5/13$, $W = (F_W \sin\theta)s = (100\text{ N})(5/13)(13\text{ m}) = 500$ N·m. Straight up, $W = (100\text{ N})(5.0\text{ m}) = 0.50$ kN·m. **19.** (a) $W = (100\text{ N})(2.0\text{ m}) = 0.20$ kN·m; (b) (1.0 cm)(1.0 cm) = (10 N)(0.25 m) = 2.5 N·m/cm^2; (c) 10 cm × 8.0 cm = 80 cm^2; (80 cm^2)(2.5 N·m/cm^2) = 0.20 kN·m. **23.** $\Delta W/\Delta t = (10^7\text{ J})/(24\text{ h} \times 3600\text{ s/h}) = 1.2 \times 10^2$ W. BMR = (90 W)/(1.5 m^2) = 60 W/m^2. **29.** KE = $\frac{1}{2}mv^2 = \frac{1}{2}(0.5 \times 10^{-3}\text{ kg})(200 \times 10^3\text{ m})^2 = 1 \times 10^7$ J. P = $\Delta KE/\Delta t = (1 \times 10^7\text{ J})/(10^{-8}\text{ s}) = 10^{15}$ W. **33.** $PE_G = mgh$; $h = (6 \times 10^9\text{ J})/(10^6\text{ kg}) \times (9.8\text{ m/s}^2) = 6 \times 10^2$ m. **39.** $Pt = W = \Delta KE = \frac{1}{2}(2 \times 10^3\text{ kg})(268\text{ m/s})^2 =$

7.18×10^7 J; $t = 8$ min. **43.** (a) $g_☾ = GM_☾/R_☾^2 = (1/100)/(1/4)^2 g_⊕ = 0.16\ g_⊕$. (b) $\Delta PE = mg\Delta h = (1000\text{ kg})(0.16) \times (9.8\text{ m/s}^2)(100\text{ m}) = 1.6 \times 10^5$ J. **45.** Let θ be the incline angle; $\tan\theta \approx \sin\theta = \frac{1}{2}\%$; the net work done by all forces acting on the train equals its ΔKE, $W = [22.7 \times 10^4\text{ N} - (35\text{ N})(1.8 \times 10^3) - (1.8 \times 10^7\text{ N})\sin\theta]s = (7.4 \times 10^4\text{ N})s$; $\Delta KE = \frac{1}{2}(45)[(4.0 \times 10^5\text{ N})/(9.8\text{ m/s}^2)][(13.4\text{ m/s})^2 - (6.7\text{ m/s})^2] = 1.237 \times 10^8$ J; hence, $W = \Delta KE$ and $s = (1.237 \times 10^8\text{ J})/(7.4 \times 10^4\text{ N}) = 1.7 \times 10^3$ m. **49.** For one car, $KE = \frac{1}{2}mv^2 = \frac{1}{2}[(7.12\text{ kN})/(9.81\text{ m/s}^2)] \times (26.67\text{ m/s})^2 = 258$ kN·m; and for the other, KE = twice that; viz., 516 kN·m. For one car on top of the mountain, $PE = mgh = (7.12\text{ kN})(33.5\text{ m}) = 238.5$ kN·m; and for the other, PE = twice that; viz., 477 kN·m. Thus, both cars have enough energy to coast over the top. **53.** $p_i = (1.00\text{ kg})v_i + 0 = p_f = (91\text{ kg})(5.0\text{ m/s})$; $v_i = 0.46$ km/s. **57.** $v_{esc} = (2GM/R)^{1/2} = [2(6.67 \times 10^{-11}\text{ N·m}^2/\text{kg}^2)(7.4 \times 10^{22}\text{ kg})/(1.74 \times 10^6\text{ m})]^{1/2} = 2.4$ km/s. It can be launched in any direction that does not intersect the Moon.
63. Taking east as positive, $p_i = (+8.0\text{ kg}) \times (0.15\text{ m/s}) - (2.0\text{ kg})(0.25\text{ m/s}) = 0.70\text{ kg·m/s} = p_f = (10.0\text{ kg})v_f$; $v_f = 0.07$ m/s. **65.** $p_i = m(15.0\text{ m/s}) + m(-10\text{ m/s}) = mv_{1f} + mv_{2f}$, $+5.0 = v_{1f} + v_{2f}$. But $v_{2f} - v_{1f} = v_{1i} - v_{2i} = 25$ m/s; hence, adding the last two equations, $2v_{2f} = 30$ m/s, $v_{2f} = 15$ m/s, and $v_{1f} = -10$ m/s. **67.** $PE_i = 0$; $E_i = 0$; hence, $\frac{1}{2}m_1v^2 + m_1gy + \frac{1}{2}m_2v^2 - m_2gy = 0$ a moment after release. $v = \sqrt{2(m_2 - m_1)gy/(m_1 + m_2)}$.
69. $m_Bv_B = (m_B + m_C)v_C$. After impact, $\frac{1}{2}(m_B + m_C)v_C^2 = (m_B + m_C)gh$, $v_C^2 = 2gh$, $v_B = [(m_B + m_C)/m_B]\sqrt{2gh}$.

CHAPTER 10

Answers to Odd-Numbered Multiple Choice Questions

1. d **3.** d **5.** b **7.** d **9.** c **11.** b **13.** b **15.** a **17.** c **19.** b

Answers to Odd-Numbered Problems

1. 6.02×10^{23} **3.** $6 \times 10^7\text{ kg/m}^3$ **5.** 49 kg **7.** 19 GJ **9.** 12.0×10^{23} **11.** 303 u **13.** 6×10^{77} **15.** 3.3×10^{28} molecules/m^3 **17.** 3.14 kg **19.** 0.46 nm **21.** 3.3 nm **23.** 1.0 kN/m **25.** 2 MPa **27.** 0.15% **29.** 0.399 8 m **31.** same **33.** 1×10^{-2} m **35.** $1.7 \times 10^2\text{ N/m}^2$ **37.** 5.0×10^2 N/m **39.** 16 J/m^2 **41.** 0.77 mm **43.** 29.5 kN **45.** 0.22% **47.** $R = 1.01$ cm, 0.173% **49.** 2 cm **51.** 0.911 MN

Solutions to Selected Problems

1. 6.02×10^{23} **5.** (13 gallons)(3.785 liters/gallon) = 49.2 liters; 1 liter = 1000 cm^3; hence, the human body contains $\approx 49 \times 10^3$ cm^3, or $\approx 49 \times 10^3$ g, or 49 kg. **9.** $C_{12}H_{22}O_{11}$: 12(12) + 22(1) + 11(16) = 342; hence, 684 g is 2 gram-moles and, therefore, 12.0×10^{23} molecules. **13.** (10^{33} g per star) × (10^{11} stars per galaxy)(10^{11} galaxies)/(10 g per mole) = number of moles = 1×10^{54} moles; (10^{54} moles)(6.02×10^{23} atoms/mole) = 6×10^{77} atoms. **17.** The mass of a molecule is 18 u, of which 2 u is hydrogen; that is, 11.1%; hence, 11.1% of 62.4 lb is 6.93 lb → 3.14 kg. **21.** 22.4 liters contain 6.0×10^{23} molecules; hence, the volume "occupied" by each one is $(22.4 \times 10^3\text{ cm}^3) \times (10^{-6}\text{ m}^3/\text{cm}^3)/6.0 \times 10^{23} = 3.73 \times 10^{-26}\text{ m}^3$; and the cube root of this is 3.3×10^{-9} m, or 3.3 nm, which is roughly 10 times the nucleus-to-nucleus separation in a solid. **23.** $F = ks$; $k = F/s = (50\text{ N})/(0.05\text{ m}) = 1.0$ kN/m. **25.** $\sigma = F/A = (200\text{ N})/(10^{-4}\text{ m}^2) = 2 \times 10^6\text{ N/m}^2 = 2$MPa. **29.** $\Delta L/L_0 = \varepsilon$; $\Delta L = (0.400\ 0\text{ m})(5.000 \times 10^{-4}) = 2.000 \times 10^{-4}$ m and $L = L_0 - \Delta L = 0.399\ 8$ m. **33.** $Y = (F/A)/(\Delta L/L_0)$; $\Delta L = (F/A)(Y/L_0) = FL_0/YA$; the mike is already on the rod and so only the weight of the *sumatori* enters; $\Delta L = (9.8\text{ m/s}^2) \times (181.4\text{ kg})(4\text{ m})/(200 \times 10^9\text{ Pa})\pi(\frac{1}{2}2 \times 10^{-3}\text{ m})^2 = 1 \times 10^{-2}$ m. **37.** $k = F/s = (10\text{ N})/(2.0 \times 10^{-2}\text{ m}) = 5.0 \times 10^2$ N/m. **41.** $F/A < 435$ MPa; $(200\text{ N})/(435\text{ MPa}) < A$; $A > 4.598 \times 10^{-7}\text{ m}^2$; $\pi R^2 > 4.598 \times 10^{-7}\text{ m}^2$; $R > 3.8 \times 10^{-4}$ m; minimum diameter is 0.77 mm. **45.** $Y = 50\text{ GPa} = (110\text{ MPa})/\varepsilon_R$; $\varepsilon_R = (110\text{ MPa})/(50\text{ GPa}) = 2.2 \times 10^{-3} = 0.22\%$. **49.** $F/A = S\gamma$; $F/S\gamma = A = (200\text{ N})/(0.6\text{ GPa}) \times (0.001) = 3.33 \times 10^{-4}\text{ m}^2 = \pi R^2$; $D = 2R = 2$ cm. **53.** $PE_e = F^2\frac{1}{2}L_0/2AY + F^2\frac{1}{2}L_0/2(2A)Y = 3F^2L_0/8AY$. **55.** From p. 341, $k = YA/L_0$, $Y = (F/A)/(\Delta L/L_0) = \sigma_R/(\Delta L/L_0)$; $PE_e = \frac{1}{2}k(\Delta L)^2 = \frac{1}{2}(YA/L_0)(\sigma_R L_0/Y)^2 = \frac{1}{2}AL_0\sigma_R^2/Y$.

CHAPTER 11

Answers to Odd-Numbered Multiple Choice Questions

1. b **3.** d **5.** a **7.** a **9.** a **11.** d **13.** b **15.** c **17.** c **19.** d

Answers to Odd-Numbered Problems

1. 3×10^4 Pa **3.** 50.7 N/cm^2 **7.** 1.0×10^4 Pa **9.** 10.3 m **11.** 1.1×10^8 Pa **13.** 4.1×10^5 Pa **15.** 10.5 **17.** 13 mN **19.** 3.9×10^3 Pa **21.** -1.1×10^4 Pa **23.** 1.1×10^5 Pa **25.** 21.5, platinum, $V = 3.00 \times 10^{-8}\text{ m}^3$ **27.** 10% is visible, 8% is above **29.** 300 cm^3 **31.** 1.1×10^5 N **33.** 72.7 kg **35.** 4.63 kN, 0.15 m **37.** 2.7×10^{-2} N **41.** 2×10^5 N **43.** $\rho_l = (F_{Wa} - F_{Wl})\rho_w/(F_{Wa} - F_{Ww})$ **45.** $16 \times 10^{-3}\text{ m}^3/\text{s}$ **47.** -27 kPa **49.** 8.8 kPa **51.** $v = \sqrt{2P_G/\rho}$ **53.** $x = 2\sqrt{yh}$ **55.** 12 m/s **57.** $v = \sqrt{2\Delta P/\rho}$ **59.** $v_p = \sqrt{2g\Delta y/[(A_p^2/A_f^2) - 1]}$ **61.** 82 m/s **63.** $v_2 = \sqrt{2gh}$, $v_3 = \sqrt{2g(h + Y)}$ **65.** $v_2 = \sqrt{2gH}$ **67.** $y = h\sin^2\theta$

Solutions to Selected Problems

3. $P_i = 5(1.013 \times 10^5\text{ N/m}^2) = 5.07 \times 10^5\text{ N/m}^2 = (5.07 \times 10^5\text{ N})/(10^4\text{ cm}^2) = 50.7$ N/cm^2. **7.** (3.0 in.)(2.54 cm/in.) = 7.62 cm; $\Delta P = \rho g\Delta h = (13.6 \times 10^3\text{ kg/m}^3) \times (9.8\text{ m/s}^2)(7.62 \times 10^{-2}\text{ m}) = 1.0 \times 10^4$ Pa. **11.** $P = \rho gh = (1.025 \times 10^3\text{ kg/m}^3)(9.8\text{ m/s}^2) \times (11 \times 10^3\text{ m}) = 1.1 \times 10^8$ Pa, or 1.6×10^4 psi. **15.** From Eq. (11.9), (10.0 g)/(0.952 g) = 10.5, and it's likely silver. **19.** $P = \rho gh = (1.0 \times 10^3\text{ kg/m}^3)(9.8\text{ m/s}^2)(0.40\text{ m}) = 3.9 \times 10^3$ Pa. **23.** At the level of the lower surface, $P = P_A + \rho g\Delta h = (1.013 \times 10^5\text{ Pa}) + (1.00 \times 10^3\text{ kg/m}^3)(9.8\text{ m/s}^2)(0.61\text{ m}) = 1.1 \times 10^5$ Pa, or 1.1 atm. **25.** From Eq. (11.9), $gm/(gm - gm_r) = (6.327 \times 10^{-3}\text{ N})/(6.327 \times 10^{-3}\text{ N} - 6.033 \times 10^{-3}\text{ N}) = 21.5$; platinum; weight of water displaced is $(6.327 \times 10^{-3}\text{ N} - 6.033 \times 10^{-3}\text{ N}) = 2.94 \times 10^{-4}$ N, equivalent to 3.00×10^{-5} kg, or $3.00 \times 10^{-8}\text{ m}^3$.
29. $P_iV_i = P_fV_f$; (228 cm Hg)(100 cm^3) = (76 cm Hg)V_f; $V_f = 300$ cm^3. **33.** mass of air displaced = mass of helium + mass of balloon; $V(1.29\text{ kg/m}^3) = V(0.178\text{ kg/m}^3) + 454$ kg; $V = 408.3\text{ m}^3$, for a mass of $\rho V = (0.178\text{ kg/m}^3)(408.3\text{ m}^3) = 72.7$ kg.
37. There are two surfaces here, so $F = 2\gamma L = 2\gamma\pi D = 2\pi(72.8 \times 10^{-3}\text{ N/m})(6.0 \times 10^{-2}\text{ m}) = 2.7 \times 10^{-2}$ N, which is the weight of a 2.8-g mass. **39.** The area of the left face is $A_a = L(L\sin\phi)$, the area of the top face is $A_c = L^2$, and that of the bottom face is $A_b = L^2\cos\phi$. So the corresponding forces are $F_a = P_aL^2\sin\phi$, $F_c = P_cL^2$, and $F_b = P_bL^2\cos\phi$. Taking the sum of the horizontal forces equal to zero, we have $F_a = F_c\sin\phi$; $P_aL^2\sin\phi = P_cL^2\sin\phi$. Thus, $P_a = P_c$. The weight of the fluid wedge is the volume, $\frac{1}{2}L(L\sin\phi)(L\cos\phi)$, times ρg; hence, the sum of the vertical forces yields $F_b = F_c\cos\phi + F_W$ or $P_bL^2\cos\phi = P_cL^2 \times \cos\phi + \frac{1}{2}\rho g L^3\sin\phi\cos\phi$ and $P_b = P_c + \frac{1}{2}\rho gL\sin\phi$. As $L \to 0$, we get $P_b = P_c$, which equals P_a. **41.** Because the pressure varies uniformly with depth, the average pressure on the wall is $\frac{1}{2}$ the pressure at the bottom. From Problem 11.1, $P_l = 3 \times 10^4$ Pa; hence, $F = (1.5 \times 10^4\text{ Pa}) \times (5\text{ m} \times 3\text{ m}) = 2 \times 10^5$ N. **43.** $F_{Wa} - F_{Ww} = \rho_W gV$; $V = (F_{Wa} - F_{Ww})/\rho_w g$; $V = (F_{Wa} - F_{Wl})/\rho_l g$; setting these equal, $(F_{Wa} - F_{Ww})/\rho_w g = (F_{Wa} - F_{Wl})/\rho_l g$; and

$$\rho_l = \frac{(F_{Wa} - F_{Wl})}{(F_{Wa} - F_{Ww})}\rho_w.$$

47. From Eq. (11.18), $P_1 - P_2 = \rho g(y_2 - y_1) = (0.68 \times 10^3\text{ kg/m}^3)(9.8\text{ m/s}^2)(-4.0\text{ m}) = -27$ kPa. $P_2 > P_1$. **51.** $v = \sqrt{2gh}$, $P_G = \rho gh$; hence, $v = \sqrt{2P_G/\rho}$. **55.** Consider a slug of gas of volume $V = AL$, where the length $L = v\Delta t$. The mass per unit time (Δt) flowing is 1.0 kg/s $= V\rho/\Delta t = Av\Delta t\rho/\Delta t = \frac{1}{4}\pi D^2\rho v$; $v = 4(1.0\text{ kg/s})/\pi D^2\rho = 12$ m/s. **63.** At the surface and the spigot using gauge pressure, $0 + \rho g(h + Y) + 0 = 0 + \rho gY + \frac{1}{2}\rho v_2^2$, hence, $v_2 = \sqrt{2gh}$. Then, between points 2 and 3, $0 + \rho gY + \frac{1}{2}\rho(2gh) = 0 + 0 + \frac{1}{2}\rho v_3^2$; $v_3 = \sqrt{2g(h + Y)}$. These are just what we expect if the water free-falls.

67. $P_2 + \frac{1}{2}\rho v_2^2 + \rho g y_2 = P_3 + \frac{1}{2}\rho v_3^2 + \rho g y_3$; $0 + \frac{1}{2}\rho v_2^2 + 0 = 0 + \frac{1}{2}\rho v_3^2 + \rho g y$; since there is no horizontal acceleration, $v_3 = v_2 \cos\theta$, $y = \frac{1}{2}(v_2^2 - v_2^2\cos^2\theta)/g$, but $\cos^2\theta + \sin^2\theta = 1$; hence, $y = (\frac{1}{2}v_2^2\sin^2\theta)/g$ and, since $v_2^2 = 2gh$, $y = h\sin^2\theta$ and it cannot exceed h. **69.** Imagine a flow tube from some distant point 1 in front of the plane at the level of the top of the wing; then $P_1 + \frac{1}{2}\rho v_1^2 + \rho g y_1 = P_\alpha + \frac{1}{2}\rho v_\alpha^2 + \rho g y_\alpha$. Now do the same thing for a tube starting at point 2 just below 1 and running to the bottom of the wing as $P_2 + \frac{1}{2}\rho v_2^2 + \rho g y_2 = P_\beta + \frac{1}{2}\rho v_\beta^2 + \rho g y_\beta$. Since $P_1 \approx P_2$, $v_1 \approx v_2$, $y_1 \approx y_2$, $P_\alpha + \frac{1}{2}\rho v_\alpha^2 + \rho g y_\alpha = P_\beta + \frac{1}{2}\rho v_\beta^2 + \rho g y_\beta$, and with $y_\alpha \approx y_\beta$, $F_L = A(P_\beta - P_\alpha) = \frac{1}{2}\rho(v_\alpha^2 - v_\beta^2)A$.

CHAPTER 12

Answers to Odd-Numbered Multiple Choice Questions

1. b **3.** a **5.** c **7.** c **9.** d **11.** c **13.** a **15.** d **17.** d **19.** b **21.** c

Answers to Odd-Numbered Problems

1. 1.8 s **3.** 1.3 Hz, 2.6π rad/s **5.** (a) 5.0 m (b) 0.064 Hz (c) 0.10 rad (d) $x = 3.1$ m **9.** 4.9×10^2 m/s^2 **11.** T = 4.4 s **13.** $\varepsilon = \pi$ **15.** $z = +0.50$ m, $z = -0.50$ m, $z = 0$ **17.** $k = 2.6$ N/m **19.** $k = 784$ N/m, $f = 3.2$ Hz **21.** $f_0 = (1/2\pi)\sqrt{(k_1 + k_2)/m}$ **23.** 0.16 Hz **25.** $L = 24.8$ m **29.** $g = 4\pi^2\Delta L/T^2$ **31.** 3.5 Hz **33.** $T = 2\pi\sqrt{b\rho/\rho_w g}$ **35.** $v_b = 0.86$ km/s **37.** 1.2 T **39.** $T = 2\pi\sqrt{m/k}$ **43.** 500 nm **45.** $\lambda = 3.40$ m **49.** 12 m **51.** 1.3×10^2 m/s **53.** 1.08×10^3 N

57.

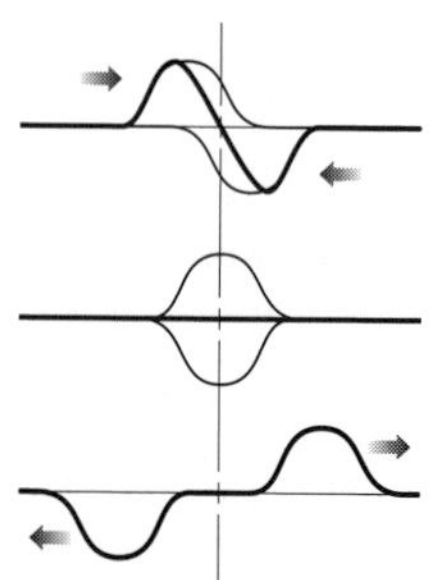

Fig. AN 57.

61. 6.3 m/s^2 **63.** 100 m/s, 314 rad/s, 0.0200 s, 2.00 m **65.** $v = 1.0 \times 10^2$ m/s **67.** $F_T = 1.9 \times 10^2$ N **69.** $y = -0.7\,A$, $y = -0.7\,A$ **71.** $v_{max} = \sqrt{gL}$, $v = \sqrt{g\Delta y + MgL/m}$

Solutions to Selected Problems

1. $T = 1/f = 1/[(33\frac{1}{3}\text{ rpm})/(60\text{ s/min})] = 1.8$ s. **5.** By comparison with Eq. (13.3), (a) $A = 5.0$ m (b) $\omega = 2\pi f = 0.40$ rad/s; $f = 0.064$ Hz (c) $\varepsilon = 0.10$ rad (d) $x = 3.1$ m. **9.** $a_0 = -A(2\pi f)^2$; dropping the sign, $a_0 = (0.005\text{ m})4\pi^2(50\text{ Hz})^2 = 4.9 \times 10^2$ m/s^2. **13.** $x = A\cos\theta = A\cos(\omega t + \varepsilon) = A\cos(2\pi f t + \varepsilon)$; hence, at $t = 0$, $x = -3 = A\cos\varepsilon$, and since $A = 3$, $\cos\varepsilon = -1$ and $\varepsilon = \pi$. **17.** $f = (1/2\pi)\sqrt{k/m}$; $k = (2\pi f)^2 m = 2.6$ N/m. **21.** Since $x = x_1 = x_2$ and $F = F_1 + F_2$; hence, $kx = k_1x_1 + k_2x_2$ and $k = k_1 + k_2$. Now, simply substitute k into Eq. (12.12). **25.** From Eq. (12.17), $T = 2\pi\sqrt{L/g}$; hence, $L = gT^2/4\pi^2 = 24.8$ m. **29.** $k\Delta L = mg$; hence, $m/k = \Delta L/g$ and, from Eq. (12.13), $T = 2\pi\sqrt{\Delta L/g}$; hence, $g = 4\pi^2\Delta L/T^2$. **33.** The additional buoyant force if the block is pushed down a distance y is $-acy\rho_w g$, which is the restoring force $-ky$, so the motion is SHM, since $F \propto -y$ and $k = ac\rho_w g$. Here, $m = abc\rho$, so $T = 2\pi\sqrt{(abc\rho)/(ac\rho_w g)} = 2\pi\sqrt{(b\rho)/(\rho_w g)}$. **37.** From Eq. (12.17), $T = 2\pi\sqrt{L/g}$, and if $L' = L + 50\%\ L = 1.50\ L$, $T' = 2\pi\sqrt{L'/g} = \sqrt{1.50}\ T = 1.2\ T$. **41.** $x = L\cos\phi + R\cos\theta$; $\cos\phi = 1 - \sin^2\phi$, but $L\sin\phi = R\sin\theta$; $x = R\cos\theta + L\sqrt{1 - (R/L)^2\sin^2\theta}$. Although the first term corresponds to SHM, the presence of the second term negates that. **45.** $v = f\lambda$; $\lambda = (1498\text{ m/s})/(440\text{ Hz}) = 3.40$ m. **49.** The round-trip distance is $2x = vt = (330\text{ m/s})(70\text{ ms}) = 23.1$ m, $x = 12$ m. **51.** $v = \sqrt{F_T/\mu} = \sqrt{(10\text{ N})/(0.59 \times 10^{-3}\text{ kg/m})} = 1.3 \times 10^2$ m/s. **53.** $F_T = v^2\mu = 1.08 \times 10^3$ N. **59.** Using the trigonometric identity $\sin(\alpha \mp \beta) = \sin\alpha\cos\beta \mp \cos\alpha\sin\beta$, where $\alpha = k(x - vt)$ and $\beta = \mp\frac{1}{2}\pi$, we get $y = \mp A\cos\frac{2\pi}{\lambda}(x - vt)$. **63.** $v = \sqrt{F_T/\mu} = 100$ m/s; $\omega = 2\pi f = 314$ rad/s; $v = f\lambda$, $\lambda = 2.00$ m; $T = 1/f = 0.0200$ s. **67.** $m = F_W/g = (0.20\text{ N})/(9.81\text{ m/s}^2) = 0.0204$ kg; $\mu = 1.7 \times 10^{-3}$ kg/m; $v = \sqrt{F_T/\mu}$; $F_T = v^2\mu = 1.9 \times 10^2$ N, or 42 lb. **69.** The phase is $2\pi[(t/0.01) - (5.0/40)] = 2\pi[(t/0.01) - (1/8)] = (200\pi t - \frac{1}{4}\pi)$; hence, $y = A\sin(200\pi t - \frac{1}{4}\pi)$ is the displacement of the bead. At $t = 0$, $y = A\sin(-\frac{1}{4}\pi) = -0.7\,A$, and at $t = 0.01$ s, which is one period later, $y = A\sin(2\pi - \frac{1}{4}\pi) = -0.7\,A$.

CHAPTER 13

Answers to Odd-Numbered Multiple Choice Questions

1. a **3.** d **5.** e **7.** d **9.** d **11.** b **13.** c **15.** a **17.** b **19.** a

Answers to Odd-Numbered Problems

1. $T = 2.27$ ms, $\lambda = 0.780$ m **3.** 3.3×10^{-4} m **5.** 2.91 **7.** 4.0×10^{-2} W/m^2, E = 4.0 MJ **9.** 0.79 m **11.** 1.9 ms **13.** 40 dB **15.** 10 dB **17.** 0 dB **19.** 4.8 dB **21.** 5.0×10^{-5} W/m^2 **23.** 51 dB **25.** 89.8 dB **27.** 75 dB **29.** $R = 2.2$ km **31.** 200 m **33.** $\beta = 20\log_{10}(P/P_0)$ **35.** (b) displacement and pressure are $\frac{1}{4}\lambda$ out-of-phase **37.** 25.8 dB **39.** 996 Hz or 1004 Hz **41.** 1.0 Hz **43.** $\lambda_6 = \frac{1}{6}$ m **45.** $f_1 = 100$ Hz **49.** zero **51.** intensity increases by a factor of 4 times **53.** 466 Hz **55.** 0.782 m/s **57.** air, $\lambda = 1.4$ m; string, 66 cm **59.** 2 Hz **61.** 1.75 kHz **63.** $\Delta f = 14$ Hz **65.** 210 N **67.** 1.8 kHz **69.** 344 m/s **71.** $v = 340$ m/s **73.** $f_O - f_s = 2v_t f_s/(v - v_t)$ **75.** $2A\cos[\frac{1}{2}(\omega_1 - \omega_2)t]$ is a cosinusoidally modulated amplitude

Solutions to Selected Problems

1. $T = 1/f = 1/440\text{ Hz} = 2.27$ ms. $\lambda = v/f = (343\text{ m/s})/(440\text{ Hz}) = 0.780$ m. **5.** $v = f\lambda = (1000\text{ Hz})\lambda = 344$ m/s; $\lambda = 0.344$ m; 2.91 waves per meter. **11.** $\tau = \frac{1}{2}T = 1/2f = 1/2(261.6\text{ Hz}) = 1.9$ ms. **15.** Ten times the power means 10 times the intensity; therefore $\Delta\beta = 10\log_{10}10 = 10$ dB. **19.** $\Delta\beta = 10\log_{10}3 = 4.8$ dB. **23.** The increase over one voice is $\Delta\beta = 10\log_{10}25 = 14$ dB; hence, each singer puts out 65 dB − 14 dB = 51 dB. **25.** $I = (1.2\text{ W})/4\pi R^2 = 0.955$ mW/m^2; $\beta = 10\log_{10}0.955 \times 10^9 = 89.8$ dB. **27.** $\beta = 10\log_{10}(56 \times 10^{-5}\text{ W/m}^2)/(10^{-12}\text{ W/m}^2) = 87$ dB. The intensity drops with the inverse square of the distance; $(5.0\text{ m})^2/(20\text{ m})^2 = I/(56 \times 10^{-5}\text{ W/m}^2)$; at 20.0 m, $I = 3.5 \times 10^{-5}$ W/m^2; and there $\beta = 75$ dB. **31.** −40 dB means a change of −4 bel or 10^{-4}; hence, from the Inverse Square Law, $R^2/(2.0\text{ m})^2 = 1/10^{-4}$ of whatever the original intensity was. $R = 200$ m. **35.** (a) N/m^2 = (N/m^2)(1/m)(m); (b) the displacement and pressure are $\frac{1}{4}\lambda$, or 90°, out-of-phase; (c) $P_0 = \frac{2\pi}{\lambda}BA$; $v = \sqrt{B/\rho}$; $P_0 = \frac{2\pi}{\lambda}v^2\rho A$. The Bulk Modulus relates pressure change to volume change in the gas, and that volume change, in turn, is related to the displacement of the molecules. P is a pressure change from ambient, and so we can expect it to be associated with B and displacement. **39.** The beat frequency is 4 Hz; hence, the wire is vibrating at *either* 996 Hz or 1004 Hz. **43.** $\frac{1}{2}\lambda = L = \frac{1}{2}$ m, $\lambda = 1.0$ m; $\lambda_6 = 1/6$ m. **47.** $f_N = (N/2L) \times \sqrt{F_T/\mu}$; but $\mu = m/L$; hence, $f_N = (\frac{1}{2}N) \times \sqrt{F_T/L^2(m/L)}$, which equals the desired expression. **51.** The addition of a 180°, or $\frac{1}{2}\lambda$, phase shift will bring the waves into phase and they will now reinforce one another, creating a peak pressure 2 times as great and an intensity 4 times that of either speaker alone. **55.** $f_O = f_s(v + v_O)/(v + v_s)$; here $v_s = 0$ and $v_O = v(f_O - f_s)/f_s = (344\text{ m/s})(1/440) = 0.782$ m/s. **59.** $f \propto \sqrt{F_T}$; hence, the increased frequency is given by $f/(440\text{ Hz}) = \sqrt{1.010\,F_T/F_T}$; $f = 442$ Hz and the beat frequency is 2 Hz. **63.** $f_1/f_1' = 200\text{ Hz}/f_1' = (331.5\text{ m/s})/(331.5 + 0.60 \times 40) = 0.932$; $f_1' = 214$ Hz; $\Delta f = 14$ Hz. **67.** $v = \sqrt{Y/\rho} = 3516$ m/s; $v = f\lambda$, $\lambda = 2L = 2.00$ m; $f = v/2L = (3516\text{ m/s})/(2.00\text{ m}) = 1.8$ kHz. **71.** There is a tiny residual signal because the paths are different; hence, the amplitudes are slightly different and the direct and reflected waves cancel, but not completely. The angle between these two beams is $\tan\theta = 1.146/2.5$, $\theta = 24.62°$; the speaker-table distance is $1.146/\sin 24.62° = 2.75$ m. The direct beam travels 5.00 m, the reflected beam travels 5.50 m, and there is cancellation; the path difference (0.50 m) must be an odd multiple of $\frac{1}{2}\lambda$. Speed is $\approx$344 m/s, and since $f = 340$ Hz, $\lambda \approx 1$ m; hence, 0.50 m cannot be $3\lambda/2$ or $5\lambda/2$ and must be $\frac{1}{2}\lambda$, ergo $\lambda = 1.00$ m and $v =$ 340 m/s. **75.** $y = y_1 + y_2 = A \sin(\omega_1 t) + A \sin(\omega_2 t)$. Using the identity $\sin \alpha + \sin \beta = 2 \sin \frac{1}{2}(\alpha + \beta) \cos \frac{1}{2}(\alpha - \beta)$, we get the desired result. Remember $\omega_1 \approx \omega_2$; there is a relatively high-frequency sinusoidal carrier oscillating at the average frequency $\frac{1}{2}(\omega_1 + \omega_2)$, and this is multiplied by an amplitude term $2A \cos[\frac{1}{2}(\omega_1 - \omega_2)t]$. This amplitude varies slowly in time, and so the modulated carrier has maxima whenever $\cos[\frac{1}{2}(\omega_1 - \omega_2)t] = \pm 1$; hence, beats occur at twice that frequency, namely, at $(\omega_1 - \omega_2)$ or $(f_1 - f_2)$.

CHAPTER 14

Answers to Odd-Numbered Multiple Choice Questions

1. b **3.** a **5.** c **7.** d **9.** a **11.** b **13.** a **15.** a **17.** b **19.** e **21.** a

Answers to Odd-Numbered Problems

1. 37°C **3.** 55.6 K **5.** 10×10^{6}°C $\approx 18 \times 10^{6}$°F **7.** 927°F **9.** $\Delta L = 25 \times 10^{-6}$ m **11.** 5.2 mm **13.** 0.54 m **15.** 1.008 cm^2 **17.** 10.002 m **19.** 0.51 cm^3 **21.** $\Delta H = -1.5$ mm **23.** 0.79×10^{-6} m^3 **25.** 7.9×10^3 kg/m^3 **27.** 2.005 9 s, ran fast **31.** 9.3 MPa **33.** 1.0 MPa **35.** 1.1×10^3 cm^3 **37.** $T_f = 199$ K **39.** (a) vapor (b) liquid (c) vapor (d) vapor (e) vapor **41.** 2.992×10^{-26} kg **43.** 3.01×10^{26} molecules **45.** 174 K **47.** N = 27×10^{18} **49.** 3.0 m/s **51.** 478.03 m/s **53.** 1.19×10^{22} **55.** 1.1 m^3 **57.** 43 kg **61.** 76 kJ **63.** $v_{rms} = (3P/\rho)^{1/2}$ **65.** 189 cm^3 **67.** 4.3 atm **69.** 0.427%

Solutions to Selected Problems

1. 37°C **5.** 10 000 000 K $\rightarrow$ 10 000 273°C $\approx 10 \times 10^{6}$°C $\approx 18 \times 10^{6}$°F. **9.** $\alpha = 25 \times 10^{-6}$ K^{-1}; hence, $\Delta L = 25 \times 10^{-6}$ m. **13.** $\Delta L = \alpha L_0 \Delta T = (12 \times 10^{-6}$ K$^{-1}) \times$ (1280 m)(35 K) = 0.54 m. **17.** $\Delta L = \alpha L_0 \Delta T = (12.15 \times 10^{-6}$ K^{-1})(10.000 m) $\times$ (16.0 K) = 1.9×10^{-3} m. $L = 10.002$ m. 2 mm is an appreciable error, but it's not likely to trouble a carpenter. **21.** $\Delta V = \beta V_0 \Delta T = (182 \times 10^{-6}$ K^{-1})(800.0 $\times 10^{-6}$ m^3) $\times$ (-95 K) = -1.383×10^{-5} m^3 for the mercury. $\Delta V = \beta V_0 \Delta T = (26 \times 10^{-6}$ K$^{-1}) \times$ (800.0 $\times 10^{-6}$ m^3)(-95 K) = -1.98×10^{-6} m^3 for the glass. Hence, the total $\Delta V = -1.185 \times 10^{-5}$ m^3. $\Delta V = A\,\Delta H = \pi r^2 \Delta H$; $\Delta H = -1.5$ mm. **25.** $\rho_i = m/V_i$; $\rho_f = m/(V_i + \Delta V)$; $\rho_i V_i = \rho_f(V_i + \Delta V) = \rho_f V_i(1 + \beta \Delta T)$; $\rho_i/(1 + \beta\Delta T) = \rho_f =$ (7.85×10^3 kg/m^3)/[(1 + (36 $\times 10^{-6}$ K$^{-1}) \times$ (-20 K)] = 7.9×10^3 kg/m^3. **29.** $\rho_0 = m/V_0$; $\rho = m/(V_0 + \Delta V)$; $\rho_0 V_0 = \rho(V_0 + \Delta V) = \rho V_0(1 + \beta\Delta T)$; $\rho_0/(1 + \beta\Delta T) = \rho$: from the Binomial Theorem, $(1 + x)^n \approx (1 + nx)$ for $x \ll 1$; here, $n = -1$, and so $\rho \approx \rho_0(1 - \beta\Delta T)$. **31.** Stress = $Y\alpha\Delta T$ = (25 GPa)(10 $\times 10^{-6}$ K^{-1})(37 K) = 9.3 MPa. Concrete is strong in compression (Table 10.3), so it's more likely to buckle than break. **33.** $P_i V_i = P_f V_f$; $P_f = 10\,P_i =$ 1.0 MPa. **37.** No change in P; $V_i/T_i = V_f/T_f$; (300 cm^3)/(298 K) = (200 cm^3)/T_f; $T_f = 199$ K. **41.** $2H \rightarrow 2(1.008$ u); $0 \rightarrow$ 15.999 u; (18.015 u)(1.660 6 $\times 10^{-27}$ kg/u) = 2.992×10^{-26} kg. **45.** $PV = nRT$; (202.6 $\times 10^3$ Pa)(10.0 $\times 10^{-3}$ m^3) = (1.40 mol) $\times$ (8.314 J/mol·K)T; $T = 174$ K. **49.** $v_{av} =$ (1.0 m/s + 2.0 m/s + 3.0 m/s + 4.0 m/s + 5.0 m/s)/5 = 15/5 m/s = 3.0 m/s. **53.** $PV = Nk_BT$; T = 98.6°F = 37°C $\rightarrow$ 310 K; PV/k_BT = (101.46 kPa)(500 $\times 10^{-6}$ m^3)/ (1.380 7 $\times 10^{-23}$ J/K)(310 K) = $N = 1.19 \times 10^{22}$ molecules, which is the same as the number of stars there are in the entire Universe. **57.** $n = PV/RT$ = (19.4 $\times 10^3$ Pa)(2000 m^3)/(8.314 J/mol·K)(218 K) = 21.4 $\times 10^3$ moles. Each mole has a mass of 0.002 kg; hence, the total mass is (0.002 kg/mol)(21.4 $\times 10^3$ mol) = 43 kg. **61.** $(\text{KE})_{av} = (3/2)k_BT$ = (3/2)(1.380 662 $\times 10^{-23}$ J/K)(273.15 K) = 5.66 $\times 10^{-21}$ J/molecule. 0.50 m^3 = 500 liter; (500 liter)/(22.4 liter/mol) = 22.3 mol; (22.3 mol)(6.022 $\times 10^{23}$ molecules/mol) $\times$ (5.66 $\times 10^{-21}$ J/molecule) = 76 kJ. **65.** To find the pressure on the gas, 3.90 cm Hg = 5.199 6 kPa, P_i = (101.05 $-$ 5.199 6) kPa = 95.85 kPa; $P_i V_i/T_i = P_f V_f/T_f$; (95.85 kPa) $\times$ (214 cm^3)/(293.15 K) = (101.32 kPa)V_f/ (273 K); $V_f = 189$ cm^3. **69.** $v_{rms} = \sqrt{3k_BT/m}$; hence, the ratio of the *rms*-speeds varies as the inverse of the square roots of the masses. The masses are 238 u + 6(18.998 u) = 351.99 u and 235 u + 6(18.998 u) = 348.99 u. $(v^{235}_{rms} - v^{238}_{rms})/v^{235}_{rms}$ = [(348.99 u)$^{-1/2}$ $-$ (351.99 u)$^{-1/2}$]/(348.99 u)$^{-1/2}$ = 0.427%.

CHAPTER 15

Answers to Odd-Numbered Multiple Choice Questions

1. e **3.** c **5.** c **7.** b **9.** a **11.** c **13.** a **15.** b **17.** b **19.** c

Answers to Odd-Numbered Problems

1. 63 kJ **3.** 116 W **5.** 9×10^5 J **7.** 17 K **9.** 0.68 kJ **11.** 0.68 kg **13.** 1.2×10^8 J **15.** 113 g **17.** 0.921 kg **19.** W = 4.148 J $\approx$ 1 cal **21.** 58°C **23.** +0.43 K **25.** 1.9 h **27.** 27 kcal **29.** 0.74 MJ **31.** c = 0.65 kJ/kg·K **33.** 1.8 h **35.** 12.7 g **37.** c = 443 J/kg·K **39.** 41.0 kJ **41.** 3.5 MJ **43.** 0.230 kg vaporizes **45.** 0.17 MJ **47.** 0.029 kg **49.** 0.125 kg **51.** T_f = 24°C **53.** 2.1×10^5 J **55.** 2.65 h **57.** 7.4×10^{-20} J **59.** 0.068 kg **61.** 354 m/s **63.** 1×10^2 g **65.** T_f = 0°C **67.** 0°C **69.** 2.3 m **71.** 0.36 MJ, T_f = 1234 K

Solutions to Selected Problems

1. 15 kcal = 63 kJ. **5.** (7 kcal/g)(30 g) = 210 kcal = 879 kJ, or to 1 significant figure, 9×10^5 J. **9.** $Q = cm\Delta T$ = (0.060) $\times$ (1.0 kcal/kg)(0.030 kg)(-90 K) = 0.162 kcal = 162 cal = 0.68 kJ. **13.** m = (3785 $\times 10^{-6}$ m^3)(0.68 $\times 10^3$ kg/m^3) = 2.574 kg; (2.574 kg)(48 MJ/kg) = 124 MJ = 1.2 $\times 10^8$ J. **17.** $-$(910 J/kg·K)m(-173 K) = (4186 J/kg·K)(4.95 kg)(7.00 K); m(157 430 J/kg) = 145 044.9 J; m = 0.921 kg. **21.** $Q/t = cm\Delta T/t$; 1.5 liters/min = 1.5 kg/min = 0.025 kg/s; 4000 W = (4186 J/kg·K)(0.025 kg/s)ΔT; ΔT = 38 K; T_f = 58°C. **25.** $Q = cm\Delta T$ = (4186 J/kg·K)(1.00 kg)(80 K) = 334 880 J; (334 880 J)/(50 J/s) = 6697.6 s = 1.9 h. **29.** 150 lb $\rightarrow$ 68.04 kg; $Q = cm\Delta T$; ΔT = 33.89°C $-$ 37°C = -3.11 K; Q = (3474.4 J/kg·K)(68.04 kg)(3.11 K) = 0.74 MJ. **33.** The average specific heat capacity of the body is 3.5 kJ/kg·K (0.83 kcal/kg·C°), and normal temperature is 37°C. From Eq. (19.1), $Q = cm\Delta T$ = (3.5 kJ/kg·K)(70 kg)(6 K) = 1470 kJ will raise the temperature 6 K. Since 200 kcal/h = 837.2 kJ/h, it will take (1470 kJ)/(837.2 kJ/h) = 1.8 h. **37.** The tension in the rope is g(6 kg) = 58.8 N, and this is the force exerted (via friction) on the rope by the rod, and it does it over a distance of $2\pi R(240)$ = 22.62 m. The work thus done is (58.8 N)(22.62 m) = 1330 J, and this result equals the heat-in: $Q = cm\Delta T$; 1330 J = c(0.250 kg)(12.0 K); c = 443 J/kg·K, which is what we might expect for some sort of steel. **41.** L_v = 2336 $\times 10^3$ J/kg; $Q = mL_v$ = (1.5 kg)(2336 $\times 10^3$ J/kg) = 3.5 MJ. **45.** $Q = cm\Delta T + mL_v$ = (138 J/kg·K) $\times$ (0.50 kg)(630 K $-$ 240 K) + (0.50 kg) $\times$ (296 $\times 10^3$ J/kg) = 26 910 J + 148 000 = 0.17 MJ. **49.** $-m$(334 000 J/kg) = (0.50 kg)(4186 J/kg·K)(0° $-$ 20°C); m = 0.125 kg. **53.** $Q/t = \lambda_T A\Delta T/d$ = (18 Cal·cm/ m^2·h·C°)(1.4 m^2)(4°C)/(2.0 cm) = 50.4 Cal/h = 0.014 kcal/s = 0.059 kJ/s = 0.059 kW. Therefore, 2.1 $\times 10^5$ J per hour. **57.** 1 mol = 18 g; at STP there are 6.02 $\times 10^{23}$ molecules/mol; to vaporize 1 mol requires $Q = mL_v$ = (0.018 kg)(2492 kJ/kg) = 44.86 kJ. Hence, per molecule the energy is (44.86 $\times 10^3$ J)/(6.02 $\times 10^{23}$ molecules/ mol) = 7.4 $\times 10^{-20}$ J. **61.** $Q = mL_f$; mL_f = (23 $\times 10^3$ J/kg)m; to which must be added $Q = cm\Delta T$ = (130 J/kg·K)m $\times$ (327 K $-$ 23 K) = (39 520 J/kg)m; the total is then (62 520 J/kg)$m = \frac{1}{2}mv^2$; v = 354 m/s. **65.** $-Q_{out} = Q_{in}$; if it dropped to zero, the calorimeter would provide an amount of heat equal to (1.00 kcal/kg·K)(0.398 kg)(5.1 K) + (0.093 kcal/kg·K)(0.102 kg)(5.1 K) = 2.078 kcal; the amount of heat needed to melt the ice is (0.040 5 kg)(80 kcal/kg) = 3.24 kcal; hence, not all the ice melts (!), and T_f = 0°C. **69.** $Q/t = k_{TB}A\Delta T/d_B = k_{TA}A\Delta T/d_A$; $k_{TB}/d_B = k_{TA}/d_A$; k_{TB}/k_{TA} = 0.60/ 0.026 = 23 = d_B/d_A; d_B = 2.3 m. **73.** In the first set-up, $Pt = c_w m\Delta T + P_l t$ and, in the second, $P't = c_w m'\Delta T + P_l t$; note that the mean temperature is the same in both cases, and so P_l is the same. Thus, solve for P_l in both equations and set them equal. $P - c_w m\Delta T/t = P' - c_w m'\Delta T/t$; and $(P - P')t = c_w \Delta T(m - m')$.

CHAPTER 16

Answers to Odd-Numbered Multiple Choice Questions

1. b **3.** d **5.** d **7.** b **9.** e **11.** b **13.** c **15.** a **17.** d **19.** b

Answers to Odd-Numbered Problems

1. +250 J **3.** −450 J **5.** +500 J **7.** 519 J **9.** $W = \frac{1}{4}P_i V_i$ **11.** $W = \frac{1}{2}P_i V_i$ **13.** +12 J **15.** 5.5% **17.** decrease T_L **19.** 3.5 kW **21.** 2.68 J/K **25.** +75.6 J/K **27.** $\Delta S_f =$ 1.22 kJ/K, $\Delta S =$ 12 J/K **29.** +9.8 kJ **31.** 0.22 g **33.** $T_f = -78$°C **35.** (a) 0.038 4 MPa (b) 207 K **37.** $W = nRT \ln(P_i/P_f)$ **41.** impossible **43.** 399 W **45.** $Q_H/W_i = 6.5$, pay for 1 unit of work and get 6.5 units of heat **47.** 0.69 W **49.** $T_H =$ 547 K **51.** $e = 62.5\%$, irreversible **53.** +0.014 kJ/K **55.** $W = 6.0$ MJ, $e = 67\%$ **59.** $T_i/T_f = 0.49$ **61.** 5.8 J/K

Solutions to Selected Problems

1. From Eq. (16.2), $\Delta U = Q - W =$ (+500 J) − (+250 J) = +250 J. **5.** Since $\Delta V = 0$, $W = 0$, and from Eq. (16.1), $Q = \Delta U = +500$ J. **9.** $W = \frac{1}{2}(\frac{1}{2}P_i) \times (2V_i - V_i) = \frac{1}{4}P_i V_i$. **13.** The work done is the area under the curve, which is $\frac{1}{2}$(3.0 MPa) × (2.0 × 10^{-6} m^3) + (4.0 × 10^{-6} m^3) × (2.0 MPa) + (1.0 MPa)(1.0 × 10^{-6} m^3) = 3.0 J + 8.0 J + 1.0 J = 12 J, and this is positive, since the gas expanded and it did work on the surroundings. **17.** $e_c = 1 - (T_L/T_H) = (T_H - T_L)/T_H$; increasing T_H increases both the numerator and the denominator; decreasing T_L increases the numerator only and therefore has the more profound effect. **21.** $\Delta S =$ (1.00 kJ)/(373 K) = 2.68 J/K. **25.** $\Delta S = Q/T = mL_f/T =$ (0.500 kg)(205 kJ/kg)/(1356 K) = +75.6 J/K. **29.** $W = P\Delta V =$ (0.30 MPa) × (0.50 m^3)(+0.065 m) = +9.8 kJ. **33.** $P_i V_i^\gamma = P_f V_f^\gamma$, where we assume the rapid expansion means an adiabatic process; final pressure is 1 atm; we don't have initial volume, so let's work per-unit-volume, (4.5 atm)(1 m^3) = (1 atm) V_f^γ; $V_f = 4.5^{1/\gamma} = 2.93$ m^3 per m^3, where $\gamma = 1.4$. To find T_f, use Ideal Gas Law; $P_i V_i/T_i = P_f V_f/T_f$; (4.5 atm)(1 m^3)/(300 K) = (1.0 atm)(2.93 m^3)/T_f; $T_f =$ 195 K, or −78°C. Notice that we didn't need to know the actual initial volume. **37.** From Eq. (16.5), $W = nRT \ln(V_f/V_i)$, but $P_i V_i = P_f V_f$; hence, $W = nRT \ln(P_i/P_f)$. **41.** The maximum efficiency from Eq. (16.9) requires the temperatures in kelvins; $T_H \rightarrow$ 144°F = 62.2°C or 355.2 K; $T_L \rightarrow$ 5°F = −15°C or 258 K; $e_c = 1 -$ (258/335.2) = 23%. As compared to the general expression for efficiency, Eq. (16.10), e = (26 kJ)/(102 kJ) = 25.4%. The engine exceeds the efficiency of a Carnot engine and is therefore impossible. **45.** η = heat-in/work-in = $Q_L/W_i = 5.5$. Moreover, $Q_H = Q_L + W_i = 5.5\ W_i + W_i = 6.5\ W_i$; hence, the ratio we want $Q_H/W_i = 6.5$. You pay for 1 unit of work and get 6.5 units of heat into the house, which compares very well with electrical heating (6.5 times better), where 100% of the electrical energy is converted into heat and you have to pay for each unit of energy that comes into the house. **49.** $e_c = 1 - T_L/T_H = 42.2\%$ and $T_L =$ 273.4 K. $T_H =$ 547 K. **53.** Since the masses are equal, $\Delta T = \pm 4.0$ K; thus, $Q = mc\Delta T =$ (20 kg) (4.186 kJ/kg·K)(±4.0 K) = ±334.88 kJ. Average temperatures are 38°C and 34°C; ΔS = (−334.88 kJ)/(311 K) + (+334.88 kJ)/(307 K) = (−1.077 kJ/K) + (+1.091 kJ/K) = +0.014 kJ/K. **57.** Using Eq. (16.7), $P_i V_i^\gamma = P_f V_f^\gamma$, we have $P_i/P_f = (V_f/V_i)^\gamma$. $P_i V_i = nRT_i$ and $P_f V_f = nRT_f$ and, dividing these, $P_i/P_f = T_i V_f/T_f V_i$; getting rid of the pressures, $(V_f/V_i)^\gamma = T_i V_f/T_f V_i$, from which Eq. (16.19) follows immediately. **61.** From Problem 60, $\Delta S = nR \ln(V_f/V_i) =$ (1.0 mol)(8.314 4 J/mol·K) ln 2 = 5.8 J/K.

CHAPTER 17

Answers to Odd-Numbered Multiple Choice Questions

1. c **3.** b **5.** c **7.** a **9.** a **11.** d **13.** b **15.** c **17.** d **19.** e

Answers to Odd-Numbered Problems

1. 5.7 × 10^{-18} kg **3.** ±1.1 × 10^{-5} C **5.** 2.3 N **7.** 5.3 μC **9.** zero **11.** 8.0 μN **13.** zero **15.** 0.58 N **17.** 8.3 N, θ = 49° **19.** 6.6 N, θ = 74° **21.** 0.11 μC **23.** $q = -0.34Q$, $a = 0.41$, where Q and q are separated by ad **25.** 8.0 × 10^2 N/C **27.** −1/10 N/C due east **29.** 5.6 × 10^{-11} N/C straight down **31.** 5.1 × 10^{11} N/C **33.** 7.8 × 10^4 N/C **35.** 6.4 × 10^2 N/C straight upward **37.** $E = \sigma/\varepsilon_0$ **39.** 4.4 nC **41.** 9.6 × 10^{10} m/s^2 **43.** $E = R\sigma/r\varepsilon$ **45.** $E = \sigma/\varepsilon_0$ **47.** 2.6 × 10^{15} m/s^2 **49.** $E = Q/4\pi R^2\varepsilon_0$ within the space, zero beyond the shells **51.** $E = rQ/4\pi\varepsilon_0 R^3$ **53.** 1.4 × 10^{21} N/C

Solutions to Selected Problems

1. N = (10^{-6} C)/(−1.6 × 10^{-19} C) = 6.2 × 10^{12} electrons. Each has a mass of 9.109 × 10^{-31} kg; hence, the mass increase is 5.7 × 10^{-18} kg. **5.** $F = kqq/r^2$ = (9.0 × 10^9 N·m^2/C^2)(1.6 × 10^{-19} C)^2/(1.0 × 10^{-14} m)^2 = 2.3 N. **9.** By symmetry, the net force is zero. **15.** $F_3 = F_{31} + F_{32}$ = (9 × 10^9 N·m^2/C^2)(−12.5 × 10^{-6} C) × (−10.0 × 10^{-6} C)/(3.0 m)^2 + (9 × 10^9 N·m^2/C^2)(−5.0 × 10^{-6} C)(−10.0 × 10^{-6} C)/(1.0 m)^2 = 0.125 N + 0.450 N = 0.58 N, in the increasing x-direction. **17.** See Fig. AN 17. $F_{12} = kq_1q_2/r_{12}^2 = 3.6$ N; $F_{13} = kq_1q_3/r_{13}^2 = 4.5$ N; $F_{14} = kq_1q_4/r_{14}^2 = 1.8$ N; F_{x1} = (1.8 N) + (4.5 N)4/5 = 5.4 N; F_{y1} = −(3.6 N) − (4.5 N)3/5 = −6.3 N; $F_1 = \sqrt{5.4^2 + 6.3^2} = 8.3$ N; θ = tan^{-1}(6.3/5.4) = 49°. **23.** The repulsive force between the two known charges is $F = kQ2Q/d^2$; put a negative charge $-q$ on the line connecting the two positive charges and between them at a distance ad from $+Q$ and $(1 - a)d$ from $+2Q$. The attraction between $+Q$ and $-q$ will cancel the repulsive force on $+Q$ when $k2Q^2/d^2 = kqQ/(ad)^2$ or when $a^2 = q/2Q$; similarly $+2Q$ will be in equilibrium when $k2Q^2/d^2 = kq2Q/[(1 - a)d]^2$, that is, $(1 - a)^2 = q/Q$; hence, $(1 - \sqrt{q/2Q}) = \sqrt{q/Q}$; $1 - \sqrt{q}/\sqrt{2Q} = \sqrt{q}/\sqrt{Q}$; $1 = \sqrt{q}(1 + \sqrt{2})/(\sqrt{2}\sqrt{Q})$; $q = 2Q/(1 + \sqrt{2})^2 = -0.34\ Q$ and $a = 0.41$. **27.** $E = F/q$ = (2.0 nN)/(−20 nC) = −1/10 N/C; due east. **31.** $E = kq_e/r^2$ = (9.0 × 10^9 N·m^2/C^2)(1.6 × 10^{-19} C)/(5.3 × 10^{-11} m)^2 = 5.1 × 10^{11} N/C. **35.** The horizontal field cancels, leaving only twice the vertical contribution from each charge: $E = 2(kq/r^2)\cos\theta$ = 2[(9.0 × 10^9 N·m^2/C^2)(+50 × 10^{-9} C)/(1.0 m)^2](cos 45°) = 6.4 × 10^2 N/C, straight upward.

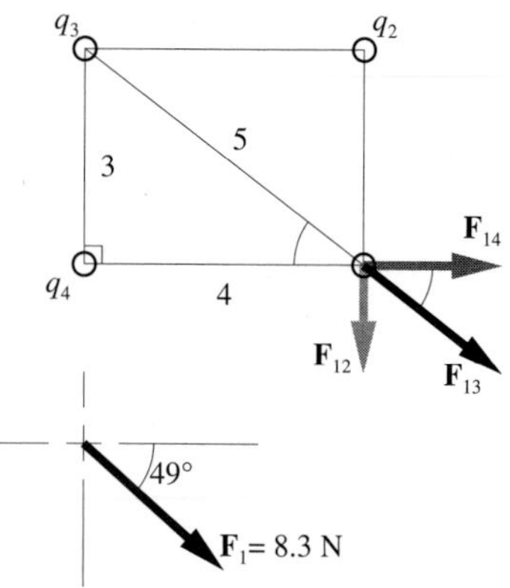

Fig. AN 17.

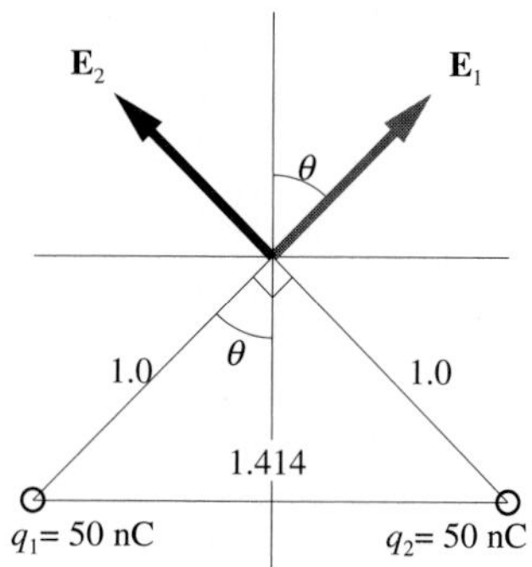

Fig. AN 35.

37. See Fig. AN 37. The field is parallel to the curved side and perpendicular to the endface; since $E = 0$ inside the metal, $EA = \sigma A/\varepsilon_0$ and $E = \sigma/\varepsilon_0$. **41.** $F = qE = ma$; $a = qE/m$ = (1.6 × 10^{-19} C)(20 × 10^3 N/C)/(3.35 ×10^{-26} kg) = 9.6 × 10^{10} m/s^2. **45.** See Fig. AN 45. Use a very short perpendicular Gaussian cylindrical surface. Assume that E and σ are constant over the tiny area of the cylinder (both may vary from region to region, depending on the shape of the conductor). Embed one endface within the conductor, where $E = 0$. Hence, $EA = \sigma A/\varepsilon_0$ and $E = \sigma/\varepsilon_0$. **49.** See Fig. AN 49. Surround the inner sphere with a spherical Gaussian surface of radius R ranging in the gap between the shells. The field is radial and $EA = E4\pi R^2 = Q/\varepsilon_0$; hence, $E = Q/4\pi R^2\varepsilon_0$, within the space. Now draw a Gaussian surface encompassing the large sphere anywhere beyond it. Again, if there is

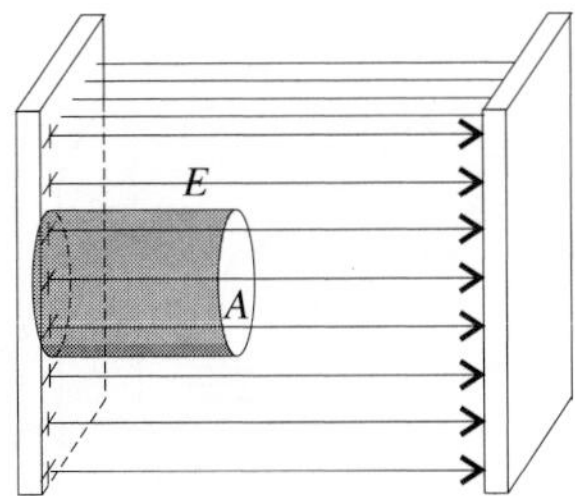

Fig. AN 37.

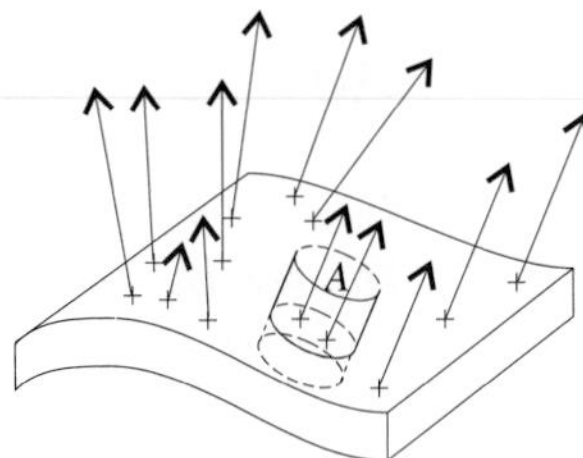

Fig. AN 45.

a field, by symmetry it would have to be radial, but now the net charge enclosed is zero and the field is therefore zero.

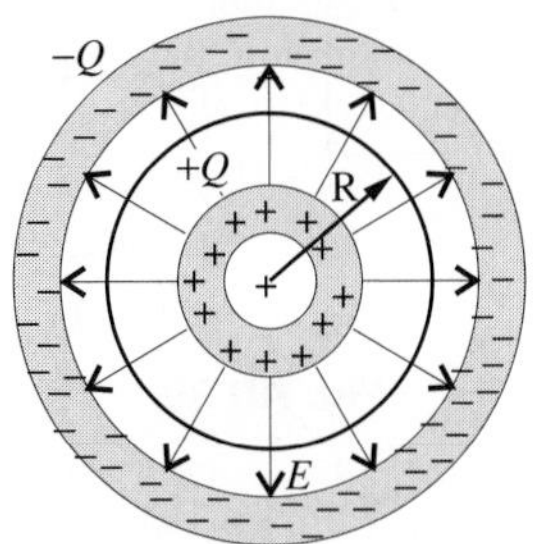

Fig. AN 49.

53. $E = kq_e/r^2 = (9 \times 10^9\ \text{N·m}^2/\text{C}^2) \times (1.6 \times 10^{-19}\ \text{C})/(1.0 \times 10^{-15}\ \text{m})^2 = 1.4 \times 10^{21}\ \text{N/C}$.

CHAPTER 18

Answers to Odd-Numbered Multiple Choice Questions

1. b **3.** c **5.** b **7.** d **9.** d **11.** d **13.** c **15.** d **17.** b **19.** b

Answers to Odd-Numbered Problems

1. -25.0×10^{-7} J **3.** ± 0.10 V **5.** V = 0 **7.** -3.6×10^5 V **9.** $\Delta\text{KE} = +500$ eV, $\Delta\text{PE}_\text{E} = -500$ eV **11.** $\Delta\text{KE} = 4.8 \times 10^{-15}$ J, 1.0×10^8 m/s **13.** 1.4×10^{-17} J **15.** 1.3×10^2 V **17.** 5.58×10^{-12} V **19.** $+0.23$ J **21.** 11 MV/m **23.** $V = -8.3$ kV, $E = 0$ **25.** $Q = 1.5 \times 10^{-10}$ C **27.** $Q = 1 \times 10^{-5}$ C **29.** $R = 90$ mm **31.** 5.0 μF **33.** 8.9×10^{-10} F **35.** 2.5 pF **37.** 110 pF **39.** 0.99 μF **41.** 10 pF **43.** 1.4 nJ **45.** 0.01 F/m^2 **47.** 7×10^{-4} C/m^2 **49.** 1.5 nJ/m **51.** 9 pF **53.** 13 μF **55.** 3.6×10^{-10} C **57.** 1.0 V **59.** 2.2 nJ **61.** 24 nF **63.** 1.4 nJ

Solutions to Selected Problems

1. $\Delta W = \Delta\text{PE}_\text{E} = q\Delta V = (-25.0 \times 10^{-9}\ \text{C})(100\ \text{V}) = -25.0 \times 10^{-7}$ J. This is the work done against the field. The negative sign means the charge has work done on it by the field. It moves from 0 to 100 V and so in the opposite direction to the field. That's the direction a negative charge moves in spontaneously, propelled by the field. **5.** A charge of $-Q$ is induced on the inner surface of the outer sphere. At a point outside, the potential is equivalent to that due to a point charge of $+Q$ and a point charge of $-Q$, both at the center. Hence, at P, $V = 0$. **13.** At the surface, $V = k_0Q/R = (9.00 \times 10^9\ \text{N·m}^2/\text{C}) \times (1.00 \times 10^{-9}\ \text{C})/(0.10\ \text{m}) = 90$ V. $W = q_e\Delta V = (1.6 \times 10^{-19}\ \text{C})(90\ \text{V}) = 1.4 \times 10^{-17}$ J. **19.** The potential at the point is $V = k_0Q_1/r_1 + k_0Q_2/r_2 = (9.000 \times 10^9\ \text{N·m}^2/\text{C}^2)(+10\ \mu\text{C})/(4.0\ \text{m}) + (9.000 \times 10^9\ \text{N·m}^2/\text{C}^2)(-25\ \mu\text{C})/(5.0\ \text{m}) = -22.5$ kV. $W = q\Delta V = (-10\ \mu\text{C})(-22.5\ \text{kV} - 0) = +0.23$ J. **23.** Each face has a diagonal of $\sqrt{2}$ m, and so there is a tilted rectangle $\sqrt{2}$ m by 1.0 m passing through the center and having four spheres at its corners. The diagonal of that square is $\sqrt{3}$ m long and half of that, $\frac{1}{2}\sqrt{3}$, is the corner-to-center distance. Hence, $V = 8kQ/r = -8.3$ kV. And $E = 0$. **25.** $Q = CV = (100 \times 10^{-12}\ \text{F})(1.5\ \text{V}) = 1.5 \times 10^{-10}$ C. **31.** $C = Q/V = (20\ \mu\text{C})/(4.0\ \text{V}) = 5.0\ \mu$F. **35.** $C = \varepsilon A/d$; $A = \pi R^2 = \pi(0.25 \times 10^{-2}\ \text{m})^2 = 1.96 \times 10^{-5}\ \text{m}^2$; $C = (4.8)(8.85 \times 10^{-12}\ \text{F/m}) \times (1.96 \times 10^{-5}\ \text{m}^2)/(0.33 \times 10^{-3}\ \text{m}) = 2.5$ pF. **41.** The 12-pF and 4.0-pF capacitors are in series and so equivalent to 3.0 pF, which, in turn, is in parallel with 7.0 pF. Hence, $C = 10$ pF. **45.** The membrane is like a rolled-up parallel-plate capacitor, so $C = \varepsilon A/d$; $C/A = \varepsilon/d = 7\varepsilon_0/d = 7(8.85 \times 10^{-12}\ \text{C}^2/\text{N·m}^2)/(6\ \text{nm}) = 0.01\ \text{F/m}^2$. **49.** $\Delta\text{PE}_\text{E}/L = \frac{1}{2}CV^2/L = \frac{1}{2}(3 \times 10^{-7}\ \text{F/m}) \times (0.1\ \text{V})^2 = 1.5$ nJ/m. **53.** The two 6.0-μF capacitors are in series (yielding 3.0 μF), as are the two 5.0-μF capacitors (yielding 2.5 μF). Now everything is in parallel, so 3.0 μF + 7.5 μF + 2.5 μF = 13 μF. **55.** The 9.0-pF capacitor is in parallel with the battery, and so the voltage is 12 V. All the capacitors are in parallel; hence, $C = 30$ pF. $Q = CV = (30\ \text{pF})(12\ \text{V}) = 3.6 \times 10^{-10}$ C. **63.** The diagram shows successive simplifications of the circuit. The equivalent capacitance is 20 pF; hence, $\text{PE}_\text{E} = \frac{1}{2}CV^2 = \frac{1}{2}(20\ \text{pF})(12\ \text{V})^2 = 1.4$ nJ.

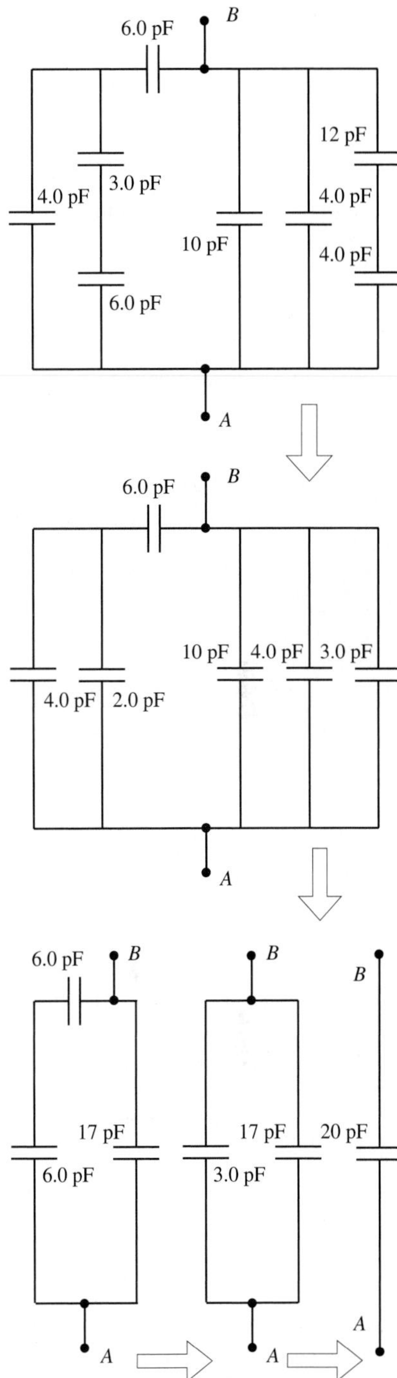

Fig. AN 63.

CHAPTER 19

Answers to Odd-Numbered Multiple Choice Questions

1. c **3.** d **5.** c **7.** e **9.** b **11.** b **13.** a **15.** c **17.** d **19.** c **21.** a

Answers to Odd-Numbered Problems

1. 10 μA **3.** 96.5 A **5.** 3.6×10^2 C **7.** 6.2×10^{14} **9.** 120 C **11.** 1.5×10^3 cells **13.** 36 kC **15.** 27 h **17.** (a) A and B (b) A and E (c) A and D and C (d) A and D (e) A and F when the headlights are on **19.** 20 cells in series form a row, and there are 20 rows in parallel **21.** 27 h **23.** 0.11 MC **25.** 10 Ω **27.** $50 \times 10^9\ \Omega$ **29.** if it's 1 m on a side **33.** $1.6 \times 10^{-8}\ \Omega\text{·m}$ **35.** 1.9 mm **37.** (a) 0.12 Ω (b) 0.10 Ω **39.** 0.1 A **41.** 0.1 W

43. 0.16 A **45.** 2.2 kW **47.** 0.017 V/m **49.** 1×10^3 °C **51.** 26 Ω **53.** 220° C **55.** 87 kW **57.** (a) 2.0×10^{-4} Ω (b) 2.4×10^{-5} V (c) 2.9×10^{-7} W **59.** 5.2 kW **61.** 30 W **63.** 1.3 W, 12 Ω **65.** 12 W **67.** $\Delta R/R_0 = 2.7$, poor: 12

Solutions to Selected Problems

1. $I = \Delta Q/\Delta t = (10\ \mu C)/(1.0\ s) = 10\ \mu A$. **7.** $I = \Delta Q/\Delta t$; $\Delta Q = (1.0\ mA)(0.10\ s) =$ 0.10 mC. The number of protons is $(0.10 \times 10^{-3}\ C)/(1.602 \times 10^{-19}\ C) = 6.2 \times 10^{14}$. **11.** (220 V)/(0.15 V per cell) = 1466.7 cells = 1.5×10^3 cells. Salt water, because of the presence of ions, is a much better conductor. **15.** Number of coulombs passing per second is $(I/A)A = (1.0\ MA/m^2)(1.00 \times 10^{-6}\ m^2) =$ 1.0 C/s; or 6.24×10^{18} electrons/s; hence, $(6.02 \times 10^{23}$ electrons$)/(6.24 \times 10^{18}$ electrons/s$)$ = 96 507 s = 27 h. **19.** First, do the voltage, (9.0 V)/(0.45 V) = 20; we need 20 cells in series forming one row. To get the current, we must have (440 mA)/(22 mA) = 20 such rows in parallel. **25.** $R = V/I = (100\ V)/(10\ A) = 10\ \Omega$. **29.** If the cube is 1 m on a side, $R = \rho$. **33.** $R = \rho L/A$; $\rho = AR/L = 1.6 \times 10^{-8}\ \Omega \cdot m$. **37.** (a) From Eq. (19.8), $R = 0.10\ \Omega + (0.10\ \Omega)(0.005\ 0\ K^{-1}) \times (30\ K) = 0.12\ \Omega$ (b) since the coefficient is essentially zero, $R = 0.10\ \Omega$. **41.** $P = IV = (16 \times 10^{-3}\ A)(9\ V) = 0.144\ W =$ 0.1 W. **47.** From Eq. (19.6) for a uniform E-field, $V = Ed$, so we must find V; since $V = IR$, we need R; $R = \rho L/A =$ $(1.7 \times 10^{-8}\ \Omega \cdot m)(1.0\ m)/(1.0 \times 10^{-6}\ m^2) =$ 0.017 Ω; $V = (1.0\ A)(0.017\ \Omega) = 0.017\ V$; $E = V/d = (0.017\ V)/(1.0\ m) = 0.017\ V/m$. **51.** $R_f = R_0(1 + \alpha_0 \Delta T) = (10\ \Omega) \times (1 + 0.003\ 9\ K^{-1} \times 400\ K) = 25.6\ \Omega =$ 26Ω. **55.** The current entering the factory must be $I = P/V = (45 \times 10^3\ W)/(110\ V) =$ 409.09 A; the total power loss in the cables is $I^2R = (409.09\ \Omega)^2(2 \times 0.25\ \Omega/\text{mile} \times 0.50\ \text{mile}) = 41.84$ kW; hence, the net power supplied must be 45 kW + 41.8 kW = 87 kW. **59.** $P = (12\ A)(500\ V) = 6.0$ kW. 86%(6.0 kW) = 5.2 kW = 6.9 hp. **63.** $P = IV = (1/3\ A)(3.9\ V) = 1.3\ W$. $P = I^2R =$ $1.3\ W = (1/3\ A)^2R$; $R = 12\ \Omega$. **69.** See Fig. AN 69. From the diagram $R/(T' + T) = R_0/T'$; and so $R = R_0(1 + T/T')$; comparing this expression with Eq. (19.8), we find that T is equivalent to ΔT and $\alpha_0 = 1/T'$.

CHAPTER 20

Answers to Odd-Numbered Multiple Choice Questions

1. d **3.** d **5.** c **7.** c **9.** a **11.** b **13.** d **15.** c **17.** a **19.** b **21.** b

Answers to Odd-Numbered Problems

1. 1.5 A, goes out **3.** 0.15 A, P doubles 1.8 W to 3.6 W **5.** 1 Ω **7.** 5.0 A **11.** 2 Ω **15.** 5.0 Ω **17.** 5 Ω **19.** 9 A through 17 Ω, 6 A through 3 Ω, 3 A through 4 Ω and 2 Ω **21.** 1 A through 4 Ω, 6/9 A through 3 Ω, 3/9 A through 6 Ω **23.** 10 Ω **25.** $R_s =$

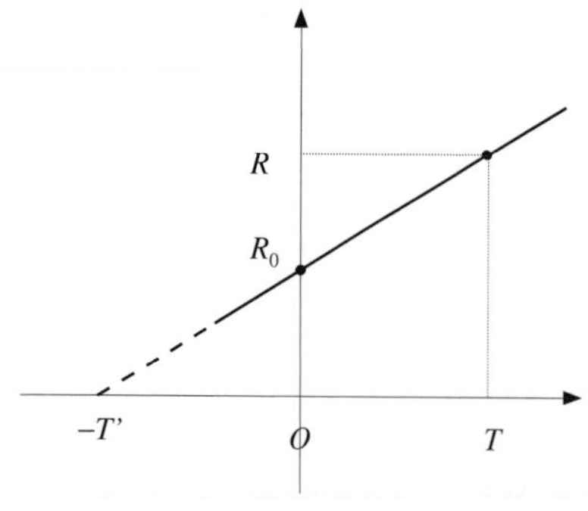

Fig. AN 69.

0.05 Ω **27.** 4.0 s **29.** 6.0 μA, 2.0 s **31.** $R_e = 1.8\ \Omega$ **33.** 2 A through 3 Ω and 1 Ω, 5 A through 2 Ω, 2 A through 4 Ω and 1 Ω **35.** 4 Ω, 95 W **37.** 24 W, C is 1.5 V above A **39.** 0.89 A, 8.6 V **41.** 27 W **43.** $R_s = 0.100\ \Omega$ **45.** 0.12 A, 0.60 ms **47.** 1.6 μA **49.** 3 A, 3 A, 6 A **51.** $I_1 =$ 2 A, $I_2 = 2/3$ A, $I_3 = 4/3$ A **53.** 4 V **55.** 32 V **57.** $I_1 = 2.0$ A, $I_2 = 1.0$ A, $I_3 = 1.0$ A **59.** 8 μC **61.** $R_1 = 46\ \Omega$, $\mathscr{E}_1 = 14$ V **63.** 28 Ω **65.** (a) zero (b) 36 μC **67.** $R = 4.0\ \Omega$, $V = 30$ V, $I_1 = 8.0$ A, $I_2 = I_6 = 2.0$ A, $I_3 = 1.0$ A, $I_4 = I_5 = 5.0$ A **69.** A at +14 V, E at +29 V, F at +12 V, D, C, B, and G at zero. **71.** 42/5 Ω.

Solutions to Selected Problems

3. $R_e = 80\ \Omega$; $I = V/R_e = 0.15$ A. After S is closed $R_e = 40\ \Omega$, and $I = 0.30$ A. $P = IV$, so the power doubles from 1.8 W to 3.6 W. **7.** Each lamp draws $I = P/V = (100\ W)/(100\ V) = 1.0$ A; hence, the source must provide 5.0 A. **11.** 6 Ω in parallel with 3 Ω is 2 Ω. **17.** The only resistance between A and B is 5 Ω. **21.** The 3.0-Ω and 6.0-Ω resistors in parallel equal 2.0 Ω, and that's in series with the 4.0-Ω resistor; hence, $R_e =$ 6.0 Ω. $I = V/R_e = (6.0\ V)/(6.0\ \Omega) = 1.0$ A. This current passes through the 4.0-Ω resistor and splits 6 parts out of 9, that is, 6/9 A, through the 3.0-Ω resistor and 3/9 A through the 6.0-Ω resistor. **25.** $(50\ \Omega)(0.001\ I) = (0.999\ I)R_s$; $R_s = 0.05\ \Omega$. **29.** $I_i = V/R =$ $(12.0\ V)/(2.0\ M\Omega) = 6.0\ \mu A$; $RC =$ $(2.0\ M\Omega)(1.0\ \mu F) = 2.0$ s. **31.** 3 Ω and 9 Ω are in parallel, and that's in parallel with the 18-Ω resistor yielding 2 Ω, which is in parallel with 3 Ω and 6 Ω for a total of 1 Ω; hence, $R_e = 1.8\ \Omega$. **37.** See Fig. AN 37. The equivalent resistance is 6.0 Ω; the battery provides 2.0 A and 24 W. Working backwards, the currents split as shown. Going from A to B, there is a rise of $\frac{1}{2}$ V, and from B to C, another rise of 1 V; hence, C is $1\frac{1}{2}$ V above A. **39.** See Fig. AN 39. The equivalent resistance is 10.14 Ω. The current from the battery is 0.89 A, and terminal voltage is 9.0 V − (0.89 A)(0.50 Ω) = 8.6 V. **43.** The coil carries 1.00 mA when the meter carries 1.00 A, and so the shunt carries a maximum of 0.999 A; $(1.00 \times 10^{-3}\ A)r = (0.999\ A)R_s$, $R_s = 0.100\ \Omega$.

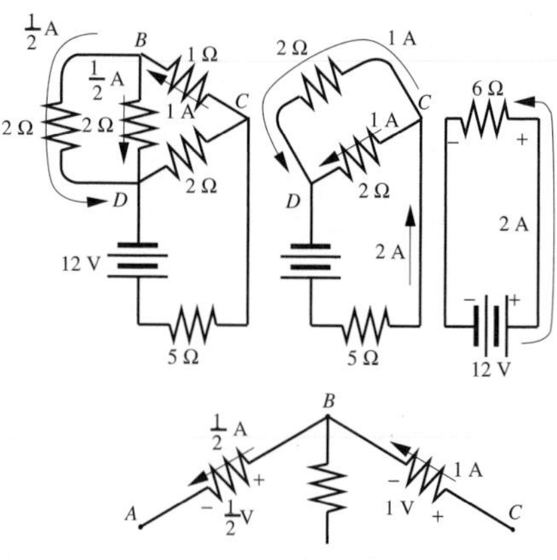

Fig. AN 37.

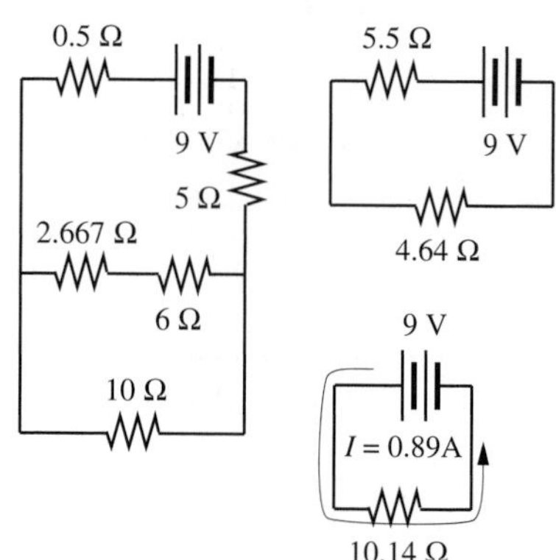

Fig. AN 39.

45. See Fig. AN 45. $I_i = V/R = (12\ V)/(100\ \Omega) = 0.12$ A; $RC = (100\ \Omega)(6.0\ \mu F) =$ 0.60 ms.

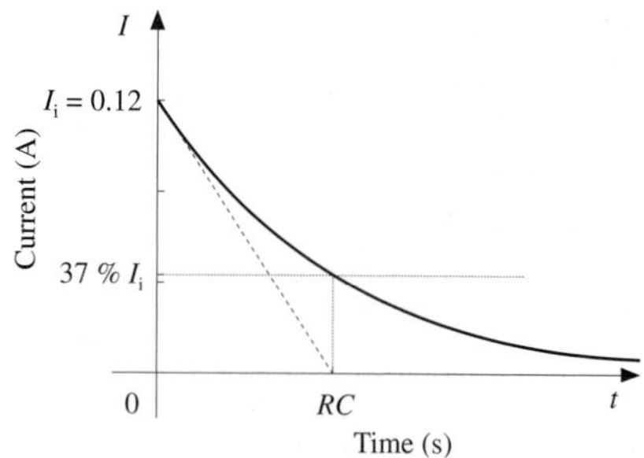

Fig. AN 45.

51. See Fig. AN 51. At A, $I_1 = I_2 + I_3$ around loop A-B-C-D-A going clockwise: $-3I_2 - 3I_2 + 20\ V - 8I_1 = 0$. Around loop A-B-C-A, clockwise: $-3I_2 - 3I_2 + 3I_3 = 0$. Around loop A-C-D-A, clockwise: $-3I_3 + 20\ V - 8I_1 = 0$. Hence, $I_1 = 2$ A, $I_2 = 2/3$ A, $I_3 = 4/3$ A. **55.** The drop across the top branch is 16 V; with that across the middle branch the current through it must be 1.0 A; hence, the current in the bottom branch, from

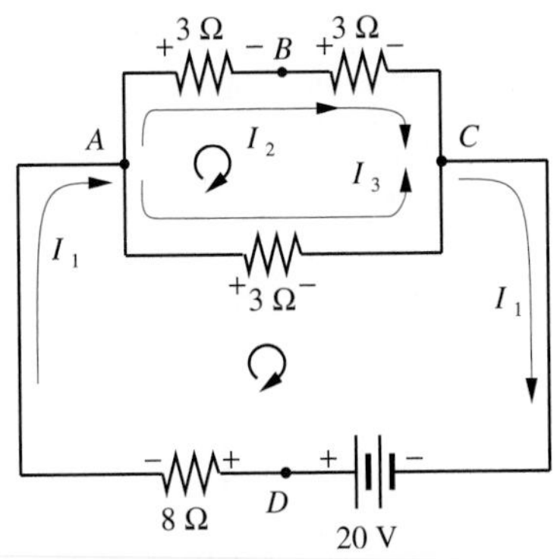

Fig. AN 51.

the Node Rule, is 1.0 A. The drop across the two bottom resistors is 16 V; hence, the voltmeter must read 32 V. **59.** The equivalent resistance in the steady state is 12 Ω; hence, the current is 1 A. The voltage across C is (1 A)(4 Ω) = 4 V, and since $CV = Q$, $Q = 8\ \mu$C. **63.** See Fig. AN 63. $I_1 = 200$ mA, $V_{AB} = +2.4$ V; hence, around the first loop clockwise: $-0.200R + 12 - 2.4 - 4 = 0$, $R = 28\ \Omega$.

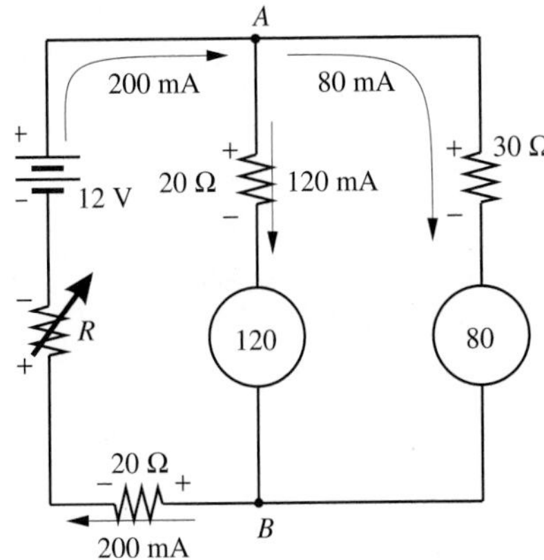

Fig. AN 63.

71. Place a source of, say, 18 V, across the group. Then solve for the six branch currents: here, 6/7 A goes through both 12-Ω resistors, 3/7 A goes through the central 6-Ω resistor, and 9/7 A passes through each outer 6-Ω resistor. Thus, the source provides 15/7 A, and so the equivalent resistance is 42/5 Ω, independent of the source.

CHAPTER 21

Answers to Odd-Numbered Multiple Choice Questions

1. d **3.** c **5.** c **7.** b **9.** b **11.** a **13.** a **15.** b **17.** c **19.** c **21.** a

Answers to Odd-Numbered Problems

3. 4.0×10^{-6} T **5.** 4.0 mT **7.** 0.10 mT **9.** 63 μT **11.** 0.63 nT **13.** 0.20 A **15.** 2.5 mT **17.** 1.3×10^{-11} T clockwise looking toward the source **19.** 500 A **21.** 24×10^{-7} T **23.** 31° W of N **25.** $B_z = \mu_0 IR^2/2z^3$ **27.** $\approx 6 \times 10^9$ A **29.** $B = \mu_0 NI/2\pi r$, same for single wire **35.** $B = \mu_0 i$ inside, $B = 0$ outside **37.** $+x$-direction **39.** 1.3 T **43.** 1.8×10^7 m/s **47.** $-y$-direction **49.** 1.00 A **51.** 0.31 mN **53.** 0.019 N·m **55.** 3.8×10^{18} m/s^2 **57.** circle, $R = 14$ mm **59.** $p = qBR$ **63.** 0.5 m/s **65.** $\tau_f = 0.75\ \tau_i$ **67.** 2.0×10^{-5} N, repulsive **69.** $f = qB/2\pi m$

Solutions to Selected Problems

3. From Eq. (21.2), $B = (1.26 \times 10^{-6}$ T·m/A)(10 A)/$2\pi(0.50$ m) $= 4.0 \times 10^{-6}$ T. **7.** $V = IR$; $I = (12$ V)/(1.2 Ω) $= 10$ A; $B = \mu_0 I/2\pi r = 0.10$ mT. **11.** $B_z = \mu_0 I/2R = 0.63$ nT. **15.** $V = IR$, $I = 0.20$ A; $B_z \approx \mu_0 nI = 2.5$ mT. **19.** $B = \mu_0 I/2\pi r$; $I = 500$ A. **23.** $B = \mu_0 I/2\pi r = 0.30 \times 10^{-4}$ T, west—due to the current; the net field is at an angle of $\tan^{-1}0.30/0.50$, or 31° west of north. **27.** We take a single circular equatorial current loop circulating with a radius of 2.3×10^3 km—it would be more realistic to assume several such loops distributed over the spherical core. We know that the field drops off as in Fig. 21.19b and the formula given. At a distance of 2.78R, it's down by a factor of about 26. $B = \mu_0 I/2R$; 0.6×10^{-4} T $= (1.26 \times 10^{-6}$ T·m/A)$I/2(2.3 \times 10^6$ m); $I = 0.22 \times 10^9$ A, but this distance is to be 6.3×10^6 m away along the z-axis, so the current should be roughly 26 times larger, or $I \approx 6 \times 10^9$ A. Since there is a south magnetic pole in the north, the current circulates from east to west. **29.** Using Ampère's Law around a circular path of radius R within the torus, $\Sigma B_\parallel \Delta l = \mu_0 \Sigma I$. If we take the field to be axial (a solenoid bent around on itself), $B2\pi r = \mu_0 NI$; $B = \mu_0 NI/2\pi r$. The field varies inversely with r—the density of the windings is not the same on the inner and outer surfaces. For a single wire, $B = \mu_0(NI)/2\pi r$, which is identical to the coil. **33.** See Fig. AN 33. For a finite sheet, the field lines would loop around the sheet counterclockwise looking into the current (along the minus z-axis). The sheet is infinite so there cannot be a preferred direction other than parallel to it—no $\pm y$-component to the field. Moreover, there's no reason for the field to be any different at different distances from the sheet because, being infinite, the sheet "looks" the same as seen from any point above it. In effect, the notion of "distance from" loses its traditional meaning in this idealized situation. Going around the Ampèrean loop in the figure, $\Sigma B_\parallel \Delta l = \mu_0 \Sigma I$; $Bl + Bl = \mu_0 il$; and $B = \frac{1}{2}\mu_0 i$. **37.** In the $+x$-direction. **41.** Using $F = qvB$; $B = F/qv$; 1 T = 1 N/C·m/s = 1 (N·m)s/C·m^2 = 1 (J/C)s/m^2 = 1 V·s/m^2. **45.** $qE = qvB$; $v = E/B$. **49.** $F = IlB$; $I = F/lB = 1.00$ A. **53.** $\tau = NIAB \sin\phi = (20)(1.5$ A)$(1.3 \times 10^{-3}$ m^2) $\times$ (0.90 T)(0.529 9) $= 0.019$ N·m **55.** $F =$

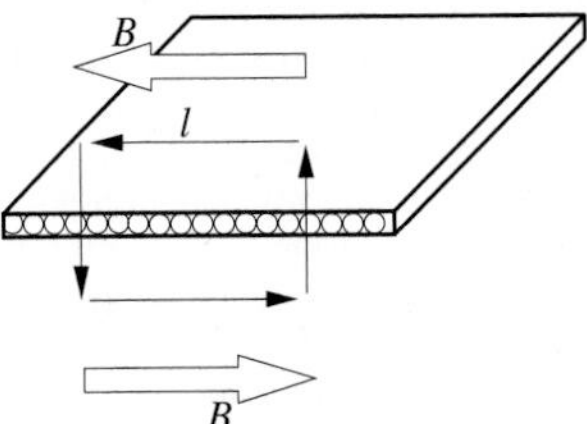

Fig. AN 33.

$qvB \sin\theta = ma$; $a = (1.60 \times 10^{-19}$ C)(5.0 $\times 10^6$ m/s)(5.0 T)0.866/(9.11 $\times 10^{-31}$ kg) $= 3.8 \times 10^{18}$ m/s^2. **59.** $qvB = mv^2/R$; $qB = mv/R$; $p = qBR$. **63.** $V = vBl$; $v = V/Bl = (1.0\ \mu$V)/(0.50 mT)(4 mm) $= 0.5$ m/s. **65.** $\tau = NIAB$; hence, $\tau_i/\tau_f = I_i/I_f = I_i/0.75I_i$; $\tau_f = 0.75\tau_i$. **69.** $qvB = mv^2/R$, $qB = mv/R$, $v = qRB/m$; if f is the frequency, $1/f$ is the period, which equals $2\pi R/v$; hence, $f = v/2\pi R$ and $f = qB/2\pi m$.

CHAPTER 22

Answers to Odd-Numbered Multiple Choice Questions

1. e **3.** c **5.** a **7.** a **9.** c **11.** c **13.** b **15.** d **17.** d **19.** b **21.** b

Answers to Odd-Numbered Problems

1. 3.0 μWb **3.** 1.2 Wb/m^2 **5.** 0.1 T, south **7.** 0.50 V **9.** 75 ms **11.** 8.0 V **13.** 0.60 V **15.** bulb would not light **17.** 90 μV **19.** $I = Blv/R$, clockwise, constant for a time l/v **21.** 4.1 μC **23.** 3.5 mV, counterclockwise **25.** P $= (Blv)^2/R$ **27.** $B = RK\theta/NA$ **29.** 0.83 mT **31.** 1.6 V **33.** 39 V **35.** 100 V, $f = 60.000$ Hz **37.** $L = 0.26$ H **39.** 0.80 mWb **41.** $N = 2.5 \times 10^3$ **43.** 8.4×10^2 **45.** 5.0 H **47.** 2.5 H **49.** 0.632 **51.** 1.2 H **53.** 21 mT **55.** 12 mV **57.** $\mathscr{E} = (1.3$ kV) sin $200\pi t$ **59.** 0.15 s **61.** 79 mJ/m **63.** (a) 6.0 A (b) 12 ms (c) 3.8 A **65.** (a) 2.0 A (b) 4.0 s (c) $\approx$20 s **69.** 7×10^{16} J, 7×10^8 gallons of gasoline **71.** $\Delta I/\Delta t = (V - IR)/L$, 0.89 kA/s never reaches 1.0 A **73.** 21 V

Solutions to Selected Problems

1. $\Phi_M = B_\perp A = (1.2$ mT)(0.002 5 m^2) $= 3.0\ \mu$Wb. **5.** The final field minus the initial field is the change in the field and that's 0.1 T, south. **9.** From Eq.(22.3), $\Delta t = 150 \times (0.030$ Wb)/(60 V) $= 75$ ms. **13.** $\mathscr{E} = vBl = (200$ m/s)(0.050 mT)(60 m) $= 0.60$ V. **19.** As soon as the leading edge of the loop starts cutting field lines, there is an emf = Blv, which produces a counterclockwise current $I = Blv/R$. The emf is constant (for a time l/v) until the trailing edge of the loop enters the field. It produces an oppositely directed emf (the flux no longer changes), and the net emf goes to zero. It remains zero until the leading edge emerges from the field at a time $3l/v$ after it entered, whereupon the emf jumps to Blv in the opposite direction. A current $I = Blv/R$ appears clockwise and remains constant for a time l/v. **23.** From Eq. (22.1), $N\Delta\Phi_M = 1(0.500\text{ T})\cos 30° \times (1.6 \times 10^{-3}\text{ m}^2) = 6.928 \times 10^{-4}\text{ T·m}^2$; from Eq.(22.3), emf = 3.5 mV. Counterclockwise. **27.** $I = V/R$; $\Delta Q = N\Delta\Phi_M/R = NBA/R = K\theta$, $B = RK\theta/NA$. **33.** For each length $\mathscr{E} = vBl = (22\text{ m/s})(0.35\text{ T})(0.10\text{ m}) = 0.77$ V. Since there are 2 lengths per turn and 25 turns, 50(0.77 V) = 39 V. **37.** $N\Phi_M = LI$; $L = (500)(0.002\,0\text{ Wb})/(3.8\text{ A}) = 0.26$ H. **41.** $N\Phi_M = LI$; $N = (2.5\text{ H})(1.80\text{ A})/(1.80\text{ mWb}) = 2.5 \times 10^3$. **45.** $\mathscr{E} = -L\dfrac{\Delta I}{\Delta t}$, $L = -(10\text{ V})/(2.0\text{ A/s}) = 5.0$ H (forget the sign, since we don't know if the current is increasing or decreasing). **49.** 1 − 0.368 = 0.632. **55.** $\mathscr{E} = \frac{1}{2}Br^2 2\pi f = \frac{1}{2}(0.050\text{ mT} \times \cos 40°)(5.0\text{ m})^2 2\pi(240/60) = 12$ mV. **59.** $\mathscr{E} = -L\dfrac{\Delta I}{\Delta t}$; $\Delta t = (1.5\text{ H})(10\text{ A})/(100\text{ V}) = 0.15$ s. **63.** (a) $I = V/R = (150\text{ V})/(25\ \Omega) = 6.0$ A; (b) $L/R = (300\text{ mH})/(25\ \Omega) = 12$ ms; (c) 0.632(6.0 A) = 3.8 A. **67.** From Eq. (22.9), $L \approx \mu N^2 A/l = An^2 l\mu$ where An^2 is a constant, so ΔL depends on the change in μl. Once the shaft is inserted, part of the region of the coil (d long) changes from μ_0 to μ; hence, $\mu d + \mu_0(l - d)$ is the new value, $\mu_0 l$ the original value. The difference $d(\mu - \mu_0)$ is what we want; hence

$$\Delta L \approx d(\mu - \mu_0)\frac{N^2 A}{l^2}$$

71. From Eq. (22.11), $\Delta I/\Delta t = (V - IR)/L$, which (in the limit as $\Delta t \to 0$) is also the instantaneous rate of change of current, or the slope of the curve in Fig. 22.21 at any instant. At $t = 0$, $I = 0$ and $\Delta I/\Delta t = V/L = 2.4$ kA/s. At $t = L/R$, $I = 0.63V/R$ and $\Delta I/\Delta t = (0.37\ V)/L = 0.89$ kA/s. The maximum current is $V/R = 0.60$ A, so it never reaches 1.0 A.

CHAPTER 23

Answers to Odd-Numbered Multiple Choice Questions

1. b **3.** d **5.** e **7.** c **9.** b **11.** e **13.** a **15.** b

Answers to Odd-Numbered Problems

1. (a) zero (b) 84.9 V **3.** $i = (14\text{ A}) \times \sin 2\pi(50\text{ Hz})t$ **5.** 141 V **7.** 0.833 A **9.** 113 W **11.** 0.13 W **13.** 2.00 Ω **15.** 22 μF **17.** $f = 11$ Hz **19.** 94 Ω **21.** $f = 100$ Hz **23.** one resistor is probably shorted internally **25.** 20 A, 120 V **27.** 15 W, 240 W **31.** 2.12 A **33.** (a) something is wrong with it (b) new capacitor is shorted internally **35.** $P_m = 2P_{av}$ **37.** 10.1 Ω **39.** 219 Ω **41.** 1.30 kΩ **43.** $\theta = 26.6°$, voltage leads current **45.** 0.25 A **47.** $\theta = -63°$ **49.** $R = 1.7$ kΩ **51.** 500 Ω **53.** 5.1×10^{-9} F **55.** 52.8 mH **57.** 100:1 **59.** $V_s = 30$ V, $I_s = 4.0$ A **61.** 5.0 A, $\theta = 51°$ **63.** $V_L = 302$ V, $V = 477$ V **65.** 27 μF **69.** (a) 20:1 (b) $I_p = 0.11$ A **71.** 98.3% **73.** 20:1

Solutions to Selected Problems

3. $I_m = 1.414(10\text{ A}) = 14$ A; $i = (14\text{ A})\sin 2\pi(50\text{ Hz})t$. **7.** $P_{av} = 100$ W $= IV$; $I = (100\text{ W})/(120\text{ V}) = 0.833$ A. **9.** $P_{av} = IV = (1.50\text{ A})(75.0\text{ V}) = 113$ W. **13.** $X_C = 1/2\pi fC$; $X_{Ci}/X_{Cf} = f_f/f_i = 100$; hence, the final reactance is 2.00 Ω. **17.** $X_L = 2\pi fL$; $f = (10\ \Omega)/2\pi(0.15\text{ H}) = 10.6\text{ Hz} = 11$ Hz. **21.** $v = V_m 0.951$; $0.951 = \sin 2\pi f(0.002\,00\text{ s})$; $f = 100$ Hz. **25.** $1/R_e = 1/(10\ \Omega) + 1/(15\ \Omega)$, $R_e = 6.0\ \Omega$; $I = V/R_e = 20$ A. The voltmeter reads 120 V. **29.** $X_L = 2\pi fL$; $V = -L\Delta i/\Delta t$, so $L = -V\Delta t/\Delta i$ and 1 H = 1 V·s/A; hence, reactance has units of (1/s)(H) = V/A = Ω. **33.** (a) $X_C = 1/2\pi fC = 132.6\ \Omega$; hence, $I = V/X_C = (120\text{ V})/(132.6\ \Omega) = 0.90$ A. The current should be about 0.9 A and, instead, the ammeter reads 0.25 A. Something is wrong with that capacitor. (b) The fuse would blow if the new capacitor was shorted internally—it would short out the first capacitor. **37.** $Z = \sqrt{R^2 + X^2} = 10.1\ \Omega$. **39.** $Z = \sqrt{R^2 + X^2} = 219\ \Omega$. **43.** $\tan\theta = X/R$; $\theta = 26.6°$, and voltage leads current. **47.** $\tan\theta = (200\ \Omega)/(100\ \Omega) = 2.0$; $\theta = -63°$, this is a capacitive reactance. **51.** $Z = 2500\ \Omega - 2000\ \Omega = 500\ \Omega$. **53.** $C = 1/4\pi^2 f_0^2 L = 5.1 \times 10^{-9}$ F—such values are not unusual. **57.** 100:1. **61.** $2\pi f = 1257\text{ s}^{-1}$; $X_L = 2\pi fL = 37.7\ \Omega$; $Z = \sqrt{R^2 + X^2} = 48.18\ \Omega$; $I = V/Z = 5.0$ A; $\tan\theta = X_L/R = 1.257$, $\theta = 51°$, voltage leads current. **63.** $2\pi f = 377\text{ s}^{-1}$; $X_L = 2\pi fL = 150.8\ \Omega$; $Z = \sqrt{R^2 + X^2} = 238.7\ \Omega$; $I = V_R/R = (370\text{ V})/(185\ \Omega) = 2.00$ A. $V_L = IX_L = 302$ V. $\tan\theta = X_L/R$; $\theta = 39.18°$; $V = V_R/\cos\theta = 477$ V. **67.** $Z = \sqrt{R^2 + X_C^2}$; $I = V_i/Z$; $V_0 = IX_C$; $V_0 = V_i X_C/Z$; $V_0/V_i = X_C/Z = (1/2\pi fC)/\sqrt{R^2 + X_C^2} = 1/\sqrt{1 + (2\pi fRC)^2}$. **69.** (a) $V_p N_s = V_s N_p$; $N_p/N_s = V_p/V_s = 20{:}1$; (b) $2\pi f = 377\text{ s}^{-1}$; $X_L = 2\pi fL = 1.131$ kΩ; $I_p = V/X_L = 0.11$ A. **73.** $Z_p = V_p/I_p$; $Z_s = V_s/I_s$; $Z_p/Z_s = (V_p/V_s) \times (I_s/I_p)$ but, as we saw earlier, $V_p N_s = V_s N_p$ and $N_p I_p = N_s I_s$; hence, $Z_p = Z_s(N_p/N_s)^2$. $\sqrt{Z_p/Z_s} = N_p/N_s = 20{:}1$.

CHAPTER 24

Answers to Odd-Numbered Multiple Choice Questions

1. c **3.** b **5.** a **7.** b **9.** a **11.** d **13.** c **15.** d **17.** b **19.** e **21.** c

Answers to Odd-Numbered Problems

1. $12.6 \times 10^6\text{ m}^{-1}$ **3.** 10.00×10^3 m **7.** 20 V/m **11.** radiowaves, 100 s **13.** 1.50×10^7 m **15.** 5 μH **17.** 1.50×10^6 Hz **19.** $1.2 \times 10^{-9}\text{ J/m}^2\text{·s}$ **21.** 6.30×10^4 J **23.** 620 nm, 484 THz, 3.20×10^{-19} J, orange-red light **27.** −20 V/m **29.** $B_0 = 6.7 \times 10^{-7}$ T **31.** 14.5 **33.** 53 W/m² **35.** $E_{max} = 1.00 \times 10^4$ eV, 1.24×10^{-10} m **37.** (a) 628 nm (b) 4.78×10^{14} Hz (c) positive x-direction **39.** (a) 0.600 m (b) $1.0 \times 10^6\text{ J/m}^3$

Solutions to Selected Problems

1. $k = 2\pi/\lambda = 2\pi/(500\text{ nm}) = 12.6 \times 10^6\text{ m}^{-1}$. **5.** See Fig. AN 5. At $x = \lambda/4$, $kx = (2\pi/\lambda) \times (\lambda/4) = \pi/2$.

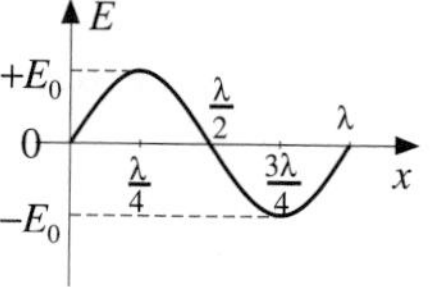

Fig. AN 5.

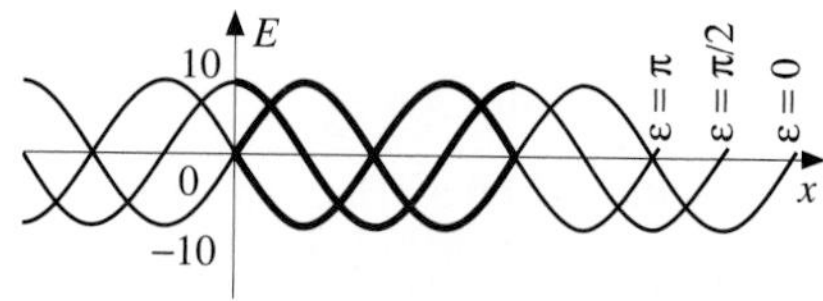

Fig. AN 9.

9. See Fig. AN 9. $E = 10\sin(kx + \varepsilon)$ mark off the axis in intervals of $\lambda/4$. Notice that shifting the phase by 90° results in a cosine function. **13.** $c = f\lambda_0$, $\lambda_0 = c/f = (3.00 \times 10^8\text{ m/s})/(20.0\text{ Hz}) = 1.50 \times 10^7$ m. **17.** $f = c/\lambda = (3.00 \times 10^8\text{ m/s})/(200\text{ m}) = 1.50 \times 10^6$ Hz. **21.** Irradiance is energy divided by area divided by time: $E/At = I$, $E = (1.05 \times 10^3\text{ W/m}^2)(1.00\text{ m}^2)(60.0\text{ s}) = 6.30 \times 10^4$ J. **23.** $1\text{ eV} = 1.602 \times 10^{-19}$ J; $E = hf = 3.20 \times 10^{-19}$ J; $f = 484$ THz; $\lambda_0 = c/(484\text{ THz}) = 620$ nm; orange-red light. **27.** $E(0,0) = 20\cos(0 - 0 + \pi) = -20$; $E(0, T/4) = 20\cos(0 - 2\pi fT/4 + \pi) = 20\cos(+\pi/2) = 0$; $E(0, T) = 20\cos(0 - 2\pi fT + \pi) = 20\cos(-\pi) = -20$ V/m. **31.** $\Delta l = c\,\Delta t = (3.00 \times 10^8\text{ m/s}) \times (30 \times 10^{-15}\text{ s}) = 9.0 \times 10^{-6}$ m. Now dividing this result by the wavelength yields the number of waves: $(9.0 \times 10^{-6}\text{ m})/(620 \times 10^{-9}\text{ m}) = 14.5$. **37.** (a) $k = 10^7 = 2\pi/\lambda$, $\lambda = 628$ nm. (b) $f = c/\lambda = (3.00 \times 10^8\text{ m/s})/(628 \times 10^{-9}\text{ m}) = 4.78 \times 10^{14}$ Hz. (c) It moves in the positive x-direction. **39.** (a) $l = c\Delta t = (3.00 \times 10^8\text{ m/s})(2.00 \times 10^{-9}\text{ s}) = 0.600$ m. (b) The volume of one pulse is $(0.600\text{ m})(\pi R^2) = 2.945 \times 10^{-6}\text{ m}^3$; hence, $(3.0\text{ J})/(2.945 \times 10^{-6}\text{ m}^3) = 1.0 \times 10^6\text{ J/m}^3$.

CHAPTER 25

Answers to Odd-Numbered Multiple Choice Questions

1. c **3.** d **5.** e **7.** a **9.** d. **11.** d **13.** d **15.** d **17.** c **19.** e **21.** d

Answers to Odd-Numbered Problems

1. 6.25% **3.** 30° **5.** ≈90° **7.** 10° **9.** 12 m **11.** 12.5 cm **13.** it misses the mirror **15.** 6.0 m **17.** unchanged **19.** $H = hd/(d + s_0)$ **23.** $L = d/\tan 2\phi$ **25.** 1.24×10^8 m/s **27.** 1.1 **29.** 2.25×10^8 m, 2.21 ×

10^8 m/s **31.** 1.3 **35.** 566 THz, 353 nm **37.** $\theta_c = 49.8°$ **39.** $\theta_c = 33.7°$ **41.** 2.2 mm, 4.2×10^3 waves **43.** $n_i = 1.57$ **45.** 2.16 cm **47.** 0.355 mm **49.** 98° **51.** $(\overline{AB} - L)/\lambda_0$, $(\overline{AB}/\lambda_0) + L(1/\lambda - 1/\lambda_0)$, 2000π **53.** 0.885 m

Solutions to Selected Problems

1. $(1/\lambda_r)^4/(1/\lambda_v)^4 = (\lambda_v/\lambda_r)^4 = 6.25\%$. **9.** The statue is 16 m from the point of incidence, and since the ray-triangles are similar, 4 m:16 m as 3 m:Y and $Y = 12$ m. **15.** See Fig. AN 15. $h = \frac{1}{2}$m, $d = 1.0$ m, $D = 11.0$ m, find H. $h/d = H/(d + D)$, $h = H/(1.0 + D)$, $\frac{1}{2}(1.0 + D) = H = \frac{1}{2}(1.0 + 11.0) = 6.0$ m.

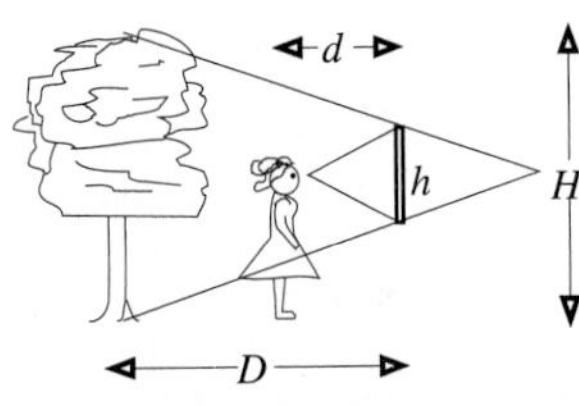

Fig. AN 15.

17. You still see the same fraction of your total body. **21.** See Fig. AN 21. There will be two images arising from single reflections from each of the single mirrors, and there will be a third image due to the double reflections, one from each mirror. The central image is *not* reversed in its handedness—you hold up a right hand, and diagonally across from it, the image will hold up its right hand.

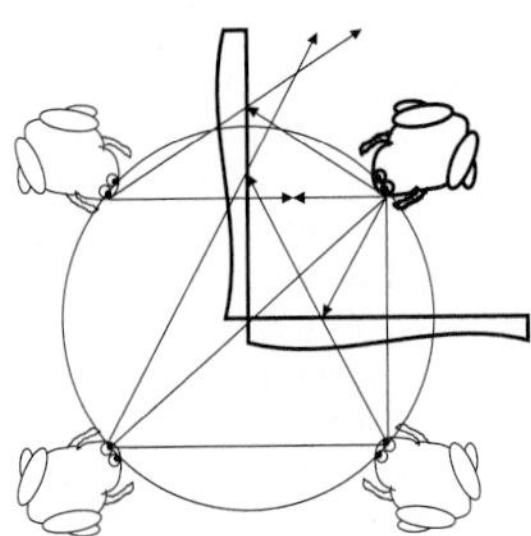

Fig. AN 21.

23. See Fig. AN 23. Construct a normal at M_2; the angle it makes with $\overline{SM_2}$ call α. $\angle M_1AM_2 = 45° + 2\phi$, hence $(45° + 2\alpha) + (45° + 2\phi) = 180°$, and so $\phi = 45° - \alpha$ and $\angle M_1SM_2 = 90° - 2\alpha = 2\phi$, thus $\tan 2\phi = d/L$; hence, $L = d/\tan 2\phi$. **29.** $l = vt = (c/n)t = (3.00 \times 10^8 \text{ m/s}) \times (1.00 \text{ s})/1.333 = 2.25 \times 10^8$ m. $n = 1.36 = c/v$; $v = c/n = 2.21 \times 10^8$ m/s. **31.** $1.00 \sin 55° = n \sin 40°$; $n = 1.27$ or 1.3. **35.** $\lambda = (1.06\ \mu\text{m})/2 = 0.530\ \mu\text{m} = 530$ nm, hence $f = 566$ THz. $(1.06\ \mu\text{m})/3 = 0.353\ \mu\text{m} = 353$ nm in the near-UV. **41.** The photo shows the trail of the pulse as it moves through the water during the 10-ps exposure time. $v = c/n = (3.0 \times 10^8 \text{ m/s})/1.36 = 2.2 \times 10^8$ m/s; the pulse moves $v\,\Delta t = (2.2 \times 10^8 \text{ m/s})(10 \times 10^{-12} \text{ s}) =$

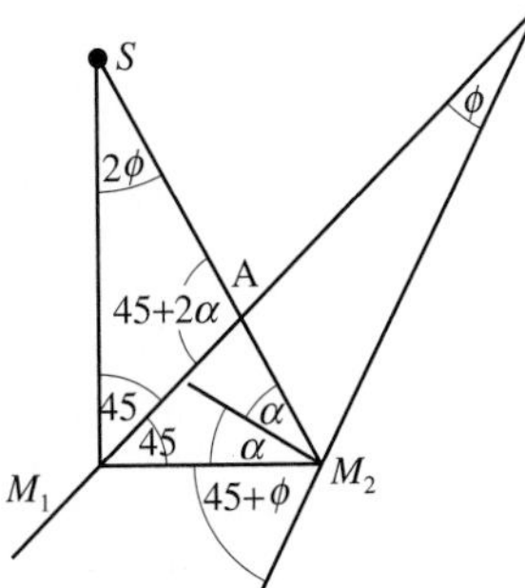

Fig. AN 23.

2.2 mm during the exposure and is about 2.2 mm long to begin with. $(2.2 \times 10^{-3} \text{ m})/(530 \times 10^{-9} \text{ m}) = 4.2 \times 10^3$ waves. **43.** $\tan\theta_i = 15/20$, $\theta_i = 36.87°$; from Snell's Law $n_i \sin 36.87° = 1.00 \sin 70°$, $n_i = 1.57$. **47.** The glass will change the depth of the object from d_R to d_A, where $d_A/d_R = 1.00/1.55$; but $d_R = 1.00$ mm; hence, $d_A = 0.645$ mm, and the camera must be raised $1.00 \text{ mm} - 0.645 \text{ mm} = 0.355$ mm. **51.** The number of waves in vacuum is $\overline{AB}/\lambda_0$. With the glass in place, there are $(\overline{AB} - L)/\lambda_0$ waves in vacuum and an additional L/λ waves in glass for a total of $(\overline{AB}/\lambda_0) + L(1/\lambda - 1/\lambda_0)$. The difference in number is $L(1/\lambda - 1/\lambda_0)$, giving a phase shift $\Delta\phi$ of 2π for each wave; hence, $2\pi L(1/\lambda - 1/\lambda_0) = 2\pi L(n/\lambda_0 - 1/\lambda_0) = 2\pi L/2\lambda_0 = 2000\pi$. **53.** See Fig. AN 53. $d_{A1}/d_{R1} = 1.50/1.33$; $d_{R1} = 1.00$ m; $d_{A1} = 1.1278$ m: $d_{R2} = d_{A1} + 0.02$ m; $d_{A2}/d_{R2} = 1.00/1.50$; $d_{A2} = 1.327\,8(1.00/1.50) = 0.885$ m.

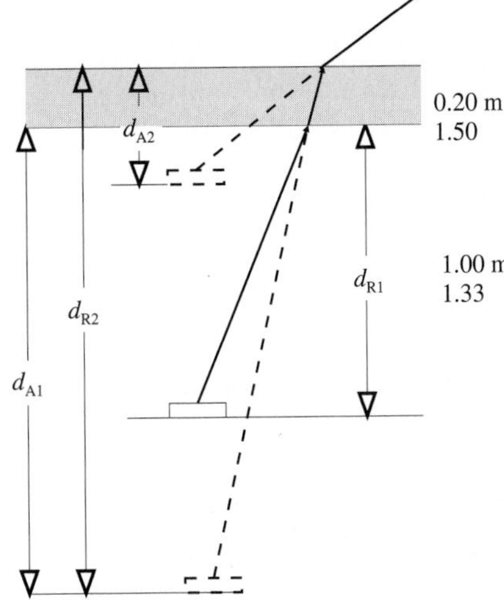

Fig. AN 53.

CHAPTER 26

Answers to Odd-Numbered Multiple Choice Questions

1. b **3.** b **5.** d **7.** b **9.** a **11.** e **13.** b **15.** d **17.** a **19.** c **21.** e

Answers to Odd-Numbered Problems

1. 100 cm **3.** 0.86 m **5.** 2.25 **7.** $s_i = 150$ cm **9.** $s_i = 0.38$ m; real, inverted, and minified **11.** $s_i = -240$ mm, virtual **13.** $f/2$ **15.** 37 mm **17.** 0.13 m **19.** 3 cm **21.** $f = -20$ cm, 40 cm in dia. **23.** $f = 10$ cm, will help **25.** $f = +4.0$ cm, yes **27.** $\mathfrak{D} = +2.94$ D **29.** $s_i = -5.0$ m **31.** $L = 6.0$ cm **33.** 1.03 m, 33× **39.** (a) −3.8 cm (b) 1st—real, inverted, magnified; after jump—virtual, right-side-up, magnified **41.** $f = 6.5 \times 10^2$ mm **43.** (a) $\mathfrak{D} = -1.02$ D (b) $s_o = 30.9$ cm **45.** The greater is s_i, the greater is M_T **47.** (a) 200× (b) $f_E = 25.4$ mm, $f_O = 8.0$ mm (c) $s_O = 8.4$ mm **49.** R = 0.08 m **51.** $f_w = 4f_a$ **53.** 3.2 m, set should be upside down **55.** $\mathfrak{D} = 5.0$ D **57.** −13 cm **59.** 1.33 m **61.** (a) −0.58 m (b) 150 cm **63.** virtual, right-side-up, minified (0.7 times), 0.14 m behind mirror **65.** 2.0×10^{-6} m **67.** R = −0.060 m **69.** $M_T = -0.14$, real, inverted, and minified **73.** $f = -2.5$ m **75.** −1.0 m

Solutions to Selected Problems

1. From Eq. (26.4), $\sigma \approx h^2/2R$, $\frac{1}{2}\text{cm} \approx (10 \text{ cm})^2/2R$; $R = 100$ cm. **3.** $1/f = 0.58 \times (2 \text{ m}^{-1})$; $f = 0.86$ m. **5.** $1/f = (n - 1) \times [1/(2.00 \text{ m}) - 1/\infty]$; $1/(1.60 \text{ m}) = (0.500 \text{ m}^{-1})(n - 1)$; $n = 2.25$. **9.** Given the focal length and the object-distance, we want the image-distance that follows from the Gaussian lens formula: $1/s_o + 1/s_i = 1/f$; $1/s_i = 1/f - 1/s_o = 2.66 \text{ m}^{-1}$; $s_i = 0.38$ m. The object is three focal lengths away from the lens and the image is real, inverted, and minified. **13.** The diameter is 25 mm; hence, f-number = (50 mm)/(25 mm) = $f/2$. **17.** $M_A = d_n\mathfrak{D}$, $2.0 = (0.254 \text{ m})\,\mathfrak{D}$, $\mathfrak{D} = 7.87$ D and $f = 0.13$ m. **21.** For the second lens $s_o = -10$ cm and $s_i = +20$ cm; hence, $1/f = 1/(-10 \text{ cm}) + 1/(20 \text{ cm})$; $f = -20$ cm. $M_T = -(20 \text{ cm})/(-10 \text{ cm}) = +2$, and the image is 40 cm in diameter. **25.** Applying Eq. (26.19) to two lenses at a time, $1/f = 1/10 + 1/(-20) + 1/5.0$; $f = +4.0$ cm. Since it's positive, it can form real images. **29.** The lens makes an object at a distance s_o, corresponding to infinity, appear to be at the far-point (s_i), where it can be seen clearly. $1/(-5.0 \text{ m}) = 1/\infty + 1/s_i$; $s_i = -5.0$ m. **33.** Putting the long focal length (+1.00 m) first, we'll need a tube of length $f_O + f_E = 1.00 \text{ m} + 0.030 \text{ m} = 1.03$ m. $M_A = -(1.00 \text{ m})/(0.030 \text{ m}) = 33\times$. **37.** $n_1\overline{HI} + n_1\overline{IJ} = n_m\overline{AB} + n_m\overline{BC} + n_1\overline{CD} + n_1\overline{DE} + n_m\overline{EF} + n_m\overline{FG}$, now divide both sides by n_m and you are back to the form of the first derivation provided $n_1/n_m \to n_1$. **41.** $M_T = y_i/y_o = -s_i/s_o$; $s_i = (-2.0 \text{ m}) \times (-0.024 \text{ m})/(0.05 \text{ m}) = 0.96$ m. $1/(2.0 \text{ m}) + 1/(0.96 \text{ m}) = 1/f$; $f = 6.5 \times 10^2$ mm. **47.** (a) 20 × 10 = 200×. (b) $f_E = (254 \text{ mm})/M_{AE} = 25.4$ mm, $f_O = -(160 \text{ mm})/(-20) = 8.0$ mm. (c) $1/s_o = 1/f_O - 1/s_i = 1/(8.0 \text{ mm}) - 1/(L + f_O) = 1/(8.0 \text{ mm}) - 1/(168 \text{ mm})$; $s_o = 8.4$ mm. **51.** From Eq. (26.7)

$1/f_w = (n_l/n_w - 1)/(n_l - 1)f_a = (1.5/1.33 - 1)/(1.5 - 1)f_a = 0.125/0.5f_a$; $f_w = 4f_a$. **55.** With her glasses on, $2.0\text{ D} = 1/(0.80\text{ m} - 0.02\text{ m}) + 1/s_i$, which gives the distance from the eyeglass plane at which she can see clearly; that is, $(s_i + 2.0\text{ cm})$ is her new near-point. The prescribed focal length of her new spectacles must satisfy the expression $1/f = 1/(0.254\text{ m} - 0.02\text{ m}) + 1/s_i$. Combining these two equations; $1/f - 4.274\text{ D} = 2.0\text{ D} - 1.282\text{ D}$; and $\mathcal{D} = 5.0\text{ D}$. **59.** $1/(2.00\text{ m}) + 1/(4.00\text{ m}) = 1/f$; $f = 1.33\text{ m}$. **63.** $1/(0.20\text{ m}) + 1/s_i = -2/(1.00\text{ m})$; $s_i = -0.14\text{ m}$. $M_T = -(-0.14\text{ m})/(0.20\text{ m}) = +0.7$. The image is virtual, minified (0.7 times life-size), and 0.14 m behind the mirror. **67.** For the image to be magnified and erect, the mirror must be concave and the image virtual; $s_o = 0.015\text{ m}$; $M_T = 2 = -s_i/(+0.015\text{ m})$; $s_i = -0.030\text{ m}$ and, hence, $1/f = 1/(0.015\text{ m}) - 1/(0.030\text{ m})$; $f = 0.030\text{ m}$ and $f = -R/2$; $R = -0.060\text{ m}$. **71.** Take another look at your solution of Problem 68; $M_T(s_o - f) = -f$; hence, $s_o = (-f + fM_T)/M_T$, which equals what we want. Furthermore, $s_i = fs_o/(s_o - f)$, and on substituting in for s_o, we do indeed get the desired expression. **75.** The mirror projects a real, inverted, life-size image onto the empty socket. $s_o = 2f = s_i = 1.0\text{ m} = 2(-R/2)$; $R = -1.0\text{ m}$.

CHAPTER 27

Answers to Odd-Numbered Multiple Choice Questions

1. c **3.** a **5.** b **7.** a **9.** c **11.** b **13.** c **15.** b **17.** a

Answers to Odd-Numbered Problems

1. 1327 W/m^2 **3.** 100 W/m^2 **5.** 188 W/m^2 **7.** 40 W/m^2 **9.** 53° **11.** 0.19 **13.** 1.501 **15.** 39° **17.** ±6.33 mm **19.** 419 nm **21.** 9.76 mm **23.** 1.84×10^{-7} m **25.** 107 nm, 322 nm, 537 nm, etc. **27.** $y_{m'} = sm'\frac{1}{2}\lambda/a$ **29.** 10 km **31.** 0.346 m **33.** $\Delta x = \lambda_f/2\alpha$ **35.** 550 nm **37.** 96.6 **39.** $\theta_m = \sin^{-1}(\sin\theta_i \pm m\lambda/a)$ **41.** 1474 nm, 737.0 nm, 491.3 nm, 368.5 nm **43.** 0.53° **45.** 12 cm **47.** Δy doubles **49.** 2.6 μm **51.** 1.32×10^{-7} rad, 7.57×10^{-6} degrees, 2.72×10^{-2} s **53.** 64.5 km **55.** 0.12 mm **57.** 11×10^4 Hz **59.** 0.597 mm **61.** 4 fringes, 2N **63.** 1.11

Solutions to Selected Problems

1. $I = \frac{1}{2}c\varepsilon_0 E_0^2 = \frac{1}{2}(2.998 \times 10^8\text{ m/s}) \times (8.854 \times 10^{-12}\text{ C}^2/\text{N}\cdot\text{m}^2)(1000\text{ V/m})^2$; hence, $I = 1327\text{ W/m}^2$. **5.** $I = I_1\cos^2\theta = (250\text{ W/m}^2)\cos^2 30 = 188\text{ W/m}^2$. **9.** $\tan\theta_p = n_t/n_i = 1.3/1.00$, $\theta_p = 53°$. **13.** $\tan\theta_p = n_t/n_i$; $n_t = \tan 56.33° = 1.501$. **17.** According to Eq. (27.4), the first minimum occurs when $m' = \pm1$. Hence, $a\sin\theta_1 = \pm\frac{1}{2}\lambda$; $\theta_1 \approx \pm\frac{1}{2}\lambda/a = \pm\frac{1}{2} \times (632.8 \times 10^{-9}\text{ m})/(0.100 \times 10^{-3}\text{ m}) = \pm3.16 \times 10^{-3}$ rad, since $y = s\theta_1 = (2.00\text{ m})(\pm3.16 \times 10^{-3}\text{ rad}) = \pm6.33$ mm. **21.** $y_4 = s4\lambda/a = (1.00\text{ m})4(487.99 \times 10^{-9}\text{ m})/(0.200 \times 10^{-3}\text{ m}) = 9.76 \times 10^{-3}$ m. **25.** There will be a 180° phase shift due to the reflections, and so the film must be a quarter-wavelength thick or odd multiples thereof. $d = (m + \frac{1}{2})\lambda_0/2n_f$; for $m = 0$, $d = \lambda_0/4n_f = (580\text{ nm})/4(1.35) = 107$ nm; for $m = 1$, $d = 322$ nm; for $m = 2$, $d = 537$ nm, etc. **29.** From Eq. (27.7), $y_1 = s\lambda/a = sc/fa = (20 \times 10^3\text{ m})(3.0 \times 10^8\text{ m/s})/(1.0 \times 10^6\text{ Hz})(600\text{ m}) = 1.0 \times 10^4$ m. **33.** The film increases half a wavelength, $\lambda_f/2$. Using the results of the previous problem, we obtain $x_{m+1} - x_m = \Delta x = \lambda_f/2\alpha$. **35.** Each arm is traversed twice, so a motion of $\lambda/2$ causes a single fringe-pair to shift past; hence, $92\lambda/2 = 2.53 \times 10^{-5}$ m and $\lambda = 550$ nm. **37.** If the cell has a length L and is traversed twice, then the change in optical path length with and without air is $2Ln_a - 2L$, and the number of waves changed is $2L(n_a - 1)/\lambda_0$, each corresponding to a fringe-pair, and so there are 96.6 of them. **41.** There will be a relative phase shift on reflection of $\frac{1}{2}\lambda$; hence, the longest wave not reflected is $550.0\text{ nm} = \frac{1}{2}\lambda_f$, or $\lambda_f = 1100$ nm, and so $\lambda_0 = (1100\text{ nm})1.340 = 1474$ nm in the IR. When $550.0\text{ nm} = \lambda_f$, $\lambda_0 = (550.0\text{ nm}) \times 1.340 = 737.0$ nm. Similarly, when $550.0\text{ nm} = 3\lambda_f/2 = 366.6$ nm or $\lambda_0 = 491.3$ nm, there is no reflection. And, finally, when $2\lambda_f = 550.0$ nm, $\lambda_f = 275.0$ nm; and $\lambda_0 = 368.5$ nm in the UV, which is not reflected. **45.** From Eq. (27.13), $\lambda = (20\text{ cm})\sin 36.87° = 12$ cm. **49.** From Eq. (27.8), $a = 2(550 \times 10^{-9}\text{ m})/0.422\,6 = 2.6 \times 10^{-6}$ m. **53.** $\theta_a = 1.22\lambda/D = 1.22(550 \times 10^{-9}\text{ m})/(4.00 \times 10^{-3}\text{ m}) = 1.68 \times 10^{-4}$ rad; hence, $L/(3.844 \times 10^8\text{ m}) = 1.68 \times 10^{-4}$ rad; $L = 64.5$ km. **57.** The wavelength should be less than the size of the target or else diffraction will be troublesome. $v = f\lambda$, $f > (330\text{ m/s})/(0.003\,0\text{ m}) = 11 \times 10^4$ Hz. **61.** The angular width of the diffraction maximum is obtained from Eq. (27.13), $\sin\theta_1 = 1\lambda/D = (500 \times 10^{-9}\text{ m})/(0.10 \times 10^{-3}\text{ m}) = 0.005$; $\theta_1 = 0.286\,5°$, and twice this value is 0.573 0°. The angular width of one of Young's fringes is $\theta = \lambda/a = (500 \times 10^{-9}\text{ m})/(0.20 \times 10^{-3}\text{ m}) = 2.5 \times 10^{-3}$ rad $= 0.143\,2°$; hence, 4.0 fringes will exist. And that might have been obvious from the fact that $a = 2D$ and the number of fringes is $2N$.

CHAPTER 28

Answers to Odd-Numbered Multiple Choice Questions

1. d **3.** c **5.** b **7.** a **9.** e **11.** e **13.** a **15.** d **17.** c

Answers to Odd-Numbered Problems

5. 90 s, Earth-based observer; 45 s, astronaut **7.** 601 s **9.** 0.499c **11.** 0.436 m **13.** 64.34 m **15.** 500 m **17.** c, classically 2c **19.** 0.999 9c ≈ c **21.** $v = 0.404\,8$c **23.** 0.436c **25.** $v = 0.9$c **27.** 4.45 ns **31.** 73.8×10^3 y **33.** 36.3° **35.** 2.9×10^{-19} kg·m/s **37.** KE = 1000 MeV, $v = 0.866\,0$c **39.** 1.0 MeV **41.** $E_0 = 8.988 \times 10^{10}$ J, $E = 9.173 \times 10^{10}$ J, $KE = 1.853 \times 10^9$ J **43.** 2.2 MeV **45.** 1.000×10^{-15} kg **47.** 3.26 MeV **49.** $f = 3.653 \times 10^{20}$ Hz **51.** $(\gamma - 1)/\gamma$, 20% **53.** $v = 0.101$ m/s for both, when $\gamma \approx 1$ **57.** $v = 0.967$c, $v = 2.4$c (classically) **59.** 0.106 0c **61.** 0.866 **63.** 3.438×10^{20} Hz

Solutions to Selected Problems

1. $t_\parallel = \dfrac{2L}{c}\dfrac{1}{(1 - \beta^2)} \to \dfrac{2L\sqrt{1 - \beta^2}}{c(1 - \beta^2)} = \dfrac{2L}{c}\dfrac{1}{\sqrt{1 - \beta^2}}$. **5.** $\Delta t_M = \Delta t_S/\sqrt{1 - v^2/c^2} = 2(45\text{ s}) = 90$ s for the Earth-based observer; 45 s for the astronaut. **9.** $\Delta t_M = \Delta t_S/\sqrt{1 - v^2/c^2}$; $60.0\text{ s} = (52.0\text{ s})/\sqrt{1 - v^2/c^2}$; $\sqrt{1 - v^2/c^2} = 0.866\,7$, $v = 0.499$c. **13.** $L_M = L_S\sqrt{1 - v^2/c^2} = (2\text{ mi}) \times (5280\text{ ft/mi})(0.020\,00) = 211.1\text{ ft} = 64.34$ m. **17.** Let S' be the frame of our laboratory, such that one photon moves right at $v_{PO'} = c$. The other photon in S moves to the left; that is, S' moves right ($+x$-direction) with respect to S at $v_{O'O} = c$. The speed of one photon with respect to the other is $v_{PO} = (c + c)/(1 + cc/c^2) = c$. Classically: 2c. **19.** Put the electron at rest in S, and have the proton move in S', which is approaching S at -0.992c (that is, moving left). The proton travels at -0.981c (also moving left). $v_{PO} = [(-0.981c) + (-0.992c)]/[1 + (0.981c)(0.992c)/c^2] = -1.973c/1.973\,2 \approx -c$. **23.** $L_M = 0.900L_S = L_S/\gamma$, $\gamma = 1/0.900 = 1.111$, $1.2345 = 1/(1 - \beta^2)$, $\beta^2 = (\gamma^2 - 1)/\gamma^2 = 0.190$, $v = 0.436$c. **31.** $\gamma \approx 1 + \frac{1}{2}\beta^2 = 1 + \frac{1}{2}8.585 \times 10^{-13}$; $\Delta t_M = \gamma\Delta t_S = (1 + \frac{1}{2}8.585 \times 10^{-13})\Delta t_S = \Delta t_S + (\frac{1}{2}8.585 \times 10^{-13})\Delta t_S$, and we want $(\frac{1}{2}8.585 \times 10^{-13})\,\Delta t_S = 1.00$ s; hence, $\Delta t_S = 2.330 \times 10^{12}\text{ s} \approx 73.8 \times 10^3$ y. **35.** $p = \gamma mv = 1.155(1.675 \times 10^{-27}\text{ kg}) \times (0.50c) = 2.9 \times 10^{-19}$ kg·m/s. **37.** $E = KE + E_0 = 2E_0$, $KE = E_0 = 1000$ MeV; $E = 2000\text{ MeV} = \gamma mc^2 = \gamma(1000\text{ MeV})$, $\gamma = 2 = 1/\sqrt{1 - v^2/c^2}$, $v = 0.866\,0$c. **41.** $E_0 = mc^2 = 8.988 \times 10^{10}$ J; $E = \gamma mc^2 = 1.021(8.988 \times 10^{10}\text{ J}) = 9.173 \times 10^{10}$ J; $E = KE + E_0$, $KE = E - E_0 = 1.853 \times 10^9$ J. **43.** 938.272 MeV + 939.566 MeV = 1877.83 MeV; this result minus 1875.6 MeV is 2.2 MeV. **47.** 2(1876.12 MeV) − 2809.41 MeV − 939.566 MeV = 3.26 MeV. **51.** $(\gamma mv - mv)/\gamma mv = (\gamma - 1)/\gamma$, $\gamma = 1.25$, $(\gamma - 1)/\gamma = 20\%$. **55.** $E^2 = E_0^2 + (pc)^2$, $4E_0^2 = E_0^2 + (pc)^2$, $3E_0^2 = p^2c^2$, $p = \sqrt{3}\,mc$. **57.** KE = 1.50 Mev, $E = (1.50\text{ MeV} + 0.511\text{ MeV}) = \gamma mc^2 = \gamma E_0 = \gamma(0.511\text{ MeV})$, $\gamma = 3.935$, $\beta^2 = (\gamma^2 - 1)/\gamma^2 = 0.935$, $v = 0.967$c. $KE_c = \frac{1}{2}mv^2 = 2.40 \times 10^{-13}$ J, $v = 2.4$c. **61.** $KE = E_0$, $E = KE + E_0 = 2E_0$, $\gamma mc^2 = 2mc^2$, $\gamma = 2 = 1/\sqrt{1 - v^2/c^2}$, $4 - 1 = 4\beta^2$, $\beta = \sqrt{3}/2 = 0.866$. **63.** $E^2 = E_0^2 + (pc)^2$, $(pc)^2 = (1.000\text{ MeV} +$

$0.511\ \text{MeV})^2 - (0.511\ \text{MeV})^2$, $p = 1.422$ MeV/c, which is the momentum of the electron. For the photon $E = pc = 1.422\ \text{MeV} = hf$, $f = 3.438 \times 10^{20}$ Hz.

Chapter 29

Answers to Odd-Numbered Multiple Choice Questions

1. d **3.** b **5.** b **7.** c **9.** b **11.** a **13.** b **15.** c **17.** a

Answers to Odd-Numbered Problems

1. 1.0978×10^{27} **3.** 35.5 g of Cl **5.** 3.55 g **7.** Ba, 25.6 g; Cl, 13.2 g **9.** 4.90×10^{-14} kg **11.** 0.11 nm **13.** $\theta = 8.54°$ **15.** 0.21 nm **17.** $\lambda = 486.2$ nm **19.** $\lambda_\infty = 364.6$ nm **21.** $\lambda_\alpha = 656.3$ nm **23.** 2.28×10^{-10} m **27.** $B = 8.7 \times 10^{-4}$ T **29.** 2.82×10^8 m/s **31.** 0.237 nm at 25.0° **33.** 12 **35.** 1.6×10^{-14} m **37.** 397 nm **41.** $\Delta\lambda = \pm 0.001$ nm

Solutions to Selected Problems

3. (965 A)(100.0 s) = 96.5 kC = 1 faraday; hence, 1 mole of each is liberated; 35.5 g of Cl. **7.** The valence of barium is 2 and that of chlorine is 1. Hence, for Ba, $m = (2.00\ \text{A}) \times (5.00\ \text{h})(3600\ \text{s/h})(137\ \text{g})/(96.5\ \text{kC})2 = 25.6$ g; for Cl, $m = (2.00\ \text{A})(5.00\ \text{h}) \times (3600\ \text{s/h})(35.5\ \text{g})/(96.5\ \text{kC})1 = 13.2$ g. **11.** $d = 1(0.090\ \text{nm})/(2 \sin 25°) = 0.11$ nm. **15.** $\lambda = 2(0.28\ \text{nm}) \sin 22° = 0.21$ nm. **19.** $1/\lambda = (1.097 \times 10^7\ \text{m}^{-1})(1/4 - 1/\infty)$; and $\lambda_\infty = 364.6$ nm. **21.** $1/\lambda = (1.097 \times 10^7\ \text{m}^{-1})(1/4 - 1/9)$; and $\lambda_\alpha = 656.3$ nm. **25.** $\tan\theta \approx \theta = v_y/v_x = a_y t/v_x = (eE/m_e)t/(E/B)$ where $t = L/v_x$; hence, $\theta \approx eLB^2/Em_e$. **31.** $d = 1\lambda/(2 \sin 25.0°) = 0.237$ nm at 25.0°. **33.** The largest angle is 90°; $2(1.22 \times 10^{-10}\ \text{m}) \sin 90° = m \times (0.020\,0 \times 10^{-9}\ \text{m})$; $m = 12.2$, so to nearest integer $m = 12$. **37.** $1/(390\ \text{nm}) = (1.097 \times 10^7\ \text{m}^{-1})(1/2^2 - 1/n^2)$; and $n = 7.8$, so we must use $n = 7$; $1/\lambda = (1.097 \times 10^7\ \text{m}^{-1})(1/2^2 - 1/7^2)$, and $\lambda = 397$ nm. **39.** $Y = y + R \tan\theta$; $\tan\theta = v_y/v_x = eLB^2/Em_e$. Using Eq. (29.2), we have $y = eL^2B^2/2m_eE$ and $Y = (e/m_e)\dfrac{B^2L}{2E}(L + 2R)$. **41.** $m\lambda = 2d \sin\theta$; $m\Delta\lambda = 2\,\Delta d \sin\theta$; $\Delta\lambda = \pm 0.001$ nm.

CHAPTER 30

Answers to Odd-Numbered Multiple Choice Questions

1. c **3.** a **5.** b **7.** b **9.** c **11.** d **13.** b **15.** c **17.** b

Answers to Odd-Numbered Problems

1. $I = 76.9$ W/m^2, P = 108 W **3.** 9.4 μm, IR **5.** 4.1×10^{15} Hz **9.** 496 nm **11.** 0.07 W/m^2 **13.** 4.61×10^{14} Hz **15.** 1.89 eV **17.** 1.46×10^{-4} nm **19.** 180° **21.** 35 keV **23.** −1.51 eV **25.** $f = 4.57 \times 10^{14}$ Hz, $\lambda = 656$ nm **27.** 476 pm **29.** $n = 137\,467$ **31.** $T = 4.46 \times 10^3$ K **33.** 0.663×10^6 m/s **35.** 4.4×10^{14} Hz, 6.8×10^{-7} m **37.** 5.82×10^{-11} m **39.** 59.6° **41.** 2.19×10^6 m/s **43.** −54.4 eV **45.** $f_n = m_e k^2 Z^2 e^4/2\pi n^3\hbar^3$ **49.** 5.54×10^{-8} eV

Solutions to Selected Problems

1. $P/A = \varepsilon\sigma(T^4 - T_e^4) = (0.97)(5.670\,3 \times 10^{-8}\ \text{W/m}^2{\cdot}\text{K}^4)(306^4 - 293^4) = I = 76.9\ \text{W/m}^2$. P = 108 W. **5.** From Eq. (30.4), $\lambda_{max}T = 0.002\,898$ m·K; $\lambda_{max} = 72$ nm; $f_{max} = 4.1 \times 10^{15}$ Hz. **9.** $E = hf = hc/\lambda$, $\lambda = hc/E = 496$ nm. **13.** KE ≈ 0 is the threshold condition; $f_0 = c/\lambda = 4.61 \times 10^{14}$ Hz. **17.** $\Delta\lambda = \lambda_s - \lambda_i = \dfrac{h}{m_e c}(1 - \cos\theta) = (0.002\,426\ \text{nm})(1 - \cos 20°) = 1.46 \times 10^{-4}$ nm. **23.** $n = 1$ is the ground state, $n = 2$ the first excited state, $n = 3$ the second excited state; $E_3 = -(13.6\ \text{eV})/3^2 = -1.51$ eV. **27.** $n = 3$; from Eq. (30.17), $r = 3^2(0.052\,9\ \text{nm}) = 476$ pm. **31.** Taking $\lambda_{max} \approx 650$ nm, $\lambda_{max}T = 0.002\,898$ m·K; $T = 4.46 \times 10^3$ K. **35.** In order to use $hf_0 = \phi$, we first need to find ϕ; $hf = KE_{max} + \phi$; $\phi = 2.929 \times 10^{-19}$ J; $f_0 = \phi/h = 4.4 \times 10^{14}$ Hz; $\lambda_0 = 6.8 \times 10^{-7}$ m. **39.** $0.060\,0(20.0\ \text{pm}) = (2.426\ \text{pm})(1 - \cos\theta)$; $\cos\theta = 0.505$; $\theta = 59.6°$. **43.** From Eq's. (30.18) and (30.19), $E_1 = (-13.6\ \text{eV})Z^2 = -54.4$ eV. **47.** For x-component, $p_i - p_s \cos\theta = p_e \times \cos\phi$; for y-component, $p_s \sin\theta = p_e \sin\phi$; square and add these two, noting that $\cos^2\phi + \sin^2\phi = 1$; (A) $p_i^2 + p_s^2 - 2p_ip_s \cos\theta = p_e^2$. For the electron (B), $E_e^2 = c^2p_e^2 + m_e^2c^4$. Conservation of Energy is (C) $E_i + m_ec^2 = E_s + E_e$. Now rewrite this equation as $E_e = E_i + m_ec^2 - E_s$ and square it and plug it into (B); also multiply (A) by c^2 and plug it into (B) to get $(E_i + m_ec^2 - E_s)^2 = c^2(p_i^2 + p_s^2 - 2p_ip_s\cos\theta) + m_e^2c^4$. Using $E = pc$ for photons and simplifying, this equation becomes $1/E_s - 1/E_i = (1/m_ec^2)(1 - \cos\theta)$, and since $E = hf = hc/\lambda$ for photons

$$\lambda_s - \lambda_i = \frac{h}{m_e c}(1 - \cos\theta) \qquad [30.13]$$

49. $\Delta E = (-3.40\ \text{eV}) - (-13.6\ \text{eV}) = 10.2$ eV; the photon of momentum $p = E/c$, Eq. (28.16), imparts that momentum to the atom; the atom's recoil energy is then $p^2/2M = E^2/2Mc^2 = (10.2\ \text{eV})^2/2(939 \times 10^6\ \text{eV/c}^2)c^2 = 5.54 \times 10^{-8}$ eV.

CHAPTER 31

Answers to Odd-Numbered Multiple Choice Questions

1. b **3.** d **5.** a **7.** d **9.** d **11.** c **13.** d **15.** a

Answers to Odd-Numbered Problems

1. 5.5×10^{-36} m **3.** 2.4×10^{-11} m **5.** 0.147 nm **7.** 150 V **9.** $l = 0$, $m_l = 0$; $l = 1$, $m_l = -1, 0, +1$; $l = 2$, $m_l = -2, -1, 0, +1, +2$ **11.** K has 2, L has 8, M has 18, N has 32 **13.** $\hbar/2L$ **15.** 5×10^{-27} m/s **17.** 5×10^{-27} J, 3×10^{-8} eV **19.** 75 MeV, 9.8% **21.** 0.27 nm **23.** $p \approx E/c$ **25.** $v_p = \frac{1}{2}v$ **27.** 12.1 eV **29.** 2.86 eV **31.** cannot exist **33.** $1s^2{\cdot}2s^22p^6{\cdot}3s^23p^5$ **35.** 2.64×10^{-29} m **37.** 1.16×10^6 m **39.** $R = r + (1.79 \times 10^{-13}\ \text{m})/r$

Solutions to Selected Problems

1. $p = h/\lambda$, $\lambda = h/p = h/mv = 5.5 \times 10^{-36}$ m. **5.** $\lambda = h/p = h/\sqrt{2m\,KE} = 0.147$ nm. **9.** $n = 3$; $n - 1 = 2$; hence, $l = 2, 1$, or 0. m_l goes from −1 to +1; that is, when

$l = 0$, $m_l = 0$
$l = 1$, $m_l = -1, 0, +1$
$l = 2$, $m_l = -2, -1, 0, +1, +2$

13. $\Delta x = L$, $\Delta p_x \approx \frac{1}{2}\hbar/\Delta x \approx \frac{1}{2}\hbar/L$. **17.** $\Delta t = 10$ ns; $\Delta E = \frac{1}{2}\hbar/\Delta t = (5.272\,9 \times 10^{-35}\ \text{J·s})/(10\ \text{ns}) = 5 \times 10^{-27}\ \text{J} = 3 \times 10^{-8}$ eV. **21.** $E_0 \gg$ KE; hence, classical form is okay; $p = \sqrt{2m_e KE}$, KE = $(20\ \text{eV})(1.60 \times 10^{-19}\ \text{J/eV})$; $p = 2.414\,6 \times 10^{-24}$ kg·m/s; $\lambda = h/p = 2.7 \times 10^{-10}\ \text{m} = 0.27$ nm. **25.** $v_p = E/p = (p^2/2m)/p = p/2m$; $v_p = \frac{1}{2}v$. **29.** From Eq. (30.19), $\Delta E = (-13.6\ \text{eV})/5^2 - (-13.6\ \text{eV})/2^2 = 2.86$ eV. **33.** $1s^2{\cdot}2s^22p^6{\cdot}3s^23p^5$ **37.** $\Delta z \Delta p_z \geq \frac{1}{2}\hbar$; $m\Delta v_z \approx \frac{1}{2}\hbar/\Delta z$; $\Delta v_z \approx 0.578\,8 \times 10^6$ m/s. After 2.00 s, $\Delta z = \Delta v_z t = 1.16 \times 10^6$ m. **39.** $\Delta z = 2r$; $\Delta p_z \approx \frac{1}{2}\hbar/\Delta z \approx \frac{1}{2}\hbar/2r$; $m\Delta v_z \approx \frac{1}{2}\hbar/2r$; $v_z \approx \frac{1}{2}\hbar/2mr$; in the time τ it takes to reach the observation screen, the beam spreads a distance $Z' = v_z\tau$. To find τ, we find v_y; KE = $(3/2)k_BT$; $\frac{1}{2}mv_y^2 = (3/2)k_B(1600\ \text{K})$; $v_y^2 = 3k_B(1600\ \text{K})/m$; $v_y = 450.1$ m/s. $\tau = (1.000\ \text{m})/(450.1\ \text{m/s}) = 2.221$ ms. $Z' = (\frac{1}{2}\hbar/2mr) \times (2.221\ \text{ms}) = (1.79 \times 10^{-13}\ \text{m})/r$; and $R = r + Z' = r + (1.79 \times 10^{-13}\ \text{m})/r$.

CHAPTER 32

Answers to Odd-Numbered Multiple Choice Questions

1. d **3.** b **5.** b **7.** d **9.** b **11.** a **13.** d **15.** b **17.** a **19.** e

Answers to Odd-Numbered Problems

1. 111 **3.** 7 protons, 8 neutrons **5.** Fr, Pb, Ag, Pr **7.** 13 043.8 MeV/c^2, 13.043 8 GeV/c^2 **9.** $9.988\,352 \times 10^{-27}$ kg **11.** 92.906 38 u **13.** 15 fm **15.** 27 **17.** 64.749 MeV **19.** 6.474 9 MeV/nucleon **21.** 15.663 MeV **23.** no **25.** 3.9 times **29.** 8.736 MeV **31.** Ca-41, 8.36 MeV; Ca-42, 11.48 MeV **33.** 5.40 MeV **35.** ${}^{27}_{13}\text{Al} + {}^{4}_{2}\alpha \rightarrow {}^{30}_{15}\text{P} + {}^{1}_{0}\text{n}$ **37.** ${}^{147}_{62}\text{Sm} \rightarrow {}^{143}_{60}\text{Nd} + {}^{4}_{2}\alpha$ **39.** 17.346 MeV **41.** 4.1×10^{-9} s^{-1} **43.** $t_{1/2} = 12$ h **45.** 4.28×10^{-4} decays/y **47.** 5.5 d **49.** 90.1 s **51.** $N/N_0 = 4.97 \times 10^{-16}$ **53.** 5.700 74 MeV **55.** 1.191 MeV **57.** 5.7×10^{12} Bq **59.** 3×10^{10} Bq **61.** 2.8×10^{10} atoms **63.** 0.18 MBq/m^3 **65.** 24.685 4 MeV

Solutions to Selected Problems

5. Fr, Pb, Ag, and Pr. **9.** (5.603 051 GeV/c^2) ×

$(1.782\ 663 \times 10^{-36}$ kg/eV)$(1.000\ 000 \times 10^{9}$ eV/GeV) $= 9.988\ 352 \times 10^{-27}$ kg. **13.** $R = R_0A^{1/3} = (1.2$ fm)(6.17) $= 7.4$ fm. $2R = 15$ fm. **19.** If $E_B = 64.749$ MeV, $E_B/A = 6.474\ 9$ MeV/nucleon. **21.** $\Delta m = 1.008\ 665$ u $+ 15.003\ 065$ u $- 15.994\ 915$ u $= 0.016\ 815$ u $\rightarrow 15.663$ MeV. **27.** Mass of C-12 atom is (12 kg)/$(6.022 \times 10^{26}$ atoms/kmole) $= 1.99 \times 10^{-26}$ kg; dividing by 12 yields 1 u $= 1.66 \times 10^{-27}$ kg. **31.** For Ca-41: (40.962 278 u) $-$ (1.008 665 u) $-$ (39.962 591 u) $= -0.008\ 978$ u. For Ca-42: (41.958 618 u) $-$ (1.008 665 u) $-$ (40.962 278 u) $= -0.012\ 325$ u. One must add 8.36 MeV to remove a neutron from Ca-41 and 11.48 MeV to remove one from Ca-42. **37.** ${}^{147}_{62}\text{Sm} \rightarrow {}^{4}_{2}\alpha + X$; $X = {}^{143}_{60}X$, which is neodymium-143. **41.** $\lambda = 0.693/t_{1/2} = 0.693/(167 \times 10^{6}\text{ s}) = 4.1 \times 10^{-9}\text{ s}^{-1}$. **45.** $\lambda = 0.693/t_{1/2} = 4.28 \times 10^{-4}\text{ y}^{-1}$. **49.** $\tau = 1/\lambda = 130$ s; $t_{1/2} = 0.693/\lambda = 90.1$ s. **51.** $\lambda = 0.693/(1.18\text{ min}) = 0.587\text{ min}^{-1}$. $N/N_0 = \exp[(-0.587\text{ min}^{-1})(60.0\text{ min})] = 4.97 \times 10^{-16}$. **55.** Using atomic masses; initial $-$ final mass; (4.002 603 u) $+$ (14.003 074 u) $-$ (16.999 131 u) $-$ (1.007 825 u) $= -0.001\ 279$ u; α must supply 1.191 MeV. **59.** $t_{1/2} = 1.236 \times 10^{6}$ s; from Eq. (32.12), $\lambda = 5.609 \times 10^{-7}\text{ s}^{-1}$; $R_0 = \lambda N_0 = 3 \times 10^{10}$ Bq. **63.** (444 MBq)/(50 weeks)(5 d/week)(10 m^3/d) $=$ 0.18 MBq/m^3. **65.** 4(1.007 276 u) $-$ (4.001 506 u) $-$ 2(0.000 548 580 u) $=$ 0.026 501 u; or 24.685 4 MeV. **67.** $N = N_0 \exp(-\lambda t)$; $N_0/2 = N_0 \exp(-\lambda t_{1/2})$; $\exp(\lambda t_{1/2}) = 2$; $\ln 2 = \lambda t_{1/2}$.

CHAPTER 33

Answers to Odd-Numbered Multiple Choice Questions

1. d **3.** e **5.** c **7.** c **9.** a **11.** a **13.** d **15.** a **17.** c **19.** b

Answers to Odd-Numbered Problems

1. (a) electromagnetic (b) strong (c) weak (d) strong **3.** (a) baryon number and strangeness (b) lepton number (c) charge and strangeness (d) strangeness and baryon number **5.** no **7.** no, electron-lepton number **9.** yes **11.** no, weakly **13.** not forbidden, 33.9 MeV **15.** 37.7 MeV, strangeness **17.** 1877 MeV **19.** 0.660 GeV **21.** 10^{30} **23.** $E_0 = hc/4\pi R$ **25.** 70.45 MeV **27.** -535.4 MeV, no—must have appreciable KE. **29.** 2.42×10^{22} Hz, 1.24×10^{-14} m, 5.34×10^{-20} kg·m/s **31.** $c\bar{u}$ **33.** $\Lambda^0 =$ uds **35.** $d\bar{s} \rightarrow u\bar{d} + d\bar{u}$ **37.** 2.47×10^{20} Hz **39.** $\approx 10^{15}$ GeV/c^2 **41.** 8×10^{27} K

Solutions to Selected Problems

1. (a) Only the electromagnetic force affects photons. (b) Since there's no strangeness, this must be a strong-force decay. (c) Λ^0 has strangeness, so this must be a weak-force decay. (d) With no strangeness, this must be a strong-force decay. **5.** No. It does not conserve baryon number—the $p\bar{p}$ pair can cancel, leaving only one nucleon on the right. Also, it doesn't conserve angular momentum. **9.** Yes. It conserves electromagnetic charge, spin, and B. There is no meson number to worry about. Because of the strangeness of K°, its decay will be controlled by the weak force. **13.** Not forbidden. (139.6 MeV) $-$ (105.7 MeV) $=$ 33.9 MeV. **17.** $2(m_pc^2) =$ 2(938.3 MeV) $=$ 1877 MeV. **21.** If 1 proton takes an average of 10^{30} y to decay, in a group of $N = 10^{30}$ protons one should decay within 1 y. $\tau = 10^{30}$ y $= 1/\lambda$; $\lambda = 10^{-30}$ fractional decays/y; $\lambda N = 1$. **25.** 2(938.3 MeV) $-$ (939.6 MeV) $-$ (938.3 MeV) $-$ (139.6 MeV) $=$ -140.9 MeV; each proton must bring in 70.45 MeV. **29.** $E = hf$; $f = E/h =$ $2.417\ 99 \times 10^{22}$ Hz $= 2.42 \times 10^{22}$ Hz; $\lambda = c/f = 1.239\ 8 \times 10^{-14}$ m $= 1.24 \times 10^{-14}$ m; $p = h/\lambda = 5.34 \times 10^{-20}$ kg·m/s. **33.** It's made up of three quarks (no antiquarks). To get the strangeness, we need $S = -1$ or one s-quark that has a $Q = -1/3$, so we must add a net charge of $+1/3$. One u and one d will do it. $\Lambda^0 =$ uds. **37.** $E = hf = 2m_ec^2 =$ 1.637×10^{-13} J; $f = 2.47 \times 10^{20}$ Hz. **41.** $E = (3/2)k_BT$; $T = 2E/3k_B = 8 \times 10^{27}$ K.

Credits

Chapter 2: p. 28, Figure 2.2 adapted from Eugene Hecht, *Physics in Perspective.* © 1980 by Addison-Wesley Publishing Company, Inc. Reprinted by permission. **Chapter 3: p. 61**, Figure 3.3 from *Road & Track,* December 1977, p. 49. Reprinted by permission. **p. 70**, Excerpt: From Associated Press News Release, July 1971. Reprinted by permission. **p. 77**, Figure 3.12 adapted from Eugene Hecht, *Physics in Perspective,* © 1980 by Addison-Wesley Publishing Company, Inc. Reprinted by permission. **Chapter 4: p. 95**, Figure 4.4 adapted from Eugene Hecht, *Physics in Perspective.* © 1980 by Addison-Wesley Publishing Company, Inc. Adapted by permission. **Chapter 5: p. 127**, Figure 5.2 adapted from Eugene Hecht, *Physics in Perspective.* © 1980 by Addison-Wesley Publishing Company, Inc. Adapted by permission. **p. 137**, Figure 5.15 from Eugene Hecht, *Physics in Perspective.* © 1980 by Addison-Wesley Publishing Company, Inc. Reprinted by permission. **Chapter 7: p. 214**, Table 7.1: Adapted from *Scientific American,* August 1992. **Chapter 9: p. 305**, Figure 9.19: Adapted from Eugene Hecht, *Physics in Perspective.* (Adapted from pg. 100). © 1980 by Addison-Wesley Publishing Company, Inc. Reprinted by permission. **Chapter 10: p. 325**, Figure 10.3: From Eugene Hecht, *Physics in Perspective.* © 1980 by Addison-Wesley Publishing Company, Inc. Reprinted by permission. **Chapter 11: p. 370**, Figure 11.22: From Eugene Hecht, *Physics in Perspective.* © 1980 by Addison-Wesley Publishing Company, Inc. Reprinted by permission. **p. 371**, Figure 11.23: From Eugene Hecht, *Physics in Perspective,* © 1980 by Addison-Wesley Publishing Company, Inc. Reprinted by permission. **p. 391**, Figure Q10: Adapted from Eugene Hecht, *Physics in Perspective.* © 1980 by Addison-Wesley Publishing Company, Inc. Reprinted by permission. **Chapter 16: p. 565**, Figure 16.19 from Eugene Hecht, *Physics in Perspective.* © 1980 by Addison-Wesley Publishing Company, Inc. Reprinted by permission. **Chapter 18: p. 627**, Figure 18.2 from Eugene Hecht, *Physics in Perspective.* © 1980 by Addison-Wesley Publishing Company, Inc. Reprinted by permission. **Chapter 19: p. 683**, Figure Q15 from *Newsday,* Wednesday, July 29, 1987. Reprinted by permission of Los Angeles Times Syndicate International. **Chapter 21: p. 729**, Figure 21.11 adapted from Eugene Hecht, *Physics in Perspective.* © 1980 by Addison-Wesley Publishing Company, Inc. Reprinted by permission. **p. 750**, Figure 21.37 adapted from Eugene Hecht, *Physics in Perspective.* © 1980 by Addison-Wesley Publishing Company, Inc. Reprinted by permission. **Chapter 23: p. 828**, Figure Q15 reprinted, with permission, from *49 Easy-to-Build Electronic Projects,* by Robert M. Brown and Tom Kneitel. Copyright 1981 by TAB Books, a Division of McGraw-Hill Inc., Blue Ridge Summit, PA 17294-0850, (1-800-233-1128). **Chapter 28: p. 1003**, Figure 28.18 adapted from W. Bertozzi, 1964, *American Journal of Physics, 32,* p. 551. Adapted by permission. **p. 1007**, Figure 28.19 adapted from R. Kollarits, 1972, *American Journal of Physics, 40,* p. 1125. Adapted by permission. **Chapter 32: p. 1104**, Figure 32.1 from A.P. Arya, *Elementary Modern Physics* (Figure 11.5), © 1974 by Addison-Wesley Publishing Company, Inc. Reprinted by permission. **p. 1105**, Figure 32.3 data from R. Hofstadter, *Annual Reviews of Nuclear Science, 7,* 231, 1957. **p. 1108**, Figure 32.6 data from G. W. Farwell and H. E. Wegner, 1954, *Physical Review, 95,* 1212.

Photo Credits

Chapter 1: **1**, J. Wray; **5**, National Optical Astronomy Observatories; **7**, Brooks/Cole; **8**, The Bryn Mawr College Archives; **9**, (top) Fotomas Index, London; (bottom) E. H.; **10**, (two at top) Stock, Boston, (two at bottom) Photo Researchers, Inc.; **12**, (top) Focus on Sports, (bottom) National Bureau of Standards; **13**, (two at top), Stock, Boston, (two at bottom) Photo Researchers, Inc.; **14**, (top) Stock, Boston, (bottom) Photo Researchers, Inc.; **16**, E. H.; **18**, (left) Stephen Frisch/Stock, Boston, (right) J. Anthony Tyson/AT&T Bell Labs

Chapter 2: **23**, Brooks/Cole; **24**, Gift of Collection Société Anonyme/Yale University Art Gallery; **28**, Ana S. Arias; **30**, © The Harold E. Edgerton 1992 Trust; **39**, Canadian Forces Photo

Chapter 3: **57**, Brooks/Cole; **59**, Duomo Photography; **71**, Vandystadt/Photo Researchers, Inc.; **73**, Biblioteca Ambosiana, Milan, Italy; **76**, Peticolas/Megna/Fundamental Photographs; **77**, Dawson Jones/Stock, Boston

Chapter 4: **89**, Brooks/Cole; **90**, Burndy Library **91**, NASA; **92**, Gendre, Paris, France; **93**, Biblioteca Ambrosiana, Milan, Italy; **96**, E. H.; **103**, © The Harold E. Edgerton 1992 Trust; **104**, Rick Stewart/Allsport USA; **109**, (top) E. H., (bottom) Richards/PhotoEdit; **111**, (top) Hewlett-Packard; (bottom) Yoram Lehmann; **112**, Mickey Gibson/Animals, Animals; **118**, U.S. Air Force; **124**, Tracy Lee Didas/U.S. Navy

Chapter 5: **125**, Brooks/Cole; **130**, Gerard Vandystadt/Photo Researchers, Inc.; **132**, K. Johnson/Focus on Sports; **134**, Joseph Sohm/Stock, Boston; **137**, Bob Martin/Allsport; **141**, James Sugar/Black Star; **142**, Gerard Vandystadt/Photo Researchers, Inc.; **144**, David Yarrow/Allsport; **145**, Six Flags Great Adventure; **147**, Brooks/Cole; **148**, Leonard Harris/Stock, Boston; **151**, (top) David Madison/Duomo Photography, (bottom) Gerard Vandystadt/Allsport USA; **152**, Gerard Vandystadt/Allsport USA; **155**, (left) E. H., (right) E. H.; **156**, E. H.

Chapter 6: **163**, Brooks/Cole; **165**, Canadian Forces Photo; **169**, Larry Molmud/Mucking Otis Press; **170**, Armed Forces Institute of Pathology; **171**, SKA; **173**, E. H.; **179**, Vanguard Racing Sailboats; **184**, Bob Daemmrich/Stock, Boston; **187**, (left) Bob Daemmrich/Stock, Boston, (right) E. H.; **190**, Hsinhai News Agency; **192**, E. H.; **193**, (top) E. H., (bottom) Central Scientific Company

Chapter 7: **207**, Brooks/Cole; **208**, (top) SKA, (bottom) Jet Propulsion Laboratory; **209**, Burndy Library; **213**, Central Scientific Company; **214**, James Sugar/Black Star; **217**, James Sugar; **221**, Uwe Fink/Dept. of Planetary Sciences, University of Arizona; **222**, Jet Propulsion Laboratory; **226**, NASA; **230**, Raymond E. Arvidson/Washington University

Chapter 8: **239**, Brooks/Cole; **255**, (top) Central Scientific Company, (bottom) Richard Megna/Fundamental Photographs; **261**, (right) NASA, (left) National Optical Astronomy Observatories, (middle) National Optical Astronomy Observatories, (bottom) Ricoh Corporation; **262**, Eadweard Muybridge; **264**, Jet Propulsion Laboratory; **265**, Gordon Garradd/Photo Researchers, Inc.; **266**, Sargent Welch

Chapter 9: **277**, Brooks/Cole; **287**, James Sugar/Black Star; **288**, Burndy Library; **292**, (top) Rick Rockman/Duomo Photography, (bottom) Marty Stouffer/Animals, Animals; **294**, Robert Bloomberg; **295**, Dorothy Littell/Stock, Boston; **296**, NASA; **300**, The Harold E. Edgerton 1992 Trust; **304**, (left) Erik Anderson/Stock, Boston, (right) The Harold E. Edgerton 1992 Trust; **306**, The Harold E. Edgerton 1992 Trust; **308**, Central Scientific Company

Chapter 10: **317**, Brooks/Cole; **318**, (top) Robert Nichols/Black Star, (bottom) SKA; **319**, E. H.; **320**, E. H.; **321**, Department of Energy; **325**, E. H.; **326**, (top) E. H., (bottom, left) E. H., (bottom, right) E. H.; **327**, E. H.; **330**, E. H.; **332**, Detroit Testing Laboratory; **333**, Elisabeth Weiland/Photo Researchers, Inc.; **334**, George Holton/Photo Researchers, Inc.; **335**, (top) E. H., (bottom) J. Gross/Photo Researchers, Inc.; **337**, Michael Fogden/Photo Researchers, Inc.; **339**, Bill Gallery/Stock, Boston; **341**, (top) E. H., (middle) E. H., (bottom) E. H.; **343**, Smithsonian Institute; **344**, E. H.; **348**, E. H.; **349**, E. H.

Chapter 11: **353**, Brooks/Cole; **355**, E. H.; **356**, E. H.; **358**, Central Scientific Company; **363**, (top, left) Victor Lorian, M.D., Bronx-Lebanon Hospital Center, (top, right) Sargent Welch, (bottom) The British Library; **364**, Jim Mejuto/FPG International, Inc.; **367**, E. H.; **371**, Frank Siteman/Stock, Boston; **372**, (top, left) Carl Purcell/Photo Researchers, Inc., (top, right), NASA, (bottom) Matson Lines; **373**, G. I. Bernard/Animals, Animals; **374**, E. H.; **375**, E. H.; **376**, Focus on Sports; **377**, (top) E. H., (bottom) Ted Thai/Time Magazine; **385**, (top) Peter Miller/Photo Researchers, Inc., (bottom) NASA; **386**, (left) NASA, (right) Peter Bradshaw/Imperial College, courtesy M. Van Dyke/Stanford University; **387**, Lockheed-Georgia; **388**, Historical Pictures Collection/Stock Montage, Inc.; **390**, E. H.; **391**, E. H.; **392**, (top) E. H., (bottom) Sargent Welch

Chapter 12: **401**, Brooks/Cole; **402**, E. H.; **417**, Robert Bloomberg; **418**, (left) E. H., (middle and right) Thomas Rossing/Northern Illinois University; **419**, MTS Systems Corp.; **420**, (left) Ealing Corporation and Dr. S. Miller, Jr., (middle) Wide World, (right) UPI/Bettmann Newsphotos; **431**, S. Kaplan, M.D.; **432**, Klein Associates; **433**, Photo Researchers, Inc.; **438**, C. F. Quate/Stanford University

Chapter 13: **441**, Brooks/Cole; **445**, Dohrn/Photo Researchers, Inc.; **449**, E. H.; **457**, E. H.; **459**, E. H.; **463**, NASA; **465**, David Ball/The Picture Cube; **467**, Educational Development Center; **475**, Central Scientific Company

Chapter 14: **481**, Brooks/Cole; **486**, (top) Smithsonian Institution, (bottom) J. F. Allen/University of St. Andrews, Scotland; **488**, (top, a, b, c) E. H., (bottom) Wide World; **492**, (left) E. H., (middle) E. H., (right) Central Scientific Company; **493**, E. H.; **502**, Richard Farrell/Photo Researchers, Inc.; **504**, Smithsonian Institution

Chapter 15: **513**, Brooks/Cole; **514**, D. Brown, R. Evans/University of Miami, Rosentiel School of Marine & Atmospheric Science; **515**, Smithsonian Institution; **516**, The Royal Institution, London; **518**, Smithsonian Institution; **520**, E. H.; **529**, Ken Eward/Photo Researchers, Inc.; **533** (from left) Blair Seitz/Photo Researchers; Philippa Scott/Photo Researchers; Stephen Frisch/Stock, Boston; Martin Dohrn/Science Photo Library; **535**, (top) NASA, (bottom) Robin Shanabarger/Amber, a Raytheon Company; **537**, (right) NASA; (left) Claus Lotscher/Peter Arnold, Inc.; **539**, NASA; **540**, E. H.

Chapter 16: **549**, Brooks/Cole; **561**, Michael Manzelli/Discover; **564**, Jean-Loup Charmet; **571**, Niels Bohr Library; **578**, Peter Arnold, Inc.

Chapter 17: **587**, Brooks/Cole; **588**, E. H.; **589**, E. H.; **592**, Discover Magazine/© 1992; **594**, NASA; **595**, Courtesy of the Coulomb Family; **596**, Burndy Library; **598**, Mark C. Burnette/Stock Boston; **601**, E. H.; **604**, Harold M. Waage/Princeton University; **605**, (left) Pat Manly/Third Coast Stock Source; **609**, Harold M. Waage/Princeton University; **610**, (top, bottom) Harold M. Waage/Princeton University; **611**, Harold M. Waage/Princeton University; **612**, (left) E. H., (right) Science Museum, Boston; **620**, E. H.; **621**, Science Museum, Boston

Chapter 18: **625**, Brooks/Cole; **626**, Courtesy, Ohio Edison; **629**, Wide World Photos, Inc.; **633**, Science Museum, Boston; **635**, Ira Wyman; **636**, Zephyr Services; **638**, Andy Freeberg/Discover Magazine; **639**, Lawrence Berkeley Laboratory; **640**, Burndy Library; **642**, (left) E. H., (right) Jennifer Hecht; **644**, E. H.; **648**, Goivaux Communication/Phototake; **651**, Lawrence Livermore National Laboratory

Chapter 19: **661**, Brooks/Cole; **662**, Physics International, a ROCK-COR subsidiary; **664**, Photographie Bulloz, Paris; **666**, (top) E. H., (bottom) University of California, Irvine; **668**, E. H.; **671**, AIP Niels Bohr Library, E. Scott Barr Collection; **674**, E. H.; **676**, E. H.; **678**, IBM Research; **679**, © Werner H. Müller/Peter Arnold, Inc.; **682**, E. H.

Chapter 20: **689**, Brooks/Cole; **690**, (left) Circus World Museum, (right) MTS Systems Corporation; **698**, Photo Researchers, Inc.; **703**, E. H.

Chapter 21: **721**, Brooks/Cole; **722**, E. H.; **723**, (left) R. P. Blakemore and N. Blakemore, (center, right) Astuko Kobayashi-Kirschvink/California Institute of Technology; **724**, E. H.; **725**, E. H.; **726**, (left, right) E. H.; **728**, General Electric Corporate Research and Development; **730**, Central Scientific Company; **731**, (left, right) Richard Megna/Fundamental Photographs; **732**, Burndy Library; **733**, (top) Hank Morgan/Photo Researchers, Inc., (bottom) Burndy Library; **735**, Richard Megna/Fundamental Photographs; **737**, Richard Megna/Fundamental Photographs; **737**, Richard Megna/Fundamental Photographs; **738**, (top) General Electric Research and Development Center, (bottom) Burndy Library; **739**, E. H.; **743**, E. H.; **744**, (top) E. H., (bottom left, right) E. H.; **745**, (left) L. A. Frank/NASA, (right) Lee Snyder/Geophysical Institute, University of Alaska; **748**, E. H.; **755**, J. F. Allen/University of St. Andrews, Scotland

Chapter 22: **763**, Brooks/Cole; **768**, Central Scientific Company; **779**, The Royal Institution, London; **780**, E. H.; **781**, E. H.

Chapter 23: **795**, Brooks/Cole; **796**, Wellcome Institute Library, London; **800**, E. H.; **814**, E. H.; **815**, E. H.; **816**, E. H.; **818**, T. J. Watson Research Center, IBM; **822**, E. H.; **824**, T. J. Watson Research Center, IBM

Chapter 24: **833**, Brooks/Cole; **834**, Smithsonian Institution; **851** (top) NASA, (bottom) E. H.; **852**, (left) Robin Shanabarger/Amber, a Raytheon Company, (right) E. H.; **854**, Jet Propulsion Laboratory, California Institute of Technology; **856** (left) Burndy Library, (right) Jim Wilson/Research Systems Inc.; **857**, Custom Medical Stock Photo; **858**, AT&T Archives

Chapter 25: **863**, Brooks/Cole; **868**, E. H.; **871**, (left, right) E. H.; **879**, E. H.; **880**, E. H.; **881**, CIRCON ACMT; **882**, Custom Medical Stock Photo; **883**, E. H.; **884**, E. H.; **885**, Fritz Goro/Life Magazine © Time-Warner; **886**, The Trustees, National Gallery, London; **887**, Cortauld Institute Galleries, London; **890**, AT&T Bell Labs

Chapter 26: **893**, Brooks/Cole; **895**, Optical Society of America; **904**, E. H.; **909**, E. H.; **910**, E. H.; **912**, E. H.; **914**, E. H.; **925**, NASA; **927**, Chris Hildreth/Cornell University Photo; **928**, E. H.; **929**, E. H.

Chapter 27: **939**, Brooks/Cole; **940**, E. H.; **945**, E. H.; **948**, (top, left and right) E. H. (bottom) E. H.; **949**, E. H.; **950**, (left, center, right) E. H.; **951**, Central Scientific Company; **954**, (top, bottom) Kodansha; **956**, E. H.; **957**, Central Scientific Company; **958**, (left, center, right) E. H.; **960**, (top, bottom) E. H.; **961**, E. H.; **965**, PASCO Scientific; **966**, (top) E. H., (bottom) Jakub Klinger/Klinger Educational Products Corp.; **967**, (top, bottom) E. H.; **968**, Benjamin J. Pernick; **969**, E. H.; **971**, ©Smith, *Principles of Holography,* by permission of John Wiley & Sons, Inc.; **973** (left, right) E. H.

Chapter 28: **979**, Brooks/Cole; **983**, Kent Reno/SKA, **995**, Ping-Kang Hsuing & Robert H. P. Dunn/Carnegie Mellon University

Chapter 29: **1015**, Brooks/Cole; **1018**, (top) E. H., (bottom) Central Scientific Company; **1019**, Cavendish Laboratory, University of Cambridge; **1020**, Cavendish Laboratory, University of Cambridge; **1022**, (top, bottom) The Metropolitan Museum of Art, Purchase, 1871; **1025**, E. H.; **1026**, (top, bottom) Smithsonian Institution; **1027**, E. H.; **1031**, Stuart Kenter Assoc.; **1032**, Stuart Kenter Assoc.; **1034**, E. H.

Chapter 30: **1043**, Brooks/Cole; **1044**, PhotoDisc; **1048**, Agency International de Physique Solvag/AIP Neils Bohr Library; **1049**, Courtesy, German Information Center; **1053**, E. H.; **1057**, AIP Emilio Segre Visual Archives; **1062**, Warren Nagourney; **1064**, Laserscope

Chapter 31: **1073**, Brooks/Cole; **1074**, (top) French Embassy, Press & Information Division, (bottom) Photo Researchers, Inc.; **1075**, (left, right) Educational Development Center, Inc.; **1077**, AIP Neils Bohr Library; **1079**, (left) Courtesy, David Sarnoff Research Center, (right) A. Tonomura/Hitachi Advanced Research Center; **1081**, A. J. G. Hey/University of Southampton; **1083**, National Optical Astronomy Observatories; **1090**, AIP Neils Bohr Library, Archives for History of Quantum Physics; **1093**, AIP Neils Bohr Library, Fankuchen Collection; **1094**, AIP Emilio Segre Visual Archives; **1095**, (left) McConnell Brain Imaging Centre, Montreal Neurological Institute, (right) Dan McCoy/Rainbow

Chapter 32: **1101**, Brooks/Cole; **1102**, Photo Researchers, Inc.; **1104**, Argonne National Laboratory; **1107**, (top, bottom) Dan McCoy/Rainbow; **1110**, AIP Meggers Gallery of Nobel Laureates; **1115**, EG&G Mound Applied Technology; **1118**, S. M. Larson and S. J. Goldsmith/Nuclear Medicine Service, Memorial Sloan-Kettering Cancer Center; **1122**, A. Agelarakis/Adelphi University; **1123**, (top) Argonne National Laboratory; **1125**, from Armin Hermann, *The New Physics,* Munich: Heinz Moos Verlag, 1979; **1126**, *N. Y. Herald Tribune;* **1127**, *Pittsburgh Post-Gazette,* AIP Emilio Serge Visual Archives; **1129**, (top) painting by Gary Sheahan, photo courtesy Argonne National Laboratory, (bottom) U. S. Council for Energy Awareness; **1130**, (top) U. S. Navy, (bottom) UPI/Bettmann Archive; **1133**, Lawrence Livermore National Laboratory

Chapter 33: **1137**, Brooks/Cole; **1139**, Fermilab Visual Media Services; **1142**, Fermilab Visual Media Services; **1157**, Brooks/Cole; **1158**, Karl Lattrell/University of Michigan; **1161**, E. H.

Index

USEFUL PHYSICAL DATA

Standard Temperature and Pressure (STP)		0°C = 273.15 K 1 atm ≡ 101.325 kPa
Water		
Density, relative (4°C)	ρ_w	1.000×10^3 kg/m³
Heat of fusion	L_f	333.7 kJ/kg
Heat of vaporization	L_v	2259 kJ/kg
Specific heat capacity	c	4.186 kJ/kg·K
Index of refraction	n_w	1.33
Standard acceleration due to Earth's gravity		9.806 65 m/s²
Speed of sound in air (20°C)		343 m/s
Speed of sound in air (STP)		331 m/s
Density of dry air (STP)		1.29 kg/m³
Molecular mass of air		28.98 g/mol

ASTROPHYSICAL DATA

	Earth	Moon	Sun
Mass	5.975×10^{24} kg	7.35×10^{22} kg	1.987×10^{30} kg
Mean radius	6.371×10^6 m	1.74×10^6 m	6.96×10^8 m
Mean density	5.52×10^3 kg/m³	3.33×10^3 kg/m³	1.41×10^3 kg/m³
Orbital period about galactic center			200×10^6 y
Orbital period	365 d 5 h 48 min	27.3 d	
Mean distance from Sun	1.50×10^{11} m		
Mean distance from Earth		3.85×10^8 m	
Surface gravitational acceleration	9.81 m/s²	1.62 m/s²	274 m/s²
Surface pressure	1.013×10^5 Pa		
Magnetic moment	8.0×10^{22} A·m²		
Surface temperature	≈287 K	125 K–375 K	5.8×10^3 K
Power output			3.85×10^{26} W
Period of rotation	23 h 56 min 4.1 s	27.3 d	~26 d

THE GREEK ALPHABET

Alpha	Α	α	Nu	Ν	ν
Beta	Β	β	Xi	Ξ	ξ
Gamma	Γ	γ	Omicron	Ο	o
Delta	Δ	δ	Pi	Π	π
Epsilon	Ε	ϵ	Rho	Ρ	ρ
Zeta	Ζ	ζ	Sigma	Σ	σ
Eta	Η	η	Tau	Τ	τ
Theta	Θ	θ	Upsilon	Υ	υ
Iota	Ι	ι	Phi	Φ	ϕ
Kappa	Κ	κ	Chi	Χ	χ
Lambda	Λ	λ	Psi	Ψ	ψ
Mu	Μ	μ	Omega	Ω	ω